CHAPTER QUESTIONS AND ANSWERS

The questions at the end of each chapter are provided to help you test your knowledge and increase your understanding of biochemistry. Since they are intended to help you learn the material, their construction does not always conform to principles for simple knowledge assessment. Specifically, you will sometimes be expected to draw on your knowledge of several areas to answer a single question, and some questions may take longer to answer than the average time allowed on certain national examinations. Occasionally, you may disagree with the answer. If this occurs, we hope that after you read the commentary on the question, you will see the point and your insight into the biochemical problem will be increased.

The *q*uestion *t*ypes conform to those commonly used in objective examinations by medical school departments of biochemistry. For each question, the *q*uestion *t*ype is given in parentheses after its number; for example, 5.(QT2) indicates that question 5 is of type 2. Keys for the question types are given below:

QT1: Choose the one *best* answer

QT2: Answer the question according to the following key:

A. If 1, 2, and 3 are correct
B. If 1 and 3 are correct
C. If 2 and 4 are correct
D. If only 4 is correct
E. If all four are correct

QT3: Answer the question according to the following key:

A. If A is greater than B
B. If B is greater than A
C. If A and B are equal or nearly equal

QT4: Answer the question according to the following key:

A. If the item is associated with A only
B. If the item is associated with B only
C. If the item is associated with both A and B
D. If the item is associated with neither A nor B

QT5: Match the numbered statement or phrase with one of the lettered options given above.

TEXTBOOK OF BIOCHEMISTRY

BOOKS OF RELATED INTEREST

The National Medical Series for Independent Study

The Basic Science Group
Anatomy
Biochemistry
Histology & Embryology
Microbiology
Pathology
Pharmacology
Physiology

The Clinical Studies Group
Medicine
Obstetrics & Gynecology
Pediatrics
Preventive Medicine & Public Health
Psychiatry
Surgery

TEXTBOOK OF BIOCHEMISTRY

With Clinical Correlations

SECOND EDITION

Edited by

Thomas M. Devlin, Ph.D.

Professor and Chairman
Department of Biological Chemistry
Hahnemann University School of Medicine
Philadelphia, Pennsylvania

A WILEY MEDICAL PUBLICATION

JOHN WILEY & SONS New York / Chichester / Brisbane / Toronto / Singapore

Library of Congress Cataloging in Publication Data:

Main entry under title:
Textbook of biochemistry.
(The National medical series for independent study)
(A Wiley medical publication)
Includes bibliographies and index.
1. Biological chemistry. 2. Clinical biochemistry.
I. Devlin, Thomas M. II. Series. III. Series: Wiley medical publication. [DNLM: 1. Biochemistry.
QU 4 T355]
QP514.2.T4 1986 612′.015 85-26318
ISBN 0-471-81462-8

Printed in the United States of America

10 9 8 7 6 5

To

Bonnie, Cathy, Marjorie, Mark, and Steven

Contributors

Paul F. Agris, Ph.D.
Professor
Division of Biological Sciences and Department of Medicine
College of Arts and Sciences and Medical School
University of Missouri—Columbia
Columbia, Missouri

Stelios Aktipis, Ph.D.
Professor
Department of Biochemistry and Biophysics
Loyola University of Chicago
Stritch School of Medicine
Maywood, Illinois

Carol N. Angstadt, Ph.D.
Assistant Professor
Department of Biological Chemistry
Hahnemann University School of Medicine
Philadelphia, Pennsylvania

William M. Awad, Jr., M.D., Ph.D.
Professor
Departments of Medicine and Biochemistry
University of Miami School of Medicine
Miami, Florida

James Baggott, Ph.D.
Associate Professor
Department of Biological Chemistry
Hahnemann University School of Medicine
Philadelphia, Pennsylvania

Stephen G. Chaney, Ph.D.
Associate Professor
Department of Biochemistry and Nutrition
School of Medicine
University of North Carolina at Chapel Hill
Chapel Hill, North Carolina

Joseph G. Cory, Ph.D.
Professor and Chairman
Department of Biochemistry
College of Medicine
University of South Florida
Tampa, Florida

David W. Crabb, M.D.
Assistant Professor
Departments of Medicine and Biochemistry
Indiana University School of Medicine
Indianapolis, Indiana

Thomas M. Devlin, Ph.D.
Professor and Chairman
Department of Biological Chemistry
Hahnemann University School of Medicine
Philadelphia, Pennsylvania

John E. Donelson, Ph.D.
Professor
Department of Biochemistry
University of Iowa School of Medicine
Iowa City, Iowa

Robert H. Glew, Ph.D.
Associate Professor
Department of Biochemistry
School of Medicine
University of Pittsburgh
Pittsburgh, Pennsylvania

Robert A. Harris, Ph.D.
Professor and Associate Chairman
Department of Biochemistry
Indiana University School of Medicine
Indianapolis, Indiana

Ulrich Hopfer, M.D., Ph.D.
Professor
Department of Developmental Genetics and Anatomy
School of Medicine
Case Western Reserve University
Cleveland, Ohio

J. Denis McGarry, Ph.D.
Professor
Departments of Internal Medicine and Biochemistry
Southwestern Medical School
University of Texas Health Science Center at Dallas
Dallas, Texas

Alan H. Mehler, Ph.D.
Professor
Department of Biochemistry
Howard University School of Medicine
Washington, D.C.

Karl H. Muench, M.D.
Professor
Departments of Medicine and Biochemistry
University of Miami School of Medicine
Miami, Florida

Merle S. Olson, Ph.D.
Professor and Chairman
Department of Biochemistry
University of Texas Health Science Center at San Antonio
San Antonio, Texas

Francis J. Schmidt, Ph.D.
Associate Professor
Department of Biochemistry
College of Agriculture and School of Medicine
University of Missouri—Columbia
Columbia, Missouri

Richard M. Schultz, Ph.D.
Professor and Chairman
Department of Biochemistry and Biophysics
Loyola University of Chicago
Stritch School of Medicine
Maywood, Illinois

Nancy B. Schwartz, Ph.D.
Professor
Departments of Pediatrics and Biochemistry
University of Chicago
Chicago, Illinois

Frank Ungar, Ph.D.
Professor
Department of Biochemistry
Medical School
University of Minnesota
Minneapolis, Minnesota

John F. Van Pilsum, Ph.D.
Professor
Department of Biochemistry
Medical School
University of Minnesota
Minneapolis, Minnesota

Marilyn S. Wells, M.D.
Adjunct Associate Professor
Department of Medicine
University of Miami School of Medicine
Chief, Spinal Cord Injury Service
Veterans Administration Medical Center
Miami, Florida

J. Lyndal York, Ph.D.
Professor
Department of Biochemistry
University of Arkansas for Medical Sciences
Little Rock, Arkansas

Preface

The purposes of the second edition of the *Textbook of Biochemistry with Clinical Correlations* remain the same as the first edition: to present a clear discussion of the biochemistry of mammalian cells; to relate the biochemical events at the cellular level to the physiological processes occurring in the whole animal; and to cite examples of deviant biochemical processes in human diseases. The topics were selected to cover the essential areas of mammalian biochemistry and physiological chemistry, and they were sequenced to permit efficient development of a knowledge of biochemistry. A principal concern in the preparation of the text was to meet the needs of a course in biochemistry for medical students, but the content and depth of presentation are appropriate for a first course in biochemistry for upper-level undergraduates and graduate students.

The entire text was reviewed and edited for the second edition and significant additions of new material, clarifications, and some deletions were made in every chapter; some chapters were rewritten entirely by new contributors. The topics of protein structure, metabolic control, molecular biology, and genetic engineering, in particular, were updated because of the rapid developments in these areas. As with the previous edition, each contributor included up-to-date information that had been substantiated as being significant but avoided observations that were so new that time had not allowed adequate evaluation.

The first six chapters cover the areas of protein and enzyme chemistry, membrane structure, and bioenergetics, setting the stage for discussion of the metabolism of the principal cellular components. Presentation of metabolic interrelationships at the cellular and tissue levels is followed by chapters on the major hormones. The biochemical events in the transmittal of genetic information and the mechanism of phenotypic expression and its control, as well as recent developments in recombinant DNA research and genetic engineering, are discussed. The remaining chapters cover essential aspects of physiological chemistry, including the biochemistry of selected tissues, iron and heme metabolism, gas transport, regulation of pH, and digestion and absorption of foodstuffs. The textbook concludes with a discussion of nutrition from a biochemical perspective.

In order to emphasize the relevancy of the topics to disease problems, each chapter includes selected clinical correlations relating normal biochemical events to pathophysiological states in humans. The correlations are intended to describe the aberrant biochemistry of the disease state rather than specific case reports. In some instances the same clinical condition is presented in different chapters, each from a different perspective. Clinical correlations are presented as separate entities in order not to interrupt the flow of the biochemistry discussion. All pertinent biochemical information is presented in the main text, and an understanding of the material does not require a reading of the correlations. In some chapters clinical discussions are part of the principal text because of the close relationship of some topics to clinical conditions.

As with any textbook, selection of the material to be presented was a difficult problem. Much of our knowledge of intracellular chemical events is derived from studies of single cells such as bacteria and yeast; in fact, for some topics we have very sketchy knowledge about the events in mammalian cells. In some sections it has been necessary to discuss information derived from bacterial systems as a model for what may occur in the complex mammalian tissue.

Two new features of the second edition are the questions at the end of each chapter and an appendix containing a discussion of selected topics of organic chemistry. The questions are in multiple choice format and represent types found in national certifying examinations. Each question has an annotated answer, with references to the page in the textbook covering the content of the question. The questions cover a range of topics in each chapter. The appendix, "Review of Organic Chemistry," was designed as a reference for the nomenclature of organic groups and compounds and some chemical reactions; it is not intended as a comprehensive review. The reader might find it valuable to become familiar with the overall content and then use the appendix as a reference for specific topics when reading related sections in the main text.

The individual contributors were requested to prepare their chapters for a teaching textbook. The book is not intended as a compendium of biochemical facts or a review of the current literature, but each chapter contains sufficient detail on the subject to make it useful as a resource. Each contributor was requested not to refer to specific researchers; our apologies to those many biochemists who rightfully should be acknowledged for their outstanding research contributions to the field of biochemistry. Each chapter contains a bibliography that can be used as an entry point to the primary research literature.

Our experience with the first edition reinforced our commitment to a multicontributor textbook as the best approach to have the most accurate and current presentation of biochemistry. Each author is involved actively in teaching biochemistry in a medical or graduate school and has an active research interest in the field in which he or she has written. Thus each has the perspective of the classroom instructor and the experience to select the topics and determine the emphasis required for students in a course of biochemistry. Every contributor, however, brings to the book an individual writing style, leading to some differences in presentation from chapter to chapter. It was decided that this should not be an impediment to the students' ability to understand the material, as students are accustomed to learning from a variety of resources. Redundancies and inconsistencies in content have been kept to a minimum. Some repetition of topics in different chapters was retained because the topics were considered of such importance that reiteration would be helpful to the reader.

In any project, one person must accept the responsibility for the final product. The decisions concerning the selection of topics and format, reviewing the drafts, and responsibility for the final checking of the book were entirely mine. I would welcome comments, criticisms, and suggestions from the faculty and students who use this textbook. It is our hope that this work will be of value to those embarking on the exciting experience of learning biochemistry for the first time and for those who are returning to a topic in which the information is expanding so rapidly.

THOMAS M. DEVLIN

Acknowledgments

This project would never have been accomplished without the encouragement and participation of many people. The contributors received the support of associates and students in the preparation of their chapters, and, for fear of omitting someone, it was decided not to acknowledge individuals by name. To everyone who gave time unselfishly and shared in the objective and critical evaluation of the text, we extend our sincerest thanks. In addition, every contributor has been influenced by former teachers and colleagues, various reference resources, and, of course, the research literature of biochemistry; we are deeply indebted to these many sources of inspiration.

As editor I extend a very special thanks to all the contributors for accepting the challenge of preparing the chapters, for sharing ideas and making recommendations to improve the book, for accepting so readily suggestions to modify their contributions, and for cooperating throughout the period of preparation.

My personal and deep appreciation goes to three friends and colleagues who have been of immeasurable value to me during the preparation of the second edition. My gratitude goes to Dr. James Baggott, who patiently allowed me to use him as a sounding board for ideas and who unselfishly shared with me his suggestions and criticisms of the text, and to Dr. Carol Angstadt, who reviewed many of the chapters and who gave me valuable suggestions for improvements. A very special note of appreciation is extended to Dr. Francis Vella, who carefully read every part of the manuscript, making many excellent suggestions and corrections, and who reviewed all of the galleys. To each I extend my deepest gratitude.

I also wish to extend my deepest appreciation to the staff of John Wiley & Sons who worked with me during the preparation of the second edition. I am indebted to Andrew Ford, Vice President, who encouraged me to undertake the project and who gave me his unqualified support. Special recognition is extended to Linda Turner, Editor, who was always available to answer my questions and who made many valuable suggestions to expedite the preparation of the book; to Beryl Matshiqi, Editorial Assistant, who handled many of the details during preparation of the manuscripts; to Bruce Williams, Editorial Supervisor, who again, as in the first edition, carefully reviewed the manuscripts; to Margery Carazzone, Production Manager, whose talents were exemplified in the transition from manuscript to book; to Denise Watov, Production Supervisor, who with unlimited patience directed the review of galleys and pages. A special thanks is due to the Illustration Department of John Wiley & Sons and the many staff members who assisted in the completion of the book.

I am deeply indebted to my own staff, including Joanne Addario, Lalise Blain, and Beth McAndrews, who faithfully and efficiently completed the multitude of little chores involved in this project. I am particularly grateful to Beverly Lyman who assisted me in the review of galleys and pages. To each I extend my heartfelt appreciation.

Finally, a very special thanks to a supportive and considerate family, particularly to my wife Marjorie, who had the foresight to encourage me to undertake this project and who created an environment in which I could devote the many hours required for the preparation of this textbook. To all my deepest and sincerest thanks.

THOMAS M. DEVLIN

Contents

TEXTBOOK OF BIOCHEMISTRY

1

Eucaryotic Cell Structure

THOMAS M. DEVLIN

1.1 CELLS AND CELLULAR COMPARTMENTS

By a process not entirely understood and in a time span that is difficult to comprehend, elements such as carbon, hydrogen, oxygen, nitrogen, and phosphorus combined, dispersed, and recombined to form a variety of molecules until a combination was achieved that was capable of replicating itself. With continued evolution and the formation of ever more complex molecules, the environment around some of these self-replicating molecules was enclosed by a membrane. This development gave these molecules a significant advantage in that they could control to some extent their own environment. Life as we know it had evolved. With life occurred the establishment of a unit of space, which is a cell. With the passing of time a diversity of cells evolved, and their chemistry and structure became more complex. Many of the chemical reactions occurring in cells can be duplicated in test tubes, but the challenge of biochemical research is to unravel the mechanisms behind the organized and controlled manner in which cells carry out these reactions.

One of the most important structures of all cells is the limiting outer membrane, termed the plasma membrane, which delineates the intracellular and extracellular environments. The plasma membrane separates the variable and potentially hostile environment outside the cell from the relatively constant milieu within the cell and is the communication link between the cell and its surroundings. It is the plasma membrane that delineates the space occupied by a cell.

Based on both microscopic and biochemical differences, living cells are divided into two major classes, the procaryotes and the eucaryotes. Procaryotic cells, which include bacteria, blue-green algae, and rickettsiae, lack extensive intracellular anatomy (Figure 1.1). Intracellular structures of cells are due to the presence of macromolecules or membrane systems, which can be visualized in a microscope under appropriate conditions. The deoxyribonucleic acid (DNA) of procaryotes is often segregated into discrete masses, but it is not surrounded by a membrane or envelope. The

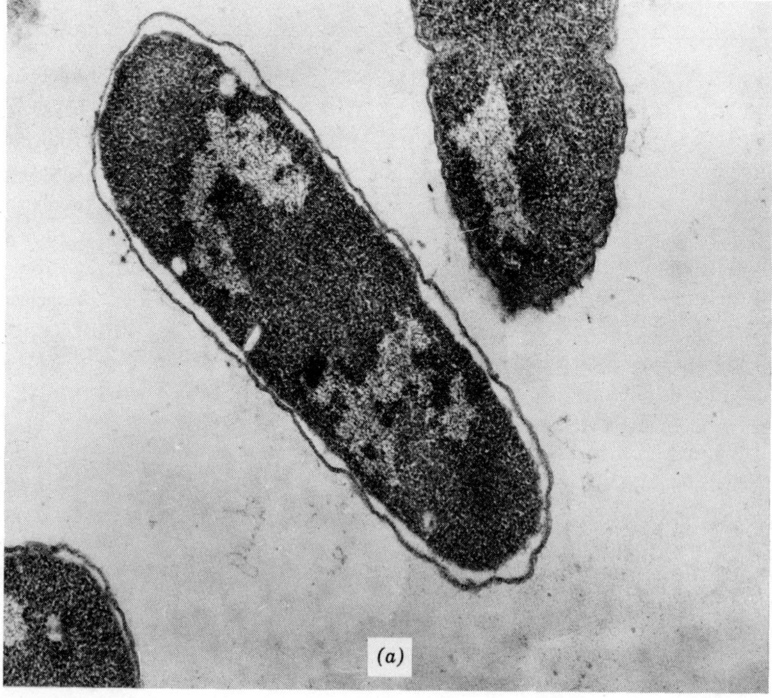

Figure 1.1
Cellular organization of procaryotic and eucaryotic cells.
(a) An electron micrograph of an E. coli, *a representative procaryote; approximate magnification 30,000×. There is little apparent intracellular organization and no cytoplasmic organelles. The chromatin is condensed into a nuclear zone but is not surrounded by a membrane. Procaryotic cells are smaller than eucaryotic cells.*
Photograph generously supplied by Dr. M. E. Bayer.

plasma membrane is often observed to be invaginated, but there are no definable subcellular organized bodies in procaryotes.

In contrast to the procaryotes, eucaryotic cells, which include yeasts, fungi, plant, and animal cells, have a well-defined membrane surrounding a central nucleus and a variety of intracellular structures and organelles. The intracellular membrane systems establish a number of distinct subcellular compartments permitting a unique degree of subcellular specialization. By compartmentalization of the cell, different chemical reactions requiring different environments can occur simultaneously. As an example, reactions requiring different ionic conditions or reactants can occur in different compartments of the same cell because the membranes surrounding a cellular compartment can control the environment in the compartment by regulating the movement of substances in and out.

Membranes are lipid in nature, and many compounds that are soluble in a hydrophobic lipid environment will concentrate in cellular mem-

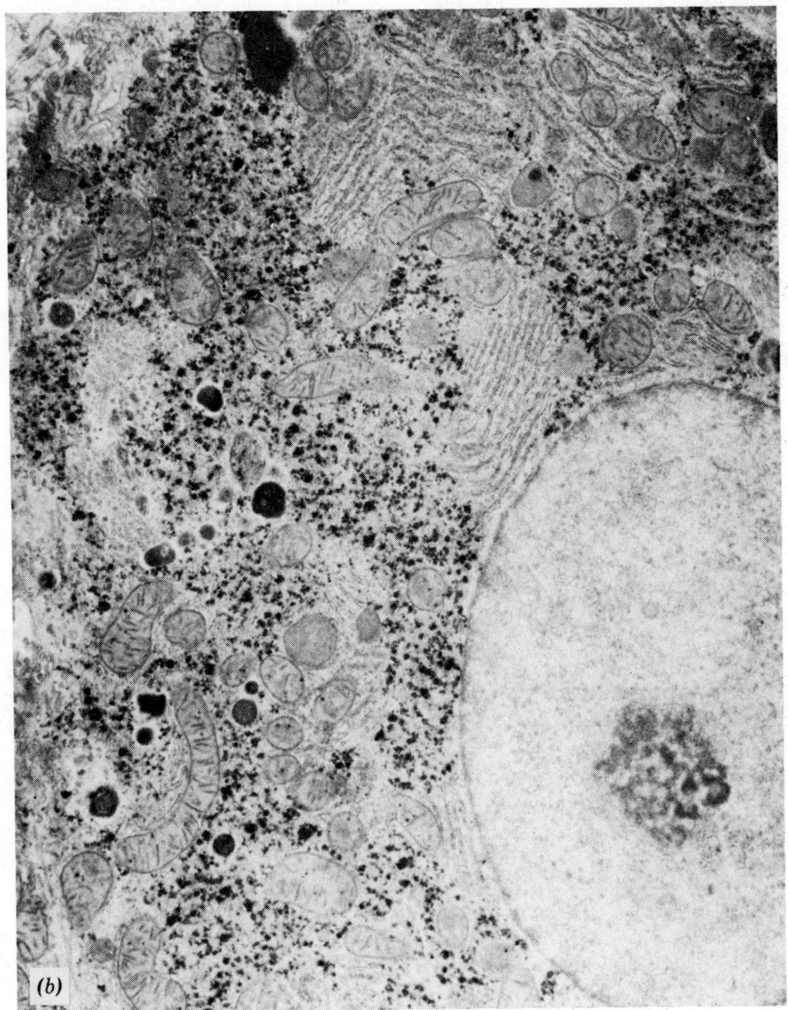

***Figure 1.1* (Continued)**
(b) An electron micrograph of a thin section of a rat liver cell (hepatocyte), a representative eucaryotic cell; approximate magnification 7500×. Note the distinct nuclear membrane, the different membrane bound organelles or vesicles, and the extensive membrane systems. The various membranes create a variety of intracellular compartments.
Photograph reprinted with permission of Dr. K. R. Porter from Porter, K. R., and Bonneville, M. A. *Fine Structure of Cells and Tissues*, Philadelphia: Lea and Febiger, 1972.

branes. Many biochemical reactions occur in specific membranes of the cell; thus, the extensive membrane systems of the eucaryotic cell create an additional environment that the cell can use for its diverse functions.

Besides the structural variations, there are significant differences in the chemical composition and biochemical activities between procaryotic and eucaryotic cells. Some of the major differences between the cell types are the lack in procaryotic cells of a class of proteins, termed histones, which in eucaryotic cells complex with DNA; major structural differences in the ribonucleic acid-protein complexes involved in the biosynthesis of proteins; differences in transport mechanisms across the plasma membrane; and a host of differences in enzyme content.

Even though there are structural and biochemical differences between the cell types, the many similarities are equally striking. The emphasis throughout this textbook is on the chemistry of eucaryotic cells, particularly mammalian cells, but much of our knowledge of the biochemistry of living cells has come from studies of procaryotic and from nonmammalian eucaryotic cells. The basic chemical components and many of the fundamental chemical reactions of all living cells are very similar. The availability of certain cell populations, for example, bacteria in contrast to human liver, has led to an accumulation of knowledge about some cells; in fact, in some areas of biochemistry our knowledge is derived exclusively from studies of procaryotes. The universality of many biochemical phenomena, however, permits many extrapolations from bacteria to humans.

Before we dissect and reassemble the complexities of mammalian cells and tissues in the following chapters, it is appropriate to review some of the chemical and physical characteristics of the environment in which the various biochemical phenomena occur. This is because it is important to recognize the constraints placed by the environment on the activities carried out by cells.

1.2 CELLULAR ENVIRONMENT—WATER AND SOLUTES

All biological cells contain essentially the same chemical components, which are listed in Table 1.1. There are differences in the concentration of specific components in different cells; in eucaryotic cells, there are even differences between intracellular compartments. It is also considered that there are microenvironments created by the cellular macromolecules in which the composition is distinct from the surrounding areas. All biological cells depend on the environment for nutrients, which are requisites for

TABLE 1.1 Chemical Components of Biological Cells

Component	*Range of Molecular Weights*
H_2O	18
Inorganic ions Na^+, K^+, Cl^-, SO_4^{2-}, HCO_3^-, Ca^{2+}, Mg^{2+}, etc.	23–100
Small organic molecules Carbohydrates, amino acids, lipids, nucleotides	100–700
Macromolecules Proteins, polysaccharides, nucleic acids	50×10^3–1×10^9

replacement of components, for growth, and for their energy requirements. The composition of the external environment can vary significantly and cells have a variety of mechanisms to cope with these variations. In addition, the different intracellular compartments also have different biochemical and chemical compositions. The one common characteristic of the different environments is the presence of water. Water, the solvent in which the substances required for the cell's existence are dissolved or suspended, has an important role in the well-being of all cells. The unique physicochemical properties of water make life possible.

Structure of Water

Two hydrogen atoms share their electrons with an unshared pair of electrons of an oxygen atom to form the water molecule. This deceptively simple molecule, however, has a number of unusual properties. The oxygen nucleus has a stronger attraction for the shared electrons than the hydrogen, and the positively charged hydrogen nuclei are left with an unequal share of electrons, creating a partial positive charge on each hydrogen and a partial negative charge on the oxygen. The bond angle between the hydrogens and the oxygen is 104.5°, making the molecule electrically asymmetric and producing an electric dipole (Figure 1.2). Water molecules interact because the positively charged hydrogens on one molecule are attracted to the negatively charged oxygen atom on another, with the formation of a weak bond between the two water molecules, as in Figure 1.3*a*. This bond is termed a hydrogen bond. Five molecules of water form a tetrahedral structure (Figure 1.3*b*), since each oxygen can share its electrons with four hydrogens and each hydrogen with another oxygen. In solid water, ice, a tetrahedral lattice structure is formed. It is the hydrogen bonding between molecules that gives ice its crystalline structure. Some of these bonds are broken as ice is transformed to liquid water. Each hydrogen bond is relatively weak compared to a covalent bond, but the large number between molecules in liquid water is the reason for the stability of water. Even at 100°C liquid water contains a significant number of hydrogen bonds, which accounts for its high heat of vaporization; in the transformation from a liquid to a vapor state the hydrogen bonds are disrupted.

Water molecules also hydrogen-bond to different chemical structures. Hydrogen bonding also occurs between other molecules and even within a molecule wherever an electronegative oxygen or nitrogen comes in close proximity to a hydrogen covalently bonded to another electronegative group. Some representative hydrogen bonds are presented in Figure 1.4. Intramolecular hydrogen bonding occurs extensively in large macromolecules such as proteins and nucleic acids and is the basis for their structural stability.

Breaking and forming hydrogen bonds occurs more rapidly than covalent bonds because of their low energy; it is estimated that hydrogen bonds in water have a half-life of less than 10^{-10} s. Liquid water actually has a definite structure due to the hydrogen bonding between molecules, but the structure is in a dynamic state as the hydrogen bonds break and reform. A similar dynamic interaction also occurs with substances present in the liquid, which are capable of hydrogen bonding. Many models have been proposed for the structure of liquid water, but none adequately explains all of its properties.

Figure 1.2
Structure of a water molecule.
The H—O—H bond angle is 104.5°, and both hydrogens carry a partial positive and the oxygen a partial negative charge, creating a dipole.

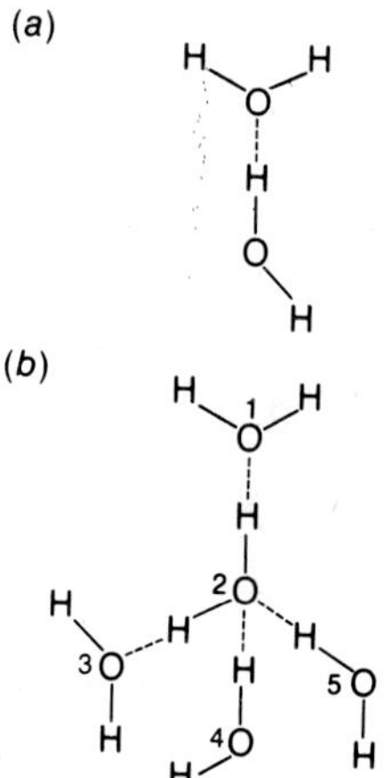

Figure 1.3
(a) Hydrogen bonding, as indicated by the dashed lines, between two water molecules. (b) Tetrahedral hydrogen bonding of five water molecules. Water molecules 1, 2, and 3 are in the plane of the page, 4 is below, and 5 is above.

Water as a Solvent

The polar nature of the water molecule and the ability to form hydrogen bonds are the basis for its unique solvent properties. Polar molecules are

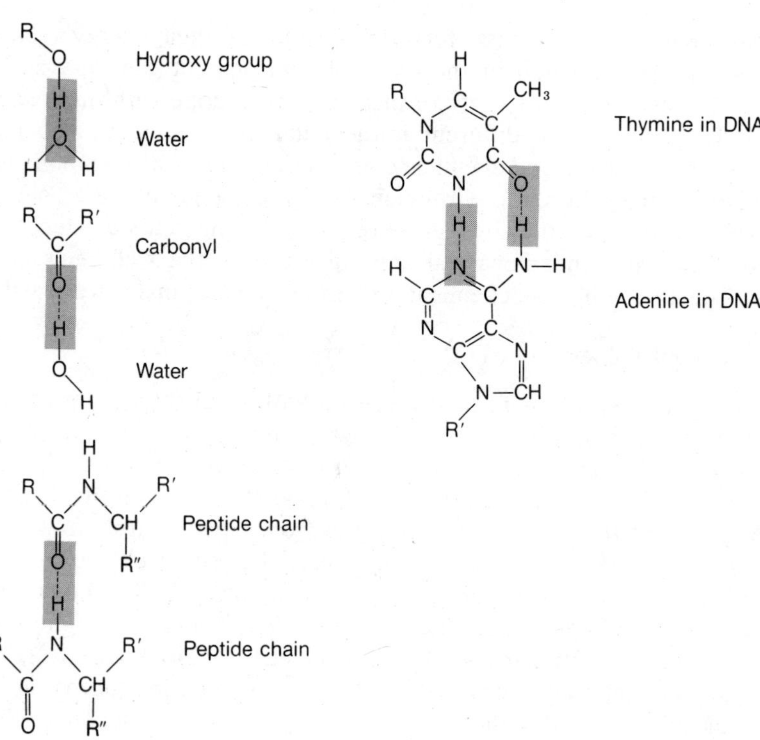

Figure 1.4
Representative hydrogen bonds of importance in biological systems.

readily dispersed in water. Salts in which the crystal lattice is held together by the attraction of the positive and negative groups dissolve in water because the electrostatic forces in the crystal can be overcome by the attraction of the charges to the dipole of water. NaCl is an example wherein the electrostatic attraction of the Na^+ and Cl^- is overcome by interaction of Na^+ with the negative charge on the oxygen, and the Cl^- with the positive charge on the protons. A shell of water surrounds the individual ions. The number of weak charge–charge interactions between water and the ions is sufficient to separate the two charged ions.

Many organic molecules containing nonionic but weakly polar groups are also soluble in water because of the attraction of the groups to water molecules. Sugars and alcohols are readily soluble in water for this reason. Amphipathic molecules, that is, compounds containing both polar and nonpolar groups, will also disperse in water if the attraction of the polar group for water can overcome possible hydrophobic interactions of the nonpolar portions of the molecules. Very hydrophobic molecules, such as compounds containing long hydrocarbon chains, however, will not readily disperse in water but interact with one another to exclude the polar water molecules.

Electrolytes

Substances that dissociate in water into a cation (positively charged ion) and an anion (negatively charged ion) are classified as electrolytes. The presence of these charged ions, partially prevented from interacting with one another because of the attraction of water molecules to the individual ions, facilitates the conductance of an electrical current through an aqueous solution. Sugars or alcohols, which readily dissolve in water but do not carry a charge or dissociate into species with a charge, are classified as nonelectrolytes.

Salts of the alkali metals (e.g., Li, Na, K), when dissolved in water at low concentrations, dissociate totally; at high concentrations there is an increased potential for interaction of the anion and cation. For biological systems in which the concentrations are low, it is customary to consider that such compounds are totally dissociated. The salts of the organic acids, for example, sodium lactate, also dissociate totally. The anion that is formed reacts to a limited extent with a proton to form the undissociated acid (Figure 1.5). These salts are also electrolytes. It is important to remember that when such salts are dissolved in water the individual ions are present in solution rather than the undissociated salt. If a solution has been prepared with several different salts (e.g., NaCl, K_2SO_4, and Na acetate) the original molecules do not exist as such in the solution, only the ions (e.g., Na^+, K^+, SO_4^{2-}, and acetate$^-$).

(1) CH_3—CHOH—COONa ⟶ Na^+ + CH_3—CHOH—COO^-
Na lactate → Lactate ion

(2) CH_3—CHOH—COO^- + H^+ ⇌ CH_3—CHOH—COOH
Lactate ion → Lactic acid

Figure 1.5
Reactions occurring when sodium lactate is dissolved in water.

Many acids, however, when dissolved in water do not totally dissociate but rather establish an equilibrium between the undissociated compound and two or more ions. An example is lactic acid, an important metabolic intermediate, which partially dissociates into a lactate anion and a H^+ as follows:

$$CH_3\text{—CHOH—COOH} \rightleftharpoons CH_3\text{—CHOH—COO}^- + H^+$$

Because of their partial dissociation, however, such compounds have a lower capacity to carry an electrical charge on a molar basis when compared to a compound that dissociates totally; they are termed weak electrolytes.

Dissociation of Weak Electrolytes

In the partial dissociation of a compound, such as lactic acid, represented by HA, the concentration of the various species where A^- represents the dissociated anion can be determined from the equilibrium equation

$$K'_{\text{eq}} = \frac{[H^+][A^-]}{[HA]}$$

The brackets indicate the concentration of each component in moles per liter.

The activities of each species rather than concentration should be employed in the equilibrium equation, but most of the compounds of concern to biochemists are present in low concentrations, and the value of the activity approaches that for the concentration; the equilibrium constant is indicated as K'_{eq} to indicate that it is an apparent equilibrium constant. The K'_{eq} is a function of the temperature of the system, increasing with increasing temperatures. The degree of dissociation of an electrolyte will depend on the affinity of the anion for a proton; if the weak dipole forces of water interacting with the anion and cation are stronger than the electrostatic forces between the H^+ and anion, there will be a greater degree of dissociation. From the dissociation equation above it is apparent that if the degree of dissociation of a substance is small, K'_{eq} will be a small number, but if the degree of dissociation is large, the number will be large. Obviously, for compounds that dissociate totally, a K'_{eq} cannot be determined because at equilibrium there is no remaining undissociated solute.

Dissociation of Water

Water also dissociates as follows:

$$\text{HOH} \rightleftharpoons H^+ + OH^-$$

The proton that dissociates will interact with the oxygen of another water molecule forming the hydronium ion, H_3O^+. For convenience, the proton will not be presented as H_3O^+, even though this is the chemical species actually present. At 25°C the value of K'_{eq} is very small and is in the range of about 1.8×10^{-16}:

$$K'_{eq} = 1.8 \times 10^{-16} = \frac{[H^+][OH^-]}{[H_2O]}$$

With such a small K'_{eq} there is nearly an insignificant dissociation of water, and the concentration of water, which is 55.5 M, will be essentially unchanged. The equation can be rewritten as follows:

$$K'_{eq} \cdot [H_2O] = [H^+][OH^-]$$

The value of $K'_{eq} \times [55.5]$ equals the product of H^+ and OH^- concentrations and is termed the ion product of water. The value at 25°C is 1×10^{-14}. In pure water the concentration of H^+ equals OH^-, and substituting $[H^+]$ for $[OH^-]$ in the equation above, the $[H^+]$ is 1×10^{-7} M. Similarly, the $[OH^-]$ is also 1×10^{-7} M. The equilibrium among H_2O, H^+, and OH^- exists in dilute solutions containing dissolved substances; if the dissolved material alters either the H^+ or OH^- concentration, such as on addition of an acid or base, a concomitant change in the other ion must occur.

Using the equation for the ion product, the $[H^+]$ or $[OH^-]$ can be calculated if the concentration of one of the ions is known.

The importance of the hydrogen ion in biological systems will become apparent in subsequent chapters. For convenience $[H^+]$ is usually expressed in terms of pH, calculated as follows:

$$pH = \log \frac{1}{[H^+]}$$

In pure water the concentration of hydrogen ion and hydroxy ion are both 1×10^{-7} M, and the pH = 7.0. The OH^- ion concentration can be expressed in a similar fashion as the pOH. For the equation describing the dissociation of water, $1 \times 10^{-14} = [H^+][OH^-]$, and taking the negative logarithm of both sides, the equation becomes 14 = pH + pOH. Table 1.2 presents the relationship between pH and H^+ concentration.

TABLE 1.2 Relationships between $[H^+]$ and pH and $[OH^-]$ and pOH

$[H^+]$ (M)	*pH*	*$[OH^-]$ (M)*	*pOH*
1.0	0	1×10^{-14}	14
0.1 (1×10^{-1})	1	1×10^{-13}	13
1×10^{-2}	2	1×10^{-12}	12
1×10^{-3}	3	1×10^{-11}	11
1×10^{-4}	4	1×10^{-10}	10
1×10^{-5}	5	1×10^{-9}	9
1×10^{-6}	6	1×10^{-8}	8
1×10^{-7}	7	1×10^{-7}	7
1×10^{-8}	8	1×10^{-6}	6
1×10^{-9}	9	1×10^{-5}	5
1×10^{-10}	10	1×10^{-4}	4
1×10^{-11}	11	1×10^{-3}	3
1×10^{-12}	12	1×10^{-2}	2
1×10^{-13}	13	0.1 (1×10^{-1})	1
1×10^{-14}	14	1.0	0

The pH of different biological fluids is presented in Table 1.3. At pH 7, the approximate pH of body fluids, H^+ ion is 0.000,000,1 M (1×10^{-7} M), whereas the concentrations of other cations are between 0.001 and 0.10 M. An increase in H^+ ion concentration to only 0.000,001 (1×10^{-6}) has a marked effect on cellular activities and is deleterious to the continued existence of life; a detailed discussion of the mechanisms by which the body maintains the intra- and extracellular pHs is presented in Chapter 23.

TABLE 1.3 pH of Some Biological Fluids

Fluid	*pH*
Blood plasma	7.4
Interstitial fluid	7.4
Intracellular fluid	
Cytosol (liver)	6.9
Lysosomal matrix	5.5–6.5
Gastric juice	1.5–3.0
Pancreatic juice	7.8–8.0
Human milk	7.4
Saliva	6.4–7.0
Urine	5.0–8.0

Acids and Bases

The definitions of an acid and a base proposed by Lowry and Brønsted are the most convenient in considering biological systems. An *acid* is a *proton donor* and a *base* is a *proton acceptor*. HCl and H_2SO_4 are strong acids because they dissociate totally and OH^- ion is a base because it will accept a proton shifting the equilibrium,

$$OH^- + H^+ \rightleftharpoons H_2O$$

When a strong acid and OH^- are combined, the H^+ from the acid and OH^- interact, participating in an equilibrium with H_2O; since the ion product for water is so small, a neutralization of the H^+ and OH^- occurs.

In dilute solutions strong acids dissociate totally; the anions formed, as an example Cl^- from HCl, are not classed as bases because they do not associate with protons in solution. When an organic acid, such as lactic acid, is dissolved in water, however, it dissociates only partially, establishing an equilibrium between the acid, an anion, and a proton as follows:

$$\text{Lactic acid} \rightleftharpoons \text{lactate}^- + H^+$$

The acid is considered to be a weak acid and the anion a base because it will accept a proton and reform the acid. The combination of a weak acid and the base that is formed on dissociation is referred to as a conjugate pair; examples are presented in Table 1.4. Ammonium ion (NH_4^+) is an acid because it is capable of dissociating to yield a H^+ and ammonia, NH_3, an uncharged species, which is the conjugate base. H_3PO_4 is an acid and PO_4^{3-} is a base, but $H_2PO_4^-$ and HPO_4^{2-} can be classified as either base or acid, depending on whether the phosphate group is accepting or donating a proton.

The tendency of a conjugate acid to dissociate can be evaluated from the K'_{eq}; as indicated above, the smaller the value of the K'_{eq}, the less the tendency to give up a proton and the weaker the acid, the larger a K'_{eq}, the greater the tendency to dissociate a proton, and the stronger the acid.

A convenient method of stating the K'_{eq} is in the form of pK', which is defined as

$$pK' = \log \frac{1}{K'_{eq}}$$

Note the similarity of this definition with the definition of pH; as with pH and $[H^+]$, the relationship between pK' and K'_{eq} is an inverse one, and the smaller the K'_{eq}, the larger the pK'. K'_{eq} and pK's for representative conjugate acids of importance in biological systems are presented in Table 1.5.

TABLE 1.4 Some Conjugate Acid–Base Pairs of Importance in Biological Systems

Proton Donor		*Proton Acceptor*
CH_3—CHOH—COOH (lactic acid)	$\rightleftharpoons$	$H^+ + CH_3$—CHOH—COO^- (lactate)
CH_3—CO—COOH (pyruvic acid)	$\rightleftharpoons$	$H^+ + CH_3$—CO—COO^- (pyruvate)
HOOC—CH_2—CH_2—COOH (succinic acid)	$\rightleftharpoons$	$2H^+ + {}^-OOC$—CH_2—CH_2—COO^- (succinate)
${}^+H_3NCH_2$—COOH (glycine)	$\rightleftharpoons$	$H^+ + {}^+H_3N$—CH_2—COO^- (glycinate)
H_3PO_4	$\rightleftharpoons$	$H^+ + H_2PO_4^-$
$H_2PO_4^-$	$\rightleftharpoons$	$H^+ + HPO_4^{2-}$
HPO_4^{2-}	$\rightleftharpoons$	$H^+ + PO_4^{3-}$
Glucose 6-PO_3H^-	$\rightleftharpoons$	H^+ + glucose 6-PO_3^{2-}
H_2CO_3	$\rightleftharpoons$	$H^+ + HCO_3^-$
NH_4^+	$\rightleftharpoons$	$H^+ + NH_3$
H_2O	$\rightleftharpoons$	$H^+ + OH^-$

TABLE 1.5 Apparent Dissociation Constant and p*K*′ of Some Compounds of Importance in Biochemistry

Compound		$K'_{eq}M$	p*K*′
Acetic acid	$(CH_3—COOH)$	1.74×10^{-5}	4.76
Alanine	$(CH_3—CH(NH_3^+)—COOH)$	4.57×10^{-3}	2.34 (COOH)
		2.04×10^{-10}	9.69 (NH_3^+)
Citric acid	$(HOOC—CH_2—COH(COOH)—CH_2—COOH)$	8.12×10^{-4}	3.09
		1.77×10^{-5}	4.75
		3.89×10^{-6}	5.41
Glutamic acid	$(HOOC—CH_2—CH_2—CH(NH_3^+)—COOH)$	6.45×10^{-3}	2.19 (COOH)
		5.62×10^{-5}	4.25 (COOH)
		2.14×10^{-10}	9.67 (NH_3^+)
Glycine	$(CH_2(NH_3^+)—COOH)$	4.57×10^{-3}	2.34 (COOH)
		2.51×10^{-10}	9.60 (NH_3^+)
Lactic acid	$(CH_3—CHOH—COOH)$	1.38×10^{-4}	3.86
Pyruvic acid	$(CH_3—CO—COOH)$	3.16×10^{-3}	2.50
Succinic acid	$(HOOC—CH_2—CH_2—COOH)$	6.46×10^{-5}	4.19
		3.31×10^{-6}	5.48
Glucose 6-PO_3H^-		7.76×10^{-7}	6.11
H_3PO_4		1×10^{-2}	2.0
$H_2PO_4^-$		2.0×10^{-7}	6.7
HPO_4^{2-}		3.4×10^{-13}	12.5
H_2CO_3		1.70×10^{-4}	3.77
NH_4^+		5.62×10^{-10}	9.25
H_2O		1×10^{-14}	14.0

A special case of a weak acid of importance to mammalian cells is that of carbonic acid. Carbon dioxide, when dissolved in water, is involved in the following equilibrium reactions:

$$CO_2 + H_2O \rightleftharpoons H_2CO_3 \rightleftharpoons H^+ + HCO_3^-$$

H_2CO_3 is a relatively strong acid with a pK' of 3.77. It is, however, in constant equilibrium with physically dissolved CO_2. In an aqueous system in contact with an air phase, dissolved CO_2 is also in equilibrium with CO_2 in the air phase. A change in any of the components in the aqueous phase will cause a shift in both equilibria, for example, increasing CO_2 causes an increase in H_2CO_3, which shifts the equilibrium of the dissociation reaction increasing the H^+. Thus CO_2 is considered a part of the conjugate acid and is in the acid component of the equilibrium equation as follows:

$$K'_{eq} = \frac{[H^+][HCO_3^-]}{[H_2CO_3 + CO_2]}$$

Including the CO_2 in the denominator lowers the K'_{eq} and the value of pK' is 6.1. The actual amount of undissociated H_2CO_3 is less than 1/200 of the CO_2 content and is normally neglected in calculations. It is common practice to refer to the dissolved CO_2 as the conjugate acid.

The equation is thus written as follows:

$$K'_{eq} = \frac{[H^+][HCO_3^-]}{[CO_2]}$$

Water is considered a very weak acid, with pK' of 14 at 25°C.

Henderson–Hasselbalch Equation

Changing the concentration of any one component in the equilibrium reaction necessitates a concomitant change in every component. An increase in $[H^+]$ will decrease the concentration of conjugate base with an equivalent increase in the conjugate acid. This relationship is conveniently expressed by rearranging the equilibrium equation and solving for H^+, as shown for the following dissociation:

$$\text{Conjugate acid} \rightleftharpoons \text{conjugate base} + H^+$$

$$K'_{eq} = \frac{[H^+]\,[\text{conjugate base}]}{[\text{conjugate acid}]}$$

Rearranging the equation by dividing through by $[H^+]$ and K'_{eq} leads to

$$\frac{1}{[H^+]} = \frac{1}{K'_{eq}} \cdot \frac{[\text{conjugate base}]}{[\text{conjugate acid}]}$$

Taking the logarithm of both sides gives

$$\text{Log}\,\frac{1}{[H^+]} = \log \frac{1}{K'_{eq}} + \log \frac{[\text{conjugate base}]}{[\text{conjugate acid}]}$$

Since $\text{pH} = \log 1/[\text{H}]^+$ and $\text{p}K' = \log 1/K'_{eq}$, the equation becomes

$$\text{pH} = \text{p}K' + \log \frac{[\text{conjugate base}]}{[\text{conjugate acid}]}$$

This equation, developed by Henderson and Hasselbalch, is a convenient way of viewing the relationship between the pH of a solution and the relative amounts of base and acid present. Figure 1.6 is a plot of the ratios of conjugate base to conjugate acid on a logarithmic scale against the pH for several weak acids. In all cases when the ratio is 1 : 1, the pH equals

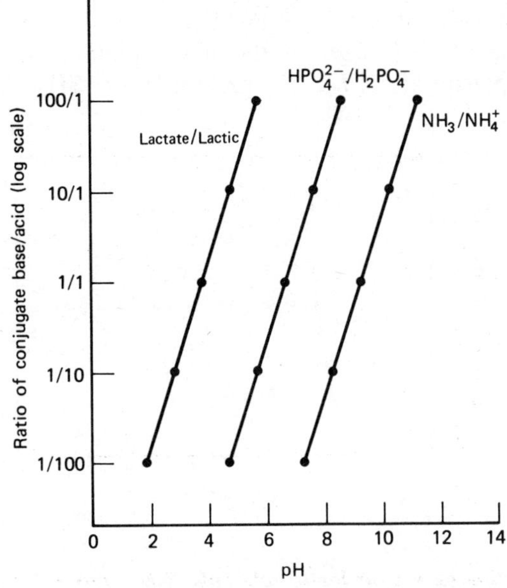

Figure 1.6
Ratio of conjugate base/acid as a function of the pH.
When the ratio of base/acid is 1, the pH equals the pK′ of the weak acid.

the pK' of the acid because log 1 = 0 and pH = pK'. If the pH is one unit less than the pK', the [base]/[acid] ratio = 1/10, and if the pH is one unit above the pK', the [base]/[acid] ratio = 10/1.

Buffers and Buffering

When NaOH is added to a solution of a weak acid, the ratio of [conjugate base]/[conjugate acid] will change because the OH^- will be neutralized by the existing H^+ to form H_2O. Decreasing the H^+ will cause a further dissociation of the weak acid to comply with requirements of its equilibrium reaction. The amount of weak acid dissociated will be nearly equal (usually considered to be equal) to the OH^- added. Thus the decrease in the amount of conjugate acid is equal to the amount of conjugate base that is formed. Titration curves of several weak acids are presented in Figure 1.7. When 0.5 equiv OH^- are added, 50% of the weak acid is dissociated and the [acid]/[base] ratio is 1.0, and the pH at this point is equal to the pK' of the acid. The shapes of the individual titration curves are similar but displaced due to the differences in pKs. As OH^- ion is added, initially there is a rather steep rise in the pH, but between 0.1 and 0.9 equiv OH^-, the pH change is only ~2. Thus a large amount of OH^- is added with a relatively small change in pH. This is called buffering, and defined as the ability of a solution to resist a change in pH when acid or base is added.

The best buffering range for a conjugate pair is in the pH range near the pK' of the weak acid. Starting from a pH one unit below to a pH one unit above the pK', ~82% of a weak acid in solution will dissociate, and therefore an amount of base equivalent to about 82% of the original acid can be neutralized with a change in pH of 2. The maximum buffering range for a conjugate pair is considered to be between 1 pH unit above and below the pK'. A weak acid such as lactic acid with pK' = 3.86 is an effective buffer in the range of pH 3–5 but has little buffering ability at a pH of 7.0. The $HPO_4^{2-}/H_2PO_4^-$ pair with pK' = 6.7, however, is an effective buffer at this pH. Thus at the pH of the cell's cytosol (about 7.0), the lactate–lactic acid pair is not an effective buffer but the phosphate system is.

The buffering capacity also depends on the concentrations of the acid and base pair. The higher the concentration of conjugate base, the more added H^+ with which it can react, and the more conjugate acid the more added OH^- can be neutralized by the dissociation of the acid. A case in point is blood plasma at pH 7.4. The pK' for $HPO_4^{2-}/H_2PO_4^-$ of 6.7 would

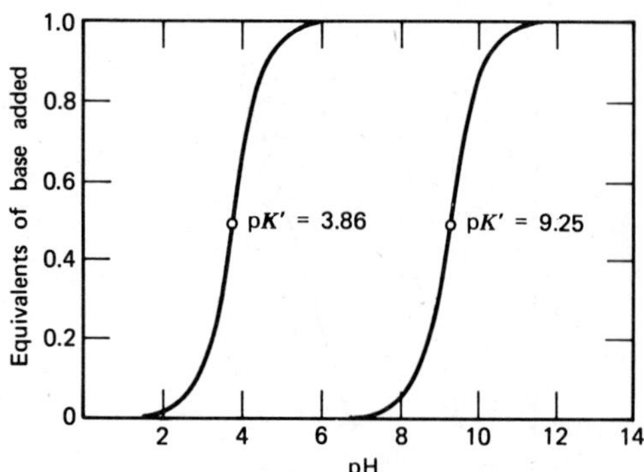

Figure 1.7
Acid–base titration curves for lactic acid (pK' 3.86) and NH_4^+ (pK' 9.25).
At the value of pH equal to the respective pK's, there will be an equal amount of the acid and base for each conjugate pair.

suggest that this conjugate pair would be an effective buffer; the concentration of the phosphate pair, however, is low compared to the HCO_3^-/CO_2 system with a pK' of 6.1, which is present at a 20-fold higher concentration and accounts for most of the buffering capacity. In considering the buffering capacity both the pK' and the concentration of the conjugate pair must be taken into account. Most organic acids are relatively unimportant as buffers in cellular fluids because their pKs are more than several pH units from the pH of the cell, and their concentrations are too low in comparison to such buffers as the various phosphate pairs and the HCO_3^-/CO_2 system.

The importance of pH and buffers in biochemistry and clinical medicine will become apparent, particularly when reading Chapters 2, 4, and 23. Figure 1.8 presents some typical problems using the Henderson–Hasselbalch equation, and Clin. Corr. 1.1 is a representative problem encountered in clinical practice.

(1) Calculate the ratio of $HPO_4^{2-}/H_2PO_4^-$ (pK = 6.7) at pH 5.7, 6.7, and 8.7.

Solution:

$$\text{pH} = \text{p}K + \log [HPO_4^{2-}]/[H_2PO_4^-]$$

$$5.7 = 6.7 + \log \text{ of ratio; rearranging}$$

$$5.7 - 6.7 = -1 = \log \text{ of ratio}$$

The antilog of −1 = 0.1 or 1/10. Thus, $HPO_4^{2-}/H_2PO_4^-$ = 1/10. Using the same procedure, the ratio at pH 6.7 = 1/1 and at pH 8.7 = 100/1.

(2) If the pH of blood is 7.1 and the HCO_3^- concentration is 8 mM, what is the concentration of CO_2 in blood (pK' for HCO_3^-/CO_2 = 6.1)?

Solution:

$$\text{pH} = \text{p}K + \log [HCO_3^-]/[CO_2]$$

$$7.1 - 6.1 + \log 8 \text{ mM}/[CO_2]\text{; rearranging}$$

$$7.1 \quad 6.1 = 1 = \log 8 \text{ mM}/[CO_2].$$

The antilog of 1 = 10. Thus, 10 = 8 mM/$[CO_2]$, or $[CO_2]$ = 8 mM/10 = 0.8 mM.

(3) At a normal blood pH of 7.4, the sum of $[HCO_3^-] + [CO_2]$ = 25.2 mM. What is the concentration of HCO_3^- and CO_2 (pK' for HCO_3^-/CO_2 = 6.1)?

Solution:

$$\text{pH} = \text{p}K + \log [HCO_3^-]/[CO_2]$$

$$7.4 = 6.1 + \log [HCO_3^-]/[CO_2]\text{; rearranging}$$

$$7.4 - 6.1 = 1.3 = \log [HCO_3^-]/[CO_2].$$

The antilog of 1.3 is 20. Thus, $[HCO_3^-]/[CO_2]$ = 20. Given, $[HCO_3^-] + [CO_2]$ = 25.2, solve these two equations for $[CO_2]$ by rearranging the first equation:

$$[HCO_3^-] = 20\,[CO_2]$$

Substituting in the second equation,

$$20\,[CO_2] + [CO_2] = 25.2$$

or

$$CO_2 = 1.2 \text{ mM}$$

Then substituting for CO_2, 1.2 + $[HCO_3^-]$ = 25.2, and solving, $[HCO_3^-]$ = 24 mM.

Figure 1.8
Typical problems of pH and buffering.

CLIN. CORR. 1.1 BLOOD BICARBONATE CONCENTRATION IN A METABOLIC ACIDOSIS

Blood buffers in the normal adult control the blood pH at a value of about 7.40; if the pH should drop much below 7.35, the condition is referred to as an acidosis. A blood pH of near 7.0 could lead to serious consequences and possibly death. Thus in acidosis, particularly that caused by a metabolic change, it is important to monitor the acid–base parameters of the person's blood. The values of interest to the clinician include the pH and HCO_3^- and CO_2 concentrations. The normal values for these are pH = 7.40, $[HCO_3^-]$ = 24.0 mM, and $[CO_2]$ = 1.20 mM.

The blood values of a patient with a metabolic acidosis were pH = 7.03, and $[CO_2]$ = 1.10 mM. What is the patient's blood HCO_3^- concentration and how much of the normal HCO_3^- concentration has been used in buffering the acid causing the condition?

1. The Henderson–Hasselbalch equation is

$$\text{pH} = \text{p}K' + \log [HCO_3^-]/[CO_2]$$

The pK' for $[HCO_3^-]/[CO_2]$ is 6.10. Substitute given values in the equation.

2. $7.03 = 6.10 + \log [HCO_3^-]/1.10 \text{ mM}$

or

$$7.03 - 6.10 = 0.93 = \log [HCO_3^-]/1.10 \text{ mM}$$

The antilog of 0.93 = 8.5; thus,

$$8.5 = [HCO_3^-]/1.10 \text{ mM}$$

or

$$[HCO_3^-] = 9.4 \text{ mM}$$

3. Since the normal value of $[HCO_3^-]$ is 24 mM, there has been a decrease of 14.6 mmol of HCO_3^- per liter of blood in this patient.

It should be noted that if much more HCO_3^- is lost, a point would be reached when this important buffer would be unavailable to buffer any more acid in the blood and the pH would drop rapidly. In Chapter 24 there is a detailed discussion of the causes and compensations that occur in such conditions.

1.3 ORGANIZATION AND COMPOSITION OF EUCARYOTIC CELLS

In many cases the membranes of eucaryotic cells observed by electron microscopy (Figure 1.9) define specific subcellular organelles, such as the nucleus, mitochondria, or lysosomes. These are self-contained units and can be isolated essentially intact from cells and tissues. Other cellular membranes are part of a tubule-like network throughout the cell, enclosing an interconnecting space or cisternae. Such is the case for the endoplasmic reticulum and the Golgi complex. On disruption of the cell, these membrane systems are disrupted, but the membranes reform into small vesicles encapsulating a portion of the isolation medium.

The semipermeable nature of cellular membranes prevents the ready diffusion of many molecules, particularly electrolytes, from one side to another. Very specific transport mechanisms in some membranes for the translocation of both charged and uncharged molecules allow the membrane to modulate the concentrations of the transported substances in various cellular compartments. In addition, macromolecules such as enzymes and nucleic acids do not move readily through biological membranes. Because of the controls exerted by cellular membranes, the fluid

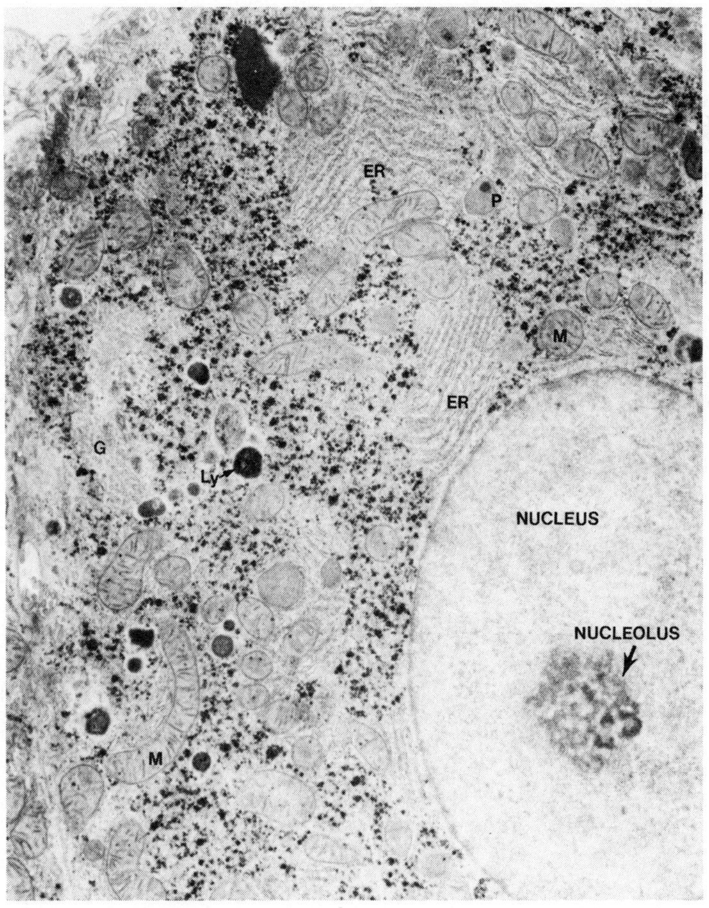

Figure 1.9
An electron micrograph of a rat liver cell labeled to indicate the major structural components of eucaryotic cells.
Note the number and variety of subcellular organelles and the network of interconnecting membranes enclosing channels, that is, cisternae. All eucaryotic cells are not as complex in their appearance, but most contain the major structures shown in the figure.
Photograph reprinted with permission of Dr. K. R. Porter.

matrix of each of the various cellular compartments has a distinctive chemical composition. Partitioning of activities and components in the membrane compartments and organelles has a number of advantages for the economy of the cell, including the sequestering of substrates and cofactors, where they are required, and adjustments of pH and ionic composition for maximum activity of a biological process.

The activities and composition of many of the cellular structures and organelles have been defined by histochemical staining techniques of intact tissues and cells and isolation of the individual organelles and membranes. The various cellular components can be separated by differential centrifugation, following disruption of the plasma membrane. A number of techniques are available to alter the plasma membrane to permit the release of subcellular components, including osmotic shock of cell suspensions, homogenization of tissues, where shearing forces cleave the plasma membrane, and chemical disruption with the use of detergents. In an appropriate isolation medium the cellular membrane systems can be separated from one another by centrifugation because of differences in size and density of the organelles. A general outline for the separation of cell fractions is presented in Figure 1.10. Techniques are available to isolate relatively pure cellular fractions from most mammalian tissues.

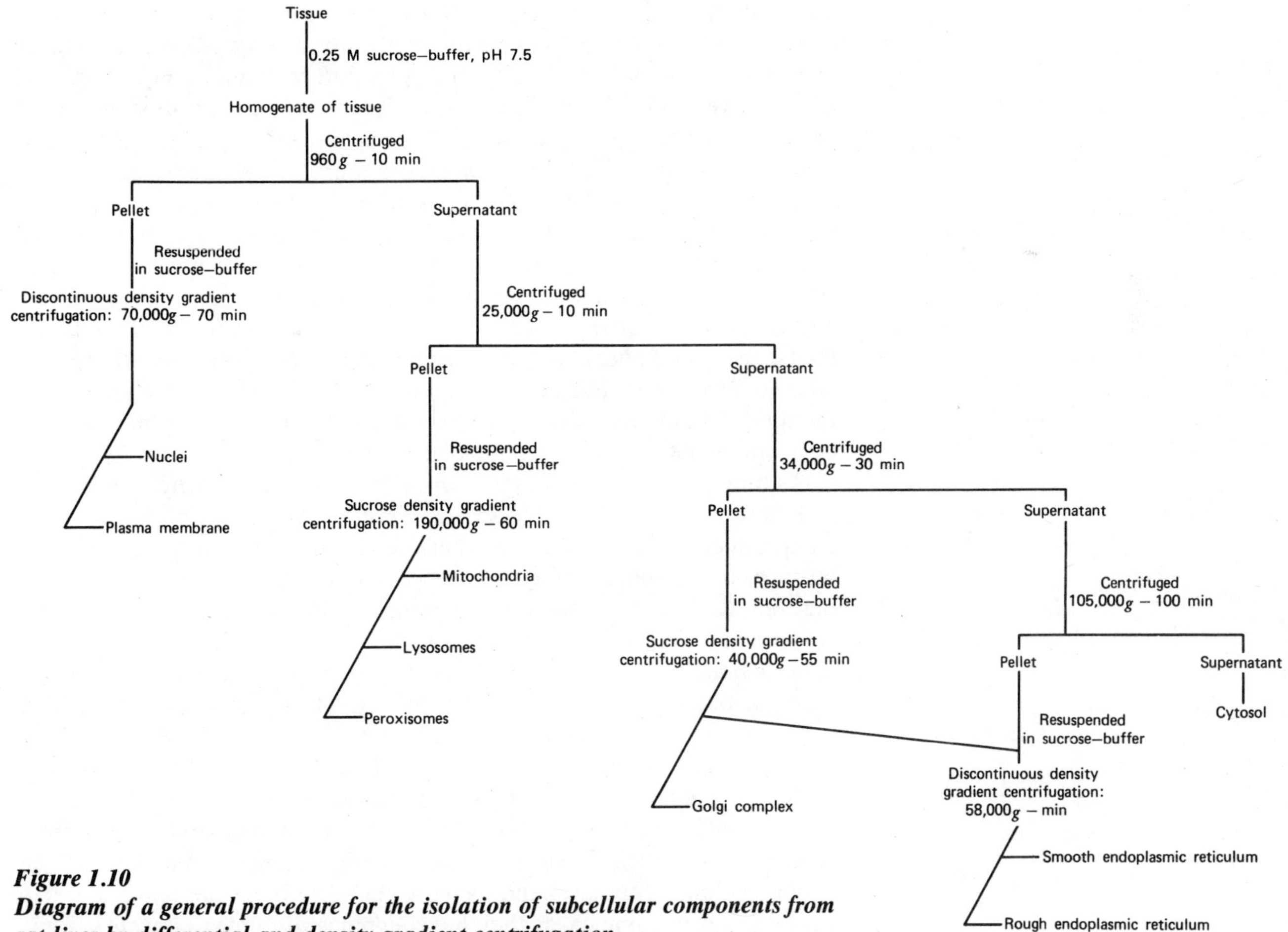

Figure 1.10
Diagram of a general procedure for the isolation of subcellular components from rat liver by differential and density gradient centrifugation.
This is an oversimplified scheme; to simplify the diagram, the composition of the various buffers and of the density gradients are not included. The gravitational fields and times are given only to indicate the range of each required for the isolation. There are steps in the procedure, particularly the washing of pellets, that have also been omitted.
Details of the procedure are found in the article by S. Fleischer and M. Kervina, Subcellular fractionation of rat liver. *Methods Enzymol.* 31:6, 1974.

In many instances the isolated structures and cellular fractions appear to retain the chemical and biochemical characteristics of the structure in situ. But biological membrane systems are very sensitive structures, subject to damage even under very mild conditions and alterations can occur during isolation which can lead to a change in the composition of the structure. The slightest damage to a membrane can alter significantly its permeability properties allowing substances that might normally be excluded to traverse the membrane barrier. In addition, many proteins are only loosely associated with the lipid membrane and easily dissociate when the membrane is damaged.

Not unexpectedly, there are differences in the structure, composition, and activities of cells from different tissues due to the diverse functions of the tissues. The major biochemical activities of the cellular organelles and membrane systems, however, are fairly constant from tissue to tissue. Thus, results of a study of biochemical pathways in liver are often applicable to other tissues. The differences between cell types are usually in the specialized activities that are distinctive to a particular tissue.

Composition of Cells

Each cellular compartment contains an aqueous fluid or matrix in which are dissolved various ions, small molecular weight organic molecules, different proteins and nucleic acids. It has not been possible to determine exactly the ionic composition of the matrix of every cell organelle; however, it is certain that each has a distinctly different ionic composition and pH. Figure 1.11 is a presentation of the ionic composition of blood plasma, interstitial fluid, and overall intracellular fluid. The major extracellular cation is Na^+, with a concentration of ~140 meq/liter; there is little Na^+ in intracellular fluids. K^+ is the major intracellular cation. Mg^{2+} is present both in extra- and intracellular compartments at concentrations much lower than Na^+ and K^+. The major extracellular anions are Cl^- and HCO_3^- with lower amounts of phosphate and sulfate. Most proteins have a negative charge at pH 7.4 (Chapter 2), thus are anions at the pH of tissue fluids. Inorganic phosphate, organic phosphate compounds, and proteins are the major intracellular anions. The total anion concentration equals the total cation concentration in the different fluids, in that there cannot be a significant deviation from electroneutrality.

As indicated previously, the intracellular ionic composition presented in Figure 1.11 is an average of the composition of all the intracellular compartments. Major differences between the matrix compositions of various organelles and the cytosol have been demonstrated. As an example, the free Ca^{2+} concentration in the cytosol is in the range of 0.0001 meq/liter but is much higher in the cisternae of the sarcoplasmic reticulum and in mitochondria. These two membrane systems in muscle actually control the cytosolic concentration of Ca^{2+} by actively sequestering the cellular Ca^{2+}; release of Ca^{2+} into the cytosol triggers a number of biochemical reactions and contraction of muscle fibrils.

The concentration of most small molecular weight organic molecules, such as sugars, organic acids, amino acids, and phosphorylated intermediates, all of which serve as substrates for enzymatic reactions in the various cellular compartments, is considered to be in the range of 0.01–1.0 mM, but in some cases they have significantly lower concentrations, depending on the cell and individual organelle. Coenzymes, organic molecules required for the activity of some enzymes, are in the same range of concentration. In contrast to inorganic ions, substrates are present in relatively low concentrations when considering their overall cellular concentrations but can be increased by localization in a specific organelle or cellular microenvironment.

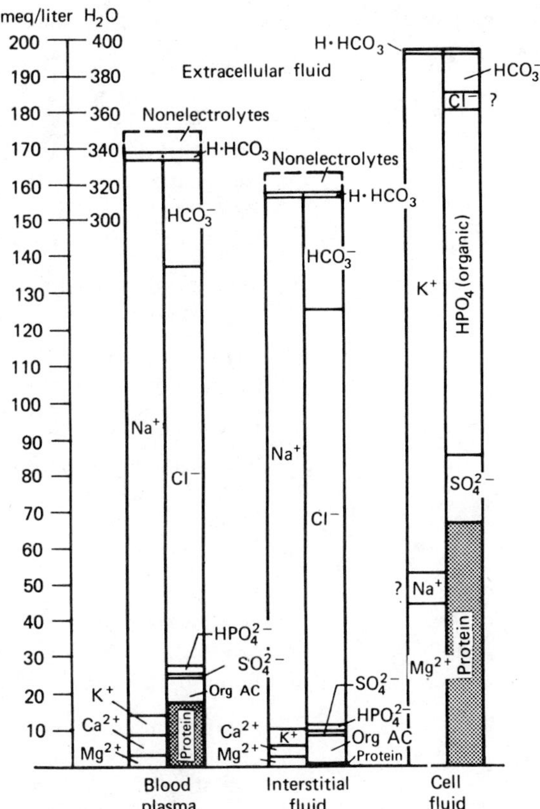

Figure 1.11
Diagram showing major chemical constituents of three fluid compartments.
Height of left half of each column indicates total concentration of cations; that of right half, concentrations of anions. Both are expressed in meq/liter of fluid. Note that chloride and sodium values in cell fluid are questioned. It is probable that, at least in muscle, the cytosol contains some sodium but no chloride.
Modified from Gamble. From Magnus I. Gregersen, in *Medical Physiology*, 11th ed., Philip Bard, ed., Mosby, St. Louis, MO. 1961, p. 307.

It is not very meaningful to determine the molar concentration of individual proteins in cells. In many cases they are localized with specific structures or in combination with other proteins to create a functional unit. It is in a restricted compartment of the cell that the individual proteins carry out their role, whether structural, catalytic, or regulatory.

1.4 FUNCTIONAL ROLE OF SUBCELLULAR ORGANELLES AND MEMBRANES

The subcellular localization of various metabolic pathways will be described throughout this textbook. In some cases an entire pathway is located in a single cellular compartment, but many metabolic sequences are divided between two locations, with the intermediates in the pathway moving or being translocated from one cell compartment to another. In general, the organelles have very specific functions. A specific enzymatic activity of an individual organelle, such as mitochondria or lysosomes, can be used as an identifying characteristic during isolation.

The following describes briefly some of the major roles of the eucaryotic cell structures indicated in Figure 1.9; the descriptions are not inclusive but rather are meant to indicate the level of complexity and

TABLE 1.6 Summary of Major Eucaryotic Cell Compartments and Their Functions

Compartment	*Major Functions*
Plasma membrane	Transport of ions and molecules Recognition Receptors for small and large molecules Cell morphology and movement
Nucleus	DNA synthesis and repair RNA synthesis
Nucleolus	RNA processing and ribosome synthesis
Endoplasmic reticulum	Membrane synthesis Synthesis of proteins and lipids for cell organelles and for export Lipid synthesis Detoxication reactions
Golgi apparatus	Modification and sorting of proteins for incorporation into organelles and for export Export of proteins
Mitochondria	Energy conservation Cellular respiration Oxidation of carbohydrates and lipids Urea and heme synthesis Control of cytosolic [Ca^{2+}]
Lysosomes	Cellular digestion: hydrolysis of proteins, carbohydrates, lipids, and nucleic acids
Peroxisomes	Oxidative reactions involving molecular O_2 Utilization of H_2O_2
Microtubules and microfilaments	Cell cytoskeleton Cell morphology Cell motility Intracellular movements
Cytosol	Metabolism of carbohydrates, lipids, amino acids, and nucleotides Protein synthesis

organization of the cell. Further details of some cellular components are presented in later chapters during discussions of their function in metabolism. Table 1.6 summarizes the major functions of cellular compartments.

Plasma Membrane

The limiting membrane of every cell has a unique role in the maintenance of a cell's integrity. One surface is in contact with the variable external environment and the other with the relatively constant environment of the cell's cytoplasm. As will be discussed in Chapter 5, the two sides of the plasma membrane, as well as all other cellular membranes, have different chemical compositions and functions. A major role of the plasma membrane is to permit the entrance of some substances but exclude many others. It contains the mechanism for phagocytosis and pinocytosis and the unique chemical structures for cell recognition. The plasma membrane with the cytoskeletal elements is involved in the shape of the cell and in cellular movements. Through this membrane cells communicate; the membrane contains multiple specific receptor sites for chemical signals, such as hormones, released by other cells. The inner surface of the plasma membrane is also the site for attachment of a variety of enzymes involved in different metabolic pathways. Plasma membranes from a variety of cells have been isolated and studied extensively; details of their structure and biochemistry are presented in Chapter 5.

Nucleus and Nucleolus

The early microscopists divided the cell into a nucleus, the largest membrane-bound compartment, and the cytoplasm, the remainder of the cell. The nucleus is surrounded by two membranes, termed the perinuclear envelope, with the outer membrane being continuous with membranes of the endoplasmic reticulum. The nucleus contains a subcompartment, clearly seen in electron micrographs, termed the nucleolus. The vast majority of cellular deoxyribonucleic acid (DNA) is located in the nucleus in the form of a DNA–protein complex termed chromatin. Chromatin is organized into chromosomes. DNA is the repository of the cell's genetic information. The importance of the nucleus in cell division and for controlling the phenotypic expression of genetic information is well established. The biochemical reactions involved in the replication of DNA during mitosis and the repair of DNA following damage (Chapter 17), the transcription of the information stored in DNA into a form that can be translated into proteins of the cell (Chapter 18), are contained in the nucleus. Transcription of DNA involves the synthesis of ribonucleic acid (RNA), which is processed following synthesis into a variety of forms. Part of this processing occurs in the nucleolus, which is very rich in RNA. The nucleus may synthesize some of the proteins required for nuclear function, but this activity is small in comparison to the very active protein synthetic activity of the cytosol and endoplasmic reticulum.

Endoplasmic Reticulum

The cytoplasm of most eucaryotic cells contains a network of interconnecting membranes enclosing channels, that is, cisternae, that thread from the perinuclear envelope of the nucleus to the plasma membrane. This extensive subcellular structure, termed the endoplasmic reticulum, consists of membrane structures with a rough appearance in some areas and smooth in other places. The rough appearance is due to the presence of ribonucleoprotein particles, that is, ribosomes, attached on the cytoplasmic side of the membrane. Smooth endoplasmic reticulum does not contain ribosomal particles. During cell fractionation the endoplasmic reticulum network is disrupted, with the membrane resealing into small vesicles referred to as microsomes, which can be isolated by differential centrifugation. Microsomes per se do not occur in cells.

The major function of the ribosomes on the rough endoplasmic reticulum is the biosynthesis of proteins for export to the outside of the cell and enzymes to be incorporated into cellular organelles such as the lysosomes. The endoplasmic reticulum also contains enzymes involved in the biosynthesis of steroid hormones and a variety of oxidative and transferase reactions required for removal of toxic substances. The endoplasmic reticulum also has a role with the Golgi apparatus in the formation of other cellular organelles such as lysosomes and peroxisomes.

Golgi Apparatus

Only in the last few decades has the biochemical role of the Golgi apparatus been clearly defined. This network of flattened smooth membranes and vesicles is responsible for the secretion to the external environment of a variety of proteins synthesized on the endoplasmic reticulum. Golgi membranes catalyze the transfer of glycosyl groups to proteins, a chemical modification required for transport of proteins across the plasma membrane. In addition, the complex is a major site of new membrane formation. Membrane vesicles are formed from the Golgi apparatus in which various proteins and enzymes are encapsulated which can be secreted from the cell after an appropriate signal. The digestive enzymes synthesized by the pancreas are stored in intracellular vesicles formed by the Golgi apparatus and released by the cell when needed in the digestive process. The role in membrane synthesis also includes the formation of intracellular organelles such as lysosomes and peroxisomes.

Mitochondria

Mitochondria have been studied extensively because of their role in cellular energy metabolism. The studies have been facilitated by the ease with which mitochondria can be isolated in a relatively intact state from tissues. In electron micrographs mitochondria appear as spheres, rods, or filamentous bodies; they are usually about 0.5–1 μm in diameter and up to 7 μm in length. The internal matrix is surrounded by two membranes, distinctively different in appearance and biochemical function. The inner membrane convolutes into the matrix of the mitochondrion to form cristae and contains numerous small spheres attached by stalks on the inner surface. The outer and inner membranes contain distinctly different sets of enzymes, which are used for identification of mitochondria during isolation. The components of the respiratory chain and the mechanism for ATP synthesis are part of the inner membrane, described in detail in Chapter 6. In addition to membrane-bound enzymes, the space between the two membranes and the internal matrix also contain a variety of enzymes. Major metabolic pathways involved in the oxidation of carbohydrates, lipids, and amino acids, and special biosynthetic pathways involving urea and heme synthesis are located in the mitochondrial matrix space. The outer membrane is relatively permeable, but the inner membrane is highly selective and contains a number of transmembrane transport systems. The inner membrane contains a specific transporter to move Ca^{2+} into and out of the matrix of the mitochondria and it is proposed that mitochondria have a role in the maintenance of cytoplasmic Ca^{2+} levels.

Mitochondria also contain a specific DNA, containing genetic information for some of the mitochondrial proteins, and the biochemical equipment for limited protein synthesis. The presence of this biosynthetic capacity indicates the unique role that mitochondria have in their own destiny.

Lysosomes

Digestion of a variety of substances inside the cell occurs in the structures designated as lysosomes. These cellular organelles have a single limiting membrane, capable of maintaining a lower pH in the lysosomal matrix than in the cytosol. Encapsulated in lysosomes is a group of enzymes (hydrolases), which catalyze the hydrolytic cleavage of carbon–oxygen, carbon–nitrogen, carbon–sulfur, and oxygen–phosphorus bonds in proteins, lipids, carbohydrates, and nucleic acids. As in the process of digestion in the lumen of the gastrointestinal system, the enzymes of the lysosome are able to split molecules into simple low molecular weight compounds, which can be utilized by the metabolic pathways of the cell. The enzymes of the lysosome have a common characteristic in that each is most active when the pH of the medium is acidic, i.e. pH 5. The relationship between pH and enzyme activity is discussed in Chapter 4. The pH of the cytosol is close to neutral, pH 7.0, and the lysosomal enzymes have little activity at this pH. Thus for the lysosomal enzymes to carry out the digestion of various substances the intralysosomal pH must be significantly lower than the cytosolic pH. A partial list of the enzymes present in lysosomes is presented in Table 1.7.

The enzyme content of lysosomes of different tissues varies and depends apparently on specific needs of individual tissues to digest different substances. Intact lysosomes isolated from other cellular components do not catalyze the hydrolysis of substrates until the membrane is disrupted, demonstrating that the lysosomal membrane is a barrier to the ready access of cellular components to the interior of the lysosome. The lysosomal membrane can be disrupted by various treatments, leading to a release of the lysosomal enzymes, which can then react with their individual substrates. The activities of the lysosomal enzymes are termed "latent" to indicate the need to disrupt the membrane to determine the activity (Figure 1.12). Disruption of the membrane in situ can lead to cellular

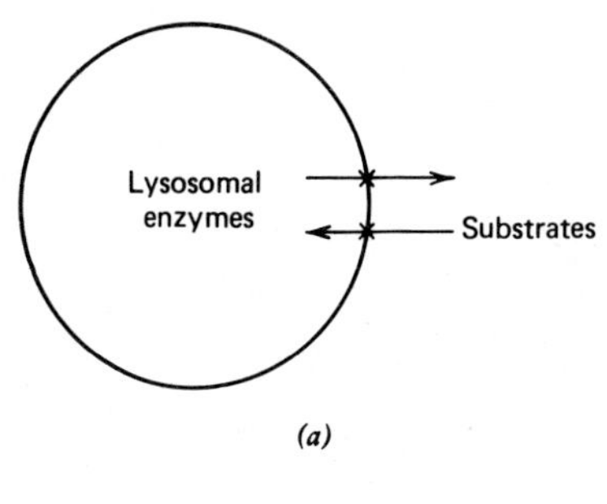

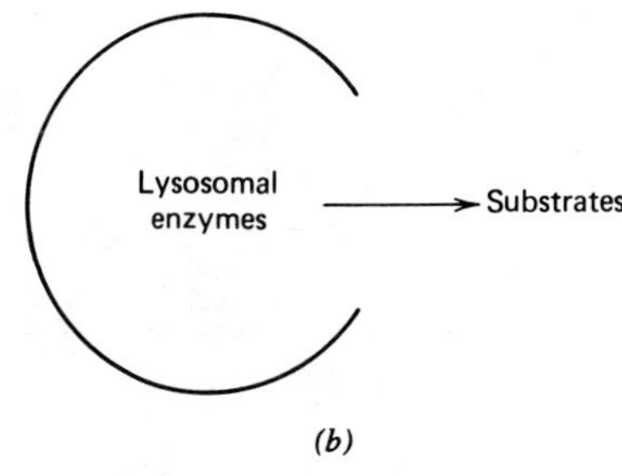

Figure 1.12
Latency of lysosomal enzymes
(a) Intact lysosomes: enzymes inactive with external substrates.
(b) Disrupted lysosomes: enzymes active with external substrates.
When the membrane of the lysosome is intact, substrates that are external will not react with the intralysosomal enzymes. Disruption of the membrane by physical or chemical means leads to a release of the enzymes and the hydrolysis of external substrates.

TABLE 1.7 Representative Lysosomal Enzymes and Their Substrates

Type of Substrate and Enzyme	*Specific Substrate*
Polysaccharide-hydrolyzing enzymes	
α-Glucosidase	Glycogen
α-Fucosidase	Membrane fucose
β-Galactosidase	Galactosides
α-Mannosidase	Mannosides
β-Glucuronidase	Glucuronides
Hyaluronidase	Hyaluronic acid and chondroitin sulfates
Arylsulfatase	Organic sulfates
Lysozyme	Bacterial cell walls
Protein-hydrolyzing enzymes	
Cathepsins	Proteins
Collagenase	Collagen
Elastase	Elastin
Peptidases	Peptides
Nucleic acid-hydrolyzing enzymes	
Ribonuclease	RNA
Deoxyribonuclease	DNA
Lipid-hydrolyzing enzymes	
Esterase	Fatty acid esters
Phospholipase	Phospholipids
Phosphatases	
Phosphatase	Phosphomonoesters
Phosphodiesterase	Phosphodiesters

digestion; a variety of pathological conditions have been attributed to release of lysosomal enzymes, including arthritis, allergic responses, several muscle diseases, and drug-induced tissue destruction (Clin. Corr. 1.2).

Lysosomes are involved in the normal digestion of both intra- and extracellular substances that must be removed by the cell. By the process of endocytosis, external material is taken into the cell and encapsulated in a membrane-bound vesicle. Formed foreign substances such as microorganisms are engulfed by the cell membrane by the process of phagocytosis, and extracellular fluid containing suspended material is taken up by pinocytosis. In both processes the vesicle containing the external material interacts with a lysosome to form a cystolic organelle containing both the material to be digested and the enzymes capable of carrying out the digestion. These vacuoles are identified microscopically by their size and often by the presence of partially formed structures in the process of being digested. Lysosomes in which the enzymes are not as yet involved in the digestive process are termed primary lysosomes, whereas secondary lysosomes are organelles in which digestion of material is under way. The latter are also referred to as digestive vacuoles and will vary in size and appearance. The general sequence of digestion of extracellular substances is represented in Figure 1.13.

Cellular constituents are synthesized and degraded continuously and lysosomes have the responsibility of digesting the cellular debris. The dynamic synthesis and degradation of cellular substances includes proteins and nucleic acids, as well as structures such as mitochondria and the endoplasmic reticulum. During the normal self-digestion process, that is autolysis, cellular substances are encapsulated within a membrane vesicle that fuses with a lysosome to complete the degradation. The overall process is termed autophagy and is also represented in Figure 1.13.

CLIN. CORR. **1.2**
LYSOSOMAL ENZYMES AND GOUT

The catabolism of purines, nitrogen-containing heterocyclic compounds found in nucleic acids, leads to the formation of uric acid which is excreted normally in the urine (see Chapter 13 for details). Gout is an abnormality in which there is an overproduction of purines leading to excessive uric acid production. A consequence of the disease is an increase in uric acid in blood and deposition of urate crystals in the joints. Uric acid is not very soluble. Some of the clinical symptoms of gout can be attributed to the damage done by the urate crystals. The crystals are phagocytosed by cells in the joint and accumulate in digestive vacuoles that contain the lysosomal enzymes. The crystals cause physical damage to the vacuoles, causing a release of the lysosomal enzymes into the cytosol of the cell. The released hydrolytic enzymes cause release of substances from the cell and autolysis. The consequences are clinical manifestations in the joint, including inflammation, pain, swelling, and increased temperature.

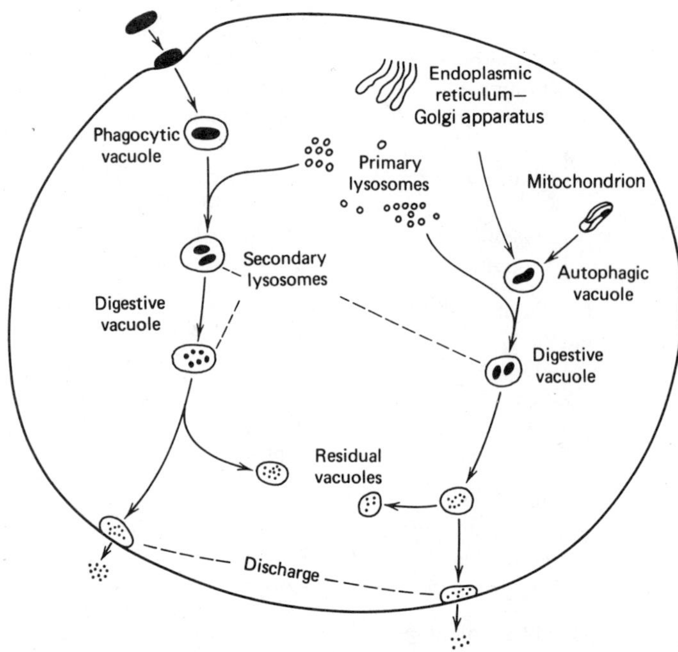

Figure 1.13
Diagrammatic representation of the role of lysosomes in the intracellular digestion of substances internalized by phagocytosis (heterophagy) and of cellular components (autophagy).
In both processes the substances to be digested are enclosed in a membrane vesicle, which is followed by interaction with either a primary or secondary lysosome.

The products of the normal lysosomal digestive process are able to diffuse across the lysosomal membrane and be reutilized by the cell. Indigestible material, however, accumulates in vesicles referred to as residual bodies; the contents of these vesicles are removed from the cell by exocytosis, where the membrane of a vesicle interacts with the plasma membrane. In some cases, however, the residual bodies, which contain a high concentration of lipid, persist for long periods of time. The lipid is oxidized and a pigmented substance, which is chemically heterogeneous and contains polyunsaturated fatty acids and protein, accumulates in the cell. This material, termed lipofuscin, has been called the "age pigment" or the "wear and tear pigment" because it accumulates in cells of older individuals. It occurs in all cells but particularly in neurons and muscle cells and has been implicated in the aging process.

Lysosomal enzymes in some cells also participate in the secretory process by hydrolyzing specific bonds in precursor protein molecules, leading to the formation of active proteins that are secreted from the cell. Under controlled conditions lysosomal enzymes are secreted from the cell for the digestion of extracellular material; an extracellular role for some lysosomal enzymes has been demonstrated in connective tissue and prostate gland and in the process of embryogenesis in which they have a role in programmed cell death.

The absence of specific lysosomal enzymes has been demonstrated in a number of genetic diseases; in these cases there is an accumulation in the cell of specific cellular components that cannot be digested. Lysosomes of the affected cell become enlarged with undigested material, which interferes with normal cellular processes. A discussion of lysosomal storage diseases is presented in Chapter 10.

$$(1)\ 2H_2O_2 \longrightarrow 2H_2O + O_2$$

$$(2)\ RH_2 + H_2O_2 \longrightarrow R + 2H_2O$$

Figure 1.14
Reactions catalyzed by catalase.

Peroxisomes

Most eucaryotic cells of mammalian origin, as well as of protozoa and plants, have a defined cellular organelle which contains several enzymes that either produce or utilize hydrogen peroxide. Frequently referred to as microbodies, the designation peroxisome is now more widely accepted. The organelles are small (0.3–1.5 μm in diameter), spherical or oval in shape, with a granular matrix and in some cases a crystalline inclusion termed a nucleoid. Peroxisomes contain enzymes that oxidize D-amino acids, uric acid, and various 2-hydroxy acids using molecular O_2 with the formation of hydrogen peroxide (H_2O_2). The enzyme catalase, also present in the peroxisome, catalyzes the conversion of H_2O_2 to water and oxygen; it can also catalyze the oxidation by H_2O_2 of various compounds (Figure 1.14). By having both the peroxide-producing and utilizing enzymes in one cellular compartment, the cell protects itself from the toxicity of hydrogen peroxide.

Peroxisomes also contain enzymes involved in lipid metabolism, particularly the oxidation of very long-chain fatty acids and the synthesis of glycerolipids and glycerol ether lipids (plasmalogens) (see Chapter 10). Clinical Correlation 1.3 describes Zellweger syndrome in which there is an absence of peroxisomes.

The peroxisomes of different tissues contain different complements of enzymes, and the peroxisome content of cells can vary depending on different cellular conditions.

CLIN. CORR. 1.3
ZELLWEGER SYNDROME AND THE ABSENCE OF PEROXISOMES

Zellweger syndrome is a rare autosomal recessive condition with a progressive degeneration of the liver, kidney, and brain. The absence of peroxisomes in the liver and kidney and structural changes in mitochondria have been observed in patients. It has been reported that fibroblasts of a person with Zellweger syndrome had a decreased level of several enzymes involved in the synthesis of plasmalogens; these enzymes are found in peroxisomes. In other afflicted patients, it was found that tissue levels of plasmalogens were reduced and there were elevated levels of serum long-chain fatty acids (C_{24} and C_{26}); both observations would be expected if peroxisomes were absent. The ability to determine the level of peroxisomal enzyme levels in fibroblasts of skin and amniotic fluid should permit early diagnosis of this disease.

N. S. Datta, G. N. Wilson, and A. K. Hajra. *New Engl. J. Med.* **311,** 1080 (1984).

Cytoskeleton

Eucaryotic cells contain both microtubules and actin filaments (microfilaments), which are part of the cytoskeletal network. The cytoskeleton has a role in maintenance of cellular morphology, intracellular transport, cell motility, and cell division. The microtubules consist of a polymer of the protein tubulin, which can be rapidly assembled and disassembled depending on the needs of the cell. A very important cellular filament occurs in striated muscle and is responsible for muscular contraction (see Chapter 22). Microtubules and microfilaments do not have a specific role in cellular metabolism but are important in other aspects of the cell.

Cytosol

The least complex in structure, but not in chemistry, is the remaining matrix or cytosol of the cell. It is here that many of the multiplicity of chemical reactions of metabolism occur and where substrates and cofactors of various enzymes interact. Even though there is no apparent structure to the cytoplasm, the high protein content precludes the matrix from being a truly homogeneous mixture of soluble components. Many reactions may be localized in selected areas of the cell, where the conditions of substrate availability are more favorable for the reaction. The actual physicochemical state of the cytosol is poorly understood. A major role of the cytosol is to support the synthesis of proteins catalyzed by the rough endoplasmic reticulum by supplying cofactors and enzymes. In addition, the cytosol contains free ribosomes, often in a polysome form, for synthesis of intracellular proteins.

Studies with isolated cytosol suggest that many reactions are catalyzed by soluble enzymes, but in the intact cell some of these enzymes may be loosely attached to one of the many membrane structures and are released upon cell disruption.

Conclusion

The eucaryotic cell is a complex structure whose purpose is to replicate itself when necessary, maintain an intracellular environment to permit a myriad of complex reactions to occur as efficiently as possible, and to protect itself from the hazards of its surrounding environment. The cells of multicellular organisms also participate in maintaining the well-being of the whole organism by exerting influences on each other to maintain all tissue and cellular activities in balance. Thus, as we dissect the separate chemical components and activities of cells, it is important to keep in mind the concurrent and surrounding activities, constraints, and influences. Only by bringing together all the separate parts, that is, reassembling the puzzle, will we appreciate the wonder of a living cell.

BIBLIOGRAPHY

Water and Electrolytes

Dick, D. A. T. *Cell water*. Washington, D.C.: Butterworths, 1966.

Eisenberg, D., and Kauzmann, W. *The structures and properties of water*. Fairlawn, N.J.: Oxford University Press, 1969.

Morris, J. G. *A biologist's physical chemistry*. Reading, Mass.: Addison-Wesley, 1968.

Stillinger, F. H. Water revisited. *Science* 209:451, 1980.

Cell Structure

Alberts, B., Bray, D., Lewis, J., Raff, M., Roberts, K., and Watson, J. D. *Molecular biology of the cell*. New York: Garland, 1983.

Alterman, P. L., and Katz, D. D. (eds.). *Cell biology*. Bethesda, Md.: Federation of American Societies of Experimental Biology, 1975.

Fawcett, D. W. *The cell: its organelles and inclusions*. Philadelphia: W. B. Saunders, 1966.

Finian, J. B., Coleman, R., and Mitchell, R. H. *Membranes and their cellular functions*, 2nd ed. London: Blackwell, 1978.

Flickinger, C. J., Brown, J. C., Kutchai, H. C., and Ogilvie, J. W. *Medical cell biology*. Philadelphia: W. B. Saunders, 1979.

Hers, H. G., and Van Hoof, F. (eds.). *Lysosomes and storage diseases*. New York: Academic, 1973.

Loewy, A., and Siekevitz, P. *Cell structure and function*, 2nd ed. New York: Holt, Rinehart & Winston, 1970.

Porter, K. R., and Bonneville, M. A. *Fine structure of cells and tissues*. Philadelphia: Lea & Febiger, 1972.

QUESTIONS

J. BAGGOTT AND C. N. ANGSTADT

Question Types are described inside the front cover.

A. Eucaryotic cell — C. Both
B. Procaryotic cell — D. Neither

1. (QT4) DNA is distributed uniformly throughout the cell.
2. (QT4) Contains histones.
3. (QT4) Lacks organized subcellular structures.
4. (QT2) Factors responsible for the polarity of the water molecule include:
   1. differences in electron affinity between hydrogen and oxygen.
   2. the tetrahedral structure of liquid water.
   3. the magnitude of the H—O—H bond angle.
   4. the ability of water to hydrogen bond to various chemical structures.
5. (QT2) Hydrogen bonds form only between electronegative atoms such as oxygen or nitrogen and a hydrogen atom bonded *to:*
   1. sulfur.
   2. carbon.
   3. hydrogen.
   4. an electronegative atom.
6. (QT1) Which of the following is *least* likely to be soluble in water?
   A. Nonpolar compound
   B. Weakly polar compound
   C. Strongly polar compound
   D. Weak electrolyte
   E. Strong electrolyte
7. (QT1) Which of the following is both a Brønsted acid and a Brønsted base in water?
   A. $H_2PO_4^-$
   B. H_2CO_3
   C. NH_3
   D. NH_4^+
   E. Cl^-

Refer to the following information for Questions 8 and 9.

A. Pyruvic acid — $pK' = 2.50$
B. Acetoacetic acid — $pK' = 3.6$
C. Lactic acid — $pK' = 3.86$
D. β-Hydroxybutyric acid — $pK' = 4.7$
E. Propionic acid — $pK' = 4.86$

8. (QT5) Which weak acid will be 91% neutralized at pH 4.86?
9. (QT5) Assuming that the sum of [weak acid] + [conjugate base] is identical for buffer systems based on the acids listed above, which has the greatest buffer capacity at pH 4.86?
10. (QT1) All of the following subcellular organelles can be isolated essentially intact *except:*
   A. nuclei.
   B. mitochondria.
   C. lysosomes.
   D. peroxisomes.
   E. endoplasmic reticulum.
11. (QT2) Biological membranes may:
   1. prevent free diffusion of ionic solutes.
   2. release proteins when damaged.
   3. contain specific systems for the transport of uncharged molecules.
   4. be sites for biochemical reactions.

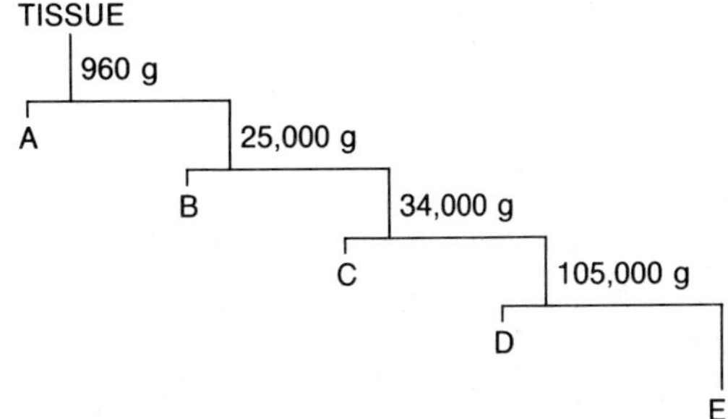

12. (QT1) In separating subcellular components of rat liver by differential centrifugation as shown schematically above:
   A. fraction A contains the Golgi apparatus.
   B. fraction B contains the mitochondria.
   C. fraction C contains the endoplasmic reticulum.
   D. fraction D contains the plasma membrane.
   E. fraction E contains the peroxisomes.
13. (QT1) Analysis of the composition of the major fluid compartments of the body shows that:
   A. the major blood plasma cation is K^+.
   B. the major interstitial fluid cation is K^+.
   C. one of the major intracellular anions is Cl^-.
   D. one of the major intracellular anions is phosphate.
   E. plasma, interstitial fluid, and the cell fluid are all very similar in ionic composition.

A. Mitochondria — C. Both
B. Nucleus — D. Neither

14. (QT4) Contains DNA.
15. (QT4) Enclosed by a single membrane.
16. (QT4) Connected to the plasma membrane by a network of membranous channels.
17. (QT2) Lysosomes may:
   1. combine with phagocytic vacuoles to become digestive vacuoles.
   2. combine with autophagic vacuoles to become digestive vacuoles.
   3. secrete their enzymes from the cell for digestion of extracellular material.
   4. combine with peroxisomes in order to add oxidative capabilities to their catalytic armamentarium.

ANSWERS

1. D The DNA of eucaryotes is confined to the nucleus and mitochondria; the DNA of procaryotes is segregated into discrete masses, the nuclear zone (pp. 2–3).
2. A Histones are unique to eucaryotes; they form complexes with the DNA (p. 4).
3. B Eucaryotic cells typically contain mitochondria, a nucleus, and other organized structures; procaryotic cells do not (p. 3).
4. B 1 and 3 true. Water is a polar molecule because the bonding electrons are attracted more strongly to oxygen than to hydrogen. The bond angle gives rise to asymmetry of the charge distribution; if water were linear, it would not be a dipole (p. 5). 2 and 4 are consequences of water's structure, not factors responsible for it.
5. D Only 4 true. Only hydrogen atoms *bonded to* one of the electro-

negative elements (O, N, F) can form hydrogen bonds (p. 5). A hydrogen participating in hydrogen bonding must have an electronegative element on both sides of it.

6. A In general, compounds that interact with the water dipoles are more soluble than those that do not. Thus ionized compounds and polar compounds tend to be soluble. Nonpolar compounds prefer to interact with one another rather than with polar solvents such as water (p. 6).

7. A $H_2PO_4^-$ can donate a proton to become HPO_4^{2+}. It can also accept a proton to become H_3PO_4. B and D are Brønsted acids; C is a Brønsted base. Cl^- in water is neither (p. 9).

8. C If weak acid is 91% neutralized, 91 parts are present as conjugate base and 9 parts remain as the weak acid. Thus the conjugate base/acid ratio is 10/1. Substituting into the Henderson–Hasselbalch equation, $4.86 = pK + \log 10/1$, and solving for pH gives the answer (p. 11).

9. E The buffer capacity of any system is maximal at pH = p*K* (p. 12). Buffer concentration also affects buffer capacity, but in this case concentrations are equal.

10. E Gentle disruption of cells will not destroy A–D. The tubelike endoplasmic reticulum, however, is disrupted and forms small vesicles. These vesicles, not the original structure from which they were derived, may be isolated (p. 14).

11. E All statements are true. 1, 2, and 3 (p. 14); 4 (p. 4).

12. B Fraction A: nuclei and plasma membrane. Fraction B: mitochondria, lysosomes, peroxisomes. Fraction C: Golgi apparatus. Fraction D: microsomes. Fraction E: cytosol (p. 15, Figure 1.10).

13. D A, B, and E: plasma and interstitial fluid are similar; cell fluid is strikingly different. Na^+ is the major cation of plasma and interstitial fluid. C and D: phosphate and protein are the major intracellular anions; most chloride is extracellular (p. 17, Figure 1.11).

14. C Most of the DNA is in the nucleus, but the mitochondria also contain DNA (pp. 19–20).

15. D Both the mitochondria and the nucleus are surrounded by double membranes (p. 19–20).

16. B This describes only the nucleus (p. 21).

17. A 1, 2, and 3 describe the activities of lysosomes (p. 22). Peroxisomes contain peroxide-producing enzymes and catalase, which destroys peroxide. This arrangement is thought to protect the rest of the cell from the damaging effects of hydrogen peroxide (p. 23).

2

Proteins I: Composition and Structure

RICHARD M. SCHULTZ

2.1 FUNCTIONAL ROLES OF PROTEINS IN HUMANS

Proteins perform a surprising variety of essential functions in mammalian organisms. These functions may be grouped into two classes: dynamic and structural (static). Dynamic functions of proteins include transport, metabolic control, contraction, and catalysis of chemical transformations. In their structural functions, proteins provide the matrix for bone and connective tissue, giving structure and form to the human organism.

One of the important groups of dynamic proteins is the enzymes. Enzymes act to catalyze chemical reactions. Almost all of the thousands of different chemical reactions that occur in living organisms and involve covalent bond formation or cleavage require a specific enzyme catalyst for the reaction to occur at a rate compatible with life. Thus the characteristics and functions of any cell are based on the particular chemistry of the cell, which in turn is generated by the specific enzyme makeup of the cell. Genetic traits are expressed through the synthesis of enzymes, which catalyze the chemical reactions that establish the trait to be expressed.

Another dynamic function for proteins is in transport. Particular examples discussed in greater detail in this text are hemoglobin and myoglobin, which transport oxygen in blood and in muscle, respectively. Transferrin transports iron in blood. Other important proteins act to transport hormones in blood from their site of synthesis to their site of action. Many drugs and toxic compounds are transported bound to proteins.

Proteins can also function in a protective role. For example, the immunoglobulins and interferon are proteins that act against bacterial or viral infection. Fibrin is a protein that is formed where required to stop the loss of blood on injury to the vascular system.

Many hormones are proteins. Protein hormones include insulin, thyrotropin, somatotropin (growth hormone), luteinizing hormone, and follicle stimulating hormone. There are many diverse protein-type hormones that have a low molecular weight (<5,000), referred to as *peptides*. In general, the term protein is used for molecules composed of over 50 component amino acids and the term peptide is used for molecules of less than 50 amino acids. Important peptide hormones include adrenocorticotropin, antidiuretic hormone, glucagon, and calcitonin.

Some proteins have roles in contractile mechanisms. Of particular importance are the proteins myosin and actin, which function in muscle contraction.

Other proteins are active in the control and regulation of gene transcription and translation. These include the histone proteins closely associated with DNA, the repressor proteins that control gene expression, and the proteins that form a part of the ribosomes.

Whereas the above proteins are "dynamic" in their function, other proteins have structural, "brick-and-mortar" roles. This group of proteins includes collagen and elastin, which form the matrix for bone and ligaments and provide structural strength and elasticity to the organs and the vascular system. α-Keratin has an essential structural role in epidermal tissue.

It is obvious that an understanding of both the normal functioning and the pathology of the mammalian organism requires a clear understanding of the structure and properties of the proteins.

2.2 AMINO ACID COMPOSITION OF PROTEINS

Proteins Are Polymers of α-Amino Acids

It is amazing that all the different types of proteins are initially synthesized as polymers of only 20 amino acids, known as the common amino acids. The *common amino acids are defined as those amino acids for which a specific codon exists in the DNA genetic code.* There are 20 amino acids for which DNA codons are known. The process of the reading of the DNA code, resulting in the polymerization of amino acids of a specific sequence into proteins based on the DNA code, is the basis of a later chapter (Chapter 19). In this chapter we will discuss only the protein product of this genetically controlled synthetic process (Figure 2.1).

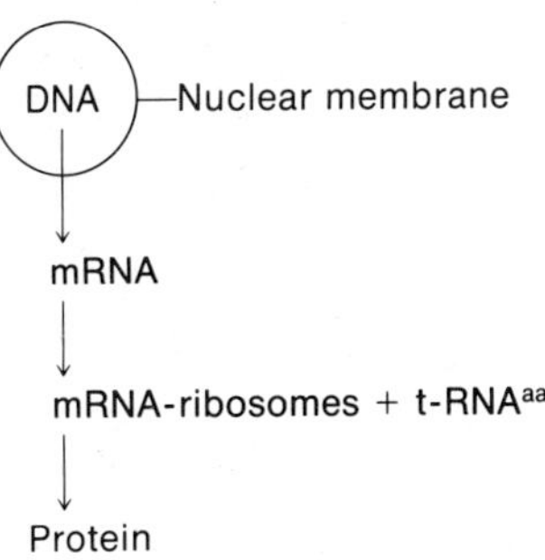

Figure 2.1
Genetic information is transmitted from a DNA sequence through RNA into the amino acid sequence of a protein.
DNA, deoxyribonucleic acid; mRNA, messenger ribonucleic acid; tRNA^{aa}, aminoacyl transfer-ribonucleic acid (see Chapters 17–20).

In addition to the common amino acids, derived amino acids are found in proteins. *Derived amino acids in proteins are formed from one of the common amino acids,* usually by an enzyme-facilitated reaction, *after the common amino acid has been incorporated into a protein structure.* An example of a derived amino acid is cystine (see the section entitled "Derived Amino Acid: Cystine" below). Other derived amino acids are desmosine and isodesmosine found in the protein elastin (Chapter 3), hydroxyproline and hydroxylysine found in collagen (Chapter 3), and γ-carboxyglutamate found in prothrombin.

General Structure of the Common Amino Acids

The common amino acids have the general structure depicted in Figure 2.2. They contain in common a central (*alpha*) carbon atom to which a carboxylic acid group, an amino group, and a hydrogen atom are covalently bonded. In addition, the *alpha* (α)-carbon atom binds a side chain group, designated R, that is different for each of the 20 common amino acids.

In the structure depicted in Figure 2.2 the ionized form for a common amino acid that is present in solution at pH 7 is shown. The α-amine is protonated and in its ammonium ion form; the carboxylic acid group is in its unprotonated or carboxylate form.

$$\overset{+}{N}H_3-\underset{\underset{R}{|}}{\overset{\overset{COO^-}{|}}{C}}-H$$

Figure 2.2
General structure of the common amino acids.

Side Chain Structures of the Common Amino Acids

The structures for the common amino acids are shown in Figure 2.3. In the category of alkyl amino acids are glycine, alanine, valine, leucine, and isoleucine. Glycine has the simplest structure, with R = H. Alanine contains a methyl (CH_3—) R group and valine an isopropyl R group (Figure 2.4). The leucine and isoleucine R groups are butyl alkyl chains that are structural isomers of each other. In leucine the branching methyl group in the isobutyl side chain occurs on the *gamma* (γ)-carbon of the amino acid. In isoleucine the butyl side chain is branched at the β-carbon.

In the category of aromatic amino acids are phenylalanine, tyrosine, and tryptophan. In phenylalanine the R group contains a benzene ring, tyrosine contains a phenol group, and the tryptophan R group contains a heterocyclic structure known as an indole. In the three aromatic amino acids the aromatic moiety is attached to the α-carbon through a methylene (—CH_2—) carbon (Figure 2.3).

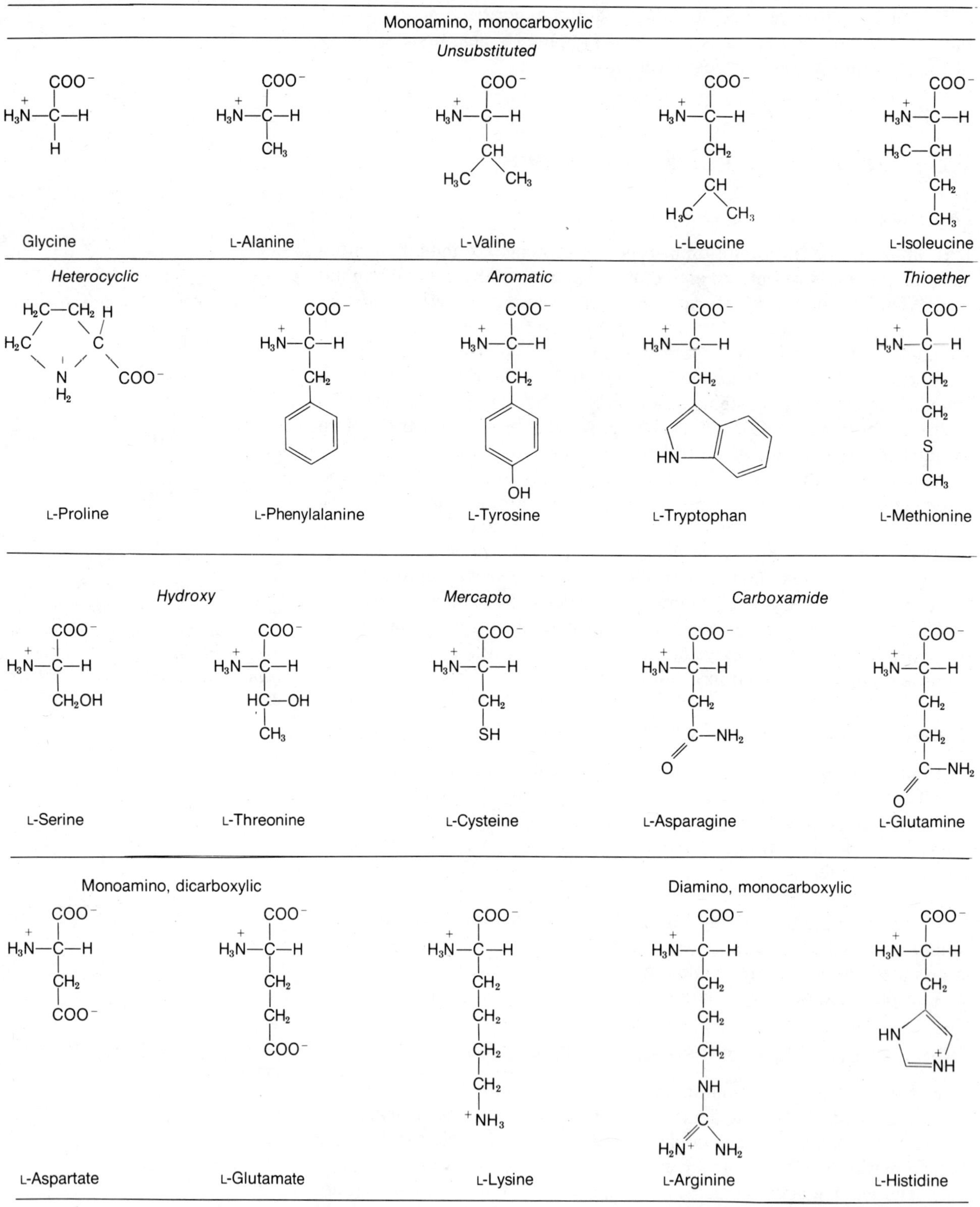

Monoamino, monocarboxylic

Unsubstituted

Glycine | L-Alanine | L-Valine | L-Leucine | L-Isoleucine

Heterocyclic

L-Proline

Aromatic

L-Phenylalanine | L-Tyrosine | L-Tryptophan

Thioether

L-Methionine

Hydroxy

L-Serine | L-Threonine

Mercapto

L-Cysteine

Carboxamide

L-Asparagine | L-Glutamine

Monoamino, dicarboxylic

L-Aspartate | L-Glutamate

Diamino, monocarboxylic

L-Lysine | L-Arginine | L-Histidine

Figure 2.3
Structures of the common amino acids.

The two sulfur-containing common amino acids are cysteine and methionine. In cysteine the R group is thiolmethyl ($HSCH_2$—). In methionine the side chain is a methyl ethyl thiol ether ($CH_3SCH_2CH_2$—).

There are two hydroxy (alcohol)-containing common amino acids, serine and threonine. In serine the side chain is a hydroxymethyl moiety ($HOCH_2$—). In threonine an ethanol structure is connected to the α-carbon of the amino acid through the 1 position of the ethanol, resulting in a secondary alcohol structure for R (CH_3CHOH—).

In proline the side chain group, R, is unique in that it incorporates the α-amino group in the side chain. Thus proline is more accurately classified as an α-imino acid, since its α-amine is a secondary amine, rather than a primary amine. In subsequent sections we discuss the ramifications of the incorporation of the α-amino nitrogen into a five-membered ring that constrains the rotational freedom around the N—C bond in proline to a specific rotational angle. This has important structural consequences for the protein structures in which proline participates.

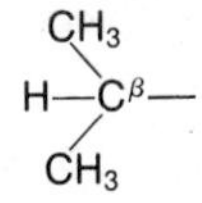

Isopropyl R group of valine

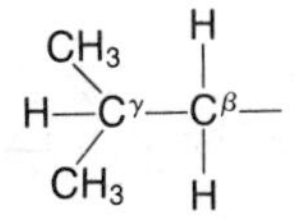

Isobutyl R group of leucine

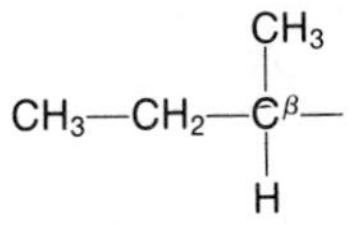

Isobutyl R group of isoleucine

Figure 2.4
Alkyl side chain groups of valine, leucine, and isoleucine.

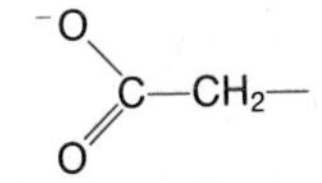

The categories of amino acids discussed so far contain uncharged side chain R groups at physiological pH. The next category of amino acids, the dicarboxylic monoamino acids, contain a negatively charged carboxylate R group at pH 7. In aspartate the side chain carboxylic acid group is separated by a single methylene carbon (—CH_2—) from the α-carbon (Figure 2.5). This differs for glutamic acid (Figure 2.5), in which the γ-carboxylic acid group is separated by two methylene (—CH_2—CH_2—) carbon atoms from the α-carbon of the generalized structure (Figure 2.2).

Dibasic monocarboxylic acid structures are present in lysine, arginine, and histidine (Figure 2.6). In these structures, the R group contains a nitrogen or nitrogens that may be protonated to form a positively charged side chain. In lysine the side chain is simply *N*-butyl amine. In arginine the side chain group contains a guanidino group separated from the α-carbon by three methylene carbons. Both the guanidinium group of arginine and the amino group of lysine are predominantly protonated at physiological pH (pH ~7) and are in their charged forms. In histidine the side chain R group contains a five-membered heterocyclic structure known as an imidazole group. The pK_a of the imidazole group of histidine is approximately 6.0 in water, and physiological solutions will contain relatively high concentrations of both the basic (imidazole) and acidic (imidazolium) forms of the histidine side chain at physiological pH (see Section 2.3).

Aspartate R group

Glutamate R group

Figure 2.5
Side chain groups of aspartate and glutamate.

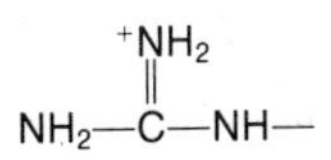

The last two common amino acids are glutamine and asparagine. These two amino acids contain an amide moiety in their side chain R group. Glutamine and asparagine may be considered structural analogs of glutamic acid and aspartic acid, respectively, with their side chain carboxylic acid groups amidated. However, DNA codons exist for glutamine and asparagine separate from those for glutamic acid and aspartic acid. The amide side chains of glutamine and asparagine cannot be protonated and are uncharged in the range of physiological pH.

Polar and Apolar Properties of the Amino Acid R Groups

It is important to have an appreciation for the relative hydrophobicity of the amino acid side chains in order to understand the role played by the different amino acids in protein structure and function. The more hydrophobic (nonpolar) amino acids are phenylalanine, leucine, isoleucine, valine, alanine, methionine, tryptophan, and cysteine. A majority of the side chain R groups from the more hydrophobic amino acids are folded into the interior of the protein molecule away from the water of solvation on the surface. However, a significant number of nonpolar residues will

Guanidinium group (charged form) of arginine

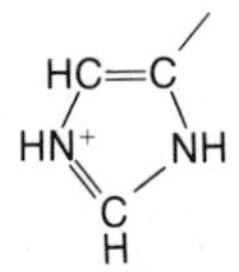

Imidazolium group of histidine

Figure 2.6
Guanidinium and imidazolium groups of arginine and histidine.

TABLE 2.1 Some Examples of Biologically Active Peptides

Amino Acid Sequence	*Name*	*Function*
1 3 pyroGlu-His-Pro(NH_2)[a]	Thyrotropin releasing factor	Secreted by hypothalamus and causes pituitary gland to release thyrotropic hormone
1 9 H-Cys-Tyr-Phe-Gln-Asn-Cys-Pro-Arg-Gly(NH_2)[b] S ——— S	Vasopressin (antidiuretic hormone)	Secreted by pituitary gland and causes kidney to retain water from urine
1 5 H-Tyr-Gly-Gly-Phe-Met-OH	Methionine enkephalin	Opiatelike peptide found in brain that inhibits sense of pain
1 10 pyroGlu-Gly-Pro-Trp-Leu-Glu-Glu-Glu-Glu-Glu- 11 17 Ala-Tyr-Gly-Trp-Met-Asp-Phe(NH_2)[a,c] | SO_3	Little gastrin (human)	Hormone secreted by mucosal cells in stomach and causes parietal cells of stomach to secrete acid
1 10 H-His-Ser-Gln-Gly-Thr-Phe-Thr-Ser-Asp-Tyr- 11 20 Ser-Lys-Tyr-Leu-Asp-Ser-Arg-Arg-Ala-Gln- 21 29 Asp-Phe-Val-Gln-Trp-Leu-Met-Asn-Thr-OH	Glucagon (bovine)	Pancreatic hormone involved in regulating glucose metabolism
1 8 H-Asp-Arg-Val-Tyr-Ile-His-Pro-Phe-OH	Angiotensin II (horse)	Pressor or hypertensive peptide; also stimulates release of aldosterone from adrenal gland
1 9 H-Arg-Pro-Pro-Gly-Phe-Ser-Pro-Phe-Arg-OH	Plasma bradykinin (bovine)	Vasodilator peptide
1 10 H-Arg-Pro-Lys-Pro-Gln-Phe-Phe-Gly-Leu-Met(NH_2)	Substance P	Neurotransmitter

[a] The NH_2-terminal Glu is in the pyro form in which its γ-COOH is covalently joined to its α-NH_2 via amide linkage; the COOH-terminal amino acid is amidated and thus also not free.
[b] Cysteine-1 and cysteine-6 are joined to form a cyclo-hexa structure within the nonapeptide.
[c] The Tyr-12 is sulfonated on its phenolic side chain OH.

remain on the surface exposed to the water solvent. These hydrophobic side chain groups on the surface appear generally dispersed and isolated among polar and ionized surface residues. The clustering of two or more nonpolar side chain groups may occur in small regions of the protein surface and are usually associated with a function of the protein such as providing a hydrophobic binding site for substrate or ligand molecules to the protein. Amino acids of intermediate polarity include glycine, tyrosine with its polar —OH group, threonine, serine, and proline. These amino acid side chains are found both in the interior and on the solvent protein interface in significant proportions. In contrast, the polar amino acids glutamine and asparagine and the amino acids containing charged R groups at pH 7 (lysine, arginine, histidine, glutamate, and aspartate) are predominantly found on the surface in globular proteins where the charge is stabilized by the water solvent. The rare positioning of a charged side chain into the interior of a globular protein is usually correlated with an essential structural or functional role for the "buried" charged side chain group within the nonpolar interior of the protein.

Amino Acids Are Polymerized into Peptides and Proteins

The polymerization of the 20 common amino acids into polypeptide chains within cells is catalyzed by enzymes and requires RNA and ribosomes to occur (Chapter 19). Chemically, the polymerization of amino acids into protein is a dehydration reaction. The chemical rationale of the reaction is shown in Figure 2.7. The figure shows that the α-carboxyl group of an amino acid with side chain R_1 may be covalently joined to the α-NH_2 group of the amino acid with side chain R_2 by the elimination of a molecule of water to form a type of amide bond known as the peptide bond. The dipeptide (two amino acid residues joined by a single peptide bond) can then form a second peptide bond through its terminal carboxylic acid group to the α-amine of a third amino acid (R_3), generating a tripeptide (Figure 2.7). Repetition of this stepwise dehydration process will generate a polypeptide or protein of specific amino acid sequence (R_1—R_2—R_3—R_4 . . . R_n). The specific amino acid sequence of a natural polypeptide is determined from the genetic information (Chapter 19). *The amino acid sequence of the polypeptide chains in a protein is known as the primary structure of the protein. It is the primary structure (amino acid sequence) that gives a protein its physical properties and causes a polypeptide chain to fold into a unique structure giving it a characteristic function and role.*

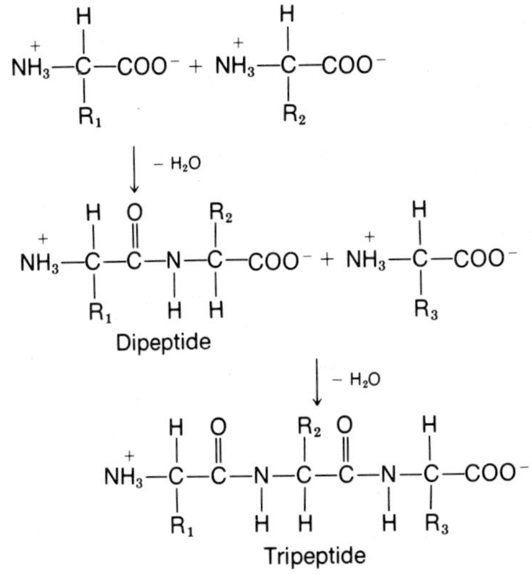

One of the largest natural polypeptide chains in humans is found in procollagen (a precursor of the structural protein collagen) which contains approximately 1,200 amino acid residues per polypeptide chain. On the other end of the spectrum there are many small peptides with less than 10 amino acids that perform important biochemical and physiological functions in humans (Table 2.1).

Some remarks are pertinent with regard to the format used to write primary structures, as exemplified by Table 2.1. Primary structures are commonly written and numbered from their NH_2-terminal end toward their COOH-terminal end. Accordingly, for thyrotropin-releasing factor the glutamic acid residue written on the left of the sequence is the NH_2-terminal amino acid of the tripeptide and is designated amino acid residue 1 in the sequence. The proline is the COOH-terminal amino acid in the structure and is designated the third amino acid residue in the sequence. The defined direction of the polypeptide chain is from Glu → Pro (NH_2-terminal amino acid to COOH-terminal amino acid).

The standard three-symbol abbreviations for the common amino acids are used in Table 2.1. These abbreviations, given in Table 2.2, will be used almost exclusively henceforth. The abbreviations of glutamic acid (Glu) and aspartic acid (Asp) should not be confused with those for glutamine (Gln) and asparagine (Asn). In some cases the experimentalist is not able to differentiate between Gln and Glu or between Asn and Asp in a primary structure. This is because the side chain amide groups are hydrolyzed in Asn and Gln to the free carboxylic acids by the chemical procedures often utilized in the determination of the amino acid sequence of a polypeptide chain (see Section 2.4). In these cases the experimentalist will depict Gln or Glu by Glx, and Asn or Asp by Asx.

Amino Acids Have an Asymmetric Center

The common amino acids with the general structure in Figure 2.2 have four substituents (R, H, COO^-, NH_3^+) covalently bonded to the α-carbon in the α-amino acid structure. A carbon atom with four different substi-

Dipeptide

Tripeptide

Figure 2.7
Peptide bond formation.

TABLE 2.2 The Three-Letter Symbols for the Amino Acids

Amino Acid	*Abbreviation*
Alanine	Ala
Arginine	Arg
Aspartic acid	Asp
Asparagine	Asn
Cysteine	Cys
Glycine	Gly
Glutamic acid	Glu
Glutamine	Gln
Histidine	His
Isoleucine	Ile
Leucine	Leu
Lysine	Lys
Methionine	Met
Phenylalanine	Phe
Proline	Pro
Serine	Ser
Threonine	Thr
Tryptophan	Trp
Tyrosine	Tyr
Valine	Val

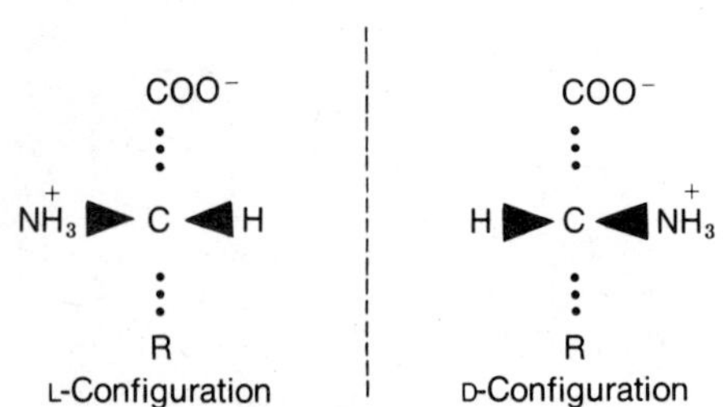

Figure 2.8
Absolute configuration of an amino acid.

tuents arranged in a tetrahedral configuration is asymmetric and exists in two enantiomeric forms. Thus each of the amino acids (*except glycine,* in which R = H and thus two of the four substituents on the α-carbon are hydrogen) exhibits optical isomerism.

The absolute configuration for an amino acid is depicted in Figure 2.8 using the Fischer projection to show the direction in space of the tetrahedrally arranged α-carbon substituents. The α-COO^- is directed up and back behind the plane of the page, and the side chain group, R, is directed down and also back behind the plane of the page. The α-H and α-NH_3^+ groups are directed toward the reader. An amino acid held in this way projects its α-NH_3^+ group either to the left or right of the α-carbon. By convention, if the α-NH_3^+ is projected to the left, the amino acid has an L absolute configuration. Its optical enantiomer, with α-NH_3^+ projected toward the right, has a designated D absolute configuration. In mammalian proteins only amino acids of L configuration are found.

As the amino acids in proteins are asymmetric, the proteins into which the amino acids are polymerized also exhibit asymmetric properties.

Derived Amino Acid: Cystine

A derived amino acid found in many protein structures is cystine. Cystine is formed by the oxidation of two cysteine thiol side chain residues, which are joined to form a disulfide covalent bond (Figure 2.9). The resulting disulfide amino acid is the derived amino acid cystine. Within proteins covalent disulfide links of cystine formed from cysteines, separated from one another in the primary structure, have an important role in stabilizing the folded conformation of proteins. Cystines are formed after the free SH-containing cysteines are incorporated into the protein's primary structure and after the protein has folded.

Two cysteines $\xrightarrow{-2H}$ Cystine

Figure 2.9
Cystine bond formation.

2.3 CHARGE AND CHEMICAL PROPERTIES OF AMINO ACIDS AND PROTEINS

Acid–Base Properties of the Common Amino Acids and Proteins

An understanding of proteins requires a knowledge of the ionizable side chain groups of the common amino acids. These ionizable groups common to proteins and amino acids are shown in Table 2.3.

The acid forms of the respective ionizable groups are on the left of the equilibrium sign in the table, and their respective conjugate bases are on the right of the equilibrium sign. Characteristic of the acid forms is that nitrogen-containing groups are positively charged, whereas acid forms that contain oxygen and sulfur atoms are neutral. In contrast, the base forms of the nitrogen-containing groups are uncharged, while the oxygen and sulfur-containing base forms are negatively charged.

The amino acids whose R groups contain nitrogen atoms (Lys and Arg) are known as the basic amino acids, since their side chains have high pK'_a values and function as good bases. They are usually in their acid forms and positively charged at physiological pH. The amino acids whose side chains contain a carboxylic acid group have relatively low pK'_a values and are called the acidic amino acids. They are predominantly in their base forms and are negatively charged at physiological pH. Proteins in which the ratio (Σ Lys + Σ Arg)/(Σ Glu + Σ Asp) is greater than 1 are referred to as basic proteins. Proteins in which the above ratio is less than 1 are referred to as acidic proteins.

TABLE 2.3 Characteristic pK_a Values for the Common Acid Groups in Proteins

Where Acid Group Is Found	*Acid Form*		*Base Form*	*Approximate pK_a Range for Group*
NH_2-terminal residue in peptides, lysine	$R—NH_3^+$ (Ammonium)	⇌	$R—NH_2 + H^+$ (Amine)	7.6–10.6
COOH-terminal residue in peptides, glutamate, aspartate	R—COOH (Carboxylic acid)	⇌	$R—COO^- + H^+$ (Carboxylate)	3.0–5.5
Arginine	$R—NH—\overset{+}{C}(\cdots NH_2)(\cdots NH_2)$ (Guanidinium)	⇌	$R—NH—C(=NH)—NH_2 + H^+$ (Guanidino)	11.5–12.5
Cysteine	R—SH (Thiol)	⇌	$R—S^- + H^+$ (Thiolate)	8.0–9.0
Histidine	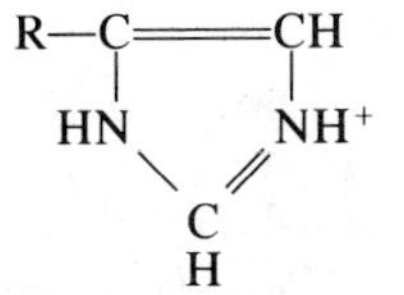 R—C=CH, HN, NH^+, C–H ring (Imidazolium)	⇌	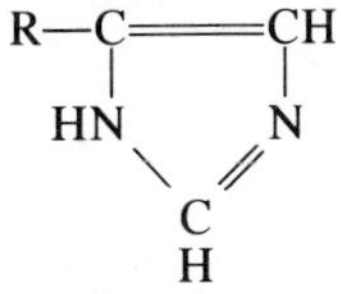 R—C=CH, HN, N ring $+ H^+$ (Imidazole)	6.0–7.0
Tyrosine	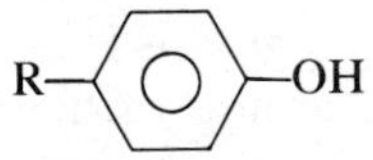 R—C₆H₄—OH (Phenol)	⇌	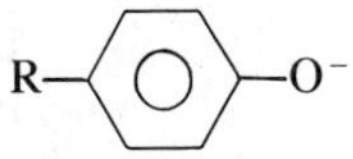 $R—C_6H_4—O^-$ (Phenolate)	9.5–10.5

Ionized Forms of Amino Acids and Proteins; Definition of pI

From a knowledge of the pK'_a for each of the ionizable acid groups in an amino acid or protein and the Henderson–Hasselbalch equation (Figure 2.10), the ionic form of the molecule can be calculated at a given pH. This is an important calculation since a change in the ionization of a protein with pH will give a molecule different functional properties at different pH values.

For example, an enzyme may require a catalytically essential histidine imidazole in its basic form for catalytic activity. If the pK'_a of the catalytically essential histidine in the enzyme is 6.0, at pH 6.0 one-half of the enzyme molecules will be in the active basic (imidazole) form and one-half in the inactive acid (imidazolium) form. Accordingly, the enzyme will exhibit 50% of its potential activity. At pH 7.0, the pH is one unit above the imidazolium pK'_a and the ratio of [imidazole]/[imidazolium] is 10 : 1 (Table 2.4). Based on this ratio, the enzyme will exhibit $10/(10 + 1) \times 100 = 91\%$ of its maximum potential activity.

$$\text{pH} = \text{p}K_a + \log \frac{[\text{conjugate base}]}{[\text{conjugate acid}]}$$

or

$$\text{pH} - \text{p}K_a = \log \frac{[\text{conjugate base}]}{[\text{conjugate acid}]}$$

Figure 2.10
Henderson–Hasselbalch equation. (For a more detailed discussion of this equation, refer to Chapter 1.)

Titration of a Monoamino Monocarboxylic Acid

An understanding of a protein's acid and base forms and their relation to charge is more clearly obtained after following the titration of the ionizable groups for the simple case of an amino acid. As presented in Figure 2.11, leucine contains an α-COOH with $pK'_a = 2.4$ and an α-NH_3^+ with $pK'_a = 9.6$. At pH 1.0 the predominant ionic form (form I) of leucine will have a formal charge of +1 and migrate toward the cathode in an electrical field. The addition of base in an amount equal to one-half of the moles of leucine present in the solution will half-titrate the α-COOH group of the leucine (i.e., $[COO^-]/[COOH] = 1$). The pH of the solution after the

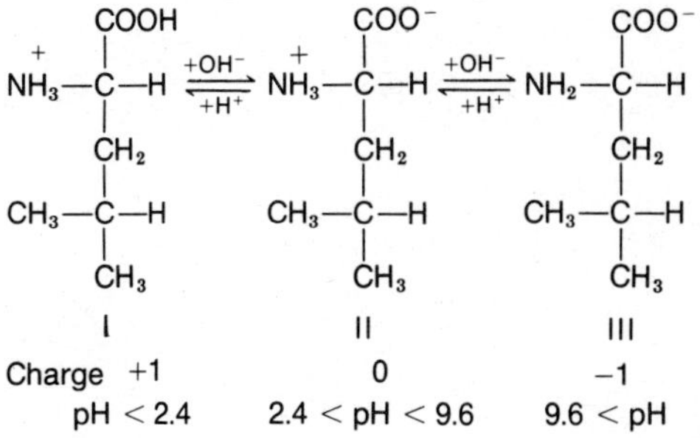

$$\underset{\text{I}}{\overset{+}{N}H_3—C(COOH)(H)—CH_2—CH(CH_3)—CH_3} \underset{+H^+}{\overset{+OH^-}{\rightleftharpoons}} \underset{\text{II}}{\overset{+}{N}H_3—C(COO^-)(H)—CH_2—CH(CH_3)—CH_3} \underset{+H^+}{\overset{+OH^-}{\rightleftharpoons}} \underset{\text{III}}{NH_2—C(COO^-)(H)—CH_2—CH(CH_3)—CH_3}$$

	I	II	III
Charge	+1	0	−1
	pH < 2.4	2.4 < pH < 9.6	9.6 < pH

Figure 2.11
Ionic forms of leucine.

TABLE 2.4 Relationship Between the Difference of pH and Acid pK'_a and the Ratio of the Concentrations of Base to Its Conjugate Acid

pH–pK'a (Difference Between pH and pK'a)	*Ratio of Concentration of Base to Conjugate Acid*
0	1
1	10
2	100
3	1000
−1	0.1
−2	0.01
−3	0.001

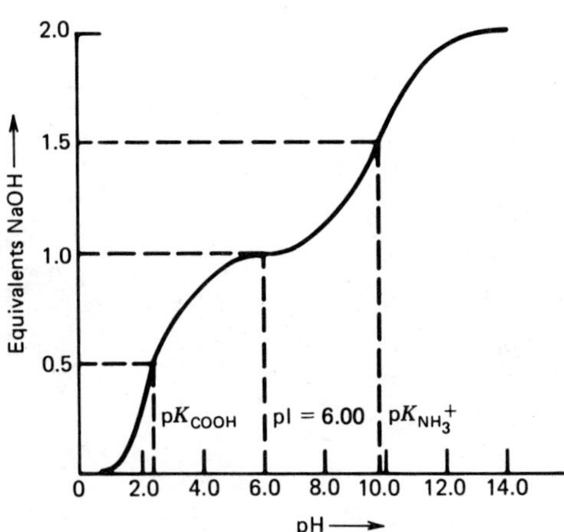

Figure 2.12
Titration curve for leucine.

addition of the 0.5 equiv of base is equal to the pK'_a of the α-COOH of the leucine (Figure 2.12).

Addition of 1 equiv of base will completely titrate the α-COOH. In the predominant form (form II), the negatively charged α-COO^- and positively charged α-NH_3^+ cancel each other and the net charge on this ionic form is zero. Form II is the zwitterion form of leucine. The *zwitterion form* is that ionic form in which the positive charge from positively charged ionized groups is exactly equal to the negative charge from negatively ionized groups of the molecule. Accordingly, the net charge on a zwitterion molecule is zero, and a zwitterion molecule will not migrate toward either the cathode or anode in an electric field.

The further addition of a 0.5 equiv of base to the zwitterion form of leucine will half-titrate the α-NH_3^+ group. At this point in the titration, the ratio of [α-NH_2]/[α-NH_3^+] is equal to 1, and pH = pK'_a for NH_3^+ (Figure 2.12).

The addition of a further 0.5 equiv of base (total of 2 full equiv; Figure 2.12) will completely titrate the α-NH_3^+ group. The solution pH is greater than 11.5, and the predominant species has a formal negative charge of −1 (form III).

It is useful to calculate the exact pH at which an amino acid is electrically neutral and in its zwitterion form. This pH is known as the *isoelectric pH* for the molecule, and the symbol is *pI*. The pI value is a constant of a particular compound at specific conditions of ionic strength and temperature. For simple molecules, such as amino acids, the pI is the average of the two pK'_a values that form the boundaries of the zwitterion form. Leucine has only two ionizable groups, and the pI is calculated as follows:

$$\text{pI} = \frac{\text{p}K'_a\text{COOH} + \text{p}K'_a\text{NH}_3^+}{2} = \frac{2.4 + 9.6}{2} = 6.0$$

At pH > 6.0 leucine will assume a partial negative charge that formally rises at high pH to a full negative charge of −1 (form III) (Figure 2.11). At pH < pI, leucine will have a partial positive charge until at very low pH it will have a formal charge of +1 (form I) (Figure 2.11). The partial charge at any pH can be calculated from the Henderson–Hasselbalch equation or from extrapolation from the titration curve of Figure 2.12.

Titration of a Monoamino Dicarboxylic Acid

A more complicated example of the relationship between molecular charge and pH is the example of glutamic acid. Its ionized forms and titration curve are shown in Figures 2.13 and 2.14. In glutamic acid the

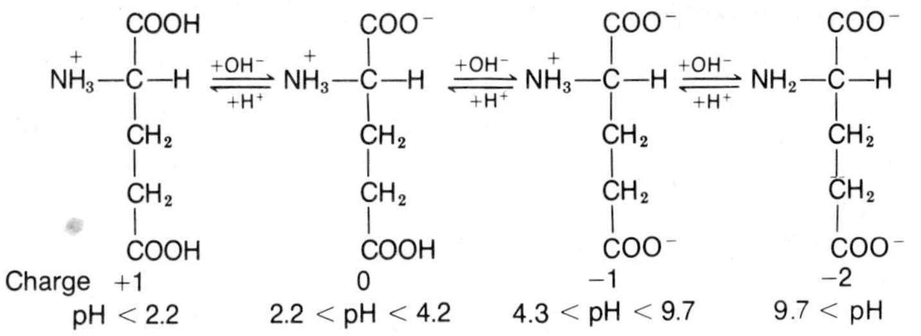

$$\overset{+}{N}H_3-\underset{\underset{\underset{\underset{COOH}{|}}{CH_2}}{|}}{\overset{\overset{COOH}{|}}{C}}-H \underset{+H^+}{\overset{+OH^-}{\rightleftharpoons}} \overset{+}{N}H_3-\underset{\underset{\underset{\underset{COOH}{|}}{CH_2}}{|}}{\overset{\overset{COO^-}{|}}{C}}-H \underset{+H^+}{\overset{+OH^-}{\rightleftharpoons}} \overset{+}{N}H_3-\underset{\underset{\underset{\underset{COO^-}{|}}{CH_2}}{|}}{\overset{\overset{COO^-}{|}}{C}}-H \underset{+H^+}{\overset{+OH^-}{\rightleftharpoons}} NH_2-\underset{\underset{\underset{\underset{COO^-}{|}}{CH_2}}{|}}{\overset{\overset{COO^-}{|}}{C}}-H$$

(each α-carbon bears CH_2-CH_2 in the side chain)

Charge	+1	0	−1	−2
	$pH < 2.2$	$2.2 < pH < 4.2$	$4.3 < pH < 9.7$	$9.7 < pH$

Figure 2.13
Ionic forms of glutamic acid.

α-COOH $pK'_a = 2.2$, the γ-COOH $pK'_a = 4.3$, and the α-NH_3^+ $pK'_a = 9.7$. The zwitterion form is generated after 1.0 equiv of base is added to the low pH form, and the isoelectric pH (pI) is calculated from the average of the two pK'_a values that form the boundaries of the zwitterion form:

$$pI = \frac{2.2 + 4.3}{2} = 3.25$$

Accordingly, at values above pH 3.25 the molecule will assume a net negative charge until at high pH the molecule will have a net charge of −2. At $pH < 3.25$ glutamic acid will be positively charged, and at extremely low pH it will have a net positive charge of +1.

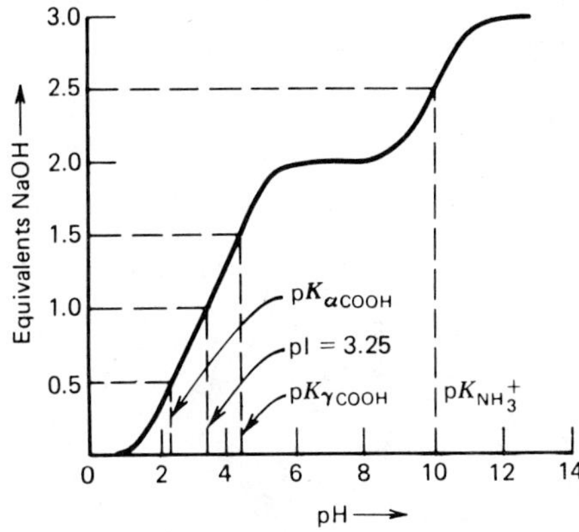

Figure 2.14
Titration curve for glutamic acid.

Common Charge Properties of Amino Acids and Proteins

An analysis of the charge forms present in the other common amino acids shows that the relationship found between pH and the respective pI constants for leucine and glutamate is generally true. That is, *at a solution pH less than the pI of the amino acid, the amino acid is positively charged. At a pH greater than the pI, the amino acid is negatively charged.* The degree of positive or negative charge is a function of the difference between the pH and the pI value of the amino acids and is calculable for an amino acid by the Henderson–Hasselbalch relationship.

Proteins contain multiple ionizable side chain groups and the pI value characteristic of a protein will depend on the relative concentrations of the different acid and basic R groups. As a protein contains many ionizable residues, calculation of its isoelectric pH from pK'_a values would be difficult. Accordingly, the pI values for proteins are almost always experimentally measured by determining the pH value in which the protein does not move in an electrical field. The pI values found for some representative proteins are given in Table 2.5.

As with the amino acids, *at a pH greater than the pI, the protein will have a net negative charge. At a pH less than the pI, the protein will have a positive net charge. The magnitude of the net charge of a protein will increase as a function of the difference between pH and pI.* For example, human plasma albumin contains 585 amino acid residues of which there are 61 glutamates, 36 aspartates, 57 lysines, 24 arginines, and 16 histidines. The albumin pI = 4.9, at which pH the net charge is zero. At pH 7.5 the imidazolium groups of histidine have been partially titrated and albumin has a formal negative charge of −10. At pH 8.6 additional groups have been titrated to their basic form, and the formal net charge is approximately −20. At pH 11 the net charge is approximately −60. On the acid side of the pI value, at pH 3, the approximate net charge on the albumin molecule in solution is +60.

TABLE 2.5 pI Values for Some Representative Proteins

Protein	*pI*
Pepsin	ca.1
Human serum albumin	4.9
α_1-Lipoprotein	5.5
Fibrinogen	5.8
Hemoglobin A	7.1
Ribonuclease	7.8
Cytochrome c	10.0
Thymohistone	10.6

$pH > pI$, then protein charge negative
$pH < pI$, then protein charge positive

Figure 2.15
Relationship between solution pH, protein pI, and protein charge.

Recorder
+
−
Optical device
Buffer solution
Buffer solution
Protein solution

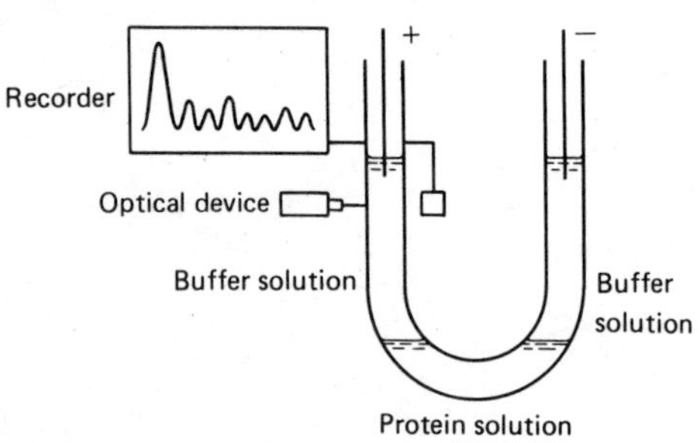

Figure 2.16
Classical electrophoresis apparatus.

Separation of Amino Acids and Proteins Based on pI Values

The techniques of electrophoresis, isoelectric focusing, and ion-exchange chromatography are some of the more important techniques for the study of biological molecules based on charge.

In *electrophoresis,* an ampholyte (protein, peptide, or amino acid) in a solution buffered at a particular pH is placed in an electric field. Depending on the relationship of the buffer pH to the pI of the molecule, the molecule will either move toward the cathode (−) or the anode (+), or remain stationary (pH = pI).

An example of a classical apparatus for protein electrophoresis is shown in Figure 2.16. The apparatus consists of a U-tube in which is placed a protein solution, followed by a buffer solution carefully layered over the protein solution. The migration of the protein is observed with an optical device that measures changes in the refractive index of the solution as the protein migrates toward the anode (Figure 2.16). This apparatus historically led to the separation and operational classification of the proteins in human plasma. For the plasma protein separation, the solution is buffered at pH 8.6, which is at a pH substantially above the pI of the major plasma proteins. The proteins are negatively charged and move toward the positive pole. The peaks in order of their rate of migration, which is related to the order of their pI values, are those of albumin, α_1-,

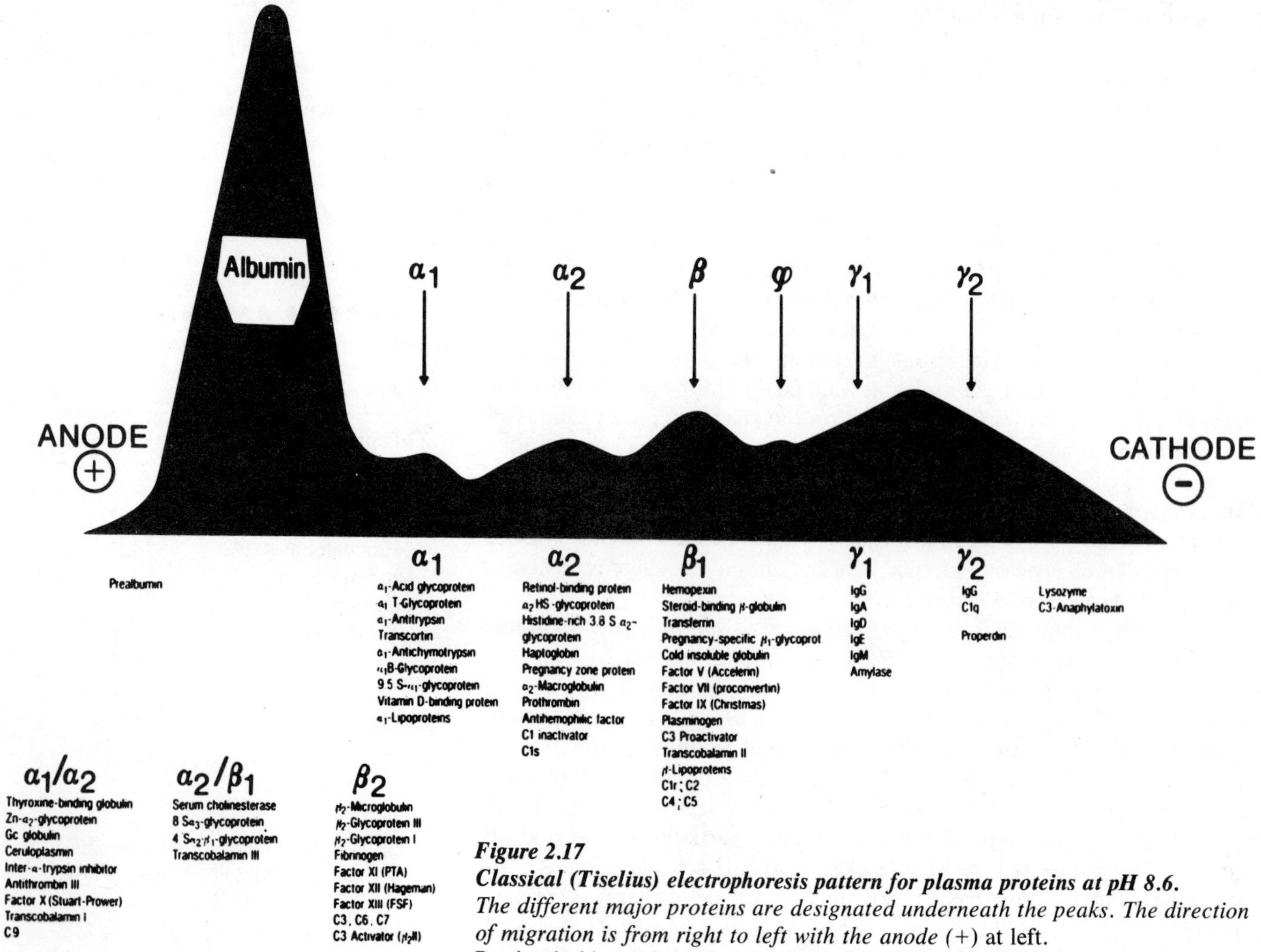

Albumin
α1
α2
β
φ
γ1
γ2
ANODE
+
CATHODE
−
Prealbumin
α1
α1-Acid glycoprotein
α1 T-Glycoprotein
α1-Antitrypsin
Transcortin
α1-Antichymotrypsin
α1B-Glycoprotein
9.5 S-α1-glycoprotein
Vitamin D-binding protein
α1-Lipoproteins
α2
Retinol-binding protein
α2 HS-glycoprotein
Histidine-rich 3.8 S α2-glycoprotein
Haptoglobin
Pregnancy zone protein
α2-Macroglobulin
Prothrombin
Antihemophilic factor
C1 inactivator
C1s
β1
Hemopexin
Steroid-binding β-globulin
Transferrin
Pregnancy-specific β1-glycoprot
Cold insoluble globulin
Factor V (Accelerin)
Factor VII (proconvertin)
Factor IX (Christmas)
Plasminogen
C3 Proactivator
Transcobalamin II
β-Lipoproteins
C1r; C2
C4; C5
γ1
IgG
IgA
IgD
IgE
IgM
Amylase
γ2
IgG
C1q
Properdin
Lysozyme
C3-Anaphylatoxin
α1/α2
Thyroxine-binding globulin
Zn-α2-glycoprotein
Gc globulin
Ceruloplasmin
Inter-α-trypsin inhibitor
Antithrombin III
Factor X (Stuart-Prower)
Transcobalamin I
C9
α2/β1
Serum cholinesterase
8 Sα3-glycoprotein
4 Sα2·β1-glycoprotein
Transcobalamin III
β2
β2-Microglobulin
β2-Glycoprotein III
β2-Glycoprotein I
Fibrinogen
Factor XI (PTA)
Factor XII (Hageman)
Factor XIII (FSF)
C3, C6, C7
C3 Activator (β2II)

Figure 2.17
Classical (Tiselius) electrophoresis pattern for plasma proteins at pH 8.6.
The different major proteins are designated underneath the peaks. The direction of migration is from right to left with the anode (+) at left.
Reprinted with permission from K. Heide, H., Haupt, and H. G. Schwick, in *The Plasma Proteins,* 2nd ed., Vol. III, F. W. Putnam, ed., Academic Press, New York, 1977, p. 545.

α_2-, and β-globulins, fibrinogen, and γ_1- and γ_2-globulins. Some of these peaks represent tens to hundreds of individually different plasma proteins that have a similar migration rate to the anode at pH 8.6. Their rate of migration under these experimental conditions is widely used for purposes of their identification and classification (Figure 2.17). (A more detailed discussion of some of the significant proteins is given in Chapter 3.)

More sophisticated procedures for electrophoresis use polymer gels, starch, or paper as support. The inert supports are saturated with buffer solution, a sample of the proteins to be examined is placed on the support, an electric field is applied across the buffered support, and the proteins migrate in the support toward a charged pole.

A commonly used high-resolution polymeric support is a polyacrylamide cross-linked gel. With the polyacrylamide gel technique, the seven peaks observed in the U-tube electrophoresis may be resolved into a greater number of distinct bands (Figure 2.18). A common *criterion for purity* of a protein is the observation of a single sharp band for a protein in a polyacrylamide gel electrophoresis experiment.

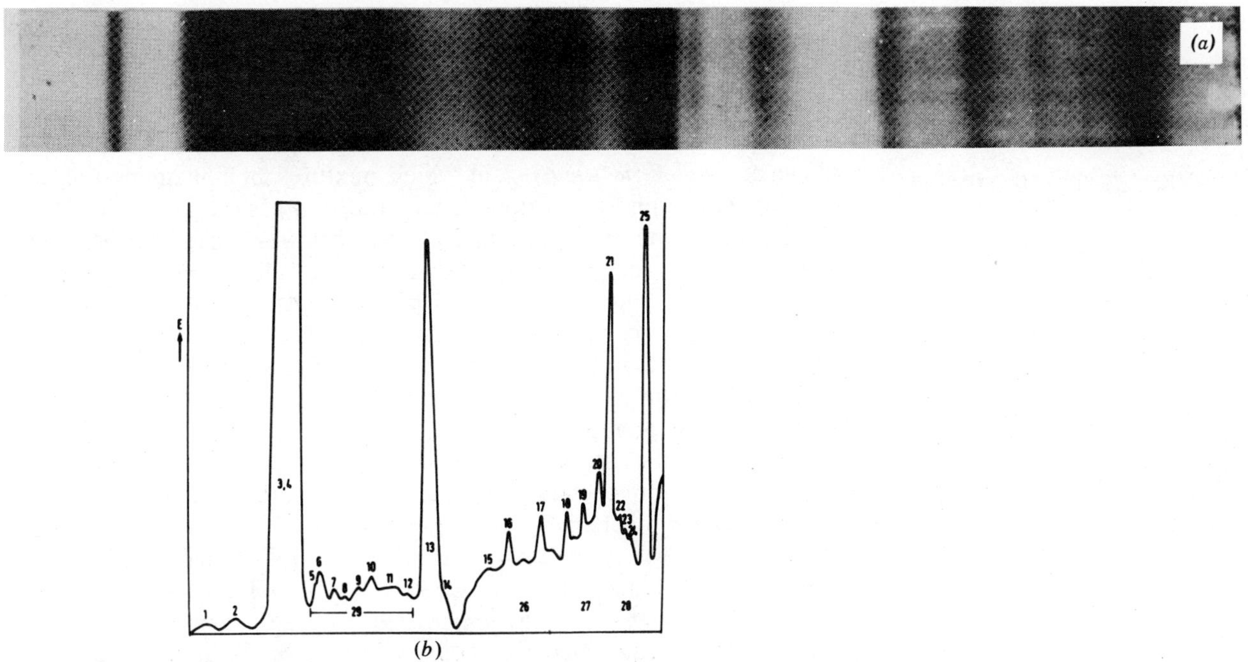

1, Prealbumin; 2, acid α_1-glycoprotein; 3, albumin; 4, α_1-antitrypsin; 5,7, Gc-globulins; 6, α_1 HS-glycoprotein, α_1-antichymotrypsin; 8, unknown; 9, unknown; 10, ceruloplasmin; 11, unknown; 12, hemopexin; 13, transferrin; 14, inter-α_1 trypsin inhibitor; 15, β_1-A-globulin; 16–19, haptoglobin polymers type 2–2; 20, β-glycoprotein; 21, α_2-macroglobulin; 22–24, haptoglobin polymers type 2–2; 25, β-lipoprotein; 26, IgA; 27, IgG; 28, IgM; 29, α_1-lipoprotein.

Figure 2.18
(a) Polyacrylamide gel electrophoresis of human serum proteins with anode (+) at left. (b) Densitometer trace of plasma proteins in polyacrylamide gel in an electrophoresis experiment similar to (a).
Figure *a* from K. Heide, H. Haupt, and H. G. Schwick, in *The Plasma Proteins*, 2nd ed., Vol. III, F. W. Putnam, ed., Academic Press, New York, 1977, p. 545; Figure *b* from R. C. Allen, *J. Chromatog.* 146:1, 1978. Reprinted with permission.

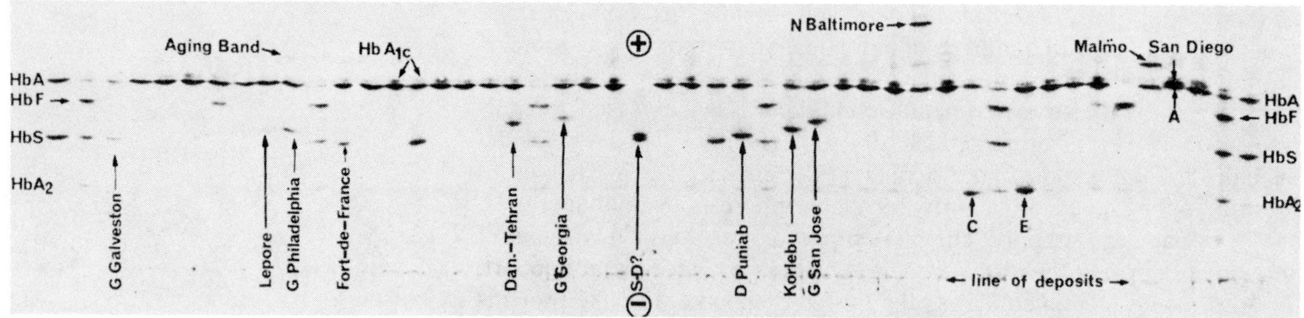

Figure 2.19
Isoelectric focusing on polyacrylamide slab gels with ampholyte gradient between pH 6.0-8.0 of normal (HbA) and abnormal human hemoglobins.
Hb F is the normal hemoglobin in fetal blood, Hb A_2 is a normal hemoglobin found in small amounts in adults, Hb S is sickle cell hemoglobin. Other abnormal hemoglobins contain changes in amino acid sequence from the normal that lead to their separation by this technique. These abnormal hemoglobins are most often named for the city from which the patient with the abnormal protein was first diagnosed.
Reprinted with permission from P. Basset, Y. Beuzard, M. C. Garel, and J. Rosa, *Blood* 51:971, 1978.

$R{-}CH_2{-}COO^-$

Negatively charged ligand: carboxymethyl

$R{-}N^+H(C_2H_5)_2$

Positively charged ligand: diethylamino

Figure 2.20
Two examples of charged ligands used in ion-exchange chromatography.

A type of electrophoresis with extremely high resolution is the technique of *isoelectric focusing* in which mixtures of polyamino-polycarboxylic acid ampholytes with a defined range of pIs are used to establish a pH gradient across an applied electric field. A charged protein will migrate through the pH gradient in the electric field until it reaches a pH region in the gradient equal to its pI value. At this point the protein becomes stationary in the electric field and may be visualized or eluted from the column in preparative quantities (Figure 2.19). Proteins that differ by as little as 0.0025 in their pI values can be separated on the appropriate pH gradient.

Separation of proteins by *ion-exchange resins* in a chromatography column is a third important technique for the separation and characterization of proteins by charge. Ion-exchange resins are prepared of insoluble materials (agarose, polyacrylamide, cellulose, glass) that contain negatively charged ligands (e.g., $—CH_2COO^-$, $—C_3H_6SO_3^-$) or positively charged ligands (e.g., diethylamino) (Figure 2.20) covalently attached to the insoluble resin. Negatively charged resins bind cations strongly and are known as *cation-exchange resins*. Similarly, positively charged resins bind anions strongly and are referred to as *anion-exchange resins*. The degree of retardation of a protein or amino acid by a resin will depend on the magnitude of the charge on the protein at the particular pH of the experiment. Molecules of the same charge as the resin are eluted first in a single band, followed by proteins with an opposite charge to that of the resin, in an order based on the protein's charge density (Figure 2.21). In situations where it is difficult to remove a molecule from the resin because of the strength of the attractive interaction between the bound molecule and resin, systematic changes in pH or in ionic strength may be used to weaken the interaction.

For example, an increasing pH gradient in the eluent buffer through a cation-exchange resin with cationic proteins bound will reduce the difference between the solution pH and the respective molecular pI values. This decrease between pH and pI reduces the magnitude of the net charge on the proteins and thus decreases the strength of interaction between the proteins and the resin.

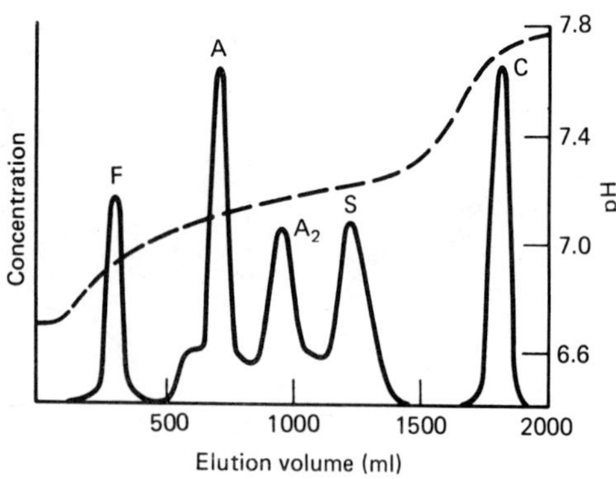

Figure 2.21
Example of ion-exchange chromatography.
Elution diagram of an artificial mixture of hemoglobins F, A, A_2, S, and C on Carboxymethyl-Sephadex C-50.
From A. M. Dozy and T. H. J. Huisman, *J. Chromatog.*, 40:62, 1969.

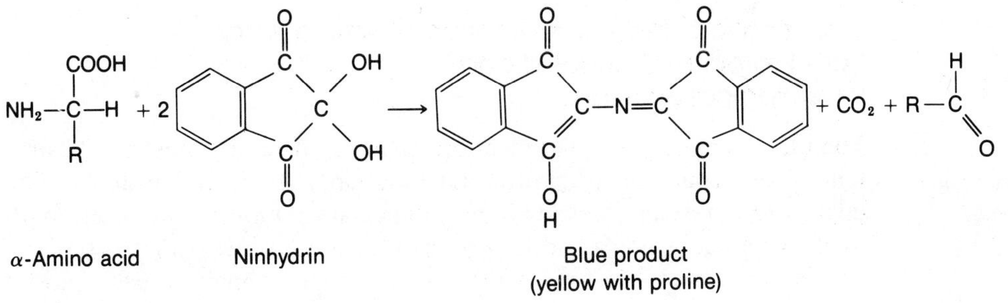

Figure 2.22
Reaction of an amino acid with ninhydrin.

An increased gradient of ionic strength in the eluting buffer will also decrease the strength of charge interactions.

Some Chemical Reactions of the Amino Acids

The amino group in amino acids reacts with *ninhydrin* on heating to give a blue compound (Figure 2.22). As the reaction is stoichiometric, the amount of an amino acid present in a solution can be quantified by the light absorption at wavelength 570 nm. Proline, an imino acid, also reacts with ninhydrin, but gives a yellow color with a maximum absorbance at 440 nm.

Fluorescamine reacts with amino acids in a 1 : 1 stoichiometry at room temperature to give a fluorescent product (Figure 2.23), the concentration of which can be determined with a spectrofluorometer. The sensitivity of the fluorescamine assay is 10 to 100 times greater than the ninhydrin assay. With current commercial instruments, the ninhydrin assay can routinely measure 3×10^{-9} moles of an amino acid, the fluorescamine procedure can measure 10^{-10} to 10^{-11} moles of an amino acid.

A number of specific chemical reactions can be used to quantify specific amino acid side chains in a solution of free amino acids or in a denatured protein. The *Sakaguchi reaction* (α-naphthol and sodium hypochlorite in alkaline solution) is used to colorimetrically quantify the guanidine side chain in arginine. *Ehrlich's reagent* specifically gives a colorimetric assay for the indole group of tryptophan. The *Pauly reagent* gives a red color with the imidazole side chain of histidine and the phenol side chain of tyrosine.

Ellman's reagent [5,5′-dithiobis-(2-nitrobenzoic acid)] reacts with the cysteine thiol side chain at pH 8.0 (Figure 2.24) to give a thionitrobenzoic acid product that has a strong absorbance at 412 nm.

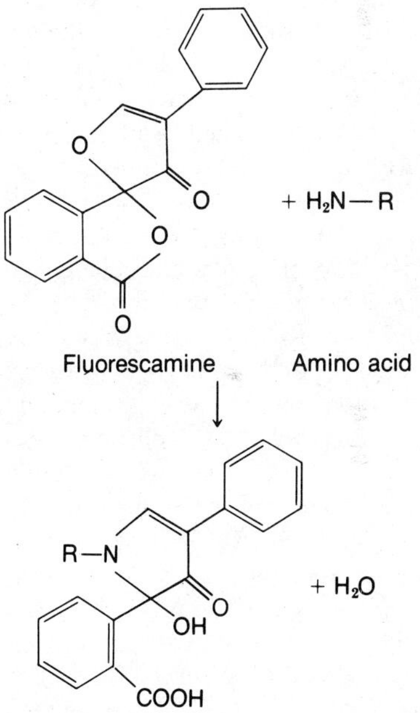

Figure 2.23
Reaction of fluorescamine with an α-amino acid to product a fluorescent product.

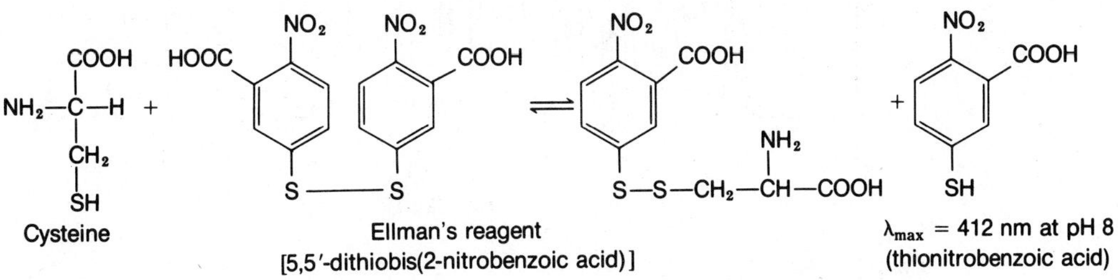

Figure 2.24
Reaction of cysteine with Ellman's reagent.

CLIN. CORR. **2.1**
THE USE OF AMINO ACID ANALYSIS IN THE DIAGNOSIS OF DISEASE

There are a number of clinical disorders in which a high concentration of amino acids are found in plasma and/or urine. An abnormally high concentration of an amino acid in urine is called an *aminoaciduria*. *Phenylketonuria* is a metabolic defect in which patients are lacking sufficient amounts of the enzyme phenylalanine hydroxylase, which catalyzes the transformation of phenylalanine to tyrosine. As a result, large concentrations of phenylalanine, phenylpyruvate, and phenyllactate accumulate in the plasma and urine. Phenylketonuria occurs clinically in the first few weeks after birth, and if the infant is not placed on a special diet, severe mental retardation will occur. *Cystinuria* is a genetically transmitted defect in the membrane transport system for cystine and the basic amino acids (lysine, arginine, and the derived amino acid ornithine) in epithelial cells. Large amounts of these amino acids are excreted in urine. Other symptoms of this disease may arise from the formation of renal stones composed of cystine precipitated within the kidney. *Hartnup disease* is a genetically transmitted defect in epithelial cell transport of neutral-type amino acids (monoamino monocarboxylic acids), and high concentrations of these amino acids are found in the urine. The physical symptoms of the disease are primarily caused by a deficiency of tryptophan. These symptoms may include a pellagra-like rash (nicotinamide is partly derived from tryptophan precursors) and cerebellar ataxia (irregular and jerky muscular movements) due to the toxic effects of indole derived from the bacterial degradation of unabsorbed (untransferred) tryptophan present in large amounts in the gut. *Fanconi's syndrome* is a generalized aminoaciduria correlated with a hypophosphatemia and a high excretion of glucose.

Identification and Quantification of Amino Acids from Protein Hydrolysates and Physiological Solutions

The identification and quantification of the amino acids present in a protein is important for structural and functional studies of proteins. The quantification of amino acids in physiological solutions is important both for biochemical studies and for the diagnosis of disease (Clin. Corr. 2.1). A common procedure for quantitative amino acid analysis uses cation-exchange chromatography to separate the amino acids which are then reacted with ninhydrin, fluorescamine, dansyl chloride (Section 2.4), or similar chromophoric or fluorophoric reagents to quantitate the separated amino acids.

Cation-exchange chromatography is carried out at a pH below the pI of the common amino acids. Accordingly, the amino acids with higher pI values are retarded more strongly by the cation-exchange resin. There is also some retention of amino acids due to hydrophobic interactions with the resin, but the charge interaction is the primary cause for the separation of the amino acids by the resin. Table 2.6 gives the pI and elution time for the common amino acids obtained from a particular laboratory in chromatography over a sulfonic acid cation-exchange resin in a gradient of buffer pH values.

Amino acids can also be separated on a chromatographic column that solely uses the hydrophobic differences between the amino acid side chain groups (Figure 2.25). The chromatography column uses closely packed insoluble resin beads coated with hydrophobic alkyl groups through which the amino acids are eluted in a mixed organic–aqueous solvent under high pressure. The chromatography procedure is known as reverse phase *high performance liquid chromatography* (HPLC) (see Section 2.8). The amino acids are derivatized with chromogenic or fluorogenic

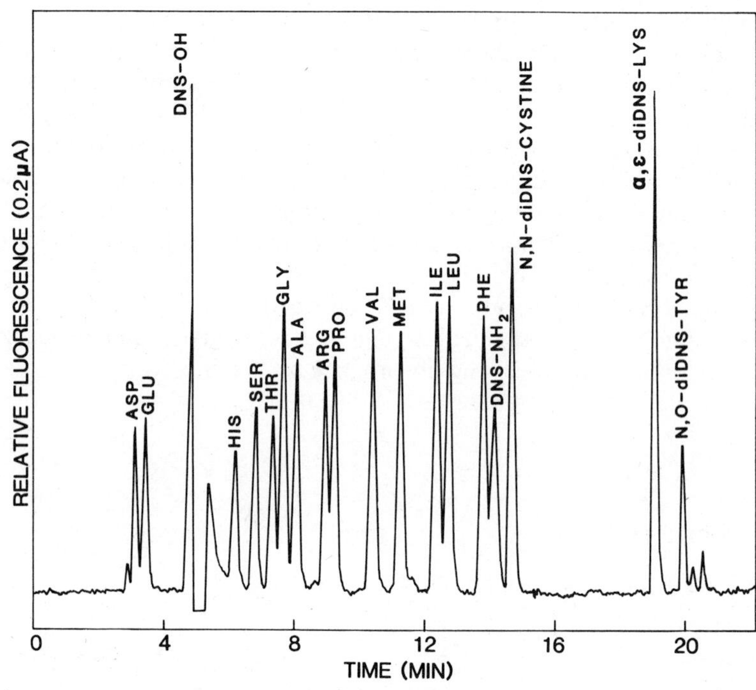

Figure 2.25
Separation of amino acids as dansyl (DNS) derivatives by reverse-phase HPLC. Reprinted with permission from M. W. Hunkapiller, J. E. Strickler, and K. J. Wilson, *Science* 226:304, 1984. Copyright 1984 by the AAAs.

reagents either before or after their placement on the column. With some types of derivatization, amino acids can be identified at concentrations as low as 0.5×10^{-12} moles ($\frac{1}{2}$ pmol).

2.4 PRIMARY STRUCTURE OF PROTEINS

Techniques Used to Determine the Amino Acid Sequence of a Protein

Knowledge of the primary structure (amino acid sequence) of a protein is required for an understanding of the relationship of a protein's structure to its function on a molecular level.

In the determination of primary structure, first the number of polypeptide chains in the protein must be ascertained. To begin with, the protein is denatured (see Section 2.6) and then treated with a reagent, such as dansyl chloride, that forms a covalent bond with the NH_2-terminal α-amino groups of each polypeptide chain within the protein (Figure 2.26). The tagged protein is then hydrolyzed to its constituent amino acids. Typical conditions for complete protein hydrolysis are 6 N HCl, at 110°C, for 18–36 h, in a sealed tube under vacuum. The vacuum prevents degradation of oxidation-sensitive amino acid side chains by oxygen in the air. Analysis of the amino acid hydrolysate by paper or other chromatographic procedures separates the dansyl-labeled NH_2-terminal amino acids, which are identified chromatographically against standards (see Figures 2.25 and 2.27).

After determining the number of polypeptide chains by identification of their NH_2-terminal amino acid, the chains must be separated in order that each may be sequenced. Since the chains may be covalently joined by the disulfide bonds of cystine, these bonds may have to be broken (Figure

TABLE 2.6 pI Values and Elution Time of Amino Acids From a Cation-Exchange Resin[a]

Amino Acid	*Elution Time (min)*	*pI*
Asp	9.7	2.97
Thr	11.5	6.53
Ser	12.5	5.68
Glu	14.8	3.22
Gly	19.8	5.97
Ala	21.4	6.02
Cys	24.5	5.02
Val	27.9	5.97
Met	34.3	5.75
Ile	37.1	6.02
Leu	38.0	5.98
Tyr	40.2	5.65
Phe	41.0	5.48
His	44.3	7.58
Lys	52.2	9.74
Arg	60.0	10.76

SOURCE: Data from J. R. Benson, *Methods Enzymol.*, **47**, 19 (1977).

[a] Amino acid analysis utilized a 0.32×15 cm bed of Durrum DC-5A resin. Eluent linear flow rate was 2.4 cm min^{-1}. Column temperature was 45°C from 0 to 22 min and 65°C after 22 min. Four buffers (0.2 M buffer component) at pH 3.25 (25 min), pH 4.25 (8 min), pH 5.25 (4 min), and pH 10.0 (28 min) were applied sequentially for the time duration indicated.

Dansyl chloride + Polypeptide $\xrightarrow{-HCl}$ Dansyl protein $\xrightarrow[6\ N\ HCl]{\sim 110°C}$ N-Terminal dansyl amino acid + free amino acids

Figure 2.26
Reaction of a polypeptide with dansyl chloride.

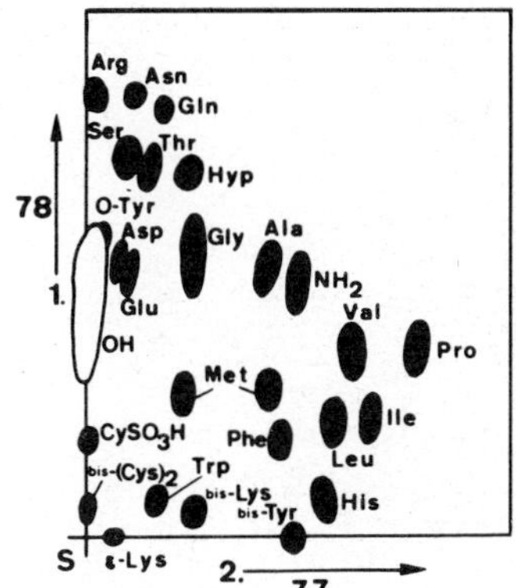

Figure 2.27
Identification of dansyl amino acids by thin-layer chromatography.
Solvent 1: water-90% formic acid (100:1.5 v/v); solvent 2: benzene-glacial acetic acid (9:1 v/v) on polyamide paper (Cheng-Chin Trading Co.).
According to D. R. Woods and K. T. Wang, *Biochem. Biophys. Acta*, 133:369, 1967, from A. Niederwieser in *Methods Enzymol.* 25:60, 1972.

2.28). The individual chains are then separated by molecular exclusion chromatography (see Section 2.8), ion-exchange chromatography, hydrophobic chromatography under high pressure (HPLC), and/or electrophoresis techniques.

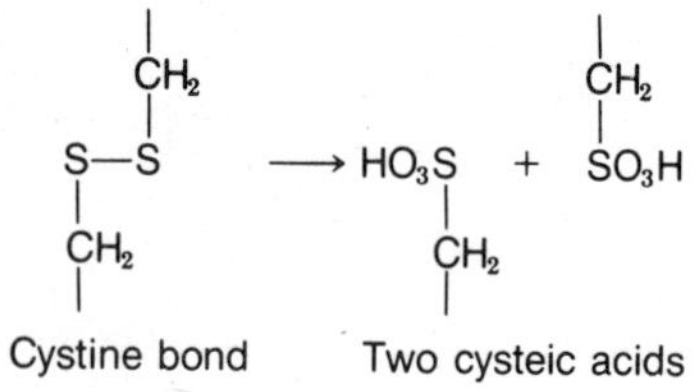

Polypeptide chains are most commonly sequenced by the Edman reaction (Figure 2.29). In the *Edman reaction,* the polypeptide chain to be sequenced is reacted with phenylisothiocyanate, which, like the dansyl chloride, forms a covalent bond to the NH_2-terminal amino acid of the chain. However, in this derivative, acidic conditions catalyze an intramolecular cyclization that results in the cleavage of the NH_2-terminal amino acid from the polypeptide chain as a phenylthiohydantoin derivative. This NH_2-terminal amino acid derivative may be separated chromatographically and identified against standards. The polypeptide chain minus the NH_2-terminal amino acid is then isolated, and the Edman reaction is repeated to identify the next NH_2-terminal amino acid of the chain. This series of reactions can theoretically be repeated until the sequence of the entire polypeptide chain is determined.

The repetition of Edman reactions under favorable conditions can be carried out for 30 or 40 amino acids into the polypeptide chain from the NH_2-terminal end. At this point in the analysis, impurities generated from incomplete reactions in the reaction series make further Edman cycles unfeasible. Since most polypeptide chains in proteins contain more than 30 or 40 amino acids, they have to be hydrolyzed into smaller fragments and sequenced in sections.

Both enzymatic and chemical methods are used to break polypeptide chains into smaller polypeptide fragments. For example, the enzymes trypsin and chymotrypsin are *proteolytic enzymes* that are commonly used for partial hydrolysis of polypeptide chains in sequencing. The enzyme trypsin preferentially catalyzes the hydrolysis of the peptide bond on the α-COOH side of the basic amino acid residues of lysine and arginine within polypeptide chains. Chymotrypsin hydrolyzes peptide bonds on the α-COOH side of amino acid residues with large apolar side chains.

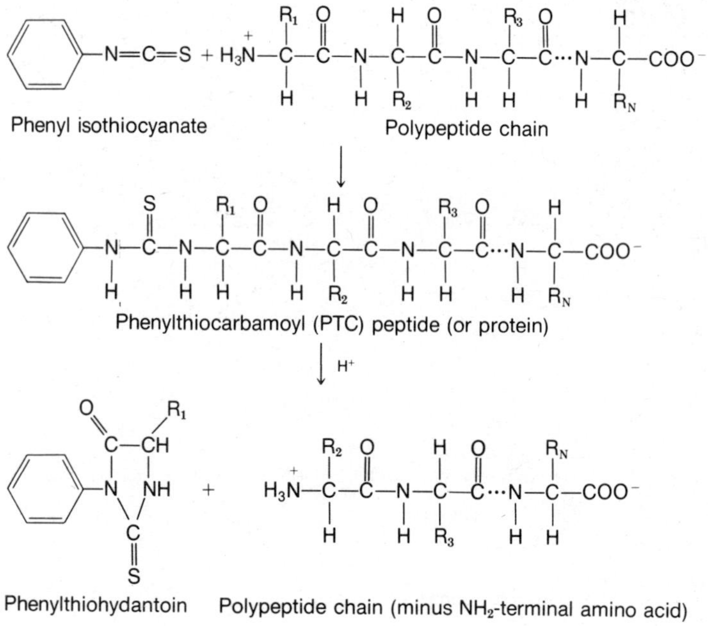

Cystine bond → Two cysteic acids

$$-CH_2-S-S-CH_2- \longrightarrow -CH_2-SO_3H + HO_3S-CH_2-$$

Figure 2.28
Oxidation of cystine bonds.

Phenyl isothiocyanate + Polypeptide chain

$$C_6H_5-N{=}C{=}S + H_3\overset{+}{N}-CH(R_1)-CO-NH-CH(R_2)-CO-NH-CH(R_3)-CO\cdots NH-CH(R_N)-COO^-$$

$$\downarrow$$

Phenylthiocarbamoyl (PTC) peptide (or protein)

$$C_6H_5-NH-CS-NH-CH(R_1)-CO-NH-CH(R_2)-CO-NH-CH(R_3)-CO\cdots NH-CH(R_N)-COO^-$$

$$\downarrow H^+$$

Phenylthiohydantoin + Polypeptide chain (minus NH_2-terminal amino acid)

$$H_3\overset{+}{N}-CH(R_2)-CO-NH-CH(R_3)-CO\cdots NH-CH(R_N)-COO^-$$

Figure 2.29
Edman reaction.

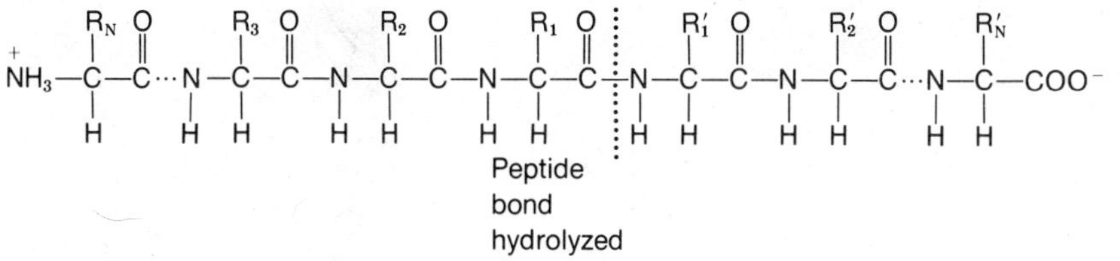

$$^{+}NH_3-\underset{H}{\overset{R_N}{C}}-\overset{O}{\overset{\|}{C}}\cdots\underset{H}{N}-\underset{H}{\overset{R_3}{C}}-\overset{O}{\overset{\|}{C}}-\underset{H}{N}-\underset{H}{\overset{R_2}{C}}-\overset{O}{\overset{\|}{C}}-\underset{H}{N}-\underset{H}{\overset{R_1}{C}}-\overset{O}{\overset{\|}{C}} \vdots \underset{H}{N}-\underset{H}{\overset{R_1'}{C}}-\overset{O}{\overset{\|}{C}}-\underset{H}{N}-\underset{H}{\overset{R_2'}{C}}-\overset{O}{\overset{\|}{C}}\cdots\underset{H}{N}-\underset{H}{\overset{R_N'}{C}}-COO^-$$

Peptide bond hydrolyzed

R_1	Reagent
Phe, Tyr, or Trp	Chymotrypsin
Arg, Lys	Trypsin
Met	Cyanogen bromide
Trp	*o*-Iodosobenzoic acid
Glu	Staphylococcus aureus endoprotease V8

Figure 2.30
Specificity of some polypeptide cleaving reagents.

The chemical reagent cyanogen bromide specifically cleaves peptide bonds on the carboxyl side of methionine residues within polypeptide chains (Figure 2.30). Thus, to establish the amino acid sequence of a large polypeptide chain, the chain is subjected to partial hydrolysis by one of the specific cleaving reagents, the polypeptide segments are separated, and the amino acid sequence of each of the small segments is determined by the Edman reaction.

To order the sequenced peptide segments correctly into the complete sequence of the original polypeptide, a sample of the original polypeptide must be subjected to a *second partial hydrolysis* by a specific hydrolytic reagent different from that used initially. The sequence of this second group of polypeptide segments gives overlapping regions for the first group of polypeptide segments. This leads to the ordering of the initially sequenced peptide segments (Figure 2.31).

Trypsin digests of a protein are often used as an analytical tool for protein identification. Figure 2.32 shows the trypsin digest of hemoglobin A and hemoglobin S chromatographed in one dimension and electrophoresed in the other dimension. Close examination of the patterns shows that they are identical except for peptide 4, which contains the genetically determined amino acid substitution that causes sickle cell anemia. The chromatography pattern of such an enzymic digest is known as the protein's "fingerprint."

Primary Structure of Insulin

Study of the primary structure of the protein hormone insulin elucidates chemical events that occur in the biosynthesis of insulin from precursor forms in the pancreatic islet cells. Insulin is produced initially in a form known as proinsulin. The proinsulin molecule is a substrate for an enzyme

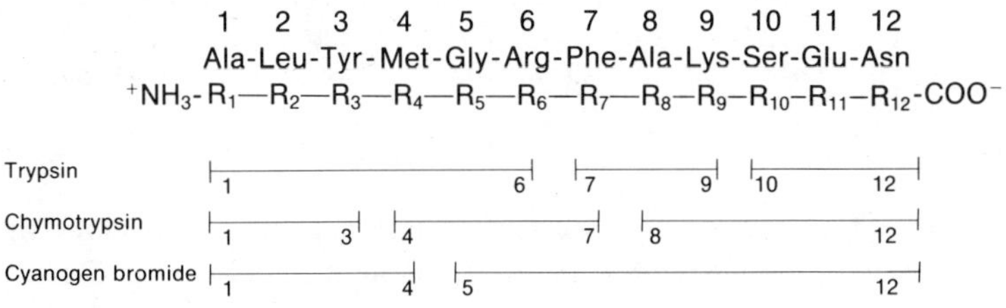

1 2 3 4 5 6 7 8 9 10 11 12
Ala-Leu-Tyr-Met-Gly-Arg-Phe-Ala-Lys-Ser-Glu-Asn
$^{+}NH_3-R_1-R_2-R_3-R_4-R_5-R_6-R_7-R_8-R_9-R_{10}-R_{11}-R_{12}-COO^-$

Reagent	Fragments
Trypsin	1–6, 7–9, 10–12
Chymotrypsin	1–3, 4–7, 8–12
Cyanogen bromide	1–4, 5–12

Figure 2.31
The ordering of peptide fragments from overlapping sequences produced by specific proteolysis of a peptide.

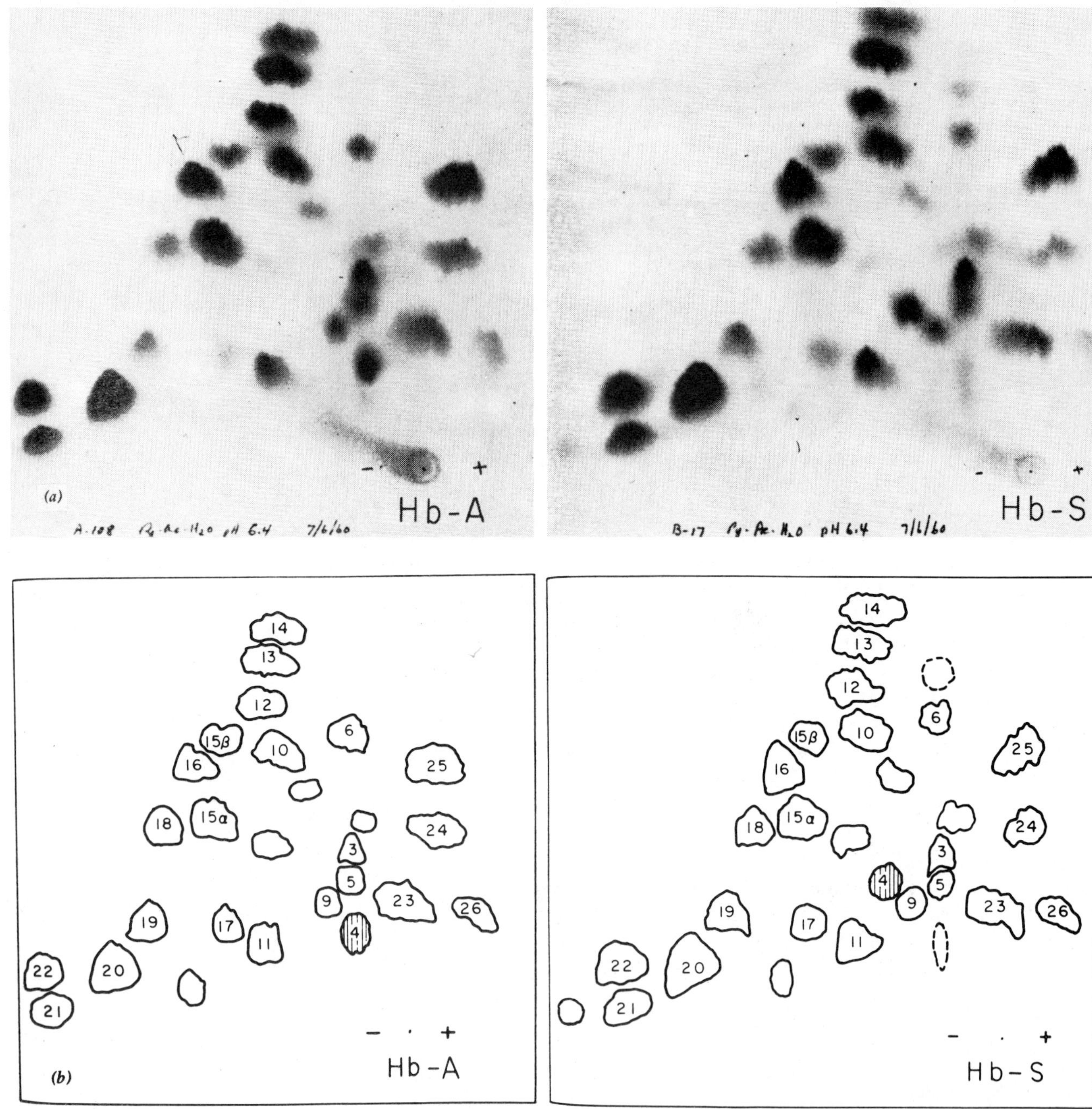

Figure 2.32
Trypsin digest of Hb A (normal) and Hb S (sickle cell hemoglobin).
(a) "Fingerprint" of digest visualized after two-dimensional chromatography. (b) Tracing of spots for peptides observed above. Dashed line tracing indicates spots that become visible only on heating. Spot 4 contains mutated amino acid that give Hb S its sickle-cell properties; otherwise all other peptides are identical.
Reprinted with permission from C. Baglioni, *Biochim. Biophys. Acta* 48:392, 1961.

that catalyzes the hydrolysis of two intrachain peptide bonds in the proinsulin, resulting in the cleavage of a 35-amino acid segment (the C-peptide) from within the polypeptide chain (Figure 2.33). The product of these proteolytic hydrolyses is active insulin, which consists of two small polypeptide chains that are covalently held together by disulfide bonds of cystine residues (Figure 2.33).

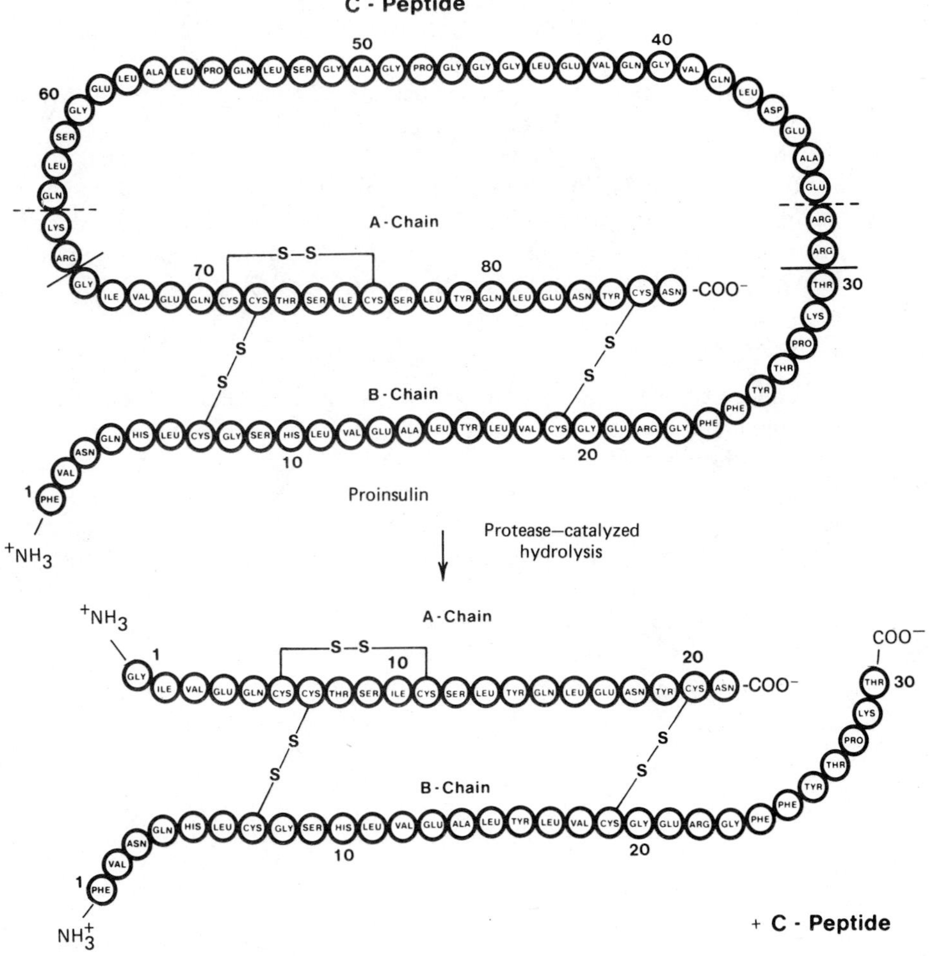

Figure 2.33
Amino acid sequence of human proinsulin and insulin.
The B-peptide extends from Phe at position 1 to Thr at position 30, the C-peptide from Arg at position 31 to Arg at position 65, and the A-peptide from Gly at position 66 to Asn at position 86. Cystine bonds from positions 7 to 72, 19 to 85, and 71 to 76 are found in proinsulin.
From G. Bell et al., *Nature* 282:525, 1979.

The primary structures of proinsulins contain from 78 (dog) to 86 amino acids (human, horse, rat). Synthesis of the primary structure of proinsulin is followed by the folding of the polypeptide chain into its native three-dimensional structure (see Section 2.6) with formation of its three intra-chain disulfide (cystine) bonds. These disulfide bonds are the same as those shown in the primary structure of its product protein, in which two of these cystine bonds become *inter*-chain bonds, whereas in proinsulin they were *intra*-chain bonds.

The C-peptide hydrolyzed from the proinsulin molecule is further processed in the pancreatic islet cells by peptidase enzymes that act to hydrolyze a dipeptide from the COOH-terminal and a second dipeptide from NH_2-terminal ends of the C-peptide. Stoichiometric amounts of the modified C-peptide and insulin are secreted into blood.

The essential or nonessential function of particular amino acids in the primary structure of the active insulin is studied by the comparison of the primary structure for active insulins from different animal species. This comparison shows that residues 8, 9, and 10 of the A chain and residue 30

TABLE 2.7 Variation in Positions A8, A9, A10, and B30 of Insulin

Species	*A8*	*A9*	*A10*	*B30*
Human	Thr	Ser	Ile	Thr
Cow	Ala	Ser	Val	Ala
Pig	Thr	Ser	Ile	Ala
Sheep	Ala	Gly	Val	Ala
Horse	Thr	Gly	Ile	Ala
Dog	Thr	Ser	Ile	Ala
Chicken[a]	His	Asn	Thr	Ala
Duck[a]	Glu	Asn	Pro	Thr

[a] Positions 1 and 2 of B chain are both Ala in chicken and duck; whereas in the other species in the table, position 1 is Phe and position 2 is Val in B chain.

CLIN. CORR. **2.2**
DIFFERENCES IN THE PRIMARY STRUCTURE OF INSULINS USED IN THE TREATMENT OF DIABETES MELLITUS

Both pig (porcine) and beef (bovine) insulins are commonly used in the treatment of human diabetics. Because of the differences in amino acid sequence from the human insulin, some diabetic individuals will have an initial allergic response to the injected insulin as their immunological system recognizes the insulin as foreign, or develop an insulin resistance due to a high anti-insulin antibody titer at a later stage in treatment. However, the number of diabetics who have a deleterious immunological response to pig and cow insulins is small; the great majority of human diabetics can utilize the nonhuman insulins without immunological complication. The compatibility of the cow and pig insulins in humans is due to the small number of changes and the conservative nature of the changes between the amino acid sequences of the insulins. These changes in primary structure do not significantly perturb the three-dimensional structure of the insulins from that of the human insulin. Pig insulin is usually more acceptable than cow insulin in insulin-reactive individuals because it is more similar in sequence to human insulin (see Table 2.7).

of the B chain may be varied without dramatically affecting the structure and corresponding physiological role of the insulin molecule (Table 2.7). The other amino acid residues of the primary structure do not vary as significantly among species, and must therefore be more critical to the structural and related functional properties of insulin.

An amino acid substituted by an alternative amino acid of similar polarity (e.g., Val for Ile in position 10 of insulin) in a primary structure is designated a *conservative type* of variation (see Clin. Corr. 2.2). A *nonconservative type* of change occurs on the substitution of an amino acid by another of dramatically different polarity (see Clin. Corr. 2.3).

Cytochrome c

Cytochrome c is the most thoroughly investigated primary structure. It is a single-chain protein of 104 amino acids in mammals and is essential to the mitochondrial electron transport system of a cell. Cytochrome c has been sequenced from over 67 plant and animal species, from yeast and fungus to the human. As with insulin, these studies indicate which residues in the primary structure are essential to the function of the protein. The invariant residues in the sequences obtained from the different species represent only 28 amino acids of the total 104.

Based on the differences in primary structure among the various cytochrome c's, an evolutionary classification of the species from which the

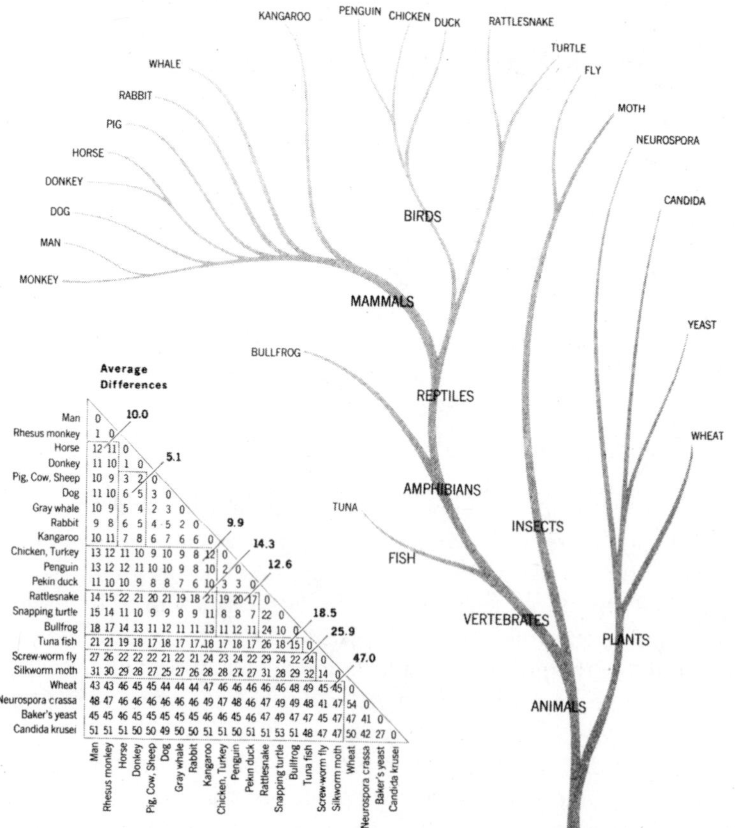

Figure 2.34
Difference in amino acid sequences among cytochrome c's from different species. *Family tree is drawn based on the calculation from number of amino acid differences.*
Reprinted with permission from R. S. Dickerson and J. Geis, *The Structure and Action of Proteins,* W. A. Benjamin, Inc., Menlo Park, 1969, p. 65.

TABLE 2.8 Evolution of Species Based on Cytochrome c Sequences

Species Compared	*Number of Residues Changed*	*Divergence of Species in Millions of Years*
Human–chicken	13	280 (assumed)
Human–tuna	21	490 (calculated)
Human–moth	31	750 (calculated)
Human–yeast	44	1180 (calculated)

SOURCE: Table from E. Margoliash and A. Schejter, *Adv. Protein Chem.* **21,** 113 (1966).

cytochrome c was obtained can be made that is similar to the phylogenetic classification based on morphological criteria (Figure 2.34 and Table 2.8).

2.5 HIGHER LEVELS OF PROTEIN ORGANIZATION

The *primary structure* of a protein, discussed in preceding sections, refers to the covalent structure of a protein. It includes the amino acid sequence of the protein and the location of disulfide (cystine) bonds. Higher levels of protein organization refer to noncovalently generated conformational properties of the primary structure. These higher levels of protein conformation and organization are customarily defined as the secondary, tertiary, and quaternary structures of a protein. The *secondary structure* refers to the conformation of the polypeptide chain in the protein. The polypeptide chain in this context refers to the covalently interconnected atoms of the peptide bonds and *alpha*-carbon linkages that string the amino acid residues of the protein together. Side chain (R) groups are not included at the level of secondary structure. For example, secondary structures of polypeptide chains may form noncovalently generated conformations that are helical (i.e., α helix). The *tertiary structure* refers to the total three-dimensional structure of the polypeptide units of the protein. It includes the conformational relationships in space of the side chain groups to the polypeptide chain and the geometric relationship of distant regions of the polypeptide chain to each other. The *quaternary structure* refers to the structure and interactions of the noncovalent association of discrete polypeptide subunits into a multisubunit protein. Not all proteins have a quaternary structure.

Proteins generally assume a unique secondary, tertiary, and quaternary conformation for their particular amino acid sequence, known as the native conformation. The folding of the primary structure into the native conformation occurs spontaneously under the influence of noncovalent forces. This unique conformation is that of the lowest Gibbs free energy kinetically accessible to the polypeptide chain(s) for the particular conditions of ionic strength, pH, and temperature of the solvent in which the folding process occurs. Cystine bonds are made after the folding of the polypeptide chain occurs, and act to stabilize covalently the native conformation.

The higher levels of protein organization are individually discussed in the following sections.

Secondary Structure of Proteins

The conformation of a polypeptide chain is based on the rotational angles assumed about the covalent bonds that interconnect the amino acids. These are the bonds in one amino acid between (1) the nitrogen and

CLIN. CORR. 2.3
A NONCONSERVATIVE MUTATION OCCURS IN SICKLE CELL ANEMIA

Hemoglobin S (Hb S) is a variant form of the normal adult hemoglobin in which a nonconservative substitution occurs in the sixth position of the β-polypeptide chain of the normal hemoglobin (Hb A_1) sequence. Whereas in Hb A_1 this position is taken by a glutamic acid residue, in Hb S the position is occupied by a valine. Consequently, individuals with Hb S have replaced a polar side chain group on the molecule's outside surface with a nonpolar hydrophobic side chain group (a nonconservative mutation). Through hydrophobic interactions with this nonpolar valine residue, Hb S in its deoxy conformation will polymerize with other molecules of deoxy-Hb S, leading to a precipitation of the hemoglobin within the red blood cell. The precipitation of the hemoglobin gives the red blood cell a sickle shape, an instability that results in a high rate of hemolysis and a lack of elasticity in diffusion through the small capillaries, which become clogged by the abnormal red blood cells.

Only individuals homozygous for Hb S exhibit the disease. Individuals heterozygous for Hb S have approximately 50% Hb A_1 and 50% Hb S in their red blood cells and do not exhibit symptoms of the sickle cell anemia disease unless under extreme hypoxia.

It is of interest that individuals heterozygous for Hb S have a resistance to the malaria parasite, which spends a part of its life cycle in the red blood cell. This is a factor selecting for the Hb S gene in malarial regions of the world and the reason for the high frequency of this homozygous lethal gene in the human genetic pool. It is known that approximately 10% of American blacks are heterozygous for Hb S, and 0.4% of American blacks are homozygous for Hb S and exhibit sickle cell anemia.

Figure 2.35
Polypeptide chain showing ϕ, ψ, and peptide bond for residue R_i within chain.
δ^+ and $\delta-$ indicate partial charge due to delocalization of bonding electrons between carbonyl oxygen and peptide bond nitrogen (see text and Figure 2.36).

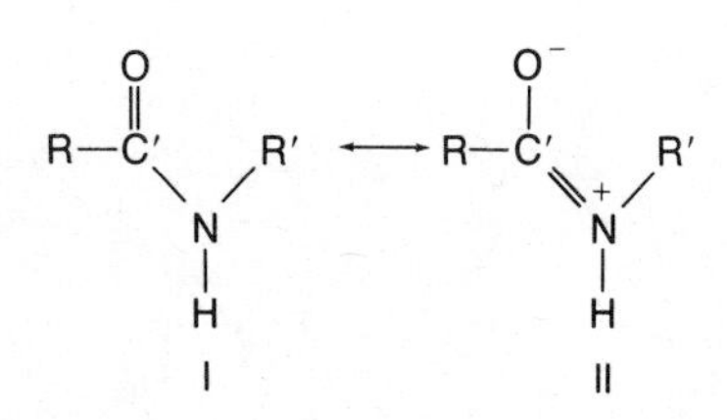

Figure 2.36
Peptide bond resonance structures.

α-carbon, and (2) the α-carbon and the carbonyl carbon, and (3) the peptide bond between the carbonyl carbon and the nitrogen of the adjacent amino acid.The first of these is designated the phi (ϕ) bond and the second the psi (ψ) bond for an amino acid residue in a polypeptide chain (Figure 2.35).

Figure 2.36 shows that the peptide bond can be depicted by two resonance configurations. In structure I, a double bond is located between the carbonyl carbon and carbonyl oxygen, and the carbonyl carbon to nitrogen bond ($C'—N_\alpha$) is depicted as a single bond. In the electronic isomer, structure II, the carbonyl carbon to nitrogen bond is a formal double bond. Peptide bonds are a resonance hybrid of the two electron isomer

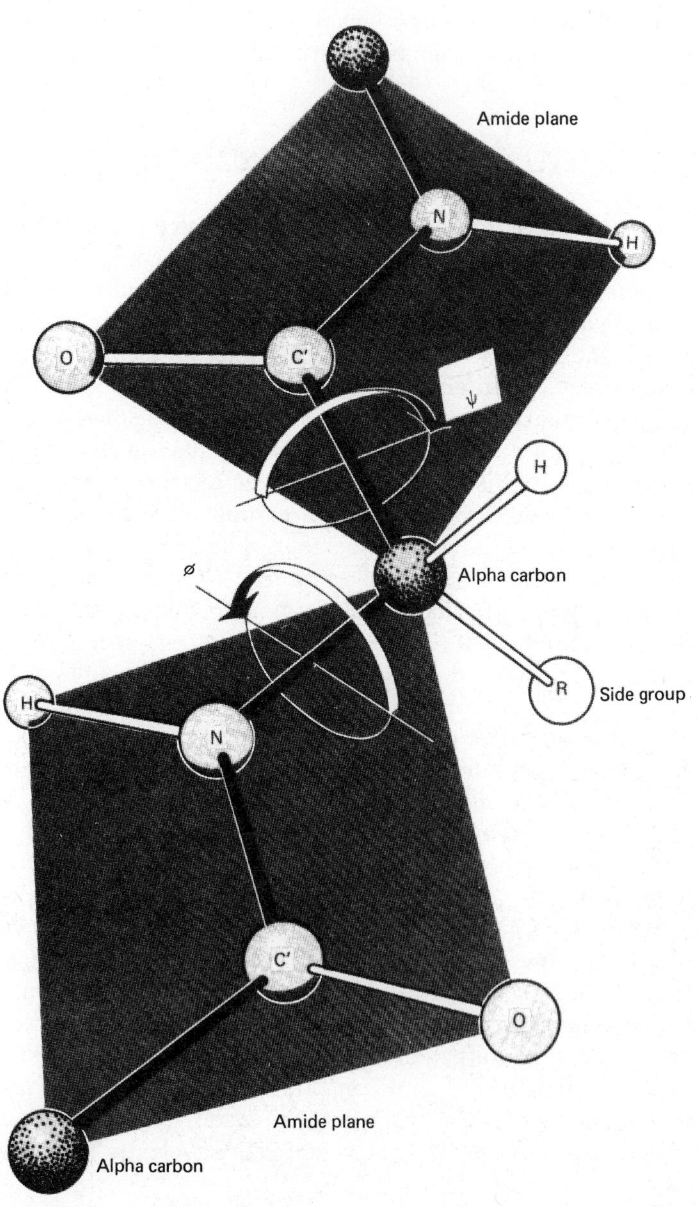

Figure 2.37
Rotation about ϕ and ψ bonds but not about peptide bonds within a polypeptide chain.
The carbonyl carbon (C'), carbonyl O, peptide nitrogen (N_α) and H lie in a plane generated from the double bond character of the $C'—N_\alpha$ peptide bond.
Reprinted with permission from R. E. Dickerson and J. Geis, *The Structure and Action of Protein*, W. A. Benjamin, Inc., Menlo Park, 1969, p. 25.

representations of Figure 2.36. The carbonyl carbon to nitrogen bond ($C'—N_\alpha$) thus contains a substantial double-bond character. The consequence of the double-bond nature of the $C'—N_\alpha$ bond is that free rotation about this bond does not normally occur at physiological temperature. Accordingly, rotations occur in only two of the three bonds (i.e. the ϕ and ψ bonds) contributed by each amino acid residue within a polypeptide chain (Figure 2.37).

Regular secondary structure conformations in segments of a polypeptide chain occur when all the ϕ bond angles in that polypeptide segment are equal to each other, and all the ψ bond angles are equal. For example, when the ϕ angles $= -57°$ and the ψ angles $= -47°$, a polypeptide segment will be in an α helix conformation. When $\phi = -139°$ and $\psi = +135$, an antiparallel β-structure conformation is found.

The α-helix and β-structure conformations for polypeptide chains are generally the most thermodynamically stable of the regular secondary structures. However, particular amino acid sequences of a primary structure in a protein may support regular conformations of the polypeptide chain other than α-helical or β-structure. Thus, whereas α-helical and β-structures are found most commonly, the actual conformation is dependent on the particular physical properties generated by the sequence present in the polypeptide chain and the solution conditions in which the protein is dissolved. In addition, in most proteins there are significant regions of "random" structure in which the ϕ and ψ angles are not equal.

Proline will break α-helical conformations of the polypeptide chain since the pyrrolidine side chain group of proline sterically interacts with the side chain group from the amino acid preceding it in the polypeptide sequence when the preceding amino acid has a ψ angle of $-47°$ as required for the α-helical structure. This repulsive steric interaction will prevent formation of α-helical structures in regions of a polypeptide chain in which proline is found.

As shown in Figure 2.38, the helical structures of polypeptide chains are geometrically defined by the number of amino acid residues per 360° turn of the helix (n) and the distance between α-carbons of adjacent amino acids measured parallel to the axis of the helix (d). The helix pitch (p), defined by the equation below, measures the distance between repeating turns of the helix on a line drawn parallel to the helix axis.

$$p = d \times n$$

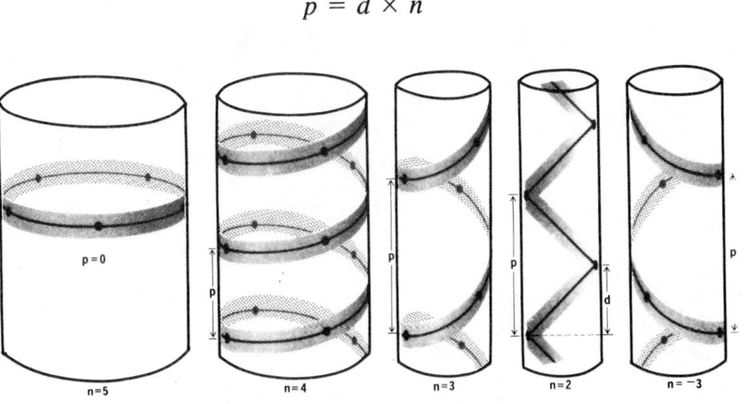

Figure 2.38
Helix parameters. Definition of pitch (p) and number of residues per turn of helix (n).
The rise per residue along the helix would be p/n (see equation in text). Each circle on line represents an α-carbon from an amino acid residue.
Reprinted with permission from R. E. Dickerson and J. Geis, *The Structure and Action of Proteins*, W. A. Benjamin, Inc., Menlo Park, 1969, p. 26.

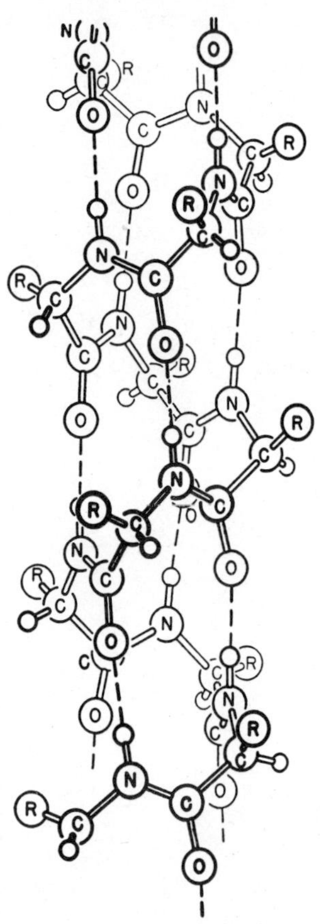

Figure 2.39
An α helix.
Reprinted with permission from L. Pauling, *The Nature of the Chemical Bond,* 3rd ed., Cornell University Press, Ithaca, N.Y., 1960.

α-Helical Structure

A polypeptide chain in an *α-helical conformation* is shown in Figure 2.39. Characteristic of the α-helical conformation are 3.6 amino acid residues per 360° turn ($n = 3.6$), $d = 1.5$ Å, and $p = 5.4$ Å.

The peptide bonds in the α-helix are directed parallel to the axis of the helix. In this geometry each peptide forms a hydrogen bond to the peptide bond of the *fourth* amino acid above and the peptide bond of the *fourth* amino acid below in the primary structure. The distance between the hydrogen donor atom and the hydrogen acceptor atom in the α-helix is 2.9 Å. Also, the donor atom, acceptor atom, and hydrogen atom in the α-helix peptide hydrogen bond are colinear, that is, they lie on a straight line. This is an optimum geometry and distance for maximum hydrogen bond strength (see Section 2.6).

The side chain groups in the α-helix conformation are perpendicularly projected from the axis of the helix on the outside of the spiral structure generated by the polypeptide chain. Due to the characteristic 3.6 residues per turn of the α-helix, every third and fourth R group in the amino acid sequence of the helix comes close to the other. If these R groups have the same charge sign or are branched at their β-carbon (valine, isoleucine), their interactions will destabilize the helix structure.

The α-helix can form its spiral in either a left-handed sense or right-handed sense, giving the helical structure asymmetric properties and a correlated property of optical activity in solution. In the structure shown, the more stable right-handed α-helix is depicted.

β-Structure

A polypeptide chain in a β-structure conformation is shown in Figure 2.40. In β-structure, segments of a polypeptide chain are in an extended helix with $n = 2$, $d = 3.5$ Å, and $p = 6.95$ Å. The helix of one polypeptide chain or segment is hydrogen-bonded to *other helices* in β-structure conformations. The polypeptide segments in a β-structure are aligned either in a parallel or antiparallel direction to its neighboring chains (Figure 2.41). Large numbers of polypeptide chains interhydrogen bonded in a β-type structure give a pleated sheet appearance (see Figure 2.42). In this structure the side chain groups are projected above and below the planes generated by the hydrogen-bonded polypeptide chains.

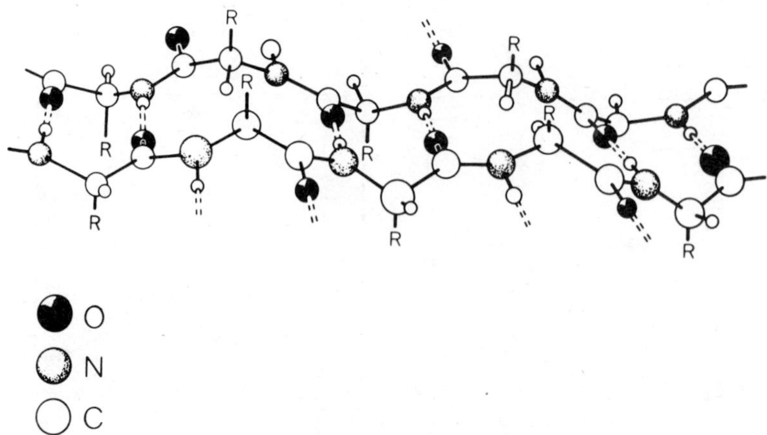

Figure 2.40
Two polypeptide chains in a β structure conformation.
Additional polypeptide chains may be added to generate more extended structure.
Reprinted with permission from A. Fersht, *Enzyme Structure and Mechanism,* Freeman, San Francisco, 1977, p. 10.

Tertiary Structure of Proteins

The tertiary structure of a protein refers to the total three-dimensional structure of the protein, including the geometric relationship between distant segments of the primary structure and the relationship of the side chain group with respect to each other in three-dimensional space. As an example of a protein's tertiary structure, the structure for cytochrome c is shown in Figure 2.43. In this figure the protein's side chain groups are not shown in order to make the general polypeptide conformation clear. Accordingly, the ribbon diagrammatically shows the relationship in space of the polypeptide segments with respect to each other, with the α-carbons of the amino acid residues in the primary structure indicated by circles. Also shown are the side chain groups of Met-80 and His-18 that form coordinate bonds to a heme group.

The heme is a nonprotein moiety that strongly associates with the polypeptide chain of cytochrome c and is essential for the function of the protein in electron transport. A nonprotein moiety that associates with a polypeptide chain of a protein in its functional state is called a *prosthetic group*. A protein without its normal or characteristic prosthetic group is referred to as the *apoprotein*.

It is the tertiary structure that brings together the Met-80, His-18, and other residues widely separated in the primary structure to form the heme binding site of the protein.

The tertiary structure of cytochrome c conforms to the general rules of folded proteins discussed previously (Section 2.2). The hydrophobic side chains are generally in the interior of the structure, away from the water interface. Ionized amino acid side chains are found on the outside of a protein structure, where they are stabilized by water of solvation.

Within the protein structure (not shown) are buried water molecules, noncovalently associated, in specific arrangements. In general, noncovalently bound water molecules typically comprise 5–60% of a protein's mass by weight.

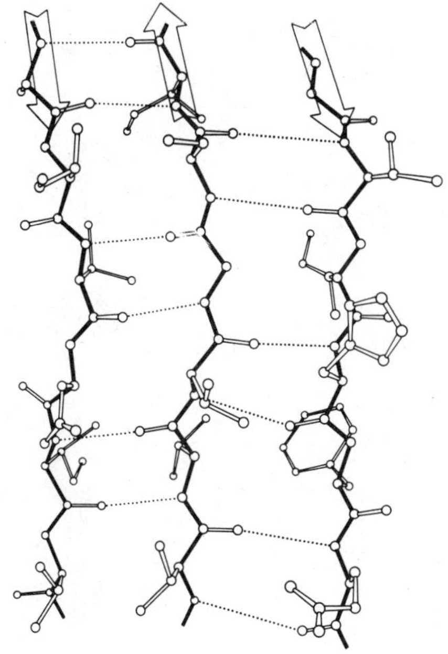

Figure 2.41
An example of antiparallel β-structure (residues 93–98, 28–33, and 16–21 of Cu, Zn superoxide dismutase).
Dotted line shows hydrogen bonds between carbonyl oxygens and peptide nitrogens; arrows show direction of polypeptide chains from N-terminal to C-terminal. In the pattern characteristic of antiparallel β-structure, pairs of closely spaced interchain hydrogen bonds alternate with widely spaced pairs.
Reprinted with permission from J. S. Richardson, *Advan. Protein Chem.* 34:168, 1981.

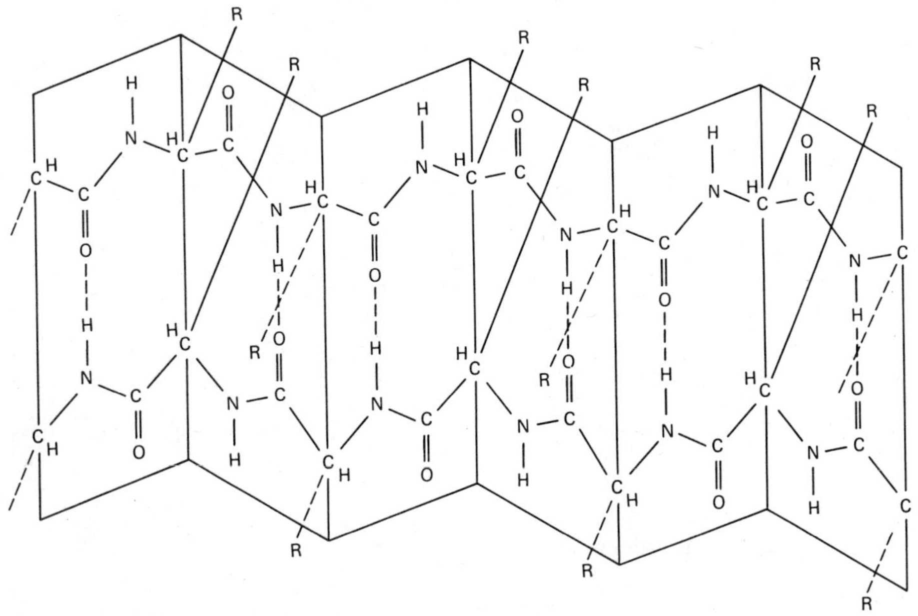

Figure 2.42
β-Pleated sheet structure between two polypeptide chains.
Additional polypeptide chains may be added above and below to generate more extended structure.

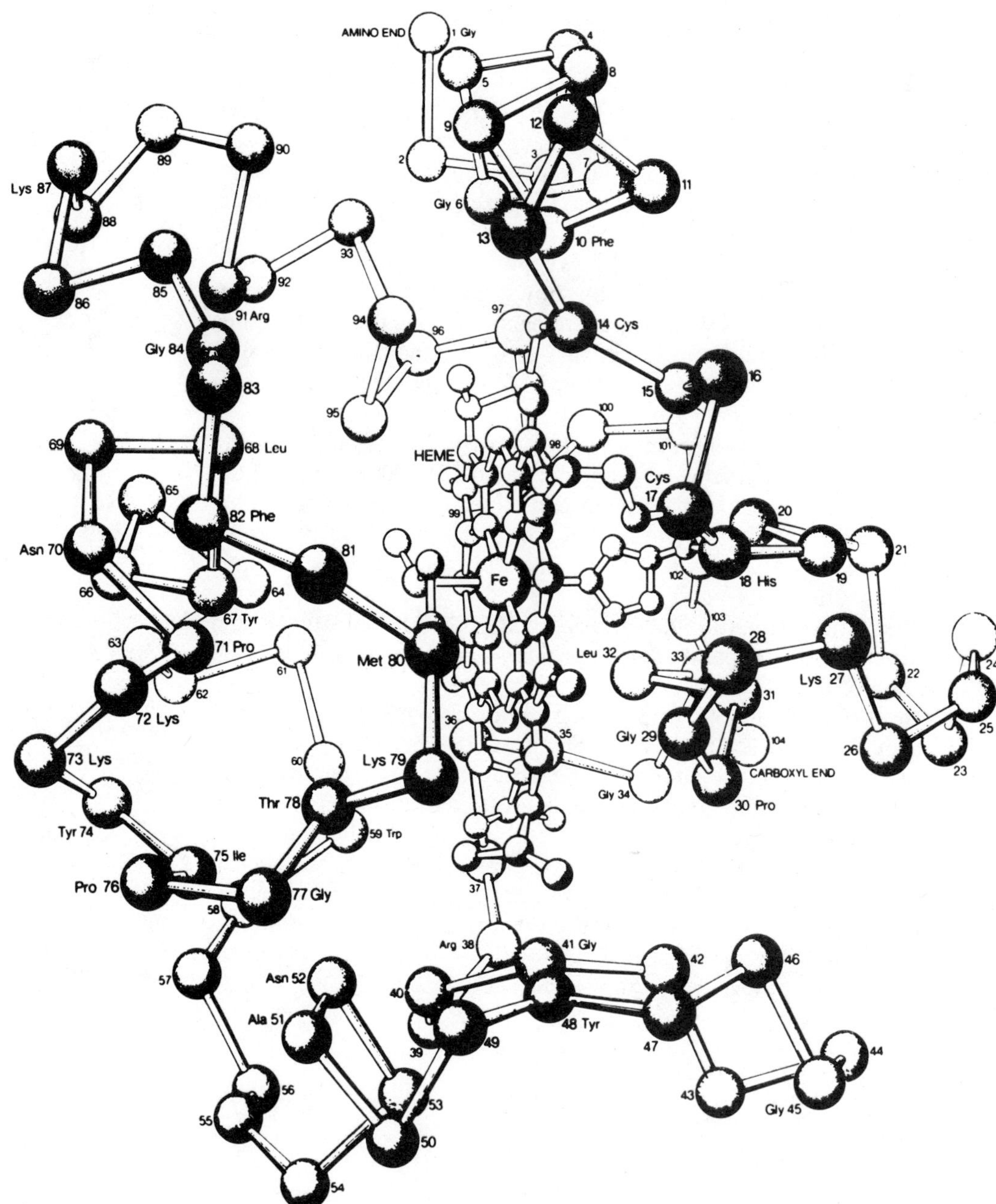

Figure 2.43
Three-dimensional structure of the protein cytochrome c.
Only α-carbons are shown for each of the amino acids in the polypeptide chain except for a methionine and histidine, which are liganded to the prosthetic group in the center of the structure. Amino acids are numbered from NH_2-terminal end (at top) to COOH-terminal end at lower right.
Reprinted with permission from R. E. Dickerson and R. Timbovich, *Enzymes* 11:407, 1975.

A long polypeptide strand of a protein will often fold into multiple compact semi-independent folded regions or *domains,* each domain having a characteristic spherical geometry with a hydrophobic core and polar outside very much like the tertiary structure of a whole globular protein. In fact, on separation of a domain unit from its polypeptide chain by hydrolysis of the polypeptide chain at a point just outside the domain's structure, the isolated domain's primary structure will potentially have an ability to fold by itself into its native domain conformation in isolation.

The domains of a multidomain protein are often interconnected by a segment of the polypeptide chain lacking regular secondary structure. Alternatively, the dense spherical folded regions can be separated by a cleft or less dense region in the tertiary structure of the protein (Figures 2.44).

The different domains within a protein can exhibit a relative motion with respect to each other, and these motions can have a functional purpose. The enzyme hexokinase (Figure 2.45), which catalyzes the phosphorylation of a glucose molecule by adenosine triphosphate, has the glucose substrate-binding site at a cleft between two domains. When the glucose molecule binds in the active site cleft, the surrounding domains close over the substrate to trap it for phosphorylation (Figure 2.45). In enzymes with more than one substrate or allosteric effector sites (Chapter 4), the different binding sites are often located in different domains. In multifunctional proteins, the different domains can each perform a different task.

A protein folds into a unique conformation for a particular sequence. Even though each native structure is unique, a comparison of tertiary structures of different proteins shows similar patterns in the arrangement

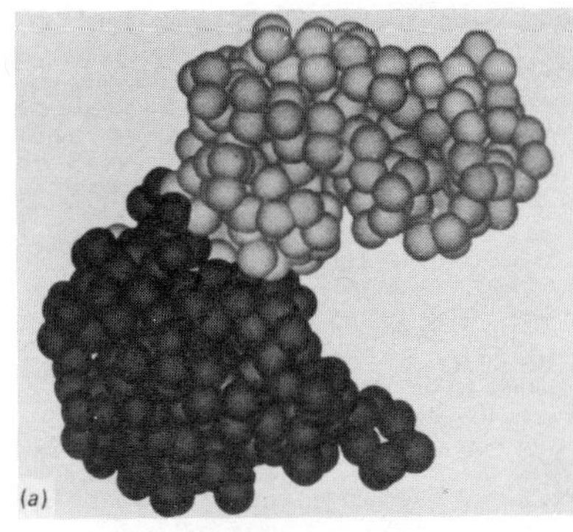

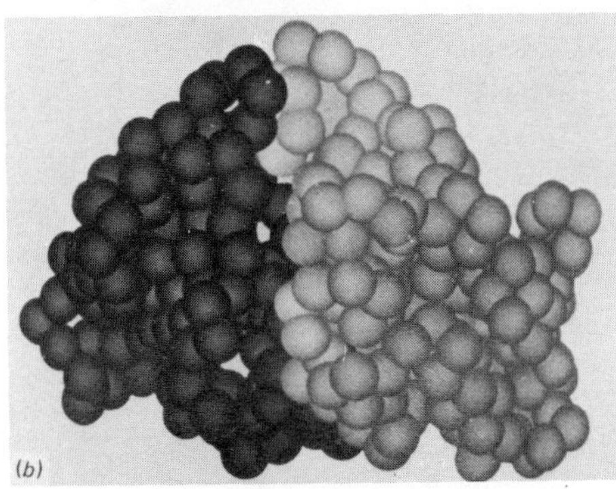

Figure 2.44
Globular domains within proteins.
(a) Phosphoglycerate kinase has two domains with a relatively narrow neck in between. (b) Elastase has two tightly associated domains separated by a narrow cleft. Each sphere in space-filling drawing represents the alpha-carbon position for an amino acid within the protein structure.
Reprinted with permission from J. S. Richardson, *Advan. Protein Chem.* 34:168, 1981.

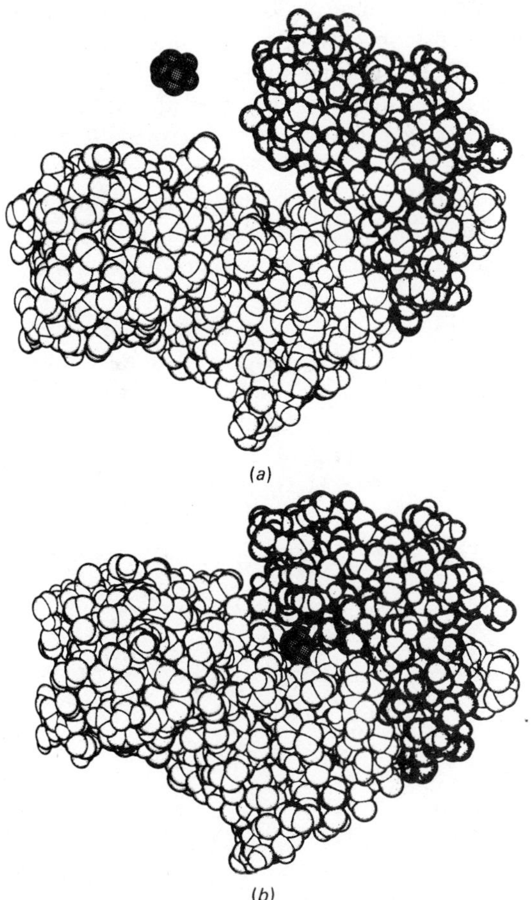

Figure 2.45
Drawings of* (a) *unliganded form of hexokinase and free glucose and* (b) *the conformation of hexokinase with glucose bound.
In this space-filling drawing each circle represents the van der Waals radii of an atom in the structure. Glucose is crosshatched, and each domain is differently shaded.
Reprinted with permission from W. S. Bennett and R. Huber, *CRC Rev. Biochem.* 15:291, 1984. Copyright CRC Press, Inc., Boca Raton, FL.

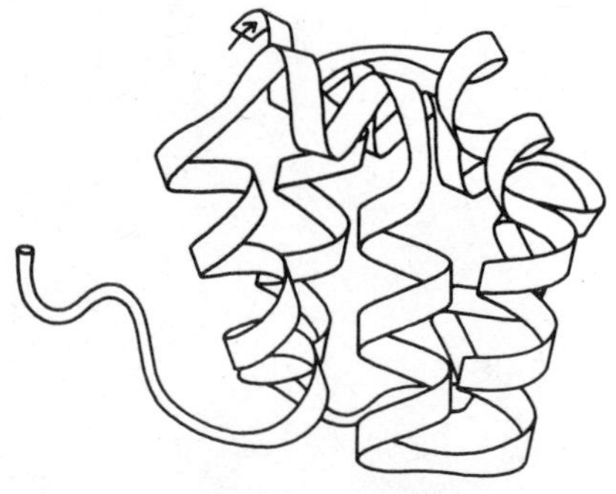

Figure 2.46
An example of an all α-folded domain.
In this drawing and the drawings that follow (Figures 2.47–2.49), only the outline of the polypeptide chain is shown. β-structure strands are shown by arrows with the direction of the arrow showing the N→C terminal direction of the chain; lightning bolts represent disulfide bonds, and circles represent metal ion cofactors (when present).
Reprinted with permission from J. S. Richardson, *Advan. Protein Chem.* 34:168, 1981.

of the secondary structure in the tertiary structures of domains. Thus proteins unrelated by function, exact sequence, or evolution will generate similar patterns of arrangement of their secondary structure. A classification system has emerged for secondary structure patterns that places most proteins or domains within proteins into four categories. An *all-α-structure* for domain 2 of the enzyme lysozyme is shown in Figure 2.46. Other examples of an all-α-structure are found in myoglobin and the subunits of hemoglobin, whose structures are extensively discussed in Chapter 3. In this folding pattern, sections of α-helices are joined by smaller segments of polypeptide chains that allow the helices to fold back upon themselves to form a spherical or globular mass.

An example of an *α/β structure* is shown by triose phosphate isomerase (Figure 2.47) in which the β-structure strands (designated by arrows) are wound into a supersecondary structure called a β barrel. Each β-structure strand in the interior of the β barrel is interconnected by α-helical regions of the polypeptide chain on the outside of the molecule. A similar supersecondary structure arrangement is found for domain 1 of pyruvate kinase (Figure 2.47). A different type of α/β structure is seen in domain 1 of lactate dehydrogenase and domain 2 of phosphoglycerate kinase (Figure 2.48) in which the interior polypeptide sections participate in a classical twisted β structure. The β-structure segments are again each joined by α-helical regions positioned on the outside of the molecule to give a characteristic α/β folding pattern.

An *all-β-structure* pattern is observed in Cu, Zn superoxide dismutase, in which the antiparallel β structure strands form a tertiary pattern called a Greek key β barrel (Figure 2.49). A similar pattern is seen in the folding of each of the domains of the immunoglobulins, discussed in Chapter 3. Concanavalin A (Figure 2.49) also shows an all-β-structure in which the segments of antiparallel β-structure assume a β barrel pattern called a "jellyroll."

Cytochrome c (Figure 2.43) is an example of a fourth category of supersecondary structure folding patterns generally found for small cystine-rich or metal-rich proteins. There are currently approximately 100 different domains from the sampling of known protein structures that can be fitted into these four patterns of supersecondary structure.

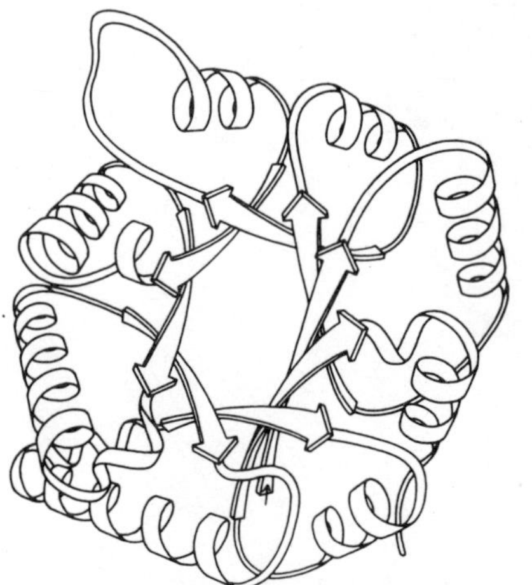

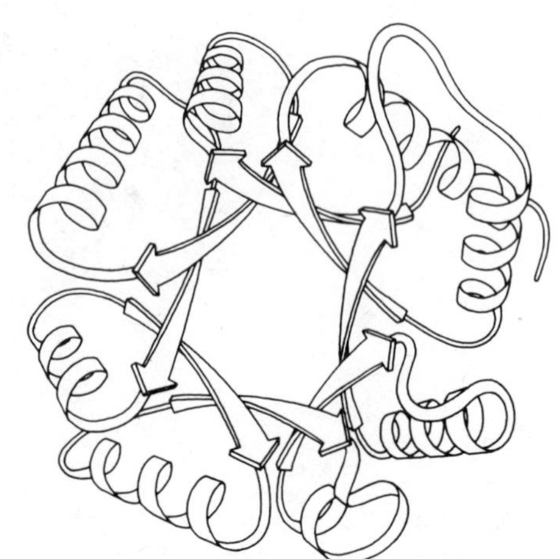

Figure 2.47
Examples of α/β-folded domains in which β-structural strands form a β-barrel in the center of the domain (see legend to Figure 2.46).
Reprinted with permission from J. S. Richardson, *Advan. Protein Chem.* 34:168, 1981.

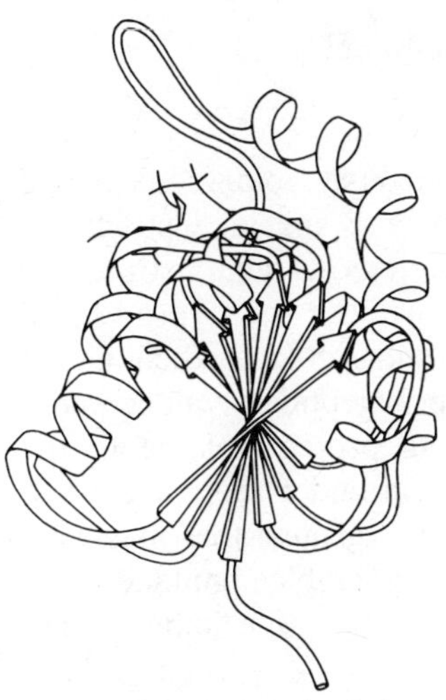

Lactate Dehydrogenase domain 1

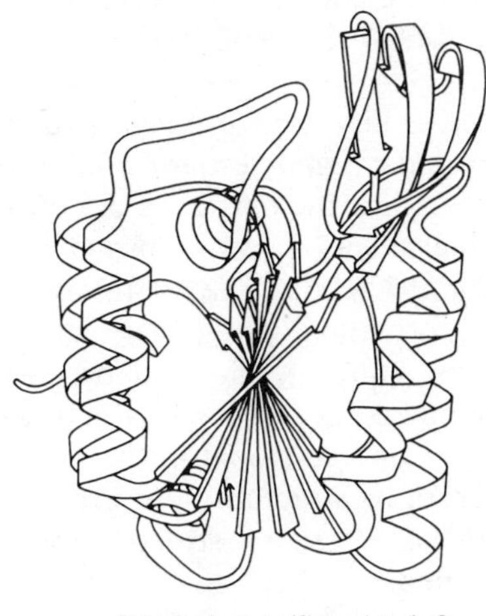

Phosphoglycerate Kinase domain 2

Figure 2.48
Examples of α/β-folded domains in which β-structure strands are in the form of a classical twisted β-sheet (see legend to Figure 2.46).
Reprinted with permission from J. S. Richardson, *Advan. Protein Chem.* 34:168, 1981.

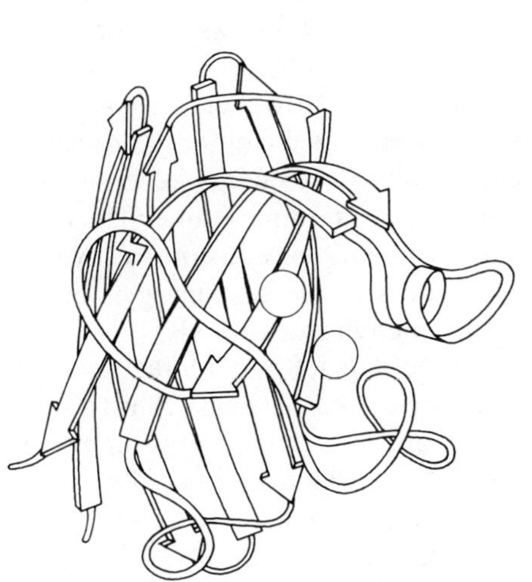

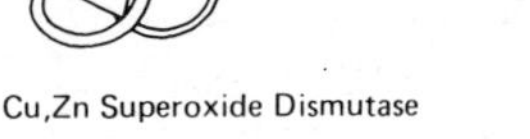

Cu,Zn Superoxide Dismutase

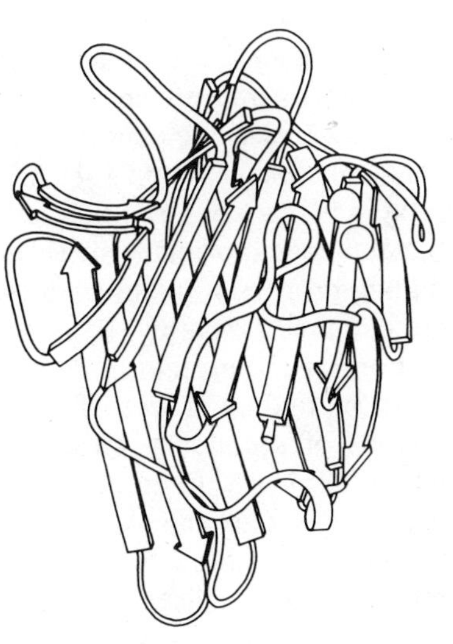

Concanavalin A

Figure 2.49
Examples of all β-folded domains (see legend to Figure 2.46).
Reprinted with permission from J. S. Richardson, *Advan. Protein Chem.* 34:168, 1981.

Quaternary Structure of Proteins

The quaternary structure of a protein is the arrangement or conformation of polypeptide chain units in a multichain protein. The subunits in a quaternary structure must be in *noncovalent* association. Thus the enzyme α-chymotrypsin contains three polypeptide chains covalently joined together by interchain disulfide bonds into a single covalent unit. Although chymotrypsin contains three polypeptide chains, it is not considered to have a quaternary structure. The protein myoglobin is composed of a single polypeptide chain and contains no quaternary structure. However, hemoglobin contains four polypeptide chains held together *noncovalently* in a specific conformation (see Chapter 3). Thus hemoglobin has a quaternary structure. The enzyme aspartate transcarbamylase (see Chapter 13) has a quaternary structure comprised of 12 polypeptide subunits. The poliovirus protein coat contains 60 polypeptide subunits, and the tobacco mosaic virus protein has 2,120 polypeptide subunits held together noncovalently in a specific structural arrangement.

2.6 FOLDING OF PROTEINS FROM RANDOMIZED TO UNIQUE STRUCTURES: PROTEIN STABILITY

The ability of a primary structure to spontaneously fold to its native secondary and tertiary conformation, without any special instructions other than the existence of noncovalent interactions, is demonstrated by experiments in which proteins are denatured without the hydrolysis of peptide bonds. These proteins, on standing, will refold to their native conformation. The experiments show that a polypeptide sequence contains sufficient physical properties to promote protein folding to the unique conformation characteristic of the protein under the correct solvent conditions and in the presence of prosthetic groups that may be a part of its structure. Quaternary structures also assemble spontaneously, after the tertiary structure of the individual polypeptide subunits are formed.

It may appear surprising that a protein folds into a single unique conformation from all the possible *a priori* rotational conformations available around single bonds in the primary structure of a protein. For example, the α chain of hemoglobin contains 141 amino acids in which there are 4 to 9 single bonds per amino acid residue around which free rotation can occur. If each bond about which free rotation occurs has two or more stable rotamer conformations accessible to it, then there are a minimum of 4^{141}–9^{141} possible conformations for this chain. However, only a single conformation is found for the α chain. The folded conformation of a protein is that conformation of the lowest Gibbs free energy accessible to the amino acid primary structure within a specified time frame. Thus the folding process is under both thermodynamic and kinetic control. A discrete number of pathways to the native structure of lowest energy conformation for the protein in its native solution environment exist.

In the folding process short-range interactions initiate folding of small regions of secondary structure in the polypeptide strand. Short-range interactions are the noncovalent interactions that occur between a side chain (R) group and the polypeptide chain to which it is covalently attached. Particular side chain R groups have a propensity to promote the formation of α-helices, β-structure, and sharp turns or bends (β turns) in the polypeptide strand. The interaction of a particular side chain group with those of its nearest neighbors in the polypeptide sequence form a grouping that determines the secondary structure into which the section of the polypeptide strand folds. Sections of the polypeptide strand, called *initiation sites,* will thus spontaneously fold into small regions of secondary structure. Medium- and long-range interactions between different initiation sites then stabilize a folded tertiary structure for the polypeptide chain. Disulfide cystine bonds are formed between cysteines in the primary structure after the protein has folded correctly.

Since it is noncovalent forces that act on the primary structure to cause a protein to fold into a unique conformational structure and then stabilize the native structure against denaturation processes, it is of importance to understand the properties of these forces. Some of the important properties of these noncovalent forces are discussed in the section that follows.

Noncovalent Forces That Lead to Protein Folding and Contribute to a Protein's Stability

Noncovalent forces are weak bonding forces of bonding strength of 1–7 kcal/mol (4–29 kJ/mol). This may be compared to the strength of covalent bonds that have a bonding strength of at least 50 kcal/mol. The noncovalent bonding forces are just higher then the average kinetic energy of molecules at 37°C (0.6 kcal/mol) (Table 2.9). However, the large number of individually weak noncovalent contacts within a protein add up to a large energy factor that is a net thermodynamic force favoring protein folding.

TABLE 2.9 Bond Strength for Typical Bonds Found in Protein Structures

Bond Type	*Bond Strength (kcal/mol)*
Covalent bonds	>50
Noncovalent bonds	0.6–7
Hydrophobic bond (i.e., 2 benzyl side chain groups of Phe)	2–3
Hydrogen bond	1–7
Ionic bond (low dielectric environment)	1–6
van der Waals	<1
Average energy of kinetic motion (37°C)	0.6

Hydrophobic Bonding Forces

The most important of the noncovalent forces that will cause a randomized polypeptide conformation to lose rotational freedom and fold into its native structure are hydrophobic bonding forces. It is important to realize that the strength of a hydrophobic bond is not due to a high intrinsic attraction between nonpolar groups, but rather to the properties of the water solvent in which the nonpolar groups are dissolved.

A nonpolar residue dissolved in water induces in the water solvent a solvation shell in which water molecules are highly ordered. When two nonpolar groups come together on the folding of a polypeptide chain, the surface area exposed to solvent is reduced and a part of the highly ordered water in the solvation shell is released to bulk solvent. Accordingly, the entropy of the water (i.e., net disorder of the water molecules in the system) is increased. The increase in entropy (disorder) is a thermodynamically favorable process, and is the driving force causing apolar moieties to come together in aqueous solvent. A favorable free energy change of approximately 2 kcal/mol for the association of two phenylalanine side chain groups in water is due to this favorable solvent entropy gain.

Calculations show that in the folding of a randomized conformation into a regular secondary conformation such as an α-helix or β-structure, approximately one-third of the ordered water of solvation about the unfolded polypeptide is lost to bulk solvent. This is an approximate driving force favoring folding in a typical globular protein of 0.5–0.9 kcal/mol per peptide residue. It is calculated that an additional one-third of the original solvation shell is lost when a protein already folded into a secondary structure then folds into a tertiary structure. The tertiary folding brings different segments of folded polypeptide chains into close proximity with the release of water of solvation between the polypeptide chains.

Hydrogen Bonds

A second important noncovalent force in proteins is hydrogen bonding. Hydrogen bonds are formed when a hydrogen atom covalently bonded to an electronegative atom is shared with a second electronegative atom. The atom to which the hydrogen atom is covalently bonded is designated the donor atom. The atom with which the hydrogen atom is shared is the hydrogen acceptor atom. Typical hydrogen bonds found in proteins are shown in Figure 2.50. It has been previously shown that α-helical and β-structure conformations are extensively hydrogen bonded.

The strength of a hydrogen bond is at a first approximation dependent on the distance between the donor and acceptor atoms. High bonding energies occur when the donor and acceptor atoms are between 2.7 and 3.1 Å apart. Of lesser importance to bonding strength than the distance requirement, but still of some importance, is the dependence of hydrogen bond strength on geometry. Bonds of higher energy are geometrically colinear, with donor, hydrogen, and acceptor atoms lying in a straight line. The dielectric constant of the medium around the hydrogen bond may also be reflected in the bonding strength. Typical hydrogen bond strengths in proteins are 1–7 kcal/mol.

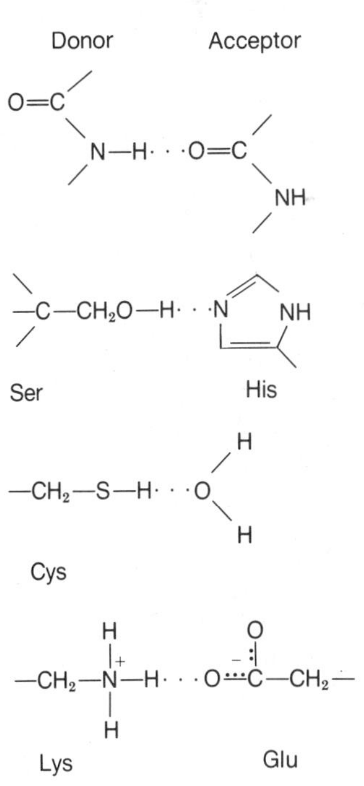

Figure 2.50
Some common hydrogen bonds found in proteins.

Although hydrogen bonds do contribute to the thermodynamic stability of a folded protein's conformation, their formation within a native protein structure may not be as major a driving force for folding as we might at first believe. This is because peptide groups and other hydrogen-bonding groups in proteins form hydrogen bonds to the water solvent in the denatured state, and these bonds must be broken before the protein folds. The energy required to break the hydrogen bonds to water must be subtracted from the energy gained from the formation of the new hydrogen bonds between atoms in the folded protein in the calculation of the net thermodynamic contribution of hydrogen bonding to the folding.

$$\Delta E_{el} \sim \frac{Z_A \cdot Z_B \cdot \varepsilon^2}{D \cdot r_{ab}}$$

Figure 2.51
Strength of electrostatic interactions.

Electrostatic Bonds

Electrostatic interactions between charged groups are of importance to particular protein structures and in the binding of charged ligands and substrates to proteins. Electrostatic forces can be repulsive or attractive depending on whether the interacting charges are of the same or opposite sign. The strength of an electrostatic force (ΔE_{el}) is directly dependent on the charge (Z) for each ion, and is inversely dependent on the dielectric constant (D) of the solvent and the distance between the charges (r_{ab}) (Figure 2.51).

Water has a high dielectric constant ($D = 80$), and ionic charge interactions in water are relatively weak in comparison to electrostatic interactions in the interior of a protein, where the dielectric constant ($D = 2$–40) is approximately a factor of 2 to 40 times lower than in water. Consequently, the strength of an electrostatic interaction in the interior of a protein, where the dielectric constant is low, may be of significant energy. However, most charged groups of proteins are on the surface of the protein where they do not strongly interact with other charged groups from the protein due to the high dielectric constant of the water solvent, but are stabilized by hydrogen bonding and polar interactions to the water. These water interactions are the driving force leading to the placement of most ionic groups of a protein on the outside of the protein structure, where they can make energetically favorable contacts with the solvent.

$$E = -\frac{A}{r_{ab}^{6}} + \frac{B}{r_{ab}^{12}}$$

Figure 2.52
Van der Waals forces.

Van der Waals–London Dispersion Forces

Van der Waals and London dispersion forces are a fourth type of weak noncovalent force of great importance in protein structure. This force has an attractive term (A) dependent on the 6th power of the distance between two interacting atoms (r_{ab}), and a repulsive term (B) dependent on the 12th power of r_{ab} (Figure 2.52). The A term contributes at its optimum distance an attractive force of <1 kcal/mol per atomic interaction. This attractive component is due to the induction of complementary partial charges or dipoles in the electron density of adjacent atoms when the electron orbitals of the two atoms approach to a close distance. The repulsive component (term B) of the van der Waals force predominates at closer distances than the attractive force when the electron orbitals of the adjacent atoms begin to overlap. This type of repulsion is commonly called steric hindrance.

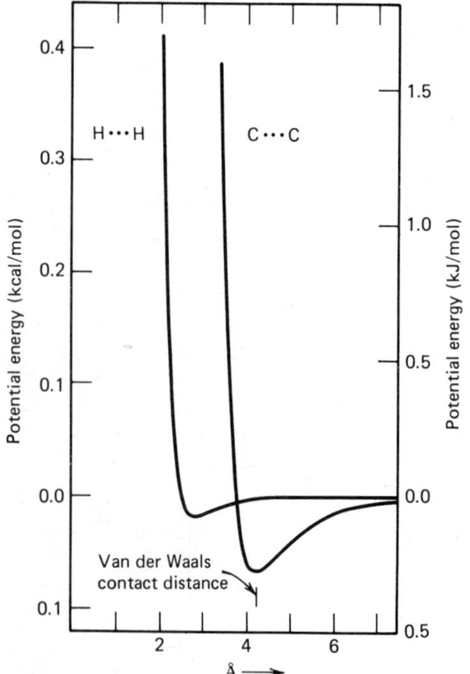

Figure 2.53
Van der Waals-London dispersion interaction energies between two hydrogen atoms and two (tetrahedral) carbon atoms.
Negative energies are favorable and positive energies unfavorable.
Redrawn from A. Fersht, *Enzyme Structure and Mechanism*, W. H. Freeman, San Francisco, 1977, p. 228.

The distance of maximum favorable interaction between two atoms is known as the van der Waals contact distance, which is equal to the sum of the van der Waals radii for the two atoms (Figure 2.53). Some van der Waals radii for atoms commonly found in proteins are given in Table 2.10. While a London dispersion–van der Waals interaction between any two atoms in a protein is usually less than 1 kcal/mol, the total number of these weak interactions in a protein molecule is in the thousands. Thus the sum of the attractive and repulsive van der Waals–London dispersion forces are extremely important to protein folding and stability.

The van der Waals contact distances of 2.8–4.1 Å are longer than hydrogen bond distances of 2.6–3.1 Å, and at least twice as long as

TABLE 2.10 Covalent Bond Radii and van der Waals Radii for Selected Atoms

Atom	Covalent Radius (Å)	van der Waals Radius (Å)[a]
Carbon (tetrahedral)	0.77	2.0
Carbon (aromatic)	0.69 along=bond 0.73 along—bond	1.70
Carbon (amide)	0.72 to amide N 0.67 to oxygen 0.75 to chain C	1.50
Hydrogen	0.33	1.0
Oxygen (—O—)	0.66	1.35
Oxygen (=O)	0.57	1.35
Nitrogen (amide)	0.60 to amide C 0.70 to hydrogen bond H 0.70 to chain C	1.45
Sulfur, diagonal	1.04	1.70

SOURCE: *CRC Handbook of Biochemistry and Molecular Biology*, 3rd ed., Sect. D, Vol. II, G. D. Fasman, ed., 1976, p. 221.

[a] The van der Waals contact distance is the sum of the two van der Waals radii for the two atoms in proximity.

normal covalent bond distances of 1.0–1.6 Å between C, H, N, and O atoms. Inasmuch as the latter bonds are shorter than the van der Waals contact distance, a repulsive van der Waals force must be overcome in forming hydrogen bonds and covalent bonds between atoms. The energy to overcome van der Waals repulsive force must form a part of the energy of activation for hydrogen bond and covalent bond formation.

A special type of interaction (π electron to π electron) occurs when two aromatic rings approach each other with the plane of their aromatic rings overlapping (Figure 2.54). This type of interaction can result in a noncovalent attractive force of up to 6 kcal/mol.

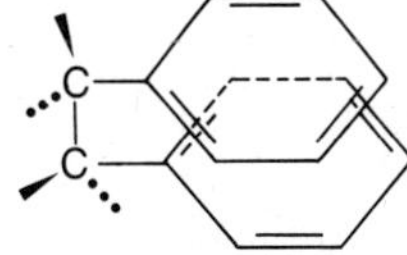

Figure 2.54
π Electron–π electron interactions between two aromatic rings.

Denaturation of Proteins

Denaturation occurs in a protein upon the loss of its native secondary, tertiary, and quaternary structures. In the denatured state the higher structural levels of the native conformation are randomized or scrambled. However, the primary (covalent) structure is not necessarily broken in denaturation.

The loss of protein function, such as a catalytic activity in an enzyme, may occur with small modifications of the protein's native conformation that do not lead to a complete scrambling of the secondary, tertiary, and quaternary levels of conformation. Denaturation is thus correlated with the loss of a protein's function; but loss of a protein's function, which may be due to only a small conformational change, is not necessarily synonymous with denaturation.

Even though the conformational differences between denatured and native structure are substantial, the free energy difference between the denatured structure and the native structure of a protein may in some cases be as low as the free energy of a single noncovalent bond. Thus the loss of a single structurally essential hydrogen bond, or electrostatic or hydrophobic interaction can lead to denaturation of a folded structure. A change in the stability of a noncovalent bond leading to denaturation can in turn be caused by a change in pH, ionic strength, and temperature, which affect the strength of noncovalent bonds. The presence of pros-

$$\text{Amino acids} \xrightarrow{\text{rate of synthesis}} [\text{protein}] \xrightarrow{\text{rate of denaturation}} \text{protein digest}$$

Figure 2.55
Steady-state concentration of a protein is due to both its rate of synthesis and denaturation.

thetic groups, cofactors, and the substrates of a protein may also affect the stability of the native conformation. These later ligands, which may have a role in the function of the protein, often also act to stabilize the native conformation of a protein.

The statement that the breaking of a single noncovalent bond in a protein can cause denaturation may appear to conflict with the observation, discussed in Section 2.4, that the amino acid sequence for a protein can often be extensively varied without loss of the native structure and related functional ability of the protein. The key to the resolution of the apparent conflict, between the extensive variability of amino acid sequence present in many proteins and the possible ease of denaturation, is the word "essential." Many noncovalent interactions in a protein are apparently not essential to the protein's overall thermodynamic stability. However, the substitution or modification of an *essential* amino acid residue of a protein that provides a critical noncovalent interaction will dramatically affect the stability of a native protein structure relative to a denatured conformation.

The concentration of a protein in vivo is under the influence of processes that both control the rate of the protein's synthesis and control the protein's rate of degradation (Figure 2.55). Therefore, an understanding of the processes that control protein degradation may be as important as an understanding of the process of protein synthesis. It is believed that the inherent denaturation rate for a protein, in many cases, is the rate-determining step in a protein's degradation. Enzymes and cellular organelles that participate in the digestion of proteins appear to "recognize" denatured protein conformations and digest these denatured conformations at faster rates than proteins in their native conformation.

In experimental situations, denaturation of a protein can often be achieved by addition of urea or detergents (sodium dodecyl sulfate, guanidine hydrochloride) that act to weaken hydrophobic bonding in proteins. Thus these reagents stabilize the denatured state and shift the equilibrium toward the denatured form of the protein. Addition of strong base, acid, or organic solvent, or heating to temperatures above 60°C are also common ways to denature a protein.

2.7 DYNAMIC ASPECTS OF PROTEIN STRUCTURE

X-Ray diffraction analysis of crystalline proteins (see Section 2.8) show that the amino acid residues within the interior of a globular protein are closely packed. The arrangement of amino acid residues appears so dense in the x-ray diffraction that in most regions of the interior the addition of a single methylene group to a side chain would prevent the side chain from fitting into the structure. However, there are observed regions of "defects" in the packing that indicate that "holes" exist in the structure that give the protein space for flexibility. In addition, the diffraction from some regions of a protein may not be well defined, indicating that the structure of this region, even in the crystalline state, is flexible and not well ordered. These pieces of evidence with data from NMR and fluorescence spectroscopy (see Section 2.8) show that in many proteins regions of flexibility and motion exist even in the interior. The concept that each

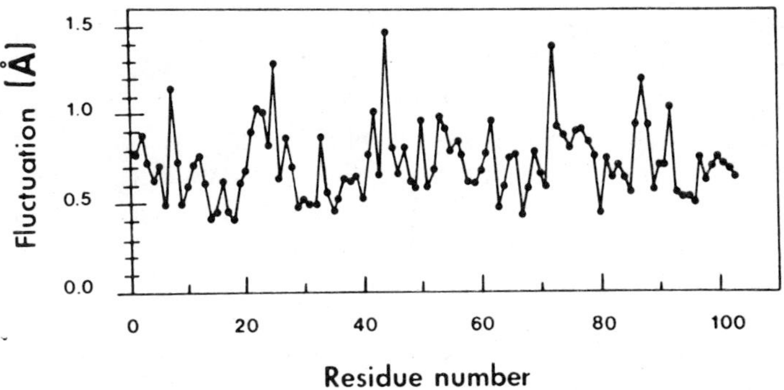

Figure 2.56
Calculated fluctuations (in angstroms) for each of the atoms in the primary structure of cytochrome c.
Reprinted with permission from J. A. McCammon and M. Karplus, *Acco. Chem. Res.* 16:187, 1983. Copyright 1983, American Chemical Society.

of the atoms in a protein structure is in constant rapid jiggling movement, such as molecules within a fluid, is an important aspect of globular protein structure. Theoretical calculations made for particular protein structures indicate that the average atom within a protein may be oscillating over a distance of 0.9 Å on a picosecond (10^{-12} s) time scale. Some atoms or groups of atoms will be moving smaller distances and others larger distances than this calculated average (Figure 2.56).

The net movement of any atom or segment of a polypeptide chain represents the sum of forces due to rapid, local jiggling and slower elastic movements of covalently attached groups of atoms. These movements within the closely packed interior of a protein molecule are large enough to allow buried tyrosine side chain rings to flip in the interior of some globular proteins.

The functional significance of the internal motions within proteins is not known in most cases. However, in the case of the proteolytic digestive enzyme trypsin, a comparison has been made between the atomic oscillations in the active enzyme and its inactive precursor trypsinogen. The trypsinogen molecule has a higher conformational disorder, correlated with a higher degree of internal atomic mobility than trypsin. It is believed that the extent of atomic flexibility in the substrate binding region of the trypsinogen structure is a factor that decreases the ability of the protein to effectively bind and catalytically transform substrate to product. On conversion of trypsinogen to trypsin, the loss in atomic movements at the substrate binding site enhances the ability of the enzyme to act as a catalyst. This gives a functional significance to the relatively high degree of atomic motion for particular regions of the trypsinogen molecule in the control of the protein's activity.

2.8 METHODS FOR THE STUDY OF HIGHER LEVELS OF PROTEIN STRUCTURE AND ORGANIZATION

X-Ray Diffraction Technique for the Determination of Protein Structure

The most important of the techniques for the study of a protein's secondary, tertiary, and quaternary structure is x-ray diffraction. The technique requires a protein in the crystalline form, although valuable information has also been obtained with fiber diffraction of noncrystalline materials that have a high degree of order. This latter type of x-ray diffraction has been especially important in the determination of fibrous protein structures such as that of collagen (Chapter 3).

Normally soluble globular proteins are crystallized from aqueous solutions of high salt concentration or on addition of a miscible organic reagent. Protein molecules associate into a crystal lattice as a subunit of the geometric building units of crystal structure known as the unit cell. The crystal also includes a significant amount of water of solvation that is noncovalently incorporated into the crystalline structure.

A beam of x-rays is diffracted by the electrons around each of the atomic nuclei in the crystal, with an intensity proportional to the number of electrons around the nucleus. With the highest resolution now available for protein structure determinations, the electron diffraction from C, N, O, and S atoms can be observed. However, the diffraction from hydrogen atoms is not clearly observed due to the low number of electrons around the hydrogen nuclei.

The intensity of the diffracted x-ray beam is observed with a photographic plate or collected by an electronic device. Whereas the intensity and the angles of deflection are discernible by such data collection, the phases of the diffracted beam are not directly determined. The solution of phase for the diffracted ray commonly requires the placement of heavy atoms in the protein molecule. If heavy atoms of high electron density (such as iodine, mercury, or lead) are placed in the molecule, their position in the unit cell of the crystal can be easily located. Usually at least two different heavy atom crystalline structures, in which the crystalline structure of the protein *is unperturbed by the heavy atom modification,* must be studied before the phase problem is solved.

Once the phases of the diffraction intensities are known from the heavy atom crystal structure, three-dimensional electron density maps can be calculated. Initially a few hundred reflections are obtained to construct a low resolution electron density map at about 6 Å. For example, in one of

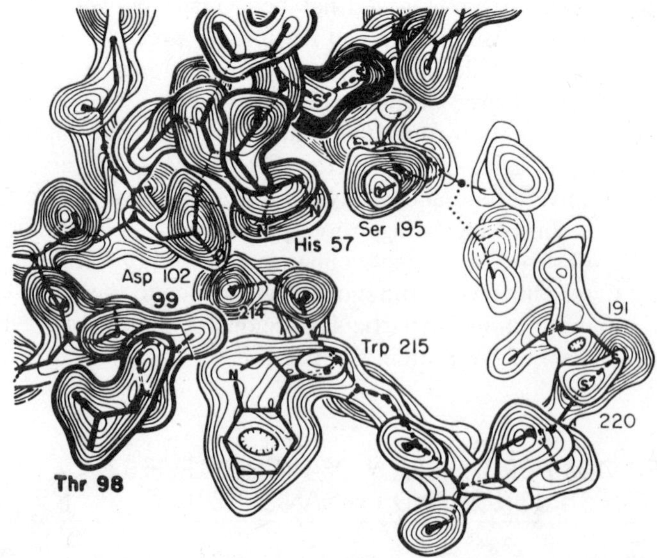

Figure 2.57
Electron-density map at 1.9Å resolution of active site region of proenzyme (trypsinogen) form of trypsin.
The catlytically active residues in trypsin (Asp-102, His-57, and Ser-195) are superposed on the map.
Reprinted with permission from T. Kossiakoff et al., *Biochemistry,* 16:654, 1977. Copyright 1977, American Chemical Society.

the first protein crystallographic structures, 400 reflections were utilized to obtain a 6-Å map of the protein myoglobin. At this level of resolution it is possible to locate clearly the molecule within the unit cell of the crystal and study the overall packing of the subunits in a protein with a quaternary structure. A trace of the polypeptide chain of an individual protein molecule may be made with difficulty. However, utilizing the low resolution structure as a base, further reflections may be used to obtain higher resolution maps. For myoglobin, whereas 400 reflections were utilized to obtain the 6-Å map, 10,000 reflections were needed for a 2-Å map, and 17,000 reflections for an extremely high resolution 1.4-Å map. A two-dimensional slice through a three-dimensional electron density map of the protein trypsinogen is shown in Figure 2.57. The known primary structure of the protein is fitted to the electron density pattern (Figure 2.57). The process of aligning a protein's primary structure to the electron density pattern until the best fit is obtained is known as refinement.

In order to observe changes in a protein's conformation that may occur on the binding of an inhibitor, activator, or substrate molecule of a protein, a difference electron density map may be computed. In this procedure, the diffraction pattern is obtained for the crystalline protein with the ligand bound and is substracted from the electron density pattern obtained for the protein without the ligand bound. The resulting difference map shows the changes that occur on the binding of the ligand (Figure 2.58).

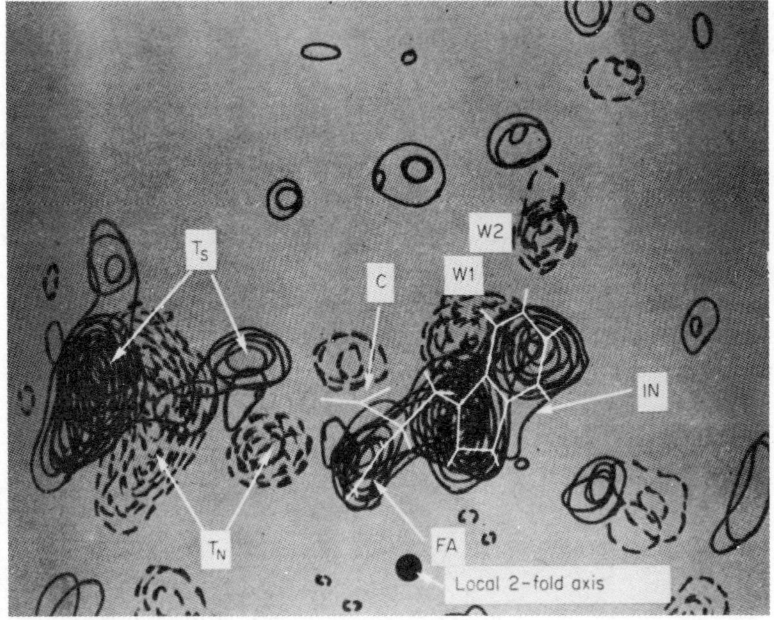

Figure 2.58
Difference electron density map between the structures of the enzyme α-chymotrypsin with substrate analog (N-formyl-L-tryptophan) bound to the enzyme active site and native α-chymotrypsin, at 2.5 Å resolution.
The smooth contours (positive density) represent electron density present in the enzyme: substrate complex and not in the native enzyme; the dashed contours (negative density) represent density present in native enzyme and not in the enzyme: substrate complex of the enzyme. Density arising from N-formyl-L-tryptophan is at IN (indole) and FA (formylamido) with structure superimposed. Negative density at W1 and W2 is due to water molecules displaced by substrate binding. The negative density T_n is the position of the carboxy terminal of Tyr-146 in the native enzyme, which moves to position T_s in the enzyme: substrate complex.
Reprinted with permission from T. A. Steitz, R. Henderson, and D. M.Blow, *J. Mol. Biol.* 46:337, 1969. Copyrighted by Academic Press, Inc. (London) Ltd.

Whereas x-ray diffraction has provided extensive knowledge on protein structure, it should be emphasized that an x-ray derived structure provides incomplete evidence for a protein's mechanism of action. The x-ray-determined structure is an average structure of a molecule in which atoms are normally undergoing rapid fluctuations in solution (see Section 2.7). In any one case, the average crystalline structure determined by x-ray diffraction may not be the active structure of a particular protein in solution. A second important consideration is that it currently takes at least a day to collect data in order to determine a structure. On this time scale, the structures of reactive enzyme–substrate complexes, intermediates, and transition states of enzyme proteins are not observed. Rather, these mechanistically important structures must be inferred from the static pictures of an *inactive* form of the protein or from complexes with inactive analogs of the normally reactive substrates of the protein.

Spectroscopy

Ultraviolet Light Spectroscopy

The side chain groups of tyrosine, phenylalanine, tryptophan, and cystine, as well as the peptide bonds in proteins, can absorb ultraviolet light. The efficiency of light energy absorption for each of these different types of absorbing chromophores is related to a molar extinction coefficient (ε), which has a characteristic value for each type of chromophoric group.

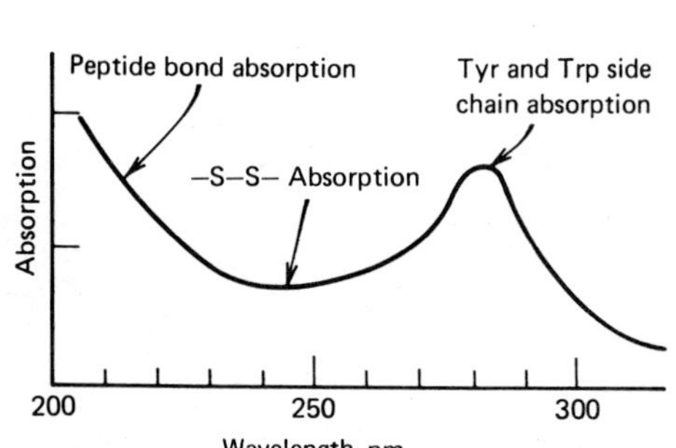

Figure 2.59
Ultraviolet absorption spectrum of the globular protein α-chymotrypsin.

A typical protein ultraviolet spectrum is shown in Figure 2.59. The absorbance between 260 and 300 nm is due to phenylalanine R groups, tyrosine R groups, and tryptophan R groups. The molar extinction coefficients for these chromophoric amino acids are plotted in Figure 2.60. When the tyrosine side chain is ionized at high pH (tyrosine R group $pK_a \simeq 10$), the absorbance for tyrosine is shifted to higher wavelength (red-shifted) and its molar absorptivity is increased (Figure 2.60).

The peptide bond absorbs in the far-ultraviolet (180–230 nm). A peptide bond in a helix conformation interacts with the electrons of other peptide bonds above and below it in the spiral conformation to create an

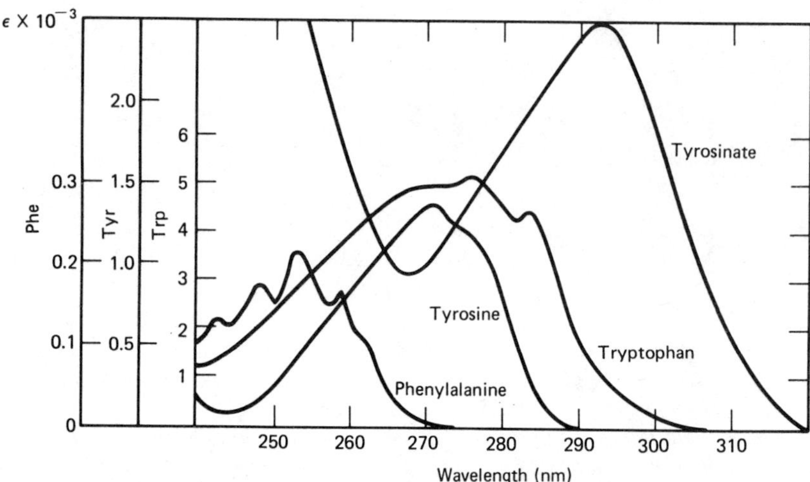

Figure 2.60
Ultraviolet absorption for the aromatic chromophores in Phe, Tyr, Trp, and tyrosinate.
Note differences in extinction coefficients on left axis for the different chromophores.
From A. d'Albis and W. B. Gratzer, in *Companion to Biochemistry,* A. T. Bull, J. R. Lagmado, J. O. Thomas, and K. F. Tipton, eds., Longmans, London, 1974, p. 170.

exciton system in which electrons are delocalized. The result is a shift of the absorption maximum from that of an isolated peptide bond to either a lower or higher wavelength (Figure 2.61).

Thus ultraviolet spectroscopy can be used to study changes in a protein's secondary and tertiary structure. As a protein is denatured (helix unfolded), differences are observed in the absorption characteristics of the peptide bonds between 180 and 230 nm due to the disruption of the exciton system. In addition, the absorption maximum for an aromatic chromophore appears at a lower wavelength in an aqueous environment than in a nonpolar environment.

The molar absorbancy of a chromophoric substrate or ligand will often change on binding to a protein. This change in the binding molecule's extinction coefficient can be used to measure its binding constant. Changes in chromophore extinction coefficients during enzyme catalysis of a chemical reaction can often be used to obtain the kinetic parameters for the reaction.

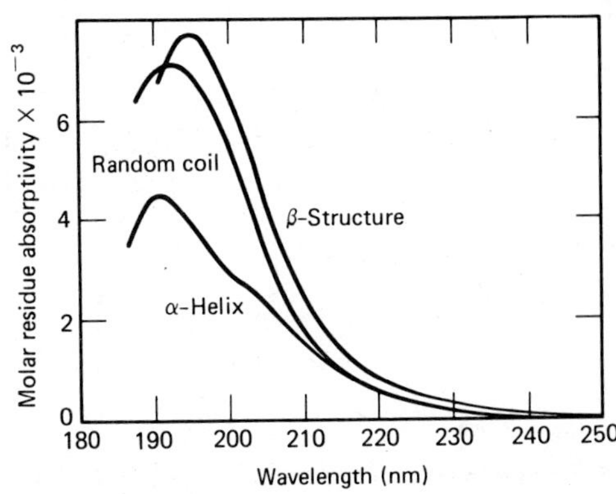

Figure 2.61
Ultraviolet absorption of the peptide bonds of a polypeptide chain in α-helical, random-coil, and antiparallel β-structure conformations.
From A. d'Albis and W. B. Gratzer, in *Companion to Biochemistry*, A. T. Bull, J. R. Lagmado, J. O. Thomas, and K. F. Tipton, eds., Longmans, London, 1970, p. 175.

Fluorescence Spectroscopy

The energy of an excited electron produced by light absorption can be lost by a variety of mechanisms. Most commonly the excitation energy is dissipated as thermal energy in a collision process. In some chromophores the excitation energy is dissipated by fluorescence.

The fluorescent emission is always at a longer wavelength of light (lower energy) than the absorption wavelength of the fluorophore. This is because vibrational energy levels formed in the excited electron state during the excitation event are lost during the time it takes the fluorescent event to occur (Figure 2.62).

In the presence of a second chromophore that can absorb light energy at the wavelength of a fluorophore's emission, the fluorescence that is normally emitted may not be observed. Rather, the fluorescence energy can be transferred to the second molecule. The acceptor molecule, in

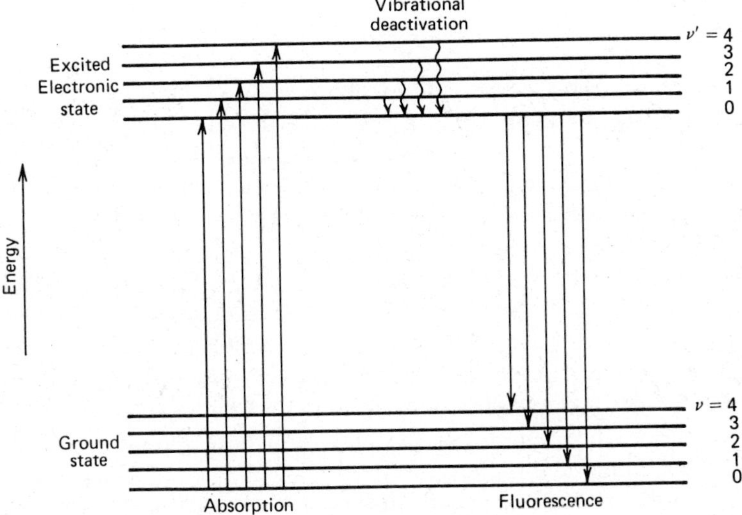

Figure 2.62
Absorption and fluorescence electronic transitions.
Excitation is from the zero vibrational level in the ground state to various higher vibrational levels in the excited state. Fluorescence is from the zero vibrational level in the excited electronic state to various vibrational levels in the ground state.
From A. d'Albis, and W. B. Gratzer, in *Companion to Biochemistry,* A. T. Bull, J. R. Lagmado, J. O. Thomas, and K. F. Tipton, eds., Longmans, London, 1970, p. 166.

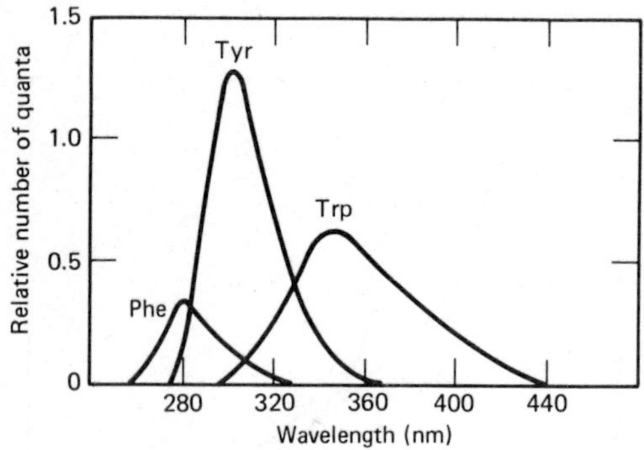

Figure 2.63
Characteristic fluorescence of aromatic groups in proteins.
From A. d'Albis and W. B. Gratzer, in *Companion to Biochemistry,* A. T. Bull, J. R. Lagmado, J. O. Thomas, and K. F. Tipton, eds. Longmans, London, 1970, p. 478.

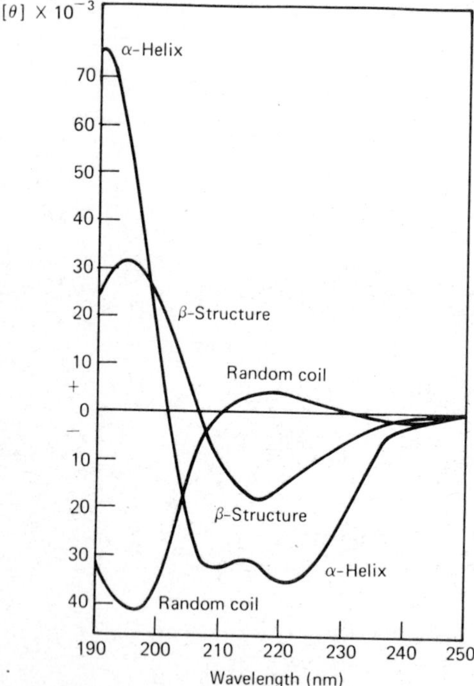

Figure 2.64
Circular dichroism spectra for polypeptide chains in α-helical, β structure, and random-coil conformations.
From A. d'Albis and W. B. Gratzer, in *Companion to Biochemistry,* A. T. Bull, J. R. Lagmado, J. O. Thomas, and K. F. Tipton, eds., Longmans, London, 1970, p. 190.

turn, can then either emit its own characteristic fluorescence or lose its excitation energy by an alternative process. If the acceptor molecule loses its excitation energy by a nonfluorescent process, it is acting as a *quencher* of the donor molecule's fluorescence. The efficiency of excitation transfer is dependent on the distance and the orientation between donor and acceptor molecules as well as the degree of overlap between the emission wavelengths of the donor molecule and the absorption wavelengths characteristic of the acceptor molecule.

The fluorescence emission spectra for phenylalanine, tyrosine, and tryptophan side chains are shown in Figure 2.63. A comparison of their emission and absorption spectra (Figure 2.63) show that the emission wavelengths for phenylalanine overlaps with the absorption wavelengths for tyrosine. In turn, the emission wavelengths for tyrosine overlap with the absorption wavelengths for tryptophan. Because of the overlap in emission and absorption wavelengths, primarily only the tryptophan fluorescence is observed in proteins that contain all three of these types of amino acids.

Excitation energy transfers occur over distances up to 80 Å, which are typical diameter distances in folded globular proteins. When a protein is denatured the distances between donor and acceptor groups become greater. The increased distance between donor and acceptor groups decreases the efficiency of energy transfer. Accordingly, an increase in the intrinsic fluorescence of the tyrosines and/or phenylalanines to the protein's emission spectrum will be observed with denaturation.

Since excitation transfer processes in proteins are distance- and orientation-dependent, the fluorescence yield is dependent on the conformation of the protein. As such, fluorescence is a highly sensitive tool with which to study protein conformation and changes in a protein's conformation related to its function.

Common prosthetic groups in enzyme proteins such as NADH and pyridoxal phosphate (see chapters on metabolism) are fluorophores. The changes in fluorescence yields from enzymes that contain these prosthetic moieties can be used to follow the chemical reactions catalyzed by the enzymes.

Optical Rotatory Dispersion and Circular Dichroism Spectroscopy

Optical rotation is caused by differences in the *refractive index* encountered by the clockwise and counterclockwise vector components of a beam of polarized light in a solution containing an asymmetric solute. *Circular dichroism* is caused by differences in the *light absorption* between the clockwise and counterclockwise component vectors of a beam of polarized light. In proteins the aromatic amino acids of asymmetric configuration give an optical rotation and circular dichroism. Also the polypeptide chains in regular helical conformation will form either a right-handed or left-handed direction spiral conformation. These two spiral conformations are not superimposable, and they generate a significant optical rotation and circular dichroism.

Circular dichroism spectra for different conformations of the polypeptide chain are shown in Figure 2.64. Because of the differences, circular dichroism is a sensitive assay for the amount and type of secondary structure in a protein.

Nuclear Magnetic Resonance (NMR)

The nucleus of the atomic isotopes ^{1}H, ^{13}C, ^{15}N, ^{19}F, and ^{31}P have a characteristic spin quantum number of one-half and in a magnetic field will absorb energy. The resonance frequency of the absorption, the pattern exhibited, and the time constants for relaxation of the excited spin state are dependent on the environment and structure into which the excited nucleus is incorporated. Accordingly, the NMR spectrum of individual atomic nuclei in a protein structure will give structural and functional information on the nucleus and its microenvironment within the protein.

Separation and Characterization of Proteins Based on Molecular Weight or Size

Ultracentrifugation: Definition of Svedberg Coefficient

A protein subjected to centrifugal force will move in the direction of the force at a velocity dependent on the protein's mass. The rate of movement can be measured with the appropriate optical system, and from the measured rate of movement the sedimentation coefficient (s) calculated in Svedberg units (units of 10^{-13} s). In the equation, which can be used to calculate a sedimentation coefficient for a molecule (Figure 2.65), γ is the measured velocity of protein movement, ω the angular velocity of the centrifuge rotor, and r the distance from the center of the tube in which the protein is placed to the center of rotation. Sedimentation coefficients between 1 and 200 Svedberg units (S) have been found for proteins (Table 2.11).

$$s = \frac{\gamma}{\omega^2 r}$$

Figure 2.65
Equation for calculation of the Svedberg coefficient.

Equations have been derived to relate the sedimentation coefficient to the molecular weight for a protein. One of the more simple equations is shown in Figure 2.66, in which R is the gas constant, T the temperature, s the sedimentation coefficient, D the diffusion coefficient of the protein, $\bar{v}$ the partial specific volume of the protein, and ρ the density of the solvent. The quantities D and $\bar{v}$ must be measured in independent experiments. In addition, *the equation assumes a spheroidal geometry for the protein*. In view of the fact that the assumption of a spheroidal geometry may not be true for any particular case, and independent measurements of D and $\bar{v}$ are difficult, often only the sedimentation coefficient for a molecule is reported. The magnitude of the protein's sedimentation coefficient will give a relative value that can be used in a generally qualitative way to characterize a protein's molecular weight.

$$\text{Molecular weight} = \frac{RTs}{D(1 - \bar{v}\rho)}$$

Figure 2.66
An equation relating the Svedberg coefficient to molecular weight.

Molecular Exclusion Chromatography

A porous gel in the form of small insoluble beads is commonly used to separate proteins by size in column chromatography. Small protein mole-

TABLE 2.11 Svedberg Coefficients for Some Plasma Proteins of Different Molecular Weights

Protein	S_{20}, $\times 10^{-13}$ *cm/s dyn*[a]	*Mol Wt*
Lysozyme	2.19	15,000–16,000
Albumin	4.6	69,000
Immunoglobulin G	6.6–7.2	153,000
Fibrinogen	7.63	341,000
C1q (factor of complement)	11.1	410,000
α_2-Macroglobulin	19.6	820,000
Immunoglobulin M	18–20	1,000,000
Factor VIII of blood coagulation	23.7	1,120,000

SOURCE: *CRC Handbook of Biochemistry and Molecular Biology*, 3rd ed., Sect. A, Vol. II, G. D. Fasman, ed., 1976, p. 242.

[a] S_{20}, $\times 10^{-13}$ is sedimentation coefficient in Svedberg units, referred to water at 20°C, and extrapolated to zero concentration of protein.

cules can penetrate the pores of the gel and will have a larger solvent volume through which to travel in the column than large proteins, which are sterically excluded from the pores. Accordingly, a protein mixture will be separated by size, the larger proteins eluted first, followed by the smaller proteins, which are retarded by their accessibility to a larger solvent volume (Figure 2.67).

As with ultracentrifugation, an assumption must be made as to the geometry of the unknown protein, and nonspheroid proteins will give anomalous molecular weights when compared to standard proteins of spherical conformation.

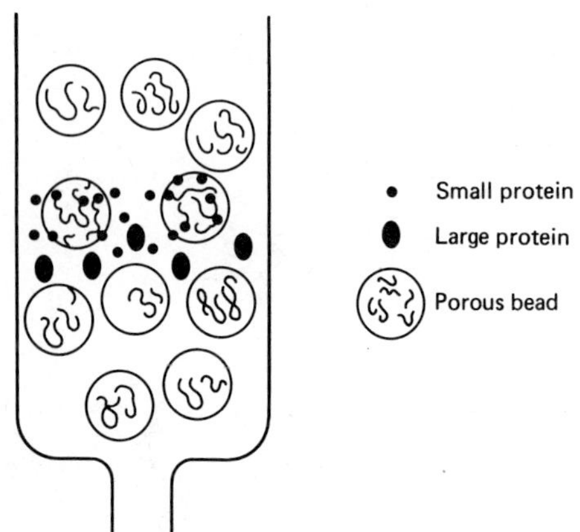

Figure 2.67
Molecular exclusion chromatography.
The small protein can enter the porous gel particles and will be retarded on the column with respect to the larger protein that cannot enter the porous gel particles.

Polyacrylamide Gel Electrophoresis in the Presence of a Detergent

If a charged detergent is added to an electrophoresis buffer and a protein is electrophoresed on a sieving support, a separation of proteins occurs based on protein size but not charge. A detergent commonly used in protein electrophoresis based on size is sodium dodecyl sulfate (SDS). The dodecyl sulfates are amphiphilic 12- carbon alkyl sulfate molecules that act to denature the protein and form a charged micelle about the denatured molecule. The inherent charge of the native protein is obliterated by the charged micelle layer of SDS, and each protein–SDS-solubilized aggregate has an identical charge per unit volume due to the charge characteristics of the SDS micelle. The negatively charged micelle particles will move through an electrophoresis gel toward the anode (+ pole). A common gel for SDS electrophoresis is cross-linked polyacrylamide. In the migration toward the positive pole, polyacrylamide acts as a molecular sieve and the protein–micelle complexes are separated by size. As the proteins are denatured, artifacts caused by nonspheroid shapes of protein native structures will not be significant (Figure 2.68).

It should be realized that the detergents dissociate quaternary structure into its constituent subunits, and only the molecular weights of covalent subunits are determined by this method.

High Performance Liquid Chromatography

In high performance liquid chromatography (HPLC), a liquid solvent containing a mixture of components to be identified or purified is passed through a column densely packed with a small-diameter insoluble beadlike resin. In column chromatography, the smaller and more tightly packed the resin beads, the greater the resolution of the separation technique. In this technique, the resin is so tightly packed that in order to overcome the resistance the liquid must be pumped through the column at high pressure. Therefore, HPLC uses precise high pressure pumps with metal plumbing and columns rather than glass and plastics as used in gravity chromatography. The resin beads can be coated with charged groupings to separate compounds by ion exchange or with hydrophobic groupings to retard nonpolar molecules passed through the resin. In hydrophobic chromatography nonpolar compounds can be eluted from the hydrophobic beads in aqueous eluents containing various percentages of an organic reagent. The higher the percentage of organic solvent in the eluent, the faster the nonpolar component is eluted from the hydrophobic resin. This latter type of chromatography over nonpolar resin beads is called reverse-phase high performance liquid chromatography (see Figure 2.25 for data showing separation of dansyl amino acids by reverse-phase HPLC). Alternatively, the beads can be porous in order to separate large molecules from smaller ones on the basis of size by molecular exclusion chromatography. HPLC separations are carried out with extremely high resolution and reproducibility in column retention times.

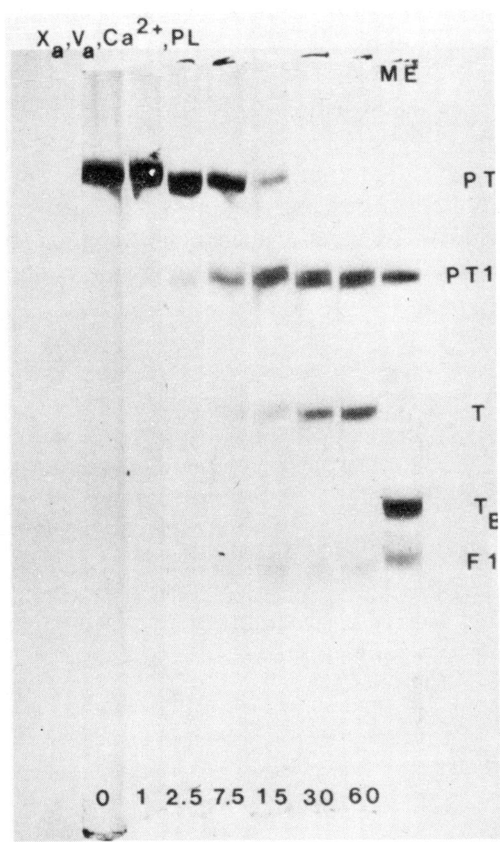

Figure 2.68
Example of sodium dodecyl sulfate (SDS)- polyacrylamide gel elecrophoresis.
Time course of activation of the coagulation enzyme thrombin as followed by SDS-electrophoresis: PT band, prothrombin; PT1, prethrombin (intermediate activation product); T, active thrombin; T_b *and F1 are degradiative products of the thrombin enzyme.*
Reprinted with permission from J. Rosing et al., *J. Biol. Chem.* 225:274, 1980.

Separation of Proteins by Affinity Chromatography

Proteins have a high affinity for their substrates, prosthetic groups, membrane receptors, specific noncovalent inhibitors, and specific antibodies made against them. These high affinity compounds can be covalently attached to an insoluble resin and utilized to purify a protein in column chromatography. In a mixture of compounds eluted through the resin, the protein of interest will be selectively retarded.

General Approach to Protein Purification

A protein must be purified prior to a meaningful characterization of its chemical composition, structure, and function. As living cells contain thousands of genetically distinct proteins, the purification of a single protein from a mixture of cellular molecules may be difficult.

The first task in the purification of a protein is the development of a facile assay for the protein. A protein assay, whether it utilizes the rate of a substrate's transformation to a product, an antibody–antigen reaction, or a physiological response in an animal assay system, must in some way give a quantitative measurement of activity per unit weight for a particular sample containing the protein. This quantity is known as the sample's specific activity. The purpose of the purification is to increase a sample's specific activity to the value equal to the specific activity expected for the pure protein.

A typical protocol for purification of a soluble cellular protein may first involve the disruption of the cellular membrane, followed by a differential centrifugation in a density gradient to isolate the protein activity from subcellular particles and high molecular weight aggregates. A further purification step may utilize selective precipitation by addition of inorganic salts (salting out) or addition of miscible organic solvent to the solution containing the protein. Final purification will include a combination of techniques previously discussed, which include methods based on molecular charge (Section 2.3), molecular size, and affinity chromatography.

BIBLIOGRAPHY

Protein Structure

Anfinsen, C. B., and Scheraga, H. A. Experimental and theoretical aspects of protein folding. *Adv. Protein Chem.* 29:205, 1975.

Bajaj, M., and Blundell, T. Evolution and the tertiary structure of proteins. *Annu. Rev. Biophys. Bioeng.* 13:453, 1984.

Chothia, C. Principles that determine the structure of proteins. *Annu. Rev. Biochem.* 53:537, 1984.

Chou, P. Y., and Fasman, G. D. Empirical prediction of protein conformation. *Annu. Rev. Biochem.* 47:251, 1978.

Dickerson, R. E., and Geis, I. *The structure and action of proteins.* New York: Harper & Row, 1969.

Neurath, H., and Hill, R. L. (eds.). *The proteins,* 3rd ed., vols. 1–4. New York: Academic, 1975–79.

Richards, F. M., and Richmond, T. Solvents, interfaces and protein structure, in *Molecular interactions and activity in proteins,* Ciba Foundation Symposium 60 (new series). Amsterdam: Excerpta Medica, 1978, p. 23.

Richardson, J. S. The anatomy and taxonomy of protein structure. *Adv. Protein Chem.* 34:168, 1981.

Amino Acid Analysis and Sequencing

Hirs, C. H. W., and Timosheff, S. N. (eds.). *Methods Enzymol.* 47, 1977.

Hunkapiller, M. W., Strickler, J. E., and Wilson, K. J. Contemporary methodology for protein structure determination. *Science* 226:304, 1984.

Motion in Folded Proteins

Huber, R. Conformational flexibility and its functional significance in some protein molecules. *Trends Biochem. Sci.* 4:271, 1979.

McCammon, J. A., Gelin, B. R., and Karplus, M. Dynamics of folded proteins. *Nature.* 267:585, 1977.

McCammon, J. A., and Karplus, M. The dynamic picture of protein structure. *Acct. Chem. Res.* 16:187, 1983.

Wüthrich, K., and Wagner, G. Internal motion in globular proteins. *Trends Biochem. Sci.* 3:227, 1978.

Protein Spectroscopy

D'Albis, A., and Gratzer, W. B. Electronic spectra and optical activity of proteins, in A. T. Bull, J. R. Lagnado, J. O. Thomas, and K. F. Tipton (eds.), *Companion to biochemistry.* London: Longmans, 1974.

Multivolume General References on Proteins: Colowick, S. P., and Kaplan, N. O. (general eds.), *Methods in Enzymol.* New York: Academic Press; and Anfinsen, C. B., Edsall, J. T., and Richards, F. M. (eds.). *Advances in Protein Chemistry.* New York: Academic Press.

QUESTIONS

J. BAGGOTT AND C. N. ANGSTADT

*Q*uestion *T*ypes are described inside the front cover.

Refer to the following structure for Questions 1 and 2.

```
Gly-Ser-Cys-Glu-Asp-Asn-Cys-Arg
         |               |
         S ------------- S
```

1. (QT2) The peptide shown above:
   1. has arginine in position 1 of the sequence.
   2. contains a derived amino acid.
   3. is basic.
   4. consists primarily of amino acids that are either of intermediate polarity or charged at pH 7.
2. (QT1) The charge on the peptide shown above is about:
   A. −2 at pH > 13.5
   B. −1 at pH ~ 11.5
   C. +1 at pH ~ 6.5
   D. +2 at pH ~ 5.5
   E. 0 at pH ~ 4.5

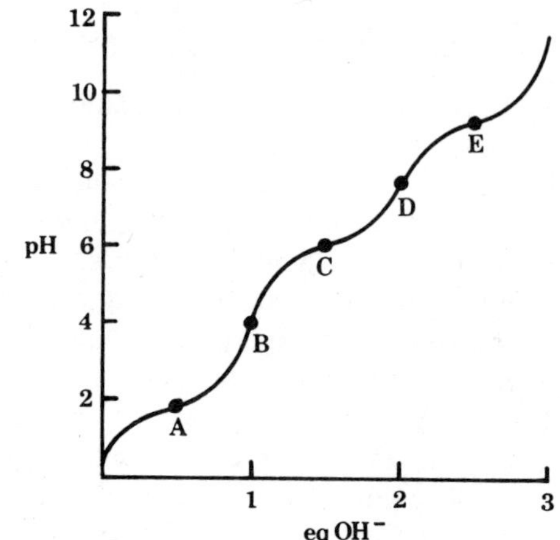

3. (QT2) The figure on page 72 shows the titration curve of one of the common amino acids. From this curve we can conclude:
   1. at point B the amino acid is zwitterionic.
   2. point D corresponds to the pK of an ionizable group.
   3. the amino acid contains two carboxyl groups.
   4. at point E the amino acid has a net negative charge.
4. (QT2) Which of the following can be used for a quantitative determination of amino acids in general?
   1. Ellman's reagent
   2. Fluorescamine
   3. The Sakaguchi reaction
   4. Ninhydrin
5. (QT3) A. Freedom of rotation of the peptide bond in an alpha helix.
   B. Freedom of rotation of the peptide bond in a "random" structure.
6. (QT2) Which of the following has quaternary structure?
   1. cytochrome c
   2. insulin
   3. α-chymotrypsin
   4. hemoglobin

Refer to the drawing for Questions 7 and 8.

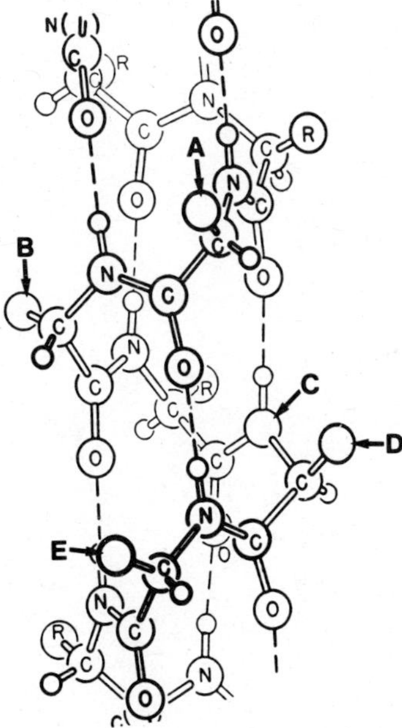

7. (QT2) When group E contains a negatively charged carboxyl function, the structure is destabilized by:
   1. aspartate at position D.
   2. glutamate at position B.
   3. alanine at position A.
   4. proline at position D.
8. (QT1) The properties of atom C are essential to which stabilizing force in the structure?
   A. hydrogen bonding
   B. steric effects
   C. ionic attraction
   D. disulfide bridge
   E. none of the above
9. (QT1) When protein subunits combine to form a quaternary structure, all of the following interactions may arise *except:*
   A. disulfide bond formation.
   B. hydrogen bonding.
   C. hydrophobic interaction.
   D. electrostatic bonding.
   E. Van der Waals forces.
10. (QT3) A. Contribution of hydrophobic bonding to the thermodynamic stability of proteins.
    B. Contribution of hydrogen bonding to the thermodynamic stability of proteins.
11. (QT3) A. Number of x-ray reflections necessary for a low resolution electron density map.
    B. Number of x-ray reflections necessary for a high resolution electron density map.
12. (QT3) A. Molar absorptivity of tyrosine at pH 7.
    B. Molar absorptivity of tyrosine at pH 11.
13. (QT3) A. Wavelength of maximum absorption by a fluorophore.
    B. Wavelength of maximum emission by the same fluorophore.
14. (QT2) Proteins may be separated according to size by:
    1. isoelectric focusing.
    2. polyacrylamide gel electrophoresis in the presence of sodium dodecyl sulfate (SDS).
    3. ion-exchange chromatography.
    4. molecular exclusion chromatography.

A. Primary structure
B. Secondary structure
C. Tertiary structure
D. Quaternary structure
E. Random conformation

15. (QT5) All ϕ angles are equal and all ψ angles are equal.
16. (QT5) May bring distant segments of a single polypeptide chain into close juxtaposition.
17. (QT5) Unaffected by binding of a charged detergent, such as sodium dodecyl sulfate (SDS).
18. (QT5) Circular dichroism spectrum exhibits a maximum in the 210–220-nm region.
19. (QT5) Exemplified by the β structure (pleated sheet).

Answers

1. C 2 and 4 true. 1: The convention is to write the N-terminal to the left. Numbering begins at the N-terminal, so glycine is in position 1 (p. 33). 2: Cystine, formed by joining two cysteine residues through a disulfide bridge, is a derived amino acid (p. 34). 3: The peptide contains two acidic amino acids, glutamate and aspartate, and only one basic amino acid, arginine, so it is acidic (p. 34). 4: Only cysteine is nonpolar; all the others are of intermediate polarity or are charged (p. 32).
2. E At pH 4.5 the peptide is in the following ionic state: The N-terminal amino group is +1, the side chain carboxyls of glutamate and aspartate each average about -0.5 (since this pH is about at their pKs), the side chain of arginine is +1, and its terminal carboxyl group is -1. The sum is zero (pp. 35, 36, and 37).
3. D Only 4 true. The axes of this titration curve are reversed from the presentation in the text. The abscissa shows that three ionizable groups are present. The pKs, where the groups are 50% titrated, are at points A (pH ~ 2), C (pH ~ 6.5), and E (pH ~ 9.5). Histidine is the only common amino acid with these pKs. At point B, its net charge is $-1 + 1 + 1 = +1$. At point E, the net charge is $-1 + 0 + 0.5 = -0.5$ (p. 35–37).
4. C 2 and 4 true. Ellman's reagent reacts only with cysteine, and the

Sakaguchi reaction is for arginine. 2 and 4 react with all amino acids (p. 41).

5. C The peptide bond has partial double bond character and does not rotate freely at physiological temperatures (p. 51).
6. D Only 4 true. Quaternary structure consists of a specific noncovalent association of subunits having their own tertiary structures (p. 57). Cytochrome c is a single polypeptide chain (p. 48). Insulin (p. 47) and chymotrypsin (p. 57) are multichain proteins, with the chains joined by disulfide bridges.
7. C 2 and 4 true. Like charges in the third or fourth position in either direction from the designated position destabilize the helix due to charge repulsion. Thus aspartate at position D is harmless, whereas glutamate at position A or B would destabilize. Alanine has a small side chain. Proline cannot fit into an α helix at all (p. 51–52).
8. A Atom C is an amide nitrogen. The attached hydrogen atom participates in hydrogen bonding (p. 52). Hydrogen bonds contribute to the stability of the structure (p. 59).
9. A Quaternary structure is stabilized exclusively by noncovalent interactions. Disulfide bonds are covalent (p. 57).
10. A Hydrophobic bonding is a major contributor to the thermodynamic stability of proteins (p. 59).
11. B More reflections must be analyzed to achieve higher resolution (p. 65). In fact, the resolution is related to the third power of the number of reflections analyzed.
12. B The pK of tyrosine's phenolic —OH group is about 10 (p. 35). In the dissociated state, tyrosine's molar absorptivity increases, and the wavelength of maximum absorption changes (p. 66).
13. B Fluorescence emission is always less energetic than the radiation required for excitation. This is because of vibrational loss of energy from the excited molecules. Longer wavelengths are less energetic (p. 67).
14. C 2 and 4 true. 2 and 4 separate on the basis of size (p. 70). 1 and 3 separate on the basis of charge (p. 40).
15. B This statement is a definition of secondary structure (p. 51).
16. C This is a consequence of folding into a compact structure (p. 53).
17. A SDS binding produces an extended conformation of a polypeptide chain due to charge repulsion, but no peptide bonds are broken (p. 70).
18. E Any protein containing α helix or β structure will exhibit a minimum in this range. Random coils exhibit a small maximum (p. 68, Figure 2.64).
19. B β structure is an important type of secondary structure (p. 52).

3

Proteins II: Physiological Proteins

RICHARD M. SCHULTZ

3.1 OVERVIEW

In this chapter, examples of proteins with important biological functions are presented as models from which you can gain insight into how a protein's structure relates to its functions on the molecular and atomic levels. This insight is essential for the proper understanding of biochemical processes carried out by proteins and the malfunction of proteins in disease. The explanation of molecular function on the atomic level is an ultimate object of protein studies, and there exist a few cases in which current knowledge may be approaching a basic understanding. Among the proteins chosen for discussion are some of the most investigated and most physiologically interesting proteins, such as hemoglobin, myoglobin, and the immunoglobulins. However, it should be clearly stated that even for these most investigated proteins, the nature of the structure–function relationship remains an extremely active area for research and even controversy.

Thus, each example discussed in detail will exemplify principles of protein molecular structure in relationship to function. The proteins chosen serve as examples of different protein types. For example, Section 3.2 describes hemoglobin and myoglobin, which are examples of globular proteins. *Globular proteins* are spherically shaped, water-soluble proteins with a varied range of molecular weights and degree of regular secondary structure. In Section 3.3, a group of physiologically important proteins known as the *immunoglobulins* or *antibodies* are discussed. These are examples of multidomain globular-type proteins.

The *plasma lipoproteins,* discussed in Section 3.4, perform an essential role in the transport and metabolism of lipids in the human. Their structure and properties may serve as models for other important groups of lipoproteins, such as those found in biological membranes.

In Section 3.5, the *glycoproteins* are described as a class in order to give you some knowledge of the structures of this important protein type.

In Section 3.6, examples from the class of *fibrous proteins* are described, stressing the example of collagen. These proteins differ from the globular type of protein in their relatively low water solubility, higher amount of secondary structure, elongated "rodlike" shape, high tensile strength, and unusual covalent cross-links. They provide a structural matrix for the organs and tissues of the mammalian organism.

Since the protein examples presented in this chapter are primarily used for insight into the structural and conformational properties of different protein types, their biosynthesis and many interesting and important aspects of their physiological functions are not addressed. Later chapters describe these aspects of the protein examples. The role of hemoglobin in gas transport and in pH regulation within blood is discussed in Chapter 23. The molecular structures of the carbohydrate portions of glycoproteins and their biosynthesis are covered in Chapter 8 (Carbohydrate Metabolism). The metabolism of plasma lipoproteins is described in Chapter 9 (Lipid Metabolism). In Chapter 19, the biosynthesis of collagen and collagen fibrils is described as an example of posttranslational modification of proteins; and in Chapter 21, other aspects of connective tissue structure involving collagen and elastin are discussed.

3.2 HEMOGLOBIN AND MYOGLOBIN

The hemoglobins are globular proteins, in high concentration in red blood cells, that bind oxygen in the lungs and transport the oxygen in blood to the tissues and cells around the capillary beds of the vascular system. On returning to the lungs from the capillary beds, hemoglobins act to transport CO_2 and protons. In this section the structural and molecular aspects of the hemoglobin and myoglobin molecules are described. The physiological role of these proteins in gas transport is discussed in Chapter 23.

Forms of Human Hemoglobin

A hemoglobin molecule consists of four polypeptide chains of two different primary structures. In the common form of human adult hemoglobin, Hb A_1, two chains of one kind are designated the α chains, and the second two chains of the same kind are designated the β chains. The polypeptide chain composition of Hb A_1 is therefore $\alpha_2\beta_2$. The α-polypeptide chain contains 141 amino acids, and the β-polypeptide chain contains 146 amino acids.

While Hb A_1 is the major form of hemoglobin in the adult human, other forms of hemoglobin predominate in the blood of the human fetus. Most early forms of human hemoglobin contain two of the same α chains found in Hb A_1, but their second kind of chain in the tetramer molecule differs in amino acid sequence from that of the β chain of adult Hb A_1 (Table 3.1). A minor form of adult hemoglobin, Hb A_2, comprises about 2% of normal adult hemoglobin and is composed of two α chains and two chains designated *delta* (δ) (Table 3.1).

TABLE 3.1 Chains of Human Hemoglobin

Primary Source	*Symbol*	*Chains*
Adult	Hb A_1	$\alpha_2\beta_2$
Adult	Hb A_2	$\alpha_2\delta_2$
Fetus	Hb F	$\alpha_2\gamma_2$
Embryo (early fetus)	Hb Gower-2	$\alpha_2\varepsilon_2$

Myoglobin

Myoglobin (Mb) is an O_2-carrying protein that binds and releases O_2 with changes in the oxygen concentration in the cytoplasm of muscle cells. In contrast to hemoglobin, which has four polypeptide chains and four O_2 binding sites, myoglobin is composed of only a single polypeptide chain and a single O_2 binding site.

A comparison between myoglobin and hemoglobin is instructive in that myoglobin is a model for what occurs when a single protomer molecule acts alone without the interactions exhibited among the four O_2 binding sites in the more complex tetramer molecule of hemoglobin.

Heme Prosthetic Group

The four polypeptide chains in hemoglobin and the single polypeptide chain of myoglobin each contain a heme prosthetic group. As defined in Chapter 2, a prosthetic group is a nonpolypeptide moiety that forms a part of a protein in its native functional state. A protein without its prosthetic group is designated an apoprotein.

The heme is a porphyrin molecule containing an iron atom in its center. The type of porphyrin found in hemoglobin and myoglobin (protoporphyrin IX) contains two propionic acid, two vinyl, and four methyl side-chain groups attached to the pyrrole rings of the porphyrin structure (Figure 3.1). The iron atom is in the ferrous (+2 charge) oxidation state in functional hemoglobin and myoglobin.

The ferrous atom in the heme can form five or six ligand bonds, depending on whether or not O_2 is bound to the protein. Of the five or six bonds, four are to the pyrrole nitrogens of the porphyrin. Since all the pyrrole rings of the porphyrin lie in a common plane, the four ligand bonds from the porphyrin to the iron atom at its center will also have a tendency to lie in the plane of the porphyrin ring. This is especially true for six-coordinate ferrous iron in the oxy form of hemoglobin. (In a later section is described how the five-coordinate bond ferrous atom of deoxyhemoglobin sits out of the plane of the porphyrin rings by about 0.6 Å.) The fifth and the potentially sixth ligand bonds to the ferrous atom of the

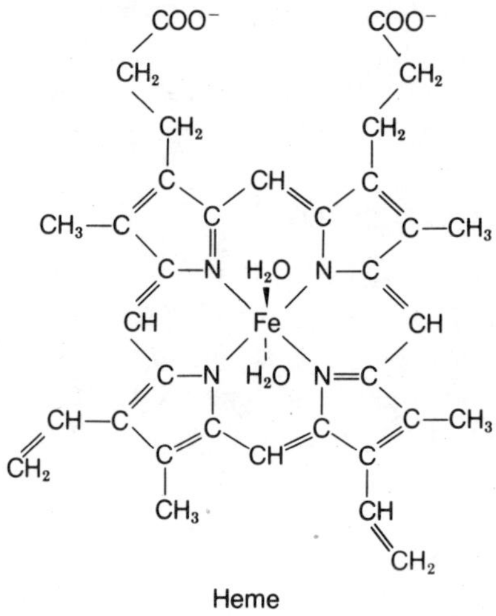

Figure 3.1
Structure of heme.

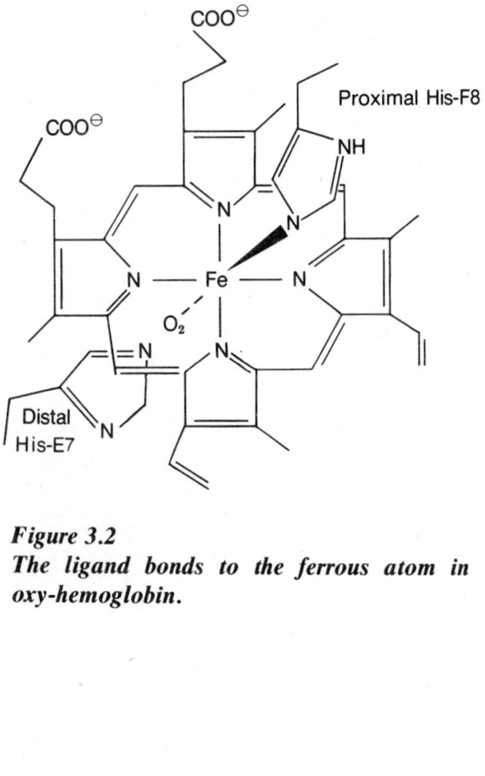

Figure 3.2
The ligand bonds to the ferrous atom in oxy-hemoglobin.

heme are directed along an axis perpendicular to the plane of the porphyrin ring (Figure 3.2).

The fifth coordinate bond of the ferrous atom in each of the hemes is to a nitrogen of a histidine imidazole. This histidine is designated the *proximal histidine* in the hemoglobin and myoglobin structures (Figures 3.2 and 3.3).

In each of the polypeptide chains with O_2 bound, the O_2 molecule forms a sixth coordinate bond to the ferrous atom. In this bonded position the O_2 is placed between the ferrous atom to which it is liganded and a second histidine imidazole, designated the *distal histidine,* in hemoglobin and myoglobin structures. In deoxyhemoglobin, the sixth coordinate position (O_2 binding position) of the ferrous atom is unoccupied.

The porphyrin part of the heme is positioned within a hydrophobic pocket formed in each of the polypeptide chains. In the heme pocket x-ray crystal diffraction structures show that approximately 80 interactions are provided by approximately 18 amino acids to the heme. Most of these noncovalent interactions are between apolar side chains of amino acids and the nonpolar regions of the porphyrin. As discussed in Section 2.6, the driving force for these interactions is the release of water of solvation on association of the hydrophobic heme with the apolar residues of the heme pocket of the protein. In myoglobin additional noncovalent interactions are made between the negatively charged propionate groups of the heme and positively charged arginine and histidine R groups of the protein. However, in hemoglobin chains a difference in the amino acid sequence in this region of the heme binding site leads to the stabilization of the porphyrin propionates by interaction with an uncharged histidine imidazole and with water molecules of the solvent toward the outer surface of the molecule.

X-Ray Crystallographic Structures of Hemoglobin and Myoglobin

Information on the structure of the deoxy and oxy forms of hemoglobin and myoglobin has primarily come from x-ray diffraction crystallography. In fact, sperm whale myoglobin was the first globular protein whose three-dimensional structure was determined by the technique of x-ray crystallography. The determination of the structure of myoglobin was

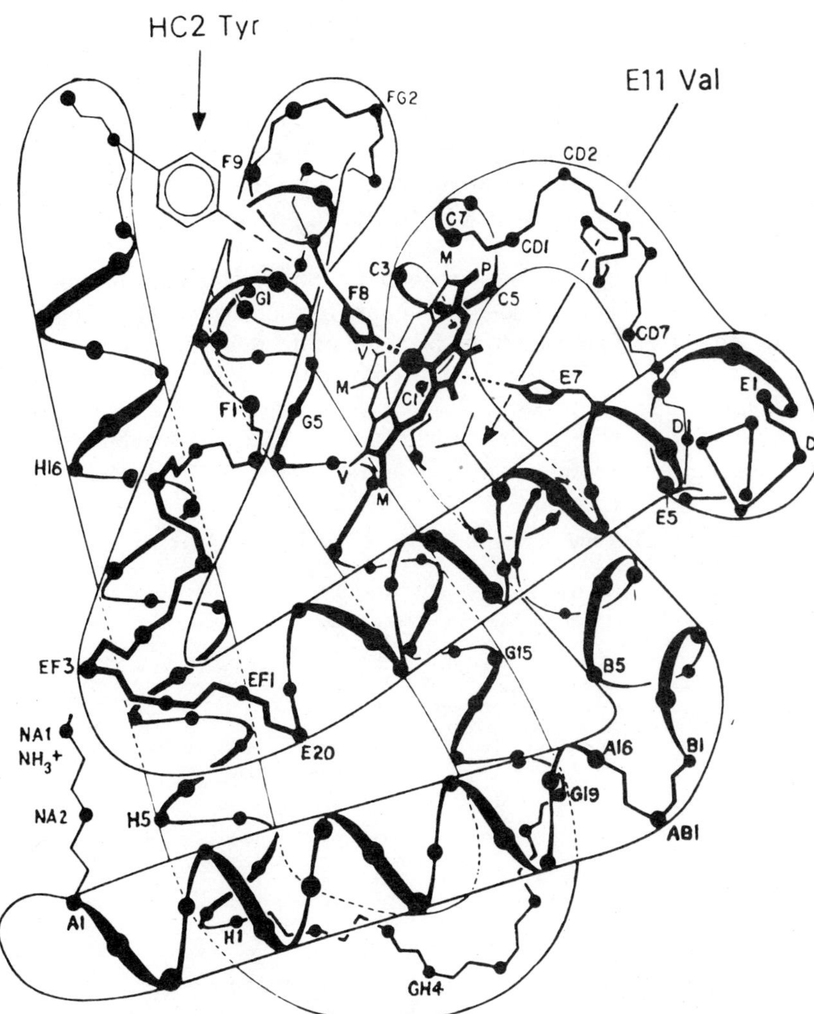

Figure 3.3
Secondary and tertiary structure characteristics of chains of hemoglobin.
The proximate His-F8, distal His-E7, and Val-E11 side chains are shown. The other amino acids of the polypeptide chain are represented by their α-carbon positions only; the letters M, V, and P refer to the methyl, vinyl, and propionate side chains of the heme.
Reprinted with permission from M. Perutz, *Brit. Med. Bull.* 32: 195, 1976.

soon followed by the x-ray structure of the more complex horse hemoglobin molecule. Some of the high resolution x-ray diffraction structures currently available are listed in Table 3.2. In this table the met forms of hemoglobin refer to structures in which the iron atoms in the hemes are in a ferric (+3 charge) oxidation state, rather than the functional ferrous (+2 charge) oxidation state. In the methemoglobins the ferric atom is six-coordinate as in normal oxyhemoglobin, with the sixth coordinate position occupied by a water molecule in methemoglobin at the position where O_2 binds in native oxyhemoglobins. The mechanism of cooperative associations of O_2, discussed below, is based on model building from these x-ray structures.

TABLE 3.2 High Resolution X-Ray Crystallographic Structures for Normal Hemoglobins

Species	*Derivative*	*Resolution* (Å)
Human A	Deoxy	2.5
	Deoxy	1.74[a]
	Oxy	2.1[b]
	CO	2.8
Human F	Deoxy	2.5
Horse	Deoxy	2.8
	Met (aquo)	2.0
	Met (cyano)	2.8

SOURCE: Data from M. F. Perutz, *Annu. Rev. Biochem.* **48,** 327 (1979) unless otherwise noted.

[a] G. Fermi, M. F. Perutz, B. Shaanan, and R. Fourme, *J. Mol. Biol.* **175,** 159 (1984).

[b] B. Shaanan, *J. Mol. Biol.* **171,** 31 (1983).

Primary, Secondary, and Tertiary Structures of Myoglobin and the Individual Hemoglobin Chains

The amino acid sequences for the polypeptide chain of myoglobin found in 23 different animal species have been determined. These myoglobins contain 153 amino acids in their polypeptide chains, of which 83 positions are shown to be invariant in the comparison of the amino acid sequences. Only 15 of these invariant residues in the myoglobin sequence are identical to the invariant residues of the currently sequenced mammalian hemoglobin chains. It should be noted, however, that the changes in residues at particular positions in the primary structure are, in the great majority of cases, varied in a conservative manner (Table 3.3). Since myoglobin is active as a monomer unit, many of its surface positions are formed to interact with water and prevent another molecule of myoglobin from associating. In contrast, the surface residues of the individual chains in hemoglobin are designed to provide hydrogen bonds and nonpolar contacts with other subunits in the hemoglobin quaternary structure. The proximal and distal histidines are, of course, preserved in the sequences of all the polypeptide chains. Other invariant residues are in the hydrophobic heme pocket and form essential nonpolar contacts with the heme that stabilizes the heme–protein complex. In addition, particular prolines in the sequence, which act to break some of the helical sections to allow the chain to fold back upon itself, are predominantly retained in most of the chains.

While there is a surprising variability in amino acid sequence among the different polypeptide chains, to a first approximation the secondary

TABLE 3.3 Amino Acid Sequences of Human Hemoglobin Chains and of Sperm Whale Myoglobin[a]

	NA 1	2	3	A 1	2	3	4	5	6	7	8	9	10	11	12	13	14	15	A 16	AB 1	B 1	2	3	4	5	6
MYOGLOBIN	val	.....	leu	ser	glu	gly	glu	trp	gln	leu	val	leu	his	val	trp	ala	lys	val	glu	ala	asp	val	ala	gly	his	gly
HEMOGLOBIN Horse α	val	.....	leu	ser	ala	ala	asp	lys	thr	asn	val	lys	ala	ala	trp	ser	lys	val	gly	gly	his	ala	gly	glu	tyr	gly
β	val	gln	leu	ser	gly	glu	glu	lys	ala	ala	val	leu	ala	leu	trp	asp	lys	val	asn	.....	.....	glu	glu	glu	val	gly
Human α	val	.....	leu	ser	pro	ala	asp	lys	thr	asn	val	lys	ala	ala	trp	gly	lys	val	gly	ala	his	ala	gly	glu	tyr	gly
β	val	his	leu	thr	pro	glu	glu	lys	ser	ala	val	thr	ala	leu	trp	gly	lys	val	asn	.....	.....	val	asp	glu	val	gly
γ	gly	his	phe	thr	glu	glu	asp	lys	ala	thr	ilu	thr	ser	leu	trp	gly	lys	val	asn	.....	.....	val	glu	asp	ala	gly
δ	val	his	leu	thr	pro	glu	glu	lys	thr	ala	val	asn	ala	leu	trp	gly	lys	val	asn	.....	.....	val	asp	ala	val	gly

	7	8	9	10	11	12	13	14	15	16	C 1	2	3	4	5	6	7	CD 1	2	3	4	5	6	7	8	D 1
MYOGLOBIN	gln	asp	ilu	leu	ilu	arg	leu	phe	lys	ser	his	pro	glu	thr	leu	glu	lys	phe	asp	arg	phe	lys	his	leu	lys	thr
HEMOGLOBIN Horse α	ala	glu	ala	leu	glu	arg	met	phe	leu	gly	phe	pro	thr	thr	lys	thr	tyr	phe	pro	his	phe	.....	asp	leu	ser	his
β	gly	glu	ala	leu	gly	arg	leu	leu	val	val	tyr	pro	trp	thr	gln	arg	phe	phe	asp	ser	phe	gly	asp	leu	ser	gly
Human α	ala	glu	ala	leu	glu	arg	met	phe	leu	ser	phe	pro	thr	thr	lys	thr	tyr	phe	pro	his	phe	.....	asp	leu	ser	his
β	gly	glu	ala	leu	gly	arg	leu	leu	val	val	tyr	pro	trp	thr	gln	arg	phe	phe	glu	ser	phe	gly	asp	leu	ser	thr
γ	gly	glu	thr	leu	gly	arg	leu	leu	val	val	tyr	pro	trp	thr	gln	arg	phe	phe	asp	ser	phe	gly	asn	leu	ser	ser
δ	gly	glu	ala	leu	gly	arg	leu	leu	val	val	tyr	pro	trp	thr	gln	arg	phe	phe	glu	ser	phe	gly	asp	leu	ser	ser

	2	3	4	5	6	7	E 1	2	3	4	5	6	7	8	9	10	11	12	13	14	E 15	16	17	18	19	20
MYOGLOBIN	glu	ala	glu	met	lys	ala	ser	glu	asp	leu	lys	lys	his	gly	val	thr	val	leu	thr	ala	leu	gly	ala	ilu	leu	lys
HEMOGLOBIN Horse α	.....	.....	.....	.....	.....	gly	ser	ala	gln	val	lys	ala	his	gly	lys	lys	val	ala	asp	gly	leu	thr	leu	ala	val	gly
β	pro	asp	ala	val	met	gly	asn	pro	lys	val	lys	ala	his	gly	lys	lys	val	leu	his	ser	phe	gly	glu	gly	val	his
Human α	.....	.....	.....	.....	.....	gly	ser	ala	gln	val	lys	gly	his	gly	lys	lys	val	ala	asp	ala	leu	thr	asn	ala	val	ala
β	pro	asp	ala	val	met	gly	asn	pro	lys	val	lys	ala	his	gly	lys	lys	val	leu	gly	ala	phe	ser	asp	gly	leu	ala
γ	ala	ser	ala	ilu	met	gly	asn	pro	lys	val	lys	ala	his	gly	lys	lys	val	leu	thr	ser	leu	gly	asp	ala	ilu	lys
δ	pro	asp	ala	val	met	gly	asn	pro	lys	val	lys	ala	his	gly	lys	lys	val	leu	gly	ala	phe	ser	asp	gly	leu	ala

	EF 1	2	3	4	5	6	7	8	F 1	2	3	4	F 5	6	7	8	9	FG 1	2	3	4	5	G 1	2	3	4
MYOGLOBIN	lys	lys	gly	his	his	glu	ala	glu	leu	lys	pro	leu	ala	gln	ser	his	ala	thr	lys	his	lys	ilu	pro	ilu	lys	tyr
HEMOGLOBIN Horse α	his	leu	asp	asp	leu	pro	gly	ala	leu	ser	asp	leu	ser	asn	leu	his	ala	his	lys	leu	arg	val	asp	pro	val	asn
β	his	leu	asp	asn	leu	lys	gly	thr	phe	ala	ala	leu	ser	glu	leu	his	cys	asp	lys	leu	his	val	asp	pro	glu	asn
Human α	his	val	asp	asp	met	pro	asn	ala	leu	ser	ala	leu	ser	asp	leu	his	ala	his	lys	leu	arg	val	asp	pro	val	asn
β	his	leu	asp	asn	leu	lys	gly	thr	phe	ala	thr	leu	ser	glu	leu	his	cys	asp	lys	leu	his	val	asp	pro	glu	asn
γ	his	leu	asp	asp	leu	lys	gly	thr	phe	ala	gln	leu	ser	glu	leu	his	cys	asp	lys	leu	his	val	asp	pro	glu	asn
δ	his	leu	asp	asn	leu	lys	gly	thr	phe	ser	gln	leu	ser	glu	leu	his	cys	asp	lys	leu	his	val	asp	pro	glu	asn

	5	6	7	8	G 9	10	11	12	13	14	15	16	17	18	19	GH 1	2	3	4	5	6	H 1	2	H 3	4	5
MYOGLOBIN	leu	glu	phe	ilu	ser	glu	ala	ilu	ilu	his	val	leu	his	ser	arg	his	pro	gly	asn	phe	gly	ala	asp	ala	gln	gly
HEMOGLOBIN Horse α	phe	lys	leu	leu	ser	his	cys	leu	leu	ser	thr	leu	ala	val	his	leu	pro	asn	asp	phe	thr	pro	ala	val	his	ala
β	phe	arg	leu	leu	gly	asn	val	leu	ala	leu	val	val	ala	arg	his	phe	gly	lys	asp	phe	thr	pro	glu	leu	gln	ala
Human α	phe	lys	leu	leu	ser	his	cys	leu	leu	val	thr	leu	ala	ala	his	leu	pro	ala	glu	phe	thr	pro	ala	val	his	ala
β	phe	arg	leu	leu	gly	asn	val	leu	val	cys	val	leu	ala	his	his	phe	gly	lys	glu	phe	thr	pro	pro	val	gln	ala
γ	phe	lys	leu	leu	gly	asn	val	leu	val	thr	val	leu	ala	ilu	his	phe	gly	lys	glu	phe	thr	pro	glu	val	gln	ala
δ	phe	arg	leu	leu	gly	asn	val	leu	val	cys	val	leu	ala	arg	asn	phe	gly	lys	glu	phe	thr	pro	gln	met	gln	ala

	6	7	8	9	10	11	12	13	14	15	16	17	18	19	20	H 21	22	23	2	HC 1	2	3	4	5
MYOGLOBIN	ala	met	asn	lys	ala	leu	glu	leu	phe	arg	lys	asp	ilu	ala	ala	lys	tyr	lys	glu	leu	gly	tyr	gln	gly
HEMOGLOBON Horse α	ser	leu	asp	lys	phe	leu	ser	ser	val	ser	thr	val	leu	thr	ser	lys	tyr	arg						
β	ser	tyr	gln	lys	val	val	ala	gly	val	ala	asn	ala	leu	ala	his	lys	tyr	his						
Human α	ser	leu	asp	lys	phe	leu	ala	ser	val	ser	thr	val	leu	thr	ser	lys	tyr	arg						
β	ala	tyr	gln	lys	val	val	ala	gly	val	ala	asn	ala	leu	ala	his	lys	tyr	his						
γ	ser	trp	gln	lys	met	val	thr	gly	val	ala	ser	ala	leu	ser	ser	arg	tyr	his						
δ	ala	tyr	gln	lys	val	val	ala	gly	val	ala	asn	ala	leu	ala	his	lys	tyr	his						

SOURCE: Based on diagram in R. E. Dickerson and I. Geis, *The Structure and Function of Proteins,* Harper & Row, New York, 1969, p. 52.

[a] Residues that are identical are enclosed in box. A, B, C . . . designate different helices of tertiary structure (see text).

and tertiary structures of each of the polypeptide chains of hemoglobin and myoglobin appear to be almost identical (Figure 3.4). The significant differences in the physiological properties between the α, β, γ, and δ chains in the hemoglobins and the single polypeptide chains of myoglobin are, therefore, due to rather small specific changes in their structures. The similarity in tertiary structure, resulting from widely varied amino acid sequences, show that the same tertiary structure for a protein can be arrived at in many different ways.

The x-ray crystallographic structures show that each of the polypeptide chains are composed of multiple α-helical regions that are broken by turns of the polypeptide chain, allowing the protein to fold into a spheroidal shape (Figure 3.4). Approximately 70% of the residues in the protein participate in the α-helical secondary structure, which generates seven helical segments in the α chain and eight helical segments in the β chain. These latter eight helical regions are commonly lettered A through H, starting from the A helix at the NH_2-terminal end, and the interhelical regions designated as AB, BC, CD, . . . , GH, respectively. The nonhelical region that lies between the NH_2-terminal end and the A helix is designated the NA region; and the region between the COOH-terminal end and the H helix is designated the HC region (Figure 3.3). This naming system allows discussion of particular residues that have similar functional and structural roles in each of the hemoglobin and myoglobin

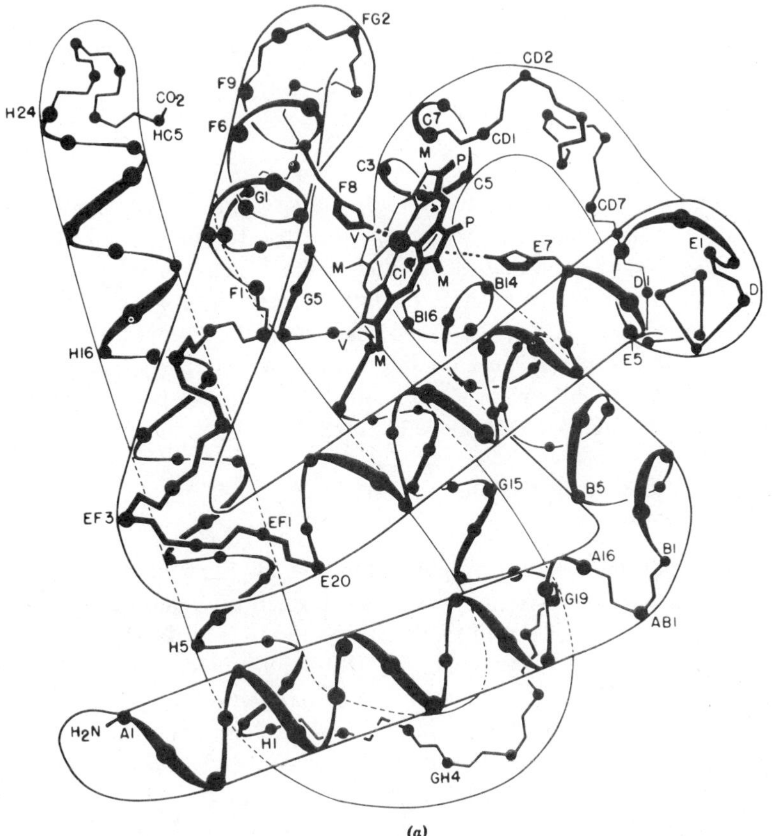

Figure 3.4
Comparison of the conformation of myoglobin (a) and β chain of Hb A_1 (b).
The overall structures are very similar, except at the NH_2-terminal and COOH-terminal ends.
Reprinted with permission from A. Fersht, *Enzyme Structure and Mechanism,* W. H. Freeman, San Francisco, Calif., 1977, pp. 12 and 13.

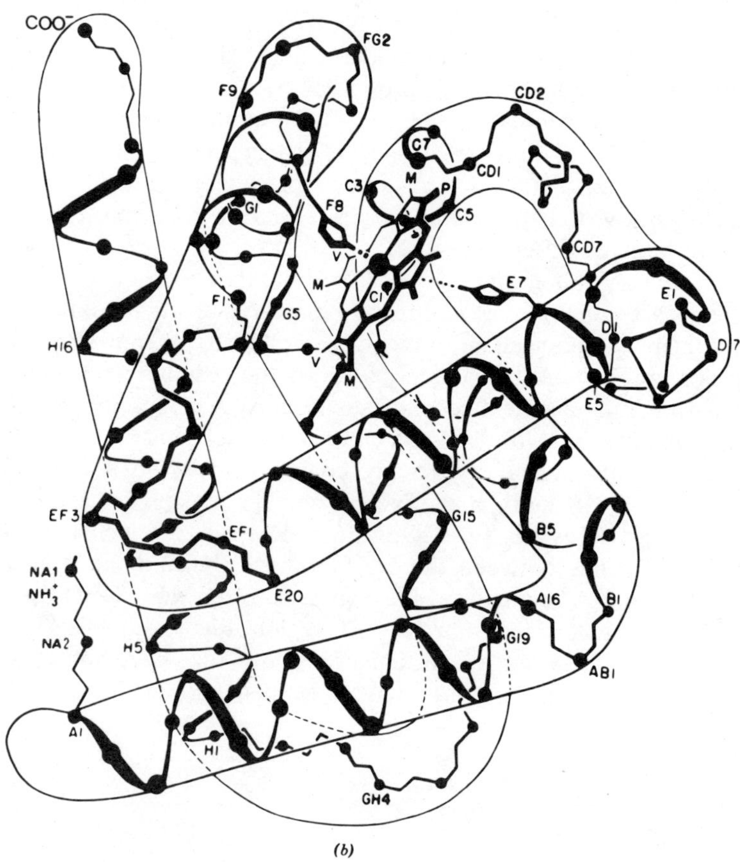

Figure 3.4 **(*Continued*)**

chains and allows for deletions and additions of amino acids made necessary because the chains are of different lengths. The proximal histidine in all chains is residue F8 (eighth residue of the F helix).

The α chain differs from the β chain by the deletion of one amino acid residue from the NA segment, one residue in the CD region, and five residues in the D helix. The α chain contains two "extra" residues in the AB region that are absent in the β chain. There are 80 amino acid differences between the two sequences (Table 3.3).

A Single Equilibrium Defines O_2 Binding to Myoglobin

Since myoglobin contains a single O_2 binding site per molecule, the association of oxygen to myoglobin is characterized by a simple equilibrium constant [equations (3.1) and (3.2)]. In equation (3.2) $[MbO_2]$ is the solution concentration of oxymyoglobin, [Mb] the concentration of deoxymyoglobin, and $[O_2]$ the concentration of oxygen, expressed in units of moles per liter. The equilibrium constant, K_{eq}, will also have the units of moles per liter. As for any true equilibrium constant, the value of K_{eq} is dependent on pH, ionic strength, and temperature.

$$\mathrm{Mb} + \mathrm{O_2} \underset{}{\overset{K_{eq}}{\rightleftharpoons}} \mathrm{MbO_2} \quad \textbf{(3.1)}$$

$$K_{eq} = \frac{[\mathrm{Mb}][\mathrm{O_2}]}{[\mathrm{MbO_2}]} \quad \textbf{(3.2)}$$

Since oxygen is a gas, it is more convenient to express O_2 concentration in terms of the pressure of oxygen in units of torr (1 torr is equal to the pressure of 1 mmHg at 0°C and standard gravity). In equation (3.3) this transfer of units has been made, with P_{50} the equilibrium constant and Po_2 the concentration of oxygen, now expressed in units of torr.

$$P_{50} = \frac{[Mb] \cdot Po_2}{[MbO_2]} \tag{3.3}$$

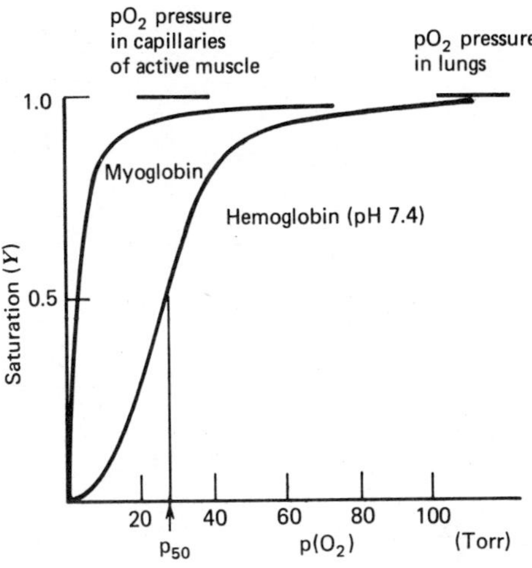

Figure 3.5
Oxygen binding curves for myoglobin and hemoglobin.

An oxygen saturation curve is used to characterize the properties of an oxygen binding protein. In this type of plot the fraction of oxygen binding sites in solution that contain oxygen [Y, equation (3.4)] is plotted on the ordinate vs the Po_2 (oxygen concentration) on the abscissa. The Y value is simply defined for myoglobin by equation (3.5). Substitution into equation (3.5) of the value of $[MbO_2]$ obtained from equation (3.3) and then dividing through by [Mb], results in equation (3.6), which shows the dependence of Y on the value of the equilibrium constant P_{50}, and the oxygen concentration. It is seen from equations (3.3) and (3.6) that the value of P_{50} is equal to the oxygen concentration, Po_2, when $Y = 0.5$ (50% of the available sites occupied). Hence the designation of the equilibrium constant by the subscript 50.

$$Y = \frac{\text{number of binding sites occupied}}{\text{total number of binding sites in solution}} \tag{3.4}$$

$$Y = \frac{[MbO_2]}{[Mb] + [MbO_2]} \tag{3.5}$$

$$Y = \frac{Po_2}{P_{50} + Po_2} \tag{3.6}$$

A plot of equation (3.6) of Y vs Po_2 generates an oxygen saturation curve for myoglobin in the form of a rectangular hyperbola (Figure 3.5).

A simple algebraic manipulation of equation (3.6) leads to equation (3.7). Taking the logarithm of both sides of equation (3.7) results in equation (3.8), which is known as the Hill equation. A plot of log $(Y/1 - Y)$ vs log Po_2, according to equation (3.8), yields a straight line with a slope equal to 1 for myoglobin (Figure 3.6). This is called the Hill plot, and the slope (n_h) is referred to as the Hill coefficient [see equation (3.9)].

$$\frac{Y}{1 - Y} = \frac{Po_2}{P_{50}} \tag{3.7}$$

$$\log \frac{Y}{1 - Y} = \log Po_2 - \log P_{50} \tag{3.8}$$

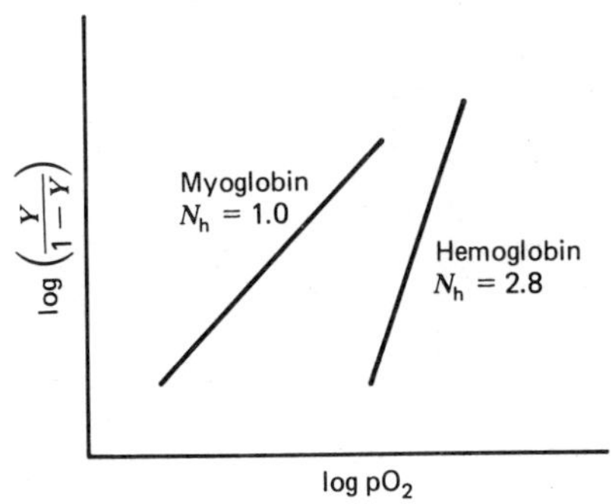

Figure 3.6
Hill plots for myoglobin and hemoglobin Hb A_1.

Binding of O_2 to Hemoglobin Is More Complex

Whereas myoglobin has a single O_2 binding site per molecule, hemoglobins contain a quaternary structure of four polypeptide chains, each with a heme binding site for O_2. The binding of the 4 O_2 in hemoglobin is found to be positively *cooperative,* so that the binding of the first O_2 to deoxyhemoglobin facilitates the binding of O_2 to the other subunits in the molecule. Conversely, the dissociation of the first O_2 from fully oxygenated hemoglobin, $Hb(O_2)_4$, will make easier the dissociation of O_2 from the other subunits of the tetramer molecule.

Based on the cooperativity in oxygen association and dissociation, the oxygen saturation curve for hemoglobin differs from that previously derived for myoglobin. A plot of Y vs Po_2 for hemoglobin follows a *sigmoid* line, indicating cooperativity in oxygen association (Figure 3.5). A plot of the Hill equation [equation (3.9)] gives a value of the slope (n_h) equal to 2.8.

$$\log \frac{Y}{1 - Y} = n_h \log Po_2 - \text{constant} \tag{3.9}$$

The meaning of the Hill coefficient to cooperative O_2 association can be quantitatively evaluated as presented in Table 3.4. A parameter known as the cooperativity index, R_x, is calculated, which shows the ratio of Po_2 required to change Y from a value of $Y = 0.1$ (10% of sites filled) to a value of $Y = 0.9$ (90% of sites filled) for designated Hill coefficient values found experimentally. In the case of myoglobin, $n_h = 1$, and an 81-fold change in oxygen concentration is required to change from $Y = 0.1$ to $Y = 0.9$. In hemoglobin, where positive cooperativity is observed, $n_h = 2.8$, and *only a 4.8-fold change in oxygen concentration* is required to change the fractional saturation from 0.1 to 0.9.

Molecular Mechanism of O_2 Cooperativity

The x-ray diffraction data on deoxyhemoglobin show that the ferrous atoms sit out of the plane of their porphyrins by about 0.4–0.6 Å. This is thought to be due to two factors. The electronic configuration of the five-coordinated ferrous atom in deoxyhemoglobin has a slightly larger radius than the distance from the center of the porphyrin to each of the pyrrole nitrogens. Accordingly, the iron can be placed in the center of the porphyrin only with some distortion of the most stable porphyrin conformation. Probably a more important consideration is that if the iron atom sits in the plane of the porphyrin, the proximal His-F8 imidazole will interact unfavorably with atoms of the porphyrin. The strength of this unfavorable steric interaction would, in part, be due to conformational constraints on the His-F8 and the porphyrin in the deoxyhemoglobin conformation that energetically forces the approach of the His-F8 toward the porphyrin to a particular path (Figure 3.7). These constraints will become less significant in the oxy conformation of hemoglobin.

TABLE 3.4 Relationship Between Hill Coefficient (n_h) and Cooperativity Index (R_x)

n_h	R_x	*Observation*
0.5	6560	Negative substrate cooperativity
0.6	1520	
0.7	533	
0.8	243	
0.9	132	
1.0	81.0	Noncooperativity
1.5	18.7	
2.0	9.0	
2.8	4.8	Positive substrate cooperativity
3.5	3.5	
6.0	2.1	
10.0	1.6	
20.0	1.3	

SOURCE: Based on Table 7.1 in A. Cornish-Bowden, *Principles of Enzyme Kinetics*, Butterworths Scientific Publishers, London and Boston, 1976.

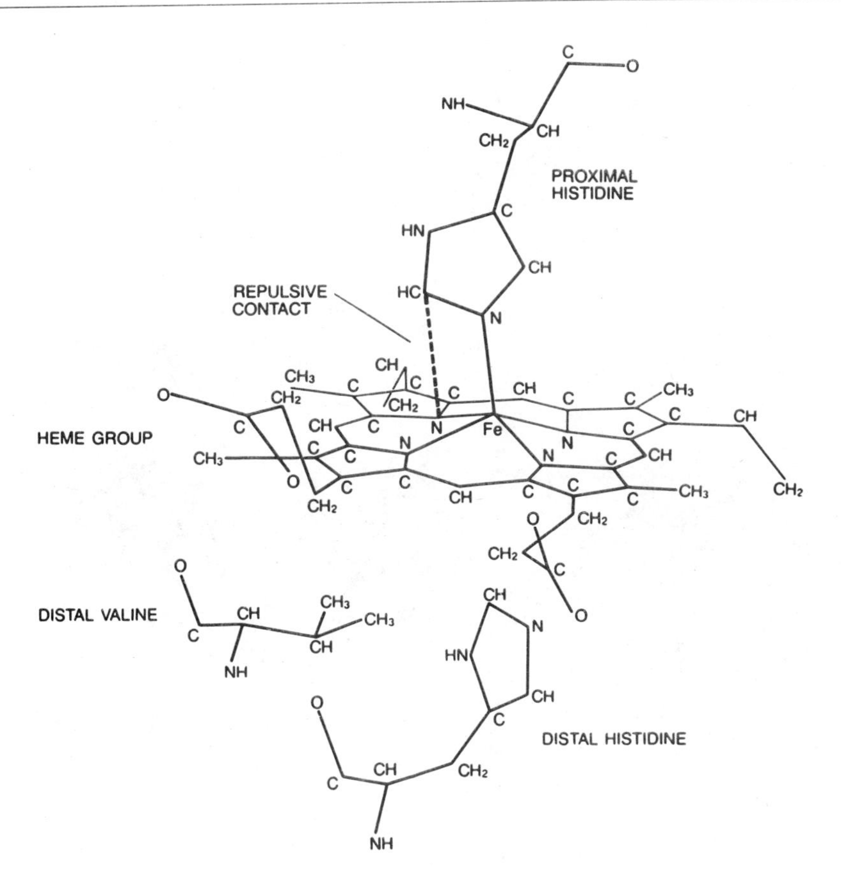

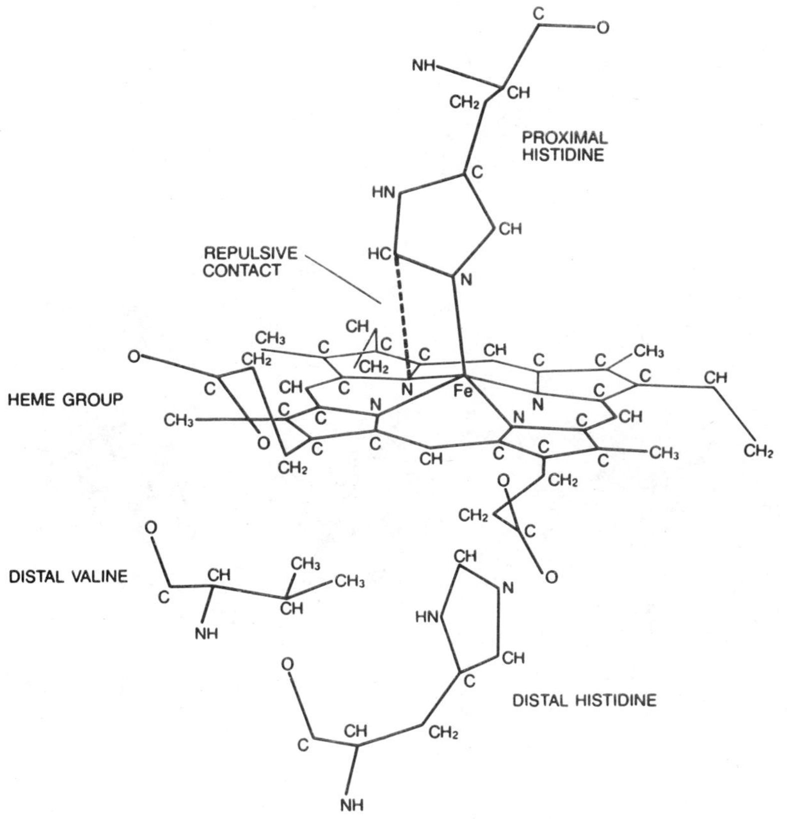

Figure 3.7
Steric hindrance between proximal histidine and porphyrin in deoxyhemoglobin.
From M. Perutz, *Sci. Am.*, 239:92, 1978 by Scientific American, Inc. All rights reserved.

The conformation with the iron atom out of the plane of the porphyrin is unstrained and energetically favored for the five-coordinate ferrous atom. However, when O_2 binds to the sixth coordinate position of the iron, this conformation becomes strained. A more energetically favorable conformation for the O_2 liganded iron is one in which the iron atom is within the plane of the porphyrin structure.

On the binding of O_2 to a ferrous atom into its sixth coordinate position, the favorable free energy of bond formation is used to overcome the repulsive interaction between the His-F8 and porphyrin, and the ferrous atom moves into the plane of the porphyrin ring. This is the most thermodynamically stable position for the now six-bonded iron atom; one axial ligand is on either side of the plane of the porphyrin ring, and the steric repulsion of one of the axial ligands with the porphyrin is balanced by the repulsion of the second axial ligand on the opposite side when the ferrous atom is in the center. If the iron atom is displaced from the center, the steric interactions of the two axial ligands with the porphyrin in the deoxy conformation are unbalanced, and the stability of the unbalanced structure will be lower than that of the equidistant conformation. Also, the radius of the iron atom with six ligands is reduced so that it can just fit into the center of the porphyrin without distortion of the porphyrin conformation.

Since the steric repulsion between the porphyrin and the His-F8 must be overcome on O_2 association, the binding of the first O_2 to hemoglobin is characterized by a relatively low affinity constant. However, when an O_2 association does occur to the first heme in a deoxyhemoglobin molecule,

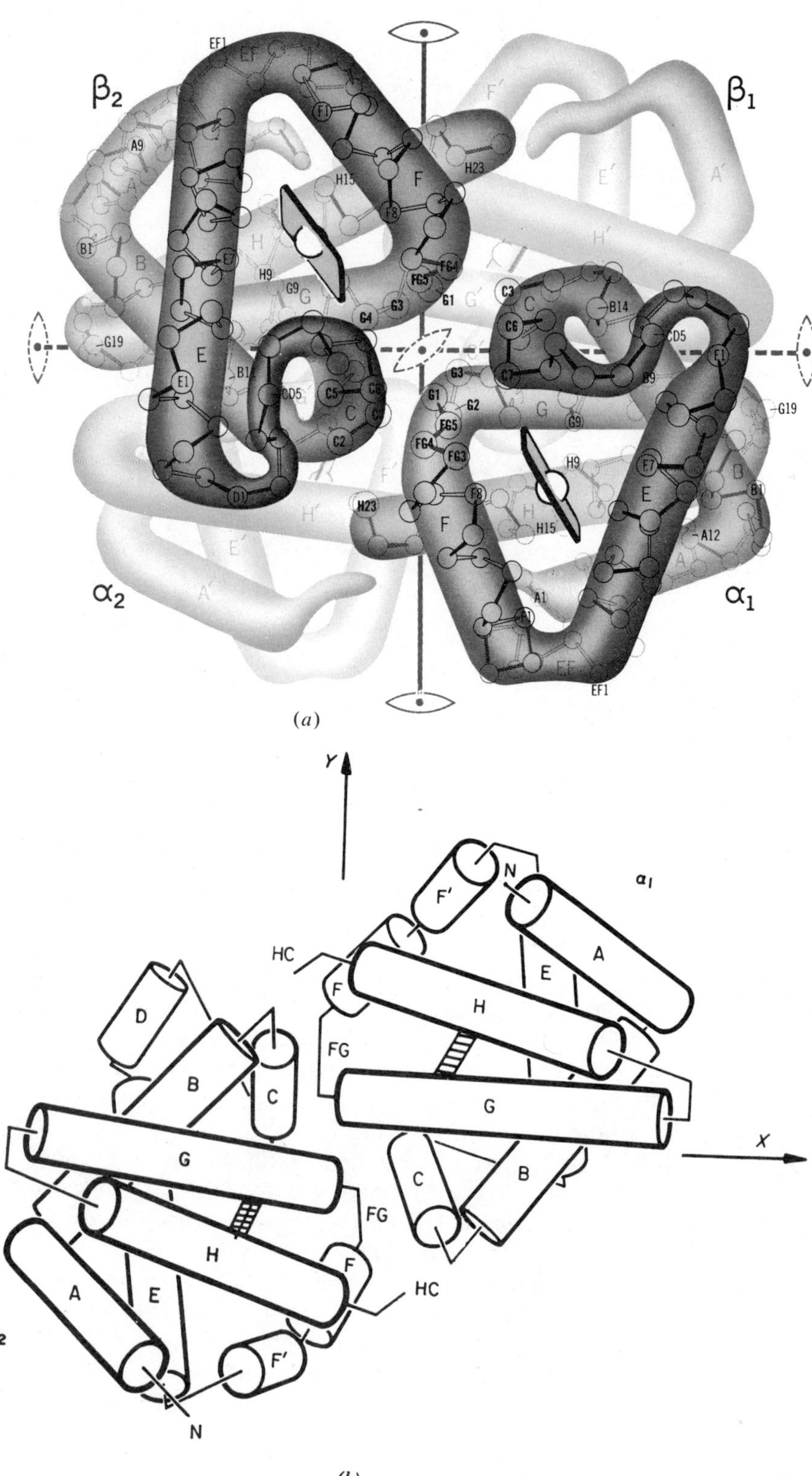

Figure 3.8

(a) Quaternary structure of hemoglobin showing the $\alpha_1\beta_2$ interface contacts between FG corners and C helix. (b) Cylinder representation of α_1 and β_2 subunits in hemoglobin molecule showing α_1 and β_2 interface contacts between FG corner and C helix, viewed from opposite side of X–Y plane from in (a).

(*a*) Figure reprinted with permission from R. E. Dickerson and I. Geis, *The Structure and Action of Proteins*, W. A. Benjamin, Inc., Menlo Park, 1969, p. 56.

(*b*) Figure reprinted with permission from J. Baldwin and C. Chothia, *J. Mol. Biol.* 129:175, 1979. Copyright by Academic Press Inc. (London) Ltd.

the change in the iron atom position from above the plane of the porphyrin into the center of the porphyrin triggers a conformational change in the whole hemoglobin molecule. The change in conformation results in a greater affinity of O_2 to the other heme sites after the first O_2 has bound. This conformation change is thought to occur as described below.

The conformation of the deoxyhemoglobin appears to be stabilized by noncovalent interactions of the quaternary structure at the interface between α and β subunits in which the FG corner of one subunit noncovalently binds to the C helix of the adjacent subunit (Figure 3.8). In addition, ionic interactions stabilize the deoxy-quaternary conformation of the protein (Figure 3.9). These particular noncovalent interactions of the deoxy conformation are now *destabilized* on the binding of O_2 to one of the heme subunits of a deoxyhemoglobin molecule. The binding of O_2 pulls the Fe^{+2} atom into the porphyrin plane, which correlated with the movement of the His-F8 toward the porphyrin moves the F helix of which the His-F8 is a part. The movement of the F helix, in turn, moves the FG corner of its subunit, destabilizing the FG noncovalent interaction with the C helix of the adjacent subunit at an $\alpha_1\beta_2$ or $\alpha_2\beta_1$ subunit interface (Figures 3.8 and 3.10).

The FG to C intersubunit contacts are thought to act as a "switch," because they apparently can exist in two different arrangements with different modes of contact between FG and C. On the binding of O_2 to deoxyhemoglobin, the movement of the FG corner in the subunit to which the O_2 is bound forces the disruption of noncovalent interactions that stabilize the deoxy conformation at the FG to C interface, and the intersubunit contacts switch to their alternative position. The switch in noncovalent interactions between the two positions involves a relative movement of FG and C in adjacent subunits of about 6 Å. In the second position of the "switch," *the tertiary conformation of the subunits participating in the FG to C intersubunit contact are less constrained and change to a new tertiary conformation* (oxy conformation). This second conformation allows the His-F8 to approach their porphyrins, on O_2 association, with a less significant steric repulsion than in the deoxy conformation of the hemoglobin subunits (Figure 3.10). An O_2 molecule can bind to the empty hemes in the less constrained oxy conformation more easily than to the original subunit conformation held by the intersubunit FG to C interaction in the deoxy conformation.

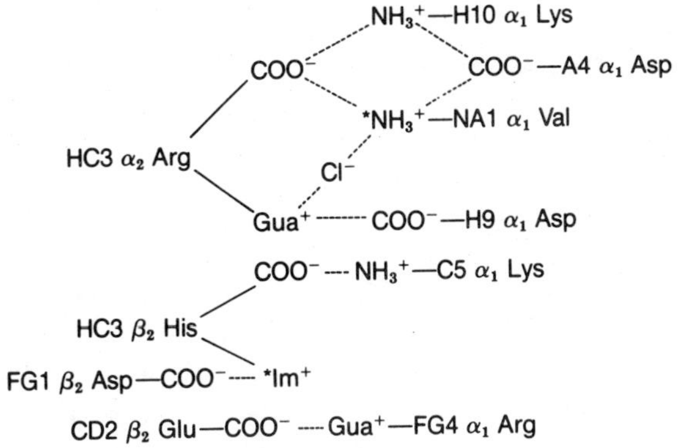

Figure 3.9
Salt bridges between subunits in deoxyhemoglobin.
Im⁺ is imidazole; Gua⁺ is quanidine; starred residues account for approximately 60% of the alkaline Bohr effect.
Redrawn from M. Perutz, *Brit. Med. Bull.*, 32:195, 1976.

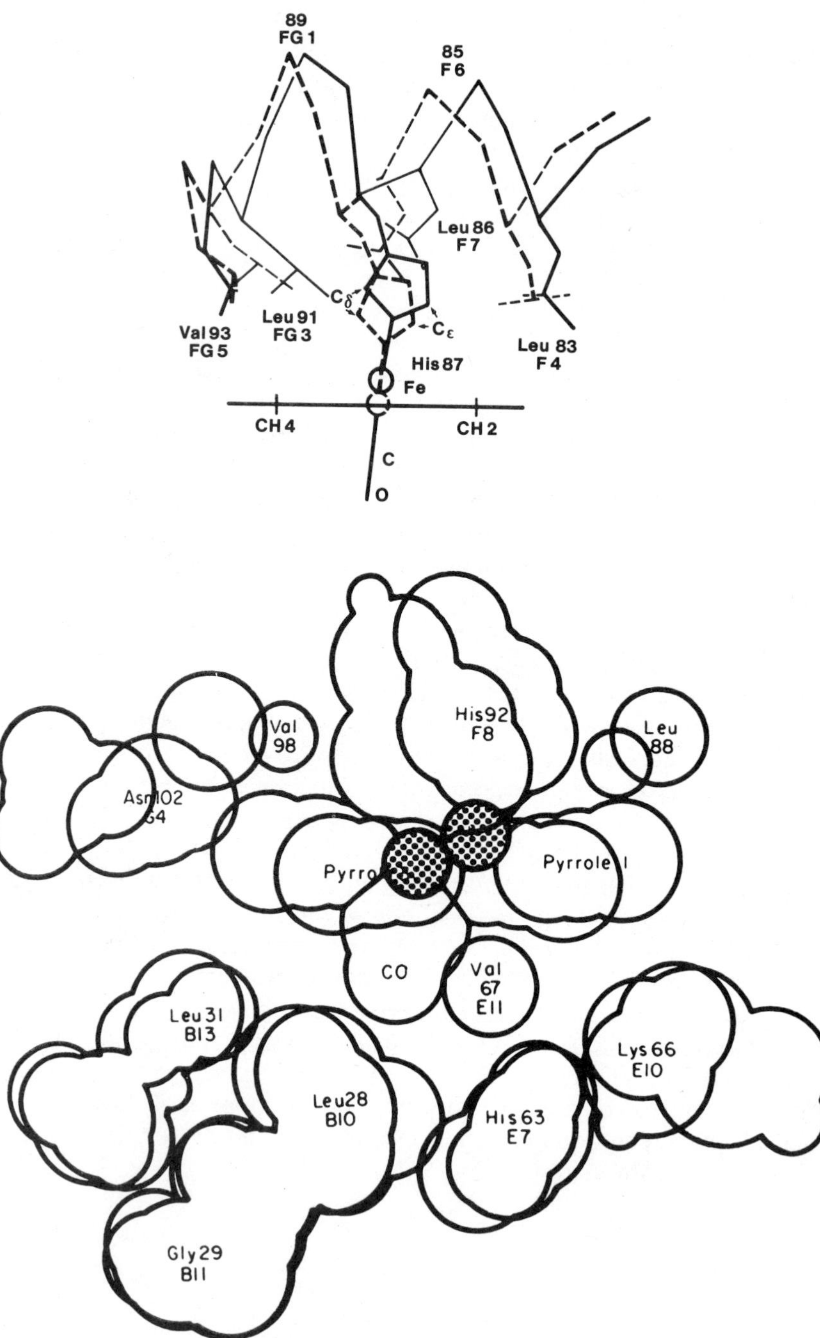

Figure 3.10
Stick and space-filling diagrams drawn by computer graphics showing movements of residues in heme environment on transition from deoxyhemoglobin to oxyhemoglobin.
(a) Dotted line outlines position of polypeptide chain and His-F8 in carbon monoxyhemoglobin, which is a model for oxyhemoglobin. Solid line outlines the same for deoxyhemoglobin. Position of iron atom shown by circle. Movements are for α subunit. (b) Similar movements in β subunit using space-filling diagram shown. Residue labels centered in density for the deoxyconformation.
Reprinted with permission from J. Baldwin and C. Chotia, *J. Mol. Biol.* 129:175, 1979. Copyrighted by Academic Press Inc. (London) Ltd.

In addition, the Val-E11 in the deoxy conformation of the β subunits is at the entrance to the O_2 binding site, where it sterically impedes O_2 association to the heme (see Figure 3.3). In the oxy conformation the heme in the β subunits appears to move approximately 1.5 Å further into the heme binding site of the protein, changing the geometric relationship of the O_2 binding site in the heme to the Val-E11 side chain, so that the Val-E11 no longer sterically interferes with O_2 binding. This appears to be an important additional factor that results in a higher affinity of O_2 to the oxy conformation of the β chain than to the deoxy conformation.

The deoxy conformation of hemoglobin is referred to as the "tense" or T conformational state. The oxyhemoglobin conformational form is referred to as the "relaxed" or R conformational state. On the binding of the initial oxygen to the heme subunits of the tetramer molecule, the molecular conformation is pushed from the T to the R conformational state. The affinity constant of O_2 is greater for the R state hemes than the T state by a factor of 150 to 300, depending on the solution conditions.

Molecular Mechanism of the Bohr Effect

The equilibrium expression for oxygen association to hemoglobin includes a term that indicates the participation of hydrogen ion in the equilibrium.

$$\text{Hb} + 4\text{O}_2 \rightleftharpoons \text{Hb(O}_2)_4 + x\text{H}^+ \qquad \textbf{(3.10)}$$

Equation (3.10) shows that the R form is more acidic, and protons dissociate when the hemoglobin is transformed to the R form. The number of equivalents of protons that dissociate per mole of hemoglobin depends on the pH of the solution and the concentration of other factors that can bind to hemoglobin, such as Cl^- and diphosphoglycerate (see Chapter 23). At pH 7.4, the value of x may vary from about 1.8 to 2.8, depending on the solution conditions. This production of protons at alkaline pH (pH > 6), when deoxyhemoglobin is transformed to oxyhemoglobin, is known as the alkaline Bohr effect.

The protons are derived from the partial dissociation of acid residues with pK_a values within ± 1.5 pH units of the solution pH, that change from a higher to lower pK_a on the transformation of the T to R conformation of the hemoglobin. For example, the HC3 His-146(β) in the deoxy conformation is predominantly in its imidazolium form (positively charged acid form), which is stabilized by a favorable interaction with the negatively charged side chain of the FG1 Asp-94(β) (Figure 3.9). This ion-pair interaction makes it more difficult to remove the imidazolium proton, and thus raises the pK_a of the imidazolium to a higher value than normally found for a free imidazolium ion in solution, where a stabilization by a proximate negatively charged group does not normally occur. However, on conversion of the protein to the R conformation, the strength of this ionic interaction is broken and the imidazolium assumes a lower pK_a. The decrease in the histidine's pK_a at blood pH results in the conversion of some of the acid form of the histidine to its conjugate base (imidazole) form, with the dissociation of free protons that form a part of the Bohr effect. Breakage of this ion pair with release of protons accounts for 50% of the protons released on conversion to the R conformation. Other acid groups in the protein contribute the additional protons due to changes in their pK_a to lower values.

The Bohr effect may fit the definition of an allosteric mechanism. An *allosteric mechanism* is a common process in protein molecules in which substrate association is influenced by the binding of other molecules that are not direct substrates of the protein. In an allosteric process there must be separate binding sites on the protein for substrate (e.g., O_2 in the case of hemoglobin) and effector (inhibitor or activator) molecules that exert allosteric control. As the effector molecule's binding site by definition is distinctly separate from that of the substrate's binding site, the effector molecule acts to increase or decrease the affinity of the substrate at the substrate binding site by either causing a conformational change or stabilizing a particular conformation of the protein. With regard to the Bohr effect, it is evident that proton binding sites exist in the hemoglobin molecule to which the binding of protons, the effector "molecules," thermodynamically stabilizes and thus increases the concentration of the T form with respect to the R form. The binding of a proton to form the ion-pair site between the His-146(β) and Asp-94(β) is one such interaction, for example, that favors the T conformation. By increasing the ratio T/R on stabilizing the T conformation, the binding of protons to their effector sites is correlated with a poorer affinity of hemoglobin for oxygen.

The Bohr effect has important physiological consequences. Cells metabolizing at high rates, with high requirements for molecular oxygen, produce carbonic acid and lactic acid, which act to increase the hydrogen ion concentration in the cell's environment. As the increase in hydrogen ion concentration forces the equilibrium of equation (3.10) to the left, from the higher O_2 affinity conformation (R) to the lower affinity conformation (T), an increased amount of oxygen is dissociated from the hemoglobin molecule.

3.3 STRUCTURE AND MECHANISM OF ACTION OF ANTIBODY MOLECULES

Antibody molecules are immunoglobulins produced by an organism in response to the invasion of foreign compounds, such as proteins, carbohydrates, and nucleic acid polymers. The antibody molecule noncovalently associates to the foreign substance, initiating a process by which the foreign substance is eliminated from the organism.

Materials that elicit antibody production in an organism are called *antigens*. An antigen may contain multiple *antigenic determinants,* which are small regions of the antigen molecule that specifically elicit the production of antibody to which the antigen binds. In proteins, for example, an antigenic determinant may comprise only six or seven amino acids of the total protein.

A *hapten* is a small molecule that cannot alone elicit the production of antibodies specific to it. However, when covalently attached to a larger molecule it can act as an antigenic determinant and elicit antibody synthesis. Whereas the hapten molecule requires attachment to a larger molecule to elicit the synthesis of antibody, when detached from its carrier, the hapten will retain its ability to bind strongly to antibody.

It is estimated that a human individual can potentially produce at least 1×10^6 different antibody structures. However, all the antibodies have a similar basic structure. The elucidation of this structure has been accomplished to a great extent from studies of immunoglobulin primary structures. In addition, electron microscopy and recently obtained x-ray diffraction data have immeasurably added to our knowledge of the three-dimensional structure of the antibody molecule.

Structural studies of proteins require pure homogeneous preparations. Such a sample of an antibody protein is extremely difficult to prepare because of the wide diversity of antibody molecules present in any one organism. Accordingly, immunoglobulins for structural studies have historically been obtained from individuals who suffer from multiple myeloma. In this disease there is a loss of control of antibody synthesis in a single type of antibody-synthesizing cell by which large amounts of antibody of a single distinct structure is produced. This antibody is identical in its general structure to normal antibodies and can be easily purified in large amounts. Homogeneous antibodies are now obtained by the monoclonal hybridoma technique.

Primary Structure of Antibody (Immunoglobulin) Molecules

Antibody molecules are composed of units of four polypeptide chains, of which two have an identical primary structure and another two have an identical primary structure. In the most common immunoglobulin, IgG, the two larger chains have approximately 440 amino acids (mol wt 50,000). These chains are designated *heavy chains,* or H chains. The smaller chains contain about one-half the number of amino acids of the H chain and are designated the *light chains,* or L chains (mol wt 25,000). The four polypeptide chains are covalently interconnected by disulfide bonds (Figure 3.11).

In the other classes of immunoglobulins (see Table 3.5) the H chains have a slightly higher molecular weight than those of the IgG class. There is a small amount of carbohydrate (2–12%, depending on immunoglobulin class) attached to the H chain, and thus the antibodies may be classified as glycoproteins (see Section 3.5).

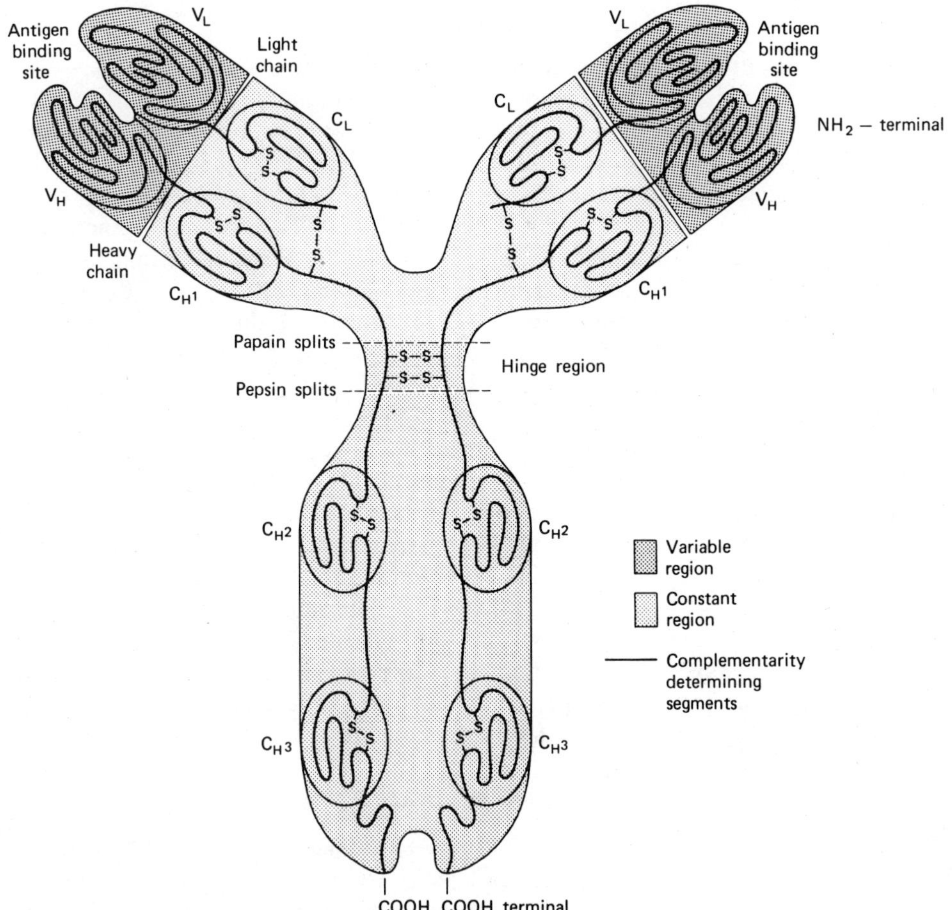

Figure 3.11
Diagrammatic structure for IgG.
The light chains (L) are divided into domains V_L (variable amino acid sequence) and C_L (constant amino acid sequence). The heavy chains (H) are divided into domains V_H (variable amino acid sequence) and C_H1, C_H2, and C_H3. The antigen binding sites are V_H-V_L. "Hinge" polypeptides interconnect the domains. The positions of inter- and intrachain cystine bonds are shown.
From C. R. Cantor and P. R. Schimmel, *Biophysical Chemistry,* Part I, W. H. Freeman, San Francisco, 1980. Reprinted with permission of Mr. Irving Geis, N.Y.

TABLE 3.5 Immunoglobulin Classes

Classes of Immunoglobulin	*Approximate Mol Wt*	*H Chain Isotype*	*Carbohydrate by Weight (%)*	*Concentration in Serum (mg/100 ml)*
IgG	150,000	γ, 53,000	2–3	600–1800
IgA	170,000–720,000[a]	α, 64,000	7–12	90–420
IgD	160,000	δ, 58,000	—	0.3–40
IgE	190,000	ε, 75,000	10–12	0.01–0.10
IgM	950,000[a]	μ, 70,000	10–12	50–190

[a] Forms polymer structures of basic structural unit.

In the three-dimensional antibody structure, each H chain is associated with an L chain such that the NH_2-terminal ends of both chains are near each other. Since the L chain is one-half the size of the H chain, only the NH_2-terminal half of the H chain (~214 amino acids in the IgG class) is associated with the L chain. This is diagrammatically shown in Figure 3.11.

Constant and Variable Regions of Primary Structure

A comparison of the amino acid sequences of antibody molecules elicited toward different antigens shows that there are regions of exact homology and regions of high variability. In particular, the sequences of the NH_2-terminal one-half of the L chains and the NH_2-terminal one-quarter of the larger H chains are highly variable. These NH_2-terminal segments are designated the variable (V) regions. Within the V region certain segments are observed to be even more variable than other segments and are termed "hypervariable" regions. Three hypervariable regions of between 5 and 7 residues in the NH_2-terminal region of the L chain and three or four hypervariable regions of between 6 and 17 residues in the NH_2-terminal region of the H chain are commonly found.

In contrast, a comparison of the amino acid sequences from different antibodies shows that the COOH-terminal three-quarters of the H chains and the COOH-terminal one-half of the L chains are mostly homologous in sequence. These regions of the polypeptide sequences are named the constant (C) regions of primary structure.

It is the C regions of H chains that determine the class to which the antibody belongs (see below). In addition, the C region provides for binding of complement proteins (Clin. Corr. 3.1) and the site necessary for antibodies to cross the placental membrane. The V regions determine the antigen specificity of the antibody molecule.

Different Classes of Immunoglobulins

The C regions of H chains within a particular class of immunoglobulins are homologous but differ significantly from the amino acid sequence of the C regions of the H chains of other antibody classes. These differences are responsible for the different physical characteristics of the immunoglobulin classes.

In some cases the H chain sequence promotes polymerization of the antibody molecules. Thus $(LH)_2$ antibody units of the IgA class are sometimes found in covalently linked dimeric forms of the structure $[(LH)_2]_2$. Similarly, IgM antibodies are pentamers of the basic $(LH)_2$ covalent structure, giving it the formula $[(LH)_2]_5$. The different H chains designated γ, α, μ, δ, and ε are found in the IgG, IgA, IgM, IgD, and IgE classes, respectively (Table 3.5; Clin. Corr. 3.2).

CLIN. CORR. 3.1 THE COMPLEMENT PROTEINS

The complement proteins are composed of at least 11 distinct proteins in plasma. They are triggered by the association of IgG or IgM binding to antigens on the outer cell membrane of invading bacterial cells, protozoa, or tumor cells. On initiation by the immunoglobulin binding event, the 11 complement proteins become activated and associate with the cell membrane, causing a lysis of the membrane and cellular death.

Many of the complement proteins are precursors of proteolytic enzymes that are present in a nonactive form prior to activation. On their activation to active enzymes during the complementation process, they will in turn activate a succeeding protein of the pathway by the hydrolysis of a specific peptide bond in the second protein. The inactive forms of enzymes are referred to as proenzymes or zymogens. The activation of enzymes by specific proteolysis (i.e., hydrolysis of a specific peptide bond in its primary structure) is an important general method for activating extracellular enzymes. For example, the enzymes that catalyze blood clot formation, induce fibrinolysis of blood clots, and digest dietary proteins in the gut are all activated by a specific proteolysis catalyzed by a second enzyme.

The *classical* complementation reaction is initiated by the binding of IgG or IgM to cell surface antigens. On association to a cellular antigen the exposure of a complement binding site in the antibody's F_c region occurs and causes the binding of the C1 complement proteins, which are a protein complex composed of three individual proteins: C1q (mol wt 400,000), C1r (mol wt 180,000), and C1s (mol wt 86,000). The C1r and C1s proteins undergo a conformational change and become active enzymes on association with the immunoglobulin on the cell surface. The activated C1 complex (C1a) catalytically hydrolyzes a peptide bond in complement proteins C2 (mol wt 117,000) and in C4 (mol wt 206,000), which then form a complex that also associates to the cell surface. The now active C2–C4 complex has a proteolytic activity that hydrolyzes a peptide bond in complement protein C3 (mol wt 180,000). Activated C3 protein binds to the cell surface, and the activated C2–C4–C3 complex activates protein C5 (mol wt 180,000). Activated protein C5 will associate with complement proteins C6 (mol wt 110,000), C7 (mol wt 100,000), C8 (mol wt 163,000), and six molecules of complement protein C9 (mol wt 79,000). This multiprotein complex binds to the cell surface and initiates membrane lysis.

The mechanism is a cascade type in which amplification of the trigger event occurs. In summary, activated C1 can activate multiple molecules of C4–C2–C3, and each activated C4–C2–C3 complex can in turn activate many molecules of C5 to C9.

The series of reactions in the classical complementation pathway is summarized in the scheme below, where a and b designate the proteolytically modified proteins and a line above a protein indicates an enzyme activity.

$$\text{IgG or IgM} \xrightarrow{\text{C1q,C1r,C1s}} \overline{\text{C1a}} \xrightarrow{\text{C2,C4}}$$

$$\text{C4b} \cdot \overline{\text{C2a}} \xrightarrow{\text{C3}}$$

$$\text{C4b} \cdot \overline{\text{C2a}} \cdot \text{C3b} \xrightarrow{\text{C5,C6,C7,C8,6C9}}$$

$$\text{C5b} \cdot \text{C6} \cdot \text{C7} \cdot \text{C8} \cdot \text{6C9}$$

There is an "alternative pathway" for C3 complement activation, initiated by aggregates of IgA or by bacterial polysaccharide in the absence of immunoglobulin binding to cell membrane antigens. This alternative pathway involves the proteins properdin (mol wt 184,000), C3 proactivator convertase (mol wt 24,000), and C3 proactivator (mol wt 93,000).

CLIN. CORR. 3.2 FUNCTIONS OF THE DIFFERENT ANTIBODY CLASSES

The IgA class of immunoglobulins are primarily found in the extravascular secretions—bronchial, nasal, and intestinal mucus secretions; tears; milk; and colostrum. As such, these immunoglobulins are the initial defense against invading viral and bacterial antigens prior to their entry in plasma or other internal space.

The IgM class are primarily found in plasma. They are the first of the antibodies to act in significant quantity on the introduction of a foreign antigen into a host's plasma. IgM antibodies can promote phagocytosis of microorganisms by macrophage and polymorphonuclear leukocytes and are also potent activators of complement (see Clin. Corr. 3.1). IgM can be found in many external secretions, but at levels lower than those of IgA.

The IgG class is found in high concentration in plasma. Its response to foreign antigens takes a longer period of time than that of IgM. However, at its maximum concentration it is present in significantly higher concentrations than the IgM antibodies. Like IgM antibodies, IgG antibodies can promote phagocytosis by phagocytic cells in plasma and can activate complement.

The normal biological functions of the IgD and IgE classes of immunoglobulins are not known. However, it is clear that the IgE antibodies play an important role in allergic responses, which are a cause of anaphylactic shock, hay fever, and asthma.

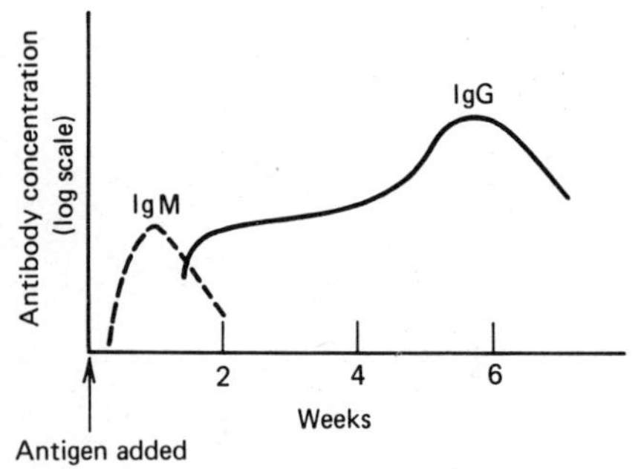

Figure 3.12
Time course of specific antibody IgM and IgG response to added antigen.
Based on figure in L. Stryer, *Biochemistry*, W. H. Freeman, San Francisco, 1975, p. 733.

Although IgG is the major immunoglobulin in plasma, the biosynthesis of a specific IgG antibody in significant concentrations after exposure to a new antigen takes about 10 days (Clin. Corr. 3.3). In the absence of an initially high concentration of IgG to a specific antigen, antibodies of the IgM class, which are synthesized at faster rates, will associate with the antigen and serve as the first line of defense against the foreign antigen until the large quantities of IgG are produced (Figure 3.12; Clin. Corr. 3.3).

Two types of L chain constant sequences are synthesized, either of which is found combined with the five classes of H chains. The chains are designated the *lambda* (λ) chain and the *kappa* (κ) chain.

Existence of Two Antigen Binding Sites Per Antibody Molecule

The NH_2-terminal V regions of the L and H chain pairs comprise an antigen binding site. As the basic antibody structure contains two LH pairs, there are two antigen binding sites per antibody molecule. Clear evidence that an antigen binding site exists in the LH pair NH_2-terminal region was obtained by chemical techniques. In these experiments an antibody is hydrolyzed by the proteolytic enzyme papain at a single peptide bond located in the hinge peptide of each H chain (see Figure 3.13). On papain hydrolysis of this peptide bond, the antibody molecule is cleaved into three products, two of which are identical and comprise the NH_2-terminal segments of the H chain associated with the full L chain (Figure 3.13). *These NH_2-terminal H · L fragments can bind antigens with a similar affinity as does the whole antibody molecule.* The two NH_2-terminal fragments are designated the Fab (antigen binding) fragments. The second type of product from the papain hydrolysis is the COOH-terminal half of the H chains bound covalently in a single covalent fragment designated the Fc (crystallizable) fragment. The Fc fragment cannot bind antigen. It is thus clear that there are two antigen binding sites per antibody molecule present in the NH_2-terminal H · L fragments, which comprise the variable (V) amino acid sequences of each of the chains.

The valency of 2 for each antibody molecule allows each antibody to bind to two different antigens. This property of bivalency facilitates the agglutination and precipitation of antigen molecules by allowing the antibodies to form an interconnected matrix of antigens and antibodies.

The L chain can be dissociated from its H chain segment within the Fab fragment by oxidation of disulfide bonds, followed by chromatography. The dissociation of the L and H chains in this way eliminates antigen binding. Accordingly, each antigen binding site must be formed from components of both the L and H variable regions acting together.

In support of these conclusions, the x-ray crystallographic structure of an Fab fragment with a hapten associated has been obtained (Figure 3.14). The structure shows that the hypervariable sequences of the V regions are specifically utilized in forming the antigen binding site (Figure 3.14). The sequence of the hypervariable regions apparently give a unique three-dimensional conformation for each antibody that makes it specific to the antigenic determinant with which it associates.

The strength of association between antibody and antigen is due to noncovalent forces (see Chapter 2). The complementarity of the structures of the antigenic determinant and antigen binding site within the antibody results in extremely high equilibrium affinity constants, between 10^5 and 10^{10} M^{-1} (strength of 7-14 kcal/mol) for this noncovalent association.

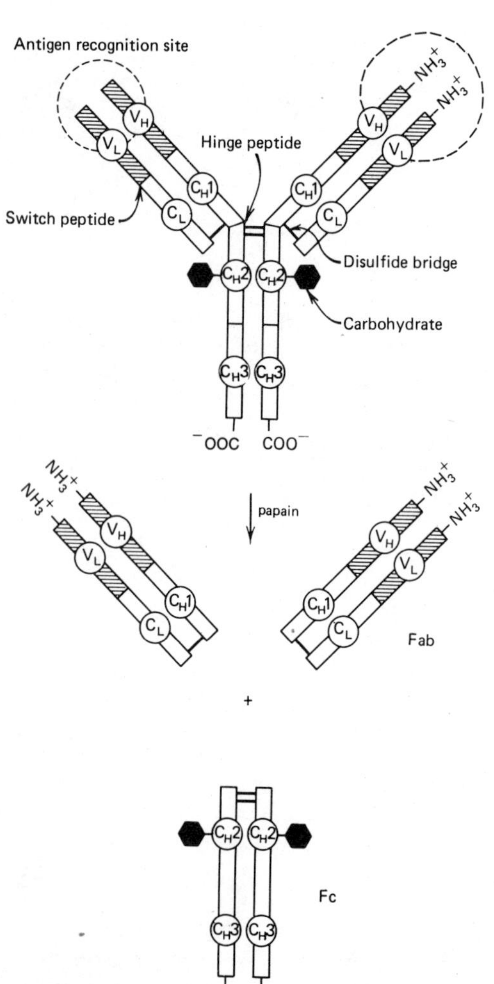

Figure 3.13
Hydrolysis of IgG into two Fab and one Fc fragments by papain, a proteolytic enzyme.

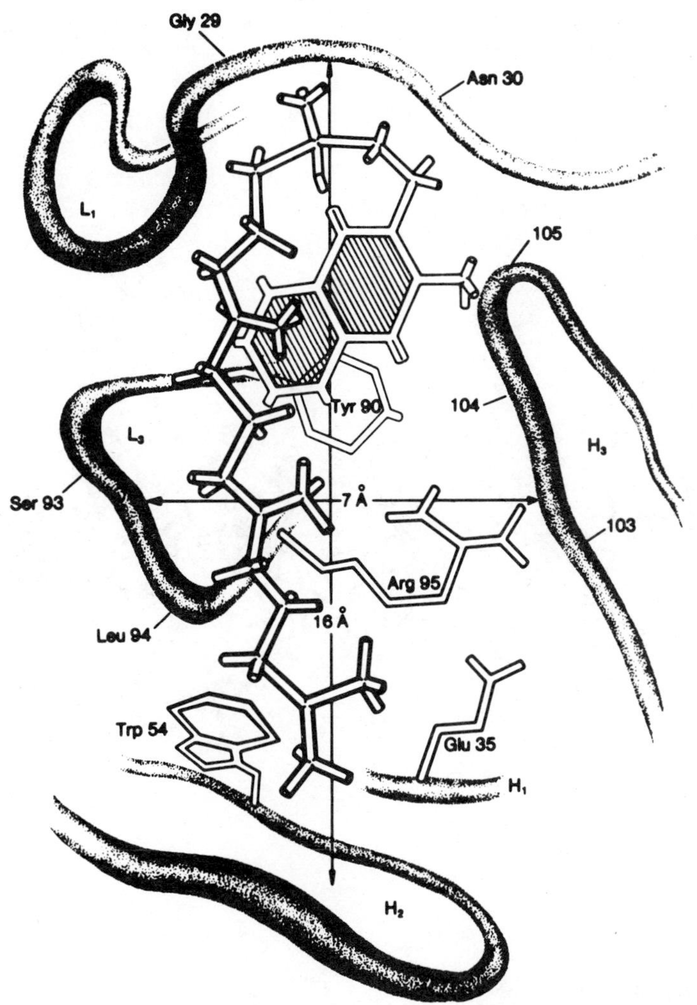

Figure 3.14
Structure of antigen (vitamin K_1-OH) bound to variable region of antibody.
Hypervariable regions designated L_1 and L_3 in light chains and H_1, H_2, and H_3 in heavy chains are shown in van der Waals contact with antigen.
Reprinted with permission from L. M. Amzel, R. J. Poljak, F. Saul, J. M. Varga, and F. F. Richards, *Proc. Natl. Acad. Sci. USA,* 71:1427, 1974.

CLIN. CORR. 3.3 IMMUNIZATION

An immunizing vaccine may consist of killed bacterial cells, inactivated viruses, killed parasites, a nonvirulent form of live bacterium related to a virulent bacterium, or a denatured bacterial toxin. The introduction of such a preparation into a human being can lead to protection against the virulent forms of the infective or toxic agents that contain the same antigen. This is because the antigens in the nonvirulent material not only cause the differentiation of lymphoid cells into cells that produce antibody toward the foreign antigen, but also cause the differentiation of some lymphoid cells into memory cells. Memory cells do not excrete antibody, but place antibodies to the antigen into their outer membrane where they act as future sensors for the antigen. These memory cells are like a long-standing radar for the potentially virulent antigen. On the reintroduction of the antigen at a later time, the binding of the antigen to the cell surface antibody in the memory cells stimulates the memory cell to divide into antibody-producing cells as well as new memory cells. This mechanism reduces the time for antibody production that is required on introduction of an antigen, and increases the concentration of antigen-specific antibody produced. It is the basis for the protection provided by immunization.

Homologous Three-Dimensional Domains of Antibody Structure

We have previously discussed the constant, variable, and hypervariable regions of primary structure found by comparing the amino acid sequences from different antibody molecules. An even closer examination of the primary structure, in which the amino acid sequences of segments of a molecule are compared with other segments within the same molecule, shows a repeating pattern of amino acid sequences within each of the chains. Analysis of the intrachain homology indicates that approximately 110 amino acids form a primary structure unit that is repeated twice in an L chain and four times in an H chain (see Figure 3.11). Each of these repeating sequences or units are cyclized by an intrachain cystine formed from two cysteines near the two ends of this sequence unit. The periodic similarities in sequence, even into the NH_2-terminal V regions, suggests a mechanism of antibody evolution in which current antibody genes have evolved by gene duplication of a primordial gene that coded for a protein 110 amino acids in length. Mutations in the duplicated DNA sequence over time resulted in the different but analogous regions seen in the present antibody protein.

Based on the presence of analogous regions in the amino acid sequence, it was proposed that each of the analogous segments may have a similar tertiary conformation. X-ray diffraction studies confirm this contention. The x-ray data dramatically show periodic domains in the H and L chains with a similar three-dimensional conformation as predicted from the study of the primary structure (Figure 3.15*b* and *c*). The globular

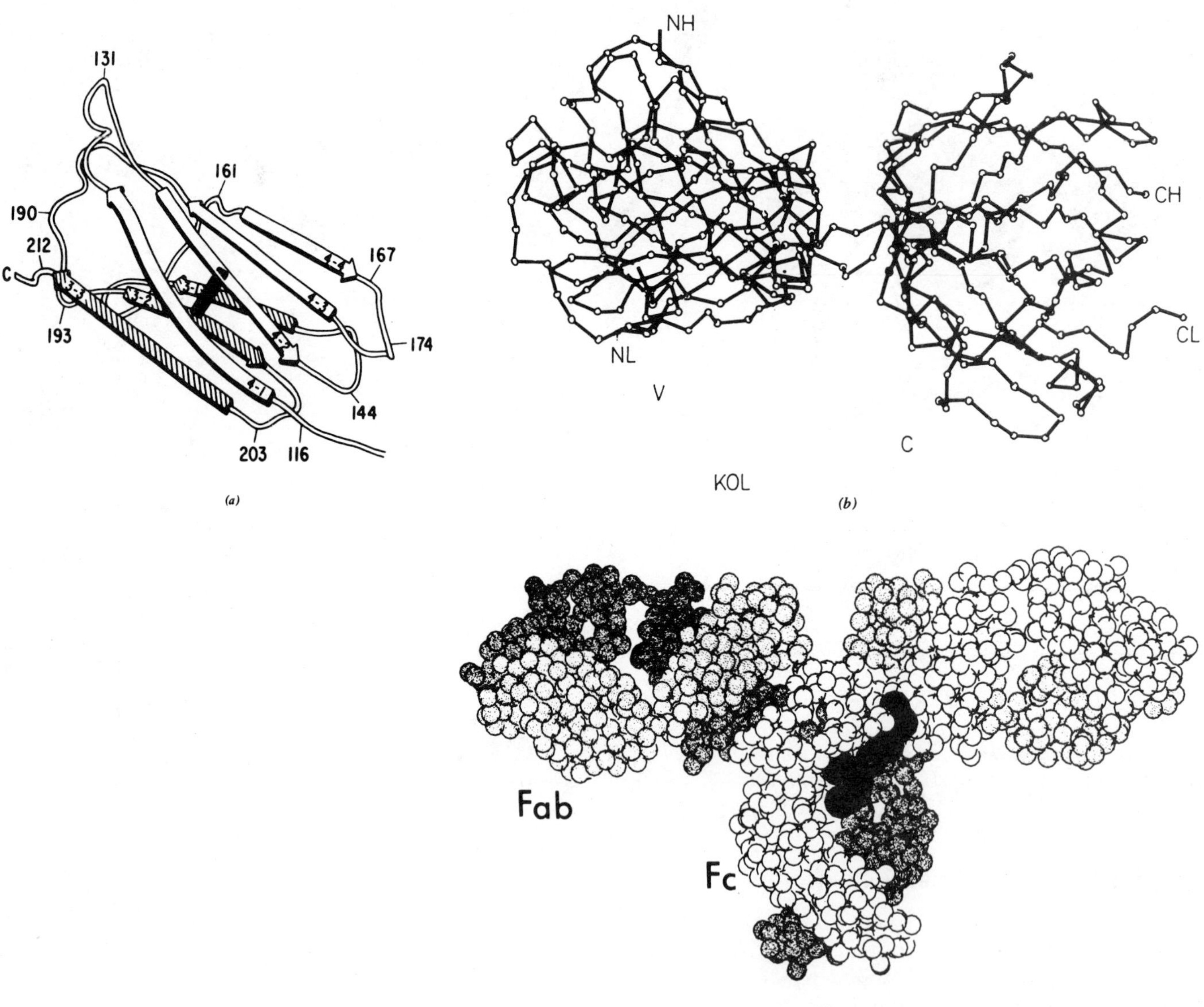

Figure 3.15
(a) Schematic diagram of folding of a C_L domain, showing β-pleated sheet structure. (b) The α-carbon (○) structure of the Fab fragment of IgG Kol showing the V_L-V_H and C_H1-C_L domains interconnected by hinge polypeptide, (c) Space-filling model of an IgG molecule. One of the H chains is in white, and the other H chain is in dark grey. The two L chains are lightly shaded. The black spheres represent the carbohydrate attached to the protein.

Reprinted with permission. Figure *a* from Edmundson *et al., Biochemistry*. Copyright 1975 American Chemical Society. Figure *b* from R. Huber, J. Deisenhofer, P. M. Coleman, M. Matsushima, and W. Palm, in *The Immune System*. 27th Mosbach Colloquium, Springer-Verlag, Berlin, 1976, p. 26. Figure *c* from Silverston et al., *Proc. Natl. Acad. Sci. USA* 74:5140, 1977.

domains are formed from the winding of the polypeptide chain back and forth upon itself forming two β-barrel structures (Figure 3.15*a,b*) that run roughly parallel to each other. In three dimensions the two sheets form a structure that appears like a double blanket with a bulky hydrophobic core (Figure 3.15*b* and *c*). In the NH_2-terminal domain a shallow crevice is opened into the center of the hydrophobic core that binds to antigen. The hypervariable sequences are present in this NH_2-terminal domain crevice, standing out as different in conformation from the rest of the folded domain. Each of the hypervariable regions extends from and interconnects strands participating in the β-barrel structure. The hypervariable loops associate noncovalently in the three dimensional structure of the V_L and V_H domains to form the binding site for the antigen (see Figure 3.14).

The x-ray structure of isolated crystalline Fab fragments, Fc fragments, and intact immunoglobulin molecules show a variability in the spatial arrangements between domains due to a rotational freedom about the hinge region polypeptide segments that interconnect the domains (Figure 3.16). On antigen binding conformational changes of the binding domain cause new interdomain relationships that "open" functional sites in the constant domains of the H chain that, for example, can facilitate the binding of the complement proteins on association of antigen to the immunoglobulin (see Clin. Corr. 3.1).

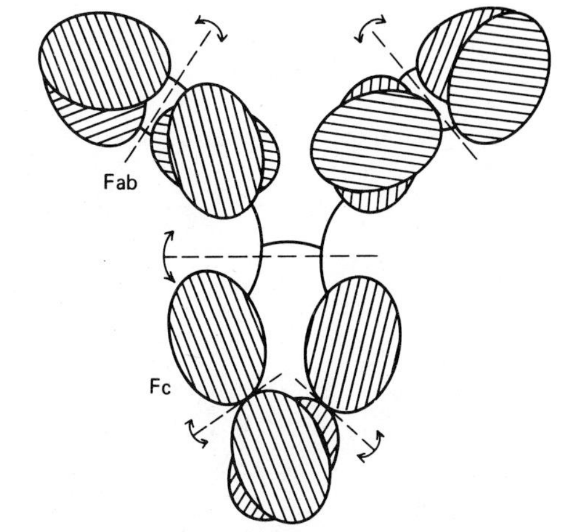

Fab

Fc

Figure 3.16
Rotations of domain regions with respect to each other about hinge polypeptides.
Reprinted with permission from R. Huber, *Trends Biochem. Sci.* 4:270, 1979.

3.4 PLASMA LIPOPROTEINS

A lipoprotein is a multicomponent complex of protein and lipids of characteristic density, molecular weight, size, and chemical composition. These complexes of protein and lipids are held together by noncovalent forces. While a certain typical chemical composition and molecular weight exists for each type of lipoprotein complex, there may exist no *exact* stoichiometry among the components of the complex.

Lipoprotein complexes serve a wide variety of functions in cellular membranes and in the transport and metabolism of lipids. In plasma, for example, lipoproteins transport lipids from the sites of their absorption into blood to the various tissues of the organism that utilize lipids. The plasma lipoproteins are the most extensively characterized of the lipoproteins, and will be described in this section. Reference should be made to Chapter 9 for a description of the structure and metabolism of particular lipids.

Classification of the Plasma Lipoproteins

The classification of the plasma lipoproteins is difficult as the physical and chemical characteristics of these complexes are often heterogeneous. However, the most popular system for classification of plasma lipoprotein particles is based on the criterion of density.

Four hydrated density classes of plasma lipoproteins are now widely recognized in normal fasting humans (Table 3.6). These are the high density lipoproteins (HDL) (d = 1.063–1.210), the low density lipoproteins (LDL) (d = 1.019–1.063), the intermediate density lipoproteins (IDL or LDL_1) (d = 1.006–1.019), and the very low density lipoproteins (VLDL) (d = 0.95–1.006) (see Clin. Corr. 3.4). In addition, lipid particles with small amounts of protein that are less dense than the VLDL appear in plasma after a fatty meal. These are the chylomicrons with $d < 0.95$.

The density of a lipoprotein particle is determined from the density of the solution into which the lipoprotein floats in an ultracentrifugation

TABLE 3.6 Hydrated Density Classes of Plasma Lipoproteins

Lipoprotein Fraction	*Density (g/ml)*	*Flotation Rate (S_f) (Svedberg Units)*	*Molecular Weight*	*Particle Diam (Å)*
HDL	1.063–1.210	—	HDL_2, 4×10^5	70–130
			HDL_3, 2×10^5	50–100
LDL (or LDL_2)	1.019–1.063	0–12	2×10^6	200–280
IDL (or LDL_1)	1.006–1.019	12–20	4.5×10^6	250
VLDL	0.95–1.006	20–400	5×10^6–10^7	250–750
Chylomicrons	<0.95	>400	10^9–10^{10}	10^3–10^4

SOURCE: Data from A. K. Soutar and N. B. Myant, in *Chemistry of Macromolecules,* IIB, R. E. Offord, ed., University Park Press, Baltimore, Md., 1979.

experiment or from its flotation rate. The flotation rate for a lipoprotein particle under standard conditions (NaCl solution $d = 1.063$) is denoted by the term S_f (Table 3.6).

Plasma lipoproteins migrate with the α- and β-globulin fractions in electrophoresis (Figure 3.17).

Composition of the Plasma Lipoproteins

The lipid fraction of the plasma lipoproteins contains significant amounts of triacylglycerols, phospholipids, free cholesterol, and cholesterol esterified with long-chain fatty acids, and other lipids present in small amounts (Table 3.7). The percent lipid content within each of the density classes is variable within extremes that set limits for each of the classes.

Generalizations that can be made from the chemical composition data of Tables 3.6 and 3.7 are (1) the size and molecular weight of the complex, (2) the percent triacylglycerol in the complex, and (3) the lipid : protein ratio of the complex—all *decrease* with *increasing* lipoprotein density.

The cholesterol content is highest in the LDL component (45%) (see Clin. Corr. 3.4).

Correlated with the decreasing lipid : protein ratio with increasing lipoprotein density, the percent protein in lipoprotein complexes varies from a low value of 2% in chylomicrons to a high value of 50% in HDL.

If the protein components of a lipoprotein are separated from the lipid components by extraction of the lipid with an organic solvent, the isolated proteins (apolipoproteins) can be shown by immunological and chemical characterization to be composed of at least seven distinct types (Table 3.8). In some of the apolipoprotein types only a single distinct protein is currently known, while other types are composed of several distinct proteins.

The initially discovered apolipoproteins from the plasma lipoproteins were named apolipoprotein A (ApoA) for the major protein of the HDL fraction, and apolipoprotein B (ApoB) for the major protein isolated from the LDL fraction. ApoB is also the major protein in the IDL and VLDL fractions (Table 3.8). A third apolipoprotein (ApoC) is found predominantly in IDL and VLDL. Other chemically and immunologically distinct apolipoproteins are ApoD and ApoE. ApoD is found in the HDL, where it is a minor component. This protein is sometimes designated $ApoA_3$, based on a common A designation for proteins isolated from the HDLs. More recently, the existence of two additional immunologically distinct plasma apolipoproteins have been reported, ApoF and ApoG. Apoproteins F and G are believed to be minor protein components in certain subfractions of HDL (Clin. Corr. 3.5.).

CLIN. CORR. 3.4 HYPERLIPOPROTEINEMIAS

Hyperlipoproteinemia is the clinical finding of an overabundance of particular classes of plasma lipoproteins. The hyperlipoproteinemias are classified (types I–V) according to the abnormal pattern observed in plasma lipoprotein gel electrophoresis.

Type I hyperlipoproteinemia is a rare disorder in which only the chylomicron fraction is abnormally increased in gel electrophoresis. This type of hyperlipoproteinemia appears in patients with a genetic deficiency of the enzyme lipoprotein lipase in adipose tissue. These patients cannot metabolize lipid normally, and increased amounts of triacylglycerol are found in the blood with a corresponding higher level of chylomicrons. This condition leads to hepatosplenomegaly.

Type II hyperlipoproteinemia (also referred to as hyper-β-lipoproteinemia or hypercholesterolemia) is characterized by an increased β-lipoprotein (LDL; type IIA) or both an increased β-lipoprotein and pre-β-lipoprotein (VLDL and LDL; type IIB) bands in plasma electrophoresis. These types of hyperlipoproteinemia exist in both a hereditary form and in an acquired form. Patients with type II hyperlipoproteinemia have an increased risk of atherosclerosis, especially of the coronary arteries, and they have high amounts of cholesterol in plasma.

Type III hyperlipoproteinemia is a rare disorder that is characterized by the appearance of an abnormal lipoprotein in plasma, which is referred to as "floating β-lipoprotein." This abnormal lipoprotein appears as a broad band between the normal β-lipoprotein and pre-β-lipoprotein in plasma electrophoresis. The disease has a hereditary form, although type III sometimes also appears as a secondary effect to hypothyroidism. Patients with type III hyperlipoproteinemia show the early appearance of atherosclerosis and have an increased risk of occlusive vascular disease.

Type IV hyperlipoproteinemia (also referred to as pre-β-hyperlipoproteinemia,

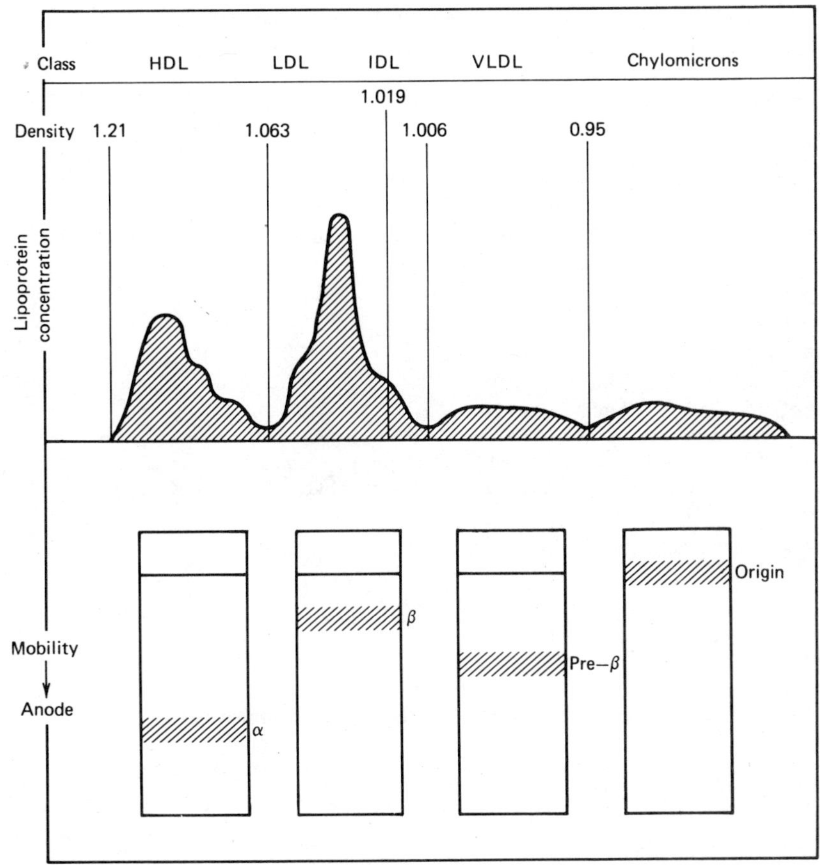

Figure 3.17
Correspondence of plasma lipoprotein density classes with electrophoresis mobility in a plasma electrophoresis.
In the upper diagram the ultracentrifugation schlieren pattern is shown. At the bottom, electrophoresis on a paper support shows the mobilities of major plasma lipoprotein classes with respect to α and β globulin electrophoresis bands.
Reprinted with permission from A. K. Soutar and N. B. Myant, in *Chemistry of Macromolecules,* IIB, R.E. Offord, ed., University Park Press, Baltimore, Md., 1979.

hypertriglyceridemia) is characterized by an increased pre-β-lipoprotein band in electrophoresis due to increased amounts of VLDL. It is thought that this type of hyperlipoproteinemia can exist in a hereditary form. Type IV is often found in conjunction with overweightness in an individual characterized by larger than normal size fat cells in the adipose tissue. Type IV is also often found in patients with a hereditary tendency toward diabetes mellitus. A reversible form of type IV is often found in individuals who imbibe excessive amounts of alcohol. Patients with type IV hyperlipoproteinemia have a high incidence of coronary artery disease and peripheral vascular disease.

Type V hyperlipoproteinemia is a rare disease that is sometimes expressed secondarily to diabetes mellitus, chronic pancreatitis, hepatopathy, and nephropathy. It is characterized by increased chylomicron and pre-β-lipoprotein bands in electrophoresis.

An abnormal lipoprotein, designated *lipoprotein X,* that electrophoreses with the β-lipoprotein band and has a density of 1.040–1.045, occurs in patients with a deficiency of the enzyme lecithin : cholesterol acyltransferase (LCAT) and in patients with cholestasis. The apoprotein of lipoprotein X consists primarily of albumin with one or two of the ApoC proteins.

There is epidemiological evidence that human coronary disease has a lower incidence in individuals with higher relative levels of HDL or α-lipoproteins.

TABLE 3.7 Chemical Composition of the Different Plasma Lipoprotein Classes

			Percent Composition of Lipid Fraction			
Lipoprotein Class	*Total Protein (%)*	*Total Lipid (%)*	*Phospholipids*	*Esterified Cholesterol*	*Unesterified Cholesterol*	*Triacylglycerols*
HDL_2[a]	40–45	55	35	12	4	5
HDL_3[a]	50–55	50	20–25	12	3–4	3
LDL	20–25	75–80	15–20	35–40	7–10	7–10
IDL	15–20	80–85	22	22	8	30
VLDL	5–10	90–95	15–20	10–15	5–10	50–65
Chylomicrons	1.5–2.5	97–99	7–9	3–5	1–3	84–89

SOURCE: Data from A. K. Soutar and N. B. Myant, in *Chemistry of Macromolecules,* IIB, R. E. Offord, ed., University Park Press, Baltimore, Md., 1979.

[a] Subclasses of HDL.

CLIN. CORR. 3.5 HYPOLIPOPROTEINEMIAS

A-β-lipoproteinemia is a genetically transmitted disease that is characterized by an absence of chylomicrons, VLDL, and LDL due to an inability to synthesize apolipoprotein B.

Tangier disease, an α-lipoproteinemia, is a rare autosomal recessive inherited disease in which the HDL is 1 to 5% of its normal value. The clinical features are due to the accumulation of cholesterol in the lymphoreticular system, which may lead to hepatomegaly and splenomegaly. In this disease the plasma cholesterol and phospholipids are greatly reduced.

Deficiency of the enzyme lecithin : cholesterol acyltransferase is a rare disease that results in the production of lipoprotein X (see Clin. Corr. 3.4). Also characteristic of this disease is the decrease in the α-lipoprotein and pre-β-lipoprotein bands, with the increase in the β-lipoprotein (lipoprotein X) in electrophoresis.

TABLE 3.8 Apolipoprotein of the Human Plasma Lipoproteins (Values in Percent of Total Protein Present)[a]

Apolipoprotein	*HDL_2*	*HDL_3*	*LDL*	*IDL*	*VLDL*	*Chylomicrons*
ApoA-I	85	70–75	Trace	0	0–3	0–3
ApoA-II	5	20	Trace	0	0–0.5	0–1.5
ApoD	0	1–2	—	—	0	1
ApoB	0–2	0	95–100	50–60	40–50	20–22
ApoC-I	1–2	1–2	0–5	<1	5	5–10
ApoC-II	1	1	0.5	2.5	10	15
ApoC-III	2–3	2–3	0–5	17	20–25	40
ApoE	Trace	0–5	0	15–20	5–10	5
ApoF	Trace	Trace	—	—	—	—
ApoG	Trace	Trace	—	—	—	—

SOURCE: Data from A. K. Soutar and N. B. Myant, in *Chemistry of Macromolecules*, IIB, R. E. Offord, ed., University Park Press, Baltimore, Md., 1979; G. M. Kostner, *Adv. Lipid Res.* **20,** 1 (1983).

[a] Values show variability from different laboratories.

TABLE 3.9 Physical Characteristics of Plasma Apolipoproteins

Apolipoprotein	*Molecular Weight*	*Number of Amino Acids*	*Calc. Secondary Structure, Lipid-Free*
ApoA-I	28,300	243	55% α-helix (~70% α-helix in lipoprotein) ~10% β-sheet ~35% disordered
ApoA-II	17,380	Dimer of two 77 amino acid chains	49% α-helix (69% with phospholipid)
ApoB	8,000–550,000	—	In lipoprotein: ~25% α-helix 37% β-sheet 37% disordered
ApoC-I	6,600	57	55% α-helix (73% with phospholipid)
ApoC-II	8,800	78	23% α-helix (59% with phospholipid)
ApoC-III	8,700	79	22% α-helix (70% with phospholipid)
ApoD	20,000	—	—
ApoE	33,000	—	?66%
ApoF	30,000	—	—
ApoG	72,000	—	—

SOURCE: Data from results summarized by A. K. Soutar and N. B. Myant, in *Chemistry of Macromolecules* IIB, R. E. Offord, ed., University Park Press, Baltimore, Md., 1979; J. D. Morrisett, R. L. Jackson, and A. M. Gotto, Jr., *Annu. Rev. Biochem.*, **44,** 183 (1975); L. C. Smith, H. J. Pownall, and A. M. Gotto, Jr., *Annu. Rev. Biochem.*, **47,** 751 (1978); J. T. Sparrow and A. M. Gotto, Jr., *CRC Crit. Rev. Biochem.* **13,** 87 (1983); and G. M. Kostner, *Adv. Lipid Res.* **20,** 1 (1983).

Within the A and C classes of apolipoproteins several distinct proteins have been identified. These proteins are distinguished either by their COOH-terminal amino acid residue or by a consecutive numbering system [i.e., A-I, A-II, C-I, C-II, etc. (Table 3.8)]. These polypeptide chains are present in defined ratios within each lipoprotein. For example, ApoA-I and ApoA-II are normally present in HDL_3 in an approximate ratio of 3 : 1.

The molecular weights of the apolipoproteins vary from 6,000 (ApoC-I) to 72,000 (ApoG). ApoB is relatively insoluble even in detergents, and reports on its molecular weight are highly variable due to aggregation of the protein and to the susceptibility of this apoprotein to proteolytic hydrolysis.

The apolipoproteins have significant amounts of α-helical structure (Table 3.9), as determined by spectrophotometric methods (see Chapter 2). Many of the α-helical regions have an amino acid sequence in which every third or fourth residue is either nonpolar or ionic. Thus a nonpolar side and an ionic side are formed along the longitudinal axis of the helix, giving it two distinct regions of polarity (Figure 3.18). Model building suggests that the ionic side of the helix forms a binding site for the polar heads of phospholipids and the hydrophobic side for the fatty acyl chains of lipids within the lipoprotein molecule.

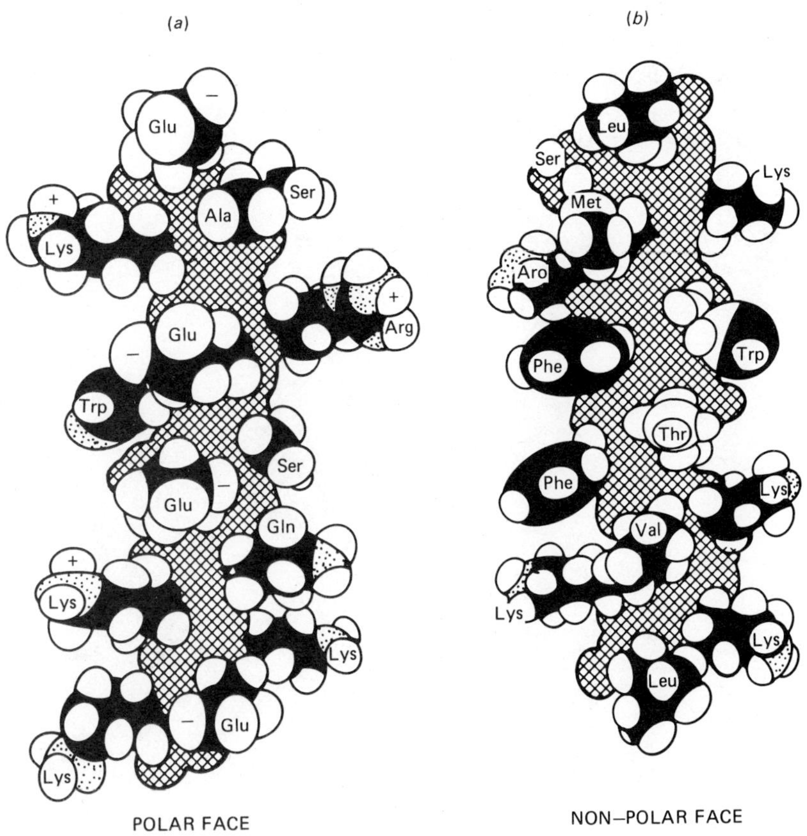

Figure 3.18
Illustration showing side chain groups of a helical segment of apolipoprotein C-1 between residues 32 and 53.
The polar face (a) shows ionizable acid residues in the center and basic residues at the edge. On the other side of the helix (b), the hydrophobic residues form a nonpolar longitudinal face.
Reprinted with permission from J. T. Sparrow and A. M. Gotto, Jr., *CRC Crit. Rev. Biochem.* 13:87, 1983. Copyright, CRC Press, Inc., Boca Raton, FL.

Structure of Lipoprotein Molecules

The diameter range for HDL is normally 95–130 Å and for LDL, 230–280 Å. The other classes are less homogeneous in their molecular size, and their diameters are accordingly variable.

The structure of lipoprotein molecules has been investigated with a wide range of methods, including electron microscopy, x-ray diffraction scattering, chemical and enzymatic modification, and spectrophotometric techniques. In addition, calculations based on the physical properties of the individual components of the lipoprotein classes have led to significant insights into their structure. However, these studies have not yet yielded a definitive structure for plasma lipoproteins, and the structure discussed below may be considered speculative.

Calculations based on the volume occupied by the chemical components of a lipoprotein molecular complex show that the sum of the volume of the protein *plus* amphoteric lipid (e.g., phosphatidylcholine and cholesterol) components *in any of the plasma lipoprotein density classes* is just sufficient to occupy an outer shell approximately 20 Å thick. The calculation thereby supports a structure of the lipoprotein in which the amphoteric components, which include the protein components, phosphatidylcholine and unesterified cholesterol, are on the outside of the complex forming an outer sphere 20 Å thick, with their apolar parts facing inside and their polar and charged parts facing outside toward the water solvent. The nonpolar neutral lipids are inside the complex in contact with the nonpolar ends of the amphoteric molecules of the outer shell (Figure 3.19). The calculations show that this model is possible for complexes in all density classes, irrespective of their particle size. Furthermore, the model nicely explains the decrease in neutral lipid to protein ratio with

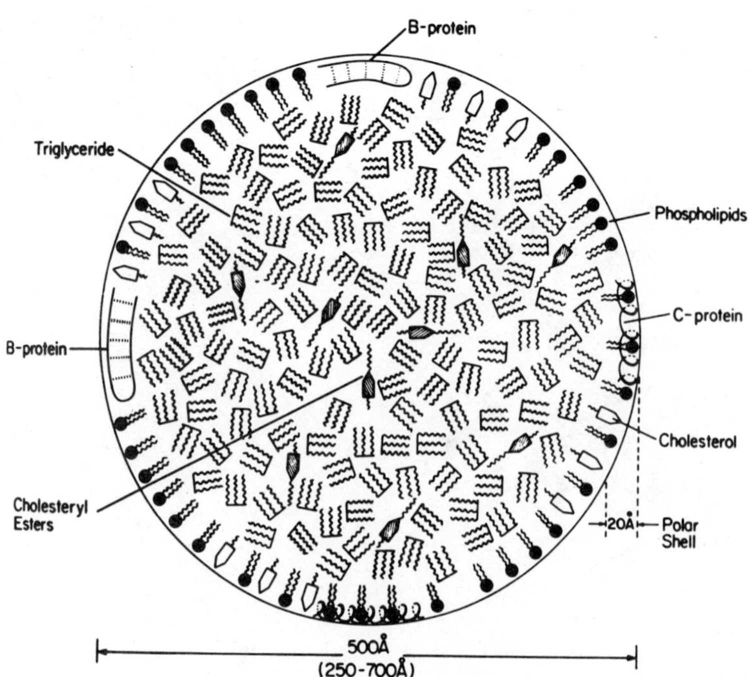

Figure 3.19
Generalized structure for the plasma lipoproteins.
This structure is drawn for VLDL. Note that protein, phospholipids, and cholesterol form a shell of 20 Å thickness on the outside.
Reprinted with permission from I. R. Morrisett, R. L. Jackson, and A. M. Gotto, *Biochim. Biophys. Acta* 472:93, 1977.

changes in size of the complex (Table 3.7). The data from x-ray scattering, electron microscopy, and other techniques generally support this overall structure. More detailed, but currently speculative, structural models for plasma lipoproteins have been put forth based on the current experimental data.

3.5 GLYCOPROTEINS

Glycoproteins are molecules composed of covalently joined protein and carbohydrate. The carbohydrate is attached to the polypeptide chains of the protein in a series of reactions that are enzymatically catalyzed after the protein component is synthesized.

The functions of glycoproteins in the human are currently of great interest. Glycoproteins in cell membranes apparently have an important role in the group behavior of cells and other important biological functions of the membrane. Glycoproteins form a major part of the mucus that is secreted by epithelial cells, where they perform an important role in lubrication and in the protection of tissues lining the body's ducts. Many other proteins secreted from cells into extracellular fluids are glycoproteins. These proteins include hormone proteins found in blood, such as follicle stimulating hormone, luteinizing hormone, and chorionic gonadotropin; and plasma proteins such as the orosomucoids, ceruloplasmin, plasminogen, prothrombin, and the immunoglobulins (Clin. Corr. 3.6).

Carbohydrate Composition of Glycoproteins

The percent of carbohydrate within the glycoproteins is highly variable. Some glycoproteins such as IgG contain low amounts of carbohydrate (4%). A human red cell membrane glycoprotein, glycophorin, has been found to contain 60% carbohydrate (Figure 3.20). Human ovarian cyst glycoprotein is composed of 70% carbohydrate, and human gastric glycoprotein is 82% carbohydrate.

The carbohydrate can be distributed fairly evenly along the polypeptide chain of the protein component or concentrated into defined regions of the polypeptide chain. For example, in human glycophorin A the carbohydrate is found in the NH_2-terminal half of the polypeptide chain that lies on the outside of the cellular membrane.

The carbohydrate attached at one or at multiple points along a polypeptide chain usually contains less than 12 to 15 sugar residues. In some cases the carbohydrate component consists of only a single sugar moiety, as in the submaxillary gland glycoprotein (single *N*-acetyl-α-D-galactosaminyl residue) and in some types of mammalian collagens (single α-D-galactosyl residue). In general, glycoproteins contain sugar residues in the D form, except for L-fucose, L-arabinose, and L-iduronic acid.

A glycoprotein from different animal species often has an identical primary structure in the protein component, but a variable carbohydrate component. This heterogeneity of a given protein may even be true within a single organism. For example, pancreatic ribonuclease is found in an A and a B form. The two forms have an identical amino acid sequence and a similar kinetic specificity toward substrates, but differ significantly in their carbohydrate composition.

Functional glycoproteins may also be found in different stages of "completion." In many cases the addition of carbohydrate units is a multienzyme-catalyzed process, which occurs over a relatively long time period. In consequence, "immature" glycoproteins may be found in biological media. These are glycoproteins in an intermediate stage of their

CLIN. CORR. 3.6 FUNCTIONS OF GLYCOPROTEINS

Glycoproteins participate in a large number of normal and disease-related functions of clinical relevance. For example, many of the proteins in the outer cellular membranes are glycoproteins. Such proteins within the cell surface are antigens, which determine the blood antigen system (A, B, O), and the histocompatibility and transplantation determinants of an individual. Immunoglobulin antigenic sites and viral and hormone receptor sites in cellular membranes are often glycoproteins. In addition, the carbohydrate portions of glycoproteins in membranes provide a surface code for cellular identification by other cells and for contact inhibition in the regulation of cell growth. As such, changes in the membrane glycoproteins can be correlated with tumorigenesis and malignant transformation in cancer.

Most of the important plasma proteins, except albumin, are glycoproteins. These plasma proteins include the blood-clotting proteins, the immunoglobulins, and many of the complement proteins. Some of the protein hormones, such as follicle-stimulating hormone (FSH) and thyroid-stimulating hormone (TSH), are glycoproteins. The important structural protein collagen contains carbohydrate. The proteins found in the mucus secretions are carbohydrate-containing proteins, where they perform a role in lubrication and in the protection of epithelial tissue. The antiviral protein interferon is a glycoprotein.

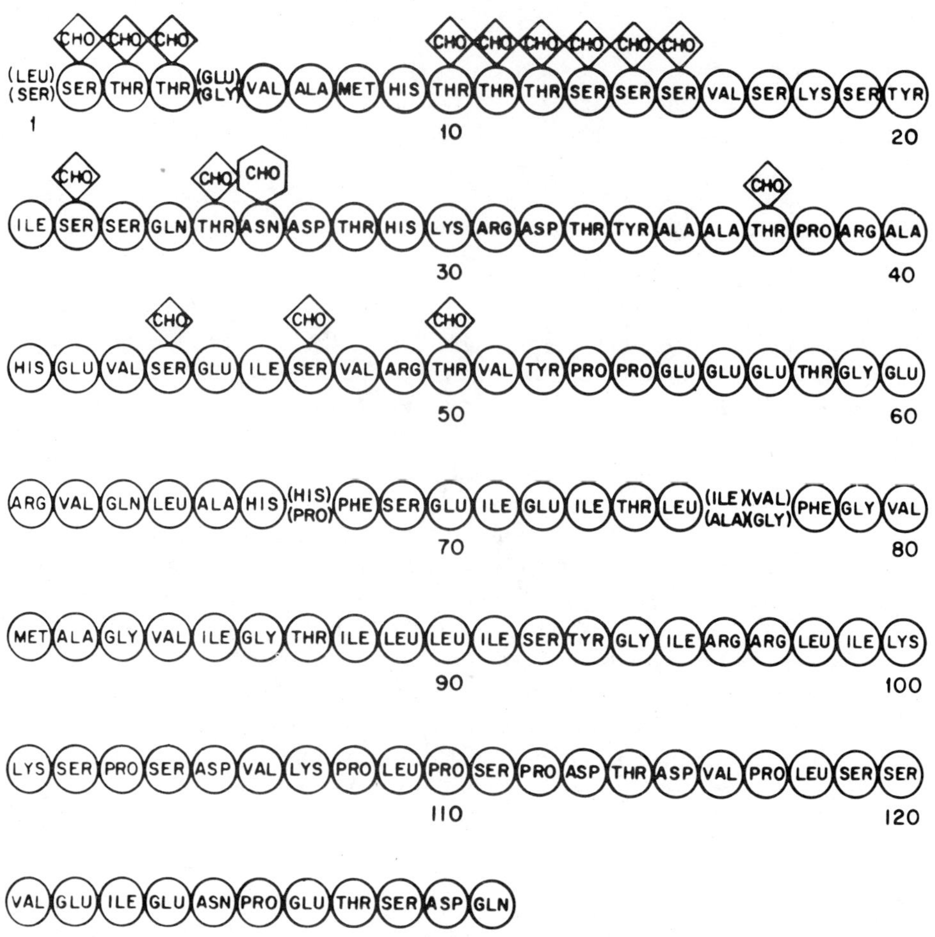

Figure 3.20
Primary structure of human red blood cell membrane glycophorin A.
O-Glycosidic carbohydrate bonds to Ser and Thr are indicated by diamonds and N-glycosidic bonds to Asn by hexagons. Residues 71–91 lie in the membrane; the NH_2-terminal end with carbohydrate attached is on the outside of the cell (residues 1-70) and the COOH-terminal end (residues 91–131) on the inside of the cell. Reprinted with permission from M. Tomita and V. T. Marchesi, *Proc. Natl. Acad. Sci., USA* 72:2964, 1975.

carbohydrate biosynthesis. (Biosynthesis of the carbohydrate portions of glycoproteins is described in Chapter 8.)

Types of Carbohydrate–Protein Covalent Linkages

Different types of covalent linkages join the sugar moieties and protein in a glycoprotein.

Type I linkages involve asparagine side chain R groups and the sugar residue *N*-acetyl-D-glucosamine (Figure 3.21). The enzyme that catalyzes the formation of type I linkages only adds *N*-acetyl-D-glucosamine to an asparagine within the sequence Asn-X-Thr (where X is any of the common amino acids). This enzymatically recognized sequence is known as the *sequon* for the formation of a type I glycoprotein linkage. The specificity for the Asn-X-Thr sequon in Asn glycosylation will limit the number of carbohydrate residues linked to asparagine in any one glycoprotein. For example, in the plasma protein human orosomucoid there are nine asparagine residues, of which only five appear in the type I sequon and are glycosylated. Bence-Jones protein is derived from an immunoglobulin L

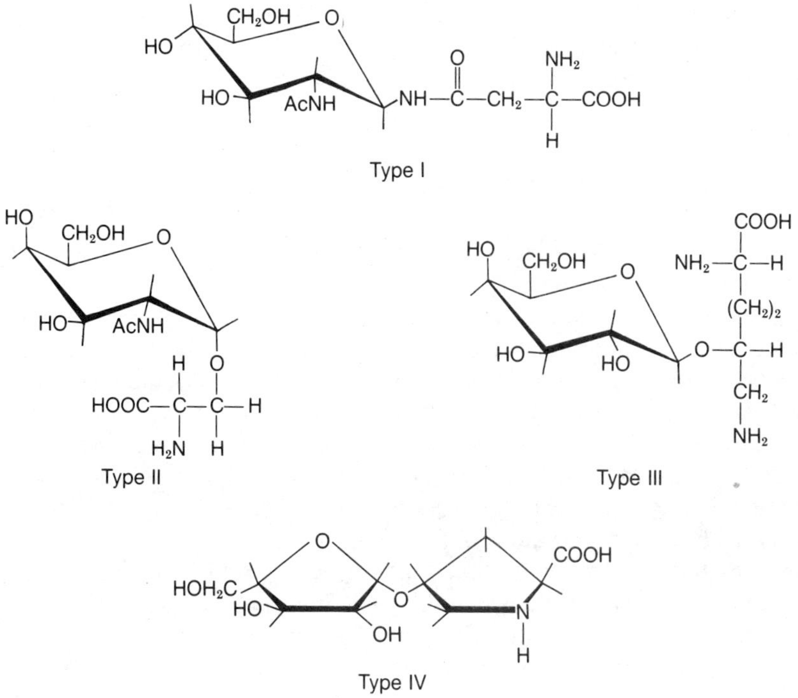

Figure 3.21
Examples of glycosidic linkages to amino acids in proteins.
Type I is an N*-glycosidic linkage through an amide nitrogen of Asn; type II is an* O*-glycosidic linkage through the OH of Ser (R = H) or Thr (R = CH_3); type III is an* O*-glycosidic linkage to the 5-OH of 5-hydroxylysine; type IV is an* O*-glycosidic linkage through 4-OH of 4-hydroxyproline.*
Diagrams modified from R. D. Marshall in *Chemistry of Macromolecules*, IIB, R. E. Offord, ed., University Park Press, Baltimore, Md., 1979, p. 12.

chain, and is found in large quantities secreted in the urine of patients with multiple myeloma. The Bence-Jones proteins do not usually contain carbohydrate. However, if an asparagine appears in an Asn-X-Thr sequon in the V region of the polypeptide chain (see Section 3.3), glycosylation will occur on this sequon.

Type II linkages involve glycosidic carbohydrate linkages to serine or threonine alcoholic R groups (Figure 3.21). This type of linkage is found in glycoproteins in mucus secretions, including those from the submaxillary glands, the epithelial cells of the gastrointestinal tract, the respiratory tract, and the female genital tract. No particular sequon is required for enzymes forming the type II bond. Type II β-D-xylopyranosyl bonds occur in the proteoglycans of connective tissue and in the important protein heparin secreted by mast cells into blood where it acts as an anticoagulant.

Type III linkages made to 5-hydroxylysine (Figure 3.21) are found in some mammalian collagens (see Section 3.6), and in the serum complement protein C1q (see Clin. Corr. 3.1). The 5-hydroxylysine is a derived amino acid (Chapter 2) formed from lysines after incorporation of the lysine into the polypeptide chains.

Other types of covalent sugar : protein linkages are *type IV,* made to 4-hydroxyproline side chains (4-hydroxyproline is also a derived amino acid); *type V,* made to cysteine side chains; and *type VI,* made to NH_2-terminal amino groups. These linkages are not widely distributed in the glycoproteins of mammals, although type V linkages are found in some glycoproteins obtained from the red blood cell membrane, and type VI are found in glycosylated adult hemoglobin, Hb_{1c} (Clin. Corr. 3.7). Type IV linkages have been found only in higher plants.

CLIN. CORR. 3.7 GLYCOSYLATED HEMOGLOBIN, HbA_{1c}

A glycosylated hemoglobin, designated HbA_{1c}, is formed spontaneously in the red blood cell by combination of the NH_2-terminal amino groups of the hemoglobin β chain and glucose. The aldehydic function of the glucose first forms a Schiff base with the NH_2-terminal amino group,

$$\text{—N=}\underset{\text{H}}{\text{C}}\text{—}\overset{\text{OH}}{\underset{\text{H}}{\text{C}}}\text{—}$$

which then rearranges to a more stable amino ketone linkage,

$$\underset{\text{H}}{\text{—N}}\text{—}\overset{\text{H}}{\underset{\text{H}}{\text{C}}}\text{—}\overset{\text{O}}{\overset{\|}{\text{C}}}\text{—}$$

by a spontaneous (nonenzymatic) reaction known as the Amadori rearrangement. The concentration of HbA_{1c} is dependent on the concentration of glucose in the blood, and in prolonged hyperglycemia may rise to 12% or more of the total hemoglobin. Patients with diabetes mellitus will tend to have high concentrations of glucose and therefore high amounts of HbA_{1c}. The changes in the concentration of HbA_{1c} in diabetic patients can be used to follow the effectiveness of treatment for the diabetes.

3.6 COLLAGEN AND OTHER FIBROUS PROTEINS

In a previous section hemoglobin and myoglobin were described as examples of globular proteins. Characteristic of the globular proteins are a spheroid shape, varied molecular weights, relatively high water solubility, and a variety of functional roles such as catalysis, transport, and physiological and genetic control. In contrast, characteristics of the fibrous proteins are a rodlike shape, a high amount of secondary structure, and a structural rather than a chemical role in the mammalian organism.

The fibrous proteins form the structural matrix of mammalian bone, ligaments, and skin. These proteins have an unusually high tensile strength. For example, the protein collagen can withstand an amazingly high longitudinal stretching force without tearing. Other fibrous proteins are constituents of muscle, where they can interdigitate in a dynamic way that leads to muscle contraction. Another of the fibrous proteins can stretch and reform like rubber, and provides the elasticity of the arteries and the ligaments.

Characteristic of many of the fibrous proteins is that their polypeptide chains are in a helical conformation and each of these chains is wound around one or two other polypeptide chains in a similar helical conformation to form a supercoiled structure, which is then meshed into an amorphous matrix of protein or polysaccharide. An exception to this structural generalization may be elastin, for which the three-dimensional structure is not known.

Collagen

Distribution of Collagen in Humans

Collagen is found in significant amounts in all tissues and organs, where it provides a framework for their form and lends structural strength. In contrast, collagen also serves as the cornea of the eye, in which the collagen fibers are stacked in a crosswise fashion in stacked arrays that act to transmit light with a minimum of scattering. The percentage of collagen by weight for some representative tissues and organs is 4% of liver, 10% of lung, 12 to 24% of the aorta, 50% of cartilage, 23% of whole cortical bone, 68% of the cornea, and 72% of skin. Collagen is the most abundant protein in the human organism (Clin. Corr. 3.8).

CLIN. CORR. 3.8 SYMPTOMS OF DISEASES OF ABNORMAL COLLAGEN SYNTHESIS

Collagen is present in virtually all tissues and is the most abundant protein in the body. Certain organs depend heavily upon normal collagen structure in order to function physiologically. Abnormal collagen synthesis or structure causes dysfunction of cardiovascular organs (aortic and arterial aneurysms and heart valve malfunction), bone (fragility and easy fracturing), skin (poor healing and unusual distensibility), joints (hyper-mobile joints, arthritis), and eyes (dislocation of the lens). Examples of diseases caused by abnormal collagen synthesis include Marfan's disease, Ehlers-Danlos syndrome, osteogenesis imperfecta, and scurvy.

Collagen is synthesized by a variety of specialized cells in the organs and tissues. These include the fibroblasts of connective tissue and tendons, osteoblasts in bone, chondroblasts in embryonic cartilage, and odontoblasts in teeth. The biosynthesis of collagen is described in Chapter 19.

The collagen fibrils in the tissues are usually imbedded in an amine-containing polysaccharide or a glycoprotein matrix with a gel-like character, known as the *ground substance*.

Amino Acid Composition of Collagen

The amino acid composition of collagen is very different from that found for a typical globular protein (Table 3.10). Its amino acid composition is high in glycine (~33% of the total), proline (~10%), the derived amino acid hydroxyproline (~10%), and the derived amino acid 5-hydroxylysine (~1%). The predominant form of hydroxyproline (Hyp) is the 4-OH deriv-

TABLE 3.10 Comparison of the Amino Acid Content of Human Skin Collagen (Type I) and Mature Elastin with That for Two Typical Globular Proteins[a]

Amino Acid	*Collagen (Human Skin)*	*Elastin (Mammalian)*	*Ribonuclease (Bovine)*	*Hemoglobin (Human)*
COMMON AMINO ACIDS		PERCENT OF TOTAL		
Ala	11	22	8	9
Arg	5	0.9	5	3
Asn			8	3
Asp	5	1	15	10
Cys	0	0	0	1
Glu	7	2	12	6
Gln			6	1
Gly	33	31	2	4
His	0.5	0.1	4	9
Ile	1	2	3	0
Leu	2	6	2	14
Lys	3	0.8	11	10
Met	0.6	0.2	4	1
Phe	1	3	4	7
Pro	13	11	4	5
Ser	4	1	11	4
Thr	2	1	9	5
Trp	2	1	9	2
Tyr	0.3	2	8	3
Val	2	12	8	10
DERIVED AMINO ACIDS				
Cystine	0	0	7	0
3-Hydroxyproline	0.1	—	0	0
4-Hydroxyproline	9	1	0	0
5-Hydroxylysine	0.6	0	0	0
Desmosine and isodesmosine	0	1	0	0

[a] Boxed numbers emphasize important differences in amino acid composition between the fibrous proteins (collagen and elastin) and typical globular proteins.

ative (Figure 3.22), although some types of collagen contain small amounts of the 3-OH form of Hyp. A small amount of carbohydrate is found in collagen covalently bonded through the 5-OH of 5-hydroxylysines. Formation of the derived amino acids and addition of the carbohydrate occurs intracellularly by an enzyme-catalyzed process prior to secretion of the collagen from its synthesizing cell (see Chapter 19).

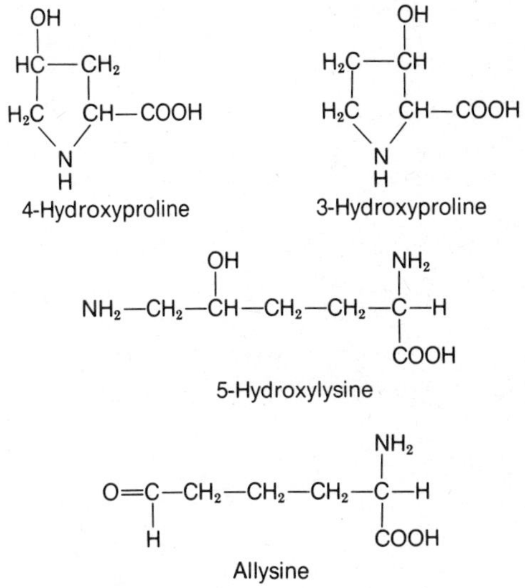

Amino Acid Sequence of Collagen

The molecular unit of collagen in collagen fibrils contains three polypeptide chains. This three polypeptide chain molecule is often referred to as *tropocollagen*. In keeping with the more recent terminology, it is simply designated *collagen* in this text. In the collagen of some tissue types, each of the three polypeptide chains has an identical amino acid sequence (e.g., type II collagen found in cartilage; type III collagen found in cartilage, scar, and soft tissue; type IV collagen found in basement membrane). In the collagen classified as type I, two of the chains are identical in sequence and the amino acid sequence of the third chain differs slightly (e.g., type I collagen found in bone, tendons, soft tissue, and scar tissue). For the latter collagen, the two identical chains are designated $\alpha 1$ chains and the third nonidentical chain, $\alpha 2$. Molecules of type V collagen may also consist of heterochains (e.g., $\alpha 1(V)\alpha 2(V)\alpha 3(V)$; see Table 3.11) of different sequence.

4-Hydroxyproline

3-Hydroxyproline

5-Hydroxylysine

Allysine

Figure 3.22
Derived amino acids found in collagen.
Carbohydrate is attached to 5-OH in 5-hydrosylysine by a type III glycosidic linkage (see Figure 3.21).

TABLE 3.11 Classification of Collagen Types

Type	*Chain Designations*	*Tissue Found*	*Characteristics*
I	$[\alpha 1(I)]_2\alpha 2(I)$	Bone, skin, tendons, scar tissue, heart valve, intestinal and uterine wall	Low carbohydrate; <10 hydroxylysines per chain; 2 types of polypeptide chains
II	$[\alpha 1(II)]_3$	Cartilage, vitreous	10% carbohydrate; >20 hydroxylysines per chain
III	$[\alpha 1(III)]_3$	Blood vessels, newborn skin, scar tissue, intestinal and uterine wall	Low carbohydrate; high hydroxyproline and Gly; contains Cys
IV	$[\alpha 1(IV)]_3$ $[\alpha 2(IV)]_3$	Basement membrane, lens capsule	High 3-hydroxyproline; >40 hydroxylysines per chain; low Ala and Arg; contains Cys; high carbohydrate (15%)
V	$[\alpha 1(V)]_2\alpha 2(V)$ $[\alpha 1(V)]_3$ $\alpha 1(V)\alpha 2(V)\alpha 3(V)$	Cell surfaces or exocytoskeleton; widely distributed in low amounts	High carbohydrate, relatively high glycine, and hydroxylysine
VI	—	aortic intima, placenta, kidney, and skin in low amounts	Relatively large globular domains in telopeptide region; high Cys and Tyr; mol wt relatively low (~160,000); equimolar amounts of hydroxylysine and hydroxyproline

The different types of collagen (types I to VI) are characterized by differences in their physical properties due to the distinct differences in their amino acid sequence and in their percent carbohydrate (Table 3.11). Additional collagen types up through type X have been reported recently.

The amino acid sequence of the polypeptide strands in the collagen molecule is quite different from the sequences of the globular proteins. It is found for example in Type I collagen that, except for a small segment of 15 to 25 amino acids on the NH_2- and COOH-terminal ends (these segments are known as the telopeptides), *glycine is found every third amino acid in the sequence*. Furthermore, *the sequences Gly-Pro-Y and Gly-X-Hyp* (where X and Y are any of the amino acids) *are each repeated more than 100 times* in the polypeptide sequence of a chain. These two triplet sequences will thus encompass over 600 amino acids within a collagen chain of approximately 1,000 amino acids.

Structure of Collagen

Polypeptides that contain only the amino acid proline (polyproline polymer) can be synthesized in the laboratory. This polyproline polypeptide will spontaneously assume a regular secondary structure in aqueous solution in which the polypeptide chain is in a tightly twisted extended helix with three residues per turn of the helix ($n = 3$), $d = 2.9$ Å, and $p = 8.7$ Å. This helix is designated the *polyproline type II helix*. The helix conformation found for polyproline has the same characteristics as the helix found for each of the polypeptide chains in the native collagen, in which the primary structure contains a proline or hydroxyproline at approximately every third position within the amino acid sequence.

Since the same helix is found in the polypeptide chains in collagen as in polyproline, the thermodynamic forces leading to the formation of this helix must be based on the properties of proline. In proline the ϕ angle is part of a five-membered ring, which constrains it to an angle compatible with the collagen helix. In addition, the hydrodynamic solvation properties of Pro and Hyp stabilize the collagen helix conformation.

In the polyproline type II helix, each peptide bond in the polypeptide chain is positioned in a plane that is perpendicular to the axis of the helix. In this geometry the peptide carbonyl groups are pointed in a direction where they can form strong *interchain* hydrogen bonds to a parallel polypeptide chain. This is in contrast to the orientation of the peptide bond in the α helix conformation, in which the peptide bond is directed in a plane parallel to the axis of the α helix and forms *intrachain* hydrogen bonds with other peptide bonds in the same polypeptide chain.

Three polypeptide chains, each in a polyproline type II helix conformation, are wound about each other in a defined way (Figure 3.23) to form the superhelix structure of the collagen molecule. This superhelix forms because glycine residues, with a small side chain group (R = H), appear at every third position in the amino acid sequence of each of the polypeptide chains. Since the helix has three amino acids per turn of the helix ($n = 3$), the glycines are positioned along the same side of the helix and thus form an apolar longitudinal edge along the outside of the helix of the polypeptide chain. This allows each of the three strands to come close to one another (because the glycine side chain is small) in a regular pattern, with the supercoil structure stabilized by the hydrophobic interstrand interactions between the apolar edges (Figure 3.23). The interchain hydrogen bonds between peptide groups further stabilize the supercoil structure. A larger side chain group than glycine would sterically prevent the adjacent strands from coming together in the superhelix structure. Although the superhelical structure predominates, there may be regions along the longitudinal axis at which the helix breaks and a globular domain is formed. This is particularly true of type IV collagen.

The three polypeptide chains that form a supercoiled structure are initially joined together by disulfide bonds formed between chains in the COOH-terminal region of procollagen (biosynthetic precursor form of collagen; see Chapter 19). This region is hydrolyzed from the polypeptide chains during the final step in the synthesis of collagen. The supercoiled type I collagen molecule is approximately 15 by 3,000 Å and has a molecular weight of 300,000.

Formation of Covalent Cross-Links in Collagen

An enzyme present in the collagen fiber (see Chapter 19) acts on some of the lysines in collagen to catalyze the oxidative deamination of the ε-amino groups to a δ-aldehyde (Figure 3.24). The resulting derived amino acid, containing an aldehydic side chain R group, is known as *allysine*. This aldehyde function will spontaneously undergo nucleophilic addition reactions with the ε-amino groups of other (nontransformed) lysines, and the δ-carbon of other allysine side chains to form covalent interchain bonds (Figure 3.24). These covalent bonds may be to adjacent collagen molecules as well as between strands within the same collagen molecule. The covalent cross-links formed with allysines are essential to the stability of the fibril structure. If the covalent cross-links are not formed, the collagen fibrils are easily dissociated and degraded.

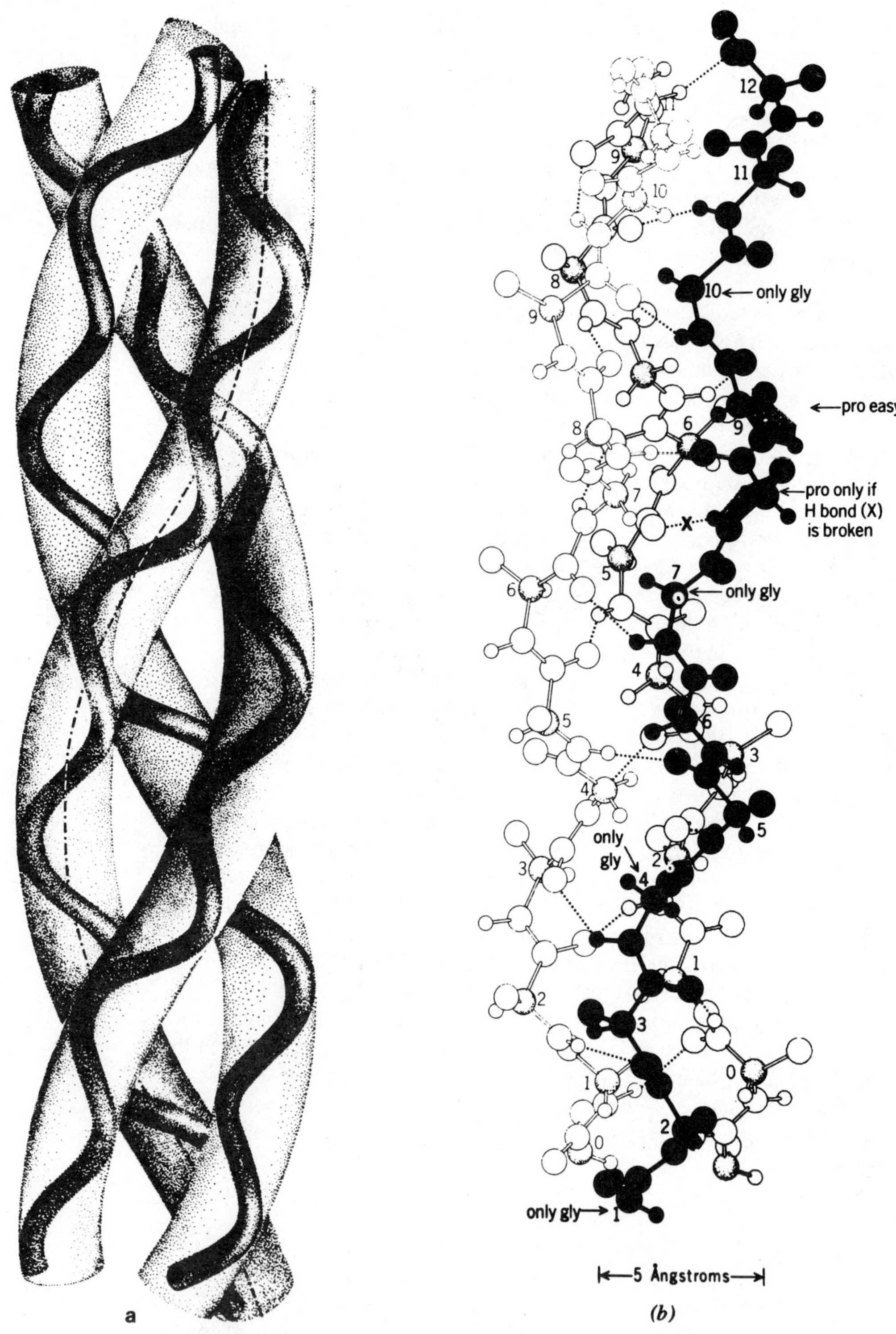

Figure 3.23
Diagram of collagen demonstrating the necessity for glycine in every third residue to allow the different chains to be in close proximity in the structure. All α carbons are numbered and proposed hydrogen bonds are shown by dotted lines.
(a) Ribbon model for supercoiled structure of collagen with each of the individual chains in a polyproline type II helix. (b) More detailed model of supercoiled conformation.
Reprinted wtih permission from R. E. Dickerson and I. Geis, *The Structure and Actions of Proteins,* W. A. Benjamin, Inc., Menlo Park, 1969, pp. 41, 42.

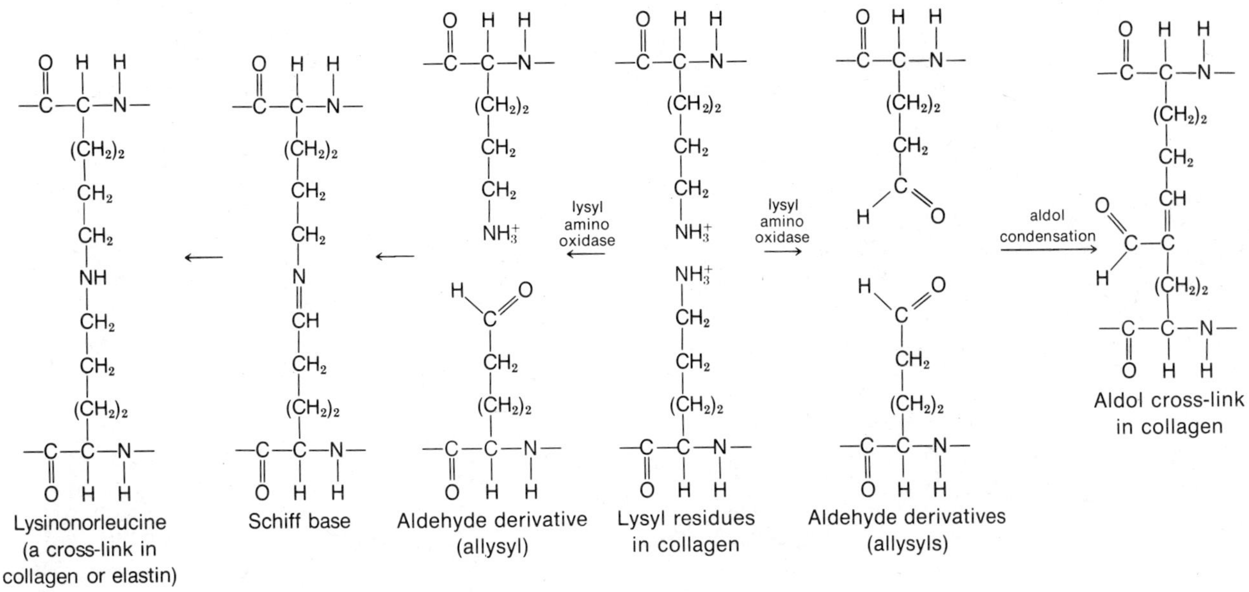

Lysinonorleucine (a cross-link in collagen or elastin) ← Schiff base ← Aldehyde derivative (allysyl) ← lysyl amino oxidase — Lysyl residues in collagen — lysyl amino oxidase → Aldehyde derivatives (allysyls) — aldol condensation → Aldol cross-link in collagen

Figure 3.24
Covalent cross-links formed in collagen from allysine.

Tropoelastin

The protein elastin gives tissues and organs the capacity to stretch without tearing. Elastin is found in high concentrations in ligaments, lungs, walls of arteries, and skin.

The amino acid composition of elastin shows it to be approximately one-third glycine, 10–14% proline, and relatively rich in hydrophobic amino acids (Table 3.10). In contrast to collagen, only small amounts of hydroxyproline are found. Elastin strands do not form a polyproline-type helix. Physical and spectral evidence indicates that in contrast to the helical rod-shaped structure of the collagen molecules and other fibrous proteins, elastin is conformationally a randomly coiled structure in which the amino acid residues in the structure are highly mobile. For example, NMR evidence shows that individual residues in the chain tumble freely in three dimensions over a time scale of 10^{-7} s. The highly mobile, kinetically free, though extensively cross-linked structure (see desmosine cross-links below) give the protein the property of a rubberlike elasticity.

An extracellular enzyme acts on particular lysine R groups in elastin, in a similar way as in collagen, to convert these lysines to allysines. In particular, the allysine residues appear to be generated where lysines appear in the sequence *Lys*-Ala-Ala-*Lys*-and-*Lys*-Ala-Ala-Ala-*Lys* in the primary structure. When the pair of lysines or allysines in this sequence from adjacent polypeptide chains comes close, three allysines and one lysine residue can react to form the heterocyclic structure of desmosine or isodesmosine that covalently cross-links polypeptide chains in elastin fibers (Figure 3.25).

α-Keratin

α-Keratin is found in the epidermal layer of the skin, in nails, and in hair. The conformation of each of the polypeptide chains in α-keratin is α-helical.

The analysis of the primary structure of the α-keratin polypeptide chain shows repetitive segments that consist of seven amino acids, of

Desmosine

This allysyl is shifted to 2 position in isodesmosine

Figure 3.25
Desmosine and isodesmosine covalent cross-link in elastin.

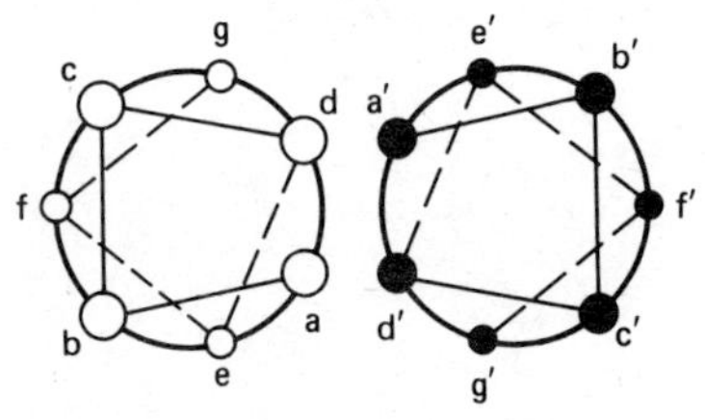

Figure 3.26
Interaction of an apolar edge of two chains in α-helical configuration as in α-keratin and in tropomyosin.
Interaction of apolar a-d′ and d-a′ residues of two α helices aligned parallel in an NH_2-terminal (top) to COOH-terminal direction is presented.
From A. D. McLachlan and M. Stewart, *J. Mol. Biol.* 98:293, 1975.

Figure 3.27
Amide cross-link formed between Gln and Lys in α-keratin.

which the first and fourth are hydrophobic and the fifth and seventh are polar and often of opposite charge. The seven-amino acid sequence can be represented by a-b-c-d-e-f-g, where a and d contain hydrophobic side chain groups, and e and g contain polar or ionized side chain groups. Since a seven-amino acid segment represents approximately two complete turns in the α-helix ($n = 3.6$), the apolar residues a and d will align to form an apolar edge along one side of the α-helix (Figure 3.26), similar to the way in which every third residue of glycine in a collagen helix forms an apolar edge. The apolar edge interacts with similar edges on other chains of α-keratin to form a superhelical structure that contains two or three polypeptide chains. The polar residues in each seven-membered repetitive sequence, which form polar edges on the individual strands, act to stabilize the multistrand supercoiled structure. The supercoil helix of α-keratin is also stabilized by covalent cross-links formed between glutamine side chains and lysine ε-NH_2 groups (Figure 3.27).

The supercoiled molecular structures (protofibrils) form part of a larger assembly in which the protofibrils are wound around one another to produce an even larger structure, designated a microfibril. The microfibril is 70 Å in diameter and may be composed of nine protofibrils (Figure 3.28). α-Keratin contains a high concentration of cysteine that acts to form disulfide cross-links between polypeptide strands within the fibril structure and with an amorphous protein (keratohyalin) that forms a matrix of high cysteine content in which the α-keratin microfibrils are imbedded. An analogy may be made to reinforced concrete, with the α-keratin serving as the metal rods and the keratohyalin as the cement. The cysteine content is higher in the keratin of nails and hair than that of epidermal tissue.

Tropomyosin

Tropomyosin is a component of the thin filament of muscle tissue (Figure 3.29). An analysis of its primary structure shows a type of repeating seven-membered amino acid sequence in terms of apolar and polar side chain R groups $(a\text{-}b\text{-}c\text{-}d\text{-}e\text{-}f\text{-}g)_n$ with residues a and d being apolar and residues e and g ionized. Two polypeptide strands with this repetitive α-

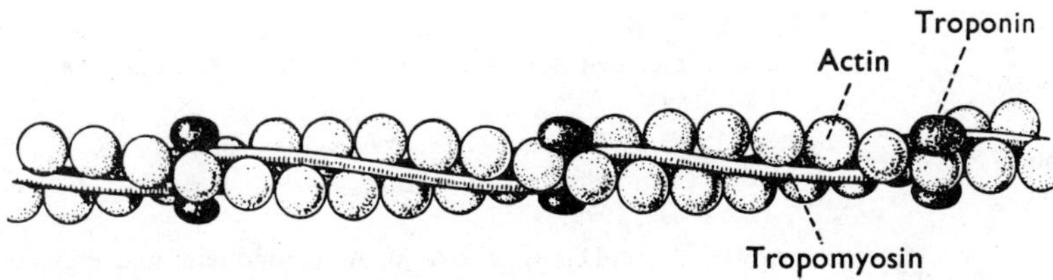

Figure 3.29
Structure of the thin filament of muscle with tropomyosin strands (containing two polypeptide chains) showing association to actin monomers and troponin.
Reprinted with permission from S. Ebashi, M. Endo, and I. Ohtsuki, *Q. Rev. Biophys.* 2:351, 1969.

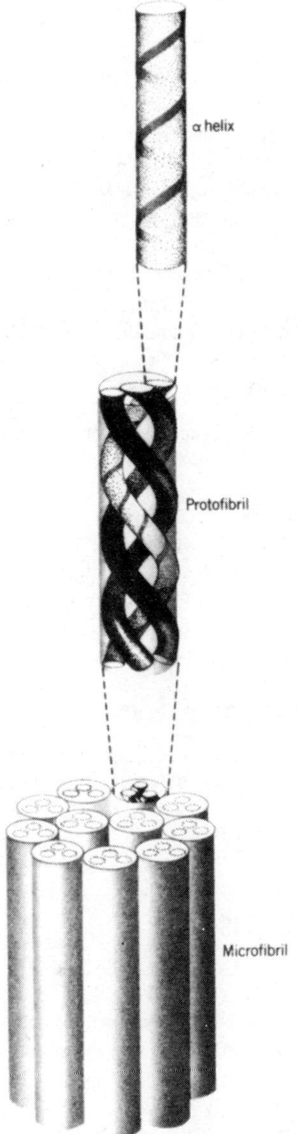

Figure 3.28
Assembly of polypeptides into supercoils and microfibrils of α-keratin.
Reprinted with permission from R. E. Dickerson and I. Geis, *The Structure and Action of Proteins,* W. A. Benjamin, Inc., Menlo Park, 1969, p. 37.

helical sequence are wound around each other into a left-handed superhelical coil.

The tropomyosin molecule is 400 Å long, with the 400 Å length divided into approximately 14 regions of 28 Å. Within each of these regions there is a zone with a high concentration of positively charged side chain groups, a zone with a high concentration of negatively charged side chain groups, and a zone with a high concentration of apolar side chain groups. These zones are thought to provide binding sites for the other protein components of muscle (actin and troponin) that associate with tropomyosin in a regular pattern (Figure 3.29).

Summary of Fibrous Protein Structures

In collagen, α-keratin, and tropomyosin, multiple polypeptide chains with a highly regular secondary structure, are found wound around each other in a supercoil conformation. In turn, the coils are wound or aligned in fibers that are stabilized by covalent cross-links in collagen and α-keratin. The amino acid sequence of the polypeptide chains are repetitive, generating edges on the cylindrical surface of these helical-shaped molecules that stabilize hydrophobic interactions between strands in the supercoiled array. In addition, collagen and α-keratin are immersed in a matrix of polysaccharide or high cysteine-containing protein that cements these fibers into the matrix of the animal tissue or organ to which they give structure and strength.

BIBLIOGRAPHY

Hemoglobin

Baldwin, J., and Chothia, C. Haemoglobin: the structural changes related to ligand binding and its allosteric mechanism. *J. Mol. Biol.* 129:175, 1979.

Dickerson, R. E., and Geis, I. *Hemoglobin: Structure, function, evolution and pathology*. Menlo Park, Calif: Benjamin/Cummings Publ. Co., 1983.

Fermi, G., Perutz, M. F., Shaanan, B., and Fourme, R. The crystal structure of human deoxyhaemoglobin at 1.74 Å resolution. *J. Mol. Biol.* 175:159, 1984.

Perutz, M. Structure and mechanism of hemoglobin. *Br. Med. Bull.* 32:195, 1976.

Perutz, M. Hemoglobin structure and respiratory transport. *Sci. Am.* 239(6):92, 1978.

Shaanan, B. Structure of human oxyhaemoglobin at 2.1 Å resolution. *J. Mol. Biol.* 171:31, 1983.

Antibodies

Kabat, E. A. The structural basis of antibody complementarity. *Adv. Protein Chem.* 32:1, 1978.

Novotny, J., Bruccoleri, R., Newell, J., Murphy, D., Haber, E., and Karplus, M. Molecular anatomy of the antibody binding site, *J. Biol. Chem.* 258:14,433, 1983.

Oi, V. T., Vuong, T. M., Hardy, R., Reidler, J., Dangl, J., Herzenberg, L. A., and Stryer, L. Correlation between segment flexibility and effector function of antibodies. *Nature* 307:136, 1984.

Poljak, R. J. Correlations between three dimensional structure and function of immunoglobins. *CRC Crit. Rev. Biochem.* 5:45, 1978.

Glycoproteins

Clamp, J. R. Structure and function of glycoproteins, in F. W. Putnam (ed.), *The plasma proteins,* vol. II. New York: Academic, 1975, p. 163.

Marshall, R. D. Structures and functions of glycoproteins in R. E. Offord (ed.), *Chemistry of macromolecules IIB*. Baltimore: University Park Press, 1979, p. 1.

Sharon, N., and Lis, H. Glycoproteins, in H. Neurath and R. L. Hill (eds.), *The proteins,* 3rd ed., vol. 5. New York: Academic Press, 1982, pp. 1–144.

Plasma Lipoproteins

Kostner, G. M. Apolipoproteins and lipoproteins of human plasma: Significance in health and in disease. *Adv. Lipid Res* 20:1, 1983.

Scanu, A. M., Byrne, R. E., and Mihovilovic, M. Functional roles of plasma high density lipoproteins. *CRC Crit. Rev. Biochem.* 13:109, 1983.

Souter, A. K., and Myant, N. B. Plasma lipoproteins, in R. E. Offord (ed.), *Chemistry of macromolecules IIB*. Baltimore: University Park Press, 1979, p. 55.

Sparrow, J. T., and Gotto, Jr., A. M. Apolipoprotein/lipid interactions: Studies with synthetic polypeptides. *CRC Crit. Rev. Biochem.* 13:87, 107, 1983.

Fibrous Proteins

Bornstein, P., and Traub, W. The chemistry and biology of collagen, in H. Neurath and R. L. Hill (eds.), *The proteins,* 3rd ed., vol. IV. New York: Academic, 1979, p. 411.

Miller, A. Structure and function of fibrous proteins, in R. E. Offord (ed.), *Chemistry of macromolecules IIA*. Baltimore: University Park Press, 1979, p. 171.

Miller, E. J. Chemistry of collagens and their distribution, in K. A. Piez and A. H. Reddi (eds.), *Extracellular matrix biochemistry,* New York: Elsevier, 1984, pp. 41–81.

Piez, K. A. Molecular and aggregate structures of the collegens, in K. A. Piez and A. H. Reddi (eds.), *Extracellular matrix biochemistry,* Amsterdam: Elsevier Science Publishing Co., 1984, pp. 1–39.

Smillie, L. B. Structure and function of tropomyosins from muscle and non-muscle sources. *Trends Biochem. Sci.* 4:151, 1979.

QUESTIONS

J. BAGGOTT AND C. N. ANGSTADT

*Q*uestion *T*ypes are described inside the front cover.

1. (QT2) Hemoglobin and myoglobin *both:*
   1. bind heme in a hydrophobic pocket.
   2. are highly α-helical.
   3. bind one molecule of heme per globin chain.
   4. consist of subunits designed to provide hydrogen bonds to an nonpolar interaction with other subunits.
2. (QT2) Hemoglobin, but not myoglobin, when it binds oxygen, exhibits:
   1. a sigmoid saturation curve.
   2. a Hill coefficient of 1.
   3. positive cooperativity.
   4. a cooperativity index of 81.
3. (QT1) All of the following are believed to contribute to the stability of the deoxy or T conformation of hemoglobin *except:*
   A. a salt bridge involving specific valyl and arginyl residues.
   B. the larger ionic radius of the six-coordinated ferrous ion as compared to the five-coordinated ion.
   C. steric interaction of His-F8 with the porphyrin ring.
   D. interactions between the FG corner of one subunit and the C helix of the adjacent subunit.
   E. a valyl residue that tends to block O_2 from approaching the hemes of the β-chains.
4. (QT3)
   A. p*K* of HC3 His-146 (β) in oxyhemoglobin.
   B. p*K* of HC3 His-146 (β) in deoxyhemoglobin.
5. (QT2) Haptens:
   1. can function as antigens.
   2. strongly bind to antibodies specific for them.
   3. may be macromolecules.
   4. can act as antigenic determinants.

   A. H chains of immunoglobulins C. Both
   B. L chains of immunoglobulins D. Neither
6. (QT4) C regions determine the class to which the antibody belongs.
7. (QT4) Contain variable regions and hypervariable regions.
8. (QT1) Study of the papain hydrolysis products of an antibody indicates:
   A. antibodies are bivalent.
   B. the products have decreased affinity for antigens.
   C. each antibody molecule is hydrolyzed into many small peptides.
   D. the hypervariable sequences are in the hinge region of the intact molecule.
   E. None of the above is true.

   A. HDL
   B. LDL
   C. IDL
   D. VLDL
   E. Chylomicrons
9. (QT5) Highest protein : lipid ratio.
10. (QT5) Highest percentage triglyceride.
11. (QT2) The apolipoproteins:
    1. contain α-helical regions in which one side of the helix is ionic and the other is nonpolar.
    2. are of only three classes: A, B, and C.
    3. A and C can be subdivided into distinct types of proteins.
    4. of different classes occur in the same ratios (i.e., apo A:apo B) in all types of plasma lipoprotein.
12. (QT2) Current views of plasma lipoprotein structure are consistent with which of the following?

1. Protein occupies a position in the outer shell of the complex.
2. Phospholipid occupies a position in the outer shell of the complex.
3. The ratio protein : neutral lipid decreases with increasing lipoprotein size.
4. Small plasma lipoproteins are denser than larger ones.

13. (QT1) Glycoproteins:
   A. are never functional unless the carbohydrate moiety is complete.
   B. always consist of more protein than carbohydrate.
   C. are generally intracellular; secretory proteins seldom contain carbohydrate.
   D. are N-glycosylated at specific amino acid sequences called sequons.
   E. can possess covalent sugar–protein linkages only at amino acid residues with alcoholic functional groups in their side chains.

   A. Proline C. Both
   B. Glycine D. Neither

14. (QT4) Important in the structure of collagen.
15. (QT4) Important in the structure of α-keratin.
16. (QT2) Structural features common to collagen, α-keratin, and tropomyosin include:
   1. superhelical coiling.
   2. disulfide bridges to neighboring proteins.
   3. repeating sequences of amino acids.
   4. a large percentage of α-helix.

ANSWERS

1. A 1, 2, and 3 true. 1: See p. 78. 2: See p. 71. 3: Hemoglobin has four chains and four oxygen binding sites, whereas myoglobin has one chain and one oxygen binding site. Each oxygen binding site is a heme (pp. 77, 78). 4: Only hemoglobin is designed to form a quaternary structure; myoglobin is structured to interact with water and to prevent association with other myoglobin molecules (p. 80).
2. B 1 and 3 true. 1: See p. 83, Figure 3.5. 2 and 3: Myoglobin has a Hill coefficient of 1; hemoglobin's Hill coefficient of 2.8 indicates positive cooperativity (p. 83). 4: A cooperativity index of 81 indicates noncooperativity; hemoglobin's lower value of 4.8 reflects cooperative oxygen binding (p. 84).
3. B Six-coordinated ferrous ion has a smaller ionic radius than the five-coordinated species and just fits into the center of the porphyrin ring without distortion (p. 85).
4. B HC3 His-146 (β) is a major contributor to the Bohr effect. Thus its pK will be lower (it will be a stronger acid) in oxyhemoglobin (p. 89).
5. C 2 and 4 true. Haptens are small molecules and cannot alone elicit antibody production; thus they are not antigens. They can act as antigenic determinants if covalently bound to a larger molecule, and free haptens may bind strongly to the antibodies thereby produced (p. 90).
6. A There are significant differences among the C regions of the H chains of the different antibody classes (p. 92).
7. C See p. 92. These regions form the antigen binding sites and differ among immunoglobulins of differing specificity.
8. A In these hydrolysis experiments, three fragments are produced: two identical Fab fragments, each of which binds antigen with an affinity similar to that of the whole antibody molecule, and one Fc fragment, which does not bind antigens (p. 94).
9. A The plasma lipoproteins are listed in the order of most dense (HDL) to least dense (chylomicrons) (Table 3.6, p. 98). Since protein is denser than lipid—this is common knowledge to anyone who has washed dishes—HDL must contain the highest percentage protein and have the highest protein : lipid ratio (p. 99).
10. E See Table 3.7, p. 99. The physiological function of chylomicrons is transport of dietary triglyceride from the intestine to other tissues.
11. B 1 and 3 true. 1: This structure gives the α-helix the ability to participate in polar interactions on one side and nonpolar interactions on the other (p. 101 and Figure 3.18). 2, 3: There are at least seven types of apoproteins: Types A and C are further subdivided (Apo A-I, A-II, etc.) (p. 107 and Table 3.8, p. 100). 4: Different classes of lipoproteins contain characteristic types of apoproteins (Table 3.8, p. 100).
12. E 1, 2, 3, and 4 true. Protein and phospholipid together are believed to form the outer shell of the plasma lipoproteins, surrounding an unorganized interior composed of neutral lipid. The structure is similar to that of a micelle (pseudomicellar). The surface area of the lipoprotein increases as the square of its radius, whereas the volume increase is proportional to the cube of the radius. Thus with increasing particle size the nonpolar portion (neutral lipid) increases faster than the polar (protein and polar lipid) (p. 102 and Figure 3.19).
13. D A: The carbohydrate moiety is often incomplete; this may not affect physiological activity. B: Some, but not all, glycoproteins are more than 50% carbohydrate. C: Secreted proteins are commonly glycoproteins; all the plasma proteins except serum albumin are glycoproteins. E: N-linked (to asparagine) and O-linked (to serine or threonine) carbohydrate moieties occur in various glycoproteins (p. 104).
14. C The high proline content constrains the individual polypeptide chains of collagen, favoring formation of the polyproline II type of helix (p. 109). The presence of glycine, with its small side chain, at every third position in collagen is what permits three polyproline II helices to form a superhelix. Larger side chains would result in too much crowding (p. 109).
15. D α-keratin is predominantly α-helical (p. 111). The presence of proline would be incompatible with an α-helix. Glycine is not known to have any special role in α-keratin.
16. B 1 and 3 true. 1: Tropomyosin contains a superhelix consisting of two α-helices (p. 113). The α-keratin superhelix contains two or three α-helices (p. 112). See also answer 14 above. 2: Disulfide bridges appear to be important only in α-keratin (p. 112). 3: In collagen gly-pro-Y and gly-X-hyp are very common, and gly is at every third residue in the sequence (p. 108; see also answer 14 above). α-Keratin contains repetitive sequences of seven amino acids, giving each helix an apolar edge (p. 112). Tropomyosin also shows this pattern (p. 112).

4

Enzymes: Classification, Kinetics, and Control

J. LYNDAL YORK

4.1 GENERAL CONCEPTS

Enzymes are proteins evolved by the cells of living organisms for the specific function of catalyzing chemical reactions. Enzymes increase the rate at which reactions approach equilibrium. *Rate* is defined as the change in the amount (moles, grams) of starting materials or products per unit time. The enzyme triggers the increased rate by acting as a catalyst. A true catalyst increases the rate of a chemical reaction, but is not itself changed in the process. The enzyme may become temporarily covalently bound to the molecule being transformed during intermediate stages of the reaction, but in the end the enzyme will be regenerated in its original form as the product is released.

There are two important characteristics of catalysts in general and enzymes in particular that should not be forgotten. The first is that the enzyme is not changed by entering into the reaction. The second is that the enzyme does not change the equilibrium constant of the reaction, it simply increases the rate at which the reaction approaches equilibrium. Therefore, a catalyst is responsible for increasing the rate but not changing the thermodynamic properties of the system with which it is interacting. In biological systems, a catalyst is necessary because at the temperature and pH of the human body reactions would not occur at a rate sufficient to support rapid muscular activity, nerve impulse generation, and all the other processes required to support life.

At this point, we need to define several terms before entering into a discussion of the mechanism of enzyme action. An *apoenzyme* is the protein part of the enzyme minus any cofactors or prosthetic groups that may be required for the enzyme to be functionally active. The apoenzyme is therefore catalytically inactive. Not all enzymes require cofactors or prosthetic groups to be active. The *cofactors* are those small organic or inorganic molecules that the enzyme requires for its activity. For example, lysine oxidase is a copper-requiring enzyme. Copper in this case is loosely bound but is required for the enzyme to be active. The *prosthetic group* is similar to the cofactor but is tightly bound to the apoenzyme. For example, in the cytochromes, the heme prosthetic group is very tightly bound and requires strong acids to disassociate it from the cytochrome. The addition of cofactor or prosthetic group to the apoprotein yields the *holoenzyme,* which is the active enzyme. The molecule the enzyme acts upon to form product is called the *substrate.* Since most reactions are reversible, the products of the forward reaction will become the substrates of the reverse reaction. Enzymes have a great deal of specificity. For example, glucose oxidase will oxidize glucose but not galactose. This specificity resides in a particular region on the enzyme surface called the *substrate binding site,* which is a particular arrangement of chemical groups on the enzyme surface that is specially formulated to bind a specific substrate. The substrate binding site may have integrated within it the *active site.* In some cases the active site may not be within the substrate binding site but may be contiguous to it in the primary sequence. In other cases the active site lies in distant regions of the primary sequence but is brought adjacent to the substrate binding site by folding of the tertiary structure. The active site contains the machinery, in the form of particular chemical groups, that is involved in catalyzing the reaction under consideration. The chemical groups involved in both binding of substrate and catalysis are often part of the side chains of the amino acids of the apoenzyme.

In some cases variant forms or isoenzymes (isozymes) are found. These isoenzymes are electrophoretically distinguishable, but they all catalyze the same chemical reaction.

In some enzymes there is another region of the molecule, the *allosteric site,* that is not at the active site or substrate binding site, but is somewhere else on the molecule. The allosteric site is the site where small molecules bind and effect a change in the active site or the substrate binding site. The binding of a specific small organic molecule at the allosteric site causes a change in the conformation of the enzyme, and that conformational change may cause the active site to become either more active or less active. It may cause the binding site to have a greater affinity for substrate, or it may actually cause the binding site to have less affinity for substrate. Such interactions are involved in the regulation of the activity of enzymes and are discussed in more detail on page 144.

4.2 CLASSIFICATION OF ENZYMES

The International Union of Biochemistry has established a system whereby all enzymes are placed into one of six major classes. Each class is then subdivided into several subclasses, which are further subdivided. A number is assigned to each class, subclass, and sub-subclass so that an enzyme is assigned a four-digit number as well as a name. The fourth digit identifies a specific enzyme. For example, alcohol : NAD oxidoreductase is assigned the number 1.1.1.1. because it is an oxidoreductase, the electron donor is an alcohol and the acceptor is the coenzyme NAD. Notice that in naming an enzyme, the substrates are stated first, followed by the reaction type to which the ending *ase* is affixed. The trivial name of the enzyme 1.1.1.1. is alcohol dehydrogenase. Many common names persist but are not very informative. For example, "aldolase" does not tell much about the substrates, although it does identify the reaction type. We will use the trivial names that are recognized by the I.U.B.

Each of the six major enzyme classes will be briefly described in the following paragraphs.

Class 1. Oxidoreductases

These enzymes catalyze oxidation and reduction reactions. For example, alcohol : NAD oxidoreductase catalyzes the oxidation of an alcohol to an aldehyde. This enzyme removes two electrons as two hydrogens from the alcohol to yield an aldehyde, and in the process, the two electrons that were originally in the carbon–hydrogen bond of the alcohol are transferred to the NAD^+, which then becomes reduced as shown in Figure 4.1. In addition to the alcohol and aldehyde functional groups, *dehydro-*

$$\mathrm{R{-}\overset{\displaystyle H}{\underset{\displaystyle H}{C}}{-}O{-}H} + \mathrm{NAD^+} \rightleftharpoons \mathrm{R{-}\overset{\displaystyle O}{\overset{\|}{C}}{-}H} + \mathrm{NADH} + \mathrm{H^+}$$

Figure 4.1
The alcohol dehydrogenase catalyzed oxidation of ethanol.

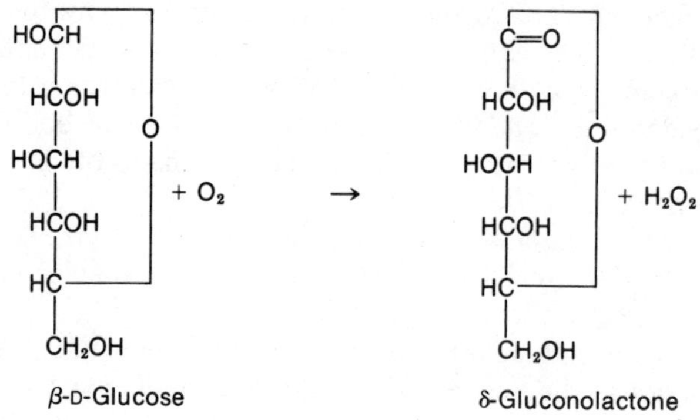

Figure 4.2
The glucose oxidase catalyzed oxidation of glucose.

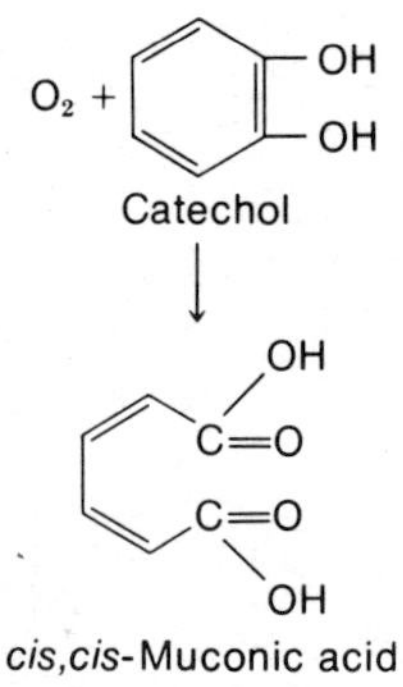

Figure 4.3
Oxygenation of catechol.

genases also act on the following functional groups as electron donors: $—CH_2—CH_2—$, $—CH_2—NH_2$, $—CH{=}NH$, as well as the nucleotides NADH and NADPH.

Other major subclasses of the oxidoreductases are summarized as follows:

Oxidases transfer two electrons from the donor to oxygen, resulting usually in hydrogen peroxide formation. For example, glucose oxidase catalyzes the reaction shown in Figure 4.2. In the case of cytochrome oxidase, H_2O rather than H_2O_2 is the product.

Oxygenases catalyze the incorporation of both atoms of oxygen into a single substrate. Catechol oxygenase catalyzes the reaction shown in Figure 4.3.

Hydroxylases incorporate one atom of molecular oxygen into the substrate; the second oxygen appears as water. The steroid hydroxylases typify this reaction type as shown in Figure 4.4.

Peroxidases utilize hydrogen peroxide rather than oxygen as the oxidant. NADH peroxidase catalyzes the reaction

$$NADH + H^+ + H_2O_2 \rightleftharpoons NAD^+ + 2H_2O$$

Catalase is unique in that hydrogen peroxide serves as both donor and acceptor. Catalase functions in the cell to detoxify hydrogen peroxide:

$$H_2O_2 + H_2O_2 \rightleftharpoons O_2 + 2H_2O$$

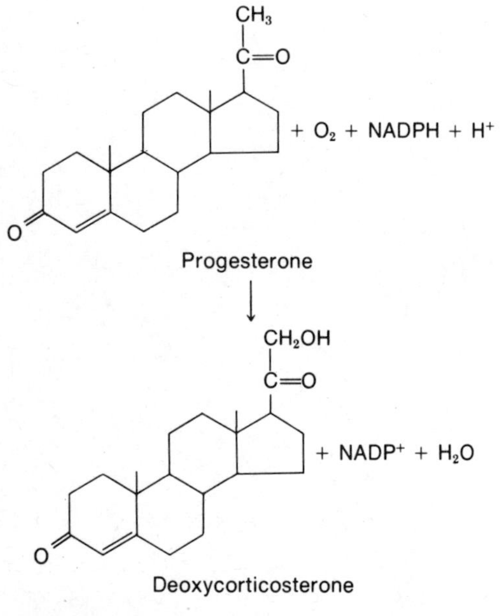

Figure 4.4
Hydroxylation of progesterone by oxygen.

Class 2. Transferases

These enzymes are involved in transferring functional groups between donors and acceptors. The amino, acyl, phosphate, one-carbon, and glycosyl groups are the major moieties that are transferred.

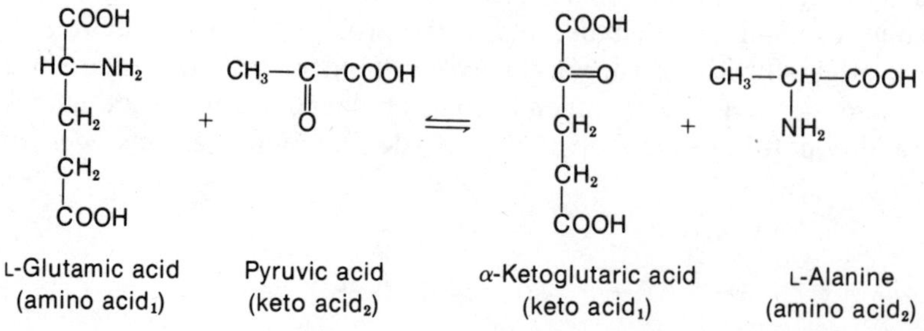

Figure 4.5
Examples of a reaction catalyzed by an aminotransferase.

Aminotransferases (*transaminases*) transfer the amino group from one amino acid to a keto acid acceptor, resulting in the formation of a new amino acid and a new keto acid (Figure 4.5).

Kinases are the phosphorylating enzymes that catalyze the transfer of the phosphoryl group from ATP or another nucleoside triphosphate, to alcohol or amino group acceptors. For example, glucokinase catalyzes the reaction:

Adenosine triphosphate (ATP) + α-D-Glucopyranose ⇌ Glucose 6-phosphate + Adenosine diphosphate (ADP)

The synthesis of glycogen depends upon *glucosyltransferases,* which catalyze the transfer of an activated glucosyl residue to a glycogen primer. The phosphoester bond in uridine diphosphoglucose is labile, which allows the glucose to be transferred to the growing end of the glycogen primer, that is,

UDP-glucose + Glycogen primer → Glycogen extended by one glucosyl unit + UDP

It should be noted that although a polymer is synthesized, the reaction is not of the ligase type, which we discuss in Class 6 below, as is the formation of protein from activated amino acids.

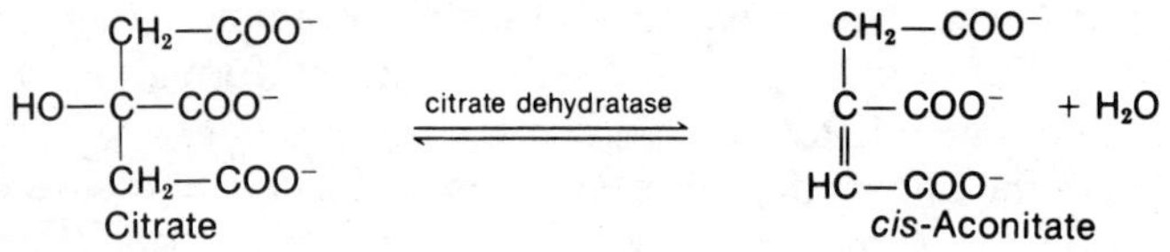

$$\begin{array}{c} CH_2—COO^- \\ | \\ HO—C—COO^- \\ | \\ CH_2—COO^- \\ \text{Citrate} \end{array} \xrightleftharpoons{\text{citrate dehydratase}} \begin{array}{c} CH_2—COO^- \\ | \\ C—COO^- \\ \| \\ HC—COO^- \\ cis\text{-Aconitate} \end{array} + H_2O$$

Figure 4.6
The dehydration of citrate.

Class 3. Hydrolases

This group of enzymes can be considered as a special class of the transferases in which the donor group is transferred to water. The generalized reaction involves the hydrolytic cleavage of C—O, C—N, O—P, and C—S bonds. The cleavage of the peptide bond is a good example of this reaction:

$$R_1—\overset{O}{\overset{\|}{C}}—NH—R_2 + H_2O \longrightarrow R_1—\overset{O}{\overset{\|}{C}}—O^- + H_3\overset{+}{N}—R_2$$

The proteolytic enzymes are a special class of hydrolases called *peptidases.*

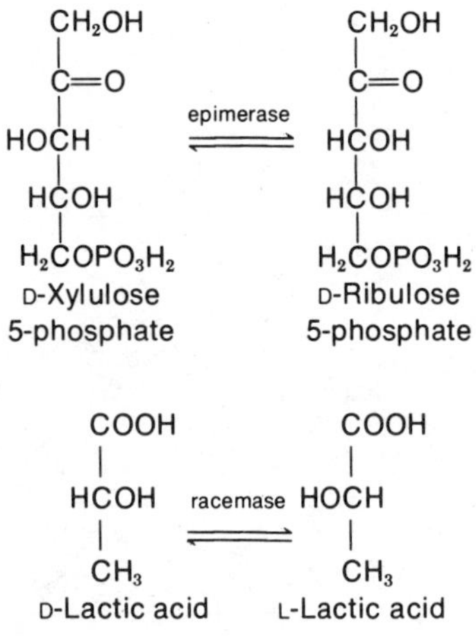

Class 4. Lyases

Lyases are enzymes which add or remove the elements of water, ammonia, or CO_2.

The *decarboxylases* remove the element of CO_2 from α- or β-keto acids or amino acids:

$$R—\overset{O}{\overset{\|}{C}}—\overset{O}{\overset{\|}{C}}—O^- + H^+ \longrightarrow R—\overset{O}{\overset{\|}{C}}—H + CO_2$$

The *dehydratases* remove the elements of H_2O in a dehydration reaction. Citrate dehydratase converts citrate to *cis*-aconitate (Figure 4.6).

$$\begin{array}{c} CH_2OH \\ | \\ C{=}O \\ | \\ HOCH \\ | \\ HCOH \\ | \\ H_2COPO_3H_2 \\ \text{D-Xylulose} \\ \text{5-phosphate} \end{array} \xrightleftharpoons{\text{epimerase}} \begin{array}{c} CH_2OH \\ | \\ C{=}O \\ | \\ HCOH \\ | \\ HCOH \\ | \\ H_2COPO_3H_2 \\ \text{D-Ribulose} \\ \text{5-phosphate} \end{array}$$

$$\begin{array}{c} COOH \\ | \\ HCOH \\ | \\ CH_3 \\ \text{D-Lactic acid} \end{array} \xrightleftharpoons{\text{racemase}} \begin{array}{c} COOH \\ | \\ HOCH \\ | \\ CH_3 \\ \text{L-Lactic acid} \end{array}$$

Figure 4.7
Examples of reactions catalyzed by an epimerase and a racemase.

Class 5. Isomerases

This is a very heterogeneous group of enzymes that catalyze isomerizations of several types. These include *cis–trans,* keto–enol, and aldose–ketose interconversions. Isomerases that catalyze inversion at asymmetric carbons are either *epimerases* or *racemases* (Figure 4.7). *Mutases* involve the intramolecular transfer of a group such as the phosphoryl. The transfer need not be direct but can involve a phosphorylated enzyme as an intermediate. An example is phosphoglycerate mutase, which catalyzes the conversion of 2-phosphoglycerate to 3-phosphoglycerate as shown in Figure 4.8.

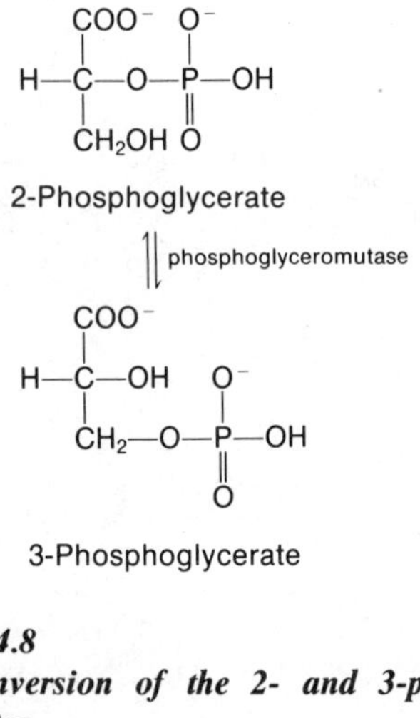

$$\begin{array}{c} COO^- \quad O^- \\ | \qquad\quad | \\ H—C—O—P—OH \\ | \qquad\quad \| \\ CH_2OH \quad O \\ \text{2-Phosphoglycerate} \end{array}$$

$$\Big\Updownarrow \text{phosphoglyceromutase}$$

$$\begin{array}{c} COO^- \\ | \\ H—C—OH \quad O^- \\ | \qquad\qquad | \\ CH_2—O—P—OH \\ \| \\ O \\ \text{3-Phosphoglycerate} \end{array}$$

Figure 4.8
Interconversion of the 2- and 3-phosphoglycerates.

Class 6. Ligases

Since to ligate means to bind, these enzymes are involved in synthetic reactions where two molecules are joined at the expense of an ATP "high-energy phosphate bond." The use of "synthetase" is reserved for this particular group of enzymes. The formation of amino acyl tRNAs, acyl coenzyme A, glutamine, and the addition of CO_2 to pyruvate, are reactions catalyzed by ligases. Pyruvate carboxylase is a good example of a ligase enzyme. The reaction is presented in Figure 4.9. The two substrates bicarbonate and pyruvate are ligated to form a four-carbon keto acid.

TABLE 4.1 Summary of the Enzyme Classes and Major Subclasses

1. Oxidoreductases	2. Transferases
Dehydrogenases	Transaldolase and transketolase
Oxidases	Acyl-, methyl-, glucosyl-, and phosphoryl-transferase
Reductases	Kinases
Peroxidases	Phosphomutases
Catalase	
Oxygenases	
Hydroxylases	
3. Hydrolases	4. Lyases
Esterases	Decarboxylases
Glycosidases	Aldolases
Peptidases	Hydratases
Phosphatases	Dehydratases
Thiolases	Synthases
Phospholipases	Lyases
Amidases	
Deaminases	
Ribonucleases	
5. Isomerases	6. Ligases
Racemases	Synthetases
Epimerases	Carboxylases
Isomerases	
Mutases (not all)	

$$\text{ATP} + \text{HCO}_3^- + \begin{matrix}\text{COOH}\\ | \\ \text{C=O}\\ | \\ \text{CH}_3\end{matrix} \xrightarrow{\text{biotin}} \begin{matrix}\text{COOH}\\ | \\ \text{C=O}\\ | \\ \text{CH}_2\\ | \\ \text{COOH}\end{matrix} + \text{ADP} + \text{P}_i$$

Pyruvate — Oxaloacetate

Figure 4.9
The reaction catalyzed by pyruvate carboxylase.

The six enzyme classes and most of the important subclass members are compiled in Table 4.1. The accepted trivial names are used for members of the subclass.

4.3 KINETICS

Basic Chemical Kinetics

Since enzymes affect the rate of chemical reactions, it is important to understand basic chemical kinetics and how kinetic principles apply to enzyme-catalyzed reactions. *Kinetics* is a study of the rate of change of the initial state of reactants and products to the final state of reactants and products. The term *velocity* is often used rather than rate. Velocity is expressed in terms of change in the concentration of substrate or product per unit time, whereas rate refers to changes in total quantity (moles, grams) per unit time. Biochemists tend to use the two terms interchangeably.

The velocity of a reaction $A \rightarrow P$ is determined from a progress curve or velocity profile of a reaction. The progress curve can be determined by following the disappearance of reactants or the appearance of product at several different times. Such a curve is shown in Figure 4.10, where product appearance is plotted against time. The slope of tangents to the progress curve yields the instantaneous velocity at that point in time. The initial velocity represents an important parameter in the assay of enzyme concentration, as we learn later. Notice that the velocity constantly changes as the reaction proceeds to equilibrium, and becomes zero at equilibrium. Mathematically, the velocity is expressed as

$$\text{Velocity} = v = \frac{-d[\text{A}]}{dt} = \frac{d[\text{P}]}{dt} \tag{4.1}$$

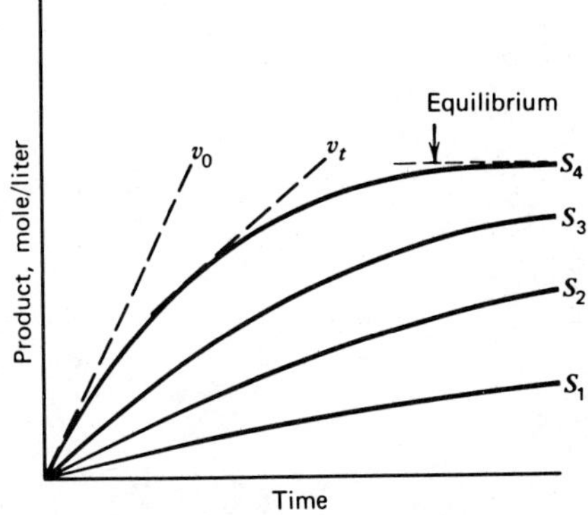

Figure 4.10
Progress curves for an enzyme-catalyzed reaction.
The initial velocity (v_0) of the reaction is determined from the slope of the progress curve at the beginning of the reaction. The initial velocity increases with increasing substrate concentration (S_1 through S_4) but reaches a limiting value that is characteristic of each enzyme. The velocity at any time, t, is denoted as v_t.

and represents the change in concentration of reactants or products per unit time.

The Rate Equation

Determination of the velocity of a reaction reveals nothing about the stoichiometry of the reactants and products or about the mechanism of the reaction. What is needed is an equation that relates the experimentally determined initial velocity to the concentration of reactants. Such a relation is the velocity or rate equation. In the case of the reaction A → P, the velocity equation is

$$\frac{-d[\mathrm{A}]}{dt} = v = k[\mathrm{A}]^n \qquad \textbf{(4.2)}$$

That is to say, the observed initial velocity will depend on the starting concentration of A to the nth power multiplied by a proportionality constant k. The latter is known as the *rate constant*. The exponent n is usually an integer from 1 to 3 that is required to satisfy the mathematical identity of the velocity expression.

Characterization of Reactions Based on Order

Another term that is useful in describing a reaction is the *order* of reaction. Empirically the order is determined as the sum of the exponents on each concentration term in the rate expression. In the case under discussion the reaction is *first order,* since the velocity depends on the concentration of A to the first power, $v = k[\mathrm{A}]^1$. In a reaction such as A + B → C, if the order with respect to A and B is 1, that is, $v = k[\mathrm{A}]^1[\mathrm{B}]^1$, overall the reaction is second order. It should be noted that the order of reaction is independent of the stoichiometry of the reaction, that is, if the reaction were third order, the rate expression could be either $v = k[\mathrm{A}][\mathrm{B}]^2$ or $v = k[\mathrm{A}]^2[\mathrm{B}]$, depending on the order in A and B. Since the velocity of the reaction is constantly changing as the reactant concentration changes, it is obvious that first-order reaction conditions would not be ideal for assaying an enzyme-catalyzed reaction because one would have two variables, the changing substrate concentration and the unknown enzyme concentration.

If the differential first-order rate expression (4.2) is integrated, one obtains

$$k_1 \cdot t = 2.3 \log \left[\frac{[\mathrm{A}]}{[\mathrm{A}] - [\mathrm{P}]}\right] \qquad \textbf{(4.3)}$$

where [A] is the initial reactant concentration and [P] is the concentration of product formed at time t. The first-order rate constant k_1 has the units of reciprocal time. If the data shown in Figure 4.10 were replotted as log [P] vs time for any one of the substrate concentrations, a straight line would be obtained whose slope is equal to $k_1/2.303$. The rate constant k_1 should not be confused with the rate or velocity of the reaction.

Many biological processes proceed under first-order conditions. The clearance of many drugs from the blood by peripheral tissues is a first-order process. A specialized form of the rate equation can be used in these cases. If we define $t_{1/2}$ as the time required for the concentration of the reactants or the blood level of a drug to be reduced by one-half the initial value, then equation (4.3) reduces to

$$k_1 \cdot t_{1/2} = 2.3 \log \left[\frac{1}{1 - \frac{1}{2}}\right] = 2.3 \log 2 = 0.69 \qquad \textbf{(4.4)}$$

or

$$t_{1/2} = \frac{0.69}{k_1} \quad \textbf{(4.5)}$$

Notice that $t_{1/2}$ is not one-half the time required for the reaction to be completed. The term $t_{1/2}$ is referred to as the *half-life* of the reaction.

Many *second-order* reactions that involve water or any one of the reactants in large excess can be treated as pseudo-first-order reactions. In the case of the hydrolysis of an ester,

$$R{-}\overset{\overset{O}{\|}}{C}{-}O{-}CH_3 + H_2O \rightleftharpoons R{-}\overset{\overset{O}{\|}}{C}{-}OH + CH_3OH$$

the second-order rate expression is

$$\text{velocity} = v = k_2[\text{ester}]^1[H_2O]^1 \quad \textbf{(4.6)}$$

but since water is in abundance (55.5 M) compared to the ester (10^{-3}–10^{-2}M), the system obeys the first-order rate law (4.2), and the reaction appears to proceed as if it were a first-order reaction. Those reactions in the cell that involve hydration, dehydration, or hydrolysis are pseudo-first order.

The rate expression for the *zero-order* reaction is $v = k_0$. Notice that there is no concentration term for reactants; therefore, the addition of more reactant does not augment the rate. The disappearance of reactant or the appearance of product proceeds at a constant velocity irrespective of reactant concentration. The units of the rate constant are concentration per unit time. Zero-order reaction conditions only occur in catalyzed reactions where the concentration of reactants is large enough to saturate all the catalytic sites. Under these conditions the catalyst is operating at maximum velocity, and all catalytic sites are filled; therefore, addition of more reactant cannot increase the rate.

Reversibility of Reactions

Although most chemical reactions are reversible, some directionality may be imposed on particular steps in a metabolic pathway through rapid removal of the end product by subsequent reactions in the pathway.

In the case of decarboxylation reactions

$$R{-}\overset{\overset{O}{\|}}{C}{-}\overset{\overset{O}{\|}}{C}{-}OH \longrightarrow R{-}\overset{\overset{O}{\|}}{C}{-}H + CO_2\uparrow$$

where carbon dioxide is liberated, the reaction is irreversible from a practical standpoint because CO_2 is a gas and can diffuse away from the reaction site; therefore the reaction proceeds in the forward direction by mass action. Many ligase reactions involving the nucleoside triphosphates result in release of pyrophosphate. These reactions are rendered irreversible by the hydrolysis of the pyrophosphate to 2 mol inorganic phosphate, P_i. Schematically,

$$\text{A} + \text{B} + \text{ATP} \longrightarrow \text{A—B} + \text{AMP} + \text{P—P}$$
$$\text{P—P} + H_2O \longrightarrow 2P_i$$

The conversion of the "high-energy" pyrophosphate to inorganic phosphate imposes irreversibility on the system by virtue of the thermodynamic stability of the products.

For those reactions that are reversible, the equilibrium constant for the reaction

$$A + B \underset{}{\overset{K_{eq}}{\rightleftharpoons}} C$$

is

$$K_{eq} = \frac{[C]}{[A][B]} \tag{4.7}$$

and can be expressed in terms of the rate constants of the forward and reverse reactions:

$$A + B \underset{k_2}{\overset{k_1}{\rightleftharpoons}} C$$

where

$$\frac{k_1}{k_2} = K_{eq} \tag{4.8}$$

Equation (4.8) shows the relationship between thermodynamic and kinetic quantities. K_{eq} is a thermodynamic expression of the state of the system, while k_1 and k_2 are kinetic expressions that are related to the speed at which that state is reached.

Enzyme Kinetics

Terminology

Enzyme activity is usually expressed in units of micromoles (μmol) of substrate converted to product per minute under specified assay conditions. One *standard unit* of enzyme activity (U) is an amount of activity that catalyzes the transformation of one μmol/minute. The *specific activity* of an enzyme preparation is defined as the number of enzyme units per milligram of protein (μmol/min/mg protein or U/mg protein). This expression, however, does not indicate whether the sample tested contains only the enzyme protein; during an enzyme purification the value will increase as contaminating protein is removed. The *catalytic constant,* or *turnover number,* for an enzyme is equal to the units of enzyme activity per mol of enzyme (μmol/min/mol of enzyme). In cases where the enzyme has more than one catalytic center, the catalytic constant is often given on the basis of the particle weight of the subunit rather than the molecular weight of the entire protein. The Commission on Enzyme Nomenclature of the International Union of Biochemistry has recommended that enzyme activity be expressed in units of mol/second, instead of μmol/minute, to conform with the rate constants used in chemical kinetics. A new unit, the *Katal* (abbreviated kat), is proposed where one kat denotes the conversion of one mol substrate/sec. Activity can be expressed, however, as millikatals (mkat), microkatals (μkat), and so forth. The specific activity and catalytic constant can also be expressed in this unit of activity.

The catalytic constant or turnover number allows a direct comparison of relative catalytic ability between enzymes. For example, the constants for catalase and α-amylase are 5×10^6 and 1.9×10^4, respectively, indicating that catalase is about 2,500 times more active than amylase.

The *maximum velocity* V_{max} is the velocity obtained under conditions of substrate saturation of the enzyme under a given set of conditions of pH, temperature, and ionic strength.

Interaction of Enzyme and Substrate

The initial velocity of an enzyme-catalyzed reaction is dependent on the concentration of substrate as shown in Figure 4.10. As the concentration of substrate is increased (S_1–S_4), the initial velocity increases until the enzyme is completely saturated with the substrate. If one plots the initial velocities obtained at given substrate concentrations (Figure 4.11), a rectangular hyperbola is obtained. The same type of curve will be obtained for the binding of oxygen to myoglobin as a function of increasing oxygen pressure. In general, the rectangular hyperbola will be obtained for any process that involves an interaction or binding of reactants or other substances at a specific but limited number of sites. The velocity of the reaction reaches a limiting maximum at the point at which all the available sites are saturated. The curve in Figure 4.11 is referred to as the *substrate saturation curve* of an enzyme-catalyzed reaction and reflects the fact that the enzyme has a specific binding site for the substrate. Obviously the enzyme and substrate must interact in some way if the substrate is to be converted to products. Initially there is formation of a complex between the enzyme and substrate:

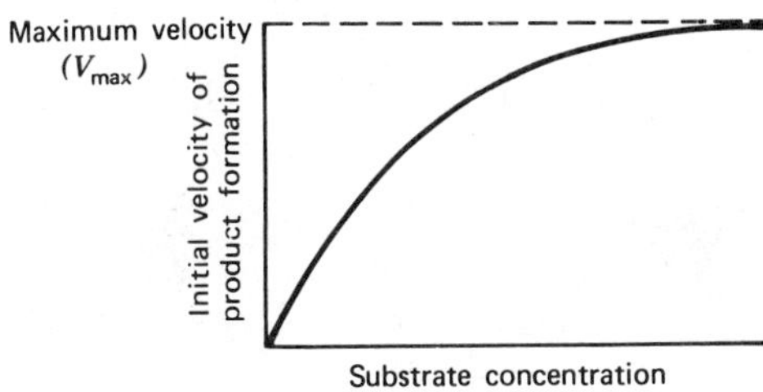

Figure 4.11
Plot of velocity vs substrate for an enzyme-catalyzed reaction.
Initial velocities are plotted against the substrate concentration at which they were determined. The curve is a rectangular hyperbola, which asymptotically approaches the maximum velocity possible with a given amount of enzyme.

$$E + S \underset{k_2}{\overset{k_1}{\rightleftharpoons}} ES \qquad \textbf{(4.9)}$$

The rate constant for formation of this ES complex is defined as k_1, and the rate constant for disassociation of the ES complex is defined as k_2. So far, we have described only an equilibrium binding of enzyme and substrate. The acual chemical event in which bonds are made or broken occurs in the ES complex. The conversion of substrate to products then occurs from the ES complex with a rate constant k_3. Therefore, equation (4.9) is transformed to

$$E + S \underset{k_2}{\overset{k_1}{\rightleftharpoons}} ES \xrightarrow{k_3} E + \text{products} \qquad \textbf{(4.10)}$$

Equation (4.10) is a general statement of the mechanism of enzyme action. The equilibrium between E and S can be expressed as an affinity constant, K_a, only if the rate of the chemical phase of the reaction, k_3, is small compared to k_2; then $K_a = k_1/k_2$. We earlier used K_{eq} to describe chemical reactions. In enzymology the association or affinity constant K_a is preferred.

The initial velocity of an enzyme-catalyzed reaction is not only dependent on the amount of substrate present, but also on the enzyme concentration. Figure 4.12 shows progress curves for increasing concentrations of enzyme, where there is enough substrate to saturate the enzyme at all levels. The initial velocity doubles as the concentration of enzyme doubles. At the lower concentrations of enzyme, equilibrium is reached more slowly than at higher concentrations, but the final equilibrium position is the same.

From our discussion thus far, we can conclude that the velocity of an enzyme reaction is dependent upon both substrate and enzyme concentration.

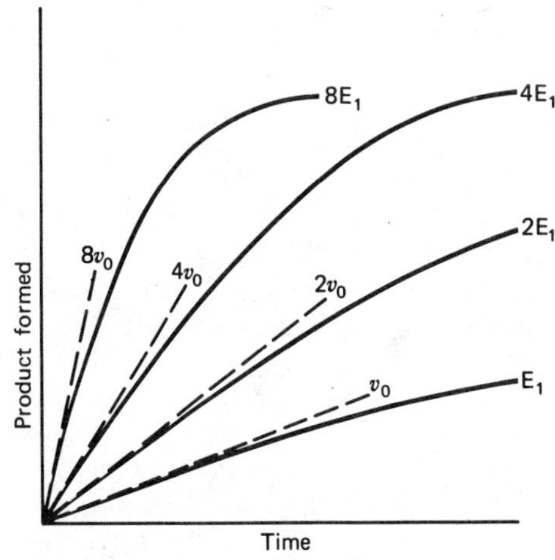

Figure 4.12
Progress curves at variable concentrations of enzyme and saturating levels of substrate.
The initial velocity (v_0) doubles as the enzyme concentration doubles. Since the substrate concentrations are the same, the final equilibrium concentrations of product will be identical in each case; however, equilibrium will be reached at a slower rate in those assays containing small amounts of enzyme.

Formulation of the Michaelis–Menten Equation

It should be recalled that in the discussion of chemical kinetics, rate equations were developed so that the velocity of the reaction could be expressed in terms of the substrate concentration. This philosophy also holds for enzyme-catalyzed reactions where the ultimate goal is to develop a relationship that will allow the velocity of a reaction to be correlated with the amount of enzyme. First, a rate equation must be developed that will relate the velocity of the reaction to the substrate concentration.

In the development of this rate equation, which is known as the *Michaelis–Menten* equation, three basic assumptions are made. The first is that the ES complex is in a steady state. That is, during the initial phases of the reaction, the concentration of the ES complex remains constant, even though many molecules of substrate are converted to products via the ES complex. The second assumption is that under saturating conditions all of the enzyme is converted to the ES complex, and none is free. The third assumption is that if all the enzyme is in the ES complex, then the rate of formation of products will be the maximum rate possible, that is,

$$V_{max} = k_3[\mathrm{ES}] \quad \textbf{(4.11)}$$

If one then writes the equilibrium expression for the formation and breakdown of the ES complex as

$$K_m = \frac{k_2 + k_3}{k_1} \quad \textbf{(4.12)}$$

then the rate expression can be obtained after suitable algebraic manipulation as

$$\text{Velocity} = v = \frac{V_{max} \cdot [\mathrm{S}]}{K_m + [\mathrm{S}]} \quad \textbf{(4.13)}$$

The complete derivation of this equation is at the end of this section. The two constants in this rate equation, V_{max} and K_m, are unique to each enzyme under specific conditions of pH and temperature. For those enzymes in which $k_3 \ll k_2$, K_m becomes the reciprocal of the enzyme–substrate binding constant, that is,

$$K_m = \frac{1}{K_a}$$

and the V_{max} reflects the catalytic phase of the enzyme mechanism as suggested by equation (4.11). In other words, in this simple model the activity of the enzyme can be separated into two phases: binding of substrate followed by chemical modification of the substrate. This biphasic nature of enzyme mechanism is reinforced in the clinical example discussed in Clin. Corr. 4.1.

CLIN. CORR. 4.1
A CASE OF GOUT DEMONSTRATES TWO PHASES IN THE MECHANISM OF ENZYME ACTION

The partitioning of the Michaelis–Menten model of enzyme action into two phases: binding, followed by chemical modification of substrate, is illustrated by studies on a family with hyperuricemia and gout. The patient excreted three times the normal amount of uric acid per day and had markedly increased levels of 5′-phosphoribosylpyrophosphate (PRPP) in his red blood cells. Assays in vitro revealed that his red cell PRPP synthetase activity was increased threefold. The pH optimum and the K_m of the enzyme for ATP and ribose 5-phosphate were normal, but V_{max} was increased threefold! This increase was not because of an increase in the amount of enzyme; immunologic testing with a specific antibody to the enzyme revealed similar titers in the normal and in the patient's red cells. This finding demonstrates that the binding of substrate as reflected by K_m and the subsequent chemical event in catalysis, which is reflected in V_{max}, are separate phases of the overall catalytic process. This situation holds only for those enzyme mechanisms in which $k_2 >> k_3$.

Significance of K_m

The concept of K_m may appear to have no physiological or clinical relevance. The truth is quite the contrary. As discussed in Section 4.9, all valid enzyme assays performed in the clinical laboratory are based on knowledge of the K_m values for each substrate.

In terms of physiological control of glucose and phosphate metabolism, two hexokinases have evolved, one with a high K_m and one with a low K_m for glucose. Together, they contribute to maintaining steady-state levels of blood glucose and phosphate, as discussed in more detail on page 280.

In general K_m values are found to be near concentrations of substrate found in the cell. Perhaps enzymes have evolved substrate binding sites with affinities comparable to in vivo levels of their substrates. Occasionally, mutation of the enzyme binding site occurs, or a different form of an enzyme (isoenzyme) with an altered K_m is expressed. Either one of these events can result in an abnormal physiology. An interesting example

CLIN. CORR. 4.2
THE PHYSIOLOGICAL EFFECT OF CHANGES IN ENZYME K_m VALUE

The unusual sensitivity of Orientals to alcoholic beverages may have a biochemical basis. In the case of Japanese and Chinese, much less ethanol is required to produce the vasodilation that results in facial flushing and tachycardia than is required to achieve the same effect in Europeans. The physiological effects arise from the acetaldehyde generated by liver alcohol dehydrogenase.

$$CH_3CH_2OH + NAD^+ \rightleftharpoons CH_3CHO + H^+ + NADH$$

The acetaldehyde formed is normally removed by a mitochondrial aldehyde dehydrogenase that converts it to acetate. In individuals of Monogoloid origins, the

given in Clin. Corr. 4.2 is the case of the expression of only the atypical form of aldehyde dehydrogenase in people of Asiatic origin.

Notice that if one allows the initial velocity, v_0, to be equal to $\frac{1}{2}V_{max}$ in equation (4.13), K_m will become equal to [S]:

$$\tfrac{1}{2}V_{max} = \frac{V_{max} \cdot [S]}{K_m + [S]}$$

$$K_m + [S] = \frac{2V_{max} \cdot [S]}{V_{max}}$$

$$K_m = [S]$$

normal electrophoretically fast-migrating form of the aldehyde dehydrogenase is missing. This form has a low K_m for acetaldehyde. In these people, only the slow-migrating, high K_m (lower affinity) form of the enzyme is present. This form of the enzyme allows a higher steady-state level of acetaldehyde in the blood after alcohol consumption, which could account for the increased sensitivity to alcohol.

Therefore, from a substrate saturation curve the numerical value of the K_m can be derived by graphical analysis, as shown in Figure 4.13. In other words, the K_m is equal to the substrate concentration that will give half the maximum velocity.

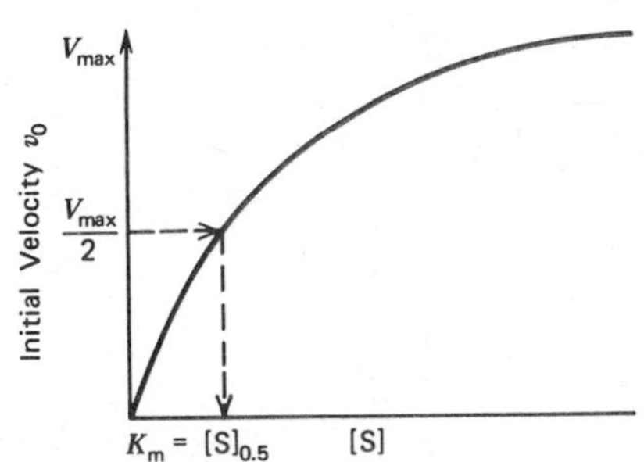

Figure 4.13
Graphic estimation of K_m for the v vs [S] plot.
K_m is the substrate concentration at which the enzyme has half-maximal activity.

Linear Form of the Michaelis–Menten Equation

In practice the determination of K_m from the substrate saturation curve is not very accurate, because V_{max} is approached asymptotically. If one takes the reciprocal of equation (4.13) and separates the variables into a format consistent with the equation of a straight line ($y = mx + b$), then

$$\frac{1}{v_0} = \frac{K_m}{V_{max}} \times \frac{1}{[S]} + \frac{1}{V_{max}}$$

A plot of the reciprocal of the initial velocity vs the reciprocal of the initial substrate concentration yields a line whose slope is K_m/V_{max} and whose y intercept is $1/V_{max}$. Such a plot is shown in Figure 4.14. It is often easier to obtain the K_m from the intercept on the X axis, which is $-1/K_m$.

This linear form of the Michaelis–Menten equation is often referred to as the *Lineweaver–Burk* or double reciprocal plot. Its advantage is that statistically significant values of K_m and V_{max} can be obtained directly with six to eight data points.

Derivation of the Michaelis–Menten Equation

The generalized statement of the mechanism of enzyme action is

$$E + S \underset{k_2}{\overset{k_1}{\rightleftharpoons}} ES \xrightarrow{k_3} E + P \qquad \textbf{(4.10)}$$

If we assume that the rate of formation of the ES complex is balanced by its rate of breakdown (the steady-state assumption), then we can write

$$v_{formation} = k_1[S]\,[E]$$

and

$$v_{breakdown} = k_2[ES] + k_3[ES] = [ES](k_2 + k_3)$$

If we set the rate of formation equal to the rate of breakdown, then

$$k_1[S]\,[E] = [ES](k_2 + k_3)$$

After dividing both sides of the equation by k_1, we have

$$[S]\,[E] = [ES]\left[\frac{k_2 + k_3}{k_1}\right] \qquad \textbf{(4.14)}$$

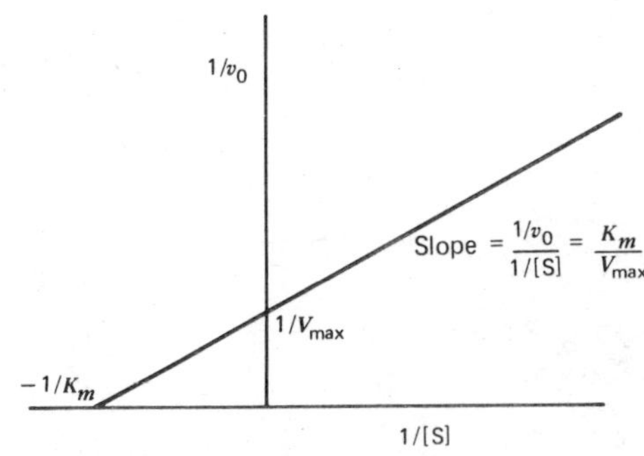

Figure 4.14
Determination of K_m and V_{max} from the Lineweaver–Burk double reciprocal plot.
Plots of the reciprocal of the initial velocity vs the reciprocal of the substrate concentration used to determine the initial velocity yield a line whose x intercept is $-1/K_m$.

If we now define the ratio of the rate constants $(k_2 + k_3)/k_1$ as K_m, the Michaelis constant, and substitute it into equation (4.14), then

$$[S]\,[E] = [ES]K_m \qquad \textbf{(4.15)}$$

Since [E] is equal to the free enzyme, we must express its concentration in terms of the total enzyme added to the system minus any enzyme in the [ES] complex, that is,

$$[E] = ([E_t] - [ES])$$

Upon substitution of the equivalent expression for [E] into equation (4.15) we have

$$[S]\,([E_t] - [ES]) = [ES]K_m$$

Dividing through by [S] yields

$$[E_t] - [ES] = \frac{[ES]K_m}{[S]}$$

and dividing through by [ES] yields

$$\frac{[E_t]}{[ES]} - 1 = \frac{K_m}{[S]} \quad \text{or} \quad \frac{[E_t]}{[ES]} = \frac{K_m}{[S]} + 1 = \frac{K_m + [S]}{[S]} \qquad \textbf{(4.16)}$$

We now need to obtain an alternative expression for $[E_t]/[ES]$, since [ES] cannot be measured easily, if at all. When the enzyme is saturated with substrate all the enzyme will be in the ES complex, and none will be free ($[E_t] = [ES]$), and the velocity observed will be the maximum possible; therefore, $V_{max} = k_3[E_t]$ [See equation (4.11).] When $[E_t]$ is not equal to [ES], $v = k_3[ES]$. From these two expressions we can obtain the ratio of $[E_t]/[ES]$, that is,

$$\frac{[E_t]}{[ES]} = \frac{V_{max}/k_3}{v/k_3} = \frac{V_{max}}{v} \qquad \textbf{(4.17)}$$

Substituting this value of $[E_t]/[ES]$ into equation (4.16) yields a form of the Michaelis–Menten equation:

$$\frac{V_{max}}{v} = \frac{K_m + [S]}{[S]}$$

or

$$v = \frac{V_{max}[S]}{K_m + [S]}$$

Enzyme-Catalyzed Reversible Reactions

As has been indicated previously, enzymes do not alter the equilibrium constant of a reaction; consequently, in a reaction

$$S \underset{k_2}{\overset{k_1}{\rightleftharpoons}} P$$

the direction of flow of material, either in the forward direction or the reverse direction, will depend on the concentration of S relative to P and the equilibrium constant of the reaction. Since enzymes catalyze the forward as well as the reverse reaction, a problem may arise if the product has an affinity for the enzyme, which is similar to that of the substrate. In this case the product can easily rebind to the active site of the enzyme and will compete with the substrate for that site. In such cases the product inhibits the reaction as the concentration of product increases. The Lineweaver–Burk plot will not be linear in those cases where the enzyme is susceptible to product inhibition. If the subsequent enzyme in the metabolic pathway has great affinity for the product and removes it, then product inhibition may not occur.

Product inhibition in a metabolic pathway provides a limited means of controlling or modulating the flux of substrates through the pathway. As the end product of the pathway increases, each intermediate will also increase via mass action. If one or more enzymes in the pathway are particularly sensitive to product inhibition, the output of the end product of the pathway will be suppressed.

Reversibility of a pathway or a particular enzyme-catalyzed reaction is dependent upon the rate of product removal and the turnover numbers of the enzymes in the pathway. If the enzymes in the pathway have high turnover numbers and the end product is a gas or is quickly removed, then the pathway may be physiologically unidirectional.

Multisubstrate Reactions

Most enzymes utilize more than one substrate, or they act upon one substrate plus a coenzyme and generate one or more products. In any case, a K_m must be determined for each substrate and coenzyme involved in the reaction when establishing an enzyme assay.

Mechanistically, enzyme reactions are divided into two major categories, ping-pong or sequential. There are many variations on these major mechanisms.

The *ping-pong mechanism* can be diagrammatically outlined as follows:

$$E + A \longrightarrow EA \xrightarrow{\overset{P_1}{\uparrow}} E' \xrightarrow{\overset{B}{\downarrow}} E'B \longrightarrow P_2 + E$$

in which substrate A reacts with E to produce product P_1, which is released before the second substrate B will bind to the modified enzyme E′. B is then converted to product P_2 and the enzyme regenerated. A good example of this mechanism is the transaminase catalyzed reaction (page 440) in which the α-amino group of amino acid$_1$ (A) is transferred to the enzyme and the newly formed keto acid$_1$ is released (P_1) followed by the binding of the acceptor keto acid$_2$ (B) and release of amino acid$_2$ (P_2). This reaction is schematically outlined in Figure 4.15.

In the *sequential mechanism,* if the two substrates A and B can bind in any order, it is a *random mechanism;* if the binding of A is required before B can be bound, then it is an *ordered mechanism*. In either case the reaction is bimolecular, that is, both A and B must be bound before reaction occurs. Examples of both these mechanisms can be found among the dehydrogenases in which the second substrate is the coenzyme (NAD, FAD, etc., page 119). The release of products may or may not be ordered in either mechanism.

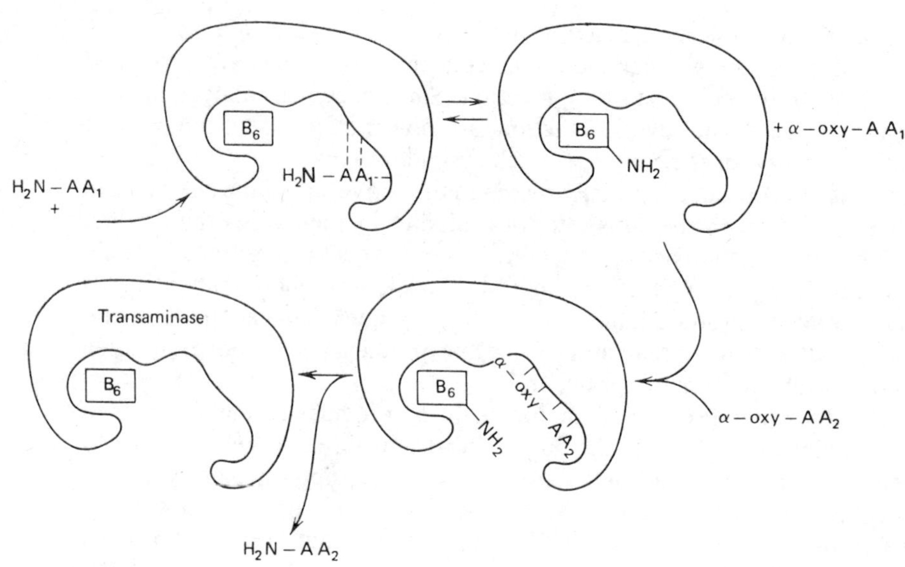

Figure 4.15
Schematic representation of the transaminase reaction mechanism—an example of a ping-pong mechanism.
Enzyme-bound vitamin B_6 coenzyme accepts the α-amino group from the first amino acid (AA_1), which is then released from the enzyme as an α-keto acid. The acceptor α-keto acid (AA_2) is then bound to the enzyme, and the bound amino group is transferred to it, forming a new amino acid which is then released from the enzyme. The terms "oxy" *and* "keto" *are used interchangeably.*

CLIN. CORR. **4.3**
MUTATION OF A COENZYME BINDING SITE RESULTS IN CLINICAL DISEASE

Cystathioninuria is a genetic disease in which the enzyme γ-cystathionase is either deficient or inactive. Cystathionase catalyzes the reaction:

Cystathionine ⟶ cysteine + α-ketobutyrate

Deficiency of the enzyme leads to accumulation of cystathionine in the plasma and mental retardation.

Since cystathionase is a pyridoxal phosphate-dependent enzyme, vitamin B_6 was administered to patients whose fibroblasts contained material that cross-reacts with antibody against cystathionase. Many responded to B_6 therapy with a fall in plasma levels of cystathionine. Such patients had the apoenzyme. In one particular patient the activity of the enzyme was undetectable in fibroblast homogenates but increased to 31% of normal with the addition of 1 mM pyridoxal phosphate to the assay mixture. It is thought that the K_m for pyridoxal phosphate binding to the enzyme was increased because of a mutation in the binding site. Activity is partially restored by increasing the concentration of coenzyme. Apparently these patients require a higher steady-state concentration of coenzyme to maintain any γ-cystathionase activity.

4.4 COENZYMES: STRUCTURE AND FUNCTION

Coenzymes function with the enzyme in the catalytic process. Often the coenzyme has an affinity for the enzyme that is similar to that of the substrate; consequently, the coenzyme can be considered a second substrate. In other cases, the coenzyme is covalently bound to the enzyme and functions at or near the active site in the catalytic event. There are other examples of enzymes where the role of the coenzyme falls between these two extremes.

Several, but not all, of the coenzymes are synthesized from the B vitamins. Vitamin B_6, pyridoxine, requires little modification to be transformed to the active coenzyme, pyridoxal phosphate (page 440). Clin. Corr. 4.3 points out the importance of the coenzyme binding site and how alterations in this site cause metabolic dysfunction.

In contrast to vitamin B_6, niacin requires major alteration by the mammalian cell before it is capable of acting as a coenzyme. This metabolic interconversion is outlined on page 518.

The structure and function of the coenzymes of only two B vitamins, niacin and riboflavin, and of ATP will be discussed in this chapter. The structure and function of coenzyme A (page 522), thiamine (page 224), biotin, and vitamin B_{12} are included in those chapters dealing with enzymes dependent upon the given coenzyme for activity.

Adenosine Triphosphate

Adenosine triphosphate (ATP) often functions as a second substrate but can also serve as a cofactor in modulating the activity of specific enzymes. This compound is so pivotal that its structure and function will be

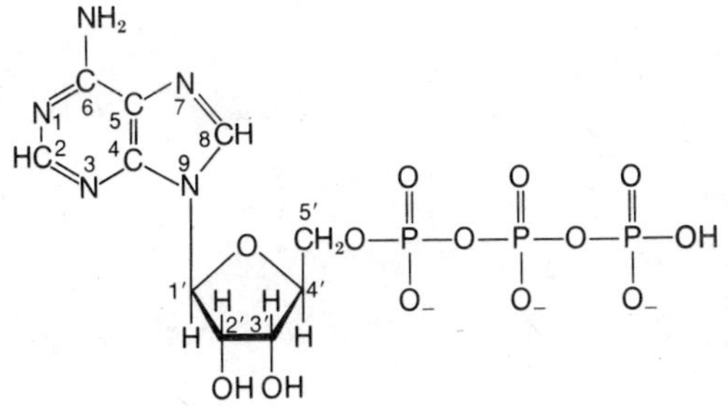

Figure 4.16
Adenosine triphosphate (ATP).

introduced here. ATP (Figure 4.16) can be synthesized de novo in all mammalian cells.

The nitrogenous heterocyclic ring is adenine. To the adenine is affixed a ribosyl 5′-triphosphate. The functional end of the molecule is the reactive triphosphate which is shown in the ionization state found in the cell. As a cosubstrate ATP is utilized by the kinases for the transfer of the terminal phosphate to various acceptors. A typical example is the glucokinase-catalyzed reaction:

$$\text{Glucose} + \text{ATP} \longrightarrow \text{glucose 6-phosphate} + \text{ADP}$$

ADP is adenosine diphosphate. The combination of adenine plus ribose is adenosine (page 492).

ATP has an additional role, other than cosubstrate: in a number of specific enzyme reactions it serves as a modulator of the activity of the enzyme. These particular enzymes have binding sites for ATP, occupancy of which changes the affinity or reactivity of the enzyme toward its substrates. Mechanistically, ATP is acting as an allosteric effector in these cases (page 144).

Coenzymes of Niacin

Niacin is pyridine-3-carboxylic acid. It is converted to two major coenzymes which are involved in the oxidoreductase class of enzymes. These coenzymes are NAD, nicotinamide adenine dinucleotide, and NADP, nicotinamide adenine dinucleotide phosphate. There are dehydrogenases that function with NADP as coenzyme but not with NAD. The reverse is also true and some enzymes function with either coenzyme. Such an arrangement allows for specificity and control over dehydrogenases that reside in the same subcellular compartment.

Structurally, NAD is composed of adenosine and *N*-ribosyl-nicotinamide linked through a pyrophosphate linkage between the 5′-hydroxyls of the two ribosyl moieties (Figure 4.17). NADP differs structurally from NAD in having an additional phosphate esterified to the 2′-hydroxyl of the adenosine moiety.

Both coenzymes function as intermediates in the transfer of two electrons between an electron donor and an acceptor. The donor and acceptor need not be involved in the same metabolic pathway. In other words, the reduced form of these nucleotides acts as a common "pool" of electrons that arise from many oxidative reactions and can be used for various reductive reactions.

The adenine, ribose, and pyrophosphate components of NAD are involved in the binding of NAD to the enzyme. Enzymes requiring NADP have a specific cationic region in their NADP binding site that is posi-

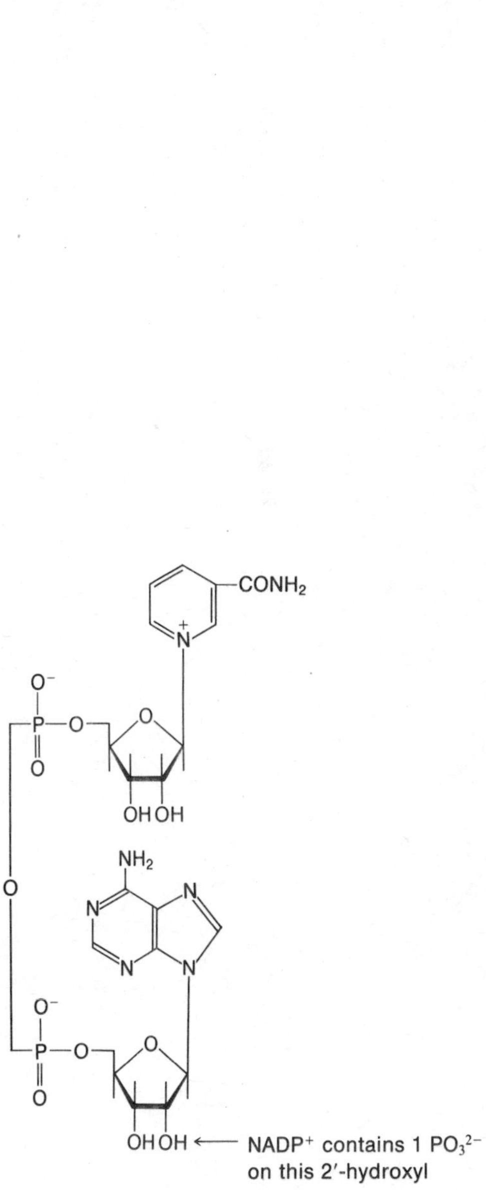

NADP$^+$ contains 1 PO_3^{2-} on this 2′-hydroxyl

Figure 4.17
Nicotinamide adenine dinucleotide (NAD$^+$).

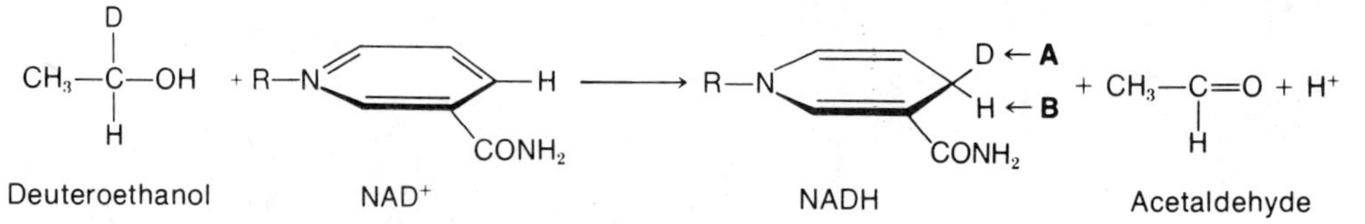

$$CH_3-\underset{H}{\overset{D}{C}}-OH + NAD^+ \longrightarrow NADH + CH_3-\underset{H}{C}=O + H^+$$

Deuteroethanol NAD+ NADH Acetaldehyde

Figure 4.18
Transfer of deuterium from deuterated ethanol to NAD⁺.

tioned so as to form an ionic bond with the 2′-phosphate of NADP. This enhances binding of NADP in preference to NAD in these particular enzymes.

The nicotinamide portion of the molecule is involved in reversibly accepting and donating two electrons at a time. It is the active center of the coenzyme. In the oxidation of deuterated ethanol by alcohol dehydrogenase, NAD^+ accepts two electrons and one deuterium from the ethanol, and the other hydrogen is released as a proton (Figure 4.18).

The specific binding of NAD^+ to the enzyme surface confers a chemically recognizable "topside" and "bottom side" to the planar nicotinamide. The former is known as the **A** face and the latter the **B** face. In the case of alcohol dehydrogenase, the proton or the deuterium ion which serves as a tracer is added to the **A** face. Other dehydrogenases utilize the **B** face. The particular effect just described demonstrates how enzymes are able to induce specificity into chemical reactions by virtue of the asymmetric binding of coenzymes and substrates.

Coenzymes of Riboflavin

The two coenzyme forms of riboflavin are FMN, riboflavin 5′-phosphate, and FAD (flavin adenine dinucleotide). The vitamin riboflavin consists of the heterocyclic ring, isoalloxazine (flavin) connected through N-10 to the alcohol ribitol as shown in Figure 4.19.

FMN has a phosphate esterified to the 5′-hydroxyl of riboflavin. FAD is structurally analogous to NAD in having adenosine linked through a pyrophosphate linkage to a heterocyclic ring, in this case riboflavin.

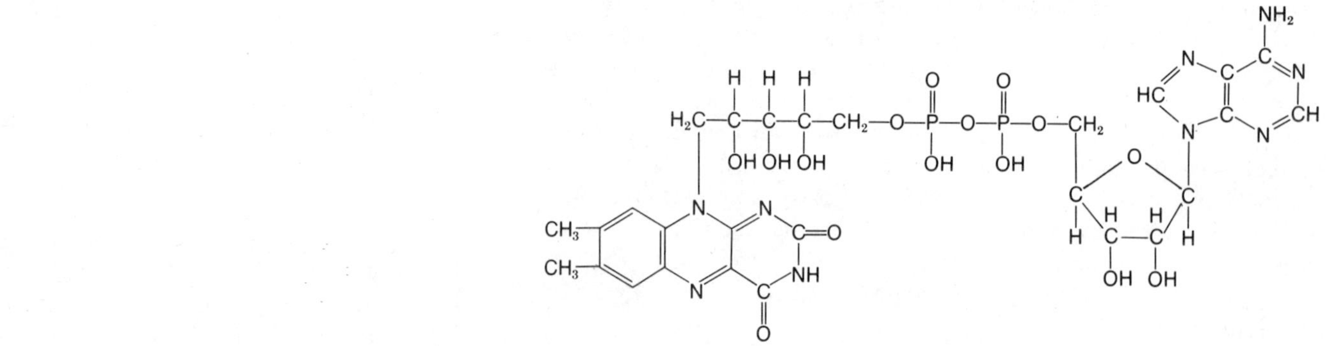

Flavin adenine dinucleotide (FAD)

FAD and FMN function in oxidation–reduction reactions by accepting and donating two electrons through the isoalloxazine ring. A typical example of FAD participation in an enzyme reaction is the oxidation of succinate to fumarate by succinate dehydrogenase (page 232).

In some cases, these coenzymes are one-electron acceptors, which lead to flavin semiquinone formation (a free radical).

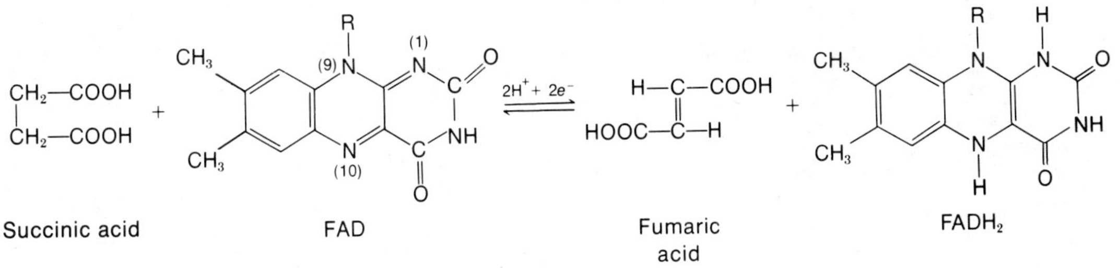

Succinic acid + FAD $\underset{}{\overset{2H^+ + 2e^-}{\rightleftharpoons}}$ Fumaric acid + $FADH_2$

There is a tendency for the flavin coenzymes to be bound much tighter to their enzymes than are the niacin coenzymes, and thus they may often function as prosthetic groups rather than as cofactors.

Metals as Cofactors

Metals are not coenzymes in the sense of FAD and NAD, but are required as cofactors in approximately two-thirds of all enzymes. There are two major areas in which metals participate in enzyme reactions—through their ability to act as *Lewis acids* and through various modes of *chelate* formation. Chelates are organometallic coordination complexes. A good example of a chelate is the complex between iron and porphyrin to form a heme (page 862).

Those metals that act as Lewis acid catalysts are found among the transition metals like Zn, Fe, Mn, and Cu, which have empty *d* electron orbitals that can act as electron sinks. The alkaline earth metals such as K and Na do not possess this ability.

A good example of a metal functioning as a Lewis acid is found in carbonic anhydrase. Carbonic anhydrase is a zinc enzyme that catalyzes the reaction

$$CO_2 + H_2O \rightleftharpoons H_2CO_3$$

The first step can be visualized as the in situ generation of a proton and a hydroxyl group from water:

$$ENZ—Zn^{2+} + H_2O \rightleftharpoons ENZ—Zn^{2+}---{}^{-}OH + H^+$$

The proton and hydroxyl are subsequently added to the carbon dioxide and carbonic acid is released. The reactions are presented in a stepwise

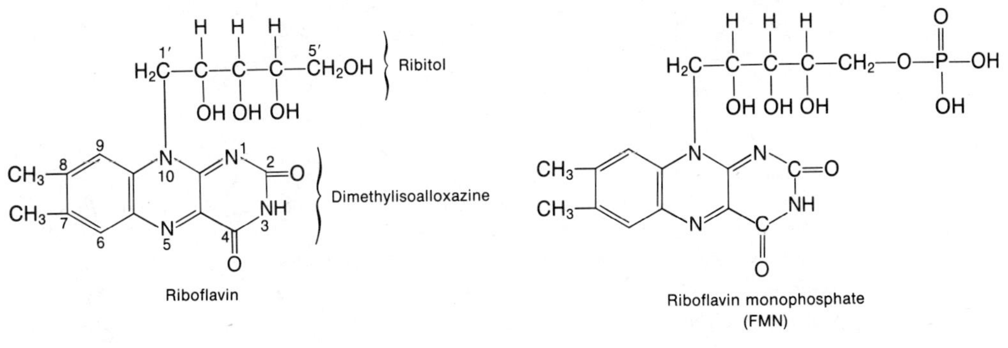

Riboflavin (Ribitol; Dimethylisoalloxazine)

Riboflavin monophosphate (FMN)

Figure 4.19
Riboflavin and riboflavin monophosphate.

fashion for clarity. Actually, the reactions may occur in a concerted fashion, that is, all at one time.

$$\text{ENZ—Zn}^{2+}\text{---}^{-}\text{O—H} + \text{H}^+ + \text{O=C=O} \rightleftharpoons \text{ENZ—Zn}^{2+}\text{---}^{-}\text{O—H} \;(\text{O=C=O}) + \text{H}^+ \rightleftharpoons$$

$$\text{ENZ—Zn}^{2+}\text{---O(H)—C(=O)—OH} \rightleftharpoons \text{ENZ—Zn}^{2+} + \text{H}_2\text{CO}_3$$

The metal can also promote catalysis by binding substrate at the site of the bond cleavage. In carboxypeptidase, the carbonyl oxygen atom is chelated to the zinc. The resulting flow of electrons from the carbonyl carbon to the electropositive metal increases the susceptibility of the peptide bond to cleavage by nucleophiles such as water or carboxylate. This is schematically shown in Figure 4.20.

Role of the Metal as a Structural Element

The functioning of a metal as a Lewis acid requires chelate formation. In addition, various modes of chelation occur between metal, enzyme, and substrate that are structural in nature, but in which no acid catalysis occurs.

In several of the kinases, creatine kinase being the best example, the true substrate is not ATP but Mg^{2+}-ATP (Figure 4.21).

In this case, the magnesium does not interact directly with the enzyme. It may serve to neutralize the negative charge density on ATP and facilitate binding to the enzyme. Ternary complexes of this conformation are

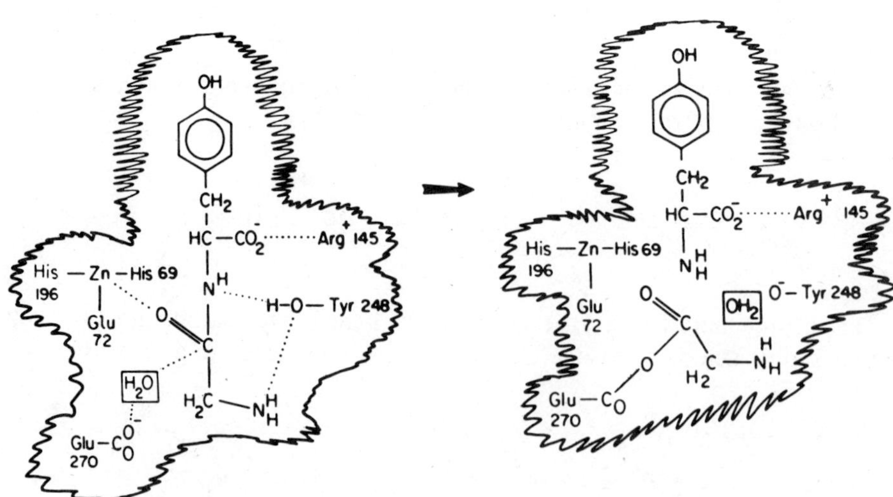

Figure 4.20
The role of zinc in carboxypeptidase A.
Enzyme-bound zinc polarizes the peptide carbonyl making the carbonyl carbon more positive and susceptible to nucleophilic attack by glutamic acid residue 270 in the active site. The end result is cleavage of the peptide bond, release of a new amino-terminal peptide and covalent addition of the remainder of the substrate to the enzyme through anhydride linkage. The latter is then released by water hydrolysis.

From W. Lipscomb, J. Hartsuck, F. Qurocho and G. Reeke, Jr., *Proc. Natl. Acad. Sci. USA* 64:39, 1969. Reprinted with permission.

known as "substrate-bridged" complexes and can be schematically represented as Enz—S—M. A hypothetical scheme for the binding of Mg–ATP and glucose in the active site of hexokinase is shown in Figure 4.22. All the kinases except muscle pyruvate kinase and phospho*enol* pyruvate carboxykinase are substrate-bridged complexes.

In pyruvate kinase Mg^{2+} serves to chelate the ATP to the enzyme as shown in Figure 4.23. The absence of the metal cofactor results in failure of the ATP to bind to the enzyme. Enzymes of this class are "metal-bridged" ternary complexes, Enz—M—S. All metalloenzymes are of this type. *Metalloenzymes* are enzymes containing a tightly bound transition metal such as Zn^{2+} or Fe^{2+}. Several enzymes catalyzing enolization and elimination reactions are metal-bridge complexes.

In addition to the role of binding enzyme and substrate, metals may also bind directly to the enzyme to stabilize it in the active conformation or perhaps to induce the formation of a binding site or active site. Not only do the strongly chelated metals like Mn^{2+} play a role in this regard, but the weakly bound alkali metals (Na^+, K^+) are also important. In pyruvate kinase, K^+ has been found to induce an initial conformation change, which is necessary, but not sufficient, for the ternary complex formation. Upon substrate binding, K^+ induces a second conformational change to the catalytically active ternary complex as indicated in Figure 4.23. In more general terms it is thought that Na^+ and K^+ stabilize the active conformation of the enzyme, but are passive from the catalytic standpoint.

Role of Metals in Oxidation and Reduction

The iron-sulfur enzymes, often referred to as *nonheme iron proteins*, are a unique class of metalloenzymes in which the active center consists of one or more clusters of sulfur-bridged iron chelates. These are of greater prominence in bacterial and plant systems than in mammalian cells. In

Figure 4.21
Mg^{2+}-ATP.

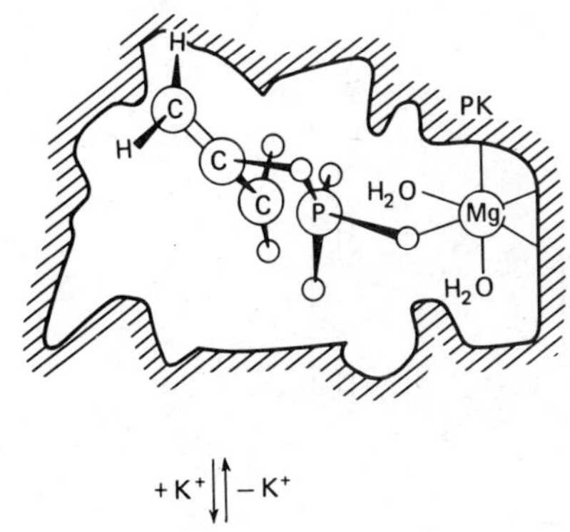

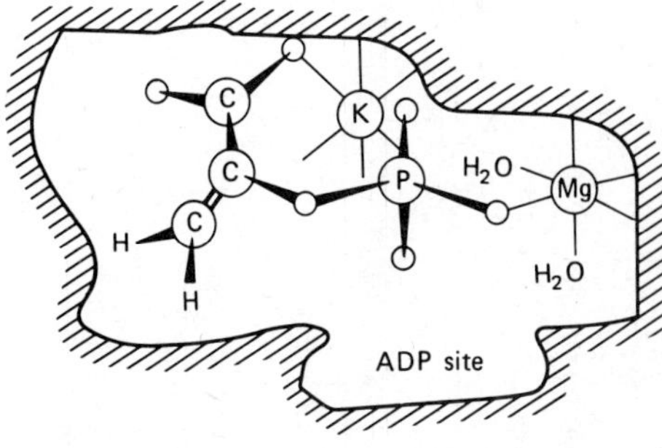

Figure 4.23
Model of the role of potassium ion in the active site of pyruvate kinase.
*Pyruvate kinase catalyzes the reaction: phospho*enol*pyruvate + ADP → ATP + pyruvate. Initial binding of K^+ induces conformational changes in the kinase, which result in increased affinity for phospho*enol*pyruvate. In addition, K^+ orients the phospho*enol*pyruvate in the correct position for transfer of its phosphate to the second substrate, which is ADP. Magnesium coordinates the substrate to the enzyme active site.*

Modified, with permission, from A. S. Mildvan, *Ann. Rev. Biochem.* 43:365, 1974. © Annual Reviews, Inc.

Figure 4.22
Model of the role of magnesium as a substrate-bridged complex in the active site of the kinases.
In hexokinase the terminal phosphate of ATP is transferred to glucose, yielding glucose 6-phosphate. Magnesium coordinates with the ATP to form the true substrate and in addition may labilize the terminal P—O bond of ATP to facilitate transfer of the phosphate to glucose. There are specific binding sites in the active site for glucose (upper left) as well as the adenine and ribose moieties of ATP.

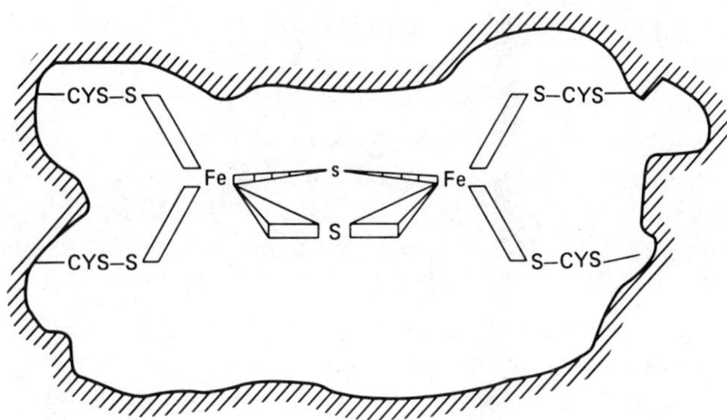

Figure 4.24
The iron binding site of adrenodoxin.
Two iron atoms are chelated to the protein via cysteine sulfhydryl groups. The two iron atoms are bridged by sulfides, which are released as hydrogen sulfide upon acidification of the protein. A formal valence state cannot be assigned to the iron atoms because they are magnetically coupled.

Reprinted, with permission, from W. Orme-Johnson in H. Sigel, ed., *Metal Ions in Biological Systems,* 7:129, 1979 Marcel Dekker, Inc.

mammalian systems succinate dehydrogenase (page 232), NADH dehydrogenase, and adrenodoxin are good representatives of this group of proteins. The structure of the iron chelate in these nonheme iron proteins is represented in Figure 4.24.

In these proteins the bridging sulfide is released as H_2S on acidifying the enzyme. Cysteine thiol groups from the enzyme hold the bridged binuclear iron complex in the enzyme. These particular enzymes have reasonably low reducing potentials (E_0') and function in electron transfer reactions. Adrenodoxin functions in the activation of oxygen in the steroid 11β- and 18-hydroxylases as a cosubstrate. It is not an enzyme.

The cytochromes, which are heme iron proteins, also function as cosubstrates for their respective reductases (page 248). The iron in the

$$\text{ENZ}\begin{matrix}\diagup Cu^{2+}\\ \diagdown Cu^{2+}\end{matrix} + \text{Ascorbate} \rightleftharpoons \text{ENZ}\begin{matrix}\diagup Cu^{+}\\ \diagdown Cu^{+}\end{matrix} + \text{Dehydroascorbate} + 2H^+$$

$$\text{ENZ}\begin{matrix}\diagup Cu^{+}\\ \diagdown Cu^{+}\end{matrix} + O_2 \rightleftharpoons \text{ENZ}\begin{matrix}\diagup Cu^{2+}\ {}^-O\\ \diagdown Cu^{2+}\ {}^-O\end{matrix}$$

$$\text{ENZ}\begin{matrix}\diagup Cu^{2+}\ {}^-O\\ \diagdown Cu^{2+}\ {}^-O\end{matrix} + \underset{\text{Dopamine}}{(HO)_2C_6H_3\text{–}CH_2\text{–}CH_2NH_2} \rightleftharpoons \text{ENZ}\begin{matrix}\diagup Cu^{2+}\\ \diagdown Cu^{2+}\end{matrix} + \underset{\text{Norepinephrine}}{(HO)_2C_6H_3\text{–}CH(OH)\text{–}CH_2NH_2} + H_2O$$

Figure 4.25
Role of copper in the activation of molecular oxygen by dopamine hydroxylase.
The normal cupric form of the enzyme is not reactive with oxygen but on reduction by the cosubstrate, ascorbate, generates the very reactive enzyme bound superoxide anion, $O_2^{-}\cdot$, in the presence of oxygen. The superoxide anion then reacts with dopamine to generate the product, norepinephrine, and the inactive cupric enzyme.

hemes of the cytochromes undergoes reversible one-electron transfers. In addition, the heme is bound to the enzyme through coordination of an amino acid side chain to the iron of the heme. Thus, in the cytochromes the metal serves not only a structural role, but also participates in the chemical event.

The last role that we will discuss here is the role of metals, specifically copper and iron, in activation of molecular oxygen. Copper is an active participant in several oxidase and hydroxylase enzymes. For example, dopamine β-hydroxylase catalyzes the introduction of one oxygen atom from O_2 into dopamine to form norepinephrine as shown in Figure 4.25. It is thought that the active form of the enzyme contains two atoms of cuprous ion that react with O_2 to form the reactive species $O_2^{-}\cdot$, superoxide anion, which then attacks the dopamine. In other metalloenzymes other species of "active" oxygen are generated.

4.5 INHIBITION OF ENZYMES

Mention was made earlier of product inhibition of enzyme activity and how an entire pathway could be controlled or modulated by this mechanism (page 131). In addition to inhibition by the immediate product, products of other enzymes can also inhibit or even activate a particular enzyme. Much of current drug therapy is based on inhibition of specific enzymes with a substrate analog. Therefore, it is important to discuss inhibition in more detail. Basically, there are three major classes of inhibitors: competitive, noncompetitive, and uncompetitive.

Competitive Inhibitors

Competitive inhibitors are defined as inhibitors whose action can be reversed by increasing amounts of substrate. Competitive inhibitors are usually enough like the substrate structurally that they bind at the substrate binding site and compete with the substrate for the enzyme. Once bound, the enzyme cannot convert the inhibitor to products. Increasing substrate concentrations will displace the reversibly bound inhibitor by the law of mass action. A competitive inhibitor need not be structurally related to the substrate.

In the succinate dehydrogenase reaction, malonate is structurally similar to succinate and is a competitive inhibitor (Figure 4.26).

COO^-–CH_2–CH_2–COO^- Succinate

COO^-–CH_2–COO^- Malonate

Figure 4.26
Substrate and inhibitor of succinate dehydrogenase.

Since the substrate and inhibitor are competing for the same site on the enzyme, the K_m for the substrate shows an apparent increase in the presence of inhibitor. This can be seen in a double-reciprocal plot as a shift in the X intercept $(-1/K_m)$ and in the slope of the line (K_m/V_{max}). If we first establish the velocity at several levels of substrate and then repeat the experiment with a given but constant amount of inhibitor at various substrate levels, two different straight lines will be obtained as shown in Figure 4.27. As can be seen, the V_{max} does not change; hence the intercept on the Y axis remains the same. In the presence of inhibitor, the X intercept is no longer the negative reciprocal of the true K_m, but of $K_{m_{app}}$ where

$$K_{m_{app}} = K_m \cdot \left[1 + \frac{[I]}{K_I}\right]$$

Thus the inhibitor constant, K_I, can be determined from the concentration of inhibitor [I] used and the K_m, which was obtained from the X intercept of the line, obtained in the absence of inhibitor.

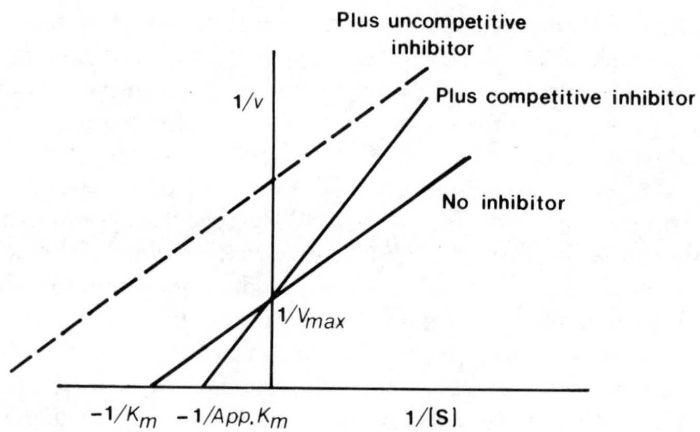

Figure 4.27
Double reciprocal plots for competitive and uncompetitive inhibition.
A competitive inhibitor binds at the substrate binding site and effectively increases the K_m *for the substrate. An uncompetitive inhibitor causes an equivalent shift in both* V_{max} *and* K_m*, resulting in a line parallel to that given by the uninhibited enzyme.*

Noncompetitive Inhibitors

A *noncompetitive* inhibitor binds at a site other than the substrate binding site. The inhibition is not reversed by increasing concentrations of substrate. Both binary, EI, and ternary, EIS, complexes form, both of which are catalytically inactive and are therefore, dead-end complexes. The noncompetitive inhibitor behaves as though it were removing active enzyme from the solution, resulting in a decrease in V_{max}. This effect is seen graphically in the double-reciprocal plot (Figure 4.28), where K_m does not change but V_{max} does change. Inhibition can often be reversed by exhaustive dialysis of the inhibited enzyme provided that the inhibitor has not reacted covalently with the enzyme. This case is considered under the irreversible inhibitors and is discussed below.

The *uncompetitive* inhibitor binds only with the ES form of the enzyme in the case of a one-substrate enzyme. The result is an apparent equivalent change in K_m and V_{max}, which is reflected in the double reciprocal

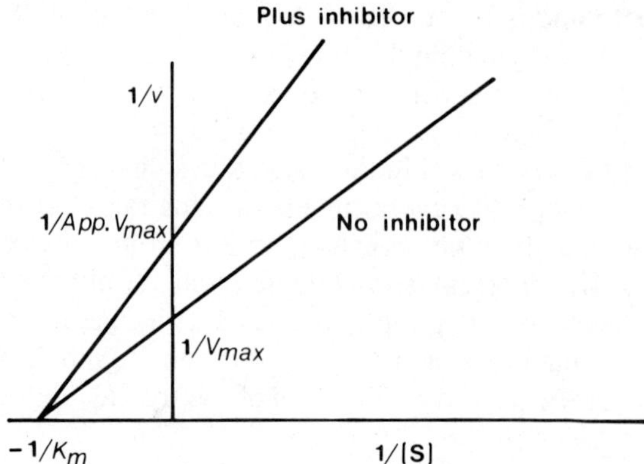

Figure 4.28
Double reciprocal plot for an enzyme subject to reversible noncompetitive inhibition.
A noncompetitive inhibitor binds at a site other than the substrate binding site; therefore, the effective K_m *does not change, but the apparent* V_{max} *decreases.*

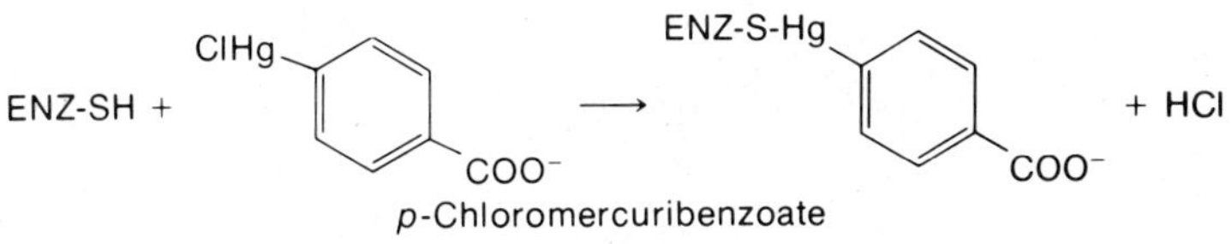

$$\text{ENZ-SH} + \text{ClHg-}C_6H_4\text{-}COO^- \longrightarrow \text{ENZ-S-Hg-}C_6H_4\text{-}COO^- + \text{HCl}$$

p-Chloromercuribenzoate

Figure 4.29
Enzyme inhibition by a covalent modification of an active center cysteine.

plot as a line parallel to that of the uninhibited enzyme (Figure 4.27). In the case of multisubstrate enzymes the interpretation is complex and will not be considered further.

Irreversible Inhibitors

In cases of covalent modification of the binding site or the active site, inhibition will not be reversed by dialysis unless the linkage is chemically labile like that of an ester or thioester. The active site thiol in glyceraldehyde 3-phosphate dehydrogenase reacts with *p*-chloromercuribenzoate to form a mercuribenzoate adduct of the enzyme as shown in Figure 4.29. Such adducts are not reversed by dialysis or by addition of substrate. Double reciprocal plots show a characteristic pattern for noncompetitive inhibition (Figure 4.28).

Drugs and Enzyme Inhibition

Most if not all of modern drug therapy is based on the concepts of enzyme inhibition that were covered in the previous section.

Drugs are designed with a view toward inhibiting a specific enzyme in a specific metabolic pathway. This application is most easily appreciated with the antiviral, antibacterial, and antitumor drugs, which are administered to the patient under conditions of limited toxicity. Such toxicity to the patient is often unavoidable because, with the exception of cell wall biosynthesis in bacteria, there are few critical metabolic pathways that are unique to tumors, viruses, or bacteria. Hence, drugs that will kill these organisms will often kill the host cells. The one characteristic that can be taken advantage of is the comparatively short generation time of the undesirable organisms. They are much more sensitive to antimetabolites and in particular those that inhibit enzymes involved in replication. *Antimetabolites* are compounds with some structural deviation from the natural substrate. In the chapters on metabolism, numerous examples of antimetabolites will be brought to your attention. Here we will present only a few examples which illustrate the concept.

Sulfa Drugs

Modern chemotherapy had its beginning in these compounds whose general formula is R—SO_2—NHR′. Sulfanilamide is the simplest member of the class and is an antibacterial agent because of its competition with *p*-aminobenzoic acid, which is required for bacterial growth. Structures of these compounds are shown in Figure 4.30.

$$H_2N\text{-}C_6H_4\text{-}SO_2NH_2 \qquad H_2N\text{-}C_6H_4\text{-}COOH$$

Sulfanilamide   p-Aminobenzoic Acid

Figure 4.30
Structure of p-aminobenzoic acid and sulfanilamide, a competitive inhibitor.

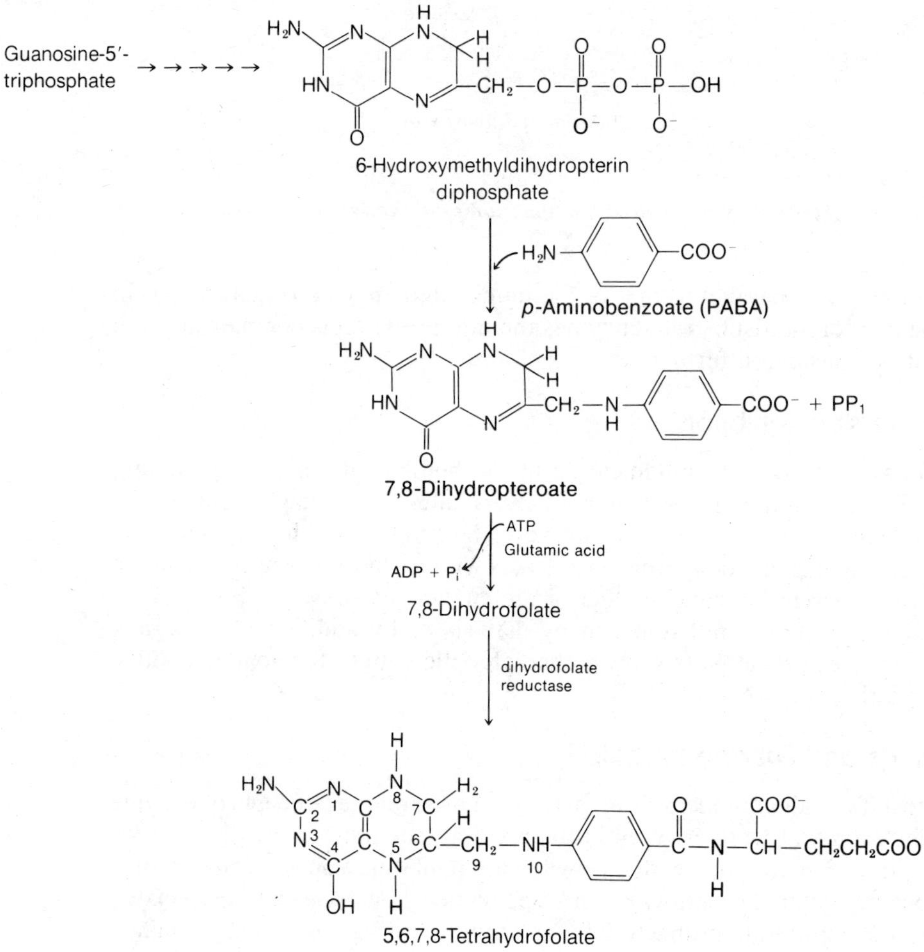

Guanosine-5′-triphosphate → → → → → 6-Hydroxymethyldihydropterin diphosphate

↓ H_2N–C_6H_4–COO^- *p*-Aminobenzoate (PABA)

7,8-Dihydropteroate + PP_i

↓ ATP, Glutamic acid → ADP + P_i

7,8-Dihydrofolate

↓ dihydrofolate reductase

5,6,7,8-Tetrahydrofolate

Figure 4.31
The route of synthesis of folate in bacteria.
p-Aminobenzoate (PABA) is a required cofactor for the synthesis of dihydrofolate. Sulfanilamide is a competitive inhibitor of PABA and is incorporated into a metabolically inactive, 7,8-dihydropteroylsulfonamide.

It is now known that bacteria cannot absorb folic acid, a required vitamin, from the host, but must synthesize it. The synthesis of folate involves the series of reactions shown in Figure 4.31.

Since sulfanilamide is a structural analog of *p*-aminobenzoate, the enzyme dihydropteroate synthetase is tricked into making a dihydropteroate containing sulfanilamide that cannot be converted to folate. Thus the bacterium is starved of the required folate and cannot grow or divide. Since man requires folate from external sources, the sulfanilamide is not harmful at the doses that will kill bacteria.

Methotrexate

The biosynthesis of purines and pyrimidines, heterocyclic bases employed in the synthesis of RNA and DNA, requires folic acid, which serves as a coenzyme in the transfer of one-carbon units from various amino acid donors (page 476).

Methotrexate (Figure 4.32) is a structural analog of folate. It has been used with great success in childhood leukemia. Its mechanism of action is based on competition with dihydrofolate for the dihydrofolate reductase. It binds 1,000-fold better than the natural substrate and is a powerful competitive inhibitor of the enzyme. This being the case, the synthesis of

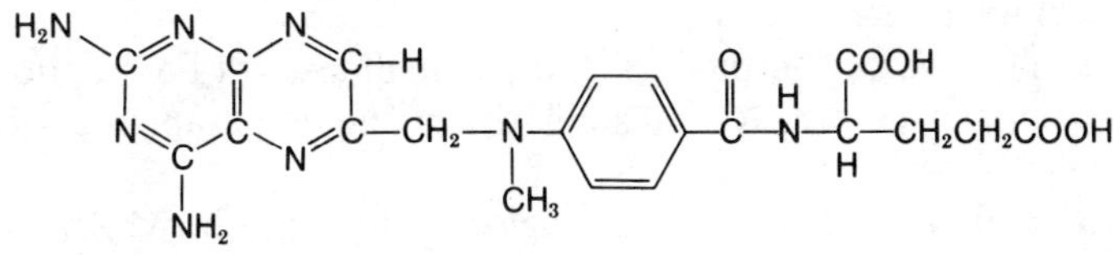

Figure 4.32
Methotrexate (4-amino-N^{10}-methyl folic acid).

thymidine monophosphate stops in the presence of methotrexate because of failure of the one-carbon metabolic system. Since cell division is dependent on thymidine monophosphate as well as the other nucleotides, the leukemia cell cannot multiply. One problem is that rapidly dividing human cells such as those in bone marrow are sensitive to the drug for the same reasons. Also, prolonged usage stimulates the tumor cells to produce larger amounts of the reductase, thus becoming resistant to the drug.

Nonclassical Antimetabolites

These compounds are also known as suicide substrates or active site directed inhibitors. The compounds are constructed so that they have an affinity for the active site, but in addition have a chemically reactive group that will form a covalent adduct with a reactive amino acid in the active site of the enzyme. Thus the compounds are specific for a particular enzyme and the inhibition is irreversible. For example, the compound shown in Figure 4.33 is a suicide substrate for the dihydrofolate reductase because the compound structurally resembles dihydrofolate and is specifically bound at the active site where the reactive benzylsulfonyl fluoride is positioned so as to react with a serine hydroxyl in the substrate binding site. Covalent binding of this suicide substrate to the enzyme prevents binding of the normal substrate and leads to inhibition of the enzyme.

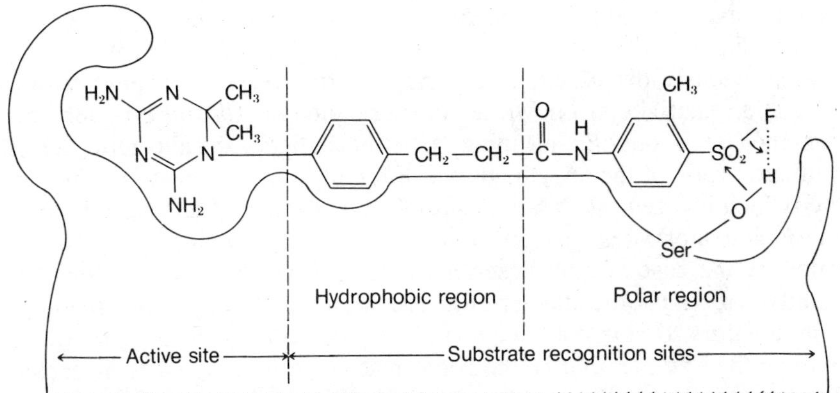

Figure 4.33
Suicide substrate inactivation of tetrahydrofolate reductase.
The suicide substrate, a substituted dihydrotriazine, structurally resembles dihydrofolate and binds specifically to the dihydrofolate site on dihydrofolate reductase. The triazine portion of the suicide substrate resembles the pterin moiety and therefore, binds to the active site. The ethylbenzene group binds to the hydrophobic site normally occupied by the p-aminobenzoyl group. The reactive end of the suicide substrate contains a reactive sulfonyl fluoride that forms a covalent linkage with a serine hydroxyl on the enzyme surface. Thus the suicide substrate irreversibly inhibits the enzyme by blocking access of dihydrofolate to the active site.

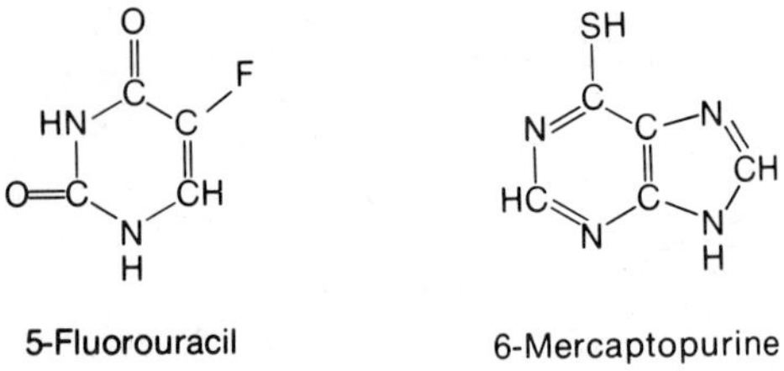

Figure 4.34
Structures of two antimetabolites.

Other Antimetabolites

Two other analogs of the purines and pyrimidines will be mentioned in order to emphasize the structural similarity of chemotherapeutic agents to normal substrates.

Fluorouracil (Figure 4.34) is an analog of thymine in which the ring bound methyl is substituted by fluorine. The deoxynucleotide of this compound is an irreversible inhibitor of the enzyme thymidylate synthetase.

6-Mercaptopurine (Figure 4.34) is an analog of hypoxanthine and therefore of adenine and guanine. 6-Mercaptopurine is a broad-spectrum antimetabolite because of its competition in most reactions involving adenine and guanine or their derivatives.

The antimetabolites discussed have been related to purine and pyrimidine metabolism. However, the general concepts developed here can be applied to any enzyme or metabolic pathway.

4.6 ALLOSTERIC CONTROL OF ENZYME ACTIVITY

Allostery and Cooperativity

Although the substrate binding site and the active site of an enzyme are well-defined structures, the activity of many enzymes can be modulated by ligands acting in ways other than as competitive or noncompetitive inhibitors. A *ligand* is any molecule that is bound to a macromolecule; the term is not limited to small organic molecules such as ATP, but is extended to low molecular weight proteins. Ligands can be activators, inhibitors, or even the substrates of enzymes. Those ligands that cause a change in enzymatic activity, but are unchanged as a result of enzyme action, are referred to as *effectors, modifiers,* or *modulators.* Most of the enzymes subject to modulation by ligands are rate-determining enzymes in metabolic pathways. In order to appreciate the mechanisms of control of metabolic pathways, the principles governing the allosteric and cooperative behavior of individual enzymes must be understood.

In addition to the substrate binding site and the active site, which we have previously discussed, those enzymes that respond to modulators have additional site(s) known as allosteric site(s). *Allosteric* is derived from the Greek root *allo,* meaning "the other"; hence the allosteric site is a unique region of the enzyme that is different from the substrate binding site (which is often the catalytic site). The existence of allosteric inhibitor or activator sites distinct from the substrate binding site is illustrated by the case of gout described in Clin. Corr. 4.4. The modulating ligands that bind at the allosteric site are known as the allosteric effectors or modulators. Binding of the allosteric effector causes an allosteric transition in the enzyme, that is, the conformation of the enzyme changes, so that the affinity for the substrate or other ligand changes. Positive (+) allosteric effectors increase the enzyme affinity for substrate or other ligand. The reverse is true for negative allosteric effectors. The allosteric

CLIN. CORR. **4.4**
A CASE OF GOUT DEMONSTRATES THE DIFFERENCE BETWEEN AN ALLOSTERIC SITE AND THE SUBSTRATE BINDING SITE

The realization that allosteric inhibitory sites are separate from activator sites as well as from the substrate binding and the catalytic sites is illustrated by a study of a gouty patient whose red blood cell PRPP levels are increased. It was found that the patient's PRPP synthetase had a normal K_m, V_{max} and sensitivity to activation by inorganic phosphate. However, it displayed reduced sensitivity to the normal allosteric inhibitors AMP, ADP, GDP, and 2,3-diphosphoglycerate (2,3-DPG). The increased PRPP levels and hyperuricemia arose because endogenous end products of the pathway (ATP, GTP) were not able to control the activity of the synthase as is normally the case. Apparently mutation had resulted in a change in the inhibitory site (I) or in the coupling mechanism between the I and the catalytic site (C). The cosubstrate binding sites (S) and the phosphate activator (A) site are shown for reference.

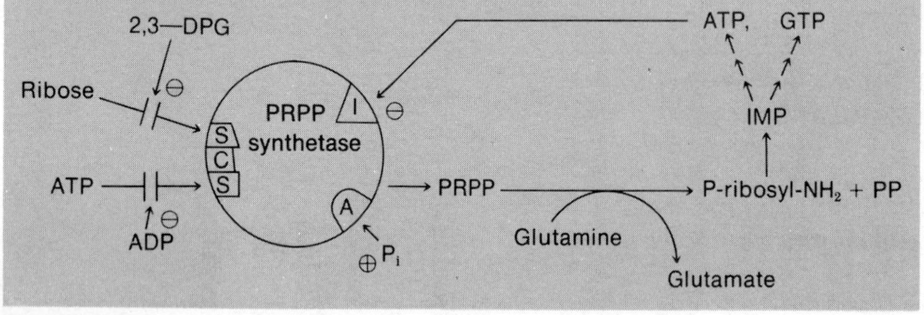

site at which the positive effector binds is referred to as the activator site. The negative effector binds at an inhibitory site.

Allosteric enzymes are divided into two classes based on the effect of the allosteric effector on the K_m and V_{max}. In the K class the effector alters the K_m but not the V_{max}, whereas in the V class the effector alters the V_{max} but not the K_m. K class enzymes give double-reciprocal plots like those given by competitive inhibitors (Figure 4.27) and V class enzymes give double-reciprocal plots like those of noncompetitive inhibitors (Figure 4.28). However, it is inappropriate to use the terms competitive and noncompetitive with allosteric enzyme systems because the mechanism of the effect of an allosteric inhibitor on a V or K enzyme is quite different from the mechanism of a simple competitive or noncompetitive inhibitor. For example, in the K class the inhibitor binds at an allosteric site, which then affects the affinity of the substrate binding site for the substrate, whereas in simple competitive inhibition the inhibitor competes with substrate for the substrate binding site. In the V class enzymes, positive and negative allosteric modifiers increase or decrease the rate of breakdown of the ES complex to products, that is, the catalytic rate constant, k_3, is affected and not the substrate binding constant. There are a few enzymes in which both K_m and V_{max} are affected.

In theory a monomeric enzyme can undergo an allosteric transition in response to a modulating ligand. In practice only two monomeric allosteric enzymes have been found, ribonucleoside diphosphate reductase and pyruvate-UDP-*N*-acetylglucosamine transferase. Most allosteric enzymes are oligomeric, that is, they consist of several subunits. The identical subunits are designated as *protomers*. Each protomer may consist of one or more polypeptide chains. As a consequence of the oligomeric nature of allosteric enzymes, binding of ligand to one protomer can affect the binding of ligands on other protomers in the oligomer. Such ligand effects are referred to as homotropic interactions. The transmission of the homotropic effects between protomers is one aspect of cooperativity, considered in detail later in this chapter. Substrate influencing substrate, activator influencing activator, or inhibitor influencing inhibitor binding are homotropic interactions. Homotropic interactions are almost always positive.

A *heterotropic interaction* is defined as the effect of one ligand on the binding of a different ligand. For example, the effect of an allosteric inhibitor on the binding of substrate or the effect of an allosteric inhibitor on the binding of an allosteric activator are heterotropic interactions. Heterotropic interactions can be either positive or negative and can occur in monomeric allosteric enzymes. Both heterotropic and homotropic effects, in an oligomeric enzyme, are mediated by cooperativity between subunits.

Based on the foregoing descriptions of allosteric enzymes, two models are pictured in Figure 4.35. In panel *a* a model for a monomeric enzyme is shown, and in panel *b* a model for an oligomeric enzyme consisting of two protomers is visualized. In both models heterotropic interactions can occur between the activator and substrate sites. In model *b*, homotropic interactions can occur between the activator sites or between the substrate sites.

Kinetics of Allosteric Enzymes

As a consequence of the interaction between the substrate site, the activator site, and the inhibitor site, a characteristic sigmoid or S-shaped curve, as shown in Figure 4.36*a* (curve *A*), is obtained in [S] vs v_0 plots of allosteric enzymes. Negative allosteric effectors move the curve toward higher substrate concentrations and enhance the sigmoidicity of the curve. If we

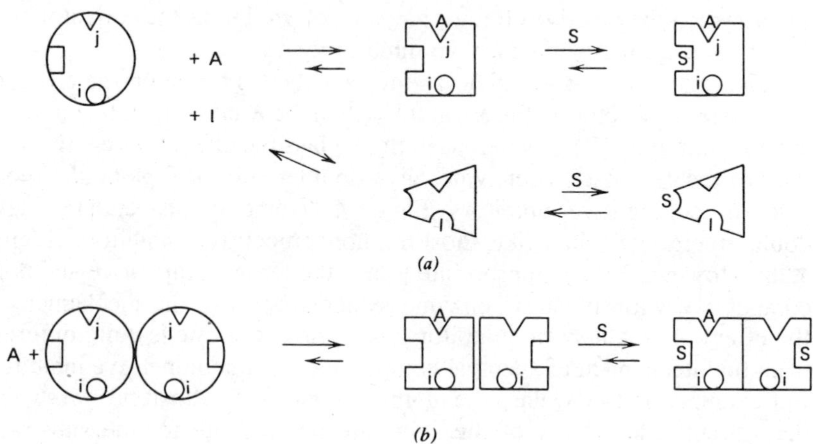

Figure 4.35
Models of allosteric enzyme systems.
(a) A model of a monomeric enzyme. Binding of a positive allosteric effector, A, to the activator site, j, induces a new conformation to the enzyme, one that has a greater affinity for the substrate. Binding of a negative allosteric effector to the inhibitor site, i, results in an enzyme conformation having a decreased affinity for substrate. (b) A model of a polymeric allosteric enzyme. Binding of the positive allosteric effector, A, at the j site causes an allosteric change in the conformation of the protomer to which the effector binds. This change in the conformation is transmitted to the second promoter through cooperative protomer–protomer interactions. The affinity for the substrate is increased in both protomers. A negative effector decreases the affinity for substrate of both protomers.

use $\frac{1}{2}V_{max}$ as a guideline, it can be seen from Figure 4.36 that a higher concentration of substrate would be required to achieve $\frac{1}{2}V_{max}$ in the presence of a negative effector (curve *C*) than is required in the absence of negative effector (curve *A*). In the presence of a positive modulator (curve *B*), $\frac{1}{2}V_{max}$ can be reached at a lower substrate concentration than is required in the absence of the positive modulator (curve *A*). Positive modulators shift the v_0 vs [S] plots toward the hyperbolic plots observed in Michaelis–Menten kinetics.

From the viewpoint of metabolic control, allosteric enzymes allow fine control of the activity of individual enzymes through small fluctuations in the level of substrate. Often the in vivo concentration of substrate corresponds with the sharply rising segment of the sigmoid [S] vs v_0 plot; conse-

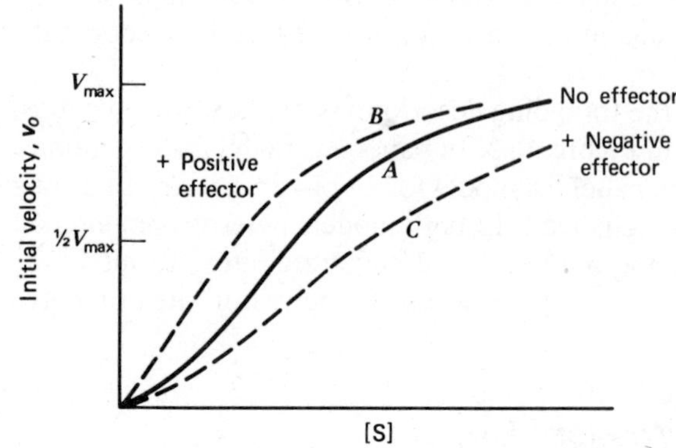

Figure 4.36
The kinetic profile of a K class allosteric enzyme.
The enzyme shows sigmoid S vs v_0 plots. Negative effectors shift the curve to the right resulting in an increase in K_m. Positive effectors shift the curve to the left and effectively lower the apparent K_m. V_{max} is not changed.

quently, large changes in enzyme activity are effected by small changes in substrate concentration. (See Figure 4.36.) It is also possible to "turn the enzyme off" with small amounts of a negative allosteric effector by its shifting the apparent K_m to higher values, values which are far above the in vivo level of substrate. Notice that at a given in vivo concentration of substrate the initial velocity, v_0, is decreased in the presence of a negative effector (compare curves *A* and *C*).

Cooperativity

The realization that most allosteric enzymes are oligomeric and also have sigmoid [S] vs v_0 plots led to the concept of cooperativity to explain the interaction between ligand sites in oligomeric enzymes. Cooperativity is defined as the influence that the binding of ligand to one protomer has on the binding of ligand to a second protomer of an oligomeric protein. It should be emphasized that kinetic mechanisms other than cooperativity can also produce sigmoid v_0 vs [S] plots; consequently, sigmoidicity is not diagnostic of cooperativity in a v_0 vs [S] plot. We have previously discussed cooperativity in terms of the binding of oxygen to hemoglobin and will now expand the concept.

The relationship between allosterism and cooperativity has been confused in many recent texts. The conformational changes occurring in a given protomer in response to ligand binding at an allosteric site is an allosteric effect. Cooperativity generally involves a change in conformation of an effector-activated protomer, which in turn transforms an adjacent protomer into a new conformation with an altered affinity for the effector ligand or for a second ligand. The conformation change may be induced by an allosteric effector or it may be induced by substrate, as it is in the case of hemoglobin. In hemoglobin the oxygen binding site on each protomer corresponds to the substrate site on an enzyme rather than to an allosteric site. Therefore, the oxygen-induced conformational change in the hemoglobin protomers is technically not an allosteric effect, although some authors identify it as such. It is a homotropic cooperative interaction. Those who consider the oxygen-induced changes in hemoglobin to be "allosteric" are using the term in a much broader sense than the original definition allows; however, "allosteric" is now used by many to describe any ligand-induced change in the tertiary structure of a protomer.

It should be emphasized that one can have an allosteric effect in the absence of any cooperativity. For example, in alcohol dehydrogenase, conformational changes can be demonstrated in each of the protomers upon the addition of positive allosteric effectors, but the active site of each protomer is completely independent of the other and there is no cooperativity between protomers, that is, induced conformational changes in one protomer are not transmitted to adjacent protomers.

In an attempt to describe mathematically experimentally observed ligand saturation curves, several models of cooperativity have been proposed. The two most prominent models are the *concerted* and the *sequential-induced fit*.

Although the *concerted model* is rather restrictive, most of the nomenclature associated with allosterism and cooperativity arose from this model. The model proposes that the enzyme exists in only two states, the T (tense or taut) and the R (relaxed). The T and R states are in equilibrium. Activators and substrates favor the R state and shift the preexisting equilibrium toward the R state by the law of mass action. Inhibitors favor the T state. A conformational change in one protomer causes a corresponding change in all protomers. No hybrid states occur. The model is diagrammed in Figure 4.37*a*. Although the model accounts for the kinetic behavior of many enzymes, it cannot account for negative cooperativity.

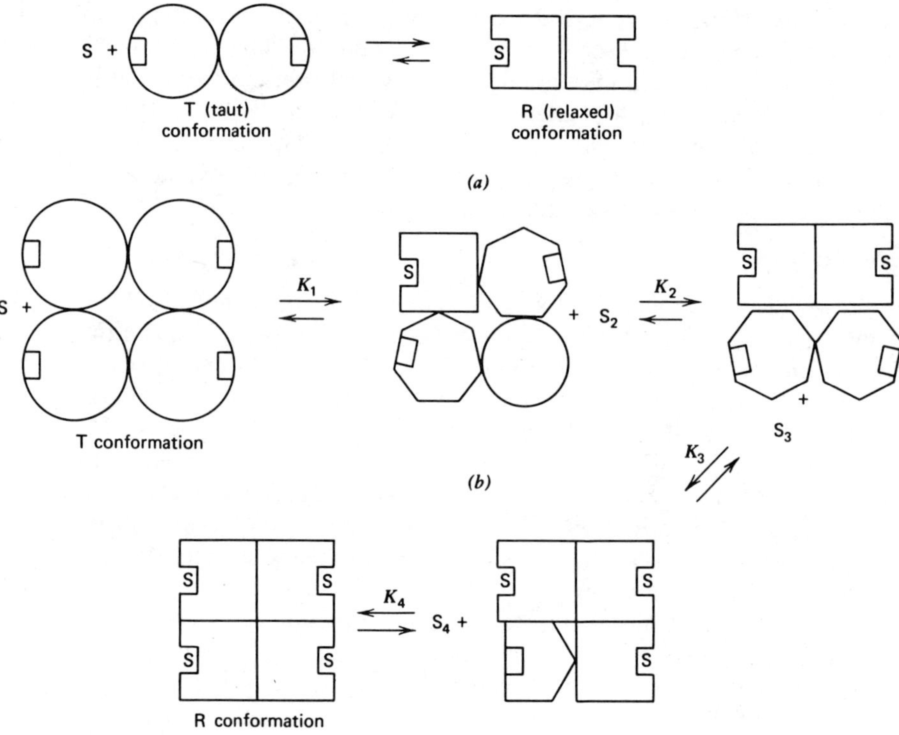

Figure 4.37
Models of cooperativity.
(a) The concerted model. The enzyme exists in only two states, the T, tense or taut, and the R, relaxed, conformation. Substrates and activators have a greater affinity for the R state and inhibitors for the T state. Ligands shift the equilibrium between the T and R states. (b) The sequential induced fit model. Binding of a ligand to any one subunit induces a conformational change in that subunit. This conformational change is transmitted partially to adjoining subunits through subunit–subunit interaction. Thus the effect of the first ligand bound is transmitted cooperatively and sequentially to the other subunits (protomers) in the oligomer resulting in a sequential increase or decrease in ligand affinity of the other protomers. The cooperativity may be either positive or negative, depending upon the ligand.

The *sequential-induced fit* model proposes that ligand binding induces a conformational change in a protomer. A corresponding conformational change is then partially induced in an adjacent protomer contiguous with the protomer containing the bound ligand. The effect of ligand binding is sequentially transmitted through the oligomer, giving rise to an increasing or decreasing affinity for the ligand by contiguous protomers as suggested by the scheme in Figure 4.37*b*. In this model numerous hybrid states occur giving rise to the cooperativity and the sigmoid $[S]$ vs v_0 plots. Both positive and negative cooperativity can be accommodated by the model. A positive modulator induces a conformation in the protomer, which has an increased affinity for the substrate. A negative modulator induces a different conformation in the protomer, one that has a decreased affinity for substrate. Both effects are cooperatively transmitted to adjacent protomers. For the V class enzymes the same reasoning applies, but the effect is on the catalytic event (k_3) rather than on K_m.

Regulatory Subunits

In the foregoing discussion the allosteric site has been considered to reside on the same protomer as the catalytic site. Furthermore, all protomers were considered to be identical. There are several very important enzymes in which a distinct regulatory protomer exists. These regulatory subunits have no catalytic function per se, but their binding with the catalytic protomer modulates the activity of the catalytic subunit through an induced conformational change.

Ribonucleotide reductase converts ribonucleoside diphosphates to deoxyribonucleotides for DNA biosynthesis (page 510). The active site is generated at the interface of two different subunits, B_1 and B_2. Both B_1 and B_2 are catalytically inactive by themselves. The allosteric sites reside on B_1; however, B_2 does not contain the active site in and of itself. The active site is generated at the interface between polymerized B_1 and B_2, as suggested in Figure 4.38, ATP is a positive effector, which binds to the ℓ allosteric site, and the end product of the enzyme, dATP, is a negative effector, which competes with ATP at the ℓ site. This particular arrangement of subunits to form the catalytic site is not common.

Completely separate regulatory subunits are observed in aspartate transcarbamoylase. This enzyme is involved in pyrimidine biosynthesis and catalyzes the transfer of the carbamoyl group from carbamoyl phosphate to the α-amino group of aspartate:

$$\text{Carbamoyl phosphate} + \text{aspartate} \longrightarrow \text{carbamoylaspartate} + P_i$$

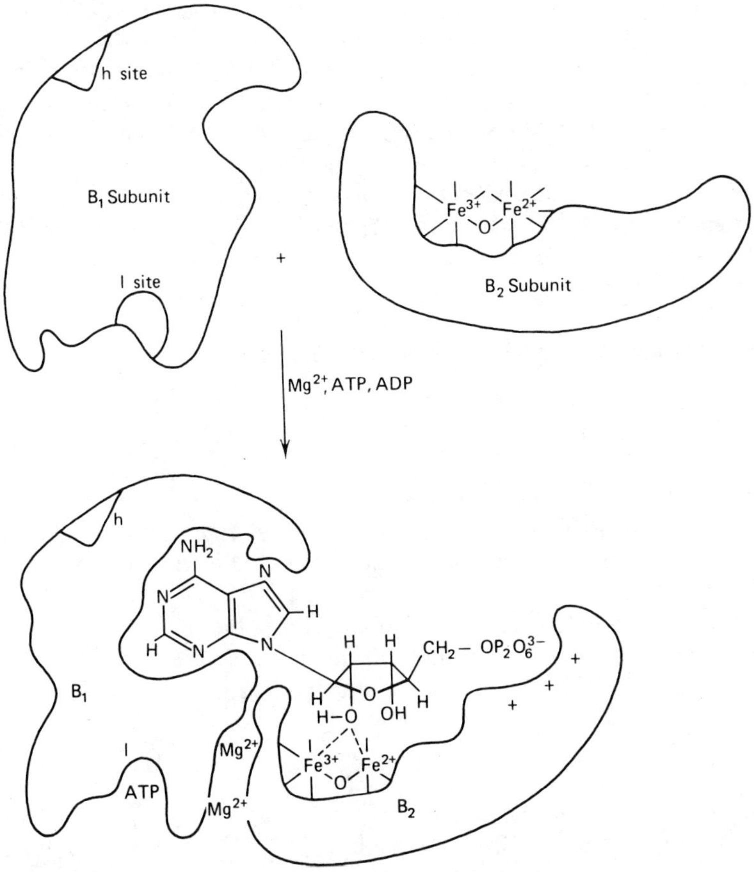

Figure 4.38
Model of an allosteric enzyme with a separate regulatory subunit.
The enzyme is ribonucleotide reductase. B_1 is the regulatory subunit. ATP binds at the "ℓ site" and is a positive effector. B_2 is the catalytic subunit but only when combined with the B_1 subunit. B_2 contains the iron cluster, which is involved in reduction of the 2'-hydroxyl of the ribose of the ribonucleoside diphosphates, which in this example is ADP. dATP is a negative modifier that competes at the "ℓ" site for ATP. Other modifiers such as dTTP bind at the "h" site and are either positive or negative modulators depending on the substrate.

The enzyme is a complex consisting of two active catalytic (C) and three regulatory (R) subunits. The end product of the pyrimidine pathway, cytidine triphosphate (CTP), binds to the R subunits as a negative effector, whereas ATP functions as a positive effector.

Two other very important enzymes, adenylate cyclase and protein kinase, have regulatory subunits. They are schematically depicted in Figures 4.62 and 4.64. These two particular enzymes are discussed in detail on page 169.

4.7 ENZYME SPECIFICITY: THE ACTIVE SITE

Enzymes are the most specific catalysts known, both from the viewpoint of the substrate as well as the type of reaction performed on the substrate. Specificity inherently resides in the substrate binding site, which lies on the enzyme surface. The tertiary structure of the enzyme is folded in such a way as to create a region that has the correct molecular dimensions, the appropriate topology, and the optimal alignment of counterionic groups and hydrophobic regions to accommodate a specific substrate. The tolerances in the active site are so small that usually only one isomer of a diastereomeric pair will bind. For example, D-amino acid oxidase will bind only D-amino acids and not L-amino acids. Some enzymes show absolute specificity for substrate. Others have broader specificity and will accept several different analogs of a specific substrate. For example, hexokinase catalyzes the phosphorylation of glucose, mannose, fructose, glucosamine, and 2-deoxyglucose, but not all at the same rate. Glucokinase, on the other hand, is specific for glucose.

The specificity of the reaction catalyzed rests in the active site and the particular arrangement of amino acids that participate in the bond-making and bond-breaking phase of catalysis. The mechanism of catalysis is discussed in Section 4.8.

Complementarity of Substrate and Enzyme

Various models have been proposed to explain the substrate specificity of enzymes. The first proposal was the "lock-and-key" model (Figure 4.39), in which a negative impression of the substrate is considered to exist on the enzyme surface. The substrate fits to this binding site just as a key fits into the proper lock or a hand into the proper sized glove. This model gives a rigid picture of the enzyme and cannot account for the effects of allosteric ligands.

A more flexible model of the binding site is the *induced fit* model. In this model, the binding site and certainly the active site are not fully formed. The essential elements of the binding site are present to the

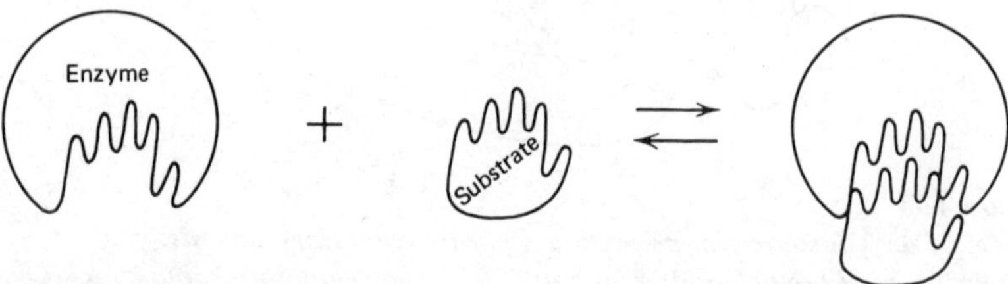

Figure 4.39
Lock and key model of the enzyme binding site.
The enzyme contains a negative impression of the molecular features of the substrate, thus allowing specificity of the enzyme for a particular substrate.

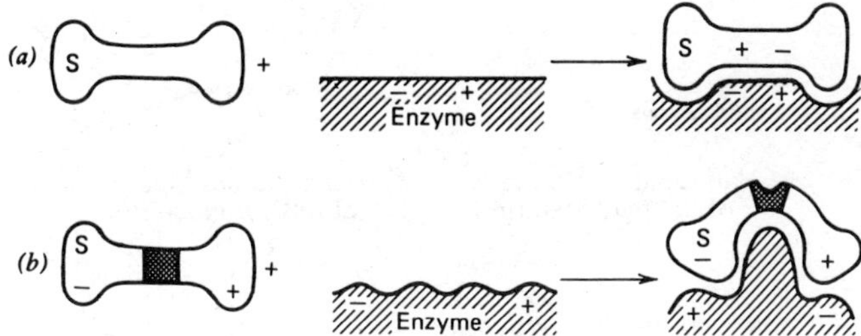

Figure 4.40
Models for induced fit and substrate strain.
(a) Approach of substrate to the enzyme induces the formation of the active site. (b) Substrate strain, induced by substrate binding to the enzyme, contorts normal bond angles and "activates" the substrate.
Reprinted, with permission, from D. Koshland *Annu. Rev. of Biochem.* 37:374, 1968. Copyrighted by Annual Reviews, Inc.

extent that the correct substrate can position itself properly in the nascent binding site. The interaction of the substrate with the enzyme induces a conformational change in the enzyme, resulting in the formation of a strongly binding site and the repositioning of the appropriate amino acids to form the active site. There is excellent x-ray evidence for the correctness of this model in the enzyme carboxypeptidase A. A diagram of the induced fit model is shown in Figure 4.40*a*.

The concept of *induced fit* combined with substrate strain accounts for more of the experimental observations concerning enzyme action than do

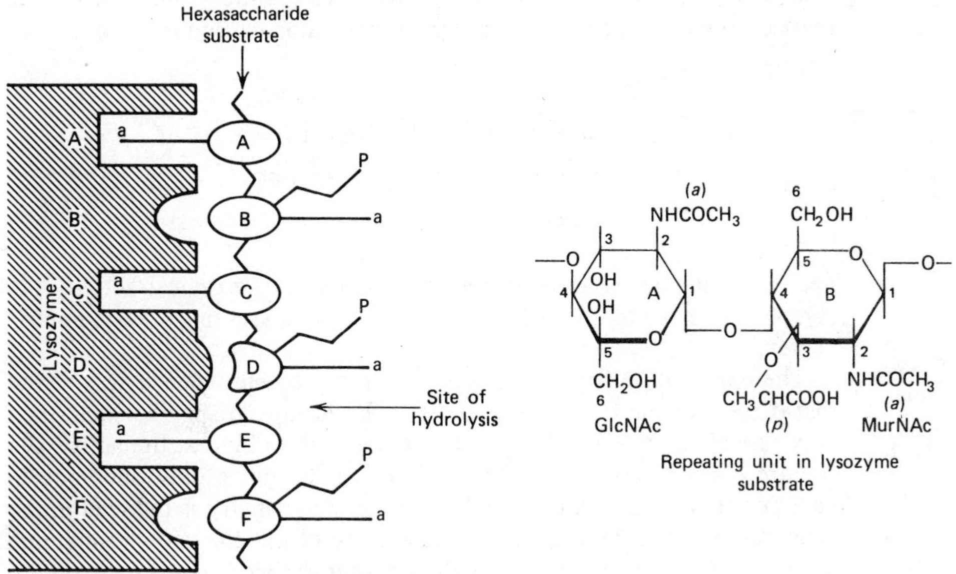

Figure 4.41
Hexasaccharide binding at the active site of lysozyme.
In the model substrate pictured, the ovals represent individual pyranose rings of the repeating units of the lysozyme substrate shown to the right in the figure. Ring D is strained by the enzyme to the half-chair conformation and hydrolysis occurs between the D and E rings. Six subsites on the enzyme bind substrate. Alternate sites are specific for acetamido groups (a) but are unable to accept the lactyl (P) side chains, which occur on the N-acetylmuramic acid residues. Thus the substrate can bind to the enzyme in only one orientation.
Reprinted, with permission from T. Imoto, L. N. Johnson, A. C. T. North, et al. *The Enzymes,* 3rd, P. Boyer, ed. 7:713, 1972. Academic Press.

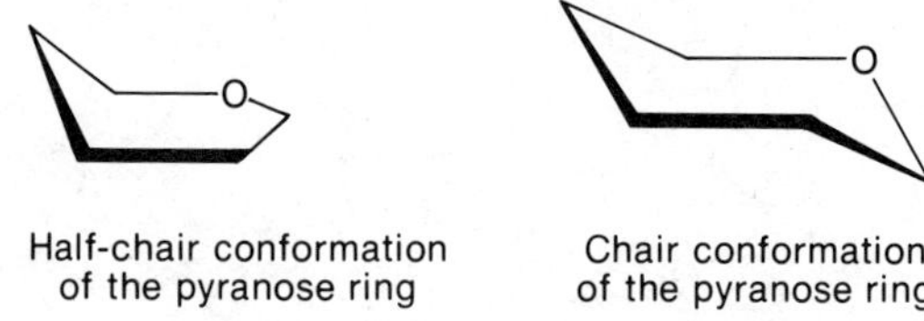

Figure 4.42
Two possible conformations of glucose.

other models. In this model (Figure 4.40*b*), the substrate is "strained" toward product formation as a result of an induced conformational transition of the enzyme. A good example of enzyme-induced substrate strain is that of lysozyme (Figure 4.41) where the conformation of the sugar residue "D" at which bond breaking occurs is strained from the stable chair to the unstable half-chair conformation upon binding. These conformations of glucose are shown in Figure 4.42.

The concept of substrate strain is useful in explaining the role of the enzyme in increasing the rate of reaction. This effect is considered in Section 4.8.

Asymmetry of the Binding Site

Not only are enzymes able to distinguish between isomers, but they are able to distinguish between two equivalent atoms in a symmetrical molecule. For example, the enzyme glycerol kinase is able to distinguish between configurations of H and OH on carbon-2 in the symmetric substrate glycerol, so that only the asymmetric product L-glycerol 3-phosphate is formed. These prochiral substrates have two identical substituents and two additional but dissimilar groups on the same carbon ($C_{aa'bd}$).

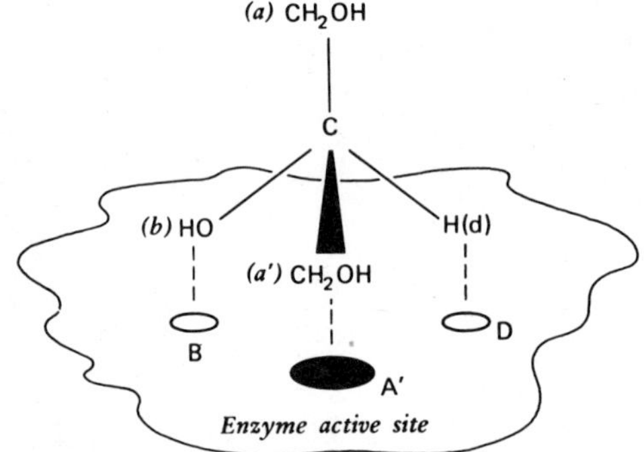

Figure 4.43
Three-point attachment of a symmetrical substrate to an asymmetric enzyme binding site.
Glycerol kinase by virtue of dissimilar binding sites for the —H and —OH group of glycerol binds only the a′ hydroxymethyl group to the active site. Thus, only one stereoisomer results from the kinase reaction, the L-glycerol 3-phosphate.

$$\begin{array}{ll} (a)\ \mathrm{CH_2OH} & \mathrm{CH_2OH} \\ \quad\ \ \mathrm{CHOH}\ (b,d) \xrightarrow[\text{ATP}]{\text{glycerol kinase}} & \mathrm{HOCH} \\ (a')\ \mathrm{CH_2OH} & \mathrm{CH_2OPO_3} \\ \quad\ \ \text{Glycerol} & \text{L-Glycerol 3-phosphate} \end{array}$$

Prochiral substrates are substances that possess no optical activity but can be converted to chiral compounds, that is, possessing an asymmetric center.

The explanation for this enigma was forthcoming when it was found that if the enzyme binds the two dissimilar groups at specific sites, then only one of the two similar substituents is able to bind at the active site (Figure 4.43). Thus, the enzyme is able to recognize only one specific configuration of the symmetrical molecule. Asymmetry is induced in the substrate by the asymmetric binding surface of the enzyme. A minimum of three different binding sites on the enzyme surface is required to distinguish between identical groups on a prochiral substrate.

4.8 MECHANISM OF CATALYSIS

All chemical reactions have a potential energy barrier that must be overcome before reactants can be converted to products. In the gas phase the reactant molecules can be given enough kinetic energy by heating so that collisions result in product formation. The same is true with solutions. However, a well-controlled body temperature of 37°C does not allow temperature to be increased to accelerate the reaction, and 37°C is not warm enough to provide the reaction rates required for fast-moving species of animals. Enzymes employ other means of overcoming the barrier to reaction, and these will be discussed after some useful definitions are covered.

A comparison of the enzyme diagrams for catalyzed and noncatalyzed reactions is shown in Figure 4.44. The energy barrier represented by the uncatalyzed curve in Figure 4.44 is a measure of the *activation* energy, E_a. The reaction coordinate is simply the pathway in terms of bond stretching between reactants and products. At the apex of the energy barrier is the activated complex known as the *transition state,* Ts. The transition state represents the reactants in their activated state. In this state reactants are in an intermediate stage along the reaction pathway and cannot be identified as starting material or products. For example, in the hydrolysis of ethyl acetate:

$$CH_3{-}\overset{\overset{\displaystyle O}{\|}}{C}{-}O{-}CH_2{-}CH_3 \xrightarrow{H_2O} CH_3{-}CH_2{-}OH + CH_3{-}\overset{\overset{\displaystyle O}{\|}}{C}{-}OH$$

the Ts might look like

$$\left[\begin{array}{c} O^- \\ | \\ CH_3{-}C{-}{-}{-}{-}O{-}{-}{-}{-}CH_2{-}CH_3 \\ \vdots \\ O \\ / \quad \diagdown \\ H \qquad H \end{array}\right]$$

The transition state complex can break down to products or go back to reactants. The Ts is not an intermediate and cannot be isolated!

Notice that in the case of the enzyme-catalyzed reaction (Figure 4.44) the energy of the reactants and products is no different than in the uncatalyzed reaction. Enzymes do not change the thermodynamics of the system but they do change the pathway for reaching the final state.

As noted on the energy diagram, there may be several plateaus or valleys on the energy contour for an enzyme reaction. At these points

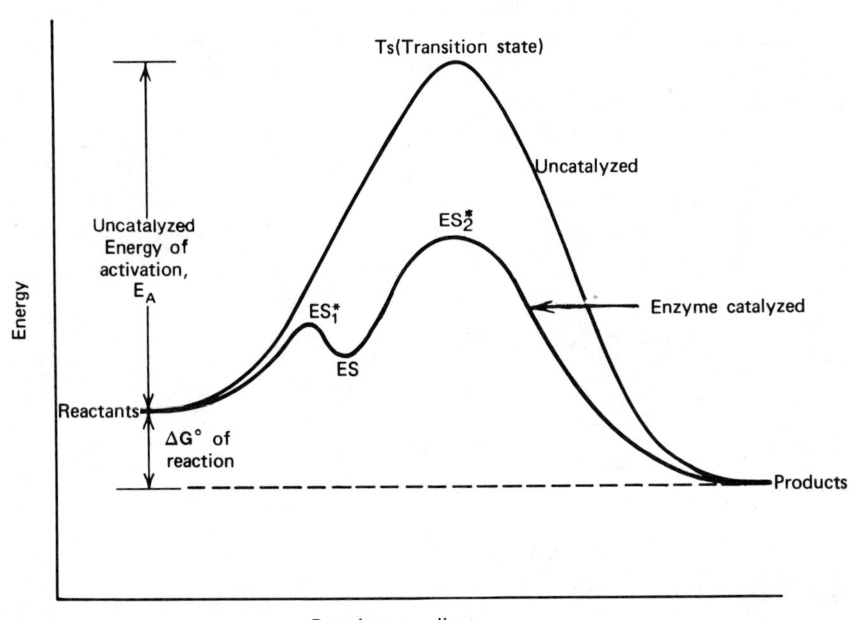

Figure 4.44
Energy diagrams for catalyzed vs noncatalyzed reactions.
The overall energy difference between reactants and products is the same in catalyzed and noncatalyzed reactions. The enzyme-catalyzed reaction proceeds at a faster rate because the energy of activation is lowered.

metastable intermediates exist. An important point is that each valley may be reached with the heat input available in a 37°C system. In other words, the enzyme allows the energy barrier to be scaled in increments. The Michaelis–Menten, ES, complex is not the transition state, but may be found in one of the valleys. This is the case because in the ES complex, the substrates are properly oriented and the substrate may be "strained," and therefore the bonds to be broken lie further along the reaction coordinate.

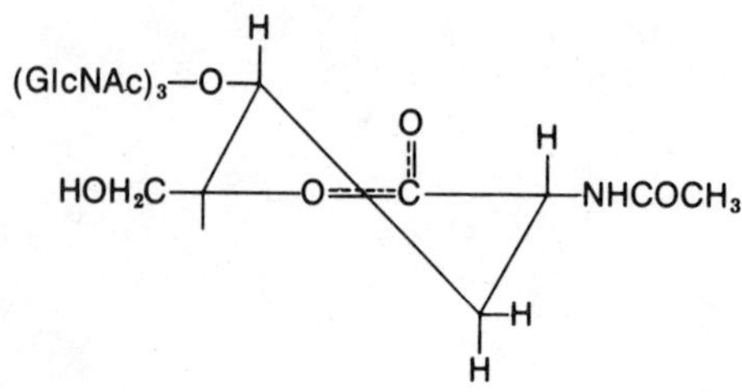

Figure 4.45
A transition state analog(tetra-N-acetylchitotetrose-δ-lactone) of ring D of the substrate for lysozyme.

If our concepts of the transition state are correct, one would expect that compounds designed to resemble closely the transition state would bind much more tightly to the enzyme than does the natural substrate. This has proven to be the case. In such substrate analogs one finds affinities 10^2 to 10^5 times greater than those for substrate. These compounds are called *transition state* analogs and are potent enzyme inhibitors. Previously, lysozyme was discussed in terms of substrate strain, and mention was made of the conversion of sugar ring D from a chair to a strained half-chair conformation. Synthesis of a transition state analog in the form of the δ-lactone of tetra-*N*-acetylchitotetrose (Figure 4.45), which has a distorted half-chair conformation, followed by binding studies, showed that this transition state analog was bound 6×10^3 times better than the normal substrate.

Factors Contributing to Decrease in Activation Energy

Enzymes are able to enhance the rates of reaction by a factor of 10^9 to 10^{12} times that of the noncatalyzed reaction. Most of this rate enhancement can be accounted for by four processes: acid–base catalysis, substrate strain, covalent catalysis, and entropy effects.

Acid–Base Catalysis

Specific acids and bases are H^+ and OH^-, respectively. Free protons and hydroxide ions are not encountered in most enzyme reactions and then only in some metal-dependent enzymes (page 135).

General acids and bases are important in enzymology. A *general* acid or base is any substance that is weakly ionizable. In the physiological pH range, the protonated form of histidine is the most important general acid and its conjugate base an important general base. These structures are shown in Figure 4.46. Other acids are the thiol —SH, tyrosine —OH and the ε-amino group of lysine. Other bases are carboxylic acid anions and the conjugate bases of the general acids.

Consideration of the mutarotation of glucose illustrates the principle of acid–base catalysis. If one dissolves pure α-D-glucose in water, over several days an equilibrium mixture of 34% α-D-glucose and 66% β-D-glucose forms. The conversion of the α anomer to the β anomer proceeds

^+H_3N—CHCOO$^-$ / CH_2 / HN NH$^+$ ⇌ ^+H_3N—CHCOO$^-$ / CH_2 / HN N: + H^+

Histidine as a general acid — Histidine as a base

Figure 4.46
Acid and base forms of histidine.

through intermediate formation of the open-chain aldehyde. Upon reforming the pyranose ring, there is no stereospecificity in the orientation of the resulting hydroxyl on carbon-1 (the anomeric carbon), and the ring can reclose with the hydroxyl in either the α or β orientation. See center reaction sequence in Figure 4.47.

General acids increase the rate of mutarotation by protonating the pyranose oxygen which labilizes the C—O bond to yield the aldehyde. The rate of mutarotation is increased 1,000-fold over the uncatalyzed reaction. General bases increase the rate another 1,000-fold over the acid rate by generating an alkoxide ion from the hydroxyl on C_1 which quickly opens the pyranose ring. These concepts are diagrammed in Figure 4.47. Both the acid and base catalyzed reactions yield the open chain form of glucose. Again, reclosure to the pyranose is not stereospecific, and a mixture of anomers is obtained.

These concepts of acid–base catalysis have been extended to the enzyme active site where specific amino acid functional groups are implicated as acid and base catalysts. Ribonuclease is a good example of the role of acid and base catalysis at the enzyme active site. Ribonuclease (RNase) cleaves the RNA chain at the 3′-phosphodiester linkage of pyrimidine nucleotides with an obligatory formation of a cyclic 2′,3′-phosphoribose on a pyrimidine nucleotide as intermediate. In the mechanism outlined in Figure 4.48, His-119 acts as a general acid to protonate

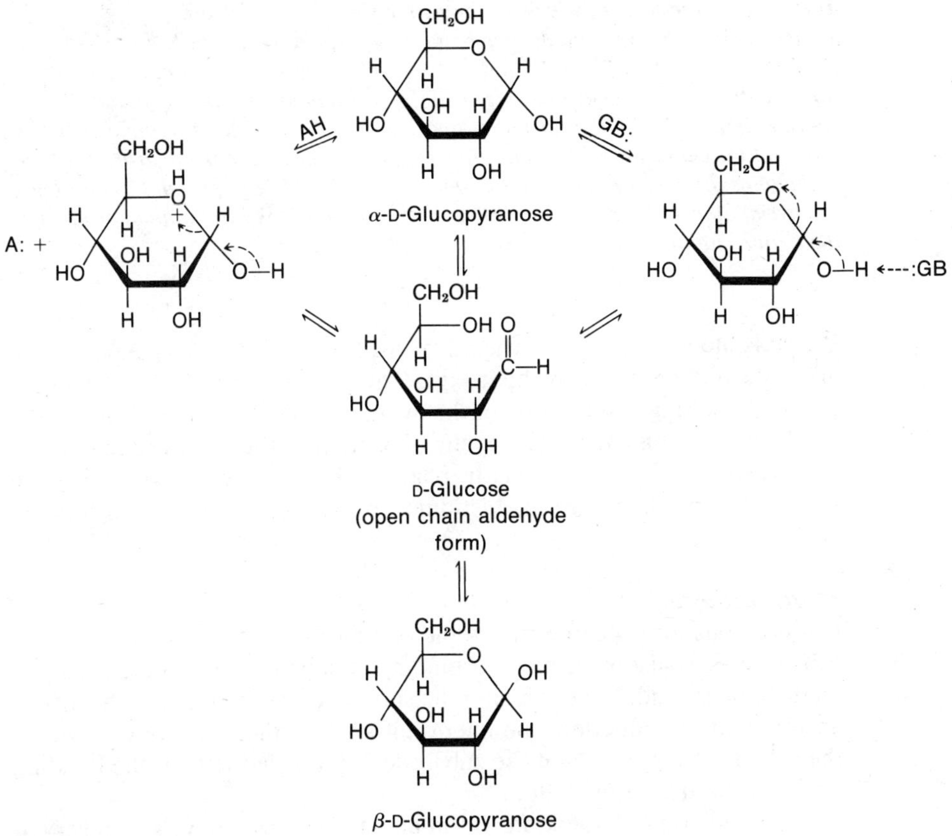

Figure 4.47
Mutarotation of glucose.
The center series of structures illustrate the uncatalyzed reaction. The structure on the left represents an intermediate in acid catalysis and the one on the right an intermediate in base catalysis. The open chain form of glucose is a common intermediate in both the catalyzed and uncatalyzed pathways.

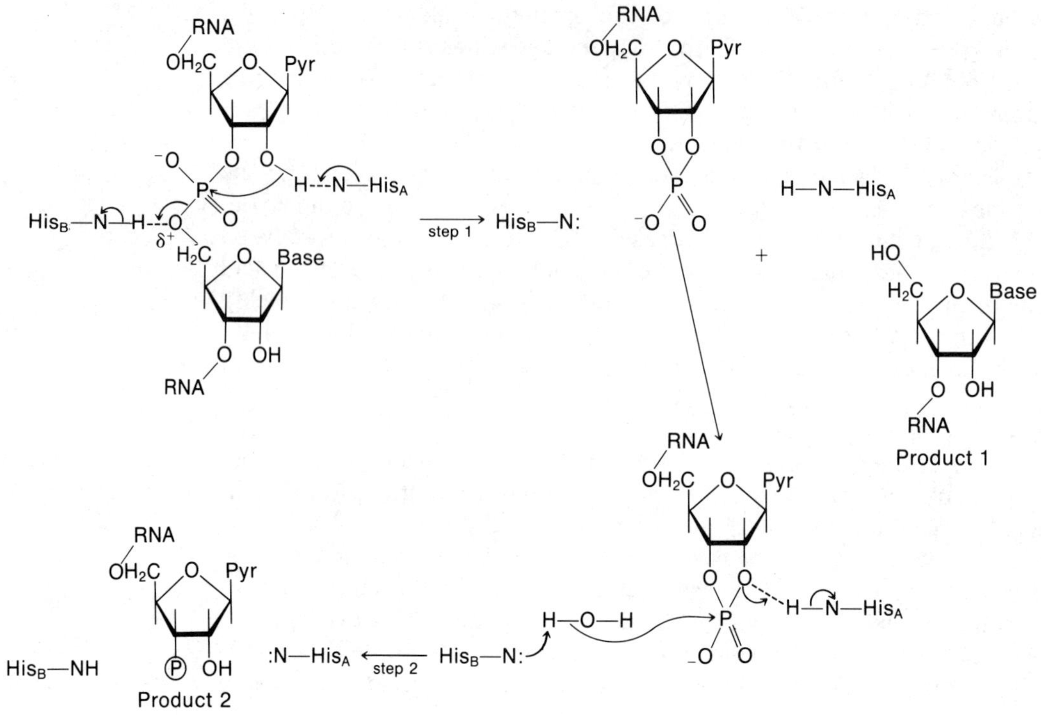

Figure 4.48
Role of acid and base catalysis in the active site of ribonuclease.
RNase cleaves the phosphodiester bond in a pyrimidine loci in RNA. HisA and HisB are histidine residues 12 and 119, respectively, at the ribonuclease active site that function as acid and base catalysts in enhancing the formation of an intermediate 2′,3′-cyclic phosphate and release of a shorter fragment of RNA (product 1). These same histidines then play a reverse role in the hydrolysis of the cyclic phosphate and release of the other fragment of RNA (product 2) that ends in a pyrimidine nucleoside 3′-phosphate. As a result of the formation of product 2, the active site of the enzyme is regenerated.

the phosphodiester bridge, whereas His-12 acts as a base in generating an alkoxide on the ribose-3′-hydroxyl. The latter then attacks the phosphorus, resulting in formation of the cyclic phosphate and breakage of the RNA chain at this locus. The cyclic phosphate is then cleaved in phase 2 by a reversal of the reactions in phase 1, but with water replacing the leaving group. The active site histidines revert to their original protonated state.

Substrate Strain

Our previous discussion of this topic related to induced fit of enzymes to substrate. It is also possible that binding of substrate to a preformed site on the enzyme induces strain into the substrate. Irrespective of the mechanism of strain induction, the important point is that the energy level of the substrate is raised, and the substrate is propelled toward the bonding found in the transition state.

A combination of substrate strain and acid–base catalysis is observed in the mechanism of lysozyme action (Figure 4.49). Ring D of the hexasaccharide substrate upon binding to the enzyme is strained to the half-chair conformation. General acid catalysis by active site glutamic acid promotes the unstable half-chair into the transition state. The carbonium ion formed in the transition state is stabilized by a negatively charged aspartate. Breakage of the glycosidic linkage between rings D and E

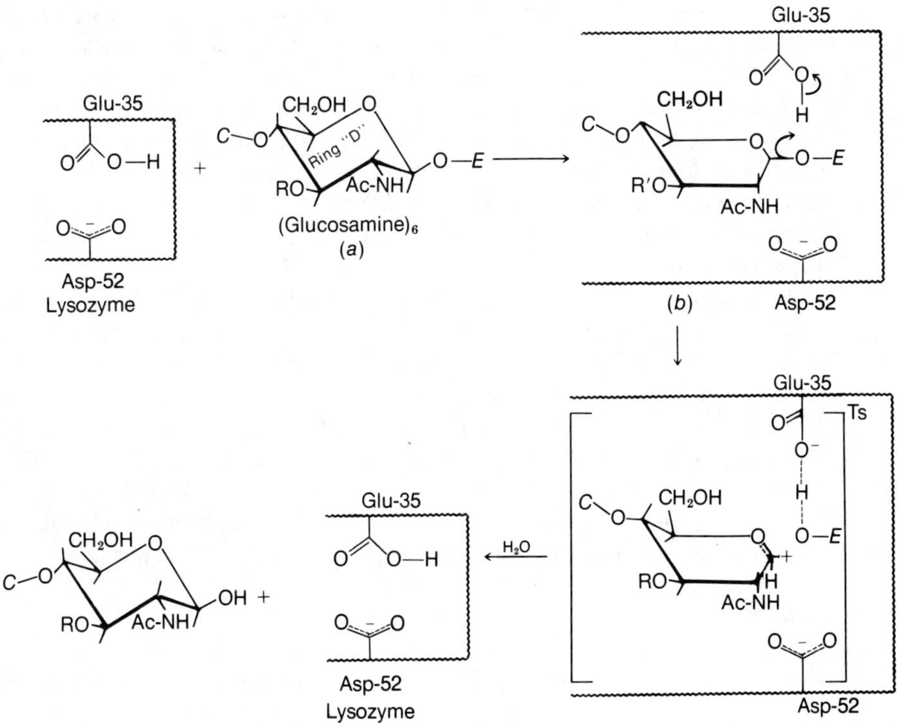

Glu-35
Lysozyme
Asp-52
CH_2OH
Ring "D"
Ac-NH
(Glucosamine)$_6$
(*a*)
(*b*)
Ts
H_2O

Figure 4.49
A mechanism for lysozyme action–substrate strain.
The binding of the stable chair (a) conformation of the substrate to the enzyme generates the strained half-chair conformation (b) in the ES complex. In the transition state, acid-catalyzed hydrolysis of the glycosidic linkage by an active site glutamic acid residue generates a carbonium ion on the D ring, which relieves the strain generated in the initial ES complex and results in collapse of the transition state to products.

relieves the strained transition state by allowing rings D and E to return to the stable chair conformation.

Covalent Catalysis

In covalent catalysis, the attack of a nucleophilic (negatively charged) or electrophilic (positively charged) group in the enzyme active site upon the substrate results in covalent binding of the substrate to the enzyme as an intermediate in the reaction sequence. Also, enzyme-bound coenzymes often form covalent bonds with the substrate. For example, in the transaminases, the amino acid substrate forms a Schiff's base with enzyme-bound pyridoxal phosphate (page 440). Evidence is now accumulating that some oxidoreductases, utilizing FAD as coenzyme, form intermediate covalent adducts of substrate and FAD. For example, in the oxidation of alcohols the scheme shown in Figure 4.50 has been proposed.

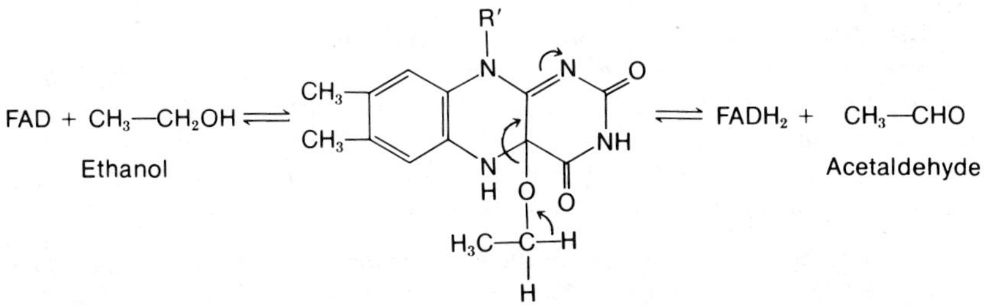

FAD + CH_3—CH_2OH ⇌ [adduct] ⇌ $FADH_2$ + CH_3—CHO
Ethanol — Acetaldehyde

Figure 4.50
Proposed covalent substrate coenzyme complex in aldehyde dehydrogenase.
The enzyme catalyzes a nucleophilic addition of the alcohol to the 4a position of the coenzyme FAD followed by collapse of this intermediate, as indicated by the arrows, to products.

In all cases of covalent catalysis, the enzyme- or coenzyme-bound substrate is more labile than the original substrate. The enzyme–substrate adduct represents one of the valleys on the energy profile (Figure 4.44).

The serine proteases, such as trypsin, chymotrypsin, and thrombin, are good representatives of the covalent catalytic mechanism. The name "serine protease" arises from the fact that serine is involved in the active site of all these enzymes. Acylated enzyme has been isolated in the case of chymotrypsin. Covalent catalysis is assisted by acid–base catalysis in these particular enzymes.

In chymotrypsin the attacking nucleophile is generated by His-57, which pulls a proton from the hydroxyl of Ser-195. The resulting alkoxide attacks the carbonyl carbon of the peptide bond, releasing the amino-terminal end of the protein and forming an acylated enzyme (through Ser-195). The acylated enzyme is then cleaved by reversal of the reaction sequence, but with water as the nucleophile rather than Ser-195. This mechanism is outlined in Figure 4.51. It was formerly believed that aspartic acid 102 increased the basicity of histidine 57 through a "charge relay," but the quantitative significance of this effect has been questioned.

Entropy Effect

Entropy is a thermodynamic term, S, which defines the extent of disorder in a system. At equilibrium, the entropy is maximal. For example, in

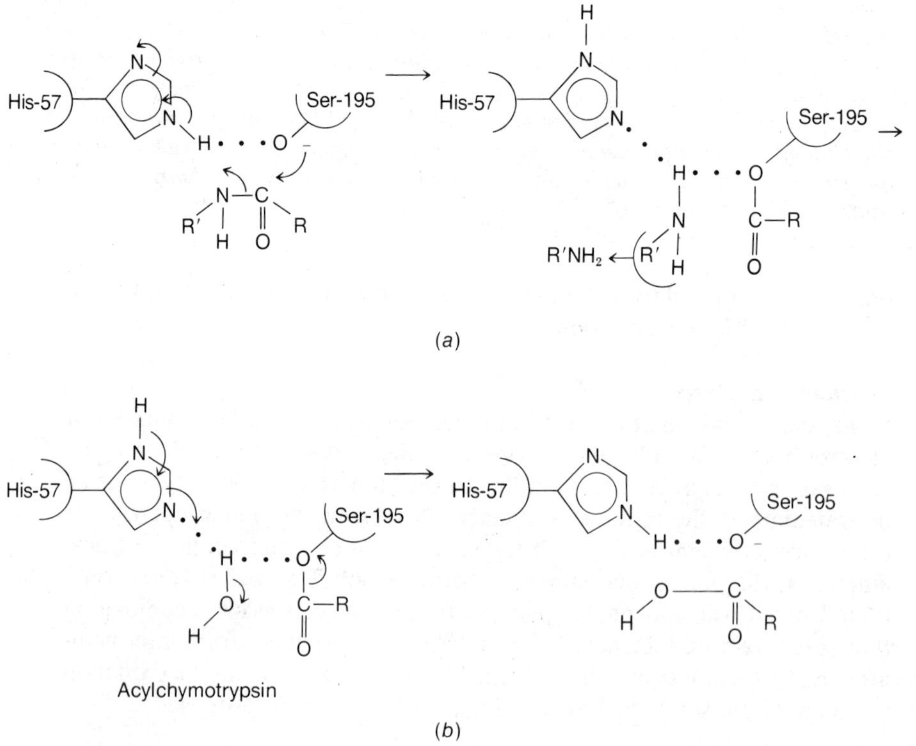

His-57 Ser-195 → His-57 Ser-195 →

R′ H O R R′NH_2 R′ H R

(*a*)

His-57 Ser-195 → His-57 Ser-195

H O C—R Acylchymotrypsin H R

(*b*)

Figure 4.51
Covalent catalysis in the active site of chymotrypsin.
Through acid-catalyzed nucleophilic attack, the stable amide linkage of the peptide substrate is converted into an unstable acylated enzyme. The latter is hydrolyzed in the rate-determining step. The new amino-terminal peptide is released concomitant with formation of the acylated enzyme. Scheme (a) represents the acylation step and (b) the deacylation step.

Modified, with permission, from D. M. Blow, J. J. Birktoft, and B. S. Hartley, *Nature* 221:339, 1969. Copyright, 1969 Macmillan Journal Ltd.

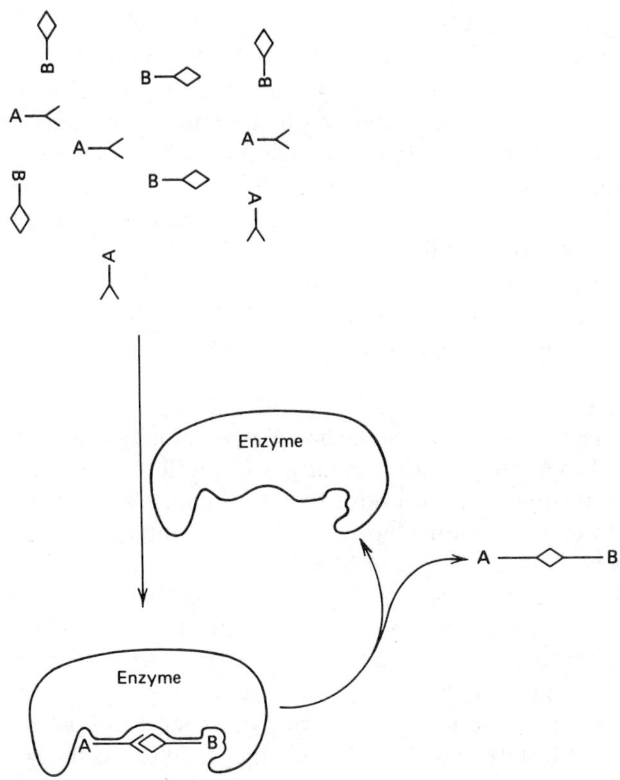

Figure 4.52
Diagrammatic representation of the entropy effect.
Substrates in dilute solution are concentrated and oriented on the enzyme surface so as to enhance the rate of reaction.

solution two reactants A—< and B—◇ exist in many different orientations. The chances of A—< and B—◇ coming together with the correct geometric orientation and with enough energy to react is small at 37°C and in dilute solution. However, if an enzyme with two high-affinity binding sites for A—< and B—◇ is introduced into the dilute solution of these reactants, as suggested in Figure 4.52, A—< and B—◇ will be bound to the enzyme in the correct orientation for the reaction to occur. They will be bound with the correct stoichiometry, and the effective concentration of the reactants will be increased on the enzyme surface—all of which will contribute to an increased rate of reaction.

Once correctly positioned on the enzyme surface, and as a result of binding, the substrates may be "strained" toward the transition state. At this point the substrates have been "set up" for acid–base and/or covalent catalysis. Thus, the proper orientation and the nearness of the substrate with respect to the catalytic groups, which has been dubbed the "proximity effect," contributes 10^3–10^4-fold to the rate enhancement observed with enzymes. It has been estimated that the decrease in entropy contributes a factor of 10^3 to the rate enhancement.

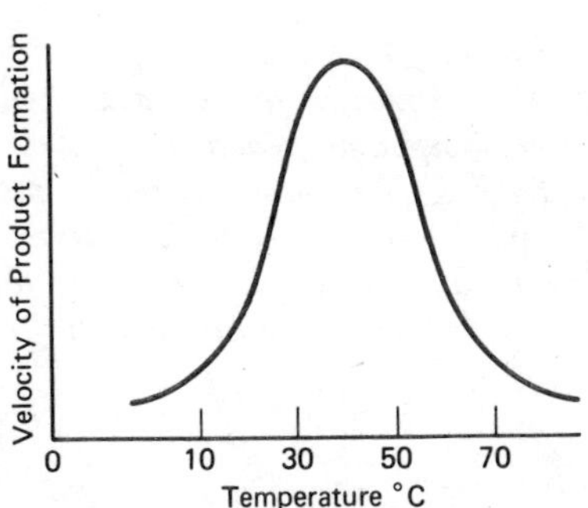

Figure 4.53
Temperature dependence of a typical mammalian enzyme.
To the left of the optimum the rate is low because the environmental temperature is too low to provide enough kinetic energy to overcome the energy of activation. To the right of the optimum, the enzyme is inactivated by heat denaturation.

Environmental Effects on Catalysis

A number of external parameters, including pH, temperature, and salt concentration, affect the enzyme activity. These effects are probably not important in vivo, under normal conditions, but are very important in setting up enzyme assays in vitro to measure enzyme activity in a patient's plasma or tissue sample.

CLIN. CORR. 4.5
THERMAL LABILITY OF G6PD RESULTS IN A HEMOLYTIC ANEMIA

Glucose-6-phosphate dehydrogenase is an important enzyme in the red cell for the maintenance of the red cell membrane (Chapter 21). A deficiency or inactivity of this enzyme leads to a hemolytic anemia. In other cases, a variant enzyme is present that normally has sufficient activity to maintain the membrane, but under conditions of oxidative stress fails. A reasonably frequent mutation of this enzyme leads to a protein with normal kinetic constants but a decreased thermal stability. This condition is especially critical to the red cell, since it is devoid of protein-synthesizing capacity and cannot renew enzymes as they denature. The end result is a greatly decreased lifetime for those red cells which have an unstable G6P dehydrogenase. These red cells are also susceptible to drug-induced hemolysis.

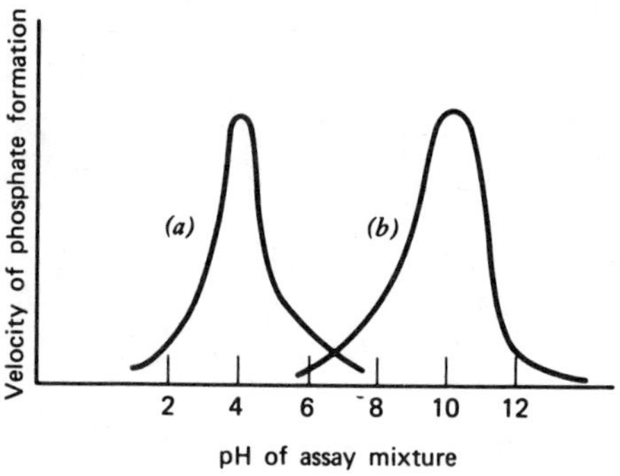

Figure 4.54
The pH dependence of (a) acid and (b) alkaline phosphatase reactions.
In each case the optimum represents the ideal ionic state for binding of enzyme and substrate and the correct ionic state for the amino acids involved in the catalytic event.

Temperature Dependence

Plots of velocity vs temperature for most enzymes reveal a bell-shaped curve with an optimum between 40 and 45°C for mammalian enzymes, as indicated in Figure 4.53. Above this temperature, heat denaturation of the enzyme occurs, while at lower temperatures there is not enough heat to overcome the energy barrier, even for the catalyzed reaction. Between 0 and 40°C, most enzymes show a twofold increase in activity for every 10° rise in temperature. Under conditions of hypothermia most enzyme reactions are depressed, which accounts for the decreased oxygen demand of living organisms at low temperature. Mutation of an enzyme to a thermolabile form can have serious consequences as discussed in Clin. Corr. 4.5.

pH Effects

Nearly all enzymes show a bell-shaped pH–velocity profile, but the maximum (pH optimum) varies greatly with different enzymes. Alkaline and acid phosphatases are both found in man, but their pH optima are greatly different, as shown in Figure 4.54. Obviously, neither functions at maximal activity at the pH of blood.

The bell-shaped curve and its position on the X axis are dependent upon the particular ionized state of the substrate that will be optimally bound to the enzyme. This in turn is related to the ionization of specific amino acids that constitute the substrate-binding site. In addition, those amino acids that are involved in catalyzing the reaction must be in the correct charge state to be functional in the catalytic event. For example, if aspartic acid is involved in catalyzing the reaction, the pH optimum may be in the region of 4.5 at which the α-carboxyl of aspartate ionizes, whereas if the ε-amino of lysine is the catalytic group, the pH optimum may be around pH 9.5, the pK_a of the ε-amino group. Studies of the pH dependence of enzymes are useful for suggesting which amino acid(s) may be operative in the catalytic event in the active site.

Clin. Corr. 4.6 points out the physiological effect of a mutation leading to a change in the pH optimum of a physiologically important enzyme. Such a mutated enzyme may function on the shoulder of the pH-rate profile, but not be optimally active, even under normal physiological conditions. Then when an abnormal condition such as alkalosis (observed in vomiting) or acidosis (observed in pneumonia and often in surgery) occurs, the enzyme activity may disappear because the pH is inappropriate. The point is that under normal conditions, the enzyme may be active enough to meet normal requirements, but under physiological stress in vivo environmental conditions may change so that the enzyme is less active and cannot meet its metabolic obligations.

4.9 CLINICAL APPLICATIONS OF ENZYMES

The principles of enzymology outlined in previous sections find practical application in the clinical laboratory in the measurement of plasma or tissue enzyme activities and concentrations of substrates in the sick individual.

The rationale for measuring plasma enzyme activities is based on the premise that changes in activities reflect changes that have occurred in a specific tissue or organ. Plasma enzymes are of two types: one is present in the highest concentration, is specific to plasma, and has a functional role; the other is normally present at very low levels and plays no functional role in the plasma. The former includes the enzymes associated with blood coagulation (thrombin), fibrin dissolution (plasmin), and processing of chylomicrons (lipoprotein lipase).

In disease of tissues and organs, the nonplasma-specific enzymes are most important. Normally, the plasma levels of these enzymes are low to absent. An insult in the form of any disease process may cause changes in cell membrane permeability or increased cell death, resulting in release of intracellular enzymes into the plasma. In cases of permeability change, those enzymes of lower molecular weight will appear in the plasma first. The greater the concentration gradient between intra- and extracellular levels, the more rapidly the enzyme diffuses out. Cytoplasmic enzymes will appear in the plasma before mitochondrial enzymes, and of course the greater the quantity of tissue damaged, the greater the increase in the plasma level. The nonplasma-specific enzymes will be cleared from the plasma at varying rates, which depend upon the stability of the enzyme and its susceptibility to the reticuloendothelial system.

In the diagnosis of specific organ involvement in a disease process it would be ideal if enzymes unique to each organ could be identified; however, this is unlikely, since the metabolism of various organs is very similar. Alcohol dehydrogenase of the liver and acid phosphatase of the prostate are useful for specific identification of disease in these organs. Other than these two examples, there are few enzymes that are tissue- or organ-specific. However, the ratio of various enzymes does vary from tissue to tissue. This fact, combined with a study of the kinetics of appearance and disappearance of particular enzymes in plasma, allows a diagnosis of specific organ involvement to be made. Figure 4.55 illustrates the time dependence of the plasma activities of enzymes released from the myocardium following a heart attack. Such profiles allow one to establish when the attack occurred and whether treatment is effective.

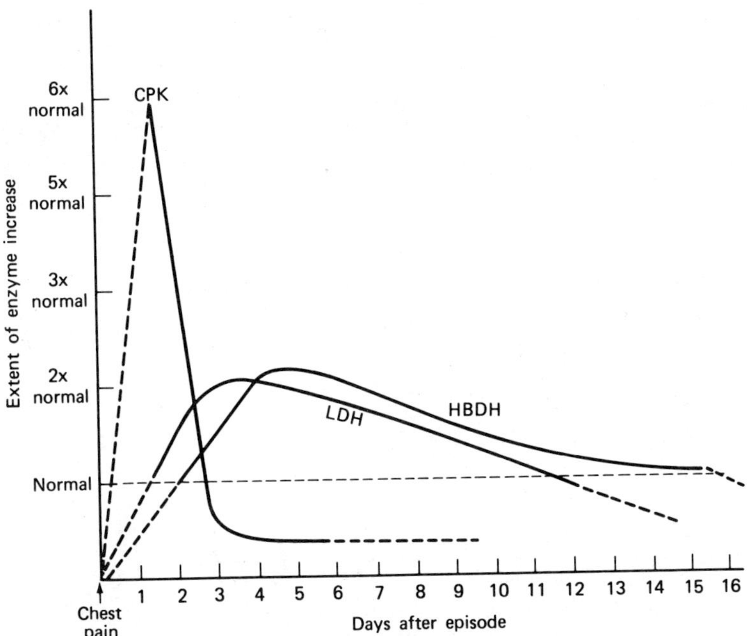

Figure 4.55
Kinetics of release of cardiac enzymes into serum following a myocardial infarction.
Creatine phosphokinase, CPK; lactic dehydrogenase, LDH; α-hydroxybutyric dehydrogenase, HBDH. Such kinetic profiles allow one to determine where the patient is with respect to the infarct and recovery. Note: CPK rises sharply but briefly; HBDH rises slowly but persists.

Reprinted, with permission, from *Diagnostic Enzymes,* E. L. Coodley, page 61. 1970, Lea and Febiger.

CLIN. CORR. 4.6
THE EFFECT OF CHANGES IN THE pH OPTIMUM OF ALCOHOL DEHYDROGENASE

In addition to the change in aldehyde dehydrogenase isoenzyme composition in Orientals (Clin. Corr. 4.2), a change in the alcohol dehydrogenase isoenzymes is also observed.

Alcohol dehydrogenase has three separate genetic loci for the production of three polypeptide chains-α, β, and γ. Now two alleles have been found in the loci responsible for the β chain such that two β chains are produced, β_1 and β_2. The correlation of ethnic background and percentage of the β_2 chain in alcohol dehydrogenase is shown in the table.

Nationality	*Percent of Total β Chain as β_2*
White American	5–10
Swiss	20
Japanese	85

The alcohol dehydrogenase containing a high percentage of the β_2 chain has a shift in pH optimum from the normal pH of 10 to 8.5. The end result of the subtle change in pH dependence is an increase in activity of alcohol dehydrogenase because of the shift in pH dependence toward more physiological pH values. Although more acetaldehyde is produced after ethanol ingestion, this factor in and of itself may not be sufficient to account for the sensitivity of Orientals to alcohol. Perhaps the increased production of acetaldehyde by the atypical alcohol dehydrogenase coupled with decreased use by the atypical aldehyde dehydrogenase result in the enhanced sensitivity to ethanol in these people.

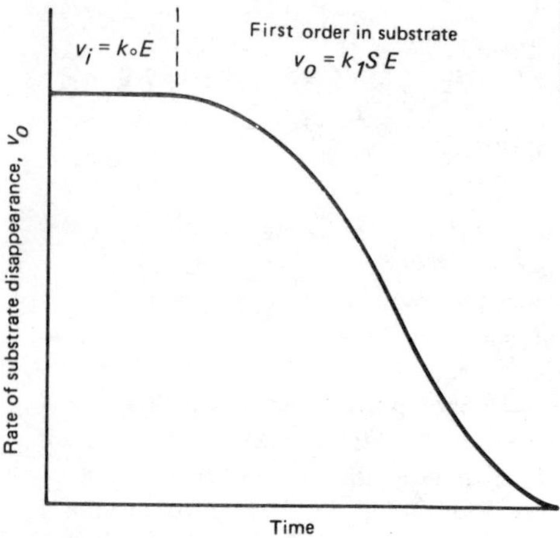

Figure 4.56
Relation of substrate concentration to order of the reaction.
When the enzyme is completely saturated, the kinetics are zero order with respect to substrate and are first order in enzyme, that is, the rate depends only on enzyme concentration. When the substrate level falls below saturating levels, the kinetics are first order in both substrate and enzyme and are therefore second order, that is, the observed rate is dependent upon both enzyme and substrate.

Clinical Correlation 4.7 demonstrates how diagnosis of a specific enzyme defect led to a rational clinical treatment that restored the patient to health.

Studies of the kinetics of appearance and disappearance of plasma enzymes require a valid enzyme assay. The establishment of a good assay is based on good temperature and pH control, as well as saturating levels of all substrates, cosubstrates, and cofactors. In order to accomplish the latter, the K_m must be known for those particular conditions of pH, ionic strength, and so on, that are to be used in the assay. You will recall that K_m is the substrate concentration at half-maximal velocity (V_{max}). To be assured that the system is saturated, the substrate concentration is generally increased 5- to 10-fold over the K_m. The importance of saturation of the enzyme with substrate is that only under these conditions is the reaction zero order. This fact is emphasized in Figure 4.56. Only under zero-order conditions are changes in velocity proportional to enzyme concentration alone. Under first-order conditions in substrate, the velocity is dependent upon both the substrate and the enzyme concentration.

Clinical Correlation 4.8 demonstrates the importance of determining if the assay conditions accurately reflect the amount of enzyme actually present. Clinical laboratory assay conditions are optimized routinely for the properties of the normal enzyme and may not reflect levels of mutated enzyme. The pH dependence and/or the K_m for substrate and cofactors may drastically change in a mutated enzyme.

Under optimal conditions a valid enzyme assay will reflect a linear dependence of velocity and amount of enzyme. This can be tested by determining if the velocity of the reaction doubles when the plasma sample size is doubled, while keeping the total volume of the assay constant, as demonstrated in Figure 4.57.

Coupled Assays

Enzymes that employ the coenzymes NAD, NADP, and FAD are easy to measure because of the optical properties of NADH, NADPH, and FAD. The absorption spectra of NADH and FAD in the ultraviolet and visible

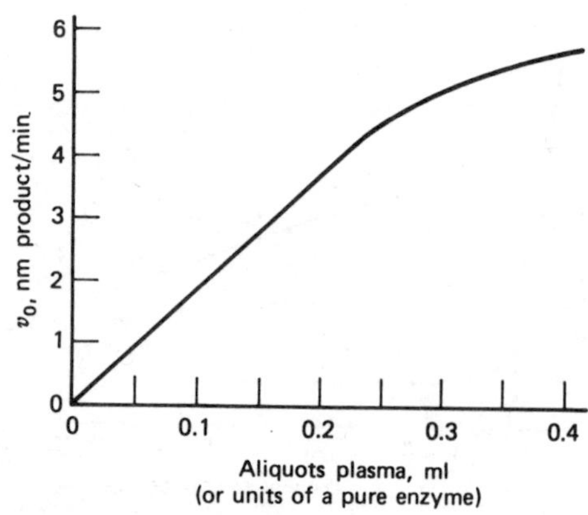

Figure 4.57
Assessing the validity of an enzyme assay.
The line shows what is to be expected for any reaction where the concentration of substrate is held constant and the aliquots of enzyme increased. In this particular example linearity between the initial velocity observed and the amount of enzyme, whether pure or in a plasma sample, is only observed up to 0.2 ml of plasma or 0.2 units of pure enzyme. If one were to measure the velocity with an aliquot of plasma greater than 0.2 ml, the actual amount of enzyme present in the sample would be underestimated.

CLIN. CORR. 4.7
IDENTIFICATION AND TREATMENT OF AN ENZYME DEFICIENCY

Enzyme deficiencies usually lead to increased accumulation of specific intermediary metabolites in plasma and hence in urine. The recognition of the intermediates that are accumulating is useful in pinpointing possible enzyme defects. After the enzyme deficiency is established, metabolites that normally occur in the pathway but are distal to the block may be supplied exogenously in order to overcome the metabolic effects of the enzyme deficiency. For example, in hereditary orotic aciduria there is a double enzyme deficiency in the pyrimidine biosynthetic pathway leading to accumulation of orotic acid.

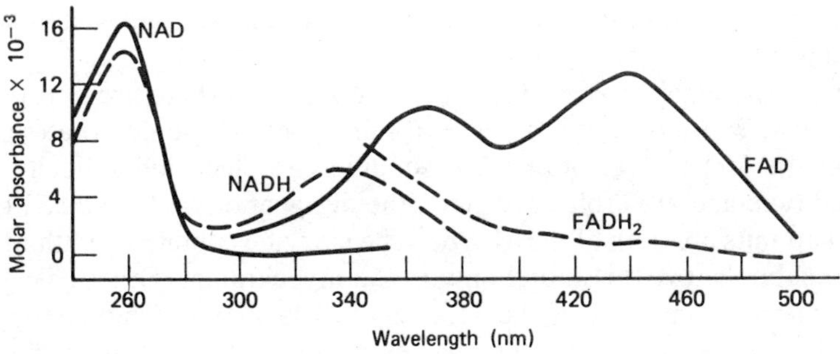

Figure 4.58
Absorption spectra of niacin and flavin coenzymes.
The reduced form of NAD (NADH) absorbs strongly at 340 nm. The oxidized form of flavin coenzymes absorbs strongly at 450 nm. Thus, one can follow the rate of reduction of NAD^+ *by observing the increase in the absorbance at 340 nm and the formation of* $FADH_2$ *by following the decrease in absorbance at 450 nm.*

light regions are shown in Figure 4.58. Oxidized FAD absorbs strongly at 450 nm, while NADH has maximal absorption at 340 nm. The concentration of both FAD and NADH is related to their absorption of light at the respective absorption maximum by the Beer–Lambert relation

$$A = \varepsilon \cdot c \cdot l$$

where l is the pathlength of the spectrophotometer cell in centimeters (usually 1 cm), ε is the absorbance of a molar solution of the substance being measured at a specific wavelength of light, A is the absorbance read off the spectrophotometer, and c is the concentration. Absorbance is the log of transmittance (I_0/I). ε is a constant which varies from substance to substance; its value can be found in a handbook of biochemistry. In an optically clear solution, the concentration c can be found after a determination of the absorbance A is made.

Many enzymes do not employ either NAD or FAD but do generate products that can be utilized by a NAD- or FAD-linked enzyme. For example, glucokinase catalyzes the reaction

$$\text{Glucose} + \text{ATP} \rightleftharpoons \text{glucose 6-phosphate} + \text{ADP}$$

Both ADP and G6P are difficult to measure directly; however, the enzyme glucose 6-phosphate dehydrogenase catalyzes the reaction,

$$\text{G6P} + \text{NADP}^+ \rightleftharpoons \text{6-phosphogluconolactone} + \text{NADPH} + \text{H}^+$$

Thus by adding an excess of the enzyme G6P dehydrogenase and NADP to the assay mixture, the velocity of production of G6P by glucokinase is proportional to the rate of reduction of NADP, which can be measured directly in the spectrophotometer.

Isoenzymes: Clinical Application

Isozymes (or *isoenzymes*) are enzymes that catalyze the same reaction but migrate differently on electrophoresis. Their physical properties may also be different, but not necessarily. The most common mechanism for the formation of isozymes involves the arrangement of subunits arising from two different genetic loci in different combinations to form the active polymeric enzyme. The isozymes that have been studied for clinical appli-

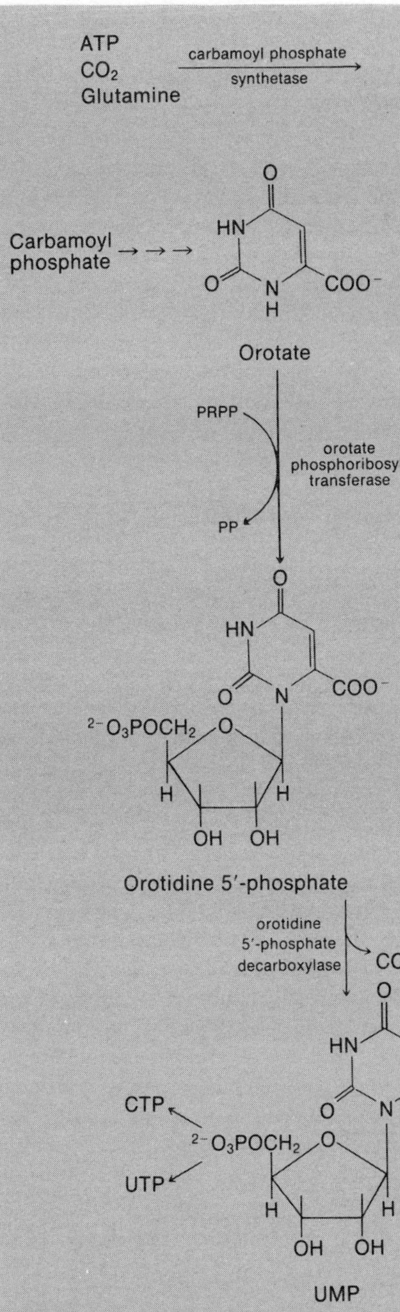

ATP, CO_2, Glutamine → (carbamoyl phosphate synthetase)

Carbamoyl phosphate → → →

Orotate

↓ PRPP → PP (orotate phosphoribosyltransferase)

Orotidine 5′-phosphate

↓ (orotidine 5′-phosphate decarboxylase) → CO_2

UMP

→ CTP, UTP

Both orotate phosphoribosyltransferase and orotidine 5′-phosphate decarboxylase are deficient. The pyrimidine nucleotides, CTP and TTP, are required for cell division, particularly in erythropoiesis. The patients are pale, weak, and fail to thrive. Administration of the missing pyrimidines as uridine and cytidine promotes growth, general well-being, and also decreases orotic acid excretion. The latter occurs because the TTP and CTP formed from the supplied nucleosides repress the carbamoyl phosphate synthetase, the committed step, by feedback inhibition.

CLIN. CORR. 4.8 AMBIGUITY IN THE ASSAY OF MUTATED ENZYMES

Structural gene mutations leading to the production of enzymes with changes in K_m are frequently observed. The K_m may be either increased or decreased, depending on the mutation. A case in point is a patient with hyperuricemia and gout, whose red blood cell hypoxanthine-guanine-phosphoribosyltransferase (HGPRT) showed little activity in assays in vitro. This enzyme is involved in the salvage of purine bases and catalyzes the reaction

$$\text{Hypoxanthine} + \text{PRPP} \longrightarrow \text{inosine monophosphate} + \text{PP}$$

where PRPP = phosphoribosyl pyrophosphate.

The absence of HGPRT activity results in a severe neurological disorder known as Lesch–Nyhan syndrome (page 501), yet this patient did not have the clinical signs of this disorder. Furthermore, immunological testing with a specific antibody to the enzyme revealed as much cross-reacting material in the patient's red blood cells as in normal controls. The conclusion was that the enzyme was being produced but was inactive in the assay in vitro. Additional experimentation revealed that by increasing the substrate concentration in the assay, full activity was measurable in the patient's red cell hemolysates. This anomaly is explained as a mutation in the substrate binding site of HGPRT, leading to an increased K_m. Neither the substrate concentration in the assay nor in the red blood cells was high enough to bind to the enzyme. This case reinforces the point that an accurate enzyme determination is dependent upon zero-order kinetics, that is, the enzyme being saturated with substrate.

Why doesn't the patient have the Lesch–Nyhan syndrome? Most likely because the cells in his nervous tissue are not affected by the mutation. This is possible since the red cell originates from a different stem cell line than the cells of the nervous system.

cations are lactate dehydrogenase, creatine kinase, and alkaline phosphatase.

Creatine kinase (page 814) occurs as a dimer. There are two types of subunits, M (muscle type) and B (brain type). These designations arise from the fact that in brain both subunits are electrophoretically of the same type and are arbitrarily given the designation B. In skeletal muscle the subunits are both of the M type. The isozyme containing both M and B type subunits (MB) is found only in the myocardium. Other tissues contain variable amounts of the MM and BB isozymes. The isozymes are numbered beginning with the species migrating the fastest to the anode thus, CPK_1 (BB), CPK_2 (MB), and CPK_3 (MM).

Lactate dehydrogenase is a tetrameric enzyme, but only two distinct subunits have been found: those designated H for heart (myocardium) and M for muscle. These two subunits are combined in five different ways. The lactate dehydrogenase isozymes, subunit compositions, and major location are as follows:

Type	*Composition*	*Location*
LDH_1	HHHH	Myocardium and RBC
LDH_2	HHHM	Myocardium and RBC
LDH_3	HHMM	Brain and kidney
LDH_4	HMMM	
LDH_5	MMMM	Liver and skeletal muscle

As an illustration of how measurement of amounts of isozymes and kinetic analysis of plasma enzyme activities are useful in medicine, activities of some CPK and LDH isozymes are plotted in Figure 4.59 as a function of time after infarction. After damage to heart tissue the cellular breakup releases CPK_2 into the blood within the first 6–18 h after an infarct, but LDH release lags behind the appearance of CPK by 1 to 2 days. Normally the activity of the LDH_2 isozyme is higher than that of LDH_1; however, in the case of infarction the activity of LDH_1 becomes

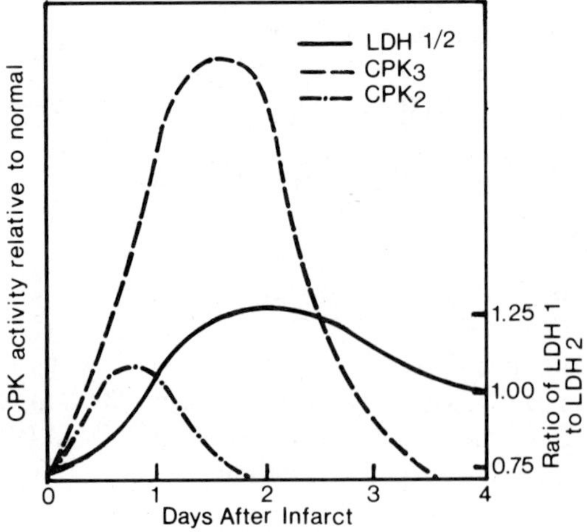

Figure 4.59
Characteristic changes in serum CPK and LDH isozymes following a myocardial infarction.
CPK_2 (MB) isozyme increases to a maximum within 1 day of the infarction. CPK_3 lags behind CPK_2 by about 1 day. The total LDH level increases more slowly. The increase of LDH_1 and LDH_2 within 12–24 h coupled with an increase in CPK_2 is diagnostic of myocardial infarction.

greater than LDH_2, at about the time CPK_2 levels are back to baseline (48–60 h). Figure 4.60 shows the fluctuations of all five LDH isozymes after an infarct. The increased ratio of LDH_2 and LDH_1 can be seen in the 24-h tracing. The LDH isozyme "switch" coupled with increased CPK_2 is diagnostic of myocardial infarct (MI) in virtually 100% of the cases. Increased activity of LDH_5 is an indicator of liver congestion. Thus secondary complications of heart failure can be monitored.

Formerly, plasma levels of the transaminases SGOT (serum glutamate-oxaloacetate transaminase also called aspartate transaminase or AST) and SGPT (serum glutamate-pyruvate transaminase also called alanine transaminase or ALT) were followed; however, these enzymes have much less specificity and predictive accuracy in diagnosing MI and liver disease. The rationale for assaying these two enzymes is that liver and heart contain high levels of both enzymes, but liver contains more GPT than GOT and the reverse is true in heart.

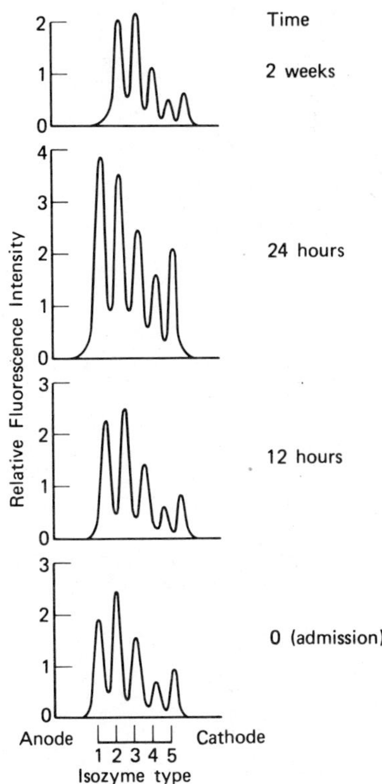

Figure 4.60
Tracings of densitometer scans of LDH isozymes at time intervals following a myocardial infarction.
As can be seen total LDH increases and LDH_1 becomes greater than LDH_2 between 12 and 24 h. Increases in LDH_5 is diagnostic of a secondary congestive liver involvement. After electrophoresis on agarose gels the LDH activity is assayed by measuring the fluorescence of the NADH formed in the LDH catalyzed reaction.
Courtesy of Dr. A. T. Gajda, Clinical Laboratories, The University of Arkansas for Medical Science.

Enzymes as Therapeutic Agents

In a few cases enzymes have been used as drugs in the therapy of specific medical problems. Streptokinase is an enzyme mixture prepared from a streptococcus. It is useful in clearing blood clots that occur in the lower extremities. Streptokinase activates the fibrinolytic proenzyme plasminogen that is normally present in plasma. The activated enzyme is plasmin. Plasmin is a serine protease like trypsin that attacks fibrin, cleaving it into several soluble components.

Asparaginase therapy is used for some types of adult leukemia. Tumor cells have a nutritional requirement for asparagine and must scavenge it from the host's plasma. By administering asparaginase i.v., the host's plasma level of asparagine is markedly depressed, which results in depressing the viability of the tumor.

Most enzymes do not have a long half-life in blood; consequently, unreasonably large amounts of enzyme are required to keep the therapeutic level up. Work is now in progress to enhance enzyme stability by coupling enzymes to solid matrices and implanting these materials in areas that are well perfused. In the future, enzyme replacement in individuals that are genetically deficient in a particular enzyme may be feasible.

4.10 REGULATION OF ENZYME ACTIVITY

Our discussion up to this point has centered upon the chemical and physical characteristics of individual enzymes, but physiologically we must be concerned with the integration of many enzymes into a metabolic pathway and the interrelationship of the products of one pathway with the metabolic activity of other pathways. For example, dietary glucose can be either converted to glycogen, fat, some nonessential amino acids or oxidized to carbon dioxide. In each case, glucose is converted to a different end product through a specific metabolic pathway involving several enzymes each of which is unique to the type of reaction catalyzed. After eating, there is an abundance of glucose in the system, but it is not diverted in equal amounts to each of the end products mentioned. Rather, there are very tightly controlled homeostatic mechanisms, which work to maintain a constant blood glucose level, utilize the glucose needed for energy production, maintain the glycogen stores, and, if any excess glucose remains, convert it to fat. The point is that all metabolic pathways are not operating at maximum capacity at all times. In fact many pathways may be shut down during certain phases in the life cycle of a cell. If this were not the case, wild, uncontrolled and uneconomical growth of the cell would occur.

Control of metabolic regulation of a pathway occurs through modulation of the enzymatic activity of one or more key enzymes in the pathway. Although the overall catalytic efficiency of a metabolic pathway is dependent upon the activity of all the individual enzymes in the pathway, the pathway can be controlled by one *rate-limiting* enzyme in the pathway. Usually this rate-controlling enzyme is the first enzyme that can be identified as unique to that particular pathway. The chemical reaction that is unique to a metabolic pathway is referred to as the *committed step*. For example, in the de novo synthesis of purines, the committed step is the reaction catalyzed by the PRPP amidotransferase, which in this case is also the rate-controlling enzyme. The rate-limiting enzyme is not necessarily the enzyme associated with the committed step. The substrate of the amidotransferase, PRPP, is also used as substrate by the pyrimidine biosynthetic pathway; hence the enzyme PRPP synthase, which produces PRPP, does not catalyze the committed step in the biosynthesis of purines because it occurs before the branch point in the two pathways (see page 523).

The activity of the enzyme associated with the committed step or with the rate-limiting enzyme can be regulated in a number of ways. First, the absolute amount of the enzyme can be regulated either by substrate or hormone stimulation of the de novo synthesis of more enzyme. Hormones can also suppress the de novo synthesis of enzyme. Second, the activity of the enzyme can be modulated by activators, inhibitors and by covalent modification through mechanisms previously discussed. Finally, the activity of a pathway can be regulated by partitioning the pathway from its initial substrate and by controlling access of the substrate to the enzymes of the pathway. This is referred to as *compartmentation*. We will now consider each of these general mechanisms of control in more detail.

Compartmentation

Generally anabolic and catabolic pathways are segregated into different organelles in order to maximize the cellular economy. There would be no point to the oxidation of fatty acids occurring at the same time and in the same compartment as biosynthesis of fatty acids. If such occurred, a futile cycle would exist. By maintaining fatty acid biosynthesis in the cytoplasm and oxidation in the mitochondria, control can be exerted by regulating transport of common intermediates across the mitochondrial membrane. For example, coenzyme A derivatives of fatty acids cannot diffuse across the mitochondrial membrane but are transported by a specific transport system. If the metabolic situation requires fatty acid biosynthesis rather than fatty acid oxidation, there could be hormonal or other control over the mitochondrial membrane fatty acid transport system such that it is depressed during fatty acid biosynthesis, but activated when fatty acid oxidation is required for cellular energy.

Table 4.2 contains a compilation of some of the important enzymes, metabolic pathways, and their intracellular distribution.

Control of Enzyme Levels

As indicated earlier, the velocity of any reaction is dependent upon the amount of enzyme present. Many rate-controlling enzymes are present in very low concentrations. More enzyme may be synthesized or existing rates of synthesis repressed through hormonally instituted activation of the mechanisms controlling gene expression. For example, insulin is an

TABLE 4.2 Intracellular Location of Major Enzymes and Metabolic Pathways[a]

Cytoplasm	Glycolysis: hexose monophosphate pathway; glycogenesis and glycogenolysis; fatty acid synthesis; purine and pyrimidine catabolism; peptidases; aminotransferases; amino acyl synthetases
Mitochondria	Tricarboxylic acid cycle; fatty acid oxidation; amino acid oxidation; fatty acid elongation; urea synthesis; electron transport and coupled oxidative phosphorylation
Lysosomes	Lysozyme; acid phosphatase; hydrolases, including proteases, nucleases, glycosidases, arylsulfatases, lipases, phospholipases and phosphatases
Endoplasmic reticulum (microsomes)	NADH- and NADPH-cytochrome c reductases; cytochrome b_5 and cytochrome P_{450} related mixed function oxidases; glucose 6-phosphatase; nucleoside diphosphatase, esterase, β-glucuronidase, and glucuronyltransferase; protein synthetic pathways; phosphoglyceride and triacylglycerol synthesis; steroid synthesis and reduction
Golgi	Galactosyl- and glucosyltransferase; chondroitin sulfotransferase; 5′-nucleotidase; NADH-cytochrome c reductase; glucose 6-phosphatase
Peroxisomes	Urate oxidase; D-amino acid oxidase; α-hydroxy acid oxidase; catalase; long chain fatty acid oxidation
Nucleus	DNA and RNA biosynthetic pathways

[a] NADH-cytochrome b_5 reductase has been found in endoplasmic reticulum, Golgi, outer mitochondrial membrane, and in the nuclear envelope. Several of the enzymes noted in the table are common to one or more of the membranous organelles.

anabolic hormone that induces the synthesis of increased amounts of glucokinase, phosphofructokinase, pyruvate kinase, and glycogen synthetase, but represses synthesis of several key gluconeogenic enzymes. The detailed mechanism of these effects are not known in mammalian systems; however, the general concepts of the regulation of eucaryotic gene expression are discussed in Chapter 20.

In some instances substrate can repress the synthesis of enzyme. For example, glucose represses the de novo synthesis of pyruvate carboxykinase. This enzyme is the rate-limiting enzyme in the conversion of pyruvate to glucose. In other words, if there is plenty of glucose available there is no point in synthesizing glucose at the expense of amino acids which are the alternative source of pyruvate; consequently, this pathway is repressed by the effect of its end product, glucose, on the synthesis of the carboxykinase enzyme. This effect of glucose may be mediated via insulin and is not direct feedback inhibition.

Many rate-controlling enzymes have relatively short half-lives, for example, that of pyruvate carboxykinase is 5 h. Teleologically this is reasonable because it provides a mechanism for effecting much larger fluctuations in the activity of a pathway than would be possible by inhibition or activation of existing levels of enzyme.

Regulation by Modulation of Activity

Regulation at the gene level is long-term. Short-term regulation occurs through modification of the activity of existing levels of enzyme by means of mechanisms we will now consider.

During various phases of the cell cycle, specific metabolic pathways are turned on or off, depending on the special requirements of a given phase of the cell cycle for a particular product. For example, there is no point to continued production of deoxyribonucleotides at all times during a cell's life, but only during replicative phases; consequently, during non-replicative phases, the concentration of deoxyribonucleotide builds up to such an extent that the ribonucleoside diphosphate reductase is inhibited by the end products of the pathway. This type of control is referred to as *feedback inhibition*. The inhibition may take the form of competitive inhibition or allosteric inhibition. In any case, the apparent K_m may be raised above the in vivo levels of substrate, and the reaction ceases or decreases in velocity.

In addition to feedback within the pathway, feedback on other pathways also occurs. This is referred to as *cross-regulation*. In cross-regulation a product of one pathway serves as an inhibitor or activator of an enzyme occurring early in another pathway as depicted in Figure 4.61. A good example, which will be considered in detail in Chapter 13, is the cross-regulation of the production of the four deoxyribonucleotides for DNA synthesis so that approximately equal amounts of each are produced.

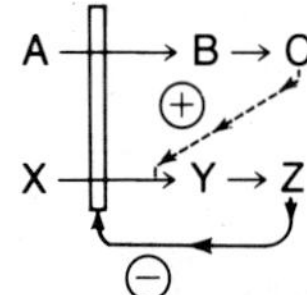

Figure 4.61
Model of feedback inhibition and cross-regulation.
The open bar indicates inhibition, and the broken line indicates activation. The product Z cross-regulates the production of C by its inhibitory effect on the enzyme responsible for the conversion of A to B in the A → B pathway. C in turn cross-regulates the production of Z. Z inhibits its own formation by feedback inhibition of the conversion of X to Y.

In addition to the inhibition and activation mechanisms just discussed, another very important mechanism of control of enzyme activity is that of reversible covalent modification, which will now be discussed in some detail.

Regulation by Covalent Modification

The first example of this mechanism of regulation was glycogen phosphorylase, in which the interconvertible a (active) and b (inactive) forms were recognized to be phosphorylated and dephosphorylated proteins, respectively.

Other examples of reversible covalent modification include acetylation–deacetylation, adenylylation–deadenylylation, uridylylation–deuridylylation, and methylation–demethylation.

The phosphorylation–dephosphorylation scheme is most common and will be considered in detail. There are four different modes of phosphorylation based upon the cofactor requirement. These are the cAMP dependent, the non-cAMP-dependent, the calcium-dependent, and the double-stranded RNA-dependent protein kinases. The cAMP dependent phosphorylation can be considered as characteristic of the other types of covalent modification. Details of the mechanisms will be covered in the chapters on metabolism and regulation.

cAMP-Dependent Phosphorylation

The phosphorylation of an enzyme occurs as the end result of a cascade of reactions initiated by the binding of a hormone such as epinephrine to a specific extracellular membrane receptor on the target cell. Such binding activates the adenylate cyclase on the intracellular surface of the plasma membrane through an induced conformational change. Adenylate cyclase catalyzes the cyclization of ATP to cAMP (Figure 4.62), which then allosterically activates cAMP-dependent protein kinases.

The cAMP activates the protein kinase by combining with its regulatory (R) subunits, resulting in the release of active catalytic subunits (C) as outlined in Figure 4.63. cAMP is referred to as the "second messenger," since it transmits the signal from the hormone, the first messenger, to the intracellular protein kinases, which then effect the intracellular response to the hormone. The hormone does not enter the cell to produce its ultimate metabolic effect. The steroid hormones operate by a different mechanism which will be discussed on page 569.

Figure 4.62
Epinephrine stimulation of adenyl cyclase.
Activation of the adenyl cyclase residing on the intracellular side of the cell membrane by extracellular epinephrine results in a rise in the intracellular levels of cyclic adenosine monophosphate (cAMP). The mechanism of hormone stimulation of the cyclase is described in Figure 4.65.

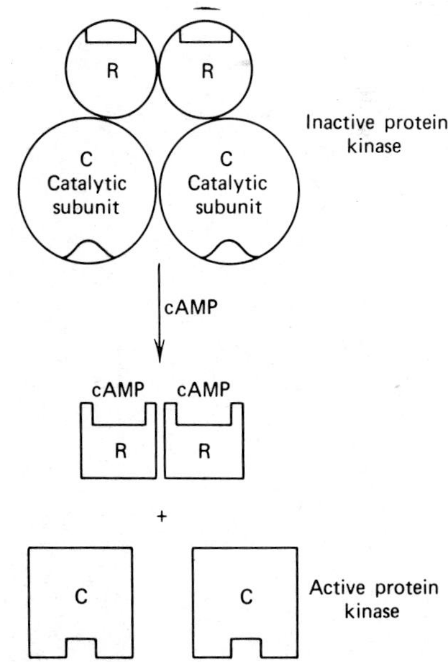

Figure 4.63
cAMP activation of protein kinase.
The enzyme is tetrameric with two catalytic and two regulatory subunits. Binding of cAMP to the regulatory subunits (R) results in dissociation of the complex and activation of the catalytic subunits (C). The active protein kinase is rather nonspecific with respect to the protein substrate.

The activated cAMP-dependent protein kinases can phosphorylate various other inactive protein kinases and activate them. The latter are kinases that phosphorylate specific enzymes. This is the point of specificity in the hormone stimulation of adenylate cyclase. As suggested by Figure 4.64, the phosphorylated form of a particular enzyme resulting from the protein kinase reaction may be activated or inhibited (or rendered less active). If enzyme Y is inactive and enzyme X is active in the phosphorylated form, stimulation by epinephrine will decrease the production of C and increase the production of A from B. The case in which B is glycogen will be considered in detail in Chapter 7, but the point should be made that the mechanism outlined provides a means for hormonal control of the synthesis and degradation of glycogen.

Additional control of such systems arises from the presence of specific phosphoprotein phosphatase enzymes that dephosphorylate a given enzyme, such as enzyme Y shown in Figure 4.64. The activity of the phosphatases is in turn controlled by substrates and products of the pathway. The detailed mechanism of this kind of control will be discussed in Chapter 7.

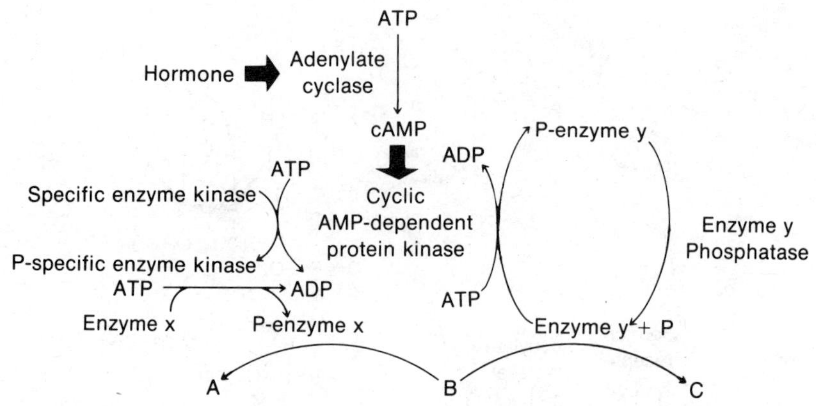

Figure 4.64
Generalized model for the nonsteroid hormone control of enzyme activity. *Stimulation of the hormone receptor increases intracellular cAMP, which activates protein kinases. The phosphorylated enzymes resulting from the protein kinase action may be either active (enzyme x) or inactive (enzyme y). The phosphorylated enzymes are converted to dephosphoenzymes by specific phosphatases, thus establishing the state before hormone binding to the receptor.*

Control of cAMP Levels

For effective regulation of any system, there must be an "off" as well as an "on" switch. cAMP can be visualized as the "on" switch which activates a number of kinases, and, as we will learn later, causes the de novo synthesis of some specific enzymes (page 774). cAMP levels are controlled in two ways. As the hormone diffuses away from the membrane receptor, guanosine triphosphate (GTP) on the regulatory subunit of the adenylate cyclase is hydrolyzed to GDP. The GDP binds tightly to the regulatory protein and induces a conformational change which results in dissociation of regulatory and catalytic subunits of adenylate cyclase. As a result of the dissociation of the catalytic and regulatory subunits the catalytic subunit assumes an inactive conformation. These effects are summarized in Figure 4.65. Binding of additional hormone stimulates the release of the GDP and binding of GTP, which activates the adenylate cyclase. Cytoplasmic levels of cAMP also are controlled by a cAMP-dependent phosphodiesterase that catalyzes the reaction

$$\text{cAMP} + H_2O \longrightarrow \text{AMP}$$

thus removing the activator of the protein kinase cascade. This phosphodiesterase is subject to control by caffeine and related purines. These substances inhibit the diesterase, resulting in a stimulatory effect by allowing cAMP levels to remain above steady state concentrations.

Calmodulin

It now appears that calcium is a "second messenger" that functions through activation of a regulatory subunit of a number of cAMP-independent protein kinases. Calmodulin is a term coined for the "calcium-dependent regulatory protein." In the presence of micromolar amounts of calcium, calmodulin undergoes a conformational change, which is transmitted cooperatively to the catalytic subunit of a number of specific enzyme kinases, such as phosphorylase kinase and glycogen synthase kinase. Calmodulin is a calcium-dependent protein allosteric activator. In phosphorylase kinase, the α and β subunits are phosphorylated by a cAMP-dependent protein kinase, whereas the δ subunit is identical to

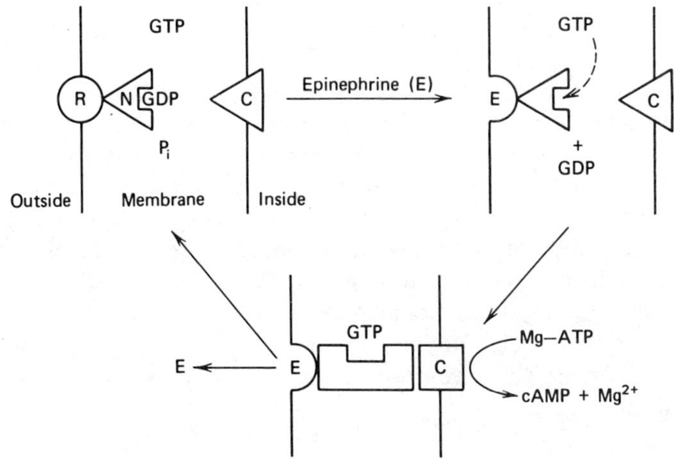

Figure 4.65
Model of hormone activation of adenylate cyclase.
The hormone receptor (R) is bound to the regulator (N) that has bound to it guanosine diphosphate (GDP). The GDP serves as a "stop" signal in the absence of hormone. Binding of a hormone like epinephrine triggers the release of GDP, which allows GTP to bind. GTP serves as an effector that initiates conformational changes in the regulator subunit so that adenylate cyclase catalytic subunit (C) is activated. Diffusion of epinephrine from the receptor allows the hydrolysis of GTP to GDP resulting in reversal of the conformation of adenylate cyclase to the inactive form.

calmodulin. The γ subunit is the catalytic subunit whose activity is regulated both by cAMP and calcium levels as suggested in Figure 4.66. Although the scheme shown in Figure 4.66 illustrates the control of phosphorylase kinase in particular through both cAMP and calcium via calmodulin, the mechanism may be much more general because it allows for control by two independent mediators, which would result in fine tuning of the metabolic pathway.

The cAMP phosphodiesterase also contains calmodulin as the regulatory protein. Calcium increases phosphodiesterase activity; hence the

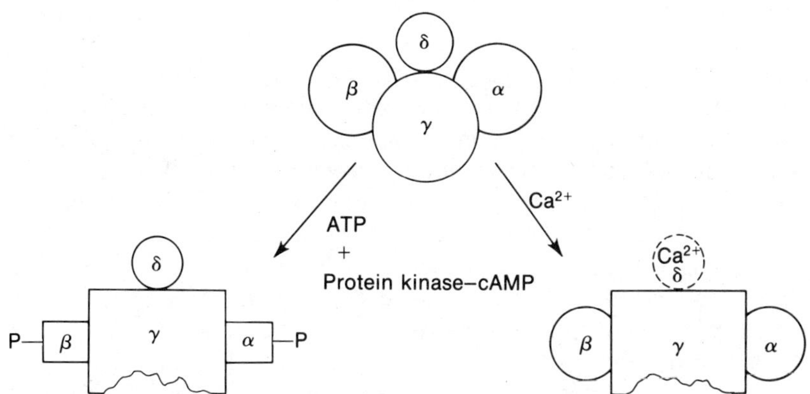

Figure 4.66
Modulation of the phosphorylase kinase reaction by cAMP-dependent protein kinase and calcium activated calmodulin.
The catalytic subunit is γ. The α and β subunits are phosphorylated by protein kinase, which results in activation. Another route of activation is through the δ subunit, which is calmodulin. Calcium binding to calmodulin activates the catalytic subunit of phosphorylase kinase.

intracellular cAMP level is controlled indirectly by the flux of calcium in the cell.

Calcium and calmodulin function is much broader than would be indicated by our discussion of the phosphorylase kinase and phosphodiesterase enzymes. In muscle the intracellular calcium level is controlled by an ATP dependent "pump," which responds indirectly to nervous stimulation. Stimulation increases the intracellular calcium levels. The calcium binds to calmodulin, which then activates all those enzymes that are responsive to it, including the calcium pump. Calcium acting through calmodulin has multifunctional regulatory capability affecting such diverse processes as cell mobility and motility, excitation–contraction, cytoplasmic streaming, chromosome movement, axonal flow, and neurotransmitter release.

4.11 AMPLIFICATION OF REGULATORY SIGNALS

In biological systems many of the signal molecules, such as hormones, that function in interorgan communication are very low in concentration but have tremendous effects on the target organ. The signal generated by the binding of very small amounts of hormone is multiplied many-fold inside the cell through a process of biological amplification. The mechanism of this amplification involves a cascade of reactions whereby the activation of the initial enzyme in the cascade activates a second proenzyme, and it in turn activates a third proenzyme, and so on. Since catalytic proteins are involved at each step in the cascade, the initial signal can be increased many times in terms of amounts of the final product generated.

Phosphorylase Cascade

A beautiful example of amplification is the epinephrine-stimulated phosphorolysis of glycogen. The left-hand series of reactions in Figure 4.64 is a good model for this system, in which B = glycogen, A = glucose 1-phosphate, and the enzyme X is phosphorylase. Levels of epinephrine of the order of 10^{-10} mol/g of muscle will stimulate the formation of 25×10^{-6} moles of glucose 1-phosphate per minute per gram of muscle. This is an amplification factor of 250,000. If we look at the initial stages of the amplification, we find that the epinephrine raises the steady-state cAMP concentration only three- to fivefold. This may at first appear to be a rather small amplification, but we must remember that the hormone is binding to only a limited number of receptor sites on the cell surface, which only affects the activation of an equivalent number of adenylate cyclase molecules. In addition, phosphodiesterases rapidly hydrolyze the cAMP formed, therefore we are not able to measure the full extent of amplification. Theoretical considerations have shown that the logarithm of the concentration of effector needed to activate 50% of the ultimate target enzymes is inversely proportional to the number of steps in the cascade. Since there are three steps between cAMP and the activation of phosphorylase, it has been calculated that only a 1% increase in the steady-state level of cAMP would produce a 50% activation of phosphorylase; consequently, three- to fivefold increase in cAMP is theoretically sufficient to activate all the phosphorylase.

In general the factors that limit the extent of amplification in a system involving a catalytic enzyme cascade are the relative amounts and the turnover numbers of each enzyme in the cascade. If each step in the cascade is bicyclic, phosphorylation-dephosphorylation, for example,

Factor XII
Collagen/platelet factor ⟹ ↓ Factor XI
XII_a ⟹ ↓ Factor IX
XI_a ⟹ ↓
IX_a
Factor X — Factor VIII
↓ ⟸ VIII_a-IX_a
X_a
Factor V_a ↓ Prothrombin
V_a-X_a ⟹
Thrombin
⇓
Fibrinogen ⟶ fibrin + fibrinopeptides

Figure 4.67
The intrinsic pathway of blood coagulation.
Each factor is a proenzyme and each activated factor is a specific peptidase except for V_a *and* VIII_a*, which are cofactors. Unlike the phosphorylase cascade, activation of each subsequent enzyme in the cascade involves hydrolytic removal of peptide segments from the inactive factor. The end product of the cascade, thrombin, is inactivated by the plasma protein antithrombin III. This plasma inhibitor provides a means of regulation of the coagulation cascade. The double arrows indicate the catalyzed phases of the reaction.*

then the ratio of the forward rate (activation) to the backward rate (deactivation) is important. If the activated enzyme is inactivated by inhibitors, as occurs in the blood coagulation cascade discussed below, then the relative rates of activation and inactivation become important.

It is interesting that the concentrations of the enzymes in the phosphorylase cascade are of the ratio expected for an amplification system. In fast twitch muscles, the molar ratios of cAMP-dependent protein kinase, phosphorylase kinase, and phosphorylase are 1 : 10 : 240. In slow twitch muscles the ratios are 1 : 0.15 : 25. Thus the absolute concentration as well as the ratio of the enzymes in a cascade are altered to provide the chemical response appropriate to the functioning of a particular tissue.

Blood Coagulation

Another cascade that is of great biological importance occurs in the coagulation of blood and involves a mechanism other than phosphorylation–dephosphorylation for control. Unlike the phosphorylase cascade, the coagulation cascade is unidirectional, that is, there is no mechanism for converting activated components into inactive forms, which can be reactivated. The intrinsic coagulation scheme is schematically outlined in Figure 4.67.

In this cascade, there is an initial release of a protease that activates a proprotease, and so on, through five cycles of activation as indicated by the double arrows in Figure 4.67. Amplification in this system is of the order of 10^6. Such large amplification factors allow the biological end response to be sensitive to very minute amounts of the inciting agent.

BIBLIOGRAPHY

Boyer, P. D. (ed.). *The enzymes,* 3rd ed. (13 vols.). New York: Academic, 1970–1976.

Nord, F. F. (ed.). *Advances in enzymology.* New York: Wiley-Interscience, issued annually.

Snell, E. E. (ed.). *Annual reviews of biochemistry.* Palo Alto: Annual Reviews, Inc. (issued annually).

Weber, G. (ed.). *Advances in enzyme regulation,* vols. 1–6. Elmsford, N.Y.: Pergamon, 1963–1979.

QUESTIONS

J. BAGGOTT AND C. N. ANGSTADT

*Q*uestion *T*ypes are described inside the front cover.

1. (QT1) The reaction

$$\begin{array}{c} CH_2\text{—}COOH \\ | \\ HO\text{—}C\text{—}COOH \\ | \\ CH_2\text{—}COOH \end{array} + \text{CoA-SH} \rightleftharpoons$$

$$HOOC\text{—}CH_2\text{—}\overset{\overset{\displaystyle O}{\|}}{C}\text{—}COOH + CH_3\text{—}\overset{\overset{\displaystyle O}{\|}}{C}\text{—}S\text{—}CoA + H_2O$$

is catalyzed by:
A. an oxidoreductase.
B. a transferase.
C. a hydrolase.
D. a lyase.
E. a ligase.

2. (QT2) Although enzymic catalysis is reversible, a given reaction may appear irreversible:
1. if the products are thermodynamically far more stable than the reactants.
2. under initial velocity conditions.
3. if a product is rapidly removed from the system.
4. at high enzyme concentrations.

3. (QT2) Metal ions may:
1. serve as Lewis acids in enzymes.
2. participate in oxidation–reduction processes.
3. stabilize the active conformation of an enzyme.
4. form chelates with the substrate, with the chelate being the true substrate.

A. Metal cofactor | C. Both
B. Nonmetal cofactor | D. Neither

4. (QT4) Can serve as a one-electron acceptor/donor.
5. (QT4) Enzyme can be generally expected to exhibit activity in the absence of the cofactor.
6. (QT2) Which of the following pairs can be distinguished on the basis of Michaelis–Menten inhibition patterns?
1. Competitive–noncompetitive
2. Competitive–irreversible
3. Noncompetitive–uncompetitive
4. Noncompetitive–irreversible

A. Michaelis–Menten kinetics | C. Both
B. Allosteric kinetics with cooperative interaction | D. Neither

7. (QT4) Lineweaver–Burk plot is useful for determining K_m and V_{max}.
8. (QT4) Inhibitor may increase the apparent K_m.
9. (QT4) Small changes in $[S]$ produce the largest changes in v when $[S] = K_m$.
10. (QT2) Drugs that act as enzyme inhibitors:
1. may function as competitive inhibitors.
2. are clinically useful only when directed against an enzyme unique to a cell that is to be killed.
3. may serve as irreversible inhibitors.
4. must be harmless to the patient.

11. (QT2) Enzymes may be specific with respect to:
1. chemical identity of the substrate.
2. optical activity of product formed from a symmetrical substrate.
3. type of reaction catalyzed.
4. which of a pair of optical isomers will react.

12. (QT2) Which of the following necessarily result(s) in formation of an enzyme–substrate intermediate?
1. Substrate strain
2. Acid–base catalysis
3. Entropy effects
4. Covalent catalysis

13. (QT1) An enzyme with histidyl residues that participate in both general acid and general base catalysis would be most likely to have a pH-activity profile resembling:
A. Curve A
B. Curve B
C. Curve C
D. Curve D
E. Curve E

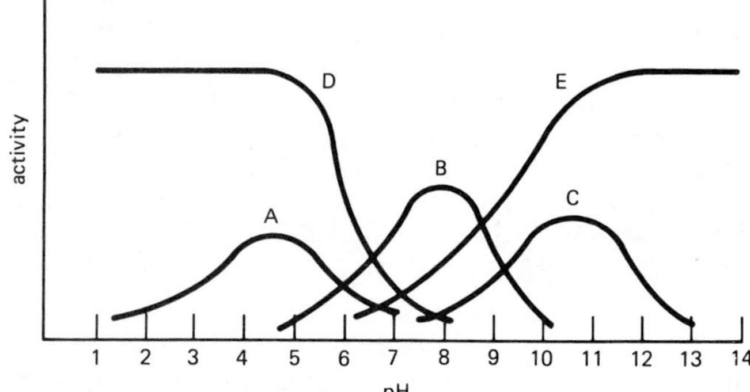

14. (QT1)

$$\begin{array}{c} \text{Cpd 3} \\ \downarrow\uparrow B \\ \text{Cpd 1} \xrightarrow{A} \text{Cpd 2} \underset{}{\overset{C}{\rightleftharpoons}} \text{Cpd 4} \xrightarrow{D} \text{Cpd 5} \overset{E}{\rightleftharpoons} \text{Cpd 6} \end{array}$$

In the reaction sequence above, the best point for controlling production of compound 6 is reaction:
A. A
B. B
C. C
D. D
E. E

15. (QT2) If the plasma activity of an intracellular enzyme is abnormally high:
1. the rate of removal of the enzyme from plasma may be depressed.
2. tissue damage may have occurred.
3. the enzyme may have been activated.
4. determination of the isozyme distribution may yield useful information.

16. (QT2) Features of all enzyme cascades include:
1. circulating hormone enters the cell, where it binds to an intracellular protein.
2. participating enzymes may be activated, inactivated, or reactivated indefinitely.
3. counterregulation.
4. signal amplification.

17. (QT2) Types of physiological regulation of enzyme activity include:
1. covalent modification.
2. changes in rate of synthesis of the enzyme.
3. allosteric activation.
4. competitive inhibition.

ANSWERS

1. D This is an unusually complicated lyase reaction, since secondary reactions are involved. It is a lyase because it removes a group (the acetyl group) with formation of a double bond (the C=O bond of the four-carbon product, oxalacetate). The common name of this enzyme is citrate synthase (p. 222).

2. A 1, 2, and 3 true. 1: Such a system is theoretically reversible, but it would be difficult to reverse it in practice (p. 125). 2: At the beginning of a reaction (initial velocity conditions), product concentration is so low that the reverse reaction is insignificant. 3: If a product is removed, it is no longer available for the reverse reaction (p. 125). 4: Is false because enzymes merely catalyze reactions; they do not alter the equilibrium, no matter what their concentration (p. 118).

3. E All of the statements are true (p. 135).

4. C The most familiar example of a metal cofactor serving as a one-electron acceptor–donor is any of the cytochromes (p. 139). Other examples exist. Flavins may also participate in one-electron transfers (p. 134).

5. D Not all enzymes have cofactors, but when they do, the cofactor is generally essential for activity (p. 118).

6. A 1, 2, and 3 true. Competitive inhibitors increase the apparent K_m, noncompetitive and irreversible inhibitors both decrease V_{max}, and uncompetitive inhibitors decrease the apparent K_m (pp. 139–141, Figures 4.27 and 4.28).

7. A The Lineweaver–Burk plot is a rearrangement of the Michaelis–Menten equation into a form that makes graphical evaluation of K_m and V_{max} easier. It is not applicable to systems that cannot be described by the Michaels–Menten equation (p. 129).

8. C Competitive inhibitors do this in systems described by Michaelis–Menten kinetics (pp. 139–140, Figure 4.27), and negative allosteric effectors affect K class allosteric enzymes in this way (p. 145, p. 146, Figure 14.36).

9. B For Michaelis–Menten enzymes, the v vs [S] curve is steepest at the lowest [S]. For allosteric enzymes, the steepest part of the curve is in the neighborhood of the [S] that corresponds to $v = \frac{1}{2} V_{max}$, that is, K_m (p. 146).

10. B 1 and 3 true. Drugs may serve as competitive inhibitors, such as sulfanilamide (p. 141), or as irreversible inhibitors, such as fluorouracil (p. 144). Pathways unique to pathogenic bacteria, viruses, and so on, are rare, so drugs are often developed that are merely *less* harmful to the host than the target cell (because of differences in cell permeability, metabolic rate, etc.) (p. 141).

11. E 1, 2, 3, and 4 true. Enzymes are specific for the substrate and the type of reaction (p. 150). The asymmetry of the binding site generally permits only one of a pair of optical isomers to react, and only one optical isomer is generated when a symmetric substrate yields an asymmetric product (p. 152).

12. D Only 4 true. All enzyme-catalyzed reactions involve an enzyme–substrate complex. There is always at least one transition state involved, but only in covalent catalysis is a covalent bond between enzyme and a portion of the substrate involved (p. 157).

13. B A group must be in the correct ionization state to act catalytically. For a histidyl group to serve as a general acid and a general base (as it does in chymotrypsin), the pH must be compatable with *both* ionization states of histidine. Since the pK of the histidyl side chain is about 6.8, the maximum activity is likely to be near that pH. Chymotrypsin's pH optimum is in the 7–9 range (p. 158).

14. D Control of reaction A would control production of Cpds 3 and 6. Reaction B is not on the direct route. Reaction C is freely reversible, so it does not need to be controlled. Reaction D is irreversible; if it were not controlled, Cpd 5 might build up to toxic levels (p. 166).

15. E 1, 2, 3, and 4 true. Intracellular enzymes may appear in abnormal amounts when tissues are damaged. Different tissues have characteristic distributions of isozymes. Choices 1 and 3 are theoretically possible, but are less common (p. 164).

16. D Only 4 true. 1: Epinephrine binds to extracellular receptors to activate adenylate cyclase (p. 168), and blood coagulation is entirely extracellular (p. 173). 2: The enzymes of blood coagulation are irreversibly inhibited (p. 173). 3: Counterregulation occurs with glycogen synthesis and breakdown (p. 169) but not with blood coagulation. 4: Amplification of a small signal by activating a sequence of enzymes (catalysts) is the purpose of a cascade (p. 172).

17. E 1, 2, 3, and 4 true. 1: Covalent modification includes zymogen activation and phospho–dephospho protein conversions (p. 168). 2: Enzyme levels may be controlled (p. 166). 3: Allosteric activation is common (p. 168). 4: End products of a reaction or reaction sequence may inhibit their own formation by competitive inhibition (p. 168).

5

Biological Membranes: Structure and Membrane Transport

THOMAS M. DEVLIN

5.1 OVERVIEW

All biological membranes, whether from eucaryotic or procaryotic cells, have the same classes of chemical components, a similarity in structural organization, and a number of properties in common. There are, of course, major differences in the specific lipid, protein, and carbohydrate components but not in the physicochemical interaction of these molecules in the membrane. Biological membranes have a trilaminar appearance when viewed by electron microscopy (Figure 5.1), with two dark bands on each side of a light band. The overall width of the various mammalian membranes is 7–10 nm; some membranes, however, have significantly smaller widths. Differences are also observed in both the size and the density of membranes depending on the staining technique employed in preparing tissue sections, making it difficult to determine the significance of the differences in width. Intracellular membranes are usually thinner than the plasma membrane. In addition, many membranes do not appear symmetrical, with the inner dense layer often thicker than the outer dense layer; as discussed below there is a chemical asymmetry of the membrane, with some components only on one or the other side. With the development of sophisticated techniques for preparation of tissue samples and staining, including negative staining and freeze fracturing, the various surfaces of membranes have been viewed; at the molecular level the surfaces are not smooth but dotted with globular-shaped components protruding from the membrane. Data from electron microscopic evaluation of membranes have been useful in developing current concepts on the molecular structure of membranes.

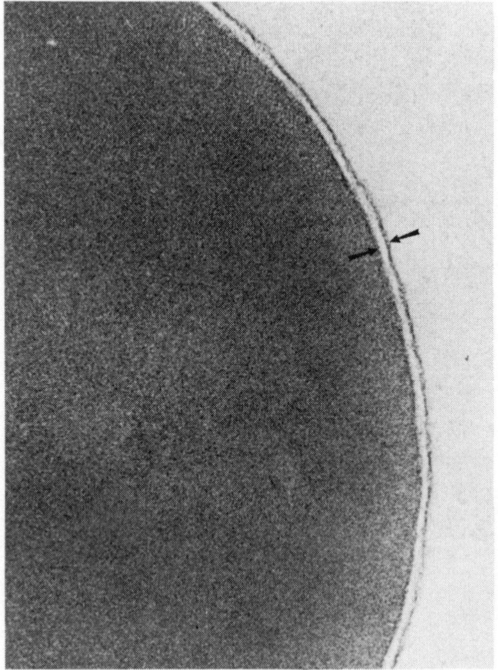

Figure 5.1
Electron micrograph of the erythrocyte plasma membrane showing the trilaminar appearance.
A clear space separates the two electron dense lines. Electron microscopy has demonstrated that the inner dense line is frequently thicker than the outer line. Magnification about 150,000×.
Courtesy of Dr. J. D. Robertson.

Even though valuable in defining structure, electron micrographs present a very static picture of membranes. Membranes are very dynamic structures with a movement that permits the cell as well as subcellular structures in eucaryotic cells to adjust their shape and to move. The chemical components of membranes, that is, lipids and protein, are ideally suited for the dynamic role of membranes. Membranes are visualized as essentially a semistructured, organized sea of lipid in a fluid state in which the various components are able to move. The lipid membrane in the fluid state is a nonaqueous compartment of the cell in which components can interact. Thus, in considering the role of membranes in the various activities of the cell, we should remember their dynamic state.

Cellular membranes control the composition of the space they enclose not only by their ability to exclude a variety of molecules but also because of the presence of selective transport systems permitting the movement of specific molecules from one side to the other. By controlling the translocation of substrates, cofactors, ions, and so on, from one compartment to another, membranes modulate the concentration of substances, thereby exerting an influence on metabolic pathways. The plasma membrane of eucaryotic cells also has a role in cell–cell recognition, maintenance of the shape of the cell and in cell locomotion. The site of action of many hormones and metabolic regulators is on the plasma membrane (Chapter 16), where there are specific recognition sites, and the information to be imparted to the cell by the hormone or regulator is transmitted by the membrane component to the appropriate metabolic pathway by a series of intracellular intermediates, termed second messengers.

The discussion that follows is directed primarily to the chemistry and function of membranes of mammalian cells but the basic observations and activities described are applicable to all biological membranes.

5.2 CHEMICAL COMPOSITION OF MEMBRANES

Lipids and proteins are the two major components of all membranes but the amount of each varies greatly between different membranes (Figure 5.2). The percent of protein ranges from about 20% in the myelin sheath to over 70% in the inner membrane of the mitochondria. Intracellular membranes have a high percentage of protein because of the greater enzymatic activity of these membranes. Membranes also contain a small amount of various polysaccharides in the form of glycoprotein and glycolipid; there is no free carbohydrate in membranes.

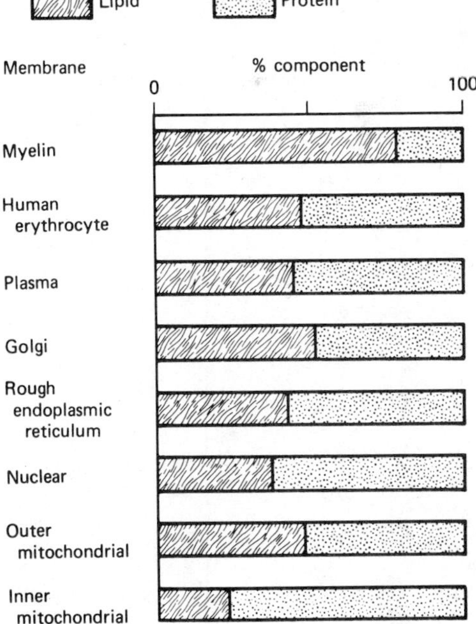

Figure 5.2
Representative values for the percentage of lipid and protein in various cellular membranes.
Values are for rat liver, except for the myelin and human erythrocyte plasma membrane. Values for liver from other species, including human, indicate a similar pattern.

Lipids of Membranes

The three major lipid components of membranes are phosphoglycerides, sphingolipids, and cholesterol; individual cellular membranes also contain small quantities of other lipids, such as triacylglycerol and diol derivatives (see Appendix, "Review of Organic Chemistry," for a discussion of the chemistry of lipids).

The percentage of each of the major classes varies significantly in different membranes and is presumably related to the specific roles of the individual membranes. This is discussed in more detail below.

Phosphoglycerides

Phosphoglycerides, also referred to as glycerophospholipids, have a glycerol molecule as the basic component to which phosphoric acid is esteri-

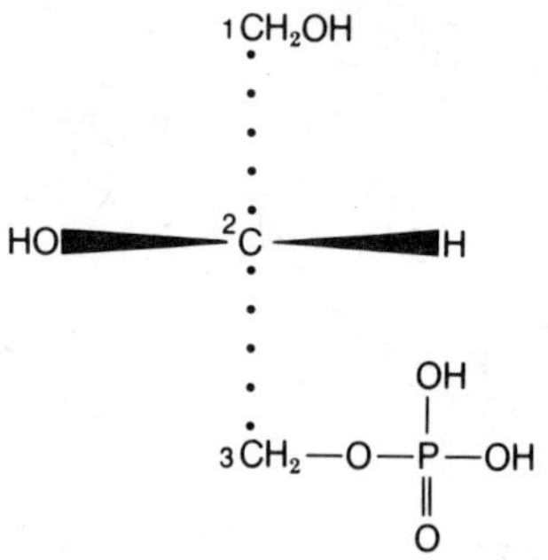

Figure 5.3
Stereochemical configuration of L-glycerol 3-phosphate (sn-glycerol 3-phosphate).
The H and OH attached to C-2 are above and C-1 and C-2 are below the plane of the page.

X
|
O
|
$O{=}P{-}O^-$
|
O
|
$^1CH_2{-}^2CH{-}^3CH_2$

Polar head

Hydrophobic tails

O | O; $C{=}O$ | $C{=}O$; $(CH_2)_{14}$–CH_3 | $(CH_2)_7$–$CH{=}CH$–$(CH_2)_7$–CH_3

Figure 5.4
Structure of phosphoglyceride.
Long-chain fatty acids are esterified at C-1 and C-2 of the L-glycerol 3-phosphate. X can be a H (phosphatidic acid) or one of several alcohols presented in Figure 5.5.

fied at the α-carbon (Figure 5.3) and two long-chain fatty acids are esterified at the remaining carbons (Figure 5.4). Even though glycerol does not contain an asymmetric carbon, the α-carbons are not stereochemically identical. Esterification of a phosphate to an α-carbon makes the molecule asymmetric. The naturally occurring phosphoglycerides are designated by the stereospecific numbering system (sn) as presented in Figure 5.3 and also discussed in Chapter 10, Section 10.2.

Phosphatidic acid, 1,2-diacylglycerol 3-phosphate, is the parent compound of a series of phosphoglycerides, where different hydroxyl-containing compounds are also esterified to the phosphate groups. The major compounds attached by the phosphodiester bridge to glycerol are choline, ethanolamine, serine, glycerol, and inositol. These structures are presented in Figure 5.5. *Phosphatidylethanolamine* (also called ethanolamine phosphoglyceride and the trivial name *cephalin*) and *phosphatidylcholine* (choline phosphoglyceride or *lecithin*) are the two most common phosphoglycerides in membranes (Figure 5.6). *Phosphatidylglycerol phosphoglyceride* (Figure 5.7) (or diphosphatidylglycerol or *cardiolipin*) contains two phosphatidic acids linked by a glycerol and is found nearly exclusively in the inner membrane of mitochondria and in bacterial membranes.

The hexahydroxy alcohol *inositol* is esterified to the phosphate in *phosphatidylinositol* (Figure 5.8); this compound should be differentiated from a class of lipids termed glycosylacylglycerols which contain a sugar in glycosidic linkage with the 3-hydroxyl group of a diacylglycerol. 4–Phospho- and 4,5-bisphosphoinositol phosphoglycerides (Figure 5.8) are found in various tissues, particularly myelin. Turnover of membrane phosphoinositol phosphoglycerides has been observed following activation of specific membrane receptors and has been implicated in controlling intracellular Ca^{2+} levels (see page 600).

Phosphoglycerides contain two fatty acyl groups esterified to carbons 1 and 2 of the glycerol; some of the major fatty acids found in phosphoglycerides are presented in Table 5.1. A saturated fatty acid is usually found on carbon-1 of the glycerol and an unsaturated fatty acid on carbon-2.

Choline	$HO{-}CH_2{-}CH_2{-}N^+(CH_3)_3$
Ethanolamine	$HO{-}CH_2{-}CH_2{-}NH_2$
Serine	$HO{-}CH_2{-}CH(NH_2){-}C(=O){-}OH$
Glycerol	$HO{-}CH_2{-}CH(OH){-}CH_2{-}OH$
Inositol	(cyclohexane ring with HO, OH, H substituents)

Figure 5.5
Structures of the major alcohols esterified to phosphatidic acid to form the phosphoglycerides.

TABLE 5.1 Major Fatty Acids in Phosphoglycerides

Common Name	*Systematic Name*	*Structural Formula*
Myristic acid	*n*-Tetradecanoic	$CH_3{-}(CH_2)_{12}{-}COOH$
Palmitic acid	*n*-Hexadecanoic	$CH_3{-}(CH_2)_{14}{-}COOH$
Palmitoleic acid	*cis*-9-Hexadecenoic	$CH_3{-}(CH_2)_5{-}CH{=}CH{-}(CH_2)_7{-}COOH$
Stearic acid	*n*-Octadecanoic	$CH_3{-}(CH_2)_{16}{-}COOH$
Oleic acid	*cis*-9-Octadecenoic acid	$CH_3{-}(CH_2)_7{-}CH{=}CH{-}(CH_2)_7{-}COOH$
Linoleic acid	*cis*-*cis*-9,12-Octadecadienoic	$CH_3{-}(CH_2)_3{-}(CH_2{-}CH{=}CH)_2{-}(CH_2)_7{-}COOH$
Linolenic acid	*cis,cis,cis*-9,12,15-Octadeca-trienoic	$CH_3{-}(CH_2{-}CH{=}CH)_3{-}(CH_2)_7{-}COOH$
Arachidonic acid	*cis,cis,cis,cis*-5,8,11,14-Icosatetraenoic	$CH_3{-}(CH_2)_3{-}(CH_2{-}CH{=}CH)_4{-}(CH_2)_3{-}COOH$

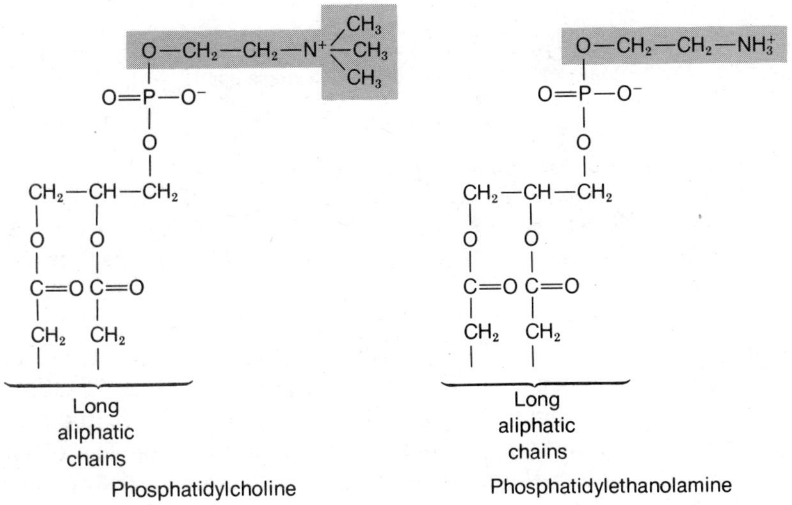

Long aliphatic chains

Phosphatidylcholine

Long aliphatic chains

Phosphatidylethanolamine

Figure 5.6
Structures of the two most common phosphoglycerides, phosphatidylcholine and phosphatidylethanolamine.

Long aliphatic chains

Long aliphatic chains

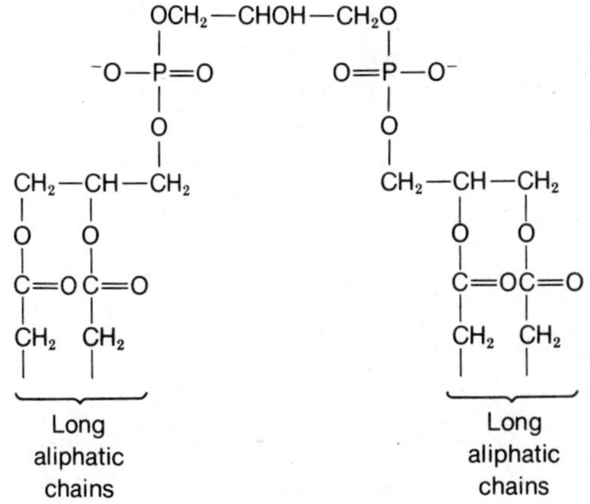

Figure 5.7
Diphosphatidylglycerol (cardiolipin).

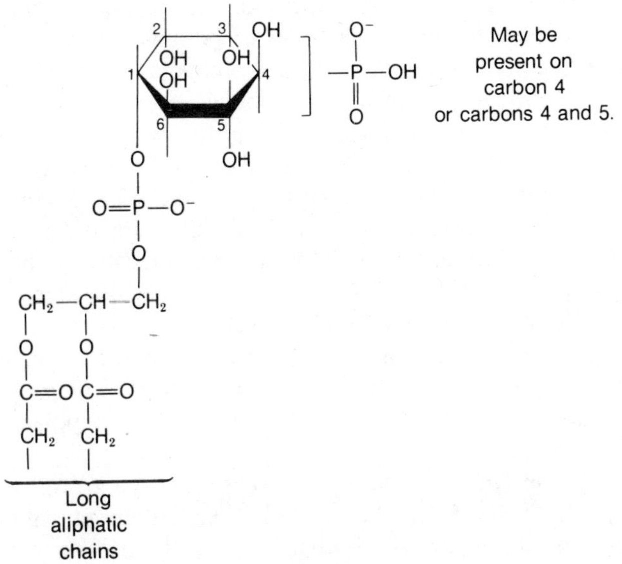

May be present on carbon 4 or carbons 4 and 5.

Long aliphatic chains

Figure 5.8
Phosphatidylinositol.
Phosphate groups are also found on C-4 or C-4 and C-5 of the inositol. The additional phosphate groups increase the charge on the polar head of this phosphoglyceride.

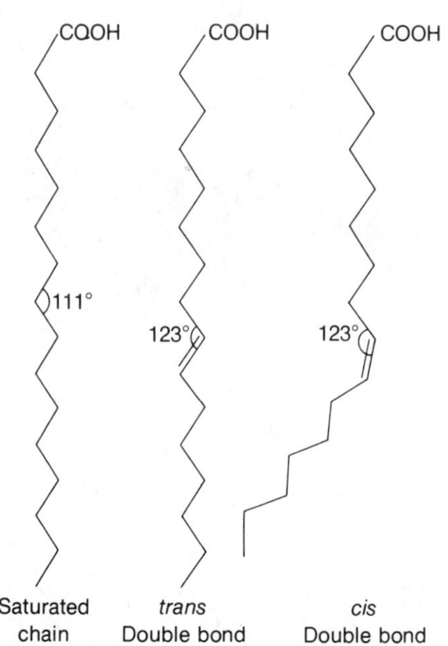

Figure 5.9
Conformation of fatty acyl groups in phospholipids.
The saturated and unsaturated fatty acids with trans *double bonds are straight chains in their minimum energy conformation, whereas a chain with a* cis *double bond has a bend. The* trans *double bond is rare in naturally occurring fatty acids.*

Polar head

$\overset{+}{N}H_3$–CH_2–CH_2–O–P(=O)(–O^-)–O–

CH_2–CH–CH_2

Hydrophobic tails

sn-1: O–CH=CH–$(CH_2)_{13}$–CH_3

sn-2: O–C(=O)–$(CH_2)_6$–CH=CH–$(CH_2)_7$–CH_3

Figure 5.10
Ethanolamine plasmalogen.
Note the ether linkage of the aliphatic chain on C-1 of glycerol.

The nomenclature for the different phosphoglycerides does not specify a specific compound because of the variety of possible fatty acid substitutions. Phosphatidylcholine usually contains palmitic or stearic in the sn-1 position and an 18-carbon unsaturated fatty acid, oleic, linoleic or linolenic, on the sn-2-carbon. Phosphatidylethanolamine also contains palmitic or oleic on sn-1 but one of the longer chain polyunsaturated fatty acids, that is, arachidonic, on the sn-2 position.

A saturated fatty acid is a straight chain, as is a fatty acid with an unsaturation in the *trans* position. The presence of a *cis* double bond, however, creates a kink in the hydrocarbon chain (Figure 5.9). A straight-chain diagram, as shown in Figures 5.4 and 5.9, does not adequately represent the chemical configuration of a long-chain fatty acid. Actually, there is a high degree of coiling of the hydrocarbon chain in a phosphoglyceride that is disrupted by the presence of a double bond. As described in Section 5.3, the presence of unsaturated fatty acids has a marked effect on the physicochemical state of the membrane.

Another group of phosphoglycerides is the *plasmologens* in which a long aliphatic chain is attached in ether linkage to the glycerol as presented in Figure 5.10. Plasmalogens containing ethanolamine (ethanolamine plasmalogen) and choline (choline plasmalogen) esterified to the phosphate are particularly abundant in nervous tissue and heart.

The phosphoglycerides contain both a polar end, referred to as the head group, due to the charged phosphate and the substitutions on the phosphate, and a nonpolar tail due to the hydrophobic hydrocarbon chains of the fatty acyl groups. These polar lipids are *amphipathic,* that is, they contain both polar and nonpolar groups. The polar groups are charged at pH 7.0 with a negative charge due to the ionization of the phosphate group (pK ~2) and the charges from the groups esterified to the phosphate (Table 5.2). Choline and ethanolamine phosphoglycerides are zwitterions at pH 7.0, with both a negative charge from the phosphate

TABLE 5.2 Predominant Charge on Phosphoglycerides and Sphingomyelin at pH 7.0

Lipid	*Phosphate Group*	*Base*	*Net Charge*
Phosphatidylcholine	−1	+1	0
Phosphatidylethanolamine	−1	+1	0
Phosphatidylserine	−1	+1, −1	−1
Phosphatidylglycerol	−1	0	−1
Diphosphatidylglycerol (cardiolipin)	−2	0	−2
Phosphatidylinositol	−1	0	−1
Sphingomyelin	−1	+1	0

and a positive charge on the nitrogen. Phosphatidylserine has two negative charges, one on the phosphate and one on the carboxyl group of serine, and a positive charge on the α-amino group of serine, with a net negative charge of 1 at pH 7.0. In contrast, the phosphoglycerides containing inositol and glycerol have only a single negative charge on the phosphate; the 4-phospho- and 4,5-bisphosphoinositol derivatives are very polar compounds with additional negative charges on the phosphate groups.

Every tissue and respective cellular membrane has a distinctive composition of phosphoglycerides. Not only are there differences in the classes of phosphoglycerides, but there are definite patterns in the fatty acid composition of the individual phosphoglycerides between tissues. There appears to be some degree of specificity for particular fatty acids in the individual tissues. There is a greater variability in the fatty acyl groups of different tissues in a single species than in the fatty acid composition of the same tissue in a variety of species. In addition, the fatty acid content of the phosphoglycerides can vary, depending on the physiological or pathophysiological state of the tissue.

Sphingolipids

The amino alcohols sphingosine (D-4-sphingenine) and dehydrosphingosine (Figure 5.11) serve as the basis for another series of membrane lipids, the sphingolipids. On the amino group of sphingosine, a saturated or unsaturated long chain fatty acyl group is present in amide linkage. This compound, termed a *ceramide* (Figure 5.12), with two nonpolar tails is similar in structure to the diacylglycerol portion of phosphoglycerides. Various substitutions are found on the hydroxyl group at position 1 of the ceramides. The sphingomyelin series has phosphorylcholine esterified to the 1-hydroxyl (Figure 5.13) and is the most abundant sphingolipid in mammalian tissues. The similarity of this structure to the choline phosphoglyceride is apparent, and they have many properties in common; note that the sphingomyelins are amphipathic compounds. It has been a common practice to classify the sphingomyelin series and the phosphoglycerides in one class of compounds, termed phospholipids. The sphingomyelin of myelin contains predominantly the longer chain fatty acids, with carbon lengths of 24; as with phosphoglycerides, there is a specific fatty acid composition of the sphingomyelin, depending on the tissue.

The *glycosphingolipids* do not contain phosphate and have a sugar attached by a β-glycosidic linkage to the 1-hydroxyl group of the sphingosine in a ceramide. One subgroup is the *cerebrosides,* which contain either a glucose or galactose attached to a ceramide and are referred to as

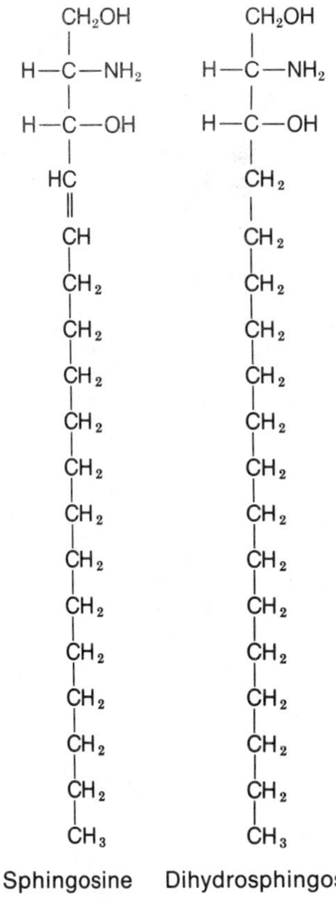

Sphingosine (D-4-sphingenine) Dihydrosphingosine (D-sphinganine)

Figure 5.11
Structures of sphingosine and dihydrosphingosine.

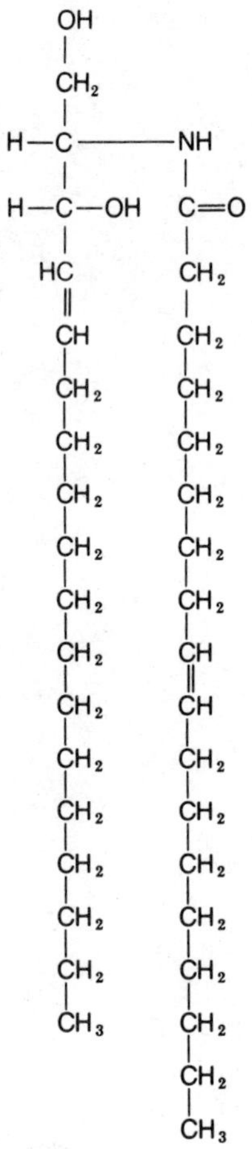

Figure 5.12
Structure of a ceramide.

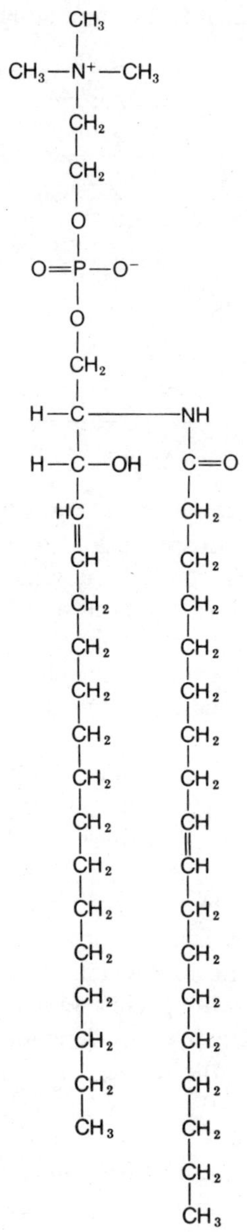

Figure 5.13
Structure of a choline containing sphingomyelin.

glucocerebrosides or galactocerebrosides, respectively (Figure 5.14). Cerebrosides are neutral compounds. Galactocerebrosides are found predominantly in brain and nervous tissue, whereas the small quantities of cerebrosides in nonneural tissues usually contain glucose. The specific galactocerebroside, phrenosine, contains a 2-hydroxy 24-carbon fatty acid. Galactocerebrosides may contain a sulfate group esterified on the 3 position of the sugar. They are called sulfatides (Figure 5.15). Cerebrosides and sulfatides usually contain very long-chain fatty acids with 22 to 26 carbon atoms.

In place of monosaccharides, neutral glycosphingolipids often have 2 (dihexosides), 3 (trihexosides), or 4 (tetrahexosides) sugar residues attached to the 1-hydroxyl group of sphingosine. Diglucose, digalactose, *N*-acetylglucosamine, and *N*-acetyldigalactosamine are the usual sugars.

The most complex group of glycosphingolipids is the *gangliosides,* which contain oligosaccharide head groups with one or more residues of

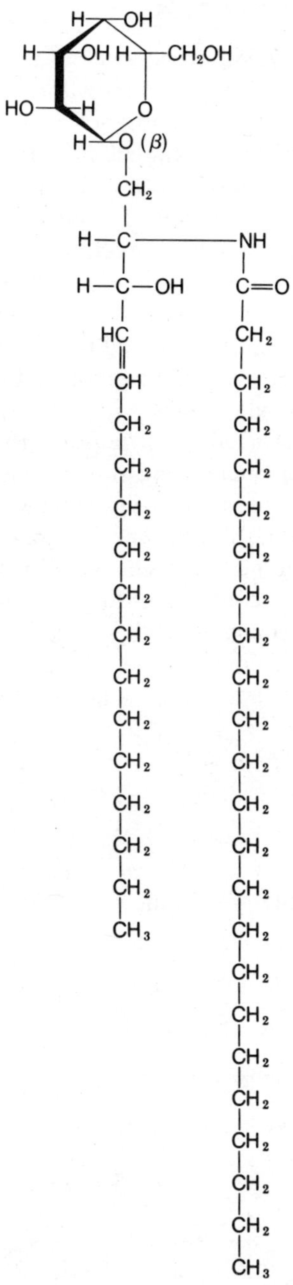

Figure 5.14
Structure of a galactocerebroside containing a C_{24} fatty acid.

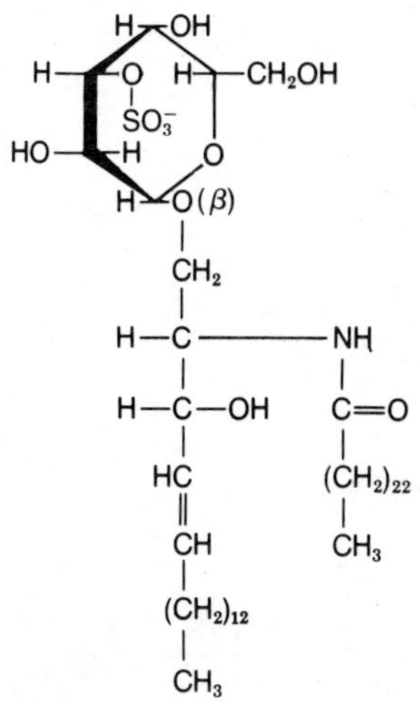

Figure 5.15
Structure of a sulfatide.

sialic acid; these are amphipathic compounds with a negative charge at pH 7.0. The gangliosides represent 5–8% of the total lipids in brain, and some 20 different types have been identified differing in the number and relative position of the hexose and sialic acid residues, which form the basis of their classification. A detailed description of the nomenclature and structures of the gangliosides is presented in Chapter 10.

Cholesterol

The third major lipid present in membranes is cholesterol. As presented in Figure 5.16 cholesterol contains four fused rings, which makes it a planar structure, a polar hydroxyl group at carbon-3, and an eight-member branched hydrocarbon chain attached to the D ring at position 17. Cholesterol is a compact hydrophobic molecule.

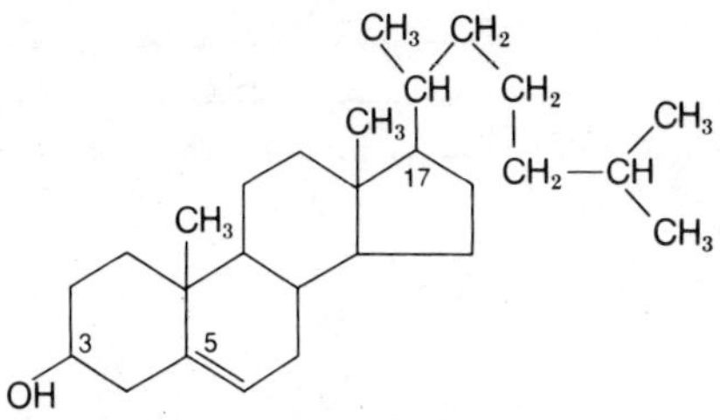

Figure 5.16
Structure of cholesterol.

Distribution of Membrane Lipids

There are large quantitative differences between the classes of lipids and individual lipids in various cell membranes. Figure 5.17 presents the lipid composition of various cellular membranes.

There is a resemblance among animal species in the lipid composition of the same intracellular membrane of cells in a specific tissue, such as liver mitochondria of rat and humans. The plasma membrane exhibits the greatest variation in percentage composition because the amount of cholesterol is affected by the nutritional state of the animal. Plasma membranes have the highest concentration of neutral lipids and sphingolipids; the myelin membranes of axons of neural tissue are rich in sphingolipids, with a high proportion of glycosphingolipids. Intracellular membranes primarily contain phosphoglycerides with little sphingolipids or cholesterol. When comparing intracellular structures, the membrane lipid composition of mitochondria, nuclei, and rough endoplasmic reticulum are similar, with the Golgi membrane being somewhere between the other intracellular membranes and the plasma membrane. As indicated previously, cardiolipin is found nearly exclusively in the inner mitochondrial membrane. The choline containing lipids, phosphatidylcholine and sphingomyelin are predominant, with ethanolamine phosphoglyceride second. The constancy of composition of the various membranes indicates the relationship between the lipids and the specific functions of the individual membranes.

Proteins of Membranes

Membrane proteins are classified on the basis of the ease of removal from isolated membrane fractions. *Peripheral* (or extrinsic) proteins are easily isolated by treatment of the membrane with salt solutions of low or high

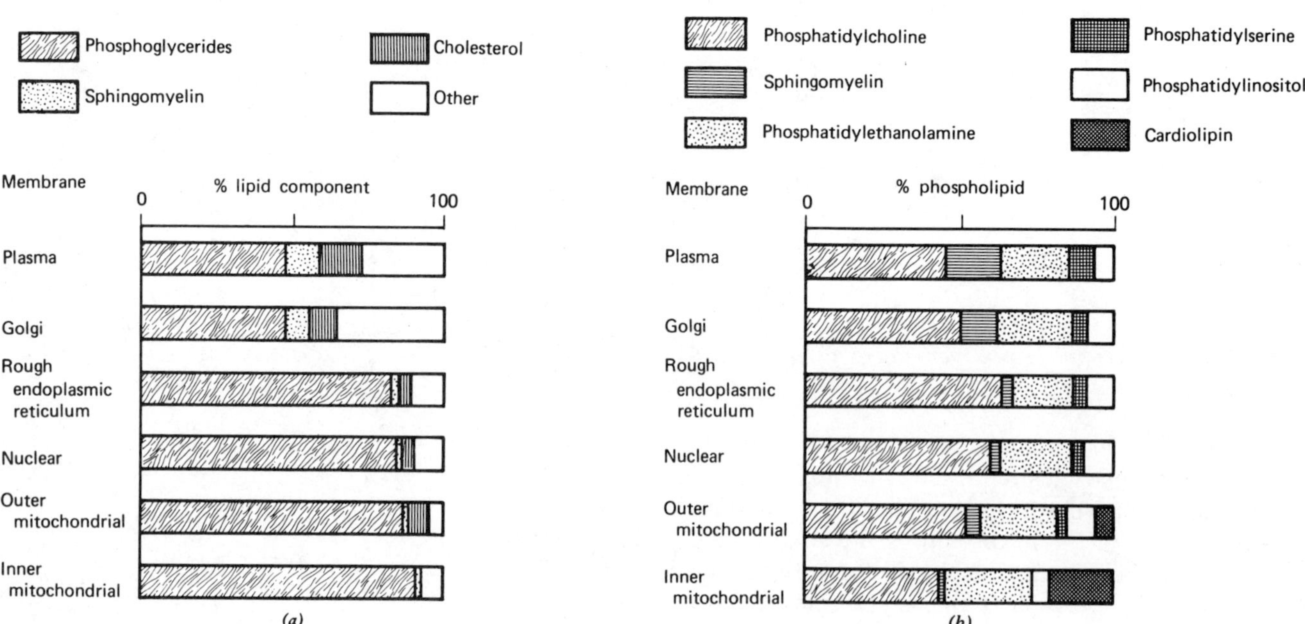

Figure 5.17
Lipid composition of cellular membranes isolated from rat liver.
(a) Amount of major lipid components as percentage of total lipid. The area labeled "Other" includes mono-, di-, and triacylglycerol, fatty acids, and cholesterol esters. (b) Phospholipid composition as a percentage of total phospholipid.
Values from R. Harrison and G. G. Lunt, *Biological Membranes,* Wiley, New York, 1975.

ionic strength, or extremes of pH, and the name is used to imply a physical location on the surface of the membrane. Peripheral proteins, many with specific enzymatic activity, are usually soluble in water and free of lipids. *Integral* (or intrinsic) proteins require rather drastic treatment, such as use of detergents or organic solvents, to be extracted from the membrane. They usually contain tightly bound lipid, which if removed leads to denaturation of the protein and loss of biological function. Removal of the integral protein leads to disruption of the membrane, whereas peripheral proteins can be removed with little or no change in the integrity of the membrane.

Of particular value in studying the chemistry and structure of integral proteins has been the use of sodium dodecyl sulfate (SDS), a detergent that dissociates the lipid-protein complex and solubilizes the protein permitting separation and analysis. The integral proteins studied have sequences of hydrophobic amino acids, which could create domains with a high degree of hydrophobicity in the tertiary structure of the protein. These hydrophobic regions of the protein interact with the hydrophobic hydrocarbons of the lipids stabilizing the protein–lipid complex.

A special class of integral proteins are the proteolipids, which are hydrophobic lipoproteins soluble in chloroform and methanol but insoluble in water. Proteolipids are present in many membranes but are particularly abundant in myelin, where they represent about 50% of the membrane protein component. An example is lipophilin, a major lipoprotein of brain myelin that contains over 65% hydrophobic amino acids and covalently bound fatty acids.

Another class of integral membrane proteins is the glycoproteins; plasma membrane of cells contain a number of different glycoproteins, each with its own unique carbohydrate content.

The complexity, variety, and interaction of membrane proteins with lipids are just being resolved. Many of the proteins are enzymes located within or on the cellular membranes. Membrane proteins also have a role in transmembrane movement of molecules and in many cells, such as neurons and erythrocytes, specific proteins have a structural role to maintain the integrity of the cell. Thus individual membrane proteins can have a catalytic, transport, structural, or recognition role, and it is not surprising to find a high protein content in a membrane being correlated with complexity and variety of function of the membrane.

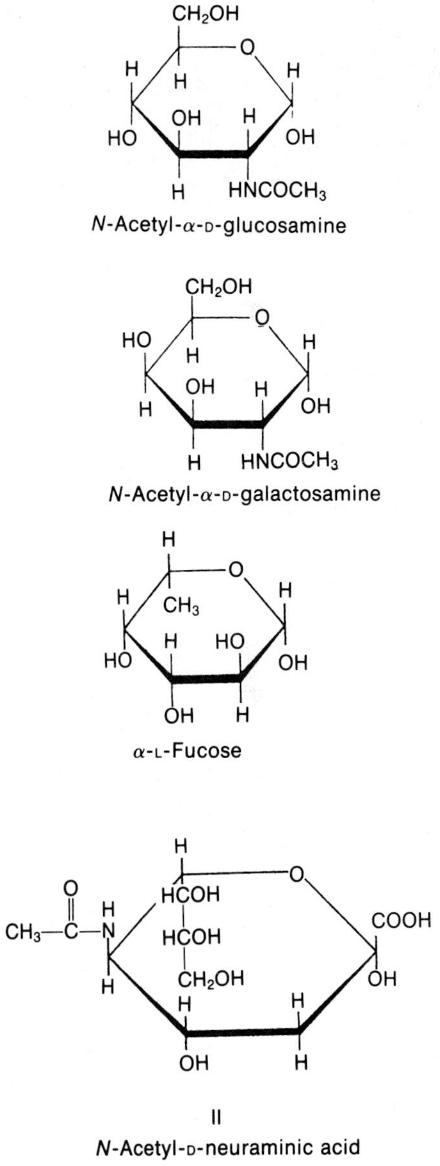

Figure 5.18
Structures of some membrane carbohydrates.

Carbohydrates of Membranes

Carbohydrates present in membranes are exclusively in the form of oligosaccharides covalently attached to proteins to form glycoproteins and to a lesser amount to lipids to form glycolipids. The sugars found in glycoproteins and glycolipids include glucose, galactose, mannose, fucose, *N*-acetylgalactosamine, *N*-acetylglucosamine, and sialic acid (see Figure 5.18 and the Appendix, "Review of Organic Chemistry," for structures). Details of the structures of glycoproteins and glycolipids are given on pages 344 and 417, respectively. It has been found that most, if not all, of the carbohydrate is on the exterior side of the plasma membrane or the luminal side of the endoplasmic reticulum. Proposed roles for membrane carbohydrates include cell–cell recognition, adhesion, and receptor action.

5.3 MOLECULAR STRUCTURE OF MEMBRANES

Micelles, Membranes, and Liposomes

The basic structural characteristic of all membranes is derived from the physicochemical properties of the major lipid components, the phos-

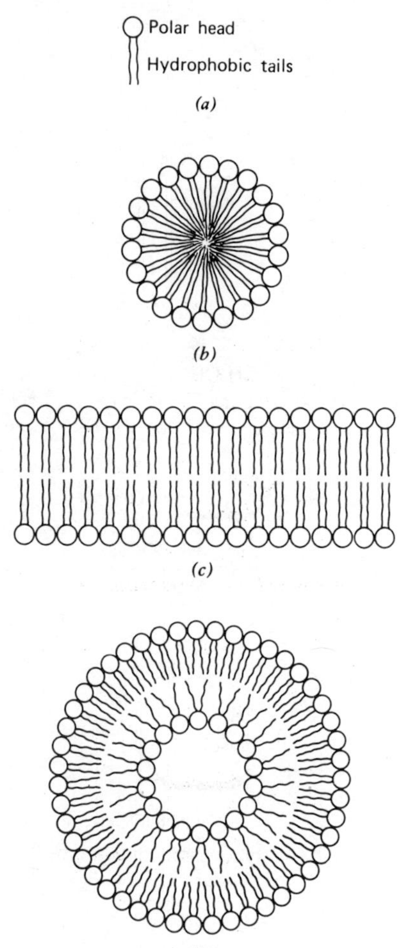

Figure 5.19
Representations of the interactions of phospholipids in an aqueous medium.
(a) Representation of an amphipathic lipid. (b) Cross-sectional view of the structure of a micelle. (c) Cross-sectional view of the structure of lipid bilayer. (d) Cross section of a liposome. Each structure has an inherent stability due to the hydrocarbon chains and the attraction of the polar head groups to water.

phoglycerides and sphingolipids. These amphipathic compounds, with a hydrophilic head and a hydrophobic tail (Figure 5.19*a*), react in a unique fashion in an aqueous system because of their very low solubility in water. Under proper conditions, these lipid molecules will come together to form spheres, termed *micelles,* with the hydrophobic tails interacting to exclude water and with the charged polar head groups on the outside. This is shown in Figure 5.19*b*. The specific concentration of lipid required for micelle formation is referred to as the *critical micelle concentration.* Micelles with a single lipid or a mixture of lipids can be made. The formation of the micelle depends also on the temperature of the system and, if a mixture of lipids are used, on the ratio of concentrations of the different lipids in the mixture. The micelle structure is very stable because of the hydrophobic interaction of the hydrocarbon chains and the attraction of the polar groups to water. As discussed in Chapter 24, Section 24.6, micelles are important in the digestion of lipids.

Depending on the conditions, amphipathic lipids will interact to form a bimolecular leaf structure with two layers of lipid in which the polar head groups are at the interface between the aqueous medium and the lipid and the hydrophobic tails interact to form an environment that excludes water (Figure 5.19*c*). This bilayer conformation is the basic lipid structure of all biological membranes.

Lipid bilayers are extremely stable structures held together by noncovalent interactions of the hydrocarbon chains of the acyl groups and the ionic interactions of the charged head groups with water. Hydrophobic interactions of the hydrocarbon chains lead to the smallest possible area for water to be in contact with the chains, and water is essentially excluded from the interior of the bilayer. If disrupted, the bilayers have a tendency to self-seal because the hydrophobic groups will seek to establish a structure in which there is the least contact with water of the hydrocarbon chains, a condition that is most thermodynamically favorable. A lipid bilayer will close in on itself, forming a spherical vesicle separating the external space from an internal compartment. These vesicles are termed *liposomes*. Because the individual lipid–lipid interactions have low energies of activation, the lipids in a bilayer have a circumscribed mobility, breaking and forming interactions with surrounding molecules but not readily escaping from the lipid bilayer (Figure 5.19*d*).

Individual phospholipid molecules can readily exchange places with neighboring molecules, which leads to rapid lateral diffusion in the plane of the membrane (see Figure 5.20). In addition, the fatty acyl chains can rotate around the carbon–carbon bonds; in fact, there is a greater degree of rotation nearer the methyl end, leading to greater motion at the center of the lipid bilayer. Individual lipid molecules cannot migrate readily from one monolayer to the other, a process termed *flip-flop*. Thus the lipid bilayer has not only an inherent stability but also a fluidity in which individual molecules can move rapidly in their own monolayer but do not exchange with adjoining monolayer. In artificial bilayer membranes composed of different lipids, the components will be randomly distributed.

Artificial membrane systems have been studied extensively as a means to determine the properties of biological membranes. A variety of techniques are available to prepare liposomes, using synthetic phospholipids and lipids extracted from natural membranes. Depending on the procedure, unilamellar vesicles and multilamellar vesicles (vesicles within vesicles) of various sizes (20 nm–1 μm in diameter) can be prepared. Figure 5.19*d* contains a representation of the structure of a liposome. The interior of the vesicle is an aqueous environment, and it is possible to prepare liposomes with different substances entrapped. Thus the external and internal environments of the liposome can be manipulated and studies conducted on a variety of properties of these synthetic membranes, including their ability to exclude molecules, their interaction with various substances, and their stability under different conditions. Na^+, K^+, Cl^-, and most polar molecules do not readily diffuse across the lipid bilayer of liposomes, whereas the membrane presents no barrier to water. Lipid-soluble nonpolar substances such as triacylglycerol and undissociated organic acids readily diffuse into the membrane remaining in the hydrophobic environment of the hydrocarbon chains. Proteins, both synthetic and those isolated from cell membranes, have been incorporated into liposomes to mimic the natural membrane. Membrane-bound enzymes and proteins involved in translocating ions have been isolated from various tissues and incorporated into the membrane of liposomes for evaluation of the proteins function. With the liposome it is easier to manipulate the various parameters of the membrane system and, thus, study the catalytic activity free of possible interfering reactions that are present in the cell membrane. Liposomes have also been used in drug therapy (Clin. Corr. 5.1).

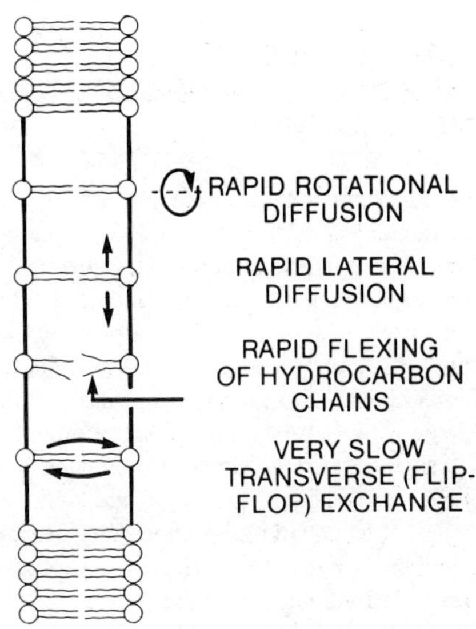

Figure 5.20
Mobility of lipid components in membranes.

Biological Membranes

Based on evidence from physicochemical, biochemical, and electron microscopic investigations, knowledge of the structure of biological membranes has evolved. The basic structure is a bimolecular leaf arrangement of lipids in which the phosphoglycerides, sphingolipids, and cholesterol are oriented so that the hydrophobic portions of the molecules interact to minimize their interactions with water or other polar groups. The polar head groups of the amphipathic compounds are at the interface with the aqueous environment. This arrangement of lipids is the same as that in synthetic phospholipid liposomes. A major problem to resolve, however, has been to explain the interaction of the integral and peripheral membrane proteins with the lipid bilayer. A number of models for the structure of biological membranes have been suggested dating back to one by H. Davson and J. Danielli in 1935, which was refined in later years by J. D. Robertson. In the early 1970s, G. L. Nicolson and S. J. Singer proposed the mosaic model for membranes in which it was suggested that proteins are on the surface as well as in the lipid bilayer. Some proteins could span the lipid bilayer with their polar groups in contact with the aqueous surroundings on both sides and hydrophobic portions interacting with the lipids in the interior of the membrane. This model has been extensively refined and is referred to as the *fluid mosaic* model to indicate the movement of both lipids and proteins in the membrane. Figure 5.21 is a pictorial representation of a biological membrane as proposed by Singer and Nicolson. The proposed structure accounts for many properties of mammalian membranes, but it continues to undergo modification and refinement.

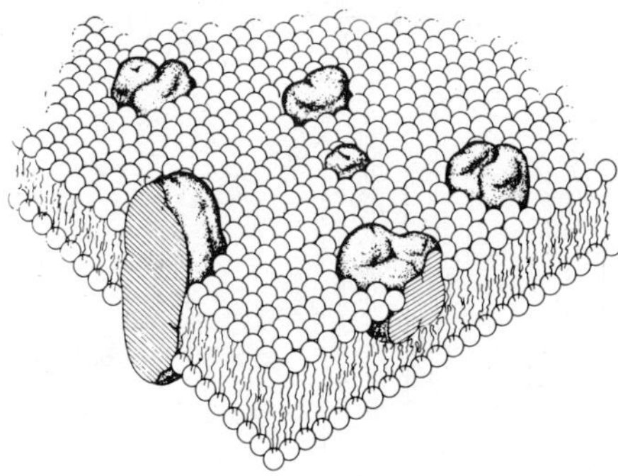

Figure 5.21
The fluid mosaic model of biological membranes.
The membrane consists of a fluid phospholipid bilayer with globular integral proteins penetrating the bilayer.
Reproduced with permission from S. J. Singer and G. L. Nicolson, *Science* 175:720, 1972. Copyrighted 1972 by the American Association for the Advancement of Science.

Membrane Lipids and Proteins

The physicochemical properties of the lipid bilayer explain many membrane properties, including the fluidity of the membrane, the flexibility of cellular membranes, which permits changes of shape and form, their self-sealing properties, and the impermeability of many substances. It is now recognized that the membrane lipids may not be randomly distributed in the monolayers, but that there are islands of lipids associated together or in contact with a specific protein. The individual lipids are not, however, immobilized is these islands but rapidly exchange with molecules in the surrounding area of the membrane.

An important difference between the Nicolson–Singer model and earlier models is that the lipid bilayer is discontinuous, with proteins embedded in the hydrophobic portion of the bilayer. Many membrane proteins span the bilayer, with portions protruding on each side. Hydrophobic interaction between the lipid and hydrophobic domains of these integral proteins prevents these proteins from being readily removed, and their extraction leads to disruption of the membrane. Obviously, there are significant differences in the strengths of interaction of different integral proteins and the lipid bilayer. Proteins are also loosely bound to the membrane by interaction between charges on the protein and the charged groups on the surface of the membrane. These are the peripheral proteins, easily removed by mild treatment with little damage to the membrane.

Even though the model would suggest that proteins are randomly distributed throughout and on the membrane, evidence from a variety of sources supports a high degree of functional organization with definite restrictions on the localization of some proteins. As an example, proteins participating in electron transport in the inner membrane of mitochondria function in consort and are organized into a functional unit both laterally and transversely in the membrane. Membrane proteins also have a definite orientation within and across the membrane. Integral proteins with enzymatic activity may have their catalytic site on either the inner or outer surface. This orientation is established during the biosynthesis of the membrane and remains unchanged. The actual location of specific proteins on the surface of plasma membranes is also controlled. Cells lining the lumen of the kidney nephron have specific plasma membrane enzymes on the luminal surface but not on the contraluminal surface of the cell; the enzymes restricted to a particular region of the membrane are located to meet the specific functions of these cells. Another restriction is that specific peripheral proteins are bound to only one side. Thus there is a high degree of molecular organization of biological membranes that is not apparent from the diagrammatic models.

CLIN. CORR. 5.1 LIPOSOMES AS CARRIERS OF DRUGS AND ENZYMES

A major obstacle in the use of many drugs is the lack of tissue specificity in the action of the drug. Drugs administered orally or intravenously often lead to the drug acting on many tissues and not exclusively on a target organ and is the basis for their toxicity. In addition, some drugs are metabolized rapidly and their period of effectiveness is relatively short. Liposomes have been prepared with drugs and even enzymes encapsulated inside and used as carriers for these substances to target organs. Liposomes prepared from purified phospholipids and cholesterol are nontoxic and biodegradable. Attempts have been made to prepare liposomes for interaction at a specific target organ. Some drugs have a longer period of effectiveness when administered encapsulated in liposomes. It is hoped, as our understanding of the structure and chemistry of cellular membranes increases, that it will be possible to prepare liposomes with a high degree of tissue specificity so that drugs and perhaps even enzyme replacement can be carried out with this technique. (For a brief review of the topic, see G. Gregoriadis, *New Engl. J. Med.* 295:704, 1976.)

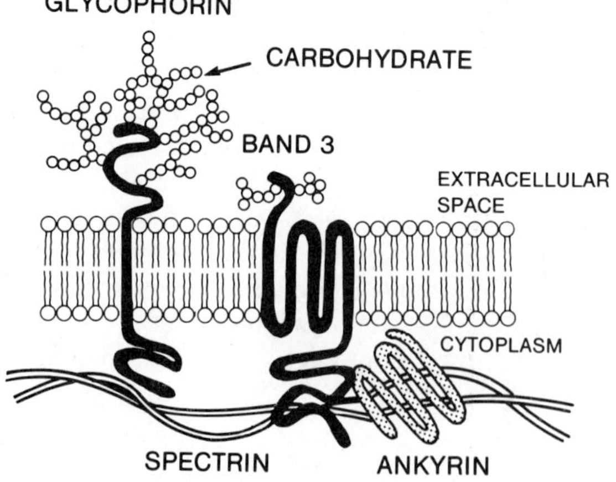

Figure 5.22
Schematic diagram of the erythrocyte membrane.
The diagram indicates the relationship of four membrane-associated proteins with the lipid bilayer. Glycophorin is a glycoprotein which contains 131 amino acids but whose function is unknown. Band 3, so designated because of its mobility in electrophoresis, contains over 900 amino acids and is involved in interacting with ankyrin and possibly in the facilitated diffusion of Cl^- *and* HCO_3^- *(see Section 5.4). Ankyrin and spectrin are part of the cytoskeleton and are peripheral membrane proteins. Ankyrin binds to band 3 and spectrin is anchored to the membrane by ankyrin.*

The structure of the plasma membrane of the human erythrocyte is under active investigation because of the ease with which the membrane can be purified from other cellular components. The structure of the membrane proteins, the composition of the lipid bilayer, and the interaction of the various components are being evaluated by chemical analysis and with a variety of physicochemical techniques. Results of these studies can be applied to other membranes. Figure 5.22 is a simplified representation of the current thinking of the placement of a few of the proteins in the membrane.

Asymmetry of the Membrane

In contrast to the random distribution of lipids between the outer and inner lipid monolayers of liposomes, there is an asymmetric distribution of lipid components across biological membranes. Each layer of the bilayer has a different composition with respect to individual phosphoglycerides and sphingolipids. The asymmetric distribution of lipids in erythrocyte membranes is presented in Figure 5.23. Similar results have been observed for other membranes, including those from human tissues. Sphingomyelin is predominantly in the outer layer, whereas phosphatidylethanolamine is predominantly in the inner lipid layer. In contrast, cholesterol is equally distributed on both sides of the membrane.

It is now considered that the asymmetry of the lipids is established during the biosynthesis of the membrane. Transverse movement from one side to the other (i.e., flip-flop movement) of the phosphoglycerides and sphingolipids is very slow and is measured in days or weeks. The slow rate of transverse movement is not unexpected, considering how unfavorable in thermodynamic terms it is to push or pull the hydrophilic polar head group of a phospholipid through the hydrophobic interior of a membrane and then reorient the group on the opposite side. The asymmetry of lipids in the erythrocyte membrane is an example of how slow the transverse movement of membrane lipids is. The mature erythrocyte has a lifetime of about 120 days, during which time there is no new membrane synthesis or even significant repair. Even so, there appears to be little mixing of the phospholipids between the molecular layers. Individual lipids, however, do exchange with lipids in the cell matrix, as well as with lipids of other membranes. Specific mechanisms to maintain both the composition and asymmetry of lipids in membranes must exist.

As discussed above, there is also a definite orientation of proteins in the membrane; the carbohydrates of glycoproteins are on the external side or that opposite the cytoplasmic face. Individual membrane proteins are also constantly being removed and replaced by newly synthesized proteins, thus participating in the dynamic steady-state of cellular constituents.

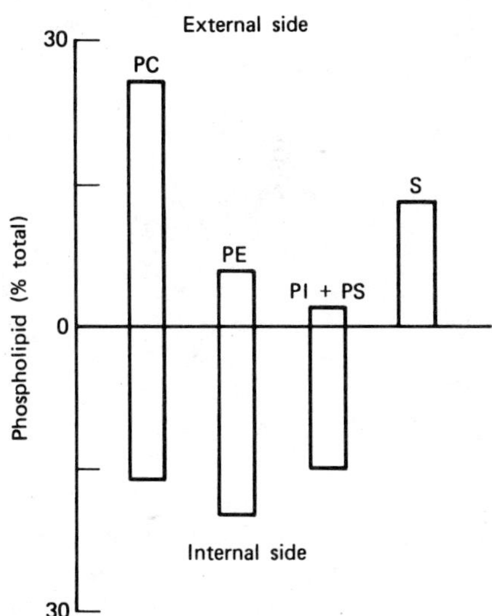

Figure 5.23
Distribution of phospholipids between inner and outer layers of the rat erythrocyte membrane.
Values are percentage of each phospholipid in the membrane. Abbreviations: PC, phosphatidylcholine; PE, phosphatidylethanolamine; PI, phosphatidylinositol; PS, phosphatidylserine; S, sphingomyelin.
Data from W. Renooij, L. M. G. van Golde, R. F. A. Zwaal, and L. L. M. van Deenen, *Eur. J. Biochem.* 61:53, 1976.

Membrane Fluidity

The interactions among the different lipids and between lipids and proteins are very complex and dynamic. There is a fluidity in the lipid portion of the membrane in which both the lipids and proteins move. The degree of fluidity is dependent on the temperature and composition of the membrane. At low temperatures, the lipids are in a gel–crystalline state, with the lipids restricted in their mobility. As the temperature is increased, there is a phase transition into a liquid–crystalline state, with an increase in fluidity. With liposomes prepared from a single pure phospholipid, the phase transition temperature, Tm, is rather precise; but with liposomes prepared from a mixture of lipids, the Tm becomes less precise because individual clusters of lipids may be in either the gel–crystalline or

CLIN. CORR. 5.2 ABNORMALITIES OF CELL MEMBRANE FLUIDITY IN DISEASE STATES

Changes in membrane fluidity can control the activity of membrane-bound enzymes, membrane functions such as phagocytosis, and cell growth. A major factor in controlling plasma membrane fluidity is the concentration of cholesterol. Higher organisms and mammals have a significant concentration of cholesterol in their membranes, which presumably has a major role in controlling the fluidity of the lipid bilayer. With increasing cholesterol content membranes become less fluid on their outer surface but more fluid in the hydrophobic core. Individuais with spur cell anemia have an increased cholesterol content of the red cell membrane. This condition occurs in severe liver disease such as cirrhosis of the liver in alcoholics. Erythrocytes have a spiculated shape and are destroyed prematurely in the spleen. The cholesterol content is increased 25–65%, and the fluidity of the membrane is decreased. The erythrocyte membrane requires a high degree of fluidity for its function and any decrease would have serious effects on the cell's physiological role of oxygen transport. The increased plasma membrane cholesterol in other cells leads to an increase in intracellular membrane cholesterol, which also affects the fluidity of other cellular membranes. Individuals with abetalipoproteinemia have an increase in sphingomyelin content and a decrease in phosphotidylcholine, thus causing a decrease in fluidity. The ramifications of these changes in fluidity are still not understood, but it is presumed that, as techniques for the measurement and evaluation of cellular membrane fluidity improve, some of the pathological manifestations in disease states will be explained on the basis of changes in membrane structure and function. (For further discussion see R. A. Cooper, Abnormalities of cell membrane fluidity in the pathogenesis of disease, *New Engl. J. Med.* 297:371, 1977.)

the liquid–crystalline state. The Tm is not precise for biological membranes because of their heterogeneous chemical composition. Interactions between specific lipids and between lipids and proteins also lead to variations in the gel–liquid state throughout the membrane and differences in fluidity in different areas of the membrane.

The specific composition of the individual biological membranes leads to differences in fluidity. Phosphoglycerides containing short-chain fatty acids will increase the fluidity as does an increase in unsaturation of the fatty acyl groups. The *cis*-double bond in an unsaturated fatty acid of phospholipid leads to a kink in the hydrocarbon chain, preventing the tight packing of the chains, and creates pockets in the hydrophobic areas. It is assumed that these spaces, which will also be mobile due to the mobility of the hydrocarbon chains, are filled with water molecules and small ions. Cholesterol with its flat stiff ring structure reduces the coiling of the fatty acid chain and decreases fluidity. Consideration has been given to the potential clinical significance of high blood cholesterol on the fluidity of cell membranes (see Clin. Corr. 5.2.). Ca^{2+} directly decreases the fluidity of a number of membranes because of its interaction with the negatively charged phospholipids, which reduces repulsion between the polar groups and increases the packing of lipid molecules. Ca^{2+} causes aggregation of lipids into clusters, which also reduces membrane fluidity.

Fluidity at different levels within the membrane also varies. The hydrocarbon chains of the lipids have a motion, which produces a fluidity in the hydrophobic core. The central area of the bilayer is occupied by the ends of the hydrocarbon chains and is more fluid than the areas closer to the two surfaces, where there are more constraints due to the stiffer portions of the hydrocarbon chains. Cholesterol makes the membrane more rigid toward the periphery because it does not reach into the central core of the membrane.

Individual lipids and proteins can move rapidly in a lateral motion along the surface of the membrane. However, electrostatic interactions of polar head groups, hydrophobic interactions of cholesterol with selected phospholipids or glycolipids, and protein–lipid interactions all lead to constraints on the movement. Thus there may be lipid domains in which lipids move together, such as an island floating in a sea of lipid.

Integral membrane proteins also move in the lipid environment, as demonstrated by the fusion of human and rat cells after antigenic membrane proteins on cells of each species were labeled with a different antibody marker. The markers permitted localization of the two different proteins on the membrane. Immediately following fusion of the cells, proteins on the membranes of the human and rat cells were segregated in different hemispheres of the new cell, but within 40 min the two groups of proteins were evenly distributed over the new cell membrane. This could occur only if there had been lateral diffusion of the proteins on the membranes. Movement of protein is slower than that of lipids. There is very little transverse motion of proteins in membranes. Movement of membrane proteins may be restricted by other membrane proteins, matrix proteins, or cellular structural elements such as microtubules or microfilaments to which they may be attached.

Evidence is accumulating that the fluidity of cellular membranes can change in response to changes in diet or physiological state. Their content of fatty acid and cholesterol is modified by a variety of factors. In addition, pharmacological agents may have a direct effect on membrane fluidity. It is now considered that some of the actions of anesthetics, which induce sleep and muscular relaxation, may be due to their effect on membrane fluidity of specific cells. A number of structurally unrelated compounds induce anesthesia, but their common feature is lipid solubility. Anesthetics increase membrane fluidity in vitro.

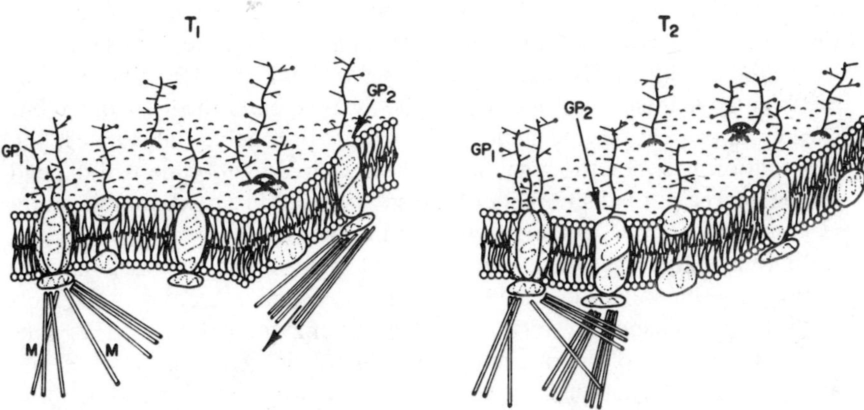

Figure 5.24
A modified version of the fluid mosaic model of biological membranes to indicate the mobility of membrane proteins.
T_1 and T_2 represent different points in time. Some integral proteins (GP_2) are free to diffuse laterally in the plane of the membrane directed by the cytoskeletal components, whereas others (GP_1) may be restricted in their mobility.
Reproduced with permission from J. L. Nicolson, *Int. Rev. Cytol.* 39:89, 1974.

Thus cellular membranes are in a constantly changing state, with not only movement of proteins and lipids laterally on the membrane but with molecules moving into and out of the membrane. A variety of forces, including hydrophobic and electrostatic interactions, are involved in maintaining the basic structural characteristics. The membrane creates a number of microenvironments, from the hydrophobic portion of the core of the membrane to the interface with the surrounding environments. It is difficult to express in words or pictures the very fluid and dynamic state, in that neither captures the time-dependent changes that occur in the structure of biological membranes. Figure 5.24 attempts to illustrate the structural and movement aspects of cellular membranes.

5.4 MOVEMENT OF MOLECULES THROUGH MEMBRANES

The lipid nature of biological membranes severely restricts the type of molecules that will readily diffuse from one side to another. Substrates carrying a charge, whether inorganic ions or charged organic molecules, will not diffuse at a significant rate because of the attraction of these molecules to water molecules and the exclusion of the charged species by the hydrophobic environment of the lipid membrane. The diffusion rate of such molecules, however, is not zero but may be too slow to accommodate a cellular need to move these molecules across the membrane. Where there is a need to move a nondiffusible substance across a particular cellular membrane, specific mechanisms are available for its translocation. Transport mechanisms are available for metabolic substrates such as carbohydrates and amino acids and for inorganic ions such as K^+, Na^+, Cl^-, and HCO_3^-. Movement through the membrane by diffusion involves three major steps: (1) the solute must leave the aqueous environment on one side and enter the membrane; (2) the solute must traverse the membrane; and (3) the solute must leave the membrane to enter a new environment on the opposite side (Figure 5.25). Each step involves an equilibrium of the solute between two states, and a variety of thermodynamic and kinetic constraints control the eventual equilibrium established for the

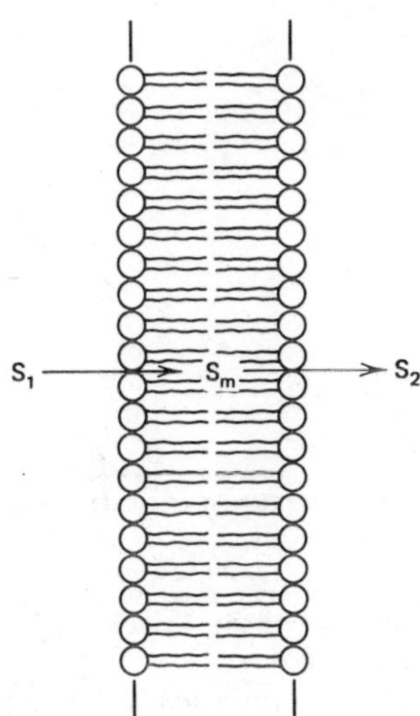

Figure 5.25
Diffusion of a solute molecule through a membrane.

concentrations of the substance on the two sides of the membrane and the rate at which it can attain the equilibrium. For simple diffusion of a hydrophilic solute with strong interaction with water molecules, the solute must have the shell of water stripped away to enter the lipid milieu but regains it on leaving the membrane. For hydrophobic substances, the distribution between the aqueous phase and lipid membrane will depend on the degree of lipid solubility of the substance; very lipid-soluble materials will essentially dissolve in the membrane.

In mammalian systems, the transport mechanism for various substances involve intrinsic membrane proteins, which interact with the molecule to be transported very much like an enzyme reacts with its substrate. Evaluation of the rate of transport is considered in the same terms, as described in Chapter 4, concerning enzyme catalyzed reactions, and the substances translocated are referred to as substrates. Some cellular transport systems move substances against a concentration gradient, that is, move from a lower to a higher concentration, which is possible only if there is an expenditure by the cell of some form of energy.

The following discussion describes the mechanisms by which molecules cross various cellular membranes; examples of specific systems will be described for illustrative purposes, but throughout this book individual systems are described in the context of specific metabolic processes.

Diffusion Across Cellular Membranes

The rate of diffusion of a solute is directly proportional to its lipid solubility and diffusion coefficient in lipids; the latter is a function of the size and shape of the substance. Uncharged lipophilic molecules, for example, fatty acids and steroids, diffuse relatively rapidly but water-soluble substances, for example, sugars and inorganic ions, diffuse very slowly. Water diffuses readily through biological membranes. It has been proposed that membranes have pores, that is, protein-lined channels through the membrane, whose size excludes most molecules except water; but definitive evidence for such pores is lacking. Movement of water may occur via the gaps in the hydrophobic environment created by the random movement of the fatty acyl chains of the lipids. Water and other small polar molecules can move into these transitory spaces and equilibrate across the membrane from one gap to another.

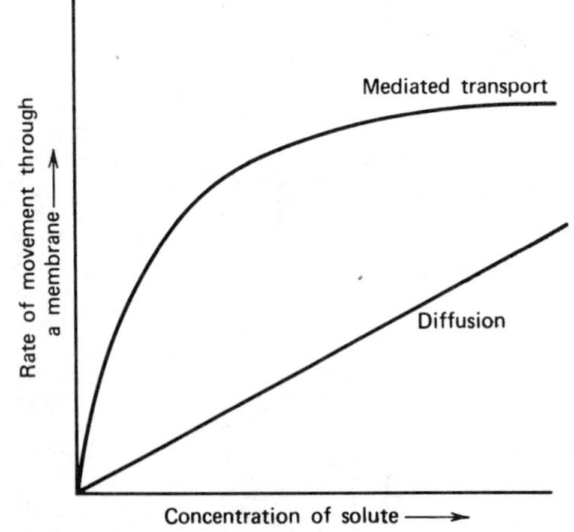

Figure 5.26
Kinetics of movement of a solute molecule through a membrane.
The initial rate of diffusion is directly proportional to the concentration of the solute. In mediated transport, the rate will reach a V_{max} *when the carrier is saturated.*

The direction of movement of solutes by diffusion is always from a higher to a lower concentration and the rate is described by Fick's first law of diffusion:

$$J = -D\left(\frac{\delta c}{\delta x}\right)$$

where J = net amount of substance moved per time, D = the diffusion coefficient, and $\delta c/\delta x$ = the chemical gradient of the substance. As the concentration of solute on one side of the membrane is increased, there will be an increasing initial rate of diffusion as illustrated in Figure 5.26. A net movement of molecules from one compartment to another will continue until the concentration in each is at a chemical equilibrium. After equilibrium is attained, there will be a continued exchange of solute molecules from one side to another, but no net accumulation on one side can occur because this would recreate a concentration gradient.

Mediated Transport

Movement of a number of different substances through different cell membranes is facilitated by the presence of specific transport systems.

Mechanisms are available for the movement of inorganic anions and cations (e.g., Na^+, K^+, Ca^{2+}, HPO_4^{2-}, Cl^-, and HCO_3^-) and uncharged and charged organic compounds (e.g., amino acids and sugars). Note that all cellular membranes do not have the same capability to move all substances; as an example, the plasma membrane has a mechanism to move K^+ and Na^+, which is not present in other cellular membranes. The transport systems of mammalian cells involve integral membrane proteins with a high degree of specificity for the substances transported. These proteins or protein complexes have been designated by a variety of names including *transporter, translocase, translocator, porter,* and *permease* or termed *transporter system, translocation mechanism,* and *mediated transport system* to name a few. For some, the term pump is applied, but this is not a very descriptive term for membrane transport systems. The designations above are used interchangeably, but for convenience we will use transporter or translocase in referring to the actual protein involved in the translocation of a substance. In many cases the mediated movement of a substance can be measured, but the actual protein involved has not been studied; in these cases it is safer to refer to them as systems or mechanisms.

Membrane transporters have a number of characteristics in common. Each facilitates the movement of a molecule or molecules through the lipid bilayer at a rate which is significantly faster than can be accounted for by simple diffusion. If S_1 is the solute on side 1 and S_2 on side 2, then the transporter promotes an equilibrium to be established as follows:

$$[S_1] \rightleftharpoons [S_2]$$

where the brackets represent the concentration of solute. If the transporter (T) is included in the equilibrium the reaction is

$$[S_1] + T \rightleftharpoons [S{-}T] \rightleftharpoons [S_2] + T$$

If there is no energy input by the system, the concentration on both sides of the membrane will be equal at equilibrium, but if there is an expenditure of energy, a concentration gradient can be established, which will depend on the thermodynamic properties of the system. Note the similarity of the role of the transporter to that of an enzyme, which increases the rate of a chemical reaction but does not determine the final equilibrium.

Table 5.3 lists the major characteristics of membrane transport systems. Mediated transport systems like enzyme-catalyzed reactions demonstrate saturation kinetics; as the concentration of the substance to be translocated increases, the initial rate of transport increases but reaches a maximum when the substance saturates the protein transporter on the membrane. A plot of solute concentration against initial rate of transport is hyperbolic, as presented in Figure 5.26. Simple diffusion does not demonstrate saturation kinetics. Constants such as V_{max} and K_m can be calcu-

TABLE 5.3 Characteristics of Membrane Transport Systems

Passive Mediated	*Active Mediated*
1. Saturation kinetics	1. Saturation kinetics
2. Specificity for solute transported	2. Specificity for solute transported
3. Can be inhibited	3. Can be inhibited
4. Solute moves down concentration gradient	4. Solute can move against concentration gradient
5. No expenditure of energy	5. Requires coupled input of energy

lated for transporters as is done in studying an enzyme. Also, as with enzymes, transporters catalyze movement of a solute in both directions across the membrane.

Most transporters have a high degree of structural and stereo-specificity for the substance transported. An example is the mediated transport system for D-glucose in the erythrocyte, where the K_m is 10 times larger for D-galactose than for D-glucose and for L-glucose it is 1,000 times larger. The transporter has essentially no activity with D-fructose or disaccharides. For many of the translocase systems competitive and noncompetitive inhibitors have been found. Structural analogs of the substrate inhibit competitively and reagents that react with specific groups on proteins are noncompetitive inhibitors.

The properties of saturation kinetics, substrate specificity, and inhibitability of transporters are characteristics in common with enzymes, but this information does not explain how the transporter actually facilitates the movement of a molecule across a distance in space. In considering mediated transport systems, therefore, we need to expand the equation above and consider four aspects: (1) *recognition* by the transporter of the appropriate solute from a variety of solutes in the aqueous environment, (2) *translocation* of the solute across the membrane, (3) *release* of the solute by the transporter, and (4) *recovery* of the transporter to its original condition to accept another solute molecule. This sequence of equilibrium reactions is presented in Figure 5.27, where T_1 and T_2 represent the location of the binding site on the transporter on different sides of the membrane.

Recognition:	$S_1 + T_1 \rightleftharpoons S{-}T_1$
Transport:	$S{-}T_1 \rightleftharpoons S{-}T_2$
Release:	$S{-}T_2 \rightleftharpoons T_2 + S_2$
Recovery:	$T_2 \rightleftharpoons T_1$

Figure 5.27
Reactions involved in mediated transport across a biological membrane.
S_1 and S_2 are the solutes on side 1 and 2 of the membrane, respectively; T_1 and T_2 are the binding sites on the transporter on side 1 and side 2, respectively.

The first step, recognition, can be readily explained on the same basis as that described for recognition of a substrate by an enzyme. The presence of very specific binding sites on the protein permits the transporter to recognize the correct structure of the solute to be translocated. The second step, translocation, is not understood. Models have been proposed based on studies on different transporters, but none has received universal support. A reasonable model (Figure 5.28) is one in which the protein transporter creates a link or channel between the environments on each side of the membrane with access through the channel being controlled by a gating mechanism (porter) in order to control which solutes can move into the channel. The transporter could have receptor sites to which the solute attaches. After association of the solute and translocase a conformational change of the protein could move the solute molecule a short distance, perhaps only 2 or 3 Å, but into the new environment of the opposite side of the membrane. In this manner, it is not necessary for the transporter physically to move the molecule the whole distance across the membrane. Earlier suggestions for the translocation step included the possibility of a diffusible or rotating carrier, but both are improbable considering that translocases are integral membrane proteins. Integral proteins do not diffuse transversely, that is, flip-flop in the membrane. A diffusion model is feasible for low molecular weight lipid soluble compounds such as ionophores (Section 5.7) or perhaps the prostaglandins (Chapter 10, Section 10.5), which can facilitate the movement of inorganic ions across membranes.

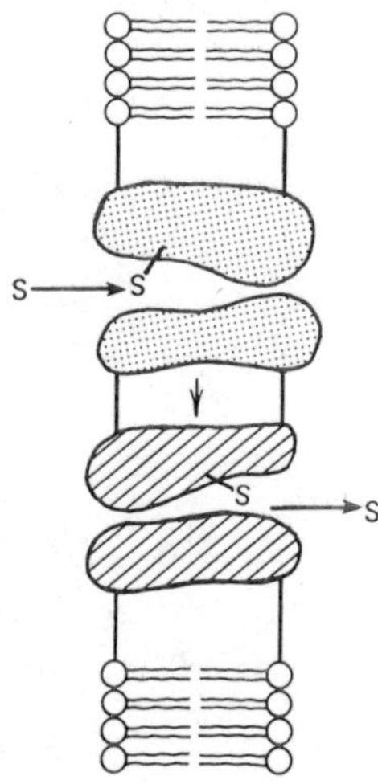

Figure 5.28
Model for a mediated transport system in a biological membrane.
The model is based on the concept of a gated channel in which conformational changes in the transporter move the bound solute a short distance but into the environment of the other side of the membrane. Once moved, the solute is released from the transporter.

Release of the solute can occur readily based on simple equilibrium considerations if the concentration of solute is lower in the new compartment than on the initial side of binding. This does not require the dissociation constant of transporter–solute complex to change. For those translocases that move a solute against a concentration gradient it is possible to envision release of the solute at the higher concentration if the affinity for the solute by the transporter is decreased. A change in the conformation of the transporter could decrease the affinity. An alternate mechanism is to alter the solute chemically while attached to the translocase, so that it is a different molecule with a lower affinity for the transporter. Examples of both of these are described below.

Finally, in some way the transporter must return to its original state. If a conformational change has occurred, there must be a return to the original conformation.

The discussion above has centered on the movement of a single solute molecule by the transporter. Actually systems are known that move two molecules simultaneously in one direction (*symport* mechanisms), two molecules in opposite directions (*antiport* mechanism), as well as a single molecule in one direction (*uniport* mechanism) (Figure 5.29). Some transporters move charged substances such as K^+, Na^+, and organic ions in which there is no direct and simultaneous movement of an ion of the opposite charge. The transporter creates a charge separation across the membrane, and the mechanism is termed *electrogenic*. If a counterion is moved to balance the charge, the mechanism is called neutral or electrically silent.

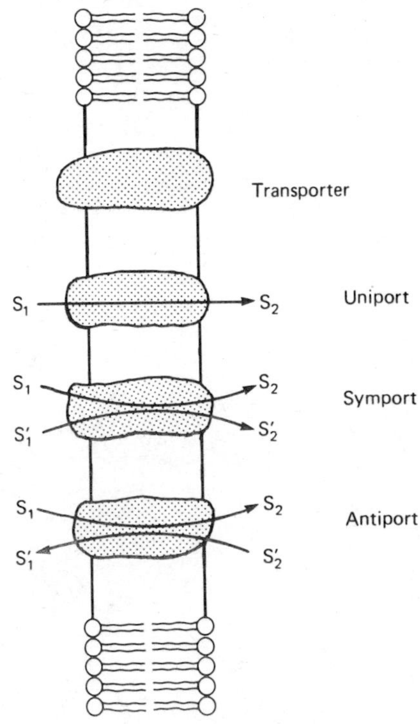

Figure 5.29
Uniport, symport, and antiport mechanisms for translocation of substances.
S and S' represent different molecules.

Energetics of Transport Systems

The change in free energy when an uncharged molecule moves from a concentration of C_1 to a concentration of C_2 on the other side of a membrane is given by equation (5.1)

$$\Delta G' = 2.3RT \log \frac{C_2}{C_1} \tag{5.1}$$

When $\Delta G'$ is negative, that is, there is release of free energy, the movement of solute will occur without the need for a driving force. When $\Delta G'$ is positive, as would be the case if C_2 is larger than C_1, then there needs to be an input of energy to drive the transport. For a charged molecule (e.g., Na^+) both the electrical potential and concentrations of solute are involved in calculating the change in free energy as in equation (5.2)

$$\Delta G' = 2.3RT \log \frac{C_2}{C_1} + ZF\Psi \tag{5.2}$$

where Z is the charge of the species moving, F is the Faraday (23.062 kcal V^{-1} mol^{-1}), and Ψ is the difference in electrical potential in volts across the membrane. The electrical component is the membrane potential and $\Delta G'$ is the electrochemical potential.

A passive transport system is one in which $\Delta G'$ is negative, that is, free energy is released, and the movement of solute occurs spontaneously. When $\Delta G'$ is positive, coupled input of energy from some source is required for movement of the solute and the process is called active transport. Several different forms of energy are available for driving active transport systems, including hydrolysis of adenosine triphosphate (ATP) to adenosine diphosphate (ADP), and the electrochemical gradient of the Na^+ ions or of H^+ ions across the membrane. In the first the chemical energy released on hydrolysis of a pyrophosphate bond drives the reaction, whereas in the latter the electrochemical gradient is dissipated to transport the solute.

Transport systems that can maintain very large concentration gradients are present in various membranes. An example is the plasma membrane transport system which maintains the Na^+ and K^+ gradients. One of the most striking examples of an active transport system is that present in the parietal cells of gastric glands which are responsible for secretion of HCl into the lumen of the stomach (see Chapter 24, page 922). The pH of plasma is about 7.4 (4×10^{-8} M H^+), and the luminal pH of the stomach can reach 0.8 (0.15 M H^+). The cells transport H^+ against a concentration gradient of $1 \times 10^{6.6}$. Assuming there is no electrical component, the energy for H^+ secretion under these conditions can be calculated from equation 5.1 and is 9.1 · kcal/mol of HCl.

5.5 PASSIVE MEDIATED TRANSPORT SYSTEMS

Passive mediated transport, also referred to as facilitated diffusion, translocates solutes through cell membranes without the expenditure of metabolic energy (see Table 5.3). As with nonmediated diffusion the direction of flow is always from a higher to a lower concentration. The distinguishing differences between measurements of simple diffusion and passive-mediated transport are the demonstration of saturation kinetics, a structural specificity for the class of molecule moving across the membrane and specific inhibition of solute movement.

The plasma membrane of many mammalian cells, but not all, has a passive mediated transport system for D-glucose. Most of our knowledge about this system is derived from studies of erythrocytes, particularly from humans. The physiological direction of movement is into the cell because the extracellular level of glucose is about 5 mM and most cells metabolize glucose rapidly thus maintaining low intracellular concentrations. Transport is by a uniport mechanism, which demonstrates saturation kinetics and is inhibitable. The system is most active with D-glucose, but D-galactose, D-mannose, and D-arabinose, and several other D-sugars as well as glycerol can be translocated by the same transporter. The L-isomers are not transported. It has been proposed that the β-D-glucopyranose is transported with carrier interaction at the hydrogens on at least C1, C3, and C6 of the sugar. The affinity of the erythrocyte carrier for D-glucose is highest with a K_m of about 6.2 mM, whereas for the other sugars the K_ms are much higher. The carrier has a very low affinity for D-fructose, which precludes the carrier from having any role in cellular uptake of fructose. A separate carrier for fructose has been proposed. With isolated erythrocytes, glucose will move either into or out of the erythrocyte, depending on the direction of the experimentally established concentration gradient, which demonstrates the reversibility of the system.

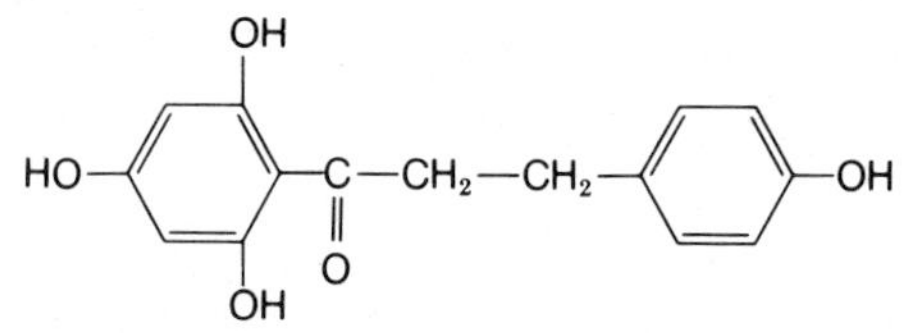

Figure 5.30
Inhibitors of passive mediated transport of D-glucose in erythrocytes.

Several sugar analogues as well as phoretin and 2,4,6-trihydroxyacetophenone (Figure 5.30) are competitive inhibitors; reagents that react irreversibly with proteins are noncompetitive inhibitors. Treatment of cells from some tissues with the hormone insulin increases the V_{max} of glucose uptake; the effect of insulin is apparently indirect rather than directly on the carrier. The alteration in the carrier by the hormone demonstrates the probable importance of the carrier in maintenance of adequate intracellular levels of glucose.

Another passive mediated transport system in erythrocytes involves the movement of the anions Cl^- and HCO_3^- by an antiport mechanism (Figure 5.31). In this mechanism Cl^- moves in one direction and simultaneously a HCO_3^- in the opposite direction; the direction of movement depends on the concentration gradients of the ions across the membrane. The transporter has an important role in adjusting the erythrocyte HCO_3^- concentrations in arterial and venous blood (see Chapter 23) and is a part of the mechanism for removing CO_2 from the body. The integral membrane protein designated band 3, because of its position in SDS polyacryl-

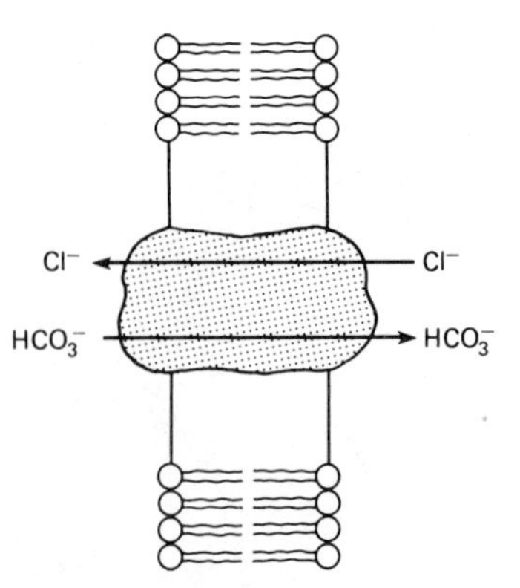

Figure 5.31
The passive anion antiport mechanism for movement of Cl^- and HCO_3^- across the erythrocyte plasma membrane.

amide gel electrophoresis, is considered to be the Cl^--HCO_3^- antiporter.

The inner mitochondrial membrane contains several antiport systems for the exchange of anions between the cytosol and mitochondrial matrix. These include (1) a transporter for exchange of ADP and ATP; (2) a transporter for exchange of phosphate and OH^-; (3) a dicarboxylate carrier that catalyzes an exchange of malate for phosphate; and (4) a translocator for exchange of aspartate and glutamate (Figure 5.32). The relationship of these translocases and energy coupling are discussed in Chapter 6, Section 6.5. In the absence of an input of energy these transporters will catalyze a passive exchange of metabolites down their concentration gradient to achieve a thermodynamic equilibrium of all intermediates. As an antiport mechanism, a concentration gradient of one compound can drive the movement of the other solute. The systems, however, can also couple to the mitochondrial energy transducing system, and the anions are moved against their concentration gradient. The ability of these translocases to function as either a passive mediated transport system or as an active mediated transport system is unique.

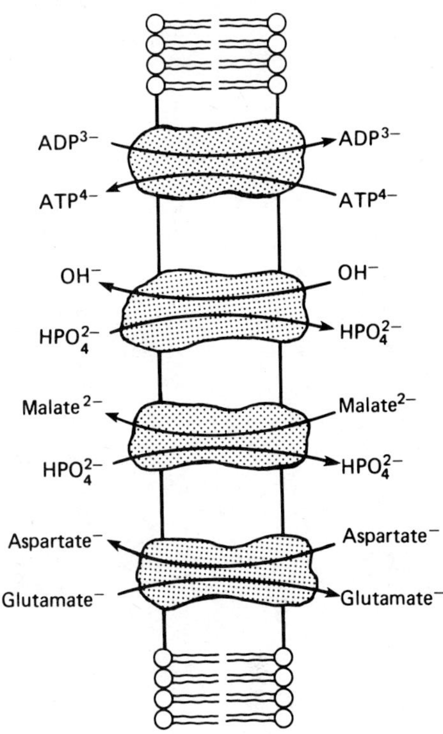

Figure 5.32
Representative anion transport systems in liver mitochondria.
Note that each is an antiport mechanism. Several other transport systems are known and are discussed in Chapter 6.

The ATP–ADP translocase has been extensively studied; it is very specific for ATP and ADP and their deoxyribose derivatives, dATP and dADP, but does not transport adenosine monophosphate (AMP) and other nucleotides. The protein responsible for the translocation has been isolated and is a dimer containing two subunits of 30,000 daltons each. It has been estimated that the transporter represents 12% of the total protein in heart mitochondria. The protein is very hydrophobic and can exist in two conformations. Atractyloside and bongkrekic acid (Figure 5.33) are specific inhibitors, each apparently reacting with a different conformation of the protein. The mitochondrial membrane potential can drive the movement of the nucleotides by this translocator, but in the absence of the potential it will function as a passive mediated transporter.

It is sometimes difficult to differentiate passive mediated transport from simple diffusion, but specific inhibition is good evidence of a carrier; this has been the case for the anion carriers of mitochondria, which have been differentiated on the basis of specific inhibitors.

Atractyloside

Bongkrekic acid

Figure 5.33
Structure of two inhibitors of the ATP/ADP transport system of liver mitochondria.

5.6 ACTIVE MEDIATED TRANSPORT SYSTEMS

A variety of mediated transport systems requiring the utilization of energy to move solutes against their concentration gradient are present in cellular membranes. Active transport systems, sometimes referred to as pumps, have the same three characteristics as passive transporters, that is, saturation kinetics, substrate specificity, and inhibitability, but in addition these systems require a coupled input of energy (see Table 5.3). If the energy source is removed or inhibited, the transport system will not function. Inhibition of ATP synthesis leads to the inhibition of many transport systems, suggesting ATP as the primary energy source for active transport. Direct utilization of ATP, however, does not occur in all active transport systems. Several systems utilize the electrochemical gradient of Na^+ across the membrane. As indicated below in the discussion on the active mediated transport of glucose, which utilizes the transmembrane Na^+ gradient in a symport mechanism, metabolic energy in the form of ATP is required for maintenance of the Na^+ gradient but not directly for moving a glucose molecule; inhibition of ATP synthesis, however, leads to a dissipation of the Na^+ electrochemical gradient, which in turn decreases transport activity utilizing the gradient. This is visualized in Figure 5.34.

Active Transport System for Na^+ and K^+: The Na^+-K^+-ATPase

For years a major research effort has been directed toward an explanation of the cellular mechanism for maintenance of the Na^+ and K^+ gradients across the plasma membrane of cells. All mammalian cells contain a Na^+-K^+ antiport system, which utilizes the direct hydrolysis of ATP for movement of ions. Knowledge of this transporter has developed along two paths: (1) from studies of a membrane enzyme which catalyzes ATP hydrolysis and has a requirement for Na^+ and K^+ ions, and (2) from measurements of Na^+ and K^+ movements across intact plasma membranes. It is now accepted that the transporter and the Na^+- K^+-ATPase are one and the same.

All mammalian membranes catalyze the reaction

$$\text{ATP} \xrightarrow[\text{Mg}^{2+}]{\text{Na}^+ + \text{K}^+} \text{ADP} + \text{inorganic phosphate}$$

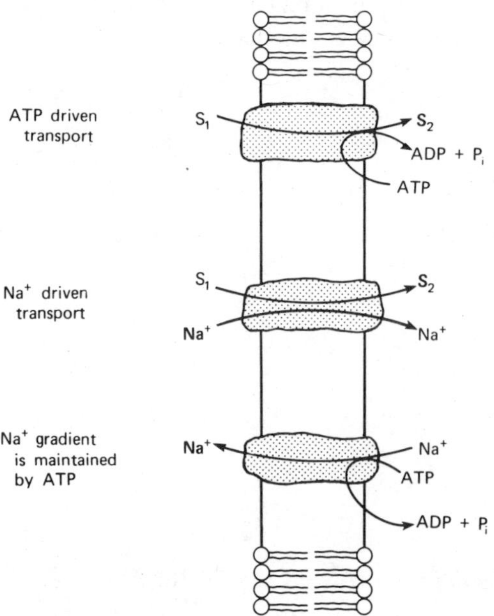

Figure 5.34
Involvement of metabolic energy (ATP) in active mediated transport systems. *The chemical energy released on the hydrolysis of ATP to ADP and inorganic phosphate is used to drive the active transport of various substances, including Na^+. The transmembrane concentration gradient of Na^+ is also used for the active transport of substances.*

The enzyme has a requirement for both Na^+ and K^+ ions as well as Mg^{2+}, which is normally required for ATP-requiring reactions. The level of the ATPase in plasma membranes correlates with the Na^+-K^+ transport activity; excitable tissue such as muscle and nerve have a high capacity of both the Na^+-K^+ transport system and the Na^+-K^+-ATPase as do cells actively involved in the movement of Na^+ ion such as those in the salivary gland and kidney cortex. The protein responsible for the Na^+-K^+-transporting ATPase activity is an oligomer containing two subunits of about 95,000 daltons each and two subunits of about 40,000 daltons each with a total mol wt of 270,000. The smallest subunits are glycoproteins, and the complex has the characteristics of a typical integral membrane protein. Figure 5.35 is a schematic diagram of the Na^+-K^+ transporter. The enzymatic activity has a requirement for phospholipids indicating its close relationship to membrane function. During the hydrolysis of ATP, the larger subunit is phosphorylated on a side chain of a specific aspartic acid forming a β-aspartyl phosphate. Phosphorylation of the protein requires Na^+ and Mg^{2+} but not K^+, whereas dephosphorylation of the protein requires K^+ but not Na^+ or Mg^{2+}. The isolated enzyme has an absolute requirement for Na^+, but K^+ can be replaced with $NH_4{}^+$ or Rb^+. Several conformations of the protein complex have been observed. A possible sequence of reactions for the enzyme is presented in Figure 5.36.

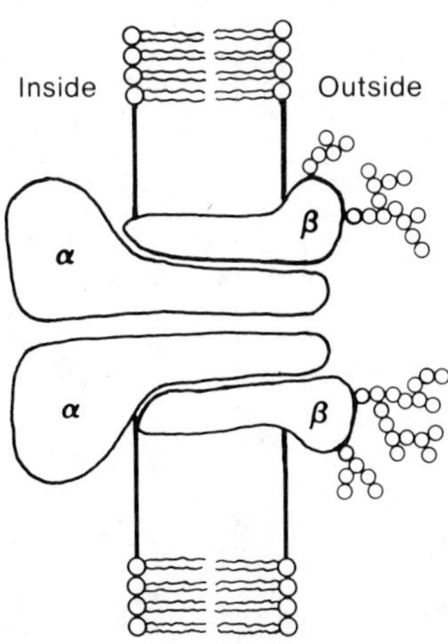

Figure 5.35
Schematic drawing of the Na^+-K^+-transporting ATPase of plasma membranes.

Of particular significance to its physiological role as a transporter, the enzyme is inhibited by a series of cardiotonic steroids. These pharmacological agents, which include digitalis, increase the force of contraction of heart muscle by altering the excitability of the tissue which is a function of the Na^+-K^+ concentration across the membrane. Ouabain (Figure 5.37) is one of the most active Na^+-K^+-transporting ATPase inhibitors of the series and its action has been studied extensively. The site of binding of ouabain is on the smaller subunit of the enzyme complex and at some distance from the ATP binding site on the larger monomer.

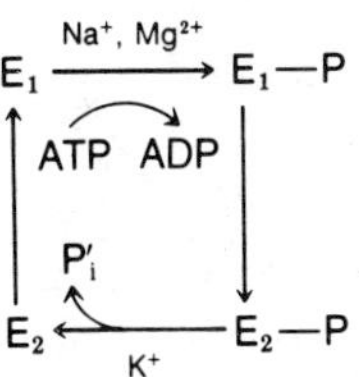

Figure 5.36
Proposed sequence of reactions and intermediates in hydrolysis of ATP by the Na^+, K^+-ATPase.
E_1 and E_2 are different conformations of the enzyme. Phosphorylation of the enzyme requires Na^+ and Mg^{2+} and dephosphorylation involves K^+.

Studies of the transporter activity have been facilitated by use of erythrocyte ghosts, which are intact erythrocyte preparations free of hemoglobin. By carefully adjusting the tonicity of the medium, erythrocytes will swell with breaks in the phospholipid bilayer, permitting the leaking from the cell of cytosolic material, including hemoglobin. The cytosol can be replaced with a defined medium by readjusting the tonicity so that the membrane reseals, trapping the isolation medium inside. In this manner the intracellular ionic and substrate composition and even protein content can be altered. With erythrocyte ghosts the intra- and extracellular Na^+ and K^+ can be manipulated as well as ATP or inhibitor content. With such preparations it has been demonstrated that movement of Na^+ and K^+ is an antiport vectorial process, with Na^+ moving out and K^+ moving into the cell. The ATP binding site on the protein is on the inner surface of the membrane in that hydrolysis occurs only if ATP, Na^+, and Mg^{2+} are inside the cell. K^+ is required externally for dephosphorylation internally of the protein. Ouabain inhibits the translocation of Na^+ and K^+ but only if it is present externally. The protein apparently spans the membrane. The actual number of translocase molecules on an erythrocyte has been estimated by binding studies of radiolabeled ouabain. It is estimated that there are between 100 and 200 molecules per erythrocyte, but the number is significantly larger for other tissues.

ATP hydrolysis by the translocase occurs only if Na^+ and K^+ are translocated, demonstrating that the enzyme is not involved in dissipation of energy in a useless activity. For each ATP hydrolyzed three ions of Na^+ are moved out of the cell but only two ions of K^+ in, which leads to an increase in external positive charges. The electrogenic movement of Na^+ and K^+ is part of the mechanism for the maintenance of membrane potential in a variety of tissues. Even though the energetics of

Figure 5.37
Structure of ouabain, a cardiotonic steroid, which is a potent inhibitor of the Na^+, K^+-ATPase and of active Na^+ and K^+ transport.

the system dictate that it functions in normal conditions in only one direction, the translocase can be reversed by adjusting the Na^+-K^+ levels; a small net synthesis of ATP has been observed when transport is forced to run in the reverse direction. Obviously, under physiological conditions translocation does not occur in the opposite direction.

A hypothetical model for the movement of Na^+ and K^+ is presented in Figure 5.38. It is proposed that the protein goes through a series of conformational changes during which the Na^+ and K^+ are moved short distances. During the transition a change in the affinity of the binding protein for the cations can occur such that there is a decrease in affinity constants, resulting in the release of the cation into a milieu where the concentration is higher than that from which it was transported.

As an indication of the importance of this enzyme, it has been estimated that the Na^+-K^+-transporting ATPase uses 60–70% of the ATP synthesized by cells such as nerve and muscle and may utilize about 35% of all ATP generated in a resting individual. A recent report suggests that some obese individuals may have a deficiency in the Na^+-K^+ transport system and thus do not require the same dietary intake for energy purposes.

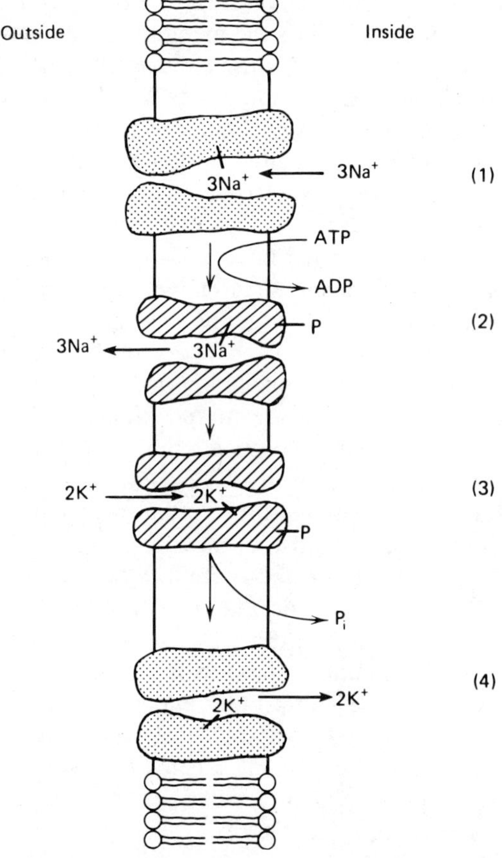

Figure 5.38
A hypothetical model for the translocation of Na^+ and K^+ across the plasma membrane by the Na^+, K^+-ATPase.
(1) Transporter in conformation 1 picks up Na^+. (2) Transporter in conformation 2 translocates and releases Na^+. (3) Transporter in conformation 2 picks up K^+. (4) Transporter in conformation 1 translocates and releases K^+.

Ca^{2+} Translocation in the Sarcoplasmic Reticulum

The sarcoplasmic reticulum of muscle contains an ATP-dependent transport system for Ca^{2+}. Muscle contraction–relaxation cycles are regulated by the cytosolic levels of Ca^{2+} (see Chapter 21, Section 21.1), and the sarcoplasmic reticulum modulates this Ca^{2+} concentration. A Ca^{2+}-dependent ATPase activity has been isolated from the reticulum which has many of the characteristics of the Na^+-K^+-ATPase. The protein apparently spans the membrane and is phosphorylated on an aspartyl residue during the translocation reaction. Two Ca^{2+} ions are translocated for each ATP hydrolyzed. The transport system can move Ca^{2+} against concentration gradients of over 1000 : 1 thereby maintaining very low cytosolic Ca^{2+} levels ($\sim 1 \times 10^{-6}$ M). It has been estimated that the Ca^{2+} translocase may represent 80% of the integral membrane protein of the sarcoplasmic reticulum and occupy a third of the surface area.

Na^+-Dependent Symport Systems

The mechanisms described above for the active transport of cations involved the hydrolysis of ATP as the driving force. Cells have another energy source in the form of the electrochemical gradient of Na^+ ion which is utilized to move sugars and amino acids actively. A symport translocation system involving the simultaneous movement of both a Na^+ ion and another molecule in the same direction is present in the plasma membrane of cells of the kidney tubule and intestinal epithelium; other tissues may also contain similar transport systems. The general mechanism is presented in Figure 5.39. The diagram represents the transport of D-glucose driven by the movement of Na^+ ion down its concentration gradient. Note that in the transport of the sugar no hydrolysis of ATP occurs. The Na^+-dependent glucose transporter is poorly understood but its activity can be distinguished from passive facilitated diffusion of glucose by differences in inhibitor effects and substrate specificity. There is an absolute requirement for Na^+, and in the process of translocation one Na^+ is moved with each glucose molecule. It can be considered that Na^+ is moving by passive facilitated transport down its electrochemical gradient. It is obligatory that the transporter also translocates a glucose with the Na^+ ion. In the transport the electrochemical gradient of Na^+ ion is dissipated and glucose can be translocated against its concentration gradient. Unless the Na^+ ion gradient is continuously regenerated, the movement of glucose will cease. The Na^+ gradient is maintained by the Na^+-K^+ transport system described above and also represented in Figure 5.39. Thus, metabolic energy in the form of ATP is indirectly involved in glucose transport because it is utilized to maintain the Na^+ ion gradient. Inhibition of energy metabolism and a subsequent decrease in ATP will alter the Na^+ ion gradient and inhibit glucose uptake. Ouabain, the inhibitor of the Na^+-K^+ transporter, inhibits uptake of glucose by preventing the cell from maintaining the Na^+-K^+ gradient. It can be calculated that each glucose molecule requires only one-third of an ATP to be translocated because 3 Na^+ ions are translocated for the hydrolysis of each ATP in the Na^+-K^+-transporting ATPase.

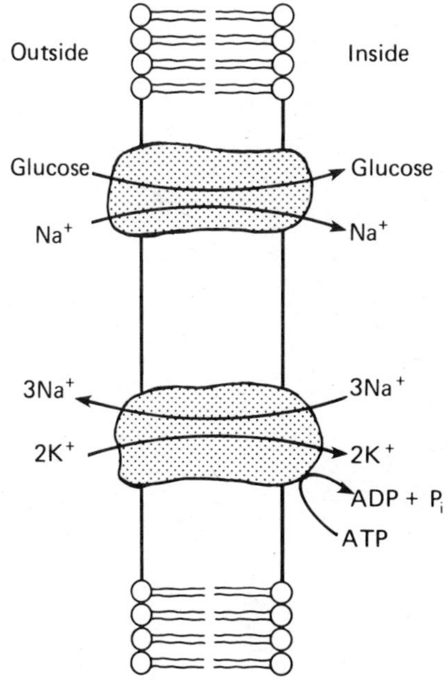

Figure 5.39
Na^+-dependent symport transport of glucose in the plasma membrane.

Amino acids are also translocated by the luminal epithelial cells of the intestines by Na^+ dependent pathways similar to the Na^+-dependent glucose transporter. At least four different translocases have been identified: (1) one for neutral amino acids such as alanine, valine, and leucine; (2) one for basic amino acids, including lysine and arginine; (3) one for the acidic amino acids, aspartate and glutamate; and (4) one for the amino acids proline and glycine.

Utilization of the Na^+ gradient as a means to drive the transport of other ions has been reported, including a symport mechanism in the small intestines for the uptake of Cl^- with Na^+ and an antiport mechanism for the secretion of Ca^{2+} out of the cell driven by the simultaneous uptake of Na^+.

It has been proposed that the chemical mechanism for the symport movement of molecules utilizing the Na^+ ion gradient involves a cooperative interaction of the Na^+ ion and the other molecule, for example, glucose, translocated on the protein. A conformational change of the protein occurs following association of the two ligands, which moves them the necessary distance to bring them into contact with the cytosolic environment. The dissociation of the Na^+ ion from the transporter because of the low Na^+ ion concentration inside the cell leads to a return of the protein to its original conformation, a decrease in the affinity for the other ligand, and a release of the ligand into the cytosol.

Group Translocation—A Special Case of Active Transport

As discussed previously, a major hurdle for any active transport system is the release of the transported molecule from the binding site after translocation. If the affinity of the transporter for the translocated molecule does not change there can not be movement against a concentration gradient. In the active transport systems previously described it is believed that there is a change in the affinity for the substance by the transporter by a conformational change of the protein. An alternate mechanism for the release of the substrate is the chemical alteration of the molecule after translocation but before release from the transporter leading to a new compound with a lower affinity for the transporter. The γ-glutamyl cycle for the transport of amino acids across the plasma membrane of some tissues is an example where the substrate is altered during transport and released into the cell as a different molecule. The reactions of the transport mechanism are presented in Figure 5.40. The pathway involves the enzyme γ-glutamyltransferase, which is membrane-bound and catalyzes a transpeptidation reaction, leading to the formation of a dipeptide

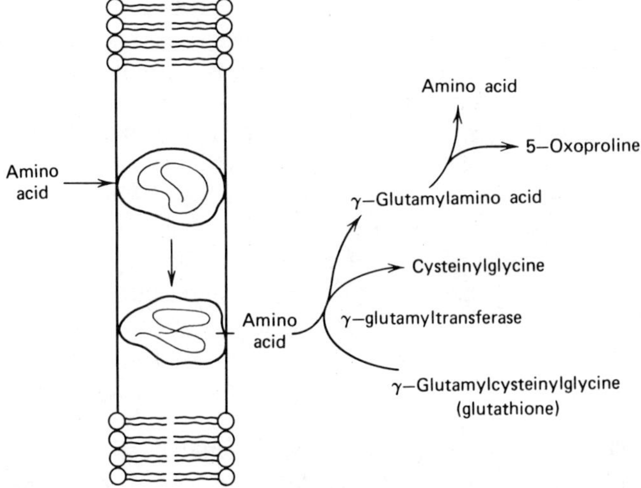

Figure 5.40
The γ-glutamyl cycle.
Represented are the key reactions involved in the group translocation of amino acids across liver cell plasma membranes. The continued uptake of amino acids requires the constant resynthesis of glutathione via a series of ATP requiring reactions described in Chapter 12, page 484.

involving the amino acid transported. The amino acid transported is the substrate to which the γ-glutamyl residue of glutathione (Figure 5.41) is transferred. The new dipeptide is not part of the chemical gradient for the amino acid across the membrane. The γ-glutamyl derivative is then hydrolyzed by a separate enzyme, not on the membrane, leaving the free amino acid and oxoproline. The process is termed group translocation.

The pathway is active in many tissues but some doubt has been raised about its physiological significance in that individuals have been identified with a genetic absence of the γ-glutamyltransferase activity without any apparent difficulty in amino acid transport. It is possible, of course, that

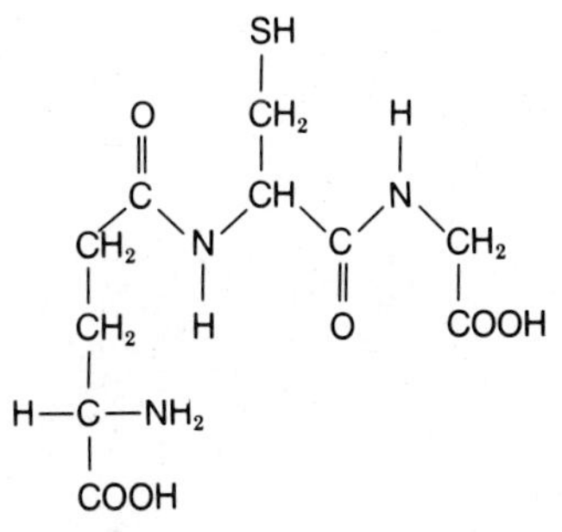

Figure 5.41
Glutathione (γ-glutamylcysteinylglycine).

TABLE 5.4 Major Transport Systems in Mammalian Cells[a]

Substance Transported	*Mechanism of Transport*	*Tissues*
Sugars		
Glucose	Passive	Most tissues
	Active symport with Na^+	Small intestines and renal tubular cells
Fructose	Passive	Intestines and liver
Amino acids		
Amino acid specific transporters	Active symport with Na^+	Intestines, kidney, and liver
All amino acids except proline	Active group translocation	Liver
Specific amino acids	Passive	Small intestine
Other organic molecules		
Cholic acid, deoxycholic acid, and taurocholic acid	Active symport with Na^+	Intestines
Organic anions, e.g., malate, α-ketoglutarate, glutamate	Antiport with counterorganic anion	Mitochondria of liver
ATP–ADP	Antiport transport of nucleotides; can be active transport	Mitochondria of liver
Inorganic ions		
Na^+	Passive	Distal renal tubular cells
Na^+-H^+	Active antiport	Proximal renal tubular cells and small intestines
Na^+-K^+	Active transport-ATP driven	Plasma membrane of all cells
Ca^{2+}	Active-ATP driven	All cells
H^+-K^+	Active transport	Parietal cells of gastric mucosa secreting H^+
Cl^-/HCO_3^- (perhaps other anions)	Mediated-antiport	Erythrocytes and many other cells

[a] The transport systems are only indicative of the variety of transporters known; others responsible for a variety of substances have been proposed. Most systems have been studied in only a few tissues and their localization may be more extensive than indicated.

cells may have several alternate methods for the transport of amino acids and are not dependent on only one mechanism.

All the amino acids except proline can be transported by this group translocation process. The energy for transport comes from the hydrolysis of a peptide bond in glutathione. For the system to continue, glutathione must be resynthesized, which requires the expenditure of three ATP molecules. Thus for each amino acid translocated, three ATPs are required. Recall that the expenditure of only one-third of an ATP is required for each amino acid transported in the Na^+-dependent translocase system. This group translocation is an expensive energetic mechanism for the transport of amino acids.

Summary of Transport Systems

The foregoing has presented the major mechanisms for the movement of molecules across cellular membranes, particularly the plasma membrane. Mitochondria also contain several active transport mechanisms utilizing the pH gradient, that is, a hydrogen ion gradient, developed across the inner membrane. These will be presented in Chapter 6. Bacteria have a number of transport systems analogous to those observed in mammalian cells including passive mediated transporters, one involving a H^+ ion gradient and group translocation.

Table 5.4 summarizes some of the characteristics of the major transport systems found in mammalian cells. (See Clin. Corr. 5.3.)

CLIN. CORR. 5.3 DISEASES DUE TO LOSS OF MEMBRANE TRANSPORT SYSTEMS

A number of pathological conditions that are due to an alteration in the transport system for specific cellular components have been defined. Some of these are discussed in the appropriate sections describing the metabolism of the intermediates. Individuals have been observed with a decrease in glucose uptake from the intestinal tract due presumably to a loss of the specific glucose transport mechanism. Fructose malabsorption syndromes have been observed, which are also due to an alteration in activity of the facilitated transport system for fructose. In Hartnup's disease there is a decrease in the transport of neutral amino acids in the epithelial cells of the intestines and of renal tubules. Little is known concerning possible changes of transport activities in tissues such as muscle, liver, and brain, but it has been suggested that there may be a number of pathological states due to the loss of specific transport mechanisms.

5.7 IONOPHORES

An interesting class of antibiotics of bacterial origin has been discovered, which facilitates the movement of monovalent and divalent inorganic ions across biological and synthetic lipid membranes. These molecules are not large macromolecules such as proteins but are relatively small molecular weight compounds (up to several thousand daltons); the class of compounds are called ionophores. The possible presence of similar molecules in mammalian tissues has been reported. Ionophores are divided into two major groups: (1) mobile carriers are those ionophores which apparently diffuse back and forth across the membrane carrying the ion from one side of the membrane to the other, and (2) ionophores which apparently form a channel that transverses the membrane and through which ions can diffuse. With both types, ions are translocated by a passive-mediated transport mechanism. The ionophores which diffuse back and forth across the membrane are more affected by the changes in the fluidity of the membrane than those that form a channel. Some major ionophores are listed in Table 5.5.

TABLE 5.5 Major Ionophores

Compound	*Major Cations Transported*	*Action*
Valinomycin	K^+ or Rb^+	Uniport, electrogenic
Nonactin	NH_4^+, K^+	Uniport, electrogenic
A 23187	$Ca^{2+}/2H^+$	Antiport, electroneutral
Nigericin	K^+/H^+	Antiport, electroneutral
Monensin	Na^+/H^+	Antiport, electroneutral
Gramicidin	H^+, Na^+, K^+, Rb^+	Forms channels
Alamethicin	K^+, Rb^+	Forms channels

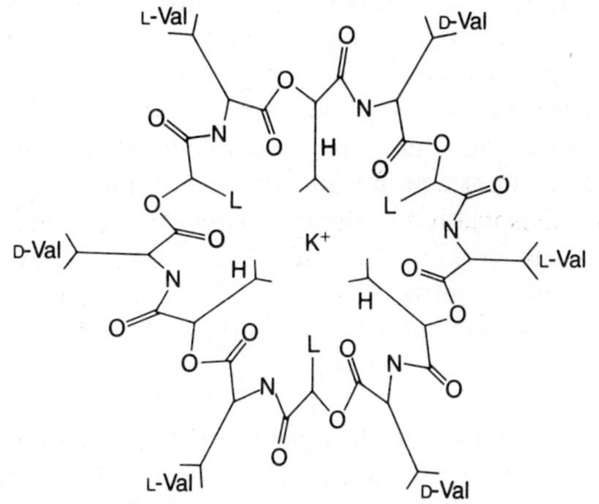

Figure 5.42
The structure of the valinomycin–K^+ complex.
Abbreviations: D-*Val* = D-*valine;* L-*Val* = L-*valine; L* = L-*lactate; and H* = D-*hydroxyisovalerate.*

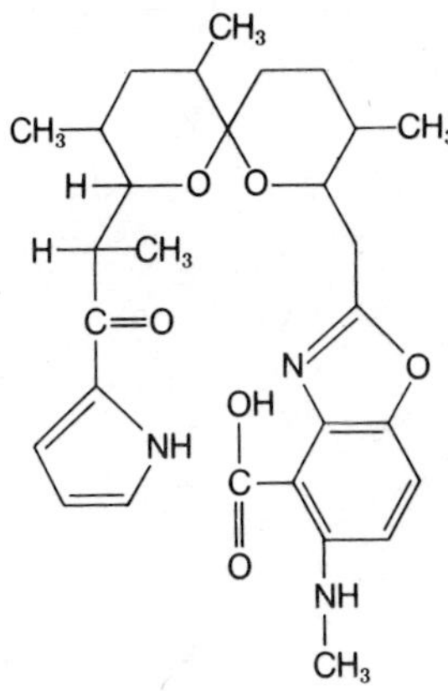

Figure 5.43
Structure of A23187, a Ca^{2+} ionophore.

Each ionophore has a definite ion specificity; valinomycin, whose structure is given in Figure 5.42, has an affinity for K^+ 1,000 times greater than that for Na^+, and the antibiotic A23187 (Figure 5.43) translocates Ca^{2+} 10 times more actively than Mg^{2+}. Several of the diffusion type ionophores that have been studied in detail have a common structural characteristic being cyclic structures, shaped like a doughnut. The metal ion is coordinated to several oxygens in the core of the ionophore, and the periphery of the molecule consists of hydrophobic groups. The interaction of the ionophore leads to a chelation of the ion, stripping away its surrounding water shell and encompassing the ion by a hydrophobic shell. The ionophore–ion complex is soluble in the lipid membrane and freely diffuses across the membrane. Since the interaction of ion and ionophore is an equilibrium reaction, a steady state develops in the concentration of ions on both sides of the membrane. The specificity of the ionophore is due in part to the size of the pore into which the ion fits and to the attraction of the ionophore for the ion in competition with water molecules.

Valinomycin transports K^+ by a uniport mechanism and is thus electrogenic, that is, it can create an electrochemical gradient. It carries a positive charge in the form of the K^+ ion across the membrane. Nigericin functions as a antiporter having a free carboxyl group, which when dissociated can pick up a K^+ ion, leading to a neutral molecule. Thus on diffusion back through the membrane it transports a proton; the overall mechanism is electrically neutral, with a K^+ ion exchanging for a H^+ ion. These mechanisms are shown in Figure 5.44.

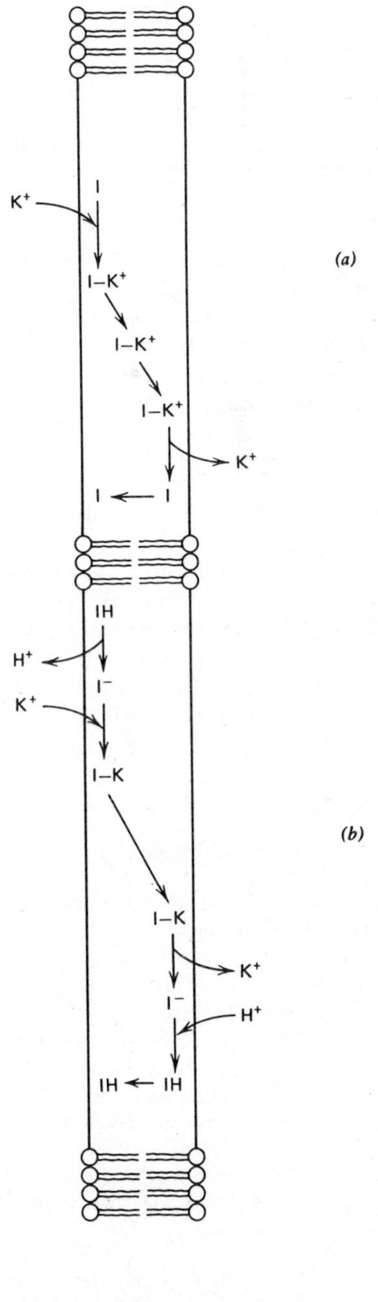

Figure 5.44
Proposed mechanism for the ionophoretic activities of valinomycin and nigericin.
(a) Transport by valinomycin. (b) Transport by nigericin. I represents the ionophore. The valinomycin–K^+ complex is positively charged and translocation of K^+ is electrogenic leading to the creation of a charge separation across the membrane. Nigericin translocates K^+ in exchange for a H^+ across the membrane and the mechanism is electrically neutral.
Diagram adopted from B. C. Pressman, *Annu. Rev. Biochem.* 45:501, 1976.

The other major type of ionophore apparently creates a pore in the membrane. The principal example is gramicidin A. These ionophores show a lower degree of selectivity toward ions, in that the ions are essentially diffusing through a hole in the membrane. Evidence suggests that two molecules of gramicidin A form a channel in the membrane and that the dimer is in constant equilibrium with the free monomer form. By the association and dissociation of the monomers in the membrane, channels can be formed and broken; the rate of interaction of two molecules of gramicidin A would control the rate of ion flux. The structure of the molecules suggests that polar peptide groups line the channel and that hydrophobic groups are on the periphery of the channel interacting with the lipid membrane.

The antibiotic ionophores have been a valuable experimental tool in studies involving ion translocation in biological membranes and for the manipulation of the ionic compositions of cells. There have been reports that proteolipids, prostaglandins, and perhaps other lipids present in mammalian tissues may function as ionophores.

BIBLIOGRAPHY

General

Andreoli, T. E., Hoffman, J. F., and Fanestil, D. D. (eds.). *Physiology of membrane disorders*. New York: Plenum, 1978.

Bittar, E. E. (ed.). *Membrane structure and function,* vols. 1–3. New York: Wiley, 1980.

Houslay, M. D., and Stanley, K. K. *Dynamics of biological membranes*. New York: Wiley, 1982.

Quinn, P. J. *The molecular biology of cell membranes*. Baltimore: University Park Press, 1976.

Weissmann, G., and Claiborne, R. (eds.). *Cell membranes,* New York: H. P. Publishing Co., 1975.

Membrane Structure

Finean, J. B., and Michell, R. H. (eds.). *Membrane structure*. Amsterdam; Elsevier, 1981.

Marchesi, V. T., Furthmayr, H., and Tomita, M. The red cell membrane. *Annu. Rev. Biochem.* 45:667, 1976.

Martonosi, A. N. (ed). *The enzymes of biological membranes; membrane structure and dynamics,* vol. 1. New York: Plenum, 1985.

Quinn, P. J., and Chapman, D. The dynamics of membrane structure. *CRC Crit. Rev. Biochem.* 8:1, 1980.

Shinitzky, M., and Henkart, P. Fluidity of cell membranes—current concepts and trends. *Int. Rev. Cyto.* 60:121, 1979.

Singer, S. J., and Nicolson, G. L. The fluid mosaic model of the structure of cell membranes. *Science* 175:720, 1972.

Stubbs, C. D. Membrane fluidity: Structure and dynamics of membrane lipids, in P. N. Campbell and R. D. Marshall (eds.), *Essays in biochemistry,* vol. 19, 1983, page 1.

Wallach, D. F. H. *Membrane molecular biology of neoplastic cells.* Amsterdam: Elsevier, 1975.

Transport Processes

Christensen, H. N. *Biological transport,* 2nd ed. Reading, Mass.: W. A. Benjamin, 1975.

MacLennan, D. H., and Holland, P. C. The Ca^{2+}-dependent ATPase of sarcoplasmic reticulum. *Annu. Rev. Biophys. Bioenerg.* 4:377, 1975.

Martonosi, A. (ed.). *The enzymes of biological membranes: membrane transport,* vol. 3, New York: Plenum, 1976.

Skou, J. C., and Norby, D. G. (eds.). *Na^{+},K^{+}-ATPase: structure and kinetics*. New York: Academic, 1979.

Wilson, D. B. Cellular transport mechanisms. *Annu. Rev. Biochem.* 47:933, 1978.

QUESTIONS

C. N. ANGSTADT AND J. BAGGOTT

*Q*uestion *T*ypes are described inside front cover.

1. (QT2) Cell membranes typically:
   1. contain phospholipids.
   2. have both intrinsic and extrinsic proteins.
   3. have some cholesterol.
   4. contain free carbohydrate such as glucose.
2. (QT2) According to the *fluid mosaic* model of a membrane:
   1. proteins may be embedded in the lipid bilayer.
   2. transverse movement (flip-flop) of a protein in the membrane is thermodynamically favorable.
   3. individual proteins may be removed and resynthesized.
   4. proteins are distributed symmetrically in the membrane.
3. (QT2) Characteristics of a mediated transport system include:
   1. the ability to bind an appropriate solute specifically.
   2. release of the transporter from the membrane following transport.
   3. a mechanism for translocating the solute from one side of the membrane to the other.
   4. release of the solute only if the concentration on the new side is lower than that on the original side.
4. (QT1) The group translocation type of transport system:
   A. does not require metabolic energy.
   B. involves the transport of two different solute molecules simultaneously.

C. has been demonstrated for fatty acids.
D. results in the alteration of the substrate molecule during the transport process.
E. uses ATP to maintain a concentration gradient.

A. Passive mediated transport C. Both
B. Active mediated transport D. Neither

5. (QT4) Require(s) a transporter that specifically binds a solute.
6. (QT4) Can transport a solute against its concentration gradient.
7. (QT4) Glucose can be transported by ________.
8. (QT1) The transport system that maintains the Na^+ and K^+ gradients across the plasma membrane of cells:
   A. involves an enzyme that is an ATPase.
   B. is a symport system.
   C. moves Na^+ either into or out of the cell.
   D. is an electrically neutral system.
   E. in the membrane, hydrolyzes ATP independently of the movement of Na^+ and K^+.
9. (QT2) A mediated transport system would be expected to:
   1. show a continuously increasing initial rate of transport with increasing substrate concentration.
   2. exhibit structural and/or stereospecificity for the substance transported.
   3. be slower than that of a simple diffusion system.
   4. establish a concentration gradient across the membrane if there is an expenditure of energy.

A. Lecithins C. Both
B. Sphingomyelins D. Neither

10. (QT4) Contain a fatty acid in an amide linkage
11. (QT4) Amphipathic
12. (QT4) Contain phosphate

A. Cerebroside C. Both
B. Ganglioside D. Neither

13. (QT4) Incorporate(s) an oligosaccharide containing sialic acid
14. (QT4) Belong(s) to the class of neutral glycosphingolipids

The answers to Questions 15 and 16 are based on the following figure:

15. (QT5) Represents a passive mediated antiport system
16. (QT5) Could represent the Na^+-driven uptake of glucose

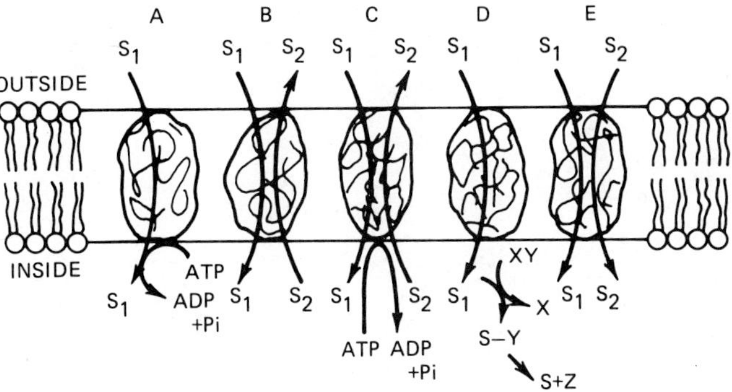

17. (QT2) An ionophore:
   1. may diffuse back and forth across a membrane.
   2. may form a channel across a membrane through which an ion may diffuse.
   3. may catalyze electrogenic mediated transport of an ion.
   4. requires the input of metabolic energy for mediated transport of an ion.

ANSWERS

1. A 1, 2, and 3 correct (Figures 5.2, 5.18). 4: All carbohydrate in membranes is in the form of glycoproteins and glycolipids (p. 187).
2. B 1 and 3 correct (Figure 5.21). 2: Transverse motion of proteins is even less than that of lipids (p. 192). 3: Membranes, like other cellular components, are synthesized and degraded (p. 191). 4: Both proteins and lipids are distributed asymmetrically (p. 191).
3. B 1 and 3 correct. 2: Recovery of the transporter to its original condition is one of the characteristics of mediated transport (p. 196). 4: Active transport, movement against a gradient, is also mediated transport.
4. D In eucaryotic cells, amino acids are transported by group translocation in which they are converted to a γ-glutamyl amino acid during transport (Figure 5.40). A,E: It is an active system with the ATP used to resynthesize the intermediate, glutathione. B,C: The system transports a single amino acid at a time (p. 204).
5. C Specific binding by the transporter is a characteristic of mediated systems (p. 196).
6. B Transportation against a gradient requires the input of energy (p. 199).
7. C In small intestines and renal tubular cells, transport is a Na^+-dependent, active symport; for most other tissues, transport is passive (Table 5.4).
8. A The Na^+-K^+-transporter is the Na^+-K^+-ATPase. It is an antiport, vectorial (Na^+ out), electrogenic ($3Na^+$, $2K^+$) system. ATP hydrolysis is not useless (p. 201).
9. C 2 and 4 correct. 1, 2: Mediated transport systems show saturation kinetics and substrate specificity (p. 195). 3: The purpose of the transporter is to aid the transport of water-soluble substances across the lipid membrane (p. 196). 4: This is a characteristic of active transport that is a mediated transport.
10. B Phosphatidylcholines (lecithins) contain two fatty acids in ester linkages (Figures 5.6, 5.13).
11. C Both have the hydrophobic fatty acid (and sphingosine for B) portions and the very hydrophilic phosphorylcholine portions (Figures 5.6, 5.13).
12. C A sphingomyelin is the only type of sphingolipid-containing phosphate (Figures 5.6, 5.13).
13. B Cerebrosides have only a single sugar (p. 183).
14. A Cerebrosides are neutral; no phosphate; uncharged sugar. Gangliosides, by virtue of the presence of sialic acid, are acidic. *Note:* Sulfatides, which are acidic, are derived from cerebrosides but are not, themselves, classified as cerebrosides (p. 184).
15. B The figure is a modified composite of Figures 5.29, 5.32, 5.34, and
16. E 5.39. All systems are mediated. A: An active uniport. B: A passive antiport; e.g., Cl^--HCO_3^-. C: An *active* antiport; e.g., Na^+-K^+-ATPase. D: A group translocation representing a change in S_1 during transport. E: A symport system; in this case, S_1 could be glucose and S_2 Na^+.
17. A 1, 2, and 3 correct. 1, 2: These are the two major types of ionophores (p. 206). 3: Valinomycin transports K^+ by a uniport mechanism (p. 207). There are also antiport systems that are electroneutral. 4: Ionophores transport by passive mediated mechanisms (p. 206).

6

Bioenergetics and Oxidative Metabolism

MERLE S. OLSON

6.1 ENERGY-PRODUCING AND ENERGY-UTILIZING SYSTEMS

Living cells are composed of a complex intricately regulated system of energy-producing and energy-utilizing chemical reactions. Metabolic reactions involved in energy generation sequentially break down ingested or stored macromolecular fuels such as carbohydrate, lipid, or protein in what are termed catabolic pathways. Catabolic reactions usually result in the conversion of large complex molecules to smaller molecules (ultimately CO_2 and H_2O), usually result in the production of storable or conservable energy, and often require the consumption of oxygen during this process. Such reactions are accelerated during periods of fuel deprivation or stress to an organism.

Energy-utilizing reactions are necessary to maintain, to reproduce, or to perform various necessary, and in many instances tissue-specific, cellular functions, for example, nerve impulse conduction and muscle contraction. Metabolic pathways in a cell, which are involved in the biosynthesis of various macromolecules, are termed anabolic pathways. Anabolic reactions usually result in the synthesis of large, complex molecules from smaller precursors and usually require the expenditure of energy. Such reactions are accelerated during periods of relative energy excess, during periods when there occurs a ready availability of precursor molecules or during periods of growth or regeneration of cellular material.

The ATP Cycle

The relationship between the energy-producing and energy-utilizing functions of the cell is illustrated in Figure 6.1. Energy may be derived from the oxidation of appropriate metabolic fuels such as carbohydrate, lipid, or protein. The proportion of each of these metabolic fuels which may be utilized as an energy source depends on the tissue and the dietary and hormonal state of the organism. For example, the mature erythrocyte and the adult brain in the fed state use only carbohydrate as a metabolic fuel, while the liver of a diabetic or fasted mammal primarily is metabolizing lipid or fat. Energy may be consumed during the performance of various energy-linked (work) functions, some of which are indicated in Figure 6.1. Again the proportion of energy expended or utilized depends largely

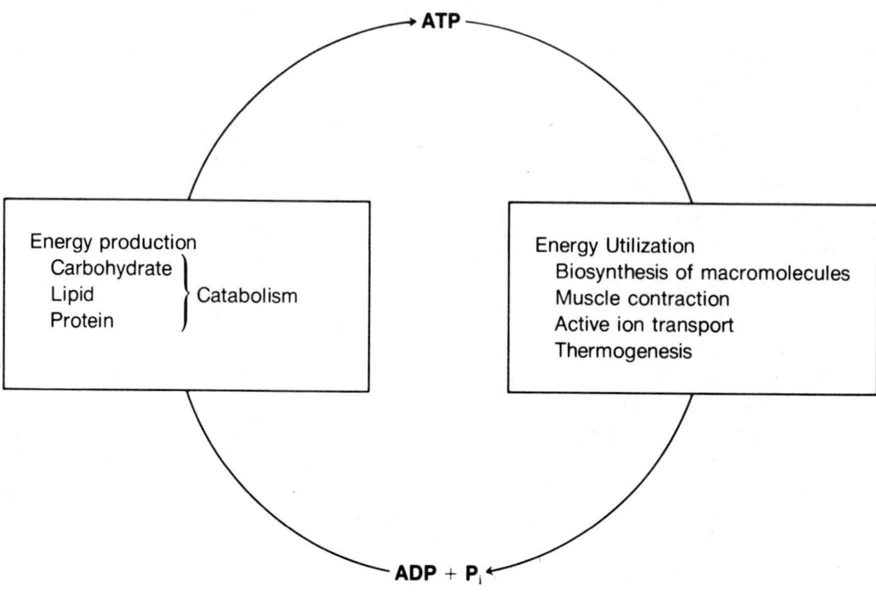

Figure 6.1
The relationship between energy production and energy utilization.

Adenine

D-Ribose

Adenosine 5′-triphosphate

H_2O

P_i

Adenosine 5′-diphosphate

Figure 6.2
Structure of ATP and ADP.

on the tissue and the physiological state of that tissue, for example, the liver and the pancreas are tissues primarily involved in biosynthetic and secretory work functions, while cardiac and skeletal muscle primarily are involved in converting metabolic energy into mechanical energy during the muscle contraction process.

The essential linkage between the energy-producing and the energy-utilizing pathways is maintained by the nucleoside triphosphate, adenosine 5′-triphosphate (ATP) (Figure 6.2). ATP is a purine (adenine) nucleotide in which the adenine ring is attached in a glycosidic linkage to D-ribose. Three phosphoryl groups are esterified to the 5 position of the ribose moiety in what are termed phosphoanhydride bonds. The two terminal phosphate groups (i.e., β and γ), which are involved in the phosphoric acid anhydride bonding, are designated as *energy-rich or high energy bonds*. Synthesizing ATP as a result of a catabolic process or consuming ATP in some type of energy–linked cellular function alternately involves the formation and either the hydrolysis or transfer of the terminal phosphate group of ATP. Because ATP is an anionic species, the physiological form of this nucleotide is thought to be chelated with a divalent metal cation such as magnesium. ADP also can chelate magnesium, but the affinity of the metal cation for ADP is considerably less than for ATP. While adenine nucleotides are involved intimately in the process of energy generation or conservation, various nucleoside triphosphates, including ATP are involved actively in transferring an energy component into biosynthetic processes. As indicated in Figure 6.3, the guanine nucle-

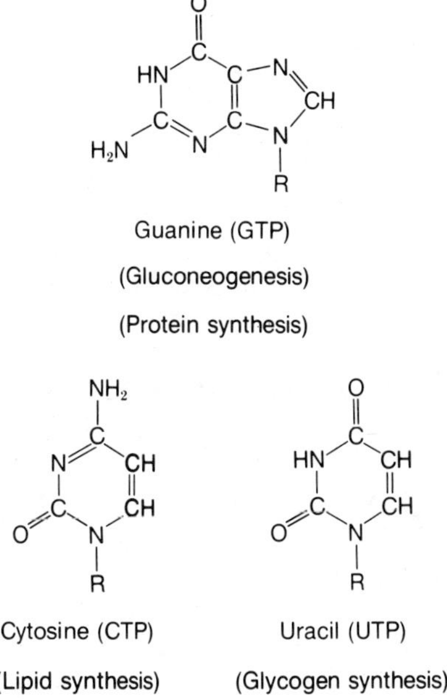

Figure 6.3
Structures of purine and pyrimidine bases involved in various biosynthetic pathways.

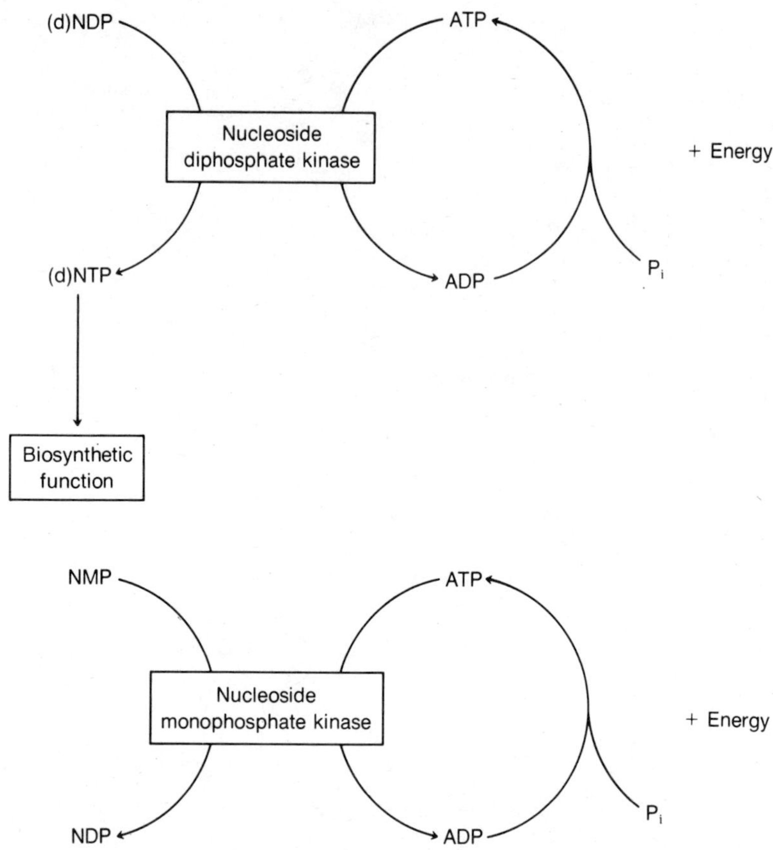

Figure 6.4
Nucleoside disphosphate kinase and nucleoside monophosphate kinase reactions.
N = any purine or pyrimidine base; (d) indicates a deoxyribonucleotide.

otide GTP serves as the source of energy input into the processes of gluconeogenesis and protein synthesis, whereas UTP (uracil) and CTP (cytosine) are utilized in glycogen synthesis and lipid synthesis, respectively. The energy in the terminal phosphate bonds of ATP may be transferred to the other nucleotides, using either the nucleoside diphosphate kinase or the nucleoside monophosphate kinase as illustrated in Figure 6.4. Two nucleoside diphosphates can be converted to a nucleoside triphosphate and a nucleoside monophosphate in various nucleoside monophosphate kinase reactions, such as the adenylate kinase reaction as indicated in Figure 6.5. A consequence of the action of these types of enzymes is that the terminal energy-rich phosphate bonds of ATP may be transferred to the appropriate nucleotides and utilized in a variety of biosynthetic processes.

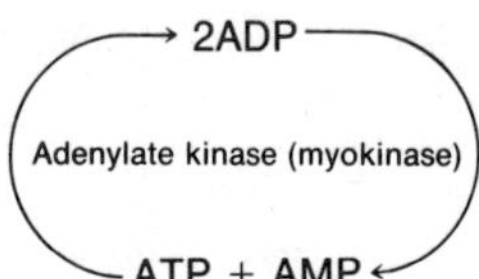

Figure 6.5
The adenylate kinase (myokinase) reaction.

6.2 THERMODYNAMIC RELATIONSHIPS AND ENERGY-RICH COMPONENTS

Because living cells are capable of the interconversion of different forms of energy and may exchange energy with their surroundings, it is helpful to review certain laws or principles of thermodynamics. Knowledge of these principles will facilitate a perception of how energy-producing and energy-utilizing metabolic reactions are permitted to occur within the same cell and how an organism is able to accomplish various work functions. The first law of thermodynamics indicates that energy can neither be created nor destroyed. This law of energy conservation stipulates that although energy may be converted from one form to another, the total energy in a system must remain constant. For example, the chemical energy that is available in a metabolic fuel such as glucose may be converted in the process of glycolysis to another form of chemical energy, ATP.

In skeletal muscle chemical energy involved in the energy-rich phosphate bonds of ATP may be converted to mechanical energy during the process of muscle contraction. It has been demonstrated that the energy involved in an osmotic electropotential gradient of protons across the mitochondrial membrane may be converted to chemical energy by using such a gradient to drive ATP synthesis.

In order to discuss the second law of thermodynamics the term entropy must be defined. Entropy (which is usually designated by the symbol S) is a measure or indicator of the degree of disorder or randomness in a system. Entropy also can be viewed as the energy in a system that is unavailable to perform useful work. All processes, whether chemical or biological, tend to progress toward a situation of maximum entropy. Equilibrium in a system will result when the randomness or disorder (entropy) is at a maximum. However, it is nearly impossible to quantitate entropy changes in systems that may be useful to study in biochemistry, and such systems are rarely at equilibrium. For the sake of simplicity and its inherent utility in these types of considerations, the quantity termed free energy is employed.

Free Energy

The free energy (denoted by the letter G) of a system is that portion of the total energy in a system that is available for useful work and may be further defined by the equation

$$\Delta G = \Delta H - T\,\Delta S$$

In this expression for a system proceeding toward equilibrium at a constant temperature and pressure, ΔG is, of course, the change in free energy, ΔH is the change in enthalpy or the heat content, T is the absolute temperature, and ΔS is the change in entropy of the system. It can be deduced from this relationship that at equilibrium $\Delta G = 0$. Furthermore, any process that exhibits a negative free energy change will occur spontaneously, since energy is given off, and this type of process is called an exergonic reaction. A process that exhibits a positive free energy change will not occur spontaneously; energy from some other source must be applied to this process to allow it to proceed toward equilibrium, and this type of process is termed an endergonic reaction. It should be noted that the change in free energy in a biochemical process is the same regardless of the pathway or mechanism employed to attain the final state. Whereas the rate of a given reaction depends on the free energy of activation, the magnitude of the ΔG is not related to the rate of the reaction. The change in free energy for a chemical reaction is related to the equilibrium constant of that reaction. For example, an enzymatic reaction may be described as

$$\mathrm{A} + \mathrm{B} \rightleftharpoons \mathrm{C} + \mathrm{D}$$

And an expression for the equilibrium constant may be written as

$$K_{eq} = \frac{[\mathrm{C}][\mathrm{D}]}{[\mathrm{A}][\mathrm{B}]}$$

The free energy change (ΔG) at a constant temperature and pressure is defined as

$$\Delta G = \Delta G^\circ + RT \ln \frac{[C][D]}{[A][B]}$$

where ΔG is the free energy change; ΔG° is the standard free energy change, which is a constant for each individual chemical reaction; the reactants and products in the reaction are present at concentrations of 1.0 M; R is the gas constant, which is 1.987 cal mol^{-1} K^{-1} or 8.134 J mol^{-1} K^{-1}, depending upon whether the resultant free energy change is expressed in calories (cal) or joules (J) per mole; and T is the absolute temperature in degrees Kelvin.

Because at equilibrium $\Delta G = 0$, the expression reduces to

$$\Delta G^\circ = -RT \ln K_{eq}$$

or

$$\Delta G^\circ = -2.3\ RT \log K_{eq}$$

Hence if the equilibrium constant for a reaction can be determined, the standard free energy change (ΔG°) for that reaction also can be calculated. The relationship between ΔG° and K_{eq} is illustrated in Table 6.1. When the equilibrium constant of a reaction is less than unity, the reaction is endergonic, and the ΔG° is positive. When the equilibrium constant is greater than 1, the reaction is exergonic, and the ΔG° is negative.

TABLE 6.1 Tabulation of Values of K_{eq} and ΔG°

K_{eq}	ΔG° $(kcal \cdot M^{-1})$
10^{-4}	5.46
10^{-3}	4.09
10^{-2}	2.73
10^{-1}	1.36
1	0
10	−1.36
10^{2}	−2.73
10^{3}	−4.09
10^{4}	−5.46

During any consideration of the energy-producing and energy-utilizing metabolic pathways in cellular systems it is important to understand that the free energy changes characteristic of individual enzymatic reactions in an entire pathway are additive, for example,

$$A \longrightarrow B \longrightarrow C \longrightarrow D$$

$$\Delta G^\circ_{A \to D} = \Delta G^\circ_{A \to B} + \Delta G^\circ_{B \to C} + \Delta G^\circ_{C \to D}$$

Although any given enzymatic reaction in a sequence may have a characteristic positive free energy change, as long as the sum of all the free energy changes is negative, the pathway will proceed spontaneously.

Another way of expressing this principle is that enzymatic reactions with positive free energy changes may be coupled to or driven by reactions with negative free energy changes associated with them. This is an important point because in a metabolic pathway such as the glycolytic pathway various individual reactions either have positive ΔG°s or ΔG°s that are close to 0. On the other hand, there are other reactions that have large and negative ΔG°s, which drive the entire pathway. The crucial consideration is that the sum of the ΔG°s for the individual reactions in a pathway must be negative in order for such a metabolic sequence to be thermodynamically feasible. Also, it is important to remember that, as for all chemical reactions, individual enzymatic reactions in a metabolic pathway or the pathway as a whole would be facilitated if the concentrations of the reactants (substrates) of the reaction exceed the concentrations of the products of the reaction.

Caloric Value

During the complete stepwise oxidation of glucose, one of the primary metabolic fuels in cellular systems, a large quantity of energy is available. The free energy released during the oxidation of glucose, whether this

TABLE 6.2 Free Energy Changes and Caloric Values Associated with Various Metabolic Fuels

Compound	*Mol Wt*	$\Delta G°$ (*kcal/mol*)	*Caloric Value* (*kcal/g*)
Glucose	180	−686	3.81
Lactate	90	−326	3.62
Palmitate	256	−2,380	9.30
Tripalmitin	809	−7,510	9.30
Glycine	75	−234	3.12

oxidation is performed in an instrument used for the combustion of such substances, called a calorimeter, or whether the oxidation occurs in a living functioning cell, is illustrated in the following equation:

$$C_6H_{12}O_6 + 6O_2 \longrightarrow 6CO_2 + 6H_2O \qquad \Delta G° = -686{,}000 \text{ cal/mol}$$

When this process is performed under aerobic conditions in most types of cells, there exists a potential to conserve less than half of this "available" energy in the form of ATP. The enzymatic machinery in cellular systems is capable of synthesizing 38 molecules of ATP during the complete oxidation of glucose. The $\Delta G°$s for the oxidation of other metabolic fuels are listed in Table 6.2. Carbohydrates and proteins (amino acids) have a caloric value of 3–4 kcal/g, while lipid (i.e., the long-chain fatty acid palmitate or the triglyceride tripalmitin) exhibits a caloric value nearly three times greater. The reason that more energy can be derived from lipid than from carbohydrate or protein relates to the average oxidation state of the carbon atoms in these substances.

Carbon atoms in carbohydrate are considerably more oxidized (or less reduced) than those in lipid (see Figure 6.6). Hence during the sequential breakdown of these fuels nearly three times as many reducing equivalents (a reducing equivalent is defined as a proton plus an electron, i.e., $H^+ + e^-$) can be extracted from lipid than from carbohydrate. Reducing equivalents may be utilized for ATP synthesis in the mitochondrial energy transduction sequence.

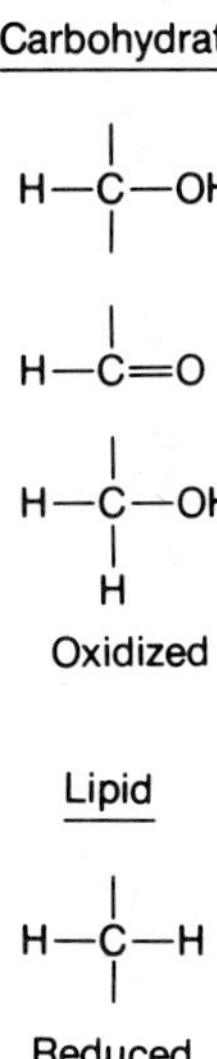

Figure 6.6
Oxidation states of typical carbon atoms in carbohydrates and lipids.

High-Energy Compounds

The two terminal phosphoryl groups in the ATP molecule contain energy-rich or high-energy bonds. What this description is intended to convey is that the free energy of hydrolysis of such an energy-rich phosphoanhydride bond is much greater than would be obtained for a simple phosphate ester. High-energy is not synonymous with stability of the bonding arrangement in question, nor does high-energy refer to the energy required to break such bonds. The concept of the high-energy compound does imply that the products of the hydrolytic cleavage of the energy-rich bond are in more stable forms than the original compound. As a rule simple phosphate esters (low-energy compounds) exhibit negative $\Delta G°$s of hydrolysis in the range 1–3 kcal/mol, whereas high-energy bonds have negative $\Delta G°$s in the range 5–15 kcal/mol. Simple phosphate esters such as glucose 6-phosphate and glycerol 3-phosphate are examples of low-energy compounds. Table 6.3 lists various types of energy-rich compounds with approximate values for their $\Delta G°$s of hydrolysis.

There are various reasons why certain compounds or bonding arrangements are energy-rich. First, the products of the hydrolysis of an energy-rich bond may exist in more resonance forms than the precursor molecule. The more possible resonance forms in which a molecule can exist

TABLE 6.3 Examples of Energy-Rich Compounds

Type of Bond	*ΔG° of Hydrolysis (kcal/mol)*	*Example*
Phosphoric acid anhydrides	−7.3	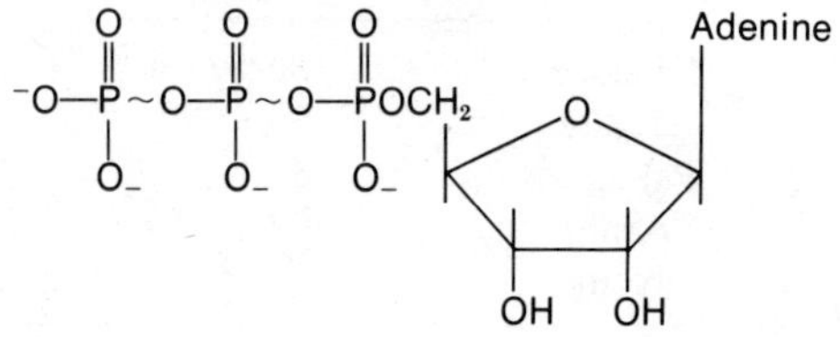 ATP
	−11.9	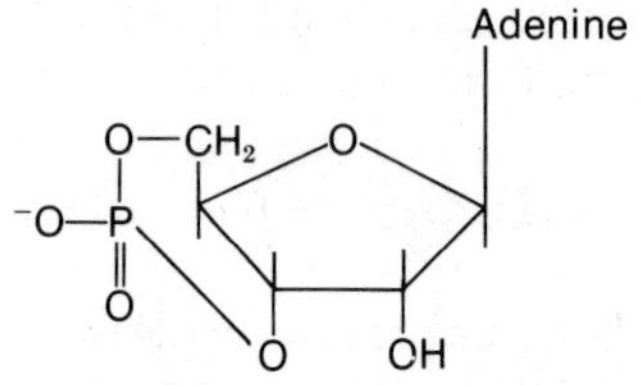 3′5′ cyclic AMP
Phosphoric-carboxylic acid anhydrides	−10.1	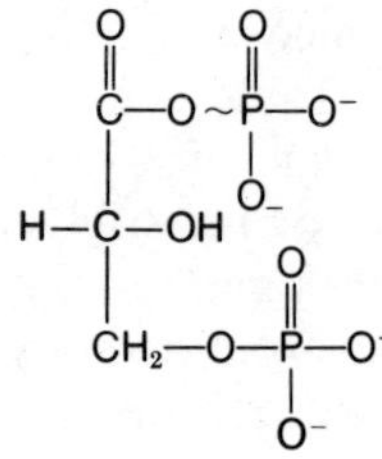 1,3-Diphosphoglycerate
	−10.3	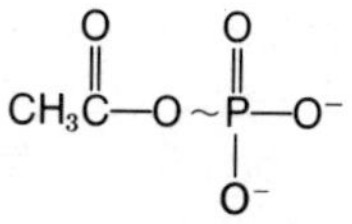 $CH_3C(=O)$—O~PO_3^{2-} Acetyl phosphate
Phosphoguanidines	−10.3	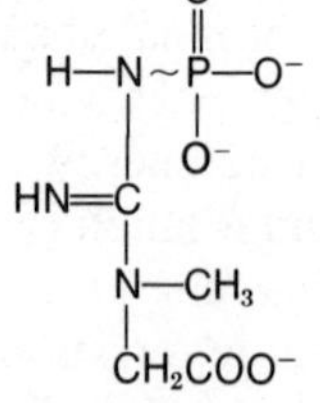 Creatine phosphate
Enol phosphates	−14.8	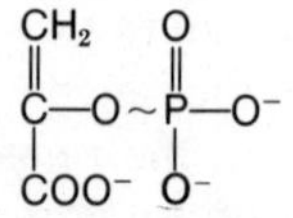 Phospho*enol*pyruvate
Thiolesters	−7.7	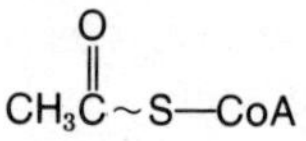 $CH_3C(=O)$~S—CoA Acetyl CoA

tend to stabilize that molecule. The resonance forms for inorganic phosphate can be written as

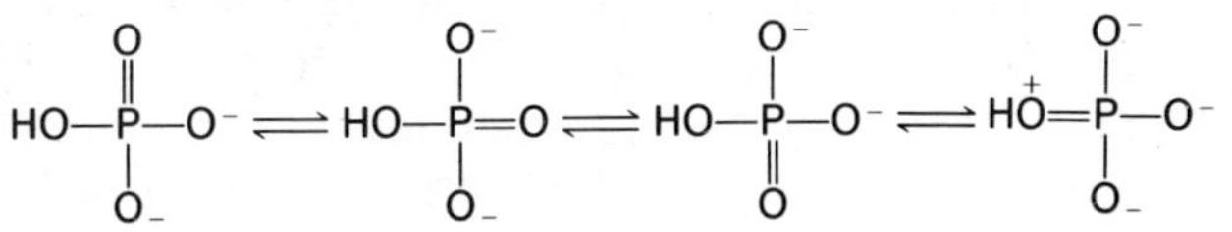

$$\mathrm{HO{-}P(=O)(O^-){-}O^-} \rightleftharpoons \mathrm{HO{-}P(O^-)(O^-){=}O} \rightleftharpoons \mathrm{HO{-}P(O^-)(=O){-}O^-} \rightleftharpoons \mathrm{H\overset{+}{O}{=}P(O^-)(O^-){-}O^-}$$

Less resonance forms may be written for ATP or a compound such as pyrophosphate than for inorganic phosphate.

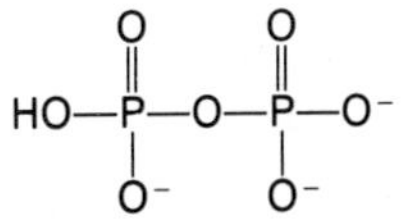

Second, many high-energy bonding arrangements have groups of similar electrostatic charge located in close proximity to each other in such compounds. Because like charges tend to repulse one another, the hydrolysis of the energy-rich bond alleviates this situation and, again, lends stability to the products of hydrolysis. Third, the hydrolysis of certain energy-rich bonds results in the formation of an unstable compound, which may isomerize spontaneously to form a more stable compound. The hydrolysis of phospho*enol*pyruvate is an example of this type of compound. The $\Delta G°$ of the isomerization reaction is considerable, and the final product, in this case pyruvate, is much more stable. Finally, if a product of the hydrolysis of a high-energy bond is an undissociated acid, the dissociation of the proton from the acidic function and its subsequent buffering may contribute to the overall $\Delta G°$ of the hydrolytic reaction. In general, any property or process that lends stability to the products of hydrolysis of a compound tends to confer a high-energy character to that compound.

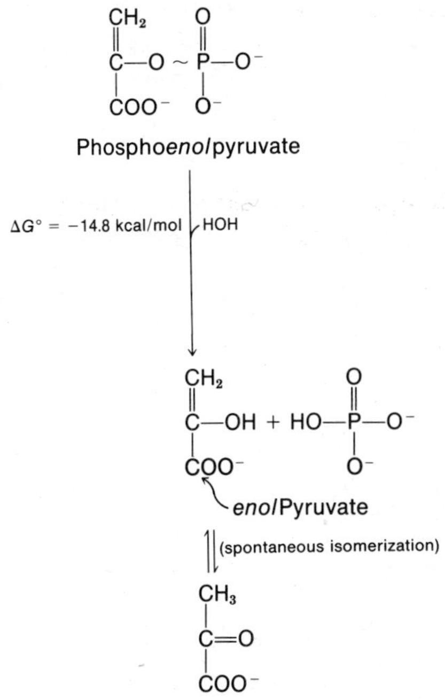

The high-energy character of 3′,5′-cyclic AMP has been attributed to the fact that the phosphoanhydride bonding character in this compound is strained as it bridges the 3′ and 5′ positions on the ribose. Further, the energy-rich character of thiol ester compounds such as acetyl CoA or succinyl CoA results from the relatively acidic character of the thiol function. Hence acetyl CoA is nearly equivalent to an anhydride bonding arrangement rather than a simple thioester.

$$\mathrm{HO{-}P(=O)(O^-){-}O{-}P(=O)(O^-){-}O^-}$$

Pyrophosphate

$$\mathrm{CH_2{=}C(COO^-){-}O{\sim}P(=O)(O^-){-}O^-}$$

Phospho*enol*pyruvate

$\Delta G° = -14.8$ kcal/mol, HOH

$$\mathrm{CH_2{=}C(COO^-){-}OH} + \mathrm{HO{-}P(=O)(O^-){-}O^-}$$

*enol*Pyruvate

(spontaneous isomerization)

$$\mathrm{CH_3{-}C(=O){-}COO^-}$$

Pyruvate (stable form)

Determination of Free Energy Changes

The $\Delta G°$ of hydrolysis of the terminal phosphate of ATP is difficult to determine by simply utilizing the K_{eq} of the hydrolytic reaction because of the position of the equilibrium.

$$\text{ATP} + \text{HOH} \rightleftharpoons \text{ADP} + \text{P}_i + \text{H}^+$$

However, the $\Delta G°$ of hydrolysis of ATP may be determined indirectly because of the additive nature of free energy changes. Hence, the free energy of hydrolysis of ATP can be determined by adding the $\Delta G°$ of an ATP-utilizing reaction such as hexokinase to the $\Delta G°$ of the reaction that cleaves the phosphate from the product of the hexokinase reaction, glucose 6–phosphate, as indicated below:

$$\text{Glucose} + \text{ATP} \xrightleftharpoons{\text{hexokinase}} \text{glucose 6-phosphate} + \text{ADP} + \text{H}^+ \qquad \Delta G° = -4.0 \text{ kcal/mol}$$

$$\text{Glucose 6-phosphate} + \text{HOH} \xrightleftharpoons{\text{glucose 6-phosphatase}} \text{glucose} + \text{P}_i \qquad \Delta G° = -3.3 \text{ kcal/mol}$$

$$\Sigma \quad \text{ATP} + \text{HOH} \rightleftharpoons \text{ADP} + \text{P}_i + \text{H}^+ \qquad \Delta G° = -7.3 \text{ kcal/mol}$$

Free energies of hydrolysis for other energy-rich compounds may be determined in a similar fashion.

1,3-Diphosphoglycerate $\xrightarrow{\text{phosphoglycerate kinase}}$ 3-phosphoglycerate (ADP → ATP)

Phospho*enol*pyruvate $\xrightarrow{\text{pyruvate kinase}}$ pyruvate (ADP → ATP)

(*a*)

Creatine $\xleftrightarrow{\text{creatine kinase}}$ creatine phosphate (ATP → ADP)

$\Delta G^\circ = +1.5$ kcal/mol

(*b*)

Glucose $\xrightarrow{\text{hexokinase}}$ glucose 6-phosphate (ATP → ADP)

$\Delta G^\circ = -4.0$ kcal/mol

(*c*)

Figure 6.7
Examples of reactions involved in the transfer of "high-energy" phosphate.

Transfer of High-Energy Bond Energies

Energy-rich compounds are capable of transferring various groups from the parent (donor) compound to an acceptor compound in a thermodynamically feasible fashion as long as an appropriate enzyme is present to facilitate the transfer. The energy-rich intermediates in the glycolytic pathway such as 1,3-diphosphoglycerate and phospho*enol*pyruvate can transfer their high-energy phosphate moieties to ATP in the phosphoglycerate kinase and pyruvate kinase reactions, respectively (Figure 6.7*a,b*). The ΔG°s of these two reactions are -4.5 and -7.5 kcal/mol, respectively, and hence the transfer of "high-energy" phosphate is thermodynamically possible, and ATP synthesis is the result. ATP can transfer its terminal high-energy phosphoryl groups to form either compounds of relatively similar high-energy character (i.e., creatine phosphate in the creatine kinase reaction) or compounds that are of considerably lower energy, such as glucose 6-phosphate formed in the hexokinase reaction (Figure 6.7*c*).

The major point of this discussion is that phosphate, or for that matter other transferable groups, can be transferred from compounds that contain energy-rich bonding arrangements to compounds that have bonding characteristics of a lower energy in thermodynamically permissible enzymatic reactions. This principle is a major premise of the interaction between energy-producing and energy-utilizing metabolic pathways in living cells.

6.3 SOURCES AND FATES OF ACETYL COENZYME A

Most of the major energy-generating metabolic pathways of the cell eventually result in the production of the two-carbon unit acetyl coenzyme A (CoA). As illustrated in Figure 6.8, the catabolic breakdown of ingested or stored carbohydrate in the glycolytic pathway, long-chain fatty acids resulting from the lipolysis of triglycerides in the β-oxidation sequence, or certain amino acids resulting from proteolysis following transamination or deamination and subsequent oxidation, provide precursors for the formation of acetyl CoA.

The structure of acetyl CoA is shown in Figure 6.9. This complex coenzyme is composed of β-mercaptoethanolamine, the vitamin pantothenic acid and the adenine nucleotide, adenosine 3′-phosphate 5′-

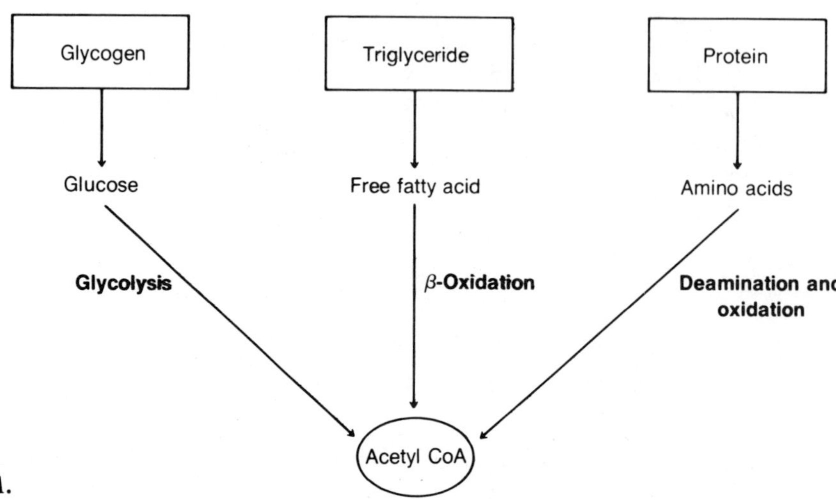

Figure 6.8
General precursors of acetyl CoA.

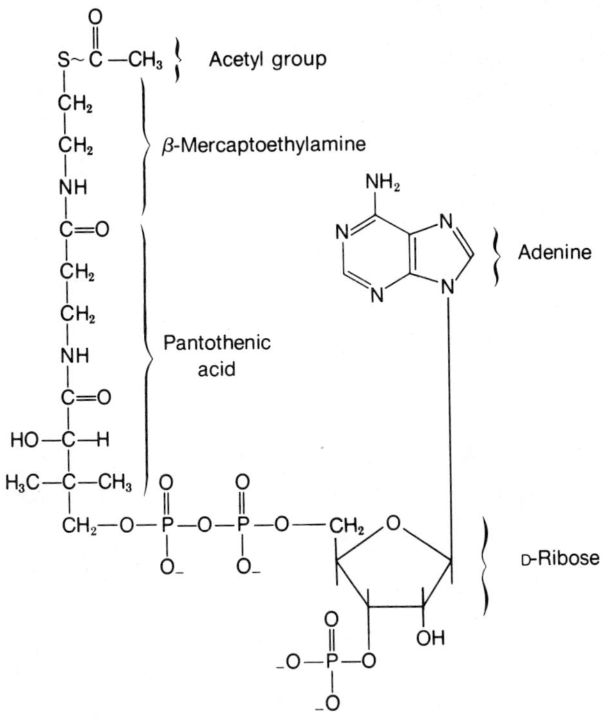

Figure 6.9
Structure of acetyl coenzyme A.

diphosphate. Normally coenzyme A exists as the reduced thiol, but in the presence of oxidizing agents two coenzyme A molecules can exist in a disulfide linkage, for example, CoA—S—S—CoA. This form of CoA is termed oxidized CoA. Coenzyme A is involved in a variety of acyl group transfer reactions, where CoA alternately serves as the acceptor, then the donor, of the acyl function. Various metabolic pathways involve only acyl CoA derivatives, for example, β-oxidation of fatty acids and branched-chain amino acid degradation. Specific information concerning the nutritional aspects of the vitamin pantothenic acid will be detailed in Chapter 26. Like many other nucleotide species, coenzyme A derivatives are not freely transported across cellular membranes. This property has necessitated the evolution of certain transport or shuttle mechanisms by which various intermediates or groups can be transferred across the membranes of the cell. Such acyl transferase reactions for acetyl groups and long-chain acyl groups will be discussed in Chapter 9. Finally, as indicated previously, the thiolester linkage in acyl CoA derivatives is an energy-rich bond, and hence these compounds can serve as effective donors of acyl groups in acyl transferase reactions. Also, in order to synthesize an acyl CoA derivative, such as in the acetate thiokinase reaction, a high-energy bond of ATP must be expended.

$$\text{Acetate} + \text{CoASH} + \text{ATP} \xrightarrow{\text{acetate kinase}} \text{acetyl CoA} + \text{AMP} + \text{PP}_i$$

As was mentioned above, the β-oxidation of fatty acids is a primary source of acetyl CoA in many tissues. Whereas a more detailed description of the mobilization, transport, and oxidation of fatty acids is presented in Chapter 9, it is important to note that the products of the β-oxidation sequence are acetyl CoA and reducing equivalents (i.e., NADH + H^+). In certain tissues (e.g., cardiac muscle) and under somewhat special metabolic conditions in other tissues (e.g., in the brain of an individual during prolonged starvation) acetyl CoA for energy generation may be derived from the ketone bodies acetoacetate and β-hydroxybutyrate.

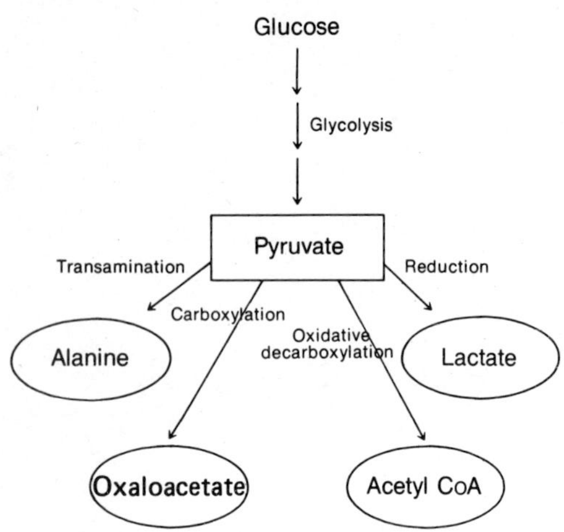

Figure 6.10
Metabolic fates of pyruvate.

Metabolic Sources and Fates of Pyruvate

During aerobic glycolysis (Chapter 7) glucose or other monosaccharides are converted to pyruvate, and hence in the presence of oxygen pyruvate is the end product per se of this cytosolic pathway. Also the degradation of amino acids such as alanine, serine, and cysteine results in the production of pyruvate (Chapter 12).

Pyruvate has a variety of metabolic fates, depending upon the tissue and the metabolic state of that tissue. The major types of reactions in which pyruvate participates are indicated in Figure 6.10. The oxidative decarboxylation of pyruvate in the pyruvate dehydrogenase reaction is discussed in this chapter; the other reactions in which pyruvate is involved are discussed in Chapter 7.

Pyruvate Dehydrogenase

Pyruvate is converted to acetyl CoA by the pyruvate dehydrogenase multienzyme complex.

$$\text{Pyruvate} + \text{NAD}^+ + \text{CoASH} \xrightarrow{\text{pyruvate dehydrogenase}}$$
$$\text{acetyl CoA} + \text{CO}_2 + \text{NADH} + \text{H}^+ \qquad \Delta G^\circ = -8 \text{ kcal/mol}$$

This enzyme is located exclusively in the mitochondrial compartment and is present in high concentrations in tissues such as cardiac muscle and kidney. Because of the large negative ΔG° of this reaction, under physiological conditions, the pyruvate dehydrogenase reaction is essentially irreversible, and this fact is the primary reason that a net conversion of fatty acid carbon to carbohydrate cannot occur, for example,

$$\text{Acetyl CoA} \not\rightarrow \text{pyruvate}$$

Molecular weights of the multienzyme complex derived from kidney, heart, or liver range from 7 to 8.5 × 10^6. The mammalian pyruvate dehydrogenase enzyme complex consists of three different types of catalytic subunits:

Number of Subunits/ Complex	*Type*	*Molecular Weight*	*Subunit Structure*
20 or 30[a]	Pyruvate dehydrogenase	154,000	$\alpha_2\beta_2$ Tetramer
60	Dihydrolipoyl transacetylase	52,000	Identical
5–6	Dihydrolipoyl dehydrogenase	110,000	α_2 Dimer

[a] Depending on source.

The structure of the pyruvate dehydrogenase complex derived from *E. coli* (particle weight, 4.6 × 10^6) is somewhat different from that of the mammalian enzyme. Electron micrographs of the bacterial enzyme complex (Figure 6.11) indicate that the transacetylase, which consists of 24 identical polypeptide chains (mol wt = 64,500), forms the cubelike core of the complex (white spheres in the model shown in Figure 6.9). Twelve pyruvate dehydrogenase dimers (black spheres; mol wt = 90,500) are distributed symmetrically on the 12 edges of the transacetylase cube. Six dihydrolipoyl dehydrogenase dimers (grey spheres; mol wt = 56,000) are distributed on the six faces of the cube. Five different coenzymes or

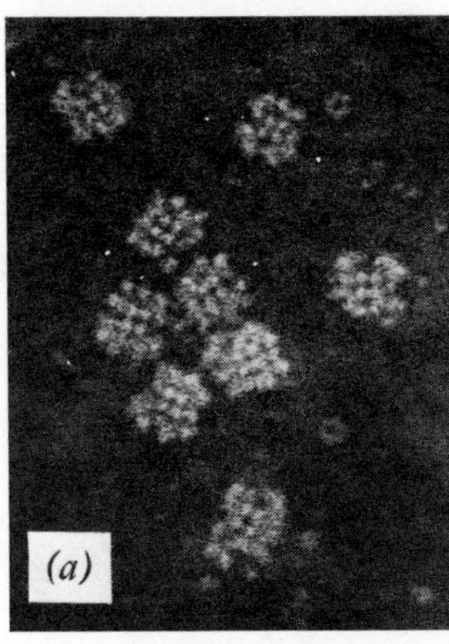

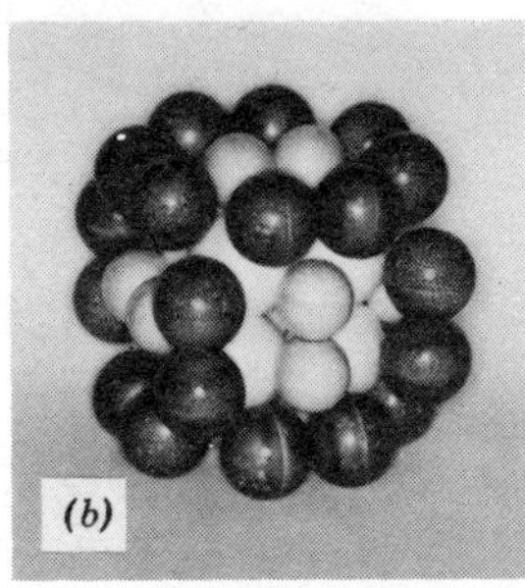

Figure 6.11
The pyruvate dehydrogenase complex from E. coli.
(a) Electron micrograph. (b) Molecular model. The enzyme complex was negatively stained with phosphotungstate. (×*200,000*)
Courtesy of Dr. Lester J. Reed, University of Texas, Austin.

prosthetic groups are involved in the pyruvate dehydrogenase reaction (Table 6.4 and Figure 6.12). The mechanism of the pyruvate dehydrogenase reaction occurs as illustrated in Figure 6.13.

Because of the active participation of thiol groups in the catalytic mechanism of the enzyme, agents which either oxidize or complex with thiol groups are strong inhibitors of the enzyme complex. Arsenite is an example of such an inhibitor.

Two types of regulation of the pyruvate dehydrogenase complex have been elucidated. First, it has been demonstrated that two of the products of the pyruvate dehydrogenase reaction, acetyl CoA and NADH, inhibit the complex in a competitive fashion. Second, the pyruvate dehydro-

TABLE 6.4 Function of Coenzymes and Prosthetic Groups of the Pyruvate Dehydrogenase Reaction

Coenzyme or Prosthetic Group	*Location*	*Function*
Thiamin pyrophosphate	Bound to pyruvate dehydrogenase	Reacts with substrate, pyruvate
Lipoic acid	Covalently attached to a lysine residue on the dihydrolipoyl transacetylase	Accepts acetyl group from thiamine pyrophosphate
Coenzyme A	Free in solution	Accepts acetyl group from lipoamide group on the transacetylase
Flavin adenine dinucleotide (FAD)	Tightly bound to dihydrolipoyl dehydrogenase	Accepts reducing equivalents from reduced lipoamide group
Nicotinamide adenine dinucleotide	Free in solution	Terminal acceptor of reducing equivalents from the reduced flavoprotein

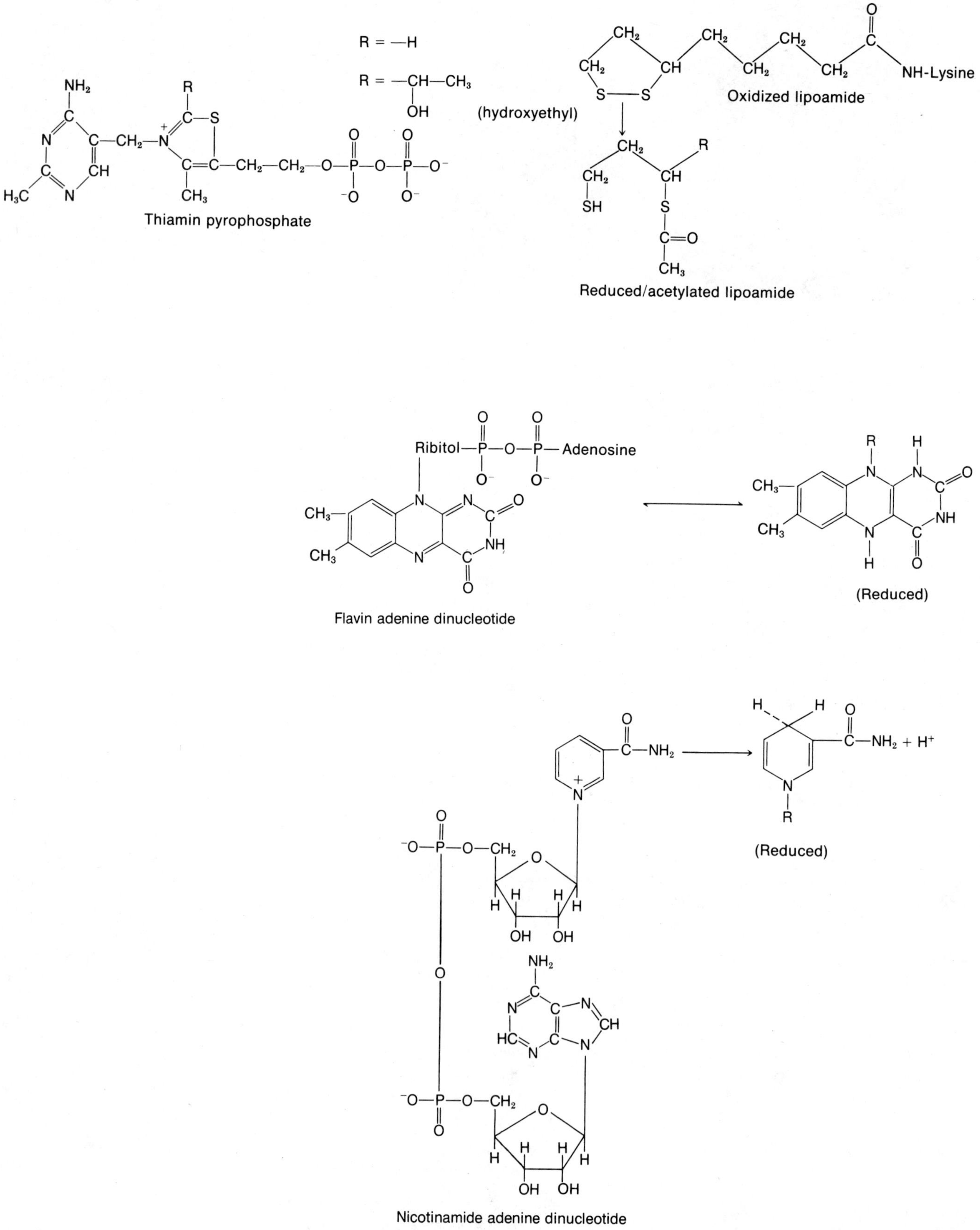

Figure 6.12
Structures of the coenzymes involved in the pyruvate dehydrogenase reaction. See also Figure 6.9.

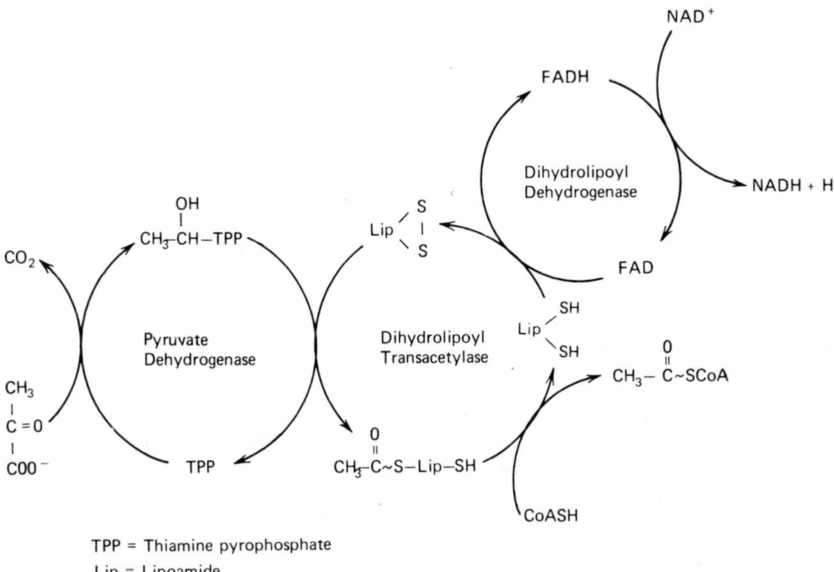

Figure 6.13
The mechanism of the pyruvate dehydrogenase reaction; the pyruvate dehydrogenase mulitenzyme complex.

genase complex exists in two forms: (1) an active, dephosphorylated complex, and (2) an inactive, phosphorylated complex. The inactivation of the complex is accomplished by a Mg^{2+}–ATP-dependent protein kinase, which is tightly bound to the enzyme complex. The reactivation of the complex is accomplished by a phosphoprotein phosphatase, which dephosphorylates the complex in a Mg^{2+}- and Ca^{2+}-dependent reaction. Three separate serine residues on the α subunit of the pyruvate dehydrogenase are phosphorylated by the protein kinase, but the phosphorylation of only one of these sites is related to the activity of the complex. The differential regulation of the pyruvate dehydrogenase kinase and phosphatase is the key to the regulation of the pyruvate dehydrogenase complex. The essential features of this complex regulatory system are illustrated in Figure 6.14. Not only can acetyl CoA and NADH, the products of the pyruvate dehydrogenase reaction, inhibit the dephospho- (active) form of the enzyme, but these two compounds stimulate the protein kinase reaction, leading to an interconversion of the complex to its inactive form. In addition, free CoASH and NAD^+ inhibit the protein kinase. Hence with any increase of the mitochondrial NADH/NAD^+ or acetyl CoA/CoA ratio, such as during rapid β-oxidation of fatty acids, pyruvate dehydrogenase will be inactivated by the kinase reaction. Also the substrate of the enzyme complex, pyruvate, is a potent inhibitor of the protein kinase, and therefore in the presence of elevated tissue pyruvate levels the kinase will be inhibited and the complex maximally active. Finally, it has been demonstrated that insulin administration can activate pyruvate dehydrogenase in adipose tissue, and catecholamines, such as epinephrine, can activate pyruvate dehydrogenase in cardiac tissue. The mechanisms of these hormonal effects are not well understood, but alterations of the intracellular distribution of calcium, such that the phosphoprotein phosphatase reaction is stimulated in the mitochondrial compartment, may be involved in these effects. These hormonal effects are not mediated directly by alterations in the tissue cAMP levels because the pyruvate dehydrogenase protein kinase and phosphatase are cAMP-independent or insensitive. (See Clin. Corr. 6.1.)

CLIN. CORR. 6.1
PYRUVATE DEHYDROGENASE DEFICIENCY

A variety of disorders in pyruvate metabolism has been detected in children. Some of these defects have been shown to involve deficiencies in each of the different component catalytic or regulatory subunits of the pyruvate dehydrogenase multienzyme complex. Children detected with a pyruvate dehydrogenase deficiency usually exhibit elevated serum levels of lactate, pyruvate, and alanine, which produce a chronic lactic acidosis. Such patients frequently exhibit severe neurological defects, and in most situations this type of enzymatic defect results in death. The diagnosis of pyruvate dehydrogenase deficiency is usually made by assaying the enzyme complex and/or its various enzymatic subunits in cultures of skin fibroblasts taken from the patient. In certain instances patients respond to dietary management in which a ketogenic diet is administered and carbohydrates are minimized.

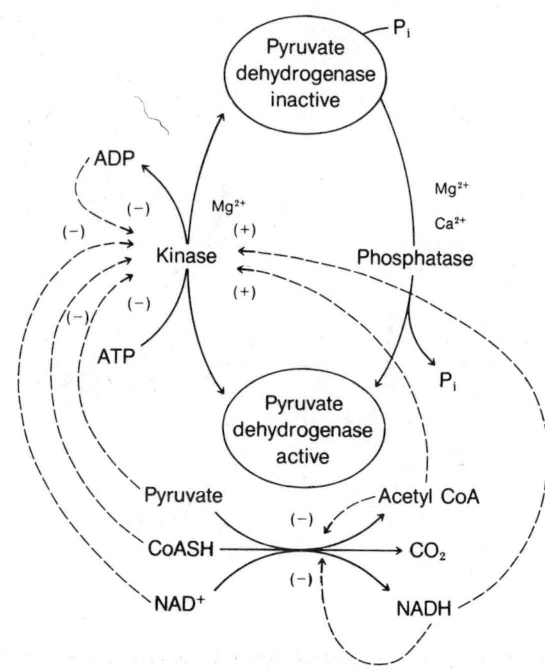

Figure 6.14
The regulation of the pyruvate dehydrogenase multienzyme complex.

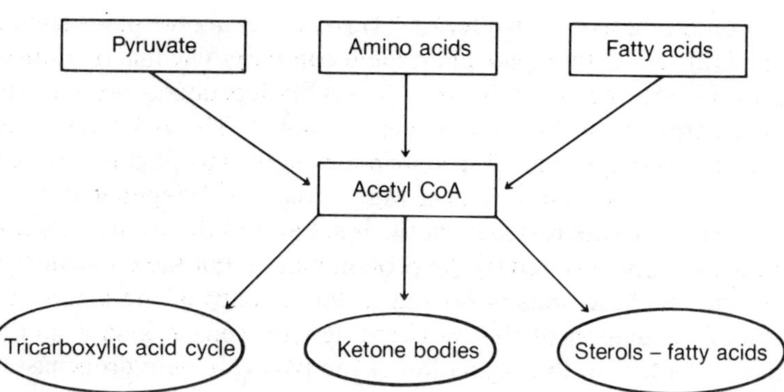

Figure 6.15
Sources and fates of acetyl CoA.

Metabolic Fates of Acetyl CoA

The various fates of acetyl CoA generated in the mitochondrial compartment include: (1) complete oxidation of the acetyl group in the tricarboxylic acid cycle for energy generation; (2) in the liver, conversion of an excess of acetyl CoA into the ketone bodies, acetoacetate and β-hydroxybutyrate; and (3) transfer of the acetyl units to the cytosol with subsequent biosynthesis of such complex molecules as sterols (Chapter 10) and long-chain fatty acids (Chapter 9) (see Figure 6.15).

6.4 THE TRICARBOXYLIC ACID CYCLE

The primary metabolic fate of acetyl CoA produced in the various energy-generating catabolic pathways of most cells is its complete oxidation in a cyclic series of oxidative reactions termed the *tricarboxylic acid cycle*. This metabolic cycle also is commonly referred to as the citric acid cycle or the Krebs cycle after Sir Hans Krebs who postulated the essential features of this pathway in 1937. Various investigators defined many of the enzymes and di- and tricarboxylic acid intermediates in this pathway, but it was Krebs who pieced together these components in his formulation of the "Krebs cycle." Although certain of the cycle enzymes are found in the cytosol, the primary location of enzymes of the tricarboxylic acid cycle is in the mitochondrion. This type of distribution is appropriate because the pyruvate dehydrogenase multienzyme complex and the fatty acid β-oxidation sequence, the two primary sources for generating acetyl CoA, are located in the mitochondrial compartment. Also, one of the primary functions of the tricarboxylic acid cycle is to generate reducing equivalents, which are utilized to generate energy, that is, ATP, in the electron transport–oxidative phosphorylation sequence, another process contained exclusively in the mitochondrion (see Figure 6.16). Mitochondrial energy transduction is discussed in Section 6.7.

The individual enzymatic reactions of the tricarboxylic acid cycle are illustrated in Figure 6.17. Figure 6.16 illustrates the essential process involved in the Krebs cycle. The substrate or input into the cycle is the two-carbon unit acetyl CoA, and the products of a complete turn of the cycle are $2CO_2$ plus one high-energy phosphate bond (as GTP) and four reducing equivalents (i.e., $3NADH + H^+$ and $1FADH_2$).

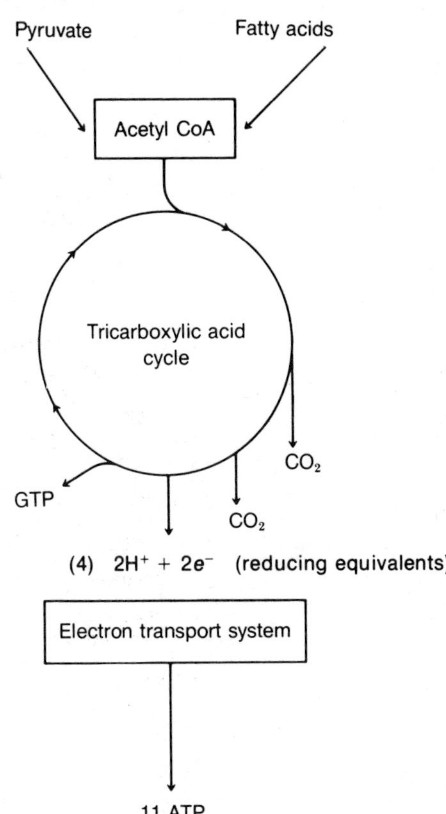

Figure 6.16
General description of mitochondrial ATP synthesis.

Individual Reactions of the Tricarboxylic Acid Cycle

The initial step of the cycle is catalyzed by the enzyme citrate synthase. This is a highly exergonic reaction and essentially commits acetyl groups toward citrate formation or oxidation in the Krebs cycle. As shown below the citrate synthase reaction involves the condensation of the acetyl moiety and the α-keto function of the dicarboxylic acid oxaloacetate. Citrate synthase is an enzyme with mol wt 100,000 and exists in the mitochondrial matrix.

$$\underset{\text{Acetyl CoA}}{\overset{*}{C}H_3\overset{*}{C}(=O)\text{—SCoA}} + \underset{\text{Oxaloacetate}}{O{=}C(COO^-)\text{—}CH_2COO^-} \longrightarrow \underset{\text{Citroyl-SCoA}}{\left[HO\text{—}C(COO^-)(\overset{*}{C}H_2\overset{*}{C}(=O)\text{—SCoA})(CH_2COO^-) \right]} \overset{H_2O}{\rightleftharpoons} \underset{\text{Citrate}}{HO\text{—}C(COO^-)(\overset{*}{C}H_2\overset{*}{C}OO^-)(CH_2COO^-)} + \text{CoASH}$$

The equilibrium of this reaction is far toward citrate formation with a $\Delta G°$ near -9 kcal/mol. The citroyl-SCoA intermediate in this reaction is not released from the enzyme during the reaction and is thought to remain bound to the catalytic site on citrate synthase. It has been estimated that the citrate synthase reaction is considerably displaced from equilibrium under in situ conditions, which makes this step a primary candidate for regulatory modulation. It has been proposed that this enzyme is regulated (inhibited) by ATP, NADH, succinyl CoA, and long-chain acyl CoA derivatives on the basis of experiments performed with the purified enzyme, but none of these effects has been proven to be operative in intact metabolic systems under physiological conditions.

It is most probable that the primary regulator of the citrate synthase reaction is the availability of its two substrates, acetyl CoA and oxaloacetate. It is important to note the many important fates and effects of

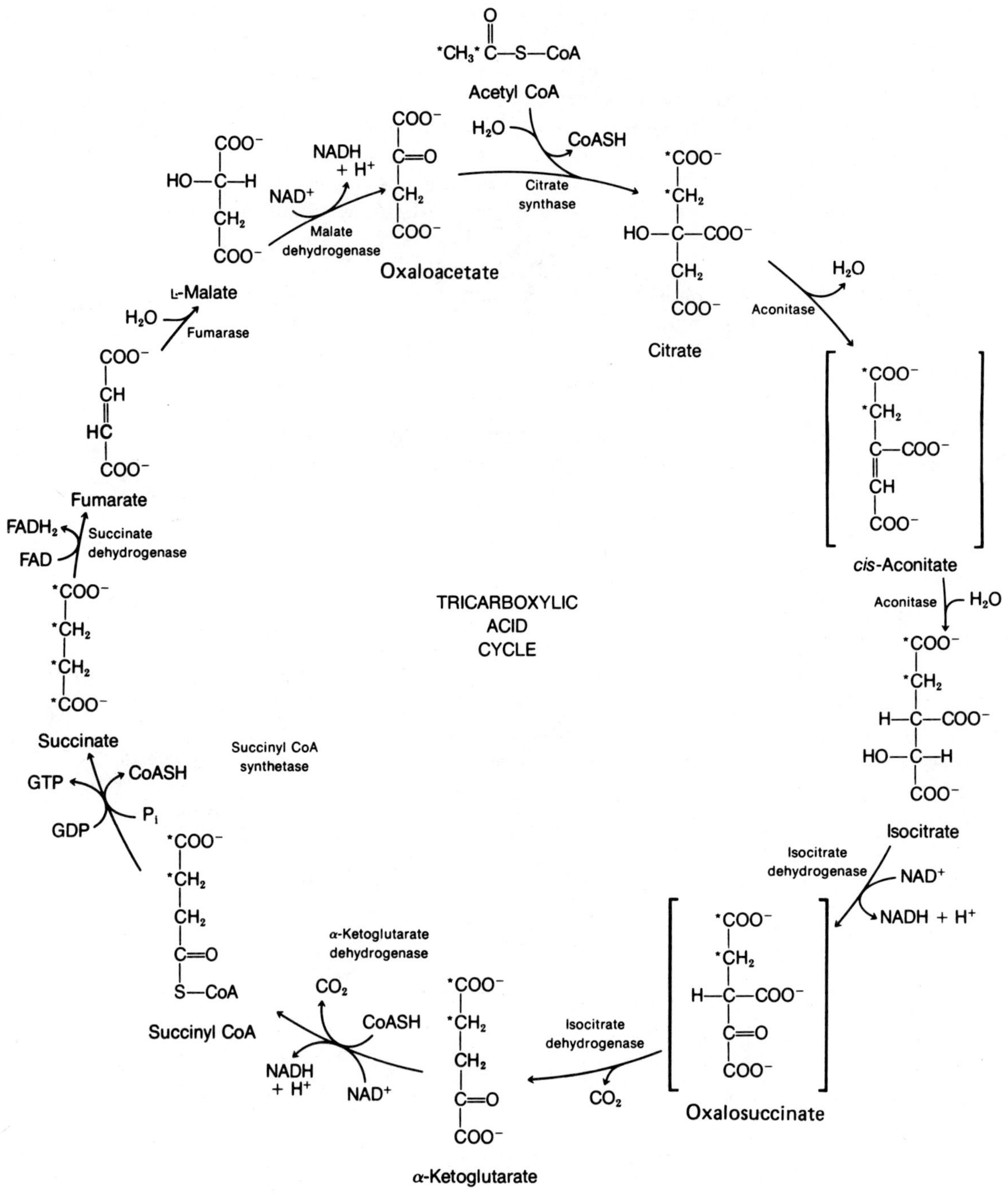

Figure 6.17
The tricarboxylic acid cycle.

citrate in energy and biosynthetic metabolism. Figure 6.18 depicts the involvement of citrate as a regulatory effector of other metabolic pathways and as a source of carbon and reducing equivalents for various synthetic purposes (see Chapters 7 and 9 for further details).

Citrate synthase can react with monofluoroacetyl CoA to form monofluorocitrate, which is a potent inhibitor of the next step in the tricarboxylic acid cycle, the aconitase reaction. In fact, whether monofluorocitrate is synthesized in situ as a result of fluoroacetate poisoning or administered experimentally, a nearly complete block of tricarboxylic acid cycle activity is observed.

Citrate is converted to isocitrate in the aconitase reaction.

$$\text{Citrate} \underset{}{\overset{H_2O}{\rightleftharpoons}} [\textit{cis}\text{-Aconitate}] \overset{H_2O}{\rightleftharpoons} \text{Isocitrate}$$

Citrate | *cis*-Aconitate | Isocitrate

Again, this reaction involves the generation of an enzyme-bound intermediate, *cis*-aconitate. At equilibrium there exists 90% citrate, 3% *cis*-aconitate, and 7% isocitrate, hence the equilibrium of aconitase lies toward citrate formation. Although the aconitase reaction does not require cofactors, it requires ferrous (Fe^{2+}) iron in its catalytic mechanism. There is evidence that this Fe^{2+} may be involved in an iron-sulfur center, which is an essential component in the hydratase activity of aconitase.

Isocitrate dehydrogenase catalyzes the first dehydrogenase reaction in the tricarboxylic acid cycle. Isocitrate is converted to α-ketoglutarate in an oxidative decarboxylation reaction, which likely occurs via an enzyme-bound intermediate, oxalosuccinate. In this step of the cycle the initial (of two) CO_2 is produced and the initial (of three) NADH + H^+ is generated. The isocitrate dehydrogenase involved in the tricarboxylic acid cycle in mitochondria from mammalian tissues requires NAD^+ as the oxidized acceptor of reducing equivalents.

$$\text{Isocitrate} \xrightleftharpoons{NAD^+ \quad NADH + H^+} [\text{Oxalosuccinate}] \xrightleftharpoons{CO_2} \alpha\text{-Ketoglutarate}$$

Isocitrate | Oxalosuccinate | α-Ketoglutarate

Mitochondria also possess an isocitrate dehydrogenase that requires $NADP^+$ as the oxidized coenzyme. The $NADP^+$-linked enzyme may also be found in the cytosol, where it is probable that it is involved in providing

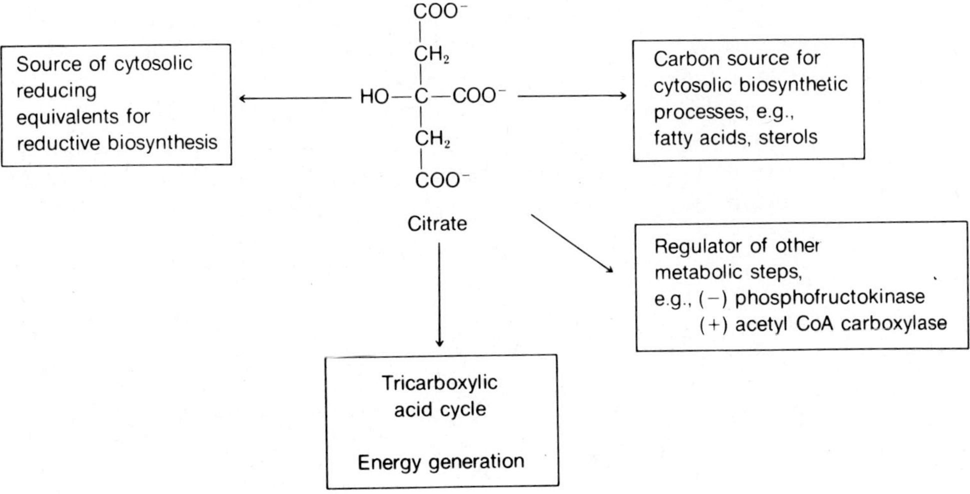

Figure 6.18
Fates and functions of citrate.

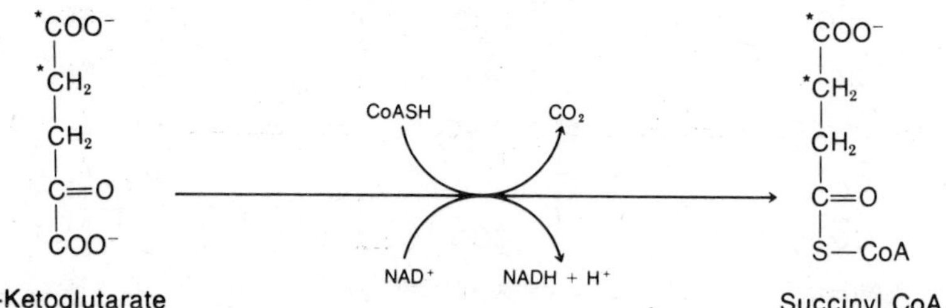

Figure 6.19
The α-ketoglutarate dehydrogenase reaction.

reducing equivalents for cytosolic reductive processes. The equilibrium of this reaction lies strongly toward α-ketoglutarate formation with a $\Delta G°$ of nearly -5 kcal/mol. The NAD^+-linked isocitrate dehydrogenase has a mol wt 380,000 and consists of eight identical subunits. The reaction requires a divalent metal cation (e.g., Mn^{2+} or Mg^{2+}) in the decarboxylation of the β position of the oxalosuccinate. The NAD^+-linked isocitrate dehydrogenase is stimulated by ADP and in some cases AMP and is inhibited by ATP and NADH. Hence under high energy conditions (i.e., high ATP/ADP + P_i and high $NADH/NAD^+$ ratios) the NAD^+-linked isocitrate dehydrogenase of the tricarboxylic acid cycle is inhibited. During periods of low energy, on the other hand, the activity of this enzyme is stimulated in order to accelerate energy generation in the tricarboxylic acid cycle.

The conversion of α-ketoglutarate to succinyl CoA is catalyzed by the α-ketoglutarate dehydrogenase multienzyme complex (Figure 6.19). This enzyme complex is nearly identical to the pyruvate dehydrogenase complex in terms of the reactions catalyzed and some of its structural features. Again, thiamine pyrophosphate, lipoic acid, CoASH, FAD, and NAD^+ participate in the catalytic mechanism. The multienzyme complex consists of the α-ketoglutarate dehydrogenase, the dihydrolipoyl transsuccinylase and the dihydrolipoyl dehydrogenase as the three catalytic subunits. The equilibrium of the α-ketoglutarate dehydrogenase reaction lies strongly toward succinyl CoA formation with a $\Delta G°$ of -8 kcal/mol. In this reaction the second molecule of CO_2 and the second reducing equivalent (i.e., NADH + H^+) of the tricarboxylic acid cycle are produced. The other product of this reaction, succinyl CoA is an example of an energy-rich thiolester compound similar to acetyl CoA. Unlike the pyruvate dehydrogenase complex the α-ketoglutarate dehydrogenase is not regulated by a protein kinase-mediated phosphorylation reaction. The nucleoside triphosphates, ATP and GTP, NADH, succinyl CoA and Ca^{2+} have been shown to inhibit this enzyme complex.

It is at the level of α-ketoglutarate in the Krebs cycle where intermediates may leave this oxidative pathway to be reductively aminated in the glutamate dehydrogenase reaction (Figure 6.20). This mitochondrial enzyme converts α-ketoglutarate to glutamate in the presence of NADH or

$$NH_4^+ + \underset{\alpha\text{-Ketoglutarate}}{COO^- - C(=O) - CH_2 - CH_2 - COO^-} \xrightleftharpoons[]{NAD(P)H \;\; NAD(P)^+} \underset{\text{Glutamate}}{COO^- - CH(NH_3^+) - CH_2 - CH_2 - COO^-}$$

Figure 6.20
The glutamate dehydrogenase reaction.

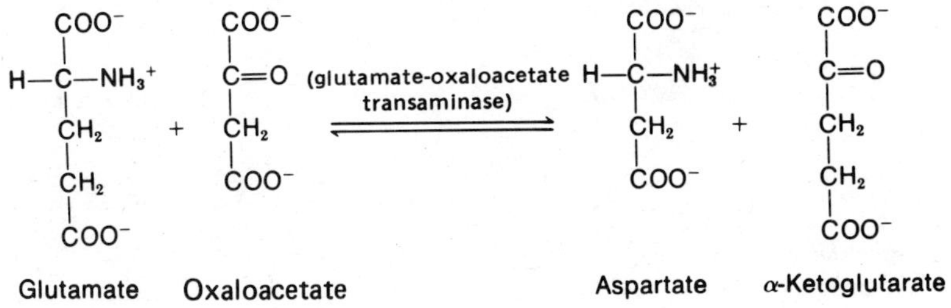

$$
\begin{array}{c} COO^- \\ | \\ H-C-NH_3^+ \\ | \\ CH_2 \\ | \\ CH_2 \\ | \\ COO^- \end{array}
+
\begin{array}{c} COO^- \\ | \\ C=O \\ | \\ CH_2 \\ | \\ COO^- \end{array}
\xrightleftharpoons{\text{(glutamate-oxaloacetate transaminase)}}
\begin{array}{c} COO^- \\ | \\ H-C-NH_3^+ \\ | \\ CH_2 \\ | \\ COO^- \end{array}
+
\begin{array}{c} COO^- \\ | \\ C=O \\ | \\ CH_2 \\ | \\ CH_2 \\ | \\ COO^- \end{array}
$$

Glutamate Oxaloacetate Aspartate α-Ketoglutarate

Figure 6.21
Glutamate-oxaloacetate transaminase, an example of a transamination reaction.

NADPH and ammonia. Using various transamination reactions (Figure 6.21) the amino group thus incorporated into glutamate is transferred to a variety of other amino acids. These enzymes and the relevance of the incorporation or release of ammonia into or from α-keto acids will be discussed further in Chapters 11 and 12.

The energy-rich character of the thiolester linkage of succinyl CoA is conserved in a substrate-level phosphorylation reaction in the next step of the tricarboxylic acid cycle. The succinyl CoA synthetase or succinate thiokinase reaction converts succinyl CoA to succinate and in mammalian tissue results in the phosphorylation of GDP to GTP (Figure 6.22). This reaction is freely reversible with a $\Delta G° = -0.7$ kcal/mol and the catalytic mechanism involves an enzyme–succinyl phosphate intermediate.

$$\text{Succinyl CoA} + P_i + \text{Enz} \longleftrightarrow \text{Enz—succinyl phosphate} + \text{CoASH}$$

$$\text{Enz—succinyl phosphate} \longleftrightarrow \text{Enz—phosphate} + \text{succinate}$$

$$\text{Enz—phosphate} + \text{GDP} \longleftrightarrow \text{Enz} + \text{GTP}$$

The enzyme is phosphorylated on the 3 position of a histidine residue during the succinyl CoA synthetase reaction. Hence in this step of the tricarboxylic acid cycle a high-energy bond is conserved as GTP. Because of the presence of the nucleoside diphosphokinase discussed earlier in this chapter, this GTP may be converted to ATP.

Succinyl CoA represents a metabolic branch point in that intermediates may enter or exit the Krebs cycle at this point (see Figure 6.23). Succinyl CoA may be formed either from α-ketoglutarate in the Krebs cycle or from methylmalonyl CoA in the final steps of the breakdown of odd-chain length fatty acids or the branched-chain amino acids valine and isoleucine. The metabolic fates of succinyl CoA include its conversion to succinate in the succinyl CoA synthetase reaction of the Krebs cycle and its condensation with glycine to form δ-aminolevulinate in the δ-aminolevulinate synthetase reaction, which is the initial reaction in porphyrin biosynthesis (see Chapter 22).

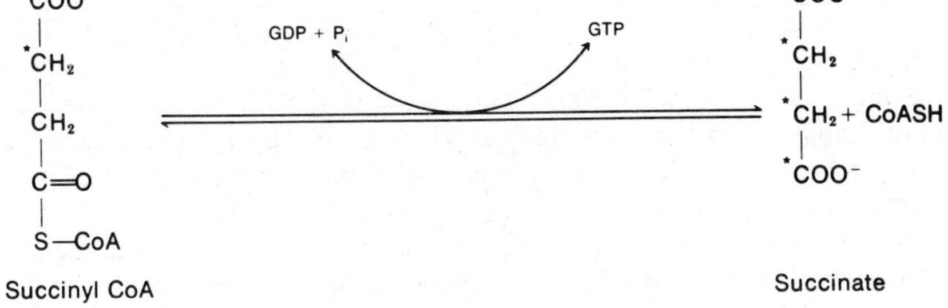

$$
\begin{array}{c} {}^{*}COO^- \\ | \\ {}^{*}CH_2 \\ | \\ CH_2 \\ | \\ C=O \\ | \\ S-CoA \end{array}
\xrightleftharpoons[]{\text{GDP} + P_i \;\curvearrowright\; \text{GTP}}
\begin{array}{c} {}^{*}COO^- \\ | \\ {}^{*}CH_2 \\ | \\ {}^{*}CH_2 \\ | \\ {}^{*}COO^- \end{array} + \text{CoASH}
$$

Succinyl CoA Succinate

Figure 6.22
The succinyl CoA synthetase reaction.

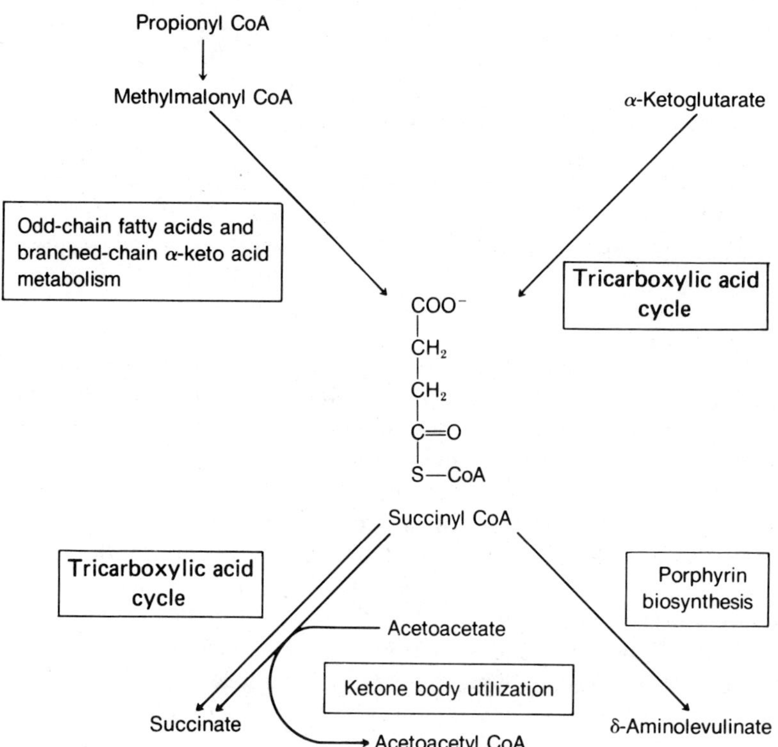

Figure 6.23
Sources and fates of succinyl CoA.

Succinate ⇌ (FAD → $FADH_2$) Fumarate

Figure 6.24
The succinate dehydrogenase reaction.

Succinate Malonate

Maleate

Figure 6.25
Structures of succinate, a tricarboxylic acid cycle intermediate; malonate, a cycle inhibitor; and maleate, a compound not involved in the cycle.

Succinate is oxidized to fumarate in the succinate dehydrogenase reaction of the Krebs cycle (Figure 6.24). Succinate dehydrogenase is tightly bound to the inner mitochondrial membrane and is composed of two subunits with mol wt 70,000 and 30,000. The 70,000 mol wt subunit contains the substrate binding site, the covalently bound (to a lysine residue) FAD, 4 nonheme iron atoms, and 4 acid-labile sulfur atoms, whereas the 30,000 mol wt subunit contains 4 nonheme irons and 4 acid-labile sulfur atoms. It is thought that this enzyme is a typical example of an iron-sulfur protein in which the nonheme iron of succinate dehydrogenase undergoes valence changes (e.g., $Fe^{2+} \rightarrow Fe^{3+}$) during the removal of electrons and protons from succinate and the subsequent transfer of these reducing equivalents through the covalently bound FAD to the mitochondrial electron transfer chain at the coenzyme Q–cytochrome b level.

Succinate dehydrogenase is strongly inhibited by malonate and oxaloacetate and is activated by ATP, inorganic phosphate, and succinate. Malonate inhibits succinate dehydrogenase competitively with respect to succinate. This inhibitory characteristic of malonate is due to the very close structural similarity between malonate and the substrate succinate (Figure 6.25). Malonate is used experimentally as a very effective inhibitor of the Krebs cycle in complex metabolic systems. In fact, the ability of malonate to inhibit the cycle was used by Krebs as evidence for the cyclic nature of this oxidative metabolic pathway.

Fumarate is hydrated to form L-malate in the next step in the tricarboxylic acid cycle by the enzyme fumarase (Figure 6.26). Fumarase is a tetramer with a mol wt of 200,000 and is stereospecific for the trans form of the substrate (the cis form, maleate, is not a substrate Figure 6.25), and the product of the fumarase reaction is only L-malate. The fumarase reaction is freely reversible under physiological conditions.

The final reaction in the Krebs cycle is the malate dehydrogenase reaction in which the final (of three) reducing equivalents as NADH + H^+ are removed from the cycle intermediates (Figure 6.27).

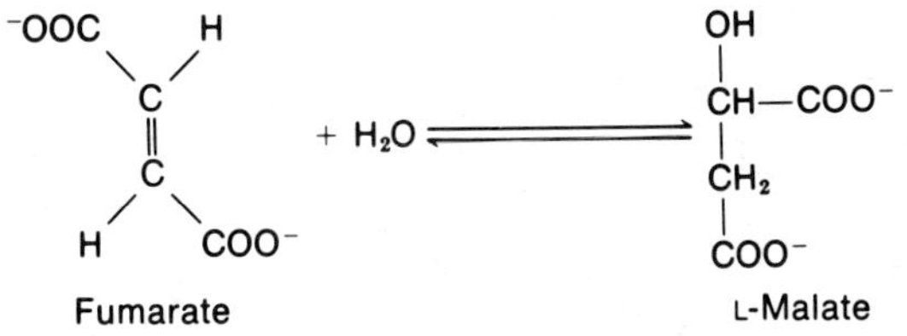

Figure 6.26
The fumarase reaction.

The equilibrium of the malate dehydrogenase reaction lies far toward L-malate formation, because $\Delta G° = +7.0$ kcal/mol. Thus the malate dehydrogenase reaction is an endothermic reaction when considered in the forward direction of the Krebs cycle. However, the citrate synthase reaction and other reactions of the cycle pull malate dehydrogenase toward oxaloacetate formation by removing oxaloacetate. Additionally, NADH produced in the various cycle NAD-linked dehydrogenases is oxidized rapidly to NAD^+ in the mitochondrial respiratory chain.

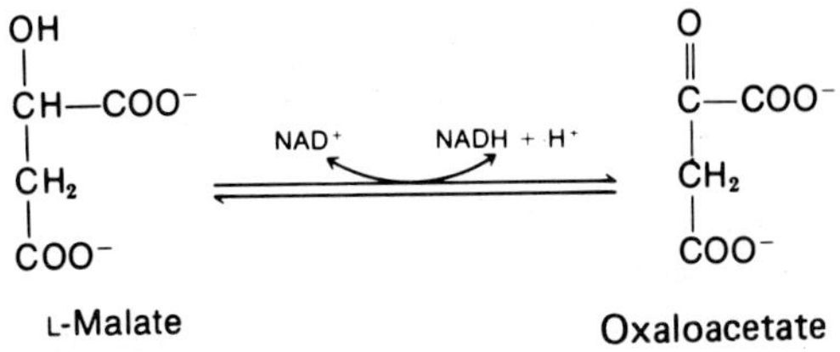

Figure 6.27
The malate dehydrogenase reaction.

Energy Yield

In summary the tricarboxylic acid cycle (Figure 6.16) serves as a terminal oxidative pathway for most metabolic fuels. Two-carbon moieties as acetyl CoA are taken into the cycle and are oxidized completely to CO_2 and H_2O. During this process 4 reducing equivalents (3 as NADH + H^+ and 1 as $FADH_2$) are produced, which are used subsequently for energy generation. As is discussed later in this chapter, oxidation of each NADH + H^+ results in the formation of 3 ATP molecules in the mitochondrial respiratory chain oxidative phosphorylation sequence, while the oxidation of the $FADH_2$ formed in the succinate dehydrogenase reaction yields 2 ATPs. Also, a high-energy bond is formed in the succinyl CoA synthetase reaction. Hence, the net yield of ATP or its equivalent (i.e., GTP) for the complete oxidation of an acetyl group in the Krebs cycle is 12.

During the complete oxidation of glucose to CO_2 and H_2O there is a net formation of (1) 2 ATPs per glucose in the conversion of glucose to two molecules of pyruvate; (2) 6 ATPs per glucose as a result of the translocation and subsequent oxidation in the mitochondrial compartment of 2NADH + H^+ formed in the glyceraldehyde 3-phosphate dehydrogenase reaction of the glycolytic pathway; and (3) 30 ATPs per glucose from the oxidation of the 2 pyruvate molecules in the pyruvate dehydrogenase reaction and subsequent conversion of 2 acetyl CoAs to CO_2 and H_2O in the tricarboxylic acid cycle. Hence, the net ATP yield during the complete oxidation of glucose to $6CO_2$ plus $6H_2O$ is 38 ATPs.

Regulation of the Tricarboxylic Acid Cycle

A variety of factors is involved in the regulation of the activity of the tricarboxylic acid cycle. First, the supply of acetyl units, whether they are derived from pyruvate (i.e., carbohydrate) or fatty acids, is a crucial factor in determining the rate of the Krebs cycle. Regulatory influences on the pyruvate dehydrogenase complex (discussed previously in this chapter) have an important effect on the activity of the cycle. Likewise any control exerted on the processes of transport and β-oxidation of fatty acids would be an effective determinant of the Krebs cycle activity.

Second, because the primary dehydrogenase reactions of the Krebs cycle are dependent upon a continuous supply of both NAD^+ and FAD, their activities are very stringently controlled by the mitochondrial respiratory chain, which is responsible for oxidizing the NADH and $FADH_2$ produced as a result of substrate oxidation in the cycle. Because the activity of the respiratory chain is coupled obligatorily to the generation

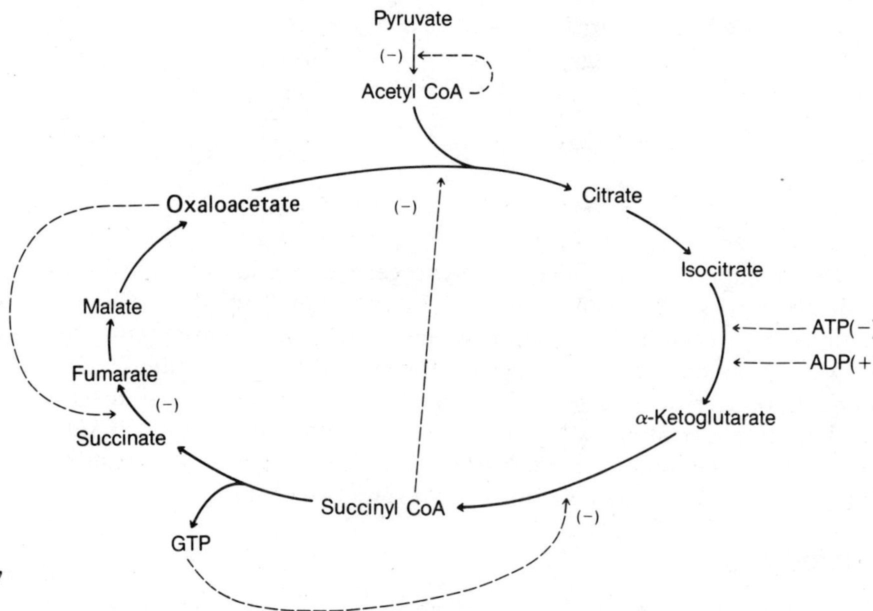

Figure 6.28
Representative examples of the regulatory interactions in the tricarboxylic acid cycle.

of ATP in the oxidative phosphorylation sequence of reactions, the activity of the Krebs cycle is very much dependent upon a "respiratory control," which is strongly affected by the availability of ADP + P_i and oxygen. Hence an inhibitory agent or metabolic condition which might interrupt the supply of oxygen, the continuous supply of ADP or the source of reducing equivalents (e.g., substrate for the cycle) would shut down cycle activity. This type of control of the cycle is generally referred to as the "coarse control" of the cycle. There are, of course, a variety of postulated effector-mediated regulatory interactions between various intermediates or nucleotides and the individual enzymes of the cycle, which may serve to exert a fine control on the activity of the cycle. Some illustrations of these interactions are shown in Figure 6.28 and have also been noted during the discussions of individual enzymes of the Krebs cycle. It must be stressed that the physiological relevance of many of these types of individual regulatory interactions has not been firmly established in intact metabolic systems.

6.5 STRUCTURE AND COMPARTMENTATION OF THE MITOCHONDRIAL MEMBRANES

Because the metabolic pathways for the oxidation of pyruvate, the end product of glycolysis, and fatty acids are located in mitochondria, a major portion of the energy-generating capacity of most cells resides in the mitochondrial compartment of the cell. The number of mitochondria in various tissues reflects the physiological function of the tissue and determines its capacity to perform aerobic metabolic functions. For example, the erythrocyte has no mitochondria and hence does not possess the capacity to generate energy, using oxygen as a terminal electron acceptor. On the other hand, cardiac tissue is a highly aerobic tissue, and it has been estimated that about half of the cytoplasmic volume of cardiac cells is composed of mitochondria. The liver is another tissue that is highly dependent upon aerobic metabolic processes for its various functions, and it has been estimated that mammalian hepatocytes contain between 800 and 2,000 mitochondria per cell. Mitochondria exist in a variety of different shapes, depending upon the cell type from which they are derived. As can be seen in Figure 6.29 mitochondria from liver are nearly spherical in shape, whereas those found in cardiac muscle are oblong or cylindrical.

Inner and Outer Mitochondrial Membranes

Mitochondria are composed of two membranes, an outer membrane and a highly invaginated inner membrane (see Figure 6.30). The outer mitochondrial membrane is thought to be a rather simple membrane, which is composed of about 50% lipid and 50% protein, with relatively few enzy-

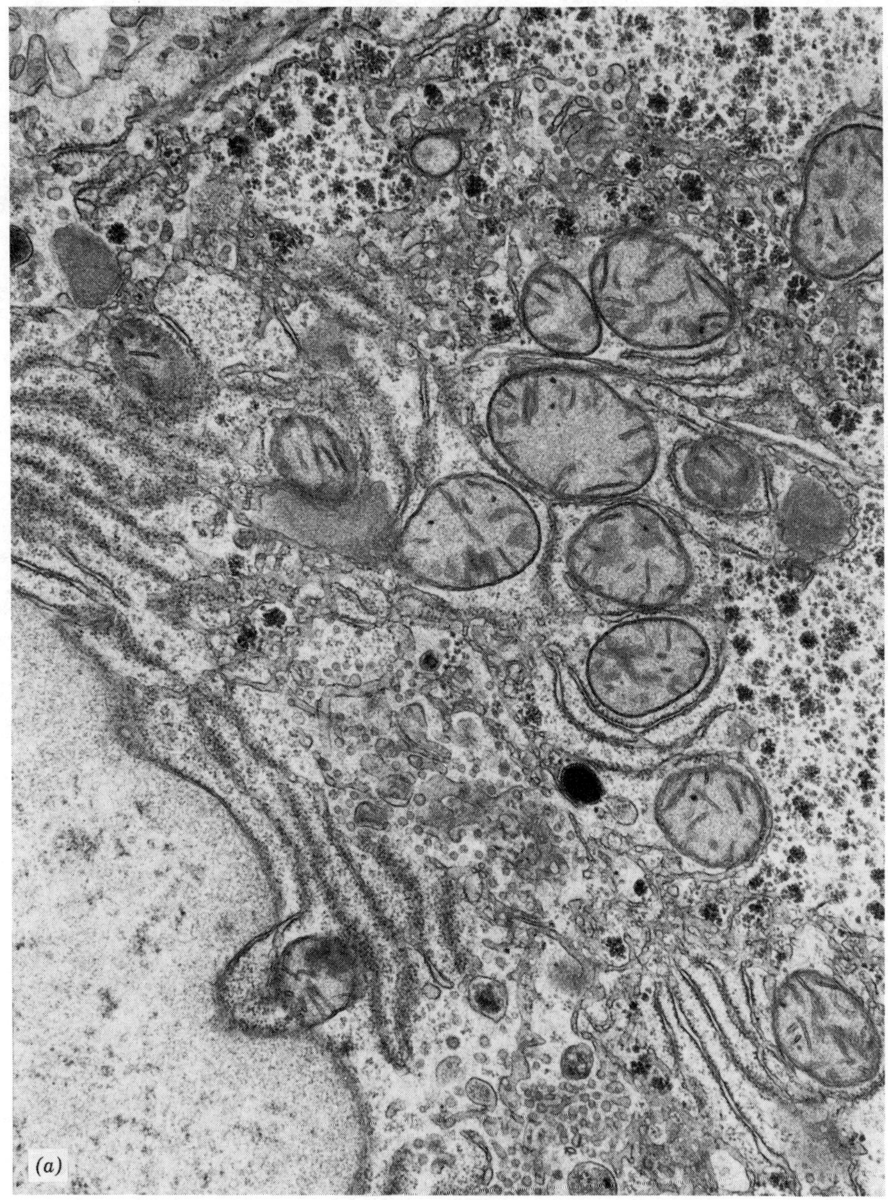

Figure 6.29
Electron micrographs of mitochondria in (a) hepatocytes from rat liver and (b) muscle fibers from rabbit heart.

Magnification 39,600×, courtesy of Dr. W. B. Winborn, Department of Anatomy, The University of Texas Health Science Center at San Antonio, and the Electron Microscopy Laboratory, Department of Pathology, The University of Texas Health Science Center at San Antonio.

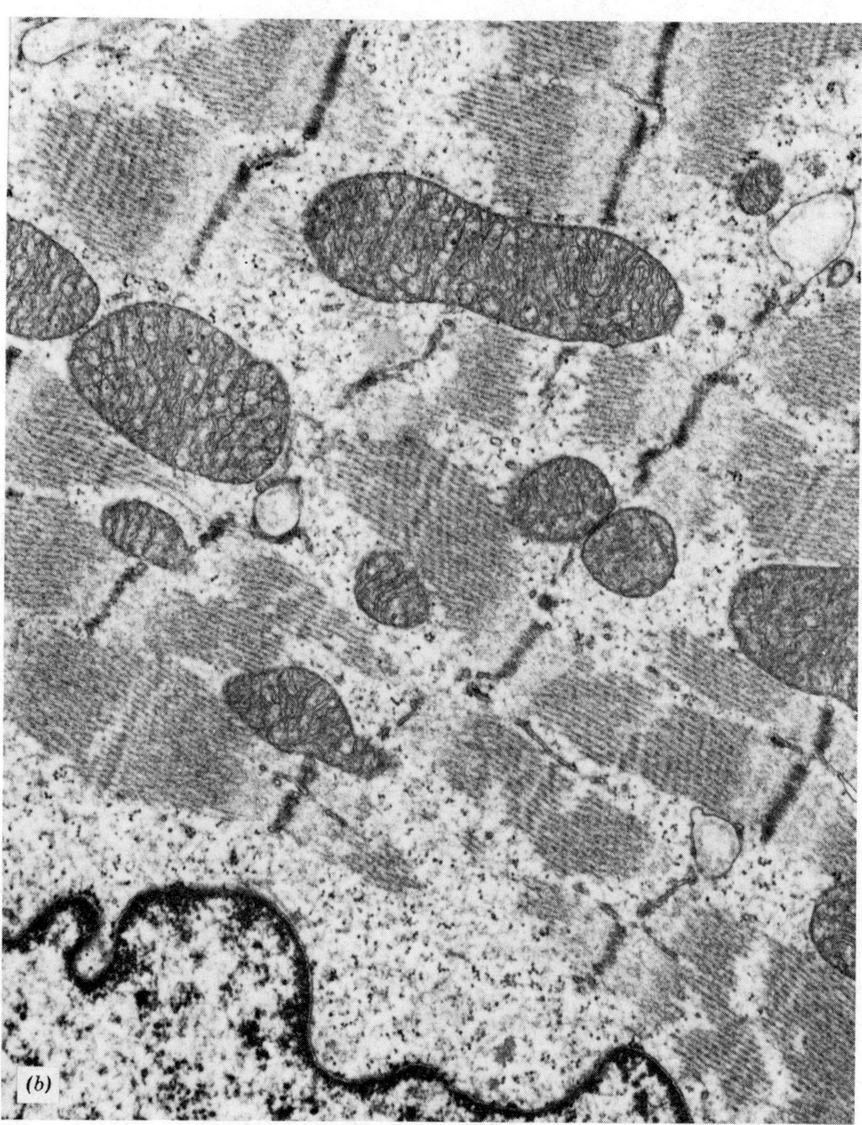

Figure 6.29 (Continued)

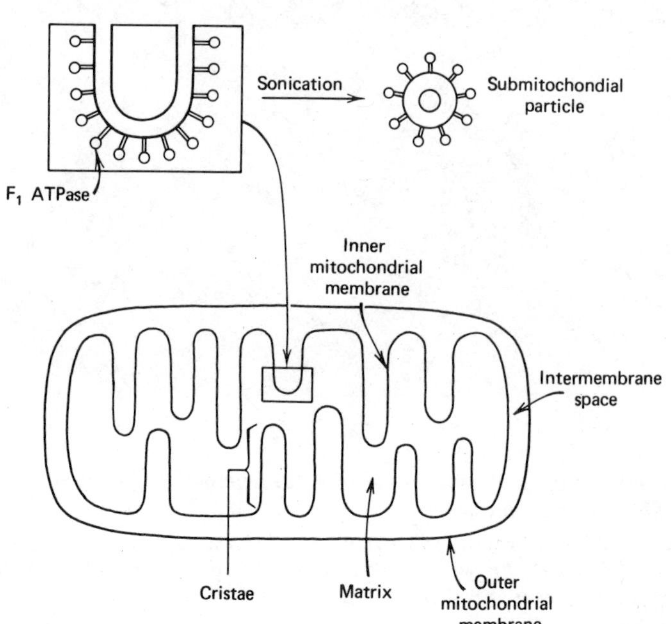

Figure 6.30
Diagram of the various submitochondrial compartments.

matic or transport functions. Table 6.5 defines some of the enzymatic components of the outer membrane.

The inner membrane is structurally and functionally much more complex than the outer membrane. Roughly 80% of the inner membrane is protein. The inner membrane contains most of the enzymes involved in electron transport and oxidative phosphorylation, various dehydrogenases, and several transport systems, which are involved in transferring substrates, metabolic intermediates, and adenine nucleotides between the cytosol and the mitochondrial matrix (Table 6.5).

Some of the enzymatic components associated with the inner mitochondrial membrane are only loosely associated with the membrane, whereas other enzymatic components are either tightly bound or are actual structural elements of the membrane. Hence there is a wide variability in the extent to which physical (ultrasonic irradiation or freezing and thawing), chemical (organic solvent or detergent treatment), or enzymatic (protease or lipase) treatments remove, release, or inactivate the enzymes associated with the inner membrane.

Experimental procedures have been developed that allow separation of inner from outer mitochondrial membranes. As indicated in Figure 6.31, the outer membrane may be stripped off and isolated, using digitonin (a detergent), osmotic shock, or ultrasonic irradiation followed by density-gradient ultracentrifugation. The resulting inner membrane plus matrix fraction is referred to as a mitoplast. The mitochondrial matrix may be released from the mitoplast, by treatment with a nonionic detergent or vigorous sonication. Once the various subcompartments of the mitochondrion have been separated, analyses may be performed to determine the location of the various characteristic marker enzymes, some of which are listed in Table 6.5. Enzymatic markers have been used effectively to detect the presence of mitochondria or even a particular portions of mitochondria in membrane preparations of diverse derivation.

TABLE 6.5 Enzymatic Composition of the Various Mitochondrial Subcompartments

Outer Membrane	*Intermembrane Space*	*Inner Membrane*	*Matrix*
Monoamine oxidase	Adenylate kinase	Succinate dehydrogenase	Pyruvate dehydrogenase
Kynurenine hydroxylase	Nucleoside diphosphate kinase	F_1-ATPase	Citrate synthase
Nucleoside diphosphate kinase		NADH dehydrogenase	Isocitrate dehydrogenase
Phospholipase A		β-Hydroxybutyrate dehydrogenase	α-Ketoglutarate dehydrogenase
Fatty acyl CoA synthetases		Cytochromes b, c_1, c, a, a_3	Aconitase
NADH : cytochrome c reductase (rotenone-insensitive)		Carnitine : acyl CoA transferase	Fumarase
Choline phosphotransferase		Adenine nucleotide translocase	Succinyl CoA synthetase
		Mono-, di-, and tricarboxylate translocase	Malate dehydrogenase
		Glutamate-aspartate translocase	Fatty acid β-oxidation system
			Glutamate dehydrogenase
			Glutamate-oxaloacetate transaminase
			Ornithine transcarbamoylase

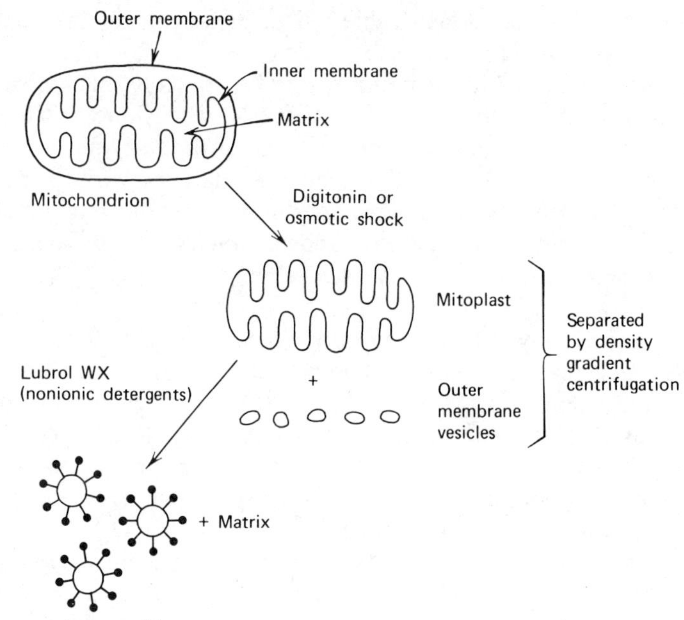

Figure 6.31
Separation of mitochondrial membranes.

Mitochondrial Transport Systems

Whereas the outer membrane presents little or no permeability barrier to substrate or nucleotide molecules of interest in energy metabolism, the inner membrane has very restricted limitations on the types of substrates, intermediates, and nucleotide species that may be transported across into the matrix compartment.

Figure 6.32 depicts various transport systems that have been described in mitochondria. Some of these transporters are well characterized, but others are not. The primary responsibility of these transport functions is

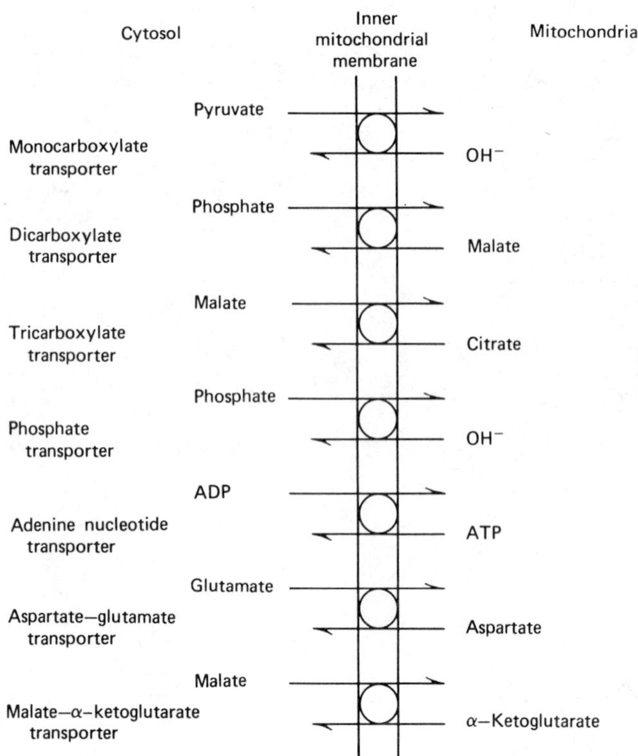

Figure 6.32
Mitochondrial metabolite transporters.

to facilitate the selective movement of various substrates, intermediates, and nucleotides back and forth across the inner mitochondrial membrane from the cytosol to the mitochondrial matrix. By virtue of these transporters, various substrates and other molecules can be accumulated in the mitochondrial matrix since the transporters can facilitate the movement of the substrate against a concentration gradient. The importance of the mitochondrial transporter systems derives from the involvement of the substances transported in a variety of mitochondrial metabolic processes.

Transport of Reducing Equivalents

The various nucleotides involved in cellular oxidation–reduction reactions (e.g., NAD^+, NADH, $NADP^+$, NADPH, FAD, and $FADH_2$) and coenzyme A and its derivatives are not permeable to the inner mitochondrial membrane. Hence, for example, in order to transport reducing equivalents (e.g., protons and electrons) from the cytosol to the mitochondrial matrix or vice versa, "substrate shuttle mechanisms" involving the reciprocal transfer of reduced and oxidized members of various oxidation–reduction couples are used to accomplish the net transfer of reducing equivalents across the membrane. Two examples of how this transfer of reducing equivalents from the cytosol to the mitochondria occurs are shown in Figure 6.33. The malate–aspartate shuttle and the α-glycerol phosphate shuttle are employed in various tissues to translocate reducing equivalents from the cytosol, where they are generated, to the mitochondrial compartment, where they are oxidized to yield energy. The operation of such substrate shuttles requires that the appropriate enzymes are localized on the correct side of the membrane and that appropriate trans-

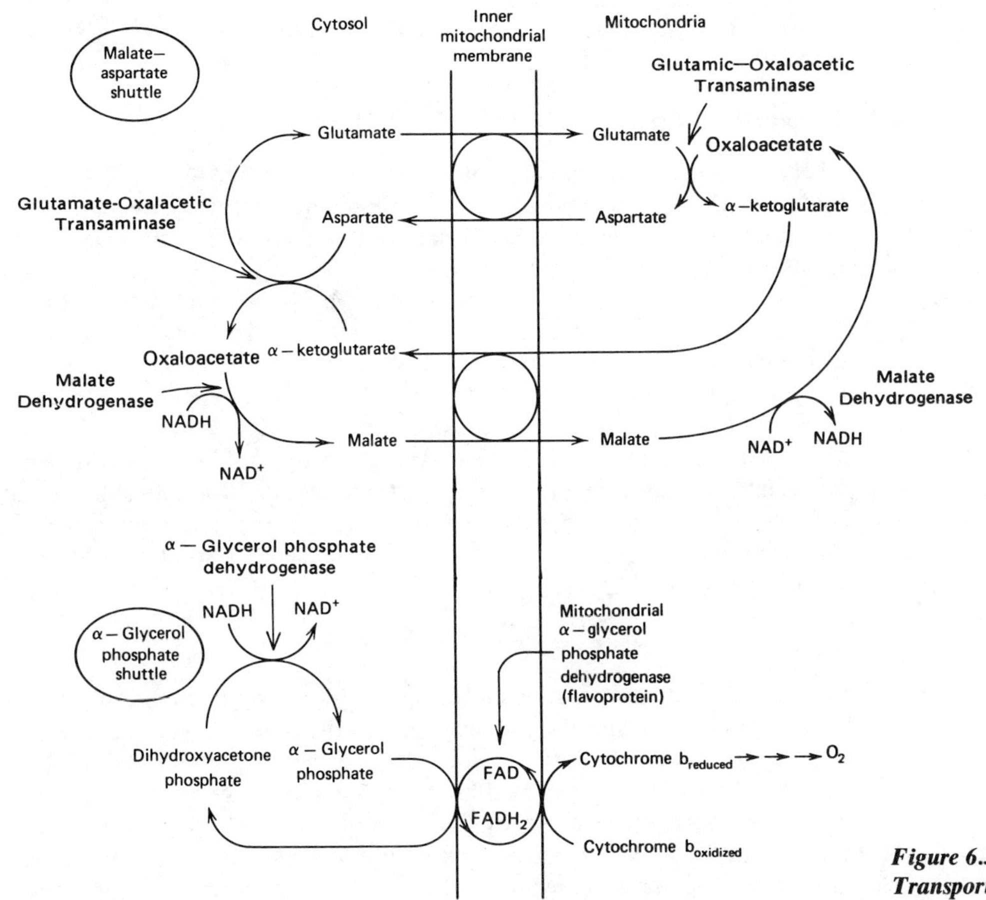

Figure 6.33
Transport shuttles for reducing equivalents.

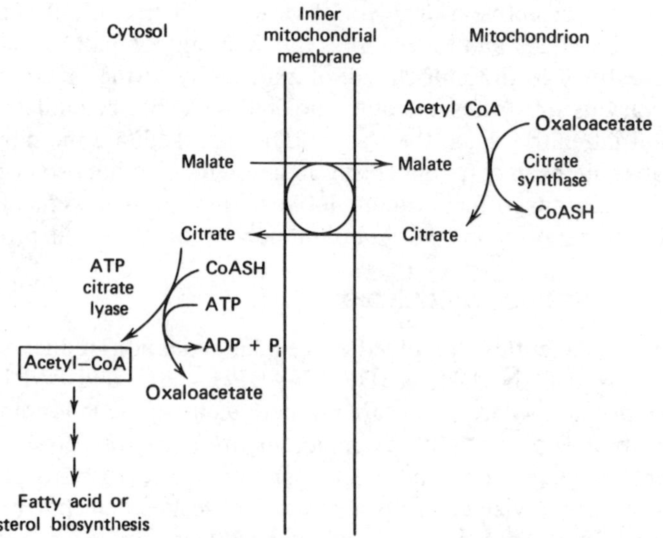

Figure 6.34
The export of intramitochondrially generated citrate to the cytosol to serve as a source of acetyl CoA for biosynthesis of fatty acids or sterols.

porters or translocases are present on/in the membrane to shuttle the various intermediates. In this regard the operation of the malate-aspartate shuttle depends on the fact that NADH, NAD^+, and oxaloacetate are not permeable to the inner mitochondrial membrane, on the distribution of malate dehydrogenase and aspartate aminotransferase on both sides of the inner mitochondrial membrane and on the existence of membrane transporters which allow the exchange of intramitochondrial aspartate for cytosolic glutamate and cytosolic malate for intramitochondrial α-ketoglutarate.

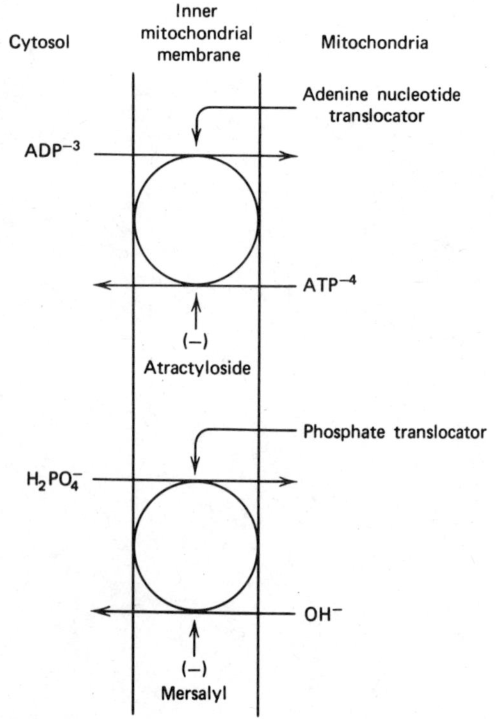

Figure 6.35
The adenine nucleotide and phosphate translocators.

Transport of Acetyl Units

Acetyl CoA is an impermeable substance but it can transfer the 2-carbon fragment (the acetyl group) from the mitochondrial compartment to the cytosol, where acetyl moieties are required for fatty acid or sterol biosynthesis, as illustrated in Figure 6.34.

Intramitochondrial acetyl CoA is converted to citrate in the citrate synthase reaction of the Krebs cycle. Subsequently the citrate is exported to the cytosol on the tricarboxylate transporter in exchange for a dicarboxylic acid such as malate. Cytosolic citrate may be cleaved to acetyl CoA and oxaloacetate at the expense of an ATP molecule in the ATP : citrate lyase reaction, which is discussed in Chapter 9. Substrate shuttle mechanisms are also involved in the movement of appropriate substrates and intermediates in both directions across the inner mitochondrial membrane in the liver during periods of active gluconeogenesis and ureogenesis (see Chapters 7 and 11).

Adenine Nucleotide Transport

Adenine nucleotides are transported across the inner mitochondrial membrane by the very specific adenine nucleotide translocator. Nucleotide species such as the guanine, uridine, or cytosine nucleotides are neither exchanged across the inner membrane on the adenine nucleotide specific translocator nor transported by a comparable carrier specific for non-adenine nucleotides. As indicated in Figure 6.35 cytosolic ADP, which is formed during energy-consuming reactions, is exchanged for mitochondrial ATP, which is generated in the process of oxidative phosphorylation. At pH 7 ADP has three negative charges and ATP has four, so that a 1 : 1 exchange of ADP : ATP would cause a charge imbalance across the membrane. Hence the ADP for ATP exchange across the mitochondrial membrane is an electrogenic process, which requires that in the end the charge imbalance must be compensated for by the movement of a proton or another charged species. An adenine nucleotide carrier has been isolated due to its capacity to bind very tightly to atractyloside, a specific inhibitor of the carrier. The carrier protein is a dimer with subunit mol wt 30,000. It is unlikely that the rate of transport of adenine nucleotides across the mitochondrial membrane is ever limiting to the overall process of mitochondrial ATP synthesis. Further, it has been observed that low concentrations of long-chain fatty acyl CoA derivatives inhibit (i.e., K_i = 1 μM) the transport of ATP and ADP in isolated liver mitochondria. However, experimental results performed under in vivo conditions in intact liver cells indicate that there occurs little, if any, inhibition of the adenine nucleotide transporter under metabolic conditions in which a large concentration of long-chain fatty acyl CoA accumulates.

There is also a transporter that transports cytosolic inorganic phosphate into the mitochondrial matrix in exchange for negatively charged hydroxyl ions in an electroneutral exchange (see Figure 6.35). This phosphate transport may also be accomplished in a proton-compensated mechanism, for example, phosphate and protons are transported in a 1 : 1 ratio. Phosphate transport is strongly inhibited by the compound mersalyl and various mercurial reagents.

Mitochondrial Calcium Transport

Finally, mitochondria from most tissues possess a transport system capable of translocating calcium across the mitochondrial inner membrane. It is difficult to overestimate the importance of the distribution of cellular calcium pools in different cell functions, such as muscle contraction, neural transmission, and hormone action and secretion. Calcium exists in distinct pools in the cell. The cytosol, mitochondria, endoplasmic reticulum, nuclei, and the Golgi membranes have their own pools of intracellular calcium. Some of the intracellular calcium is bound to nucleotides, metabolites, or membrane ligands, while a portion of the intracellular calcium is free in solution. A gradient of calcium exists from outside to inside a cell. Estimates of intracellular (e.g., cytosolic) calcium range from 10^{-6} M in the liver to 10^{-7}–10^{-8} M in the heart and skeletal muscle, whereas extracellular calcium likely is at least two orders of magnitude greater than this. Total intramitochondrial calcium has been estimated to be ~10^{-4} M but the free calcium concentration in the mitochondrion may be in the range of 10^{-5}–10^{-6} M. Hence processes involved in the alternate sequestering and release of an intracellular store of calcium can greatly influence intracellular calcium pools and various cell functions. Mitochondria have been known to accumulate rather large quantities of calcium at the expense of ATP hydrolysis, respiration, or an electrochemical gradient. Mitochondrial calcium transport is inhibited by low concentrations of lanthanides (trivalent metal cations) and by a compound called ruthenium red. Magnesium can compete with calcium for the carrier in certain types of mitochondria. The current view is that there is a specific carrier in the inner mitochondrial membrane, which is likely a glycoprotein (Figure 6.36). The mitochondrial calcium carrier exhibits saturation kinetics, has a high affinity for calcium, and is highly specific for calcium. Permeant counterions such as phosphate or acetate stimulate calcium transport and allow the metal cation to be retained by the mitochondria.

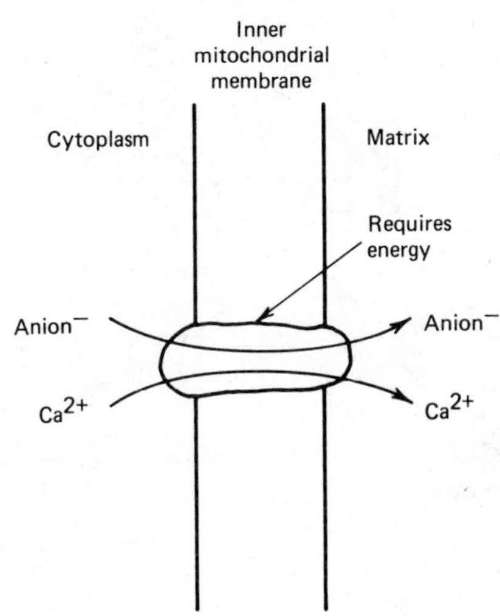

Figure 6.36
Mitochondrial calcium carrier.
The energy requirement can be met from ATP, ΔpH, or membrane potential.

Most interesting is the finding that certain hormones may affect intracellular calcium distribution (e.g., epinephrine or vasopressin) as part of the mechanism for the mediation of the hormone response. Various cytosolic protein kinases such as those involved in glycogen metabolism are calcium-sensitive.

In summary, the inner mitochondrial membrane possesses a variety of transport systems that are involved in the movement of nucleotides, substrates, metabolites, and metal cations into and out of the mitochondrial compartment. An understanding of these transport functions is essential in order to understand complex cellular metabolic pathways and their regulation. (See Clin. Corr. 6.2.)

CLIN. CORR. 6.2 MITOCHONDRIAL MYOPATHIES

Diseases that involve defects in various metabolic functions of muscle have been described. Clinically, patients with myopathies complain of weakness and cramping of the affected muscles; infants have difficulty feeding and crawling; severe fatigue results from minimal exertion; and there is usually evidence of muscle wasting. On the basis of electron microscopic examination and enzymatic characterization of muscle biopsy material, many myopathies have been found that have a primary lesion in mitochondrial function.

Deficiencies in mitochondrial transport functions (i.e., carnitine: palmitoyl CoA transferase) and in components of the mitochondrial electron transport chain (NADH dehydrogenase, cytochrome b, cytochrome a,a_3, or the mitochondrial ATPase) have been described. In many mitochondrial myopathies large paracrystalline inclusions have been found within the mitochondrial matrix. (See the accompanying figure.) It is not known whether this crystalline material is inorganic or organic in composition. In certain mitochondrial myopathies electron transport is only loosely coupled to ATP production; in other cases these processes exhibit normal tight coupling.

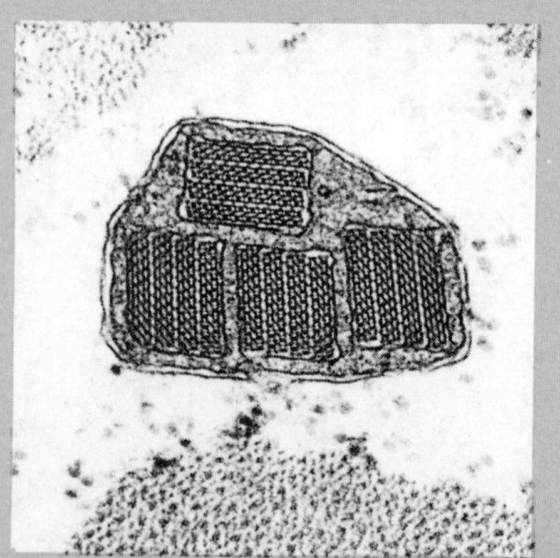

Example of paracrystalline inclusions in mitochondria from muscles of ocular myopathic patients.

Magnification 36,000×. Courtesy of Dr. D. N. Landon, Institute of Neurology, University of London.

6.6 ELECTRON TRANSFER

During the enzymatic reactions involved in glycolysis, fatty acid oxidation, and the tricarboxylic acid cycle, reducing equivalents are derived from the sequential breakdown of the initial metabolic fuel. In the case of glycolysis, NADH is produced in the glyceraldehyde 3-phosphate dehydrogenase reaction, and this reducing equivalent must be either reoxidized in the cytosol (e.g., by lactate dehydrogenase) or transported to the mitochondrial matrix via one of the substrate shuttle mechanisms in order to realize the maximum energy yield from the oxidation of glucose. In the case of fatty acid oxidation and in the tricarboxylic acid cycle, reducing equivalents as both NADH and $FADH_2$ are produced in the mitochondrial matrix. In order to transduce this reducing power into utilizable energy, mitochondria have a system of electron carriers in or associated with the inner mitochondrial membrane, which convert reducing equivalents in the presence of oxygen into utilizable energy by synthesizing ATP. This process is called electron transport, and, as will be seen later, NADH and $FADH_2$ oxidation in this process results in the production of 3 and 2 mol ATP/mol reducing equivalent transferred to oxygen, respectively.

Oxidation–Reduction Reactions

Prior to the presentation of a description of the many components and the mechanism of the electron transport sequence, it is important to discuss some basic information concerning oxidation–reduction reactions. The mitochondrial electron transport system is little more than a sequence of linked oxidation–reduction reactions, for example,

$$AH_2 + B \rightleftharpoons A + BH_2$$

or

$$\text{Electron donor} \rightleftharpoons \text{electron} + \text{electron acceptor}$$

Oxidation–reduction reactions occur when there is a transfer of electrons from a suitable electron donor (the reductant) to a suitable electron acceptor (the oxidant). In some oxidation–reduction reactions only electrons are transferred from the reductant to the oxidant (i.e., electron transfer between cytochromes),

$$\text{Cytochrome c } (Fe^{2+}) + \text{cytochrome a } (Fe^{3+}) \rightleftharpoons \text{cytochrome c } (Fe^{3+}) + \text{cytochrome a } (Fe^{2+})$$

whereas in other types of reactions, both electrons and protons (hydrogen atoms) are transferred (i.e., electron transfer between NADH and FAD).

$$NADH + H^+ + FAD \rightleftharpoons NAD^+ + FADH_2$$

The oxidized and the reduced forms of the compounds or groups operating in oxidation–reduction-type reactions are referred to as redox couples or pairs. The facility with which a given electron donor (reductant) gives up its electrons to an electron acceptor (oxidant) is expressed quantitatively as the oxidation–reduction potential of the system. An oxidation–reduction potential is measured in volts as an electromotive force of a half-cell made up of both members of an oxidation–reduction couple when compared to a standard reference half-cell (usually the hydrogen electrode reaction) (see Figure 6.37). The potential of the standard hydrogen reference electrode is set by convention at 0.0 V at pH 0.0. However, when this standard potential is corrected for pH 7.0 the reference electrode potential becomes −0.42 V. The oxidation–reduction potentials for a variety of important biochemical reactions have been determined and are tabulated in Table 6.6.

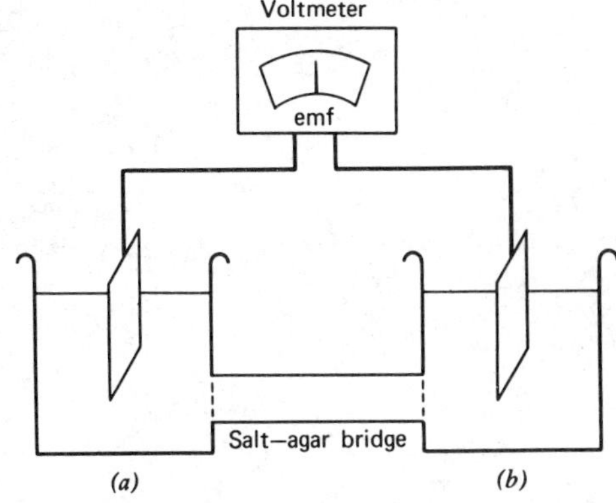

Figure 6.37
A system for determining oxidation–reduction potentials in half-cell reactions.
(a) Sample half-cell; Red:H + Ox ⇌ Ox:H + Red, where Red = reductant and Ox = oxidant (both initially present at 1 M). (b) Reference half-cell; $2H^+ + 2e^- \rightleftharpoons H_2$*, where emf = 0.0 V(1 M* H^+ *in solution and* H_2 *gas at 1 atm).*

An important concept is indicated in this listing of oxidation–reduction potentials. The reductant of an oxidation–reduction pair with large negative oxidation–reduction potential will give up its electrons more readily than pairs with smaller negative or positive redox potentials. On the other hand, a strong oxidant (e.g., characterized by a large positive potential) has a very high affinity for electrons.

The Nernst equation characterizes the relationship between the standard oxidation–reduction potential of a particular redox pair (E_0'), the observed potential (E), and the ratio of the concentrations of the oxidant and reductant in the system:

$$E = E_0' + \frac{2.3\,RT}{nF}\log\frac{[\text{oxidant}]}{[\text{reductant}]}$$

E is the observed potential with all concentrations at 1 M.

E_0' is the standard potential at pH 7.0.

TABLE 6.6 Standard Oxidation–Reduction Potentials for Various Biochemical Reactions

Oxidation–Reduction System	*Standard Oxidation–Reduction Potential* E_0' (V)
Acetate + $2H^+ + 2e^- \rightleftharpoons$ acetaldehyde	−0.60
$2H^+ + 2e^- \rightleftharpoons H_2$	−0.42
Acetoacetate + $2H^+ + 2e^- \rightleftharpoons$ β-hydroxybutyrate	−0.35
$NAD^+ + 2H^+ + 2e^- \rightleftharpoons$ NADH + H^+	−0.32
Acetaldehyde + $2H^+ + 2e^- \rightleftharpoons$ ethanol	−0.20
Pyruvate + $2H^+ + 2e^- \rightleftharpoons$ lactate	−0.19
Oxaloacetate + $2H^+ + 2e^- \rightleftharpoons$ malate	−0.17
Coenzyme $Q_{ox} + 2e^- \rightleftharpoons$ coenzyme Q_{red}	+0.10
Cytochrome b (Fe^{3+}) + $e^- \rightleftharpoons$ cytochrome b (Fe^{2+})	+0.12
Cytochrome c (Fe^{3+}) + $e^- \rightleftharpoons$ cytochrome c (Fe^{2+})	+0.22
Cytochrome a (Fe^{3+}) + $e^- \rightleftharpoons$ cytochrome a (Fe^{2+})	+0.29
$\frac{1}{2}O_2 + 2H^+ + 2e^- \rightleftharpoons H_2O$	+0.82

R is the gas constant of 8.3 J deg^{-1} mol^{-1}.

T is the absolute temperature in degrees Kelvin.

n is the number of electrons being transferred.

F is the Faraday of 96,500 J V^{-1}.

When the observed potential is equal to the standard potential, a potential is defined which is referred to as the midpoint potential. At the midpoint potential the concentration of the oxidant is equal to that of the reductant. Knowing the standard oxidation–reduction potentials of a diverse variety of biochemical reactions allows one to predict the direction of electron flow or transfer when more than one redox pair is linked together by the appropriate enzyme which causes a reaction to occur. For example, as shown in Table 6.6 the NAD^+/NADH pair has a standard potential of −0.32 V, and the pyruvate/lactate pair possesses a potential of −0.19. This means that electrons will flow from the NAD^+/NADH system to the pyruvate/lactate system as long as the enzymatic component (lactate dehydrogenase) is present, for example,

$$\text{Pyruvate} + \text{NADH} + \text{H}^+ \rightleftharpoons \text{lactate} + \text{NAD}^+$$

Hence in the mitochondrial electron transfer system electrons or reducing equivalents are being produced in NAD- and FAD-linked dehydrogenase reactions, which have standard potentials at or close to that of NAD^+/NADH and are passed through the electron transfer chain, which has as its terminal acceptor the oxygen/water couple.

Free Energy Changes in Redox Reactions

Oxidation–reduction potential differences between two redox pairs are similar to free energy changes in a chemical reaction, in that both quantities depend on the concentration of the reactants and of the products of the reaction. Because of this similarity the following relationship can be expressed:

$$\Delta G^{\circ\prime} = -nF\,\Delta E_0'$$

Using this expression the free energy change for electron transfer reactions can be readily calculated if the potential difference between two oxidation–reduction pairs is known. Hence for the mitochondrial electron transfer process in which electrons are transferred between the NAD^+/NADH couple ($E_0' = -0.32$ V) and the $\frac{1}{2}O_2/H_2O$ couple ($E_o' = +0.82$ V) the free energy change for this process can be calculated:

$$\Delta G^\circ = -nF\,\Delta E_0' = -2 \times 23.062 \times 1.14\ \text{V}$$

$$\Delta G^\circ = -52.6\ \text{kcal/mol}$$

where 23.062 is the Faraday in kcal V^{-1} and n is the number of electrons transferred; for example, in the case of NADH $\rightarrow O_2$, $n = 2$. Thus the free energy available from the potential span between NADH and oxygen in the electron transfer chain is capable of generating more than enough energy to synthesize three ATPs per two reducing equivalents or two electrons transported to oxygen. Additionally, because of the negative sign of the free energy available in the process of mitochondrial electron transfer, this process is exergonic and will proceed spontaneously provided that the necessary enzymatic components are present.

Major Components of the Mitochondrial Electron Transport Chain

Before cataloging the mechanistic details of the mitochondrial electron transport chain it is necessary to describe the various components that participate in the transfer of electrons in this system. The major enzymes or proteins functioning as electron transfer components involved in the mitochondrial electron transfer system are as follows:

1. NAD^+-linked dehydrogenases
2. Flavin-linked dehydrogenases
3. Iron–sulfur proteins
4. Cytochromes

NAD-Linked Dehydrogenases

The initial stage in the mitochondrial electron transport sequence consists of the generation of reducing equivalents in the tricarboxylic acid cycle, the fatty acid β-oxidation sequence, and various other dehydrogenase reactions. The NAD-linked dehydrogenase reactions of these pathways reduce NAD^+ to NADH while converting the reduced member of an oxidation–reduction couple to the oxidized form, for example, for the isocitrate dehydrogenase reaction

$$\text{Isocitrate} + NAD^+ \rightleftharpoons \alpha\text{-ketoglutarate} + CO_2 + NADH + H^+$$

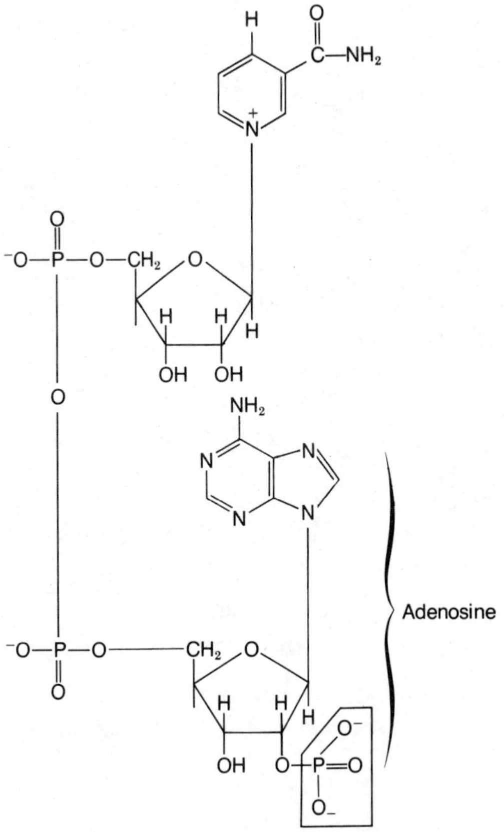

Two nicotinamide nucleotides are involved in various metabolic reactions, NAD and NADP (Figure 6.38). NADP has a phosphate esterified to the 2 position of the ribose in the adenosine portion of the dinucleotide. Each NAD(P)-linked dehydrogenase catalyzes a stereospecific transfer of the reducing equivalent from the substrate to the nucleotide:

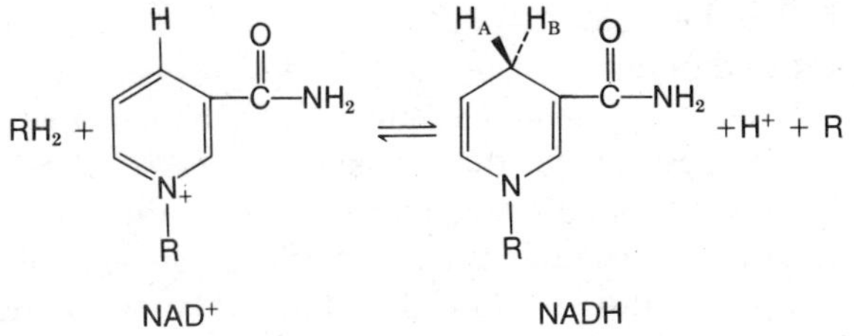

$$RH_2 + NAD^+ \rightleftharpoons NADH + H^+ + R$$

NAD(P)-linked dehydrogenases are either A-specific or B-specific in that the transfer of hydrogen occurs between either the oxidized or reduced metabolite and the A-side (projecting out from the plane of the pyridine ring) or the B-side (below the plane of the ring). Table 6.7 lists examples of the stereospecificity of $NAD(P)^+$-linked dehydrogenases.

TABLE 6.7 The Stereospecificity of NAD(P)-Linked Dehydrogenases

NAD(P)-Linked Dehydrogenase	*Specificity*
Alcohol dehydrogenase	A
Malate dehydrogenase	A
Lactate dehydrogenase	A
Isocitrate dehydrogenase ($NADP^+$)	A
Hydroxyacyl CoA dehydrogenase	B
Glyceraldehyde 3-phosphate dehydrogenase	B
Glucose 6-phosphate dehydrogenase ($NADP^+$)	B

Adenosine

Figure 6.38
Structure of nicotinamide adenine dinucleotide phosphate: $NADP^+$.

Once formed, NAD(P)H is released from the primary dehydrogenase and serves as the substrate for the mitochondrial electron transport system. NADPH is not a substrate for the mitochondrial respiratory chain but is used in the reductive biosynthetic reactions of such processes as fatty acid and sterol synthesis. When $NAD(P)^+$ is converted to NAD(P)H, there is a characteristic change in the absorbant and fluorescent properties of these nucleotides, which occurs as a result of the reduction of $NAD(P)^+$. As seen in Figure 6.39, the reduced form of the nicotinamide nucleotide [NAD(P)H] has an absorbance maximum at 340 nm, not present in the oxidized $NAD(P)^+$ form. Further, when the reduced form of the nicotinanide nucleotide is excited by light at 340 nm a fluorescence emission maximum is seen at 465 nm. These absorbant and fluorescent properties of the nicotinamide nucleotides have been employed extensively in developing assays for dehydrogenase reactions and have been utilized to monitor the oxidation–reduction state of a tissue or a preparation of intact mitochondria. With an appropriate spectrophotometer (e.g., a dual wavelength), capable of measuring small absorbancy changes in turbid cell or mitochondrial suspensions, the relative changes in the oxidized/reduced nicotinamide nucleotides may be determined as a function of the metabolic condition of the cell or subcellular suspension (e.g., changes in substrate, oxygen concentration, or on drug or hormone additions). This type of spectrophotometric technique and more sophisticated techniques, in which a light guide is used to direct a beam of excitation light to the surface of an intact organ or tissue, and another light guide is employed to pick up the reflected fluorescence emission at a longer wavelength, have been valuable tools in understanding the very complicated relationships that exist between the mitochondrial respiratory chain and the metabolic characteristics of various tissues.

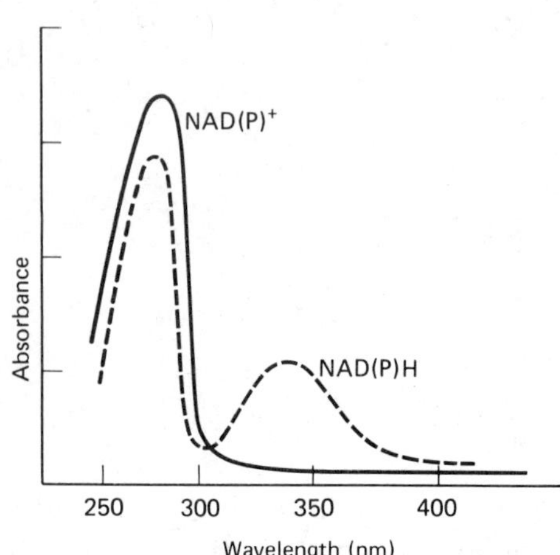

Figure 6.39
Absorbance properties of NAD^+ and NADH.

Another effective method for monitoring the oxidation–reduction state of the cytosolic or the mitochondrial compartments is to measure the oxidized and reduced members of various redox couples in tissue extracts, in the bathing solution of a tissue or in the effluent perfusate of an isolated, perfused organ. Because lactate dehydrogenase is exclusively a cytosolic enzyme the pyruvate/lactate ratio in the tissue or organ perfusate should accurately reflect the cytosolic NAD^+/NADH ratio under a variety of metabolic conditions. In a like manner the β-hydroxybutyrate dehydrogenase is exclusively mitochondrial, and hence the acetoacetate/β-hydroxybutyrate ratio should reflect the oxidation–reduction state of the mitochondrial NAD^+/NADH system. If the acetoacetate/β-hydroxybutyrate ratio and the equilibrium constant for β-hydroxybutyrate dehydrogenase are known, the NAD^+/NADH ratio under any condition can be calculated:

$$\text{Acetoacetate} + \text{NADH} + \text{H}^+ \rightleftharpoons \beta\text{-hydroxybutyrate} + \text{NAD}^+$$

$$K_{eq} = \frac{[\beta\text{-hydroxybutyrate}][\text{NAD}^+]}{[\text{acetoacetate}][\text{NADH}][\text{H}^+]}$$

Flavin-Linked Dehydrogenases

The second type of oxidation–reduction reaction essential to a discussion of mitochondrial electron transport employs a flavin (e.g., derived from riboflavin 5′-phosphate) as the electron acceptor in the reaction. These reactions are catalyzed by a group of flavin-linked dehydrogenases. The two flavins commonly utilized in oxidation–reduction reactions are FAD (flavin adenine dinucleotide) and FMN (flavin mononucleotide) (Figure 6.40).

Among the flavin-containing enzymes that have been described, five play an essential role in energy metabolism in mammalian mitochondria

TABLE 6.8 Various Flavin-Linked Dehydrogenases

Enzyme	*Function*	*Flavin Nucleotide*
Succinate dehydrogenase	Tricarboxylic acid cycle	FAD
Dihydrolipoyl dehydrogenase	Component in pyruvate and α-ketoglutarate dehydrogenase complexes	FAD
NADH dehydrogenase	Electron transport chain	FMN
Electron-transferring flavoprotein	Electron transport chain	FAD
Acyl CoA dehydrogenase	Fatty acid β-oxidation	FAD
D-Amino acid oxidase	Amino acid oxidation	FAD
Monoamine oxidase	Oxidation of monoamines	FAD

(Table 6.8). In the discussion of the pyruvate and α-ketoglutarate dehydrogenase multienzyme complexes, the final reaction catalyzed by this complex involved the flavoprotein enzyme, dihydrolipoyl dehydrogenase, which accepts electrons via a bound FAD moiety from reduced lipoamide groups on the transacylase subunit and transfers these reducing equivalents to NAD^+. Also, in the tricarboxylic acid cycle, succinate dehydrogenase is a flavin-linked enzyme, which oxidizes succinate to fumarate and converts FAD to $FADH_2$. The first dehydrogenation reaction in β-oxidation of fatty acids is catalyzed by acyl CoA dehydrogenase, another flavin-linked enzyme. Finally, oxidation of NADH in the mitochondrial respiratory chain is catalyzed by a FMN-containing enzyme, the NADH dehydrogenase, and the reducing equivalents are then transferred to another flavoprotein called the electron-transferring flavoprotein.

The flavins FAD and FMN either may be bound very tightly with noncovalent bonding (i.e., with dissociation constants in the range of 10^{-10} M) to their respective enzymes, as is the case with the NADH dehydrogenase, or they may be bound covalently to the enzyme (i.e., to a histidine residue), as is the case with succinate dehydrogenase. Flavoproteins may be classified into two groups: (1) the dehydrogenases in which the reduced flavin is reoxidized by electron carriers other than oxygen (e.g., coenzyme Q, other flavins, or chemicals such as ferricyanide, methylene blue, or phenazine methosulfate), and (2) the oxidases in which the flavin may be reoxidized using as the electron acceptor molecular oxygen, O_2, yielding hydrogen peroxide, H_2O_2, as the product. The H_2O_2 may then be broken down to water and oxygen by the enzyme catalase,

$$2H_2O_2 \xrightarrow{\text{catalase}} 2H_2O + O_2$$

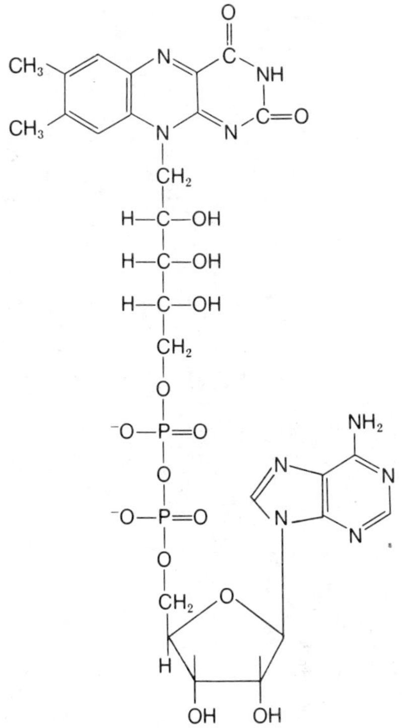

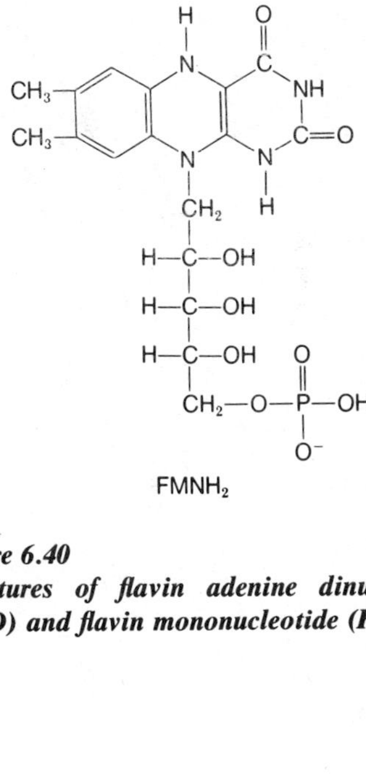

Figure 6.40
Structures of flavin adenine dinucleotide (FAD) and flavin mononucleotide (FMN).

Iron–Sulfur Proteins

A number of flavin-linked enzymes have nonheme iron (i.e., an iron-sulfur center) involved in the catalytic mechanism. In these enzymes the iron is converted from the oxidized (Fe^{3+}) form to the reduced (Fe^{2+}) form during the transfer of reducing equivalents on and off the flavin moiety. Both succinate dehydrogenase and NADH dehydrogenase contain iron-sulfur centers. The iron component of the iron-sulfur center is bound in various arrangements to cysteine residues in the protein and to acid-labile sulfur, for example, $Fe_4S_4Cys_4$; $Fe_2S_2Cys_4$; $Fe_1S_0Cys_4$. Iron-sulfur proteins are found in abundance in all species from the simplest microorganism to the mammal. Certain flavin-linked enzymes (xanthine oxidase) have one or two molybdenum atoms associated with their catalytic mechanism. The tightly bound molybdenum undergoes a valence change during the transfer of electrons $Mo^{+6} \longrightarrow Mo^{+5}$.

Cytochromes

Organisms that require oxygen (i.e., aerobic organisms) in their energy-generating functions possess various cytochromes that are involved in electron-transferring systems. Cytochromes are a class of proteins characterized by the presence of an iron-containing heme group covalently bound to the protein. Unlike the heme group in hemoglobin or myoglobin in which the heme iron remains in the Fe^{2+} state, the iron in the heme of a cytochrome alternately is oxidized (Fe^{3+}) or reduced (Fe^{2+}) as the cytochrome transfers electrons toward oxygen in the electron transport chain.

The cytochromes of mammalian mitochondria were designated as a, b, and c on the basis of the α band of their absorption spectrum and the type of heme group (see Figure 6.41). Cytochrome c is a small protein (104 amino acid residues) with mol wt = 13,000. Amino acid sequences of cytochrome c from a great many species have been described and show

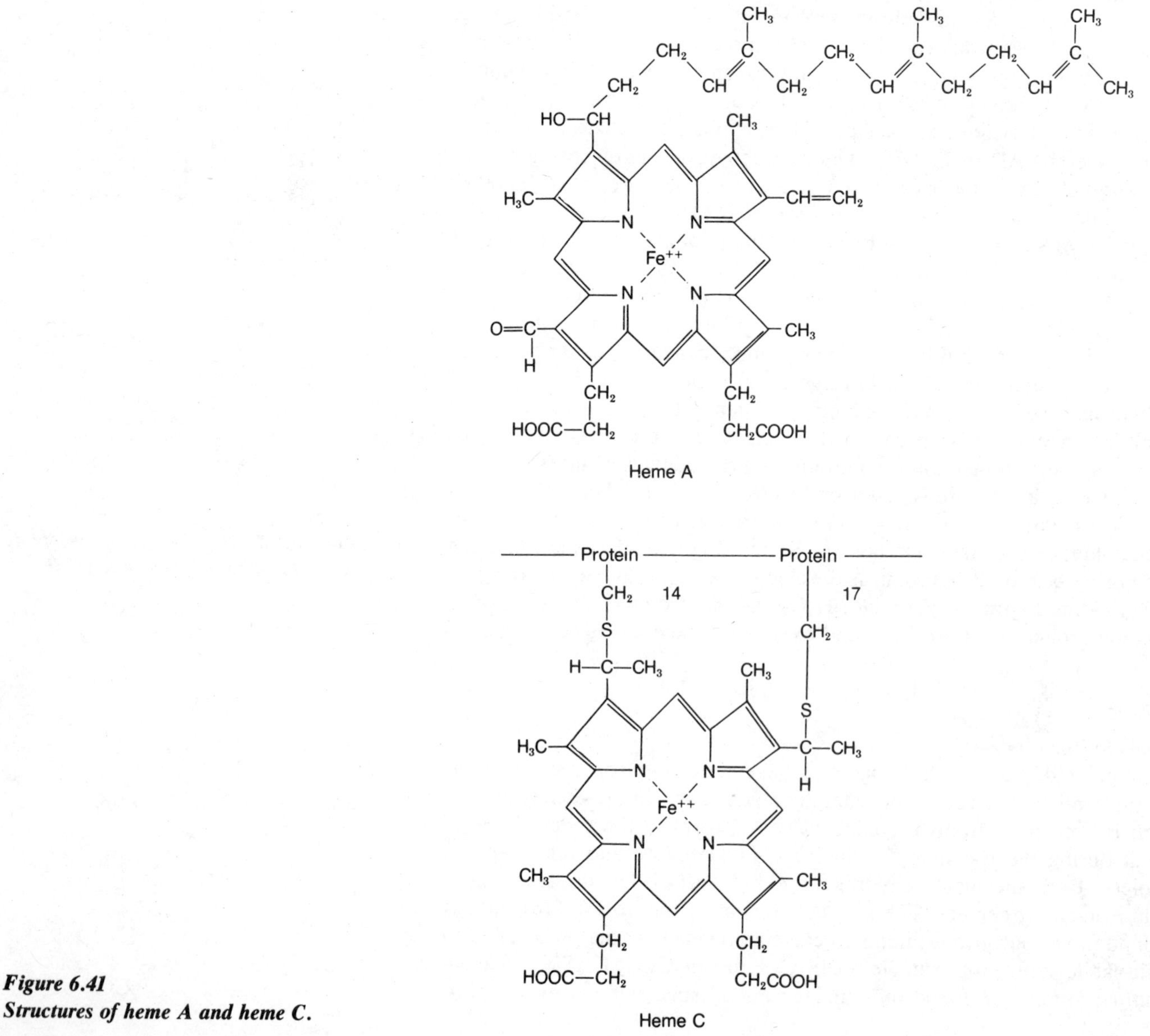

Heme A

Heme C

Figure 6.41
Structures of heme A and heme C.

that 20 out of 104 amino acid residues are invariant. The iron of the heme group in cytochrome c is coordinated between the four nitrogens of the tetrapyrrole structure of the porphyrin group, whereas the fifth and sixth coordination positions are occupied by the methionine residue at position 80 and histidine residue at position 18 of the protein (Figure 6.42). The fact that all six coordination positions are filled in most of the cytochromes prohibits oxygen from binding directly to the iron and prevents such respiratory inhibitors as cyanide, azide, and carbon monoxide from binding to most cytochromes. The notable exception is cytochrome a_3, which is involved in the terminal step in the mitochondrial electron transport chain. The heme group in cytochrome c is attached to the protein, not only by the fifth and sixth coordination positions of the heme iron, but also by the vinyl side chains of the protoporphyrin IX structure, from which both heme a and c are derived. These vinyl side chains are reduced by the addition of H—S across the vinyl group and the resulting sulfhydryl linkages attach the heme group to cysteine residues at positions 14 and 17 in cytochrome c. Hence the heme group is covalently linked to the protein as well as being coordinated through the Fe^{2+} group in the heme.

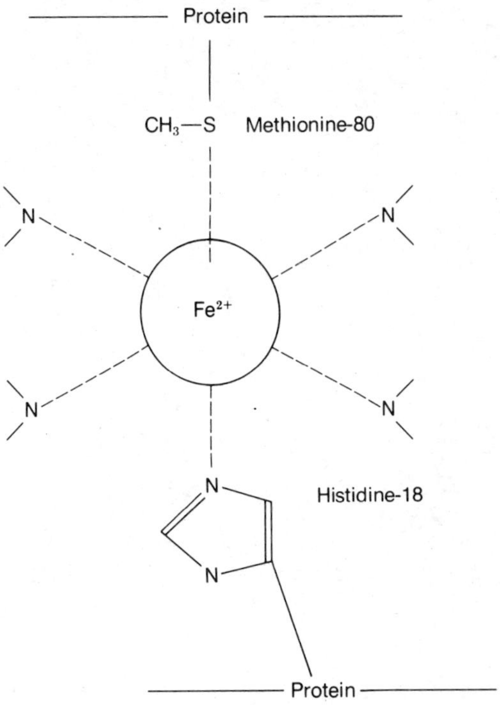

Protein

CH_3—S Methionine-80

N N Fe^{2+} N N

N Histidine-18 N

Protein

Figure 6.42
The six coordination positions of cytochrome c.

Coenzyme Q

Coenzyme Q, also called ubiquinone, is neither a nucleotide species nor a protein but a lipophilic electron carrier. Like the pyridine nucleotides and to a certain extent cytochrome c, coenzyme Q serves as a "mobile" electron transport component which operates between the various flavin-linked dehydrogenases, for example, NADH dehydrogenase, succinate dehydrogenase, or fatty acyl CoA dehydrogenase, and cytochrome b of the electron transport chain. As shown in Figure 6.43, the quinone portion of the coenzyme Q molecule is alternately oxidized and reduced by the addition of 2 reducing equivalents, for example, 2 protons (H^+) and 2 electrons (e^-). The number (n) of isoprene units in the side chain of coenzyme Q varies between 6 and 10, depending upon the source of the coenzyme Q. The side chain renders the coenzyme Q lipid-soluble and facilitates the accessibility of this electron carrier to the lipophilic portions of the inner mitochondrial membrane, where the enzymatic aspects of the mitochondrial electron transfer chain are localized.

Mitochondrial Electron Transport Chain

The various electron-transferring proteins and other carriers that comprise the mitochondrial electron transfer chain are arranged in a sequential pattern in the inner mitochondrial membrane. Reducing equivalents are extracted from substrates in the tricarboxylic acid cycle, the fatty acid β-oxidation sequence, and indirectly from glycolysis, and passed sequentially through the electron transport chain to molecular oxygen. The current arrangement of the mitochondrial electron transport carriers is illustrated in Figure 6.44. Electrons or reducing equivalents are fed into the electron transport chain at the level of NADH or coenzyme Q from the primary NAD^+- and FAD-linked dehydrogenase reactions and are transported to molecular oxygen through the cytochrome chain. This electron transport system is set up so that the reduced member of one redox couple is oxidized by the oxidized member of the next component in the system:

$$\text{NADH} + \text{H}^+ + \text{FMN} \rightleftharpoons \text{FMNH}_2 + \text{NAD}^+$$

or

$$\text{Cytochrome b } (\text{Fe}^{2+}) + \text{cytochrome } c_1\ (\text{Fe}^{3+}) \rightleftharpoons \text{cytochrome b } (\text{Fe}^{3+}) + \text{cytochrome } c_1\ (\text{Fe}^{2+})$$

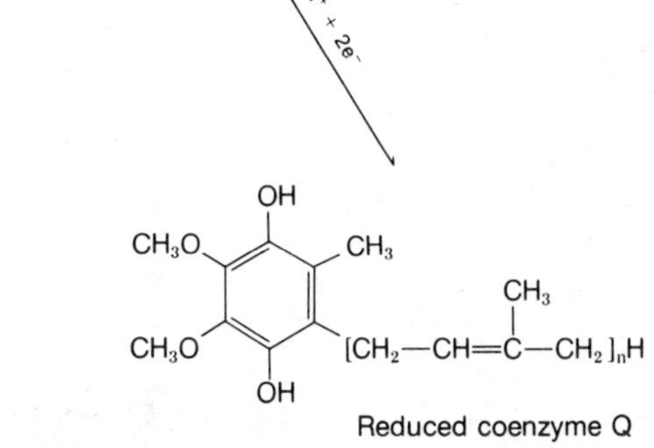

O

CH_3O CH_3

CH_3

CH_3O $[CH_2—CH{=}C—CH_2]_nH$

O

Oxidized coenzyme Q

$2H^+ + 2e^-$

OH

CH_3O CH_3

CH_3

CH_3O $[CH_2—CH{=}C—CH_2]_nH$

OH

Reduced coenzyme Q

Figure 6.43
The oxidation and reduction of coenzyme Q.

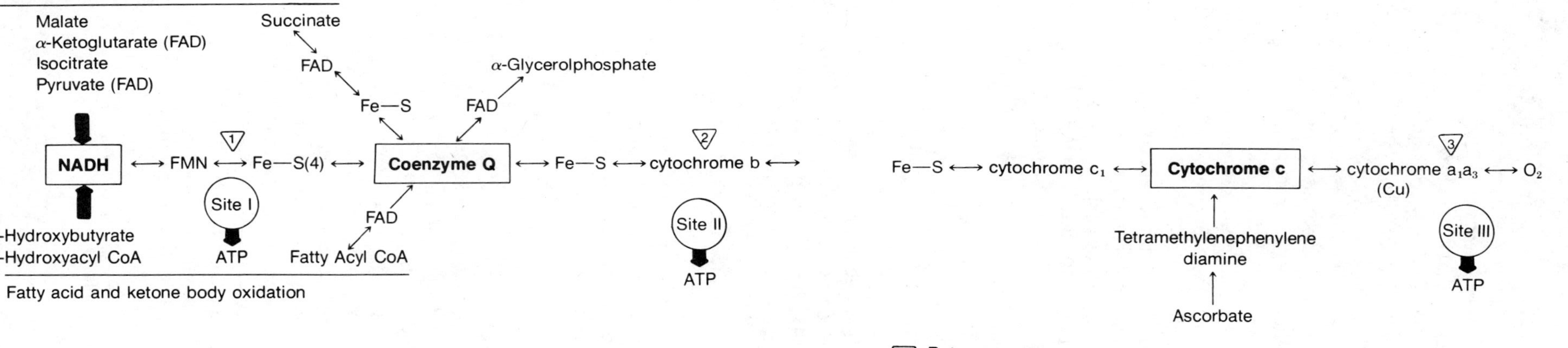

Figure 6.44
Mitochondrial electron transport chain.

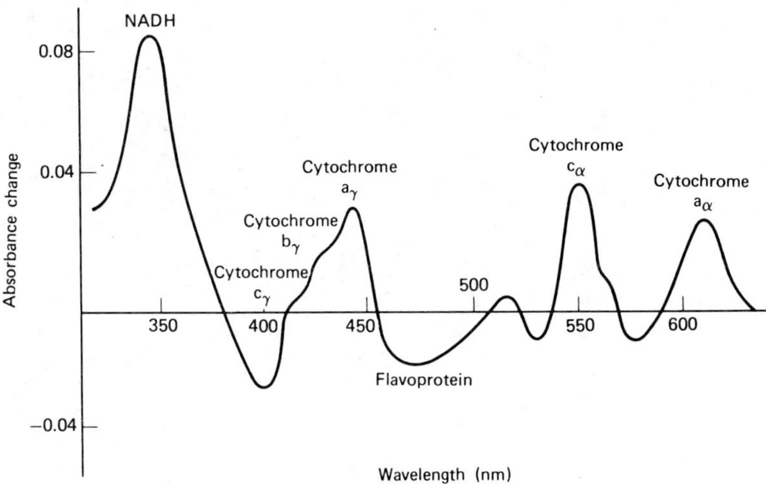

Figure 6.45
Difference spectra of liver mitochondrial suspensions (oxidized/reduced).

It should be noted that the electron transfer reactions from NADH through coenzyme Q transfer two electrons, whereas the reactions between coenzyme Q and oxygen involving the various cytochromes are one-electron transfer reactions.

The various components of the respiratory chain have characteristic absorption spectra which can be visualized in suspensions of isolated mitochondria or submitochondrial particles using a dual beam spectrophotometer. The different absorption bands are shown in Figure 6.45. One of the light beams of the spectrophotometer was passed through a suspension of liver mitochondria, which was maintained under fully reduced conditions (e.g., substrate plus no oxygen), and the other beam was passed through an identical suspension in the presence of oxygen. Hence the resulting spectrum is a difference spectrum of the reduced minus the oxidized states of the mitochondrial respiratory chain.

During the transfer of electrons from the NADH/NAD$^+$ couple ($E_0{}' = -0.32$) to molecule oxygen ($E_0{}' = +0.82$) there occurs an oxidation–reduction potential decrease of 1.14 V. As shown in Figure 6.46, this drop in potential occurs in discrete steps as reducing equivalents or electrons are passed between the different segments of the chain. There is at least a 0.3 V decrease in potential between each of the three coupling or phosphorylation sites. A potential drop of 0.3 V is more than sufficient to

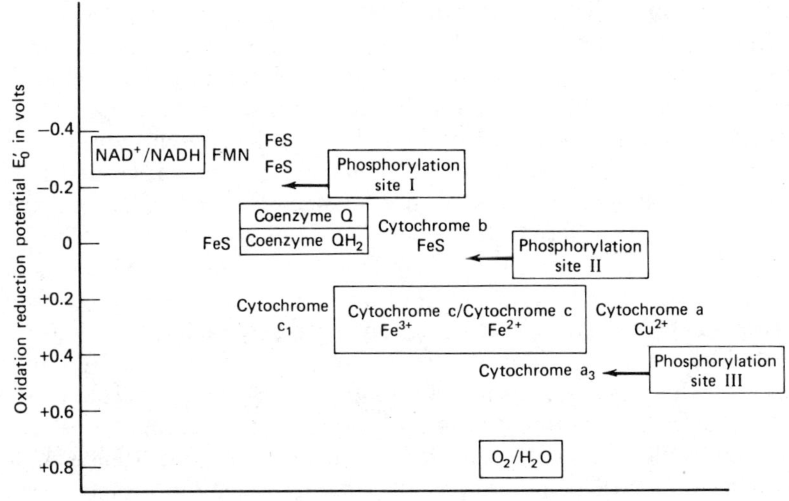

Figure 6.46
Oxidation–reduction potentials of the mitochondrial electron transport chain carriers.

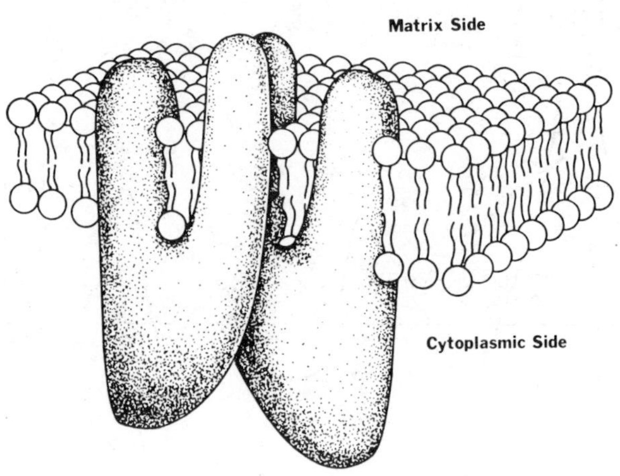

Figure 6.47
Model of cytochrome c oxidase dimer in the mitochondrial inner membrane.
Reprinted with permission from T. G. Frey, M. J. Costello, B. Karlsson, J. C. Haselgrove, and J. S. Leigh, *J. Mol. Biol. 162*:113, 1982.

CLIN. CORR. **6.3**
CYANIDE POISONING

Inhalation of hydrogen cyanide gas or ingestion of potassium cyanide causes a rapid and extensive inhibition of the mitochondrial electron transport chain at the cytochrome oxidase step. Cyanide is one of the most potent and rapidly acting poisons known. Cyanide binds to the Fe^{3+} in the heme of the cytochrome a,a_3 component of the terminal step in the electron transport chain and prevents oxygen from reacting with cytochrome a,a_3. Mitochondrial respiration and energy production cease, and cell death occurs rapidly. Death due to cyanide poisoning occurs from tissue asphyxia, most notably of the central nervous system. If cyanide poisoning is not lethal, an individual who has been exposed to cyanide is given various nitrites that convert oxyhemoglobin to methemoglobin, which merely involves converting the Fe^{2+} of hemoglobin to Fe^{3+} in methemoglobin. Methemoglobin (Fe^{3+}) competes with cytochrome a,a_3 (Fe^{3+}) for cyanide, forming a methemoglobin-cyanide complex. Administration of thiosulfate causes the cyanide to react with the enzyme rhodanese forming thiocyanate.

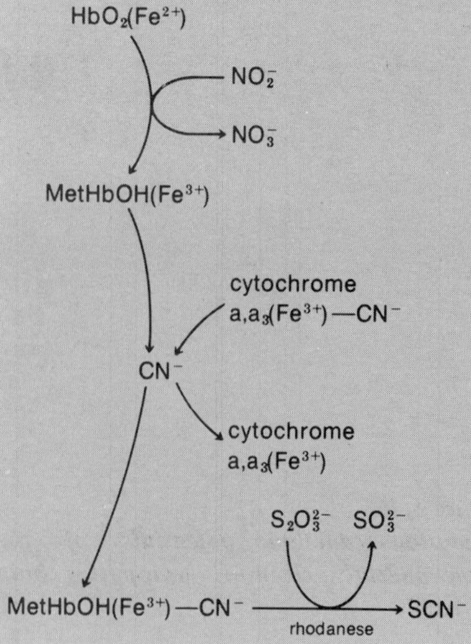

accommodate the synthesis of a high-energy phosphate bond such as occurs in ATP synthesis, for example,

$$\Delta E_0' = 0.3 \text{ V}$$
$$\Delta G° = -nF\,\Delta E_0'$$
$$\Delta G° = -2 \times 23.062 \times 0.3$$
$$\Delta G° = -13.8 \text{ kcal/mol}$$

Various components of the electron transport chain are located asymmetrically in the mitochondrial membrane. An example of this asymmetric localization is shown in Figure 6.47. Cytochrome c oxidase, which catalyzes the terminal step in the electron transfer chain, consists of a dimer that spans the membrane between the matrix and the intermembrane space or the cytosol. Cytochrome c binds to the oxidase from the cytosolic side of the membrane, whereas oxygen binds from the matrix side of the membrane during the electron transferring event.

Inhibitors of Electron Transfer

The illustration of the mitochondrial respiratory chain shown in Figure 6.45 indicates that a number of chemical compounds are capable of specifically inhibiting electron flow in the chain at different points. The fish poison rotenone (Figure 6.48) and the barbiturate amytal (Figure 6.48) inhibit the electron transfer chain at the level of the flavoprotein, NADH dehydrogenase. Hence electrons or reducing equivalents derived from NAD^+-linked dehydrogenases are not oxidized by a rotenone-inhibited respiratory chain, whereas those derived from flavin-linked dehydrogenases are freely oxidized. The antibiotic antimycin A (Figure 6.48) inhibits electron transfer at the level of cytochrome b, whereas the terminal step in the respiratory chain catalyzed by cytochrome oxidase is inhibited by cyanide, azide, or carbon monoxide. (See Clin. Corr. 6.3.) These later three compounds merely combine with the oxidized heme iron (Fe^{3+}) in cytochromes a and a_3 in order to prevent the reduction of this heme iron by electrons derived from reduced cytochrome c. Hence ingestion or injections of respiratory chain inhibitors leads to a blockage of electron transfer and impairment of the normal energy generating function

Rotenone

Amytal

Antimycin A

Figure 6.48
Structures of various respiratory chain inhibitors.

of the mitochondrial electron transport chain, and if the exposure to such an inhibitor is prolonged, death of the organism would result.

Reversal of Electron Transport

It should be pointed out that the various events in the mitochondrial electron transport system and the closely coupled reactions or processes in the oxidative phosphorylation sequence are reversible, provided an appropriate amount of energy is supplied to drive the system. In mitochondrial systems, reducing equivalents derived from succinate can be transferred to NADH with the concomitant hydrolysis of ATP (see Figure 6.49). Electron transport across the other two phosphorylation sites can be reversed in a similar fashion.

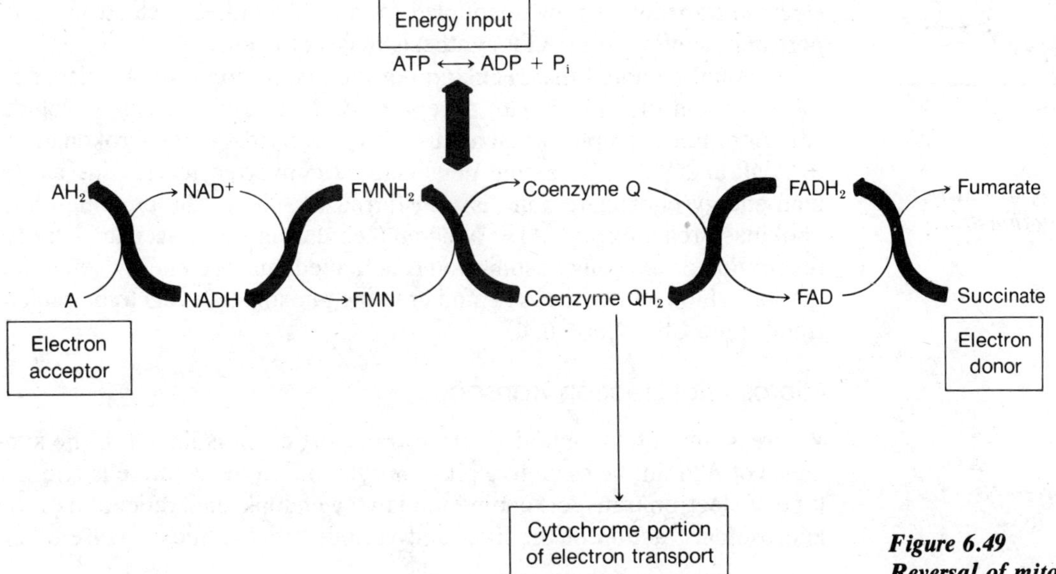

Figure 6.49
Reversal of mitochondrial electron transfer.

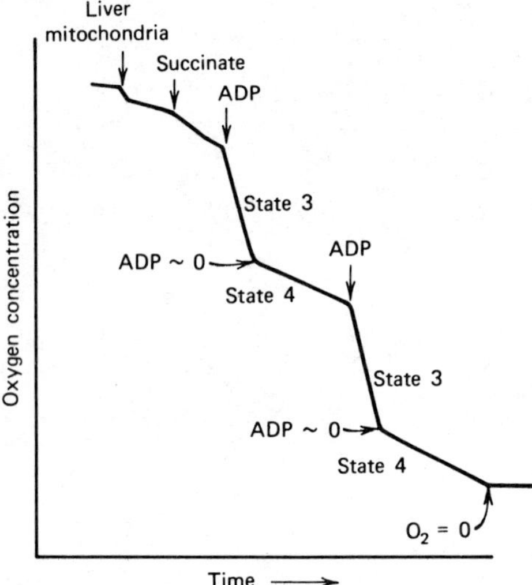

Figure 6.50
A demonstration of the coupling of electron transport to oxidative phosphorylation in a suspension of liver mitochondria.
State 3/state 4 = respiratory control ratio.

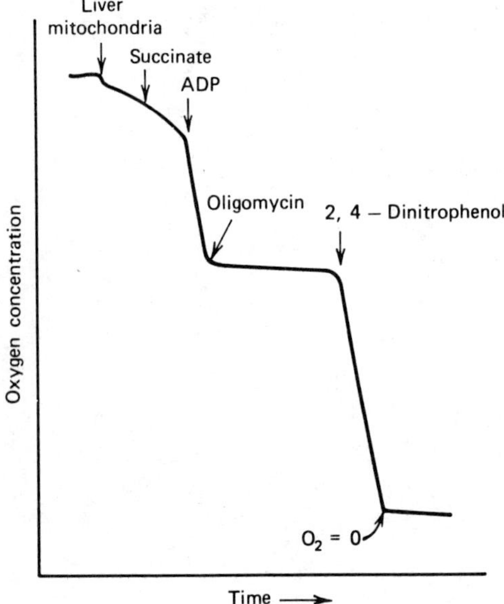

Figure 6.51
Inhibition and uncoupling of oxidative phosphorylation in liver mitochondria.

Coupling of Oxidative Phosphorylation to Electron Transport

The obligatory tight coupling between the electron-transferring reactions and the reactions in oxidative phosphorylation can best be illustrated in the experiment shown in Figure 6.50. Mitochondrial electron transport monitored by measuring the rate of oxygen consumption by a suspension of liver mitochondria can occur only at a rapid rate, following the addition of an oxidizable substrate (the electron donor) and ADP (a phosphate acceptor) plus inorganic phosphate. The ''active'' state in the presence of substrate and ADP has been designated state 3 and is a situation in which there occurs rapid electron transfer, oxygen consumption, and rapid synthesis of ATP. Following the conversion of all of the added ADP to ATP the rate of electron transfer subsides back to the rate observed prior to ADP addition. Hence respiration is tightly coupled to ATP synthesis and this relationship has been termed respiratory control or phosphate acceptor control. The ratio of the active (state 3) to the resting (state 4) rates of respiration is referred to as the respiratory control ratio and is a measure of the ''tightness'' of the coupling between electron transfer and oxidative phosphorylation. Damaged mitochondrial preparations and preparations to which various uncoupling compounds (see below) have been added exhibit low respiratory control ratios, indicating that the integrity of the mitochondrial membrane is required for tight coupling.

The effect of uncouplers and inhibitors of the electron transport–oxidative phosphorylation sequence is illustrated in Figure 6.51. Following the addition of ADP, which initiates a rapid state 3 rate of respiration, an inhibitor of the oxidative phosphorylation sequence (actually the mitochondrial ATPase), oligomycin, is added. Oligomycin stops ATP synthesis, and because the processes of electron transport and ATP synthesis are coupled tightly, respiration or electron transport is inhibited nearly completely. Following the inhibition of both oxygen consumption and ATP synthesis, the addition of an uncoupler of these two processes such as 2,4-dinitrophenol,

NO_2

O_2N— —OH

Dinitrophenol

causes a rapid initiation of oxygen consumption. Because respiration or electron transport is now uncoupled from ATP synthesis, electron transport may continue but ATP synthesis may not occur.

It should be noted that regulation of the respiration rate of a tissue by the provision of a phosphate acceptor, ADP, is a normal physiological situation. For example, when a muscle is exercised, ATP is broken down to ADP and P_i, and creatine phosphate is converted to creatine as the high-energy phosphate bond is transferred to ATP in the creatine phosphokinase reaction. As ADP accumulates during the muscular activity, respiration or oxygen consumption is activated, and the energy generated in this fashion allows the ATP and creatine phosphate levels to be replenished. (See Clin. Corr. 6.4.)

Microsomal Electron Transport

Whereas the mitochondrial electron transport chain is linked to the synthesis of ATP in the oxidative phosphorylation sequence there is another type of electron transport chain found in the endoplasmic reticulum or the microsomal fraction of the liver and various other tissues (Figure 6.52).

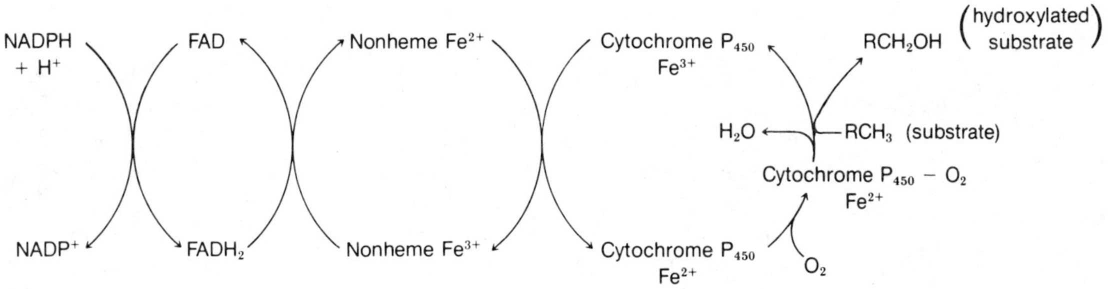

NADPH + H^+ → $NADP^+$; FAD → $FADH_2$; Nonheme Fe^{2+} → Nonheme Fe^{3+}; Cytochrome P_{450} Fe^{3+} → Cytochrome P_{450} Fe^{2+}; O_2 → Cytochrome P_{450} – O_2 Fe^{2+}; RCH_3 (substrate) → RCH_2OH (hydroxylated substrate); H_2O

Figure 6.52
Microsomal electron transport.

This electron transport chain exhibits several important differences which distinguish it from its mitochondrial counterpart. First, microsomal electron transport utilizes NADPH as the substrate or source of reducing equivalents for the initial flavin-linked reaction in this sequence, for example, the NADPH-cytochrome P_{450} reductase. Second, the microsomal system contains a cytochrome that is unique to this membranous system as it is not found in mitochondria. Cytochrome P_{450} was named as such because the reduced form of the cytochrome has an absorption band at 450 nm. During the terminal reaction of this electron transport sequence the Fe^{2+} of cytochrome P_{450} reacts with molecular oxygen to form a cytochrome P_{450} Fe^{2+}–O_2 complex. Thereafter, one of the oxygen atoms is incorporated into the substrate in a hydroxylation reaction, while the other oxygen is reduced to water. This electron transport chain is nonphosphorylating, since ATP is not synthesized during the transfer of electrons. The primary purpose of this system is in the hydroxylation of various drugs (e.g., phenobarbital), steroids or sterols, fatty acids, polycyclic hydrocarbons, and some amino acids. In fact the synthesis of components in the microsomal electron transport sequence is induced in the liver upon administration of certain drugs or other substrates of this system.

6.7 OXIDATIVE PHOSPHORYLATION

One of the most vexing problems that has confronted biochemists during the past three decades is the delineation of the mechanism of oxidative phosphorylation. Despite the countless man-years of experimental consideration that have been expended on this problem, a precise description of the mechanism by which energy derived from the passage of electrons sequentially along the electron transport chain is transduced into the chemical energy involved in the phosphoanhydride bonds of ATP is not available. Many hypotheses for the mechanism of oxidative phosphorylation have been tested, and three general theories have emerged as reasonable proposals, and one of these theories is now widely accepted.

Chemical Coupling Hypothesis

The chemical-coupling hypothesis for oxidative phosphorylation was developed in the early 1950s. This mechanism was based upon an analogy with the mechanism for substrate level phosphorylation observed in the glyceraldehyde 3-phosphate dehydrogenase reaction of the glycolytic pathway. In this reaction glyceraldehyde 3-phosphate is oxidized and a high-energy phosphoric–carboxylic acid anhydride bond is generated in

CLIN. CORR. 6.4 HYPOXIC INJURY

Acute hypoxic tissue injury has been studied in a variety of human tissues. The occlusion of one of the major coronary arteries during a myocardial infarction produces a large array of biochemical and physiological sequelae. When a tissue is deprived of its oxygen supply, the mitochondrial electron transport–oxidative phosphorylation sequence is inhibited, resulting in the decline of cellular levels of ATP and creatine phosphate. As cellular ATP levels diminish, anaerobic glycolysis is activated in an attempt to maintain normal cellular functions. Glycogen levels are rapidly depleted and lactic acid levels in the cytosol increase, reducing the intracellular pH. Hypoxic cells in such an energetic deficit begin to swell as they can no longer maintain their normal intracellular ionic environments. Mitochondria swell and begin to accumulate calcium, which may be deposited in the matrix compartment as calcium phosphate. The cell membranes of swollen cells become more permeable, leading to the leakage of various soluble enzymes, coenzymes, and other cell constituents from the cell. As the intracellular pH falls, damage occurs to lysosomal membranes, which release various hydrolytic proteases, lipases, glucosidases, and phosphatases into the cell. Such lysosomal enzymes begin an autolytic digestion of cellular components.

Cells that have been exposed to short periods of hypoxia can recover, without irreversible damage, upon reperfusion with an oxygen-containing medium. The exact point at which hypoxic cell damage becomes irreversible is not precisely known.

the product of the reaction, 1,3-diphosphoglycerate. An enzyme-bound high-energy intermediate is generated in this reaction, which is utilized to form the intermediate high-energy compound 1,3-diphosphoglycerate and ultimately to form ATP in the next reaction in the glycolytic pathway, that of phosphoglycerate kinase (see Chapter 7). Another example of a substrate level phosphorylation reaction, which was defined in the 1960s, is the succinyl CoA synthetase reaction of the tricarboxylic acid cycle. In this reaction the high-energy character of succinyl CoA is converted to the phosphoric acid anhydride bond in GTP with the intermediate participation of a high-energy, phosphorylated histidine moiety on the enzyme. Originally it was thought (incorrectly) that this phosphohistidine was a high-energy intermediate in the oxidative phosphorylation sequence. Because of these types of substrate level phosphorylation reactions it was proposed that the mechanism of mitochondrial energy transduction involved a series of high-energy intermediates that were generated in the mitochondrial membrane as a consequence of electron transport:

$$AH_2 + B + I \longrightarrow A \sim I + BH_2$$

$$A \sim I + P_i \longrightarrow I \sim P_i + A$$

$$I \sim P_i + ADP \longrightarrow ATP + I$$

In this representation A and B are electron carriers, whereas I is a hypothetical ligand which participates in the formation initially of a high-energy compound with the respiratory carrier ($A \sim I$) and thereafter with inorganic phosphate to form a phosphorylated high-energy intermediate ($I \sim P_i$). The phosphorylated high-energy intermediate is then utilized to form ATP in the ATP synthetase reaction. Uncouplers of oxidative phosphorylation were proposed to act by hydrolyzing the nonphosphorylated high-energy intermediate prior to the incorporation of phosphate into the system. Oligomycin was suggested to inhibit the incorporation of phosphate into ATP. The strongest argument for this type of mechanism was its basic simplicity, while its primary detraction is the fact that none of the proposed high-energy intermediates have ever been defined or isolated. Hence it is believed that such intermediates may not actually exist.

Conformational Coupling Hypothesis

A second proposal for the mechanism of oxidative phosphorylation is the conformational-coupling hypothesis. This hypothesis has an analogy in the process of muscle contraction in which ATP hydrolysis is used to drive conformational changes in myosin head groups which result in the disruption of cross-bridges to the actin thin filaments. The conformation coupling hypothesis suggests that a consequence of electron transport in the inner mitochondrial membrane is the induction of a conformational change in a membrane protein. ATP is synthesized by a mechanism which allows the membrane protein in its high-energy conformation to revert to its low-energy or random state, with the resultant formation of ATP from ADP and P_i. Hence the high-energy state of the membrane protein is transduced into the bond energy of the γ-phosphate group of ATP (see Figure 6.53).

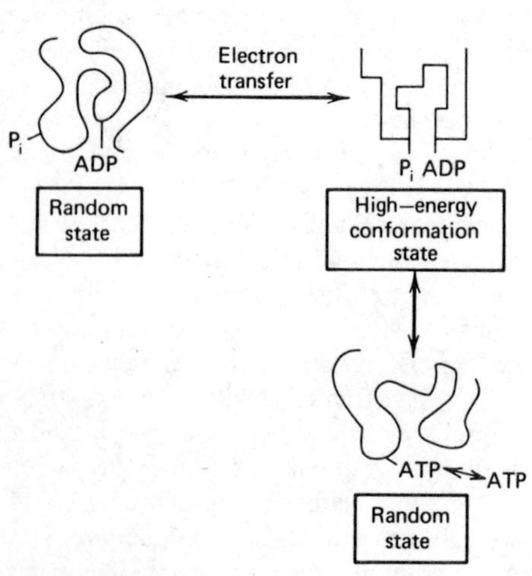

Figure 6.53
The conformational coupling hypothesis.

There are various experimental observations which indicate that mitochondrial membrane proteins undergo conformational state changes during the process of active electron transport. However, there is relatively little evidence demonstrating conclusively that such conformational changes are actually involved in the mechanism of ATP synthesis.

Chemiosmotic Coupling Hypothesis

Finally, the chemiosmotic-coupling hypothesis originally proposed by Peter Mitchell has gained widespread appreciation as a mechanism for energy transduction in mitochondria, as well as other biological systems. Mitchell's original proposition of the chemiosmotic theory of oxidative phosphorylation compared the energy-generating systems in biological membranes to a common storage battery. Just as energy may be stored in batteries because of the separation of positive and negative charges in the different components of the battery, energy may be generated as a consequence of the separation of charges in complex membranous systems. The chemiosmotic hypothesis (Figure 6.54) suggests that an electrochemical or proton gradient is established across the inner mitochondrial membrane during electron transport. This proton gradient is formed by pumping protons from the mitochondrial matrix side of the inner membrane to the cytosolic side of the membrane. Once there is a substantial electrochemical gradient established, the subsequent dissipation of the gradient is coupled to the synthesis of ATP by the mitochondrial ATPase. The chemiosmotic hypothesis requires that the electron transport carriers and the F_1-ATPase are localized in such a fashion in the inner mitochondrial membrane that protons are pumped out of the matrix compartment during the electron transport phase of the process, and protons are pumped or allowed back through the membrane during the ATP synthetase aspect of the process.

Uncouplers, which are usually relatively lipophilic weak acids, act to dissipate the proton gradient by transporting protons through the membrane from the intermembrane space to the matrix, essentially short-circuiting thc normal flow of protons through the ATP synthetic portion of the system. One of the strongest arguments supporting the chemiosmotic hypothesis is that ATPases can be purified, incorporated into artificial membrane vesicles, and are able to synthesize ATP when an electrochemical gradient is established across the membrane. In recent years a considerable experimental effort has been expended to purify the various components of the mitochondrial ATPase. It has been determined that proton-translocating ATPases are present and may be purified from a variety of mammalian tissues, bacteria, and yeast. The ATPase is a multi-

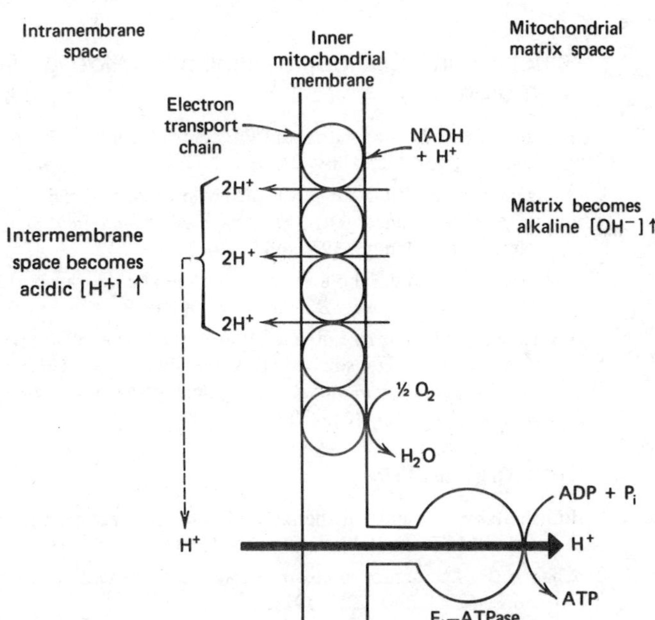

Figure 6.54
The chemiosmotic coupling hypothesis.

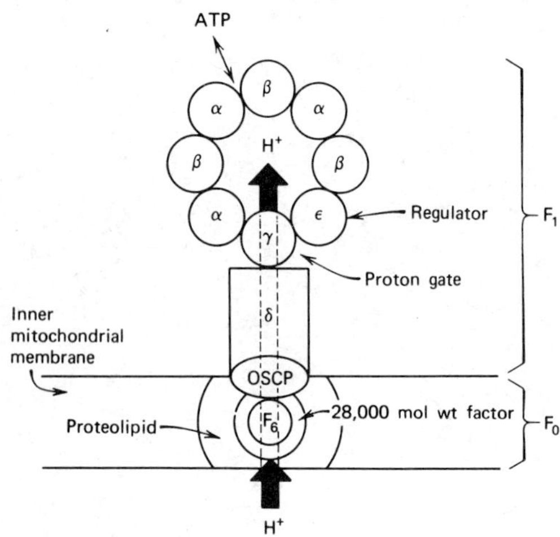

Figure 6.55
A model for the mitochondrial F_1 F_0-ATPase.

component complex with a suggested molecular weight of 480,000–500,000 (Figure 6.55). These ATPases can be incorporated into artificial membranes and can catalyze ATP synthesis. The ATPase complex consists of a water-soluble portion called F_1 and a hydrophobic portion called F_0. The F_1 consists of five nonidentical subunits (α, β, γ, δ, and ε) with a subunit stoichiometry of $\alpha_3\beta_3\gamma\delta\varepsilon$ and a molecular weight of 350,000–380,000. Nucleotide binding sites of the enzyme have been localized on the α and β subunits. The γ subunit has been proposed to function as a gate to the proton translocating activity of the complex, while the δ subunit has been suggested to be necessary for the attachment of the F_1 to the membrane. The ε subunit has been proposed to be involved in regulating the F_1-ATPase. The F_0 portion of the ATPase consists of three or four nonidentical subunits and is an integral part of the membrane from which the ATPase is derived. When the purified F_0 portion of the ATPase is incorporated into an artificial membrane, it renders the membrane permeable to protons. In addition the F_0 contains a subunit called the oligomycin-sensitivity-conferring protein which, as the name implies, causes the ATPase complex to show sensitivity to the inhibitory action of oligomycin.

While the chemiosmotic-coupling hypothesis for oxidative phosphorylation is widely accepted in principle as the mechanism for energy transduction in various biological systems, there are various questions that remain unanswered concerning the exact mechanism by which this important biochemical process occurs. For instance, what is the mechanism by which protons are pumped out of the mitochondrial matrix during electron transport? What is the stoichiometry of protons pumped per ATP synthesized? What is the mechanism by which protons are pumped back into the matrix "through" the F_1-ATPase? Is there a high-energy intermediate involved at some point in the ATP synthetic process?

BIBLIOGRAPHY

Energy-Producing and Energy-Utilizing Systems

Atkinson, D. E. *Cellular energy metabolism and its regulation.* New York: Academic, 1972.

Bock, R. M. Adenine nucleotides and properties of pyrophosphate compounds, in P. D. Boyer, H. Lardy, and K. Myrback (eds.), *The enzymes,* 2nd ed., vol. 2. New York: Academic, 1960, pp. 3–38.

Lipmann, F. Metabolic generation and utilization of phosphate bound energy, *Advan. Enzymol.* 1:99, 1941.

Sources and Fates of Acetyl CoA

Denton, R. M., and Halestrap, A. Regulation of pyruvate metabolism in mammalian tissues. *Essays Biochem.* 15:37, 1979.

Reed, L. J. Multienzyme complexes. *Acct. Chem. Res.* 7:40, 1974.

Reed, L. J., Pettit, F. H., and Yeaman, S. J. Pyruvate dehydrogenase complex: structure, function, and regulation, in P. A. Srere and R. W. Estabrook (eds.), *Microenvironments and metabolic compartmentation. New York: Academic, 1978, pp. 305–315.*

The Tricarboxylic Acid Cycle

Hansford, R. G. Control of mitochondrial substrate oxidation. *Curr. Topics Bioenerg.* 10:217, 1980.

Krebs, H. A. The history of the tricarboxylic acid cycle, *Perspect. Biol. Med.* 14:154, 1980.

Lowenstein, J. M. (ed.). *Citric acid cycle: control and compartmentation.* New York: Marcel Dekker, 1969.

Srere, P. M. The enzymology of the formation and breakdown of citrate. *Advan. Enzymol.* 43:57, 1975.

Structure and Compartmentation of Mitochondrial Membranes

Bygrave, F. L. Mitochondria and the control of intracellular calcium. *Biol. Rev.* 53:43, 1978.

Klingenberg, M. The ATP–ADP carrier in mitochondrial membranes, in A. N. Martinosi (ed.), *The enzymes of biological membranes.* New York: Plenum, 1976, pp. 383–438.

LaNoue, K. F., and Schoolwerth, A. C. Metabolite transport in mitochondria. *Annu. Rev. Biochem.* 48:871, 1979.

Williamson, J. R. The role of anion transport in the regulation of metabolism, in R. W. Hanson and M. A. Mehlman (eds.), *Gluconeogenesis: its regulation in mammalian species.* New York: Wiley-Interscience, 1976, pp. 165–220.

Electron Transfer

Baltsheffsky, H., and Baltsheffsky, M. Electron transport phosphorylation. *Annu. Rev. Biochem.* 43:871, 1974.

Chance, B. The nature of electron transfer and energy coupling reactions, *FEBS Lett.* 23:3, 1972.

Oxidative Phosphorylation

Boyer, P. D., Chance, B., Ernster, L., Mitchell, P., Racker, E., and Slater, E. C. Oxidative phosphorylation and photophosphorylation. *Annu. Rev. Biochem.* 46:955, 1977.

Fillingame, R. H. The proton-translocating pumps of oxidative phosphorylation. *Annu. Rev. Biochem.* 49:1079, 1980.

Kagawa, Y., Sone, N., Hirata, H., and Yooshida, M. Structure and function of H^+-ATPases. *J. Bioenerget. Biomemb.* 11:39, 1979.

Mitchell, P. Keilin's respiratory chain concept and its chemiosmotic consequences. *Science* 206:1148, 1979.

Racker, E. From Pasteur to Mitchell: a hundred years of bioenergetics. *Fed. Proc.* 39:210, 1980.

QUESTIONS

J. BAGGOTT AND C. N. ANGSTADT

*Q*uestion *T*ypes are described inside the front cover.

1. (QT1) At 37°C, −2.303RT = −1.42 kcal/mol. For the reaction A ⇌ B, if $\Delta G°$ = −7.1 kcal/mol, what is the equilibrium ratio of B/A?
   A. 10,000,000/1
   B. 100,000/1
   C. 1000/1
   D. 1/1000
   E. 1/100,000
2. (QT1) A bond may be "high-energy" for any of the following reasons *except:*
   A. products of its cleavage are more resonance stablized than the original compound.
   B. the bond is unusually stable, requiring a large energy input to cleave it.
   C. electrostatic repulsion is relieved when the bond is cleaved.
   D. a cleavage product may be unstable, tautomerizing to a more stable form.
   E. the bond may be strained.
3. (QT2) The active form of pyruvate dehydrogenase is favored by the influence of which of the following on pyruvate dehydrogenase kinase?
   1. Low $NADH/NAD^+$
   2. Low acetyl CoA/CoASH
   3. High [pyruvate]
   4. Low $[Ca^{2+}]$
4. (QT2) At which of the following enzyme-catalyzed steps of the tricarboxylic acid cycle do(es) net incorporation of the elements of water into an intermediate of the cycle occur?
   1. Citrate synthase
   2. Succinyl CoA synthase
   3. Fumarase
   4. Aconitase
5. (QT2) The freely reversible reactions of the tricarboxylic acid cycle include:
   1. the citrate synthase reaction.
   2. the isocitrate dehydrogenase reaction.
   3. the α-ketoglutarate dehydrogenase reaction.
   4. the succinyl CoA synthase reaction.
6. (QT2) Which of the following tricarboxylic acid cycle intermediates may be added or removed by other metabolic pathways?
   1. Oxalosuccinate
   2. α-ketoglutarate
   3. Isocitrate
   4. Succinyl CoA
7. (QT2) Regulation of tricarboxylic acid cycle activity *in vivo* may involve the concentration(s) of:
   1. acetyl CoA.
   2. ATP.
   3. ADP.
   4. oxygen.
8. (QT2) The mitochondrial membrane contains a transporter for:
   1. NADH.
   2. acetyl CoA.
   3. GTP.
   4. ATP.
9. (QT1) Which line of the accompanying table correctly describes the indicated properties of *both* the malate shuttle and the α-glycerophosphate shuttle?

Table for Question 9

Property	*Malate Shuttle*	*α-Glycerophosphate Shuttle*
A. Location	Inner mitochondrial membrane	Outer mitochondrial membrane
B. ATP generated per cytoplasmic NADH	3	2
C. Transporter	Malate dehydrogenase	α-Glycerophosphate dehydrogenase
D. Species transported	Malate	α-Glycerophosphate
E. Matrix electron acceptor	Oxaloacetate	Cytochrome b

A. NAD　　C. Both
B. NADP　　D. Neither

10. (QT4) Sterospecific hydrogen transfer.
11. (QT4) Reduced form is the usual source of reducing equivalents for anabolic processes.
12. (QT4) Irradiation with light of 300 nm wavelength causes the reduced form to emit fluorescence at 465 nm.
13. (QT4) Reduced form is an electron donor for microsomal electron transport but not for mitochondrial respiratory chain electron transport.
14. (QT1) If rotenone is added to the mitochondrial electron transport chain:
    A. the P/O ratio of NADH is reduced from 3/1 to 2/1.
    B. the rate of NADH oxidation is diminished to 2/3 of its initial value.
    C. succinate oxidation remains normal.
    D. oxidative phosphorylation is uncoupled at site 1.
    E. electron flow is inhibited at site II.
15. (QT1) If cyanide is added to tightly coupled mitochondria that are actively oxidizing succinate:
    A. subsequent addition of 2,4-dinitrophenol will cause ATP hydroylsis.
    B. subsequent addition of 2,4-dinitrophenol will restore succinate oxidation.
    C. electron flow will cease, but ATP synthesis will continue.

D. electron flow will cease, but ATP synthesis can be restored by subsequent addition of 2,4-dinitrophenol.
E. subsequent addition of 2,4-dinitrophenol *and* the phosphorylation inhibitor, oligomycin, will cause ATP hydrolysis.

16. (QT2) The heme iron of which of the following is bound to the protein by only one coordination linkage?
1. Cytochrome a
2. Cytochrome a_3
3. Cytochrome b
4. Cytochrome P_{450}

A. Substrate level phosphorylation
B. Oxidative phosphorylation
C. Both
D. Neither

17. (QT4) Occur(s) in mitochondria.
18. (QT4) High-energy intermediate compound has been found.
19. (QT4) ATP synthesis is linked to dissipation of a proton gradient.

ANSWERS

1. B $\Delta G^\circ = -2.3RT \log K$. $\log 100{,}000 = 5$. Substitution gives $\Delta G^\circ = -7.1$ (p. 216).
2. B A "high-energy" bond is so designated because it has a high free energy of hydrolysis. This could arise for reasons A, C, D, or E. High-energy does not refer to a high energy of formation (bond stability) (p. 217).
3. A 1, 2, and 3 true. NADH and acetyl CoA activate pyruvate dehydrogenase kinase, thus inactivating pyruvate dehydrogenase. Pyruvate inhibits the kinase, favoring the active dehydrogenase. High Ca^{2+} favors the active dehydrogenase but by activating the phosphatase (p. 226, Figure 6.14).
4. B 1 and 3 true. 1 and 3 clearly incorporate water, whereas 4 merely removes and then adds water (p. 228, Figure 6.17). In 2, a thioester bond is cleaved, just as it is in 1. The mechanism, however, involves a molecule of inorganic phosphate, not water; the phosphate is ultimately incorporated into GTP (p. 231).
5. D Only 4 true. 1 is irreversible due to cleavage of the thioester link, a high-energy bond. In 2 and 3, CO_2 is released. In 4 there are high-energy compounds on both sides of the reaction, namely GTP and succinyl CoA (p. 231).
6. C 2 and 4 true. 2 can be formed from glutamate, and 4 can be formed from methylmalonyl CoA (p. 231).
7. E 1, 2, 3, and 4 true. 1 is the substrate (p. 227). 2 inhibits isocitrate dehydrogenase, and 3 activates it (p. 230, p. 234, Figure 6.12). The cycle requires oxygen to oxidize NADH and ADP to be converted to ATP (respiratory control) (p. 234).
8. D Only 4 true. Reducing equivalents from NADH are shuttled across the membrane, as is the acetyl group of acetyl CoA, but NADH and acetyl CoA themselves cannot cross (p. 239, Figure 6.33 and 6.34). Of the nucleotides, only ATP and ADP are transported. The translocator is inhibited by atractyloside (p. 240).
9. B A: Both shuttles operate across the inner membrane. C: Two transporters are used by the malate shuttle, the malate α-ketoglutarate antiport and the aspartate-glutamate antiport. D: α-Glycerophosphate is not translocated; only reducing equivalents are. E: Oxaloacetate is a reaction product. NAD is the electron acceptor (p. 239, Figure 6.33).
10. C NAD- and NADP-linked dehydrogenases both exhibit specificity for the A or the B side of the pyridine ring (p. 245).
11. B NADPH is not a substrate for mitochondrial electron transport (p. 246).
12. D Fluorescence excitation of the reduced pyridine ring occurs in a wavelength range where it absorbs light, about 340 nm. 300 nm is an absorbance minimum (p. 246, Figure 6.39).
13. B Microsomal electron transport uses NADPH; mitochondrial respiratory chain electron transport uses only NADH (p. 255).
14. C Rotenone inhibits at the level of NADH dehydrogenase (site 1), preventing all electron flow and all ATP synthesis from NADH. Flavin-linked dehydrogenases feed in electrons below site 1 and are unaffected by site 1 inhibitors (p. 252, p. 250, Figure 6.44).
15. A Cyanide inhibits electron transport at site III, blocking electron flow throughout the system. In coupled mitochondria, ATP synthesis ceases too. Addition of an uncoupler permits the mitochondrial ATPase (which is normally driven in the synthetic direction) to operate, and it catalyzes the favorable ATP hydrolysis reaction unless it is inhibited by a phosphorylation inhibitor such as olgomycin (p. 254).
16. C 2 and 4 true. Fe^{2+} has six coordination positions. In heme, four are filled by the porphyrin ring. In cytochromes a and b, the other two are filled by the protein. But in cytochromes a_3 and P_{450}, one position must be left vacant to provide an oxygen-binding site (pp. 249, 255).
17. C Substrate level phosphorylation is catalyzed by succinate thiokinase (p. 231).
18. A Enzyme-bound high-energy intermediates of substrate level phosphorylation have been found (p. 255), but none has been found for oxidative phosphorylation (p. 256).
19. B Current thinking is that the energy required for ATP synthesis comes from a proton gradient rather than from a chemical intermediate or a high-energy conformational state (p. 257).

7

Carbohydrate Metabolism I: Major Metabolic Pathways and Their Control

ROBERT A. HARRIS

7.1 OVERVIEW

GLYCOGEN

Glycogenolysis ⬇ ⬆ Glycogenesis

GLUCOSE

Glycolysis ⬇ ⬆ Gluconeogenesis

LACTATE

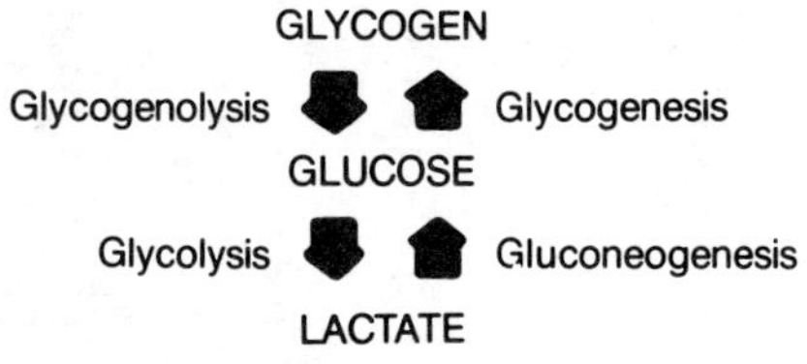

Figure 7.1
Relationship of glucose to the major pathways of carbohydrate metabolism.

The major pathways of carbohydrate metabolism either begin or end with glucose (Figure 7.1). This chapter will describe the utilization of glucose as a source of energy, the formation of glucose from noncarbohydrate precursors, the storage of glucose in the form of glycogen for later use, and the subsequent release of glucose from this storage form for use by cells. A thorough understanding of the pathways and the details of their metabolic regulation is necessary because of the important role played by glucose in the body. Glucose is the major form in which the carbohydrate coming from the intestinal tract is presented to the cells of the rest of the body. Glucose is the only fuel used to any significant extent by a few specialized cells, and it is the major fuel used by the brain. Indeed, glucose is so important to these specialized cells and the brain that several of the major tissues of the body work together to ensure a continuous supply of this essential substrate. Of importance to the practicing physician, glucose metabolism is defective in two very common metabolic diseases, obesity and diabetes, which in turn are contributing factors in the development of a number of major medical problems, including atherosclerosis, hypertension, small vessel disease, kidney disease, and blindness.

The discussion begins with glycolysis, a pathway that can be used by all cells of the body to extract part of the chemical energy inherent in the glucose molecule. This pathway also converts glucose to pyruvate and thus sets the stage for the complete oxidation of glucose to CO_2 and H_2O. The de novo synthesis of glucose, that is, gluconeogenesis, is considered next. It is a function of the liver and kidneys and can be conveniently discussed following glycolysis because gluconeogenesis will seem, without careful examination of the pathway, to be simply the reverse of the glycolytic pathway. In contrast to glycolysis, which produces ATP, gluconeogenesis requires ATP and is therefore an energy-requiring process. The consequence is that only some of the enzyme-catalyzed steps can be common to both the glycolytic and gluconeogenic pathways. Indeed, additional enzyme-catalyzed steps and even mitochondria become involved to make the overall process of gluconeogenesis exergonic. (Note on a confusing point: Gluconeogenesis is an energy-requiring process, that is, it requires ATP, but in order to occur the overall process has to be exergonic and is exergonic because of the ATP-driven steps.) Regulation of the rate-limiting and key enzyme-catalyzed steps will be stressed throughout the chapter. This will be particularly true for glycogen synthesis (glycogenesis) and glycogen degradation (glycogenolysis). Many cells store glycogen for the purpose of having glucose available for later use. The liver is less selfish, storing glycogen not for its own use, but rather for the maintenance of blood glucose levels to help ensure that other tissues of the body, especially the brain, have an adequate supply of this important substrate. Regulation of the synthesis and degradation of glycogen has been extensively studied and now serves as a model for our current understanding of how hormones work and how other metabolic pathways may be regulated. This subject will be emphasized because it contributes much to our understanding of the diabetic condition, starvation, and how tissues of the body respond to stress, severe trauma, and injury. See the Appendix, "Review of Organic Chemistry," for a discussion of the nomenclature and chemistry of the carbohydrates.

7.2 GLYCOLYSIS

Importance of the Glycolytic Pathway

The Embden–Meyerhof or glycolytic pathway represents an ancient process, possessed by all cells of the human body, in which anaerobic degradation of glucose to lactate occurs. This is one example of anaerobic fermentation, a term used to refer to pathways by which organisms extract chemical energy from high-energy fuels in the absence of molecular oxygen. Glycolysis represents an emergency energy-yielding pathway, capable of yielding 2 moles of ATP from a mole of glucose in the absence of molecular oxygen (Figure 7.2). This means that when the oxygen supply is shut off to a tissue, ATP levels can still be maintained for at least a short period of time by glycolysis. Many examples could be given, but the capacity to turn to glycolysis as a source of energy is particularly important to the human being at birth. With the exception of the brain, circulation of blood decreases to most parts of the body of the neonate during delivery. The brain is not normally deprived of oxygen during delivery, but other tissues must depend upon glycolysis for their supply of ATP until circulation returns to normal and oxygen becomes available once again. This conserves oxygen for use by the brain, illustrating one of many mechanisms that have evolved to assure survival of brain tissue in times of stress. Glycolysis also sets the stage for aerobic oxidation of carbohydrate in cells. Oxygen is not necessary for glycolysis, and the presence of oxygen can indirectly suppress glycolysis, a phenomenon called the Pasteur effect, that is considered in a later section. Nevertheless, glycolysis can and does occur in cells with an abundant supply of molecular oxygen. Provided that the cells also contain mitochondria, the end product of glycolysis in the presence of oxygen becomes pyruvate rather than lactate. Pyruvate can then be completely oxidized to CO_2 and H_2O by enzymes housed within the mitochondria. The overall process of glycolysis *plus* the subsequent mitochondrial processing of pyruvate to CO_2 and H_2O has the following equation:

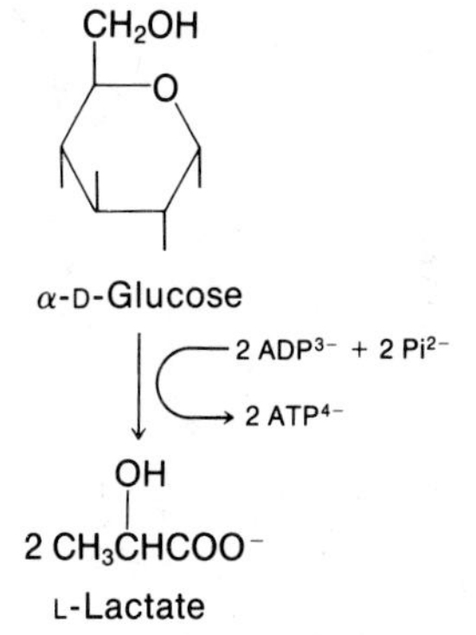

$$\text{D-Glucose } (C_6H_{12}O_6) + 6O_2 + 38ADP^{3-} + 38P_i^{2-} + 38H^+ \longrightarrow 6CO_2 + 6H_2O + 38ATP^{4-}$$

CH₂OH

α-D-Glucose

2 ADP³⁻ + 2 Pi²⁻

2 ATP⁴⁻

OH

2 CH₃CHCOO⁻

L-Lactate

Figure 7.2
Overall balanced equation for the sum of the reactions of the glycolytic pathway.

It should be noted that much more ATP is produced in the complete oxidation of glucose to CO_2 and H_2O than in the conversion of glucose to lactate. This has important consequences, which are considered in detail later. The important point at the moment is that, in order for glucose to be completely oxidized to CO_2 and H_2O, it must first be converted to pyruvate by glycolysis. This makes glycolysis a preparatory pathway for aerobic metabolism of glucose, as shown in Figure 7.3.

The importance of glycolysis as a preparatory pathway is best exemplified by the brain. This tissue has an absolute need for glucose and processes most of it via the glycolytic pathway. The pyruvate obtained is then oxidized completely to CO_2 and H_2O in brain mitochondria. Approximately 120 g of glucose is used by the adult human brain each day in order to meet its extraordinary need for ATP. The brain makes extensive use of glycolysis as a means of "preparing" the carbon of glucose for

D-Glucose $\xrightarrow{\text{glycolysis}}$ 2 pyruvate $\xrightarrow{\text{PDH}}$ 2 acetyl CoA

2 L-lactate　　$2CO_2$　　TCA → $4CO_2$

No O_2 requirement for glycolysis

O_2 requirement for pyruvate dehydrogenase (PDH) plus TCA cycle activity

Figure 7.3
Glycolysis is a preparatory pathway for aerobic metabolism of glucose.
TCA refers to the tricarboxylic acid cycle (see Figure 6.17).

complete oxidation. In contrast, glycolysis but with lactate as the end product, is the major mechanism of ATP production in a number of other tissues. Red blood cells (erythrocytes) lack mitochondria and therefore are unable to convert pyruvate to CO_2 and H_2O. The cornea, lens, and regions of the retina have a limited blood supply and also lack mitochondria (because mitochondria would absorb and scatter light) and likewise depend on glycolysis as the major mechanism for ATP production. Kidney medulla, testis, leukocytes, and white muscle fibers are almost totally dependent upon glycolysis as a source of ATP, again because these tissues have relatively few mitochondria. Combined, the tissues that are dependent primarily upon glycolysis for ATP production consume about 40 g of glucose per day in the normal human adult.

In any listing of the importance of glycolysis, it is impossible to ignore alcoholic fermentation. The overall balanced equation for the most common type is given in Figure 7.4. This pathway plays an important role in the making of "good brews," one of those things that make life worth living. This pathway is found in many yeast and certain bacteria but, somewhat surprisingly, the human body also produces a significant amount of ethanol. Most of the production, if not all, may be accounted for by microorganisms present in the intestinal tract. (There is a theory, however, that some of us have a better intestinal flora for the production of ethanol than others!) The pathway of alcoholic fermentation involves the same enzyme-catalyzed reactions as the glycolytic pathway, with the exception that lactate is not the end product. Rather than being reduced to lactate, pyruvate is decarboxylated to give acetaldehyde, which is reduced to ethanol to complete the pathway.

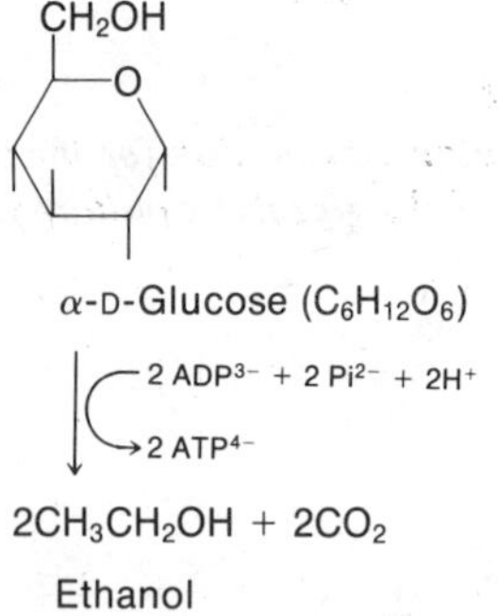

Figure 7.4
Overall balanced equation for ethanol production by alcoholic fermentation.

The major dietary sources of glucose are indicated in Chapter 24 in the discussion of the enzymes of digestion. Recall that starch is the storage form of glucose in plants and that it contains α-[1 → 4] glucosidic linkages along with α-[1 → 6] branches. Glycogen is the storage form of glucose in animal tissues and contains the same sort of glucosidic linkages and branches. It is important to distinguish between endogenous and exogenous sources of glucose. Exogenous refers to that which we eat and digest in the intestinal tract, whereas endogenous refers to that which is stored or synthesized in our tissues. Exogenous starch or glycogen is hydrolyzed within the lumen of the intestinal tract with the production of glucose, whereas stored glycogen endogenous to our tissues is converted to glucose or glucose 6-phosphate by enzymes present within the cells of these tissues. The disaccharides, which are important sources of glucose in our diet, include milk sugar (lactose) and grocery store sugar (sucrose). The hydrolysis of these sugars by enzymes of the brush border of the intestinal tract is discussed on page 929. Glucose can be used as a source of energy to satisfy the needs of the cells of the intestinal tract. However, these cells are not designed to depend upon glucose to any great extent, most of their energy requirement being met by glutamine catabolism. Most of the glucose passes through the cells of the intestinal tract into the blood, where it goes by way of the portal blood and the general circulation to be used by other cells of the body. The first major tissue to have an opportunity to remove it from the portal blood is the liver. When blood glucose is too high, the liver removes glucose from the blood by the glucose-consuming processes of glycogenesis and glycolysis. When blood glucose is too low, the liver supplies the blood with glucose by the glucose-producing processes of glycogenolysis and gluconeogenesis. The liver is also the first organ exposed to the blood flowing from the pancreas and therefore "sees" the highest concentrations of the hormones released from this endocrine tissue—glucagon and insulin. These are important hormonal regulators of blood glucose levels, in part because of their regulatory effects upon enzyme-catalyzed steps in the liver.

Overview of What Happens to Glucose in Various Cells

After penetrating the plasma membrane by mediated transport, glucose is metabolized mainly by glycolysis in red blood cells (Figure 7.5*A*). Since red blood cells lack mitochondria, the end product of glycolysis is lactate, which is released from the cells back into the blood plasma. Glucose used by the pentose phosphate pathway (see Chapter 8) in red blood cells provides NADPH, necessary in these cells primarily to keep glutathione in the reduced state. Reduced glutathione, in turn, plays an important role in the destruction of organic peroxides and hydrogen peroxide by the reaction catalyzed by glutathione peroxidase (Figure 7.6). NADPH is absolutely required for the reduction of oxidized glutathione (GSSG) back to reduced glutathione (GSH) by glutathione reductase. Peroxides cause irreversible oxidative damage to membranes, DNA, and numerous other cellular components and must, therefore, be destroyed.

The brain, like red blood cells, takes up glucose by mediated transport in an insulin-independent manner (Figure 7.5*B*). Glycolysis in the brain yields pyruvate, which is then oxidized completely to CO_2 and H_2O, as discussed above. The pentose phosphate pathway is also quite active in these cells, generating part of the NADPH needed for reductive synthesis and the maintenance of glutathione in the reduced state. Muscle and heart cells readily utilize glucose (Figure 7.5*C*), and transport of glucose into both is dependent upon the presence of insulin in the blood. Once taken up by these cells, glucose can be utilized by glycolysis to give pyruvate and lactate. Again pyruvate can be further utilized by the pyruvate dehydrogenase complex and the TCA cycle within the mitochondria to provide considerable energy in the form of ATP. Muscle and heart cells, in contrast to the other cells just considered, are capable of synthesizing significant quantities of glycogen. The synthesis and degradation of glycogen are extremely important processes in these cells. Adipose tissue also accumulates glucose by an insulin-dependent mechanism (Figure 7.5*D*). Pyruvate, as in other cells, is generated by glycolysis and can be oxidized by the pyruvate dehydrogenase complex to give acetyl CoA within adipocytes. However, instead of being completely oxidized to CO_2 and H_2O for the production of ATP, the acetate moiety of acetyl CoA is used primarily for de novo fatty acid synthesis in this tissue. Generation of NADPH by the pentose phosphate pathway is an important process in adipose tissue, considerable quantities of NADPH being necessary for the reductive steps of fatty acid synthesis. Adipose tissue also has the capacity for glycogenesis and glycogenolysis, but these processes are much more limited in this tissue than in muscle and heart. The final tissue to be men-

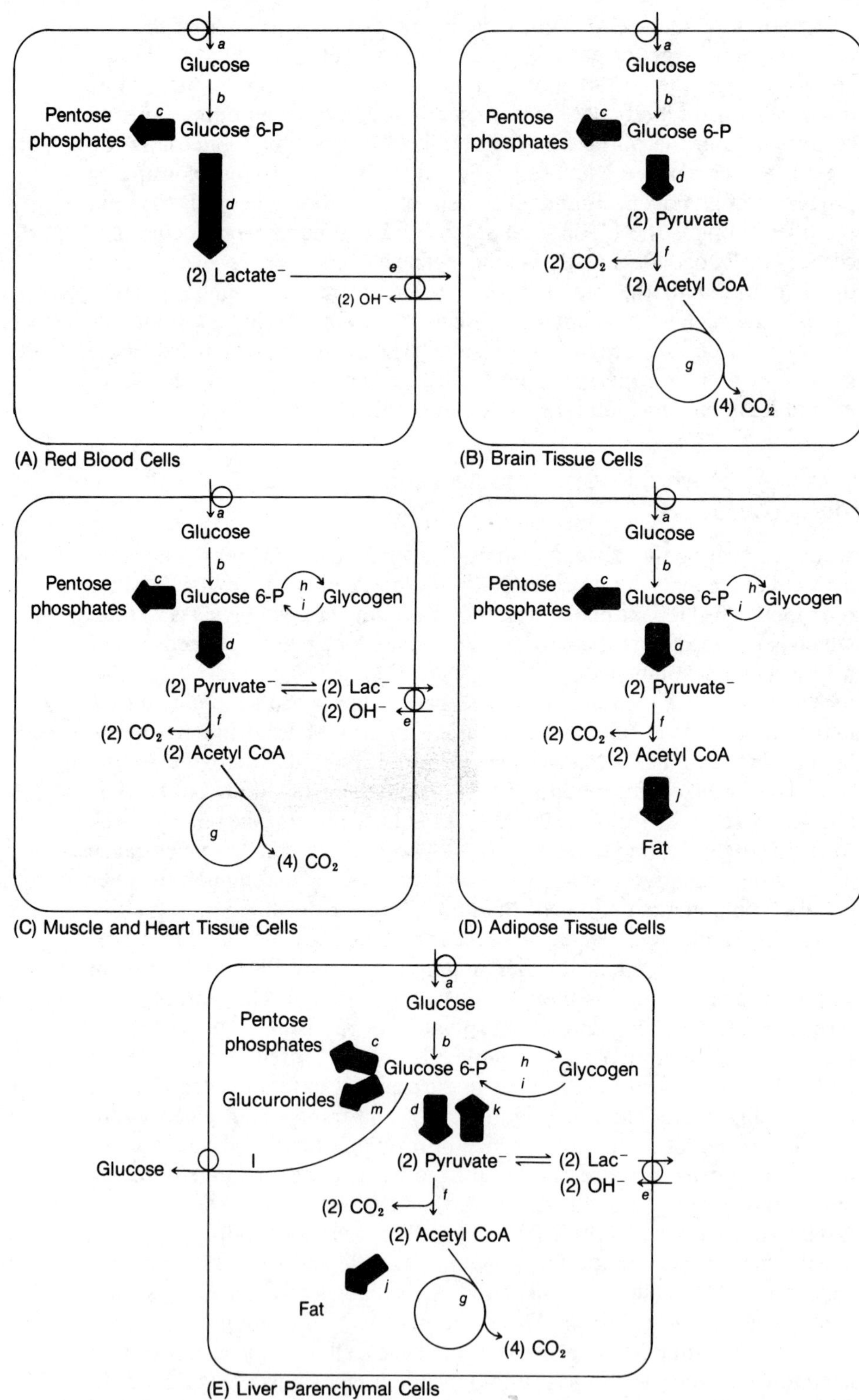

Figure 7.5
Overviews of the major ways in which glucose is metabolized within cells of selected tissues of the body.
a, glucose transport into the cell; b, glucose phosphorylation by hexokinase; c, the pentose phosphate pathway; d, glycolysis; e, lactate transport out of the cell in exchange for hydroxide ion; f, pyruvate decarboxylation by pyruvate dehydrogenase; g, TCA cycle; h, glycogenesis; i, glycogenolysis; j, lipogenesis; k, gluconeogenesis; l, hydrolysis of glucose 6-phosphate and release of glucose from the cell into the blood; m, formation of glucuronides (drug and bilirubin detoxification by conjugation) by the glucuronic acid pathway.

tioned is the liver, the cells of which are involved in the greatest number of ways with glucose metabolism (Figure 7.5*E*). Uptake of glucose by the liver occurs independent of insulin by means of a high-capacity transport system. Glucose is used rather extensively by the pentose phosphate pathway for the production of NADPH, which is needed for reductive synthesis, maintenance of glutathione in the reduced state, and numerous reactions catalyzed by microsomal enzyme systems (see Chapter 6). A quantitatively less important but nevertheless vital function of the pentose phosphate pathway is the provision of ribose phosphate, required for the synthesis of nucleotides such as ATP and those found in DNA and RNA. Glucose is also used for glycogen synthesis, making glycogen storage an important feature of the liver. Glucose can also be used in the glucuronic acid pathway, important in drug and bilirubin detoxification (see Chapter 21). The liver also has significant capacity for glycolysis, the pyruvate produced being used as a source of acetyl CoA for complete oxidation by the TCA cycle and for the synthesis of fat by the process of de novo fatty acid synthesis. In contrast to the other tissues discussed above, the liver is unique in that it also has the capacity to convert three-carbon precursors, such as lactate, pyruvate, and alanine, into glucose by the process of gluconeogenesis. The glucose produced can then be used to meet the need for glucose of other cells of the body.

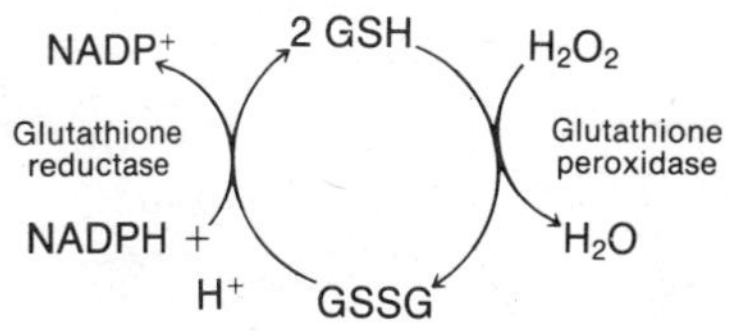

NADP$^+$ — 2 GSH — H_2O_2

Glutathione reductase — Glutathione peroxidase

NADPH + H^+ — GSSG — H_2O

Figure 7.6
Destruction of hydrogen peroxide is dependent upon reduction of glutathione by NADPH generated by the pentose phosphate pathway.

7.3 THE GLYCOLYTIC PATHWAY

Glucose is combustible and will burn in a test tube to yield heat and light but, of course, no ATP. Cells use some 30 steps to take glucose to CO_2 and H_2O, a seemingly inefficient process, since it can be done in a single step in a test tube. However, side reactions and some of the actual steps used by the cell to "burn" glucose to CO_2 and H_2O lead to the conservation of a significant amount of energy in the form of ATP. In other words, ATP is produced by the controlled "burning" of glucose in the cell, glycolysis representing only the first few steps, shown in Figure 7.7, in the overall process.

Glycolysis can be conveniently pictured as occurring in three major stages (also see Figure 7.7):

Priming stage

$$\text{D-Glucose} + 2\text{ATP}^{4-} \rightarrow \text{D-fructose 1,6-bisphosphate}^{4-} + 2\text{ADP}^{3-} + 2\text{H}^+$$

Splitting stage

$$\text{D-Fructose 1,6-bisphosphate}^{4-} \rightarrow 2\ \text{D-glyceraldehyde 3-phosphate}^{2-}$$

Oxidoreduction–phosphorylation stage

$$2\ \text{D-Glyceraldehyde 3-phosphate}^{2-} + 4\text{ADP}^{3-} + 2\text{P}_i^{2-} + 2\text{H}^+ \rightarrow 2\ \text{L-lactate}^- + 4\text{ATP}^{4-}$$

Sum

$$\text{D-Glucose} + 2\text{ADP}^{3-} + 2\text{P}_i^{2-} \rightarrow 2\ \text{L-lactate}^- + 2\text{ATP}^{4-}$$

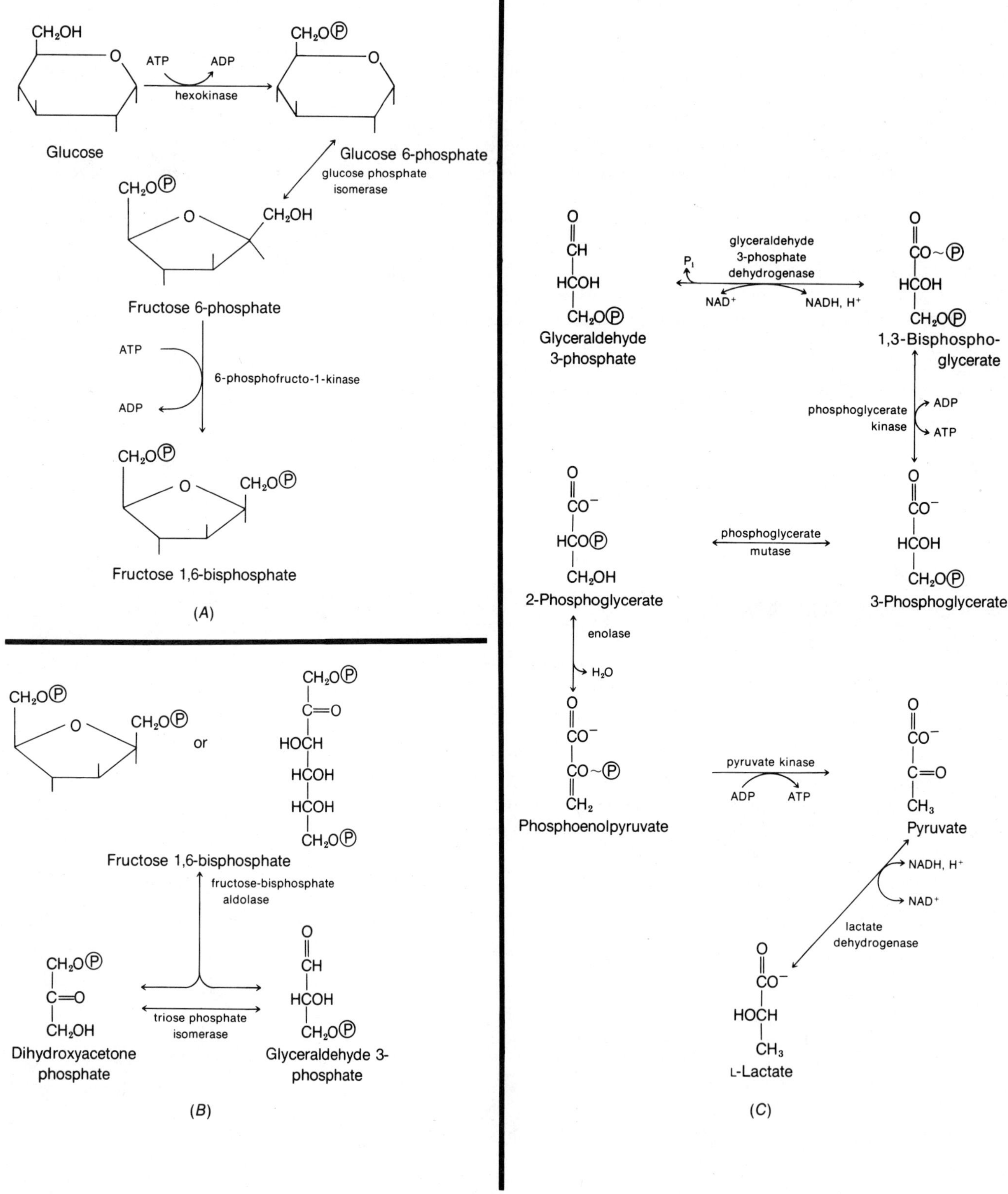

Figure 7.7
The glycolytic pathway, divided into its three stages.
The symbol Ⓟ refers to the phosphoryl group PO_3^{2-}; ~ indicates a high-energy phosphate bond. (A) Priming stage. (B) Splitting stage. (C) Oxidoreduction-phosphorylation stage.

The priming stage involves the input of two molecules of ATP with the conversion of glucose into a molecule of fructose 1,6-biphosphate. ATP is "invested" in the priming stage of glycolysis. However, ATP beyond that invested is gained from the subsequent completion of the glycolytic process. The splitting stage "splits" the six-carbon molecule fructose 1,6-bisphosphate into two molecules of glyceraldehyde 3-phosphate. In the oxidoreduction–phosphorylation stage two molecules of glyceraldehyde 3-phosphate are converted into two molecules of lactate with the production of four molecules of ATP. The sum reaction for the overall process of glycolysis comes to the generation of two molecules of lactate and two molecules of ATP at the expense of one molecule of glucose.

Priming Stage

Hexokinase catalyzes the first step of the glycolytic pathway (see Figure 7.7*A* and Step 1). Although this reaction consumes ATP, it gets glycolysis off to a good start by trapping glucose in the form of glucose 6-phosphate within the cytosol of the cell where all of the glycolytic enzymes are located. Phosphate esters of charged, hydrophilic compounds do not readily penetrate cell membranes. The phosphorylation of glucose with ATP is a thermodynamically favorable reaction, requiring the use of one high-energy phosphate bond. It is an irreversible reaction under the conditions that exist in cells and represents, therefore, a way to synthesize glucose 6-phosphate. However, it is *not,* by the reverse reaction, a way to synthesize ATP or to hydrolyze glucose 6-phosphate to give glucose. Hydrolysis of glucose 6-phosphate is accomplished by a completely different reaction, catalyzed by the enzyme glucose 6-phosphatase:

$$\text{Glucose 6-phosphate}^{2-} + H_2O \longrightarrow \text{glucose} + P_i^{2-}$$

This reaction is thermodynamically favorable in the direction written and cannot be used under conditions existing within biological cells for the synthesis of glucose 6-phosphate from glucose. (A common mistake is to notice that ATP and ADP are involved in the reaction catalyzed by hexokinase but not to notice that they are *not* involved in the reaction catalyzed by glucose 6-phosphatase.) Glucose 6-phosphatase is an important enzyme in liver, functioning to produce free glucose from glucose 6-phosphate in the last step of both gluconeogenesis and glycogenolysis, but it plays no role in the glycolytic pathway.

The next reaction is a readily reversible step of the glycolytic pathway, catalyzed by the enzyme phosphoglucoisomerase (Step 2). This step is not subject to regulation and, since it is readily reversible, functions in both glycolysis and gluconeogenesis.

6-Phosphofructo-1-kinase (also referred to as phosphofructokinase-1) catalyzes the next reaction of the glycolytic pathway, an ATP-dependent phosphorylation of fructose 6-phosphate to give fructose 1,6-bisphosphate (Step 3). This is the favorite enzyme of most students of biochemistry, being subject to regulation by a score of effectors and considered the rate-limiting enzyme of the glycolytic pathway. The reaction is irreversible under intracellular conditions, that is, it represents a way to produce fructose 1,6-bisphosphate but not a way to produce either ATP or fructose 6-phosphate by the reverse reaction. This reaction utilizes the second ATP needed to "prime" glucose, thereby completing the first stage of glycolysis.

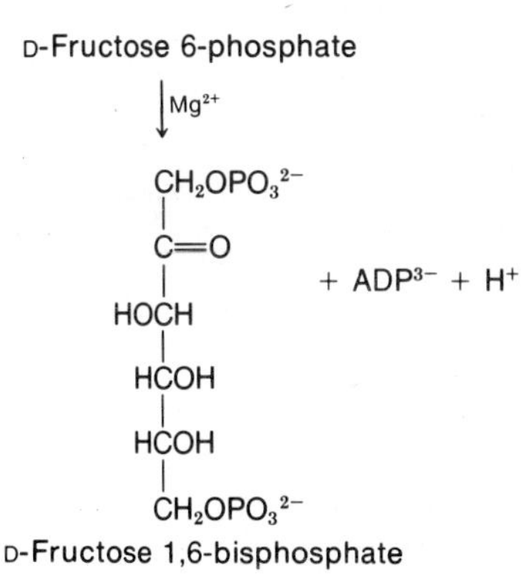

Splitting Stage

Fructose 1,6-bisphosphate aldolase catalyzes the next step of the glycolytic pathway (see Figure 7.7*B*), cleaving fructose 1,6-bisphosphate into a

CH_2OH (ring) $+ ATP^{4-} \xrightarrow{Mg^{2+}}$ $CH_2OPO_3^{2-}$ (ring) $+ ADP^{3-} + H^+$

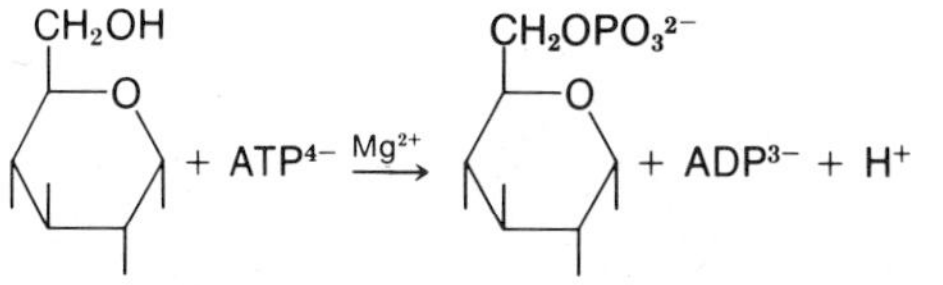

α-D-Glucose α-D-Glucose 6-phosphate

Step 1

$CH_2OPO_3^{2-}$ (ring) $\rightleftharpoons$ CH_2OH–C=O–HOCH–HCOH–HCOH–$CH_2OPO_3^{2-}$

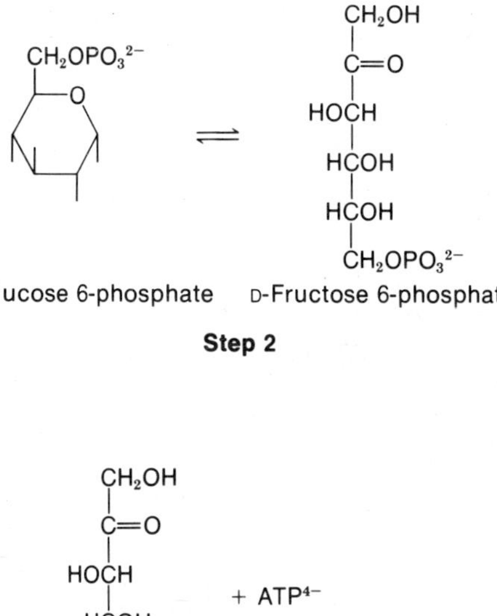

α-D-Glucose 6-phosphate D-Fructose 6-phosphate

Step 2

CH_2OH–C=O–HOCH–HCOH–HCOH–$CH_2OPO_3^{2-}$ $+ ATP^{4-}$

D-Fructose 6-phosphate

$\downarrow Mg^{2+}$

$CH_2OPO_3^{2-}$–C=O–HOCH–HCOH–HCOH–$CH_2OPO_3^{2-}$ $+ ADP^{3-} + H^+$

D-Fructose 1,6-bisphosphate

Step 3

$CH_2OPO_3^{2-}$ | C=O | HOCH | HCOH | HCOH | $CH_2OPO_3^{2-}$

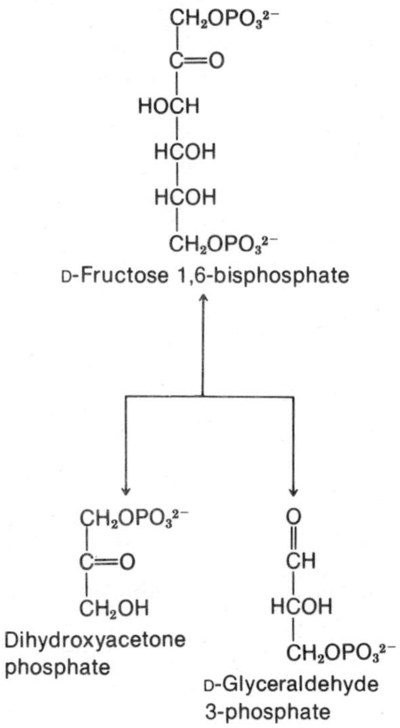

D-Fructose 1,6-bisphosphate

$CH_2OPO_3^{2-}$ | C=O | CH_2OH

Dihydroxyacetone phosphate

O=CH | HCOH | $CH_2OPO_3^{2-}$

D-Glyceraldehyde 3-phosphate

Step 4

CH_2OH | C=O | $CH_2OPO_3^{2-}$

Dihydroxyacetone phosphate

⇅

O=CH | HCOH | $CH_2OPO_3^{2-}$

D-Glyceraldehyde 3-phosphate

Step 5

O=CH | HCOH | $CH_2OPO_3^{2-}$ + NAD^+ + P_i^{2-}

D-Glyceraldehyde 3-phosphate

⇅

O=$COPO_3^{2-}$ | HCOH | $CH_2OPO_3^{2-}$ + NADH + H^+

1,3-Bisphospho-D-glycerate

Step 6

molecule each of dihydroxyacetone phosphate and glyceraldehyde 3-phosphate (Step 4). This is a reversible reaction, the enzyme being called aldolase because the overall reaction is a variant of an aldol cleavage in one direction and an aldol condensation in the other. Triose phosphate isomerase then catalyzes the reversible interconversion of dihydroxyacetone phosphate and glyceraldehyde 3-phosphate to complete the splitting stage of glycolysis (Step 5). With the transformation of dihydroxyacetone phosphate into glyceraldehyde 3-phosphate, the net conversion of one molecule of glucose into two molecules of glyceraldehyde 3-phosphate has been accomplished.

Oxidoreduction–Phosphorylation Stage

The first reaction of the last stage of glycolysis (Figure 7.7*C*) is catalyzed by the enzyme glyceraldehyde 3-phosphate dehydrogenase (Step 6). This reaction is of considerable interest, not so much because of the regulation of the enzyme involved nor because this complex reaction is reversible under intracellular conditions, but rather because of what is accomplished in a single enzyme-catalyzed step. In this reaction an aldehyde (glyceraldehyde 3-phosphate) is oxidized to a carboxylic acid with the reduction of NAD^+ to NADH. Besides producing NADH, however, the reaction also produces a high-energy phosphate compound (1,3-bisphosphoglycerate), which is a mixed anhydride of a carboxylic acid and phosphoric acid. 1,3-Bisphosphoglycerate has a large negative free energy of hydrolysis, enabling it to participate in a subsequent reaction that yields ATP. The overall reaction catalyzed by glyceraldehyde 3-phosphate dehydrogenase can be visualized as the coupling of a very favorable exergonic reaction with a very unfavorable endergonic reaction on the surface of the enzyme. The exergonic reaction can be thought of as being composed of a half-reaction in which an aldehyde (glyceraldehyde 3-phosphate) is oxidized to a carboxylic acid (1,3-bisphosphoglycerate), which is then coupled with a half-reaction in which NAD^+ is reduced to NADH:

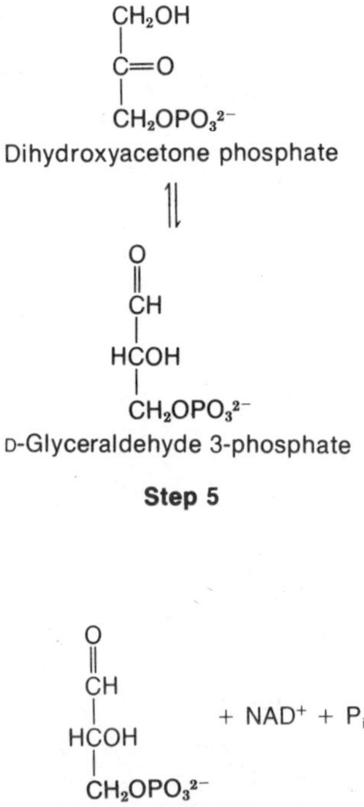

$$R{-}\overset{O}{\overset{\|}{C}}H + H_2O \longrightarrow R{-}\overset{O}{\overset{\|}{C}}OH + 2H^+ + 2e^-$$

$$NAD^+ + 2H^+ + 2e^- \longrightarrow NADH + H^+$$

The overall reaction (sum of the half-reactions) is quite exergonic, with the aldehyde being oxidized to a carboxylic acid and NAD^+ being reduced to NADH:

$$R{-}\overset{O}{\overset{\|}{C}}H + NAD^+ + H_2O \longrightarrow R{-}\overset{O}{\overset{\|}{C}}OH + NADH + H^+,\ \Delta G^{\circ\prime} = -10.3\ \text{kcal/mol}$$

The endergonic component of the reaction corresponds to the formation of a mixed anhydride between the carboxylic acid and phosphoric acid:

$$R{-}\overset{O}{\overset{\|}{C}}OH + P_i^{2-} \longrightarrow R{-}\overset{O}{\overset{\|}{C}}{-}OPO_3^{2-} + H_2O,\ \Delta G^{\circ\prime} = +11.8\ \text{kcal/mol}$$

The overall reaction involves coupling of the endergonic and exergonic components to give an overall standard free energy change of +1.5 kcal/mol.

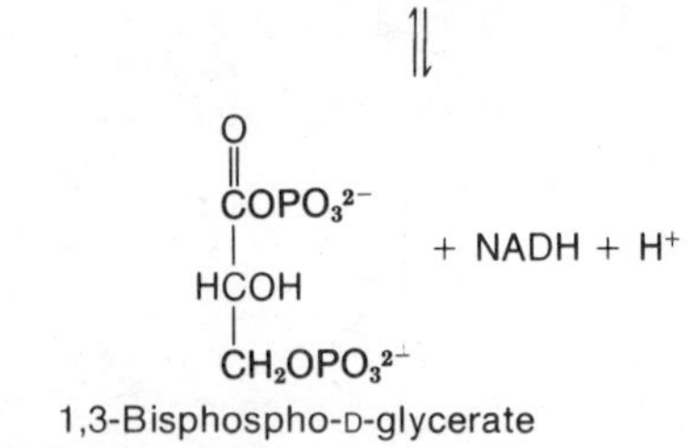

$$\text{Sum: } R{-}\overset{O}{\overset{\|}{C}}H + NAD^+ + P_i^{2-} \longrightarrow R{-}\overset{O}{\overset{\|}{C}}OPO_3^{2-} + NADH + H^+,\ \Delta G^{\circ\prime} = +1.5\ \text{kcal/mol}$$

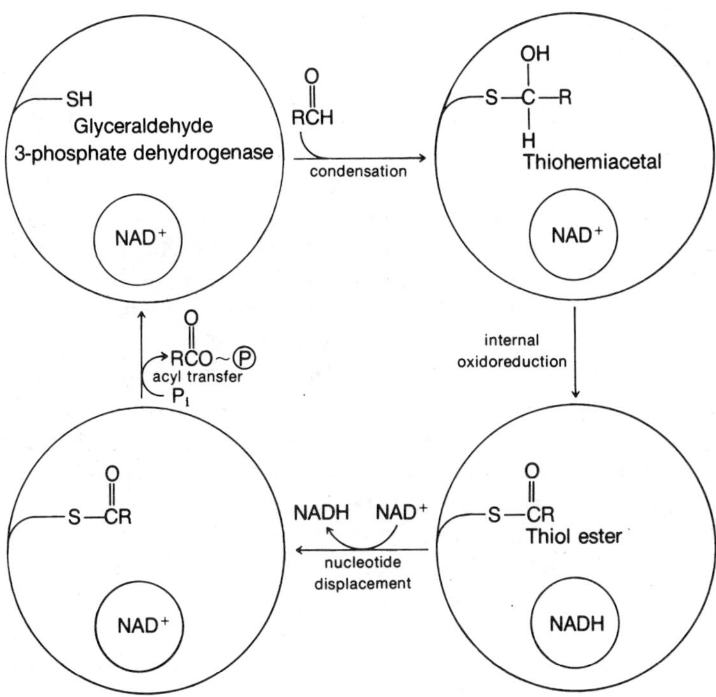

Figure 7.8
Mechanism of action of glyceraldehyde 3-phosphate dehydrogenase.
Large circle represents the enzyme; small circle, the binding site for NAD^+; $R\overset{\overset{O}{\|}}{C}H$, glyceraldehyde 3-phosphate; −SH, the sulfhydryl group of the cysteine residue located at the active site; and ~ P, the high-energy phosphate bond of 1,3-bisphosphoglycerate.

The reaction is freely reversible under intracellular conditions and is used in both the glycolytic and gluconeogenic pathways. The proposed mechanism for the enzyme-catalyzed reaction is shown in Figure 7.8. Glyceraldehyde 3-phosphate reacts with a sulfhydryl group of a cysteine residue of the enzyme to generate a thiohemiacetal. An internal oxidation–reduction reaction takes place on the surface of the enzyme in which the bound NAD^+ is reduced to NADH and the thiohemiacetal is oxidized to give a high-energy thiol ester. Exogenous NAD^+ then replaces the bound NADH and the high-energy thiol ester reacts with inorganic phosphate to form the mixed anhydride and regenerate the free sulfhydryl group. The mixed anhydride then dissociates from the enzyme. It should be noted that, in contrast to the exergonic and endergonic components of the reactions discussed above, a carboxylic acid (RCO_2H) is not considered to be an intermediate in the actual reaction mechanism. Instead, the enzyme uses the strategy of generating a high-energy thiol ester, which can be readily converted into another high-energy compound, a mixed anhydride of carboxylic and phosphoric acids.

The reaction catalyzed by glyceraldehyde 3-phosphate dehydrogenase requires NAD^+ and produces NADH. Since the cytosol of cells has only a limited amount of NAD^+, it is imperative for continuous glycolytic activity that the NADH be converted back (turned over) to NAD^+. Without turnover of NADH, glycolysis will stop for want of NAD^+. The options that cells have for accomplishing the regeneration of NAD^+ are considered in detail in a later section of this chapter (see page 274).

The next reaction, catalyzed by the enzyme phosphoglycerate kinase, produces ATP from the high-energy compound 1,3-bisphosphoglycerate (Step 7). This is the first site of ATP production in the glycolytic pathway.

O
‖
$COPO_3^{2-}$
|
HCOH + ADP^{3-}
|
$CH_2OPO_3^{2-}$
1,3-Bisphospho-D-glycerate

⇅ Mg^{2+}

O
‖
CO^-
|
HCOH + ATP^{4-}
|
$CH_2OPO_3^{2-}$
3-Phospho-D-glycerate

Step 7

Since two ATPs were "invested" for each glucose molecule in the priming stage [one at the hexokinase-catalyzed step (1) and one at the 6-phosphofructo-1-kinase-catalyzed step (3)], and since two molecules of 1,3-bisphosphoglycerate are produced from each glucose, all of the ATP "invested" in the priming stage is recovered in this step of glycolysis. Since ATP production occurs in the forward direction and ATP utilization in the reverse direction, it is somewhat surprising that the reaction is freely reversible and can be used in both the glycolytic and gluconeogenic pathways. The reaction provides a means for the generation of ATP in the glycolytic pathway but, when needed for glucose synthesis, can also be used in the reverse direction for the synthesis of 1,3-bisphosphoglycerate at the expense of ATP. The glyceraldehyde 3-phosphate dehydrogenase-phosphoglycerate kinase system is an example of substrate-level phosphorylation, a term used to refer to a process in which a substrate participates in an enzyme-catalyzed reaction that yields ATP or GTP. Substrate-level phosphorylation stands in contrast to oxidative phosphorylation in which electron transport by the respiratory chain of the mitochondrial inner membrane is used to provide the energy necessary for ATP synthesis (see Chapter 6). Note, however, that the combination of the reactions catalyzed by glyceraldehyde 3-phosphate dehydrogenase and phosphoglycerate kinase accomplishes the coupling of an oxidation (an aldehyde goes to a carboxylic acid) to a phosphorylation.

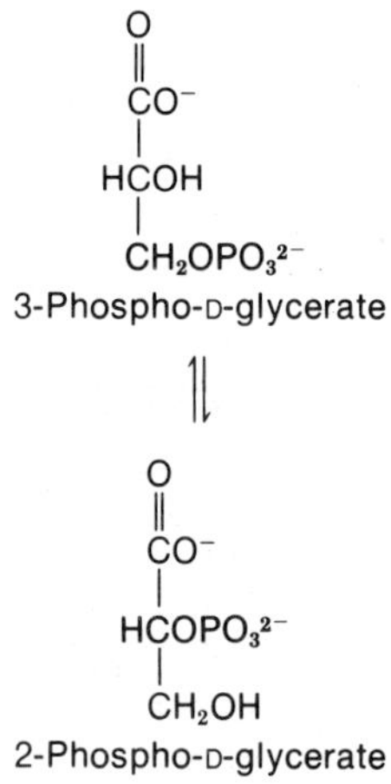

$$\begin{array}{c} O \\ \| \\ CO^- \\ | \\ HCOH \\ | \\ CH_2OPO_3^{2-} \end{array}$$

3-Phospho-D-glycerate

$$\rightleftharpoons$$

$$\begin{array}{c} O \\ \| \\ CO^- \\ | \\ HCOPO_3^{2-} \\ | \\ CH_2OH \end{array}$$

2-Phospho-D-glycerate

Step 8

Phosphoglycerate mutase catalyzes the step in which 3-phosphoglycerate is converted to 2-phosphoglycerate (Step 8). This is a freely reversible reaction in which 2,3-bisphosphoglycerate functions as an obligatory intermediate at the active site of the enzyme (E):

$$\text{E} + \text{2,3-bisphosphoglycerate} \rightleftharpoons \text{E-phosphate} + \text{2 phosphoglycerate}$$

$$\text{E-phosphate} + \text{3-phosphoglycerate} \rightleftharpoons \text{E} + \text{2,3-bisphosphoglycerate}$$

$$\text{Sum: 3-Phosphoglycerate} \rightleftharpoons \text{2-phosphoglycerate}$$

2,3-Bisphosphoglycerate is synthesized by a reaction catalyzed by another enzyme, 2,3-bisphosphoglycerate mutase:

$$\begin{array}{c} O \\ \| \\ COPO_3^{2-} \\ | \\ HCOH \\ | \\ CH_2OPO_3^{2-} \end{array} \rightleftharpoons \begin{array}{c} O \\ \| \\ CO^- \\ | \\ HCOPO_3^{2-} \\ | \\ CH_2OPO_3^{2-} \end{array} + H^+$$

1,3-Bisphospho-D-glycerate   2,3-Bisphospho-D-glycerate

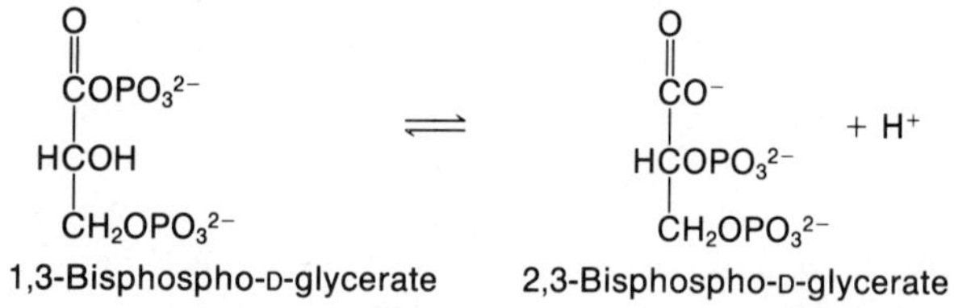

The mutase is quite unusual in that it is a *bifunctional* enzyme, serving also as a phosphatase that converts 2,3-bisphosphoglycerate to 3-phosphoglycerate and inorganic phosphate. Most cells contain minute quantities of 2,3-bisphosphoglycerate since it is only needed in catalytic amounts for the reaction catalyzed by phosphoglycerate mutase. Note that E-phosphate in the phosphoglycerate mutase reaction scheme cannot be generated without 2,3-bisphosphoglycerate. Red blood cells represent a special case, in which 2,3-bisphosphoglycerate (usually abbreviated DPG) accumulates to high concentrations and functions as a physiologically important allosteric effector of the association of oxygen with hemoglobin (see Chapter 23). From 15 to 25 percent of the glucose converted to lactate in red blood cells goes by way of the "DPG shunt" (Figure 7.9). Glucose catabolized by the shunt generates no net ATP since the reaction catalyzed by the phosphoglycerate kinase reaction is bypassed.

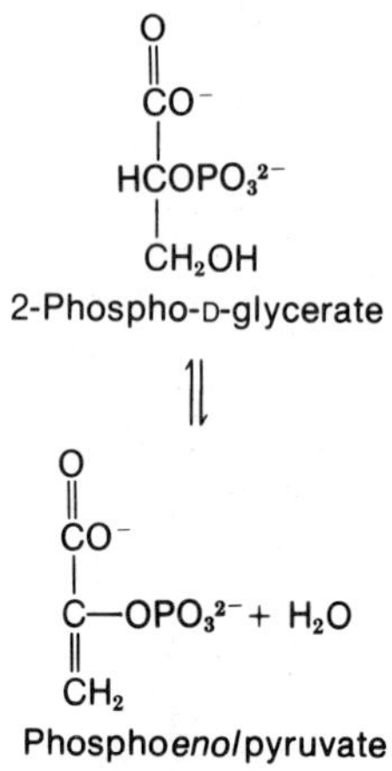

$$\begin{array}{c} O \\ \| \\ CO^- \\ | \\ HCOPO_3^{2-} \\ | \\ CH_2OH \end{array}$$

2-Phospho-D-glycerate

$$\rightleftharpoons$$

$$\begin{array}{c} O \\ \| \\ CO^- \\ | \\ C{-}OPO_3^{2-} \\ \| \\ CH_2 \end{array} + H_2O$$

Phospho*enol*pyruvate

Step 9

Enolase catalyzes the elimination of water from 2-phosphoglycerate to form phospho*enol*pyruvate in the next reaction (Step 9). This is a

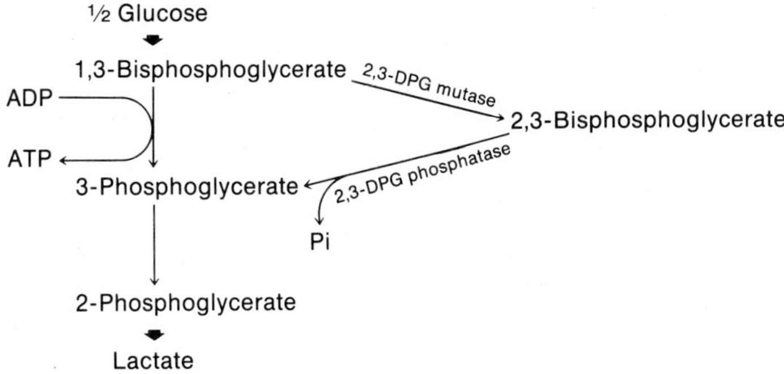

Figure 7.9
The 2,3-diphosphoglycerate (2,3-DPG) shunt consists of reactions catalyzed by the bifunctional enzyme, 2,3-DPG synthase/phosphatase.

remarkable reaction from the standpoint that a high-energy phosphate compound is generated from one of markedly lower energy level. The standard free energy change ($\Delta G^{\circ\prime}$) for the hydrolysis of phospho*enol*pyruvate is −14.8 kcal/mol, a value strikingly greater than that of 2-phosphoglycerate (−4.2 kcal/mol). Although the reaction catalyzed by the enzyme is freely reversible, a large change in the distribution of energy occurs as a consequence of the action of enolase upon 2-phosphoglycerate.

The next step of the glycolytic sequence is catalyzed by pyruvate kinase (Step 10). This enzyme accomplishes substrate level phosphorylation; that is, the synthesis of ATP with the conversion of the high-energy compound phospho*enol*pyruvate into pyruvate. The reaction is not reversible under intracellular conditions. It constitutes a way to synthesize ATP, but in contrast to the reaction catalyzed by phosphoglycerate kinase is not reversible under conditions that exist in cells and is not a reaction that can be used for the synthesis of phospho*enol*pyruvate when needed for glucose synthesis.

The last step of the glycolytic pathway is an oxidoreduction reaction catalyzed by lactate dehydrogenase (Step 11). Pyruvate is reduced in this reaction to give L-lactate, whereas NADH is oxidized to give NAD^+. This is a freely reversible reaction and the only one of the body in which L-lactate participates, that is, the only reaction that can result in L-lactate formation or L-lactate utilization. It should be noted that NADH generated by glyceraldehyde 3-phosphate dehydrogenase is converted back to NAD^+ by lactate dehydrogenase (see Figure 7.7), the major option used by cells under anaerobic conditions for the regeneration of cytosolic NAD^+.

O
‖
CO^-
|
C—OPO_3^{2-} + ADP^{3-} + H
‖
CH_2
Phospho*enol*pyruvate

↓ + Mg^{2+}

O
‖
CO^-
|
C=O + ATP^{4-}
|
CH_3
Pyruvate

Step 10

Stoichiometries of the Glycolytic Pathway

An examination of the overall glycolytic pathway will show that there is a perfect coupling between the generation of NADH and its utilization (Figure 7.7). Two molecules of NADH are generated at the level of glyceraldehyde 3-phosphate dehydrogenase and two molecules of NADH are utilized at the level of lactate dehydrogenase in the overall conversion of one molecule of glucose into two molecules of lactate. NAD^+, a soluble molecule present in the cytosol, is available in only limited amounts to participate in the glycolytic pathway. It is essential, therefore, that NAD^+ be regenerated from NADH for the glycolytic pathway to continue unabated. The NAD^+ reacts at the level of glyceraldehyde 3-phosphate dehydrogenase to produce NADH, which diffuses through the cytosol

O
‖
CO^-
|
C=O + NADH + H^+
|
CH_3
Pyruvate

⇅

O
‖
CO^-
|
HOCH + NAD^+
|
CH_3
L-Lactate

Step 11

until it makes contact with lactate dehydrogenase, which, if a molecule of pyruvate is also available, forms lactate with the regeneration of NAD^+. The overall reaction catalyzed by the combined actions of these two enzymes is the conversion of glyceraldehyde 3-phosphate, pyruvate, and inorganic phosphate into lactate and 1,3-bisphosphoglycerate.

$$\text{D-Glyceraldehyde 3-phosphate} + NAD^+ + P_i \longrightarrow \text{1,3-bisphospho-D-glycerate} + NADH + H^+$$

$$\text{Pyruvate} + NADH + H^+ \longrightarrow \text{L-lactate} + NAD^+$$

$$\text{Sum:}\quad \text{D-Glyceraldehyde 3-phosphate} + \text{pyruvate} + P_i \longrightarrow \text{1,3-bisphosphoglycerate} + \text{L-lactate}$$

This perfect coupling of reducing equivalents in the glycolytic pathway only has to occur under conditions of anaerobiosis, or in cells that lack mitochondria. With the availability of oxygen and mitochondria, reducing equivalents in the form of NADH generated at the level of glyceraldehyde 3-phosphate dehydrogenase can be "shuttled" into the mitochondria for the synthesis of ATP. When this occurs, the end product of glycolysis becomes pyruvate. Two shuttle systems are known to exist for the transport of reducing equivalents from the cytosolic space to the mitochondrial matrix space (mitosol). The mitochondrial inner membrane is not permeable to NADH; therefore, NADH cannot penetrate directly across the mitochondrial inner membrane to gain access to the NADH dehydrogenase of the mitochondrial electron transfer chain.

Shuttle Pathways

The glycerol phosphate shuttle is shown in Figure 7.10*A*; the malate–aspartate shuttle in Figure 7.10*B*. All tissues that have mitochondria appear also to have the capability of "shuttling" reducing equivalents from the cytosol to the mitosol. The relative proportion of the activities of the two shuttles varies from tissue to tissue, with liver making greater use of the malate–aspartate shuttle, whereas some muscle cells may be more dependent on the glycerol phosphate shuttle. The shuttle systems are irreversible, that is, they represent mechanisms for moving reducing equivalents into the mitosol, but not mechanisms for moving mitochondrial reducing equivalents into the cytosol.

The transport of aspartate out of the mitochondria in exchange for glutamate is the irreversible step in the malate–aspartate shuttle. The mitochondrial inner membrane has a large number of transport systems (see Chapter 6), but lacks one which is effective for oxaloacetate. For this reason oxaloacetate transaminates with glutamate to produce aspartate, which then exits irreversibly from the mitochondrion in exchange for glutamate. The aspartate entering the cytosol transaminates with α-ketoglutarate to give oxaloacetate and glutamate. The oxaloacetate accepts the reducing equivalents of NADH and becomes malate. Malate then penetrates the mitochondrial inner membrane, where it is oxidized by the mitochondrial malate dehydrogenase. This produces NADH within the mitosol and regenerates oxaloacetate to complete the cycle. The overall balanced equation for the sum of all the reactions of the malate–aspartate shuttle is simply,

$$NADH_{cytosol} + H^+_{cytosol} + NAD^+_{mitosol} \longrightarrow NAD^+_{cytosol} + NADH_{mitosol} + H^+_{mitosol}$$

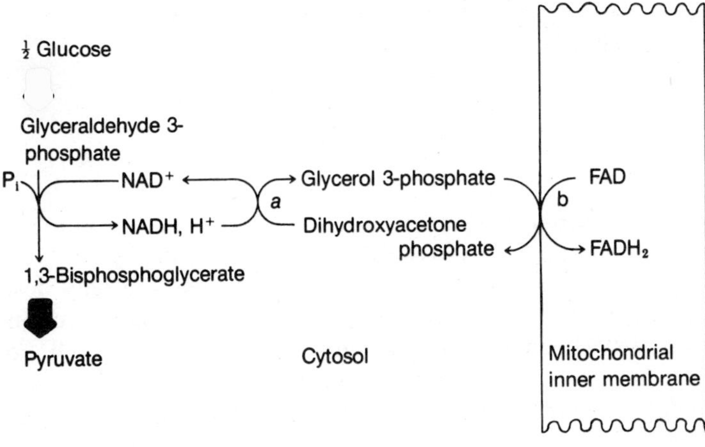

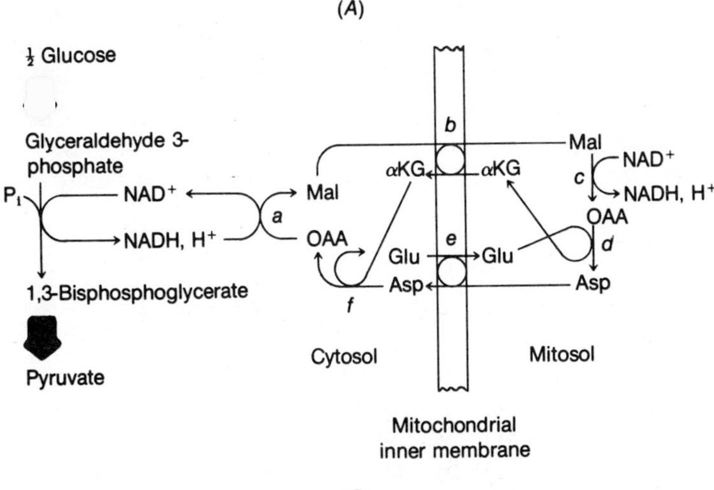

Figure 7.10
Shuttles for the transport of reducing equivalents from the cytosol to the mitochondrial electron-transfer chain.
(A) Glycerol phosphate shuttle; a, cytosolic glycerol 3-phosphate dehydrogenase oxidizes NADH; b, glycerol 3-phosphate dehydrogenase of the outer surface of the mitochondrial inner membrane reduces FAD. (B) Malate-aspartate shuttle: a, cytosolic malate dehydrogenase reduces oxaloacetate (OAA) to malate; b, dicarboxylic acid antiport of the mitochondrial inner membrane catalyzes electrically silent exchange of malate for α-ketoglutarate (α-KG); c, mitochondrial malate dehydrogenase produces intramitochondrial NADH; d, mitochondrial aspartate aminotransferase transaminates glutamate and oxaloacetate; e, glutamate-aspartate antiport of the mitochondrial inner membrane catalyzes electrogenic exchange of glutamate for aspartate; f, cytosolic aspartate aminotransferase transaminates aspartate and α-ketoglutarate.

The glycerol phosphate shuttle is simpler, in the sense that fewer reactions are involved, but it should be noted that $FADH_2$ is generated as the end product within the mitochondrial inner membrane, rather than NADH within the mitosolic compartment. The irreversible step of the shuttle is catalyzed by the mitochondrial glycerol 3-phosphate dehydrogenase. The active site of this enzyme is exposed on the cytosolic surface of the mitochondrial inner membrane, making it unnecessary for glycerol 3-phosphate to penetrate completely into the mitosol for oxidation. The overall balanced equation for the sum of the reactions of the glycerol phosphate shuttle is

$$NADH_{cytosol} + H^+ + FAD_{inner\ membrane} \longrightarrow NAD^+_{cytosol} + FADH_{2\,inner\ membrane}$$

Alcohol Oxidation

We should not get trapped into thinking, because of Figure 7.10, that the glycerol phosphate and malate–aspartate shuttles are only designed to handle the NADH generated by glycolysis. These shuttles are important in using NADH generated in the cytosol in other ways as well. For example, the first step of alcohol (i.e., ethanol) metabolism is its oxidation to acetaldehyde with the production of NADH by the enzyme alcohol dehydrogenase.

$$\underset{\text{Ethanol}}{CH_3CH_2OH} + NAD^+ \longrightarrow \underset{\text{Acetaldehyde}}{CH_3\overset{\overset{O}{\|}}{C}H} + NADH + H^+$$

This enzyme is located almost exclusively in the cytosol of liver parenchymal cells. The acetaldehyde generated is able to traverse the mitochondrial inner membrane for oxidation by a mitosolic aldehyde dehydrogenase.

$$CH_3\overset{\overset{O}{\|}}{C}H + NAD^+ \longrightarrow CH_3\overset{\overset{O}{\|}}{C}O^- + NADH + 2H^+$$

The NADH generated by the last step can be used directly by the mitochondrial electron transfer chain. However, the NADH generated by cytosolic alcohol dehydrogenase cannot be used directly, and must be oxidized back to NAD^+ by one of the shuttles. Thus, the capacity of human beings to oxidize alcohol is dependent on the ability of their liver to transport reducing equivalents from the cytosol to the mitosol by these shuttle systems.

Glucuronide Formation

Another situation in which the shuttles play an important role has to do with the formation of water-soluble glucuronides of bilirubin and various drugs. Conjugation with glucuronic acid occurs so that these compounds can be eliminated from the body in the aqueous media of urine and bile. In this process UDP-glucose (for structure of this activated form of glucose, see discussion of glycogen synthesis on page 311) is oxidized to UDP-glucuronic acid (structure on page 339).

$$\text{UDP-D-glucose} + 2NAD^+ + H_2O \longrightarrow \text{UDP-D-glucuronic acid} + 2NADH + 2H^+$$

In a reaction that occurs primarily in the liver, the "activated" glucuronic acid molecule is then transferred to a nonpolar, acceptor molecule, such as some compound (e.g., a drug) foreign to the body:

$$\text{UDP-D-glucuronic acid} + \text{R-OH} \longrightarrow \text{R-}O\text{-glucuronic acid} + \text{UDP}$$

Excess NADH generated by the first reaction has to be eliminated from the cytosol for this process to continue, and, of course, the shuttles are called into play for this purpose. Since ethanol oxidation and drug conju-

CLIN. CORR. 7.1
ALCOHOL AND BARBITURATES

Acute alcohol intoxication causes increased sensitivity of an individual to the general depressant effects of barbiturates. This drug combination is very dangerous, normal prescription doses of barbiturates having potentially lethal consequences in the presence of ethanol. In addition to the

gation are properties of the liver, the two of them occurring together may overwhelm the combined capacity of the shuttles. A good thing to remember—and to tell patients—is not to mix the intake of pharmacologically active compounds and the consumption of alcohol (see Clin. Corr. 7.1).

Energetics of NADH Oxidation

The mitosolic NADH formed as a consequence of malate–aspartate shuttle activity can be used in the presence of oxygen by the mitochondrial respiratory chain for the production of three molecules of ATP by oxidative phosphorylation:

$$NADH_{mitosol} + H^+ + \tfrac{1}{2}O_2 + 3ADP + 3P_i \longrightarrow NAD^+_{mitosol} + 3ATP + H_2O$$

In contrast, the $FADH_2$ obtained by glycerol phosphate shuttle activity is worth only two ATPs:

$$FADH_{2\,inner\ membrane} + \tfrac{1}{2}O_2 + 2ADP + 2P_i \longrightarrow FAD_{inner\ membrane} + 2ATP + H_2O$$

Without the intervention of these shuttle systems, the conversion of one molecule of glucose to two molecules of lactate by glycolysis results in the *net* formation of two molecules of ATP. Two molecules of ATP are used in the priming stage to set glucose up so that it can be cleaved. However, subsequent steps then yield four molecules of ATP so that the overall net production of ATP by the glycolytic pathway is two molecules of ATP. Biological cells have only a limited amount of ADP and inorganic phosphate. Flux through the glycolytic pathway is also dependent, therefore, upon an adequate supply of these substrates. Consequently, the ATP generated has to be used, that is, turned over, in normal work-related processes in order for glycolysis to occur. The equation for the use of ATP for any work-related process is simply

$$ATP^{4-} \longrightarrow ADP^{3-} + P_i^{2-} + H^+ + \text{"work"}$$

When this equation is added to that given above for glycolysis, excluding the work accomplished, the overall balanced equation for the glycolytic process when coupled to some ATP-utilizing work performance becomes

$$\text{D-Glucose} \longrightarrow 2\ \text{lactate}^- + 2H^+$$

If the ATP is not utilized for performance of work glycolysis will stop for want of ADP and/or inorganic phosphate. Thus glycolytic activity is dependent on the turnover of ATP to ADP and P_i, just as it is dependent on the turnover of NADH to NAD^+.

Inhibitors of the Glycolytic Pathway

The best known inhibitors of the glycolytic pathway include 2-deoxyglucose, sulfhydryl reagents, and fluoride. 2-Deoxyglucose causes inhibition at the first step, that is, at the reaction catalyzed by hexokinase. 2-Deoxyglucose serves as a substrate for this enzyme, being converted to the 6-phosphate ester by the reaction shown in Figure 7.11. Like glucose 6-phosphate, 2-deoxyglucose 6-phosphate is an effective inhibitor of the reaction catalyzed by hexokinase, but unlike glucose 6-phosphate, 2-deoxyglucose 6-phosphate will not function as a substrate for the reaction

depressant effects of both ethanol and barbiturates on the central nervous system, ethanol inhibits the metabolism of barbiturates and prolongs the time barbiturates remain effective in the body. Hydroxylation of barbiturates by the endoplasmic reticulum of the liver is inhibited by ethanol. This reaction, catalyzed by the NADPH-dependent cytochrome P_{450} system, results in water soluble derivatives of the barbiturates which are eliminated readily from the circulation by the kidneys. Blood levels of barbiturates remain high when ethanol is present, causing increased central nervous system depression.

Surprisingly, the alcoholic when sober is less sensitive to barbiturates. Chronic ethanol consumption apparently causes adaptive changes in the sensitivity of the central nervous system to barbiturates. It also results in the induction of the enzymes of liver endoplasmic reticulum involved in drug hydroxylation reactions. Consequently, the sober alcoholic is able to metabolize barbiturates more rapidly. This sets up the following scenario. A sober alcoholic has trouble falling asleep, even after taking several sleeping pills, because his liver has increased capacity to hydroxylate the barbiturate contained in the pills. In frustration he consumes more pills and then alcohol. Sleep results, but may be followed by respiratory depression and death because the alcoholic, although less sensitive to barbiturates when sober, remains sensitive in the presence of alcohol.

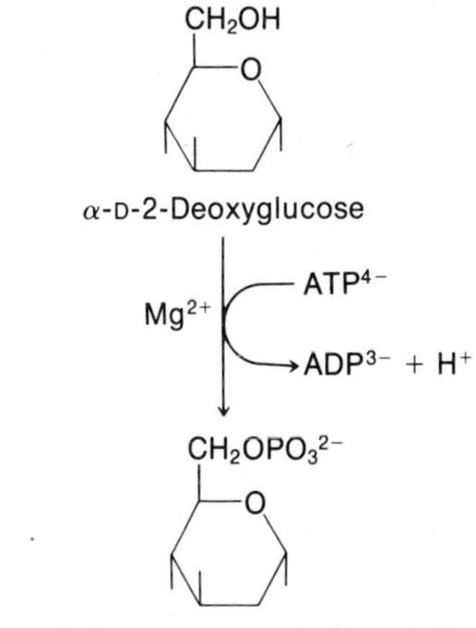

Figure 7.11
Hexokinase catalyzes the phosphorylation of 2-deoxyglucose.

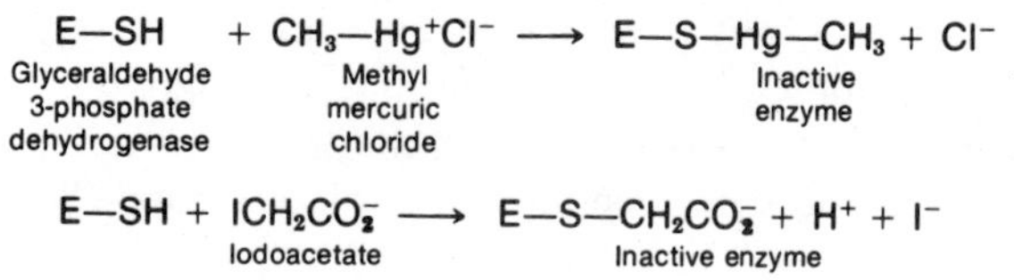

$$\underset{\text{Glyceraldehyde 3-phosphate dehydrogenase}}{E{-}SH} + \underset{\text{Methyl mercuric chloride}}{CH_3{-}Hg^+Cl^-} \longrightarrow \underset{\text{Inactive enzyme}}{E{-}S{-}Hg{-}CH_3} + Cl^-$$

$$E{-}SH + \underset{\text{Iodoacetate}}{ICH_2CO_2^-} \longrightarrow \underset{\text{Inactive enzyme}}{E{-}S{-}CH_2CO_2^-} + H^+ + I^-$$

Figure 7.12
Mechanism responsible for inactivation of glyceraldehyde 3-phosphate dehydrogenase by sulfhydryl reagents.

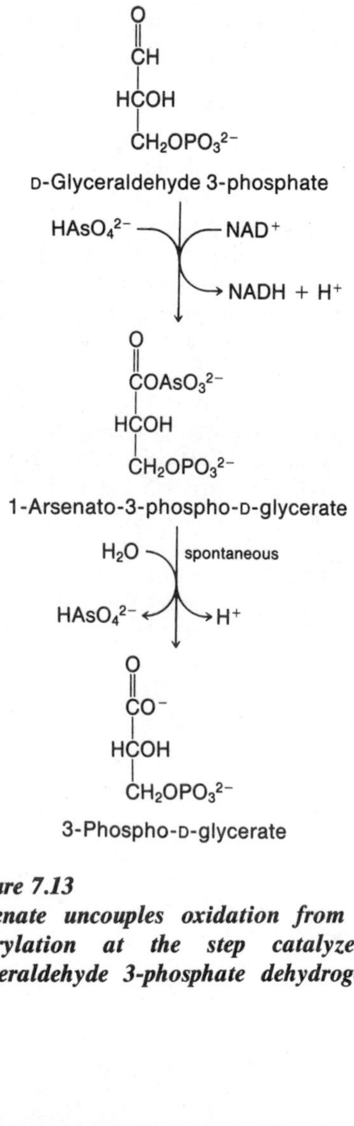

D-Glyceraldehyde 3-phosphate ($O{=}CH{-}HCOH{-}CH_2OPO_3^{2-}$)

$HAsO_4^{2-}$, NAD^+ → $NADH + H^+$

1-Arsenato-3-phospho-D-glycerate ($O{=}COAsO_3^{2-}{-}HCOH{-}CH_2OPO_3^{2-}$)

H_2O (spontaneous) → $HAsO_4^{2-}$ + H^+

3-Phospho-D-glycerate ($O{=}CO^-{-}HCOH{-}CH_2OPO_3^{2-}$)

Figure 7.13
Arsenate uncouples oxidation from phosphorylation at the step catalyzed by glyceraldehyde 3-phosphate dehydrogenase.

catalyzed by phosphoglucoisomerase. Deoxyglucose 6-phosphate inhibition of hexokinase prevents glucose from being phosphorylated, resulting in the inhibition of glycolysis at the very first step.

Sulfhydryl reagents bring about an inhibition at the level of glyceraldehyde 3-phosphate dehydrogenase. As discussed above this enzyme has a cysteine residue at the active site, the sulfhydryl group of which reacts with glyceraldehyde 3-phosphate to give a thiohemiacetal. Sulfhydryl reagents are usually mercury-containing compounds or alkylating compounds, such as iodoacetate, which readily react with the sulhydryl group of glyceraldehyde 3-phosphate dehydrogenase to prevent the formation of the thiohemiacetal (see Figure 7.12).

Fluoride is a potent inhibitor of enolase. Mg^{2+} and inorganic phosphate are believed to form an ionic complex with fluoride ion, which is responsible for inhibition of the enzyme, apparently by interfering with the combination of the enzyme with its substrate (a Mg^{2+}-2-phosphoglycerate complex).

Arsenate has important effects on the glycolytic pathway and can be toxic. In the sense that it does not prevent flux through glycolysis, arsenate is not an inhibitor of the process. However, by causing arsenolysis at the step catalyzed by glyceraldehyde 3-phosphate dehydrogenase, arsenate prevents net synthesis of ATP by the pathway. Arsenate looks a lot like inorganic phosphate and is able to substitute for inorganic phosphate in enzyme-catalyzed reactions. The result, in the case of glyceraldehyde 3-phosphate dehydrogenase, is the formation of a mixed anhydride of arsenic acid and the carboxyl group of 3-phosphoglycerate (Figure 7.13). 1-Arsenato 3-phosphoglycerate is unstable, undergoing spontaneous hydrolysis to give 3-phosphoglycerate and inorganic arsenate. Hence, glycolysis continues unabated in the presence of arsenate, but 1,3-bisphosphoglycerate is not formed, resulting in the loss of the capacity to synthesize ATP at the step catalyzed by phosphoglycerate kinase. The consequence is that net ATP synthesis does not occur when glycolysis is carried out in the presence of arsenate, the ATP invested in the priming stage being only balanced by the ATP generated in the pyruvate kinase step. This means that in the presence of arsenate, glycolysis does not generate the ATP required to meet the energy needs of a cell. This, along with the fact that arsenolysis also interferes with ATP formation by oxidative phosphorylation, makes arsenate a toxic compound (see Clin. Corr. 7.2).

CLIN. CORR. 7.2 ARSENIC POISONING

Most forms of arsenic are toxic, but the trivalent form (arsenite as AsO_2^-) is much more toxic than the pentavalent form (arsenate or $HAsO_4^{2-}$). Less ATP is produced whenever arsenate substitutes for inorganic phosphate in biological reactions. Arsenate competes for inorganic

7.4 REGULATION OF THE GLYCOLYTIC PATHWAY

Depending somewhat on the tissue under consideration, the regulatory enzymes of the glycolytic pathway are commonly considered to be hexokinase, 6-phosphofructo-1-kinase, and pyruvate kinase. A summary of the important regulatory features of these enzymes is presented in Figure

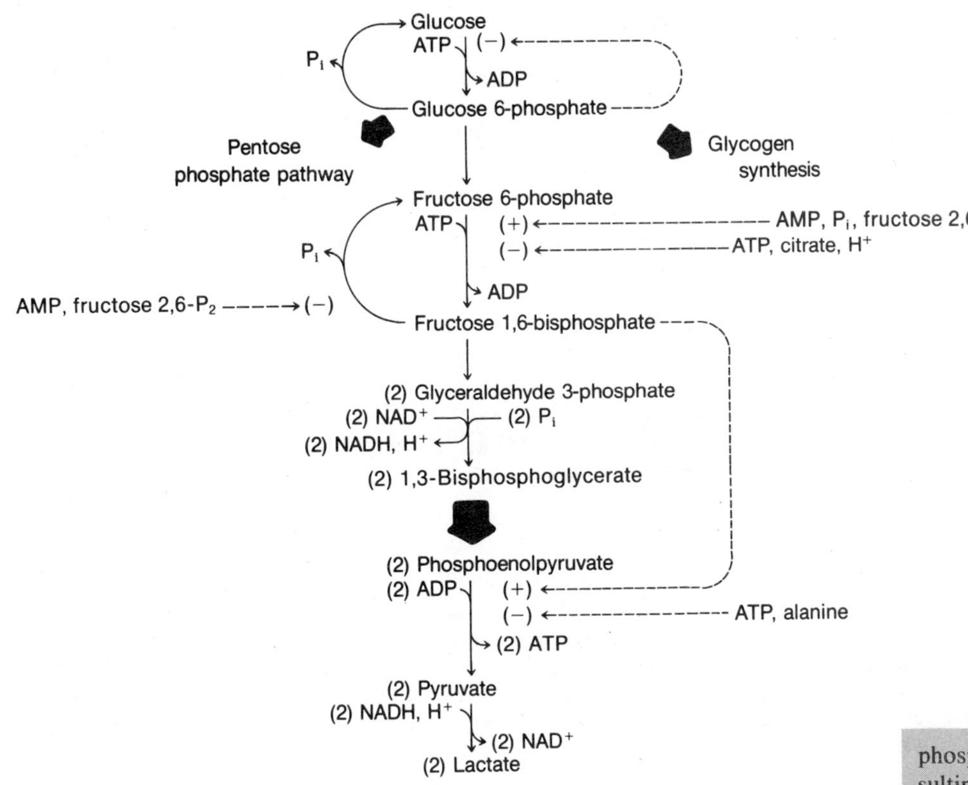

Figure 7.14
Important regulatory features of the glycolytic pathway.
Because of differences in isoenzyme distribution, not all tissues of the body have all of the regulatory mechanisms shown here.

7.14. A regulatory enzyme is defined as an enzyme that is subject to control by either allosteric effectors or covalent modification. Both mechanisms are used by cells to control the most important of the regulatory enzymes. A regulatory enzyme can often be identified by determining whether the concentrations of the substrates and products within a cell indicate that the reaction catalyzed by the enzyme is close to equilibrium. An enzyme that is not subject to regulation will catalyze a "near-equilibrium" reaction, whereas a regulatory enzyme will catalyze a "nonequilibrium reaction" under intracellular conditions. This makes sense because flux through the step catalyzed by a regulatory enzyme is restricted because of controls imposed upon that enzyme. Whether an enzyme-catalyzed reaction is near equilibrium or nonequilibrium can be determined by comparing the established equilibrium constant for the reaction with the mass–action ratio as it exists within a cell. The equilibrium constant for the reaction A + B → C + D is defined as

$$K_{eq} = \frac{[C][D]}{[A][B]}$$

where the brackets indicate the concentrations at equilibrium. The mass–action ratio is calculated in a similar manner, except that the steady-state (ss) concentrations of reactants and products within the cell are used in the equation:

$$\text{Mass–action ratio} = \frac{[C]_{ss}[D]_{ss}}{[A]_{ss}[B]_{ss}}$$

phosphate-binding sites on enzymes, resulting in the formation of arsenate esters, which are unstable. Arsenite works by a completely different mechanism, involving the formation of a stable complex with enzyme bound lipoic acid:

(dithiol lipoyl–enzyme, SH SH, $(CH_2)_4C(=O)$ – enzyme) + AsO_2^- + H^+

↓

(cyclic S–As(OH)–S complex, $(CH_2)_4C(=O)$ – enzyme) + H_2O

For the most part arsenic poisoning is explained by inhibition of those enzymes which require lipoic acid as a coenzyme. These include pyruvate dehydrogenase, α-ketoglutarate dehydrogenase, and branched-chain α-keto acid dehydrogenase. Chronic arsenic poisoning from well water contaminated with arsenical pesticides or through the efforts of a murderer is best diagnosed by determining the concentration of arsenic in the hair or fingernails of the victim. About 0.5 mg of arsenic would be found in a kilogram of hair from a normal individual. The hair of a person chronically exposed to arsenic could have 100 times as much.

TABLE 7.1 Apparent Equilibrium Constants and Mass–Action Ratios for the Reactions of Glycolysis and Gluconeogenesis in Liver

Reaction Catalyzed by	*Reaction in the Pathway of*		*Apparent Equilibrium Constant* (K'_{eq})	*Mass–Action Ratios*	*Considered Near-Equilibrium Reaction?*
	Glycolysis	*Gluconeogenesis*			
Glucokinase	Yes	No	2×10^3	0.02	No
Glucose 6-phosphatase	No	Yes	850 M	120 M	No
Phosphoglucoisomerase	Yes	Yes	0.36	0.31	Yes
6-Phosphofructo-1-kinase	Yes	No	1×10^3	0.09	No
Fructose 1,6-bisphosphatase	No	Yes	530 M	19 M	No
Aldolase	Yes	Yes	13×10^{-5} M	12×10^{-7} M	Yes[a]
Glyceraldehyde 3-phosphate dehydrogenase + phosphoglycerate kinase	Yes	Yes	2×10^3 M^{-1}	0.6×10^3 M^{-1}	Yes
Phosphoglycerate mutase	Yes	Yes	0.1	0.1	Yes
Enolase	Yes	Yes	3.0	2.9	Yes
Pyruvate kinase	Yes	No	2×10^4	0.7	No
Pyruvate carboxylase + phospho*enol*pyruvate carboxykinase	No	Yes	7.0 M	1×10^{-3} M	No

[a] Reaction catalyzed by aldolase appears to be out of equilibrium by two orders of magnitude. However, in vivo concentrations of fructose 1,6-bisphosphate and glyceraldehyde 3-phosphate are so low (μM concentration range) that significant enzyme binding of both metabolites is believed to occur. Although only the total concentration of any metabolite of a tissue can be measured, only that portion of the metabolite that is not bound should be used in the calculations of mass–action ratios. This is usually not possible, introducing uncertainty in the comparison of in vitro equilibrium constants to in vivo mass–action ratios.

If the mass–action ratio is approximately equal to the K_{eq}, the enzyme is said to be active enough to catalyze a near-equilibrium reaction and the enzyme is not considered subject to regulation. When the mass–action ratio is considerably different from the K_{eq}, the enzyme is said to catalyze a nonequilibrium reaction and usually will be found subject to regulation by one or more mechanisms. Mass–action ratios and equilibrium constants are compared for the glycolytic enzymes of liver in Table 7.1. The reactions catalyzed by glucokinase (liver isoenzyme of hexokinase), 6-phosphofructo-1-kinase, and pyruvate kinase in the intact liver are considered far enough from equilibrium to indicate that these enzymes are "regulatory" in this tissue.

Hexokinase

Different isoenzymes of hexokinase are found in different tissues of the body. The hexokinase isoenzymes found in most tissues have a low K_m for glucose (<0.1 mM) relative to its concentration in blood (~5 mM) and are strongly inhibited by the product of the reaction, glucose 6-phosphate. The latter is an important regulatory feature because it prevents hexokinase from tying up all of the inorganic phosphate of a cell in the form of phosphorylated hexoses (see Clin. Corr. 7.3). Thus the reaction catalyzed by hexokinase is not at equilibrium within cells that contain this enzyme because of the inhibition imposed by glucose 6-phosphate. Liver parenchymal cells are unique in that they contain glucokinase, an isoenzyme of hexokinase with strikingly different kinetic properties from the other hexokinases. This isoenzyme catalyzes the same reaction, that is, an ATP-dependent phosphorylation of glucose, but has a much higher K_m for glucose and is not subject to product inhibition by glucose 6-phosphate. The high K_m of glucokinase for glucose contributes to the capacity of the liver to "buffer" blood glucose levels. Glucose equilibrates readily across the plasma membrane of the liver, the concentration within the

CLIN. CORR. 7.3 FRUCTOSE INTOLERANCE

Patients with hereditary fructose intolerance are deficient in the liver aldolase responsible for splitting fructose 1-phosphate into dihydroxyacetone phosphate and glyceraldehyde. Consumption of fructose by these patients results in the accumulation of fructose 1-phosphate and depletion of inorganic phosphate and ATP in the liver. The reactions involved are those catalyzed by fructokinase and the enzymes of oxidative phosphorylation:

Fructose + ATP ⟶ fructose 1-phosphate + ADP

ADP + P_i + "energy provided by electron transfer chain" ⟶ ATP

Net: P_i + fructose ⟶ fructose 1-phosphate

Tying up inorganic phosphate in the form of fructose 1-phosphate makes it impossible for liver mitochondria to generate ATP by oxidative phosphorylation. ATP levels fall precipitously, making it also impossible for the liver to carry out its normal work functions. Damage results to the liver reflecting that of the blood. Since the K_m of glucokinase for glucose (~10 mM) is considerably greater than normal blood glucose concentrations (~5 mM), any increase in glucose concentration leads to a proportional increase in the rate of glucose phosphorylation by glucokinase (see Figure 7.15). Likewise, any decrease in glucose concentration leads to a proportional decrease in the rate of glucose phosphorylation. The result is that the liver uses glucose at a significant rate only when blood glucose levels are greatly elevated. This buffering effect of liver glucokinase on blood glucose levels would not occur if glucokinase had the low K_m for glucose characteristic of other hexokinases and was, therefore, completely saturated at physiological concentrations of glucose (see Figure 7.15). On the other hand, a low K_m form of hexokinase is a good choice for tissues such as the brain in that it allows phosphorylation of glucose even when blood and tissue glucose concentrations are dangerously low.

In spite of the fact that glucokinase is not subject to inhibition by glucose 6-phosphate, the reaction catalyzed by glucokinase is not near-equilibrium under the intracellular conditions of liver cells (Table 7.1). Part of the explanation lies in the rate restriction imposed by the high K_m of glucokinase for glucose. Another important factor is that the activity of glucokinase is opposed in liver by that of glucose 6-phosphatase. Like glucokinase, this enzyme has an unusually high K_m (3 mM) with respect to the normal intracellular concentration (about 0.2 mM) of its primary substrate, glucose 6-phosphate. The result is that flux through this enzyme-catalyzed step is almost directly proportional to the intracellular concentration of glucose 6-phosphate. As shown in Figure 7.16, the combined action of glucokinase and glucose 6-phosphatase constitutes a futile cycle, that is, the sum of their reactions is simply the hydrolysis of ATP to give ADP and P_i without the performance of any work. It turns out that when blood glucose concentrations are about 5 mM, the activity of glucokinase is almost exactly balanced by the opposing activity of glucose 6-phosphatase. The result is that no net flux occurs in either direction. This futile cycling between glucose and glucose 6-phosphate is wasteful of ATP but, combined with the process of gluconeogenesis, contributes significantly to the "buffering" action of the liver on blood glucose levels. Furthermore, it provides a mechanism for preventing glucokinase from tying up all of the inorganic phosphate of the liver (see Clin. Corr. 7.3). An accumulation of glucose 6-phosphate in the liver does not inhibit glucokinase activity because glucose 6-phosphate is not an effective inhibitor of this isoenzyme of hexokinase. The effectiveness of glucokinase is nullified, however, because an increase in glucose 6-phosphate concentration increases the rate of glucose 6-phosphate hydrolysis by glucose 6-phosphatase. This helps prevent glucose 6-phosphate from accumulating faster than it can be used by other metabolic processes of the liver.

Glucokinase is an inducible enzyme. This means that under various physiological conditions the *amount* of the enzyme either increases or decreases. Induction of the synthesis of an enzyme and the opposite—

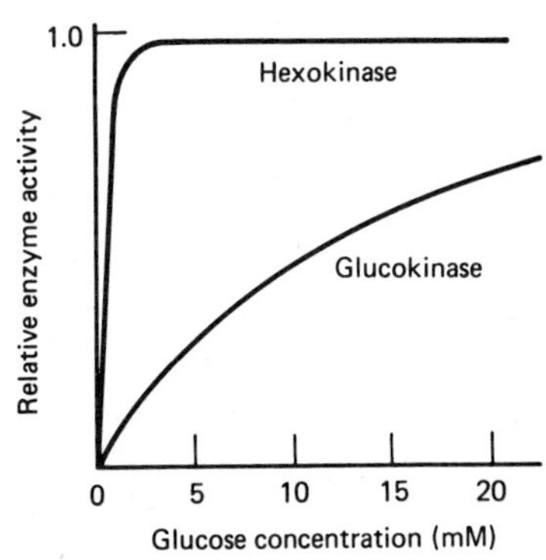

Figure 7.15
Comparison of the substrate saturation curves for hexokinase and glucokinase.

cells in large part because they are unable to maintain normal ion gradients by means of the ATP-dependent cation pumps. The cells swell and eventually lose their internal contents by osmotic lysis.

Although patients with fructose intolerance are particularly sensitive to fructose, humans in general have a limited capacity to handle this sugar. The capacity of the normal liver to phosphorylate fructose greatly exceeds its capacity to split fructose 1-phosphate. This means that fructose use by the liver is poorly controlled and that excessive fructose could deplete the liver of inorganic phosphate and ATP. Fructose was actually tried briefly in hospitals as a substitute for glucose with patients being maintained by parenteral nutrition. The rationale was that fructose would be a better source of calories than glucose because fructose utilization is relatively independent of the insulin status of a patient. Delivery of large amounts of fructose by intravenous feeding was soon found to result in severe liver damage. Similar attempts have been made to substitute sorbitol and xylitol for glucose. These sugars also tend to deplete the liver of ATP and, like fructose, should be used for parenteral nutrition with caution and only under special circumstances, if at all.

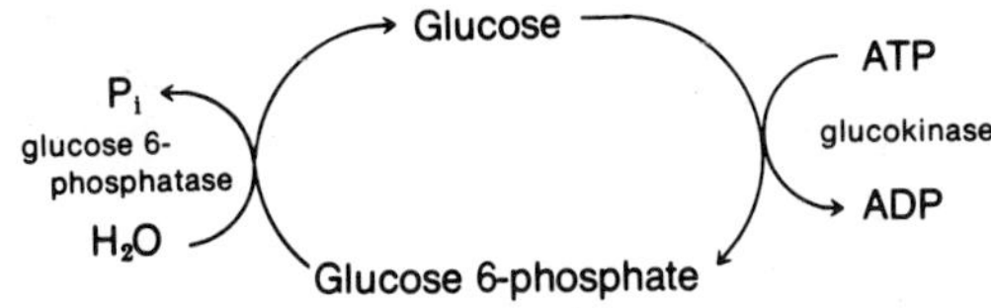

Figure 7.16
Phosphorylation of glucose followed by dephosphorylation constitutes a futile cycle in parenchymal cells of the liver.

CLIN. CORR. 7.4 DIABETES MELLITUS

Diabetes mellitus is a chronic disease characterized by derangements in carbohydrate, fat, and protein metabolism. Two types are recognized clinically—the juvenile-onset or insulin-dependent type (see Clin. Corr. 14.7) and the maturity-onset or insulin-independent type (see Clin. Corr. 14.9).

The oral glucose tolerance test is commonly used for the diagnosis of diabetes. It consists of determining the blood glucose level in the fasting state and at intervals of 30–60 min for 2 h or more after consuming a 100 g carbohydrate meal. In a normal individual blood glucose returns to normal levels within 2 h after ingestion of the carbohydrate meal. In the diabetic patient, blood glucose will reach a higher level and remain elevated for longer periods of time, depending upon the severity of the disease. However, many factors may contribute to an abnormal glucose tolerance test. The patient must have consumed a high carbohydrate diet for the preceding 3 days, presumably to allow for induction of enzymes of glucose-utilizing pathways, for example, glucokinase, fatty acid synthase, and acetyl CoA carboxylase. In addition, almost any infection (colds, etc.) and less well-defined "stress," possibly by effects on the sympathetic nervous system, can result in (transient) abnormalities of the glucose tolerance test. Because of problems with the glucose tolerance test, elevation of the fasting glucose level should probably be the sine qua non for the diagnosis of diabetes. Glucose uptake by cells of insulin-sensitive tissues, that is, muscle and adipose, is decreased in the diabetic state. Insulin is required for glucose uptake by these tissues, and the diabetic patient either lacks insulin or has developed "insulin resistance" in these tissues. Resistance to insulin is an abnormality of the insulin receptor or in subsequent steps mediating the metabolic effects of insulin. Parenchymal cells of the liver do not require insulin for glucose uptake. Without insulin, however, the liver has diminished enzymatic capacity to remove glucose from the blood. This is explained in part by decreased glucokinase activity plus the loss of insulin's action on key enzymes of glycogenesis and the glycolytic pathway.

repression of the synthesis of an enzyme—are relatively slow control processes, usually requiring several hours before significant changes are realized. As long as insulin is also present, the amount of glucokinase in the liver tends to reflect how much glucose is being delivered to the liver via the portal vein. In other words, a person consuming large meals rich in carbohydrate will have greater amounts of glucokinase in the liver than one who is not. The liver in which glucokinase has been induced can make a greater contribution to the lowering of elevated blood glucose levels. The exact details of how induction of glucokinase takes place at the gene level are not known, but it is clear that insulin is involved. The absence of insulin makes the liver of the diabetic patient deficient in glucokinase, in spite of high blood glucose levels, and this is one of the reasons why the liver of the diabetic has less blood glucose "buffering" action (see Clin. Corr. 7.4).

6-Phosphofructo-1-kinase

Much evidence suggests that 6-phosphofructo-1-kinase is the rate-limiting enzyme and most important regulatory site of glycolysis in most tissues. Usually we think of the first step of a pathway as the most logical choice for the rate-limiting step. Notice, however, that 6-phosphofructo-1-kinase catalyzes the first *committed* step of the glycolytic pathway. The phosphoglucoisomerase catalyzed reaction is reversible, and most cells can use glucose 6-phosphate for glycogen synthesis and in the pentose phosphate pathway. The reaction catalyzed by 6-phosphofructo-1-kinase commits the cell to the metabolism of glucose by glycolysis and is, therefore, a logical site for the step of the pathway that is rate limiting and subject to the greatest degree of regulation by allosteric effectors. ATP, citrate, and hydrogen ions (low pH) are the most important negative allosteric effectors, whereas AMP, fructose 2,6-bisphosphate, and inorganic phosphate (P_i) are the most important positive allosteric effectors (Figure 7.14). Through their actions as strong inhibitors or activators of 6-phosphofructo-1-kinase, these compounds signal different rates of glycolysis in response to changes in (1) the energy state of the cell (ATP, AMP, and P_i), (2) the internal environment of the cell (hydrogen ions), (3) the availability of alternate fuels such as fatty acids and ketone bodies (citrate), and (4) the insulin-to-glucagon ratio in the blood (fructose 2,6-bisphosphate). Evidence for the physiological importance of these effectors comes in part from application of the crossover theorem to the glycolytic pathway.

Crossover Theorem and Regulation of 6-Phosphofructo-1-kinase by ATP, AMP, and Inorganic Phosphate

For the hypothetical pathway $A \rightarrow B \rightarrow C \rightarrow D \rightarrow E \rightarrow F$ . . . , the crossover theorem proposes that an inhibitor that partially inhibits the conversion of C to D will cause a "crossover" in the metabolite profile between C and D. This means that when the steady-state concentrations of the intermediates in the presence and absence of an inhibitor are compared, the concentrations of the intermediates before the site of inhibition should increase in response to the inhibitor, whereas those after the site should decrease. Crossover plots are constructed by setting the concentrations of all intermediates without some effector of the pathway equal to 100%. The concentrations of the intermediates observed in the presence of the effector are then expressed as percentages of these values. The expected result with a negative effector is shown in Figure 7.17*a*. The effect of returning the perfused rat heart from an anoxic condition to a

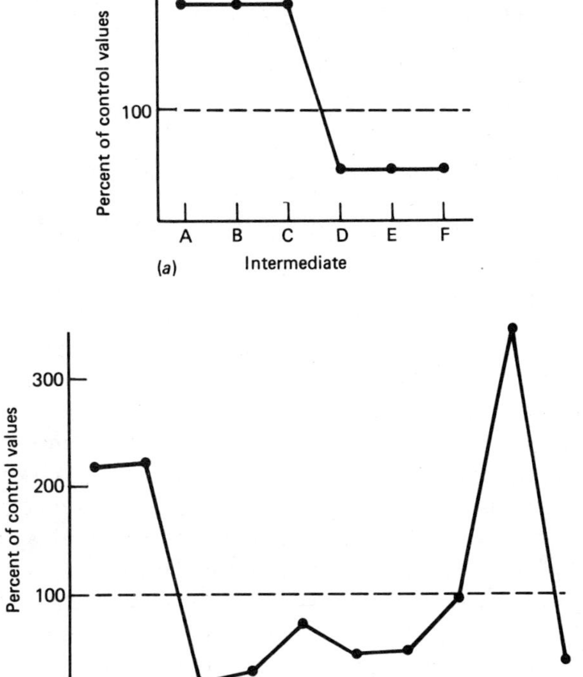

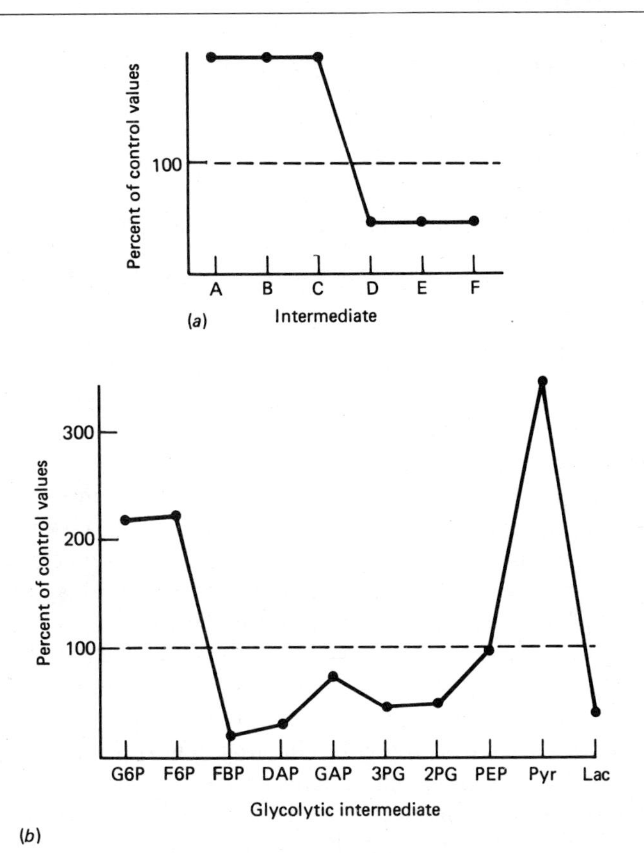

Figure 7.17
Crossover analysis is used to locate sites of regulation of a metabolic pathway.
(a) Theoretical effect of an inhibitor of the C to D step in the pathway of $A \rightarrow B \rightleftharpoons C \rightarrow D \rightleftharpoons E \rightarrow F$. *Steady-state concentrations of all intermediates of the pathway without the inhibitor present are arbitrarily set equal to 100%. Steady-state concentrations of all intermediates when the inhibitor is present are then expressed as percentages of the control values. (b) Effect of oxygen on the relative steady-state concentrations of the intermediates of the glycolytic pathway in the perfused rat heart. The changes in concentrations of metabolites of hearts perfused without oxygen caused by subsequent perfusion with oxygen (95%* O_2, *5%* CO_2) *are recorded as percentages of the anoxic values. Oxygen strongly inhibits glucose utilization and lactate production under such conditions. The dramatic increase in pyruvate concentration occurs as a consequence of greatly increased utilization of cytosolic NADH by the shuttle systems. Abbreviations: G6P, glucose 6-phosphate; F6P, fructose 6-phosphate; FBP, fructose 1,6-bisphosphate; DAP, dihydroxyacetone phosphate; GAP, glyceraldehyde 3-phosphate; 3PG, 3-phosphoglycerate; 2PG, 2-phosphoglycerate; PEP, phospho*enol*pyruvate; Pyr, pyruvate; and Lac, lactate.*
From J.R. Williamson, *J. Biol. Chem.* 241: 5026, 1966.

well-oxygenated state is also shown (Figure 7.17*b*). This transition with the perfused rat heart is known to establish new steady-state concentrations of the glycolytic intermediates, the flux being much greater through the glycolytic pathway in the absence of oxygen. Under experimental conditions used, the perfused hearts consumed glucose at rates some 20 times greater in the absence than in the presence of oxygen. This example illustrates what is known as the Pasteur effect, defined as the inhibition of glucose utilization and lactate accumulation by the initiation of respiration (oxygen consumption). This is readily understandable on a thermodynamic basis, the complete oxidation of glucose to CO_2 and H_2O yielding

CLIN. CORR. 7.5
LACTIC ACIDOSIS

This problem is characterized by elevated blood lactate levels, usually greater than 5 mM, along with decreased blood pH and bicarbonate concentrations. Lactic acidosis is the most commonly encountered form of metabolic acidosis and can be the consequence of overproduction of lactate, underutilization of lactate, or both. Lactate production is normally balanced by lactate utilization, with the result that lactate is usually not present in the blood at concentrations greater than 1.2 mM. All tissues of the body have the capacity to produce lactate by anaerobic glycolysis, but most tissues do not produce large quantities because much more ATP can be gained by the complete oxidation of the pyruvate produced by glycolysis. However, all tissues respond with an increase in lactate generation when oxygenation is inadequate. A decrease in ATP resulting from reduced oxidative phosphorylation allows the activity of 6-phosphofructo-1-kinase to increase. These tissues have to rely on anaerobic glycolysis for ATP production under such conditions and this results in lactic acid production. A good example is muscle exercise, which can deplete the tissue of oxygen and cause an overproduction of lactic acid. Tissue hypoxia occurs, however, in all forms of shock, during convulsions, and in diseases involving circulatory and pulmonary failure.

The major fate of lactate in the body is either complete combustion to CO_2 and H_2O or conversion back to glucose by the process of gluconeogenesis. Both require oxygen. Decreased oxygen availability, therefore, increases lactate production and decreases lactate utilization. The latter can also be decreased by liver diseases, ethanol, and a number of other drugs. Phenformin, a drug which was once used to treat the hyperglycemia of insulin-independent diabetes, was well-documented to induce lactic acidosis in certain patients.

Bicarbonate is usually administered in an attempt to control the acidosis associated with lactic acid accumulation. The key to successful treatment, however, is to find and eliminate the cause of the overproduction and/or underutilization of lactic acid and most often involves the restoration of circulation of oxygenated blood.

much more ATP than anaerobic glycolysis:

$$\text{Glycolysis:}\quad \text{D-Glucose} + 2\text{ADP}^{3-} + 2\text{P}_i^{2-} \longrightarrow 2\ \text{L-lactate}^- + 2\text{ATP}^{4-}$$

$$\text{Complete Oxidation:}\quad \text{D-Glucose} + 6\text{O}_2 + 38\text{ADP}^{3-} + 38\text{P}_i^{2-} + 38\text{H}^+ \longrightarrow 6\text{CO}_2 + 6\text{H}_2\text{O} + 38\text{ATP}^{4-}$$

ATP is used by a cell only to meet its metabolic demand, that is, to provide the necessary energy for the work processes (metabolic demand) inherent to that cell. Since so much more ATP is produced from glucose in the presence of oxygen, much less glucose has to be consumed to meet the metabolic demand of the cell. The "crossover" at the conversion of fructose 6-phosphate to fructose 1,6-bisphosphate argues that oxygen imposes an inhibition at the level of 6-phosphofructo-1-kinase. This can be readily rationalized on the basis that ATP is a well-recognized inhibitor of 6-phosphofructo-1-kinase, and more ATP can be generated in the presence of oxygen than in the absence. However, ATP levels do not change greatly between these two conditions (in the experiment of Figure 7.17*b*, ATP increased from 4.7 μmol/g wet weight in the absence of oxygen to 5.6 μmol/g wet weight in the presence of oxygen). Since 6-phosphofructo-1-kinase is severely inhibited at concentrations of ATP (2.5–6 mM) normally present in cells, such a small difference in ATP concentration cannot account completely for the change in flux through 6-phosphofructo-1-kinase. However, much greater changes, percentagewise, occur in the concentrations of AMP and P_i, both positive allosteric effectors of 6-phosphofructo-1-kinase. The changes that occur in the steady-state concentrations of AMP and P_i when oxygen is introduced into the system are exactly what might have been predicted, that is, the levels of both go down dramatically. These changes result in less 6-phosphofructo-1-kinase activity, greatly suppressed glycolytic activity, and account in large part for the Pasteur effect. AMP levels automatically go down in a cell when ATP levels increase. Although this is not intuitively obvious, the reason is simple. The sum of the adenine nucleotides in a cell, that is, ATP + ADP + AMP, is nearly constant under most physiological conditions, but the relative concentrations are such that the ATP concentration is always much greater than the AMP concentration. Furthermore, the adenine nucleotides are maintained in equilibrium in the cytosol through the action of an enzyme called nucleoside monophosphokinase, which catalyzes the reaction 2ADP $\rightleftharpoons$ ATP + AMP. The equilibrium constant (K'_{eq}) for this reaction is given by

$$K'_{eq} = \frac{[\text{ATP}][\text{AMP}]}{[\text{ADP}]^2}$$

Since this reaction is "near-equilibrium" under intracellular conditions, the concentration of AMP is given by

$$[\text{AMP}] = \frac{K'_{eq}[\text{ADP}]^2}{[\text{ATP}]}$$

Due to the fact that intracellular [ATP] >> [ADP] >> [AMP], a small decrease in [ATP] causes a substantially greater percentage increase in [ADP]; and, since [AMP] is related to the square of the [ADP], an even greater percentage increase in [AMP]. Because of this relationship, a small decrease in ATP concentration is amplified into a much larger (percentage) change in AMP concentration. This makes AMP an excellent signal of the energy status of the cell and allows it to function as an important allosteric effector of 6-phosphofructo-1-kinase activity. Furthermore, AMP influences in yet another way the effectiveness of the reaction catalyzed by 6-phosphofructo-1-kinase. An enzyme called fructose 1,6-bisphosphatase catalyzes an irreversible reaction, which opposes that of 6-phosphofructo-1-kinase:

$$\text{Fructose 1,6-bisphosphate} + H_2O \longrightarrow \text{fructose 6-phosphate} + P_i$$

This enzyme sits "cheek by jowl" with 6-phosphofructo-1-kinase in the cytosol of many cells. Together they catalyze a futile cycle (ATP → ADP + P_i + "heat"), and, at the very least, they decrease the "effectiveness" of one another. The AMP concentration is a perfect signal of the energy status of the cell—not only because AMP activates 6-phosphofructo-1-kinase but also because AMP *inhibits* fructose 1,6-bisphosphatase. The result is that a small decrease in ATP concentration triggers, via the increase in AMP concentration, a large increase in the net conversion of fructose 6-phosphate into fructose 1,6-bisphosphate. This increases the glycolytic flux by increasing the amount of substrate available for the splitting stage. In cells containing hexokinase, it also results in greater phosphorylation of glucose because a decrease in fructose 6-phosphate automatically causes a decrease in glucose 6-phosphate, which, in turn, results in less inhibition of hexokinase activity.

The decrease in lactate production in response to the onset of respiration is another feature of the Pasteur effect that can be readily explained. The most important factor is the decreased glycolytic flux caused by oxygen; however, secondary factors include competition between lactate dehydrogenase and the mitochondrial pyruvate dehydrogenase complex for pyruvate, as well as competition between lactate dehydrogenase and the shuttle systems for NADH. For the most part, lactate dehydrogenase loses the competition in the presence of oxygen.

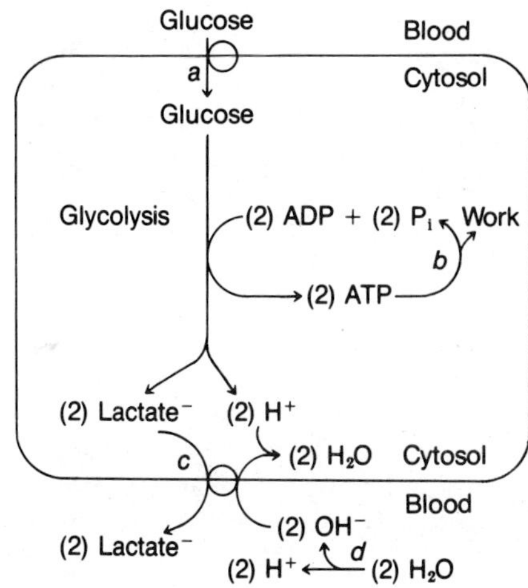

Figure 7.18
Unless lactate formed by glycolysis is released from the cell, the intracellular pH is decreased as a consequence of the accumulation of intracellular lactic acid.
The low pH decreases 6-phosphofructo-1-kinase activity so that further lactic acid production by glycolysis is shut off. a, glucose transport into the cell; b, all work performances which convert ATP back to ADP and P_i; c, lactate-hydroxide ion antiport; d, ionization of water in the blood to give hydroxide ions for exchange with lactate.

Regulation of 6-Phosphofructo-1-kinase by Intracellular pH

It would be natural to suspect that lactate, as the end product of glycolysis, would inhibit the rate-limiting enzyme of the glycolytic pathway. It does not. However, hydrogen ions, the other glycolytic end product, do inhibit 6-phosphofructo-1-kinase. As shown in Figure 7.18, glycolysis in effect generates lactic acid, and the cell must dispose of it as such. This accounts for why excessive glycolysis in the body lowers blood pH and leads to an emergency medical situation termed *lactic acidosis* (see Clin. Corr. 7.5). Plasma membranes of cells appear to contain either an antiport for lactate and hydroxide ions or a symport for lactate and hydrogen ions (experimentally they are difficult to distinguish). Regardless of the exact mechanism, lactic acid is released from the cell into the bloodstream. This ability to transport lactic acid out of the cell is a defense mechanism, preventing the pH from getting so low that everything becomes pickled (see Clin. Corr. 7.6). The sensitivity of 6-phosphofructo-1-kinase to hydrogen ions is also part of this mechanism. Hydrogen ions are able to shut off glycolysis, the process responsible for decreasing the pH. Note that transport of lactic acid out of a cell requires that the blood be available to the cell in order to carry this "end product" away. When blood flow to a group of cells is inadequate, for example, in heavy exercise of a skeletal muscle or an attack of angina pectoris in the case of the heart, hydrogen ions cannot escape from the cells fast enough. Yet the need for ATP within such cells, because of the lack of oxygen, may partially override

CLIN. CORR. **7.6**
PICKLED PIGS AND MALIGNANT HYPERTHERMIA

In patients with malignant hyperthermia, a variety of agents, especially the widely used general anesthetic halothane, will produce a dramatic rise in body temperature, metabolic and respiratory acidosis, hyperkalemia, and muscle rigidity. This genetic abnormality occurs in about 1 in 15,000 children and 1 in 50,000–100,000 adults. Death often results the first time a susceptible person is anesthetized. Onset occurs within minutes of drug exposure and the hyperthermia must be recognized immediately. Packing the patient in ice is effective and should be accompanied by measures to combat acidosis. The drug dantrolene is also effective.

A phenomenon similar, if not identical, to malignant hyperthermia is known to occur in pigs. Pigs with this problem, called porcine stress syndrome, respond poorly to stress. This genetic disease usually manifests itself as the pig is being shipped to market. Pigs with the syndrome can be identified by exposure to halothane, which triggers the same response seen in patients with malignant hyperthermia. The meat of pigs that have died as a result of the syndrome is pale, watery, and of very low pH (i.e., nearly pickled).

Muscle is considered the site of the primary lesion in both malignant hyperthermia and porcine stress syndrome. In response to halothane the skeletal muscles become rigid and generate heat and lactic acid. Although much experimental work has been conducted, the biochemical basis for the increased heat production remains obscure. Heat produced by glycolytic activity and muscle contraction is not believed sufficient to explain the dramatic increase in body temperature. Uncontrolled futile cycling in which ATP hydrolysis is greatly accelerated has been suggested to be involved:

$$\text{ATP} + H_2O \longrightarrow \text{ADP} + P_i + \text{heat}$$

Indeed halothane has been shown to accelerate futile cycling at the level of 6-phosphofructo-1-kinase/fructose 1,6-bisphosphatase in muscles of pigs with porcine stress syndrome. Perhaps one of these regulatory enzymes will be found defective with respect to allosteric effector control in patients with malignant hyperthermia. There is also evidence that the sarcoplasmic reticulum of such patients may be defective and that the anesthetic triggers inappropriate release of Ca^{2+} from the sarcoplasmic reticulum. This could result in uncontrolled stimulation of a number of heat-producing processes, that is, myosin ATPase, glycogenolysis, glycolysis, and cyclic uptake and release of Ca^{2+} by mitochondria and sarcoplasmic reticulum.

CLIN. CORR. 7.7 ANGINA PECTORIS AND MYOCARDIAL INFARCTION

Chest pain associated with reversible myocardial ischemia is termed angina pectoris. The pain is the result of an imbalance between demand for and supply of blood flow to cardiac muscles and is most commonly caused by coronary artery obstructive disease. The patient experiences a heavy squeezing pressure or ache substernally, often radiating to either the shoulder and arm or occasionally to the jaw or neck. Attacks occur with exertion, last from 1 to 15 min, and are relieved by rest. The coronary arteries involved are obstructed by atherosclerosis (i.e., lined with characteristic fatty deposits) or less commonly narrowed by spasm. Myocardial infarction occurs if the ischemia persists long enough to cause severe damage (necrosis) to the heart muscle. In myocardial infarction, tissue death occurs and the characteristic pain is longer lasting, and often more severe.

Nitroglycerin and other nitrates are frequently prescribed to relieve the pain caused by the myocardial ischemia of angina pectoris. These drugs can be used prophylactically, enabling patients to participate in activities that would otherwise precipitate an attack of angina. Nitroglycerin may work in part by causing dilation of the coronary arteries, improving oxygen delivery to the heart and washing out lactic acid. Probably more important is the effect of nitrates on the peripheral circulation. Nitrates relax smooth muscle, causing vasodilation throughout the body. This reduces arterial pressure and allows blood to accumulate in the veins. The result is decreased return of blood to the heart, which reduces the volume of blood the heart has to pump, which reduces the energy requirement of the heart. In addition, the heart empties itself against less pressure, which also spares energy. The overall effect is a lowering of the oxygen requirement of the heart, bringing it in line with the oxygen supply via the diseased coronary arteries.

The coronary artery bypass operation is used in severe cases of angina that cannot be controlled by medication. In this operation veins are removed from the leg and interposed between the aorta and coronary arteries of the heart. The purpose is to bypass the portion of the artery diseased by atherosclerosis and provide the affected tissue with a greater blood supply. Remarkable relief from angina can be achieved by this operation, with the patient being able to return to normal productive life in some cases.

the inhibition of 6-phosphofructo-1-kinase by hydrogen ions. The unabated accumulation of hydrogen ions then results in pain, which, in the case of skeletal muscle, can be relieved by simply terminating the exercise. In the case of the heart, rest or pharmacologic agents that increase blood flow or decrease the need for ATP within the myocytes may be effective (see Clin. Corr. 7.7).

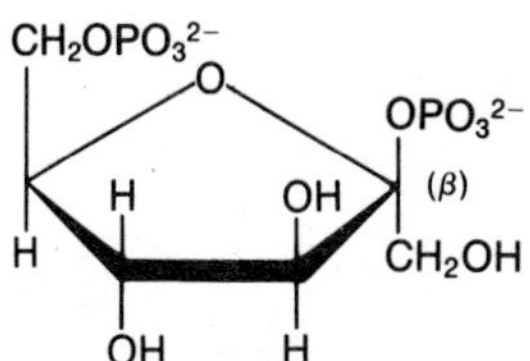

Figure 7.19
Structure of fructose 2,6-bisphosphate.

Regulation of 6-Phosphofructo-1-kinase by Alternate Fuels that Increase Intracellular Citrate

Many tissues prefer to use fatty acids and ketone bodies as oxidizable fuels in place of glucose. Most of these tissues have the capacity to use glucose but actually prefer to oxidize fatty acids and ketone bodies. This unselfish act helps preserve glucose for those tissues, such as brain, that are absolutely dependent upon glucose as an energy source. The mechanism responsible for this preference is relatively simple. Oxidation of both fatty acids and ketone bodies elevates the levels of cytosolic citrate, which inhibits 6-phosphofructo-1-kinase. The result is decreased glucose utilization by the tissue when fatty acids or ketone bodies are available.

Hormonal Control of 6-Phosphofructo-1-kinase by cAMP and Fructose 2,6-bisphosphate

Fructose 2,6-bisphosphate, a recently discovered compound, has been clearly established as the most important regulatory molecule of 6-phosphofructo-1-kinase in liver. The structure of this compound is given in Figure 7.19. This compound is probably present in all tissues, but its exact role in regulation of glycolysis is only understood for liver. Fructose 2,6-bisphosphate behaves exactly like AMP in that it functions as a positive allosteric effector of 6-phosphofructo-1-kinase and as a negative allosteric effector of fructose 1,6-bisphosphatase. Indeed, without the presence of this compound, glycolysis could not occur in the liver because 6-phosphofructo-1-kinase would have insufficient activity and fructose 1,6-bisphosphatase would have too much activity for net conversion of fructose 6-phosphate to fructose 1,6-bisphosphate. Fructose 2,6-bisphosphate was not discovered until 1980 because of its very unusual acid lability. Most phosphate esters of sugars are stable in dilute acid, and dilute acid is used routinely for the extraction of phosphorylated metabolites from tissues for identification and quantitation. The presence of fructose 2,6-bisphosphate in tissues was missed largely because this compound is readily hydrolyzed to fructose 6-phosphate and inorganic phosphate in such extracts. It was finally isolated from neutral extracts of tissues by investigators who became convinced that some factor responsible for 6-phosphofructo-1-kinase regulation must be present in tissues. This working hypothesis led to the discovery of the most important regulatory mechanism for glycolysis in liver and thereby greatly increased our understanding of the molecular mechanism of hormone action in this tissue.

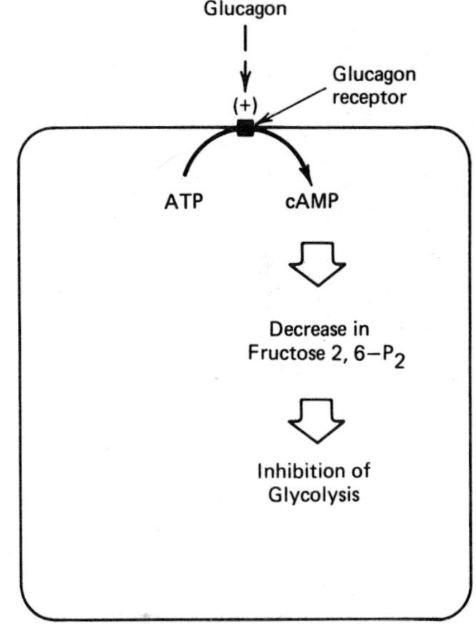

Figure 7.20
Overview of the mechanism responsible for glucagon inhibition of hepatic glycolysis.
The glucagon receptor-adenylate cyclase complex (■) is an intrinsic component of the plasma membrane.

The fundamental role of fructose-2,6-bisphosphate in hormonal control of hepatic glycolysis is already well understood. Figure 7.20 gives a brief overview of the mechanism involved. Understanding this mechanism requires an appreciation of the role of cAMP (Figure 7.21) as the "second messenger" of hormone action. As discussed in more detail in Chapters 14 and 16, glucagon is released from the α cells of the pancreas and circulates in the blood until it comes in contact with glucagon receptors located on the outer surface of the liver plasma membrane (Figure 7.20). Binding of glucagon to these receptors is sensed by adenylate cyclase, an enzyme located on the inner surface of the the plasma membrane, stimulating it to convert cytosolic ATP into cytosolic cAMP and

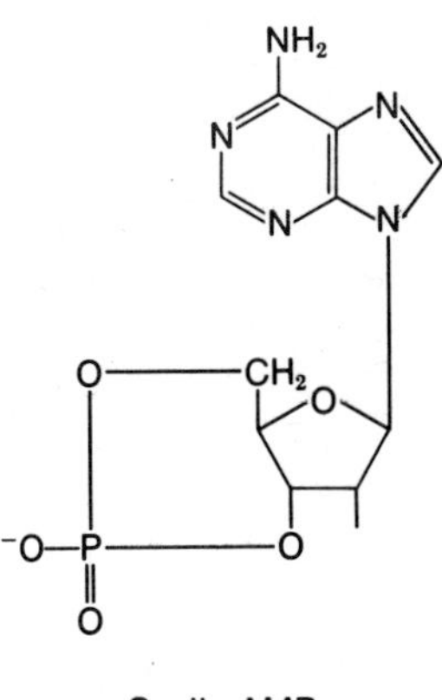

Figure 7.21
Structure of cyclic AMP.

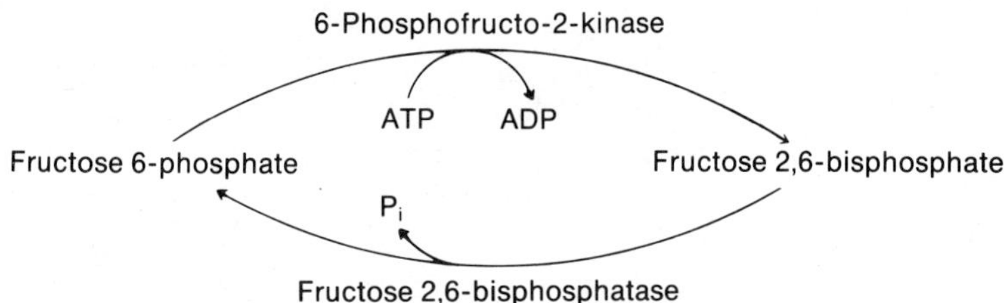

Figure 7.22
Reactions involved in the formation and degradation of fructose 2,6-bisphosphate.

pyrophosphate. Cyclic AMP triggers a series of intracellular events, the details of which are discussed below, that result ultimately in a decrease in fructose 2,6-bisphosphate levels. A decrease in this compound makes 6-phosphofructo-1-kinase less effective but makes fructose 1,6-bisphosphatase more effective, thereby severely restricting flux from fructose 6-phosphate to fructose 1,6-bisphosphate in the glycolytic pathway.

Note that fructose 2,6-bisphosphate is not an intermediate of the glycolytic pathway. As shown in Figure 7.22, fructose 2,6-bisphosphate is produced from fructose 6-phosphate by an enzyme called 6-phosphofructo-2-kinase (also referred to as phosphofructokinase-2). We now have two "phosphofructokinases" to contend with, one producing an intermediate (fructose 1,6-bisphosphate) of the glycolytic pathway and the other producing an important allosteric effector (fructose 2,6-bisphosphate) of the first enzyme. Before the discovery of fructose 2,6-bisphosphate, 6-phosphofructo-1-kinase could be called phosphofructokinase or even PFK among friends. Now we must carefully distinguish between these two very important enzymes.

Fructose 2,6-bisphosphate is destroyed in cells by being converted back to fructose 6-phosphate by fructose 2,6-bisphosphatase (Figure 7.22). This is a simple hydrolysis, with no ATP or ADP being involved, in contrast to fructose 2,6-bisphosphate synthesis. Of interest is the fact that a *bifunctional* enzyme carries out both the synthesis and degradation of fructose 2,6-bisphosphate. You may recall that a bifunctional enzyme is responsible for the synthesis and degradation of 2,3-bisphosphoglycerate (see page 272) and you will find in Chapter 9 that a multifunctional enzyme (fatty acid synthase) catalyzes numerous reactions during the process of fatty acid synthesis. Because of its bifunctional nature, the combined name of 6-phosphofructo-2-kinase/fructose 2,6-bisphosphatase is used to refer to the enzyme that makes and degrades fructose 2,6-bisphosphate. As mentioned above, cAMP is responsible for regulation of fructose 2,6-bisphosphate levels in the liver. How is this possible when the same enzyme carries out both the synthesis and degradation of the molecule? The answer is that a mechanism exists whereby cAMP is able to inactivate the kinase function and, at the same time, activate the phosphatase function of this bifunctional enzyme.

cAMP is an activator of an enzyme called cAMP-dependent protein kinase. This enzyme—in its inactive state—consists of two regulatory subunits of mol wt 85,000 each plus two catalytic subunits of mol wt 40,000 each. cAMP binds only to the regulatory subunits. Binding of cAMP causes conformational changes in the regulatory subunits which sets the catalytic subunits free. The catalytic subunits are active only after being dissociated from the regulatory subunit by this action of cAMP. The liberated protein kinase then catalyzes the phosphorylation of specific serine residues of the polypeptide chains of several different enzymes (Figure 7.23).

Phosphorylation of an enzyme can more conveniently be abbreviated as

$$\boxdot + \text{ATP} \longrightarrow \odot\text{–P} + \text{ADP}$$

$$\text{HN–CH(CH}_2\text{OH)–C(=O)} + ATP^{4-} \xrightarrow{Mg^{2+}} \text{HN–CH(CH}_2\text{OPO}_3^{2-}\text{)–C(=O)} + ADP^{3-} + H^+$$

Peptide chain Peptide chain

Figure 7.23
Enzymes subject to covalent modification are usually phosphorylated on specific serine residues.

where ⊡ and ⊙–P are used to indicate the dephosphorylated and phosphorylated enzymes, respectively. The circle and square symbols are used because phosphorylation of enzymes subject to regulation by covalent modification causes a change in their conformation, which affects the active site. It turns out that the change in conformation due to phosphorylation greatly increases the catalytic activity of some enzymes but greatly decreases the catalytic activity of others. It depends on the enzyme involved. Only a few of the known enzymes are subject to this type of regulation, called covalent modification. Regardless of whether phosphorylation or dephosphorylation activates the enzyme, the active form of the enzyme is called the "a" form and the inactive form the "b" form. Likewise, regardless of the effect of phosphorylation on catalytic activity, the action of cAMP-dependent protein kinase is always opposed by that of a phosphoprotein phosphatase, which catalyzes the reaction of

$$\odot\text{–P} + H_2O \longrightarrow \boxdot + P_i$$

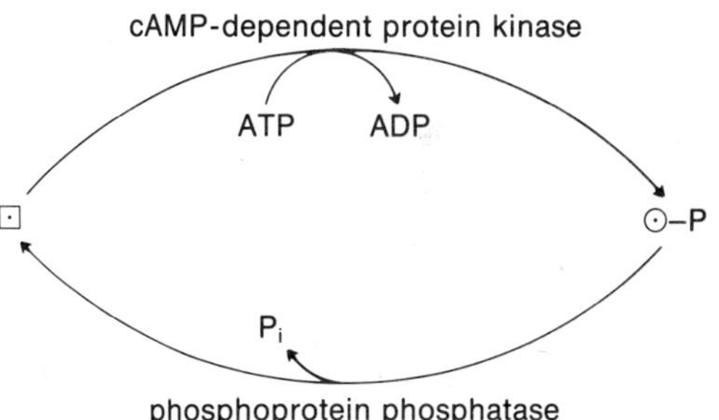

Figure 7.24
General model for the mechanism responsible for regulation of enzymes by phosphorylation/dephosphorylation.
The symbols ⊡ and ⊙–P indicate that different conformational and activity states of the enzyme are produced as a result of phosphorylation/dephosphorylation.

Putting these together creates a cyclic control system (see Figure 7.24), such that the ratio of phosphorylated enzyme to dephosphorylated enzyme is a function of the relative activities of the cAMP-dependent protein kinase and the phosphoprotein phosphatase. If the kinase has greater activity than the phosphatase, more enzyme will be in the phosphorylated mode—and vice versa. Since the activity of an interconvertible enzyme (i.e., an enzyme subject to covalent modification) is determined by whether it is in the phosphorylated or dephosphorylated mode, the relative activities of the kinase and phosphatase determine the amount of a particular enzyme which is in the catalytically active state.

As discussed above, most enzymes are either turned on or off by phosphorylation but, in the case of 6-phosphofructo-2-kinase/fructose 2,6-bisphosphatase, advantage is taken of the bifunctional nature of the enzyme. Phosphorylation causes inactivation of the active site responsible for synthesis of fructose 2,6-bisphosphate but activation of the active site responsible for hydrolysis of fructose 2,6-bisphosphate. Dephosphorylation of the enzyme has the opposite effects (Figure 7.25). A sensitive mechanism has evolved, therefore, to set the intracellular concentration of fructose 2,6-bisphosphate in response to changes in blood levels of glucagon (Figure 7.26). Increased levels of glucagon cause

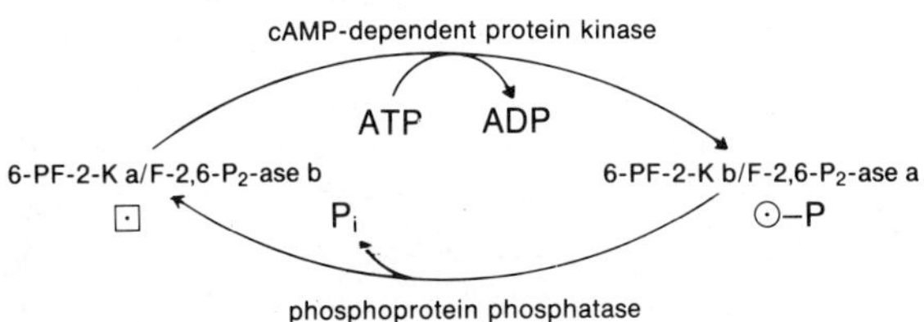

Figure 7.25
Mechanism responsible for covalent modification of the bifunctional enzyme 6-phosphofructo-2-kinase/fructose 2,6-bisphosphatase.
Name of the enzyme is abbreviated as 6-PF-2-K/F-2,6-P_2. Letters a and b indicate the active and inactive forms of the enzymatic activities respectively.

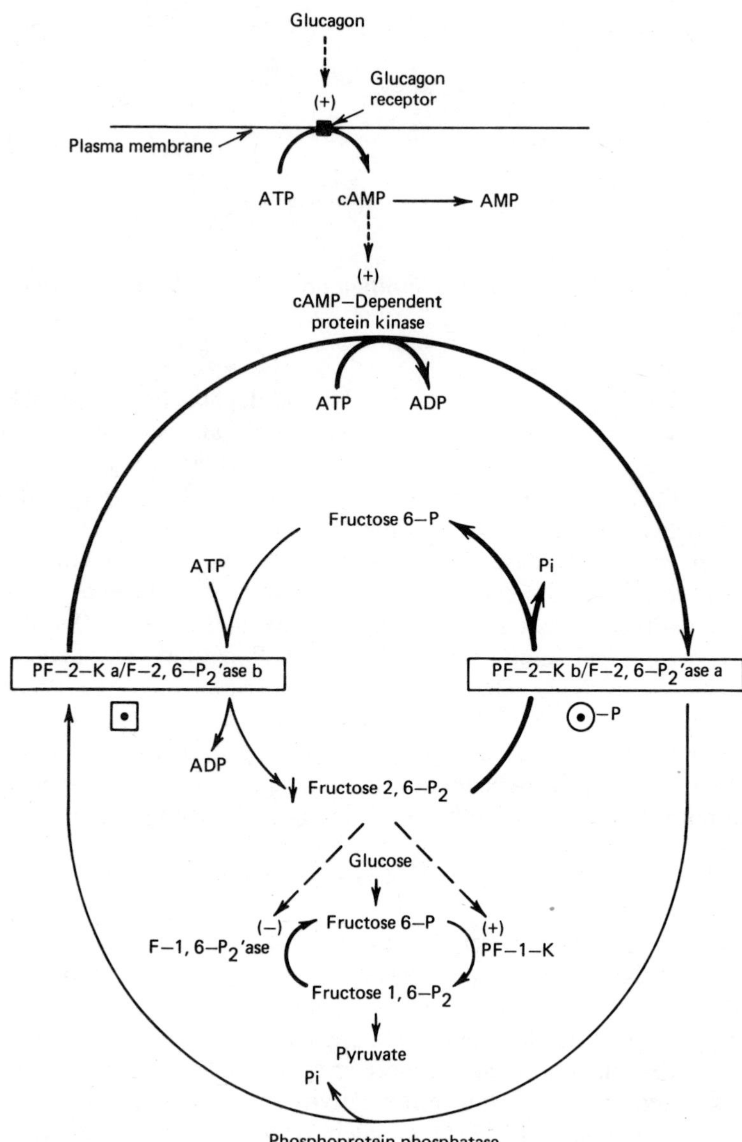

Figure 7.26
Mechanism responsible for glucagon inhibition of hepatic glycolysis via cAMP-mediated decrease in fructose 2,6-bisphosphate concentration.
The glucagon receptor-adenylate cyclase complex (■) is an intrinsic component of the plasma membrane. The (+) and (−) symbols indicate activation and inhibition of the designated enzymes, respectively. The heavy arrows indicate the reactions that predominate in the presence of glucagon. The small arrow (↓) in front of fructose 2,6-bisphosphate indicates a decrease in concentration of this compound in response to glucagon.

an increase in intracellular levels of cAMP. The second messenger activates cAMP-dependent protein kinase, which, in turn, phosphorylates 6-phosphofructo-2-kinase/fructose 2,6-bisphosphatase. The latter event inhibits fructose 2,6-bisphosphate synthesis and promotes its degradation. The resulting decrease in fructose 2,6-bisphosphate makes 6-phosphofructo-1-kinase less effective and fructose 1,6-bisphosphatase more effective. The overall result is inhibition of glycolysis at the level of the conversion of fructose 6-phosphate to fructose 1,6-bisphosphate. Decreased levels of glucagon in the blood result in less cAMP in the liver because adenylate cyclase is less active and the cAMP that had accumulated is converted to AMP by the action of cAMP phosphodiesterase (Figure 7.27). Loss of the cAMP signal results in inactivation of cAMP-dependent protein kinase and a corresponding decrease in the rate of phosphorylation of 6-phosphofructo-2-kinase/fructose 2,6-bisphosphatase by cAMP-dependent protein-kinase. A phosphoprotein phosphatase removes phosphate from the bifunctional enzyme to produce active 6-phosphofructo-2-kinase and inactive fructose 2,6-bisphosphatase. Fructose 2,6-bisphosphate can now accumulate to a higher steady-state concentration and, by activating 6-phosphofructo-1-kinase and inhibiting fructose 1,6-bisphosphatase, greatly increase the rate of glycolysis. It

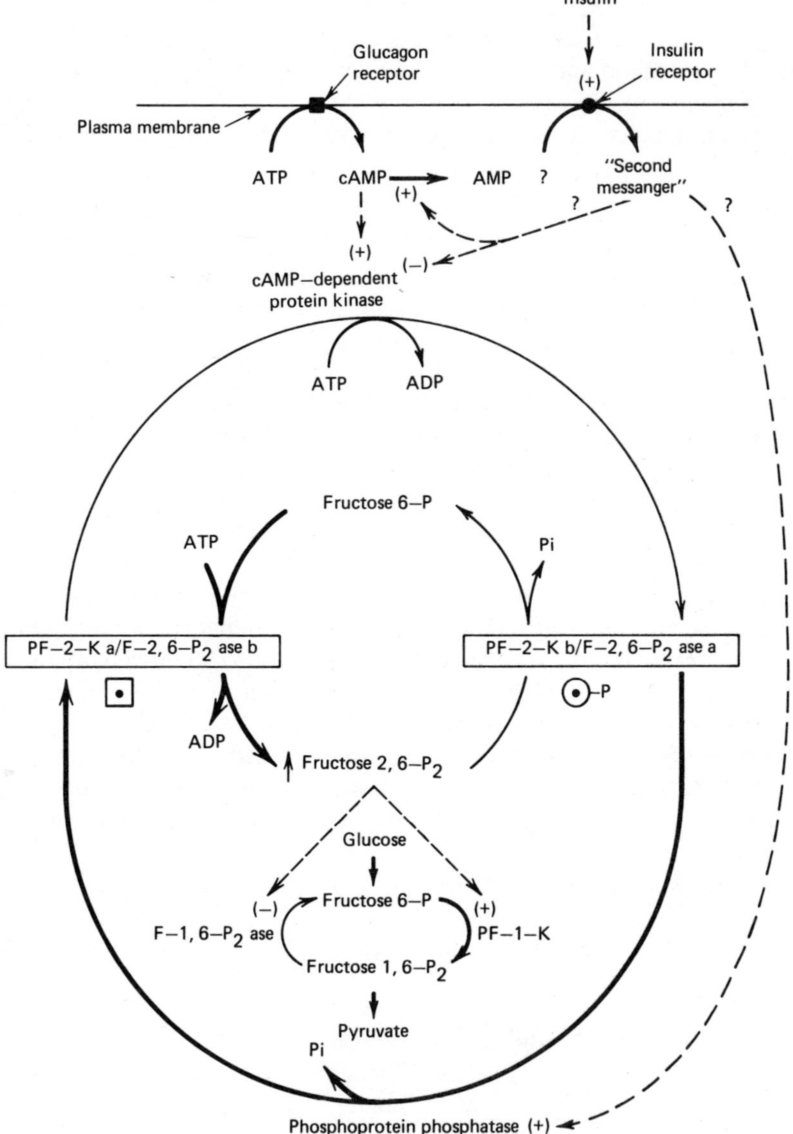

Figure 7.27
Mechanism responsible for accelerated rates of hepatic glycolysis when the concentration of glucagon is low and that of insulin is high in the blood.
See legend to Figure 7.26 for the meaning of symbols and heavy arrows. The insulin receptor complex (●) is an intrinsic component of the plasma membrane. The small arrow in front of fructose 2,6-bisphosphate indicates an increase in concentration of this compound in response to an increased insulin signal and decreased glucagon signal. The question marks indicate that the details of the mechanism of action of insulin are unknown at this time.

CLIN. CORR. 7.8
PYRUVATE KINASE DEFICIENCY HEMOLYTIC ANEMIA

Mature erythrocytes are absolutely dependent upon glycolytic activity for ATP production. ATP is needed for the ion pumps, especially the Na^+, K^+–ATPase, which maintain the biconcave disk shape of erythrocytes, a characteristic that helps erythrocytes slip through the capillaries as they deliver oxygen to the tissues. Without ATP the cells swell and lyse. Anemia due to excessive erythrocyte destruction is referred to as hemolytic anemia. Pyruvate kinase deficiency is rare, but is by far the most common genetic defect of the glycolytic pathway known to cause hemolytic anemia. Although most pyruvate kinase-deficient patients have 5–25% of normal red blood cell pyruvate kinase levels, flux through the glycolytic pathway is restricted severely, resulting in markedly lower ATP concentrations. The expected crossover of the glycolytic intermediates is observed, that is, those intermediates proximal to the pyruvate kinase-catalyzed step accumulate, whereas pyruvate and lactate concentrations decrease. Normal ATP levels are observed in reticulocytes of patients with this disease. Although deficient in pyruvate kinase, these "immature" red blood cells have mitochondria and can generate ATP by oxidative phosphorylation. Maturation of reticulocytes into red blood cells results in the loss of mitochondria and complete dependence on glycolysis for ATP production. Since glycolysis is defective, the mature cells are lost rapidly from the circulation. Anemia results because the cells cannot be replaced rapidly enough by erythropoiesis.

should be apparent from this discussion that glucagon is an extracellular signal that stops the liver from using glucose, whereas fructose 2,6-bisphosphate is an intracellular signal that promotes glucose utilization by this tissue.

The role of insulin in regulation of fructose 2,6-bisphosphate levels is poorly understood. Although it is clear that this hormone opposes the action of glucagon, exactly how insulin works after binding to the plasma membrane remains to be established. One current hypothesis is presented in Figure 7.27. The idea is that insulin binding may promote the formation of an intracellular messenger, much like glucagon promotes the formation of its intracellular messenger, cAMP. Obvious enzyme targets that this hypothetical "second messenger" might influence include cAMP phosphodiesterase, cAMP-dependent protein kinase, and phosphoprotein phosphatase (Figure 7.27). Regardless of the exact mechanism of action of insulin, glucagon and insulin clearly act in opposition to one another, and the insulin-to-glucagon ratio of the blood must determine intracellular levels of fructose 2,6-bisphosphate and, therefore, the rate of glycolysis.

Pyruvate Kinase

Pyruvate kinase is another regulatory enzyme of glycolysis (see Clin. Corr. 7.8). However, as with hexokinase, the reaction catalyzed by pyruvate kinase has to be considered a secondary site of regulation of glycolysis. This enzyme is drastically inhibited by physiological concentrations of ATP, so much so that its potential activity is never fully realized under physiological conditions. The isoenzyme found in liver is greatly activated by fructose 1,6-bisphosphate, thereby linking regulation of pyruvate kinase to what is happening at the level of 6-phosphofructo-1-kinase. Thus, if conditions favor increased flux through 6-phosphofructo-1-kinase, the level of fructose 1,6-bisphosphate increases and acts as a feed-forward activator of pyruvate kinase. The liver enzyme is also subject to covalent modification, being active in the dephosphorylated state and inactive in the phosphorylated state (Figure 7.28). Inactivation of pyruvate kinase by phosphorylation is a function of cAMP-dependent protein kinase in the liver. Glucagon inhibition of hepatic glycolysis and stimulation of hepatic gluconeogenesis are explained in part by the elevation of cAMP levels caused by this hormone. This aspect is explored more thoroughly under the section of this chapter on gluconeogenesis, and in Chapter 14.

Pyruvate kinase, like glucokinase, is induced to higher steady-state concentrations in the liver by the combination of high carbohydrate intake and high insulin levels. This is a major reason why the liver of the well-fed

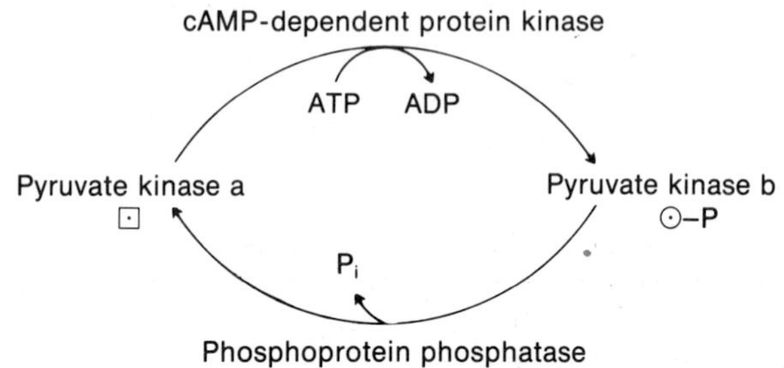

Figure 7.28
Glucagon acts via cAMP to cause the phosphorylation and inactivation of hepatic pyruvate kinase.

individual has much greater capacity for utilizing carbohydrate than a fasting or diabetic person (see Clin. Corr. 7.4).

7.5 GLUCONEOGENESIS

Importance of Glucose Synthesis

The net synthesis or formation of glucose from a large variety of noncarbohydrate substrates is termed gluconeogenesis. This includes the use of various amino acids, lactate, pyruvate, propionate and glycerol, as sources of carbon for the pathway (see Figure 7.29). Glucose is also synthesized from galactose and fructose. Glycogenolysis, that is, the formation of glucose or glucose 6-phosphate from glycogen, should be carefully differentiated from gluconeogenesis; glycogenolysis refers to

$$\text{Glycogen or (glucose)}_n \longrightarrow n \text{ molecules of glucose}$$

and thus does not correspond to de novo or new synthesis of glucose, the hallmark of the process of gluconeogenesis.

The capacity to synthesize glucose is crucial for the survival of humans and other animals. Blood glucose levels have to be maintained to support metabolism of those tissues that use glucose as their primary substrate (see Clin. Corr. 7.9). This includes brain, red blood cells, kidney medulla, lens and cornea of the eye, testis, and a number of other tissues. Gluconeogenesis enables the maintenance of blood glucose levels long after all dietary glucose has been absorbed and completely oxidized.

The Cori and Alanine Cycles

Two important cycles between tissues are recognized in gluconeogenesis. The Cori cycle and the alanine cycle, given in Figure 7.30, consist of gluconeogenesis in the liver followed by glycolysis in a peripheral tissue. The purpose of both is to provide a mechanism for continuously supplying glucose to tissues that are dependent on it as their primary energy source. The cycles are only functional, however, between the liver and tissues that do not completely oxidize glucose to CO_2 and H_2O. In order to participate in these cycles, the peripheral tissue must release either alanine or lactate as the end product of glycolysis. The type of recycled three-carbon intermediate is the major difference between the Cori cycle and the alanine cycle, carbon returning to the liver in the form of lactate in the Cori cycle but in the form of alanine in the alanine cycle. Another difference is that the NADH generated by glycolysis in the alanine cycle

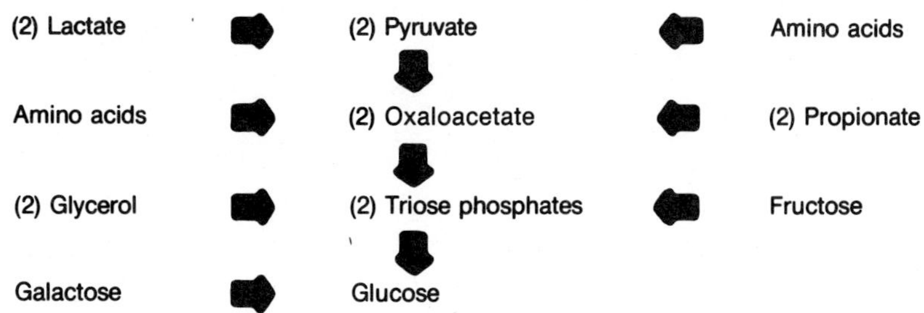

Figure 7.29
Abbreviated pathway of gluconeogenesis, illustrating the major substrate percursors for the process.

CLIN. CORR. 7.9
HYPOGLYCEMIA AND PREMATURE INFANTS

Premature and small-for-gestational-age neonates have a greater susceptibility to hypoglycemia than full-term, appropriate-for-gestational-age infants. Several factors appear to be involved. Children in general are more susceptible than adults to hypoglycemia, simply because they have larger brain to body weight ratios and the brain utilizes disproportionately greater amounts of glucose than the rest of the body. Newborn infants have a limited capacity for ketogenesis, apparently because the transport of long-chain fatty acids into liver mitochondria of the neonate is poorly developed. Since ketone body use by the brain is directly proportional to the circulating ketone body concentration, the neonate is unable to spare glucose to any significant extent by using ketone bodies. The consequence is that the neonate's brain is almost completely dependent upon glucose obtained from liver glycogenolysis and gluconeogenesis.

The capacity for hepatic glucose synthesis from lactate and alanine is also limited in newborn infants. This is because the rate-limiting enzyme phospho*enol*pyruvate carboxykinase is present in very low amounts during the first few hours after birth. Induction of this enzyme to the level required to prevent hypoglycemia during the stress of fasting requires several hours. Premature and small-for-gestational-age infants are believed to be more susceptible to hypoglycemia than normal infants because of smaller stores of liver glycogen. Fasting depletes their glycogen stores more rapidly, making these neonates more dependent on gluconeogenesis than normal infants.

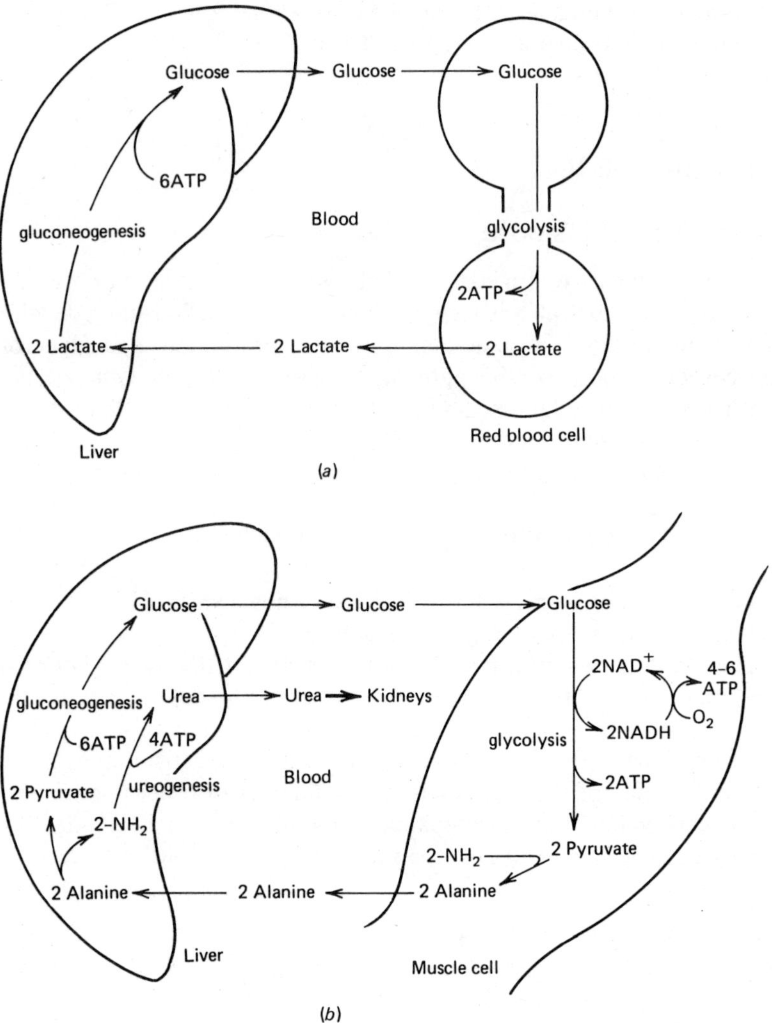

Figure 7.30
Relationship between gluconeogenesis in the liver and glycolysis in the rest of the body.
(a) Cori cycle. (b) Alanine cycle.

cannot be used to reduce pyruvate to lactate. In tissues that have mitochondria, the electrons of NADH can be transported into the mitochondria by the malate–aspartate shuttle or the glycerol phosphate shuttle for the synthesis of ATP by oxidative phosphorylation:

$$NADH + H^+ + \tfrac{1}{2}O_2 + 3ADP + 3P_i \longrightarrow NAD^+ + 3ATP$$

or

$$FADH_2 + \tfrac{1}{2}O_2 + 2ADP + 2P_i \longrightarrow FAD + 2ATP$$

The consequence is that six to eight molecules of ATP can be formed per glucose molecule in peripheral tissues that participate in the alanine cycle. This stands in contrast to the Cori cycle, in which only two ATPs per glucose are produced. Inspection of Figure 7.30*a* will reveal that the overall stoichiometry for the Cori cycle is

$$6ATP_{\text{liver}} + 2(ADP + P_i)_{\text{red blood cells}} \longrightarrow 6(ADP + P_i)_{\text{liver}} + 2ATP_{\text{red blood cells}}$$

The six ATPs are needed in the liver to provide the energy necessary for glucose synthesis. The alanine cycle also transfers the energy equivalent of a high-energy phosphate from liver to peripheral tissues and, because of the 6 to 8 ATPs produced per glucose, is an energetically more efficient cycle. However, as shown in Figure 7.30*b*, the participation of alanine in the cycle presents the liver with amino nitrogen, which must be disposed of as urea. In terms of ATP, urea synthesis is expensive (4 ATP molecules per urea molecule). The concurrent need for urea synthesis results in more ATP being needed per glucose molecule synthesized in the liver. The overall stoichiometry for the alanine cycle, as presented in Figure 7.30*b*, is then

$$10ATP_{liver} + 6\text{–}8(ADP + P_i)_{muscle} + O_{2\ muscle} \longrightarrow 10(ADP + P_i)_{liver} + 6\text{–}8ATP_{muscle}$$

Note that the last equation makes the point that, in contrast to the Cori cycle, oxygen and mitochondria are required in the peripheral tissue for participation in the alanine cycle.

Liver was used as the example in Figure 7.30 because it is the most important gluconeogenic tissue. The kidneys, on a wet weight basis, have about the same capacity for the process. However, the liver is the largest organ in the body, exceeding the combined weight of the kidneys by a factor of 4, and thus contributes much more to the maintenance of blood glucose levels by gluconeogenesis. Certain muscle fibers may have the capacity for limited gluconeogenesis. Since the adult human has 18 times more muscle mass than liver, glucose synthesis in muscle may eventually be shown to be quantitatively important. However, muscle tissue lacks glucose 6-phosphatase, the enzyme that catalyzes the last step of the gluconeogenic pathway. Thus, any gluconeogenesis occurring in muscle should be pictured as taking place in order to help replenish glycogen stores in this tissue, rather than for the production of free glucose for the maintenance of blood sugar levels.

Pathway Responsible for Glucose Synthesis

Gluconeogenesis from lactate is an ATP-requiring process with the overall equation of

$$2\ \text{L-Lactate}^- + 6ATP^{4-} \longrightarrow \text{glucose} + 6ADP^{3-} + 6P_i^{2-} + 4H^+$$

Many of the enzymes of the glycolytic pathway are common to the gluconeogenic pathway but, it is obvious from the overall equation for glycolysis,

$$\text{Glucose} + 2ADP^{3-} + 2P_i^{2-} \longrightarrow 2\ \text{L-lactate}^- + 2ATP^{4-}$$

that additional reactions have to be involved. Also, as pointed out in the discussion of glycolysis, certain steps of this pathway are irreversible under intracellular conditions and are replaced by irreversible steps of the gluconeogenic pathway. The reactions of gluconeogenesis from lactate are given in Figure 7.31. The initial step is the conversion of lactate to pyruvate by lactate dehydrogenase. NADH is generated and is also needed for a subsequent step in the pathway. Pyruvate cannot be converted to phospho*enol*pyruvate by reversing the step used in glycolysis because the reaction catalyzed by pyruvate kinase is irreversible under intracellular conditions. Pyruvate is converted into the high-energy phosphate compound phospho*enol*pyruvate by the coupling of two reactions

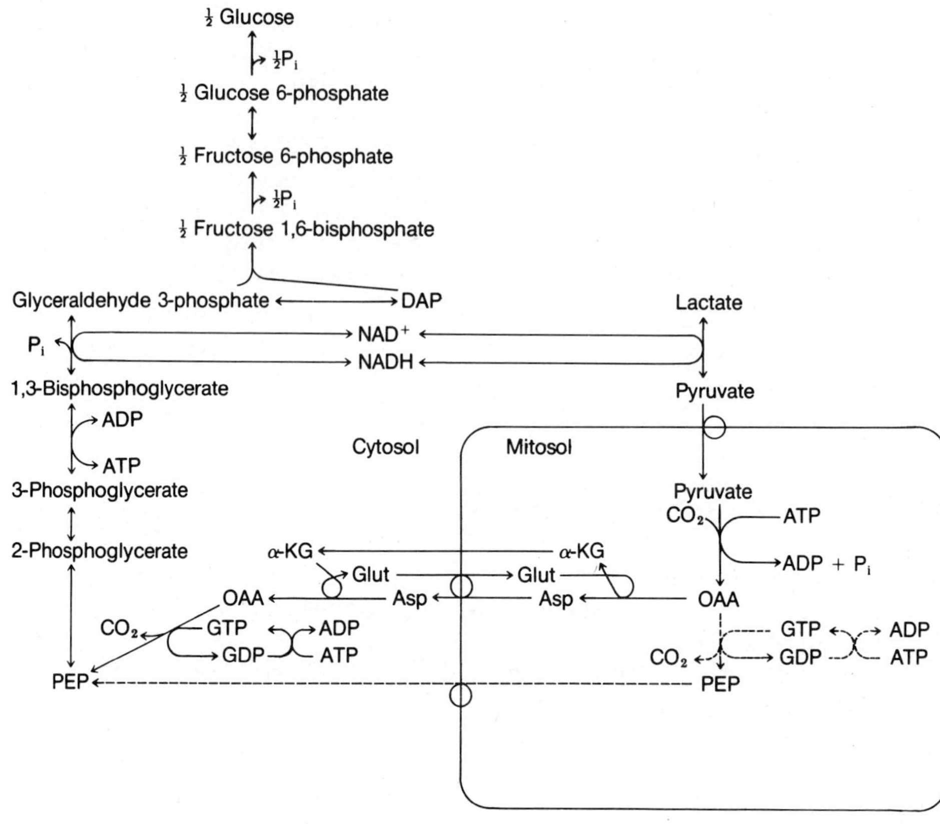

Figure 7.31
Pathway of gluconeogenesis from lactate.
The involvement of the mitochondrion in the process is indicated in the figure. Dashed arrows refer to an alternate route which employs mitosolic phosphoenolpyruvate carboxykinase rather than the cytosolic isoenzyme. Abbreviations: OAA, oxaloacetate; α-KG, α-ketoglutarate; PEP, phosphoenolpyruvate; and DAP, dihydroxyacetone phosphate.

requiring high-energy phosphate compounds (an ATP and a GTP). The first is catalyzed by pyruvate carboxylase and the second by phospho*enol*pyruvate carboxykinase (see Figure 7.32).

Since the GTP required for the phospho*enol*pyruvate carboxykinase catalyzed reaction is equivalent to an ATP through the action of nucleoside diphosphate kinase (GDP + ATP $\rightleftharpoons$ GTP + ADP), and since CO_2 and HCO_3^- readily equilibrate by the action of carbonic anhydrase ($CO_2 + H_2O \rightleftharpoons H_2CO_3 \rightleftharpoons H^+ + HCO_3^-$), the sum of these reactions is

$$\text{Pyruvate}^- + 2\text{ATP}^{4-} \longrightarrow \text{phospho}\textit{enol}\text{pyruvate}^{3-} + 2\text{ADP}^{3-} + 2\text{P}_i^{2-} + 4\text{H}^+$$

Whereas the conversion of phospho*enol*pyruvate to pyruvate by the enzyme pyruvate kinase yields the cell one molecule of ATP, the conversion of pyruvate into phospho*enol*pyruvate by the combination of pyruvate carboxylase and phospho*enol*pyruvate carboxykinase costs the cell 2 ATP molecules.

Now for some interesting details. As shown in Figure 7.31, the conversion of cytosolic pyruvate into cytosolic phospho*enol*pyruvate requires the participation of the mitochondrion. Pyruvate carboxylase is housed within the mitochondrion, making these particles mandatory for glucose synthesis. There are two routes that oxaloacetate can then take to glu-

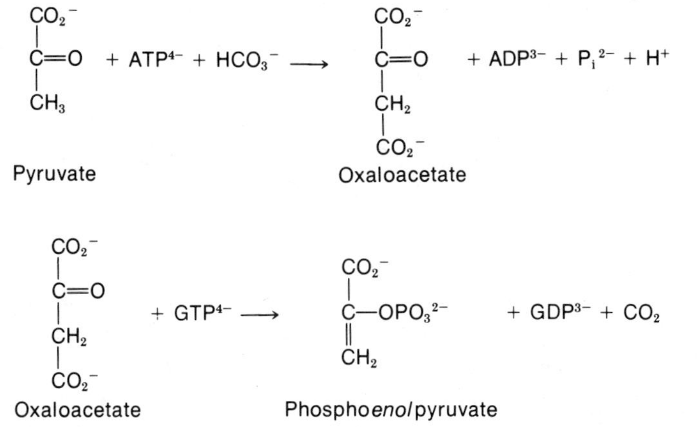

$$\begin{array}{c} CO_2^- \\ | \\ C{=}O \\ | \\ CH_3 \end{array} + ATP^{4-} + HCO_3^- \longrightarrow \begin{array}{c} CO_2^- \\ | \\ C{=}O \\ | \\ CH_2 \\ | \\ CO_2^- \end{array} + ADP^{3-} + P_i^{2-} + H^+$$

Pyruvate   Oxaloacetate

$$\begin{array}{c} CO_2^- \\ | \\ C{=}O \\ | \\ CH_2 \\ | \\ CO_2^- \end{array} + GTP^{4-} \longrightarrow \begin{array}{c} CO_2^- \\ | \\ C{-}OPO_3^{2-} \\ \| \\ CH_2 \end{array} + GDP^{3-} + CO_2$$

Oxaloacetate   Phospho*enol*pyruvate

Figure 7.32
Energy requiring steps involved in phospho*enolpyruvate *formation from pyruvate.
*Reactions are catalyzed by pyruvate carboxylase and phospho*enolpyruvate *carboxykinase, respectively.*

cose—and both are important in human liver. This happens because phospho*enol*pyruvate carboxykinase occurs in both the cytosolic and mitosolic compartments. The simplest pathway to follow is the one involving the mitochondrial phospho*enol*pyruvate carboxykinase. In this case, oxaloacetate is simply converted within the mitochondrion into phospho*enol*pyruvate, which then traverses the mitochondrial inner membrane in search of the rest of the enzymes of the gluconeogenic pathway. The second pathway would also be simple if oxaloacetate could traverse the mitochondrial inner membrane to reach the cytosolic phospho*enol*pyruvate carboxykinase; however, as already discussed with respect to the malate–aspartate shuttle (Figure 7.10), oxaloacetate per se cannot escape from the mitochondrion. Thus, the trick is again used, as in the malate–aspartate shuttle (Figure 7.10*B*), of converting oxaloacetate into aspartate that traverses the mitochondrial inner membrane by way of the aspartate–glutamate antiport. Aspartate is converted back to oxaloacetate in the cytosol by transamination with α-ketoglutarate.

The steps from phospho*enol*pyruvate to fructose 1,6-bisphosphate are already familiar, being just the reverse of steps of the glycolytic pathway. Note that the NADH generated by lactate dehydrogenase is utilized by the reaction catalyzed by glyceraldehyde 3-phosphate dehydrogenase.

6-Phosphofructo-1-kinase catalyzes an irreversible step in the glycolytic pathway and cannot be used for the conversion of fructose 1,6-bisphosphate to fructose 6-phosphate. A way around this problem is offered by the enzyme fructose 1,6-bisphosphatase, which catalyzes the irreversible reaction shown in Figure 7.33. Note that ATP and ADP are not involved and that this reaction can be used to yield fructose 6-phos-

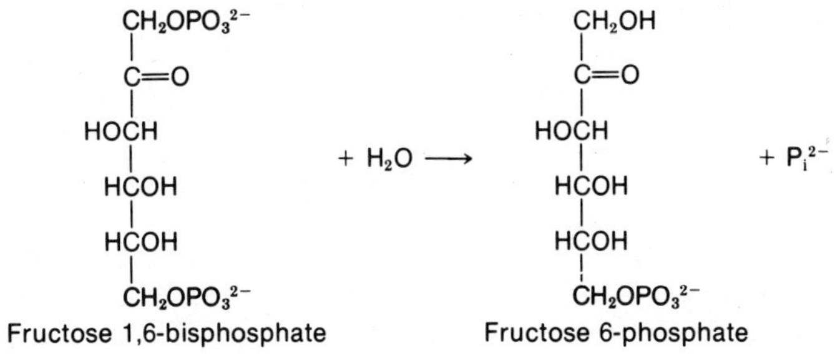

$$\begin{array}{c} CH_2OPO_3^{2-} \\ | \\ C{=}O \\ | \\ HOCH \\ | \\ HCOH \\ | \\ HCOH \\ | \\ CH_2OPO_3^{2-} \end{array} + H_2O \longrightarrow \begin{array}{c} CH_2OH \\ | \\ C{=}O \\ | \\ HOCH \\ | \\ HCOH \\ | \\ HCOH \\ | \\ CH_2OPO_3^{2-} \end{array} + P_i^{2-}$$

Fructose 1,6-bisphosphate   Fructose 6-phosphate

Figure 7.33
Reaction catalyzed by fructose 1,6-bisphosphatase.

$$\underset{\alpha\text{-D-Glucose 6-phosphate}}{CH_2OPO_3^{2-}\text{(pyranose ring)}} + H_2O \longrightarrow \underset{\alpha\text{-D-Glucose}}{CH_2OH\text{(pyranose ring)}} + P_i^{2-}$$

Figure 7.34
Reaction catalyzed by glucose 6-phosphatase.

phate, but since it is irreversible, cannot be used in glycolysis to yield fructose 1,6-bisphosphate.

The reaction catalyzed by phosphoglucoisomerase is freely reversible and functions in both the glycolytic and gluconeogenic pathways. However, glucose 6-phosphatase has to be used instead of glucokinase for the last step. Glucose 6-phosphatase catalyzes an irreversible reaction under intracellular conditions (Figure 7.34). It should be noted again that nucleotides do not have a role in this reaction and that the function of this enzyme is to generate glucose, not to convert glucose into glucose 6-phosphate. Glucose 6-phosphatase is unique among the enzymes required for gluconeogenesis. It is a membrane-bound enzyme, housed within the endoplasmic reticulum, with its active site available for glucose 6-phosphate hydrolysis on the cisternal surface of the tubules (see Figure 7.35). A translocase for glucose 6-phosphate is required to move glucose 6-phosphate from the cytosol to its site of hydrolysis within the endoplasmic reticulum. A genetic defect in either the translocase or the phosphatase interferes with gluconeogenesis and results in massive accumulation of glycogen in the liver. This will be discussed later in our consideration of glycogen metabolism (Section 7.6).

Special Features of Gluconeogenesis

The pathway for gluconeogenesis from lactate (Figure 7.31) illustrates that, regardless of whether carbon exits the mitochondrion in the form of phospho*enol*pyruvate or aspartate, there is a perfect equivalence between the NADH generated by lactate dehydrogenase and the NADH used by glyceraldehyde 3-phosphate dehydrogenase. This is tidy—none is left over and no extra is required. Consider pyruvate, however, as a gluconeogenic substrate (Figure 7.36). Pyruvate should, one would think, just follow the same pathway as the pyruvate generated from lactate by lactate

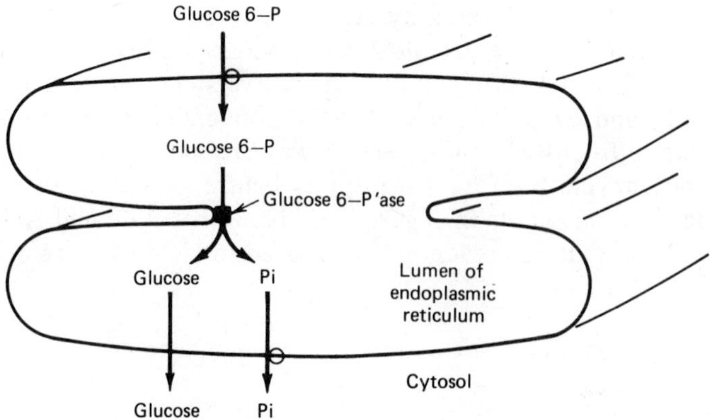

Figure 7.35
Glucose 6-phosphate is hydrolyzed by glucose 6-phosphatase (■) located on the cisternal surface of the endoplasmic reticulum.
Two transporters (○) are involved; one moves glucose 6-phosphate into the lumen, and the other moves inorganic phosphate back to the cytosol. The membrane is freely permeable to glucose.

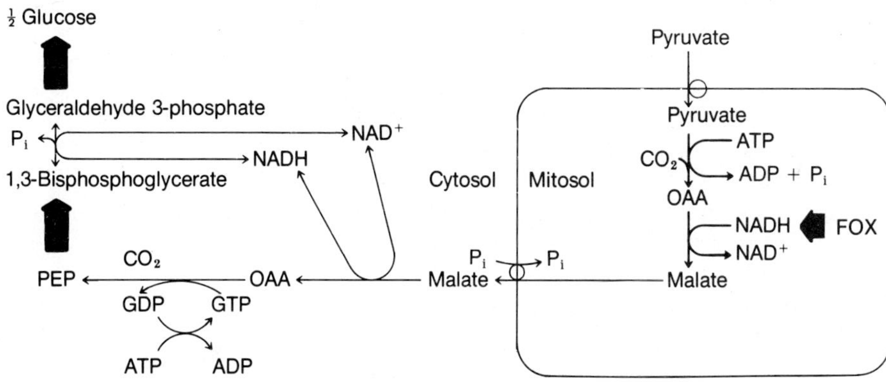

Figure 7.36
Pathway of gluconeogenesis from pyruvate.
Abbreviations are as in Figure 7.31. Large arrows indicate portions of the pathway which are identical to those given in Figure 7.31. FOX refers to the process of fatty acid oxidation.

dehydrogenase as in Figure 7.31. The NADH needed by the glyceraldehyde 3-phosphate dehydrogenase-catalyzed reaction, however, would not be generated by such a pathway. The problem is solved nicely, as shown in Figure 7.36, by having the carbon exit from the mitochondrion as malate. Thus, pyruvate penetrates into the mitochondrion where it is carboxylated to oxaloacetate. Oxaloacetate is reduced to malate by mitosolic malate dehydrogenase, using NADH generated by fatty acid oxidation. Egress of malate from the mitochondrion occurs by means of the malate–P_i antiport, which provides, by way of the cytosolic malate dehydrogenase, the oxaloacetate to be used as carbon for gluconeogenesis and the NADH needed at the level of glyceraldehyde 3-phosphate dehydrogenase.

All amino acids except leucine and lysine can supply carbon for the net synthesis of glucose by gluconeogenesis. The details of the pathways of amino acid catabolism are covered in Chapter 12. For our purposes here, it is very important to note that if the catabolism of an amino acid can yield either net pyruvate or net oxaloacetate formation, then net glucose synthesis can occur from that amino acid. As shown in Figures 7.31 and 7.36 oxaloacetate is an intermediate in gluconeogenesis and pyruvate is readily converted to oxaloacetate by the action of pyruvate carboxylase. The abbreviated pathway, given in Figure 7.29, illustrates how amino acid catabolism fits with the process of gluconeogenesis. The catabolism of amino acids feeds carbon into the tricarboxylic acid cycle at more than one point. As long as net synthesis of a tricarboxylic acid cycle intermediate occurs as a consequence of the catabolism of a particular amino acid, net synthesis of oxaloacetate will follow. Reactions that "fill up" the tricarboxylic acid cycle with intermediates, that is, lead to the net synthesis of tricarboxylic acid cycle intermediates, are called anaplerotic reactions. Such reactions support gluconeogenesis because they provide for the net synthesis of oxaloacetate. The reactions catalyzed by pyruvate carboxylase and glutamate dehydrogenase are good examples of anaplerotic reactions (anaplerosis):

$$\text{Pyruvate}^- + \text{ATP}^{4-} + \text{HCO}_3^- \longrightarrow \text{oxaloacetate}^{2-} + \text{ADP}^{3-} + \text{P}_i^{2-} + \text{H}^+$$

$$\text{Glutamate}^- + \text{NAD(P)}^+ \longrightarrow \alpha\text{-ketoglutarate}^{2-} + \text{NAD(P)H} + \text{NH}_4^+ + \text{H}^+$$

On the other hand, the glutamate-oxaloacetate transaminase reaction is not an anaplerotic reaction,

$$\alpha\text{-Ketoglutarate} + \text{aspartate} \rightleftharpoons \text{glutamate} + \text{oxaloacetate}$$

because net synthesis of a tricarboxylic acid cycle intermediate is not accomplished (note presence of an intermediate of the tricarboxylic acid cycle on both sides of the equation).

Glucose Synthesis from Amino Acids

Gluconeogenesis from amino acids imposes an additional nitrogen load upon the liver, as pointed out in the description of the alanine cycle in Figure 7.30*b*. Since the liver has to convert this nitrogen into urea, there is a close relationship between urea synthesis and glucose synthesis from amino acids. This relationship is illustrated in Figure 7.37 for alanine, the most important gluconeogenic amino acid. Two alanine molecules are shown being transaminated to give two molecules of pyruvate, which enter the mitochondrion where each is used by a separate pathway, one yielding malate plus NH_4^+ and the other aspartate. The latter two nitrogen-containing compounds should be recognized as the primary substrates for the urea cycle. The aspartate leaves the mitochondrion and becomes part of the urea cycle after reacting with citrulline. The carbon of aspartate is released from the urea cycle in the form of fumarate, the latter then being converted to malate by cytosolic fumarase. Both this malate and the malate exiting from the mitochondria are converted to glucose by the action of the cytosolic enzymes of the gluconeogenic pathway. As

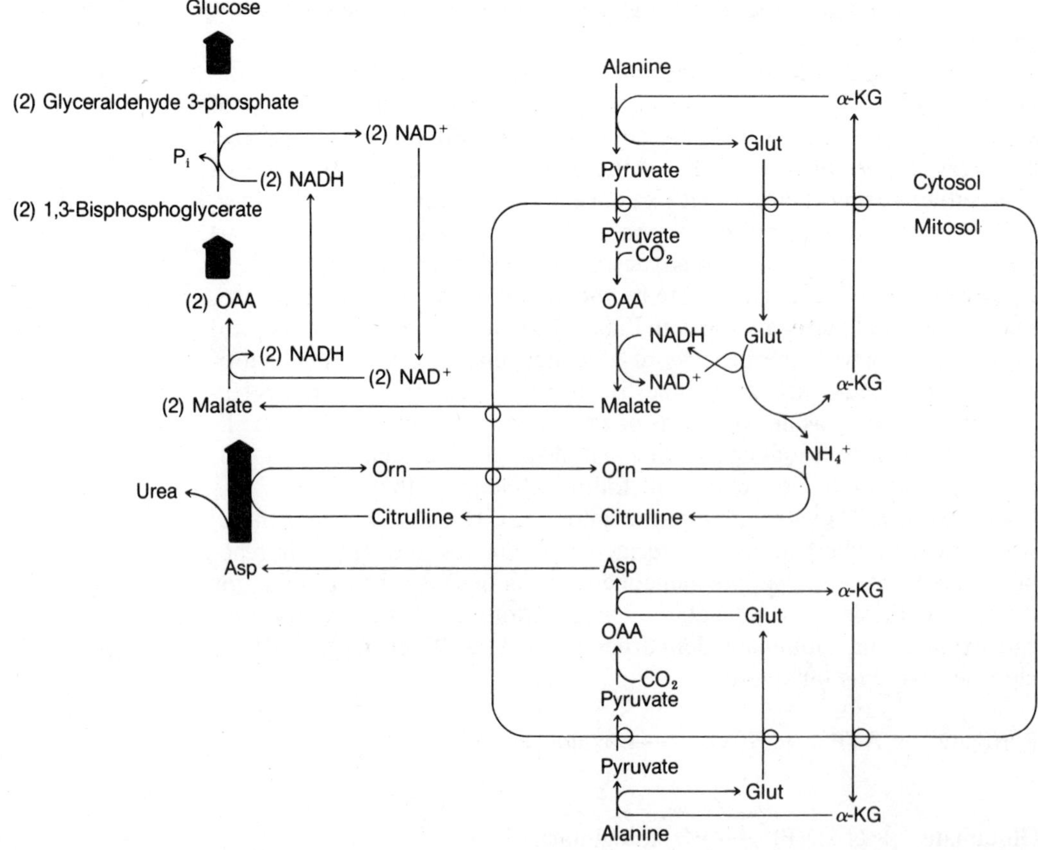

Figure 7.37
Pathway of gluconeogenesis from alanine and its relationship to urea synthesis.

shown in Figure 7.37, a balance is achieved between the reducing equivalents (NADH) generated and those required in both the cytosolic and mitosolic spaces.

Leucine and lysine are the only amino acids that cannot function as carbon sources for the net synthesis of glucose. These are the only amino acids that are only ketogenic and not also glucogenic. As shown in Table 7.2, all other amino acids are classified as glucogenic, or at least both glucogenic and ketogenic. Glucogenic amino acids give rise to the net synthesis of either pyruvate or oxaloacetate, whereas glucogenic-ketogenic amino acids also yield the ketone body acetoacetate, or at least acetyl CoA which is readily converted into ketone bodies. Acetyl CoA is the end product of lysine metabolism, and acetoacetate and acetyl CoA are the end products of leucine metabolism. In the human and other animals, no pathway exists for converting acetoacetate or acetyl CoA into pyruvate or oxaloacetate. It may not be immediately obvious why acetyl CoA cannot be used for net synthesis of glucose but remember that the reaction catalyzed by pyruvate dehydrogenase complex is irreversible:

TABLE 7.2 The Glucogenic and Ketogenic Amino Acids

Glucogenic	*Ketogenic*	*Both*
Glycine	Leucine	Threonine
Serine	Lysine	Isoleucine
Valine		Phenylalanine
Histidine		Tyrosine
Arginine		Tryptophan
Cysteine		
Proline		
Hydroxyproline		
Alanine		
Glutamate		
Glutamine		
Aspartate		
Asparagine		
Methionine		

$$\text{Pyruvate} + \text{NAD}^+ + \text{CoASH} \longrightarrow \text{acetyl CoA} + \text{NADH} + \text{CO}_2$$

meaning this reaction cannot be used to synthesize pyruvate from acetyl CoA. It might be argued that oxaloacetate is generated from acetyl CoA by way of the Tricarboxylic Acid cycle:

$$\text{Acetyl CoA} \longrightarrow \text{citrate} \xrightarrow{\text{TCA}} 2\text{CO}_2 + \text{oxaloacetate}$$

However, this is a fallacious argument because oxaloacetate must react with acetyl CoA to give citrate by way of citrate synthase:

$$\begin{aligned} \text{Acetyl CoA} + \text{oxaloacetate} &\longrightarrow \text{citrate} + \text{CoA} \\ \text{Citrate} &\xrightarrow{\text{TCA}} 2\text{CO}_2 + \text{oxaloacetate} \\ \text{Sum:}\quad \text{Acetyl CoA} &\longrightarrow 2\text{CO}_2 + \text{CoA} \end{aligned}$$

The point is that, although students of biochemistry have tried every conceivable way in the laboratory and on examinations, it turns out to be impossible for animals to synthesize net oxaloacetate or glucose from acetyl CoA.

Glucose Synthesis from Fat

This lack of an anaplerotic pathway from acetyl CoA also means that in general it is impossible to synthesize glucose from fatty acids. Most fatty acids found in the human body are of the straight-chain variety with an even number of carbon atoms. Their catabolism by fatty acid oxidation (FOX) followed by ketogenesis or complete oxidation to CO_2 can be abbreviated as given in Figure 7.38. Since acetyl CoA and other intermediates of even numbered fatty acid oxidation cannot be converted to oxaloacetate or any other intermediate of gluconeogenesis, it is impossible to synthesize glucose from fatty acids. An exception to this general rule applies to fatty acids with methyl branches (e.g., phytanic acid, obtained as a breakdown product of chlorophyll; see discussion of Refsum's disease, Clin. Corr. 9.4) and fatty acids with an odd number of carbon atoms. The catabolism of such compounds yields propionyl CoA:

$$\text{Fatty acid with an odd number } (n) \text{ of carbon atoms} \longrightarrow \frac{(n-3)}{2}\text{ acetyl CoA} + 1\text{ propionyl CoA}$$

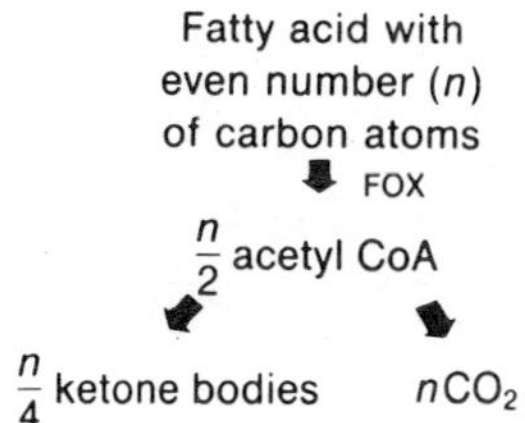

Figure 7.38
Overview of the catabolism of fatty acids to ketone bodies and CO_2.

CO_2^-
CH_2
CH_3
Propionate
CoA, ATP → PP_i, AMP (propionyl CoA synthetase)
O
‖
CSCoA
CH_2
CH_3
Propionyl CoA
CO_2, ATP → P_i, ADP (propionyl CoA carboxylase (biotin))
O
‖
CSCoA
H—C—CO_2^-
CH_3
(*S*)-Methylmalonyl CoA
↓ methylmalonyl CoA racemase
O
‖
C—SCoA
H—C—CH_3
CO_2^-
(*R*)-Methylmalonyl CoA
↓ methylmalonyl CoA mutase (vitamin B_{12})
O
‖
CSCoA
CH_2
CH_2
CO_2^-
Succinyl CoA
↓
$\frac{1}{2}$ Glucose

Figure 7.39
Pathway of gluconeogenesis from propionate.
The large arrow refers to steps of the tricarboxylic acid cycle (see Figure 6.17) plus steps of lactate-gluconeogenesis (see Figure 7.31).

CH_2—O—C(=O)—R
R′—C(=O)—O—CH
CH_2—O—C(=O)—R″

Figure 7.40
General structure of triacylglycerol.
R—C(=O)— refers to long-chain acyl groups of the molecule.

Propionate is a good precursor for gluconeogenesis, generating oxaloacetate by the anaplerotic pathway shown in Figure 7.39. Although not indicated in the figure, all of the steps involved in anaplerosis from propionate take place within the mitochondrion, and carbon has to exit from the mitochondrion in the form of malate in order to balance the reducing equivalent stoichiometry during gluconeogenesis. Propionate is also produced in the catabolism of valine and isoleucine and the conversion of cholesterol into bile acids.

It is sometimes loosely stated that fat *cannot* be converted into carbohydrate (glucose) by the liver. In a sense this is certainly true, that is, fatty acid metabolism, with the exception of fatty acids with branched chains or an odd number of carbon atoms, cannot give rise to net synthesis of glucose. However, the term "fat" is usually used to refer to triacylglycerols which are composed of three *O*-acyl groups combined with 1 glycerol molecule (Figure 7.40). Hydrolysis of this molecule of fat yields three fatty acids and glycerol, the latter compound being an excellent substrate for gluconeogenesis as shown in Figure 7.41. Phosphorylation of glycerol by glycerol kinase produces glycerol 3-phosphate, which can be converted back into fat by esterification with long-chain acyl CoA esters. However, of immediate concern is that glycerol 3-phosphate can be converted by glycerol 3-phosphate dehydrogenase into dihydroxyacetone phosphate, an intermediate of the gluconeogenic pathway (see Figure 7.31). As indicated in Figure 7.41, the last stage of glycolysis can compete with the gluconeogenic pathway and convert dihydroxyacetone phosphate into lactate (or into pyruvate for subsequent complete oxidation to CO_2 and H_2O).

Glucose Synthesis from Other Sugars

Fructose

Humans consume considerable quantities of fructose in the form of sucrose, and much of the fructose obtained by sucrose hydrolysis in the small bowel is converted into glucose in the liver. Like glucose, fructose is phosphorylated in the liver by a special ATP-linked kinase (Figure 7.42). Phosphorylation of fructose occurs in the 1 position to yield fructose 1-phosphate (see Clin. Corr. 7.3). A special aldolase then cleaves fructose 1-phosphate but not fructose 1,6-bisphosphate to give 1 molecule of dihydroxyacetone phosphate and 1 molecule of glyceraldehyde. The latter compound can be reduced to glycerol and used by the same pathway given for glycerol in the previous figure. The 2 molecules of dihydroxyacetone phosphate obtainable from one molecule of fructose can then be converted to glucose by enzymes of the gluconeogenic pathway or, alternatively, into pyruvate or lactate by the last stage of glycolysis. In analogy to glycolysis, the conversion of fructose into lactate is termed fructolysis.

Fructose is also generated in the body of man for an interesting purpose. The major energy source of spermatozoa is fructose, formed from glucose by cells of the seminal vesicles by the pathway given in Figure 7.43. Note that an NADPH-dependent reduction of glucose to sorbitol is followed by an NAD^+-dependent oxidation of sorbitol to fructose. Fructose is secreted from the seminal vesicles in a fluid that becomes part of the semen. Although the fructose concentration in human semen can exceed 10 mM, tissues that come in contact with semen utilize fructose poorly, allowing this substrate to be conserved to meet the energy demands of spermatozoa in their search for ova. Spermatozoa contain mitochondria and thus can metabolize fructose completely to CO_2 and H_2O by the combination of fructolysis and TCA cycle activity. The mitochondria of sperm are unique. They are the only mitochondria known to contain lactate dehydrogenase. In all other cells this enzyme is confined to the cytosol. This enables sperm mitochondria to oxidize lactate obtained by fructolysis and makes shuttle systems for the transport of reducing equivalents into the mitosol unnecessary.

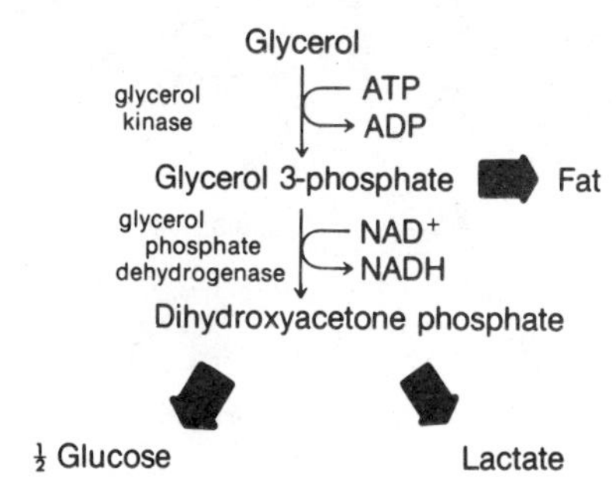

Figure 7.41
Pathway of gluconeogenesis from glycerol, along with competing pathways.
Large arrows indicate steps of the glycolytic and gluconeogenic pathways that have been given in detail in Figure 7.7 and 7.31, respectively. The large arrow pointing to fat refers to the synthesis of triacylglycerols and glycerophospholipids.

Galactose

Milk sugar or lactose constitutes an important source of galactose in the human diet. Glucose formation from galactose follows the pathway shown in Figure 7.44. The role of UDP-glucose as a recycling intermediate in the overall process of converting galactose into glucose should be noted. The absence of the enzyme galactose 1-phosphate uridylyltransferase accounts for most cases of galactosemia (see Clin. Corr. 8.3).

Mannose

Mannose is found in our diet, but in very limited quantities. Fortunately, its pathway of metabolism is also short and simple. It is first phosphorylated by hexokinase and then converted into fructose 6-phosphate by mannose phosphate isomerase:

$$\text{D-Mannose} + \text{ATP} \longrightarrow \text{D-mannose 6-phosphate} + \text{ADP}$$

$$\text{D-Mannose 6-phosphate} \rightleftharpoons \text{D-fructose 6-phosphate}$$

The latter compound can then be used in either the glycolytic pathway or the gluconeogenic pathway.

Cost of Glucose Synthesis

The synthesis of glucose is costly in terms of ATP. At least 6 molecules of ATP are required for the synthesis of 1 molecule of glucose from 2 molecules of lactate. The ATP needed by the liver cell for glucose synthesis is provided in large part by fatty acid oxidation. Metabolic conditions under which the liver is required to synthesize glucose generally favor increased availability of fatty acids in the blood. These fatty acids find their way to the liver mitochondria where they are oxidized to ketone bodies with the concurrent production of large amounts of ATP. This ATP is used to support the energy requirements of gluconeogenesis, regardless of the substrate being used as the carbon source for the process.

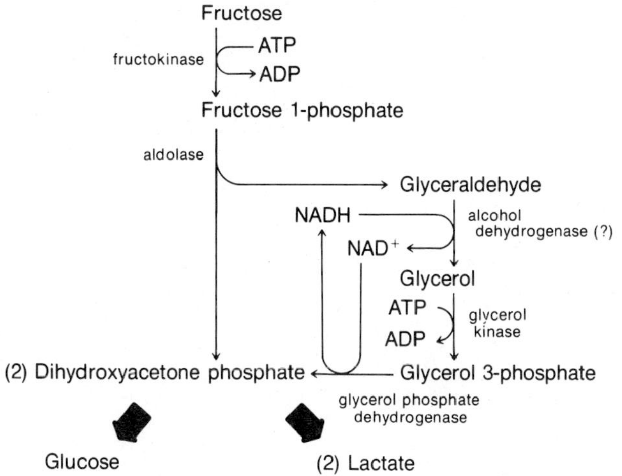

Figure 7.42
Pathway of glucose formation from fructose, along with the competing pathway of fructolysis.
Large arrows indicate steps of the glycolytic and gluconeogenic pathways that have been given in detail in Figures 7.7 and 7.31, respectively.

$$\text{D-Glucose} + \text{NADPH} + \text{H}^+ \longrightarrow \begin{matrix} CH_2OH \\ | \\ HCOH \\ | \\ HOCH \\ | \\ HCOH \\ | \\ HCOH \\ | \\ CH_2OH \\ \text{D-Sorbitol} \end{matrix} + \text{NADP}^+$$

$$\text{D-Sorbitol} + \text{NAD}^+ \longrightarrow \text{D-fructose} + \text{NADH} + \text{H}^+$$

Figure 7.43
The pathway responsible for the formation of sorbitol and fructose from glucose.

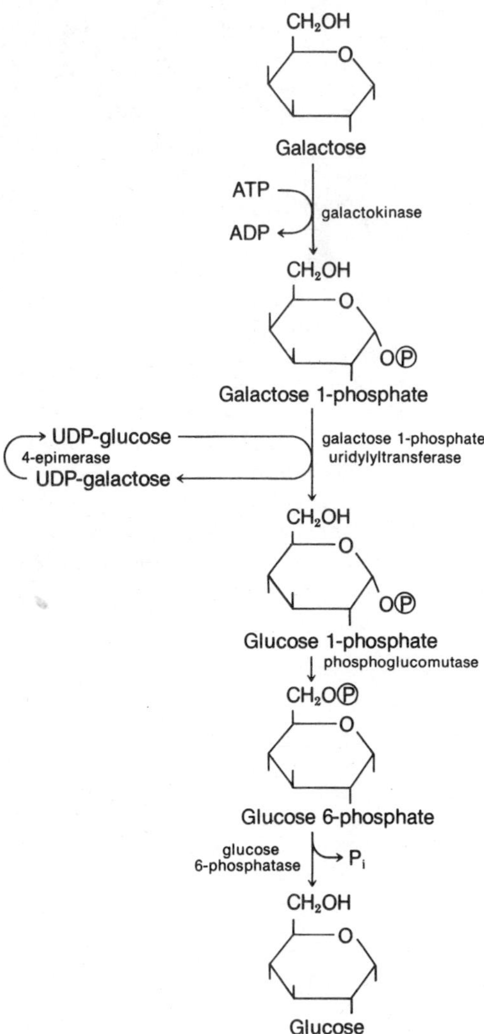

Figure 7.44
Pathway of glucose formation from galactose.

Regulation of the Gluconeogenic Pathway

The sites of regulation of the gluconeogenic pathway are apparent from the mass action ratios and equilibrium constants in Table 7.1, and are further indicated in Figure 7.45. Those enzymes that are used to "get around" the irreversible steps of glycolysis are primarily involved in regulation of the pathway, that is, pyruvate carboxylase, phospho*enol*pyruvate carboxykinase, fructose 1,6-bisphosphatase, and glucose 6-phosphatase. Considering the regulation of hepatic gluconeogenesis is almost the same as considering the regulation of hepatic glycolysis, which was discussed in some detail in earlier sections of this chapter. Inhibition of glycolysis at its chief regulatory sites, or repressing the synthesis of the enzymes involved at these sites (glucokinase and pyruvate kinase), greatly increases the effectiveness of the opposing gluconeogenic enzymes. Turning on gluconeogenesis is accomplished in large part, therefore, by shutting off glycolysis. Fatty acid oxidation does more than just supply ATP for the process. It actually promotes glucose synthesis. First of all, it increases the steady-state concentration of mitochondrial acetyl CoA, a positive allosteric effector of the mitochondrial enzyme pyruvate carboxylase. Second, the increase in acetyl CoA and in pyruvate carboxylase activity results in a greater synthesis of citrate, a negative effector of 6-phosphofructo-1-kinase. A secondary effect of inhibition of 6-phosphofructo-1-kinase is a decrease in fructose 1,6-bisphosphate concentration, an activator of pyruvate kinase. This decreases the flux of phospho*enol*pyruvate to pyruvate by pyruvate kinase, and increases the effectiveness of the combined efforts of pyruvate carboxylase and phospho*enol*pyruvate carboxykinase in the conversion of pyruvate to phospho*enol*pyruvate. An increase in ATP levels with the consequential decrease in AMP levels would favor gluconeogenesis by way of inhibition of 6-phosphofructo-1-kinase and pyruvate kinase and activation of fructose 1,6-bisphosphatase (see Figure 7.45 and the discussion of the regulation of glycolysis, page 282). A shortage of oxygen for respiration, a shortage of fatty acids for oxidation, or any inhibition or uncoupling of oxidative phosphorylation would be expected to cause the liver to turn from gluconeogenesis to glycolysis.

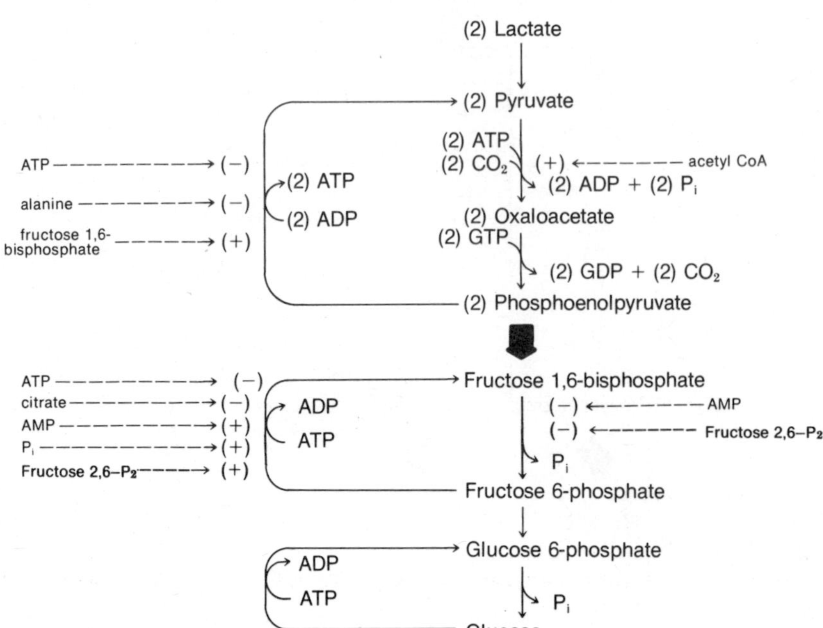

Figure 7.45
Important allosteric regulatory features of the gluconeogenic pathway.

Hormonal control of gluconeogenesis is a matter of regulating the supply of fatty acids to the liver and, in addition, regulating the enzymes of both the glycolytic and gluconeogenic pathways. Glucagon increases plasma fatty acids by promoting lipolysis in adipose tissue, an action which is opposed by insulin. The greater availability of fatty acids caused by glucagon results in more fatty acid oxidation by the liver which, as discussed above, promotes glucose synthesis. Insulin, on the other hand, has the opposite effect. Glucagon and insulin also regulate gluconeogenesis by influencing the state of phosphorylation of hepatic enzymes subject to covalent modification. As discussed in detail previously, pyruvate kinase of the glycolytic pathway is active in the dephosphorylated mode and inactive in the phosphorylated mode (see Figure 7.14). Glucagon activates adenylate cyclase to produce cAMP, which activates cAMP-dependent protein kinase, which, in turn, phosphorylates and inactivates pyruvate kinase. Inactivation of this glycolytic enzyme stimulates the opposing pathway (gluconeogenesis) by blocking the futile conversion of phospho*enol*pyruvate back to pyruvate. Glucagon also stimulates gluconeogenesis at the conversion of fructose 1,6-bisphosphate to fructose 6-phosphate by decreasing the concentration of fructose 2,6-bisphosphate present in the liver. Recall from our previous discussion of the regulation of glycolysis (page 286) that fructose 2,6-bisphosphate is a powerful allosteric activator of 6-phosphofructo-1-kinase and a powerful allosteric inhibitor of fructose 1,6-bisphosphatase. Glucagon, again working via its second messenger cAMP, lowers fructose 2,6-bisphosphate levels by stimulating the phosphorylation of the bifunctional enzyme 6-phosphofructo-2-kinase/fructose 2,6-bisphosphatase. Phosphorylation of this enzyme inactivates the site (kinase moiety) that makes fructose 2,6-bisphosphate from fructose 6-phosphate but activates the site (phosphatase moiety) that hydrolyzes fructose 2,6-phosphate back to fructose 6-phosphate. The consequence of a glucagon-induced fall in fructose 2,6-bisphosphate levels is that 6-phosphofructo-1-kinase becomes less active while fructose 1,6-bisphosphatase become more active (Figure 7.45). The overall effect is an increased conversion of fructose 1,6-bisphosphate to fructose 6-phosphate and a corresponding increase in the rate of gluconeogenesis. Insulin has effects opposite to those of glucagon—but the nature of the intracellular messenger formed in response to insulin is not known.

Glucagon and insulin also have long-term effects upon the levels of hepatic enzymes involved in glycolysis and gluconeogenesis. A high glucagon : insulin ratio in the blood increases the capacity for gluconeogenesis and decreases the capacity for glycolysis in the liver. A low glucagon : insulin ratio has the opposite effects. In addition to the short-term or acute mechanisms discussed above, this is accomplished by induction and repression of the synthesis of key enzymes of the pathways. Thus the glucagon : insulin ratio in the blood increases when gluconeogenesis is needed. This serves to signal the induction within the liver of the synthesis of greater quantities of phospho*enol*pyruvate carboxykinase, glucose 6-phosphatase, and various aminotransferases. The same signal causes the repression of the synthesis of glucokinase and pyruvate kinase. The opposite response occurs when glucose synthesis is not needed, that is, when a low glucagon : insulin ratio prevails because of maintenance of high blood glucose levels by glucose input from the gastrointestinal tract.

It is impossible to leave gluconeogenesis without saying something about the effects of alcohol on the process (see Clin. Corr. 7.10). Ethanol inhibits gluconeogenesis by the liver. Ethanol is oxidized primarily in the liver with the production of a large load of reducing equivalents that must be transported into the mitochondria by the malate–aspartate shuttle.

CLIN. CORR. 7.10
HYPOGLYCEMIA AND ALCOHOL INTOXICATION

Consumption of alcohol, especially by an undernourished person, can cause hypoglycemia. The same effect can result from drinking alcohol after strenuous exercise. In both cases the hypoglycemia results from the inhibitory effects of alcohol on hepatic gluconeogenesis and thus occurs under circumstances of hepatic glycogen depletion. The problem is caused by the NADH produced during the metabolism of alcohol. The liver simply cannot handle the reducing equivalents provided by ethanol oxidation fast enough to prevent metabolic derangements. The extra reducing equivalents block the conversion of lactate to glucose and promote the conversion of alanine into lactate, resulting in considerable lactate accumulation in the blood. Since lactate has no place to go, lactic acidosis (see Clin. Corr. 7.5) can develop.

Low doses of alcohol cause impaired motor and intellectual performance; high doses have a depressant effect, which can lead to stupor and anesthesia. Low blood sugar can contribute to these undesirable effects of alcohol. What is more, a patient may be thought to be simply inebriated when in fact he is suffering from hypoglycemia, with irreversible damage to the central nervous system. Alcohol-induced hypoglycemia has also been suggested to be a causative factor in the fetal alcohol syndrome. This syndrome, characterized by growth retardation and abnormalities of the central nervous system, is seen in the offspring of alcoholic and heavy-drinking women, and may be produced by even "social" drinking during pregnancy.

This excess NADH in the cytosol creates problems for liver gluconeogenesis because it forces the equilibrium of the lactate dehydrogenase- and malate dehydrogenase-catalyzed reactions in the directions of lactate and malate formation, respectively:

$$\underset{\text{Ethanol}}{CH_3CH_2OH} + NAD^+ \longrightarrow \underset{\text{Acetaldehyde}}{CH_3\overset{\overset{\displaystyle O}{\|}}{C}H} + NADH + H^+$$

$$\text{Pyruvate} + NADH + H^+ \longrightarrow \text{lactate} + NAD^+$$

$$\text{Sum:}\quad \text{Ethanol} + \text{pyruvate} \longrightarrow \text{acetaldehyde} + \text{lactate}$$

or

$$\text{Oxaloacetate} + NADH + H^+ \longrightarrow \text{malate} + NAD^+$$

$$\text{Sum:}\quad \text{Ethanol} + \text{oxaloacetate} \longrightarrow \text{acetaldehyde} + \text{malate}$$

In the presence of ethanol there is no shortage of NADH for the gluconeogenic pathway at the level of glyceraldehyde 3-phosphate dehydrogenase; however, forcing the equilibrium of lactate dehydrogenase and malate dehydrogenase as shown above inhibits glucose synthesis because pyruvate and oxaloacetate are no longer available in sufficient concentrations for the reactions catalyzed by pyruvate carboxylase and phospho*enol*pyruvate carboxykinase, respectively. The take home message is: Don't drink while synthesizing glucose!

7.6 GLYCOGENOLYSIS AND GLYCOGENESIS

Significance

Glycogenolysis refers to the intracellular breakdown of glycogen; glycogenesis to the intracellular synthesis of glycogen. We will be concerned here mainly with these processes in muscle and liver because of their greater quantitative importance in these tissues. However, it should be appreciated that these processes are of some importance in almost every tissue of the body.

The liver has tremendous capacity for storing glycogen. In the well-fed human the liver glycogen content can account for as much as 10% of the wet weight of this organ. Muscle stores less when expressed on the same basis—a maximum of only 1–2% of its wet weight. However, since the average person has more muscle than liver, there is about twice as much total muscle glycogen as liver glycogen.

Muscle and liver glycogen stores serve completely different roles. Muscle glycogen is present to serve as a fuel reserve for the synthesis of ATP within that tissue, whereas liver glycogen functions as a glucose reserve for the maintenance of blood glucose concentrations. Liver glycogen levels vary greatly in response to the intake of food, accumulating to high levels shortly after a meal and then decreasing slowly as it is mobilized to help maintain a nearly constant blood glucose level (see Figure 7.46). Liver glycogen reserves in the human are called into play between meals and to an even greater extent during the nocturnal fast. In both man and the rat, the store of glycogen in the liver lasts somewhere between 12 and 24 h during fasting, depending greatly, of course, upon whether the individual under consideration is caged or running wild.

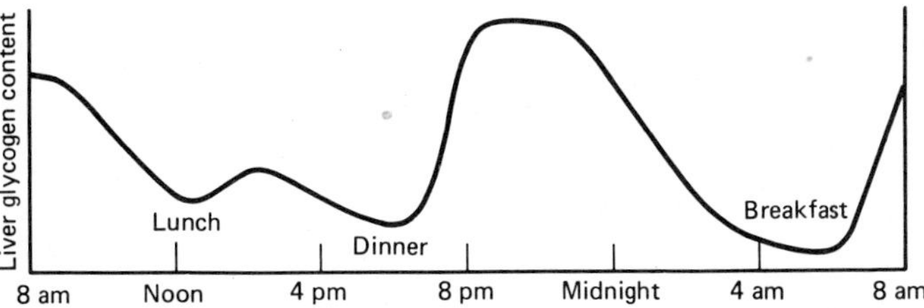

Figure 7.46
Variation of liver glycogen levels between meals and during the nocturnal fast.

Glycogen in muscle is used within this tissue when needed as a source of ATP for increased muscular activity. Most of the glucose of the glycogen molecule is consumed within muscle cells without the formation of free glucose as an intermediate. However, because of a peculiarity of glycogen catabolism to be discussed below, about 8% of muscle glycogen is converted into free glucose within the tissue. Some of this glucose is released into the bloodstream, but most gets metabolized by the glycolytic pathway (Figure 7.5) in the muscle. Since muscle cells lack glucose 6-phosphatase, and since most of the free glucose formed during glycogen breakdown is further catabolized, muscle glycogen is not of quantitative importance in the maintenance of blood glucose levels. Muscle glycogen levels vary much less than liver glycogen levels in response to food intake. The processes of glycogenesis and glycogenolysis within the liver work to "buffer" blood glucose levels, but this is not an important role of these processes in muscle. Exercise of a muscle, that is, increased mechanical work, is what triggers the mobilization of muscle glycogen. It serves to supply the ATP needed for the performance of work. The yield of ATP and the fate of the carbon of glycogen depend upon whether a "white" or "red" muscle is under consideration. Red muscle fibers are supplied with a rich blood flow, contain large amounts of myoglobin, and are packed with mitochondria. Glycogen mobilized within these cells is converted into pyruvate, which, because of the availability of O_2 and mitochondria, can be converted into CO_2 and H_2O. In contrast, white muscle fibers have a poorer blood supply and fewer mitochondria. Glycogenolysis within this tissue supplies substrate for glycolysis, with the end product being primarily lactate. White muscle fibers have enormous capacity for glycogenolysis and glycolysis, much more than red muscle fibers. Since their glycogen stores are limited, however, muscles of this type can only function at full capacity for relatively short periods of time. Breast muscle and the heart of chicken are good examples of white and red muscles, respectively. The heart has to beat continuously and is therefore blessed with many mitochondria and a rich supply of blood via the coronary arteries. The heart stores glycogen to be used when a greater work load is imposed. The breast muscle of the chicken, in contrast to the heart, is not continuously carrying out work. Its important function is to enable the chicken to fly rapidly for short distances, as in fleeing from predators (or amorous roosters). Because glycogen can be mobilized so rapidly, these muscles are designed for maximal activity for a relatively short period of time. Although it was easy to point out readily recognizable white and red muscles in the chicken, most skeletal muscles of the human body are composed of a mixture of red and white fibers in order to provide for both rapid and sustained muscle activity. The distribution of white and red muscle fibers in cross sections of a human skeletal muscle can be readily shown by using special staining procedures (see Figure 7.47).

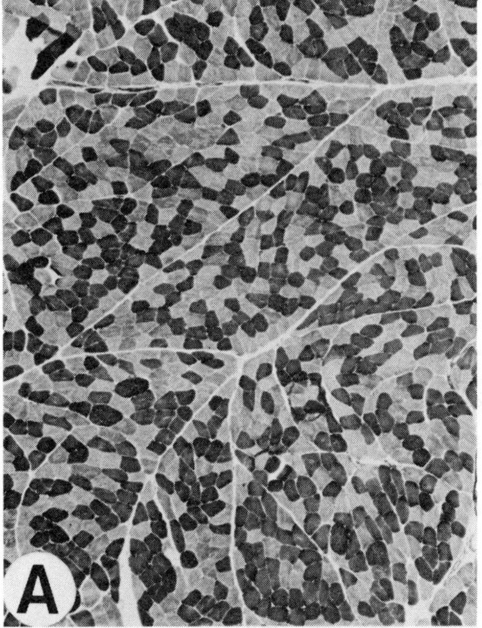

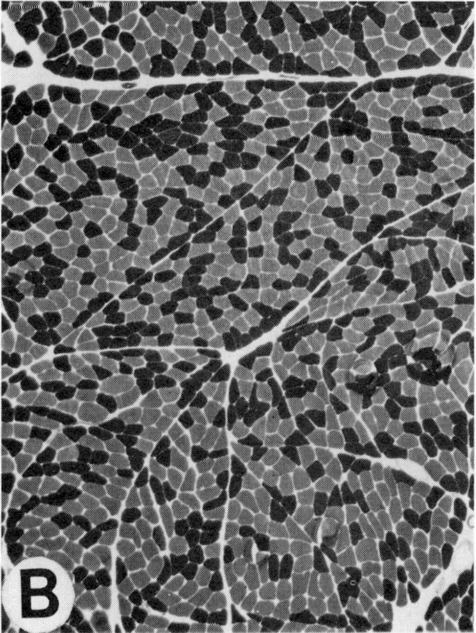

Figure 7.47
Cross section of human skeletal muscle showing red and white muscle fibers.
Sections were stained for NADH diaphorase activity in A; for ATPase activity at pH 9.4 in B. The red fibers are dark and the white fibers are light in A; vice versa in B.
Pictures provided by Dr. Michael H. Brooke of the Jerry Lewis Neuromuscular Research Center, St. Louis, Mo.

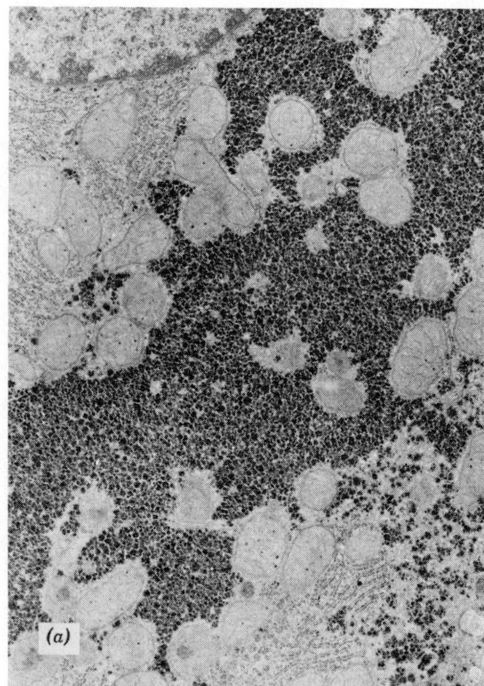

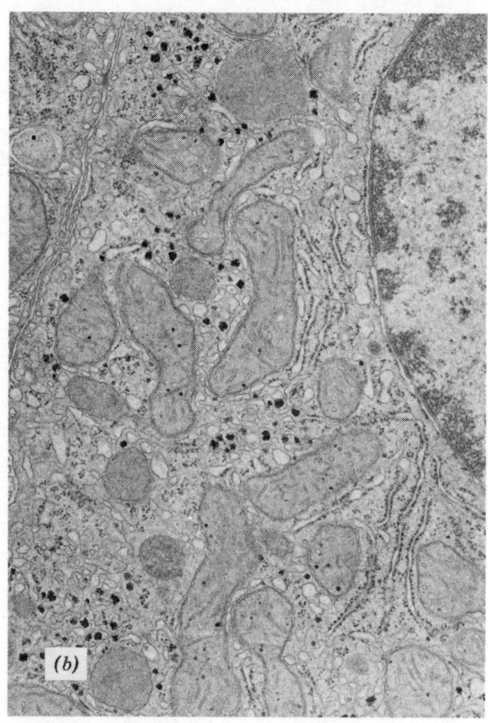

Figure 7.48
Electron micrographs showing glycogen granules (darkly stained material) in the liver of a well-fed rat (a) and the relative absence of such granules in the liver of a rat starved for 24 h (b).
Micrographs provided by Dr. Robert R. Cardell of the Department of Anatomy at the University of Cinncinnati.

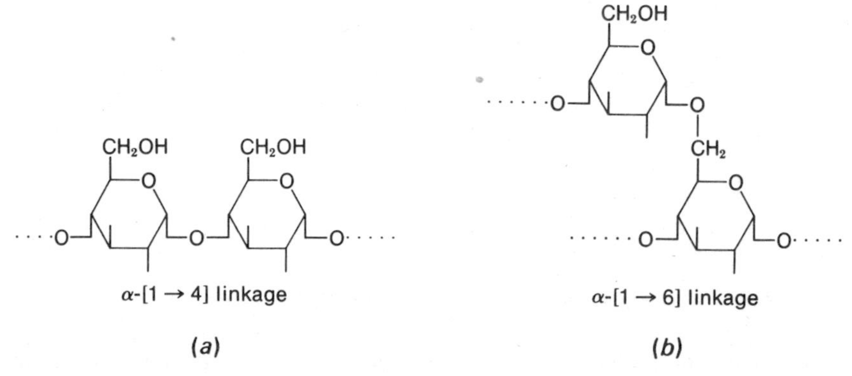

α-[1 → 4] linkage

(a)

α-[1 → 6] linkage

(b)

Figure 7.49
Two types of linkage between glucose molecules are present in glycogen.

Glycogen granules are abundant in the liver of the well-fed animal but are virtually absent from the liver of the 24-hour-fasted animal (Figure 7.48). Heavy exercise causes the same loss of glycogen granules in muscle fibers. These granules of glycogen correspond to clusters of glycogen molecules, the molecular weights of which can approach 2×10^7. Glycogen is composed entirely of glucosyl residues, the majority of which are linked together by α-[1 → 4] glucosidic linkages (Figure 7.49). Branches also occur in the glycogen molecule, however, because of frequent α-[1 → 6] glucosidic linkages (Figure 7.49). A limb of the glycogen "tree" (see Figure 7.50) is characterized by branches at every fourth glucosyl residue within the more central core of the molecule. These branches occur much less frequently in the outer regions of the molecule. An interesting question, which we shall attempt to answer below, is why this polymer is constructed by the cell with so many intricate branches and loose ends? Glycogen certainly stands in contrast to proteins and nucleic acids in this regard but, of course, it is a storage form of fuel and never has to catalyze a reaction nor convey information within a cell.

Pathway of Glycogen Degradation

The first step of glycogen degradation is catalyzed by the enzyme glycogen phosphorylase (see Figure 7.51). This enzyme catalyzes the phosphorolysis of glycogen, a reaction in which the elements of inorganic phosphate are used in the cleavage of an α-[1 → 4] glucosidic bond to yield glucose 1-phosphate. This always occurs at a terminal, nonreducing end of a glycogen molecule:

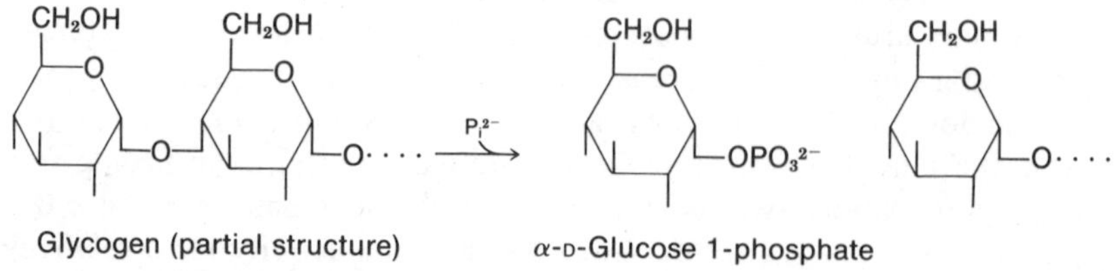

$\xrightarrow{P_i^{2-}}$

Glycogen (partial structure) α-D-Glucose 1-phosphate

The reaction catalyzed by glycogen phosphorylase should be carefully distinguished from that catalyzed by α-amylase, the enzyme responsible for glycogen (and starch) degradation in the gut (see Chapter 24, page 929). Alpha-amylase acts by simple hydrolysis, using the elements of water rather than inorganic phosphate to cleave α-[1 → 4] glucosidic bonds. Since a molecule of glycogen may contain up to 100,000 glucose residues, its structure is usually abbreviated (glucose)$_n$. The reaction catalyzed by the enzyme glycogen phosphorylase can then be written as

$$(\text{Glucose})_n + P_i^{2-} \longrightarrow (\text{glucose})_{n-1} + \alpha\text{-D-glucose 1-phosphate}^{2-}$$

The next step of glycogen degradation is catalyzed by phosphoglucomutase:

$$\text{Glucose 1-phosphate} \rightleftharpoons \text{glucose 6-phosphate}$$

This is a near-equilibrium reaction under intracellular conditions, allowing it to function in both glycogen degradation and synthesis. It has the interesting feature of having a reaction mechanism analogous to that catalyzed by phosphoglyceromutase (page 272) in that a bisphosphate compound is an obligatory intermediate:

$$\text{E-P} + \text{glucose 1-phosphate} \rightleftharpoons \text{E} + \text{glucose 1,6-bisphosphate}$$

$$\text{E} + \text{glucose 1,6-bisphosphate} \rightleftharpoons \text{E-P} + \text{glucose 6-phosphate}$$

$$\text{Sum:}\quad \text{glucose 1-phosphate} \rightleftharpoons \text{glucose 6-phosphate}$$

As with phosphoglyceromutase, a catalytic amount of the bisphosphate compound must be present for the reaction to occur. It is produced in small quantities for this specific purpose by an enzyme called phosphoglucokinase:

$$\text{Glucose 6-phosphate} + \text{ATP} \longrightarrow \text{glucose 1,6-bisphosphate} + \text{ADP}$$

The next enzyme involved in glycogenolysis depends on the tissue under consideration (see Figure 7.51). In liver the glucose 6-phosphate produced by glycogenolysis would be primarily hydrolyzed by glucose 6-phosphatase to give free glucose:

$$\text{Glucose 6-phosphate}^{2-} + H_2O \longrightarrow \text{glucose} + P_i^{2-}$$

Lack of this enzyme or of the translocase that transports glucose 6-phosphate into the endoplasmic reticulum (see page 298) results in type I glycogen storage disease (see Clin. Corr. 7.11). The overall balanced equation for the removal of one glucosyl residue from glycogen in the liver by glycogenolysis is then

$$(\text{Glucose})_n + H_2O \longrightarrow (\text{glucose})_{n-1} + \text{glucose}$$

In other words, glycogenolysis in the liver involves phosphorolysis but, because the phosphate ester is cleaved by a phosphatase, the overall reaction adds up to be hydrolysis of glycogen. It should be noted that no ATP is used or formed in the process of glycogenolysis.

In peripheral tissues the glucose 6-phosphate generated by glycolysis would be used by the glycolytic pathway, which would lead primarily to the generation of lactate in white muscle fibers and primarily to the complete oxidation of the glucose to CO_2 and H_2O in red muscle fibers. Since no ATP had to be invested to produce the glucose 6-phosphate obtained from glycogen, the overall equation for glycogenolysis followed by glycolysis is

$$(\text{Glucose})_n + 3\text{ADP}^{3-} + 3P_i^{2-} + H^+ \longrightarrow (\text{glucose})_{n-1} + 2\ \text{lactate}^{-1} + 3\text{ATP}^{4-}$$

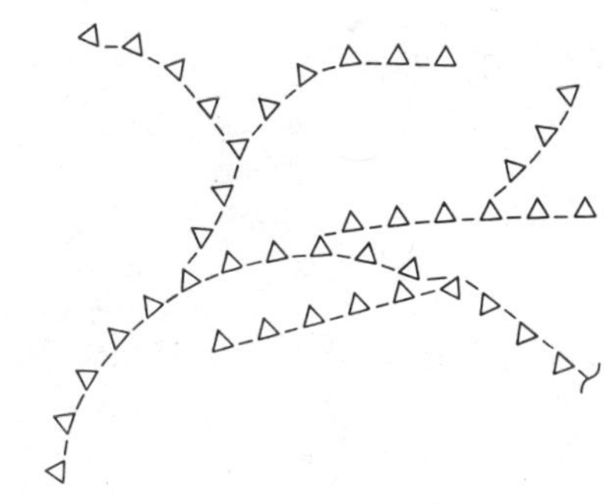

Figure 7.50
The branched structure of glycogen.

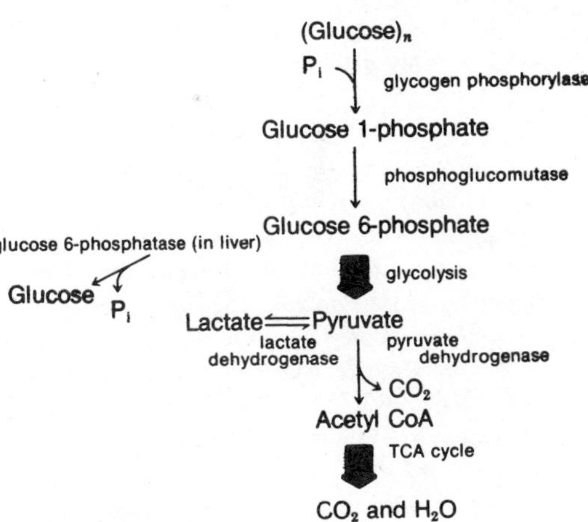

Figure 7.51
Glycogenolysis and the fate of glycogen degraded in liver versus its fate in peripheral tissues.

Up to this point in our consideration of glycogenolysis we have been able to ignore a rather messy feature caused by all the branches that exist in the glycogen molecule. The first enzyme involved in glycogen degradation, glycogen phosphorylase, is specific for α-[1 → 4] glucosidic linkages. It does not even like to go near α-[1 → 6] linkages. Indeed it stops attacking α-[1 → 4] glucosidic linkages four glucosyl residues from an α-[1 → 6] branch point. A glycogen molecule that has been degraded to the limit by phosphorylase looks like a well-trimmed hedge and is called the phosphorylase-limit dextrin. The action within cells of a "debranching" enzyme is what allows glycogen phosphorylase to continue to degrade glycogen. The "debranching" enzyme is a *bifunctional* enzyme that catalyzes two reactions necessary for the debranching of glycogen. The first is a 4-α-D-glucanotransferase activity in which a strand of three glucosyl residues is removed from a four-glucosyl residue branch of the glycogen molecule (see Figure 7.52). The strand remains covalently attached to the enzyme until it can be transferred to a free 4-hydroxyl of a glucosyl residue at the end of the same or an adjacent glycogen molecule (see Figure 7.52). The result is a longer amylose chain with only one glucosyl residue remaining in [1 → 6] linkage. This linkage is broken hydrolytically by the other enzyme action of the "debranching" enzyme, that is, its amylo-α-[1,6]-glucosidase activity:

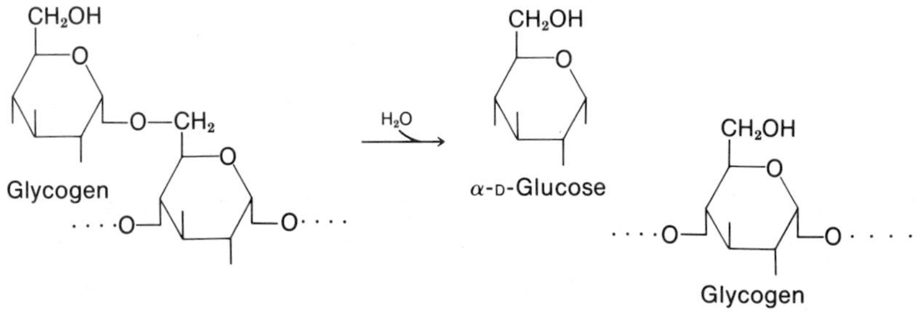

Glycogen $\xrightarrow{H_2O}$ α-D-Glucose + Glycogen

The cooperative and repetitive action of phosphorylase and debranching enzyme results in almost complete phosphorolysis and/or hydrolysis of the glycogen molecule. The average molecule of glycogen yields about 12 molecules of glucose 1-phosphate by the action of phosphorylase for every molecule of free glucose produced by the action of the debranching enzyme.

There is another pathway for glycogen degradation which is quantitatively not very important. Major problems result, however, when this pathway is defective in an individual. As discussed in Clin. Corr. 7.11, a glucosidase of lysosomes degrades glycogen which has entered, perhaps inadvertently, into these organelles during normal turnover of intracellular components.

Pathway of Glycogen Synthesis

The pathway involved in glycogen synthesis is given in Figure 7.53. The first reaction is already familiar, being catalyzed by glucokinase in hepatic tissue and hexokinase in peripheral tissues:

$$\text{Glucose} + \text{ATP} \longrightarrow \text{glucose 6-phosphate} + \text{ADP}$$

The next enzyme involved, phosphoglucomutase, was discussed in relation to glycogen degradation, although this reversible reaction was written in the opposite direction:

$$\text{Glucose 6-phosphate} \longrightarrow \text{glucose 1-phosphate}$$

CLIN. CORR. **7.11**
GLYCOGEN STORAGE DISEASES

There are a number of well-characterized glycogen storage diseases, all due to inherited defects of one or more of the enzymes involved in the synthesis and degradation of glycogen. The liver is usually the tissue most affected, but heart and muscle glycogen metabolism can also be defective.

VON GIERKE'S DISEASE

The most common glycogen storage disease, referred to as type I or von Gierke's disease, is caused by a deficiency of liver, intestinal mucosa, and kidney glucose 6-phosphatase. Thus, diagnosis by small bowel biopsy is possible. Patients with this disease can be further subclassified into those lacking the glucose 6-phosphatase enzyme per se (type 1a) and those lacking the glucose 6-phosphatase translocase (type 1b) (see Figure 7.35). A genetic abnormality in glucose 6-phosphate hydrolysis occurs in only about 1 person in 200,000 and is transmitted as an autosomal recessive trait. Clinical manifestations include fasting hypoglycemia, lactic acidemia, hyperlipidemia, and hyperuricemia with gouty arthritis. The fasting hypoglycemia is readily explained as a consequence of the glucose 6-phosphatase deficiency, the enzyme required to obtain glucose from liver glycogen and gluconeogenesis. The liver of these patients does release some glucose by the action of the glycogen debrancher enzyme. The lactic acidemia occurs because the liver can not use lactate effectively for glucose synthesis. In addition, the liver inappropriately produces lactic acid in response to glucagon. This hormone should trigger glucose release without lactate production; however, the opposite occurs because of the lack of glucose 6-phosphatase. Hyperuricemia results from increased purine degradation in the liver; hyperlipidemia because of increased availability of lactic acid for lipogenesis and chronic lipid mobilization from the adipose tissue.

POMPE'S DISEASE

Type II glycogen storage disease or Pompe's disease is caused by the absence of α-1,4-glucosidase (or acid maltase), an enzyme normally found in lysosomes. The absence of this enzyme leads to the accumulation of glycogen in virtually every tis-

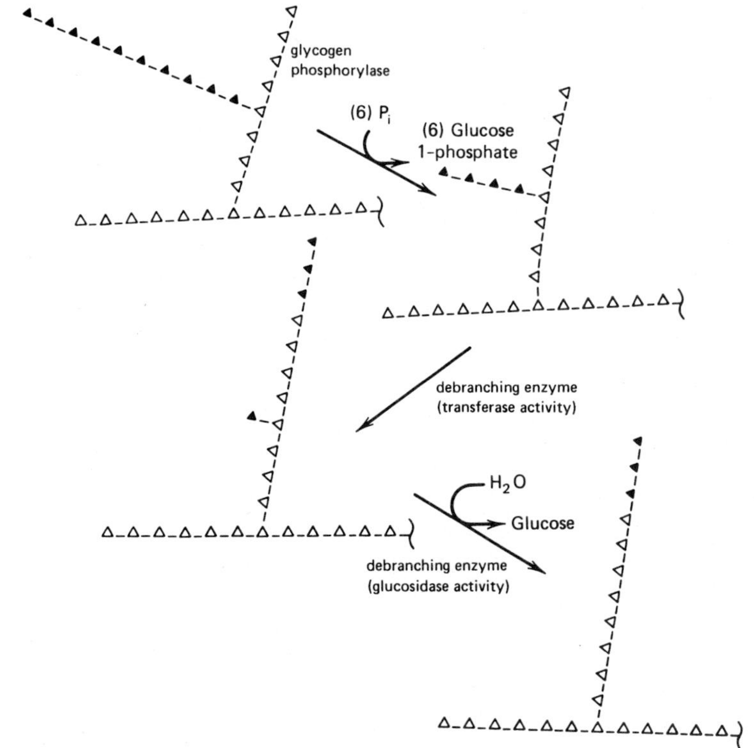

Figure 7.52
Action of the glycogen debranching enzyme.

Glucose

hexokinase: ATP → ADP

Glucose 6-phosphate

phosphoglucomutase

Glucose 1-phosphate

glucose 1-phosphate uridylyltransferase: UTP → PP_i

CH_2OH, O, O, O, O—P—O—P—O—CH_2, OH, OH, O, HN, O, N, O

UDP-glucose

glycogen synthase: $(Glucose)_n$ → $(Glucose)_{n+1}$

UDP

Figure 7.53
Pathway of glycogen synthesis.

sue. This is somewhat surprising, but lysosomes take up glycogen granules and become defective with respect to other functions if they lack the capacity to destroy the granules. Because other synthetic and degradative pathways of glycogen metabolism are intact, metabolic derangements such as those in von Gierke's disease are not seen. The reason for extralysosomal glycogen accumulation is unknown. Massive cardiomegaly occurs and death results at an early age from heart failure.

CORI'S DISEASE

Also called type III glycogen storage disease, Cori's disease is caused by a deficiency of the glycogen debrancher enzyme. Glycogen accumulates because only the outer branches can be removed from the molecule by phosphorylase. Hepatomegaly occurs, but diminishes with age. The clinical manifestations are similar to but much milder than those seen in von Gierke's disease, because gluconeogenesis is unaffected, and hypoglycemia and its complications are less severe.

McARDLE'S DISEASE

Also called type V glycogen storage disease, McArdle's disease is caused by an absence of muscle phosphorylase. Patients suffer from painful muscle cramps and are unable to perform strenuous exercise, presumably because muscle glycogen stores are not available to the exercising muscle. Thus, the normal increase in plasma lactate (released from muscle) following exercise is absent. The muscles are probably damaged because of inadequate energy supply and glycogen accumulation. Release of muscle enzymes creatine phosphokinase and aldolase and of myoglobin is common.

A unique reaction found in the next step, involves the formation of UDP-glucose by the action of glucose 1-phosphate uridylyltransferase:

$$\text{Glucose 1-phosphate} + \text{UTP} \longrightarrow \text{UDP-glucose} + \text{PP}_i$$

This reaction generates an "activated" glucosyl residue, which can be used to build the glycogen molecule. The formation of UDP-glucose is made energetically favorable and the reaction is irreversible by the subsequent hydrolysis of pyrophosphate by pyrophosphatase:

$$\text{PP}_i^{4-} + \text{H}_2\text{O} \longrightarrow 2\text{P}_i^{2-}$$

Glycogen synthase, utilizing glycogen and UDP-glucose as substrates, then catalyzes the transfer of the activated glucosyl moiety to the glycogen molecule so that a new glucosidic bond is formed between the hydroxyl group of carbon-1 of the activated sugar and carbon-4 of a glucosyl residue of the growing glycogen chain. The reducing end of glucose (carbon-1 of glucose is an aldehyde that can reduce other compounds) is always added to a nonreducing end of the glycogen chain. Note that the glycogen molecule, regardless of its size, theoretically has only one free reducing end tucked away within the core. Also note that UDP, *not* UMP, is the product of the reaction catalyzed by glycogen synthase. UDP can be converted back to UTP by the action of nucleoside diphosphate kinase:

$$\text{UDP} + \text{ATP} \longrightarrow \text{UTP} + \text{ADP}$$

Glycogen synthase is very specific, that is, it will create chains of glucose molecules with α-[1 → 4] linkages but will not participate in the formation of α-[1 → 6] branches. Its action alone would only produce amylose, the straight-chain polymer of glucose with α-[1 → 4] linkages. Once an amylose chain of at least 10 residues has been formed, a "branching" enzyme comes into play. Its name is 1,4-α-glucan branching enzyme because it removes a block of glucosyl residues from a growing chain and transfers it to another chain to produce an α-[1 → 6] linkage (see Figure 7.54). The last peculiarity to be mentioned is that the new branch has to be introduced at least four glucosyl residues from an adjacent branch point. Thus the creation of the highly branched structure of glycogen requires the concerted efforts of glycogen synthase and the branching enzyme. The overall balanced equation for glycogen synthesis by the pathway just outlined is

$$(\text{Glucose})_n + \text{glucose} + 2\text{ATP} \longrightarrow (\text{glucose})_{n+1} + 2\text{ADP} + 2\text{P}_i$$

As noted above, the combination of glycogenolysis and glycolysis yields only three molecules of ATP per glucosyl residue:

$$(\text{Glucose})_n + 3\text{ADP} + 3\text{P}_i \longrightarrow (\text{glucose})_{n-1} + 2\text{ lactate} + 3\text{ATP}$$

Thus the combination of glycogen synthesis plus glycogen degradation to lactate actually yields the cell only 1 ATP, that is, the sum of the last two equations is

$$\text{Glucose} + \text{ADP} + \text{P}_i \longrightarrow 2\text{ lactate} + \text{ATP}$$

It should be realized, however, that glycogen synthesis and degradation are carried out during different time frames in a cell. For example, white muscle fibers synthesize glycogen at rest when glucose is plentiful and

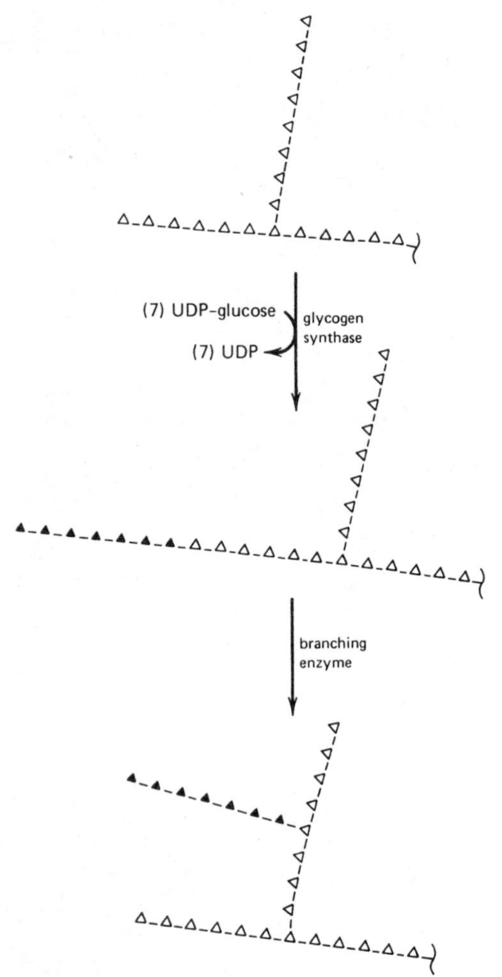

Figure 7.54
Action of the glycogen branching enzyme.

ATP for muscle contraction is not needed. Glycogen is then used during periods of exertion. Although in such terms glycogen storage is not a very efficient process, it provides cells with a fuel reserve that can be quickly and efficiently mobilized.

Special Features of Glycogen Degradation and Synthesis

Since glycogen is such a good fuel reserve, it is obvious why we synthesize and store glycogen in liver and muscle. But why store glucose as glycogen? Why not store our excess glucose calories entirely as fat instead of glycogen? The answer is at least threefold: (1) we do store fat, some of us lots of it, but fat cannot be mobilized as rapidly in muscle as glycogen; (2) fat cannot be used as a source of energy in the absence of oxygen; and (3) fat cannot be converted to glucose by any pathway of the human body in order to maintain blood glucose levels for use by tissues such as the brain. Why not just pump glucose into cells and store it as free glucose until needed? Why waste so much ATP making a polymer out of glucose? The problem is that glucose is osmotically active. It would cost ATP to "pump" glucose into a cell, regardless of the mechanism, and glucose would have to reach concentrations of 400 mM in liver cells to match the "glucose reserve" provided by the usual liver glycogen levels. Unless balanced by the outward movement of some other osmotically active compound, the accumulation of such concentrations of glucose would cause the uptake of considerable water and the osmotic lysis of the cell. Assuming the molecular mass of a glycogen molecule is of the order of 10^7 daltons, 400 mM glucose is in effect stored at an intracellular glycogen concentration of 0.01 μM. Storage of glucose as glycogen, therefore, creates absolutely no osmotic pressure problem for the cell.

Another interesting feature about glycogen is that a primer is needed for its synthesis. No template is required, but like DNA synthesis, a primer is necessary. Glycogen itself is the usual primer, that is, glycogen synthesis usually takes place by the addition of glucosyl units to glycogen "core" molecules, which are almost invariably present in the cell. The outer regions of the glycogen molecule get removed and resynthesized much more rapidly than the inner core. Glycogen within a cell is frequently sheared by the combined actions of glycogen phosphorylase and debranching enzyme but is seldom ever obliterated before glycogen synthase and branching enzyme rebuild the molecule. This is a good time to point out why nature has evolved such an elaborate mechanism for creating and disposing of the branched structure of glycogen. In other words, why is glycogen a branched molecule with only one real beginning (the reducing end) and many branches terminating with nonreducing glucosyl units? The answer is that this gives numerous sites of attack for glycogen phosphorylase on a mature glycogen molecule and the same number of sites for glycogen synthase to add glucosyl units. If cells synthesized amylose, that is, an unbranched glucose polymer, there would only be one nonreducing end per molecule. The result would be that glycogen degradation and synthesis would surely be much slower processes. As it is, glycogen phosphorylase and glycogen synthase are usually found in tight association with glycogen granules in a cell, as though they exist in the branches of the glycogen tree with ready access to a multitude of nonreducing sugars at the ends of its limbs.

We digressed, however, from the problem of a need of a primer for glycogen synthesis. Perhaps as a consequence of the great number of nonreducing ends, glycogen synthase has a very low K_m for very large glycogen molecules. However, the K_m gets larger and larger as the glycogen molecule gets smaller and smaller. This phenomenon is so pronounced that it is clear that glucose, at its physiological concentration, could never function as a primer. This led to the notion that glycogen must be immortal, that is, some glycogen must be handed down from one cell generation to the next in order for glycogen to be synthesized. Although immortality is attractive, it is now thought that one or more proteins probably function as primers for glycogen synthesis. The hydroxyl groups of serine or threonine residues of certain proteins may become glycosylated and then serve as a nucleus for the synthesis of glycogen. Hydrolytic breakage of one of the resulting branches of the glycogen tree rooted in the side of the protein could then provide additional primer for the synthesis of glycogen, independent of any further need for such a protein. Alas, glycogen is probably not immortal.

If glycogen synthase becomes more efficient as the glycogen molecule gets bigger, we ought to worry somewhat about how synthesis of this ball of sugar is curtailed. Fat cells have an almost unlimited capacity to pack away fat—but then fat cells do not have to do anything else. Muscle cells participate in mechanical activity and liver cells carry out many processes other than glycogen synthesis. Even in the face of excess glucose, there has to be a way to limit the intracellular accumulation of glycogen. It turns out that glycogen itself inhibits glycogen synthase, the regulatory enzyme involved in glycogen synthesis. Inhibition of the enzyme by glycogen is complicated, and will not be discussed until we have had a chance to present more details with respect to the regulatory enzymes involved in glycogen metabolism.

As stressed in the clinical correlations given in this text, biochemistry has made many significant contributions and is relevant without question to modern clinical medicine. However, clinical medicine also contributes greatly to our understanding of the biochemistry of a number of complex biochemical processes. A case in point is the enzymes actually involved in glycogen synthesis and degradation. At one time the enzyme glycogen phosphorylase was believed responsible for both the synthesis and degradation of glycogen. This enzyme is responsible for glycogen degradation in the cell but can be readily assayed in the test tube in the direction of glycogen synthesis:

$$(\text{Glucose})_n + \alpha\text{-D-glucose 1-phosphate} \longrightarrow (\text{glucose})_{n+1} + P_i$$

The K'_{eq} for this reaction is approximately unity, and since one of the substrates $(\text{glucose})_n$ looks almost identical to one of the products $(\text{glucose})_{n+1}$ to both us and the enzyme, the direction of flux is determined by the ratio of the steady-state concentrations of P_i and glucose 1-phosphate. Measurements of the concentrations of these components under intracellular conditions make it clear that the reaction is nonequilibrium and that net flux can only be in the direction of net degradation of glycogen. In other words, the intracellular concentration of inorganic phosphate is always much greater than that of glucose 1-phosphate. However, it really did not become clear that glycogen phosphorylase is involved only in glycogen degradation until studies were conducted with a patient who presented with a rare glycogen storage disease, now known as McArdle's disease. The enzyme glycogen phosphorylase of skeletal muscle is missing in this type of glycogenosis (see Clin. Corr. 7.11). Nevertheless, skeletal muscles of patients with McArdle's disease are loaded with glycogen. This observation made it apparent that another enzyme had to be involved in the synthesis of glycogen, and helped lead to the isolation and characterization of glycogen synthase.

Regulation of Glycogen Synthesis and Degradation

Regulatory Enzymes of the Pathways

Glycogen synthase and glycogen phosphorylase are the regulatory enzymes of glycogen synthesis and degradation, respectively. Both catalyze nonequilibrium reactions, and both are subject to control by allosteric effectors and covalent modification. The details of the mechanisms responsible for the regulation of these enzymes will be considered first. This will be followed by an overview of how hormones control glycogen metabolism in muscle and liver at the level of these enzymes. A review of the material presented previously (page 168) on the regulation of enzymes by covalent modification and the role of cAMP as a "second messenger" of hormone action may be helpful to you.

Regulation of Glycogen Phosphorylase

The mechanisms responsible for the regulation of glycogen phosphorylase are summarized in a rather formidable fashion in Figure 7.55. The enzyme is subject to allosteric activation by AMP and allosteric inhibition by glucose and ATP. Although these effectors are considered to be of some physiological significance in the regulation of glycogen metabolism, effector control of phosphorylase has to be considered "primitive" by contrast with its very elaborate control by covalent modification. Phosphorylase exists in an "a" form, which is active, and a "b" form, which is inactive. These forms of the enzyme are interconverted by the actions of phosphorylase kinase and phosphoprotein phosphatase (Figure 7.55). A conformational change caused by phosphorylation transforms the enzyme

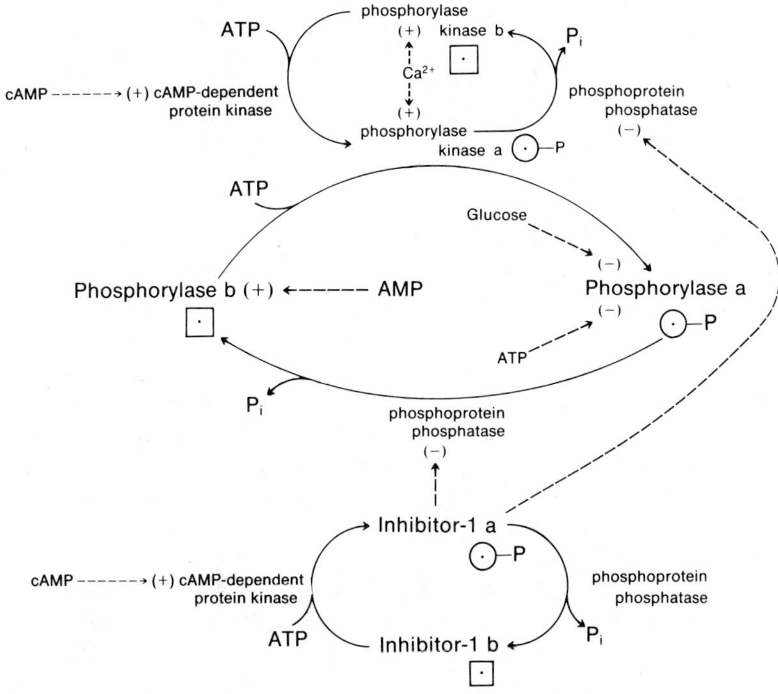

Figure 7.55
Regulation of glycogen phosphorylase by covalent modification.
Note that phosphorylation converts glycogen phosphorylase, phosphorylase kinase, and phosphatase inhibitor-1 from their inactive (b) forms to their active (a) forms.

into a more active catalytic state. Phosphorylase b has some catalytic activity and can be greatly activated by AMP. This allosteric effector has little activating effect, however, on the already active phosphorylase a. Hence the covalent modification mechanism can be bypassed by the allosteric mechanism or vice versa. Although there is still some uncertainty, phosphorylase appears to be composed of two identical 95,000 mol wt subunits. Serine-14, counting from the amino terminus, is phosphorylated on both subunits by phosphorylase kinase. Thus, two ATP molecules are converted to two ADP molecules in the conversion of phosphorylase b to phosphorylase a. Such details are usually ignored for the sake of simplicity in attempts (Figure 7.56) to present the overall mechanism involved in the regulation of an enzyme subject to covalent modification.

Phosphorylase kinase is responsible for the phosphorylation and activation of phosphorylase (Figure 7.55). Moreover, phosphorylase kinase itself is also subject to regulation by a cyclic phosphorylation/dephosphorylation mechanism. Cyclic AMP-dependent protein kinase is responsible for phosphorylation and activation of phosphorylase kinase; phosphoprotein phosphatase in turn is responsible for dephosphorylation and inactivation of phosphorylase kinase. Phosphorylase kinase is a large enzyme complex, composed of four subunits with four molecules of each subunit in the complex ($\alpha_4\beta_4\gamma_4\delta_4$). The α and β subunits are phosphorylated in the transition from the inactive (b) form to the active (a) form of the enzyme. Cyclic AMP-dependent protein kinase does not interact with phosphorylase directly—this protein kinase can only exert an effect on phosphorylase via its ability to phosphorylate phosphorylase kinase. Thus, a bicyclic system is required for the activation of phosphorylase in response to cAMP-mediated signals.

The δ subunit of phosphorylase kinase also plays an important role in regulation of the activity of this enzyme as well as the overall control of glycogen metabolism. The δ subunit is identical to the Ca^{2+}-binding regulatory protein, calmodulin—meaning Ca^{2+}-modulating protein. Calmodulin is not unique to phosphorylase kinase. It is found in cells as the free molecule and is also bound to other enzyme complexes. Calmodulin functions as a Ca^{2+} receptor in the cell, responding to changes in intracellular Ca^{2+} concentration and affecting the relative activities of a number of enzyme systems. For example, the binding of Ca^{2+} to the calmodulin subunit of phosphorylase kinase changes the conformation of the complex, making the enzyme more active with respect to the phosphorylation of phosphorylase. Note in Figure 7.55 that Ca^{2+} is indicated as an activator of both phosphorylase kinase a and phosphorylase kinase b. This means that maximum activation of phosphorylase kinase requires both the phosphorylation of specific serine residues of the enzyme and the interaction of Ca^{2+} with the calmodulin subunit of the enzyme. At least in part because of intracellular effects exerted via calmodulin, Ca^{2+} functions as an important "second messenger" of hormone action, as will be discussed in detail below.

It is obvious that activation of phosphorylase kinase by phosphorylation and Ca^{2+} will have a substantial effect on the activity of glycogen phosphorylase. It is equally obvious that turning off phosphoprotein phosphatase could achieve the same thing. But what would really provide ultimate control for the activation of phosphorylase would involve the simultaneous turning off of phosphoprotein phosphatase and turning on of phosphorylase kinase, and vice versa, for the inactivation of the enzyme. Since phosphoprotein phosphatase also acts on phosphorylase kinase, turning off phosphoprotein phosphatase would also achieve greater activation of phosphorylase kinase. It appears from recent work that such a mechanism exists for this sort of reciprocal relationship, with just a slight twist to make it interesting. Cells of many tissues contain a protein that inhibits phosphoprotein phosphatase. This protein is called phosphatase inhibitor-1. It also is subject to covalent modification by cAMP-dependent protein kinase and phosphoprotein phosphatase, as shown in Figure 7.55. Only the a form (phosphorylated form) of phosphatase inhibitor-1 will inhibit phosphoprotein phosphatase. It should be noted that cAMP, by completely indirect mechanisms, causes the activation of phosphorylase kinase and the inhibition of phosphoprotein phosphatase—making it possible theoretically to activate glycogen phosphorylase completely (see Figure 7.55).

It is interesting that the phosphorylated form of phosphatase inhibitor-1 is converted back to its dephosphorylated form by the enzyme it inhibits—the phosphoprotein phosphatase! However, the inhibitor cleverly does not inhibit its own dephosphorylation, just the dephosphorylation of phosphorylase kinase a and phosphorylase a. In contrast to many other interconvertible enzymes that become phosphorylated on serine residues, inhibitor-1 becomes phosphorylated on a threonine residue. This may account in part for why it can inhibit the action of phosphoprotein phosphatase against other phosphorylated enzymes and yet serve itself as a substrate for phosphoprotein phosphatase.

Now if you have not already given up because of the complexities of the glycogen phosphorylase regulatory system, note that there is a good reason for the existence of the bicyclic control system for the phosphorylation of phosphorylase plus the additional control on its dephosphorylation. This provides a tremendous amplification mechanism. Think about it with relation to Figure 7.55. One molecule of epinephrine or one molecule of glucagon can cause, by the activation of adenylate cyclase, the formation of many molecules of cAMP. cAMP can then activate cAMP-dependent protein kinase, which, in turn, can cause the activation of many molecules of phosphatase inhibitor-1 as well as the activation of many molecules of phosphorylase kinase. In turn, phosphorylase kinase can cause the phosphorylation of many molecules of glycogen phosphorylase—which in turn can cause the phosphorolysis of many glucosidic bonds of glycogen. This mechanism provides an elaborate amplification system in which the signal provided by just a few molecules of hormone can be amplified into production of an enormous number of glucose 1-phosphate molecules. If each step represents an amplification factor of 100, then a total of four steps would result in an amplification of 100 million! This system is so rapid, in large part because of the amplification system, that all of the stored glycogen of white muscle fibers could be completely mobilized within just a few seconds.

Regulation of Glycogen Synthase

Glycogen synthase is the regulatory enzyme involved in glycogen synthesis. It has to be active for glycogen synthesis and inactive during glycogen degradation. The combination of the reactions catalyzed by glycogen synthase, glycogen phosphorylase, glucose 1-phosphate uridylyltransferase, and nucleoside diphosphate kinase adds up to a futile cycle with the overall equation: $ATP \rightarrow ADP + P_i$. Hence glycogen synthase needs to be turned off when glycogen phosphorylase is turned on, and vice versa.

The primitive allosteric mechanism of glucose 6-phosphate activation of glycogen synthase might be of physiological significance under some circumstances. However, as with glycogen phosphorylase, this mode of control is integrated with regulation by covalent modification (see Figure 7.56). Glycogen synthase is known to exist in two forms. One is designated the D form because this form of the enzyme is dependent on the

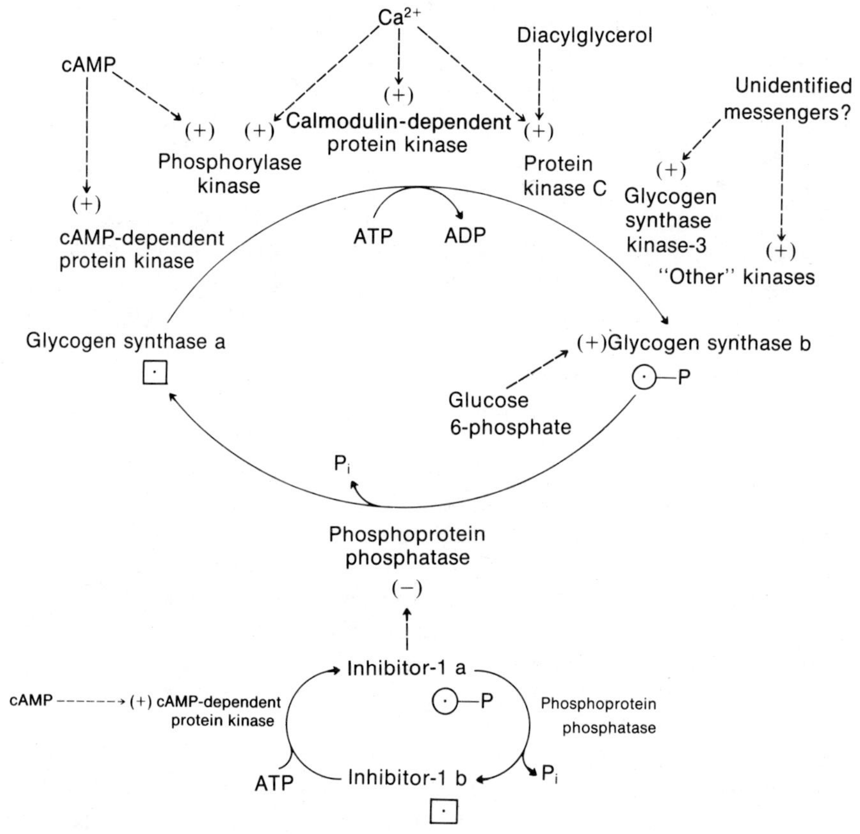

Figure 7.56
Regulation of glycogen synthase by covalent modification.
Note that phosphorylation converts glycogen synthase from its active (a) form to its inactive (b) form, whereas phosphorylation converts phosphatase inhibitor-1 from its inactive (b) form to its active (a) form. Each protein kinase named in the figure is capable of phosphorylating glycogen synthase independent of the presence of the other indicated enzymes.

presence of glucose 6-phosphate for activity. The other is designated the I form because this form of the enzyme is active in the absence of glucose 6-phosphate. These are old names for the two forms of the enzyme, used before the enzyme was established to be subject to covalent modification. The D form corresponds to the b or inactive form of the enzyme, the I form to the a or active form of the enzyme. Phosphorylation of this enzyme can be catalyzed by several different kinases, which, in turn, are regulated by several different second messengers of hormone action, including cAMP, Ca^{2+}, diacylglycerol, and perhaps some yet to be identified compounds (Figure 7.56). Each of the protein kinases shown in Figure 7.56 is capable of catalyzing the phosphorylation and at least partial inactivation of glycogen synthase. Although glycogen synthase is a simple tetramer (α_4) with only one subunit type of mol wt 85,000, this subunit can be phosphorylated on at least seven different serine residues! At last count, eight different protein kinases have been identified that could phosphorylate glycogen synthase at one or more specific sites. This stands in striking contrast to glycogen phosphorylase, which is phosphorylated only by phosphorylase kinase and only at one specific site. Also in contrast to glycogen phosphorylase, the phosphorylation of glycogen synthase results in inactivation rather than activation of the enzyme. Likewise, phosphorylation of phosphatase inhibitor-1 by cAMP-dependent protein kinase results in inhibition of phosphoprotein phosphatase, which prevents reactivation of glycogen synthase. Since cAMP plays such an important role in the regulation of glycogen phosphorylase (Figure 7.55), it should be immediately appreciated that cAMP is an extremely important intracellular signal for reciprocally controlling glycogen synthase and glycogen phosphorylase. An increase in intracellular cAMP signals the activation of glycogen phosphorylase by the two different mechanisms, as shown in Figure 7.55, and signals the inactivation of glycogen synthase by the same two mechanisms, as shown in Figure 7.56. Note, however, that the regulation of glycogen synthase by cAMP is not bicyclic. cAMP-dependent protein kinase directly phosphorylates glycogen synthase, bypassing the need for phosphorylase kinase. Exactly why cAMP-dependent protein kinase directly phosphorylates glycogen synthase, thereby eliminating the need for bicyclic control of glycogen synthase and losing some of the amplification factor important in the regulation of glycogen phosphorylase, is not known. This is particularly surprising in view of the fact that phosphorylase kinase is one of the enzymes capable of phosphorylating glycogen synthase (Figure 7.56). This may be important because phosphorylase kinase, in contrast to cAMP-dependent protein kinase, is sensitive to regulation by Ca^{2+}. Thus, both cAMP and Ca^{2+} influence the phosphorylation state and, therefore, the activity state of the two regulatory enzymes of glycogen metabolism. Furthermore, two cAMP-independent, Ca^{2+}-activated protein kinases have been identified that also may have physiological significance. One of these has been named calmodulin-dependent protein kinase and the other protein kinase C. Both enzymes phosphorylate glycogen synthase, but neither enzyme can use glycogen phosphorylase as substrate. Protein kinase C requires phospholipid, diacylglycerol, and Ca^{2+} for full activity. There is considerable interest in protein kinase C because tumor-promoting agents called phorbol esters have been found to mimic diacylglycerol as activators of this enzyme. This and other findings suggest that diacylglycerol may be an important "second messenger" of hormone action, acting via protein kinase C to regulate numerous cellular processes. Another protein kinase with excellent activity toward glycogen synthase is called glycogen synthase kinase-3 (Figure 7.56). It has no activity toward glycogen phosphorylase and does not appear subject to regulation by either cAMP or Ca^{2+}. Several "other" protein kinases that phosphorylate glycogen synthase have also been isolated (Figure 7.56). Whether "second messenger" systems also exist in cells to regulate glycogen synthase kinase-3 and as well as the "other" protein kinases is not known at this time.

From the information given above, it should be apparent that the role of cAMP in the regulation of glycogen metabolism is rather well understood. The roles of Ca^{2+} and Ca^{2+}-activated protein kinases are beginning to be understood. We also have much to learn with respect to the physiological importance of several other protein kinases. Furthermore, several phosphoprotein phosphatases and another phosphatase inhibitor have been characterized. Exactly how all of the known protein kinases, phosphoprotein phosphatases, and phosphatase inhibitors should be inserted into our overall picture of the control of glycogen metabolism (Figures 7.55 and 7.56) is unclear at this time.

Effector Control of Glycogen Metabolism

Referred to above as a "primitive" regulatory mechanism, effector control is clearly important under some physiological conditions. For example, certain muscles under anaerobic conditions have been shown to mobilize their glycogen stores rapidly without marked conversion of phosphorylase b into phosphorylase a or glycogen synthase a into glycogen synthase b. Presumably this is accomplished by effector control in which ATP levels decrease, causing less inhibition of phosphorylase; glucose 6-phosphate levels decrease, causing less activation of glycogen synthase; and AMP levels increase, causing activation of phosphorylase. This enables the muscle to keep working, for at least a short period of time, by using the ATP produced by glycolysis of the glucose 6-phosphate obtained from glycogen.

Proof that effector control can operate has also been obtained in studies of a special strain of mice that are deficient in muscle phosphorylase kinase. Phosphorylase b in the muscle of such mice cannot be converted into phosphorylase a. Nevertheless, heavy exercise of these mice results in depletion of muscle glycogen, presumably because of stimulation of phosphorylase b by effectors.

Negative Feedback Control of Glycogen Synthesis by Glycogen

As mentioned previously (see page 313), glycogen is able to exert feedback control over its own formation. The portion of glycogen synthase in the active (a) form decreases as glycogen accumulates in a particular tissue. The mechanism is not well understood, but glycogen may make the a form of glycogen synthase a better substrate for one or more of the protein kinases, or, alternatively, glycogen may inhibit dephosphorylation of glycogen synthase b by phosphoprotein phosphatase. Either of these mechanisms would account for the shift in the steady state in favor of glycogen synthase b that occurs in response to glycogen accumulation.

Phosphorylase a Functions as a "Glucose Receptor" in the Liver

Consumption of a carbohydrate-containing meal results in an increase in blood and liver glucose, which, in turn, signals an increase in glycogen synthesis in the latter tissue. The mechanism involves glucose stimulation of insulin release from the pancreas and subsequent effects of this hormone on hepatic glycogen phosphorylase and glycogen synthase. This cannot be the entire story, however, because hormone-independent mechanisms appear to be important in the liver (Figure 7.57). Direct

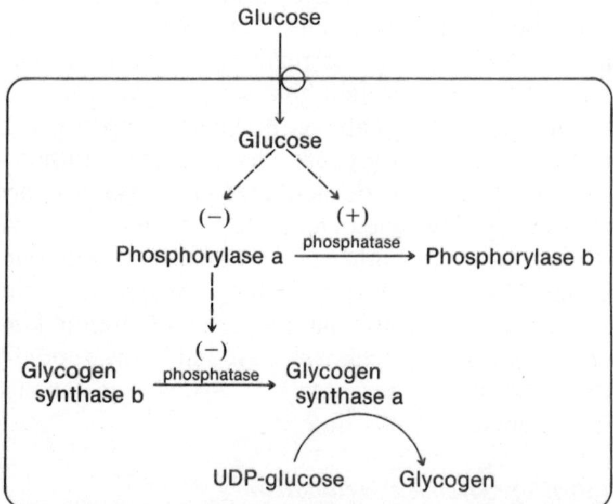

Figure 7.57
Overview of the mechanism responsible for glucose stimulation of glycogen synthesis in the liver.

inhibition of phosphorylase a by glucose is probably of some importance. Moreover, binding of glucose to phosphorylase makes the a form of phosphorylase a better substrate for dephosphorylation by phosphoprotein phosphatase. This has led to the hypothesis that phosphorylase a can function as a glucose receptor in the liver. The basic idea is that binding of glucose to phosphorylase a promotes the inactivation of phosphorylase a, with the overall result being inhibition of glycogen degradation by glucose. This "negative feedback" control of glycogenolysis by glucose would not necessarily promote glycogen synthesis. However, there also is experimental evidence that phosphorylase a is an inhibitor of the dephosphorylation of glycogen synthase b by phosphoprotein phosphatase. This inhibition is lost once phosphorylase a has been converted to phosphorylase b (Figure 7.57). In other words, phosphoprotein phosphatase can turn its attention to glycogen synthase b following the dephosphorylation of phosphorylase a. Thus, as a result of the interaction of glucose with phosphorylase a, phosphorylase becomes inactivated, glycogen synthase becomes activated, and glycogen is synthesized rather than degraded in the liver. Phosphorylase a can serve this function of "glucose receptor" in liver because the concentration of glucose in liver always reflects the blood concentration of glucose. This is not true for extrahepatic tissues. Liver cells have a very high-capacity transport system for glucose and a high K_m enzyme for glucose phosphorylation (glucokinase). Cells of extrahepatic tissues as a general rule have low-capacity transport systems for glucose and low K_m enzymes for glucose phosphorylation (hexokinase) that maintain intracellular glucose at concentrations too low for phosphorylase a to function as a "glucose receptor."

Glucagon Stimulates Glycogen Degradation in the Liver

Glucagon is released from α cells of the pancreas in response to low glucose levels in the blood. One of glucagon's primary jobs during periods of low food intake (fasting or starvation) is to mobilize liver glycogen, that is, stimulate glycogenolysis, in order to ensure that adequate blood glucose is available to meet the needs of glucose-dependent tissues. Glucagon circulates in the blood until it interacts with glucagon receptors such as those located on the plasma membrane of liver cells (Figure 7.58). Binding of glucagon to these receptors activates adenylate cyclase and triggers the cascades that result in activation of glycogen phosphorylase and inactivation of glycogen synthase by the mechanisms given in Figures 7.55 and 7.56, respectively. Glucagon also inhibits the use of glucose by glycolysis at the level of 6-phosphofructo-1-kinase and pyruvate kinase by the mechanisms given in Figures 7.26 and 7.28, respectively. The net result of these effects of glucagon, all mediated by the second messenger cAMP and covalent modification, is a very rapid increase in blood glucose levels. Hyperglycemia might be expected but is avoided because less glucagon is released from the pancreas as blood glucose levels increase.

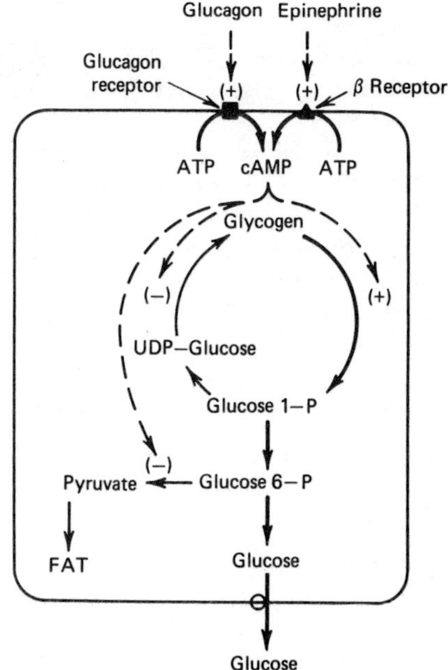

Figure 7.58
Cyclic AMP mediates the stimulation of glycogenolysis in liver by glucagon and β-agonists.
The glucagon receptor-adenylate cyclase complex (■), β-adrenergic receptor-adenylate cyclase complex (▲), and glucose transporter (○) are intrinsic components of the plasma membrane.

Epinephrine Stimulates Glycogen Degradation in the Liver

Epinephrine is released into the blood from the chromaffin cells of the adrenal medulla in response to stress. This hormone is our "fright, flight or fight" hormone, preparing the body for either combat or escape.

Epinephrine mobilizes liver glycogen by at least three different mechanisms, but the physiologically most important mechanism is not clearly established for humans. One mechanism involves epinephrine stimulation of glucagon release from the α cells of the pancreas. Glucagon then travels by way of the blood to mobilize liver glycogen as discussed above. Epinephrine can also interact directly with receptors in the plasma membrane of the liver cells to activate adenylate cyclase (Figure 7.58). The

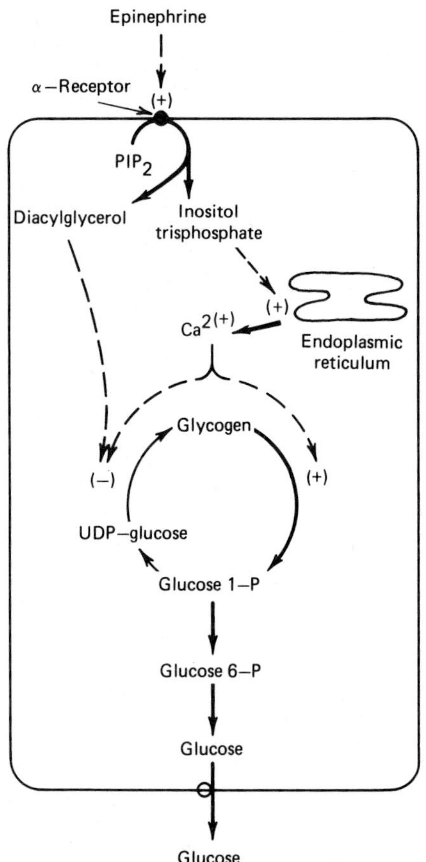

Figure 7.59
Inositol trisphosphate and Ca^{2+} mediate the stimulation of glycogenolysis in liver by α-agonists.
The α-adrenergic receptor-phospholipase C complex (●) and glucose transporter (○) are intrinsic components of the plasma membrane. Although not indicated, phosphatidylinositol 4,5-bisphosphate (PIP_2) is also a component of the plasma membrane.

resulting increase in cAMP has the same effect as that caused by glucagon. The binding site for epinephrine on the plasma membrane, which is in communication with adenylase cyclase, is called the β-adrenergic receptor. Although there are considerable species differences and the picture is a bit hazy for the human at this time, the plasma membrane of liver cells also has another binding site for epinephrine, called the α-adrenergic receptor. Interaction of epinephrine with α-adrenergic receptors leads to the formation of inositol trisphosphate and diacylglycerol (Figure 7.59). These compounds are second messengers, produced in the plasma membrane by the action of a phospholipase-C on phosphatidylinositol 4,5-bisphosphate by the reaction shown in Figure 7.60. Formation of these compounds occurs in response to epinephrine binding to α-adrenergic receptors, much like cAMP formation occurs in response to epinephrine binding to β-adrenergic receptors. Inositol trisphosphate stimulates the release of Ca^{2+} from intracellular stores such as the endoplasmic reticulum (Figure 7.59). As previously discussed for Figure 7.55, the increase in Ca^{2+} activates phosphorylase kinase, which in turn activates glycogen phosphorylase. Likewise, as previously discussed for Figure 7.56, Ca^{2+}-mediated activation of phosphorylase kinase, calmodulin-dependent protein kinase, and protein kinase C, as well as diacylglycerol-mediated activation of protein kinase C, may all be important for inactivation of glycogen synthase.

The consequences of all of the mechanisms described for epinephrine are the same—increased release of glucose into the blood from the glycogen stored in the liver. This makes more blood glucose available to tissues that are called upon to meet the challenge of the stressful situation that triggered the release of epinephrine from the adrenal medulla.

Epinephrine Stimulates Glycogen Degradation in Skeletal Muscle

Epinephrine also stimulates glycogen degradation in skeletal muscle. This tissue lacks glucagon but has β-adrenergic receptors. Cyclic AMP, produced in response to epinephrine stimulation of adenylate cyclase via β-adrenergic receptors (Figure 7.61), signals the concurrent activation of glycogen phosphorylase and inactivation of glycogen synthase by the mechanisms given previously in Figures 7.55 and 7.56, respectively. This does not lead, however, to glucose release into the blood from this tissue. In contrast to liver, skeletal muscle lacks glucose 6-phosphatase, and in this tissue cAMP does not inhibit glycolysis. Thus, the role of epinephrine

Phosphatidylinositol 4,5-bisphosphate

H_2O

Diacylglycerol

Inositol trisphosphate

Figure 7.60
Phospholipase C cleaves phosphoinositol 4,5-bisphosphate to produce diacylglycerol and inositol trisphosphate.
R—C(=O)— refers to long-chain acyl groups of the molecules. Ⓟ refers to phosphate (PO_4^{2-}) groups.

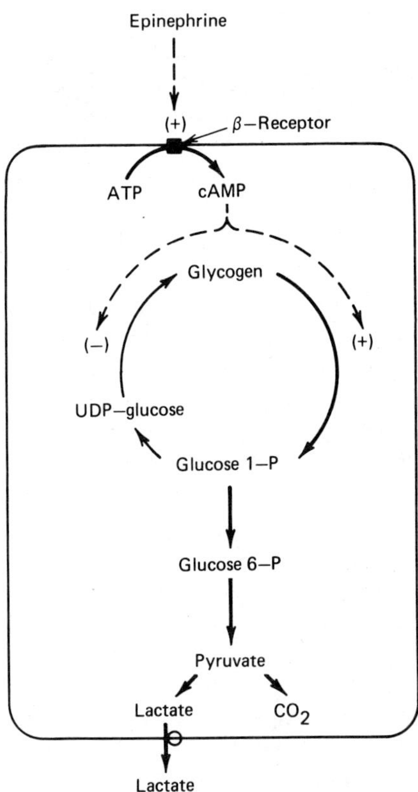

Figure 7.61
Cyclic AMP mediates the stimulation of glycogenolysis in muscle by β-agonists. *The β-adrenergic receptor-adenylate cyclase complex (■) is an intrinsic component of the plasma membrane.*

on glycogen metabolism in skeletal muscle is to make more substrate (glucose 6-phosphate) available for glycolysis. ATP generated by glycolysis can then be used to meet the metabolic demand imposed upon skeletal muscle by the stress that triggered epinephrine release.

Neural Control of Glycogen Degradation in Skeletal Muscle

Nervous excitation of muscle activity is mediated via changes in intracellular Ca^{2+} concentrations (Figure 7.62). The nerve impulse causes membrane depolarization which, in turn, causes Ca^{2+} release from the sarcoplasmic reticulum into the sarcoplasm of muscle cells. This release of Ca^{2+} triggers muscle contraction, whereas reaccumulation of Ca^{2+} by the sarcoplasmic reticulum causes relaxation. The same change in Ca^{2+} concentration effective in causing muscle contraction (from 10^{-8} to 10^{-6} M) also greatly affects the activity of phosphorylase kinase. As Ca^{2+} concentrations increase there is more muscle activity and a greater need for ATP. The activation of phosphorylase kinase by Ca^{2+} leads to the subsequent activation of glycogen phosphorylase and perhaps the inactivation of glycogen synthase. The result is that more glycogen is converted to glucose 6-phosphate so that more ATP can be produced to meet the greater energy demand of muscle contraction.

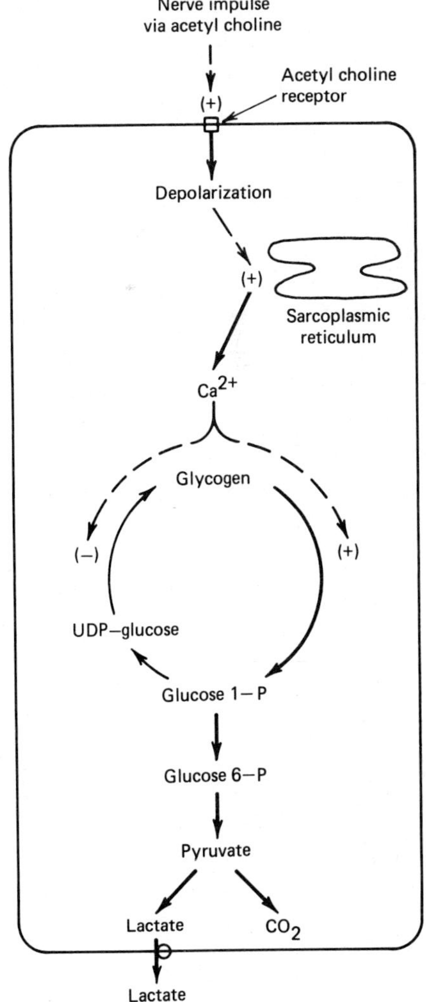

Figure 7.62
Ca^{2+} mediates the stimulation of glycogenolysis in muscle by nervous excitation.

Synergistic Effects of Epinephrine and Neural Signals on Glycogen Degradation in Skeletal Muscle

As has been discussed for Figures 7.55 and 7.56, phosphorylase kinase can be activated both by phosphorylation of the α and β subunits and by interaction of Ca^{2+} with the δ subunit. These two different mechanisms for phosphorylase kinase activation provide independent routes to the activation of phosphorylase by epinephrine (Figure 7.61) and neural signals (Figure 7.62). However, maximum activation of phosphorylase kinase requires phosphorylation of the enzyme plus the binding of Ca^{2+} to the enzyme. This provides a mechanism useful to the skeletal muscle cell. For example, if a muscle is not signaled to contract in response to a particular stressful situation, that muscle will require no additional burst of ATP production from glycolysis and there will have been no reason for epinephrine to have triggered degradation of glycogen. Epinephrine serves to make the muscle more sensitive to changes in intracellular Ca^{2+} concentration. The release of epinephrine into the blood in response to stress triggers the cascade that causes the conversion of phosphorylase kinase b to its a form in skeletal muscle. However, this will only partially activate phosphorylase because phosphorylase kinase a is not fully active without Ca^{2+}. If nervous stimulation triggers work to be done in the muscle, intracellular Ca^{2+} levels will increase, phosphorylase kinase a will be maximally activated, phosphorylase will be maximally converted to its a form, and appropriate amounts of glycogen will be mobilized to meet the metabolic demand of the stimulated tissue.

Insulin Stimulates Glycogen Synthesis in Muscle and Liver

An increase in blood glucose signals the release of insulin from β cells of the pancreas. Insulin circulates in the blood, serving as a first messenger to inform several tissues that excess glucose is present. Insulin receptors,

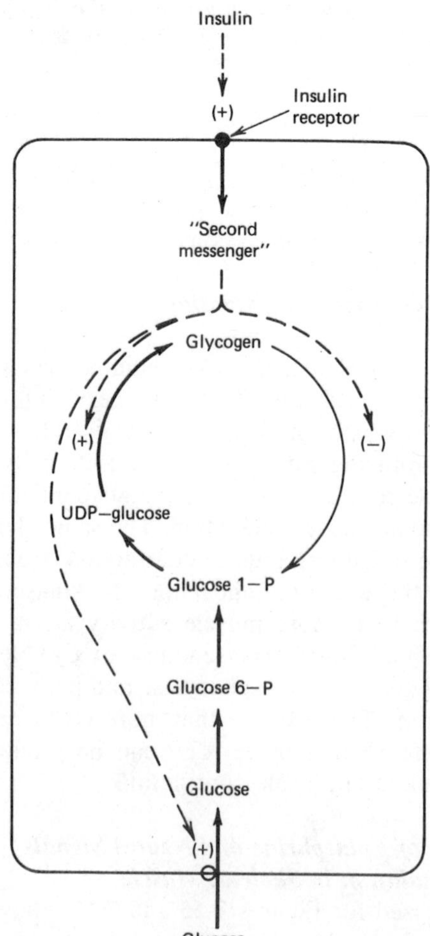

Figure 7.63
Insulin acts via unidentified second messenger system to promote glycogen synthesis in muscle.

located only on the plasma membranes of insulin-responsive cells, are thought to respond to insulin binding by producing an intracellular second messenger that promotes glucose use within these tissues. The pancreas responds to a decrease in blood glucose with less release of insulin but greater release of glucagon. These hormones have opposite effects on glucose utilization by the liver, thereby establishing the pancreas as a fine-tuning device that prevents dangerous fluctuations in blood glucose levels.

Insulin increases glucose utilization rates in part by promoting glycogenesis and inhibiting glycogenolysis in muscle (Figure 7.63) and liver (Figure 7.64). Insulin stimulation of glucose transport at the plasma membrane is important for these effects in muscle but not liver. Hepatocytes have a high-capacity, insulin-insensitive transport system, whereas muscle cells are equipped with a low capacity system that requires insulin for maximum rates of glucose uptake. Insulin further promotes glycogen accumulation in both tissues by activating glycogen synthase. The insulin "second messenger", assuming there is one, can be pictured as having effects opposite to those of cAMP. Recall from Figures 7.55 and 7.56 that cAMP promotes phosphorylation of glycogen phosphorylase and glycogen synthase by activating protein kinases and inhibiting phosphoprotein phosphatase. The putative insulin "second messenger" promotes dephosphorylation of these enzymes, presumably by inhibiting protein kinases and/or activating phosphoprotein phosphatase. The details of the mechanism are not known, and even the existence of the insulin "second messenger" remains to be established. As shown in Figure 7.65, evidence

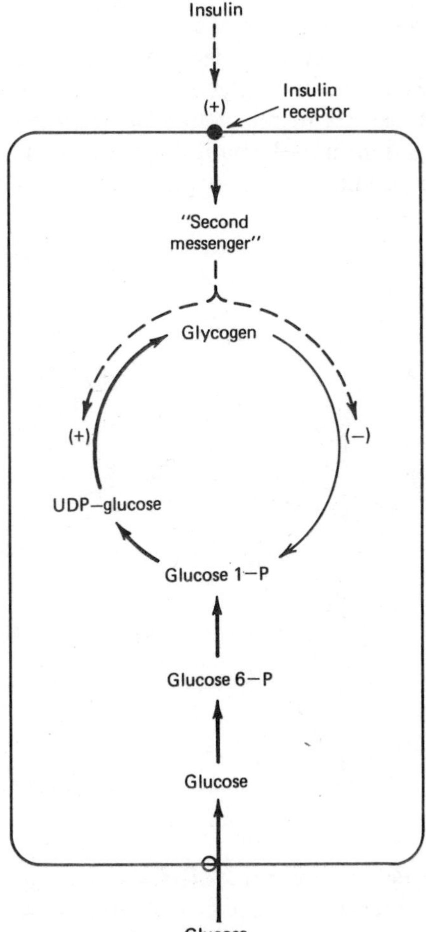

Figure 7.64
Insulin acts via unidentified second messenger system to promote glycogen synthesis in liver.

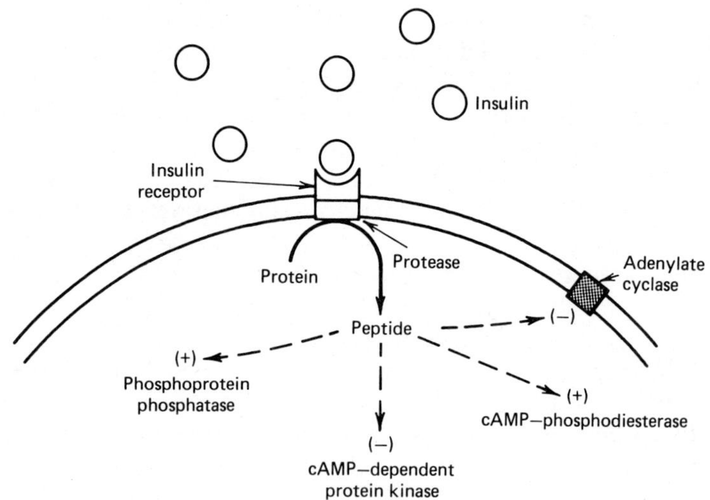

Figure 7.65
The second messenger of insulin action may be a peptide generated by a protease located in the plasma membrane.

has been presented that the messenger may be a peptide and that an insulin-dependent protease responsible for generating the peptide may be associated with the insulin receptor. The peptide has not been characterized, but extracts believed to contain the peptide have been reported to inhibit cAMP-dependent protein kinase and adenylate cyclase as well as activate cAMP-dependent phosphodiesterase and phosphoprotein phosphatase (Figure 7.65). Such effects would oppose the actions of hormones such as glucagon and epinephrine and would cause an increase in glycogen synthesis and a decrease in glycogen degradation.

A second hypothesis for insulin action has emerged recently as a result of detailed investigations into the nature of the insulin receptor. This work suggests that one subunit of the insulin receptor is actually a protein

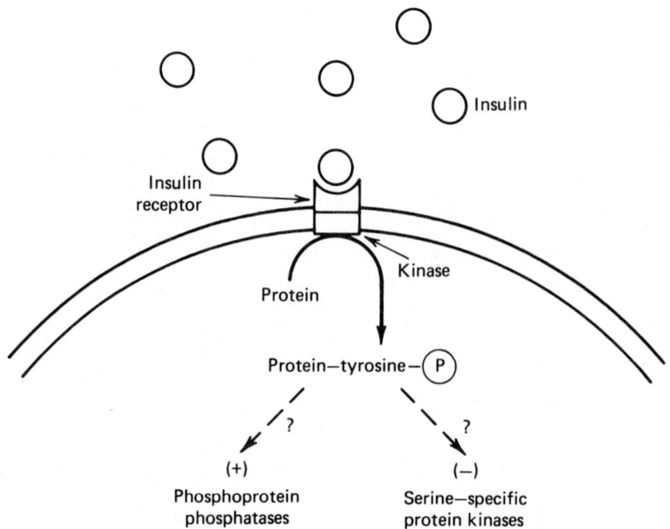

Figure 7.66
A tyrosine-specific protein kinase may mediate the action of insulin.
Ⓟ refers to a phosphate (PO_4^{2-}) group attached covalently to a tyrosine residue of the protein.

kinase that uses ATP to phosphorylate specific tyrosine (but not serine) residues of acceptor protein molecules. The emerging picture, outlined in Figure 7.66, is that binding of insulin to its receptor activates this tyrosine-specific kinase to phosphorylate tyrosine residues of an intracellular enzyme. Although this enzyme has not been identified, we can readily visualize how such an event might be part of a cascade that would culminate in dephosphorylation of glycogen synthase and glycogen phosphorylase.

That much progress has been made and that definition of the mechanism of action of insulin finally seems imminent, make this a most exciting time for all students of biochemistry. Moreover, recent evidence suggests that factors that stimulate normal cell growth and viruses that cause malignant transformations may also act in part via tyrosine-specific protein kinases. We can expect, therefore, that future investigations in this area of basic research may contribute significantly to the eventual solution of our greatest health problems, that is, atherosclerosis, cancer, and diabetes.

BIBLIOGRAPHY

Arion, W. J., Lange, A. J., Walls, H. E., and Ballas, L. M. Evidence for the participation of independent translocases for phosphate and glucose 6-phosphate in the microsomal glucose 6-phosphatase system. *J. Biol. Chem.* 255:10396, 1980.

Berridge, M. J. Review article: Inositol trisphosphate and diacylglycerol as second messengers. *Biochem. J.* 220:345, 1984.

Brooke, M. H., and Kaiser, K. K. The use and abuse of muscle histochemistry. *Ann. N.Y. Acad. Sci.* 228:121, 1974.

Claus, T. H., El-Maghrabi, M. R., Regen, D. M., Stewart, H. B., McGrane, M., Kountz, P. D., Nyfeler, F., Pilkis, J., and Pilkis, S. J. The role of fructose 2,6-bisphosphate in the regulation of carbohydrate metabolism. *Curr. Top. Cell. Regul.* 23:57, 1984.

Cohen, P. *Control of enzyme activity, 2nd* ed. New York: Wiley, 1983.

Denton, R. M., and Pogson, C. I. *Metabolic regulation.* New York: Wiley, 1976.

Geelen, M. J. H., Harris, R. A., Beynen, A. C., and McCune, S. A. Short-term hormonal control of hepatic lipogenesis. *Diabetes* 29:1006, 1980.

Greene, H. L., Slonin, A. E., and Burr, I. M. Type I glycogen storage disease: a metabolic basis for advances in treatment, in L. A. Barness (ed.), *Advances in pediatrics,* vol. 26, Chicago: Year Book Publishers, 1979, p. 63.

Hanson, R. W., Mehlman, M. A. (eds.). *Gluconeogenesis, its regulation in mammalian species.* New York: Wiley, 1976.

Hers, H. G., and Hue, L. Gluconeogenesis and related aspects of glycolysis. *Annu. Rev. Biochem.* 52:617, 1983.

Houslay, M. D. The search for a molecular mechanism for the action of insulin. *Biochem. Educ.* 12:49, 1984.

Ingebritsen, T. S., and Cohen, P. Protein phosphatases: Properties and role in cellular regulation. *Science (Wash. D.C.)* 221: 331, 1983.

Isselbacher, K. J., Adams, R. D., Braundwald, E., Petersdorf, R. B., and Wilson, J. D. (eds.). *Harrison's principles of internal medicine,* 9th ed. New York: McGraw-Hill, 1980.

Joseph, S. K. Inositol trisphosphate: An intracellular messenger produced by Ca^{2+} mobilizing hormones. *Trends Biochem. Sci.* 9:420, 1984.

Lieber, C. S. The metabolism of alcohol. *Sci. Am.* 234:25, 1976.

Metzler, D. E. *Biochemistry, the chemical reactions of living cells,* New York, Academic Press, 1977.

Mitchell, G., Heffron, J. J. A., and van Rensburg, A. J. J. A halothane-induced biochemical defect in muscle of normal and malignant hyperthermia-susceptible landrace pigs. *Anesth. Analg.* 59:250, 1980.

Nelson, T. E., and Flewellen, E. H. Current concepts: the malignant hyperthermia syndrome. *New Engl. J. Med.* 309:416, 1983.

Newsholme, E. A., and Leech, A. R., *Biochemistry for the medical sciences.* New York: Wiley, 1983.

Newsholme, E. A., and Start, C. *Regulation in metabolism.* New York: Wiley, 1973.

Nishizuka, Y. Protein kinases in signal transduction. *Trends. Biochem. Sci.* 9:163, 1984.

Roach, P. J. Hormonal control of glycogen metabolism, in H. Rupp (ed.), *Regulation of heart function: Basic concepts and clinical applications.* New York: Thieme-Stratton Inc., 1985.

Roach, P. J. Principles of the regulation of enzyme activity, in L. Goldstein and D. M. Prescott (eds.), *Cell biology, a comprehensive treatise,* vol. IV. New York: Academic Press, 1980.

Stanbury, J. B., Wyngaarden, J. B., and Fredrickson, D. S. (eds.). *The metabolic basis of inherited disease,* 4th ed. New York: McGraw-Hill, 1978.

Stanley, C. A., Anday, E. K., Baker, L., and Delivoria-Papadopolous, M. Metabolic fuel and hormone responses to fasting in newborn infants. *Pediatrics* 64:613, 1979.

QUESTIONS

J. BAGGOTT AND C. N. ANGSTADT

*Q*uestion *T*ypes are described inside the front cover.

1. (QT1) In glycolysis ATP synthesis is catalyzed by:
   A. hexokinase.
   B. 6-phosphofructo-1-kinase.
   C. glyceraldehyde-3-phosphate dehydrogenase.
   D. phosphoglycerate kinase.
   E. none of the above.

2. (QT2) NAD^+ can be regenerated in the cytoplasm if NADH reacts with:
   1. pyruvate.
   2. dihydroxyacetone phosphate
   3. oxaloacetate
   4. the flavin bound to NADH dehydrogenase

   A. Glucokinase C. Both
   B. Hexokinase D. Neither

3. (QT4) K_m is well above normal blood glucose concentrations.
4. (QT4) Found in muscle.
5. (QT4) Inhibited by glucose 6-phosphate.
6. (QT1) 6-Phosphofructo-1-kinase can be inhibited by all of the following *except:*
   A. ATP at high concentrations.
   B. citrate.
   C. AMP.
   D. low pH.
7. (QT2) Which of the following supports gluconeogenesis:
   1. α-ketoglutarate + aspartate $\rightarrow$ glutamate + oxaloacetate
   2. pyruvate + ATP + HCO_3 $\rightarrow$ oxaloacetate + ADP + P_i + H^+
   3. acetyl CoA + oxaloacetate + H_2O $\rightarrow$ citrate + CoA
   4. glutamate + NAD^+ $\rightarrow$ α-ketoglutarate + NADH + NH_4^+

   A. Cori cycle C. Both
   B. Alanine cycle D. Neither

8. (QT4) Involves only tissues with aerobic metabolism (i.e., mitochondria and O_2).
9. (QT4) Three-carbon compounds arising from glycolysis are converted to glucose at the expense of energy from fatty acid oxidation.
10. (QT2) The uncontrolled production of NADH from NAD^+ during ethanol metabolism blocks gluconeogenesis from:
    1. pyruvate.
    2. oxaloacetate.
    3. glycerol.
    4. galactose.
11. (QT2) Gluconeogenic enzymes include:
    1. pyruvate carboxylase.
    2. fructose-1,6-bisphosphatase.
    3. phospho*enol*pyruvate carboxykinase.
    4. phosphoglucomutase.

    A. Glucose production from pyruvate C. Both
    B. Glucose production from glycogen D. Neither

12. (QT4) Consumes ATP.
13. (QT4) Glucose 6-phosphatase is involved.

    A. Synthesis of glycogen C. Both
    B. Breakdown of glycogen D. Neither

14. (QT4) Phosphoglucomutase.
15. (QT4) Glucose 1-phosphate uridylyltransferase.
16. (QT4) Chain of glucosyl residues is transferred.
17. (QT4) Stimulated by epinephrine.
18. (QT2) Phosphorylation activates:
    1. glycogen phosphorylase.
    2. inhibitor-1.
    3. phosphorylase kinase.
    4. protein kinase.
19. (QT2) AMP activates:
    1. 6-phosphofructo-1-kinase.
    2. protein kinase.
    3. glycogen phosphorylase.
    4. hexokinase.

ANSWERS

1. D A and B use ATP; both catalyze irreversible reactions. C synthesizes 1,3-bisphosphoglycerate. D synthesizes ATP in the forward direction; the reaction is reversible (p. 271).
2. A 1, 2, and 3 true. 1 may be converted to lactate. 2 and 3 are the cytoplasmic acceptors for shuttle systems. 4 is mitochondrial (p. 273).
3. A Blood glucose is ~5 mM. K_m of glucokinase is ~10 mM. K_m of hexokinase is <0.1 mM (p. 281).
4. B Hexokinases are widely distributed. Glucokinase is hepatic (p. 280).
5. B Glucose 6-phosphate inhibition is an important control of hexokinase (p. 280).
6. C AMP is an allosteric regulator that relieves inhibition by ATP (p. 282).
7. C 2 and 4 true. 1: α-Ketoglutarate and oxaloacetate both give rise to glucose; interconversion of one to the other accomplishes nothing (p. 300). 3: Citrate ultimately gives rise to oxaloacetate, losing two carbons in the process; again nothing is gained (p. 301). 2 is on the direct route of conversion of pyruvate to glucose. 4 converts an amino acid to a compound that is converted to oxaloacetate (p. 299). *Net* synthesis of a TCA intermediate is required to support gluconeogenesis.
8. B If alanine is the end product of glycolysis, NADH must be reoxidized aerobically (p. 294, Figure 7.30B).
9. C Lactate or alanine is transported to the liver for glucose synthesis. The liver's major energy source is fatty acid (p. 303).
10. A 1, 2, and 3 true. High NADH/NAD^+ converts pyruvate to lactate and oxaloacetate to malate (p. 306). It prevents oxidation of glycerol 3-phosphate to dihydroxyacetone phosphate (p. 302, Figure 7.41). Gluconeogenesis from galactose is not affected by the redox state of the cell (p. 303, Figure 7.44).
11. A 1, 2, and 3, along with glucose 6-phosphatase, are the gluconeogenic enzymes (p. 296). 4 is on the pathway between glucose and glycogen (p. 309), which is not part of gluconeogenesis (p. 293).
12. A ATP is required for gluconeogenesis but not for glycogenolysis (pp. 303, 309).
13. C Both processes produce glucose 6-phosphate, which must be hydrolyzed (p. 309). Some free glucose is, however, produced by the debranching enzyme (p. 310).
14. C The enzyme interconverts glucose 6-phosphate and glucose 1-phosphate (pp. 309, 310).
15. A The enzyme synthesizes UDP-glucose (p. 312).
16. C Synthesis involves "branching" enzyme (p. 312), and breakdown requires "debranching" enzyme (p. 310).
17. B Epinephrine activates glycogen phosphorylase and inactivates glycogen synthase via Ca^{2+} or the cAMP-regulated cascade (p. 321, Figure 7.59 and 7.61).
18. A 1, 2, and 3 are activated by phosphorylation (p. 315, Figure 7.55). Protein kinase is not a phospho–dephospho enzyme (p. 316).
19. B 1 and 3 true. 1 (p. 285) and 3 (p. 315) are allosterically activated by AMP. 2 is activated by cAMP (p. 316). 4 is controlled by glucose 6-phosphate.

8

Carbohydrate Metabolism II: Special Pathways

NANCY B. SCHWARTZ

8.1 OVERVIEW

In addition to the catabolism of glucose for the specific purpose of energy production in the form of ATP, several other pathways involving sugar metabolism exist in cells. One, the pentose phosphate pathway, is particularly important in animal cells. As will be discussed, this pathway does not operate instead of glycolysis and the tricarboxylic acid cycle, but rather it functions side by side with them for production of reducing power and pentose intermediates. Thus, the metabolic significance of this pathway, known also as the hexose monophosphate shunt or the 6-phosphogluconate pathway, is not to obtain energy from the oxidation of glucose in animal tissues. In fact, starting with glucose 6-phosphate, no ATP is generated, nor is any required. The pentose phosphate pathway is rather a multifunctional pathway whose primary purpose is to generate reducing power in the form of NADPH. It has previously been mentioned that the fundamental distinction between NADH and NADPH in most biochemical reactions is that NADH is oxidized by the respiratory chain to produce ATP, whereas NADPH serves as a hydrogen and electron donor in reductive biosynthetic reactions. The enzymes involved in this pathway are located in the cytosol, indicating that the oxidation that occurs is not dependent on mitochondria or the tricarboxylic acid cycle. Another important function is to convert hexoses into pentoses, particularly ribose 5-phosphate. This five-carbon sugar or its derivatives are components of ATP, CoA, NAD, FAD, RNA, and DNA. The pentose phosphate pathway also catalyzes the interconversion of 3-, 4-, 6-, and 7-carbon sugars, some of which can enter the glycolytic sequence. In order to fulfill another function, which will not be discussed further, the pentose phosphate pathway may be modified to participate in the formation of glucose from CO_2 in photosynthesis.

There are specific pathways for synthesis and degradation of monosaccharides, oligosaccharides, and polysaccharides (other than glycogen), and a profusion of chemical interconversions, whereby one sugar can be changed into another. All of the monosaccharides, and the oligo- and polysaccharides synthesized from the monosaccharides, originate from glucose. The interconversion reactions by which one sugar is changed into another may occur directly or at the level of nucleotide-linked sugars. In addition to their important role in sugar transformation, nucleotide sugars are the obligatory activated form for saccharide synthesis. Monosaccharides are also often found as components of more complex macromolecules like oligo- and polysaccharides, glycoproteins, glycolipids, and proteoglycans. Certain oligosaccharides, covalently linked to protein or lipid, form the major structural components of bacterial cell walls, and include peptidoglycan, teichoic acids, and lipopolysaccharides. Chitin, a linear homopolymer of *N*-acetylglucosamine residues, is the predominant organic structural component of the exoshells of the invertebrates, as well as most fungi, many algae, and some yeast. Cellulose, an unbranched polymer of glucose residues, is the major structural component in the plant kingdom. In higher animals some of the complex carbohydrate molecules are also predominantly structural elements found in ground substance filling the extracellular space in tissues and as components of cell membranes. Increasingly though, more dynamic functions for these complex macromolecules, such as recognition markers and as determinants of biological specificity, are being discovered. The discussion of complex carbohydrates in this chapter is limited to the chemistry and biology of those complex carbohydrates found in animal tissues and fluids. See the Appendix, "Review of Organic Chemistry," for a discussion of the nomenclature and chemistry of the carbohydrates.

8.2 PENTOSE PHOSPHATE PATHWAY

Reaction Sequence of Pentose Phosphate Pathway

The oxidative pentose phosphate pathway provides a means for cutting the carbon chain of a sugar molecule one carbon at a time. However, in contrast to glycolysis and the tricarboxylic acid cycle, the operation of this pathway does not occur as a consecutive set of reactions leading directly from glucose 6-phosphate to six molecules of CO_2. For simplification, the overall pathway can be visualized as occurring in two stages. In the first, hexose is decarboxylated to pentose. The two oxidation reactions that lead to formation of NADPH also occur in this stage. The pathway may continue further, and, by a series of transformations, six molecules of pentose may undergo rearrangements to yield five molecules of hexose. It is these various transformations that give the pentose phosphate pathway its characteristic complexity. To understand this pathway, it is necessary to examine each reaction individually.

Formation of Pentose Phosphate

The first reaction of the pentose phosphate pathway (Figure 8.1) is the enzymatic dehydrogenation of glucose 6-phosphate at C-1 to form 6-phosphogluconolactone and NADPH. The enzyme catalyzing this reaction is glucose 6-phosphate dehydrogenase, the first enzyme found to be specific for $NADP^+$. Special interest in this enzyme stems from the severe

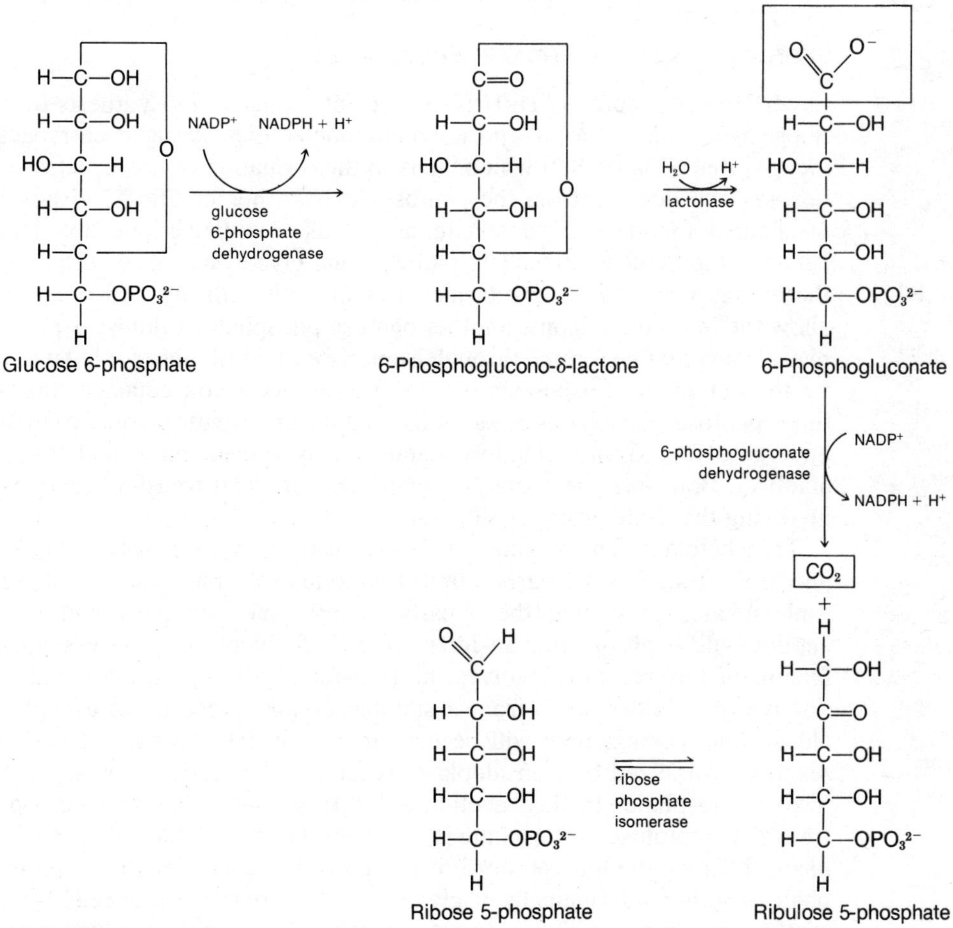

Figure 8.1
Formation of pentose phosphate.

CLIN. CORR. 8.1
GLUCOSE-6-PHOSPHATE DEHYDROGENASE: GENETIC DEFICIENCY OR PRESENCE OF GENETIC VARIANTS IN ERYTHROCYTES

When certain seemingly harmless drugs, such as antimalarials, are administered to susceptible patients, an acute hemolytic anemia may result. Susceptibility to drug-induced hemolytic disease may be due to a deficiency of glucose 6-phosphate dehydrogenase activity in the erythrocyte, and was one of the early indications that genetic deficiencies of this enzyme exist. This enzyme, which catalyzes the coupled oxidation of glucose 6-phosphate to 6-phosphogluconate and the reduction of $NADP^+$, is particularly important, since the pentose phosphate pathway is the only known pathway of CO_2 production in the red cell. For example, red cells with the relatively mild A-type of glucose 6-phosphate dehydrogenase deficiency are able to oxidize glucose at a normal rate when the demand for NADPH is normal. However, if the rate of NADPH oxidation is increased, the enzyme-deficient cells are unable to increase adequately the rate of glucose oxidation and carbon dioxide production. Additionally, cells lacking glucose 6-phosphate dehydrogenase do not reduce enough $NADP^+$ to maintain glutathione in its reduced state. Reduced glutathione is necessary for the integrity of the erythrocyte membrane, thus rendering enzyme-deficient red cells more susceptible to hemolysis by a wide range of compounds. Therefore, the basic abnormality in glucose 6-phosphate deficiency is the formation of mature red blood cells that have diminished glucose 6-phosphate dehydrogenase activity. This enzymatic deficiency, which is usually undetected until administration of certain drugs, illustrates the interplay of heredity and environment on the production of disease. Enzyme absence is only one of several abnormalities that can affect enzyme activity, and others have been detected independent of drug administration. Increasingly, these variants can be distinguished from one another by certain clinical and biochemical differences.

anemia that may result from the absence of glucose 6-phosphate dehydrogenase in erythrocytes or from presence of one of several genetic variants of the enzyme (Clin. Corr. 8.1). Formation of the intermediate product of this reaction, a lactone, is freely reversible. Although the lactone is unstable and hydrolyzes spontaneously, a specific gluconolactonase causes a more rapid ring opening and ensures that the reaction goes to completion. The overall equilibrium of these two reactions lies far in the direction of NADPH maintaining a high $[NADPH]/[NADP^+]$ ratio within cells. A second dehydrogenation and decarboxylation is catalyzed by 6-phosphogluconate dehydrogenase, a Mg^{2+}-dependent enzyme. The pentose phosphate, ribulose 5-phosphate, and a second molecule of NADPH are produced. The final step in synthesis of ribose 5-phosphate is the isomerization of ribulose 5-phosphate by phosphopentose isomerase. Like a similar reaction in the glycolytic pathway, this ketose-aldose isomerization proceeds through an enediol intermediate.

These first reactions resulting in decarboxylation may be considered to be the most important, since all the oxidation reactions leading to production of NADPH occur in this early part of the pathway. Under certain metabolic conditions, the pentose phosphate pathway may end at this point, with the utilization of NADPH for reductive biosynthetic reactions and ribose 5-phosphate as a precursor for nucleotide synthesis. The overall equation may be written as

$$\text{Glucose 6-phosphate} + 2\text{NADP}^+ + H_2O \longrightarrow \text{ribose 5-phosphate} + 2\text{NADPH} + 2H^+ + CO_2$$

Interconversions of Pentose Phosphates

In certain cells more NADPH is needed for reductive biosynthesis than ribose 5-phosphate for incorporation into nucleotides. A sugar rearrangement system (Figure 8.2) which leads to the formation of triose, tetrose, hexose, and heptose from the pentoses exists, thus creating a disposal mechanism for ribose 5-phosphate, as well as providing a reversible link between the pentose phosphate pathway and glycolysis, since certain of these sugars are common intermediates of both pathways. In order to allow the interconversions, another pentose phosphate, xylulose 5-phosphate, must first be formed through isomerization of ribulose 5-phosphate by the action of phosphopentose epimerase. As a consequence, these three pentose phosphates exist as an equilibrium mixture, and can then undergo further transformations catalyzed by transketolase and transaldolase. Both enzymes catalyze chain cleavage and transfer reactions involving the same group of substrates.

Transketolase, an enzyme involving thiamine pyrophosphate (TPP) and Mg^{2+}, transfers a 2-carbon unit from xylulose 5-phosphate to ribose 5-phosphate, producing the 7-carbon sugar sedoheptulose and glyceraldehyde 3-phosphate, an intermediate of glycolysis. The essential feature of this reaction involves the transfer of a C_2 group, designated "active glycolaldehyde," from a suitable donor ketose to an acceptor aldose (another example will be encountered later). A further transfer reaction, catalyzed by transaldolase, results in the recovery of the first hexose phosphate. In this reaction a 3-carbon unit (dihydroxyacetone) from sedoheptulose 7-phosphate is transferred to glyceraldehyde 3-phosphate, forming the tetrose, erythrose 4-phosphate, and fructose 6-phosphate, another intermediate of glycolysis. This reaction proceeds by a mechanism similar to that of fructose bisphosphate aldolase (Chapter 7), except that transaldolase is unable to react with or form free dihydroxyacetone or its phosphate.

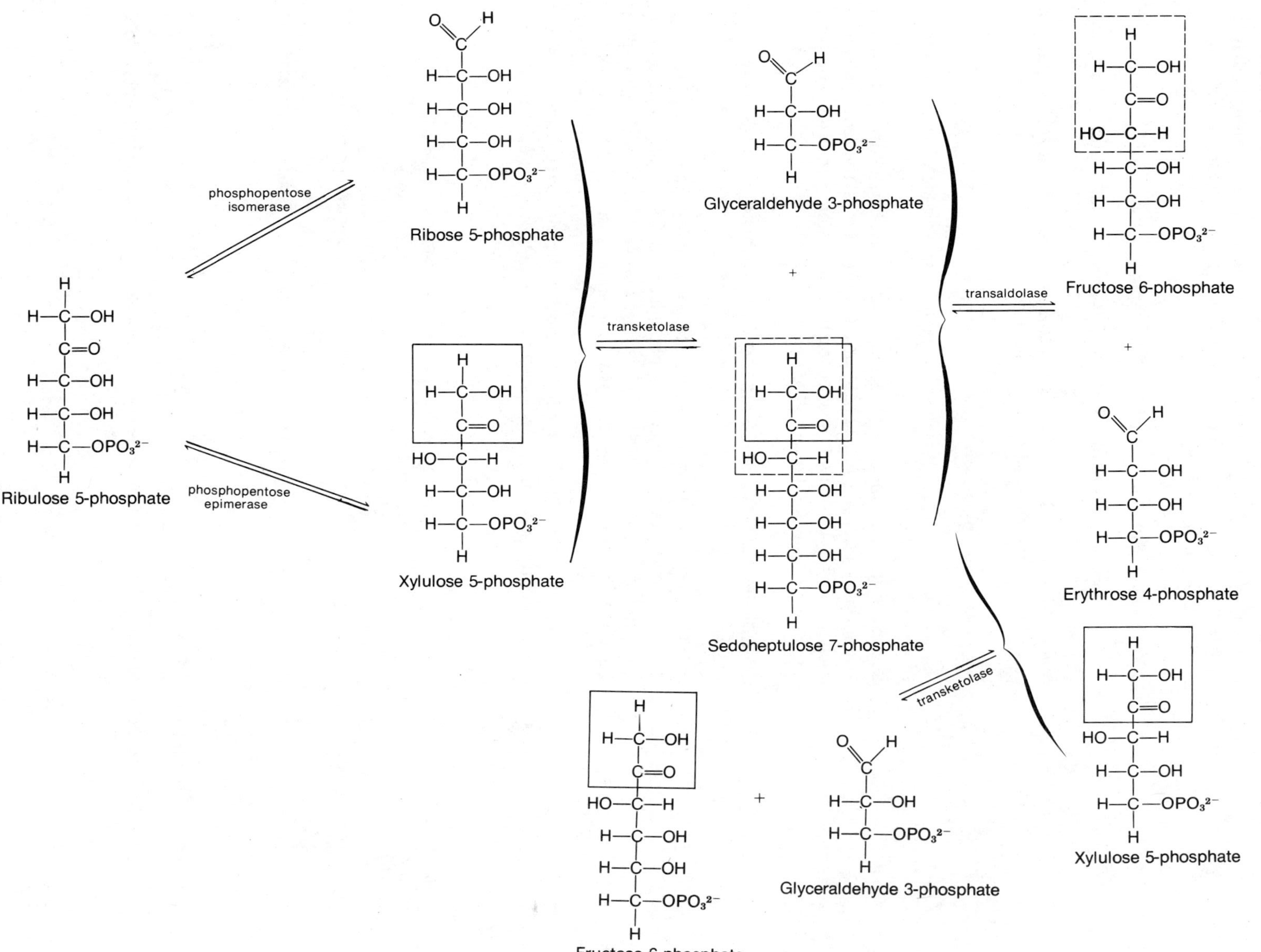

Figure 8.2
Interconversions of pentose phosphates.

In a third reaction, specific to this pathway, transketolase catalyzes the synthesis of fructose 6-phosphate and glyceraldehyde 3-phosphate from erythrose 4-phosphate and a second molecule of xylulose 5-phosphate. In this case, the 2-carbon unit is transferred from xylulose 5-phosphate to an acceptor 4-carbon sugar, now forming two glycolytic intermediates. The sum of these reactions is

$$\text{2 Xylulose 5-phosphate + ribose 5-phosphate} \rightleftharpoons \text{2 fructose 6-phosphate + glyceraldehyde 3-phosphate}$$

Since xylulose 5-phosphate is derived from ribose 5-phosphate, the net reaction starting from ribose 5-phosphate is

$$\text{3 Ribose 5-phosphate} \rightleftharpoons \text{2 fructose 6-phosphate + glyceraldehyde 3-phosphate}$$

Therefore, excess ribose 5-phosphate, whether it arises from the initial oxidation of glucose 6-phosphate or from the degradative metabolism of nucleic acids, is effectively scavenged by conversion to intermediates that can enter the carbon flow of glycolysis.

Complete Oxidation of Glucose 6-Phosphate to CO_2

In certain cells and tissues, like lactating mammary gland, a pathway for complete oxidation of glucose 6-phosphate to CO_2, with concomitant reduction of $NADP^+$ to NADPH, also prevails (Figure 8.3). By a complex sequence of reactions, the ribulose 5-phosphate produced by the pentose phosphate pathway is recycled into glucose 6-phosphate by transketolase, transaldolase, and certain enzymes of the gluconeogenic pathway. Hexose continually enters this system, and CO_2 evolves as the only car-

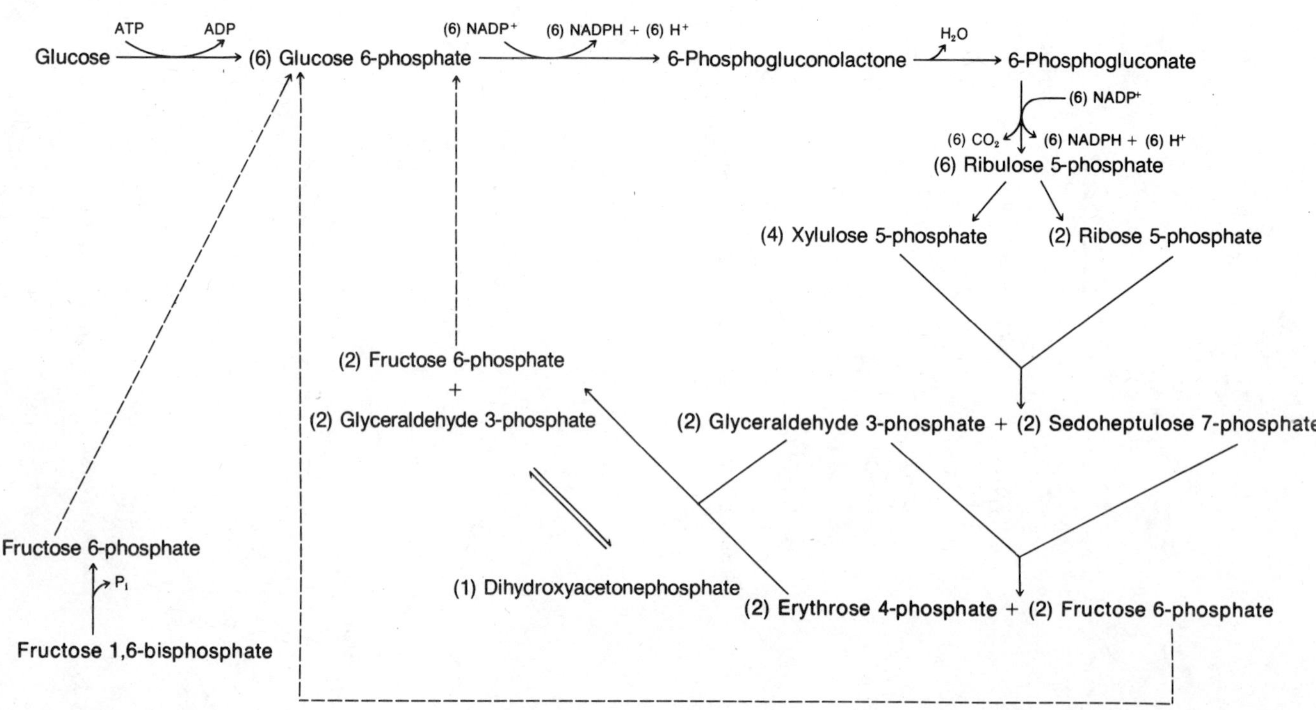

Figure 8.3
Pentose phosphate pathway.

bon compound. A balanced equation for this process would involve the oxidation of six molecules of glucose 6-phosphate to six molecules of ribulose 5-phosphate and $6CO_2$. This represents essentially the first part of the pentose phosphate pathway and results in transfer of 12 pairs of electrons to $NADP^+$, the requisite amount for total oxidation of 1 glucose to $6CO_2$. The remaining six molecules of ribulose 5-phosphate are then rearranged by the pathway described above to regenerate five molecules of glucose 6-phosphate. The overall equation can be written as

$$\text{6 Glucose 6-phosphate} + 12NADP^+ + 7H_2O \longrightarrow \text{5 glucose 6-phosphate} + 6CO_2 + 12NADPH + 12H^+ + P_i$$

The net reaction is therefore

$$\text{Glucose 6-phosphate} + 12NADP^+ + 7H_2O \longrightarrow 6CO_2 + 12NADPH + 12H^+ + P_i$$

Significance of Pentose Phosphate Pathway

The pentose phosphate pathway serves several purposes, including a mechanism for synthesis and degradation of sugars other than hexoses, particularly pentoses necessary for nucleotides and nucleic acids, and other glycolytic intermediates. Most important, however, is the ability to synthesize NADPH, which plays a unique role in biosynthetic reactions. The direction of flow and path taken by glucose 6-phosphate after entry into the pentose phosphate pathway is determined largely by the needs of the cell for NADPH or sugar intermediates. The situation in which more NADPH than ribose 5-phosphate is required has already been examined and results in a continuation of the pathway, leading to complete oxidation of glucose 6-phosphate to CO_2 and resynthesis of glucose 6-phosphate from ribulose 5-phosphate. Alternatively, if more ribose 5-phosphate than NADPH is required, glucose 6-phosphate is converted to fructose 6-phosphate and glyceraldehyde 3-phosphate by the glycolytic pathway. Two molecules of fructose 6-phosphate and one molecule of glyceraldehyde 3-phosphate are then converted into three molecules of ribose 5-phosphate by a reversal of the transaldolase and transketolase reactions.

The distribution of the pentose phosphate pathway in the tissues of the body is consistent with its functions. As previously mentioned, it is present in erythrocytes for production of NADPH, which in turn is used to generate reduced glutathione, essential for maintenance of normal red cell structure. It is also active in tissues such as liver, mammary gland, testis, and adrenal cortex, which are active sites of fatty acid or steroid synthesis, processes that require the reducing power of NADPH. In contrast, in mammalian striated muscle, where little fatty acid or steroid synthesis occurs, there is no direct oxidation of glucose 6-phosphate through the pentose phosphate pathway. Rather, all catabolism proceeds via glycolysis and the TCA cycle. In some other tissues like liver, 20–30% of the CO_2 may arise from the pentose phosphate pathway, and the balance between glycolysis and the pentose phosphate pathway depends on the metabolic requirements of the cell.

8.3 SUGAR INTERCONVERSIONS AND NUCLEOTIDE SUGAR FORMATION

In preceding discussions, the general principles of carbohydrate metabolism, specifically those involving glucose, were considered. Now we shall examine certain aspects of the metabolism of other monosaccharides, oligosaccharides, and polysaccharides. Most of the monosaccharides found in biological compounds derive from glucose. The most common reactions for sugar transformations in mammalian systems are summarized in Figure 8.4.

Common Reactions in Carbohydrate Metabolism: Isomerization and Phosphorylation

The formation of some saccharides may occur directly, starting from glucose via modification reactions, such as the conversion of glucose 6-phosphate to fructose 6-phosphate by phosphoglucose isomerase in the glycolytic pathway. A similar aldose-ketose isomerization catalyzed by phosphomannose isomerase results in synthesis of mannose 6-phosphate.

Internal transfer of a phosphate group on the same sugar molecule from one hydroxyl to another is a modification that has been previously described. Glucose 1-phosphate, resulting from enzymatic phosphorolysis of glycogen, is converted to glucose 6-phosphate for entry into the glycolytic pathway by phosphoglucomutase. Galactose may also be phosphorylated directly to galactose 1-phosphate by a galactokinase and mannose to mannose 6-phosphate by a mannokinase; the latter may equilibrate with fructose 6-phosphate as discussed above. Similarly, free fructose, an important dietary constituent, may be phosphorylated in the liver to fructose 1-phosphate by a special fructokinase. However, no mutase exists to interconvert fructose 1-phosphate and fructose 6-phosphate, nor can phosphofructokinase synthesize fructose 1,6-bisphosphate from fructose 1-phosphate. Rather, a specific aldolase cleaves fructose 1-

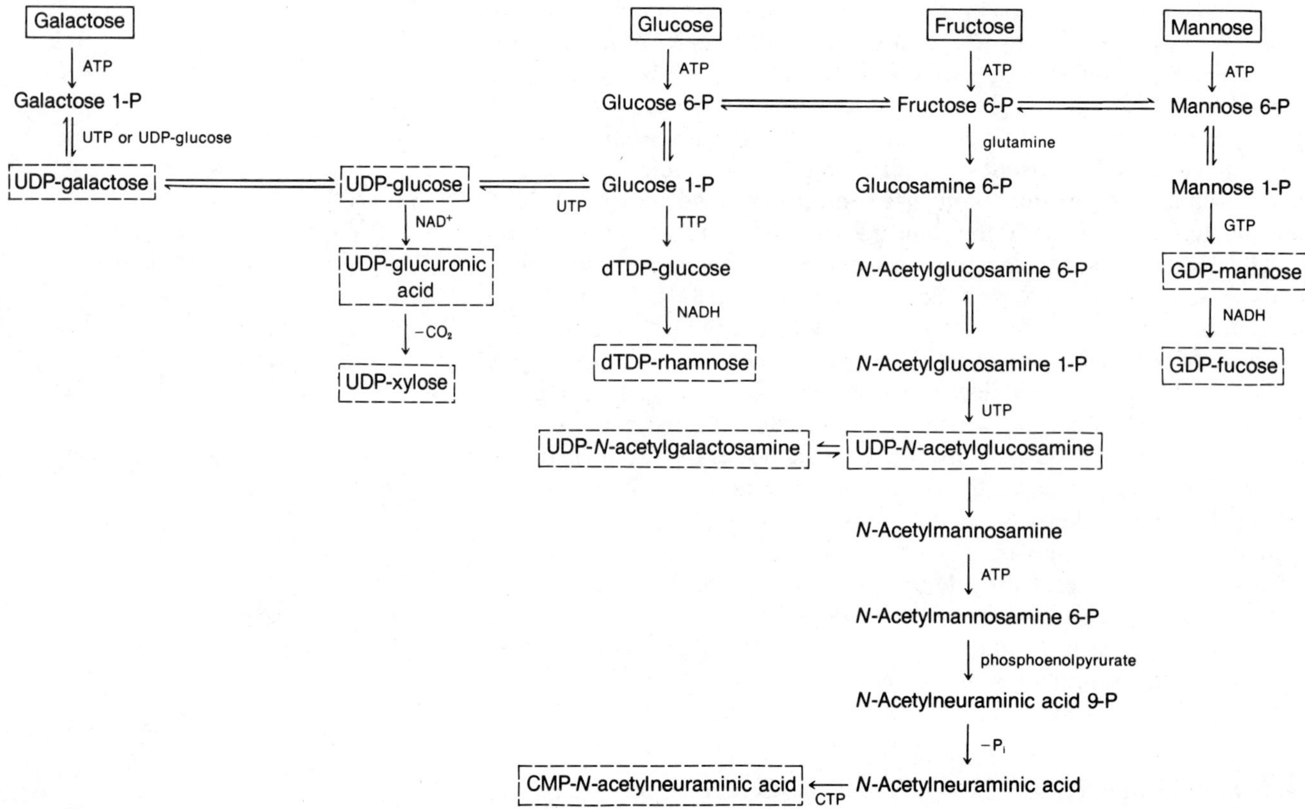

Figure 8.4
Pathways of formation of nucleotide-sugars and interconversions of some hexoses.

phosphate to dihydroxyacetone phosphate (DHAP), which enters the glycolytic pathway directly, and glyceraldehyde which must first be reduced to glycerol, phosphorylated, and then reoxidized to DHAP. Lack of this aldolase may lead to fructose intolerance (see Clin. Corr. 8.2).

Formation of Nucleotide-Sugars

Most other sugar transformation reactions require the prior conversion into *nucleotide*-linked sugars. Nucleotides are phosphoric acid esters of nucleosides; nucleosides are pentosyl derivatives of a purine or pyrimidine base. Formation of nucleoside diphosphate sugar involves the reaction of hexose 1-phosphate and nucleoside triphosphate, catalyzed by a pyrophosphorylase. These reactions are readily reversible. However, in vivo pyrophosphate is rapidly hydrolyzed irreversibly by an inorganic pyrophosphatase, thereby driving the synthesis of nucleotide-sugars. These reactions are summarized as follows:

$$\text{NTP} + \text{sugar 1-phosphate} \rightleftharpoons \text{NDP-sugar} + \text{PP}_i$$
$$\text{PP}_i + \text{H}_2\text{O} \longrightarrow 2\text{Pi}$$
$$\text{NTP} + \text{sugar 1-phosphate} + \text{H}_2\text{O} \longrightarrow \text{NDP-sugar} + 2\text{P}_i$$

A common nucleotide sugar involved in synthesis of glycogen and presumably certain glycoproteins is UDP-glucose. It is synthesized from glucose 1-phosphate and UTP in a reaction catalyzed by UDP-glucose pyrophosphorylase. Pyrophosphate derives from the terminal two phosphoryl groups of UTP.

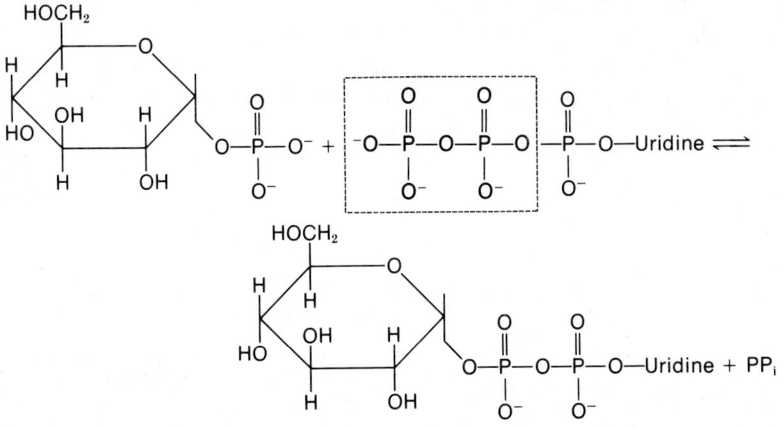

HOCH₂ ... O—P(=O)(O⁻)—O⁻ + ⁻O—P(=O)(O⁻)—O—P(=O)(O⁻)—O—P(=O)(O⁻)—O—Uridine ⇌

HOCH₂ ... O—P(=O)(O⁻)—O—P(=O)(O⁻)—O—Uridine + PP_i

Nucleoside diphosphate-sugars contain two phosphoryl bonds with a large negative ΔG of hydrolysis, which contribute to the energized character of these compounds as glycosyl donors in further transformation and transfer reactions, as well as conferring specificity on the enzymes catalyzing these reactions. This is discussed in greater detail in Section 8.4. UDP usually serves as the glycosyl carrier in higher animals; however, ADP, GDP, and CMP have also been shown to act as carriers in other reactions. As previously mentioned, many of the sugar transformation reactions occur only at the level of nucleotide-sugars. Some of these modification reactions involving nucleotide-sugars include epimerization, oxidation, decarboxylation, reduction, and rearrangement.

Epimerization

One of the most common types of reaction in carbohydrate metabolism is epimerization. For example, the reversible conversion of glucose to ga-

CLIN. CORR. 8.2
ESSENTIAL FRUCTOSURIA AND FRUCTOSE INTOLERANCE: DEFICIENCY OF FRUCTOKINASE AND FRUCTOSE 1-PHOSPHATE ALDOLASE, RESPECTIVELY

Fructose may account for 30 to 60% of the total carbohydrate intake of mammals. It is predominantly metabolized by a specific fructose pathway. The first enzyme in this pathway, fructokinase, is deficient in essential fructosuria. This disorder is a benign asymptomatic metabolic anomaly, which appears to be inherited as an autosomal recessive. Biochemically, following intake of fructose, blood levels appear unusually high; however, 80 to 90% of fructose is eventually metabolized. In contrast, hereditary fructose intolerance is characterized by severe hypoglycemia after ingestion of fructose. Prolonged ingestion in young children may lead to a chronic condition or death. In this disorder fructose 1-phosphate aldolase is deficient, and fructose 1-phosphate accumulates intracellularly.

CLIN. CORR. **8.3**
GALACTOSEMIA: INABILITY TO TRANSFORM GALACTOSE INTO GLUCOSE

The reactions of galactose are of particular interest because in man they are subject to genetic defects resulting in the hereditary disorder galactosemia. When the defect is present, individuals are unable to metabolize the galactose derived from lactose (milk sugar) to glucose metabolites, often with resultant cataract formation, growth failure, or eventual death from liver damage. The genetic disturbance is expressed as a cellular deficiency of either galactokinase, causing a relatively mild disorder, or of galactose 1-phosphate uridylyltransferase, resulting in a more severe disease. Since the alternate enzyme for forming UDP-galactose directly from galactose 1-phosphate does not become active until later years in life, infants with galactosemia are unable to metabolize galactose by either pathway. The development of this second enzyme does account for the fact that galactosemic patients tend to tolerate galactose as they mature. Thus, if maintained on a galactose-free diet during infancy, serious damage may be avoided.

CLIN. CORR. **8.4**
PENTOSURIA: DEFICIENCY OF XYLITOL DEHYDROGENASE

The glucuronic acid oxidation pathway presumably is not essential to human carbohydrate metabolism, since individuals in whom the pathway is blocked suffer no ill effects. This metabolic variation, called idiopathic pentosuria, results from reduced activity of NADP-linked xylitol dehydrogenase, the enzyme that catalyzes the reduction of xylulose to xylitol. Hence affected individuals excrete large amounts of pentose into the urine especially following intake of glucuronic acid.

lactose in animals occurs by epimerization of UDP-glucose to UDP-galactose, by UDP-glucose epimerase. The UDP-galactose that participates in the epimerization reaction is an important intermediate in the metabolism of free galactose, which is derived from the hydrolysis of lactose in the intestinal tract, and may be formed in either of two ways. Galactose can first be phosphorylated by galactokinase and ATP to yield galactose 1-phosphate:

$$\text{Galactose} + \text{ATP} \rightleftharpoons \text{galactose 1-phosphate} + \text{ADP}$$

An enzyme, galactose 1-phosphate uridylyltransferase, transforms galactose 1-phosphate into UDP-galactose by displacing glucose 1-phosphate from UDP-glucose:

$$\text{UDP-glucose} + \text{galactose 1-phosphate} \rightleftharpoons \text{UDP-galactose} + \text{glucose 1-phosphate}$$

A hereditary disorder, galactosemia, results from the absence of this uridylyltransferase (Clin. Corr. 8.3). Alternatively, UDP-galactose can be formed directly from galactose 1-phosphate and UTP, catalyzed by UDP-galactose pyrophosphorylase:

$$\text{Galactose 1-phosphate} + \text{UTP} \rightleftharpoons \text{UDP-galactose} + \text{PP}_i$$

However, this alternative pathway for galactose 1-phosphate metabolism only develops in later life.

An NAD^+-dependent galactose 4-epimerase converts an equatorial hydroxyl at C-4 in glucose to an axial one in galactose in the epimerization of these two sugars (Figure 8.5). Most likely, NAD^+ accepts the hydrogen atom at C-4, and a 4-keto intermediate is formed. Inversion of the hydroxyl group then occurs when NADH transfers its hydrogen to the

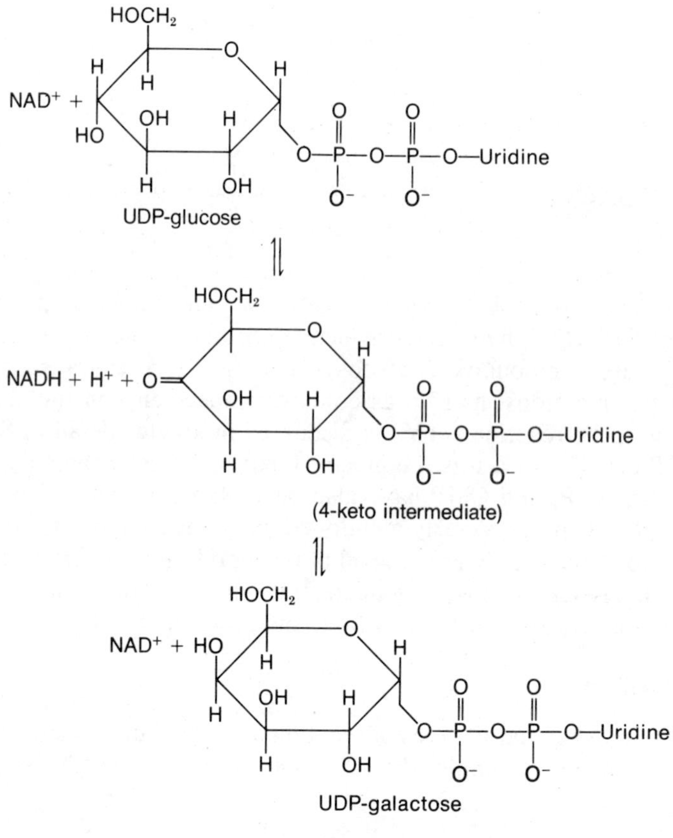

NAD⁺ + UDP-glucose ⇌ NADH + H⁺ + (4-keto intermediate) ⇌ NAD⁺ + UDP-galactose

Figure 8.5
Epimerization of UDP-glucose to UDP-galactose.

other side of C-4. A combination of these reactions allows an efficient transformation of galactose derived from the diet into glucose 1-phosphate, which can then be further metabolized by previously described pathways. Alternatively, the 4-epimerase can operate in the reverse direction when UDP-galactose is needed for biosynthesis. Other 4-epimerases are known such as those that convert UDP-*N*-acetylglucosamine to UDP-*N*-acetylgalactosamine, or UDP-xylose to UDP-arabinose. Presumably these operate by a mechanism similar to the UDP-galactose 4-epimerase.

More recent work has shown that epimerization reactions are not exclusively restricted to nucleotide-linked sugars but may also occur at the polymer level. Thus D-mannuronic acid is epimerized to L-guluronic acid after incorporation into alginic acid (a polyglycosyluronate compound produced by seaweed and certain bacteria), and D-glucuronic acid is epimerized to L-iduronic acid after incorporation into heparin and dermatan sulfate, which is discussed later.

Glucuronic Acid Metabolism

Oxidation and reduction interconversions also result in formation of many additional sugars. One of the most important is formation of glucuronic acid, which serves as a precursor of L-ascorbic acid in those animals which synthesize vitamin C, but is converted to L-xylulose, the ketopentose excreted by humans with essential pentosuria (Clin. Corr. 8.4), and participates in physiological processes of detoxification by production of glucuronide conjugates (Clin. Corr. 8.5). Glucuronic acid is formed by oxidation of UDP-glucose catalyzed by UDP-glucose dehydrogenase.

UDP-glucose + $2NAD^+$ ⟶ UDP-glucuronic acid + 2NADH + $2H^+$

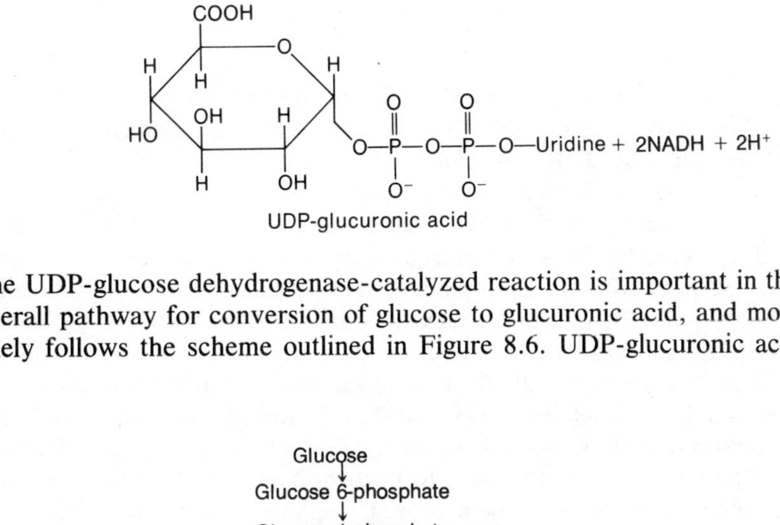

The UDP-glucose dehydrogenase-catalyzed reaction is important in the overall pathway for conversion of glucose to glucuronic acid, and most likely follows the scheme outlined in Figure 8.6. UDP-glucuronic acid

CLIN. CORR. 8.5 GLUCURONIC ACID: PHYSIOLOGICAL SIGNIFICANCE OF GLUCURONIDE FORMATION

The biological significance of glucuronic acid extends to its ability to conjugate with certain endogenous and exogenous substances, forming a group of compounds collectively termed glucuronides in a reaction catalyzed by UDP-glucuronyltransferase. Conjugation of a compound with glucuronic acid produces a strongly acidic compound that is more water-soluble at physiological pH than the precursor and therefore may alter its metabolism, transport, or excretion properties. Glucuronide formation is important in several processes, including drug detoxification, steroid excretion, and bilirubin metabolism. An example of glucuronide conjugation in the latter process is presented. Bilirubin is the major metabolic breakdown product of heme, the prosthetic group of hemoglobin. The central step in excretion of bilirubin is conjugation with glucuronic acid by UDP-glucuronyltransferase. The development of this conjugating mechanism occurs gradually and may take several days to 2 weeks after birth to be active in humans. Although not from a single cause, "physiological jaundice of the newborn" may result from the inability of the neonatal liver to form bilirubin glucuronide at a rate comparable to that of bilirubin production. A defect in glucuronide synthesis has been found in a mutant strain of Wistar ("Gunn") rats, due to a deficiency of UDP-glucuronyltransferase and results in hereditary hyperbilirubinemia. In human beings a similar defect is found in congenital familial nonhemolytic jaundice (Crigler–Najjar syndrome). Patients with this condition are also unable to conjugate foreign compounds efficiently with glucuronic acid.

Glucose
↓
Glucose 6-phosphate
↓
Glucose 1-phosphate
↓
UDP-Glucose
↓
UDP-Glucuronic acid
↓
D-Glucuronic acid 1-phosphate
↓
D-Glucuronic acid

Figure 8.6
Biosynthesis of D-glucuronic acid.

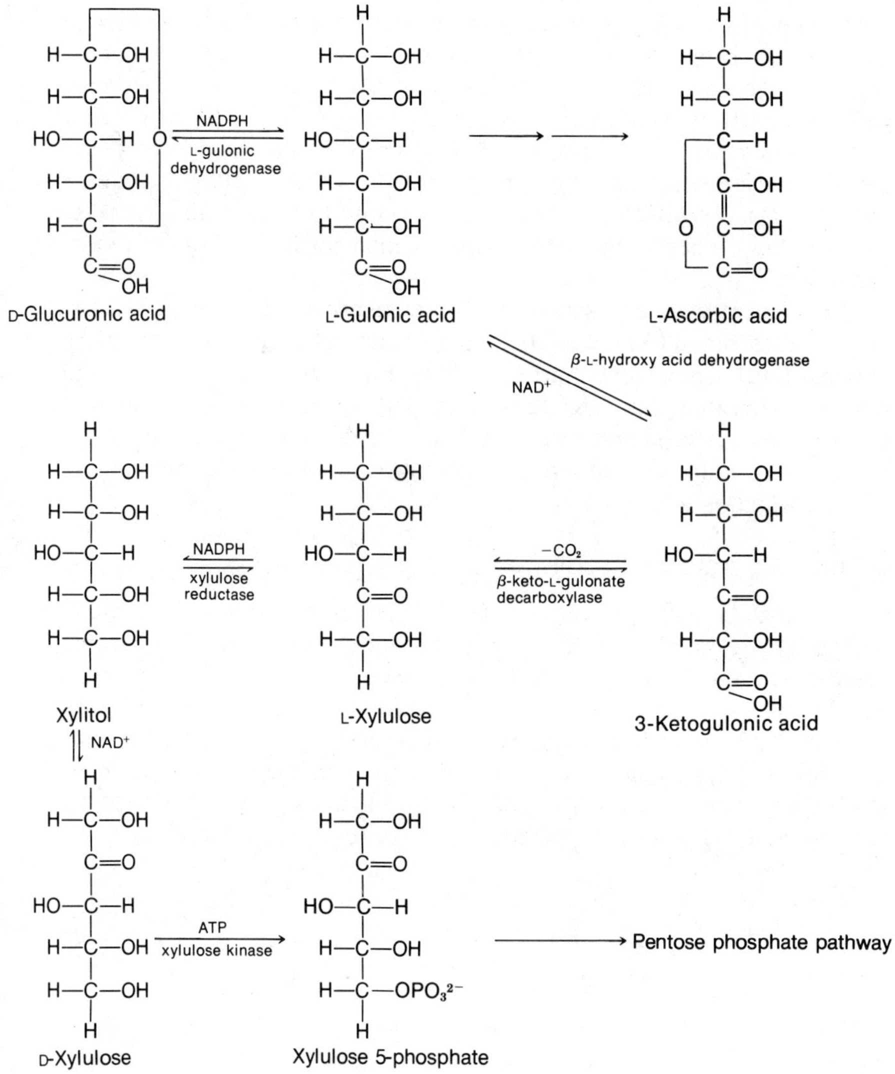

Figure 8.7
Glucuronic acid oxidation pathway.

may then be epimerized to UDP-galacturonic acid. In a similar manner, GDP-mannose is oxidized to GDP-mannuronic acid, which may undergo epimerization to GDP-guluronic acid.

Following its formation, free glucuronic acid is further metabolized by reduction with NADPH to L-gulonic acid (Figure 8.7).Gulonic acid is converted by a two-step process through L-gulonolactone to L-ascorbic acid (vitamin C) in plants and most higher animals. Man, other primates, and the guinea pig lack the enzyme that converts L-gulonolactone to L-ascorbic acid and therefore must satisfy their needs for ascorbic acid by its ingestion. Gulonic acid may also be oxidized to 3-ketogulonic acid and decarboxylated to L-xylulose. L-Xylulose is in turn converted by reduction to xylitol, reoxidized to D-xylulose and phosphorylated with ATP and an appropriate kinase to xylulose 5-phosphate. The latter compound may then re-enter the pentose phosphate pathway described previously. This complex catabolic pathway for glucuronic acid represents another shunt pathway for oxidation of glucose. It should be noted that, in contrast to other pathways of carbohydrate metabolism in which only phosphate esters participate, these reactions also involve free sugars or sugar acids. Evidence suggests that this pathway operates in adipose tissue, and its activity may be increased in tissue from starved or diabetic animals; however, the regulation and extent to which these reactions proceed has not been adequately evaluated.

Decarboxylation, Oxidoreduction, and Transamination

Decarboxylation, which is an important mechanism for degrading sugars one carbon atom at a time, has been previously encountered in the major metabolic pathways. The only known decarboxylation of a nucleotide-sugar is the conversion of UDP-glucuronic acid to UDP-xylose, necessary for synthesis of proteoglycans (Section 8.6). UDP-xylose is a potent inhibitor of UDP-glucose dehydrogenase, which oxidizes UDP-glucose to UDP-glucuronic acid (Figure 8.4). Thus the level of these nucleotide-sugar precursors is regulated by this sensitive feedback mechanism.

Deoxyhexoses and dideoxyhexoses are also synthesized, while the sugars are attached to nucleoside diphosphates, by a multistep process. For example, L-rhamnose is synthesized from glucose by a series of oxidation–reduction reactions starting with dTDP-glucose and yielding dTDP-rhamnose, catalyzed by oxidoreductases. Presumably, similar reactions account for the synthesis of GDP-fucose from GDP-mannose, as well as for various dideoxyhexoses.

Formation of amino sugars, often found as constituents of antibiotics, occurs by transamidation. For example, synthesis of glucosamine 6-phosphate occurs by the reaction of fructose 6-phosphate with glutamine.

Fructose 6-phosphate + Glutamine (NH_2—C(=O)$(CH_2)_2$—CH(NH_3^+)—COO^-) —transamidase→ Glucosamine 6-phosphate + Glutamate ($^-OOC(CH_2)_2$—CH(NH_3^+)—COO^-)

Glucosamine 6-phosphate may then be N-acetylated, forming N-acetylglucosamine 6-phosphate, followed by isomerization to N-acetylglucosamine 1-phosphate. This latter sugar is converted to UDP-N-acetylglucosamine by reactions similar to those of UDP-glucose synthesis.

UDP-N-acetylglucosamine —4-epimerase→ UDP-N-acetylgalactosamine

UDP-N-acetylglucosamine, a precursor of glycoprotein synthesis, may be epimerized to UDP-N-acetylgalactosamine, necessary for proteoglycan synthesis. This first reaction in hexosamine synthesis, the fructose 6-phosphate : glutamine transamidase reaction, is under negative feedback control by UDP-*N*-acetylglucosamine; thus, the synthesis of both nucleotide sugars is regulated (Figure 8.4). This regulation is meaningful in certain tissues such as skin, in which this pathway may involve up to 20% of the glucose flux.

Synthesis of Sialic Acid

Another product derived from UDP-*N*-acetylglucosamine is CMP-*N*-acetylneuraminic acid, one of a family of 9-carbon sugars, called sialic acids (Figure 8.8). The first reaction in this complex pathway involves epimerization of UDP-*N*-acetylglucosamine by a 2-epimerase to *N*-acetylmannosamine, concomitant with elimination of UDP. Since the monosaccharide product is no longer bound to nucleotide, this epimerization is clearly different from those previously encountered. Most likely, this 2-epimerase reaction proceeds by a *trans* elimination of UDP, with formation of the unsaturated intermediate, 2-acetamidoglucal. In mammalian tissues *N*-acetylmannosamine is phosphorylated with ATP to *N*-

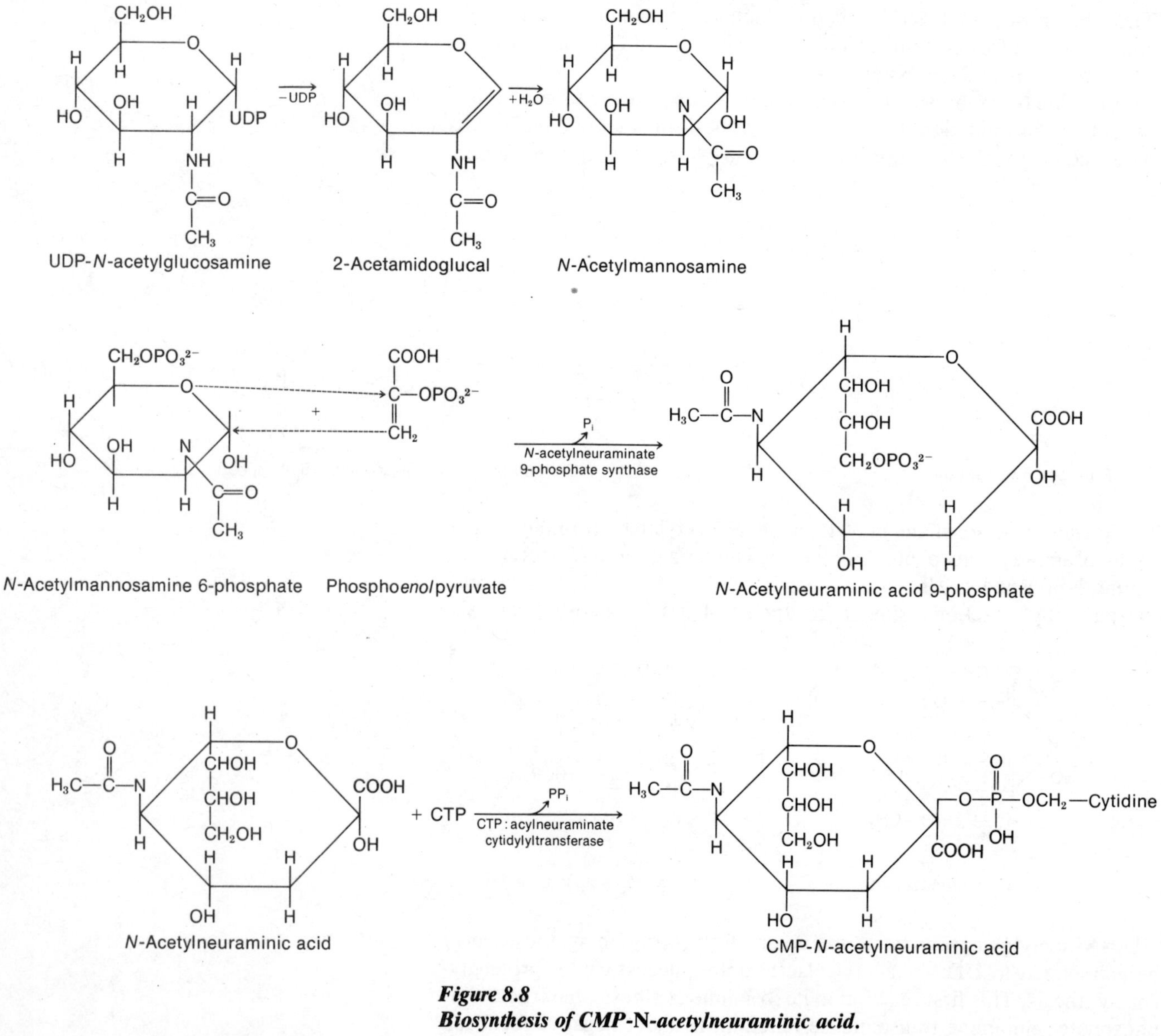

UDP-*N*-acetylglucosamine —(−UDP)→ 2-Acetamidoglucal —(+H_2O)→ *N*-Acetylmannosamine

N-Acetylmannosamine 6-phosphate + Phospho*enol*pyruvate —(*N*-acetylneuraminate 9-phosphate synthase; P_i)→ *N*-Acetylneuraminic acid 9-phosphate

N-Acetylneuraminic acid + CTP —(CTP : acylneuraminate cytidylyltransferase; PP_i)→ CMP-*N*-acetylneuraminic acid

Figure 8.8
Biosynthesis of CMP-N-acetylneuraminic acid.

acetylmannosamine 6-phosphate, which then condenses with phospho*enol*pyruvate to form *N*-acetylneuraminic acid 9-phosphate. This product is cleaved by a phosphatase and activated by CTP to form the CMP derivative, CMP-*N*-acetylneuraminic acid. This is an unusual nucleotide sugar containing only one phosphate group, and is formed by a reaction that is irreversible. *N*-Acetylneuraminic acid is a precursor of other sialic acid derivatives, some of which evolve by modification of *N*-acetyl to *N*-glycolyl or *O*-acetyl after incorporation into glycoprotein.

8.4 BIOSYNTHESIS OF COMPLEX CARBOHYDRATES

In complex carbohydrate-containing molecules, sugars are linked to other sugars by glycosidic bonds, which are formed by specific glycosyltransferases. Energy is required for synthesis of a glycosidic bond and is made available through the use of nucleotide-sugars as donor substrates. A glycosyltransferase reaction proceeds by donation of the glycosyl unit from the nucleotide derivative to the nonreducing end of an acceptor sugar. The nature of the bond formed is

$$\underset{\text{(donor)}}{\text{Nucleoside diphosphate—glycose}} + \underset{\text{(acceptor)}}{\text{glycose}_2} \xrightarrow{\text{glycosyltransferase}}$$

$$\underset{\text{(glycoside)}}{\text{glycosyl}_1\text{—}O\text{-glycose}_2} + \text{nucleoside diphosphate}$$

determined by the specificity of an individual glycosyltransferase, which is unique for the sugar acceptor, the sugar transferred, and the linkage formed. Thus polysaccharide synthesis is controlled by a nontemplate mechanism (see Chapter 17) in which genes code for specific glycosyltransferases.

At least 40 different glycosidic bonds have been identified in mammalian oligosaccharides and about 15 additional ones in connective tissue polysaccharides. The number of possible linkages is even greater and arises both from the diversity of monosaccharides covalently bonded and from the formation of both α and β linkages, with each of the available hydroxyl groups on the acceptor saccharide. The large and diverse number of molecules that can be generated suggests that oligosaccharides have the potential for great informational content. In fact, it is known that the specificity of many biological molecules is determined by the nature of the composite sugar residues. For example, the specificity of the major blood types is determined by sugars (see Clin. Corr. 8.6). *N*-Acetylgalactosamine is the immunodeterminant of blood type A and galactose of blood type B. Removal of *N*-acetylgalactosamine from type A erythrocytes, or of galactose from type B erythrocytes, will convert both to type O erythrocytes. Increasingly, other examples of sugars as determinants of specificity for cell surface receptor and lectin interactions, targeting of cells to certain tissues, and survival or clearance from the circulation of certain molecules, are being recognized.

All the glycosidic bonds that have been identified in biological compounds are degraded by specific hydrolytic enzymes, glycosidases. In addition to being valuable tools for the structural elucidation of oligosaccharides, recent interest in this class of enzymes stems from the fact that many genetic diseases of complex carbohydrate metabolism result from defects in glycosidases (see Clin. Corrs. 8.7 and 8.8).

CLIN. CORR. 8.6 BLOOD GROUP SUBSTANCES

The surface of the human erythrocyte is covered with a complex mosaic of specific antigenic determinants, many of which are saccharides. There are about 100 blood group determinants, belonging to 15 independent human blood group systems. The most widely studied are the antigenic determinants of the ABO blood group system and the closely related Lewis system. From the study of these systems, a definite correlation was established between gene activity as it relates to specific glycosyltransferase synthesis and oligosaccharide structure. The genetic variation is achieved through specific glycosyltransferases responsible for synthesis of the heterosaccharide determinants. For example, the *H* gene codes for a fucosyltransferase, which adds fucose to a peripheral galactose in the heterosaccharide precursor. A and B blood group specificities arise by addition of either *N*-acetylgalactosamine or galactose to the nonreducing terminal galactose of the *H*-specific oligosaccharide. The Lewis (*Le*) gene codes for another fucosyltransferase, which adds fucose to a peripheral *N*-acetylglucosamine residue in the precursor. Absence of the *H* gene gives rise to the Le^a specific determinant, whereas in the absence of both the *H* and *Le* genes, the interaction product responsible for the Le^b specificity is found. The elucidation of the structures of these oligosaccharide determinants represents a milestone in carbohydrate chemistry. This knowledge has proved useful not only for medicine and biology, but also for legal and historical purposes. For example, tissue dust containing complex carbohydrates has been used in serological analysis to establish the blood group of Tutankhamen and his probable ancestral background.

CLIN. CORR. 8.7
ASPARTYLGLYCOSYLAMINURIA: ABSENCE OF 4-L-ASPARTYLGLYCOSAMINE AMIDOHYDROLASE

A group of human inborn errors of metabolism involving storage of glycolipids, glycopeptides, mucopolysaccharides, and oligosaccharides exists. These diseases are caused by defects in glycosidase activity, which prevents the catabolism of oligosaccharides. The disorders involve accumulation in tissues and urine of compounds derived from incomplete degradation of the oligosaccharides, and may be accompanied by skeletal abnormalities, renal, hepatic, or cardiovascular defects, or severe mental retardation. One disorder resulting from a defect in catabolism of asparagine-*N*-acetylglucosamine-linked oligosaccharides is aspartylglycosylaminuria. A deficiency in the enzyme 4-L-aspartylglycosylamine amidohydrolase allows the accumulation of aspartylglucosamine-linked structures (see the accompanying table).

Other disorders have been described involving accumulation of oligosaccharides derived from both glycoproteins and glycolipids, which may share common oligosaccharide structures (see table). Examples of genetic diseases include mannosidosis (L-mannosidase), G_{M2} gangliosidosis variant O (Sandhoff-Jatzkewitz disease; β-*N*-acetylhexosaminidases A and B), and G_{M1} gangliosidosis (β-galactosidase). (See also Chapter 10.)

Enzymic Defects in Degradation of Asn-GlcNAc Type Glycoproteins[a]

Disease	*Deficient Enzyme*[b]
Aspartylglycosylaminuria	4-L-Aspartylglycosylamine amidohydrolase (1)
Mannosidosis	α-Mannosidase (3)
G_{M2}-Gangliosidosis variant O (Sandhoff–Jatzkewitz disease)	β-*N*-Acetylhexosaminidases (A and B) (4)
G_{M1}-Gangliosidosis	β-Galactosidase (5)
Mucolipidosis I and II	Sialidase (6)

[a] A typical Asn-GlcNAc oligosaccharide structure.

```
NeuAc -(6)α- Gal -(5)β- GlcNAc -(4)β-\
                                      Man
NeuAc ---α--- Gal ---β--- GlcNAc ---β-/    \ α
                                        (3) \
                                            Man -β- GlcNAc -(2)β- GlcNAc -(1)- Asn
                                            /    |                  |
                                           /   GlcNAc              Fuc
NeuAc ---α--- Gal ---β--- GlcNAc ---β--- Man α
                                   (4) /
NeuAc -(6)α- Gal -(5)β- GlcNAc ---β---/
```

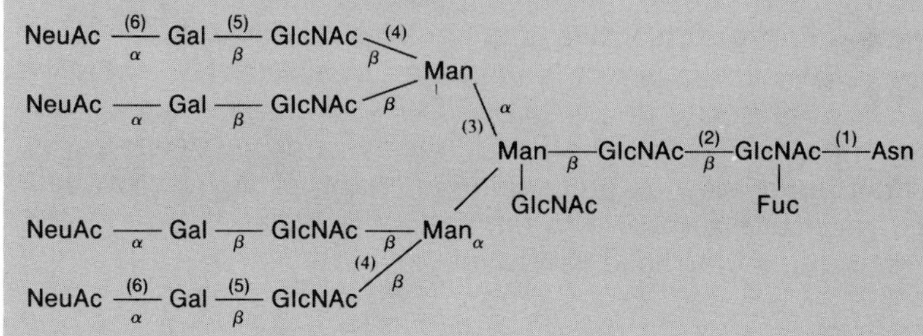

[b] The numbers in parentheses refer to the enzymes that hydrolyze those bonds.

8.5 GLYCOPROTEINS

Glycoproteins have been restrictively defined as conjugated proteins containing as a prosthetic group one or more saccharides lacking a serial repeat unit and bound covalently to a peptide chain. This definition excludes proteoglycans, which are discussed in later sections. At present, the structures of relatively few oligosaccharide components have been completely elucidated. The microheterogeneity of glycoproteins, arising from incomplete synthesis or partial degradation, makes structural analyses extremely difficult. However, certain generalities about the structure of glycoproteins have emerged.

Structure of Glycoprotein

The covalent linkage of sugars to the peptide chain is a central part of glycoprotein structure, and only a limited number of bonds are found. The three major types of glycopeptide bonds, as shown in Figure 8.9, are *N*-glycosyl to asparagine (Asn), *O*-glycosyl to serine (Ser), or threonine (Thr) and *O*-glycosyl to 5-hydroxylysine. The latter linkage, representing the carbohydrate side chains of either a single galactose or the disaccharide glucosylgalactose covalently bonded to hydroxylysine, is generally confined to the collagens. The other two linkages occur in a wide variety of glycoproteins. Of the three major types, only the *O*-glycosidic linkage to serine or threonine is labile to alkali cleavage. By this procedure two types of oligosaccharides (simple and complex) are released. Examination of the simple class from porcine submaxillary mucins reveals some general structural features. A core structure exists, consisting of galactose (Gal) linked β-(1 → 3) to *N*-acetylgalactosamine (GalNAc) *O*-glycosidically linked to serine or threonine residues. Residues of L-fucose (Fuc), sialic acid (NeuAc), and another *N*-acetylgalactosamine are found at the nonreducing periphery of this class of glycopeptides. The general structure of this type of glycopeptide is as follows:

$$\begin{array}{ccccc} \text{GalNAc} \xrightarrow{1,3} & \text{Gal} & \xrightarrow{1,3} & \text{GalNAc} & \longrightarrow O\text{-Ser/Thr} \\ & \uparrow{\scriptstyle 1,2} & & \uparrow{\scriptstyle 2,6} & \\ & \text{Fuc} & & \text{NeuAc} & \end{array}$$

More complex heterosaccharides are also linked to peptides via serine or threonine and are exemplified by the blood group substances. The study of these determinants has shown how complex and variable these structures are, as well as how the oligosaccharides of cell surfaces are assembled and how that assembly pattern is genetically determined. An example of how oligosaccharide structures on the surface of red blood cells determine blood group specificity is presented in Clin. Corr. 8.6.

Certain common structural features of the oligosaccharide N-glycosidically-linked to asparagine have also emerged. These glycoproteins commonly contain a core structure consisting of mannose (Man) residues linked to N-acetylglucosamine (GlcNAc) in the following structure:

$$(\text{Man})_n \xrightarrow[1,4]{} \text{Man} \xrightarrow[1,4]{} \text{GlcNAc} \xrightarrow[1,4]{} \text{GlcNAc} \longrightarrow \text{Asn}$$

Synthesis of Glycoprotein

In contrast to the synthesis of O-glycosidically linked glycoproteins, which involves the sequential action of a series of glycosyltransferases, the synthesis of N-glycosidically linked peptides, involves a somewhat

> **CLIN. CORR. 8.8**
> **HEPARIN AS AN ANTICOAGULANT**
>
> Heparin is a naturally occurring sulfated polysaccharide that is used to reduce the clotting tendency of patients. Both in vivo and in vitro heparin prevents the activation of clotting factors, but does not act directly on the clotting factors. Rather, it is thought that the anticoagulant activity of heparin is brought about by the binding interaction of heparin with an inhibitor of the coagulation process. Presumably, heparin binding induces a conformational change in the inhibitor that generates a complementary interaction between the inhibitor and the activated coagulation factor, thereby preventing the factor from participating in the coagulation process. The inhibitor that interacts with heparin directly is antithrombin III, a plasma protein inhibitor of serine proteases. In the absence of heparin, antithrombin III slowly (10–30 min) combines with several clotting factors, yielding complexes devoid of proteolytic activity. In the presence of heparin, inactive complexes are formed within a few seconds.

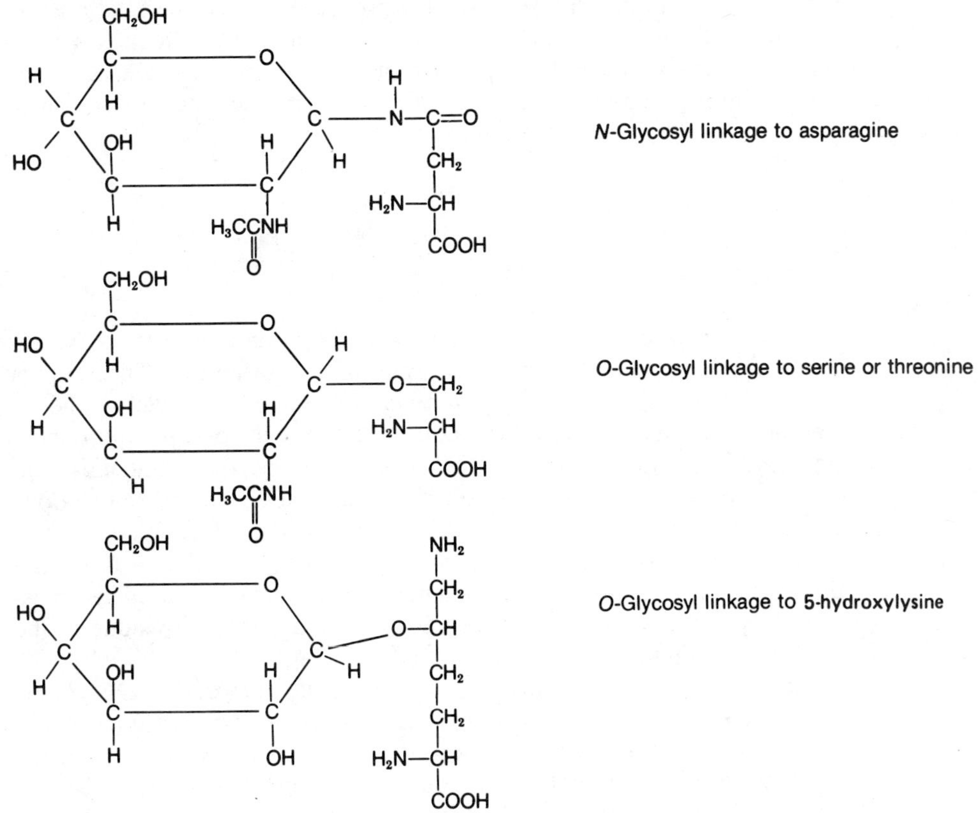

N-Glycosyl linkage to asparagine

O-Glycosyl linkage to serine or threonine

O-Glycosyl linkage to 5-hydroxylysine

Figure 8.9
Structure of three major glycopeptide bonds.

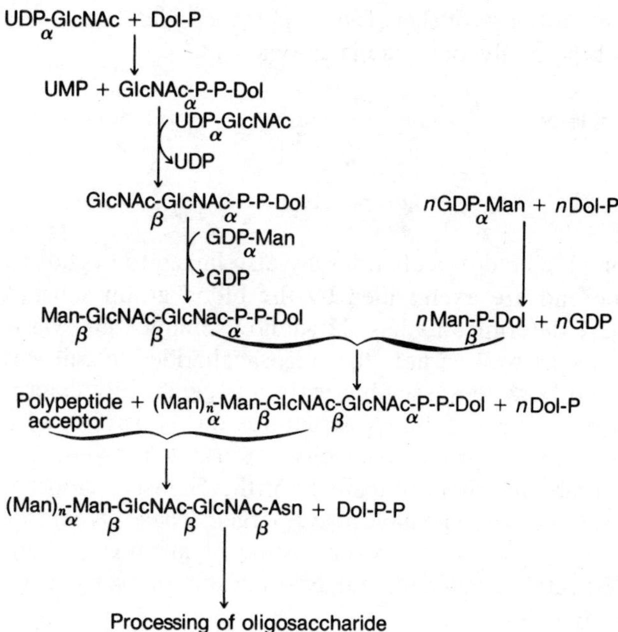

Figure 8.10
Biosynthesis of the oligosaccharide core in asparagine-*N*-acetylgalactosamine-linked glycoproteins.
Abbreviation: Dol, dolichol.

different and more complex mechanism (Figure 8.10). The common core is preassembled as a lipid-linked oligosaccharide prior to incorporation into the polypeptide. Similar assembly processes for synthesis of precursor units followed by transfer *en bloc* have been reported in bacterial cell wall synthesis, but are uncommon in mammalian heterosaccharide synthesis. During synthesis, the oligosaccharide intermediates are bound to derivatives of dolichol phosphate.

$$(CH_2{=}\overset{\overset{\displaystyle CH_3}{|}}{C}{-}CH{=}CH)_n{-}CH_2{-}\overset{\overset{\displaystyle CH_3}{|}}{CH}{-}CH_2{-}CH_2O{-}PO_3H_2$$

Dolichol phosphate

Dolichols are a class of polyprenols (C_{80}–C_{100}) containing 16 to 20 isoprene units, in which the final isoprene unit is saturated. These lipids participate in two types of reactions in core oligosaccharide synthesis. The first reaction involves formation of *N*-acetylglucosaminylpyrophosphoryldolichol with release of UMP from the respective nucleotide sugars. The second *N*-acetylglucosamine and mannose transferase reactions proceed by sugar transfer from the nucleotide without formation of intermediates. The subsequent addition of a mannose unit also occurs via a dolichol-linked mechanism. In the final step, the oligosaccharide is transferred from the dolichol pyrophosphate to an asparagine residue in the polypeptide chain.

After synthesis of the specific core region, the oligosaccharide chains are completed by the action of glycosyltransferases without further participation of lipid intermediates. Recent evidence suggests that extensive "processing," involving the addition and subsequent removal of certain glycosyl residues, occurs during the course of synthesis of asparagine-*N*-acetylglucosamine-linked glycoproteins. The pathways of "oligosaccharide processing" are currently under investigation in certain enveloped viruses (VSV and Sindbis) and in some secretory glycoprotein systems. It is thought that oligosaccharides destined to become simple or oligomannoside-type glycoproteins probably undergo more limited processing, whereas the more complex glycoproteins undergo more extensive processing and elongation. It is not yet understood why some glycoproteins become one type or another, but it may be of evolutionary significance.

Initiation of glycosylation by addition of the core saccharides may occur while the nascent peptide is still bound to ribosomes or soon after completion. Processing and elongation reactions then take place as the peptide moves through the rough endoplasmic reticulum to the Golgi. Once elongation is complete by addition of external sugars within the Golgi apparatus, the glycoprotein migrates toward the plasma membrane within a vesicle. The membrane of the transport vesicle fuses with the plasma membrane, and secretory glycoproteins are extruded, while internal membrane glycoproteins remain part of the plasma membrane.

Just as the synthesis of oligosaccharides requires specific glycosyltransferases, the degradation requires specific glycosidases. Exoglycosidases remove sugars sequentially from the nonreducing end, exposing the substrate for the subsequent glycosidase. The absence of a particular glycosidase prevents the action of the next enzyme, resulting in cessation of catabolism and accumulation of the product (Clin. Corr. 8.7). There is also evidence for the presence of endoglycosidases with broader specificity. However, the sequence of action of endo- and exoglycosidases in the catabolism of glycoproteins is not well understood. The primary degradation process occurs in lysosomes, but there are also specific microsomal glycosidases involved in the processing of glycoproteins during synthesis.

8.6 PROTEOGLYCANS

In addition to glycoproteins, which usually contain proportionally less carbohydrate than protein by weight, there is another class of complex macromolecules, containing 95% or more carbohydrate. The properties of these compounds resemble polysaccharides more than proteins. To distinguish these compounds from other glycoproteins, they are referred to as proteoglycans and their carbohydrate chains as glycosaminoglycans. An older name, mucopolysaccharides, is still in use, especially in reference to the group of storage diseases, mucopolysaccharidoses, which result from an inability to degrade these molecules.

The proteoglycans are high molecular weight polyanionic substances consisting of many different glycosaminoglycan chains linked covalently to a protein core. Although six distinct classes of glycosaminoglycans are now recognized, certain features are common to all classes. The long heteropolysaccharide chains are made up largely of disaccharide repeating units, in which one sugar is a hexosamine and the other a uronic acid. Other common constituents of glycosaminoglycans are sulfate groups, linked by ester bonds to certain monosaccharides or by amide bonds to the amino group of glucosamine. An exception, hyaluronic acid, is not sulfated, and as yet has not been shown to exist covalently attached to protein. The carboxyl and sulfate groups contribute to the nature of glycosaminoglycans as highly charged polyanions. Both their electrical charge and macromolecular structure aid in their biological role as lubricants and support elements in connective tissue. These proteoglycans form solutions with high viscosity and elasticity by absorbing large volumes of water. This allows them to act in stabilizing and supporting fibrous and cellular elements of tissues, as well as contributing to the maintenance of water and salt balance in the body.

Hyaluronic Acid

Among the glycosaminoglycans, hyaluronic acid is very different from the other five types. As previously mentioned, it is unsulfated, is not covalently complexed with protein, and is the only glycosaminoglycan not limited to animal tissue, but is also produced by bacteria. It is nevertheless classified as a glycosaminoglycan because of its structural similarity to these other polymers. It consists solely of repeating disaccharide units of *N*-acetylglucosamine and glucuronic acid (Figure 8.11). Although hyaluronic acid has the least complex chemical structure of all the glycosaminoglycans, the chains may reach molecular weights of 10^5 to 10^7. The large molecular weight, polyelectrolyte character, and large volume of water it occupies in solution all contribute to the properties of hyaluronic acid as a lubricant and shock absorbant. Hence, it is found predominantly in synovial fluid, vitreous humor, and umbilical cord.

Chondroitin Sulfates

The most abundant glycosaminoglycans in the body are the chondroitin sulfates. The individual polysaccharide chains are attached to specific serine residues in a protein core of approximate molecular weight of 200,000 through a tetrasaccharide linkage region.

$$\text{GlcUA} \xrightarrow[1,3]{} \text{Gal} \xrightarrow[1,3]{} \text{Gal} \xrightarrow[1,4]{} \text{Xyl} \longrightarrow O\text{-Ser}$$

The characteristic repeating disaccharide units of *N*-acetylgalactosamine and glucuronic acid are covalently attached to this linkage region (Figure

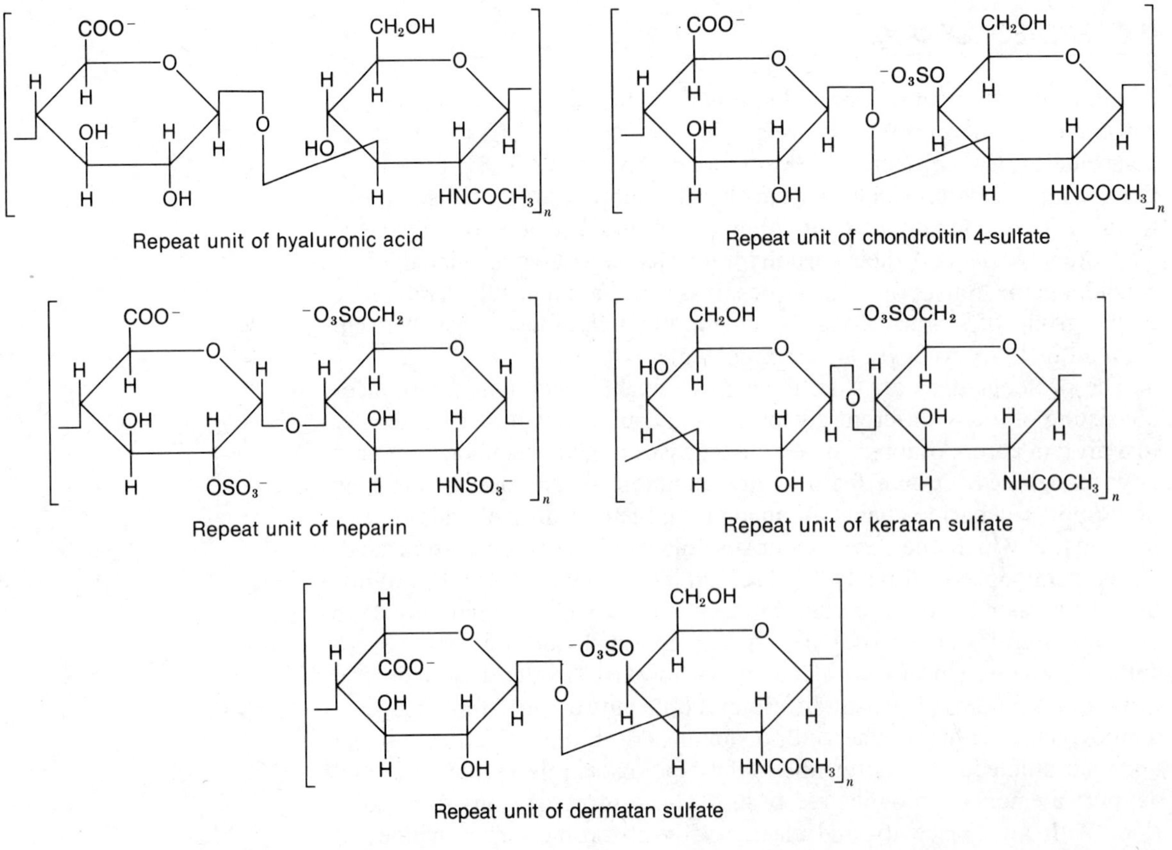

Repeat unit of hyaluronic acid

Repeat unit of chondroitin 4-sulfate

Repeat unit of heparin

Repeat unit of keratan sulfate

Repeat unit of dermatan sulfate

Figure 8.11
Major repeat units of glycosaminoglycan chains.

8.11). The disaccharides may be sulfated in either the 4 or 6 position of N-acetylgalactosamine. Each polysaccharide chain contains between 30 and 50 such disaccharide units, corresponding to molecular weights of 15,000 to 25,000. An average chondroitin sulfate proteoglycan molecule has approximately 100 chondroitin sulfate chains attached to the protein core, giving rise to a molecular weight of 1.5 to 2×10^6. Proteoglycan preparations are, however, extremely heterogeneous, differing in length of protein core, degree of substitution, and distribution of polysaccharide chains, length of chondroitin sulfate chains, and degree of sulfation. Chondroitin sulfate proteoglycans have also recently been shown to aggregate noncovalently with hyaluronic acid, forming much larger structures. They appear to exist in vivo in this aggregated form in the ground substance of cartilage, and have also been isolated from tendons, ligaments, and aorta.

Dermatan Sulfate

Dermatan sulfate differs from chondroitin 4- and 6-sulfates in that its predominant uronic acid is L-iduronic acid, although D-glucuronic acid is also present in variable amounts. The glycosidic linkages are the same in position and configuration as in the chondroitin sulfates. with average polysaccharide chains of molecular weights of 2 to 5×10^4. The physiological function of dermatan sulfate is poorly understood. Unlike the chondroitin sulfates, dermatan sulfate is antithrombic like heparin, but in contrast to heparin, it shows only minimal whole blood anticoagulant and blood lipid-clearing activities. As a connective tissue macromolecule, dermatan sulfate is found in skin, blood vessels, and heart valves.

Heparin and Heparan Sulfate

Heparin differs from other glycosaminoglycans in a number of important respects. Glucosamine and D-glucuronic acid or L-iduronic acid form the characteristic disaccharide repeat unit, as in dermatan sulfate (Figure 8.11). In contrast to the glycosaminoglycans in ground substance, heparin contains α-glycosidic linkages. Almost all glucosamine residues are bound in sulfamide linkages, but a small number of glucosamine residues are N-acetylated. The sulfate content of heparin, although variable, approaches 2.5 sulfate residues per disaccharide unit in preparations with the highest biological activity. In addition to N-sulfate and O-sulfate on C-6 of glucosamine, heparin may also contain sulfate on carbon-3 of the hexosamine and C-2 of the uronic acid. Unlike the other glycosaminoglycans, which are predominantly extracellular components, heparin is an intracellular component of mast cells. Heparin is known as an anticoagulant and lipid-clearing agent; however, the natural physiological role of this polysaccharide still remains unclear (see Clin. Corr. 8.8).

Heparan sulfate contains a similar disaccharide repeat unit but has more N-acetyl groups, fewer N-sulfate groups, and a lower degree of O-sulfate groups. Heparan sulfate appears to be extracellular in distribution and has been isolated from blood vessel walls, amyloid, and brain. Most recently, it has been shown to be an integral and ubiquitous component of the cell surface.

Keratan Sulfate

More than any of the other glycosaminoglycans, keratan sulfate is characterized by molecular heterogeneity. The polysaccharide is composed principally of a repeating disaccharide unit of N-acetylglucosamine and galactose, with no uronic acid in the molecule (Figure 8.11). Sulfate content is variable, with ester sulfate present on C-6 of both galactose and hexosamine. Two types of keratan sulfate have been distinguished, which differ in their overall carbohydrate content and tissue distribution. Both contain as additional monosaccharides, mannose, fucose, sialic acid, and *N*-acetylgalactosamine. Keratan sulfate I, isolated from cornea, is linked to protein by an *N*-acetylglucosamine–asparaginyl bond, typical of glycoproteins. Keratan sulfate II, isolated from cartilage, is attached to protein through *N*-acetylgalactosamine in *O*-glycosidic linkage to either serine or threonine. Skeletal keratan sulfates are often found covalently attached to the same core protein as are the chondroitin sulfate chains.

Biosynthesis of Proteoglycans

Most of the problems in studies on the biosynthesis of proteoglycans are similar to those previously encountered in the formation of other glycoproteins. The polysaccharide chains are assembled by the sequential action of a series of glycosyltransferases, which catalyze the transfer of a monosaccharide from a nucleotide sugar to an appropriate acceptor, either the nonreducing end of another sugar or a polypeptide. Since the biosynthesis of the chondroitin sulfates is most thoroughly understood, this pathway will be discussed as the prototype for glycosaminoglycan formation (Figure 8.12).

In analogy with the mechanisms established for glycoprotein biosynthesis, the formation of the core protein of the chondroitin sulfate proteoglycan is the first step in this process. However, whether the initiation

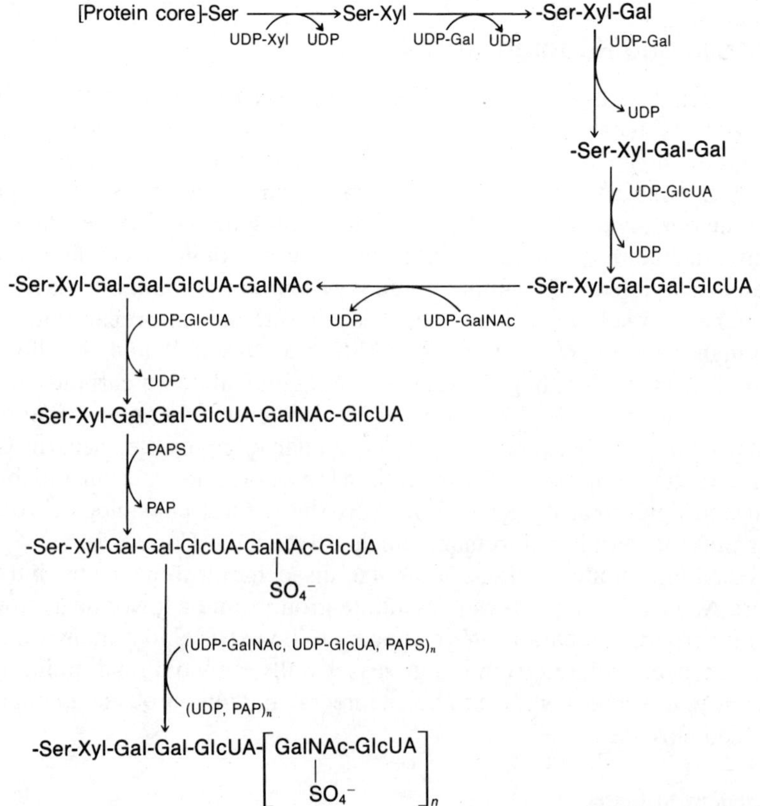

Figure 8.12
Synthesis of chondroitin sulfate proteoglycan.
Abbreviations: Xyl, xylose; Gal, galactose; GluUA, glucuronic acid; GalNAc, N-*acetylgalactosamine; PAPS, phosphoadenosine phosphosulfate.*

of polysaccharide chains precedes completion of the peptide chain and release from the ribosomes has not been determined. The polysaccharide chains are assembled by six different glycosyltransferases. Strict substrate specificity is required for completion of the unique tetrasaccharide linkage region. Polymerization then results from the concerted action of two glycosyltransferases, an *N*-acetylgalactosaminyltransferase and a glucuronosyltransferase, which alternately add the two monosaccharides, forming the characteristic repeating disaccharide units. Sulfation of *N*-acetylgalactosamine residues in either the 4 or 6 position apparently occurs along with chain elongation. The sulfate donor in these reactions, as in other biological systems, is 3′-phosphoadenosine 5′-phosphosulfate (PAPS), which is formed from ATP and sulfate in two steps (Figure 8.13).

Synthesis of the other glycosaminoglycans requires additional transferases specific for the sugars and linkages found in these molecules. Completion of these glycosaminoglycans often involves modifications in addition to O-sulfation, including epimerization, acetylation, and *N*-sulfation. Interestingly, the epimerization of D-glucuronic acid to L-iduronic acid occurs after incorporation into the polymer chain and is coupled with the process of sulfation.

It still remains unclear what fundamental central mechanisms determine the quantity as well as the qualitative nature of the proteoglycans synthesized. It appears that different proteoglycans are synthesized because of the presence and strict substrate specificity of the enzymes and the formation of specific acceptor proteins in a cell. Two mechanisms by which the levels of nucleotide sugars may be regulated have previously been described. The first specific reaction in hexosamine synthesis, the fructose 6-phosphate : glutamine transamidase reaction (Figure 8.4), is subject to feedback inhibition by UDP-*N*-acetylglucosamine, which is in equilibrium with UDP-*N*-acetylgalactosamine. Hence, synthesis of both proteoglycans and glycoproteins is regulated by the same mechanism. More specific to proteoglycan synthesis, the levels of UDP-xylose and UDP-glucuronic acid are stringently controlled by the inhibition by UDP-xylose of the UDP-glucose dehydrogenase conversion of UDP-glucose to UDP-glucuronic acid (Figure 8.4). Since xylose is the first sugar added

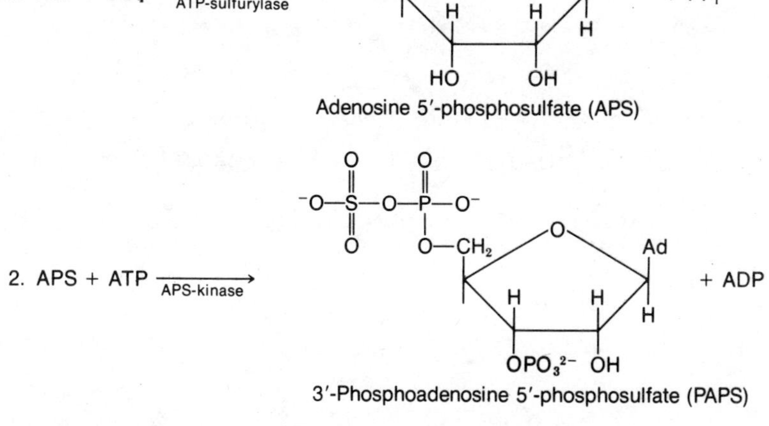

1. ATP + SO_4^{2-} $\xrightarrow{\text{ATP-sulfurylase}}$ Adenosine 5′-phosphosulfate (APS) + PP_i

2. APS + ATP $\xrightarrow{\text{APS-kinase}}$ 3′-Phosphoadenosine 5′-phosphosulfate (PAPS) + ADP

Figure 8.13
Biosynthesis of PAPS.

during synthesis of chondroitin sulfate, dermatan sulfate, heparin, and heparan sulfate, the earliest effect of decreased core protein synthesis would be accumulation of UDP-xylose. This sensitive regulatory mechanism may be responsible for maintaining a balance between synthesis of the protein and polysaccharide moieties of these complex macromolecules.

Proteoglycans, like glycoproteins, are presumably degraded by the sequential action of proteases and glycosidases, as well as deacetylases and sulfatases. Much of the information about metabolism and degradation of proteoglycans has been derived from the study of mucopolysaccharidoses (Clin. Corr. 8.9). This group of human genetic disorders is characterized by accumulation in tissues and excretion in urine of oligosaccharide products derived from incomplete breakdown of the proteoglycans, due to a deficiency of one or more lysosomal hydrolases. In the diseases for which the biochemical defect has been identified, it has been shown that a product accumulates with a nonreducing terminus that would have been the substrate for the deficient enzyme.

CLIN. CORR. 8.9 MUCOPOLYSACCHARIDOSES

A group of human genetic disorders characterized by excessive accumulation and excretion of the oligosaccharides of proteoglycans exists, collectively called mucopolysaccharidoses. These disorders result from a deficiency of one or more lysosomal hydrolases responsible for the degradation of dermatan and/or heparan sulfate. The enzymes lacking in specific mucopolysaccharidoses that have been identified are presented in the accompanying table. Presumably, for complete sequential degradation of the glycosaminoglycans, additional enzymes are required to hydrolyze bonds (4) *N*-acetyl *β*-galactosaminidase, (7) *α*-glucosaminidase, and (8) *N*-acetylglucosamine sulfatase, for which deficiency diseases remain unknown.

Although the chemical basis for this group of disorders is similar, their mode of inheritance as well as clinical manifestations may vary. Hurler's syndrome and Sanfilippo's syndrome are transmitted as autosomal recessives, whereas Hunter's disease is sex-linked. Both Hurler's syndrome and Hunter's disease are characterized by skeletal abnormalities and mental retardation, which in severe cases may result in early death. In contrast, in the Sanfilippo syndrome, the physical defects are relatively mild, while the mental retardation is severe. Collectively, the incidence for all mucopolysaccharidoses is 1 per 30,000 births.

In addition to those listed in the table, some others are less well understood. Morquio syndrome involves impaired degradation of keratan sulfate, presumably due to a deficiency of a 4-sulfatase; multiple sulfatase deficiency involves the deficiency of at least six sulfatases; and "I-cell" disease involves a marked decrease in several hydrolase enzymes.

These disorders are amenable to prenatal diagnosis, since the pattern of metabolism by affected cells obtained from amniotic fluid is strikingly different from normal. Furthermore, a promising approach for correcting certain of these disorders is enzyme or gene replacement therapy.

Enzyme Defects in the Mucopolysaccharidoses

Disease	*Accumulated Products*[a]	*Deficient Enzyme*[b]
Hunter	Heparan sulfate Dermatan sulfate	Iduronate sulfatase (1)
Hurler + Scheie	Heparan sulfate Dermatan sulfate	*α*-L-Iduronidase (2)
Maroteaux–Lamy	Dermatan sulfate	*N*-Acetylgalactosamine (3) sulfatase
Mucolipidosis VII	Heparan sulfate Dermatan sulfate	*β*-Glucuronidase (5)
Sanfilippo A	Heparan sulfate	Heparan sulfamidase (6)
Sanfilippo B	Heparan sulfate	*N*-Acetylglucosaminidase (9)

[a] Structures of dermatan sulfate and heparan sulfate.

Dermatan sulfate: —IdUA —(2) α— GalNAc —(4) β— GlcUA —(5) β— GalNAc —β—
(IdUA: (1) OSO_3H; GalNAc: (3) OSO_3H; GalNAc: OSO_3H)

Heparan sulfate: —IdUA —(2) α— GlcN —(7) α— GlcUA —(5) β— GlcNAc —(9) α—
(IdUA: (1) OSO_3H; GlcN: (6) OSO_3H; GlcNAc: (8) OSO_3H)

[b] The numbers in parentheses refer to the enymes that hydrolyze those bonds.

BIBLIOGRAPHY

Dutton, G. J. (ed.). *Glucuronic acid, free and combined.* New York: Academic, 1966.

Ginsburg, V., and Robbins, P. (eds.). *Biology of carbohydrates,* vol. 1. New York: Wiley, 1981.

Horecker, B. L. *Pentose metabolism in bacteria.* New York: Wiley, 1962.

Hughes, R. C. *Glycoproteins.* London: Chapman and Hall, 1983.

Lennarz, W. J. (ed.). *The biochemistry of glycoproteins.* New York: Plenum, 1980.

Margolis, R. V., and Margolis, R. K. (eds.). *Complex carbohydrates of nervous tissue.* New York: Plenum, 1979.

Sharon, N. *Complex carbohydrates—their chemistry, biosynthesis and functions.* Reading, Mass.: Addison-Wesley, 1975.

Stanbury, J. B., Wyngaarden, J. B., Fredrickson, D. S., Goldstein, J. L., and Brown, M. S. (eds.). *The metabolic basis of inherited disease.* 5th ed. New York: McGraw-Hill, 1983.

Walborg, E. F., Jr. (ed.). *Glycoproteins and glycolipids in disease processes,* in ACS Symposium Series 80. Washington, D.C.: American Chemical Society, 1978.

QUESTIONS

C. N. ANGSTADT AND J. BAGGOTT

*Q*uestion *T*ypes are described inside the front cover.

1. (QT1) [NADPH]/[$NADP^+$] is maintained at a high level in cells primarily by:
   A. lactate dehydrogenase.
   B. the combined actions of glucose 6-phosphate dehydrogenase and gluconolactonase.
   C. the action of the electron transport chain.
   D. shuttle mechanisms such as the α-glycerophosphate dehydrogenase shuttle.
   E. the combined actions of transketolase and transaldolase.
2. (QT1) Transketolase:
   A. transfers a two-carbon fragment to an aldehyde acceptor.
   B. transfers a three-carbon kctone-containing fragment to an acceptor.
   C. converts the ketose sugar ribulose 5-phosphate to ribose 5-phosphate.
   D. is part of the irreversible oxidative phase of the pentose phosphate pathway.
   E. converts two five-carbon sugar phosphates to fructose 6-phosphate and erythrose 4-phosphate
3. (QT1) If a cell requires more NADPH than ribose 5-phosphate:
   A. only the first phase of the pentose phosphate pathway would occur.
   B. glycolytic intermediates would flow into the reversible phase of the pentose phosphate pathway.
   C. there would be sugar interconversions but no net release of carbons from glucose 6-phosphate.
   D. the equivalent of the carbons of glucose 6-phosphate would be released as $6CO_2$.
   E. only part of this need could be met by the pentose pathway, and the rest would have to be supplied by another pathway.
4. (QT2) Which of the following interconversions of monosaccharides (or derivatives) do(es) *not* require a nucleotide-linked sugar intermediate?
   1. Galactose 1-phosphate to glucose 1-phosphate
   2. Glucose 6-phosphate to mannose 6-phosphate
   3. Glucose to glucuronic acid
   4. D-Glucuronic acid to L-iduronic acid
5. (QT1) Fructose:
   A. unlike glucose, cannot be catabolized by the glycolytic pathway.
   B. in the liver, enters directly into glycolysis as fructose 6-phosphate.
   C. must be isomerized to glucose before it can be metabolized.
   D. is converted to a UDP-linked form and then epimerized to UDP-glucose.
   E. catabolism in liver uses fructokinase and a specific aldolase that recognizes fructose 1-phosphate.
6. (QT1) Galactosemia:
   A. is a genetic deficiency of a uridylyltransferase that exchanges galactose 1-phosphate for glucose on UDP-glucose.
   B. necessitates lifelong abstinence from galactose.
   C. is not apparent at birth but symptoms develop in later life.
   D. is an inability to form galactose 1-phosphate.
   E. would be expected to interfere with the use of fructose as well as galactose because the deficient enzyme is common to the metabolism of both sugars.
7. (QT2) Glucuronic acid:
   1. enhances the water-solubility of compounds to which it is conjugated.
   2. as a UDP-derivative can be decarboxylated to a component used in proteoglycan synthesis.
   3. is a precursor of ascorbic acid in many mammalian species but not in humans.
   4. formation from glucose is under feedback control by a UDP-linked intermediate.
8. (QT2) The conversion of fructose 6-phosphate to glucosamine 6-phosphate:
   1. is a transamination reaction with glutamate as the nitrogen donor.
   2. is under feedback inhibition by UDP-*N*-acetylglucosamine.
   3. requires that fructose 6-phosphate first be linked to a nucleotide.
   4. is a first step in the formation of *N*-acetylated amine sugars.
9. (QT2) *N*-Acetylneuraminic acid:
   1. is a sialic acid.
   2. is activated by conversion to CMP-*N*-acetylneuraminic acid.
   3. is derived from UDP-*N*-acetylglucosamine.
   4. formation includes a condensation of *N*-acetylmannosamine 6-phosphate with phospho*enol*pyruvate.
10. (QT2) Roles for the complex carbohydrate moiety of glycoproteins include:
    1. determinant of blood type.
    2. cell surface receptor specificity.
    3. determinant of the rate of clearance from the circulation of certain molecules.
    4. template for the synthesis of glycosaminoglycans.
11. (QT3) A. Percentage of carbohydrate by weight in glycoproteins
    B. Percentage of carbohydrate by weight in proteoglycans
12. (QT2) In glycoproteins, the carbohydrate portion may be linked to the protein by:
    1. an *N*-glycosyl bond to asparagine.
    2. an *O*-glycosyl bond to lysine.
    3. an *O*-glycosyl bond to serine.
    4. noncovalent bonds.

    A. *O*-linked glycoproteins C. Both
    B. *N*-linked glycoproteins D. Neither

13. (QT4) Core structure is assembled on dolichol phosphate before transfer to the protein.
14. (QT4) Core structure typically contains fucose and sialic acid.
15. (QT4) Processing and elongation of the core structure occurs in the rough endoplasmic reticulum and Golgi apparatus.
16. (QT2) Glycosaminoglycans:
    1. are the carbohydrate portion of proteoglycans.
    2. contain large segments of a repeating unit typically consisting of a hexosamine and a uronic acid.
    3. are polyanions.
    4. exist as at least six different classes.

    A. Hyaluronic acid
    B. Chondroitin sulfate
    C. Dermatan sulfate
    D. Heparin
    E. Keratan sulfate
17. (QT5) Differs from other glycosaminoglycans in being predominantly intracellular rather than extracellular.
18. (QT5) Only glycosaminoglycan not covalently linked to protein.
19. (QT5) Only glycosaminoglycan that is not sulfated.
20. (QT2) Proteoglycan:
    1. specificity is determined, in part, by the action of glycosyltransferases.
    2. synthesis is regulated, in part, by UDP-xylose inhibition of the conversion of UDP-glucose to UDP-glucuronic acid.
    3. synthesis involves sulfation of carbohydrate residues by PAPS.
    4. degradation is catalyzed in the cytosol by nonspecific glycosidases.

Answers

1. B Although the glucose 6-phosphate dehydrogenase reaction, specific for NADP, is reversible, hydrolysis of the lactone assures that the overall equilibrium lies far in the direction of NADPH. A, C, D: These all use NAD, not NADP. E: These enzymes are part of the pentose phosphate pathway but catalyze freely reversible reactions that do not involve NADP (p. 331).
2. A Both reactions catalyzed by transketolase are of this type. B, E: Describe transaldolase. C. Describes an isomerase. D. Transketolase is part of the reversible phase of the pentose phosphate pathway that also allows glycolytic intermediates to be converted to pentose sugars, if necessary (p. 332).
3. D A, C, D, E: Glucose 6-phosphate yields ribose 5-phosphate + CO_2 in the oxidative phase. If this is multiplied by six, the six ribose 5-phosphates can be rearranged to five glucose 6-phosphates by the second, reversible phase. B: If more ribose 5-phosphate than NADPH were required, the flow would be in this direction to supply the needed pentoses (p. 334).
4. C 2, 4 correct. 1: Occurs via an epimerase at the UDP-galactose level. 2: The glucose and mannose phosphates are both in equilibrium with fructose 6-phosphate by phosphohexose isomerases. 3: This oxidation of glucose is catalyzed by UDP-glucose dehydrogenase. 4: Although most epimerizations involve nucleotide-linked sugars, this one does not, occurring after the glucuronic acid is incorporated into heparin or dermatan sulfate (p. 336).
5. E A, C, E: Fructokinase produces fructose 1-phosphate. Since this cannot be converted to fructose 1,6-bisphosphate, a specific aldolase cleaves it to dihydroxyacetone phosphate and glyceraldehyde. The first product is a glycolytic intermediate; the second requires modification to enter glycolysis. D: Glucose and fructose are not epimers (p. 336).
6. A B, C: Galactosemics can tolerate galactose in later life because an alternate enzyme for galactose 1-phosphate conversion to UDP-galactose is expressed then. E: Fructose metabolism does not use the uridylyltransferase that is deficient in galactosemia (p. 338).
7. E All four correct. 1: Enhancing water-solubility is a major physiological role for glucuronic acid, for example, bilirubin metabolism. 2, 4: Decarboxylation of UDP-glucuronic acid gives UDP-xylose, which is a potent inhibitor of the oxidation of UDP-glucose to the acid. 3: The reduction of D-glucuronic acid to L-gulonic acid leads to ascorbate as well as xylulose 5-phosphate for the pentose phosphate pathway (p. 339, Figure 8.6).
8. C 2, 4 correct. 1, 3: This conversion is a trans*amidation* of the amide nitrogen of glutamine and does not involve nucleotide intermediates. 2, 4: Glucosamine 6-phosphate is acetylated. UDP-*N*-acetylglucosamine is formed, and the UDP-derivative can be epimerized to the galactose derivative. It also is a feedback inhibitor of the transamidase reaction, thus controlling formation of the nucleotide sugars (p. 341).
9. E All four correct. 2, 4: This complex pathway begins with an unusual epimerization of UDP-*N*-acetylglucosamine to free *N*-acetylmannosamine that is phosphorylated and then condensed with phospho*enol*pyruvate. 3: Unlike most sugars, the activated form of *N*-acetylneuraminic acid is a CMP rather than a UDP-derivative (p. 342).
10. A 1, 2, 3 correct. Because of the diversity possible with oligosaccharides, they play a significant role in determining the specificity of many biological molecules. 4: The synthesis of complex carbohydrates is not template directed but determined by the specificity of individual enzymes (p. 343).
11. B The term "proteoglycans" is reserved for species that contain at least 95% carbohydrate (p. 347).
12. B 1, 3 correct. 2: *O*-linkage occurs to hydroxylysine in collagen not to lysine. 4: Carbohydrates are covalently linked to protein (p. 344).
13. B Synthesis of *O*-linked glycoproteins involves the sequential addition to the *N*-acetylgalactosamine linked to serine or threonine (p. 345).
14. A Core also contains galactose and *N*-acetylgalactosamine; core structure of *N*-linked carbohydrates contains mannose and *N*-acetylglucosamine (p. 344).
15. C Both types of sugar require processing of the core structure (p. 347).
16. E All four correct. 1: These are quite different from the carbohydrate of glycoproteins. 2: This is a major distinction from glycoproteins, which, by definition, do not have a serial repeating unit. 3: The anionic character contributed by carboxyl and sulfate (another common feature) groups is important to the biological function (p. 348).
17. D (p. 349).
18. A Classified as a glycosaminoglycan because of its structural similarity to the others (p. 348).
19. A Both heparin and heparan sulfate are sulfated (p. 348).
20. A 1, 2, 3 correct. 1: Strict substrate specificity of the enzymes is important in determining the type and quantity of proteoglycans synthesized. Formation of specific protein acceptors for the carbohydrate is also important. 2: Both xylose and glucuronic acid levels are controlled by this; xylose is the first sugar added in the synthesis of four of the six types. 3: This is necessary for the formation of all proteoglycans (hyaluronic acid is not part of a proteoglycan). 4: Degradation is lysosomal; deficiencies of one or more lysosomal hydrolases leads to accumulation of proteoglycans in the mucopolysaccharidoses (p. 350).

Lipid Metabolism I: Utilization and Storage of Energy in Lipid Form

J. DENIS McGARRY

9.1 OVERVIEW

As the human body builds and renews its structures, obtains and stores energy, and performs its various functions, there are numerous circumstances in which it is essential to use molecules or parts of molecules that do not associate with water. This property of being nonpolar and hydrophobic is largely supplied by the substances classed as lipids. Most of these are molecules that contain or are derived from fatty acids. In the early stages of biochemical research these substances were not studied as intensively as other body constituents, largely because the techniques for studying aqueous systems were easier to develop. This benign neglect led to early assumptions that the lipids were relatively inert and their metabolism was of lesser importance than that of carbohydrates, for instance.

As the methodology for investigation of lipid metabolism developed, however, it soon became evident that fatty acids and their derivatives had at least two major roles in the human body. On the one hand, the oxidation of fatty acids was shown to be a major means of metabolic energy production, and it became clear that their storage in the form of triacylglycerols was more efficient and quantitatively more important than storage of carbohydrates as glycogen. On the other hand, as details of the chemistry of biological structures were elucidated, hydrophobic structures were found to be largely composed of fatty acids and their derivatives. Thus the major separation of cells and subcellular structures into separate aqueous compartments is accomplished by the use of membranes whose hydrophobic characteristics are largely supplied by the fatty acid moieties of complex lipids. These latter compounds contain constituents other than fatty acids and glycerol. They frequently have significant covalently-bound hydrophilic moieties, notably carbohydrates in the glycolipids and organic phosphate esters in the phospholipids.

In addition to these two major functions of lipids, energy production and structure building, there are several other quantitatively less important roles, which are nonetheless of great functional significance. These include the use of the surface active properties of some complex lipids for specific functions, such as maintenance of lung alveolar integrity and solubilization of nonpolar substances in body fluids. In addition, several classes of lipids, the steroid hormones and the prostaglandins, have highly potent and specific physiological roles in control of metabolic processes. The interrelationships of some of the processes involved in lipid metabolism are outlined in Figure 9.1.

Since the metabolism of fatty acids and triacylglycerols is so crucial to proper functioning of the human body, imbalances and deficiencies in these processes lead to significant pathological processes, and disease states related to fatty acid and triacylglycerol metabolism include some of the major clinical problems to be encountered by physicians, for instance, ketoacidosis, obesity, and abnormalities in transport of lipids in blood. In addition, some unique deficiencies have been found, such as Refsum's

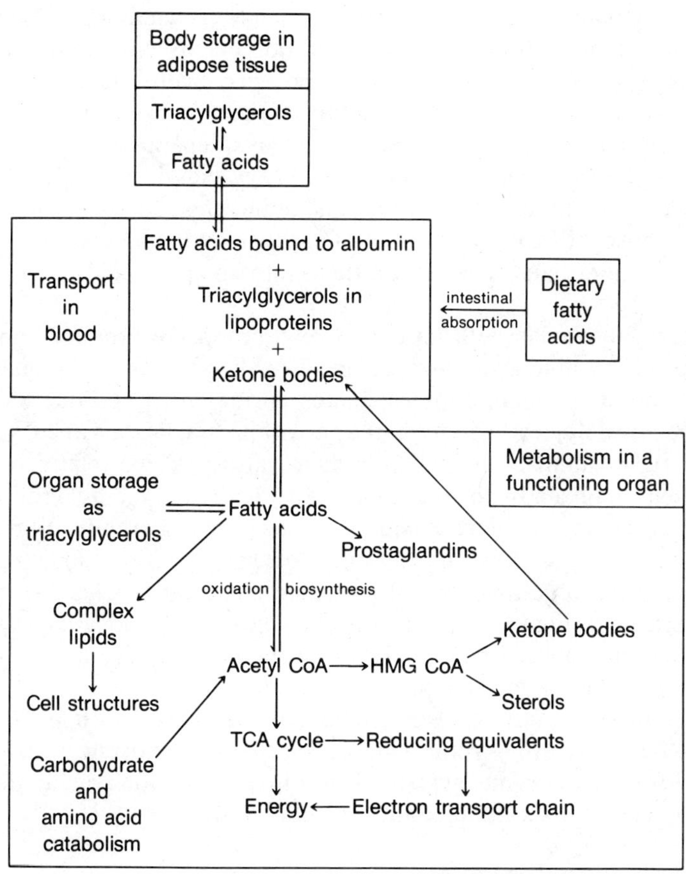

Figure 9.1
The metabolic interrelationships of fatty acids in the human body.

disease and familial hypercholesterolemia, which have helped to elucidate some pathways in lipid metabolism.

In this chapter on lipid metabolism we will be primarily concerned with the structure and metabolism of the fatty acids themselves and of their major storage form, the triacylglycerols. After a discussion of the structures of the more important fatty acids in the human body we describe how they are supplied to the human metabolic machinery from the diet or by biosynthesis. Since their storage as triacylglycerols is a major process we next discuss how this storage is accomplished and how the fatty acids themselves are mobilized and transported throughout the body to sites where they are needed. The central process of energy production from fatty acids is then discussed, and finally we introduce some concepts about the role and metabolism of polyunsaturated fatty acids.

See the Appendix, "Review of Organic Chemistry," for a discussion of the nomenclature and chemistry of lipids and Chapter 24 for a presentation of the digestion and absorption of lipids.

9.2 THE CHEMICAL NATURE OF FATTY ACIDS AND ACYLGLYCEROLS

The Structure of Fatty Acids

Fatty acids consist of an alkyl chain with a terminal carboxyl group, and the simplest configuration is a completely saturated straight chain. The basic formula is $CH_3—(CH_2)_n—COOH$. The fatty acids of importance for humans have relatively simple configurations, although fatty acids in some organisms are occasionally quite complex, containing cyclopropane rings or extensive branching. Unsaturation occurs commonly in human fatty acids, with up to six double bonds per chain, and the bonds are almost always of the cis configuration. If there is more than one double bond per molecule, these bonds are always separated by a methylene ($—CH_2—$) group. The most common fatty acids in biological systems have an even number of carbon atoms, although some organisms do synthesize those with an odd number of carbons. Human beings can use the latter for energy and incorporate them into complex lipids to a minimal degree.

A few fatty acids with an α-hydroxyl group are produced and used structurally by humans. However, more oxidized forms are normally produced only as metabolic intermediates during energy production or for specific physiological activity in the case of prostaglandins and thromboxanes. Higher animals, including humans, also produce relatively simple branched-chain acids, the branching being limited to methyl groups along the chain at one or more positions. These are apparently produced to contribute specific physical properties to some secretions and structures. For instance, large amounts of branched-chain fatty acids, particularly isovaleric acid (Figure 9.2), occur in the lipids of echo-locating structures in marine mammals. The elucidation of the role of these lipids in sound focusing should be fascinating.

$$CH_3—CH(CH_3)—CH_2—COOH$$

Figure 9.2
Isovaleric acid.

The bulk of the fatty acids in the human body have 16, 18, or 20 carbon atoms, but there are several with longer chains that occur principally in the lipids of the nervous system. These include nervonic acid, and a 22-carbon acid with six double bonds (Figure 9.3).

$$CH_3—(CH_2)_7—CH{=}CH—(CH_2)_{13}—COOH$$

Nervonic acid

$$CH_3—(CH_2—CH{=}CH)_6—(CH_2)_2—COOH$$

All-*cis*-4,7,10,13,16,19-docosahexaenoic acid

Figure 9.3
Long-chain fatty acids.

Fatty Acid Nomenclature

The most abundant fatty acids have common names that have been accepted for use in the official nomenclature. Examples are given in Table 9.1 along with the official systematic names. The approved abbreviations consist of the number of carbon atoms followed, after a colon, by the number of double bonds. Carbon atoms are numbered with the carboxyl carbon as number 1, and the double bond locations are designated by the number of the carbon atom on the carboxyl side of it. These designations of double bonds are in parentheses after the rest of the symbol. See Table 9.1 for examples.

Structure of Acylglycerols

Fatty acids occur primarily as esters of glycerol, as shown in Figure 9.4, when they are stored for future utilization. Compounds with one (mono-

TABLE 9.1 Fatty Acids of Importance to Humans

Numerical Symbol	*Structure*	*Trivial Name*	*Systematic Name*
16:0	$CH_3—(CH_2)_{14}—COOH$	Palmitic	Hexadecanoic
16:1(9)	$CH_3—(CH_2)_5—CH{=}CH—(CH_2)_7—COOH$	Palmitoleic	*cis*-9-Hexadecenoic
18:0	$CH_3—(CH_2)_{16}—COOH$	Stearic	Octadecanoic
18:1(9)	$CH_3—(CH_2)_7—CH{=}CH—(CH_2)_7—COOH$	Oleic	*cis*-9-Octadecenoic
18:2(9,12)	$CH_3—(CH_2)_3—(CH_2—CH{=}CH)_2—(CH_2)_7—COOH$	Linoleic	*cis,cis*-9,12-Octadecadienoic
18:3(9,12,15)	$CH_3—(CH_2—CH{=}CH)_3—(CH_2)_7—COOH$	Linolenic	*cis,cis,cis*-9,12,15-Octadecatrienoic
20:4(5,8,11,14)	$CH_3—(CH_2)_3—(CH_2—CH{=}CH)_4—(CH_2)_3—COOH$	Arachidonic	*cis,cis,cis,cis*-5,8,11,14-Icosatetraenoic

acylglycerols) or two (diacylglycerols) acids esterified are present only in relatively minor amounts and occur largely as metabolic intermediates in the biosynthesis and degradation of glycerol-containing lipids. The bulk of the fatty acids in the human body exist as triacylglycerols, in which all three hydroxyl groups on the glycerol are esterified with a fatty acid. Historically, these compounds have been termed *neutral fats* or *triglycerides,* and these terms are still in common usage. However, there are other types of "neutral fats" in the body, and the term "triglyceride" is chemically incorrect and should no longer be used. The same can be said for the terms "monoglyceride" and "diglyceride."

The distribution of various fatty acids in the different positions of the glycerol moiety of triacylglycerols in the body at any given time is the result of a number of factors, some of which are not completely understood. Suffice it to say that the fatty acid pattern varies with the time, diet, and anatomical location of the triacylglycerol. Compounds with the same fatty acid in all three positions are rare and the usual case is for a complex mixture.

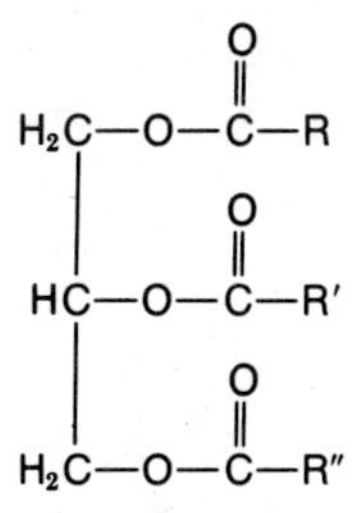

$H_2C{-}O{-}C({=}O){-}R$ / $HC{-}O{-}C({=}O){-}R'$ / $H_2C{-}O{-}C({=}O){-}R''$

Triacylglycerol

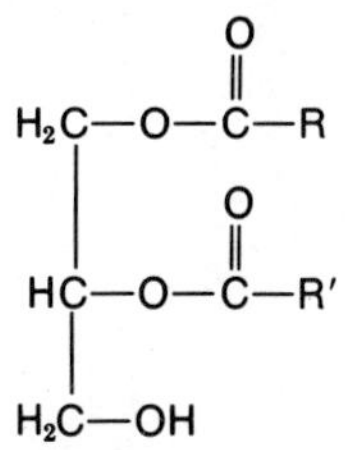

$H_2C{-}O{-}C({=}O){-}R$ / $HC{-}O{-}C({=}O){-}R'$ / $H_2C{-}OH$

Diacylglycerol

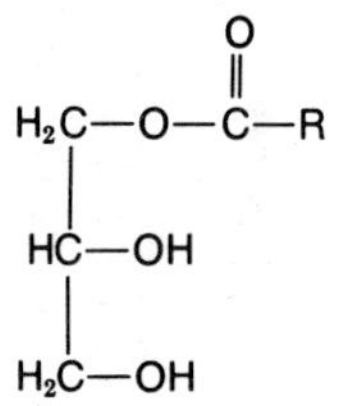

$H_2C{-}O{-}C({=}O){-}R$ / $HC{-}OH$ / $H_2C{-}OH$

1-Monoacylglycerol

$H_2C{-}OH$ / $HC{-}O{-}C({=}O){-}R$ / $H_2C{-}OH$

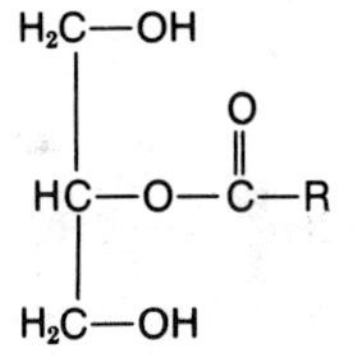

2-Monoacylglycerol

Figure 9.4
Acylglycerols.

Physical and Chemical Properties

Certainly one of the most prominent and significant properties of fatty acids and triacylglycerols is their lack of affinity for water. The long hydrocarbon chains have negligible possibility for hydrogen bonding, and the acids, whether unesterified or in a complex lipid, have a much greater tendency to associate with each other or with other hydrophobic structures, such as sterols and the hydrophobic side chains of amino acids, than they do with water or polar organic compounds. It has been calculated that the van der Waals-London forces between closely packed, relatively long-chain fatty acid moieties in lipids can approach the strength of a covalent bond. This hydrophobic character is essential for construction of complex biological structures and the separation of aqueous compartments as described in Chapter 5. It is also essential for use in biological surface active molecules, as in the intestinal tract.

Of major significance is the fact that the hydrophobic nature of triacylglycerols and their relatively reduced state make them efficient compounds for storing energy. Three points deserve emphasis. First, on a weight basis pure triacylglycerols yield nearly $2\frac{1}{2}$ times the amount of ATP on complete oxidation that pure glycogen does. Second, the triacylglycerols can be stored as pure lipid without associated water, whereas glycogen is quite hydrophilic and binds about twice its weight of water when stored in tissues. Thus the equivalent amount of metabolically recoverable energy stored as hydrated glycogen would weigh about four times as much as if it were stored as triacylglycerols. Third, the average 70-kg person stores about 100 g of carbohydrate as liver glycogen and 250 g as muscle glycogen. This represents about 1,400 kcal of available energy, barely enough to sustain bodily functions for 24 hours of fasting. By contrast, a normal complement of fat stores will provide sufficient energy to allow several weeks of survival during total food deprivation.

The bulk of the fatty acids in the lipids of the human body are either saturated or contain only one double bond. Consequently, although they are readily catabolized by appropriate enzymes and cofactors, they are fairly inert chemically. This is an added advantage of their use for energy storage. However, the smaller amounts of the more highly unsaturated fatty acids in the tissues are much more susceptible to oxidation. Some possible biological consequences of this oxidation are discussed later in this chapter.

9.3 SOURCES OF FATTY ACIDS

Both diet and biosynthesis supply the fatty acids needed by the human body for energy and for construction of hydrophobic parts of biomolecules. Excess amounts of protein and carbohydrate obtained in the diet are readily converted to fatty acids and stored as triacylglycerols.

Dietary Supply

A great proportion of the fatty acids utilized by humans is supplied in their diet. Various animal and vegetable lipids are ingested, hydrolyzed at least partially by digestive enzymes, and absorbed through the intestinal wall to be distributed through the body, first in the lymphatic system and then in the bloodstream. These processes are extensively discussed in Chapter 24. To some extent, then, dietary supply governs the composition of the fatty acids in the body lipids. On the other hand, metabolic processes in the tissues of the normal human body can modify the dietary fatty acids, and/or those that are synthesized in these tissues, to produce almost all the various structures that are needed. For this reason, with one exception, the actual composition of the fatty acids supplied in the diet is relatively unimportant. This one exception involves the need for appropriate proportions of the relatively highly unsaturated fatty acids and particularly relates to the fact that many higher mammals, including humans, are unable to produce fatty acids with double bonds very far toward the methyl end of the molecule, either during de novo synthesis or by modification of dietary acids. Despite this inability, certain polyunsaturated acids with double bonds within the last seven linkages toward the methyl end are essential for some specific functions. Although all the reasons for this need are not yet elucidated, certainly one is that some of these acids are precursors of prostaglandins, highly active oxidation products (see Chapter 10).

In humans a dietary precursor is essential for two series of fatty acids. These are the linoleic series and the linolenic series (Figure 9.5).

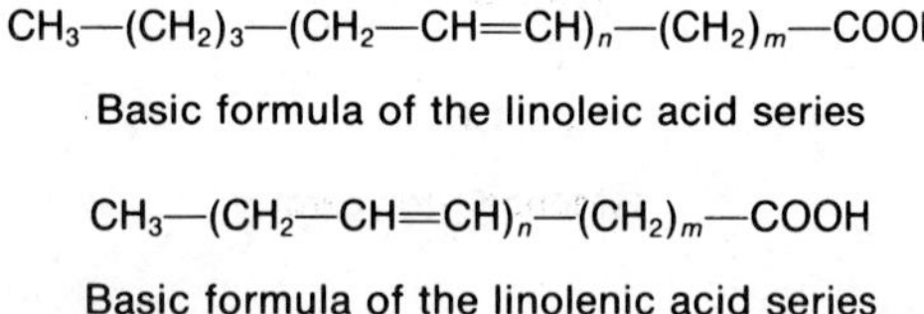

Figure 9.5
Linoleic and linolenic acid series.

Biosynthesis of Palmitate, the "Stem" Fatty Acid

Besides dietary supply, the second major source of fatty acids for humans is their biosynthesis from small-molecule intermediates, which can be derived from metabolic breakdown of sugars, of some amino acids, and of other fatty acids. In a majority of instances the saturated, straight-chain 16-carbon acid, palmitic acid, is first synthesized, and all other fatty acids are made by modification of palmitic acid. Acetyl CoA is the direct source of all carbon atoms for this synthesis, and the fatty acids are made by sequential addition of two-carbon units to the activated carboxyl end of the growing chain. In mammalian systems the sequence of reactions is carried out by fatty acid synthetase.

Fatty acid synthetase is a fascinating enzyme complex that is still being intensively studied. In bacteria it is a complex of several proteins whereas in mammalian cells it is a single multifunctional protein. For the most part its function is to form palmitate, but in some circumstances this pathway can be altered to produce other short-chain fatty acids. Some of the details of these modifications are discussed in later paragraphs, but first we will outline the basic scheme for synthesis of palmitate.

Either acetyl CoA or butyryl CoA is the priming unit for fatty acid synthesis, and the methyl end of these primers becomes the methyl end of palmitate. The addition of the rest of the two-carbon units requires further activation of the methyl carbon of acetyl CoA by carboxylation to malonyl CoA. However, the CO_2 added in this process is lost when the

$$CH_3-\overset{O}{\overset{\|}{C}}-SCoA + HCO_3^- + ATP \xrightarrow[\text{carboxylase}]{\text{acetyl CoA}} {}^-OOC-CH_2-\overset{O}{\overset{\|}{C}}-SCoA + H_2O + ADP + P_i$$

Figure 9.6
Acetyl CoA carboxylase reaction.

condensation of malonyl CoA to the growing chain occurs, so the carbons in the palmitate chain originate only from the acetyl CoA.

Formation of Malonyl CoA

The metabolic process that commits acetyl CoA to fatty acid synthesis is its carboxylation to malonyl CoA by the enzyme acetyl CoA carboxylase (Figure 9.6). This reaction is similar in a number of ways to the carboxylation of pyruvate, which starts the process of gluconeogenesis. The reaction requires energy from ATP and uses dissolved bicarbonate as the source of CO_2. As in the case of pyruvate carboxylase, the first step in this reaction is the formation of activated CO_2 on the biotin moiety of the acetyl CoA carboxylase using the energy from ATP. This is then transferred to the acetyl CoA.

Acetyl CoA carboxylase catalyzes the committed step in the process of fatty acid synthesis and is thus an essential control point. The enzyme can be isolated in an inactive protomeric state, and these protomers aggregate to active polymers upon addition of citrate in vitro. In vitro studies also have demonstrated that palmitoyl CoA inhibits the active enzyme. The action of these two effectors is very logical; increased synthesis of fatty acids to store energy being desirable when citrate is in high concentration, and decreased synthesis being necessary if high levels of the product accumulate. However, the degree to which these regulatory mechanisms actually operate in vivo is unknown.

Acetyl CoA carboxylase is also controlled by a cAMP-mediated phosphorylation–dephosphorylation mechanism in which the phosphorylated enzyme is less active than the dephosphorylated one. There is evidence suggesting that the phosphorylation is promoted by glucagon and that the presence of the active form is fostered by insulin. These effects of hormone-mediated phosphorylation are probably separate from the allosteric effects of citrate and palmitoyl CoA (see Table 9.2).

In longer-term effects the rate of synthesis of acetyl CoA carboxylase is regulated. More enzyme is produced by animals on high-carbohydrate or fat-free diets, and fasting or high-fat diets decrease the rate of enzyme synthesis.

Reaction Sequence for the Synthesis of Palmitic Acid

The first step catalyzed by the fatty acid synthetase is the transacylation of the primer molecule, either acetyl CoA or butyryl CoA, to a 4′-phosphopantetheine moiety on a protein constituent of the enzyme complex in bacteria. This protein is acyl carrier protein (ACP), and its phosphopantetheine unit is identical with that in coenzyme A. The mammalian enzyme also contains a phosphopantetheine unit. Six or seven two-carbon units are then added sequentially to the enzyme complex until the palmitate molecule is completed. After each addition of a two-carbon unit a series of reductive steps takes place. The reaction sequence starting with an

TABLE 9.2 Regulation of Fatty Acid Synthesis

Enzyme		*Regulatory Agent*	*Effect*
		PALMITATE BIOSYNTHESIS	
Acetyl CoA carboxylase	Short term	Citrate	Allosteric activation
	Short term	C_{16}–C_{18} acyl CoAs	Allosteric inhibition
	Short term	Insulin	Stimulation
	Short term	Glucagon	Inhibition
	Short term	cAMP-mediated phosphorylation	Inhibition
	Short term	Dephosphorylation	Stimulation
	Long term	High-carbohydrate diet	Stimulation by increased enzyme synthesis
	Long term	Fat-free diet	Stimulation by increased enzyme synthesis
	Long term	High-fat diet	Inhibition by decreased enzyme synthesis
	Long term	Fasting	Inhibition by decreased enzyme synthesis
	Long term	Glucagon	Inhibition by decreased enzyme synthesis
Fatty acid synthetase		Phosphorylated sugars	Allosteric activation
		High carbohydrate diet	Stimulation by increased enzyme synthesis
		Fat-free diet	Stimulation by increased enzyme synthesis
		High-fat diet	Inhibition by decreased enzyme synthesis
		Fasting	Inhibition by decreased enzyme synthesis
		Glucagon	Inhibition by decreased enzyme synthesis
		BIOSYNTHESIS OF FATTY ACIDS OTHER THAN PALMITATE	
Fatty acid synthetase		High ratio of $\frac{\text{methylmalonyl CoA}}{\text{malonyl CoA}}$	Increased synthesis of methylated fatty acids
		Thioesterase cofactor	Termination of synthesis with short-chain product
Stearoyl CoA desaturase		Various hormones	Stimulation of unsaturated fatty acid synthesis by increased enzyme synthesis
		Dietary polyunsaturated fatty acids	Decreased activity

acetyl CoA primer and leading to butyrl-ACP is as presented in Figure 9.7.

The next round of synthesis is initiated by transfer of the growing fatty acid chain from the 4′-phosphopantetheine moiety of ACP to the functional SH group of β-ketoacyl-ACP synthase (analogous to reaction 3a). This liberates the SH group of ACP for acceptance of a second malonyl unit from malonyl CoA (reaction 2) and allows reactions 3b–6 to generate hexanoyl-ACP. The process is repeated five more times at which point palmitoyl-ACP is acted upon by a thioesterase with the production of free

$$(1)\quad CH_3{-}\overset{\overset{\displaystyle O}{\|}}{C}{-}SCoA + ACPSH \xrightarrow[\text{acetyltransferase}]{\text{(acyl carrier protein)}} CH_3{-}\overset{\overset{\displaystyle O}{\|}}{C}{-}SACP + CoASH$$

$$(2)\quad {}^{-}OOC{-}CH_2{-}\overset{\overset{\displaystyle O}{\|}}{C}{-}SCoA + ACPSH \xrightarrow[\text{malonyltransferase}]{\text{(acyl carrier protein)}} {}^{-}OOC{-}CH_2{-}\overset{\overset{\displaystyle O}{\|}}{C}{-}SACP + CoASH$$

$$(3)\quad (a)\ CH_3{-}\overset{\overset{\displaystyle O}{\|}}{C}{-}SACP + Enz{-}SH \xrightarrow[\text{synthase}]{\beta\text{-ketoacyl-(acyl carrier protein)}} CH_3{-}\overset{\overset{\displaystyle O}{\|}}{C}{-}S{-}Enz + ACPSH$$

$$(b)\ CH_3{-}\overset{\overset{\displaystyle O}{\|}}{C}{-}S{-}Enz + {}^{-}OOC{-}CH_2{-}\overset{\overset{\displaystyle O}{\|}}{C}{-}SACP \xrightarrow[\text{synthase}]{\beta\text{-ketoacyl-(acyl carrier protein)}} CH_3{-}\overset{\overset{\displaystyle O}{\|}}{C}{-}CH_2{-}\overset{\overset{\displaystyle O}{\|}}{C}{-}SACP + CO_2 + Enz{-}SH$$

$$(4)\quad CH_3{-}\overset{\overset{\displaystyle O}{\|}}{C}{-}CH_2{-}\overset{\overset{\displaystyle O}{\|}}{C}{-}SACP + NADPH + H^+ \xrightarrow[\text{reductase}]{\beta\text{-ketoacyl-(acyl carrier protein)}} CH_3{-}\overset{\overset{\displaystyle OH}{|}}{CH}{-}CH_2{-}\overset{\overset{\displaystyle O}{\|}}{C}{-}SACP + NADP^+$$

$$(5)\quad CH_3{-}\overset{\overset{\displaystyle OH}{|}}{CH}{-}CH_2{-}\overset{\overset{\displaystyle O}{\|}}{C}{-}SACP \xrightarrow[\text{dehydratase}]{\beta\text{-hydroxyacyl-(acyl carrier protein)}} CH_3{-}CH{=}CH{-}\overset{\overset{\displaystyle O}{\|}}{C}{-}SACP + H_2O$$

$$(6)\quad CH_3{-}CH{=}CH{-}\overset{\overset{\displaystyle O}{\|}}{C}{-}SACP + NADPH + H^+ \xrightarrow[\text{reductase}]{\text{enoyl-(acyl carrier protein)}} CH_3{-}CH_2{-}CH_2{-}\overset{\overset{\displaystyle O}{\|}}{C}{-}SACP + NADP^+$$

Figure 9.7
Reactions catalyzed by fatty acid synthetase.

palmitic acid (Figure 9.8). Note that at this stage the sulfhydryl groups of ACP and β-ketoacyl-ACP synthase are both free so that another cycle of fatty acid synthesis can begin.

$$CH_3{-}(CH_2)_{14}{-}\overset{\overset{\displaystyle O}{\|}}{C}{-}SACP + H_2O \xrightarrow{\text{thioesterase}} CH_3{-}(CH_2)_{14}{-}COO^- + ACPSH$$

Figure 9.8
Release of palmitic acid from fatty acid synthetase.

Mammalian Fatty Acid Synthetase

The reaction sequence given above is fairly well established as the basic pattern for fatty acid biosynthesis in living systems. However, the details of the reaction mechanisms are still far from clear and may vary among species. The enzyme complex termed fatty acid synthetase catalyzes all these reactions, but its structure and properties vary considerably. The enzymes in *E. coli* are dissociable, and the reaction sequence was worked out with that organism. This sequence has been confirmed in mammalian systems, but the enzyme complex itself has not been dissociated. Some investigators postulate that the mammalian synthetase is composed of two possibly identical subunits, each of which is a multienzyme polypeptide. Even among mammalian species and tissues there are certainly variations.

Despite the gaps in present knowledge, it appears likely that the growing fatty acid chain is continually bound to the enzyme complex and is sequentially transferred between the 4′-phosphopantetheine group of ACP and the sulfhydryl group of a cysteine residue on β-ketoacyl-ACP synthase during the condensation reaction (reaction 3) (see Figure 9.9).

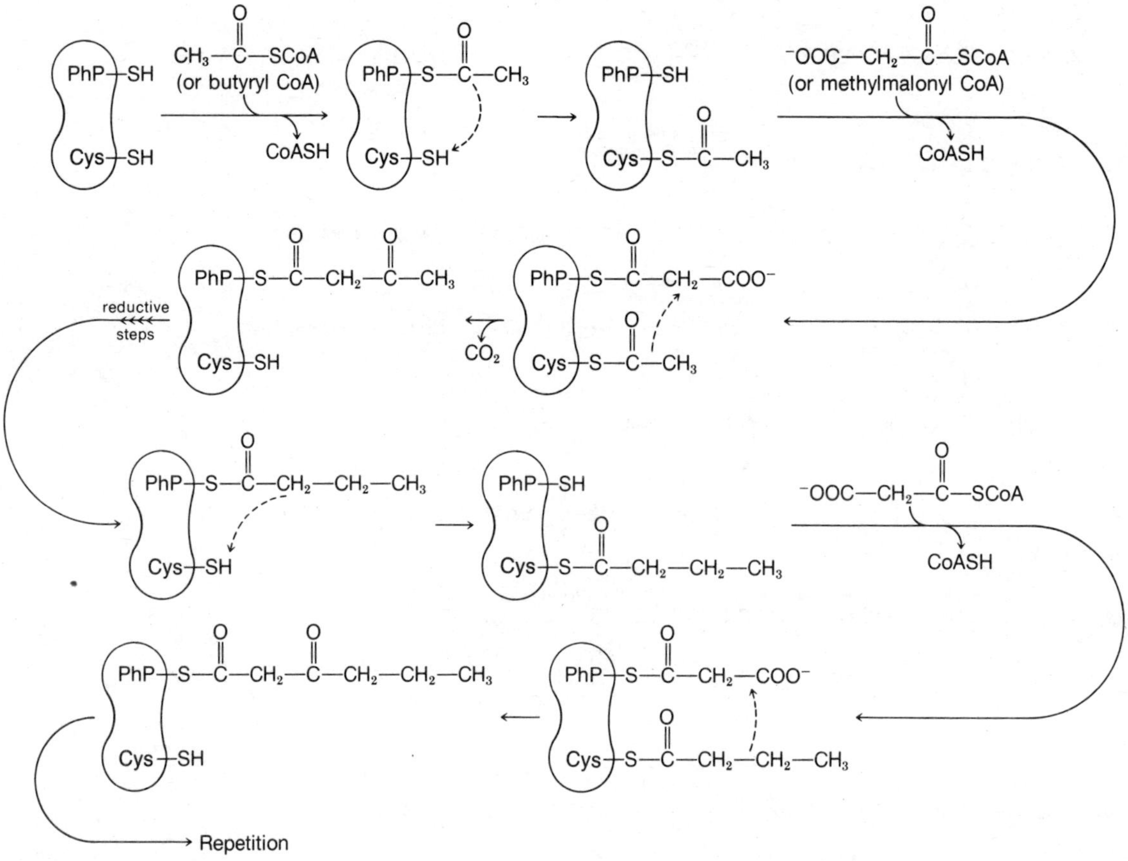

PhP = 4′-phosphopantetheine on acyl carrier protein
Cys = enzyme protein cysteine

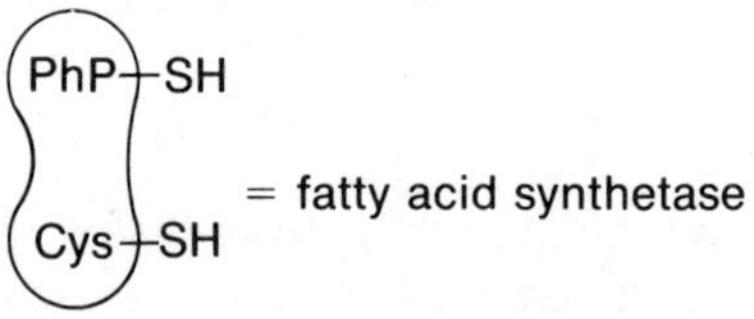

PhP–SH / Cys–SH = fatty acid synthetase

Figure 9.9
Proposed mechanism of elongation reactions taking place on mammalian fatty acid synthetase.

It is also probable that an intermediate acylation to an enzyme serine takes place when acyl CoA units add to the enzyme-bound ACP in the transacylase reactions.

There is suggestive evidence that some short-term regulation of the fatty acid production is carried out by control of the activity of fatty acid synthetase, but this is yet to be established firmly. An allosteric stimulation of phosphorylated sugars has been proposed and a reversible control mechanism by addition or removal of the 4′-phosphopantetheine cofactor is suggested. However, most control of palmitate biosynthesis through fatty acid synthetase probably occurs by controlling the rate of synthesis and degradation of the enzyme. The agents and conditions which do this are given in Table 9.2. They are logical in terms of balancing an efficient utilization of the various biological energy substrates.

Stoichiometry of Fatty Acid Biosynthesis

If acetyl CoA is the primer for palmitate biosynthesis, the overall reaction is

$$CH_3\text{—}\overset{\displaystyle O}{\overset{\|}{C}}\text{—SCoA} + 7\,^{-}OOC\text{—}CH_2\text{—}\overset{\displaystyle O}{\overset{\|}{C}}\text{—SCoA} + 14NADPH + 14H^+ \longrightarrow$$

$$CH_3\text{—}(CH_2)_{14}\text{—}COO^- + 7CO_2 + 14NADP^+ + 8CoASH + 6H_2O$$

To calculate the energy needed for the overall conversion of acetyl CoA to palmitate, we must add the ATP used in formation of malonyl CoA:

$$7CH_3\text{—}\overset{\displaystyle O}{\overset{\|}{C}}\text{—SCoA} + 7CO_2 + 7ATP \longrightarrow 7\,^{-}OOC\text{—}CH_2\text{—}\overset{\displaystyle O}{\overset{\|}{C}}\text{—SCoA} + 7ADP + 7P_i$$

Then the overall stoichiometry for conversion of acetyl CoA to palmitate is

$$8CH_3\text{—}\overset{\displaystyle O}{\overset{\|}{C}}\text{—SCoA} + 7ATP + 14NADPH + 14H^+ \longrightarrow$$

$$CH_3\text{—}(CH_2)_{14}\text{—}\overset{\displaystyle O}{\overset{\|}{C}}\text{—}O^- + 8CoASH + 7ADP + 7P_i + 6H_2O + 14NADP^+$$

Subcellular Localization and Sources of Substrates

Fatty acid synthetase and acetyl CoA carboxylase are found primarily in the cytosol, and palmitate biosynthesis occurs largely in that subcellular compartment. However, mammalian tissues must use special processes to ensure an adequate supply of acetyl CoA and NADPH for this synthesis.

Specifically, the major source of acetyl CoA is the pyruvate dehydrogenase reaction inside the mitochondria. Since mitochondria are not readily permeable to acetyl CoA, a bypass mechanism moves it to the cytosol for palmitate biosynthesis. This mechanism, outlined in Figure 9.10,

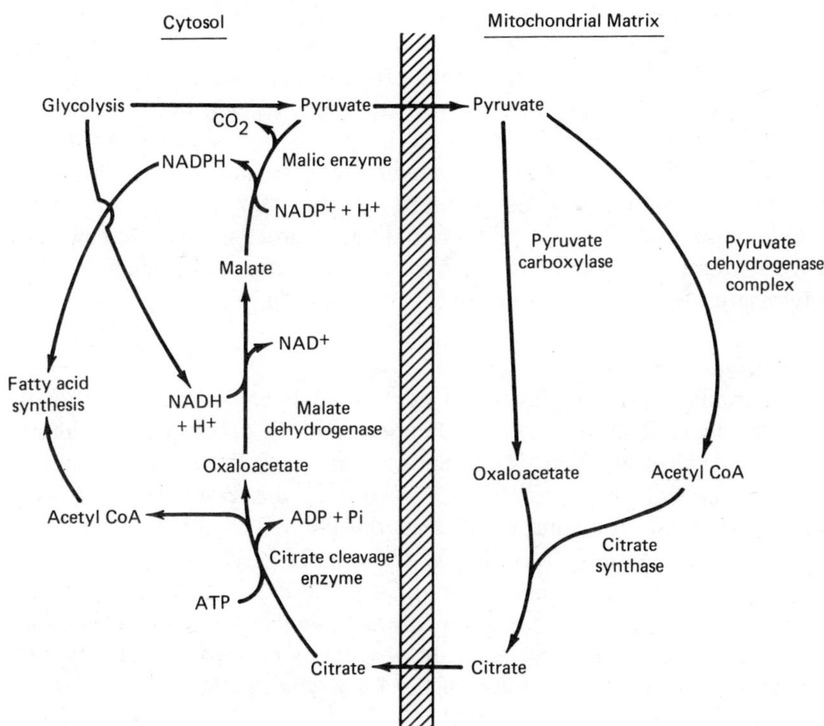

Figure 9.10
Mechanism for transfer of acetyl CoA from mitochondria to cytosol for fatty acid biosynthesis.

takes advantage of the facts that citrate does exchange freely from the mitochondria to the cytosol and that an enzyme exists in the cytosol to convert citrate to acetyl CoA and oxaloacetate. When there is an excess of citrate for the tricarboxylic acid cycle, citrate will pass into the cytosol and supply acetyl CoA for fatty acid biosynthesis. Citrate cleavage enzyme catalyzes the cleavage which requires a mole of ATP:

$$\text{Citrate} + \text{ATP} + \text{CoA} \xrightarrow[\substack{\text{citrate}\\ \text{cleavage}\\ \text{enzyme}}]{} \text{acetyl CoA} + \text{ADP} + \text{P}_i + \text{oxaloacetate}$$

This mechanism has other advantages because CO_2 and NADPH for synthesis of palmitate can be produced from excess cytoplasmic oxaloacetate. As shown in Figure 9.10, the process produces NADPH from NADH, which was formed during glycolysis, by the sequential action of NAD-linked malate dehydrogenase and NADP-linked malic enzyme (malate : NADP oxidoreductase-decarboxylating). The products are pyruvate and CO_2. The cycle is completed by return of the pyruvate to the mitochondrion where it can be carboxylated to regenerate oxaloacetate, as has been described in the process of gluconeogenesis.

In sum, one NADH can be converted to NADPH for each acetyl CoA transferred from mitochondria to the cytosol, each transfer requiring one ATP. The transfer of the 8 acetyl CoA used for each molecule of palmitate can thus supply 8 NADPH. Since palmitate biosynthesis requires 14 NADPH per mole, the other 6 must be supplied from the cytosolic pentose phosphate pathway. This stoichiometry is, of course, hypothetical. The in vivo relationships are complicated by the fact that the transport of citrate and the other di- and tricarboxylic acids across the inner mitochondrial membrane takes place by several one-for-one exchange mechanisms. The actual flow rates are probably controlled by a composite of the concentration gradients of several of these exchange systems.

Synthesis of Other Fatty Acids from Palmitate

The human body can synthesize all of the fatty acids it needs except for the essential, polyunsaturated fatty acids. These syntheses involve a variety of enzyme systems in a number of locations, and the palmitic acid produced by fatty acid synthetase is modified by three processes: elongation, desaturation, and hydroxylation. In this section we discuss the process of elongation, the initial mechanism of desaturation, and the hydroxylation of brain fatty acids that are destined for incorporation into sphingolipids in nerve tissue. An α-oxidation process, involved in fatty acid degradation, and the more elaborate desaturation schemes producing polyunsaturated fatty acids are outlined in later sections.

Elongation Reactions

In mammalian systems elongation of fatty acids can occur either in the endoplasmic reticulum or in mitochondria, and the processes are slightly different in these two loci. In the endoplasmic reticulum the sequence of reactions is similar to that which occurs in the cytosolic fatty acid synthetase in that the source of two-carbon units is again malonyl CoA, and NADPH provides the reducing power. The preferred substrate for elongation in most cases is palmitoyl CoA, but in contrast to the system for de novo fatty acid synthesis, the intermediates in subsequent reactions are CoA esters, suggesting that the process is carried out by separate enzymes rather than a complex of the fatty acid synthetase type. It appears that in most tissues this elongation system in the endoplasmic reticulum almost exclusively converts palmitate to stearate. However, brain contains one or more additional elongation systems, which synthesize the longer chain acids (up to 24 C) that are needed for the brain lipids. These other systems also use malonyl CoA as substrate.

The elongation system in mitochondria is different from that in the endoplasmic reticulum in that acetyl CoA is the source of the added two-carbon units and both NADH and NADPH serve as reducing agents (Figure 9.11). Note that this system operates by simple reversal of the opposing pathway of fatty acid β-oxidation (see Section 9.6) with the exception that NADPH-linked enoyl-CoA reductase (last step of elongation) replaces FAD-linked acyl CoA dehydrogenase (first step in β-oxidation). The process has little activity with acyl CoA substrates of 16 carbons or longer, suggesting that it serves primarily in the elongation of shorter-chain species.

R—CH₂—C(=O)—SCoA

CH_3—C(=O)—SCoA → CoASH (β-ketothiolase)

R—CH_2—C(=O)—CH_2—C(=O)—SCoA

NADH + H^+ → NAD^+ (β-hydroxyacyl CoA dehydrogenase)

R—CH_2—CH(OH)—CH_2—C(=O)—SCoA

→ H_2O (enoyl CoA hydratase)

R—CH_2—CH=CH—C(=O)—SCoA

NADPH + H^+ → $NADP^+$ (enoyl CoA reductase)

R—CH_2—CH_2—CH_2—C(=O)—SCoA

Figure 9.11
Mitochondrial elongation of fatty acids.

Formation of Monoenoic Acids by Stearoyl CoA Desaturase

In higher animals desaturation of fatty acids occurs in the endoplasmic reticulum, and the oxidizing system used to introduce cis double bonds is significantly different from the main fatty acid oxidation process in mitochondria. The systems in endoplasmic reticulum have sometimes been termed "mixed function oxidases" because the terminal enzymes simultaneously oxidize two substrates. In the case of fatty acid desaturation one of these substrates is NADPH and the other is the fatty acid. The electrons from NADPH are transferred through a specific flavoprotein reductase and a cytochrome to "active" oxygen so that it will then oxidize the fatty acid. Although the complete mechanism is not worked out, this latter step may involve a hydroxylation. The three components of the system are the desaturase enzyme, cytochrome b_5, and NADPH-cytochrome b_5 reductase. The overall reaction is

$$\text{R—CH}_2\text{—CH}_2\text{—(CH}_2)_7\text{—COOH} + \text{NADPH} + \text{H}^+ + \text{O}_2 \longrightarrow \text{R—CH=CH—(CH}_2)_7\text{—COOH} + \text{NADP}^+ + 2\text{H}_2\text{O}$$

As noted before, the enzyme specificity is such that the R group must contain at least six carbon atoms. The two main products in most organs are palmitoleic and oleic acids.

The control mechanisms that govern the conversion of the palmitate product of fatty acid synthetase to unsaturated fatty acids are largely unexplored. One of the most important considerations is the control of the proportions of the unsaturated fatty acids available for a proper maintenance of physical state of stored triacylglycerols and membrane phospholipids. A critical committed step in the formation of unsaturated fatty acids from palmitate is the introduction of the first double bond by stearoyl CoA desaturase. The activity of this enzyme and its synthesis are controlled by both dietary and hormonal mechanisms. Increasing the amounts of polyunsaturated fatty acids in the diet of experimental animals decreases the activity of stearoyl CoA desaturase in liver, and insulin, triiodothyronine, and hydrocortisone cause its induction.

Formation of Hydroxy Fatty Acids in Nerve Tissue

There are apparently two different processes that produce α-hydroxy fatty acids in higher animals. One occurs in the mitochondria of many tissues and acts on relatively short-chain fatty acids. This is discussed in

Section 9.6. The second process has so far been demonstrated only in tissues of the nervous system where it produces long-chain fatty acids with a hydroxyl group on carbon-2. These are needed for the structure of some myelin lipids. The specific case of α-hydroxylation of lignoceric acid to cerebronic acid has been studied. These enzymes preferentially use C_{22} and C_{24} fatty acids and show characteristics of the "mixed function oxidase" systems, requiring molecular oxygen and reduced NAD or NADP. This synthesis may be closely coordinated with the biosynthesis of the sphingolipids, which contain the hydroxylated fatty acids.

Fatty Acids Formed by Modification of Fatty Acid Synthetase Function

The schemes outlined in previous sections, which utilize palmitate synthesized by fatty acid synthetase and modify it by further enzymatic action, account for the great bulk of fatty acid biosynthesis in the human body, particularly that involved in energy storage. However, there are a number of special instances where smaller amounts of different fatty acids are needed for specific structural or functional purposes, and these acids are produced by modifications of the process carried out by fatty acid synthetase. Two examples are the production of fatty acids shorter than palmitate in mammary glands and the synthesis of branched-chain fatty acids in certain secretory glands.

Recent work has shown that milk produced by many animals contains varying amounts of fatty acids with shorter chain lengths than palmitate. The amounts produced by the mammary gland apparently vary with species and especially with the physiological state of the animal. This is probably true of humans, although most investigations have been carried out with rats, rabbits, and various ruminants. The same fatty acid synthetase that produces palmitate synthesizes the shorter chain acids when the linkage of the growing chain with the acyl carrier protein is split before the full 16-carbon chain is completed. This hydrolysis is caused by soluble thioesterases whose activity is under hormonal control.

As noted in an earlier section, there are relatively few branched-chain fatty acids in higher animals, and, until recently, their metabolism has been studied mostly in primitive species such as *Mycobacteria,* where they are present in greater variety and amount. It is now known that simple branched-chain fatty acids are synthesized by tissues of higher animals for specific purposes, such as the production of waxes in sebaceous glands and avian preen glands and the elaboration of structures in the echo-locating systems of porpoises.

The majority of branched-chain fatty acids in higher animals are simply methylated derivatives of the saturated, straight-chain acids, and they are synthesized by fatty acid synthetase. When methylmalonyl CoA is used as a substrate instead of malonyl CoA, a methyl side chain is inserted in the fatty acid, and the reaction is as follows:

$$CH_3{-}(CH_2)_n{-}\overset{\overset{\displaystyle O}{\|}}{C}{-}SACP + HOOC{-}\overset{\overset{\displaystyle CH_3}{|}}{CH}{-}\overset{\overset{\displaystyle O}{\|}}{C}{-}SCoA \longrightarrow$$

$$CH_3{-}(CH_2)_n{-}\overset{\overset{\displaystyle O}{\|}}{C}{-}\overset{\overset{\displaystyle CH_3}{|}}{CH}{-}\overset{\overset{\displaystyle O}{\|}}{C}{-}SACP + CO_2 + CoASH$$

The regular reduction steps then follow. Apparently these reactions occur in many tissues normally at a rate several orders of magnitude slower than the utilization of malonyl CoA to produce palmitate. However, it has been suggested that the proportion of branched-chain fatty acids synthesized is largely governed by the relative availability of the two precursors, and an increase in branching can occur by decreasing the ratio of malonyl CoA to methylmalonyl CoA. A malonyl CoA decarboxylase capable of causing this decrease occurs in many tissues. It has also been suggested that increased levels of methylmalonyl CoA in pathological situations, such as vitamin B_{12} deficiency, can lead to excessive production of branched-chain fatty acids.

Production of Fatty Alcohols

As discussed in Chapter 10, many phospholipids contain fatty acid chain moieties in ether linkage rather than ester linkage. The biosynthetic precursors of these ether-linked chains are fatty alcohols (Figure 9.12) rather than fatty acids. These alcohols are formed in higher animals by a two-step, NADPH-linked reduction of fatty acyl CoAs in the endoplasmic reticulum. In organs that produce relatively large amounts of ether-containing lipids, the concurrent production of fatty acids and fatty alcohols is probably closely coordinated.

$$CH_3{-}(CH_2)_n{-}CH_2OH$$

Figure 9.12
Fatty alcohol.

9.4 STORAGE OF FATTY ACIDS AS TRIACYLGLYCEROLS

Most tissues in the human body can convert fatty acids to triacylglycerols by a common sequence of reactions, but liver and adipose tissue carry out this process to the greatest extent. The latter organ is a specialized connective tissue, which is designed for the synthesis, storage, and hydrolysis of triacylglycerols, and this is the main mechanism that the human body has for relatively long-term energy storage. We are concerned here with white adipose tissue as opposed to brown adipose tissue, which occurs in much lesser amounts and has other specialized functions. The triacylglycerols are stored as liquid droplets in the cytoplasm, but this is not "dead storage," since they turn over with an average half-life of only a few days. Thus, in a homeostatic situation there is continuous synthesis and breakdown of triacylglycerols in adipose tissue. Some storage also occurs in skeletal and cardiac muscle, but this is only for local consumption.

Triacylglycerol synthesis in the liver is used primarily for production of blood lipoproteins, although the products can serve as energy sources for other liver functions. The required fatty acids may come from the diet, from adipose tissue via blood transport, or from liver biosynthesis. The acetyl coenzyme A for biosynthesis is principally derived from glucose catabolism.

Biosynthesis of Triacylglycerols

Triacylglycerols are synthesized in most tissues from activated fatty acids and a phosphorylated three-carbon product of glucose catabolism (see Figure 9.13). The latter can be either glycerol 3-phosphate or dihydroxyacetone phosphate. Glycerol phosphate is formed either by reduction of dihydroxyacetone phosphate produced in glycolysis or by phosphorylation of glycerol. It is important to note that there is little or no glycerol kinase in white adipose tissue, so in that particular organ glycerol phosphate must be supplied from glycolytic intermediates. The fatty acids are

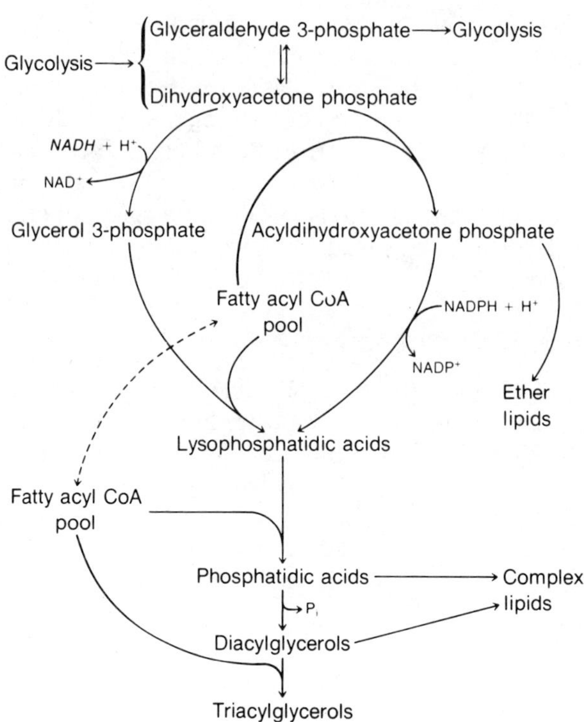

Figure 9.13
Alternative pathways for biosynthesis of triacylglycerols from dihydroxyacetone phosphate.

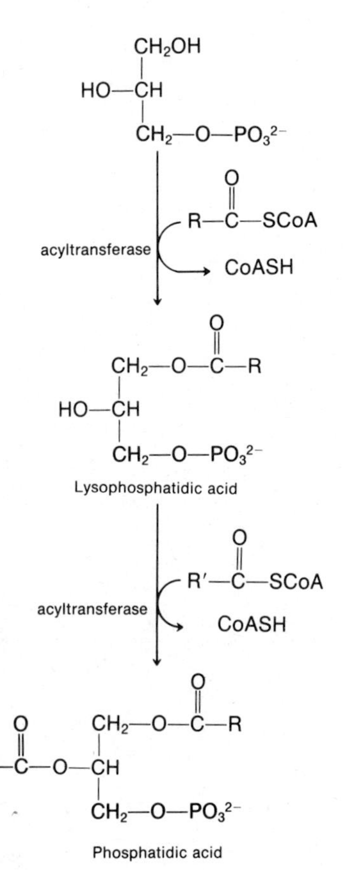

Figure 9.14
Synthesis of phosphatidic acid from glycerol 3-phosphate.

activated by conversion to their coenzyme A esters in the following reaction:

$$R{-}\overset{O}{\overset{\|}{C}}{-}O^- + ATP + CoASH \xrightarrow[\text{synthetase}]{\text{acyl CoA}} R{-}\overset{O}{\overset{\|}{C}}{-}SCoA + AMP + PP_i + H_2O$$

This is a two-step reaction with an acyl adenylate as intermediate and is driven by hydrolysis of the pyrophosphate to inorganic phosphate.

The synthesis of triacylglycerols from the phosphorylated three-carbon fragments involves formation of phosphatidic acid, which is a key intermediate in the synthesis of other lipids as well (see Chapter 10). This may be formed by two sequential acylations of glycerol 3-phosphate, as shown in Figure 9.14. Alternatively, dihydroxyacetone phosphate may be directly acylated at carbon 1 followed by reduction at carbon 2. The resultant lysophosphatidic acid can then be further esterified, as illustrated in Figure 9.15. If phosphatidic acid from either of these routes is to be used for synthesis of triacylglycerols, the phosphate group is next hydrolyzed by phosphatidate phosphatase to yield diacylglycerols. The latter are then acylated to triacylglycerols, as shown in Figure 9.16.

There is at least one tissue, the intestinal mucosa, in which the synthesis of triacylglycerols does not require formation of phosphatidic acid as described above. A major product of intestinal digestion of lipids is 2-monoacylglycerols, which are absorbed as such into the mucosa cells. An enzyme in these cells catalyzes the acylation of these monoacylglycerols with acyl coenzyme A to form 1,2-diacylglycerols, which then can be further acylated as shown above.

The degree of specificity of the acylation reactions in all the steps above is still quite controversial. Analysis of fatty acid patterns in triacylglycerols from various human tissues shows that the distribution of dif-

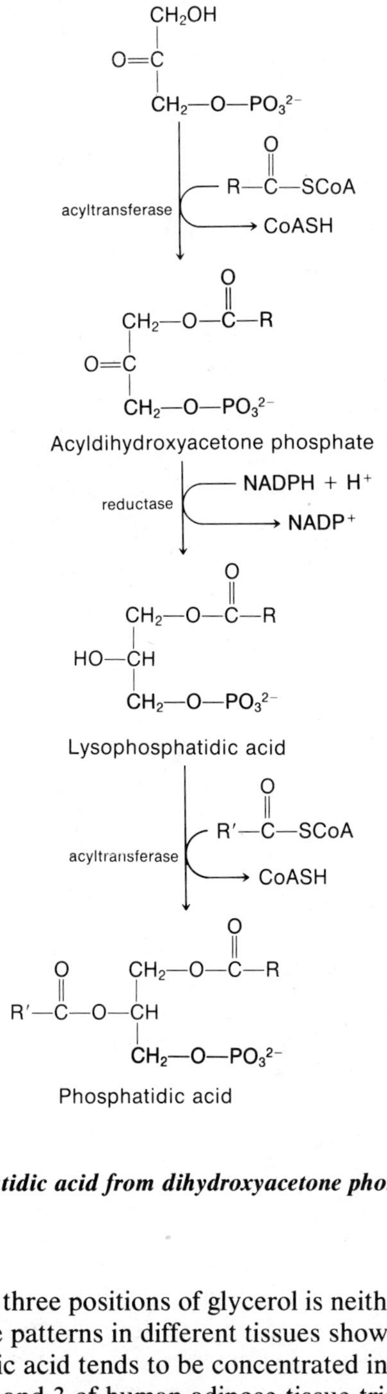

Figure 9.15
Synthesis of phosphatidic acid from dihydroxyacetone phosphate.

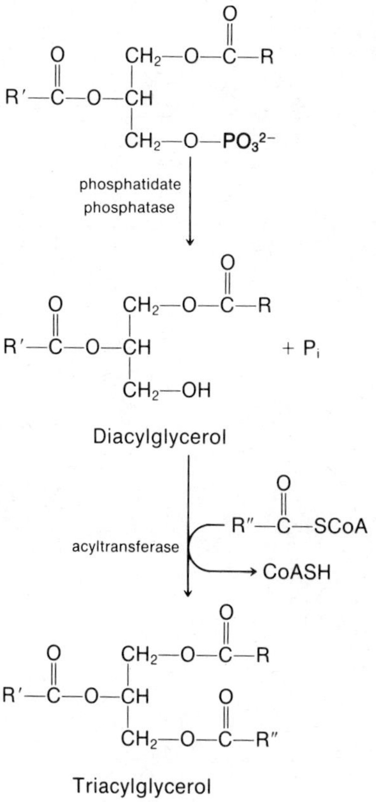

ferent acids on the three positions of glycerol is neither random nor absolutely specific. The patterns in different tissues show some characteristic tendencies. Palmitic acid tends to be concentrated in position 1 and oleic acid in positions 2 and 3 of human adipose tissue triacylglycerols. (Position 3 is the one from which phosphate was removed in hydrolysis of phosphatidic acid.) The two main factors that determine the localization of a given fatty acid to a given position on glycerol are the specificity of the acyltransferase involved and the relative availability of the different fatty acids in the fatty acyl CoA pool. Other factors are probably involved also, but their relative importance is yet to be determined.

Mobilization of Triacylglycerols

The first step in recovering stored fatty acids for energy production is the hydrolysis of triacylglycerols. A variety of lipases catalyze this reaction, the sequence of hydrolysis from the three positions on glycerol depending upon the specificities of the particular lipases involved.

Figure 9.16
Synthesis of triacylglycerol from phosphatidic acid.

CLIN. CORR. **9.1**
OBESITY

When storage of energy as triacylglycerols in adipose tissue becomes abnormal and excessive deposition occurs, the condition is called obesity. In some cases the increased body weight is obvious, but in other patients it is difficult to assess the exact point at which excessive triacylglycerol storage needs to be corrected. Statistics do indicate, however, that obesity is correlated with shorter life span and increased incidence of other health problems.

Obesity is a symptom rather than a disease, and many factors lead to it and need to be considered when trying to correct it. The basic cause is, of course, an excess intake of calories as food and drink over and above the energy requirements of the individual for the period involved. Many possible purely biochemical causes for obesity can be hypothesized, such as inadequate lipases to mobilize the triacylglycerols from adipose tissue, but the human body has so many compensating mechanisms to ensure sufficient energy supply to its tissues that such simple causes rarely, if ever, occur. An appetite-controlling center, or centers, in the hypothalamus has been conclusively demonstrated, but the exact details of the peripheral mechanisms that control food intake are yet to be elucidated.

Treatment of obesity is a significant medical problem, although it is frequently treated by nonmedical practitioners. The plethora of available fad and weight-reducing programs shows vividly the economic incentives involved. Responsible treatment for obesity should involve a careful assessment of the individual patient to evaluate the relative importance of the various possible contributing factors. The most logical approach is then to decrease energy input below the needs of the obese individual until excess weight is lost. This decreased dietary intake should be programmed to provide a balance of constituents and a sufficient supply of essential nutrients. A program of increased physical activity within the capabilities of the individual should hasten the process. The use of various weight-reducing drugs is a very controversial alternative.

$$\begin{array}{l} CH_2{-}O{-}\overset{O}{\overset{\|}{C}}{-}R \\ R'{-}\overset{O}{\overset{\|}{C}}{-}O{-}CH \\ CH_2{-}O{-}\overset{O}{\overset{\|}{C}}{-}R'' \end{array} + 3H_2O \xrightarrow[\text{lipases}]{} \begin{array}{l} CH_2{-}OH \\ HO{-}CH \\ CH_2{-}OH \end{array} + R{-}\overset{O}{\overset{\|}{C}}{-}OH + R'{-}\overset{O}{\overset{\|}{C}}{-}OH + R''{-}\overset{O}{\overset{\|}{C}}{-}OH$$

The lipases in adipose tissue are, of course, the key enzymes for release of the major energy stores. The lipase which removes the first fatty acid is a carefully controlled enzyme, which is sensitive to a variety of circulating hormones. This control of triacylgylcerol hydrolysis must be balanced with the process of triacylglycerol synthesis described in the previous section to assure adequate energy stores and avoid obesity (see Clin. Corr. 9.1). The fatty acids and glycerol produced by the adipose tissue lipases are released to the circulating blood, where the fatty acids are bound by serum albumin and transported to tissues for use. The glycerol returns to the liver, where it is converted to dihydroxyacetone phosphate and enters the glycolytic or gluconeogenic pathways.

9.5 METHODS OF INTERORGAN TRANSPORT OF FATTY ACIDS AND THEIR PRIMARY PRODUCTS

The energy available in fatty acids needs to be distributed throughout the body from the site of fatty acid absorption, biosynthesis, or storage to the functioning tissues that consume them. This transport is closely integrated with the transport of other lipids, especially cholesterol. Since these transport systems appear intimately involved in the pathological processes leading to atherosclerosis, they are being intensively studied, but many important questions are still unanswered.

The human body uses three types of substances as vehicles to transport lipid-based energy: (1) chylomicrons and other plasma lipoproteins in which triacylglycerols are carried in protein-coated lipid droplets, the latter also containing other lipids; (2) fatty acids bound to serum albumin; and (3) the so-called "ketone bodies," principally acetoacetate and β-hydroxybutyrate. These three vehicles are used in varying proportions to carry the energy in the bloodstream via three routes. The first is transport of dietary fatty acids as chylomicrons throughout the body from the intestine after absorption. The second is the transport of lipid-based energy processed by or synthesized in the liver and distributed either to adipose tissue for storage or to other tissues for utilization; in this case they use "ketone bodies" and plasma lipoproteins other than chylomicrons. The third is transport of energy released from storage in adipose tissue to the rest of the body in the form of fatty acids which are bound to serum albumin.

Forms in Which Lipid-Based Energy Is Transported in Blood

The proportions of energy being transported in any one of the modes outlined above varies considerably with metabolic and physiological state. At any one time, the largest amount of lipid in blood is in the form of triacylglycerols in the various lipoproteins. However, the fatty acids bound to albumin are utilized and replaced very rapidly so the total energy transport for a given period of time by this mode may be very significant.

Plasma Lipoproteins

The plasma lipoproteins are synthesized both in the intestine and in the liver and are a heterogeneous group of lipid–protein complexes composed of various types of lipids and apoproteins (see page 97 for detailed discussion of structure). The two most important categories for delivery of lipid-based energy are the chylomicrons and the VLDL, since they contain relatively large amounts of triacylglycerols. Chylomicrons are formed in the intestine and function in the absorption and transport of dietary fat. The exact precursor–product relationships between the other types of plasma lipoproteins are yet to be completely defined, as are the roles of the various protein components. It seems clear, however, that the liver synthesizes VLDL and that the fatty acids from the triacylglycerols are taken up by adipose tissue and other tissues. In the process the VLDL are converted to LDL. The role, if any, of HDL in transport of lipid-based energy is yet to be clarified. All of these lipoproteins are integrally involved in transport of other lipids, especially cholesterol. The lipid components can interchange to some extent between different classes of lipoprotein, and some of the apoproteins probably have functional roles in modifying enzyme activity during exchange of lipids between plasma lipoproteins and tissues. Other apoproteins serve as specific recognition sites for cell surface receptors. Such interaction constitutes the first step in receptor-mediated endocytosis of certain lipoproteins. Studies of rare genetic abnormalities have been helpful in elucidating the roles of some of these apoproteins (see Clin. Corr. 9.2).

Fatty Acids Bound to Serum Albumin

Serum albumin acts as a carrier of a number of substances in the blood, some of the most important being fatty acids. These acids are of course, water insoluble in themselves, but when they are released into the plasma during triacylglycerol hydrolysis they are quickly bound to albumin. This protein has a number of binding sites for fatty acid, two of them bind with high affinity. At any one time the proportion of sites on albumin actually loaded with fatty acids is far from complete, but the rate of turnover is high, so binding by this mechanism constitutes a major route of energy transfer.

Ketone Bodies

The third mode of transport of lipid-based energy-yielding molecules is in the form of small water-soluble molecules, principally acetoacetate and β-hydroxybutyrate (Figure 9.17), which are produced primarily by the liver during the oxidation of fatty acids. The reactions involved in their formation and utilization will be discussed in a later section of this chapter. Under certain conditions, these substances can reach excessive concentrations in blood, leading to ketosis and acidosis. When this occurs, some spontaneous decarboxylation of acetoacetate to acetone occurs. This led early investigators to call the group of soluble products ''ketone

$$CH_3{-}\overset{\overset{O}{\|}}{C}{-}CH_2{-}\overset{\overset{O}{\|}}{C}{-}OH$$

Acetoacetic acid

$$CH_3{-}\overset{\overset{OH}{|}}{CH}{-}CH_2{-}\overset{\overset{O}{\|}}{C}{-}OH$$

β-Hydroxybutyric acid

Figure 9.17
Structures of ketone bodies.

CLIN. CORR. 9.2 GENETIC ABNORMALITIES IN LIPID-ENERGY TRANSPORT

Diseases that affect the transport of lipid-based energy frequently result in abnormally high plasma triacylglycerols, cholesterol, or both. They are classified as hyperlipidemias. Some of them are genetically transmitted, and presumably they result from the alteration or lack of one or more proteins involved in the production or processing of plasma lipids. The nature and function of all of these proteins is yet to be determined, so the elucidation of exact causes of the pathology in most of these diseases is still in the early stages. However, in several cases a specific protein abnormality has been associated with altered lipid transport in the patients' plasma.

In the extremely rare disease analbuminemia, there is an almost complete lack of serum albumin. Despite the many functions of this protein, the symptoms of the disease are surprisingly mild. Lack of serum albumin effectively eliminates the transport of fatty acids unless they are esterified in acylglycerols or complex lipids. However, since patients with analbuminemia do have elevated plasma triacylglycerol levels, presumably the deficiency in lipid-based energy transport caused by the absence of albumin to carry fatty acids is filled by increased use of plasma lipoproteins to carry triacylglycerols.

A more serious genetic defect is the absence of lipoprotein lipase. The major problem here is the inability to process chylomicrons after a fatty meal. Although pathological fat deposits do occur throughout the body, if patients are put on a low-fat diet they respond reasonably well.

Another rare but more severe disease, abetalipoproteinemia, is caused by defective synthesis of apoprotein B, an essential component in the formation of chylomicrons and VLDL. Under these circumstances the major pathway for transporting lipid-based energy from the diet to the body is unavailable. Chylomicrons, VLDL, and LDL are absent from the plasma and fat absorption is deficient or nonexistent. There are other serious symptoms, including neuropathy and red cell deformities whose etiology is less clear.

bodies.'' In fact, these substances are continually produced by the liver and, to a lesser extent, by the kidney. Skeletal and cardiac muscle then utilize them to produce ATP. Nervous tissue, which obtains almost all of its energy from glucose if it is available, is unable to take up and utilize the fatty acids bound to albumin for energy production. However, it can use β-hydroxybutyrate when glucose supplies are insufficient.

Mechanism for Transfer of Fatty Acids Between Blood and Tissues

Lipid-based energy distributed as fatty acids bound to albumin or as ''ketone bodies'' is readily taken up by various tissues for oxidation and production of ATP. However, the energy in fatty acids stored or circulated as triacylglycerols is not directly available, but rather the latter compounds must be enzymatically hydrolyzed to release the fatty acids and glycerol. There are two types of lipases involved in this hydrolysis: (1) lipoprotein lipases, which hydrolyze triacylglycerols in the plasma lipoproteins, and (2) so-called ''hormone-sensitive triacylglycerol lipase,'' which initiates hydrolysis of triacylglycerols in adipose tissue and the release of fatty acids and glycerol into the plasma.

Lipoprotein lipases are located on the surface of the endothelial cells of capillaries and possibly of adjoining tissue cells. They hydrolyze fatty acids from the 1 and/or 3 position of tri- and diacylglycerols when the latter are present in VLDL or chylomicrons. One of the lipoprotein apoproteins must be present to activate the process. The fatty acids that are released are either bound to serum albumin or taken up by the tissue. The monoacylglycerol products may either pass into the cells or be further hydrolyzed by serum monoacylglycerol hydrolase.

A completely distinct type of lipase controls the mobilization of fatty acids from the triacylglycerols stored in the adipose tissue. One of them is hormonally controlled by a cAMP-mediated mechanism. There are a number of lipase activities in the tissue, but the enzyme attacking triacylglycerols initiates the process. Two other lipases then rapidly complete the hydrolysis of mono- and diacylglycerols, releasing fatty acids to the plasma where they are bound to serum albumin (see Table 9.3).

TABLE 9.3 Regulation of Triacylglycerol Metabolism

Enzyme	*Regulatory Agent*	*Effect*
	TRIACYLGLYCEROL MOBILIZATION	
''Hormone-sensitive'' lipase	''Lipolytic hormones,'' e.g., epinephrine, glucagon, ACTH, etc.	Stimulation by cAMP-mediated phosphorylation of relatively inactive enzyme
	Insulin	Inhibition
	Prostaglandins	Inhibition
Lipoprotein lipase	Lipoprotein apoprotein C-II	Activation
	Insulin	Activation
	TRIACYLGLYCEROL BIOSYNTHESIS	
Phosphatidate phosphatase	Steroid hormones	Stimulation by increased enzyme synthesis

9.6 UTILIZATION OF FATTY ACIDS FOR ENERGY PRODUCTION

The fatty acids that arrive at the surface of tissues are taken up by the cells and can be used for energy production. This process occurs primarily inside the mitochondria and is intimately integrated with the processes of energy production from other sources. The energy-rich intermediates produced from fatty acids are the same as those obtained from sugars, that is, NADH and $FADH_2$, and the final stages of the oxidation process are exactly the same as for carbohydrates, that is, the metabolism of acetyl CoA by the TCA cycle and the production of ATP in the mitochondrial electron transport system.

The degree of utilization of fatty acids for energy production varies considerably from tissue to tissue and depends to a significant degree upon the metabolic status of the body, whether it is fed or fasted, exercising, and so on. For instance, nervous tissue apparently oxidizes fatty acids to a minimal degree if at all, but cardiac and skeletal muscle depend heavily on fatty acids as a major energy source. During prolonged fasting most tissues are able to use fatty acids or ketone bodies for their energy requirements.

β-Oxidation of Straight-Chain Saturated Fatty Acids

For the most part, fatty acids are oxidized by a mechanism that is similar to, but not identical with, a reversal of the process of palmitate synthesis described earlier in this chapter. That is, two-carbon fragments are removed sequentially from the carboxyl end of the acid after steps of dehydrogenation, hydration, and oxidation to form a β-keto acid, which is split by thiolysis. These processes take place while the acid is activated in a thioester linkage to the 4′-phosphopantetheine of coenzyme A.

Activation with Coenzyme A

The first step in oxidation of a fatty acid must therefore be its activation to a fatty acyl CoA. This is the same reaction described for synthesis of triacylglycerols in Section 9.4 and occurs in the endoplasmic reticulum or the outer mitochondrial membrane.

Fatty acids occurring inside the mitochondria can also be activated to a limited extent. This process is analogous to the extramitochondrial one, except that it is dependent on energy from guanine nucleotides instead of adenine nucleotides. The physiological significance of this mitochondrial process is not yet clear.

Transport of Fatty Acyl CoAs into the Mitochondria

Since most of the fatty acyl CoAs are formed outside the mitochondria while the oxidizing machinery is inside the inner membrane, which is impermeable to coenzyme A and its derivatives, the cell has a major logistical problem. An efficient shuttle system overcomes this problem by using carnitine as the carrier of acyl groups across the membrane. The steps involved are outlined in Figure 9.18. There are enzymes on both sides of the inner mitochondrial membrane that transfer the fatty acyl group between coenzyme A and carnitine according to the equation:

$$CH_3{-}(CH_2)_n{-}\overset{\overset{\displaystyle O}{\|}}{C}{-}SCoA + (CH_3)_3{-}\overset{+}{N}{-}CH_2{-}\overset{\overset{\displaystyle OH}{|}}{CH}{-}CH_2{-}COOH \underset{\substack{\text{carnitine}\\ \text{palmitoyl}\\ \text{transferase}}}{\rightleftharpoons} CH_3{-}(CH_2)_n{-}\overset{\overset{\displaystyle O}{\|}}{C}{-}O{-}\overset{\overset{\displaystyle (CH_3)_3N^+CH_2}{|}}{CH}{-}CH_2{-}COOH + CoASH$$

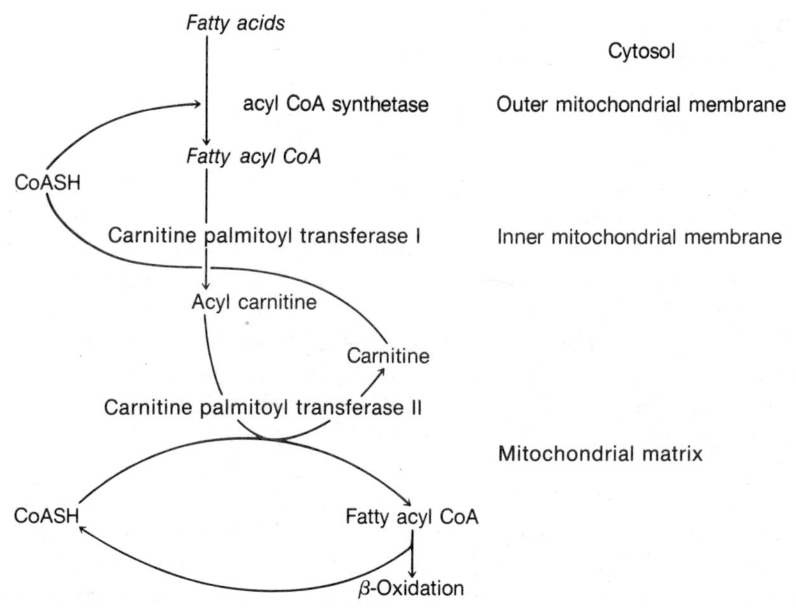

Figure 9.18
Mechanism for transfer of fatty acids from the cytosol through the mitochondrial membranes for oxidation.

On the outer surface, the acyl group is transferred to carnitine in a reaction catalyzed by carnitine palmitoyltransferase I (CPT I). The acyl carnitine exchanges across the inner mitochondrial membrane with free carnitine. The latter becomes available on the inner surface when the fatty acyl group is transferred back to coenzyme A under the influence of carnitine palmitoyltransferase II (CPT II). This process functions primarily in the mitochondrial transport of fatty acyl CoAs with chain lengths of C_{12}–C_{18}, and genetic abnormalities in the system lead to muscle pathology (see Clin. Corr. 9.3). By contrast, shorter-chain fatty acids can cross the inner mitochondrial membrane directly and become activated to their CoA derivatives in the matrix compartment; that is, their oxidation is carnitine independent.

β-Oxidation Reaction Sequence

Once the fatty acyl groups have been transferred back to coenzyme A at the inner surface of the inner mitochondrial membrane they can be oxidized by a group of acyl CoA dehydrogenases in this membrane that remove hydrogens and form enoyl CoA with a trans double bond between carbons 2 and 3. The several dehydrogenases have different specificities for chain length of the acyl CoA oxidized, and the hydrogen acceptor is a flavoprotein. The reaction is

$$CH_3\text{—}(CH_2)_n\text{—}CH_2\text{—}CH_2\text{—}\overset{O}{\overset{\|}{C}}\text{—SCoA} + \text{FAD-protein} \xrightarrow[\text{acyl CoA dehydrogenase}]{} CH_3\text{—}(CH_2)_n\text{—}CH{=}CH\text{—}\overset{O}{\overset{\|}{C}}\text{—SCoA} + \text{FADH}_2\text{-protein}$$

As is the case in the TCA cycle, the enzyme-bound flavoproteins transfer electrons through several other flavoproteins to ubiquinone in the electron transport scheme and only 2 ATP can be obtained for each double bond formed.

CLIN. CORR. **9.3**
GENETIC DEFICIENCIES IN CARNITINE OR CARNITINE PALMITOYL TRANSFERASE

There are several known diseases that are caused by genetic abnormalities in the carnitine system for transporting fatty acids across the inner mitochondrial membrane. They cause deficiencies either in the level of carnitine or in the functioning of the carnitine palmitoyl transferase enzymes. In both cases it is difficult to correlate the degree of the deficiency with the severity of the symptoms.

The clinical symptoms of carnitine deficiency, in particular, seem to vary greatly. They can range from mild, recurrent muscle cramping to severe weakness and death. The deficiency has been classified into two categories; that which is a generalized systemic carnitine deficiency and that which appears to be limited to deficiency only in muscles. The etiology of both disorders remains obscure. Pathological accumulation of triacylglycerols in muscles usually occurs, since fatty acids are inefficiently transported into mitochondria for oxidation. The degree of muscle wasting is variable and results in unpredictable blood carnitine levels, since carnitine can be released into the blood as a result of muscle tissue breakdown. Carnitine therapy has proved effective in some cases, as has the replacement of normal dietary fat by triacylglycerols containing medium chain length fatty acids whose oxidation is carnitine independent.

The deficiency of carnitine palmitoyl transferase is usually a less serious situation and the symptoms are limited to recurrent muscle cramps and aches following strenuous exercise and/or fasting. Muscle mitochondria from patients with this disease have been studied, and the data show that the deficiency can occur at the level of carnitine palmitoyltransferase I or carnitine palmitoyltransferase II or both. Strangely, triacylglycerols do not accumulate in muscles in this disease.

The second step in β-oxidation is hydration of the trans double bond to an L-3-hydroxyacyl CoA.

$$CH_3{-}(CH_2)_n{-}CH{=}CH{-}\overset{O}{\overset{\|}{C}}{-}SCoA + H_2O \xrightarrow[\text{enoyl CoA hydratase}]{} CH_3{-}(CH_2)_n{-}\overset{OH}{\overset{|}{C}H}{-}CH_2{-}\overset{O}{\overset{\|}{C}}{-}SCoA$$

This reaction is stereospecific, in that the L isomer is the product when the trans double bond is hydrated. The stereospecificity of the oxidative pathway is governed by the next enzyme, which is specific for the L isomer as its substrate.

$$CH_3{-}(CH_2)_n{-}\overset{OH}{\overset{|}{C}H}{-}CH_2{-}\overset{O}{\overset{\|}{C}}{-}SCoA + NAD^+ \xrightarrow[\text{L-}\beta\text{-hydroxyacyl CoA dehydrogenase}]{} CH_3{-}(CH_2)_n{-}\overset{O}{\overset{\|}{C}}{-}CH_2{-}\overset{O}{\overset{\|}{C}}{-}SCoA + NADH + H^+$$

The final step is the cleavage of the two-carbon fragment by a thiolase, which, like the preceding two enzymes, has relatively broad specificity with regard to chain length of the acyl group being oxidized.

$$CH_3{-}(CH_2)_n{-}\overset{O}{\overset{\|}{C}}{-}CH_2{-}\overset{O}{\overset{\|}{C}}{-}SCoA + CoASH \xrightarrow[\beta\text{-ketothiolase}]{} CH_3{-}(CH_2)_n{-}\overset{O}{\overset{\|}{C}}{-}SCoA + CH_3{-}\overset{O}{\overset{\|}{C}}{-}SCoA$$

In the overall process then, an acetyl CoA is produced and the acyl CoA product is ready for the next round of oxidation starting with acyl CoA dehydrogenase.

As yet it has been impossible to show conclusively that any of the enzymes in the β-oxidation scheme are control points, although under rather rigid in vitro conditions some apparently have slower maximum rates of reaction than others. It is generally assumed that control is exerted by the availability of substrates and cofactors and by the rate of processing of the acetyl CoA product by the TCA cycle. Mitochondria contain several acyl dehydrogenases with different chain length specificities and these enzymes presumably oxidize the products of adjacent enzymes as rapidly as they are produced. One way in which substrate availability is controlled is by regulation of the shuttle mechanism that transports fatty acids into the mitochondria, a phenomenon of central importance in the regulation of hepatic ketone body production.

Stoichiometry of β-Oxidation Energy Yield

Each set of oxidations resulting in production of a two-carbon fragment yields, in addition to the acetyl CoA, 1 reduced flavoprotein and 1 NADH. In the oxidation of palmitoyl CoA seven such cleavages take place, and in the last cleavage 2 acetyl CoAs are formed. The products of β-oxidation of palmitate are thus 8 acetyl CoAs, 7 reduced flavoproteins, and 7 NADH.

Each of the reduced flavoproteins can yield 2 ATP and each of the NADH can yield three when processed through the electron transport chain, so the reduced nucleotides yield 35 ATP per palmitoyl CoA. As described earlier in Chapter 6, the oxidation of each acetyl CoA through the TCA cycle yields 12 ATP, so the eight 2-carbon fragments from a palmitate molecule produce 96 ATP. However, two ATP equivalents (1 ATP going to 1 AMP) were used to activate palmitate to palmitoyl CoA. Therefore, each palmitic acid entering the cell from the action of lipoprotein lipase or from its combination with serum albumin can yield 129 ATP per mole by complete oxidation.

Comparison of the β-Oxidation Scheme with Palmitate Biosynthesis Reactions

In living metabolic systems the reactions in a catabolic pathway are sometimes quite similar to those in a reversal of the corresponding anabolic pathway, but there are significant differences which provide for separate control of the two schemes. This is true of the palmitate biosynthetic scheme and the scheme for β-oxidation of fatty acids. The critical differences between these two pathways are outlined in Table 9.4. This comparison illustrates some basic mechanisms for separation of metabolic pathways. These include separation by subcellular compartmentation (β-oxidation occurring inside the mitochondria and palmitate biosynthesis in the cytosol), and use of different cofactors (NADPH in biosynthesis; FAD and NAD^+ in oxidation).

Oxidation by Other Processes

The β-oxidation scheme described in the previous sections accounts for the bulk of energy production from fatty acids in the human body. However, it is clear that these reactions must be supplemented by a few other mechanisms so that all fatty acids that are ingested can be oxidized. The principal modifications are those required to oxidize odd-chain fatty acids and unsaturated fatty acids, and those which catalyze α- and ω-oxidation.

TABLE 9.4 Comparison of Schemes for Biosynthesis and β-Oxidation of Palmitate

Parameter	*Biosynthesis*	*β-Oxidation*
Subcellular localization	Primarily cytosolic	Primarily mitochondrial
Phosphopantetheine-containing active carrier	Acyl carrier protein	Coenzyme A
Nature of small carbon fragment added or removed	Carbons 1 and 2 of malonyl CoA after initial priming	Acetyl CoA
Nature of oxidation–reduction coenzyme	NADPH	FAD when saturated chain dehydrogenated, NAD^+ when hydroxy acid dehydrogenated
Stereochemical configuration of β-hydroxy intermediates	D-β-Hydroxy	L-β-Hydroxy
Energy equivalents yielded or utilized in interconversion of palmitate ↔ acetyl CoA	7ATP + 14NADPH = 49ATP equiv	$7FADH_2$ + 7NADH − 2ATP = 33ATP equiv

α-Oxidation occurs at carbon-2 instead of carbon-3 as occurs in the β-oxidation scheme. ω-Oxidation occurs at the methyl end of the fatty acid molecule. Partial oxidation of fatty acids with cyclopropane ring structures probably occurs in humans, but the mechanisms are not worked out.

Odd-Chain Fatty Acids

The oxidation of fatty acids with an odd number of carbons proceeds exactly as described above, but the final product is a molecule of propionyl CoA (Figure 9.19). In order that this compound can be further oxidized, it undergoes carboxylation, molecular rearrangement, and conversion to succinyl CoA. These reactions are identical with those described in Chapter 12 for the metabolism of propionyl CoA when it is formed as a product of the metabolic breakdown of some amino acids.

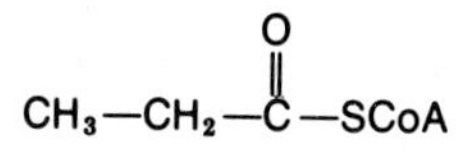

$CH_3—CH_2—C(=O)—SCoA$

Figure 9.19
Propionyl CoA.

Unsaturated Fatty Acids

The many unsaturated fatty acids in the diet are readily available for the production of energy by the human body. However, in several respects the structures encountered in these dietary acids may differ from those required by the specificity of the enzymes in the β-oxidation pathway. One problem is that the naturally occurring unsaturated fatty acids are almost all of the *cis* configuration, whereas those produced during β-oxidation have the *trans* structure (Figure 9.20). The enzyme that hydrates double bonds in the β-oxidation process will hydrate *cis* double bonds, but it forms the D-β-hydroxyacyl CoA. In the next step the dehydrogenase that forms the β-ketoacyl structure is stereospecific for the L configuration of the β-hydroxy substrate (Figure 9.20). However, a racemase enzyme exists in mitochondria, which catalyzes interconversion of D and L isomers. The L isomer is constantly oxidized by the β-hydroxyacyl CoA dehydrogenase, and thus the D isomer is continually converted to the L form and metabolized.

A second problem in β-oxidation of unsaturated fatty acids is that during the process of sequential excision of two-carbon fragments the double bond is sometimes encountered between carbons 3 and 4 instead of between carbons 2 and 3 as required for the β-oxidation scheme enzymes. This problem is overcome by another enzyme, which converts the *cis* bond between carbons 3 and 4 to a trans bond between carbons 2 and 3. The regular scheme can then proceed.

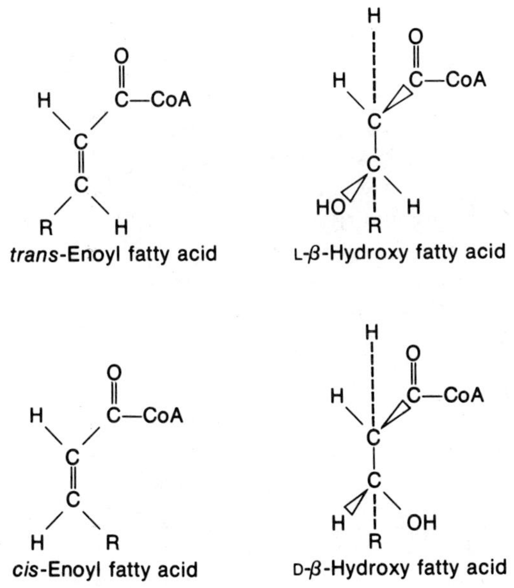

trans-Enoyl fatty acid

L-β-Hydroxy fatty acid

cis-Enoyl fatty acid

D-β-Hydroxy fatty acid

Figure 9.20
Geometric isomers of fatty acids.

$$CH_3{-}(CH_2)_4{-}CH{=}CH{-}CH_2{-}CH{=}CH{-}CH_2{-}(CH_2)_4{-}CH_2{-}CH_2{-}\overset{O}{\overset{\|}{C}}{-}SCoA \quad \text{Linoleoyl CoA}$$

β-Oxidation → $3CH_3{-}\overset{O}{\overset{\|}{C}}{-}SCoA$

$$CH_3{-}(CH_2)_4{-}CH{=}CH{-}CH_2{-}CH{=}CH{-}CH_2{-}\overset{O}{\overset{\|}{C}}{-}SCoA \quad \text{A } cis\text{-3-enoyl CoA}$$

Enoyl CoA isomerase ↓

$$CH_3{-}(CH_2)_4{-}CH{=}CH{-}CH_2{-}CH_2{-}CH{=}CH{-}\overset{O}{\overset{\|}{C}}{-}SCoA \quad \text{A } trans\text{-2-enoyl CoA}$$

β-Oxidation → $2CH_3{-}\overset{O}{\overset{\|}{C}}{-}SCoA$

$$CH_3{-}(CH_2)_4{-}CH{=}CH{-}\overset{O}{\overset{\|}{C}}{-}SCoA \quad \text{A } cis\text{-2-enoyl CoA}$$

Enoyl CoA hydratase ↓

$$CH_3{-}(CH_2)_4{-}C(OH)(H){-}CH_2{-}\overset{O}{\overset{\|}{C}}{-}SCoA \quad \text{A D-}\beta\text{-hydroxyacyl CoA}$$

β-Hydroxyacyl CoA racemase ↓

$$CH_3{-}(CH_2)_4{-}C(H)(OH){-}CH_2{-}\overset{O}{\overset{\|}{C}}{-}SCoA \quad \text{An L-}\beta\text{-hydroxyacyl CoA}$$

β-Oxidation ↓

$$4CH_3{-}\overset{O}{\overset{\|}{C}}{-}SCoA$$

Figure 9.21
Oxidation of linoleoyl CoA.

Both of these problems are encountered in the oxidation of linoleoyl coenzyme A, which is outlined in Figure 9.21.

α-Oxidation

As noted in the earlier discussion of fatty acid biosynthesis, there are several mechanisms for hydroxylation of fatty acids. The one discussed previously is for α-hydroxylation of the long-chain acids needed for the synthesis of sphingolipids. In addition, there are systems in other tissues which hydroxylate the α-carbon of shorter chain acids in order to start their oxidation. The sequence is as follows:

$$CH_3{-}(CH_2)_n{-}CH_2{-}\overset{O}{\overset{\|}{C}}{-}OH \longrightarrow CH_3{-}(CH_2)_n{-}\overset{OH}{\overset{|}{C}H}{-}\overset{O}{\overset{\|}{C}}{-}OH \longrightarrow$$

$$CH_3{-}(CH_2)_n{-}\overset{O}{\overset{\|}{C}}{-}\overset{O}{\overset{\|}{C}}{-}OH \longrightarrow CH_3{-}(CH_2)_n{-}\overset{O}{\overset{\|}{C}}{-}OH + CO_2$$

These hydroxylations probably occur in the endoplasmic reticulum and mitochondria and involve the "mixed function oxidase" type of mechanism discussed previously, because they require molecular oxygen, reduced nicotinamide nucleotides and specific cytochromes. These reactions are particularly important in oxidation of methylated fatty acids (see Clin. Corr. 9.4).

ω-Oxidation

Another minor pathway for fatty acid oxidation also involves hydroxylation and occurs in the endoplasmic reticulum of many tissues. In this case the hydroxylation takes place on the methyl carbon at the other end of the molecule from the carboxyl group or on the carbon next to the methyl end. It also uses the "mixed function oxidase" type of reaction requiring cytochrome P_{450}, O_2, and NADPH as well as the necessary enzymes. The hydroxylated fatty acid can be further oxidized to a dicarboxylic acid, and the β-oxidation can proceed from either end of the molecule. This process probably occurs primarily with medium-chain length fatty acids, and the degree to which it contributes to fatty acid oxidation depends on the tissue and its metabolic state. The overall reactions are

$$CH_3{-}(CH_2)_n{-}\overset{\displaystyle O}{\overset{\|}{C}}{-}OH \longrightarrow HO{-}CH_2{-}(CH_2)_n{-}\overset{\displaystyle O}{\overset{\|}{C}}{-}OH \longrightarrow$$

$$HO{-}\overset{\displaystyle O}{\overset{\|}{C}}{-}(CH_2)_n{-}\overset{\displaystyle O}{\overset{\|}{C}}{-}OH$$

Ketone Body Formation and Utilization

As noted previously, the so-called ketone bodies, which are the most water-soluble form of lipid-based energy, consist mainly of acetoacetic acid and β-hydroxybutyric acid. The latter is a reduction product of the former. β-Hydroxybutyryl CoA and acetoacetyl CoA are intermediates near the end of the β-oxidation sequence, and it was initially presumed that enzymatic removal of coenzyme A from these compounds was the main route for production of the free acids. However, more definitive studies indicated that β-oxidation proceeds completely to acetyl CoA production without accumulation of any intermediates, and that acetoacetate and β-hydroxybutyrate are formed subsequently from acetyl CoA by a separate mechanism.

Formation of Acetoacetate

The primary site for the formation of ketone bodies is the liver, with lesser activity occurring in the kidney. The entire process takes place within the mitochondrial matrix and begins with the condensation of two acetyl CoA molecules to form acetoacetyl CoA (Figure 9.22). The enzyme involved, β-ketothiolase, is probably an isozyme of that which catalyzes the reverse reaction as the last step of β-oxidation. Acetoacetyl CoA then condenses with another molecule of acetyl CoA to form β-hydroxy-β-methylglutaryl coenzyme A (HMG CoA). Cleavage of HMG CoA then yields acetoacetic acid and acetyl CoA.

Formation of β-Hydroxybutyrate and Acetone

In mitochondria a proportion of acetoacetate is reduced to β-hydroxybutyrate depending upon the intramitochondrial $NADH/NAD^+$ ratio.

CLIN. CORR. 9.4 REFSUM'S DISEASE

Although the use of the α-oxidation scheme is a relatively minor one in terms of total energy production, it is significant in the metabolism of dietary fatty acids which are methylated. A principal one of these is phytanic acid,

$$\begin{array}{l} CH_3 \\ | \\ CH{-}CH_3 \\ | \\ (CH_2)_3 \\ | \\ CH{-}CH_3 \\ | \\ (CH_2)_3 \\ | \\ CH{-}CH_3 \\ | \\ (CH_2)_3 \\ | \\ CH{-}CH_3 \\ | \\ CH_2 \\ | \\ COOH \end{array}$$

Phytanic acid

a metabolic product of phytol, which occurs as a constituent of chlorophyll. Phytanic acid is a significant constituent of milk lipids and animal fats, and normally it is metabolized by an initial α-hydroxylation followed by dehydrogenation and decarboxylation. β-Oxidation cannot occur initially because of the presence of the 3-methyl group, but it can proceed after the decarboxylation. The whole reaction produces three molecules of propionyl CoA, three molecules of acetyl CoA, and one of isobutyryl CoA.

In a rare genetic disease called Refsum's disease, the patients lack the α-hydroxylating enzyme and accumulate large quantities of phytanic acid in their tissues and serum. This leads to serious neurological problems for reasons which are still obscure.

$$2CH_3{-}\overset{O}{\overset{\|}{C}}{-}SCoA \xrightarrow{\beta\text{-ketothiolase}} CH_3{-}\overset{O}{\overset{\|}{C}}{-}CH_2{-}\overset{O}{\overset{\|}{C}}{-}SCoA + CoASH$$

$$CH_3{-}\overset{O}{\overset{\|}{C}}{-}CH_2{-}\overset{O}{\overset{\|}{C}}{-}SCoA + CH_3{-}\overset{O}{\overset{\|}{C}}{-}SCoA + H_2O \xrightarrow{\text{HMG—CoA synthase}} HOOC{-}CH_2{-}\underset{CH_3}{\overset{OH}{C}}{-}CH_2{-}\overset{O}{\overset{\|}{C}}{-}SCoA + CoASH$$

$$HOOC{-}CH_2{-}\underset{CH_3}{\overset{OH}{C}}{-}CH_2{-}\overset{O}{\overset{\|}{C}}{-}SCoA \xrightarrow{\text{HMG—CoA lyase}} HOOC{-}CH_2{-}\overset{O}{\overset{\|}{C}}{-}CH_3 + CH_3{-}\overset{O}{\overset{\|}{C}}{-}SCoA$$

Figure 9.22
Pathway of acetoacetate formation.

$$CH_3{-}\overset{O}{\overset{\|}{C}}{-}CH_2{-}\overset{O}{\overset{\|}{C}}{-}O^- + NADH + H^+ \xrightarrow[\text{dehydrogenase}]{\beta\text{-hydroxybutyrate}} CH_3{-}\overset{OH}{\overset{|}{C}H}{-}CH_2{-}\overset{O}{\overset{\|}{C}}{-}O^- + NAD^+$$

Note that the product of this reaction is D-β-hydroxybutyrate, whereas β-hydroxybutyryl CoA formed during the course of β-oxidation is of the L configuration.

β-Hydroxybutyrate dehydrogenase is tightly associated with the inner mitochondrial membrane and, because of its high activity in liver, the concentrations of substrates and products of the reaction are maintained close to equilibrium. Thus, the ratio of β-hydroxybutyrate to acetoacetate in blood leaving the liver can be taken as a reflection of the mitochondrial $NADH/NAD^+$ ratio.

A certain amount of acetoacetate is continually undergoing slow, spontaneous nonenzymatic decarboxylation to acetone.

$$CH_3{-}\overset{O}{\overset{\|}{C}}{-}CH_2{-}\overset{O}{\overset{\|}{C}}{-}O^- + H_2O \longrightarrow CH_3{-}\overset{O}{\overset{\|}{C}}{-}CH_3 + H_2CO_3$$

Under normal conditions acetone formation is negligible, but when pathological accumulations of acetoacetate occur, as for example in severe diabetic ketoacidosis (see Clin. Corr. 9.5), the amount of acetone in blood can be sufficient to cause it to be detectable in a patient's breath.

As seen from Figure 9.23, the pathway leading from acetyl CoA to HMG CoA also operates in the cytosolic space of the liver cell (indeed, this applies to essentially all tissues of the body). However, in this compartment HMG CoA lyase is absent and the HMG CoA formed is used for the purposes of cholesterol biosynthesis (see Chapter 10). What distinguishes liver from nonhepatic tissues is its high complement of intramitochondrial HMG CoA synthase, thus providing an enzymological basis for the primacy of this organ in ketone body production.

CLIN. CORR. 9.5
DIABETIC KETOACIDOSIS

Diabetic ketoacidosis (DKA) is a common illness among patients with insulin-dependent diabetes mellitus. Although mortality rates have declined, they are still in the range of 6–10%. The condition is triggered by severe insulin deficiency coupled with glucagon excess and is frequently accompanied by concomitant elevation of other stress hormones, such as epinephrine, norepinephrine, cortisol, and growth hormone. The major metabolic derangements are marked hyperglycemia, excessive ketonemia, and ketonuria. Blood concentrations of acetoacetic plus β-hydroxybutyric acids as high as 20 mM are not uncommon. Because these are relatively strong acids (pK ~ 3.5), the situation results in life-threatening metabolic acidosis.

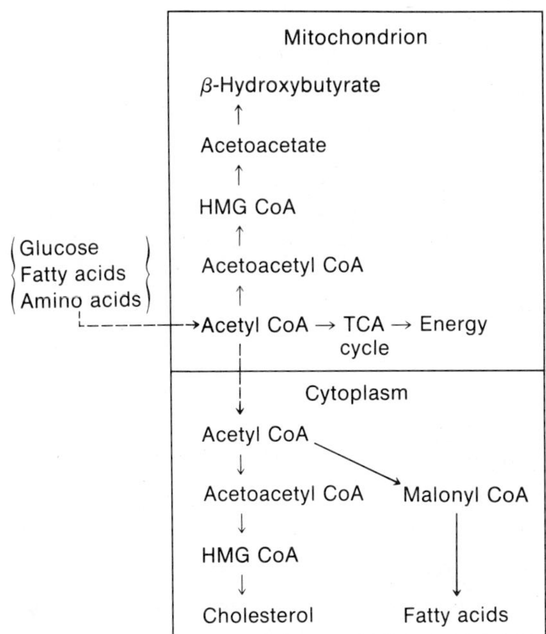

Figure 9.23
Interrelationships of ketone bodies with lipid, carbohydrate, and amino acid metabolism in liver.

The massive accumulation of ketone bodies in the blood in DKA stems from a greatly accelerated hepatic production rate such that the capacity of nonhepatic tissues to use them is exceeded. In biochemical terms the initiating events are identical with those operative in the development of starvation ketosis; that is, increased glucagon : insulin ratio → elevation of liver [cAMP] → decreased [malonyl CoA] → deinhibition of CPT I → activation of fatty acid oxidation and ketone production (see text for details). However, in contrast to physiological ketosis, where insulin secretion from the pancreatic β cells limits free fatty acid (FFA) availability to the liver, this restraining mechanism is absent in the diabetic individual. As a result, plasma FFA concentrations can reach levels as high as 3–4 mM, which drive hepatic ketone production at maximal rates.

Correction of DKA requires rapid treatment that will be dictated by the severity of the metabolic abnormalities and the associated tissue water and electrolyte imbalance. Insulin is essential. It lowers the plasma glucagon level, antagonizes the catabolic effects of glucagon on the liver, inhibits the flow of ketogenic and gluconeogenic substrates (FFA and amino acids) from the periphery, and stimulates glucose uptake in target tissues.

Utilization of Ketone Bodies

Acetoacetate and β-hydroxybutyrate produced by the liver serve as excellent fuels of respiration for a variety of nonhepatic tissues, such as cardiac and skeletal muscle, particularly when glucose is in short supply (starvation) or inefficiently used (insulin deficiency). But since under these conditions the same tissues can readily use free fatty acids (whose blood concentration rises as insulin levels fall) as a source of energy, a nagging question for many years was why the liver should produce ketone bodies in the first place. The answer emerged in the late 1960s with the recognition that during prolonged starvation in humans the ketone bodies replace glucose as the major fuel of respiration for the central nervous system, which has a low capacity for fatty acid oxidation. Also noteworthy is the fact that during the neonatal period of development, acetoacetate and β-hydroxybutyrate serve as important precursors for cerebral lipid synthesis.

The mechanism for use of ketone bodies requires that acetoacetate first be reactivated to its CoA derivative. This is accomplished by a mitochondrial enzyme present in most nonhepatic tissues (but absent from liver) that uses succinyl CoA as the source of the coenzyme. The reaction is depicted in Figure 9.24. Through the action of β-ketothiolase, acetoacetyl CoA is then converted into acetyl CoA, which in turn enters the tricarboxylic acid cycle with the production of energy. Mitochondrial β-hydroxybutyrate dehydrogenase reconverts β-hydroxybutyrate into acetoacetate as the concentration of the latter is decreased.

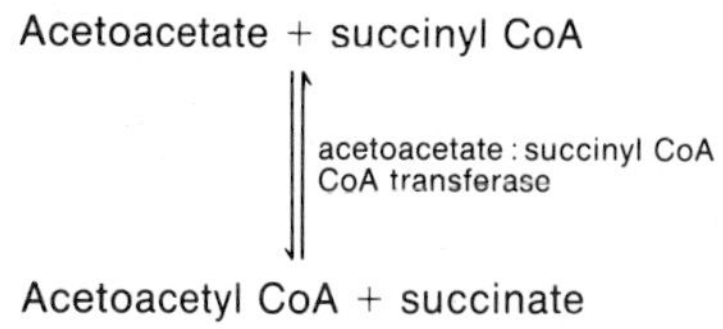

Figure 9.24
Initial step in the utilization of acetoacetate by nonhepatic tissues.

Ketotic States

Under normal feeding conditions, the hepatic production of acetoacetate and β-hydroxybutyrate is minimal and the concentration of these compounds in the blood is very low (<0.2 mM). However, with food deprivation ketone body synthesis is greatly accelerated, and the circulating level of acetoacetate plus β-hydroxybutyrate may rise to the region of 3–5 mM.

This is a normal response of the body to a shortage of carbohydrate and, as alluded to above, subserves a number of crucial roles. In the early stages of fasting, use of the ketone bodies by heart and skeletal muscle conserves glucose for support of the central nervous system. With more prolonged starvation, the increased blood concentration of acetoacetate and β-hydroxybutyrate ensures their efficient uptake by the brain, thereby further sparing glucose consumption.

In contrast to the physiological ketosis of starvation, certain pathological conditions, most notably diabetic ketoacidosis (Clin. Corr. 9.5), are characterized by excessive accumulation of ketone bodies in the blood (up to 20 mM). The hormonal and biochemical factors operative in the overall control of hepatic ketone body production are discussed in detail in Chapter 14.

Role of Peroxisomes in Fatty Acid Oxidation

Most of the oxidation of fatty acid probably occurs in mitochondria, but recent experimental evidence has led to the hypothesis that significant oxidation of fatty acids also takes place in the peroxisomes of liver, kidney, and other tissues. Peroxisomes are a class of subcellular organelles with distinctive morphological and chemical characteristics. Their initial distinguishing characteristic was a high content of the enzyme catalase, and it has been suggested that peroxisomes may function in a protective role against oxygen toxicity. Two lines of evidence suggest that they may also be involved in lipid catabolism. First, the analogous structures in plants, glyoxysomes, are capable of oxidizing fatty acids, and, second, a number of drugs used clinically to decrease triacylglycerol levels in patients cause a marked increase in histologically detectable peroxisomes. Subsequently, it has been shown conclusively that liver peroxisomes, isolated by differential centrifugation, do oxidize fatty acids, and do possess most of the enzymes needed for the β-oxidation process.

The mammalian peroxisomal fatty acid oxidation scheme, which is similar to that in plant glyoxysomes, differs from the mitochondrial β-oxidation system in three important respects: first, the initial dehydrogenation is accomplished by a cyanide-insensitive oxidase system, as shown in Figure 9.25. The hydrogen peroxide is then eliminated by catalase, and the remaining steps are the same as in the mitochondrial system. Second, there is evidence that the peroxisomal and mitochondrial enzymes are slightly different and that the specificity in peroxisomes is for somewhat longer chain length. Third, although rat liver mitochondria will oxidize a molecule of palmitoyl CoA to eight molecules of acetyl CoA, the β-oxidation system in peroxisomes from the same organ will not proceed beyond the stage of octanoyl CoA (C_8). The possibility is thus raised that one function of peroxisomes is to shorten the chains of relatively long-chain fatty acids to a point at which β-oxidation can be completed in mitochondria.

$$CH_3{-}(CH_2)_n{-}CH_2{-}CH_2{-}\overset{O}{\overset{\|}{C}}{-}SCoA$$

flavoprotein → flavoprotein-H_2; O_2 → H_2O_2

$$CH_3{-}(CH_2)_n{-}CH{=}CH{-}\overset{O}{\overset{\|}{C}}{-}SCoA$$

Figure 9.25
Initial step in peroxisomal fatty acid oxidation.

9.7 CHARACTERISTICS, METABOLISM, AND FUNCTIONAL ROLE OF POLYUNSATURATED FATTY ACIDS

In recent years there has been considerable renewed interest in elucidating the specific physiological roles of the polyunsaturated fatty acids at the biochemical level. This is due to some extent to the results of initial studies which suggested that a diet in which the proportion of polyunsaturated to saturated fatty acids was relatively high could help to lower blood cholesterol levels in some patients. The relationship between these diet modifications and the development of atherosclerosis, if any, is not simple, but the initial reports did tend to spur interest in the polyunsaturated fatty acids.

It is important to emphasize that the essential fatty acids mentioned in Section 9.3 are polyunsaturated but that most of the individual polyunsaturated fatty acids need not be specifically supplied in the diet. All fatty acids with three or more double bonds, without regard to their position in the chain, are polyunsaturated and most of these can be synthesized by human tissues. The essential acids refer only to the linoleic and the linolenic series, which have double bonds near the methyl end of the chain, and which cannot be synthesized by humans. The degree to which these acids really are essential for humans is still to be determined, although clearly reproducible deficiency states can be produced in rats and some other animals by carefully controlled diets. The need for the linoleic acid series is clearer since the discovery of the prostaglandins, which are derived from arachidonic acid, one of the linoleic series. The need for the linolenic acid series is very obscure, although the 22-carbon hexaenoic acid derivative of it shown in Section 9.2 is concentrated in some membranes of nerve and retina. A deficiency syndrome for linolenate has yet to be produced in any animal except rainbow trout, but this is possibly due to extremely efficient mechanisms for conservation of linolenate and its derivatives in the body.

A major role of all polyunsaturated fatty acids seems to be to produce the proper fluidity in biological membranes. As described in Chapter 10, the various phospholipids have variable amounts of polyunsaturated fatty acids as constituents, and it has been conclusively demonstrated that lower organisms can alter the fatty acid patterns in their membrane phospholipids to maintain proper fluidity under changing conditions, such as temperature alterations. This can be done by increasing the proportion of fatty acids with a few double bonds in them or by increasing the degree of unsaturation of the fatty acids.

Metabolic Modifications of Unsaturated Fatty Acids

The human body can synthesize a variety of polyunsaturated fatty acids by the elongation and desaturation reactions described in Section 9.3. The stearoyl CoA desaturase introduces an initial double bond between carbons 9 and 10 in a saturated fatty acid, and then double bonds can be introduced just beyond carbons 4, 5, or 6. Desaturation at carbon-8 probably occurs also in some tissues. The positions of these desaturations are shown in Figure 9.26. The relative specificities of the various enzymes are still to be elucidated completely, but it seems likely that elongation and desaturation can occur in either order. The conversion of linolenic acid to all *cis*-4, 7, 10, 13, 16, 19-docosahexaenoic acid in brain is a specific example of such a sequence.

$CH_3—(CH_2—CH{=}CH)_3—CH_2—CH_2—CH_2—CH_2—CH_2—CH_2—CH_2—COOH$

Linolenic acid

↓ "Δ^6-desaturase"

$CH_3—(CH_2—CH{=}CH)_3—CH_2—CH{=}CH—CH_2—CH_2—CH_2—COOH$

↓ elongation

$CH_3—(CH_2—CH{=}CH)_3—CH_2—CH{=}CH—CH_2—CH_2—CH_2—CH_2—CH_2—COOH$

↓ "Δ^5-desaturase"

$CH_3—(CH_2—CH{=}CH)_3—CH_2—CH{=}CH—CH_2—CH{=}CH—CH_2—CH_2—CH_2—COOH$

↓ elongation

$CH_3—(CH_2—CH{=}CH)_3—CH_2—CH{=}CH—CH_2—CH{=}CH—CH_2—CH_2—CH_2—CH_2—CH_2—COOH$

↓ "Δ^4-desaturase"

$CH_3—(CH_2—CH{=}CH)_3—CH_2—CH{=}CH—CH_2—CH{=}CH—CH_2—CH{=}CH—CH_2—CH_2—COOH$

All-*cis*-4,7,10,13,16,19-docosahexaenoic acid

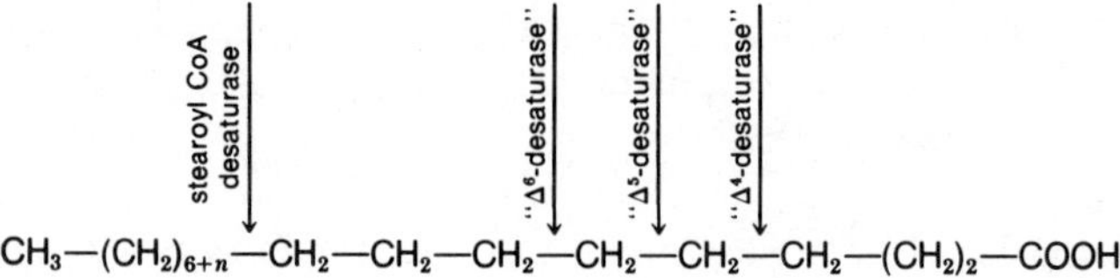

Figure 9.26
Positions in the fatty acid chain where desaturation can occur in the human body. *There must always be at least six single bonds in the chain toward the methyl end of the molecule just beyond the bond being desaturated.*

The polyunsaturated fatty acids, particularly arachidonic acid, are the precursors of the highly active prostaglandins and thromboxanes. A number of different classes of prostaglandins are formed depending on the precursor fatty acid and the sequence of various oxidations which convert the acids to the active compounds. A detailed discussion of these substances and their formation is given in Chapter 10.

Autooxidation of Polyunsaturated Fatty Acids

Polyunsaturated fatty acids in living systems have a significant potential for autooxidation, a process that may have important physiological and/or pathological consequences. This is the set of reactions that causes rancidity in fats and the curing of linseed oil in paints.

The basis behind the process is that the methylene carbon between any two double bonds in the polyunsaturated fatty acids is quite susceptible to hydrogen abstraction and free radical formation. Once this abstraction occurs the reactions can take place in any sequence, and many of the reactive breakdown products can contribute to further oxidation. Oxygen can attach to acids from which hydrogen has been abstracted, forming free radicals which can then react with another lipid molecule, leading to abstraction of hydrogen from the second molecule. The products of this reaction are a lipid hydroperoxide in the first molecule and a new free radical in the molecule attacked. The lipid hydroperoxide molecules break up, forming dialdehydes, the most prominent being malondialdehyde. This product can cause cross-linking between various types of molecules, such linkages leading to cytotoxicity, mutagenicity, membrane breakdown, and enzyme modification. Malondialdehyde also polymerizes with itself and other tissue breakdown products, forming an insoluble pigment, lipofuscin, which accumulates in some aging tissues.

Possible Autooxidation Initiators

A number of external agents can initiate autooxidation in vitro. The extent to which they can lead to such reactions in vivo in humans is undetermined. Various types of radiation, including sunlight, and environmental pollutants such as oxides of nitrogen and carbon tetrachloride are examples of such external agents. The detoxification mechanisms for CCl_4 in the liver use cytochrome P_{450} and generate transient free radicals. The latter can initiate lipid autooxidation and lead to carcinogenesis. Metabolism of the herbicide paraquat, sometimes used for marijuana control, produces superoxide anions, which can also initiate fatty acid autooxidation.

It is quite possible that autooxidation can be initiated without the need for an external agent. Theoretically at least, the enzymes involved in various oxidative processes can produce singlet oxygen and transient partial reduction products of oxygen (superoxide anion, hydrogen peroxide, and the hydroxyl radical), any or all of which could potentially lead to lipid free radicals and/or lipid hydroperoxides. For instance, under proper circumstances rat liver microsomes cause extensive formation of lipid peroxides in vitro, presumably initiated by an enzyme-bound reactive form of iron. A number of enzymes such as xanthine oxidase, superoxide dismutase, and lipoxidases can initiate lipid peroxide formation in vitro.

Possible Protective Mechanisms in Vivo

Recent evidence suggests that in special circumstances the partial reduction products of oxygen which can potentially initiate lipid autooxidation may actually be produced for beneficial purposes, for example, by leukocytes in killing bacteria; however, under most conditions the human body utilizes potent mechanisms to ensure against accumulation of these substances. Three types of enzymes, the catalases, peroxidases, and superoxide dismutases, seem primarily designed to destroy them rapidly and keep tissue levels of their substrates negligible. An additional safeguard is the presence of scavenging molecules in the body, which interact with any free radicals produced, but do not in turn produce self-propagating chain reactions. The principal one present in humans is α-tocopherol, vitamin E. Evidence for its protective role in vivo is purely circumstantial.

BIBLIOGRAPHY

Dietschy, J. M., Gotto, A. M., Jr., and Ontko, J. A. (eds.). *Disturbances in lipid and lipoprotein metabolism.* Bethesda, Md.: American Physiological Society, 1978.

Foster, D. W., and McGarry, J. D., The metabolic derangements and treatment of diabetic ketoacidosis. *N. Engl. J. Med.* 309:159, 1983.

Gurr, M. I., and James, A. T. *Lipid biochemistry, an introduction,* 3rd ed. London: Chapman and Hall, 1980.

IUPAC-IUB Commission on Biochemical Nomenclature. The nomenclature of lipids. *Biochem. J.* 171:21, 1978.

Lech, J. J. (chmn.). Symposium on control of endogenous triglyceride metabolism in adipose tissue and muscle. *Fed. Proc.* 36:1984, 1977.

McGarry, J. D., and Foster, D. W., Regulation of hepatic fatty acid oxidation and ketone body production. *Annu. Rev. Biochem.* 49:395, 1980.

Nilsson-Ehle, P., Garfinkel, A. S., and Schotz, M. C. Lipolytic enzymes and plasma lipoprotein metabolism. *Annu. Rev. Biochem.* 49:667, 1980.

Robinson, A. M., and Williamson, D. H., Physiological roles of ketone bodies as substrates and signals in mammalian tissues. *Physiol. Rev.* 60:143, 1980.

Scow, R. O., Blanchette-Mackie, E. J., and Smith, L. C. Transport of lipid across capillary endothelium. *Fed. Proc.* 39:2610, 1980.

Volpe, J. J. Lipid metabolism: fatty acid and cholesterol biosynthesis. *Adv. Mod. Nutr.* 37, 1978.

Wakil, S. J., Stoops, J. K., and Joshi, V. C., Fatty acid synthesis and its regulation. *Annu. Rev. Biochem.* 52:537, 1983.

QUESTIONS

C. N. ANGSTADT AND J. P. BAGGOTT

*Q*uestion *T*ypes are described inside front cover.

1. (QT2) Fatty acids occurring in humans most commonly:
   1. are straight chain but may have some methyl branches.
   2. have double bonds present in trans configuration.
   3. contain an even number of carbons.
   4. do not contain more than 16 carbons.
2. (QT1) Triacylglycerols:
   A. would be expected to be good emulsifying agents.
   B. yield about the same amount of ATP on complete oxidation as would an equivalent weight of glycogen.
   C. are stored as hydrated molecules.
   D. in the average individual, represent sufficient energy to sustain life for several weeks.
   E. are generally negatively charged molecules at physiological pH.
3. (QT1) In humans, fatty acids:
   A. can be synthesized from excess dietary carbohydrate or protein.
   B. are not required at all in the diet.
   C. containing double bonds cannot be synthesized.
   D. must be supplied entirely by the diet.
   E. other than palmitate, must be supplied in the diet.
4. (QT2) Acetyl CoA carboxylase:
   1. undergoes protomer–polymer interconversion during its physiological regulation.
   2. requires biotin.
   3. is inhibited by cAMP-mediated phosphorylation.
   4. content in a cell responds to changes in fat content in the diet.
5. (QT2) In the synthesis of palmitate:
   1. the addition of malonyl CoA to fatty acid synthase elongates the growing chain by three carbons.
   2. a β-keto residue on the 4′-phosphopantetheine moiety is reduced to a saturated residue by NADPH.
   3. palmitoyl CoA is released from the synthase.
   4. transfer of the growing chain from ACP to another —SH must precede the addition of the next malonyl CoA.

6. (QT2) Citrate stimulates fatty acid synthesis by:
   1. allosterically activating acetyl CoA carboxylase.
   2. providing a mechanism to transport acetyl CoA from the mitochondria to the cytosol.
   3. participating in a pathway that ultimately produces CO_2 and NADPH in the cytoplasm.
   4. participating in the production of ATP.

   A. Fatty acid elongation in mitochondria　C. Both
   B. Fatty acid elongation in cytoplasm　D. Neither
7. (QT4) Malonyl CoA is the source of carbons.
8. (QT4) Preferred substrates are fatty acyl CoAs shorter than 16 carbons.
9. (QT1) Fatty acid synthase:
   A. synthesizes only palmitate.
   B. yields an unsaturated fatty acid by skipping a reductive step.
   C. produces hydroxy fatty acids in nerve tissue.
   D. can stop with the release of a fatty alcohol instead of an acid.
   E. can produce a branched-chain fatty acid if methyl malonyl CoA is used as a substrate.
10. (QT2) Which of the following events is/are usually involved in the synthesis of triacylglycerols in adipose tissue?
    1. Addition of a fatty acyl CoA to a diacylglycerol
    2. Addition of a fatty acyl CoA to a lysophosphatide
    3. Hydrolysis of phosphatidic acid by a phosphatase
    4. Glycerol kinase reaction
11. (QT2) Plasma lipoproteins:
    1. are not the only carriers of lipid-based energy in the blood.
    2. usually have a nonpolar core containing triacylglycerols and cholesterol esters.
    3. do not generally include free (unesterified) fatty acids.
    4. include chylomicrons generated in the intestine.
12. (QT1) Lipoprotein lipase:
    A. is an intracellular enzyme.
    B. is stimulated by cAMP-mediated phosphorylation.
    C. functions to mobilize stored triacylglycerols from adipose tissue.
    D. is stimulated by one of the apoproteins present in VLDL.
    E. readily hydrolyzes three fatty acids from a triacylglycerol.
13. (QT1) A deficiency of carnitine might be expected to interfere with:
    A. β-oxidation.
    B. ketone body formation from acetyl CoA.
    C. palmitate synthesis.
    D. mobilization of stored triacylglycerols from adipose tissue.
    E. uptake of fatty acids into cells from the blood.
14. (QT2) β-Oxidation of fatty acids:
    1. has the potential to generate ATP even if acetyl CoA is not subsequently oxidized.
    2. is controlled primarily by allosteric effectors.
    3. can use odd-chain and unsaturated fatty acids as substrates.
    4. uses $NADP^+$.
15. (QT1) Ketone bodies:
    A. are formed by removal of coenzyme A from the corresponding intermediate of β-oxidation.
    B. are synthesized from cytoplasmic β-hydroxy-β-methyl glutaryl coenzyme A (HMGCoA).
    C. are excellent energy substrates for liver.
    D. include both β-hydroxybutyrate and acetoacetate, the ratio reflecting the intramitochondrial $NADH/NAD^+$ ratio in liver.
    E. form when β-oxidation is interrupted.
16. (QT2) The high glucagon : insulin ratio seen in starvation:
    1. promotes mobilization of fatty acids from adipose stores.
    2. stimulates β-oxidation by inhibiting the production of malonyl CoA.
    3. leads to increased concentrations of ketone bodies in the blood.
    4. produces a condition that results in an increased utilization of ketone bodies by the brain.
17. (QT2) Polyunsaturated fatty acids:
    1. cannot be synthesized by humans.
    2. are important in determining fluidity of membranes.
    3. have no known functions other than as membrane components.
    4. are quite susceptible to autooxidation.

ANSWERS

1. B 1, 3 correct. 2: Most naturally occurring double bonds are *cis,* an important factor in β-oxidation of unsaturated fatty acids. 4: 18- and 20-carbon fatty acids are very common (p. 357).
2. D A, C, E. Triacylglycerols are neutral, hydrophobic molecules with no hydrophilic portion and, therefore, are not emulsifying agents and are stored anhydrously. B: Their more reduced state, compared to carbohydrates, makes them more energy-rich (p. 359).
3. A It is important to realize that triacylglycerol is the ultimate storage form of excess dietary intake. B–E: We can synthesize most fatty acids, including those with double bonds, except for the essential fatty acids, linoleic and linolenic (p. 360).
4. E All four correct. 1: Acetyl CoA carboxylase shifts between its protomeric (inactive) and polymeric (active) forms under the influence of a variety of regulatory factors. 3: Since cAMP increases at times when energy is *needed,* it is consistent that a process that uses energy would be inhibited. 4: Long-term control is related to enzyme synthesis and responds appropriately to dietary changes (Table 9.2, p. 361).
5. C 2, 4 correct. 1: Splitting CO_2 from malonyl CoA is the driving force for the condensation reaction so the chain grows two carbons at a time. 3: Palmitate is released as the free acid; the conversion to the CoA ester is by a different enzyme (p. 363). 4: It is important to realize that only ACP binds the incoming malonyl CoA so it must be freed before another addition can be made (p. 363).
6. A 1, 2, 3, correct. 1: Table 9.2. 2: Acetyl CoA is generated primarily in mitochondria but does not cross the membrane readily. 3: Oxaloacetate generated by citrate cleavage enzyme, when converted to malate, yields CO_2 and NADPH by the malic enzyme (Figure 9.10). 4: Citrate *consumes* ATP when acted upon by citrate cleavage enzyme (p. 366).
7. B The cytoplasmic system is very similar to fatty acid synthase except that the enzymes are not part of a multienzyme complex (p. 366).
8. A The role of mitochondrial fatty acid elongation seems to be to elongate short-chain fatty acids; the cytoplasmic system is most active with palmitate (p. 366).
9. E E. This is much slower than reaction with malonyl CoA, but it is significant. A: In certain tissues, for example, mammary glands, shorter-chain products are formed. B–D: These products are all formed by other processes. Reactions proceeding on a multienzyme complex generally do not "stop" at intermediate steps (p. 367).
10. A 1, 2, 3 correct. 4: Does not occur to any significant extent in adipose tissue. The sequential addition of fatty acyl CoAs to

glycerol 3-phosphate forms lysophosphatidic acid, then phosphatidic acid whose phosphate is removed before the addition of the third fatty acyl residue (p. 369).

11. E All four correct. 1, 3: Fatty acids bound to serum albumin and ketone bodies are other sources. 2: All lipoproteins (Section 9.5) have this same general structure, a nonpolar core surrounded by a more polar shell.

12. D A–C: These are characteristics of hormone-sensitive lipase. E: It generally requires more than one lipase to hydrolyze all of the fatty acids (p. 374, Table 9.3).

13. A Carnitine functions in transport of fatty acyl CoA esters formed in cytosol into the mitochondria (p. 375).

14. B 1, 3 correct. 1, 4: It is important to realize that β-oxidation, itself, generates $FADH_2$ and NADH, which can be reoxidized to generate ATP. 2: Carnitine transport to provide the substrate and reoxidation of reduced cofactors control β-oxidation. 3: β-Oxidation is a general process requiring only minor modifications to oxidize nearly any fatty acid in the cell (p. 376, Table 9.4).

15. D A, E: β-Oxidation proceeds to completion; ketone bodies are formed by a separate process. B, C: Ketone bodies are formed, but not used, in liver mitochondria; cytosolic HMGCoA is a precursor of cholesterol (p. 381).

16. E All four correct. High glucagon: insulin ratio results in cAMP-mediated phosphorylations that activate hormone-sensitive lipase and inhibit acetyl CoA carboxylase. Both of these, as well as other events, promote ketone body formation by greatly increasing acetyl CoA production in mitochondria, thereby assuring efficient uptake and utilization by brain (p. 383).

17. C 2, 4 correct. 1: Humans can introduce additional double bonds between carbon 9 and the carboxyl end but not toward the methyl end of the molecule. 2, 3: Maintaining proper membrane fluidity is a major role; some polyunsaturated fatty acids are precursors of prostaglandins and thromboxanes. 4: They can form free radicals that initiate a sequence of events (p. 385).

10

Lipid Metabolism II: Pathways of Metabolism of Special Lipids

ROBERT H. GLEW

10.1 OVERVIEW

Lipid is a general term that describes substances that are relatively water-insoluble and extractable by nonpolar solvents. The complex lipids of humans fall into one of two broad categories: the nonpolar lipids, such as triacylglycerols and cholesterol esters, and the polar lipids, which are amphipathic in that they contain both a hydrophobic domain and a hydrophilic region in the same molecule. This chapter will discuss the two major subdivisions of the polar lipids, the *phospholipids* and the *sphingolipids*. The hydrophobic and hydrophilic domains are bridged by a glycerol moiety in the case of glycerophospholipids and by sphingosine in the case of sphingomyelin and the glycosphingolipids. In terms of location, triacylglycerol is confined largely to storage sites in adipose tissue, whereas the polar lipids occur primarily in cellular membranes. Membranes generally contain about 40% of their dry weight as lipid and 60% as protein, the two being held together by noncovalently bonded interactions. The detailed models of membrane structure and their functions are discussed in Chapter 5.

The processes of cell–cell recognition, phagocytosis, contact inhibition, and rejection of transplanted tissues and organs are all phenomena of medical significance that involve highly specific recognition sites on the surface of the plasma membrane. The synthesis of these complex glycosphingolipids that appear to play a role in these important biological events will be described.

The glycolipids are worthy of study because the ABO antigenic determinants of the blood groups are primarily glycolipid in nature. In addition, various sphingolipids are the storage substances that accumulate in the liver, spleen, kidney, or nervous tissue of persons suffering from certain genetic disorders called sphingolipidoses. In order to understand the basis of these enzyme-deficiency states, a knowledge of the relevant chemical structures involved is required.

A very important lipid is *cholesterol*. This chapter describes the pathway of cholesterol biosynthesis and its regulation and shows how cholesterol functions as a precursor to the bile salts and steroid hormones. Also described is the role of high density lipoprotein (HDL) and lecithin : cholesterol acyltransferase (LCAT) in the management of plasma cholesterol.

Finally, this chapter describes the metabolism and function of two pharmacologically powerful classes of hormones derived from arachidonic acid, namely the prostaglandins, the thromboxanes and the leukotrienes. See the Appendix, "Review of Organic Chemistry," for a discussion of the nomenclature and chemistry of lipids.

Figure 10.1
Stereospecific numbering of glycerol.

Figure 10.2
Structures of some common polar head groups of phospholipids.

10.2 PHOSPHOLIPIDS

The two principal classes of acylglycerolipids are the triacylglycerols and glycerophospholipids. They are referred to as glycerolipids because the core of these compounds is provided by the 3-carbon polyol, glycerol.

The two primary alcohol groups of glycerol are not stereochemically identical and in the case of the phospholipids, it is usually the same hydroxyl group that is esterified to the phosphate residue. The stereospecific numbering system is the best way to designate the different hydroxyl groups. In this system, when the structure of glycerol is drawn in the Fischer projection with the C-2 hydroxyl group projecting to the left of the page, the carbon atoms are numbered as shown in Figure 10.1. When the stereospecific numbering (sn) system is employed, the prefix "sn-" is used before the name of the compound. Glycerophospholipids usually contain a sn-glycerol 3-phosphate moiety. Although each contains the glycerol moiety as a fundamental structural element, the neutral triacylglycerols and the charged, ionic phospholipids have very different physical properties and functions. Let us now consider the structure and function of the various phospholipids.

Structure of Phospholipids

The phospholipids are polar, ionic lipids composed of 1,2-diacylglycerol and a phosphodiester bridge that links the glycerol backbone to some base, usually a nitrogenous one, such as choline, serine, or ethanolamine (Figures 10.2 and 10.3). The most abundant phospholipids in human tissues are phosphatidylcholine (also called lecithin), phosphatidylethanolamine, and phosphatidylserine. Note that C-2 of the phospholipids represents an asymmetric center (Figure 10.3). At physiologic pH, phosphatidylcholine and phosphatidylethanolamine have no net charge and exist as dipolar zwitterions, whereas phosphatidylserine has a net charge of −1, causing it therefore to be an acidic phospholipid. Note that phosphatidylethanolamine (PE) is related to phosphatidylcholine in that trimethylation of PE produces lecithin. Most phospholipids contain more than one kind of fatty acid per molecule, so that a given class of phospholipids from any tissue actually represents a family of molecular species. Phosphatidylcholine (PC) contains mostly palmitic acid (16 : 0) or stearic acid (18 : 0) in the sn-1 position and primarily the unsaturated 18-carbon fatty acids oleic, linoleic, or linolenic in the sn-2 position. Phosphatidylethanolamine has the same saturated fatty acids as PC at the sn-1 position but contains more of the long-chain polyunsaturated fatty acids—namely, 18 : 2, 20 : 4, and 22 : 6—at the sn-2 position.

Phosphatidylinositol is an acidic phospholipid that occurs in mammalian membranes (Figure 10.4). Phosphatidylinositol is rather unusual because it often contains almost exclusively stearic acid in the sn-1 position and arachidonic acid (20 : 4) in the sn-2 position.

Another phospholipid comprised of a polyol polar head group is phosphatidylglycerol. Phosphatidylglycerol (Figure 10.4) occurs in relatively large amounts in mitochondrial membranes and is a precursor of cardiolipin. Phosphatidylglycerol and phosphatidylinositol both carry a formal charge of −1 at neutral pH and are therefore acidic lipids.

The phospholipids that have been discussed so far contain only *O*-acyl residues attached to glycerol. *O*-(1-Alkenyl) substituents occur at C-1 of the sn-glycerol moiety of phosphoglycerides in combination with an *O*-acyl residue esterified to the C-2 position; compounds in this class are known as plasmalogens (Figure 10.5). Relatively large amounts of ethanolamine plasmalogen (also called plasmenylethanolamine) occur in myelin with lesser amounts in heart muscle where choline plasmalogen is abundant.

An unusual plasmalogen called "platelet activating factor" (PAF) is a major mediator of hypersensitivity, acute inflammatory reactions and anaphylactic shock. In hypersensitive individuals, cells of the polymorphonuclear (PMN) leukocyte family (basophils, neutrophils, and eosinophils) macrophages, and monocytes are coated with IgE molecules that are specific for a particular antigen (e.g., ragweed pollen, bee venom). Subsequent reexposure to the antigen and formation of antigen–IgE complexes on the surface of the aforementioned inflammatory cells provokes the synthesis and release of PAF, whose structure is shown in Figure 10.6. PAF is a choline plasmalogen containing an acetyl residue instead of

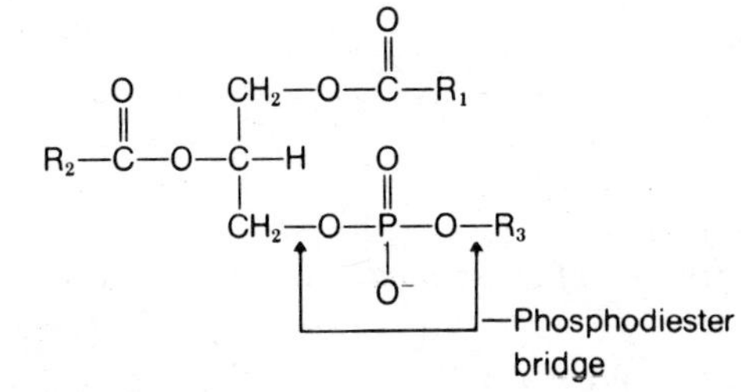

Phosphodiester bridge

Figure 10.3
Generalized structure of a phospholipid where R_1 and R_2 represent the aliphatic chains of fatty acids, and R_3 represents some polar head group.

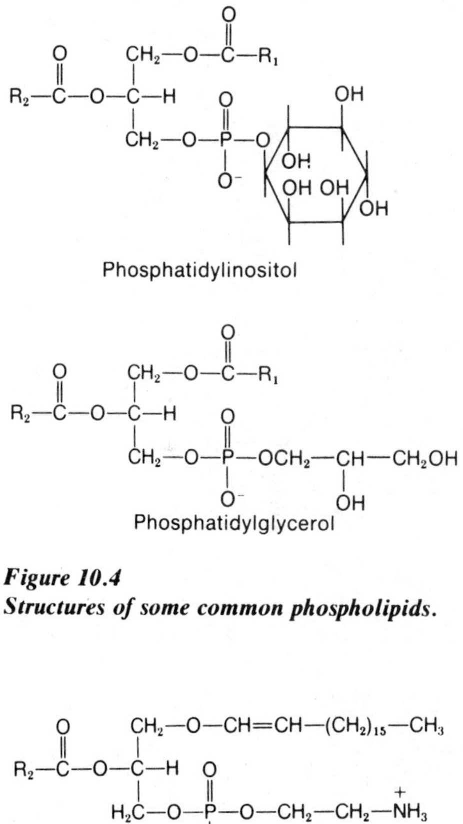

Phosphatidylinositol

Phosphatidylglycerol

Figure 10.4
Structures of some common phospholipids.

Figure 10.5
Structure of ethanolamine plasmalogen.

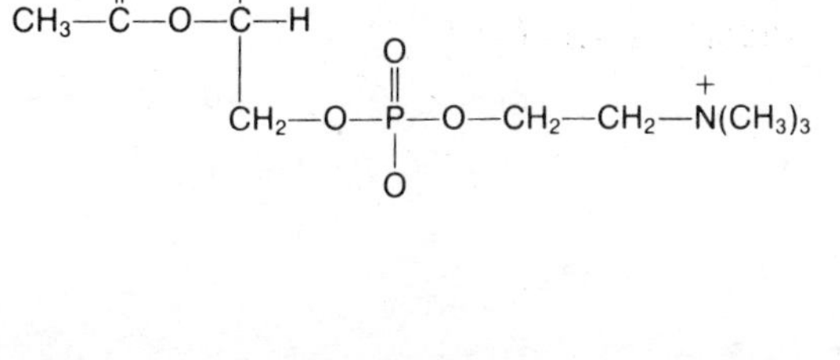

Figure 10.6
Structure of platelet activating factor (PAF).

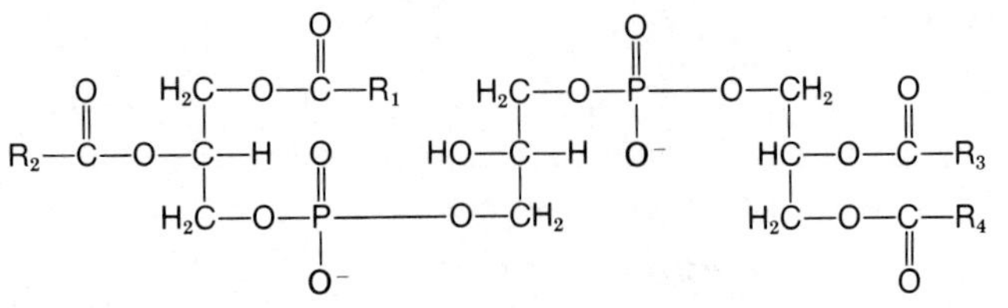

Figure 10.7
Structure of cardiolipin.

a long-chain fatty acid in position 2 of the glycerol moiety. PAF is not stored; it is synthesized and released when PMNs are stimulated. Platelet aggregation, cardiovascular and pulmonary changes, edema, hypotension, and PMN cell chemotaxis are affected by PAF.

Cardiolipin, a very acidic (charge, −2) phospholipid, is composed of two molecules of phosphatidic acid linked together covalently through a molecule of glycerol. Cardiolipin is found primarily in the inner membrane of mitochondria and in bacterial membranes (Figure 10.7).

Functions of Phospholipids

Although present in body fluids such as plasma and bile, the phospholipids are found in highest concentration in the various cellular membranes where they perform many different functions. Following is a description of some of the important functions of phospholipids.

As discussed in the introduction to this chapter, a major function of the phospholipids is to serve as structural components of membranes of the cell surface and subcellular organelles. For example, nearly half the mass of the erythrocyte membrane is comprised of various phospholipids. The amphipathic character of the phospholipids allows them to self-associate through hydrophobic or van der Waals interactions between the long-chain fatty acyl moieties in adjacent molecules in such a way that the polar head groups project outward toward water where they can interact with protein molecules.

Phospholipids also play a role in activating certain enzymes. β-Hydroxybutyrate dehydrogenase, a mitochondrial enzyme imbedded in the inner membrane of that organelle, catalyzes the reversible abstraction of electrons from β-hydroxybutyrate (see page 381). The enzyme has an absolute requirement for phosphatidylcholine. Phosphatidylserine and phosphatidylethanolamine cannot substitute for phosphatidylcholine in activating the enzyme.

Normal lung function depends upon a constant supply of an unusual phospholipid called dipalmitoyllecithin (Figure 10.8) in which the lecithin molecule contains palmitic acid (16 : 0) residues in both the sn-1 and sn-2 positions. More than 80% of the phospholipid in the extracellular liquid layer that lines alveoli of normal lungs is contributed by dipalmitoyllecithin. This particular phospholipid—called surfactant–is produced by type II epithelial cells and prevents atelectasis at the end of the expiration phase of breathing (Figure 10.9). This lipid derives its name from its powerful capacity to decrease the surface tension of the aqueous surface layer of the lung. Lecithin molecules that do not contain two residues of

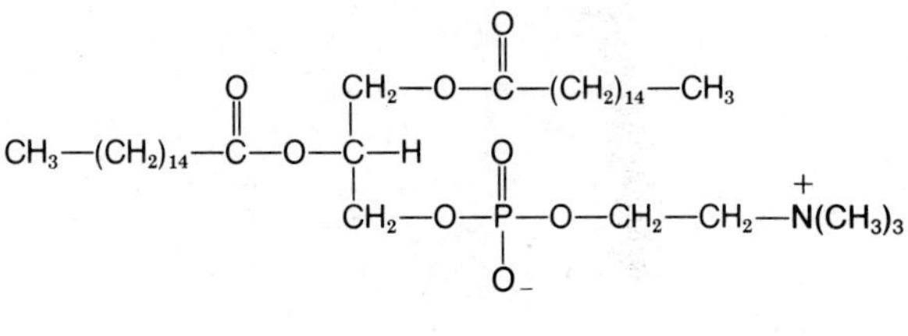

Figure 10.8
Structure of dipalmitoyllecithin.

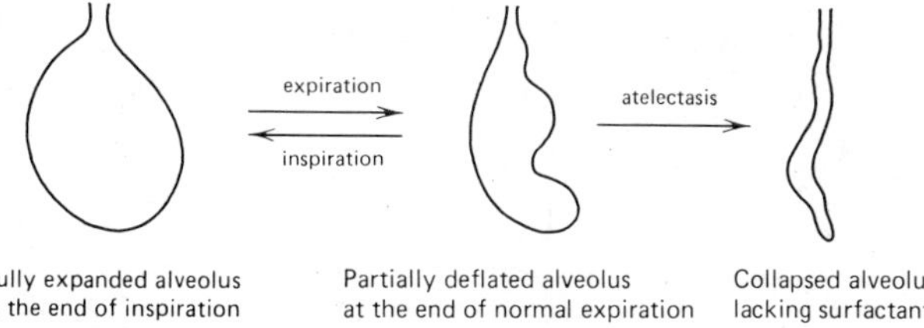

Figure 10.9
Role of surfactant in preventing atelectasis.

palmitic acid are not effective in lowering the surface tension of the fluid layer lining alveoli.

During the third trimester–before the twenty-eighth week of gestation–the fetal lung is synthesizing primarily sphingomyelin. Normally, at this time, glycogen that has been stored in epithelial type II cells is converted to fatty acids and then to dipalmitoyllecithin. During lung maturation there is a good correlation between the increase in lamellar inclusion bodies that represent the intracellular pulmonary surfactant (phosphatidylcholine) storage organelles, called lamellar bodies, and the simultaneous decrease in glycogen content of type II pneumocytes. At the twenty-fourth week of gestation the type II granular pneumocytes appear in the alveolar epithelium, and within a few days they produce their typical osmiophilic lamellar inclusion bodies. The number of type II cells increases until the 32nd week at which time surface active agent appears in the lung and amniotic fluid. Surface tension decreases when the inclusion bodies increase in the type II cells. In the few weeks before term one can perform screening tests on amniotic fluid to detect newborns that are at risk for RDS (respiratory distress syndrome) (Clin. Corr. 10.1). These tests are useful in timing elective deliveries, in applying vigorous preventive therapy to the newborn infant and to determine if the mother should be treated with a glucocorticoid drug such as beta-methasone to accelerate maturation of the fetal lung.

Respiratory failure due to an insufficiency in surfactant can also occur in adults whose type II cells or surfactant-producing pneumocytes have been destroyed as an adverse side effect of the use of immunosuppressive medications or chemotherapeutic drugs.

The detergent properties of the phospholipids, especially phosphatidylcholine, play an important role in bile where they function to solubilize cholesterol. An impairment in phospholipid production and secretion into bile can result in the formation of cholesterol and bile pigment gallstones.

Phosphatidylinositol and phosphatidylcholine also serve as donors of arachidonic acid for the synthesis of prostaglandins, thromboxanes, leukotrienes and related compounds.

Biosynthesis of Phospholipids

Phosphatidic Acid Synthesis

L-α-Phosphatidic acid (commonly called phosphatidic acid) and sn-1,2-diglyceride (1,2-diacyl-sn-glycerol) are common intermediates in the pathways of phospholipid and triacylglycerol biosynthesis (Figure 10.10). Furthermore, the biosynthesis of triacylglycerols proceeds by way of a pathway comprised of enzymes shared by the pathway of phospholipid synthesis (see Chapter 9). Essentially all cells are capable of synthesizing phospholipids to some degree (except mature erythrocytes), whereas triacylglycerol biosynthesis occurs only in the liver, adipose tissue, and the intestine. In most tissues, the pathway for phosphatidic acid synthesis begins with α-glycerophosphate (sn-glycerol 3-phosphate), and there are two sources of this triose phosphate. The most general source of α-glycerophosphate, particularly in adipose tissue, is from reduction of the

CLIN. CORR. **10.1**
RESPIRATORY DISTRESS SYNDROME

The respiratory distress syndrome (RDS) is a major cause of neonatal morbidity and mortality in many countries. It accounts for approximately 15–20% of all neonatal deaths in Western countries and somewhat less in the developing countries. The disease affects only premature babies and its incidence varies directly with the degree of prematurity. Premature babies develop RDS because of immaturity of their lungs, resulting from a deficiency of pulmonary surfactant. The maturity of the fetal lung can be predicted antenatally by measuring the lecithin/sphingomyelin (L/S) ratio in the amniotic fluid. The mean L/S ratio in normal pregnancies increases gradually with gestation until about 31 or 32 weeks when the slope rises sharply. The ratio of 2.0 that is characteristic of the term infant at birth is achieved at the gestational age of about 34 weeks. In terms of predicting pulmonary maturity, the critical L/S ratio is 2.0 or greater. The risk of developing RDS when the L/S ratio is less than 2.0 has been worked out: for an L/S ratio of 1.5–1.9, the risk is approximately 40%, and for a ratio less than 1.5 the calculated risk of developing RDS is about 75%.

Although the L/S ratio in amniotic fluid is still widely used to predict the risk of RDS, the results are unreliable if the amniotic fluid specimen has been contaminated by blood or meconium obtained during a complicated pregnancy. In recent years the determination of saturated palmitoylphosphatidylcholine (SPC) has been found to be more specific and a more sensitive predictor of the RDS than the L/S ratio. There are rapid techniques for the isolation of saturated dipalmitoylphosphatidylcholine that can be used to assay the surfactant content in amniotic fluid.

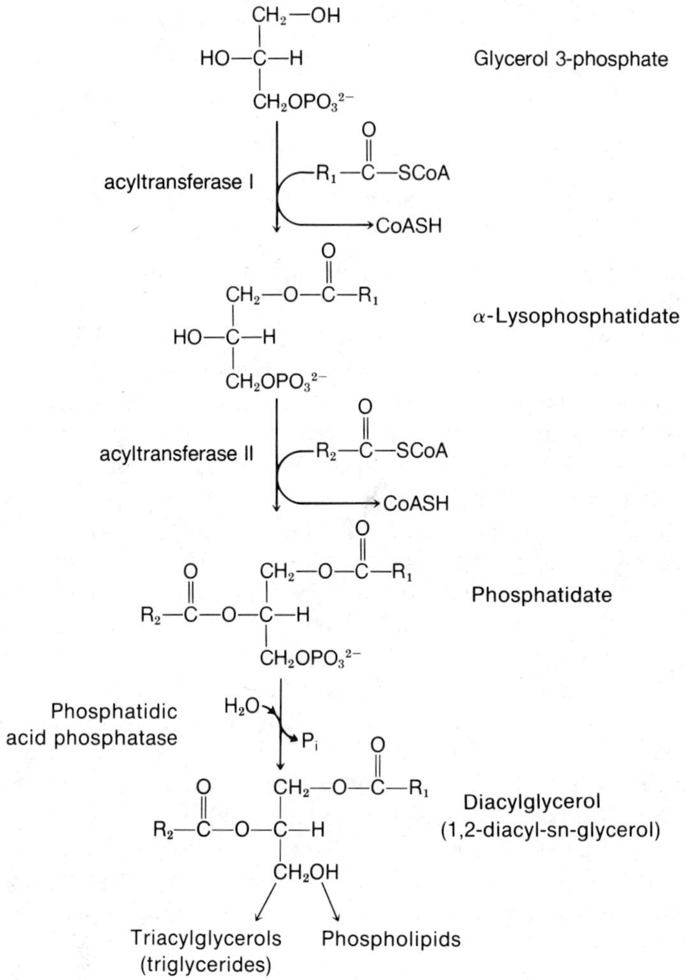

$CH_2{-}OH$ / $HO{-}C{-}H$ / $CH_2OPO_3^{2-}$ Glycerol 3-phosphate

acyltransferase I: $R_1{-}C(=O){-}SCoA$ → CoASH

$CH_2{-}O{-}C(=O){-}R_1$ / $HO{-}C{-}H$ / $CH_2OPO_3^{2-}$ α-Lysophosphatidate

acyltransferase II: $R_2{-}C(=O){-}SCoA$ → CoASH

$CH_2{-}O{-}C(=O){-}R_1$ / $R_2{-}C(=O){-}O{-}C{-}H$ / $CH_2OPO_3^{2-}$ Phosphatidate

Phosphatidic acid phosphatase: H_2O → P_i

$CH_2{-}O{-}C(=O){-}R_1$ / $R_2{-}C(=O){-}O{-}C{-}H$ / CH_2OH Diacylglycerol (1,2-diacyl-sn-glycerol)

Triacylglycerols (triglycerides) Phospholipids

Figure 10.10
Phosphatidic acid biosynthesis from glycerol 3-phosphate and the role of phosphatidic acid phosphatase in the synthesis of phospholipids and triacylglycerols.

glycolytic intermediate, dihydroxyacetone phosphate, in the reaction catalyzed by α-glycerophosphate dehydrogenase:

$$\text{Dihydroxyacetone phosphate} + \text{NADH} + \text{H}^+ \rightleftarrows \text{glycerol 3-phosphate} + \text{NAD}^+$$

A few specialized tissues, including the liver, kidney, and intestine, derive α-glycerophosphate by means of the glycerol kinase reaction:

$$\text{Glycerol} + \text{ATP} \xrightarrow{Mg^{2+}} \text{glycerol 3-phosphate} + \text{ADP}$$

The next two steps in phosphatidic acid biosynthesis involve stepwise transfer of long-chain fatty acyl groups from the activated donor, fatty acyl CoA. The first acyltransferase (I) is called glycerol phosphate : acyltransferase and attaches predominantly saturated fatty acids and oleic acid to the sn-1 position to produce 1-acylglycerol phosphate or α-lysophosphatidic acid. The second enzyme (II), 1-acylglycerol phosphate : acyltransferase, catalyzes the acylation of the sn-2 position, usually with an unsaturated fatty acid (Figure 10.10). The high-energy, highly reactive donor of acyl groups is the coenzyme A thioester derivative of the long-chain fatty acids.

The specificity of the two acyltransferases does not always match the fatty acid asymmetry that we find in the phospholipids of a particular cell. Remodeling reactions discussed below function to modify the fatty acid composition at the C-1 and C-2 positions of the glycerol phosphate backbone.

Cytoplasmic phosphatidic acid phosphatase (also called phosphatidic acid phosphohydrolase) hydrolyzes phosphatidic acid (1,2-diacylglycerophosphate) that is generated on the endoplasmic reticulum, thereby yielding sn-1,2-diacylglycerol that serves as the branch point in triacylglycerol and phospholipid synthesis (Figure 10.10).

Phosphatidic acid can also be formed by a second pathway that begins with dihydroxyacetone phosphate (DHAP). The DHAP pathway is usually an alternative supportive route used by some tissues to produce phosphatidic acid (see Chapter 9).

Biosynthesis of Specific Phospholipids

The major pathway for the biosynthesis of phosphatidylcholine (lecithin) involves the sequential conversion of choline to phosphocholine, CDP-choline and phosphatidylcholine. In this pathway, the phosphocholine polar head group is activated using CTP, according to the following reactions. Free choline is first phosphorylated by ATP in a reaction catalyzed by choline kinase (Figure 10.11). Choline is a dietary requirement for most mammals. Phosphocholine in turn is converted to CDP-choline at the expense of CTP in the reaction catalyzed by phosphocholine cytidylyltransferase. Note that inorganic pyrophosphate (PP_i) is a product of this reaction resulting from attack by the phosphoryl residue of phosphocholine on the internal α-phosphorus atom of CTP. The high energy pyrophosphoryl bond in CDP-choline is very unstable and reactive such that the phosphocholine moiety can be transferred readily to the nucleophilic center provided by the OH group at position 3 of 1,2-diacylglycerol in the reaction catalyzed by choline phosphotransferase shown in Figure 10.12. This is the principal pathway for the synthesis of dipalmitoyllecithin in the lung.

The rate-limiting step for phosphatidylcholine biosynthesis is the cytidylyltransferase reaction that forms CDP-choline (Figure 10.11). This enzyme is regulated by a novel mechanism that involves exchange of the enzyme between the cytosol and the endoplasmic reticulum. The cytosolic form of cytidylyltransferase is inactive and appears to function as a reservoir of enzyme; binding of the enzyme to the membrane results in activation. Translocation of cytidylyltransferase from the cytosol to the

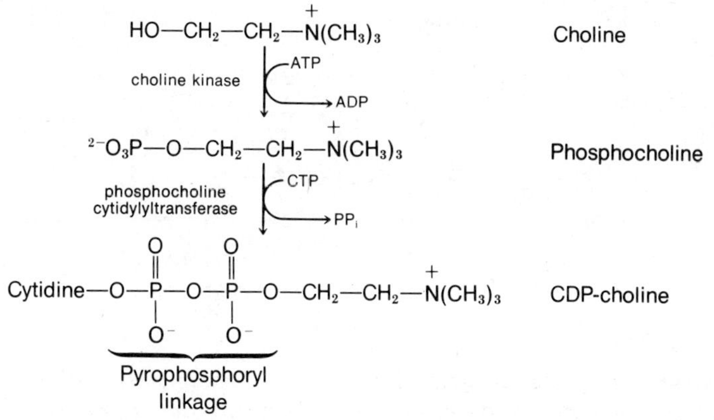

$HO-CH_2-CH_2-\overset{+}{N}(CH_3)_3$ Choline

choline kinase: ATP → ADP

$^{2-}O_3P-O-CH_2-CH_2-\overset{+}{N}(CH_3)_3$ Phosphocholine

phosphocholine cytidylyltransferase: CTP → PP_i

$$\text{Cytidine}-O-\underset{O^-}{\overset{O}{\overset{\|}{\underset{|}{P}}}}-O-\underset{O^-}{\overset{O}{\overset{\|}{\underset{|}{P}}}}-O-CH_2-CH_2-\overset{+}{N}(CH_3)_3$$ CDP-choline

Pyrophosphoryl linkage

Figure 10.11
The biosynthesis of CDP-choline from choline.

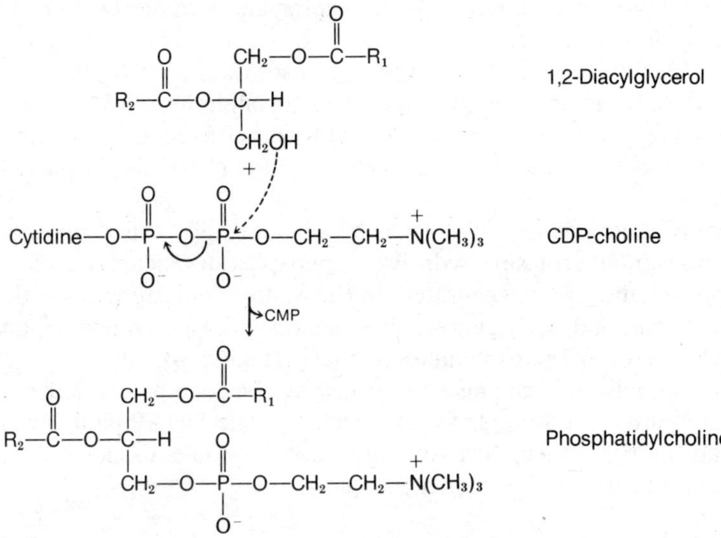

1,2-Diacylglycerol

CDP-choline

CMP

Phosphatidylcholine

Figure 10.12
The choline phosphotransferase reaction.

endoplasmic reticulum is regulated by cAMP and fatty acids. Reversible phosphorylation in which a cAMP-dependent kinase phosphorylates the enzyme causes it to be released from the membrane, rendering it inactive. Subsequent dephosphorylation will cause the cytidylyltransferase to rebind to the membrane and become active. Fatty acyl CoAs activate the enzyme by promoting binding to the endoplasmic reticulum.

In liver only, phosphatidylcholine can also be formed by repeated methylation of the phospholipid phosphatidylethanolamine. Phosphatidylethanolamine *N*-methyltransferase, a microsomal enzyme, catalyzes the transfer of methyl groups–one at a time–from S-adenosylmethionine (AdoMet) to phosphatidylethanolamine to produce phosphatidylcholine (Figure 10.13). It is not known if one or more enzymes are involved in the conversion of phosphatidylethanolamine to phosphatidylcholine.

The primary pathway for phosphatidylethanolamine synthesis in liver and brain involves the microsomal enzyme ethanolamine phosphotransferase that catalyzes the reaction shown in Figure 10.14. This enzyme is particularly abundant in liver. CDP-ethanolamine is formed through the reaction catalyzed by ethanolamine kinase:

$$\text{Ethanolamine} + \text{ATP} \xrightarrow{Mg^{2+}} \text{phosphoethanolamine} + \text{ADP}$$

and the phosphoethanolamine cytidylyltransferase reaction:

$$\text{Phosphoethanolamine} + \text{CTP} \xrightarrow{Mg^{2+}} \text{CDP-ethanolamine} + \text{PP}_i$$

Liver mitochondria can also generate phosphatidylethanolamine by decarboxylation of phosphatidylserine; however, this is thought to repre-

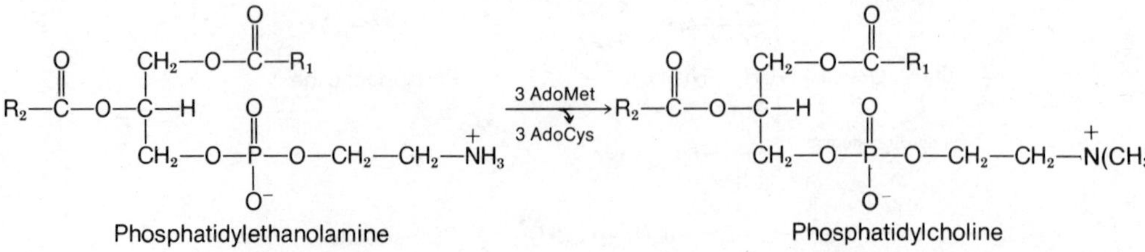

3 AdoMet

3 AdoCys

Phosphatidylethanolamine

Phosphatidylcholine

Figure 10.13
Biosynthesis of phosphatidylcholine from phosphatidylethanolamine and S-adenosylmethionine (AdoMet); S-adenosylhomocysteine (AdoCys).

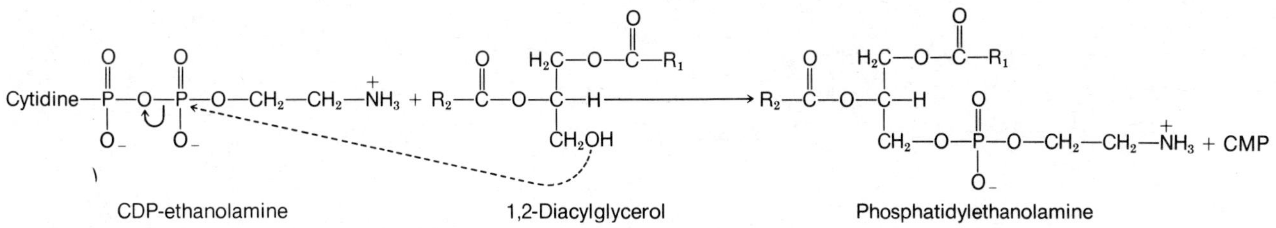

Cytidine—P(=O)(O⁻)—O—P(=O)(O⁻)—O—CH_2—CH_2—$\overset{+}{N}H_3$ + R_2—C(=O)—O—C(H)(—H_2C—O—C(=O)—R_1)—CH_2OH ⟶ R_2—C(=O)—O—C(H)(—H_2C—O—C(=O)—R_1)—CH_2—O—P(=O)(O⁻)—O—CH_2—CH_2—$\overset{+}{N}H_3$ + CMP

CDP-ethanolamine 1,2-Diacylglycerol Phosphatidylethanolamine

Figure 10.14
Biosynthesis of phosphatidylethanolamine from CDP-ethanolamine and diacylglycerol; the reaction is catalyzed by ethanolamine phosphotransferase.

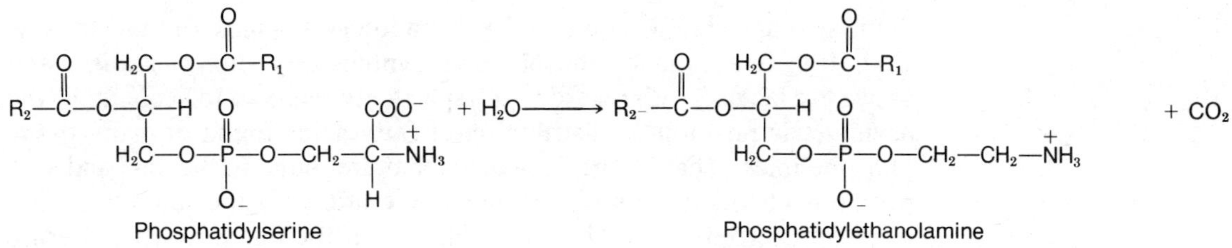

Phosphatidylserine + H_2O ⟶ Phosphatidylethanolamine + CO_2

Phosphatidylserine Phosphatidylethanolamine

Figure 10.15
Formation of phosphatidylethanolamine by the decarboxylation of phosphatidylserine.

sent only a minor pathway in phosphatidylethanolamine synthesis (Figure 10.15).

The major source of phosphatidylserine in mammalian tissues is provided by the "base-exchange" reaction shown in Figure 10.16 in which the polar head group of phosphatidylethanolamine is exchanged for the amino acid serine; since there is no net change in the number or kinds of bonds, this reaction is reversible and has no requirement for ATP or any other high-energy compound. The reaction is initiated by attack on the phosphodiester bond of phosphatidylethanolamine by the hydroxyl group of serine.

Phosphatidylinositol is made via CDP-diacylglycerol and free myo-inositol (Figure 10.17) in a reaction catalyzed by microsomal phosphatidylinositol synthase.

Remodeling Reactions: Role of Phospholipases in Phospholipid Synthesis

The activities of two phospholipases, phospholipase A_1 and phospholipase A_2, occur in many tissues, and each plays a role in the formation of specific phospholipid structures containing the proper kinds of fatty acids

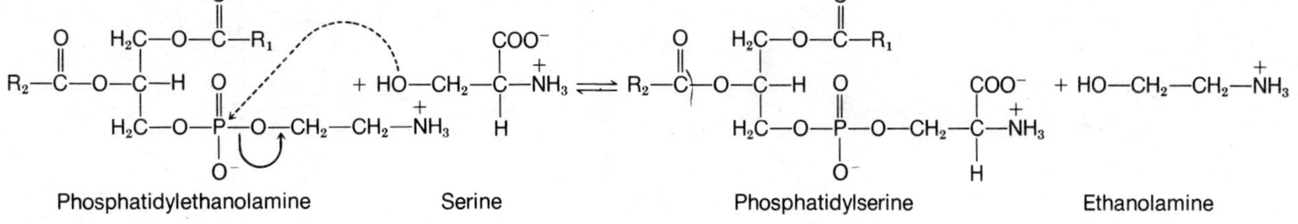

Phosphatidylethanolamine + HO—CH_2—CH(COO^-)—$\overset{+}{N}H_3$ ⇌ Phosphatidylserine + HO—CH_2—CH_2—$\overset{+}{N}H_3$

Phosphatidylethanolamine Serine Phosphatidylserine Ethanolamine

Figure 10.16
Biosynthesis of phosphatidylserine from serine and phosphatidylethanolamine by "base exchange."

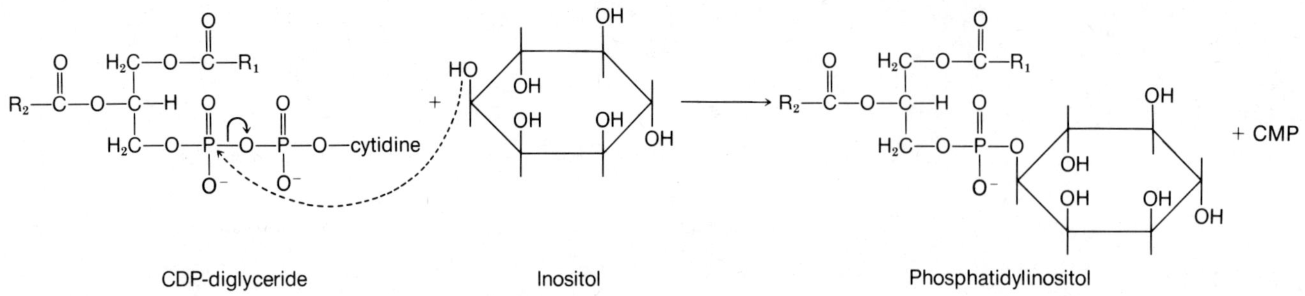

Figure 10.17
Biosynthesis of phosphatidylinositol.

in the sn-1 and sn-2 positions. It has been found that most of the fatty acyl CoA transferases and phospholipid synthesizing enzymes discussed above are lacking with regard to the specificity required to account for the asymmetric position or distribution of fatty acids found in many tissue phospholipids. That is, the fatty acids that are found in the sn-1 and sn-2 positions of the various phospholipids are often not the same ones that were transferred to the glycerol backbone in the initial acyl transferase reactions of the phospholipid biosynthesis pathways. Phospholipases A_1 and A_2 catalyze the reactions indicated in Figure 10.18 where X represents the polar head group of a phospholipid. The products of the action of phospholipases A_1 and A_2 are called lysophosphatides.

For example, if it becomes necessary for a cell to remove some undesired fatty acid, such as stearic acid from the sn-2 position of phosphatidylcholine, and replace it by a more unsaturated one like arachidonic acid, then this can be accomplished by the action of phospholipase A_2 followed by a reacylation step. The insertion of arachidonic acid into the 2 position of sn-2-lysophosphatidylcholine can then be accomplished by one of two means; either by direct acylation from arachidonyl CoA (Figure 10.19) or from some other arachidonic acid-containing phospholipid by an exchange-type reaction (Figure 10.20). Remodeling by direct acylation would be accomplished by an arachidonic acid-specific acyl CoA transacylase. The lysolecithin exchange reaction is catalyzed by lysolecithin : lecithin acyltransferase (LLAT) (Figure 10.20). Note that, since there is no change in either the number or nature of the bonds involved in products and reactants, there is no ATP requirement for this acylation reaction. Reacylation of lysophosphatidylcholine is the major route for the remodeling of phosphatidylcholine.

Lysophospholipids, particularly sn-1-lysophosphatidylcholine, can also serve as sources of fatty acid in the remodeling reactions. For example, the following diagram summarizes the remodeling reactions that

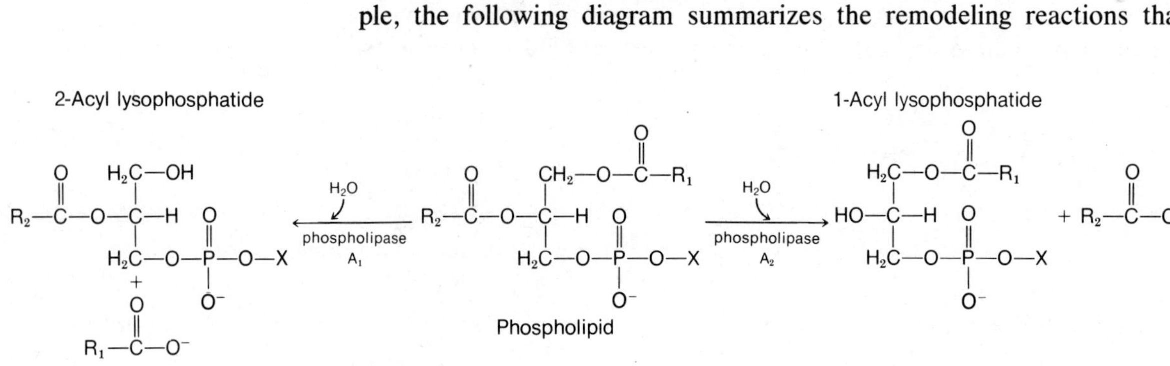

Figure 10.18
Reactions catalyzed by phospholipase A_1 and phospholipase A_2.

$$\underset{\text{Lysophosphatidylcholine}}{\begin{array}{l}CH_2{-}O{-}\overset{O}{\overset{\|}{C}}{-}R_1\\ HO{-}C{-}H\\ CH_2{-}O{-}\underset{O^-}{\overset{O}{\overset{\|}{P}}}{-}O{-}CH_2{-}CH_2{-}\overset{+}{N}(CH_3)_3\end{array}} \xrightarrow[\text{CoASH}]{\text{arachidonyl S-CoA}} \underset{\text{Phosphatidylcholine}}{\begin{array}{l}\qquad\qquad\quad CH_2{-}O{-}\overset{O}{\overset{\|}{C}}{-}R_1\\ R_2{-}\overset{O}{\overset{\|}{C}}{-}O{-}C{-}H\\ \qquad\qquad\quad CH_2{-}O{-}\underset{O^-}{\overset{O}{\overset{\|}{P}}}{-}O{-}CH_2{-}CH_2{-}\overset{+}{N}(CH_3)_3\end{array}}$$

Figure 10.19
Synthesis of phosphatidylcholine by reacylation of lysophosphatidylcholine where $R_2{-}\overset{O}{\overset{\|}{C}}{-}O{-}$ ***represents arachidonic acid. This reaction is catalyzed by acyl-CoA: 1-acylglycerol-3-phosphocholine O-acyltransferase.***

$$\underset{\text{Lysophosphatidylcholine}}{\begin{array}{l}CH_2{-}O{-}\overset{O}{\overset{\|}{C}}{-}R_1\\ HO{-}C{-}H\\ CH_2{-}O{-}\underset{O^-}{\overset{O}{\overset{\|}{P}}}{-}O{-}CH_2{-}CH_2{-}\overset{+}{N}(CH_3)_3\end{array}} + \underset{\text{Phosphatidylethanolamine}}{\begin{array}{l}\qquad\qquad\quad CH_2{-}O{-}\overset{O}{\overset{\|}{C}}{-}R_1\\ R_2{-}\overset{O}{\overset{\|}{C}}{-}O{-}C{-}H\\ \qquad\qquad\quad CH_2{-}O{-}\underset{O^-}{\overset{O}{\overset{\|}{P}}}{-}O{-}CH_2{-}CH_2{-}\overset{+}{N}H_3\end{array}} \rightleftharpoons$$

$$\underset{\text{Phosphatidylcholine}}{\begin{array}{l}\qquad\qquad\quad CH_2{-}O{-}\overset{O}{\overset{\|}{C}}{-}R_1\\ R_2{-}\overset{O}{\overset{\|}{C}}{-}O{-}C{-}H\\ \qquad\qquad\quad CH_2{-}O{-}\underset{O^-}{\overset{O}{\overset{\|}{P}}}{-}O{-}CH_2{-}CH_2{-}\overset{+}{N}(CH_3)_3\end{array}} + \underset{\text{Lysophosphatidylethanolamine}}{\begin{array}{l}CH_2{-}O{-}\overset{O}{\overset{\|}{C}}{-}R_1\\ HO{-}C{-}H\\ CH_2{-}O{-}\underset{O^-}{\overset{O}{\overset{\|}{P}}}{-}O{-}CH_2{-}CH_2{-}\overset{+}{N}H_3\end{array}}$$

Figure 10.20
Formation of phosphatidylcholine by lysolecithin exchange, where $R_2{-}\overset{O}{\overset{\|}{C}}{-}O{-}$ ***represents arachidonic acid.***

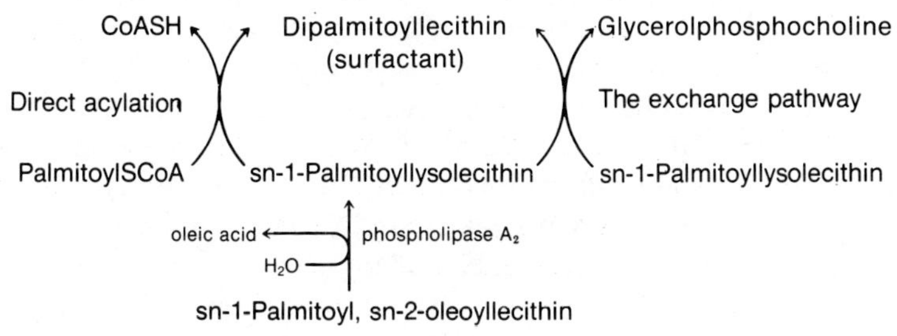

Figure 10.21
Two pathways for the biosynthesis of dipalmitoyllecithin from sn-1 palmitoyllysolecithin.

might be involved in the synthesis of dipalmitoyllecithin (surfactant) from 1-palmitoyl-2-oleoylphosphatidylcholine. Note that sn-1-palmitoyl lysolecithin is the source of palmitic acid in the acyltransferase exchange reaction (Figure 10.21).

10.3 CHOLESTEROL

Structure and Function of Cholesterol

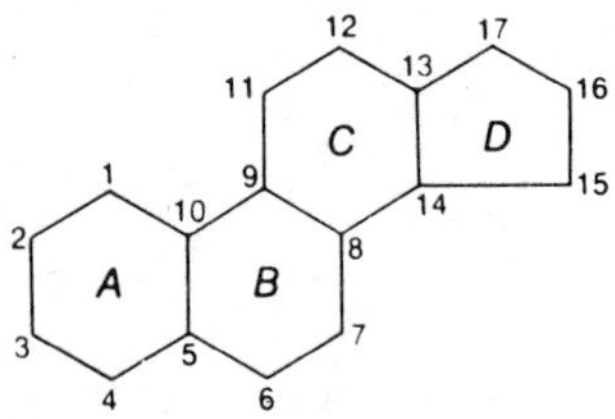

Cholesterol is an alicyclic compound whose structure includes (1) the perhydrocyclopentanophenanthrene nucleus with its four fused rings, (2) a single hydroxyl group at C-3, (3) an unsaturated center between carbon atoms 5 and 6, (4) an eight-membered branched hydrocarbon chain attached to the D ring at position 17, and (5) a methyl group (designated C-19) attached at position 10 and another methyl group (designated C-18) attached at position 13. (See Figures 10.22 and 10.23).

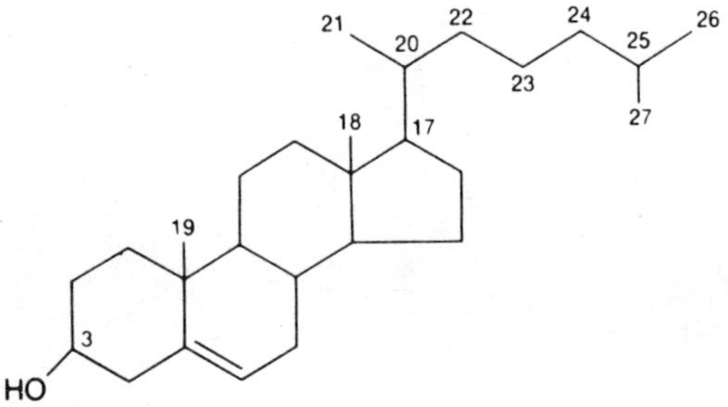

Figure 10.22
The cyclopentanophenanthrene ring.

In terms of physical properties, cholesterol is a lipid with very low solubility in water; at 25°C, the limit of solubility is approximately 0.2 mg/100 ml, or 4.7 μM. The actual concentration of cholesterol in plasma of healthy people is usually 150–200 mg/dl: on a milligram basis, this value is almost twice the normal concentration of blood glucose. The very high solubility of cholesterol in blood is due to the presence of proteins called plasma lipoproteins (mainly LDL and VLDL) that have the ability to bind and thereby solubilize large amounts of cholesterol (see page 97).

Figure 10.23
Structure of cholesterol (cholest-5-en-3β-ol).

Actually, only about 30% of the total circulating cholesterol occurs free as such; approximately 70% of the cholesterol in plasma lipoproteins exists in the form of cholesterol esters where some long-chain fatty acid, usually linoleic acid, is attached by an ester bond to the OH—group on carbon-3 of the A ring. The presence of the long-chain fatty acid residue enhances the hydrophobicity of cholesterol (Figure 10.24).

Cholesterol is also abundant in bile where the normal concentration is 390 mg/dl. In contrast to the finding of predominantly cholesterol esters in plasma, only 4% of the cholesterol in bile is esterified to some long-chain fatty acid. Bile does not contain appreciable amounts of any of the lipoproteins and the solubilization of free cholesterol is achieved in part by the detergent property of phospholipids present in bile that are produced in the liver. A chronic disturbance in phospholipid metabolism in the liver can result in the deposition of cholesterol-rich gallstones. Bile salts, which are derivatives of cholesterol, also aid in keeping cholesterol in solution in bile.

In the clinical laboratory total cholesterol is estimated by the Liebermann–Burchard reaction. The proportions of free and esterified cholesterol can be determined by gas–liquid chromatography.

Cholesterol, which can be derived from the diet or manufactured de novo in virtually all the cells of humans, plays a number of important roles. It is the major sterol in humans and a component of virtually all cell surfaces and intracellular membranes. Cholesterol is especially abundant in the myelinated structures of the brain and central nervous system but is present in small amounts in the inner membrane of the mitochondrion. In contrast to the situation in plasma, most of the cholesterol in cellular membranes occurs in the free, unesterified form.

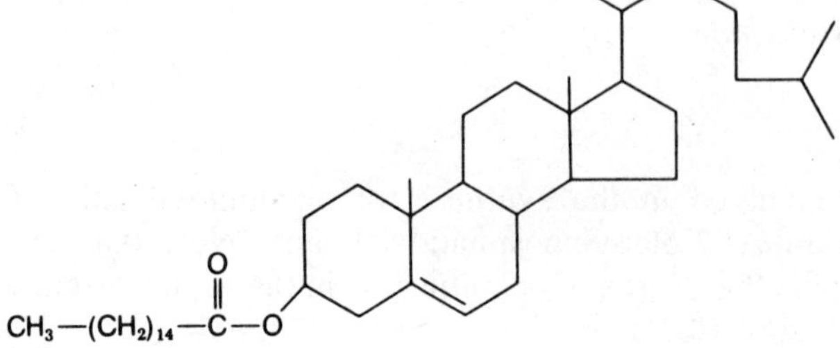

$CH_3—(CH_2)_{14}—\overset{O}{\overset{\|}{C}}—O$

Figure 10.24
Structure of cholesterol (palmitoyl-) ester.

The second role of cholesterol is as the immediate precursor of the bile acids that are synthesized in the liver and that function to facilitate the absorption of dietary triacylglycerols and fat-soluble vitamins (Chapter 24). It is important to realize that the ring structure of cholesterol cannot be metabolized to CO_2 and water in humans. The route of excretion of cholesterol is by way of the liver and gallbladder through the intestine in the form of bile acids.

The third physiological role of cholesterol is as the precursor of the various steroid hormones (Chapter 15). Progesterone is the 21-carbon keto steroid sex hormone secreted by the corpus luteum of the ovary. The metabolically powerful corticosteroids of the adrenal cortex are derived from cholesterol; these include deoxycorticosterone, corticosterone, cortisol, and cortisone. The mineralocorticoid aldosterone is derived from cholesterol in zona glomerulosa tissue of the cortex of the adrenal gland. Cholesterol also serves as the precursor to the female steroid hormones, the estrogens (e.g., estradiol) in the ovary and to the male steroids (e.g., testosterone) in the testes.

Although all of the steroid hormones are structurally related to and biochemically derived from cholesterol, they have widely different physiological properties that relate to spermatogenesis, pregnancy, lactation and parturition, mineral balance, and energy (amino acids, carbohydrate and fat) metabolism. Their metabolism and functions will be discussed in Chapter 15.

The hydrocarbon skeleton of cholesterol is also found in the plant sterols, for example ergosterol, a precursor to vitamin D (Figure 10.25). Ergosterol is converted in the skin by ultraviolet irradiation to vitamin D_3 (cholecalciferol). Vitamin D_3 is involved in calcium and phosphorus metabolism. (Chapter 26).

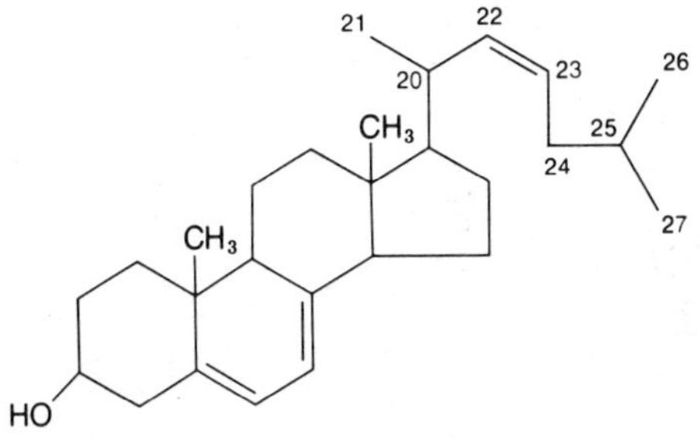

21 22 20 23 26 CH_3 25 24 27 CH_3 HO

Figure 10.25
Structure of ergosterol.

Cholesterol Biosynthesis

Although de novo biosynthesis of cholesterol occurs in virtually all cells, this capacity is greatest in liver, intestine, adrenal cortex, and reproductive tissues, including ovaries, testes, and placenta. From an inspection of its structure it is apparent that cholesterol biosynthesis will require a source of carbon atoms and considerable reducing power to generate the numerous carbon–hydrogen and carbon–carbon bonds. All of the carbon atoms of cholesterol are derived from acetate. Reducing power in the form of NADPH is provided mainly by enzymes of the hexose monophosphate shunt, specifically, glucose 6-phosphate dehydrogenase and 6-phosphogluconate dehydrogenase (page 331). It should be remembered that for each glucose molecule oxidized by way of the hexose monophosphate shunt, 2 equiv of NADPH are produced. With regard to the requirement for high-energy bond containing compounds, the pathway of cholesterol synthesis is driven in large part by the hydrolysis of the high-energy thioester bonds of acetyl CoA and the high-energy phosphoanhydride bonds of ATP. Realizing that cholesterol biosynthesis occurs in the cytoplasm of cells, let us consider the individual steps in the pathway.

Formation of Mevalonic Acid from Acetate

The first compound unique to the pathway of cholesterol biosynthesis is mevalonic acid. Mevalonic acid is derived from the two-carbon precursor acetyl CoA that is located at the hub of the pathways of fat, carbohydrate, and amino acid metabolism. Acetyl CoA can be obtained from several sources: (1) the β-oxidation of long-chain fatty acids (Chapter 9); (2) the oxidation of ketogenic amino acids such as leucine and isoleucine (Chapter 12); and (3) the pyruvate dehydrogenase reaction that links glycolysis

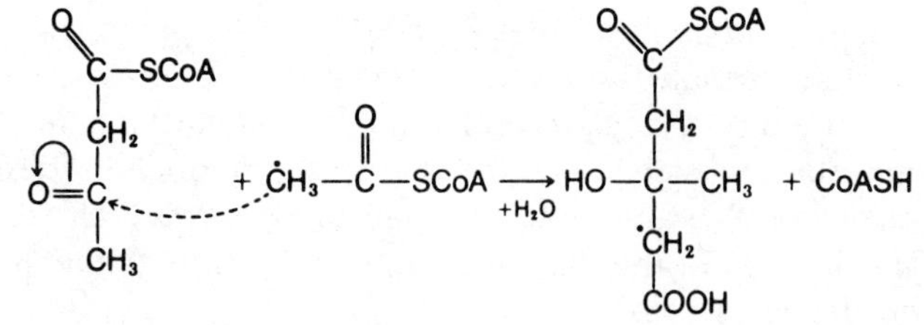

$$\underset{\text{Acetoacetyl CoA}}{CH_3-CO-CH_2-CO-SCoA} + \underset{\text{Acetyl CoA}}{CH_3-\overset{O}{\overset{\|}{C}}-SCoA} \xrightarrow[+H_2O]{} \underset{\text{HMG CoA}}{HOOC-CH_2-C(OH)(CH_3)-CH_2-CO-SCoA} + CoASH$$

Figure 10.26
The HMG CoA synthase reaction.

and the TCA cycle (Chapter 6). In addition, free acetate can be activated to its thioester derivative at the expense of ATP by the enzyme acetokinase, which is also referred to as acetate thiokinase:

$$ATP + CH_3COO^- + CoASH \longrightarrow CH_3-\overset{O}{\overset{\|}{C}}-SCoA + AMP + PP_i$$

The first two steps in the pathway of cholesterol synthesis are shared by the pathway that also produces ketone bodies (Chapter 9). Two molecules of acetyl CoA condense to form acetoacetyl CoA in a reaction catalyzed by acetoacetyl CoA thiolase (acetyl CoA : acetyl CoA acetyltransferase):

$$CH_3-\overset{O}{\overset{\|}{C}}-SCoA + CH_3-\overset{O}{\overset{\|}{C}}-SCoA \longrightarrow CH_3-\overset{O}{\overset{\|}{C}}-CH_2-\overset{O}{\overset{\|}{C}}-SCoA + CoASH$$

Note that the formation of the carbon–carbon bond in acetoacetyl CoA in this reaction is favored energetically by the cleavage of a thioester bond and the generation of free coenzyme A (CoASH).

The next step introduces a third molecule of acetyl CoA into the cholesterol pathway and forms the branched-chain compound 3-hydroxy-3-methylglutaryl CoA (HMG CoA) (Figure 10.26). This condensation reaction is catalyzed by HMG CoA synthase (3-hydroxy-3-methylglutaryl CoA : acetoacetyl CoA lyase). Liver parenchymal cells contain two isoenzyme forms of HMG CoA synthase; one is found in the cytosol and is involved in cholesterol synthesis, while the other has a mitochondrial location and functions in the pathway that forms ketone bodies (Chapter 9). In the HMG CoA synthase reaction, an aldol condensation occurs between the methyl carbon of acetyl CoA and the β-carbonyl group of acetoacetyl CoA with the simultaneous hydrolysis of the thioester bond of acetyl CoA. Note that the thioester bond in the original acetoacetyl CoA substrate molecule remains intact.

HMG CoA can also be formed from the oxidative degradation of the branched-chain amino acid leucine, which proceeds through the intermediates 3-methylcrotonyl CoA and 3-methylglutaconyl CoA (Chapter 12).

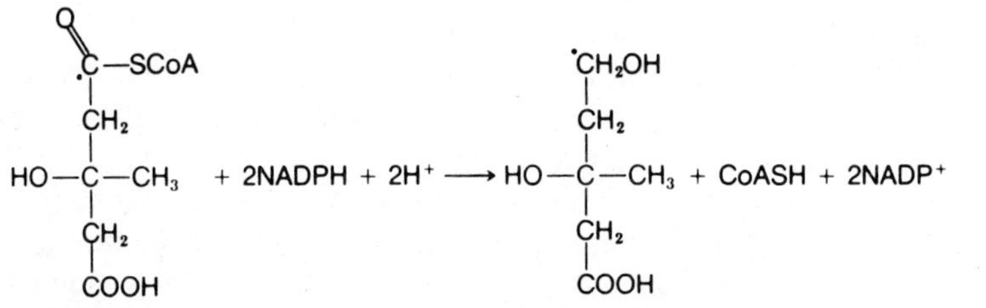

$$\underset{\text{HMG CoA}}{HOOC-CH_2-C(OH)(CH_3)-CH_2-CO-SCoA} + 2NADPH + 2H^+ \longrightarrow \underset{\text{Mevalonic acid}}{HOOC-CH_2-C(OH)(CH_3)-CH_2-CH_2OH} + CoASH + 2NADP^+$$

Figure 10.27
The HMG CoA reductase reaction.

The step that produces the unique compound mevalonic acid from HMG CoA is catalyzed by the important microsomal enzyme HMG CoA reductase (mevalonate : $NADP^+$ oxidoreductase) that has an absolute requirement for NADPH as the reductant (Figure 10.27). Note that this reductive step (1) consumes two molecules of NADPH from the pentose phosphate pathway, (2) results in the hydrolysis of the thioester bond of HMG CoA, and (3) generates a primary alcohol residue in mevalonate. This reduction reaction is irreversible and produces R-(+) mevalonate, which contains six carbon atoms.

Conversion of Mevalonic Acid to Farnesyl Pyrophosphate

The various reactions involved in the conversion of mevalonate to farnesyl pyrophosphate are described below and summarized in Figure 10.28.

The stepwise transfer of the terminal γ-phosphate group from two molecules of ATP to mevalonate (A) to form 5-pyrophosphomevalonate (B) are catalyzed by mevalonate kinase (enzyme I) and phosphomevalonate kinase (enzyme II). The next step affects the decarboxylation of 5-pyrophosphomevalonate and generates Δ^3-isopentenyl pyrophosphate (D); this reaction is catalyzed by pyrophosphomevalonate decarboxylase. In this ATP-dependent reaction in which ADP, P_i, and CO_2 are produced it is thought that decarboxylation–dehydration proceeds by way of the triphosphate intermediate, 3-phosphomevalonate 5-pyrophosphate (C). Next, isopentenyl pyrophosphate is converted to its allylic isomer 3,3-dimethylallyl pyrophosphate (E) in a reversible reaction catalyzed by isopentenyl pyrophosphate isomerase. The condensation of 3,3-dimethylallyl pyrophosphate (E) and 3-isopentenyl pyrophosphate (D) generates geranyl pyrophosphate (F).

The stepwise condensation of three 5-carbon isopentenyl units to form the 15-carbon unit farnesyl pyrophosphate (G) is catalyzed by one enzyme, a cytoplasmic prenyl transferase called geranyl transferase.

Formation of Cholesterol from Farnesyl Pyrophosphate via Squalene

The last steps in cholesterol biosynthesis involve the "head-to-head" fusion of two molecules of farnesyl pyrophosphate to form squalene and finally the cyclization of squalene to yield cholesterol. The reaction that

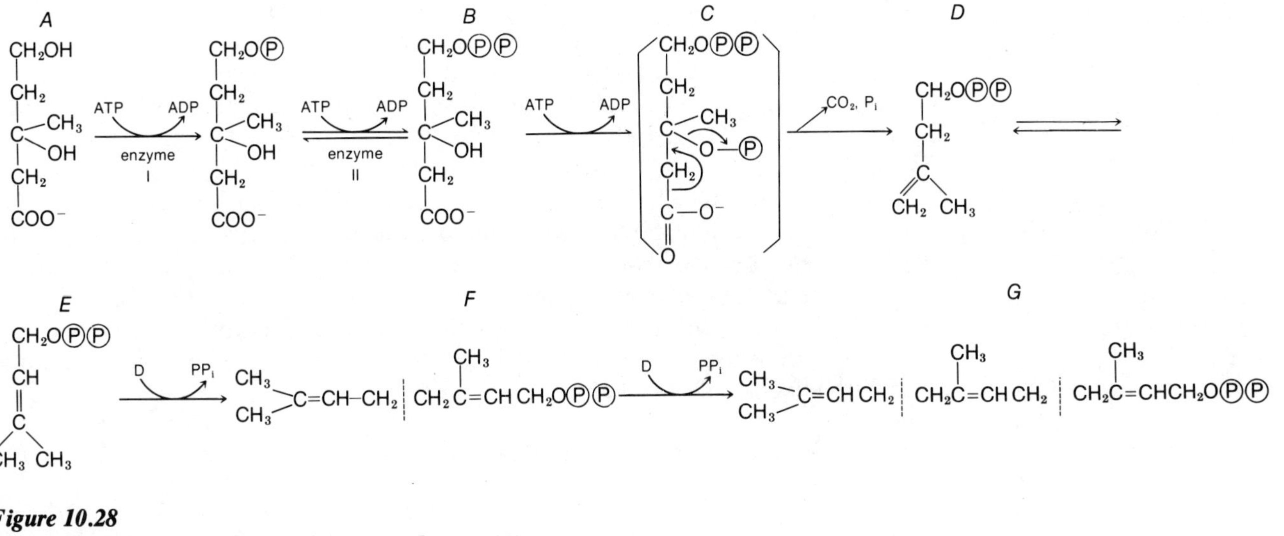

Figure 10.28
Formation of farnesyl-PP (F) from mevalonate (A).
The dotted lines divide the molecules into isoprenoid-derived units. O is 3-isopentenyl pyrophosphate.

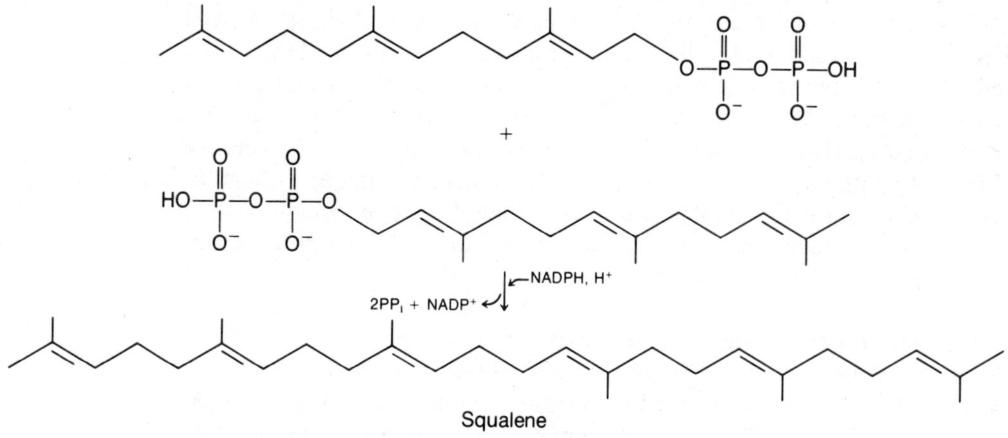

Figure 10.29
Formation of squalene from two molecules of farnesyl pyrophosphate.

produces the 30-carbon molecule of squalene from two 15-carbon farnesyl pyrophosphate moieties (Figure 10.29) is unlike the previous carbon–carbon bond-forming reactions in the pathway (Figure 10.28).

In this reaction catalyzed by the microsomal enzyme squalene synthetase, two pyrophosphate groups are released, with loss of a hydrogen atom from one molecule of farnesyl pyrophosphate and replacement by a hydrogen from NADPH. Several different intermediates probably occur between farnesyl pyrophosphate and squalene. By rotation about carbon–carbon single bonds, the conformation of squalene indicated in Figure 10.30 can be obtained. Note the similarity of the overall shape of the compound to cholesterol. Observe also that squalene is devoid of oxygen atoms.

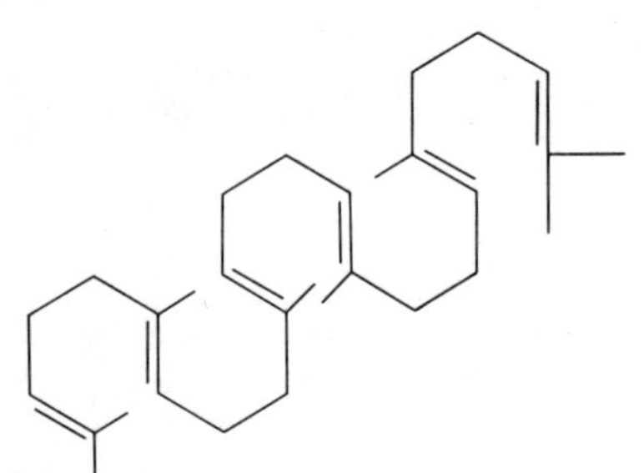

Figure 10.30
Structure of squalene, C_{30}.

Cholesterol biosynthesis from squalene proceeds through the intermediate lanosterol, which contains the fused tetracyclic ring system and an 8-carbon side chain:

$$\text{Squalene} \longrightarrow \text{squalene 2,3-epoxide} \longrightarrow \text{lanosterol}$$

The many carbon–carbon bonds that are formed during the cyclization of squalene are generated in a concerted fashion as indicated in Figure 10.31. Note that the —OH group of lanosterol projects above the plane of the A ring; this is referred to as the β orientation. Groups that extend down below the ring in a trans relationship to the —OH group are designated as α (alpha) by a dotted line. During this reaction sequence a hydroxyl group is added to C-3, a methyl group undergoes two 1,2 shifts, and a proton is eliminated. The oxygen atom is derived from molecular oxygen. The reaction is catalyzed by a microsomal enzyme system called squalene oxidocyclase that appears to be composed of at least two enzymes, an epoxidase or monoxygenase and a cyclase (lanosterol cyclase).

The cyclization process is initiated by epoxide formation between what will become C-2 and C-3 of cholesterol, the epoxide being formed at the expense of NADPH:

$$\text{Squalene} + O_2 + \text{NADPH} + H^+ \rightarrow \text{squalene 2,3-epoxide} + H_2O + \text{NADP}^+$$

This reaction is catalyzed by the monoxygenase or epoxidase component of squalene oxidocyclase. Hydroxylation at C-3 by way of the epoxide

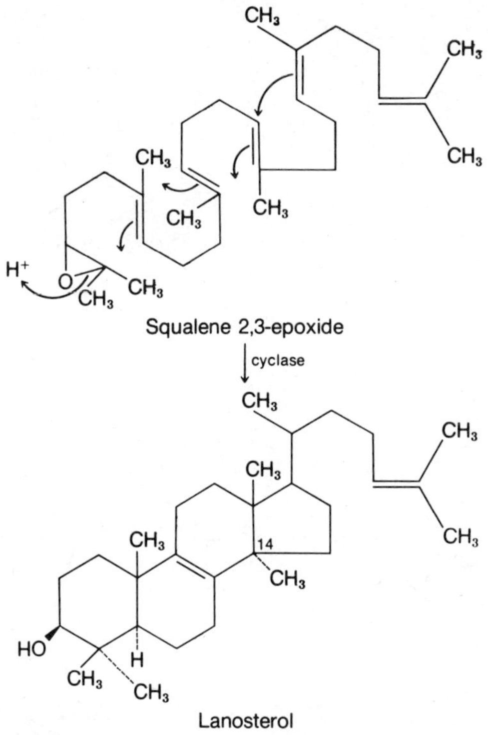

Squalene 2,3-epoxide

cyclase

Lanosterol

Figure 10.31
Conversion of squalene 2,3-epoxide to lanosterol.

intermediate triggers the cyclization of squalene to form lanosterol as shown in Figure 10.31. In the process of cyclization, two hydrogen atoms and two methyl groups migrate to neighboring positions.

The transformation of lanosterol to cholesterol involves many steps and a number of different microsomal enzymes. These steps include (1) removal of the methyl group at C-14, (2) removal of the two methyl groups at C-4, (3) migration of the double bond from C-8 to C-5, and (4) reduction of the double bond between C-24 and C-25 in the side chain. (See Figure 10.32).

Regulation of Cholesterol Biosynthesis

The cholesterol pool of the body is derived from two sources: absorption of dietary cholesterol and biosynthesis de novo, primarily in the liver and the intestine. When the amount of dietary cholesterol is reduced, cholesterol synthesis is increased in the liver and intestine to satisfy the needs of other tissues and organs. Cholesterol synthesized de novo is transported from the liver and intestine to peripheral tissues in the form of lipoproteins. These two tissues are the only ones that can manufacture apolipoprotein B, the protein component of the cholesterol transport proteins LDL and VLDL. Most of the apolipoprotein B is secreted into the circulation as VLDL, which is converted into LDL by removal of triacylglycerol and the apolipoprotein C components, probably in peripheral tissues and the liver. In contrast, when the quantity of dietary cholesterol increases, cholesterol synthesis in the liver and intestine is almost totally suppressed. Thus, the rate of de novo cholesterol synthesis is inversely related to the amount of dietary cholesterol taken up by the body.

The primary site for control of cholesterol biosynthesis is HMG CoA reductase, which catalyzes the step that produces mevalonic acid. This is the committed step and the rate-limiting reaction in the pathway of cholesterol biosynthesis (Figure 10.33). Cholesterol effects feedback inhibition of its own synthesis by inhibiting the activity of preexisting HMG

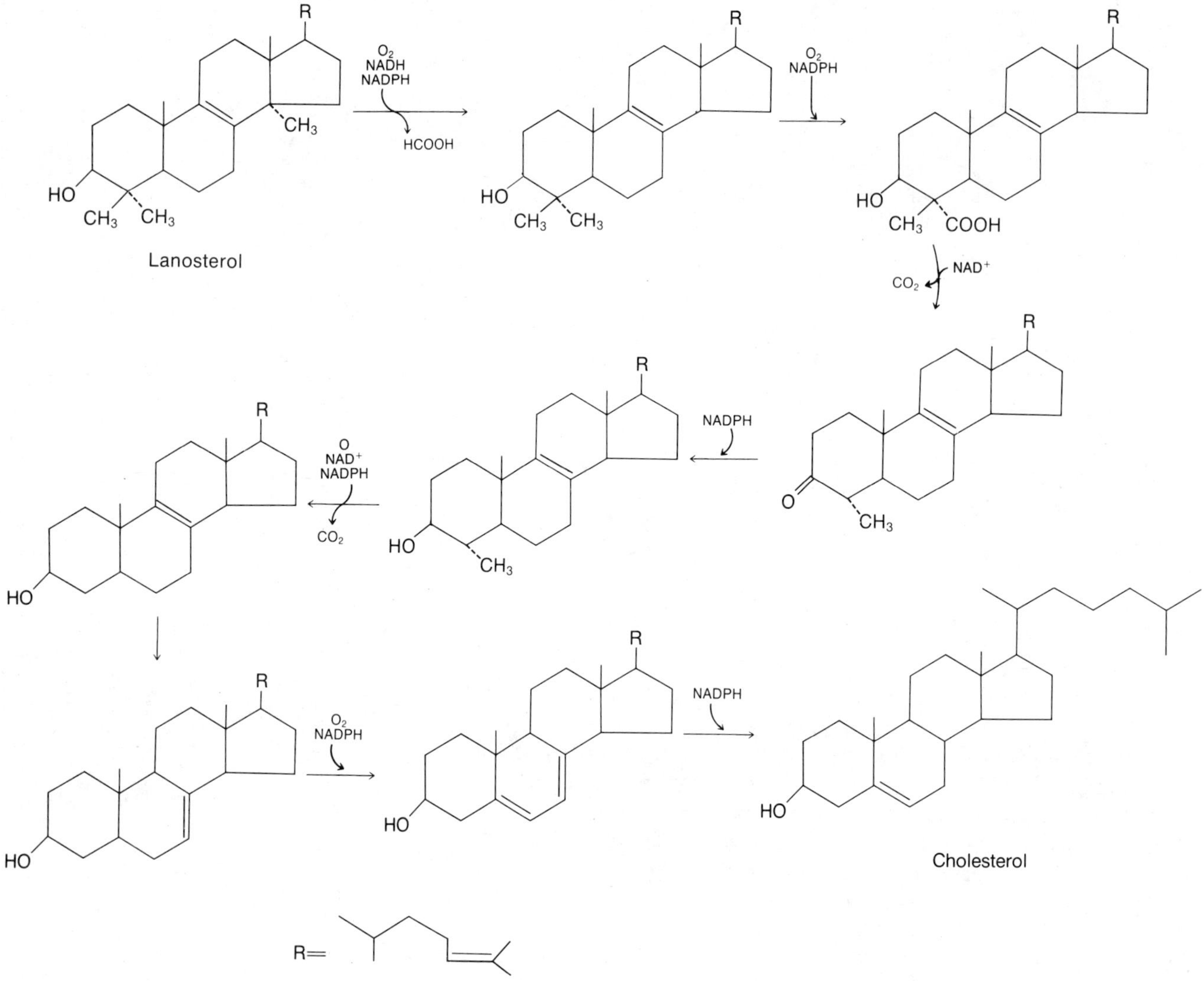

Lanosterol

O_2 NADH NADPH HCOOH

O_2 NADPH

NAD⁺ CO_2

NADPH

O NAD⁺ NADPH CO_2

O_2 NADPH

NADPH

Cholesterol

R=

Figure 10.32
Conversion of lanosterol to cholesterol.

CoA reductase and also by promoting rapid inactivation of the enzyme by mechanisms that remain to be elucidated. The dietary cholesterol that suppresses HMG CoA reductase activity and cholesterol synthesis emerges from the intestine in the form of chylomicrons.

In a normal healthy adult on a low cholesterol diet about 1,300 mg of cholesterol is returned to the liver each day for disposal. This cholesterol comes from (1) cholesterol reabsorbed from the gut by means of the enterohepatic circulation and (2) HDL that carries cholesterol to the liver from peripheral tissues. The liver disposes of cholesterol in one of three ways: (1) excretion in bile as free cholesterol and after conversion to bile salts—each day, about 250 mg of bile salts and 550 mg of cholesterol are lost from the enterohepatic circulation, (2) esterification and storage in the liver as cholesterol esters, and (3) incorporation into lipoproteins (VLDL and LDL) and secretion into the circulation. On a low-cholesterol diet, the liver will synthesize about 800 mg of cholesterol per day to replace bile salts and cholesterol lost from the enterohepatic circulation in the feces.

The mechanism of suppression of cholesterol biosynthesis by LDL-bound cholesterol involves specific LDL receptors that project from the

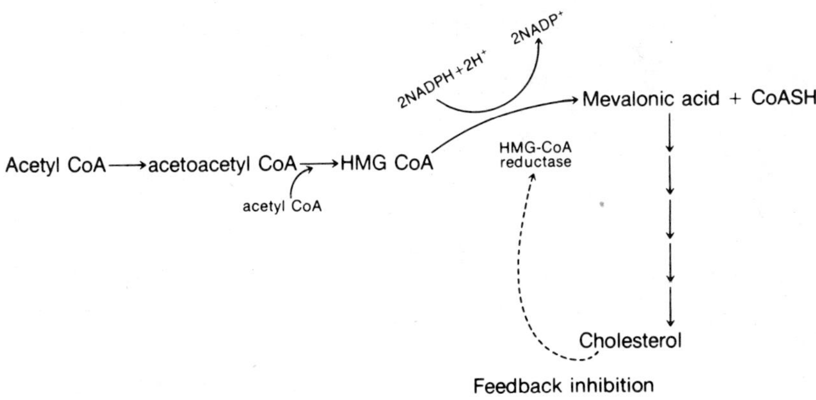

Figure 10.33
Summary of the pathway of cholesterol synthesis indicating feedback inhibition of HMG CoA reductase by cholesterol.

surface of human cells. The first step of the regulatory mechanism involves the binding of the lipoprotein LDL to these LDL receptors, thereby extracting the LDL particles from the blood. The binding reaction is characterized by its saturability, high affinity, and high degree of specificity. The receptor recognizes only LDL and VLDL, the two plasma lipoproteins that contain apolipoprotein B. Once LDL binds to the cell surface LDL receptor, the cholesterol-charged lipoprotein is endocytosed in the form of vesicles called endosomes. This process is termed *receptor-mediated endocytosis*. The next step involves the fusion of the endosome with a lysosome that contains numerous hydrolytic enzymes, including proteases and cholesterol esterase. Eventually the LDL receptor separates from LDL and returns to the cell surface. Inside the lysosome the cholesterol ester component of LDL is hydrolyzed by lysosomal cholesterol esterase to produce free cholesterol and a molecule of long-chain fatty acid. Free cholesterol then diffuses into the cytoplasm where, by some unknown mechanism, it inhibits the activity of microsomal HMG CoA reductase and suppresses the synthesis of HMG CoA reductase enzyme. There is evidence that cholesterol acts at the level of DNA and the protein synthesis apparatus to decrease the rate of synthesis of HMG CoA reductase. At the same time, microsomal fatty acyl CoA : cholesterol acyltransferase (ACAT) is activated by cholesterol, thereby promoting the formation of cholesterol esters, principally cholesterol oleate. The accumulation of intracellular cholesterol esters eventually inhibits the replenishment of LDL receptors on the cell surface, a phenomenon called *down regulation,* thereby blocking further uptake and accumulation of cholesterol by the cell.

Patients with familial (genetic) hypercholesterolemia suffer from accelerated atherosclerosis (Clin. Corr. 10.2) and have a defect in this regulatory system. In such individuals, referred to as receptor-negative, cells lack functional LDL receptors on the surfaces. As a result, there is no binding of LDL to the cell, cholesterol is not transferred into the cell, cholesterol synthesis is not inhibited, and the cholesterol content of the blood increases.

In specialized tissues such as the adrenal gland and ovary, the cholesterol derived from LDL serves as a precursor to the steroid hormones made by these organs, such as cortisol and estradiol, respectively. In the liver, cholesterol extracted from LDL and HDL is converted into bile salts that function in intestinal fat digestion.

CLIN. CORR. **10.2**
ATHEROSCLEROSIS

Atherosclerosis is a complex and chronic disease involving the gradual accumulation of lipids, collagen, elastic fibers, and proteoglycans in the arterial wall. Since cholesterol esters and cholesterol are major components of atherosclerotic lesions, the interaction of the cholesterol-carrying lipoproteins in plasma with the cells of the arterial wall seem to be important. An increased level of total plasma cholesterol and an increase in the major cholesterol-carrying lipoprotein, LDL, are associated with an increased risk of developing atherosclerotic cardiovascular disease because the cholesterol of atherosclerotic plaques is derived from LDL. There is also considerable evidence that an elevated level of plasma triacylglycerol is also a risk factor for ischemic heart disease. On the other hand, high concentrations of HDL seem to decrease the risk for ischemic heart disease.

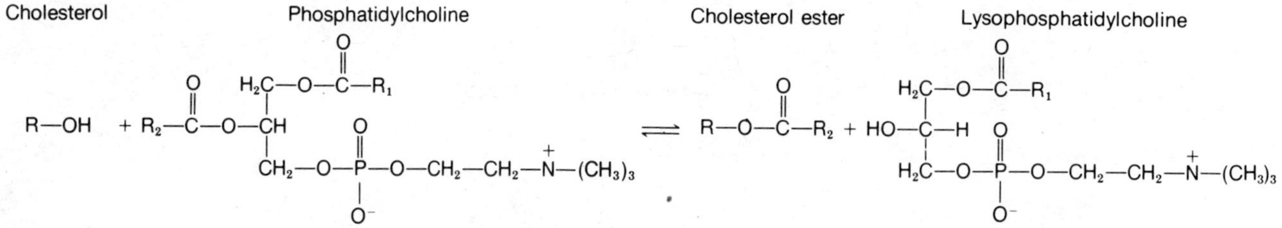

Cholesterol + Phosphatidylcholine ⇌ Cholesterol ester + Lysophosphatidylcholine

$$R{-}OH + R_2{-}\overset{O}{\overset{\|}{C}}{-}O{-}CH(CH_2{-}O{-}\overset{O}{\overset{\|}{C}}{-}R_1)(CH_2{-}O{-}\overset{O}{\overset{\|}{P}}(O^-){-}O{-}CH_2{-}CH_2{-}\overset{+}{N}{-}(CH_3)_3) \rightleftharpoons R{-}O{-}\overset{O}{\overset{\|}{C}}{-}R_2 + HO{-}CH(CH_2{-}O{-}\overset{O}{\overset{\|}{C}}{-}R_1)(CH_2{-}O{-}\overset{O}{\overset{\|}{P}}(O^-){-}O{-}CH_2{-}CH_2{-}\overset{+}{N}{-}(CH_3)_3)$$

Figure 10.34
The lecithin : cholesterol acyltransferase (LCAT) reaction where R—OH = cholesterol.

Dynamics of Plasma Cholesterol

Plasma cholesterol is in a dynamic state, entering the blood complexed with lipoproteins that keep the lipid in solution and leaving the blood as tissues remove cholesterol from these lipoproteins or degrade them intracellularly. Cholesterol occurs in plasma lipoproteins in two forms: as free cholesterol and esterified to some long-chain fatty acid. From 70 to 75% of plasma cholesterol is esterified to long-chain fatty acids. It is the free, unesterified form of cholesterol that exchanges readily between different lipoproteins and the plasma membranes of cells.

HDL and the enzyme lecithin : cholesterol acyltransferase (LCAT) play important roles in the transport and elimination of cholesterol from the body. LCAT catalyzes the freely reversible reaction shown in Figure 10.34, which transfers the fatty acid in the sn-2 position of phosphatidylcholine to the 3-hydroxyl of cholesterol. LCAT is a plasma enzyme produced mainly by the liver. The actual substrate for LCAT is cholesterol contained in HDL. The LCAT–HDL system functions to protect cells, especially their plasma membranes, from the damaging effects of excessive amounts of free cholesterol. Cholesterol ester generated in the LCAT reaction diffuses into the core of the HDL particle where it is then transported from the tissues and plasma to the liver, the latter being the only organ capable of metabolizing and excreting cholesterol. Thus, by this mechanism, referred to as the *reverse transport of cholesterol,* LCAT acting on HDL provides a vehicle for transporting cholesterol from peripheral tissues to the liver.

Bile Acids and the Excretion of Cholesterol

The bile acids are the end products of cholesterol metabolism. Primary bile acids are those that are synthesized in hepatocytes directly from cholesterol. The most abundant bile acids in human beings are derivatives of cholanic acid (Figure 10.35). The most common bile acids in human beings are cholic acid and chenodeoxycholic acid (Figure 10.36). The primary bile acids (1) are composed of 24 carbon atoms, (2) are made in liver parenchymal cells, (3) contain two or three hydroxyl groups, and (4) have a side chain that ends in a carboxyl group that is ionized at pH 7.0 (hence the name bile salt). The carboxyl group of the primary bile acids is often conjugated via an amide bond to either glycine (NH_2—CH_2—COOH) or taurine (NH_2—CH_2—CH_2—SO_3H) to form glycocholic or taurocholic acid, respectively. The structure of glycocholic acid is shown in Figure 10.37.

When the primary bile acids undergo further chemical reactions by microorganisms in the gut, they give rise to secondary bile acids that also

Figure 10.35
Structure of cholanic acid.

possess 24 carbon atoms. Examples of secondary bile acids are deoxycholic acid and lithocholic acid, which are derived from cholic acid and chenodeoxycholic acid, respectively, by the removal of one hydroxyl group (Figure 10.36).

The changes in the cholesterol molecule that occur during its transformation into bile acids include (1) epimerization of the 3β-hydroxyl group; (2) reduction of the Δ^5 double bond; (3) introduction of hydroxyl groups at C-7 (chenodeoxycholic acid) or at C-7 and C-12 (cholic acid); and (4) conversion of the C-27 side chain into a C-24 carboxylic acid by elimination of a propyl equivalent.

Bile acids formed in liver parenchymal cells are secreted into the bile canaliculi, which are specialized channels formed by adjacent hepatocytes. Bile canaliculi unite with bile ductules, which in turn come together to form bile ducts. The bile acids are then carried to the gallbladder for storage and ultimately to the small intestine where they are excreted. The capacity of the liver to produce bile acids is insufficient to meet the physiological demands, so the body relies upon an efficient enterohepatic circulation that carries the bile acids from the intestine back to the liver several times each day. The primary bile acids, after removal of the glycine or taurine residue in the gut, are reabsorbed by an active transport process from the intestine, primarily in the ileum, and returned to the liver by way of the portal vein. Bile acids that are not reabsorbed are acted upon by bacteria in the gut and converted into secondary bile acids: a portion of secondary bile acids, primarily deoxycholic acid and lithocholic acid, are reabsorbed passively in the colon and returned to the liver where they are secreted into the gallbladder. Hepatic synthesis normally produces 0.2–0.6 g of bile acids per day to replace those lost in the feces. The gallbladder pool of bile acids is 2–4 g. Because the enterohepatic circulation recycles 6–12 times each day, the total amount of bile acids absorbed per day from the intestine corresponds to 12–32 g.

In terms of function the bile acids are significant in medicine for several reasons:

1. They represent the only significant way in which cholesterol can be excreted. The carbon skeleton of cholesterol is not oxidized to CO_2 and water in human beings but is excreted in bile via the gallbladder and intestine as free cholesterol and after conversion to bile acids.
2. They prevent the precipitation of cholesterol out of solution in the gallbladder. Bile acids and phospholipids function to solubilize cholesterol in bile.
3. They act as emulsifying agents to prepare dietary triacylglycerols for attack by pancreatic lipase. Bile acids may also play a direct role in activating pancreatic lipase. (See Chapter 24.)
4. They facilitate the absorption of fat-soluble vitamins, particularly vitamin D, from the intestine.

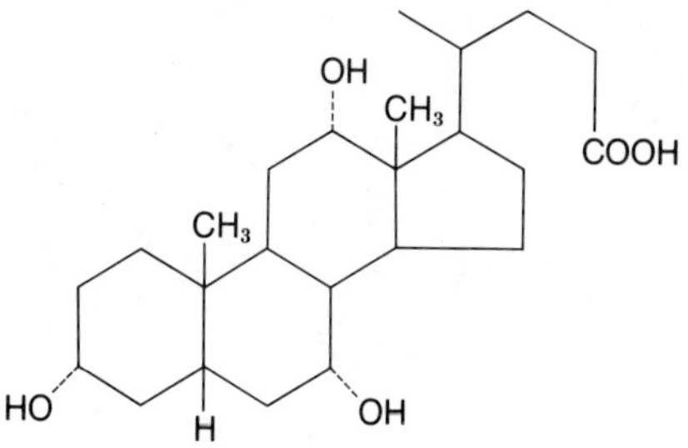

Cholic acid

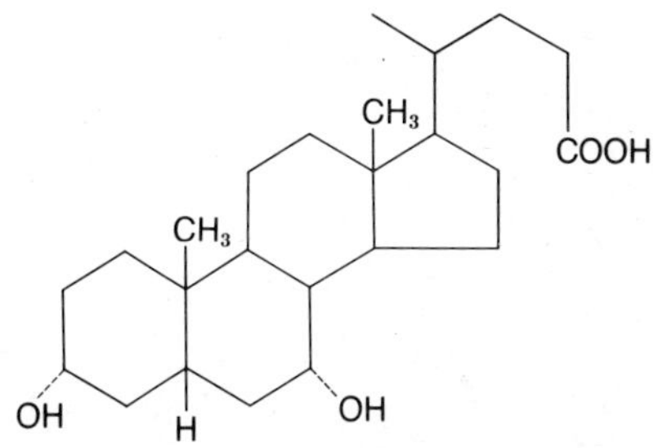

Chenodeoxycholic acid

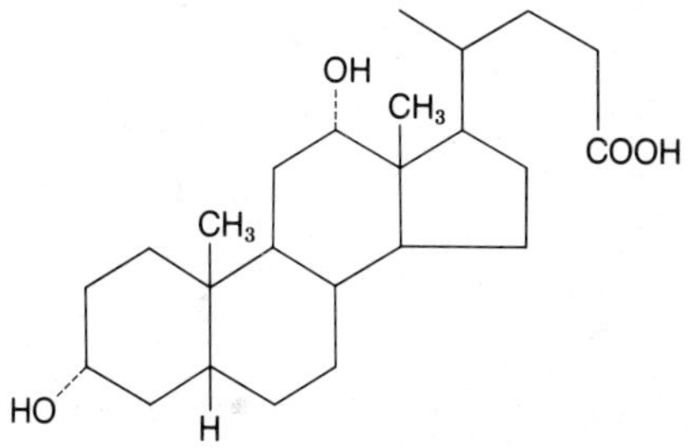

Deoxycholic acid

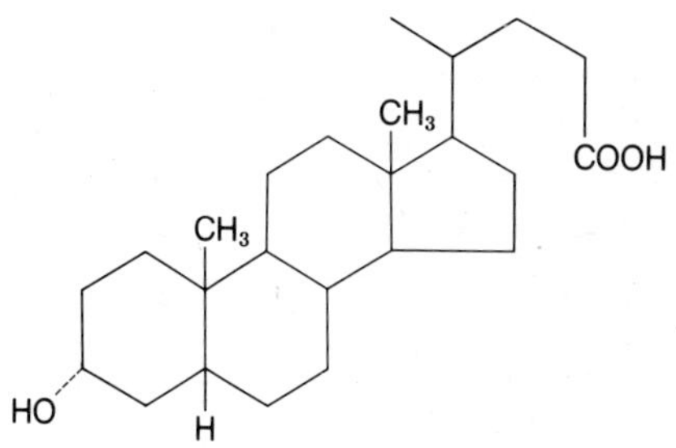

Lithocholic acid

Figure 10.36
Structures of some common bile acids.

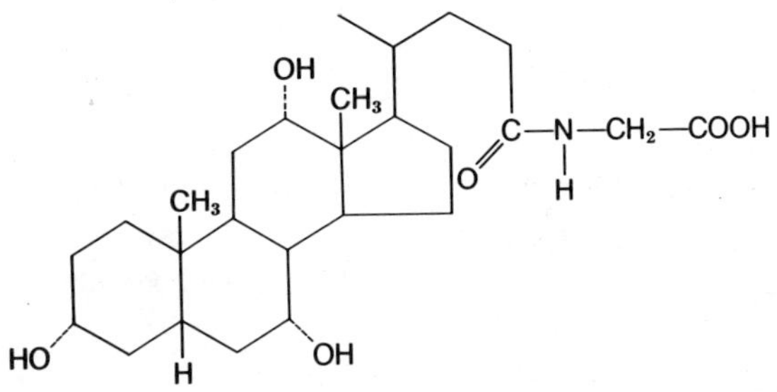

Figure 10.37
Structure of glycocholic acid, a conjugated bile acid.

10.4 SPHINGOLIPIDS

Structure and Biosynthesis of Sphingosine and Ceramide

Sphingolipids are complex lipids whose core structure is provided by the long-chain amino alcohol sphingosine (Figure 10.38). Another common name for sphingosine is 4-sphingenine and the formal name for sphingosine is *trans*-1,3-dihydroxy-2-amino-4-octadecene. We shall learn that sphingosine (1) possesses two asymmetric carbon atoms (positions C-2 and C-3); of the four possible optical isomers, naturally occurring sphingosine is of the D-erythro form: (2) the Δ^4 carbon–carbon double bond has the trans configuration: (3) the primary alcohol group at C-1 is a nucleophilic center that forms covalent bonds with sugars to form sphingomyelin; (4) the amino group at C-2 always bears a long-chain (usually C_{20}–C_{26}) fatty acid in amide linkage; and (5) the secondary alcohol at C-3 is never derivatized and is always free. It is useful to appreciate the structural similarity of a part of the sphingosine molecule to the glycerol moiety of the acyl glycerols. When one views the structure of sphingosine from another perspective, the similarity between carbons 1, 2, and 3 of sphingosine and glycerol becomes apparent (Figure 10.38). Note that both glycerol and sphingosine have nucleophilic groups, hydroxyl or amino, at positions C-1, C-2, and C-3.

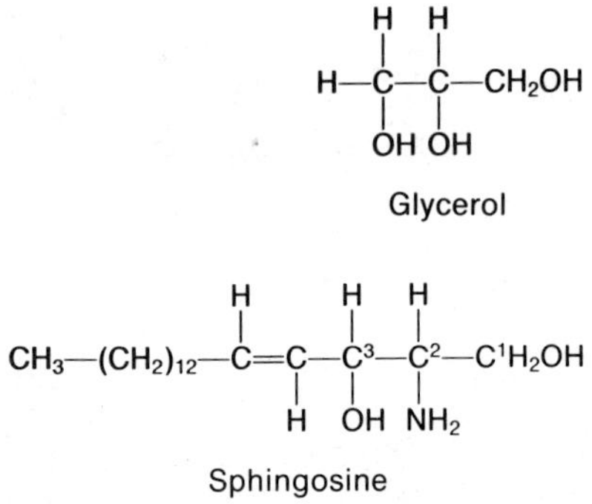

H—C(H)(OH)—C(H)(OH)—CH_2OH

Glycerol

CH_3—$(CH_2)_{12}$—C(H)=C(H)—C^3(H)(OH)—C^2(H)(NH_2)—C^1H_2OH

Sphingosine

Figure 10.38
Comparison of the structures of glycerol and sphingosine (trans-1,3-dihydroxy-2-amino-4-octadecene).

The sphingolipids occur in blood and nearly all of the tissues of human beings. However, the highest concentrations of sphingolipids are found in the white matter of the central nervous system. Various sphingolipids are components of the plasma membrane of practically all cells.

Sphingosine is synthesized by way of sphinganine (dihydrosphingosine) in two steps from the precursors L-serine and palmitoyl CoA: serine is the source of C-1, C-2 and the amino group of sphingosine, while palmitic acid provides the remaining carbon atoms. The condensation of serine with palmitoyl CoA is catalyzed by a pyridoxal phosphate-dependent enzyme serine palmitoyltransferase and the driving force for the reaction is provided by both the cleavage of the reactive, high-energy, thioester bond of palmitoyl CoA and the release of CO_2 from serine (Figure 10.39). The next step involves the reduction of the carbonyl group in 3-keto dihydrosphingosine with reducing equivalents being derived from NADPH to produce sphinganine (also called dihydrosphingosine) (Figure 10.40). The insertion of the double bond into sphinganine to produce sphingosine occurs at the level of ceramide (see below).

Sphingosine as such with its free amino group does not occur naturally. The fundamental building block or core structure of the natural sphingolipids is ceramide. Ceramide is the long-chain fatty acid amide derivative of sphingosine. Most often the acyl group is behenic acid, a saturated C_{22} fatty acid, but other long-chain acyl groups can be used. The long-chain fatty acid is attached to the 2-amino group of sphingosine through an

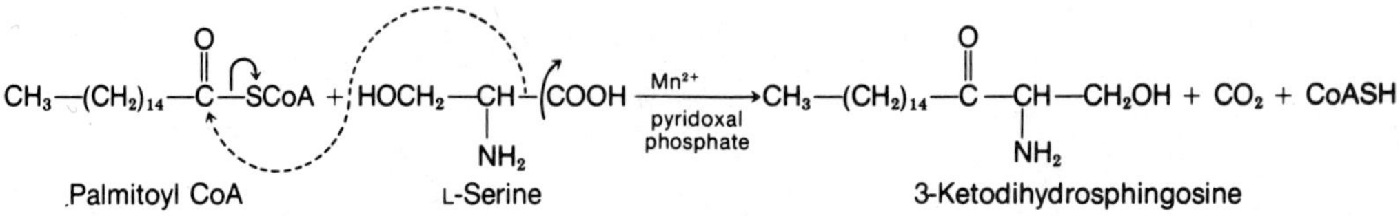

CH_3—$(CH_2)_{14}$—C(=O)—SCoA + $HOCH_2$—CH(NH_2)—COOH $\xrightarrow[\text{pyridoxal phosphate}]{Mn^{2+}}$ CH_3—$(CH_2)_{14}$—C(=O)—CH(NH_2)—CH_2OH + CO_2 + CoASH

Palmitoyl CoA L-Serine 3-Ketodihydrosphingosine

Figure 10.39
Formation of 3-ketodihydrosphingosine from serine and palmitoyl CoA.

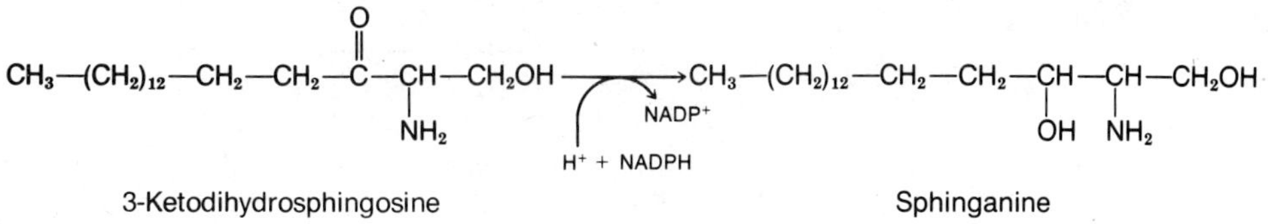

$$CH_3{-}(CH_2)_{12}{-}CH_2{-}CH_2{-}\overset{O}{\overset{\|}{C}}{-}\underset{NH_2}{\underset{|}{CH}}{-}CH_2OH \xrightarrow[H^+ + NADPH]{NADP^+} CH_3{-}(CH_2)_{12}{-}CH_2{-}CH_2{-}\underset{OH}{\underset{|}{CH}}{-}\underset{NH_2}{\underset{|}{CH}}{-}CH_2OH$$

3-Ketodihydrosphingosine　　　　Sphinganine

Figure 10.40
Conversion of 3-ketodihydrosphingosine to sphinganine.

amide bond (Figure 10.41). There are two long-chain hydrocarbon domains in the ceramide molecule; these hydrophobic regions are responsible for the lipoidal character of the sphingolipids.

Ceramide is synthesized from dihydrosphingosine and a molecule of long-chain fatty acyl CoA by a microsomal enzyme with dihydroceramide as an intermediate that is then oxidized by dehydrogenation at C-4 and C-5 (Figure 10.42). Free ceramide is not a component of membrane lipids but rather is an intermediate in the biosynthesis and catabolism of glycosphingolipids and sphingomyelin. Figure 10.43 contains, in diagrammatic form, the structures of the prominent sphingolipids of humans.

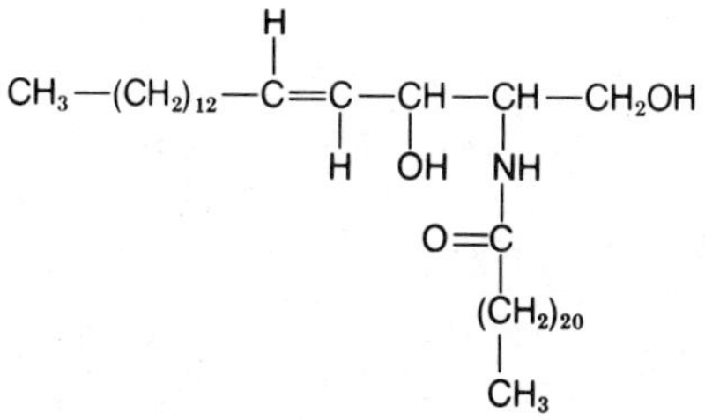

$$CH_3{-}(CH_2)_{12}{-}\overset{H}{\overset{|}{C}}{=}\underset{H}{\underset{|}{C}}{-}\underset{OH}{\underset{|}{CH}}{-}\underset{NH{-}C(=O){-}(CH_2)_{20}{-}CH_3}{\underset{|}{CH}}{-}CH_2OH$$

Figure 10.41
Structure of a ceramide (N-acylsphingosine).

Sphingomyelin

Sphingomyelin is one of the principal structural lipids of the membranes of nervous tissue. It is the only sphingolipid that is a phospholipid. In sphingomyelin the primary alcohol group at C-1 of sphingosine is esterified to choline through a phosphodiester bridge of the kind that occurs in the acyl glycerophospholipids and the amino group of sphingosine is attached to a long-chain fatty acid by means of an amide bond. Sphingomyelin is therefore a ceramide phosphocholine. It contains one negative and one positive charge so that it is neutral at physiological pH (Figure 10.44).

The most common fatty acids in sphingomyelin are palmitic, stearic, lignoceric (a C_{24}, saturated fatty acid), and nervonic acid [24:1, $CH_3{-}(CH_2)_7CH{=}CH{-}(CH_2)_{13}{-}COOH$]. The sphingomyelin of myelin contains predominantly longer chain fatty acids, mainly lignoceric and

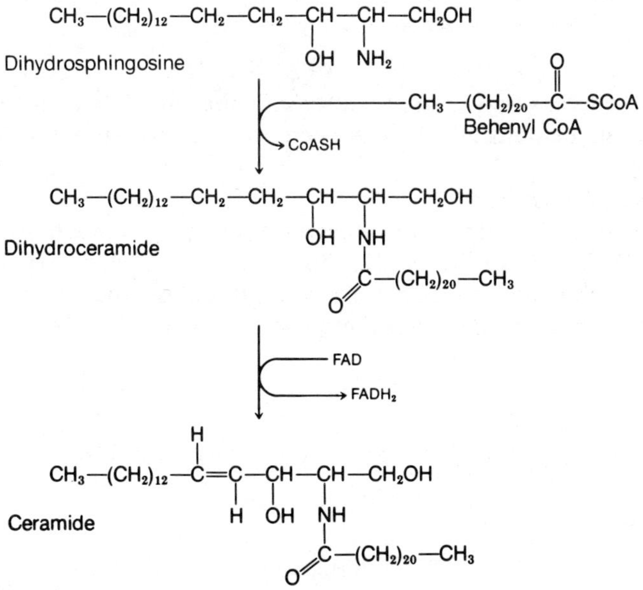

Dihydrosphingosine
$$CH_3{-}(CH_2)_{12}{-}CH_2{-}CH_2{-}\underset{OH}{\underset{|}{CH}}{-}\underset{NH_2}{\underset{|}{CH}}{-}CH_2OH$$

$$\downarrow \; CH_3{-}(CH_2)_{20}{-}\overset{O}{\overset{\|}{C}}{-}SCoA \text{ (Behenyl CoA)} \rightarrow CoASH$$

Dihydroceramide
$$CH_3{-}(CH_2)_{12}{-}CH_2{-}CH_2{-}\underset{OH}{\underset{|}{CH}}{-}\underset{NH{-}C(=O){-}(CH_2)_{20}{-}CH_3}{\underset{|}{CH}}{-}CH_2OH$$

$$\downarrow \; FAD \rightarrow FADH_2$$

Ceramide
$$CH_3{-}(CH_2)_{12}{-}\overset{H}{\overset{|}{C}}{=}\underset{H}{\underset{|}{C}}{-}\underset{OH}{\underset{|}{CH}}{-}\underset{NH{-}C(=O){-}(CH_2)_{20}{-}CH_3}{\underset{|}{CH}}{-}CH_2OH$$

Figure 10.42
Formation of ceramide from dihydrosphingosine.

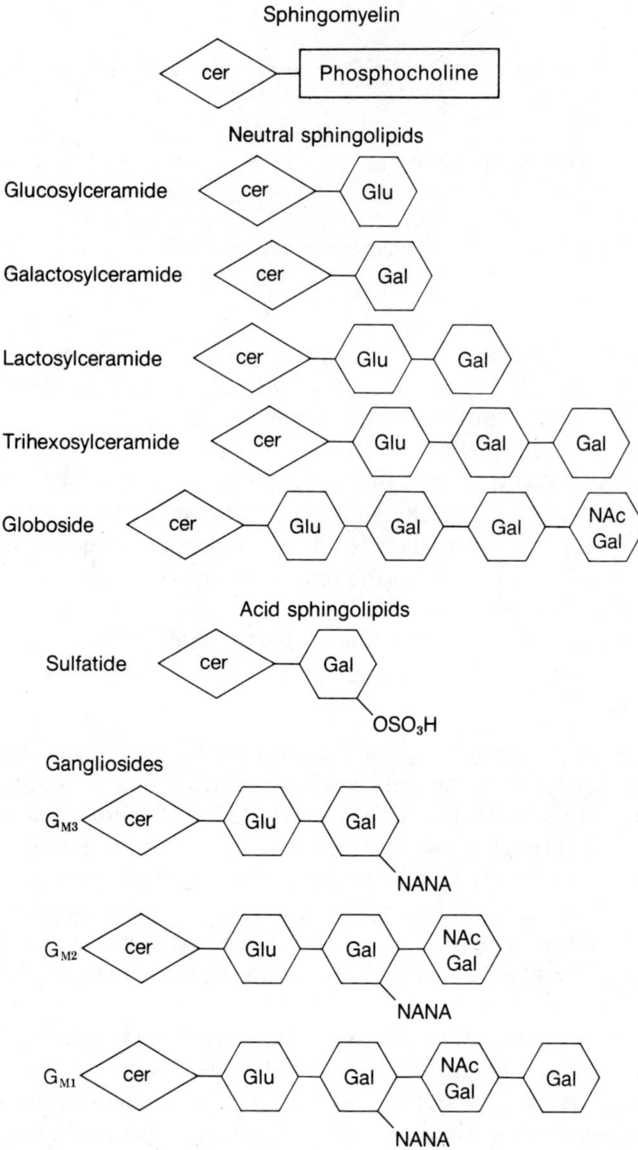

Figure 10.43
Structures of some common sphingolipids in diagrammatic form, where cer = ceramide, Glu = glucose, Gal = galactose, NAcGal = N-acetylgalactosamine, and NANA = N-acetylneuraminic acid (sialic acid).

nervonic, whereas that of grey matter contains largely stearic acid. Excessive accumulations of sphingomyelin occur in Niemann–Pick disease.

Sphingomyelin Synthesis

The conversion of ceramide to sphingomyelin involves the transfer of the phosphocholine group from CDP-choline to the primary, C-1 hydroxyl group of ceramide: this reaction is catalyzed by the enzyme CDP-choline : ceramide cholinephosphotransferase (Figure 10.45).

$$CH_3{-}(CH_2)_{12}{-}\overset{H}{\overset{|}{C}}{=}\underset{\overset{|}{H}}{C}{-}\underset{\overset{|}{OH}}{CH}{-}\underset{\overset{|}{NH{-}C(=O){-}(CH_2)_{16}{-}CH_3}}{CH}{-}CH_2{-}O{-}\overset{O}{\overset{\|}{P}}(\text{-}O^-){-}O{-}CH_2{-}CH_2{-}\overset{+}{N}(CH_3)_3$$

Figure 10.44
Structure of sphingomyelin.

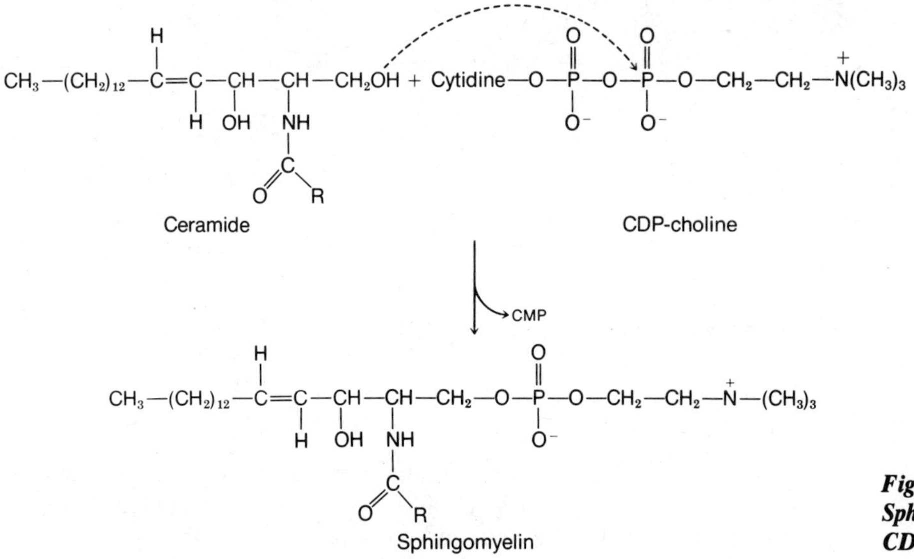

$$CH_3-(CH_2)_{12}-CH=CH-CH(OH)-CH(NHCOR)-CH_2OH + Cytidine-O-P(O)(O^-)-O-P(O)(O^-)-O-CH_2-CH_2-\overset{+}{N}(CH_3)_3$$

Ceramide CDP-choline

$$\downarrow \rightarrow CMP$$

$$CH_3-(CH_2)_{12}-CH=CH-CH(OH)-CH(NHCOR)-CH_2-O-P(O)(O^-)-O-CH_2-CH_2-\overset{+}{N}-(CH_3)_3$$

Sphingomyelin

Figure 10.45
Sphingomyelin synthesis from ceramide and CDP-choline.

Carbohydrate-Containing Sphingolipids

The principal glycosphingolipid classes are cerebrosides, sulfatides, globosides, and gangliosides. In the glycolipid class of compounds the polar head group is attached to sphingosine via the glycosidic linkage of a sugar molecule rather than a phosphate ester bond, as is the case in the phospholipids.

Cerebrosides

The cerebrosides are a group of ceramide monohexosides. The two most common cerebrosides of 1-β-glycosylceramides encountered in medicine are galactocerebroside and glucocerebroside. Unless specified otherwise, the term cerebroside usually refers to galactocerebroside. Galactocerebroside is also called "galactolipid." In Figure 10.46 note that the monosaccharide units are attached at C-1 of the sugar moiety to the C-1 position of ceramide, and the anomeric configuration of the glycosidic bond between ceramide and hexose in both galactocerebroside and glucocerebroside is β (beta). The largest amount of galactocerebroside in healthy people is found in the brain. Moderately increased amounts of galactocerebroside accumulate in the white matter in Krabbe's disease, also called globoid leukodystrophy, due to a deficiency in the lysosomal enzyme galactocerebrosidase.

Glucocerebroside (glucosyl ceramide) is not normally a structural component of membranes and is an intermediate in the synthesis and degradation of more complex glycosphingolipids. However, hundredfold increases in the glucocerebroside content of spleen and liver occur in the genetic lipid storage disorder called Gaucher's disease, which results from a deficiency of lysosomal glucocerebrosidase (See Figure 10.47).

Galactocerebroside and glucocerebroside are synthesized from ceramide and the activated nucleotide sugars UDP-galactose and UDP-

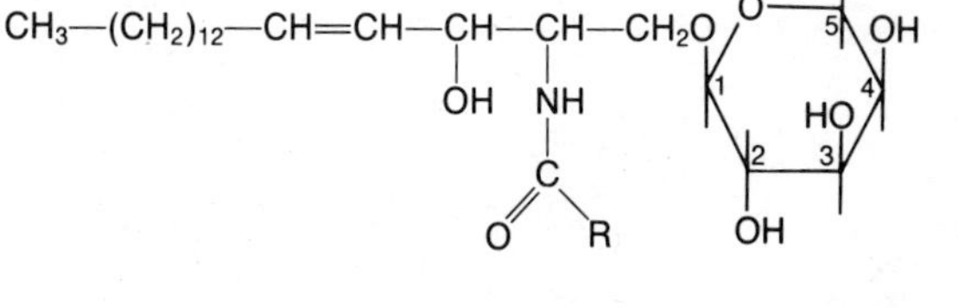

$$CH_3-(CH_2)_{12}-CH=CH-CH(OH)-CH(NHCOR)-CH_2O-[\text{galactose: } {}^6CH_2OH,\ 1,\ 2\text{-}OH,\ 3,\ 4\text{-}HO,\ 5\text{-}OH]$$

Figure 10.46
Structure of galactocerebroside (galactolipid).

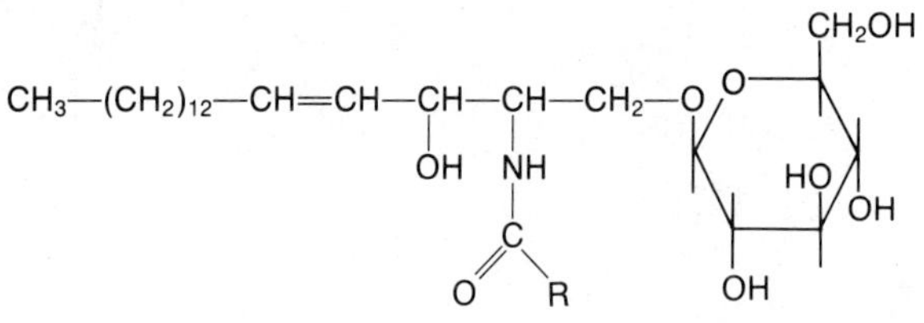

$$CH_3-(CH_2)_{12}-CH=CH-\underset{OH}{CH}-\underset{NH-C(=O)-R}{CH}-CH_2-O-\text{[glucose ring: }CH_2OH\text{, HO, OH, OH]}$$

Figure 10.47
Structure of glucocerebroside.

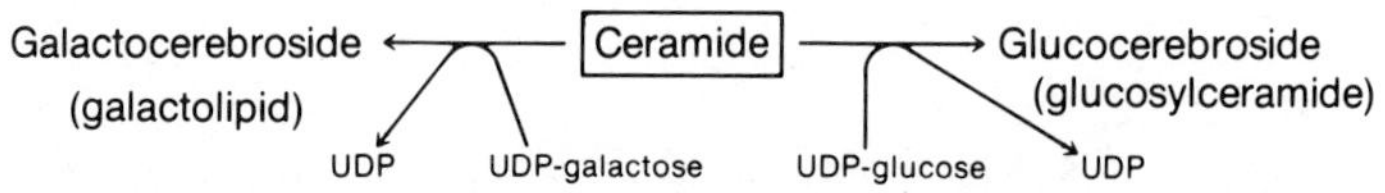

Galactocerebroside (galactolipid) ← [UDP-galactose → UDP] — Ceramide — [UDP-glucose → UDP] → Glucocerebroside (glucosylceramide)

Figure 10.48
Synthesis of galacto- and glucocerebroside.

glucose, respectively. The enzymes that catalyze these reactions, glucosyl and galactosyl transferases, are associated with the endoplasmic reticulum (Figure 10.48). Alternatively, in some tissues, the synthesis of glucocerebroside (glucosylceramide) proceeds by way of glucosylation of sphingosine:

$$\text{Sphingosine} + \text{UDP-glucose} \xrightarrow[\text{glucosyltransferase}]{} \text{glucosylsphingosine} + \text{UDP}$$

followed by fatty acylation:

$$\text{Glucosylsphingosine} + \text{stearoyl CoA} \longrightarrow \text{glucocerebroside} + \text{CoASH}$$

Sulfatide

Sulfatide, or sulfogalactocerebroside as it is sometimes called, is a sulfuric acid ester of galactocerebroside. Galactocerebroside 3-sulfate is the major sulfolipid in brain and accounts for approximately 15% of the lipids of white matter. (See Figure 10.49).

Galactocerebroside sulfate is synthesized from galactocerebroside and "activated sulfate" or PAPS (3′-phosphoadenosine 5′-phosphosulfate) in a reaction catalyzed by microsomal sulfotransferase:

$$\text{Galactocerebroside} + \text{PAPS} \longrightarrow \text{PAP} + \text{galactocerebroside 3-sulfate}$$

The structure of PAPS is indicated in Figure 10.50. Large quantities of sulfatide accumulate in the tissues of the central nervous system in metachromatic leukodystrophy due to a deficiency in a specific sulfatase.

Globosides: Ceramide Oligosaccharides

This family of compounds represents cerebrosides that contain two or more sugar residues, usually galactose, glucose, or *N*-acetylgalactos-

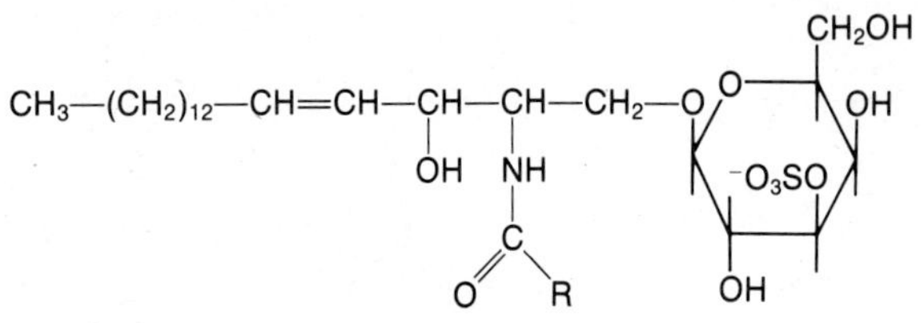

$$CH_3-(CH_2)_{12}-CH=CH-\underset{OH}{CH}-\underset{NH-C(=O)-R}{CH}-CH_2-O-\text{[galactose ring: }CH_2OH\text{, OH, }{}^{-}O_3SO\text{, OH]}$$

Figure 10.49
Structure of galactocerebroside sulfate (sulfolipid).

amine. The ceramide oligosaccharides are neutral compounds and contain no free amino groups.

Lactosylceramide is a component of the erythrocyte membrane (Figure 10.51).

Another prominent globoside is ceramide trihexoside or ceramide galactosyllactoside: ceramide-β-glc(4 $\leftarrow$ 1)-β-gal-(4 $\leftarrow$ 1)-α-gal. Note that the terminal galactose residue of this globoside has the α-anomeric configuration. Ceramide trihexoside accumulates in the kidneys of patients with Fabry's disease who are deficient in lysosomal α-galactosidase A activity.

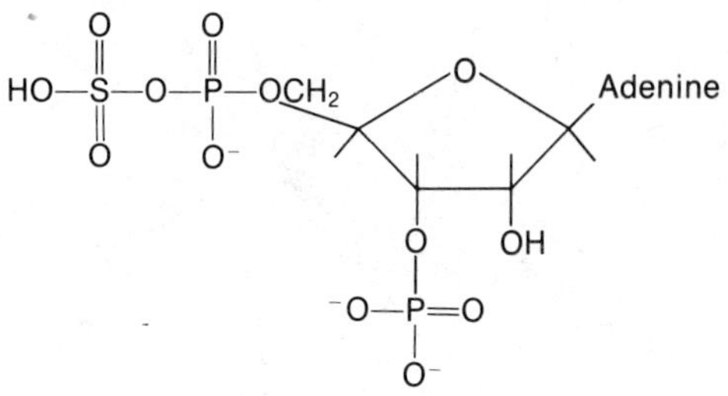

Figure 10.50
Structure of PAPS (3'-phosphoadenosine-5'-phosphosulfate).

Gangliosides

The name ganglioside was adopted for the class of sialic acid-containing glycosphingolipids that are highly concentrated in the ganglion cells of the central nervous system, particularly in the nerve endings. The central nervous system is unique among human tissues because more than half of the sialic acid is in ceramide-lipid bound form, with the remainder of the sialic acid occurring in the oligosaccharides of glycoproteins. Lesser amounts of gangliosides are contained in the surface membranes of the cells of most extraneural tissues where they account for less than 10% of the total sialic acid.

Neuraminic acid (abbreviated Neu) is present in gangliosides, glycoproteins, and mucins. The amino group of neuraminic acid occurs most often as the N-acetyl derivative, and the resulting structure is called N-acetylneuraminic acid or sialic acid, commonly abbreviated NANA. (See Figure 10.52).

The hydroxyl group on C-2 occurs most often in the α-anomeric configuration and the linkage between NANA and the oligosaccharide ceramide

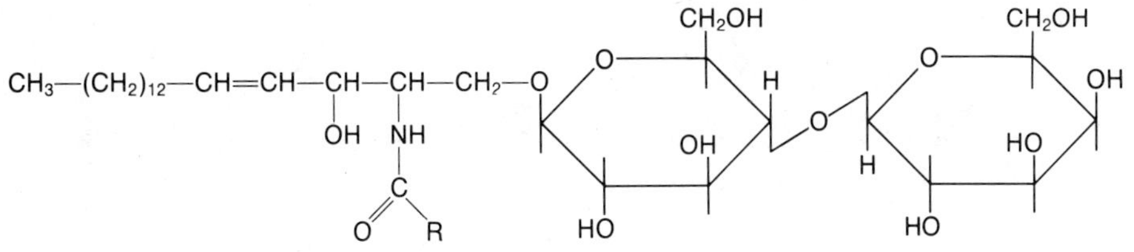

Figure 10.51
Structure of ceramide-β-glc-(4 ← 1)-β-gal-(4 ← 1) (lactosylceramide).

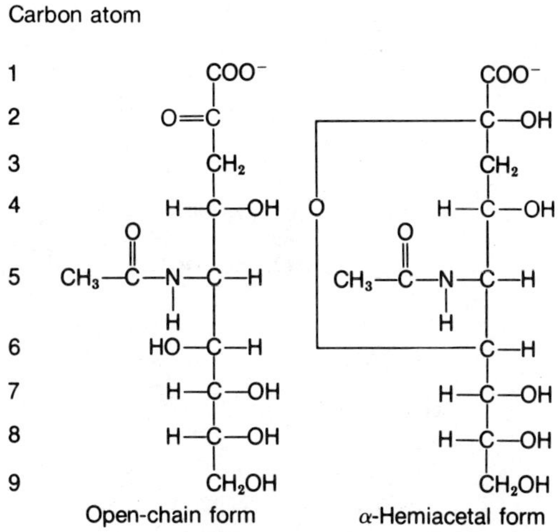

Figure 10.52
Structure of N-acetylneuraminic acid (NANA).

always involves the —OH group on position-2 of *N*-acetylneuraminic acid.

The structures of some of the common gangliosides are indicated in Table 10.1. The principal gangliosides in brain are G_{M1}, G_{D1a}, G_{D1b}, and G_{T1}. Nearly all of the gangliosides of man are derived from the family of compounds originating with glucosylceramide.

With regard to the nomenclature of the sialoglycosylsphingolipids, the letter G refers to the name ganglioside. The subscripts M, D, T, and Q indicate mono-, di-, tri-, and quatra(tetra)-sialic acid-containing gangliosides. The numerical subscripts 1, 2, and 3 designate the carbohydrate sequence that is attached to ceramide as indicated as follows: 1, Gal-GalNAc-Gal-Glc-ceramide; 2, GalNAc-Gal-Glc-ceramide; and 3, Gal-Glc-ceramide. Consider the nomenclature of the Tay–Sachs ganglioside; the designation G_{M2} denotes the ganglioside structure shown in Table 10.1.

A specific ganglioside on intestinal mucosal cells mediates the action of cholera toxin. Cholera toxin is a protein of mol w 84,000 that is secreted by the pathogen *Vibrio cholerae*. The toxin stimulates the secretion of

TABLE 10.1 The Structures of Some Common Gangliosides

Code Name	*Chemical Structure*
G_{M3}	Galβ → 4Glcβ → Cer 3 (on Gal) ↑ αNANA
G_{M2}	GalNAcβ → 4Galβ → 4Glcβ → Cer 3 (on Gal) ↑ αNANA
G_{M1}	Galβ → 3GalNAcβ → 4Galβ → 4Glcβ → Cer 3 (on Gal) ↑ αNANA
G_{D1a}	Galβ → 3GalNAcβ → 4Galβ → 4Glcβ → Cer 3 (on terminal Gal) ; 3 (on internal Gal) ↑ ; ↑ αNANA ; αNANA
G_{D1b}	Galβ → 3GalNAcβ → 4Galβ → 4Glcβ → Cer 3 (on internal Gal) ↑ αNANA8 ← αNANA
G_{T1a}	Galβ → 3GalNAcβ → 4Galβ → 4Glcβ → Cer 3 (on terminal Gal) ; 3 (on internal Gal) ↑ ; ↑ αNANA8 ← αNANA ; αNANA
G_{T1b}	Galβ → 3GalNAcβ → 4Galβ → 4Glcβ → Cer 3 (on terminal Gal) ; 3 (on internal Gal) ↑ ; ↑ αNANA ; αNANA8 ← αNANA
G_{Q1b}	Galβ → 3GalNAcβ → 4Galβ → 4Glcβ → Cer 3 (on terminal Gal) ; 3 (on internal Gal) ↑ ; ↑ αNANA8 ← αNANA ; αNANA8 ← αNANA

chloride ions into the gut lumen, resulting in the severe diarrhea characteristic of cholera. Two kinds of subunits, A and B, comprise the cholera toxin; there is one copy of the A subunit (28,000 daltons) and six copies of the B subunit (~12,000 daltons each). After binding to the cell surface membrane through a domain on the B subunit, the active subunit A passes into the cell, where it activates adenylate cyclase on the inner surface of the membrane. The cAMP that is generated then stimulates chloride ion transport and produces diarrhea. The choleragenoid domain, as the B subunits are called, binds to the gangliosidc G_{M1} that has the structure shown in Table 10.1.

Gangliosides are also thought to be receptors for other toxic agents, such as tetanus toxin, and certain viruses, such as the influenza viruses. There is also speculation that gangliosides may play an informational role in cell–cell interactions by providing specific recognition determinants on the surface of cells.

The gangliosides are also of medical interest for there are several lipid storage disorders that involve the accumulation of sialic acid-containing glycosphingolipids. The two most common gangliosidoses involve the storage of the gangliosides G_{M1} (G_{M1} gangliosidosis) and G_{M2} (Tay–Sachs disease).

G_{M1} gangliosidosis is an autosomal recessive metabolic disease characterized by impaired psychomotor function, mental retardation, hepatosplenomegaly, and death within the first few years of life. The massive cerebral and visceral accumulation of G_{M1} ganglioside is due to a profound deficiency of β-galactosidase.

THE SPHINGOLIPIDOSES: LYSOSOMAL STORAGE DISEASES AND THE CATABOLIC PATHWAY OF THE SPHINGOLIPIDS

The various sphingolipids are normally degraded within lysosomes of phagocytic cells, particularly the histiocytes or macrophages of the reticuloendothelial system located primarily in the liver, spleen, and bone marrow. Degradation of the sphingolipids by visceral organs begins with the engulfment of the membranes of white cells and erythrocytes that are rich in lactosylceramide (Cer-Glc-Gal) and hematoside (Cer-Glc-Gal-NANA). In the brain, the majority of the cerebroside-type lipids are gangliosides. Particularly during the neonatal period, ganglioside turnover in the central nervous system is extensive so that glycosphingolipids are rapidly being broken down and resynthesized. The pathway of sphingolipid catabolism is summarized in Figure 10.53. Note that among the various sphingolipids that comprise this pathway, there occurs a sulfate ester (in sulfolipid or sulfogalactolipid); *N*-acetylneuraminic acid groups (in the gangliosides); an α-linked galactose residue (in ceramide trihexoside); several β-galactosides (in galactocerebroside and the ganglioside G_{M1}); the ganglioside G_{M2}, which terminates in a β-linked *N*-acetylgalactosamine unit; and glucocerebroside, which is composed of a single glucose residue attached to ceramide through a β linkage. We also see that the phosphodiester bond in sphingomyelin is broken to produce ceramide, which is turn is converted in sphingosine by the cleavage of an amide bond to a long-chain fatty acid. This overall pathway of sphingolipid catabolism is composed of a series of enzymes that cleave specific bonds in the compounds that comprise the pathway: these enzymes include α- and β-galactosidases, a β-glucosidase, a neuraminidase, hexosaminidase, a sphingomyelin-specific phosphodiesterase (sphingomyelinase), a sulfate

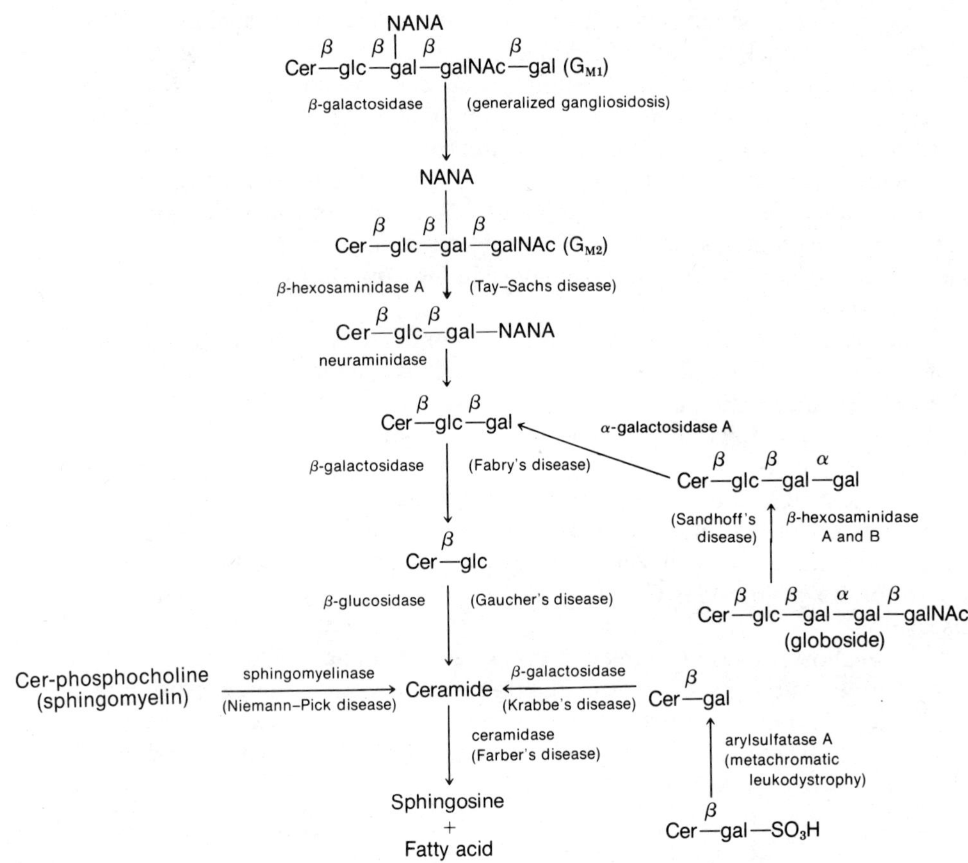

Figure 10.53
Summary of the pathways for the catabolism of sphingolipids by lysosomal enzymes.
The genetically determined enzyme deficiency diseases are indicated in the parentheses.

esterase (sulfatase), and a ceramide-specific amidase (ceramidase). The following statements summarize the important features of the sphingolipid catabolic pathway: (1) all the reactions take place within the lysosome; that is, the enzymes of the pathway are contained in lysosomes; (2) the enzymes are hydrolases; therefore, one of the substrates in each reaction is water; (3) the pH optimum of each of the hydrolases is in the acid range, pH 3.5–5.5; (4) most of the enzymes are relatively stable and occur as isoenzymes. For example, hexosaminidase occurs in two forms: hexosaminidase A (HexA) and hexosaminidase B (HexB); (5) the hydrolases of the sphingolipid pathway are glycoprotein in character and often occur firmly bound to the lysosomal membrane; and (6) the pathway is composed of a series of compounds that are related to the nearest compound in the pathway and which differ by only one sugar molecule, a sulfate group, or a fatty acid residue. That is, the substrates are converted to products by the sequential, stepwise removal of constituents such as sugars and sulfate, by hydrolytic, irreversible reactions.

In most cases, the pathway of sphingolipid catabolism functions smoothly, and all of the various complex glycosphingolipids and sphingomyelin are degraded to the level of their basic building blocks, namely, sugars, sulfate, fatty acid, phosphocholine, and sphingosine. However, when the activity of one of the hydrolytic enzymes in the pathway is markedly reduced in the tissues of a person due to a genetic, inborn error, then the substrate for the defective or missing enzyme accumulates and deposits within the lysosomes of the tissue responsible for the catabolism of that lipid. For most of the reactions in Figure 10.53, patients have been identified who lack the enzyme that normally catalyzes that reaction. These disorders are called sphingolipidoses. Table 10.2 summarizes the individual diseases that comprise the sphingolipidoses.

One can generalize about some of the common features of lipid storage diseases: (1) usually only a single sphingolipid accumulates in the involved organs; (2) the ceramide portion is shared by the various storage lipids; (3) the rate of biosynthesis of the accumulating lipid is normal; (4) a catabolic enzyme is missing in each of these disorders; and (5) the extent of the enzyme deficiency is the same in all tissues.

The diagnosis of a given sphingolipidosis can be made from a biopsy of the involved organ, usually bone marrow, liver, or brain, or on morphologic grounds on the basis of the highly characteristic appearance of the storage lipid within lysosomes. Biochemical methods involving enzyme assays are also widely used to confirm the diagnosis of a particular lipid storage disease. Of great practical value is the fact that, for most of the diseases, peripheral leukocytes and culture skin fibroblasts express the relevant enzyme deficiency and can be used as a source of enzyme for diagnostic purposes. In some cases (e.g., Tay–Sachs disease) serum, and even tears, have been used as a source of enzyme for the diagnosis of a lipid storage disorder. Because the sphingolipid storage diseases for the

TABLE 10.2 Sphingolipid Storage Diseases of Man

Disorder	*Principal Signs and Symptoms*	*Principal Storage Substance*	*Enzyme Deficiency*
1. Tay–Sachs disease	Mental retardation, blindness, cherry red spot on macula, death between second and third year	Ganglioside G_{M2}	Hexosaminidase A
2. Gaucher's disease	Liver and spleen enlargement, erosion of long bones and pelvis, mental retardation in infantile form only	Glucocerebroside	Glucocerebrosidase
3. Fabry's disease	Skin rash, kidney failure, pains in lower extremities	Ceramide trihexoside	α-Galactosidase A
4. Niemann–Pick disease	Liver and spleen enlargement, mental retardation	Sphingomyelin	Sphingomyelinase
5. Globoid leukodystrophy (Krabbe's disease)	Mental retardation, absence of myelin	Galactocerebroside	Galactocerebrosidase
6. Metachromatic leukodystrophy	Mental retardation, nerves stain yellowish brown with cresyl violet dye (metachromasia)	Sulfatide	Arylsulfatase A
7. Generalized gangliosidosis	Mental retardation, liver enlargement, skeletal involvement	Ganglioside G_{M1}	G_{M1} ganglioside: β-galactosidase
8. Sandhoff–Jatzkewitz disease	Same as 1; disease has more rapidly progressing course	G_{M2} ganglioside, globoside	Hexosaminidase A and B
9. Fucosidosis	Cerebral degeneration, muscle spasticity, thick skin	Pentahexosylfucoglycolipid	α-L-Fucosidase

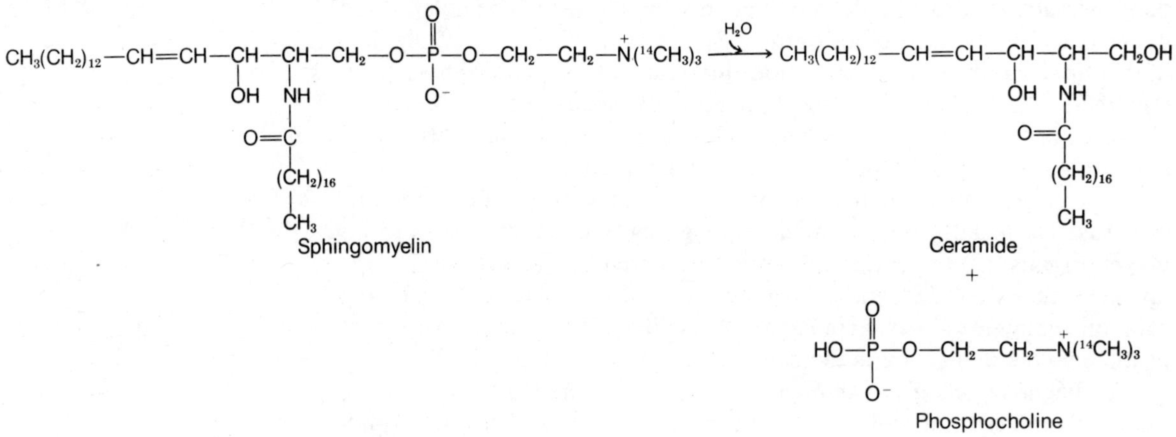

$$CH_3(CH_2)_{12}-CH{=}CH-\underset{OH}{CH}-\underset{NH-C(=O)-(CH_2)_{16}-CH_3}{CH}-CH_2-O-\overset{O}{\underset{O^-}{P}}-O-CH_2-CH_2-\overset{+}{N}(^{14}CH_3)_3 \xrightarrow{H_2O} CH_3(CH_2)_{12}-CH{=}CH-\underset{OH}{CH}-\underset{NH-C(=O)-(CH_2)_{16}-CH_3}{CH}-CH_2OH + HO-\overset{O}{\underset{O^-}{P}}-O-CH_2-CH_2-\overset{+}{N}(^{14}CH_3)_3$$

Sphingomyelin → Ceramide + Phosphocholine

Figure 10.54
The sphingomyelinase reaction.

most part are recessive in terms of their hereditary mode of transmission, and disease occurs only in homozygotes with a defect in both chromosomes, enzyme assays can also be used to identify carriers or heterozygotes. Let us consider some representative examples of the use of enzyme assays for diagnostic purposes.

In Niemann–Pick disease, the deficient enzyme is sphingomyelinase, which normally catalyzes the reaction shown in Figure 10.54. Sphingomyelin, radiolabeled in the methyl groups with carbon-14, provides a useful substrate for determining sphingomyelinase activity. Extracts of white blood cells from healthy, appropriate controls, will hydrolyze the labeled substrate and produce the water-soluble product, [^{14}C]phosphocholine. Extraction of the final incubation medium with an organic solvent such as chloroform will result in radioactivity in the upper, aqueous phase; the unused, lipidlike substrate sphingomyelin will be found in the chloroform phase. On the other hand, if the white blood cells were derived from a patient with Niemann–Pick disease, then after incubation with labeled substrate and extraction with chloroform, little or no radioactivity (i.e., phosphocholine) would be found in the aqueous phase and the diagnosis of sphingomyelinase deficiency or Niemann–Pick disease would be confirmed.

Because many hospitals or medical centers do not have access to radioactivity counting devices, it is fortunate that many of the lysosomal hydrolases have broad or versatile substrate specificity in that they will hydrolyze not only the true, natural sphingolipid substrate but also nonphysiologic, unnatural synthetic substrates that can be measured colorimetrically and fluorometrically. For example, the sulfatase deficiency in metachromatic leukodystrophy can be demonstrated, using the artificial substrate, *p*-nitrocatechol sulfate. The assay, called the arylsulfatase determination, is indicated in Figure 10.55. In alkaline medium, *p*-nitrocatechol is red. Thus the amount of red color generated by incubating the enzyme extract with nitrocatechol sulfate can be quantitated with the aid

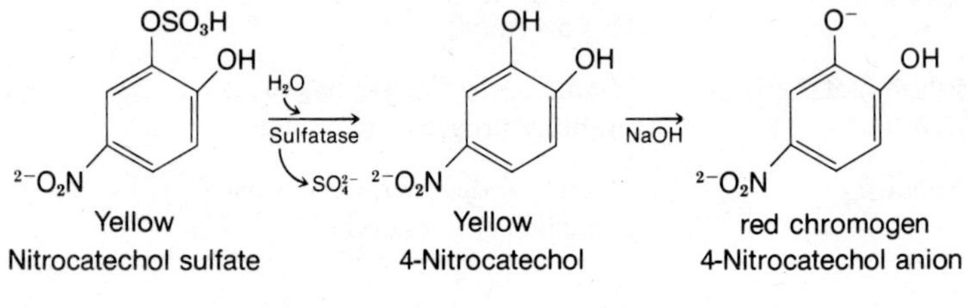

Nitrocatechol sulfate (Yellow; OSO_3H, OH, $^{2-}O_2N$) $\xrightarrow[\text{Sulfatase}]{H_2O}$ 4-Nitrocatechol (Yellow; OH, OH, $^{2-}O_2N$) + SO_4^{2-} $\xrightarrow{NaOH}$ 4-Nitrocatechol anion (red chromogen; O^-, OH, $^{2-}O_2N$)

Figure 10.55
The arylsulfatase assay.

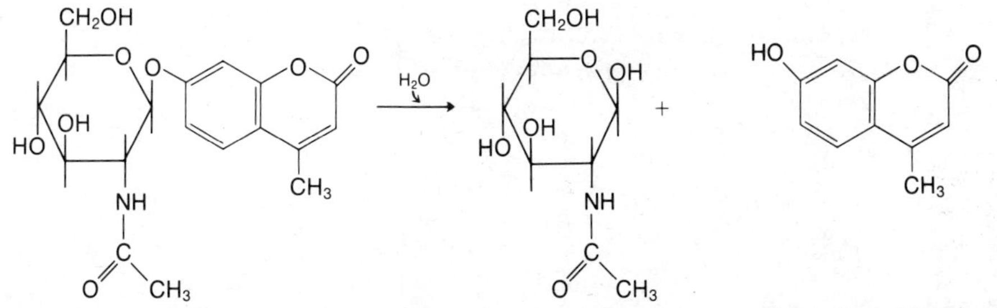

Figure 10.56
The β-hexosaminidase reaction.

of a spectrophotometer and is proportional to the level of sulfatase activity. The utility of the enzyme assay is due to the fact that sulfolipid sulfatase will act on the *p*-nitrocatechol sulfate.

Another disease that can be diagnosed by use of an artificial substrate is Tay–Sachs disease. Tay–Sachs disease is the most common form of G_{M2} gangliosidosis. In this fatal disorder the ganglion cells of the cerebral cortex are swollen and the lysosomes are engorged with the acidic lipid, G_{M2} ganglioside. This results in a loss of ganglion cells, proliferations of glial cells and demyelination of peripheral nerves. The pathognomonic finding is a cherry-red spot on the macula caused by swelling and necrosis of ganglion cells in the eye. In Tay–Sachs disease, the commercially available artificial substrate 4-methylumbelliferyl-β-*N*-acetylglucosamine is used to confirm the diagnosis. The compound is recognized and hydrolyzed by hexosaminidase A, the deficient lysosomal hydrolase, to produce the intensely fluorescent product 4-methylumbelliferone (Figure 10.56). Unfortunately, the diagnosis may be confused by the presence of hexosaminidase B in tissue extracts and body fluids. This enzyme is not deficient in the Tay–Sachs patient and will hydrolyze the test substrate, thereby confusing the interpretation of results. The problem is usually resolved by taking advantage of the relative heat lability of hexosaminidase A and the heat stability of hexosaminidase B. The tissue extract or serum specimen to be tested is first heated at 55°C for 1 h and then assayed for hexosaminidase activity. The amount of heat-labile activity is a measure of hexosaminidase A, and this value is used in making the diagnosis of Tay–Sachs disease.

Enzyme assays of serum or extracts of tissues, peripheral leukocytes and fibroblasts have proven useful in heterozygote detection. Once carriers of a lipid storage disease have been identified, or if there has been a previously affected child in a family, the pregnancies at risk for these diseases can be monitored. All nine of these lipid storage disorders are transmitted as recessive genetic abnormalities. In all but one the allele is carried on an autosomal chromosome. Fabry's disease is linked to the X chromosome. In all of these conditions statistically one of four pregnancies will be homozygous (or hemizygous in Fabry's disease), two fetuses will be carriers, and one will not be involved at all. The enzyme assay procedures have been used to detect affected fetuses and carriers in utero, using cultured fibroblasts obtained by amniocentesis as a source of enzyme.

There is no therapy for the sphingolipidoses; the role of medicine at the present time is prevention through genetic counseling based upon enzymologic assays of the type discussed above. A discussion of the diagnosis of Gaucher's disease is presented in Clin. Corr. 10.3.

CLIN. CORR. 10.3 DIAGNOSIS OF GAUCHER'S DISEASE IN AN ADULT

Gaucher's disease is an inherited disease of cell membrane catabolism that results in deposition of glucocerebroside in macrophages of the reticuloendothelial system. Because of the large numbers of macrophages in the spleen, splenomegaly and its sequelae (thrombocytopenia or anemia) are the most common signs of the disease.

A 61-year-old nurse (H.V.) of Ashkenazi Jewish ancestry was examined for persistent epigastric pain. A slightly enlarged spleen was detected, which was also seen with a liver–spleen scan. Leukocyte hydrolase assays were performed and showed low levels of β-glucosidase and glucocerebrosidase activity (see accompanying Table) in the range for patients with Gaucher's disease. A year later, the spleen appeared smaller by scan, and this raised questions about the diagnosis of Gaucher's disease. Repeat leukocyte hydrolase assays confirmed the low levels detected in the past. The patient then underwent bone marrow aspiration to obtain macrophages. The macrophages had the appearance of Gaucher cells, and electron micrographs showed the typical storage deposits of glucocerebroside.

Glucocerebrosidase and β-Glucosidase Activities in Leukocytes from the Propositus

Source of Leukocytes	*β-Glucosidase Activity*[a] *[nmol/(hr · mg protein)]*	*Glucocerebrosidase Activity*[b] *[nmol/(hr · mg protein)]*
Controls (n = 25)[c]		
Range	6.50–13.8	4.45–7.67
Mean	8.42	5.59
Patients with confirmed type 1 Gaucher's disease		
A.A.	1.69	1.51
P.T.	1.65	0.94
T.P.	1.20	0.70
Propositus (H.V.)		
February, 1979	1.00[12][d]	0.80[14]
April, 1980	0.95[11]	0.04[17]

[a] Determined using the artificial, fluorogenic substrate 4-methylumbelliferyl-β-D-glucopyranoside in the presence of 1.2% (w/v) sodium taurocholate under conditions that measure relative glucocerebrosidase activity.

[b] Determined using authentic, radiolabeled glucocerebroside as the substrate.

[c] n, number of individual samples.

[d] The number in brackets represents the value for the propositus expressed as a percentage of the mean value for the controls.

10.5 PROSTAGLANDINS AND THROMBOXANES

Structure and Biosynthesis

The prostaglandins were discovered through their effects on smooth muscle, specifically their ability to promote the contraction of intestinal and uterine muscle and the lowering of blood pressure. Although complexity of their structures and the diversity of their sometimes conflicting functions sometimes create a sense of frustration, the potent pharmacological effects of the prostaglandins have afforded them an important place in human biology and medicine. With the exception of the red blood cell, the prostaglandins are produced and released by nearly all mammalian cells and tissues; they are not confined to specialized cells as insulin is to the pancreas. Furthermore, unlike most other hormones, the prostaglandins are not stored in cells but instead are synthesized and released immediately.

There are three major classes of primary prostaglandins, the A, E, and F. series. The structures of the more common prostaglandins A, E, and F are shown in Figure 10.57. They are all related to prostanoic acid (Figure

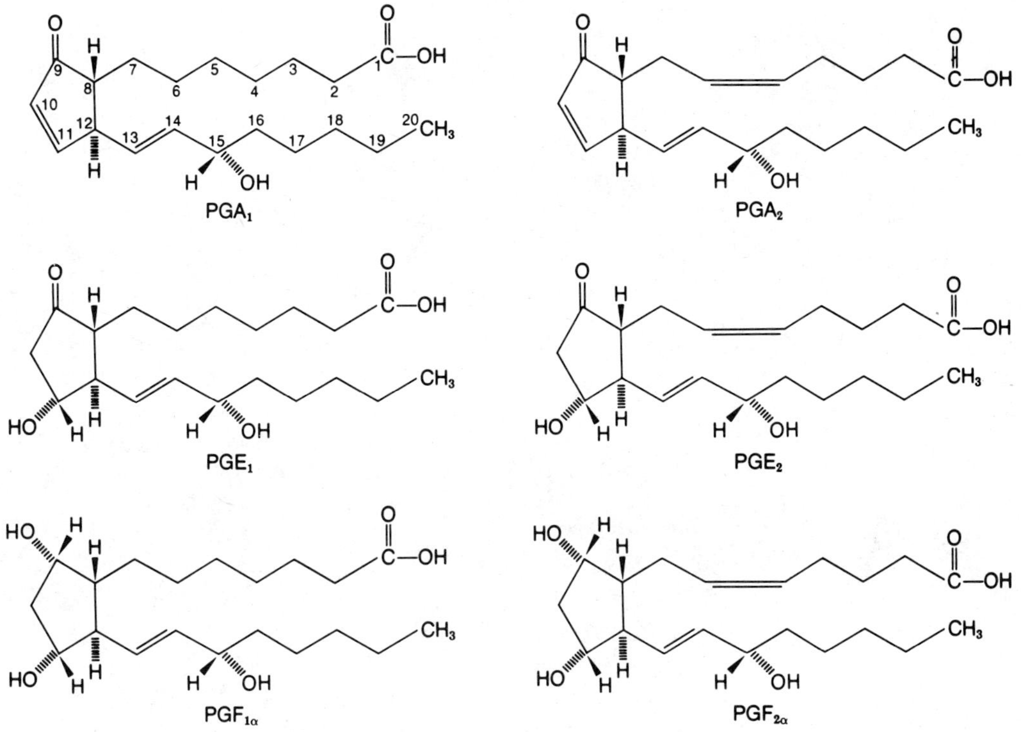

Figure 10.57
Structures of the major prostaglandins.

10.58). Note that the prostaglandins contain a multiplicity of functional groups; for example, PGE_2 contains a carboxyl group, a β-hydroxyketone, a secondary alkylic alcohol and two carbon–carbon double bonds. The three classes (A, E, and F) are distinguished on the basis of the functional groups about the cyclopentane ring: the E type is a β-hydroxyketone, the F series are 1,3-diols, and those in the A series are α,β-unsaturated ketones. The subscript numerals 1, 2, or 3, refer to the number of double bonds in the side chains. The subscript "α" refers to the configuration of the C-9 hydroxyl group: an α-hydroxyl group projects "down" from the plane of the ring.

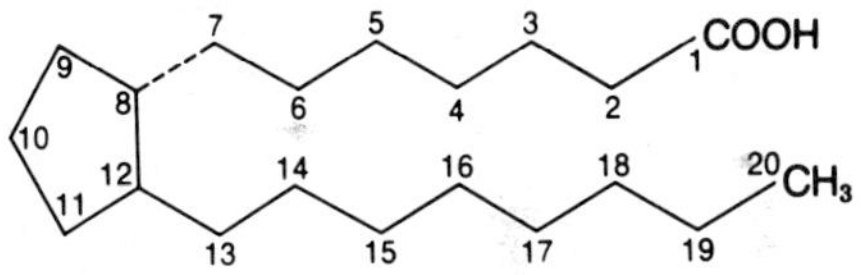

Figure 10.58
Structure of prostanoic acid.

The most important dietary precursor of the prostaglandins is linoleic acid (18 : 2) which is an essential fatty acid. In adults linoleic acid is ingested daily in amounts of about 10 g. Only a very minor part of this total intake is converted by elongation and desaturation in the liver to arachidonic acid and to some extent also to dihomo-γ-linoleic acid. Since the total daily excretion of prostaglandins and their metabolites is only about 1 mg, it is clear that the formation of prostaglandins is a quantitatively unimportant pathway in the overall metabolism of fatty acids. At the same time, however, the metabolism of prostaglandins is completely dependent on a regular and constant supply of linoleic acid. When the diet is deficient in linoleic acid, then there is decreased production of prostaglandins.

The immediate precursors to the prostaglandins are 20-carbon polyunsaturated fatty acids containing 3, 4, and 5 carbon–carbon double bonds. During their transformation into various prostaglandins they are cyclized and take up oxygen. Dihomo-γ-linoleic acid (C_{20}-Δ8,11,14) is the precursor to PGE_1 and $PGF_{1\alpha}$; arachidonic acid (C_{20}-Δ5,8,11,14) is the precursor to PGE_2 and $PGF_{2\alpha}$; and eicosopentaenoic acid (C_{20}-Δ5,8,11,14,17) is the precursor to PGE_3 and $PGF_{3\alpha}$. (See Figure 10.59).

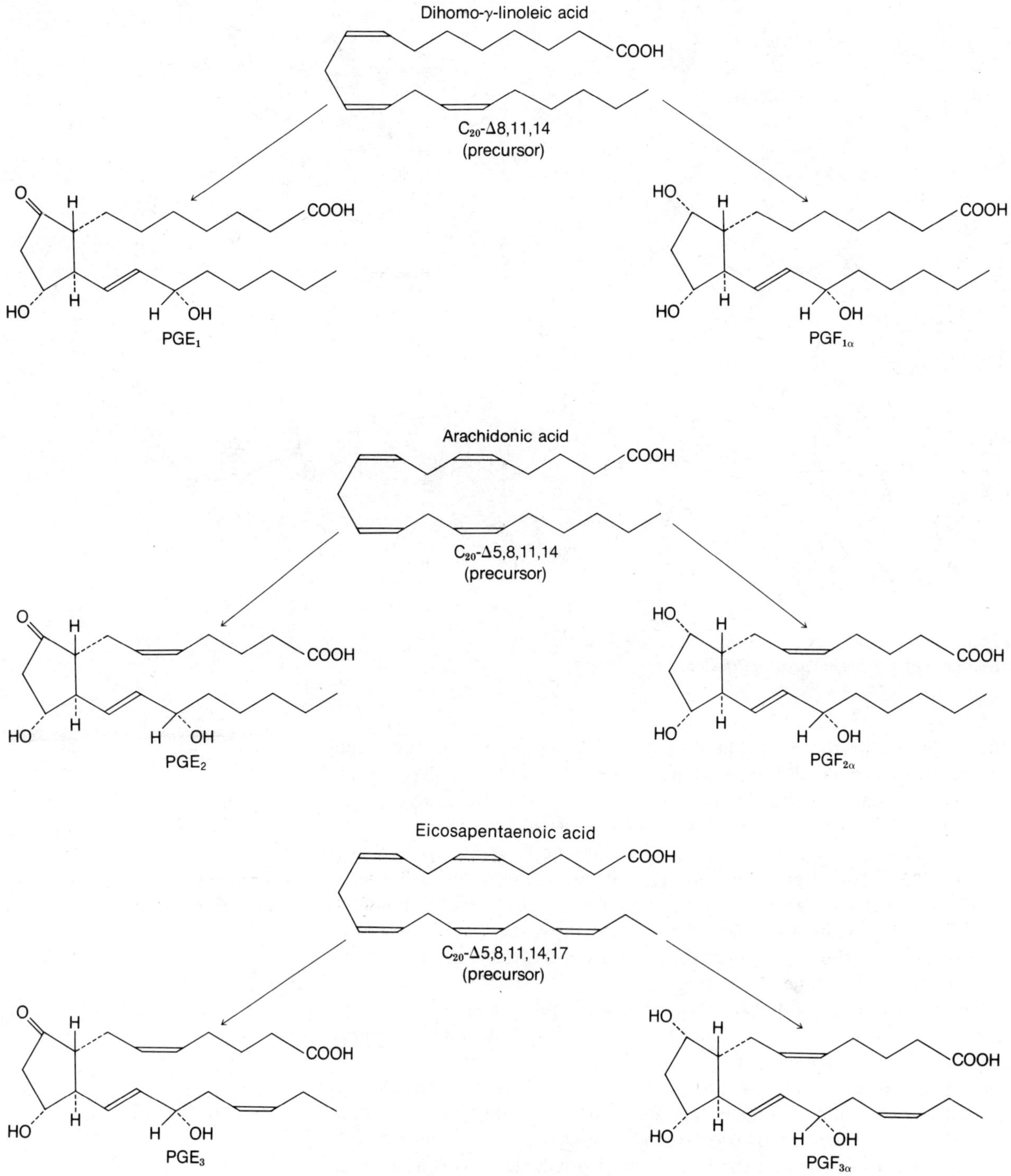

Dihomo-γ-linoleic acid

COOH

C_{20}-Δ8,11,14
(precursor)

PGE$_1$

PGF$_{1\alpha}$

Arachidonic acid

COOH

C_{20}-Δ5,8,11,14
(precursor)

PGE$_2$

PGF$_{2\alpha}$

Eicosapentaenoic acid

COOH

C_{20}-Δ5,8,11,14,17
(precursor)

PGE$_3$

PGF$_{3\alpha}$

Figure 10.59
Synthesis of E and F prostaglandins from fatty acid precursors.

Compounds of the 2-series derived from arachidonic acid are the principal prostaglandins in man and are of the greatest significance biologically. Thus, one should focus attention primarily on the metabolism of arachidonic acid.

The central enzyme system in prostaglandin biosynthesis is the prostaglandin synthase complex, which catalyzes the oxidative cyclization of polyunsaturated fatty acids. The major pathway of prostaglandin biosyn-

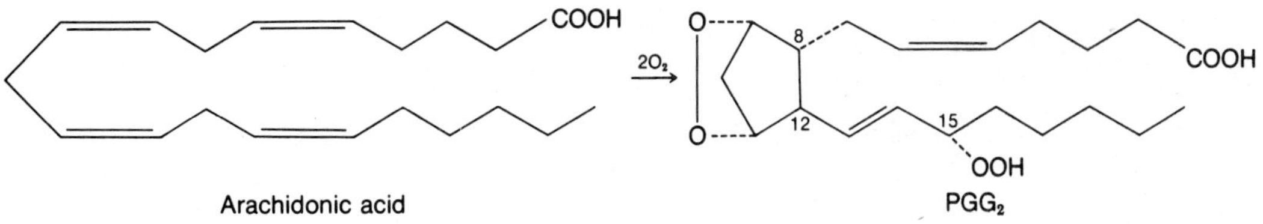

Figure 10.60
The cyclooxygenase reaction.

thesis will be illustrated, using arachidonic acid as substrate. Arachidonic acid is derived from membrane phospholipids by the action of the hydrolase phospholipase A_2. This esterolytic cleavage step is important because it is the rate-limiting step in prostaglandin synthesis and because certain agents that stimulate prostaglandin production act by stimulating the activity of phospholipase A_2. Cholesterol esters containing arachidonic acid may also serve as a source of arachidonic acid substrate. The first step catalyzed by the microsomal cyclooxygenase component of the prostaglandin synthase complex involves the cyclization of carbon atoms C-8 to C-12 of arachidonic acid to form the cyclic endoperoxide 15-hydroperoxide, PGG_2, and the reaction requires 2 molecules of oxygen as shown in Figure 10.60. The 15-endoperoxide PGG_2 is then converted to prostaglandin H_2 (PGH_2) by a reduced glutathione (GSH)-dependent peroxidase that catalyzes the reaction shown in Figure 10.61. The details of the additional steps leading to the individual prostaglandins remain to be elucidated. The reactions that cyclize polyunsaturated fatty acids are found in the membranes of the endoplasmic reticulum. The major pathways of prostaglandin biosynthesis are summarized in Figure 10.62. The formation of the primary prostaglandins of the D, E, and F series and of thromboxanes or prostacyclin (PGI_2) is mediated by different specific enzymes, whose presence varies depending upon the cell type and tissue. This results in a degree of tissue specificity as to the type and quantity of prostaglandin produced. Thus, in the kidney and spleen PGE_2 and $PGF_{2\alpha}$ are the major prostaglandins produced. In contrast, blood vessels produce mostly PGI_2. In the heart PGE_2, $PGF_{2\alpha}$, and PGI_2 are formed in about equal amounts. Thromboxane A_2 (TXA_2) is the main prostaglandin endoperoxide formed in platelets.

The prostaglandins have a very short half-life. Soon after release they are rapidly taken up by cells and inactivated. The lungs appear to play an important role in inactivating prostaglandins.

Thromboxanes, mentioned above, are the highly active metabolites of the PGG_2- and PGH_2-type prostaglandin endoperoxides that have the cyclopentane ring replaced by a six-membered oxygen-containing (oxane) ring. The term thromboxane is derived from the fact that these compounds have a thrombus-forming potential. The microsomal enzyme thromboxane A synthetase that is abundant in lung and platelets catalyzes

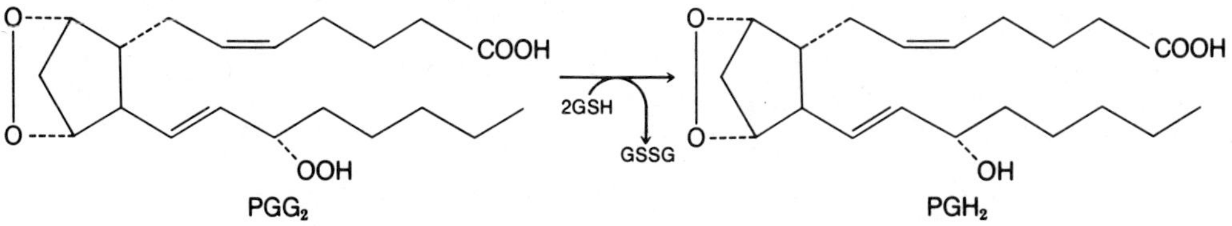

Figure 10.61
Conversion of PGG_2 to PGH_2.

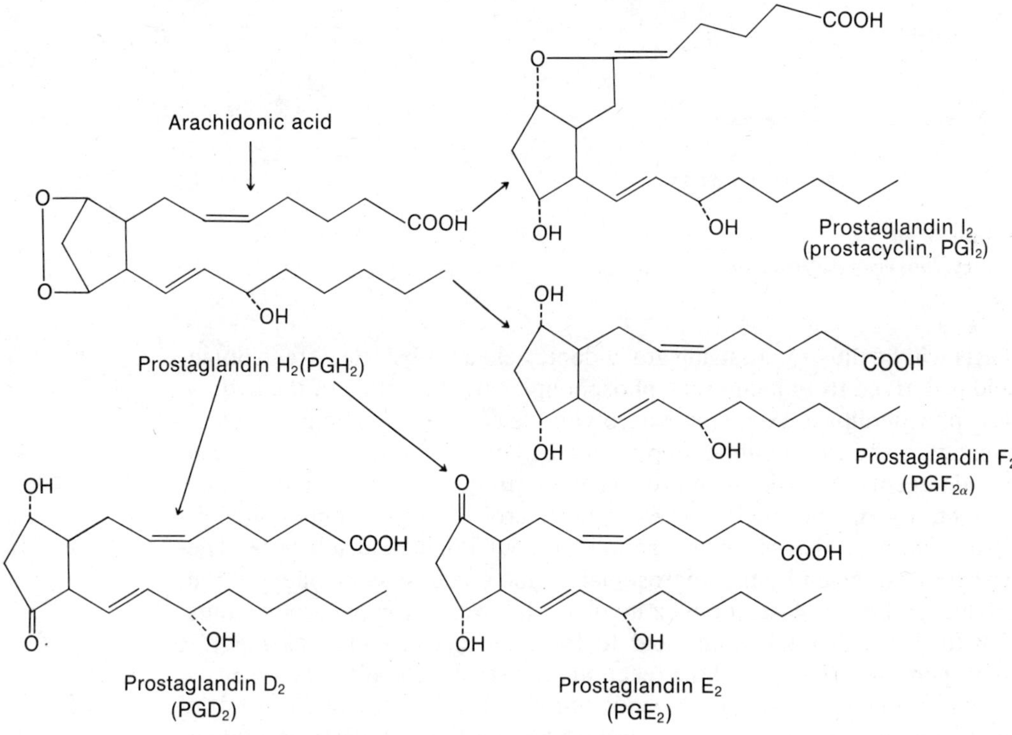

Arachidonic acid

Prostaglandin H_2(PGH_2)

Prostaglandin I_2 (prostacyclin, PGI_2)

Prostaglandin $F_{2\alpha}$ ($PGF_{2\alpha}$)

Prostaglandin D_2 (PGD_2)

Prostaglandin E_2 (PGE_2)

Figure 10.62
Major routes of prostaglandin biosynthesis.

the conversion of endoperoxide PGH_2 to TXA_2. The half-life of TXA_2 is very short in water ($t_{1/2}$, 1 min) as the compound is transformed rapidly into biologically inactive thromboxane B_2 (TXB_2) by the reaction shown in Figure 10.63.

Inhibitors of Cyclooxygenase

Clinically there are two types of drugs that affect prostaglandin metabolism and are therapeutically useful. First, there are the nonsteroidal, anti-inflammatory agents such as aspirin (acetylsalicylic acid), indomethacin,

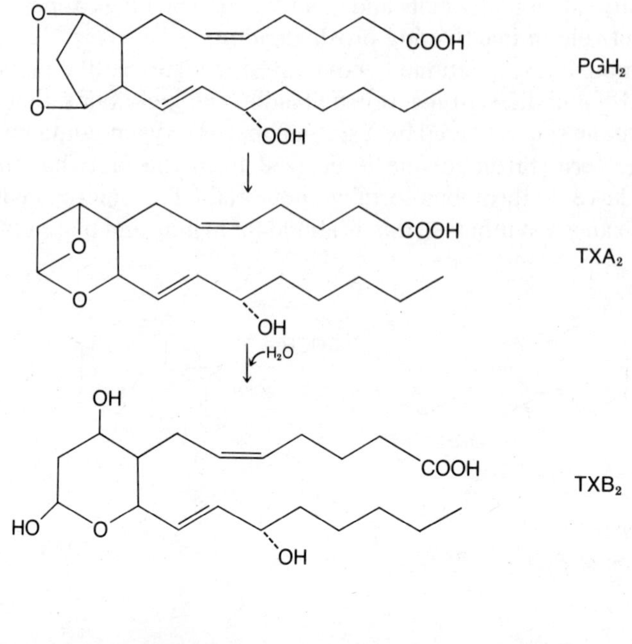

PGH_2

TXA_2

H_2O

TXB_2

Figure 10.63
Synthesis of TXB_2 from PGH_2.

and phenylbutazone, which block prostaglandin production by irreversibly inhibiting the enzyme cyclooxygenase. In the case of aspirin, inhibition occurs presumably by acetylation of the enzyme. These drugs are not without their undesirable side effects: aplastic anemia can result from phenylbutazone therapy. The second group, the steroidal antiinflammatory drugs like hydrocortisone, prednisone, and betamethasone, appear to act by blocking prostaglandin release by inhibiting phospholipase A_2 activity so as to interfere with mobilization of arachidonic acid, the substrate for cyclooxygenase. (See Figure 10.64)

The factors that govern the biosynthesis of prostaglandins are poorly understood, but, in general, prostaglandin release seems to be triggered following hormonal or neural excitation or after muscular activity. For example, histamine stimulates an increase in the prostaglandin concentration in gastric perfusates. Also, prostaglandins are released during labor and after cellular injury (e.g., platelets exposed to thrombin, lungs irritated by dust).

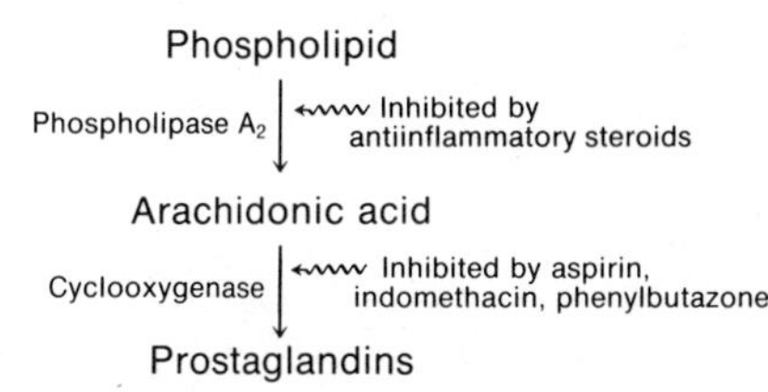

Figure 10.64
Site of action of inhibitors of prostaglandin synthesis.

Physiological Effects of the Prostaglandins

Inflammation

Prostaglandins appear to be one of the natural mediators of inflammation. Inflammatory reactions most often involve the joints (rheumatoid arthritis), skin (psoriasis), and eyes, and inflammation of these sites is frequently treated with corticosteroids that inhibit prostaglandin synthesis. Administration of the prostaglandins PGE_2 and PGE_1 induce the signs of inflammation that include redness and heat (due to arteriolar vasodilation), and swelling and edema resulting from increased capillary permeability.

Pain and Fever

PGE_2 in amounts that alone do not cause pain, prior to administration of the autocoids, histamine and bradykinin, enhance both the intensity and the duration of pain caused by these two agents. It is thought that pyrogen activates the prostaglandin biosynthetic pathway, resulting in the release of PGE_2 in the region of the hypothalamus where body temperature is regulated. Aspirin, which is an antipyretic drug, acts by inhibiting cyclooxygenase.

Reproduction

The prostaglandins have been used extensively as drugs in the reproductive area. PGE_2 and PGF_2 have been used to induce parturition and for the termination of an unwanted pregnancy. There is also evidence that the PGE series of prostaglandins may play some role in infertility in males.

Gastric Secretion and Peptic Ulcer

Synthetic prostaglandins have proven to be very effective in inhibiting gastric acid secretion in patients with peptic ulcers. The inhibitory effect of PGE compounds appears to be due to inhibition of cAMP formation in gastric mucosal cells. Prostaglandins also accelerate the healing of gastric ulcers.

Regulation of Blood Pressure

Prostaglandins play an important role in controlling blood vessel tone and arterial pressure. The vasodilator prostaglandins, PGE, PGA, and PGI_2, lower systemic arterial pressure, thereby increasing local blood flow and decreasing peripheral resistance. There is hope that the prostaglandins may eventually prove useful in the treatment of hypertension.

Ductus Arteriosus and Congenital Heart Disease

PGE_2 functions in the fetus to maintain the patency of the ductus arteriosus prior to birth. There are two clinical applications of prostaglandin biochemistry in this area. First, if the ductus remains open after birth, closure can be hastened by administration of the cyclooxygenase inhibitor indomethacin. In other situations it may be desirable to keep the ductus open. For example, in the case of infants born with congenital abnormalities where the defect can be corrected surgically, infusion of prostaglandins will maintain blood flow through the ductus over this interim period.

Platelet Aggregation and Thrombosis

Certain prostaglandins, especially PGI_2, inhibit platelet aggregation, whereas PGE_2 and TXA_2 promote this clotting process. TXA_2 is produced by platelets and accounts for the spontaneous aggregation that occurs when platelets contact some foreign surface, collagen or thrombin. Endothelial cells lining blood vessels release PGI_2 and may account for the lack of adherence of platelets to the healthy blood vessel wall.

10.6 LIPOXYGENASE AND THE OXY-EICOSATETRAENOIC ACIDS

In addition to cyclooxygenase, which directs polyunsaturated fatty acids into prostaglandins, there exists another equally important arachidonic acid oxygenating enzyme, called lipoxygenase. The products of the lipoxygenase reaction, which arise by addition of hydroperoxy groups to arachidonic acid, are designated hydroperoxy-eicosatetraenoic acids (HPETEs). Figure 10.65 shows the conversion of arachidonic acid to 5-HPETE. Thus, in contrast to the cyclooxygenase component of prostaglandin endoperoxide synthetase, which catalyzes the bis-dioxygenation of unsaturated fatty acids to endoperoxides, lipoxygenase catalyzes

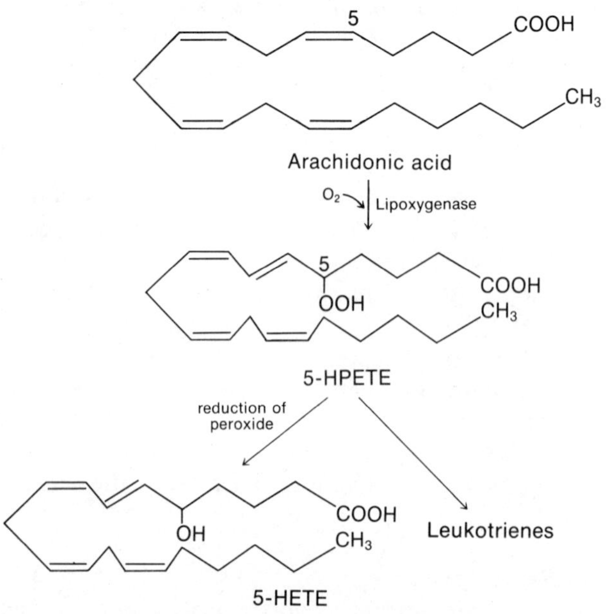

Figure 10.65
The lipoxygenase reaction and the role of 5-hydroperoxy eicosatetraenoic acid (5-HPETE) as the precursor of 5-hydroxy-eicosatetraenoic acid (5-HETE).

the monodioxygenation of unsaturated fatty acids to allylic hydroperoxides. Hydroperoxy substitution of arachidonic acid by lipoxygenases in humans may occur at position 5, 8, 9, 11, 12, or 15. 5-HPETE is the major lipoxygenase product in basophils, polymorphonuclear (PMN) leukocytes, and macrophages; 12-HPETE predominates in platelets, pancreatic endocrine islet cells, and glomerular cells; 15-HPETE is the principal lipoxygenase product in reticulocytes, eosinophils and T-lymphocytes. Specific stimuli or signals determine which type of lipoxygenase product a given type of cell produces.

The HPETE-hydroperoxides themselves are not hormones, but instead are highly reactive, unstable intermediates that are converted either to the analogous alcohol (hydroxy fatty acid) by reduction of the peroxide moiety or to leukotrienes. Figure 10.66 shows how 5-HPETE serves as a precursor to the hydroxy fatty acids (e.g., LTB_4) and the leukotrienes, thus emphasizing that this particular HPETE occurs at an important branch point in the lipoxygenase pathway.

The peroxidative reduction of 5-HPETE to the stable 5-hydroxy-eicosatetraenoic acid (5-HETE) is illustrated in Figure 10.65. Note that the carbon–carbon double bonds in 5-HETE occur at positions 6, 8, 11, and 14, that these double bonds are unconjugated, and that the geometry of the double bonds is trans, cis, cis, cis, respectively.

Leukotrienes are derived from the unstable precursor 5-HPETE by a reaction that generates an epoxide called leukotriene A_4 (LTA_4). As indi-

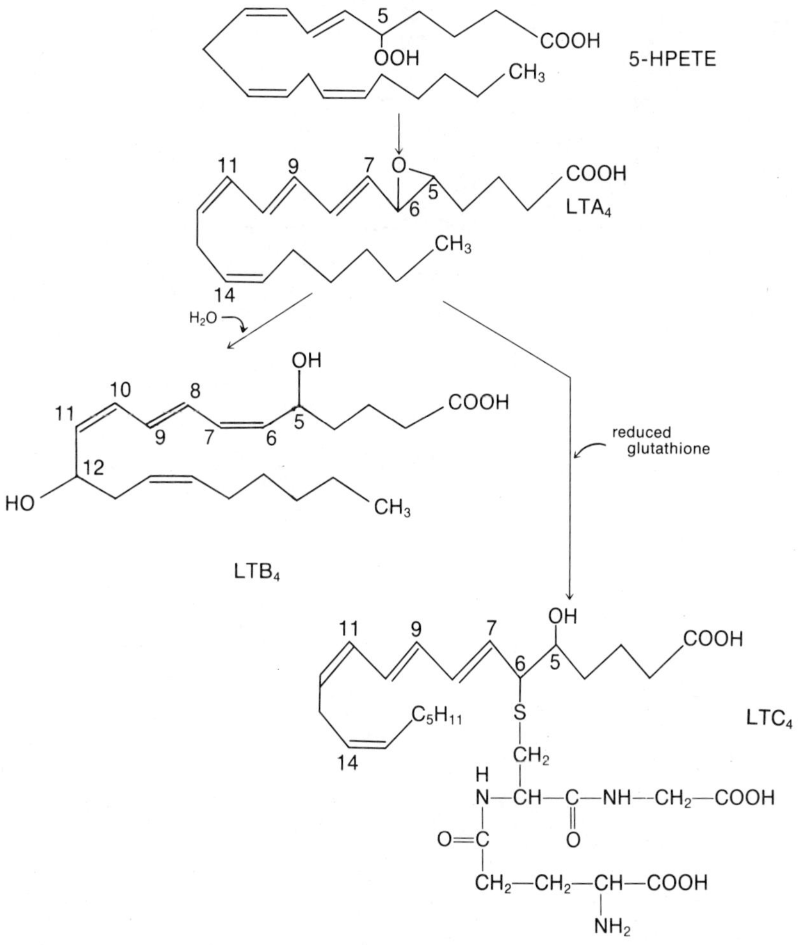

5
COOH
OOH
CH_3
5-HPETE

11 9 7 O
5
6
COOH
LTA_4
CH_3
14
H_2O

OH
10 8
11
COOH
5
9 7 6
12
HO
CH_3
LTB_4

reduced glutathione

OH
11 9 7
COOH
6 5
C_5H_{11}
S
LTC_4
CH_2
14
H
N—CH—C—NH—CH₂—COOH
O=C O
CH₂—CH₂—CH—COOH
NH_2

Figure 10.66
Conversion of 5-HPETE to LTB_4 and LTC_4 through the intermediate of LTA_4.

cated in Figure 10.66, LTA_4 occurs at a branch point; it can be converted either to 5, 12-dihydroxyeicosatetraenoic acid (designated leukotriene B_4 or LTB_4) or to the leukotrienes LTC_4 and LTD_4.

Conversion of 5-HPETE to the diol LTB_4, shown in Figure 10.66, is catalyzed by a cytosolic enzyme in a reaction initiated by the addition of water to the double bond between carbon atoms 11 and 12.

The diversion of LTA_4 to leukotrienes LTC_4, LTD_4, and LTE_4 requires the participation of reduced glutathione that opens the epoxide ring in LTA_4 to produce LTC_4 (Figure 10.66). Sequential removal of glutamic acid and glycine residues by specific peptidases yields the leukotrienes LTD_4 and LTE_4 (Figure 10.67). The subscript 4 denotes the total number of carbon–carbon double bonds.

The leukotrienes persist for as long as 4 hours in the body, but little is known about the mechanisms that degrade or eliminate them. The biological actions of the thionyl peptides LTC_4, and LTD_4, and LTE_4 comprise what has been referred to for decades as the slow-reacting substance of anaphylaxis (SRS-A).

In general, the HETEs (especially 5-HETE) and LTB_4 are involved mainly in regulating neutrophil and eosinophil function: They mediate chemotaxis, stimulate adenylate cyclase, and induce PMNs to degranulate and release lysosomal hydrolytic enzymes. In contrast, the leukotrienes LTC_4 and LTD_4 are humoral agents that promote smooth muscle contraction, constriction of pulmonary airways, trachea and intestine, and changes in capillary permeability (edema).

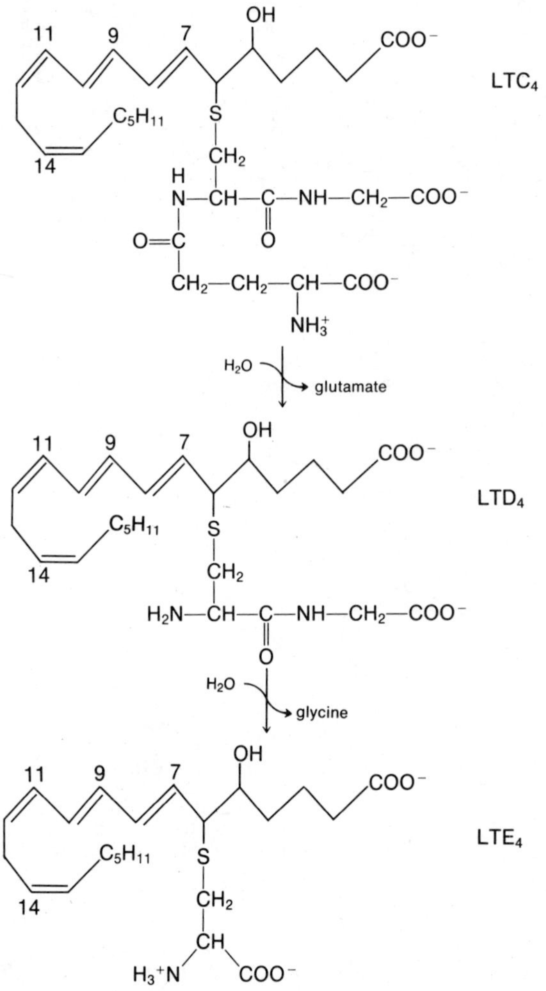

Figure 10.67
Conversion of LTC_4 to LTD_4 and LTE_4.

The hydroxyeicosatetraenoic acids that comprise the lipoxygenase pathway are potent mediators of processes involved in allergy (hypersensitivity) and inflammation. The initial allergic event, namely the binding of IgE antibody to receptors on the surface of the mast cell, causes the release of substances, including leukotrienes, that are referred to as *mediators of immediate hypersensitivity*. LTC_4, LTD_4, and LTE_4 are much more potent than histamine in contracting nonvascular smooth muscles of bronchi and intestine. LTD_4 increases the permeability of the microvasculature. LTB_4 and the mono-HETEs stimulate migration (chemotaxis) of eosinophils and neutrophils, making them the principal mediators of PMN-leukocyte infiltration in inflammatory reactions.

Although a potential fertile ground for the application of pharmacologic agents, to date, therapeutic use of lipoxygenase inhibitors has been prevented by their toxicity or lack of specificity.

BIBLIOGRAPHY

Phospholipid Metabolism

Holub, B. J., and Kuksis, A. Metabolism of molecular species of diacylglycerophospholipids. *Adv. Lipid Res.* 16:1, 1978.

Pelech, S. L., and Vance, D. E. Regulation of phosphatidylcholine biosynthesis. *Biochim. Biophys. Acta* 779:217, 1984.

Lung Surfactant

Shelly, S. A., Kovacevic, M., Paciga, J. E., and Balis, J. U. Sequential changes of surfactant phosphatidylcholine in hyaline-membrane disease of the newborn, *New Engl. J. Med.* 300:112, 1979.

Torday, J., Carson, L., and Lawson, E. E. Saturated phosphatidylcholine in amniotic fluid and prediction of the respiratory distress syndrome. *New Engl. J. Med.* 300:1013, 1979.

Cholesterol Synthesis

Ahrens, E. H. Dietary fats and coronary heart disease: Unfinished business. *Lancet* 1345, 1979.

Brown, M. S., and Goldstein, J. L. How LDL receptors influence cholesterol and atherosclerosis. *Sci. Am.* Nov. p. 58, 1984.

Brown, M. S., and Goldstein, J. L. Multivalent feedback regulation of HMG CoA reductase, a control mechanism coordinating isoprenoid synthesis and cell growth. *J. Lipid Res.* 21:505, 1980.

Eisenberg, S. High density lipoprotein metabolism. *J. Lipid Res.* 25:1017, 1984.

Lipid Research Clinics Program. The lipid research clinics coronary primary prevention trial results. I. Reduction in incidence of coronary heart disease. *J. Am. Med. Assoc.* 251:351, 1984.

Bile Acids

Gibbons, G. F., Mitropoulos, K. A., and Myant, N. B. *Biochemistry of cholesterol.* New York: Elsevier Biomedical Press, 1982.

Hanson, R. F., and Pries, J. M. Synthesis and enterohepatic circulation of bile salts. *Gastroenterology* 73:611, 1977.

Sphingolipids and the Sphingolipidoses

Glew, R. H., and Peters, S. P. *Practical enzymology of the sphingolipidoses.* New York: Alan R. Liss, 1977.

Stanbury, J. B., Wyngaarden, J. B., Fredrickson, D. S., Goldstein, J. L., and Brown, M. S. *The metabolic basis of inherited disease,* 5th ed. New York: McGraw-Hill, 1983.

Prostaglandins, Thromboxanes, and Leukotrienes

Curtis-Prior, B. P. *Prostaglandins, an introduction to their biochemistry, physiology and pharmacology.* New York: North-Holland, 1976.

Johnson, M., Carey, F., and McMillan, R. M. Alternate pathways of arachidonate metabolism: Prostaglandins, thromboxane, and leukotrienes, in P. N. Campbell and R. D. Marshall (eds.), *Essays in biochemistry.* New York: Academic Press, 1983.

Piper, P. J. Formation and action of leukotrienes. *Phys. Rev.* 64:744, 1984.

Samuelsson, B. Prostaglandins and thromboxanes, in R. O. Greep (ed.), *Recent progress in hormone research,* vol. 34. New York: Academic Press, 1978, p. 239.

QUESTIONS

C. N. ANGSTADT AND J. BAGGOTT

*Q*uestion *T*ypes are described inside the front cover.

A.

$$\mathrm{CH_2{-}O{-}\overset{O}{\overset{\|}{C}}{-}(CH_2)_{14}CH_3}$$
$$\mathrm{CH_3(CH_2)_{16}{-}\overset{O}{\overset{\|}{C}}{-}O{-}CH}$$
$$\mathrm{CH_2{-}O{-}\overset{O}{\overset{\|}{C}}{-}(CH_2)_{14}CH_3}$$

B.

$$\mathrm{CH_2{-}O{-}\overset{O}{\overset{\|}{C}}{-}(CH_2)_7CH{=}CH{-}(CH_2)_5{-}CH_3}$$
$$\mathrm{CH_3(CH_2)_3(CH_2{-}CH{=}CH)_4{-}(CH_2)_3{-}\overset{O}{\overset{\|}{C}}{-}O{-}CH}$$
$$\mathrm{CH_2{-}O{-}\overset{O}{\overset{\|}{C}}{-}(CH_2)_{14}CH_3}$$

C.

$$\mathrm{CH_2{-}O{-}CH{=}CH(CH_2)_{15}CH_3}$$
$$\mathrm{R{-}\overset{O}{\overset{\|}{C}}{-}O{-}CH}$$
$$\mathrm{CH_2{-}O{-}\overset{O}{\overset{\|}{P}}(O^-){-}O{-}(CH_2)_2{-}\overset{+}{N}H_3}$$

D.

$$\mathrm{CH_2{-}O{-}\overset{O}{\overset{\|}{C}}{-}R_1}$$
$$\mathrm{R_2{-}\overset{O}{\overset{\|}{C}}{-}O{-}CH}$$
$$\mathrm{CH_2{-}O{-}\overset{O}{\overset{\|}{P}}(O^-){-}O{-}(CH_2)_2{-}\overset{+}{N}(CH_3)_3}$$

E.

$$\mathrm{CH_2{-}O{-}\overset{O}{\overset{\|}{C}}{-}R_1 \quad CH_2{-}O{-}\overset{O}{\overset{\|}{P}}(O^-){-}O{-}CH_2}$$
$$\mathrm{R_2\overset{O}{\overset{\|}{C}}{-}O{-}CH \qquad HO{-}C{-}H \qquad HC{-}O{-}\overset{O}{\overset{\|}{C}}{-}R_3}$$
$$\mathrm{CH_2{-}O{-}\overset{O}{\overset{\|}{P}}(O^-){-}O{-}CH_2 \qquad R_4\overset{O}{\overset{\|}{C}}{-}O{-}CH_2}$$

1. (QT5) A plasmalogen.
2. (QT5) A cardiolipin.
3. (QT5) An acylglycerol that would likely be liquid at room temperature.
4. (QT2) Roles of various phospholipids include:
   1. a surfactant function in the lung.
   2. activation of certain membrane enzymes.
   3. a detergent function in bile.
   4. cell–cell recognition.
5. (QT1) Which of the following represents a correct group of enzymes involved in triacylglycerol synthesis in ADIPOSE tissue?
   A. choline phosphotransferase, glycerol kinase, phosphatidic acid phosphatase.
   B. choline phosphotransferase, glycerol phosphate:acyl transferase, α-glycerolphosphate dehydrogenase.
   C. glycerol phosphate:acyl transferase, α-glycerolphosphate dehydrogenase, phosphatidic acid phosphatase.
   D. glycerol phosphate:acyl transferase, α-glycerolphosphate dehydrogenase, glycerol kinase.
   E. α-glycerolphosphate dehydrogenase, glycerol kinase, phosphatidic acid phosphatase.
6. (QT1) Phospholipases A_1 and A_2:
   A. have no role in phospholipid synthesis.
   B. are responsible for the initial insertion of fatty acids in sn-1 and sn-2 positions during synthesis.
   C. are responsible for base-exchange in the interconversion of phosphatidylethanolamine and phosphatidylserine.
   D. hydrolyze a phosphatidic acid to a diglyceride.
   E. remove a fatty acid in an sn-1 or sn-2 position so it can be replaced by another in phospholipid synthesis.
7. (QT2) In the biosynthesis of cholesterol:
   1. 3-hydroxy-3-methyl glutaryl CoA (HMG CoA) is synthesized by cytosolic HMG CoA synthase.
   2. HMG CoA reductase catalyzes the rate-limiting step.
   3. the conversion of mevalonic acid to farnesyl pyrophosphate requires more than 1 mol of ATP/mole of mevalonic acid.
   4. the conversion of squalene to lanosterol is initiated by formation of an epoxide.
8. (QT2) The cholesterol present in LDL (low-density lipoproteins):
   1. binds to a cell receptor and diffuses across the cell membrane.
   2. when it enters a cell, suppresses the cell's cholesterol synthesis by inhibiting HMG CoA reductase.
   3. once in the cell is converted to cholesterol esters by LCAT (lecithin-cholesterol acyl transferase).
   4. once it has accumulated in the cell, inhibits the replenishment of LDL receptors.

9. (QT1) Primary bile acids:
   A. are any bile acids that are found in the intestinal tract.
   B. are any bile acids reabsorbed from the intestinal tract.
   C. are synthesized in the intestinal tract by bacteria.
   D. are synthesized in hepatocytes directly from cholesterol.
   E. are converted to secondary bile acids by conjugation with glycine or taurine.
10. (QT2) A ganglioside may contain which of the following?
    1. One or more sialic acids
    2. Glucose or galactose
    3. A ceramide structure
    4. Phosphate
11. (QT1) Sphingomyelins differ from the other sphingolipids in that they are:
    A. not based on a ceramide core.
    B. acidic rather than neutral at physiological pH.
    C. the only types containing *N*-acetylneuraminic acid.
    D. the only types that are phospholipids.
    E. not amphipathic.
12. (QT2) The degradation of sphingolipids:
    1. occurs by hydrolytic enzymes contained in lysosomes.
    2. is a sequential, stepwise removal of constituents.
    3. is inhibited in the types of diseases known as sphingolipidoses (lysosomal storage diseases).
    4. is catalyzed by enzymes that are specific for a type of linkage rather than for a particular compound.
13. (QT2) Structural features that are common to all prostaglandins include:
    1. 20 carbons.
    2. an internal ring structure.
    3. at least one double bond.
    4. a peroxide group at carbon 15.
14. (QT2) The prostaglandin synthase complex:
    1. contains both a cyclooxygenase and a peroxidase component.
    2. is inhibited by anti-inflammatory steroids.
    3. produces PGH_2.
    4. uses as substrate the pool of free arachidonic acid in the cell.
15. (QT1) Thromboxane A_2:
    A. is a long-lived prostaglandin.
    B. is an inactive metabolite of PGE_2.
    C. is the major prostaglandin produced in all cells.
    D. does not contain a ring structure.
    E. is synthesized from the intermediate PGH_2.
16. (QT2) Hydroperoxy eicosatetraenoic acids (HPETEs):
    1. are derived from arachidonic acid by a lipoxygenase reaction.
    2. are mediators of hypersensitivity reactions.
    3. are intermediates in the formation of leukotrienes.
    4. are relatively stable compounds (persist for as long as 4 hours).

ANSWERS

1. C Only one with an ether instead of an ester link at sn-1. D is a phosphatidylcholine (p. 393).
2. E Two phosphatidic acids connected by glycerol (p. 394).
3. B Note the two unsaturated fatty acids. A: With all saturated fatty acids, would likely be solid at room temperature.
4. A 1, 2, 3 correct. 1: Especially dipalmitoyllecithin (p. 394). 2: For example, β-hydroxybutyrate dehydrogenase (p. 394). 3: Solubilize cholesterol (p. 395). 4: This function appears to be associated with complex glycosphingolipids (p. 392).
5. C Glycerol kinase is not present in adipose tissue, which must rely on the α-glycerolphosphate dehydrogenase; choline phosphotransferase is involved in the synthesis of phosphatidylcholines, not triacylglycerols (p. 395).
6. E Phospholipases A_1 and A_2, as their names imply, hydrolyze a fatty acid from a phospholipid and so are part of phospholipid degradation. They are also important in synthesis, however, in assuring the asymmetric distribution of fatty acids that occurs in phospholipids (p. 399).
7. E All four correct (p. 403). Remember that cholesterol biosynthesis is cytosolic; mitochondrial biosynthesis of HMG CoA leads to ketone body formation.
8. C 2, 4 correct. 1: The LDL binds to the cell receptor and is endocytosed and then degraded in lysosomes to release cholesterol. 3: LCAT is a plasma enzyme; the tissue enzyme that forms cholesterol esters is ACAT (p. 409).
9. D The intestinal tract contains a mixture of primary and secondary bile acids, both of which can be reabsorbed. Secondary bile acids are formed by bacteria in the intestine by chemical reactions, such as the removal of the 7-hydroxyl group (p. 410).
10. A 1, 2, 3 correct. The glycosphingolipids do not contain phosphate. Ceramide is the base structure from which the glycosphingolipids are formed (p. 417).
11. D Sphingomyelins are not glycosphingolipids. They are formed from ceramides, are amphipathic, and are neutral. C is the definition of gangliosides (p. 414).
12. E All four correct. 4: Many of the sphingolipids share the same types of bonds, for example, a β-galactosidic bond, and one enzyme, for example, β-galactosidase, will hydrolyze it whenever it occurs (p. 419, Figure 10.53).
13. A 1, 2, 3 correct. This is so whether or not you include thromboxane A_2 as a prostaglandin (in TXA_2 the ring is 6-membered rather than 5). 4 is true only of the intermediate of synthesis, PGG_2 (Figures 10.59–10.63).
14. B 1, 3 correct. 2: Anti-inflammatory steroids inhibit the release of the precursor fatty acid by phospholipase A_2. 4: Arachidonic acid is not free in the cell but is part of the membrane phospholipids (p. 426).
15. E TXA_2 is very active, has a very short half-life, contains a 6-membered ring, and is the main prostaglandin in platelets but not all tissues (p. 427).
16. B 1, 3 correct. 2–4: HPETEs, themselves, are not hormones but highly unstable intermediates that are converted to either the HETEs (mediators of hypersensitivity) or leukotrienes (p. 430).

11

Amino Acid Metabolism I: General Pathways

ALAN H. MEHLER

11.1 OVERVIEW

Amino acids are a group of molecules of major importance that contain a common chemical structure R—CH(NH_3^+)COO^-. As written, this structure represents two stereoisomers because of the asymmetry of the α-carbon atom; except where specifically noted, all of the statements in this chapter apply to the L isomers. The primary amino acids share the function of being polymerized with each other to form proteins, and some of these compounds are interconverted by metabolic reactions. Each naturally occurring amino acid, however, is a unique compound with individual biological functions and metabolism. In this chapter, the metabolic processes that affect amino acids collectively will be described; the metabolism of the individual compounds will be presented in the following chapter.

Amino acids occur mainly as constituents of proteins, which make up most of the dry weight of the human body. Many of the amino acids incorporated into human proteins are the compounds absorbed by the intestine from digests of dietary protein. However, since the composition of dietary protein does not correspond precisely to the needs of any individual, some of the amino acids in human proteins are synthesized from other dietary constituents; the carbon, hydrogen, and oxygen may be derived from other amino acids or from carbohydrate or fat precursors, but the nitrogen comes almost entirely from ingested amino acids.

Feeding experiments have shown that hydrolysates of proteins serve as well as the intact polymers. Subsequent studies with artificial mixtures of pure amino acids led to the grouping of these compounds as essential or nonessential (Table 11.1). That several amino acids needed for protein can be produced by human metabolism is the basis for listing those amino acids as nonessential. It must be understood that a deficiency of these amino acids in the diet can be overcome only when sufficient amounts of precursors are available to permit them to be formed as needed. With one exception, those amino acids listed as essential cannot be synthesized by the human body because our cells lack the necessary biosynthetic enzymes. A net synthesis of arginine apparently can meet the needs of adult humans, but normal growth requires an exogenous supply of this amino acid. In some of the enzyme deficiency diseases to be discussed in Chapter 12, needs for other amino acids, tyrosine and cysteine, must be met by the diet.

TABLE 11.1 Dietary Requirements of Amino Acids

Essential	*Nonessential*
Arginine[a]	Alanine
Histidine	Aspartate
Isoleucine	Cysteine
Leucine	Glutamate
Lysine	Glycine
Methionine[b]	Proline
Phenylalanine[c]	Serine
Threonine	Tyrosine
Tryptophan	
Valine	

[a] Arginine is synthesized by mammalian tissues, but the rate is not sufficient to meet the need during growth.
[b] Methionine is required in large amounts to produce cysteine if the latter is not supplied adequately by the diet.
[c] Phenylalanine is needed in larger amounts to form tyrosine if the latter is not supplied adequately by the diet.

The ordinary diets of developed countries contain more than adequate amounts of both essential and nonessential amino acids. Therefore, the categories are of practical importance only in disease, when specific supplements are administered, or in designing diets for people in certain areas. Even when syntheses are not apparently necessary to meet nutritional needs, the reactions described in Chapter 12 constantly redistribute the chemical elements of the metabolic systems involved and, thus, are of great theoretical relevance in interpreting studies on human and animal metabolism.

The amino acids listed as essential or nonessential are only a small fraction of those that occur naturally. These lists include only those amino acids that must be present wherever proteins are synthesized. The 18 compounds are activated and incorporated into proteins (Chapter 19). In addition, two derivatives, glutamine from glutamic acid and asparagine from aspartic acid, are synthesized for subsequent incorporation into proteins, increasing the list of primary building blocks for proteins to 20. Asparagine survives digestion and is absorbed from the intestine like other amino acids, whereas glutamine is mainly hydrolyzed to glutamate. All of the many other amino acids found in specific proteins are derivatives made by so-called post-translational modifications of one or more of the 20 primary building blocks after they have been assembled into polypeptide chains; none of these compounds, such as hydroxyproline, is required in the diet.

Since the amino acids comprise a complex group of compounds, the biochemistry of these substances can be considered from several points of view. The nutritional requirements and the demands of protein synthesis have already been mentioned. In Chapter 12, the metabolism of each amino acid will be considered from the point of view of degradation of the portion not needed for protein synthesis and also from the overlapping point of view of use of amino acids for the biosynthesis of other essential cellular materials. Another way of looking at amino acid metabolism is from the point of view of nitrogen balance.

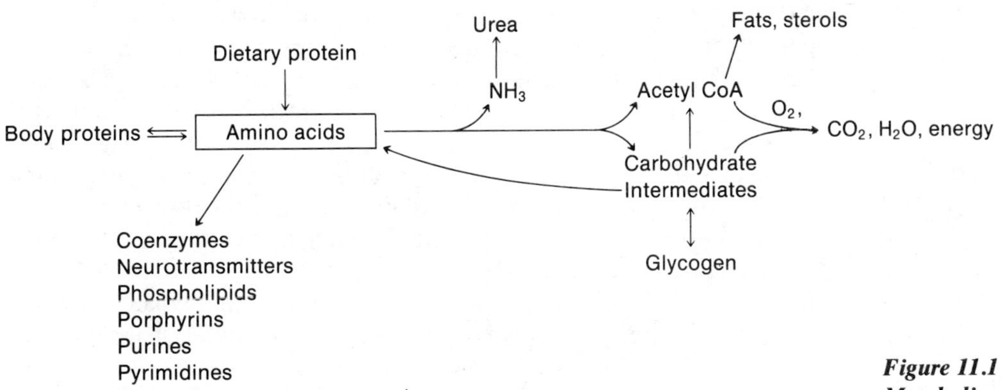

Figure 11.1
Metabolic relationships of amino acids.

The amount of nitrogen in each individual is regulated to stay relatively constant except during growth, when the amount must be increased in proportion to growth. There is no storage form for nitrogen reserves; only a small part of this nitrogen exists in the form of free amino acids or other compounds that can be used for synthesis of amino acids. Therefore, for maintenance of optimum body structure, an adequate supply of amino acids must be eaten frequently. If for any reason the supply of protein in the diet is insufficient, the need to synthesize specific proteins for vital physiological functions results in a redistribution of amino acids among proteins. For example, hemoglobin is degraded to the extent of almost 1% a day as red blood cells die, and under normal circumstances the degradation is balanced by resynthesis. In a deficiency of amino acids, relatively less hemoglobin is synthesized because a degree of anemia is more tolerable than a deficiency of certain other proteins.

Amino acids are substrates for many essential biosynthetic processes as shown in the outline of Figure 11.1; these will be discussed in Chapter 12. However, normal diets include a large excess of amino acids over the amount needed for synthesis of proteins or other cell constituents, and most excess amino acids are degraded to products that are either oxidized for energy or stored as fat and glycogen. In each case the nitrogen is liberated as ammonia. Some ammonia is reused in the synthesis of amino acids, some is used in other biosynthetic reactions, some is excreted in the urine, but the largest part is converted to urea, which is excreted by the kidneys. Urea synthesis occurs mainly in the liver, which is also the site of most of the biosynthesis of nonessential amino acids and a large part of the degradation of all amino acids.

11.2 GENERAL REACTIONS OF AMINO ACIDS

Transamination

Quantitatively, the most important reaction of amino acid metabolism is the transfer of the amino group to an α-keto acid. This reaction, transamination, is involved in the synthesis of nonessential amino acids, in the degradation of most amino acids, and in the exchange of amino groups; transamination has been demonstrated in vivo for all of the primary amino acids except lysine and threonine. The best studied enzyme of the family of transaminases (also called aminotransferases) is the aspartate transaminase of the cytosol; a similar but different enzyme for the reaction has been found in mitochondria. Most transaminases specifically require α-ketoglutarate or glutamate as one of the reacting pair; the specificities of various transaminases for the other substrate (amino or keto

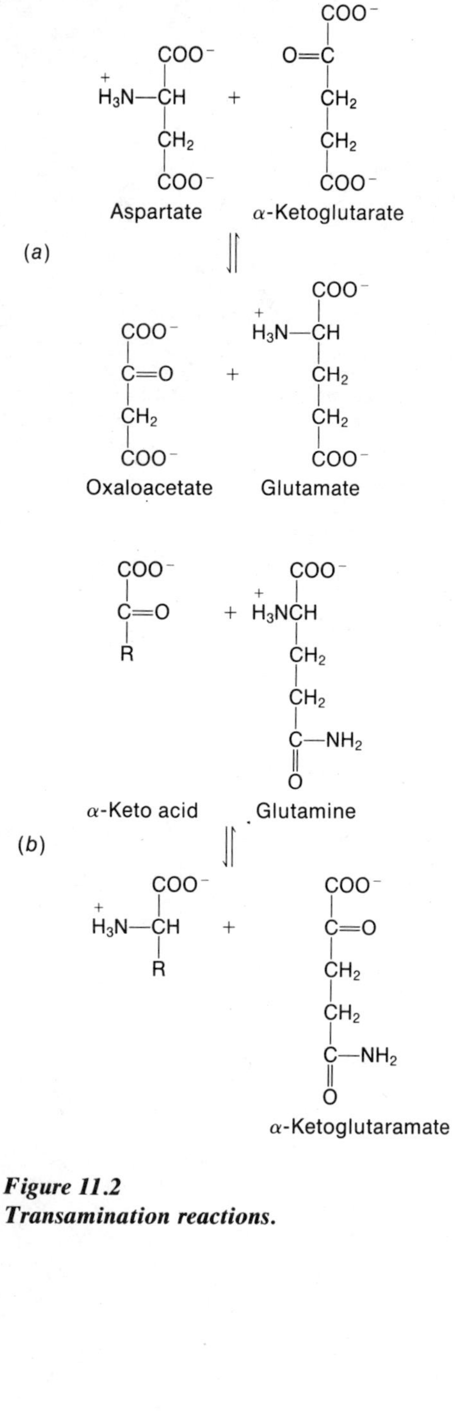

(a)

Aspartate + α-Ketoglutarate ⇌ Oxaloacetate + Glutamate

(b)

α-Keto acid + Glutamine ⇌ (amino acid) + α-Ketoglutaramate

Figure 11.2
Transamination reactions.

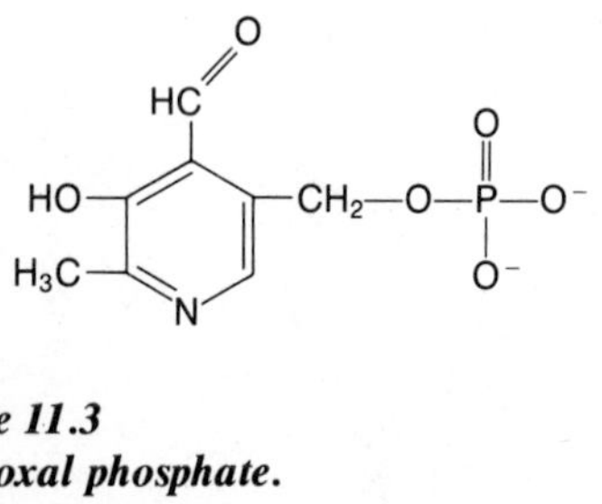

acid) may be strict or broad. In each case an amino acid and its corresponding keto acid are equilibrated with α-ketoglutarate and glutamate as shown in Figure 11.2*a*, and the equilibrium constant is near 1. The actual direction taken thus depends upon the concentrations of the four reactants, which are determined by the other cellular processes that produce or consume these compounds. A few transaminases use glutamine instead of glutamate as the amino donor and produce α-ketoglutaramate (Figure 11.2*b*). One of the first discoveries that resulted from the introduction of ^{15}N as a tracer in biochemistry by Schoenheimer and his students was that amino groups of many amino acids of proteins are rapidly labeled in vivo when the isotope is administered in a single compound. This led to the concept of "dynamic equilibrium," meaning a steady state in which not only the proteins are constantly degraded and resynthesized but also the atoms of the amino acids are exchanged. Transaminases are responsible for an active redistribution of amino groups among amino acids in vivo.

All transaminases require a coenzyme, pyridoxal phosphate (Figure 11.3). The coenzyme is bound to the protein by ionic forces involving the pyridine ring and the phosphate group. In addition, the aldehyde group forms a Schiff's base, an aldimine, with an ε-amino group of a lysine residue. A series of steps illustrated in Figure 11.4 takes advantage of the versatility of the bound coenzyme. In the reaction with an appropriate amino acid, the carbon of the aldimine is transferred to the amino group of the substrate. The ability of the pyridine ring to serve as an electron sink facilitates the dissociation of the α-hydrogen of the amino acid and permits a tautomerization in which the double bond of the aldimine shifts to form a ketimine; this is hydrolyzed to liberate the keto acid and the enzyme is left bearing pyridoxamine as its coenzyme. A similar series of reactions with the other keto acid results in the formation of a new amino acid while the enzyme reverts to its original state.

Pyridoxal phosphate is used as the coenzyme of a large number of enzymes that catalyze many different kinds of reactions. These will be described later. In each case, it is likely that the chemical mechanism is initiated by the same sort of aldimine formation used by transaminases, which is followed by an electronic rearrangement, but the different enzymes are able to labilize bonds other than that of the α-hydrogen.

Glutamate Dehydrogenase

The transfer of amino groups from one carbon chain to another is an efficient way to maintain proper ratios of the various amino acids while individual compounds are being synthesized or degraded. In addition to this means of redistribution, a mechanism is needed to dispose of surplus amino groups when ingestion of excess protein results in a surfeit of amino acids that must be metabolized and, in some circumstances, a mechanism is needed to use ammonia to increase the total amino acid concentration. The major mechanism for net use or production of amino acids is the reaction catalyzed by glutamate dehydrogenase.

$$\text{Glutamate} + \text{NAD}^+\text{(P)} \rightleftharpoons \alpha\text{-ketoglutarate} + \text{ammonia} + \text{NAD(P)H} + \text{H}^+$$

Figure 11.3
Pyridoxal phosphate.

In vitro glutamate dehydrogenase uses either NAD or NADP. At neutral pH, the equilibrium of the reaction lies to the side of glutamate synthesis. It should be noted that the equilibrium constant is a value derived from a theoretical consideration of a hypothetical situation with arbitrarily selected concentrations of reagents. Under physiological conditions in which other reactions compete for the products of the reaction (especially

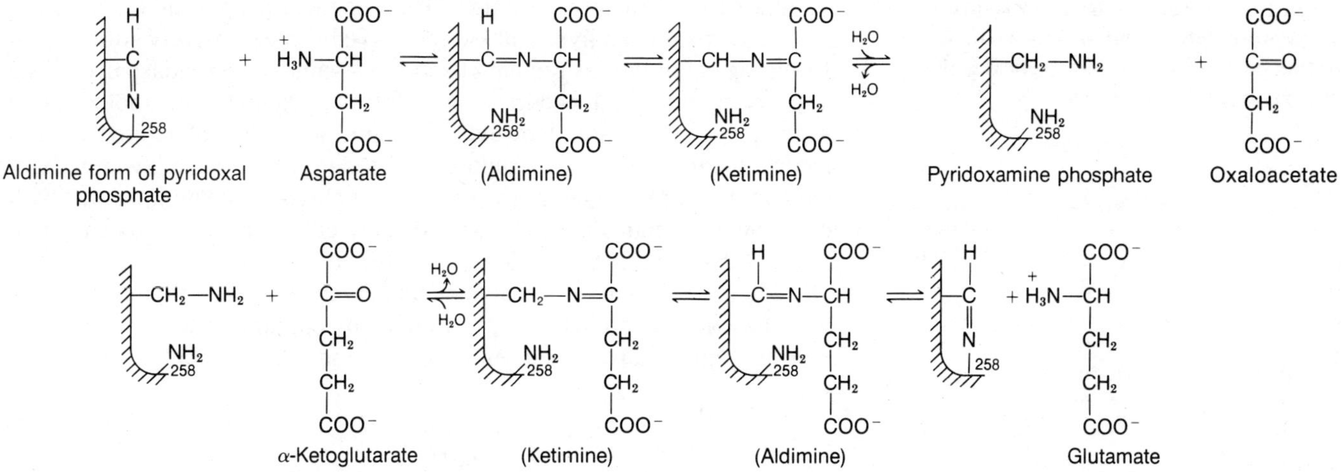

Aldimine form of pyridoxal phosphate + Aspartate ⇌ (Aldimine) ⇌ (Ketimine) ⇌ Pyridoxamine phosphate + Oxaloacetate

Pyridoxamine phosphate + α-Ketoglutarate ⇌ (Ketimine) ⇌ (Aldimine) ⇌ Aldimine form of pyridoxal phosphate + Glutamate

Figure 11.4
Mechanism of enzymatic transamination represented in schematic form.

NADH oxidase and α-ketoglutarate dehydrogenase, Chapter 6, and ammonia conversion to urea, page 443), an efficient oxidation of glutamate can be catalyzed. It should be noted that the equilibrium constant for this reaction is more complicated than most, even when the H^+ term is eliminated, because the number of products is greater than the number of reactants:

$$K_{eq} = \frac{[\alpha\text{-ketoglutarate}][\text{NAD(P)H}][\text{NH}_3]}{[\text{glutamate}][\text{NAD}^+(\text{P})]}$$

The efficient removal of any one of the products can shift the equilibrium concentrations of the other reactants; lowering the concentration of a second product has a compounding effect. The facts that most of the coenzyme in vivo is in the oxidized form, that ammonia is removed from cells to give a very low steady-state concentration and that intracellular glutamate occurs at more than 10^{-3}M concentrations create conditions that favor oxidative deamination of glutamate. Like many reactions of biochemistry, however, the concentrations are poised so that glutamate can be formed when excess ammonia is produced. A major source of ammonia in the liver is from bacterial metabolism in the intestine and transport through the portal system.

Glutamate dehydrogenase from bovine liver, the best studied representative of this ubiquitous enzyme family, is a complex enzyme composed of six identical subunits of M_r 56,000. It is subject to extensive allosteric control by a diverse group of substances; of these, GTP and ATP inhibit, whereas GDP and ADP activate. The less active form of the enzyme demonstrates a weak alanine dehydrogenase activity. These properties suggest a teleological rationale; when amino acids are needed for energy production, the activity of the dehydrogenase is greatest, and when nucleoside triphosphate levels are high, the activity is decreased. Inhibition by several steroid hormones and by thyroxine has been described but not established as having physiological significance.

Amino Acid Oxidases

For many years two enzymes have been known to oxidize D- and L-amino acids, respectively, to the corresponding α-keto acids and ammonia with molecular oxygen as the electron acceptor. In these reactions the oxygen

TABLE 11.2 Relative Rates of Oxidation by Representative Amino Acid Oxidases Acting on the Appropriate Optical Isomer of a Series of Amino Acids

	L-*Amino Acid Oxidase*[a]	D-*Amino Acid Oxidase*[a]
Alanine	3.5	100
Arginine	0	0
Aspartate	0	0
Cysteine	0	0
Glutamate	0	0
Glycine	0	15
Histidine	—	20
Isoleucine	9	90
Leucine	100	20
Lysine	0	0
Methionine	87.5	110
Phenylalanine	32	90
Proline	77	300
Serine	0	50
Threonine	0	30
Tryptophan	35.7	45
Valine	5.3	65

[a] The L-amino acid oxidase was purified from rat kidney and the D-amino acid oxidase was from sheep kidney.

is reduced to hydrogen peroxide. The chemical properties of these enzymes, purified from liver and kidney, have been extensively studied, but the biological functions remain obscure. D-Amino acid oxidase uses FAD as its coenzyme and occurs in amounts sufficient to metabolize large quantities of substrate. However, D-amino acids are of limited natural occurrence in mammals. L-Amino acid oxidase has a much lower activity and contains FMN. These enzymes do not attack some amino acids with ionizable side chains but otherwise display broad specificity as shown in Table 11.2. L-Amino acid oxidase can also attack α-hydroxy acids, and D-amino acid oxidase oxidizes glycine and sarcosine (*N*-methyl glycine). In vitro these flavoproteins can substitute many oxidants, such as methylene blue, for molecular oxygen, and the reaction can be seen to proceed in steps:

$$\text{Amino acid} + \text{flavoprotein} + H_2O \longrightarrow \alpha\text{-keto acid} + NH_3 + \text{flavoprotein } H_2$$

$$\text{flavoprotein } H_2 + X \longrightarrow \text{flavoprotein} + XH_2$$

When the oxidant is O_2, the product of its reduction is H_2O_2; in liver and kidney this potentially destructive material is decomposed by a very active enzyme, catalase, which catalyzes the reaction

$$2H_2O_2 \longrightarrow 2H_2O + O_2$$

Superficially the overall reaction catalyzed by the amino acid oxidases appears to resemble the complex of reactions initiated by glutamate dehydrogenase and followed by the mitochondrial oxidation of NADH. An important difference is that in the latter case the oxidation is coupled with oxidative phosphorylation and the generation of ATP.

11.3 REACTIONS OF AMMONIA

In addition to forming the α-amino groups of amino acids, ammonia is incorporated into several other metabolites. The mechanisms of formation of the purine ring, the amino group of the pyrimidine cytosine, and amino sugars all involve the intermediate formation of glutamine. As seen in Figure 11.5, glutamine is synthesized from glutamate and ammonia by the enzyme glutamine synthetase. This amino acid amide is an important compound because it is one of the 20 primary amino acids used for protein synthesis. In addition, in the kidney, glutamine supplies the bulk of the ammonia excreted. Thus the ammonia is transported in the blood as a nontoxic, nonionized amide that is used in regulation of urinary pH by release of ammonia in a simple hydrolysis of the amide group catalyzed by the enzyme glutaminase (Figure 11.6) followed by the removal of the amino group by glutamate dehydrogenase. Blood levels of glutamine normally exceed those of any other amino acid. Ammonia released by degradation of amino acids in the liver is made available as glutamine to cells in other organs for the synthesis of purines, pyrimidines, and other compounds.

Glutamine also donates the amide group to form its lower homologue, asparagine, from aspartate. Although there appears to be no need for energy to drive a reaction in which the substrates and products have similar groups, that is, a carboxyl group and a carboxamide, the synthesis

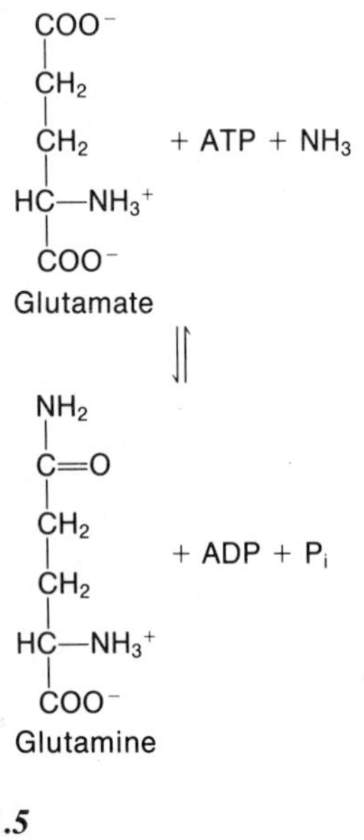

COO^-—CH_2—CH_2—HC—NH_3^+—COO^- + ATP + NH_3
Glutamate

⇅

NH_2—$C{=}O$—CH_2—CH_2—HC—NH_3^+—COO^- + ADP + P_i
Glutamine

Figure 11.5
Reaction catalyzed by glutamine synthetase.

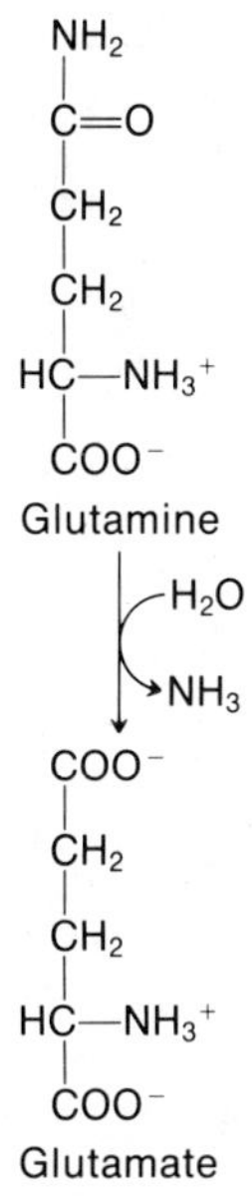

NH_2 | $C{=}O$ | CH_2 | CH_2 | $HC{-}NH_3^+$ | COO^-

Glutamine

$\xrightarrow{H_2O \quad NH_3}$

COO^- | CH_2 | CH_2 | $HC{-}NH_3^+$ | COO^-

Glutamate

Figure 11.6
Reaction catalyzed by glutaminase.

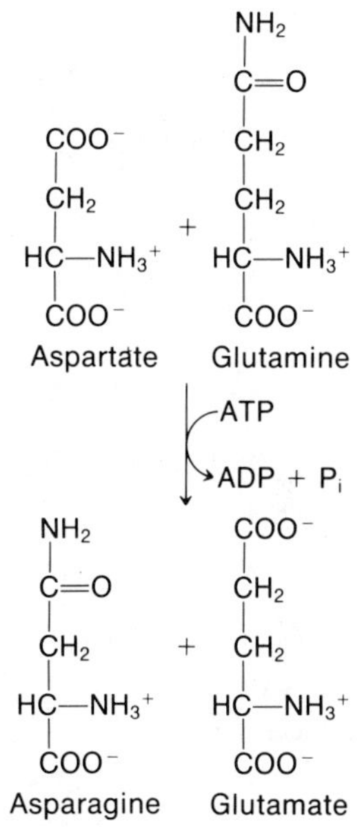

COO^- | CH_2 | $HC{-}NH_3^+$ | COO^- + NH_2 | $C{=}O$ | CH_2 | CH_2 | $HC{-}NH_3^+$ | COO^-

Aspartate + Glutamine

$\xrightarrow{ATP \quad ADP + P_i}$

NH_2 | $C{=}O$ | CH_2 | $HC{-}NH_3^+$ | COO^- + COO^- | CH_2 | CH_2 | $HC{-}NH_3^+$ | COO^-

Asparagine + Glutamate

Figure 11.7
Synthesis of asparagine.

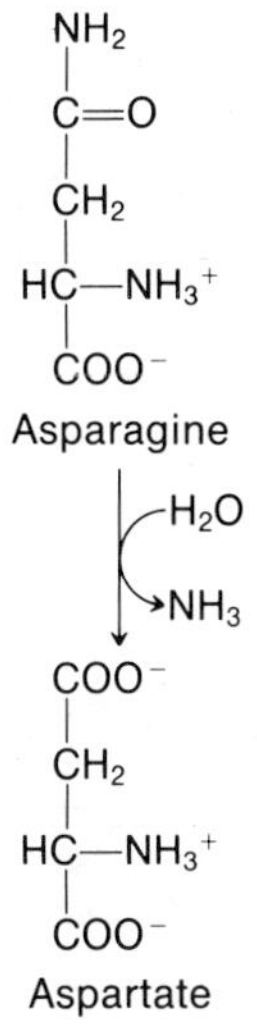

NH_2 | $C{=}O$ | CH_2 | $HC{-}NH_3^+$ | COO^-

Asparagine

$\xrightarrow{H_2O \quad NH_3}$

COO^- | CH_2 | $HC{-}NH_3^+$ | COO^-

Aspartate

Figure 11.8
Reaction catalyzed by asparaginase.

of asparagine requires ATP as shown in Figure 11.7. The ATP is needed to activate the β-carboxyl group of aspartate. In general it is not possible to alter carboxyl groups under biological conditions until they have been activated by reaction with a nucleoside triphosphate.

Asparagine is not known to have any function in mammals other than incorporation into proteins. In general animal cells appear to be able to synthesize enough asparagine for their own needs. Some rapidly dividing leukemic cells have little or no ability to produce asparagine and depend upon asparagine taken from blood to synthesize their proteins; this is the basis for the therapeutic use of asparaginase (Figure 11.8) to treat leukemic patients, depriving the neoplastic cells of asparagine by lowering the serum concentration drastically.

11.4 THE UREA CYCLE

Normal adults are in nitrogen balance; that is, the amount of nitrogen ingested is balanced by the excretion of an equivalent amount (Chapter 25). About 80% of the excreted nitrogen is in the form of urea (Figure 11.9). The discovery in 1828 of a laboratory synthesis of this simple compound was momentous in the history of science because it demonstrated for the first time that compounds characteristic of living organisms can also be made by the methods of chemistry. The elucidation of the relatively complex method by which our bodies produce urea had a similar impact on scientific thought in developing the concept of metabolic cycles. The urea cycle (also the Krebs, Krebs-Henseleit, or ornithine cycle) is a true cycle in which the carrier molecule, ornithine, is regenerated with the same atoms in its skeleton after the formation of each molecule of urea.

Amino acids are metabolized extensively throughout the body, but most urea synthesis occurs in the liver. The importance of the liver in

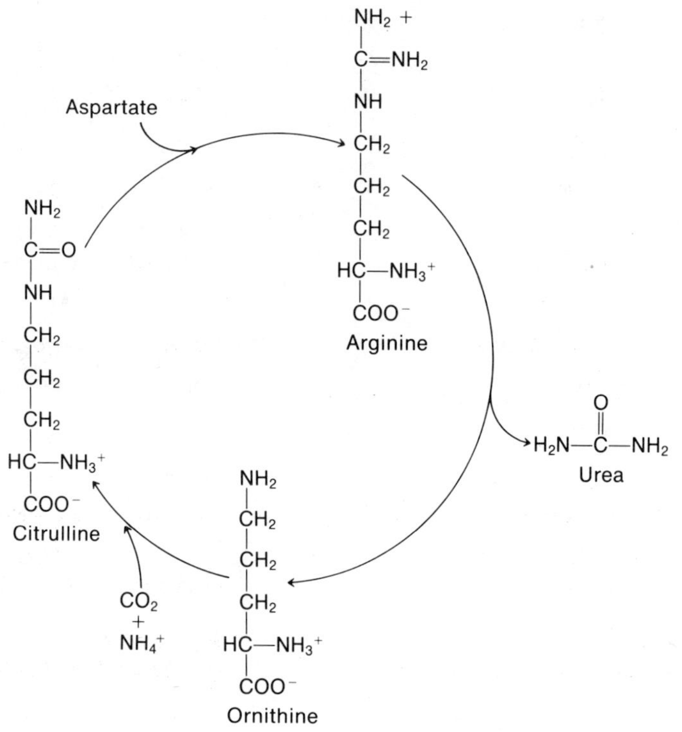

Figure 11.9
Concept of the urea cycle.

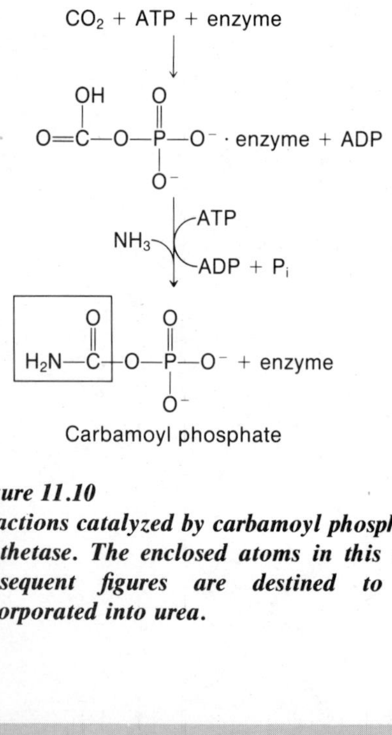

Figure 11.10
Reactions catalyzed by carbamoyl phosphate synthetase. The enclosed atoms in this and subsequent figures are destined to be incorporated into urea.

CLIN CORR. **11.1**
HYPERAMMONEMIA AND HEPATIC COMA

Hyperammonemia, an increase in the concentration of ammonia in the blood over the normal levels, may result in coma. The normal range is 30–60 μM (part of the var-

nitrogen metabolism is further illustrated in Clin. Corr. 11.1. The concept of the urea cycle was derived from observations on urea synthesis in liver slices. Only a few compounds were found to cause a major increase in the rate of urea synthesis and were effective in catalytic amounts; that is, the extra synthesis of urea exceeded the amounts added of arginine, ornithine, or citrulline (Figure 11.9). The explanation of the stimulation by these three compounds on the basis of "paper chemistry" has been confirmed and elaborated by the identification and characterization of each of the enzymes involved. Thus, the urea cycle is established as a process, not merely a theory.

Carbamoyl Phosphate Formation

The formation of citrulline from ornithine requires the addition of elements of ammonia and CO_2. These are joined in a preliminary reaction in which a phosphate group from ATP is also used to produce a relatively stable compound, carbamoyl phosphate.

In the reaction of carbamoyl phosphate synthetase, one molecule of ATP activates CO_2 to form enzyme-bound carboxy phosphate, which reacts with ammonia to form carbamate. A second molecule of ATP reacts with the enzyme-bound intermediate in a kinase reaction to form carbamoyl phosphate, shown in Figure 11.10. Actually, two distinct carbamoyl phosphate synthetases are found in mammalian livers. The one that participates in the urea cycle, carbamoyl phosphate synthetase I, is part of the mitochondrial matrix and is inactive in the absence of an N-acylated glutamate as an allosteric activator. The physiological activator is N-acetylglutamate. A cytosolic carbamoyl phosphate synthetase II participates in the biosynthesis of pyrimidines (Chapter 13) and does not require an acylated glutamate for activity.

iation in normal values is attributed to differences obtained with various methods for determining blood ammonia). Loss of consciousness can also be produced by mechanical injury to the brain, anoxia, and hypoglycemia. It is not fully established how each of these conditions alters the normal function of the cerebral cortex. A general mechanism could be impaired neuronal function because of low ATP levels. Ammonia does cross the blood-brain barrier and might affect the formation of glutamine, decreasing both glutamate and ATP, but this is probably of only minor significance. More important is the shift of the glutamate dehydrogenase equilibrium, which results in the depletion of α-ketoglutarate to a point at which the tricarboxylic acid cycle cannot supply enough electrons for the normal production of ATP through oxidative phosphorylation.

Hyperammonemia is usually caused by inability of the patient to form urea at a sufficient rate to maintain tolerable levels of ammonia. Inherited deficiency of individual urea cycle enzymes has been implicated in many cases. A delay in development of urea cycle enzymes has been observed to cause a transient hyperammonemia in newborns. Ammonia is not cleared from the blood efficiently in cases in which the normal blood flow through the liver is reduced. Cirrhosis of the liver can result in the development of collateral circulation of blood from the portal system directly to the posterior vena cava, bypassing the liver. In some cases this type of bypass is deliberately produced surgically as a portal-caval shunt.

In acute cases of hyperammonemia, removal of excess ammonia by dialysis is effective therapy. In cases involving circulatory bypass of the liver, it is important to restrict the ingestion of protein. Treatments have been developed and shown to be effective over long periods of time that are based on excretion of nitrogen in forms other than urea. The two reactions (see accompanying Figure) used for this purpose were previously known as detoxification mechanisms for benzoate and phenylacetate. Benzoate and phenylacetate are efficiently converted to conjugates with glycine and glutamine, respectively, and the conjugates are rapidly excreted in the urine. The reactions forming these amides, formerly thought to be of importance only as means to eliminate the foreign organic acids, can be initiated deliberately by administration of benzoate and phenylacetate to cause stoichiometric amounts of nitrogen to be eliminated as hippuric acid and phenylacetylglutamine. In Chapter 12, reactions are described that show how glycine can be formed in large quantities from major metabolite pools, and in this chapter it is seen that glutamine is also available as needed as long as anaplerotic reactions maintain the supply of α-ketoglutarate. With careful management, these reactions can replace urea formation and prevent ammonia toxicity. An additional treatment that minimizes ammonia toxicity is substitution in the diet of the α-keto analogues for some amino acids; this reduces the total intake of nitrogen and diverts some nitrogen from an excretory fate to essential amino acids for protein synthesis.

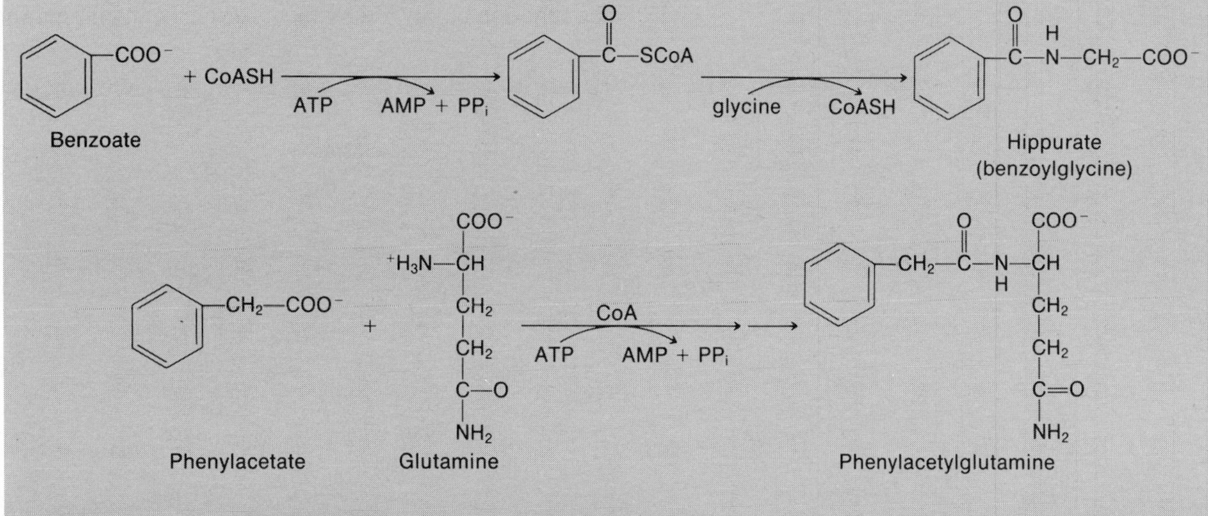

Benzoate + CoASH —(ATP → AMP + PP_i)→ C_6H_5–C(=O)–SCoA —(glycine → CoASH)→ C_6H_5–C(=O)–NH–CH_2–COO^- Hippurate (benzoylglycine)

Phenylacetate (C_6H_5–CH_2–COO^-) + Glutamine (^+H_3N–CH(COO^-)–CH_2–CH_2–C(=O)–NH_2) —CoA (ATP → AMP + PP_i)→ → Phenylacetylglutamine (C_6H_5–CH_2–C(=O)–NH–CH(COO^-)–CH_2–CH_2–C(=O)–NH_2)

Detoxification reactions used as alternatives to the urea cycle.

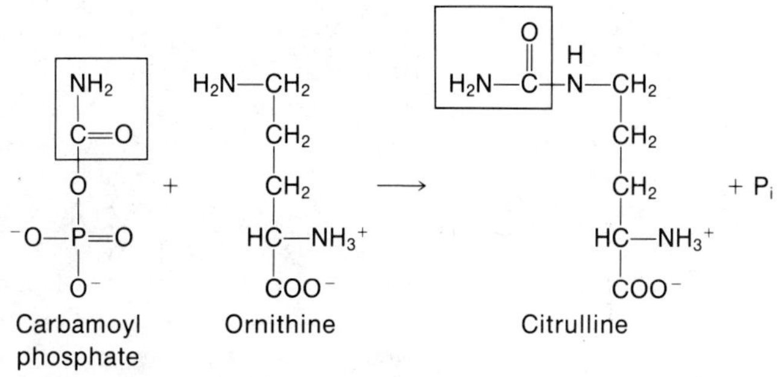

Carbamoyl phosphate + Ornithine ⟶ Citrulline + P_i

Figure 11.11
Reaction catalyzed by ornithine transcarbamoylase.

Citrulline Synthesis

Citrulline is formed by the transfer of the carbamoyl group from its phosphoric acid anhydride to the δ-amino group of ornithine (Figure 11.11). The enzyme ornithine transcarbamoylase is also mitochondrial. Citrulline diffuses from mitochondria to the cytosol, where the rest of the urea cycle occurs.

Formation of Argininosuccinate

In two steps citrulline is converted to arginine, which has a similar structure but with a nitrogen in place of the ureido oxygen. The new nitrogen atom comes from aspartate. The transfer involves a condensation, as shown in Figure 11.12, to form argininosuccinate. Since the ureido group is very stable, the condensation requires activation by ATP. The reaction probably proceeds in steps analogous to the formation of active derivatives of fatty acids (Chapter 9) and amino acids (Chapter 19). In each case an intermediate compound containing AMP is formed with the release of inorganic pyrophosphate. The activated molecule is then transferred to its acceptor without dissociating from the enzyme. The reaction is reversible, but no partial reactions have been observed, leading to the conclusion that everything is bound firmly to the enzyme until the reaction is complete. The reaction proceeds strongly in the direction of synthesis of argininosuccinate because of the hydrolysis of the inorganic pyrophosphate.

Formation of Arginine

The formation of arginine is catalyzed by argininosuccinate lyase (Figure 11.13). The reaction catalyzed by this enzyme is an elimination; the nitro-

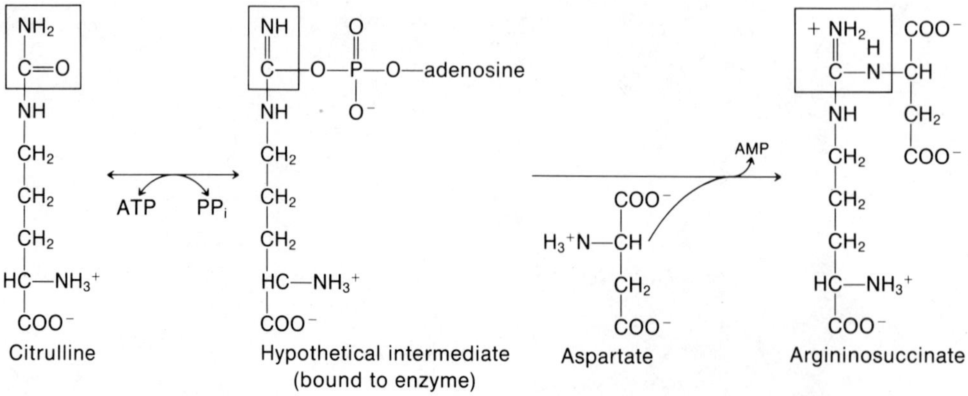

Citrulline + ATP ⇌ Hypothetical intermediate (bound to enzyme) + PP_i; Hypothetical intermediate + Aspartate ⟶ Argininosuccinate + AMP

Figure 11.12
Reactions catalyzed by argininosuccinate synthetase.

gen of what was originally aspartate is eliminated along with a proton from the 4-carbon dicarboxylic compound, leaving a double bond. The unsaturated product is fumarate, a tricarboxylic acid cycle intermediate (Chapter 6). The nitrogen atom that is eliminated becomes one of two equivalent atoms of the guanidinium group of arginine.

Argininosuccinate → Arginine + Fumarate

Figure 11.13
Formation of arginine by argininosuccinate lyase.

Formation of Urea

The urea cycle is completed by the hydrolysis of arginine to ornithine and urea, a reaction catalyzed by arginase (Figure 11.14). Arginase is a tetrameric protein that depends upon easily dissociable Mn^{2+} for its activity. This enzyme is found in brain, kidney, and other organs in which the production of ornithine from dietary arginine may have more physiological importance than formation of urea. The ornithine that is released in this reaction in the urea cycle is the identical molecule that was used to initiate the series of reactions that resulted in the synthesis of urea and, after transport into mitochondria, reacts with carbamoyl phosphate to start another cycle. The urea diffuses into the blood from which it is cleared by the kidneys.

Arginine + H_2O → Urea + Ornithine

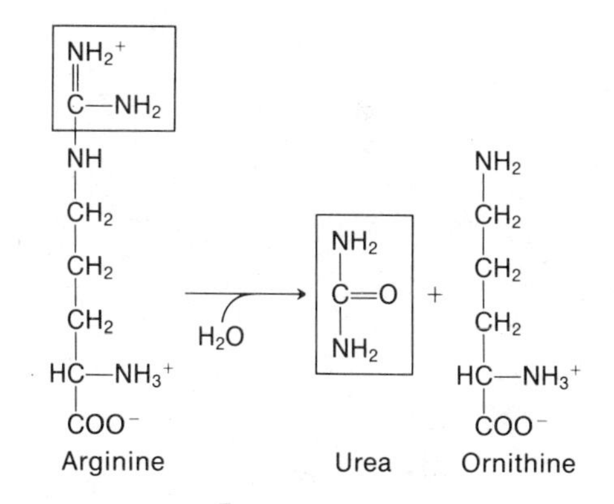

Figure 11.14
Hydrolysis of arginine by arginase.

The biological formation of urea is much more complicated than the chemical synthesis, which can be as simple as heating ammonium cyanate, NH_4CNO. The urea cycle uses the energy of ATP instead of high temperature and, equally important, it uses the ornithine structure as a handle for the enzymes that build the urea molecule and enables each step to be integrated in a smooth process that maintains the optimum physiological concentration of ammonia. The outline in Figure 11.15 shows the essential simplicity of the cycle in which complex enzymes participate.

Urea has no role in animal metabolism other than as an end product that is excreted. An enzyme, urease, present in plants and microorganisms hydrolyzes urea to CO_2 and two equivalents of ammonia. A portion of the urea synthesized by the urea cycle diffuses into the intestine and is degraded by the urease of intestinal bacteria to produce ammonia that is recycled to the liver. This is of significance only in cases of kidney failure, when a large amount of urea may pass through the intestine. Urease from the jackbean is renowned in biochemistry as the first enzyme to be shown to form crystals.

The urea cycle is considered as a mechanism for elimination of excess ammonia as urea. The mechanism, however, requires that one of the nitrogen atoms be donated by aspartate. In the synthesis of carbamoyl phosphate two equivalents of ATP lose their terminal phosphate groups, and another ATP is degraded to AMP and pyrophosphate in the synthesis

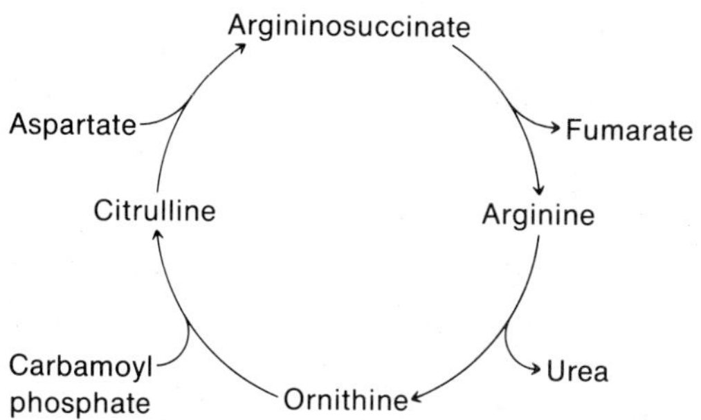

Figure 11.15
Outline of the urea cycle.

of citrulline. The overall chemical balance of the biosynthesis of urea is as follows:

$$
\begin{aligned}
NH_3 + CO_2 + 2\ ATP &\longrightarrow \text{carbamoyl phosphate} + 2\ ADP + P_i \\
\text{Carbamoyl phosphate} + \text{ornithine} &\longrightarrow \text{citrulline} + P_i \\
\text{Citrulline} + ATP + \text{aspartate} &\longrightarrow \text{argininosuccinate} + AMP + PP_i \\
\text{Argininosuccinate} &\longrightarrow \text{arginine} + \text{fumarate} \\
\text{Arginine} &\longrightarrow \text{urea} + \text{ornithine} \\
\hline
\text{Sum: } 2NH_3 + CO_2 + 3\ ATP &\longrightarrow \text{urea} + 2\ ADP + AMP + PP_i + 2P_i
\end{aligned}
$$

The reaction has a large negative free energy derived largely from the cleavage of the pyrophosphate bonds of ATP and the increased number of ionized groups in the split products (15 in the products compared with 12 in the three molecules of ATP). Therefore, the cycle is not reversible. In effect, four equivalents of ATP are required for each turn of the cycle since two phospho anhydride bonds must be used to regenerate ATP from AMP.

The essential aspartate is generated almost entirely by transamination of oxaloacetate by glutamate, which is thereby converted to α-ketoglutarate. Since oxaloacetate and α-ketoglutarate are both intermediates in the tricarboxylic acid cycle, the two cycles discovered by Krebs are metabolically linked (Figure 11.16 illustrates this relationship). The fumarate produced in the urea cycle is also a tricarboxylic acid cycle intermediate.

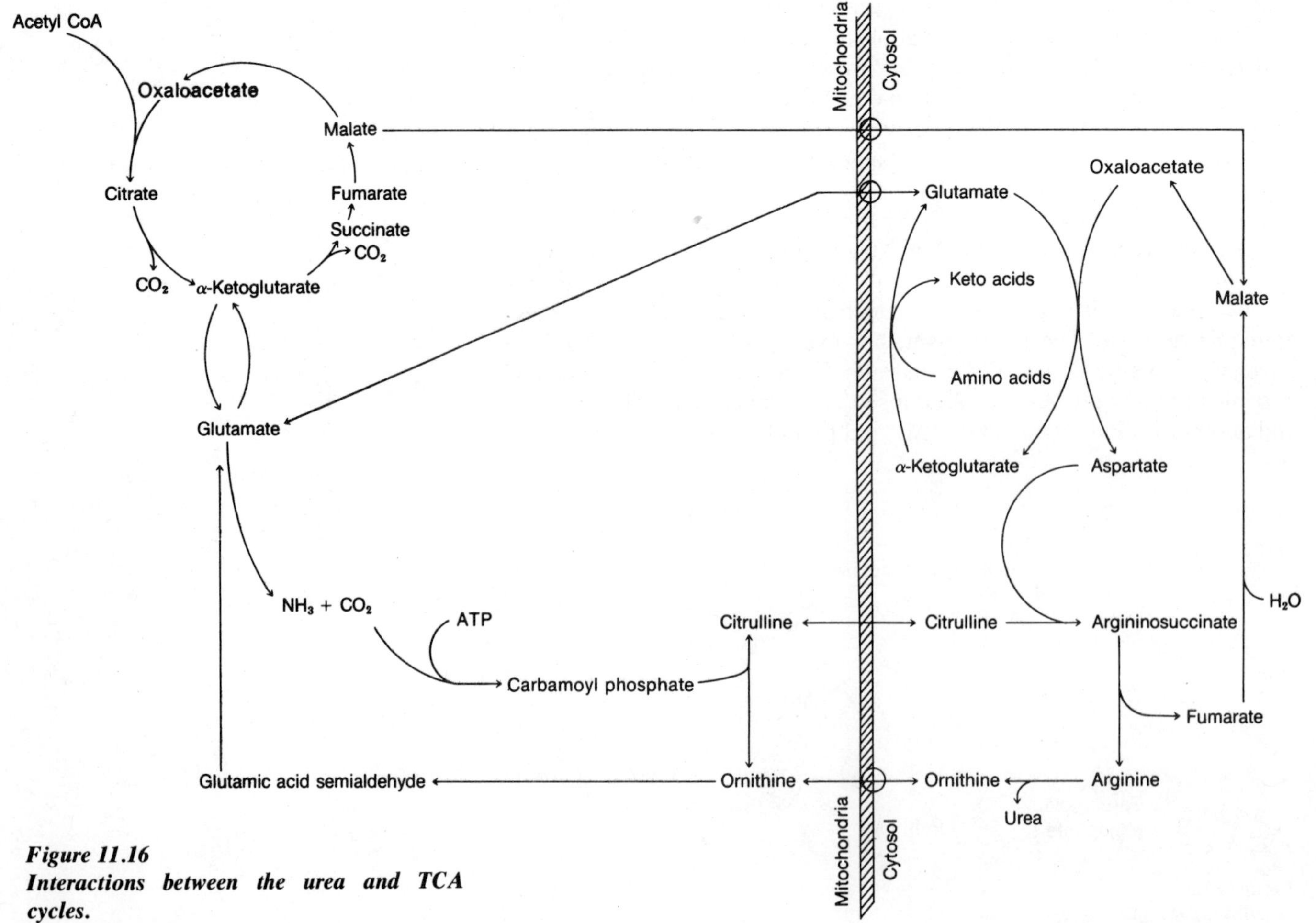

Figure 11.16
Interactions between the urea and TCA cycles.

A theoretical sequence of reactions can regenerate the aspartate from these intermediates and ammonia:

$$\begin{array}{rl} \alpha\text{-ketoglutarate} + NH_3 + NADH & \longrightarrow \text{glutamate} + NAD^+ + H_2O \\ \text{fumarate} + H_2O & \longrightarrow \text{malate} \\ \text{malate} + NAD^+ & \longrightarrow \text{oxaloacetate} + NADH \\ \text{oxaloacetate} + \text{glutamate} & \longrightarrow \text{aspartate} + \alpha\text{-ketoglutarate} \\ \hline \text{Sum: fumarate} + NH_3 & \longrightarrow \text{aspartate} \end{array}$$

In living cells the dicarboxylic acids are equilibrated with tricarboxylic acid cycle intermediates and pass rapidly between the cytosol and mitochondria.

11.5 REGULATION OF THE UREA CYCLE

The urea cycle operates only to eliminate excess ammonia. The excess comes mainly from ingested amino acids not used promptly for the synthesis of proteins; only a few nitrogenous compounds comprising a small percentage of nitrogen turnover, such as purines, pyrimidines, creatine, and nicotinic acid, are not degraded to ammonia. The general nutritional status does not influence the degradation of excess amino acids; whether the carbon skeletons from amino acids are used immediately to produce ATP or are stored as fat and glycogen, the amino groups are released as ammonia, which must be eliminated. In nitrogen balance the amount of urea produced represents "metabolic wear and tear," a poorly understood combination of cell death and regeneration and protein turnover with partial, but never complete, reuse of amino acids. These metabolic processes persist even during protein deficiency in a high-fat, high-carbohydrate diet so that urea production never ceases. The rate, however, is precisely regulated to the need to eliminate potentially toxic ammonia. Interference with this process results in serious diseases, described in Clin. Corr. 11.2.

Gross control of the urea cycle is accomplished by altering the amounts of the enzymes of the cycle. The levels of the individual enzymes in laboratory animals have been seen to change 10–20-fold in response to extreme changes in diet, and similar changes presumably occur in humans. During extensive starvation, when muscle proteins are broken down to amino acids to be oxidized for energy, the importance of removing ammonia is accompanied by elevated levels of urea cycle enzymes in the liver.

Fine control of the urea cycle appears to be applied primarily at the synthesis of carbamoyl phosphate. Carbamoyl phosphate synthetase is inactive in the absence of its allosteric activator, N-acetylglutamate. This activator is synthesized by a liver enzyme from acetyl CoA and glutamate (Figure 11.17) and is hydrolyzed by a specific deacylase. The steady-state concentration of N-acetylglutamate is determined by the concentrations of the substrates, acetyl CoA and glutamate, and by the concentration of arginine, which is an activator of N-acetylglutamate synthetase.

An additional control of the urea cycle is effected by the concentration of the intermediates of the cycle. Although the cycle is theoretically perfect in that the ornithine is regenerated completely in each turn, in real life there are reactions that convert the intermediates to other products. A normal diet supplies enough arginine to maintain the necessary concentration of ornithine; an enzyme system in the intestinal mucosa synthesizes ornithine from glutamate (Chapter 12).

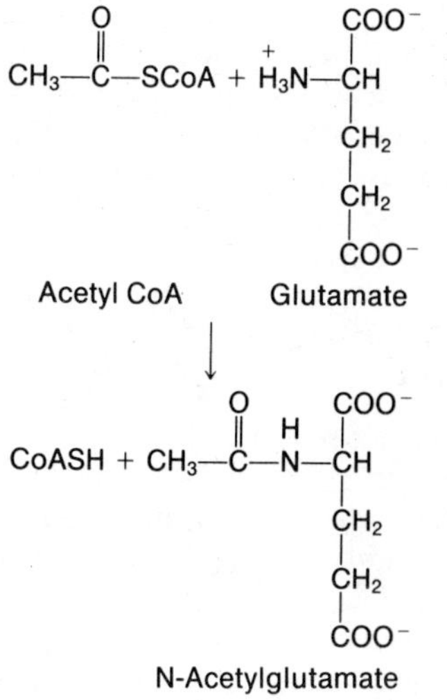

$$CH_3-\overset{O}{\overset{\|}{C}}-SCoA + H_3\overset{+}{N}-CH(COO^-)-CH_2-CH_2-COO^-$$

Acetyl CoA Glutamate

$$\downarrow$$

$$CoASH + CH_3-\overset{O}{\overset{\|}{C}}-\overset{H}{N}-CH(COO^-)-CH_2-CH_2-COO^-$$

N-Acetylglutamate

Figure 11.17
Reaction catalyzed by acetylglutamate synthetase.

CLIN CORR. 11.2 DEFICIENCIES OF UREA CYCLE ENZYMES

Even though the normal elimination of nitrogen and the prevention of ammonia toxicity require the function of the urea cycle, infants born with total deficiency of one or more enzymes of the cycle survive at least for several days. Many of the enzyme deficiencies that have been identified are partial and are reported as so much percent of normal activity. Analogy with mutations in other enzymes suggests that the enzymes are likely to have altered K_m values in many cases, rather than altered V_{max}, but there is as yet little information on characterizing the human mutations.

Cases are known of deficiencies of each of the urea cycle enzymes. Interruption of the cycle at each point affects nitrogen metabolism differently because some of the intermediates can diffuse from hepatocytes, accumulate in the blood, and pass into the urine. Therefore, the symptomatology, prognosis, and treatment differ for the various enzyme deficiencies. In general these are severe diseases with high incidences of mental retardation, seizures, coma, and early death.

N-ACETYLGLUTAMATE SYNTHETASE DEFICIENCY

The importance of an activator for carbamoyl phosphate synthetase was demonstrated by the finding of hyperammonemia and general hyperaminoacidemia in a newborn whose liver contained no detectable ability to synthesize N-acetylglutamate. In this case normal nitrogen metabolism was maintained by a low protein diet and the administration of carbamoyl glutamate, an analogue of N-acetylglutamate that is also an activator of carbamoyl phosphate synthetase.

CARBAMOYL PHOSPHATE SYNTHETASE DEFICIENCY

Hyperammonemia has been observed in infants with 0–50% of the normal level of carbamoyl phosphate synthetase synthesis in their livers. Treatment with benzoate and phenylacetate has been effective in maintaining such infants. Low protein diets supplemented with arginine have been attempted on the hypothesis that activation of N-acetylglutamate synthesis by arginine would indirectly stimulate the low level of carbamoyl phosphate synthetase to provide enough of its product to sustain the urea cycle. In general this deficiency is associated with mental retardation, which may be an indirect consequence of the periods of uncontrolled hyperammonemia.

ORNITHINE TRANSCARBAMOYLASE DEFICIENCY

The most common deficiency disease involving urea cycle enzymes is lack of ornithine transcarbamoylase. Early death can be prevented by removal of excess ammonia and prevention of further accumulation by the same sort of diet effective with carbamoyl phosphate synthetase deficiency. The occasional finding of normal development in treated patients supports the idea that the mental retardation usually associated with this deficiency is caused by the excess ammonia before adequate therapy.

Genetic analysis of ornithine transcarbamoylase deficiency indicates that the gene is located on the X chromosome. Therefore, males generally are more seriously affected than females, who are often asymptomatic as heterozygotes. In addition to ammonia and amino acids appearing in the blood in increased amounts, orotic acid also increases, presumably because carbamoyl phosphate that cannot be used to form citrulline diffuses into the cytosol, where it condenses with aspartate, ultimately becoming orotate (Chapter 13). This suggests that normal orotate synthesis is limited by the cytosolic carbamoyl phosphate synthetase II and that the production of excess carbamoyl phosphate intended for urea synthesis bypasses the normal control mechanism.

ARGININOSUCCINATE SYNTHETASE DEFICIENCY

The inability to continue the urea cycle by condensing citrulline with aspartate results in accumulation of citrulline in the blood and excretion in the urine. In some cases a majority of the nitrogen excreted may be as citrulline. In addition to restricted nitrogen intake, therapy for this disease requires specific supplementation with arginine for protein synthesis and for the formation of creatine (Chapter 12) and ornithine, which also has essential functions other than its role in urea synthesis (Chapter 12).

ARGININOSUCCINATE LYASE DEFICIENCY

Impaired ability to split argininosuccinate to form arginine resembles argininosuccinate synthetase deficiency in that the substrate, in this case argininosuccinate, is excreted in large amounts. For undetermined reasons, the severity of symptoms in this disease varies greatly so that it is hard to evaluate the effect of therapy. Nevertheless, dietary restriction of nitrogen, alternative excretion of nitrogen by administration of benzoate and phenylacetate, and supplementation of arginine appear to be useful.

ARGINASE DEFICIENCY

Arginase deficiency is a rare disease that causes many abnormalities in the development and function of the central nervous system. In this condition arginine accumulates and is excreted. In addition precursors of arginine and products of arginine metabolism (Chapter 12) may also be excreted. Unexpectedly, some urea is also excreted; this has been attributed to a second type of arginase found in the kidney. Feeding low nitrogen diets including essential amino acids but excluding arginine, or in some cases using the keto analogue in place of essential amino acids, has been used successfully; the addition of sodium benzoate has also been effective.

BIBLIOGRAPHY

See bibliography for Chapter 12.

QUESTIONS

C. N. ANGSTADT AND J. BAGGOTT

*Q*uestion *T*ypes are described inside the front cover.

1. (QT1) Amino acids considered nonessential for humans are:
   A. those not incorporated into protein.
   B. not necessary in the diet if sufficient amounts of precursors are present.
   C. the same for adults as for children.
   D. the ones made in specific proteins by post-translational modifications.
   E. generally not provided by the ordinary diet.
2. (QT2) Aminotransferases:
   1. usually require α-ketoglutaramate or glutamine as one of the reacting pair.
   2. require pyridoxal phosphate as an essential cofactor for the reaction.
   3. catalyze reactions that result in a net use or production of amino acids.
   4. catalyze freely reversible reactions.
3. (QT2) The production of ammonia in the reaction catalyzed by glutamate dehydrogenase:
   1. requires the participation of NAD^+ or $NADP^+$.
   2. does not proceed through a Schiff's base intermediate.
   3. is favored by removal of the products of the reaction.
   4. may be reversed to consume ammonia if it is present in excess.
4. (QT2) The net synthesis of aspartate using ammonia as the source of nitrogen would be expected to involve:
   1. glutamate dehydrogenase.
   2. a transamination reaction.
   3. oxaloacetate.
   4. L-amino acid oxidase.
5. (QT1) L-Amino acid oxidase:
   A. catalyzes an oxidation coupled to the production of ATP.
   B. is present in large amounts in normal cells.
   C. in vivo, catalyzes a reaction producing H_2O_2.
   D. uses pyridoxal phosphate as its coenzyme.
   E. transfers the amino group of an amino acid to an acceptor molecule.
6. (QT2) Glutamine:
   1. amide nitrogen represents a nontoxic transport form of ammonia.
   2. is a major source of ammonia for urinary excretion.
   3. is used in the synthesis of asparagine in humans.
   4. is an intermediate in the synthesis of purines and pyrimidines.

   A. Aspartate
   B. Arginine
   C. Carbamoyl phosphate
   D. Citrulline
   E. Ornithine
7. (QT5) Urea is formed by a hydrolysis of this compound.
8. (QT5) As part of the urea cycle, synthesized in the cytosol but used in the mitochondria.
9. (QT5) May be synthesized in both mitochondria and cytosol by different isozymes.
10. (QT2) In the formation of urea from ammonia by the urea cycle:
    1. aspartate supplies one of the nitrogens found in urea.
    2. part of the large negative free energy change of the process may be attributed to the hydrolysis of pyrophosphate.
    3. fumarate is produced.
    4. genetic deficiency of any one of the enzymes can lead to hyperammonemia.
11. (QT2) Urea production, as a process to eliminate ammonia:
    1. is not necessary during periods of starvation.
    2. fluctuates as amounts of urea cycle enzymes change in response to changing diets.
    3. because it is cyclic, is unaffected by changing concentrations of intermediates.

4. has its primary control at the level of carbamoyl phosphate synthesis.

12. (QT1) Carbamoyl phosphate synthetase I:
   A. is a flavoprotein.
   B. is controlled primarily by feedback inhibition.
   C. is unresponsive to changes in arginine.
   D. requires acetyl glutamate as an allosteric effector.
   E. requires ATP as an allosteric effector.

13. (QT1) Asparagine:
   A. is formed in proteins after the protein has been synthesized.
   B. is an essential amino acid in humans.
   C. is an intermediate in the urea cycle.
   D. is a primary source of ammonia in the kidney.
   E. from external sources is a requirement for certain abnormal cells.

14. (QT1) Pyridoxal phosphate:
   A. is covalently bound to its enzyme.
   B. facilitates labilization of bonds by acting as an electron-attracting entity.
   C. forms a Schiff's base with amino acids.
   D. all of the above.
   E. none of the above.

15. (QT1) If glutamate labeled with ^{14}C in the α-carbon atom and ^{15}N is included in the diet, the ratio of ^{14}C to ^{15}N in aspartate derived from the reactions of the citric acid cycle and transamination will most likely be:
   A. identical to the ratio in the glutamate fed.
   B. greater than the ratio in the glutamate fed.
   C. lower than the ratio in the glutamate fed.
   D. depending on circumstances, higher or lower than the ratio in the glutamate fed.

ANSWERS

1. B A: All of the 20 common amino acids are incorporated into protein. B, E: Although most of our supply of nonessential amino acids comes from the diet, we can make them if necessary, given the precursors. C: Arginine is not believed to be required for adults (Section 11.1).
2. C 2, 4 correct. 1: Most mammalian aminotransferases use glutamate or α-ketoglutarate. 3: One amino acid is converted into another amino acid; there is neither net gain nor net loss (Section 11.2).
3. E All four correct. 1, 2: The cofactor is a pyridine nucleotide, and the enzyme recognizes both NAD^+ and $NADP^+$ (Note: Production of NH_4^+ requires the oxidized form). 3, 4: The reaction is freely reversible and is usually pulled in the direction of ammonia production by removal of products, but it can easily reverse to meet changing cellular conditions (p. 440).
4. A 1, 2, 3 correct. 1: Glutamate dehydrogenase is necessary to convert the ammonia to an organic compound, glutamate. 2, 3: A transamination from glutamate to oxaloacetate produces aspartate. 4: The L-amino acid oxidase reaction is essentially irreversible in the direction of producing ammonia (p. 440).
5. C A, D: The reduced flavoprotein is reoxidized directly by O_2, not by the electron transport system. B: It is present in rather small amounts. E: Free ammonia is produced (p. 441).
6. E All four correct. It is in the form of the amide nitrogen of glutamine that much of amino acid nitrogen is made available in a nontoxic form (p. 442).
7. B This is the direct production of urea (p. 447).
8. E Ornithine transcarbamoylase, which utilizes ornithine, is mitochondrial; ornithine is regenerated in the cytosol by arginase (Section 11.4).
9. C Carbamoyl phosphate synthetase I, used in the urea cycle, is mitochondrial; CPSII is cytoplasmic and leads to pyrimidines (p. 444).
10. E All four correct. 1, 2, 3: One of the nitrogens is supplied as aspartate, with its carbons being released as fumarate. This reaction is physiologically irreversible because of the hydrolysis of pyrophosphate. 4: Since this is the main pathway for disposal of ammonia, any defect leads to hyperammonemia (p. 445).
11. C 2, 4 correct. 1, 2: Regardless of the fate of the amino acid carbon chain, it is necessary to remove ammonia so this process never ceases although its rate may fluctuate depending on how many amino acid carbon chains are being produced. 3: This would be true only if the cycle were completely closed but, in fact, intermediates, especially ornithine, can be diverted to other processes. 4: The primary control in humans is on CPSI (p. 447).
12. D B, C, D: The primary control is by the allosteric effector, N-acetylglutamate. Synthesis of the effector, and therefore activity of CPSI, is increased in the presence of arginine. E: ATP is a substrate (p. 449).
13. E This is the basis for the therapeutic use of asparaginase in leukemia. A, B: Amides are formed from their corresponding acids before incorporation into protein. C, D: In mammals, asparagine has no known function other than protein synthesis (p. 442).
14. D A, C: Pyridoxal phosphate forms a Schiff's base (covalent bond) with first an ε-amino of lysine in its enzyme and then with an amino acid substrate. B: The ability of the pyridine ring to attract electrons is the basis of its action (p. 440).
15. B ^{15}N is diluted, without an accompanying dilution of the carbon label, by the actions of glutamate dehydrogenase and other transaminases than that using aspartate. Therefore, the oxaloacetate formed from the labeled carbon receives nitrogen with relatively less ^{15}N than originally present. Since glutamate is the most efficient precursor of oxaloacetate, it is unlikely that transamination would transfer nitrogen from glutamate more rapidly than this substrate is oxidized (p. 440).

12

Amino Acid Metabolism II: Metabolism of the Individual Amino Acids

ALAN H. MEHLER

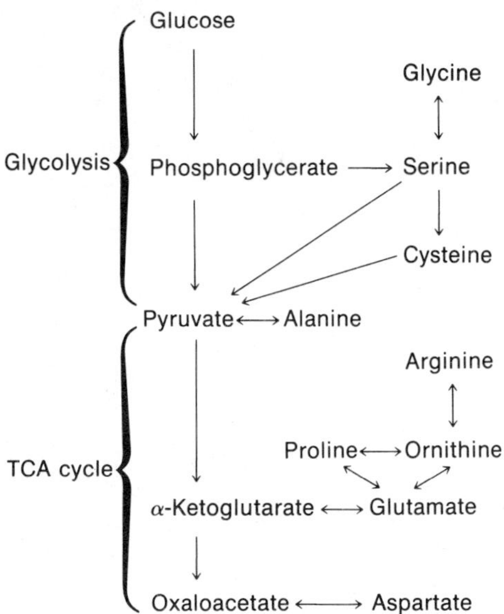

Figure 12.1
Interconversion of amino acids and intermediates of carbohydrate metabolism.

12.1 OVERVIEW

The metabolism of amino acids is a major part of all intermediary metabolism. Since transamination reactions are freely reversible, the nonessential amino acids alanine, aspartate, and glutamate, which feed into glycolytic and tricarboxylic acid cycle reactions, are synthesized, when needed, from glycolytic and tricarboxylic acid cycle intermediates. Proline and arginine (via ornithine) participate indirectly in these processes through glutamate; serine and glycine are interconverted and related to glycolysis in both anabolic and catabolic reactions. These relationships are outlined in Figure 12.1.

Each of the amino acids is metabolized by an individual pathway, although some steps may be shared, notably the transamination and oxidation of the branched-chain amino acids valine, leucine, and isoleucine. The subsequent steps even in this group of similar molecules are completely different for each compound.

An important aspect of amino acid metabolism is the dependence on coenzymes derived from the vitamins folic acid, B_{12}, and biotin for certain of the reactions in addition to the more widely used pyridoxal phosphate and nicotinamide nucleotides. Some of these reactions and their abnormally accumulated substrates are sensitive indicators of deficiency diseases involving both enzymes and coenzymes.

Individual amino acids serve not only to form other amino acids but also as precursors for many other essential groups of diverse compounds. These include the porphyrins, phospholipids, catecholamines and other hormones and neurotransmitters, a large number of methylated compounds, the sulfate esterified to both endogenous and foreign molecules, the nicotinic acid of the pyridine nucleotides, purines, pyrimidines, the pigment melanin, the phosphagen creatine, carnitine, the polyamines spermine and spermidine, and many other compounds, some of whose functions are not yet known. This long list is not intended to be discouraging but is to emphasize the importance of each of the amino acids in intermediary metabolism. An understanding of the interrelationships, controls, and physiological significance of the amino acids requires that each be recognized as an important metabolite in its own right. The amount of information presented in outlining this area might appear to be excessive for the nonspecialist to assimilate, but familiarity with the broad scope of amino acid metabolism is a precondition for seeking precise information from references when problems are encountered involving this part of biochemistry.

$CH_3{-}CH(OH){-}CH_2{-}COO^-$
β-Hydroxybutyrate

NAD ⇄ NADH + H^+

$CH_3{-}C(=O){-}CH_2{-}COO^-$
Acetoacetate

→ CO_2

$CH_3{-}C(=O){-}CH_3$
Acetone

Figure 12.2
Ketone bodies.
The primary product of fatty acid metabolism, acetoacetate, gives rise to β-hydroxybutyrate and acetone when it accumulates during ketosis.

12.2 GLUCOGENIC AND KETOGENIC AMINO ACIDS

Before the detailed metabolism of the amino acids was studied by isolation of enzymes and identification of intermediates, overall relationships of amino acids to the processes of fat and carbohydrate metabolism were established through physiological experiments. By measuring variables such as blood glucose, liver glycogen, and circulating ketone bodies (Figure 12.2), it was seen that some amino acids increase the total amount of glucose in animals, other increase the fat or ketone bodies associated with fatty acid oxidation, and some do both. This evidence led to a classification of amino acids as glucogenic, ketogenic, or both (page 300). The reaction pathways described in the following sections provide explanations for the fates of each of the amino acids derived from proteins. In each case, the reactions are specific for the L isomer of the amino acid.

Although D-amino acids are not found in normal diets, when these are administered to animals in some cases they are metabolized like the L isomers after oxidation to the common α-keto analogue by D-amino acid oxidase. The keto acids may be transaminated to form the natural isomers or may be further metabolized by the pathways described in this chapter.

12.3 METABOLISM OF INDIVIDUAL AMINO ACIDS

Alanine

The only known functions for alanine are incorporation into proteins and participation in transamination. The ability of glutamate dehydrogenase to use alanine or pyruvate in place of glutamate or α-ketoglutarate is of doubtful significance in mammalian metabolism. A large amount of nitrogen is transported from muscle and other peripheral tissues to the liver. This physiological process uses pyruvate produced by glycolysis to accept nitrogen from other amino acids in the formation of alanine, which is converted back to pyruvate in the liver where it participates in gluconeogenesis. Since the liver supplies glucose to other tissues, pyruvate and alanine constitute a shuttle mechanism for carrying nitrogen to be reutilized or converted to urea.

Branched-Chain Amino Acids: Valine, Leucine, and Isoleucine

Valine, leucine, and isoleucine bear nonpolar side chains, each with a methyl group branch. They are essential in the diet of all higher animals, which totally lack the relevant biosynthetic enzymes. The physical and chemical similarities are emphasized by the finding of homologous proteins in various organisms in which these branched-chain compounds replace each other in certain positions without greatly altering the functional properties of the proteins (conservative substitution, Chapter 2). The biosynthetic pathways in plants and microorganisms involve some common steps, and the initial reactions in mammalian catabolism are at some steps carried out by the same enzymes in the case of all three of these amino acids. Subsequent steps, however, are entirely different and yield different products. Thus, valine is glucogenic, leucine is ketogenic, and isoleucine is both.

Where the branched-chain amino acids are present in excess over what is needed for protein synthesis, they are transaminated with α-ketoglutarate to form the corresponding branched-chain α-keto acids. This first step in degradation can be deficient, as seen in Clin. Corr. 12.1. The deaminated products can be considered as higher homologues of pyruvate, and they are oxidized by a complex of enzymes very similar to those that oxidize pyruvate and α-ketoglutarate. The oxidation products include CO_2, NADH that feeds electrons into the electron transport system, and the branched-chain acyl CoAs. These analogues of fatty acyl CoA are oxidized by specific dehydrogenases as if they were unbranched to form the corresponding α–β unsaturated compounds. These reactions are illustrated in Figure 12.3. At this point, the products derived from valine and isoleucine follow one pathway (Figure 12.4), whereas the leucine pathway (Figure 12.5) is quite different.

The unsaturated compounds derived from valine (methylacrylyl CoA) and isoleucine (tiglyl CoA) are hydrated like the unbranched fatty acyl CoA thioesters to form β-hydroxyisobutyryl CoA and α-methyl-β-hydroxyisobutyryl CoA, respectively. The 4-carbon product from valine at

CLIN. CORR. 12.1
DISEASES OF METABOLISM OF BRANCHED-CHAIN AMINO ACIDS

The long series of specific reactions used to degrade the three branched-chain amino acids offer many potential sites for interference with the use of these compounds. Diseases caused by metabolic deficiencies in these pathways are not common, but several different conditions have been identified that have given new insight into the specific reactions for the metabolism of the individual branched-chain amino acids and into their physiological roles. In general these diseases are detected because of acidosis in newborns or young children, and correct identification of the primary cause is necessary for appropriate therapy and prognosis.

Very rare instances have been reported of hypervalinemia and hyperleucine–isoleucinemia. These cases do not provide sufficient information to understand the basic pathology or to design an effective therapy. It has been suggested that the two types of disease might indicate the existence of a specific transaminase for valine and one for leucine and isoleucine. Alternatively, rare mutations could alter the specificity of a single enzyme to restrict its specificity.

The most common cause of abnormality in the metabolism of branched-chain amino acids is deficiency of the branched-chain keto acid dehydrogenase complex. There are several variations according to the severity of the deficiency and the enzyme component involved, but all patients with this disease excrete the α-keto acids and corresponding hydroxyacids and other side products; an unidentified excretory product is responsible for a characteristic odor that gives the name Maple Syrup Urine disease to the group. A few cases respond to high doses of thiamine. A large percentage of the cases show serious mental retardation, ketoacidosis, and short life span, but treatment with diets to reduce the ketoacidemia seems to be effective in some cases. A few cases have been reported of diseases caused by deficiency of enzymes in later reactions of branched-chain amino acids. These include a block at the oxidation of isovaleryl CoA and accumulation of isovalerate, β-methylcrotonyl CoA carboxylase deficiency as judged from excretion of β-methylcrotonylglycine and β-hydroxyisovalerate (an abnormal product of hydration of β-methylcrotonate), deficiency of β-hydroxy-β-methyl glutaryl CoA lyase and deficiency of the β-ketothiolase that splits α-methyloacetoacetyl CoA (with no defect in acetoacetate cleavage). In the latter condition, development is normal and symptoms appear to be related only to episodes of ketoacidosis.

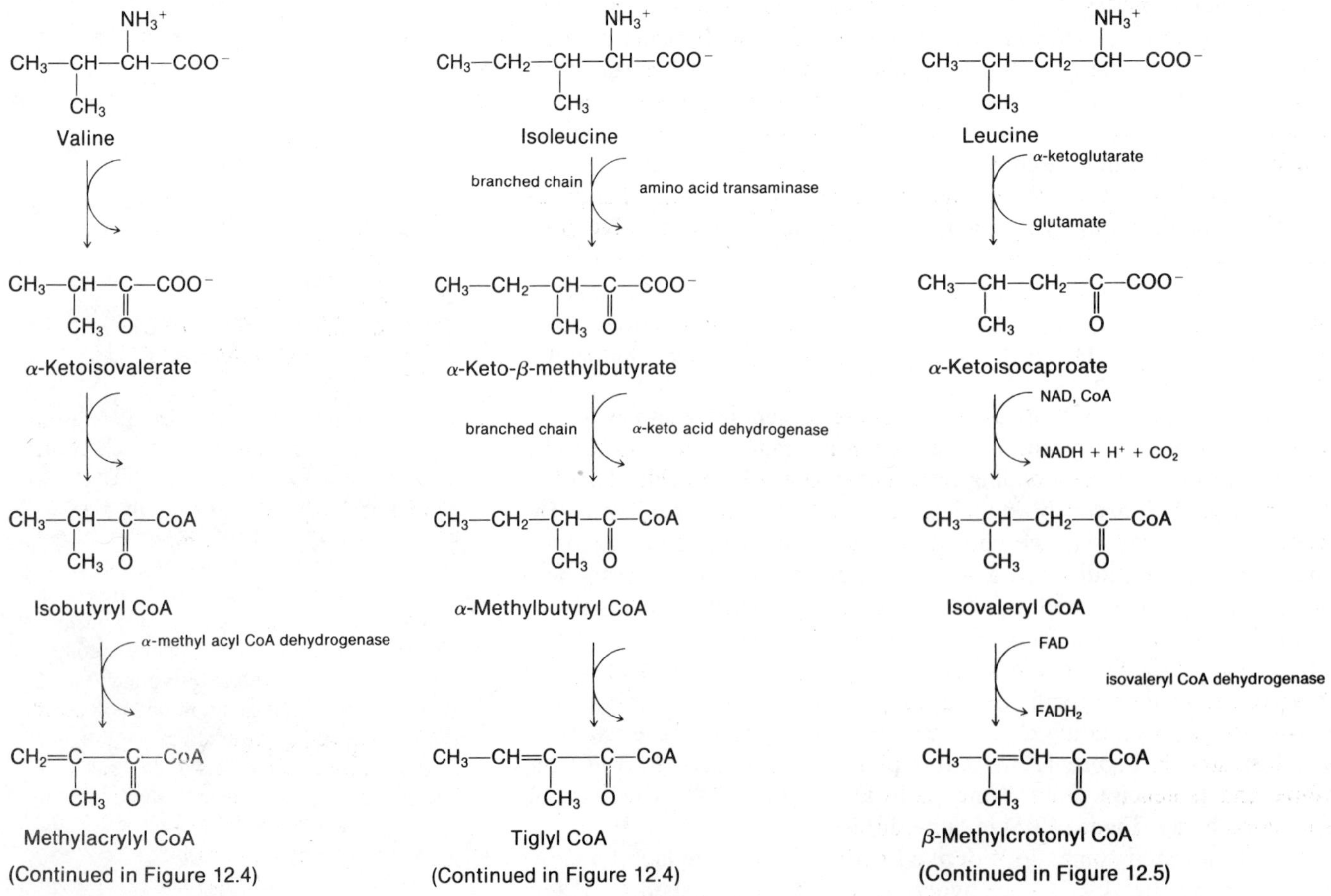

Valine: $CH_3-CH(CH_3)-CH(NH_3^+)-COO^-$

Isoleucine: $CH_3-CH_2-CH(CH_3)-CH(NH_3^+)-COO^-$

Leucine: $CH_3-CH(CH_3)-CH_2-CH(NH_3^+)-COO^-$

↓ branched chain amino acid transaminase (α-ketoglutarate → glutamate)

α-Ketoisovalerate: $CH_3-CH(CH_3)-C(=O)-COO^-$

α-Keto-β-methylbutyrate: $CH_3-CH_2-CH(CH_3)-C(=O)-COO^-$

α-Ketoisocaproate: $CH_3-CH(CH_3)-CH_2-C(=O)-COO^-$

↓ branched chain α-keto acid dehydrogenase (NAD, CoA → NADH + H⁺ + CO_2)

Isobutyryl CoA: $CH_3-CH(CH_3)-C(=O)-CoA$

α-Methylbutyryl CoA: $CH_3-CH_2-CH(CH_3)-C(=O)-CoA$

Isovaleryl CoA: $CH_3-CH(CH_3)-CH_2-C(=O)-CoA$

↓ α-methyl acyl CoA dehydrogenase (isobutyryl CoA); isovaleryl CoA dehydrogenase (FAD → $FADH_2$)

Methylacrylyl CoA: $CH_2=C(CH_3)-C(=O)-CoA$

(Continued in Figure 12.4)

Tiglyl CoA: $CH_3-CH=C(CH_3)-C(=O)-CoA$

(Continued in Figure 12.4)

β-Methylcrotonyl CoA: $CH_3-C(CH_3)=CH-C(=O)-CoA$

(Continued in Figure 12.5)

Figure 12.3
Common reactions in the degradation of branched-chain amino acids.

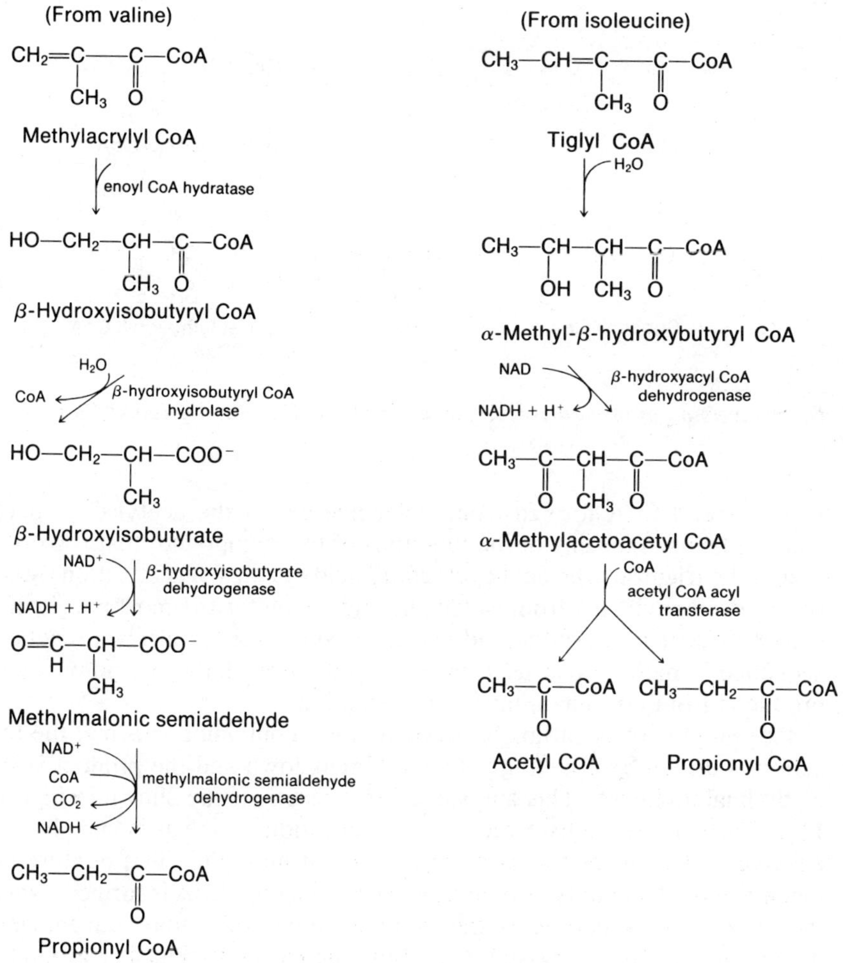

(From valine)

$CH_2=C(CH_3)-C(=O)-CoA$

Methylacrylyl CoA

↓ enoyl CoA hydratase

$HO-CH_2-CH(CH_3)-C(=O)-CoA$

β-Hydroxyisobutyryl CoA

↓ H_2O → CoA; β-hydroxyisobutyryl CoA hydrolase

$HO-CH_2-CH(CH_3)-COO^-$

β-Hydroxyisobutyrate

↓ NAD^+ → $NADH + H^+$; β-hydroxyisobutyrate dehydrogenase

$O=C(H)-CH(CH_3)-COO^-$

Methylmalonic semialdehyde

↓ NAD^+, CoA → CO_2, NADH; methylmalonic semialdehyde dehydrogenase

$CH_3-CH_2-C(=O)-CoA$

Propionyl CoA

(From isoleucine)

$CH_3-CH=C(CH_3)-C(=O)-CoA$

Tiglyl CoA

↓ H_2O

$CH_3-CH(OH)-CH(CH_3)-C(=O)-CoA$

α-Methyl-β-hydroxybutyryl CoA

↓ NAD → $NADH + H^+$; β-hydroxyacyl CoA dehydrogenase

$CH_3-C(=O)-CH(CH_3)-C(=O)-CoA$

α-Methylacetoacetyl CoA

↓ CoA; acetyl CoA acyl transferase

$CH_3-C(=O)-CoA$ Acetyl CoA + $CH_3-CH_2-C(=O)-CoA$ Propionyl CoA

Figure 12.4
Terminal reactions in degradation of valine and isoleucine.

this point is hydrolyzed from the CoA, then the hydroxyl group is oxidized by a dehydrogenase to an aldehyde, methylmalonic semialdehyde. A new thioester is formed when the aldehyde is oxidized by an NAD-requiring dehydrogenase and during the oxidation the original carboxyl group is lost as CO_2. This complex reaction results in the formation of propionyl CoA. The 5-carbon intermediate from isoleucine is oxidized as its CoA derivative to α-methylacetoacetyl CoA, which is cleaved by thiolase to form two thioesters, acetyl CoA and propionyl CoA.

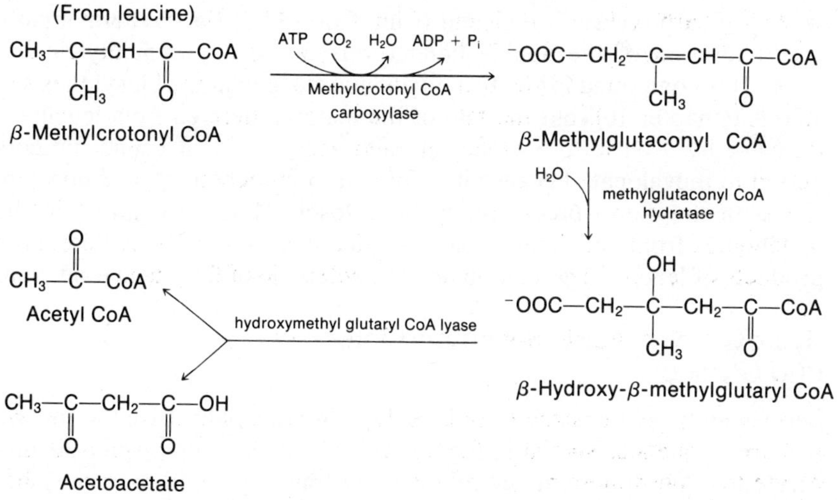

(From leucine)

$CH_3-C(CH_3)=CH-C(=O)-CoA$

β-Methylcrotonyl CoA

→ ATP, CO_2, H_2O → ADP + P_i; Methylcrotonyl CoA carboxylase

$^-OOC-CH_2-C(CH_3)=CH-C(=O)-CoA$

β-Methylglutaconyl CoA

↓ H_2O; methylglutaconyl CoA hydratase

$^-OOC-CH_2-C(OH)(CH_3)-CH_2-C(=O)-CoA$

β-Hydroxy-β-methylglutaryl CoA

← hydroxymethyl glutaryl CoA lyase

$CH_3-C(=O)-CoA$ Acetyl CoA + $CH_3-C(=O)-CH_2-C(=O)-OH$ Acetoacetate

Figure 12.5
Terminal reactions of leucine degradation.

CLIN. CORR. 12.2 DISEASES OF PROPIONATE AND METHYLMALONATE METABOLISM

The three enzymes shown in Figure 12.6 have all been implicated in the production of ketoacidosis, which has serious consequences if not controlled. Propionate is formed in the degradation of valine, isoleucine, methionine, threonine, the side chain of cholesterol, and odd-chain fatty acids; therefore, each of these could be implicated as a source of the accumulated acid but quantitatively the amino acids appear to be the main precursors since decreasing or eliminating protein from the diet has an immediate effect in minimizing acidosis.

A defect in propionyl CoA carboxylase results in accumulation of propionate in body fluids. Secondary accumulation of other compounds results in formation of propionylglycine, methylcitrate (formed by propionyl CoA condensing with oxaloacetate as an abnormal reaction of citrate synthetase), and tiglate. The inability of propionyl CoA to be metabolized normally results in increased amounts being diverted to alternative pathways, oxidation like long chain fatty acids to β-hydroxypropionate, and incorporation of the 3-carbon chain into fatty acids in place of an acetyl group to form odd-chain fatty acids. The extent of these reactions is very limited, however, and they do not serve effectively to relieve the acidosis. In one case administration of large amounts of biotin, the coenzyme of the carboxylase, was reported to produce beneficial effects. This suggests that more than one defect results in decreased propionyl CoA carboxylase.

A group of children have been found to suffer from acidosis caused by high levels of methylmalonate, which is normally not detectable in blood. Analyses of enzymes from liver taken at autopsy or from cultured fibroblasts have established that some cases were due to deficiency of methylmalonyl CoA mutase, but these were further divided into two groups on the basis of the enzyme studies. One group was unable to convert methylmalonyl CoA to succinyl CoA under any conditions, but extracts from cells of another group of cases carried out the conversion when the coenzyme adenosylcobalamin was added. Clearly, patients with a structural defect in the enzyme cannot metabolize methylmalonate, but some patients

$$\underset{\text{Propionyl CoA}}{CH_3-CH_2-\overset{O}{\overset{\|}{C}}-CoA} \xrightleftharpoons[\text{carboxylase}]{CO_2 \;\; \text{propionyl CoA}} \underset{\text{D-Methylmalonyl CoA}}{{}^-OOC-\overset{CH_3}{\overset{|}{CH}}-\overset{O}{\overset{\|}{C}}-CoA}$$

$$\updownarrow \text{ methylmalonyl CoA racemase}$$

$$\underset{\text{Succinyl CoA}}{{}^-OOC-CH_2-CH_2-\overset{O}{\overset{\|}{C}}-CoA} \xrightleftharpoons[\text{mutase}]{\text{methylmalonyl CoA}} \underset{\text{L-Methylmalonyl CoA}}{{}^-OOC-\underset{CH_3}{\underset{|}{CH}}-\overset{O}{\overset{\|}{C}}-CoA}$$

Figure 12.6
Interconversion of propionyl CoA, methylmalonyl CoA, and succinyl CoA.

The acetyl CoA formed from isoleucine enters the acetyl CoA pool and can be used for any of the functions of this compound, that is oxidation in the tricarboxylic acid cycle, fatty acid synthesis, acetylations, and so on. Propionyl CoA from isoleucine and valine for the most part is not oxidized further as a fatty acid but is the substrate for a novel reaction sequence, which is also used in the metabolism of the propionyl CoA produced from odd-chain fatty acid oxidation.

Propionyl CoA is carboxylated by a biotin-containing enzyme; the reaction is driven by coupling with ATP breakdown and the product is D-methylmalonyl CoA. This and subsequent reactions are shown in Figure 12.6. Further metabolism requires racemization to the L isomer by an enzyme that labilizes the α-hydrogen so that an equilibrium mixture of racemic (equal amounts of D- and L-) methylmalonyl CoA is formed. Only the L isomer is a substrate for the methylmalonyl CoA mutase, an enzyme with a vitamin B_{12} coenzyme, that shifts the carboxy–thioester group to the methyl carbon atom to form the straight-chain succinyl CoA. This compound is metabolized as if it were produced during the operation of the tricarboxylic cycle. Deficiency of any of the three enzymes that together convert propionate to succinyl CoA causes disease, as discussed in Clin. Corr. 12.2. The three carbon atoms derived from valine and isoleucine may be oxidized completely to CO_2, but they can be incorporated into carbohydrate by oxidation of succinate to oxaloacetate and the reactions of gluconeogenesis (Chapter 7).

The unsaturated β-methylcrotonyl CoA produced from leucine is carboxylated by another specific biotin-containing enzyme. An unusual acidosis is caused by hydrolysis of the thioester to free β-methylcrotonate when the carboxylase is deficient (Clin. Corr. 12.1). In the normal pathway hydration of the double bond forms β-hydroxy-β-methylglutaryl CoA. This compound is identical to an intermediate in the biosynthesis of sterols (Chapter 10), but the fate of the material derived from leucine is cleavage by a lyase to acetoacetate and acetyl CoA; it cannot be converted to mevalonate because it is formed in mitochondria and does not mix with the sterol precursor in the cytosol. Thus, in contrast to the metabolites from valine and isoleucine that can form carbohydrate, both products of leucine degradation are characteristic of fatty acid oxidation.

Hydroxyamino Acids: Serine and Threonine, and Glycine

Serine can be synthesized from glycolytic intermediates by either of two reaction sequences shown in Figure 12.7: (1) oxidation of 3-phosphoglycerate to 3-phosphopyruvate followed by transamination to 3-phospho-

with defects in handling vitamin B_{12} do respond to massive doses of the vitamin. Other cases of methylmalonic aciduria suffer from a more fundamental inability to use vitamin B_{12} that results in deficiency in methylcobalamin, the coenzyme of methionine salvage, as well as adenosylcobalamin deficiency. A recent report described the sudden deterioration of an adolescent who had been an honor student. Within a year her mental ability dropped to a first-grade level, she developed an unsteady gait, and other signs and symptoms appeared including large amounts of homocysteine and methylmalonate in the urine. Treatment with massive amounts of vitamin B_{12} eliminated the biochemical abnormalities and restored both mental and physical functions, although after 3 months she still showed evidence of myelopathy.

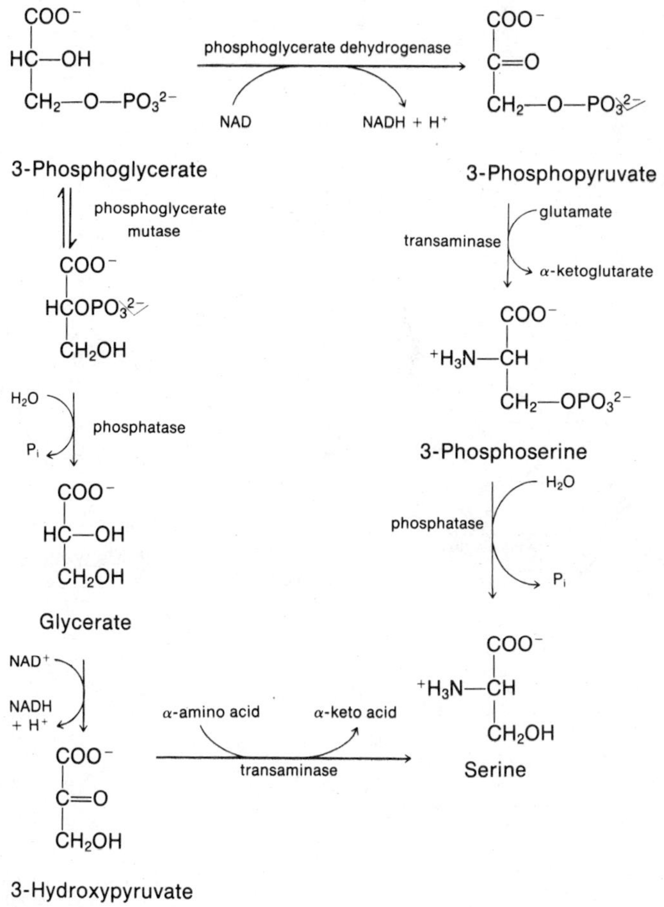

3-Phosphoglycerate (COO^-, HC—OH, $CH_2—O—PO_3^{2-}$) —phosphoglycerate dehydrogenase (NAD → NADH + H^+)→ 3-Phosphopyruvate (COO^-, C=O, $CH_2—O—PO_3^{2-}$)

3-Phosphopyruvate —transaminase (glutamate → α-ketoglutarate)→ 3-Phosphoserine (COO^-, $^+H_3N—CH$, $CH_2—OPO_3^{2-}$)

3-Phosphoserine —phosphatase (H_2O → P_i)→ Serine (COO^-, $^+H_3N—CH$, CH_2OH)

3-Phosphoglycerate ⇅ phosphoglycerate mutase → COO^-, $HCOPO_3^{2-}$, CH_2OH —phosphatase (H_2O → P_i)→ Glycerate (COO^-, HC—OH, CH_2OH)

Glycerate (NAD^+ → NADH + H^+)→ 3-Hydroxypyruvate (COO^-, C=O, CH_2OH) —transaminase (α-amino acid → α-keto acid)→ Serine

Figure 12.7
Formation of serine from intermediate metabolites of glycolysis.

serine and hydrolysis to the free amino acid or (2) hydrolysis of 2-phosphoglycerate to glycerate then oxidation to hydroxypyruvate and transamination to serine. As will be discussed below, serine is also interconvertible with glycine so that multiple routes are available to produce this amino acid as it is needed. In contrast, threonine is an essential amino acid in animals.

When serine is used to produce energy, the principal pathway is conversion to pyruvate by a pyridoxal phosphate requiring enzyme, serine dehydratase, that eliminates water and NH_3 from the amino acid (Figure 12.8). A similar reaction forms α-ketobutyrate from threonine. Oxidation of α-ketobutyrate, probably by the pyruvate dehydrogenase complex, produces propionyl CoA. A novel pathway of threonine oxidation is an oxidative decarboxylation to aminoacetone and CO_2; oxidation of this product leads to pyruvate or lactate.

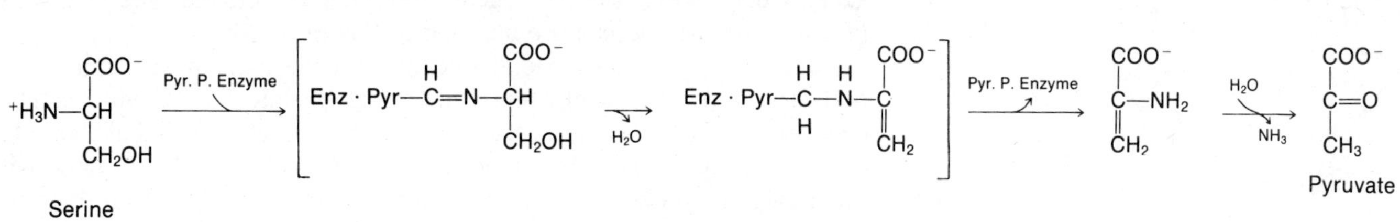

Serine (COO^-, $^+H_3N—CH$, CH_2OH) —Pyr. P. Enzyme→ [Enz · Pyr—CH=N—CH(COO^-)(CH_2OH) → (H_2O) Enz · Pyr—CH₂—NH—C(COO^-)=CH_2] —Pyr. P. Enzyme→ COO^-, C—NH_2, =CH_2 —(H_2O → NH_3)→ Pyruvate (COO^-, C=O, CH_3)

Figure 12.8
Reaction of serine dehydratase, a pyridoxal phosphate requiring enzyme.

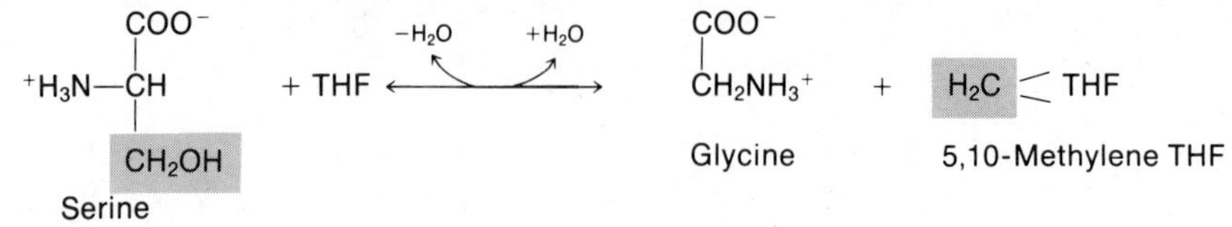

$$^{+}H_3N{-}CH(COO^-){-}CH_2OH \text{ (Serine)} + THF \xrightleftharpoons[+H_2O]{-H_2O} CH_2NH_3^{+}{-}COO^- \text{ (Glycine)} + H_2C{<}THF \text{ (5,10-Methylene THF)}$$

Figure 12.9
Interconversion of serine and glycine.

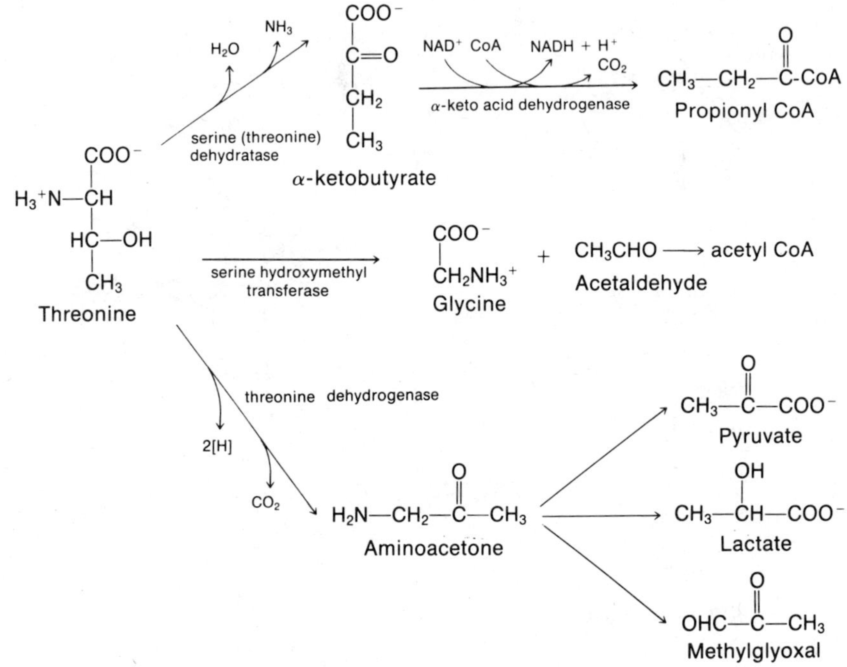

Threonine ($H_3^{+}N{-}CH(COO^-){-}HC{-}OH{-}CH_3$):

- serine (threonine) dehydratase (H_2O in, NH_3 out) → α-ketobutyrate ($COO^-{-}C{=}O{-}CH_2{-}CH_3$) → [$NAD^+$, CoA → NADH + H^+, CO_2; α-keto acid dehydrogenase] → $CH_3{-}CH_2{-}C(=O){-}CoA$ Propionyl CoA
- serine hydroxymethyl transferase → $COO^-{-}CH_2NH_3^{+}$ Glycine + CH_3CHO Acetaldehyde → acetyl CoA
- threonine dehydrogenase (2[H], CO_2) → $H_2N{-}CH_2{-}C(=O){-}CH_3$ Aminoacetone → $CH_3{-}C(=O){-}COO^-$ Pyruvate; $CH_3{-}CH(OH){-}COO^-$ Lactate; $OHC{-}C(=O){-}CH_3$ Methylglyoxal

Figure 12.10
Outline of threonine metabolism.

$COO^-{-}CH_2{-}NH_3^{+}$ Glycine → (O_2 in; NH_3, H_2O_2 out) → $COO^-{-}C(=O){-}H$ Glyoxalate → (NAD^+ → NADH + H^+) → $COO^-{-}COO^-$ Oxalate

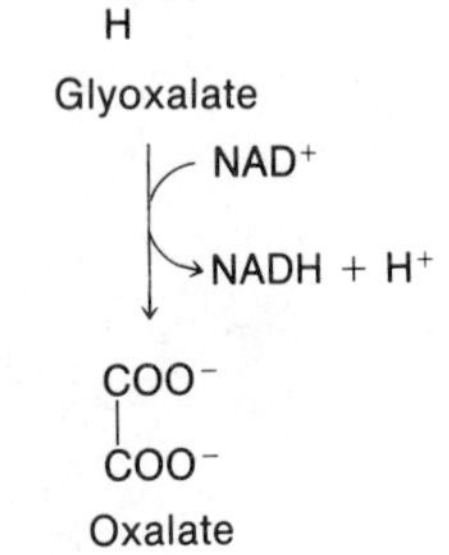

Figure 12.11
Oxidation of glycine.

The formation of glycine from serine is a reversible reaction in which a pyridoxal phosphate enzyme, serine hydroxymethyl transferase, uses another coenzyme, tetrahydrofolic acid (THF), as a substrate. In the transfer reaction shown in Figure 12.9, carbon 3 of serine, which is at the oxidation level of formaldehyde, is transferred to the acceptor where it bridges the nitrogen atoms at positions 5 and 10. This product, 5,10-methylenetetrahydrofolate, can transfer the 1-carbon element back to glycine, which is thus in equilibrium with serine. It will be shown later that the methylene group is part of a pool of 1-carbon fragments at several levels of oxidation. A significant amount of serine is used in the formation of phospholipids.

Threonine is also a substrate for serine hydroxymethyl transferase. In producing glycine from threonine, no acceptor is needed for the other fragment, which is released as free acetaldehyde. An aldehyde dehydrogenase using both NAD and CoA converts the acetaldehyde efficiently to acetyl CoA, making the cleavage of threonine irreversible. The pathways of threonine metabolism are outlined in Figure 12.10.

Glycine is oxidized by two distinct enzyme mechanisms. D-Amino acid oxidase attacks glycine efficiently to form glyoxalate. This product can be converted back to glycine by transamination but is also oxidized further to oxalate (Figure 12.11). Oxalate is of major importance in the formation of renal calculi, some of which are mainly precipitates of calcium oxalate. The α-carbon of glycine can also enter the 1-carbon pool through the action of an enzyme that contains pyridoxal phosphate and uses both NAD^+ and tetrahydrofolate (see Figures 12.34–12.36). The products are ammonia, CO_2, NADH, and 5,10-methylenetetrahydrofolate. Since the reaction is reversible, it provides a mechanism for the synthesis of glycine and also explains why the α- and β-carbon atoms of isotopically labeled serine and the α-carbon atom of glycine are rapidly randomized in vivo. The metabolic importance of this reaction is seen in Clin. Corr. 12.3.

Glycine is incorporated intact as a constituent of purines (Chapter 13). It also contributes to the synthesis of δ-aminolevulinate en route to porphyrin formation (Chapter 22). Creatine, discussed later in this chapter, is a substituted glycine. Other substituted glycines to be discussed later are sarcosine and betaine.

Glutamate, Proline, Ornithine, and Arginine

In Chapter 11 a major role was described for glutamate in the incorporation of ammonia into amino acids, transferring amino groups, and eliminating nitrogen. Glutamate is also a key intermediate in the metabolism of other amino acids that share with it a common carbon skeleton. The nonessential amino acids proline and arginine are built directly from the 5-carbon chain of glutamate.

Proline

The cyclization of glutamate to a five-membered ring at neutral pH requires prior modification of the γ-carboxyl group. The pyrrolidine ring of proline is more reduced than the direct cyclization product, 5-oxoproline (pyrrolidone carboxylate, page 484). The pathway brackets the ring closure with reduction steps. Glutamate is activated in a reaction with ATP; the putative γ-carboxyl phosphate does not dissociate from the protein but is reduced to the corresponding aldehyde by NADH. The resulting glutamic semialdehyde is an equilibrium mixture composed of only a trace of an open chain and mainly the cyclic Schiff base, Δ^1-pyrroline-5-carboxylate. A second reduction by NADPH saturates the ring and gives a stable product, proline, as seen in Figure 12.12.

The derivatives of proline found in collagen and a few other proteins, 3- and 4-hydroxyprolines, are formed by mixed-function oxygenases from proline residues incorporated in the polypeptide chains (Chapter 19). During growth and development there is extensive turnover of the structural proteins, and both proline and the hydroxyprolines are degraded.

Catabolism of proline superficially appears to reverse the synthetic pathway, but the conversion to glutamate involves different enzymes that catalyze different reactions. Oxidation of proline is carried out by a mitochondrial enzymes (probably a flavoprotein by analogy with a similar bacterial activity) that transfers electrons to the electron transport system in forming Δ^1-pyrroline 5-carboxylate. The further oxidation of pyrroline carboxylate is a dehydrogenation by an enzyme nonspecific for NAD^+ and $NADP^+$ to form glutamate with no involvement of phosphate. Like all oxidations of aldehydes to free carboxyl groups, this reaction is effectively irreversible. 4-Hydroxyproline is oxidized by an oxidase similar to that for proline but not the same. The second oxidation, however, is catalyzed by the same dehydrogenase involved in proline oxidation. The product, 4-hydroxy-2-ketoglutarate, is split by an aldolase to pyruvate and glyoxylate. The structures of these compounds are shown in Figure 12.13. Little is known about the oxidation of 3-hydroxyproline.

Ornithine

Δ^1-Pyrroline-5-carboxylate is in equilibrium with the open-chain aldehyde that participates in a transamination reaction to form ornithine. In mammals the conversion of glutamate to ornithine has been demonstrated

CLIN. CORR. 12.3
NONKETOTIC HYPERGLYCINEMIA

Nonketotic hyperglycinemia is characterized by severe mental deficiency, and many patients do not survive infancy. The name of this very serious disease is meant to distinguish it from ketoacidosis in abnormalities of branched-chain amino acid metabolism in which, for unknown reasons, the glycine level in the blood is also elevated. Deficiency of glycine decarboxylase has been demonstrated in homogenates of liver from several patients, and isotopic studies in vivo have confirmed that this enzyme is not active in the patients. The reason that this apparently minor metabolic reaction is of major biological significance is unknown. Vigorous measures to reduce the glycine levels [infusion of 5-formyltetrahydrofolate, administration of sodium benzoate (see Clin. Corr. 11.1), and exchange transfusion] fail to alter the course of the disease.

$$^-OOC—CH_2—CH_2—CH(NH_3^+)—COO^-$$
Glutamate

ATP → ADP + P_i; NADH + H^+ → NAD^+ (uncharacterized enzymes)

$$O=C(H)—CH_2—CH_2—CH(NH_3^+)—COO^-$$
Glutamate semialdehyde

spontaneous

Δ^1-Pyrroline-5-carboxylate

NADPH + H^+ → $NADP^+$ (pyrroline-5-carboxylate reductase)

Proline

Figure 12.12
Synthesis of proline from glutamate.

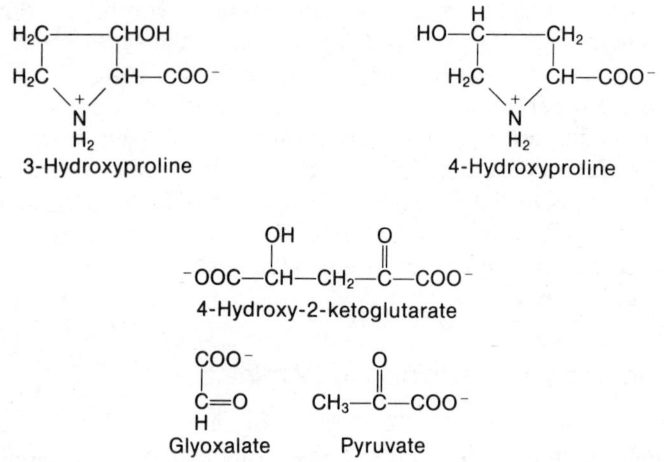

3-Hydroxyproline

4-Hydroxyproline

$^{-}OOC—CH(OH)—CH_2—C(=O)—COO^{-}$

4-Hydroxy-2-ketoglutarate

Glyoxalate

$CH_3—C(=O)—COO^{-}$

Pyruvate

Figure 12.13
Hydroxyprolines and metabolic products.

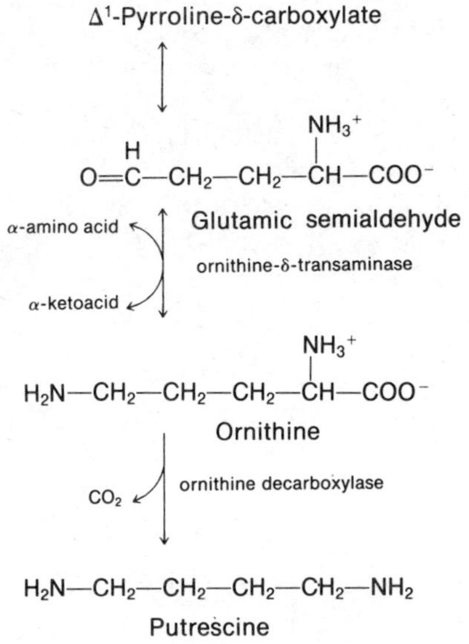

Δ^1-Pyrroline-δ-carboxylate

$O{=}C(H)—CH_2—CH_2—CH(NH_3^+)—COO^-$

Glutamic semialdehyde

α-amino acid → α-ketoacid (ornithine-δ-transaminase)

$H_2N—CH_2—CH_2—CH_2—CH(NH_3^+)—COO^-$

Ornithine

CO_2 (ornithine decarboxylase)

$H_2N—CH_2—CH_2—CH_2—CH_2—NH_2$

Putrescine

Figure 12.14
Formation and decarboxylation of ornithine.

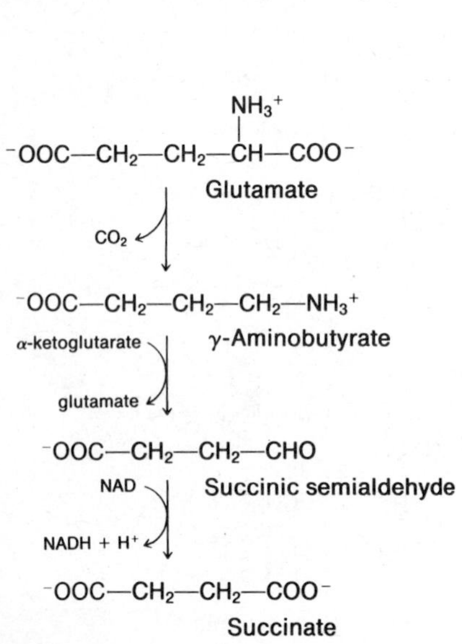

only in intestinal mucosa, not in liver (Figure 12.14). The existence of a common intermediate makes it possible for ornithine and proline to be interconverted without involving the formation of glutamate.

In addition to its participation in urea cycle reactions, arginine is the donor in a transamidinase reaction. The physiological acceptor is glycine, which is converted to guanidinoacetate. This compound is an intermediate in the synthesis of creatine (page 479).

Ornithine is prominent in metabolism primarily in relationship to arginine, as a precursor and product in the urea cycle, or simply as a product when arginine is degraded by arginase in tissues other than liver or when the transamidinase reaction produces guanidinoacetate. Since most of the arginine ingested is not required for protein synthesis, a large amount of ornithine is produced and started on a degradative pathway by ornithine transaminase (Clin. Corr. 12.9). Another very important function for ornithine is the formation of putrescine by decarboxylation (Figure 12.14). The putrescine is required for the synthesis of polyamines, spermidine, and spermine. The exact function of polyamines is still uncertain, but it is clear that initiation of cell division depends upon ornithine decarboxylase.

γ-Aminobutyrate

Removal of the α-carboxyl group of glutamate leaves γ-aminobutyrate (GABA). The glutamate decarboxylase reaction is prominent in the brain, where GABA is an important neurotransmitter. GABA is converted to succinic semialdehyde by transamination and then oxidized to succinate (Figure 12.15).

$^{-}OOC—CH_2—CH_2—CH(NH_3^+)—COO^-$

Glutamate

CO_2

$^{-}OOC—CH_2—CH_2—CH_2—NH_3^+$

γ-Aminobutyrate

α-ketoglutarate → glutamate

$^{-}OOC—CH_2—CH_2—CHO$

Succinic semialdehyde

NAD → NADH + H^+

$^{-}OOC—CH_2—CH_2—COO^-$

Succinate

Figure 12.15
Metabolism of γ-aminobutyrate (GABA).

Sulfur Amino Acids

In the human adult methionine can serve as the sole source of sulfur, but on a normal diet a larger amount of sulfur is ingested as the more abundant cysteine. The synthesis of cysteine requires that the thioether of methionine be made available as a sulfhydryl group. The mechanism of exposing the sulfur is the same that is used to make methionine the principal source of methyl groups that are transferred to many diverse acceptors in reactions of great physiological importance and also to enable the side chain of methionine to be used for other biosynthetic reactions.

The activation of methionine, described in Figure 12.16, is a unique reaction in that the sulfur of the thioether becomes a sulfonium atom by the addition of a third carbon atom, the 5′-carbon of the ribose of ATP. In this reaction all of the phosphates of ATP are eliminated, one as inorganic orthophosphate and the others as inorganic pyrophosphate. The product, *S*-adenosylmethionine, is reactive because of the positive charge that reverts to a neutral thioether when any of its three substituents is lost. Enzymatic reactions are known that specify transfer to appropriate acceptors of either the methyl group, a 3- or 4-carbon fragment from the rest of the methionine structure, or the adenosyl group, in each case leaving the other two substituents in the thioether.

The most prominent reactions of *S*-adenosyl methionine are methyl transferases. These are for the most part presented as reactions of the methyl acceptors. In each case, as illustrated in Figure 12.17 with nicotinamide, the methyl donor becomes *S*-adenosyl homocysteine. Hydrolysis of the latter yields adenosine and homocysteine. There are three principal routes of metabolism of homocysteine; the distribution of homocysteine among these pathways is determined by the physiological needs of the organism. If cysteine is needed, the homocysteine condenses with serine to form cystathionine in a reaction catalyzed by an enzyme with pyridoxal phosphate as a coenzyme. Another pyridoxal phosphate enzyme, cystathionase, cleaves this thioester to cysteine, α-ketobutyrate, and ammonia (Figure 12.18). The sum of cystathionine synthase and cystathionase reactions is transsulfuration. If, however, methionine is in short supply, homocysteine is remethylated by N^5-methyltetrahydrofolate or betaine. When both sulfur amino acids are present in adequate amounts, the same enzyme, cystathionase, appears to be responsible for the activity called homocysteine desulfhydrase that hydrolyzes homocysteine to α-ketobutyrate, ammonia, and H_2S. *S*-Adenosylmethionine activates cystathionine synthase and thus regulates the competing pathways of homocysteine metabolism. Defects of enzymes involved in the metabolism of cystathionine result in accumulation of methionine, homocysteine, or cystathionine (Clin. Corr. 12.4).

S-Adenosylmethionine can be decarboxylated by a specific enzyme. The decarboxylated product, *S*-adenosylmethylthiopropylamine, is the donor of the 3-carbon fragment that is used in the formation of polyamines. The acceptor of the propylamino group is the diamine putrescine produced by decarboxylation of ornithine and the first condensation product is spermidine. A second transfer of a propylamino residue produces the symmetrical polyamine, spermine. Spermine is a highly cationic molecule that binds tightly to nucleic acids and other polyanions. It has been found to stimulate a variety of reactions in vitro and appears to be essential for the function of a topoisomerase (Chapter 19). This may explain the early increase of ornithine decarboxylase activity in cell division, but the precise biological functions of the polyamines remain to be determined.

Although methionine is an essential amino acid, it is resynthesized to a large extent by salvage pathways. 5,10-Methylenetetrahydrofolate produced from serine or glycine or by reduction of other 1-carbon derivatives of tetrahydrofolate can be reduced by NADH to 5-methyltetrahydrofolate. The only known reaction to use 5-methyltetrahydrofolate is the methyltransferase with homocysteine as the methyl acceptor to regenerate methionine. The reaction requires cobalamin, the coenzyme form of vitamin B_{12}. The salvage reaction in which betaine is the methyl donor appears to be a simpler reaction.

The sulfur-containing product of polyamine synthesis, methylthioadenosine, is not wasted. The steps in the salvage pathway have not yet all been identified, but a phosphorylase is known to replace the adenine with a phosphate. Isotopic evidence shows that the sulfur, the methyl carbon, and the ribose chains are all incorporated into methionine,

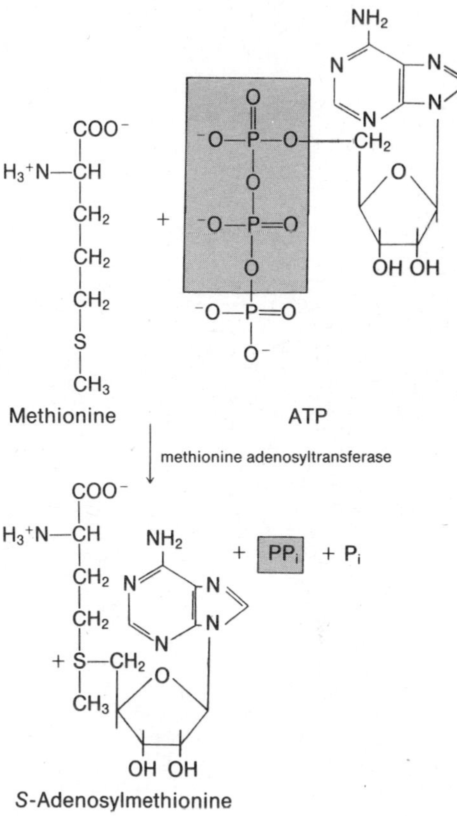

Methionine + ATP → (methionine adenosyltransferase) → *S*-Adenosylmethionine + PP_i + P_i

Figure 12.16
Synthesis of S-adenosylmethionine.

S-Adenosylmethionine + Nicotinamide → (*N*-methyltransferase) → *S*-Adenosylhomocysteine + *N*′-Methylnicotinamide

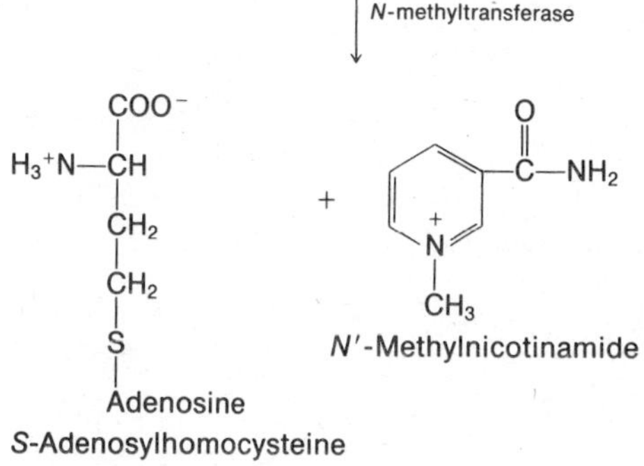

Figure 12.17
A representative methylation reaction.
N′-Methylnicotinamide is the major product excreted in the breakdown of NAD^+; S-adenosylhomocysteine is a product of all methyl transfers from S-adenosylmethionine.

$$\underset{\text{Homocysteine}}{CH_2{-}SH,\ CH_2,\ HC{-}NH_3^+,\ COO^-} + \underset{\text{Serine}}{CH_2OH,\ HC{-}NH_3^+,\ COO^-} \xrightarrow[H_2O]{\text{cystathionine synthase}} \underset{\text{Cystathionine}}{CH_2{-}S{-}CH_2,\ CH_2,\ C{-}NH_3^+,\ HC{-}NH_3^+,\ COO^-,\ COO^-} \xrightarrow[NH_3]{\text{cystathionase}} \underset{\alpha\text{-Ketobutyrate}}{CH_3,\ CH_2,\ C{=}O,\ COO^-} + \underset{\text{Cysteine}}{SH,\ CH_2,\ HC{-}NH_3^+,\ COO^-}$$

Figure 12.18
Biosynthesis of cysteine.
The enzyme cystathionine synthase is sometimes named β-cystathionine synthase, and cystathionase is sometimes designated γ-cystathionase. The Greek letters are used to distinguish these enzymes from similar activities in bacteria that participate in a reversal of this pathway, making homocysteine (and methionine) from cysteine.

presumably through formation of α-keto-γ-methylthiobutyrate, which is transaminated to regenerate methionine. It should be noted that this process, pictured in Figure 12.19, was initiated by activation of methionine and cannot give a net synthesis of the essential amino acid.

Obviously, a transaminase that makes methionine can also degrade it via the same α-keto-γ-methylthiobutyrate. A degradative sequence that starts with this transamination is probably responsible for the formation of toxic mercaptans, including methanethiol and H_2S. Normally the amounts of these compounds are negligible, but they become significant in methionine toxicity, caused by excessive ingestion or decreased transsulfuration during liver disease.

Cysteine is a major source of sulfur for the body. Sulfur occurs at several oxidation levels, from sulfide to sulfate. Reports of a direct desulfhydration of cysteine, analogous to dehydration of serine, has been claimed, but at this time the best evidence indicates that H_2S (or S^{2-}) is derived indirectly from cysteine in a series of reactions catalyzed by cystathionase. Very little of this toxic material is produced in normal metabolism. The sulfur of cysteine is made available indirectly for detoxi-

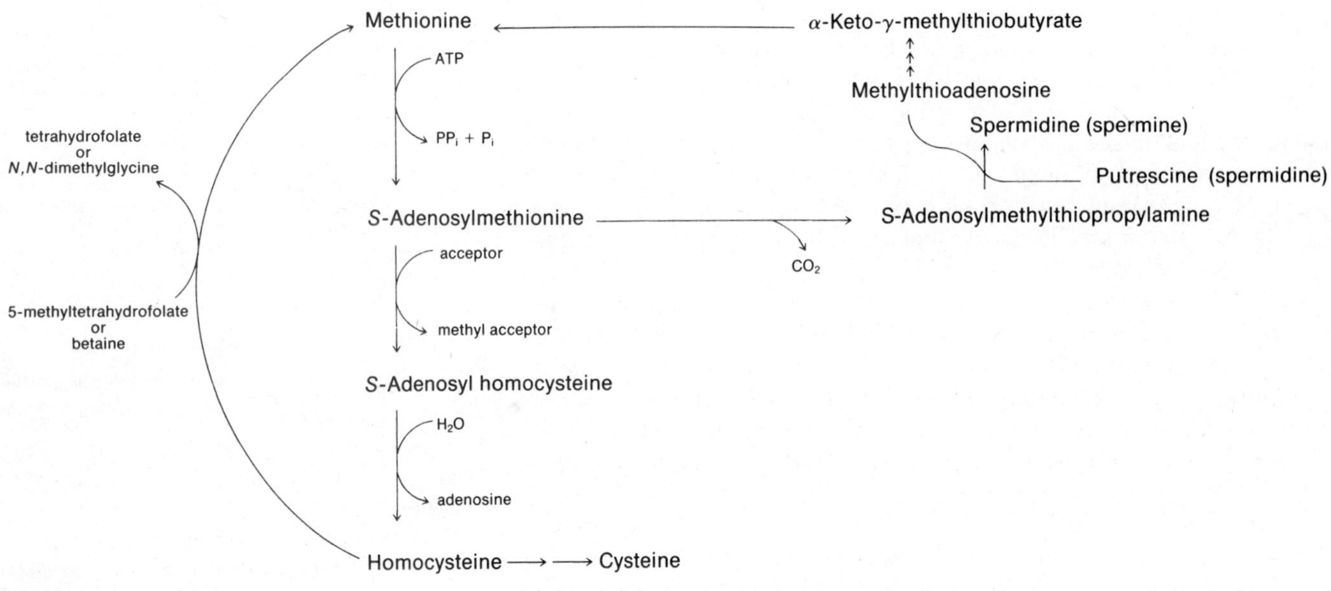

Figure 12.19
Uses and resynthesis of methionine.

CLIN. CORR. 12.4 DISEASES OF SULFUR AMINO ACIDS

TRANSSULFURATION DEFECTS

Congenital deficiency of any of three enzymes involved in transsulfuration results in accumulation of sulfur-containing amino acids. Hypermethioninemia has been attributed to a deficiency of methionine adenosyltransferase. A total lack of this enzyme would eliminate most methylation reactions and has never been found. The partial deficiencies are probably manifestations of K_m mutants; that is, the patients have enzymes that require higher than normal concentrations of methionine for saturation so that a steady state is established in which methionine at a high concentration permits a rate of formation of *S*-adenosylmethionine adequate for the various functions illustrated in Figure 12.19. The accumulation of methionine in these patients has no other consequence so the condition is benign. This is of significance in making a differential diagnosis since hypermethioninemia also occurs in cystathionine synthase deficiency (see below), in severe liver disease, and in tyrosinemia.

Deficiency of cystathionine synthase causes homocysteine to accumulate, and this causes even higher levels of methionine to be formed by remethylation. Many minor products of these amino acids are formed and excreted as a result of these accumulations. No mechanism has been established to explain why the accumulation of homocysteine should lead to pathological changes, but the deficiency of the synthase is associated with many abnormalities. The lens of the eye frequently is dislocated some time after age 3 years, and other ocular abnormalities are often seen. Osteoporosis and other skeletal abnormalities develop during childhood. Mental retardation is often the first indication of this deficiency. Thromboembolism and vascular occlusion may occur at any age. Rational attempts to treat this biochemical lesion are complex because of the variety of symptoms. Restriction of methionine intake and feeding of betaine (or its precursor, choline) have been found to lower the homocystine level. In a number of cases significant improvement has been obtained by feeding pyridoxine (vitamin B_6), but other cases are nonresponsive to the vitamin. This suggests that the deficiency can be caused by more than one type of gene mutation; one type may affect the K_m for pyridoxal phosphate and others may alter the K_m for other substrates, V_{max}, or the amount of enzyme.

In contrast to the severe manifestations of cystathionine synthase deficiency, lack of cystathionase does not seem to cause any clinical abnormalities other than accumulation of cystathionine and excretion of this compound in the urine. It is worth noting that the first discovery of cystathioninuria was in a mental patient, and subsequent searches for additional cases tended to concentrate on the mentally retarded. Even so, a majority of reported cases are mentally normal and the retarded cases include cases with endocrine disorders, phenylketonuria, and explanations other than cystathioninuria. A large majority of reported cases are responsive to pyridoxine. The amount of cysteine synthesized in the various deficiencies of enzymes of transsulfuration is not known, but supplementation is not necessary except when a low methionine diet is used in cases of hypermethioninemia.

DISEASES INVOLVING CYSTINE

Two clinical conditions feature the disulfide cystine. Cystinuria is a disease of defective membrane transport of cystine and the basic amino acids that results in increased renal excretion of these compounds. Since extracellular sulfhydryl compounds are quickly oxidized to disulfides, cysteine in the blood and urine exists as cystine. The low solubility of cystine results in the formation of calculi, the serious feature of this disease. Treatment is limited to attempts to remove stones or to prevent precipitation by diet, including large amounts of water, or drugs to make soluble derivatives of cystine.

A much more serious disease is cystinosis in which cystine accumulates in lysosomes. The mechanism of normal transport is unknown. The stored cystine forms crystals in many cells, with most serious loss of function of the kidneys, usually causing death within 10 years.

Four cases have been reported in which a mixed disulfide of cysteine and β-mercaptolactate is excreted. Two of the reports describe severely retarded individuals, and two concern individuals with normal development. It is accepted that the abnormal metabolite is derived from β-mercaptopyruvate, but why this intermediate accumulates to be reduced and excreted together with cysteine remains a mystery.

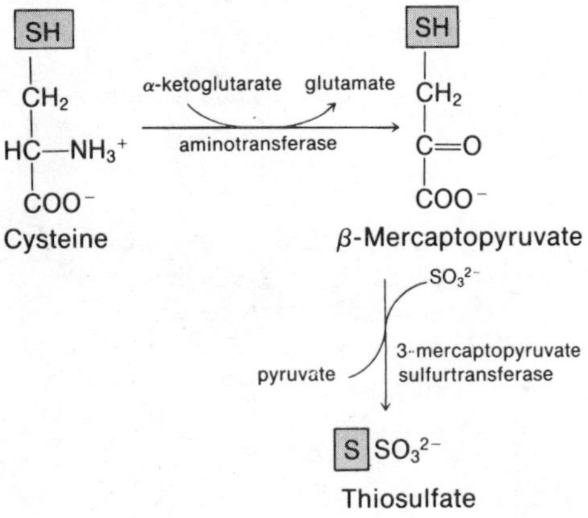

Figure 12.20
Formation and use of β-mercaptopyruvate.

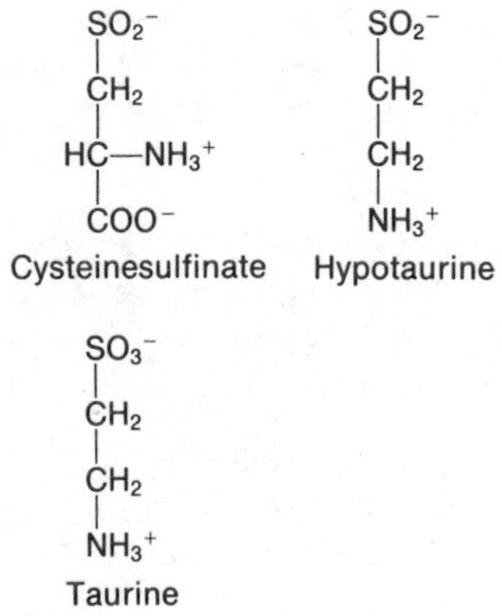

Figure 12.21
Taurine and its precursors.

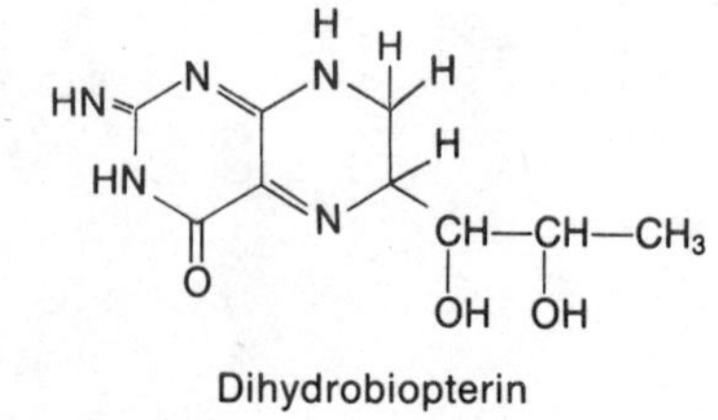

Figure 12.22
Coenzyme of aromatic amino acid hydroxylation.
The quinonoid form is produced in the oxygenation reaction and is reduced to the coenzyme by a dehydrogenase using NADH.

fication reactions and for incorporation into proteins such as ferredoxin. A sulfur donor produced by cystathionase is thiocysteine.

$$\text{Cystine} + H_2O \longrightarrow$$
$$\text{pyruvate} + NH_3 + HS—S—CH_2—CH(NH_3^+)—COO^-$$
Thiocysteine

Rhodanese catalyzes the transfer of sulfur from thiocysteine (and other sulfane donors, such as thiosulfate) to many acceptors; the best known of these is cyanide, which is converted to the much less toxic thiocyanate.

$$\text{Thiocysteine} + CN^- \longrightarrow \text{cysteine} + SCN^-$$

Glutamate aminotransferase and other aminotransferases are able to use cysteine as an amino donor. The resulting 3-mercaptopyruvate is the substrate of a sulfur transferase. In the absence of an acceptor elemental sulfur is formed, but with acceptors such as sulfite and cyanide, the transfer reaction shown in Figure 12.20 predominates.

A major pathway of cysteine metabolism is oxidation by an oxygenase to cysteinesulfinate. The enzyme has the requirements of a mixed-function oxygenase, Fe^{2+} and a reduced nicotinamide nucleotide, but nevertheless inserts two atoms of oxygen from O_2 into the product. Cysteinesulfinate and compounds derived from it react spontaneously to give several products, so it has been difficult to evaluate the biochemical pathways using this intermediate. At this time it seems likely that two reactions account for most of its metabolism. Decarboxylation yields hypotaurine, which is rapidly oxidized to taurine shown in Figure 12.21. An alternative pathway of cysteinesulfinate metabolism starts with transamination, which should give β-sulfinylpyruvate, but only sulfite and pyruvate are found. Sulfite oxidase, an enzyme containing molybdenum and cytochrome b_5, catalyzes the main production of sulfate from bisulfite.

$$HSO_3^- + O_2 + H_2O \xrightarrow{\text{sulfite oxidase}} SO_4^{2-} + H_2O_2 + H^+$$

Although much of the sulfate is excreted, some is also used as a component of many structural elements; some proteins, polysaccharides, and lipids are sulfated. In addition many compounds are excreted as sulfate esters. Metabolites of the steroidal sex hormones and exogenous phenolic compounds are prominent among the urinary sulfates. Each of these materials is made by transfer of sulfate from a nucleotide carrier, 3′-phosphoadenosine-5′-phosphosulfate (PAPS), which is synthesized in two steps:

$$SO_4^{2-} + ATP \xrightarrow{\text{ATP sulfurase, } Mg^{2+}}$$
$$\text{adenosine-5'-phosphosulfate(AMPS)} + PP_i$$

$$\text{Adenosine-5'-phosphosulfate} + ATP \xrightarrow[Mg^{2+}]{\text{AMPS phosphokinase}}$$
$$\text{3'-phosphoadenosine-5'-phosphosulfate} + ADP$$

Sulfate incorporated into this nucleotide is also transferred to some unknown acceptor to form taurine. The physiological role of taurine is still conjectural, but the high concentrations found in the brain suggest a more important function than its use by the liver for conjugation with bile acids. An alternative route for taurine synthesis starts with cysteamine, but the source of this precursor appears to be limited to the breakdown of CoA; a direct decarboxylation of cysteine has not been observed. Children maintained for long periods on parenteral nutrition have been found to have decreased plasma levels of taurine and to produce abnormal electroretinograms; these conditions were corrected by addition of taurine to the solutions that were administered intravenously. It is not yet known at which stages in human development or under which clinical conditions dietary taurine is necessary.

The vitamins thiamine, biotin, and lipoic acid contain sulfur but do not contribute measurably to the pool of sulfur metabolites.

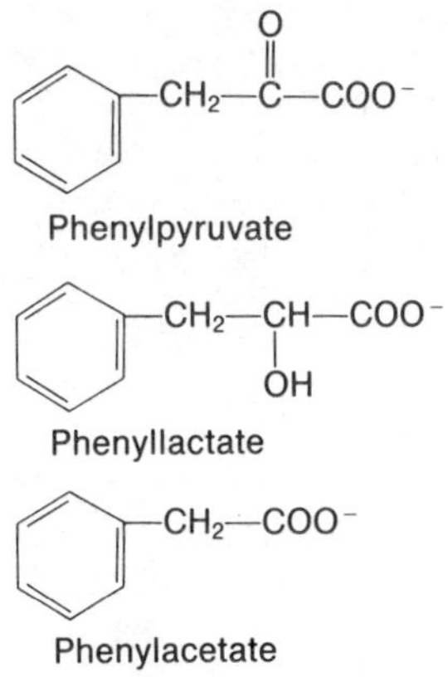

C_6H_5—CH_2—C(=O)—COO^-

Phenylpyruvate

C_6H_5—CH_2—CH(OH)—COO^-

Phenyllactate

C_6H_5—CH_2—COO^-

Phenylacetate

Figure 12.23
Minor products of phenylalanine metabolism.

Phenylalanine and Tyrosine

Under normal circumstances almost all of the phenylalanine that is degraded undergoes only a single metabolic conversion—hydroxylation at position 4 to form tyrosine. The hydroxylation is a mixed-function oxygenation that uses as cosubstrate a specific coenzyme, tetrahydrobiopterin (Figure 12.22). This relative of folic acid serves as the coenzyme for several hydroxylases. When the hydroxylation reaction is deficient (Clin. Corr. 12.5), excess phenylalanine accumulates in the blood and equilibrates with its transamination product, phenylpyruvate. This compound is in part excreted and also is converted to phenyllactate and phenylacetate by reduction and oxidation, respectively; these products are also excreted in the urine (Figure 12.23).

Tyrosine is degraded by a major pathway in terms of the quantity of amino acid metabolized, but the so-called minor pathways have great physiological importance. The major pathway, shown in Figure 12.24, occurs in the liver, where it begins with transamination to *p*-hydroxyphenylpyruvate. The aminotransferase is a prominent "inducible" enzyme, which is increased many fold in animals treated with both tyrosine and glucocorticoid hormones. The oxidation of *p*-hydroxyphenylpyruvate is very complex. A single enzyme adds the atoms of O_2 to the α-carbon atom of the side chain and to the ring carbon to which the side chain is attached. The enzyme contains Fe^{2+} and is protected by ascorbic acid, but a biological role of vitamin C has not been established in this system. The oxidation of the carbonyl group to a carboxyl is accompanied by decarboxylation of the original acid group. To accommodate the new hydroxyl group in the ring, the shortened side chain is shifted to an adjacent carbon atom. The product is homogentisate. Another iron-containing oxygenase cleaves the aromatic ring by adding oxygen atoms to carbons 1 and 2; the *cis*-configuration of the double bond in the open-chain compound is the basis for the designation maleylacetoacetate. Further degradation requires the isomerization of the double bond to the trans configuration of fumarylacetoacetate; the isomerase appears to have a specific requirement for glutathione. Hydrolysis yields fumarate, a tricarboxylic acid cycle intermediate, and acetoacetate, a ketone body. Phenylalanine and tyrosine are, therefore, both glycogenic and ketogenic. Several genetic diseases are associated with this series of reactions (Clin. Corr. 12.6).

Several prominent physiological agents are produced by a sequence of reactions (Figure 12.25) initiated by hydroxylation of tyrosine to 3,4-dihydroxyphenylalanine (dopa). The hydroxylase is very similar to the enzyme that forms tyrosine from phenylalanine; Fe^{2+} is essential for the activity and the cosubstrate is tetrahydrobiopterin. The decarboxylation of dopa by a pyridoxal phosphate enzyme yields the corresponding amine, which is formed in specific regions of the brain where it functions

CLIN. CORR. **12.5**
PHENYLKETONURIA

Phenylketonuria (PKU) is the most prominent disease caused by a deficiency of an enzyme of amino acid metabolism. The name comes from the excretion of phenylpyruvic acid (a phenylketone) in the urine. An oxidation product of phenylpyruvate, phenylacetate, is also excreted and gives the urine a "mousey" odor. Phenylpyruvate is detected as a green color produced with ferric chloride, but other compounds also give similar colors. Therefore, the routine screening legally required in many states is an estimation of phenylalanine in the blood; over 20 mg per 100 ml is considered positive. The requirement for this laboratory analysis to be performed on all infants is a reflection of the high incidence of the genetic deficiencies that interfere with the hydroxylation of phenylalanine and the belief that mental retardation associated with the condition can be prevented by dietary control.

Classical PKU is a deficiency of phenylalanine hydroxylase. Genetic analysis shows that the deficiency is caused by an autosomal recessive gene; thus, the enzyme is produced normally by genes on two chromosomes, one from each parent, and the presence of one defective gene results in a heterozygus individual whose liver produces less than normal amounts of the enzyme. When the normal pathway of phenylalanine metabolism in an offspring of two heterozygus parents is blocked by this deficiency (in most cases there is no enzyme detected by either catalytic or immunochemical methods), the concentration of phenylalanine increases in all body fluids and "minor" products accumulate and are excreted. In some cases there are severe neurological symptoms and very low IQ. These are generally attributed to toxic effects of phenylalanine, possibly on the transport and metabolism of other aromatic amino acids in the brain. It should be noted that there is great variation in the symptoms of children with this enzyme deficiency. Another characteristic of the condition is light color of skin, eyes, and internal tissues where melanin is normally accumulated. This suggests that tyrosine deficiency is also characteristic; that is, dietary tyrosine is not sufficient to compensate for lack of endogenous production. There is a surprising dearth of literature on the effect of tyrosine supplementation in PKU, although it has been suggested that intermittent restriction of protein synthesis because of lack of tyrosine could account for the failure of the brain to develop normally. The conventional treatment is to feed infants a synthetic diet low in phenylalanine for about 4–5 years and to restrict proteins in the diet for several more years or for life.

The diagnosis of phenylketonuria is complicated by the increase of phenylalanine in the blood (phenylalaninemia) by several mechanisms. About 3% of infants with high levels of phenylalanine have normal hydroxylase but are defective in either the synthesis or reduction of the essential cofactor, biopterin. Since this cofactor is also necessary for the synthesis of neurotransmitters, central nervous functions are more seriously affected and treatment at this time includes administration of 5-hydroxytryptophan and dopa to overcome the deficiency of precursors of serotonin and catecholamines.

Great caution must be observed in interpreting screening tests for phenylalanine. In a few cases the increase in blood levels indicative of a serious disease did not occur during the first several days postpartum. On the other hand, there are many cases of benign transient neonatal hyperphenylalaninemia. Defects in tyrosine metabolism also result in accumulation of phenylalanine.

Phenylalanine
tetrahydrobiopterin dihydrobiopterin
O_2 H_2O
phenylalanine hydroxylase
Tyrosine
α-ketoglutarate
glutamate
p-Hydroxyphenylpyruvate
CO_2 O_2
p-hydroxyphenylpyruvate oxidase
Homogentisate
O_2
homogentisate oxidase
Maleylacetoacetate
maleylacetoacetate isomerase
Fumarylacetoacetate
fumarylacetoacetate hydrolase
Fumarate + Acetoacetate

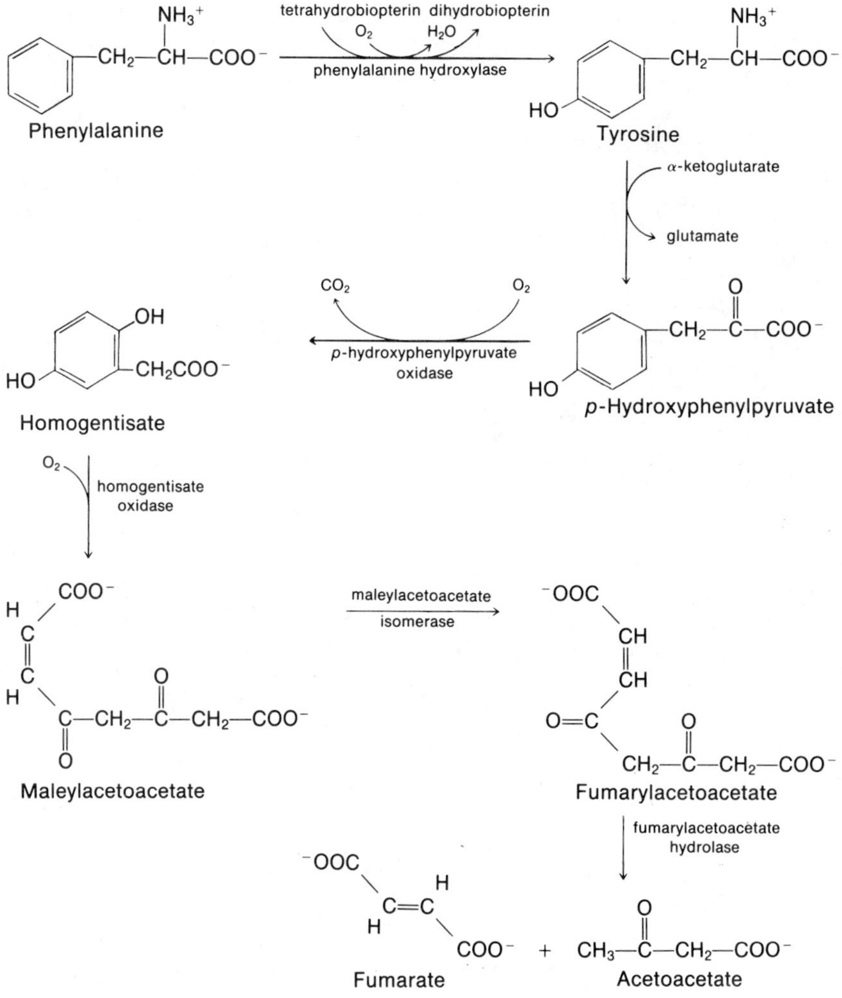

Figure 12.24
Synthesis and degradation of tyrosine.

CLIN. CORR. **12.6**
DISORDERS OF TYROSINE METABOLISM

TYROSINEMIAS

The absence or deficiency of cytosolic tyrosine transaminase (tyrosine aminotransferase, TAT) is responsible for the accumulation and excretion of tyrosine and metabolites including *N*-acetyltyrosine, *p*-hydroxyphenylpyruvate, *p*-hydroxyphenyllactate, *p*-hydroxyphenylacetate, and tyramine. Since *p*-hydroxyphenylpyruvate, the presumed precursor of some of the other products, is also the product of the transaminase, it is likely that these products come from mitochondrial transaminases or oxidases in extrahepatic tissues. The disease is characterized by eye and skin lesions, and most

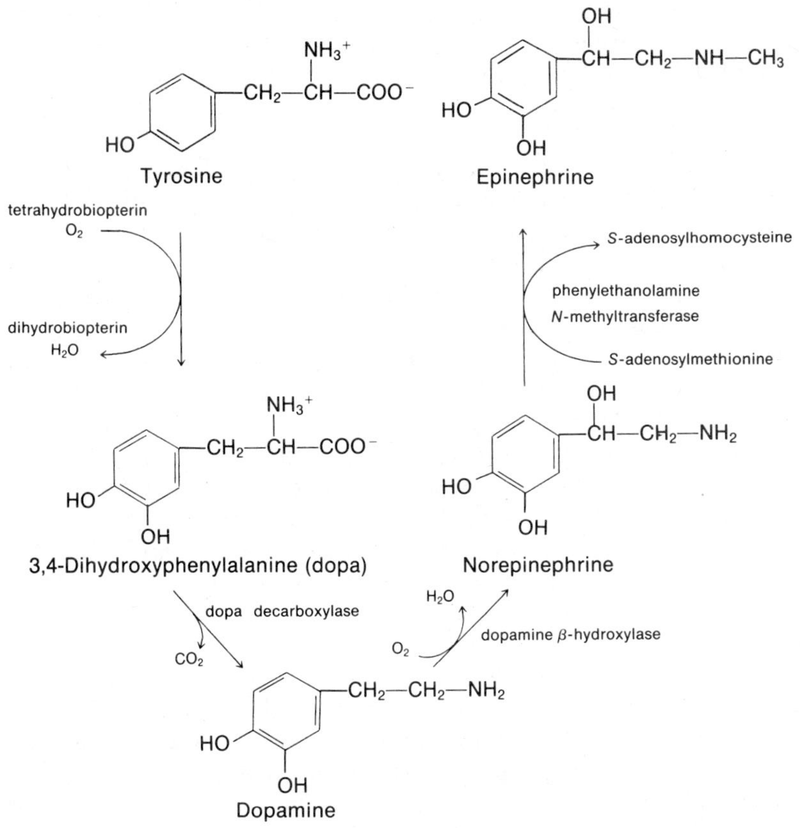

Tyrosine

tetrahydrobiopterin
O_2
dihydrobiopterin
H_2O

3,4-Dihydroxyphenylalanine (dopa)

dopa decarboxylase
CO_2

Dopamine

O_2
H_2O
dopamine β-hydroxylase

Norepinephrine

S-adenosylmethionine
phenylethanolamine N-methyltransferase
S-adenosylhomocysteine

Epinephrine

Figure 12.25
Synthesis of the catecholamines.

but not all of the cases reported have been mentally retarded. This condition is called oculocutaneous or type II tyrosinemia.

Type I, hepatorenal tyrosinemia, is a more serious disease involving liver failure, renal tubular dysfunction, rickets, and polyneuropathy in addition to excretion of tyrosine, other amino acids, and other metabolites. All of this appears to be caused by a deficiency of fumarylacetoacetate hydrolase.

Additional diseases are associated with lack of other enzymes of tyrosine degradation. Deficiency of *p*-hydroxylphenylpyruvate oxidase is believed to be responsible for neonatal tyrosinemia, which is usually a temporary condition and in some cases responds to ascorbic acid, given on the hypothesis that this compound protects the enzyme from substrate inhibition.

The very different consequences of deficiencies at various points in tyrosine metabolism show the necessity of analyzing all of the factors that might be relevant and avoiding a simplistic explanation. Interruption of the pathway at homogentisate oxidation causes this compound to accumulate without any other metabolic effects. Homogentisate is not an intrinsically reactive compound, and it does not alter any enzyme before it is excreted. In contrast, in the absence of its hydrolase, fumarylacetoacetate accumulation causes maleylacetoacetate to accumulate also and this is chemically reactive, especially combining with sulfhydryl compounds. Another toxic compound, succinylacetone, has been suggested as a secondary product of fumarylacetoacetate metabolism that might be responsible for some of the biochemical lesions in this disease.

ALBINISM

Skin and hair color are controlled by an unknown number of genetic loci in humans; in mice 147 genes have been identified in color determination. It is not surprising, therefore, that skin color exists in infinite variations and also that formation of pigment can be interfered with in many ways. Many conditions have been described in which the skin has little or no pigment, but the chemical basis is not established for any except classical albinism. In this condition the enzyme tyrosinase is deficient and melanin is not formed. Lack of pigment in the skin makes albinos sensitive to sunlight, which may cause carcinoma of the skin in addition to burns; lack of pigment in the eyes causes photophobia. Lack of eye pigment does not imply impaired eyesight; a description of albinos among American Indians in 1699 indicated that their vision at night was superior to normal.

ALCAPTONURIA

The first condition to be identified as an "inborn error of metabolism" was alcaptonuria. People deficient in homogentisate oxidase excrete almost all ingested tyrosine as homogentisic acid in their urine. This hydroquinone is colorless, but on standing it autooxidizes to the corresponding quinone, which polymerizes to form an intensely dark color. Concern about the dark urine is the only consequence of this condition early in life. Homogentisate is slowly oxidized to pigments that are deposited in bones, connective tissue, and various organs, presumably as a function of the metabolic inertness of tissues that allow foreign material to accumulate over long periods of time. This generalized pigmentation is called ochronosis because of the ochre color seen in the light microscope. Pigment deposition is thought to be responsible for the arthritis that develops in many alcaptonuric individuals, especially in males.

The analysis of alcaptonuria by Garrod that first indicated its genetic basis as an autosomal recessive deficiency condition includes an unusual historical description of the condition. This is of great value in appreciating the iatrogenic suffering that can be inflicted by physicians who act on the basis of inadequate information and false assumptions.

CLIN. CORR. 12.7 PARKINSON'S DISEASE

Usually in people over the age of 60 years but occasionally in much younger people, tremors may develop that gradually interfere with motor function of various muscle groups. This condition is named Parkinson's disease for the physician who described "shaking palsy" in 1817. The primary cause has not been identified, and there may be more than one etiological agent. The defect, however, is established as degeneration of cells in certain small areas of the brain, substantia nigra and locus coeruleus. Cells in these nuclei produce dopamine as a neurotransmitter, and the amount of dopamine released decreases as a function of the number of surviving cells. Symptomatic relief, often dramatic, is obtained by increasing the availability of dopa, the precursor of dopamine. A number of clinical problems were recognized when dopa (L-dopa, levodopa) became available for treating the large number of people suffering from Parkinson's disease. Side effects included nausea, vomiting, hypotension, cardiac arrhythmias and varied central nervous symptoms. These were explained as effects of dopamine produced outside the central nervous system. Administration of analogues of dopa that inhibit dopa decarboxylase and are unable to cross the blood-brain barrier has been effective both in decreasing the side effects and increasing the effectiveness of the dopa. It should be emphasized that interactions of the many neurotransmitters are very complex, that degeneration of cells continues after treatment, and that elucidation of the major biochemical abnormality has not yet led to complete control of Parkinson's disease.

as a neurotransmitter. Decreased production of dopamine is the cause of Parkinson's disease (Clin. Corr. 12.7). In the adrenal medulla, dopamine is also made from tyrosine and is further hydroxylated on the β-carbon of the side chain by a copper-containing enzyme located in chromaffin granules. Some of the product, norepinephrine, is stored in secretory granules until the cells are stimulated to release the hormone into the blood. The larger part of the norepinephrine is methylated by a relatively nonspecific phenylethanolamine N-methyltransferase to become epinephrine, which is also stored in the chromaffin granules. Epinephrine and norepinephrine are hormones known collectively as catecholamines. Norepinephrine is stored in vesicles in the termini of axons to be released for synaptic transmission of nerve impulses. They affect the physiological functions and metabolism of most organs very rapidly, at least in some cases by stimulating the synthesis of cyclic AMP.

An oxidation that closely resembles that carried out by tyrosine hydroxylase is carried out by a copper-containing enzyme, tyrosinase. This is an oxygenation that does not use a cofactor but that uses the presumed product of the hydroxylation, dopa, as the hydrogen donor in a mixed-function oxygenation so that the material that accumulates is dopaquinone (Figure 12.26). This is a very reactive molecule that cyclizes and condenses to form melanin. Melanin is a family of high molecular weight polymers that may include cysteine; melanin granules are very insoluble and very dark in color. Melanin is concentrated in special cells called melanocytes in the skin and also in the choroid plexus, the retina and ciliary body of the eye and in the substantia nigra of the brain. Inability to form melanin is the basis of albinism (Clin. Corr. 12.6).

The hormone thyroglobulin contains iodinated aromatic amino acids derived from tyrosine. These are produced by posttranslational modification of the protein in the thyroid gland (Chapter 15).

Aspartate

Aspartate is a metabolically active compound because of its interconversion with the C-4 dicarboxylic acids of the tricarboxylic acid cycle after transamination. It is important as a precursor of asparagine, a primary amino acid in protein synthesis, as is aspartate itself. The other reactions of aspartate are as amino donors in urea synthesis (page 446) and in purine synthesis (pages 498 and 500) and as a precursor of pyrimidine rings through the formation of carbamoyl aspartate in the cytosol (page 507).

Tryptophan

The metabolism of tryptophan does not resemble that of any other metabolite. Its principal degradative pathway occurs in the liver and leads to the formation of nicotinic acid, usually classified as a vitamin, and many

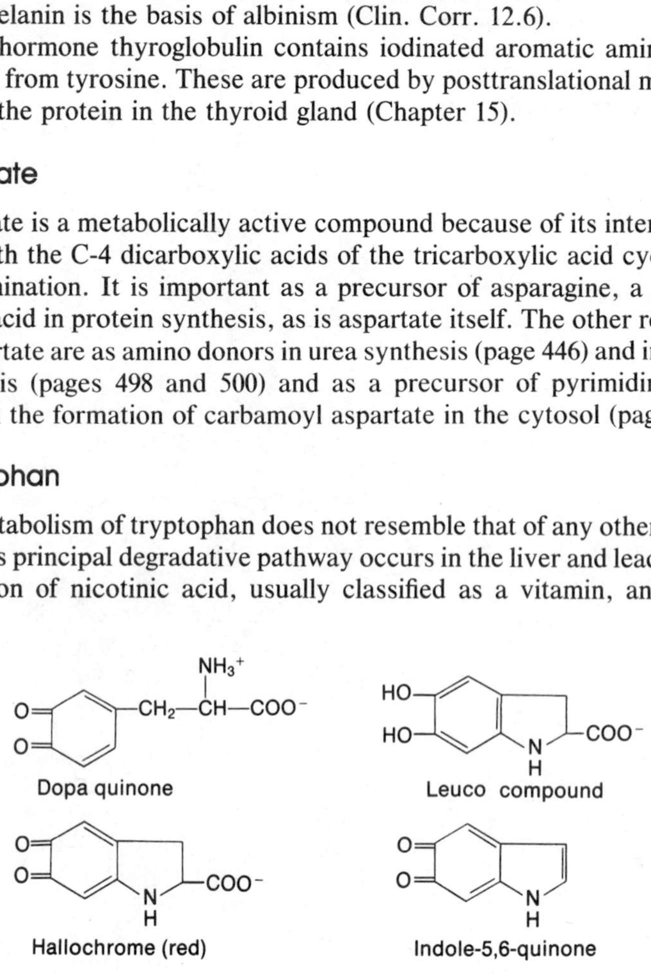

Dopa quinone

Leuco compound

Hallochrome (red)

Indole-5,6-quinone

Figure 12.26
Some intermediates in the formation of melanin by tyrosinase.

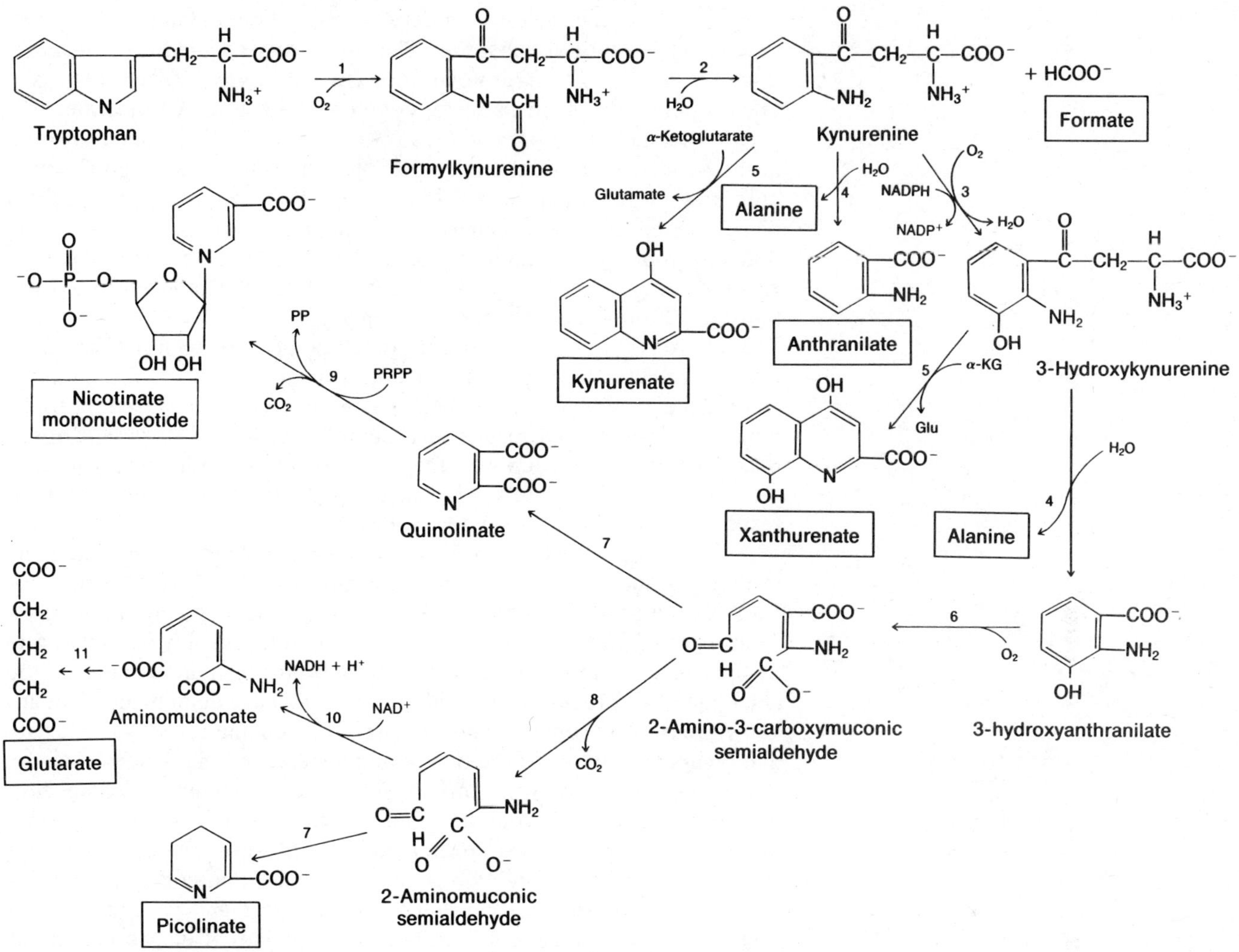

Figure 12.27
Metabolism of tryptophan.
The enzymes indicated by bold numbers are (1) tryptophan oxygenase; (2) kynurenine formamidase; (3) kynurenine hydroxylase; (4) kynureninase; (5) transaminase; (6) 3-hydroxyanthranilate oxidase; (7) spontaneous nonenzymatic reaction; (8) picolinate carboxylase; (9) quinolinate phosphoribosyl transferase; (10) aldehyde dehydrogenase; (11) complex series of reactions including reduction and deamination to α-ketoadipate and further metabolism as described for lysine (p. 474). The compounds named in boxes are end products or familiar metabolites of energy metabolism.

byproducts that accumulate under normal circumstances. Tryptophan metabolism also leads to a physiologically active amine.

The initial reaction in the main pathway (Figure 12.27) is an oxygenation that opens the five-membered ring of the indole nucleus of tryptophan to form formylkynurenine. The enzyme, tryptophan oxygenase (tryptophan pyrrolase, tryptophan dioxygenase), contains an iron porphyrin and is one of the most studied inducible mammalian enzymes. Administration of large amounts of tryptophan causes an increase of more than a factor of 10 in the level of tryptophan oxygenase by protecting the enzyme against degradation. High levels of the adrenal glucocorticoid hormones (cortisone, dihydrocortisone, and synthetic analogues) cause similar increases by stimulating enzyme synthesis.

The oxidation and cleavage of the ring occur without the formation of a detectable intermediate. Formylkynurenine is hydrolyzed by a constitutive enzyme that acts on a variety of aromatic formamides but most rapidly on its natural substrate. The formate is handled as a 1-carbon fragment by the tetrahydrofolate system (page 476). Kynurenine is a branch point; the main pathway continues with a mixed-function oxygenase that includes FAD and uses NADH or NADPH as the cosubstrate in the synthesis of 3-hydroxykynurenine. One side branch splits off the bulk of the side chain as alanine through the action of the pyridoxal phosphate enzyme kynureninase, and another side branch removes the α-amino group by transamination (also using pyridoxal phosphate), but the expected keto group forms a Schiff base with the aromatic amine to form the stable aromatic compound kynurenate.

3-Hydroxykynurenine is also a branch point. The main pathway now uses the kynureninase that caused a branch earlier to remove alanine but to produce 3-hydroxyanthranilate. The branch is again caused by transamination, which also results in a quinoline ring by cyclization of the presumed carbonyl group with the amine, to form xanthurenic acid. The quinoline compounds are significant components of normal urine and are responsible for part of its yellow color.

A third oxygenase cleaves 3-hydroxyanthranilate to an unstable intermediate, 2-amino-3-carboxymuconic semialdehyde. This enzyme uses Fe^{2+}. In the absence of a competing enzyme, the unstable intermediate cyclizes to a Schiff base in a first-order reaction, probably limited in rate by the isomerization of the double bond to bring the amino group near the aldehyde. The product is a pyridine dicarboxylate, quinolinate. For unknown reasons in some species, an enzyme, picolinic carboxylase, exists that competes with the formation of quinolinate, and in other species, for equally mysterious reasons, this enzyme appears in animals that develop diabetes. It decarboxylates the intermediate to one that cyclizes more rapidly than its precursor to form picolinate. Most of the decarboxylated material is caught by a dehydrogenase, however, that converts the aldehyde to an acid and leads through α-ketoadipate and glutaryl CoA to acetoacetyl CoA (see lysine metabolism, page 474). Thus, to the extent that tryptophan contributes to energy metabolism it is ketogenic in forming acetoacetyl CoA as well as glucogenic because of the alanine produced in the early steps.

Nicotinic acid is not formed in mammals as a free compound but as a product of a concerted reaction in which quinolinate is decarboxylated in the act of forming a nucleotide, nicotinate mononucleotide, as described in Chapter 13. At this point it should be noted that this biosynthetic pathway is unusual in containing a nonenzymatic step, the formation of quinolinate, which may be regulated by diversion of an intermediate to an oxidative degradation by a completely enzymatic route.

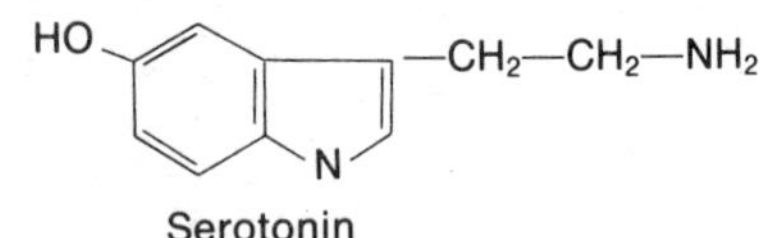

Figure 12.28
Structure of serotonin.

Serotonin (Figure 12.28) is an important neurotransmitter, but it probably has additional physiological functions since it is found in many organs outside the central nervous system, especially in mast cells and platelets, and it causes contraction of smooth muscle in arterioles and bronchioles. This amine may act as a transmitter in the gastrointestinal tract to evoke release of peptide hormones. The formation of serotonin is very similar to the synthesis of the catecholamines. Tryptophan is hydroxylated at carbon-5 by a mixed-function oxygenase that uses tetrahydrobiopterin as a cosubstrate. The resulting 5-hydroxytryptophan is decarboxylated by a pyridoxal phosphate enzyme to give serotonin, 5-hydroxytryptamine.

Histidine

The catabolism of histidine is of greater importance in 1-carbon metabolism than in energy production. The principal pathway of histidine degradation leads to glutamate by a series of reactions of different types but

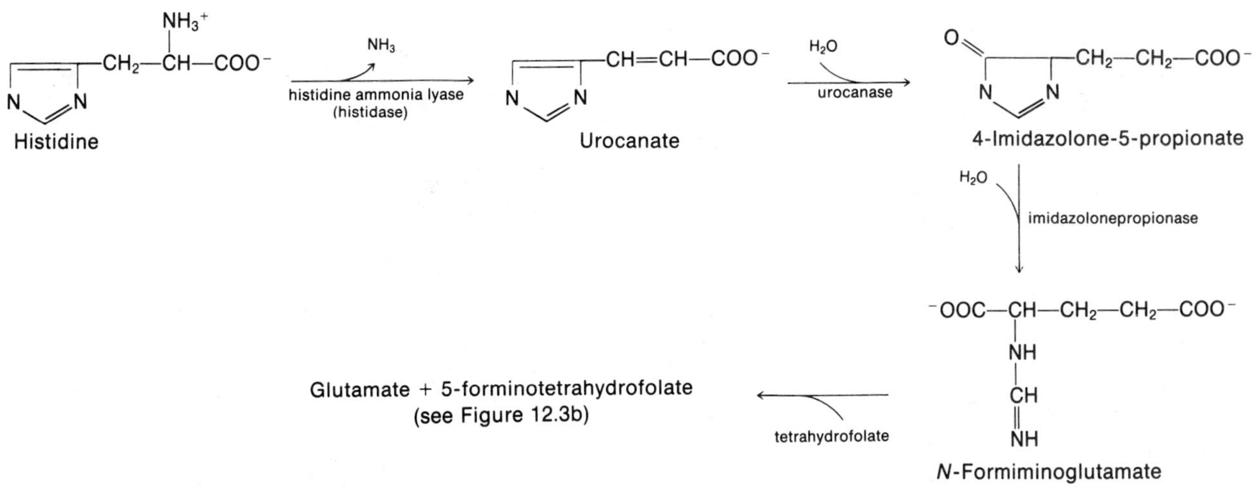

Figure 12.29
Degradation of histidine.

without oxidation. An obligate step requires tetrahydrofolate. Minor pathways in terms of the percentage of histidine consumed are also of physiological importance.

The first step in histidine degradation is the elimination of ammonia with the formation of a double bond. This kind of reaction is known to occur with other amino acids in bacteria and plants, but the histidase reaction is unique in higher animals although elimination reactions do occur in the transfer of amino groups after condensation of aspartate in the synthesis of arginine and purines. A disease associated with deficiency of histidase is described in Clin. Corr. 12.8. The unsaturated urocanate is hydrated to 4-imidazolone-5-propionate. Histidase and urocanase have been well characterized only as bacterial enzymes that catalyze the same reactions that occur in liver. The bacterial enzymes have no organic cofactors but contain altered amino acids, α-ketobutyrate at the former *N*-terminus of urocanase and a tentatively identified dehydroalanine in histidase; each of these participates in the catalytic mechanism. Hydrolysis of imidazolonepropionate opens the five-membered ring to form *N*-formiminoglutamate. The only reaction in which this compound participates is a transfer of the formimino group, a 1-carbon fragment bearing a nitrogen atom, to N-5 of tetrahydrofolate, leaving glutamate. The medical significance of this reaction is illustrated in Clin. Corr. 12.11. These reactions are illustrated in Figure 12.29.

The decarboxylation of histidine to give the corresponding amine, histamine (Figure 12.30), occurs in various organs, especially in mast cells. Histamine is secreted in the physiological control of such diverse functions as production of acid by the gastric mucosa and the dilation and constriction of specific blood vessels. Excess reaction to histamine causes the symptoms of asthma and various allergic reactions.

Two unusual peptides containing histidine, carnosine and anserine, are found in muscle. They are both derivatives of β-alanine. This compound is produced as a degradation product of pyrimidines (Chapter 13) but can be synthesized from malonic semialdehyde, which is derived from propionate. Transfer to the α-amino group of histidine of a β-alanyl residue from an anhydride with AMP, with no carrier such as CoA or tRNA, forms carnosine; a similar reaction with 1-methylhistidine forms anserine. The methylation of carnosine by *S*-adenosylmethionine has been demonstrated in vitro but is of doubtful significance in vivo. The functions of these peptides in muscle is conjectural; a role has been proposed for carnosine in olfactory function. The origin of methyl histidine is un-

CLIN. CORR. 12.8 HISTIDINEMIA

Histidinemia, an elevated level of histidine in the blood, was first detected as a false positive in the screening of the urine of infants with ferric chloride. The green color was given by an analogue of phenylpyruvate, imidazolepyruvate, formed by transamination of histidine. The accumulation of histidine is due to a deficiency of histidase. A convenient assay for this enzyme uses skin, which produces urocanate as a constituent of sweat; urocanase and the other enzymes of histidine catabolism in liver are not found in skin.

Most reported cases of histidinemia have shown impaired mental development and a variety of other developmental problems, but there is no clear pattern of response to high levels of histidine or lack of normal metabolites derived from this amino acid. No treatment has been shown to have beneficial effects.

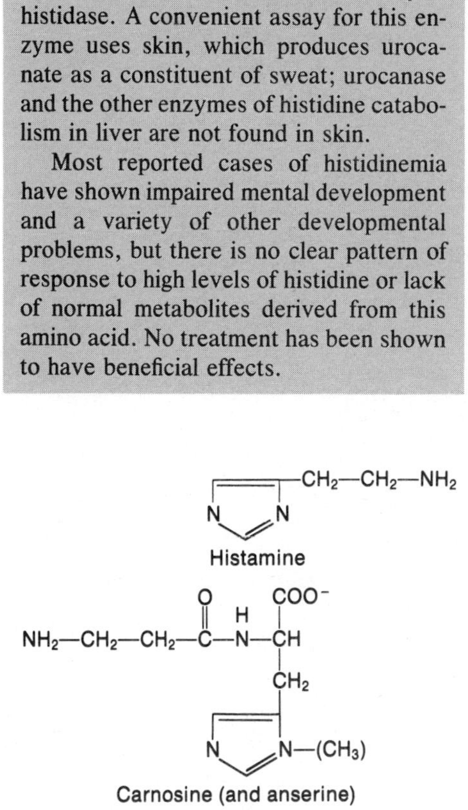

Figure 12.30
Some derivatives of histidine.

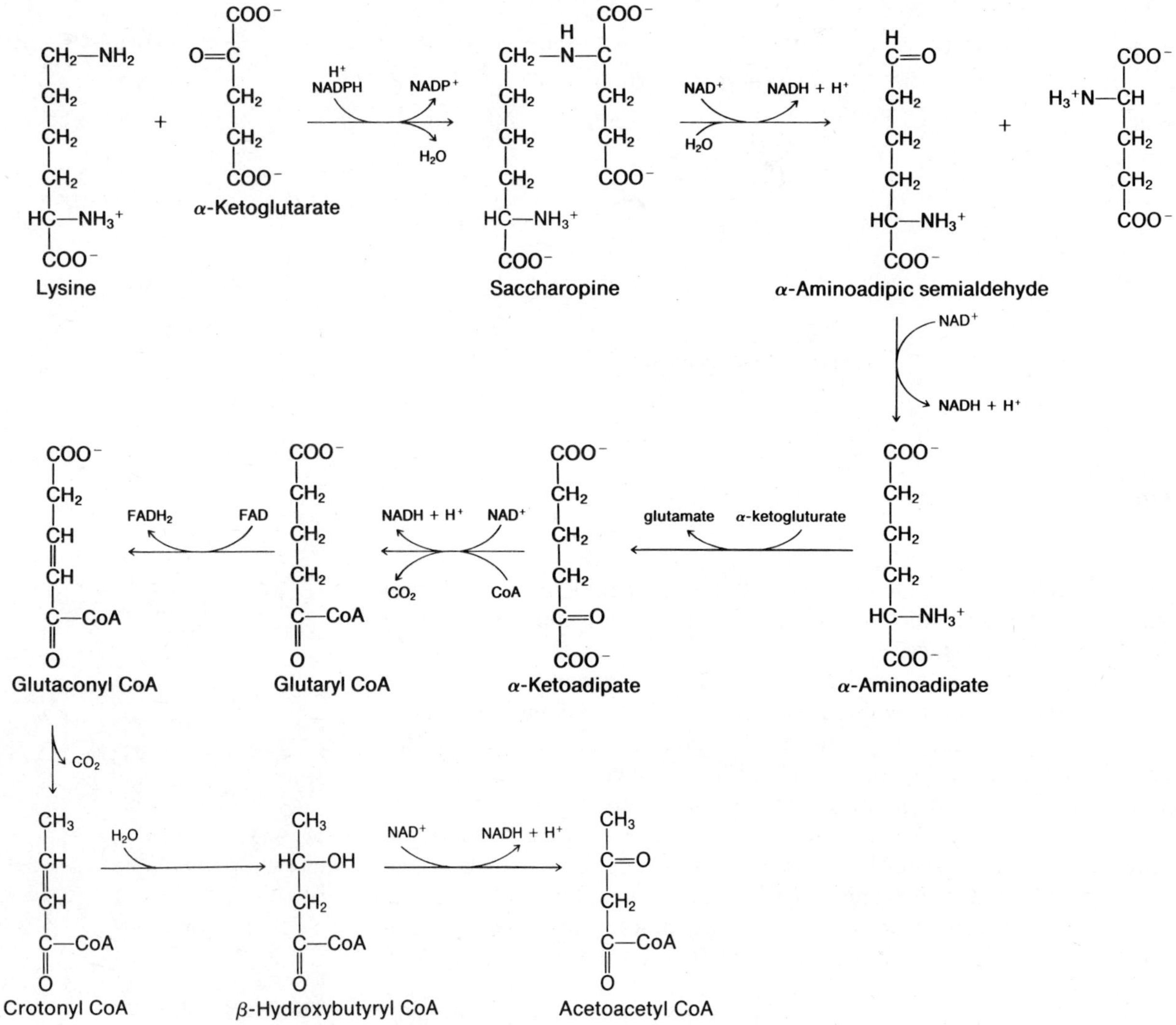
Lysine
α-Ketoglutarate
Saccharopine
α-Aminoadipic semialdehyde
α-Aminoadipate
α-Ketoadipate
Glutaryl CoA
Glutaconyl CoA
Crotonyl CoA
β-Hydroxybutyryl CoA
Acetoacetyl CoA
NADPH
NADP+
NAD+
NADH + H+
H2O
glutamate
α-ketogluturate
FADH2
FAD
CO2
CoA

Figure 12.31
Principal pathway of lysine degradation.

known, but its occurrence in certain proteins suggests that it is formed in a posttranslational modification of histidine in a polypeptide, and the free amino acid is liberated when the protein is degraded.

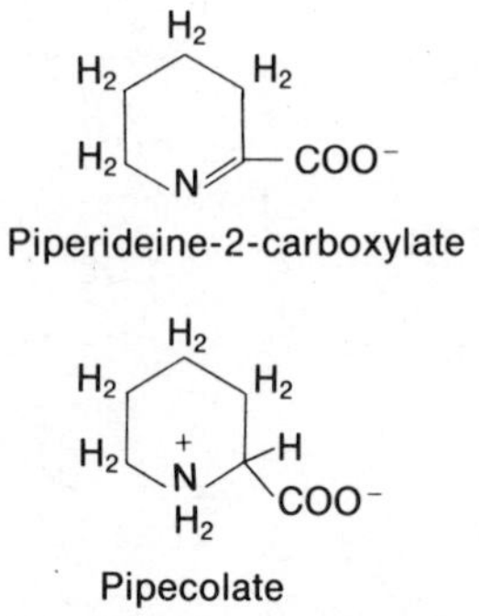
Piperideine-2-carboxylate
Pipecolate

Figure 12.32
Minor products of lysine metabolism.

Lysine

Lysine is one of two essential amino acids whose α-amino group does not equilibrate with the body pool of amino groups; the other is threonine. The amino group of lysine is transferred to other amino acids, but the reverse does not occur. The α-keto acid corresponding to lysine, α-keto-ε-aminocaproic acid, cannot replace lysine in the diet; this compound cyclizes by Schiff base formation and exists mainly as Δ^1-piperideine-2-carboxylic acid. Derivatives of lysine in which the ε-amino group is blocked, such as ε-*N*-acetyllysine and ε-*N*-methyllysine, are converted to lysine in vivo by hydrolytic or oxidative enzymes found in the liver and kidney.

Most degradation of lysine occurs by a unique pathway shown in Figure 12.31 in which a secondary amine is formed between the ε-amino group and the carbonyl group of α-ketoglutarate. The product, saccharopine, is formed by an enzyme that reduces the hypothetical Schiff base with NADPH. Normally saccharopine does not accumulate but is oxidized by another dehydrogenase that splits the linkage on the other side of the bridge nitrogen. The sum of the reduction and oxidation reactions is effectively a transamination yielding glutamate and α-aminoadipic semialdehyde. The latter, of course, can also form a Schiff base but with the double bond on the side of the nitrogen atom away from the carboxylate. This compound can be oxidized by another dehydrogenase to become α-aminoadipate. A conventional transamination converts this higher homologue of glutamate to the corresponding α-ketoadipate. In a reaction analogous to the oxidation of α-ketoglutarate to succinyl CoA, glutaryl CoA is formed. Another oxidation introduces a double bond, forming glutaconyl CoA, which is decarboxylated to crotonyl CoA. This unsaturated fatty acyl CoA is an intermediate in the normal oxidation of fatty acids, and subsequent reactions of the material derived from lysine are those of fatty acid oxidation leading to acetoacetyl CoA.

Although lysine does not participate in transamination, the α-amino group is removed by a dehydrogenase and Δ^1-piperideine-2-carboxylate is formed. This reaction is of minor importance in that it cannot compensate for deficiency in the saccharopine pathway. The ring is saturated by reduction with a nicotinamide nucleotide to form pipecolate, shown in Figure 12.32. Following the reduction, an oxidation produces an unsaturation on the other side of the nitrogen to reform the lysine passing through the saccharopine pathway at α-aminoadipate semialdehyde. Abnormalities associated with lysine metabolism are discussed in Clin. Corr. 12.9.

Carnitine Formation from Lysine

For reasons that are not understood, some of the ε-amino groups of lysine residues in many proteins are methylated to mono-, di-, or trimethyllysine by an *N*-methyl transferase that uses *S*-adenosylmethionine as the methyl donor. Speculation that the methylation plays a regulatory role in the function of such proteins as histones is supported by the observation that the methyl groups are turned over rather rapidly by oxidation to formaldehyde. Some lysyl residues in proteins are also acetylated. Probably the same enzyme that demethylates lysyl residues in proteins removes the methyl groups from free mono- and dimethyllysine, which are produced when the methylated proteins are degraded in normal protein turnover. The trimethyllysine that is liberated by proteolysis has an unusual metabolic role as the precursor of carnitine. The biosynthesis of carnitine in mammals is absolutely dependent upon the formation of trimethyllysine residues in proteins. Free lysine is not methylated, and the intermediates in carnitine formation are derived exclusively from trimethyllysine.

Trimethyllysine is oxidized by a specific oxygenase that requires as cosubstrate α-ketoglutarate. The resulting β-hydroxytrimethyllysine is a substrate for serine hydroxymethyl transferase, very much like threonine is, and is converted to γ-butyrobetaine aldehyde and glycine. The aldehyde is oxidized by a specific NAD-requiring dehydrogenase to form γ-butyrobetaine. Subsequent hydroxylation at the β position to form carnitine is catalyzed by an enzyme with the same requirements as for the earlier hydroxylation of trimethyllysine and that may be the same enzyme. These reactions are illustrated in Figure 12.33. The role of carnitine in fatty acid transport across the inner mitochondrial membrane is described in Chapter 9.

CLIN. CORR. 12.9 DISEASES INVOLVING LYSINE AND ORNITHINE

LYSINE

Several cases have been described in which lysine levels of the bloods are elevated. The nature of these cases is very poorly understood. Several of the patients were severely retarded, but some had normal mental and physical development. One authentic case is the basis for a classification of periodic hyperlysinemia attributed to a partial deficiency of the dehydrogenase that produces saccharopine. Administration of lysine or a diet with unrestricted protein triggers in an unknown way ammonia intoxication with vomiting and abnormal EEG and coma. Persistent hyperlysinemia has been observed in a small heterogeneous series. In some cases α-*N*-acetyllysine is excreted; in some pipecolic acid has been found, and in one case saccharopine has found in blood and cerebrospinal fluid.

ORNITHINE

In contrast to the inconsistent findings in patients with hyperlysinemia, elevated ornithine levels are generally due to a deficiency of the aminotransferase that enables ornithine to be oxidized through glutamate and the TCA cycle. A well-defined clinical entity, gyrate atrophy of the choroid and retina, characterized by progressive loss of vision leading to blindness by the fourth decade, is caused by a deficiency of ornithine transaminase (ornithine-δ-aminotransferase). The mechanism of the specific pathological changes in the eye is not known, but apparently ornithine itself or a metabolic product is directly involved since progression of the disease is halted by restriction of arginine intake and/or administration of pyridoxine, which reduce the level of ornithine in body fluids.

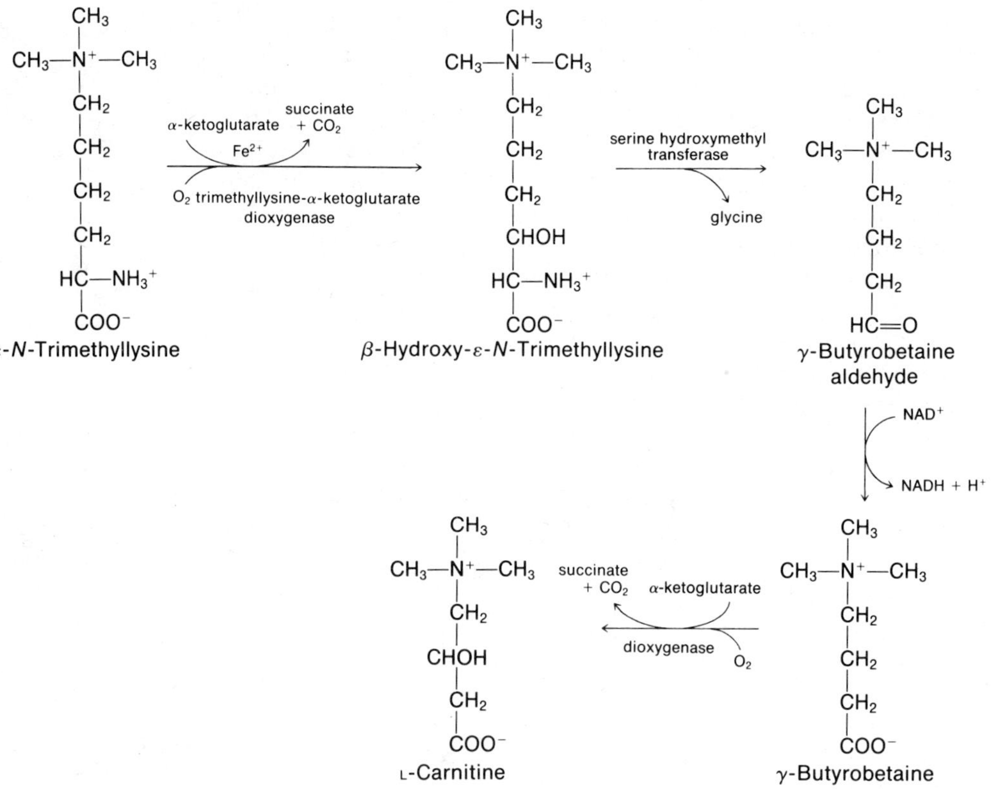

ε-N-Trimethyllysine → (α-ketoglutarate, O_2, Fe^{2+} → succinate + CO_2; trimethyllysine-α-ketoglutarate dioxygenase) → β-Hydroxy-ε-N-Trimethyllysine → (serine hydroxymethyl transferase → glycine) → γ-Butyrobetaine aldehyde → (NAD^+ → $NADH + H^+$) → γ-Butyrobetaine → (α-ketoglutarate, O_2 → succinate + CO_2; dioxygenase) → L-Carnitine

Figure 12.33
Biosynthesis of carnitine.

12.4 FOLIC ACID AND ONE-CARBON METABOLISM

One-carbon metabolism is a term applied to reactions in which a chemical group built around a single carbon atom is transferred from one compound to another. The groups that correspond to derivatives of methanol, formaldehyde, and formate are transferred mainly by enzymes that use the cofactor tetrahydrofolate. The most reduced 1-carbon compound, methane, is inert in animal metabolism, and the most oxidized molecule, carbon dioxide, is produced by a variety of decarboxylases and is used in a similarly diverse group of reactions not related to the rest of 1-carbon metabolism.

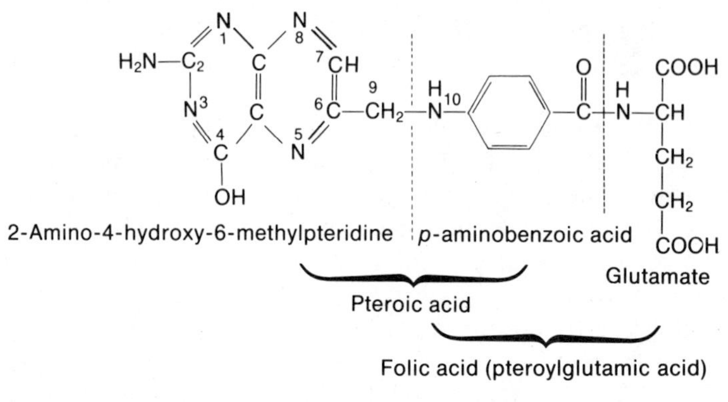

2-Amino-4-hydroxy-6-methylpteridine | p-aminobenzoic acid | Glutamate

Pteroic acid

Folic acid (pteroylglutamic acid)

Figure 12.34
Components of folic acid.

CLIN. CORR. 12.10
FOLIC ACID DEFICIENCY

The 100–200 μg of folic acid required daily by an average adult are easily obtained from conventional Western diets. Deficiency of folic acid is not uncommon, however. It may result from limited diets, especially when food is cooked at high temperatures for long periods, which destroys the vitamin. Intestinal diseases, notably coeliac disease, often are characterized by folic acid deficiency caused by malabsorption.

In contrast to pernicious anemia, a fairly common disease caused by decreased ability to absorb vitamin B_{12} (see Chapter 22), specific inability to absorb folic acid has been found rarely. Another contrast is the appearance of pernicious anemia late in life because of failure of the gastric mucosa to produce intrinsic factor, whereas folate deficiency is usually seen in newborns. Both conditions, however, are characterized by megaloblastic anemia, a disease in which developing blood cells are much larger than normal. Of the small series of patients studied, some

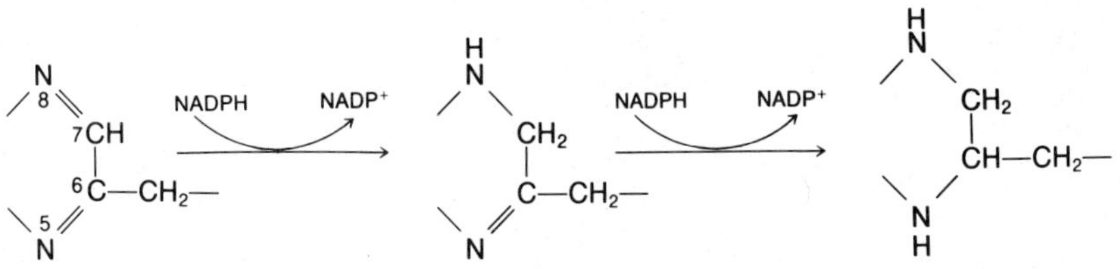

Folate   Dihydrofolate   Tetrahydrofolate

NADPH NADP⁺ NADPH NADP⁺

Figure 12.35
Reduction of folate by dihydrofolate reductase.

The coenzyme tetrahydrofolate is derived from the vitamin folic acid (Figure 12.34) that is absorbed from the intestine (Clin. Corr. 12.10). The vitamin is reduced in two steps catalyzed by the same enzyme, named for the second step dihydrofolate reductase (Figure 12.35). Deficiency of this enzyme and other enzymes of 1-carbon metabolism are discussed in Clin. Corr. 12.11. Actually there are several species of tetrahydrofolate with up to seven glutamate residues in γ-amide linkage. Until now no functional differences have been able to be attributed to these species, and attention will be focused on the pteroic acid portion of the molecule where interactions with substrates occur.

One-carbon elements at the oxidation levels of formaldehyde and formate are combined with tetrahydrofolate to produce the structures illustrated in Figure 12.36. Free formaldehyde adds spontaneously to tetrahydrofolate and forms a bridge between nitrogen atoms 5 and 10, 5,10-methylenetetrahydrofolate. The major donor of the 1-carbon bridge of 5,10-methylenetetrahydrofolate is serine in the formation of glycine. Oxidation of glycine also forms this bridge compound.

5,10-Methylenetetrahydrofolate can be both oxidized and reduced by dehydrogenases using pyridine nucleotide coenzymes. Reduction breaks

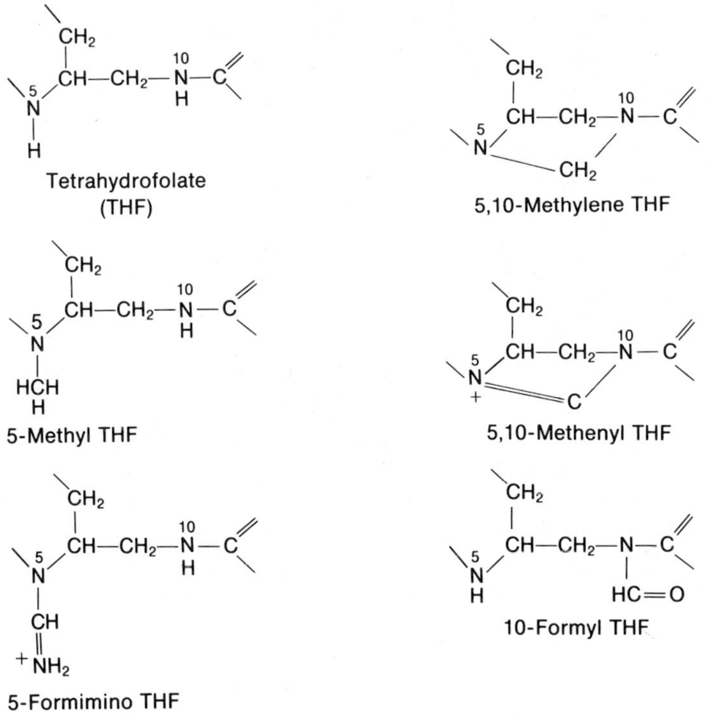

Tetrahydrofolate (THF)   5,10-Methylene THF

5-Methyl THF   5,10-Methenyl THF

5-Formimino THF   10-Formyl THF

Figure 12.36
Active center of tetrahydrofolate and its one-carbon derivatives.

were responsive to large doses of oral folate, whereas one required parenteral administration; these observations indicate that absorption of the vitamin must be an active process but the nature of the defect(s) is not yet known. Besides the anemia, mental and other central nervous symptoms are seen in patients with folate deficiency, and all respond to continuous therapy although permanent damage appears to be caused by delayed or inadequate treatment.

A classical experiment was carried out by a physician, apparently serving as his own experimental subject, to study the human requirements for folic acid. In this study the diet consisted only of foods boiled repeatedly to extract the water-soluble vitamins to which vitamins (and minerals) were added, omitting folic acid. On this regimen, symptoms attributable to folate deficiency did not appear for 7 weeks (altered appearance of blood cells and formiminoglutamate excretion were seen only at 13 weeks) and serious symptoms (irritability, forgetfulness, macrocytic anemia) appeared only after 4 months on a totally deficient diet. The neurological symptoms were alleviated within 2 days after folic acid was added to the diet; as expected from the time required for erythropoiesis, the blood picture became normal more slowly.

The occurrence of folic acid in essentially all natural foods makes deficiency difficult, and apparently a normal person accumulates more than adequate reserves of this vitamin. For pregnant women the situation is very different. The needs of the fetus for normal growth and development include constant, uninterrupted supplies of coenzymes (in addition to amino acids and other cell constituents). In the complex circumstances of human gestation it is not clearly established that nutritional deficiency of folic acid is responsible for irreversible damage, but it is clear that antifolate drugs, such as methotrexate, can cause abortion and congenital abnormalities.

CLIN. CORR. 12.11
DISEASES OF FOLATE METABOLISM

DIHYDROFOLATE REDUCTASE DEFICIENCY

Although folic acid derivatives in all tissues that have been examined are mainly reduced to the level of THF, these compounds are easily oxidized so that some significant fraction of the absorbed vitamin must be reduced in order to function as a coenzyme. In a few cases symptoms of folate deficiency were found to be due to a deficiency of dihydrofolate reductase (which catalyzes both steps of the conversion of folate to THF). An effective treatment for such cases is parenteral administration of 5-formyl THF, the most stable of the reduced folates.

METHYLENE TETRAHYDROFOLATE REDUCTASE DEFICIENCY

In the few cases of central nervous system abnormality attributed to a deficiency of methyleneTHF reductase, an accompanying biochemical abnormality is homocystinuria. A younger sibling of a patient was found to have homocystinuria, the same reductase level as the patient (about 18% of normal), but no symptoms. It is not known whether symptoms develop late or whether some people function adequately without this enzyme activity. The lowered enzyme activity decreases the amount of 5-methylTHF that is formed so that the source of methyl groups for the salvage of homocysteine is limiting. Administering large amounts of folic acid reverses the biochemical abnormalities and, in at least one case, the neurological disorder.

FORMIMINOTRANSFERASE DEFICIENCIES

A series of patients with widely divergent presentations has shown deficiencies in the transfer of the formimino group from formiminoglutamate (FIGLU) to THF. The patients excrete varying amounts of FIGLU; some respond to administration of large doses of folate, but others do not. Histidine loading generally increases the excretion of FIGLU, and a low histidine diet decreases the excretion. No clear mechanism is known, whereby a deficiency of formiminotransferase should cause pathological changes. Speculation on possible mechanisms includes distortion of the normal distribution of 1-carbon forms of folate because of different amounts of transferase and cyclodeaminase activity and hypothetical effects of one or more excessively accumulated forms on other reactions involving folate.

TETRAHYDROFOLATE METHYLTRANSFERASE DEFICIENCY

A single patient has been described who suffered from severe central nervous system maldevelopment and signs of folate deficiency, all attributable to a deficiency of THF methyltransferase. Other patients with similar deficiencies have been found to lack vitamin B_{12}, but treatment with cobalamin did not help this patient although the anemia responded partially to folic acid therapy. The favored explanation for these findings is the "methyl trap hypothesis," which proposes that conversion of folate to methylTHF is an irreversible blind alley when the transferase is less active than normal. Since the patient had 30% of the normal methyltransferase activity, this was adequate to prevent accumulation of homocysteine, but it is possible that the activity in vivo depends upon higher than normal levels of methylTHF. The hypothesis proposes that additional vitamin is needed to maintain other folate-dependent reactions by replacing the coenzyme lost by the drain into methyl THF.

the bond to N-10 and forms 5-methyltetrahydrofolate. Oxidation, in contrast, produces 5,10-methenyltetrahydrofolate, which is reversibly hydrolyzed to 10-formyltetrahydrofolate. Two other mechanisms are of importance in forming bound 1-carbon units at this level of oxidation: (1) free formate is activated by ATP and forms 10-formyltetrahydrofolate and (2) the formimino group attached to glutamate in the catabolism of histidine is transferred to N-5 and in a second step the carbon forms the cyclic 5,10-methenyl compound while eliminating ammonia. Both the transfer reaction (formiminotransferase) and the elimination reaction (cyclodeaminase) are associated with a single enzyme.

Both the methenyl compound and 10-formyltetrahydrofolate are 1-carbon donors in specific reactions. This is seen in the synthesis of purines (Chapter 13) in which the former compound is the specific donor of C-8 and the latter the donor of C-2. The methyl group associated with tetrahydrofolate is used only in the resynthesis of methionine from homocysteine. The methyl group of methionine is the source of most other methyl groups in the body through transfer reactions of *S*-adenosylmethionine.

The methyl group of thymine is formed by transfer of the 1-carbon unit of 5,10-methylenetetrahydrofolate to deoxyuridine 5-phosphate in a unique reaction in which the coenzyme reduces the 1-carbon group at the same time that it joins the pyrimidine ring. The coenzyme is liberated as dihydrofolate (Chapter 13) and must be reduced to the tetrahydro state in order to continue to function in 1-carbon metabolism. Inhibition of dihydrofolate reductase by folic acid analogues (methotrexate, amethopterin) is the basis for the cytotoxic action of these agents that are used widely in the chemotherapy of leukemias and other neoplasms.

12.5 NITROGENOUS DERIVATIVES OF AMINO ACIDS

Creatine and Creatinine

The energy for driving most biochemical reactions is derived from the hydrolysis of ATP. Both ATP and ADP are powerful effectors of many enzymes. Therefore, the concentrations of adenine nucleotides must be carefully regulated and cannot be increased to provide a reservoir of energy. To permit bursts of activity, a large amount of phosphate is stored as phosphocreatine, which has about the same free energy of hydrolysis as ATP; creatine kinase catalyzes the transfer of phosphate from ATP to form creatine phosphate during periods of rest and regenerates ATP when it is needed.

Creatine is synthesized from three amino acids: glycine, arginine, and methionine (Figure 12.37). The transfer of the amidine group from arginine to glycine produces guanidinoacetate and ornithine. Creatine is formed by addition of a methyl group from *S*-adenosylmethionine. The formation of creatine accounts for more utilization of *S*-adenosylmethionine than all other transfers of methyl groups combined.

The reactivity of the phosphoguanidine group is responsible for a nonenzymatic cyclization of creatine in which the carboxyl group displaces the phosphate (Figure 12.38). The cyclic compound is called creatinine.

NH_3^+—CH_2—COO^- (Glycine) + NH_2—$\overset{+}{C}$=NH_2—NH—CH_2—CH_2—CH_2—$HCNH_3^+$—COO^- (Arginine) —transamidinase→ NH_2—$\overset{+}{C}$=NH_2—NH—CH_2—COO^- (Guanidinoacetate) + NH_2—CH_2—CH_2—CH_2—$HCNH_3^+$—COO^- (Ornithine)

Guanidinoacetate —*S*-adenosylmethionine→

$$NH_2\text{—}\overset{\overset{+}{NH_2}}{\overset{\|}{C}}\text{—}\underset{CH_3}{\underset{|}{N}}\text{—}CH_2\text{—}COO^-$$ (Creatine) + Adenosylhomocysteine

Figure 12.37
Synthesis of creatine.

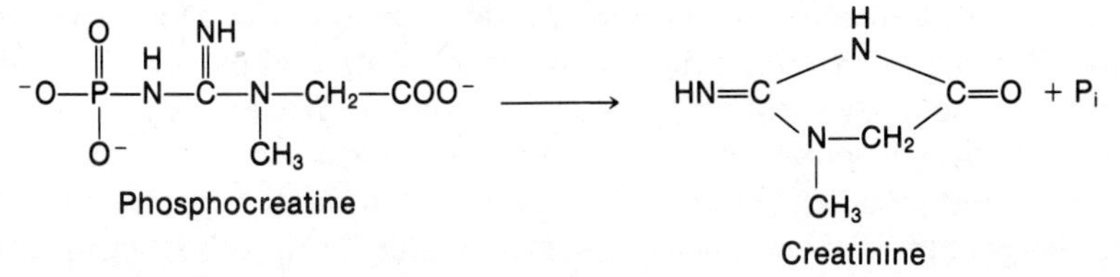

$$^{-}O-\overset{O}{\overset{\|}{P}}(O^{-})-\overset{H}{N}-\overset{NH}{\overset{\|}{C}}-N(CH_3)-CH_2-COO^{-} \longrightarrow \text{Creatinine} + P_i$$

Phosphocreatine → Creatinine

Figure 12.38
Spontaneous reaction forming creatinine.

Since the amount of creatine phosphate is roughly proportional to the muscle mass of the body, the spontaneous formation of creatinine is characteristic of each individual and proceeds at a constant rate from day to day. All of the creatinine produced is excreted in the urine, where it is the most accurate measure of the time during which the urine was produced. Thus, measurement of creatinine in 24-h urine samples is carried out to ensure the accuracy of the collection.

Choline

Choline (Figure 12.39) is a highly methylated compound that plays a central role in nerve conduction and also is a component of the phospholipid lecithin. Isotopic evidence demonstrated long ago that the 2-carbon skeleton of choline is derived from serine, but the details of the conversion are not well established. A decarboxylation of the serine in a phosphotidylserine is generally considered to initiate a minor pathway and exogenous ethanolamine is rapidly assimilated. However, little is known about the formation of free ethanolamine or the relative amounts of methyl transfer to ethanolamine and its derivatives. It is clear that three methyl groups are transferred from *S*-adenosylmethionine to make the quaternary ammonium group of choline.

Choline is acetylated by acetyl CoA in nervous tissue to form the neurotransmitter acetylcholine. Cholinesterase hydrolyzes the ester to acetate and choline. Choline is oxidized in a succession of reactions to the corresponding aldehyde and carboxylic acid, named betaine aldehyde and betaine, respectively. Betaine is an effective methyl donor and participates in a minor variant of the salvage pathway to reform methionine from homocysteine. Only one of the three methyl groups can be transferred; the other methyl groups are removed as formaldehyde by oxidation by flavoproteins, giving *N*-methylglycine (sarcosine) then glycine.

Amines and Polyamines

As described earlier, decarboxylation of certain amino acids produces the corresponding amines, and some of these are combined to form polyamines. The amines that act as extracellular effectors, such as neurotransmitters, must be metabolized quickly to limit responses and allow

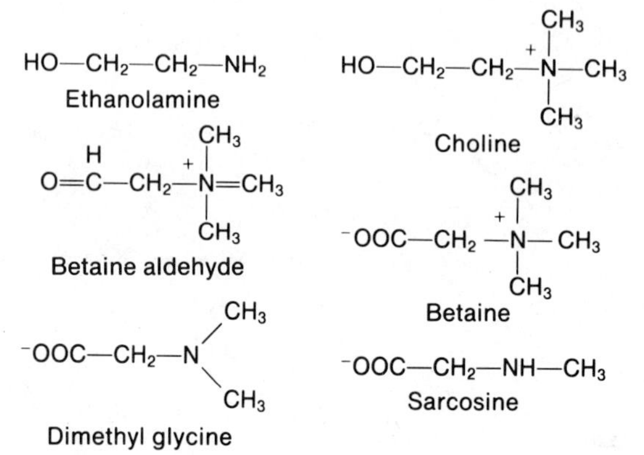

$HO-CH_2-CH_2-NH_2$
Ethanolamine

$HO-CH_2-CH_2-\overset{+}{N}(CH_3)_3$
Choline

$O{=}CH-CH_2-\overset{+}{N}(CH_3)_3$
Betaine aldehyde

$^{-}OOC-CH_2-\overset{+}{N}(CH_3)_3$
Betaine

$^{-}OOC-CH_2-N(CH_3)_2$
Dimethyl glycine

$^{-}OOC-CH_2-NH-CH_3$
Sarcosine

Figure 12.39
Choline and related compounds.

other messages to be effective. In the nervous system a physiological mechanism recovers a large part of the amines released from nerve endings by reuptake back into the cells that produced them. The part that is not taken up is inactivated by two mechanisms, oxidation of the amino group or methylation of the phenolic hydroxyl groups.

In most cells an enzyme located in the outer membrane of mitochondria oxidizes amines to the corresponding aldehydes as follows:

$$RCH_2NH_2 + O_2 + H_2O \longrightarrow RCHO + NH_3 + H_2O_2$$

This is one type of *monoamine oxidase*. It contains covalently bound FAD and uses only O_2 as an electron acceptor. In addition to primary amines, secondary and tertiary amines with methyl groups added to the nitrogen also serve as substrates. In the serum a very different monoamine oxidase oxidizes only primary amines but with little specificity; in some mammalian species the so-called monoamine oxidase also attacks typical diamine oxidase substrates, including histamine and spermine. The serum enzymes contain copper and a cofactor previously thought to be pyridoxal phosphate but recently identified as pyrrolo quinoline quinone (PQQ), depicted in Figure 12.40; these enzymes catalyze the same reaction as the mitochondrial enzyme. *Diamine oxidase* has been purified from kidney and also shown to contain copper and a cofactor that is probably PQQ (the purified enzyme has a pink color, similar to some of the serum monoamine oxidases). It is sometimes referred to as *histaminase* for its only known physiological substrate.

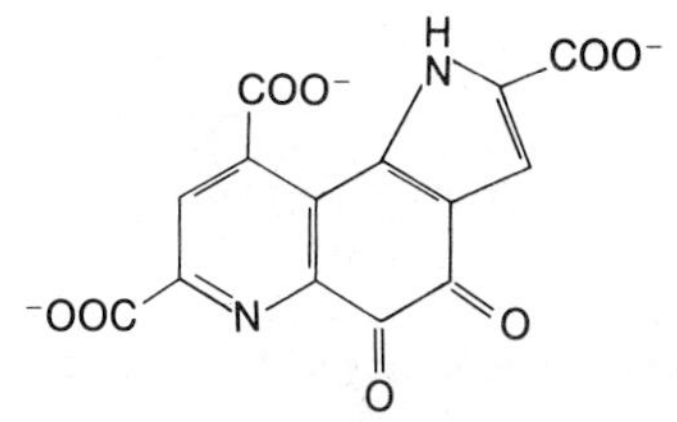

Figure 12.40
Pyrrolo quinoline quinone.
The biosynthesis of this recently discovered cofactor has not been reported.

Methylation at N-1 of the imidazole ring by an enzyme specific for histamine inactivates most of the histamine released; the methylhistamine produced is also a substrate of diamine oxidase. For the most part the aldehydes produced by amine oxidases are oxidized by aldehyde dehydrogenases to the corresponding carboxyl compounds, but a small part is sometimes found as the alcohol, presumably the result of reduction of the aldehyde by a pyridine nucleotide and alcohol dehydrogenase. Thus major products of histamine metabolism are methylhistamine, imidazoleacetate, and methylimidazoleacetate (Figure 12.41). In addition, some hista-

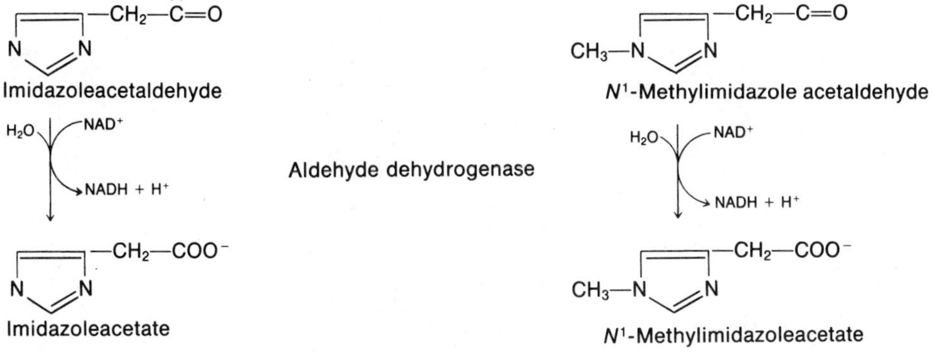

Figure 12.41
Metabolism of histamine.

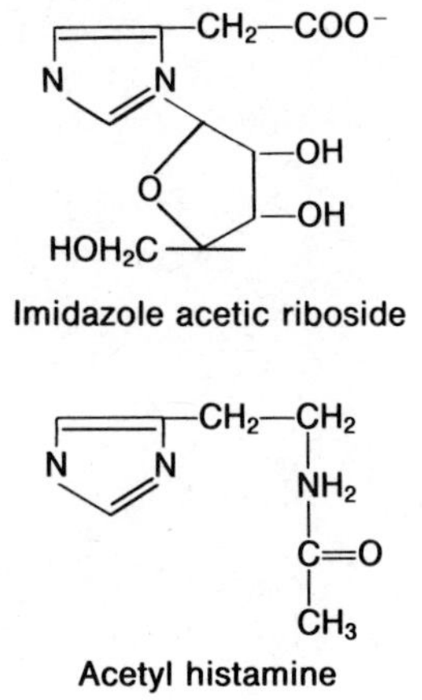

Figure 12.42
Additional metabolites of histamine.

mine is acetylated (Figure 12.42). An unusual derivative was isolated from the urine of animals given large doses of histamine; the imidazoleacetate was conjugated with ribose. The nucleoside is formed in a reaction with 5-phosphoribosyl-1-pyrophosphate catalyzed by a liver enzyme that liberates both ortho- and pyrophosphate.

Methylation of norepinephrine and similar compounds is catalyzed by catechol-*O*-methyl transferase (COMT). This enzyme uses *S*-adenosylmethionine as the donor for a great variety of substrates. The presence of methyl groups does not affect the action of amine oxidases, and COMT does not require an amine in the substrate, so that the major degradative pathways of the catechol amines converge to give vanillylmandelic acid as the major product. In addition to the other products shown in Figure 12.43, sulfate esters and glucuronides of the catechols are also excreted.

Spermine and spermidine are substrates for the serum copper–pyridoxal phosphate oxidase. Both primary amino groups of spermine are converted to aldehydes, but only the propionylamine of spermidine is

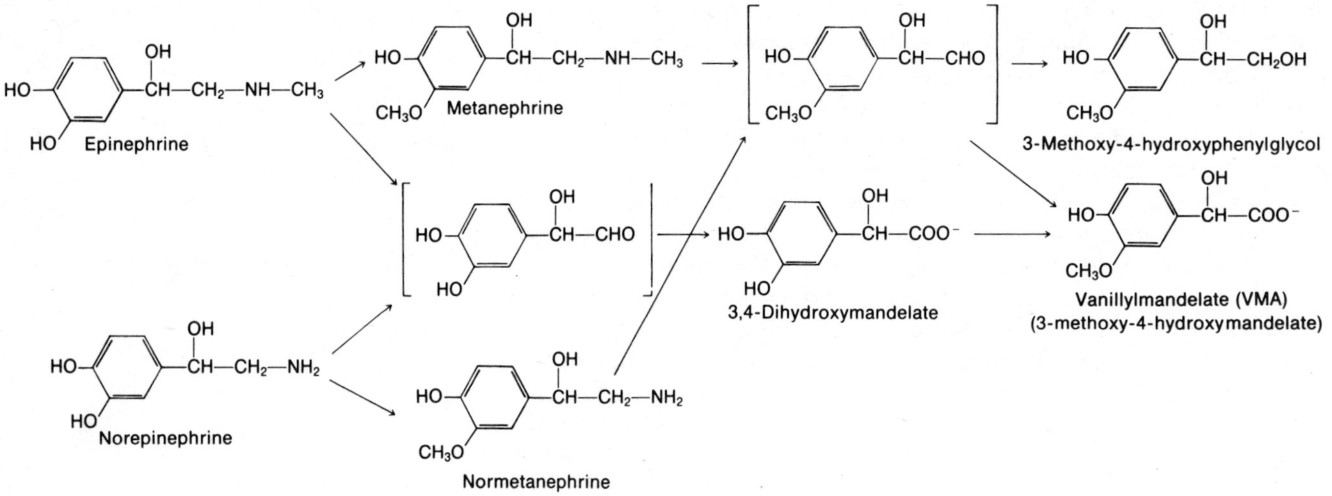

Figure 12.43
Metabolism of the catecholamines.
Each of the compounds named is excreted in the urine. Epinephrine, norepinephrine, and their methyl derivatives are excreted mainly as conjugates with sulfate or glucuronic acid.

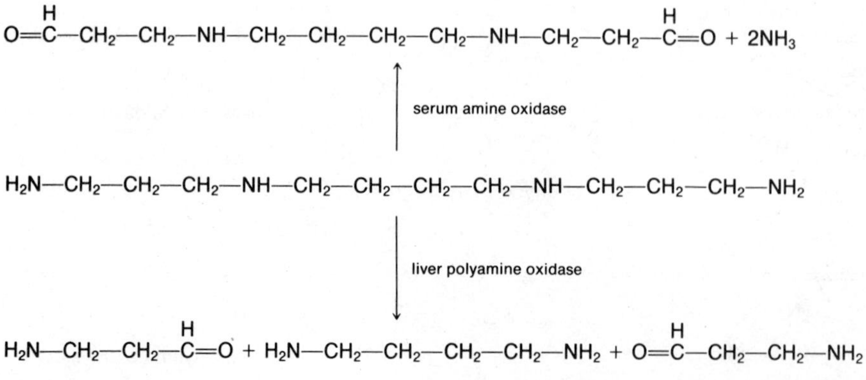

Figure 12.44
Oxidation of spermine.

oxidized. In liver a flavoprotein oxidizes spermine internally to produce 3-aminopropionaldehyde and spermidine, which is oxidized similarly to form putrescine. The oxidative reactions are illustrated in Figure 12.44. Liver also contains two types of acetylase, nuclear enzymes that form N^1 and N^8-acetyl-spermidine and an inducible cytosolic enzyme that forms exclusively N^1-acetylspermidine. Both enzymes also acetylate spermine, but only the nuclear enzymes acetylate the basic histone proteins. N^1-acetyl spermidine and N^1-acetyl spermine are substrates for the liver polyamine oxidase and also for a hydrolytic deacylase. The reason for acetylating and deacetylating polyamines is unknown.

A unique reaction produces a new amino acid by transfer of the butyl-amino residue of spermidine to a single lysine residue in a specific protein, the initiation factor of protein synthesis eIF-4D. After the amino-butyl group is transferred, it is hydroxylated by an iron-dependent enzyme, presumably a mixed-function oxygenase. The modified amino acid, N^1-(4-amino-2-hydroxybutyl)-lysine, is named hypusine (Figure 12.45).

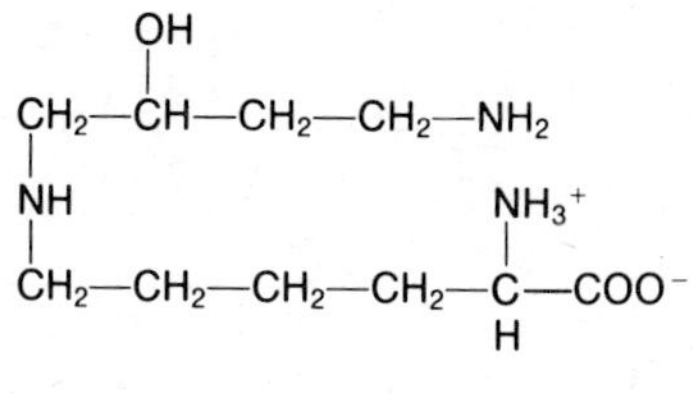

OH
CH_2—CH—CH_2—CH_2—NH_2
NH NH_3^+
CH_2—CH_2—CH_2—CH_2—C—COO^-
H

Figure 12.45
Structure of hypusine.

12.6 GLUTATHIONE

The tripeptide glutathione, γ-glutamylcysteinylglycine (Figure 12.46), is a major constituent of cells and serves several independent functions. Recent insight into the metabolism of glutathione suggests several potential clinical applications, discussed in Clin. Corr. 12.12. One group of functions is based on the reducing powers of the sulfhydryl group, which is the reason the accepted abbreviation of glutathione is GSH. Since the primary sulfhydryl compound synthesized, cysteine, is toxic in high concentrations, possibly because of the effects of degradation products, chelation of metals, or other properties of this reactive molecule, a derivative that is better tolerated is used as the reservoir of reducing power. This is graphically illustrated in red blood cells, in which GSH is present in high concentration and serves to reduce methemoglobin back to hemoglobin. The primary reductants are the pentose phosphate shunt reactions that produce NADPH, which is used by glutathione reductase to reduce the disulfide of oxidized glutathione (GSSG):

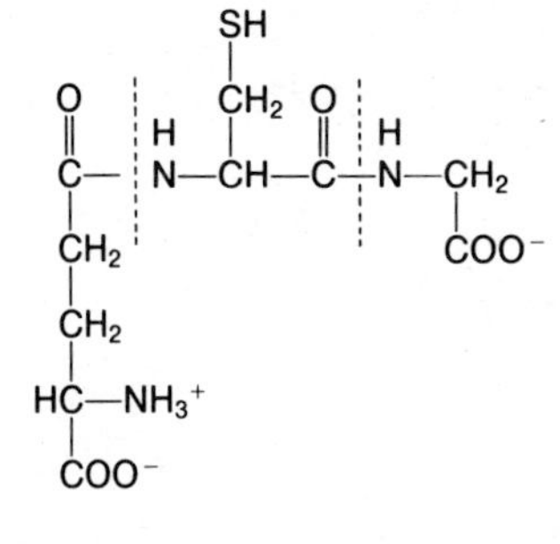

SH
O CH_2 O
H H
C— N—CH—C—N—CH_2
CH_2 COO^-
CH_2
HC—NH_3^+
COO^-

Figure 12.46
Structure of glutathione (γ-glutamyl-cysteinylglycine).

$$2\ GSH + 2\ MetHb \longrightarrow GSSG + 2\ Hb + 2\ H^+$$

$$NADPH + H^+ + GSSG \longrightarrow NADP^+ + 2\ GSH$$

The steady state within cells generally maintains a ratio of about 100/1 of GSH to GSSG.

Toxic amounts of peroxides and free radicals are produced in vivo by irradiation and other mechanisms; these are scavenged by glutathione peroxidase, a selenium-containing enzyme:

$$2\ GSH + H_2O_2 \longrightarrow GSSG + 2H_2O$$

The sulfhydryl group of GSH also participates in a disulfide interchange reaction that rearranges disulfide bonds in proteins until the thermodynamically most stable structure is formed. The sulfhydryl group is also involved in the cofactor function of GSH with several enzymes, including glyoxalase, maleylacetoacetate isomerase, and prostaglandin PGE_2 synthetase.

Glutathione is synthesized from free amino acids in two steps: first the γ-carboxyl group of glutamate is activated by ATP and forms an amide with the amino group of cysteine, then a similar activation of the carboxyl

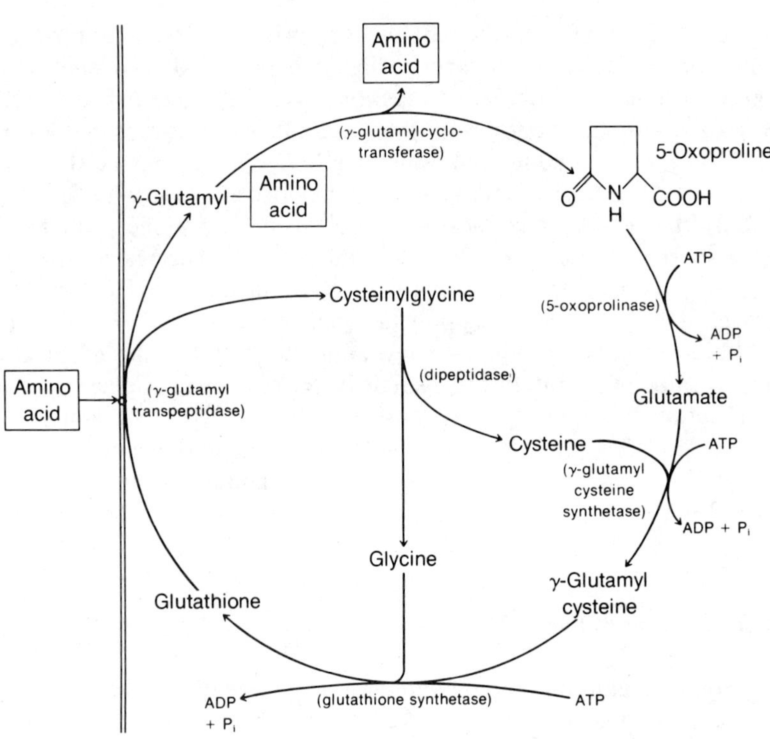

Figure 12.47
The γ-glutamyl cycle.

group of the cysteinyl residue of the dipeptide permits condensation with glycine. Although glutathione is present in high concentration relative to free amino acids in the cytosol, isotopic evidence shows that it is rapidly turned over; that is, it is constantly being broken down and resynthesized. An explanation of this metabolic activity is the γ-glutamyl cycle (Figure 12.47), a mechanism that appears to be used by some cells to transport amino acids across the cell membrane.

The enzyme γ-glutamyl transpeptidase is associated with cell membranes and forms isopeptides of glutamate and various free amino acids, releasing cysteinyl glycine. Both peptides are brought into the cytosol, where the cysteinyl glycine is converted to free amino acids as a result of action of a dipeptidase. The γ-glutamyl peptide is split by a specific enzyme, cyclotransferase, which releases the other amino acid while converting the glutamate residue to a ring compound 5-oxoproline (also called pyrrolidone carboxylic acid). To open this ring at neutral pH requires energy; an enzyme 5-oxoprolinase couples hydrolysis of ATP with hydrolysis of the cyclic amide to form glutamate, ADP, and P_i. The sum of the reactions is the transport of an amino acid at the expense of three molecules of ATP. This appears to be a needless extravagance but may be necessary for rapid, high capacity transport in the kidney and other organs for certain amino acids, especially cysteine and glutamine. The physiological significance of the cycle is not clear, but the high level of activity, even when the system is incomplete (Clin. Corr. 12.12) suggests an important function.

Glutathione is a substrate for a group of enzymes, glutathione *S*-transferases, that add its sulfur to a great variety of acceptor molecules. Acceptors include halogenated and nitro compounds, organophosphates (including insecticides), allylic compounds, epoxides (including metabolites of carcinogens), and more. The conjugation is usually considered to be

$$RX + GSH \longrightarrow RSG + HX$$

CLIN. CORR. 12.12 CLINICAL PROBLEMS AND POSSIBILITIES RELATED TO GLUTATHIONE

DEFICIENCY OF ENZYMES OF γ-GLUTAMYL CYCLE

Decreased ability to form glutathione has been associated with hemolytic disease and, in some cases, neurological and mental abnormalities. Two different deficiencies result in excretion of 5-oxoproline. An obvious cause is that oxoproline cannot be hydrolyzed, which applies only to a benign condition since neither acidosis nor lack of glutamate production in the γ-glutamyl cycle strains the capacity of the body to maintain normal pH or glutamate. However, an indirect mechanism initiated by lack of glutathione synthetase causes much more 5-oxoproline to be accumulated so that acidosis affects the central nervous system. In this disease low levels of glutathione are insufficient to provide normal feedback inhibition of γ-glutamylcysteine synthetase so that large amounts of this dipeptide are formed. γ-Glutamylcysteine is a good substrate for γ-glutamylcyclotransferase and cannot be used appreciably to make GSH, so 5-oxoproline is formed at almost the rate of formation of γ-glutamylcysteine. Besides administration of bicarbonate to control the acidosis, experimental approaches to therapy include giving inhibitors for γ-glutamyltranspeptidase to maintain higher intracellular concentrations of GSH and use of an inhibitor of γ-glutamylcysteine synthetase, buthionine sulfoximine, a structural analogue of γ-glutamate derivatives (see accompanying figure).

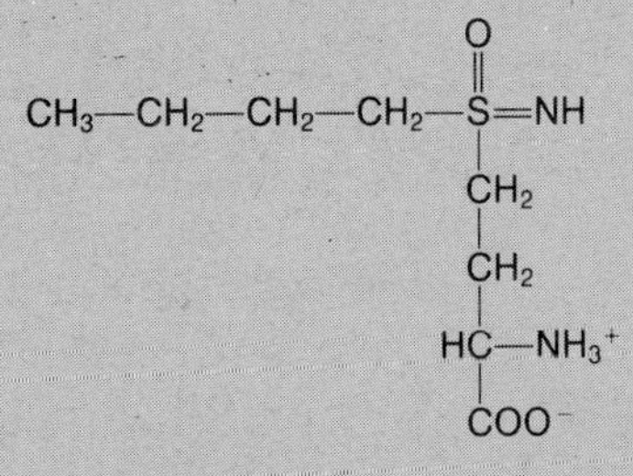

$$CH_3{-}CH_2{-}CH_2{-}CH_2{-}\overset{\overset{\displaystyle O}{\|}}{S}{=}NH$$
$$|$$
$$CH_2$$
$$|$$
$$CH_2$$
$$|$$
$$HC{-}NH_3^+$$
$$|$$
$$COO^-$$

Buthionine sulfoximine.

Deficiency of γ-glutamyl cysteine synthetase results in low levels of glutathione. For many years patients with this disease show no abnormality except hemolytic anemia, but later in life (in the third decade for two patients) serious neurologic defects develop rapidly. In contrast, two patients deficient in γ-glutamyl transpeptidase were mentally deficient and excreted glutathione.

CONTROL OF GLUTATHIONE LEVELS

Animal experiments offer possibilities for both increasing and decreasing cellular GSH, which might be useful in the future for treating different conditions. Two ways to increase GSH are administration of the intermediate γ-glutamylcysteine and giving glutathione esters. Increased intracellular cysteine has been accomplished by administration of 2-oxothiazolidine-4-carboxylate, an analogue of 5-oxoproline that is a good substrate for 5-oxoprolinase. Hydrolysis of the analogue should form 5-carboxycysteine, which would be expected to decompose spontaneously to cysteine and CO_2. Depletion of glutathione is brought about by inhibitors of γ-glutamyl cycle enzymes. Buthionine sulfoximine acts as an analogue of glutamate and binds to γ-glutamylcysteine synthetase but does not inhibit glutamine synthetase as analogues with shorter alkyl groups do (e.g., ethionine sulfoxime).

Why should depletion of GSH be desirable? Again, in animal studies the administration of buthionine sulfoximine has been shown to increase the sensitivity of tumors to radiation. In some situations chemotherapy might also be made more effective by lowering GSH. Certain parasites, including trypanosomes, contain marginal levels of GSH and, therefore, they may be more sensitive to inhibitors of GSH synthesis than human cells. Some drugs are inactivated by conjugation with GSH and may be more effective when the GSH concentration is lowered. The role of GSH in removal of H_2O_2 suggests that catalase-deficient organisms may become less virulent when GSH is decreased. Other reactions of GSH including metabolism of prostaglandins and leukotrienes may also be modulated for therapeutic purposes by adjusting the GSH level.

Besides possible use in γ-glutamylcysteine synthetase deficiency, increasing GSH might be useful in protecting normal cells against radiation, free radicals, and other toxic materials.

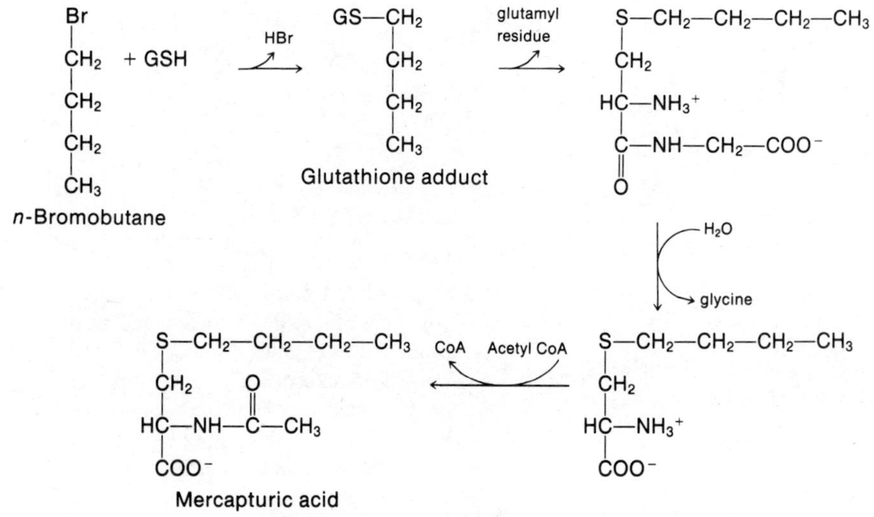

Figure 12.48
Formation of a mercapturic acid.

a detoxication and the conjugate is degraded by the enzymes of the γ-glutamyl cycle, first losing the γ-glutamyl residue by transpeptidation, then the glycyl by hydrolysis. The resulting cysteinyl derivative is acetylated with acetyl CoA to become a mercapturic acid, which is excreted in the urine, along with intermediate compounds of this pathway (Figure 12.48). The fates of the glutathione conjugates vary greatly as a function of the chemical structure of the acceptors.

BIBLIOGRAPHY

General

Boyer, P. *The enzymes,* 3rd ed. Volumes 4–9 and 11–13. New York: Academic, 1970.

Cooper, A. J. L. Biochemistry of sulfur-containing amino acids. *Ann. Rev. Biochem.* 52:187, 1983.

Friedhoff, A. J. *Catecholamines and behavior.* New York: Plenum, 1975.

Meister, A. *Biochemistry of the amino acids,* 2nd ed. New York: Academic, 1965.

Urea Cycle

Grisolia, S., Baguena, R., and Mayor, F. *The urea cycle.* New York: Wiley, 1976.

Holmes, F. L. Hans Krebs and the discovery of the ornithine cycle. *Fed. Proc.* 39:216, 1980.

Polyamines

Tabor, C. W., and Tabor, H. Polyamines. *Ann. Rev. Biochem.* 53:749, 1984.

Williams-Ashman, H. G., and Canellakis, Z. N. Polyamines in mammalian biology and medicine. *Perspect. Biol. Med.* 22:421, 1979.

Transport

Meister, A., and Anderson, M. E. Glutathione. *Ann. Rev. Biochem.* 52:711, 1983.

Disorders of Amino Acid Metabolism

Rosenberg, L. E., and Scriver, C. R. Disorders of amino acid metabolism, in P. K. Bondy and L. E. Rosenberg (eds.), *Metabolic control and disease,* 8th ed. Philadelphia: W. B. Saunders, 1980, p. 583.

Stanbury, J. B., Wyngaarden, J. B., Fredrickson, D. O., Goldstein, J. L., and Brown, M. S. *The metabolic basis of inherited disease,* 4th ed. New York: McGraw-Hill, 1978; 5th ed., 1983.

Wellner, D., and Meister, A. A survey of inborn errors of amino acid metabolism and transport. *Ann. Rev. Biochem.* 50:911, 1981.

QUESTIONS

C. N. ANGSTADT AND J. BAGGOTT

Question Types are described inside front cover.

1. (QT1) Pyruvate and alanine are components of a shuttle that involves:
   A. hepatic and renal gluconeogenesis.
   B. hepatic gluconeogenesis and transport of muscle nitrogen to liver as alanine.

C. transport of alanine to muscle to supply pyruvate.
D. the production of alanine for use in protein synthesis in most peripheral tissues.
E. transport of alanine between cytosol and mitochondria of liver.

2. (QT2) The branched-chain amino acids:
   1. are essential in the diet.
   2. differ in that one is glucogenic, one is ketogenic, and one is classified as both.
   3. are catabolized in a manner that bears a resemblance to β-oxidation of fatty acids.
   4. are oxidized by a dehydrogenase complex to branched-chain acyl CoAs one carbon shorter than the parent compound.

A. Serine C. Both
B. Threonine D. Neither

3. (QT4) Substrate(s) for a hydroxymethyltransferase that produces glycine.
4. (QT4) Can be synthesized from an intermediate of glycolysis.
5. (QT2) Glycine:
   1. is oxidized to glyoxylate by D-amino acid oxidase.
   2. if labeled in its α-carbon can lead to a serine labeled in its α- and/or β-carbons.
   3. is a contributor to the pool of 1-carbon tetrahydrofolate compounds.
   4. apart from its catabolism through serine or incorporation into proteins has no other known functions.

A. Proline C. Both
B. Ornithine D. Neither

6. (QT4) May be formed from or converted to glutamic semialdehyde or its cyclic Schiff base, Δ^1-pyrroline-5-carboxylate.
7. (QT4) Play(s) a major role in the urea cycle.
8. (QT2) Which of the following is/are a product of decarboxylation of an amino acid.
   1. Guanidinoacetate
   2. Putrescine
   3. Spermidine
   4. γ-Aminobutyrate (GABA)
9. (QT2) *S*-Adenosylmethionine:
   1. contains a positively charged sulfur (sulfonium) that facilitates the transfer of substituents to suitable acceptors.
   2. yields homocysteine when used as a methyl donor.
   3. participates in the formation of the polyamine, spermine.
   4. generates H_2S by transsulfuration.
10. (QT1) In humans, sulfur of cysteine may participate in all of the following *except* the:
    A. conversion of cyanide to less toxic thiocyanate.
    B. formation of thiosulfate.
    C. formation of urinary sulfur products.
    D. donation of the sulfur for methionine formation.
    E. formation of PAPS.
11. (QT2) An inability to generate tetrahydrobiopterin might be expected to:
    1. inhibit the normal degradative pathway of phenylalanine.
    2. lead to albinism.
    3. reduce formation of the neurotransmitter, serotonin.
    4. reduce the body's ability to transfer 1-carbon fragments.
12. (QT1) Both tyrosine aminotransferase and tryptophan oxygenase are enzymes that can be induced by adrenal glucocorticoids. This is reasonable because:
    A. tyrosine and tryptophan are precursors of physiologic amines.
    B. glucocorticoids work by inducing enzymes.
    C. tryptophan is the precursor of nicotinic acid needed for NAD synthesis.
    D. tyrosine is the precursor of catecholamines in the adrenal gland.
    E. these two enzymes initiate the major catabolic pathways in the liver of tyrosine and tryptophan.
13. (QT2) Histidine:
    1. unlike most amino acids, is not converted to an α-keto acid when the amino group is removed.
    2. is a contributor to the tetrahydrofolate 1-carbon pool.
    3. decarboxylation produces a physiologically active amine.
    4. forms a peptide with β-alanine.
14. (QT2) Lysine as a nutrient:
    1. may be replaced by its α-keto acid analogue.
    2. produces acetoacetyl CoA in its catabolic pathway.
    3. is methylated by *S*-adenosylmethionine.
    4. is the only one of the common amino acids that is a precursor of carnitine.
15. (QT2) In folic acid-dependent 1-carbon metabolism:
    1. the formation of the methyl group of thymine involves a reduction in the process of transfer of the 1-carbon group.
    2. the major donor of the 1-carbon group of 5,10-methylene tetrahydrofolate is serine.
    3. carbons at different oxidation levels may be interconverted by suitable enzymes.
    4. the only acceptor for the methyl form is homocysteine in a reaction dependent on a derivative of vitamin B_{12}.
16. (QT2) In the catabolism and excretion of physiologically active amines:
    1. mitochondrial monoamine oxidase oxidizes primary and methylated amines to aldehydes.
    2. methylation may occur.
    3. conjugation with sulfate and/or glucuronate may occur.
    4. opening of a ring, if present, is common.
17. (QT1) Glutathione does all of the following *except* to:
    A. participate in the transport of amino acids across cell membranes.
    B. scavenge peroxides and free radicals.
    C. form sulfur conjugates for detoxication of compounds.
    D. exist in the cell primarily in the oxidized state.
    E. act as a cofactor for some enzymes.

ANSWERS

1. B Peripheral tissues, especially muscle, collect amino group nitrogen that is eventually transferred to pyruvate by transamination to produce alanine. Alanine, in the liver, is converted back to pyruvate and then to glucose. The glucose is transported back to muscle to once again produce pyruvate through glycolysis (p. 455).
2. E All four correct. 2–4: Although their catabolism is similar, the end products are different because of the differences in the branching. After transamination, the α-keto acids are oxidized by a dehydrogenase complex in a fashion similar to pyruvate dehydrogenase. The similarity to β-oxidation comes in steps like oxidation to an α–β unsaturated CoA, hydration of the double bond, and oxidation of a hydroxyl to a carbonyl (p. 455).
3. C Both are substrates for serine hydroxymethyltransferase, which produces glycine, although with serine the reaction is reversible and with threonine is irreversible (p. 460).
4. A Serine can be synthesized from either 2- or 3-phosphoglycerate, but threonine is an essential amino acid (p. 458).
5. A 1, 2, 3 correct. 1: Glycine is neither a D- nor an L-amino acid but is a substrate for this enzyme. 2, 3: Serine and glycine are inter-

convertible via 5,10-methylene tetrahydrofolate, and glycine, itself, also produces this compound, resulting in a randomization of isotopic label among these carbons. 4: Glycine is a precursor of purines, creatine, and heme (p. 460).

6. C Ornithine and proline are interconvertible because they both give rise to and are formed from glutamic semialdehyde. They both enter the TCA cycle as glutamate for the same reason (p. 461).

7. B Ornithine is both a substrate and product of the urea cycle leading to or being formed from arginine (p. 462).

8. C 2, 4 correct. 1: Product of transamidination of arginine. 2, 3: Decarboxylation of ornithine produces putrescine, which is a precursor of the *polyamine* spermidine. 4: GABA is the decarboxylation product of glutamate (p. 462).

9. A 1, 2, 3 correct. 1, 2: The reactive, positively charged sulfur reverts to a neutral thioether when the methyl group is transferred to an acceptor. The product, *S*-adenosylhomocysteine, is hydrolyzed to homocysteine. 3: Decarboxylation of *S*-adenosylmethionine generates the 3-carbon amine fragment that is transferred to putrescine. 4: Transsulfuration refers to the combined action of cystathionine synthase and cystathionase transferring methionine's sulfur to serine to yield cysteine (p. 463).

10. D A, B: Transamination to β-mercaptopyruvate with subsequent formation of thiosulfate and/or conversion of cystine to thiocysteine allows transfer of the sulfur to detoxify cyanide. C, E: SO_4^{2-}, the most oxidized form of sulfur found physiologically, is either excreted or activated as PAPS for use in detoxifying phenolic compounds or in biosynthesis. D: Methionine is the source of sulfur for cysteine (via homocysteine), but the reverse is not true in humans (p. 464).

11. B 1, 3 correct. 1, 3: Tetrahydrobiopterin is a necessary cofactor for phenylalanine, tyrosine, and tryptophan hydroxylases. The first catalyzes the major pathway of phenylalanine catabolism, and the third leads to the formation of serotonin. Catecholamine formation, catalyzed by the second enzyme, would also be deficient. 2: Albinism stems from a deficiency of tyrosinase, which, while giving the same product as tyrosine hydroxylase, is not a tetrahydrobiopterin-requiring enzyme. 4: One-carbon fragments are transferred from either *S*-adenosylmethionine or the tetrahydrofolate 1-carbon pool (p. 467).

12. E Although all of the statements are true, only E offers a suitable rationale. Tyrosine and tryptophan both yield a glucogenic fragment (fumarate, alanine) upon catabolism. Glucocorticoids are secreted in response to low blood glucose or stress (p. 467, 470).

13. E All four correct. 1: Elimination of ammonia from histidine leaves a double bond (urocanate) unlike both transamination and oxidative deamination reactions. 2: A portion of the ring is released as 5-formiminotetrahydrofolate. 3, 4: Histamine; carnosine (p. 472).

14. C 2, 4 correct. 1: Lysine does not participate in transamination (L-amino acid oxidase reaction is not reversible), probably in part because the α-keto acid exists as a cyclic Schiff's base. 3, 4: Free lysine is not methylated, but lysyl residues in a protein are methylated in a posttranslational modification. Intermediates of carnitine synthesis are derived from trimethyllysine liberated by proteolysis (p. 474).

15. E All four correct. 1: 5,10-methylene tetrahydrofolate is reduced to a methyl in the process of transfer to dUMP, releasing the carrier as dihydrofolate. 3: This is probably the most important point to remember about the tetrahydrofolate 1-carbon pool. 4: This is an important consideration in vitamin B_{12} deficiency. Once converted to methionine, this methyl is now available for transmethylation reactions in general (p. 476).

16. A 1, 2, 3 correct. 1–3: Oxidation to aldehydes and/or methylation are extremely common. Frequently the products are conjugated to enhance solubility in urine. 4: Modification, rather than cleavage, of aromatic or imidazole rings is the norm (p. 481).

17. D Most of the functions of glutathione listed are dependent upon the sulfhydryl group (—SH), and glutathione reductase helps to maintain the ratio of GSH : GSSG at about 100 : 1 (p. 483).

13

Purine and Pyrimidine Nucleotide Metabolism

JOSEPH G. CORY

13.1 OVERVIEW

Purine and pyrimidine nucleotides are critically important metabolites that participate in many cellular functions. These functions range from serving as the monomeric precursors of the nucleic acids, to serving as energy stores, effectors, group transfer agents, and mediators of hormone action. The nucleotides are formed in the cell de novo from amino acids, ribose, formate, and CO_2. The de novo pathway for the synthesis of the nucleotides requires a relatively high input of energy. To compensate for this, most cells have very efficient "salvage" pathways by which the preformed purine or pyrimidine bases can be reutilized.

Because of the manner in which nucleotides are synthesized and "salvaged," the purines and pyrimidines occur primarily as nucleotides in the cell. The concentrations of free bases or free nucleosides under normal conditions are exceedingly small. The levels of nucleotides in the cell are very finely regulated by a series of allosterically controlled enzymes in the pathway. Nucleotides are the regulators of these reactions.

Ribonucleotides (at the diphosphate level) serve as the precursors of the deoxyribonucleotides. While the concentrations of ribonucleotides in the cell are in the millimolar range, the concentrations of deoxyribonucleotides are in the micromolar range. DNA replication requires that there be sufficient quantities of the deoxyribonucleoside triphosphates. To facilitate this, the activities of several of the enzymes involved with deoxyribonucleotide metabolism increase just prior to the replication of DNA in the cell.

There are several diseases or syndromes that result from defects in the metabolic pathways for the synthesis of nucleotides either de novo or by salvage or for the degradation of the nucleotides. These include gout, the Lesch–Nyhan syndrome, orotic aciduria, and immunodeficiency diseases. Since nucleotides are obligatory for DNA and RNA synthesis in dividing cells, the metabolic pathways involving the synthesis of nucleotides have been the sites at which many antitumor agents have been directed.

It should be kept in mind during the reading of this chapter on the metabolism of purine and pyrimidine nucleotides that the metabolism discussed has been limited exclusively to mammalian cells. In certain instances there are major differences between the bacterial and mammalian cells in nucleotide synthesis, degradation, and regulation.

13.2 METABOLIC FUNCTIONS OF NUCLEOTIDES

All types of cells (mammalian, bacterial, and plant) contain a wide variety of nucleotides and their derivatives. Some of these nucleotides occur in relatively high concentrations (millimolar range) in the cells. The reason for the large number of nucleotides and their derivatives in the cell is that they are involved in many metabolic processes that must be carried out for normal cellular growth and function.

These functions include the following:

1. *Role in Energy Metabolism:* As we have already seen, ATP is the main form of chemical energy available to the cell. Quantitatively, ATP is generated in cells by oxidative phosphorylation and substrate-level phosphorylation. ATP is utilized to drive metabolic reactions, as a phosphorylating agent, and is involved in such processes as muscle

contraction, active transport, and maintenance of cell membrane integrity. As a phosphorylating agent, ATP serves as the phosphate donor for the generation of the other nucleoside 5′-triphosphates (e.g., GTP, UTP, CTP).

2. *Monomeric Units of Nucleic Acids:* The nucleic acids, DNA and RNA, are composed of monomeric units of the nucleotides. In the reactions in which the nucleic acids are synthesized, the nucleoside 5′-triphosphates are the substrates and are linked in the polymer through 3′,5′-phosphodiester bonds with the release of pyrophosphate.
3. *Physiological Mediators:* More recently recognized functions of nucleotides and their derivatives involve those in which the nucleotides or nucleosides serve as mediators of key metabolic processes. The role of cAMP as a "second messenger" in epinephrine- and glucagon-mediated control of glycogenolysis and glycogenesis has already been discussed. The importance of cGMP as a mediator of cellular events has also been recognized. ADP has been shown to be very critical for normal platelet aggregation and hence blood coagulation. Adenosine has been shown to cause dilation of coronary blood vessels and therefore may be important in the regulation of coronary blood flow.
4. *Components of Coenzymes:* Coenzymes such as NAD, FAD, and coenzyme A are important metabolic constituents of cells and are involved in many metabolic pathways.

   NAD is a coenzyme (cosubstrate) that is involved in oxidation–reduction reactions. This coenzyme contains AMP as part of the molecule. While the AMP moiety is not directly involved in the electron transfer, it is critical for the binding of the coenzyme to the particular enzyme.

   FAD likewise is a coenzyme involved in electron transfer reactions. AMP is also part of this molecule and serves the same function as it does in NAD, that is, the binding of the coenzyme to the enzyme.

   Coenzyme A functions as an acyl group transfer agent (e.g., acetyl CoA, palmityl CoA). In this coenzyme, the nucleotide present is adenosine 3′,5′-diphosphate. The 3′,5′-ADP moiety is not involved in binding the acyl groups (which are actually bound as a thiol ester), but appears to be critical for the binding of the coenzyme to the appropriate enzyme.
5. *Activated Intermediates:* The nucleotides also serve as carriers of "activated" intermediates required for a variety of reactions. A compound such as UDP-glucose is a key intermediate in the synthesis of glycogen and glycoproteins. GDP-mannose, GDP-fucose, UDP-galactose, and CMP-sialic acid are all key intermediates in reactions in which sugar moieties are transferred for the synthesis of glycoproteins. CTP is utilized to generate CDP-choline, CDP-ethanolamine, and CDP-diacylglycerols, which are involved in phospholipid metabolism. Other activated intermediates include *S*-adenosylmethionine (SAM) and 3′-phosphoadenosine 5′-phosphosulfate (PAPS). *S*-Adenosylmethionine is a methyl donor in reactions involving the methylation of the sugar and base moieties of RNA and DNA and in the formation of compounds such as phosphatidylcholine from phosphatidylethanolamine, carnitine from lysine, and so on. *S*-Adenosylmethionine also provides the aminopropyl groups for the synthesis of spermine from ornithine. PAPS is used as the sulfate donor to generate sulfated biomolecules such as the proteoglycans and the sulfatides.
6. *Allosteric Effectors:* Many of the regulated steps of the metabolic pathways are controlled by the intracellular concentrations of nucleotides.

Many examples have already been discussed in the previous chapters, and the roles of nucleotides in the regulation of mammalian nucleotide metabolism will be discussed in this chapter.

Occurrence in Cells

The principal form of purine and pyrimidine compounds found in cells is the 5′-nucleotide derivative. In normally functioning cells the nucleotide of highest concentration is ATP. Depending on the cell type, the concentrations of the nucleotides vary greatly. For example, in the red cell the adenine nucleotides far exceed the other nucleotides, which are barely detectable. In the liver cells and other tissues a complete spectrum of the mono-, di-, and triphosphates are found along with UDP-glucose, UDP-glucuronic acid, NAD^+, NADH, and so on. The presence of the free bases, nucleosides or 2′- and 3′-nucleotides in the acid-soluble fraction of the cell represents degradation products of either the endogenous or exogenous nucleotides or nucleic acids. The presence of the so-called minor bases is due to the degradation of nucleic acids.

The ribonucleotide concentration in the cell is in the millimolar range while the concentration of deoxyribonucleotides in the cell is in the micromolar range. As a specific example, the ATP concentration in Ehrlich tumor cells is 3,600 pmol/10^6 cells, while the dATP concentration in these cells is only 4 pmol/10^6 cells. The deoxyribonucleotide levels, however, are subject to major fluctuations during the cell cycle, in contrast to the ribonucleotide levels, which remain relatively constant.

In normal cells the total concentrations of the nucleotides are fixed within rather narrow limits, although the concentration of the individual components can vary. That is, the total concentration of adenine nucleotides (AMP, ADP, ATP) is constant, although there is a variation in the ratio of ATP to AMP + ADP, depending on the energy state of the cells. The basis for this "fixed concentration" is that the synthesis of nucleotides is one of the most finely regulated pathways occurring in the cell, as is discussed later.

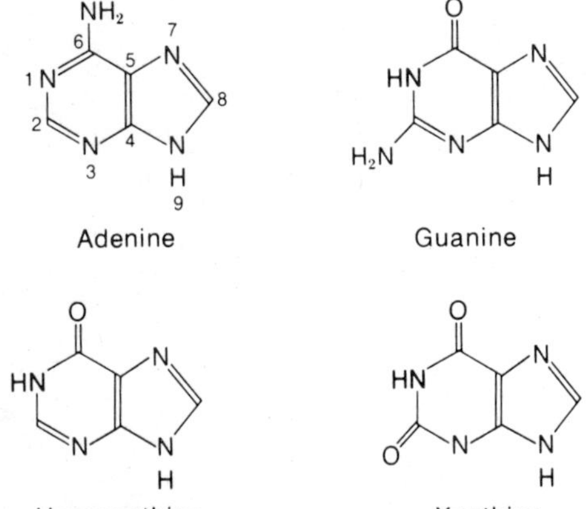

Figure 13.1
Purine bases.

13.3 CHEMISTRY OF NUCLEOTIDES

Quantitatively, the major purine derivatives found in the cell are those of adenine and guanine. Other purine bases encountered are hypoxanthine and xanthine (Figure 13.1). Nucleoside derivatives of these molecules will contain either ribose or 2-deoxyribose linked to the purine ring through a β-N-glycosidic bond at N-9. Ribonucleosides contain ribose, while deoxyribonucleosides contain deoxyribose as the sugar moiety (Figure 13.2). Nucleotides are phosphate esters of the purine nucleosides (Figure 13.3). 3′-Nucleotides such as adenosine 3′-monophosphate (3′-AMP) may occur in cells as a result of nucleic acid degradation.

In normally functioning cells, the tri- and diphosphates of the nucleosides are found to a greater extent than the monophosphates, nucleosides, or free bases.

The pyrimidine nucleotides found in highest concentrations in the cell are those containing uracil, cytosine, and thymine. The structures of the bases are presented in Figure 13.4. Uracil and cytosine nucleotides are the major pyrimidine components of RNA, whereas cytosine and thymine are the major pyrimidine components of DNA. As with purine derivatives, the pyrimidine nucleosides or nucleotides contain either ribose or 2-deoxyribose. The sugar moiety is linked to the pyrimidine in a β-N-glycosidic bond at N-1. The nucleosides of the pyrimidines are uridine,

Adenosine   Deoxyadenosine

Figure 13.2
Adenosine and deoxyadenosine.

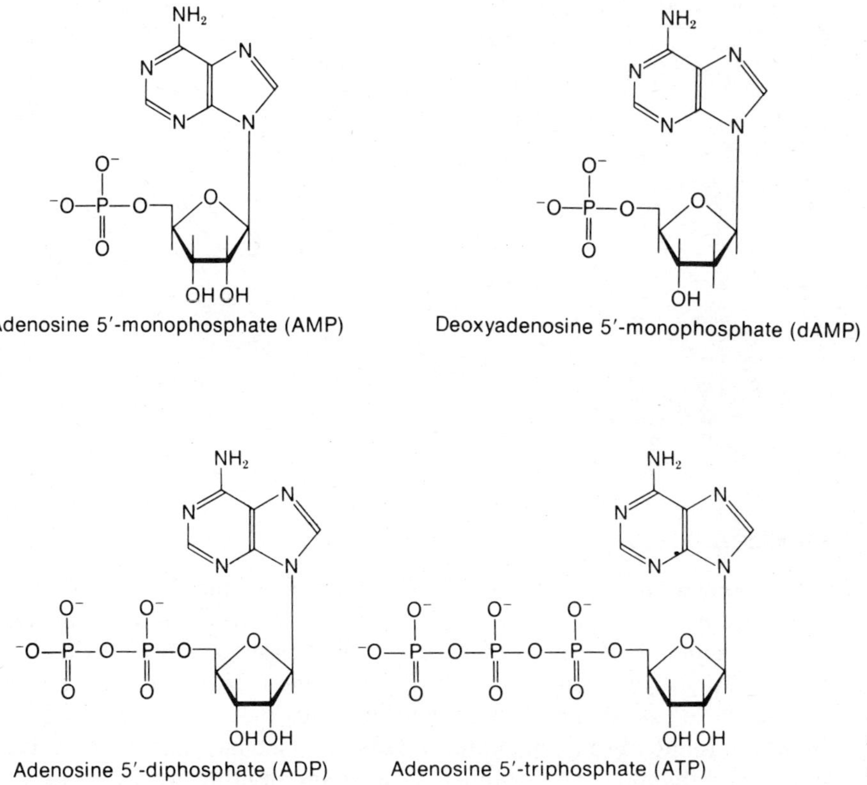

Adenosine 5′-monophosphate (AMP)

Deoxyadenosine 5′-monophosphate (dAMP)

Adenosine 5′-diphosphate (ADP)

Adenosine 5′-triphosphate (ATP)

Figure 13.3
Adenosine nucleotides.

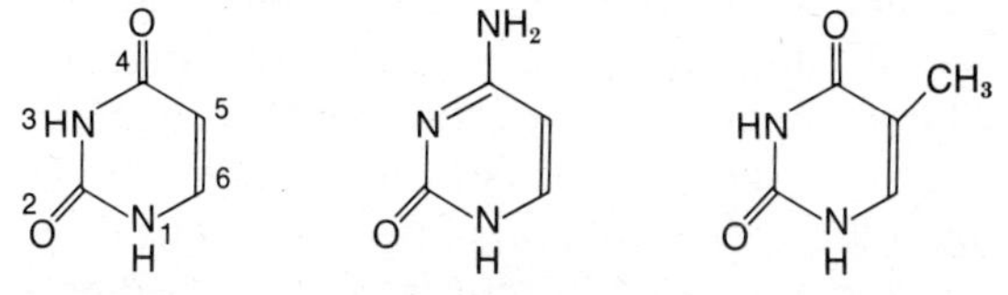

Uracil

Cytosine

Thymine

Figure 13.4
Pyrimidine bases.

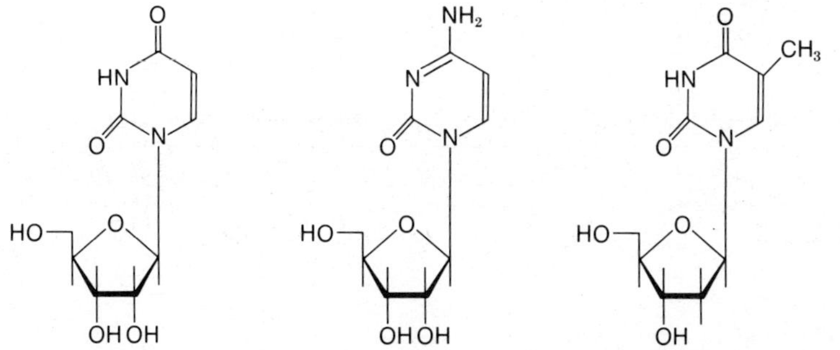

Uridine

Cytidine

Thymidine

Figure 13.5
Pyrimidine nucleosides.

cytidine and thymidine (Figure 13.5). The phosphate esters of the pyrimidine nucleosides are UMP, CMP, and TMP. In the cell the major pyrimidine derivatives found are the tri- and diphosphates (Figure 13.6).

The symbols and abbreviations for the bases, nucleosides, and nucleotides are summarized in Table 13.1. See the appendix, "Review of Organic Chemistry," for a summary of the nomenclature and chemistry of the purines and pyrimidines.

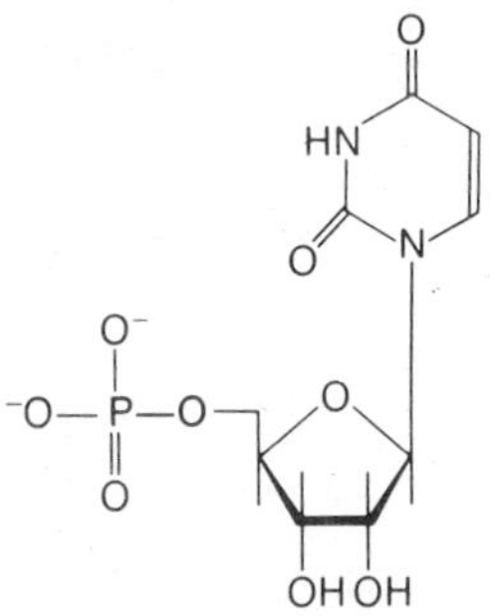

Uridine 5′-monophosphate (UMP)

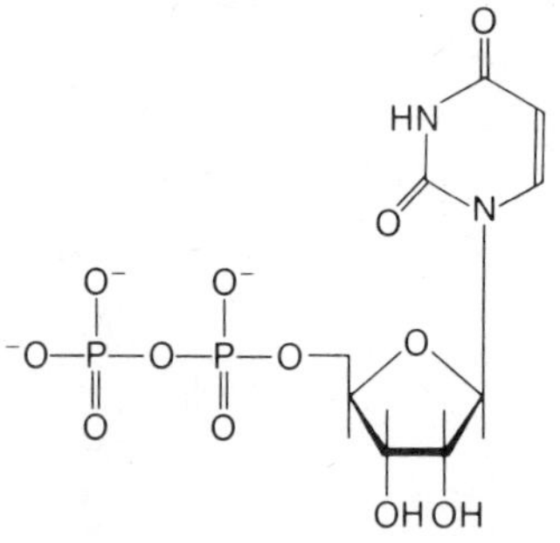

Uridine 5′-diphosphate (UDP)

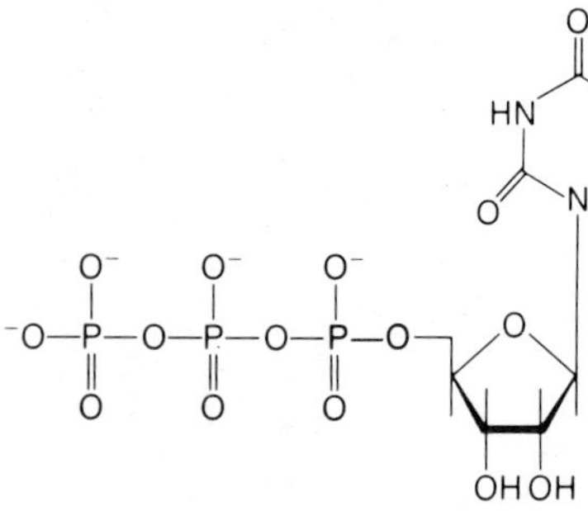

Uridine 5′-triphosphate (UTP)

Figure 13.6
Uracil nucleotides.

Modified Bases

The modified bases are formed by alteration of the purine or pyrimidine ring only *after* the parent base has been incorporated into the nucleic acids. Examples of some of these modified bases are 6-methylaminopurine, 7-methylguanine, and 5-methylcytosine (Figure 13.7). The term "minor base" is frequently used and indicates only that these modified bases are found in small quantities relative to adenine, guanine, cytosine, uracil, and thymine. The modified bases found in urine are a direct measure of the turnover of nucleic acids in the cells.

An additional modified nucleoside is also found as a constituent of tRNA. This modified nucleoside is called pseudouridine (Figure 13.8) and is unusual in that this nucleoside contains a C-glycosidic bond rather than a N-glycosidic bond as shown in the structure for pseudouridine. As with the modified bases, pseudouridine is formed only after UMP has been incorporated into the RNA. The level of pseudouridine in the urine is also an excellent measure of tRNA turnover. Ribothymidine is found as a minor component of tRNA. This derivative is formed by the methylation of uracil at C-5 by *S*-adenosylmethionine as a component of tRNA rather than by the direct incorporation of a ribothymidine nucleotide into this RNA species.

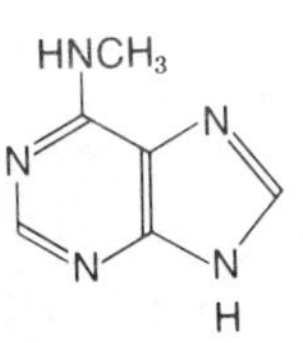

6-Methylaminopurine

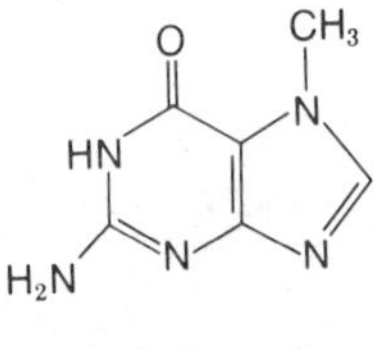

7-Methylguanine

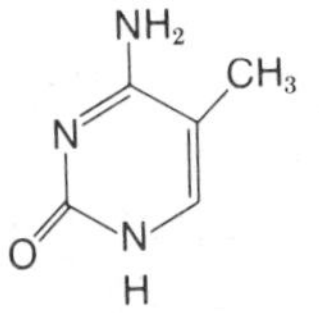

5-Methylcytosine

Figure 13.7
Modified bases.

TABLE 13.1 Symbols for Bases, Nucleosides, and Nucleotides

Compound	*Abbreviations*
Adenine	Ade
Cytosine	Cyt
Guanine	Gua
Thymine	Thy
Uracil	Ura
Adenosine	Ado
Cytidine	Cyd
Guanosine	Guo
Thymidine (2′-deoxythymidine)	dThd
Uridine	Urd
Adenosine 5′-mono-, di-, and triphosphate	AMP, ADP, and ATP
Cytidine 5′-mono-, di-, and triphosphate	CMP, CDP, and CTP
Guanosine 5′-mono-, di-, and triphosphate	GMP, GDP, and GTP
Thymidine 5′-mono-, di-, and triphosphate	TMP, TDP, and TTP
Uridine 5′-mono-, di-, triphosphate	UMP, UDP, and UTP

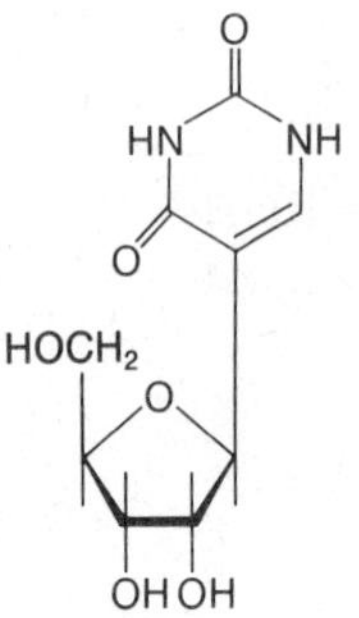

Pseudouridine

Figure 13.8
Pseudouridine.

Properties of Nucleotides

Cellular components containing either the purine or pyrimidine bases can be easily detected because of the strong absorption of uv light by these compounds. The purine bases, nucleosides and nucleotides have stronger absorptions than the pyrimidines and their derivatives. The molar extinction coefficients (a measure of the light absorption at a specific wavelength of a compound) and λ_{max} for these are given in Table 13.2. The wavelength of light at which maximum absorption occurs varies with the particular base component, but in most cases the uv maximum is close to 260 nm. The uv spectrum for each of the nucleoside or nucleotide derivatives responds differently to changes in pH. The strong uv absorptions and the differences due to the specific structure of the base moiety provide the basis for sensitive methods in assaying these compounds both qualitatively and quantitatively. For example, the deamination of cytosine nucleosides or nucleotides to the corresponding uracil derivatives causes a marked shift in λ_{max} from 271 to 262 nm, which is easily determined. Because of the high molar extinction coefficients of the purine and pyrimidine bases and their high concentrations in the nucleic acids, a solution of RNA or DNA at a concentration of 1 mg/ml would have an absorbance at 260 nm of ~20, whereas a typical protein at a concentration of 1 mg/ml would have an absorbance at 280 nm of ~1. Consequently, the nucleic acids are easily detected at low concentrations.

TABLE 13.2 Spectrophotometric Constants for Purine and Pyrimidine Nucleosides

Nucleoside	*Molar Extinction Coefficient* × 10^{-3}	λ_{max} *pH 7*
Adenosine	15.4	259
Guanosine	13.7	253
Cytidine	8.9	271
Uridine	10.0	262
Thymidine	10.0	262

The N-glycosidic bond of the purine and pyrimidine nucleosides and nucleotides are stable to alkali. However, the stability of this bond to acid hydrolysis differs markedly. The N-glycosidic bond of purine nucleosides and nucleotides is easily hydrolyzed by dilute acid at elevated temperatures (e.g., 60°C) to yield the free purine base and the sugar or sugar phosphate. On the other hand, the N-glycosidic bond of uracil, cytosine, and thymine nucleosides and nucleotides is very stable to acid treatment. Strong conditions, such as perchloric acid (60%) and 100°C, will cause the release of the free pyrimidine but with the complete destruction of the sugar moiety. The N-glycosidic bond of the pyrimidine nucleosides and nucleotides containing dihydrouracil is labile to mild acid treatment.

Because of the highly polar phosphate group, the purine and pyrimidine nucleotides are considerably more soluble in aqueous solutions than are their nucleosides and free bases. In general, the nucleosides are more soluble than the free purine or pyrimidine bases.

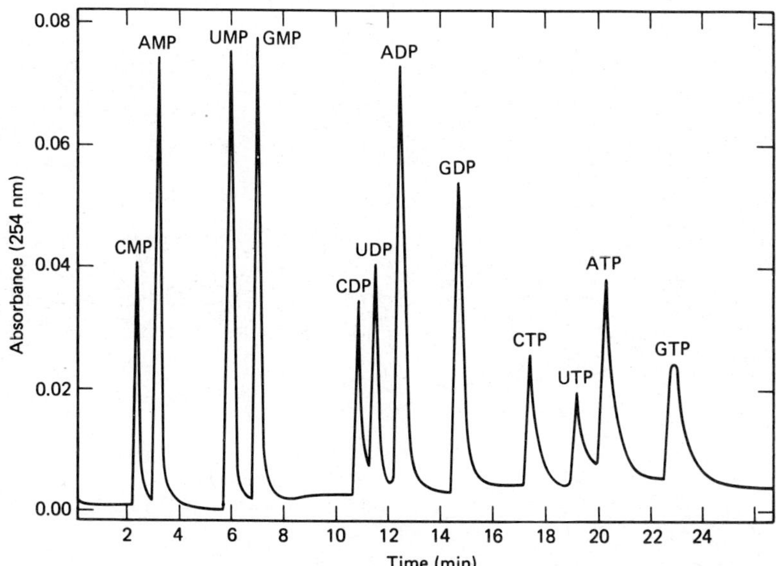

Figure 13.9
Separation of nucleotides by high pressure liquid chromatography.
A mixture of the 12 ribonucleotides was injected onto a Partisil SAX column (25 × 0.46 cm). A linear gradient was developed using 0.01 M ammonium phosphate, pH 2.77 and 0.5 M ammonium phosphate, pH 4.8 at a rate of 5%/min. The solvent flow rate was 2 ml/min.

The purine and pyrimidine bases and their nucleoside and nucleotide derivatives can be easily separated by a variety of techniques. These methods include: paper chromatography; thin-layer chromatography, utilizing plates with cellulose or ion-exchange resins; electrophoresis; and ion-exchange column chromatography. The most recent advance in the separation of the purine and pyrimidine components involves the use of high pressure liquid chromatography (HPLC); nanomole quantities of these components are easily separated in a brief period of time. Figure 13.9 shows the separation by HPLC of a mixture containing AMP, ADP, ATP, GMP, GDP, GTP, UMP, UDP, UTP, CMP, CDP, and CTP in 25 min. The development of this instrument and the high resolution columns (anion and cation exchange and reverse phase) has allowed rapid determination of nucleoside and nucleotide pools under a variety of cellular conditions.

13.4 METABOLISM OF PURINE NUCLEOTIDES

The purine ring is synthesized de novo in mammalian cells utilizing amino acids as carbon and nitrogen donors, and formate and CO_2 as carbon donors.

The numbering system for the purine ring is shown in Figure 13.10 with the sources for the various carbon and nitrogen atoms indicated. From the sources of carbon and nitrogen which make up the purine ring it is very evident that amino acids play an important role in nucleotide metabolism. It has been shown that glutamine and aspartate levels can influence the rate of purine nucleotide synthesis in tumor cells. Many of the reactions that are required for the de novo synthesis utilize the hydrolysis of ATP to drive the particular reaction. The overall set of reactions leading to the synthesis of a purine nucleotide is, therefore, expensive in terms of ATP required.

De Novo Synthesis

All the enzymes involved with purine nucleotide synthesis and degradation are found in the cytosol of the cell. However, not all cells are capable of de novo purine nucleotide synthesis. The reactions leading to the de novo synthesis of purine nucleotides are as follows.

1. *Formation of N-Glycosidic Bond* (N-9 of purine ring and C-1 of ribose)

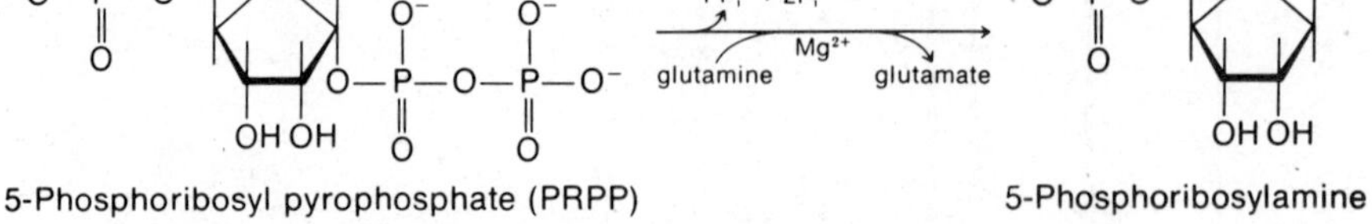

5-Phosphoribosyl pyrophosphate (PRPP) —(glutamine → glutamate; $PP_i \rightarrow 2P_i$; Mg^{2+})→ 5-Phosphoribosylamine

This reaction is catalyzed by the enzyme PRPP amidotransferase, which is the committed step in this pathway and, as we will see later, the major regulated step.

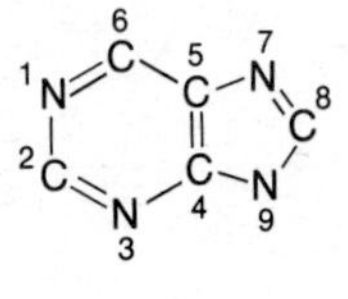

N-3 and N-9 from amide N of glutamine
C-4, C-5, and N-7 from glycine
C-2 and C-8 from formate via H_4folate
C-6 from CO_2
N-1 from aspartate

Figure 13.10
Sources of carbon and nitrogen atoms in purine ring.

In the formation of 5-phosphoribosylamine from PRPP there is inversion at carbon-1, giving rise to the β configuration of the N-glycosidic bond in purine nucleotides.

2. *Addition of Glycine* (C-4, C-5, and N-7 of purine ring)

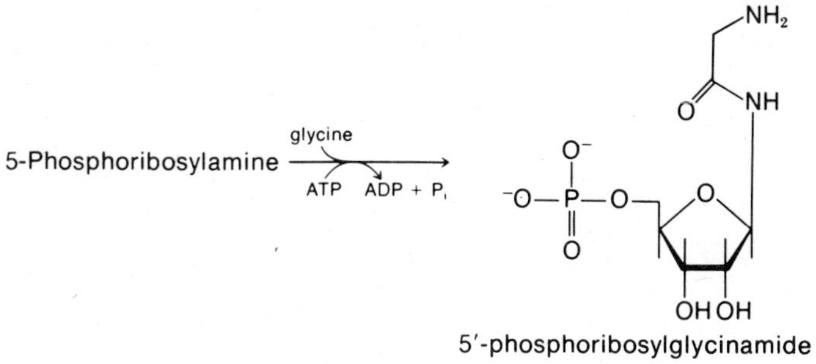

5-Phosphoribosylamine —(glycine, ATP → ADP + P_i)→ 5′-phosphoribosylglycinamide

This reaction is catalyzed by phosphoribosylglycinamide synthetase and requires ATP as the high energy source to drive the reaction.

3. *Introduction of Formyl Group via H_4folate* (Figure 13.11) (C-8 of purine ring).

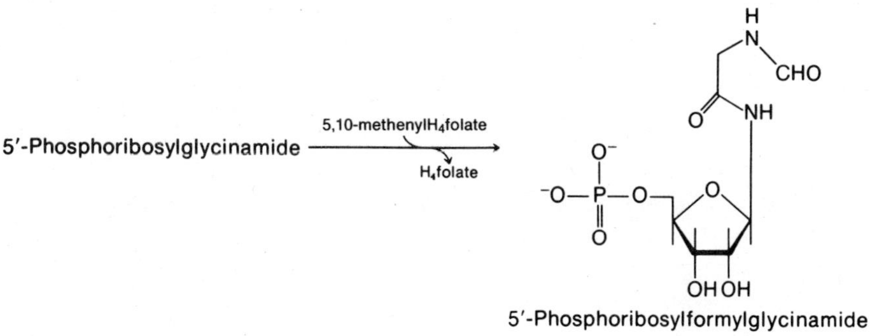

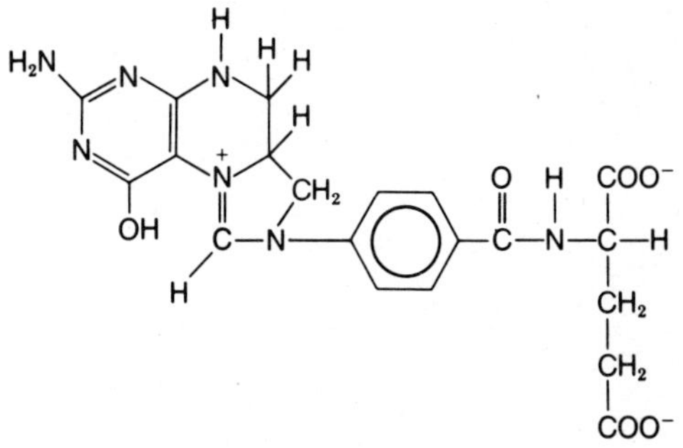

5′-Phosphoribosylglycinamide —(5,10-methenylH_4folate → H_4folate)→ 5′-Phosphoribosylformylglycinamide

Figure 13.11
5,10-Methenyl H_4folate.

This reaction is catalyzed by phosphoribosylglycinamide formyltransferase, which requires 5,10-methenylH_4folate as the one-carbon donor.

4. *Addition of Nitrogen from Glutamine* (N-3 of purine ring)

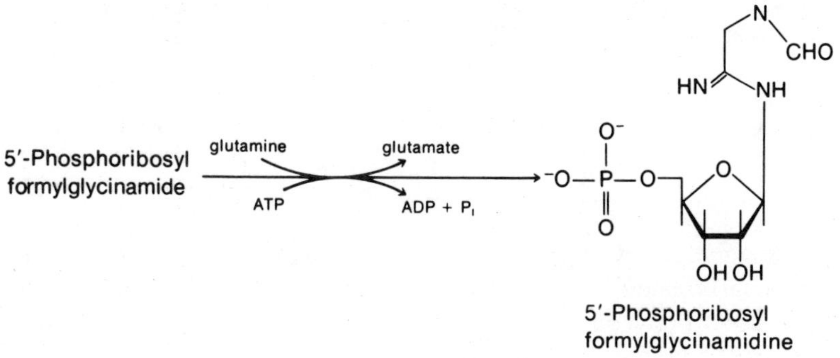

5′-Phosphoribosyl formylglycinamide —(glutamine → glutamate, ATP → ADP + P_i)→ 5′-Phosphoribosyl formylglycinamidine

This reaction is catalyzed by phosphoribosylformylglycinamide synthetase. ATP hydrolysis provides the energy for this reaction.

5. *Ring Closure to Form Imidazole Ring*
The enzyme for this step is phosphoribosylaminoimidazole synthetase, which again requires ATP hydrolysis to close the ring.

5′-Phosphoribosyl formylglycinamidine → (ATP → ADP + P) → 5′-Phosphoribosyl-5-aminoimidazole

6. *Addition of CO_2* (C-6 of purine ring)

5′-Phosphoribosyl-5-aminoimidazole → (CO_2) → 5′-Phosphoribosyl 5-aminoimidazole-4-carboxylic acid

The enzyme that catalyzes this reaction is phosphoribosylaminoimidazole carboxylase. This reaction is not inhibited by avidin, indicating that this enzyme is not a biotin-requiring carboxylase.

7. *Addition of NH_2 Group from Aspartate* (N-1 of purine ring)

5′-Phosphoribosyl 5-aminoimidazole-4-carboxylic acid → (aspartate; ATP → ADP + P_i) → 5′-Phosphoribosyl 5-aminoimidazole-4-N-succinocarboxamide

The enzyme catalyzing this reaction is phosphoribosylaminoimidazole-succinocarboxamide synthetase. ATP is required for the formation of the amide bond between the amino group of aspartate and the carboxylate group of the imidazole derivative.

8. *Cleavage of N—C Bond of Aspartate*
Adenylosuccinase is the enzyme that cleaves the N—C bond, in effect, to transfer the amino group to the imidazole derivative. Adenylosuccinase is the same enzyme that is used in the conversion of adenylosuccinate to AMP and fumarate later in the pathway.

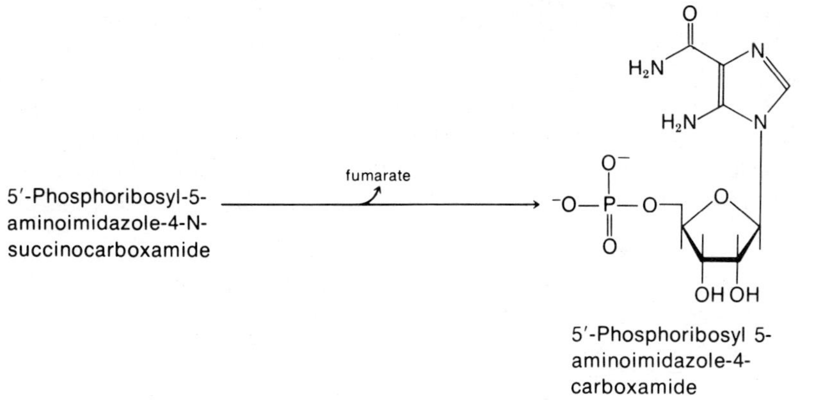

5′-Phosphoribosyl-5-aminoimidazole-4-N-succinocarboxamide → (fumarate) → 5′-Phosphoribosyl 5-aminoimidazole-4-carboxamide

9. *Introduction of Formate via H_4 Folate* (Figure 13.12) (C-2 of purine ring)

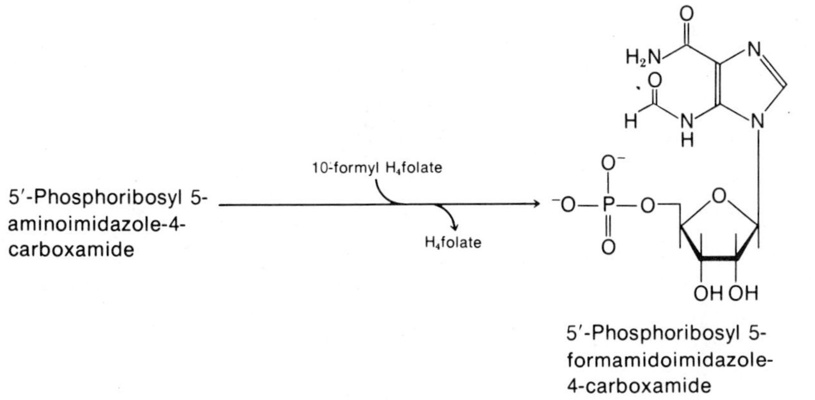

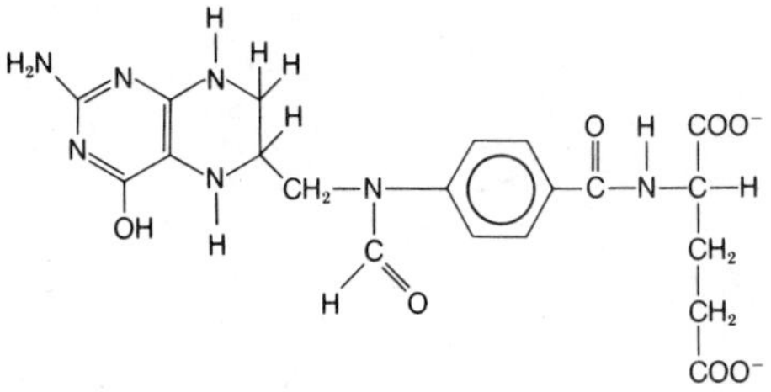

5′-Phosphoribosyl 5-aminoimidazole-4-carboxamide + 10-formyl H_4folate → H_4folate + 5′-Phosphoribosyl 5-formamidoimidazole-4-carboxamide

Figure 13.12
10-Formyl H_4folate.

The enzyme catalyzing this reaction is phosphoribosylaminoimidazole carboxamide formyltransferase. N^{10}-Formyl H_4folate is the cosubstrate in this reaction and is the one-carbon donor.

10. *Ring Closure to Form IMP*

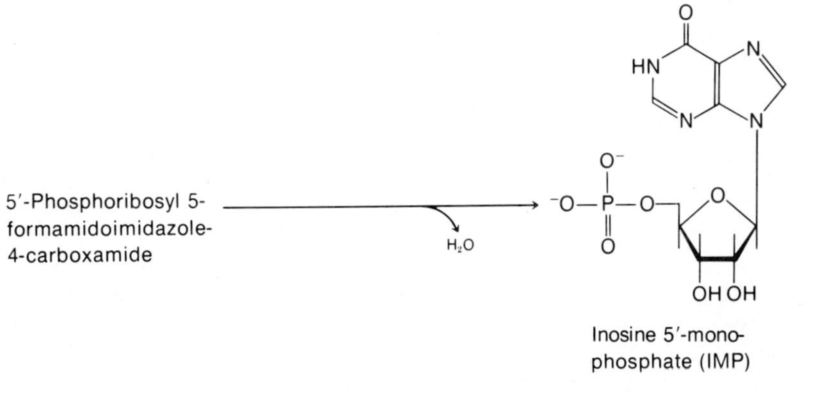

5′-Phosphoribosyl 5-formamidoimidazole-4-carboxamide → H_2O + Inosine 5′-monophosphate (IMP)

In the final reaction of the de novo synthesis of the purine ribonucleotide, IMP is formed from phosphoribosylformamidoimidazole carboxamide by inosinicase.

This series of 10 reactions leads to the synthesis of the ribonucleotide IMP. However, IMP is not found to any extent under normal conditions in the cell either as a component of the nucleotide pool or as a constituent of the nucleic acids. IMP is converted to the adenine and guanine nucleotides found in the nucleotide pool and nucleic acids by the pathways shown in Figure 13.13.

From these reactions it is clear that the conversion of IMP to AMP and GMP does not occur randomly. It is seen that the conversion of IMP to GMP requires ATP as the energy source, while the conversion of IMP to AMP requires GTP as the energy source. Therefore, when there is suffi-

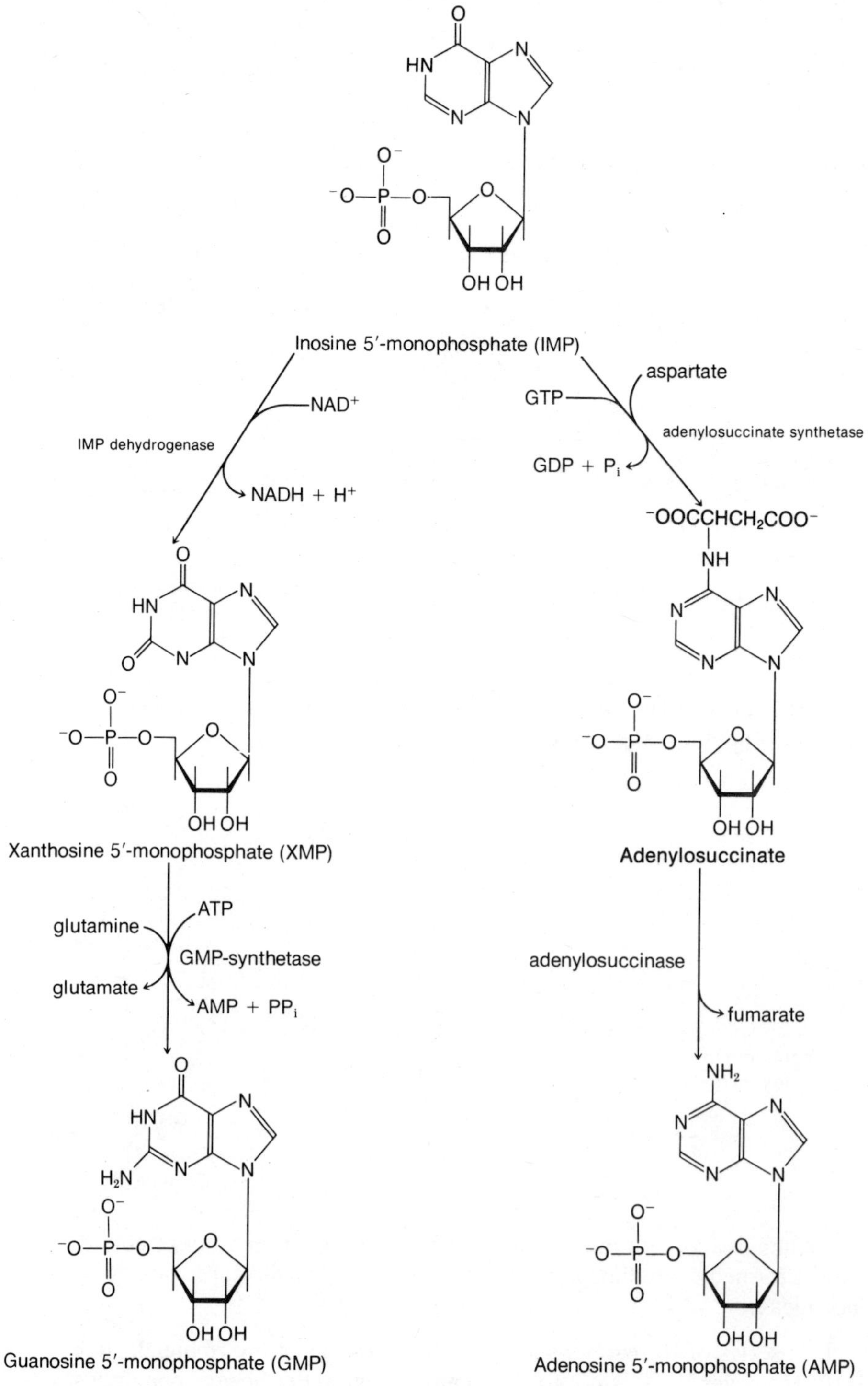

Inosine 5′-monophosphate (IMP)

IMP dehydrogenase: NAD^+ → $NADH + H^+$

GTP, aspartate → GDP + P_i (adenylosuccinate synthetase)

$^-OOCCHCH_2COO^-$

Xanthosine 5′-monophosphate (XMP)

Adenylosuccinate

glutamine, ATP → glutamate, AMP + PP_i (GMP-synthetase)

adenylosuccinase → fumarate

Guanosine 5′-monophosphate (GMP)

Adenosine 5′-monophosphate (AMP)

Figure 13.13
Formation of AMP and GMP from IMP branch point.

cient ATP in the cell, IMP will be converted to GMP, and conversely, when there is sufficient GTP in the cell IMP will be converted to AMP.

Regulation of de Novo Purine Nucleotide Synthesis

As is frequently the case, the committed step of a metabolic pathway is the step that is regulated. Such is the case for the de novo synthesis of purine nucleotides. The reaction catalyzed by PRPP amidotransferase, in which 5-phosphoribosylamine is formed, is the committed step for purine nucleotide synthesis. This reaction is strongly regulated by IMP, GMP, and AMP. It has been found that PRPP amidotransferase from human placenta exists in two forms with molecular weights of 133,000 and 270,000. The enzyme activity is correlated with the smaller form. In the presence of 5′-nucleotides the small active form is converted to the large inactive form, whereas PRPP causes the large form to shift to the active form. This can be viewed as presented on the right.

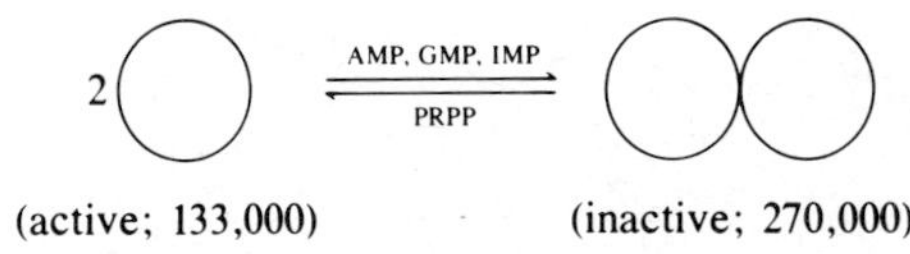

The placental enzyme appears to have at least two effector binding sites. One site specifically binds the oxypurine nucleotides (IMP and GMP), while the other site binds the aminopurine nucleotide (AMP). The simultaneous binding of an oxypurine nucleotide and an aminopurine nucleotide results in a synergistic inhibition of the enzyme.

The enzyme PRPP amidotransferase displays hyperbolic kinetics (Figure 13.14) with respect to glutamine. With respect to the second substrate, PRPP, the enzyme shows sigmoidal kinetics. The presence of nucleotides further exaggerates the sigmoidal nature of the v (velocity) vs [PRPP] plot. The intracellular concentrations of glutamine are approximately equal to the K_m of the enzyme for glutamine. On the other hand, the PRPP levels in the cell, which do fluctuate, can be 10 to 100 times less than the K_m of the enzyme for PRPP. Consequently, the PRPP concentrations in the cell play an important role in controlling de novo purine nucleotide synthesis.

Between the formation of 5-phosphoribosylamine and IMP, there are no known regulated steps. However, there is regulation at the branch point of IMP to GMP and IMP to AMP. The two enzymes which utilize IMP at this branch point, IMP dehydrogenase and adenylosuccinate synthetase, have similar K_ms for IMP. AMP is a competitive inhibitor (with respect to IMP) of adenylosuccinate synthetase, while GMP is a competitive inhibitor of IMP dehydrogenase. Two levels of control are therefore in effect at the IMP branch point. GTP serves as an energy source for the adenylosuccinate synthetase reaction, while AMP is a competitive inhibitor of this step; and ATP serves as the energy source in the conversion of XMP to GMP, while GMP acts as an inhibitor of XMP formation.

The regulation of purine nucleotide synthesis is summarized in Figure 13.15.

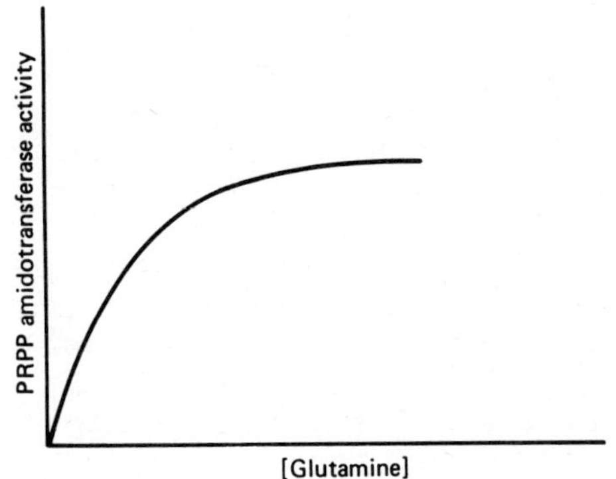

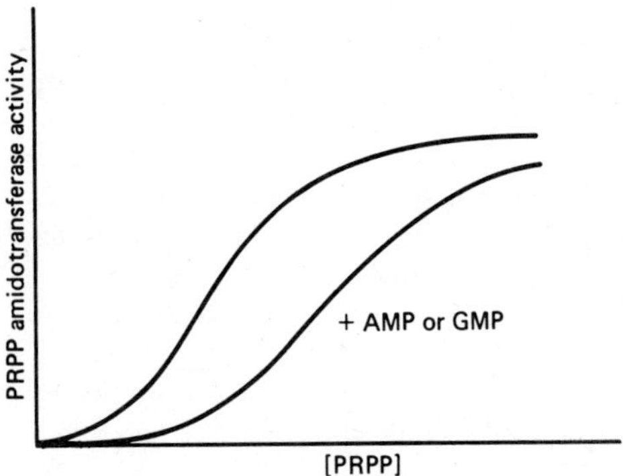

Figure 13.14
PRPP-amidotransferase activity as a function of glutamine or PRPP concentrations.

Salvage Pathways for the Purine Bases

The efficiency of cellular metabolism under normal conditions is expressed by the presence of the so-called "salvage pathways" for the purine bases. In the metabolic pathway for the de novo synthesis of the purine nucleotides, a great deal of energy in the form of ATP is required to synthesize the purine nucleotides. In the "salvage" pathways the preformed bases (from exogenous sources or from the turnover of nucleic acids) can be reutilized, resulting in a considerable energy saving for the cell. There are two distinct enzymes involved. The reactions are as follows.

$$\text{Guanine} + \text{PRPP} \longrightarrow \text{GMP} + \text{PP}_i$$

$$\text{Hypoxanthine} + \text{PRPP} \longrightarrow \text{IMP} + \text{PP}_i$$

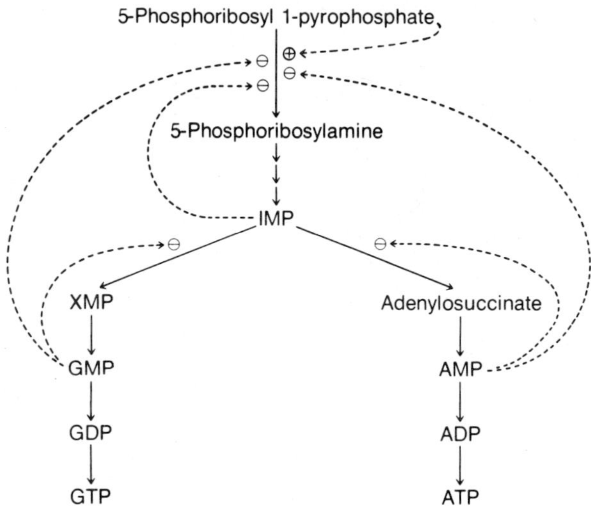

Figure 13.15
Regulation of purine nucleotide synthesis.
The solid arrows indicate enzyme-catalyzed reactions. The dashed arrows indicate the regulated steps (⊕, activation; ⊖, inhibition).

The enzyme that catalyzes both these reactions is hypoxanthine-guanine phosphoribosyltransferase (HGPRTase) and requires Mg^{2+}. HGPRTase is regulated by the presence of IMP or GMP. IMP and GMP are competitive inhibitors with respect to PRPP in the HGPRTase reaction. The competitive nature of the inhibition implies that high concentrations of PRPP can overcome the regulation of this metabolic step.

$$\text{Adenine} + \text{PRPP} \longrightarrow \text{AMP} + \text{PP}_i$$

The enzyme that catalyzes this reaction is adenine phosphoribosyltransferase (APRTase) and also requires Mg^{2+}. AMP, the product of the reaction catalyzed by APRTase, is an inhibitor of this reaction.

The source of adenine utilized in the APRTase reaction appears to be mainly from the synthesis of polyamines. For each molecule of spermine synthesized, two molecules of 5′-methylthioadenosine are generated. 5′-Methylthioadenosine is then degraded to 5-methylthioribose-1-phosphate and adenine via the 5′-methylthioadenosine phosphorylase-catalyzed reaction. The adenine base is salvaged through the APRTase reaction, whereas the carbon skeleton of 5-methylthioribose-1-phosphate is utilized in a reaction sequence that regenerates methionine.

These reactions are important not only because they conserve energy, but also because they permit cells such as erythrocytes to form nucleotides from the bases. The erythrocyte, for example, does not have PRPP amidotransferase and hence cannot synthesize 5-phosphoribosylamine, the first unique metabolite in the pathway of purine nucleotide synthesis. As a consequence, the red cell must depend on the purine phosphoribosyltransferases to replenish the nucleotide pools.

The importance of these reactions is further demonstrated in the situation in which the HGPRTase activity is markedly depressed. Such a deficiency results in the Lesch–Nyhan syndrome (Clin. Corr. 13.1), which is characterized clinically by hyperuricemia, mental retardation, and self-mutilation.

CLIN. CORR. 13.1
LESCH–NYHAN SYNDROME

The Lesch–Nyhan syndrome is characterized clinically by hyperuricemia, excessive uric acid production, and neurological problems, which may include spasticity, mental retardation, and self-mutilation. This disorder is associated with a very severe or complete deficiency of HGPRTase activity. It is an inherited defect that affects only males. In a study of the available patients, it was observed that if the HGPRTase activity was less than 2% of normal, mental retardation was present, and if the activity was less than 0.2% of normal, the self-mutilation aspect was expressed. This defect leads not only to the overproduction and excretion of uric acid, but also to increased excretion of hypoxanthine.

There appear to be several variant forms of the HGPRTase-deficient patients. In one form there appeared to be a complete lack of HGPRTase activity, and yet by titration with antibodies prepared against the purified enzyme from normal erythrocytes, the Lesch–Nyhan patient had an equivalent amount of immunoprecipitable protein. In another form, the enzyme appeared to be unstable. "Young" red cells separated from the "old" red cells by density centrifugation had much higher (although still markedly reduced compared to normal) HGPRTase activity than did the "old" red cells. In still another case, a child was studied who had all the clinical manifestations of the Lesch–Nyhan syndrome, but had normal levels of HGPRTase activity when determined in the laboratory. Further kinetic analysis showed that this variant was a "K_m mutant." When assayed under saturating levels of PRPP, the level of activity was comparable to normal. When assayed at the concentration of PRPP which would approximate the intracellular concentration of PRPP, the activity was diminished to the range found in Lesch–Nyhan patients.

As discussed previously, the role of HGPRTase is to catalyze reactions in which hypoxanthine and guanine are converted to nucleotides. The hyperuricemia and excessive uric acid production which occur in patients with the Lesch–Nyhan syndrome are easily explained by the lack

Interconversion of Purine Nucleotides

Along with the very fine control exhibited in the cell for the de novo synthesis of purine nucleotides, there are enzymes that can be used to balance the levels of guanine and adenine nucleotides. As discussed earlier, IMP can be converted by one pathway to GMP and by a different pathway to AMP. There is no known direct pathway for the conversion of GMP to AMP or AMP to GMP. However, these purine nucleotides can be redistributed to meet the cellular needs through the conversion of GMP and AMP back to IMP. These reactions are carried out by separate enzymes, each under separate controls. The pathways are summarized in Figure 13.16. The reductive deamination of GMP to IMP by GMP reductase is activated by GTP and inhibited by XMP.

XMP is a competitive inhibitor of human GMP reductase, having a $K_i = \sim 0.2\ \mu M$. Because of this low K_i, the concentration of XMP in the cell could influence the conversion of GMP to IMP. On the other hand, GTP is a nonessential activator of GMP reductase and serves to lower the K_m of the enzyme with respect to GMP and to increase the V_{max}. The activity of AMP deaminase (5′-AMP aminohydrolase), which specifically catalyzes the deamination of AMP to yield IMP, is activated by K^+, and ATP and is inhibited by inorganic phosphate, GDP, and GTP. In the absence of K^+ ions the v vs [AMP] curve is sigmoidal. The presence of K^+ ions is not required for maximum activity but rather acts as a positive allosteric effector to reduce the apparent K_m for AMP.

Degradation of Purine Nucleotides

The purine nucleotides, nucleosides, and bases funnel through a common pathway for the degradation of these biomolecules. The end product of purine degradation in man is uric acid. The catabolic pathways are as shown in Figure 13.17. The enzymes involved in the degradation of the nucleic acids and the nucleotides and nucleosides vary in specificity. The nucleases show specificity toward either RNA or DNA and also toward the bases and position of cleavage of the 3′,5′-phosphodiester bonds. The nucleotidases range from those with relatively high specificity, such as 5′-AMP nucleotidase, to those with broad specificity, such as the acid and alkaline phosphatases, which will hydrolyze any of the 3′- or 5′-nucleotides. AMP deaminase is specific for AMP. Adenosine deaminase is much less specific, since not only adenosine, but also 2′-deoxyadenosine and many other 6-aminopurine nucleosides are deaminated by this enzyme.

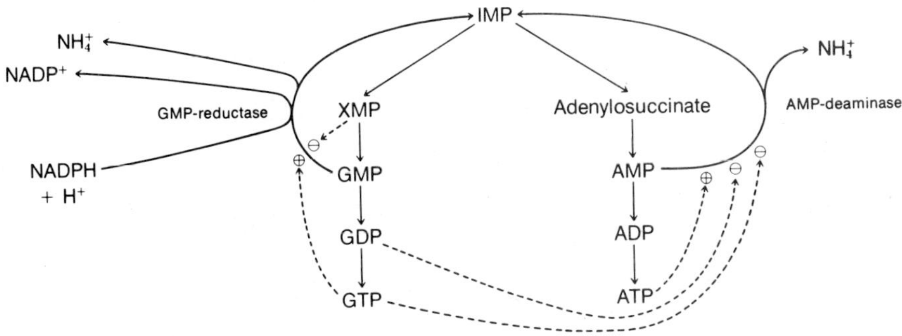

Figure 13.16
Interconversion of purine nucleotides.
The solid arrows represent enzyme-catalyzed reactions. The dashed lines represent steps of regulation (⊕, activation; ⊖, inhibition).

of HGPRTase activity. The hypoxanthine and guanine are not salvaged leading to increased intracellular pools of PRPP and decreased levels of IMP or GMP. Both these factors promote the de novo synthesis of purine nucleotides without regard for the proper regulation of this pathway.

It is not understood why a severe defect in this salvage pathway leads to the neurological problems. The adenine phosphoribosyltransferase activity in these patients is normal or in fact elevated. With this salvage enzyme, presumably the cellular needs for purine nucleotides could be met via the pathway,

$$\text{Adenine} \xrightarrow{\text{PRPP}} \text{AMP} \longrightarrow \text{IMP} \longrightarrow \text{GMP}$$

if the cell's de novo pathway were not functioning. The normal tissue distribution of HGPRTase activity perhaps could explain the neurological symptoms. The brain (frontal lobe, basal ganglia, and cerebellum) has 10 to 20 times the enzyme activity found in liver, spleen, or kidney and from 4 to 8 times that found in the erythrocytes. Individuals who have primary gout with excessive uric acid formation and hyperuricemia do not display the neurological problems. It is argued that on this basis the products of purine degradation (hypoxanthine, xanthine, and uric acid) cannot be toxic to the central nervous system (CNS). However, it is possible that these metabolites are toxic to the developing CNS or that the lack of the enzyme leads to an imbalance in the concentrations of the purine nucleotides at critical times during development.

Perhaps the lesion which leads to the neurological problems is not related to the HGPRTase deficiency. It will require further study to understand the relationship between these neurological defects and the decreased levels of HGPRTase.

Treatment of Lesch–Nyhan patients with allopurinol will decrease the amount of uric acid formed, relieving some of the problems caused by sodium urate deposits. However, since the Lesch–Nyhan patient has a marked reduction in HGPRTase activity, hypoxanthine and guanine are not salvaged, PRPP is not consumed, and consequently the de novo synthesis of purine nucleotides is not shut down. There is no known treatment for the neurological problems. These patients usually die from kidney failure, resulting from the high sodium urate deposits.

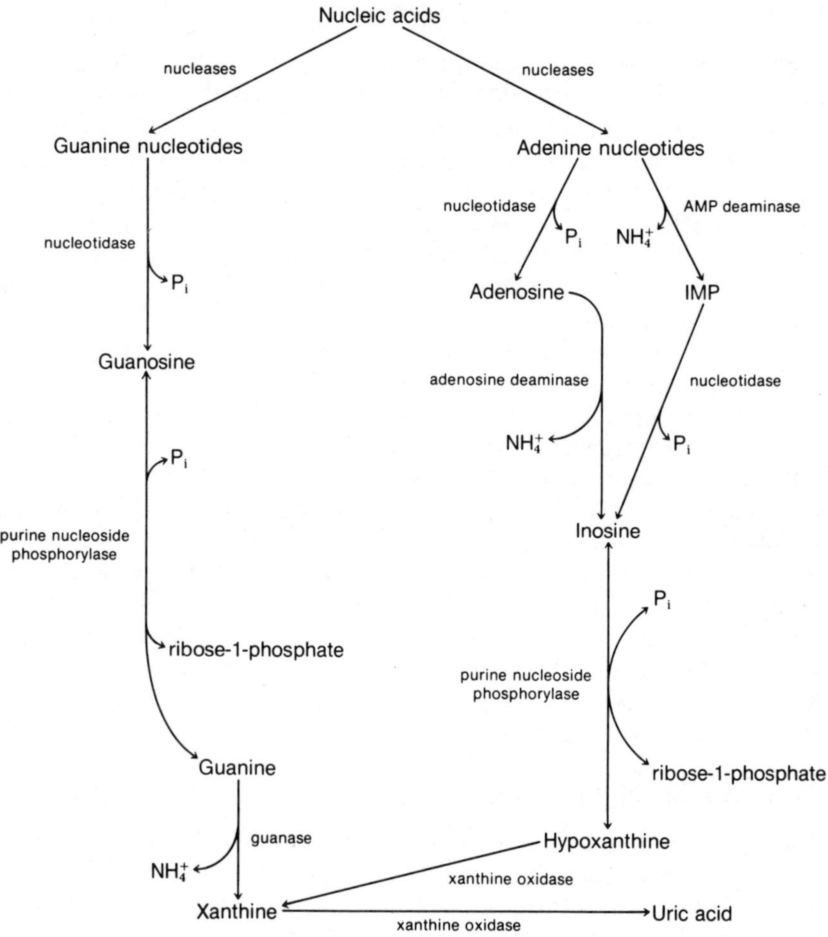

Figure 13.17
Degradation of purine nucleotides.

A special comment should be made about purine nucleoside phosphorylase. As indicated, the reaction is readily reversible.

$$\text{Inosine} + P_i \rightleftharpoons \text{hypoxanthine} + \text{ribose 1-P}$$

or

$$\text{Guanosine} + P_i \rightleftharpoons \text{guanine} + \text{ribose 1-P}$$

or

$$\text{Xanthosine} + P_i \rightleftharpoons \text{xanthine} + \text{ribose 1-P}$$

Deoxyinosine and deoxyguanosine are also excellent substrates for purine nucleoside phosphorylase. While the equilibrium for the reactions catalyzed by purine nucleoside phosphorylase favors nucleoside synthesis, it would appear that in the cell the concentrations of the free purine and ribose 1-phosphate are too low to support nucleoside synthesis under most conditions. The main function of purine nucleoside phosphorylase in the cells seems to be its role in purine nucleoside degradation. This is supported by the conditions observed in the cases where a deficiency of purine nucleoside phosphorylase has been detected. Under these conditions there is a large buildup of the substrates (inosine, guanosine, deoxyinosine and deoxyguanosine) for purine nucleoside phosphorylase with a corresponding decrease in uric acid formation.

These enzymes may therefore be regarded as part of the degradative pathway. On the other hand, since the reaction is readily reversible, it may serve as part of the salvage pathway under certain metabolic conditions.

Deficiencies in two of the enzymes of this degradative pathway (adenosine deaminase and purine nucleoside phosphorylase) have been observed in man in two disease states. Adenosine deaminase deficiency has been associated with a severe combined immunodeficiency, and purine nucleoside phosphorylase deficiency is associated with a defective T-cell immunity and a normal B-cell immunity (Clin. Corr. 13.2).

Formation of Uric Acid

Xanthine oxidase is an enzyme which contains FAD, Fe(III), and Mo(VI). In the reaction, molecular oxygen is a substrate with H_2O_2 being

CLIN. CORR. 13.2 IMMUNODEFICIENCY DISEASES ASSOCIATED WITH DEFECTS IN PURINE NUCLEOTIDE METABOLISM

Recently two different immunodeficiency diseases have been recognized that are associated with deficiencies in the enzymes adenosine deaminase and purine nucleoside phosphorylase. Referring to Figure 13.17, it is seen that these two enzymes are involved in the degradation of purine nucleosides.

The deficiency in adenosine deaminase is associated with a severe combined immunodeficiency involving T-cell and usually B-cell dysfunction. Adenosine deaminase deficiency is not associated with the overproduction of purine nucleotides. The mechanism by which the lack of adenosine deaminase interferes with immune function is not completely understood. However, it has been shown that in a patient with adenosine deaminase deficiency there is an extremely large buildup of deoxyadenosine triphosphate in the erythrocytes examined. In fact, the dATP concentration exceeded the ATP concentration in these cells. It is therefore thought that the failure of the cells to metabolize deoxyadenosine to deoxyinosine for further conversion to hypoxanthine leads to the increased levels of dATP. dATP is known to be a very effective inhibitor of ribonucleotide reductase and consequently of DNA synthesis (cell replication). It is thought that this is the site that leads to the deficiencies in the immune system. Other suggestions have included the proposals that the elevated adenosine is toxic to the cells by virtue of its ability to increase the intracellular concentrations of cAMP or due to the inhibition of *S*-adenosyl homocysteine hydrolase, leading to increased intracellular levels of *S*-adenosyl-L-homocysteine.

The deficiency in purine nucleoside phosphorylase is associated with an impairment of T-cell function with no apparent effects on B-cell function. There is no overproduction of purine nucleotides associated with this deficiency. However, there is a marked *decrease* in uric acid formation with the corresponding increased levels of the purine nucleoside phosphorylase substrates, guanosine, deoxyguanosine, inosine, and deoxyinosine. When these various nucleosides were incubated with normal T lymphocytes in culture, deoxyguanosine was found to be the most toxic. In addition, it was found that dGTP was the major nucleotide that accumulated in the red cells from patients with purine nucleoside phosphorylase deficiency. It is suggested that dGTP, which acts as an inhibitor of CDP reductase, is the actual agent which is toxic to the development of normal T cells.

In both of these enzyme deficiencies, it is not entirely clear how these defects lead to the immune problems. However, it is clear that defects in enzymes that have been casually considered in metabolic pathways in the past, reveal their importance to normal metabolism when they are absent or severely decreased.

CLIN. CORR. **13.3**
GOUT

Primary gout is characterized by excessive uric acid due to a variety of metabolic abnormalities that lead to the overproduction of purine nucleotides via the de novo pathway. With the overproduction of uric acid, the levels of uric acid in the serum are elevated, and there are deposits of sodium urate crystals in the joints of extremities. The metabolic bases for the increased production of purine nucleotides which in turn are manifested by increased uric acid levels have been identified in several situations. Many, if not all, of the clinical symptoms associated with the overproduction of uric acid arise because uric acid is not very soluble. Formation of sodium urate crystals leads not only to the joint problems but also to renal disease. Hyperuricemia resulting from the overproduction of uric acid via the de novo pathway as opposed to hyperuricemia resulting from renal damage or increased cell death (e.g., radiation therapy) can be relatively easily distinguished. The feeding of ^{15}N-glycine to an "overproducer" will result in a marked ^{15}N enrichment of the N-7 of uric acid isolated from the urine or serum of these patients, whereas there would be little ^{15}N enrichment in the uric acid from individuals with renal problems or with increased nucleic acid degradation.

Various studies directed at determining the molecular basis in primary gout for the overproduction of purine nucleotides, and consequently of uric acid, have uncovered a diverse group of metabolic defects. While the primary metabolic defects that have been determined may be seemingly unrelated to the de novo synthesis of purine nucleotides, a common feature of these defects evolves.

The defects described in human beings include the following:

1. *PRPP-Synthetase:* Mutant forms of PRPP-synthetase have been detected, which are not subject to allosteric regulation by inorganic phosphate or to feedback inhibition by GDP and ADP. Under these conditions, the intracellular concentration of PRPP is elevated, leading to increased formation of 5-phosphoribosylamine.
2. *Partial HGPRTase Deficiency:* A characteristic of this deficiency is the overproduction of purine nucleotides. The basis for this appears to be twofold. First, the lack of HGPRTase activity decreases the amount of hypoxanthine or guanine that can be "salvaged." Consequently, the level of PRPP is increased because PRPP is not consumed via the salvage enzyme. The increased PRPP levels lead to increased PRPP amidotransferase activity. Second, the lack of salvage of hypoxanthine or guanine leads to decreased levels of IMP and GMP, which in turn act as feedback regulators of the PRPP amidotransferase step.

In both of these conditions, the common feature of the defect that leads to the overproduction of purine nucleotides is that the intracellular concentration of PRPP is elevated. These defined defects fully support the conclusion that the PRPP amidotransferase step is the rate-controlling step in purine nucleotide synthesis.

Further support comes from clinical cases of secondary gout (in consequence of another metabolic defect). A deficiency in glucose 6-phosphatase (glycogen storage disease, type I; von Gierke's) leads to increased purine nucleotide synthesis de novo. The lack of conversion of glucose 6-phosphate to glucose leads to increased hexose monophosphate shunt activity. The increased utilization of glucose 6-phosphate via the shunt results in increased ribose 5-phosphate levels and consequently increased PRPP levels. An elevation of glutathione reductase activity has also been correlated with increased uric acid levels. Glutathione reductase generates NADP$^+$, which is required to drive the first two reactions of the hexose monophosphate shunt. This will also lead to increased PRPP levels as discussed above.

These latter two examples show quite clearly that a defect in one pathway can cause major problems in a metabolic pathway that is seemingly unrelated and points to the critical nature of the interrelationships among various pathways.

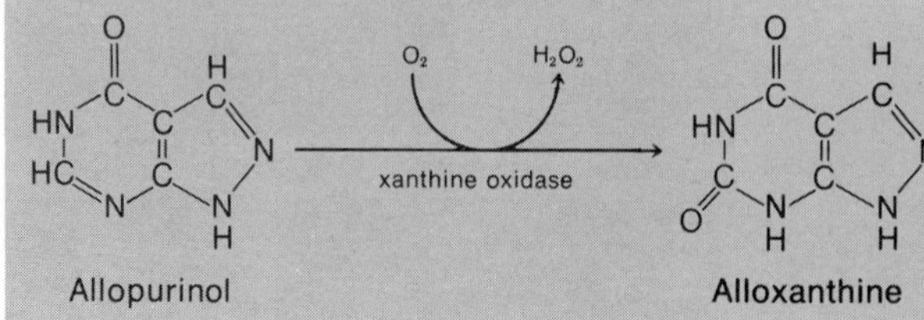

Allopurinol $\xrightarrow[\text{xanthine oxidase}]{O_2 \quad H_2O_2}$ **Alloxanthine**

generated as a product. Uric acid is the end product of purine

Hypoxanthine —xanthine oxidase (O_2 → H_2O_2)→ Xanthine —xanthine oxidase (O_2 → H_2O_2)→ Uric acid

nucleotide catabolism and is excreted in the urine. There are clinical disorders in which the serum level of uric acid is markedly elevated (hyperuricemia) and which can lead to the deposit of sodium urate crystals. This condition is known as gout (Clin. Corr. 13.3). Allopurinol is an inhibitor of xanthine oxidase; this is discussed in detail as a clinical drug in Clin. Corr. 13.3.

13.5 METABOLISM OF PYRIMIDINE NUCLEOTIDES

De Novo Synthesis

The pyrimidine ring is synthesized de novo in mammalian cells utilizing amino acids as carbon and nitrogen donors and CO_2 as a carbon donor.

The numbering system for the pyrimidine ring is shown in Figure 13.18 along with the sources for the various carbon and nitrogen atoms indicated.

As in the case of the de novo synthesis of purine nucleotides, amino acids also play an important role in the de novo synthesis of pyrimidine nucleotides. The reactions leading to the synthesis of pyrimidine nucleotides in mammalian cells are as follows.

1. *Formation of Carbamoyl Phosphate* (C-2 and N-3 of pyrimidine ring)

Glutamine + CO_2 + 2ATP → (glutamate; 2ADP, P_i) Carbamoyl phosphate (NH_2–C(=O)–OPO_3^{2-})

The enzyme that catalyzes this reaction is cytosolic carbamoyl phosphate synthetase II. It is distinct from carbamoyl phosphate synthetase I, a mitochondrial enzyme involved in the urea cycle.

2. *Addition of Aspartate* (N-1, C-4, C-5, and C-6 of pyrimidine ring)

Carbamoyl phosphate + Aspartate (COO^-–CH_2–HC–NH_3^+–COO^-) → (P_i) N-Carbamoylaspartate

This reaction is catalyzed by the enzyme aspartate carbamoyltransferase. The mammalian enzyme is not allosterically regulated, although this could be considered to be the committed step for pyrimidine synthesis.

3. *Ring Closure to Form Pyrimidine Ring*

This reaction is catalyzed by the enzyme dihydroorotase.

Probably the major treatment for primary gout involves the use of the drug allopurinol. The overall effect of allopurinol treatment is to lower the uric acid levels in vivo. It is generally reported that allopurinol is an inhibitor of xanthine oxidase. However, allopurinol is oxidized by xanthine oxidase to Alloxanthine and this product (Alloxanthine) binds tightly to the reduced form of xanthine oxidase. The dissociation constant for the binding of Alloxanthine to reduced xanthine oxidase is about 0.5 nM, which makes Alloxanthine a very effective inhibitor of xanthine oxidase.

The inhibition of xanthine oxidase by Alloxanthine decreases the formation of uric acid, while increasing the levels of hypoxanthine and xanthine excreted. This benefits the patient, since the amount of the purine degradative products will be distributed among three compounds instead of just uric acid. Hypoxanthine and xanthine are more soluble than uric acid, so that the total amount (hypoxanthine, xanthine, and uric acid) that will be soluble is increased by allopurinol treatment.

In "overproducers" who do not have a deficiency in HGPRTase, allopurinol treatment not only lowers the formation of uric acid with an increase in the excretion of hypoxanthine and xanthine, but also decreases the overall production of purine nucleotides via the de novo pathway. The metabolic basis for this appears to be due to the increased salvage of hypoxanthine and xanthine, which requires the consumption of PRPP and the subsequent formation of IMP and XMP, which can block de novo synthesis at the PRPP amidotransferase step.

N-1, C-4, C-5, and C-6 from aspartate
C-2 from CO_2
N-3 from amide N of glutamine

Figure 13.18
Sources of carbon and nitrogen atoms in pyrimidines.

N-Carbamoylaspartate → Dihydroorotate

4. *Oxidation of Dihydroorotate*

Dihydroorotate —(NAD^+ → $NADH + H^+$)→ Orotate

Dihydroorotate dehydrogenase, a flavoprotein, catalyzes the reaction in which orotic acid is formed.

5. *Addition of the Ribose 5-Phosphate* (formation of N-riboside bond) The enzyme catalyzing the reaction in which the first pyrimidine nucleotide is formed is orotate phosphoribosyltransferase. PRPP is the ribose-5-phosphate donor.

Orotate + PRPP —(→ pyrophosphate)→ Orotidine 5′-mono phosphate (OMP)

6. *Decarboxylation of OMP*

Orotidine 5′ mono phosphate (OMP) —(→ CO_2)→ Uridine 5′-mono-phosphate (UMP)

OMP-decarboxylase catalyzes this reaction. The absence of both or either of these last two enzyme activities leads to a condition termed orotic aciduria (Clin. Corr. 13.4).

CLIN. CORR. 13.4 OROTIC ACIDURIA

In a clinical condition called hereditary orotic aciduria, characterized by retarded growth and severe anemia, high levels of orotic acid are excreted. The biochemical basis for this increased production of orotic acid is the absence of either or both of the enzymes orotate phosphoribosyltransferase and OMP-decarboxylase. While this is a relatively rare disease, the understanding of the metabolic basis for it has led not only to a successful treatment, but also has confirmed the site of regulation of pyrimidine nucleotide synthesis.

When these patients are fed either cytidine or uridine there is not only a marked improvement in the hematologic manifestation but also a decrease in orotic acid formation. The biochemical basis for this can be explained as follows. Uridine or cytidine after conversion to the nucleotide by the cell bypasses the block at orotate phosphoribosyltransferase/OMP-decarboxylase to provide the rapidly growing cells, such as the erythropoietic cells, with the pyrimidine nucleotides required for RNA and DNA synthesis. In addition, the intracellular formation of UTP from these nucleosides acts as a feedback inhibitor of carbamoyl phosphate synthetase II to shut down the synthesis of orotic acid.

Although the de novo pathway for UMP synthesis requires six enzyme activities for the six steps, the activities are found on only three gene products. Carbamoyl phosphate synthetase, aspartate carbamoyl transferase, and dihydroorotase activities (pyr 1-3) are present on the same polypeptide chain (mol wt 200,000); dihydroorotate dehydrogenase activity is on a separate protein; and orotate phosphoribosyltransferase and OMP-decarboxylase activities (pyr 5,6) are on the same polypeptide (mol wt 51,000). The multifunctional enzymes (pyr 1-3 and pyr 5,6) are found in the cytosol, whereas dihydroorotate dehydrogenase is a mitochondrial enzyme. As a result of the channeling of intermediates through these enzyme systems, essentially none of the metabolites between the first step and the last step is found in the intracellular pool of the cells.

By these reactions the pyrimidine nucleotide UMP is synthesized. The formation of cytidine nucleotides proceeds from the uridine nucleotide but at the triphosphate level rather than at the monophosphate level. This reaction for the synthesis of this pyrimidine is as follows:

ATP → ADP + P_i; glutamine → glutamate

Uridine 5′-triphosphate (UTP) → Cytidine 5′-triphosphate (CTP)

CTP synthetase catalyzes this reaction. The enzyme does not have an absolute requirement for GTP, but concentrations as low as 0.2 mM GTP stimulate CTP synthetase activity 5- to 10-fold.

As indicated, carbamoyl phosphate is synthesized in the cytosol by a specific carbamoyl phosphate synthetase II. This is the only source of carbamoyl phosphate for pyrimidine synthesis in extrahepatic tissue. As is discussed in more detail in the section on regulation of pyrimidine nucleotide synthesis, carbamoyl phosphate synthetase II is the regulated enzyme of the pyrimidine nucleotide de novo pathway. However, it is quite clear that in liver, carbamoyl phosphate synthetase I can provide cytosolic carbamoyl phosphate. Under stressed physiological conditions in which there is excessive ammonia, the liver, through carbamoyl phosphate synthetase I (a component of the urea cycle) can utilize this pathway to detoxify the NH_3 by forming carbamoyl phosphate in the mitochondria. This carbamoyl phosphate passes into the cytosol and becomes a substrate for aspartate carbamoyltransferase. High levels of orotic acid have been observed to be excreted as a result of ammonia toxicity in man.

The pathway for de novo synthesis of pyrimidine nucleotides differs in two major respects (on a comparative basis) from the de novo pathway for purine nucleotides. First, in purine nucleotide synthesis the N-glycosidic bond is formed in the first committed step of the pathway. In the pyrimidine pathway, the pyrimidine ring is formed first, and then the sugar phosphate is added. Second, all the enzymes of the purine nucleotide pathway are in the cytosol. For pyrimidine synthesis one of the enzymes, dihydrooratate dehydrogenase, is found in the mitochondria, whereas the other five activities (present on only two proteins) are located in the cytosol.

Regulation of Pyrimidine Nucleotide Formation

Unlike in bacterial cells, the regulation of mammalian pyrimidine nucleotide synthesis does not occur at the aspartate carbamoyltransferase step. The regulation of the pyrimidine nucleotide synthesis in mammalian cells occurs at the level of carbamoyl phosphate synthetase II (the cytosolic enzyme), which is inhibited by UTP.

The next level of regulation of pyrimidine nucleotide synthesis is at the level of OMP-decarboxylase. UMP and, to a lesser extent, CMP are inhibitors of OMP-decarboxylase but not orotate phosphoribosyltransferase. However, since these two enzymes comprise the components of a complex, orotate phosphoribosyltransferase cannot continue to function, since there is no place to transfer the product. Under the conditions in which OMP-decarboxylase is inhibited, orotate, *not* OMP accumulates.

In a clinical condition called orotic aciduria excessive amounts of orotic acid are produced. This is caused by deficiencies in either orotate phosphoribosyltransferase or OMP decarboxylase or both enzymes (Clin. Corr. 13.4).

Another possible regulatory site is the CTP synthetase step. This enzyme shows a hyperbolic curve for a plot of velocity (v) vs [UTP]. However, in the presence of CTP, the plot of v vs [UTP] curve becomes sigmoidal. In this way the activity of CTP synthetase is depressed, preventing all of the UTP from being converted to CTP.

Salvage Pathways for Pyrimidines

Pyrimidines can be "salvaged" by conversion to the nucleotide level by reactions involving pyrimidine phosphoribosyltransferase.

The general reaction is

$$\text{Pyrimidine} + \text{PRPP} \longrightarrow \text{pyrimidine nucleoside monophosphate} + \text{pyrophosphate}$$

The enzyme from human erythrocytes has been purified and can utilize orotate, uracil, and thymine as substrates. Cytosine is not a substrate.

It should be mentioned that as uracil becomes available to the cell, competing reactions can occur. Uracil can be degraded to β-alanine or can be salvaged. Normal liver, when presented with uracil, will readily degrade it, whereas regenerating liver would convert the uracil to UMP. This is the result of the availability of PRPP, the enzyme levels, and in general the metabolic state of the animal.

The pyrimidine nucleoside kinases can also be thought of as "salvage" enzymes. The net effect of the kinase reaction is to divert the pyrimidine nucleoside from the degradative pathway to the pyrimidine nucleotide level for cellular utilization.

13.6 DEOXYRIBONUCLEOTIDE FORMATION

As indicated earlier in the chapter, the concentrations of deoxyribonucleotides are extremely low in the "resting" cell. Only at the time of DNA replication (S phase) does the deoxyribonucleotide pool increase to support the required DNA synthesis.

Reduction of Ribonucleotides

Deoxyribonucleotides are formed by the direct reduction of the 2′ position of the corresponding ribonucleotides. The reaction is strongly regulated not only by allosteric effectors (both activators and inhibitors) but also by drastic changes in the level of the enzyme catalyzing the formation of deoxyribonucleotides. This reaction occurs at the level of the nucleoside diphosphates. The general reaction can be summarized as follows:

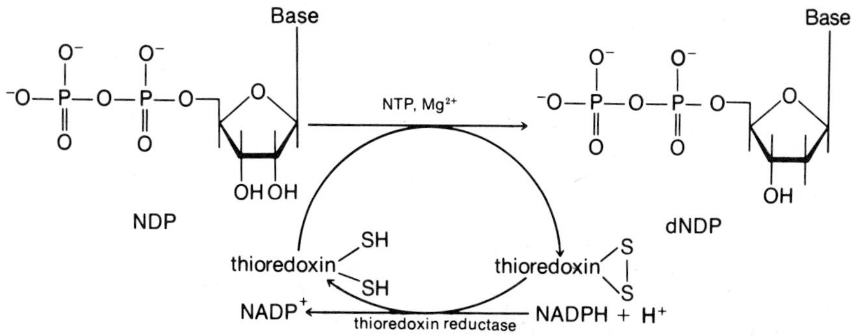

The enzyme catalyzing the formation of 2′-deoxyribonucleotides is nucleoside diphosphate reductase (ribonucleotide reductase). The reduction of a specific NDP requires a specific NTP as a positive effector of the enzyme. This reaction is also subject to regulation by other NTPs, which can serve as negative effectors. The specificity of the effectors for the various substrates is summarized in Table 13.3.

The mammalian ribonucleotide reductase enzyme consists of two nonidentical subunits neither of which alone has enzymatic activity. One of the subunits contains the effector binding site or sites, while the other subunit contains the nonheme iron. For the mammalian system it has not been completely resolved whether there is one enzyme for all four substrates, or whether there are separate enzymes or binding sites for each of the substrates. dATP is a potent inhibitor of the reduction of all four nucleoside diphosphate substrates. This fact provides the biochemical basis for the toxicity of deoxyadenosine for a variety of mammalian cells.

Thioredoxin is a small molecular weight protein (12,000 daltons), which is oxidized during the reduction of the 2′-hydroxyl group of the ribose moiety. To complete the catalytic cycle, reduced thioredoxin is regenerated by thioredoxin reductase (a flavoprotein) and NADPH.

The importance of this enzyme for DNA replication cannot be overemphasized. Ribonucleotide reductase is uniquely responsible for catalyzing the reactions in which the deoxynucleoside triphosphates are generated for DNA replication. It appears that in controlling the deoxyribonucleotide levels in the cell, at least two approaches are utilized by the cell. These are (1) the actual concentration of reductase in the cells and (2) the very strict allosteric regulation of enzyme activity by the nucleoside triphosphates.

TABLE 13.3 Effectors of Ribonucleotide Reductase Activity

Substrate	*Major Positive Effector*	*Major Negative Effector*
CDP	ATP	dATP, dGTP, dTTP
UDP	ATP	dATP, dGTP
ADP	dGTP	dATP, ATP
GDP	dTTP	dATP

5,10-methylene H_4folate

H_2folate

Deoxyuridine 5′-monophosphate (dUMP)

Thymidine 5′-monophosphate (TMP)

Figure 13.19
Synthesis of deoxythymidine nucleotide.

Figure 13.20
5,10-Methylene H_4folate.

Synthesis of Deoxythymidine Nucleotide

Deoxythymidylate is formed in a unique reaction. The enzyme thymidylate synthetase catalyzes this reaction in which a one carbon unit is not only transferred but also reduced to a methyl group (Figure 13.19). In the process 5,10-methylene H_4folate acts both as the one carbon transfer agent and as the reducing agent. As a result dihydrofolate is generated (See Figure 13.20.)

The dUMP for this reaction can arise from two different pathways. In one reaction dCMP is deaminated directly to dUMP by dCMP deaminase, whereas in the other pathway UDP is reduced to dUDP which is then converted to dUMP (Figure 13.21). From labeling studies it appears that for most cells the major source of dUMP is the step in which dCMP is deaminated to dUMP. dCMP deaminase activity in the mammalian cell far exceeds the level of UDP reductase activity.

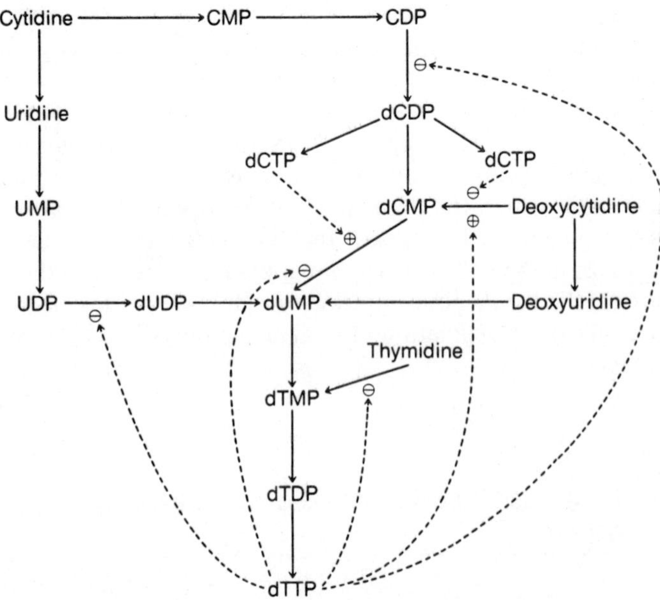

Figure 13.21
Pyrimidine interconversions.
The solid arrows represent enzyme-catalyzed reactions. The dashed lines represent the regulation that certain nucleotides have on the various steps (⊕, activation; ⊖, inhibition).

Pyrimidine Nucleotide Interconversions

Because of the cell's critical need for the deoxyribonucleotides to support DNA synthesis and hence cell replication, a series of enzymes is present that is responsible for the interconversion of the deoxyribopyrimidine nucleotides. These interconversions are summarized in Figure 13.21. The enzymes catalyzing these interconversions are strongly regulated. The control of dCDP and dUDP formation through the ribonucleotide reductase reaction has already been discussed. dCMP deaminase is activated by dCTP and inhibited by dTTP. The inhibition by dTTP can be overcome by increasing concentrations of dCTP. A v vs [dCMP] curve is sigmoidal for dCMP deaminase. The presence of dCTP shifts the curve to a hyperbolic activity curve. dTTP is also an inhibitor of thymidine kinase, an enzyme which is important in "salvaging" thymidine and deoxyuridine for the cell. It is seen that dTTP serves as a negative effector of several conversions, and this is the basis for the toxicity of high concentrations of thymidine for mammalian cells.

Degradation of Pyrimidine Nucleotides

The turnover of nucleic acids results in the release of pyrimidine nucleotides. These nucleotides are also in a steady state, and there is constant synthesis and degradation. The degradation of pyrimidine nucleotides follows the pathways shown in Figure 13.22. In these degradative pathways, the nucleotides are first converted to the nucleosides and then to the free base uracil or thymine. The conversion of the pyrimidine nucleotides to nucleosides is catalyzed by various nonspecific phosphatases. Deoxycytidylate deaminase has preference for dCMP but can utilize CMP as substrate. Cytidine and deoxycytidine are deaminated to uridine and deoxyuridine, respectively, by the nucleoside deaminase. Uridine phosphorylase catalyzes the phosphorolysis not only of uridine, but also deoxyuridine and deoxythymidine.

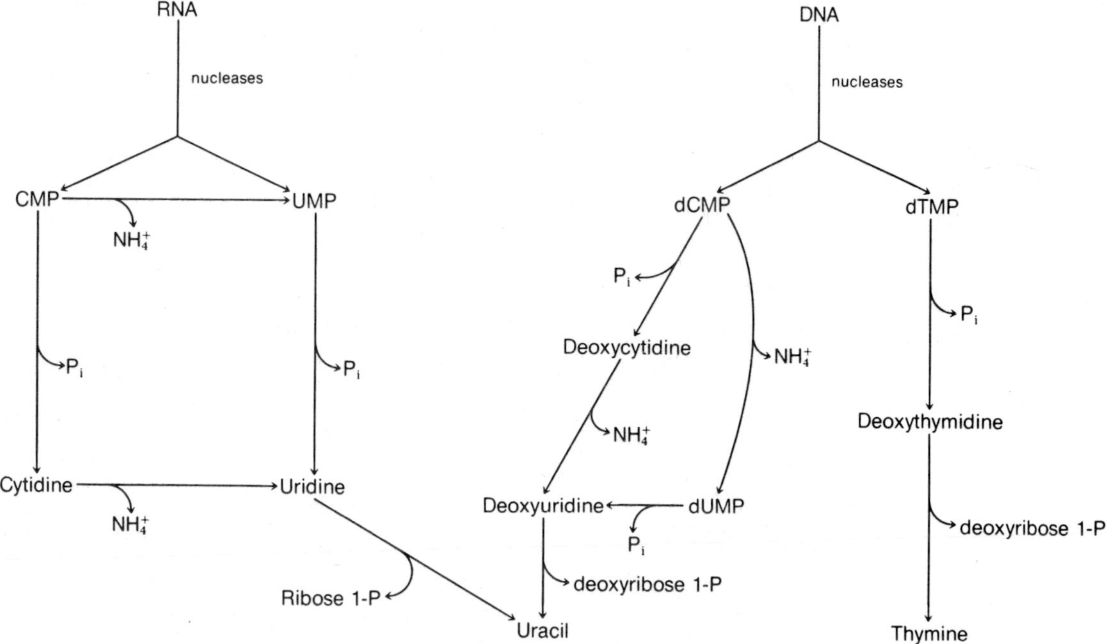

Figure 13.22
Pathways for the degradation of pyrimidine nucleotides.

It is important to note that mammalian cells have a very specific dUTPase in high concentration, which catalyzes the reaction:

$$\text{dUTP} \longrightarrow \text{dUMP} + \text{pyrophosphate}$$

It is critical for normal DNA replication that dUTP not be present in the nucleotide pool. As is shown in Chapter 17, dUTP can very effectively replace dTTP as a substrate in the DNA polymerase reaction.

The K_m of dUTP for DNA polymerase is approximately 10 μM, whereas the K_m of dUTP for dUTPase is only 1 μM. Consequently, the K_m strongly favors the dUTPase reaction. dUTPase, therefore, serves to prevent misincorporation of dUMP into the DNA, which would have other major consequences.

Uracil and thymine are then further degraded by analogous reactions, although the final products are different as shown in Figure 13.23. Uracil is degraded further to β-alanine, NH_4^+ and CO_2.

The three enzymes (dihydropyrimidine dehydrogenase, dihydropyrimidinase, and ureidopropionase) required to catalyze the degradation of uracil or thymine to their respective products are separate proteins. These enzymes, however, appear to utilize uracil or thymine and their intermediates equally well as substrates.

None of these products is unique to uracil degradation, and consequently the turnover of cytosine or uracil nucleotides cannot be estimated from the end products of this pathway. Thymine degradation proceeds to β-aminoisobutyric acid, NH_4^+ and CO_2. β-Aminoisobutyric acid is excreted in the urine of man and originates from the degradation of thymine. Increased levels of β-aminoisobutyric acid are excreted after the administration of diets rich in DNA or in cancer patients undergoing chemotherapy or radiation therapy in which large numbers of cells are being killed. It is possible therefore to estimate the turnover of DNA or thymidine nucleotides by the measurement of β-aminoisobutyrate production.

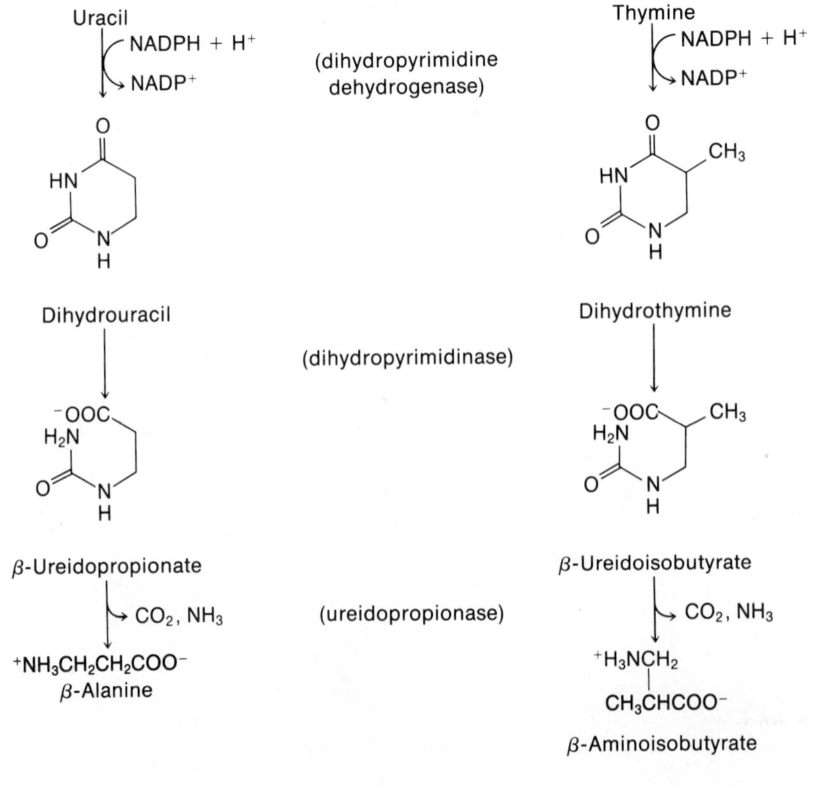

Uracil → ($NADPH + H^+$ → $NADP^+$) (dihydropyrimidine dehydrogenase) → Dihydrouracil → (dihydropyrimidinase) → β-Ureidopropionate → (ureidopropionase; CO_2, NH_3) → $^+NH_3CH_2CH_2COO^-$ β-Alanine

Thymine → ($NADPH + H^+$ → $NADP^+$) → Dihydrothymine → β-Ureidoisobutyrate → (CO_2, NH_3) → $^+H_3NCH_2CH(CH_3)COO^-$ β-Aminoisobutyrate

Figure 13.23
Degradation of uracil and thymine.

13.7 NUCLEOSIDE AND NUCLEOTIDE KINASES

As shown in the pathways for the de novo synthesis of purine and pyrimidine nucleotides, the nucleotide is synthesized as the monophosphate. Most, if not all, reactions in which the nucleotides function require that these nucleotides be at the di- or triphosphate level.

There are specific kinases to "salvage" nucleosides to nucleotides and to convert the nucleoside monophosphates to the di- and triphosphates.

Examples of these are as follows.

1. *Uridine/Cytidine Kinase*

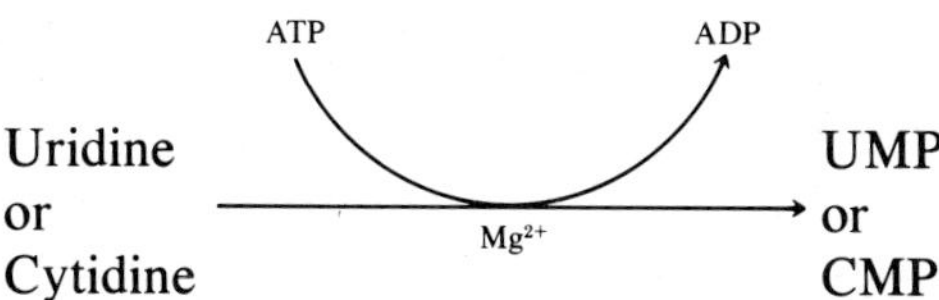

The enzyme is specific for uridine or cytidine as substrates. UTP and CTP are inhibitors of these reactions.

2. *Deoxycytidine Kinase*

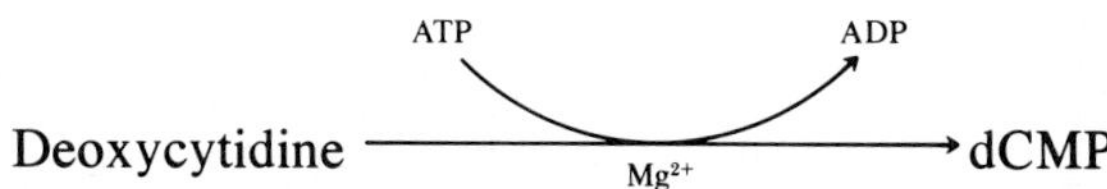

Deoxycytidine is the preferred substrate for this kinase. Cytidine, uridine, and thymidine are not substrates. However, deoxyadenosine and deoxyguanosine, although poor substrates, are phosphorylated by this enzyme. dCTP is a potent inhibitor of this reaction. dTTP will reverse the inhibition caused by dCTP.

3. *Pyrimidine Nucleoside Monophosphate Kinase*

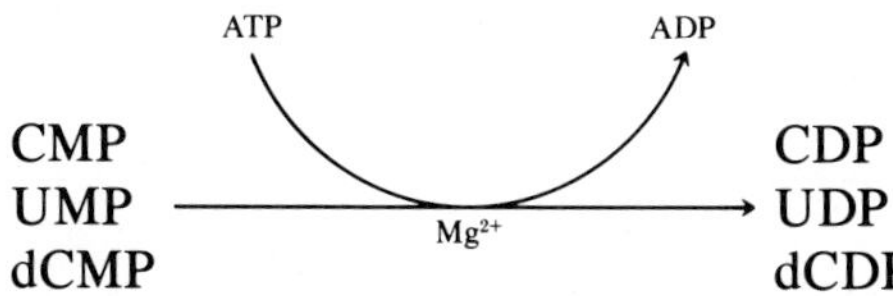

This enzyme shows specificity for the substrates CMP, UMP, and dCMP. dUMP is not a substrate.

4. *Thymidine Kinase*

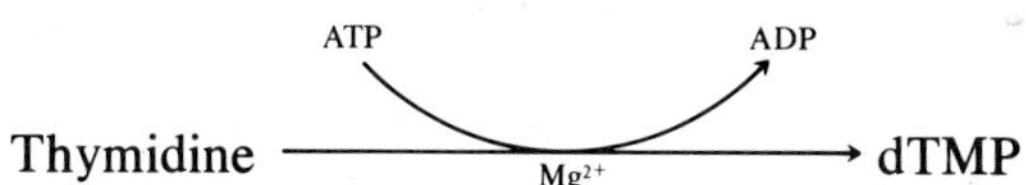

This kinase is specific for thymidine and deoxyuridine as substrates. This enzyme is elevated in rapidly growing tissues.

5. *Thymidylate Kinase*

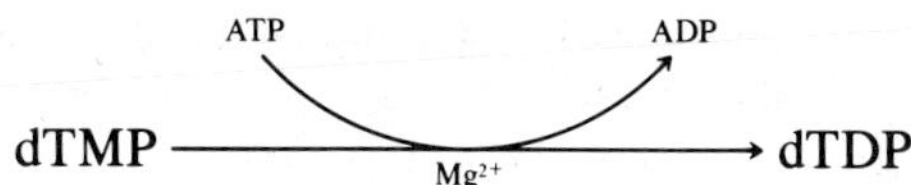

This enzyme is specific for dTMP.

6. *Adenosine Kinase*

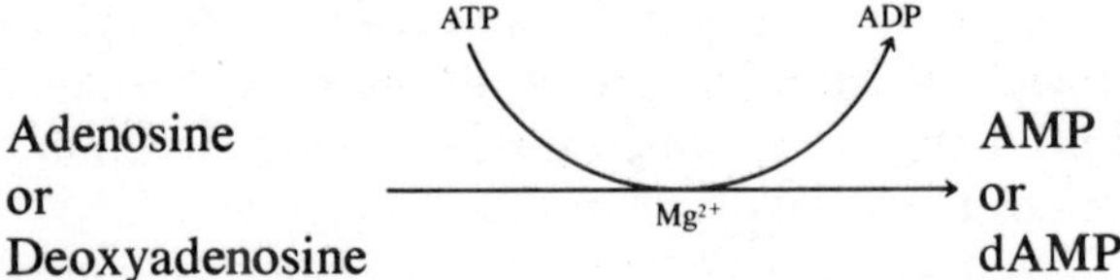

This enzyme is specific for adenosine or deoxyadenosine. Inosine is not a substrate for this enzyme.

7. *AMP Kinase*

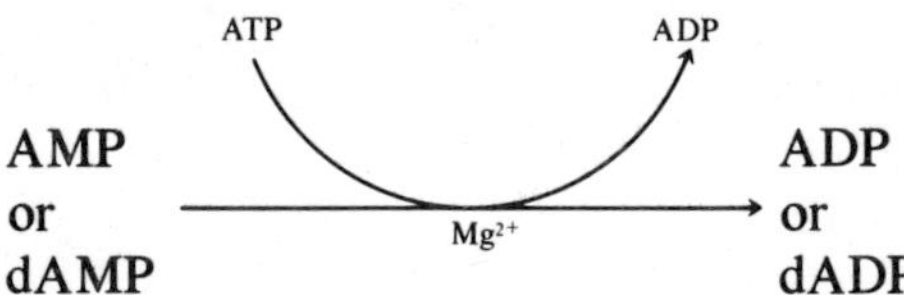

AMP kinase shows specificity for AMP. Although dAMP can be utilized as substrate, AMP is phosphorylated at a rate 10 times higher than that of dAMP.

8. *GMP Kinase*

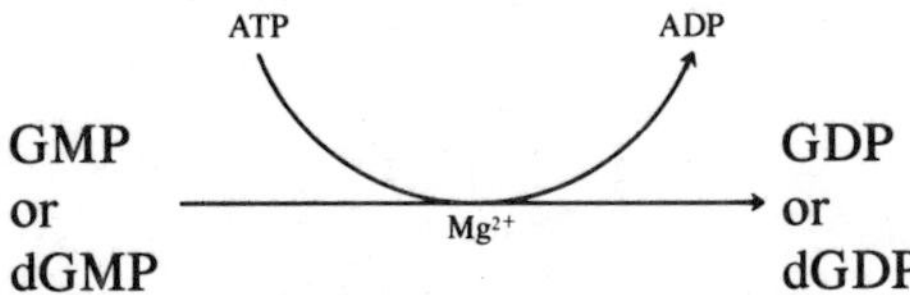

GMP kinase is distinct from AMP kinase showing specificity for GMP and dGMP.

9. *Nucleoside Diphosphokinase*

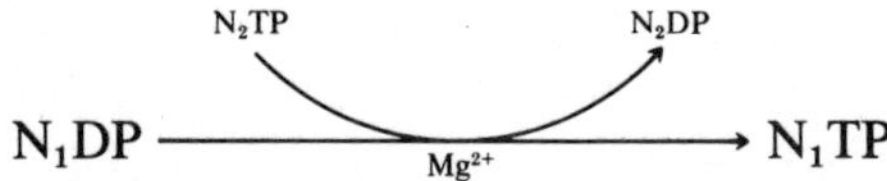

Mammalian cells contain an enzyme, nucleoside diphosphokinase, which is not specific for either the phosphate donor or phosphate acceptor in terms of either the purine or pyrimidine base or the sugar moiety. Since in most cells the concentration of ATP is the highest of the triphosphates and most easily regenerated via glycolysis or oxidative phosphorylation, ATP is probably the major phosphate donor for these reactions in intact cells.

13.8 NUCLEOTIDE METABOLIZING ENZYMES AS A FUNCTION OF THE CELL CYCLE AND RATE OF CELL DIVISION

For normal cell division to occur, essentially all components of the cell must double. The events which lead from the formation of a daughter cell through mitosis, to the completion of the processes required for its own division into two daughter cells is described by the term cell cycle. The periods of the cell cycle have been termed mitosis (M), gap 1 (G_1), synthesis (S), and gap 2 (G_2). The total period of the cell cycle will vary with the particular cell type. In many mammalian cells the periods of M, S, and G_2 are relatively constant, while G_1 varies widely causing cells to have long or short doubling times. The cell cycle is represented in Figure 13.24. In preparation for DNA replication during the S phase of the cell cycle, there is considerable synthesis of enzymes involved in nucleotide metabolism during the G_1 phase, especially during late G_1/early S. RNA and protein synthesis occur continuously, although at varying rates during G_1, S, and G_2 phases, although DNA replication occurs only during S.

The strict regulation of nucleotide synthesis requires that certain mechanisms must be available to the cell to meet the requirements for the ribonucleotides and deoxyribonucleotide precursors at the time of increased RNA synthesis and DNA replication. To meet these needs, the cell responds by increasing the levels of specific enzymes involved with nucleotide formation during very specific periods of the cell cycle.

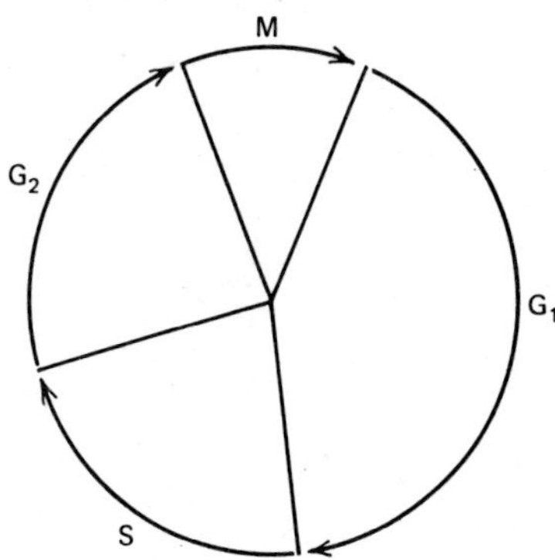

Figure 13.24
Diagrammatic representation of the cell cycle.
For a mammalian cell with a doubling time of 24 h, G_1 would last~12 h; S, 7 h; G_2, 4 h; and M, 1 h.

The enzymes involved in purine nucleotide synthesis and interconversions which are elevated during the S phase of the cell cycle are PRPP amidotransferase and IMP dehydrogenase. Adenylosuccinate synthetase and adenylosuccinase do not appear to increase.

The enzymes involved in pyrimidine nucleotide synthesis which are elevated during the S phase of the cell cycle include aspartate carbamoyltransferase, dihydroorotase, dihydroorotate dehydrogenase, orotate phosphoribosyltransferase, and CTP synthetase.

Many of the enzymes involved in the synthesis and interconversions of deoxyribonucleotides are also elevated during the S phase of the cell cycle. Included in these enzymes are ribonucleotide reductase, thymidine kinase, dCMP deaminase, thymidylate synthetase, and TMP kinase. The importance of the increases in these enzymes during S phase to DNA replication is worthy of further discussion with a specific example. As has been discussed previously, the deoxyribonucleotide pool is extremely small in "resting" cells (less than 1 μM). As a result of the increase in ribonucleotide reductase the levels of deoxyribonucleotides reach levels of 10–20 μM during DNA synthesis. However, this concentration would sustain DNA synthesis for only minutes, while complete DNA replication would require hours. Consequently, the levels of ribonucleotide reductase activity not only must increase but must be sustained during S phase in order to provide the necessary substrates for DNA synthesis.

If one looks at a population of cells as a whole (i.e., tissue) rather than as individual cells going through the cell cycle, it is observed that rapidly growing tissues such as regenerating liver, embryonic tissues, intestinal mucosal cells, and erythropoietic cells are geared toward DNA replication and RNA synthesis. These tissues will show elevated levels of those key enzymes involved with purine and pyrimidine nucleotide synthesis and interconversions and complementary decreases in the levels of the enzymes that catalyze reactions in which these precursors are degraded. Of course these changes reflect the proportion of the cells in that tissue which are in S phase.

As a result of Weber's molecular correlation concept, an understanding has evolved of the biochemical changes that occur to satisfy the proliferative life-style of tumor cells. It has been determined that gene expression has been altered to result not only in quantitative changes in enzyme levels but also qualitative changes (isozyme shifts). As a result of careful experimental study, utilizing a series of liver and kidney tumors of varying growth rates, it has been possible to categorize these biochemical changes as (1) transformation-linked (meaning that all tumors regardless of growth rate show certain increased and certain decreased enzyme levels); (2) progression-linked (alterations that correlate with the growth rate of the tumor); and (3) coincidental alterations (not connected to the malignant state). As very limited examples, the levels of ribonucleotide reductase, thymidylate synthetase, and IMP dehydrogenase increase as a function of the tumor growth rate. PRPP amidotransferase, UDP kinase, and uridine kinase are examples of enzymes whose activity is increased in all tumors, whether they are slow-growing or the most rapidly growing tumors.

It is important to point out that, while certain of the enzymes are increased in both fast-growing normal tissue (e.g., embryonic and regenerating) and tumors, the total quantitative and qualitative patterns for normal and tumor tissue can easily be distinguished.

13.9 NUCLEOTIDE COENZYME SYNTHESIS

The coenzymes NAD, FAD, and coenzyme A are synthesized in mammalian cells provided that there is a suitable source of the vitamin component (niacin, riboflavin, and pantothenic acid).

NAD Synthesis

NAD is synthesized in mammalian cells by at least three different pathways shown in Figure 13.25.

When the exogenous source of tryptophan is in excess of the amount required for protein synthesis, tryptophan can be metabolized to quinolinic acid. Quinolinic acid is utilized in a reaction with PRPP to form nicotinate mononucleotide. The enzyme catalyzing this reaction, quinolinate phosphoribosyltransferase, is found only in the liver and kidney. Therefore, this pathway is specific for these tissues.

Nicotinate reacts with PRPP to form nicotinate mononucleotide. The enzyme catalyzing this reaction is nicotinate phosphoribosyltransferase and is widely distributed in various tissues. Nicotinate mononucleotide reacts with ATP to yield nicotinate adenine dinucleotide. The enzyme catalyzing this reaction is NAD-pyrophosphorylase and is widely distributed in various tissues. Nicotinate adenine dinucleotide reacts with glutamine with the hydrolysis of ATP to yield nicotinamide adenine dinucleotide (NAD). NAD-synthetase catalyzes this reaction.

Nicotinamide reacts with PRPP to give nicotinamide mononucleotide. The enzyme which catalyzes this reaction is nicotinamide phosphoribosyltransferase. The enzyme is specific for nicotinamide and is entirely distinct from nicotinate phosphoribosyltransferase. Nicotinamide mononucleotide reacts with ATP to yield NAD. The enzyme that catalyzes this reaction is the same enzyme that catalyzes the reaction between nicotinate mononucleotide and ATP.

In nucleated cells NAD-pyrophosphorylase is located exclusively in the nucleus of the cell. However, the erythrocyte is entirely capable of synthesizing NAD from nicotinate or nicotinamide.

The intracellular concentration of pyridine nucleotides is maintained at a constant level, implying a pathway that is tightly regulated. Nicotinamide phosphoribosyltransferase appears to be the regulated enzyme in NAD synthesis. NMN, NAD^+, $NADP^+$, and NADPH are strong inhibitors of nicotinamide mononucleotide synthesis. ATP stimulates this reaction, although it is not an absolute requirement. ATP lowers the K_m for PRPP 10-fold and nicotinamide 100-fold, while increasing the V_{max}. The resulting K_m for nicotinamide approaches the intracellular concentration of this compound. Nicotinate phosphoribosyltransferase does not appear to be regulated by the end products of the pathway utilizing nicotinate as the substrate.

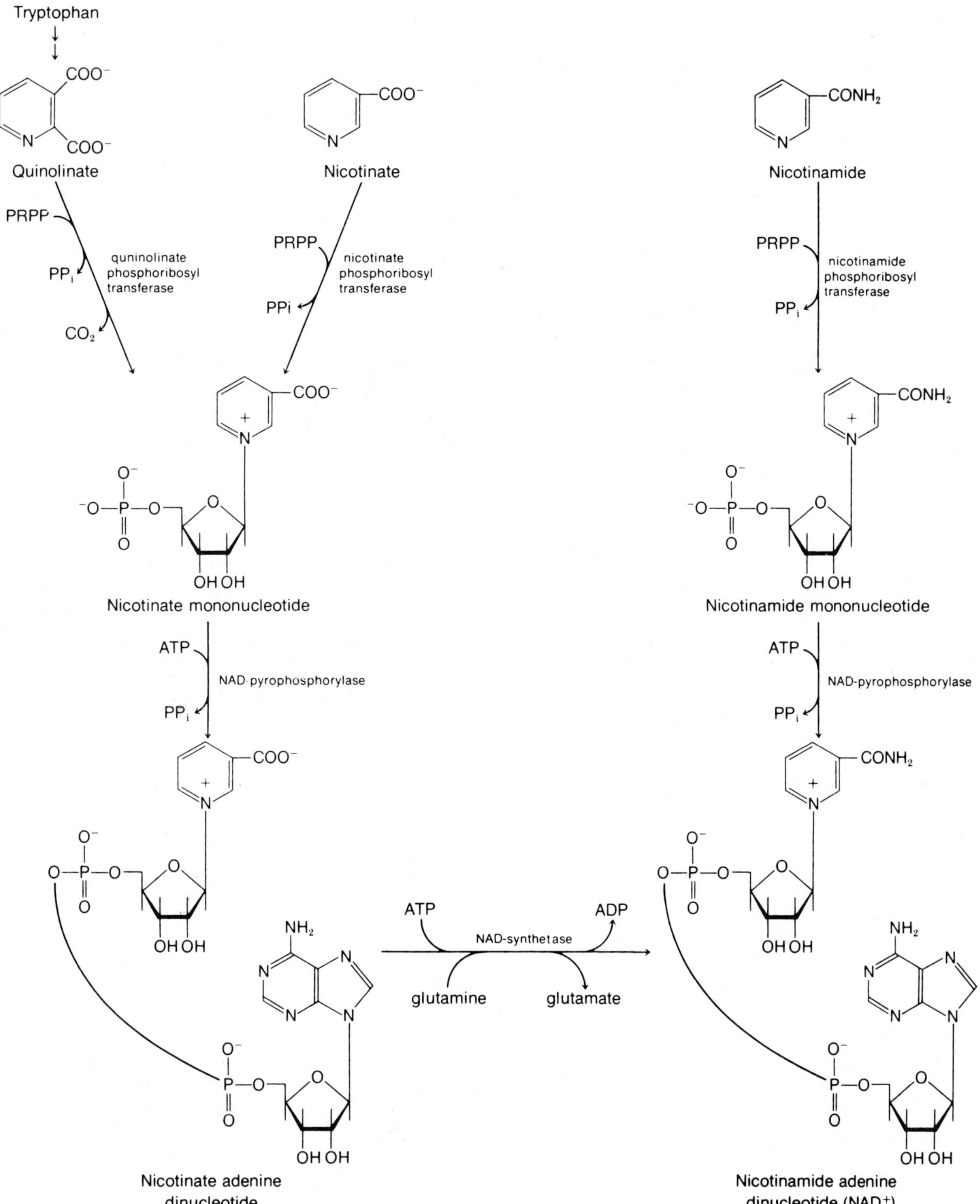

Figure 13.25
Pathways for NAD⁺ synthesis.

NAD is synthesized also by the mitochondria. However, the relative importance of the mitochondrial pathway to the pathways just described is not known.

There is considerable turnover of NAD in cells. NAD is consumed by two distinct reactions. In one reaction NAD-glycohydrolase catalyzes the conversion of NAD^+ to nicotinamide and adenosine diphosphoribose. This enzyme is located in microsomes. In the second, poly(ADP-ribose) synthetase catalyzes the polymerization of the ADP-ribose moiety of NAD^+ onto nuclear proteins with the release of nicotinamide. This enzyme is found exclusively in the nucleus and the level of activity is highest during G_2 and lowest during S phase of the cell cycle. A possible role of polyADP ribosylation in cell regulation has been proposed.

There is no known enzymatic conversion of nicotinate directly to nicotinamide. Since only the liver and kidney can utilize tryptophan for NAD synthesis, the NAD glycohydrolase reaction can be thought to provide nicotinamide to extrahepatic tissues for NAD synthesis. In some tissues nicotinate is a more efficient precursor of NAD than is nicotinamide.

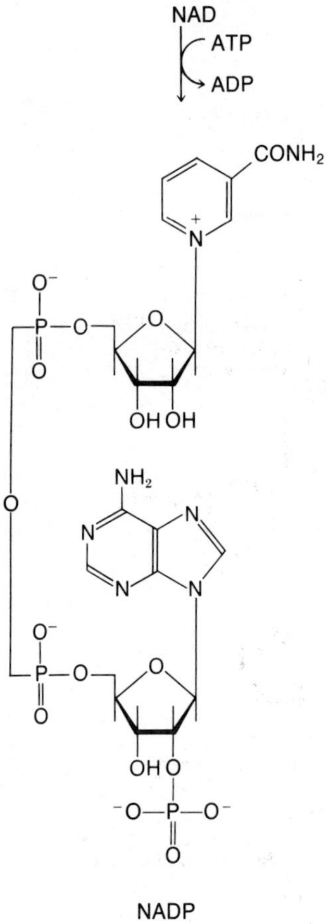

NADP Synthesis

NAD is the immediate precursor of nicotinamide adenine dinucleotide phosphate (NADP). The reaction is catalyzed by NAD-kinase as presented in Figure 13.26. NAD-kinase is found in the cytosol of the cell and NADPH, the reduced form, is a negative effector of this reaction.

NAD
ATP
ADP
$CONH_2$
N
O⁻
P—O
O
O
OH OH
NH_2
N
N
N
N
O
O⁻
P—O
O
O
OH O
⁻O—P—O⁻
O
NADP

Figure 13.26
Synthesis of $NADP^+$.

Degradation of NAD

NAD can be degraded by two pathways, although the product is nicotinamide by either pathway as shown in Figure 13.27. Nicotinamide is excreted in the urine of human beings as the N-methyl derivatives of nicotinamide and 2-pyridone-5-carboxamide. An excess of these products gives urine a bright yellow fluorescent color. It should be noted that the excretion of excess nicotinamide requires the consumption of *S*-adenosylmethionine.

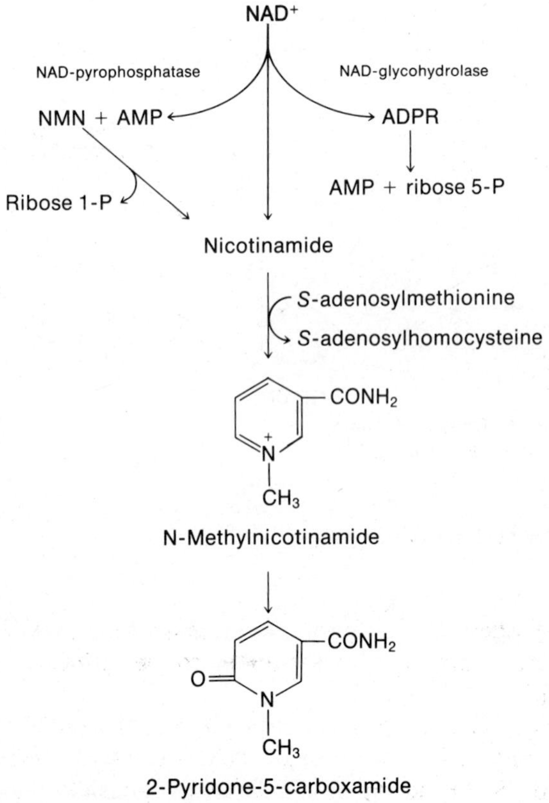

NAD^+
NAD-pyrophosphatase
NAD-glycohydrolase
NMN + AMP
ADPR
AMP + ribose 5-P
Ribose 1-P
Nicotinamide
S-adenosylmethionine
S-adenosylhomocysteine
$CONH_2$
N
CH_3
N-Methylnicotinamide
$CONH_2$
O
N
CH_3
2-Pyridone-5-carboxamide

Figure 13.27
Degradation of NAD^+.

FAD Synthesis

FAD is synthesized in a two-step reaction. Riboflavin is required from an exogenous source because human beings cannot synthesize the isoalloxazine moiety. Riboflavin is phosphorylated by ATP in a reaction catalyzed by riboflavin kinase (flavokinase) to give riboflavin phosphate. Mg^{2+} ions are the preferred divalent cation required for this kinase. GTP will partially replace ATP as the phosphorylating agent. Flavokinase is found in the cytosol of the cell of a variety of tissues such as liver, kidney, brain, spleen, and heart. Many references are made to this compound as flavin mononucleotide (FMN), although this is not a true nucleotide. FMN then reacts with ATP to yield FAD in a reaction catalyzed by FAD pyrophosphorylase. This enzyme shows an absolute requirement for ATP and Mg^{2+} ions are required. FAD-pyrophosphorylase activity has been reported to be located mainly in mitochondria. The pathway for FAD synthesis is summarized in Figure 13.28.

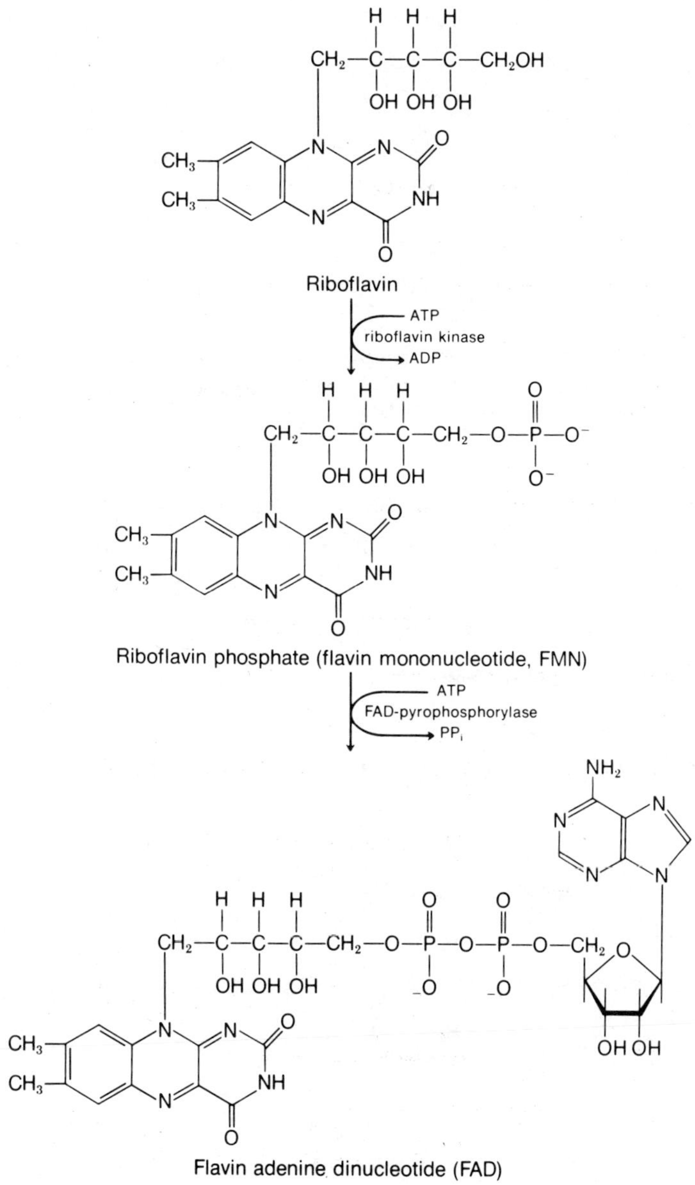

Riboflavin

ATP
riboflavin kinase
ADP

Riboflavin phosphate (flavin mononucleotide, FMN)

ATP
FAD-pyrophosphorylase
PP_i

Flavin adenine dinucleotide (FAD)

Figure 13.28
Synthesis of FAD.

Coenzyme A Synthesis

Coenzyme A is synthesized in humans beings by a series of reactions, which has an absolute requirement for an exogenous source of pantothenic acid. The pathway for the synthesis of coenzyme A in mammalian cells is shown in Figure 13.29. Pantothenic acid is phosphorylated by ATP to give 4-phosphopantothenic acid. In the next reaction, cysteine is added to provide the —SH group, which will ultimately be the ''business end'' of CoA. The α-carboxyl group of cysteine is then removed from 4-phosphopantothenoyl-L-cysteine to yield 4-phosphopantotheine. In a pyrophosphorylase reaction, ATP is then added to give dephosphocoenzyme A. The dephosphocoenzyme A is then phosphorylated at the 3′ position of the adenosine moiety to give coenzyme A. The enzymes, dephosphoCoA pyrophosphorylase and dephospho-CoA kinase, appear to exist in nature as a bifunctional enzyme complex. The enzymes copurify, with the ratio of the two activities remaining constant through many steps.

This pathway appears to be regulated at the phosphopantothenoylcysteine decarboxylase step. The product of the reaction catalyzed by this

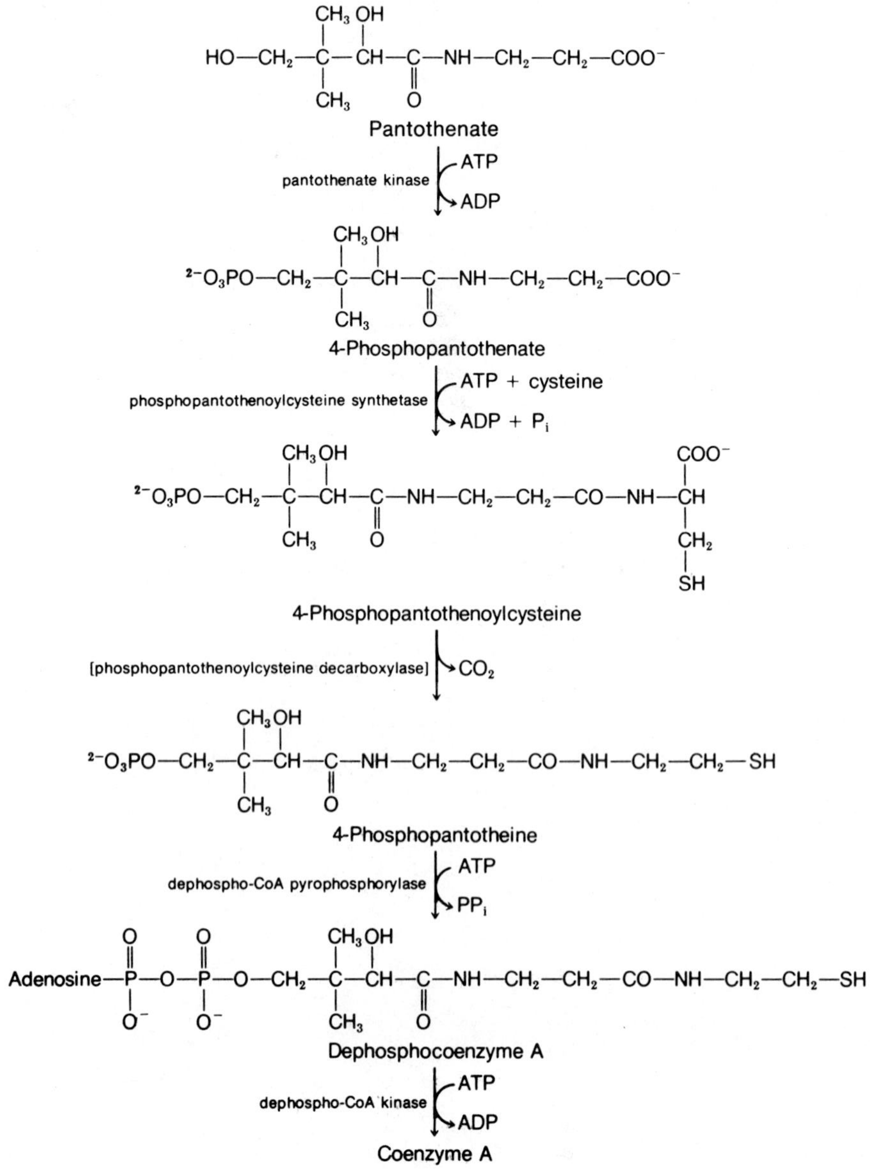

$HO{-}CH_2{-}C(CH_3)_2{-}CH(OH){-}C(=O){-}NH{-}CH_2{-}CH_2{-}COO^-$

Pantothenate

pantothenate kinase: ATP → ADP

${}^{2-}O_3PO{-}CH_2{-}C(CH_3)_2{-}CH(OH){-}C(=O){-}NH{-}CH_2{-}CH_2{-}COO^-$

4-Phosphopantothenate

phosphopantothenoylcysteine synthetase: ATP + cysteine → ADP + P_i

${}^{2-}O_3PO{-}CH_2{-}C(CH_3)_2{-}CH(OH){-}C(=O){-}NH{-}CH_2{-}CH_2{-}CO{-}NH{-}CH(COO^-){-}CH_2{-}SH$

4-Phosphopantothenoylcysteine

[phosphopantothenoylcysteine decarboxylase]: → CO_2

${}^{2-}O_3PO{-}CH_2{-}C(CH_3)_2{-}CH(OH){-}C(=O){-}NH{-}CH_2{-}CH_2{-}CO{-}NH{-}CH_2{-}CH_2{-}SH$

4-Phosphopantotheine

dephospho-CoA pyrophosphorylase: ATP → PP_i

$\text{Adenosine}{-}P(=O)(O^-){-}O{-}P(=O)(O^-){-}O{-}CH_2{-}C(CH_3)_2{-}CH(OH){-}C(=O){-}NH{-}CH_2{-}CH_2{-}CO{-}NH{-}CH_2{-}CH_2{-}SH$

Dephosphocoenzyme A

dephospho-CoA kinase: ATP → ADP

Coenzyme A

Figure 13.29
Synthesis of coenzyme A.

enzyme, 4-phosphopantotheine, is a relatively strong competitive inhibitor of this reaction.

13.10 SYNTHESIS AND UTILIZATION OF 5-PHOSPHORIBOSYL 1-PYROPHOSPHATE

The intracellular concentration of 5-phosphoribosyl 1-pyrophosphate (PRPP) plays an important role in regulating several important pathways. The synthesis and utilization of PRPP by the cell will determine the steady-state concentration of PRPP and hence the metabolic pathways that compete for PRPP.

PRPP is synthesized in the cell in the reaction catalyzed by 5-phosphoribose pyrophosphokinase (PRPP synthetase) utilizing α-ribose 5-phosphate and ATP. The reaction requires Mg^{2+} ions (Figure 13.30). The ribose 5-phosphate used in this reaction is generated from glucose 6-phosphate metabolism via the hexose monophosphate shunt or from ribose 1-phosphate (generated by phosphorolysis of nucleotides) via a phosphoribomutase reaction.

As expected for such a critical reaction, the formation of PRPP is regulated. The enzyme has an absolute requirement for inorganic phosphate ions. The velocity vs $[P_i]$ curve for PRPP-synthetase is sigmoidal and at the concentration of P_i normally found in the cell, the activity is markedly depressed because of the sigmoidal rather than hyperbolic curve.

The importance of the sigmoidal curve relative to inorganic phosphate in regulating PRPP formation, and hence the overproduction of uric acid, was shown in a "gouty" individual who had a marked increase in uric acid formation. Analysis of the patient's red cells for PRPP-synthetase revealed that this patient had a mutant form of the enzyme, which showed a hyperbolic v vs $[P_i]$ curve. At the intracellular concentration of P_i, the production of PRPP was greatly elevated. Presumably the increased levels of PRPP in other tissues lead to the overproduction of purine nucleotides and consequently of uric acid.

The levels of PRPP synthetase are elevated in cells undergoing rapid cell division and decrease to basal levels in cells that have reached confluence. The activity of PRPP synthetase is inhibited by nucleoside di- and triphosphates. ADP is the most potent inhibitor of PRPP synthetase activity in human placenta, rat liver, and mouse tumor cells. ADP is a competitive inhibitor with respect to ATP. Its inhibition constant (K_i) is less than the intracellular concentration of ADP, indicating that it can serve as a physiological effector of PRPP synthetase. 2,3-Bisphosphoglycerate is also an inhibitor of PRPP synthetase, and this is of importance in red cell metabolism dealing with the salvage pathways.

Factors that lead to increased flux of glucose 6-phosphate through the hexose monophosphate shunt pathway can result in increased intracellular levels of PRPP. Pyrroline-5-carboxylate, an intermediate in the inter-

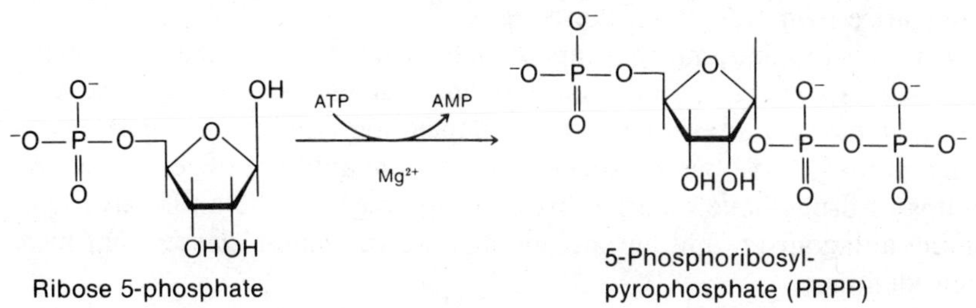

Ribose 5-phosphate

ATP AMP

Mg^{2+}

5-Phosphoribosyl-pyrophosphate (PRPP)

Figure 13.30
Synthesis of PRPP.

conversions of ornithine, glutamic acid, and proline, stimulates the hexose monophosphate shunt via the generation of $NADP^+$ in the pyrroline-5-carboxylate reductase-catalyzed reaction. This leads to elevated concentrations of PRPP.

PRPP formed in the cells is a required substrate for many key metabolic reactions depending on the cell type. These reactions and the pathways in which they are involved are as follows:

1. De novo purine nucleotide synthesis
   a. PRPP + glutamine $\longrightarrow$ 5-phosphoribosylamine + glutamate + PP_i
2. "Salvage" of purine bases
   a. PRPP + hypoxanthine (guanine) $\longrightarrow$ IMP (GMP) + PP_i
   b. PRPP + adenine $\longrightarrow$ AMP + PP_i
3. De novo pyrimidine nucleotide synthesis
   a. PRPP + orotate $\longrightarrow$ OMP + PP_i
4. "Salvage" of pyrimidine bases
   a. PRPP + uracil $\longrightarrow$ UMP + PP_i
5. NAD^+ synthesis
   a. PRPP + nicotinate $\longrightarrow$ nicotinate mononucleotide + PP_i
   b. PRPP + nicotinamide $\longrightarrow$ nicotinamide mononucleotide
   c. PRPP + quinolinate $\longrightarrow$ nicotinate mononucleotide + PP_i

In the red cell, the major reactions in which PRPP is consumed are reactions 2a and 2b and 5a and 5b. In a rapidly growing tumor cell, PRPP would be consumed by all five pathways. The direction in which PRPP would be consumed would depend on several factors, including the relative K_ms of the competing enzymes for PRPP, the availability of the second substrate, and the concentration of the effector for the particular reaction. For example, if the concentration of AMP in the cell were high, adenine phosphoribosyltransferase would be inhibited and PRPP would be consumed via hypoxanthine-guanine phosphoribosyltransferase provided hypoxanthine or guanine were present in the cell and the IMP and GMP concentrations were low.

13.11 COMPOUNDS THAT INTERFERE WITH PURINE AND PYRIMIDINE NUCLEOTIDE METABOLISM

As has been discussed, the de novo synthesis of purine and pyrimidine nucleotides is critical to normal cell replication and function. The regulation of these pathways has also been shown to be important, since disease states arise from defects in these steps.

With this in mind, many compounds have been synthesized or isolated from plants, bacteria, fungi, and so on, that are directed at relatively specific metabolic sites involved with nucleotide synthesis or interconversion and these have been screened as potential antitumor agents. Several groups of drugs have emerged from these screens. The antifolates, glutamine antagonists, and antimetabolites are the major classes that have been identified.

Glutamine Antagonists

Many reactions occur in mammalian cells in which glutamine serves as an amino donor. These amidation reactions are very critical for the synthesis of the purine ring de novo (N-3 and N-9), in the conversion of IMP to GMP, in the conversion of UTP to CTP and in the conversion of nicotinate adenine dinucleotide to NAD.

Compounds that inhibit these reactions are referred to as glutamine antagonists. Azaserine (*O*-diazoacetyl-L-serine) and 6-diazo-5-oxo-L-norleucine (DON) (Figure 13.31), which were first isolated from cultures of *Streptomyces,* are very effective inhibitors of glutamine utilization. Since azaserine and DON inactivate the enzymes irreversibly, glutamine will not reverse the effects of these two drugs. It would require that many types of metabolites such as guanine, cytidine, hypoxanthine (or adenine), and nicotinamide would have to be utilized to overcome the sites blocked by these glutamine antagonists. As expected from the multiple sites of inhibition at key steps, the glutamine antagonists are extremely toxic.

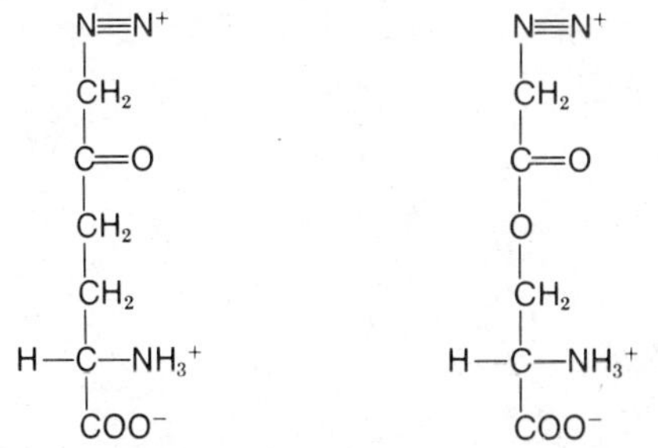

6-Diazo-5-oxo-L-norleucine: $N{\equiv}N^+{-}CH_2{-}C(=O){-}CH_2{-}CH_2{-}CH(NH_3^+){-}COO^-$

Azaserine: $N{\equiv}N^+{-}CH_2{-}C(=O){-}O{-}CH_2{-}CH(NH_3^+){-}COO^-$

Figure 13.31
Structures of glutamine antagonists.

Antifolates

Antifolates are compounds that interfere with the formation of tetrahydrofolate and dihydrofolate from folic acid (Figure 13.32). The decreased intracellular concentration of H_4folate results in the impaired formation of purine nucleotides via the de novo pathway and thymidylate via thymidylate synthetase.

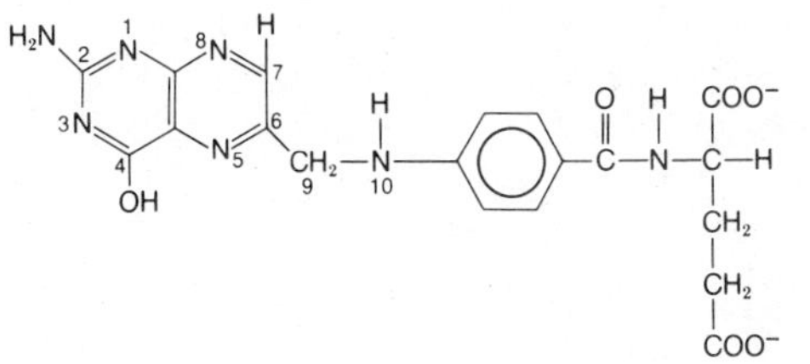

Methotrexate (MTX) (Figure 13.33) is an example of an antifolate, which is currently in use in the treatment of various forms of cancer in human beings.

Methotrexate is a close structural analog of folic acid. The differences in structure are at C-4, where an amino group replaces a hydroxyl group, and at N-10, where a methyl group replaces a hydrogen atom. The mode of action of MTX is quite specific. It inhibits dihydrofolate reductase with a K_i in the range of 10^{-10} M and thus inhibits the reactions:

$$\text{Folate} \xrightarrow[\text{MTX}]{\uparrow} H_2\text{folate} \xrightarrow[\text{MTX}]{\uparrow} H_4\text{folate}$$

Figure 13.32
Folic acid.

When tumor cells in culture are treated with MTX, the cells die. However, the cytotoxic effects of MTX can be overcome by 5-formyl-H_4folate (citrovorum factor), thymidine and hypoxanthine. Some cell lines will respond to thymidine alone, but other cell lines will require both thymidine and hypoxanthine to rescue the cells from MTX treatment. The reversal by these compounds indicates that the direct effect of MTX on the H_4folate levels leads indirectly to the lack of thymidine and purine nucleotide formation.

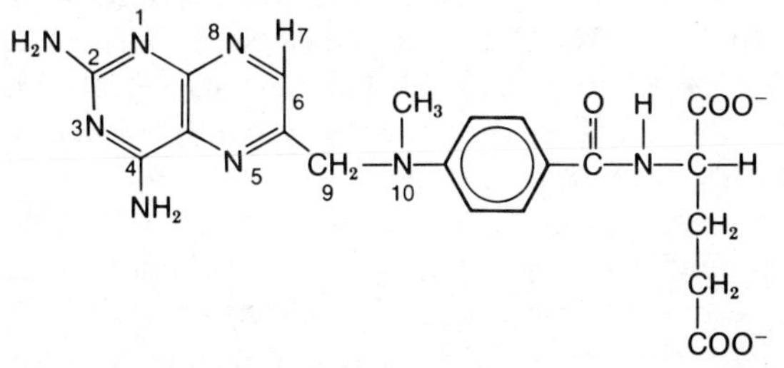

Methotrexate

Figure 13.33
Methotrexate.

Clinically, high-dose MTX therapy followed by "rescue" treatment with citrovorum factor has been successful in many situations in "curing" several forms of human cancer. Clinical trials utilizing "thymidine rescue" following high-dose MTX treatment are currently in progress.

Antimetabolites

Antimetabolites are generally structural analogs of the purine and pyrimidine bases or nucleosides which interfere with rather specific metabolic sites. Only a few of these will be discussed to show: (1) the importance of the de novo pathways to normal cell metabolism; (2) that the regulation of these pathways occurs in vivo; (3) the concept of the requirement for metabolic activation of the drugs; and (4) that the inactivation of these compounds can greatly influence their usefulness.

6-Mercaptopurine (*6-MP*) is a useful antitumor drug in man. The cytotoxic activity of this agent is related to the formation of 6-mercaptopurine ribonucleotide by the tumor cell. The tumor cell utilizing PRPP and HGPRTase converts 6-MP to its nucleotide form. 6-Mercaptopurine ribonucleoside 5′-monophosphate accumulates in the cell and serves as a negative effector of PRPP-amidotransferase, the committed step in the de novo pathway. In addition, this nucleotide acts as an inhibitor of the conversion of IMP to GMP at the IMP-dehydrogenase step and IMP to AMP at the adenylosuccinate synthetase step. Since 6-mercaptopurine is a substrate for xanthine oxidase and is oxidized to 6-thiouric acid, allopurinol can be administered to inhibit the degradation of 6-MP and potentiate the antitumor properties of 6-mercaptopurine.

Adenine arabinoside (araA) has been used as an antitumor drug and as an antiviral agent in man. The effective metabolite of araA is its triphosphate, araATP. AraA, therefore, must be metabolized through araAMP and araADP. AraATP is an inhibitor of DNA-polymerase.

Cytosine arabinoside (araC) is currently in use in the treatment of several forms of human cancer. AraC, of itself, is not active. It must be converted to the nucleoside triphosphate (araCTP) to exert its cytotoxic effects. AraCTP is a potent competitive inhibitor with respect to dCTP of DNA polymerase. It has been shown that the effectiveness of araC as an antitumor drug correlates with the absolute level of araCTP achieved in the cell and the half-life of araCTP in the tumor cells.

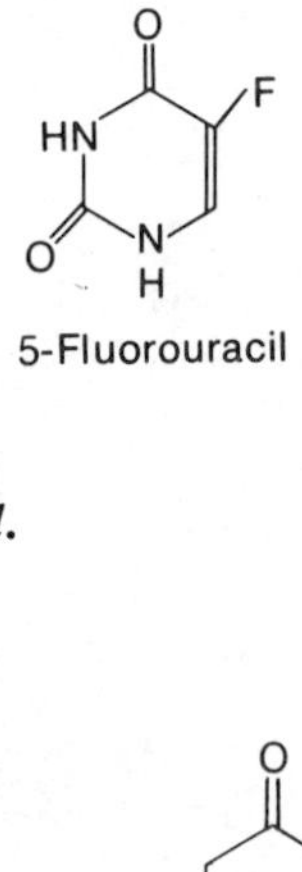

5-Fluorouracil

Figure 13.34
5-Fluorouracil.

5-Fluorouracil (FUra) (Figure 13.34) is a pyrimidine analog, which of itself has no biological activity. FUra must be activated via the salvage pathways to the nucleotide level to exert its cytotoxic effects. FUra has been found to be converted to F-deoxyuridylate (FdUMP) and to FUTP. These two metabolites are the cytotoxic agents. FdUMP is a very potent and specific inhibitor of thymidylate synthetase. In the presence of thymidylate synthetase, H_4folate and FdUMP, a ternary complex is formed, which results in the covalent binding of FdUMP to thymidylate synthetase. FdUMP, therefore, causes what amounts to a "thymineless death" for the cells. The second metabolite FUTP, is incorporated into the various RNA species of the cell. This incorporation of FUra into RNA has as its major consequence the inhibition of the maturation of 45S precursor ribosomal RNA into the 28S and 18S species. The realization that the incorporation of FUra into RNA has serious effects on normal RNA metabolism and is a factor in the cytotoxicity of this agent has been recognized only recently.

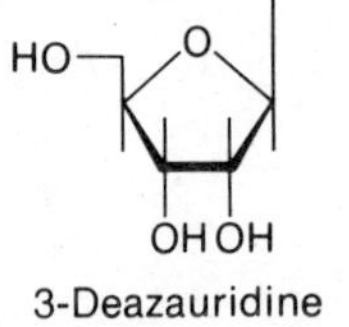

3-Deazauridine

Figure 13.35
3-Deazauridine.

The importance of these two metabolic sites to the cytotoxic action of this agent is confirmed by the fact that thymidine and uridine are both required to rescue completely the FUra-treated cells.

Deazauridine (Figure 13.35) is a pyrimidine nucleoside analog that is an effective antitumor agent. This nucleoside must be "activated" to the di- and triphosphates to exert its cytotoxic effects. 3-Deazauridine-5′-triphosphate, an active form of this antimetabolite, is a potent inhibitor of CTP synthetase. DeazaUTP is a competitive inhibitor with respect to UTP of this enzyme. DeazaUDP is an inhibitor of ribonucleotide reductase activity. The net result of the inhibition at these sites is that the cells become deficient in cytidine and deoxycytidine nucleotides, causing inhibition of both RNA and DNA synthesis. In addition, CTP is required for normal phospholipid metabolism (e.g., CDP-choline).

Hydroxyurea

Tumor cells treated with hydroxyurea show a specific inhibition of DNA synthesis with little or no effect on RNA or protein synthesis. The metabolic basis for the specific inhibition of DNA synthesis is that hydroxyurea is an inhibitor of ribonucleotide reductase, blocking the reduction of all four nucleoside diphosphate substrates (CDP, UDP, GDP, and ADP). Toxicity to this drug results from the depletion of the deoxyribonucleoside triphosphates required for DNA replication.

Although this drug is specific for the inhibition of ribonucleotide reductase, its clinical use is limited due to its rapid rate of clearance and the high drug concentration that is required for effective inhibition of this enzyme.

The above drugs are only a few of the many compounds in clinical use, but serve as examples in which the knowledge of basic biochemical pathways and mechanisms lead to the generation of effective drugs. Another important point to be made regarding many of the antimetabolites used as drugs is that they must be activated to the nucleotide level to exert their cytotoxic effects. To become activated many of the cells' normal enzymes are utilized. The activation of 6-mercaptopurine to 6-mercaptopurine ribonucleotide requires the presence of HGPRTase activity and PRPP. Resistance of tumor cells to 6-mercaptopurine could result from a deficiency of HGPRTase activity. As the cells' enzymes can be utilized to convert the antimetabolites to the active form, the cells' enzymes can also be utilized to inactivate these drugs. Two examples of this aspect are the drugs adenine arabinoside (araA) and cytosine arabinoside (araC). For araA to be an effective antiviral drug it must be converted to its nucleotide form, araATP. This takes place through the reactions

$$\text{araA} \longrightarrow \text{araAMP} \longrightarrow \text{araADP} \longrightarrow \text{araATP}$$

utilizing the cells' constitutive enzymes. On the other hand, araA is a substrate for adenosine deaminase. The product of this reaction, hypoxanthine arabinoside (araHx), is ineffective as a cytotoxic agent. Consequently, the effectiveness of araA as an antitumor or antiviral agent will depend, at least in part, on the ratio of the drug converted to araAMP and to araHx. Strategies using potent adenosine deaminase inhibitors, deoxycoformycin and erythro-9-(2-hydroxy-3-nonyl)adenine (EHNA), have been developed in an attempt to potentiate the formation of araAMP and inhibit the production of araHx.

The conversion of araC to araCTP depends on the presence of deoxycytidine kinase. Tumor cell resistance to araC could result from either a lack of deoxycytidine kinase or from an excess of deoxycytidine deaminase, which would inactivate araC to the araU derivative. Attempts to potentiate the activity of araC have involved combinations of drugs, which include inhibitors of deoxycytidine deaminase. Tetrahydrouridine is one such potent inhibitor of deoxycytidine deaminase. In fact, utilizing tetrahydrouridine, which itself has no effect on the cell, the concentration of araC required for cytotoxicity is markedly decreased.

These examples show the importance that the cells' enzymes play in drug action.

BIBLIOGRAPHY

Baliga, B. S., and Borek, E. Metabolism of thymine in tumor tissue: The origins of β-aminoisobutyric acid. *Adv. Enzyme Regul.* 13:27, 1975.

Becker, M. A., Raivio, K. O., and Seegmiller, J. E. Synthesis of phosphoribosylpyrophosphate in mammalian cells. *Adv. Enzymol.* 49:281, 1979.

Cory, J. G. Role of ribonucleotide reductase in cell division. *Pharmacol. Ther.* 21:265, 1983.

Henderson, J. F., and Patterson, A. R. P. *Nucleotide metabolism: An introduction.* New York: Academic, 1973.

Jones, M. E. Pyrimidine nucleotide biosynthesis in animals: Genes, enzymes, and regulation of UMP biosynthesis. *Annu. Rev. Biochem.* 49:253, 1980.

Stanbury, J. B., Wyngaarden, J. B., Fredrickson, D. S., Goldstein, J. L., and Brown, M. S. *The metabolic basis of inherited disease,* 5th ed. New York: McGraw-Hill, 1983.

Traut, T. W., and Loechel, S. Pyrimidine catabolism: Individual characterization of the three sequential enzymes with a new assay. *Biochemistry* 23:2533, 1984.

Tremblay, G. C., Crandall, D. E., Knott, C. E., and Alfant, M. Orotic acid biosynthesis in rat liver: Studies on the source of carbamoylphosphate. *Arch. Biochem. Biophys.* 178:264, 1977.

Weber, G. Biochemical strategy of cancer cells and the design of chemotherapy: G. H. A. Clowes Memorial Lecture. *Cancer Res.* 43:3466, 1983.

Wyngaarden, J. B. Regulation of purine biosynthesis and turnover. *Adv. Enzyme Regul.* 14:25, 1976.

QUESTIONS

C. N. ANGSTADT AND J. BAGGOTT

*Q*uestion *T*ypes are described inside front cover.

1. (QT1) Nucleotides serve all of the following roles *except:*
   A. monomeric units of nucleic acids.
   B. physiological mediators.
   C. sources of chemical energy.
   D. structural components of membranes.
   E. structural components of coenzymes.

A. [structure: cytosine, NH_2]

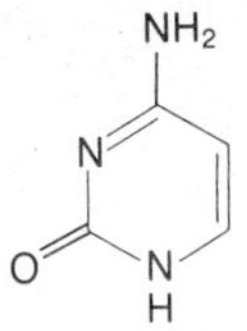

B [structure: adenine, NH_2]

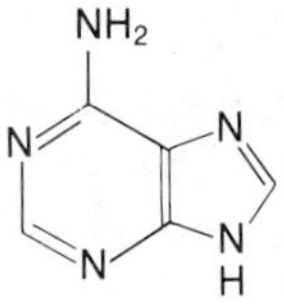

C. [structure: uridine, $HOCH_2$, HO, OH]

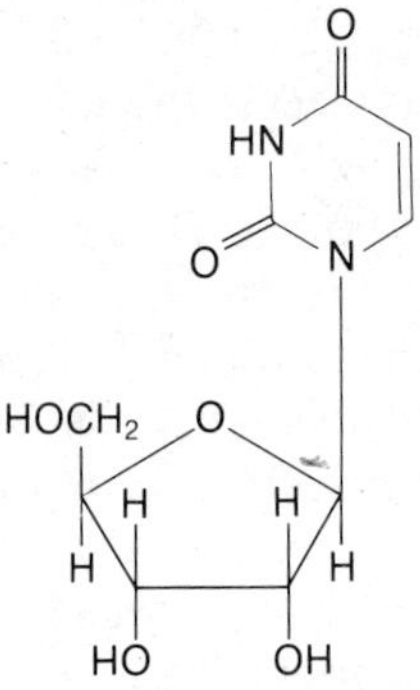

D. [structure: guanosine 5'-monophosphate]

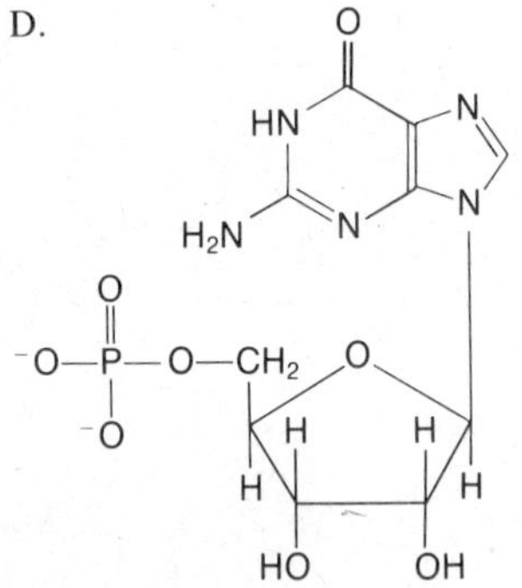

E. [structure: cytidine 5'-monophosphate]

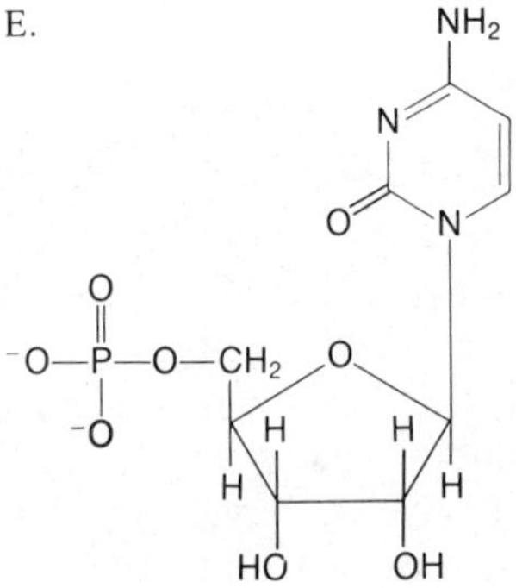

2. (QT5) Adenine
3. (QT5) A pyrimidine nucleoside
4. (QT5) CMP
5. (QT1) The term *modified base* refers to:
   A. a purine or pyrimidine attached to deoxyribose.
   B. a purine or pyrimidine that has been altered, for example, by methylation.
   C. a nitrogen-containing ring other than a purine or pyrimidine that is part of a nucleotide.
   D. a purine or pyrimidine attached to a sugar by an *O*-glycosidic bond.
   E. those purine and pyrimidine bases that can be incorporated into a growing nucleotide chain.

A. De novo synthesis of purine nucleotides — C. Both
B. De novo synthesis of pyrimidine nucleotides — D. Neither

6. (QT4) A source of nitrogen is the amide nitrogen of glutamine.
7. (QT4) PRPP is a substrate of the rate-limiting step.
8. (QT4) The two nucleotides found in RNA are formed in a branched pathway from a common intermediate.
9. (QT4) A free base is formed in the process.

10. (QT2) Which of the following are aspects of the overall regulation of de novo purine nucleotide synthesis?
   1. AMP, GMP, and IMP cause a shift of PRPP amido transferase from a small form to a large form.
   2. PRPP levels in the cell can be severalfold less than the K_m of PRPP amidotransferase for PRPP.
   3. GMP is a competitive inhibitor of IMP dehydrogenase.
   4. UMP is a competitive inhibitor of OMP-decarboxylase.
11. (QT2) The type of enzyme known as a phosphoribosyltransferase is involved in:
   1. salvage of pyrimidine bases.
   2. the de novo synthesis of pyrimidine nucleotides.
   3. salvage of purine bases.
   4. the de novo synthesis of purine nucleotides.
12. (QT1) Uric acid is:
   A. formed from xanthine in the presence of O_2.
   B. a degradation product of cytidine.
   C. deficient in the condition known as gout.
   D. a competitive inhibitor of xanthine oxidase.
   E. oxidized, in humans, before it is excreted in urine.
13. (QT2) In nucleic acid degradation:
   1. there are nucleases that are specific for either DNA or RNA.
   2. nucleotidases convert nucleotides to nucleosides.
   3. the conversion of a nucleoside to a free base is an example of a phosphorolysis.
   4. because of the presence of deaminases, hypoxanthine rather than adenine is formed.
14. (QT1) Deoxyribonucleotides:
   A. cannot be synthesized so they must be supplied preformed in the diet.
   B. are synthesized de novo using dPRPP.
   C. are synthesized from ribonucleotides by an enzyme system involving thioredoxin.
   D. are synthesized from ribonucleotides by nucleotide kinases.
   E. can be formed only by salvaging free bases.
15. (QT2) If a cell were unable to synthesize PRPP, which of the following processes would be likely to be *directly* impaired?
   1. FAD synthesis
   2. NAD synthesis
   3. Coenzyme A synthesis
   4. OMP synthesis
16. (QT2) Which of the following antitumor agents work by impairing de novo purine synthesis?
   1. Azaserine (glutamine antagonist)
   2. 5-Fluorouracil (antimetabolite)
   3. Methotrexate (antifolate)
   4. Hydroxyurea
17. (QT2) Sources of the nicotinamide portion of NAD include:
   1. *N*-Methylnicotinamide.
   2. The vitamin riboflavin.
   3. PRPP.
   4. tryptophan.

Answers

1. D Both cAMP and cGMP are physiological mediators. NAD, FAD, and coenzyme A all contain AMP as part of their structures (p. 490).
2. B Adenine is the free purine. (A is a pyrimidine).
3. C A nucleoside contains a base plus sugar but no phosphate.
4. E CMP is a pyrimidine nucleotide. (D is a purine nucleotide) (p. 492).
5. B Modified bases are formed by alteration of the purine or pyrimidine ring after the parent nucleotide has been incorporated into a nucleic acid (p. 494).
6. C Nitrogens 3 and 9 of purine nucleotides (p. 496) and nitrogen 3 of pyrimidine nucleotides (p. 507) are supplied by glutamine.
7. A The rate-limiting step of purine nucleotide synthesis is the amido transfer between glutamine and PRPP (p. 496). PRPP is used in pyrimidine nucleotide synthesis but only after a pyrimidine has been formed (p. 508).
8. A GMP and AMP are both formed from the first purine nucleotide, IMP, in a branched pathway (Figure 13.13). The pyrimidine nucleotides UMP and CTP are formed in a sequential pathway from orotic acid (p. 508).
9. B There is no free purine base at any point of the pathway, but in pyrimidine nucleotide synthesis, orotic acid is formed. It is only then that the sugar phosphate is added (p. 508).
10. A 1, 2, 3 correct. 1 is the mechanism of inhibition since the large form of the enzyme is inactive (p. 501). 2: PRPP amidotransferase shows sigmoidal kinetics with respect to PRPP so large shifts in concentration of PRPP have the potential for altering velocity (p. 501). 3 plays a major role in controlling the branched pathway of IMP to GMP or AMP (p. 501). 4: OMP is a pyrimidine nucleotide so it would not be expected to be part of the regulation of purine nucleotide synthesis.
11. A 1, 2, 3 correct. Phosphoribosyltransferases are important salvage enzymes for both purines and pyrimidines (pp. 501, 510) and are also part of the synthesis of pyrimidines since OPRT catalyzes the conversion of orotate to OMP (p. 508). In purine nucleotide synthesis, though, the purine ring is built up stepwise on ribose-5-phosphate and not transferred to it (p. 496).
12. A The xanthine oxidase reaction produces uric acid. B, E: Uric acid is an end product of purines, not pyrimidines. C: Gout is characterized by excess uric acid (p. 505).
13. E All four correct. 1: They can also show specificity toward the bases and positions of cleavage. 2: A straight hydrolysis. 3: The product is ribose-1-phosphate rather than the free sugar. 4: AMP deaminase and adenosine deaminase remove the 6-NH_2 as ammonia. The IMP or inosine formed is eventually converted to hypoxanthine (Figure 13.17).
14. C Deoxyribonucleotides are synthesized from the ribonucleoside diphosphates by nucleoside diphosphate reductase that uses thioredoxin as the direct hydrogen-electron donor (p. 511). A, B, E: There is a synthetic mechanism as just described but it is not a de novo pathway. D: Nucleotide kinases are enzymes that add phosphate to a base or nucleotide.
15. C 2, 4 correct. PRPP is a substrate in both of these processes (p. 523). Both FAD and coenzyme A contain AMP, which is supplied as ATP (p. 521).
16. B 1, 3 correct. 1: Glutamine is the source of nitrogens 3 and 9 for the purine ring. 2: 5-Fluorouracil is a pyrimidine analog not a purine analog. 3: Antifolates reduce the concentration of tetrahydrofolate compounds that are necessary for two steps of purine synthesis. 4: Hydroxyurea inhibits the reduction of ribonucleotides to deoxyribonucleotides so is not involved in de novo purine synthesis (p. 525).
17. D Only 4 correct. 1: *N*-Methylnicotinamide is the excretory form of nicotinamide. 2: Nicotinamide comes from the vitamin nicotinic acid (riboflavin-FAD). PRPP supplies the ribose-phosphate not the nicotinamide. 4: Tryptophan can be metabolized, through quinolinic acid, to nicotinate mononucleotide and eventually NAD (p. 518).

14

Metabolic Interrelationships

ROBERT A. HARRIS
DAVID W. CRABB

14.1 OVERVIEW

In this chapter the interdependence of the metabolic processes of the major tissues of the body in physiological situations will be stressed. Not all of the major metabolic pathways and processes of the body operate in

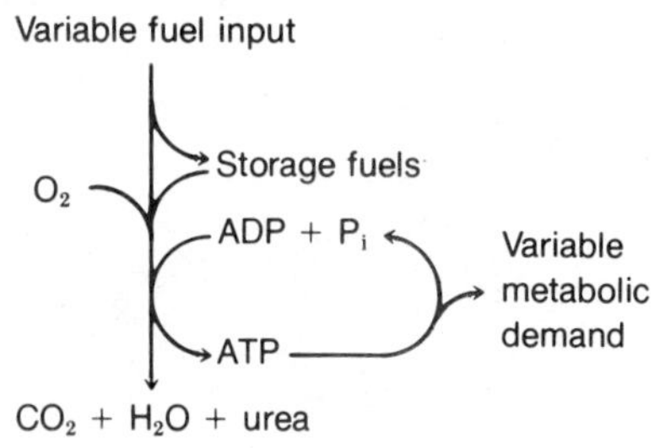

Variable fuel input

O_2 → Storage fuels

ADP + P_i ← Variable metabolic demand

ATP →

$CO_2 + H_2O$ + urea

Figure 14.1
Human beings are able to use a variable fuel input to meet a variable metabolic demand.

every tissue at any given time. Given the nutritional and hormonal status of a patient, we need to be able to say, at least qualitatively, which of the major metabolic pathways of the body are functional and how these pathways relate to one another.

The metabolic processes with which we are concerned in this chapter are glycogenesis, glycogenolysis, gluconeogenesis, glycolysis, fatty acid synthesis, fatty acid oxidation, citric acid cycle activity, ketogenesis, amino acid oxidation, protein synthesis, proteolysis, and urea synthesis. It is important to know: (1) which tissues are most active in these various processes, (2) when these processes are most or least active, and (3) how these processes are controlled and coordinated in different metabolic states.

The best way to gain an understanding of the relationships of the major metabolic pathways to one another is to become familiar with the changes in metabolism that occur during the starve–feed cycle. As shown in Figure 14.1, the starve–feed cycle allows a variable fuel consumption to meet a variable metabolic demand. Starve is a poor choice of words in this case; fast is what we mean, but the phrase fast–feed cycle brings to mind lunch at McDonald's rather than what we are trying to express. Feed refers to the intake of meals (the variable fuel input) after which we store the fuel (in the form of glycogen and fat) to be used to meet our metabolic demand while we fast. Note the participation of an ATP cycle within the starve–feed cycle (Figure 14.1). ATP functions as the energy-transferring agent in the starve–feed cycle, being like money to the cell—the payoff to the cell for going to the trouble of burning fuels to CO_2 and H_2O.

Humans have the capacity to consume food at a rate some 100 times greater than their basal caloric requirements. This allows us to survive from meal to meal without nibbling continuously between meals. We thus store the calories as glycogen and fat and consume them as needed. Unfortunately, an almost unlimited capacity to consume food is matched by an almost unlimited capacity to store it as fat. Obesity, a very common problem in this country, is the consequence of excess consumption (Clin. Corr. 14.1), whereas other forms of malnutrition are more prevalent in developing countries (Clin. Corr. 14.2 and 14.3). Every day represents a series of starve–feed cycles. It balances out quite nicely for most, that is, fuel consumption equals fuel utilization. The regulation of food consumption is complex and not well understood. The tight control needed is indicated by the calculation that eating two extra pats of butter (~100 cal) per day over caloric expenditures results in a 10 lb weight gain per year.

CLIN. CORR. **14.1**
OBESITY

Obesity is the most common nutritional problem in the United States and other affluent countries of the world. It causes a reduction in life span and is a risk factor in the development of diabetes mellitus, hypertension, osteoarthritis, gallstones, and cardiovascular diseases. Obesity is easy to explain—an obese person has eaten more than he or she required. The accumulation of massive amounts of body fat is not otherwise possible. Appreciation of this fact, however, is of little or no consolation to the obese person. For unknown reasons, the neural control of caloric intake to balance energy expenditure is abnormal. Rarely, obesity is secondary to a correctable disorder. Hypothyroidism causes puffiness and some weight gain but is hardly ever the cause of obesity. Cushing's syndrome, the result of increased levels of glucocorticoids, causes fat deposition in the face and trunk, with wasting of the limbs, and glucose intolerance. These effects are due to increased protein breakdown in muscle and conversion of the amino acids to glucose and fat. Less commonly, tumors, vascular accidents, or maldevelopment of the nervous system hunger control centers in the hypothalamus cause obesity. The most common "gland" problem causing obesity is overactivity of the salivary glands, that is, overeating.

In the most common type of obesity, the number of adipocytes of the body does not increase, they just get large as they become engorged with triacylglycerols. If obesity develops before puberty, however, an increase in the number of adipocytes can also occur. In the latter case,

14.2 STARVE–FEED CYCLE

Well-Fed State

Figure 14.2 shows what happens to glucose, amino acids, and fat obtained from food in the gut. Note the different route by which fat enters the bloodstream. Glucose and amino acids pass directly into the blood from the intestinal epithelial cells and are presented to the liver by way of the portal vein. Fat, contained in chylomicrons, is secreted by the intestinal epithelial cells into lymphatics, which drain the intestine. The lymphatics lead to the thoracic duct, which, by way of the subclavian vein, delivers chylomicrons to the blood at a site of rapid blood flow. The latter provides for rapid distribution of the chylomicrons and prevents coalescence of the fat particles.

Liver is the first tissue to have the opportunity to use dietary glucose. Glucose can be converted into glycogen by glycogenesis and into pyru-

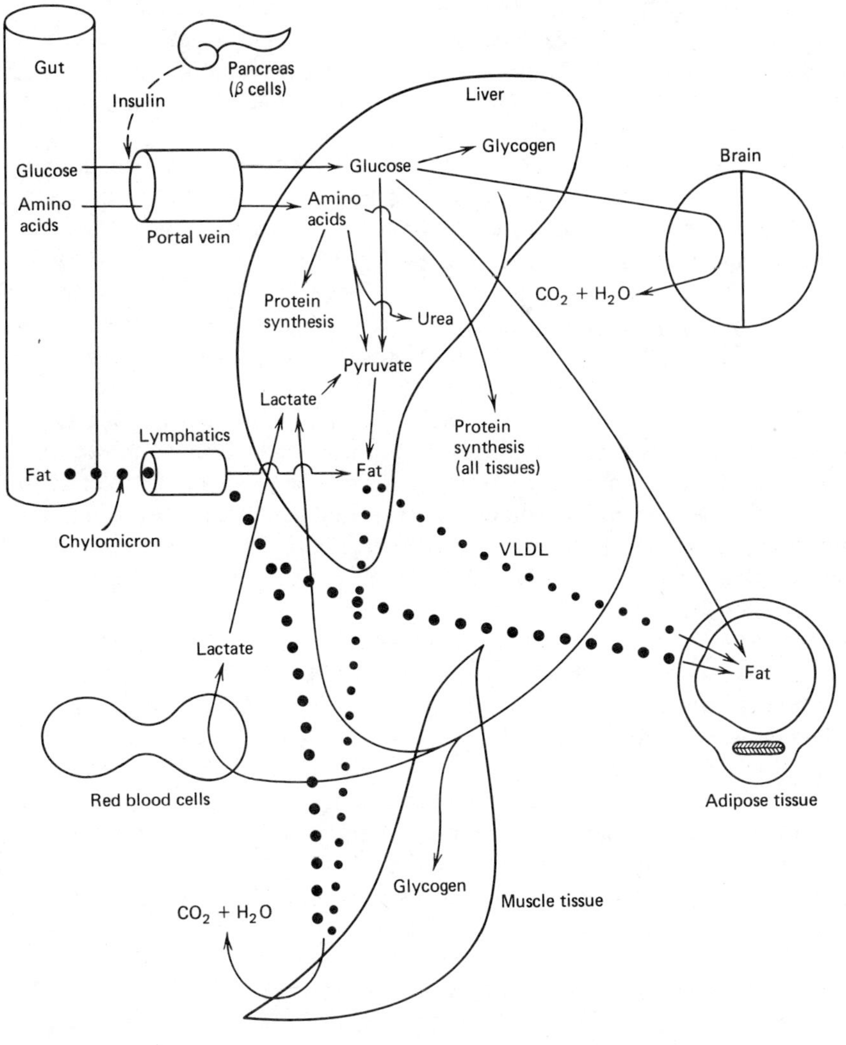

Figure 14.2
Disposition of glucose, amino acids, and fat by various tissues in the well-fed state.

vate and lactate by glycolysis or can be used in the pentose phosphate pathway for the generation of NADPH for reductive synthetic processes. Pyruvate can be oxidized to acetyl CoA, which, in turn, can be converted into fat or oxidized to CO_2 and water by the TCA cycle. Some of the glucose coming from the intestine escapes the liver and circulates to other tissues. The brain is a major user of glucose, being almost solely dependent upon this substrate for the production of ATP. Other major users of glucose include red blood cells, which can only convert glucose to lactate and pyruvate, and the adipose tissue, which converts it into fat. Muscle also has the capacity to use glucose, converting it to glycogen or using it in the glycolytic pathway and the TCA cycle pathways. A number of tissues produce lactate and pyruvate from circulating glucose by glycolysis. Lactate and pyruvate generated in peripheral tissues are taken up by the liver and converted to fat by the process of lipogenesis. In the very well-fed state, the liver uses glucose and does not engage in gluconeogenesis. Thus, the Cori cycle, which involves conversion of glucose to lactate in the peripheral tissues followed by conversion of lactate back to glucose in the liver, is interrupted in the well-fed state.

both hyperplasia (increase in cell number) and hypertrophy (increase in cell size) are contributing factors to the magnitude of the obesity.

The only effective treatment of obesity is reduction in the ingestion, absorption, or use of calories. Practically speaking, this means dieting. Unfortunately, the body compensates for decreased energy intake with reduced formation of triiodothyronine and a corresponding decrease in the basal metabolic rate. Thus, there is a biochemical basis for the universal complaint that it is far easier to gain than to lose weight. Furthermore, about 95% of people who are able to lose a significant amount of weight regain it within 1 year.

CLIN. CORR. 14.2 PROTEIN MALNUTRITION

Protein malnutrition is the most important and widespread nutritional problem among young children in the world today. The clinical syndrome, called kwashiorkor, occurs mainly in children 1 to 3 years of age and is precipitated by weaning an infant from breast milk onto a starchy, protein-poor diet. The name originated in Ghana, meaning "the sickness of the older child when the next baby is born." Common in many developing countries, it is a consequence of feeding the child a diet adequate in calories but deficient in protein. It may become clinically manifest when protein requirements are increased by infection, for example, malaria, helminth infestation, or gastroenteritis. The syndrome is characterized by poor growth, low plasma protein levels, muscle wasting, edema, diarrhea, and increased susceptibility to infection. The presence of subcutaneous fat clearly differentiates it from caloric starvation. The maintenance of fat stores is due to the high carbohydrate intake and resulting high insulin levels. In fact, the high insulin level interferes with the adaptations described for starvation. Specifically, fat is not mobilized as an energy source, ketogenesis does not take place, and there is no transfer of amino acids from the skeletal muscle to the internal organs, that is, the liver. The lack of dietary amino acids results in diminished protein synthesis in all tissues. The liver becomes enlarged and infiltrated with fat, reflecting the need for hepatic protein synthesis for the formation and release of lipoproteins from this tissue. In addition, protein malnutrition impairs the function of the gut, resulting in malabsorption of calories, protein, and vitamins, which accelerates the disease. The consequences of the disease include a permanent stunting of physical growth and poor intellectual and psychological development.

CLIN. CORR. 14.3 STARVATION

Starvation, or an overall deficit in food intake, including both calories and protein, leads to the development of a syndrome known as nutritional marasmus. Marasmus is a word of Greek origin meaning "to waste." Although not restricted to any age group, it is most common in children under 1 year of age. In developing countries early weaning of infants from breast milk is a common cause of marasmus. This may result from pregnancies in rapid succession, the desire of the mother to return to work, or switching to artificial formulas. Although the latter usually provides a complete and nutritious diet for a growing infant, economically deprived parents tend to dilute it with water to make it last longer. This practice leads to insufficient intake of calories. Likewise, diarrhea and malabsorption can develop if safe water and sterile procedures are not used.

In contrast to kwashiorkor (see Clin. Corr. 14.2), subcutaneous fat, hepatomegaly, and fatty liver are absent in marasmus because fat is mobilized as an energy source and muscle provides adequate amino acids to the liver for the synthesis of glucose and hepatic proteins. Low insulin levels allow the liver to oxidize fatty acids and to produce ketone bodies for other tissues. Ultimately, energy and protein reserves are exhausted, and the child starves to death.

Dietary protein is hydrolyzed in the intestine, the cells of which use some amino acids as an energy source. Most dietary amino acids are transported into the portal blood, but the intestine metabolizes aspartate, asparagine, glutamate, and glutamine, and releases alanine, lactate, citrulline, and proline into the portal blood.

Liver then has the opportunity to remove absorbed amino acids from the blood (Figure 14.2). The liver lets most of each amino acid pass through, unless the concentration of the amino acid is unusually high. This is especially important for the essential amino acids, needed by all tissues of the body for protein synthesis. The liver can catabolize amino acids, but the K_m values for amino acids of many of the enzymes involved is high, allowing the amino acids to be present in excess before significant catabolism can occur. In contrast, the tRNA-charging enzymes have much lower K_m values for amino acids. This ensures that as long as all the amino acids are present, protein synthesis can occur as needed for growth and protein turnover. Amino acids can be oxidized completely to CO_2 and water, or the intermediates generated can be used as substrates for lipogenesis, ketogenesis, or gluconeogenesis. Thus, excess amino acids not needed for protein synthesis are converted to ketone bodies, glucose, or fat, with the amino nitrogen converted to urea. Amino acids that escape the liver can be used for protein synthesis or for energy in other tissues. Skeletal muscle and heart muscle have a high capacity for amino acid transamination and oxidation of the resulting α-keto acids to CO_2 and water. The branched-chain amino acids (leucine, isoleucine, and valine) are handled in an interesting manner. The liver has low capacity for transamination of these amino acids but considerable capacity for oxidative decarboxylation of their corresponding α-keto acids. On the other hand, skeletal muscle has considerable capacity for transamination but is relatively deficient in the enzymes responsible for subsequent catabolism. As a consequence, much of the transamination occurs in peripheral tissues such as skeletal muscle, the α-keto acids escape into the blood, and the liver oxidizes the α-keto acid. Branched-chain amino acids are a major source of nitrogen for the production of alanine and glutamine in muscle.

When considering fat delivery to the tissues, we must differentiate between endogenous and exogenous fat (Figure 14.2). Glucose, lactate, pyruvate, and amino acids can be used to support hepatic lipogenesis. The fat formed from these substrates is released from the liver in the form of VLDL. Dietary fat is delivered to the bloodstream as chylomicrons. Both chylomicrons and VLDL circulate in the blood until they are acted upon by a special extracellular enzyme attached to the endothelial cells of the capillaries of many tissues. This enzyme, lipoprotein lipase, is particularly abundant in the capillaries in adipose tissue. It acts on both the VLDL and chylomicrons, liberating fatty acids by hydrolysis of the triacylglycerols. The fatty acids are then taken up by the adipocytes, reesterified with glycerol 3-phosphate to form triacylglycerols, and stored as large fat droplets within these cells. The glycerol 3-phosphate needed for triacylglycerol formation in adipose tissue is generated from glucose, using the first half of the glycolytic pathway to generate dihydroxyacetone phosphate, which is reduced to glycerol 3-phosphate by glycerol 3-phosphate dehydrogenase.

The β cells of the pancreas are very responsive to the influx of glucose and amino acids in the fed state. The β cells release insulin during and after eating, which is essential for the metabolism of these nutrients by liver, muscle, and adipose tissue. The role of insulin in the starve–feed cycle is discussed in more detail in Section 14.3.

Early Fasting State

Figure 14.3 shows what happens in early fasting after fuel stops coming in from the gut. Hepatic glycogenolysis for the maintenance of blood glucose is very important during this transitional period. Lipogenesis is curtailed, and the lactate, pyruvate, and amino acids that were used by that pathway are diverted into the formation of glucose. The Cori cycle is shown in Figure 14.3; glucose is produced from lactate by the liver and is then converted back to lactate by glycolysis in peripheral tissues such as red blood cells. The alanine cycle, in which carbon returns to the liver in the form of alanine rather than lactate, also becomes important as a mechanism for maintaining blood glucose levels. The catabolism of amino acids for energy is greatly diminished in the early fasting condition because less is available from the gut.

Fasting State

Figure 14.4 shows what happens in the fasting state. No fuel enters from the gut and little glycogen is left in the liver. Tissues that use glucose are completely dependent on hepatic gluconeogenesis, primarily from lactate, glycerol, and alanine. The Cori and the alanine cycles described above play important roles in supplying glucose; however, the Cori and the

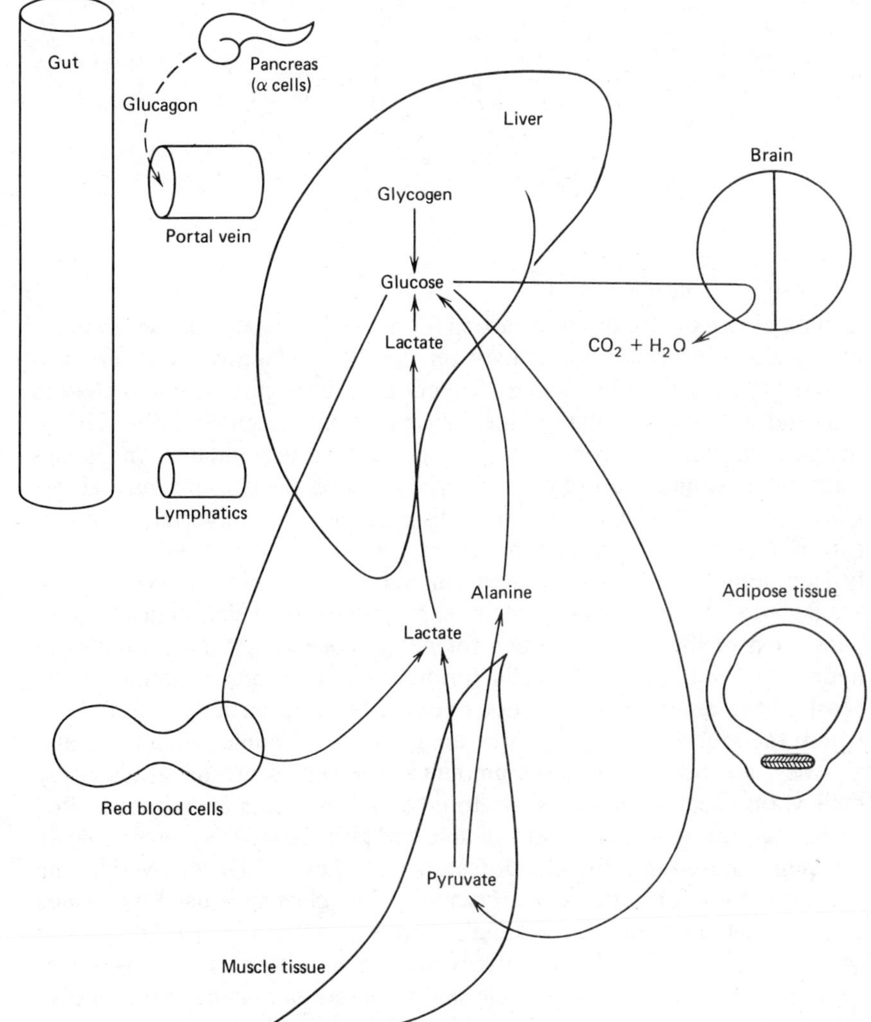

Figure 14.3
Metabolic interrelationships of the major tissues of the body in the early fasting state.

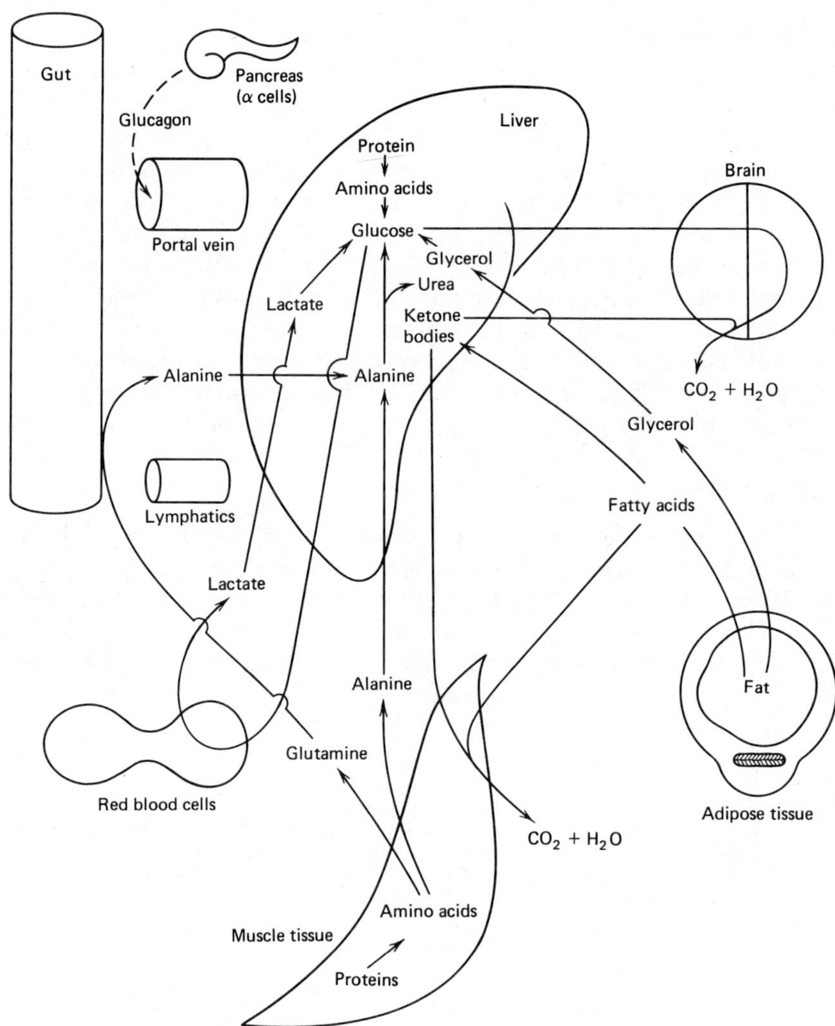

Figure 14.4
Metabolic interrelationships of the major tissues of the body in the fasting state.

alanine cycles do not provide carbon for net synthesis of glucose. In these cycles glucose formed by the liver replaces that which was converted to lactate by the peripheral tissues. The brain oxidizes glucose completely to CO_2 and water and probably does not participate in either cycle. Hence net glucose synthesis from some source of carbon is mandatory in fasting. Fatty acids cannot be used for the synthesis of glucose, because acetyl CoA obtained by fatty acid catabolism cannot be converted to three-carbon intermediates of gluconeogenesis. Glycerol, a by-product of lipolysis in adipose tissue, is an important substrate for glucose synthesis in the fasted state. However, protein, especially from skeletal muscle, supplies most of the carbon needed for net glucose synthesis. Proteins are hydrolyzed within muscle cells (proteolysis) to produce amino acids. Most of the amino acids are not released but are partially metabolized within the muscle cell. Only three amino acids—alanine, glutamine, and glycine—are released in large amounts. The others are metabolized by their various catabolic pathways to give intermediates (pyruvate and α-ketoglutarate), which can yield alanine and glutamine. These amino acids are then released into the blood, from which they can be removed by the liver or kidney for net glucose formation. The glucose is used by tissues other than kidney or cardiac and skeletal muscle, which oxidize fatty acids in preference to glucose. Pyruvate formed in other tissues by glycolysis may be taken up by muscle and released as alanine. Muscle also releases branched-chain α-keto acids to the liver, which synthesizes glucose from the keto acid of valine, ketone bodies from the keto acid of leucine, and both glucose and ketone bodies from the keto acid of isoleucine. There is evidence that much of the glutamine released from muscle is converted into alanine by the intestinal epithelium. Glutamine is partially oxidized in these cells to supply energy to meet part of the metabolic demand of this tissue, and the carbon and amino groups left over are released back into the bloodstream in part as alanine and NH_4^+. This pathway probably involves the formation of malate from glutamine via the TCA cycle and the conversion of malate to oxaloacetate, oxaloacetate to phospho*enol*pyruvate, and phospho*enol*pyruvate to pyruvate. Direct decarboxylation of malate to pyruvate is also possible. Pyruvate is then transaminated to alanine. Alanine is quantitatively the most important gluconeogenic amino acid to reach the liver in the fasting state. Glycine released from muscle is transformed in part to serine by the kidneys. Serine is subsequently converted into glucose by the liver or kidney.

The synthesis of glucose in the liver during fasting is closely linked to the synthesis of urea. Most amino acids can give up the amino nitrogen by transamination with α-ketoglutarate, forming glutamate and a new α-keto acid, which can be further oxidized. Further metabolism of glutamate provides both nitrogen compounds required for urea synthesis, ammonia from the oxidative deamination of glutamate by glutamate dehydrogenase, and aspartate from the transamination of glutamate with oxaloacetate by aspartate aminotransferase. An additional important source of ammonia is the gut mucosa, which converts glutamine to alanine and ammonia. The gut also releases precursors of ornithine such as citrulline, which may be important for urea cycle activity during fasting.

The adipose tissue is also very important in the fasting state. Because of the low insulin : glucagon ratio existing in this condition, lipolysis is greatly activated. This raises the blood levels of fatty acids, which can be used as alternative fuels to glucose by many tissues. In heart and muscle, the oxidation of fatty acids inhibits glycolysis. The brain, on the other hand, does not oxidize fatty acids because fatty acids cannot cross the blood-brain barrier. Fatty acids play a very important role in liver, providing by the β-oxidation pathway most of the ATP needed to support the energy requirements of gluconeogenesis. Very little of the acetyl CoA generated by β-oxidation in the liver is oxidized completely by the TCA cycle. The acetyl CoA formed from fatty acids is converted into ketone bodies by liver mitochondria under these conditions. The ketone bodies (acetoacetate and β-hydroxybutyrate) are released into the blood and are a source of energy for many tissues. Like fatty acids, ketone bodies are preferred by many tissues over glucose. Ketone bodies, in contrast to fatty acids, penetrate the blood–brain barrier. Once their blood concentration is high enough, ketone bodies function as a good alternative fuel for the brain. They are unable, however, to replace the need for glucose by the brain completely. Ketone bodies do decrease brain glucose utilization and also suppress proteolysis in skeletal muscle and decrease to a certain extent muscle wasting, which inevitably occurs during starvation. As long as ketone body levels are maintained at a high level by hepatic β-oxidation, there is less need for glucose, less need for gluconeogenic amino acids, and less need for using up precious muscle tissue by proteolysis.

The interrelationships between liver, muscle, and adipose tissue in supplying glucose for the brain are shown in Figure 14.4. The liver synthesizes the glucose, the muscle supplies the substrate (alanine), and the adipose tissue supplies the ATP (via fatty acid oxidation) needed for hepatic gluconeogenesis. These relationships are disrupted in Reye's syndrome (Clin. Corr. 14.4). This interaction is dependent upon a low insu-

CLIN. CORR. **14.4** REYE'S SYNDROME

Reye's syndrome is a devastating illness of children that tends to follow viral infections. It is characterized by evidence of brain dysfunction and edema (irritability, lethargy, coma) and liver dysfunction (elevated plasma-free fatty acids, fatty liver, hypoglycemia, hyperammonemia, and accumulation of short-chain organic acids). In some respects, it appears that hepatic mitochondria are specifically damaged, which impairs β-oxidation and synthesis of carbamoyl phosphate and ornithine (for ammonia detoxification) and oxaloacetate (for gluconeogenesis). On the other hand, the accumulation of organic acids has suggested that the oxidation of these compounds is defective and that the CoA esters of some of these acids may inhibit specific enzymes, such as carbamoyl phosphate synthetase I, pyruvate dehydrogenase, pyruvate carboxylase, and the adenine nucleotide transporter, all present in mitochondria. The issue has not yet been resolved. The therapy for Reye's syndrome consists of measures to reduce brain edema (administration of hypertonic solutions of mannitol, hyperventilation, and occasionally pentobarbital-induced coma) and the provision of glucose intravenously. Glucose administration prevents hypoglycemia and elicits a rise in insulin levels that may (1) inhibit lipolysis in adipose cells and (2) reduce proteolysis in muscles and the release of amino acids, which (3) reduces the deamination of amino acids to ammonia.

lin : glucagon ratio, opposite of the ratio favoring those processes that are characteristic of the well-fed state. Glucose levels are lower in the fasting condition, reducing the secretion of insulin but favoring the release of glucagon from the pancreas. In addition, fasting reduces the formation of triiodothyronine, the active form of thyroid hormone, from thyroxine. This reduces the daily basal energy requirements by as much as 25%.

Early Refed State

Figure 14.5 shows what happens soon after fuel starts coming in from the gut. Fat is metabolized as described above for the well-fed state. In contrast, glucose is poorly extracted by the liver and serves as a very poor precursor of hepatic glycogen synthesis during this period of the starve–feed cycle. Glycolysis by the liver is also slowly established, so that the liver remains in the gluconeogenic mode for a few hours after feeding. Rather than providing blood glucose, however, hepatic gluconeogenesis provides glucose units in the form of glucose 6-phosphate for glycogenesis. This means that liver glycogen is not repleted after a fast by direct synthesis from blood glucose. Rather, the incoming glucose is catabolized in peripheral tissues to lactate, which is converted in the liver to glycogen without the intervening formation of free glucose. Gluconeogenesis from specific amino acids entering from the gut may also play an important role

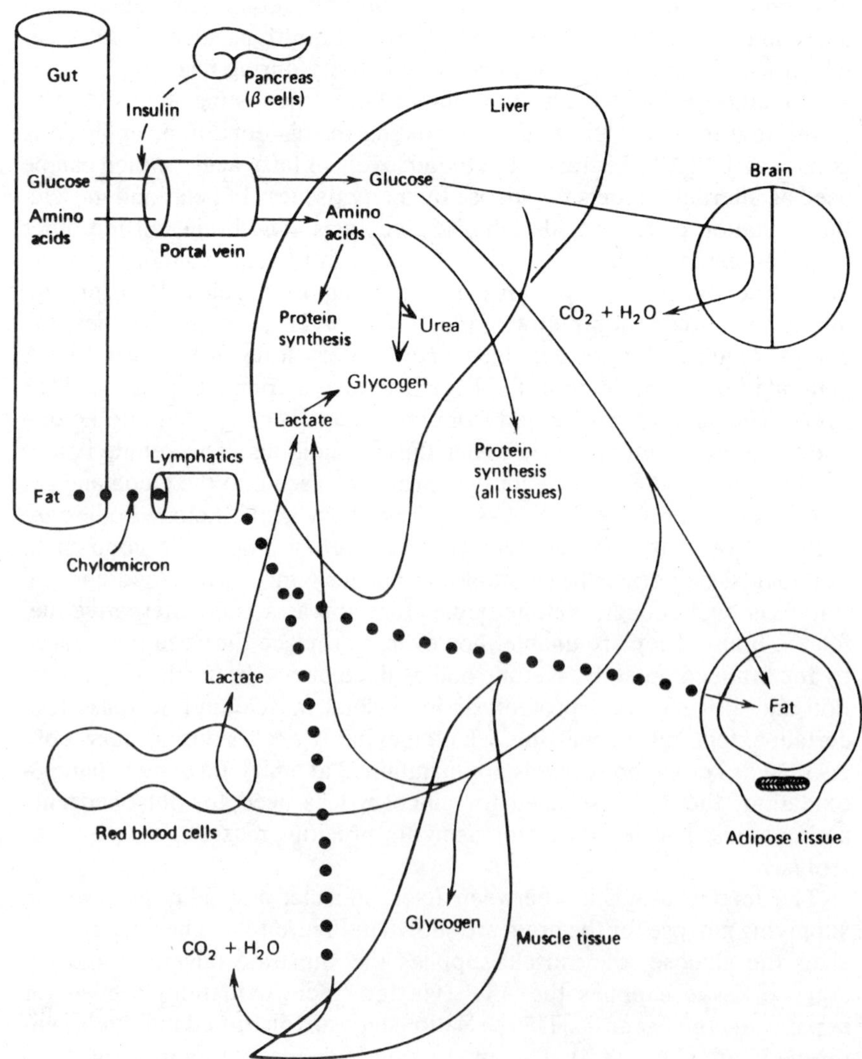

Figure 14.5
Metabolic interrelationships of the major tissues of the body in the early refed state.

in reestablishing normal liver glycogen levels. After the maintenance of the well-fed state for a few hours, the metabolic interrelationships of Figure 14.2 become established. The rate of gluconeogenesis declines, glycolysis becomes the predominant means of glucose disposal in the liver, and liver glycogen is sustained by direct synthesis from glucose.

Caloric Homeostasis

The major tissues of the body work closely together to maintain a constant availability of oxidizable fuels in the blood. This is termed caloric homeostasis, which, as illustrated in Table 14.1, means that regardless of whether a person is in the well-fed state, fasting, or starving to death, the blood level of ATP equivalent fuel does not fall below certain limits. The changes in insulin : glucagon ratio, discussed above and shown in Table 14.1, are crucial to the proper maintenance of caloric homeostasis. Note that blood glucose levels are controlled within very tight limits, whereas fatty acid concentrations in the blood can vary by an order of magnitude and ketone bodies by two orders of magnitude. The fact that glucose is maintained within tighter limits is again related to the absolute need of the brain for this substrate. If the blood glucose level falls too low (<1.5 mM), coma results from lack of ATP production, and death will follow shortly unless the situation can be rapidly corrected. On the other hand, hyperglycemia must be avoided because of the risk of hyperosmolar, hyperglycemic coma (see Clin. Corr. 14.5). Hyperglycemia also leads to the glycosylation of a number of proteins, which is postulated to be a more insidious complication of prolonged high concentrations of glucose (see Clin. Corr. 14.6).

Energy Requirements and Reserves

The average person leading a sedentary life, such as a portly professor of biochemistry, consumes daily about 200 g of carbohydrate, 70 g of pro-

TABLE 14.1 Substrate and Hormone Levels in Blood of Well-fed, Fasting, and Starving Humans[a]

Hormone or Substrate (units)	*Very Well-Fed*	*Postabsorptive 12 hours*	*Fasted 3 days*	*Starved 5 weeks*
Insulin (μU/ml)	40	15	8	6
Glucagon (pg/ml)	80	100	150	120
Insulin : glucagon ratio (μU/pg)	0.50	0.15	0.05	0.05
Glucose (mM)	6.1	4.8	3.8	3.6
Fatty acids (mM)	0.14	0.6	1.2	1.4
Acetoacetate (mM)	0.04	0.05	0.4	1.3
β-Hydroxybutyrate (mM)	0.03	0.10	1.4	6.0
Lactate (mM)	2.5	0.7	0.7	0.6
Pyruvate (mM)	0.25	0.06	0.04	0.03
Alanine (mM)	0.8	0.3	0.3	0.1
ATP equivalents (mM)	313	290	380	537

[a] Data are for normal weight subjects except for the 5-week starvation values, which are from obese subjects undergoing therapeutic starvation. ATP equivalents were calculated on the basis of the ATP yield expected on complete oxidation of each substrate to CO_2 and H_2O: 38 molecules of ATP for each molecule of glucose; 144 for the average fatty acid (oleate); 23 for acetoacetate; 26 for β-hydroxybutyrate; 18 for lactate; 15 for pyruvate; and 13 (corrected for urea formation) for alanine. Taken in part from: Ruderman, N. B., Aoki, T. T., and Cahill, G. F., Jr. Gluconeogenesis and its disorders in man, in R. W. Hanson, and M. A. Mehlman (eds.), *Gluconeogenesis, its regulation in mammalian species*. New York: Wiley, 1976, p. 515.

CLIN. CORR. 14.5 HYPEROSMOLAR, HYPERGLYCEMIC COMA

In contrast to young type I diabetic patients who can become ill with diabetic ketoacidosis within a day or so, older type II diabetic patients sometimes develop a condition called hyperosmolar, hyperglycemic coma. This is particularly common in the elderly. Hyperglycemia, perhaps worsened by failure to take insulin or hypoglycemic drugs, an infection, or a coincidental medical problem such as a heart attack, leads to urinary losses of water, glucose, and electrolytes (sodium, chloride, and potassium). This osmotic diuresis reduces the circulating blood volume, a stress that in turn worsens insulin resistance and hyperglycemia. In addition, elderly patients may be less able to sense thirst or to obtain fluids. Over the course of several days these patients can become extremely hyperglycemic (glucose > 1,000 mg/dl), dehydrated, and ultimately comatose. Ketoacidosis does not develop in these patients possibly because free fatty acids are not always elevated or because adequate insulin concentrations exist in the portal blood to inhibit ketogenesis (although it is not high enough to inhibit gluconeogenesis). Therapy is aimed at restoring water and electrolyte balance and correcting the hyperglycemia with insulin. The mortality of this syndrome is considerably higher than that of diabetic ketoacidosis.

CLIN. CORR. 14.6 HYPERGLYCEMIA AND PROTEIN GLYCOSYLATION

Glycosylation of enzymes is known to cause changes in their activity, solubility and susceptibility to degradation. In the case of hemoglobin A, glycosylation occurs by a nonenzymatic reaction between glucose and the amino-terminal valine of the β chain. A Schiff base between glucose and valine forms, followed by a rearrangement of the molecule to give a 1-deoxyfructose molecule attached to the valine. The reaction is favored by high glucose levels and the resulting protein called hemoglobin A_{1c}, is a good index of how high a person's average blood–glucose concentration has been over the previous several weeks. The concentration of this protein increases substantially in the red blood cells of an uncontrolled diabetic.

It has been proposed that increased glycosylation of proteins resulting from hyperglycemia may contribute to the medical complications caused by diabetes, for example, coronary heart disease, retinopathy, nephropathy, and neuropathy. Collagen, fibrin, and antithrombin III (an inhibitor of blood coagulation) can become glycosylated and undergo alterations in physical and enzymatic properties. It is possible that these changes favor the accelerated blood vessel damage that occurs in patients with diabetes. Likewise, increased glycosylation of the lens protein α-crystallin may contribute to the development of diabetic cataracts.

TABLE 14.2 The Energy Reserves of Humans[a]

Stored Fuel	*Tissue*	*Fuel Reserves in Units of* Grams	*Fuel Reserves in Units of* Kilocalories
Glycogen	Liver	70	280
Glycogen	Muscle	120	480
Glucose	Body fluids	20	80
Fat	Adipose	15,000	135,000
Protein	Muscle	6,000	24,000

[a] The data are for a normal subject weighing 70 kg. Carbohydrate contains 4 kcal/g; fat 9 kcal/g; protein 4 kcal/g.

tein, 60 g of fat, and, during the academic year, 100 g of ethanol and an occasional graduate student. As shown in Table 14.2, the energy reserves of an average-sized person are considerable. We tend to emphasize the details of glycogen metabolism, and the ability to mobilize glycogen rapidly is indeed very important. Table 14.2 demonstrates, however, that our glycogen reserves are minuscule with respect to our fat reserves. Fat reserves are large in obesity with little or no significant increase in glycogen reserves. The fat stores of obese subjects can weigh as much as 80 kg, adding another 585,000 kcal to their energy reserves. Protein is listed in Table 14.2 as an energy reserve, and, in the sense that it can be used to provide substrate for amino acid oxidation, protein is an energy reserve. On the other hand, protein is not inert like stored fat and glycogen. Proteins make up the muscles that allow us to move and breathe and the enzymes that carry out all the catabolic and anabolic processes. Hence it is not as dispensable as fat and glycogen and is given up by the body more reluctantly.

The Five Phases of Glucose Homeostasis

Figure 14.6 comes from the work of Cahill and his colleagues with obese patients undergoing long-term starvation for therapeutic purposes. It illustrates the effects of starvation on those processes that are used by the tissues of the body to maintain caloric homeostasis as described above. For convenience of discussion, the time period involved has been divided into five phases. Phase I is the well-fed state, in which glucose is provided by dietary carbohydrate. Once this supply is exhausted, glycogenolysis in the liver maintains blood glucose levels during phase II. As this supply of glucose starts to dwindle, hepatic gluconeogenesis from lactate, glycerol, and alanine becomes increasingly important until, in phase III, it is the major source of blood glucose. Note that all of these changes occur within just 20 or so hours of fasting, depending of course on how well fed the individual was prior to the fast, how much hepatic glycogen was present, and the sort of physical activity occurring during the fast. Several days of fasting move one into phase IV, where the dependence upon gluconeogenesis actually decreases. The explanation for this surprising phenomenon, already discussed above, is that ketone bodies have accumulated to concentrations that are high enough for them to enter the brain and meet some of the energy needs of this tissue. Phase V occurs after very prolonged starvation of extremely obese individuals. It is characterized by even less dependence of the body upon gluconeogenesis, the energy needs of almost every tissue being met to an even greater extent by either fatty acid or ketone body oxidation.

As long as ketone body concentrations are high, proteolysis will be somewhat restricted, and conservation of muscle proteins and enzymes

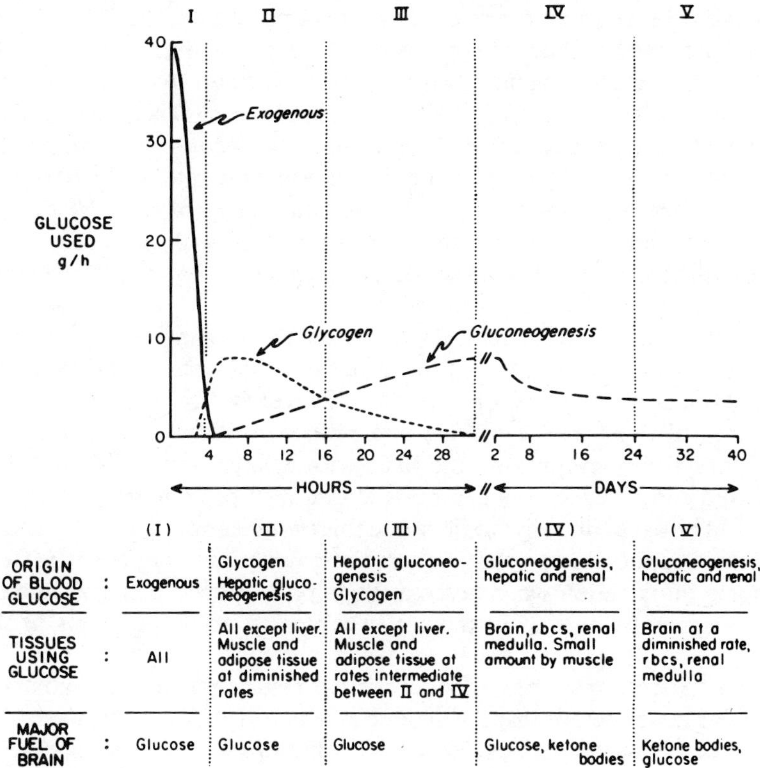

Figure 14.6
The five phases of glucose homeostasis in human beings.
From Ruderman, N. B., Aoki, T. T., and Cahill, G. F., Jr. Gluconeogenesis and its disorders in man, in R. W. Hanson, and M. A. Mehlman (eds.), *Gluconeogenesis, its regulation in mammalian species*. New York: Wiley, 1976, p. 515. Reproduced with permission.

will occur. This continues until practically all of the fat is gone as a consequence of starvation. After all of it is gone, the body has to use muscle protein. Before it is gone—you are gone (see Clin. Corr. 14.3).

14.3 MECHANISMS INVOLVED IN SWITCHING THE METABOLISM OF THE LIVER BETWEEN THE WELL-FED STATE AND THE STARVED STATE

The liver of the well-fed person is actively engaged in processes that favor the synthesis of glycogen and fat; such a liver is glycogenic, glycolytic, lipogenic, and cholesterogenic. The liver of the fasting person is quite a different organ; it is glycogenolytic, gluconeogenic, ketogenic, and proteolytic. The strategy is to store calories when food is available, but then to be able to mobilize these stores when the rest of the body is in need. The liver is switched between these metabolic extremes by a variety of regulatory mechanisms: substrate supply, allosteric effectors, covalent modification, and induction-repression of enzymes.

Substrate Supply

Because of the other, more sophisticated levels of control, the importance of substrate supply is often ignored. However, the concentration of fatty acids in the blood of the portal vein is clearly a major determinant of the rate of ketogenesis. Excess fat is not synthesized and stored unless one consumes excessive amounts of substrates that can be used for the process of lipogenesis. Glucose synthesis by the liver is also restricted by the rate at which gluconeogenic substrates (e.g., lactate, pyruvate, alanine, and glycerol) flow to the liver. Delivery of excess amino acids to the liver of the diabetic, because of the accelerated and uncontrolled proteolysis that occurs in this metabolic condition, increases the rate of gluconeogenesis and exacerbates the hyperglycemia characteristic of diabetes. On the other hand, failure to supply the liver adequately with glucogenic substrate explains some types of hypoglycemia, such as that observed during pregnancy or advanced starvation.

Another pathway that may be regulated by substrate supply is urea synthesis. Amino acid metabolism in the intestines may provide a substantial fraction of the ammonia used by the liver for urea production. The intestines also releases citrulline and proline, metabolic precursors of ornithine. This may provide the liver with a larger ornithine pool and increased capacity for urea synthesis after a high protein meal. Your pet cat could be used (literally) to illustrate this particular interaction between substrate supplies. Cats vomit, become comatose, and may even die from ammonia intoxication when given a single protein meal deficient in arginine. Arginine is needed in the cat to replenish ornithine levels in the liver.

From the examples provided above we can conclude that substrate supply is a major determinant of the rate at which virtually every metabolic process of the body operates. However, variations in substrate supply are not sufficient to account for the tremendous changes in metabolism that must occur in the starve–feed cycle. As discussed below, regulation by allosteric effectors plays an important role in the different states.

Allosteric Effectors

Figures 14.7 and 14.8 summarize the effects of negative and positive allosteric effectors believed to be important in the well-fed and starved states, respectively. As shown in Figure 14.7, glucose inactivates glycogen phosphorylase and activates glycogen synthase (indirectly by stimulating phosphoprotein phosphatase(s), see Chapter 7, page 320), thereby preventing degradation and promoting synthesis of glycogen; fructose 2,6-bisphosphate stimulates 6-phosphofructo-1-kinase and inhibits fructose 1,6-bisphosphatase, thereby stimulating glycolysis and inhibiting gluconeogenesis; fructose 1,6-bisphosphate activates pyruvate kinase, thereby stimulating glycolysis; pyruvate activates pyruvate dehydrogenase (indirectly by inhibition of pyruvate dehydrogenase kinase, see Chapter 6, page 225) and citrate activates acetyl CoA carboxylase, thereby stimulating fatty acid synthesis; and malonyl CoA inhibits carnitine palmitoyltransferase I, thereby inhibiting fatty acid oxidation.

As shown in Figure 14.8, acetyl CoA stimulates gluconeogenesis in the fasted state by activating pyruvate carboxylase (direct allosteric effect) and inhibiting pyruvate dehydrogenase (direct allosteric effect and also indirect effect via stimulation of pyruvate dehydrogenase kinase; see Chapter 7, page 304); long-chain acyl CoA esters inhibit acetyl CoA carboxylase, which, in turn, lowers the level of malonyl CoA and permits greater carnitine palmitoyltransferase I activity and fatty acid oxidation rates; NADH produced by fatty acid oxidation inhibits TCA cycle activity.

Although not shown in Figure 14.8, cAMP should be noted as an important allosteric effector. Its concentration is elevated in the liver in the starved state. Often referred to as the "hunger signal" because its concentration increases when glucose is lacking, cAMP is a positive effector

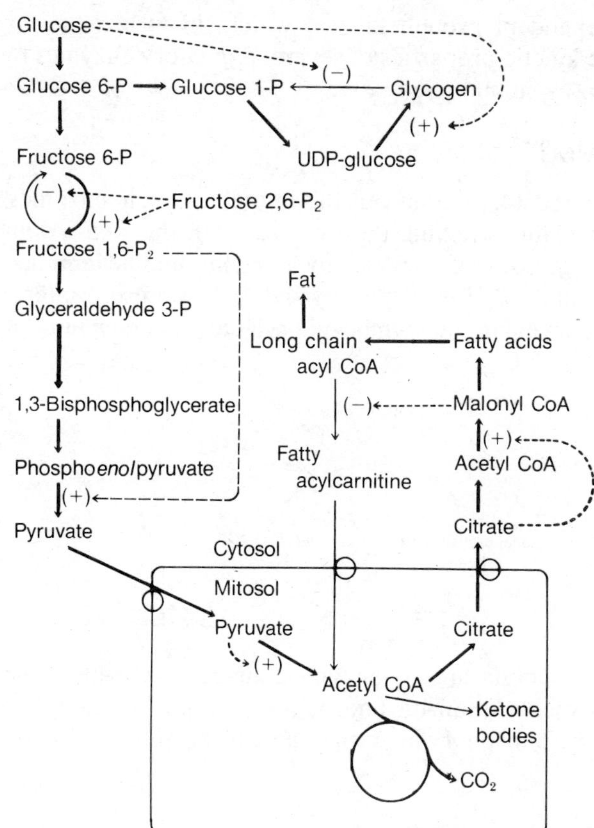

Figure 14.7
Control of hepatic metabolism in the well-fed state by allosteric effectors.

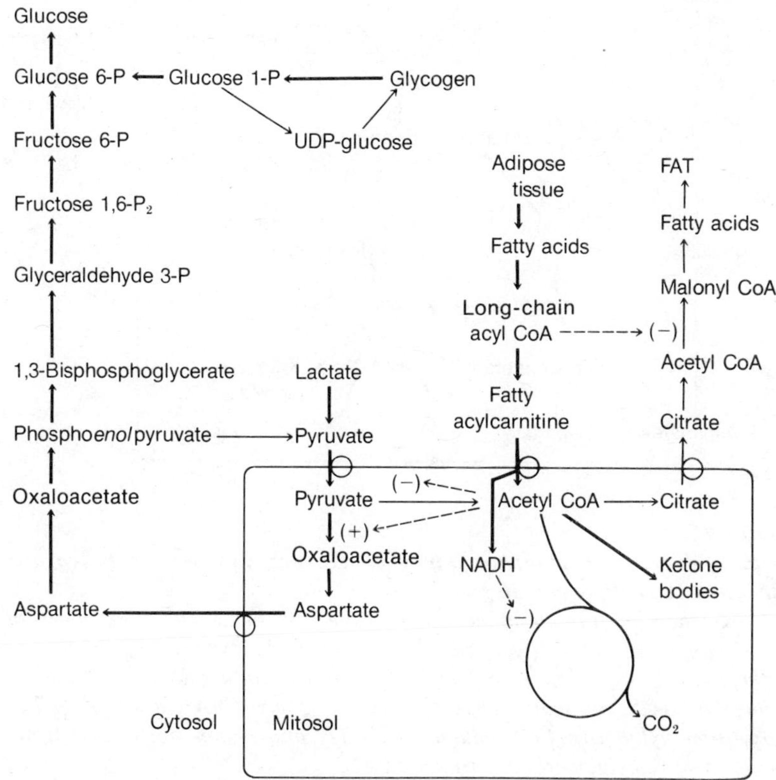

Figure 14.8
Control of hepatic metabolism in the fasting state by allosteric effectors.

of cAMP-dependent protein kinase, which, in turn, is responsible for changing the kinetic properties of several regulatory enzymes by covalent modification, as summarized next.

Covalent Modification

Figures 14.9 and 14.10 point out the interconvertible enzymes that play important roles in switching the liver between the well-fed and starved states. The regulation of enzymes by covalent modification has been discussed in Chapter 7. Recall that ⊡ and ⊙-P represent interconvertible forms of an enzyme in the nonphosphorylated and phosphorylated states, respectively.

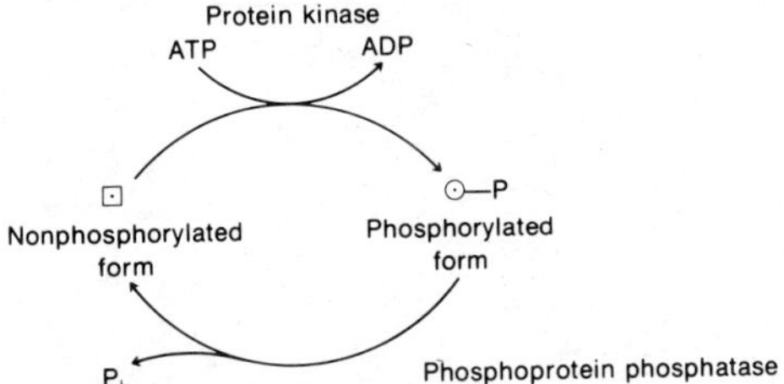

The important points are (1) enzymes subject to covalent modification undergo phosphorylation on one or more serine residues by a protein kinase; (2) the phosphorylated enzyme can be returned to the dephos-

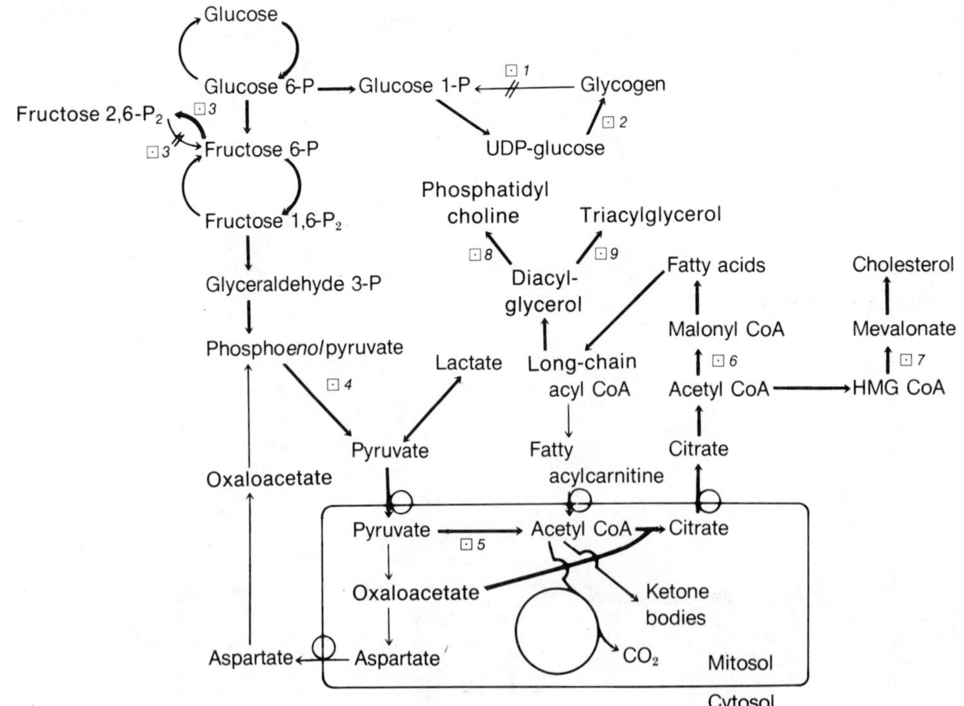

Figure 14.9
Activity and state of phosphorylation of the enzymes subject to covalent modification in the lipogenic liver.
The dephosphorylated mode is indicated by the symbol ⊡. The interconvertible enzymes numbered are 1, glycogen phosphorylase; 2, glycogen synthase; 3, 6-phosphofructo-2-kinase/fructose-2,6-bisphosphatase (bifunctional enzyme); 4, pyruvate kinase, 5, pyruvate dehydrogenase; 6, acetyl CoA carboxylase; 7, β-hydroxy-β-methylglutaryl CoA reductase; 8, CTP : phosphocholine cytidylyltransferase; and 9, diacylglycerol acyltransferase.

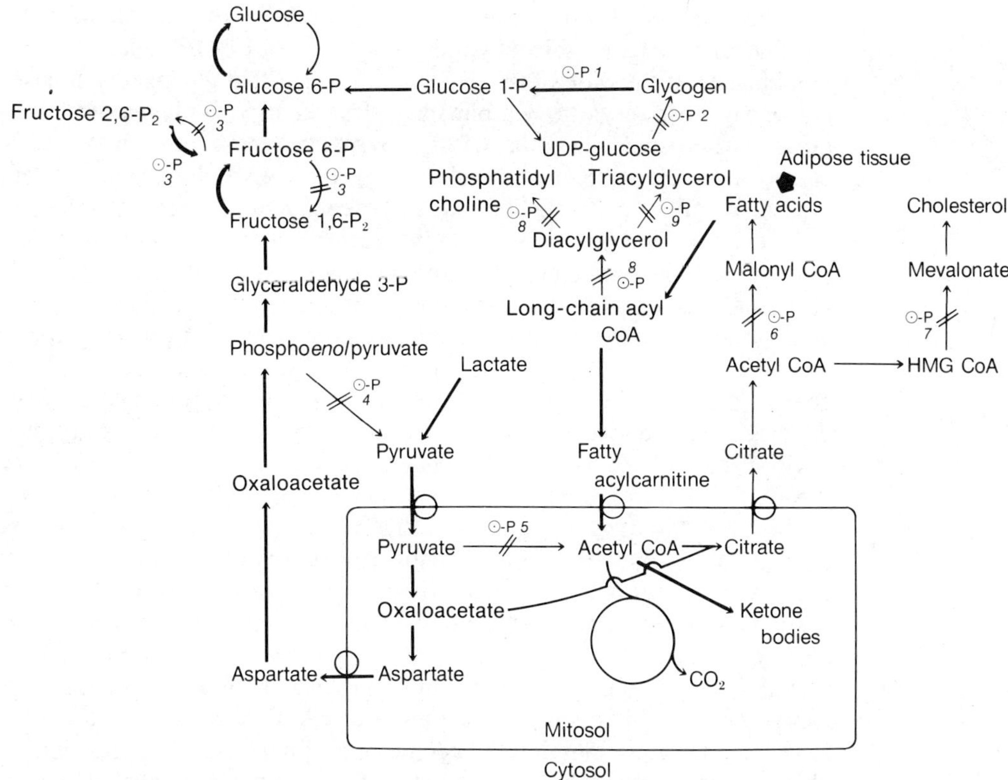

Figure 14.10
Activity and state of phosphorylation of the enzymes subject to covalent modification in the glucogenic liver.
The phosphorylated mode is indicated by the symbol ⊙—P. The numbers refer to the same enzymes as in Figure 14.9.

phorylated state by the action of the phosphoprotein phosphatase; (3) hormonal regulation via protein kinases is better understood than hormonal regulation via phosphoprotein phosphatases; (4) phosphorylation of the enzyme changes its conformation and its catalytic activity; (5) some enzymes are active only in the dephosphorylated state, others only in the phosphorylated state; (6) cAMP is the messenger that signals the phosphorylation of many, but not all, of the enzymes subject to covalent modification; (7) cAMP acts by activating cAMP-dependent protein kinase; (8) cAMP also indirectly promotes phosphorylation of interconvertible enzymes by signaling the phosphorylation of inhibitor-1, a heat-stable protein that functions as an inhibitor of phosphoprotein phosphatase when it is phosphorylated on a threonine residue by cAMP-dependent protein kinase (see Chapter 7, page 317); (9) glucagon increases cAMP levels in the liver by activating adenylate cyclase; and (10) insulin, by an unknown mechanism that probably involves formation of an intracellular messenger, opposes the action of glucagon and cAMP and thereby promotes dephosphorylation of the interconvertible enzymes.

As shown in Figure 14.9, the hepatic enzymes currently believed subject to covalent modification are in the dephosphorylated mode in the liver of the well-fed animal. Although not shown in the figure, it should be noted that phosphorylase kinase and inhibitor-1 would be in the dephosphorylated mode in the well-fed state. Insulin : glucagon ratios are high in the blood, and the cAMP levels are low in the liver in this situation. This results in a low activity for the cAMP-dependent protein kinase and, because inhibitor-1 is not phosphorylated, a high activity for phosphoprotein phosphatase. It is important to realize, however, that not all interconvertible enzymes are subject to phosphorylation by cAMP-dependent protein kinase. At the time of this writing, it is clear that glycogen synthase, glycogen phosphorylase (via phosphorylase kinase), 6-phosphofructo-2-kinase, fructose-2,6-bisphosphatase, pyruvate kinase, and acetyl CoA carboxylase are subject to regulation by the cAMP-dependent protein kinase. Whether there is a link to cAMP-dependent protein kinase for the other enzymes is a question of current research interest. Only three of the interconvertible enzymes, glycogen phosphorylase, phosphorylase kinase, and fructose-2,6-bisphosphatase, are inactive in the dephosphorylated mode. All of the other identified interconvertible enzymes (glycogen synthase, 6-phosphofructo-2-kinase, pyruvate kinase, pyruvate dehydrogenase, acetyl CoA carboxylase, β-hydroxy-β-methylglutaryl CoA reductase, CTP:phosphocholine cytidylyltransferase and diacylglycerol acyltransferase) are active. Glycogenesis, glycolysis, and lipogenesis are greatly favored as a result of placing the interconvertible enzymes in the dephosphorylated mode. On the other hand, the opposing pathways, glycogenolysis, gluconeogenesis, and ketogenesis, are inhibited.

As shown in Figure 14.10, the hepatic enzymes believed subject to covalent modification are in the phosphorylated mode in the liver of the fasting animal. Insulin:glucagon ratios are low in the blood, hepatic cAMP levels are high, and the insulin "messenger" presumably is low in this situation. This results in activation of cAMP-dependent protein kinase and, because inhibitor-1 becomes phosphorylated, inactivation of phosphoprotein phosphatase. The net effect is a much greater degree of phosphorylation of the interconvertible enzymes than in the well-fed state. In the starved state, three of the interconvertible enzymes—glycogen phosphorylase, phosphorylase kinase, and fructose-2,6-bisphosphatase—are in the active catalytic state. All the other interconvertible enzymes are inactive in the phosphorylated mode. As a result, glycogenesis, glycolysis and lipogenesis are shut down almost completely, and glycogenolysis, gluconeogenesis and ketogenesis predominate.

Two additional hepatic enzymes, phenylalanine hydroxylase and branched-chain α-keto acid dehydrogenase, have recently joined the list of enzymes whose activity is controlled by phosphorylation/dephosphorylation. These enzymes catalyze rate-limiting steps in the disposal of phenylalanine and the branched-chain amino acids (leucine, isoleucine, and valine), respectively. These enzymes are not included in Figures 14.9 and 14.10 because of special features of their control by covalent modification. Phenylalanine hydroxylase, a cytosolic enzyme, is active in the phosphorylated state, and phosphorylation is stimulated by glucagon via cAMP-dependent protein kinase. Branched-chain α-keto acid dehydrogenase, a mitochondrial enzyme, is active in the dephosphorylated state, and its activity is regulated by branched-chain α-keto acid dehydrogenase kinase and a phosphoprotein phosphatase. Phenylalanine acts as a positive allosteric effector for the phosphorylation and activation of phenylalanine hydroxylase by cAMP-dependent protein kinase. Branched-chain α-keto acids activate branched-chain α-keto acid dehydrogenase indirectly by inhibiting branched-chain α-keto acid dehydrogenase kinase. Covalent modification of these enzymes provides a very sensitive means for control of the degradation of phenylalanine and the branched-chain amino acids. The clinical experience with phenylketonuria (see Clin. Corr. 12.5) and maple syrup urine disease (see Clin. Corr. 12.1) emphasizes the importance of maintaining low blood and tissue levels of these amino acids. On the other hand, the regulatory mechanism must function to prevent depletion of the body stores of these essential amino acids. Therefore, the tissue requirements for these amino acids supersedes the phase of the starve–feed cycle in establishing the phosphorylation and activity state of these interconvertible enzymes.

Largely because it also contains enzymes subject to covalent modification, adipose tissue responds just as dramatically as liver to the starve–feed cycle. Pyruvate kinase, pyruvate dehydrogenase, acetyl CoA carboxylase, and hormone-sensitive lipase (not found in liver) are all in the dephosphorylated mode in the adipose tissue of the well-fed person. As in liver, the first three enzymes are active in the dephosphorylated mode. Hormone-sensitive lipase is like glycogen phosphorylase, that is, inactive in the dephosphorylated mode. The high insulin : glucagon ratio, the low tissue cAMP concentration, and perhaps a high insulin "messenger" level are believed to be important determinants of the phosphorylation state of the interconvertible enzymes of adipocytes. Lipogenesis within adipose tissue is favored in the well-fed state. During fasting, adipocytes quickly shut down lipogenesis and activate lipolysis. This is accomplished in large part by the phosphorylation of the enzymes described above, a consequence of the decrease in the insulin : glucagon ratio induced by fasting and the resulting increase in cAMP levels and protein kinase activity. In this manner, adipose tissue is transformed from a fat storage tissue into a source of fatty acids for other tissues.

Regulation of the metabolic processes of liver and adipose tissue by hormonal effects upon the interconvertible enzymes is clearly of great importance in the starve–feed cycle. Such effects are probably important in muscle and kidney as well, but less is known about the starve–feed cycle in these tissues. This type of control is like allosteric effectors and substrate supply, a short-term regulatory mechanism, operating on a minute-to-minute basis. Adaptive changes in enzyme activities due to changes in the absolute amounts of key enzymes of a tissue are also subject to hormonal and nutritional factors but require several hours to come into effect.

Adaptive Changes in Enzyme Levels

The adaptive change in enzyme levels is a mechanism of regulation involving changes in the rate of synthesis or degradation of key enzymes involved in metabolic processes. Whereas allosteric effectors and covalent modification affect either the K_m or V_{max} of an enzyme, this mode of regulation involves the actual quantity of an enzyme in the tissue. In other words, because of the influence of hormonal and nutritional factors on its turnover, there are more or fewer enzyme molecules present in the tissue. For example, when a person is maintained in a well-fed or overfed condition, the liver improves its capacity to synthesize fat. To be sure, this can be explained in part by increased substrate supply, as well as appropriate changes in allosteric effectors (Figure 14.7) and the conversion of the interconvertible enzymes into the dephosphorylated mode (Figure 14.9). This is not the entire story, however, because the liver also has more molecules of those enzymes that play a key role in fat synthesis (see Figure 14.11). A whole battery of enzymes is induced, including glucokinase and pyruvate kinase for faster rates of glycolysis; glucose 6-phosphate dehydrogenase, 6-phosphogluconate dehydrogenase, and malic enzyme to provide greater quantities of NADPH for reductive synthesis; and citrate cleavage enzyme, acetyl CoA carboxylase, fatty acid synthase, and Δ^9-desaturase for more rapid rates of fatty acid synthesis. All of these enzymes are present at higher levels in the well-fed state, possibly in response to the increased insulin : glucagon ratios. While these enzymes are induced, there is a decrease in the enzymes that favor glucose synthesis. Phospho*enol*pyruvate carboxykinase, glucose 6-phospha-

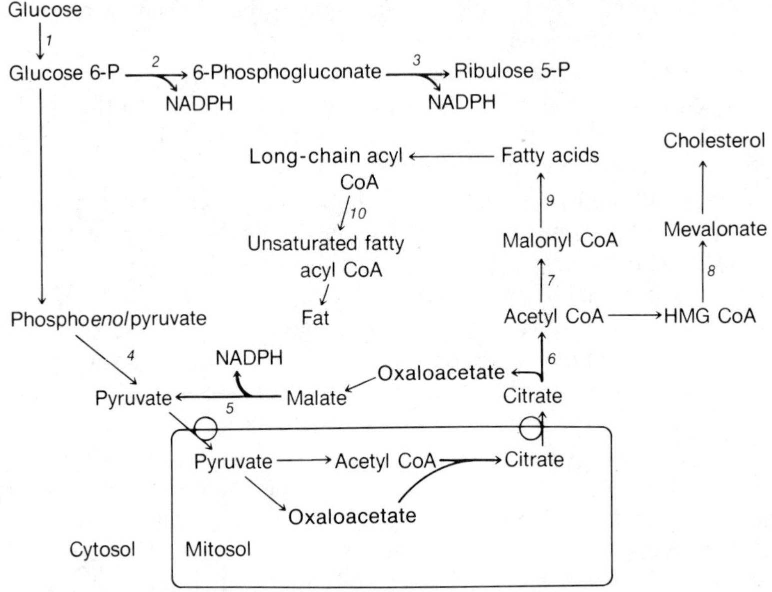

Figure 14.11
Enzymes induced in the liver of the well-fed individual.
The inducible enzymes are numbered: 1, *glucokinase;* 2, *glucose 6-phosphate dehydrogenase;* 3, *6-phosphogluconate dehydrogenase;* 4, *pyruvate kinase;* 5, *malic enzyme;* 6, *citrate cleavage enzyme;* 7, *acetyl CoA carboxylase;* 8, *β-hydroxy-β-methylglutaryl CoA reductase;* 9, *fatty acid synthase; and* 10, *Δ^9-desaturase.*

tase, and some aminotransferases are decreased in amount; that is, their synthesis is reduced or degradation increased in response to increased circulating glucose and insulin.

If a person fasts for several hours, the enzyme pattern characteristic of the liver changes dramatically (Figure 14.12). The enzymes involved in lipogenesis decrease in quantity, possibly because their synthesis is decreased or degradation of these proteins is increased. At the same time a number of enzymes favoring gluconeogenesis are induced (Figure 14.12), making the liver much more effective in synthesizing glucose for the rest of the body. In addition, the enzymes of the urea cycle are induced, possibly by the presence of higher blood glucagon levels. This permits the disposal of nitrogen, as urea, from the alanine used in gluconeogenesis.

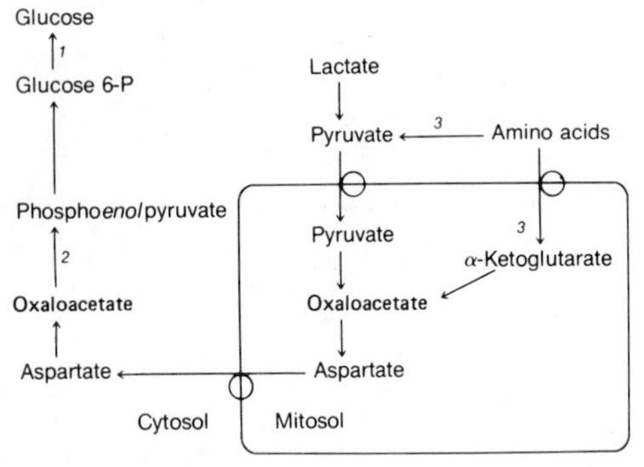

Figure 14.12
Enzymes induced in the liver of an individual during fasting.
The inducible enzymes are numbered: 1, *glucose 6-phosphatase;* 2, *phospho*enol*-pyruvate carboxykinase; and* 3, *various aminotransferases.*

These adaptive changes are clearly important in the starve–feed cycle, greatly affecting the capacity of the liver for its various metabolic processes. Although often overlooked, the adaptive changes also influence the effectiveness of the short-term regulatory mechanisms. For example, long-term starvation or uncontrolled diabetes greatly decreases the level of acetyl CoA carboxylase. Taking away long-chain acyl CoA esters that inhibit this enzyme, increasing the level of citrate that activates this enzyme, or creating conditions that activate this interconvertible enzyme by dephosphorylation will not have any effect when the enzyme is virtually absent from the tissue. Another example is afforded by the glucose intolerance of starvation. A chronically starved person, because of the absence of the key enzymes needed for glucose metabolism, cannot effectively utilize a sudden load of glucose. A glucose load, however, will set into motion the induction of the required enzymes and the reestablishment of short-term regulatory mechanisms.

14.4 METABOLIC INTERRELATIONSHIPS OF TISSUES IN VARIOUS NUTRITIONAL AND HORMONAL STATES

Many of the changes that occur in various nutritional and hormonal states of the human body are just variations on the starve–feed cycle and are completely predictable from what we have learned about the cycle. Some examples are given in Figure 14.13. Others are so obvious that a diagram is not necessary; for example, in rapid growth of a child, amino acids are directed away from catabolism and into protein synthesis. On the other hand, the changes that occur in some physiologically important situations are rather subtle and poorly understood. An example of the latter is aging, which seems to lead to a decreased "sensitivity" of the major tissues of the body to hormones. The important consequence is a decreased ability of the tissues to respond normally during the feed–starve cycle. Whether

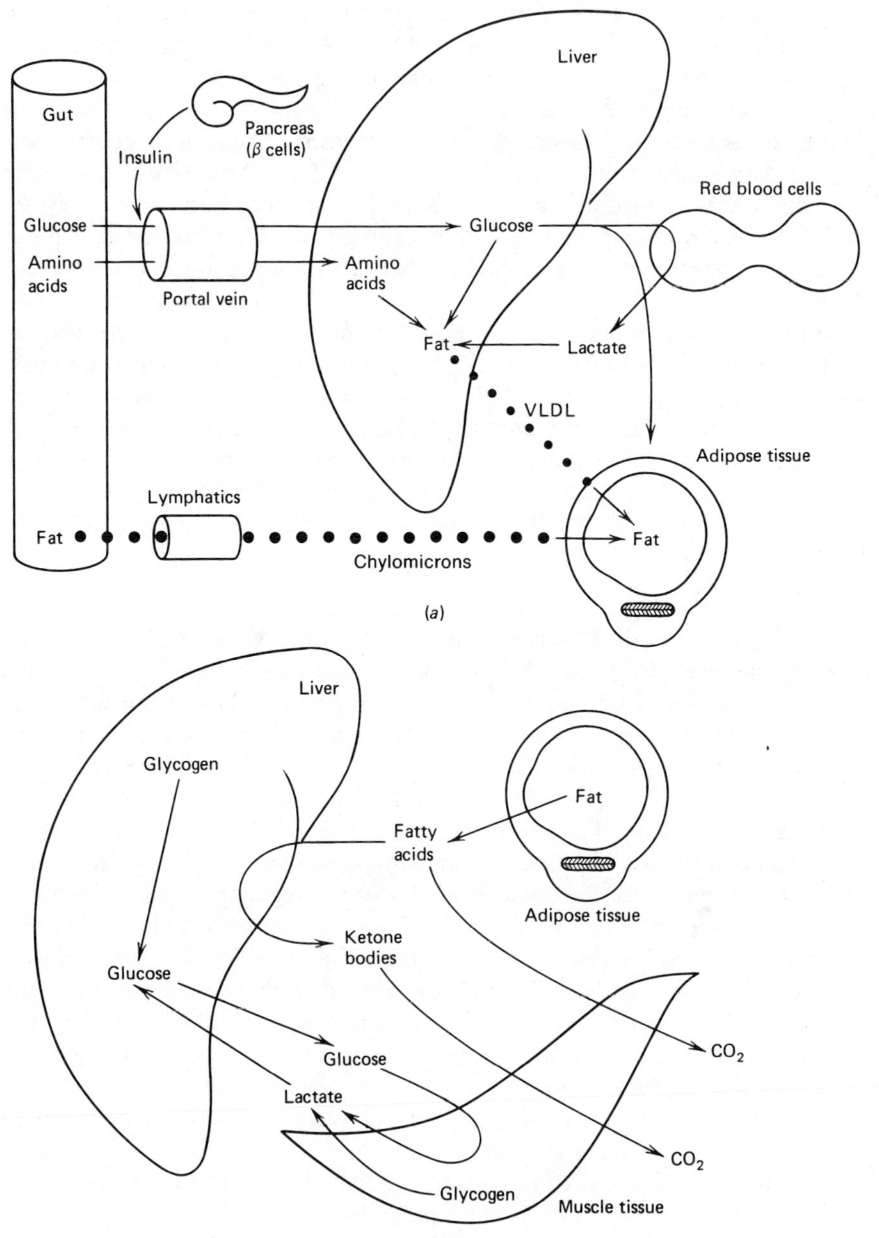

Figure 14.13
Metabolic interrelationships of tissues in various nutritional, hormonal, and disease states.
(a) Obesity. (b) Exercise. *Continued on pages 551–556.*

this is a contributing factor to or a consequence of the aging process is not known.

Obesity

Figure 14.13*a* illustrates the metabolic interrelationships prevailing much of the time in an obese person. Most of the body fat of the human is either provided by the diet or synthesized in the liver and transported to the adipose tissue for storage. Obesity is caused by a person staying in such a well-fed state that stored fuel (particularly fat) does not get used up during the fasting phase of the cycle. The body then has no option other than to accumulate fat.

Exercise

It is important to differentiate between two distinct types of exercise—aerobic and anaerobic. Aerobic exercise is exemplified by long-distance running, anaerobic exercise by sprinting or weight lifting. During anaerobic exercise there is really very little interorgan cooperation. The muscle largely relies upon its own stored glycogen and phosphocreatine. Phosphocreatine serves as a source of high energy phosphate bonds for ATP synthesis until glycogenolysis and glycolysis are stimulated. The blood vessels within these muscles are compressed during peak contraction, thus these cells are isolated from the rest of the body. Aerobic exercise is metabolically more interesting (Figure 14.13*b*). The body in the well-fed state does not store enough glucose and glycogen to provide the energy needed for running long distances. It is also known that the respiratory quotient, the ratio of carbon dioxide exhaled to oxygen consumed, falls during distance running. This indicates the progressive switch to using free fatty acids during the race. Apparently, lipolysis gradually increases as glucose stores are exhausted, and, as in the fasted state, the muscle will oxidize fatty acids in preference to glucose as the former becomes available. Unlike what happens in fasting, there is little increase in blood ketone body concentration. This may simply reflect a balance between hepatic ketone body synthesis and muscle ketone body oxidation.

Pregnancy

The fetus is another nutrient-requiring tissue (Figure 14.13*c*). It mainly uses glucose for energy, but it may also use amino acids, lactate, and ketone bodies. Fatty acids do not cross the placenta to the fetus, but maternal LDL cholesterol is an important precursor of placental steroids. During pregnancy, the starve–feed cycle is perturbed. The placenta secretes a hormonal polypeptide, placental lactogen, and two steroid hormones, estradiol and progesterone. Placental lactogen stimulates lipolysis in adipose tissue, and the steroid hormones seem to induce an insulin-resistant state. In the postprandial state, pregnant women enter the starved state more rapidly than do nonpregnant women. This results from increased consumption of glucose and amino acids by the fetus. Plasma glucose, amino acids, and insulin levels fall rapidly, and glucagon and placental lactogen levels rise and stimulate lipolysis and ketogenesis. The consumption of glucose and amino acids by the fetus may be great enough to cause maternal hypoglycemia. On the other hand, in the fed state pregnant women have increased levels of insulin and glucose and demonstrate resistance to exogenous insulin. These swings of plasma hormones and fuels are even more exaggerated in diabetic women and make control of blood glucose difficult in these patients.

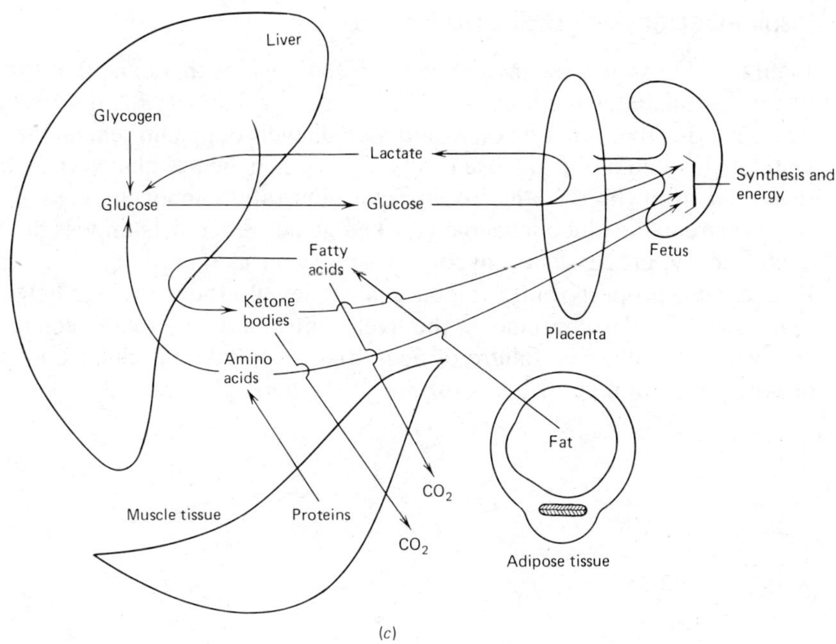

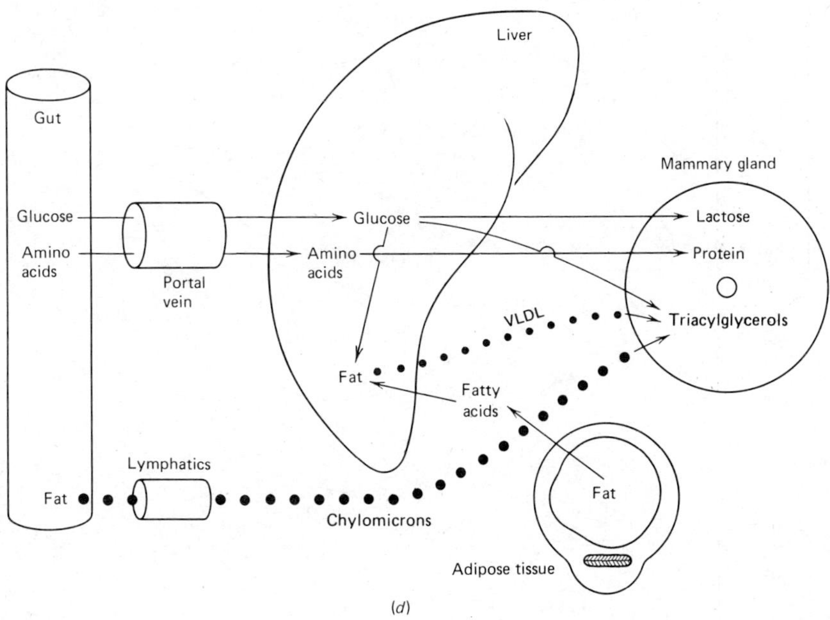

Figure 14.13 (Continued)
(c) Pregnancy. (d) Lactation.

Lactation

In late pregnancy placental hormones induce lipoprotein lipase in the mammary gland and promote the development of milk-secreting cells and ducts. During lactation (see Figure 14.13*d*) the breast utilizes glucose for lactose and triacylglycerol synthesis, as well as its major energy source. Amino acids are taken up for protein synthesis, and chylomicrons and VLDL are utilized as sources of fatty acids for triacylglycerol synthesis. If these compounds are not supplied by the diet, proteolysis, gluconeogenesis, and lipolysis must supply them, resulting eventually in maternal malnutrition and poor quality milk.

CLIN. CORR. 14.7 INSULIN-DEPENDENT DIABETES MELLITUS

Insulin-dependent diabetes mellitus was once called juvenile-onset diabetes because it usually appears in childhood or in the teens, but it is not limited to these patients. Insulin is either absent or nearly absent in this disease because of defective or absent β cells in the pancreas. Untreated, it is characterized by hyperglycemia, hyperlipoproteinemia (chylomicrons and VLDL), and episodes of severe ketoacidosis. Far from being a disease of defects in carbohydrate metabolism alone, abnormalities exist in fat and protein metabolism in such patients as well. The hyperglycemia results in part from the inability of the insulin-dependent tissues to take up plasma glucose and in part by accelerated hepatic gluconeogenesis from amino acids derived from muscle protein. The ketoacidosis results from increased lipolysis in the adipose tissue and accelerated fatty acid oxidation in the liver. Hyperchylomicronemia is the result of low lipoprotein lipase activity in adipose tissue capillaries, an enzyme dependent upon insulin for its synthesis. Although insulin does not cure the diabetes, its use markedly alters the clinical course of the disease. The injected insulin promotes glucose uptake by tissues and inhibits gluconeogenesis, lipolysis, and proteolysis. The life span of the treated diabetic is still decreased, perhaps because it remains impossible to maintain perfect control of metabolism by repeated injections of insulin.

Insulin-Dependent Diabetes Mellitus

Figure 14.13*e* shows the metabolic interrelationships that exist in insulin-dependent diabetes mellitus (see Clin. Corrs. 14.7 and 14.8). Because of defective β-cell production of insulin, blood levels of insulin remain low in spite of elevated blood glucose levels. Even when dietary glucose is being delivered from the gut, the insulin : glucagon ratio cannot increase, and the liver remains gluconeogenic and ketogenic. Since it is impossible to switch to the processes of glycolysis, glycogenesis, and lipogenesis, the liver cannot properly buffer blood glucose levels. Indeed, since hepatic gluconeogenesis is continuous, the liver contributes to hyperglycemia in the well-fed state. The failure of many tissues to take up glucose in the absence of insulin contributes further to the hyperglycemia. Accelerated

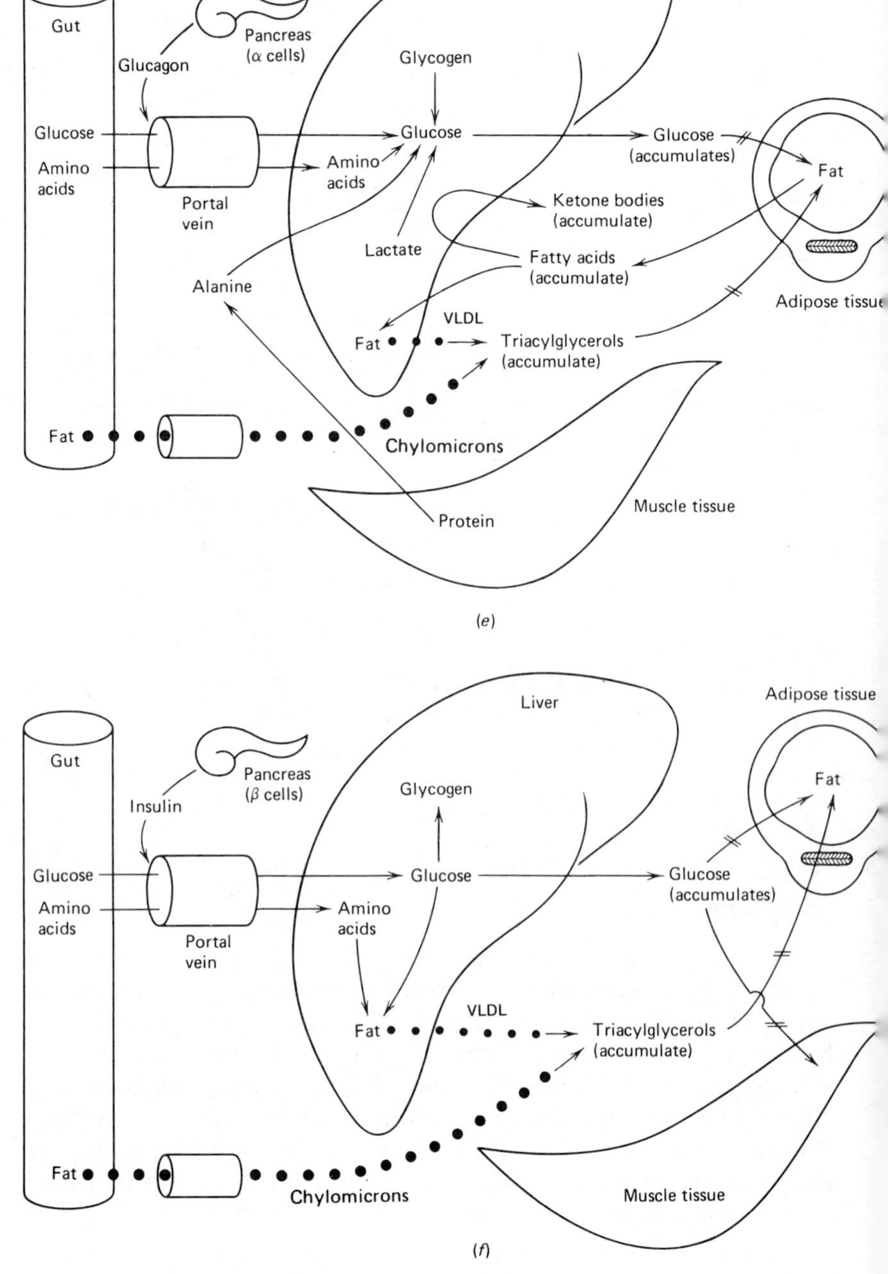

Figure 14.13 (Continued)
(e) Insulin-dependent diabetes mellitus. (f) Non-insulin diabetes mellitus.

gluconeogenesis, fueled by substrate made available by body protein degradation, maintains the hyperglycemia even in the starved state.

It may seem an enigma that hypertriglyceridemia is characteristic of this condition, since fatty acid synthesis is greatly diminished in the diabetic state. However, the low insulin : glucagon ratio results in uncontrolled rates of lipolysis in the adipose tissue. This increases blood levels of fatty acids and results in accelerated ketone body production by the liver. If the ketone bodies are not used as rapidly as they are formed, a dangerous condition, known as ketoacidosis, develops due to the accumulation of ketone bodies and hydrogen ions. Regardless of whether ketoacidosis develops, not all of the fatty acid taken up by the liver can be handled by the pathway of fatty acid oxidation and ketogenesis. The excess is esterified and directed into VLDL synthesis. Hypertriglyceridemia results because VLDL is synthesized and released by the liver more rapidly than these lipid-laden particles can be cleared from the blood by lipoprotein lipase, the activity of which is dependent upon a high insulin : glucagon ratio. The defect in lipoprotein lipase also results in hyperchylomicronemia, since lipoprotein lipase is also required for chylomicron catabolism in adipose tissue.

The most important thing to remember about the diabetic state is that every tissue continues to play the catabolic role that it was designed to play in starvation, in spite of delivery of adequate or even excess fuel from the gut. The consequence is that metabolism becomes stuck in the starve phase of the starve–feed cycle, with life-threatening consequences.

Noninsulin-Dependent Diabetes Mellitus

Figure 14.13*f* shows the metabolic interrelationships characteristic of a person suffering from noninsulin-dependent diabetes. In contrast to insulin-dependent diabetes discussed above, insulin is not absent in noninsulin-dependent diabetes (see Clin. Corr. 14.9). Indeed high levels of insulin may be observed in this form of diabetes, and the problem is primarily insulin resistance rather than lack of insulin. Insulin resistance is a poorly understood phenomenon in which the tissues fail to respond to insulin. The number or affinity of insulin receptors is reduced in some patients; others have normal insulin binding, but abnormal postreceptor responses.

The majority of patients with noninsulin-dependent diabetes mellitus are obese. Their insulin levels, which may be high, are not as high as those of a nondiabetic but similarly obese person (who also has some degree of insulin resistance). Hence, this form of diabetes is also a form of β-cell failure, and exogenous insulin will reduce the hyperglycemia. Hyperglycemia results mainly because of poor uptake of glucose by peripheral tissues. In contrast to insulin-dependent diabetes, ketoacidosis does not develop because uncontrolled lipolysis in the adipose tissue is not a feature of this disease. On the other hand, hypertriglyceridemia is characteristic of noninsulin-dependent diabetes but usually results from an increase in VLDL without hyperchylomicronemia. This is most likely explained by rapid rates of de novo hepatic synthesis of fatty acids and VLDL rather than increased delivery of fatty acids from the adipose tissue.

Stress and Injury

Stress includes injury, surgery, renal failure, burns, and infections (Figure 14.13*g*). Characteristically, blood cortisol, glucagon, catecholamines, and growth hormone levels are increased. The patient is resistant to insulin.

CLIN. CORR. 14.8 COMPLICATIONS OF DIABETES AND THE POLYOL PATHWAY

Diabetes is complicated by several disorders that may share a common pathogenesis. The lens, peripheral nerve, glomerulus, and possibly retinal capillaries contain two enzymes that constitute the polyol pathway (the term polyol refers to polyhydroxy sugars). The first is aldose reductase, an NADPH-requiring enzyme. It reduces glucose to form sorbitol. Sorbitol is further metabolized by sorbitol dehydrogenase, an NAD^+-requiring enzyme that oxidizes sorbitol to fructose. Aldose reductase has a high K_m for glucose; therefore this pathway is only quantitatively important during hyperglycemia. It is known that in diabetic animals the sorbitol content of lens, nerve, and glomerulus is elevated. There are now experimental inhibitors of the reductase that prevent the accumulation of sorbitol in these tissues and that improve some of the symptoms of nerve injury in humans. Sorbitol accumulation may damage these tissues by causing them to swell. Another metabolic disturbance in nerve is a reduction in myo-inositol content. Inositol is a component of the membrane phospholipid phosphatidyl-inositol. By unknown mechanisms, aldose reductase inhibitors also increase the level of inositol in nerve.

CLIN. CORR. 14.9 NONINSULIN-DEPENDENT DIABETES MELLITUS

Noninsulin-dependent diabetes mellitus, which accounts for 80–90% of the diagnosed cases of diabetes, is also called maturity-onset diabetes to differentiate it from insulin-dependent, juvenile diabetes. It usually occurs in middle-aged obese people. Insulin is present at near-normal to elevated levels in this form of the disease. The defect in these patients may be at the level of the insulin receptors located on the plasma membranes of normally insulin-responsive cells, that is, hepatocytes, adipocytes, and muscle cells. Noninsulin-dependent diabetes is characterized by hyperglycemia, often with hypertriglyceridemia. The ketoacidosis characteristic of the insulin-dependent disease is not observed. Increased levels of VLDL are probably the result of increased hepatic triacylglycerol synthesis stimulated by hyperglycemia and hyperinsulinemia. Obesity often precedes the development of insulin-independent diabetes and appears to be the major contributing factor. Obese patients are usually hyperinsulinemic. An inverse relationship between insulin levels and the number of insulin receptors has been established. The higher the basal level of insulin, the fewer receptors present on the plasma membranes. In addition, there are defects within insulin-responsive cells at sites beyond the receptor. The consequence is that insulin levels remain high, but glucose levels are poorly controlled because of the lack of normal responsiveness to insulin. Although the insulin level is high, it is not as high as in a person who is obese but not diabetic. In other words, there is a relative deficiency in the insulin supply from the β cells. Diet alone can most often control the disease in the obese diabetic. If the patient can be motivated to lose weight, insulin receptors will increase in number, and the postreceptor abnormalities will improve, which will increase both tissue sensitivity to insulin and glucose tolerance. The noninsulin-dependent diabetic has a less severe form of the disease and tends not to develop ketoacidosis but nevertheless develops many of the same complications as the insulin-dependent diabetic, that is, nerve, eye, kidney, and coronary artery disease.

The basal metabolic rate and blood glucose and free fatty acid levels are elevated. However, ketogenesis is not accelerated as in fasting, and perhaps as a result of this, the body is less able to preserve protein stores in muscle. It can be very difficult to reverse this protein breakdown, although now it is common to replace amino acids, glucose, and fat by infusing solutions of these nutrients intravenously. It has recently been proposed that the negative nitrogen balance of injured or infected patients is mediated by a monocyte protein, interleukin-1. This polypeptide, which is also responsible for the production of fever, is reported to increase skeletal muscle proteolysis in isolated muscle preparations. Increased production of interleukin-1 may be the common link between the diverse conditions considered to be metabolic stresses (see Clin. Corr. 14.10).

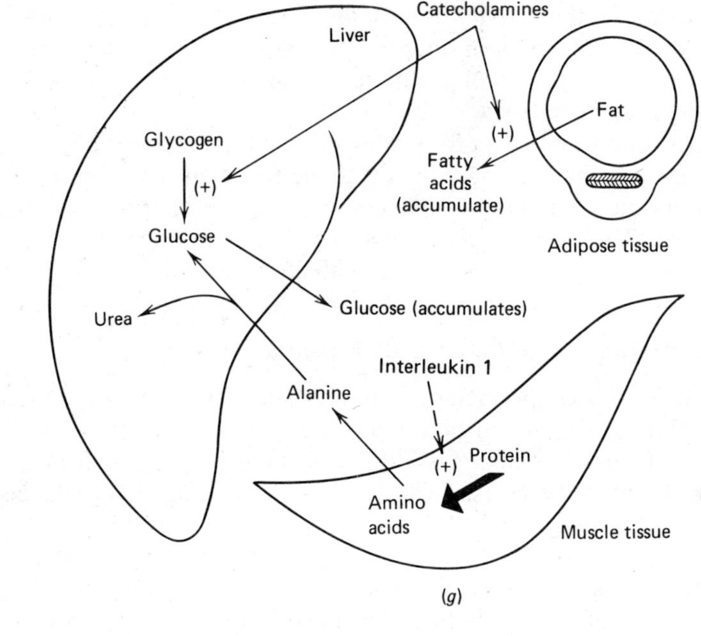

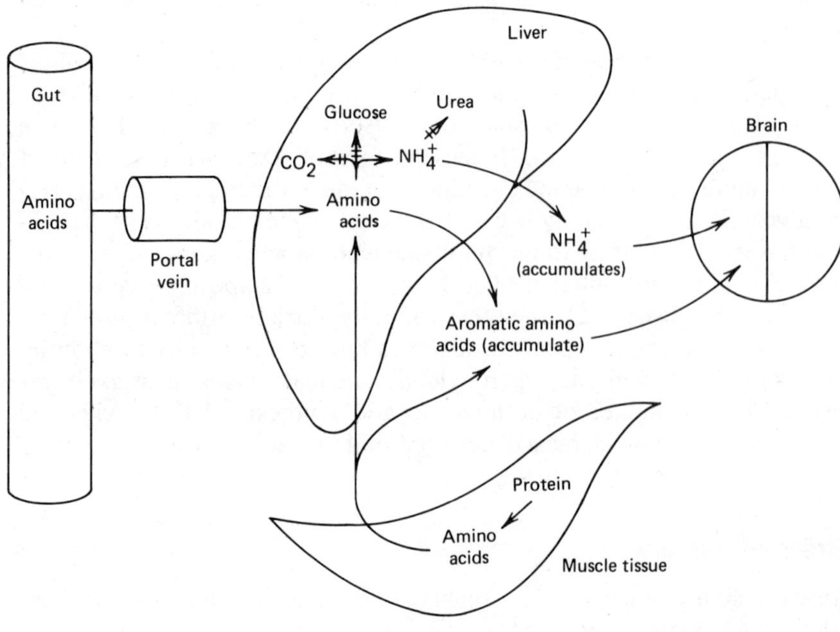

Figure 14.13 (Continued)
(g) Stress. (h) Liver disease.

Liver Disease

Since the liver is central to the body's metabolic interrelationships, liver disease is associated with major metabolic derangements (Figure 14.13*h*). The most important abnormalities are those in the metabolism of amino acids. The liver is the only organ capable of urea synthesis. In patients with advanced liver disease, the liver is unable to convert ammonia into urea rapidly enough, and the blood ammonia rises. Ammonia arises from certain enzyme reactions, such as glutaminase, glutamate dehydrogenase, and adenosine deaminase, during the metabolism of amino acids by the intestines and liver, and from the intestinal lumen, where bacteria can split urea into ammonia and carbon dioxide. Ammonia is very toxic to the central nervous system and is a major reason for the coma that sometimes occurs in patients in liver failure.

In advanced liver disease, aromatic amino acids accumulate in the blood to higher levels than branched-chain amino acids, apparently because of defective hepatic catabolism of the aromatic amino acids. This is important because aromatic amino acids and branched-chain amino acids are transported into the brain by the same carrier system. An elevated ratio of aromatic amino acids to branched-chain amino acids in liver disease results in increased brain uptake of aromatic amino acids. Increased synthesis of neurotransmitters in the brain as a consequence of increased availability of aromatic amino acids has been suggested to be responsible for some of the neurological abnormalities characteristic of liver disease. Finally, in outright liver failure, patients sometimes die of hypoglycemia because the liver is unable to maintain the blood glucose level by gluconeogenesis.

Renal Disease

Nitrogenous wastes, including urea and creatinine, accumulate in patients with renal failure (Figure 14.13*i*). This accumulation is worsened by high dietary protein intake or accelerated proteolysis. The fact that gut bacteria can split urea into ammonia and that the liver can use ammonia and α-keto acids to form nonessential amino acids has been used to control the

CLIN. CORR. 14.10 CANCER CACHEXIA

Unexplained weight loss may be a sign of malignancy, and weight loss is common in advanced cancer. Decreased appetite and food intake contribute to but do not entirely account for the weight loss. The weight loss is largely from skeletal muscle and adipose tissue, with relative sparing of visceral protein (i.e., liver, kidney, heart). Although tumors commonly exhibit high rates of glycolysis and release lactate, the energy requirement of the tumor probably does not explain weight loss, because weight loss is common with even small tumors. In addition, the presence of another energy-requiring growth, the fetus in an expecting mother, does not normally lead to weight loss. Several endocrine abnormalities have been recognized in cancer patients. They tend to be insulin resistant, have higher cortisol levels, and have a higher basal metabolic rate compared with controls matched for weight loss. Two other phenomena may contribute to the metabolic disturbances. Some tumors synthesize and secrete biologically active peptides such as ACTH, nerve growth factor, and insulin-like growth factors, which could modify the endocrine regulation of energy metabolism. It is also possible that the host response to a tumor, by analogy to chronic infection, includes release of interleukin-1, a polypeptide that induces fever and that may signal increased rates of skeletal muscle proteolysis.

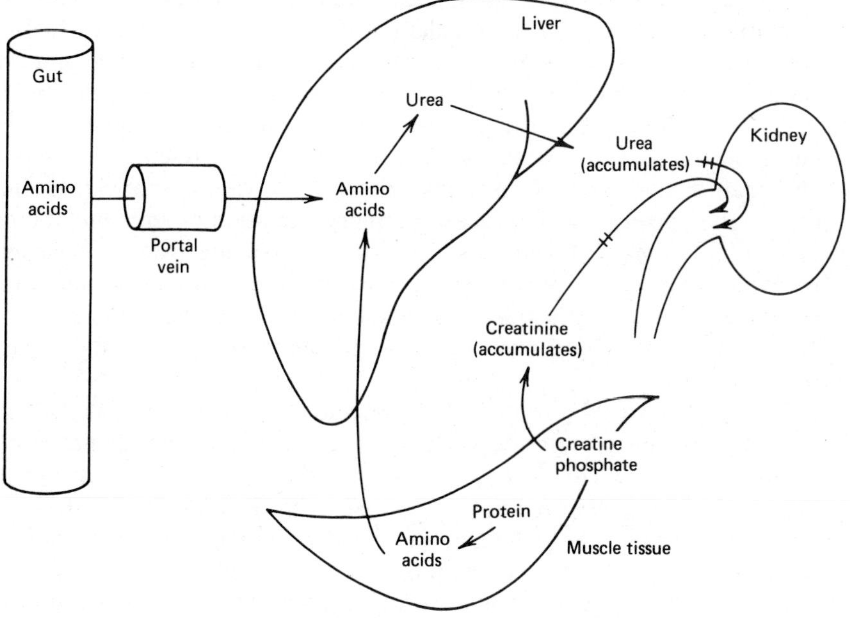

***Figure 14.13* (*Continued*)**
(*i*) *Kidney failure.*

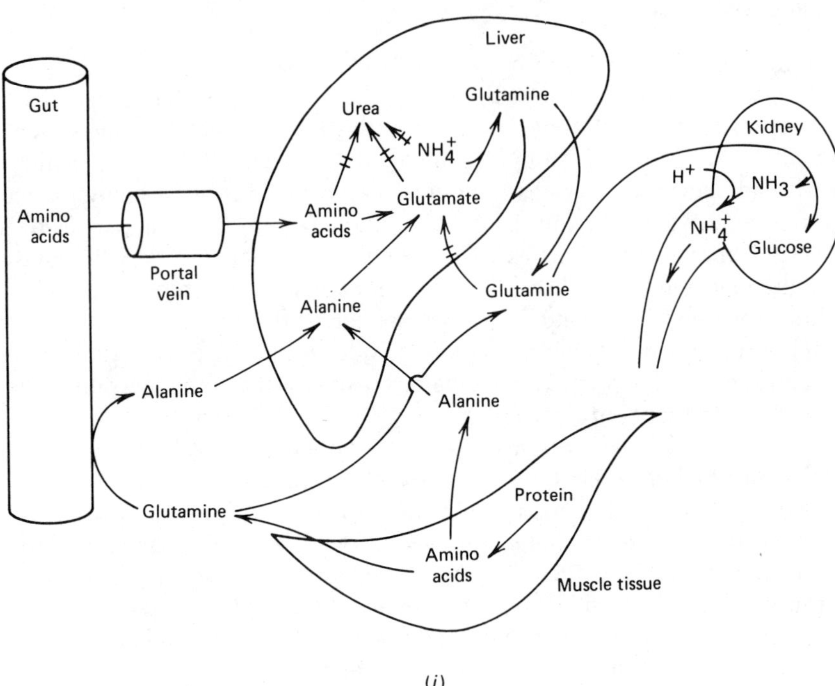

***Figure 14.13* (*Continued*)**
(j) Acidosis.

level of nitrogen wastes in renal patients. The patients are given a diet high in carbohydrate calories, and the amino acid intake is limited as much as possible to essential amino acids. Under these circumstances, the liver synthesizes nonessential amino acids from TCA acid cycle intermediates. This type of diet therapy may extend the time before the renal failure patient requires dialysis.

Acid–Base Disorders

The regulation of acid–base balance, like that of nitrogen excretion, is shared by the liver and kidney. This topic has recently become controversial. The classical view holds that metabolism of proteins generates excess hydrogen ions, which must be excreted by the kidney. The kidney is primarily responsible for the regulation of blood pH by excreting hydrogen ions, which is necessary for the reabsorption of bicarbonate and the titration of phosphate and ammonia in the tubular filtrate (see Chapter 23, page 897). Glutamine is the precursor of renal ammonia production. In chronic acidosis (see Figure 14.13*j*), the activities of renal glutaminase, glutamate dehydrogenase, and phospho*enol*pyruvate carboxykinase increase and correlate with increased urinary excretion of ammonium ions and increased renal gluconeogenesis from amino acids. The liver participates in this process by synthesizing less urea, which makes more glutamine available for the kidney. In alkalosis, urea synthesis increases in the liver, and gluconeogenesis and ammonium ion excretion by the kidney decrease.

A recent departure from this explanation has been suggested. This view argues that protein metabolism produces a large bicarbonate load not an acid load. The urea cycle is viewed as a means of consuming bicarbonate (or generating acid) by this balanced equation: $2\ NH_4^+ + 2\ HCO_3^- \rightarrow \text{urea} + 3\ H_2O + CO_2$. The liver, by modulation of the rate of urea synthesis, is primarily responsible for the regulation of blood pH. When urea synthesis in the liver is diminished, glutamine is used by the kidney as an alternate route of ammonia excretion. Thus, during acidosis, less bicarbonate is available, urea synthesis is slowed to conserve bicarbonate, and ammonia is excreted in the urine. During alkalosis, urea synthesis increases to consume bicarbonate, leaving less ammonia for glutamine synthesis and less need for renal ammonia excretion. Renal ammonia excretion therefore falls.

Regardless of which mechanism is involved, the muscle supplies glutamine to the liver for ureagenesis and to the kidney for ammonia and glucose synthesis. The partitioning of glutamine between these alternative fates appears to be determined by plasma pH, acting by affecting the activities of hepatic glutamine synthetase (favored by acidosis) and glutaminase (favored by alkalosis). High hepatic glutaminase activity favors ureagenesis; conversely, high rates of glutamine synthesis allow renal ammonia excretion.

14.5 SUMMARY

This chapter has stressed the working relationship between liver, kidney, muscle, gut, and adipose tissue in the maintenance of caloric and nitrogen homeostasis during the starve–feed cycle. The liver functions at the center of this relationship, switching from an organ primarily involved in the synthesis of glycogen and fat in the well-fed state into an organ primarily involved in the synthesis of glucose and ketone bodies in the fasting state. This dramatic transformation is brought about by the combined effects of several different regulatory mechanisms, including substrate supply, allosteric effectors, covalent modification, and induction–repression. The metabolic response of the major tissues of the body to various diseases is often analogous to that characteristic of the starve–feed cycle.

BIBLIOGRAPHY

Atkinson, D. E., and Bourke, E. The role of ureagenesis in pH homeostasis. *Trends Biochem. Sci.* 9:297, 1984.

Brownlee, M., Vlassara, H., and Cerami, A. Nonenzymatic glycosylation and the pathogenesis of diabetes complications. *Ann. Intern. Med.* 101:527, 1984.

Cahill, G. F., Jr. Diabetes mellitus: a brief overview. *Johns Hopkins Med. J.* 143:155, 1978.

Cerami, A., and Koenig, R. H. Hemoglobin A_{Ic} as a model for the development of the sequelae of diabetes mellitus. *Trends Biochem. Sci.* 3:73, 1978.

Cohen, P. *Control of enzyme activity*. New York: Wiley, 1976.

Denton, R. M., and Pogson, C. I. *Metabolic regulation*. New York: Wiley, 1976.

Foster, D. W. Banting lecture 1984. From glycogen to ketones and back. *Diabetes* 33:1188, 1984.

Geelen, M. J. H., Harris, R. A., Beynen, A. C., and McCune, S. A. Short-term hormonal control of hepatic lipogenesis. *Diabetes* 29:1006, 1980.

Gibson, D. M., and Parker, R. A. Control of HMG CoA reductase by reversible phosphorylation, in E. G. Krebs (ed.), *The enzymes* (series): *Enzyme control by phosphorylation*. New York: Academic Press, 1985 (in press).

Goldberg, A. L., Baracos, V., Rodemann, P., Waxman, L., and Dinarello, C. Control of protein degradation in muscle by prostaglandins, calcium, and leukocytic pyrogen (interleukin 1). *Federation Proc.* 43:1301, 1984.

Hers, H. G., and Hue, L. Gluconeogenesis and related aspects of glycolysis. *Annu. Rev. Biochem.* 52:617, 1983.

Ingebritsen, T. S., and Cohen, P. Protein phosphatases: properties and role in cellular regulation. *Science (Wash. D.C.)* 221:331, 1983.

Krebs, H. A. Some aspects of the regulation of fuel supply in omnivorous animals. *Adv. Enzyme Regul.* 10:387, 1972.

Krebs, H. A., Williamson, D. H., Bates, M. W., Page, M. A., and Hawkins, R. A. The role of ketone bodies in caloric homeostasis. *Adv. Enzyme Regul.* 9:387, 1971.

Larner, J. *Intermediary metabolism and its regulation*. Englewood Cliffs, N.J.: Prentice-Hall, 1971.

Newsholme, E. A., and Leech, A. R., *Biochemistry for the medical sciences*. New York: Wiley, 1983.

Newsholme, E. A., and Start, C. *Regulation in metabolism*. New York: Wiley, 1973.

Roth, J. Insulin receptors in diabetes. *Hosp. Pract.* May 1980, p. 98.

Ruderman, N. B., Aoki, T. T., and Cahill, G. F., Jr. Gluconeogenesis and its disorders in man, in R. W. Hanson, and M. A. Mehlman (eds.), *Gluconeogenesis, its regulation in mammalian species*. New York: Wiley, 1976, p. 515.

Williamson, D. H., and Whitelow, E. Physiological aspects of the regulation of ketogenesis. *Biochem. Soc. Symp.* 43:137, 1978.

QUESTIONS

C. N. ANGSTADT AND J. BAGGOTT

Question Types are described inside the front cover.

A. Well-fed state
B. Early fasting state
C. Fasting state
D. Early refed state

1. (QT5) Hepatic glycogenolysis is a primary source of blood glucose during this period.
2. (QT5) Ketone bodies supply a significant portion of the brain's fuel.
3. (QT5) The Cori cycle is interrupted since the liver is primarily in a glycolytic rather than gluconeogenic state.
4. (QT3) A. Variation in blood glucose concentration between the fed and fasted states.
   B. Variation in blood fatty acid concentration between the fed and fasted states.
5. (QT1) The fact that the K_m of aminotransferases for amino acids is much higher than that of aminoacyl-tRNA synthetases means that:
   A. at low amino acid concentrations, protein synthesis will take precedence over amino acid catabolism.
   B. the liver cannot accumulate amino acids.
   C. amino acids will undergo transamination as rapidly as they are delivered to the liver.
   D. any amino acids in excess of immediate needs for energy must be converted to protein.
   E. amino acids can be catabolized only if they are present in the diet.
6. (QT1) Branched-chain amino acids:
   A. are normally completely catabolized by muscle to CO_2 and H_2O.
   B. can be catabolized by liver but not muscle.
   C. are the main dietary amino acids metabolized by intestine.
   D. are in high concentration in blood following the breakdown of muscle protein.
   E. are a major source of nitrogen for alanine and glutamine produced in muscle.
7. (QT1) The largest energy reserve (in terms of kilocalories) in humans is:
   A. blood glucose.
   B. liver glycogen.
   C. muscle glycogen.
   D. adipose tissue triacylglycerol.
   E. muscle protein.

Use the accompanying figure (representing hours postfeeding) to answer questions 8 and 9.

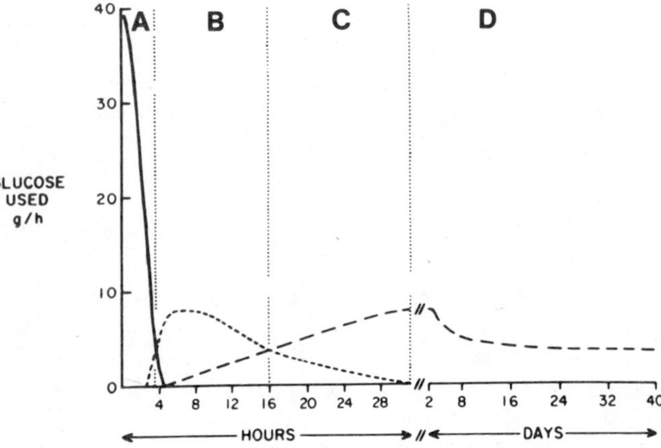

8. (QT5) the origin of most of the glucose in the period represented by ________ is liver glycogen.
9. (QT5) Ketone bodies are a significant fuel for brain as well as muscle in the period ________.
10. (QT2) Which of the following may represent control of a metabolic process by substrate availability?
    1. Increased urea synthesis after a high protein meal
    2. Rate of ketogenesis
    3. Hypoglycemia of advanced starvation
    4. Response of glycolysis to fructose 2,6-bisphosphate
11. (QT2) Which of the following would favor gluconeogenesis in the fasted state?
    1. Fructose 1,6-bisphosphate stimulation of pyruvate kinase
    2. Long-chain acyl CoA ester inhibition of acetyl CoA carboxylase
    3. Malonyl CoA inhibition of carnitine palmitoyltransferase I
    4. Acetyl CoA activation of pyruvate carboxylase
12. (QT2) Conversion of a nonphosphorylated enzyme to a phosphorylated one:
    1. usually changes its activity.
    2. may be catalyzed by a cAMP-dependent protein kinase.
    3. is favored when inhibitor-1 is phosphorylated.
    4. is more likely to occur in the well-fed than in the fasted state.
13. (QT1) Adipose tissue responds to low insulin : glucagon by:
    A. dephosphorylating the interconvertible enzymes.
    B. stimulating the deposition of fat.
    C. increasing the amount of pyruvate kinase.
    D. stimulating hormone-sensitive lipase.
    E. stimulating phenylalanine hydroxylase.
14. (QT2) Changing the level of enzyme activity by changing the number of enzyme molecules:
    1. is considerably slower than allosteric or covalent modification methods.
    2. may involve enzyme induction.
    3. may override the effectiveness of allosteric control.
    4. may be caused by hormonal influences or by changing the nutritional state.
15. (QT1) Muscle metabolism during exercise:
    A. is the same in both aerobic and anaerobic exercise.
    B. shifts from primarily glucose to primarily fatty acids as fuel during aerobic exercise.
    C. uses largely glycogen and phosphocreatine in the aerobic state.
    D. causes a sharp rise in blood ketone body concentration.
    E. uses only phosphocreatine in the anaerobic state.

A. Insulin-dependent diabetes mellitus — C. Both
B. Noninsulin-dependent diabetes mellitus — D. Neither

16. (QT4) Hepatic gluconeogenesis occurs in both the well-fed and fasted states.
17. (QT4) Hypertriglyceridemia and ketoacidosis are commonly present in the untreated state.

ANSWERS

1. B The response of glycogenolysis to fasting is rapid, and during this period there is still glycogen present. In fasting, the glycogen is depleted and in the other two states, glycogenesis would occur (p. 535).
2. C If ketone body concentration in blood is high, ketone bodies can cross the blood-brain barrier and they are a good fuel. High ketone body concentrations do not occur in the other states (p. 537).
3. A The Cori cycle involves glycolysis in peripheral tissues and gluconeogenesis in liver. This is the only state of those listed in which the liver is more likely to be carrying out glycolysis than gluconeogenesis (p. 533).
4. B Various control mechanisms keep blood glucose concentration within tight limits, but fatty acid concentration can vary by an order of magnitude from very low in the fed state (when most plasma lipid is in the form of chylomicrons and VLDL triacylglycerols) to high in the fasted state when mobilized by the low insulin : glucagon ratio (p. 539).
5. A A high K_m means that a reaction will proceed slowly at low concentration, whereas a low K_m means the reaction can be rapid under the same circumstances. Protein synthesis requires only that all amino acids be present. Unless amino acids are in high enough concentration, the liver does not catabolize them (p. 534).
6. E A, B: Muscle has high levels of the aminotransferases for branched-chain amino acids, whereas liver has high levels of enzymes for the catabolism of the branched-chain α-keto acids. C: Intestine metabolizes several dietary amino acids but not these. D, E: When branched-chain amino acids are derived from muscle protein, transamination transfers the nitrogen to alanine or glutamine, which are transported to the liver and kidney (p. 534).
7. D A: Blood glucose must be maintained but is a relatively minor reserve. B, C: Glycogen is a rapidly mobilizable reserve of energy but not a large one. E: Protein can be used for energy, but that is not its primary role. D: The caloric content of adipose tissue fat is more than 5 times as great as that of muscle protein and almost 200 times as great as that of the combined carbohydrates (Table 14.2).
8. B Glycogen is the most rapidly mobilized source of glucose in fasting but lasts for only a little more than a day. A represents dietary glucose.
9. D The decreased rate of gluconeogenesis indicates that the brain is using ketone bodies and muscle proteolysis is restricted, conserving body protein (Figure 14.6).
10. A 1, 2, 3 correct. 1: After a high protein meal, the intestine produces ammonia and precursors of ornithine for urea synthesis. 2: Ketogenesis is dependent on the availability of fatty acids. 3: This represents lack of gluconeogenic substrates. 4: Fructose 2,6-bisphosphate is an allosteric effector (activates the kinase and inhibits the phosphatase) of two enzymes controlling glycolysis (p. 541).
11. C 2, 4 correct. 1: Stimulation of pyruvate kinase stimulates glycolysis, opposing gluconeogenesis. 2, 3: Decreased activity of acetyl CoA carboxylase results in decreased synthesis of malonyl CoA and greater transport of fatty acids into mitochondria for β-oxidation, a necessary source of energy for gluconeogenesis. 4: Pyruvate carboxylase is a key gluconeogenic enzyme (p. 542).
12. A 1, 2, 3 correct. 1: Some enzymes are active when phosphorylated; for others the reverse is true. 2: This is the most common, though not only, mechanism of phosphorylation. 3: Inhibitor-1 (active when phosphorylated) inhibits phosphoprotein phosphatase. 4: In the well-fed state, insulin : glucagon is high and cAMP levels are low (p. 544).
13. D A: Low insulin : glucagon means high cAMP and, thus, high activity of cAMP-dependent protein kinase and protein phosphorylation. B, D: Phosphorylation activates hormone-sensitive lipase to mobilize fat. C: cAMP works by covalent modification of enzymes. E: This is a liver enzyme (p. 545).
14. E All four correct. 1: Adaptive changes are examples of long-term control. 2, 4: Both hormonal and nutritional effects are involved in inducing certain enzymes and/or altering their rate of degradation. 3: If there is little or no enzyme because of adaptive changes, allosteric control is irrelevant. This is important to keep in mind in refeeding a starved person (p. 547).
15. B A: Anaerobic muscle uses glucose almost exclusively; aerobic muscle uses fatty acids and ketone bodies. B: This is indicated by the drop in the respiratory quotient. D: Ketone bodies are good aerobic substrates so the blood concentration does not increase greatly. E: Phosphocreatine is only a short-term source of ATP (p. 550).
16. A Because the defect is an inability of the β cells of the pancreas to produce insulin, the insulin : glucagon ratio is always low and gluconeogenesis is stimulated. Insulin levels may be high in type B, and hyperglycemia is primarily because of poor uptake by peripheral tissues (p. 552).
17. A Hypertriglyceridemia is present in both types, although for different reasons, but ketoacidosis is common only in the insulin-dependent type, again because the low insulin : glucagon ratio results in excessive lipolysis of adipose tissue (p. 552).

15

Biochemistry of Hormones I: Hormone Receptors, Steroid and Thyroid Hormones

FRANK UNGAR

15.1 BIOCHEMICAL REGULATION

Enzyme Synthesis and Activation

The purpose of Chapters 15 and 16 is to classify by name and function those substances called *hormones* that are involved in the transmission of information from one organ to another and from cell to cell. In reading these chapters, it would be well to review those sections in Chapters 7 and 14 that discuss metabolic control.

The controlled entry of required substrates and the exclusion of noxious agents are fundamental properties of the cell. These events are controlled by the permeability characteristics of the membrane and by specific transport systems within the cell. Nuclear activity and the energy-producing systems that depend on the access of suitable substances for metabolism, in turn regulate the production of protein and lipid components of the membrane and supply the energy to maintain these same protective membrane barriers. At any instant there is an interdependence among nuclear, cellular, and membrane events, which involves a communication network of substances of varying forms and modes of action. Every action is in response to a preceding one, which sets the pattern for subsequent events. In most cases we are able to define these changes in ionic and substrate concentrations or of enzyme activities as a result of chemical or physical modifications. Many of the changes in cellular activity are modified by extracellular signals or messages and involve chemical and physical modifications of simple substances and macromolecules within the cell.

Cellular activity is modified by the flow of substrate within the cells, the turnover of enzyme protein, and, at the enzyme level, essentially five basic mechanisms:

1. Chemical modification of enzyme protein by phosphorylation and dephosphorylation, as in phosphorylase and glycogen synthetase
2. Cleavage of a covalent bond in the polypeptide chain, as in the conversion of trypsinogen to trypsin
3. Association by aggregation, or disassociation of peptide subunits, as in removal of a regulatory subunit by cAMP to activate protein kinase
4. Allosteric modification, which leads to a conformational change, such as the effect of concentration levels of AMP and ATP on phosphofructokinase and fructose 1,6-bisphosphatase
5. Cofactors, small molecules, or ions acting on enzyme complexes as prosthetic groups or as activators, as in Ca^{2+}- and calmodulin-regulated events

Examples of these mechanisms as a result of cellular responses to hormone stimulation will be seen in the following sections. Hormones are best thought of as modifying agents; they do not directly initiate or terminate enzyme activity. In addition, it is worth noting that expression of enzyme activity, once initiated, is the same irrespective of the source of stimulation, endocrine or nonendocrine. The cell can alter its activities with respect to growth and differentiation by increasing or decreasing enzyme or protein synthesis. Induction and turnover of enzymes as proposed in the Jacob–Monod model, is described in detail in Chapter 20.

The flow of this information could be controlled effectively at each of the five steps of initiation, transcription, translation, posttranslational modification and degradation, any one of which could be the possible site of hormone action (Figure 15.1). Considerable evidence places sites of hormonal control at the levels of initiation and transcription.

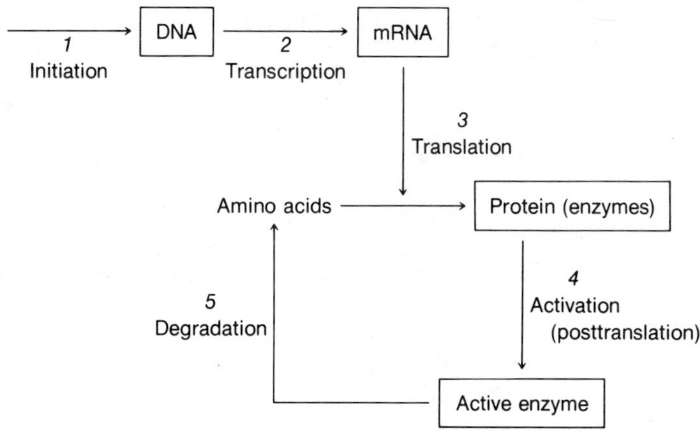

Figure 15.1
Sites of control of enzyme activity levels.

In the human a number of the regulatory systems have been defined, which aid communication within cells and between groups of cells or tissues. Whereas microorganisms have limited defenses against changes in the environmental medium, multicellular organisms are so organized that the extracellular environment itself can be controlled, and communication links in this environment over great distances can be maintained. This control consists of both neural and hormonal elements.

A number of agents involved in these regulatory systems are described. To understand their involvement in biochemical regulation we first must characterize these compounds chemically, and then describe their role with respect to hormone action at the molecular level.

Definitions

The group of compounds discussed in this and the next chapter are involved in the control and regulation of cellular activity. These agents, which have been designated as *hormones,* are listed in Table 15.1 according to the tissue of origin and primary function. The role of the hormones in metabolic activity and their interactions comprise a tightly controlled organized pattern, which in its entirety is commonly referred to as the *endocrine system.*

It is useful at this point to compare the actions of *vitamins* and *hormones* and to differentiate the two classes of compounds. Both consist of chemically heterogeneous groups of substances. *Vitamins* are organic substances present in the diet in trace amounts, which are necessary for maintaining normal growth, reproduction, and health. Compounds classified as vitamins either cannot be synthesized in the body or cannot be formed in amounts sufficient for normal needs. In the instances where biochemical role and function are well established, vitamins have been demonstrated to serve as cofactors or coenzymes and interact directly with enzyme systems.

Hormones are organic substances produced in trace amounts by specific cells and secreted directly into the bloodstream, where they circulate and travel to other parts of the body to produce a biological effect. In order to induce a biological response in a cell, the hormone must first bind as a ligand to a specific protein called a *receptor.*

The classification of substances as hormones or vitamins has been established primarily by historical precedent and may be modified as new information is acquired. It should be recognized that thiamine, riboflavin, and other substances of the vitamin B group are classified as vitamins due to their dietary requirement in human nutrition, but in plants, where these substances can be synthesized, their actions on growth and development qualify these agents as true plant hormones. Ascorbic acid, or vitamin C,

TABLE 15.1 The Principal Hormones and Their Actions

Gland of Origin	*Hormone Name (Symbol)*	*Primary Actions*
Hypothalamus	Thyrotropin releasing factor (TRF)	Release of pituitary thyrotropin (TSH)
	Gonadotropin releasing factor (LH/FSH-RF)	Release of both pituitary FSH and LH
	Somatostatin or somatropin release inhibiting factor (SRIF)	Inhibits the release of pituitary growth hormone
	Corticotropin releasing factor (CRF)	Release of ACTH
(Stored in posterior pituitary)	Vasopressin (antidiuretic hormone, ADH)	Contraction of blood vessels, kidney reabsorption of water
	Oxytocin	Stimulates uterine contraction, milk ejection
	Vasotocin	Maintains water balance (nonmammalian species)
(Median eminence)	Melanocyte-stimulating hormone (MSH)	Dispersion of pigment granules
Anterior Pituitary	Somatotropin or growth hormone (STH or GH)	Growth of body, organs, and bones
	Thyrotropin (TSH)	Size and function of thyroid
	Adrenocorticotropic hormone (ACTH)	Size and function of adrenal cortex
	Follicle-stimulating hormone (FSH)	Growth of Graafian follicle, spermatogenesis (with LH)
	Luteinizing hormone (LH); interstitial cell stimulating hormone, (ICSH)	Causes ovulation *with* FSH, formation of testosterone and progesterone in interstitial cells
	Prolactin, mammotropin (luteotropin)	Growth of mammary gland, lactation, corpus luteum function
	Lipotropin (fat-mobilizing factor)	Release and oxidation of fats from adipose tissue
Parathyroid	Parathyroid hormone	Increases blood calcium Excretion of phosphate by kidney
Parathyroid and thyroid	Calcitonin	Lowers blood calcium
Thyroid	Thyroxine (T_4) Triiodothyronine (T_3)	Growth and maturation and metabolic rate Metamorphosis
Pancreatic islets		
β cells	Insulin	Hypoglycemic factor Regulation of CHO, fats, proteins
α cells	Glucagon	Liver glycogenolysis
Adrenal medulla	Epinephrine	Liver and muscle glycogenolysis
	Norepinephrine	

TABLE 15.1 (*Continued*)

Gland of Origin	*Hormone Name (Symbol)*	*Primary Actions*
Adrenal cortex	Cortisol	Carbohydrate metabolism
	Aldosterone	Mineral metabolism
	Adrenal androgens	androgenic activity (esp. females)
Pineal gland (epiphysis)	Indoles, serotonin, and melatonin	Effects on biological rhythms and brain function
		Counteracts MSH activity
Ovaries	Estrogens	Estrous cycle, female sex properties
	Progesterone	Secretory phase (with estrogens) of uterus and mammary glands
	Relaxin	Relaxes symphysis pubis for birth
Testis	Testosterone and androgens	Male sex properties and spermatogenesis
Placenta	Placental lactogenic hormone	Growth hormone–prolactin activity
	Chorionic gonadotropin, estrogen, progesterone	Adjunct to other endocrine glands in 2nd and 3rd stages of pregnancy
Kidney	Renin	Hydrolysis of blood precursor protein to yield angiotensin
Prostate, gonads, many tissues	Prostaglandins (PG)	Many effects at membrane site of synthesis
Gastrointestinal (GI) tract	Gastrin	Stimulates parietal cell secretions
	Secretin	Stimulates pancreatic juice
	Cholecystokinin	Contraction of gallbladder
Brain	Endorphins: β-endorphin, enkephalins	Endogenous peptides which bind to morphine receptor

is a vitamin in human nutrition, and in the guinea pig as well, because an enzyme which converts gulonolactone to ascorbic acid is missing. Since the rat has this enzyme and ascorbic acid can be synthesized in its tissues, ascorbic acid is not a vitamin for this species. Vitamin D_3, or cholecalciferol, has been classified as a vitamin in human nutrition because of its low rate of conversion from a precursor sterol under limiting (no sunlight) conditions. It is now recognized that the cholecalciferol precursor, which originates in cells in the skin, is modified by liver and kidney action to a more active form, which stimulates Ca^{2+} ion absorption in the mucosal cells of the gastrointestinal (GI) tract. By virtue of its conversion by hydroxylation in a specific tissue to form a derivative that affects other cells of the body, vitamin D_3 can be considered to be a hormone.

Vitamin D_3 shares one other important attribute with substances classified as hormones. As is the case with other steroid compounds and with thyroid hormones, vitamin D_3 binds to receptors within the cell nucleus.

Response of the cell to hormone stimulation in general is considered to involve gene activation with increased protein or enzyme synthesis due to the DNA-directed synthesis of messenger RNA. A large body of evidence has accumulated both for those hormones that bind to nuclear receptors and for the hormones that bind to cell plasma membrane receptors to support the concept of nuclear activation as an integral part of hormone action.

Secretory Cells

One further distinction may be useful in differentiating endocrine cells and other cell types and their secretions (Figure 15.2). Cells are referred to as *endocrine* cells (Figure 15.2*a*) if they produce and secrete active substances (hormones) directly into the bloodstream. The secretion of the hormone insulin by the pancreatic β cell is an example. *Neuroendocrine* cells (Figure 15.2*b*) are secretory nerve cells. A substance, for example vasopressin, produced by nerve cells in the hypothalamus and secreted by the neurohypophysis directly into the bloodstream, is called a *neurohormone*. *Paracrine* cells (Figure 15.2*c*) produce substances that affect adjacent cells directly without transport into the circulating blood. Pancreatic somatostatin is an example. These substances by definition would not be considered hormones. Substances released by nerve cells (Figure 15.2*d*) into the synaptic cleft to stimulate contiguous nerve cells are called *neurotransmitters*. These substances, such as norepinephrine or acetylcholine, are secreted and inactivated locally at the synaptic cleft and therefore would not fit the classical definition of hormone. However, epinephrine and norepinephrine also are produced by the adrenal medulla and are secreted directly into the circulating blood. Since they have effects on muscle and liver carbohydrate metabolism, they traditionally have been listed as hormones.

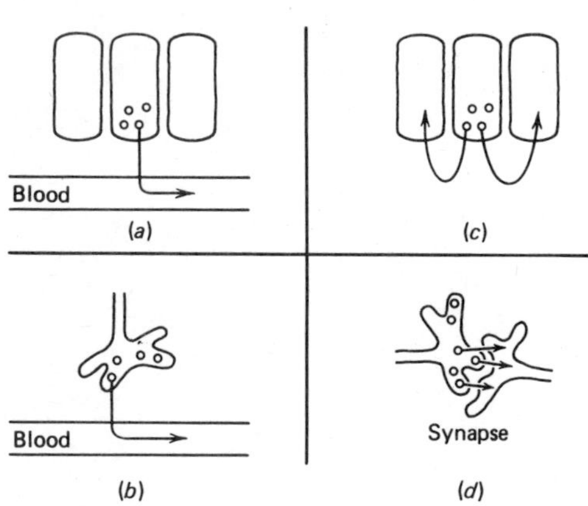

Figure 15.2
Secretory cells.
(a) Endocrine cell. (b) Neuroendocrine cell. (c) Paracrine cell. (d) Nerve cell.

Classification of cells and of their secretions into distinct groups is now considered of less importance due to the recent demonstrations that many agents, including hormones and growth factors, come from cells from different parts of the body and have apparently common actions in cell regulation and communication.

Peptide Hormone Assays

When biologically active substances, particularly if they are peptides or proteins, are present in circulating plasma in trace amounts, the tasks of isolation, characterization, and quantitation are difficult. The usual spectrophotometric or analytical procedures are not sensitive at the levels (10^{-7} to 10^{-11} M) at which most hormones circulate, so that radioactive tracers have to be employed. The isolation and assay of biologically active substances, if their chemical structure is not known or when concentrations are extremely low, depend upon procedures which by necessity are indirect and less reliable. The assessment and validity of measurement of a trace substance therefore rests upon an accumulation of data from separate, independent analyses. The use of any one assay by itself usually is not sufficient to provide precise information concerning peptides in tissues or in plasma. Therefore, whenever possible the combined use of three assay procedures—(1) bioassay, (2) radioimmunoassay, and (3) receptor assay—is used for more effective and reliable estimates of levels of biologically active peptides.

Bioassay

Bioassay is the most sensitive and is the definitive measure of a biologically active substance. The bioassay quantitates a specific biological response in an *in vivo* or *in vitro* system by measuring an amount of test substance on a log dose–response curve plotted with known amounts of authentic standards. Although the precision of the bioassay is inferior to the chemical/physical assay, the sensitivity and specificity can be very good when an appropriate biological response is chosen. Examples of well-known bioassay procedures include the lowering *in vivo* of blood glucose levels in the rat with varying doses of injected insulin; or measuring the contraction of cardiac muscle with varying doses of epinephrine added to a medium bathing the tissue *in vitro*.

Receptor Assays

Receptor assays consist of the displacement of a radioactive form of ligand by a nonlabeled ligand bound to the receptor protein of a target cell. The amount of radioactivity displaced under standardized conditions of time and temperature is proportional to the amount of added nonlabeled ligand. The displacement by nonlabeled insulin of $[^{125}I]$insulin bound to a fat cell ghost, and by nonlabeled ACTH of $[^{125}I]$ACTH bound to an adrenal cortical cell membrane are examples of receptor assays used to measure levels of hormone in the circulating plasma.

Radioimmunoassay

Radioimmunoassay is the most frequently used hormone assay due to its extreme level of sensitivity, specificity, and general versatility. It is also one of the least complicated assays to perform and thus lends itself to automated procedures capable of measurement of large numbers of tests using small sample volumes. In analogy to the receptor assay, an ^{125}I-labeled hormone is used to compete with a nonlabeled hormone for binding to an antibody. Antibodies to a protein hormone can be raised specifically when the hormone used as an antigen is injected into an animal. When the ligand of interest is not antigenic, that is, it cannot generate an antibody by itself, the substance, used as a hapten, is made antigenic by covalent linkage to a protein such as bovine serum albumin. The complex is then injected into an animal, such as a rabbit or guinea pig, to form antibodies. The procedure is thus generally applicable for the assay of a variety of proteins, peptides, hormones, vitamins, drugs, and organic substances.

It should be apparent that the three assays applied to a given amount of unknown protein may yield different quantitative results. This is due to the fact that the determinants for each assay will depend upon different parts of the protein structure with different amino acids as binding or reacting sites. The measurement of protein will vary considerably from one assay procedure to another, depending on the specificity of the antibody or of the receptor for the ligand, the presence of other competing ligands, or the loss or denaturation of part of the isolated protein ligand molecule.

15.2 THE RECEPTOR MODEL FOR HORMONE ACTION

Estradiol Binding by Uterine Tissue

The activity of a hormone at the target cell is determined by its concentration, by its binding affinity to a receptor, by the number of receptor sites occupied, and by the duration of binding. The successful demonstration of specific incorporation of a hormone by its target tissue was first achieved with the use of trace amounts of tritium-labeled estradiol of extremely high specific activity (Figure 15.3). A single injection in the rat

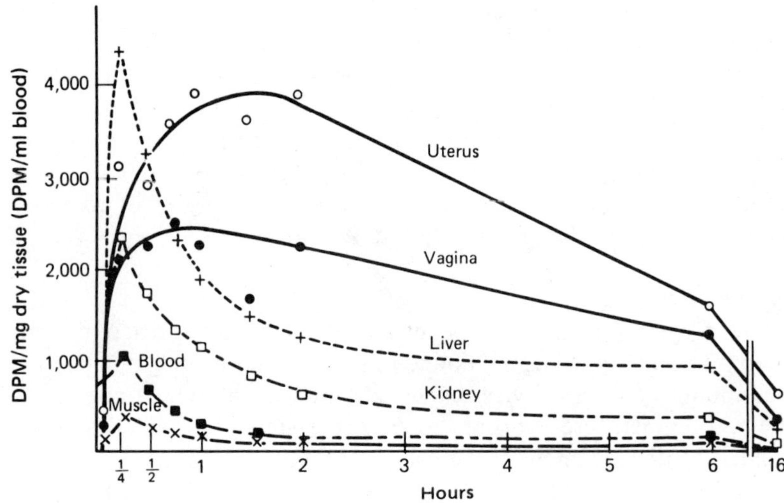

Figure 15.3
Tissue distribution of estradiol after a single subcutaneous injection of 6,7-^{3}H-estradiol.

Reproduced with permission from E. V. Jensen and H. I. Jacobson, *Recent Prog. Hormone Res.*, 18:387, 1962.

permitted the accurate measurement of as little as 1 pg(10^{-12} g) of steroid in the tissues and blood. A high concentration of radioactivity was attained very rapidly in muscle, kidney, and liver, followed by a rapid decrease which paralleled that found in the blood. In contrast, the estrogen growth responsive tissues, uterus and vagina, achieved high levels of radioactivity, which were maintained for longer periods of time. The selective concentration of hormone by its target tissue demonstrated a unique trapping mechanism in the target cell.

This avidity of the target tissue for the hormone was due to the presence of a specific protein in those cells. The term *receptor* was applied to the protein, since it could be shown that (1) the binding was specific for estradiol, and (2) there was an associated stimulation of uterine growth in response to the estrogen binding. No chemical alteration of the steroid molecule occurred as a result of estradiol binding to its receptor. To fully characterize a cell receptor, specificity of binding to the receptor by the ligand and a biological response as a result of the ligand binding must be demonstrated. A substance that competes with a hormone for binding to a receptor is described as an *agonist* if the response of the cell is the same as, or mimics, the action of the hormone. A substance that competes with a hormone for binding to a receptor is referred to as an *antagonist* if it blocks or inhibits the response of the cell to the action of the hormone.

Due to the very low concentrations of specific receptor sites, tissues can easily be overwhelmed by excess ligand, in which case the high level of nonspecific binding would mask the specific receptor binding. A basic procedure to determine the binding characteristics of receptors, the number of binding sites, different classes of receptors, and the binding affinity or association constant of a ligand utilizes the Scatchard plot, described in the next section.

Scatchard Plot

The kinetics of ligand binding to a protein has been presented in the discussion of enzyme kinetics in Chapter 4. Except for the fact that there is no product involved, the assumptions and treatment of ligand–protein interaction are no different from that for the substrate–enzyme complex.

Michaelis–Menten behavior is observed with receptor binding of ligand. Data can be arranged in a typical Michaelis plot, Lineweaver–Burk plot, or other usual modes of presentation. The equation for the Scatchard plot is derived from the law of mass action. It is presented below in the form in which it is most useful for the purpose of determining ligand–receptor data under experimental conditions.

$$[A] + [P] \rightleftharpoons [AP]$$

$$K_{eq} = \frac{[AP]}{[A][P]} = K_a = \frac{1}{K_d}$$

Assume total binding sites = unbound + bound sites. [TP] = [P] + [AP]; then unbound sites [P] = [TP] − [AP]; substitute in K_a equation for [P],

$$K_a = \frac{[AP]}{[A]([TP] - [AP])}$$

Rearranging yields

$$\frac{\text{Bound}}{\text{Free}} = \frac{[AP]}{[A]} = K_a\,([TP] - [AP])$$

The equation is now set in the form of a straight line obtained by plotting [AP]/[A] vs [AP] (Figure 15.4*a*). Note that all of the parameters necessary for plotting values for the association constant (K_a) and for the total number of binding sites [TP] contains the ligand term (A). Using a radioactive form of ligand (^{3}H-A), it is then necessary only to separate the bound (protein-bound) ligand from the unbound (free) ligand by a suitable procedure and plot the radioactive counts along the ordinate and abscissa. Complete separation is achieved by protein precipitation, adsorption of the free form, chromatography, or electrophoresis. Since receptor concentrations and hormone (ligand) concentrations are very low (10^{-8} to 10^{-12} M), radioactive ligands must be used with exceedingly high (10–50 Ci/mM) specific activity.

The slope of the line (which is negative) gives the association constant ($-K_a$). Extrapolation to the abscissa gives the value for [TP] = total number of sites. In terms of Michaelis kinetics, the slope constant when half the sites are occupied and half are unoccupied, is K_d, which is equivalent to K_m. The total number of ligand binding sites (maximum bound) is equivalent to V_{max}.

Receptor sites have the characteristics of being saturable, reversible, and specific for the ligand. Nonspecific binding of ligand has a relatively low affinity constant and nonsaturable kinetics. Therefore, specific binding is established by demonstrating high affinity and low capacity of receptor for the ligand. A correction for nonspecific binding can be made by diluting the radioactive ligand with a large excess of nonlabeled carrier ligand. Specific binding is equal to the total counts bound minus the correction factor for nonspecific binding (Figure 15.4*b*).

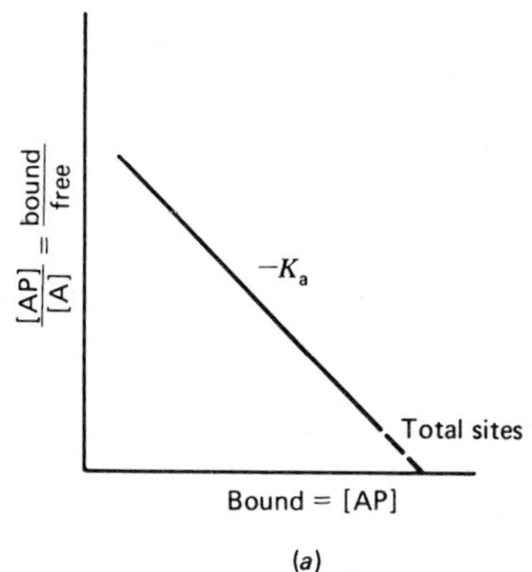

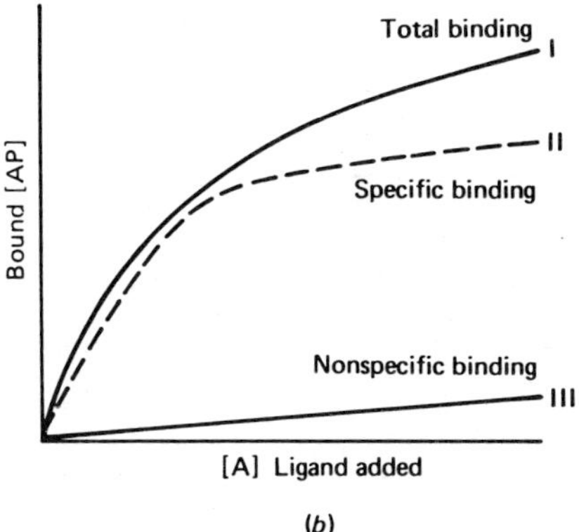

Figure 15.4
(a) Scatchard plot. A plot of the ratio of radioactivity in the protein-bound form to the free ligand form, (AP)/(A), vs (AP) on the abscissa gives the slope = $-K_a$ *(the association constant), and total sites (TP) by extrapolation to baseline. The points used to determine the straight line have been corrected for nonspecific binding by the procedure shown in (b). (b) Saturation analysis. Total radioactivity of ligand bound to protein (AP) is plotted vs radioactive ligand (A) added to give curve I. When a large excess of nonradioactive ligand is added with radioactive ligand (A) to the same amount of receptor in the same volume, curve III is obtained. Curve II, specific binding is obtained by subtraction of values of curve III from curve I.*

Steroid Hormone Receptors

When trace amounts of radioactive estradiol are administered to a female rat, most of the radioactivity is subsequently bound in the nucleus of uterine cells. Estradiol binds to a protein component to form a steroid–receptor complex, which is transferred to an acceptor site on the chromatin in the nucleus. The transfer involves both hormone and receptor being transported as a complex for nuclear binding. Each of the steroid hor-

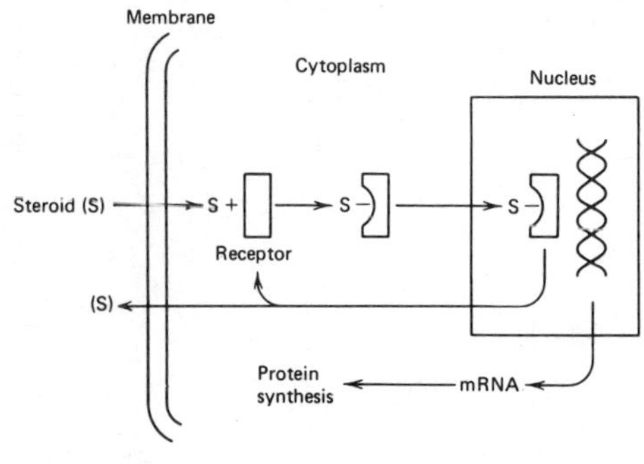

Steroid	*Target Cell Nucleus*
Estradiol	Uterus, breast, brain
Testosterone	Testes, brain
Cortisol	Liver, lymphocyte
Aldosterone	Kidney
Progesterone	Uterus, breast
Cholecalciferol	Intestinal mucosa

Figure 15.5
Receptor model for steroid hormones.

mones binds to nuclear sites in their respective target cells. Receptor proteins with high affinity binding ($K_a = 10^8$–10^{10} M^{-1}) have been found for each steroid target cell (Figure 15.5).

Vitamin D and thyroid hormones also bind as receptor complexes to nuclear sites. The high affinity binding constants ($K_a = 10^{11}$ M^{-1}) for thyroid hormone in the liver cell nucleus indicates specific binding sites are present. Nuclear concentration of active forms of vitamin D have been demonstrated by radioautography in cells of bone, intestine, and kidney, presumed target cells for vitamin D.

Steroid hormones and thyroid hormones are lipid-soluble compounds and freely permeable in the plasma cell membrane. There appears to be no barrier to their entrance or exit from tissue cells. Uptake and subsequent response to steroid hormones are determined by the ratio of occupied to unoccupied receptor sites present in the cell. Upon release of steroid, receptor may be recycled in the cytoplasm.

In contrast to lipid-soluble compounds, the cell membrane serves as a barrier to water-soluble substances as represented by the catecholamines, indole amines, peptides, and larger protein molecules. The first contact water-soluble hormones have with their target cells involves binding to receptor proteins located on the outer surface of the plasma cell membrane.

Peptide Hormone Receptors

Glucagon is representative of a peptide hormone which binds to a receptor on the plasma cell membrane. Both liver and adipose tissue cells respond to glucagon binding by stimulation of the membrane-bound adenylate cyclase (Chapter 7). An in vitro membrane preparation can be used to study simultaneously the binding reaction using [^{125}I]-glucagon and the response as a result of that binding by measuring cAMP production. In addition, receptor specificity can be studied using the fat cell, since it is a target organ for glucagon, epinephrine, ACTH, lipotropin, and growth hormone. The response of the fat cell (release of free fatty acids) to a mixture of different hormones is additive, that is, there is an increased increment of release with the addition of each hormone, indicating separate receptor proteins for each hormone. In this system receptor binding results in the activation of adenylate cyclase as a common response for each stimulus. A number of hormone receptor sites on the outer surface of the membrane can stimulate adenylate cyclase situated on the inner (cytoplasmic) surface of the membrane (Figure 15.6). Other factors resulting from receptor binding of individual peptide hormones are discussed in Chapter 16.

Plasma membrane of fat cell
Outer surface receptors (R)
Inner surface adenylate cyclase (AC)
Glucagon R_1
ACTH R_2
Growth hormone R_3
AC
ATP
cAMP

Figure 15.6
Receptor model for peptide hormones.

15.3 STEROID HORMONES

Metabolic Considerations

The steroid hormones are produced in specific cells of the adrenal cortex, the testis, the ovary, and placenta (Table 15.2). The biological actions of the steroid hormones are listed in Table 15.3. Biologically active steroids, particularly androgens and estrogens, are formed also in nonendocrine tissues from steroid precursors circulating in blood. These peripheral tissues include skin, liver, brain, mammary, and adipose tissues. The steroids are not stored in appreciable amounts in the endocrine glands but are secreted into the general circulation and distributed to all tissues of the body. There are two established carrier proteins in blood, cortisol-

TABLE 15.2 Tissue Source of Steroid Hormones

Organ	*Cell type*	*Secretion*	*Control*
Adrenal cortex	Glomerulosa	Aldosterone	Angiotensin and Na^+/K^+ ratio
	Fasciculata-reticularis	Cortisol Dehydroepiandrosterone (DHEA)	Adrenocorticotropic hormone (ACTH)
Testis	Leydig cell	Testosterone 4-Androstene-3,17-dione	Luteinizing hormone (LH)
Ovary	Follicle (theca)	Estradiol	Luteinizing hormone and follicle stimulating hormone
	Stroma	4-Androstene-3,17-dione	Luteinizing hormone
	Corpus luteum	Progesterone	Luteinizing hormone
Placenta		Progesterone	Chorionic gonadotropin (HCG)
		Estradiol	Chorionic gonadotropin
		Estriol	Chorionic gonadotropin

binding globulin (CBG, or transcortin), which binds cortisol, corticosterone, and progesterone; and the sex steroid-binding globulin (17 β-globulin), which has a high affinity for testosterone and estradiol. Steroids also bind with a low affinity to serum albumin.

Protein binding does not limit the access of the hormone into tissue cells since there is an equilibrium at all times with the free steroid form in solution. At high concentrations of steroids, the concentration of free steroids will increase in blood, as the capacity of the high affinity globulins to bind steroids is exceeded and more steroid is bound to low affinity nonspecific protein sites.

TABLE 15.3 Biological Actions of the Steroid Hormones

Class	*Hormones*	*Primary Target Tissue*	*Effects*
Glucocorticoid	*Cortisol,* cortisone, and corticosterone	Muscle Liver	Protein catabolism Gluconeogenesis
Mineralocorticoid	*Aldosterone* and 11-deoxycorticosterone	Kidney tubules	Sodium retention and potassium excretion
Androgen	*Testosterone,* 5α-dihydrotestosterone, and dehydroepiandrosterone	Reproductive organs (primary and secondary) Muscle	Spermatogenesis, secondary male characteristics, bone maturation, virilization
Estrogen	*Estradiol* and estrone	Reproductive organs (primary and secondary)	Feminization Cyclic rhythms
Progestin	*Progesterone*	Uterus	Nidation and maintenance of pregnancy

The plasma concentration of the steroid hormones at any moment represents the net difference between the rates of formation and secretion of the hormone by the endocrine gland and the rates of metabolism in liver and excretion by the kidneys. There appears to be no apparent limit in the capacity of these organs to metabolize steroids, and there is no appreciable storage in the secreting tissue. The rate of turnover of steroid hormones is rapid. The half-life of steroids in plasma ranges from 30 to 90 min. The rate of formation and secretion of a steroid hormone by the endocrine gland, therefore, is an essential control point for the biological action of the steroid hormone at its target tissue.

The liver is the primary organ for metabolizing steroid hormones. Reduced steroids are formed by the action of stereospecific dehydrogenases, using nicotinamide nucleotides as cofactors (Figure 15.7). The reduced

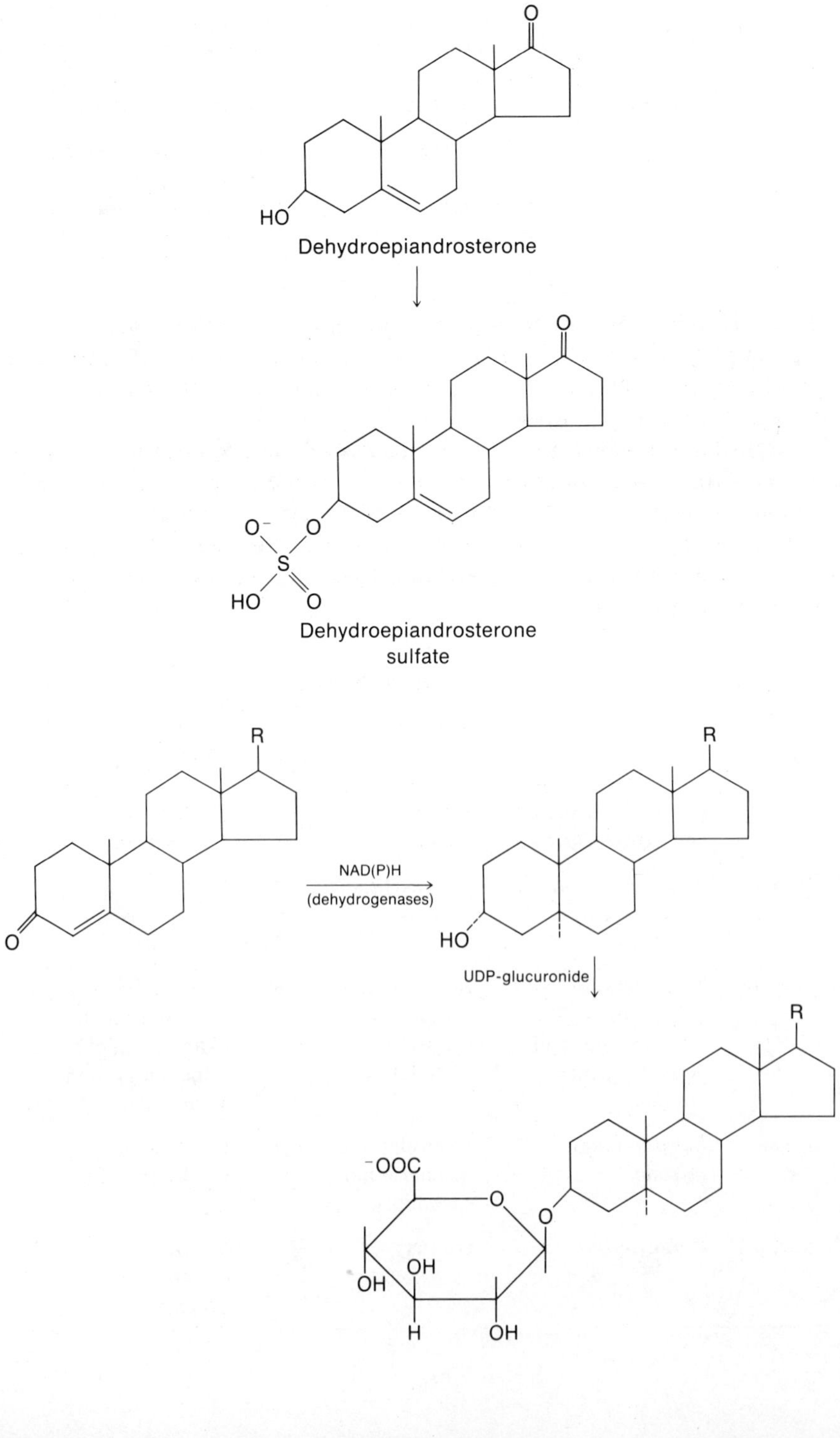

Dehydroepiandrosterone

Dehydroepiandrosterone sulfate

NAD(P)H (dehydrogenases)

UDP-glucuronide

Figure 15.7
Steroid sulfates and glucuronides.

metabolites are conjugated at the hydroxyl groups as sulfates or glucuronides in which they circulate in the blood. They are rapidly excreted into the urine, since kidney clearance is greatest for glucuronide and sulfate conjugates of steroids. Normally only trace amounts of free nonreduced steroids are found in urine; the low clearance is partly a result of their binding to plasma proteins.

The measurement of steroid hormones and their metabolites in blood and in urine is routine at most medical centers. A number of older classical chemical procedures, have been replaced by newer radioimmunoassay procedures.

Some storage of the estrogenic and progestational steroids occurs in adipose tissue. These steroids also have an appreciable enterohepatic circulation; that is, they rapidly appear in the bile in their conjugated forms and, like cholesterol, enter the gastrointestinal (GI) tract and are reabsorbed via the portal vein and taken back to the liver. In contrast, androgens and adrenocortical steroids are excreted in the reduced conjugate forms in urine. Since total steroid metabolism and clearance is relatively rapid, the amount of steroidal metabolites found in urine can approximate the secretion rate by the endocrine tissues over the period (4–24 h) of urine collection.

Steroidogenesis (Figure 15.8)

There is a common metabolic pathway for the formation of all the steroid hormones, which is initiated by the conversion of cholesterol to pregnenolone. Cleavage of the cholesterol side chain and subsequent hydroxylation reactions occur at the site of the cytochrome P_{450} system in the steroid-producing cell. Adrenal cortex mitochondria contain a P_{450} system for cholesterol side-chain cleavage, and 11β- and 18-hydroxylations. The endoplasmic reticulum contains a P_{450} system 17α-, 21-, and 19-hydroxylations. Several of the P_{450} cytochromes have now been purified, including $P_{450\,CSCC}$, (cholesterol side-chain cleavage), $P_{450,11\beta}$ (11β-hydroxylase) and $P_{450,17\alpha}$ (17α-hydroxylase). They have different specificities due to protein differences, which would account for specific hydroxylation sites on the steroid molecule. Hydroxylation at carbons 2, 7, and 16 can be catalyzed by a cytochrome P_{450} system (the drug metabolizing P_{450}) in liver microsomes; however, the major metabolic fate for steroids in the liver is reduction and conjugation.

Steroidogenesis is stimulated by pituitary trophic factors, which determine the amount of cholesterol available at the P_{450} enzyme site and the rate of side-chain cleavage. The amount of cholesterol converted to the C_{21} pregnenolone is regulated by ACTH in the adrenal cortex and by luteinizing hormone (LH) in the interstitial cells of the testis and ovary. Pregnenolone is immediately converted by hydroxylases and dehydrogenases to the appropriate steroid hormones, which are then released into the plasma.

The steroid metabolic pathway (Figure 15.8) depicts a series of reactions leading from cholesterol to pregnenolone, 17-α-hydroxypregnenolone and dehydroepiandrosterone in which Δ^5-3β-hydroxyl structure, as it occurs in cholesterol, remains intact. This, in fact, represents the pattern of steroid conversions that occurs in the fetal adrenal of the human and other species before the Δ^5-3β-hydroxysteroid dehydrogenase enzyme (Δ^5-3β-OHD) attains optimal activity. Consequently, steroid metabolites circulating in blood and those in urine at the time of parturition contain predominantly the Δ^5-3β-hydroxyl configuration.

Increased Δ^5-3β-OHD activity in the neonate and in the adult leads to the formation of Δ^4-3-ketosteroids. The oxidation of the 3β-hydroxyl to the 3-ketone requires the nicotinamide nucleotide, NAD^+, as cofactor.

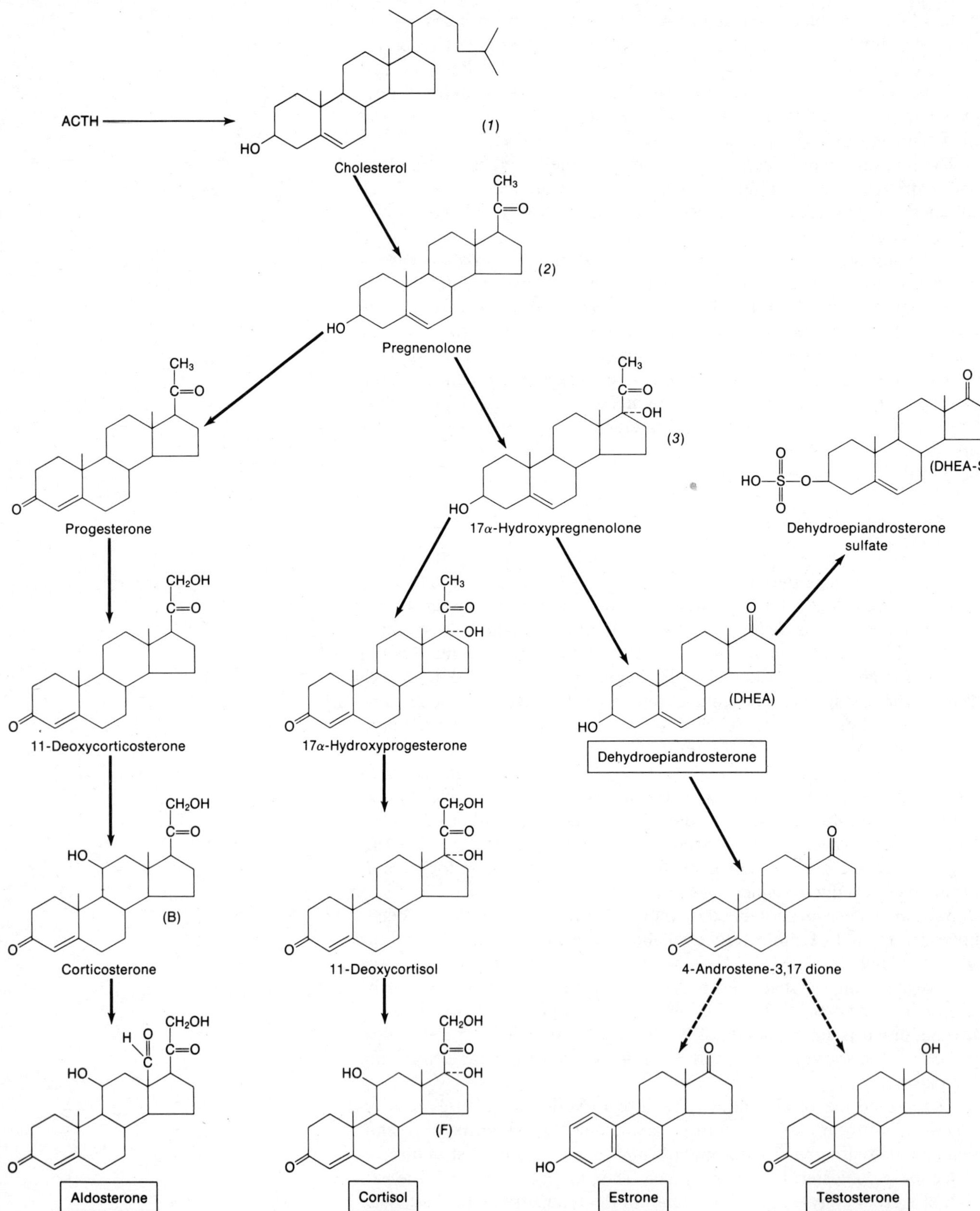

Figure 15.8
Biosynthesis of steroid hormones.
Early phase of biosynthesis includes events from ACTH stimulation to rate-limiting step at (1), cholesterol side-chain cleavage to form pregnenolone. Later phases occur at key branch points (2) and (3) and additional hydroxylations to form the steroid secretory products.

The Δ^5-3β-OHD reaction [Figure 15.8 (*2*)] converts pregnenolone to progesterone, which precedes subsequent hydroxylation reactions at C-21 and C-11 to form corticosterone. In the glomerulosa zone of the adrenal there is a C-18-hydroxylation and oxidation to a C-18-aldehyde to form aldosterone.

17α-Hydroxypregnenolone formed from pregnenolone can be oxidized to 17α-hydroxyprogesterone [Figure 15.8 (*3*)] or by a desmolase reaction in which cleavage of the side chain on C-18 occurs to form the C_{19} steroid, dehydroepiandrosterone. The hydroxylation reactions and desmolase reactions require specific cytochrome P_{450} enzymes, oxygen and the reduced NADP as cosubstrate. The relative activities of the Δ^5-3β-OHD and of the 17α-hydroxylase and desmolase at the major branch points (*2* and *3*, Figure 15.8) of the steroid pathway determine the final type and amount of steroid hormones secreted by the adrenal cortex. In the human adrenal cortex, dehydroepiandrosterone (DHEA) is converted by a sulfotransferase to DHEA $\cdot$ SO_4, which is then secreted into the blood. Major steroid hormones secreted by the human adrenal cortex consist of the C_{19}-steroid, DHEA $\cdot$ SO_4, the 17α-hydroxyl C_{21}-steroid, cortisol, the C_{21}-steroid, corticosterone, and the C-18-oxoC_{21}-steroid, aldosterone. Alternate routes in the steroid biosynthetic pathway are possible as, for example, the conversion of progesterone to 17α-hydroxyprogesterone and to 4-androstene-3, 17-dione. Changes in enzyme activities at the major branch points and for specific steroid hydroxylases can occur under a variety of abnormal adrenal conditions and as a consequence will give rise to altered steroid secretory patterns that are characteristic, and therefore diagnostic, for a number of different adrenal disease states.

In the testis and ovary the reactions involving cholesterol side-chain cleavage to pregnenolone, the formation of progesterone, 17α-hydroxylated compounds, and the C_{19} 17-ketosteroids occur as in the adrenal. In the testis, the Leydig cells respond to LH to produce the male hormone testosterone (Figure 15.9). In the ovary, androstenedione is converted to estrone by the concerted action of an enzyme complex which contains P_{450} hydroxylases for C-19 and C-2 of the steroid nucleus. The enzyme complex that forms estrogen is referred to as *aromatase*.

LH stimulation of cells of the ovarian corpus luteum produces progesterone. Progesterone secretion is cyclic; its production occurs almost entirely during the luteal phase of the menstrual cycle when the corpus luteum is active. During pregnancy, while the corpus luteum persists and after the development of the placenta, larger amounts of progesterone are secreted.

The steroid-producing capacity of the placenta is under trophic hormone control also. The hormone is called human chorionic gonadotropin (HCG), a protein with LH activity but with an amino acid sequence different from that of LH and immunologically distinct from LH. As the name implies, HCG is not derived from the anterior pituitary but is produced by cells of the placenta. During pregnancy, steroid sulfates formed in fetal tissues are converted by the placenta to estrogens. A sulfatase converts dehydroepiandrosterone sulfate to the free steroid, which is then converted to estrone by aromatase (Figure 15.9). By a similar sequence, 16α-hydroxydehydroepiandrosterone sulfate formed in the fetus is transported to the placenta, where it is hydrolyzed by the sulfatase and aromatized to estriol. Estriol is the major estrogen metabolite formed in late pregnancy.

Regulation of Steroidogenesis

The dynamic state that controls secretion of steroid hormones is the result of the interaction of a number of regulatory factors. For the adrenal

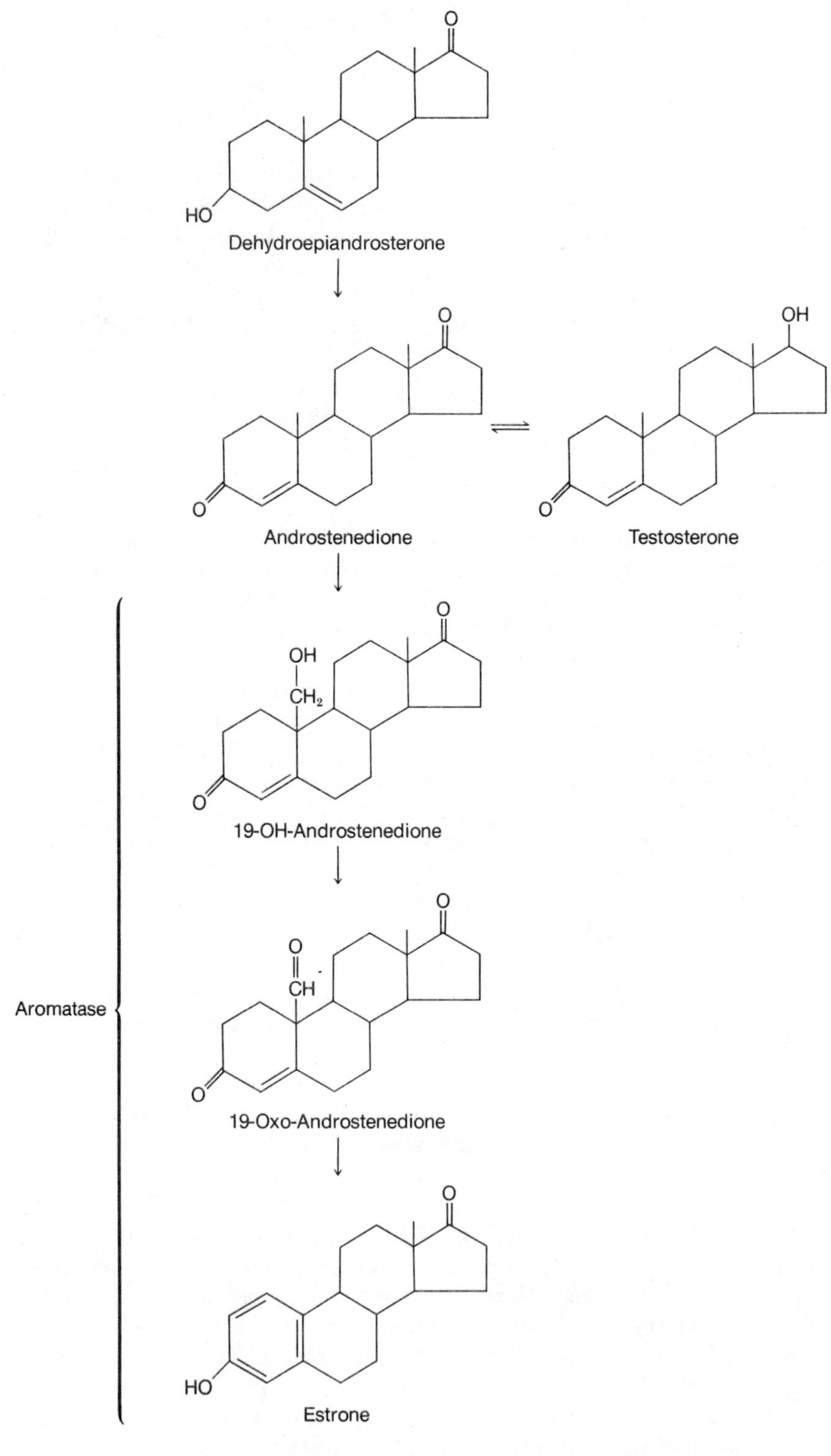

Figure 15.9
Biosynthesis of estrogen and androgen.

cortex this includes cortisol, which exerts a negative feedback effect on the hypothalamic–pituitary system; the anterior pituitary peptide adrenocorticotropic hormone (ACTH), which stimulates the adrenal cortical cell; a hypothalamic factor, corticotropin releasing factor (CRF), which stimulates the pituitary cell to secrete ACTH; and other neurotransmitter agents that regulate the secretion of CRF.

The hypothalamus serves as an integrating center which receives signals from the central nervous system and the higher cortical centers via

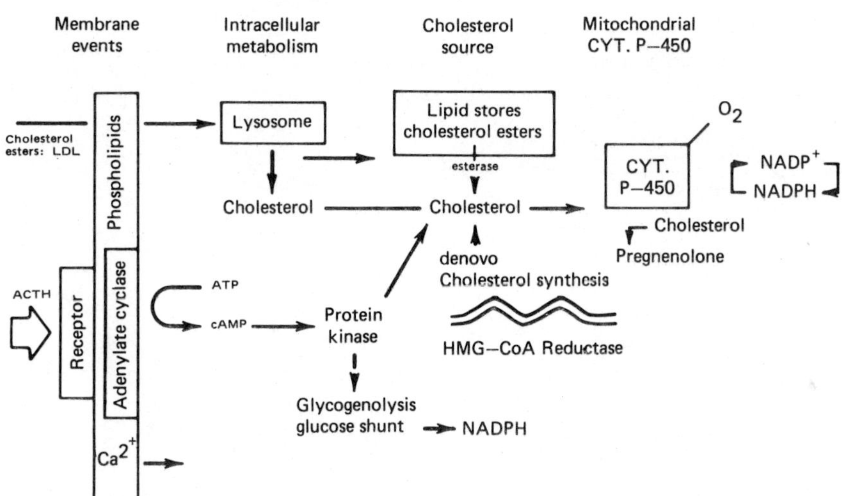

Figure 15.10
Regulation of adrenal steroidogenesis.

neurotransmitters. The composite signal results in the secretion of CRF into the portal blood system draining the hypothalamus, which enters the anterior pituitary to stimulate ACTH secretion. One form of CRF peptide contains 41 amino acids. In response to CRF, the pituitary secretes ACTH. ACTH is available for clinical use as the natural polypeptide containing 39 amino acids and as synthetic polypeptides containing the first 17–24 amino acids with complete biological activity. ACTH binds to the plasma membrane of the fasciculata-reticularis zone in the adrenal cortex to initiate a chain of events leading to the secretion of steroid hormones (Figure 15.10).

ACTH binding results in activation of adenylate cyclase to convert ATP to cAMP. With increasing doses of ACTH, there is a corresponding increase in the cellular concentration of cAMP. cAMP activates a specific protein kinase, which phosphorylates phosphorylase, and a cholesterol esterase, which hydrolyzes cholesterol esters stored in lipid droplets to the free cholesterol form. There is enhanced uptake of cholesterol from the LDL in plasma. The formation of cholesterol de novo by the adrenal HMG-CoA reductase system (Chapter 10) is increased as well. The cytoplasmic cholesterol is probably transported by a carrier protein and converted by the mitochondrial cytochrome P_{450} system to pregnenolone. Sterol carrier protein (SCP) occurs in mitochondria and cytosol of the adrenal cortex.

Functional Zonation of the Adrenal Gland

The adrenal gland can be differentiated by structure and function into three independent biologically active zones: (1) the centrally located *medulla,* (2) the cortex layers of the *fasciculata-reticularis* zones, and (3) a thin outer cortex layer of the *glomerulosa* zone beneath the adrenal capsule (Figure 15.11).

The adrenal medulla, which consists essentially of nerve tissue (chromaffin cells), responds to splanchnic nerve stimulation or hypoglycemia by increased formation of norepinephrine and epinephrine. The enzyme *N*-methyltransferase, with *S*-adenosylmethionine as the methyl donor, converts norepinephrine to epinephrine. Epinephrine and norepinephrine are secreted by the adrenal medulla in a ratio of ~4 : 1 in the human. Despite the close proximity of the medullary cells to the cortical cells which produce steroid hormones, there appears to be little physiological interaction of the catecholamines on the cortex, or for action of the steroid hormones on the medulla. There is a prominent circadian biorhythm

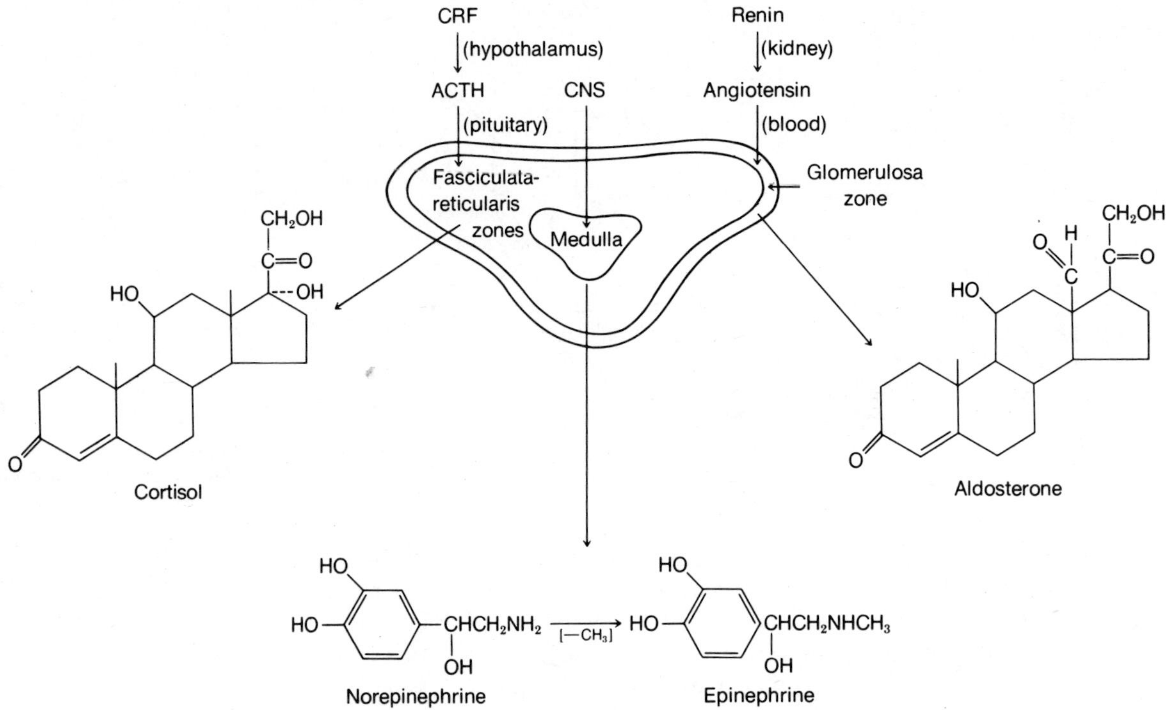

Figure 15.11
Functional zonation of the adrenal gland.

in the enzyme activities of the medulla and cortex, which can be entrained by light/dark and activity schedules. The *N*-methyltransferase activity to form epinephrine in the medulla has been related to the cortisol circadian cycle.

The cells of the fasciculata-reticularis zone of the adrenal cortex appear to comprise a morphological continuum; the cells of the fasciculata being somewhat larger, containing more lipid droplets, and are arranged in orderly columns. The fasciculata-reticularis zone, produce the C_{19} 17-ketosteroids of the adrenal. It is clear that both zones respond to ACTH stimulation; the loss in lipid droplets in the fasciculata zone with stimulation is particularly prominent. The hypertrophy of the adrenal cortex as a result of chronic ACTH stimulation is confined to the fasciculata-reticularis zones. The prominent biochemical feature of ACTH stimulation of the adrenal cortex is enhanced production of cortisol. There is increased secretion of dehydroepiandrosterone and lesser amounts of other C_{21} and C_{19} steroids as well.

The glomerulosa (capsular) zone consists of a thin layer of cells beneath the outer capsule of the adrenal gland. The most prominent biochemical feature of this zone is the production of the mineralocorticoid, aldosterone. The C-18-aldehyde group unique for aldosterone is derived from C-18-hydroxylated steroids in the glomerulosa zone. Aldosterone secretion is stimulated by sodium restriction or potassium excess, or a combination of both. The width and number of cells of the glomerulosa zone are increased with chronic sodium restriction.

Angiotensin II, a peptide containing eight amino acids, plays an important role in the regulation of aldosterone formation and secretion. The regulatory chain of events is initiated in the kidney with the secretion of renin. Renin, a kidney enzyme secreted in response to osmotic changes, anoxia, or kidney trauma, acts upon a blood precursor protein, angiotensinogen (Figure 15.12). A 10-amino acid peptide, angiotensin I, which has

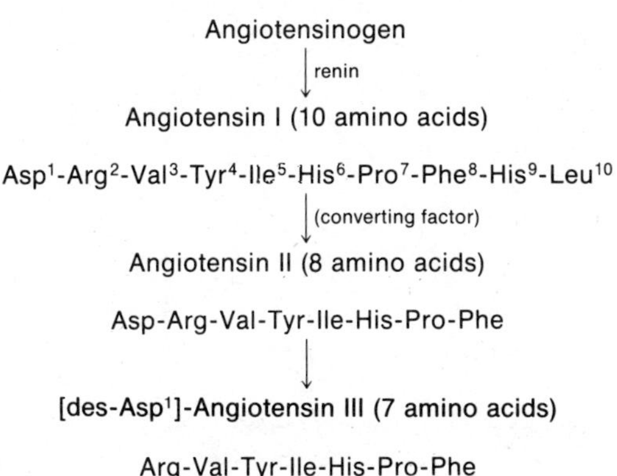

Figure 15.12
Angiotensin formation.

potent blood pressor activity, is formed. The decapeptide gives rise to the octapeptide angiotensin II by an enzyme converting factor. An active heptapeptide is also formed. Participating agents in the regulatory system comprise aldosterone, angiotensin II, renin, and sodium, among others, but their interactions at the molecular level have not been adequately resolved.

Biological Actions of Adrenal Steroids

The biological actions of corticosteroids are classified in terms of three major categories: mineralocorticoid activity, glucocorticoid activity, and antiinflammatory activity (Table 15.4).

Mineralocorticoid Activity

The most active mineralocorticoid and the physiologically effective agent produced by the glomerulosa zone is aldosterone. The action of aldosterone is at the distal convoluted tubule of the kidney to promote Na^+ reabsorption. The Na^+ retention is accompanied by a corresponding excretion of K^+ and H^+. In the absence of aldosterone there is no mechanism to retain Na^+, and the body will continue to lose Na^+, even with severe Na^+ restriction in the diet. The excretion of Na^+ (and the consequent retention of K^+ and H^+ ions) is a critical factor in adrenal insufficiency as in Addison's disease.

Glucocorticoid Activity

Cortisol is the major glucocorticoid secreted by the human adrenal cortex. The term glucocorticoid refers to steroids that affect carbohydrate metabolism. Specifically, this relates primarily to the ability of steroids to increase blood glucose levels, and to the increased deposition of glycogen in the liver. Both effects are due to enhanced liver gluconeogenesis as a consequence of steroid actions discussed below.

Glucose utilization by the muscle cell is inhibited by cortisol (Figure 15.13). This action opposes the effect of insulin to increase muscle perme-

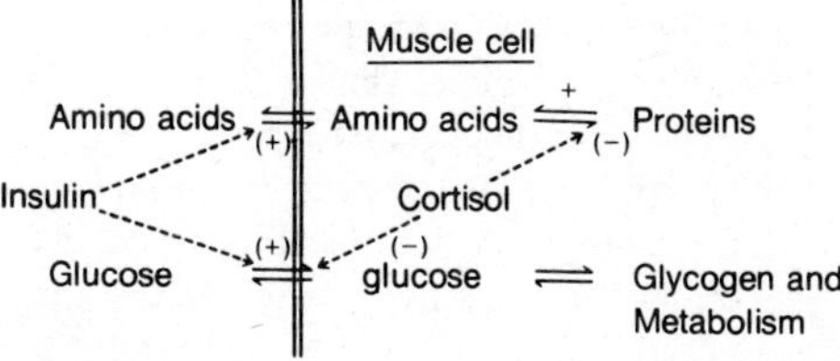

Figure 15.13
Insulin as an anabolic agent enhances (+) the incorporation of glucose and amino acids into muscle cells. Cortisol decreases (−) glucose uptake and incorporation of amino acids into protein, which leads to increased protein breakdown (catabolism).

TABLE 15.4 Biological Actions of Corticosteroids (Cortisone = 1.0)

Steroid	*Na^+ Retention*	*CHO Activity*[a]	*Antiinflammatory*
Cortisone	1.0	1.0	1.0
Cortisol	1.5	1.5	12.5
Corticosterone	2.5	0.5	0
11-Deoxycorticosterone	30.0	0	0
Aldosterone	600.0	0.3	0

[a] Effect on carbohydrate metabolism.

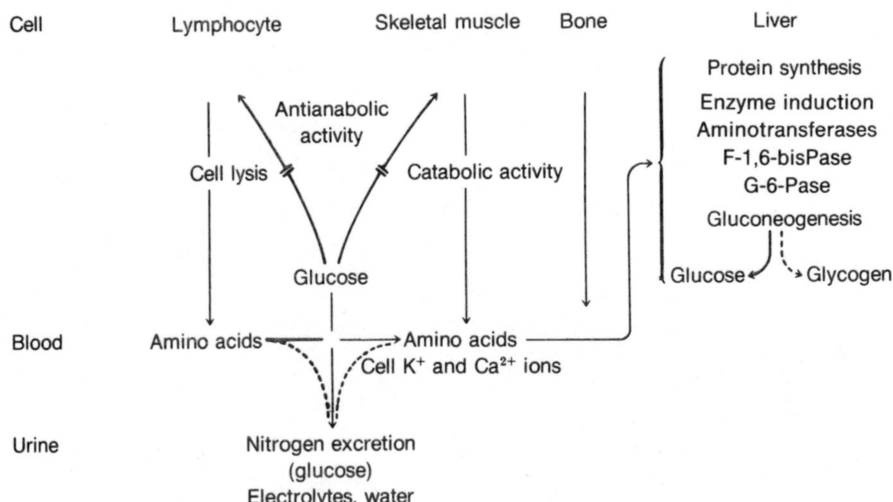

Figure 15.14
Cortisol action in peripheral tissue can be termed antianabolic and catabolic because the formation of protein is inhibited, and the breakdown of protein to amino acids is enhanced.

ability to glucose and amino acids. The antianabolic action of the cortisol results in a breakdown (catabolic action) of muscle protein, a consequence of protein turnover.

Cortisol receptor sites have been demonstrated in the nuclei of liver cells. Increased induction of the enzymes involved in transamination and in gluconeogenesis have been found in liver after cortisol administration (Figure 15.14). As a result of cortisol activity, the amino acids released from peripheral tissues are converted by aminotransferases to keto acids and via the gluconeogenic pathway to increased formation of glucose (Chapter 7). Part of this glucose is released by the liver to increase blood glucose, and part is utilized by the liver cell to form glycogen. The glycogen deposition occurs by the action of insulin on glycogen synthetase. The enzymes involved in these reactions, aminotransferases, fructose 1, 6-bisphosphatase, and glucose 6-phosphatase, are increased as a result of cortisol activity. Note that as a result of cortisol action, there is an increase in protein synthesis in the liver while protein breakdown to amino acids is occurring in peripheral muscle tissues (Clin. Corr. 15.1).

Antiinflammatory Activity

Antiinflammatory activity refers to those actions of cortisol that interfere with the normal cellular processes of inflammation, wound healing, and growth. There is a marked specificity for this activity shared only by cortisol and cortisol derivatives (Table 15.4). There are now a large number of synthetic cortisol analogs available for clinical use. These orally active steroids have greater potency than cortisol with respect to antiinflammatory activity and glucocorticoid activity, but have diminished mineralocorticoid activity.

It is on the basis of their antiinflammatory activity that the glucocorticoids have attained such widespread use as palliative agents in clinical therapy. These uses include treatment for asthma and allergies, inhibition of immune processes and tissue rejection in organ transplantation, increasing resistance to the stress of life-threatening situations, and a wide variety of disease processes.

CLIN. CORR. 15.1 BIOLOGICAL ACTIONS OF GLUCOCORTICOIDS

The term *permissive action* is used frequently to describe the physiologic role of the glucocorticoids. That is, they act in a synergistic manner with other agents to allow certain events to occur. It is not possible to define these actions at the molecular level, however, since those effects that can readily be observed in vivo cannot be reproduced in an in vitro system. This is an indication that the observed effects are the result of an indirect rather than a direct action of the steroids.

The glucocorticoid, cortisol, and the synthetic analogs of cortisol are used extensively in medicine as *antiinflammatory* agents. Since they can diminish normal but undesirable body responses to noxious stimuli or trauma, advantages are gained by their use in counteracting stressful situations and in decreasing pain and discomfort. In this respect, they are used as *palliative* agents, as the symptoms rather than the cause of the disability or disease is treated. The danger of *iatrogenic* sequelae, a result of the side effects of steroid treatment, will often set a limit on the advantages achieved by their use.

15.4 STEROID SEX HORMONES: REGULATION AND BIOLOGICAL ACTIVITY

The major function of the steroid sex hormones is the development, growth, maintenance, and regulation of the reproductive system. Sex steroids are classified according to biological activity: *androgens* are male sex hormones and are C_{19} steroids; *estrogens* are female sex hormones and are C_{18} steroids (ring A is phenolic and lacks the C-19-methyl); and *progesterone* is a C_{21} steroid which is secreted during the luteal phase of the ovarian cycle and during pregnancy.

Androgens

Androgens are produced by the Leydig cells of the testes and the adrenals in both sexes. The ovary also produces androgens in small amounts. Testosterone is the major androgen produced by the testes. About 6–7 mg/day are secreted in the adult male. Testosterone is synthesized from cholesterol by a pathway shown in the section on steroidogenesis (Figure 15.8). Although the main secretory product of the testes is testosterone, the active hormone in many tissues is not testosterone, but rather a metabolite, 5α-dihydrotestosterone (DHT).

The Leydig cells are the principal site of production of androgens made in the testes. The Sertoli cells are capable of forming androgen and also some estrogen. One function of Sertoli cells appears to be the production of androgen-binding protein (ABP), which is secreted into the seminiferous tubules. This protein specifically binds testosterone and DHT. Its presence in the seminiferous tubules is associated with the action of androgens on spermatogenesis. The production of androgen is primarily under regulation of LH, while FSH (follicle stimulating hormone) acts on the Sertoli cell to stimulate production of ABP and other factors, including, possibly, estradiol.

Androgens influence the development, maintenance, and function of the male reproductive organs and the male secondary sex characteristics. Androgens have widespread effects throughout the body as well as localized effects on specific tissues. These include the structures concerned with the formation and delivery of sperm, the low pitch of the male voice due to enlargement and thickening of the vocal cords, the male pattern of hair distribution. They have profound anabolic activity, leading to the retention of nitrogen and increased muscle and bone mass. They are responsible for bone maturation and the cessation of growth after puberty. There is an effect on the brain, leading to characteristic male sexual behavior and aggressiveness. Receptor sites for androgens have been demonstrated in brain, muscle, and other target tissues where androgen effects are known to occur.

Regulation

Receptor binding of LH by the Leydig cell leads to the stimulation of membrane-bound adenylate cyclase and the production of cAMP. As is the case for ACTH and the adrenal cortex, the result of stimulation by LH is increased cleavage of the side chain of cholesterol to form pregnenolone. Pregnenolone is immediately converted to testosterone by P_{450} enzymes of the Leydig cell endoplasmic reticulum (Figure 15.8). An increased secretion of testosterone is seen within a few minutes after the administration of the gonadotropins, LH and FSH.

Metabolism of Androgen

Both testosterone and DHT circulate in blood bound to 17β-estradiol-binding globulin, or testosterone-estrogen binding globulin (TeBG). The

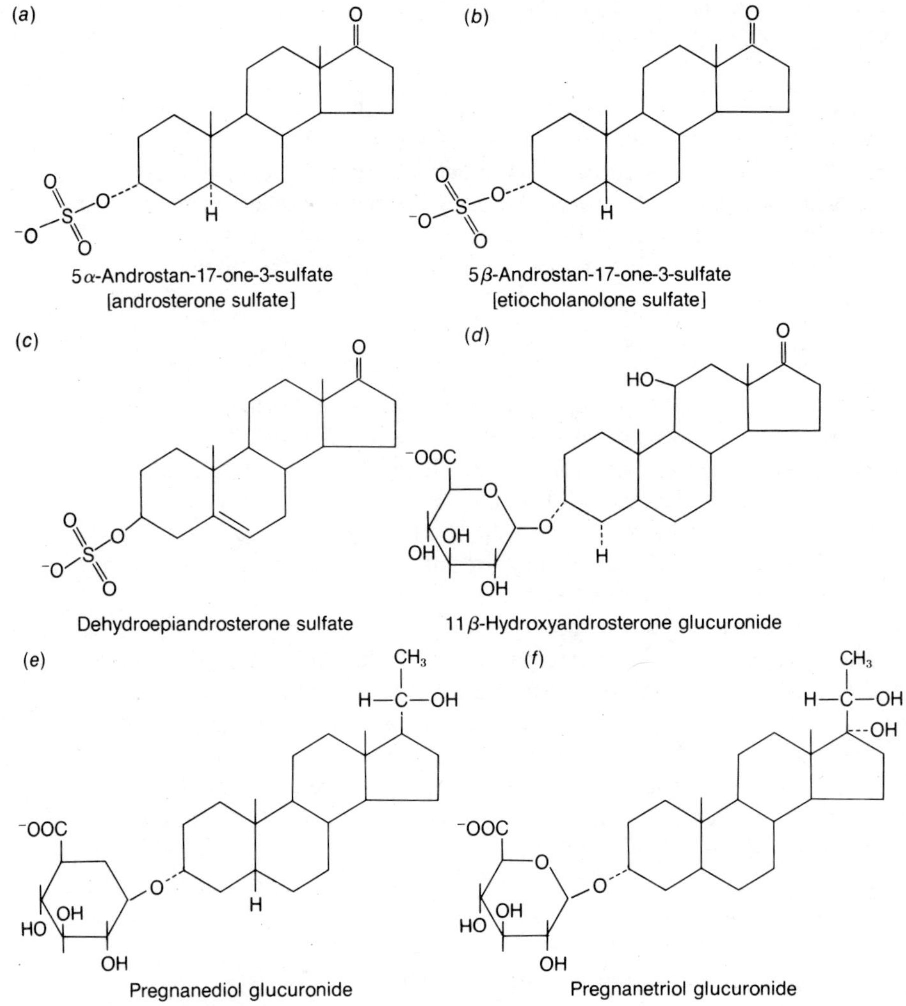

(a) 5α-Androstan-17-one-3-sulfate [androsterone sulfate]

(b) 5β-Androstan-17-one-3-sulfate [etiocholanolone sulfate]

(c) Dehydroepiandrosterone sulfate

(d) 11β-Hydroxyandrosterone glucuronide

(e) Pregnanediol glucuronide

(f) Pregnanetriol glucuronide

Figure 15.15
Steroid metabolites in urine.

liver is the major site of metabolism of testosterone. The main products of metabolism are conjugated and excreted via the kidney as sulfates and glucuronides. Figure 15.15 illustrates the conjugate structures for androsterone (*a*), a weak androgen, and etiocholanolone (*b*), an isomer of androsterone devoid of androgenic activity. Together with dehydroisoandrosterone sulfate (*c*), they comprise the principal steroids referred to as urinary 17-ketosteroids (17-KS). However, the androgens derived from the adrenal (Figure 15.8) are also precursors of the urinary 17-KS, and are secreted in amounts greater than is testosterone and other androgens derived from the ovary or testis. For this reason, urinary 17-KS reflect adrenal function to a greater extent than gonadal function. Urinary values of 17-KS are approximately one-third higher in males than females due to the testicular production of testosterone.

Estrogens

Estrogens are produced by the follicles and corpus luteum of the ovary and by the placenta in greatly increased amounts during the second and third trimesters of pregnancy. The ovary secretes both estradiol (E_2) and estrone (E_1), whereas the placenta produces these steroids and estriol (E_3) (Figure 15.16).

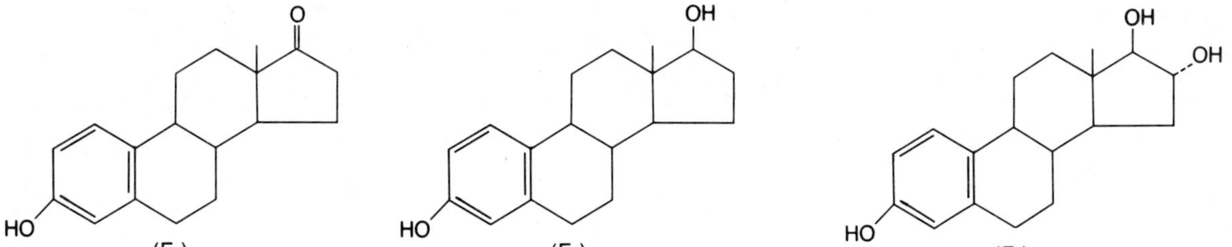

(E_1)
Estrone

(E_2)
Estradiol

(E_3)
Estriol

Figure 15.16
Major human estrogens.

Effects of Estrogens

Estrogens influence the development, maintenance, and functions of the female reproductive organs, the sexual activity cycles, and the female secondary sex characteristics. Estrogens also have general metabolic effects throughout the body. These include an increased lipogenesis in adipose tissue, which could account in part for differences in body form and in the distribution of fat in women. There is an increase in blood triacylglycerol levels and a decrease in blood cholesterol levels; the latter has been associated with relatively higher plasma lipoprotein levels of HDL compared to LDL. Estrogens increase the synthesis of a number of proteins by liver, including transferrin, ceruloplasmin, and the binding proteins, CBG and TBG. Estrogens, like androgens, have an effect on the maturation of bone and the cessation of growth after puberty. Estrogens have an effect on the cardiovascular system and on the blood-clotting mechanism by decreasing clotting time and increasing platelet aggregation. Consistent with these actions, receptor sites for estradiol have been demonstrated in the uterus, vagina, mammary gland, brain, and other target tissues where estrogen effects are known to occur.

Estrogens are responsible for the maintenance of the menstrual cycle. They have growth-promoting activities on cells of the uterus, vagina, Graafian follicles of the ovary, and the mammary gland. Estrogens induce in the uterus and mammary gland the synthesis of progesterone receptors that are necessary for these tissues to respond to progesterone. This activity of estrogen that is necessary for progesterone action is referred to as ''estrogen priming'' and is an essential feature in the growth and function of these target tissues.

A number of natural compounds has been found to have estrogenic activity. Some of these have had important clinical consequences, as described in Clin. Corr. 15.2.

Regulation

Receptors for LH have been demonstrated in the plasma membranes of the interstitial cells in the stroma and in the corpus luteum of the ovary. The interstitial cells are stimulated by LH and the granulosa cells of the Graafian follicle by FSH. The combined action of these two systems are required for the synthesis of estradiol.

Metabolism of Estrogens

Estradiol is bound to 17β-estradiol-binding globulin in the circulating plasma. Estrone and estriol are only weakly bound. Estradiol undergoes a complex metabolism in peripheral tissues and in the liver. Metabolic events include a reversible oxidation to estrone and irreversible hydroxylation at C-2 and C-16. The metabolites are conjugated with sulfuric or

CLIN. CORR. **15.2** ESTROGENS AND THE ENVIRONMENT

Estrogenic substances have assumed an importance in our society even more extensive than their role in normal reproductive physiology. They are extremely potent substances with biological effectiveness in the picomolar (10^{-12} M) range, and unlike other steroid hormones, estrogenic activity can be observed with compounds of many types of chemical structures, indeed, the presence of the phenolic (HO—⟨benzene ring⟩—R) grouping being the only relatively constant factor. Many aromatic compounds that are found in plants (including phytoestrogens) and substances produced in industry (including insecticides) have estrogenic activity.

In addition to the modified estrogenic compounds used in the formulation of the contraceptive pill (see accompanying figure) to control the reproductive cycle, synthetic compounds like diethylstilbesterol (DES II) have been widely used for a variety of purposes in humans and animals as orally active substitutes for estradiol. Insecticides, such as DDT, can be shown to have intrinsic estrogenic activity, since they will compete with estradiol for the tissue estrogen receptor site. The widespread use of DDT has been considered to be a factor that affects reproductive activity of grazing animals and in the increased fragility of eggshells and consequent decline in bird populations. Increased problems in waste disposal have added to environmental pollution in soils and water. The unusual ringed compound *Zearalenone* is an estrogen. It is a product of the mold *Fusarium* and has been used as a food supplement to promote weight increase in livestock. It can replace DES, which was used formerly for this purpose until it was withdrawn because of possible links to cancer formation. It is a fortunate circumstance that natural steroids such as estradiol are substrates for microbial degradation in the soil.

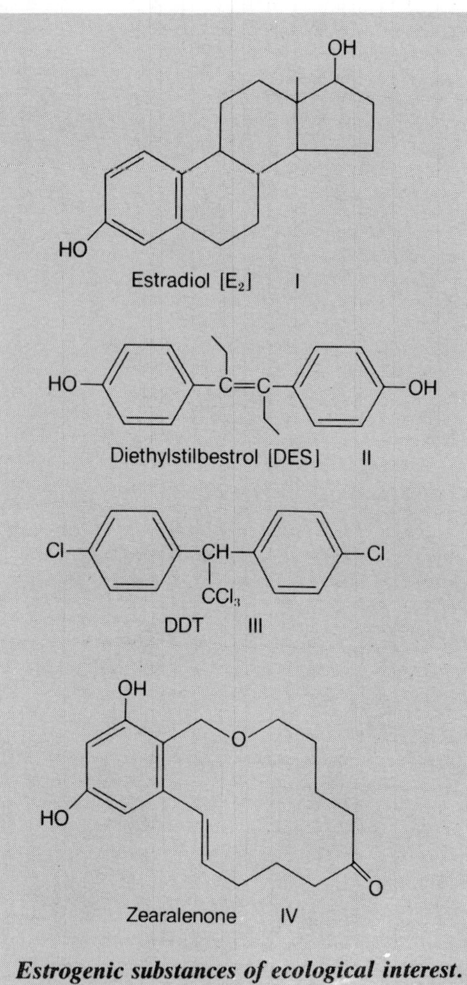

Estrogenic substances of ecological interest.

glucuronic acid and excreted either into urine or bile. Biliary metabolites may undergo further modification by action of the intestinal flora and are reabsorbed into the portal circulation.

A metabolite of estradiol, which comprises at least 20% of the total amount secreted in human, is the 2-hydroxy derivative (2,3-diol in ring A) (Figure 15.17). These derivatives, referred to as catechol estrogens, are unstable in alkali. Since alkali is ordinarily used in extraction for estrogen assays, the assays for catechol estrogens require special procedures.

The conversion to catechol estrogen occurs in a number of tissues, including the brain. A portion of estrogen is irreversibly (covalently) bound to tissue protein, and it has been proposed that these catechol forms are involved in expression of their biological activity. It is also of interest that the same enzyme, *O*-methyltransferase, which converts the 2-hydroxy to the 2-methoxy derivative (Figure 15.17), also forms the methoxy derivatives of epinephrine and norepinephrine. An interaction with and control of epinephrine action by estrogen, which is known to occur, thus may be exerted through the process of altering the rate of inactivation of these substrates by competition for the *O*-methyltransferase.

Progesterone

Progesterone is formed as an intermediate in all steroid-secreting cells. It is produced and secreted as a hormone by the corpus luteum and by the placenta. In the corpus luteum it is derived from blood and ovarian cholesterol (Figure 15.8). Production of progesterone by the corpus luteum is about 20–40 mg/day and is under the control of LH. During pregnancy maternal circulating cholesterol is the chief precursor to progesterone in the placenta, and near term the placenta secretes 200–300 mg progesterone/day.

Biological Actions of Progesterone

Progesterone is necessary for the implantation of the fertilized ovum and for the inhibition of uterine contraction in the maintenance of pregnancy. Progesterone stimulates the growth of the secretory glandular portions of the uterine and breast tissue after estrogen priming. It has a thermogenic effect of unknown origin which raises the normal body temperature by 0.4–1.4°F. Both the thermogenic effect and the large increment in progesterone secretion by the corpus luteum define the luteal phase of the menstrual cycle and are used most commonly as indicators that ovulation has occurred.

Progesterone as well as estradiol, in relatively large amounts, will inhibit ovulation by suppression of gonadotropic stimulation. Orally active, long-acting synthetic derivatives of progesterone and estradiol also will prevent ovulation in relatively small amounts. These are the active agents in the contraceptive pill. (See Clin. Corr. 16.5, p. 618.)

2-Hydroxyestradiol
[catechol estrogens]
S-adenosylmethionine (*O*-methyltransferase)
2-Methoxyestradiol

Figure 15.17
Formation of 2-methoxyestradiol.

Metabolism of Progesterone

Progesterone circulates in plasma bound to transcortin. Transcortin (CBG) actually has a higher affinity for progesterone than for cortisol. The major urinary metabolite of progesterone is pregnanediol glucuronide (Figure 15.15). Progesterone and its metabolites also have an extensive enterohepatic circulation. Urinary excretion of pregnanediol glucuronide is used as an index of production of progesterone by the corpus luteum and during pregnancy. Pregnanetriol glucuronide is a urinary metabolite of 17α-hydroxyprogesterone, which is increased in certain types of adrenal abnormality.

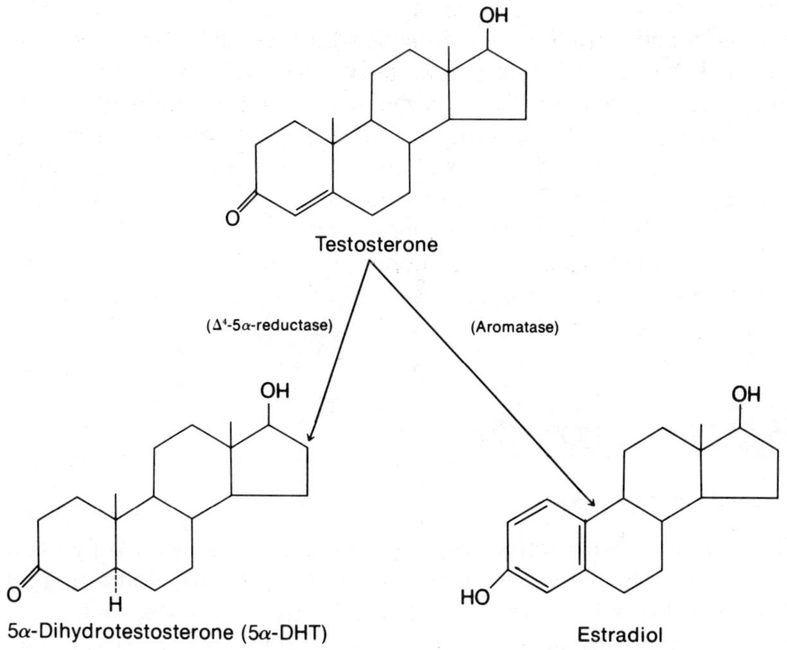

OH

O

Testosterone

(Δ^4-5α-reductase)

(Aromatase)

OH

OH

O

H

HO

5α-Dihydrotestosterone (5α-DHT)

Estradiol

Figure 15.18
Biologically active metabolites of testosterone.

Peripheral Metabolism of Androgens and Estrogens

Many tissues in the body, in addition to the endocrine glands, liver, and kidney, contain enzymes that metabolize steroids. These enzymes are primarily dehydrogenascs and hydroxylases. The conversion of testosterone to 5α-dihydrotestosterone (5α-DHT), for example, is by the action of a Δ^4-5α-reductase. The conversion of testosterone or 4-androstene-3,17-dione to estrogen is due to the action of the enzyme complex, aromatase (Figure 15.18). These reactions occur in tissues such as skin, adipose

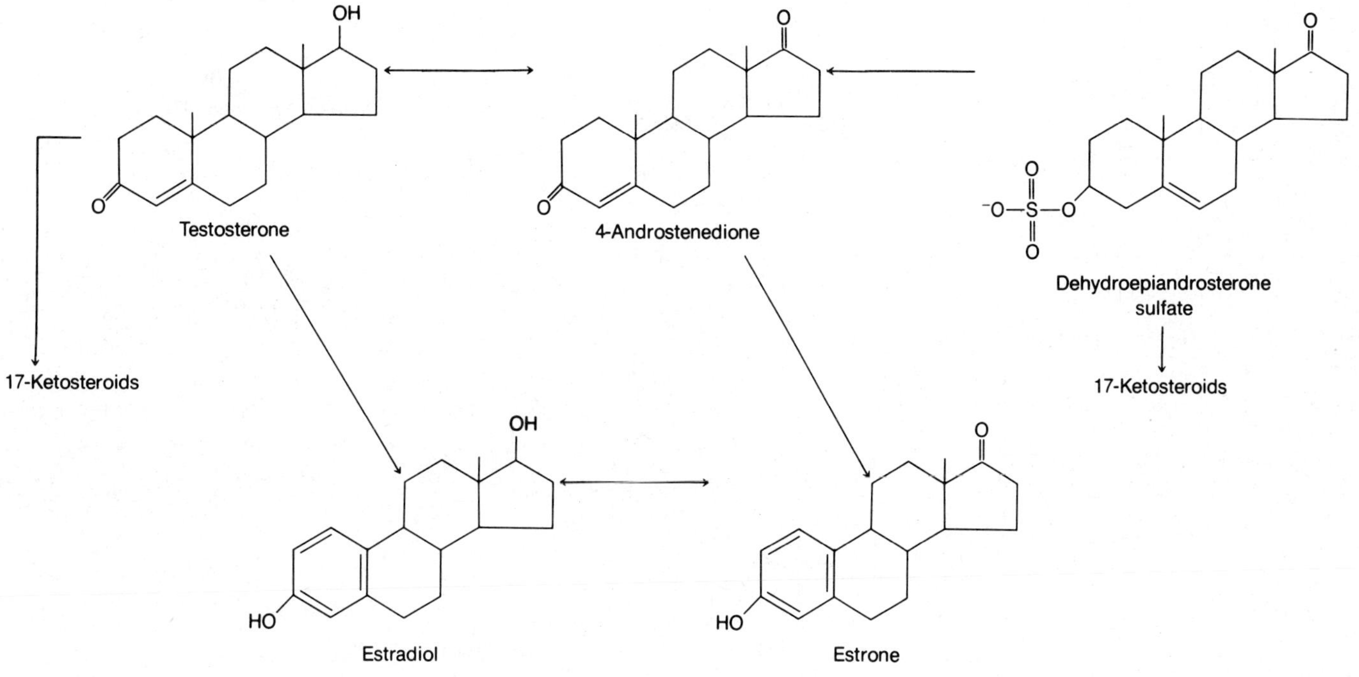

OH

O

O

Testosterone

4-Androstenedione

$^{-}$O—S—O

Dehydroepiandrosterone sulfate

17-Ketosteroids

17-Ketosteroids

OH

O

HO

HO

Estradiol

Estrone

Figure 15.19
Interaction of biologically active steroids in peripheral tissues.

tissue, and brain. Note that the two reactions which form 5α-DHT, an androgen, and estradiol, an estrogen, are irreversible. Testosterone, the principal steroid sex hormone secreted by the testes, may be considered a prohormone, since 5α-DHT is a biologically active androgen and binds avidly to testosterone receptors. 5α-DHT and estradiol formation should be considered as two distinct and essential features of testosterones biological activity. The effect of steroids on the CNS and brain tissue in sexual maturation and behavior and the role of steroids in the etiology of mammary carcinoma are two examples of clinical interest in which metabolic alterations by peripheral tissues have important consequences (Figure 15.19). (See Clin. Corr. 15.3.)

CLIN. CORR. 15.3
ESTROGEN RECEPTORS IN BREAST CANCER

Hormone treatment and endocrine ablation for the treatment of prostate and breast cancer was first introduced in the 1940s by Charles Huggins at the University of Chicago. A further advance in the clinical evaluation for the treatment of breast cancer was the application of a binding assay for estrogen-receptor interaction developed by Elwood Jensen and co-workers at the University of Chicago in the 1960s. It was reasoned that hormone-dependent tumors, those that respond to estrogen for growth, would have demonstrable cytoplasmic and nuclear receptors for estrogen. A positive binding assay using tritium-labeled estradiol and extracts of tumor biopsy tissue should differentiate those tumors that would be most likely to respond to subsequent endocrine ablation and hormone therapy. In very few instances is there a response as a result of such treatment when the estrogen-receptor assay is negative, as compared to a response in 50% of the cases that demonstrate the presence of estrogen receptors in the tumor cells.

15.5 THYROID HORMONES

Formation of T_4 and T_3

There are two major active hormones produced and secreted by the follicle cells of the thyroid gland, thyroxine (T_4) and triiodothyronine (T_3) (Figure 15.20). With respect to biological activity they are similar qualitatively, but there are important differences. Of the two, T_3 is considerably more active, and it is thought that T_4 activity is due to its conversion to T_3. On a weight basis, T_3 is 3 to 5 times more active than T_4, depending upon the bioassay used. The half-life of the thyroid hormones in the body is considerably longer than for any of the other hormones for reasons not well understood: $T_{1/2}$ for T_4 is 6 days and $T_{1/2}$ for T_3 is 2 days. Binding to plasma thyroxine-binding globulin (TBG) and other proteins is extensive with less than 0.04% of T_4 and T_3 in the free form. Metabolic clearance could partially account for the differences in half-life, but other factors must be involved. As with cortisol-binding globulin and circulating corti-

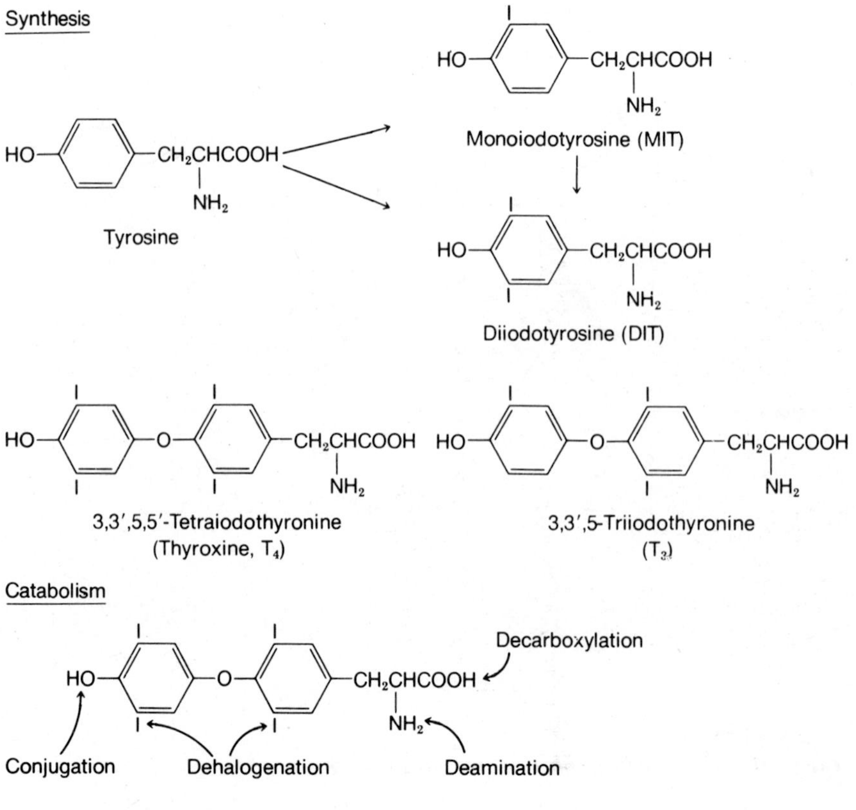

Synthesis

Tyrosine

Monoiodotyrosine (MIT)

Diiodotyrosine (DIT)

3,3′,5,5′-Tetraiodothyronine (Thyroxine, T_4)

3,3′,5-Triiodothyronine (T_3)

Catabolism

Decarboxylation

Conjugation

Dehalogenation

Deamination

Figure 15.20
Thyroid hormone metabolism.

sol (about 95% bound), the free hormones, which are in equilibrium with the bound forms, have unhindered access to the tissue cells. Other than as a storage function the physiological role for TBG (and for CBG) remains to be established. The high plasma protein-bound levels of active hormone could be due to a decrease in both kidney clearance and liver inactivation.

More T_4 is produced by the thyroid than T_3, and plasma levels for T_4 (4.5–13.0 μg/dl) are considerably higher than for T_3 (0.06–0.20 μg/dl). A unique feature of the thyroid hormones is the presence of organically bound iodine. Other proteins may contain small amounts of iodine, and the tyrosine precursors of the thyroid hormones contain some iodine; however, the bulk of the halide is maintained very efficiently in the thyroid tissue in the form of T_4 and T_3. Normal iodine plasma levels are less than 1 μg/dl and the small loss of iodide (75–100 μg/day) is usually adequately replaced in the diet, which ordinarily contains more than 200 μg/day.

Regulation of Thyroid Hormone Formation

Regulation of thyroid hormone formation is under the control of the anterior pituitary hormone, thyroid-stimulating hormone (thyrotropin, TSH). TSH formation and secretion is regulated by the hypothalamic thyrotropin releasing factor, TRF. TRF contains only three amino acids, pyroglutamyl-histidyl-prolinamide and was the first of the releasing factors identified and synthesized. Feedback inhibition by both the more active T_3 and by T_4 is exerted at the hypothalamic and pituitary levels; however, greater inhibitory effects occur at the pituitary level. Since the original isolation, TRF has been found in other tissues, and the release of prolactin as well as TSH occurs with TRF administration. This raises questions about other biological roles for TRF, for example, as a neurotransmitter, but its function as a releasing agent for TSH as a true physiological event is generally accepted.

The accumulation of I^- in the thyroid follicle exceeds that in any other tissue. The extensive uptake of I^- is referred to as the "iodide trap" and is a function of TSH activity (Figure 15.21). Iodide is oxidized by a thyroid peroxidase, resulting in organic binding to tyrosine molecules. First, a monoiodo- and then a diiodotyrosine is formed. Then the coupling of two iodotyrosines results in the formation of tetraiodo- and triiodothyronine molecules. These iodinated compounds are linked covalently in a peptide linkage to a large (M_r = 700,000) protein molecule called thyroglobulin. The reactions leading to the formation of the iodinated thyroglobulin in the colloid of the thyroid follicle are controlled by TSH through the activation of cAMP-mediated events. These reactions include the release of proteolytic enzymes from lysosomes, which hydrolyze thyroglobulin to yield the iodinated tyrosine and thyronine molecules. Approximately 100 μg of T_4 and T_3 in a ratio of 20 : 1 are released into the general circulation per day. A specific thyroid microsomal *dehalogenase* removes the iodide from the tyrosines but not from the thyronines and returns the halogen to the follicle for reutilization.

The uptake of iodide by the thyroid and the subsequent release of iodinated thyronines from thyroglobulin are both mediated through TSH stimulation. The conversion of T_4 and T_3 by thyroxine dehalogenase occurs in a number of tissues. This reaction is catalyzed by an enzyme that is different from the dehalogenase of thyroid tissue, since iodinated tyrosines are not substrates. In the liver T_4 and T_3 are further metabolized by deamination, decarboxylation, and conjugation (Figure 15.20). Both sulfate and glucuronide esters at the phenolic hydroxyl are formed in the

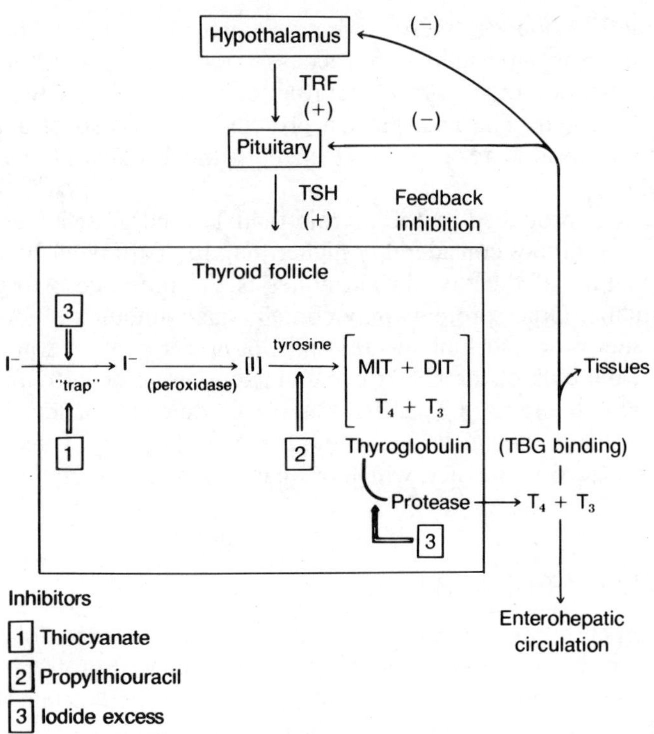

Figure 15.21
Regulation of thyroid hormone synthesis.

liver. There is an extensive enterohepatic circulation of the conjugated hormone.

Inhibition of Thyroid Hormone Formation

When the formation of the thyroid hormones is blocked or inhibited, the low circulating blood levels at the pituitary will result in increased TSH secretion as the normal feedback inhibition is diminished. Increased TSH levels will stimulate the growth of the thyroid follicles to give an enlarged thyroid gland. The abnormally large thyroid gland is called a *goiter*. Drugs that interfere with the formation of thyroid hormone, resulting in the enlargement of the thyroid gland, are called goitrogenic substances, or goitrogens.

Thiocyanates, perchlorates, and nitrates interfere with the iodide-trapping mechanism of the thyroid (Figure 15.17), resulting in decreased utilization of iodide and deficient thyroid hormone formation. The *Brassica* class (cabbage family) of plants contain large amounts of thiocyanates, and their ingestion in large amounts tends to be goitrogenic. Compounds in the thiocarbamide class, such as thiouracil and thiourea, sulfonamide compounds, and imidazole drugs, inhibit the organic binding of iodide. These compounds interfere with the oxidation reaction, which involves a peroxidase, probably by competing as substrates for the iodide.

Iodide itself has paradoxical effects on thyroid function. Below normal levels of iodide an increase in thyroid size may occur by TSH stimulation, as a result of diminished thyroid hormone formation. Euthyroid patients with goiters are found in certain regions of the world (Great Lakes, U.S.A., or Switzerland) where natural soil iodide levels are low. Increased iodide concentration interferes with TSH stimulatory effects on the thyroid follicles. Large doses of iodide are used to reduce thyroid size in certain hyperthyroid states (see Clin. Corr. 15.4).

Thyroid Hormone Function

All cells of the body, with the possible exception of adult brain and testes, are target cells for thyroid hormone. In most thyroid responsive cells T_3 binding sites have been demonstrated in the chromatin fraction of the nucleus. T_4 represents only 15% of the total iodothyronine bound to the nuclear receptor as the T_4 is converted in the cell to T_3 prior to binding. The receptor has not been well characterized. It has a high affinity ($K_d = 10^{-11}$ M) and low capacity for T_3, which would be expected for a receptor. In the normal (euthyroid) state only 15% of the total T_3 in the cell is bound, and ~50% of the available nuclear receptor sites are occupied. T_3 bound to the nuclear receptors is in equilibrium with the T_3 in the cytoplasm and circulating plasma. It is believed that increased mRNA production and increased protein synthesis are direct consequences of T_3-nuclear receptor binding, but evidence to support this concept is minimal and circumstantial. Another potential site for thyroid hormone action that has been proposed is the mitochondrion, based on the demonstration of high affinity binding. The mechanism of thyroid hormone action at the molecular level to explain the well-known physiological effects described below remains to be determined.

CLIN. CORR. 15.4 USES OF RADIOACTIVE IODIDE

When a tracer dose of radioactive form of iodine (^{123}I, $T_{1/2}$ = 8 days) is administered orally, its rate of uptake by the thyroid gland ("iodide trap") is proportional to the degree of stimulation by TSH. The rate of uptake is relatively high in the hyperfunctioning and low in the hypofunctioning gland (normal range is 5 to 50%). Similarly, the release of ^{123}I-thyronine from the thyroid to the blood can be monitored by a radioactivity scanning device 24 hours or more after the administration of a tracer dose of ^{123}I, since the release of thyroid hormone bound in thyroglobulin stored in the gland is also stimulated by TSH.

Some types of thyroid cancer can be treated as an adjunct to surgery by the administration of larger doses of ^{131}I, since the radioactivity will concentrate in the thyroid tissue as a result of a very efficient trapping mechanism. The high energy of the β-emitter will destroy thyroid tissue selectively, particularly the more active cancer cells. Regions of high uptake are referred to as *hot spots* as distinct from inactive *cold spots*. It has been the practice in the past to restrict the use of radioactive iodine in young individuals, especially in women of childbearing age.

Thermogenesis and Oxygen Consumption

Increased heat production and oxygen consumption are characteristic for most tissues responding to thyroid hormone (brain, testis, and spleen excluded). Indeed, this feature was the basis for measurement (basal metabolic rate (BMR)) of thyroid hormone activity in the human prior to the development of more specific assay procedures for T_4 and T_3. The increased O_2 demand could be accounted for in part by the stimulation of membrane-bound sodium-potassium ATPase to maintain normal ionic gradients. Since thyroid hormone is present in cold-blooded as well as warm-blooded animals, its effect on thermogenesis requires more study to be fully integrated into present concepts of mode of action. Correlation of increased oxygen consumption with a mitochondrial site of thyroid hormone action has been considered, but the evidence that associates thyroid effects with energy production (electron transport system regulation) has been disputed. The ATPase enzyme activity in some obese patients have been disputed. The Na^+-K^+-ATPase enzyme activity in some obese patients has been found to be below normal. Since there is a direct correlation between thyroid hormone levels and Na^+-K^+-ATPase activity, obesity in some patients has been suggested to be the result of a decreased energy and heat production due to diminished Na^+-K^+-ATPase activity.

Metabolic Effects

Effects of T_4 and T_3 include actions on metabolism of carbohydrates, proteins, lipids, electrolytes, and water. Thyroid hormone effects on carbohydrate metabolism involve increased intestinal absorption of glucose balanced by increased glucose utilization. The net effect is one of hyperglycemia and an abnormal glucose tolerance curve. In protein metabolism a dual effect by thyroid hormone is observed, with anabolic activity during growth and development, and catabolic activity (protein breakdown) in the hyperthyroid state. Cholesterol blood levels are high in hypothyroidism, and the high levels can be decreased with thyroid hormone administration. There is increased lipid utilization with thyroid hormone. Retention of water and electrolytes in the hypothyroid state can be reversed by thyroid administration.

Altering the thyroid hormone state in the human causes well-known changes in the central nervous system, in nerve and muscle function, in the gastrointestinal tract, and in the vascular system. The skin is a good indicator of the thyroid state. In hyperthyroidism the skin is smooth, warm, and moist as a result of vasodilation. In contrast, the skin is cold and has a rough texture due to vasoconstriction in the hypothyroid state. The characteristic accumulation of fluid and mucopolysaccharides, with the resulting puffiness (pitting edema) of the skin, gives rise to the adult hypothyroid state, myxedema.

The heart reflects the changes in thyroid state, having a slow rate and decreased blood flow in the hypothyroid condition. A major toxic effect of thyrotoxicosis is on the heart, with increased heart rate, cardiac hypertrophy, and cardiac contractility.

The thyroid hormones act in conjunction with pituitary growth hormone as the principal anabolic agents during growth and in maintaining protein stores. Synergistic effects of the two can be demonstrated on protein synthesis in the liver and on transamidinase (guanidine transferase) activity in the kidney. The important role of thyroid hormone in human development is apparent in *cretinism,* a condition brought about by thyroid deficiency during the prenatal period, resulting in a serious detriment to both mental and physical development in the growing child.

BIBLIOGRAPHY

Receptor Models

Baxter, J. D., and MacLoed, K. M. Molecular basis for hormone action, in P. K. Bondy and I. N. Rosenberg (eds.), *Metabolic control and disease*. Philadelphia: W. B. Saunders, 1980, p. 140.

Jensen, E. V., and Jacobson, H. I. Basic guides to the mechanism of estrogen action. *Rec. Prog. Horm. Res.* 18:387, 1962.

Schrader, W. T. Methods for extraction and quantification of receptors, in B. W. O'Malley and J. G. Hardeman (eds.), *Methods Enzymol.* 36:187, 1975.

Thyroid Hormones

Edelman, L. S. Thyroid thermogenesis. *N. Engl. J. Med.* 290:1303, 1974.

Frieden, E. Iodine and the thyroid hormone. *Trends Biochem. Sci.* 6:50, 1981.

Oppenheimer, J. H., Schwartz, H. L., Surks, M. I., Koerner, D. H., and Dillman, W. H. Nuclear receptors and initiation of thyroid hormone action. *Rec. Prog. Horm. Res.* 32:529, 1976.

Steroid Hormones

Dorfman, R. I., and Ungar, F. *Metabolism of steroid hormones*. New York: Academic, 1965.

Gwynne, J. T., and Straus, J. F., III. The role of lipoproteins in steroidogenesis and cholesterol metabolism in steroidogenic glands. *Endocr. Rev.* 3:299, 1982.

Lieberman, S., Greenfield, N. J., and Wolfson, A. A heuristic proposal for understanding steroidogenic processes. *Endocr. Rev.* 5:128, 1984.

Takemori, S., and Kominami, S. The role of cytochromes P_{450} in adrenal steroidogenesis. *Trends Biochem. Sci.* 9:393, 1984.

Wilson, J. D. Recent studies on the mechanism of action of testosterone. *N. Engl. J. Med.* 287:1284, 1972.

QUESTIONS

J. BAGGOTT AND C. N. ANGSTADT

*Q*uestion *T*ypes are described inside the front cover.

1. (QT1) Hormones may act:
   A. by phosphorylating or dephosphorylating enzymes.
   B. by binding to enzymes and causing subunit association or dissociation.
   C. indirectly on enzymes through a sequence of steps.
   D. as enzyme cofactors.
   E. by cleaving peptide bonds of proenzymes or zymogens.

   A. Nerve cell   C. Both
   B. Endocrine cell   D. Neither

2. (QT4) May synthesize hormones.
3. (QT4) Produce(s) a substance that affects an adjacent cell directly.
4. (QT2) Which of the following is/are secreted directly into the bloodstream?
   1. Steroid hormone
   2. Peptide hormone
   3. Neurohormone
   4. Neurotransmitter
5. (QT1) According to the receptor model for hormone action:
   A. any given hormone is bound tightly for long periods of time to most tissues of the body.
   B. no substance other than the hormone may bind to a specific receptor.
   C. specific receptors exhibit low affinity for the hormone.
   D. specific receptors are not saturable but bind more and more hormone as the hormone concentration is increased.
   E. receptors bind hormones in a manner analogous to the binding of substrates to enzymes.

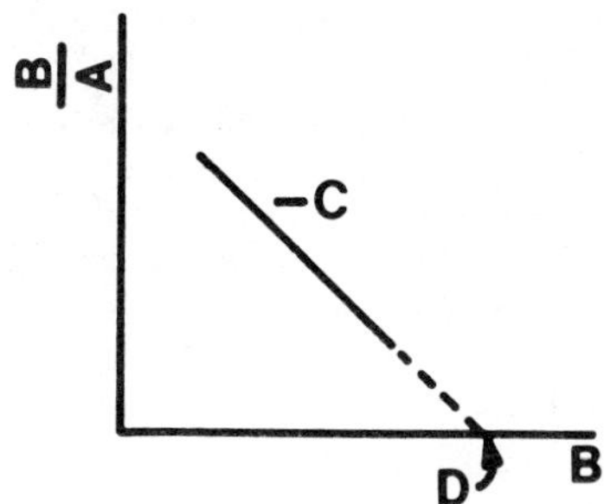

6. (QT1) The Scatchard plot, shown in the accompanying figure, could be used to determine kinetic parameters of an enzyme. Which letter in the graph corresponds to total binding sites in a Scatchard plot or V_{max} in an enzyme kinetic plot?
   A. Hormone   C. Both
   B. Vitamin   D. Neither
7. (QT4) The body synthesizes its entire requirement.
8. (QT4) Exerts its effect at a site *other* than the site of its synthesis.
9. (QT4) May bind to a nuclear receptor.
10. (QT1) Steroid hormones:
   A. are all synthesized in most tissues of the body.
   B. have long half-lives and so are stored in endocrine tissues until needed.
   C. in blood, circulate in both the free form and bound to protein.
   D. are converted to excretory forms in the tissues that synthesize them.
   E. and their metabolites are excreted only in urine.
11. (QT1) Which of the following is common to the synthesis of all steroid hormones?
   A. Cholesterol side chain cleavage.
   B. Conversion of pregnenolone to progesterone
   C. 17α-hydroxylation
   D. 21-hydroxylation
   E. formation of an aromatic ring
12. (QT2) Dehydroepiandrosterone (DHEA):
   1. is a weak androgen.
   2. is secreted by the adrenal conjugated to sulfate.
   3. is a precursor of steroid hormones in nonendocrine tissues.
   4. is converted to estrogens by the placenta during pregnancy.
13. (QT2) Chronic administration of cortisol for its anti-inflammatory properties might be expected to cause:
   1. chronic stimulation of the adrenal cortex by ACTH.
   2. elevated levels of blood glucose.
   3. stimulation of protein synthesis in muscle.
   4. a negative feedback effect on corticotropin-releasing factor (CRF).
14. (QT2) Steroid hormone production in the glomerulosa zone of the adrenal cortex is stimulated by:
   1. ACTH.
   2. excess Na^+.
   3. low glucose concentration.
   4. angiotensin II.
15. (QT2) Testosterone:
   1. is the main secretory product of the testes.
   2. may function as a prohormone.
   3. production is stimulated by luteinizing hormone (LH).
   4. is a catabolic hormone.
16. (QT1) Estrogen:
   A. effects are confined to sexual maturation and sexual characteristics.
   B. production is not related to LH and FSH levels.
   C. metabolism contributes to the 17-keto steroids of the urine.
   D. is inactivated by conversion to estrone.
   E. is synthesized from androgens by the action of an aromatase.
17. (QT2) Metabolites of steroid hormones:
   1. are frequently conjugated with either sulfate or glucuronate.
   2. no longer have an intact steroid ring system.
   3. may be present in the enterohepatic circulation.
   4. must circulate in the blood bound to 17β-globulin.
   A. T_3 C. Both
   B. T_4 D. Neither
18. (QT4) In the blood, mostly bound to thyroxine-binding proteins.
19. (QT4) Released when a hypothalamic factor stimulates the thyroid gland.
20. (QT4) Formed as part of a thyroglobulin molecule.
21. (QT4) Substrate of a broadly distributed dehalogenase.

ANSWERS

1. C Hormones do not directly initiate or terminate enzyme activity (p. 562). A, B, D, E refer to ways in which enzyme activity is modified, and they may be influenced indirectly by hormones.
2. C Nerve cells may produce neurohormones such as vasopressin. Endocrine cells may produce hormones such as insulin (p. 566).
3. A The statement describes paracrine cells. Neurotransmitter release by nerve cells also fits, however (p. 566).
4. A 1, 2, 3 correct. Hormones, by definition, are secreted directly into the blood (p. 563). Neurotransmitters act locally (p. 566).
5. E A: Hormones bind tightly for long periods to specific tissues. B: Receptors also bind agonists and antagonists. C: Specific binding is tight. D, E: Specific receptor binding is like substrate binding to an enzyme and can be saturated (p. 568).
6. D A is free ligand concentration (analogous to substrate concentration), B is bound ligand concentration (analogous to v), C is the equilibrium constant (analogous to K_m), and D is the extrapolated maximum number of binding sites (analogous to V_{max}) (p. 569). Note that the first equation on p. 569 is identical in form to equation (4.9) on p. 127.
7. A Most vitamins cannot be synthesized by the body at all, and no vitamin can be synthesized in adequate amounts (p. 563).
8. C The statement is part of the definition of "hormone" and must also be true of vitamins, since they are synthesized outside the body, but serve as enzyme cofactors (p. 563).
9. A True of steroid and thyroid hormones (p. 565).
10. C Because steroid hormones are so poorly water-soluble, they circulate largely bound to specific proteins, but small amounts are present in the free state. A: Hormone production is tissue specific. (Figure 15.2). B: Rapid turnover of the hormones is part of control. D: A hormone is produced in one tissue, exerts its effects on a different target tissue, and is frequently metabolized in still a different tissue (often liver). E: Excretion into bile is also important (p. 570).
11. A Side chain cleavage converts cholesterol to pregnenolone. B: Formation of androgens in the adrenal cortex is via 17 α-hydroxypregnenolone. C–E: Not all steroid hormones have all of these functional groups (Figure 15.8).
12. E All four correct. The adrenal cortex secretes large amounts of DHEA sulfate, which represents a potential reservoir of precursor steroid for peripheral tissues (p. 573). During pregnancy, fetal steroid sulfates are precursors for estrogens (p. 575).
13. C 2,4 correct. 1 and 4 are related to control. Cortisol exerts a

negative feedback effect on the hypothalamic–pituitary system. Depression of CRF release would also result in depression, not stimulation, of ACTH (p. 576). The effects of cortisol are to enhance protein catabolism in muscle. The amino acids generated would be used by the liver for gluconeogenesis (Figure 15.14).

14. D Only 4 correct. The glomerulosa zone produces aldosterone, which stimulates Na^+ retention when $[Na^+]$ or extracellular volume is low. 1 and 3 are factors influencing the fasiculata–reticularis zones (p. 577, Figure 15.11).

15. A 1, 2, 3 correct. The Leydig cells, which produce testosterone, are controlled by LH. As a prohormone testosterone is converted to either 5-α-dihydrotestosterone or estradiol (p. 581, 585). 4: Androgens are profoundly anabolic (p. 581).

16. E Testosterone is a precursor of estrogens, and this requires the formation of an aromatic ring by an aromatase. (Figure 15.9, p. 585). A: Estrogens have general metabolic effects as well as sexual ones. B: Their synthesis is stimulated by the combined action of LH and FSH. C: 17-Ketosteroids are primarily androgen metabolites. D: Estrone is one of the active estrogens (p. 582).

17. B 1, 3 correct. Conjugation with glucuronate or sulfate is an important mechanism for enhancing water-solubility of very hydrophobic materials, and the conjugates are excreted in bile as well as urine (p. 572). 2: Humans cannot degrade the steroid ring. 4: 17-β-globulin is the sex steroid hormone binding globulin for the active hormones not the metabolites (p. 570).

18. C Only about 0.04% of T_3 and T_4 are free (p. 586).

19. D The thyroid gland is stimulated by TSH, a pituitary hormone (p. 587, Figure 15.21).

20. C T_3 and T_4 are formed in a posttranslational modification of thyroglobulin and are subsequently released by proteolysis (p. 588, Figure 15.21).

21. B T_4 is dehalogenated to T_3 in many tissues. A specific thyroid dehalogenase attacks iodinated tyrosines but not T_3 or T_4 (p. 587).

16

Biochemistry of Hormones II: Peptide Hormones

FRANK UNGAR

16.1 PLASMA MEMBRANE RECEPTORS AND MEDIATORS OF HORMONE ACTION

Two principal types of hormone–receptor interactions were introduced in Chapter 15. In one type, lipid-soluble substances, including steroid hormones, 1,25-dihydroxycholecalciferol, and thyroid hormones, pass freely through the plasma membrane and bind to a cytoplasmic receptor. The hormone–receptor complex translocates to the chromatin of the cell nu-

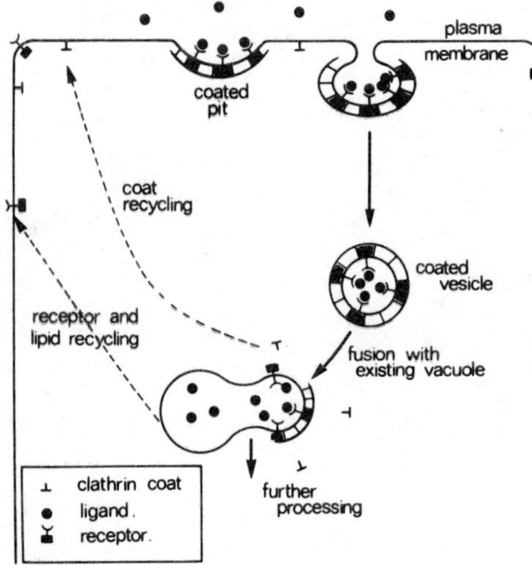

Figure 16.1
Internalization of substances through the cell membrane using coated pits and coated vesicles.
Pathway by which peptide hormones can enter the cell. Inside the cell the ingested materials, receptors and hormones, are processed by lysosomal hydrolytic enzymes. Processed molecules can then be transferred to other cell compartments.
From B. Pearse, *Trends Biochem. Sci.*, 5:26, 1980. Reproduced with permission.

cleus to initiate the DNA-directed synthesis of mRNA. The second type of hormone–receptor interaction involves water-soluble substances, which cannot permeate through the cell plasma membrane; their first contact with the cell is to bind to protein receptors on the outer membrane surface.

In order for the message to reach intracellular sites, the hormone–receptor complex can be internalized (endocytosis) as in Figure 16.1, or the information could be transferred at the membrane site (transduction) and carried by other mediators or messengers within the cell compartments. Substances that are considered to participate in the intracellular transfer of information include the calcium ion (Ca^{2+}), the prostaglandins, the cyclic nucleotides, phospholipids, such as phosphatidylinositol, and the proteins with which they interact. Examples of the participation of these mediators in the expression of hormones are presented in this chapter and diagrammed in Figure 16.2.

The plasma membrane presents a barrier to the entrance of water-soluble molecules, and within the cell the subcellular membranes restrain free access of substances. The membrane, by enfolding materials from the exterior, can internalize them (endocytosis), or by coalescing with vesicles from within, can transport materials out (exocytosis) of the cell. Cellular activity is enhanced by removing the membrane restraint by increasing its permeability to the flow of ions or organic substances. This is described in terms of creating membrane pores, channels, or gaps by which water-soluble substances can be transferred into the cell or between the cell compartments created by membranes. Hormone binding to the cell membrane or by indirectly affecting subcellular membrane permeability is in effect removing a restraint to cell activity or subparticle inter-

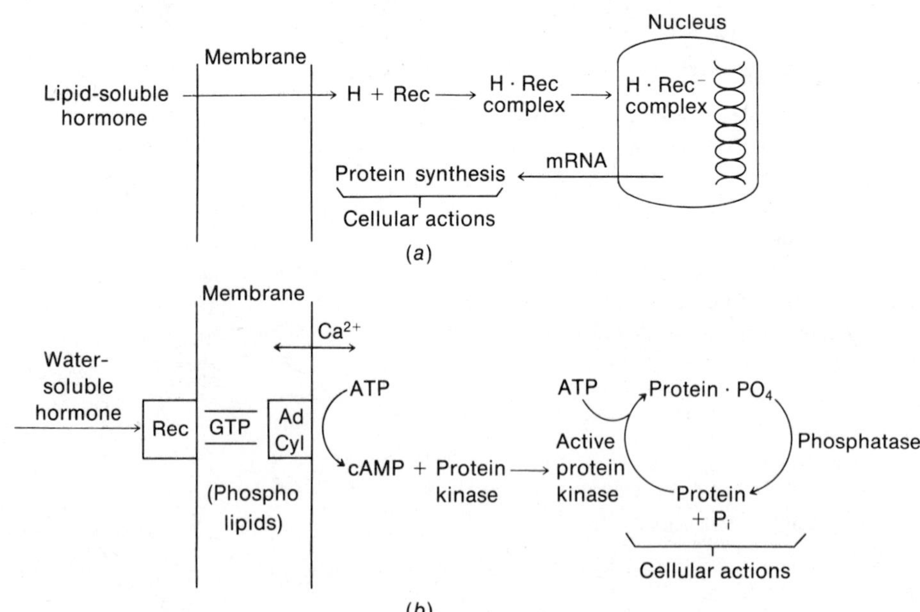

Figure 16.2
Initial interactions of hormones with target cells.
(a) Lipid-soluble hormones bind to receptors within the cell and as a hormone-receptor complex initiate DNA-directed synthesis of mRNA (examples: steroid hormones, vitamin D, thyroid hormones). (b) Water-soluble hormones (examples: glucagon, epinephrine, ACTH) bind to specific receptors on the cell plasma membrane. By activation of adenylate cyclase and protein kinase, a number of cellular processes are stimulated. See text for actions of Ca^{2+}, Mg^{2+}, GTP, phospholipids, and prostaglandins.

action. Water-soluble hormones, neurotransmitters, or other agents (lectins and growth factors) bind to a cell surface receptor as a result of a noncovalent association with a relatively high affinity to initiate a set of membrane events. As a result, there is a release of mediators (secondary messengers) that react with cytoplasmic components. Most polypeptide hormones stimulate the production of cAMP as the secondary messenger. There are important exceptions to this, however. Insulin and prolactin activity can be expressed without direct stimulation of adenylate cyclase. Their membrane interactions are described in a later section. (See Clin. Corr. 16.1.)

Ionic Fluxes in Transmembrane Signaling

Rapid changes in monovalent ionic fluxes and membrane potential may serve as a transmembrane signal in the early action of some growth factors and hormones. Cytoplasmic Ca^{2+} functions as an intermediary messenger of growth factor activity, sometimes mediated by calmodulin but others not. Surface membrane stimuli can result in a change in cellular pH. This cellular event may be the result of a Na^+/H^+ exchange at the membrane site of growth factor stimulation and is due to the difference between Na^+ and H^+ transmembrane gradients. Inhibitors of Na^+/H^+ exchange inhibit growth factor effects on this exchange and the change in intracellular pH. The exchange is not accompanied by a membrane depolarization but is an electroneutral transport system distinct from basal Na^+ permeability. This event precedes the stimulation of the ouabain-sensitive Na^+/K^+ transport that is activated by mitogenic factors. Early Na^+ influx is the common mechanism that stimulates the Na^+/K^+-ATPase system in fibroblasts, lymphocytes, and neuroblastoma cells. Growth factors, such as epithelial growth factor and hormonal agents, may utilize the Na^+/H^+ exchanger as a signal transducer to shift intracellular pH toward alkaline pHs in order to modulate the rate of important cellular processes.

The acetylcholine receptor, which alters ion fluxes, (Figure 16.3) has been purified and its subunit character fully determined. The unusually high concentration of this receptor in the electric organ of the sting ray (*Torpedo*) and eel (*Electrophorus electricus*) allowed purification of the

CLIN. CORR. 16.1 MEMBRANE AND NUCLEAR SITES OF HORMONE ACTION

The binding of a peptide hormone to a plasma membrane receptor or that of steroid and thyroid hormones to a nuclear receptor does not describe adequately the full range of hormone action normally encountered in clinical medicine. Long-terms effects, particularly those associated with regulation of cell division and tissue growth, have not been studied extensively and are poorly understood. Subsequent to membrane receptor binding, peptide hormones can be internalized by a process referred to as *receptor-mediated endocytosis,* a process of membrane enfolding to form cytoplasmic vesicles, which transport the hormone receptor to the lysosome and possibly other subcellular compartments. Plasma membrane receptors have a complex subunit structure that may involve other transport functions in addition to hormone binding. The actions of insulin, prolactin, and other hormones that have been described at the intracellular level may be a consequence of additional events other than those associated with cyclic AMP-mediated protein kinase-phosphatase interactions. Similarly, membrane associated events can be associated with administration of steroid and thyroid hormones, although the mechanisms involved cannot be described at the level of molecular interaction.

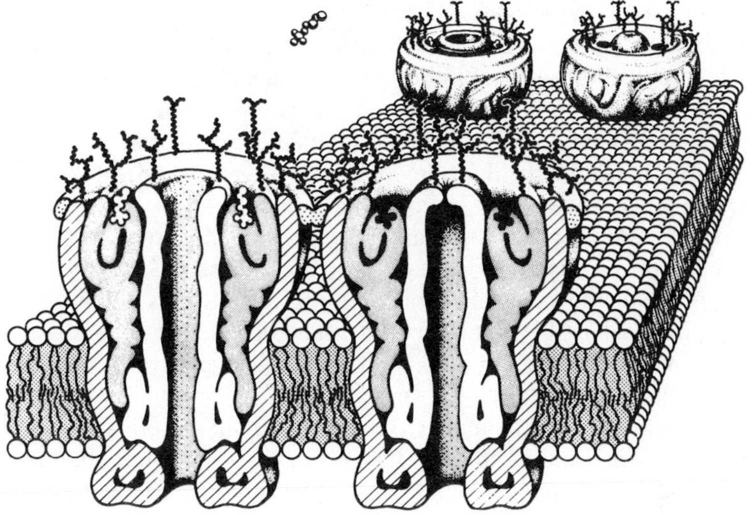

Figure 16.3
Model of the transmembrane pentameric receptor for acetylcholine.
Channel open (left), channel closed (right).
From J. Lindstrom, *Advan. Immunol.*, 27:1, 1979. Reproduced with permission.

acetylcholine receptor to homogeneity after a 500-fold purification. Due to their very low concentrations in cells, mammalian hormone receptors require many thousandsfold purification to achieve the same degree of homogeneity. The acetylcholine receptor is a pentamer of four homologous proteins of M_r 255,000, with subunits of M_r 40K, 50K, 60K, and 65K, forming a structure of $\alpha_2\beta,\gamma,\delta$. The receptor is a transmembrane protein with a carbohydrate portion on the outer (synaptic) face of the membrane and forms a channel for ion transfer. Acetylcholine is a neurotransmitter released at nerve terminals in response to an electrical signal at a nerve–muscle structure known as a motor end plate. Both the binding of acetyl choline to its receptor and the ion flux can be monitored simultaneously in the time scale of milliseconds. The receptor behaves as if it were an allosteric protein that can undergo a conformation change in a membrane environment consisting of neutral lipids, cholesterol, phospholipid and protein. Specific receptors are present across a synaptic gap in the postsynaptic membrane, which receives acetylcholine, binds and hydrolyzes it by reaction with an esterase. This interaction opens channels in the receptor to allow Na^+ and K^+ to pass through.

Cyclic Nucleotides as Second Messengers

The fact that activation of adenylate cyclase is a major route of hormone stimulation can be demonstrated by the addition of cAMP or dibutyryl cAMP to a target cell. The addition of the nucleotide will mimic most, if not all, of the actions of some hormones on cellular events. Furthermore, the addition of caffeine will enhance hormone stimulatory effects on a cell. This is consistent with the effect of methylxanthines, such as caffeine, as inhibitors of phosphodiesterase, the enzyme that converts cAMP to the inactive 5′-AMP. The participation of cyclic nucleotides in the control of cellular metabolic events is discussed throughout the textbook. The role of hormones that bind to the plasma membrane as activators of this system is described in this section.

Hormone receptor and adenylate cyclase can be isolated and purified as separate proteins from a number of different cell membranes. Cells are lysed under hypoosmotic conditions, the cell contents are removed, and the heavy tissue fragments, which sediment by centrifugation, contain intact membrane particles (ghosts). Liver cell ghosts or fat cell ghosts are commonly used as model systems. Since the hormone receptor and adenylate cyclase are present and active in the same particle, it is possible to measure both hormone binding to the receptor as well as cAMP production, as a membrane response to hormone stimulation. An unusual feature of the adenylate cyclase reaction is the stimulation of the enzyme by halides, particularly fluoride ions. The organization of the enzyme complex in the membrane is important, since the enzyme in a soluble form still retains activity, but can no longer be stimulated by hormones.

Regulation of adenylate cyclase activity to form cAMP by hormone action involves the association of three protein systems within the membrane—the specific hormone receptor, the G protein, which is actually a complex of two trimeric proteins, and adenylate cyclase (see Figure 16.4). The G protein consists of two similar proteins, G_s and G_i, each with a trimeric structure containing α, β, and γ subunits. Both G_s and G_i are activated by the binding of GTP that occurs following interaction of the appropriate hormone with the receptor. The G_s–GTP complex stimulates the adenylate cyclase, whereas the G_i–GTP complex inhibits the cyclase. The stimulatory or inhibitory effect is terminated by the hydrolysis of the GTP to GDP and phosphate, a GTPase reaction also catalyzed by the G protein, and the release of the GDP. This permits a fine control of initiation of the activity when GTP binds then termination when GTP is hydro-

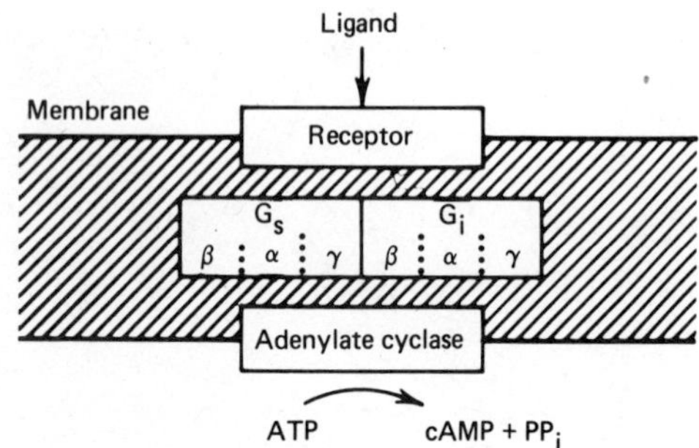

Figure 16.4
Regulation of adenylate cyclase.

lyzed. The α subunits are different in G_s and G_i, but the β subunits are identical. The β subunit is a regulatory protein and dissociates when GTP binds. The function of the γ subunit is unknown. Hormones that increase cAMP levels lead to activation of G_s and ligands such as acetyl choline and insulin, and some α-adrenergic agents are considered to cause a change in the G_i subunit. Cholera toxin, a protein (84K) secreted by *Vibrio cholerae,* is responsible for the harmful effects of cholera in the intestines. It has a structure of three peptides $\alpha\gamma\beta n$, arranged into an A (α,γ) subunit and a B (βn) subunit. The A subunit of the molecule enters the membrane of a cell and splits into A_1 and A_2 subunits. The A_1 peptide of 23 Kdal catalyzes the transfer of an ADP-ribose unit from an NAD^+ molecule to an arginine residue in the G_s subunit of the GTP-binding protein of the adenylate cyclase system in the membrane. As a result of the ADP ribosylation, the normal GTPase activity of the G protein is blocked, and the adenylate cyclase remains "on" in an active state (Figure 16.5). In an analogous manner but with a different result, the toxin of the *Pertussis* bacterium catalyzes ADP ribosylation of the inhibitory G_i subunit of the GTP binding protein that sustains the "off" or inactive form of adenylate cyclase. The action of the cholera toxin results in the extreme stimulation of adenylate cyclase to produce a constant formation of cAMP. This alters the permeability of the intestinal membrane and allows copious amounts of water and electrolytes to be lost, which accounts for the typical severity of the disease.

The binding subunit B of cholera toxin has some amino acid homology to the β subunits of the pituitary glycoprotein hormones, TSH, FSH, and LH and to HCG. (See Table 16.1.) However, this homology also extends to the serine proteases, which have no membrane function, so that amino acid homology alone may not be significant. There is no analogous fate of the α subunits of the glycoprotein hormones comparable to that of the A subunit of cholera toxin. In the A subunit interaction, NAD^+ is used to form a covalently bound ADP-ribosylated protein with the release of nicotinamide, resulting in the irreversible stimulation of adenylate cyclase. No such reaction occurs with the subunit peptides of the pituitary hormones. Although the cholera toxin will stimulate hormone-responsive tissues and provides important insight into the regulation of adenylate cyclase, it does not reflect on the manner of interaction of the pituitary hormones with their respective receptors or their subsequent mode of stimulation of adenylate cyclase.

Guanylate cyclase, which catalyzes the conversion of GTP to cGMP and PP_i, can be maximally stimulated in the presence of added Mn^{2+} ion. Mg^{2+} is required, since, similar to Mg^{2+}-ATP, the substrate form of the nucleotide is Mg^{2+}-GTP. In marked distinction to the adenylate cyclase system, guanylate cyclase is not stimulated directly by hormones. The membrane-bound enzyme can be activated by oxidative events involving membrane lipid constituents, including possibly the prostaglandins and phospholipid turnover.

In the control of cellular activity by secondary messengers involving the same cyclic nucleotides, Ca^{2+} ion, Mg^{2+} ion, and prostaglandins, the question arises as to the specificity of cell response and specialization of cellular activity. Assuming the genetic makeup of each cell is identical and it possesses a common potential, the expression of cellular activity is the result of the pattern and amounts of enzyme activities that are free to be expressed. This probably occurs early in development by removal of restraints at the level of the genome, which will vary among the different cell types.

If the pattern of response of the cell is formed early in the differentiation of the organism, at the other end of the spectrum the specificity of

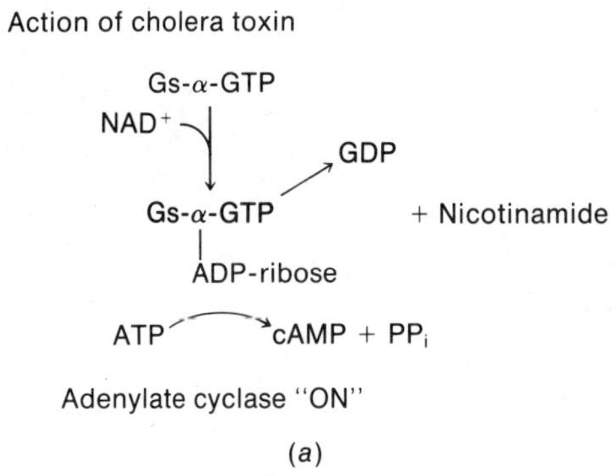

Action of cholera toxin

Gs-α-GTP
NAD^+ → GDP
Gs-α-GTP + Nicotinamide
|
ADP-ribose

ATP → cAMP + PP_i

Adenylate cyclase "ON"

(*a*)

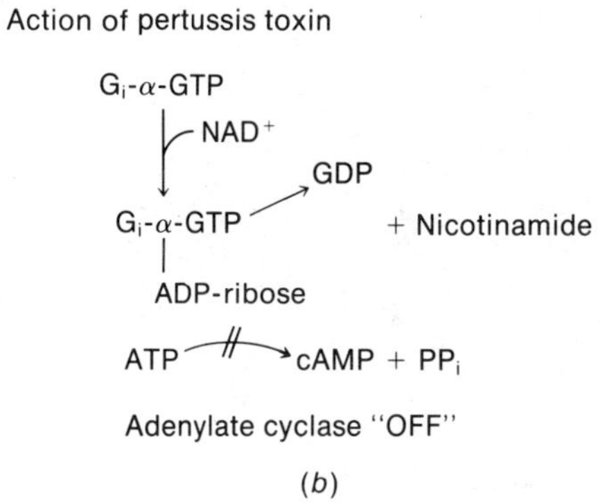

Action of pertussis toxin

G_i-α-GTP
NAD^+ → GDP
G_i-α-GTP + Nicotinamide
|
ADP-ribose

ATP ⇸ cAMP + PP_i

Adenylate cyclase "OFF"

(*b*)

Figure 16.5
Site of action of cholera and pertussis toxin 3 on the G_s and G_i complexes.

TABLE 16.1 Families of Peptide Hormones

Source	*Hormone*	*No. of Amino Acids*	*Molecular Weight*	*Special Features*
GLUCOSE REGULATION				
Pancreas				
(β cell)	Insulin	51	6K	3 Disulfide bridges
(α cell)	Glucagon	29	3.5K	Straight chain
HYPOTHALAMUS–PITUITARY				
Posterior pituitary	Oxytocin	9	1K	1 Disulfide bridge
	Vasopressin	9	1K	1Disulfide bridge
SOMATOTROPIN-LIKE				
Anterior pituitary	Growth hormone (somatotropin)(STH)	191	22K	Straight chain; 2 disulfide bridges
	Prolactin	198	22K	3 Disulfide bridges
Placenta	Placental lactogen (somatomammotropin)	191	22K	2 Disulfide bridges
ACTH/ENDORPHIN				
Anterior pituitary	Adrenocorticotropin (ACTH)	39	4.5K	Straight chain
(hypothalamus)	Melanocyte-stimulating (α-MSH)	13	1.8K	Acetyl on Ser
	β-Lipotropin	91	9.5K	Straight chain
(hypothalamus)	β-Endorphin	30	4.0K	Straight chain
GLYCOPROTEIN HORMONES				
Anterior pituitary	Thyroid-stimulating hormone (TSH)	209	27K	CHO 15–30%
	Follicle stimulating hormone (FSH)	236	32K	α subunit
	Luteinizing hormone (LH)	215	30K	β subunit
Placenta	Human chorionic gonadotropin (HCG)	231	46K	α, 92 aa; β, 139 aa
CALCIUM REGULATION				
Parathyroid	Parathyroid hormone (PTH)	84	9.5K	Straight chain
Parathyroid and thyroid	Calcitonin	32	3.6K	Straight chain

response is determined by the presence of cell receptors at the plasma membrane, which initiates those events leading to that cell response. The ability of a hormone to stimulate a cell is regulated by the type and amount of receptor present on the cell membrane. It is probable that each cell has the potential to respond to any hormone if a sufficient number of receptors are present and active on the membrane surface.

The number of receptors on the cell membrane is relatively low, on the order of 5,000 to 10,000 per cell. In most systems relatively few receptors, 10–20% of the total present, need to be occupied to produce a maximum response of the cell. The presence of "spare receptors" is considered a feature which ensures by the redundancy of the system that a given cell will respond to a hormone stimulus. The number of receptors per cell is not fixed and can be shown to vary under different physiological conditions as determined by the use of Scatchard plots.

The terms *desensitization* and *down-regulation* have been applied to a decrease in cell response to hormone stimulation. Down-regulation is observed whenever cells are exposed under chronic conditions to relatively high tissue concentrations of hormone. Explanations for the occurrence of down-regulation may vary for different hormones, for example, (1) increased number of receptor sites occupied by hormone, (2) the presence of antibodies that may block the receptor site, (3) the turnover of receptor sites from the membrane surface to the membrane interior, and (4) the decreased synthesis and turnover of receptor protein by the cell. At present there is evidence to support each of these postulates.

The concentration of a membrane receptor that is saturated by hormone levels at 10^{-8}–10^{-10} M must be correspondingly low and requires a high degree of purification that is difficult to achieve. Affinity chromatography with the hormone attached as a ligand to an inert phase is used to obtain receptor preparations from mammalian tissues that are relatively pure. Model systems for the study of hormone receptors and their activation of membrane-bound adenylate cyclase have been most successful for the glucagon receptor and the α-adrenergic receptor with epinephrine.

Prostaglandins

The molecular basis for the actions of the prostaglandins is known in only a few systems of the many in which they have been implicated. In most cases, they exert their actions within the membrane where they are formed. As a result they can modulate the effect of peptide hormones bound to their receptors. Prostaglandins can inhibit the actions of some hormones; their inhibitory actions on epinephrine stimulation of lipolysis in adipose tissue already has been cited. They can also stimulate hormone-mediated events; a slight ACTH-like effect on adrenal steroidogenesis has been observed. Some of these actions are related to the stimulation of adenylate cyclase, and others, as with epinephrine, to the inhibition of adenylate cyclase. Their exact role as modulators of hormone action with respect to membrane function and activity is yet to be determined.

At the present time it is not known whether many effects attributed to prostaglandin E and F represent true physiological events, or whether these substances are merely stable end products of a series of short-lived reactants involving unstable forms of oxygen-containing intermediates such as endoperoxides, thromboxanes, prostacyclins and leukotrienes.

Their ubiquity, their short duration of activity, and the physiological importance of the systems involved hold great promise for prostaglandins and their analogs, both as important links in the study of the complex

cAMP + [R|C] ⟶
Inactive protein kinase A

cAMP-[R] + Activated [C] protein kinase A

Figure 16.6
Activation of protein kinase A. [R] is a regulatory monomer, and [C] is the catalytic monomer of the kinase.

CLIN. CORR. 16.2
EGF AND THE erbB ONCOGENE PRODUCT

The epidermal growth factor (EGF) receptor (M_r 170,000) consists of an apoprotein (138,000) plus carbohydrate (37,000) portion in a straight chain with the amino terminal end extending to the outside and carboxyl terminal end to the inside of the cell. The receptor (Figure below) has three domains: (1) the outer carbohydrate containing portion, which contains the EGF binding site; (2) the transmembrane portion; and (3) the inner cytosol domain. When EGF (53 amino acids) binds to the receptor, a series of reactions occurs in the cell, culminating in cell division. In the process the carboxyl terminus of the receptor that has protein kinase activity phosphorylates the receptor at tyrosine residues.

The transforming growth factor (TGF-1) of the rat transformed fibroblast contains 50 amino acids with appreciable homology to EGF. Both TGF-1 and EGF contain three disulfide linkages. TGF-1 competes with EGF at the EGF receptor. The A431 cell, obtained from a human epidermoid cancer, has larger numbers (3×10^6) of EGF receptors per cell. TGF-1 competes with EGF for A431 cell receptors. Both TGF-1 and EGF will trigger autophosphorylation and will phosphorylate tyrosine residues on membrane proteins. Both will stimulate DNA synthesis in fibroblasts grown in media depleted of serum. The DNA sequence for EGF receptor is amplified by 30-fold in A431 cells.

The gene protein product derived from the avian erythroblastosis virus *erbB* has been compared to the amino acid sequence of EGF receptor peptides obtained

nature of tissue and metabolic regulation and as models for the synthesis of new classes of biologically active therapeutic agents.

Protein Kinase Activation

One of the properties common to a diverse group of peptide hormones, growth factors, and oncogenes is the activation of protein kinase activity during the course of their actions. Phosphorylation of proteins is a common form of posttranslational modification. Serine and threonine sites are preferred by many protein kinases, but a unique protein kinase is specific for phosphorylation of tyrosine. Activation of protein kinase C with phosphorylation of tyrosine has been found with a number of factors (epidermal growth factor, nerve growth factor, insulin, angiotensin, etc.). Ligand receptor interaction increases the number of tyrosine residues phosphorylated by a factor of at least 10-fold, but this increase comprises less than 1% of the total serine and threonine residues that can be phosphorylated. The protein kinase of these receptors phosphorylates tyrosine residues of various cellular proteins and histones. The identification of tyrosine-specific protein kinases as protein products of transforming factors and of oncogene activity has stimulated further interest in these enzymes. The function of gene products of oncogenes and the determination of a more specific role for proteins phosphorylated at tyrosine residues are under active investigation. These reactions may consist of a cascade of phosphorylations and dephosphorylations that comprise a network for communication between the transduction of signals at the membrane and transcription by the genome and the translation of protein products (see Clin. Corr. 16.2).

Protein Kinase A and G

The cAMP-dependent protein kinase A (PK-A) is stimulated by the binding of peptide hormones to their membrane receptors. Figure 16.6 illustrates the role of cAMP in activating the protein kinase A. A wide range of proteins serves as substrates for the enzyme, including plasma membrane proteins, enzymes involved in metabolic pathways, and histones. Phosphorylation of the serine and threonine residues are the common sites for PK-A kinase activity. There also exists a protein kinase G (PK-G), which is activated by cyclic GMP.

Protein Kinase C

Protein kinase C (PK-C) is activated by both 1,2-diacylglycerol and Ca^{2+}. Following interaction of an appropriate ligand with a membrane receptor, hydrolysis of phosphatidylinositol-4,5-bisphosphate occurs leading to formation of 1,2-diacylglycerol and inositol 1,4,5-trisphosphate (IP_3); this hydrolysis is catalyzed by phospholipase C (Figure 16.7). The 1,2-diacylglycerol stimulates the activity of protein kinase C, and the inositol trisphosphate mobilizes cellular Ca^{2+} from nonmitochondrial sources, presumably the endoplasmic reticulum or Ca^{2+} that is bound to the inner surface of the plasma membrane. Ca^{2+} is also an activator of PK-C, and thus increased Ca^{2+} levels lead to a sustained activation of PK-C. 1,2-Diacylglycerol and Ca^{2+} thus act synergistically on the PK-C. Ca^{2+} is an activator of phospholipase C. The Ca^{2+} activation of phospholipase C leads to further generation of the two intracellular mediators. Even though the intracellular Ca^{2+} may return to prestimulation levels, the PK-C once stimulated remains active as long as the 1,2-diacylglycerol is being formed from the phosphatidylinositides. Phosphatidylinositol-4,5-bisphosphate is rapidly resynthesized.

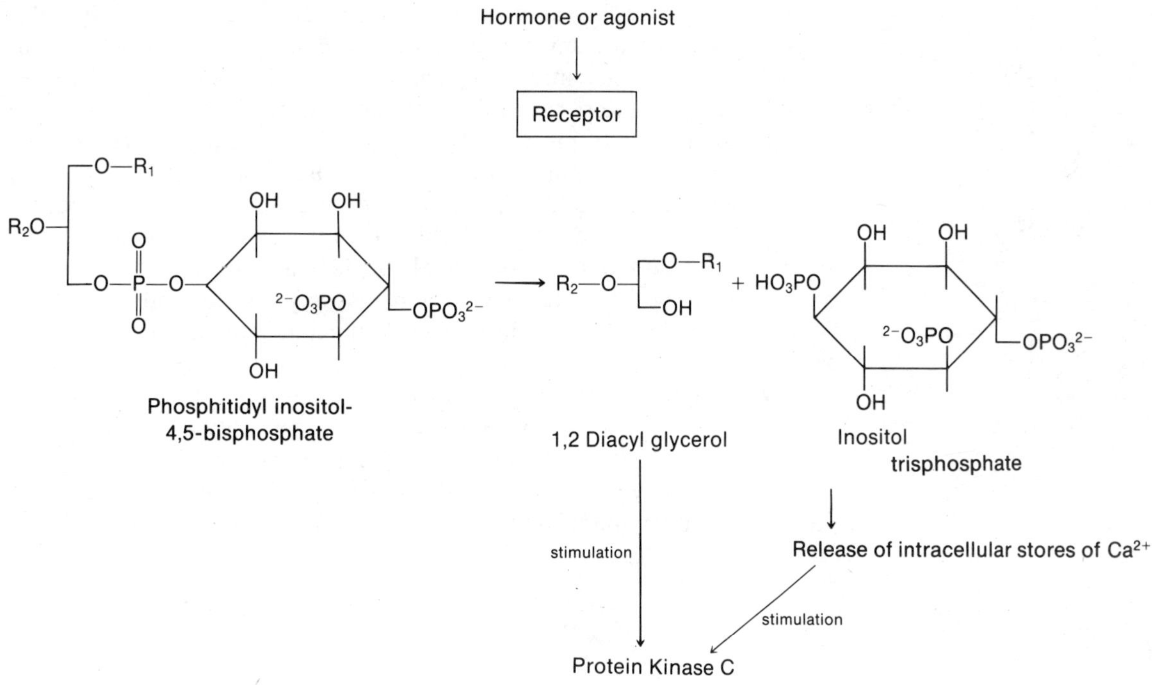

Figure 16.7
Phosphatidylinositol turnover and stimulation of protein kinase C.

Of added interest is the effect of phorbol esters on stimulating protein kinase C. Phorbol is a diterpene that in various chemical forms, such as tetradecanoyl phorbol acetate (TPA), is known to be a major promotor of tumor formation. Associated with this action is the stimulation of protein kinase C and Ca^{2+} mobilization.

Ca^{2+}-Calmodulin and Ca^{2+}-Phosphatidylserine Activated Protein Kinase

Protein kinase activation by Ca^{2+} ion and calmodulin that includes a phosphatidylserine turnover is yet another system that leads to protein phosphorylation on serine and threonine residues. Phosphodiesterase and myosin light chain kinase are examples of substrates for this kinase activity.

16.2 HORMONE REGULATION OF CALCIUM

The intracellular concentration of calcium at 10^{-6} to 10^{-7} M is maintained by an active Ca^{2+}/Mg^{2+} ATPase system, which extrudes Ca^{2+} into an extracellular environment where the $[Ca^{2+}]$ ion is 10^{-3} M. The dominant role exerted by calcium and calmodulin within the cell with respect to the regulation of a number of enzyme systems, particularly with respect to hormone action on cyclic nucleotide regulation, has been reviewed in Chapter 4, Section 4.10. Extracellular calcium is rigidly controlled as well, and this discussion describes those hormonal factors that are involved in this regulation. The plasma $[Ca^{2+}]$ level is maintained closely at 10 mg/dl by three substances: (1) parathyroid hormone (PTH), (2) calcitonin, and (3) cholecalciferol or vitamin D. The actions of these three hormones to achieve homeostasis are exerted in specific tissues of the body in a concerted fashion, with the ultimate control being maintained by the levels of blood calcium itself.

after partial hydrolysis. Six of 14 peptides have amino acid homology almost identical to the *erbB* protein. The *erbB* gene product contains the cytoplasmic protein domain of the EGF receptor but has lost the amino terminal binding site for EGF. This portion of the receptor is equivalent to the control site.

Viral genes have similar DNA counterparts to the host DNA (protooncogenes). Segments of the normal gene are replaced by fusion with viral transformed genes. These transformed genes produce a product whose function is mainly unknown. The *erbB* gene product is a part of the EGF receptor. The protein $p60^{src}$ produced by the *src* gene is a tyrosine kinase, but its function is unknown. The *ras* oncogene protein product is equivalent to the G protein that activates the membrane enzyme, adenylate cyclase. The *sis* oncogene codes for a protein that is one of the two protein chains comprising the platelet derived growth factor (PDGF) that stimulates cell division.

The close amino acid homology between a number of growth promoting agents, their similarity in activation of tyrosine specific protein kinase, and the types of correlation that have been described above hold great promise for future research prospects.

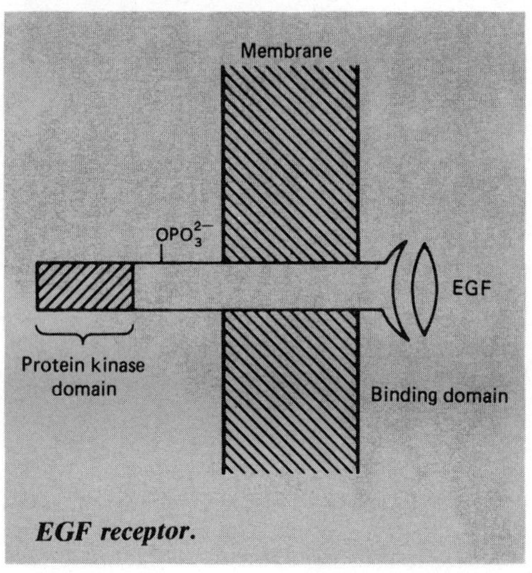

EGF receptor.

The major physiological counterion to calcium is phosphate. The solubility product of calcium phosphate is exceedingly low, and a large increase in one ion will cause a reciprocal decrease in the other; otherwise, insoluble calcium phosphate will precipitate. At pH 7.4 (that of blood) the most common form of phosphate is HPO_4^{2-}. At an acid pH (pH ~6.0) of urine $H_2PO_4^-$ is the common form excreted. The insoluble $(Ca^{2+})_3(PO_4^{3-})_2$ is found in bone and teeth in the crystalline lattice form of hydroxyapatite, $Ca_{10}(PO_4)_6(OH)_2$, which also contains Cl^-, F^-, CO_3^{2-}, Na^+, K^+, and Mg^{2+}. In the blood, one-half of the calcium is bound to plasma proteins. It is the remaining 50% of free Ca^{2+} ion that is physiologically effective. Bone contains over 99% of the body calcium and is the main reservoir to be used when calcium is required. In the young adult, dietary calcium is usually more than sufficient to maintain normal blood Ca^{2+} levels. The amount of Ca^{2+} in foods, however, is not always readily available; it may be present in an insoluble form, or it may not be absorbed by the intestinal mucosa. Normally Ca^{2+} is excreted in the feces. The urinary excretion of Ca^{2+} is usually minimal but increases when blood levels of Ca^{2+} are raised.

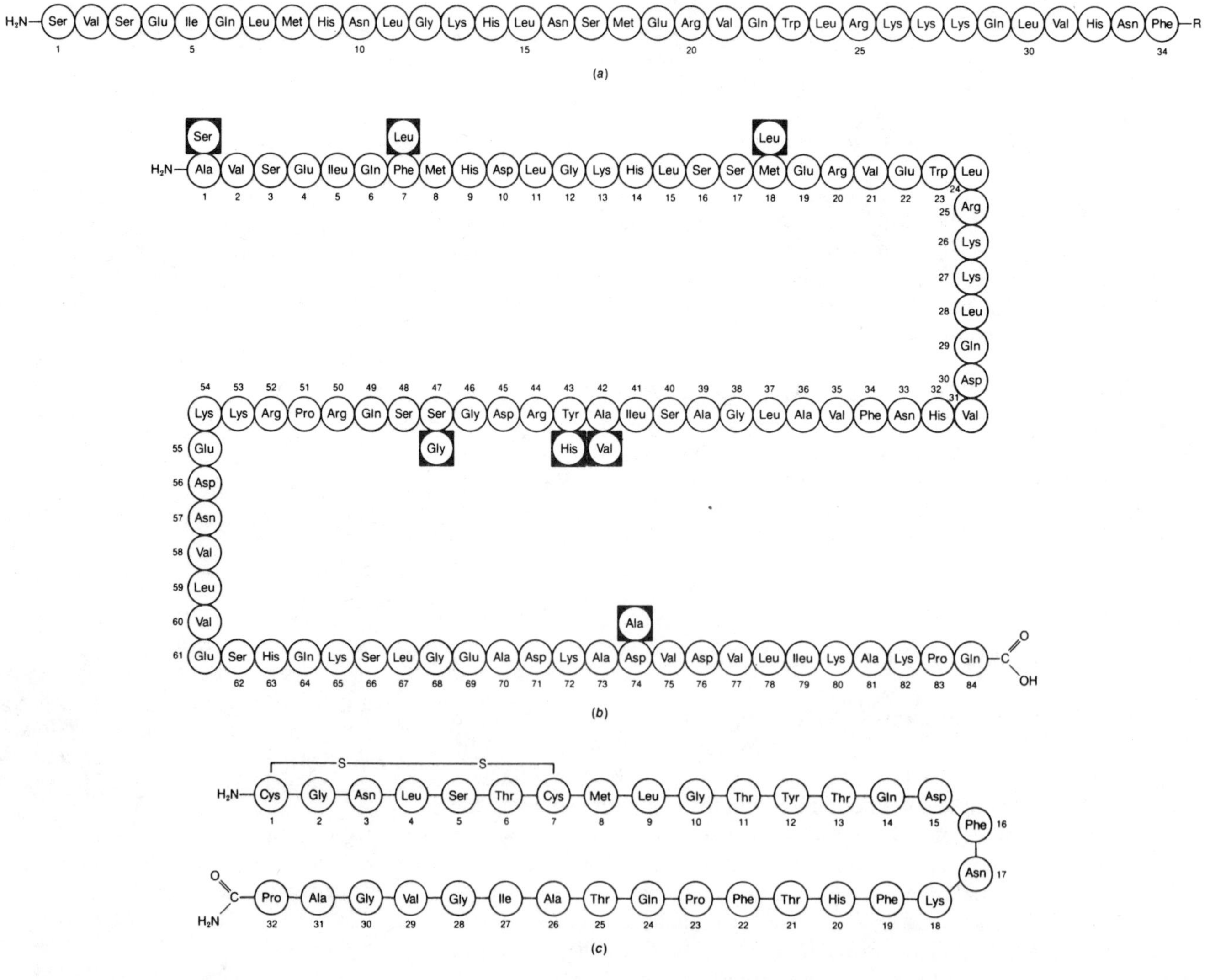

H_2N—Ser Val Ser Glu Ile Gln Leu Met His Asn Leu Gly Lys His Leu Asn Ser Met Glu Arg Val Gln Trp Leu Arg Lys Lys Lys Gln Leu Val His Asn Phe—R
1 5 10 15 20 25 30 34

(a)

Ser (1) Leu (7) Leu (18)
H_2N—Ala Val Ser Glu Ileu Gln Phe Met His Asp Leu Gly Lys His Leu Ser Ser Met Glu Arg Val Glu Trp Leu
1 2 3 4 5 6 7 8 9 10 11 12 13 14 15 16 17 18 19 20 21 22 23 24
25 Arg 26 Lys 27 Lys 28 Leu 29 Gln 30 Asp
54 53 52 51 50 49 48 47 46 45 44 43 42 41 40 39 38 37 36 35 34 33 32 31
Lys Lys Arg Pro Arg Gln Ser Ser Gly Asp Arg Tyr Ala Ileu Ser Ala Gly Leu Ala Val Phe Asn His Val
Gly (47) His (43) Val (42)
55 Glu 56 Asp 57 Asn 58 Val 59 Leu 60 Val
Ala (74)
61 Glu Ser His Gln Lys Ser Leu Gly Glu Ala Asp Lys Ala Asp Val Asp Val Leu Ileu Lys Ala Lys Pro Gln—C(=O)OH
62 63 64 65 66 67 68 69 70 71 72 73 74 75 76 77 78 79 80 81 82 83 84

(b)

S—S (1–7)
H_2N—Cys—Gly—Asn—Leu—Ser—Thr—Cys—Met—Leu—Gly—Thr—Tyr—Thr—Gln—Asp—Phe—Asn
1 2 3 4 5 6 7 8 9 10 11 12 13 14 15 16 17
H_2N—C(=O)—Pro—Ala—Gly—Val—Gly—Ile—Ala—Thr—Gln—Pro—Phe—Thr—His—Phe—Lys
32 31 30 29 28 27 26 25 24 23 22 21 20 19 18

(c)

Figure 16.8
(a) Amino acid sequence (1–34) of human parathyroid hormone compared to that of (b) bovine (white circles) and ovine (dark circles) hormones (1–84). (c) Amino acid sequence of human thyrocalcitonin.

Parathyroid Hormone and Calcitonin

The blood [Ca^{2+}] determines the rate of secretion of parathyroid hormones (PTH). Low blood [Ca^{2+}] releases the hormone contained in vesicles in the parathyroid cell. The hormone is a straight-chain polypeptide (M_r 9,500) containing 84 amino acids. It is formed as a preprohormone and is cleaved to a prohormone (90 amino acids) and to its final active form (84 amino acids), as it is transferred from the endoplasmic reticulum to Golgi apparatus to secretory vesicle. Once secreted, PTH has a relatively short half-life, measured in minutes, as is the case for most peptide hormones. Inactive fragments, which can be detected by radioimmunoassay using an antibody to the entire PTH (1–84), are found in target cell membranes or circulating in the plasma. The major portion contains the COOH end of the molecule and is biologically inactive. The smaller NH_2 end of the molecule retains some biological activity. The correspondence of biological activity of PTH to the amount of hormone as determined by radioimmunoassay may be quite variable for this reason. (See Figure 16.8.)

Calcitonin is a peptide (M_r 3,600) containing 32 amino acids in a straight chain, which is produced from a larger peptide in a prohormone form. Calcitonin is secreted from cells of the parathyroid and the thyroid gland in response to high blood levels of Ca^{2+}. The release of calcitonin leads to a gradual lowering of blood Ca^{2+} levels. This action results from increased removal (resorption) of calcium from bone. A primary action of PTH is on the calcification system of bone with a direct effect on osteolytic activity leading to bone Ca^{2+} resorption. The actions of both hormones involve cAMP in an undetermined manner in which PTH stimulates, and calcitonin inhibits, the release of bone Ca^{2+}. Thus the level of blood Ca^{2+} is the determining factor in its own regulation. High blood Ca^{2+} levels stimulate calcitonin secretion, which tends to lower [Ca^{2+}], and low blood Ca^{2+} levels stimulate PTH secretion, which tends to raise Ca^{2+} levels (see Figure 16.9).

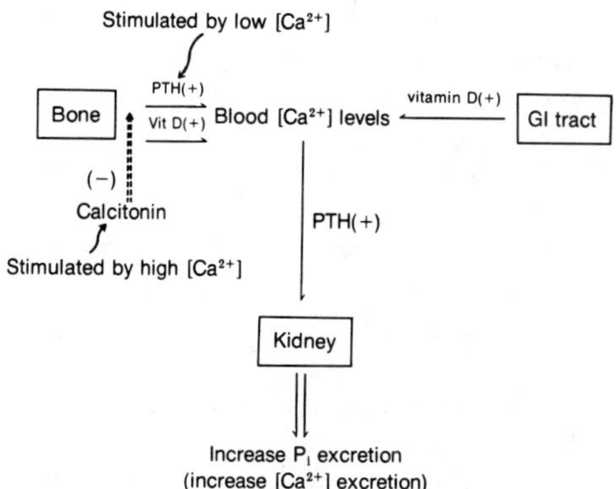

Figure 16.9
Role of parathyroid hormone (PTH), vitamin D, and calcitonin in controlling Ca^{2+} metabolism.

PTH has a direct effect on kidney to cause increased excretion of phosphate. In the absence of PTH, phosphate is reabsorbed by the kidney. The increased excretion of phosphate by kidney results in the lowering of blood phosphate levels. PTH action in the kidney also is mediated by cAMP. The major source of cAMP in urine is derived from kidney and is thought to be due primarily to the action of PTH. The administration of PTH leads to the following course of events with time: (1) there is first an early (4 h) increased excretion of urinary phosphate followed by (2) a lowering of blood phosphate levels (~8 h); (3) then there is an increase in blood Ca^{2+} levels (12–24 h), followed by (4) an increased excretion of Ca^{2+} into the urine (24+ h).

After PTH administration, the turnover and excretion of administered radioactive $^{45}Ca^{2+}$ is considerably increased. The chronic increased levels of PTH, which occurs in hyperparathyroidism, will result in loss of bone calcium and often can result in renal calculi and deposition of calcium phosphate in soft tissues. Hypoparathyroidism can occur by the removal of the parathyroid glands during thyroidectomy. In the absence of PTH, blood [Ca^{2+}] levels can fall to a level of 7 mg/dl. The low [Ca^{2+}] levels increase neuromuscular irritability. This may lead to tetany, which is the rapid uncontrolled contraction of skeletal muscle.

The stimulus of lower blood Ca^{2+} levels to increase parathyroid hormone secretion also is expressed in the kidney with the increased 1-α-hydroxylation of vitamin D to form 1,25-dihydroxycholecalciferol (Figure 16.10). This, the most active form of vitamin D, acts on the mucosal cells to increase the intestinal transport of Ca^{2+}. See Chapter 26 for details of synthesis of vitamin D. It is thought that the action of vitamin D on the mucosal cell results in the formation of a calcium-binding protein, which

1,25-Dihydroxycholecalciferol

Figure 16.10
1,25-Dihydroxycholecalciferol.

enhances the absorption of Ca^{2+} by the gut. 1,25-Dihydroxycholecalciferol is also active on bone and enhances the mobilization of calcium to the blood.

Clinical Correlation 16.3 discusses the condition of postmenopausal osteoporosis and the needs for Ca^{2+}.

> **CLIN. CORR. 16.3**
> **POSTMENOPAUSAL OSTEOPOROSIS**
>
> Osteoporosis is a skeletal disorder in which the absolute amount of bone is decreased as compared to that of a young adult, whereas the chemical composition of the remaining bone is normal. The defect can be attributed to a number of factors including age, nutrition, sex, hormonal status, physical activity, and heredity. Bone loss normally occurs in aging people and to a greater extent in females than males.
>
> The loss of bone that occurs in both sexes during senescence begins in the fourth and fifth decades. In the postmenopausal female, 25–30% have major orthopedic problems. For the more serious problem of hip fractures, the rate per 1,000 in women is 5 at age 65, and increases to 10 at age 75. With increasing age, there are decreasing amounts of vitamin D_3. Parathyroid hormone (PTH) increases and calcitonin decreases with age due to the deficiency of Ca^{2+} intake and lower Ca^{2+} plasma levels. Bone loss is increased with aging, particularly in patients with diabetes mellitus or in cases of alcoholism.
>
> Since the age-related bone loss in postmenopausal osteoporosis suggests involvement of the ovaries, estrogens have been administered to prevent some of the loss. However, the role of estrogen deficiency and ovarian function is not clear, and rapid or slow loss of bone cannot be related to estrogen levels. Estrogen receptors on bone are difficult to demonstrate. The estrogen effect may be indirect, that is, on other hormones (PTH, vitamin D_3), or via absorption and excretion of Ca^{2+} and phosphate ions. Estrogen administration, or estrogen plus progesterone, as contained in the contraceptive pill, may retard bone loss. Long-term use of these agents requires periodic monitoring of cortical bone mass (metacarpal) and gynecological examinations.
>
> Calcium deficiency appears to be an important factor in postmenopausal osteoporosis. The published RDA values of 800 mg Ca^{2+} per day is now estimated to be too low, and if it is would place some normal women in a negative Ca^{2+} balance. Factors resulting in Ca^{2+} deficiency are due to decreased Ca^{2+} absorption in the gut with low estrogen levels and greater loss of Ca^{2+} in the urine with a decreased renal function. In addition, the lack of ex-

16.3 PANCREATIC HORMONES

The levels of blood glucose are maintained relatively constant by the actions of liver to take up or release glucose. In the liver glucose is converted to glucose 6-phosphate, a key intermediate leading to a number of different metabolic pathways: (1) formation of glycogen, (2) metabolism via glycolysis, (3) oxidation by way of the pentose phosphate pathway, or (4) release to the blood as free glucose by action of glucose 6-phosphatase. The fate of the glucose 6-phosphate is determined essentially by the actions of two peptide hormones, insulin and glucagon.

High blood levels of glucose lead to the increased secretion of insulin, and low blood levels of glucose stimulate the secretion of glucagon. The action of glucagon to stimulate phosphorylase by a cAMP-dependent process has been described (Chapter 7, page 315). The cascade of events involving a series of protein phosphorylations, which is initiated by cAMP as an allosteric effector of protein kinase, leads to a breakdown of glycogen by phosphorylase *a*. An opposite effect is seen with insulin due to a shift from phosphorylase to synthase activity, leading to glycogen formation.

Increased cAMP levels lead to the phosphorylation of phosphorylase (activation), the phosphorylation of glycogen synthetase (inactivation), and the phosphorylation of a specific phosphatase inhibitor protein, which binds and inactivates the phosphatase that activates the glycogen synthetase. By the action of protein kinase A on the three proteins, there is more concerted and precise control over the regulation of glycogen breakdown. Ca^{2+} and calmodulin also exert an important control on the system since they are part of the subunit structure of phosphorylase kinase and regulate the rate of dephosphorylation of the enzyme through an action on a phosphoprotein phosphatase. Calmodulin binding to the subunits of the protein is induced by Ca^{2+}. In a similar manner, Ca^{2+}–calmodulin also exert an effect on adenylate cyclase and on phosphodiesterase activity.

Other binding sites in addition to an allosteric effector site on phosphorylase *a* for glucose have been considered. It has been proposed that an intracellular mediator of insulin, formed from the insulin–receptor complex, binds to another allosteric site on phosphorylase *a* to produce an effect similar to that of glucose; that is, removal of phosphate and the formation of the active synthase *a* form.

Insulin

Insulin monomer (M_r 6,000) consists of an A chain of 21 amino acids, and a B chain of 30 amino acids connected by two disulfide bridges (see Figure 19.15 for structure). Aggregates of 2, 4, 6, or more monomers can occur, especially at increased insulin concentrations. The element zinc found with insulin in the pancreas is associated with the aggregate forms of insulin. Insulin is derived from larger precursor peptide molecules, which are synthesized on the endoplasmic reticulum of the β cells of the pancreas. A preproinsulin with a leader signal sequence of 16 amino acids first is cleaved by proteolysis to form the proinsulin molecule, which is extruded into the lumen and stored in the Golgi system. There the proinsulin is gradually cleaved by proteolysis at two points containing arg-arg and arg-lys. The four basic amino acids are released, and a C-peptide (connecting peptide) of 30 amino acids and the two chain insulin molecule of 51 amino acids are formed. Insulin and the C-peptide are released from the vesicles due to increased glucose concentrations entering the cell. It is uncertain whether glucose or a glucose metabolite serves as the stimulus, but there is a concomitant increase of Ca^{2+} ion, which triggers insulin release.

Antibodies have been raised against the C-peptide and against insulin. Since a C-peptide is released with each insulin molecule, a radioimmunoassay for C-peptide can be used to measure endogenous insulin production. Insulin derived by injection from an exogenous source would not be accompanied by C-peptide, and the low estimates by radioimmunoassay would reflect this difference.

Purified preparations of insulin receptors have been obtained from membrane fractions of liver cells and from fat cells. Solubilized receptor preparations using detergent Triton X-100 are added to an affinity column in which insulin is covalently bound to agarose; the receptor is selectively retained and subsequently eluted with a urea solution. It has been estimated that a fat cell contains about 10,000 receptors. The dissociation constant of the insulin–receptor complex is $K_d = 10^{-10}$ M, which corresponds to the level of insulin circulating in blood.

Insulin, as well as other peptide hormones such as prolactin and FSH, are internalized by endocytosis (Figure 16.11). Radioactively labeled hormones can be localized within the cell lysosomes, which could imply a

ercise and relative inactivity of the elderly and ingestion of a high protein diet also contribute to a greater Ca^{2+} loss. Sodium fluoride at less than 60 mg per day has been used in therapy on the basis that it forms a more stable complex in the bone crystal; however, the chronic use of fluoride can often lead to abnormal bone cytology.

At this time there is no unifying hypothesis to explain osteoporosis. There is no acceptable therapy to stimulate bone formation. The process of bone loss is considered to be irreversible. Therefore, the various forms of therapy are designed to prevent further loss of bone. Present-day research on regimens for treatment consists of administration of different combinations of the following agents listed in random order: estrogen (with progesterone or other steroids); fluoride; 1,25-dihydroxycholecalciferol (or other synthetic vitamin D forms); dietary regimens of high calcium, low phosphate, low protein; and supplementary Ca^{2+} to ensure an intake above 1 g/day.

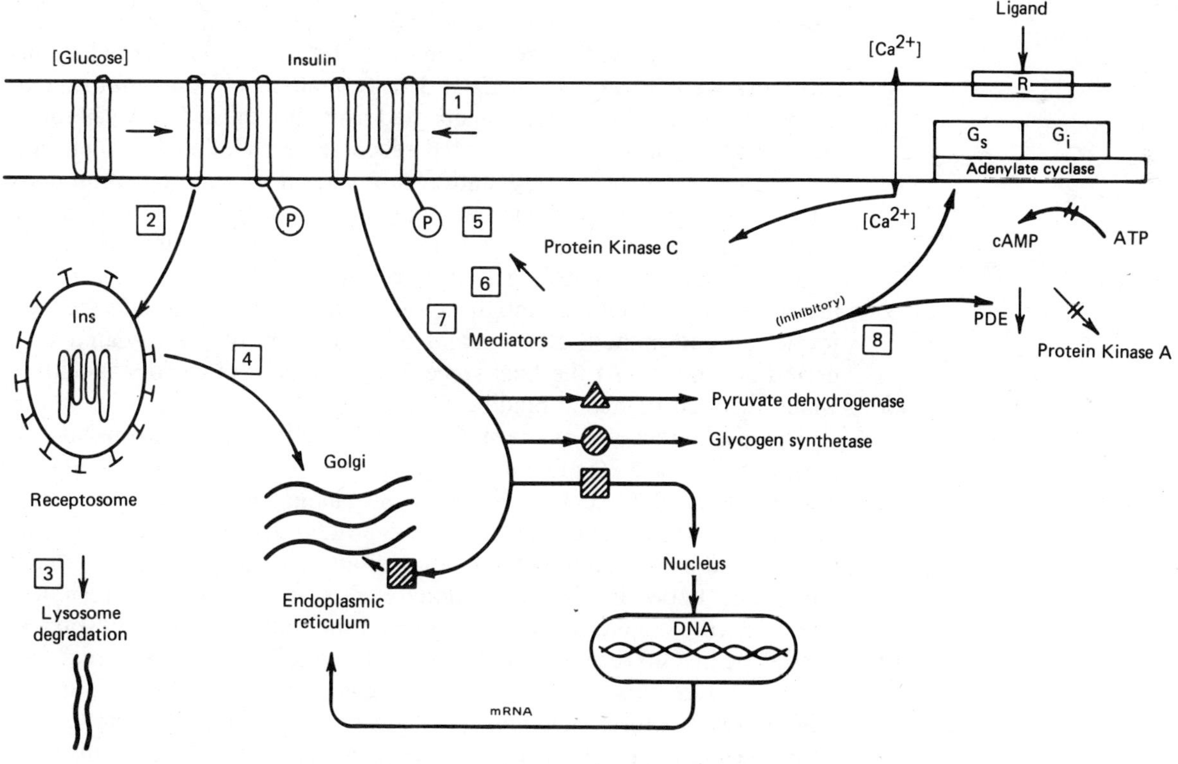

Figure 16.11
Cellular intereactions of the insulin-receptor complex.
(1) Receptor aggregation with insulin; (2) endocytosis of insulin receptor; (3) lysosomal degradation; (4) processing within Golgi; (5) autophosphorylation; (6) activation of protein kinase C; (7) release of mediators; (8) inhibition of adenylate cyclase by activation of G_i *protein or the phosphodiesterase (PDE).*

means of inactivation and breakdown by the cell. However, radioactivity has also been localized in the Golgi system, which could be physiologically significant. It is not known if the internalized hormones are biologically active. Insulin receptors at the cell membrane surface cluster and are subsequently internalized. Aggregation or cross-linking of the insulin receptor is considered an important aspect of insulin action. Fat cells are more sensitive to the actions of insulin than are liver cells. These differences in sensitivity do not appear to be due to differences in the subunit structure of the receptor. Using an electron dense complex of insulin–ferritin as ligand, a major difference is seen by electron microscopy in that insulin receptors on the liver plasma membrane occur singly, whereas insulin receptors on the membrane surface of the adipocyte occur in natural groups of two or more. The receptor clustering or internalization initiates a signal, which can be mimicked with bivalent $F(Ab)_2$ antibodies but not their univalent fragments (Fab), similar to the action stimulated by insulin. These antibodies to the insulin receptor protein, rather than from insulin itself, were obtained from patients who were resistant to the action of insulin. Some but not all insulin affects are observed with $F(Ab)_2$ binding.

The concept of *down-regulation,* decreased response to chronic exposure to high hormone levels, was first applied to the behavior of insulin. A net loss of insulin receptors occurs with insulin exposure and provides a signal to the cell interior. Mouse tumor 3T3 fibroblast cells (referred to as preadipocytes) treated with insulin for 3 days differentiate into adipocytes. Insulin receptors increase from 8,500 to 170,000 per cell during this period. By using a heavy isotope density-shift method, it was possible to measure receptor synthesis and degradation and translocation in the cells. Heavy amino acids containing over 95% of ^{15}N, ^{13}C, and ^{2}H were incorporated into newly synthesized receptors. The heavy and normal (light) receptors were separated on CsCl density gradients. Upon exposure to ligand, insulin receptors are translocated from the cell surface to an intracellular site resulting in accelerated receptor degradation and then in a lower down-regulated steady-state receptor level. Down-regulation does not alter transit time to the surface but causes a two-fold increase in the inactivation rate of surface receptors. In another study, using ^{35}S-labeled insulin receptor in the 3T3 cell system, the rate of glycosylation of the insulin receptor from the time the apoprotein was synthesized to the final form was determined. Both α and β subunits were glycosylated in a period of 3 hours. At the final stage of synthesis, sialic acid was added, which resulted in insulin-binding activity.

The insulin receptor on the surface of liver cells or fat cells consists of a tetramer of 2α and 2β subunits ($\alpha_2\beta_2$) with a molecular weight of 350,000. The α subunit (M_r 135,000) provides the binding site for insulin. The β subunit (M_r 95,000) is a transmembrane protein that has a protein kinase activity that is activated by the binding of insulin. The β unit is autophosphorylated under in vitro conditions exclusively on tyrosine residues. Under in vivo conditions, both serine and threonine are also phosphorylated. The insulin receptor, with respect to tyrosine protein kinase activity and to autophosphorylation, is analogous in behavior to receptors of other growth factors, including epidermal growth factor (EGF) and platelet-derived growth factor (PDGF). The significance in cell communication and metabolism remains to be determined, but the protein kinase activity of the β subunit may have an important role in the action of insulin on a number of key enzymes in carbohydrate metabolism. It has been suggested that as a result of insulin binding a conformational change is produced, followed by release of a small peptide from the receptor, which acts as an intracellular mediator of insulin action.

Evidence has been presented for the release of several small peptide-like fragments (M_r 500 to 2,000), referred to as mediators (Figure 16.11). One action ascribed to a mediator is the phosphorylation of a 40 S ribosomal protein. A variety of enzymes influenced by phosphorylation–dephosphorylation reactions such as pyruvate dehydrogenase and glycogen synthase may be activated by such a mediator of insulin action. Regulation by insulin of phosphatases and of kinases within the cell on the phosphorylation state of target proteins is an attractive idea that, unfortunately, lacks sufficient substantiation. These effects of insulin occur at higher concentrations (2–8 nM) than are required for metabolic effects or that occur in circulating plasma. The immediate short-term effects of insulin may also involve conformational changes that affect membrane permeability for glucose, amino acids, and ionic fluxes. Receptor autophosphorylation may be a signal for and may precede receptor endocytosis. Later effects of insulin on metabolism and growth would be due to the action of insulin or of its mediators on intracellular events.

Biological Actions of Insulin

The binding of insulin to its receptor in muscle results in rapid changes in membrane permeability to glucose and amino acids. Two classical experiments are basic to the understanding of insulin action. In the first, when intact rat diaphragm muscle is immersed for 10 s in a dilute solution containing insulin and then washed thoroughly, the treated muscle will incorporate glucose to a far greater extent than its control. In the second experiment, insulin covalently bound to a much larger ferritin molecule, which cannot enter the cell, exerts the same biological effect as soluble insulin. Both experiments demonstrate a rapid effect on membrane function by insulin, and are consistent with the concept that insulin exerts part of its biological effects while bound to a receptor on the plasma membrane without entering the cell.

Insulin is an important anabolic agent in muscle, liver, and fat cells. Its actions are also anticatabolic. It increases the synthesis of glycogen, fatty acids and triacylglycerols, and proteins. Insulin achieves these effects by making energy available through the formation of NADPH and ATP. The major source of energy in the cell is produced by glycolysis and subsequent oxidation of the intermediate products via the TCA cycle and oxidative phosphorylation. These events are promoted by the induction by insulin of key enzymes of glycolysis. With the lowering of activity levels in liver of fructose 1,6-bisphosphatase and pyruvate carboxykinase, it decreases gluconeogenesis.

There is no compelling reason to assume that the different effects exerted by insulin (pleiotropic effects) involve a single common mechanism. Indeed, a role in stimulating glucose and amino acid transport into the cell could be distinct and separate from a role of insulin as a growth-promoting agent inducing mitosis, although they clearly are related. A satisfactory explanation of insulin action will require a more comprehensive view that takes into account the observations previously discussed, including: (1) the requirement of insulin to cross-link with its receptor and the need for aggregation of receptors and endocytosis; (2) the binding of insulin to its receptor elicits the release of small molecular weight peptide fragments that lead to activation of enzymes within the cell; (3) the function of autophosphorylation of the β subunit of the insulin receptor elicited by insulin binding (the tyrosine kinase activity observed with insulin binding indicates a possible link with protein kinase C mediated events); (4) the putative role of protein kinase C for the actions of EGF, PDGF, transforming factors, and oncogenes with respect to activation of key events during growth and cell division; (5) the similarity in cellular events common to a diverse group of growth factors, including insulin, provides the means for relating short-term membrane receptor activity with long-term growth-related phenomenon.

Glucagon

Glucagon is a straight-chain polypeptide (M_r 3,500) containing 29 amino acids (Figure 16.12). A proglucagon peptide has been isolated. It is secreted by the pancreatic islet α-cell in response to low blood glucose as a stimulus. As with insulin release, a concomitant change in Ca^{2+} ion also is observed.

Glucagon and glucagonlike immunoreactive (GLI) material are also formed in cells of the small intestines and in brain tissues. A substance similar to proglucagon, glycentin, has been found in the gut. These substances cross-react with antibody to glucagon and compete with glucagon for the liver cell glucagon receptor.

The principal action of glucagon is to activate the cAMP-dependent phosphorylase system of liver, with a breakdown of glycogen to glucose, which is released into the blood circulation. In addition to its marked effect on glucose production, glucagon increases liver gluconeogenesis by increasing the liver pool of glucose precursors. It decreases pyruvate kinase and acetyl CoA carboxylase activity. In the fat cell glucagon stimulates cAMP-dependent lipase activity to convert triacylglycerols to free fatty acids and glycerol.

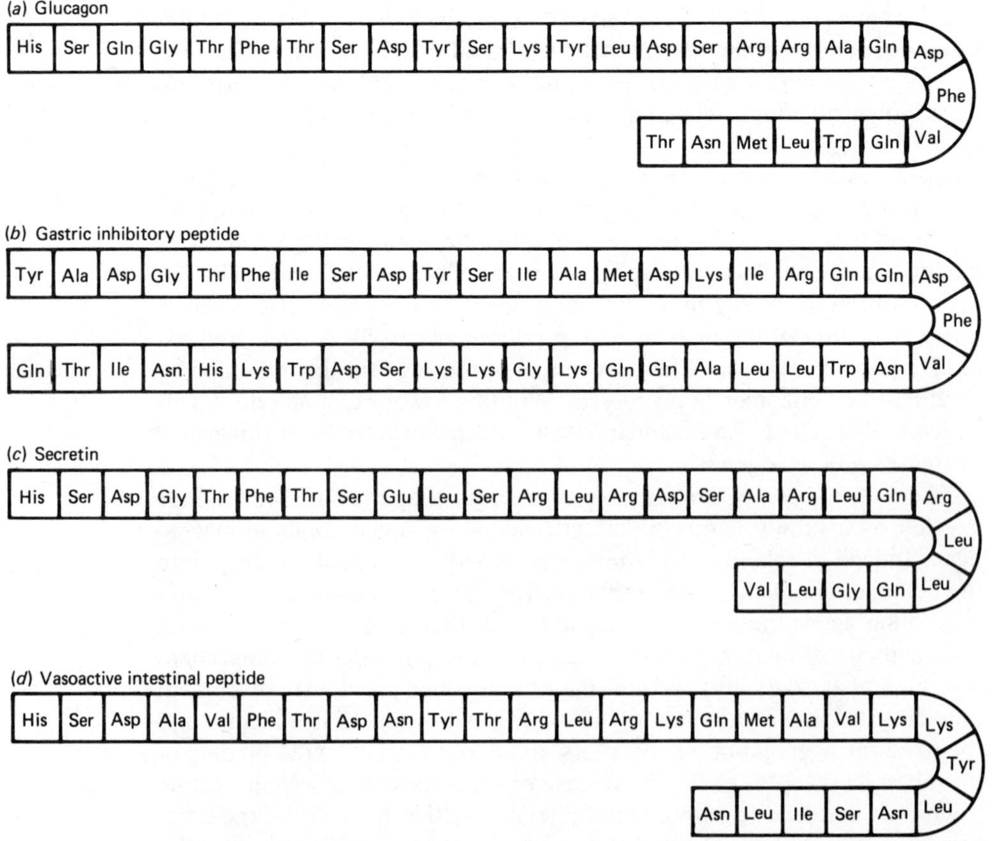

Figure 16.12
Structure of glucagon and related peptides.
Glucagon (29 amino acids) is produced by the α-cell of the islets of the pancreas. Three other peptides shown are produced by cells of the stomach (gastrin) or by cells of the small intestine (secretin) and vasoactive intestinal peptide (VIP).

Glucagon is rapidly inactivated ($t_{\frac{1}{2}}$= 5 min) within liver and kidney tissues, primarily. Since many of the metabolic effects of glucagon are opposed by insulin action, the effects of glucagon, increased breakdown of fats and increased ketogenesis, are most pronounced in patients with diabetes (deficiency of insulin) and in those suffering from starvation (when insulin levels are lowered).

The glucagon receptor has been studied extensively in liver cell and fat cell ghost preparations. Since adenylate cyclase is located on the inner surface of the membrane of the ghost cells, the association of glucagon binding to its receptor at the outer surface of the membrane with the production of cAMP at the inner surface has provided most valuable information with respect to the transduction of hormone-receptor signaling to intracellular communication. The liver cell ghost glucagon receptor system was used in studies of the regulatory role of GTP and the G protein on adenylate cyclase activity.

Other peptides that resemble glucagon in structure (Figure 16.12) include gastric inhibitory peptide, secretin, and vasoactive intestinal peptide (VIP). The occurrence of peptides such as VIP in the brain as well as in the intestine suggests other roles for these peptides that need further clarification.

Somatostatin

Somatostatin is found in the highest concentrations in the median eminence of the pituitary stalk and in specific pancreatic islet cells, as well as in the hypothalamus and intestines. Since it acts primarily in the area of its release, it has a major paracrine action in addition to other neuroendocrine or neurotransmitter functions. It is formed in the islet D cells as a larger peptide precursor which is cleaved to the tetradecapeptide by proteolysis. Somatostatin inhibits the release of glucagon and insulin by the α and β cells, respectively. The inhibition by somatostatin in the pancreas as well as in other tissues, such as the inhibition of release of somatotropin (STH) by the pituitary, cholecystokinin and vasoactive intestinal peptide (VIP) in the gut, appears to be the result of blocking Ca^{2+} entry into the hormone-secreting cells.

Other peptides besides glucagon and somatostatin with known functions found in the gastrointestinal tract include *gastrin* (17 amino acids), which stimulates gastric secretion, *secretin* (27 amino acids), which stimulates the secretion of pancreatic enzymes, and *cholecystokinin* (39 amino acids), which stimulates digestive secretions and gallbladder contractions. It is of interest that the gut contains these as well as a number of other peptides and catecholamines, indoleamines, and endorphins known to be present in the brain. Also, the gut hormone, cholecystokinin, is found in several regions of the brain; its function is unknown.

16.4 HYPOTHALAMIC–PITUITARY HORMONES

For most pituitary hormones, a large peptide precursor is formed by the protein synthesizing system in the endoplasmic reticulum. The protein contains the amino acid sequence for one or several biologically active peptides. The protein precursor may, in fact, contain several copies of the same peptide in a single protein sequence. The formation of the biologically active peptide fragments occurs by a process of proteolysis of the parent protein chain, in some cases to form different hormone or prohormone molecules, in other cases to form several copies of the same peptide hormone. Very little is known about the nature of these proteases, what determines the specificity of the cleavage site (amino acid signal sequence), and how the action of the protease is regulated. The larger peptides, identified after chromatographic separation by their immunoreactivity, are referred to as "big," "intermediate," or "little"; that is, "big ACTH" or "big" growth hormone, and so on. In the circulating plasma, peptide hormones can aggregate to form dimers or larger molecules and would be referred to as "big" insulin or "big" growth hormone. "Little" insulin would be the monomer-sized molecule (M_r 6,000).

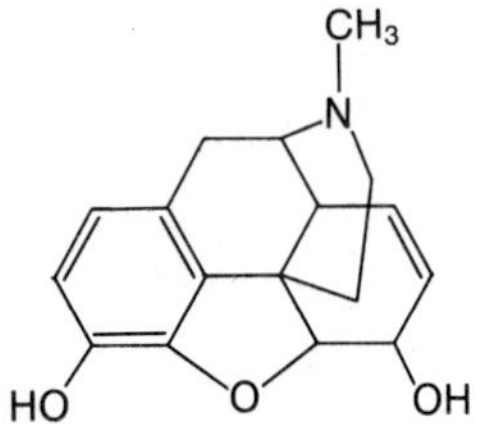

Endorphins are a class of naturally occurring peptides which bind to morphine receptors (*endo*genous m*orphin*e-like) in brain tissue. This unexpected development was a direct result of a search for natural substances that could occupy the morphine receptor sites. A physiological relevance for these peptides was indicated on the basis of their ability to mimic the actions of morphine with respect to analgesia, behavior, and withdrawal effects. For this reason the term *opioid* has been used for these peptides. The morphine derivative, naloxone (Figure 16.13), which competes for the receptor site and blocks morphine effects, is also an antagonist for the endorphins. The first peptides isolated were the enkephalins, containing five amino acids. The most active peptide, containing 30 amino acids, called β-endorphin, was first isolated from β-lipotropin.

Morphine

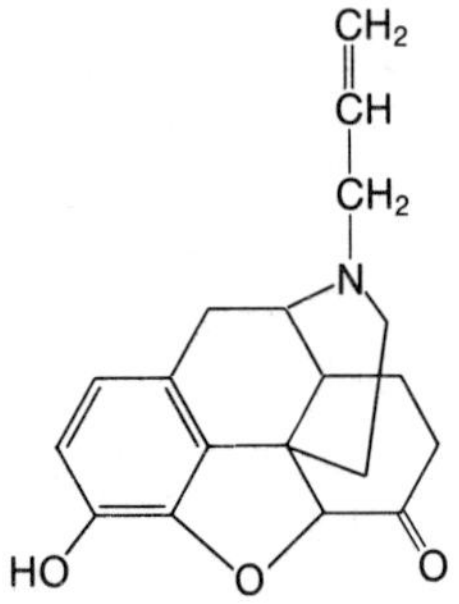

Naloxone

Figure 16.13
Morphine and naloxone.

There are no direct nerve connections from the hypothalamus to the anterior pituitary, but there is a rich blood supply from the hypophyseal portal system. When the pituitary stalk is transected, serum prolactin levels rise, while the levels of other anterior pituitary hormones fall. This demonstrates the stimulatory effect of the hypothalamus on the anterior pituitary hormone secretions and the dominant inhibitory effect on the control of prolactin secretion. (See Figure 16.14). Hypothalamic actions are mediated by chemical messengers, which are transported by the portal blood system from the median eminence of the hypothalamus to the anterior pituitary. By contrast, the posterior pituitary has direct neuronal connections to hypothalamic nuclei. The hormones vasopressin and oxytocin, which are secreted from the posterior pituitary gland, are synthesized in cells of the hypothalamic nuclei, travel down the nerve axons, and are stored in vesicles in the posterior pituitary. Other regulatory agents or hormones are produced in the hypothalamus by these peptidergic neurons, with cell bodies located in different hypothalamic nuclei. The activity of these neurons are controlled by neurotransmitters, such as dopamine, norepinephrine, and serotonin.

Hypothalamic Releasing Factors

Five hypothalamic peptides that have been isolated and characterized as to structure and function are shown in Figure 16.15. These include the thyrotropin-releasing factor (TRF), gonadotropin-releasing factor (GnRF, also known as FSH/LH-RH), growth hormone-releasing factor (GRF), growth hormone release-inhibiting factor (GRIF, also known as somatostatin), and corticotropin-releasing factor (CRF). The release of pituitary trophic hormones is inhibited by a negative feedback control by the target organ hormones acting on hypothalamic and pituitary secretory cells. A retrograde blood flow in the pituitary stalk can deliver the pituitary hormones to the median eminence and result in a "short loop" regulation of the secretion of hypothalamic releasing factors. In particular, by exerting both negative and positive feedback at the hypothalamic level, prolactin and growth hormone may regulate their own rate of production by altering the secretion of releasing factors or release-inhibiting factors.

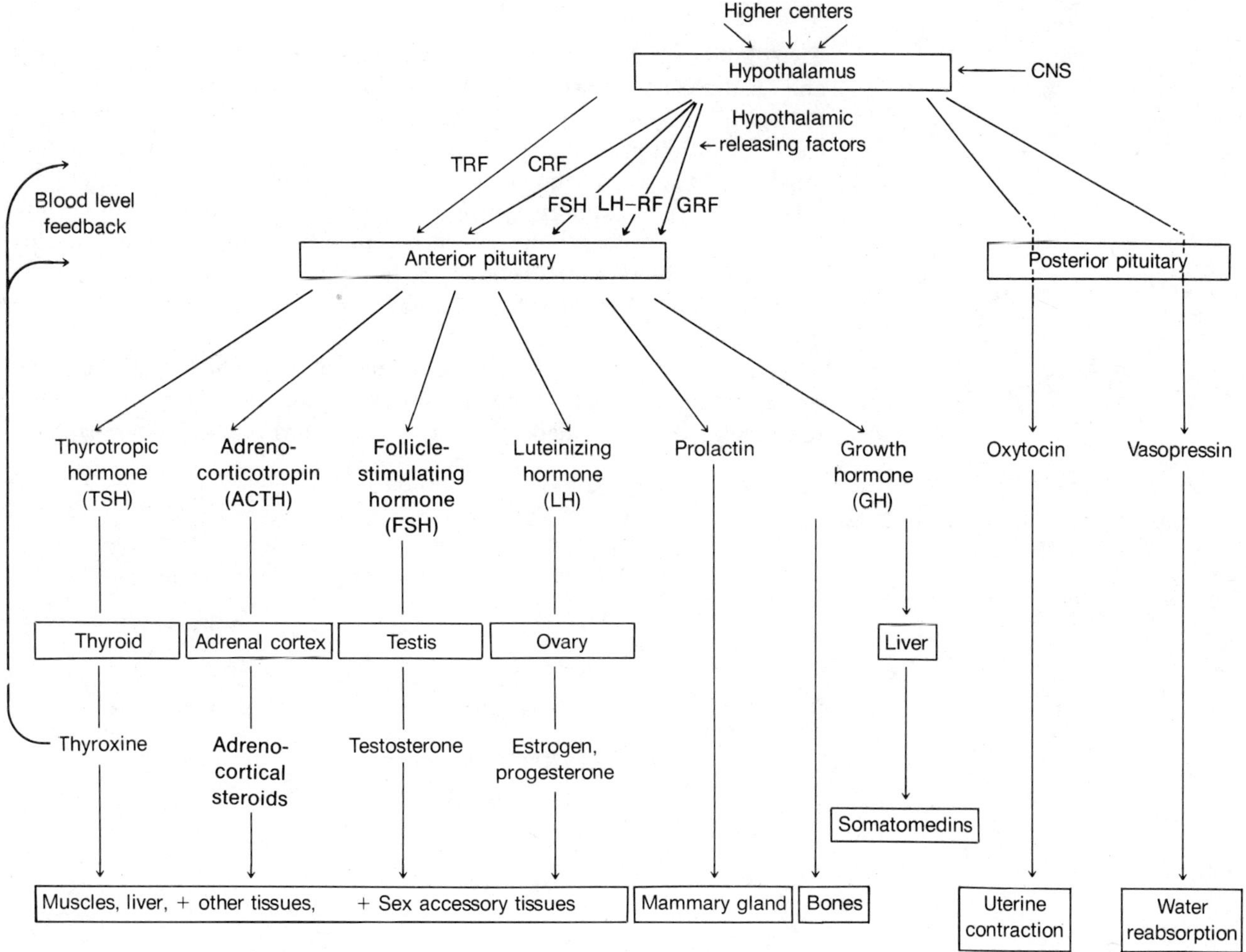

Figure 16.14
Hypothalamus-pituitary-target organ relationships.
Releasing factors from the hypothalamus stimulate pituitary trophic hormones, which stimulate target organs to secrete their respective hormones. These target organ hormones secreted into the circulation will exert a negative feedback at the level of the hypothalamus and pituitary tissues to inhibit further secretion of releasing factors and trophic hormones. Prolactin and growth hormones have a release-inhibiting factor as well as a releasing factor. Oxytocin and vasopressin are produced in the hypothalamus and are stored in the posterior pituitary.

Hypothalamic releasing factors (with known amino acid sequence)

TRF—thyrotropin-releasing factor
  pyroglu-his-proNH_2
GnRF—Gonadotropin-releasing factor (same as FSH/LH-RF)
  pyroglu-his-trp-ser-tyr-gly-leu-arg-pro-gly-glyNH_2
GRF—Growth hormone-releasing factor (same as hpGRF from pancreatic islet tumor)
  2 forms—44 amino acids and 40 amino acids
CRF—Corticotropin-releasing factor
  41 amino acids
Somatostatin (SRIF—somatotropin release-inhibiting factor)
  ala-gly-cys-lys-asn-phe phe-trp-lys-thr-phe-thr-ser-cys
  (S — S bridge linking the two cys residues)

Figure 16.15
Hypothalamic releasing factors.

Oxytocin and Vasopressin

Two of the peptides present in the posterior lobe of the pituitary, oxytocin and vasopressin, are shown in Figure 16.16, which also lists their major biological actions. Oxytocin, named for its action on uterine tissue to cause contraction of the myometrium, also has an effect on the smooth muscle of the ducts of the mammary gland to cause the expulsion of milk. Oxytocin is released in response to CNS signals in response to the suckling stimulus. Vasopressin, named for its action to increase blood pressure, is also known as the antidiuretic hormone (ADH) for its action at the distal convoluted tubule of the kidney, which results in reabsorption of water.

These two hormones have a prominent role in the history of the development of our concepts of biologically active peptides. They were the first to be isolated and purified, the first to be synthesized. A Nobel Prize was

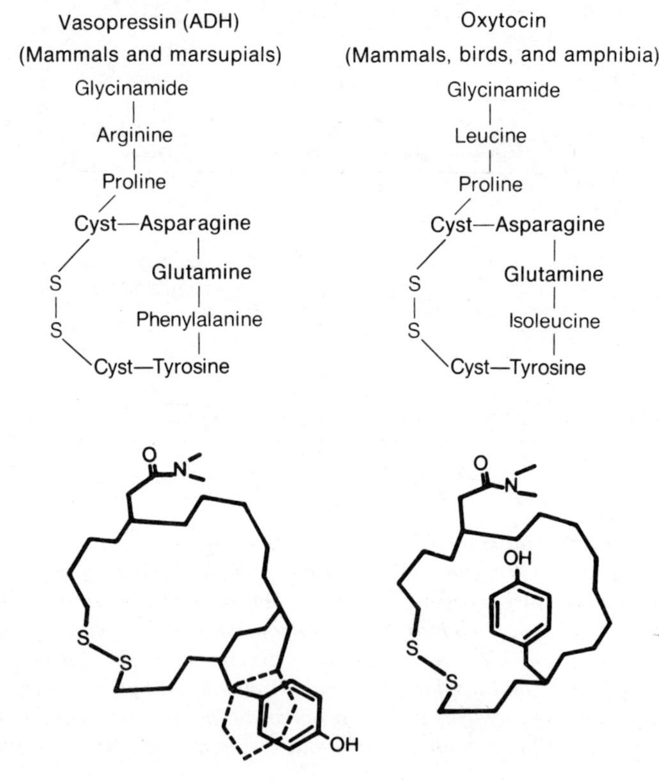

Biological Activity	*Vasopressin*	*Oxytocin*
Antidiuretic activity	++++	+
Blood pressor activity (rat)	++++	+
Uterine contraction (rat)	+	++++
Milk ejection	+	++++

++++ = very active in 10^{-9}–10^{-14} M range; + = <5% activity.

Figure 16.16
Structures and biological actions of vasopression and oxytocin.
The middle figures depict three-dimensional (solution) structures obtained from NMR data. Phenylalanine stacking with tyrosine prevents H-bonding of tyrosine to amide of asparagine that occurs in oxytocin.
From R. Walters, *Fed. Proc.*, 36:1872, 1977. Reproduced with permission.

awarded in 1954 to Vincent Du Vigneaud for these studies. It was demonstrated that a single peptide structure could have more than one biological action. A number of analogs have been synthesized. The effect on biological activity varied according to the type of amino acid substitution and its position in the molecule. Whereas the substitution of leucine (nonpolar) for arginine (basic) results in a profound change in activity (Figure 16.16), the substitution of one basic amino acid (arginine, in human ADH) for another (lysine, in ovine ADH) does not alter the type or degree of biological action. In some cases, the substitution of leucine for arginine, for example, confers new properties not observed in either vasopressin or oxytocin. This substance, called vasotocin, was first synthesized before it was subsequently found to be a natural peptide present in most nonmammalian species. The new property, maintaining water balance, is particularly important in amphibia.

Oxytocin and vasopressin are synthesized in neuronal cells of specific hypothalamic nuclei. Large peptide precursors (M_r 20,000) are formed which are cleaved to smaller (M_r 10,000) peptides called neurophysins, each specific for vasopressin and oxytocin. The complex of vasopressin–neurophysin and oxytocin–neurophysin streams down the axon of the nerve cells to the posterior lobe of the pituitary. They are stored in noncovalent association in vesicles and are released by appropriate stimuli. In response to small changes of the order of 1% in plasma [Na^+], vasopressin and its specific neurophysin molecule are released separately into the circulation.

With more sophisticated technology utilizing computer analysis of physical measurements, such as nuclear magnetic resonance (nmr), it has become possible to determine the three-dimensional structure of peptides in solution. Differences in the proton resonance patterns due to molecular position and the effect of neighboring groups can be related to alterations in amino acid substitutions and in changes in the solution environment, pH, temperature, D_2O vs H_2O, and so on. The complicated patterns of shifts in absorbancies can be analyzed by computer, and ultimately specific assignments of absorption peaks can be related to positions on the carbon–nitrogen skeleton. In this manner the three-dimensional structures in Figure 16.16 were derived for oxytocin and vasopressin. The hydrogen bonding of tyrosine to asparagine observed in oxytocin is in sharp contrast to the structure indicated for vasopressin, in which this conformation by tyrosine is prevented by the stacking effect of the aromatic rings due to the presence of phenylalanine.

On the basis of such studies chemists have gained new information by which to synthesize new analogs with predictable behavior in biological functions. Synthetic analogs of ADH are available in which the blood pressor activity is negligible and the antidiuretic activity is increased by 30-fold over the natural substance. Further progress in designing synthetic analogs of specific function are to be anticipated. It is also expected that these studies will provide important information with respect to the conformation of hormones and their receptors during the binding process.

Relationship of ACTH and Opioid Peptides

The role of ACTH in adrenal steroidogenesis is described in Chapter 15, Section 15.3. The biosynthesis of ACTH and a family of related peptides, including α-melanocyte-stimulating hormone (α-MSH) and the endorphins, are discussed in this section.

ACTH is a straight-chain peptide (M_r 4,500) containing 39 amino acids, without disulfide bridges (Figure 16.17). The mammalian ACTH structures that have been sequenced vary by only two amino acids in the

α-MSH

1	2	3	4	5	6	7	8	9	10	11	12	13
SER -Ac	TYR	SER	MET	GLU	HIS	PHE	ARG	TRY	GLY	LYS	PRO	VAL $-NH_2$

ACTH

1	2	3	4	5	6	7	8	9	10	11	12	13	14	15
SER	TYR	SER	MET	GLU	HIS	PHE	ARG	TRY	GLY	LYS	PRO	VAL	GLY	LYS

16	17	18
LYS	ARG	ARG

39	38	37	36	35	34	33	32	31	30	29	28	27	26	25	24	23	22	21	20	19
PHE	GLU	LEU	PRO	PHE	ALA	GLU	ALA	SER	GLU	ASP	GLU	ALA	GLY	ASN	PRO	TYR	VAL	LYS	VAL	PRO

Figure 16.17
Amino acid homology of α-MSH and ACTH.
α-MSH has an acetyl group on Ser[1]. It has no ACTH activity. ACTH has an α-MSH-like effect of increasing skin pigmentation. ACTH action requires the presence of amino acids 1–18. The fragment containing amino acids 18–39 is known as CLIP. It has no biological activity.

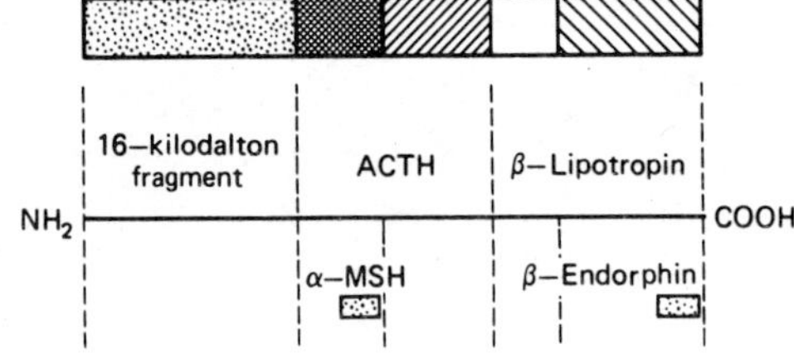

Figure 16.18
The 31K peptide-pro-ACTH/endorphin peptides.
The 31K peptide contains the fragments α-MSH (ACTH 1–13), the ACTH (1–39) molecule, β–endorphin (61–91) and β–lipotropin (1–91) molecule. The first 16K fragment is devoid of activity.

carboxyl terminal half of the molecule. The amino terminal half contains the steroidogenic activity. The molecule has been synthesized, and peptides containing the first 18 to 24 amino acids have activity. Commercial preparations containing 24 amino acids are used clinically and are equal in most respects to the natural ACTH of 39 amino acids. The carboxyl terminal half (ACTH, 18–39) is referred to as CLIP and is devoid of biological activity. Radioimmunoassays are available using antibodies raised against the ACTH (1–18), ACTH (18–39), as well as the complete molecule of ACTH (1–39). The melanocyte-stimulating hormone, α-MSH, is the amino terminal half of ACTH (1–13) with Ser[1] in the acetylated form. With the use of immunoassay and immunoprecipitation techniques, it has been possible to relate those forms to larger peptide molecules present in the anterior pituitary and median eminence, which could serve as biosynthetic precursors. With these techniques it has been possible to show that large peptides of 31kDa ("big" ACTH), 13kDa ("intermediate" ACTH), and 4.5kDa ("little" ACTH) contain amino acid sequences in common with the ACTH (1–39) molecule. Similar studies using antibodies to β-lipotropin and β-endorphin indicate a common recognition site in the 31kDa peptide. The 31kDa peptide, which reacts with both the antibodies to ACTH (1–39) and β-lipotropin contains these two separate amino acid sequences and is referred to as pro-ACTH/endorphin or proopiomelanocortin, since it is the precursor of endorphin, MSH, and ACTH. The complete analysis of this precursor peptide and the relationship of the identifiable fragments to the entire molecule is shown in Figure 16.18.

The 31kDa peptide contains a 16kDa fragment not yet identified with any known biological activity, ACTH (1–39), and the β-LPH (1–91), the carboxyl end of which contains β-endorphin (61–91). Although the latter fragment also contains the five amino acid sequence of methionine–enkephalin, it is likely that this pentapeptide, which has been found in a variety of tissues, is formed from an entirely different protein precursor.

CLIN. CORR. 16.4
HORMONAL REGULATION OF PROTEIN SYNTHESIS

Protein anabolism is increased by the actions of insulin, growth hormone, and testosterone as evidenced by the fact that blood amino acid levels are lowered by these hormones and that they produce a positive nitrogen balance. Growth hormone and testosterone, in particular, increase total body mass. Optimal growth in

The enkephalins are pentapeptides with the following structures:

Tyr-Gly-Gly-Phe-Met Met-Enkephalin

Tyr-Gly-Gly-Phe-Leu Leu-Enkephalin

These were the first of the naturally occurring peptides isolated that bind to opiate receptors in brain membrane preparations and to show morphine-like activity. It is suggested that these substances may have a neural transmitter function at the synaptic junction. The most active of the opioids isolated is the 30-amino acid compound β-endorphin.

Growth Hormone and Related Peptides

Growth hormone and prolactin are straight-chain polypeptides that are similar in size and structure and share in some of their biological properties (Table 16.1). In addition there are proteins produced by the placenta that have growth hormone and prolactinlike properties. These substances, called placental lactogens, can be differentiated from the pituitary hormones by the use of hormone-specific antibodies. Due to similarity in structure, antibodies generated for each specific hormone will cross-react to some extent with the other hormones. (See Figure 16.19.)

A variety of factors have been implicated in growth hormone regulation. Growth hormone release is regulated by a releasing factor and by GRIF (somatostatin). The highest concentration of somatostatin is found in the hypothalamus, but it has been found in other parts of the central nervous system and in other tissues, such as the pancreas. The structure of the growth hormone releasing factor (GRF) has been reported as containing 40 amino acids as well as a 44-amino acid peptide (Figure 16.15). In addition to GRF, the neurotransmitters, dopamine and norepinephrine, have known stimulatory effects. Hypoglycemia, exercise, fasting, and amino acids, particularly arginine, are some factors that stimulate the release of growth hormone. As its name implies, growth hormone is active in the regulation of a number of growth processes. The term somatotropin (STH), by which GH is also known, connotes a wider range of activity, which affects soft tissues, organs, and bones. It is well to remember that there are other agents such as insulin and thyroid hormone that also exert important stimulatory effects during the period of growth. In particular, to achieve normal growth rates thyroid hormone and growth hormone act synergistically during development. Thyroid hormone may be necessary for the optimum production of growth hormone in the pituitary. Both are involved in achieving normal rates of protein synthesis in the liver during growth, and in the induction of the enzyme transamidinase in the kidney (Clin. Corr. 16.4). In the hypophysectomized animal, growth hormone is specifically required for the formation of guanidinoacetate, which is the direct precursor of creatine (Chapter 21).

Human growth hormone has 191 amino acids (Figure 16.20) with M_r 21,500 and pI 4.9. There are two disulfide bridges in a straight chain on purification. It can form aggregates and has been assayed and isolated from plasma as a dimer. Due to the heterogeneity of circulating growth hormone and the plasma of somatomedins, the extent of immunoreactivity of plasma constituents by RIA for growth hormone is quite variable and often much lower than estimates based on radioreceptor binding assays or assays for biological activity.

Differences in peptide structure among animal species are greater for growth hormone than for the other anterior pituitary hormones. Only primate growth hormone (monkey or human) is active in the human.

the growing organism requires the presence of both growth hormone and thyroid hormone. Testosterone and estradiol have important growth promoting effects on the accessory reproductive tissues. Specific liver proteins, CBG, TBG, and metalloproteins are induced by the action of estradiol. Tyrosine aminotransferase is induced in perfused liver after the administration of glucagon, insulin, or cortisol or by a protein digest.

Synthetic analogs of testosterone have been used as protein anabolic agents in chronically ill, debilitated patients. These steroid alkyl derivatives have received considerable notoriety owing to their use by athletes seeking to increase body weight and strength while undergoing intensive training programs involving exercise and massive caloric intake. The effectiveness of anabolic steroids for increasing body mass under these conditions in a normal male remains to be established, and in addition the steroids may produce toxic effects. Because of the androgenic nature of these agents, their use by female athletes produces more obvious masculinizing effects. It is now routine practice to test for steroids as well as other drugs in the urine of athletes prior to major events.

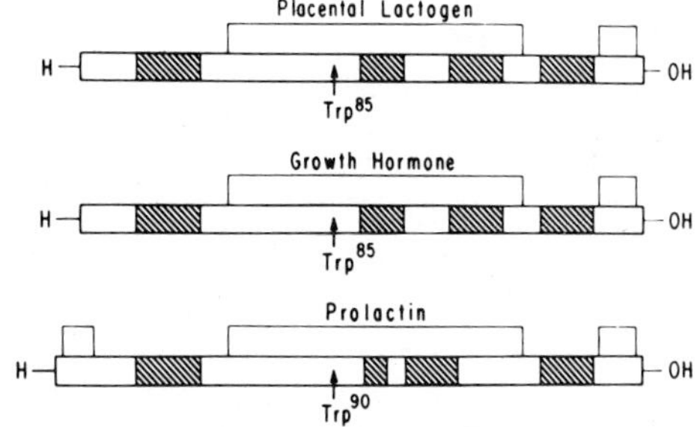

Figure 16.19
Similarity in structures of prolactin, growth hormone, and placental lactogen.
Homology indicated by position of disulfide bridges (thin lines), repeating polypeptide sequences internally, and position of tryptophan.

From H. Niall, M. Hogan, R. Sauer, I. Rosenblum, and F. Greenwood, *Proc. Natl. Acad. Sci., USA*, 68:869, 1971. Reproduced with permission.

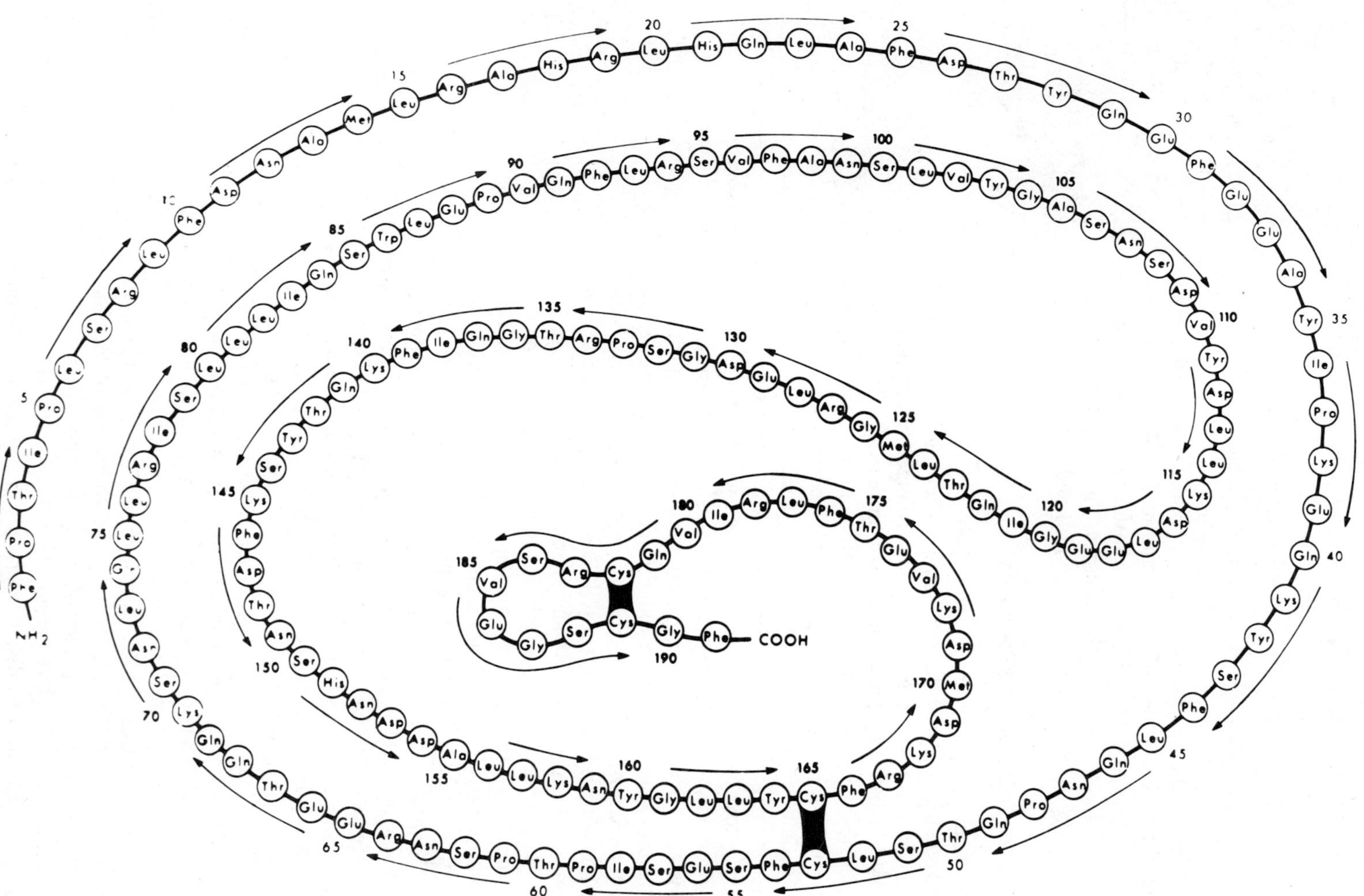

Figure 16.20
Structure of human growth hormone.
The hormone contains 191 amino acids with two disulfide bridges.
From C. H. Li and T. H. Bewley, *Proc. Natl. Acad. Sci. USA*. 73:1476, 1976. Reproduced with permission.

Bovine growth hormone is active in murine and avian species, and fragments after proteolytic procedures have some activity in humans. Different portions of the molecule that have been synthesized contain some biological activity. For these reasons it is felt that some smaller peptide fragment (an active core) may be involved in expression of biological activity.

Some of growth hormone's effects are indirect, since many of its actions are mediated through peptide intermediates called somatomedins. Three somatomedins, A, B, and C, have been characterized. The term somatomedin refers to plasma factors which mediate growth and are dependent on growth hormone for their synthesis or release. Other factors that affect growth of tissues, such as nerve growth factor and epidermal growth factor, are formed in liver but do not require stimulation by growth hormone. One group of somatomedins has isoelectric points above pH 7.5. They include SM-B, SM-C, and insulinlike growth factor (IGF-1). The structures of SM-C and IGF-1 are identical, and their biological activities appear to be similar. IGF-II is a somatomedin with a neutral pI. Receptors for the somatomedins are found in the cells of human term placenta, although no functional role for these receptors has been determined. Two types of receptors have been characterized. Type I receptor binds SM-C, SM-B, or IGF-I. It is an oligomer, probably a tetramer, with subunits of 350kDa and 140kDa linked together by disulfide bonds. Type II receptor, which binds IGF-II, appears to be a straight chain (M_r 250,000) with no disulfide bonds. The insulinlike growth factors (IGF-1 and IGF-II) have structures similar to proinsulin. Somatomedin C, like growth hormone (GH), has an effect on the hypothalamus to stimulate somatostatin release, which would inhibit further (GH) release.

One of the somatomedins, formerly referred to as *sulfation factor,* stimulates sulfate uptake by cartilage. The incorporation of $^{35}SO_4$ is used as a bioassay for growth hormone or somatomedin. Somatomedins also stimulate protein synthesis and amino acid and glucose uptake by isolated rat diaphragm muscle. Some somatomedin actions are similar while others are antagonistic to insulin activity, for example, a diabetogenic effect. The many multiple effects ascribed may be due to different forms of growth hormone produced by the pituitary, to its cleavage to active fragments, or to the formation of somatomedins by action on the liver. Specific receptors for growth hormone have been demonstrated separate from that of prolactin in liver, kidney, and in other tissues.

Prolactin

Prolactin is a hormone found throughout the animal kingdom and has numerous and diverse effects that regulate a variety of physiological functions. These activities are grouped into actions related to water and electrolyte metabolism, actions related to reproduction, actions affecting growth, developmental, and metabolic events. In mammals prolactin has osmoregulatory functions and growth-promoting effects, but the best-known activity is related to the growth of the mammary glands and the initiation of lactation. In the rodent, prolactin stimulates corpus luteum function and modulates growth and secretions of male sex accessory organs. Placental lactogen is a substance produced during pregnancy, which has attributes of both pituitary prolactin and pituitary growth hormone. An older name for this substance is *chorionic somatomammotropin*. As with growth hormone, the actions of prolactin are complicated by the fact that they are synergistic to or are modulators of other hormonal agents. Prolactin affects the release of pituitary gonadotropins. It modulates the actions of antidiuretic hormone (ADH) and aldosterone in the kidney with respect to water and electrolyte metabolism.

The control of prolactin secretion is complex. Agents that release prolactin are known, although the identity of a physiological prolactin-releasing factor (PRF) has not been established. Inhibition of prolactin release by the pituitary is the dominant feature of control exerted by the hypothalamus. One prolactin-inhibiting factor (PIF) is considered to be dopamine.

The most studied mechanism of action of prolactin is in the stimulation of lactogenesis in the mammary gland, where it acts in concert with insulin and glucocorticoids. The stimulation of growth of the ductal tissue by estrogens and the glandular alveolar tissue by progesterone precedes the lactogenic activity. Specific prolactin receptors have been found on the plasma membrane of mammary tissue. Events that have been implicated as a response of the tissue to the receptor binding of prolactin include (1) activation of a membrane associated Na^+-K^+-ATPase, (2) activation of cyclic nucleotide synthesis and the mediation of calcium-dependent events, (3) synthesis of prostaglandins, and (4) synthesis of polyamines, for example, spermidine, which affects RNA synthesis and the induction of protein synthesis during lactation.

It has been shown that prolactin can be internalized within the cell by use of autoradiography. Since part of the [^{125}I]prolactin has been found associated with the Golgi apparatus, a functional role can be postulated. However, it is not known whether the prolactin detected within the cell is biologically active.

CLIN. CORR. 16.5
CONTRACEPTIVES

The natural estrogens and progesterone inhibit the secretion of gonadotropin during pregnancy and therefore prevent further ovulation from occurring. The synthetic estrogens and progestins used in birth control pills have the same effect as the natural hormones and in that sense produce a state of pseudopregnancy. The natural hormones have a short half-life and must be injected to obtain full biological effect. There are a number of synthetic compounds that are long acting and orally active progestins. Oral contraceptives contain one of the synthetic progestins alone or in combination with an estrogen. When taken each day for 21 days, the concentrations of 4 mg progestin and 100 μg estrogen per pill will inhibit the release of FSH and LH midcycle and as a consequence inhibit ovulation. With a different regimen and with lower dosage (minipill form), some pills may achieve their birth control effects through other biological mechanisms.

The key to the success in the development of an orally active contraceptive pill was the concept of a daily ingestion of hormone in a regimen that would mimic the endogenous estrogen–progesterone changes in the normal cycling female. In this respect, the pill was used initially to normalize the periods of the menstrual cycle in the treatment of infertility, in addition to their use in contraception. The successful development of the pill in the 1960s was the result of a collaborative effort by a gynecologist (John Rock, Boston), a physiologist (Gregory Pincus, Worcester), a reproductive biologist (Min-Chueh Chang, Worcester), and organic chemists from several pharmaceutical

Pituitary Glycoprotein Hormones

The pituitary trophic hormones LH, FSH, and TSH are glycoproteins of similar structure (Table 16.1). They consist of two subunits in a noncovalent association, a 13,000-dalton α subunit and a 15,000-dalton β subunit. Both α and β subunits have carbohydrate portions containing sialic acid, mannose, galactose, fucose, and *N*-acetylhexoses. The amount of carbohydrate is variable and ranges from ~15–30% by weight for each subunit.

The amino acid composition has been determined for the α and β subunits in a number of species. Within each species there is a very good amino acid homology in the α subunit of LH, FSH, and TSH. There are marked differences in the amino acid sequence of the β subunits. The specificity in receptor binding and in biological activity of LH, FSH, and TSH resides in the structural differences of the β subunits.

The placental gonadotropin, human chorionic gonadotropin (HCG), although not of pituitary origin, is structurally and biologically very similar to the luteinizing hormone (LH). Because of its availability from pregnancy urine, HCG has been used clinically instead of the more difficult to obtain LH. HCG also contains an α subunit very similar to that of the pituitary hormones, but the β subunit is markedly different from that of FSH and TSH. The amino acid sequences of the β subunit of HCG and LH are sufficiently alike to account for the similar biological activities of the two hormones. HCG and LH can be distinguished immunologically, however.

Antibodies can be produced that are relatively specific for LH, FSH, TSH, and HCG by injection of pure trophic hormone of one species into another. The specificity of the antibody resides primarily in the differences in amino acid structure of the β subunit. Due to the similarity in amino acid sequences of the α subunit, antibodies to the entire protein molecule interact to some extent (cross-react) with the other trophic hormones. Both α and β subunits of the pituitary hormones can be obtained in pure form; indeed, they are present in the tissues where they have been isolated in subunit form to a limited extent. The separate subunits have little biological activity, but they are immunologically active. Antibodies raised against the respective pure α and β subunits have the expected specificities; the α-antibodies will react with each hormone, LH, FSH, TSH, and HCG, whereas the β-antibody is quite specific for only the one entire hormone (or its β subunit) from which it was derived. For example, there is no cross-reactivity between the antibody of the β subunit of HCG for FSH or TSH. Since there are similarities between the β subunits of LH and HCG, some cross-reactivity can be observed. Generally, the small degree of cross-reactivity of antibodies from different hormones or subunits may not present a problem for routine purposes of measurement, that is, assay for circulating concentrations of pituitary hormones in blood.

An antibody to any one α-subunit will provide the needed information as well as another. However, in certain conditions the small degree of cross-reactivity may become significant, as in the measurement of HCG in early pregnancy, the monitoring of ectopic hormone-producing tumors, or in the preparation of a vaccine against pregnancy. In each case, cross-reaction with pituitary LH would be misleading. The use of the purest form of β-HCG subunit or the use of a chemically modified β subunit as an antigen, or the preparation of monoclonal antibody to β-HCG is warranted to obtain the highest degree of specificity possible, when both LH and HCG molecules are present to a significant degree.

The interrelationships of gonadotropin stimulation and steroid hormone levels during the menstrual cycle is depicted in Figure 16.21. Refer to the regulation of steroidogenesis in the ovary discussed in Chapter 15. The effect of oral contraceptives on FSH and LH is discussed in Clin. Corr. 16.5.

concerns who synthesized suitable compounds for clinical testing.

Barrier methods of contraception, sterilization, intrauterine devices, and rhythm methods have been developed and used with varying success in different countries of the world as alternatives to the use of the pill or the administration of chemical substances, but a practical and effective oral contraceptive for the male, although sought after, has yet to be achieved.

One stated goal of the approach to population growth by family planning is to limit increases at least to a rate of 1% or less growth per year. This rate has been achieved in the United States and the western world but has not been in many countries in South America, Africa, and Asia. A contraception program that has attained some success is the use of intramuscular injection of medroxyprogesterone acetate as a depot form that can exert its contraceptive action for many months. Its potential use for effective family planning with respect to population control in many developing countries and in the developed countries worldwide has implications that are of vital concern for obvious sociological, political, and religious reasons.

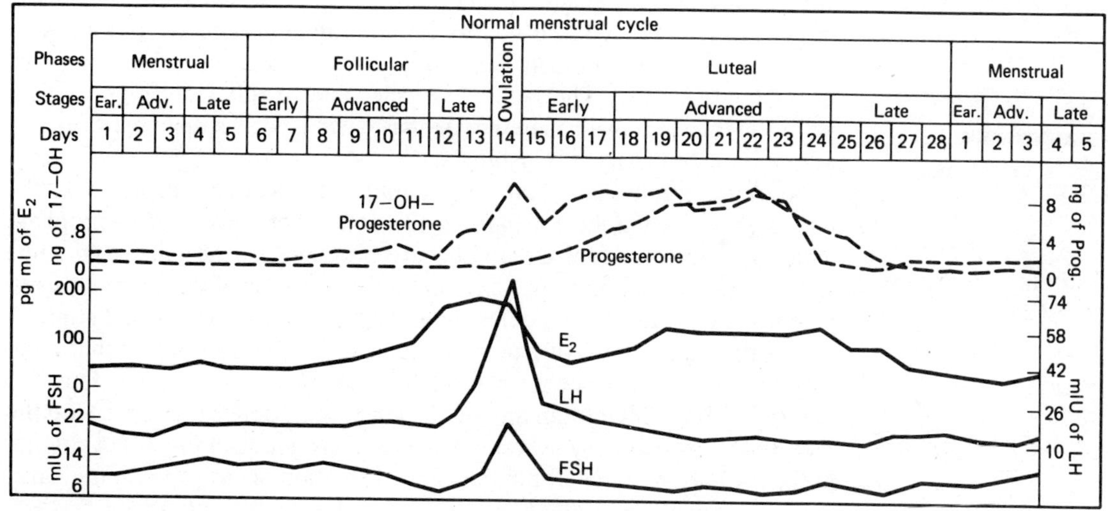

Figure 16.21
Human menstrual cycle.
The relationships among the gonadotropins, LH and FSH, and the steroid hormones, estradiol and progesterone, at the time of ovulation should be noted. Refer to Chapter 15 on ovarian steroidogenesis.

From G. T. Ross, C. M. Cargille, M. B. Lipsett et al., *Recent Prog. Hormone Res.*, 26:1, 1970.

16.5 BIOLOGICAL RHYTHMS

All animals exhibit patterns of change that have cycles of ~24 h. The term circadian rhythm (*circa*, about; *dies*, day) applies to rhythms that persist in the absence of any environmental or sensory input that are not exactly 24 h in length; they are free-running and can be compared to autonomous oscillations. In fact, however, most rhythms are entrained by external periodic signals such as light and dark, temperature changes, feeding, and activity patterns. These internal and environmental factors are referred to as zeitgebers (time keepers). If a human is shielded completely from all possible input (individuals have been isolated alone in deep mines underground with no outside stimuli or time cues for several months) their circadian rhythms run at their own frequency of slightly less than or longer than 24 h. In the human there is a multiplicity of oscillators, which may be coupled or desynchronized and free-run with different frequencies. The search for the biochemical or physiological equivalent of an internal oscillator has not been successful.

Desynchronization of rhythms can have profound physiological effects. Best known of these is the condition of "jet-lag," which occurs when many time zones are crossed in a jet plane, particularly in traveling west to east. Recovery requires several days, and usually a week is needed to reestablish normal circadian rhythms.

The endocrine system provides a good model for the study of circadian rhythms, since they have pronounced cycles with fluctuations of hormone levels that are often greater than 50% during the light/dark cycle. A well-established cycle is that of corticosteroid levels in blood or urine, which is a consequence of the secretion of hypothalamic (CRF) and pituitary (ACTH) hormones (Figure 16.22). The degree of inhibitory negative feedback exerted by cortisol will vary at different phases of the 24-h cycle. These effects have practical clinical implications with respect to the timing of the administered dose of cortisol. Furthermore, the efficacy in response to drugs in general can be altered, depending upon the time at which they are administered. The normal cortisol cycle in the human peaks in the early AM and has its lowest values in late afternoon. As a result of hypersecretion in Cushing's disease, cortisol values are high throughout the day. The lack of a cycle, due to relatively high cortisol values in the PM as well as in the AM, has been used as a diagnostic test for adrenal hyperactivity of Cushing's.

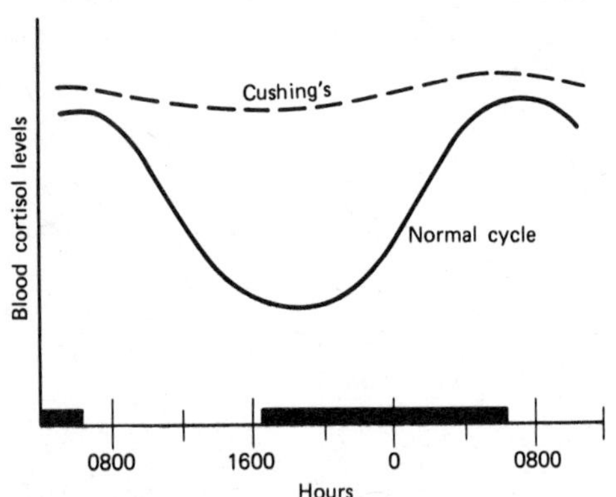

Figure 16.22
Absence of normal* AM/PM *variation in cortisol levels in Cushing's disease.

Another organ that elicits a pronounced circadian response is the pineal gland. A number of biologically active derivatives of tryptophan, including serotonin and melatonin, are produced in the pineal. In some species, as in the amphibia, external light serves as a direct stimulus to pineal activity. In other species, including the human, the pineal stimulation is through release of neurotransmitters by nerve impulses that emanate from other nerve centers, including the suprachiasmatic nucleus and the optic nerve. Melatonin and other pineal secretions have effects on the hypothalamic–pituitary system. There is a pronounced circadian rhythm in the formation of melatonin, since its formation occurs only in the dark. The sharp peak in content in the dark is due to the activity of a single enzyme, *N*-acetyltransferase, which converts serotonin to acetylserotonin (Figure 16.23). Neither serotonin formation nor the *O*-methyltransferase, which converts acetylserotonin to melatonin, is affected by light/dark cycles. In many rhythms the peak and trough of activity correspond to the temporal behavior of the animal; the nocturnal animal has a peak in corticosteroids in the PM, and the diurnal animal (human) has its corticosteroid peak in the early AM. In contrast, the melatonin peak formation

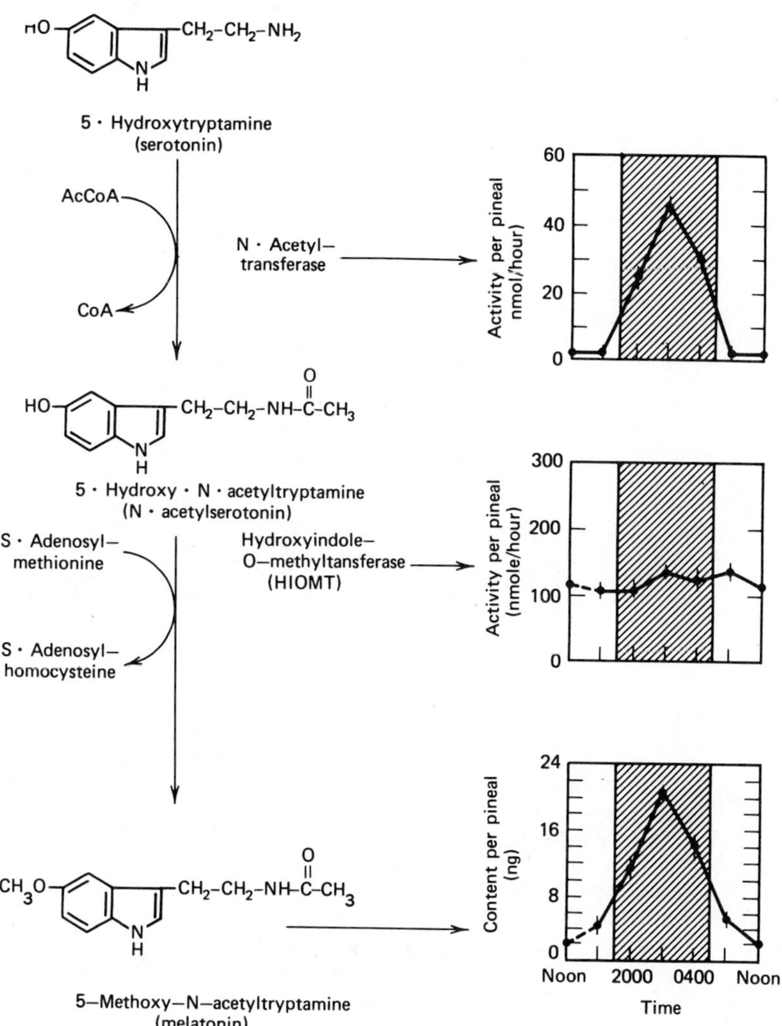

Figure 16.23
Circadian rhythm in the formation of melatonin in the pineal gland of the chick is due to the stimulation of N-acetyltransferase in the dark.
The O-methyltransferase (HIOMT) as well as the formation of precursor serotonin do not have large activity changes in the light/dark cycle.
From S. Brinkley, S. MacBride, D. Klein, and C. Ralph, *Science,* 181:273, 1974. Reproduced with permission. Copyright 1974 by the American Association for the Advancement of Science.

occurs in the dark in early AM regardless of activity periods. Pineal and melatonin activity have been implicated in the regulation of gonad and adrenal function. The pineal is considered a part of the chain that determines phase relationships, with respect to circadian rhythms of these endocrine units. It is not likely that the pineal is the basic pacemaker of the body; it could be responding to signals from higher centers, for example. It is probable that there are a number of pacemakers, since individual cells may have endogenous clocks of their own; although a pacemaker could be more dominant in one species than another in determining certain patterns of activity or behavior.

BIBLIOGRAPHY

Cell Regulation

Goldstein, J. L., Anderson, R. G. W., and Brown, M. S. Coated pits, coated vesicles, and receptor-mediated endocytosis. *Nature* 279:679, 1979.

Greengard, P. Phosphorylated proteins as physiological effectors. *Science* 199:146, 1978.

Ingebritsen, T. S., and Cohen, P. Protein phosphatases: Properties and role in cellular regulation. *Science* 221:331, 1983.

Means, A. R. and Dedham, J. R. Calmodulin: An intracellular calcium receptor. *Nature* 285:73, 1980.

Norman, A. W., Roth, J., and Orci, L. The Vitamin D endocrine system: Steroid metabolism, hormone receptors and biological response (calcium binding proteins). *Endocrine Rev.* 3:331, 1982.

Schlessinger, J. The mechanism and role of hormone induced clustering of membrane receptors. *Trends Biochem. Sci.* 5:210, 1980.

Acetylcholine Receptor

Changeux, J-P., Devillers-Thiery, A., and Chemouilli, P. Acetylcholine receptor: An allosteric protein. *Science* 223:1335, 1984.

Protein Kinases

Michell, R. H. Inositol phospholipids and cell surface receptor function. *Biochem. Biophys. Acta* 415:81, 1975.

Nishizuka, Y. Turnover of inositol phospholipids and signal transduction. *Science* 225:1365, 1984.

Schramm, M., and Selinger, Z. Message transmission: Receptor controlled adenylate cyclase system. *Science* 225:1350, 1984.

Insulin

Jacobs, S., and Cuatrecasas, P. Insulin receptor: Structure and function. *Endocrine Rev.* 2:251, 1981.

Pilch, P. F., and Czech, M. P. The subunit structure of the high affinity insulin receptor. *J. Biol. Chem.* 225:1722, 1980.

Oxytocin

Walter, R. Identification of sites in oxytocin involved in uterine receptor recognition and activation. *Fed. Proc.* 36:1872, 1977.

Prolactin

Rillema, J. A. Mechanism of prolactin action. *Fed. Proc.* 39:2593, 1980.

Growth Hormone

Martin, J. B. Functions of central nervous system neurotransmitters in regulation of growth hormone. *Fed. Proc.* 39:2902, 1980.

Paladini, A. C., Pena, C., and Retegni, L. A. The intriguing nature of the multiple actions of growth hormone. *Trends Biol. Sci.* 4:250, 1979.

Gonadotropins

Ghai, R. D., Mise, T., Pandian, M. R., and Bahl, D. P. Immunological properties of the β-subunit of human chorionic gonadotropin. *Endocrinology* 107:1556, 1980.

ACTH/Endorphins

Eipper, B. A., and Mains, R. C. Structure and biosynthesis of pro-adrenocorticotropin/endorphin and related peptides. *Endocrine Rev.* 1:1, 1980.

Biological Rhythms

Menaker, M. Symposium on physiological and biochemical aspects of circadian rhythms. *Fed. Proc.* 35:2325, 1976.

Spelsberg, T. C., and Halberg, F. Circannual rhythms in steroid receptor concentration and nuclear binding in chick oviduct. *Endocrinology* 107:1234, 1980.

QUESTIONS

J. BAGGOTT AND C. N. ANGSTADT

Question Types are described inside the front cover.

1. (QT2) Plasma membrane receptors for peptide hormones:
   1. are generally easy to isolate and purify.
   2. need not be fully saturated by the appropriate hormone to produce a maximal cell response.
   3. generally mediate the effect of the hormone by mechanisms independent of adenylate cyclase.
   4. determine the specificity of cell responses to hormones.
2. (QT1) Plasma Ca^{2+}:
   A. increases in response to calcitonin secretion.
   B. is directly proportional to plasma phosphate.
   C. increases in response to parathyroid hormone (PTH).
   D. is deposited in bone under the influence of vitamin D.
   E. is mostly bound to plasma proteins.
3. (QT2) Prostaglandins:
   1. are found in many tissues.
   2. have short half-lives.
   3. may stimulate cAMP synthesis.
   4. generally exert their effects at their sites of synthesis.
4. (QT2) Glucagon and insulin are similar in that both:
   1. have prohormone precursors.
   2. affect liver and adipose tissue metabolism.
   3. are produced by the pancreas.
   4. act by direct stimulation of adenylate cyclase.
5. (QT2) ACTH:
   1. release is influenced by CRF.
   2. contains the entire primary structure of α-MSH.
   3. arises from the same precursor as β-lipotropin.
   4. contains the entire primary structure of β-endorphin.

   A. Oxytocin   C. Both
   B. Vasopressin   D. Neither
6. (QT4) Contain(s) one disulfide bridge.
7. (QT4) Contain(s) two aromatic amino acid residues.
8. (QT4) Precursor is transferred from the hypothalamus to the pituitary gland by an intracellular route.
9. (QT4) Precursor of at least one of the endorphins.
10. (QT1) Growth hormone:
    A. is released in response to high blood glucose.
    B. is required for creatine synthesis.
    C. bears a structural resemblance to insulin.
    D. exhibits the greatest similarity among animal species of any anterior pituitary hormone.

E. affects the metabolism of a narrow range of tissues, notably liver and kidney.

11. (QT1) Peptide hormones:
    A. are synthesized as active hormones with little or no posttranslational modification required.
    B. in their active form must be free of attached carbohydrate.
    C. may have structural homologies with one another.
    D. usually bind to intracellular receptors.
    E. in their active form consist of a single polypeptide chain.
12. (QT1) The mediation of a hormonal signal by cAMP is an example of:
    A. endocytosis.
    B. transduction.
    C. exocytosis.
    D. a zeitgeber.
    E. negative feedback.

A. Parathyroid gland
B. Anterior pituitary gland
C. Posterior pituitary gland
D. Pancreas
E. None of the above

13. (QT5) Actions mediated by releasing factors from the hypothalamus.
14. (QT5) Site of conversion of cholecalciferol to it's most active metabolite.
15. (QT5) Releases a hormone having some antidiuretic activity and considerable milk ejection activity.
16. (QT5) Source of hormones that are glycoproteins.
17. (QT5) Exhibits increased *N*-acetyltransferase in the dark, leading to increased synthesis of melatonin.

ANSWERS

1. C 2 and 4 true. Most peptide hormones act through cAMP (p. 595). Receptor concentration is too low to permit easy purification. Receptor specificity determines the specificity of a cell's response, and only 10–20% need to be occupied to produce a maximum response (p. 599).
2. C Plasma Ca^{2+} is reciprocally related to plasma phosphate (p. 602). Calcitonin decreases plasma Ca^{2+} (p. 603). PTH increases Ca^{2+} via inhibition of renal phosphate reabsorption (p. 603). Vitamin D stimulates Ca^{2+} release from bone. Plasma Ca^{2+} is about 50% bound (p. 602, Figure 16.9, p. 603).
3. E All statements are true (p. 599).
4. A 1, 2, 3 true. 1: Both are derived from larger precursors (pp. 604, 608). 2: Insulin is anabolic in muscle, liver, and fat cells (p. 607); glucagon stimulates glucose production in liver and lipolysis in adipose (p. 608). 3: Both are pancreatic (p. 604). 4: Insulin is unusual among peptide hormones in being cAMP independent; glucagon acts via cAMP (pp. 595, 604, 609).
5. A 1, 2, 3 true. 1: CRF stimulates ACTH release (pp. 576, 611). 2, 3, 4: A 31K peptide is the precursor of ACTH and β-lipotropin. ACTH contains α-MSH, and β-lipotropin contains β-endorphin (p. 614, Figure 16.18).
6. C See Figure 16.16, p. 612.
7. B See Figure 16.16, p. 612. Note the effect of aromatic ring stacking on the structures of these similar peptides (p. 613).
8. C Both are transported in nerve cell axons (p. 613).
9. D β-Endorphin is part of the β-lipotropin structure (p. 614).
10. B A and B: Hypoglycemia stimulates growth hormone, which is required for formation of guanidinoacetate, the direct precursor of creatine (p. 615). C: Growth hormone resembles prolactin and placental lactogen (p. 615, Figure 16.19). D: It differs more among species (p. 615). E: It affects many tissues (p. 615).
11. C A: They are synthesized as large precursors (pp. 604, 609, 614). B and C: LH, FSH, and TSH are glycoproteins of similar structure (p. 618). Other families of related peptides exist (pp. 612, 615). D: They generally bind to plasma membrane receptors; steroid hormones bind to intracellular receptors (p. 595). E: Insulin (p. 604) and the pituitary glycoprotein hormones (p. 618) consist of two chains, although insulin is derived from a single-chain precursor.
12. B See definitions, pp. 594, 620. E: Negative feedback describes how circulating hormone levels may be controlled (p. 611).
13. B See Figure 16.14, p. 611.
14. E 1,25-dihydroxycholecalciferol is formed in the kidney (p. 603).
15. C The hormone is oxytocin (p. 612).
16. B LH, FSH, and TSH are glycoproteins (p. 618) from the anterior pituitary (p. 611).
17. E This describes the pineal gland (p. 620).

17

DNA: The Replicative Process and Repair

STELIOS AKTIPIS

17.1 BIOLOGICAL PROPERTIES OF DNA

Overview

This chapter reviews the chemical structure of DNA and examines the relationship between this structure and the biochemical function of DNA. Within this context the process of DNA replication and repair is detailed. The remaining processes through which DNA regulates the expression of biological information (i.e., transcription and translation) are the subjects of other chapters.

One of the striking aspects of natural order is the sense of unity that exists between the members of successive generations in each species. It is apparent that an almost totally stable bank of information must always be preserved and passed from one generation to the next if individual species are to maintain their identities relatively unchanged over millions of years. It is now well established that the bank of genetic information takes the form of a stable macromolecule, deoxyribonucleic acid (DNA), which serves as the carrier of genetic information in both procaryotes and eucaryotes. DNA exhibits a rare purity of function by being one of the few macromolecules known to perform, with only minor exceptions, the same basic functions across species barriers.

It is apparent that the properties of cells are to a large extent determined by their constituent proteins. Many proteins serve as indispensable structural components of the cell. Other proteins, such as enzymes and certain hormones, are functional in character and determine most of the biochemical properties of the cell. As a result, the factors that control *which proteins* a cell may synthesize, at *what quantities,* and with *which sequence* are the same factors that primarily determine the function as well as the destiny of every living cell.

It is now well recognized that DNA is the macromolecule that ultimately controls, primarily through protein synthesis, every aspect of cellular function. DNA exercises this control as suggested by the sequence

$$\circlearrowleft\text{DNA} \longrightarrow \text{RNA} \longrightarrow \text{protein}$$

The flow of biological information is clearly from one class of nucleic acid to another, from DNA to RNA, with only minor exceptions, and from there to protein. In order for this transfer of information to occur faithfully, each preceding macromolecule serves as a structure-specifying template for the synthesis of the subsequent member in the sequence.

In addition to regulating cellular expression, DNA plays an exclusive role in heredity. This role is suggested by a circular arrow engulfing DNA, which depicts DNA as a *replicon,* a molecule that can undergo self-replication. The significance of *replication* is far reaching. It permits DNA to make copies of itself as a cell divides. These copies are bestowed to the daughter cells, which can thus inherit each and every property and characteristic of the original cell.

First, the important message to be retained is that DNA ultimately determines the properties of a living cell by *regulating the expression of biological information,* primarily by the control of protein synthesis. Second, but not less importantly, it should be clear that DNA transfers biological information from one generation to the next, that is, it is essential for the *transmittance* of genetic information.

Transforming Properties

The above principles, universally accepted today, were rejected outright not long ago. In fact, prior to the 1950s the general view was that nucleic acids were substances of somewhat limited cellular importance. The first convincing suggestion that DNA is the genetic material was made during the mid-1940s. The experiment involved the *transformation* of one type of pneumococcus, surrounded by the presence of a polysaccharide capsule and referred to as the S form because of its property of forming colonies with smooth-looking cellular perimeters, to a mutant without capsule, called the R form, which forms colonies with rough-looking outlines. These two forms are genetically distinct and cannot interconvert spontaneously. The transformation experiment demonstrated that a pure extract of DNA from the S form, when incorporated into the R form of pneumococcus, conveyed to the R form the specific property of synthesizing the characteristic polysaccharide capsule. Furthermore, the bacteria transformed from the R form to the S form maintained the property of synthesizing the capsule over succeeding generations. It was thus demonstrated that DNA was the *transforming* agent, as well as the material responsible for *transmitting* genetic information from one generation to the next. Almost three-quarters of a century had to elapse from the time nucleic acids were discovered until their important biological role was generally recognized.

DNA: A Molecule with Unusual Capacity

One of the striking characteristics of DNA is that it is able to encode an enormous quantity of biological information. An undifferentiated mammalian fetal cell contains only a few picograms (10^{-12} g) of DNA. Yet this minute amount of material is sufficient to direct the synthesis of an enormous number of distinct proteins that will determine the form and biochemical behavior of a large variety of differentiated tissues in the adult animal.

The compactness with which such information is stored in DNA is unique. Even the sophisticated memory elements of contemporary computers would appear pitifully inadequate by comparison. How does DNA achieve such a supreme coding effectiveness? The answer must obviously be sought in the nature of its chemical structure. It turns out that this structure is not only consistent with the unique efficiency of DNA as a "memory bank," but also provides the basis for understanding how DNA eventually "translates" this information into proteins.

17.2 STRUCTURE OF DNA

Structurally DNA is a *polynucleotide*. A formal analogy between polynucleotides and proteins may therefore be perceived. Polynucleotides are the products of *nucleotide* condensation, just as proteins are produced by the polymerization of *amino acids*. This similarity of structures is an important element which facilitates the transfer of genetic information between these two distinct classes of macromolecules. The structure of nucleotides and their constituent purine and pyrimidine bases are examined in Chapter 13.

The base composition of DNA varies considerably among species, particularly procaryotes, which have a range of 25–75% in adenine–thymine content. This range narrows with evolution, reaching limiting values of ~45–53% in mammals.

In addition to the four common bases, adenine, guanine, thymine, and cytosine, which occur in DNA from all sources, DNA isolated from many plant and animal tissues (e.g., wheat germ, thymus gland) contains small amounts of the base 5-methylcytosine. Methylated derivatives of the ba-

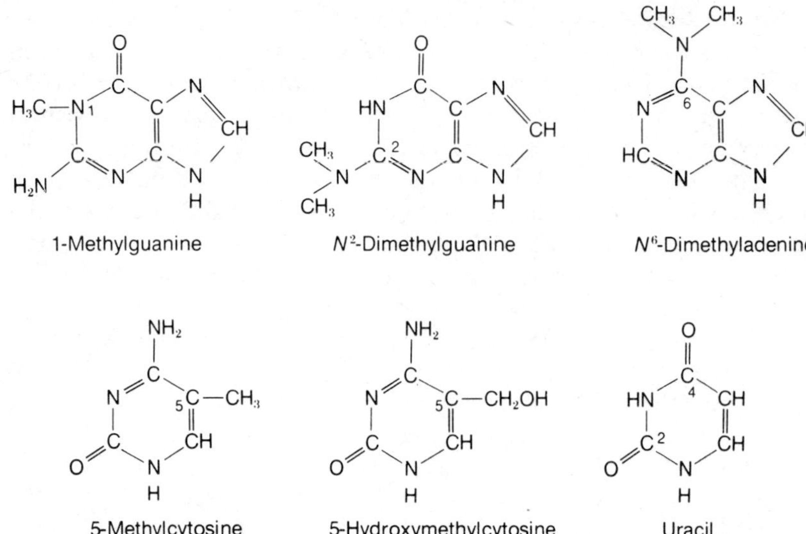

Figure 17.1
Structures of some less common bases occurring in DNA.

ses are also present in all DNA molecules examined to date. In addition, the DNA of certain bacteriophages (the T-even coliphages) contain 5-hydroxymethylcytosine in place of cytosine, and this derivative occurs in a glucosylated form. Even uracil, a base constituent of RNA, has been found in certain *Bacillus subtilis* phages, instead of thymine. The structures of some of these bases are shown in Figure 17.1.

Polynucleotides

Polynucleotides are formed by the joining of nucleotides by phosphodiester bonds. The phosphodiester bond is the formal analog of the peptide bond in proteins. It serves to join, as a result of the esterification of two of the three hydroxyl groups of phosphoric acid, two adjoining nucleotide residues. Two free hydroxyl groups are present in deoxyribose on the C-3′ and C-5′ atoms. Therefore these are the only hydroxyl groups that can participate in the formation of a phosphodiester bond. Indeed, it turns out that the nucleotide residues in DNA polynucleotides are joined together by 3′, 5′-phosphodiester bonds, as shown in Figure 17.2.

In some instances polynucleotides are linear polymers. The last nucleotide residue at each of the opposite ends of the polynucleotide chain serve as the two terminals of the chain. It is apparent that these terminals are not structurally equivalent, since one of the nucleotides must terminate at a 3′-hydroxyl group and the other at a 5′-hydroxyl group. These ends of the polynucleotides are referred to as the 3′ and the 5′ termini, and they may be viewed as corresponding to the amino and carboxyl termini in proteins. Polynucleotides also exist as cyclic structures, which contain no free terminals. Esterification between the 3′-OH terminus of a polynucleotide with its own 5′-phosphate terminus can produce a cyclic polynucleotide.

In this discussion long polymers of nucleotiaes joined by phosphodiester bonds are referred to as *polynucleotides,* in accordance with the prevailing nomenclature. A distinct name, oligonucleotide, is reserved for shorter nucleotide-containing polymers. According to formal rules of nomenclature, however, polynucleotides must be named by using roots derived from the names of the corresponding nucleotides, and using the ending *ylyl.* For example, the polynucleotide segment in Figure 17.2, in

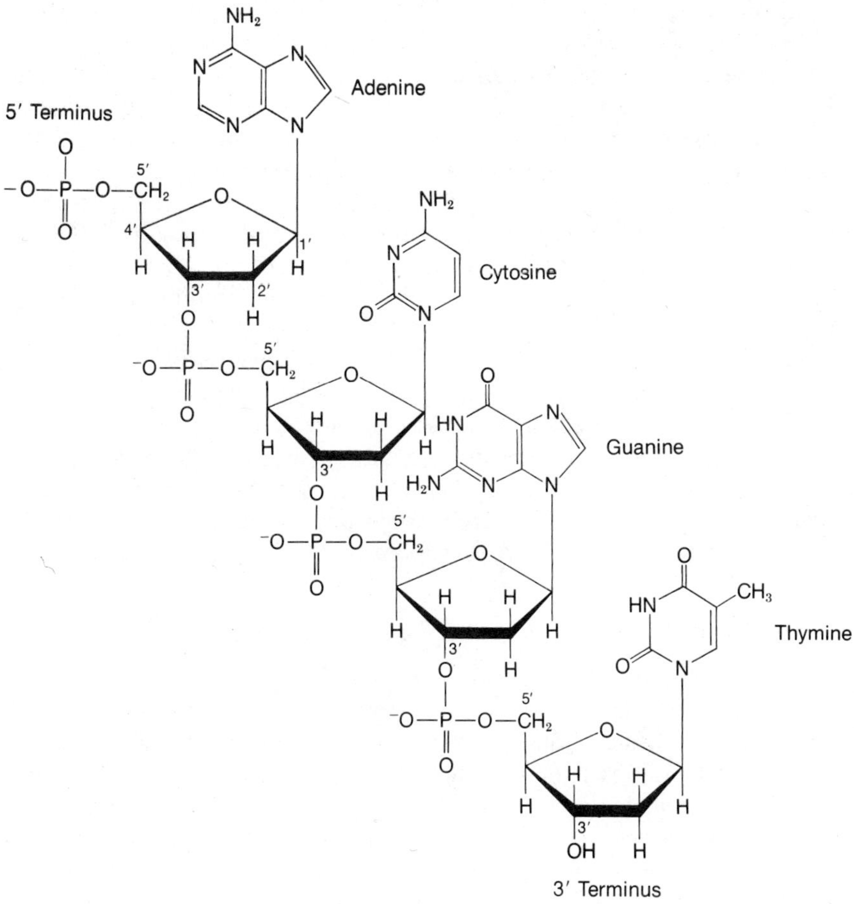

Figure 17.2
Structure of a DNA polynucleotide segment.
The example shown in this figure is a tetranucleotide, that is, an oligonucleotide consisting of four monomeric units. Although an exact polymerization size for this change in name does not exist, as a general rule a polymer containing less than 30 to 40 nucleotides is referred to as an oligonucleotide.

which the 5′ terminal is on the left of each nucleotide residue, should be named from left to right as

. . .deoxyaden*ylyl,* deoxycytid*ylyl,* deoxyguan*ylyl,* deoxythymid*ylyl*. . .

It is apparent, however, that the result of this approach is so cumbersome that abbreviations are generally preferred. For example, the oligonucleotide shown in Figure 17.2 is usually referred to as dAdCdGdT, and a polynucleotide containing only one kind of nucleotide, for example, dA, may be written as poly(dA). Oligo- and polynucleotide structures are also written out in shorthand, as shown in Figure 17.3. In every instance the sequence is written starting on the left with the nucleotide of the 5′ terminus.

DNA is made of polynucleotides, and it is the specific sequence of bases along a polynucleotide chain that determines the biological properties of the polymer. Although the structure of the nucleic acid building blocks, the bases, had been correctly known for many years, the polymeric structure initially proposed for DNA turned out to be one of the

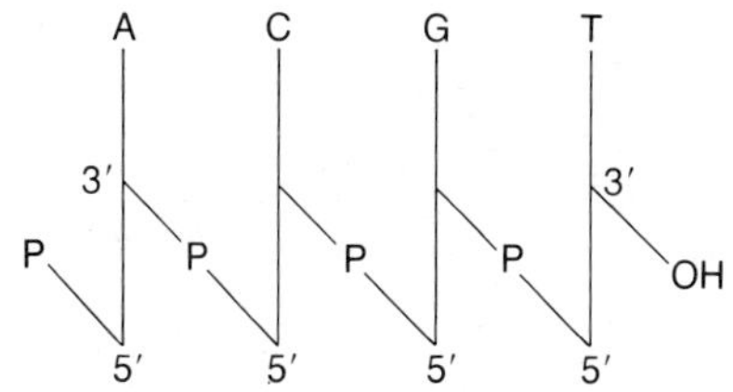

Figure 17.3
Shorthand form for structure of oligonucleotides.
The convention used in writing the structure of an oligo- or polynucleotide is a perpendicular bar representing the deoxyribose moiety, with the 5′-OH position of the sugar located at the bottom of the bar and the 3′-OH at a midway position. Bars joining the 3′ and 5′ positions represent the 3′,5′-phosphodiester bond, and the P on the left side of the perpendicular bar represents a 5′-phosphate ester. A 3′-phosphate ester is represented by placing the phosphate group on the right side of the bar. The base is indicated by its initial.

classical errors in the history of biochemistry. Experimental data obtained from what appears to have been partially degraded samples of DNA, and several other misconceptions, led to the erroneous conclusion that DNA consisted of repeating tetranucleotide units. Each tetranucleotide supposedly contained equimolar quantities of the four common bases. These impressions persisted to some degree until the late 1940s and early 1950s, when they were clearly shown to be in error. In the interim, however, these misconceptions were responsible for setting back the acceptance of the concept that the DNA of chromosomes carried genetic information. The monotonous structure of repeating tetranucleotides appeared incapable of having the versatility to encode for the enormous number of messages necessary to convey hereditary traits. Instead proteins, which can be ordered in an almost unlimited number of amino acid sequences, were favored as the most suitable candidates for a hereditary function. The transformation experiment carried out in the mid-1940s, and the subsequent finding that DNA consists of polynucleotide rather than tetranucleotide chains, were responsible for the general acceptance of the hereditary role of DNA that followed.

Hydrolysis of the Phosphodiester Bond: Nucleases

The nature of the linkage between nucleotides to form polynucleotides was elucidated primarily by the use of exonucleases, which are enzymes that hydrolyze these polymers in a selective manner. *Exonucleases* cleave the last nucleotide residue in either of the two terminals of an oligonucleotide. Oligonucleotides can thus be degraded by the stepwise removal of individual nucleotides or small oligonucleotides from either the 5′ or the 3′ terminus. Nucleases sever the bonds in one of two nonequivalent positions indicated in Figure 17.4 as proximal (p) or distal (d) to the base which occupies the 3′ position of the bond. For example, the treatment of an oligodeoxyribonucleotide with venom diesterase, an enzyme obtained from snake venom, yields deoxyribonucleoside 5′-phosphates. In contrast, treatment with a diesterase isolated from animal spleen produces deoxyribonucleoside 3′-phosphates.

It should be noted that other nucleases, which cleave phosphodiester bonds located in the interior of polynucleotides and are designated as *endonucleases,* behave similarly in this respect. For instance, DNase I cleaves only p linkages, while DNase II cleaves d linkages. The points of cleavage along an oligonucleotide chain are indicated by arrows in Figure 17.4. Some *endonucleases* have been particularly useful in the development of early methodologies for sequencing of RNA polynucleotides. More recently other endonucleases, known as *restriction endonucleases,* have provided the basis for the development of recombinant DNA techniques.

Many nucleases do not exhibit any specificity with respect to the base adjacent to the linkage that is hydrolyzed. Certain nucleases, however, act more discriminately next to specific types of bases or even specific individual bases. Restriction nucleases act only on sequences of bases specifically recognized by each restriction enzyme. Nucleases also exhibit specificities with respect to the overall structure of polynucleotides. For instance, some nucleases act on either single- or double-stranded polynucleotides, whereas others discriminate between these two types of structures. In addition, some nucleases exclusively designated as *phosphodiesterases* will act on either DNA or RNA, whereas other nucleases will limit their activity to only one type of polynucleotide. The nucleases listed in Table 17.1 illustrate some of the diverse properties of these enzymes.

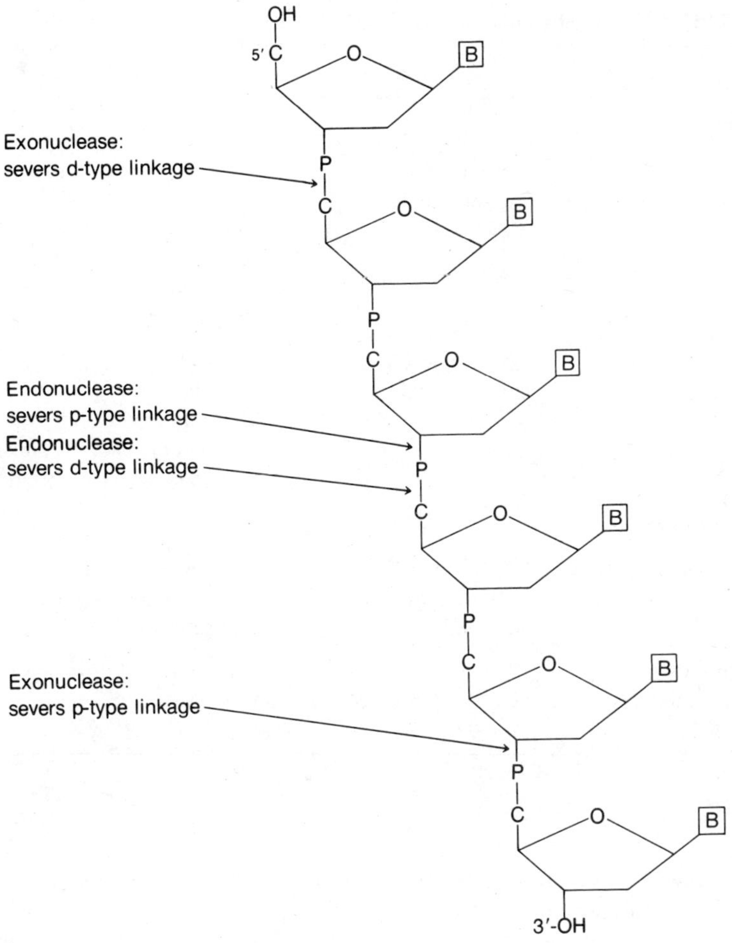

Exonuclease:
severs d-type linkage

Endonuclease:
severs p-type linkage
Endonuclease:
severs d-type linkage

Exonuclease:
severs p-type linkage

Figure 17.4
Nucleases of various specificities.
Exonucleases remove nucleotide residues from either of the terminals of a polynucleotide, depending on their specificity. Endonucleases hydrolyze interior phosphodiester bonds. Both endo- and exonucleases hydrolyze either d- or p-type linkages, as illustrated in the figure (see text for explanation of d- and p-type linkages).

Secondary Structures of DNA

As has been emphasized previously, the polypeptide chains of protein are often arranged in space in a manner that leads to the formation of *periodic* structures. For instance, in the α helix each residue is related to the next by a translation of 1.5 Å along the helix axis and a rotation of 100°. This arrangement places 3.6 amino acid residues in each complete turn of the polypeptide helix. The property of *periodicity* is also encountered with polynucleotides, which usually occur in the form of helices.

Such preponderance of helical conformations among macromolecules is not surprising. The formation of helices tends to accommodate the effects of intramolecular forces, which in a helix can be distributed at regular intervals. The precise geometry of the polynucleotide helices varies, but the helical structure invariably results from the stacking of bases along the helix axis. In many instances stacking produces helices in which the bases are more or less perpendicularly oriented along the helix and touch one another. This arrangement, which obviously leaves no free

TABLE 17.1 Specificities of Various Types of Nucleases

Enzyme	*Substrate*	*Specificity*
EXONUCLEASES		
Snake venom phosphodiesterase	DNA or RNA single-stranded only	Cleaves all type p linkages, starting with a free 3′-OH group and moving toward the 5′ terminal; releases nucleoside 5′-phosphates; has no base specificity
Bovine spleen phosphodiesterase	DNA or RNA single-stranded only	Cleaves all type d linkages, starting at the free 5′-OH and proceeding to the 3′ terminal; releases nucleoside 3′-phosphates; has no base specificity
ENDONUCLEASES		
Bovine pancreas deoxyribonuclease (DNase I)	DNA single- or double-stranded	Cleaves all type p linkages but prefers those between purine and pyrimidine bases
Calf thymus deoxyribonuclease (DNase II)	DNA single- or double-stranded	Cleaves all type d linkages randomly

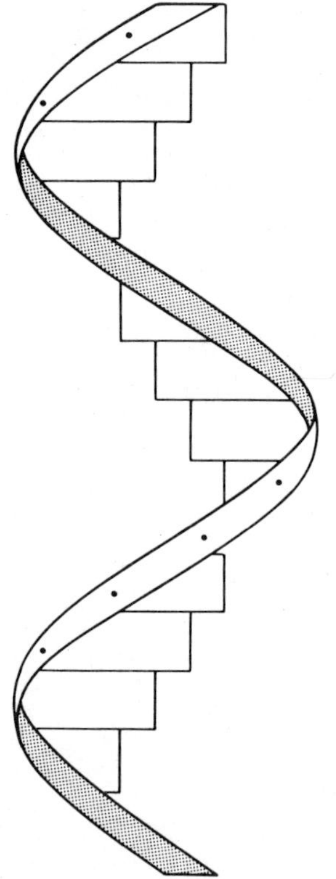

Figure 17.5
Conformation of a hypothetical, perfectly helical, single-stranded polynucleotide.
The helical band represents the phosphate backbone of the polynucleotide. The bases are shown in a side view as solid blocks in tight contact with their neighbors, above and below each base. The surfaces of the rings are in contact with each other and are not visible to the observer in the perspective from which the figure was drawn.

space between two successive neighboring bases, is illustrated in Figure 17.5. Such stacked single-stranded helices, however, are not commonly encountered in nature. Rather, as it will become apparent from the subsequent discussion, polynucleotide helices tend to associate with one another to form double helices.

Forces That Determine Polynucleotide Conformation

The hydrophobic properties of the bases are, to a large extent, responsible for forcing polynucleotides to adopt helical conformations. Examination of molecular models of the bases reveals that the edges of the rings contain polar groups (i.e., amino and hydroxyl group residues) that are able to interact with other polar groups or surrounding water molecules. The faces of the rings, however, are unable to participate in such interactions and tend to avoid any contact with water. Instead they tend to interact with one another, producing the stacked conformation. The stability of this arrangement is further reinforced by an interchange between the electrons that circulate in the π orbitals located above and below the plane of each ring.

Clearly then, single-stranded polynucleotide helices are stabilized by both *hydrophobic* as well as *stacking* interactions involving the π orbitals of the bases. The stability of the helical structures is also influenced by the potential repulsion among the charged phosphate residues of the polynucleotide backbone. These repulsive forces introduce a certain degree of rigidity to the structure of the polynucleotide. Under physiological conditions, that is, at neutral pH and relatively high concentrations of salts, the charges on the phosphate residues are partially shielded by the cations present, and the structure can be viewed as a fairly flexible coil. Under more extreme conditions the stacking of the bases is disrupted and the helix collapses. A collapsed helix is commonly described as a random coil. A conversion between a stacked helix and a random-coil conformation is depicted in Figure 17.6.

DNA Double Helix

Although certain forms of cellular DNA exist as single-stranded structures, the most widespread DNA structure is the double helix. The double helix can be visualized as resulting from the interwinding around a common axis of two right-handed helical polynucleotide strands. The two strands achieve contact through hydrogen bonds, which are formed at the hydrophilic edges of their bases. These bonds extend between purine residues in one strand and pyrimidine residues in the other, so that the two types of resulting pairs are always adenine–thymine and guanine–cytosine. A direct consequence of these hydrogen-bonding specificities is that double stranded DNA contains equal amounts of purines and pyrimidines. Examination of space-filling models clearly indicates the structural compatibility of these bases in forming linear hydrogen bonds.

This relationship between bases in the double helix is described as *complementarity*. The bases are complementary because every base of one strand is matched by a complementary hydrogen-bonding base on the other strand. For instance, for each adenine projecting toward the common axis of the double helix, a thymine must be projected from the opposite chain so as to fill exactly the space between the strands by hydrogen bonding with adenine. Neither cytosine nor guanine *fits precisely in the available space in a manner that allows the formation of hydrogen bonds across strands*. These hydrogen-bonding specificities, illustrated in Figure 17.7 ensure that the entire base sequence of one strand is complementary to that of the other strand.

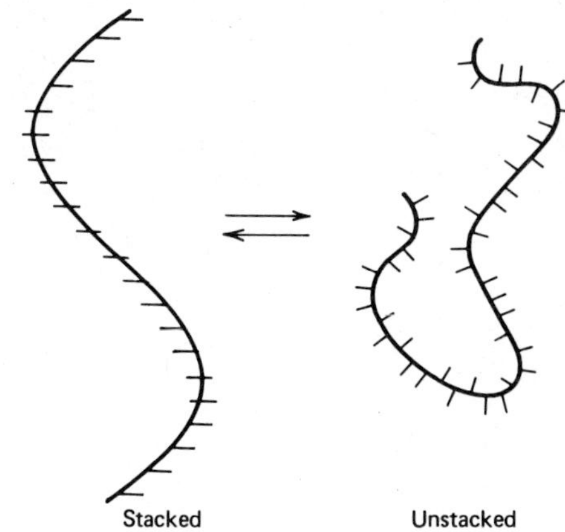

Figure 17.6
Stacked and unstacked conformations of a polynucleotide.
Stacking of the bases decreases the flexibility of a polynucleotide and tends to produce a more extended, often helical, structure.

Thymine
C-1′ of deoxyribose
Adenine
C-1′ of deoxyribose

Cytosine
C-1′ of deoxyribose
Guanine
C-1
of deoxyribose

Figure 17.7
Formation of hydrogen bonds between complementary bases in double-stranded DNA.
The interaction between polynucleotide strands is a highly selective process. The property of complementarity depends not only upon the geometric factors that allow the proper fitting between the complementary bases of the two strands, but also on the electronic specificity of interaction between complementary bases. Thus specificity of interaction between purines and pyrimidines has also been noted both in solution and in the crystal form, and it is expressed in terms of strong hydrogen bonding between monomers of adenine and uracil or monomers of guanine and cytosine. In double-stranded DNA adenine interacts instead with thymine, which is a structural analog of uracil.

The conventional double helix exists in various geometries designed as forms A, B, and C. These forms, however, share certain common characteristics. Specifically, the phosphate backbone is always located on the outside of the helix. Also, because the diesters of phosphoric acid are fully ionized at neutral pH, the exterior of the helix is negatively charged. The bases are well packed in the interior of the helix, where their faces are protected from contact with water. In this environment the strength of the hydrogen bonds that connect the bases can be maximized. The interwinding of the two polynucleotide strands produces a structure having two deep helical grooves that separate the winding phosphate backbone ridge.

However, the precise geometry of the double helix varies among the different forms. The original x-ray data obtained with highly oriented DNA fibers suggested the occurrance of a form, later designated as B, which appears to be the one commonly found in solution and in vivo (Figure 17.8). A characteristic of this form is that one of its grooves is

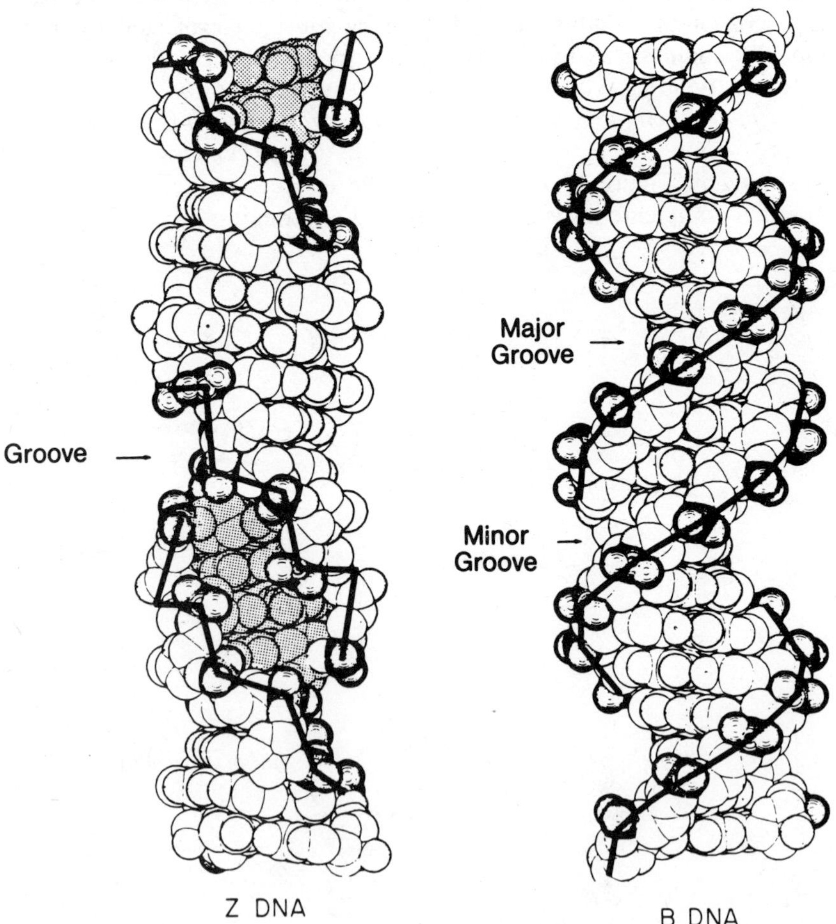

Figure 17.8
Space-filling molecular models of B and Z DNA.
Watson and Crick were the first to postulate a double-stranded model for the structure of DNA. The double helix is still referred to as the Watson and Crick model, although this structure has been substantially refined since it was proposed. B-DNA may be the most typical form of DNA occurring in the cell. Z-DNA may be present in the cell as small stretches, consisting of alternating purines and pyrimidines, incorporated between long stretches of B-DNA. The zigzag nature of the Z DNA backbone is illustrated by the heavy lines that connect phosphate residues along the chain.
Reprinted with permission from *A. Rich, J. Biomol. Struct. Dynamics* 1:1, 1983.

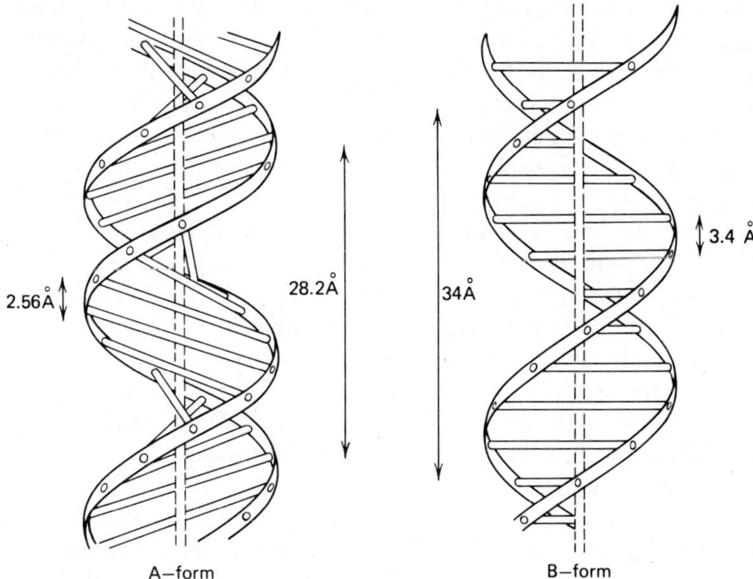

Figure 17.9
The various geometries of the DNA-double helix.
Depending on conditions, the double helix can acquire various forms of distinct geometries. In the B-form of DNA the centers of the bases are ~3.4 Å apart and produce a complete turn of a helix with a pitch of 34 Å. Such an arrangement results in a complete turn of the helix for every 10 base pairs. The diameter of the helix is 20 Å. Form C (not shown) is very similar to the B structure, with a pitch of 33 Å and 9 base pairs per turn. Form A, which is obtained from form B when the relative humidity of the fiber is reduced to 75%, differs from B in that the base pairs are not perpendicular to the helical axis but are tilted. This tilt results in a pitch of 28.2 Å and a shortening of the helix by the packing of 11 pairs per helical turn.
From W. Guschelbauer, *Nucleic Acid Structure* Berlin: Springer-Verlag, 1976.

wider than the other, and it is referred to as the major groove to distinguish it from the second, or minor, groove. The nucleotide sequence of the polynucleotides can be discerned without dissociating the double helix by looking inside these grooves. As each of the four bases has its own orientation with respect to the rest of the helix, each base always shows the same atoms through the grooves. For instance, the C-6, N-7, and C-8 of the purine rings and the C-4, C-5, and C-6 of the pyrimidine rings line up in the major groove. The minor groove is paved with the C-2 and N-3 of the purine and the C-2 of the pyrimidine rings. Forms A and C differ from B in the pitch of the base pairs relative to the helix axis as well as in other geometric parameters of the double helix, as shown in Figure 17.9 and Table 17.2.

TABLE 17.2 Nucleic Acid Helix Parameters

Family	*A*		*B*		*Z*
Environment	Crystal	Fiber	Crystal	Fiber	Crystal
Helix sense	Right	Right	Right	Right	Left
Sugar ring conformation (pucker)	C3′-*endo*	C3′-*endo*	Variable	C2′-*endo*	Alternating
Base pairs per turn	10.7	11	9.7	10	12
Rise per base pair (Å)	2.3	2.6	3.3	3.4	3.7

A new form of DNA was discovered recently, which has geometric characteristics radically different from those of the conventional forms. In this DNA, termed Z-DNA, the polynucleotide phosphodiester backbone assumes a "zig-zag" arrangement rather than the smooth conformation that characterizes other double-stranded forms. The Z-DNA structure forms a single groove as opposed to the two grooves that characterize B-DNA. Therefore, the conformation of Z-DNA may be viewed as the result of the major groove of B-DNA having "popped out" in order to form the outer convex surface of Z-DNA. This change places the stacked bases on the outer part of Z-DNA rather than in their conventional positions in the interior of the double helix. Another highly unusual property of the Z structure is that it consists of left-handed rather than right-handed helices, which characterize the conventional forms. These major structural differences between the B-DNA and the Z-DNA, which are illustrated in Figure 17.8, are partly the result of different conformations in the nucleotide residues between the two forms.

The biological function of Z-DNA is not known with certainty. Some evidence exists suggesting that Z-DNA influences gene expression and regulation. Apparently, Z-DNA is incorportated in small stretches, normally containing approximately one or two dozen nucleotide residues, in regions of the gene that regulate transcriptional activities. These stretches consist of alternating purines and pyrimidines in the sequence, which is a condition favoring the formation of the Z conformation. The Z form of DNA is stabilized by the presence of cations or polyamines and by methylation of either guanine residues in the C-8 and N-7 positions or cytosine residues in the C-5 position.

An important structural characteristic of all double-stranded DNA is that its strands are *antiparallel*. Polynucleotides are asymmetric structures with an intrinsic sense of polarity built into them. As it may be concluded from inspection of Figure 17.10, the two strands are aligned in opposite directions that is, if two adjacent bases in the same strand, for example thymine and cytosine, are connected in the $5' \rightarrow 3'$ direction, their complementary bases adenine and guanine will be linked in the $3' \rightarrow 5'$ direction (directions are defined by linking the 3′ and 5′ positions within the same nucleotide). This antiparallel alignment produces a stable association between strands to the exclusion of the alternate parallel arrangement. Just as peptide geometries and the formation of α helices determine the overall preferred conformation of proteins, the formation of hydrogen bonds between complementary bases on antiparallel polynucleotide strands leads to the formation of the double helix.

The double-stranded structure for DNA was proposed in 1953. The proposal was partly based on the results of previously available x-ray diffraction studies, which suggested that the structures of DNAs from various sources exhibited remarkable similarities. These studies also suggested that DNA had a helical structure containing two or more polynucleotides. An additional piece of evidence of central importance to the proposal was the clarification of the quantitative base composition of DNA, which was obtained independently in 1950. These results indicated the existence of molar equivalence between purines and pyrimidines, which turned out to be the essential observation suggesting the existence of complementarity between the two strands.

Stability of the DNA Structure

The same factors that stabilize single-stranded polynucleotide helices, *hydrophobic* and *stacking* forces, are also instrumental in stabilizing the double helix. The separation between the hydrophobic core of the stacked

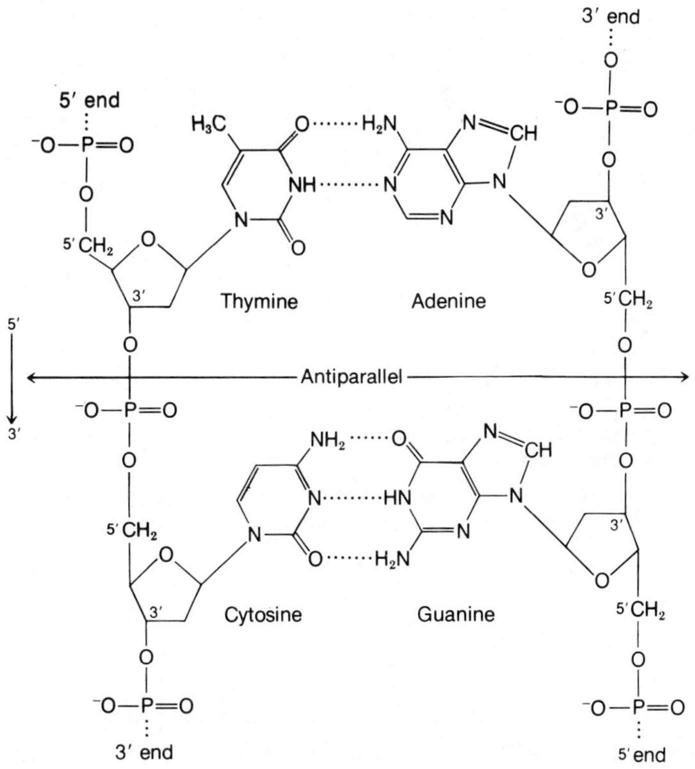

Figure 17.10
Antiparallel nature of the DNA strands.
The strands of a double-stranded DNA are arranged in such a manner that, as the complementary bases pair with one another, the two strands are aligned with opposite polarities, that is, the conventional assignment of the 5′ → 3′ direction to each of the strands suggests opposite directions. It should be noted that the geometry of the helices does not prevent a parallel alignment, but such an arrangement is not found in DNA.

bases and the hydrophilic exterior of the charged sugar-phosphate groups is even more striking in the double helix than with single-stranded helices. This arrangement, which produces substantial stabilization of the double-stranded structures over single-stranded conformations, explains the preponderance of the former. The stacking tendency of single-stranded polynucleotides may be viewed as resulting from a tendency of the bases to avoid contact with water. The double-stranded helix is by far a more favorable arrangement, as it permits the phosphate backbone to be highly solvated by water while the bases are essentially removed from the aqueous environment.

Additional stabilization of the double helix results from its extensive network of cooperative hydrogen bonding. Although this bonding per se makes only a relatively minor contribution to the free energy of stabilization of the double helix, the physiological importance of hydrogen bonds should not be underestimated. By contrast to hydrophobic forces, hydrogen bonds are highly directional and for this reason are able to provide a discriminatory function for choosing between correct and incorrect base pairs. In addition, because of their directionality, hydrogen bonds tend to orient the bases in a way that favors stacking. Therefore although hydrogen bonds make a minor contribution to the total energy of stabilization, their contribution is essential for the stability of the double helix.

In the past, the relative importance of hydrogen bonding and hydrophobic forces in stabilizing the double helix was not always appreciated.

TABLE 17.3 Effects of Various Reagents on the Stability of the Double Helix[a]

Reagent	*Adenine Solubility × 10^{-3} (in 1 M reagent)*	*Molarity Producing 50% Denaturation*
Ethylurea	22.5	0.60
Propionamide	22.5	0.62
Ethanol	17.7	1.2
Urea	17.7	1.0
Methanol	15.9	3.5
Formamide	15.4	1.9

SOURCE: Data from L. Levine, J. Gordon, and W. P. Jencks, *Biochemistry,* 2:168, 1963.

[a] The destabilizing effect of the reagents listed below on the double helix is independent of the ability of these reagents to break hydrogen bonds. Rather, the destabilizing effect is determined by the solubility of adenine. Similar results would be expected if the solubility of the other bases were examined.

However, studies on the effect of various reagents on the stability of the double helix have suggested that the destabilizing effect of a reagent is not related to the ability of the reagent to break hydrogen bonds. Rather, the stability of the double helix is determined by the solubility of the free bases in the reagent, the stability decreasing as the solubility increases. Some of these findings, summarized in Table 17.3, emphasize the importance of hydrophobic forces in maintaining the structure of double-stranded DNA.

Ionic forces also have an effect on the stability and the conformation of the double helix. At physiological pH the electrostatic *intrastrand* repulsion between negatively charged phosphates forces the double helix into a relatively rigid rodlike conformation. In addition the repulsion between phosphate groups located on opposite strands tends to separate the complementary strands. In distilled water, DNA strands will separate at room temperature; near the physiological salt concentration, cations (in addition to other charged groups, for example, the basic side chains of proteins) shield the phosphate groups and decrease repulsive forces. Therefore, the flexibility of the double helix is partially restored and its stability is enhanced.

Denaturation

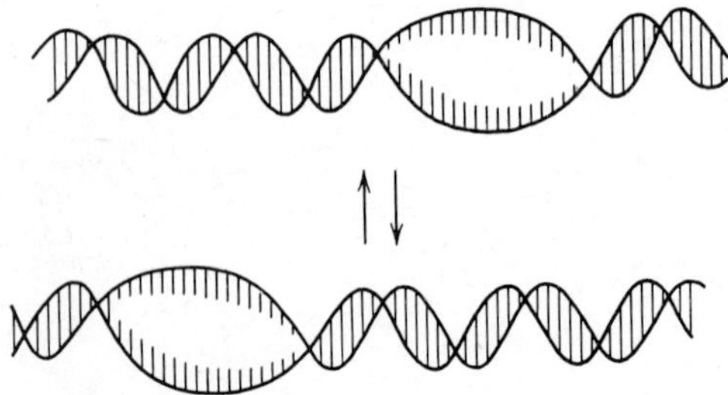

Figure 17.11
"Zipper" model for the DNA double helix. *DNA contains short sections of open-strandedness that can "move" up and down the helix.*

The double helix is stabilized by ~1 kcal per base pair. Therefore a relatively minor perturbation can produce disruption in double strandedness, provided that only a short section of the DNA is involved. As soon as the relatively few base pairs have separated, they close up again and release free energy, and then the adjacent base pairs unwind. In this manner minor disruptions of double strandedness can be propagated along the length of the double helix. Therefore, at any particular moment the large majority of the bases of the double helix remain hydrogen bonded, but all bases can pass through the single-stranded state, a few at a time. This dynamic state of the double helix is characterized by the movement of an "open-stranded" portion up and down the length of the helix, as indicated in Figure 17.11. The "dynamic" nature of this structure is an essential prerequisite for the biological function of DNA and especially the process of DNA synthesis.

Furthermore, the strands of DNA can be completely separated by increasing the temperature in solution. At relatively low temperatures a few base pairs will be disrupted, creating one or more "open-stranded bubbles." These "bubbles" form initially in sections that contain relatively higher proportions of adenine and thymine pairs. Adenine-thymine pairs are bound by two hydrogen bonds and are therefore less stable than guanine-cytosine pairs, which contain three such bonds per pair. As the temperature is raised, the size of the "bubbles" increases and eventually the thermal motion of the polynucleotides overcomes the forces that stabilize the double helix. This transformation is depicted in Figure 17.12. At even higher temperatures the strands can separate physically and acquire a random-coil conformation, as shown in Figure 17.13, referred to as *denaturation*. The process of denaturation is accompanied by a number of physical changes, including a buoyant density increase, a reduction in viscosity, a change in the ability to rotate polarized light and changes in absorbancy.

Changes in absorbancy are frequently used for following experimentally the process of denaturation. DNA absorbs in the uv region due to the heterocyclic aromatic nature of its purine and pyrimidine constitutents. Although each base has a unique absorption spectrum, all bases exhibit maxima at or near 260 nm. This property is responsible for the absorption of DNA at 260 nm. However, this absorbancy is almost 40% lower than that expected from adding up the absorbancy of each of the base components of DNA. This property of DNA, referred to as *hypochromic effect*, results from the close stacking of the bases along the DNA helices. In this special arrangement interactions between the electrons of neighboring bases produce a decrease in absorbancy. However, as the ordered structure of the double helix is disrupted at increasing temperatures, stacking interactions are gradually decreased. Therefore, a totally disordered polynucleotide, a random coil, eventually approaches an absorbance not very different from the sum of the absorbancy of its purine and pyrimidine constituents.

Slow heating of double-stranded DNA in solution is accompanied by a gradual change in absorbancy as the strands separate. However, since the interactions between the two strands are cooperative, the transition from double-stranded to random-coil conformation occurs over a narrow range of temperatures, as indicated in Figure 17.14. Before the rise of the melting curve, DNA is double stranded. In the rising section of the curve an increasing number of base pairs is interrupted as the temperature rises. Strand separation occurs at a critical temperature corresponding to the upper plateau of the curve. However, if the temperature is decreased before the complete separation of the strands, the native structure is completely restored.

The midpoint temperature, T_m, of this process, under standard conditions of concentration and ionic strength, is characteristic of the base content of each DNA. The higher the guanine-cytosine content, the higher the transition temperature between the double stranded helix and the single strands. This difference in T_ms is attributed to the increased stability of guanine-cytosine pairs, as a result of the three hydrogen bonds that connect them in DNA, in contrast to only two hydrogen bonds that connect adenine and thymine pairs.

Rapid cooling of a heated DNA solution normally produces denatured DNA, a structure that results from the reformation of some hydrogen bonds either between the separate strands or between different sections of the same strand. The latter must contain complementary base sequences. By and large denatured DNA is a disordered structure containing substantial amounts of random-coil and single-stranded regions.

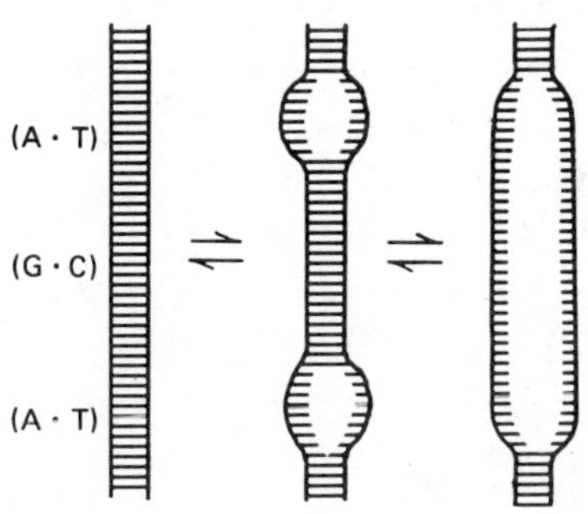

Figure 17.12
Structure of double-stranded DNA at increasing temperatures.
Disruptions of the double-stranded structure appear first in regions of relatively high adenine-thymine content. The size of these "bubbles" increases with increasing temperatures, leading to extensive disruptions in the structure of the double helix at elevated temperatures.

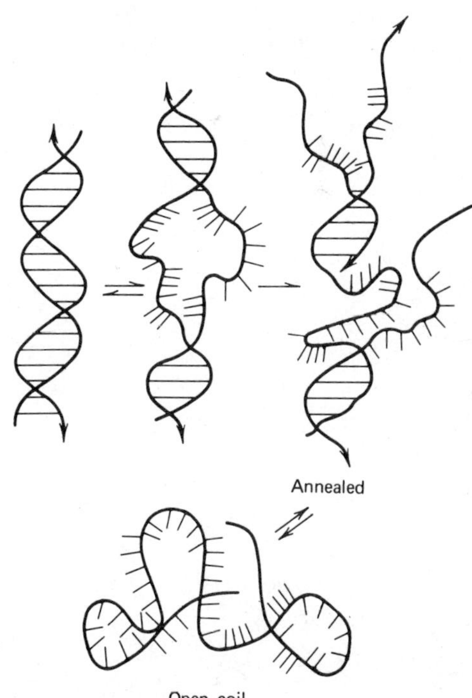

Figure 17.13
Denaturation of DNA.
At high temperatures the double-stranded structure of DNA is completely disrupted, with the eventual separation of the strands and the formation of single-stranded open coils. Denaturation also occurs at extreme pH ranges or at extreme ionic strengths.

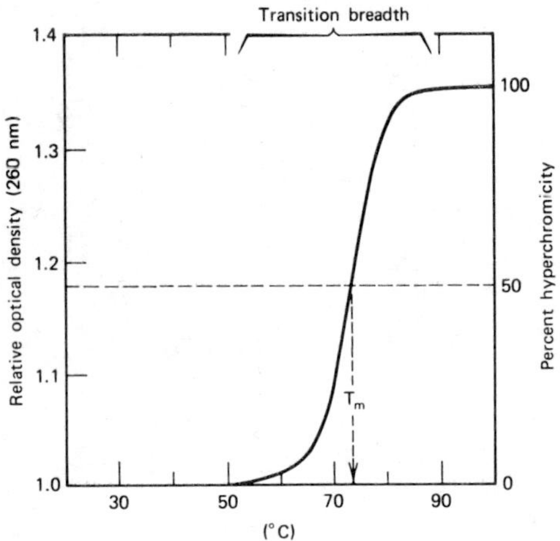

Figure 17.14
Temperature–optical density profile for DNA.
When DNA is heated, the optical density increases with rising temperature. A graph in which optical density versus temperature is plotted is called a "melting curve." Relative optical density is the ratio of the optical density at the temperature indicated to that at 25°C. The temperature at which one-half of the maximum optical density is reached is the midpoint temperature (T_m).

Redrawn from D. Freifelder, *The DNA molecule: structure and properties*. San Francisco: W. H. Freeman, 1978.

DNA can also be denatured at a pH above 11.3 as the charge on several substituents on the rings of the bases is changed preventing these groups from participating in hydrogen bonding. Alkaline denaturation is often used as an experimental tool in preference to heat denaturation to prevent breakage of phosphodiester bonds that can occur to some degree at high temperatures. Denaturation can also be induced at low ionic strengths, because of enhanced interstrand repulsion between negatively charged phosphates, as well as by various denaturing reagents, that is, compounds that weaken or break hydrogen bonds. A complete denaturation curve similar to that shown in Figure 17.4 can be obtained at a relatively low constant temperature, for instance room temperature, by variation of the concentration of an added denaturant.

Renaturation

Complementary DNA strands, separated by denaturation, can reform a double helix if appropriately treated by a process referred to as *renaturation* or *reannealing*. Renaturation depends upon the meeting of complementary DNA strands in an exact manner that can lead to the reformation of the original structure, and it is therefore a slow, concentration-dependent process. As a rule, maintaining DNA at temperatures 10–15°C below its T_m under conditions of moderate ionic strength (about 0.15 M), provides the maximum opportunity for renaturation. At lower salt concentrations, the charged phosphate groups repel one another and prevent the strands from associating. As renaturation begins, some of the hydrogen bonds formed are extended between short tracts of polynucleotides that might have been distant in the original native structure. Short sequences, consisting for example of four to six base pairs, are reiterated many times within every DNA strand. Furthermore eucaryotic DNA contains a large number of much longer nucleotide sequences reiterated many times within each genome. Such sequences provide sites for initial base pairing, which produces a partially hydrogen-bonded double helix. These randomly base-paired structures are short-lived because the bases that surround the short complementary segments cannot pair and lead to the formation of a stable fully hydrogen-bonded structure. However, once the correct bases begin to pair by chance, the double helix over the entire DNA molecule is rapidly reformed. Clearly then renaturation is a two-step process. The first step, which determines the rate of association, involves the chance meeting of two complementary sequences on different strands and it is, therefore, a second-order reaction. The rate of renaturation is thus proportional to the product of the concentrations of the two homologous dissociated strands and is expressed as $dt/dc = -kc^2$, where k is the rate constant for the association. Integration of this equation gives $C/C_o = 1/(1 + kCot)$, where C is the concentration of single-stranded DNA expressed as moles of nucleotides per liter at time t, and Co is the concentration of DNA at time zero. A plot of C/Co (which is proportional to DNA that is single stranded or of the DNA fraction that is reassociated) vs Cot can be constructed, Figure 17.15, and a Cot½ (Cot-a-half) value, which corresponds to $C/Co = 0.5$ can be determined. The Cot-a-half value is proportional to the complexity of the genome, which is equal to the molecular weight of the genome provided that the genome consists of unique nucleotide sequences. Also, as it will be apparent from subsequent discussion on eucaryotic DNA classes, that Cot curves are also useful for measuring the number of repetitive DNA classes, and the proportion of the total genome represented by those classes.

Hybridization

The self-association of complementary polynucleotide strands has also provided the basis for the development of the technique of *hybridization*.

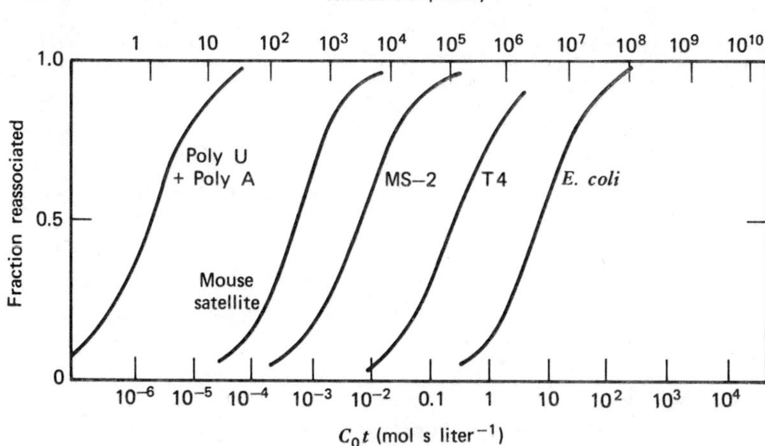

Figure 17.15
Reassociation kinetics for DNA isolated from various sources.
Each DNA is first fragmented to segments of approximately 400 nucleotides. The denatured segments are subsequently then allowed to renature. The fraction of each polynucleotide reassociated, calculated from changes in hypochromicity, is plotted against the total concentration of nucleotides multiplied by the renaturation time (C_0t). The top scale shows the kinetic complexity of each DNA sample. Whenever a DNA contains reiterated sequences, these sequences are present in the fragments at higher concentrations than they would have been if a unique sequence had been fragmented. As a result, renaturation of fragments, obtained from DNAs containing reiterated sequences, proceeds more rapidly the higher the degree of repetition. This is exemplified by the rates of renaturation of fragments obtained from the synthetic double-stranded polynucleotide poly(A)–poly(U) and mouse satellite DNA, a DNA that contains many repeated sequences. For a homogeneous DNA, which contains a distribution of different extents of reiterated sequences, kinetic complexity can be defined as the minimum length of DNA needed to contain a whole single copy of the reiterated sequence.
After R. J. Britten and D. E. Kohne, *Science*, 161:529, 1968.

This technique depends on the association between any two polynucleotide chains, which may be of the same or of different length, provided that a relationship of base complementarity exists between these chains. Hybridization can take place not only between DNA chains but also between appropriately related RNA chains as well as DNA–RNA combinations.

Appropriate techniques have been developed for measuring the *maximum amount of polynucleotide* that can be hybridized as well as the *rates of hybridization*. These techniques are among the most indispensable basic tools of contemporary molecular biology and are also specifically used for the following: (1) determining whether or not a certain sequence occurs more than once in the DNA of a particular organism; (2) demonstrating a *genetic* or *evolutionary relatedness* between different organisms; and (3) determining the number of genes transcribed in a particular mRNA. Clearly DNA : RNA hybridizations are needed for accomplishing the last goal.

As the first step, DNA to be tested for hybridization is denatured. The resulting single strands are immobilized by binding to a suitable polymer, which is then used to pack a chromatography column. DNA formed in the presence of labeled precursors, usually tritiated thymidine, is allowed to run through the column that contains the bound, unlabeled DNA. The rate at which radioactivity is retained by the column obviously equals the rate of annealing between complementary strands.

As discussed in the preceding section, measurements of such rates have established that the DNA of eucaryotic cells contain a given nucle-

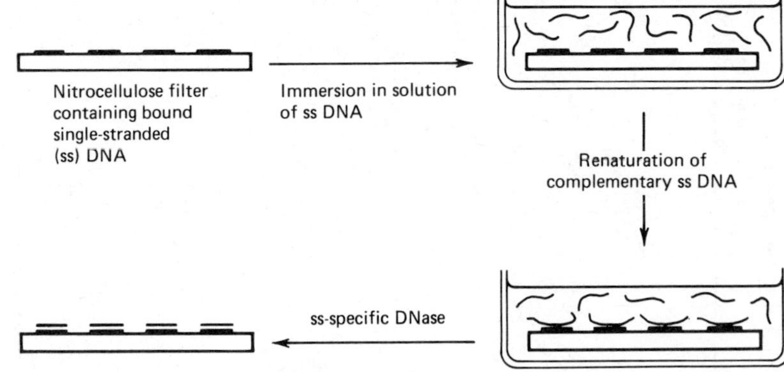

Figure 17.16
Hybridization of DNA bound to nitrocellulose filters.
These filters bind single-stranded DNA along the sugar–phosphate backbone, leaving the hydrogen-bonding bases accessible for hybridization. In contrast, double-stranded DNA is not retained by these filters. After the single-stranded DNA is fixed on the filter, the filter is treated with a reagent that prevents additional binding of single-strands. A radioactively labeled single-stranded DNA or RNA of known origin or sequence is then used as a hybridization probe. *In the final step, the filter is treated with a single-strand specific DNase that degrades single-stranded DNA leaving the double-stranded DNA intact. If the probe is retained by the filter, the complementary nucleotide sequence is identified as having been present among the sequences fixed on the filter.*

Reprinted with permission from D. Freifelder, *Molecular Biology, A Comprehensive Introduction,* Jones and Bartlett, Inc., Portola Valley, Ca., 1983.

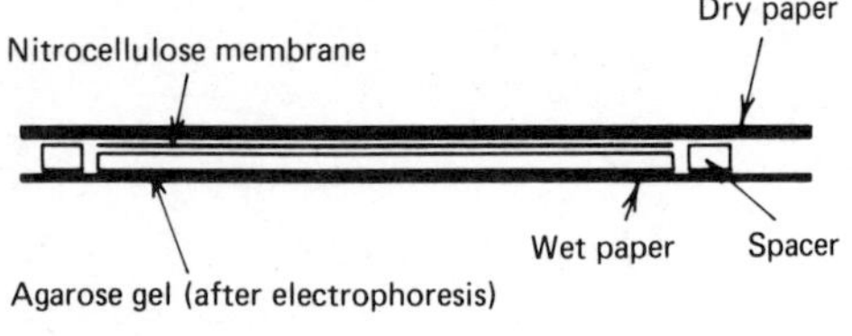

Figure 17.17
Southern hybridization.
The Southern transfer (or Southern blotting) is a technique combining gel electrophoresis, for the separation of fragments produced by partial digestion of DNA, with direct transfer of these fragments to a nitrocellulose filter where they can be hybridized. The principle of this widely used technique is exceedingly simple. DNA fragments are carried from the agarose plate into a nitrocellulose filter by the capillary action-induced flow of a buffer solution from wet filter paper to dry filter paper. More recent techniques permit the hybridization of DNA bands in situ, that is, on the agarose gel used for separation without prior transfer to nitrocellulose. In the Southern method, radioactive RNA of known sequence is used for DNA hybridization. The reverse technique, in which electrophoresis is performed with RNA and DNA is used as the radioactive probe, is referred to as Northern blotting.

Reprinted with permission from E. M. Bradbury and W. McLean, *DNA, Chromatin and Chromosomes,* Blackwell Sci. Pub. Ltd. Oxford, 1981.

otide sequence reiterated a number of times. The principle of this determination is simple. For a DNA of a given size the rate of annealing depends on the frequency with which two complementary segments can collide with each other. Therefore, the larger the number of reiterated sequences in a given DNA, the greater is the chance that a particular collision will result in the formation of annealed polynucleotides. On this basis the extent to which annealing takes place within a unit of time can be used to determine the number of reiterated sequences in the DNA.

Determinations of the maximum amount of DNA that can be hybridized have been used to establish homologies between the DNA of different species. This is possible because the base sequences of the DNA in each organism are unique for this organism. Therefore the annealed helices represent the same unique sequences of DNA even if the individual annealed strands originate from different cells. On this basis annealing can be used to compare the degree to which DNAs isolated from different species are related to one another. Consequently, the observed homologies serve as indices of *evolutionary relatedness* and have been particularly useful for defining *phylogenies* in procaryotes. "Hybridization" studies between DNA and RNA have, in addition, provided very useful information about the biological role of DNA, particularly the mechanism of transcription.

In recent years hybridization techniques using membrane filters, usually made of nitrocellulose, have found increasing application (Figure 17.16). In general, hybridization can be quantitated by either measuring *the amount of hybrid in equilibrium or the rate of hybrid formation* under conditions in which one nucleic acid is present in large excess. The approach used for the latter determination is analogous to the Cot procedure and when it is used for DNA-RNA hybridization and RNA is present in excess it is referred to as the Rot method, or the Dot method when DNA is in excess.

A variant of filter hybridization, known as the *Southern transfer,* can be used for identifying the location of specific genes (Figure 17.17). Since a gene sequence represents a very small percentage of total DNA, the gene must be separated from the remaining DNA and amplified before hybridization.

Finally, the principle of *hybridization* has also served as the basis for the development of a technique that has permitted the construction of precise physical maps of DNA genes. This technique depends on the direct visualization under the electron microscope of single-stranded loops in the structures of artificially formed double-stranded DNA molecules known as *heteroduplexes*. The principle of this technique is simple. Heteroduplexes are constructed by hybridization of two complementary DNA strands. One of these strands, however, is selected on the basis that, as the result of a known mutation, it misses the gene being mapped. As is apparent from Figure 17.18, the complementary strands of the heteroduplex pair perfectly throughout the length of the molecule, with one important exception. Across from the position of the missing gene in the mutant strand the complementary strand forms a clearly visible loop. The position of the loop identifies the location of the deleted gene.

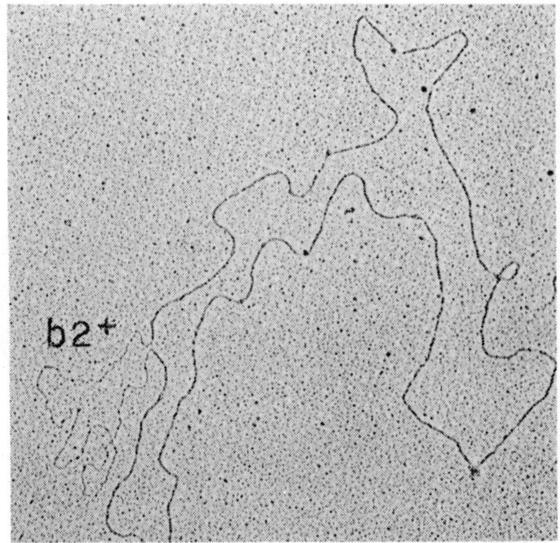

Figure 17.18
Heteroduplex formation in bacteriophage λ. *Electron micrograph of a heteroduplex DNA molecule constructed from complementary strands of bacteriophage λ and a bacteriophage λ deletion mutant (bacteriophage λβ2). In λβ2 a segment of DNA has been deleted, producing, at the site of deletion, a loop labeled* $b2^+$.

Reprinted with permission from B. C. Westmoreland et al., *Science,* 163:1343, 1969. Copyrighted 1969 by the American Association for the Advancement of Science.

17.3 TYPES OF DNA STRUCTURE

The subject of DNA structure has been treated so far as though DNA were a "generic" substance, that is, only the essential features common to all DNAs have been presented. In fact, the specific structural features of DNA vary, depending on the origin and the function of each DNA molecule. DNAs differ in size, conformation, and topology.

Size of DNA

The size of DNA varies from a few thousand base pairs for the DNA of the small viruses, to millions for the chromosomal DNA of bacteria, and to billions for the chromosomal DNA of animals. Several types of expressions are commonly used to describe DNA size, including number of base pairs, molecular weights, the length of the strands, and even the actual weight of DNA. The units used in these expressions, however, can be easily interconverted, taking into account that a 1-million-mol-wt DNA contains approximately 1,500 base pairs which comprise a macromolecular segment of 0.5 nm length. Also, since DNA is a macromolecule, DNA weight can be converted to molecular weight by division with the average molecular weight of a DNA nucleotide pair.

As is apparent from Table 17.4, the amount of DNA per cell increases as the complexity of the cellular function increases. It should be noted

TABLE 17.4 The DNA Cell Content of Some Species

Type of Cell	*Organism*	*DNA/Cell (pg)*[a]
Phage	T4	2.4×10^{-4}
Bacterium	*E. coli*	4.4×10^{-3}
Fungi	*N. crassa*	1.7×10^{-2}
Avian erythrocyte	Chicken	2.5
Mammalian leukocyte	Human	3.4

SOURCE: B. Lewin, *Gene expression,* vol. 2, 2nd ed. New York: Wiley, 1980, p. 958.

[a] pg = picograms.

that although mammalian cells contain some of the highest amounts of DNA per cell, some amphibian, fish, and plant cells may contain even higher amounts. In fact, lung fish cells contain more than 40 times the amount of DNA in human cells, but such extraordinary amounts of DNA reflect a reiteration of nucleotide sequneces within the DNA macromolecule and do not represent an actual increase in the size of DNA in terms of unique sequences. But aside from these minor irregularities, the size of the DNA of higher cells is very large indeed. The DNA contained within a single human cell, if it were stretched end to end, would be about one meter long. This suggests that the polynucleotides are exquisitely packed in order to fit within the minute dimensions of the cell nucleus.

Because of their extraordinary length, relative to the total mass, DNA molecules are extremely sensitive to shearing forces that develop during ordinary laboratory manipulations. Even careful pipetting may shear a DNA molecule. In addition, during the process of isolation it is difficult to prevent with absolute confidence the disruption of some phosphodiester bonds by contaminating endonucleases (nicking). For these reasons the precise size of DNA, especially that of the higher species, could not be determined until special handling techniques were developed, both for the isolation of DNA and the measurement of its molecular weight.

Techniques for Determining DNA Size

In any event, devising suitable methods for the measurement of the molecular size of DNA has been a scientific challenge. The classical methods for determining size in proteins, such as light scattering, sedimentation diffusion, sedimentation equilibrium, or osmometry proved to be unsuitable for measuring the molecular weight of even relatively small DNAs. For instance, because of the great mass of DNA the sedimentation coefficient of the macromolecule is so high that centrifuges could not be run slowly enough to yield useful data with existing methodology. Instead custom-tailored methods had to be devised. Equilibrium centrifugation in a density gradient (usually a concentrated cesium chloride solution), electron microscopy, and electrophoresis in agarose gels are among the principle methods providing reliable information about the molecular weights of various DNAs. Electron microscopy provides a measure of the length of DNA strands. Molecular weights can be calculated from known values of the mass per unit length. DNA can be visualized under the electron microscope if it is first coated with protein and a metal film. Determination of molecular weights by electrophoresis depends on the molecular-sieving effect of porous agarose gels. Over a limited range of molecular weights the mobility of DNA is directly proportional to the logarithm of the molecule's weight. The range of the method is further extended by appropriate adjustments in the density of the agarose gels, which leads to changes in mobility.

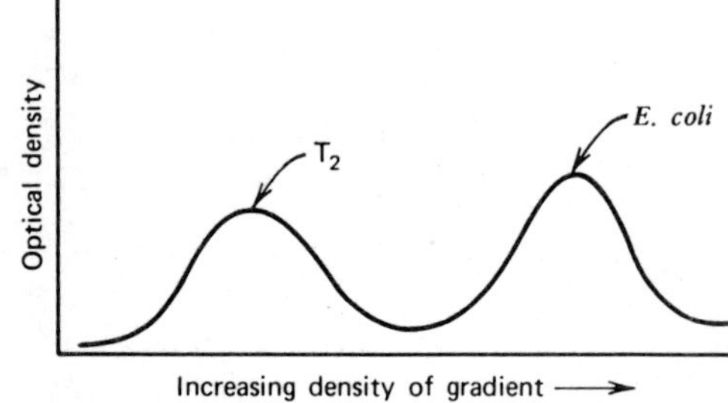

Figure 17.19
Equilibrium gradient centrifugation of DNA. *The DNA macromolecules travel into the increasingly dense regions of the gradient driven by centrifugal forces. The macromolecules equilibrate as soon as they reach an area of the gradient of density equal to their own. For example, bacteriophage T2 DNA and* E. coli *DNA can be resolved into two distinct bands. The width of the bands at equilibrium is related to the molecular weight of DNA.*

In order to determine the molecular weight of DNA by equilibrium centrifugation a small portion of a DNA solution to be analyzed is layered on top of a gradient in a centrifuge tube. Upon centrifugation, the molecules of DNA sediment to equilibrium through the gradient. Under these conditions a homogeneous high molecular weight DNA will form a Gaussian band centered at a position in the gradient that corresponds to the density of the macromolecule. Molecules with different densities are resolved into a series of bands that sediment independently of one another, as shown in Figure 17.19. A relationship can be demonstrated between the width of the bands at equilibrium and the molecular weights, permitting the determination of accurate molecular weights.

$$\begin{matrix} 3'\ \mathrm{HO}\text{———}\mathrm{P}\ 5' \\ 5'\ \mathrm{P}\text{———}\mathrm{OH}\ 3' \end{matrix} \xrightarrow[\text{phosphatase}]{\text{alkaline}} \begin{matrix} 3'\ \mathrm{HO}\text{———}\mathrm{OH}\ 5' \\ 5'\ \mathrm{HO}\text{———}\mathrm{OH}\ 3' \end{matrix} \xrightarrow[\text{kinase}\ [\gamma\text{-}^{32}\mathrm{P}]\mathrm{ATP}]{\text{polynucleotide}} \begin{matrix} 3'\ \mathrm{HO}\text{———}{}^{32}\mathrm{P}\ 5' \\ 5'\ {}^{32}\mathrm{P}\text{———}\mathrm{OH}\ 3' \end{matrix}$$

Figure 17.20
End-group labeling procedure.
The 5′ terminals on the opposite ends of DNA are labeled with ^{32}P by treatment with alkaline phosphatase and esterification of the resulting 5′-hydroxyl groups with [γ-^{32}P]ATP.

A biochemical method based on the labeling of the terminals of a macromolecule has been used successfully for determining molecular weights in proteins. In this case the DNA is treated with the enzyme alkaline phosphatase, which converts the 5′-phosphate nucleotide terminals of double-stranded DNA to the corresponding hydroxyl groups. These terminals are then esterified, using [γ-^{32}P]ATP with the enzyme polynucleotide kinase. The free 5′ terminus of each polynucleotide chain becomes labeled as shown in Figure 17.20. The labeled DNA is then analyzed by zonal centrifugation and detected from both its absorbancy at 260 nm and ^{32}P, counting as indicated in Figure 17.21. The molecular weight is calculated from the ratio of the amount of ^{32}P to the absorbancy, both measured at the coinciding peaks of the bands.

The above methods have permitted the determination of DNA molecular weights with an accuracy of at least 10%, but the usefulness of each method is limited within certain molecular weight ranges. Electrophoresis is most suitable for molecular weights in the range between 1.5×10^5 and 1.5×10^7. This range can be extended upward to 2×10^8 by electron microscopy. The most versatile method, however, is equilibrium centrifugation, the range of which extends approximately between 2×10^5 and 10^9. The high range of the method is limited because of the effect of shear forces on larger molecules. Therefore, because even bacterial DNAs often have molecular weights in excess of 10^9, it is apparent that none of the above methods can be used for very large DNAs. For DNA molecules of mol wt 10^{10} a specifically designed low shear viscometric method, described in Figure 17.22 has been developed. This method, known as viscoelastic retardation, is based on mildly stretching long DNA molecules by hydrodynamic shear forces. Once these forces are removed, the DNA molecules can relax back to their normal unstressed conformation. The relaxation time is related to, and can be used to determine with accuracy, the molecular weight of DNA molecules of the size found in eucaryotic chromosomes.

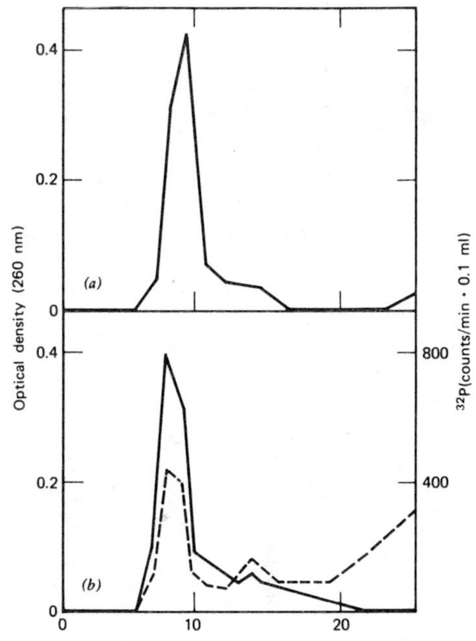

Figure 17.21
Zonal centrifugation profiles of denatured T7 DNA treated by the end-group labeling procedure.
Sedimentation is from right to left. (a) Untreated DNA. (b) DNA treated by the end-group labeling procedure. Zonal centrifugation is performed on a sucrose density gradient and should be distinguished from density gradient centrifugation. The latter is an equilibrium centrifugation with the macromolecules reaching equilibrium at regions within the tube at which their density equals the density of the environment. With zonal centrifugation the macromolecules move continuously until they reach the bottom of the tube or until the centrifuge is stopped. The molecular weight is calculated from the ratio of the amount of ^{32}P (dotted line) to the optical density (solid line) at the peak of the curve.

Redrawn from C. C. Richardson, *J. Mol. Biol.*, 15:49, 1966.

Linear and Circular DNA

The DNA of several small viruses occurs in the form of typical linear double-stranded helices of equal size. In addition, certain DNAs have naturally occurring interior single-stranded breaks. The breaks found in natural bacteriophage molecules result mostly from broken phosphodiester bonds, although occasionally a deoxyribonucleoside may be missing. The DNA of coliphage T5 consists of one intact strand and a complementary strand, which is really four different well-defined complementary fragments ordered perfectly along the intact strand. A similar regularity in the points of strand breaks is noted with a few other DNAs, for example, *Pseudomonas aeruginosa* phage B3, but generally interior breaks seem to be randomly distributed along the strands. The overall structure of the

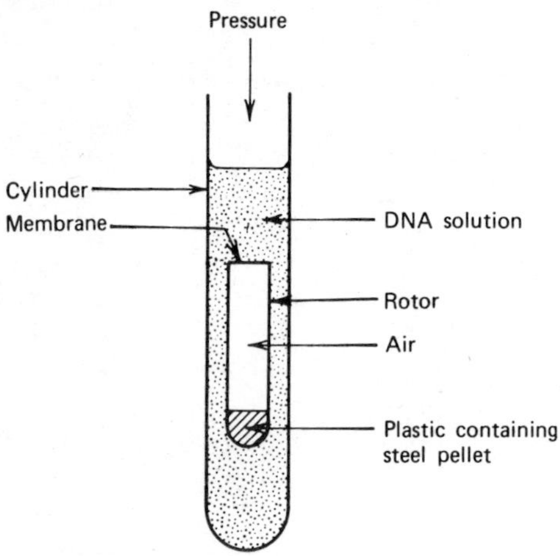

Figure 17.22
Viscoelastic retardation technique for the measurement of the molecular weight of large DNA molecules.
An appropriately constructed cylinder holds the solution to be measured. A free-floating rotor, which can maintain a suspended position by adjusting the pressure exerted on the surface of the solution, is inserted into the tube. The rotor is magnetically rotated, but once the magnetic field is removed the tube does not come to a complete stop. Instead, after slowing down it reverses direction before coming to a final stop. This reversal can be used to measure the relaxation time of DNA. In practice, cells are lysed in the measuring tube so as to avoid breakage of DNA caused by a transfer of DNA solution. Incubation with detergent at 65°C inactivates nucleases and insures the intactness of the resulting DNA strands. Proteins bound to DNA are removed by addition of the proteolytic enzyme pronase. The technique measures the size of the largest DNA molecule in the mixture rather than the average molecular weight of the molecules present.
Redrawn from D. Freifelder, *The DNA molecule: structure and properties*. San Francisco: W. H. Freeman, 1978. Copyrighted, 1978.

double helix is maintained because the breaks that occur in one strand are generally in different locations from breaks in the complementary strand.

Double-Stranded Circles

Most naturally occurring DNA molecules exist in circular form. In some instances circular DNA exists even as interlocked circles. Provided that suitable precautions are taken to avoid shearing the DNA, the circular form can be isolated intact and observed by electron microscopy. The circular structure results from the circularization of a linear DNA by formation of a phosphodiester bond between the 3′ and 5′ terminals of a linear polynucleotide.

The circular nature of DNA of the small phage ϕX174 was first suspected from studies that showed no polynucleotide ends were available for reactions with exonucleases. Sedimentation studies also revealed that endonuclease cleavage yielded one rather than two polynucleotides. These suspicions were later confirmed by observation with electron microscopy.

During the early 1960s, after workable methods for avoiding the shear of large molecules were developed, the circular nature of the DNA chromosome of *E. coli* was demonstrated by the use of autoradiography techniques. Soon it became apparent that many other DNAs (e.g., those of mitochondria, chloroplasts, bacterial plasmids, and mammalian viruses) also existed as closed circles. Obviously the strands of a circular DNA cannot be irreversibly separated because they exist as intertwined closed circles. The absence of 3′ or 5′ termini apparently provides an evolutionary advantage because it endows the circular DNA with complete resistance toward exonucleases, which act by hydrolyzing the phosphodiester bond of terminal nucleotides only. Thus circularity may be a protective mechanism against cellular exonucleases, which insures the longevity of DNA.

The DNA of some bacteriophages exists in a linear double-stranded form, which has the tendency to circularize when it enters the host cell. The linear DNA form of bacteriophage λ of *E. coli,* for instance, has single-stranded 5′ terminals of 20 nucleotides each. These terminals have complementary sequences, so that an *open circle* structure can be formed when the linear λ molecule acquires a circular shape, which allows the overlap of these complementary sequences. Subsequently, the enzyme *DNA ligase,* which forms phosphodiester bonds between properly aligned polynucleotides, joins the 3′- and 5′-terminal residues of each strand and transforms the DNA into a covalently *closed circle,* as illustrated in Figure 17.23.

Single-Stranded DNA

With the exception of a few small bacteriophages (e.g., ϕX174, G4) that can acquire a single-stranded form, most circular as well as linear DNAs exist as double-stranded helices. The single-stranded nature of the nonreplicative form of ϕX174 DNA was first suspected in the 1950s when it was discovered that the base composition of this DNA did not conform to the base *equivalence* rules, that is, for this DNA A $\neq$ T and G $\neq$ C. The single-stranded nature of this structure was also confirmed by the observation that the amino groups of the bases reacted rapidly with formaldehyde, which indicated that the bases were exposed. Furthermore, electron micrographs of ϕX174 indicated that single-stranded DNA appears more "kinky" and less thick than the double-stranded form. It may be noted that the discovery of the single-stranded circular form of ϕX174 actually preceded the identification of the replicative double-stranded form, which has a normal complementary base composition.

DNA Topology—Superhelices

The double-stranded circular DNAs, with few apparent exceptions, possess an intriguing topological characteristic. The circular structure contains twists, which are referred to as *supercoils* and can actually be visualized by electron microscopy. In order to understand the origin of these twists, it may be helpful to consider two possible approaches by which, in principle, linear DNA can be converted to a circular molecule. Circular DNA may be formed by bringing together, and joining by a phosphodiester bond, the free terminals of linear DNA. If no other manipulations are used, the resulting circular DNA will be *relaxed;* that is, the circular molecule will have the thermodynamically favored structure of the linear double helix (B-DNA), which accommodates one complete turn of the helix for a unit of length of polynucleotide consisting of approximately 10 base pairs. However, if before sealing the circle, one DNA terminus is held steady while the other terminus is rotated one or more full turns in a direction that unwinds the double helix, the resulting structure will be strained. This strained structure, which is characterized by a deficiency of turns, is known as negative *superhelical* DNA. The strain produced by this deficit of turns can be accommodated by the disruption of hydrogen bonds and the opening of the double helix over a small region of the macromolecular structure. The resulting structure may be viewed as consisting of a small-stranded loop along with regions of regularly spaced relaxed double-helical turns. If, however, hydrogen bonds are not disrupted, the circular DNA will twist in a direction opposite to the one in which it was rotated in order to relieve the strain induced by the unwinding. Thus the rotational strain that was introduced before the circularization of DNA can also be accommodated by the formation of tertiary structures with visible *supercoils* (Figure 17.24). These two representations of the negative superhelix should not be viewed as two distinct types of superhelices, but rather as two manifestations of the same underlying phenomenon. In general a dynamically imposed compromise, determined by the environment and the status of circular DNA, is reached between hydrogen bond disruption and supertwisting. In practice this means that supercoiled DNA consists of *twisted structures* with *enhanced tendency* to contain regions with disrupted hydrogen bonding (bubbles).

In a circular DNA that is initially relaxed, the transient strand unwinding would tend to introduce compensating supertwists. However, if DNA is superhelical to begin with, the density of the superhelix will obviously tend to fluctuate with the "breathing" of the helix. All naturally occurring DNA molecules contain a deficit of helical turns; that is, they exist as *negative superhelices* with a superhelical density that remains remarkably constant among different DNAs. Normally one negative twist is found for every 20 turns of the helix.

If before converting a linear DNA to the corresponding circular structure one of the terminals of the linear polynucleotides is rotated in the direction of *overwinding* rather than *unwinding* the double helix, the resulting DNA will contain *positive* superhelices. *Positive* supercoils can be experimentally produced by specialized enzymes, the topoisomerases, and may be present in vivo transiently.

The notion of superhelicity is often difficult to grasp fully without examining an appropriate physical model. In the absence of a more suitable alternative, you might attempt to twist two pieces of *fully extended* thick rope past the point at which considerable resistance develops. At that point the rope would represent a positive supercoil, which, in order to be accommodated without undue strain, must be allowed to escape the fully extended conformation and acquire the form of a compact coil. This

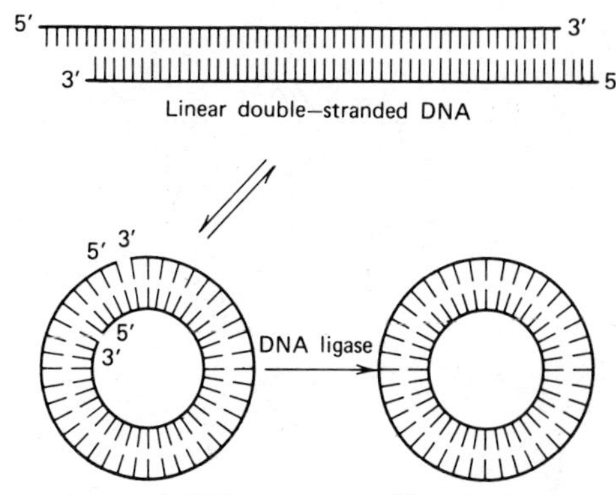

Figure 17.23
Circularization of λ DNA.
The DNA of bacteriophage λ exists in both a linear and a circular form, which are interconvertible. The circularization of λ DNA is possible because of the complementary nature of the single-stranded 5′ terminals of the linear form.

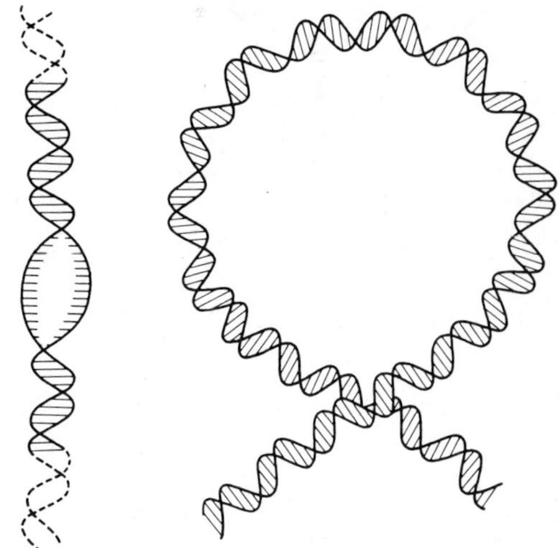

Figure 17.24
Two representations of superhelicity in DNA.
In both representations the terminals of the macromolecules must be viewed as unable to rotate freely around the axis of the double helix; if they did the structures would revert to the relaxed forms. Although the supercoiled structure on the right is a stable conformation, the structure on the left may be viewed as only a transient form. Both structures are characterized by a deficit of helical turns.

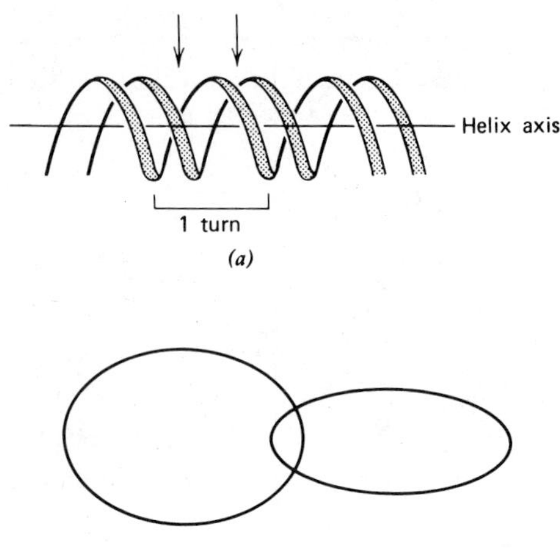

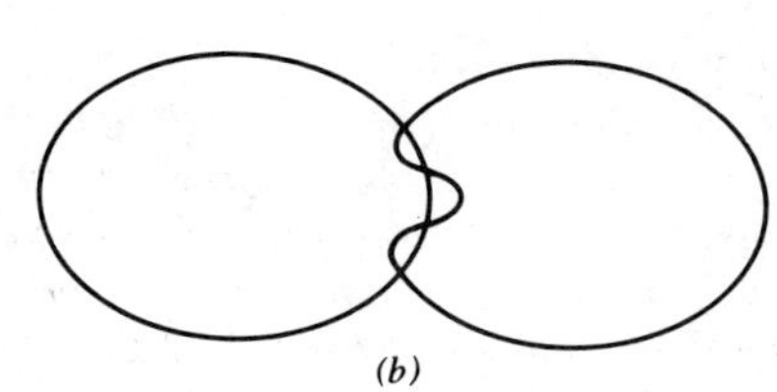

Figure 17.25
Determination of the linking number α in superhelical DNA.
(a) Side view of a schematic representation of the double helix. Note that the strands cross twice for each turn of the helix. (b) DNA circles interwound once and twice. Note that each pair of crossings is equivalent to one interwind.

model also highlights the concept that superhelicity is inseparably associated with the existence of a closed or restricted topological domain. Superhelicity in this example will be preserved only for as long as both hands grasp the rope firmly so as to maintain a closed topological system. Once this closed system is interrupted, the superhelix can unwind and acquire a relaxed form.

Geometric Description of Superhelical DNA

The conformations acquired by the interlocking rings of a closed circular complex can be formally characterized by three parameters: the linking number α, the number of helical turns β, and the number of supercoils or tertiary turns τ. These parameters are related by the equation $\alpha = \beta + \tau$. The nature of β and τ is self-explanatory. When interlocked rings are viewed with one ring held in a plane, the linking number α may be defined as the number of times one ring passes through the other. As is apparent in Figure 17.25, α can also be determined by counting the number of times the two rings appear to cross each other and dividing this number by 2. This is because for each turn of the helix of a closed complex the second strand must pass through the circle formed by the first twice when viewed perpendicular to the helix axis.

Two important conclusions can be reached from consideration of these definitions and from examination of Figure 17.26. First, it is apparent that for every *relaxed* DNA the linking number and the number of helical turns are identical. However, as will be apparent shortly, the reverse is not true. Second, DNAs with a specific linking number can acquire various different arrangements in space. In the case of superhelical DNAs different types of supercoils may be formed. However, *all conformations with the same linking number α are interconvertible without the need of breaking any covalent bonds.*

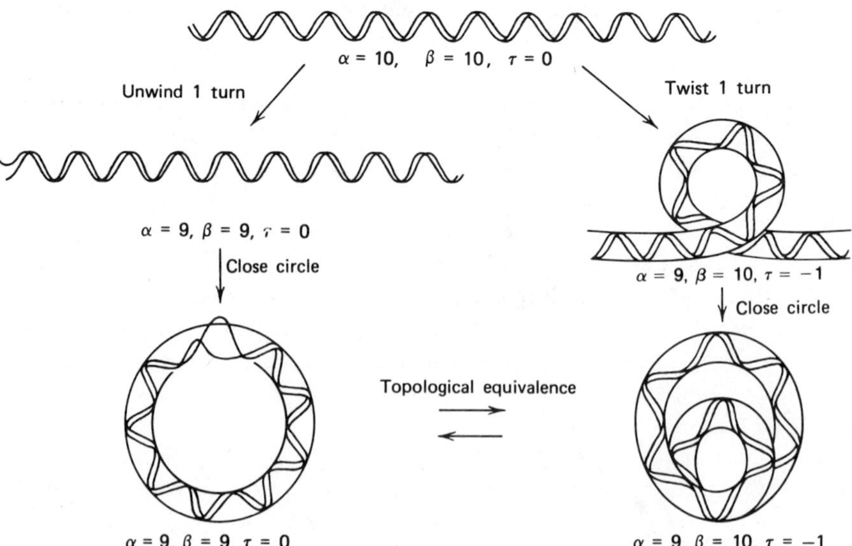

Figure 17.26
Various types of DNA superhelices.
An accurate representation of superhelical DNA structures can be made, using the number of helical turns β and the number of supercoils or tertiary turns τ along with a third parameter referred to as the linking number α as defined in the text. The figure shows ways of introducing one supercoil into a DNA segment of 10 duplex turns and the parameters of the resulting superhelices.

Redrawn with permission from C. R. Cantor and P. R. Schimmel, *Biophysical chemistry*, Part III, W. H. Freeman and Co., San Francisco, 1980. Copyrighted, 1980.

The various forms of supercoiled DNAs can be described using the α, β, and τ numbers. The mental exercise shown in Figure 17.26 illustrates how these numbers apply. It should be recalled that the turns of the typical double helix are right-handed. Therefore, if a hypothetical linear DNA duplex that is 10 turns long ($\alpha = 10$ and $\beta = 10$) is unwound by, say, one turn, the resulting structure will have the following characteristics: $\alpha = 9$ and $\beta = 9$. A *potentially* equivalent structure can be formed if instead the ends of the same hypothetical DNA are secured so that they cannot rotate and the molecule is looped in a counterclockwise manner. Since in this case untwisting is not permitted to occur, the number of helical turns remains unchanged, that is, $\beta = 10$. However, as a result of the "looping" operations, the linking number is now reduced by 1, that is, $\alpha = 9$. The structure resulting from this deliberate introduction of a loop is visibly superhelical. Furthermore, application of the equation that relates the values of α, β, and τ indicates that τ must be equal to -1, that is, the structure is a *negative* superhelix with *one* superhelical turn.

The two structures described above, $\alpha = 9, \beta = 9, \tau = 0$ and $\alpha = 9, \beta = 10, \tau = -1$ obviously have the same linking number and are therefore interconvertible without the disruption of any phosphodiester bonds. The potential equivalence of these two types of structure becomes more apparent when the ends of the polynucleotides in each structure are joined into a circle without the strands being allowed to rotate. Circularization produces an underwound circular structure and a doughnut-shaped superhelical arrangement referred to as a *toroidal* turn, which are freely interconvertible. A third equivalent structure, called an *interwound* turn, shown in Figure 17.27, can be produced by unfolding a toroidal turn along an axis which is distinct from the supercoil axis.

In summary, if the termini of a linear DNA molecule are covalently attached, a "relaxed" covalent circle results. However, if one end of the double helix is maintained in a fixed and stationary position while the other end is rotated in either direction prior to closing the circle, the resulting structure will twist in the opposite direction so as to generate a superhelical structure. For each additional complete turn of the helix, the circle will acquire one more superhelical twist in the opposite direction of the rotation in order to relieve the intensifying strain. As a result, topologically equivalent structures, such as those shown in Figure 17.26 will be created. A real superhelical DNA must exist as an equilibrium among

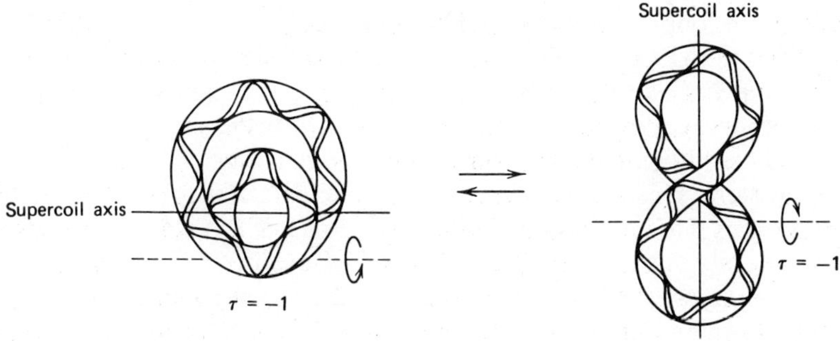

Figure 17.27
Equilibrium between two equivalent supercoiled forms of DNA.
The forms shown are freely interconvertible by unfolding the doughnut-shaped toroidal form along an axis parallel to the supercoil axis or by folding the 8-shaped interwound form along an axis perpendicular to the supercoil axis. The two forms have the same α, β, and τ numbers.
From C. R. Cantor and P. R. Schimmel, op. cit.

these forms and many other intermediate arrangements in space that have the same linking number.

Although the closed circular form of DNA is an ideal candidate for acquiring a superhelical structure, any segment of double-stranded DNA that is in some way immobilized at both of its terminals qualifies for superhelicity. This property therefore is not the exclusive province of circular DNA. Rather, any appropriately anchored DNA molecule can acquire a superhelical conformation.

The DNA of animal cells, for instance, normally associated with nuclear proteins, falls into this category. Because of the fragility and the large size of this DNA, it has been difficult to establish whether it generally consists of a single circular piece, although this may be the case. However, even in the absence of a circular structure, animal DNA can acquire a superhelical form because its association with nuclear proteins creates numerous closed topological domains. In addition, most bacterial phages, animal viruses, bacterial plasmids, and cell organelles, such as mitochondria and chloroplasts, contain superhelical DNA. The existence of negative superhelicity appears to be an important factor, promoting the packaging of DNA within the confines of the cell because supercoils generate compact structures. For instance, while the length of DNA in each human chromosome is of the order of centimeters, the condensed mitotic chromosomes that contain this DNA are only a few nanometers long. Negative superhelicity may also be instrumental in facilitating the process of localized DNA strand separation during the process of DNA synthesis.

Topoisomerases

Although much remains to be learned as to how superhelices are generated, specific enzymes known as *topoisomerases* appear to regulate the formation of superhelices. Topoisomerases act by catalyzing the concerted breakage and rejoining of DNA strands, which produces a DNA that is more or less superhelical than the original DNA. Topoisomerases are classified into type I, which break only one strand, and type II, which break both strands of DNA simultaneously. Topoisomerases I act by making a transient *single-strand break* in a supercoiled DNA duplex resulting in *relaxation of the supercoiled DNA*. In the past these enzymes were referred to as the nicking–closing enzymes. Topoisomerase I isolated from *E. coli* is also known as the *omega protein* (ω protein). Topoisomerase II acts by binding to a DNA molecule in a manner that generates two supercoiled loops, as shown in step 1 of Figure 17.28. Since one of these loops is positive and the other negative, the overall linking number of the DNA remains unchanged. In subsequent steps, however, the enzyme *nicks both strands* and passes one DNA segment through this break before resealing it. This manipulation inverts the sign of the positive supercoil, resulting in the introduction of *two negative supercoils* in each catalytic step. This reaction occurs at the expense of ATP; that is, topoisomerases II are ATPases. Obviously, reversal of this reaction removes two supercoils in each step. Therefore, topoisomerases II can either *add* or *remove* supercoils. During the reaction the enzyme remains bound to DNA by forming a covalent bond extended between a tyrosyl residue and a phosphoryl group at the incision site. This enzyme–polynucleotide bond conserves the energy of the interrupted phosphodiester bond for the subsequent repair of the nick. The cleavage sites do not consist of unique nucleotide sequences, although certain sequences are preferentially found at cleavage sites. Many topoisomerases of type II have been purified. One of these, *Gyrase,* which has been isolated from *E. coli,* has been studied quite extensively.

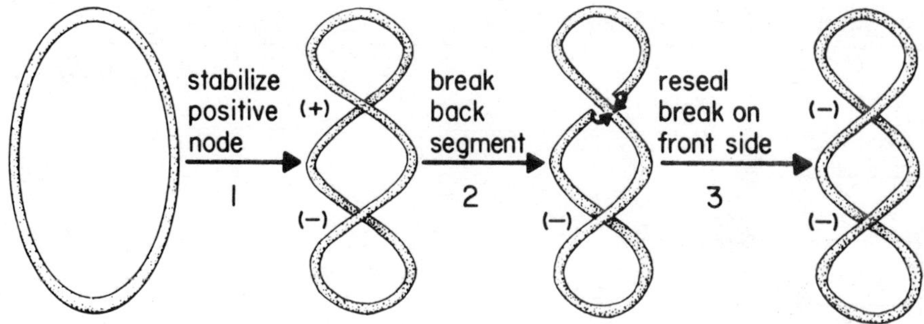

Figure 17.28
Mechanism of action of gyrase.
Gyrase, and other type II topoisomerases, can change the linking number of DNA by binding to a DNA molecule and passing one DNA segment through a reversible break formed at a different segment of the same DNA molecule. The mechanism of action of gyrase is illustrated above using as an example the conversion of a relaxed DNA molecule to a molecule that contains first two supercoils, one positive and one negative (step 1). Passage of a DNA segment through the positive supercoil shown on the rightmost part of the figure (step 3) changes the linking number, producing a molecule that contains two negative supercoils.
Reprinted with permission from P. O. Brown and N. R. Cozzarelli, *Science,* 206:1081, 1979. Copyright 1979 by the American Association for the Advancement of Science.

Separation of superhelical DNA from the corresponding relaxed or linear forms can be achieved by gel electrophoresis or by equilibrium centrifugation. With the latter method separation is achieved because the density of supercoiled DNA differs from that of the relaxed forms.

Nucleoproteins of Eucaryotes

DNA in eucaryotes is associated in the cell with various types of protein known as *nucleoprotein.* The main protein constituents of nucleoproteins are a class of highly basic proteins known as histones. *Histones,* regardless of their source, consist of five distinct types of polypeptides of different size and composition as listed in Table 17.5. The most "conserved" histones are the H4 and H3, which differ very little even between extremely diverse species; histone H4 from peas and cows are very similar, differing only by two amino acids. The H2A and H2B histones are less highly conserved, but still exhibit substantial evolutionary stability, especially within their nonbasic portions. The H1 histones are quite distinct from the inner histones. They are larger, more basic, and by far the most tissue-specific and species-specific histones. As a result of their unusually high content of the basic amino acids lysine and arginine, histones are highly polycationic and interact with the polyanionic phosphate backbone of DNA so as to produce uncharged nucleoproteins. All five histones are characterized by a central nonpolar polypeptide domain, which under appropriate conditions of ionic strength tends to form a globular structure, and N-terminal and C-terminal regions that contain most of the basic amino acids. The basic N-terminal regions of histones H2A, H2B, H3, and H4 are the major, but not the exclusive, sites of interaction with DNA. The nonpolar domains and the C-terminal regions are involved in interactions both between histones and between DNA and histones.

In addition to histones, a heterogenous group of proteins with high species, and even organ, specificity is present in nucleoproteins. These proteins, which are grouped together under the somewhat unimaginative name of *nonhistone proteins,* consist of several hundred different proteins, most of which are present in trace amounts.

TABLE 17.5 The Structure of the Five Types of Histones[a]

Name	*Structure*[b]	*Residues*	*Molecular Weight*
H4	N├──────● C	102	11,300
H3	N├────────●────┤C	135	15,300
H2A	N├─────●────┤C	129	14,000
H2B	N├──────●────┤C	125	13,800
H1	N├──────●──────────────────────┤C	~216	~21,000

SOURCE: From D. E. Olins and A. L. Olins, *Am. Sci.*, 66:704, 1978.

[a] Histones, which are highly basic polypeptides, are often classified as lysine-rich (H1), slightly lysine-rich (H2A and H2B), and arginine-rich (H3 and H4). Many of the basic amino acids are clustered on amino-terminal tails (i.e., the first 30–40 amino acids on the N side). The nonbasic mid- and carboxyl-terminal portions of the histones (C side) form globular structures that appear to be the sites of interaction between histones in the nucleohistones. The basic tail, on the other hand, interacts with DNA. The H1 class is almost twice as large and more basic than the other histones. Its globular region is nearer the N terminal.

[b] Scale ├┤ 10 amino acids.

Nucleosomes and Polynucleosomes

Nucleoproteins interacting with DNA have a *periodic* structure in which an elementary unit known as *nucleosome* is regularly repeated. Each nucleosome consists of a DNA segment associated with a histone cluster and nonhistone proteins. Each DNA-histone cluster is a disk-shaped structure about 10 nm in diameter and 6 nm in height composed of two molecules each of H2A, H2B, H3, and H4 histones. The clusters are organized as tetramers consisting of $(H3)_2$ $(H4)_2$, with an H2A–H2B dimer stacked on each face in the disk. DNA is wrapped around the octamer as a negative toroidal superhelix. For the purpose of describing the structure of nucleoproteins two distinct structures can be distinguished—the nucleosome core and the chromatosome as presented in Figure 17.29. The chromatosome constitutes the basic structural element of nucleoproteins.

The next level of organization of nucleoprotein is the *polynucleosome,* which consists of numerous nucleosomes joined by "linker" DNA the size of which differs between cell types. Usually the nucleosome core is used as the elementary unit for describing the polynucleosome, in which case linker DNA size varies anywhere from about 20 to 80 base pairs. (Linker sequences would of course be proportionally smaller if the chromatosome were to be used as the elementary unit for the polynucleosome.) Since in addition to the linker sequence 145 base pairs are wrapped around the nucleosome core, the polynucleosome has a minimum *nucleosome repeat frequency* of 165 base pairs. As a rule, repeat frequencies appear to be relatively long in transcriptionally inactive cells and are found to depend upon both the organism and the organ from which the cell is isolated. For example, chick erythrocytes have a repeat frequency of 212 base pairs. Active cells, such as yeast cells which have a frequency of 165 base pairs, generally appear to be smaller. Although nucleosomes are periodically positioned along the polynucleosome, their distribution may not be random with respect to the base sequence of DNA.

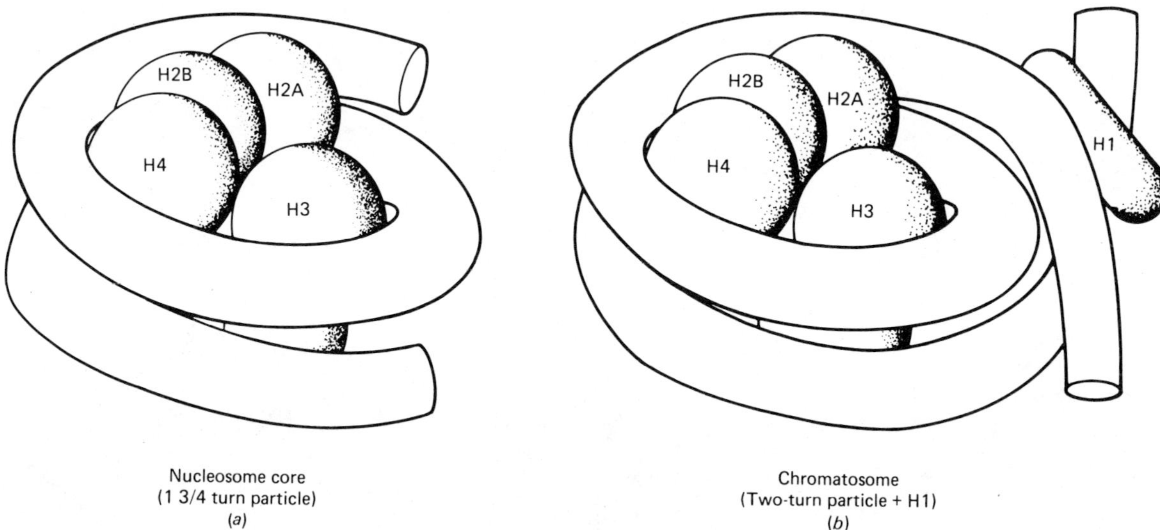

Figure 17.29
Postulated structures for the nucleosome core and the chromatosome.
The nucleosome core consists of 146 base pairs of DNA corresponding to $1\frac{3}{4}$ superhelical turns wound around a histone octamer. The chromatosome (two-turn particle) consists of 166 base pairs of DNA (two superhelical turns). The H1 subunit is retained by this particle and may be associated with it, as shown. Nucleosome particles containing less than 166 base pairs do not bind the H1 subunit.

Chromosomal Structure

Almost all the DNA of differentiated cells is present as *chromatin*. Chromatin, which fills the entire nucleus of resting cells, becomes highly condensed into distinct *chromosomes* as soon as the process of DNA replication is completed. Chromosomes consist of fibers about 300 Å wide, which at low ionic strengths can be dissociated into 100 Å wide fibers known as *nucleofilaments* (Figure 17.30).

A small amount of eucaryotic DNA is located in the mitochondria and the chloroplasts of plant cells. This DNA, which occurs in the form of small superhelices, is generally free of protein.

Nucleosomes are universally present among eucaryotic organisms and appear to be the first level of chromosomal organization beyond the DNA helix. There is little doubt that the nucleosome has a definite packaging function for chromosomal DNA.

Histones H3 and H4, as well as the central regions of H2A and H2B, apparently participate in interactions essential for maintaining chromatin structure and function. In the H1 sequence, variations are also confined within the basic N-terminal and the very basic C-terminal of the histone.

Nucleosomes have a definite packaging function for chromosomal DNA. The DNA from a human cell, which has a length of the order of 1 m, must be condensed so that it can fit within a nucleus with a diameter of approximately 10 μm. In order for this DNA to fit within the confines of the nucleus, it must be made compact by various types of sequential folding that are stabilized by histones and other proteins. In addition to the nucleofilament and the supercoiled or solenoid structures, the eucaryotic chromosome can become further condensed by the formation of 600 Å knoblike structures. These various consecutive levels of chromosomal organization are depicted in Figure 17.31. For the last stage of packaging, the supercoiled DNA is organized into separate loops, each

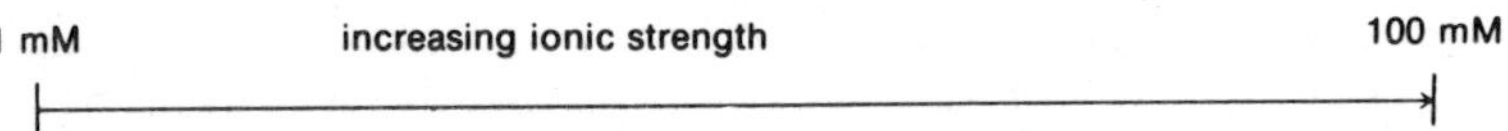

100 Å fiber

300 Å fiber or solenoid

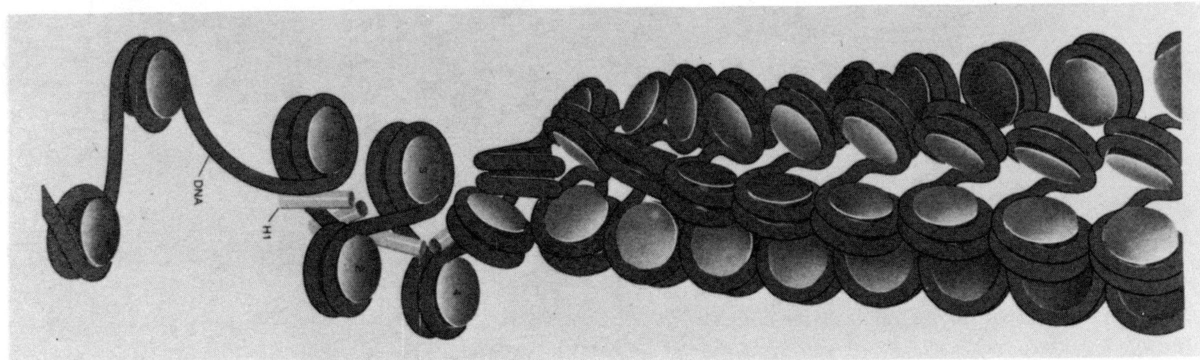

Figure 17.30
Nucleofilament structure.
Nucleofilament has the "string of beads" appearance, which corresponds to an extended polynucleosome chain. H1 histone is attached to the "linker" regions between nucleosomes, but in the resulting structure H1 molecules, associated to adjacent nucleosomes, are located close to one another. Furthermore, at higher salt concentrations polynucleosomes can be transformed into the higher order structure of the 300 Å fiber. It has been proposed that at higher ionic strengths the nucleofilament forms a very compact helical structure or a helical solenoid, as illustrated in the upper part of the figure. H1 histones appear to interact strongly with one another in this structure. In fact the organization of the 100 Å nucleofilament into the 300 Å coil or solenoid requires, and may be dependent upon, the presence of H1.

Reprinted with permission from *The nucleosome*, R. D. Kornberg and A. Klug.

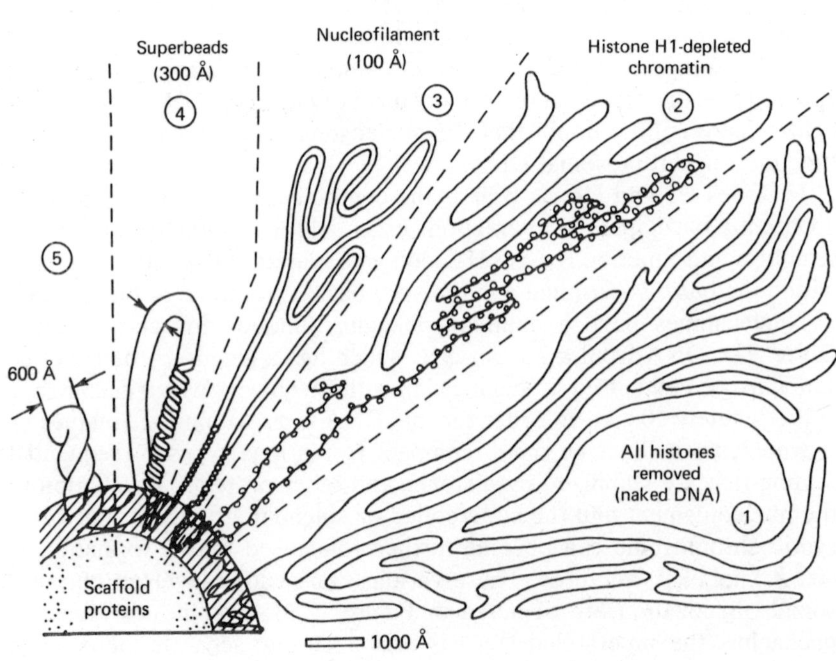

Figure 17.31
Various levels of organization of chromatin in the cell.
The "string of beads" structure of nucleofilament (section 3) is generated as a result of histone interaction with histone-free DNA (section 1). Superbeads (section 4) are formed by condensation of the nucleofilament at physiological ionic strengths. The final packaging of chromatin is in the form of the 600 Å knoblike structures depicted in section 5. Chromatin in all levels of its organization remains associated with a cellular protein scaffold.

Reprinted with permission from W. Saenger, *Principles of nucleic acid structure,* Springer-Verlag, New York, N.Y., 1983.

40–80 thousand base pairs long, which merge in the center of the chromosome within a protein-rich region referred to as the *scaffold*. The function of histones in DNA packaging may be regulated by various in vivo reactions such as methylation, acetylation, and phosphorylation. Maximum phosphorylation is observed before mitosis when chromosomes are most compact. Much remains to be done before arriving at an understanding of the control of eucaryotic transcription and replication, but it is becoming increasingly apparent that both histone and nonhistone proteins are involved in these processes. While the dissociation of histones from chromosomal DNA may be a prerequisite for transcription, nonhistone proteins may provide more finely tuned transcription controls. However, the exact manner in which nonhistone proteins interact with DNA and regulate gene expression remains to be elucidated. Finally nonhistone proteins may control gene expression during differentiation and development and may serve as sites for the binding of hormones and other regulatory molecules.

Viral DNA is almost always complexed with protein, where the function of the protein is generally one of "packaging." In essence the protein protects the DNA from mechanical damage or digestion by endonucleases by providing housing for the DNA within the tertiary structure of the protein.

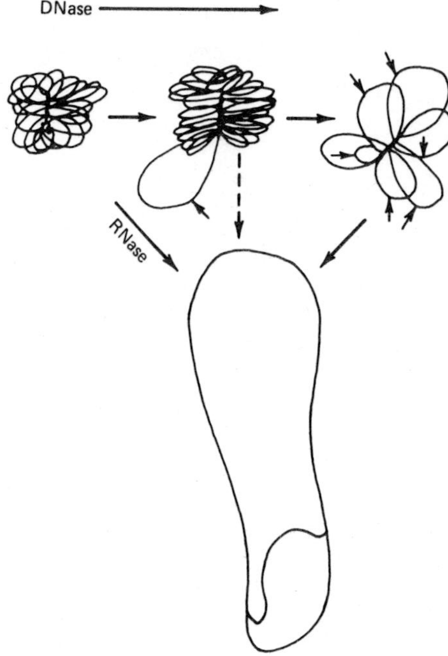

Figure 17.32
***A schematic depiction of the folded chromosome of* E. coli.**
This chromosome contains about 50 loops of supercoiled DNA organized by a central RNA scaffold. DNase relaxes the structure progressively by opening individual loops, one at a time. RNase completely unfolds the chromosome in a single step.
Redrawn from A. Worcel and E. Burgi, *J. Mol. Biol.* 71:127, 1972.

Nucleoproteins of Procaryotes

In procaryotic cells, DNA is generally present as a double-stranded circular supercoil, which is in part associated with the inner side of the plasma membrane. An abundant histonelike protein found in *E. coli* may be responsible for the presence of "beaded" chromatin fibers in procaryotes, which were long thought to lack histones. Histones together with various cations; polyamines such as spermine, spermidine, putrescine and cadaverine; RNA; and nonhistone proteins apparently account for the organization of bacterial chromosomes. For example, the chromosome of *E. coli* exists as a highly compact structure, known as *nucleoid,* which consists of a single supercoiled DNA molecule organized into 45 loops merging into a scaffold rich in protein and RNA (Figure 17.32). In procaryotic scaffolds, the loops are maintained by interactions between DNA and RNA rather than DNA-protein interactions only, as is the case with eucaryotes.

17.4 DNA STRUCTURE AND FUNCTION

Nucleotide Sequences in DNA

Overall *base composition* characterizes DNA only in a very general manner. Yet information on composition is often useful for DNA characterization. For determining this parameter, the nucleic acid is first hydrolyzed to its nucleotide components either by chemical or enzymatic means. The resulting nucleotides are separated usually by ion-exchange chromatography, and their amounts are determined spectrophotometrically. Alternatively, the composition can be indirectly estimated by equilibrium centrifugation of the DNA in a density gradient or by measurement of the melting temperature under standard conditions. In both of these techniques, the guanine–cytosine content of DNA influences in a quantitative manner the buoyant density and the thermal stability of the macromolecule, respectively.

A more specific property, which characterizes any DNA in a unique way, is its nucleotide *sequence*. Clearly the uniqueness of each DNA does not rest on its base composition but rather in the sequential arrangement of its individual bases. The direct determination of nucleotide sequences in DNA remained, until recently, an intimidating undertaking. This has been the case, in spite of the fact that the amino acid sequences of proteins and the nucleotide sequences of certain small RNAs such as tRNAs, have been accessible for many years through the application of effective but tedious methods of digestion of these macromolecules by appropriate enzymes. These enzymes, which sever macromolecules at specific sequences, can be chosen so that they yield fragments with overlapping sequences from which overall sequences can be gathered. Such approaches could not be used for sequencing DNA, partly because even the smallest DNA molecules are very large compared to proteins or small RNAs. In addition, enzymes that cleave DNA next to either a specific base or a specific sequence were not available until recently.

Palindromes

The discovery of *restriction endonucleases,* which cleave DNA chains in a specific sequence-dependent manner, has made possible the sectioning of large DNA molecules into small segments amenable to sequencing.

These highly specific bacterial enzymes act in vivo by making two cuts, one in each strand of double-stranded DNA of an invading phage, generating 3′-OH and 5′-P termini. This initial fragmentation exposes phage DNA to eventual degradation by bacterial exonucleases.

Approximately 500 restriction endonucleases have been obtained in pure form. With few exceptions, these enzymes have been found to recognize sequences of four to six nucleotides long. These sequences, known as *palindromes,* are characterized by local symmetry as illustrated by the examples listed in Table 17.6. The order of the bases is the same or nearly the same, when the two strands of the palindrome are read in opposite directions. For example, in the case of the restriction enzyme EcoR1, isolated from *E. coli,* the order of the bases is GAATTC when read from the 5′ terminus of either of the strands.

Restriction enzymes are classified into three categories: Enzymes of types I and III make cuts in the vicinity of the recognition site in a unpredictable manner; type II enzymes cleave specifically DNA within

TABLE 17.6 Examples of Sites of Cleavage of DNA by Restriction Enzymes of Various Specificities[a]

			No. of Cleavage Sites for 2 Commonly Used Substrates	
Enzyme	*Microorganism*	*Specific Sequence*	ϕX174	pBR 322
EcoR1	*E. coli*	-G↓AATT-C- -C-TTAA↓G-	25	9
Hae III	*Haemophilus aegyptus*	-GG↓CC- -CC↓GG-	11	22
Hpa II	*Haemophilus parainfluenzae*	-C↓CG-G- -G-GC↑C-	5	26
Hind III	*Haemophilus influenzae* Rd.	-A↓AGCT-T- -T-TCGA↑A-	0	1

[a] Cleavage takes place within palindromes. The cleavage sites are indicated by arrows.

```
                  ↓
EcoRI    5'.... GAATTC....3'      5'....G3'            5'AATTC....3'
         3'.... CTTAAG....5'  →   3'....CTTAA5'   +         3'G....5'
                     ↑
                      ↓
PstI     5'....CTGCAG....3'       5'....CTGCA3'            5'G....3'
         3'....GACGTC....5'   →   3'....G5'       +   3'ACGTC....5'
               ↑
                   ↓
HaeIII   5'....GGCC....3'         5'....GG3'              5'CC....3'
         3'....CCGG....5'     →   3'....CC5'      +       3'GG....5'
                 ↑
```

Figure 17.33
Types of products generated by type II restriction endonucleases.
Enzymes exemplified by EcoRI and PstI nick on both sides of the center of symmetry of the palindrome generating single-stranded stubs. Commonly used enzymes generate 5′ ends, although some produce stubs with 3′ stubs as shown for PstI. Other restriction nucleases cut across the center of symmetry of the recognition sequence, producing flush or blunt ends as exemplified by HaeIII.

the recognition sequence. The cuts, made by type II enzymes, are indicated in Table 17.5 by arrows. Examples of the products generated by these various specificities are shown in Figure 17.33.

Restriction enzymes fragment DNA very selectively. A typical bacterial DNA, which may contain about 3 million base pairs, is broken down only into a few hundred fragments. A small virus or plasmid may have no cutting sites at all or may have only a few such sites. The significance of this selectivity of restriction enzymes is that a particular enzyme generates a *unique family of fragments* for any given DNA molecule. A second enzyme acting on the same DNA obviously generates a different unique family of fragments, and so on. It is precisely this property of restriction endonucleases that has made these enzymes particularly valuable tools.

The availability of restriction enzymes for sectioning large DNA sequences and the development of new techniques for separating DNA segments which differ from one another by only a single nucleotide has made the determination of sequences a simple matter. These sequencing techniques are described in Section 17.9. Also, the availability of sufficient amounts of isolated single-copy chromosomal genes for sequencing has been enormously facilitated by the development of recombinant DNA techniques, as described in Section 17.8.

Early attempts to determine DNA sequences were limited to small DNA regions, which could be easily sectioned off from the remaining DNA. Sequences that bind selectively with various functional proteins, for example RNA polymerase and the repressor proteins, have been among the first to be determined. As a rule these sections can be separated from the remaining DNA by nuclease digestion of the complexes formed between DNA and the respective proteins. The protein protects the DNA section over which it is bound from the action of nuclease, and the protected DNA is recovered after digestion by dissociation of the protein. These studies indicated that many functional proteins and enzymes interact with DNA over regions of palindromic sequence (Figure 17.34).

Palindromes also serve as recognition sites for *methylases,* enzymes that modify the host DNA by introducing methyl groups into two bases of the palindrome. Once methylated, these palindromes cannot be recognized by the corresponding restriction enzymes, and the DNA of the host is protected from cleavage.

The new sequencing methods have made possible the determination of the complete nucleotide sequences of the DNA of many small viruses

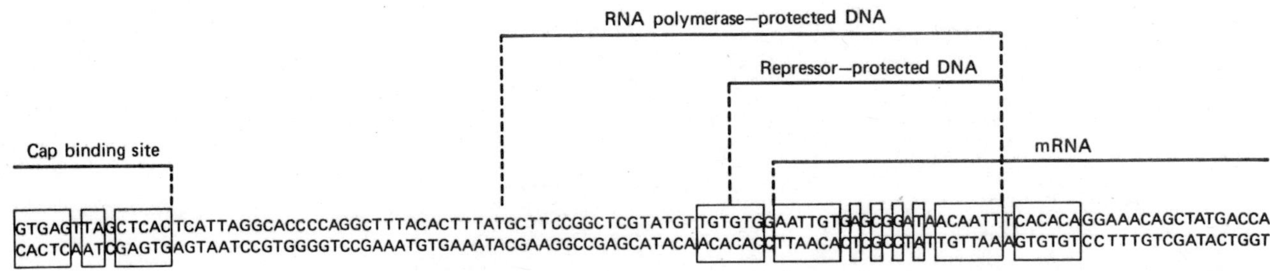

Figure 17.34
The nucleotide sequence of part of the DNA segment that controls the synthesis of the enzyme β-galactosidase in* E. coli *(the lac operon).
The binding regions of the cap protein, which acts as an activator of transcription, and of the lac repressor protein, an inhibitor of transcription, are indicated. Also shown is the region of RNA polymerase interaction. The presence of two palindromic sequences is indicated by boxes.
From C. R. Cantor and P. R. Schimmel, op. cit. Part I.

containing thousands of nucleotide residues. The effectiveness of the new methods is such that sequencing the DNA of even higher cells is now becoming a routine undertaking.

Procaryotic DNA

In procaryotes a large percentage of total chromosomal DNA codes for specific proteins. In certain bacteriophages the primary structure of DNA reveals that *structural genes,* nucleotide sequences coding for protein, do not always have distinct physical locations in DNA. Rather they frequently overlap with one another, as illustrated by the partial sequence of bacteriophage ϕX174 shown in Figure 17.35. It is believed that this type of overlap provides for the efficient and economic utilization of the limited DNA present in small procaryotes. This arrangement of genes may also be a factor in controlling the sequence in which genes are expressed.

Eucaryotic DNA

A typical mammalian DNA, with 20 times as many genes as that of *E. coli,* contains 500 times more DNA than *E. coli*. Clearly then, the DNA content of a mammalian cell is much too high on the assumption that it consists mostly of structural genes, along with some sequences used to control gene expression, as is the case with procaryotes. For example, only 10% of the DNA of the human cell may suffice for coding for the approximately 50 thousand genes that are probably present in the human genome. Determination of the complete nucleotide sequences of whole eucaryotic DNA could shed some light on the function of this excess DNA. This is an impractical task. However, the sequencing of large sections of eucaryotic DNA, for instance, structural genes and their surrounding regions, has now become a relatively easy undertaking. As a result, sequence data of eucaryotic DNA segments are accumulating rapidly. These data indicate that, in contrast to procaryotes, eucaryotic genes not only do not overlap, but with few exceptions (e.g., the genes of histones and the majority of tRNA genes) are interrupted by intervening nucleotide sequences, introns, as shown in Figure 17.36.

As a rule the sequence and the size of introns vary greatly among species, but generally these intervening segments may be 5–10 times longer than the sum of the length of the parts of the structural genes they separate. Some genes are interrupted only once, whereas others are

```
                  (PROTEIN A)..................GLU   SER   LYS   ASN   TYR   LEU   ASP   LYS   ALA   GLY   ILE   THR   THR
                     (ORIGIN OF PROTEIN K) MET   SER   ARG   LYS   ILE   ILE   LEU   ILE   LYS   GLN   GLU   LEU   LEU   LEU

(NUCLEOTIDE SEQUENCE)..............A T G A G T C G A A A A A T T A T C T T G A T A A A G C A G G A A T T A C T A C T
                                   51                  61                  71                  81                  91

ALA   CYS   LEU   ARG   ILE   LYS   SER   LYS   TRP   THR   ALA   GLY   GLY   LYS (TERMINUS OF PROTEIN A)
  LEU   VAL   TYR   GLU   LEU   ASN   ARG   SER   GLY   LEU   LEU   ALA   GLU   ASN   GLU   LYS   ILE   ARG   PRO   ILE
                                                              (ORIGIN OF PROTEIN C) MET   ARG   LYS   PHE   ASP   LEU   SER
G C T T G T T T A C G A A T T A A A T C G G A G T G G A C T G C T G G C G G A A A A T G A G A A A A T T C G A C C T A T
                   101                 111                 121                 131                 141                 151

  LEU   ALA   GLN   LEU   GLU   LYS   LEU   LEU   LEU   CYS   ASP   LEU   SER   PRO   SER   THR   ASN   ASP   SER   VAL
    LEU   ARG   SER   SER   ARG   SER   SER   TYR   PHE   ALA   THR   PHE   ARG   HIS   GLN   LEU   THR   ILE   LEU   SER
C C T T G C G C A G C T G A C G A A G C T C T T A C T T T G C G A C C T T T C G C C A T C A A C T A A C G A T T C T G T
                   161                 171                 181                 191                 201                 211

  LYS   ASN (TERMINUS OF PROTEIN K)
    LYS   THR..........(PROTEIN C CONTINUES)
C A A A A A C T..............
```

Figure 17.35
Partial nucleotide sequences of contiguous and overlapping genes of bacteriophage ϕX174.
The complete nucleotide sequence of ϕX174 is known. Only the sequence starting with nucleotide 51 and continuing to nucleotide 219 is shown in this figure. This sequence codes for the complete amino acid sequence of one of the proteins of ϕX174, protein K. A part of the same sequence, nucleotide 51 to nucleotide 133, codes for part of the nucleotide sequence of another protein, protein A (the remaining part of protein A is coded by a sequence, not shown, extending on the left beyond nucleotide 51). The remaining part of the sequence coding for protein K, which starts with nucleotide 133, also codes for part of a third protein, protein C. Similar overlaps are noted between other genes of ϕX174; for instance the sequence coding for a fourth protein, protein B, extends on the left of nucleotide 51.

Adapted with permission from M. Smith, *Am. Sci.*, 67:61, 1979. Journal of Sigma Xi, The Scientific Research Society.

highly fragmented. For instance, the conalbumin gene of the chicken, which codes for a major protein in egg white, may be divided by introns in as many as 17 distinct sections. The possibility exists that these sequences play a role in the control of gene expression. The suggestion has been made that genes subdivided by introns could, on an evolutionary

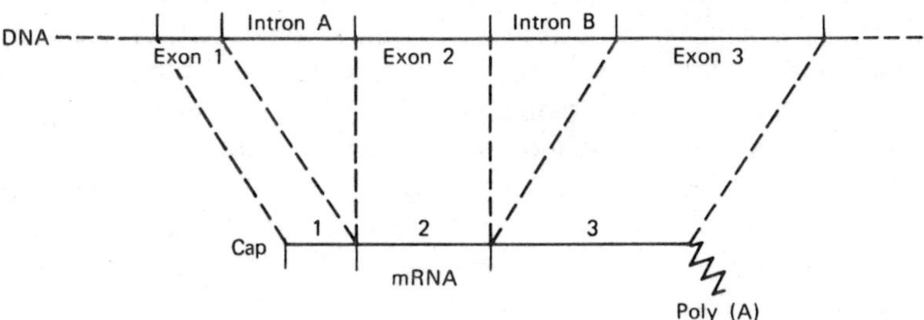

Figure 17.36
Schematic presentation of a eucaryotic gene.
The nucleotide sequences of eucaryotic genes are frequently separated by polynucleotide segments that are not present in mRNA and therefore are not translated to protein. These segments are referred to as introns. The gene is thus separated into noncontiguous segments called exons. The top horizontal line in the figure represents a part of the DNA genome of a eucaryote: the bottom line represents the mRNA produced by it. In this hypothetical example the DNA consists of two introns and three exons. The intron sequences are transcribed as hnRNA (precursor mRNA) but are not present in mature mRNA.

Redrawn from F. Crick. *Science*, 204:264, 1979. Copyrighted 1979 by the American Association for the Advancement of Science.

scale of time, be more easily shuffled to produce new gene combinations than genes that are put together as one piece.

Repeated Sequences

Until recently, the nucleotide sequences of eucaryotic DNA had been extensively studied by reassociation techniques. The more recent application of direct sequencing methods on DNA fragments, obtained by restriction endonuclease digestion, further extended the scope of these studies. As a result, in addition to obtaining the sequences of specific DNA sections, a good understanding of the complex characteristics of the primary structures of eucaryotic DNAs is now beginning to evolve.

As distinct from procaryotes, the DNA of eucaryotes contains multiple copies of certain nucleotide sequences that are repeated anywhere from a few times, for certain coding genes, to millions of times per genome for certain simple, relatively short, sequences. In addition to evidence obtained from sequencing data and DNA reassociation studies, the repetition of certain types of DNA sequences can be observed directly by electron microscopy, as in the cases of rRNA genes undergoing transcription.

Based on the number of times a sequence is repeated, three classes of sequences have been distinguished—*single copy, moderately reiterated, and highly reiterated*. These classes are defined experimentally from their rates of reassociation. Reassociation rates have also been used to define a fourth class of DNA, *inverted repeats*.

It may be recalled that the genome size of procaryotic DNA can be determined by fragmenting the DNA, denaturing the fragments and then allowing them to reassociate and form double-stranded molecules (Figure 17.15). The kinetics of reassociation obey a single second-order equation, indicating that all the sequences in the procaryotic genomes occur as single copies. When a mouse DNA was first studied by this method, unexpected results were obtained, which lead to the realization that eucaryotic DNAs contain reiterated sequences. A priori, it was assumed that since mammalian genes are about three orders of magnitude larger than *E. coli* genes, the rates of reassociation of denatured mammalian DNA would be exceedingly slow. Instead it turned out that a fraction of the mouse DNA, the highly-repetitive fraction reassociated far more rapidly than even the DNAs of small viruses (Figure 17.15). This is reasonable, since the probability that a fragment will encounter a complementary fragment leading to reassociation is proportional to the number of similar sequences repeated in the original DNA prior to fragmentation. The more reiterated the sequence, the more rapid the reassociation. Consequently, the reassociation kinetics of eucaryotic DNAs provided the first evidence for four classes or sequences. The inverted repeat and the highly repetitive sequences reassociate extremely rapidly. The unique sequences reassociate slowly, and the moderately reiterated renature at intermediate rates.

Most of the highly reiterated sequences have a distinct base composition from that of the remaining DNA. These sequences can be isolated from the total genomic DNA by shearing the DNA into segments of a few hundred nucleotides each and separating the fragments by density gradient centrifugation. These fragments are termed *satellite* DNA because after centrifugation they appear as satellites of the band of bulk DNA. For example the highly reiterated DNA sequence in the rat consisting of the repeated sequence 5′-GCACAC-3′ can be separated as satellite DNA. Other highly reiterated sequences, however, cannot be isolated by centrifugation, although they can be identified by virtue of their property of rapid reannealing. Some of the highly reiterated sequences can also be isolated by digestion of total DNA with restriction endonucleases that cleave at specific sites within the reiterated sequence.

The exact boundaries separating the various types of reiterated DNAs do not appear to have been strictly defined. Keeping this limitation in mind, the following distinctions among reiterated eucaryotic DNA types may be used.

Single Copy DNA

About half of the human genome is made up of unique nucleotide sequences, but only a small fraction of it codes for specific proteins. A part of the remaining DNA is devoted to *pseudogenes;* that is, tracts of DNA that have significant nucleotide homology to a functional gene but that contain mutations that prevent gene expression. These genes, which may be present in a frequency as high as one pseudogene for every four functional genes, significantly increase the size of eucaryotic genomes without contributing to their *expressible* genetic content. Additional DNA sequences are committed to serve as introns and as regions flanking genes. For instance, the genes producing the ε, β, γ, and δ chains of hemoglobin are located in a cluster and are separated by a noncoding sequence 400 base pairs long interspersed among these genes. The function of the remaining single copy DNA remains unclear.

Moderately Reiterated DNA

In this class of DNA, we may include copies of identical or closely related sequences that are reiterated anywhere from a few to several hundred times. These sequences are relatively long, varying between a few hundred to many thousand nucleotides before the same polynucleotide sequence is repeated. Normally single copy and moderately reiterated sequences are present on the chromosome in an orderly pattern known as the *interspersion pattern,* which consists of alternating blocks of single copy DNA and moderately reiterated DNA. For example, in the human genome sequences 600 nucleotides long, present in about 100 copies per genome, are tandemly reiterated over about 6% of the genome. Another 52% of the genome consists of reiterated sequences interspersed with single copy sequences about 2,250 nucleotides long. The remainder of the genome consists of unique sequences sparsely interspersed with repeated sequences. With the exception of some insects that have a distinctly different interspersion patterns, the human chromosome is typical of the observed pattern.

The short period of the interspersion pattern implicates the interspersed reiterated sequences in the control of transcription of the structural genes present in DNA. A role for the moderately reiterated DNA in controlling the transcription of the adjacent structural genes is plausible in view of the fact that the large majority of the structural gene sequences occur adjacent to reiterated sequences. Other moderately reiterated sequences are present as *segregated tandem arrays.* These two distinct types of arrangements of the moderately reiterated sequences appear to relate to different functions for these sequences. Tandem arrays are used for the synthesis of products that must be rapidly generated in numerous copies, such as ribosomal RNA and certain proteins of specialized function. For example, in sea urchin oocyte histone, genes are amplified so that sufficient amounts of histone are available for DNA packaging during the rapid cycles of DNA replication that follow fertilization. The genes for the five histones are arranged in tandemly repeated clusters, with each histone gene separated from its neighbor in the cluster by *spacers* that vary from about 400 to 900 nucleotides in length. These spacers are AT-rich and can therefore be separated as satellite DNA from the GC-rich DNA of the histone genes.

The arrangement of structural genes, with their introns, and the moderately repetitive segments of DNA of the interspersed and tandem array types is illustrated in Figure 17.37. In this hypothetical segment of eucaryotic DNA the single-copy and moderately repetitive sequences are indicated, which together normally account for more than 80% of the total nucleotide content of the genome.

Highly Reiterated DNA

The major part of the remaining DNA consists of sequences constructed by the repetition, many thousand times, of a nucleotide sequence that is typically shorter than 20 nucleotides. Because of the manner in which they are constructed, highly reiterated DNAs are also referred to as *simple sequence DNA*. You should draw a distinction between the terms "reiterated" and "repetitive" in describing a DNA sequence. The term *reiterated* is used to describe a unique DNA sequence, usually several hundred nucleotides long, present in multiple copies in a genome. An individual DNA sequence is termed *repetitive* if a certain, usually short, nucleotide sequence is repeated many times over the DNA sequence. Simple sequences are typically present in the DNA of most, if not all, eucaryotes. In some eucaryotes only one major type of simple sequence may be present, as for example in the rat in which the sequence 5′GCA-CAC3′ is repeated every six bases. In other eucaryotes several simple sequences are repeated up to a million times. Some considerably longer repeat units for simple sequence DNA have also been identified. For instance, in the genome of the African green monkey a 172-base pair segment has been found to be highly repeated. Determination of the nucleotide sequence of the highly repetitive DNA reveals that there are few sequence repetitions within the 172-base segment. The repeated units consist of a set of closely related but variant sequences. Because of its characteristic composition, simple sequence DNA can often be isolated as satellite DNA. The function of simple sequence DNA is not known, but this DNA appears to be concentrated in the centromers of chromosomes, and since it is not transcribed, a structural role in the organization of the eucaryotic chromosome is proposed.

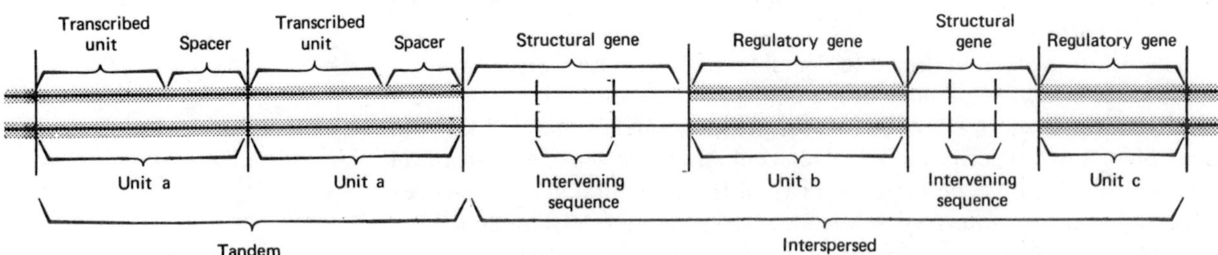

Figure 17.37
Structural genes with intervening sequences of the "interspersed" and "tandem array" types.
A hypothetical segment of eucaryotic DNA may be visualized as consisting of nonrepetitive sequences (indicated by a thin line) as well as moderately repetitive sequences (indicated by a shaded line). The latter can be of the interspersed or the tandem type. The interspersed sequences are separated by structural genes. The structural genes themselves are frequently interrupted by intervening sequences. The terms unit a, unit b, and unit c refer to distinct nucleotide sequences.

Inverted Repeat DNA

Short inverted repeats, each consisting of no more than six nucleotides, such as the palindromic sequence GAATTC, occur by chance about once for every 3,000 nucleotides. Such short repeats cannot form a stable "hair pin" structure that can be formed by longer palindromic sequences. Inverted repeat sequences that are long enough to form stable "hair pins" are not likely to occur by chance, and therefore they should be classified as a separate class of eucaryotic sequences. Experimentally they can be easily detected and quantitated by virtue of their extremely rapid rates of reassociation. In human DNA, about 2 million inverted repeats are present, with an average length of about 200 base pairs, although inverted sequences longer than 1,000 base pairs have been detected. Some of these repeats may be separated by a spacer sequence that is not part of the inverted repeat. Most inverted repeat sequences are repeated 1,000 or more times per cell.

17.5 FORMATION OF THE PHOSPHODIESTER BOND IN VIVO

The processes of enzymatic repair of certain randomly introduced changes in the chemical structure of the DNA bases and the process of DNA replication is discussed in this section.

DNA repair and particularly DNA replication are very complex processes. Although key similarities in the mechanisms of DNA replication and repair are discernible among different organisms, a considerable amount of diversity exists in terms of individual detail. This diversity further complicates any attempt to present a simplified and universally applicable model for each of these two processes. To resolve this difficulty the basic mechanistic elements of the substeps of each process are first described and subsequently integrated, using as an example the *E. coli* replication system but differences between *E. coli* and eucaryotic systems are pointed out.

DNA-Dependent DNA Polymerase

The common denominator between the processes of DNA replication and repair is the enzymatically catalyzed synthesis of DNA polynucleotide segments, which can be assembled with preexisting polynucleotides, leading to products of repair or replication. The synthesis of these polynucleotide segments is catalyzed by the enzyme DNA-dependent DNA polymerase, which in the case of *E. coli* has been isolated in three distinct forms, the polymerases I, II, and III listed in Table 17.7. The DNA polymerases are characterized by a $3' \rightarrow 5'$-exonuclease activity in addi-

TABLE 17.7 DNA Polymerase I, II, and III of *E. coli*

	Polymerase		
Properties	*I*	*II*	*III*
Molecular weight	110,000	120,000	180,000
Molecules per cell	400	100	10
Polymerization activity (turnover number)	1,000	50	15,000
Exonuclease activity $3' \rightarrow 5'$	Active	Active	Active
Exonuclease activity $5' \rightarrow 3'$	Active	Inactive	Active

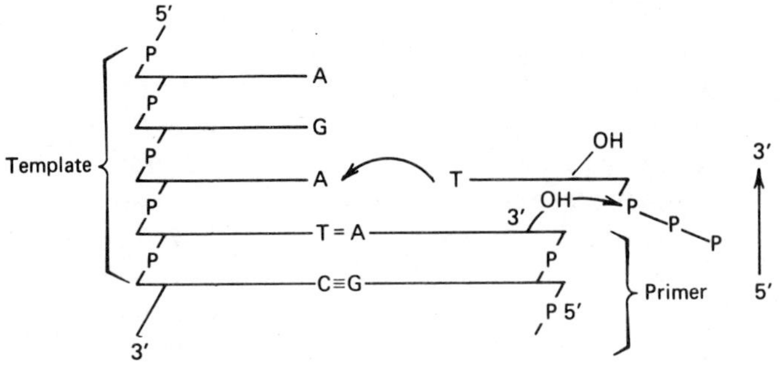

Figure 17.38
Synthetic activity of DNA polymerase.
DNA polymerase catalyzes the polymerization of nucleotides in the 5′ → 3′ direction. A phosphodiester bond is formed between a free 3′-hydroxyl group of the strand undergoing elongation (the primer) and an incoming deoxynucleoside 5′-triphosphate. Pyrophosphate is eliminated.
Redrawn from A. Kornberg, *Science* 163:1410, 1969. Copyright 1969 by the American Association for the Advancement of Science.

tion to synthetic activities. Polymerases I and III are also 5′ → 3′-exonucleases. The involvement of all these enzymatic activities in the processes of repair and replication will be apparent shortly.

The synthetic activity of DNA polymerase can be described by referring to Figure 17.38 in which two complementary DNA strands of unequal length are shown. This conformation, in which the shorter strand has a free 3′ terminus, is essential for the function of DNA polymerase. The enzyme catalyzes the addition of free 5′-deoxynucleoside triphosphates to the 3′ terminus of the short strand, the *primer*. The term *primer* applies to the initial terminus of a molecule, in this instance the 3′-polynucleotide end, onto which additional monomeric units can be added stepwise to yield the final product. The free portion of the longer complementary strand is used as a *template* to direct the condensation of selected 5′-deoxynucleotides onto the growing primer. In the present context the term *template* refers to a single strand of nucleic acid, which provides the specific information necessary for the synthesis of a complementary strand. DNA polymerase requires both a primer and a template in order to function. The primer provides a site for the polymerization to begin, and the template provides the information that determines the precise nucleotide sequence of the new polymer.

The DNA polymerase-catalyzed reaction permits the selection of 5′-deoxyribonucleoside triphosphates, one at a time, with a base complementary to that present in the corresponding position of the template. The specificity of the polymerase reaction with respect to the template is vested in the strong association of each of the bases of the template with their normal complementary partners present in the cell as free 5′-deoxyribonucleotides. Strong binding between complementary bases is apparently achieved because the bases become confined within custom-fitted cages created by appropriate hydrophobic regions of the DNA polymerase. As a result the reading of the template is extremely accurate. In addition, the fidelity of the reading is probably enhanced because the 3′ → 5′ exonuclease activity of the polymerase may be used for proofreading the bases, selected by the enzyme, for possible errors. Specifically, if a 5′-deoxyribonucleotide which is not complementary with the corresponding base on the template is erroneously condensed with the primer, the enzyme can temporarily reverse its synthetic activity and hydrolyze the phosphodiester bond formed between the primer and the erroneous base.

Thus, in effect the enzyme can retrace the path it covers and remove any erroneously introduced mismatched bases. Because of these precautions, rates of error are extremely low, generally in the order of a few mispairings per billion of added bases.

The polymerase has well-defined selectivities also in a different sense. Only the 3′ terminus of a strand can be used for priming. Therefore the enzyme can elongate a strand only in the 5′ → 3′ direction, as indicated in Figure 17.38. The 5′ terminus of the strand is rejected as a primer because the polymerase is unable to elongate a polynucleotide in the opposite 3′ → 5′ direction.

17.6 MUTATION AND REPAIR OF DNA

Mutations

One of the fundamental requirements for a structure that serves as a permanent depository of genetic information is extreme stability. Such stability is essential, at least in terms of those characteristics of the structure that code for the genetic information. Therefore a prerequisite for the structure of DNA is extreme stability in its base content and in its sequence, in which hereditary information is encoded. Yet the structure of the DNA bases is not totally exempt from gradual change. Normally, changes occur infrequently and then affect very few bases, but nevertheless they do take place. Chemical or irradiation-induced reactions may modify the structure of some bases or may disrupt phosphodiester bonds and sever the strands. Errors may also occur during the processes of replication and strand recombination, leading to the incorporation of one or more erroneous bases into a new strand. In almost every instance, however, a few cycles of DNA replication are required before a modification in the structure of a base can lead to irreversible damage, that is, DNA polymerase must use the polynucleotide initially damaged as a template for the synthesis of a complementary strand for the initial change to become permanent. As Figure 17.39 suggests, use of the damaged strand as template extends the damage from a change of a single base to a change of a complete base pair and subsequent replication perpetuates the change.

Since the properties of cells and of the organisms constructed from them ultimately depend on the DNA sequences of their genes, irreversible alterations in a few DNA base pairs can cause substantial changes in the corresponding organism. These changes, referred to as mutations, may be hidden or visible, that is, phenotypically silent or expressed. Therefore, a *mutation* may be defined as a stable change in the DNA structure of a gene, which may be expressed as a phenotypic change in the correspond-

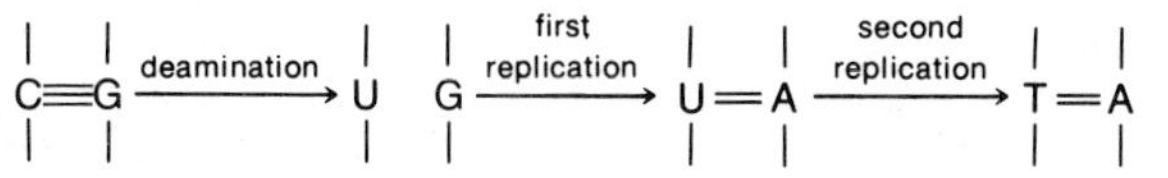

C≡G —deamination→ U G —first replication→ U=A —second replication→ T=A

Figure 17.39
Mutation perpetuated by replication.
Mutations introduced on a DNA strand, such as the replacement of a cytosine residue by a uracil residue resulting from deamination of cytosine, extend to both strands when the damaged strand is used as a template during replication. In the first round of replication uracil selects adenine as the complementary base. In the second round of replication uracil is replaced by thymine. Similar events occur when the other bases are altered.

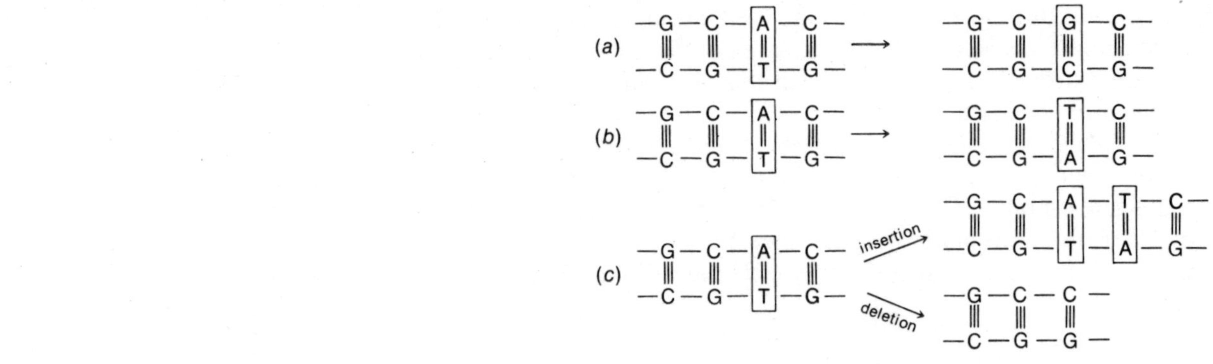

Figure 17.40
Mutations.
Mutations can be classified as transitions, transversions, and frame shift. Bases undergoing mutation are shown in boxes. (a) Transitions: A purine–pyrimidine base pair is replaced by another. This mutation occurs spontaneously, possibly as a result of adenine enolization or can be induced chemically by such compounds as 5-bromouracil or nitrous acid. (b) Transversions: A purine–pyrimidine base pair is replaced by a pyrimidine–purine pair. This mutation occurs spontaneously and is common in man. About one-half the mutations in hemoglobin are of this type. (c) Frame shift: This mutation results from insertion or deletion of a base pair. Insertions can be caused by mutagens such as acridines, proflavin and ethidium bromide. Deletions are caused by deaminating agents. Alteration of bases by these agents prevents pairing.

ing organism. Mutations may be classified, depending on their origin, into two categories: base substitutions and frame shift mutations. Base substitutions include *transitions,* substitutions of one purine–pyrimidine pair by another, and *transversions,* substitutions of a purine–pyrimidine pair by a pyrimidine–purine pair. *Frame shift* mutations, which are the most radical, are the result of either the insertion of a new base pair or the deletion of a base pair or a block of base pairs from the DNA base sequence of the gene. These changes are illustrated in Figure 17.40.

Mutagens

A more systematic coverage of the subject of mutations, especially with respect to the expression of a mutation as a change in the product of the corresponding gene, must await the detailed description of the processes of replication, transcription, and translation. In this section, the factors that cause mutations are listed, and a few examples of structural DNA changes brought about by these factors are given. Irradiation and certain chemical compounds are recognized as the main mutagens. Some rare events of incorporation of erroneous bases by DNA polymerase can also lead to mutations.

Chemical Modification of the Bases

The bases in DNA are sensitive to the action of numerous chemicals. Among them are nitrous acid (HNO_2), hydroxylamine (NH_2OH), and various alkylating agents such as dimethyl sulfate and *N*-methyl-*N'*-nitro-*N*-nitrosoguanidine. Chemical modifications of bases, brought about by these reagents are shown in Figure 17.41. The conversion of guanine to xanthine by nitrous acid has no effect on the hydrogen-bonding properties of this base. The new base, xanthine, can pair with cytosine, the normal partner of guanine. However, the conversion of either adenine to hypoxanthine or the change from cytosine to uracil disrupts the normal hydro-

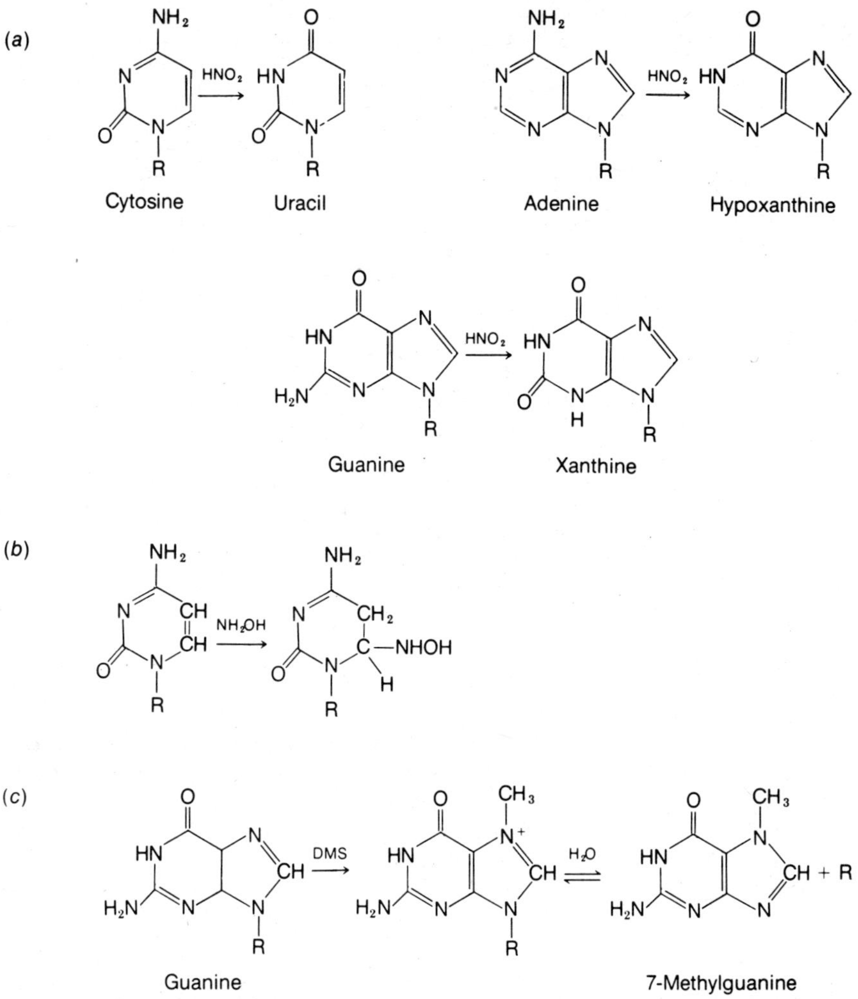

(a)

Figure 17.41
Reactions of various mutagens.
(a) Deamination by nitrous acid (HNO_2) converts cytosine to uracil, adenine to hypoxanthine, and guanine to xanthine. (b) Reaction of bases with hydroxylamine (NH_2OH) as illustrated by the action of this reagent on cytosine. (c) Alkylations of guanine by dimethyl sulfate (DMS). The formation of a quaternary nitrogen destabilizes the deoxyriboside bond and releases deoxyribose. Among the effective agents for methylation of the bases are certain nitrosoguanidines such as N-methyl-N′-nitro-N-nitrosoguanidine.

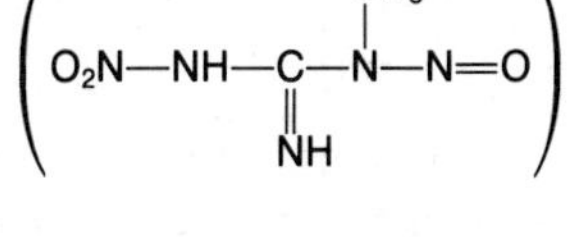

$$O_2N{-}NH{-}\underset{\underset{NH}{\|}}{C}{-}\overset{\overset{CH_3}{|}}{N}{-}N{=}O$$

gen bonding of the double helix. This is because neither hypoxanthine nor uracil can form complementary pairs with the base present in the initial double helix (Figure 17.42). Subsequent replication of the DNA extends and perpetuates these base changes. Alkylating agents may affect both the structure of the bases as well as disrupt phosphodiester bonds so as to lead to the fragmentation of the strands. In addition, certain alkylating agents can interact covalently with both strands, creating interstrand bridges.

In principle, DNA bases can also be modified as a result of spontaneous reactions in the absence of reactive chemicals. For instance, adenine

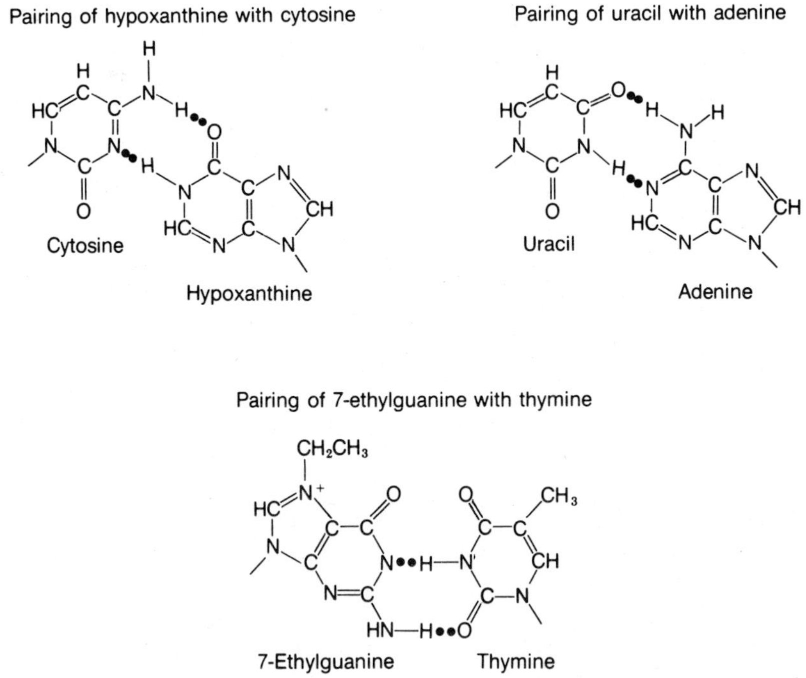

Pairing of hypoxanthine with cytosine

Cytosine

Hypoxanthine

Pairing of uracil with adenine

Uracil

Adenine

Pairing of 7-ethylguanine with thymine

7-Ethylguanine

Thymine

Figure 17.42
Chemical modifications that alter the hydrogen-bonding properties of the bases.
Hypoxanthine, obtained by deamination of adenine, has different hydrogen-bonding properties from adenine, for example, it pairs with cytosine. Similarly, uracil obtained from cytosine, has a different hydrogen-bonding specificity than cytosine and pairs with adenine. Alkylation of guanine modifies the hydrogen-bonding properties of the base.

Adenine

Hypoxanthine

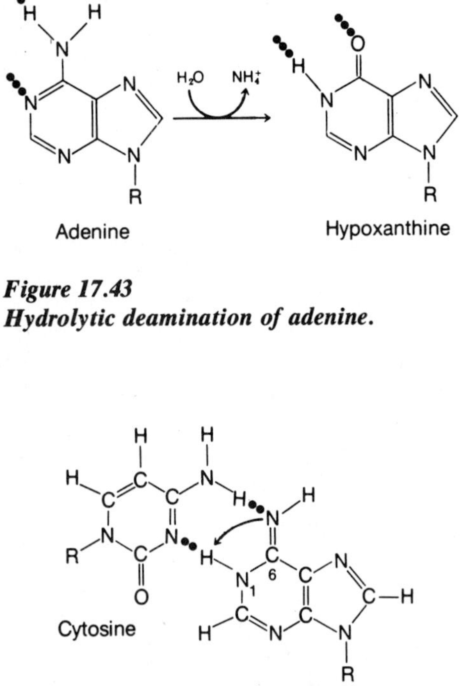

Figure 17.43
Hydrolytic deamination of adenine.

could undergo hydrolytic deamination to hypoxanthine as shown in Figure 17.43. The extent to which such spontaneous chemical reactions may take place is not established.

Radiation Damage

Both ultraviolet as well as x-ray irradiation are generally very effective means of producing mutations. The bases normally exist as keto or amino forms in equilibrium, with only minor amounts of the enol or the imino structures. Radiation energy absorbed by the bases tends to shift the equilibrium to the minor forms. The minor forms, however, cannot pair with the normal partners of the bases. For example, the enol form of adenine pairs with cytosine instead of thymine, the normal adenine partner. This atypical base pairing is shown in Figure 17.44. It has been suggested that the existence of increased amounts of the enol forms of the bases at the moment of replication increases the frequency of mutations of the newly synthesized DNA strand because the enol forms select new bases that pair with them rather than the normal hydrogen-bonding partners of the more predominant keto forms.

Exposure of DNA to high energy radiation (x-rays or γ-rays) may also bring about direct modifications in the structure of the bases. Intermediates produced by electron expulsion can be rearranged, leading to the opening of the heterocyclic rings of the bases and the disruption of phosphodiester bonds. In the presence of oxygen additional reactions take place, yielding a variety of oxidation products.

Irradiation by uv light primarily affects the pyrimidines. Activation of the ethylene bond of these bases frequently leads to a photochemical

Cytosine

Rare tautomer of adenine

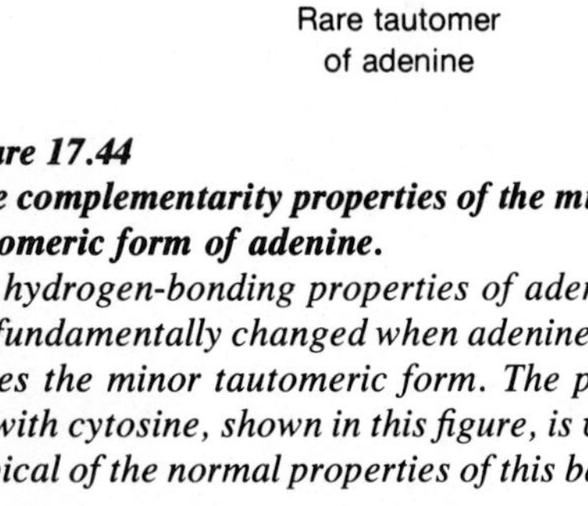

Figure 17.44
Base complementarity properties of the minor tautomeric form of adenine.
The hydrogen-bonding properties of adenine are fundamentally changed when adenine acquires the minor tautomeric form. The pairing with cytosine, shown in this figure, is very atypical of the normal properties of this base.

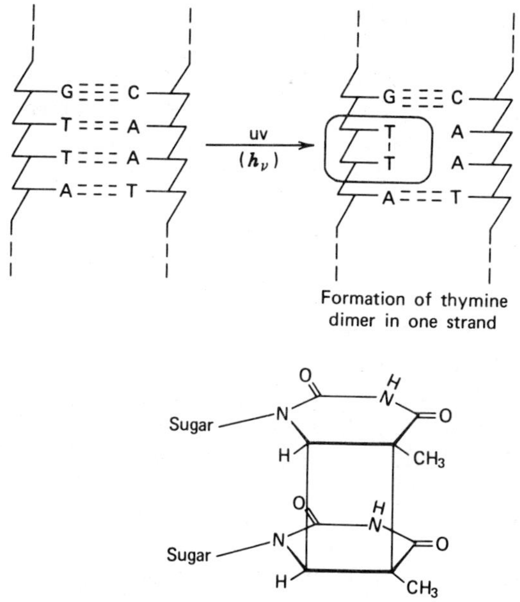

Formation of thymine dimer in one strand

Figure 17.45
Dimerization of adjacent pyrimidines in irradiated DNA.
A residue of thymine which is activated by the absorption of ultraviolet light can react with a second neighboring thymine and form a thymine dimer.

dimerization of two adjacent pyrimidines, as shown in Figure 17.45. Thymine residues are particularly susceptible to this reaction, although cytosine dimers and thymine–cytosine combinations are also produced.

DNA Polymerase Errors

When the appropriate deoxyribonucleotides are available, DNA-dependent DNA polymerase generally functions with a very high degree of fidelity. Not all DNA polymerases, however, are equally discriminating. For instance, the RNA-dependent DNA polymerase (a polymerase that uses RNA as a template) associated with a virus causing a form of leukemia in birds can make numerous errors in selecting nucleotides complementary to the bases of the template. Perhaps more than one erroneous base per 1,000 correctly chosen nucleotides might be incorporated into DNA by this low specificity polymerase. Furthermore, even the most discriminating DNA polymerases are unable to distinguish between the normal deoxyribonucleoside triphosphate substrates and other nucleotides with very similar structures.

On the basis of the tendency of DNA polymerase to accept, in place of the normal substrates, structural analogs of the common bases, certain mutations can be introduced into DNA by design. For instance, 2-aminopurine, incorporated instead of adenine into a newly synthesized DNA strand, can associate with cytosine and produce an A–T → G–C transition. A somewhat more complex example is provided by 5-bromouracil. This base, in the form of the corresponding deoxynucleoside triphosphate, can be incorporated into a strand in place of thymine. However, the equilibrium between the enol and the keto forms for these two bases allows for the formation of a somewhat higher proportion of the enol form in 5-bromouracil than in thymine. This occurs presumably because of the higher electronegative nature of the bromine atom in comparison to the corresponding methyl group in thymine. Because the enol form of 5-bromouracil pairs with guanine, as shown in Figure 17.46 the substitution of thymine by bromouracil produces an A–T → G–C transition.

Minor tautomer of 5-bromouracil

Guanine

Figure 17.46
Hydrogen-bonding properties of the minor enol form of 5-bromouracil.
The enol form of 5-bromouracil, an analog of thymine, pairs with guanine instead of adenine, the normal partner of thymine.

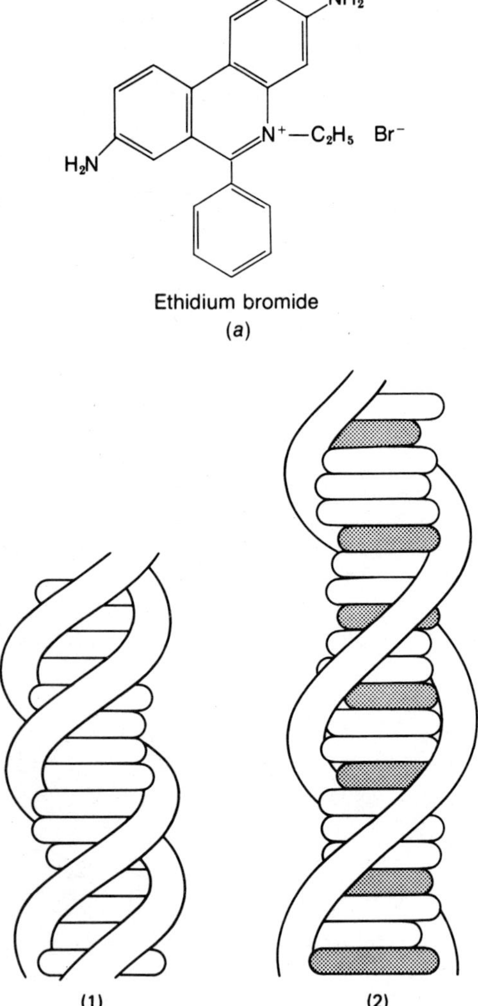

Figure 17.47
Intercalation between base pairs of the double helix.
The insertion of the planar ring system of intercalators (a) between two adjacent base pairs requires the stretching of the double helix (b). During replication this stretching apparently changes the frame used by DNA polymerase for reading the sequence of nucleotides. Consequently, newly synthesized DNA is frame-shifted. (b-1) The original DNA helix; (b-2) the helix with intercalative binding of ligands.

Reprinted with permission from S. J. Lippard, *Accts. Chem. Res.* 11:211, 1978. Copyright 1978 by the American Chemical Society.

Stretching of the Double Helix

Certain organic compounds, which are characterized by planar aromatic ring structures of appropriate size, can be inserted between base pairs in double-stranded DNA. This process is referred to as *intercalation*. During intercalation neighboring base pairs in DNA are separated to allow for the insertion of the intercalating ring system, causing an elongation of the double helix by stretching. The continuity of the base sequences in DNA is disrupted, and the reading of the bases by the DNA polymerase produces a new strand, with an additional base inserted near the site of intercalation. The resulting mutation is referred to as a *frame shift. Acridines, ethidium bromide,* and other intercalators are known to be effective *frame-shift mutagens* (Figure 17.47).

Clinical Correlation 17.1 discusses mutations and the etiology of cancer. Clinical Correlation 17.2 discusses the diagnosis of genetic disease by hybridization techniques that identify the associated mutations.

Repair of DNA

Mutations may be necessary for effective evolutionary response to environmental change. Procaryotes are able to generate new forms rapidly; this flexibility is essential for survival. Adaptability, however, is obtained at high cost, since most random mutations are not beneficial but instead produce deleterious effects.

Because of the generally harmful nature of mutations, mechanisms have also evolved that counteract the effects of these changes. In effect the rate of initial DNA damage is generally much higher than the rate of the expressed lethal or damaging mutations. Obviously the most direct means by which a potential mutation can be reversed is the repair of the affected base prior to replication. Almost all cells have the capacity to detect distortions in the structure of the DNA double helix. Such distortions may result from a permanent interruption of the hydrogen bonding between one or more base pairs which originates from a change in the structure of one of the bases. Two principal mechanisms that lead to the restoration of the normal structure of DNA are presented here; those that reverse the DNA damage and those that lead to the replacement of the damaged DNA section. *Error-prone (SOS) repair,* an additional repair process, is not a true repair mechanism; this process instead allows DNA replication to take place in a stop-gap manner until the damage is permanently repaired by another mechanism. In effect, a daughter strand is synthesized by skipping over the damaged base, and, therefore, after synthesis the daughter strand is found to be missing a base that would be normally present across from the damaged base. The missing base is added *postreplicatively* after the parental strand, and with the damaged base still present, is separated from the daughter strand. The mechanism of this process remains poorly understood.

Mechanisms that Reverse Damage

The uv light-induced formation of pyrimidine dimers can be reversed by the action of light that leads to the regeneration of the base monomers. Most cells contain enzymes, *photolyases,* that are activated by light in the 300–600-nm range and disrupt the covalent bonds that hold together the pyrimidine molecules in the dimer.

Another example of direct reversal of damage is the removal of a methyl or ethyl group from the 6 position of the enol form of a guanine residue in DNA; these alkyl groups can be removed, and the normal structure of guanine reestablished, by the action of a specific protein that accepts alkyl groups and in the process itself becomes alkylated.

CLIN. CORR. 17.1
MUTATIONS AND THE ETIOLOGY OF CANCER

Considerable progress in our understanding of the etiology of cancer has been made in recent years by our ever-increasing realization that long-term exposure to certain chemicals leads to various forms of cancer. Some experts are now suggesting that the great majority of cancers are in fact triggered by environmental factors.

Carcinogenic (cancer-causing) compounds are not only introduced into the environment by the increasing use of new chemicals in industrial applications but are also present in the form of natural products. For instance, the *aflatoxins*, produced by certain molds, and *benz[a]anthracene*, present in cigarette smoke and charcoal broiled foods, are

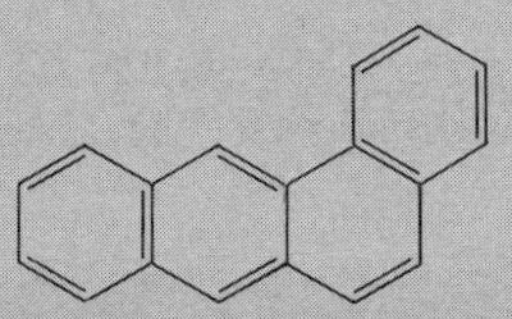

Benz[*a*]anthracene

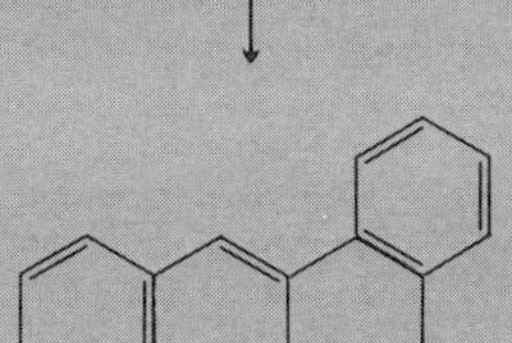

↓

5,6-Epoxide (carcinogenic)

carcinogenic. Some carcinogens act directly, while others, such as benz[*a*]anthracene, must undergo prior hydroxylation by arylhydroxylases, present mainly in the liver, before their carcinogenic potential can be expressed.

The reactivity of many carcinogenic compounds toward guanine residues results in modification of the guanine structure, usually by alkylation at the N7 position, as well as in breaks of the phosphodiester bond, events that, upon replication, lead to permanent mutations. Chemicals that produce mutations generally turn out to be carcinogenic and vice versa.

The vulnerability of DNA toward alkylating agents, and other chemicals as well, underscores the concerns expressed today by many scientists about the ever-increasing exposure of our environment to new chemicals. What is particularly distressing is that the carcinogenic potential of new chemicals released into the environment cannot be predicted with confidence even when they appear chemically innocuous toward DNA.

Until recently, tests for carcinogenicity, that is, the ability of a substance to cause cancer, required the use of large numbers of experimental animals to which high doses of the suspected carcinogen were administered over a long period of time. Such tests, which are time-consuming as well as expensive, are the only approach still available for testing carcinogenicity directly. Recently, however, a much simpler and much more inexpensive indirect test for carcinogenicity was developed. This test is based on the premise that carcinogenicity and mutagenicity are essentially manifestations of the same underlying phenomenon, the structural modification of DNA. The test measures the rate of mutation that bacteria undergo when exposed to chemicals suspected to be carcinogens.

A major criticism advanced against the test is that the assumption of an equivalence between mutagenicity and carcinogenicity is not always valid. Because of the unusually large economic implication of labeling a chemical with widespread use as a potential carcinogen, the scrutiny that is often exercised in assessing the reliability of applicable tests for labeling a chemical as a carcinogen is understandable. Yet, certain exceptions notwithstanding, the great majority of chemicals tested has reinforced the view that a good correlation exists between the tendency of a chemical to produce bacterial mutations and animal cancer. Furthermore, even the direct and very costly tests for carcinogenicity, in which large numbers of animals are used, have not completely escaped criticism. The reliability of the test has been questioned because of the relatively large doses of chemicals employed in these tests; doses essential for shortening the long-term chemical exposure of the animals to a practically manageable period of time.

In addition, the necessity of projecting data from animals, usually rodents, to humans has often been used as an argument against the validity of the test.

CLIN. CORR. 17.2 DIAGNOSIS OF GENETIC DISEASE BY SOUTHERN BLOTTING

The application of restriction enzymes, in conjunction with hybridization techniques, has been responsible for the development of methods that permit the diagnosis of certain genetic defects. Diseases, such as sickle-cell anemia, β-thalassemia, Huntington's chorea, and Duchenne's muscular dystrophy, can be diagnosed by determining variations in the nucleotide sequences of DNA obtained from the skin or the blood of adults or children and even from the amniotic fluid for in utero testing.

Southern blotting is easiest to apply for detecting diseases characterized by deletion or insertion mutations, such as certain forms of thalassemia. A gene deletion often results in the disappearance of a restriction fragment in the Southern blot. These deletions are frequently associated with α-thalassemia in Southeast Asia and with some forms of Indian thalassemia. Southern blotting is not generally suitable for picking out point mutations, but there are exceptions. Sickle-cell anemia, associated with an A → T point mutation, can be easily detected by the Southern blotting technique, because this mutation eliminates the site of interaction for the restriction enzyme Mst III. As a result, a different pattern of β-globin gene fragments are generated from the DNA of the sickle-cell homozygote than the normal cell upon digestion with this enzyme. In fact, this test, which generates fragments 1,350 base pairs long for sickle cell DNA vs 1,150 base pairs for the normal genome, is so sensitive that it can be applied directly to uncultured amniotic fluid cells.

It should be noted that the use of specific hybridization probes in identifying defective genes, as exemplified above, has general applicability. In essence, the technique requires the isolation of mRNA from a patient's cells. This is used to synthesize (via reverse transcriptase) a specific cDNA probe. Hybridization with this probe detects DNA "markers" on the patient genome that are linked to a mutant gene. Many probes have been recently isolated, associated with such genetic abnormalities as the Lesch-Nyhan syndrome and phenylketonuria, and many more are being developed.

Excision Repair

In this type of repair the damaged part of DNA is removed and replaced with the correct structure. The process of excision repair in *E. coli* is a multistep process involving the coordinated participation of several enzymes, including endonucleases, DNA polymerases and DNA ligase; this is illustrated in Figure 17.48. The repair is initiated by the recognition of the distortion of the DNA structure by an endonuclease consisting of three different proteins that are the products of genes UvrA, UvrB, and UvrC of *E. coli*. The enzyme binds near the distorted area and introduces a nick on the 5′ side of a pyrimidine dimer. Once the 5′ end is liberated, DNA polymerase I begins to remove nucleotides one at a time from the 5′ end of the nicked strand. Polymerase I is also vested with the endonuclease activity illustrated in Figure 17.49. This endonuclease activity supplements the 5′ → 3′ exonuclease action of polymerase I in removing the dimeric-pyrimidines. The resulting gap is concurrently filled by DNA polymerase I so as to restore the original structure.

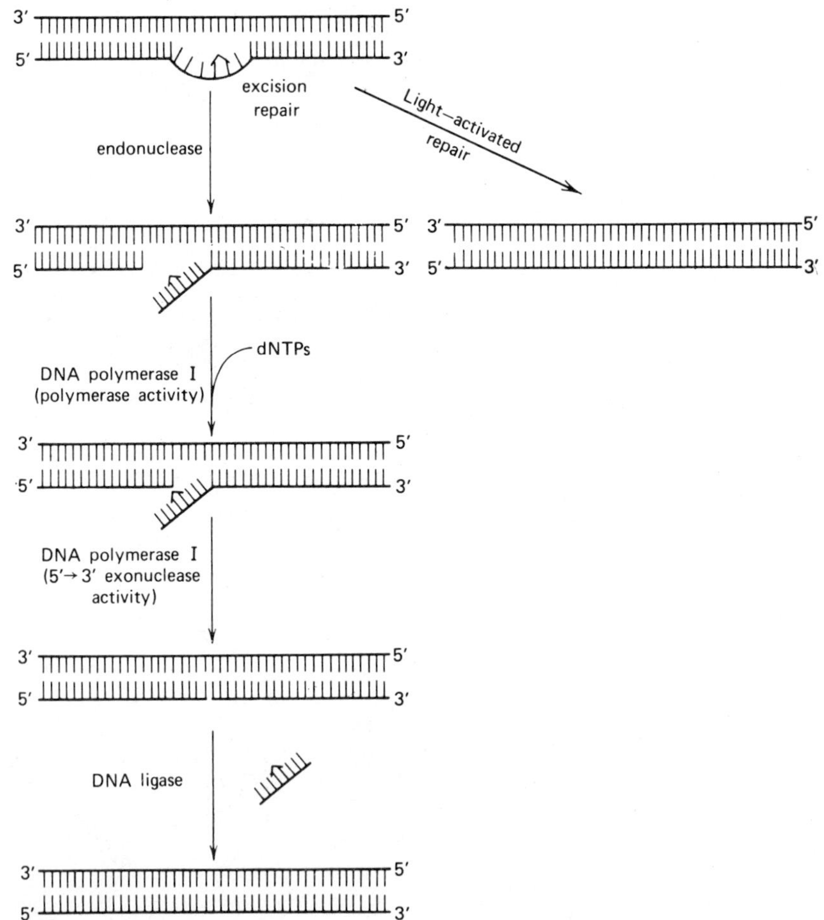

Figure 17.48
Repair of DNA damaged by ultraviolet light.
Excision repair requires the coordinated action of many enzymes. The repair begins with an incision by a specific endonuclease at a point several bases away from the distorted region of DNA that contains the dimer. DNA polymerase I elongates the 3′-OH terminus, using the other DNA strand as a template. Polymerase I also removes the DNA fragments containing the dimer. The final gap is sealed by the action of DNA ligase, and the damaged DNA region is restored to an intact state. Alternatively, the dimerization can be directly reversed by a process catalyzed by light, known as light activated repair.

The mechanism by which the DNA polymerase I fills the gap in the DNA strands undergoing repair is applicable to the processes of both DNA repair as well as DNA synthesis. The polymerase exhibits a combination of synthetic and exonucleolytic activities, which are vested in separate active sites of the enzyme. The synthetic site uses as a primer the 3′ terminus of the strand from which the pyrimidine dimer has been removed. The intact complementary strand serves as the template that directs the incorporation of the nucleotides needed to restore the original structure of the strand undergoing repair. The 3′ → 5′ exonucleolytic activity of polymerase I insures the fidelity of the synthesis as usual. At the same time the distinct 5′ → 3′ hydrolytic site of the enzyme removes nucleotides from the 5′ terminus of the strand undergoing repair. The concerted actions of polymerase I therefore close the gap created by the excision of the pyrimidine dimer until the gap is reduced to the size of a nick, that is, until a single phosphodiester bond remains open. As a result of the combined synthetic–nucleolytic action of polymerase I, the nick can move along the strand undergoing repair (nick translation) until it is finally bridged by the action of DNA ligase, as illustrated in Figure 17.50.

An additional distinct type of excision repair has recently been found to operate in T4 phage infected *E. coli* and in *M. luteus*. This process, illustrated in Figure 17.51, depends upon the action of a *N-glycosylase* that acts on the pyrimidine dimer by nicking the *N*-glycoside bond that connects one of the dimeric pyrimidines to its polynucleotide strand. DNA must subsequently be nicked on both sides of the nucleotide residue that carries the dimeric pyrimidine by *apurinic-apyrimidinic (AP) endonuclease* so that the dimeric residue can be removed. The glycosylase-AP endonuclease activity is present in a single protein induced by uv irradiation. The single nucleotidyl gap that results from the removal of the dimer can then be filled by DNA polymerase I.

Excision repair can also remove cross-links between complementary DNA strands, such as those introduced by the mustards and drugs used in cancer therapy (i.e., mitomycin D and platinum complexes). In such cases error-free repair is not possible if the cross-link extends across directly opposing bases. Clinical Correlation 17.3 discusses defects in DNA repair that are associated with human disease.

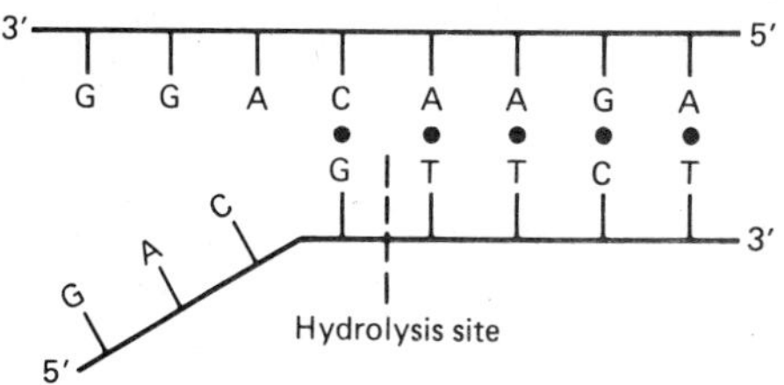

Figure 17.49
The endonuclease activity of DNA polymerase I.
The specialized endonuclease activity of polymerase I illustrated in this figure catalyzes the hydrolysis of a phosphodiester bond connecting two base pairs that follow a segment of unpaired bases which terminates in a 5′-phosphate. Note that bases located left from the hydrolysis site are generally not complementary.

Figure 17.50
The action of DNA ligase.
The enzyme catalyzes the joining of polynucleotide strands that are part of a double-stranded DNA. A single phosphodiester bond is formed between the 3′-hydroxyl and the 5′-phosphate ends of the two strands.

In E. coli cells the energy for the formation of the bond is derived from the cleavage of the pyrophosphate bond of NAD$^+$. In eucaryotic cells and bacteriophage-infected cells energy is provided by the hydrolysis of the α,β-pyrophosphate bond of ATP.

Figure 17.51
DNA repair by the combined action of N-glycosylase and AP endonuclease.
Repair is initiated by a cleavage of an N-glycosidic bond connecting the thymine dimer with ribose by a dimer-specific glycosylase, followed by cleavage of phosphodiester bonds on both sides of the dimeric thymidine residue by AP endonuclease.

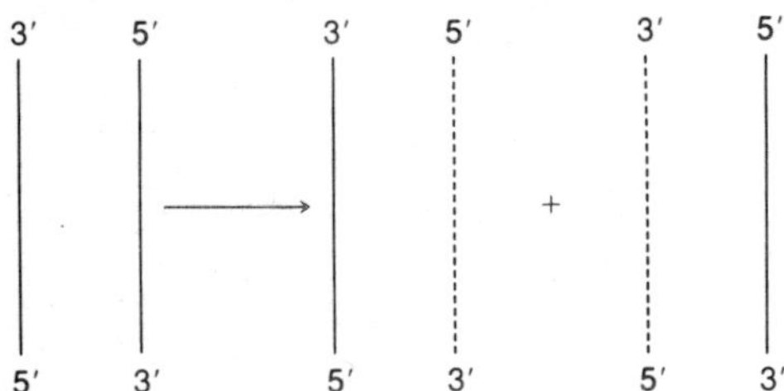

Figure 17.52
Each DNA strand serves as template for the synthesis of a new complementary strand. *Replication of DNA proceeds by a mechanism in which a new DNA strand (indicated by a dashed line) is synthesized that matches each original strand (shown by solid lines).*

CLIN. CORR. **17.3**
DISEASES OF DNA REPAIR AND SUSCEPTIBILITY TO CANCER

Defects in DNA repair may lead to a number of diseases including *xeroderma pigmentosum, hereditary retinoblastoma* and *Fanconi's anemia. Xeroderma pigmentosum* is the best understood of these conditions. Those suffering from the last affliction are particularly sensitive to the effects of sunlight. Exposure to the sun creates severe skin reactions which range initially from excessive flecking and skin ulceration to the eventual development of skin cancers. Some forms of the disease are also accompanied by various neurological abnormalities.

Normal mammalian cells can carry out excision repair of DNA. In contrast, skin cells from *xeroderma pigmentosum* patients are unable to repair the DNA damage produced by ultraviolet light. This defect originates from a reduced effectiveness of the first step of the repair mechanism, that is, the step at which an endonuclease nicks the DNA undergoing repair near a pyrimidine dimer. It has been suggested, based on indirect evidence, that in *xeroderma pigmentosum* patients this enzyme is defective. It is more likely, however, that the excision repair defect is the result of impaired access of the endonuclease to chromosomal DNA rather than a manifestation of the presence of a defective enzyme per se. If this is the case, then some yet unidentified control factors needed to expose DNA, in order that they become susceptible to the action of endonuclease, could be missing.

17.7 DNA REPLICATION

Basic Elements of the Mechanism of Replication

From the very same moment the double-stranded structure of DNA was proposed, it was apparent that this structure could serve as the basis of a mechanism for DNA replication. The complementary structure of the strands was immediately perceived as a characteristic, which, in principle, permitted each one of the strands to serve as a template for the synthesis of a new strand identical to the other strand, as suggested in Figure 17.52. A number of pathways could be easily visualized that lead to the synthesis of two new double-stranded helices, identical to one another and to the maternal double helix.

In the more than quarter-century since the double helix was proposed, the correctness of this overall scheme of replication has been solidly established. Even bacteriophages, which contain single-stranded instead of double-stranded DNA, have been shown to convert their DNA to a double-stranded form before replication. Our expanding knowledge of the character of DNA replication has also revealed that the simplicity of the basic scheme conceals, in fact, a rather complex set of more intricate substeps. A multiplicity of enzymes and protein factors participate in the process of replication. Before the synthesis of a DNA molecule can be brought to successful completion, the enzymes involved in replication must deal with a variety of topological problems.

These problems originate partly because the DNA-dependent DNA polymerase can synthesize new strands by operating *along the 5′ → 3′ direction only,* and therefore it is unable to elongate the two *antiparallel* strands of the helix in the *same macroscopic direction*. Also, the replication of DNA cannot proceed unless the complementary strands are separated at an early stage of the synthesis. Separation requires the commitment of energy for disrupting the thermodynamically favorable double-helical arrangement and the unwinding of a highly twisted double helix at extremely rapid rates. As if these difficulties were not enough, double-stranded DNA is normally a topologically closed domain which, unless properly modified, will not tolerate strand unwinding to any appreciable degree. Obviously, these multiple difficulties are enzymatically resolved before the replication of DNA can take place.

In order to ease the complexity of describing the process of DNA replication, the substeps of this process will be described first. The presentation of a unique model with universal applicability is not feasible because variations in the mode of replication have been noted among different species.

Replication Is Semiconservative

The concept that DNA strands are separable and that new strands, complementary to the preexisting strands, can be assembled from free nucleotides on each separate strand, is not new. In a macroscopic sense three possibilities by which information transfer could take place during replication were initially visualized as indicated in Figure 17.53. *Conservative* replication could in principle yield a product consisting of a double helix of the original two strands and a daughter DNA consisting of completely newly synthesized chains. A second possibility, labeled *dispersive,* would have resulted if the nucleotides of the parental DNA were randomly scattered along the strands of the newly synthesized DNA.

The synthesis of DNA eventually proved to be a *semiconservative* process. After each round of replication, the structure of paternal DNA is

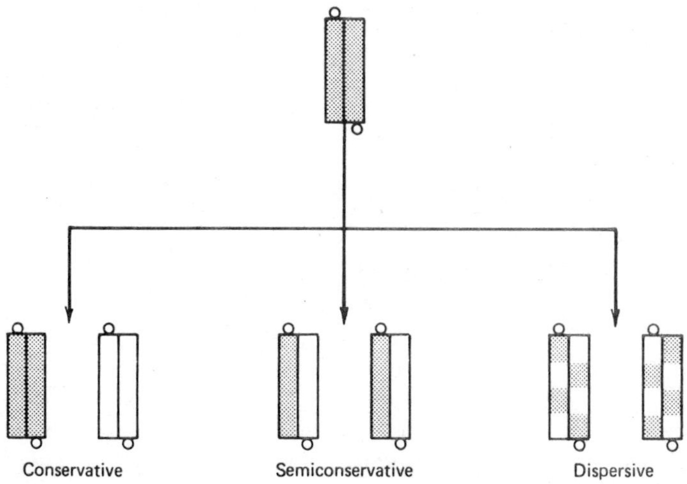

Figure 17.53
Three possible types of DNA replication.
Replication has been shown to occur exclusively according to the semiconservative model, that is, after each round of replication one of the parental strands is maintained intact, and it combines with one newly synthesized complementary strand.

One particularly intriguing aspect of diseases related to defective DNA repair is their possible relationship to carcinogenesis. Although these diseases are rare autosomal recessive conditions, the carriers of the defective genes are relatively common; carriers of *xeroderma pigmentosum* gene account for as much as 1% of the general population. An interesting finding is that not only are those who are afflicted by *xeroderma pigmentosum* susceptible to cancer but the carriers of the gene have a higher incidence of skin cancer. These and other similar findings suggest the presence of some subtle defect in DNA repair among the carriers of these genes. Further research on the predisposition of these carriers to cancer might provide some clues regarding the mechanism of carcinogenesis for some types of cancer.

found to preserve one of its own original strands combined with a newly synthesized complementary polynucleotide.

The semiconservative nature of replication was elegantly suggested by a classic experiment which allowed the physical separation and identification of the paternal and the newly synthesized strands. For this experiment *E. coli* was grown in a medium containing [^{15}N]ammonium chloride as the exclusive source of nitrogen. Several cell divisions were allowed to occur during which the naturally occurring ^{14}N in the DNA of *E. coli* was, for all practical purposes, replaced by the heavier ^{15}N isotope. The ^{14}N-containing nutrient was then added, and cells were removed at appropriate intervals. The DNA of these cells was extracted, and the ratios of ^{14}N ^{15}N content were determined by equilibrium density gradient centrifugation. The separation between ^{14}N and ^{15}N DNA was achieved based on the lower density of DNA, which contained the lighter isotope. In subsequent experiments, the newly synthesized DNA was thermally denatured and the individual strands were completely separated. The results, shown in Figure 17.54, demonstrated that daughter DNA molecules consisted of two strands with different densities, corresponding to the densities of single-stranded polynucleotides containing exclusively ^{14}N or ^{15}N.

Clearly, the synthetic activity of DNA polymerase makes it possible for the enzyme to synthesize new complementary DNA strands by using in turn each parental DNA strand as template.

A Primer Is Required

The semiconservative nature of DNA replication requires that each strand serve as a DNA polymerase template for the synthesis of a new complementary strand. The polymerase that catalyzes the primary synthetic reaction, that is, the elongation of DNA polynucleotides, is polymerase III (Table 17.7) as distinguished from polymerase I, which is primarily a repair enzyme. Polymerase III is an ATP-dependent enzyme, consisting of at least 13 discrete polypeptide units. This polymerase is unable to assemble the first few nucleotides of a new strand and needs a

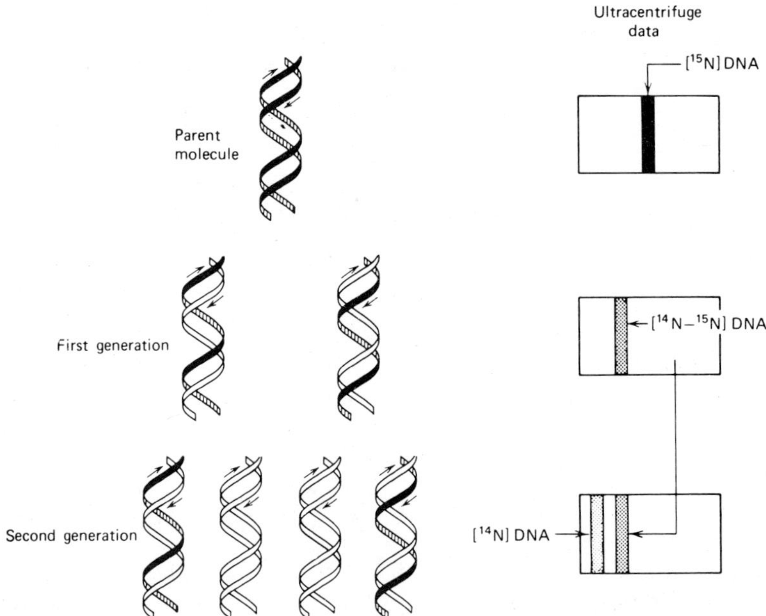

Figure 17.54
Semiconservative replication of DNA.
Schematic representation of the experiment of Meselson and Stahl that demonstrated semiconservative replication of DNA. This model of replication requires that, if the parent molecule (shown in black) contains ^{15}N, each of the molecules produced during the first generation contain ^{15}N in one strand and ^{14}N in the other. Furthermore, in the second generation two molecules must contain only ^{14}N, and two molecules must contain equal amounts of ^{14}N and ^{15}N. The results of separating DNA molecules from successive generations, shown on the right are consistent with this model.

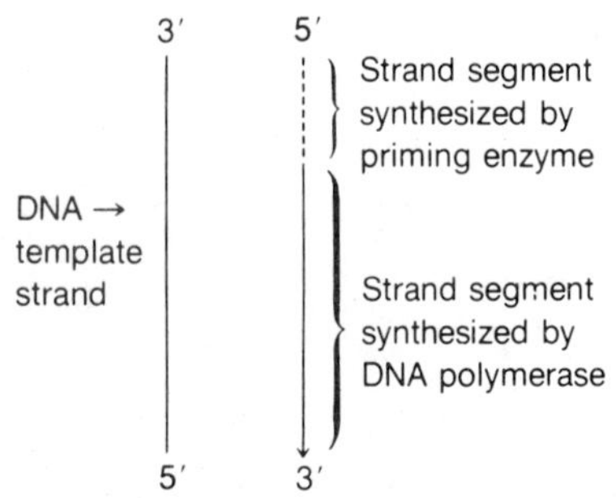

Figure 17.55
Synthesis of primer for DNA replication.
The primer (dashed line) is synthesized by specific enzymes. The existence of a primer permits new DNA (solid line) to be synthesized after which the primer is excised.

primer that varies in size between a few nucleotides in procaryotes (1–5) to about 10 nucleotides in animal cells. With few exceptions, the primer is an oligonucleotide synthesized by other enzymes, as indicated in Figure 17.55.

In some bacterial systems and phages, the priming enzyme has activity characteristic of an RNA polymerase, that is, the nucleotides condensed to form the primer are ribonucleotides rather than deoxyribonucleotides. In other bacterial systems and in some phages, initiation is achieved by the action of a *primase*. An interesting characteristic of primase is that in vitro it is not particularly discriminating between 5′-ribo and 5′-deoxyribonucleotides, both of which can be selected under appropriate conditions for the formation of the primer. Recently, the term "primase" appears to be used in a generic manner to describe all enzymes with priming activity.

Once the primers have been synthesized, the DNA polymerase can move in and take over the process of synthesis. It is not clear what signal causes a switchover from primase to DNA polymerase, although it has been suggested that a ribonuclease (RNaseH) is involved. RNaseH activity appears to be specifically directed at RNA that is hydrogen-bonded to a DNA strand.

Both Strands of DNA Serve as Templates Concurrently

In the preceding section, the events leading to the synthesis of DNA by DNA polymerase were examined and attention was directed to one of the two parental DNA strands used as template. In fact, synthetic events occur at both strands almost concurrently. This would appear to generate some problems of geometry. Specifically, if a single initiation site is considered, and if the synthesis is assumed to continue until each template is completely copied, the result of the synthesis would be the creation of *two* new double-stranded molecules. Examination of Figure 17.56 indicates that at least in the case of linear double-stranded DNA, none of these two hypothetical DNA molecules would be identical to the parental DNA.

Such an outcome is not in agreement with the actual course of DNA replication. The discrepancy can be accounted for because it is recognized that the microscopic synthesis of the new strands does not proceed uninterrupted. In fact, the synthesis occurs in a discontinuous fashion and in a manner that permits the assembly of the synthesized polynucleotide portions into appropriate complete DNA strands.

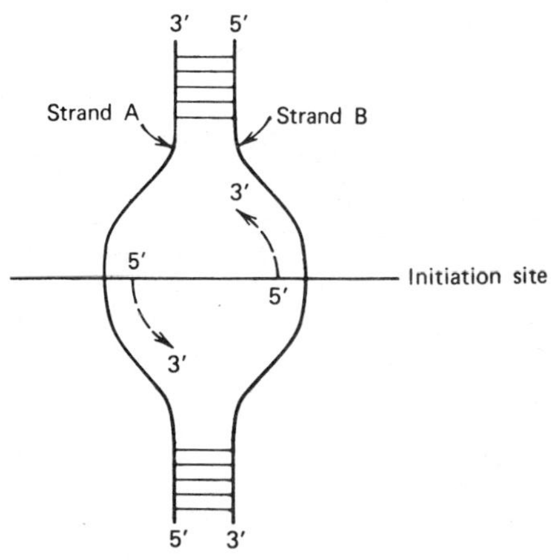

Figure 17.56
Both DNA strands serve as templates for DNA synthesis.
Each DNA must serve as a template for DNA synthesis. The new DNA can be synthesized only in the 5′ → 3′ direction. For these reasons if only a single initiation site were considered, the result of the synthesis would be the formation of two new nonidentical double-stranded DNA molecules (one above and one below the initiation origin). Also, the upper part of strand A and the lower part of strand B could not have been used as templates. More than a single initiation site is involved.

Synthesis Is Discontinuous

Examination of the overall process of DNA synthesis should now be expanded past the immediate vicinity of initiation and encompass a larger section of DNA. The attention of the reader should be focused on only one of the two parts of DNA that would be generated if the macromolecule were divided at the site of chain initiation, as indicated in Figure 17.57. In most instances, the synthesis is bidirectional, which means that the synthetic events occurring at the part of the molecule indicated by solid lines are of the same general nature as those occurring on the other site with dashed lines.

A prerequisite for the semiconservative mechanism of replication is that the two complementary strands of DNA gradually separate as the synthesis of new strands takes place. The mechanics of this separation is addressed later, but it may be apparent that as a result of separating the strands at an interior position, two topologically equivalent forks are created at the point of diversion of the two strands.

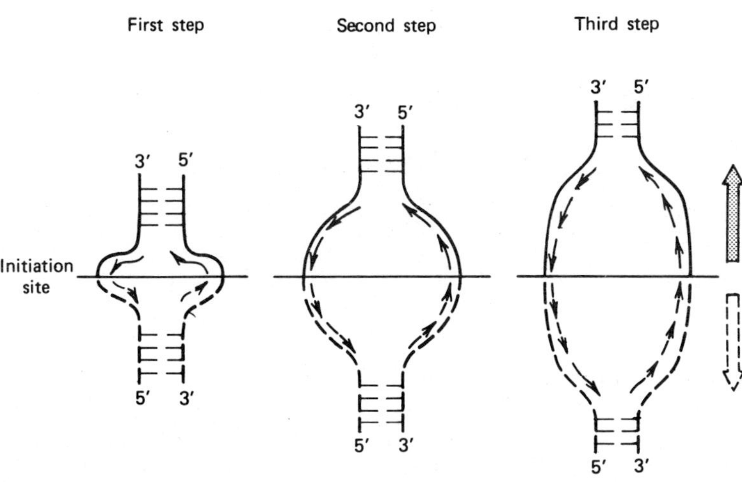

Figure 17.57
Discontinuous synthesis of DNA.
In this figure emphasis is placed on the synthetic events occurring at only one side of the initiation site (solid line). The two complementary strands of DNA separate as the discontinuous synthesis of small DNA segments takes place on both strands. After excision of the primers, the excised parts are repaired, and the segments are joined together. Although the segments are clearly synthesized in opposite directions on the two strands, the overall macroscopic impression is that the DNA grows in the single direction suggested by the solid arrow on the right.

The observation to be emphasized presently is that DNA polymerase acts in a discontinuous manner, that is, along each DNA molecule there are numerous points at which primers are formed. How these points are selected is not precisely known, but, at least in *E. coli* and some of its phages, they appear to be determined through the action of *prepriming proteins*. Prepriming proteins form, along with primases, complex assemblies known as *primosomes*. In *E. coli,* the primosome may contain, in addition to primase, six distinct proteins designated as n, n′, n″, i, dnaB, and dnaC. Among these proteins, n′ may recognize DNA structures that serve as signals for the initiation of the prepriming events. Once a signal has been recognized, the n′ protein begins to displace DNA-binding proteins using ATP. Concurrently, dnaB, which is also a DNA-dependent ATPase, in association with dnaC uses ATP to modify the secondary structure of DNA. The resulting conformational change facilitates the action of primase leading to the formation of a primer as the primosome is propelled to the next priming site. After promoting initiation at one point, primosomes may move along the template strand, assisted by their own ATPase activity. Thus, the primosome can promote the synthesis of the next polynucleotide segment and so on until a large DNA segment is completed.

The segments built by DNA polymerase upon each primer, which are known as *precursor* (*Okazaki*) *fragments* or *nascent* DNA, vary in size from 100 to 200 deoxyribonucleotides in the case of eucaryotes to about 10 times as long in the case of bacteria. Once the small segments of the new DNA strands are synthesized on both strands of a fork (upper part of Figure 17.54), the fork opens up further, and the same process of synthesis can be repeated.

Shortly after synthesis, the primer portions of the Okazaki fragments are excised by the 5′ → 3′ exonuclease activity of DNA polymerase. The DNA polymerase therefore serves both as an exonuclease as well as a repair enzyme in a manner conceptually similar to that by which photochemically damaged DNA is repaired. The nicks, which remain after repair, are closed by DNA ligase.

This discontinuous mechanism compensates for the inability of DNA polymerase to synthesize strands in the 3′ → 5′ direction. By synthesizing portions of DNA strands only in the 5′ → 3′ direction on both antiparallel strands of the parental DNA, the polymerase is able to produce *the illusion that both strands are concurrently elongated in the same macroscopic direction*. In Figure 17.57 this direction is indicated by a large solid arrow. It should be noted that the first strand synthesized, often referred to as the *leading* strand, appears with few exceptions to be synthesized *continuously*. It is the second strand, *the lagging strand, that must be synthesized discontinuously*.

Macroscopic Synthesis Is as a Rule Bidirectional

Examination of Figure 17.57 indicates that at the site of initiation of DNA synthesis two identical forks are created. Therefore two possibilities exist for the synthesis of DNA: the process may occur at only one fork and proceed in a single direction, as shown by the thick solid arrow, or alternatively it may occur at both forks and in both directions away from the starting point. The events occurring in the forks located below the starting line are simply a mirror image repetition of what occurs in the fork that is located above the line. Bidirectional replication is the exclusive mechanism of DNA synthesis by animal cells. There are exceptions to the rule of bidirectionality in a small number of phages and plasmids that replicate unidirectionally. The direction of replication can be experimentally determined by the application of autoradiographic techniques. In the case of a small linear chromosome (e.g., bacteriophage λ) each fork moves along, synthesizing new DNA, until the end of the chromosome is reached. In a circular chromosome (e.g., *E. coli*) the two forks proceed in opposite directions until they meet at a predetermined site on the other side of the chromosome, as depicted in Figure 17.58. As the two forks meet, a new copy of the parental DNA is completed and released. The average rate at which each fork moves during replication is of the order of 60,000 bases per minute at 37°C.

Strands Must Unwind and Separate

Separation of the strands of the parental DNA prior to the synthesis of new strands is a requirement because the bases of each template must be made accessible to the complementary deoxyribonucleotides from which the new strands are constructed. The overall process of separation consists of a number of enzymatically catalyzed, coordinated steps, including the local unwinding of the helix, and the nicking and rejoining of the strands necessary for the continuation of the unwinding process. Once the strands are unwound, they must be kept separate so that they can operate freely as templates.

Helicase. The fact that most DNAs are circular supertwisted molecules facilitates the unwinding of the helix. They contain a net deficit of helical turns in comparison to the corresponding "relaxed" molecules, which introduces into the double helix a tendency toward partial unwinding. This tendency is even more pronounced in regions richer in A-T pairs, which are intrinsically less stable than G-C pairs. Therefore A-T-rich regions are susceptible to local melting, that is, unwinding and partial separation of the strands of the double helix.

The cell has to resort to the services of specialized proteins to accomplish the rapid orderly unwinding of the strands. These proteins separate DNA strands in advance of the moving replication fork; for the *E. coli* system they are referred to as *helicase II* and *rep protein*. Helicases move unidirectionally along one or the other strand of the DNA and separate the strands in advance of replication. They destabilize the interaction between complementary base pairs at the expense of ATP.

Binding Proteins. Once the strands have been separated, the single-stranded regions are stabilized by specific proteins, the *binding proteins,*

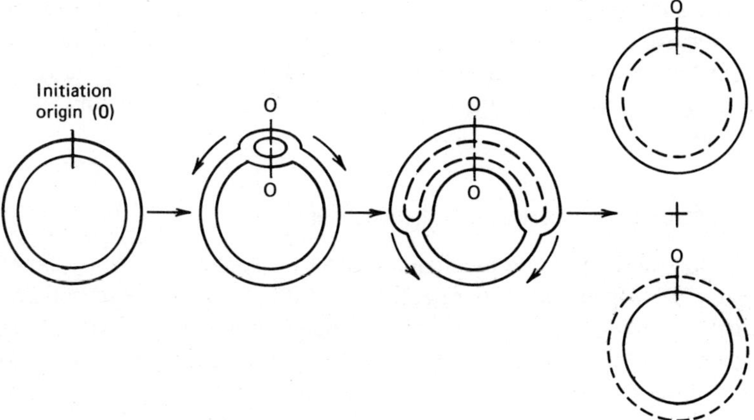

Figure 17.58
Bidirectional replication of a circular chromosome.
Replication starts at a fixed initiation origin (0), and goes in opposite directions until the replication forks meet. Newly synthesized strands are indicated by a dashed line.

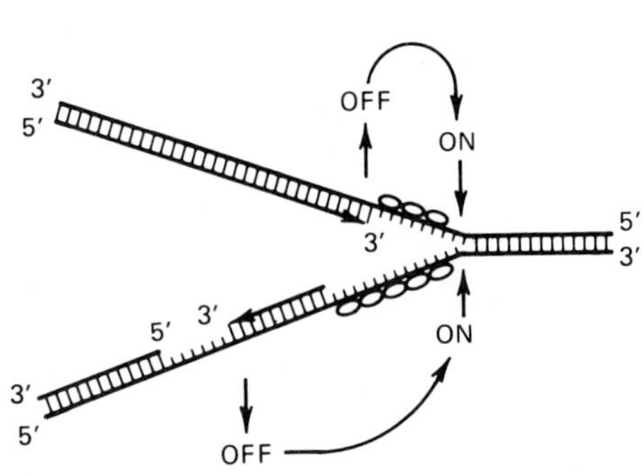

Figure 17.59
Function of single-strand binding protein in replication.
Single-strand binding proteins operate at regions of single-strandedness including the replication fork. The protein may follow the movement of the fork using the hypothetical scheme shown in this figure.
Redrawn with permission from *DNA Replication* by A. Kornberg, W. H. Freeman and Company, Copyright © 1980.

TABLE 17.8 Replication Proteins of *E. coli*

Protein	*Function*
SSB	Single-strand binding
Protein i Protein n Protein n′ Protein n″ dnaC dnaB	Primosome assembly and function
Primase	Primer synthesis
Pol III holoenzyme	Processive chain elongation
Pol I	Gap filling, primer excision
Ligase	Ligation
Gyrase gyrA gyrB	Supercoiling
rep Helicase II	Helicase
dnaA	Origin of replication

which have been identified in both procaryotic and eucaryotic systems. DNA single strands are covered by the binding proteins because of the high stability of the complexes formed between these proteins and single-stranded DNA regions. As the helicase moves in advance of the replication fork, binding proteins go on and off the DNA, with protein molecules that are displaced from one site reassociating with another (Figure 17.59). Binding proteins do not consume ATP and do not exhibit any enzymatic activities. Their role is only to keep the strands apart long enough for the priming process to occur.

Topoisomerases. Even after the local unwinding and separation of the strands is achieved, other practical problems must still be solved in order for the replication to proceed unimpeded. For the *E. coli* DNA it may be calculated that the parental double helix must unwind at a rate of about 6,000 turns per minute. These high rates would generate serious difficulties if strands were to separate over an appreciable length of DNA. The large free energy requirements of bringing about the unwinding of large regions of DNA can, however, be reduced to manageable levels by the nicking of one or both of the DNA strands near the replicating fork. Since the fork is a moving entity, the nicking must be visualized as a reversible cut-and-rejoin process, which moves along with the fork.

Nicking is indispensable for a topological reason as well. Unwinding at one of the two forks requires that the parental double helix rotates in the opposite direction to that necessary for the unwinding of the opposite fork. Furthermore, even if only one of the forks is considered and the other is ignored, the topological problem still remains unresolved. In the absence of a nick as the unwinding at one of the forks would progress, an increasing number of positive supercoils would have to be introduced into the double helix. Once the limit of the helix to accommodate the supercoils is reached, the unwinding and the replication would have to stop.

The above topological restraints can be overcome if DNA is maintained during replication in the *negative superhelical* form. This form could serve as a "sink" for the positive supercoils that can potentially be generated during replication. In *E. coli, gyrase,* a topoisomerase type II, induces the formation of negative supercoils at the expense of ATP. Topoisomerases type I, on the other hand, tend to relax supercoiled DNA. The in vivo superhelicity of DNA may be negatively regulated through a balance between topoisomerases of types I and II; that is, a diminishment of topoisomerase II activity may bring about a decrease in the amount of negative superhelicity that can be created, whereas an inhibition of topoisomerase I activity may increase it. During replication the linking number between the parental strands decreases from a large value at the beginning of replication to zero at the end of a complete round of DNA synthesis.

Basic Model for the Replication of DNA

Our understanding of the mechanism of DNA replication is still far from complete. Extensive studies in *E. coli* and its phages have permitted the proposal of a replication model that depends upon the action of a large number of proteins listed in Table 17.8. With the specific exceptions noted in the following two sections, this model may also be viewed as a basic scheme for DNA replication in other cells. Synthesis of DNA begins with the initiation of the *leading* strand that occurs at a specific site of the chromosome, referred to as the replication origin or *oriC* (Figure 17.60a). *oriC* consists of a unique sequence of 245 base pairs that is recognized by an enzyme system responsible for the initiation of bidirectional replication. This enzyme system is composed of five proteins; the precise role of

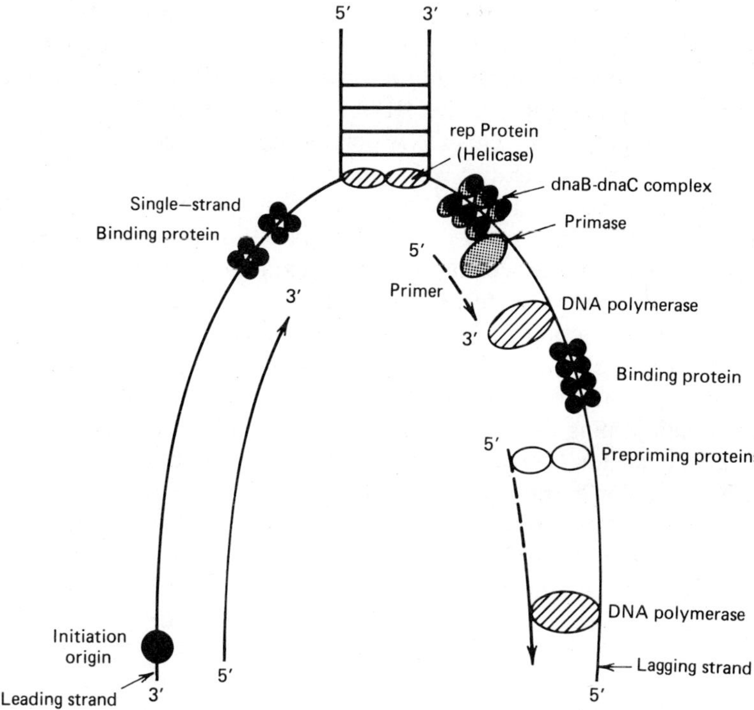

Figure 17.60
***Model for DNA replication in* E. coli.**
In this figure the initial stages of replication are depicted. The primers are subsequently removed from the newly synthesized segments of DNA at the lagging strand, and the segments are joined. Since replication is normally bidirectional, similar events take place concurrently at the other side of the initiation origin.

the individual proteins in the initiation reaction is not yet understood. There is only one *oriC* site per DNA molecule. The subsequent initiation of the synthesis of precursor (Okazaki) fragments on the *lagging* strand, which is controlled by the primosome, is a process different from the initiation of replication. Clearly, initiation of the synthesis of the leading strand at *oriC,* as opposed to the repetitive initiation of the lagging strand, requires, in addition to the formation of a primer, the unwinding of the double helix. The repeated initiations of the lagging strand occur by a different mechanism.

The initiation of the leading strand does not present the cell with serious topological problems because of the negative superhelicity initially present in the circular DNA. For the continuation of the synthesis, however, the presence of helicase activity, which in *E. coli* originates from the two enzymes designated as *helicase II* and *rep protein,* is essential. These enzymes unwind and separate the strands in each of the two forks created by the initiation event. As the helicases move in advance of each fork, two single-stranded regions are generated on parental DNA. These regions are immediately covered by single-stranded binding protein that keeps the fork open and allows DNA polymerase III to take over the elongation of primers. A short time after initiation at the leading strand a signal, uncovered on the template for the initiation of the lagging strand by the movement of helicase, directs a primosome to become attached to the template. Subsequently, primase bound to the lagging strand template moves in the same direction as helicase (a direction that is opposite to that of chain elongation) to set the stage for the discontinuous synthesis of the lagging strand. Primase, the action of which is triggered by the prepriming

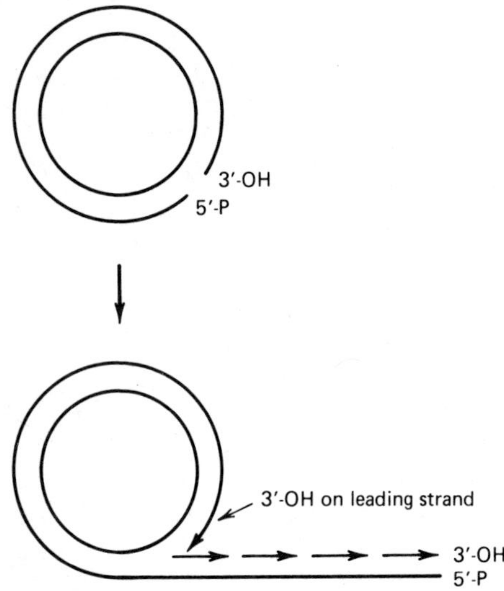

Figure 17.61
Replication by the rolling circle mechanism. *Synthesis of the leading strand occurs by elongation of the 3′-OH terminus generated by endonuclease cleavage of the DNA. As the leading strand is synthesized, the parental template directs the synthesis of the lagging strand in the form of precursor fragments.*

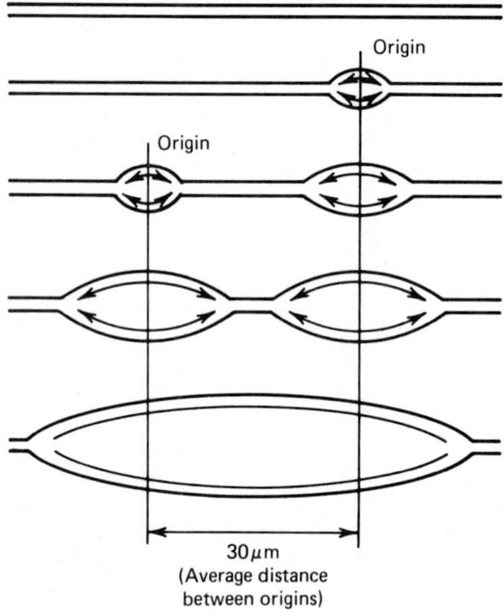

Figure 17.62
Replication of mammalian DNA.
Mammalian DNA replicates by using a very large number of replicating forks simultaneously. This mechanism serves to accelerate the process of replication, which in mammalian systems is limited by rates of fork movement that are considerably slower than those characteristic of procaryotes.
Redrawn from J. A. Huberman and A. D. Riggs, *J. Mol. Biol.* 32:327, 1968.

proteins, synthesizes a brief complementary segment of the strand. This segment serves as a primer for covalent extension of the strand synthesized by DNA polymerase III and for the formation of Okazaki fragments. It appears likely that polymerase III functions as a pair of enzymatically active protein molecules. This DNA polymerase III dimer might be associated with a primosome and with one or more helicase molecules in a multisubunit structure referred to as a *replisome*. Replisomes, which may be formed at each replicating fork, coordinate through physical association the priming and the replicative processes. It is likely that the process of replication occurs concurrently on the two strands. This could be achieved by the action of a replisome on DNA that is appropriately looped near the replicating fork. Removal of the primer portions at the 5′ end of the Okazaki fragments by DNA polymerase I, repair by the same enzyme, and joining of the repaired fragments by DNA ligase produces intact DNA strands. Termination of the synthesis occurs near the center of a 270 Kb region across from *oriC*. It is postulated that upon completion of the synthesis the newly synthesized DNA is untangled from the parental DNA by the action of topoisomerase II.

Considerable variations in this basic replication scheme are apparent in both procaryotic and eucaryotic replication, some of which are examined below.

Rolling Circle Model for Replication

DNA synthesis directed by the circular DNA of mitochondria and in some instances of bacteria and viruses occurs in a manner that initially gives rise to linear daughter DNA molecules that contain the base sequence of parental DNA repeated numerous times. These repeated linear DNAs, which are known as *concatemers,* are important intermediates in phage production and are essential for the process of bacterial mating.

The synthesis of concatemer DNA occurs by a mechanism known as *rolling circle* replication, which may be involved in the process of gene amplification. The essential aspects of continuous replication of the leading strand and discontinuous replication of lagging strand described above is present in the rolling circle model. DNA replication by the rolling circle mechanism, shown in Figure 17.61, is different from the basic scheme of DNA replication described above in two important ways. The initiation of the synthesis of the leading strand does not make use of an RNA primer; instead the initiation depends upon the nicking of one strand by a phage-encoded endonuclease that generates 3′-OH and a 5′-P termini. The synthesis of the leading strand can thus occur by elongation of the 3′-OH terminus at a fork created under the influence of helicase and single-stranded binding protein. As the leading strand is elongated by the action of polymerase III, the parental template for the synthesis of the lagging strand is displaced and begins to replicate in the usual manner, that is, via Okazaki fragments.

A second characteristic that distinguishes the rolling circle model from conventional replication is that the circular template, used for the synthesis of the leading strand, does not dissociate from the complementary strand during the synthesis. Instead the replication of the leading strand goes on beyond the length of circle-generating linear concatemeric DNA. Appropriately sized DNA molecules are generated from concatemers by specific endonuclease cleavage.

Eucaryotic Replication

Replication of the eucaryotic chromosome must allow for additional complexities, including the organization of the DNA into chromosomes and the very large size of the eucaryotic chromosomes. In eucaryotes the rates of fork, and therefore polymerase, movement do not exceed 30,000 base pairs per minute, which is considerably slower than the rates observed for *E. coli*. Based on the higher DNA content of animal cells and the lower activities of DNA polymerases in comparison to bacteria, the replication cycle of eucaryotic cells could be expected to take as long as a month to complete. In fact, however, the replication cycle is completed within hours, because compensating factors are in operation. Eucaryotic cells contain a large number of DNA polymerase molecules (in excess of 20,000) as compared to no more than a couple of dozen molecules found in each *E. coli* cell. In addition, DNA polymerase molecules initiate bidirectional synthesis, not in one, but at several initiation points along the chromosome. The DNA segments between two initiation points are termed *replicons*. Therefore, for a DNA molecule that contains 1,000 replicons, replication may proceed simultaneously at as many as 2,000 forks. At each of these forks, strands are being replicated as Okazaki fragments. Replicons vary considerably in size and may extend across as many as 40,000 nucleotide pairs. Each mammalian chromosome may use for replication as many as several hundred replicons of different sizes as indicated in Figure 17.62.

In eucaryotes DNA is present in packaged form as chromatin. Therefore DNA replication is sandwiched between two additional steps, namely a carefully ordered and incomplete dissociation of the chromatin and, post replicatively, the reassociation of DNA with the histone octamers to form nucleosomes. The synthesis of new histones occurs simultaneously with DNA replication and the reassociation of newly synthesized nucleosomes contain only newly synthesized histones, indicating that the parental histone octamers are conserved into the constituent histones. Furthermore the histone octamers appear not to dissociate from DNA completely as indicated by the observations that the newly synthesized histone octamers are associated exclusively with one of the two daughter strands rather than becoming distributed between both daughter strands (Figure 17.63).

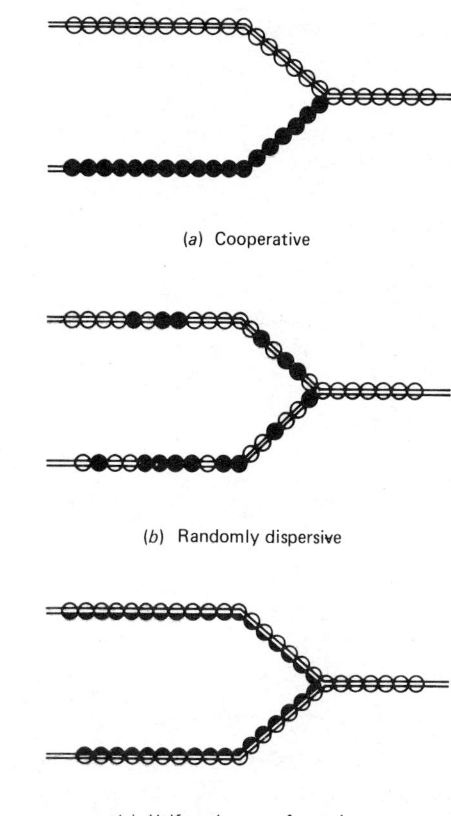

Figure 17.63
Distribution of nucleosomes at a replication fork.
Three possible mechanisms for the distribution of newly synthesized nucleosomes are depicted. The actual mechanism appears to be cooperative. (○, Parental nucleosome; ● newly synthesized nucleosomes).
Reproduced with permission from H. Weintraub, *Cell* 9:419, 1976. Copyright, MIT Press.

17.8 DNA RECOMBINATION AND RECOMBINANT DNA

DNA recombination is a general phenomenon during which two "parental" DNA molecules are spliced together, giving rise to a new DNA that contains genetic information from both parental strands. Recombination underlies many essential biological processes, including the crossing over between eucaryotic chromosomes during meiosis and the events that lead to exchange of genetic material between related DNAs. The extent to which DNA from different organisms "mixes" in nature is strictly controlled with combinations occurring only between "suitable" closely related DNA molecules. The most common example of DNA "mixing" is the *integration* of phage or plasmid DNA into the corresponding bacterial hosts. Another form of DNA exchange occurs as certain phages or animal viruses incorporate small segments of DNA of the host cell and transfer it to a recipient cell upon infection, a process termed *transduction*.

DNA Recombination

One type of recombination, known as *general recombination,* occurs between homologous DNA regions, that is, regions that are largely or completely complementary in their nucleotide sequences. However, a second type of recombination that results in the splicing of the parental DNA

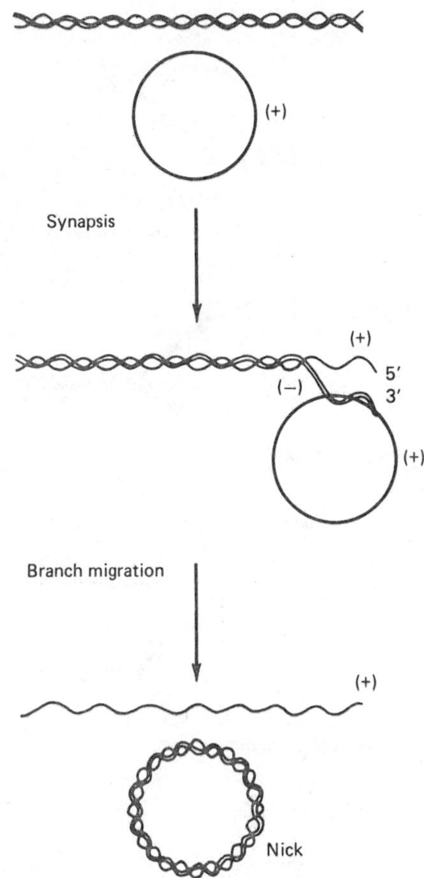

Figure 17.64
Homologous recombination.
In this example, a circular φX174 single strand is assimilated into a double-stranded (replicative form) φX174 DNA that is present in linear form. The process is initiated by a synapsis at one of the terminals of the linear duplex followed by branch migration. Both reactions require single-strand binding proteins, recA protein, and ATP.

Redrawn from A. Kornberg, *DNA replication.* 1982 supplement. New York: W. H. Freeman and Co., 1982.

strands at specific sites, referred to as *site-specific recombination,* takes place at sites characterized by limited complementary. An example of site-specific recombination is the integration of bacteriophage λ that occurs at a predetermined site on the bacterial chromosome and is characterized by the alignment of the integrated phage in a specific orientation within the *E. coli* chromosome. Bacteriophage λ and the *E. coli* chromosome have distinct recombining sites and both sites have a common sequence of 15 nucleotides. Such short sequences of homology are, as a rule, characteristic of site specific recombination. Four sites have been identified as phage attachment sites, one of which is in the homologous region that binds a phage protein with type I topoisomerase activity known as *integrase*. Integrase produces a staggered cleavage seven base pairs apart within the homologous region and catalyzes the exchange of strands at the position of the cut. The integration reaction is formally reversible but is precisely controlled with its forward and reverse steps separately regulated.

Site-specific recombinations appear to be widespread in nature and lead to rearrangements in the DNA of many genes that are referred to as *transpositions*. Transposable elements of DNA, or "jumping genes," are capable of movement from one chromosome to another or to a different site within the same chromosome. Transposable elements, which are also known as *transposons,* can modify gene expression and therefore introduce a substantial element of genetic flexibility into eucaryotic genomes. This flexibility challenges the traditional view that associates a specific gene with a particular chromosome.

Recombination that can occur at any complementary location along the length of the combining strands is referred to as *general* (*or homologous*) *recombination*. This process requires, in addition to extensive homology, that one of the parental DNA molecules be at least partially single stranded and that a free end exists at some location in either one of the parental DNA molecules. The example shown in Figure 17.64 illustrates the recombination between the single-stranded form of φX 174 viral DNA (replicative form) and a linear DNA duplex. General recombination is catalyzed by a protein, *recA*, which pairs the single-stranded DNA with the DNA duplex by placing the two molecules in homologous register in a process referred to as *synapsis*. RecA also catalyzes the migration of the branch formed during synapsis that leads to the transfer of one of the strands of the duplex onto the circular DNA to form an heteroduplex. Both reactions require ATP, but ATP hydrolysis is essential only for the formation of the heteroduplex by the process of *branch migration*. RecA protein controls the process of general recombination via its *ATPase* and *recombinase* activities that catalyze the binding of the duplex DNA and the unwinding of the double strands. RecA has no topoisomerase activity that explains why a free polynucleotide end must be present for recombination to occur. General recombination between two double-stranded DNAs may occur as indicated in Figure 17.65. RecA also exhibits a highly specific *protease* activity, activated by the presence of *unpaired* DNA strands, directed at specific regulatory proteins. Digestion of such regulatory molecules vests recA with unique properties for the coordinate regulation of a number of cellular functions that occur when DNA damage, or the interruption of DNA replication, leads to the production of single-stranded DNA segments. An example of such a process is the postreplication repair of DNA damaged by uv light or other mutagens.

Recombinant DNA

In recent years, the development of appropriate new techniques and the discovery of new enzymes, especially type II restriction endonucleases, has made possible the in vitro recombination of DNA fragments from the genomes of organisms that are totally unrelated to one another. The goal of most recombination experiments is to separate a single gene, or some other DNA element, from other genetic information present in the genome of the cell. Once separated, the DNA fragment, which is referred to as *passenger* or *foreign* DNA, must be amplified (*cloned*) so that sufficient quantities can be made available for studying the properties of the passenger DNA. From such studies the nucleotide sequence and the physicochemical properties of the passenger DNA can be determined and the organization of the genetic material and the control of its expression can be studied. For amplification (cloning) the passenger DNA must be introduced into a host cell (frequently *E. coli*) that will propagate the passenger as the host divides and multiplies (Figure 17.66). The capacity for amplification of the passenger depends strictly on coupling the passenger to a DNA molecule (*vector*) that is normally capable of promoting its own replication within the host cell. The coupling between passenger and vector DNAs is accomplished through making a recombinant DNA molecule by joining these two separate DNA segments through the use of appropriate techniques.

Practical benefits are also gained if the amplified (cloned) gene can be expressed in vitro. Technology is now available for the introduction of an intact piece of genetic information isolated from one organism, or synthesized by chemical or biochemical techniques, into DNA vectors so that the DNA that is introduced can be both replicated and expressed within a suitable host. This technology therefore permits the amplified synthesis of a polypeptide encoded by a particular gene, which normally would be available only in very limited amounts.

A boom to recombinant DNA technology was provided by the discovery of restriction endonucleases that can cut from a large DNA genome

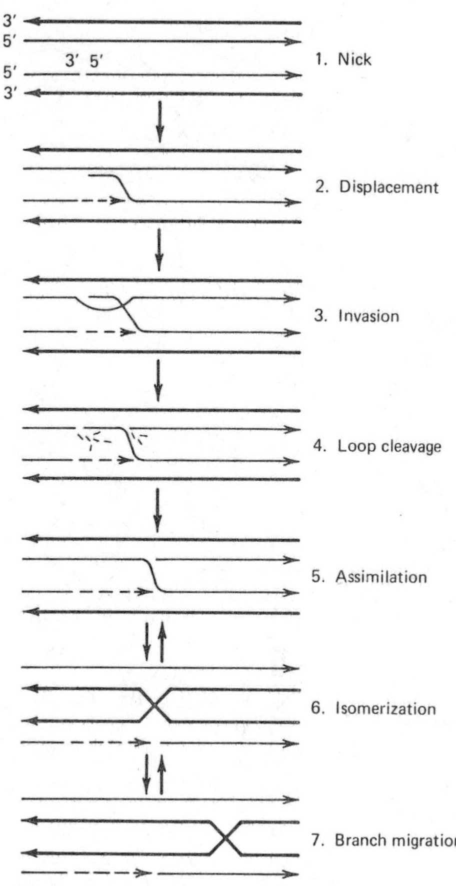

Figure 17.65
A model for general recombination between two duplex DNAs.
Recombination between two double-stranded DNA molecules may occur via the multistep process shown above. Dashed lines represent newly synthesized DNA.
Reprinted with permission from M. Meselson and C. Radding, *Proc. Natl. Acad. Sci.* 72:358, 1975.

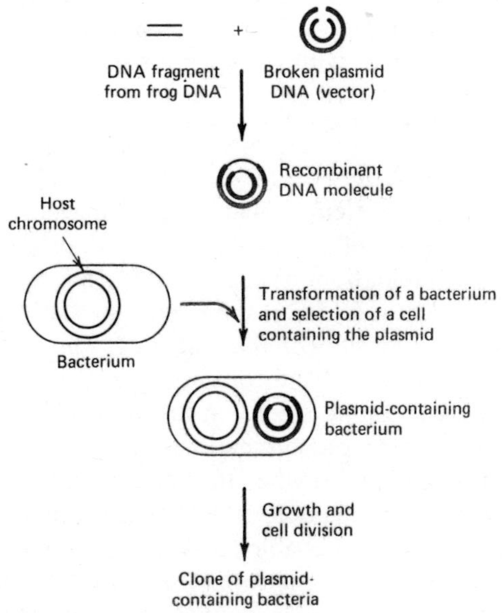

Figure 17.66
Overview of a cloning experiment.
A recombinant DNA molecule is obtained by joining a DNA fragment from frog to plasmid DNA, a small self-replicating circular DNA, which has been opened by treatment with a restriction endonuclease. The recombinant plasmid transforms a bacterium (usually E. coli*) and replicates along with the bacterial chromosome. Frog DNA is carried to all progeny bacteria.*

well-defined DNA segments that can be used as passengers. Because of the unusual and specific manner in which many restriction enzymes introduce cuts into DNA strands, the use of restriction endonucleases facilitates the recombination process between a passenger and an appropriate vector. Recombinant DNA technology provides practical means for (1) *obtaining suitable DNA passenger molecules,* (2) *selecting appropriate DNA vector molecules,* (3) *constructing the desired passenger-vector combination,* and (4) *amplifying the resulting recombinant DNA.*

Passenger DNA

Commonly, fragments suitable for cloning are produced by digestion of DNA with type II restriction endonucleases. Depending on the specific enzyme used, these nucleases give rise to passengers with either single-stranded extensions or blunt ends. In any case, the pattern of cleavage is specific and reproducible for each enzyme and depends on the location and the size of the palindromes that the enzyme recognizes. Enzymes that recognize a hexanucleotide sequence will cleave DNA into fragments several thousand nucleotides long. The large variety of available restriction enzymes allows cleavage of a desirable fragment out of a large DNA. This is accomplished by selecting the enzyme most suitable for cutting on both sides of the prospective passenger.

In some occasions the known restriction enzymes may also produce cuts within the desired passenger sequence. In such cases DNA can be fragmented into blunt-ended fragments ranging in molecular weight from a few to several million daltons by hydrodynamic shear forces, such as those generated by vigorous stirring or sonication. If the desired passenger is a gene, which might typically have a molecular weight of a million daltons, this method will generally cleave within a few gene molecules, but on average it will produce intact genes suitable for cloning. For cloning of DNA sequences that encode for a specific polypeptide, a passenger can, in addition, be synthesized from the corresponding mRNA by the use of the enzyme *reverse transcriptase* (Figure 17.67). This passenger is referred to as *copy DNA* (cDNA). Finally, a passenger that encodes for a relatively small polypeptide, with a known amino acid sequence, can be synthesized chemically.

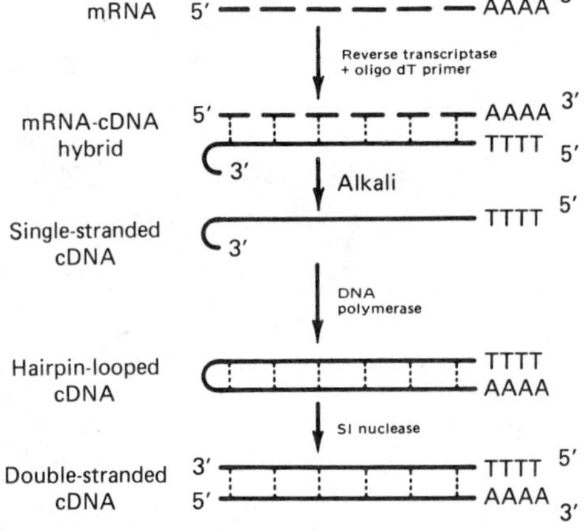

Figure 17.67
Synthesis of double-stranded cDNA from an mRNA template.
mRNA serves as a template for the synthesis of single-stranded cDNA, which can be transformed to a hairpin-looped double-stranded DNA by DNA polymerase-catalyzed extension of the 3′-terminus. Regular duplex cDNA is obtained by nicking the hairpin loop by a specific nuclease.

Cloning Vectors

Any self-replicating DNA element can in principle serve as a cloning vector for the amplification of a passenger DNA. In practice, for a vector to be suitable for recombination, it must be characterized, as a minimum, by the followed properties: (1) It must be small so that it can be manipulated in vitro without being subjected to shearing. This requirement alone eliminates chromosomal DNA from consideration. Large DNA molecules will also tend to have multiple sites that are recognized and cut by the restriction enzyme used in recombination. The process of passenger–vector recombination is simplest when the vector contains a single cut site that can be bridged by the passenger DNA. (2) The vector–passenger recombinant DNA must be amenable to introduction into an appropriate host. DNA does not normally penetrate through cellular membranes, but there are notable exceptions; that is, animal viruses readily penetrate into animal cells. (3) The vectors must carry some genetic marker that makes it possible to conveniently identify and separate the vector–passenger recombinant DNA from other DNA molecules. The vector–passenger recombinant DNA should endow the host cells into which it is incorporated with properties that easily distinguish these cells from cells free of the recombinant DNA.

Only a few viruses, such as the bacteriophages λ and M13, and certain extrachromosomal DNA elements found in bacteria and known as *plasmids,* possess all the qualities necessary for serving as effective vectors. Plasmids consist of a circular DNA of small size; that is a few thousand nucleotides long, which are present in one or more copies in many bacteria. Plasmid DNA contains genes coding for proteins that can degrade antibiotics, and as a result they confer to bacteria antibiotic resistance. This resistance serves as the genetic marker essential for selecting host cells containing the vector. Finally, plasmids can be introduced into the cells of many bacterial species that serve as hosts if the cells are appropriately treated.

One of the most commonly used vectors is a plasmid known as pBR322, which was constructed by recombinant DNA techniques to possess the most desirable characteristics for a vector. It consists of approximately 4,400 nucleotides of known sequence and carries single restriction sites for each one of four commonly used restriction enzymes (Figure 17.68). It also carries the genes that confer bacterial resistance to two antibiotics—ampicillin and tetracycline. Phage λ, also employed as a vector, has a genome of manageable size (mol wt 3.2×10^7), but its DNA is still considerably larger than that of a plasmid. It contains several palindromic sites recognized by the restriction enzymes Hind III and EcoR1. To serve as a useful vector, this phage must be modified by deletion of dispensable (nonessential) parts of its genome. The deleted parts are selected so that most of the restriction sites are excised. The resulting modified phage is referred to as an *insertion vector* if only a single restriction site remains. Phages retaining two restriction sites are referred to as *replacement vectors* (Figure 17.69). A separation of a modified vector from the corresponding passenger–vector combination is frequently based on their different molecular sizes. This size determines whether the DNA of the phage can or cannot be packaged inside the phage and therefore whether it can or it cannot be cloned. Many modified vectors have been constructed that contain antibiotic markers that allow the separation of the passenger–vector recombinant DNA from other DNA molecules.

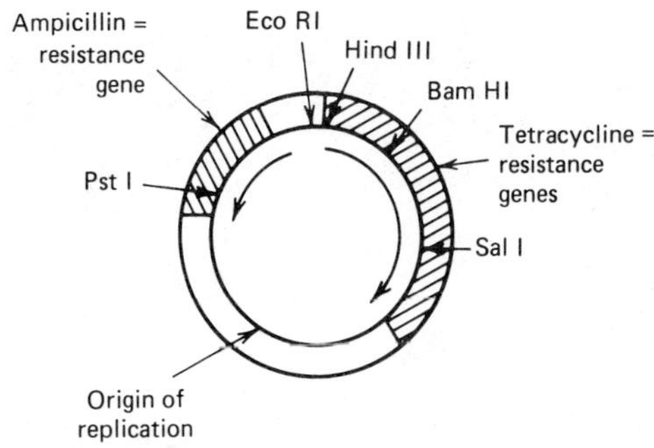

Figure 17.68
Structure of the* E. coli *plasmid pBR322.
The location of antibiotic resistance genes and the sites at which four commonly used restriction enzymes cut the duplex are indicated.

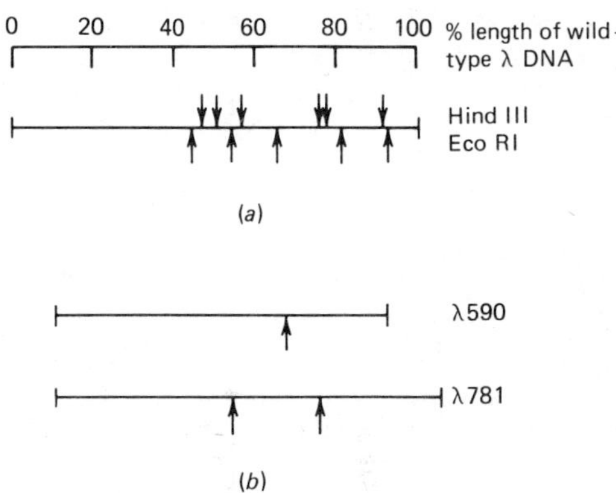

Figure 17.69
Formation of insertion and replacement vectors from λ phage.
(a) Location of EcoRI and Hind III restriction sites on the DNA of the λ phage. (b) Simplified structures for the Hind III insertion vector λ 590 and the EcoRI replacement vector λ 781.

Construction of a Recombinant DNA

The recombination of DNA segments obtained from different sources; that is, the joining between a vector and a passenger, is very much facilitated if both vector and passenger have been obtained by DNA treatment with the same type II endonuclease. It may be recalled that some of these endonucleases cleave a palindromic sequence in a manner that produces at each location cleaved two short oligonucleotide stubs that are the antiparallel complement of one another. For example, if the enzyme Hind III, isolated from a strain of *Haemophilus influenzae* is used for sectioning out a DNA passenger molecule, the passenger stubs will have the sequences shown in Figure 17.70. Use of the same enzyme for introducing a cut in the corresponding palindrome of the pBR322 plasmid will generate a linear vector with terminal single-stranded sequences homologous to those present in the passenger. This homology permits a close fitting of the two segments, which is essential for recombination, as shown in Figure 17.71.

Annealing between the segments generated by separate treatment of the two different DNAs produces a number of new DNA combinations. The key to the success of the splicing of the donor DNA into the vector is the presence, at the end of each fragment, of the complementary single-stranded polynucleotide stubs generated by digestion with Hind III or some other restriction nuclease. Treatment with ligase seals the open phosphodiester bonds and establishes a number of different recombinant DNA molecules. Among the combinations are circular DNA molecules

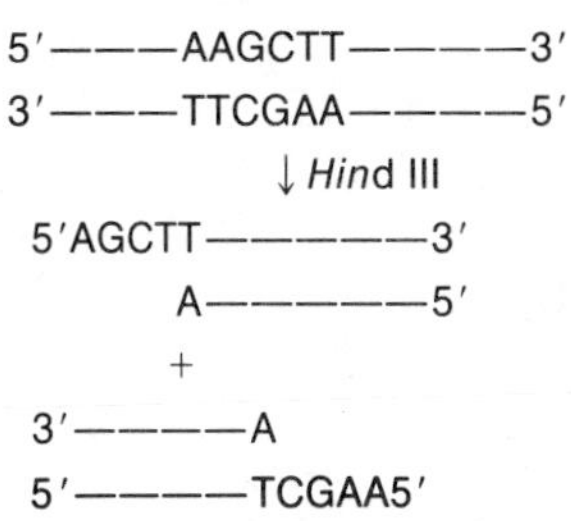

Figure 17.70
Sites of cleavage of DNA by the restriction enzyme Hind III.

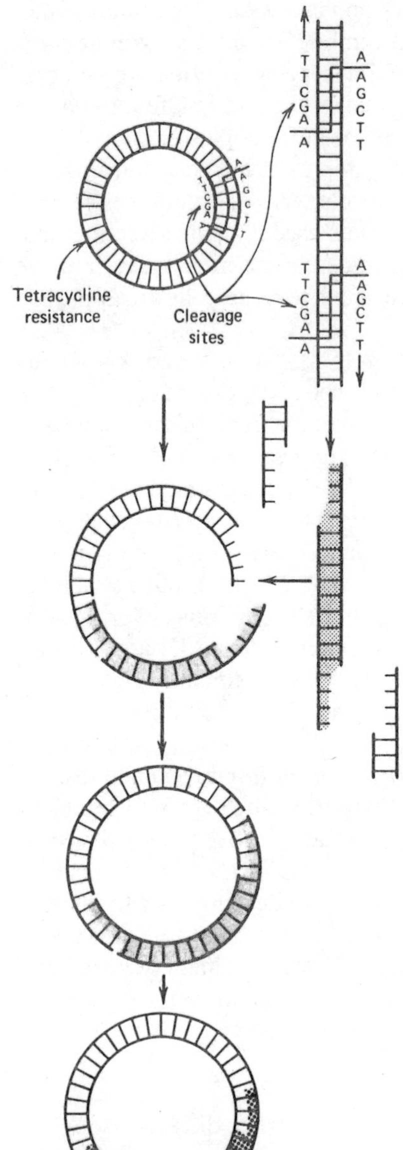

Figure 17.71
Formation of recombinant DNA.
In constructing a recombinant DNA molecule, foreign DNA is joined to a cloning vector and the recombinant is introduced into a bacterial cell for amplification. The basic steps involved are as follows. (a) The cloning vector and the foreign DNA are cleaved at homologous sites (i.e., at sites with the same nucleotide sequence) by a restriction enzyme, often selected so that it cleaves the cloning vector at a single site and the donor DNA at several sites. (b) In the sequences that are cleaved the enzyme creates staggered cuts that are homologous between the cloning vector and the donor DNA. Under annealing conditions an opened cloning vector will add a fragment of foreign DNA, and then the circle will close. The nicks in the annealed ends are then sealed by the action of ligase, which forms a stable circular DNA. (c) The recombinant DNA is taken up by a bacterial cell and replicated during cell division.

consisting of the plasmid into which one *appropriate* segment of donor DNA has been inserted. Alternatively, recombinant DNA molecules can be constructed by joining blunt-ended passenger DNA molecules with vectors present in a linear form also with blunt ends, either directly or after appropriate modification. Blunt-ended DNA molecules can be joined using a ligase that can be isolated from *E. coli* infected by T4 phage (Figure 17.72). In another approach, blunt-ended DNA molecules can be modified by the addition to their termini of single-stranded stubs of a complementary nature (Figure 17.73). For example, the 3′-OH terminal of a passenger can be extended by the formation of a homopolymer tail consisting of thymidine nucleotides (dT), whereas the corresponding terminal of the vector can be extended by the addition of a poly dA segment. In these reactions, which are catalyzed by a DNA polymerase, *terminal deoxynucleotidyl transferase,* which does not require a template, nucleotides are added stepwise to a polynucleotide chain. Finally, appropriate single-stranded stubs can also be added to a DNA segment by using

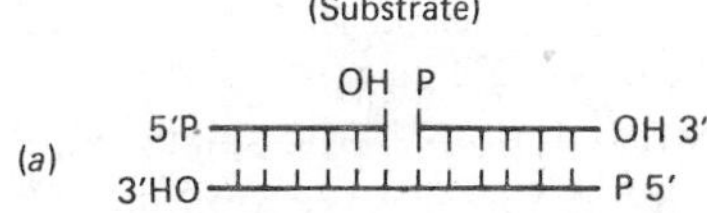

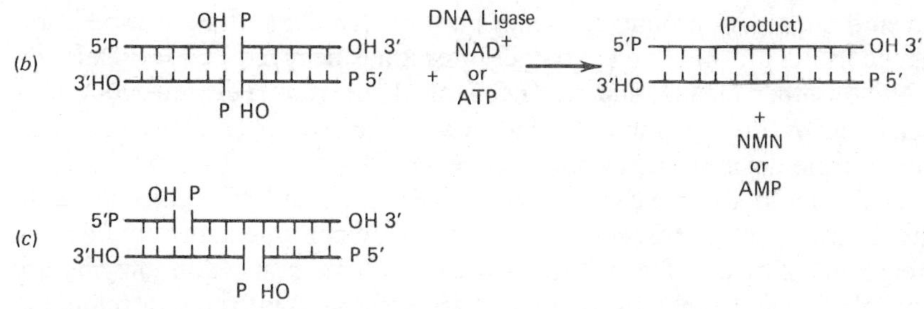

Figure 17.72
Action of DNA ligase.
(a) Sealing of a single nick. (b) Joining of two blunt-ended fragments. (c) Sealing of a "cohesive-end" DNA.

suitable oligonucleotide adaptors (Figure 17.74). These oligonucleotides, which are chemically synthesized, are designed so that they contain a palindromic sequence identical to the one present in the termini of their prospective vectors. After blunt-end ligation of adaptor molecules by T4 ligase on each end of the passenger, the "adapted" passenger is treated with the appropriate restriction endonuclease. Obviously single-stranded stubs, homologous to those generated by the vector when treated with the same enzyme, are produced that facilitate the recombination.

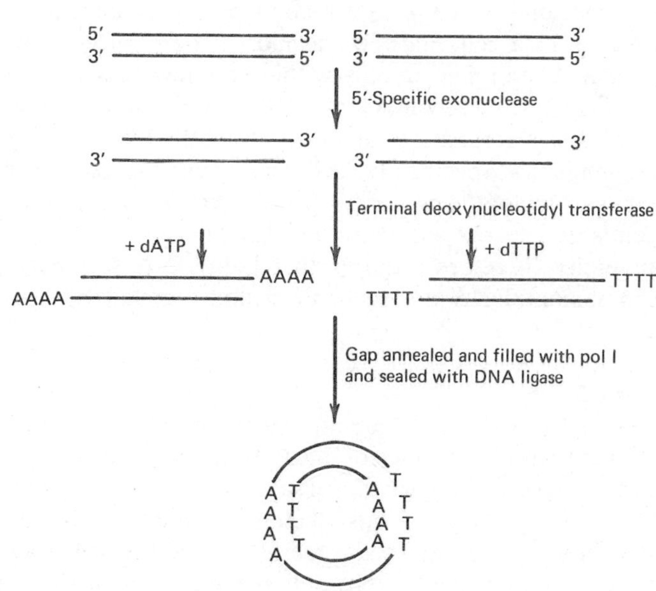

Figure 17.73
Modification of blunt-ended DNA molecules by the addition of homopolymer tails.
The enzyme terminal deoxynucleotidyl transferase catalyzes the addition of homopolymer extensions to substrates with 3'-OH terminals. Both blunt-ended and single-stranded 3'-OH terminals can be extended. The annealed structures can be repaired in vivo *and therefore they can be taken up by the host cell without prior ligation.*

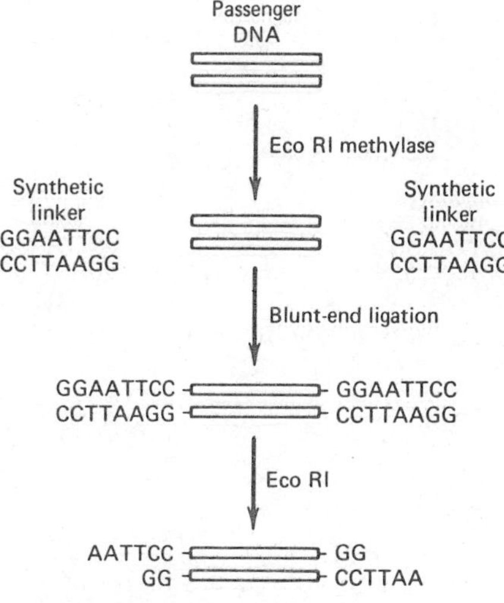

Figure 17.74
Modification of DNA molecules by the method of blunt-end ligation.
The linker used in this method is a short blunt-ended oligonucleotide that contains the recognition site for the same restriction enzyme, in this example EcoRI, which has been utilized for the cleavage of the prospective vector. After the linker is attached to the passenger by blunt-end ligation, the passenger is treated with the enzyme so as to generate cohesive ends. Those ends therefore make the passenger suitable for annealing with the vector.

Limitations in the Construction of Clones

The methods described for recombining different DNAs have almost unlimited applicability because they place few restrictions in the selection of a passenger DNA. The only major requirement for most recombination experiments is the existence of identical palindromes at or near the termini of both the passenger and the vector that are candidates for splicing. Even if such palindromic sequences are not available, they can be introduced into a prospective passenger after it has been excised from a larger DNA by other means, such as mechanical shearing. This introduction is achieved by the ligations of appropriate synthetic oligonucleotides that contain the desired palindromic sequence.

These methods give rise to very few recombinant DNA molecules with the desired vector–passenger combination. When fragments obtained by treatment of two different DNAs with the same restriction enzyme are annealed, many types of molecules are generated by random rejoining. For instance, the complementary termini of a vector, in addition to forming the desired recombinant DNA molecule, can become associated to a second vector molecule or the vector may self-anneal; that is, reform a circular structure without the incorporation of passenger. Some passenger DNA molecules will be joined together, whereas other passenger combinations may form circular structures. Appropriate enzymatic techniques are available that minimize the reactions that lead to these undesirable by-products. In addition, it is possible to enrich the polynucleotide digest obtained by restriction enzymes in the fragment that contains the passenger prior to cloning. If a method is available for easy detection of a passenger, enrichment can be achieved by electrophoresis or chromatography techniques.

An additional limitation in producing sufficient quantities of specific recombinant DNA molecules arises from the low efficiency with which a recombinant DNA molecule can normally be introduced into a host cell. On the average, only a few dozen recombinant plasmids can be introduced into each host cell, and the plasmids that are introduced can become established and replicate only within very few cells of the host. In order to obtain a desired recombinant DNA in sufficient quantities, methods are used that discriminate for those host cells that incorporate the desired recombinant. Specifically, cells containing a recombinant DNA molecule, consisting of the passenger DNA inserted into the vector, must be preferentially selected and cultured. Amplification of the proper recombinant entity therefore requires that both DNA segments present in the recombinant DNA are recognized and selected by appropriate methods.

Selection of Clones

Different strategies for the detection, selection and characterization of the recombinant DNA molecules of interest are used depending upon the specific type of vector and passenger used for the recombination. An exhaustive listing of these various strategies cannot be provided here. Instead, the principle of one of the methods used for selection will be described using as an example DNA obtained by recombination of a passenger with the plasmid pBR322, because cells that have stably incorporated this plasmid can be easily selected by growth in a medium containing either ampicillin or tetracycline. The ampicillin and tetracycline resistance genes present in this plasmid make this selection possible by generating gene products that degrade these antibiotics. A second property of the plasmid essential for the selection process is that it is subject to cleavage by the restriction endonuclease PstI that cleaves at a single site within the gene sequence coding for ampicillin resistance. Similarly the tetracycline resistance gene of this plasmid can be cleaved at a single site by the endonucleases BamHI and SalI (Figure 17.75).

Insertion of the passenger DNA into any one of the cleavage sites permits narrowing down the selection process only to those cells that carry plasmids that have incorporated a passenger. This narrowing down of the selection process, known as *insertional inactivation,* can be achieved because the insertion of a passenger within a gene coding for antibiotic resistance will inactivate the gene. Clones obtained by the insertion of foreign DNA into a pBR322 plasmid cleaved within the ampicillin resistance gene will be resistant to tetracycline but will not grow in a medium containing ampicillin. Obviously a reversal in resistance with respect to the two antibiotics will be observed if the passenger is inserted within the tetracycline gene. These properties can therefore be used to distinguish between recombinant and nonrecombinant plasmids, the latter being resistant to both antibiotics. Several other approaches based on the same principle of insertional inactivation can be used to select between cells carrying recombinant and nonrecombinant molecules.

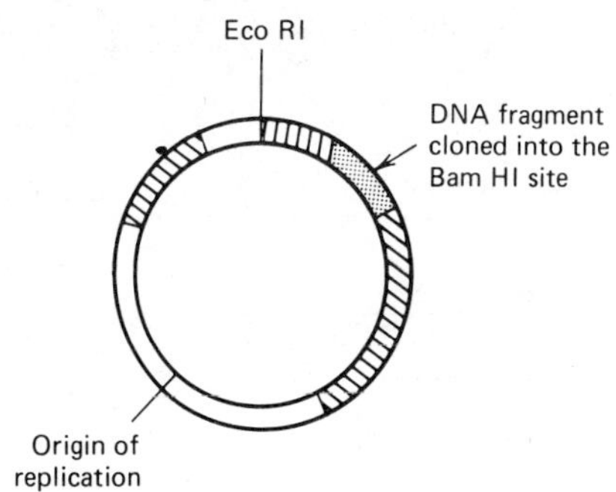

Figure 17.75
Insertional inactivation.
The structure of a recombinant plasmid derived from pBR322 in which the cloned DNA has caused insertional inactivation of the tetracycline resistance genes.

Detection and Characterization of Recombinant DNA

A method must be available for the detection and isolation of the passenger segment of the recombinant DNA. Three general approaches can be used, namely *hybridization, immunochemical techniques,* and *structural methods.* All hybridization methods depend upon the availability of easily detectable DNA or RNA polynucleotides with a sequence that is complementary (or nearly complementary) to the sequence of the cloned passenger. Such a polynucleotide, known as a *hybridization probe,* upon association with the passenger under renaturing conditions permits the easy detection of the passenger. The mRNA coding for a cDNA passenger or a chemically synthesized oligonucleotide can be used as hybridization probes. Chemical synthesis of the appropriate polynucleotide sequence can be accurately guided if, for instance, the polynucleotide coded for by the passenger is known. Identification of host cell colonies that contain recombinant DNA molecules incorporating the passenger DNA segment depends upon the synthesis of radioactively labeled probes. One method for in situ screening of cells containing the passenger molecule is shown in Figure 17.76.

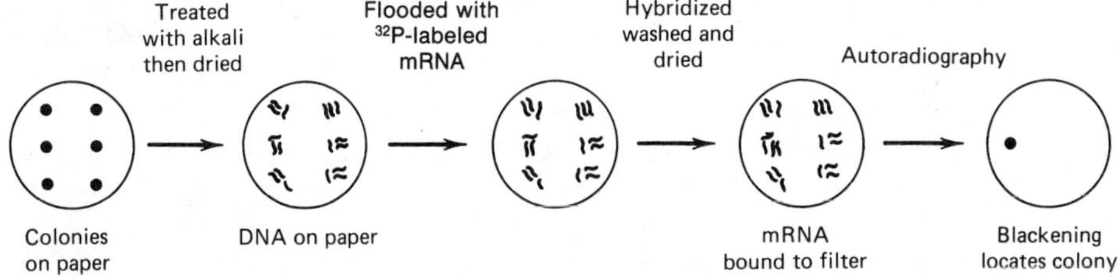

Figure 17.76
Colony hybridization.
Colonies from a reference plate are first transferred on paper, and the DNA from each colony is released by alkali treatment on this paper. Radioactively labeled mRNA hybridizes with complementary DNA, which, in this hypothetical example, is present in only one of the colonies transferred from the reference plate.
Redrawn from D. Freifelder, Molecular biology. *A comprehensive introduction to prokaryotes and eukaryotes.* New York: Van Nostrand Reinhold, New York: 1983.

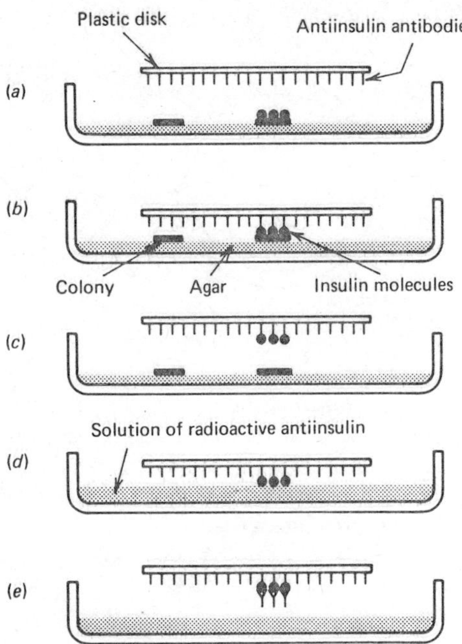

Figure 17.77
Immunodetection of antigen-expressing recombinant DNA.
In this radioimmune assay antibody to insulin is transferred to a plastic disk. Insulin molecules from an insulin-producing colony bind to the insulin antibody on the disk. Radioactive insulin antibodies are then used to identify the insulin-producing colonies.
Reproduced with permission from W. I. P. Mainwaring, J. H. Paris, J. D. Pickering, and N. H. Mann, *Nucleic acid biochemistry and molecular biology*. Oxford, London: Blackwell Scientific Publications, 1982.

Immunological techniques are based upon a similar approach, but they are dependent upon the prior synthesis of the protein coded for in the passenger and the fact that the polypeptide product has easily distinguishable antigenic properties (Figure 17.77).

In the cases in which neither immunological nor hybridization techniques are applicable, the size of the passenger and the distribution of fragments obtained upon digestion of the passenger with various restriction enzymes (restriction pattern) can be determined. Since, as a rule, the recombinant DNA is produced by restriction digestion and subsequent ligation of the fragments, it is often a simple matter to reverse the process of ligation by using the same enzyme originally used to obtain the passenger out of a larger DNA. Such *structural* methods of recombinant DNA characterization obviously depend upon prior knowledge of the size and restriction pattern of the passenger. Clinical Correlation 17.4 discusses applications of recombinant DNA technology.

17.9 NUCLEOTIDE SEQUENCING OF DNA

The determination of the sequence of a large DNA molecule often begins by cutting the DNA, by the use of appropriate restrictions endonucleases, into pieces of a more manageable size. In recent years several effective methods have been developed for the rapid sequencing of relatively large polydeoxyribonucleotides. These methods are impressively accurate and can be used for the sequencing of RNA molecules by prior conversion of the polyribonucleotide sequences of RNA to complementary polydeoxyribonucleotides by use of *reverse transcriptase*. The accuracy of sequencing methods rests on the availability of numerous restriction enzymes. Digests obtained using individual enzymes produce segments with overlapping lengths of nucleotide sequences. These sequences allow the sequencing of the DNA by use, if necessary, of multiple sets of independent sequence data. In addition, the accuracy of the sequencing methods can be further increased by determination of the sequence of the complementary strand in order to verify the original sequence.

Contemporary sequencing methods have led to determinations of the sequence of DNAs containing as many as 50,000 nucleotides. Sequences up to 500 nucleotides can now be determined in a single operation, permitting the determination of sequences at a rate of several hundred nucleotides per day. Sequences of many polynucleotide segments with important regulatory or coding functions and entire sequences of several viral genomes have been determined. Two methods are currently in use—the *enzymatic* (or *dideoxy*) method and the *chemical* method. Neither of these methods determines the nucleotide sequence of a polynucleotide directly. Instead, the terminal nucleotides of a large collection of polynucleotide segments of various sizes, all smaller than the polynucleotide undergoing sequencing, are determined. To obtain the overall sequence, the terminal nucleotides are determined in the order in which they were present in the original structure.

In the chemical method, known also as the Maxam and Gilbert method, the collection of segments used for sequencing is obtained by cleavage of the DNA undergoing sequencing by four separate chemical reactions. The reactions, which are specific for particular nucleotides, are not allowed to go to completion, thus generating polynucleotides of all possible lengths each one terminating with a different nucleotide. The position of this terminal nucleotide in the original sequence can be deduced from the *length of the specific* fragment in which the nucleotide is present as a terminal residue.

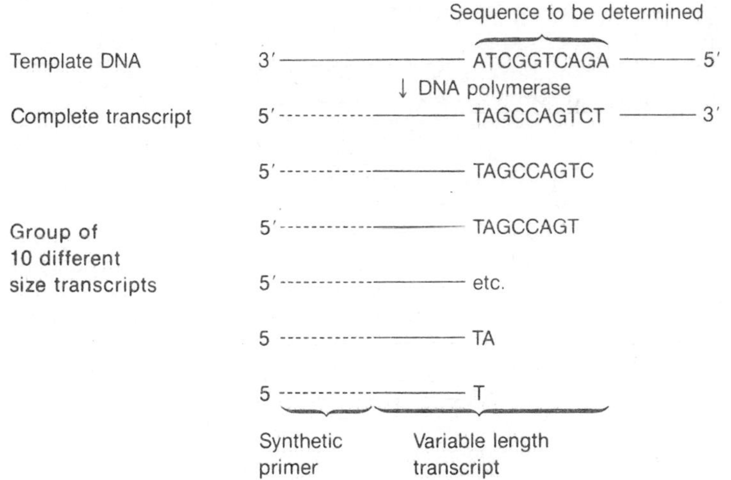

Figure 17.78
Synthesis of variable length transcripts for determination of a nucleotide sequence in DNA.
In this example a hypothetical 10 nucleotide sequence is transcribed in a manner that produces a series of variable length transcripts, 10 in all, which terminate at various nucleotides within the sequence to be determined. Use of a synthetic primer insures that all transcripts are "in register" relative to the sequence of interest.

CLIN. CORR. 17.4 RECOMBINANT DNA

Recombinant DNA techniques are extensively used today. The reason why these techniques are receiving wide application is that they offer great promise both in basic research and for practical benefit. The complexity of eucaryotic genomes makes the study of a given gene and its expression extremely difficult. Recombination techniques now allow the study of an isolated (and amplified) gene by transplantation into a bacterial system. The possibilities for practical benefit are numerous. For instance, the insertion of a synthetic gene coding for the mammalian peptide hormone somatostatin into bacteria has induced bacteria to produce the hormone, that is, the synthetic gene can be replicated, transcribed, and translated by the host. Similarly, the gene coding for mammalian proinsulin has been appropriately recombined and grown in *E. coli*. As a result, bacteria can be made to secrete proinsulin. Numerous other achievements in this area are presently taking place at a high rate. In fact, the potential for benefit from recombinant DNA techniques appears enormous. This technology may be expected to increase the supply of hormones and other very rare proteins, for example, interferon, which show promise as new tools in medicine. One form of interferon (α-interferon) shows promise for the treatment of Kaposi's sarcoma and some forms of leukemia. Another intriguing application of this interferon is its potential use, in the form of nasal spray, for preventing the common cold.

In addition, recombinant techniques can, in principle, be adapted for the synthesis of proteins that can be utilized as vaccines. Normally, in order to make a vaccine, an infectious organism must be grown in large amounts, which often proves to be a difficult or dangerous undertaking. Furthermore in some instances, for example, the hepatitis B virus, the infectious agent cannot be grown in vitro. In vitro techniques that can be used to produce only the protein responsible for the antibody response obviously have the potential for resolving such difficulties.

An analogous set of polynucleotide segments is used for the enzymatic sequencing method, the Sanger method; the segments are not generated by degradation but rather by the synthesis of polynucleotides using the enzyme DNA polymerase. Each of the segments is complementary to gradually increasing portions of the polynucleotide that is being sequenced (Figure 17.78). Knowledge of the terminal nucleotide incorporated into each one of these *variable length transcripts* permits the reconstruction of the sequence of the template that directed their synthesis.

Obviously, both the enzymatic and the chemical methods depend upon the satisfactory separation of polynucleotides that differ in length by as little as a single nucleotide residue. The development of high resolution electrophoresis in polyacrylamide gels has made this type of separation easy to achieve. If these polynucleotides are radioactively labeled, the resolved polynucleotides can be detected by autoradiography on the gel.

Both sequencing methods are in current use, with the chemical method being the most widely used. This method is easy to use, is applicable to almost any single- or double-stranded DNA fragment, and is especially convenient for determining the sequence of double-stranded fragments. In addition the method, properly modified, can be used for determining the sites of interaction between proteins and nucleic acids because chemical cleavage does not occur at nucleotide sites that are protected by proteins interacting with these sites. The enzymatic method is particularly suitable for determining sequences of extensive lengths, but it is not directly applicable to double-stranded DNA fragments. This limitation can of course be circumvented by prior separation of the strands by gel electrophoresis or by various approaches, one of which, the so-called "shot gun" DNA sequencing, has been particularly useful for genome sequencing. In this method, fragments several hundred nucleotides long, generated by random digestion of the genome, are introduced for cloning into an EcoRI restriction site of the replicative form of the M13 phage. A complete collection of these DNA fragments is then isolated as single strands from the single-stranded progeny form of the bacteriophage and sequenced. This method has the advantage of replacing the time-consuming and complex operations of fractionating endonuclease restriction digests by a convenient cloning procedure.

BIBLIOGRAPHY

Adams, R. L. P., Burdon, R. H., Campbell, A. M., Leader, D. P., and Smellie, R. M. S. *The biochemistry of nucleic acids*. Chicago: Year Book Medical Publishers, 1981.

Bradbury, M. E., Maclean, N., and Mathews, H., *DNA chromatin and chromosomes*. New York: Chapman and Hall, 1981.

Cantor, C. R., and Schimmel, P. R. *Biophysical chemistry, part I: The conformation of biological macromolecules*. San Francisco: W. H. Freeman, 1980.

Cantor, C. R., and Schimmel, P. R. *Biophysical chemistry, part III: The behavior of biological macromolecules*. San Francisco: W. H. Freeman, 1980.

Davidson, J. N., and Cohn, W. E. (eds.). *Progress in nucleic acid research and molecular biology*. New York: Academic Press (a series published regularly).

Freifelder, D. *Molecular biology: A comprehensive introduction to prokaryotes and eukaryotes*. New York: Van Nostrand Reinhold, 1983.

Hunt, T., Prentis, S., and Tooze, J. *DNA makes RNA makes protein*. New York: Elsevier Press, 1983.

Kornberg, A. *DNA replication*. San Francisco: W. H. Freeman, 1980.

Kornberg, A. *DNA replication, 1982 Supplement*. New York: W. H. Freeman, 1982.

Mainwaring, W. I. P., Paris, J. H., Pickering, J. D., and Mann, N. H. *Nucleic acid biochemistry and molecular biology*. Oxford, London: Blackwell Scientific Publications, 1982.

Mizobuchi, K., Watanabe, I., and Watson, J. D. *Nucleic acids research: Future developments*. 1983.

Rodriquez, R. L., and Tait, R. C. *Recombinant DNA Techniques: An introduction*. Reading Mass.: Addison-Wesley, 1983.

Saenger, W. *Principles of nucleic acid structure*. New York: Springer-Verlag, 1983.

Woods, R. A. *Biochemical genetics*. New York: Chapman and Hall, 1980.

QUESTIONS

C. N. ANGSTADT AND J. BAGGOTT

*Q*uestion *T*ypes are described inside the front cover.

1. (QT1) A polynucleotide is a polymer in which:
   A. the two ends are structurally equivalent.
   B. the monomeric units are joined together by phosphodiester bonds.
   C. there are at least 20 different kinds of monomers that can be used.
   D. the monomeric units are not subject to hydrolysis.
   E. purine and pyrimidine bases are the repeating units.
2. (QT1) The *best* definition of an endonuclease is an enzyme that hydrolyzes:
   A. a nucleotide from only the 3′ end of an oligonucleotide.
   B. a nucleotide from either terminal of an oligonucleotide.
   C. a phosphodiester bond located in the interior of a polynucleotide.
   D. a bond only in a specific sequence of nucleotides.
   E. a bond that is distal (d) to the base that occupies the 5′ position of the bond.
3. (QT2) Which of the following tends to favor a helical conformation of a single polynucleotide chain?
   1. Hydrophobic interactions of the rings of the purine and pyrimidine bases, which exclude water
   2. Interchange of electrons in the π orbitals of the purine and pyrimidine bases
   3. Charge–charge repulsion of phosphate residues of the polynucleotide backbone
   4. Hydrogen bonding between appropriate purine–pyrimidine pairs

Use the accompanying figure to answer the next two questions.

4. (QT5) A, B, and C represent conformations at different temperatures. Which one represents the highest temperature?

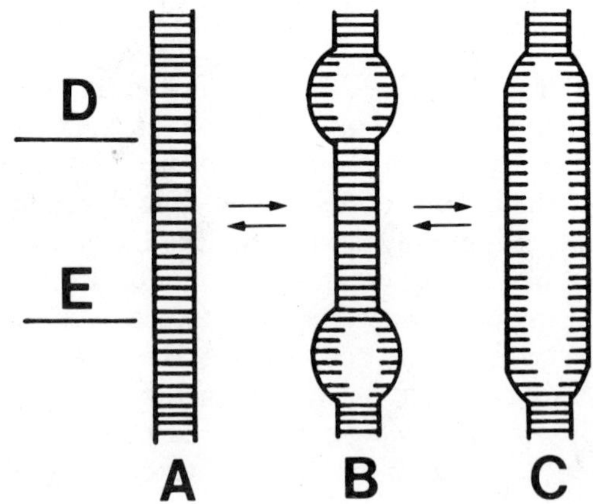

5. (QT5) Which section, D or E, has the higher content of guanine and cytosine?
   A. Annealing
   B. Viscoelastic retardation
   C. Equilibrium centrifugation
   D. Dideoxy method (Sanger procedure)
6. (QT5) A technique for determining the molecular weight of very large (>10^9 daltons) DNA
7. (QT5) A technique used in forming a recombinant DNA
8. (QT5) A technique used for determining the sequence of a DNA segment

9. (QT2) The superhelices that form in double-stranded circular DNA:
   1. may have fewer turns of the helix per unit length than does a linear double helix.
   2. are associated with a restricted topological domain.
   3. may exist in multiple conformations that are interconvertible without breaking covalent bonds.
   4. may be either formed or relaxed by enzymes called topoisomerases.

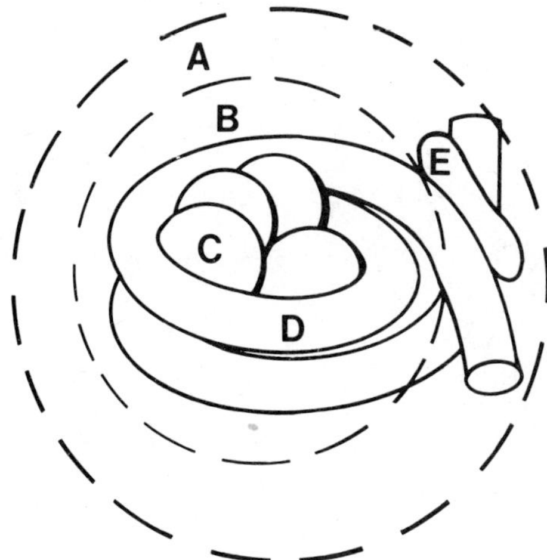

10. (QT5) A chromatosome
11. (QT5) DNA
12. (QT5) H1 class of histones
13. (QT1) A palindrome is a sequence of nucleotides in DNA that:
    A. is highly reiterated.
    B. is part of the introns of eucaryotic genes.
    C. is a structural gene.
    D. has local symmetry and may serve as a recognition site for various proteins.
    E. has the information necessary to confer antibiotic resistance in bacteria.
14. (QT2) Which of the following would result in a frame-shift mutation?
    1. Insertion of a new base pair in the DNA
    2. Substitution of a purine–pyrimidine pair by a pyrimidine–purine pair
    3. Intercalation of ethidium bromide into the nucleotide chain
    4. Deamination of cytosine to uracil

    A. DNA replication C. Both
    B. DNA excision repair D. Neither
15. (QT4) Both strands of DNA serve as templates concurrently.
16. (QT4) Require(s) both 5′ → 3′ exonuclease and polymerase activities of a DNA polymerase.
17. (QT4) Require(s) the activity of DNA ligase.

ANSWERS

1. B The structure of a polynucleotide possesses an intrinsic sense of direction that does not depend on whether a 3′-OH or 5′-OH terminal is esterified. C, E: There are only four different monomers, and the repeating unit is the base monophosphate (p. 628).
2. C Both A and B describe exonucleases. D does refer to an endonuclease but only to a specific type, a restriction endonuclease, and is therefore not a definition of the general type. E: Both endo- and exonucleases show specificity toward the bond hydrolyzed and so this is not a definition of an endonuclease (p. 630).
3. A 1, 2, 3 correct. The exclusion of water by stacking of the bases is a strong stabilizing force that is enhanced by the interaction of π orbital electrons. The repulsive forces of the phosphate groups confer a certain rigidity to the structure. 4: This is very important in holding two different polynucleotide chains together, but it is unlikely that the proper positioning would occur within a single chain (p. 632).
4. C The figure represents the process of denaturation with the extent
5. E of disruption increasing as temperature increases. Since a guanine–cytosine pair has three hydrogen bonds and an adenine–thymine pair only has two, higher temperatures are required to disrupt regions high in G-C (Figure 17.12).
6. B This is a special low shear method. Equilibrium centrifugation (C) is also a method for determining molecular weight, but it is limited to a molecular weight of 10^9 daltons or less because of the effects of shear forces on large molecules (p. 645).
7. A Self-association of complementary polynucleotide strands permits annealing of the homologous terminal sequences produced when the same restriction enzyme is used to cut two different DNA molecules (p. 640, Figure 17.71).
8. D This is just one of the two principle methods for determining sequences (p. 692).
9. E All four correct. 1 describes a negative superhelix. There may also be more turns per unit length in a positive superhelix. 2: Once a closed system is interrupted, a superhelix can unwind. 3: All conformations with the same linking number α are interconvertible without breaking covalent bonds. 4: Topoisomerase I (omega protein) from *E. coli* relaxes and gyrase (topoisomerase II) can introduce or remove superhelices (depending on the conditions) (p. 647).
10. A The nucleosome core is a discrete particle consisting of an octa-
11. E mer of specific histones with a segment of DNA wrapped around
12. C it. B: The chromatosome, the basic structural element of nucleoprotein, contains the nucleosome core with associated H1 histones. The strand depicted represents DNA; the circles, histones. The H_1 class of histones is bound to the spacer regions between nucleosomes. D represents H_4, H_3, H_{2A}, or H_{2B}, which are part of the nucleosomes (Figure 17.29).
13. D A palindrome, by definition, reads the same forward and backward. Short palindromic segments of DNA are recognized by a variety of proteins such as restriction endonucleases and CAP-binding protein. A is not likely since it would be incompatible with specific recognition. B is possible but has not been shown. C is not correct since genes are thousands of base pairs in length, whereas palindromes are short segments. E also would not be likely because palindromes are too short (p. 656).
14. B 1, 3 correct. Since the bases are read in groups of three, insertion of an additional base would shift the reading frame (p. 666). Intercalation stretches the DNA so when DNA is replicated an additional base is inserted near the intercalation site (p. 670). 2 and 4 are both examples of base substitution type of mutations.
15. A In excision repair, the damaged segment of a strand is removed so both strands are not available (p. 672, Figure 17.48).
16. C In replication, the exonuclease activity removes the primer and the space is filled by the polymerase activity (p. 678). In excision repair, the damaged portion is removed by the exonuclease activity and the space is filled by polymerase activity (p. 672).
17. C In both cases, ligase is needed to seal the newly synthesized strand to the end of the strand already in place (p. 672, 678).

18

RNA: Transcription and Posttranscriptional Modification

PAUL F. AGRIS and FRANCIS J. SCHMIDT

18.1 OVERVIEW

The primary genetic information of a eucaryotic cell is localized within the cell nucleus and encoded in the DNA. Expression of part or all of this information, a cell's functional capacity, or phenotype, usually takes the form of polypeptides, which are synthesized in the cytoplasm. Since genomic DNA is not a normal cytoplasmic component, cells require mechanisms for transferring the necessary information from the nucleus to the cytoplasmic protein-synthesizing machinery. Macromolecules that mediate the transfer of information must reflect the sequence of the purines and pyrimidines of the DNA. These macromolecules, called ribonucleic acids (RNAs), are linear polymers of ribonucleoside 5′-monophosphates. The process by which the RNA copies of selected DNA sequences are made is termed *transcription*.

Analysis of total cellular RNA reveals several distinct molecular sizes and functional families of RNA. These are presented in Table 18.1. The primary role of RNA within the cell is its involvement in protein synthesis, that is, *translation*. Messenger RNAs serve as templates, transfer RNAs serve as amino acid carriers and translators of the genetic code, and ribosomal RNAs function in ribosomes during messenger RNA binding and peptide bond formation. RNA involvement as an intermediary in translation is required; there is no specific affinity for amino acids by the purines and pyrimidines of DNA, which of course are not accessible to cytoplasmic protein synthesis in eucaryotes. Cytoplasmic RNA molecules, which contain the information for directing synthesis of a specific

TABLE 18.1 Characteristics of Cellular RNAs

Type of RNA	*Abbreviation*	*Function*	*Size and Sedimentation Coefficient*	*Site of Synthesis*	*Structural Features*
Messenger RNA Cytoplasmic	mRNA	Transfer of genetic information from nucleus to cytoplasm, or from gene to ribosome	Depends upon size of protein 1,000 to 10,000 nucleotides	Nucleoplasm	Blocked 5′ end; poly (A) tail on 3′ end; nontranslated sequences before and after coding regions; few base pairs and methylations
Mitochondrial	mt mRNA		9S to 40S	Mitochondria	
Transfer RNA Cytoplasmic	tRNA	Transfer of amino acids to mRNA ribosome complex and correct sequence insertion	65–110 nucleotides 4S	Nucleoplasm	Highly base-paired; many modified nucleotides; common specific structure
Mitochondrial	mt tRNA		3.2S to 4S	Mitochondria	
Ribosomal RNA Cytoplasmic	rRNA	Structural framework for ribosomes	28S, 5,400 nucleotides 18S, 2,100 nucleotides 5.8S, 158 nucleotides 5S, 120 nucleotides	Nucleolus Nucleolus Nucleolus Nucleoplasm	5.8S and 5S highly base-paired; 28S and 18S have some base-paired regions and some methylated nucleotides
Mitochondrial	mt rRNA		16S, 1,650 nucleotides 12S, 1,100 nucleotides	Mitochondria	
Heterogeneous nuclear RNA	hnRNA	Some are precursors to mRNA and other RNAs	Extremely variable 30S to 100S	Nucleoplasm	mRNA precursors may have blocked 5′ ends and 3′-poly(A) tails; many have base-paired loops
Small nuclear RNA	snRNA	Structural and regulatory RNAs in chromatin	100–300 nucleotides	Nucleoplasm	

polypeptide, are messenger RNAs (mRNA). The size of each mRNA is directly related to the size of the protein for which it codes. The molecules that transfer specific amino acids from soluble amino acid pools to ribosomes, and ensure the alignment of these amino acids in the proper sequence prior to peptide bond formation, are transfer RNAs (tRNA). All the tRNA molecules are approximately the same size and shape. The assembly site, or factory, for peptide synthesis involves ribosomes, complex subcellular particles containing at least three different RNA molecules called ribosomal RNAs (rRNA), and 70 to 80 ribosomal proteins.

Protein synthesis requires a close interdependent relationship between mRNA, the informational template, tRNA, the amino acid adaptor molecule, and rRNA, part of the synthetic machinery. In order for protein synthesis to occur at the correct time in a cell's life, the synthesis of mRNA, tRNA, and rRNA must be coordinated with the cell's response to the intra- and extracellular environments.

All cellular RNA is synthesized on a DNA template and reflects a portion of the DNA base sequence. Therefore, all RNA is associated with DNA at some time. The RNA involved in protein synthesis functions in the cytoplasm outside the nucleus, while some RNAs remain in the nucleus, where they have structural and regulatory roles. RNA synthesized in the mitochondria remains there and is involved in mitochondrial protein synthesis.

Although DNA is the more prevalent genetic store of information, RNA can also carry genetic information in the sequence of the bases and serves as the genome in several viruses. However, RNA is not normally found as the genome in eucaryotic or procaryotic cells. Genomic RNA is found in the RNA tumor viruses and the other small RNA viruses, such as poliovirus and reovirus.

18.2 STRUCTURE OF RNA

Components and Primary Structure

RNA has been shown to be a general constituent of eucaryotes and procaryotes. Chemically it is very similar to DNA. Although RNA is one of the more stable components within a cell, it is not as stable as DNA. Some RNAs, such as messenger RNA, are synthesized, used, and degraded, whereas others, such as ribosomal RNA, are not turned over rapidly.

RNA is an unbranched linear polymer in which the monomeric subunits are the ribonucleoside 5′-monophosphates. The purines found in RNA are *adenine* and *guanine;* the pyrimidines are *cytosine* and *uracil.* Except for uracil, which replaces thymine, these are the same bases found in DNA. Complete analysis of cellular RNA reveals that other bases are also found in low concentrations, except in transfer RNA in which they represent as much as 25% of the nucleotides. These minor nucleotides are for the most part derivatives of A, G, U, and C and are a small but significant portion of nucleotides having prominent roles in RNA metabolism and function (Table 18.2). Modification of the purines and pyrimidines occurs only after polymerization of the nucleotides into RNA. Modifications to the ribose, usually methylation of the 2′-hydroxyl position, also occur after transcription.

The monomers are connected by single phosphate groups linking the 3′-carbon of one ribose with the 5′-carbon of the next ribose. The internucleotide link, a 3′,5′ phosphodiester, forms a chain or backbone from which the bases extend (Figure 18.1). The length of natural RNA mole-

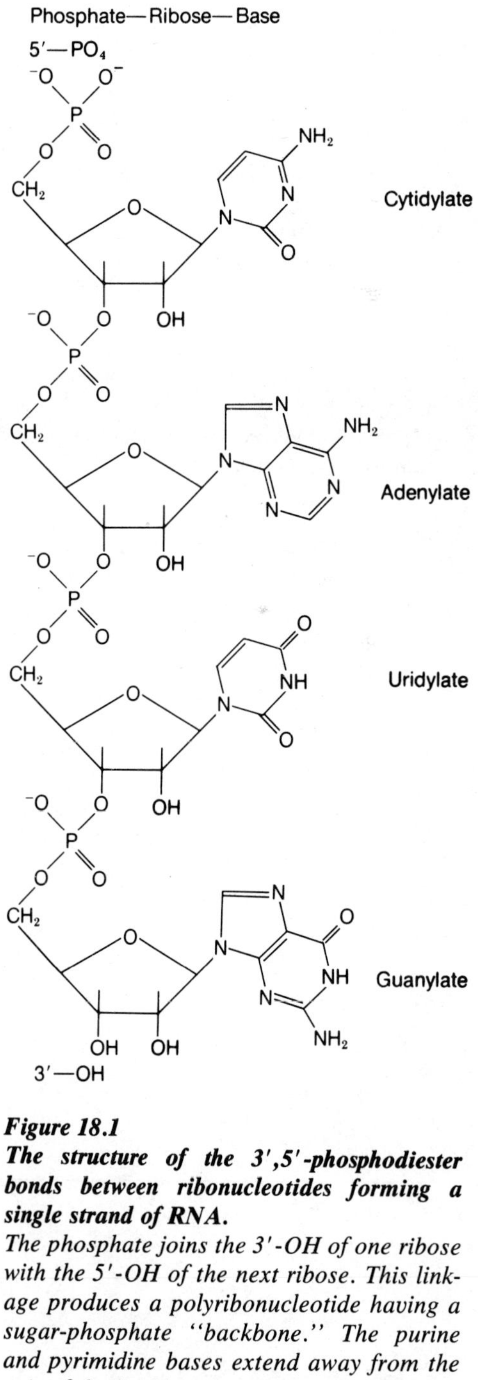

Figure 18.1
The structure of the 3′,5′-phosphodiester bonds between ribonucleotides forming a single strand of RNA.
The phosphate joins the 3′-OH of one ribose with the 5′-OH of the next ribose. This linkage produces a polyribonucleotide having a sugar-phosphate "backbone." The purine and pyrimidine bases extend away from the axis of the backbone and may pair with complementary bases to form double-helical base-paired regions.

TABLE 18.2 Some Modified Nucleosides Found in RNA

Purine Derivatives

Nucleosides with a methylated base
- 1-Methyladenosine (m^{1}A)
- 2-Methyladenosine (M^{2}A)
- 7-Methylguanosine (m^{7}G)
- 1-Methylguanosine (m^{1}G)
- N^6-Methyladenosine (m^{6}A)
- N^6,N^6-Dimethyladenosine (m$_2^6$A)
- N^2-Methylguanosine (m^{2}G)
- N^2,N^2-Dimethylguanosine (m$_2^2$G)

2′-*O*-Methylated derivatives
- 2′-*O*-Methyladenosine (Am)
- 2′-*O*-Methylguanosine (Gm)

Deaminated derivatives
- Inosine (I)
- 1-Methylinosine (m^{1}I)

Adenosine derivatives with an isopentenyl group
- N^6-(Δ^2-Isopentenyl)adenosine (i^{6}A)
- N^6-(Δ^2-Isopentenyl)-2-methylthioadenosine (ms^{2}i^{6}A)
- N^6-(4-Hydroxy-3-methylbut-2-enyl)-2-methylthioadenosine

Other nucleosides
- *N*-[9-(β-D-Ribofuranosyl)purin-6-ylcarbamoyl]threonine (t^{6}A)
- *N*-[9-(β-D-Ribofuranosyl)purin-6-yl-*N*-methylcarbamoyl]threonine (mt^{6}A)
- 7-4, 5-*cis*-dihydroxy-1-cyclopenten-3-ylaminomethyl-7-deaza-quanosine (Q), queuosine

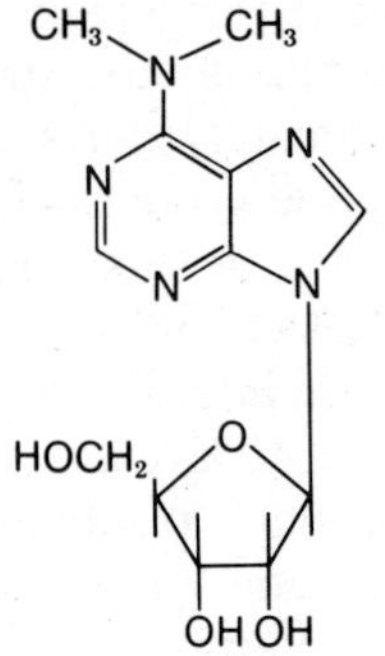

N^6,N^6-Dimethyladenosine

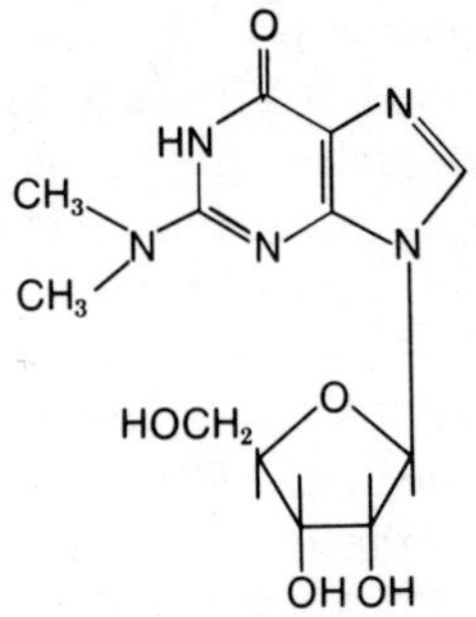

N^2,N^2-Dimethylguanosine

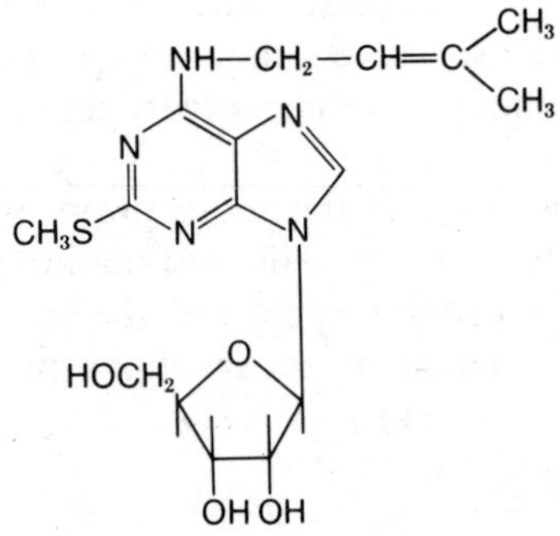

N^6-(Δ^2-Isopentenyl)-2-methylthioadenosine

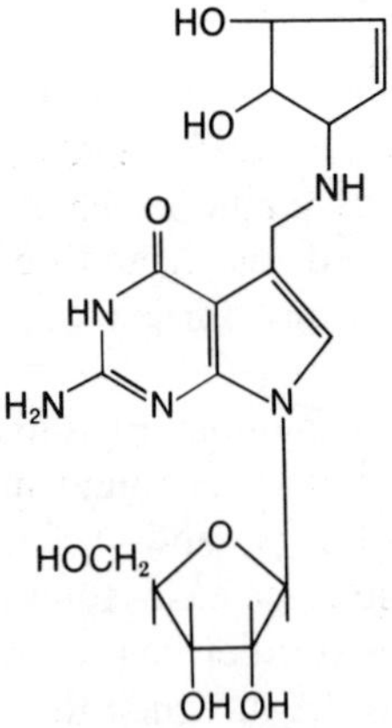

Queuosine

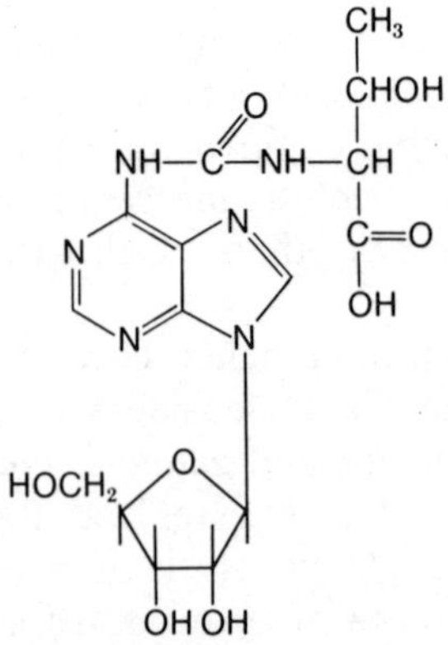

N-[9-(β-D-Ribofuranosyl)purin-6-yl carbamoyl]-L-Threonine

cules in eucaryotic cells varies from approximately 65 nucleotides to more than 6,000 nucleotides. The sequences of the bases are complementary to the base sequences of specific portions of only *one strand of DNA*. Thus, unlike the base composition of DNA, molar ratios of A + U and G + C in RNA are not equal. All cellular RNA so far examined is linear and single-stranded, but double-stranded RNA is present in some viruses.

TABLE 18.2 (*Continued*)

Pyrimidine Derivatives

Nucleosides with a methylated base
- Thymine riboside (T)
- 5-Methylcytidine (m^5C)
- 3-Methylcytidine (m^3C)
- 3-Methyluridine (m^3U)

2′-*O*-Methylated derivatives
- 2′-*O*-Methyluridine (Um)
- 2′-*O*-Methylcytidine (Cm)
- 2′-*O*-Methylpseudouridine (Ψm)
- 2′-*O*-Methylthymine riboside (Tm)

Sulfur-containing nucleosides
- 4-Thiouridine (s^4U)
- 5-Carboxymethyl-2-thiouridine methyl ester (cm^5s^2U)
- 5-Methylaminomethyl-2-thiouridine (mnm^5s^2U)
- 5-Methyl-2-thiouridine (2-thiothymidine, s^2T)
- 2-Thiocytidine (s^2C)

Other nucleosides
- Pseudouridine (Ψ)
- 5,6-Dihydrouridine (D)
- 4-Acetylcytidine (ac^4C)
- Uridine-5-oxyacetic acid (V)

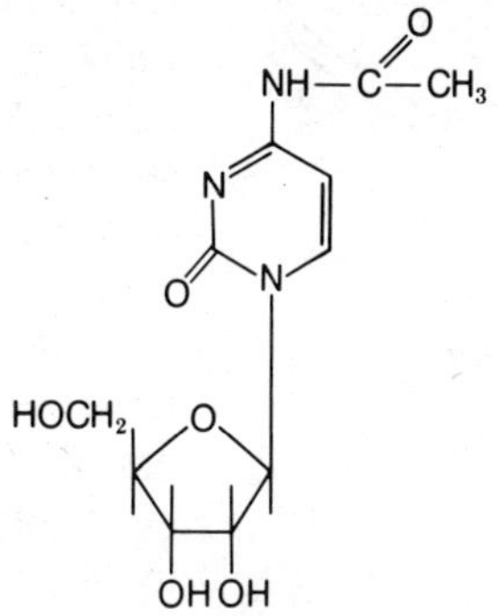

*N*⁴-Acetylcytidine

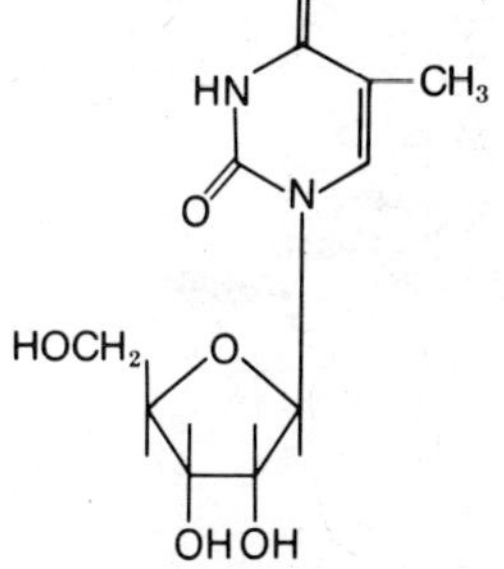

5-Methyluridine

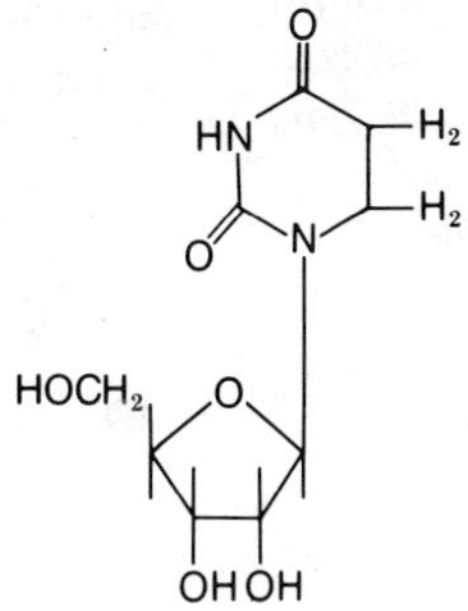

5,6-Dihydrouridine

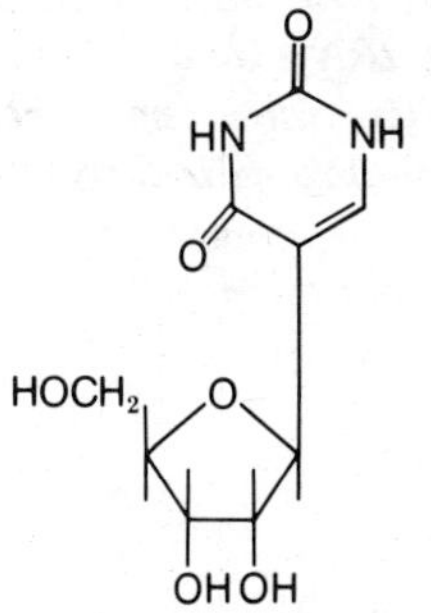

5-(β-D-Ribofuranosyl)uracil (Pseudouridine)

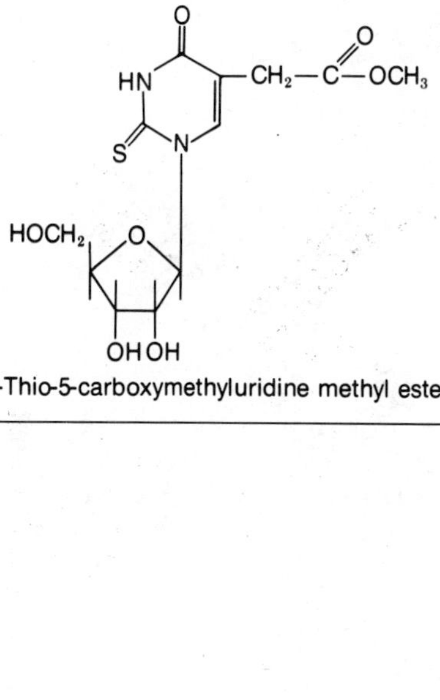

2-Thio-5-carboxymethyluridine methyl ester

Secondary Structure of RNA

RNA in solution exhibits a greater variety of structures than DNA. In low ionic strength solutions the molecules appear as extended polyelectrolyte chains. Shifting the molecules to solutions of high ionic strength causes the RNA to contract. Increasing the temperature denatures the RNA by disrupting the hydrogen-bonded base pairs and the base stacking. These changes can be monitored by measuring absorption of ultraviolet light at 260 nm, much like DNA. Since the RNA is single-stranded, the hyperchromic shifts from an ordered molecular conformation, a partial helix, to an extended and then random coil are not as high as for DNA. However, considerable helical structure exists in RNA in the absence of extensive base pairing. This helix is due to the strong base-stacking forces between A, G, and C residues. Base stacking is more important than

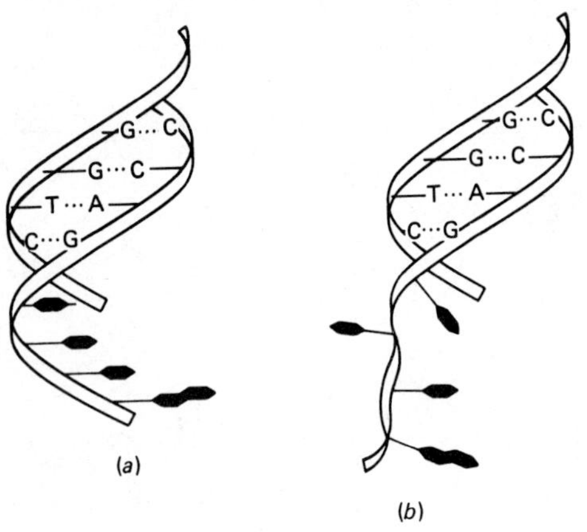

Figure 18.2
Models indicating a helical structure due to (a) base stacking in the —CCA_{OH} terminus of tRNA and (b) the lack of an ordered helix when no stacking occurs in this non-base-paired region.
Redrawn from M. Sprinzl and F. Cramer, *Prog. Nucl. Res. Mol. Biol.,* 22:9, 1979.

simple hydrogen bonding in determining inter- and intramolecular interactions. These forces act to restrict the possible conformations of an RNA molecule (Figure 18.2). Base stacking is a result of the van der Waals forces between the π electron clouds above and below the unsaturated rings of the purines and pyrimidines and the hydrophobic nature of the bases. As with DNA, the distance restriction of the phosphodiester-ribosyl backbone and the near-perpendicular angle of the β-glycosidic bond do not permit the bases to stack directly over each other. Therefore each succeeding base is offset by ~35°, forming RNA helical structures with 10 or 11 nucleotides per turn in a double helix in comparison to 10 in DNA. In single-stranded helical RNA, each succeeding base is offset by 60°. The result is a turn with six nucleotides.

A single strand of RNA may have double-helical regions formed by hydrogen bonding between complementary base sequences located within the molecule. These intermolecular duplex structures, often called "hairpins," may or may not have large unpaired loops at the end. There are considerable variations in the fine structural details of the "hairpin" structures. These variables include the length of base-paired regions and the size and number of unpaired loops (Figure 18.3). Transfer RNAs have a large proportion of their bases involved in these helical structures and are excellent examples of base stacking and hydrogen bonding in a single-stranded molecule (Figure 18.4*a*). The anticodon region in tRNA is an

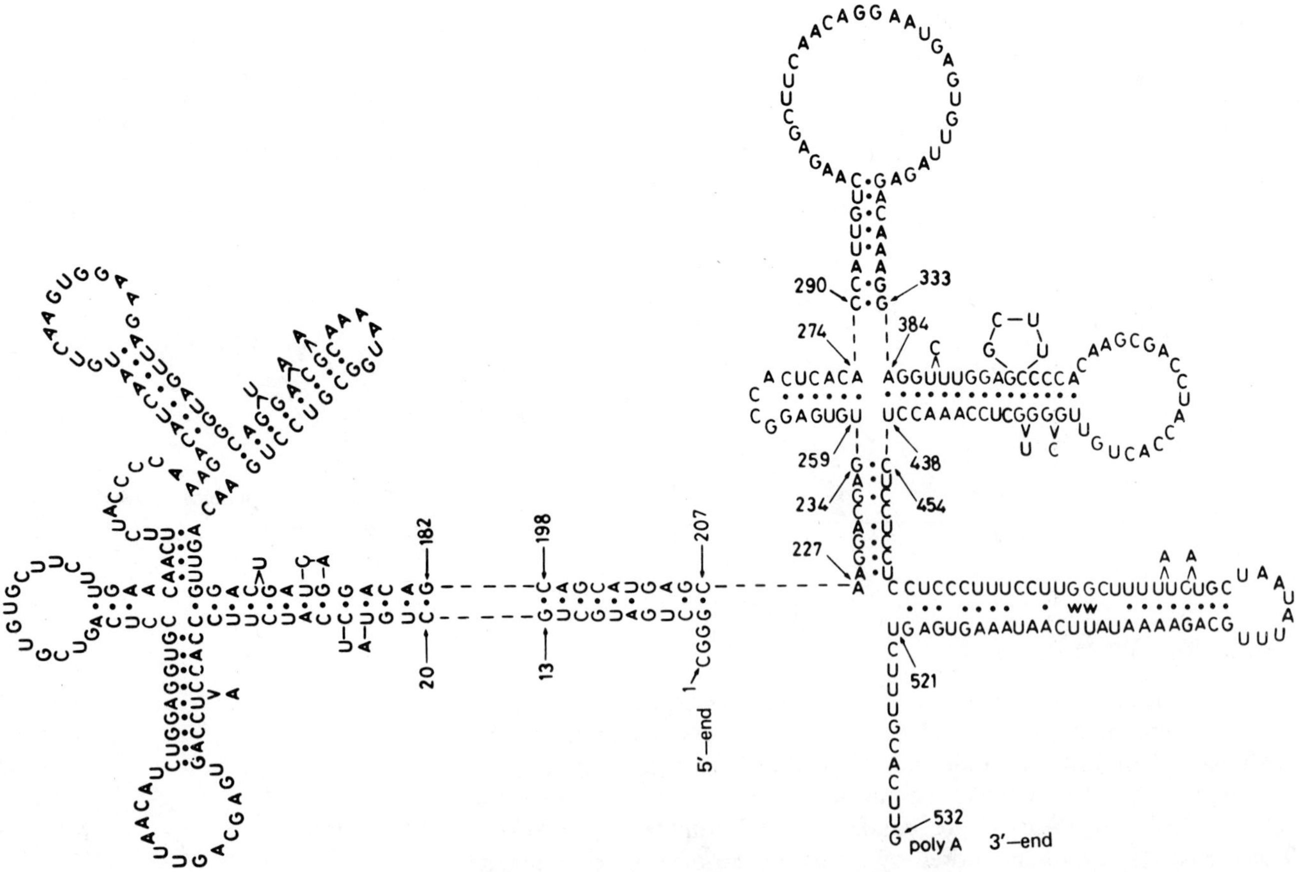

Figure 18.3
Proposed base-pairing regions in the mRNA for mouse immunoglobulin light chain.
Base-paired structures shown have free energies of at least −5 kcal. Note the variance in loop size and length of paired regions.
P. H. Hamlyn et al., *Cell,* 15:1073, 1978. Reproduced with permission.

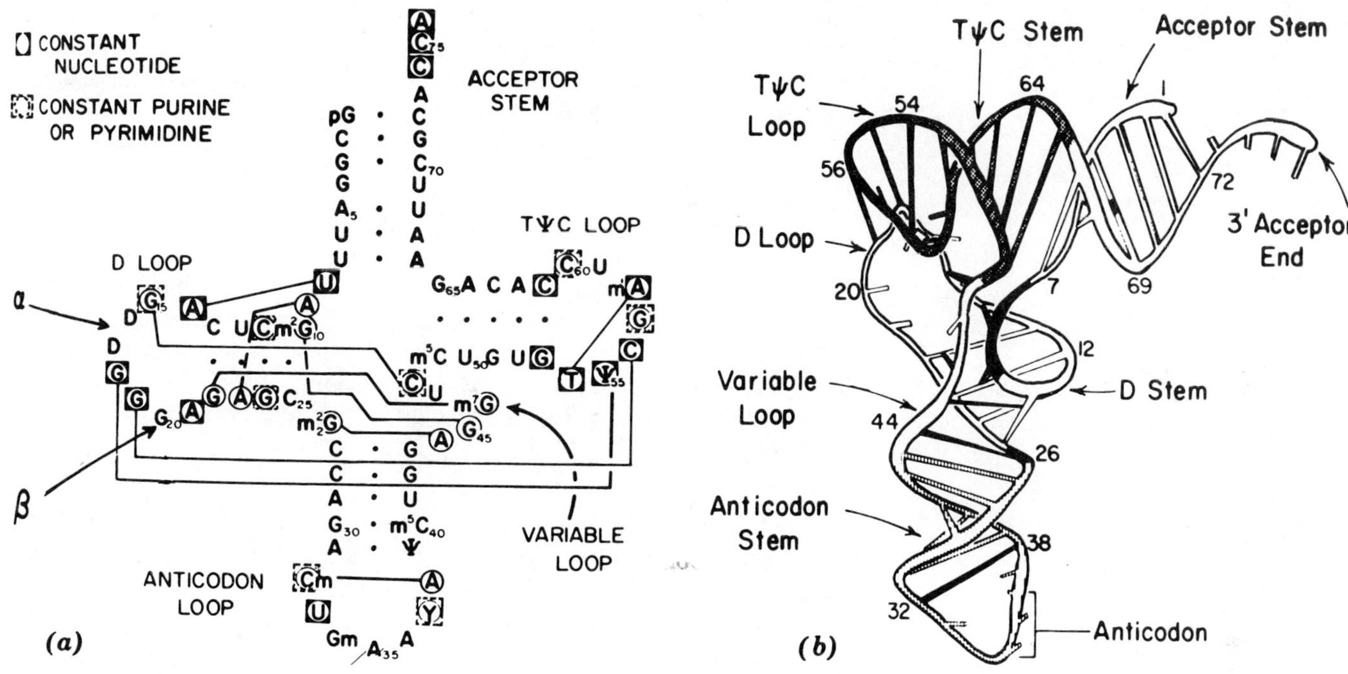

Figure 18.4
(a) Cloverleaf diagram of the two-dimensional structure and nucleotide sequence of yeast tRNA^{Phe}. Solid lines connecting circled nucleotides indicate hydrogen-bonded bases. Solid squares indicate constant nucleotides; dashed squares indicate a constant purine or pyrimidine. Insertion of nucleotides in the D loop occurs at positions α and β for different tRNAs. (b) Tertiary folding of the cloverleaf structure in (a). Hydrogen bonds are indicated by cross rungs.

From G. J. Quigley and A. Rich, *Science,* 194:797, 1976. Reproduced with permission.

unpaired, base-stacked loop of seven bases. The partial helix caused by base stacking in this loop binds, by specific base pairing, to a complementary codon in mRNA so that translation (peptide bond formation) can occur. Within the tRNA molecule itself ~60% of the bases are paired in four double-helical regions called stems. In addition, the unpaired regions have the capability to form base pairs with free bases in the same or other looped regions, thereby contributing to the molecule's tertiary structure. The role and extent of base pairing in each type of RNA is described in the following sections.

Tertiary Structure of RNA

The actual functioning structures for the RNAs are more complex than the base-stacked and hydrogen-bonded helices mentioned above. The RNAs in vivo are dynamic molecules in solution, which undergo changes in conformation during synthesis, processing, and functioning. Like DNA, RNA is always associated with one or more proteins that have functional and conformational roles. These proteins lend stability to the RNA much like the base pairing and subsequent molecular folding. Transfer RNA in solution is folded in a compact "L-shaped" conformation (Figure 18.4*b*). The arms and loops are folded in specific conformations held in position not only by traditional Watson-Crick base pairing as found in DNA, but also base interactions involving more than two nucleotides. There are interactions between bases and the phosphodiester backbone and between bases with particular regions of the sugars, especially the 2′-OH group. The folding of the tRNA molecules apparently occurs during transcription. Transfer RNA is the only biologically functional RNA molecule for which a three-dimensional structure is known. The first structural determination was accomplished by x-ray crystallography of yeast tRNA specific for phenylalanine.

18.3 TYPES OF RNA

RNA molecules in the cell can be classified by a number of different schemes, including their function, stability, and cytological localization. Traditionally, RNA species have been classified as transfer, ribosomal, and messenger RNAs; however, more recently it has been recognized that RNA molecules perform or facilitate a variety of different functions in a cell.

Transfer RNA

The tRNAs comprise ~15% of the total cellular RNA. Although synthesized in the nucleus, the tRNAs are rapidly processed and used in the cytoplasm. The primary roles of tRNA are to transport amino acids to the polyribosomes (ribosomes–mRNA complex) and to translate the genetic code in mRNA. The three nucleotides of the tRNA anticodon loop region bind to complementary nucleotide triplets of the mRNA. Transfer RNAs therefore have two primary active sites, the —CCA_{OH}, *3′-hydroxyl terminus,* to which specific amino acids are enzymatically attached covalently, and the *anticodon triplet*. These two active sites are responsible for the *adapter* function of tRNA; that is, the conversion of information encoded in a nucleic acid (DNA or mRNA) sequence into protein sequence during translation.

Although there are only 20 amino acids used in most proteins, there are at least 56 different species of tRNAs in any given cell, with each tRNA having different anticodon triplets. (Mitochondria synthesize a much smaller number of tRNAs.) There is often more than one tRNA for any particular amino acid, and these RNAs may be defined as isoacceptor tRNAs. A tRNA that binds phenylalanine would be written $tRNA^{Phe}$, whereas one for carrying tyrosine would be $tRNA^{Tyr}$.

Transfer RNAs are relatively small for nucleic acids and range in length from 65 to 110 nucleotides. This corresponds to a molecular weight range from ~22,000 to 37,300. The sedimentation coefficient for tRNAs as a group is 4S, and the term 4S RNA is often used to designate tRNA. Nucleotide sequences for tRNAs can all be drawn to conform to the general two-dimensional "cloverleaf" structure and three-dimensional "L-form," as determined by x-ray crystallography, shown in Figure 18.4.

From the nucleotide sequence and structure of the $tRNA^{Phe}$ shown in Figure 18.4, it is clear that tRNAs have several modified nucleotides as well as a high proportion of bases involved in secondary conformations, helices, and tertiary folding. Some of the modified nucleotides found in tRNA are listed in Table 18.2 and their positions in tRNAs indicated in Figure 18.5. The modified nucleotides affect tRNA structure and stability but are not essential for the formation or maintenance of tertiary conformation. The modifications do not appear to have a role in general aminoacylation, or "charging," of tRNAs, but may be involved in regulating specific and nonspecific recognition of enzymes and proteins and also specify interactions between tRNA and ribosomes.

Many structural features are common to all tRNA molecules (see Table 18.3). Seven base pairs are always in the amino acid acceptor stem, which, in functioning molecules, is terminated with the nucleotide triplet

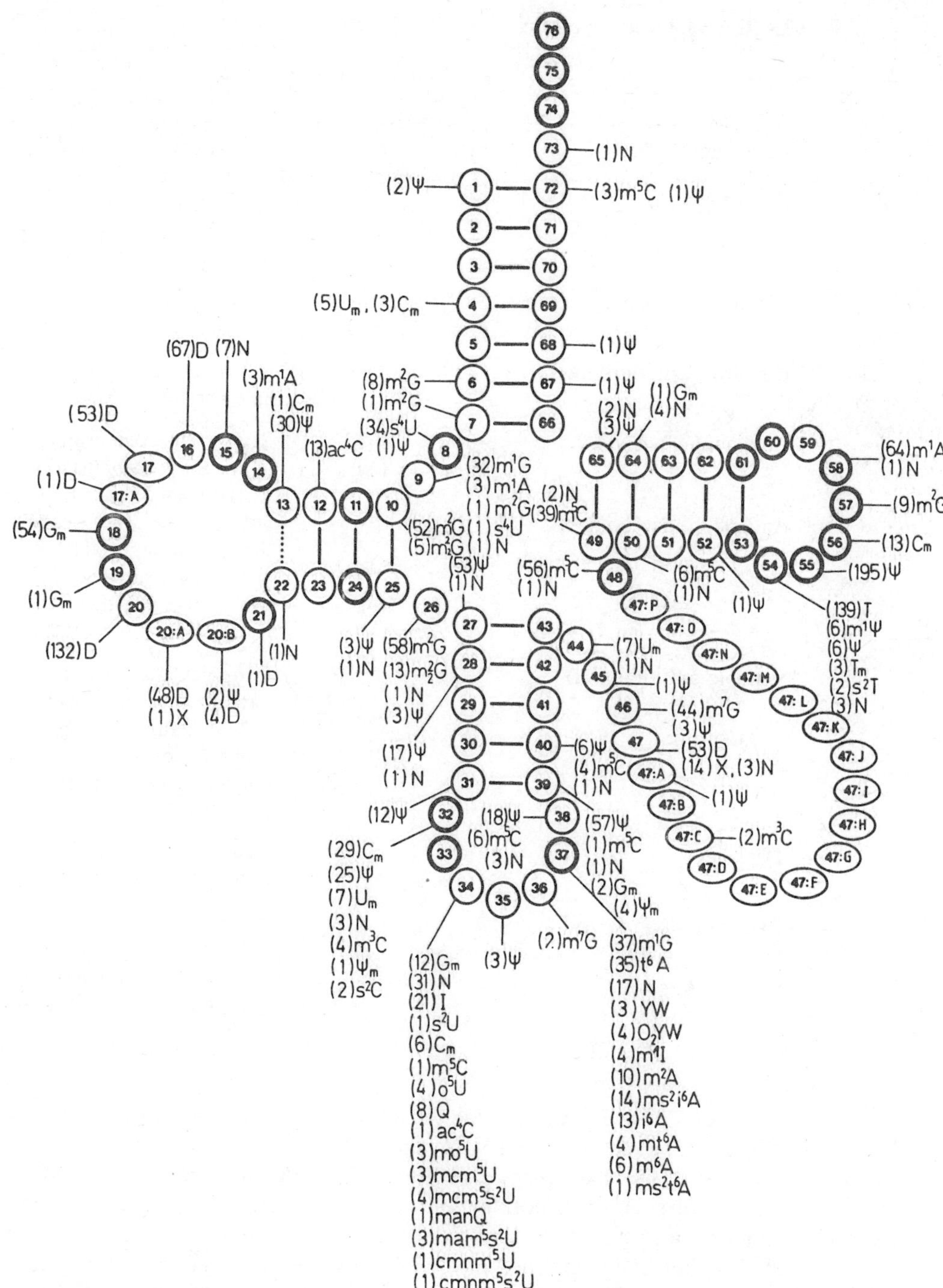

Figure 18.5

Cloverleaf secondary structure of tRNA showing positions and prevalence of modified nucleotides.

The figure depicts the cloverleaf secondary structure of tRNA, the standardized numbering of the nucleotide sequences, and those nucleotides that are always present (thick-edged circles) or commonly present (thin-edged circles) in the structure. Ovals represent nucleotides that are not present in each sequence. The location and prevalence of each of the modified nucleotides in the known sequences of 211 different tRNAs is shown as the frequency of appearance at a particular location in parentheses preceding the nucleotide abbreviation. Table 18.2 lists the full names and abbreviations of the modified nucleotides.

P. F. Agris and R. A. Kopper (eds). *The modified nucleosides of transfer RNA II*. New York: Alan R. Liss, 1983. Reproduced with permission.

TABLE 18.3 Characteristics of Regions in tRNA

Region	*Number of Nucleotides*	*Comments*
Amino acid helix	14 (7 base pairs)	Region where base mispairing occurs frequently; G · U is common; —CCA_{OH} is added posttranscriptionally
Dihydrouracil stem	6 or 8 (3 or 4 base pairs)	First and last base pairs are usually C · G
Dihydrouracil loop (loop I)	7–10	Region exhibits considerable variation
Anticodon stem	10 (5 base pairs)	Second base pair from anticodon loop is usually C · G; from 5′ end, the 3rd, 4th, and 5th bases are the anticodon; 5′ side of anticodon is always a pyrimidine; 3′ side is usually a modified purine
Anticodon loop (loop II)	7	
Variable arm (loop III)	3–21	Extremely variable in structure and often lacks a helical stem; the arm probably forms hydrogen-bonded stem region (3–7 base pairs)
TΨC stem	10 (5 base pairs)	Base pair adjacent to TΨC loop is C · G
TΨC loop (loop IV)	7	All tRNAs contain the sequence T-Ψ-C-purine at the same location in the loop; the purine is usually guanine

—CCA_{OH}. This —CCA_{OH} triplet is not base-paired. The dihydrouracil or "D" stem has three or four base pairs, while the anticodon and —TΨC— stems have five base pairs each. Both the anticodon loop and —TΨC— loop have seven nucleotides. Since the distance from the amino acid attachment to the anticodon triplet is constant, about 75 Å, differences in the number of nucleotides in different tRNAs are accounted for by the variable loop. Thus 80% of tRNAs have small variable loops of 4 or 5 nucleotides, while the others have larger loops with 13 to 21 nucleotides. Five or six nucleotides may also be incorporated at two positions in the D loop, which varies from 8 to 14 nucleotides. Several of the nucleotides are in constant positions (see Figure 18.4*a*).

Ribosomal RNA

The eucaryotic cytoplasmic ribosomes are composed of four RNA molecules and 70 to 80 proteins. These RNAs and proteins are distributed specifically between the two ribosomal subunits. The smaller subunit, the 40S particle, contains one 18S rRNA and 55% of the proteins. The large ribosomal subunit, the 60S particle, contains the remaining rRNAs and proteins. The three rRNAs in the large subunit are the 28S rRNA, the 5.8S rRNA, and the smaller 5S rRNA.

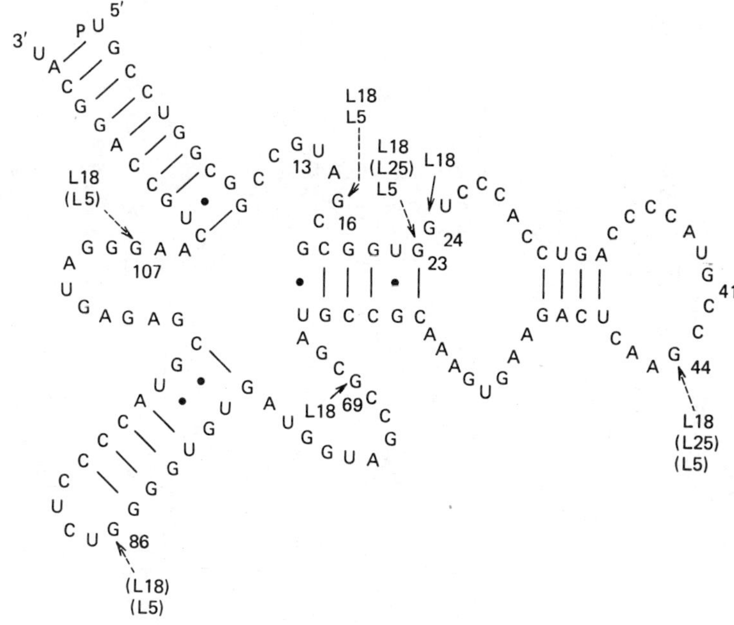

Figure 18.6
Secondary, base-paired, structure proposed for 5S rRNA.
Arrows indicate regions protected by proteins in the large ribosomal subunit.
Combined information from G. E. Fox and C. R. Woese, *Nature,* 256:505, 1975, and R. A. Garrett and P. N. Gray.

The rRNAs account for 80% of the total cellular RNA and are metabolically stable. This stability, required for repeated functioning of the ribosome, is enhanced by close association with the ribosomal proteins. Some of the ribosomal proteins bind directly to the rRNAs during transcription. The 28S, 18S, and 5.8S rRNAs are synthesized in the *nucleolar region* of the nucleus. The 5S rRNA is not transcribed in the nucleolus but rather from separate genes within the nucleoplasm. Processing of the rRNAs (see Section 18.5) includes cleavage to the functional size, limited formation of internal base pairing via hydrogen bonds, modification of particular nucleotides, and association with ribosomal proteins to form a stable tertiary conformation.

The 5S rRNA is 68% base-paired, with helical regions formed between proximal as well as distal internal complementary sequences (see Figure 18.6). The nucleotide sequences and proposed conformations for 5S rRNA have been highly conserved throughout the evolutionary scale. The length, some sequences, and most helical regions are the same for *E. coli* and human 5S rRNA. A specific function for 5S rRNA has not been described, although it is apparently required, in a structural role, for protein synthesis. A lack of 5S rRNA in a ribosome or cleavage at specific locations in the 5S rRNA nucleotide chain render the ribosome inactive.

The 5.8S rRNA, with 158 nucleotides, is closely associated, by hydrogen bonds, with the 28S rRNA. The 5.8S rRNA has considerable internal base-pairing, while the 5′ and 3′ ends are free to interact with the 28S rRNA. Like the 5S rRNA, the 5.8S rRNA nucleotide sequence has been conserved during evolution, but the 5.8S rRNA is not found in procaryotes.

The larger rRNAs contain most of the altered nucleotides found in rRNA. These are primarily methylations on the 2′ position of the ribose, giving 2′-*O*-methylribose. At least one methylation of rRNA has been directly related to bacterial antibiotic resistance (Clin. Corr. 18.1). There

CLIN. CORR. **18.1** STAPHYLOCOCCAL RESISTANCE TO ERYTHROMYCIN

Bacteria exposed to antibiotics in a clinical or agricultural setting often develop resistance to the drugs. This resistance can arise from a mutation in the target cell's DNA, which gives rise to resistant descendants. An alternative and clinically more serious mode of resistance arises when plasmids coding for antibiotic resistance proliferate through the bacterial population. These plasmids may carry multiple resistance determinants and render several antibiotics useless at the same time.

Erythromycin inhibits protein synthesis by binding to the large ribosomal subunit. *Staphylococcus aureus* can become resistant to erythromycin and similar antibiotics as a result of a plasmid-borne RNA methylase enzyme that converts a single adenosine in 23S rRNA to N^6-dimethyladenosine. Synthesis of the methylase enzyme is induced by erythromycin.

The microorganism that produces an antibiotic must also be immune to it or else it would be inhibited by its own toxic product. The organism that produces erythromycin, *Streptomyces erythreus,* itself possesses an rRNA methylase that acts at the same ribosomal site as the one from *Staphylococcus aureus.*

Which came first? It is likely that many of the resistance genes in target organisms evolved from those of producer organisms. In several cases, DNA sequences from resistance genes of the same specificity are conserved between producer and target organisms. We may therefore look on plasmid-borne antibiotic resistance as a case of "natural genetic engineering," whereby DNA from one organism (the *Streptomyces* producer) is appropriated and expressed in another (the *Staphylococcus* target).

TABLE 18.4 Ribosomal RNA Characteristics

Sedimentation Value[a]	*Molecular Weight*	*G + C (%)*	*Methylations*	*Number of Gene Copies*
Eucaryotes				
p45S (nucleolar)	4.3×10^6	70	Mostly 2′-*O*-methylribose All retained in 28S and 18S	500–1,000
p41S	3.1×10^6	~70		—
p32S	2.1×10^6	70		—
p20S	0.95×10^6	<70		—
m28S	1.7×10^6	65	9–13/1,000 NT	—
m18S	0.7×10^6	58	15–19/1,000 NT + m_2^6A	—
m5.8S	5.1×10^4	~65	?	—
m5S (nuclear)	3.9×10^4	65	None	Several hundred
Mitochondrial				
16S	5.4×10^5		Very low	1
12S	3.5×10^5		Very low	1
Procaryotes				
23S	1.1×10^6		11/1,000 NT	5–6
16S	0.6×10^6		17/1,000 Nt	5–6
5S	4×10^4	65	None	~7

[a] p, precursor form of the RNA; m, mature form of the RNA.

are a small number of N^6-dimethyladenines present in the 18S rRNA. The 28S rRNA has about 45 methyl groups and the 18S rRNA has 30 methyl groups, which may be involved in processing of the 45S precursor molecule (see Table 18.4).

In addition to the base-pairing within each 18S and 28S rRNA there is evidence for a base sequence in mRNAs, which can base-pair with the rRNA of the smaller subunit, forming a translation complex. The hinging mechanism between the two ribosomal subunits, which enables translocation and mRNA movement, is thought to involve protein–protein interactions and base-pairing between the 18S and 28S rRNAs.

Messenger RNA

The mRNAs are the direct carriers of genetic information from the nucleus to the cytoplasmic ribosomes. Each eucaryotic mRNA contains information for only one polypeptide chain, and therefore these mRNAs have been designated monocistronic, whereas in procaryotes mRNA can exist as polycistronic molecules. A cell's phenotype and functional state is related directly to the cytoplasmic mRNA content. In cells exhibiting highly active protein synthesis, such as pancreatic cells, the DNA : RNA ratio is very low due to the large amounts of mRNA and rRNA. However, in cells with low rates of protein synthesis, such as muscle cells, the DNA : RNA ratio is much higher, since there is no requirement for large quantities of mRNAs and ribosomes.

In the cytoplasm mRNAs have relatively short life spans, which in part are determined by the cell's particular needs at any given time. Some mRNAs are known to be synthesized and stored in an inactive or dormant state in the cytoplasm, ready for a quick protein synthetic response. An example of this is the unfertilized egg of the African clawed toad, *Xenopus laevis*. Immediately upon fertilization the egg undergoes rapid protein synthesis, indicating the presence of preformed mRNA and ribosomes.

Eucaryotic mRNAs have unique structural features not found in rRNA or tRNA (see Figure 18.7). These features aid in proper mRNA function-

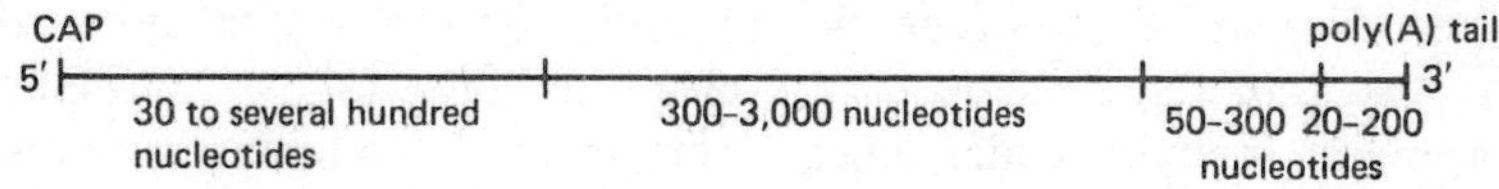

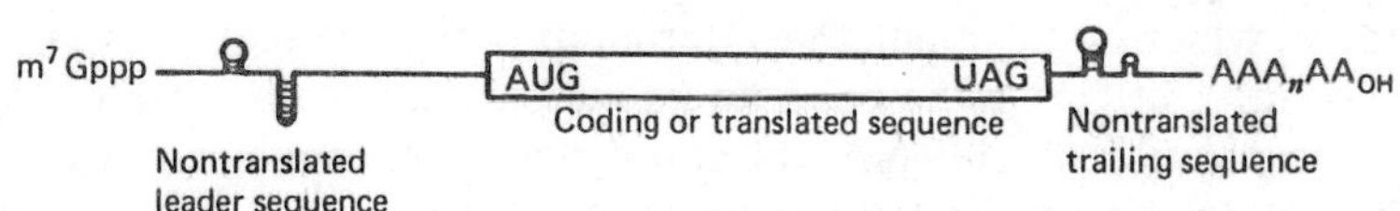

Figure 18.7
General structure for a eucaryotic mRNA.
There is a "blocked" 5′ terminus, cap, followed by the nontranslated leader containing a promoter sequence. The coding region usually begins with the initiator codon AUG and continues to the translation termination sequence UAG, UAA, or UGA. This is followed by the nontranslated trailer and a poly(A) tail on the 3′ end.

ing. Since the information within mRNA lies in the linear sequence of the nucleotides, the integrity of this sequence is extremely important. Any loss or change of nucleotides could alter the protein being translated. The translation of mRNA on the ribosomes must also begin and end at specific sequences. Structurally, starting from the 5′ terminus, there is an inverted methylated base attached via *5′-phosphate-5′-phosphate* bonds rather than the usual internucleotide 3′,5′ phosphodiester linkages between adjacent riboses. This structure, called a "cap," is a guanosine 5′-triphosphate methylated at the number 7 nitrogen of the ring ($m^7G^{5'}ppp$). The cap is attached to the first transcribed nucleotide, usually a purine, methylated on the 2′-OH of the ribose (see Figure 18.8). The cap is followed by a nontranslated or "leader" sequence to the 5′ side of the coding region. Following the leader sequence is the initiation sequence or codon, most often AUG, and then the translatable message or coding region of the molecule. At the end of the coding sequence is a termination sequence

Figure 18.8
Diagram of the "cap" structure or blocked 5′ terminus in mRNA.
The 7-methylguanosine is inverted to form a 5′-phosphate to 5′-phosphate linkage with the first nucleotide of the mRNA. This nucleotide is often a methylated purine.

signaling termination of polypeptide formation and release from the ribosome. A second nontranslated or "trailer" sequence follows, terminated by a string of adenylic acids, called a poly(A) tail, which makes up the 3′ terminus of the mRNA. This poly(A) section may vary from 20 to 200 nucleotides.

The 5′ cap structure blocks the action of RNA exonucleases and phosphatases, which could attack the 5′ terminus of the message. The cap also has a positive effect on the initiation of message translation. In the initiation of translation of a mRNA, the cap structure is recognized by a single ribosomal protein, an initiation factor (see Chapter 19). Several methylated nucleotides also occur in the internal portions of some mRNAs in addition to those on the cap and adjacent nucleotides. The majority of the internal methylations are m^6-adenosines with some m^5-cytidines.

The role for the poly(A) tail is unclear. Several possibilities exist and the actual function may include some, all, or none of these: the poly(A) may function in transport of mRNA through the nuclear membrane; it may serve as a buffer for exonucleolytic attack from the 3′ terminus; or it may have some role in translation by interacting with the ribosome or a nonribosomal protein. Although the majority of mRNAs have a poly(A) tail, some do not. For example, the messengers for some histones lack 3′-poly(A) termini.

There are two types of protein associated closely with the eucaryotic mRNAs. One type binds tightly to the 5′ cap region and may be cap specific. The second type of protein is associated with the 3′ end of the mRNAs.

RNA in Ribonucleoprotein Particles

Besides tRNA, rRNA, and mRNA, small, stable RNA species can be found in the nucleus, cytoplasm, and mitochondria. These small RNA species function as ribonucleoprotein particles (RNPs), with one or more protein subunits attached. Different RNP species, both cytoplasmic and nuclear, have been implicated in a variety of cellular functions. These functions include RNA trimming, splicing, transport, and control of translation, as well as recognition of proteins for export. The actual roles of these species are described more fully in the discussion of specific metabolic events.

RNA Species in Mitochondria

Mitochondria have their own protein-synthesizing mechanisms, including ribosomes, tRNAs, and mRNAs. The mt rRNAs, 12S and 16S, are transcribed from the mitochondrial DNA (mtDNA) as are at least 19 specific tRNAs and some mRNAs. Note that there are fewer mt tRNAs than procaryotic or cytoplasmic tRNA species; there is only one mt tRNA species per amino acid. The mt RNAs account for 4% of the total cellular RNA. They are transcribed by a mitochondrial-specific RNA polymerase. mt tRNAs are not as highly methylated as their cytoplasmic counterparts and are processed from precursor molecules smaller than their cytoplasmic counterparts. Genes for 12 tRNAs are located on the heavy mtDNA strand and 7 on the light strand. Some of the mRNAs have eucaryotic characteristics, such as 3′-poly(A) tails. A large degree of coordination exists between the nuclear and mitochondrial genomes. Most of the aminoacylating enzymes for the mt tRNAs and most of the mitochondrial ribosomal proteins are specified by nuclear genes, translated in the cytoplasm and transported into the mitochondria. Furthermore, at least some of the modified bases in mt tRNA species are synthesized by enzymes encoded in nuclear DNA.

18.4 MECHANISMS OF TRANSCRIPTION

Overview

The process by which RNA chains are made from DNA templates is called *transcription*. All known transcription reactions take the following form:

$$\text{DNA template} + n(\text{NTP}) \longrightarrow \text{pppN(pN)}_n + (n - 1)\text{PP}_i + \text{DNA template}$$

Enzymes that catalyze this reaction are designated RNA polymerases. The energetics favoring the RNA polymerase reaction are twofold: First, the 5′ α-nucleotide phosphate of the ribonucleoside triphosphate is converted from a phosphate anhydride to a phosphodiester bond with a change in free energy ($\Delta G'$) of approximately 3 kcal/mol under standard conditions. Second, the released pyrophosphate, PP_i, can be cleaved to two phosphates by other enzymatic systems. This latter reaction means that $[PP_i]$ is low and phosphodiester bond formation is favored relative to standard conditions (see Chapter 6 for a fuller discussion of metabolic coupling).

Since a DNA template is required for RNA synthesis, eucaryotic transcription takes place in the cell nucleus or mitochondrial matrix. Within the nucleus, the *nucleolus* is the site of rRNA synthesis, whereas mRNA and tRNA are synthesized in the nucleoplasm. Procaryotic transcription is accomplished on the cell's DNA, which is located in a relatively small region of the cell. In the case of procaryotic plasmids, the DNA template need not be associated with the chromosome.

Structural changes in DNA occur during its transcription. In the polytene chromosomes of *Drosophila,* transcriptionally active genes are visualized in the light microscope as puffs distinct from the condensed, inactive chromatin. Furthermore, the nucleosome patterns of active genes are disrupted so that active chromatin is more accessible to, for example, DNase attack. In procaryotes, the DNA double helix is transiently opened (unwound) as the transcription complex proceeds down the DNA.

These openings and unwindings are a manifestation of a topological necessity: If the RNA chain were copied off DNA without this unwinding, the transcription complex and growing end of the RNA chain would have to wind around the double helix once every 10 base pairs as they travel from the beginning of the gene to its end. Such a process would wrap the newly synthesized RNA chain around the DNA double helix; the problem then would be to unwind it before the RNA could be exported to the cytoplasm. Local opening and unwinding of the DNA solves this problem before it occurs by allowing transcription to proceed on a single face of the DNA. In addition, the opening of DNA base pairs during transcription allows Watson-Crick base pairing between template DNA and the newly synthesized RNA.

The process of transcription is divided into three parts: *Initiation* refers to the recognition of an active gene starting point by RNA polymerase and the beginning of the bond formation process. *Elongation* is the actual synthesis of the RNA chain, and is followed by chain *termination and release*.

Required Components for Transcription: The Template

Each cycle of transcription begins and ends with the recognition of certain sites in the DNA template. DNA sequencing of a large number of transcription start regions, called *promoters,* has shown that certain sequences occur with great regularity. These sequences are called *conserved* or *consensus sequences*. Similar considerations demonstrate that termination occurs at specific consensus-type sequences. In addition, sites within a transcript may allow premature termination of transcription. These sites can act as molecular switches affecting the continuation of synthesis of an RNA molecule.

Consensus sequences near the transcription start are found for both procaryotic and some eucaryotic promoters. In addition eucaryotic transcription has been shown in some cases to be affected by *internal* promoter elements and other sequences called *enhancers*. Enhancers are gene-specific sequences that positively affect transcription. The most extensively studied is a twofold repeat of a 72 base pair sequence in the simian virus 40 (SV40) chromosome. The remarkable observation is that the SV40 and other enhancers function at various positions and in either orientation in the SV40 DNA. In contrast to, for example, a site in which RNA polymerase initially binds DNA, enhancer sequences can stimulate transcription whether they are located at the beginning or the end of a gene. The enhancer sequence must be on the same DNA strand as the transcribed gene (genetically in a position *cis*) but can function in either orientation. Cellular proteins are known that specifically bind enhancers. The most likely hypothesis is that eucaryotic enhancers serve to bring about a structural change in the DNA template, allowing transcription to occur.

RNA Polymerase

RNA polymerase from procaryotes is a large multisubunit enzyme, consisting in *E. coli* of five subunits (Table 18.5). Two α subunits, one β subunit and one β' subunit, constitute the *core enzyme,* which is capable of faithful transcription but not of specific (i.e., correctly initiated) RNA synthesis. The addition of a fifth protein subunit, designated σ, results in the *holoenzyme* that is capable of specific RNA synthesis in vitro and in vivo. The logical conclusion, that σ is involved in the specific recognition of promoters, has been borne out by a variety of biochemical studies and

TABLE 18.5 Comparative Properties of Some RNA Polymerases

	Nuclear				
	I (A)	*II (B)*	*III (C)*	*Mitochondrial*	E. coli
High mol wt subunits[a]	195–197 117–126	240–214 140	155 138	65	160 (β') 150 (β)
Low mol wt subunits	61–51 49–44 29–25 19–16.5	41–34 29–25 27–20 19.5 19 16.5	89 70 53 49 41 32 29 19		86 (σ) 40 (α) 10 (ω)
Variable forms	2–3 types	3–4 types	2–4 types	1	1
Specialization	Nucleolar; rRNA	mRNA Viral RNA	tRNA 5S rRNA	All mtRNA	None
Inhibition by α-amanitin	Insensitive (>1 mg/ml)	Very sensitive 10^{-9}–10^{-8} M	Sensitive 10^{-5}–10^{-4} M	Insensitive, but sensitive to rifampicin	Rifampicin-sensitive

[a] Molecular weight × 10^{-3}.

is discussed below. Furthermore, specific σ factors can recognize different classes of genes. This has been shown to be the case in several systems. In *E. coli* a specific σ factor recognizes a class of promoters for genes that are turned on as a result of heat shock. In *Bacillus subtilis* specific σ factors are made that recognize genes turned on during sporulation. Some bacteriophage synthesize σ factors that allow the appropriation of the cell's RNA polymerase for transcription of the viral DNA.

The common procaryotic RNA polymerases are inhibited by the antibiotic *rifampicin* (used in treating tuberculosis), which binds to the β subunit (Figure 18.9 and Clin. Corr. 18.2). Eucaryotic nuclear RNA polymerases are inhibited differentially by the compound α-amanitin, which is synthesized by the poisonous mushroom *Amanita phalloides*. A particular concentration of amanitin can be used to inhibit synthesis of a class of RNA in vitro or in vivo. Using such experiments, three nuclear RNA polymerase classes can be distinguished. Very low concentrations of α-amanitin inhibit the synthesis of mRNA and some small nuclear RNAs (snRNAs); higher concentrations inhibit the synthesis of tRNA and other

CLIN. CORR. 18.2
ANTIBIOTICS

Rifamycin (rifampicin) is toxic to acid-fast and gram-positive bacteria and has been used to treat tuberculosis (Figure 18.9).

Actinomycin D is an effective bacteriocidal agent; however, its toxicity prevents it from becoming widely used as an antitumor compound, although it has been used successfully for Wilm's tumor.

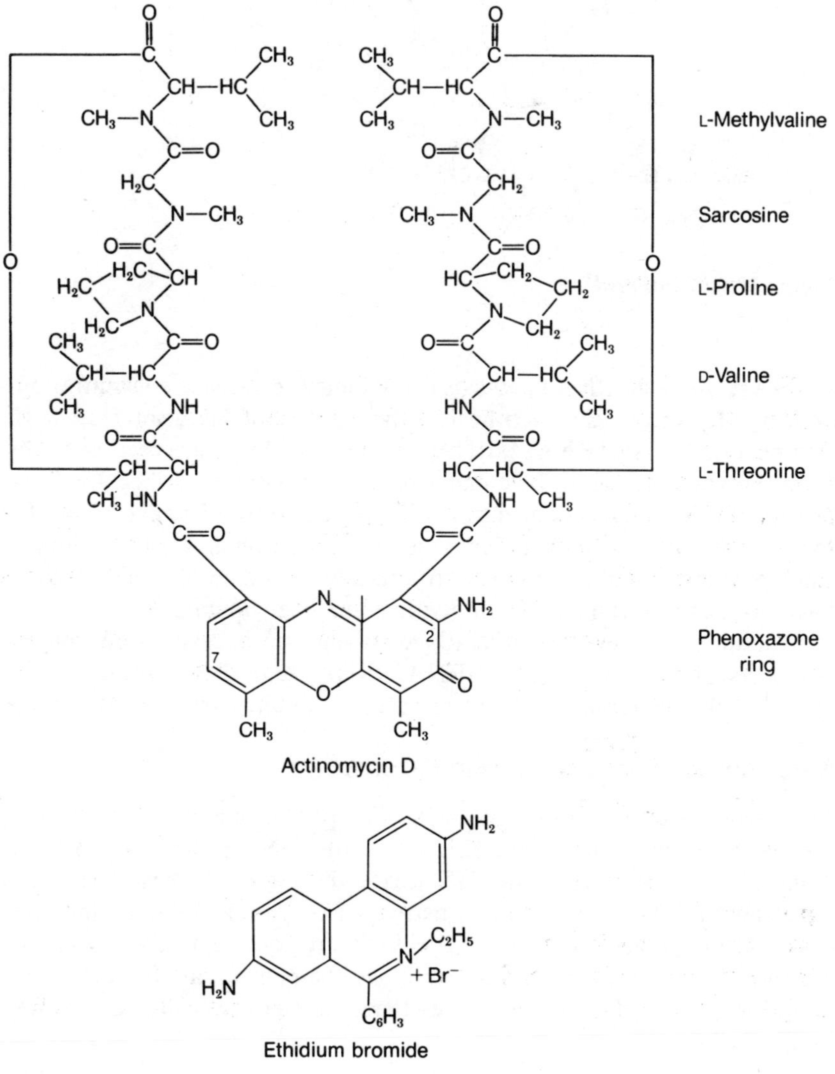

L-Methylvaline

Sarcosine

L-Proline

D-Valine

L-Threonine

Phenoxazone ring

Actinomycin D

Ethidium bromide

Figure 18.9
Inhibitors of RNA synthesis.
Actinomycin D and ethidium bromide bind to DNA, blocking transcription. Rifamycin, rifampicin and α-amanitin bind directly to RNA polymerase.

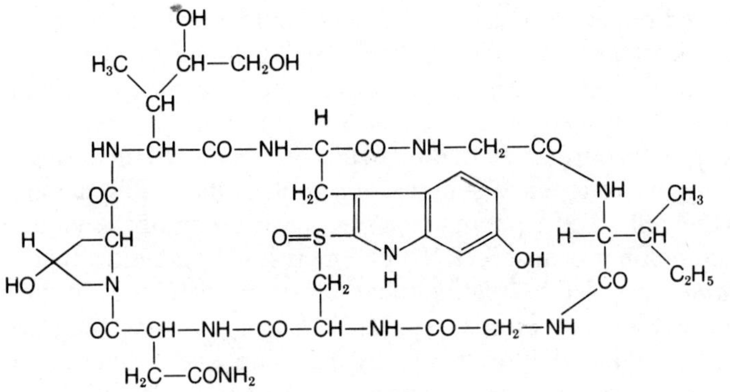

α-Amanitin

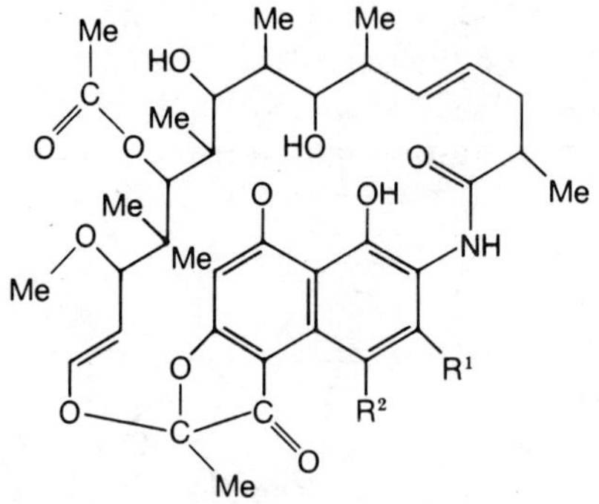

Rifamycin B $R^1 = H$. $R^2{=}O{-}CH_2{-}COOH$

Rifampicin $R^1 = {-}CH{=}N$ (piperazine ring) $N{-}CH_3$. $R^2{=}OH$

Figure 18.9 **(*Continued*)**

snRNAs, whereas rRNA synthesis is not inhibited at these concentrations of drug. Messenger RNA synthesis is the function of RNA polymerase II. Synthesis of transfer RNA, 5sRNA, and some snRNAs are carried out by RNA polymerase III. Ribosomal RNA genes are transcribed by RNA polymerase I, which is concentrated in the nucleolus. (The numbers refer to the order of elution of the enzymes from a chromatography column.) Each enzyme is highly complex structurally (Table 18.5), and functions have not been established for individual enzyme subunits.

In addition, a mitochondrial RNA polymerase is responsible for the synthesis of this organelle's mRNA, tRNA, and rRNA species. This enzyme, like bacterial RNA polymerase, is inhibited by rifampicin.

Mechanics of Transcription in Procaryotes

Procaryotic transcription begins with the specific binding of RNA polymerase to a gene's promoter (Figure 18.10). RNA polymerase holoenzyme binds to one face of the DNA extending some 45 or so base pairs upstream and 10 base pairs downstream from the RNA initiation site. Two short oligonucleotide sequences in this region are highly conserved. One sequence that is located about 10 base pairs upstream from the transcription start is the consensus sequence (sometimes called a Pribnow box):

$$\overset{*}{T}\overset{*}{A}TAA\overset{*}{T}$$

The positions marked with an asterisk are the most conserved; indeed, the last T residue is *always* found in *E. coli* promoters.

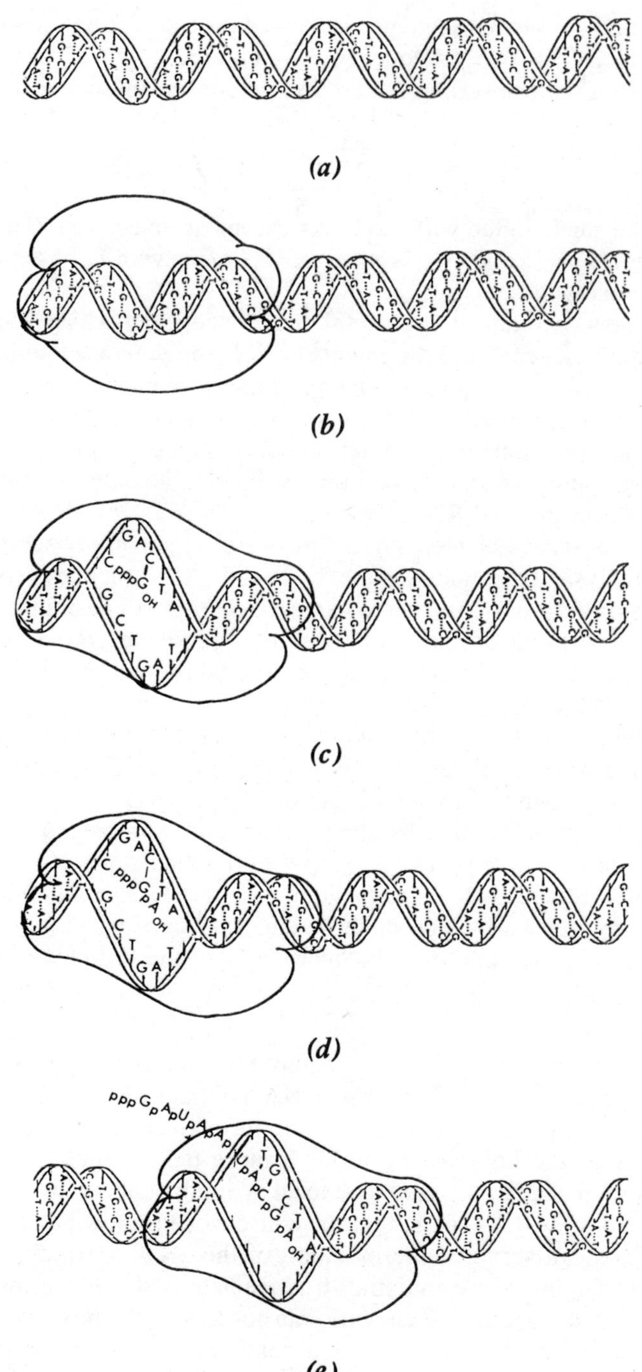

Figure 18.10
Early events in procaryotic transcription.
(a) DNA double helix. (b) Recognition: RNA polymerase with "sigma" factor binds to a DNA promoter region in a "closed" conformation. (c) Initiation: The complex is converted to an "open" conformation and the first nucleoside triphosphate aligns with the DNA. (d) The first phosphodiester bond is formed and the "sigma" factor released. (e) Elongation: Synthesis of nascent RNA proceeds with movement of the RNA polymerase along the DNA. The double helix reforms.

A second consensus sequence is located upstream from the Pribnow or "−10" box. This "−35 sequence" is centered about 35 basepairs upstream from the transcription start and its consensus sequence is

$$\overset{*}{T}\overset{*}{T}\overset{*}{G}ACA$$

where the nucleotides with asterisks are most conserved. The spacing between the "−35 and −10 sequences" is crucial with 17 base pairs being highly conserved.

Consensus sequences are useful and striking, but what difference do they make to a gene? Measurements of RNA polymerase binding affinity and initiation efficiency to various promoter sequences have shown that the most active promoters fit the consensus sequences most closely. Statistical measurements of promoter homology conform closely to the measured "strength" of a promoter; that is, its kinetic ability to initiate transcription with purified RNA polymerase.

Other conserved elements in the promoter region are observed by statistical analysis: bases flanking the "−35 and −10 sequences", bases near the transcription start, and bases located near the −16 position are weakly conserved. In some of these weakly conserved regions, RNA polymerase may require that a particular nucleotide *not* be present or that local variations in DNA helical structure be present.

Promoters for *E. coli* heat shock genes have different consensus sequences at the "−35" and "−10" homologies. This is consistent with their being recognized by a different σ factor.

An RNA transcript usually starts with a purine riboside triphosphate; that is, pppG . . . or pppA . . . , but pyrimidine starts are also known (Figures 18.10 and 18.11). The position of an RNA chain initiation differs slightly between various promoters, usually occurring from five to eight base pairs downstream from the invariant T of the Pribnow box.

Initiation

Two kinetically distinct steps are required for RNA polymerase to initiate an RNA chain. In the first step, RNA polymerase holoenzyme binds electrostatically to the promoter DNA to form a "closed complex." In the second step, the holoenzyme forms a more tightly bound "open complex," which is characterized by a local opening of about 10 base pairs of the DNA double helix. Note that the consensus Pribnow box is A-T rich; it therefore can serve as the initiation of this local unwinding. The unwound DNA binds the initiating triphosphate and then forms the first phosphodiester bond. The enzyme translocates to the next position (this is the rifampicin-inhibited step) and continues synthesis. At or a short time after the initial bond formation, σ factor is released and the enzyme is considered to be in an *elongation* mode.

Elongation

RNA polymerase continues the binding-bond formation-translocation cycle at a rate of about 40 nucleotides per second. This rate is only an average, however, and there are many examples known for which RNA polymerase pauses or slows down at particular sequences, usually inverted repeats (palindrome sequence of nucleotides). As will be discussed below, these pauses can bring about termination.

Alternatively, examples are known, especially in phage λ, in which changes in the transcription complex during the elongation phase affect subsequent termination events. These changes depend on the binding of another cellular protein (*nusA* protein) to core RNA polymerase. Failure

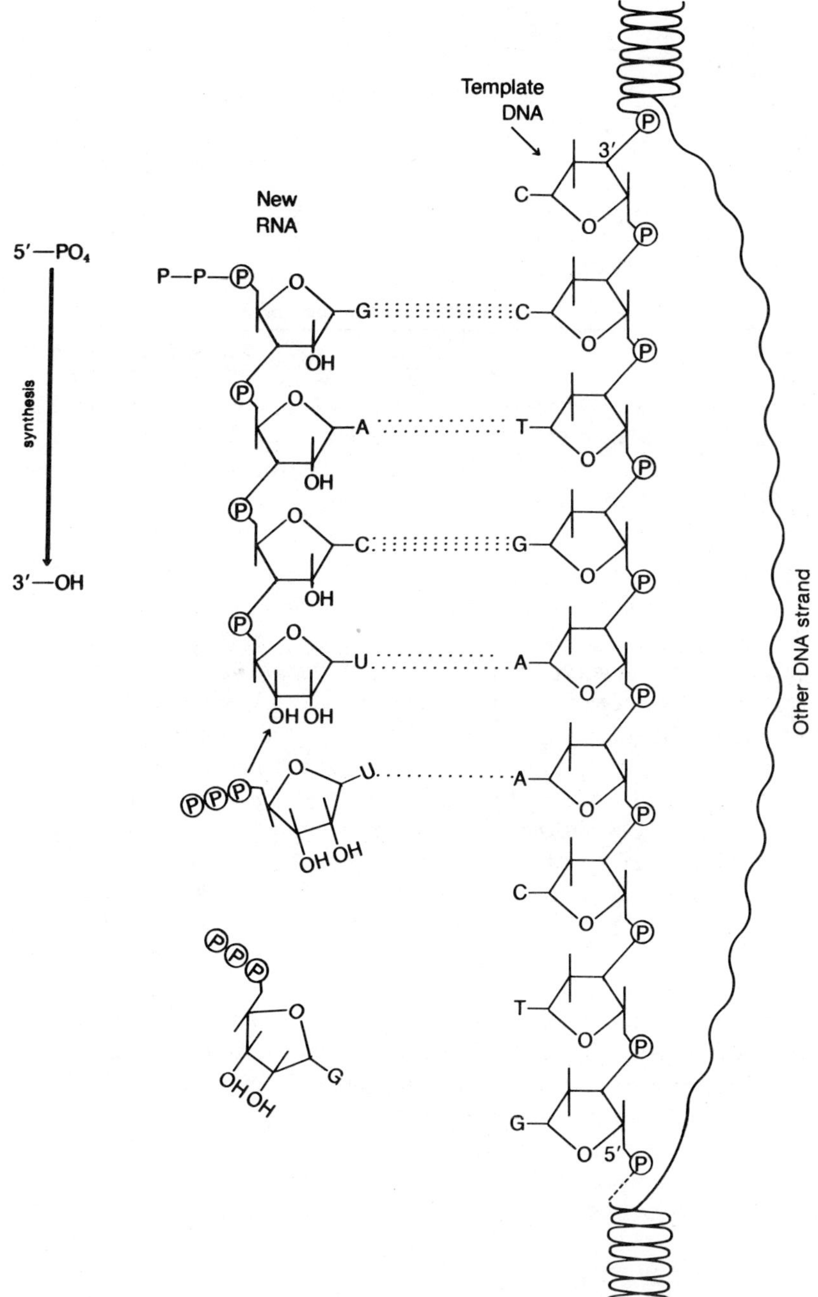

Figure 18.11
Biosynthesis of RNA showing asymmetry in transcription.
Nucleoside 5′-triphosphates align with complementary bases on one DNA strand, the template. RNA polymerase catalyzes the formation of the 3′,5′ phosphodiester links by attaching the 5′-phosphate of the incoming nucleotide to the 3′-OH of the growing nascent RNA releasing P_i*. The new RNA is synthesized from its 5′ end toward the 3′ end.*

to bind sometimes results in an increased frequency of termination and, consequently, a reduced level of gene expression.

Termination

The RNA polymerase complex recognizes the *ends* of genes as well (Figure 18.12). Transcription termination can occur in either of two modes,

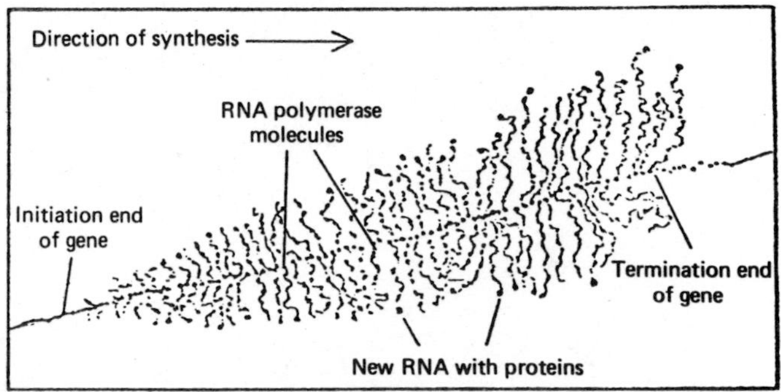

Figure 18.12
Simultaneous transcription of a gene by many RNA polymerases, depicting the increasing length of nascent RNA molecules.
Courtesy of Dr. O. L. Miller, University of Virginia. Reproduced with permission from O. L. Miller and B. R. Beatty, *J. Cell Physiol.*, 74:225, 1969.

depending on whether or not it is dependent on the protein factor rho. Terminators are thus classified as rho-independent or rho-dependent.

Rho-independent terminators are better characterized. A consensus-type sequence is involved here: a G-C rich *palindrome* (inverted repeat) precedes a sequence of 6–7 U residues in the RNA chain. As a result the RNA chain forms a stem and loop structure preceding the U sequence. The secondary structure of the stem and loop is crucial for termination; base change mutations in the stem and loop that disrupt pairing also reduce termination. Furthermore, the most efficient terminators are the most G-C rich and therefore most stable. The terminator stem and loop stabilize procaryotic mRNA against nucleolytic degradation.

Rho-dependent terminators are less well characterized, and the biochemical mechanism of rho action is still unclear. Rho factor does possess an RNA-dependent ATPase activity that is required for termination. Further, rho does seem to act at polymerase pausing sites in the DNA sequence. This has suggested that rho-dependent termination occurs as a result of competition between elongation and termination.

It is also important to remember that procaryotic ribosomes usually attach to the nascent mRNA while it is still being transcribed. This coupling between transcription and translation is important in gene control by *attenuation,* which is discussed in Chapter 20.

Eucaryotic Transcription

In contrast to procaryotic polymerases, which correctly initiate transcription on naked DNA, eucaryotic RNA polymerases require other components as well. The first requirement arises from the fact that active eucaryotic genes form an altered chromatin structure. This alteration is detectable by DNase sensitivity and is most apparent in the region upstream of the structural sequences. By analogy with bacterial RNA polymerase, we would expect this region of the gene to bind RNA polymerase II; in several cases, this has been shown. The transition from nucleosomes to an extended DNA structure is also observed in the structural portions of genes, especially in genes such as rRNA, which are very actively transcribed. On the other hand, cases are known in which other genes appear to remain organized in nucleosomes while they are being transcribed.

Promoters for mRNA Synthesis

Accurate transcription by eucaryotic RNA polymerase II usually requires the presence of two consensus sequences upstream from the mRNA start site. The first and most prominent of these, sometimes called the TATA box, has the sequence

$$\text{TATA}^{\text{A}}_{\text{T}}\text{A}^{\text{A}}_{\text{T}}$$

These nucleotides are highly conserved, much more so than in the *E. coli* "−10" consensus sequence previously described. The TATA box is centered about 25 base pairs upstream from the transcription unit. Experiments in which it was deleted suggest that it is required for efficient transcription, although weak promoters may lack it entirely.

A second region of homology is located further upstream, in which the CAAT box sequence

$$\text{GG}^{\text{T}}_{\text{C}}\text{CAATCT}$$

is found. This sequence is not as highly conserved as the TATA box, and some active promoters may not possess it.

In a few cases accurate transcription has been produced on naked DNA using highly purified RNA polymerase II and DNA-binding proteins. The DNA-binding proteins locate RNA polymerase on the template, perhaps by binding to a specific sequence and also to polymerase. Very little is known about eucaryotic mRNA termination, and the 3′ ends of mature mRNA are largely derived by processing (see below).

Internal Promoters for RNA Polymerase III

RNA polymerase III can accurately initiate synthesis of 5S or tRNA, but transcription requires the presence of an *internal* promoter on the template. This DNA sequence binds a specific protein factor. In fact the correct 5′ end of the RNA gene is not absolutely necessary—transcription begins a defined distance upstream from the internal protein-binding site. The sequence of the upstream region does determine overall promoter efficiency, perhaps by an effect on initial RNA polymerase III binding.

RNA Polymerase I and rRNA Genes

Ribosomal RNA genes in nucleolar DNA exist in tandem arrays separated by spacer regions of varying lengths. Within the tandem array single gene clusters are transcribed in an "all or none" fashion; that is, either completely packed with RNA polymerase I molecules or not transcribed at all. Each transcriptional unit begins with a promoter sequence extending about 140 base pairs upstream of the rRNA transcript. The primary RNA transcript contains three rRNA sequences in order from the 5′ terminus 28 S − 5.8 S − 18 S, in an RNA sedimenting at 45 S. A termination signal defines the 3′ end of the transcript. The 45 S RNA is subsequently processed to generate mature rRNA species (see below).

18.5 POSTTRANSCRIPTIONAL PROCESSING

The immediate product of transcription is a precursor RNA molecule, the *primary transcript,* which is modified subsequently to a mature functional molecule. Primary transcripts have longer nucleoside sequences than the final RNA, have no modifications on the bases or sugars, and thus contain only A, G, C and U residues. The primary transcript is copied from a

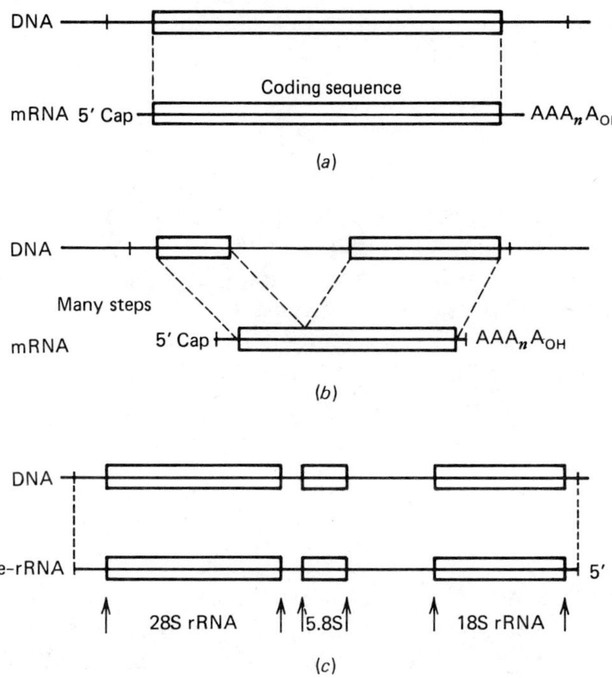

Figure 18.13
Models for transcription units in DNA showing coding regions and transcription products.
Arrows indicate cleavage points. (a) Linear information as expressed in histone genes. (b) Discontinuous information with subsequent splicing of intervening sequences as expressed in hemoglobin and ovalbumin genes. (c) Tandem information, single precursor with multiple mature RNAs: examples are rRNA and tRNA.

linear segment of DNA, a *transcriptional unit,* between specific initiation and termination sites. Transcriptional units may contain information in one or more forms: (1) information that is contiguous or without interruption; (2) information that is discontinuous, having sequences coding for a single protein or RNAs that are interrupted by unwanted nucleotide stretches; and (3) information that is in a tandem or repeated form in which information for multiple molecules is linked together and at some later stage requires separation. A gene, therefore, may not necessarily be colinear with the nucleotide or amino acid sequence of the final gene product (see Figure 18.13).

The summation of all the enzymatic reactions leading to mature functional RNA molecules from primary transcripts is called RNA processing. Processing involves a variety of events, which include base modifications, sugar modifications, pyrimidine ring rearrangements, formation of helices and tertiary conformations, additions to the 5′ terminus, additions to the 3′ terminus, specific exonucleolytic cleavages, specific endonucleolytic cleavages, complex cleavages with splicing of pieces, and formation of RNA–protein complexes. The number, type, and order of processing events is different for each group of RNA and often varies for each specific type within the groups.

Transfer RNA Processing

The proper recognition of mRNA by the tRNAs and the proper aminoacylation of the specific tRNAs are important and need to be rigorously

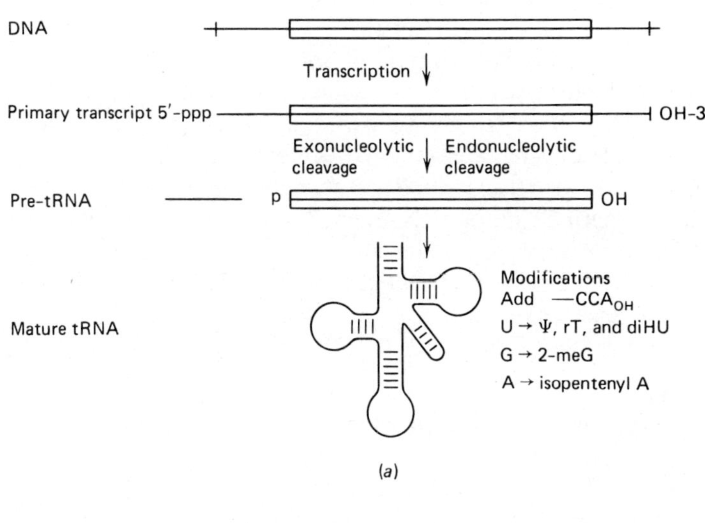

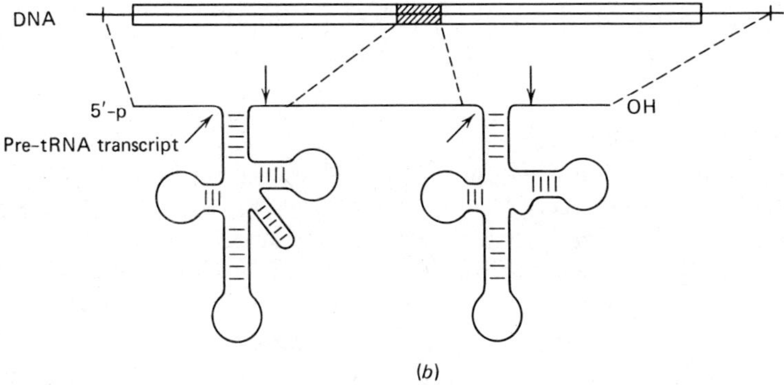

Figure 18.14
Processing of tRNA transcription products.
(a) Processing for single tRNA gene. (b) Processing for tandem tRNA genes. Arrows in (b) indicate cleavage points. Nucleoside modifications occur early in the maturation process.

controlled. Much of the specificity that results in overall tRNA function resides in the many nucleotide modifications (see Figure 18.5).

Cleavage

Primary transcripts for the tRNAs are trimmed to the proper size from both the 3′ and 5′ termini. These may be identical tRNA repeats or different tandem-linked tRNAs. The processing of tRNA precursors in *E. coli* has been well documented. Precursor tRNAs in *E. coli* include both monomeric and multimeric precursors. In the monomeric precursors such as $tRNA_I^{Leu}$, $tRNA_{III}^{Gly}$, $tRNA^{Asn}$, $tRNA^{Ile}$, $tRNA_{II}^{Glu}$, and $tRNA_I^{Asp}$ the 5′ extra nucleotides, ~15, are cleaved by the endonuclease RNase P. The 3′ end then is trimmed by an exonuclease, RNase D. In the multimeric precursors most nucleotide modifications are made prior to cleavage (see Figure 18.14). These tRNAs are separated first by RNase PC and then matured by RNase P. RNase P has counterparts in eucaryotic cells. The more fully understood procaryotic enzyme is an extremely interesting structure composed of RNA and protein. Under conditions of relatively high Mg^{2+} or polycation concentrations, the RNA alone is capable of catalyzing the cleavage of pre-tRNA, to make this the first known enzymatic activity by a biomolecule other than protein. One precursor molecule includes $tRNA_{III}^{Ser}$ and $tRNA_{II}^{Arg}$, while another has tandem repeats of $(tRNA_{III}^{Gly})^n$. In yeast, several but not all of the pre-tRNAs have extra segments of nucleotides internally adjacent to the anticodon loop. Yeast $tRNA^{Tyr}$ is transcribed from eight genes, and each transcript has 14 extra internal nucleotides that are removed. An extra internal sequence of 18 to 19 nucleotides is also transcribed and removed from yeast $tRNA^{Phe}$. Yeast cells have several clustered tRNA genes separated by only 9 nucleotides, which are cotranscribed. Similar structures occur in several eucaryotes, although extra internal (intervening) sequences for tRNAs have only been detected in yeast cells.

Additions

Each functional tRNA has at its 3′ terminus the sequence —$pCpCpA_{OH}$. In most instances this sequence is added sequentially by nucleotidyltransferase. Cells grown in the presence of actinomycin D, an antibiotic that blocks transcription, still add —CCA quickly to presynthesized tRNAs. Nucleotidyltransferase prefers ATP and CTP as substrates and always incorporates them into tRNA at a ratio of 2C/1A. —CCA_{OH} ends are found on both cytoplasmic and mitochondrial tRNAs.

Modified Nucleosides

Transfer RNA nucleotides are the most highly modified of all nucleic acids (Figure 18.5). More than 60 different modifications to the bases and ribose, requiring well over 100 different enzymatic reactions, have been found in tRNA. Many are simple, one-step methylations, but others involve multistep synthesis. Two derivatives, pseudouridine and queuosine (7-4, 5-*cis*-dihydroxy-1-cyclopenten-3-ylamino methyl-7-deazaguanosine) (see Table 18.2) actually require severing of the β-glycosidic bond to the transcribed base. One enzyme or set of enzymes produces a single site-specific modification in more than one species of tRNA molecule. Separate enzymes or sets of enzymes produce the same modifications at more than one location in tRNA. In other words, most modification enzymes are site- or nucleoside sequence-specific, not tRNA specific. All modifications are synthesized posttranscriptionally. Most are completed before the tRNA precursors have been cleaved to mature tRNA size. One noted exception is the methylation of ribose, which occurs late in processing.

Some modifications are found at extremely high frequencies in certain nucleotide sequences or locations within the tRNA structure. Ribothymidine (5-methyluridine) and pseudouridine (ψ) are almost always found together in the constant sequence GTψC of the hairpin loop proximal to the 3′ terminus, although ψ is also found elsewhere in tRNAs. Dihydrouridine is found in the loop nearest the 5′ terminus, and 7-methylguanosine is often found in the small or "extra" loop. The consistency of certain modifications hint at their possible functions in the tRNA structure.

Although first described over 20 years ago, the functions of the modifications are only just becoming evident through the use of a combination of biochemical, genetic, and biophysical techniques. Analyses of modified nucleotide function in tRNA has been limited to those steps in protein synthesis that can be assayed: aminoacylation, binding to initiation and elongation factors, ribosome binding, codon recognition and fidelity, for instance. There is direct evidence that modifications at the first position of the anticodon (the wobble base position) and immediately adjacent to the 3′ end of the anticodon are important for correct and effective recognition of the codon in mRNA.

Ribosomal RNA Processing

The primary transcript for the ribosomal RNAs (28S, 18S, and 5.8S) is a large 45S precursor in the nucleolus. This is an example of multiple RNA molecules being transcribed in tandem and the functional sections cut out and retained, while the unwanted sequences are degraded. The processing of the 45S rRNA precursor includes modification of nucleotides, cleavage, and strand folding or base pairing. The overall scheme for rRNA processing is depicted in Figure 18.15. In eucaryotic cells only 50–60% of the 4.6×10^6 dalton 45S precursor is actually retained as ribosomal rRNA.

As the 45S rRNA is transcribed, it associates with some ribosomal proteins and with the methyltransferases, which add the required methyl groups to the bases. About 80% of the total methyl groups on the mature molecule are on the 2′-OH of ribose; methylation of the riboses occurs at a different cellular site. All methylations occur on the 45S rRNA, except for the formation of N^6-dimethyladenine on the 18S rRNA. The methyl groups are retained in the mature 28S and 18S rRNAs. The methylations, which use *S*-adenosylmethionine as the donor, are sensitive to actinomycin D. Pseudouridine, ~1.2–1.8 mol %, is also present in the 45S rRNA precursor.

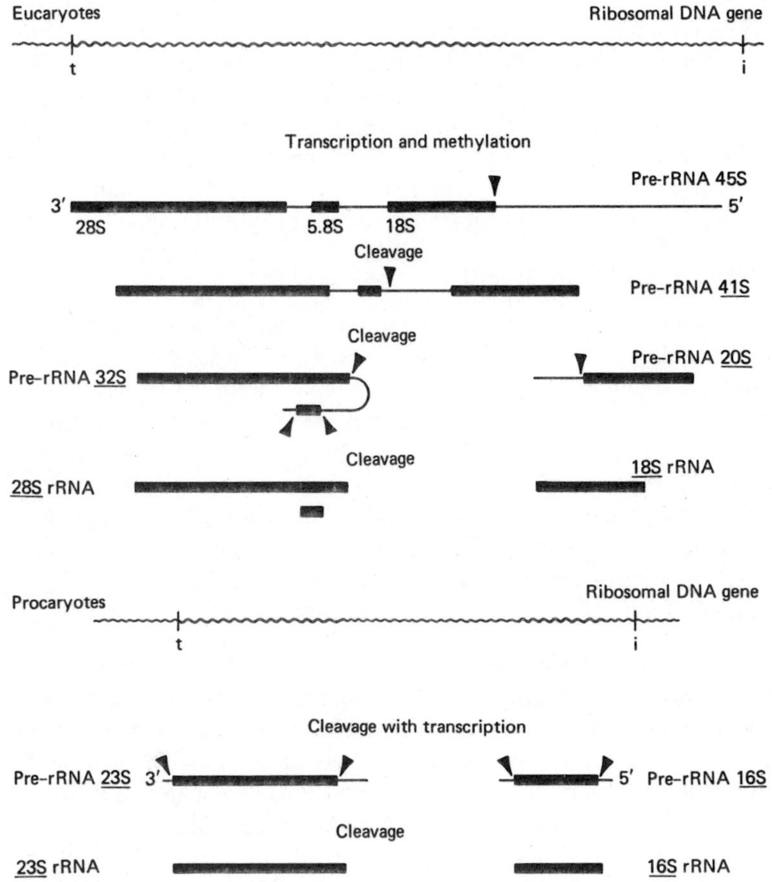

Figure 18.15
Schemes for transcription and processing of rRNAs.
Redrawn with permission from R. Perry, *Ann. Rev. Biochem.* 45:611, 1976. Copyright 1976 by Annual Reviews, Inc.

Cleavage of the 45S rRNA occurs by sequential and specific endonucleolytic attack, yielding discrete intermediate RNAs. RNase P and RNase III activities for this type of cleavage have been detected in human cells. Base compositions have been determined for the 45S, 32S, 28S, and 18S RNAs. Whereas the 45S RNA has a G + C content near 70%, the 28S and 18S rRNAs have G + C contents similar to the cellular average of 56–65%. This indicates preferential degradation of G—C-rich regions (78% G + C). The unwanted regions may be bounded by methylated sequences serving as cleavage signals (see Table 18.4). Further stabilization of the rRNAs occurs by formation of base-paired helical hairpin loops. Many of the base-paired regions are retained in 28S rRNA. The remainder are degraded, but may first serve as processing signals for cleavage or ribosomal protein binding.

The locations of the 28S and 18S rRNAs within the 45S precursor rRNA have been determined by nucleotide sequence analysis, oligonucleotide mapping, and hybridization competition experiments, using the intermediate RNAs, 45S RNA, and nucleolar DNA. The 28S rRNA originates from the 3′ end of the 45S molecule, while the 18S rRNA is processed from the middle of the 5′ half of the 45S precursor.

Processing of rRNA in procaryotes also involves cleavage of high molecular weight precursors to smaller precursor and mature molecules (see Figure 18.15). Some of the bases are modified by methylation on the ring nitrogens of the bases rather than the ribose and by the formation of pseudouridine (only 0.06–0.3 mol %). The *E. coli* genome has approximately seven rRNA transcriptional units dispersed throughout the DNA. Each contains at least one 16S, one 23S, and one 5S rRNA or tRNA sequence. Processing of the rRNA is coupled directly to transcription, so that cleavage of a large precursor primary transcript, 30S, occurs at double-helical regions yielding precursor 16S, precursor 23S, precursor 5S, and precursor tRNAs. These precursors are slightly larger than the functional molecules and only require trimming for maturation.

Messenger RNA Processing

Most eucaryotic mRNAs have distinctive structural features, which are not a consequence of RNA polymerase action. These features, added in the nucleus, include a 3′-terminal poly(A) tail, methylated internal nucleotides, and a cap or blocked methylated 5′ terminus. Cytoplasmic mRNAs are shorter than their nuclear primary transcripts, which can contain additional terminal and internal sequences. The noncoding sequences present within the coding portion of pre-mRNA molecules, but not present in cytoplasmic mRNAs, are called *intervening sequences* or *introns*. The *expressed* or retained sequences are called *exons*. The general pattern for mRNA processing is depicted in Figure 18.16.

Blocking of the 5′ Terminus

The 5′ termini of most mRNAs, including histone mRNAs, have methylated guanosines attached by three phosphates linking the 5′p of the cap nucleotide to the 5′p of the first transcribed nucleotide. The general form this takes is $m^7G^{5}ppp^{5'}N$ . . . (see Figure 18.7). The cap is attached to a purine in 75% of the blocked mRNAs and to a pyrimidine in the other 25%.

Addition of the cap structures occurs on mRNAs or pre-mRNAs, and the enzymatic addition is associated with the initiation of transcription and RNA polymerase II (B). No capping is detected during transcription with RNA polymerases I (A) or III (C). The synthesis of the cap requires several allosteric enzyme activities. The steps in mRNA capping are outlined in Figure 18.17. The second methylation is usually on the 2′-OH

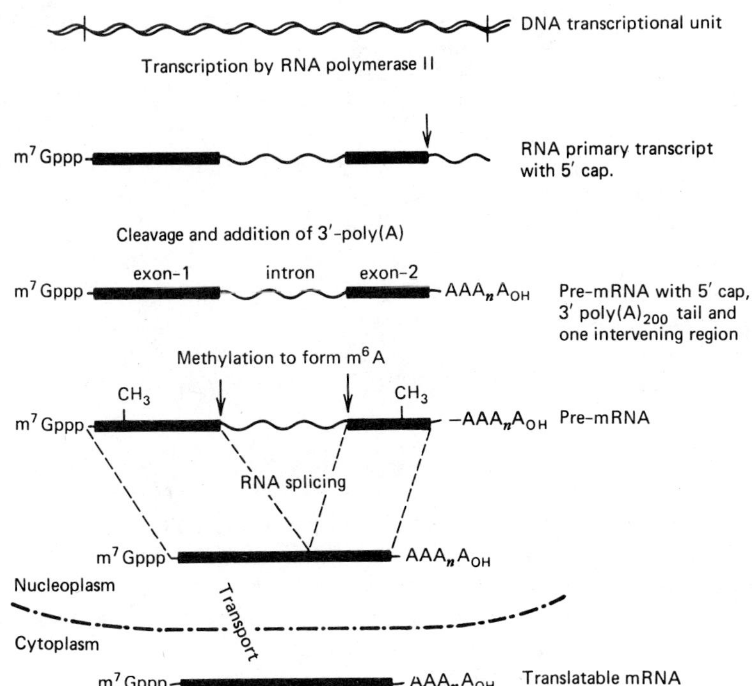

Figure 18.16
Scheme for processing mRNA.
The points for initiation and termination of transcription are indicated on the DNA. Arrows indicate cleavage points. The many proteins associated with the RNA and tertiary conformations are not shown.

1. Initiation of transcription

$$\text{pppG} + \text{pppC} \xrightarrow{\text{RNA polymerase II}} \text{pppGpC} + \text{PP}_i$$

2. Removal of inorganic phosphate to yield a 5′-diphosphate

$$\text{pppGpC} \xrightarrow{\text{nucleotide phosphohydrolase}} \text{ppGpC} + \text{P}_i$$

3. Addition of the inverted or cap guanyl residue

$$\text{pppG} + \text{ppGpC} \xrightarrow{\text{guanyltransferase}} \text{GpppGpC} + \text{PP}_i$$

4. Methylation of the terminal guanine

$$\text{GpppGpC} + S\text{-adenosylmethionine} \xrightarrow{\text{methyltransferase}} \text{m}^7\text{GpppGpC} + S\text{-adenosylhomocysteine}$$

5. Methylation of the penultimate nucleoside

$$\text{m}^7\text{GpppGpC} + S\text{-adenosylmethionine} \xrightarrow{\text{methyltransferase}} \text{m}^7\text{GpppGmpC} + S\text{-adenosylhomocysteine}$$

Figure 18.17
Reactions in synthesis of mRNA caps.
The five-step biochemical reaction pathway for addition of caps to mRNA is shown.

position. Caps are conserved during the remainder of the processing and are therefore found in cytoplasmic mRNAs.

Additions to the 3′ Terminus

A poly(A) tail is found on most cytoplasmic and many mitochondrial mRNAs. The larger nuclear RNAs, destined to become mRNAs, also have this modification. The string of adenylic acids (AMP) varies in length from 20 to 200 nucleotides. Poly(A) is synthesized in the nucleus by a poly(A) polymerase using ATP and the 3′-OH end of pre-mRNA as the substrates.

$$\text{RNA} + n\text{ATP} \xrightarrow[\text{Mg}^{2+}]{\text{poly(A) polymerase}} \text{RNA—(A)}_n + n\text{PP}_\text{i}$$

The addition of poly(A) can be inhibited by the nucleoside analog 3′-deoxyadenosine, cordycepin. This analog can be phosphorylated at the 5′ position and added by poly(A) polymerase to the end of mRNAs; however, the lack of a 3′-OH prevents further additions of AMP. Cells treated in this manner show a lack of newly synthesized mRNA transported to the cytoplasm. Separation of transcription events and poly(A) addition also has been demonstrated by treating cells with actinomycin D. Actinomycin D, at very low concentrations, blocks transcription but not the action of poly(A) polymerase, which adds AMP to the pre-mRNAs transcribed before addition of actinomycin D. Cytoplasmic addition of poly(A) to mRNAs is possible, but only in cells infected with certain DNA viruses, such as vesicular stomatitis virus (VSV) and vaccinia virus. In these situations the poly(A) polymerase is a viral enzyme. In mt mRNAs with 3′-poly(A) tails, the poly(A) is added by a mitochondrial specific poly(A) polymerase.

CLIN. CORR. **18.3** AUTOIMMUNITY IN CONNECTIVE TISSUE DISEASE

Humoral antibodies in the sera of patients with various connective tissue diseases recognize cellular RNA–protein complexes. Patients with systemic lupus erythematosus exhibit a serum antibody activity designated Sm, and those with mixed connective tissue disease exhibit an antibody designated RNP. Each antibody recognizes a distinct site on the same RNA–protein complex, U1 RNP, that now has been implicated in mRNA processing in mammalian cells. The U1-RNP complex contains U1 RNA, a 165 nucleotide sequence highly conserved among eucaryotes, that at its 5′ terminus includes a sequence complementary to intron–exon splice junctions (Figure 18.2). Addition of antibody to in vitro splicing assays inhibits splicing presumably by removal of the U1-RNP from the reaction. Sera from patients with other connective tissue diseases recognize different nuclear antigens, nucleolar proteins, and the chromosome centromere, for example. Sera of patients with myositis have been shown to recognize cytoplasmic antigens such as aminoacyl-tRNA synthetases. Although humoral antibodies have been reported to enter live cells that have F_c receptors, there is no evidence of such as part of the autoimmune disease mechanism. The immunogens of these diseases remain unknown.

Cleavage and Splicing

The primary transcript for mRNAs may contain considerable noninformational nucleotide sequences. These sequences may appear in the cytoplasmic mRNAs immediately preceding the "start" codon for transcription and after the termination codon. Most of the extra nucleotides, however, are removed during processing. Extra nucleotides transcribed beyond the poly(A) site are trimmed by endonucleolytic cleavage and there may be a few nucleotides to the 5′ side of the cap sequence. The extra 5′ end trimming has been inferred for some mRNAs by the presence of a cap followed by a pyrimidine rather than a purine nucleotide.

Noninformational or intervening sequences are interspersed throughout the informational portion of the primary transcripts in many eucaryotes and hence were present in the DNA transcriptional unit for the gene. The presence of these sequences in DNA and primary transcripts and the loss of these sequences in cytoplasmic mRNAs may occur by two mechanisms.

1. *DNA recombinations:* DNA may be rearranged prior to transcription so that widely separated DNA sequences become adjacent. An example of this is the synthesis of mouse immunoglobulins that helps explain the flexibility of the immune system. Typical DNA recombination mechanisms can be used to achieve this rearrangement (Figure 18.18). This type of control over mRNA sequences may be limited only to specialized systems, such as the synthesis of the immunoglobulins. The processing described below is probably more prominent.
2. *RNA splicing:* RNA polymerase transcribes a primary transcript containing exons (expressed sequences) and introns (intervening se-

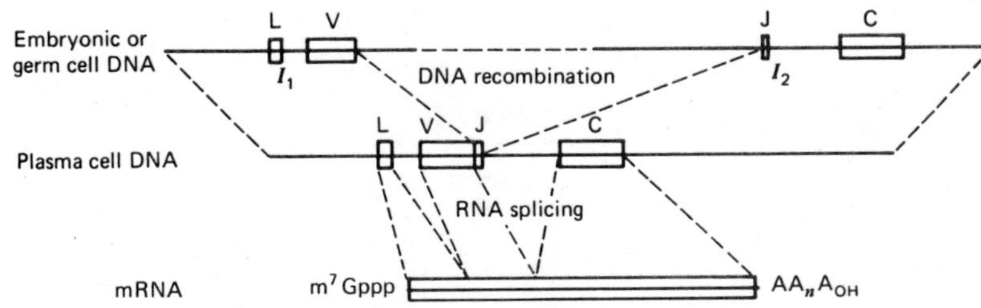

Figure 18.18
Diagram of a mechanism proposed for immunoglobulin gene rearrangements.
In embryonic or germ cells the coding sequences for the variable and constant regions of the peptide are located extremely far apart. In the differentiated cell, as indicated by nucleotide sequences of primary transcripts, these genetic regions are closer together. DNA recombination, in response to developmental pressures, may account for the genetic rearrangement. A primary transcript is made, which is subsequently spliced to form a mature mRNA. The leader sequence (L), the variable (V), and constant (C) coding regions, and the junction (J) sequence are indicated by open boxes. The regions between L and V and between J and C are intervening sequences found in both cell types and spliced out during RNA processing.
Courtesy of P. Leder.

quences). The introns are removed by endonucleolytic enzymes, and the exons are joined or spliced together forming a continuous informational sequence (see Figure 18.19). Splicing occurs in the nucleus probably after capping, methylation and poly(A) addition. Specific enzymes may be responsible for the recognition of intron–exon borders, cleavage, and ligation of the various pre-mRNAs. These mammalian enzymes have not been characterized, although a specific protein–RNA complex that is recognized by autoimmune antibodies in patients (Clin. Corr. 18.3) has been implicated. This complex of 170,000 daltons is designated U1-RNP because it contains U1 RNA, an RNA of highly conserved sequence among eucaryotes, plus a small number of proteins in the size range of 15,000–60,000 daltons. The 5′ terminal sequence of U1 RNA (see Figure 18.20) begins with a trimethylated guanosine in a cap structure. It continues with a sequence of nucleotides complementary to intron–exon splice junctions (see Figure 18.21) that may direct the enzymatic activity to the proper splice site. By virtue of this U1 RNA sequence being free of H bonding in the secondary structure and not being bound by protein in the U1 RNP complex as is most of the RNA molecule, it is free to bind at splice

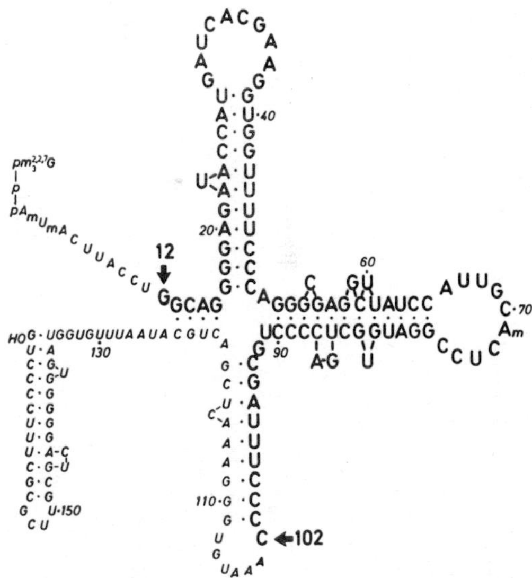

Figure 18.20
Nucleotide sequence and secondary structure of human and calf U1 RNA.
U1 RNA is found in a protein complex, U1 RNP, within the nuclei of cells from divergent organisms. The U1 RNP complex is implicated in the splicing of hnRNA to mRNA and is recognized by human autoimmune antibodies. The 5′-terminal sequence is complementary to intron–exon splice junctions. The nucleotide sequence in heavy type is required for one antibody's recognition of U1 RNP protein.
Adapted from P. F. Agris et al. *Immunol. Commun.* 13:137, 1984. By courtesy of Marcel Dekker, Inc.

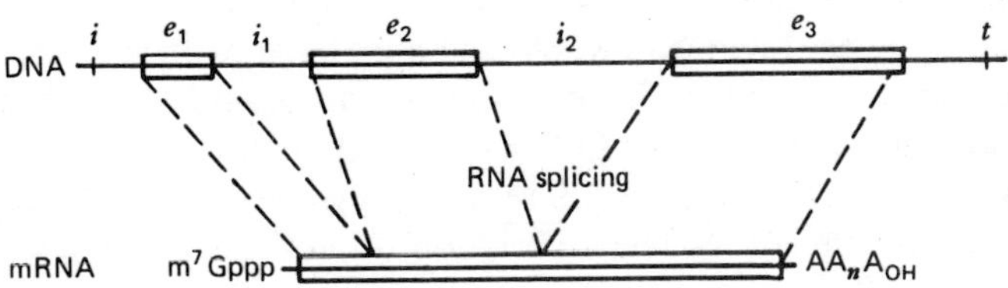

Figure 18.19
Diagram of RNA splicing showing the sequence of intervening and expressed regions in DNA.
A primary transcript is synthesized from i through t and processed to form an mRNA as indicated. The linearity of the exons is retained throughout. e_1, e_2, and e_3 refer to expressed regions, or exons. i_1 and i_2 refer to intervening sequences, or introns.

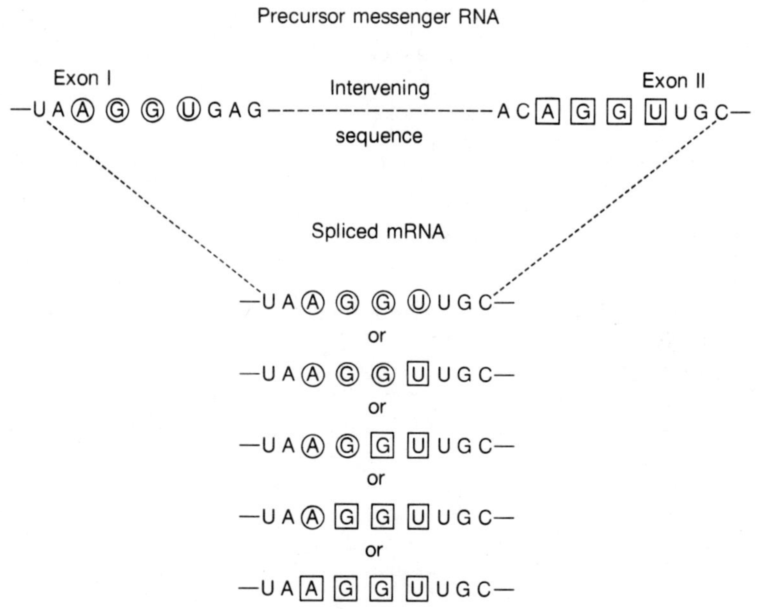

Precursor messenger RNA

Exon I — Intervening sequence — Exon II

—UA(A) (G) (G) (U)GAG----------------ACA[A] [G] [G] [U]UGC—

Spliced mRNA

—UA(A) (G) (G) (U)UGC—

or

—UA(A) (G) (G) [U]UGC—

or

—UA(A) (G) [G] [U]UGC—

or

—UA(A) [G] [G] [U]UGC—

or

—UA[A] [G] [G] [U]UGC—

Figure 18.21
An example of an RNA splicing sequence and the way in which splicing can occur at several positions while yielding the same mRNA product.
Redrawn with permission from F. Crick, *Science*, 204:265, 1979. Copyright 1979 by the American Association for the Advancement of Science.

junctions. An enzyme for yeast tRNA splicing has been isolated and requires Mg^{2+}, K^+, and ATP. The nuclease and ligase activities are independent. The yeast tRNA-splicing enzyme does not seem to work on mRNAs.

The individual exons for some proteins have been shown to contain the genetic information which corresponds to each of the structural domains within the proteins. Thus, an exon may simply code for an α helix or β-pleated sheet. The evolutionary consequences of this arrangement of genetic information is profound when we consider that exon rearrangements in DNA could facilitate construction of new proteins. However, not all protein genes with introns have exons which correspond to protein structural domains.

The sequences of bases within the introns of a given primary transcript do not have any singular features and contain no apparent genetic information. Intron lengths vary from 100 to 1,000 nucleotides for mRNAs and from 14 to 34 nucleotides for yeast tRNAs. The nucleotide sequences of the intron–exon borders have some similarities that appear to be common for several species. Introns in RNA generally begin with . . . pGpU . . . and end with . . . pApG. Those intron–exon borders sequenced have a tetranucleotide sequence (—pApGpGpU—) that allows for several splice points within the border while still retaining the correct mRNA sequence. Such a mechanism places less restriction upon the splicing enzyme and reduces mRNA loss during processing (see Figure 18.21). Procaryotes do not have the ability to process intervening sequences, and consequently most eucaryotic genes are not expressed in procaryotic hosts.

Examples of Eucaryotic Spliced mRNAs

Experiments in vitro with crude splicing systems from mammalian cells have shown that the splicing mechanism may proceed through a novel RNA structure. As depicted in Figure 18.22, the production of a 2′–5′

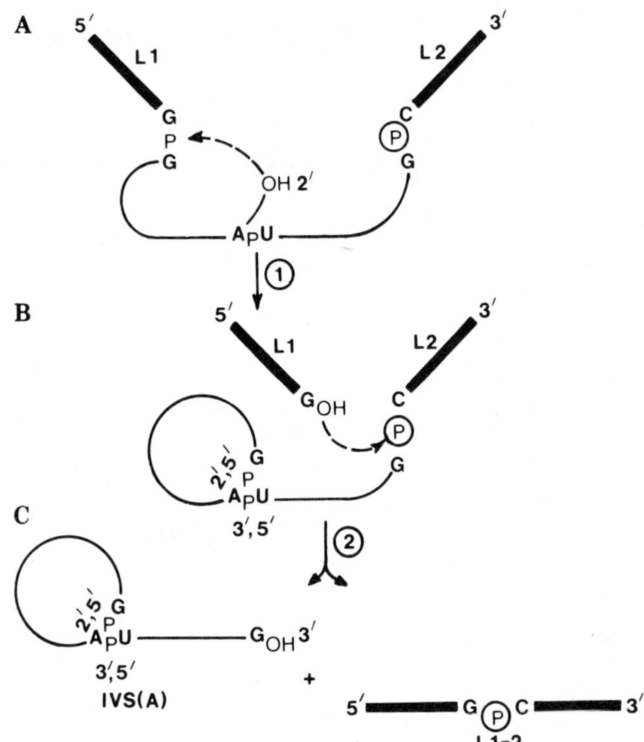

Figure 18.22
Proposed scheme for mRNA splicing to include the lariat structure.
A messenger RNA is depicted with two exons (L1 and L2, heavy lines) and intervening intron (light line). A 2′-OH of the intron sequence reacts with the 5′-phosphate of the intron's 5′-terminal nucleotide producing a 2′–5′ linkage and the lariat structure. Simultaneously, the L1-exon-to-intron phosphate ester bond is broken leaving a 3′-OH terminus on this exon free to react with the 5′-phosphate of the L2 exon, displacing the intron and creating the spliced mRNA.

phosphodiester linkage between the middle and 5′ end of the intron frees the 3′ terminus of the first exon. This also results in the formation of an intron lariat structure attached to the 5′ end of the second exon. Splicing of the 3′ end of the first exon to the 5′ terminus of the second is envisioned as the means by which the intron is completely excised and exons are joined. Mammalian hemoglobin mRNA was the first eucaryotic mRNA isolated. Whereas the globin mRNA is only 9S, the primary transcript for α-globin is 2.5 to 5 times longer than the mature mRNA. In the mouse the mRNA contains a cap, a 52-nucleotide leader sequence, the structural message for 144 amino acids, a 110-nucleotide nontranslated 3′ sequence, and a poly(A) tail. β-Globin pre-mRNAs with lengths of 5,000 and 1,500 nucleotides have been found. The half-life of the longer sequence is less than 10 minutes. Two introns have been located in the mouse, rabbit, and human β-globin primary transcripts. Although the intron sequences differ, their locations between codons 30-31 and codons 104-105 are identical for all three species. In mouse β-globin the first intron is about 646 nucleotides long and the second 116 nucleotides. The mouse α-globin gene also has two introns in positions similar to the β-globin genes (see Clin. Corr. 18.4 and 18.5 and Figure 18.23).

Immunoglobulins are made up of four polypeptide chains, two light and two heavy. The light chain has variable regions and constant regions. Both DNA rearrangements and RNA splicing are required for the expression of an immunoglobulin gene. The DNA genes are put in proximal

CLIN. CORR. 18.4
ALTERED HUMAN GLOBIN GENES

Many clinical examples of altered globin gene expression exist in humans. These are due either to malfunctioning genes or incorrect RNA processing. Abnormal globin proteins, with a single amino acid change, may be produced, following point mutations in the coding region of the gene. Over 300 single amino acid substitutions have been identified. There may also be gene fusions as well as deletions of gene segments, which can cause other globin abnormalities. Hemoglobin Lepore results from the fusion of the δ- and β-globin genes. Another example of a fusion protein is hemoglobin Kenya. Neither fusion protein occurs as a result of RNA splicing.

CLIN. CORR. 18.5
THE THALASSEMIAS

In the diseases known as the thalassemias the amount of a particular globin chain is severely reduced or absent. There is a deficiency of α-globin chains in α-thalassemia, with a resultant precipitation of β-chains. Homozygotes with α^0-thalassemia have a syndrome called *hydrops fetalis* in which death occurs prior to or shortly after birth. The genes for the α chain are on chromosome 16, while genes for many of the other globin types are on chromosome 11.

In the β-thalassemias the amount of β-globin varies. No β-globin mRNA can be detected in one type of β^0-thalassemia. In another, mRNA is present in the nucleus but not the cytoplasm. In a third type the β-globin mRNA is present but not translated, while α-globin mRNA is translated normally. β^+-thalassemia is characterized by low concentrations of cellular β-globin chains. These are translated from reduced amounts of mRNA. Both β- and δ-globins are absent in $\delta\beta^0$-thalassemia. This condition is due to a deletion from the 5′ region of the δ-globin gene to the 3′ side of the β-globin gene (Figure 18.23).

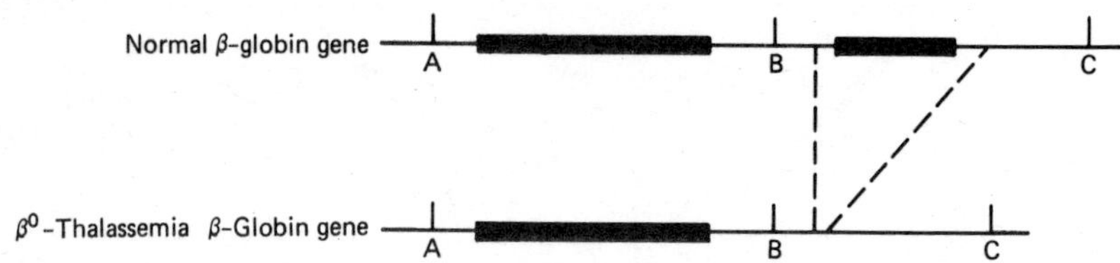

Figure 18.23
Diagram showing the deleted region of the β-globin gene in β° thalassemia.
A, B, and C indicate markers in the DNA. A 600-base pair segment is deleted from the normal gene.

positions by DNA recombination during differentiation and before transcription. However, the primary transcript still includes two introns, which are processed very efficiently. In mouse myeloma cells the globulin mRNAs are transcribed at ~20 molecules per minute, requiring 50 polymerases per gene. Each cell contains nearly 30,000 heavy chains and 40,000 light chains. Only 100 to 150 of these are found in the nucleus. The heavy chains are capped with $m^7Gppp(m^6)Ap^mAp^mC$ . . . and the light chains with $m^7GpppGp^mAp^mA$. . . . These mRNAs have a high metabolic stability with half-lives of 24 h.

Ovalbumin is synthesized in response to hormone stimulation in chickens. The ovalbumin mRNA has 1,890 nucleotides and is processed from a primary transcript of 7,900 nucleotides having a cap and poly(A) tail. The primary transcript has a 5′ leader sequence of 76 nucleotides, 7 mRNA exons (185, 51, 129, 118, 143, 156, and 1,043 nucleotides), and 7 introns (1,560, 238, 601, 411, 1,029, 323, and 1,614 nucleotides) (see Figure 18.24).

Another chicken protein, ovomucoid, is translated from a small mRNA of 800 nucleotides. However, this mRNA is processed from a primary transcript having 6,000 nucleotides and at least six intervening sequences.

Histone mRNAs are also processed prior to translation. Nuclear pre-mRNAs of about 14,000 nucleotides have been found, that do not have intervening sequences and therefore do not require splicing. The histone mRNAs have 5′ caps, $m^7G^{5'}ppp^5X^mpY$ . . ., but only some have poly(A) tails. Each histone mRNA contains noncoding leader and trailer

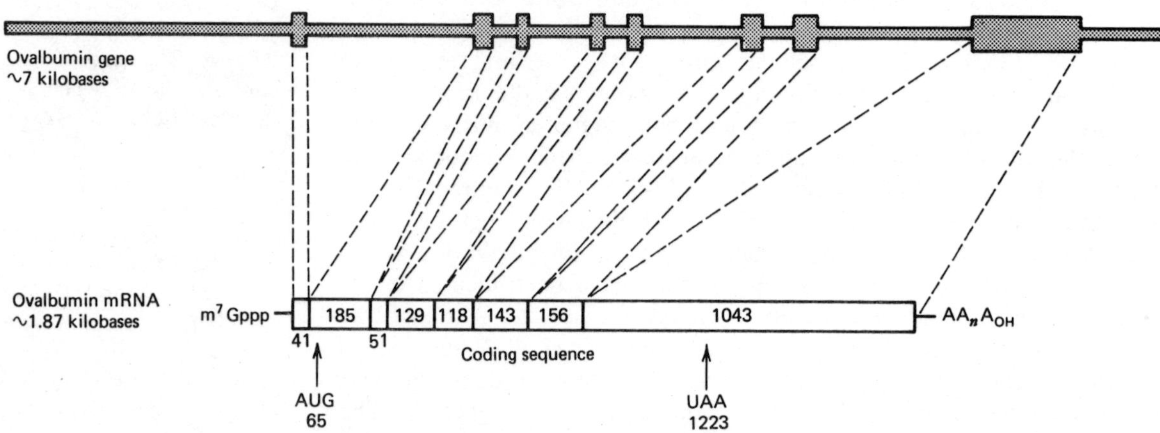

Figure 18.24
Diagram of the ovalbumin gene or primary transcript and the several RNA splicing sites required to synthesize the ovalbumin mRNA.
The two arrows indicate the translated portion of the mRNA.
Redrawn with permission from P. Chambon et al., *Miami Winter Symp.*, 16:57, 1979.

sequences. Histones are found in all eucaryotic cells and have resisted evolutionary pressures. The genes for the five major histones are clustered within the genome in a repetitive and tandemly linked arrangement. There are 30 to 40 copies of each histone gene in human cells and each gene is linked by a variable spacer region having a high percentage of A-T base pairs. Histone mRNA translation is linked, and transcription of histone genes also may be linked, to DNA replication in the S phase of the cell cycle.

18.6 NUCLEASES AND RNA TURNOVER

The different roles of RNA and DNA in genetic expression are reflected in their metabolic fates. A cell's information store (DNA) must be preserved; the results of this biological necessity are seen in the myriad DNA repair and editing systems in the nucleus. Thus, although individual stretches of nucleotides in DNA may turn over, the molecule as a whole is metabolically inert when not replicating. RNA molecules, on the other hand, are individually dispensable; they can be replaced by newly synthesized species of the same specificity. It is therefore no surprise that RNA

TABLE 18.6 Characteristics of Ribonucleases

Nuclease	*Specificity*	*Products*
Endonucleases yielding 3′-phosphates		from 5′-pApGpCpGpUpU$_{OH}$-3′
Pancreatic RNase	After pyrimidines	pApGpCp+GpUp+U$_{OH}$
T_1	After guanines	pApGp + CpGp + UpU$_{OH}$
U_1	After guanines	pApGp + CpGp + UpU$_{OH}$
T_2	After adenines	pAp + GpCpGpUpU$_{OH}$
Rat liver RNase-1	All phosphodiester bonds	pAp + 2Gp + Cp + Up + U$_{OH}$
Rat liver RNase-2	Between adjacent pyrimidines	pApGpCpGpUp + U$_{OH}$
E. coli RNase III	Double-helical structures	Cleaves rRNA precursors
Endonucleases yielding 5′-phosphates		
Rat liver alkaline RNase I	Nonspecific	pA + 2pG + pC + pU + pU$_{OH}$
E. coli RNase P also in mammalian cells	Precursor tRNA	Mature tRNA + fragment
E. coli RNase H also in mammalian cells	RNA of DNA–RNA hybrid	Nucleoside 5′-monophosphates + DNA
Exonucleases		
E. coli RNase II also in mammalian cell nuclei	Single strands, 3′ ⟶ 5′	Nucleoside 5′-monophosphates, may trim RNA precursors after endonucleases
E. coli RNase V	Single strands, 5′ ⟶ 3′ precursor mRNAs	Nucleoside 5′-monophosphates
E. coli oligoribonuclease	Short oligoribonucleotides	Nucleoside 5′-monophosphates
RNase H from RNA tumor virus	RNA of DNA · RNA hybrid 5′ ⟶ 3′ 3′ ⟶ 5′	Nucleoside monophosphates
Polynucleotide phosphorylase	Single-strand RNA 3′ ⟶ 5′	Nucleoside 5′-diphosphates
Nonspecific nucleases		
Micrococcal endonuclease	Single-strand RNA or DNA	Nucleoside 3′-monophosphates
Nuclease S1	Single-strand RNA or DNA	5′-Phosphate oligonucleotides
Venom phosphodiesterase	Exonuclease 3′ ⟶ 5′ Blocked by 3′-phosphate end	Nucleoside 5′-monophosphates
Spleen phosphodiesterase	Exonuclease 5′ ⟶ 3′	Nucleoside 3′-monophosphates

repair systems are not known. Instead, defective RNAs are removed from the cell by being degraded to nucleotides, which then are repolymerized into new RNA species.

This principle is clearest for messenger RNAs, which are classified as unstable. However, even the so-called stable RNAs turn over; for example, the half-life of tRNA species in liver is on the order of 5 days. A fairly long half-life for a eucaryotic mRNA would be 30 h.

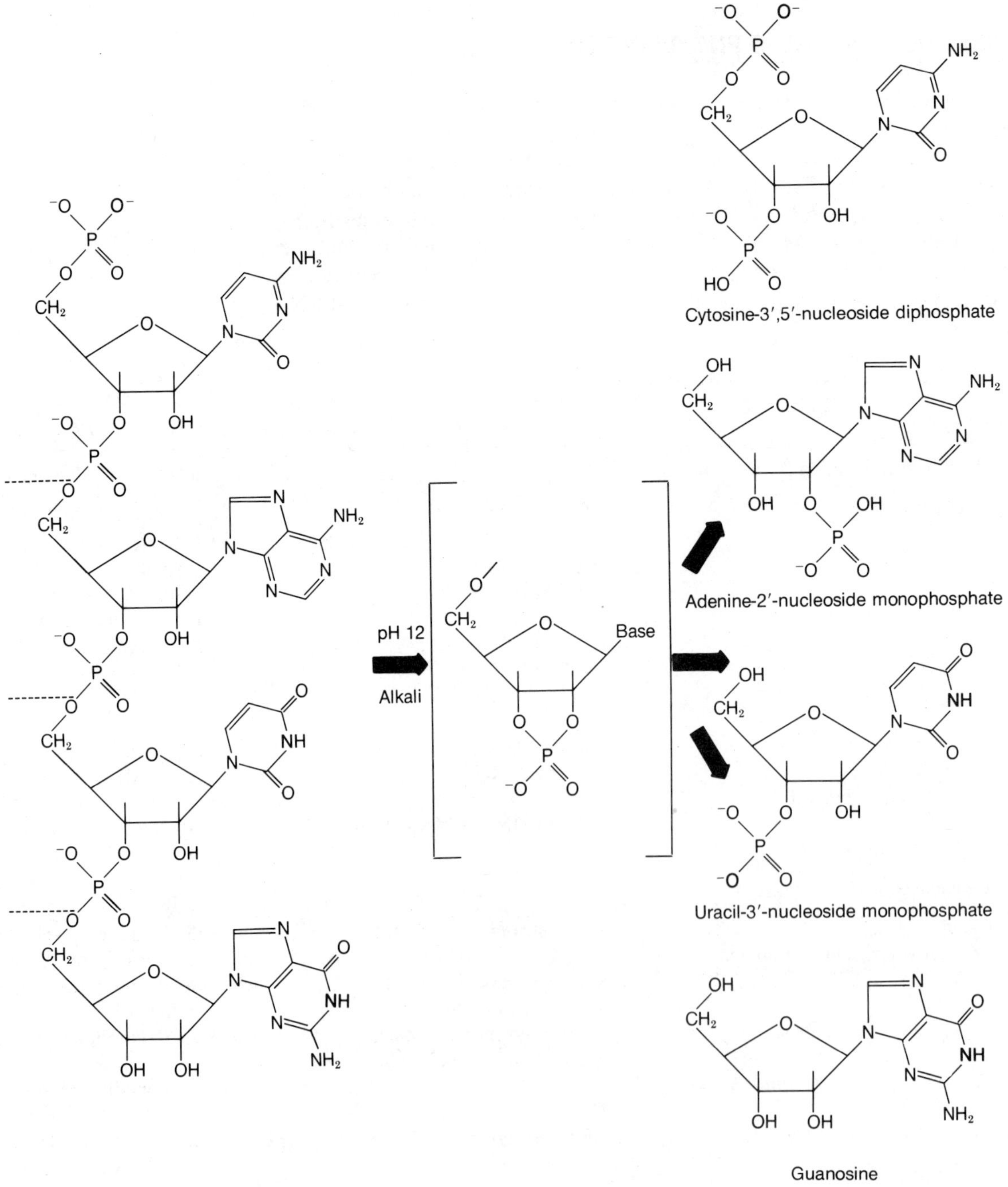

Cytosine-3′,5′-nucleoside diphosphate

Adenine-2′-nucleoside monophosphate

pH 12

Alkali

Base

Uracil-3′-nucleoside monophosphate

Guanosine

Figure 18.25
Alkaline hydrolysis of a hypothetical RNA.
Cleavage proceeds through a cyclic 2′,3′-phosphate intermediate and results in a 3′,5′-diphosphate nucleoside from the 5′ terminus, 2′- or 3′-nucleoside monophosphates from the internal residues and a nonphosphorylated nucleoside from the 3′ terminus. DNA is not subject to alkaline hydrolysis.

This removal of RNAs from the cytoplasm is accomplished by cellular ribonucleases. Messenger RNAs are at least initially degraded in the cytoplasm. The rates vary for different RNAs, raising the possibility of control by differential degradation. Lysosomes also contain large numbers of nucleases, and RNAs transported there could be degraded as well. The four common nucleotides are recycled to triphosphates and used again. Nucleosides containing modified bases do not recycle since the modifications are accomplished after the RNA is transcribed (see Section 18.5). The modified nucleosides released by degradation of tRNA or rRNA are ultimately excreted in the urine. The excretion is enhanced in human diseases exhibiting rapid cellular metabolism, division, and turnover, including cancers. Multivariant analysis of excreted nucleosides may ultimately yield clinical markers for specific disease states.

Nucleases are of several types and specificities (Table 18.6). The most useful distinction is between *exonucleases,* which degrade RNA from either the 5′ or 3′ end, and *endonucleases,* which cleave phosphodiester bonds within a molecule. The products of RNase action contain either 3′ or 5′ terminal phosphates, and both endo- and exonucleases can be further characterized by the position (5′ or 3′) at which the monophosphate created by the cleavage is located. The enzymatic mechanism of endoribonucleases that produce 3′ phosphates is best understood. The action of these enzymes, which include pancreatic RNase A, RNase T, and others, is chemically very similar to the base-catalyzed degradation of RNA (Figure 18.25).

RNA structure also affects nuclease action. Most degradative enzymes are less efficient on highly ordered RNA structure. Thus, tRNAs are preferentially cleaved in unpaired regions of the sequence. Limited digestion of other RNAs leaves base-paired regions uncleaved by the enzyme. On the other hand, many RNases involved in maturation of RNA require a defined three-dimensional structure for enzyme activity. These enzymes are discussed more fully above in the consideration of RNA processing pathways.

BIBLIOGRAPHY

Abelson, J. RNA processing and the intervening sequence problem. *Ann. Rev. Biochem.* 48:1035, 1979.

Agris, P. F., and Kopper, R. A. (eds.). *The modified nucleosides of transfer RNA II.* New York: Alan R. Liss, 1983.

Chamberlain, M. J. Bacterial DNA-dependent RNA Polymerases, in P. D. Boyer (ed.), *The enzymes,* vol. XV. New York: Academic, 1982, p. 61.

Cohn, W. E. (ed.). *Progress in nucleic acid research and molecular biology,* vol. 23. New York: Academic, 1979.

Cohn, W. E., and Volkin, E. (eds.). *Progress in nucleic acid research and molecular biology, mRNA: the relation of structure to function,* vol. 19. New York: Academic, 1976.

Davidson, J. N. *The biochemistry of the nucleic acids,* 8th ed., revised by R. L. P. Adams. New York: Academic, 1976.

Lewis, M. K., and Burges, R. R. Eucaryotic RNA polymerases, in P. D. Boyer (ed.), *The enzymes,* vol. XV. New York: Academic, 1982, p. 109.

Losick, R., and Chamberlain, M. (eds.). *RNA polymerase.* Cold Spring Harbor, N.Y.: Cold Spring Harbor Laboratory, 1976.

Rosenberg, M., and Court, D. Regulatory sequences involved in the promotion and termination of RNA transcription. *Ann. Rev. Genet.* 13:319, 1979.

Russell, T. R., Brew, K., Faber, H., and Schultz, J. (eds.). *From gene to protein: information transfer in normal and abnormal cells,* Miami Winter Symposium, vol. 16. New York: Academic, 1979.

QUESTIONS

C. N. ANGSTADT AND J. BAGGOTT

*Q*uestion *T*ypes are described inside the front cover.

1. (QT2) RNA:
   1. contains modified purine and pyrimidine bases that are formed posttranscriptionally.
   2. is usually single stranded in mammals.
   3. structures exhibit base stacking and hydrogen-bonded base pairing.
   4. usually contains about 65–100 nucleotides.

A. HnRNA
B. mRNA
C. rRNA
D. snRNA
E. tRNA

2. (QT5) Has the highest percentage of modified bases of any RNA.
3. (QT5) Stable RNA representing the largest percentage by weight of cellular RNA.
4. (QT5) Contains both a 7-methylguanosine triphosphate cap and a polyadenylate segment.
5. (QT2) In eucaryotic transcription:
   1. RNA polymerase does not require a template.
   2. different kinds of RNA are synthesized in different parts of the nucleus.
   3. consensus sequences are the only known promoter elements.
   4. phosphodiester bond formation is favored, in part, because it is accomplished by pyrophosphate hydrolysis.
6. (QT1) An enhancer:
   A. is a consensus sequence in DNA located where RNA polymerase first binds.
   B. may be located in various places in different genes.
   C. may be on either strand of DNA in the region of the gene.
   D. functions by binding RNA polymerase.
   E. stimulates transcription in both procaryotes and eucaryotes.
7. (QT1) The sigma (σ) subunit of procaryotic RNA polymerase:
   A. is part of the core enzyme.
   B. binds the antibiotic rifampicin.
   C. is inhibited by α-amanitin.
   D. must be present for transcription to occur.
   E. specifically recognizes promoter sites.

Use this schematic representation of a procaryotic gene to answer questions 8–10. Numbers refer to positions of base bairs relative to the beginning of transcription.

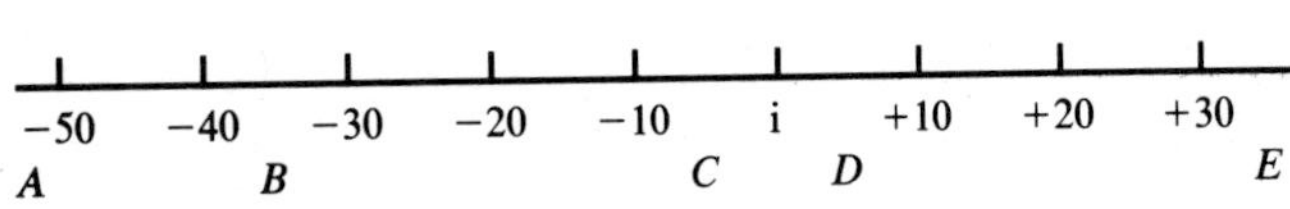

8. (QT5) Sigma (σ) factor might be released from RNA polymerase.
9. (QT5) An "open complex" should form in this region.
10. (QT5) Events beyond this region should be catalyzed by core enzyme.
11. (QT1) Termination of a procaryotic transcript:
    A. is a random process.
    B. requires the presence of the rho subunit of the holoenzyme.
    C. does not require rho factor if the end of the gene contains a G-C rich palindrome.
    D. is most efficient if there is an A-T rich segment at the end of the gene.
    E. requires an ATPase in addition to rho factor.
12. (QT2) Eucaryotic transcription:
    1. is independent of the presence of consensus sequences upstream from the start of transcription.
    2. may involve a promoter located within the region transcribed rather than upstream.
    3. requires a separate promoter region for each of the three ribosomal RNAs transcribed.
    4. often involves an alteration in chromatin structure.
13. (QT2) The primary transcript in eucaryotes:
    1. is usually longer than the functional RNA.
    2. may contain nucleotide sequences that are not present in functional RNA.
    3. will contain no modified bases.
    4. could contain information for more than one RNA molecule.
14. (QT2) The processing of transfer RNA involves:
    1. cleavage of extra bases from both the 3′ and 5′ ends.
    2. nucleotide sequence-specific methylation of bases.
    3. addition of the sequence CCA by a nucleotidyl transferase.
    4. addition of a methylated guanosine at the 5′-end.
15. (QT1) Cleavage and splicing:
    A. are features of ribosomal RNA processing.
    B. cause sequences that are widely separated in a DNA molecule to be placed next to each other.
    C. remove noninformational sequences occurring anywhere within a primary transcript.
    D. are usually the first events in mRNA processing.
    E. are catalyzed by enzymes that recognize and remove specific introns.
16. (QT3) A. Rate of turnover of transfer RNA.
    B. Rate of turnover of messenger RNA.
17. (QT2) In the cellular degradation of RNA:
    1. any of the nucleotides released may be recycled.
    2. regions of extensive base pairing are more susceptible to cleavage.
    3. endonucleases may cleave the molecule starting at either the 5′ or 3′ end.
    4. the products are nucleotides with a phosphate at either the 3′ or 5′ hydroxyl group.

ANSWERS

1. A 1, 2, 3 correct. 1: Only the four bases A, G, U, and C are incorporated during transcription. 2, 3: Although single stranded, RNA exhibits considerable secondary and tertiary structure. 4: Only tRNA would be this small; sizes can range to more than 6,000 nucleotides (Section 18.2).
2. E Modified bases seem to be very important in the three-dimensional structure of tRNA.
3. C Stability of rRNA is necessary for repeated functioning of ribosomes.
4. B These are important additions during processing that yield a functional eucaryotic mRNA (Section 18.3, Table 18.1).
5. C 2, 4 correct. 1, 2: Transcription is directed by the genetic code, generating rRNA precursors in the nucleolus and mRNA and tRNA precursors in nucleoplasm. 3: Eucaryotic transcription may have internal promoter regions as well as enhancers. 4: This is an important mechanism for driving reactions (p. 711).
6. B B, C: Enhancer sequences seem to work whether they are at the beginning or end of the gene, but they must be on the same DNA strand as the transcribed gene. D: They seem to function by structurally altering the template (p. 712).
7. E A, D, E: Sigma factor is required for correct initiation and dissociates from the core enzyme after the first bonds have been formed. Core enzyme can transcribe but cannot correctly initiate transcription. B, C: Rifampicin binds to the β subunit, and α-amanitin is an inhibitor of eucaryotic polymerases (p. 712).
8. D Sigma factor is released when, or a short time after, the initial bond is formed.

9. C The high A-T content of the Pribnow box is believed to facilitate initial unwinding.
10. E Elongation, which requires only the core enzyme, is well underway in this region (p. 714).
11. C C, D: Rho-independent termination involves secondary structure, which is stabilized by high G-C content. A, B, E: There is a rho-dependent as well as a rho-independent process. Rho is a separate protein from RNA polymerase and appears to possess ATPase activity (p. 717).
12. C 2, 4 correct. 1: RNA polymerase II activity involves the TATA and CAAT boxes. 2: RNA polymerase III uses an internal promoter. 3: RNA polymerase I produces *one* transcript, which is later processed to yield three rRNAs. 4: This is a major difference between pro- and eucaryotic transcription (p. 719).
13. E All four correct. Modification of bases, cleavage, and splicing are all important events in posttranscriptional processing to form functional molecules (p. 719).
14. A 1, 2, 3 correct. 1: The primary transcript is longer than the functional molecule. 2: The same modifications, catalyzed by a certain (set of) enzyme(s), occurs at more than one location. 3: This is a posttranscriptional modification. 4: Capping is a feature of mRNA (p. 721).
15. C A: Cleavage occurs, but splicing does not. B: This occurs by DNA recombination. D: Splicing occurs after other events. E: Specificity of cleavage is related to specific sequences at the intron–exon junctions, not to the sequence of the intron itself (p. 721).
16. B Although all RNA turns over, the rapid turnover of mRNA may be part of control (p. 731).
17. D Only 4 correct. 1: Modified bases cannot be recycled. 2: Although some enzymes of maturation may require an ordered structure, degradative enzymes are less efficient on an ordered structure. 3: An endonuclease cleaves an interior phosphodiester bond (p. 721).

19

Protein Synthesis: Translation and Posttranslational Modification

KARL H. MUENCH

19.1 THE GENETIC CODE

Information necessary for synthesis of the unique primary structure of each protein is analogous to the written human languages that consist of a sequence of letters forming words and sentences. In the genetic language, however, there are only four letters. Four nucleotidyl residues, differing only in the four purine and pyrimidine base residues, A, G, U, C (adenine, guanine, uracil, cytosine), recur in various permutations of sequence embodied in mRNA. The concept of a code arises from the need for the nucleotidyl residue sequences, hereafter designated base sequences, to be translated into amino acid residue sequences, constituting proteins. Specific contiguous base sequences define the code message for each of the 20 coded amino acids found in proteins. Thus in translation a four-letter language becomes a 20-letter language. Implicit in the analogy to language is the necessity for directional specificity. Base sequences are by convention written left to right from the 5′ terminal to the 3′ terminal, and amino acid sequences are by convention written left to right from the amino terminal to the carboxy terminal. Later we shall see that these directions in mRNA and proteins correspond.

TABLE 19.1 Sixteen Permutations of a Four-Letter Code Taken in Pairs

	Second Base			
First Base	*U*	*C*	*A*	*G*
U	UU	UC	UA	UG
C	CU	CC	CA	CG
A	AU	AC	AA	AG
G	GU	GC	GA	GC

Codons Are Three-Letter Words

The four bases taken as pairs give only $4^2 = 16$ permutations, as shown in Table 19.1. This hypothetical code of two-letter words is illustrative but would be insufficient to specify 20 different amino acids, not to mention start and stop signals. Actually, the genetic code consists of three-letter words. The four bases taken in threes give $4^3 = 64$ permutations (words), more than sufficient to designate 20 amino acids. These words of three bases are called codons. The 64 codons are customarily arranged in tabular form, as shown in Table 19.2, and comprise the genetic code. Perusal of the genetic code reveals that different amino acids have different num-

TABLE 19.2 The Genetic Code[a]

5′ Base	U	C	A	G	3′ Base
U	UUU Phe	UCU Ser	UAU Tyr	UGU Cys	U
U	UUC Phe	UCC Ser	UAC Tyr	UGC Cys	C
U	UUA Leu	UCA Ser	UAA Stop	UGA Stop	A
U	UUG Leu	UCG Ser	UAG Stop	UGG Trp	G
C	CUU Leu	CCU Pro	CAU His	CGU Arg	U
C	CUC Leu	CCC Pro	CAC His	CGC Arg	C
C	CUA Leu	CCA Pro	CAA Gln	CGA Arg	A
C	CUG Leu	CCG Pro	CAG Gln	CGG Arg	G
A	AUU Ile	ACU Thr	AAU Asn	AGU Ser	U
A	AUC Ile	ACC Thr	AAC Asn	AGC Ser	C
A	AUA Ile	ACA Thr	AAA Lys	AGA Arg	A
A	AUG Met	ACG Thr	AAG Lys	AGG Arg	G
G	GUU Val	GCU Ala	GAU Asp	GGU Gly	U
G	GUC Val	GCC Ala	GAC Asp	GGC Gly	C
G	GUA Val	GCA Ala	GAA Glu	GGA Gly	A
G	GUG Val	GCG Ala	GAG Glu	GGG Gly	G

[a] The genetic code comprises 64 codons, which are permutations of four bases taken in threes. Note the importance of sequence: three bases, each used once per triplet codon, give six permutations: ACG, AGC, GAC, GCA, CAG, and CGA, for threonine, serine, aspartate, alanine, glutamine, and arginine, respectively.

bers of codons. For example, tryptophan has only one codon, whereas arginine, leucine, and serine each have six codons. The existence of multiple codons for one amino acid has been called "degeneracy."

A useful way to review the tabular genetic code is as a composite of 16 boxes, each containing four codons. In each box of four codons the first two bases are the same. In 8 of the 16 boxes is seen a four-codon family for a single amino acid. Members of these codon families differ only in the third base. In these eight codon families two-base codons would seem adequate, and there is evidence that in some cases only the first two bases are sufficient to specify the amino acid. We shall return to this point below.

Punctuation Codons

Four of the 64 codons specify start and stop signals for protein synthesis. The start signal, AUG, also specifies the amino acid methionine. When AUG appears at the beginning of a message, methionine becomes the initial amino acid at the amino terminal of the protein to be synthesized. Because AUG is the only methionine codon, it also specifies methionine residues in the interior of the protein sequence. In contrast to the start signal, the stop signals, UAA, UAG, and UGA, specify no amino acid. For that reason they are sometimes called nonsense codons.

Codon–Anticodon Interaction

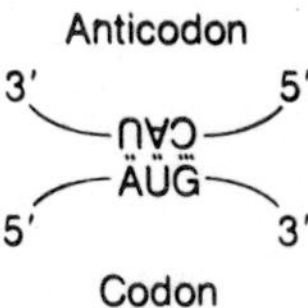

Figure 19.1
The codon–anticodon interaction for methionine.

With the partial exception of a unique codon usage in mitochondria, the genetic code is universal. We shall see that the code as deciphered from studies using bacterial components is precisely the same as that used by human beings for synthesis of all proteins. The translation of a codon is accomplished by a mechanism involving complementary anticodons. The codon-anticodon interaction for methionine is shown in Figure 19.1. The anticodons are three-base sequences in tRNA, as discussed in Chapter 18. When tRNAs are depicted in their secondary or cloverleaf structures, as shown in Figure 18.4, the anticodons are always found in the anticodon loop, remote from the amino acid acceptor stem. Anticodon sequences in tRNAs of known specificity have confirmed the genetic code. Conversely, tRNAs of unknown amino acid specificity have been identified by their anticodons. When tRNA is depicted in its L-shaped tertiary structure as in Figure 18.4, the anticodon is at one extremity of the molecule, and the amino acid residue attachment point on the 3′ terminus of the acceptor stem is at the other extremity of the molecule. Thus the structure of tRNA is ideally suited to its role as an adaptor or bridge between the base sequence in mRNA and the amino acid sequence in nascent protein.

For reasons that are not understood, mitochondria use certain codons differently. The codons involved in this unique departure of the mitochondrion from the general genetic code are shown in the table:

Codon	*General Code*	*Mitochondrial Code*
CUA	Leu	Thr
AUA	Ile	Met
UGA	Stop	Trp

Exceptions to Classical Base Pairing

Perfect adherence to Watson-Crick base pairing would demand absolute specificity of a given tRNA for a given codon. However, the degeneracy of the genetic code is not matched by a precise correspondence in multiplicity of tRNAs, and one tRNA may actually translate several different codons, all with the same bases in the first two positions. Exceptions to Watson-Crick base pairing at the third position of the codon are given in the "wobble" rules, according to which G in the third position of the codon may pair with U in the first position of the anticodon, U in the third position of the codon may pair with G in the first position of the anticodon, and U, C, or A in the third position of the codon may pair with I (hypoxanthine) in the first position of the anticodon:

Anticodon	3′	--U	--G	--I	--I	--I
Codon	5′	--G	--U	--U	--C	--A

Hypoxanthine is one of the numerous unusual bases found in tRNA. Some of these unusual bases are found in or near anticodons. The wobble rules are not sufficient to explain other non-Watson-Crick codon-anticodon interactions that do occur. For example, in the case of valine all four codons (GUU, GUC, GUA, and GUG) can be read by each of the anticodons U*AC, GAC, and IAC (where U* is the unusual base, 5-oxyacetic acid uridine). Therefore, either base pairs prohibited by wobble rules (I : G, G : G, and A : G) actually do form, or a two-letter code suffices for translation in this valine family. According to the wobble rules 31 different tRNAs would suffice to read the 64 codons. However, human (HeLa cell) mitochondria contain only 23 tRNAs, which must then be translating certain codons by the first two letters.

Deciphering the Code

In the presence of higher than physiological concentrations of magnesium ion in vitro, specific initiation components required for protein synthesis in vivo are not necessary, and the synthetic polymers, poly(A), poly(U), and poly(C), are readily translatable to produce the synthetic polypeptides, polylysine, polyphenylalanine, and polyproline, respectively. From such studies one can deduce the respective codons to be AAA, UUU, and CCC. Interesting variations and extrapolations of such experiments led historically to the complete deciphering of the genetic code. For example, a perfectly alternating copolymer, AUAUAUAUAUAUAUAU. . . or $(AU)_n$ contains two and only two codons, AUA and UAU. Furthermore, these two codons occur not at random but in a perfectly alternating sequence, a fact leading to the expectation that translation of this synthetic messenger would produce a polypeptide, Ile-Tyr-Ile-Tyr-Ile-Tyr-, and this was indeed realized. Similarly a perfect synthetic message containing the sequence $(CUG)_n$ contains the possible codons CUG for leucine, UGC for cysteine, and GCU for alanine. However, in this case any one codon, once selected, sets the reading frame and is followed by repetition of the same codon. The selection of a particular initial codon in this case is random. We shall later see that selection of the initial codon for protein synthesis in vivo is not random but specific. A special initiation codon, the start signal AUG, sets the unique reading frame for most natural mRNAs. The synthetic proteins realized from the synthetic message $(CUG)_n$ are three different homogeneous polypeptides: polyleucine, polycysteine, and polyalanine. The student should here be convinced that a perfectly regular sequence of, for example, (CUCG) would produce a single synthetic polypeptide of the sequence (Leu-Ala-Arg-Ser). These data are summarized in Table 19.3. The results show that in protein synthesis the codons are taken in their precise sequence without the omission of a single base and without the double use of any base. In the latter case

CLIN. CORR. 19.1 MISSENSE MUTATION: HEMOGLOBIN

Clinically the most important missense mutation known is the change from A to U in either the GAA or GAG codon for glutamate to give a GUA or GUG codon for valine in the sixth position of the β chain for hemoglobin. An estimated 1 of 10 American blacks are carriers of this mutation, which in its homozygous state is the basis for sickle cell disease, the most common of all hemoglobinopathies. The second most common hemoglobinopathy is hemoglobin C disease, in which a change from G to A in either the GAA or GAG codon for glutamate results in an AAA or AAG codon for lysine in the sixth position of the β chain.

CLIN. CORR. 19.2 DISORDERS OF TERMINATOR CODONS

In the disorder of hemoglobin McKees Rocks the UAU or UAC codon normally designating tyrosine in position 145 of the

TABLE 19.3 Polypeptide Products of Synthetic mRNAs[a]

mRNA	*Codon Sequence*	*Products*
—$(AU)_n$—	— AUA UAU AUA UAU —	—$(\text{Ile-Tyr})_{n/3}$—
—$(CUG)_n$—	— CUG CUG CUG CUG —	—Leu_n—
	— UGC UGC UGC UGC —	—Cys_n—
	— GCU GCU GCU GCU —	—Ala_n—
—$(CUCG)_n$—	CUC GCU CGC UCG	—$(\text{Leu-Ala-Arg-Ser})_{n/3}$—

[a] The horizontal brackets accent the reading frame.

the reading frames would be overlapping rather than sequential. Such an overlapping reading frame sequence would place restrictions on which amino acids could follow others and would allow a single base change to alter three resulting amino acid residues instead of only one. The absence of these phenomena proves the nonoverlapping nature of the genetic code.

Mutations

The genetic code provides a basis for understanding mutation. Originally a genetic term, the word *mutation* describes a change in a gene, and some mutations have been characterized in biochemical terms. For example, missense mutations involve a codon change, usually by a single base, such that the mutant codon designates a different amino acid than does the original codon. Of almost 300 mutations expressed in the β chain of human hemoglobin, the vast majority are of the missense variety. Every one of these missense mutations confirms the genetic code, because the amino acid replacement in the mutant hemoglobin is reducible to and explained by a single base change in a codon (Clin. Corr. 19.1).

Just as a codon may change to designate another amino acid, so an amino acid codon may change to a stop codon, thereby producing a terminator mutation. Two human illnesses representing terminator mutations are a polycythemia, resulting from the extremely high oxygen affinity of hemoglobin McKees Rocks, and a variety of β-thalassemia (Clin. Corr. 19.2).

When a stop codon mutates to become a sense codon, then base sequences not normally translated at the 3′ end of the mRNA are read as sequential codons until the appearance of another stop codon. The result is the appearance of a larger protein in place of the normal protein. This phenomenon is the basis for the several human disorders summarized in Table 19.4 and discussed in Clin. Corr. 19.3. Although all the data in Table 19.4 are equally consistent with UAG as the normal stop codon for

TABLE 19.4 Reverse Terminator Mutations Producing Abnormal α-Globins

Hemoglobin	*α-Codon 142*	*Amino Acid 142*	*α-Globin Length (residues)*
A	UAA	—	141
Constant Spring	CAA	Glutamine	172
Icaria	AAA	Lysine	172
Seal Rock	GAA	Glutamate	172
Koya Dora	UCA	Serine	172

β chain has mutated to the terminator codon UAA or UAG. The result is a shortening of the β chain from its normal 146 residues to 144 residues, a change that gives the hemoglobin molecule an unusually high oxygen affinity. An attempt to compensate for decreased oxygen delivery by increased red blood cell production gives the polycythemic phenotype. Another human illness resulting from a terminator mutation is a variety of β-thalassemia. The thalassemias comprise a group of disorders characterized at the molecular level by an inbalance in the stoichiometry of globin synthesis. In β^0-thalassemia no β-globin is synthesized. As a result α-globin, unable to associate with β-globin to form hemoglobin, accumulates and precipitates in erythroid cells. The precipitation damages cell membranes, causing hemolytic anemia and interfering with erythropoiesis. One variety of β^0-thalassemia results from a terminator mutation at codon 17 of the β-globin. The normal codon AAG designates a lysyl residue at β-17 but becomes the stop codon UAG in this variety of β^0-thalassemia. In contrast to the situation with hemoglobin McKees Rocks, in which the terminator mutation occurs late in the β-globin message, the terminator mutation occurs so early in the mRNA of β^0-thalassemia that no useful partial β-globin sequence can be synthesized, and β-globin is absent.

CLIN. CORR. 19.3 THALASSEMIA

The disorders summarized in Table 19.4 are forms of α-thalassemia, because the abnormally long α-globin molecules, which replace normal α-globin, are present only in small amounts. The small amounts of α-globin result either from decreased rate of synthesis or from increased rate of breakdown. As can be seen, the normal stop codon, UAA, for α-globin mutates to any of four sense codons with resultant placement of four different amino acids at position 142. Normal α-globin is only 141 residues in length, but the four abnormal α-globins are 172 residues in length, presumably because of a terminator codon in position 173.

TABLE 19.5 A Human Frameshift Mutation Produces Hemoglobin Wayne[a]

Position	137	138	139	140	141	142	143	144	145	146	147
Normal α-globin amino acid sequence	- Thr	- Ser	- Lys	- Tyr	- Arg						
Normal α-globin codon sequence	- ACP	- UC(U)	- AAA	- UAC	- CGU	- [UAA]	- GCU	- GGA	- GCC	- UCG	- GUA
Wayne α-globin codon sequence	- ACP	- UCA	- AAU	- ACC	- GUU	- AAG	- CUG	- GAG	- CCU	- CGG	- [UAG]
Wayne α-globin amino acid sequence	- Thr	- Ser	- Asn	- Thr	- Val	- Lys	- Leu	- Glu	- Pro	- Arg	

[a] The base deletion causing the frameshift is encircled. The stop codons are boxed.
P = A, G, U, or C.

α-globin, evidence from another kind of mutation, a frameshift mutation, indicates that UAA is indeed the normal stop codon. Frameshift mutations result from a single base deletion or addition and may cause the appearance of a sequence of different amino acids or the appearance or disappearance of a stop mutation. Hemoglobin Wayne contains an abnormal α-globin chain as a consequence of a frameshift mutation, illustrated in Table 19.5. A deletion of the indicated U in codon 138 causes the resulting sequential codon changes beginning at the next position, 139, such that the amino acid sequence changes. In addition, the normal stop codon UAA is now no longer read in that frame, and protein synthesis continues beyond that point until another stop codon in the new reading frame is encountered at position 148 as shown. If UAG were the normal stop codon for α-globin, as would be allowed from the data in Table 19.4, then codon 142 in Wayne α-globin would be AGG for arginine instead of AAG for lysine as actually found.

The importance of reading frame is underscored by the phenomenon of overlapping genes. Although the genetic code is not overlapping, certain viruses use the same length of DNA to encode different genes in different reading frames. In this way more information can be stored in the DNA, which for reasons of viral packaging must be limited in total amount. Thus the genome of ϕX174 is too small to code for its nine proteins without use of different and overlapping reading frames. At one point of some viral DNA all three reading frames may be used. Mammalian viruses with overlapping genes include simian virus 40, which causes tumors in apes, and the closely homologous BK virus, isolated from human sources.

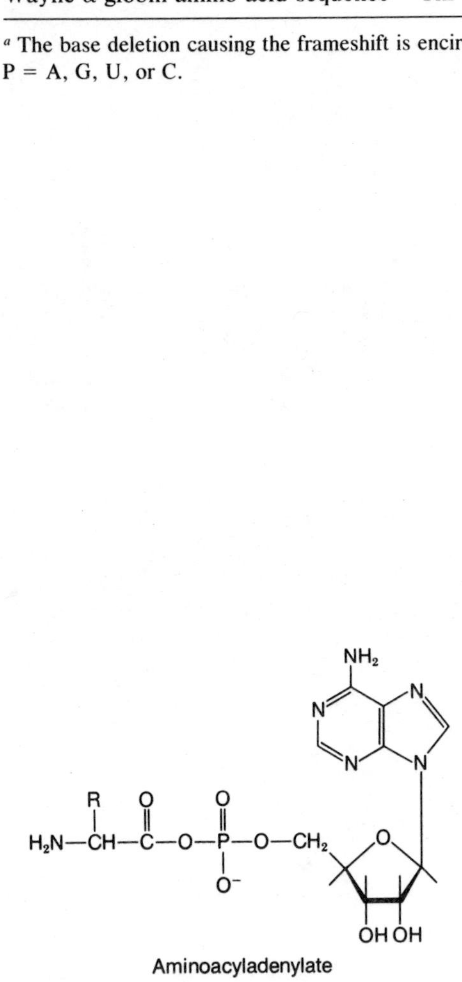

Aminoacyladenylate

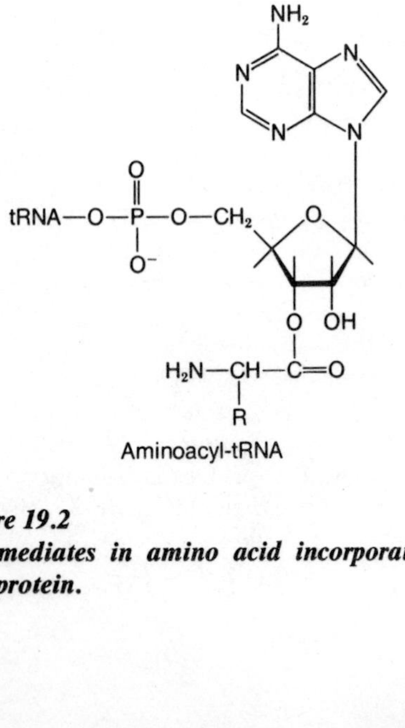

Aminoacyl-tRNA

Figure 19.2
Intermediates in amino acid incorporation into protein.

19.2 AMINOACYL–tRNA SYNTHETASES

Protein synthesis involves convergence of diverse components. A useful way to conceptualize the process as a sequence of reactions is to follow the pathway of amino acids. From the "point of view" of amino acids, the aminoacyl-tRNA synthetases catalyze the first two steps in protein synthesis. These are the activation of each amino acid as aminoacyl-AMP, a mixed acid anhydride (Figure 19.2) and the subsequent attachment of each amino acid to its cognate tRNA in ester linkage (Figure 19.2) for later transfer into polypeptide. The two steps may be written with *L*-valine, for example,

$$\text{L-Valine} + \text{ATP} + (\text{ValRS}) \rightleftharpoons \text{Val} \sim \text{AMP} \cdot (\text{ValRS}) + \text{PP}_i$$

$$\text{Val} \sim \text{AMP} \cdot [\text{ValRS}] + \text{tRNA}^{\text{Val}} \rightleftharpoons \text{Val} \sim \text{tRNA}^{\text{Val}} + \text{AMP} + (\text{ValRS})$$

Valyl-tRNA synthetase (ValRS) is entered as a reactant because of the formation of Val~AMP, not as a free intermediate but as an enzyme-bound intermediate. The sum of the two reactions is then

$$\text{L-Valine} + \text{ATP} + \text{tRNA}^{\text{Val}} \rightleftharpoons \text{Val} \sim \text{tRNA}^{\text{Val}} + \text{AMP} + \text{PP}_i$$

Viewed in this way ATP is split to AMP and PP_i with concomitant formation of one aminoacyl-tRNA. The actual equilibria of these reactions in vivo is shifted far to the product side because of cleavage of pyrophosphate by ubiquitous pyrophosphatases.

Each of the 20 coded amino acids has a specific aminoacyl-tRNA synthetase, and each aminoacyl-tRNA synthetase recognizes one or several cognate tRNAs, capable of linking only the specific amino acid. The specificity of aminoacyl-tRNA biosynthesis is as important for accurate protein biosynthesis as is the specificity of the codon–anticodon interaction. This is true because once attached to tRNA an amino acid residue has no effect on its own specific placement into protein, that placement now becoming entirely the responsibility of the codon-anticodon interaction. Thus in vitro cysteinyl-tRNA^{Cys} can be reduced with Raney nickel to alanyl-tRNA^{Cys}. The chemically modified aminoacyl-tRNA has been used as a precursor in hemoglobin biosynthesis by reticulocyte lysates, and the alanine residues occupy known positions of cysteine, but not of alanine, in the finished hemoglobin.

Mechanisms to Ensure Fidelity

Accurate protein synthesis depends absolutely on mechanisms to insure fidelity of amino acid attachment to tRNA. These mechanisms are remarkably discriminating, far more so than those of ordinary metabolic enzymes. For some amino acids the discrimination is relatively simple; for example, glycine is smaller than any other amino acid, thus easily distinguished. However, others of the 20 coded amino acids are so similar to one another as to require further mechanisms of fidelity. For example, consider the minimal structural difference, a single methylene group, between valine and isoleucine. During the synthesis of rabbit hemoglobin, even the most difficult error to prevent, the substitution of valine for isoleucine, occurs only about once in 3,000 opportunities. Such fine discrimination cannot be explained only by relative K_m and K_{cat} values in aminoacyl-tRNA synthesis. Then what is the explanation? The double-sieve mechanism provides one and is so named for explaining discrimination against larger analogs and discrimination against smaller analogs. For example, take ValRS and isoleucine as a model of discrimination against a larger analog. As shown in Figure 19.3 ValRS discriminates against isoleucine in the activation step. The relative rates of activation of valine and of isoleucine depend on their concentrations and on their K_{cat}/K_m ratios. Thus

$$\nu = \frac{K_{cat}}{K_m}[\text{E}][\text{S}]$$

at any substrate concentration, where ν is the reaction rate, [E] is concentration of free enzyme, and K_{cat} is the turnover number. The reaction rates for valine and for isoleucine can then be compared according to the equation:

$$\frac{\nu_{\text{val}}}{\nu_{\text{Ile}}} = \frac{[\text{Val}](K_{\text{cat}}^{\text{Val}}/K_{\text{m}}{}^{\text{Val}})}{[\text{Ile}](K_{\text{cat}}^{\text{Ile}}/K_{\text{m}}{}^{\text{Ile}})}$$

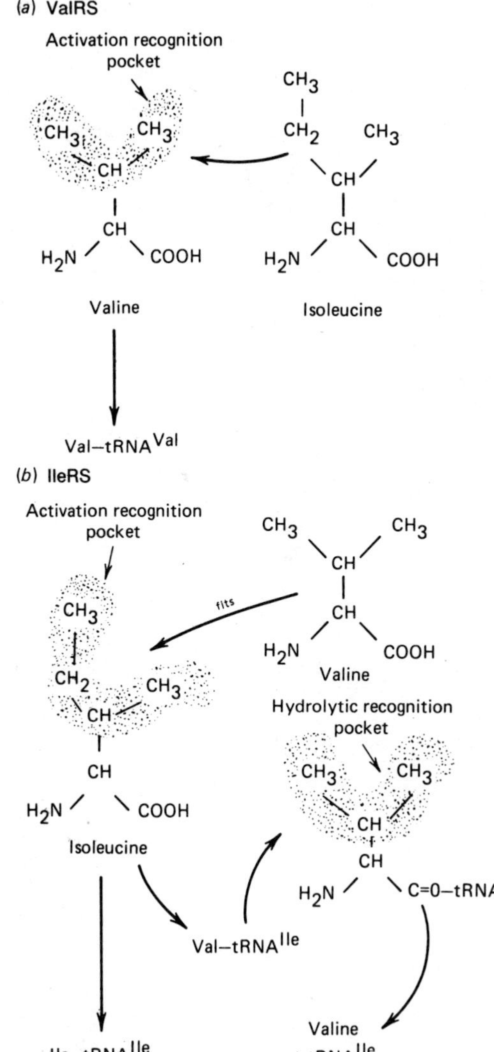

Figure 19.3
The double-sieve mechanism.
(a) ValRS discriminates against the larger analog, isoleucine, as measured by K_m and K_{cat}. (b) Val-tRNA^{Ile} forms but is immediately hydrolyzed in the hydrolytic recognition pocket, which cannot accommodate the correct product, Ile-tRNA^{Ile}.

The ratio actually measured in model bacterial systems is about 10^5, sufficient to explain the observed low mistake levels in vivo.

Now take IleRS and valine as a model of discrimination against an isosteric or a smaller analog. Although the activation site of IleRS discriminates against valine, the relative rates of valine and isoleucine activation are not different enough to explain the low, observed in vivo mistake levels. In this case significant Val-$tRNA^{Ile}$ is actually formed. However, a hydrolytic site on the IleRS surface recognizes this mismatch and rapidly hydrolyzes the ester bond to give valine and $tRNA^{Ile}$. The hydrolytic site excludes the correct product, Ile-$tRNA^{Ile}$, by steric hindrance, because the hydrolytic site is smaller than the activation site and just large enough to accommodate Val-$tRNA^{Ile}$. This editing function is analogous to the 3′ to 5′ editing function of DNA polymerase I.

19.3 PROTEIN SYNTHESIS

The needs of various cells to synthesize protein vary markedly. At one extreme are mature red blood cells, which lack the enzymes and organelles required for protein synthesis. They undergo no cell division and have a finite life span of about 120 days because of their inability to replenish necessary proteins. Other cells must maintain levels of certain enzymes and replace structural proteins to remain viable. For growing and dividing cells greater levels of protein synthesis are required. Finally, some cells synthesize relatively large amounts of protein for export. Examples are pancreatic acinar cells, which synthesize digestive enzymes such as trypsin and chymotrypsin, endocrine cells such as the insulin-producing β cells of the pancreas, and liver cells, which synthesize many proteins, including serum albumin. These cells with the highest protein synthetic activities contain large numbers of ribosomes, which lend the cytoplasm its basophilic staining qualities, seen by light microscopy.

In essence protein synthesis involves the convergence of sequence information in the form of mRNA, activated amino acids in the form of aminoacyl-tRNAs, energy in the form of GTP, and various protein factors. The process occurs on the surface of ribosomes (Figure 19.4) which are multiprotein, multi-RNA complexes that provide at least one required

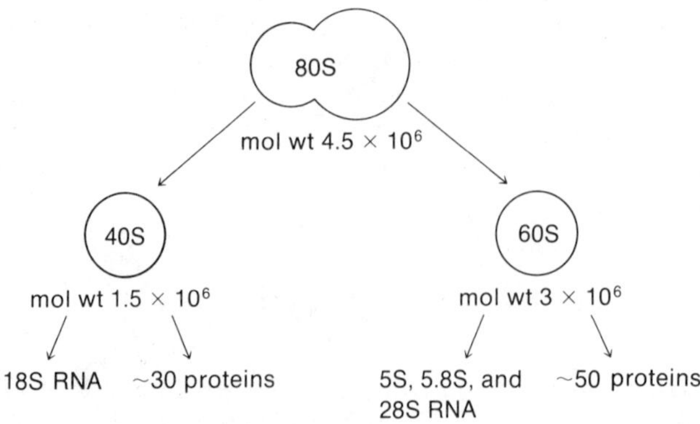

Figure 19.4
A representative eucaryotic ribosome.
Sizes of eucaryotic ribosomes vary by as much as 10% for different organisms, and the number of proteins and the structure are less well defined than those of the highly uniform procaryotic ribosomes.

enzyme. That enzyme is peptidyl-transferase, one of many proteins making up the larger ribosomal subunit and imbedded in the surface of the subunit. Peptidyl-transferase catalyzes peptide bond formation, the actual covalent linkage of one amino acid residue to another. The rate of this bond formation can be as high as 1,200 residues per minute per ribosome at 37°C. The rate of peptide bond formation per mRNA molecule is much higher because many ribosomes simultaneously translate each mRNA. The resulting polyribosomes are visible by electron microscopy, as shown in Figure 19.5, and are measurable by zone sedimentation in sucrose gradients as shown in Figure 19.6.

Structure of mRNA

In contrast to prokaryotes mRNA in eucaryotes is synthesized in the nucleus and must cross the nuclear membrane to be translated in the cytoplasm. This mRNA transport is closely related to the mRNA processing described in Chapter 18. Thus eucaryotic mRNA is complex and

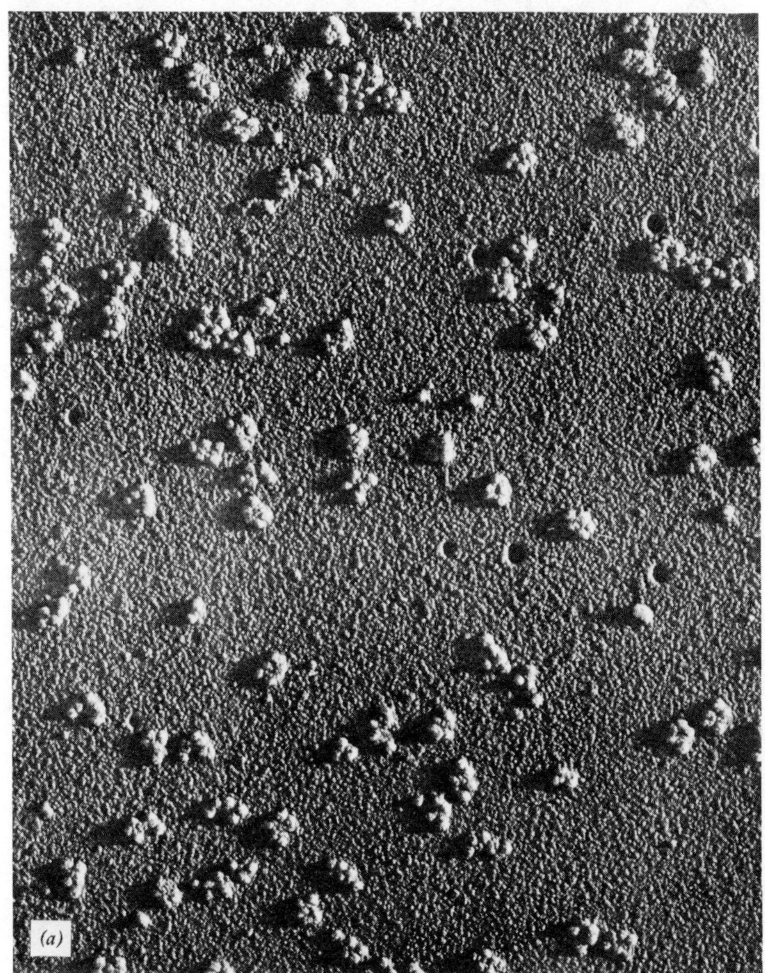

Figure 19.5
Reticulocyte polyribosomes shadowed with platinum as shown above appear as clusters of three to six ribosomes, a number consistent with the size of mRNA for a globin chain.
Further magnification after uranyl acetate staining as shown in (b) (p. 746) reveals one extraordinarily clear five-ribosome polysome with part of the mRNA visible.
Courtesy of Dr. Alex Rich, MIT.

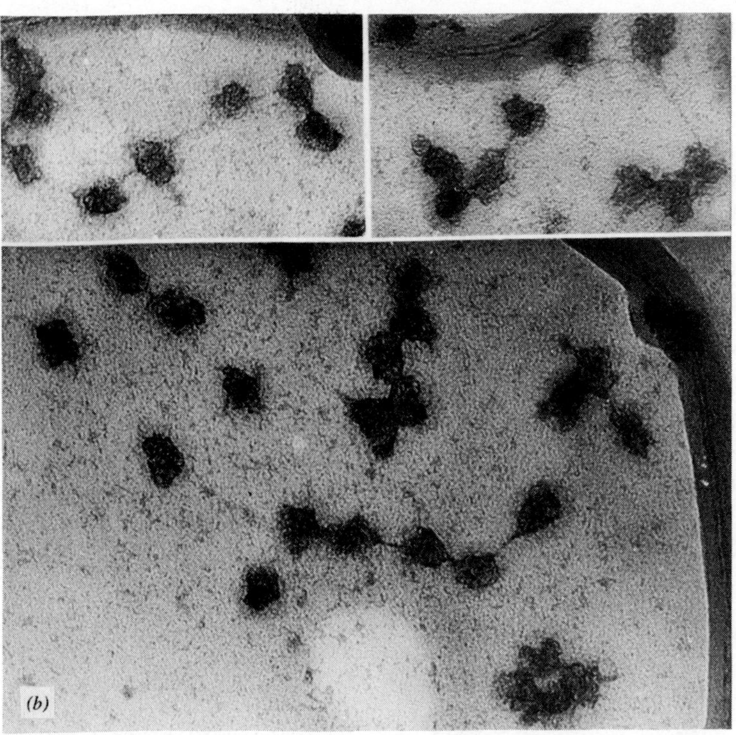

Figure 19.5 **(Continued)**

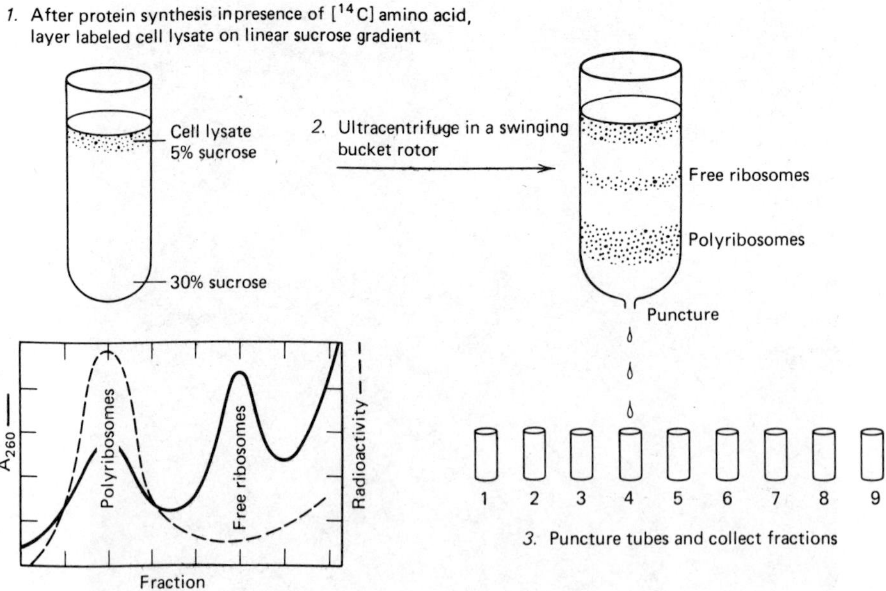

Figure 19.6
Zone sedimentation in a sucrose gradient is a technique used to demonstrate that polyribosomes, not free ribosomes, are the site of protein synthesis.
The four basic steps of the technique are indicated. Polyribosomes labeled with [^{14}C]amino acid in a cell lysate actively synthesizing protein sediment in the radioactive peak. The absorbance at 260 nm measures RNA, mostly rRNA with a prominent peak at the position of free ribosomes.

contains more than the simple base sequence required to specify a protein. To review briefly, at the 5′ end is a characteristic cap structure. After an untranslated region near the 5′ end is a ribosome binding site consisting of about 10 base residues believed to pair with a complementary sequence in ribosomal RNA. Shortly after the ribosomal binding site is the AUG codon and the base sequence for the protein, followed by a stop codon. Eucaryotic mRNAs are usually monocistronic, that is, carrying the base sequence for only a single protein. After the stop codon is a noncoding sequence before the 3′ terminus of poly A. The noncoding sequence is at least 96 base residues long in mRNA for human α-globin, as shown by the chain termination mutants already discussed: those giving hemoglobins Constant Spring, Icaria, Seal Rock, and Koya Dora (Table 19.4). This noncoding sequence must have an important normal function, because its sequence is highly conserved, being 80% homologous from rabbit to man.

Direction of Translation

Just as these words are read in sequence from left to right, so mRNA is translated in a defined direction and not in a random fashion. The defined direction of translation is such that the amino terminal of the nascent protein is synthesized first, and the carboxy terminal of the nascent protein is synthesized last. This can be demonstrated in reticulocytes synthesizing hemoglobin at artificially lowered temperature so as to decrease the rate of protein synthesis and make the process more amenable to study. In such a system radioactive amino acids are added for short, measured periods of time, and the free globin chains synthesized are then isolated for analysis. As shown in Figure 19.7 the radioactive precursor should be

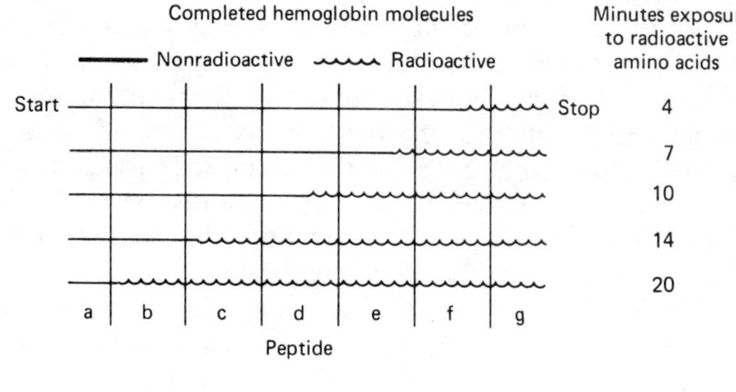

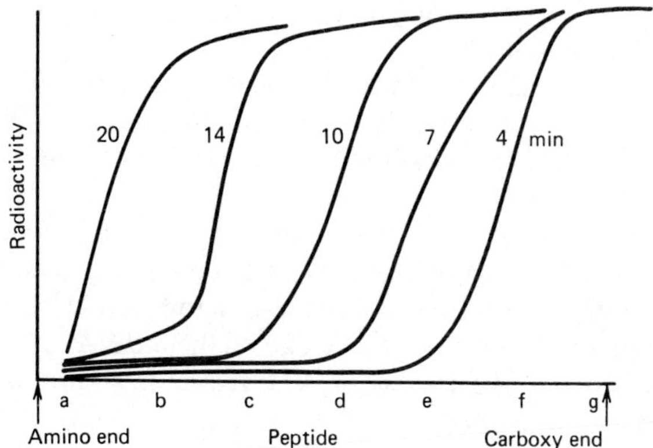

Figure 19.7
Hemoglobin synthesis starts at the amino end and stops at the carboxy end.
See text for explanation.

found in the portion of the globin chain synthesized last. After longer and longer exposure to the radioactive precursor more and more of the globin chain becomes labeled until eventually even the first part of the molecule to be synthesized is labeled with radioactivity. The peptides obtained by digesting the globin product are arranged in order of specific radioactivity. From the known sequence of the peptides in globin it can be seen that the most highly radioactive peptides, for example, peptides F and G in Figure 19.7, are always at the carboxy end of the globin. Conversely, the least radioactive peptide, peptide A in Figure 19.7, is always at the amino end of the globin. Therefore the temporal sequence of protein synthesis is from the amino to the carboxy end.

Information on the direction of protein synthesis and knowledge of the genetic code together prove that the direction of translation of mRNA must be 5′ to 3′. Thus mRNA is translated in the same direction in which it is transcribed, and the 5′ end of mRNA corresponds to the amino end of the protein for which the mRNA codes. The existence of stable, isolatable polyribosomes and the directional, sequential nature of protein synthesis imply that ribosomes do not attach and detach from mRNA at random, that a given ribosome remains bound to a given mRNA molecule until the message is completely translated, and that in the translation process there is movement of each ribosome relative to the mRNA.

Initiation of Protein Synthesis

Three operational divisions of protein synthesis provide a useful descriptive framework. They are initiation, elongation, and termination. Initiation consists of the placement of the amino-terminal amino acid and involves, first, the formation of a complex on the 40S ribosomal subunit and, second, the addition of the 60S ribosomal subunit to give the whole ribosome 80S initiation complex. The process (Figure 19.8) is intricate and requires proteins called initiation factors. Initiation factors bind reversibly to ribosomal complexes but are not ribosomal structural proteins. In eucaryotes the number and functions of all initiation factors are not completely understood, but their counterparts in procaryotes have provided useful, simpler models. Eucaryotic initiation factor 2, or eIF-2, specifically binds Met-$tRNA_i^{Met}$, a specific tRNA for initiation, with GTP and then joins with the 40S ribosomal subunit to form an entry complex. The 40S ribosomal subunit exists bound to eIF-3, which dissociates the 40S and 60S ribosomal subunits. That dissociation is required in each cycle of protein synthesis as done by each ribosome. Whereas its procaryotic counterpart is a single polypeptide chain, eIF-3 has 9–11 subunits, testifying to the greater complexity and/or to our lesser understanding of eucaryotic initiation.

Met-$tRNA_i^{Met}$ functions only in initiation. Conversely, Met-$tRNA_m^{Met}$ (note the subscript) is required for elongation and does not place methionine residues in the initiator (amino-terminal) position. Met-$tRNA_m^{Met}$ is not recognized by eIF-2.

The next step in initiation is the joining of the entry complex and mRNA to form the 40S initiation complex, with codon–anticodon pairing. This step is mediated by four more initiation factors, eIF-1, 4A, 4B, and 4E. The initial binding is at or near the 5′ cap on the mRNA, and eIF-4E is the cap-binding protein. Hydrolysis of ATP is required, but mechanism and stoichiometry remain obscure. The recognition of the start codon, AUG, by the anticodon in Met-$tRNA_i^{Met}$ (Figure 19.1) sets the reading frame. The AUG codon nearest the 5′ cap is always the start codon. Subsequent AUG codons signal methionine residues in the elongation process.

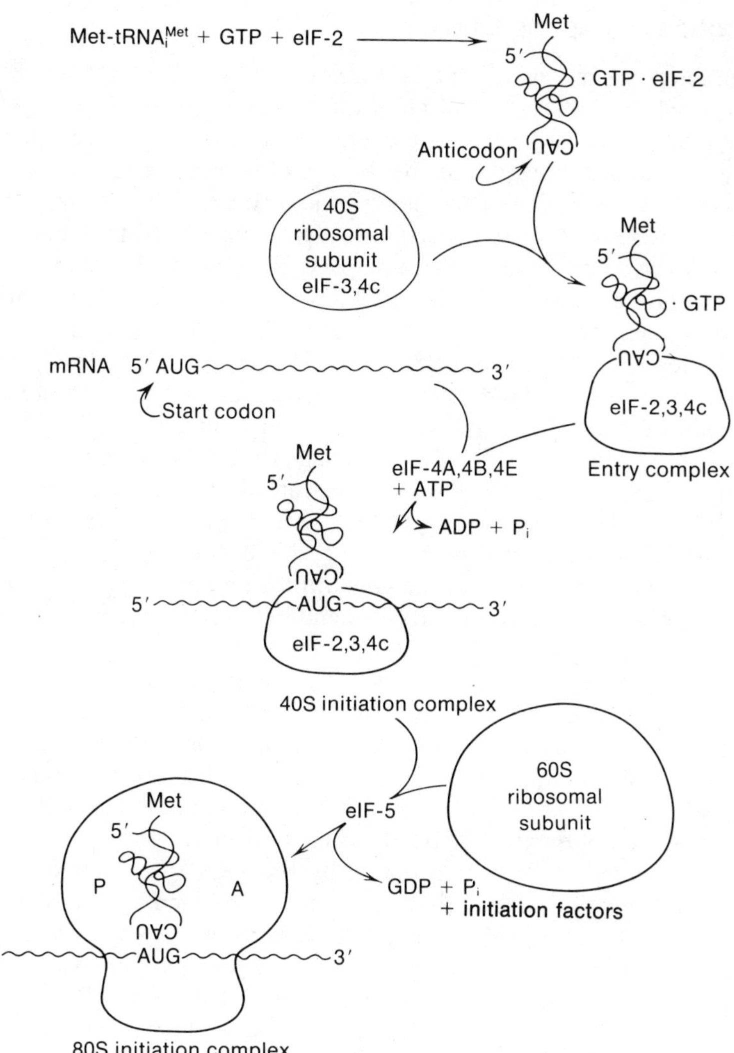

Figure 19.8
Initiation of protein synthesis.
See text for description. Shown is initiation for the first ribosome. Later entry complexes add not to "naked" mRNA but to a polyribosome.

In the final step of initiation eIF-5 mediates formation of the 80S initiation complex (Figure 19.8) by addition of the 60S subunit to the 40S initiation complex, with concomitant release of initiation factors and required breakdown of GTP to GDP and P_i. Similarly one of the procaryotic initiation factors has a ribosome-dependent GTPase activity.

Other initiation factors have been partially characterized and are found to be stimulatory, but not absolutely required, for various steps of initiation in eucaryotes. Perhaps their roles when fully elucidated will be found to relate to control.

As shown in Figure 19.8, Met-tRNA$_i^{Met}$ is bound at the P site in the 80S initiation complex. The A site is the entry position for new aminoacyl-tRNAs in elongation. The P site bears only the peptidyl-tRNA during elongation and holds the nascent chains in readiness for addition of the next amino acid residue. The two sites are experimentally differentiated and defined by sensitivity to attack by the antibiotic, puromycin, which is discussed below.

Elongation of Peptide Chain

Elongation is the second and major division of protein biosynthesis. Elongation consists of placement of all of the amino acid residues except the first one and involves three basic events, repeated for each residue: (1) binding of aminoacyl-tRNA at the A site; (2) formation of the peptide bond; and (3) movement of the new peptidyl-tRNA to the P site.

The aminoacyl-tRNA specified by the next codon in the initiation complex does not simply join to that complex. Rather, this incoming tRNA forms an entry complex with GTP and a protein called elongation factor 1. This entry complex relinquishes its aminoacyl-tRNA, as specified by the next codon, to the A site on the 80S initiation complex. With this binding the GTP is released as GDP and P_i. Elongation factor 1 forms complexes with aminoacyl-tRNAs for all amino acids, including Met-$tRNA_m^{Met}$, but does not bind Met-$tRNA_i^{Met}$, whose carriage to the ribosome is mediated only by initiation factor 2. With completion of this binding step, Met-$tRNA_i^{Met}$, and the next aminoacyl-tRNA now lie juxtaposed on the ribosome, as shown in Figure 19.9 but no peptide exists yet.

Figure 19.9
The first step of elongation is placement of the second aminoacyl-tRNA, here Leu-tRNA, in the A site, the position for peptide bond formation.
See text for description.

Peptidyltransferase, an enzyme constituting one of the proteins of the large ribosomal subunit of procaryotes and believed to occupy a similar, strategic surface location in the 60S subunit of eucaryote ribosomes, now catalyzes a nucleophilic attack by the free amino group of the second aminoacyl-tRNA on the carbonyl carbon of Met-$tRNA_i^{Met}$, to form the first peptide bond, as shown in Figure 19.10. With completion of this step, $tRNA_i^{Met}$ remains bound at the P site, ready to be discarded, and dipeptidyl-tRNA is bound at the A site, ready to move to the P site in the next step. Formation of the peptide bond requires no energy source other than the aminoacyl-tRNA here represented by Met-$tRNA_i^{Met}$. Therefore, the energy for formation of each peptide bond is provided by the peptidyl-tRNA, not by the incoming aminoacyl-tRNA. In this example, Met-Leu-tRNA will provide the energy for formation of the next peptide bond.

The final event of elongation is translocation. Dipeptidyl-tRNA in codon-anticodon register with the codon for the second amino acid (leucine in Figure 19.10) now moves to the P site under mediation of translocase, also called elongation factor 2, and $tRNA_i^{Met}$ is released simultaneously. For its function EF-2 requires GTP hydrolysis to GDP and P_i, and EF-2 has a ribosome-dependent GTPase activity. Presentation of translocation as a separate event may be misleading because translocation occurs simultaneously with the first event for the next amino acid, that is, the

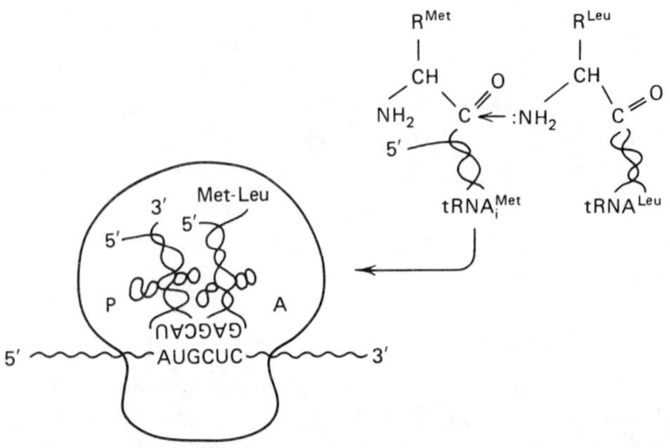

Figure 19.10
Peptidyltransferase catalyzes the second step of elongation with formation of the peptide bond.

binding of aminoacyl-tRNA. Thus, as translocation occurs, elongation factor 1 places the third aminoacyl-tRNA at the A site, as described for the second aminoacyl-tRNA. In the combined event the precise stoichiometry of GTP hydrolysis remains obscure. The large subunit of the procaryotic ribosome contains two copies of a GTPase called the A protein, which may be involved in these events. Completion of the translocation-binding step results in the complex shown in Figure 19.11. Repetition of the peptidyltransferase step and of the translocation-binding step places all remaining amino acid residues in the nascent protein.

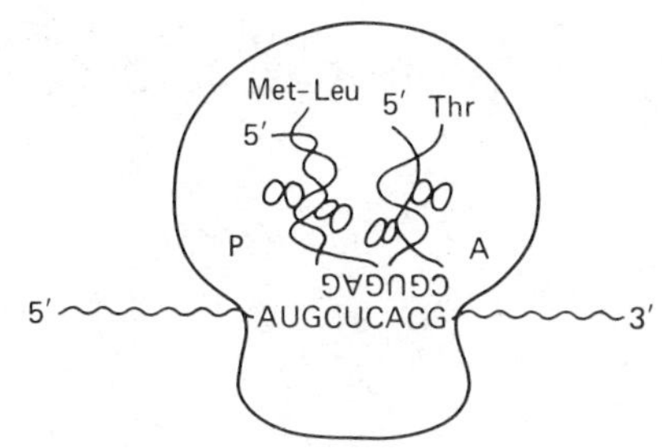

Figure 19.11
After translocation and simultaneous binding of the third aminoacyl-tRNA, here Thr-tRNA, at the A site, Met-Leu-tRNA is in the P site.
Peptide bond formation will follow.

Termination of Protein Synthesis

Termination is the final operational division of protein biosynthesis. Termination occurs after placement of the final amino acid residue at the carboxy terminal of the nascent protein and consists of release of the newly synthesized protein, release of the tRNA of the final amino acid residue, and release of mRNA and ribosomal subunits, all in response to a termination signal. The termination signal is one or more of the codons UAG, UAA, or UGA. These termination signals are recognized not by a tRNA, but by a protein-release factor, which requires GTP to bind to ribosomes and has a ribosome-dependent GTPase activity that is stimulated by termination codons. In the release reaction GTP is split to GDP and P_i. Peptidyltransferase activity is required for release to occur. The release factor allows water to substitute for the α-amino group of an amino acid in the nucleophilic attack (Figure 19.10) on the carbonyl carbon of the terminal amino acid residue. The termination codon is a specific requirement and does not simply represent a codon for which there is no tRNA. Thus the simple absence of the required aminoacyl-tRNA to translate a codon within a cistron temporarily stops protein synthesis but does not result in termination at that point.

If procaryotic protein synthesis is a correct model, the separation of ribosomal subunits during the termination process is not an accidental but a necessary part of the protein synthetic scheme, mediated by initiation factor 3, which remains bound to the smaller ribosomal subunit until the next initiation complex is formed.

Antibiotic Mechanisms

Many antibiotics interfere with protein synthesis, and this interference may be clinically useful if it is sufficiently specific for procaryotic vis-à-vis eucaryotic protein synthesis. Some of the clinically useful antibiotics that act by inhibition of protein synthesis are the tetracyclines, the aminoglycosides, chloramphenicol, and erythromycin. Tetracyclines work by binding to the procaryotic ribosome and preventing the normal binding of aminoacyl-tRNA at the A site. Streptomycin and other aminoglycosides work by binding to the small (30S) subunit of the procaryotic ribosome, distorting it, and thereby producing incorrect codon–anticodon interaction with consequent misreading of the genetic code. A particular ribosomal protein, S12, constitutes the binding site for streptomycin, and organisms with mutant forms of this protein may be resistant to or even dependent upon streptomycin. Erythromycin interferes with procaryotic protein synthesis by inhibiting translocase. Chloramphenicol specifically inhibits procaryotic peptidyltransferase. The antibiotic also inhibits mitochondrial ribosomal peptidyltransferase, this inhibition being one measure of similarity between mitochondria and procaryotes. Cycloheximide inhibits eucaryotic but not procaryotic peptidyltransferase, and thus is the converse of chloramphenicol in its activity. Of course, the antibiotic has no clinical usefulness for this reason.

Many antibiotics too toxic for clinical use are important in experimental work and have sometimes played an important role in elucidating mechanisms of various steps of protein synthesis. Puromycin is perhaps the best example because reactivity with puromycin defines whether a peptidyl-tRNA resides at the P site or at the A site of a ribosome. Puromycin has a structure closely resembling the 3′ end of aminoacyl-tRNA (Figure 19.12) and stops protein synthesis instantly by completing a peptide bond, under mediation of peptidyltransferase, with the nascent peptidyl-tRNA on the P site. The peptide, now with carboxy-terminal puromycin in covalent linkage, is immediately released from the ribosome, having no tRNA to hold it to the A site. Puromycin does not react with peptidyl-tRNA bound at the A site.

Another clinically relevant example of inhibition of protein synthesis is the inactivation of mammalian translocase, elongation factor 2, by the toxins of *Corynebacterium diphtheriae* and *Pseudomonas aeruginosa*. These protein toxins bind to the cell membrane, then enter the cell either intact (*Pseudomonas* toxin) or as an active subunit (diphtheria toxin). Once inside the cell the toxin functions as an enzyme, catalyzing the reaction.

$$\text{Translocase} + \text{NAD}^+ \xrightarrow{\text{toxin}} \text{ADPR-translocase} + \text{nicotinamide} + \text{H}^+$$

The translocase, thus covalently modified, is irreversibly inactivated under in vivo conditions. Thus the extraordinary potency of these toxins results from their catalytic rather than stoichiometric mode of action. Perhaps one molecule of toxin is sufficient to kill a cell. Consistent with this mode of action, these toxins kill cells slowly in contrast to other bacterial toxins. The site in translocase for the ADPR attachment is a modified histidine residue known as diphthamide. We shall deal later with such posttranslational modifications.

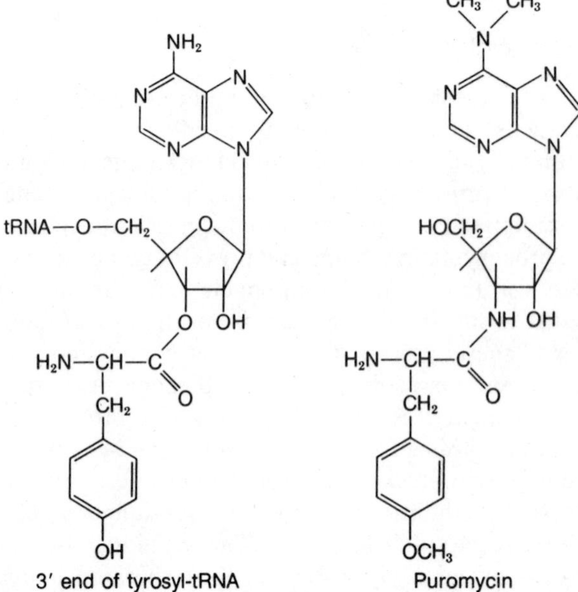

Figure 19.12
Puromycin (right) interferes with protein synthesis by functioning as an analog of aminoacyl-tRNA, here tyrosyl-tRNA (left) in the peptidyltransferase reaction.

Proteins for Export

Proteins synthesized for export are sequestered during synthesis in the rough endoplasmic reticulum, an organelle of ribosomes bound to membranes and particularly rich in protein exporting cells, as shown in Figure 19.13. From the inside of the rough endoplasmic reticulum the proteins travel in vesicles to the cell membrane, where they are extruded after fusion of the vesicle and cell membranes. The membrane-bound ribosomes of the rough endoplasmic reticulum are identical to and in equilibrium with free ribosomes. They bind by the 60S subunit to the membrane of the rough endoplasmic reticulum through the mediation of the transmembrane glycoproteins, ribophorin I and ribophorin II. However, the binding does not take place until synthesis of the protein for export has begun. At the amino terminal of such proteins are signal peptides, which identify proteins to be exported. The signal peptides are variable but always hydrophobic sequences ranging in length from 15 to 30 amino acid residues. There are a wide variety of different signal peptides, even different immunoglobulin G light chains have different signal peptides. Proteins not to be exported, for example, the α- and β-globins of hemoglobin, do not

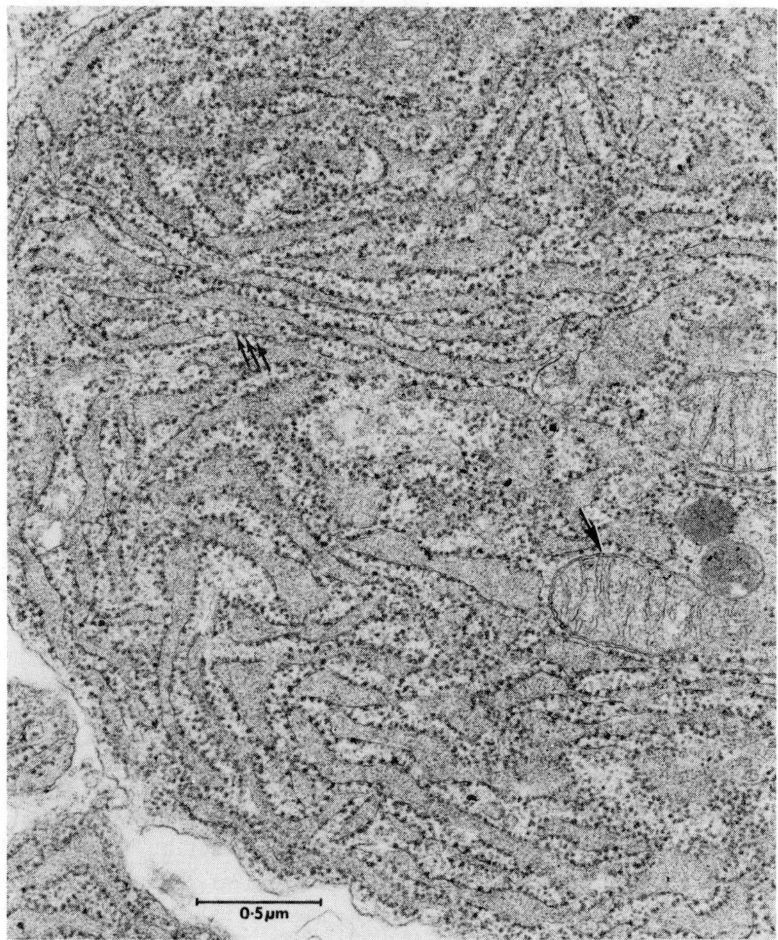

Figure 19.13
Rough endoplasmic reticulum of a plasma cell.
The three parallel arrows indicate three ribosomes among the many attached to the extensive membranes. The single arrow indicates a mitochondrion for comparison.
Courtesy of Dr. U. Jarlfors, University of Miami.

have signal peptides. As the export protein is synthesized, it is threaded through the rough endoplasmic reticulum membrane by a mechanism not fully understood but thought to involve the action of signal peptidase, an enzyme that cleaves the signal peptide from the nascent export protein before synthesis of the protein is complete. Curiously, one prominent export protein, ovalbumin, has no signal peptide but does have a hydrophobic sequence at the amino terminal. Ovalbumin competes with other proteins excreted by oviduct cells, following the same pathway. Proteins destined to become membrane proteins also may have transient signal peptides but remain imbedded in the membrane rather than undergoing excretion from the cells for reasons not well understood.

Mitochondrial Protein Synthesis

Synthesis of special proteins for export introduces the compartmentalization of protein synthesis and the involvement of organelles in that compartmentalization. The mitochondria provide an interesting example of protein synthesis occurring entirely within an organelle. However, most mitochondrial proteins are not synthesized within the mitochondrion but are synthesized in the cytoplasm on free ribosomes as directed by information in mRNA transcribed from nuclear DNA. The proteins or their larger precursors bind to the mitochondrial surface and cross the double membrane into the mitochondria with subsequent cleavage or other covalent modification and subunit assembly. However, mitochondria have DNA, which codes not only for rRNA and for tRNAs but for a few mitochondrial proteins, comprising about 10% of mitochondrial protein mass. Some mitochondrial proteins consist of subunits of nuclear origin and subunits of mitochondrial genetic origin. Human mitochondrial DNA is a circular double helix of 16,500 base pairs. Recall that tRNA genes would require less than 100 base pairs and that a protein with molecular weight 37,000 requires a gene with about 1,000 base pairs, not counting leaders and intervening sequences. Interestingly, human mitochondria use the codon UGA, a terminator in all other known systems, as a tryptophan codon in addition to the single ordinary tryptophan codon, UGG. Moreover, human mitochondria use the codon AUA, an isoleucine codon in all other systems, as a second methionine codon in addition to AUG, the initiator codon. These exceptions to usual codon usage are unique.

The mitochondria have a complete and independent protein synthetic apparatus, including RNA polymerase, ribosomes, aminoacyl-tRNA synthetases, tRNAs, and all other necessary factors. However, the tRNAs, the aminoacyl-tRNA synthetases, and the ribosomes are not the same as those in the cytoplasm. The mitochondrial ribosomes are smaller than cytoplasmic ribosomes and resemble bacterial ribosomes in that respect and in their sensitivity to inhibition of peptidyltransferase by chloramphenicol. In fact the mitochondrial protein synthetic system resembles that of bacteria in other respects. For example, as in procaryotic protein synthesis the initiator in mitochondrial protein synthesis is not Met-$tRNA_i^{Met}$, but formyl-Met-$tRNA_i^{Met}$. Recall that two different tRNAs for methionine, $tRNA_i^{Met}$ and $tRNA_m^{Met}$, are both substrates for methionyl-tRNA synthetase, giving Met-$tRNA_i^{Met}$ and Met-$tRNA_m^{Met}$. Further recall that Met-$tRNA_i^{Met}$ is recognized by initiation factor 2 but not by elongation factor 1, whereas the converse is true for Met-$tRNA_m^{Met}$. A third enzyme that recognizes the difference between $tRNA_i^{Met}$ and $tRNA_m^{Met}$ is transformylase, which catalyzes the formylation of the α-amino group of methionine in Met-$tRNA_i^{Met}$, but not in Met-$tRNA_m^{Met}$, as shown in the reaction

$$N^{10}\text{-Formyltetrahydrofolate} + \text{Met-tRNA}_i^{Met} \longrightarrow \text{tetrahydrofolate} + \text{HCO-Met-tRNA}_i^{Met}$$

Usually the formyl group or the formylmethionine residue is removed from the finished protein. Nevertheless, this is the first mention of a covalently modified amino acid in a protein, and it serves as an introduction to a discussion of posttranslational modification of polypeptides.

19.4 POSTTRANSLATIONAL MODIFICATION OF POLYPEPTIDES

The newly synthesized protein as released from the ribosome may not be in a final, functional state. To achieve function, posttranslational modifications may be required. Proteins undergo a variety of modifications to provide final structures composed of subunits, final structures containing prosthetic groups, or final structures containing covalently modified amino acid residues.

Consider first the formation of multimoric proteins, which may result either from aggregation of subunits or from cleavage of a larger precursor. Hemoglobin formation is an example of the former process. α- and β-Globins, synthesized separately with information from mRNAs transcribed from DNA in different chromosomes, associate with heme and with each other to form hemoglobin. The synthesis of α-globin, β-globin, and heme must be coordinated to produce stoichiometric amounts. The coordination requires control of protein synthesis, a control already implicit in the marked variation between cells of different types in terms of amounts and types of proteins synthesized. Much of the necessary control is exerted at the transcriptional level and is expressed by differing amounts of mRNAs in different cells or in the same cell at different times. If we regard this as gross control, fine tuning may occur in translation. An example of the complexities of such translational control is seen in reticulocytes synthesizing hemoglobin. Here in the presence of constant concentrations of mRNAs for α- and for β-globin, their balanced synthesis is dependent upon presence of heme. In its absence a protein kinase, which is not dependent on cyclic nucleotides, phosphorylates the α-subunit of eIF-2 (Figure 19.14), rendering it unable to form the 40S initiation complex. This protein kinase uses ATP and is called hemin-controlled translation repressor. In addition to decreasing globin synthesis, it decreases α-globin synthesis more than β-globin synthesis, a phenomenon suggesting different affinities of the respective mRNAs for the initiation complex.

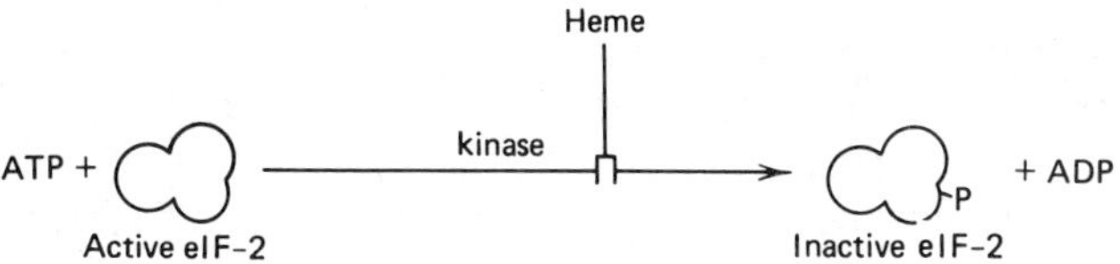

Figure 19.14
Role of heme in blockade of eIF-2 inactivation by protein kinase, termed hemin-controlled translation repressor.

Defects in coordination of heme, α-globin, and β-globin synthesis are manifest at the clinical level. For example, iron deficiency decreases heme synthesis and results in hemoglobin deficiency in red blood cells. Thalassemia has already been mentioned as comprising a class of disorders characterized by decreased or absent α- or β-globin with relative excess of the other globin, a state causing anemia from decreased red blood cell production and increased red blood cell destruction. Any of the many possible reasons for decreased or absent globins, from a gene deletion to a terminator mutation (page 741) will give the same end result, and thalassemia, with many causative defects, remains the best example of genetic heterogeneity that is understood at a molecular level.

Biosynthesis of Insulin

Perhaps the simplest mechanism for synthesis of stoichiometric quantities of protein subunits is that using a larger precursor, which is specifically cleaved to provide the subunits or peptide chains. Thus in insulin biosynthesis translation of the mRNA produces preproinsulin, in which "pre-" signifies the signal peptide and "pro-" signifies a precursor of insulin containing both the A and B peptides and the intervening C-peptide, as shown in Figure 19.15. Specific proteases within the endoplasmic reticulum or Golgi apparatus or in the associated vesicles cleave proinsulin to insulin and C-peptide, which are released into the bloodstream with a small amount of uncleaved proinsulin. In familial hyperproinsulinemia an interesting and different situation exists (Clin. Corr. 19.4).

Perhaps the importance of proinsulin synthesis resides in the requirement for proinsulin in correct disulfide bond formation. Unlike the subunits of hemoglobin, which cohere by noncovalent bonding alone, the A- and B-peptides of insulin are held together by disulfide bonds, and the A-peptide contains an intrachain disulfide bond. Although formation of disulfide bonds in proteins may occur spontaneously, correct cysteine-residue specificity requires the correct tertiary structure as dictated by the primary structure, according to the thermodynamic hypothesis. Thus when insulin itself is reduced and denatured, then gently renatured with concomitant reoxidation of thiol groups to form disulfide bonds, those bonds form in random fashion, and native insulin structure and function is not achieved. In contrast, when the same experiment is done with proinsulin, the intact total primary structure assures the correct reformation of native proinsulin, which can be subsequently cleaved to give native insulin.

CLIN. CORR. 19.4 FAMILIAL HYPERPROINSULINEMIA

Familial hyperproinsulinemia, an autosomal dominant condition, results from lack of one protease or, more probably, from a structural change in proinsulin, preventing cleavage. Approximately equal amounts of insulin and proinsulin are released into the circulation. Although affected individuals have high levels of proinsulin in the blood, they are apparently normal in terms of glucose metabolism, being neither diabetic nor hypoglycemic.

Zymogen Activation

Precursor cleavage is a mechanism to obtain active protein from totally inactive precursor. The phenomenon is classically illustrated by activation of zymogen, for example, the cleavage of chymotrypsinogen to give chymotrypsin and two inactive peptides (Chapter 24). Inappropriate activation of zymogens in the pancreas for any reason leads to autodigestion expressed clinically as acute pancreatitis. Not all zymogens yield subunit enzymes after cleavage. For example, pepsinogen and trypsinogen are activated by cleavage and removal of amino terminal peptides to leave single-chain, active proteins.

Many proteins are modified by removal or addition of terminal amino acid residues. All proteins do not have an amino-terminal methionine residue, and proteins lacking this initial residue have had it and perhaps further residues removed. Other proteins are modified by addition of amino acid residues. In some cases such addition is mediated by tRNA. For example, arginyl-tRNA is a donor of the arginine residue for the

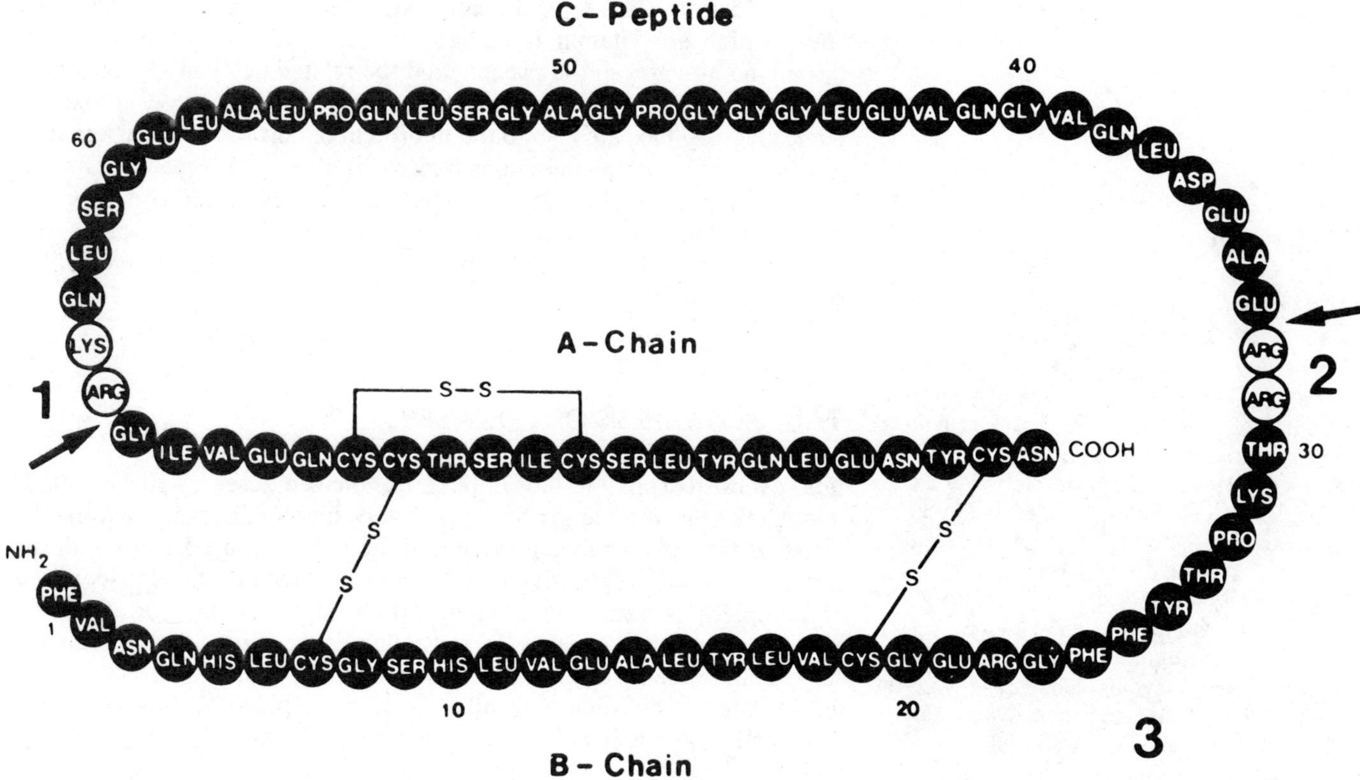

Figure 19.15
Human proinsulin.
After cleavage at the two sites indicated by arrows the arginine residues 31, 32, and 65 and the lysine residue 64 are removed to give insulin and C-peptide.
Reprinted, by permission of the *New Engl. J. Med.* 302:165, 1980.

amino terminal of some proteins. In at least one case, however, the addition is not tRNA-dependent. Thus, the enzyme tubulin-tyrosine ligase catalyzes an ATP-dependent addition of a tyrosine residue to the carboxy terminal of tubulin, a protein in microtubules.

Modification of Coded Amino Acids

Although only 20 amino acids are specifically coded by the genetic code, their modification after protein synthesis leads to a wide variety of unusual amino acid derivatives, about 120 of which have been described in various proteins taken from all varieties of life. Some examples are given in Table 19.6. The unusual amino acids are derived from all of the coded amino acids except for isoleucine and tryptophan. Some unusual amino acid residues in human proteins are of critical relevance to clinical medicine. For example, in a vitamin K-dependent reaction the first 10 glutamic acid residues from the amino terminal end of prothrombin are carboxylated to γ-carboxyglutamic acid residues as shown:

$$\begin{array}{c} CH_2\text{—}COO^- \\ | \\ CH_2 \\ | \\ \text{—CO—NH—CH—CO—NH—} \end{array} \longrightarrow \begin{array}{c} {}^-OOC\text{—}CH\text{—}COO^- \\ | \\ CH_2 \\ | \\ \text{—CO—NH—CH—CO—NH—} \end{array}$$

The modified residues chelate calcium ions, required for normal blood clotting. For the same reason vitamin K is also required for synthesis of

TABLE 19.6 Some Representative Amino Acid Derivatives Found in Proteins

Coded Amino Acid	*Derivative in Protein*
Alanine	*N*-Acetylalanine
Cysteine	*S*-Galactosylcysteine
Glutamic acid	γ-Carboxyglutamic acid
Glycine	Glycinamide
Histidine	Diphthamide
Lysine	ε-*N*-Methyllysine
Methionine	*N*-Formylmethionine
Tyrosine	Tyrosine O^4-sulfate

clotting factors VII, IX, X. In patients treated with coumarin anticoagulants, which are vitamin K antagonists, an inactive precursor of prothrombin circulates in the plasma, and the related delay in blood clotting is measured by the test known as the prothrombin time. γ-Carboxyglutamic acid residues are also found in protein of cortical bone, where they presumably play a role in binding hydroxyapatite to the organic bone matrix, again through chelation of calcium ions. These comments on covalent modification of amino acid residues serve as an introduction to a more detailed look at extensive modifications required for the biosynthesis of collagen.

CLIN. CORR. 19.5
EHLERS–DANLOS SYNDROME, TYPE IV

The Ehlers–Danlos syndrome is the name for a group of at least 10 disorders that are clinically, genetically, and biochemically distinguishable but that all share manifestations of structural weaknesses in connective tissue. The weaknesses result from defects in collagen structure. For example, type IV Ehlers–Danlos syndrome is caused by defects in type III collagen, which is particularly important in skin, arteries, and hollow organs. Manifestations may be severe, with arterial rupture, intestinal perforation, rupture of the uterus during pregnancy or labor, and easy bruisability of thin, translucent skin. The basic defect in several variants of type IV Ehlers–Danlos appears to be one of several changes in the primary structure of proα1(III) chains that slow synthesis, delay secretion, or increase rate of degradation, thereby preventing normal formation of type III collagen fibrils.

19.5 BIOSYNTHESIS OF COLLAGEN

For several reasons the biosynthesis of collagen deserves special attention. Collagen, forming the fibrous network that holds organs and tissues intact, is the most abundant protein in the human body. Its biosynthesis serves as the most extensive example in protein synthesis of a sequence of multiple, posttranslational, modification steps and of export. Unusual and important features include extracellular processing and spontaneous self-assembly of collagen molecules into fibrils. Various hereditary defects related to collagen biosynthesis cause a number of illustrative diseases and show the importance of mutation in posttranslational modification as distinct from mutations in primary structure.

A brief review of collagen structure (Figure 19.16) is essential for an understanding of its biosynthesis. Briefly, a collagen chain is a left-handed helical polypeptide of approximately 1,000 residues. Every third residue is glycine. Thus the approximate formula for a chain is $(\text{-Gly-X-Y-})_{333}$. A collagen molecule consists of three chains twisted about each other in a right-handed helix with the glycine residues at the central crossing points.

The collagen molecules aggregate in a parallel, staggered fashion to produce fibrils, which range 10–200 nm in diameter depending on the collagen type. The fibrils are visible in the extracellular matrix of connective tissue by electron microscopy, which reveals a banded structure resulting from the staggered nature of collagen molecules in the fibril, each molecule being displaced from its neighbors by 0, 1, 2, 3, or 4 axial stagger lengths of 234 $\pm$ 1 residues. In $(\text{Gly-X-Y})_{333}$ approximately 100 of the X residues are proline, and approximately 100 of the Y residues are 4-hydroxyproline. The cyclic nature of these residues limits rotation and thereby provides stability. Other amino acids in clusters of hydrophobic and charged residues occupy the other positions designated by X and Y. The different primary structures of the α chains are specified by different genes and give rise to at least eight different types of collagen, having different organ distributions. The major types of fibrillar collagen are I, II, and III. Type I is the most abundant. Type II is localized to hyaline cartilage. Nonfibrillar types of collagen, such as type IV, found in basement membranes, are so far not known to be associated with genetic disorders; we will focus on types I and III. Type I collagen consists of two α1(I) chains and one α2(I) chain, whereas type III collagen consists of three α1(III) chains. The α1(I) gene is on chromosome 17, and the α2(I) gene is on chromosome 7.

Some disorders of collagen structure are listed in Table 19.7 (Clin. Corrs. 19.5–19.11).

CLIN. CORR. 19.6
OSTEOGENESIS IMPERFECTA

Osteogenesis imperfecta is the name for a group of at least four clinically, genetically, and biochemically distinguishable disorders all characterized by multiple fractures with resultant bone deformities. Several variants result from mutations producing shortened α(I) chains. In the clearest example a deletion mutation causes absence of about 80 amino acids in the α1(I) chain. The shortened α1(I) chains are synthesized, because the mutation leaves the reading frame in register. The short α1(I) chains associate with normal α1(I) and α2(I) chains, thereby preventing normal triple helix formation, with resultant degradation of all of the chains, a phenomenon aptly named "protein suicide." Three-fourths of all of the collagen molecules formed have at least one short (defective) α1(I) chain, an amplification of the effect of a heterozygous gene defect. The amplification results from the subunit structure of collagen.

In addition to hydroxyproline, finished collagen contains hydroxylysine and glycosylated hydroxylysine. The existence of modifications in coded amino acid residues signals enzymatic mechanisms for their bio-

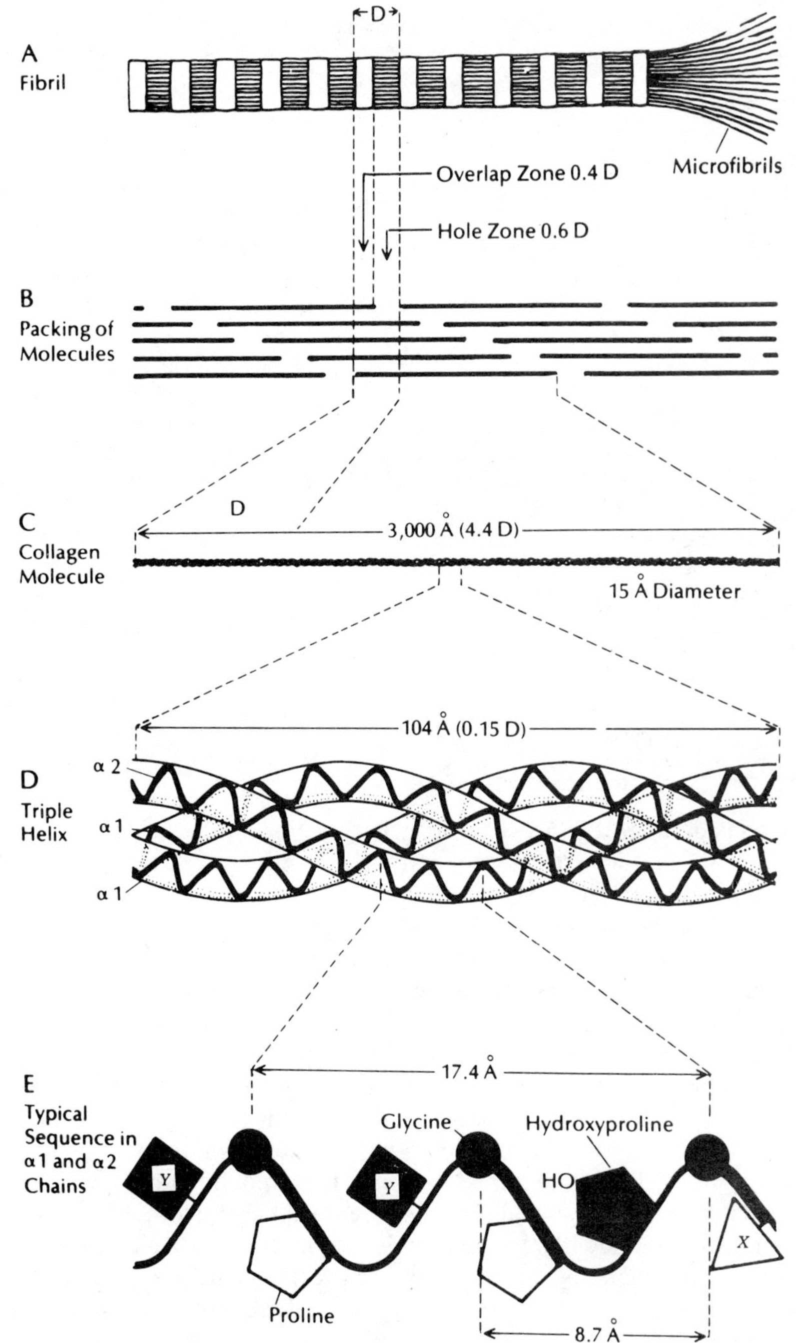

Figure 19.16
Collagen structure, illustrating the regularity of primary sequence, the left-handed α helix, the right-handed triple helix, the 300-nm molecule, and the organization of molecules in a typical fibril, within which the collagen molecules are cross-linked.
Figure by B. Tagawa. Reproduced with permission from D. J. Prockop and N. A. Guzman. *Hosp. Proct.* 12:61, 1977.

synthesis and opens possibilities for defects in those mechanisms. Such defects would be expected to cause disorders, just as do direct mutations in primary structure. More than 100 amino acid residues in each proα chain are modified in reactions requiring at least eight specific enzymes and several other enzymes.

CLIN. CORR. **19.7**
SCURVY AND HYDROXYPROLINE SYNTHESIS

In scurvy the severe ascorbic acid deficiency causes decreased hydroxyproline synthesis. Collagen containing insufficient hydroxyproline loses temperature stability and is less stable than normal collagen at body temperature. The resultant clinical manifestations are distinctive and understandable: suppression of the orderly growth process of bone in children, poor wound healing, and increased capillary fragility with resultant hemorrhage, particularly in the skin. Severe ascorbic acid deficiency leads secondarily to a decreased rate of procollagen synthesis.

CLIN. CORR. **19.8**
DEFICIENCY OF LYSYLHYDROXYLASE

In type VI Ehlers–Danlos syndrome lysylhydroxylase is deficient. As a result collagen with decreased hydroxylysine content is synthesized, and subsequent cross-linking of collagen fibrils is less stable. The clinical features include marked hyperextensibility of the skin and joints, poor wound healing, and musculoskeletal deformities. Some patients with this form of Ehlers–Danlos syndrome have a mutant form of lysylhydroxylase with a higher Michaelis constant for ascorbic acid than the normal enzyme. Accordingly, they respond to high doses of ascorbic acid.

CLIN. CORR. **19.9**
EHLERS–DANLOS SYNDROME, TYPE VII

In Ehlers–Danlos syndrome, type VII, the skin bruises easily and is hyperextensible, but the major manifestations are dislocations of major joints, such as the hips and knees. The laxity of ligaments is caused by incomplete removal of the amino-terminal propeptide. In one variant the deficiency is in procollagen aminopeptidase. The deficiency occurs in the autosomal recessive dermatosparaxis of cattle, sheep, and cats, in which skin fragility is so extreme as to be lethal. In the other variant the proα2(I) chain has a structural mutation that prevents its normal cleavage by procollagen aminopeptidase.

TABLE 19.7 Some Disorders of Collagen Structure

Disorder	*Collagen Defect*	*Clinical Manifestations*
Ehlers–Danlos IV	Decrease in type III	Arterial, intestinal, or uterine rupture; thin, easily bruised skin
Osteogenesis imperfecta	Decrease in type I	Blue sclerae, multiple fractures, and bone deformities
Scurvy	Decreased hydroxyproline	Poor wound healing, deficient growth; increased capillary fragility
Ehlers–Danlos VI	Decreased hydroxylysine	Hyperextensible skin and joints, poor wound healing, musculoskeletal deformities
Ehlers–Danlos VII	Amino terminal propeptide present	Hyperextensible, easily bruised skin; hip dislocations
Ehlers–Danlos V and cutis laxa	Decreased cross-linking	Skin and joint hyperextensibility
Marfan syndrome	Decreased cross-linking	Skeletal deformities, aortic aneurism, valvular heart disease, ectopia lentis

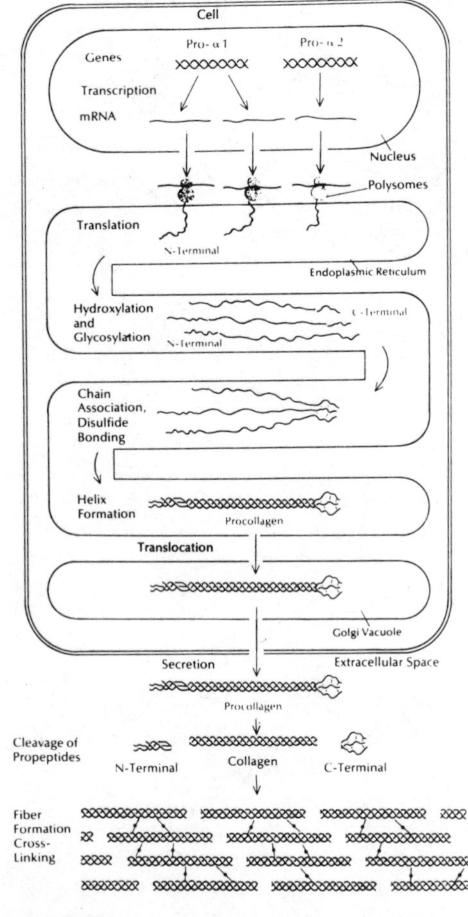

Figure 19.17
Diagram of collagen biosynthesis, showing transcription (only two genes are depicted), translation in the endoplasmic reticulum, various and extensive posttranslational modifications, helix formation, secretion into the extracellular space, cleavage of propeptides, and cross-linking to give fibrils.
Figure by B. Tagawa. Reproduced with permission from D. J. Prockop and N. A. Guzman. *Hosp. Pract.* 12:61, 1977.

Preprocollagen to Procollagen

A consideration of the steps in collagen biosynthesis (Figure 19.17) is necessary for understanding of still other disorders. The initial translation product is a procollagen chain with an amino-terminal propeptide of about 180 amino acid residues and a carboxy-terminal propeptide of about 300 amino acid residues destined not to appear in the final collagen molecule. Actually, a preprocollagen containing a signal sequence directs the transfer of the procollagen into the rough endoplasmic reticulum. The signal sequence is removed by signal peptidase. The first modification of procollagen then takes place within the rough endoplasmic reticulum. The procollagen contains cysteine residues in both terminal propeptides. Those at the carboxy terminal become involved in interchain disulfide bonds, and those at the amino terminal become involved in intrachain disulfide bonds, which are essential for subsequent triple helix formation. Whether or not the disulfide bond formation is enzymatic is unclear. Interestingly, no cysteine and no cystine residues are present in mature type I collagen. Another modification of the carboxy-terminal propeptide is the addition of a mannose-rich oligosaccharide. Another modification required for subsequent triple helix formation is hydroxylation of proline and lysine residues. Hydroxyproline is an amino acid found only in collagen. Three different enzymes are required: prolyl-4-hydroxylase, prolyl-3-hydroxylase, and lysyl hydroxylase. All require ferrous ion and ascorbic acid, and all use O_2 and α-ketoglutarate in the general reaction

$$\text{AA residue} + O_2 + \alpha\text{-ketoglutarate} \xrightarrow{\text{hydroxylase}} \text{HO-AA residue} + CO_2 + \text{succinate}$$

To be a substrate the amino acid residue must be in a particular sequence in a nonhelical region. Thus lysyl hydroxylase acts only on lysyl residues in the Y position of (Gly-X-Y). Similarly, prolyl-4-hydroxylase acts only on a prolyl residue in the Y position. Prolyl-3-hydroxylase is specific for prolyl residues in the X position but only when a hydroxyprolyl residue is already present in the Y position. The structures of hydroxylysine, 4-hydroxyproline, and 3-hydroxyproline are shown in Figure 19.18.

Further modifications of certain of the hydroxylysyl residues are required for subsequent stable helices and must be performed before the collagen molecule assumes the helical conformation. The modifications are glycosylations and require two enzymes, a galactosyltransferase and a glucosyltransferase. Both enzymes require a divalent cation, preferably Mn^{2+}. The galactosyltransferase uses UDPGal as a donor to add galactosyl residues to specific hydroxylysyl residues,

$$\begin{array}{c} CH_2NH_3^+ \\ | \\ HOCH \\ | \\ (CH_2)_2 \\ | \\ -NH-CH-CO- \end{array} + \text{UDPGal} \xrightarrow{Mn^{2+}} \begin{array}{c} CH_2NH_3^+ \\ | \\ \text{Gal}-OCH \\ | \\ (CH_2)_2 \\ | \\ -NH-CH-CO- \end{array} + \text{UDP}$$

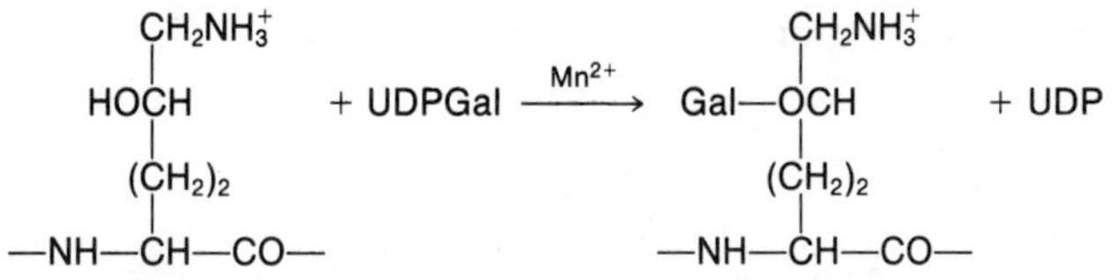

and the glucosyltransferase uses UDPG as a donor to add glucosyl residues to certain of the newly formed galactosylhydroxylsyl residues.

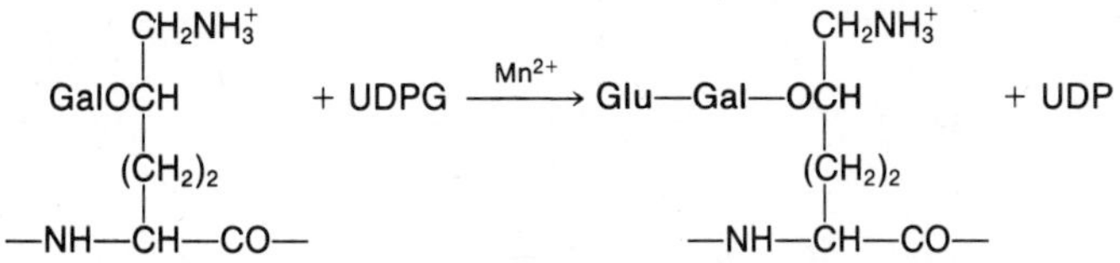

$$\begin{array}{c} CH_2NH_3^+ \\ | \\ \text{GalOCH} \\ | \\ (CH_2)_2 \\ | \\ -NH-CH-CO- \end{array} + \text{UDPG} \xrightarrow{Mn^{2+}} \begin{array}{c} CH_2NH_3^+ \\ | \\ \text{Glu}-\text{Gal}-OCH \\ | \\ (CH_2)_2 \\ | \\ -NH-CH-CO- \end{array} + \text{UDP}$$

When collagen helices do not form because of any of the defects mentioned in Clin. Corrs. 19.5–19.9, the nonfunctional collagen precursor is secreted slowly by the cells. Such a product is essentially gelatin, the familiar protein that we obtain for food by the heat denaturation of collagen. Normal helical procollagen with completed disulfide bonds, hydroxylations, and glycosylations now passes through the Golgi complex and is extruded from the cell. Further processing of this helical procollagen product is extracellular.

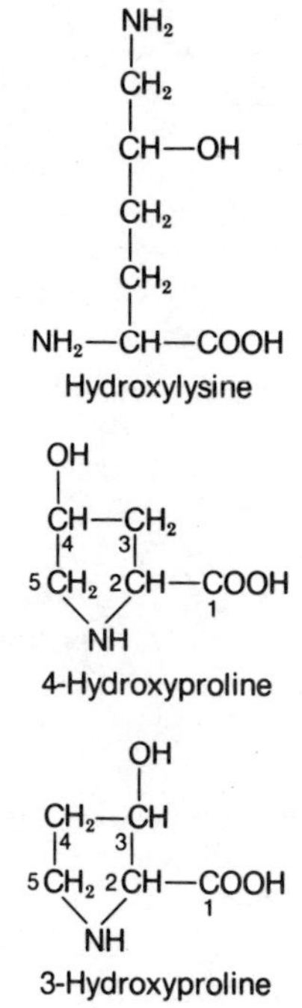

Procollagen to Collagen

The first steps of extracellular processing involve the removal of the propeptides. Two enzymes are involved: procollagen carboxypeptidase, which removes the carboxy-terminal propeptide, and procollagen aminopeptidase, which removes the amino-terminal propeptide. After normal removal of propeptides spontaneous self-assembly of collagen molecules into collagen fibrils ensues. The amino acid sequence alone provides the necessary information for correct fibril formation, and the fibrils formed are remarkably different for the different types of collagen. For example, type I and type II collagen have a totally different appearance by electron microscopy.

Although the collagen at this stage is indistinguishable by electron microscopy from mature collagen, intermolecular cross-linking is required for the necessary tensile strength of mature collagen. The first step and the only enzymatic step in cross-link formation is the oxidative deamination of certain lysyl and hydroxylysyl residues to give the corresponding aldehydes. This reaction is catalyzed by lysyl oxidase, an enzyme requiring cuprous ion and molecular oxygen.

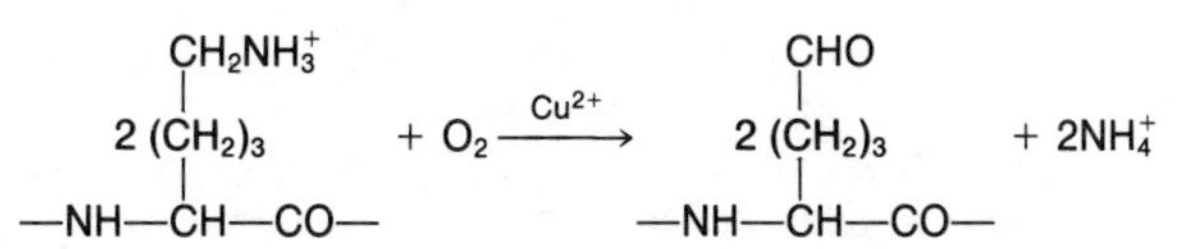

$$2\begin{array}{c} CH_2NH_3^+ \\ | \\ (CH_2)_3 \\ | \\ -NH-CH-CO- \end{array} + O_2 \xrightarrow{Cu^{2+}} 2\begin{array}{c} CHO \\ | \\ (CH_2)_3 \\ | \\ -NH-CH-CO- \end{array} + 2NH_4^+$$

$$\begin{array}{c} NH_2 \\ | \\ CH_2 \\ | \\ CH-OH \\ | \\ CH_2 \\ | \\ CH_2 \\ | \\ NH_2-CH-COOH \end{array}$$

Hydroxylysine

$$\begin{array}{l} \quad OH \\ \quad | \\ \quad CH-CH_2 \\ \quad |4 \quad 3| \\ 5\,CH_2 \;\; 2\,CH-COOH \; (1) \\ \quad \diagdown \; \diagup \\ \quad NH \end{array}$$

4-Hydroxyproline

$$\begin{array}{l} \qquad\quad OH \\ \qquad\quad | \\ \quad CH_2-CH \\ \quad |4 \quad 3| \\ 5\,CH_2 \;\; 2\,CH-COOH \; (1) \\ \quad \diagdown \; \diagup \\ \quad NH \end{array}$$

3-Hydroxyproline

Figure 19.18
Structures of hydroxylysine, 4-hydroxyproline, and 3-hydroxyproline.

CLIN. CORR. 19.10
CUTIS LAXA

In type V Ehlers–Danlos syndrome and in some forms of cutis laxa there is a deficiency in lysyl oxidase with consequent cross-linking defects in both collagen and elastin. In cutis laxa the defect is manifested in loose skin, which appears excessively wrinkled, hangs in folds, and lacks the elastic qualities of the hyperextensible skin in Ehlers–Danlos syndrome. In type V Ehlers–Danlos syndrome the skin hyperextensibility is accompanied by joint hypermobility, but the latter is limited to the digits. Copper-deficient animals have deficient cross-linking of elastin and collagen, apparently because of the requirement for cuprous ion by lysyl oxidase. A woman taking high doses of the copper-chelating drug, D-penicillamine, gave birth to an infant with an acquired Ehlers–Danlos-like syndrome, which subsequently cleared. Side effects of D-penicillamine therapy include poor wound healing and hyperextensible skin.

CLIN. CORR. 19.11
MARFAN SYNDROME

The Marfan syndrome is an autosomal dominant disorder (probably a group of disorders) characterized by skeletal, eye, and heart manifestations and often ending with cardiovascular catastrophy, especially rupture of the aorta. The organ and tissue distribution of defects have pointed to type I collagen as the basis for this disorder. Many studies have indicated a defect in cross-linking, but no defect has ever been found in lysyl oxidase, the enzyme essential for cross-linking, in Marfan patients. The explanation must lie in a structural mutation. That has been found in one patient whose $\alpha 2(I)$ chains have an insert of about 20 amino acid residues near the carboxy terminal. The insert alters the register of the mutant $\alpha 2(I)$ chain relative to the normal chains, thereby interfering with normal cross-linking.

The aldehyde groups undergo two types of linking reactions, probably neither one of which is enzymatic. One type of linking is aldol condensation, mostly between two aldehyde groups on the same α chain. The second type of linkage is Schiff base formation between aldehydes and ε-amino groups of lysyl, hydroxylysyl, and glycosylated hydroxylysyl residues, predominantly in neighboring α chains. The Schiff bases are not stable but serve as precursors to further reactions, leading to stable cross-links. For example, when a hydroxylysyl residue is the source of the aldehyde residue (hydroxyallysyl residue), the compound derived by Schiff base formation with a lysyl residue on another chain can be stabilized by Amadori rearrangement, as shown:

Lysyl residue: $(CH_2)_4$–NH_2 ; Hydroxyallysyl residue: CHO–CHOH–$(CH_2)_2$ $\xrightarrow{\text{cross-linking}}$ Schiff base: $(CH_2)_4$–N=CH–CHOH–$(CH_2)_2$ $\xrightarrow{\text{rearrangement}}$ Stable cross-link: $(CH_2)_4$–NH–CH_2–C=O–$(CH_2)_2$

Ultimately, the aldehydes derived from hydroxylysyl residues give the most stable cross-links, a point underscoring the importance of hydroxylysyl formation.

The disorders of collagen biosynthesis serve well to illustrate the consequences of different types of mutational defects: those affecting regulation of synthesis with formation of inadequate quantities of normal collagen, those with defects in procollagen primary structure, and those characterized by defects in modifying enzymes. These disorders also serve as a rich source of information on the molecular basis for genetic heterogeneity.

BIBLIOGRAPHY

Biosynthesis of Proteins

Austin, S. A., and Kay, J. E. Translational control of protein synthesis in eukaryotes. *Essays Biochem.* 18:79, 1982.

Caskey, C. T. Peptide chain termination. *Trends Biochem. Sci.* 5:234, 1980.

Clark, B. The elongation step of protein biosynthesis. *Trends. Biochem. Sci.* 5:207, 1980.

Clemens, M. Enzymes and toxins that regulate protein synthesis. *Nature* 310:727, 1984.

Fersht, A. R. Enzymic editing mechanisms and the genetic code. *Proc. R. Soc. Lond.* B212:351, 1981.

Hunt, T. The initiation of protein synthesis. *Trends Biochem. Sci.* 5:178, 1980.

Kozak, M. Comparison of initiation of protein synthesis in prokaryotes, eukaryotes, and organelles. *Microbiol. Rev.* 47:1, 1983.

Lagerkvist, U. Codon misreading: A restriction operative in the evolution of the genetic code. *Am. Sci.* 68:192, 1980.

Leader, D. P. Protein biosynthesis on membrane-bound ribosomes. *Trends Biochem. Sci.* 4:205, 1979.

Pollack, M. Pseudomonas aeruginosa exotoxin A. *N. Engl. J. Med.* 302:1360, 1980.

Uy, R., and Wold, F. Post-translational covalent modification of proteins. *Science* 198:890, 1977.

Wickner, W. Assembly of proteins into membranes. *Science* 210:861, 1980.

Special Proteins

Eyre, D. R. Collagen: molecular diversity in the body's protein scaffold. *Science* 207:1315, 1980.

Nienhuis, A. W., and Benz, E. J. Jr. Regulation of hemoglobin synthesis during the development of the red cell. *N. Engl. J. Med.* 297:1318, 1977.

Disorders of Protein Synthesis

Prockop, D. J., and Kivirikko, K. I. Heritable diseases of collagen. *N. Engl. J. Med.* 311:376, 1984.

Spritz, R. A., and Forget, B. G. The thalassemias: molecular mechanisms of human genetic disease. *Am. J. Hum. Genet.* 35:333, 1983.

QUESTIONS

J. BAGGOTT AND C. N. ANGSTADT

*Q*uestion *T*ypes are described inside the front cover.

1. (QT1) Degeneracy of the genetic code denotes the existence of:
   A. multiple codons for a single amino acid.
   B. codons consisting of only two bases.
   C. base triplets that do not code for any amino acid.
   D. different protein synthesis systems in which a given triplet codes for different amino acids.
   E. codons that include one or more of the "unusual" bases.
2. (QT2) Deletion of a single base from a coding sequence of mRNA may result in a polypeptide product with:
   1. a sequence of amino acids that differs from the sequence found in the normal polypeptide.
   2. more amino acids.
   3. fewer amino acids.
   4. a single amino acid replaced by another.
3. (QT1) During initiation of protein synthesis:
   A. methionyl-tRNA appears at the A site of the 80S initiation complex.
   B. eIF-3 and the 40S ribosomal subunit participate in forming the entry complex.
   C. eIF-2 is phosphorylated by GTP.
   D. the same methionyl-tRNA is used as is used during elongation.
   E. a complex consisting of mRNA, the 60S ribosomal subunit, and certain initiation factors is formed.
4. (QT2) Requirements for protein synthesis include:
   1. mRNA.
   2. ribosomes.
   3. GTP.
   4. 20 different amino acids in the form of aminoacyl-tRNAs.
5. (QT2) During the elongation stage of eucaryotic protein synthesis:
   1. the incoming aminoacyl-tRNA binds to the A site.
   2. a new peptide bond is synthesized by peptidyltransferase in a GTP-requiring reaction.
   3. the peptide, still bound to a tRNA molecule, is translocated to a different site on the ribosome.
   4. streptomycin can cause premature release of the incomplete peptide.
6. (QT1) Diphtheria toxin and *Pseudomonas* toxin differ from all antibiotic inhibitors of protein synthesis discussed in this chapter in that they:
   A. act catalytically.
   B. release incomplete polypeptide chains from the ribosome.
   C. inhibit translocase
   D. prevent release factor from recognizing termination signals.
   E. have no clinical usefulness.

Protein synthesis in:

A. Mitochondria	C. Both
B. Eucaryotic free polysomes	D. Neither

7. (QT4) UGA is a stop signal.
8. (QT4) Inhibited by chloramphenicol.
9. (QT4) AUG codes for initiation and for internal methionyl residues.
10. (QT4) Synthesis of proteins for export.
11. (QT2) Posttranslational modification of polypeptides can include:
    1. removal of a signal peptide.
    2. removal of one or more terminal amino acid residues.
    3. removal of a peptide from an internal region.
    4. addition of one or more terminal amino acid residues.
12. (QT2) Hydroxylation of specific prolyl residues during collagen synthesis requires:
    1. Fe^{2+}.
    2. O_2.
    3. ascorbic acid.
    4. succinate.
13. (QT1) In the formation of an aminoacyl-tRNA:
    A. ADP and P_i are products of the reaction.
    B. aminoacyl adenylate appears in solution as a free intermediate.
    C. the aminoacyl-tRNA synthetase is believed to recognize and hydrolyze incorrect aminoacyl-tRNA's it may have produced.
    D. there is a separate aminoacyl-tRNA synthetase for every amino acid appearing in the final, functional protein.
    E. there is a separate aminoacyl-tRNA synthetase for every tRNA species.
14. (QT2) During collagen synthesis, events that occur extracellularly include:
    1. disulfide bond formation.
    2. modification of lysyl residues.
    3. modification of prolyl residues.
    4. hydrolysis of N-terminal and C-terminal peptides from the triple helical procollagen.

A. Intracellular lysyl oxidation	C. Both
B. Extracellular lysyl oxidation	D. Neither

15. (QT4) O_2 is required.
16. (QT4) Cu^{2+} is required.
17. (QT4) α-Ketoglutarate is required.

ANSWERS

1. A A is the definition of degeneracy (p. 739). B and E are not known to occur, although sometimes tRNA reads only the first two bases of a triplet (wobble), and sometimes unusual bases occur in anticodons (p. 740). C denotes the stop (nonsense) codons (p. 739). D is a deviation from universality of the code, as found in mitochondria (p. 739).
2. A 1, 2, and 3 true. Deletion of a single base causes a frameshift mutation (p. 742). The frameshift would destroy the original stop codon; another one would be generated before or after the original location. In contrast, replacement of one base by another would cause replacement of one amino acid (missense mutation), unless a stop codon is thereby generated (p. 741).
3. B A: methionyl-$tRNA_i^{met}$ appears at the P site. C: phosphorylation of eIF-2 inhibits initiation. D: methionyl-$tRNA_m^{met}$ is used internally. E: mRNA associates first with the 40S subunit (p. 748).
4. E 1, 2, 3, and 4 true. See p. 744. Absence of a required aminoacyl-tRNA stops protein synthesis (p. 751).
5. B 1 and 3 true. 2: Peptide bond formation requires no energy source other than the aminoacyl-tRNA (p. 750). 4: Streptomycin inhibits formation of the procaryotic 70S initiation complex (analogous to the eucaryotic 80S complex) and causes misreading of the genetic code when the initiation complex is already formed (p. 751).
6. A These toxins catalyze the formation of an ADP ribosyl derivative of translocase, which irreversibly inactivates the translocase (p. 752). Erythromycin inhibits translocate stoichiometrically, not catalytically (p. 751). The toxins are, of course clinically useless, but so are cycloheximide and puromycin (p. 751).
7. B UGA is a stop signal in all systems except mitochondria, in which it codes for tryptophan (p. 739).

8. A Mitochondrial and procaryotic peptidyltransferases are inhibited by chloramphenicol (pp. 751, 754).
9. C See p. 739.
10. D Proteins synthesized for export are synthesized on ribosomes bound to the endoplasmic reticulum (p. 753).
11. E All statements are true. See pp. 756 for examples of each.
12. A 1, 2, and 3 required. Succinate is a product; α-ketoglutarate is the reactant (p. 760).
13. C A and B: ATP and the amino acid react to form an enzyme-bound aminoacyl adenylate; PP_i is released into the medium (p. 742). C: Bonds between a tRNA and an incorrect smaller amino acid may form but are rapidly hydrolyzed (Figure 19.3, p. 743). D: Some amino acids, such as hydroxyproline and hydroxylysine, arise by co- or posttranslational modification (p. 757). E: An aminoacyl-tRNA synthetase may recognize any of several tRNA's specific for a given amino acid (p. 743).
14. C 2 and 4 are extracellular; See p. 761. Some modification of lysyl residues also occurs intracellularly. (p. 760).
15. C Both enzymes require molecular oxygen (pp. 760, 761).
16. B The intracellular enzyme requires Fe^{2+} (pp. 760, 761).
17. A Only the intracellular process requires α-ketoglutarate (p. 760).

20

Regulation of Gene Expression

JOHN E. DONELSON

20.1 OVERVIEW

In order to survive, a living cell must be able to respond to changes in its environment. One of many ways in which cells adjust to these changes is to alter the expression of specific genes, which, in turn, affects the number of the corresponding proteins. This chapter will focus on some of the molecular mechanisms that determine when a given gene will be expressed and to what extent. The attempt to understand how the expression of genes is regulated is one of the most active areas of biochemical research today.

It makes sense for a cell to vary the amount of a given gene product that is available under different conditions. For example, the bacterium *Escherichia coli* (*E. coli*) contains genes for about 3,000 different proteins, but it does not need to synthesize all of these proteins at the same time. Therefore, it regulates the number of molecules of these proteins that are made. The classic illustration of this phenomenon, as we shall see, is the regulation of the number of β-galactosidase molecules in the cell. This enzyme converts the disaccharide lactose to the monosaccharides, glucose and galactose. When *E. coli* is growing in a medium containing glucose as the carbon source, β-galactosidase is not required and only about five molecules of the enzyme are present in the cell. When lactose is the sole carbon source, however, 5,000 or more molecules of β-galactosidase occur in the cell. Clearly, the bacteria respond to the need to metabolize lactose by increasing the synthesis of β-galactosidase molecules. If lactose is removed from the medium, the synthesis of this enzyme stops as rapidly as it began.

The complexity of eucaryotic cells means that they have even more extensive mechanisms of gene regulation than do procaryotic cells. The differentiated cells of higher organisms have a much more complicated physical structure and often a more specialized biological function as determined by the expression of their genes. For example, insulin is synthesized in specific cells of the pancreas and not in kidney cells even though the nuclei of all cells of the body contain the insulin genes. Molecular regulatory mechanisms facilitate the expression of insulin in the pancreas and prevent its synthesis in the kidney and other cells of the body. In addition, during development of the organism the appearance or disappearance of proteins in specific cell types are tightly controlled with respect to the timing and sequence of the developmental events.

As expected from the differences in complexities, far more is understood about the regulation of genes in procaryotes than in eucaryotes. However, studies on the control of gene expression in procaryotes often provide exciting new ideas that can be tested in eucaryotic systems. And, sometimes, discoveries about eucaryotic gene structure and regulation alter the interpretation of data on the control of procaryotic genes.

In this chapter several of the best studied examples of gene regulation in bacteria will be discussed, followed by some illustrations of the organization and regulation of related genes in the human genome. Finally, the use of recombinant DNA techniques to express some human genes of clinical interest will be presented.

20.2 THE UNIT OF TRANSCRIPTION IN BACTERIA: THE OPERON

The single *E. coli* chromosome is a circular double-stranded DNA molecule of about 4 million base pairs. Most of the approximately 3,000 *E. coli* genes are not distributed randomly throughout this DNA; instead, the genes that code for the enzymes of a specific metabolic pathway are clustered in one region of the DNA. In addition, genes for associated structural proteins, such as the 70 or so proteins that comprise the ribosome, are frequently adjacent to one another. The members of a set of clustered genes are usually coordinately controlled; they are transcribed together to form a "polycistronic" mRNA species that contains the coding sequences for several proteins. The term *operon* is used to describe the complete regulatory unit of a set of clustered genes. An operon includes the adjacent *structural genes* that code for the related enzymes or associated proteins, a *regulatory gene* or genes that code for regulator protein(s), and *control elements* that are sites on the DNA near the structural genes at which the regulator proteins act. Figure 20.1 shows a partial genetic map of the *E. coli* chromosome that gives the locations of the structural genes of some of the different operons.

When transcription of the structural genes of an operon increases in response to the presence of a specific substrate in the medium, the effect

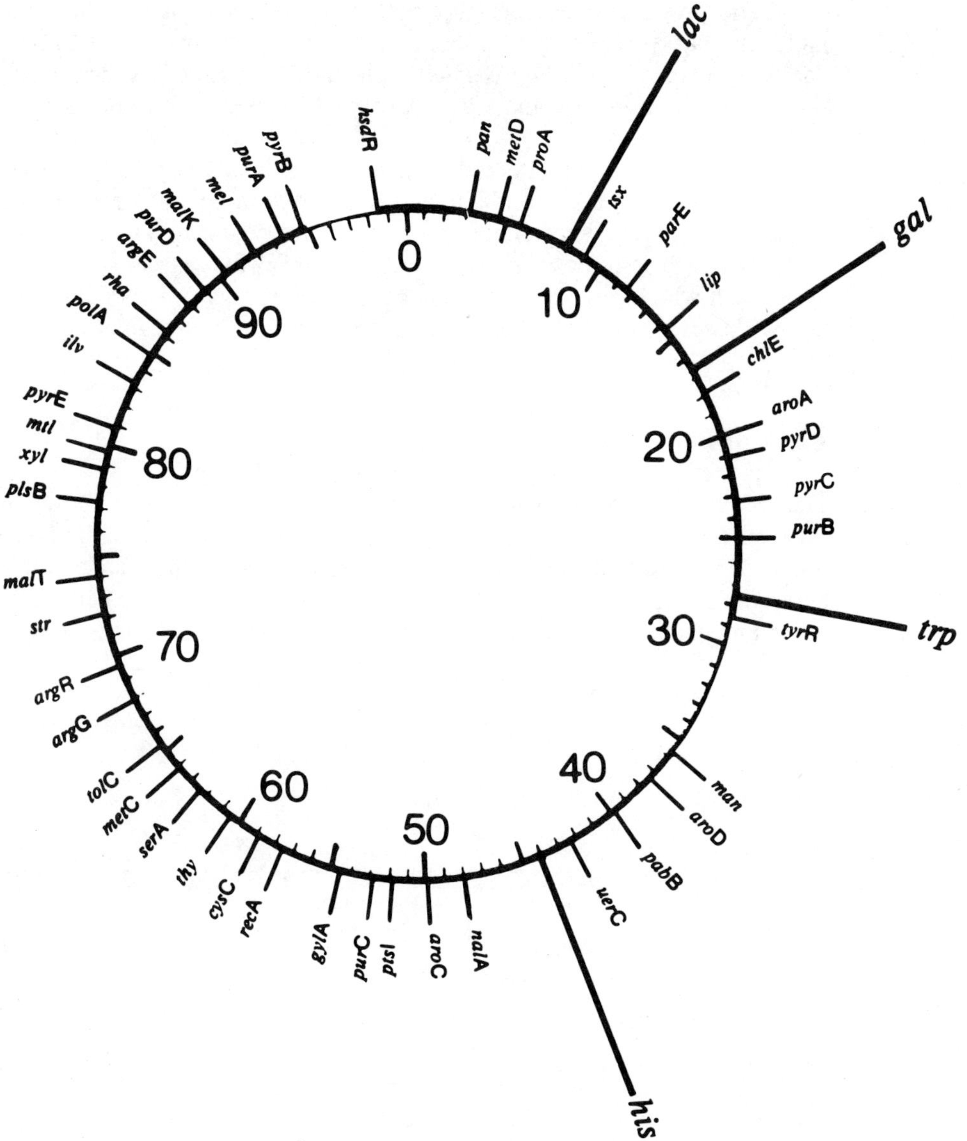

Figure 20.1
***Partial genetic map of* E. coli.**
The locations of only a few of the genes identified and mapped in E. coli *are shown here. Four operons discussed in this chapter are indicated.*

Reproduced with permission from G. S. Stent and R. Calendar, *Molecular Genetics, An Introductory Narrative*. W. H. Freeman, San Francisco: 1978, p. 289; modified from B. J. Bachmann, K. B. Low, and A. L. Taylor, *Bacteriol. Rev.*, 40:116, 1976.

is known as *induction*. The increase in transcription of the β-galactosidase gene when lactose is the sole carbon source is an example of induction. Bacteria can also respond to nutritional changes by quickly turning off the synthesis of enzymes that are no longer needed. As will be described below, *E. coli* synthesizes the amino acid tryptophan as the end product of a specific biosynthetic pathway. However, if tryptophan is supplied in the medium, the bacteria do not need to make it themselves, and the synthesis of the enzymes for this metabolic pathway is stopped. This is called *repression*. It permits the bacteria to avoid using their energy for making unnecessary and even harmful proteins.

Induction and repression are manifestations of the same phenomenon. In one case the bacterium changes its enzyme composition so that it can utilize a specific substrate in the medium; in the other it reduces the number of enzyme molecules so that it doesn't overproduce a specific metabolic product. The signal for both types of regulation is the small molecule that is a substrate for the metabolic pathway or a product of the pathway, respectively. These small molecules are called *inducers* when they stimulate induction and *corepressors* when they cause repression to occur.

Section 20.3 below will describe in detail the lactose operon, the best studied example of a set of inducible genes. Section 20.4 will present the tryptophan operon, an example of a repressible operon. The following sections will briefly describe some other operons as well as some gene systems in which physical movement of the genes themselves within the DNA, that is, gene rearrangements, plays a role in their regulation.

20.3 THE LACTOSE OPERON OF *E. COLI*

The lactose operon contains three adjacent structural genes as shown in Figure 20.2. *LacZ* codes for the enzyme β-galactosidase, which is composed of four identical subunits of 1,021 amino acids. *LacY* codes for a permease, which is a 275-amino acid protein that occurs in the cell membrane and participates in the transport of sugars, including lactose, across

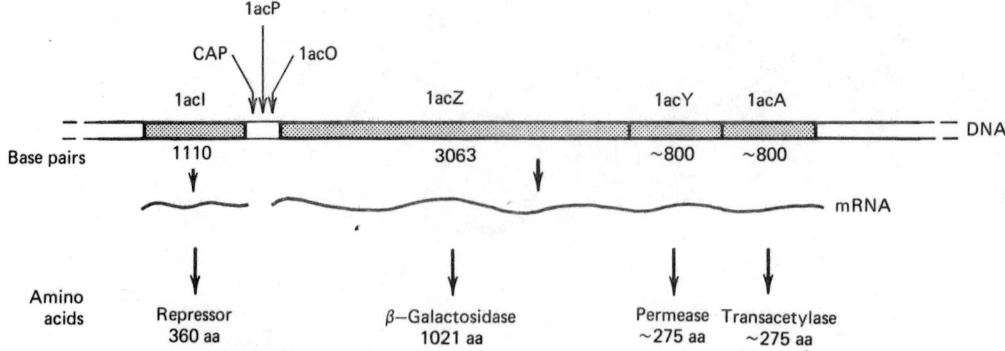

Figure 20.2
***Lactose operon of* E. coli.**
The lactose operon is composed of the lacI *gene, which codes for a repressor, the control elements of* CAP, lacP, *and* lacO; *and three structural genes,* lacZ, lacY, *and* lacA, *which code for β-galactosidase, a permease and a transacetylase, respectively. The* lacI *gene is transcribed from its own promotor. The three structural genes are transcribed from the promotor,* lacP, *to form a polycistronic mRNA from which the three proteins are translated.*

the membrane. The third gene, *lacA*, codes for β-galactoside transacetylase, a 275-amino acid enzyme that transfers an acetyl group from acetyl CoA to β-galactosides. Interestingly, of these three proteins, only β-galactosidase actually participates in a known metabolic pathway. However, the permease is clearly important in the utilization of lactose since it is involved in transporting lactose into the cell. The significance of the acetylation reaction has long been a mystery, but it may be associated with the detoxification and excretion reactions of nonmetabolized analogs of β-galactosides.

Mutations in *lacZ* or *lacY* that destroy the function of the β-galactosidase or the permease prevent the cells from breaking down lactose or acquiring it from the medium, respectively. Curiously, mutations in *lacA* that destroy the acetylase activity do not seem to have an identifiable effect on cell growth and division. Perhaps there are other related enzymes in the cell that serve as backups for this enzyme, or perhaps it has a specific unknown function that is required only under certain conditions.

A single mRNA species containing the coding sequences of all three structural genes is transcribed from a promoter that occurs just upstream from the *lacZ* gene. The induction of these three genes occurs during the initiation of their transcription. Without the inducer, transcription of the gene cluster occurs only at a very low level. In the presence of the inducer, transcription begins at the promoter, called *lacP,* and goes through all three genes to a transcription terminator located slightly beyond the end of *lacA*. Therefore, the genes are *coordinately expressed;* either all three are transcribed in unison or none is transcribed.

The presence of the three coding sequences on the same mRNA molecule means that the *relative* amounts of the three proteins is always the same under varying conditions of induction. An inducer that causes a high rate of transcription will result in a high level of all three proteins; an inducer that stimulates only a little transcription of the operon will result in a low level of the proteins. The inducer can be thought of as a molecular switch that influences synthesis of the single mRNA species for all three genes. The number of molecules of each protein in the cell may be different, but this does not reflect differences in transcription; it reflects differences in translation rates of the coding sequences or in degradation of the proteins themselves.

The lactose mRNA is very unstable; it is degraded with a half-life of about 3 min. Therefore, expression of the operon can be altered very quickly. Transcription ceases as soon as the inducer is no longer present, the existing mRNA disappears within a few minutes, and the cell stops making the proteins.

The Repressor of the Lactose Operon

The regulatory gene of the lactose operon, *lacI,* was originally identified by mutations that mapped outside the *lacZYA* region but affected the transcription of all three of these structural genes. *LacI* is now known to code for a protein whose only function is to control the transcription initiation of the three *lac* structural genes. This regulator protein is called the *lac repressor*. The *lacI* gene is located just in front of the controlling elements for the *lacZYA* gene cluster. However, it is not obligatory that a regulatory gene be physically close to the gene cluster that it regulates. In some of the other operons it is not. Transcription of *lacI* is not regulated; instead, this single gene is always transcribed from its own promoter at a low rate that is relatively independent of the cell's status. Therefore, the affinity of the *lacI* promoter for RNA polymerase seems to be the only factor involved in its transcription initiation.

The *lac* repressor is initially synthesized as monomers of 360 amino acids that associate to form a tetramer, the active form of the repressor. Usually there are about 10 tetramers per cell. The repressor has a strong affinity for a specific DNA sequence that lies between *lacP* and the start of *lacZ*. This sequence is called the *operator* and is designated *lacO*. The operator overlaps the promoter somewhat so that the presence of the repressor bound to the operator physically prevents RNA polymerase from binding to the promoter and initiating transcription.

In addition to recognizing and binding to the *lac* operator DNA sequence, the repressor also has a strong affinity for the inducer molecules of the *lac* operon. Each monomer has a binding site for an inducer molecule. The binding of inducer to the monomers causes an allosteric change in the repressor that greatly lowers its affinity for the operator sequence (Figure 20.3). In other words, when inducer molecules are bound to their sites on the repressor, a conformational change in the repressor occurs that alters the binding site for the operator. The result is that the repressor no longer binds to the operator so that RNA polymerase, in turn, can begin transcription from the promoter. A repressor molecule that is already bound to the operator when the inducer becomes available can still bind to the inducer so that the repressor–inducer complex immediately disassociates from the operator.

The study of the lactose operon has been greatly facilitated by the discovery that some small molecules fortuitously serve as inducers but are not actually metabolized by β-galactosidase. Isopropylthiogalactoside (IPTG) is one of several thiogalactosides with this property. They are called *gratuitous inducers*. They bind to the inducer sites on the repressor molecule causing the conformational change but are not cleaved by the induced β-galactosidase. Therefore, they affect the system without themselves being altered (metabolized) by it. If it were not possible to manipu-

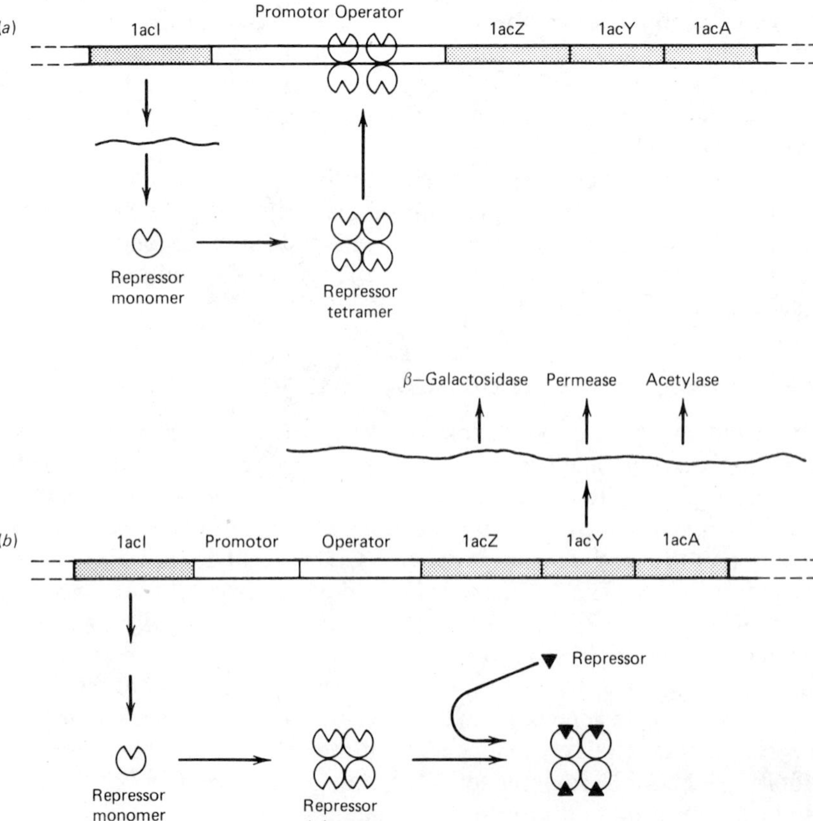

Figure 20.3
Control of the* lac *operon.
(a) *The repressor tetramer binds to the operator and prevents transcription of the structural genes.* (b) *The inducer binds to the repressor tetramer, which prevents the repressor from binding to the operator. Transcription of the three structural genes can occur from the promotor.*

late the system with these gratuitous inducers experimentally, it would have been much more difficult to reach our current understanding of the lactose operon in particular and bacterial gene regulation in general.

The product of the *lacI* gene, the repressor protein, acts in *trans;* that is, it is a diffusible product that moves through the cell to its site of action. Therefore, mutations in the *lacI* gene can exert an effect on the expression of other genes located far away or even on genes located on different DNA molecules. *LacI* mutations can be of several types. One class of mutations changes or deletes amino acids of the repressor that are located in the binding site for the inducer. These changes interfere with the interaction between the inducer and the repressor but do not affect the affinity of the repressor for the operator. Therefore, the repressor is always bound to the operator, even in the presence of inducer, and the *lacZYA* genes are never transcribed above a very low basal level. Another class of *lacI* mutations changes the amino acids in the operator binding site of the repressor. Most of these mutations lessen the affinity of the repressor for the operator. That means that the repressor does not bind to the operator and the *lacZYA* genes are always being transcribed. These mutations are called *repressor-constitutive* mutations because the *lac* genes are permanently turned on. Interestingly, a few rare *lacI* mutants actually increase the affinity of the repressor for the operator over that of wild type repressor. In these cases inducer molecules can still bind to the repressor, but they are less effective in releasing the repressor from the operator.

The repressor-constitutive mutants illustrate the features of a negative control system. An active repressor, in the absence of an inducer, shuts off the expression of the *lac* structural genes. An inactive repressor results in the constitutive, unregulated, expression of these genes. It is possible, using the recombinant DNA techniques described in Chapter 17, to introduce into constitutive *lacI* mutant cells a recombinant plasmid containing the wild type *lacI* gene (but not the rest of the *lac* operon). Therefore, these cells have one wild type and one mutant *lacI* gene and will synthesize both active and inactive repressor molecules. Under these conditions, normal wild type regulation of the lactose operon occurs. In genetic terms, the wild type induction is dominant over the mutant constitutivity. This is the main feature of a negative control system.

The Operator of the Lactose Operon

The known control elements in front of the structural genes of the lactose operon are the operator and the promoter. The operator was originally identified, like the *lacI* gene, by mutations that affected the transcription of the *lacZYA* region. Some of these mutations also result in the constitutive synthesis of lac mRNA; that is, they are operator-constitutive mutations. In these cases the operator DNA sequence has undergone one or more base pair changes so that the repressor no longer binds as tightly to the sequence. Thus, the repressor is less effective in preventing RNA polymerase from initiating transcription.

In contrast to mutations in the *lacI* gene that affect the diffusible repressor, mutations in the operator do not affect a diffusible product. They exert their influence on the transcription of *only* the three *lac* genes that lie immediately downstream of the operator on the same DNA molecule. This means that if a second *lac* operon is introduced into a bacterium on a recombinant plasmid, the operator of one operon does not influence action on the other operon. Therefore, an operon with a wild type operator will be repressed under the usual conditions, whereas in the same bacterium a second operon that has an operator-constitutive mutation will be continuously transcribed.

Operator mutations are frequently referred to as *cis-dominant* to emphasize that these mutations affect only adjacent genes on the same DNA molecule and are not influenced by the presence in the cell of other copies of the unmutated sequence. *Cis*-dominant mutations occur in DNA sequences that are *recognized* by proteins rather than sequences that *code* for the diffusible proteins. *Trans-dominant* mutations occur in genes that specify the diffusible products. Therefore, *cis*-dominant mutations also occur in promoter and transcription termination sequences, whereas *trans*-dominant mutations occur in the genes for the subunit proteins of RNA polymerase, the ribosomes, and so on.

Figure 20.4 shows the sequence of both the *lac* operator and promoter. The operator sequence has an axis of dyad symmetry. The sequence of the upper strand on the left side of the operator is nearly identical to the lower strand on the right side; only three differences occur between these inverted DNA repeats. This symmetry in the DNA recognition sequence reflects symmetry in the tetrameric repressor. It probably facilitates the tight binding of the subunits of the repressor to the operator, although this has not been definitively demonstrated. A common feature of many protein-binding or recognition sites on double-stranded DNA is a dyad symmetry in the nucleotide sequence.

The 30 base pairs that constitute the *lac* operator are an extremely small fraction of the total *E. coli* genome of 4 million base pairs and occupy an even smaller fraction of the total volume of the cell. Therefore, it would seem that the approximately 10 tetrameric repressors in a cell might have trouble finding the *lac* operator if they just randomly diffuse about the cell. Although this remains a puzzling consideration, there are factors that confine the repressor to a much smaller space than the entire volume of the cell. First, it probably helps that the repressor gene is very close to the *lac* operator. This means that the repressor doesn't have far to diffuse if its translation begins before its mRNA is fully synthesized. Second, and more importantly, the repressor possesses a low general affinity for *all* DNA sequences. When the inducer binds to the repressor, its affinity for the operator is reduced about a 1,000-fold, but its low affinity for random DNA sequences is unaltered. Therefore, all of the repressors of the cell probably spend the majority of the time in loose association with the DNA. As the binding of the inducer releases a repressor molecule from the operator, it quickly reassociates with another nearby region of the DNA. Therefore, induction redistributes the repressor on the DNA rather than generates freely diffusing repressor molecules. This confines the repressor to a smaller volume within the cell.

Another question is how does lactose enter a *lac* repressed cell in the first place if the *lacY* gene product, the permease, is repressed yet is

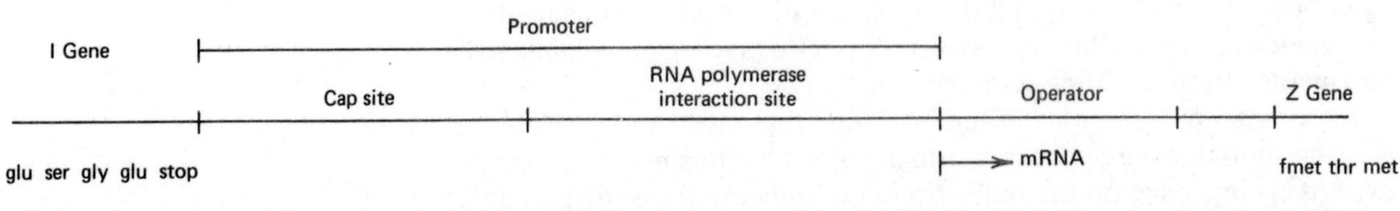

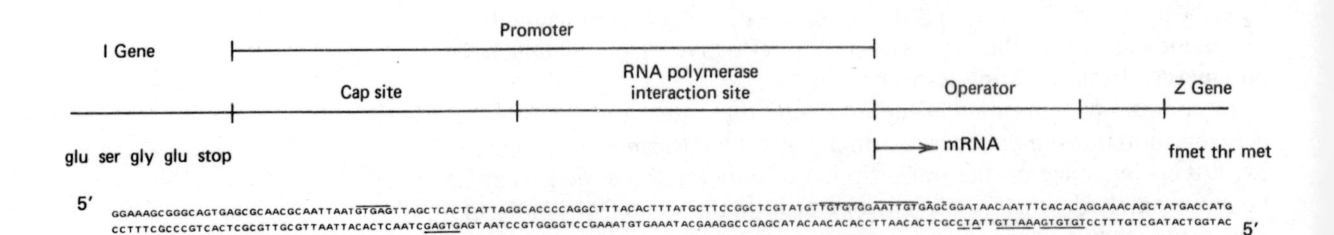

Figure 20.4
Nucleotide sequence of the control elements of the lactose operon.
The end of the I gene (coding for the lactose repressor) and the beginning of the Z gene (coding for β-galactosidase) are also shown. Lines above and below the sequence indicate symmetrical sequences within the CAP site and the operator.

required for lactose transport across the cell membrane? The answer is that even in the fully repressed state, there is a very low basal level of transcription of the *lac* operon that provides five or six molecules of the permease per cell. Perhaps this is just enough to get a few molecules of lactose inside the cell and begin the process.

An even more curious observation is that, in fact, lactose is not the natural inducer of the lactose operon as we would expect. When the repressor is isolated from fully induced cells, the small molecule bound to each repressor monomer is *allolactose* not lactose. Allolactose, like lactose, is composed of galactose and glucose, but the linkage between the two sugars is different. It turns out that a side reaction of β-galactosidase (which normally breaks down lactose to galactose and glucose) converts these two products to allolactose. Therefore, it appears that a few molecules of lactose are taken up and converted by β-galactosidase to allolactose, which then binds to the repressor and induces the operon. Further confirmation that lactose itself is not the real inducer comes from experiments that indicate that lactose binding to the purified repressor slightly *increases* the repressor's affinity for the operator. Therefore, in the induced state a small amount of allolactose must be present in the cell to overcome this ''anti-inducer'' effect of the lactose substrate.

The Promoter of the Lactose Operon

Immediately in front of the *lac* operator sequence is the promoter sequence. This sequence contains the recognition sites for two different proteins, RNA polymerase and the CAP-binding protein (Figure 20.4). The site at which RNA polymerase interacts with the DNA to initiate transcription has been identified using several different genetic and biochemical approaches. Point mutations that occur in this region frequently affect the affinity to which RNA polymerase will bind the DNA. Deletions (or insertions) that extend into this region also dramatically affect the binding of RNA polymerase to the DNA. The end points of the sequence to which RNA polymerase binds have been identified by DNase protection experiments. Purified RNA polymerase was bound to the *lac* promoter region cloned in a bacteriophage DNA or a plasmid, and this protein/DNA complex was digested with DNase I. The DNA segment protected from degradation by DNase was recovered and its sequence determined. The ends of this protected segment varied slightly with different DNA molecules but corresponded closely to the boundaries shown in Figure 20.4.

The sequence of the RNA polymerase interaction site is not composed of symmetrical elements similar to those described for the operator sequence. This is not surprising since RNA polymerase must associate with the DNA in an assymmetrical fashion for RNA synthesis to be initiated in only one direction from the binding site. However, that portion of the promoter sequence that is recognized by the CAP-binding protein does contain some symmetry. A DNA-protein interaction at this region enhances transcription of the *lac* operon as described in the next section.

The CAP-Binding Site of the Lactose Promoter

Escherichia coli prefers to use glucose instead of other sugars as a carbon source. For example, if the concentrations of glucose and lactose in the medium are the same, the bacteria will selectively metabolize the glucose and not utilize the lactose. This phenomenon is illustrated in Figure 20.5, which shows that the appearance of β-galactosidase, the *lacZ* product, is delayed until all of the glucose in the medium is depleted. Only then can lactose be used as the carbon source. This indicates that glucose inter-

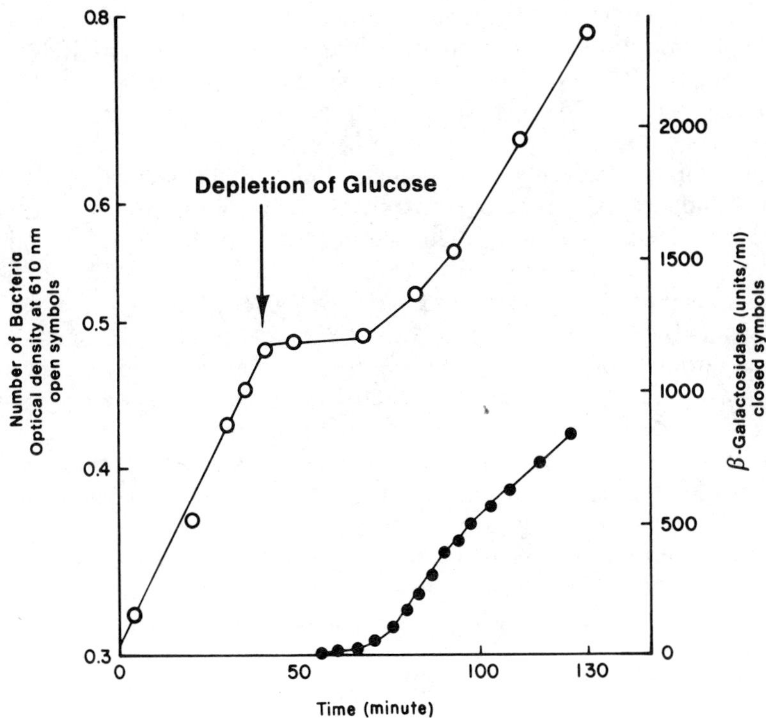

Figure 20.5
Lack of synthesis of β-galactosidase in* E. coli *when glucose is present.
The bacteria are growing in a medium containing initially 0.4 mg/ml glucose and 2 mg/ml lactose. The left-hand ordinate indicates the cell density of the growing culture, an indicator of the number of bacterial cells. The right-hand ordinate indicates the units of β-galactosidase per milliliter. Note that the appearance of β-galactosidase is delayed until the glucose is depleted.
Redrawn from W. Epstein, S. Naono, and F. Gros, *Biochem. Biophys. Res. Commun.*, 24:588, 1966.

feres with the induction of the lactose operon. This effect is called *catabolite repression* because it occurs during the catabolism of glucose and may be due to a catabolite of glucose rather than glucose itself. An identical effect is exerted on a number of other inducible operons, including the arabinose and galactose operons, that code for enzymes involved in the utilization of various substances as energy sources. It probably is a general coordinating system for turning off synthesis of unwanted enzymes whenever the preferred substrate, glucose, is present.

Catabolite repression begins in the cell when glucose lowers the concentration of intracellular cyclic AMP (cAMP). The exact mechanism by which this reduction in the cAMP level is accomplished is not known. Perhaps glucose influences either the rate of synthesis or degradation of cAMP. At any rate cAMP can bind to another regulatory protein, which has not been discussed yet, called *CAP* (for *c*atabolite *a*ctivator *p*rotein) or CRP (for *c*AMP *r*eceptor *p*rotein). CAP is an allosteric protein, and when it is combined with cAMP, it is capable of binding to the CAP regulatory site that is near the promoter of the *lac* (and other) operons. The CAP–cAMP complex exerts positive control on the transcription of these operons. Its binding to the CAP site on the DNA facilitates the binding of RNA polymerase to the promoter (Figure 20.6). Alternatively, if the CAP site is not occupied, RNA polymerase has more difficulty binding to the promoter, and transcription of the operon occurs much less efficiently. Therefore, when glucose is present, the cAMP level drops, the

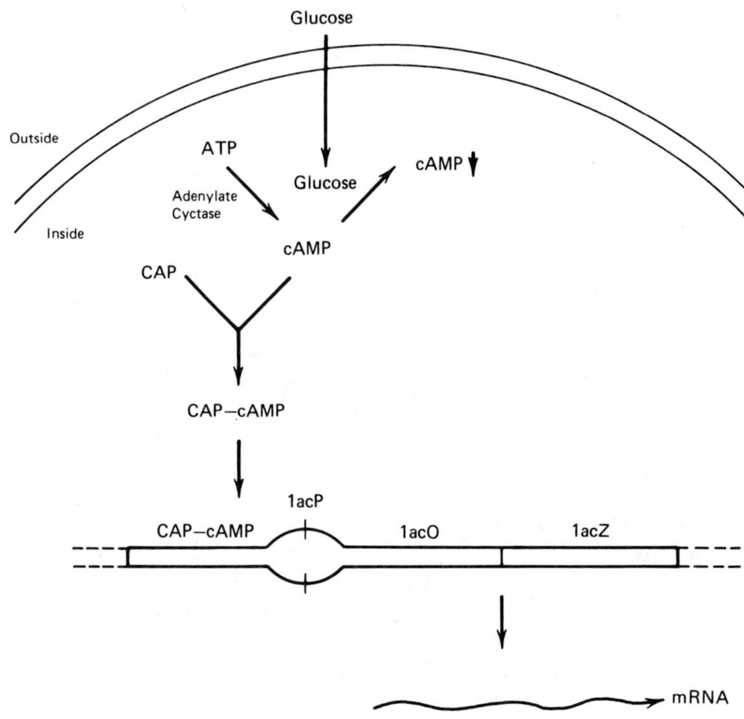

Figure 20.6
Control of* lacP *by cAMP.
A CAP–cAMP complex binds to the CAP site and enhances transcription at lacP. *Catabolite repression occurs when glucose lowers the intracellular concentration of cAMP. This reduces the amount of the CAP–cAMP complex and decreases transcription from* lacP *and from the promotors of several other operons.*

CAP–cAMP complex does not form, and the positive influence on RNA polymerase does not occur. Conversely, if glucose is absent, the cAMP level is high, a CAP–cAMP complex binds to the CAP site, and transcription is enhanced.

20.4 THE TRYPTOPHAN OPERON OF *E. COLI*

Tryptophan is essential for bacterial growth; it is needed for the synthesis of all proteins that contain tryptophan. Therefore, if tryptophan is not supplied in sufficient quantity by the medium, the cell must make it. In contrast, lactose is not absolutely required for the cell's growth; many other sugars can substitute for it, and, in fact, as we saw in the previous section, the bacterium prefers to use some of these other sugars for the carbon source. As a result, synthesis of the tryptophan biosynthetic enzymes is regulated differently than synthesis of the proteins coded for in the lactose operon.

In *E. coli* tryptophan is synthesized from chorismic acid in a five-step pathway that is catalyzed by three different enzymes as shown in Figure 20.7. The tryptophan operon contains the five structural genes that code for these three enzymes (two of which have two different subunits). Upstream from this gene cluster is a promoter where transcription begins and an operator to which binds a repressor protein that is coded for by the unlinked *trpR* gene. Transcription of the lactose operon is generally "turned off" unless it is *induced* by the small molecule inducer. The tryptophan operon, on the other hand, is always "turned on" unless it is *repressed* by the presence of a small molecule *corepressor* (a term used to

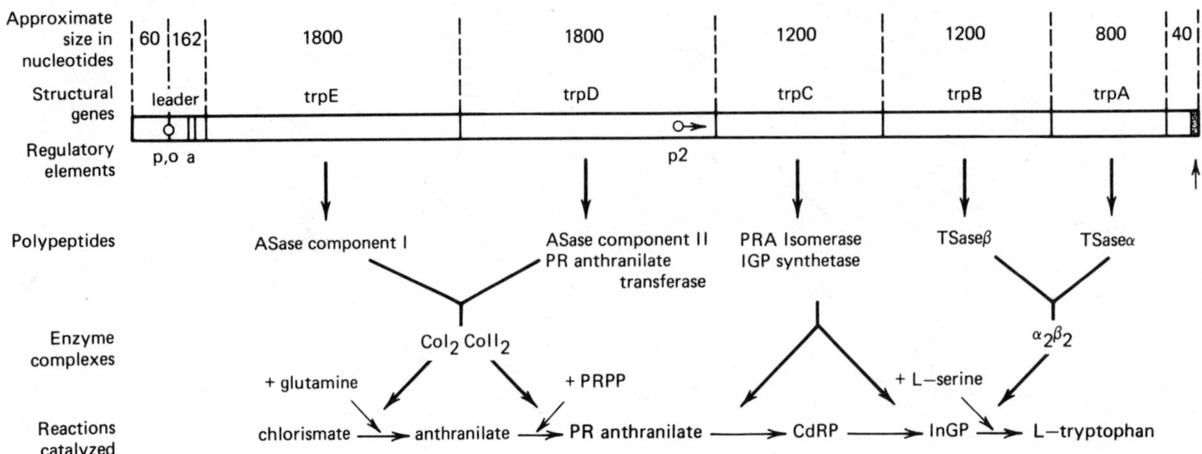

Figure 20.7
***Genes of the tryptophan operon of* E. coli.**
Regulatory elements are the primary promoter (trpP), the operator (trpO), the attenuator (trp a), the secondary internal promoter (trpP2), and the terminator (trp t). Sites of mRNA initiation are given by ⏀, sites of termination by ■. CoI and CoII signify components I and II, respectively, of the anthranilate synthetase complex; PR-anthranilate is N*-5′-phosphoribosyl-anthranilate; CdRP is 1-(o-carboxy-phenylamino)-1-deoxyribulose-5-phosphate; InGP is indole-3-glycerol phosphate; and PRPP is 5-phosphoribosyl-1-pyrophosphate.*
Redrawn from T. Platt, The Tryptophan Operon in *The operon,* eds. J. H. Miller and W. Reznikoff, Cold Spring Harbor, 1978, p. 263.

distinguish it from the repressor protein). Hence, the *lac* operon is inducible, whereas the *trp* operon is repressible. When the tryptophan operon is being actively transcribed, it is said to be *derepressed;* that is, the *trp* repressor is not preventing RNA polymerase from binding. This is mechanistically the same as an *induced* lactose operon in which the *lac* repressor is not interfering with RNA polymerase.

The biosynthetic pathway for tryptophan synthesis is regulated by mechanisms that affect both the *synthesis* and the *activity* of the enzymes that catalyze the pathway. For example, anthranilate synthetase, which catalyzes the first step of the pathway, is coded for by the *trpE* and *trpD* genes of the *trp* operon. The number of molecules of this enzyme that is present in the cell is determined by the transcriptional regulation of the *trp* operon. However, the catalytic activity of the existing molecules of the enzyme is regulated by *feedback inhibition*. This is a common short-term means of regulating the first committed step in a metabolic pathway. In this case, tryptophan, the end product of the pathway, can bind to an allosteric site on the anthranilate synthetase and interfere with its catalytic activity at another site. Therefore, as the concentration of tryptophan builds up in the cell, it begins to bind to anthranilate synthetase and immediately decreases its activity on the substrate, chorismic acid. In addition, as we shall see, tryptophan also acts as a corepressor to shut down the synthesis of new enzyme molecules from the *trp* operon. Therefore, feedback inhibition is a short-term control that has an immediate effect on the pathway, whereas repression takes a little longer but has the more permanent effect of reducing the number of enzyme molecules.

The *trp* repressor is a tetramer of four identical subunits of about 100 amino acids each. Under normal conditions about 20 molecules of the repressor tetramer are present in the cell. The repressor by itself does *not* bind to the *trp* operator. It must be complexed with tryptophan in order to bind to the operator and therefore acts in vivo only in the presence of tryptophan. This is exactly the opposite of the *lac* repressor, which binds to its operator only in the *absence* of its small molecule inducer.

Interestingly, the *trp* repressor also regulates transcription of *trpR,* its own gene. As the *trp* repressor accumulates in the cell, the repressor–tryptophan complex binds to a region upstream of this gene turning off its transcription and maintaining the equilibrium of 20 repressors per cell. In addition, the repressor–tryptophan complex represses transcription of still another gene, *aroH*. This gene is not linked to any of the other genes of the *trp* operon. However, it codes for one of three enzymes that catalyze the first steps in the common pathway of aromatic amino acid biosynthesis. Therefore, the *trp* repressor influences the level of other amino acids besides tryptophan. The genes for the other two enzymes, *aroF* and *aroG* are controlled by other regulator molecules.

Another difference from the *lac* operon is that the *trp* operator occurs entirely within the *trp* promoter rather than adjacent to it, as shown in Figure 20.8. The operator sequence is a region of dyad symmetry, and the mechanism of preventing transcription is the same as in the *lac* operon. Binding of the repressor–corepressor complex to the operator physically blocks the binding of RNA polymerase to the promoter.

Repression results in only about a 70-fold decrease in the rate of transcription initiation at the *trp* promoter. (In contrast, the basal level of the *lac* gene products is about 1,000-fold lower than the induced level.) However, the *trp* operon contains additional regulatory elements that impose further control on the extent of its transcription. One of these additional control sites is a secondary promoter, designated *trpP2,* which is located within the coding sequence of the *trpD* gene (shown in Figure 20.7). This promoter is not regulated by the *trp* repressor. Transcription from it occurs constitutively at a relatively low rate and is terminated at the same location as transcription from the regulated promoter for the whole operon, *trpP*. The resulting transcription product from *trpP2* is an mRNA that contains the coding sequences for *trpCBA,* the last three genes of the operon. Therefore, two polycistronic mRNAs are derived from the *trp* operon, one containing the coding sequences of all five structural genes and one possessing only the last three genes. Under conditions of maximum repression the basal level of mRNA coding sequence for the last three genes is about five times higher than the basal mRNA level for the first two genes.

The reason for the need of a second internal promoter is not clear, but there are several possibilities. Perhaps the best alternative comes from the observation that three of the five proteins do not contain tryptophan; only the *trpB* and *trpC* genes contain the single codon that specifies tryptophan. Therefore, under extreme tryptophan starvation, these two proteins would not be synthesized, which would prevent the pathway from being activated. However, since both of these genes lie downstream of the unregulated second promoter, their protein products will always be present at the basal level necessary to maintain the pathway.

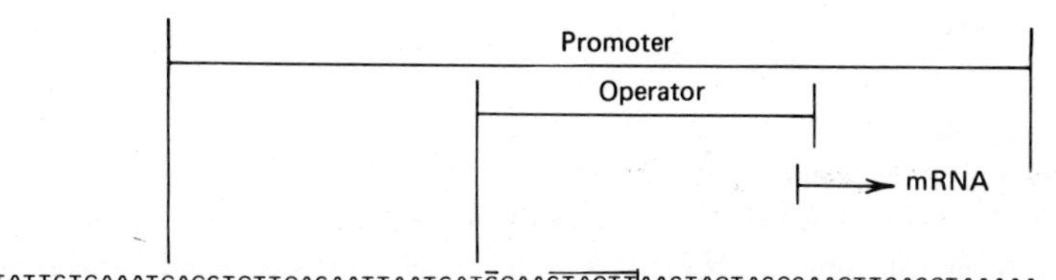

Figure 20.8
Nucleotide sequence of the control elements of the tryptophan operon.
Lines above and below the sequence indicate symmetrical sequences within the operator.

Attenuation of the Tryptophan Operon

Another important control element of the *trp* operon that is not present in the *lac* operon is the *attenuator* site (Figure 20.9). It lies within the 162 nucleotides between the start of transcription from *trpP* and the initiator codon of the *trpE* gene. Its existence was first deduced by the identification of mutations that mapped in this region and increased transcription of all five structural genes. Within the 162 nucleotides, called the *leader sequence,* are 14 adjacent codons that begin with a methionine codon and are followed by an in-phase termination codon. These codons are preceded by a canonical ribosome-binding site and could potentially specify a 14-residue leader peptide. This peptide has never been detected in bacterial cells, perhaps because it is degraded very rapidly. It has been shown that the ribosome-binding site does function properly when its corresponding DNA sequence is ligated upstream of a structural gene using recombinant DNA techniques.

The attenuator region provides RNA polymerase with a second chance to stop transcription if the *trp* enzymes are not needed by the cell. In the presence of tryptophan, it acts like a rho-independent transcription termination site to produce a short 140-nucleotide transcript. In the absence of tryptophan, it has no effect on transcription, and the entire polycistronic mRNA of the five structural genes is synthesized. Therefore, at both the operator and the attenuator, tryptophan exerts the same general influence. At the operator it participates in repressing transcription, and at the attenuator it participates in stopping transcription by those RNA polymerases that have escaped repression. It has been estimated that attenuation has about a 10-fold effect on transcription of the *trp* structural genes. When multiplied by the 70-fold effect of derepression at the operator, about a 700-fold range exists in the level at which the *trp* operon can be transcribed.

The molecular mechanism by which transcription is terminated at the attenuator site is a marvelous example of the cooperative interaction between bacterial transcription and translation to achieve the desired levels of a given mRNA. The first hints that ribosomes were involved in the mechanism of attenuation came from the observation that mutations in the gene for *trp*-tRNA synthetase (the enzyme that charges the tRNA with tryptophan) or the gene for an enzyme that modifies some bases in

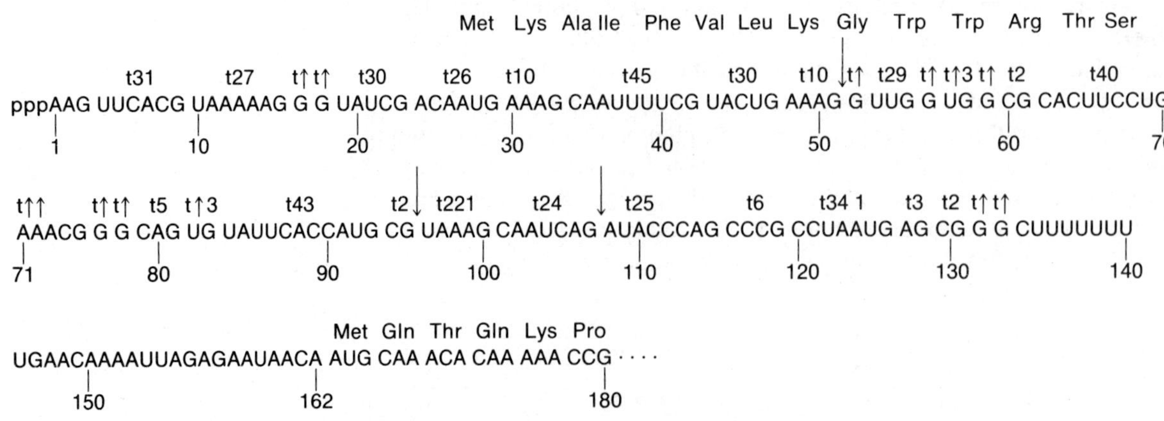

Figure 20.9
Nucleotide sequence of leader RNA from the trp operon.
The 14 amino acids of the putative leader peptide are indicated over their codons. Redrawn with permission from D. L. Oxender, G. Zurawski, and C. Yanofsky, *Proc. Natl. Acad. Sci. USA,* 76:5524, 1979.

the tRNA prevents attenuation. Therefore, a functional *trp*-tRNA must participate in the process.

The leader peptide (Figure 20.9) of 14 residues contains two adjacent tryptophans in positions 10 and 11. This is unusual because tryptophan is a relatively rare amino acid in *E. coli*. It also provides a clue about the involvement of *trp*-tRNA in attenuation. If the tryptophan in the cell is low, the amount of charged *trp*-tRNA will also be low and the ribosomes may be unable to translate through the two *trp* codons of the leader peptide region. Therefore, they will stall at this place in the leader RNA sequence.

It turns out that the RNA sequence of the attenuator region can adopt several possible secondary structures (Figure 20.10). The position of the ribosome within the leader peptide-coding sequence determines the secondary structure that will form. This secondary structure, in turn, is recognized (or "sensed") by the RNA polymerase that has just transcribed through the attenuator coding region and is now located a small distance downstream. The RNA secondary structure that forms when a ribosome is *not* stalled at the *trp* codons is a termination signal for the RNA polymerase. Under these conditions the cell does not need to make tryptophan, and transcription stops after the synthesis of a 140-nucleotide transcript, which is quickly degraded. On the other hand, the secondary structure that results when the ribosomes *are* stalled at the *trp* codons is *not* recognized as a termination signal, and the RNA polymerase continues on into the *trpE* gene. Figure 20.11 shows these different secondary structures in detail.

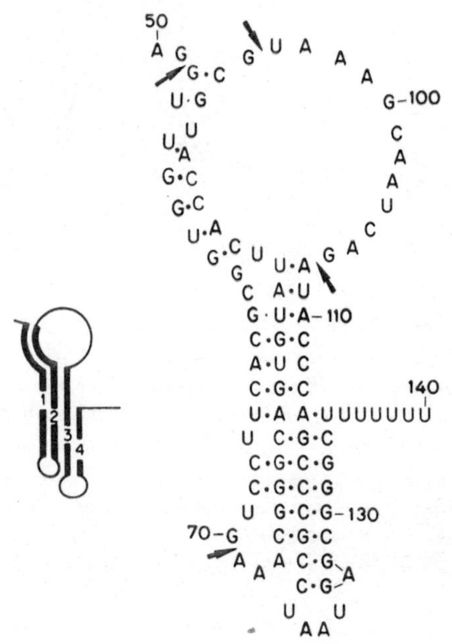

Figure 20.10
***Schematic diagram showing the proposed secondary structures in trp leader RNA from* E. coli.**

Four regions can base-pair to form three stem and loop structures. These are shown as 1-2, 2-3, and 3-4.

Reproduced with permission from D. L. Oxender, G. Zurawski, and C. Yanofsky. *Proc. Natl. Acad. Sci. USA*, 76:5524, 1979.

In the first structure in Figure 20.11, region 1 base pairs with region 2, and region 3 pairs with region 4. The two *trp* codons, UGG, occur at the beginning of the base-paired region 1. The pairing between regions 3 and 4 results in a hairpin loop because of base pairing between the Gs and Cs. This is followed by eight U residues, a common feature of sequences that occur at sites of transcription termination. As the leader RNA sequence is being synthesized in the presence of tryptophan, it is likely that a loop between regions 1 and 2 will form first so that the region 3 and 4 loop will then occur and be recognized as a signal for termination by the RNA polymerase.

A different structure occurs if region 1 is prevented from base pairing with region 2 (Figure 20.11*b*). Under these circumstances, region 2 has the potential to base pair with region 3. This region 2 and 3 hairpin ties up the complementary sequence to region 4, which now must remain single stranded. Therefore, the region 3 and 4 hairpin loop that serves as the termination signal does not form, and the RNA polymerase continues on with its transcription. Thus, for transcription to proceed past the attenuator, region 1 must be prevented from pairing with region 2. This is accomplished if the ribosome stalls in region 1 due to an insufficient amount of charged *trp*-tRNA for translation of the leader peptide to continue beyond this point (Figure 20.11). When this happens, region 1 is bound within the ribosome and cannot pair with region 2. Since regions 2 and 3 are synthesized before region 4, they, in turn, will base pair before region 4 appears in the newly transcribed RNA. Therefore, region 4 remains single stranded, the termination hairpin does not form, and RNA polymerase continues transcription into the structural genes.

The tryptophan codons occur in region 1 right at the beginning of the hairpin between regions 1 and 2. This means that if the ribosome happens to stall at an earlier codon in the leader sequence, it will have little effect on attenuation. For example, starvation for lysine, valine, or glycine would be expected to reduce the amount of the corresponding charged tRNA and stall the ribosome at that codon, but a deficiency in these

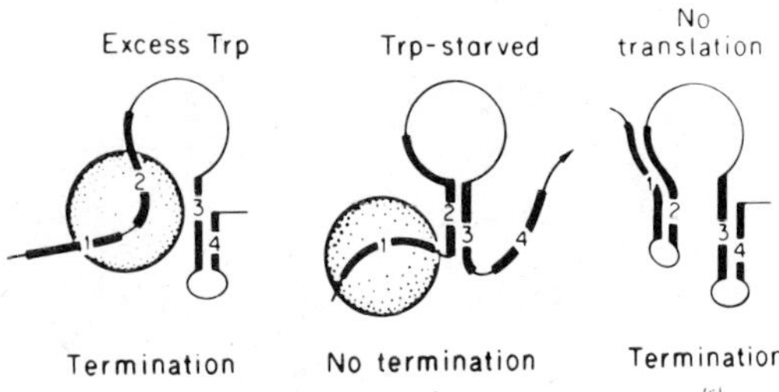

Figure 20.11
***Schematic diagram showing the model for attenuation in the trp operon of* E. coli.**
(a) *Under conditions of excess tryptophan, the ribosome (the shaded circle) translating the newly transcribed leader RNA will synthesize the complete leader peptide. During this synthesis the ribosome will mask regions 1 and 2 of the RNA and prevent the formation of stem and loop 1-2 or 2-3. Stem and loop 3-4 will be free to form and signal the RNA polymerase molecule (not shown) to terminate transcription.* (b) *Under conditions of tryptophan starvation, charged tryptophan-tRNA will be limiting, and the ribosome will stall at the adjacent trp codons in the leader peptide coding region. Because only region 1 is masked, stem and loop 2-3 will form, excluding the formation of stem and loop 3-4, which is required as the signal for transcription termination. Therefore RNA polymerase will continue transcription into the structural genes.* (c) *Under conditions in which the leader peptide is not translated, stem and loop 1-2 will form preventing the formation of stem and loop 2-3, and thereby permit the formation of stem and loop 3-4. This will signal transcription termination.*
Reproduced with permission from D. L. Oxender, G. Zurawski, and C. Yanofsky, *Proc. Natl. Acad. Sci. USA*, 76:5524, 1979.

amino acids has no effect on transcription of the *trp* operon. An exception is arginine whose codon occurs immediately after the tryptophan codons. Starving for arginine does attenuate transcription termination somewhat, probably because of ribosome stalling at this codon, but to less of an extent than a deficiency in tryptophan.

Cis-acting mutations in the attenuator region support the alternate hairpin model. Most of these mutations result in increased transcription because they disrupt base pairing in the double-stranded portion of the termination hairpin and render it less stable. Some mutations, however, increase termination at the attenuator. One of these interferes with base pairing between regions 2 and 3, allowing region 3 to be available for pairing with region 4 even when region 1 is bound to a stalled ribosome. Another mutation occurs in the AUG initiator codon for the leader peptide so that the ribosome cannot begin its synthesis.

Transcription Attenuation at Other Operons

Attenuation is a common phenomenon in bacterial gene expression; it occurs in at least six other operons that code for enzymes catalyzing amino acid biosynthetic pathways. Figure 20.12 shows the corresponding leader peptide sequences specified by each of these operons. In each case, the leader peptide contains several codons for the amino acid end product of the pathway. The most extreme case is the 16-residue leader peptide of the histidine operon that contains seven contiguous histidines. Starvation for histidine results in a decrease in the amount of charged *his*-tRNA and a dramatic increase in transcription of the *his* operon. As with the *trp* operon, this effect is diminished by mutations that interfere with the level of charged *his*-tRNA. Furthermore, the nucleotide sequence of the attenuator region suggests that ribosome stalling at the histidine codons also influences the formation of alternate hairpin loops, one of which resembles a termination hairpin followed by several U residues. In contrast to the *trp* operon, transcription of the *his* operon is regulated entirely by attenuation; it does not possess an operator that is recognized by a repressor protein. Instead, the ribosome acts rather like a positive regulator protein, similar to the cAMP–CAP complex discussed with the *lac* operon. If the ribosome is bound to (i.e., stalled at) the attenuator site, then transcription of the downstream structural genes is enhanced. If the ribosome is not bound, then transcription of these genes is greatly reduced.

Operon	Leader peptide sequence	Regulatory amino acids
his	Met-Thr-Arg-Val-Gln-Phe-Lys-His-His-His-His-His-His-His-Pro-Asp	His
pheA	Met-Lys-His-Ile-Pro-Phe-Phe-Phe-Ala-Phe-Phe-Phe-Thr-Phe-Pro	Phe
thr	Met-Lys-Arg-Ile-Ser-Thr-Thr-Ile-Thr-Thr-Thr-Ile-Thr-Ile-Thr-Thr-Gly-Asn-Gly-Ala-Gly	Thr Ile
leu	Met-Ser-His-Ile-Val-Arg-Phe-Thr-Gly-Leu-Leu-Leu-Leu-Asn-Ala-Phe-Ile-Val-Arg-Gly-Arg-Pro-Val-Gly-Gly-Ile-Gln-His	Leu
ilv	Met-Thr-Ala-Leu-Leu-Arg-Val-Ile-Ser-Leu-Val-Val-Ile-Ser-Val-Val-Val-Ile-Ile-Ile-Pro-Pro-Cys-Gly-Ala-Ala-Leu-Gly-Arg-Gly-Lys-Ala	Leu, Val, Ile

Figure 20.12
***Leader peptide sequences specified by biosynthetic operons of* E. coli.**
All contain multiple copies of the amino acid(s) synthesized by the enzymes coded for by the operon.

Transcription of some of the other operons shown in Figure 20.12 can be attenuated by more than one amino acid. For example, the *thr* operon is attenuated by either threonine or isoleucine; the *ilv* operon is attenuated by leucine, valine, or isoleucine. This effect can be explained in each case by stalling of the ribosome at the corresponding codon, which, in turn, interferes with the formation of a termination hairpin. Although it is not proven, it is possible that with some of the longer leader peptides stalling at more than one codon is necessary to achieve maximal transcription through the attenuation region.

20.5 OTHER BACTERIAL OPERONS

Many other bacterial operons have been studied and found to possess the same general regulatory mechanisms as the *lac, trp,* and *his* operons as discussed above. However, each operon has evolved its own distinctive quirks. For example, the galactose operon contains three structural genes coding for enzymes that metabolize galactose to glucose 1-phosphate. This operon is inducible. A repressor prevents transcription of the gene cluster unless galactose is present. Galactose binds to the repressor causing it to dislodge from the operator, which, in turn, allows transcription to begin. The cAMP–CAP complex greatly enhances this transcription. However, this operon has two different promoters and a different order to the operator, CAP-binding site, and the promoters as shown in Figure 20.13. The operator occurs first in the sequence followed by the CAP-binding site and the promoters. In the induced state (when galactose is present), transcription starts at promoter P1 if the cAMP–CAP complex is bound to the CAP-site. If the CAP-binding site is not occupied, however, transcription begins at a lower rate at a second promoter, P2, that is five base pairs in front of P1. These two overlapping promoters are close enough to the CAP-binding site that CAP must interact directly with RNA polymerase when transcription begins at P1. Furthermore, because the operator and the promoters are separated by the CAP site, it is possible that the repressor does not directly block binding of the RNA polymerase at the promoters. Perhaps repressor binding distorts the double helix in the vicinity of the promoters so that they are not recognized by RNA polymerase.

Another interesting group of operons are those that contain the structural genes for the 70 or more proteins that comprise the ribosome (Figure 20.14). Each ribosome contains one copy of each ribosomal protein (ex-

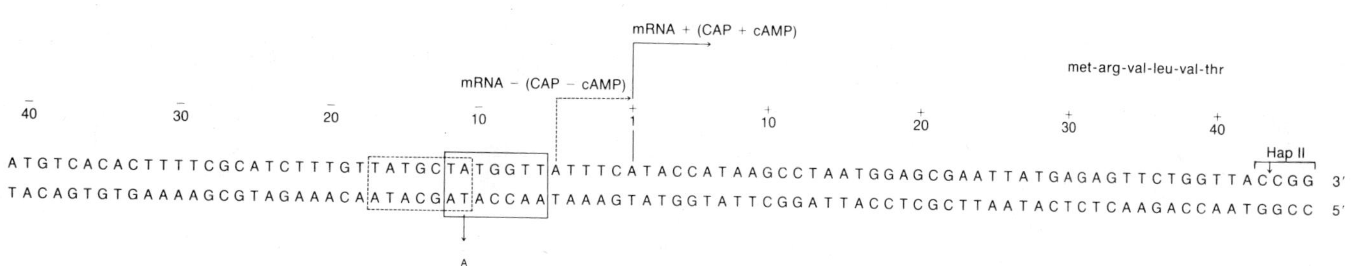

Figure 20.13
DNA sequence of the operator–promoter region of the galactose operon.
The +1 corresponds to the start of the CAP–cAMP-dependent mRNA. Two similar overlapping heptamers preceding each start site are boxed.

Operon	Regulator	Proteins specified by the operon
Spc	S8	L14-L24-L5-S14-S8-L6-L18-S5-L15-L30
S10	L4	S10-L3-L2-L4-L23-S19-L22-S3-S17-L16-L29
str	S7	S12-S7-Ef · G-EF · Tu
α	S4	S13-S11-S4-α-L17
L11	L1	L11-L1
rif	L10	L10-L7-β-β'

Figure 20.14
***Location of genes for ribosomal proteins of* E. coli.**
Genes for the protein components of the small (S) and large (L) ribosomal subunits of E. coli *are clustered on several operons. Some of these operons also contain genes for RNA polymerase subunits ($\alpha\beta\beta'$) and protein synthesis factors (EF.G and EF.Tu). At least one of the protein products of each operon usually regulates expression of that operon (see text).*

cept for protein L7/L12, which is probably present in four copies). Therefore, all 70 proteins are required in equimolar amounts, and it makes sense that their synthesis is regulated in a coordinated fashion. The characterization of this set of operons is not yet complete, but six operons, containing about half of the ribosomal protein genes, have been identified in two major gene clusters. One cluster contains four adjacent operons (*str, spc, S10, and* α), and the other two operons are near each other elsewhere in the *E. coli* chromosome. There seems to be no real pattern to the distribution of these genes among the different operons. Some operons contain genes for proteins of just one ribosomal subunit; others code for proteins of both subunits. In addition to the structural genes for the ribosomal proteins, these operons also contain genes for other (related) proteins. For example, the *str* operon contains genes for the two soluble elongation factors, EF-Tu and EF-G, as well as genes for some proteins in the 30S subunit. The α operon has genes for proteins of both the 30S and 50S subunits plus a gene for one of the subunits of RNA polymerase. The *rif* operon has genes for two other protein subunits of RNA polymerase and ribosomal protein genes.

A common theme among the six ribosomal operons is that their expression is regulated by one of their own structural gene products; that is, they are self-regulated. The precise mechanism of this self-regulation varies considerably with each operon and is not yet understood in detail. However, in at least some cases the regulation occurs at the level of *translation* not *transcription* as discussed for the other operons. After the polycistronic mRNA is made, the "regulatory" ribosomal protein determines which regions, if any, are translated. In general, the ribosomal protein that regulates expression of its own operon, or part of its own operon, is a protein that is associated with one of the ribosomal RNAs (rRNAs) in the intact ribosome. This ribosomal protein has a high affinity for the rRNA and a lower affinity for one or more regions of its own mRNA. Therefore, a competition between the rRNA and the operon's mRNA for binding with the ribosomal protein occur. As the ribosomal protein accumulates to a higher level than the free rRNA, it binds to its own mRNA and prevents the initiation of protein synthesis at one or more of the coding sequences on this mRNA (Figure 20.15). As more ribosomes are formed, the excess of this particular ribosomal protein is used up and translation of its coding sequence on the mRNA can begin again.

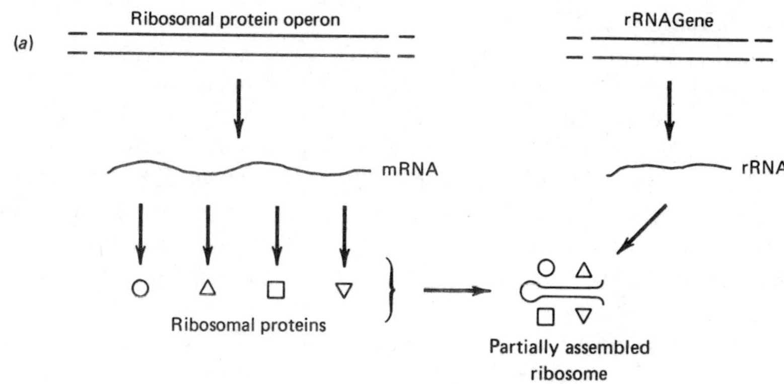

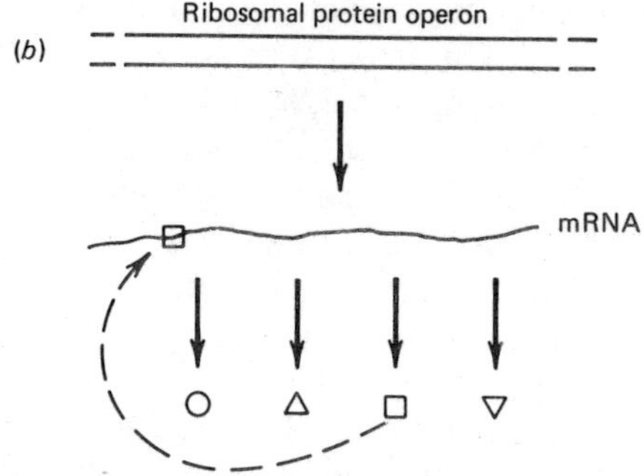

Figure 20.15
Self-regulation of ribosomal protein synthesis.
Individual ribosomal proteins bind to the polycistronic mRNA from their own operon, blocking further translation, if free rRNA is not available for the assembly of new ribosomal subunits.

The Stringent Response of Bacteria

Bacteria have several ways in which to respond molecularly to emergency situations, that is, times of extreme general stress. One of these situations is when the cell does not have a sufficient pool of amino acids to maintain protein synthesis. Under these conditions the cell invokes what is called the *stringent response*, a mechanism that reduces the synthesis of the rRNAs and tRNAs about 20-fold. This places many of the activities within the cell on hold until conditions improve. The mRNAs are less affected, but there is also about a three-fold decrease in their synthesis.

The stringent response is triggered by the presence of an uncharged tRNA in the A site of the ribosome. This occurs when the concentration of the corresponding charged tRNA is very low. The first result, of course, is that further peptide enlongation by the ribosome stops. This event causes a protein called the *stringent factor,* the product of the *relA* gene, to synthesize two small molecules, guanosine tetraphosphate (ppGpp) and guanosine pentaphosphate (pppGpp), from ATP and GTP or GDP as shown in Figure 20.16. The stringent factor is loosely associated with a few, but not all, ribosomes of the cell. Perhaps a conformational change in the ribosome is induced by the occupation of the A site by an uncharged tRNA, which, in turn, activates the associated stringent factor. The exact functions of ppGpp and pppGpp are not known. However, they seem to inhibit transcription initiation of the rRNA and tRNA genes. In addition they affect transcription of some operons more than others.

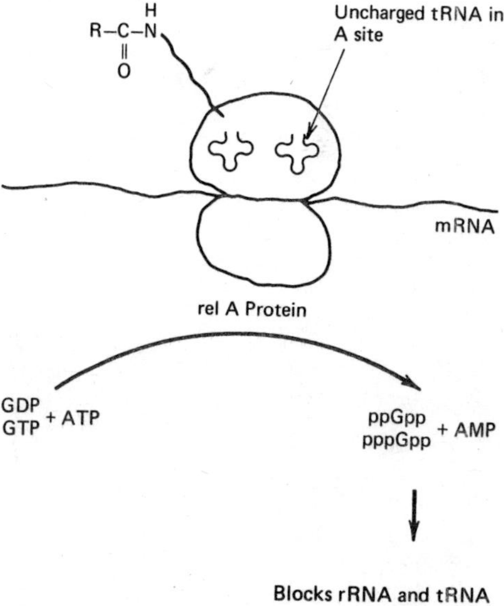

Figure 20.16
***Stringent control of protein synthesis in* E. coli.**
During extreme amino acid starvation, an uncharged tRNA in the A site of the ribosome activates the relA *protein to synthesize ppGpp and pppGpp which, in turn, are involved in decreasing transcription of the genes coding for rRNAs and tRNAs.*

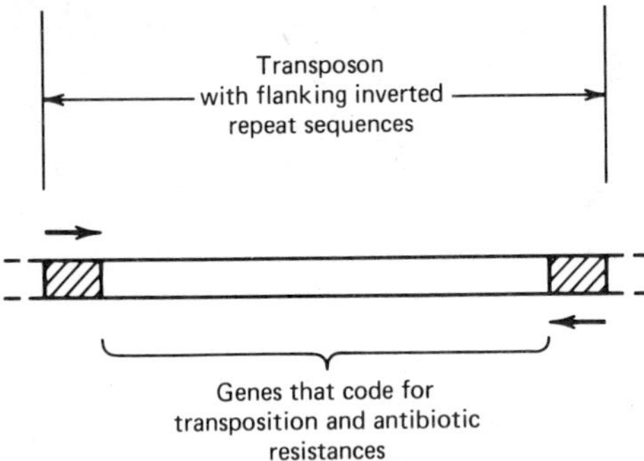

Figure 20.17
The general structure of transposons.
Transposons are relatively rare mobile segments of DNA that contain genes coding for their own rearrangement and (usually) genes that specify resistance to various antibiotics.

20.6 BACTERIAL TRANSPOSONS

So far we have only discussed the regulation of bacterial genes that are fixed in the chromosome. Their positions relative to the neighboring genes do not change. The vast majority of bacterial genes are of this type. In fact the genetic maps of *E. coli* and *Salmonella typhimurium* are quite similar, indicating the lack of much evolutionary movement of most genes within the bacterial chromosome. There is a class of bacterial genes, however, in which newly duplicated gene copies ''jump'' to another genomic site with a frequency of about 10^{-7} per generation, the same rate as spontaneous point mutations occur. The mobile segments of DNA containing these genes are called *transposable elements* or *transposons* (Figure 20.17).

Transposons were first detected as rare insertions of foreign DNA into structural genes of bacterial operons. Usually, these insertions interfere with the expression of the structural gene (into which they have inserted) and all downstream genes of the operon. This is not surprising since they can potentially destroy the translation reading frame, introduce transcription termination signals, affect the mRNA stability, and so on. A number of transposons and the sites into which they insert have now been isolated using recombinant DNA techniques and have been extensively characterized. These studies have revealed many interesting features about the mechanisms of transposition and the nature of genes located within transposons.

Different transposons vary tremendously in length. Some are a few thousand base pairs and contain only two or three genes; others are many thousands of base pairs long, possessing several genes. Several small transposons can occur within a large transposon. All transposons contain at least one gene that codes for a *transposase*, an enzyme required for the transposition event. Often they also contain genes that code for resistance to antibiotics or heavy metals. All transpositions involve the generation of an addition copy of the transposon and the insertion of this copy into another location. The original transposon copy is the same after the duplication as before; that is, the donor copy is unaffected by insertion of its duplicate into the recipient site. All transposons contain short inverted terminal repeat sequences that are essential for the insertion mechanism, and in fact these inverted repeats are often used to define the two boundaries of a transposon. The multiple target sites into which most transposons can insert seem to be fairly random in sequence; other transposons have a propensity for insertion at specific ''hot spots.'' The duplicated transposon can be located in a different DNA molecule than its donor. Frequently, transposons are found on plasmids that pass from one bacterial strain to another and are the source of a suddenly acquired resistance to one or more antibiotics by a bacterium (Clin. Corr. 20.1).

As with bacterial operons, each transposon or set of transposons has its own distinctive characteristics. It is beyond the scope of this chapter to compare different transposons so one well-characterized transposon called *Tn3* will be discussed as an example of their general properties.

CLIN. CORR. 20.1 TRANSMISSIBLE MULTIPLE DRUG RESISTANCE

Pathogenic bacteria are becoming increasing resistant to a large number of antibiotics, which is viewed with alarm by many physicians. Many cases have been documented in which a bacterial strain in a patient being treated with one antibiotic suddenly becomes resistant to that antibiotic and, simultaneously, to several other antibiotics even though the bacterial strain had never been previously exposed to these other antibiotics. This occurs when the bacteria suddenly acquire from another bacterial strain a plasmid that contains several different transposons, each containing one or more antibiotic-resistance genes.

The Tn3 Transposon

Tn3 has been cloned using recombinant DNA techniques and its complete sequence determined. Its general structure is shown in Figure 20.18. It contains 4,957 base pairs including 38 base pairs at one end that occur as an inverted repeat at the other end. Three genes are present on *Tn3*. One gene codes for the enzyme β-lactamase, which hydrolyzes ampicillin and renders the cell resistant to this antibiotic. The other two genes, *tnpA* and *tnpR*, code for a transposase and a repressor protein, respectively. The

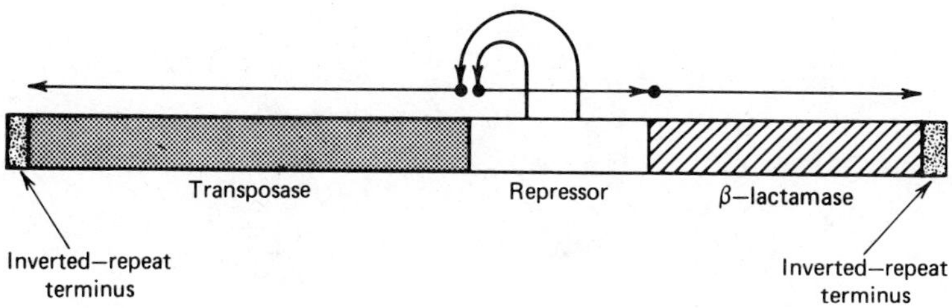

Figure 20.18
Functional components of the transposon Tn3.
Genetic analysis shows there are at least four kinds of regions: the inverted-repeat termini; a gene for the enzyme β-lactamase, which confers resistance to ampicillin and related antibiotics; a gene encoding an enzyme required for transposition (transposase); and a gene for a repressor protein that controls the transcription of the genes for transposase and for the repressor itself. The arrows indicate the direction in which DNA of various regions is transcribed.
Redrawn from S. N. Cohen and J. A. Shapiro, *Sci. Am.* 242:40, 1980. W. H. Freeman and Company, Copyright 1980.

transposase is composed of 1,021 amino acids and it binds to single-stranded DNA. Little else is known about its action, but it is thought to recognize the repetitive ends of the transposon and to participate in the cleavage of the recipient site into which the new transposon copy inserts. The *tnpR* gene product is a protein of 185 amino acids. In its role as a repressor it controls transcription of both the transposase gene and its own gene. The *tnpA* and *tnpR* genes are transcribed divergently from a 163 base-pair control region located between the two genes that is recognized by the repressor. In addition to serving as a repressor of these two genes, the *tnpR* product also participates in the recombination process that results in the insertion of the new transposon. Transcription of the ampicillin-resistance gene is not affected by the *tnpR* gene product.

Mutations in the transposase gene generally decrease the frequency of *Tn3* transposition, demonstrating its direct role in the transposition process. Mutations that destroy the repressor function of the *tnpR* product cause an increased frequency of transposition. These mutations derepress the *tnpA* gene resulting in more molecules of the transposase, which increases the formation of more transposons. They also derepress the *tnpR* gene but since the repressor is inactive, this has no effect on the system.

When a transposon, containing its terminal *inverted* repeats, inserts into a new site, it generates short (5–10 base pairs) *direct* repeats of the sequence at the recipient site that flank the new transposon. This is due to the mechanism of recombination that occurs during the insertion process, as illustrated in Figure 20.19. The first step in the process is the generation of staggered nicks at the recipient sequence. These staggered single-strand, protruding 5′ ends then join covalently to the inverted repeat ends of the transposon. The resulting intermediate then resembles two replicating forks pointing toward each other and separated by the length of the transposon. The replication machinery of the cell fills in the gaps and continues the divergent elongation of the two primers through the transposon region. This ultimately results in two copies of the transposon sequence. Reciprocal recombination within the two copies regenerates the original transposon copy at its old (unchanged) position and completes the process of forming a new copy at the recipient site that is flanked by direct repeats of the recipient sequence.

In recent years the practical importance of transposons located on plasmids has taken on increased significance for the use of antibiotics in

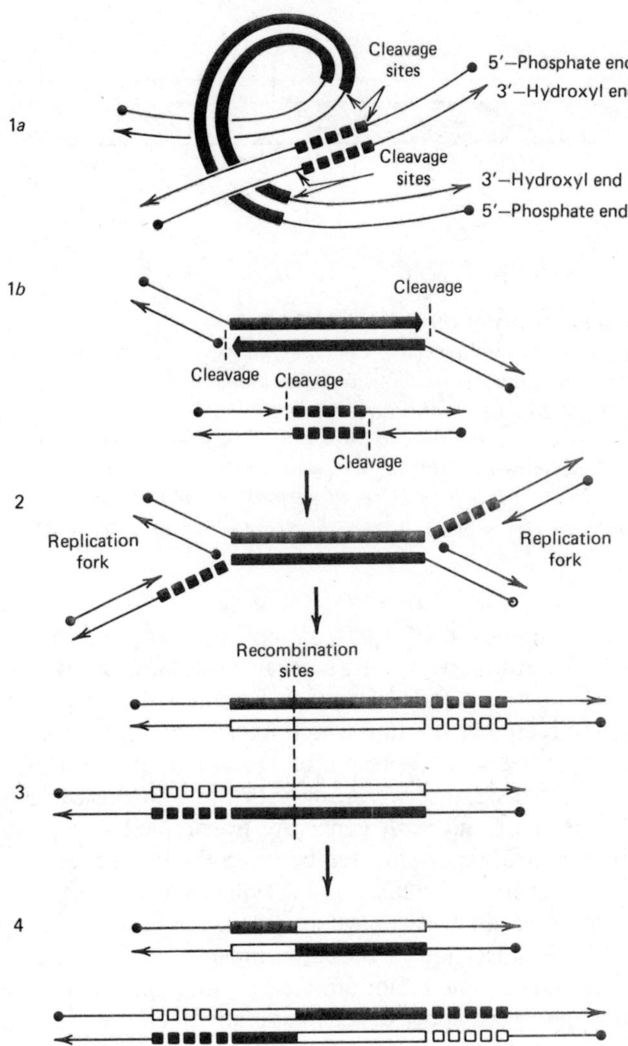

Figure 20.19
Proposed molecular pathway for transposition and chromosome rearrangements. *The donor DNA, including the transposon, is in solid bars; the recipient DNA is in squares. The pathway has four steps, beginning with single-strand cleavage (1*a*) at each end of the transposable element and at each end of the "target" nucleotide sequence to be duplicated. The cleavages expose (1*b*) the chemical groups involved in the next step: the joining of DNA strands from donor and recipient molecules in such a way that the double-stranded transposable element has a DNA-replication fork at each end (2). DNA synthesis (3) replicates the transposon (open bars) and the target sequence (□) accounting for the observed duplication. This step forms two new complete double-stranded molecules; each copy of the transposable element joins a segment of the donor molecule and a segment of the recipient molecules. (The copies of the element serve as linkers for the recombination of two unrelated DNA molecules.) In the final step (4), reciprocal recombination between copies of the transposable element inserts the element at a new genetic site and regenerates the donor molecule. The mechanism of this recombination is not known.*

Redrawn from S. N. Cohen and J. A. Shapiro. *Sci. Am.* 242:40, 1980. W. H. Freeman and Company, Copyright 1980.

the treatment of bacterial infections. Plasmids that have not been altered in the laboratory usually contain genes that facilitate their transfer from one bacterium to another. As the plasmids transfer (e.g., between different infecting bacterial strains), their transposons containing antibiotic-resistance genes are moved into new bacterial strains. Once inside a new bacterium, the transposon can be duplicated onto the chromosome and become permanently established in that cell's lineage. The result is that more and more pathogenic bacterial strains become resistant to an increasing number of antibiotics.

20.7 INVERSION OF GENES IN *SALMONELLA*

A very interesting mechanism of differential gene regulation has recently been discovered for one set of genes in *Salmonella*. Since similar control mechanisms exist for the expression of other genes in other procaryotes (e.g., the bacteriophage mu), this gene system will be briefly described as the final illustration of gene regulation in procaryotes.

Bacteria move by waving their flagella. The flagella are composed predominantly of subunits of a protein called flagellin. Many *Salmonella* species possess two different flagellin genes and express only one of these genes at a time. The bacteria are said to be in *phase 1* if they are expressing the H1 flagellin gene and in *phase 2* if they are expressing the H2 flagellin gene. A bacterial clone in one phase switches to the other phase about once every 1,000 divisions. This switch is called *phase variation,* and its occurrence is controlled at the level of transcription of the *H1* and *H2* genes.

The organization of the flagellin genes and their regulatory elements is shown in Figure 20.20. A 995-base pair segment of DNA flanked by 14 base-pair repeats is adjacent to the *H2* gene and a *rhl* gene that codes for a repressor of *H1*. The *H2* and *rhl* genes are coordinately transcribed. Therefore, when *H2* is expressed, the repressor is also made and turns off

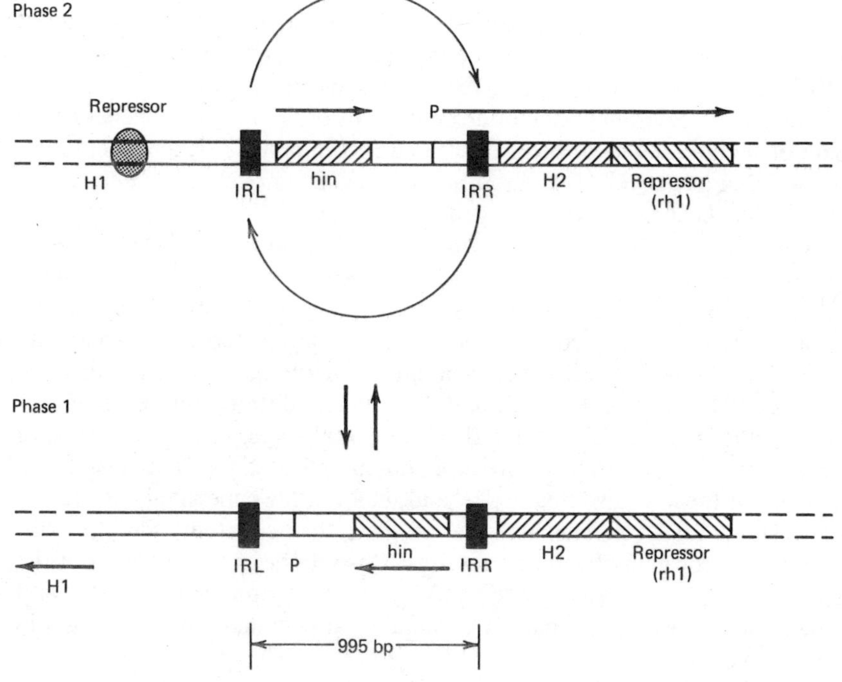

Figure 20.20
Organization of the flagellin genes.
The orientation of a 995 base pairs DNA segment flanked by 14 base pairs inverted repeats (IRL and IRR) controls the expression of the H1 and H2 flagellin genes. In phase 2, transcription initiates at promoter P within the invertable segment and continues through the H2 and rhl genes. In phase 1, the orientation is reversed so that transcription of the H2 and rhl genes does not occur.

H1 expression. When H2 protein and the repressor are not made, the *H1* gene is derepressed and H1 synthesis occurs.

The promoter for the operon containing *H2* and *rhl* lies near one end of the 995-base pair segment, just inside one copy of the 14-base pair repeats. Furthermore, this segment can undergo inversions about the 14-base pair repeats. In one orientation of the segment, the promoter is upstream of the *H2-rhl* transcription unit; in the other orientation it points toward the opposite direction so that *H2* and *rhl* are not transcribed. In addition to containing this promoter, the invertible segment of DNA possesses the *hin* gene whose product is an enzyme that catalyzes the inversion event itself. The *hin* gene seems to be transcribed constitutively at a low rate. Mutations in *hin* reduce the rate of inversion by 10,000-fold.

Therefore, phase variation is controlled by the physical inversion of the segment of DNA that removes a promoter from its position in front of the *H2-rhl* operon. When the promoter is in the opposite direction, it presumably still serves to initiate transcription, but the fate of that RNA is not known. It does not code for the *H1* that maps in this direction. That gene apparently has its own promoter that is controlled directly by the presence of the *rhl* repressor.

The inversion of the *hin* segment probably occurs via recombination between the 14-base pair inverted repeats that is similar to recombination events involved in the transposition of a transposon. In fact, transposons have been shown to invert relative to their flanking sequences in a fashion exactly analogous to the *hin* inversion. Furthermore, the amino acid sequence of the *hin* product shows considerable homology with the *tnpR* product of the *Tn3* transposon, which, as described above, participates in the integration of the transposon into a new site. Thus, it is possible, and even likely, that the two processes are evolutionarily related.

20.8 ORGANIZATION OF GENES IN MAMMALIAN DNAs

The past 10 years have seen a virtual explosion of new information about the organization, structure, and regulation of genes in eucaryotic organisms. The reason for this enormous increase in our knowledge about eucaryotic genes has been the concurrent development of recombinant DNA or "genetic engineering" techniques and the development of techniques for determining nucleotide sequences of DNA. Both of these new sets of techniques have been described in Chapter 17. Experiments undreamed of a few years ago are now routine accomplishments because of the use of these laboratory procedures.

The mammalian genome contains over 10^9 base pairs of DNA, nearly 1,000 times more DNA than the *E. coli* chromosome. Each mammalian cell contains virtually a complete copy of this genome, and all except the haploid germ line cells contain two copies. Different types of mammalian cells express widely different proteins even though each contains the same complement of genes. In addition, widely different patterns of protein synthesis occur at different developmental stages of the same type of cells. Therefore, extremely intricate and complicated mechanisms of regulation for these genes must exist, and, in fact, these mechanisms are not understood for even one mammalian gene to the extent that they are understood for many bacterial operons. Despite the great advances of the past 10 years, our understanding of gene regulation in mammals, and indeed all eucaryotes, remains fragmentary at best and most likely is still very naive.

The Size of Eucaryotic DNA

It was appreciated even before the advent of recombinant DNA methodology that eucaryotic cells, including mammalian cells, contain far more DNA than seems necessary to code for all of the required proteins. Furthermore, organisms that appear rather similar in complexity can have a several-fold difference in cellular DNA content. A housefly, for example, has about six times the cellular DNA content of a fruit fly. Some plant cells have almost 10 times more DNA than human cells. Therefore, DNA content does not always correlate with the complexity and diversity of functions of the organism.

It is difficult to obtain an accurate estimate of the number of different proteins, and therefore genes, that are found in a mammalian cell or in the entire mammalian organism. However, nucleic acid hybridization procedures have been used to determine that a maximum of 5,000–10,000 different mRNAs may be present in a mammalian cell at a given time. Most of these mRNAs code for proteins that are common to many cell types. Therefore, a generous estimate might be approximately 40,000 genes for the entire mammalian genome. If the average coding sequence is 1,500 nucleotides (specifying a 500-amino acid protein), this accounts for 6% of the genome. Controlling elements, repetitive genes for ribosomal RNAs, and so on, may account for another 5 to 10%. However, as much as 80–90% of the mammalian genome may not have a direct genetic function. This is in contrast to the bacterial genome in which virtually all of the DNA is consumed by genes and their coding elements.

Introns (Intervening Sequences) of Eucaryotic Genes

As discussed in Chapter 18, the coding sequences (*exons*) of eucaryotic genes are frequently interrupted by *intervening sequences* or *introns* that do not code for a product. These intervening sequences are transcribed into a precursor RNA species found in the nucleus and are removed by "RNA splicing" events in the processing of the precursor RNA to the mature mRNA found in the cytoplasm. The number and length of the intervening sequences in a gene can vary tremendously. Histone genes and interferon genes do not have intervening sequences; they contain a continuous coding sequence for the protein as do bacterial genes. The mammalian collagen gene, on the other hand, has over 50 different intervening sequences that collectively consume 10 times more DNA than the coding sequence for collagen. On the basis of the mammalian genes analyzed to date, it appears that most have one or more intervening sequences but that 50 introns in one gene is an extreme case. Therefore, intervening sequences of genes can account for some of the "excess" DNA present in eucaryotic genomes. The evolutionary significance of intervening sequences and their potential biological functions, if any, are the subject of much current speculation and experimentation. In many ways, they remain as big an enigma as they were when first discovered a few years ago.

20.9 REPETITIVE DNA SEQUENCES IN EUCARYOTES

Another curiosity about mammalian DNA, and the DNA of most higher organisms, is that, in contrast to bacterial DNA, it contains repetitive sequences in addition to single copy sequences. This repetitive DNA falls into two general classes–highly repetitive simple sequences and moderately repetitive longer sequences of several hundred to several thousand base pairs.

Highly Repetitive Sequences

The highly repetitive sequences range from 5 to about 100 base pairs in length and occur in tandem. Their contribution to the total genomic size is extremely variable, but in most organisms these sequences are tandemly repeated millions of times and in some organisms they consume 50% or more of the total DNA. They are concentrated at the ends of chromosomes, that is, the telomeres, and at the chromosomal centromeres. Figure 20.21 shows the three main repeat units of the highly repetitive sequences of the fruit fly, *Drosophila virilis*. Repeats of these three sequences of seven base pairs comprise 40% of the organism's DNA. They are obviously related evolutionarily since two of the repeats can be derived from the third by a single base pair change. Relatively little transcription occurs from the highly repetitive sequences, and their biological importance remains, for the most part, a mystery. The repetitive sequences that occur near the telomeres are probably required for the replication of the ends of the linear DNA molecules. The ones that occur at the centromeres might play a structural role since this is the region of attachment during chromosome pairing and segregation in mitosis and meiosis.

Moderately Repetitive Sequences

The "moderately" repetitive sequences comprise a large number of different sequences that are repeated to such different extents that it is somewhat misleading to group them under one heading. Some moderately repetitive sequences are clustered in one region of the genome; others are scattered throughout the DNA. Some moderate repeats are several thousand base pairs in length; other repeats come in a unit-size of only a hundred base pairs. Sometimes the sequence is highly conserved from one repeat to another; in other cases, different repeat units of the same basic sequence will have undergone considerable divergence. To illustrate the diversity in the structures and functions of moderately repetitive sequences, two examples from the human genome will be briefly described.

Percent of genome	Number of copies in genome	Predominant sequence
25	1×10^7	5′ -ACAAACT- 3′ 3′ -TGTTTGA- 5′
8	3.6×10^6	5′ -A[T]AAACT- 3′ 3′ -T[A]TTTGA- 5′
8	3.6×10^6	5′ -ACAAA[T]T- 3′ 3′ -TGTTT[A]A- 5′

Figure 20.21
***Main repeat units of repetitive sequences of* Drosophilia virilis.**
Approximately 41% of the genomic DNA of the fruit fly, Drosophilia virilis, *is comprised of three related repeat sequences of seven base pairs. The bottom two sequences differ from the top sequence at one base pair shown in the box.*

In mammalian cells the 18S, 5.8S, and 28S ribosomal RNAs (rRNAs) are transcribed as a single precursor transcript that is subsequently processed to yield the mature rRNAs. In humans the length of this precursor is 13,400 nucleotides, about half of which is comprised of the mature rRNA sequences. Several sequential posttranscriptional cleavage steps remove the extra sequences from the ends and the middle of the precursor RNA, releasing the mature rRNA species. The DNA that contains the corresponding rRNA genes is a moderately repetitive sequence of about 43,000 base pairs of which 30,000 base pairs are nontranscribed spacer DNA. Clusters of this entire DNA unit occur on five human chromosomes. In total there are about 280 repeats of this unit, which comprise about 0.3% of the total genome (Figure 20.22). The 5S rRNA genes are also repeated but in different clusters. The human genome contains about 2,000 repeats of the 5S rRNA genes. The need for so many rRNA genes is because the rRNAs are structural RNAs. Each transcript from the gene yields only one rRNA molecule. On the other hand, each transcript from a ribosomal *protein* gene can be repeatedly translated to give many protein molecules.

In contrast to the repetitive ribosomal RNA genes that are clustered at only a few sites, most moderately repetitive sequences in the mammalian genome do not code for a stable gene product and are interspersed with nonrepetitive sequences that occur only once or a few times in the genome. The average size of these interspersed repetitive sequences is about 300 base pairs. Almost half of these sequences are members of a general family of moderately repetitive sequences called the *Alu family* because they can be cleaved by the restriction enzyme Alu I. There are about 300,000 Alu sequences scattered throughout the human haploid genome (on the high side of being *moderately* repetitive). Individual members are related in sequence but frequently are not identical. Their average homology with a consensus sequence is about 87%

Interestingly, there is some repeat symmetry within an individual Alu sequence itself. The sequence appears to have arisen by a tandem duplication of a 130 base sequence with a 31-base pair insertion in one of the two adjacent repeats. Furthermore, some members of the *Alu* family resemble bacterial transposons in that they are flanked by short direct repeats. This does not prove that an Alu repeat can be duplicated and transposed to another site like true transposons, but it suggests that such events may occur.

The biological function, if any, of Alu sequences is not known. One interesting suggestion is that they serve as the multiple origins for the

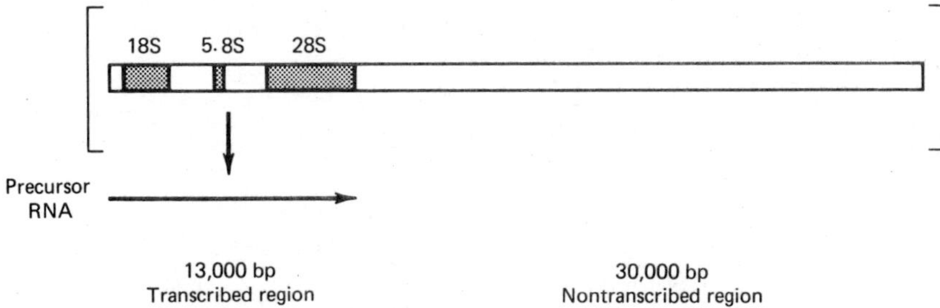

Figure 20.22
Repetitive sequence in human DNA for rRNA.
In human cells a single transcription unit of 13,000 nucleotides is processed to yield the 18S, 5.8S, and 28S rRNAs. About 280 copies of the corresponding rRNA genes are clustered on five chromosomes. Each repeat contains a nontranscribed spacer region of about 30,000 base pairs.

DNA replication during the S phase of the cell, but more of these sequences occur than seem necessary for this function. Some Alu sequences appear in the intervening sequences of some genes and are transcribed as part of large precursor RNAs in which the Alu sequences are removed during RNA splicing to form the mature mRNA. Other Alu sequences are transcribed into small RNA molecules whose function is unknown. A final point is that all mammalian genomes appear to have a counterpart to the human interspersed Alu sequence family although the size of the repeat and its distribution can vary considerably from one species to another.

20.10 GENES FOR GLOBIN PROTEINS

Most of the mammalian structural genes that have been cloned by recombinant DNA techniques specify proteins that either occur in large quantity in a specific cell type (such as the globins of the red blood cell) or after

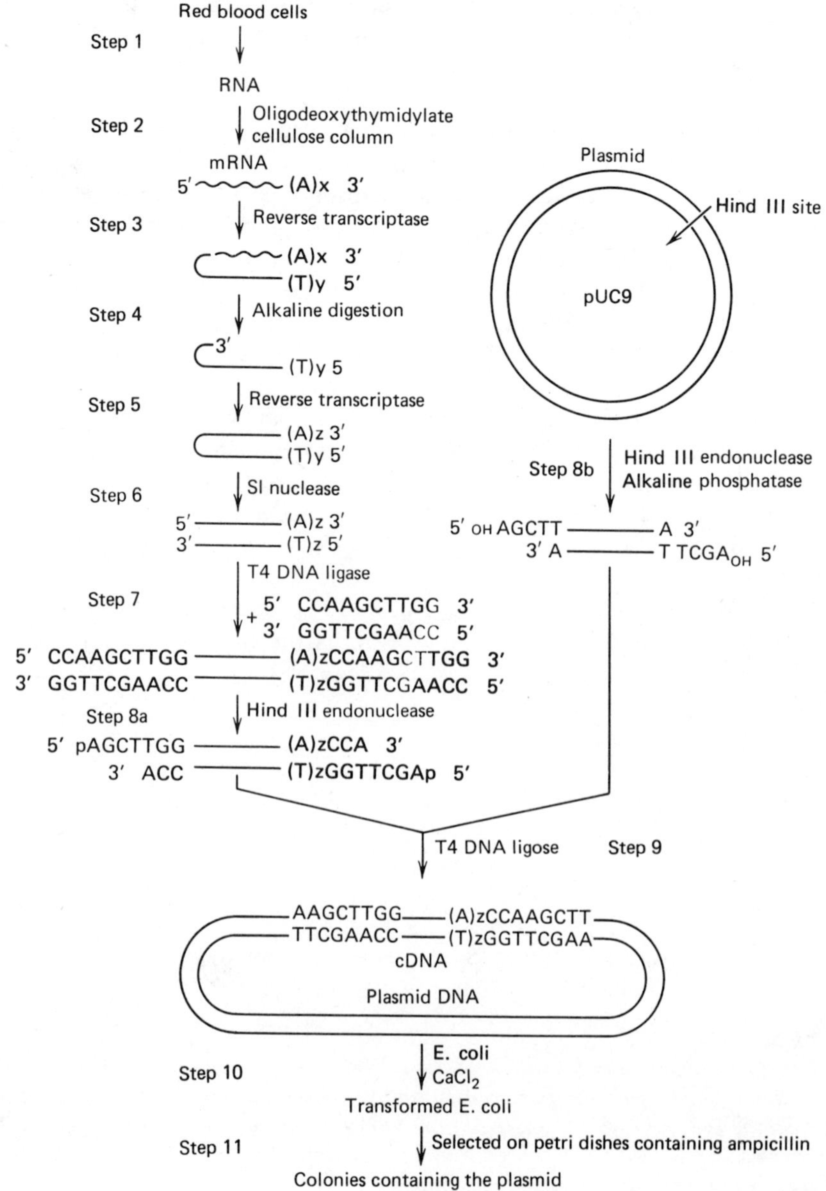

induction of a specific cell type (e.g., growth hormone or prolactin in the pituitary). As a result, more is understood about the regulation of these genes than of other genes whose protein products occur at lower levels in many different cell types. We shall discuss the organization, structure, and regulation of the related members of two gene families—the globin and the growth-hormone-like proteins.

The first step in characterizing a eucaryotic gene is usually to use recombinant DNA techniques to clone a complementary DNA (cDNA) copy of that gene's corresponding mRNA. In fact, this is the reason that the most extensively studied mammalian genes are ones that code for the major proteins of specific cells; a large fraction of the total mRNA isolated from these cells codes for the protein of interest. The main constituent of red cells in human adults is hemoglobin which is comprised of two α subunits (141 amino acids) and two β subunits (146 amino acids), each of which is complexed with a heme group. Almost all of the mRNA isolated from immature red cells (reticulocytes) codes for these two subunits of hemoglobin. Thus, the study of the structure and regulation of the globin genes began with the identification of cloned cDNA molecules that contain the coding sequences for α- and β-globin.

There are several experimental variations of the procedure for synthesizing double-stranded cDNA copies of isolated mRNA *in vitro*. As discussed in Chapter 17, several different plasmid and viral DNA vectors are available for cloning the (passenger) cDNA molecules. Figure 20.23 shows one protocol for constructing and cloning cDNAs prepared from mRNA of reticulocytes.

←

Figure 20.23
Cloning of globin cDNA.
Step 1: Total RNA is extracted from red cells. Step 2: The total RNA is passed through an oligodeoxythymidylate cellulose column. This column separates the polyadenylated mRNA (see Chapter 17) from rRNA and tRNA. The polyadenylated RNA is thought to contain significant amounts of mRNA coding for hemoglobin. Step 3: The mRNA is transcribed into cDNA using reverse transcriptase, the viral enzyme that transcribes DNA on RNA templates (see Chapter 17). Step 4: The mRNA is hydrolyzed with alkali. Step 5: The single-stranded cDNA is converted into a double-stranded DNA helix by using reverse transcriptase. Step 6: The resulting double helix contains a single-stranded hairpin loop that is removed by S1 nuclease, an enzyme that hydrolyzes single-stranded DNA. Step 7: The cDNA is now a double helix with an unknown number of A–T base pairs at one end. In order to produce cohesive ends for the introduction of this cDNA into a plasmid, a chemically synthesized decanucleotide is attached to both ends using DNA ligase. This decanucleotide contains the palindromic symmetry recognized by the Hind III restriction nuclease. Step 8a: Treatment with Hind III restriction nuclease produces a cDNA molecule with Hind III cohesive ends. Step 8b: The plasmid pUC9, which contains an ampicillin-resistance gene, is cleaved with Hind III restriction endonuclease and exposed to bacterial alkaline phosphatase, an enzyme that removes the phosphates from the cleaved 5′-terminal ends of the plasmids at the Hind III site. This prevents the cleaved plasmid from recircularizing without the insertion of the cDNA. Step 9: The linear plasmid and the cDNA molecules are mixed, and formation of circular, dimeric, "recombinant" DNA molecules is allowed to take place. DNA ligase is used to repair the breaks. Step 10: This mixture is used to transform an E. coli *strain. Step 11: Individual* E. coli *cells that took up the plasmid were selected by their ability to grow on ampicillin. The globin cDNA is confirmed by determining the nucleotide sequence of the small DNA fragment released from the plasmid DNA by Hind III restriction endonuclease; if the observed nucleotide sequences corresponded to those expected based on the known amino acid sequence of α- and β-globin, then the cDNA is identified.*

A synthetic oligonucleotide composed of 12–18 residues of deoxythymidine is hybridized to the 3′-polyadenylate tail of the mRNA and serves as a primer for reverse transcriptase, an enzyme that copies the RNA sequence into a DNA strand in the presence of the four deoxynucleoside triphosphates. The resulting RNA-DNA heteroduplex is treated with NaOH, which degrades the RNA strand and leaves the DNA strand intact. The 3′-end of the remaining DNA strand can then fold back and serve as a primer for initiating synthesis of a second DNA strand at random locations by *E. coli* DNA polymerase I. The hairpin loop is then nicked by S1 nuclease, an enzyme that cleaves single-stranded DNA but has little activity against double-stranded DNA. The 3′-ends of the resulting double-stranded cDNAs are ligated to small synthetic "linker" oligonucleotides that contain the recognition site for the restriction enzyme *HindIII*. Digestion of the resulting DNA with *HindIII* generates DNA fragments that contain HindIII-specific ends. These fragments can be ligated into the *HindIII* site of a plasmid, and when the resulting circular "recombinant" DNA species are incubated with *E. coli* in the presence of cations such as calcium or rubidium, a few molecules will be taken up by the bacteria. The incorporated recombinant DNAs will be replicated and maintained in the progeny of the original transformed bacterial cell.

The collection of cloned cDNAs synthesized from the total mRNA isolated from a given tissue or cell type is called a cDNA library, for example, a liver cDNA library or a reticulocyte cDNA library. Since most of the mRNAs of a reticulocyte code for either α- or β-globin, it is relatively easy to identify these globin cDNAs in a reticulocyte cDNA library using procedures discussed in Chapter 17. Once identified, the nucleotide sequences of the cDNAs can be determined to confirm that they do code for the known amino acid sequences of the α- and β-globins. In cases in which the amino acid sequence of the protein is not known, other procedures (usually immunological) are used to confirm the identification of the desired cDNA clone.

Among the first things to be noticed in comparing the α- and β-globin cDNA sequences with the corresponding chromosomal globin genes (which have also been cloned using recombinant DNA techniques) is that all members of both sets of genes contain two introns at approximately the same positions relative to the coding sequences (Figure 20.24). The alpha- (and alpha-like) genes have an intron of 95 base pairs between codons 31 and 32 and a second intron of 125 base pairs between codons 99 and 100. The beta- (and beta-like) genes have introns of 125–150 and 800–900 base pairs located between codons 30 and 31 and codons 104 and 105, respectively. Although the function of introns is presently unknown, they do separate the coding sequences of different functional domains of *some* proteins, including the globins. The coding region between the two globin introns specifies the region of the protein that interacts with the heme group. The final coding region (after the second intron) encodes the region of the protein that provides the interface with the opposite subunit, that is the alpha-beta protein–protein interaction. This separation of the coding sequences for functional domains of a protein by introns is not a general phenomenon, however. The positioning of introns in other genes seems to bear little relationship to the final three-dimensional structure of the encoded protein.

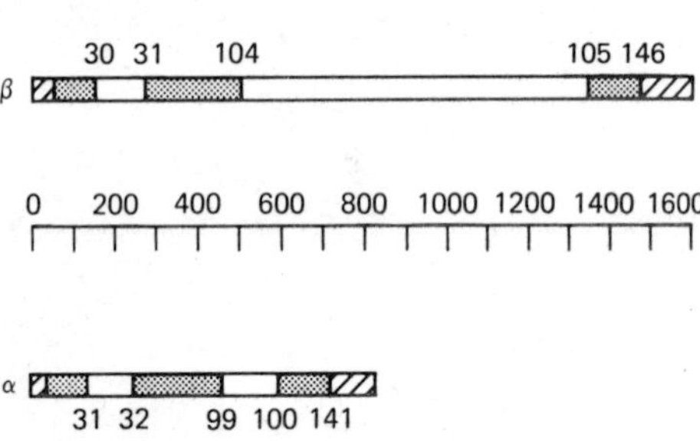

Figure 20.24
Structures of human globin genes.
Structures for the human α-like and β-like globin genes are drawn to approximate scale. Solid and open boxes represent coding (exon) and noncoding (intron) sequences, respectively. Cross-hatched boxes indicate the (5′) upstream and (3′) downstream nontranslated regions in the RNA. The α-like globin genes contain introns of approximately 95 and 125 base pairs, located between codons 31 and 32 and 99 and 100, respectively. The β-like globin genes contain introns of approximately 125–150 and 800–900 base pairs, located between codons 30 and 31 and 104 and 105, respectively.

Different alpha-like and beta-like globin subunits are synthesized at different developmental stages. For example, embryonic red cells contain a different hemoglobin tetramer than do adult red cells. These developmentally distinct subunits have slightly different amino acid sequences and oxygen affinities but are closely related to each other. In humans there are two alpha-like chains—zeta, which is expressed in the embryo during the first 8 weeks and alpha itself, which replaces zeta in the fetus and continues through adulthood. There are four beta-like chains. Epsilon and gamma are expressed in the embryo, gamma in the fetus, and delta plus beta in the adult. Adults possess 97% $\alpha_2\beta_2$, 2% $\alpha_2\delta_2$, and 1% $\alpha_2\gamma_2$, which persists from the fetus.

Each of the different globin chains is coded by at least one gene in the haploid genome. The alpha-like genes are clustered on human chromosome 16, and the beta-like genes are clustered on the short arm of chromosome 11. The gene organization within these two clusters is shown in Figure 20.25. Interestingly, the genes within a cluster are positioned relative to one another in the order of both their transcriptional direction and their developmental expression; that is, 5′ — embryonic—fetal—adult—3′.

The alpha gene cluster encompasses about 28 kb and includes one zeta gene, one zeta pseudogene, two alpha genes, and one alpha pseudogene. The pseudogenes contain sequences that are very similar to the active genes, but a determination of their nucleotide sequences has revealed that they do not code for a functional globin subunit found in the red cell. Pseudogenes are, in fact, rather common in eucaryotic genomes. In most cases they do not seem to be deleterious to the organism and probably arose via a duplication of a segment of DNA followed by evolutionary mutation. Both alpha genes are active and code for identical proteins.

The beta gene cluster contains five active genes and one pseudogene (for the gamma subunit). Of the five functional genes, two are for the gamma subunit and specify proteins that differ only at position 136, which is a glycine in the G variant and an alanine in the A variant. Only a single haploid gene exists for the epsilon, delta, and beta globin subunits.

Other mammalian species often have a different number of globin-like genes within the two clusters. For example, rabbits have only four beta-like genes, goats have seven, and mice have as many as nine. At least some of these additional genes are pseudogenes similar to the gamma pseudogene in the human genome.

Many patients have been identified who have abnormalities in hemoglobin structure or expression. Furthermore, in many cases the precise molecular defect that is responsible for these abnormalities is known. The two that have been the most extensively studied are *sickle cell anemia* and a family of diseases collectively called *thalassemias*.

Sickle Cell Anemia

A single base pair change within the coding region for the beta chain appears to be responsible for sickle cell anemia. This change occurs in the second position of the codon for position 6 of the beta chain. In the corresponding mRNA the codon, GAG, which specifies glutamate in normal beta chains, is converted to the codon, GUG, which specifies valine. The resultant hemoglobin, called hemoglobin S, has altered surface charge properties (because the polar side group of glutamate has been replaced by valine's distinctly nonpolar group), which is responsible for clinical symptoms of the sickle cell trait. Carriers of the sickle cell mutation can be detected by restriction enzyme digestion of a sample of the potential carrier's DNA followed by Southern hybridization with the β-globin cDNA as described in Clin. Corr. 20.3.

Alpha- and Beta-Thalassemias

Thalassemias are a family of related genetic diseases that occur in people who are frequently from or originate from populations living in the Mediterranean areas. If there is a reduced synthesis or a total lack of synthesis

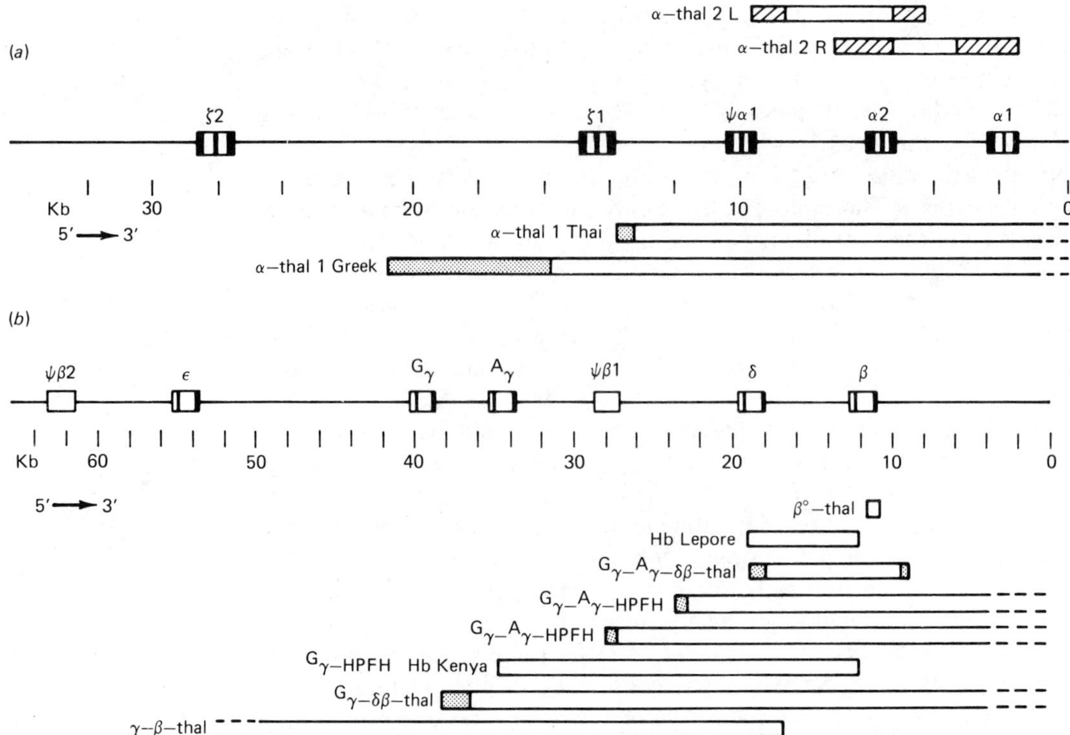

Figure 20.25
Gene organization for α-like and β-like genes of human hemoglobin.
(a) *Linkage arrangement of the human α-like globin genes and locations of deletions within the α-like gene cluster. The positions of the adult (α1, α2) and embryonic (ζ1, ζ2) α-like globin genes and the α-like pseudogene (ψα1) are shown. For each gene the black and white boxes represent coding (exon) and noncoding (intron) sequences respectively. The introns in ζ2 are assumed to exist by analogy with the other α-like genes. The locations of deletions associated with the leftward and rightward types of α-thalassemia 2 are indicated by the rectangles labeled α-thal 2 L and α-thal 2 R. The crosshatched boxes at the ends of these rectangles indicate regions of sequence homology. The breakpoints of each type of α-thalassemia 2 deletion can occur anywhere within the regions of homology. The locations of deletions associated with two cases of α-thalassemia 1 (α-thal 1 Thai and α-thal 1 Greek) are shown below the linkage map. The stippled boxes indicate uncertainty in the extent of each deletion.* (b) *Linkage arrangement of the human β-like globin genes and locations of deletions within the β-like gene cluster. The positions of the embryonic (ε), fetal ($^{G}\gamma^{A}\gamma$), and adult (δ, β) β-like globin genes and the two β-like pseudogenes (ψβ1, ψβ2) are shown. For each gene the black and white boxes represent the coding (exon) and noncoding (intron) sequences, respectively. The distribution of coding and noncoding sequences within ψβ1 and ψβ2 is not known. The locations of various deletions within the gene cluster are presented below the map. Open boxes represent areas known to be deleted; dashed lines indicate that the endpoint of the deletion has not been determined; and stippled boxes represent uncertainty in the extent of the deletions. For δβ-thalassemia and HPFH, the type of fetal globin chain that is produced ($^{G}\gamma$ and/or $^{A}\gamma$) is indicated in the name of each syndrome (for example, in $^{G}\gamma$-$^{A}\gamma$-δβ-thalassemia, the $^{G}\gamma$- and $^{A}\gamma$-globin chains are produced). The percentage of HbF observed in heterozygotes is given to the right of each deletion. An asterisk (*) indicates that the HbF is entirely of the $^{G}\gamma$-type.*

Redrawn from T. Maniatis, E. F. Fritsch, J. Lauer, and R. M. Lawn. *Annu. Rev. Genet.* 14:145, 1980. Copyright 1980, *Annu. Rev. Inc.*

of α-globin mRNA, the disease is classified as α-thalassemia; if the β-globin mRNA level is affected, it is called β-thalassemia. Thalassemias can be due to the deletion of one or more globin-like genes in either of the globin gene clusters or be caused by a defect in the transcription or processing of a globin gene's mRNA.

Since each human chromosome 16 contains two adjacent α-globin genes, a normal diploid individual has four copies of this gene. α-Thalassemic patients have been identified who are missing one to four α-globin genes. The condition in which one alpha gene is missing is referred to as *α-thal-1;* when two alpha genes are gone, the condition is *α-thal-2*. In both cases the individuals can experience mild to moderate anemia but may have no additional symptoms. When three alpha genes are missing, many more beta-chain molecules are synthesized than alpha-chain molecules resulting in the formation a globin tetramer of four beta chains, which causes *HbH disease* and accompanying anemia. When all four alpha genes are absent, the disease *hydrops fetalis* occurs, which is fatal at or before birth. Some chromosomal deletions that have been mapped in the alpha-gene cluster are shown in Figure 20.25.

The different β-thalassemias also exhibit differing degrees of severity and can be caused by a variety of defects or deletions. In one case the beta gene is present but has undergone a mutation in the codon 17, which generates a termination codon. In another case the beta gene is transcribed in the nucleus but no beta-globin mRNA occurs in the cytoplasm. Thus, a defect has occurred in the processing and/or transport of the primary transcript of the gene.

Other β-thalassemias are clearly caused by deletions within the beta gene cluster on chromosome 11, as illustrated in Figure 20.25. In some cases these deletions remove the DNA between two adjacent genes resulting in a new fusion gene. For example, in the normal person the linked delta and beta genes differ in only about 7% of their positions. In *Hb Lepore* a deletion has placed the front portion of the delta gene in register with the back portion of the beta gene. From this fusion gene a new beta-like chain is produced in which the N-terminal sequence of delta is joined to the C-terminal sequence of beta. Several variants of Hb Lepore are now known, and in each case the globin product is a composite of the delta and beta sequence, but the actual fusion junction is different.

Another fusion beta-like globin is produced in *Hb Kenya*. This deletion results in a gene product that contains the N-terminal sequence of the A_γ gene and the C-terminal sequence of the beta gene. Still another series of deletions has been found in which both the delta and beta genes are removed, causing *HPFH* (hereditary persistence of fetal hemoglobin). Frequently there are no clinical symptoms of this condition because fetal hemoglobin (α_2, γ_2) continues to be synthesized after the time at which gamma gene expression is normally turned off. (See Clin. Corr. 20.2 and 20.3.)

20.11 GENES FOR HUMAN GROWTH HORMONE-LIKE PROTEINS

Human growth hormone (somatotropin) is a single polypeptide of 191 amino acids. A larger precursor of the protein is synthesized in the somatotrophs of the anterior pituitary, and the mature form is secreted into the circulatory system. Growth hormone induces liver (and perhaps other) cells to produce other hormones called somatomedins, which are insulin-like growth factors that stimulate proliferation of mesodermal tissues such

CLIN. CORR. 20.2 PRENATAL DIAGNOSIS OF THALASSEMIA

If a fetus is suspected of being thalassemic because of its genetic background, recombinant DNA techniques can now be used to determine if one or more globin genes are missing from its genome. Fetal DNA can be easily obtained (in relatively small quantities) from amniotic fluid cells aspirated early during the second trimester of pregnancy. This DNA is digested with one of several specific restriction enzymes that places the globin genes on DNA restriction fragments that are several thousand base pairs in length. These fragments are separated by electrophoresis through an agarose gel and hybridized with radioactive cDNA for α- and/or β-globin using the Southern hybridization procedure described in Chapter 17. If one or more globin genes are missing, the corresponding restriction fragment will not be detected or its hybridization to the radioactive cDNA probe will be reduced (in the case when only one of two diploid genes is absent).

CLIN. CORR. 20.3 PRENATAL DIAGNOSIS OF SICKLE CELL ANEMIA

Sickle cell anemia can also be diagnosed using fetal DNA obtained by amniocentesis. In this case the disease is caused by a single point change that converts a glutamate codon to a valine codon in the sixth position of β-globin. In the normal β-globin gene, the sequence that specifies amino acids 5, 6, and 7 (Pro-Glu-Glu) is CCT-GAG-GAG. In a carrier of sickle cell anemia, this sequence is CCT-GTG-GAG. An A in the middle of the sixth codon has been changed to a T. The restriction enzyme *MstII* recognizes and cleaves the sequence CCT-GAG-G, which is present at this position in normal DNA but not the mutated DNA. Therefore, digestion of the fetal DNA with *MstII* followed by a Southern hybridization experiment with β-globin cDNA as the radioactive probe will reveal if this restriction site is present in one or both allelic copies of the gene. If it is absent in both copies, the fetus will likely be homozygous for the sickle trait; if it is missing in only one copy, the fetus will be heterozygous for the trait. There is increasing evidence that other genetic diseases (e.g., Huntington's disease, muscular dystrophy, and hemophilias) can be diagnosed by similar restriction fragment length polymorphisms (RFLP).

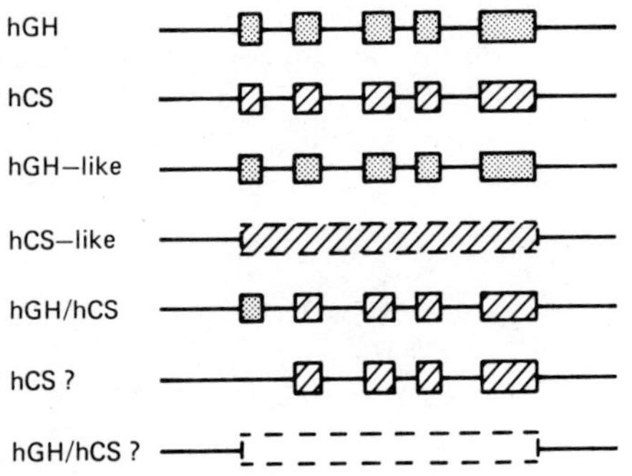

Figure 20.26
Members of the hGH gene family.
Various cloned members of the hGH family are diagrammed. Solid boxes: sequences most similar to hGH. Hatched boxes: sequences most similar to hCS. Open boxes: undefined sequences. Solid outlines: sequenced exons. Dashed outlines: sequences defined by restriction and hybridization analysis. The structure of the hCS gene, which has not been conclusively identified, is defined by analogy with hGH.

Redrawn from D. D. Moore, M. A. Conkling, and H. M. Goodman. *Cell* 29:285, 1982.

as bone, cartilage, and muscle. Infants with a deficiency in growth hormone become dwarfs, whereas those who produce too much become giants.

A closely related protein, displaying 85% homology with growth hormone, is human chorionic somatomammotropin (also called placental lactogen), which is synthesized in the placenta. The complete role of this hormone in normal fetal–maternal physiology is still unclear, but it participates in placental growth and contributes to mammary gland preparation for lactation during pregnancy.

Human growth hormone (hGH) and chorionic somatomammotropin (hCS) are an example of two very similar proteins that serve different biological roles and are synthesized in different tissues. Therefore, it was expected that their genes would also be closely related but expressed in a different tissue-specific fashion. Analysis of the cloned cDNAs and corresponding chromosomal genes for hGH and hCS has revealed that, indeed, the two DNA coding sequences are very similar. Furthermore, in addition to the hGH and hCS genes, there are at least five other closely related DNA regions in the human genome, as shown in Figure 20.26. It is not known if all of these regions are linked in the genomes.

These seven (and perhaps other) related DNA regions comprise the human growth hormone gene family. Some of these regions have been extensively characterized; others require more analysis. The hGH and hCS genes are each comprised of five coding regions (exons) interrupted by four intervening sequences (introns). The introns occur at exactly the same positions relative to the exons in the two genes. On the basis of partial analysis, the seven genes can be divided into three classes: hGH type, hCS type, and hybrids between hGH and hCS. The hGH class contains two members: the hGH gene itself and a hGH-like variant, which has a very similar but nonidentical sequence. The hCS class has three members: the hCS gene, a closely related hCS variant, and another variant that appears to have lost the first exon. The hybrid class contains at least one member in which the first exon is nearly identical to the first exon of hGH, whereas at least the next three exons are identical to the hCS cDNA. Thus, this gene may be the result of a crossover event between two other genes that occurs in or near the first intron.

Only the hGH and hCS genes are known to be transcribed. Transcription of the other related genes has not been shown. They may be pseudogenes; alternatively, they may be transcribed at a very low level or in other unsuspected tissues. Another surprising observation is that about 10% of the hGH produced in all examined pituitaries is missing 15 internal amino acids. These 15 amino acids are specified by the beginning of the second exon. Determination of the DNA sequence of the second exon revealed a sequence very similar to the upstream boundary between the first intron and second exon. This sequence occurs 15 codons in from this boundary. Therefore, the RNA splicing event that removes the first intron (between the first and second exon) may not occur with complete faithfulness. If the splicing reaction occurs at the alternative interior site 10% of the time, the shortened version of hGH would occur in the detected proportion. However, direct evidence for this alternative splicing event is not yet available. Therefore, the shortened hGH could be the result of the expression of another as yet unidentified member of the hGH gene family.

The expression of both the hGH and hCS genes is under the regulation of other hormones. Both thyroxine and cortisol stimulate increased transcription of these genes. Studies using cultured rat pituitary tumor cells reveal that these hormones act in a synergistic fashion in inducing growth hormone mRNA synthesis. Pituitary cells that have only about two molecules of growth hormone mRNA per cell can be stimulated to a level of

1,000 growth hormone mRNA molecules per cell—a 500-fold range that is comparable to the magnitude of induction of many bacterial operons.

The precise mechanism by which thryoxine and cortisol stimulate this increased transcription is not known. The regulation is clearly more complicated molecularly than the control of bacterial operon transcription. The regulatory hormones may be transported into the nucleus and either directly or in association with a binding protein affect transcription initiation of the gene. Alternatively, they may interact with other factors in the cell that in turn regulate the level of transcription. There is some evidence that DNA regulatory site for the glucocorticoid influence is upstream of the site at which transcription of the gene begins. Other evidence hints that it may also be located within the first intron. Clearly, our understanding of the mechanisms of eucaryotic gene regulation is still in its infancy, and activity in this research area will continue for many years to come.

20.12 BACTERIAL EXPRESSION OF FOREIGN GENES

Recombinant DNA techniques are now frequently used to construct bacteria that are "factories" for making large quantities of specific human proteins that are useful in the diagnosis or treatment of disease. The two examples to be illustrated here are the construction of bacteria that synthesize human insulin and human growth hormone.

Many factors must be considered in designing recombinant plasmids that contain a eucaryotic gene to be expressed in bacteria. First, the cloned eucaryotic gene cannot have any introns since the bacteria do not have the RNA-splicing enzymes that correctly remove introns from the initial transcript. Therefore, the actual eucaryotic chromosomal gene is usually not used for these experiments; instead, the cDNA or a synthetic equivalent of the coding sequence or a combination of both is placed in the bacterial plasmid.

Another consideration is that different nucleotide sequences comprise the binding sites for RNA polymerase and ribosomes in bacteria and eucaryotes. Therefore, to achieve expression of the desired protein it is necessary to insert the eucaryotic coding sequence directly behind a set of bacterial controlling elements. This has the advantage that the foreign gene is now under the regulation of the bacterial control elements, but its disadvantage is that considerable recombinant DNA manipulation is required to make the appropriate plasmid. Still other factors to be considered are that the foreign gene product must not be degraded by bacterial proteases or require modification before it is active (e.g., specific glycosylation events that the bacteria cannot perform) and must not be toxic to the bacteria. Furthermore, even when the bacteria do synthesize the desired product, it must be isolated from the 1,000 or more endogenous bacterial proteins.

Bacteria that Make Human Insulin

Insulin is produced by the β cells of the pancreatic islets of Langerhans. It is initially synthesized as preproinsulin, a precursor polypeptide that possesses an N-terminal signal peptide and an internal C-peptide of 33 amino acids that are removed during the subsequent maturation and secretion of insulin. The A-peptide (21 amino acids) and B-peptide (30 amino acids) of mature insulin are both derived from this initial precursor and are held together by two disulfide bridges. Bacteria do not have the processing enzymes that convert the precursor form to mature insulin. Therefore, the initial strategy for the bacterial synthesis of human insulin involved the production of the A chain and B chain by separate bacteria followed by purification of the individual chains and the subsequent formation of the proper disulfide linkages.

The first step was to use organic chemistry synthetic methods to prepare small single-stranded oligonucleotides (between 11 and 15 nucleotides in length) that were both complementary and overlapping with each other. When these oligonucleotides were mixed together in the presence of DNA ligase under the proper conditions, they formed a double-stranded fragment of DNA with termini equivalent to those that are formed by specific restriction enzymes (Figure 20.27). Furthermore, the sequences of the oligonucleotides were carefully chosen so that one of the two strands contained a methionine codon followed by the coding sequence of the A-chain of insulin and a termination codon. A second set of overlapping complementary oligonucleotides were prepared and ligated together to form another double-stranded DNA fragment that contained a methionine codon followed by 30 codons specifying the B-chain of insulin and a termination codon.

These two double-stranded fragments were then individually cloned at a restriction site in the β-galactosidase gene of the lactose operon that was

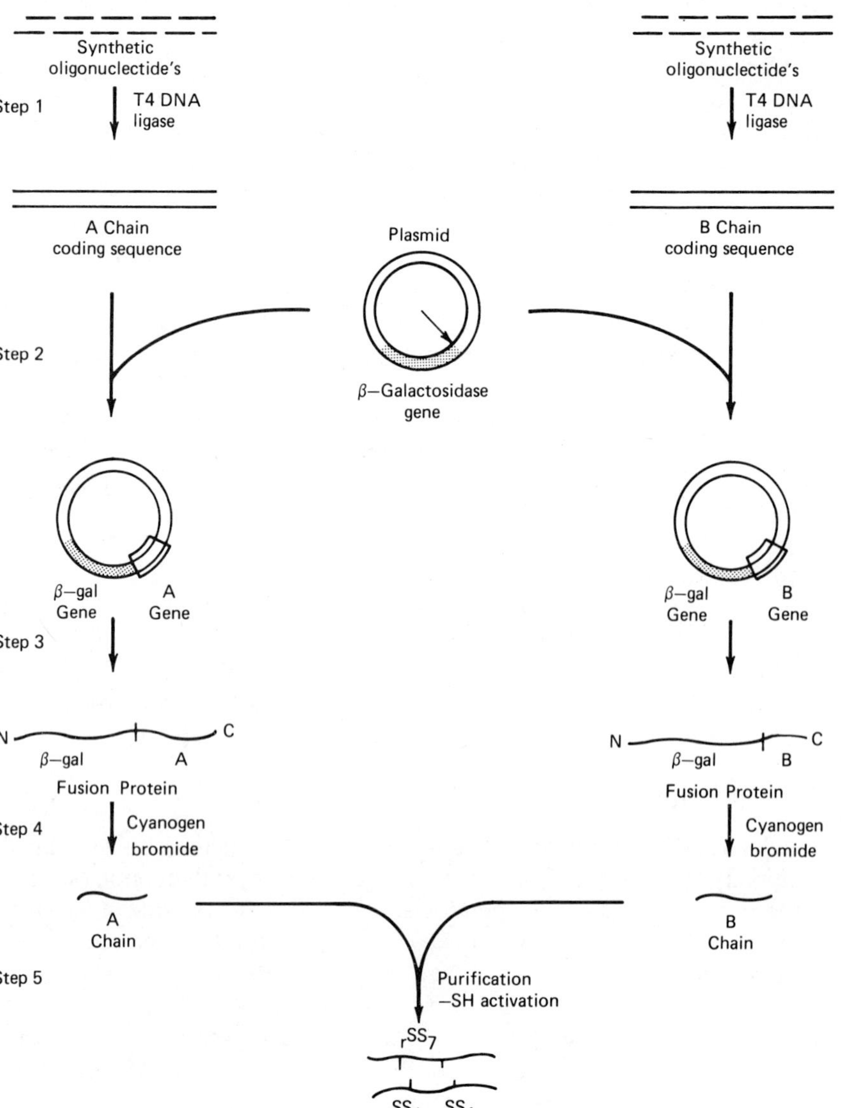

Figure 20.27
Bacterial expression of the A and B chains of human insulin.
Step 1: A series of overlapping, complementary, oligonucleotides (11 for the A chain and 18 for the B chain) were synthetically prepared and ligated together. One strand of the resulting small DNA fragments contained a methionine codon followed by the coding sequence for the A chain and B chain, respectively. Step 2: The small DNA fragments were ligated into a restriction site near the end of the β-galactosidase gene of the lactose operon that was on a plasmid. Step 3: The recombinant plasmids were introduced into E. coli *and the β-galactosidase gene induced with IPTG, an inducer of the lactose operon. A fusion protein was produced that contained most of the β-galactosidase sequence at the N-terminus and the A chain (or B chain) at the C-terminus. Step 4: Bacterial cell lysates containing the fusion protein were treated with cyanogen bromide, which cleaves peptide bonds following methionine residues. Step 5: The A and B chains were purified away from all of the other cyanogen bromide peptides using biochemical and immunological separation techniques. The —SH groups were activated and reacted to form the intra- and interchain disulfide bridges found in mature human insulin.*

Redrawn from R. Crea et al. *Proc. Natl. Acad. Sci. USA* 75:5765, 1980.

on a plasmid. These two recombinant plasmids were introduced into bacteria. The bacteria could now produce a fusion protein of β-galactosidase and the A-chain (or B-chain) that is under the control of the lactose operon. In the absence of lactose in the bacterial medium, the lactose operon is repressed and only very small amounts of the fusion protein are synthesized. Using induction with IPTG and some additional genetic tricks, the bacteria can be forced to synthesize as much as 20% of their protein as the fusion protein. The A-peptide (or B-peptide) can be released from this fusion protein by treatment with cyanogen bromide, which cleaves on the carboxyl side of methionine residues. Since neither the A- nor B-peptide contains a methionine, they will be liberated intact and can be subsequently purified to homogeneity. The final steps involve chemically activating the free —SH groups on the cysteines and mixing the activated A and B chains together in a way that the proper disulfide linkages form to generate molecules of mature human insulin.

Bacteria that Synthesize Human Growth Hormone

The strategy for generating a recombinant DNA plasmid from which bacteria can synthesize human growth hormone is somewhat different than

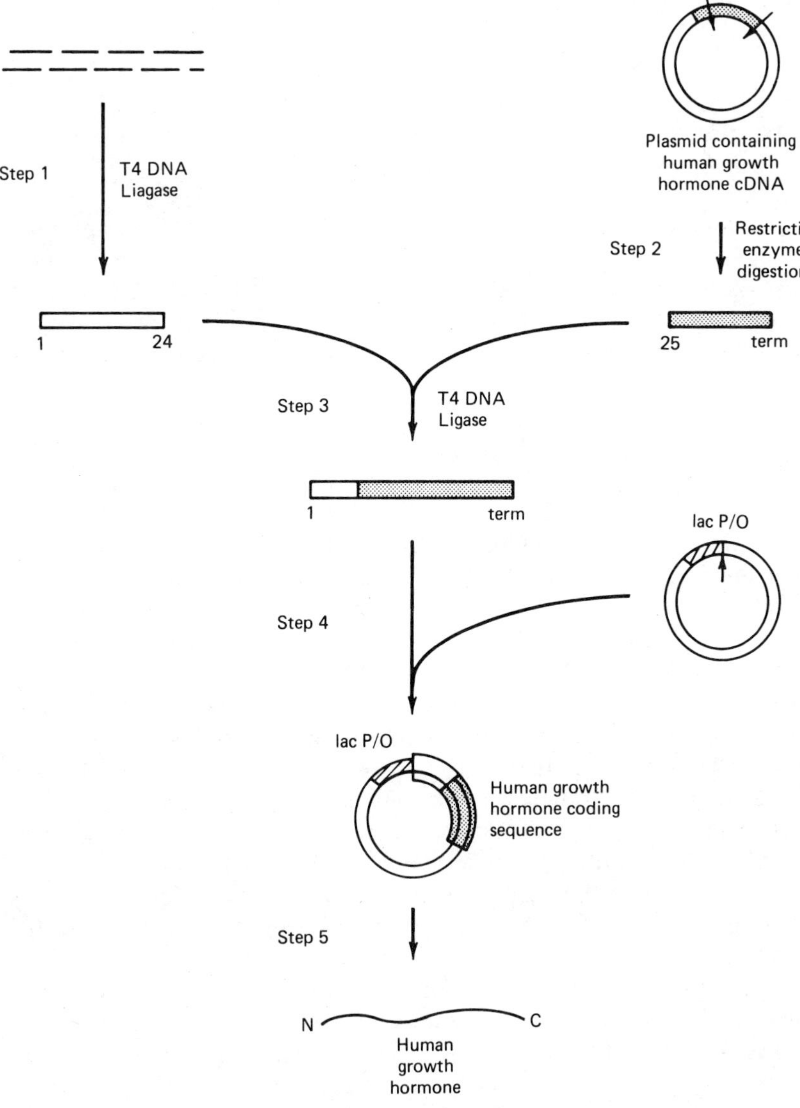

Figure 20.28
***Expression of human growth hormone in* E. coli.**
Step 1: Several overlapping, complementary, oligonucleotides were synthesized and ligated together. One strand of the resulting small DNA fragment contains the coding sequence for the first 24 amino aicds of mature *human growth hormone (after removal of the N-terminal signal peptide). Step 2: A recombinant plasmid with a full length human growth hormone cDNA (which is not expressed) is cleaved with restriction enzymes that release a fragment containing the complete growth hormone coding sequence after codon 24. Step 3: The synthetic fragment and the partial cDNA-containing fragment are ligated together to yield a new fragment containing the complete coding sequence of mature human growth hormone. Step 4: The new fragment is ligated into a restriction site just downstream from the lactose promoter–operator region cloned on a plasmid. Step 5: The resulting recombinant DNA plasmid is introduced into bacteria in which synthesis of human growth hormone can be induced with IPTG, an inducer of the lactose operon.*

for insulin synthesis. First, human growth hormone is 191 amino acids long so the total synthetic construction of the corresponding DNA coding sequence is more difficult (although certainly not impossible) than in the above insulin case. On the other hand, growth hormone is a single polypeptide so it is not necessary to deal with the production of two chains and their subsequent dimerization to form a protein with biological activity.

Because of these considerations, the growth hormone coding sequence was initially cloned into a bacterial expression plasmid using part of a cloned growth hormone cDNA and several synthetic oligonucleotides (Figure 20.28). The overlapping oligonucleotides were prepared so that, when ligated together, they would form a small double-stranded DNA containing the codon for the first 24 amino acids of mature human growth hormone. One end of this DNA fragment was designed so that the fragment could be ligated in front of a restriction fragment of growth hormone cDNA that provided the rest of the coding sequence, including the termination codon. The other end of the synthetic fragment was chosen so that the composite coding sequence could be easily inserted into a site immediately downstream of the promoter-operator-ribosome binding site of the lactose operon cloned on a plasmid. After the introduction into bacteria, the bacteria were induced with IPTG to transcribe this foreign coding region and the greatly overproduced human growth hormone subsequently purified away from the bacterial proteins.

20.13 INTRODUCTION OF THE RAT GROWTH HORMONE GENE INTO MICE

The previous section described examples of the use of bacteria to produce large quantities of a human protein that is used to treat a disease. In some cases it is now possible to microinject molecules of purified RNA or DNA directly into living cells. This provides a very powerful approach for identifying conditions under which specific genes are expressed in eucaryotic cells. One of the most dramatic illustrations of this approach is the microinjection of a chromosomal DNA fragment containing the structural gene for *rat* growth hormone into the pronuclei of fertilized *mouse* eggs. The eggs were then reimplanted into the reproductive tracts of foster mouse mothers. Some of the mice that developed from this procedure were *transgenic;* one or more copies of the microinjected growth hormone gene integrated into a host mouse chromosome at an early stage of embryo development. These foreign genes were transmitted through the germ line and became a permanent feature in the host chromosomes of the progeny (Figure 20.29).

Analysis of these transgenic mice revealed that in some cases several tandem copies of the rat growth hormone gene had integrated into a mouse chromosome; in other cases only one gene copy was present. In all cases at least some transcription occurred from the integrated gene(s), and in a few cases a dramatic overproduction of rat growth hormone resulted. In these latter cases, as much as 800 times more growth hormone was present in the transgenic mice than in normal mice, resulting in animals more than three times the size and weight of their unaffected littermates.

These results present many potential experimental possibilities for the future and raise a number of issues. One implied possibility is the use of similar growth hormone gene insertions to stimulate rapid growth of com-

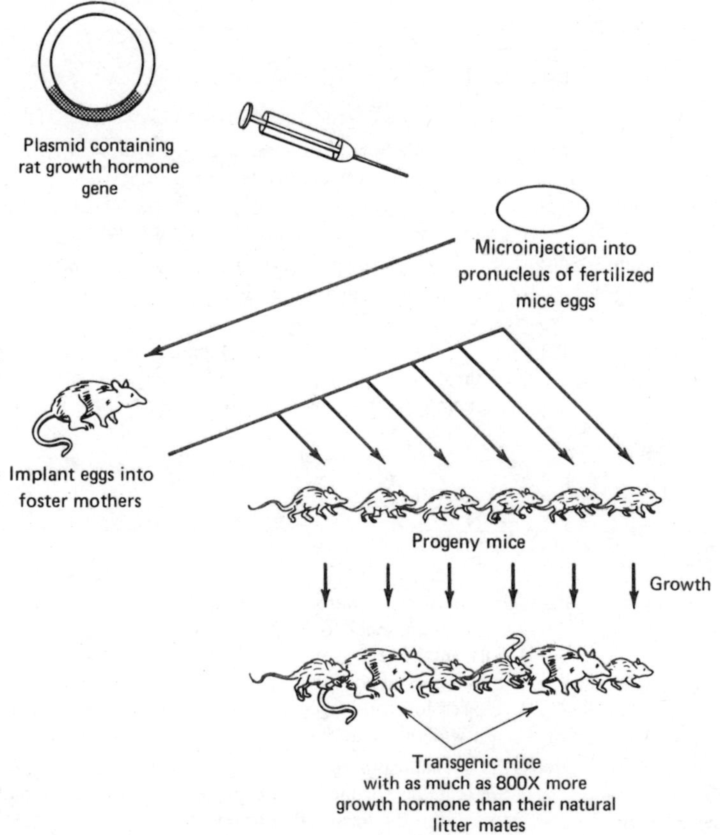

Figure 20.29
Schematic illustration of the introduction of rat growth hormone gene into mice. *Copies of a recombinant plasmid DNA containing the rat growth hormone gene were microinjected into fertilized mouse eggs that were reimplanted into foster mothers. Some of the progeny that resulted contained the foreign gene integrated into their own genome and over-expressed growth hormone, growing much larger than their normal-sized litter mates.*
From R. D. Palmiter, R. L. Brenster, R. E. Hammer, et al. *Nature* 300:611, 1982.

mercially valuable animals. This could result in a shorter production time and increased efficiency of food utilization. Another long-term possibility is the use of this approach to correct certain human genetic diseases or mimic the diseases in experimental animals so that they can be studied more carefully. One obvious human disease that is a candidate for this "gene therapy" approach is thalassemia. For example, an individual with two to three missing α-globin genes might benefit tremendously from receiving bone marrow transplants of his or her own cells that have been established in culture and microinjected with additional copies of the normal α-globin gene. At the moment this is not technically feasible, but in the future it may become possible. A final point is that insertion of normal genes into human somatic cells of a defective tissue or organ (instead of germ line cells) does not result in transmission of these genes to the progeny. This lessens the ethical considerations that enter into the design of experiments intended to alter germ line characteristics.

BIBLIOGRAPHY

Procaryotic Gene Expression

Cohen, S. N., and Shapiro, J. A. Transposable genetic elements. *Sci. Am.* 242:40, 1980.

Miller, J. H. The *lac* gene: Its role in *lac* operon control and its use as a genetic system, in J. H. Miller and W. S. Resnikoff (eds.), *The Operon.* Cold Spring Harbor Laboratory, 1978, p. 31.

Platt, T. Regulation of gene expression in the tryptophan operon of *Escherichia coli,* in J. H. Miller and W. S. Resnikoff (eds.), *The Operon.* Cold Spring Harbor Laboratory, 1978, p. 263.

Simon, M. et al. Phase variation: Evolution of a controlling element. *Science* 209:1370, 1980.

Eucaryotic Gene Expression

Brown, D. D. Gene expression in eukaryotes. *Science* 211:667, 1981.

Maniatis, T., Fritsch, E. F., Laurer, J., and Lawn, R. M. The molecular genetics of human hemoglobins. *Annu. Rev. Genet.* 14:145, 1980.

Moore, D. D., Conkling, M. A., and Goodman, H. M. Human growth hormone: A multigene family. *Cell* 29:285, 1982.

Orkin, S. H., and Kazazian, H. H., Jr. The mutation and polymorphism of the human beta-globin gene and its surrounding DNA. *Annu. Rev. Genet.* 18:131, 1984.

Palmiter, R. D., Brinster, R. L., Hammer, R. E., et al. Dramatic growth of mice that develop from eggs microinjected with metallothionein-growth hormone fusion genes. *Nature* 300:611, 1982.

QUESTIONS

J. BAGGOTT AND C. N. ANGSTADT

*Q*uestion *T*ypes are described inside the front cover.

A. Induction — C. Both
B. Repression — D. Neither

1. (QT4) Occur(s) in response to the presence of a specific low molecular weight organic compound.

2. (QT3) A. Affinity of the *lac* repressor for the *lacO* sequence when lactose is present in the growth medium of *E. coli.*
B. Affinity of the *lac* repressor for the *lacO* sequence when lactose is absent.

3. (QT2) The *E. coli lacZYA* region will be transcribed at a rate greater than the low basal rate if:
   1. there is a defect in binding of the inducer to the product of the *lacI* gene.
   2. glucose and lactose are both present in the growth medium, but there is a defect in the cell's ability to synthesize cAMP.
   3. glucose and lactose are both readily available in the growth medium.
   4. the operator has mutated so it can no longer bind repressor.

4. (QT2) An operon:
   1. includes structural genes.
   2. is expected to code for polycistronic mRNA.
   3. contains control sequences such as an operator.
   4. can have only a single promoter.

A. Tryptophan synthesis in *E. coli.* — C. Both
B. Lactose utilization by *E. coli.* — D. Neither

5. (QT4) Controlled by a repressor.
6. (QT4) Controlled by feedback inhibition.
7. (QT4) Affected by a secondary promotor.

8. (QT3) A. Basal level of mRNA for the first gene in the *E. coli* tryptophan operon.
B. Basal level of mRNA for the last gene in the *E. coli* tryptophan operon.

A. *trp* operator site — C. Both
B. *trp* attenuator site — D. Neither

9. (QT4) Site at which tryptophan availability plays a role in preventing transcription.

10. (QT1) Ribosomal operons:
    A. all contain genes for proteins of just one ribosomal subunit.
    B. all contain genes for proteins of both ribosomal subunits.
    C. all contain genes for only ribosomal proteins.
    D. expression can be regulated at the level of translation.
    E. are widely separated in the *E. coli* chromosome.

11. (QT2) Transposons:
    1. are a means for the permanent incorporation of antibiotic resistance into the bacterial chromosome.
    2. contain short inverted terminal repeat sequences.
    3. include at least one gene that codes for a transposase.
    4. code for an enzyme that synthesizes guanosine tetraphosphate and guanosine pentophosphate, which inhibit further transposition.

12. (QT1) Repetitive DNA:
A. is common in bacterial and mammalian systems.
B. is all uniformly distributed throughout the genome.
C. includes DNA that codes for rRNA.
D. consists mostly of DNA that codes for enzymes catalyzing major metabolic processes.
E. is resistant to the action of restriction endonucleases.

A. α-Globin gene — C. Both
B. β-Globin gene — D. Neither

13. (QT4) Free of introns.
14. (QT4) Defective in sickle cell anemia.
15. (QT4) Four copies are found in the normal human.
16. (QT1) In designing a recombinant DNA for the purpose of synthesizing an active eucaryotic polypeptide in bacteria all of the following should be true *except:*
    A. the eucaryotic gene may contain its usual complement of introns.
    B. the foreign polypeptide should be resistant to degradation by bacterial proteases.
    C. glycosylation of the polypeptide should be unnecessary.
    D. the foreign polypeptide should be nontoxic to the bacteria.
    E. bacterial controlling elements are necessary.

ANSWERS

1. C Induction is an increase in transcription of a structural gene in response to a low molecular weight substrate (p. 768). Repression is a decrease in transcription in response to a low molecular weight product (p. 768).
2. B The repressor protein has a high affinity for the *lacO* (operator) sequence of DNA unless the repressor has previously bound the inducer, allolactose (p. 770). The complex of inducer with repressor, however, does not bind to the operator site.
3. D Only 4 true. 1: The product of the lacI gene is the repressor protein. When this protein binds an inducer, it changes its conformation, no longer binds to the operator site of DNA, and transcription occurs at an increased rate. Failure to bind an inducer prevents this sequence. 2 and 3: In the presence of glucose catabolite repression occurs. Glucose lowers the intracellular level of cAMP. The catabolite activator protein (CAP) then cannot complex with cAMP, so there is no CAP–cAMP complex to activate transcription. The same would occur if the cell had lost its capacity to synthesize cAMP (p. 774). 4: If the operator is unable to bind repressor, the rate of transcription is greater than the basal level (p. 771).
4. A 1, 2, 3 true. 1, 2, 3: An operon is the complete regulatory unit of a set of clustered genes, including the structural genes (which are transcribed together to form a polycistronic mRNA), regulatory genes, and control elements, such as the operator (p. 767). 4: An operon may have more than one promoter, as does the tryptophan and galactose operon of *E. coli* (p. 777).
5. C Both are controlled by a repressor. A is "turned on" unless tryptophan is present, whereas B is "turned off" unless lactose is present (p. 775).
6. A Tryptophan synthesis is controlled both by control of enzyme synthesis and by regulation of enzyme activity through feedback (p. 776).
7. A The tryptophan operon has a secondary promoter (p. 777).
8. B The tryptophan operon has two promoters. The first one is controlled by the repressor, whereas the second one (located in the *trpD* gene) is not. Although the *trpP2* promoter is not very efficient, it does promote transcription, so more mRNA for the last three genes of the *trp* operon are made than for the entire *trp* operon (p. 777).
9. C At the *trp* operator site the repressor–tryptophan complex binds, preventing transcription. Tryptophan serves as a corepressor (p. 775). At the *trp* attenuator site, the presence of trp-tRNA (which can be made only if tryptophan is available) permits synthesis of leader peptide, which contains two *trp* residues. If this occurs, the mRNA on which it occurs folds into a secondary structure that causes termination of its own synthesis (p. 778). Figure 20.10 shows the secondary structure, a hairpin loop with many CG pairs in its stem, followed by an oligo-U sequence (bases 114–140). This is a standard termination sequence.
10. D A, B, C, E: The genes for half of the ribosomal proteins are in two major clusters. There is no pattern to the distribution of genes for the proteins of the two ribosomal subunits, and they are intermixed with genes for other proteins involved in protein synthesis. D: Excess ribosomal protein binds to its own mRNA, preventing initiation of further synthesis of that protein (p. 782).
11. A 1, 2, 3 true. 1: See p. 787. 2 and 3: See p. 784. 4: These guanosine phosphates are synthesized by the product of the *relA* gene; they inhibit initiation of transcription of the rRNA and tRNA genes, shutting off protein synthesis in general. This is the stringent response (p. 783).
12. C A and B: Highly repetitive and moderately repetitive DNA are found only in eucaryotes. Highly repetitive sequences tend to be clustered, as are some moderately repetitive sequences (p. 790). C: This makes sense, since many copies of these structural elements are needed (p. 791). D: Most repetitive DNA does not code for a stable gene product (p. 791). E: The *Alu* family of moderately repetitive DNA is named for the restriction endonuclease that cleaves them.
13. D Both genes contain two introns (p. 794). Introns are very common in eucaryotic genes.
14. B Sickle cell anemia is due to a genetic error in the β-chain gene (p. 795).
15. A The normal human diploid cell contains four copies of the α-globin gene and two copies of the β-globin gene (p. 797).
16. A A and C: The bacterial system has no mechanism for posttranscriptional modification of mRNA or for posttranslational (or cotranslational) modification of protein. E: Bacterial systems need bacterial promoters, etc. (p. 799).

21

Metabolism of Individual Tissues

JOHN F. VAN PILSUM

21.1 SKELETAL MUSCLE

Structure

Skeletal muscle is composed of fibers that are striated in both the longitudinal and transverse directions. The longitudinal striations are the result of the fiber containing many myofibrils, and the transverse striations are due to the fact that the composition of the myofibril varies regularly along its length. A diagrammatic representation of skeletal muscle fibers is shown in Figure 21.1. Two main striations of the myofibril have been identified by light microscopy. The A band, or anisotropic band, is birefringent; it appears light in the dark field of a polarizing microscope because the light absorbing properties of the substance in this band are unequal in all directions. This property suggests a regular geometric arrangement of the molecules in the A band. The I, or isotropic band, is not birefringent, indicating that its light properties are equal in all directions, and it contains a substance of a more random arrangement. Other bands have been detected in the myofibril by staining procedures and light microscopy. A diagrammatic representation of the structure of the myofibril is shown in Figure 21.2. The Z band appears as a narrow dark line in the middle of the I band. The H zone appears in the center of the A band. The sarcomere is defined as the length of a myofibril bounded by two adjacent Z bands. In the center of the sarcomere at the H zone is an M line. The myofibril is composed of thick and thin myofilaments, and the bands observed by light microscopy can be explained by the regular arrangement of the thick and thin myofilaments in the myofibril. The H and the I bands contain only thick or thin filaments, respectively, whereas the A band contains both thick and thin filaments. The Z line is a dense amorphous material to which the thin filaments are attached and the M line is an enlargement of the thick filaments. A regular spatial arrangement of the thick and thin filaments around each other is also known to occur. Each thick filament is surrounded by six thin filaments and each thin filament is surrounded by three thick filaments. The thick filaments are studded with

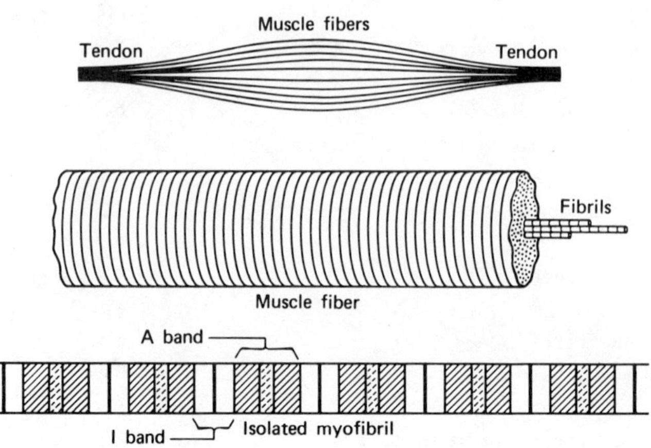

Figure 21.1
Diagrammatic representation of skeletal muscle fibers and myofibrils.

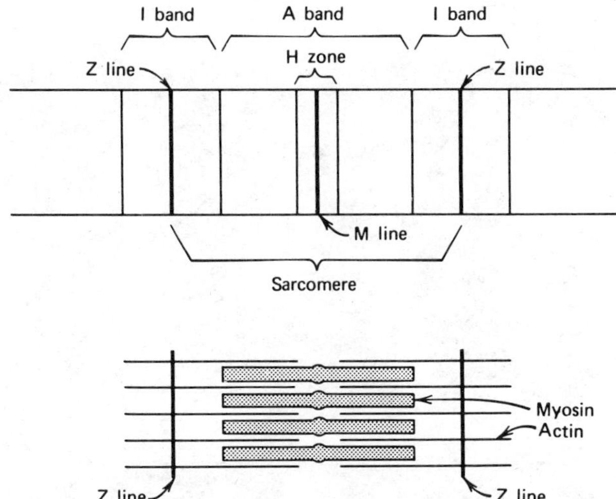

Figure 21.2
Diagrammatic representation of the structure of the myofibril.

projections that touch the thin filaments. The thick and thin filaments contain the proteins myosin and actin, respectively.

Each muscle fiber is covered with a sarcolemma, which has an outer layer of collagenous and reticular fibers that insert into the basement membrane. The sarcolemma is not a smooth membrane but has small indentations that act as a pinocytosis system for the transport of materials into the fibers. The myofibrils are imbedded in the cytoplasm of the muscle cell, called the sarcoplasm. The proteins myoglobin, myoalbumin, myogen, and droplets of triacylglycerol and glycogen are present in the sarcoplasm. Myoglobin, or muscle hemoglobin, accepts O_2 from hemoglobin in the blood; thus oxygenated myoglobin serves as a reservoir of O_2 for the muscle fibers. Myogen has the properties of the enzyme aldolase. The function of myoalbumin is not known. Each striated muscle fiber contains many nuclei and mitochondria, a Golgi apparatus, and a specialized endoplasmic reticulum called the sarcoplasmic reticulum. The sarcoplasmic reticulum is a closed tubular membrane system that extends throughout the sarcoplasm and surrounds the bundles of contractile proteins within each myofibril. A three-dimensional model of the sarcoplasmic membrane system in the rat diaphragm is shown in Figure 21.3. It is composed of longitudinal sections that run the length of each sarcomere, and the terminal cisternae portions which are observed in electron micrographs as triads at the level of the Z band. The other intracellular membrane in muscle is the transverse tubule system which occurs at the Z band, or in some muscles it lies closer to the A–I junction. The T tubule is a tubular invagination of the sarcolemma which penetrates to the interior of the muscle fiber and is considered to be extracellular space. The terminal cisternae portions of the sarcoplasmic reticulum abut the transverse system; however, the two membranes remain separated by a section of sarcoplasm.

At least two types of striated muscle fibers are found in skeletal muscle—red fibers and white fibers. Other types of fibers, intermediary between the red and white varieties, have also been detected. Red fibers have a large amount of sarcoplasm, and contain more nuclei and mitochondria than white fibers. Red fibers also contain a greater amount of myoglobin, mitochondrial iron-containing cytochromes, and more lipid droplets than white fibers. Red fibers twitch longer, are more easily tetanized, and contract more slowly than white fibers. Three types of skeletal muscle fibers have been designated based on their metabolic and physiological characteristics. The fast-twitch white muscle fibers have a low

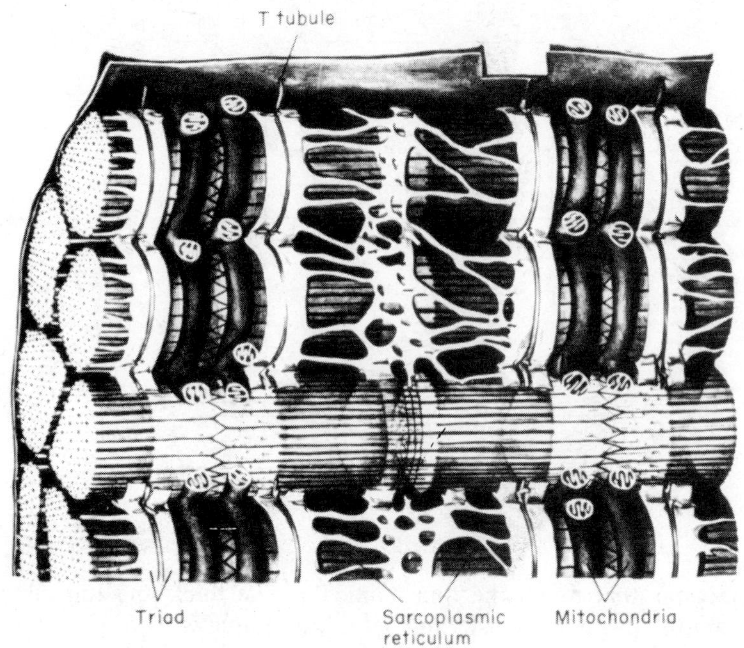

Figure 21.3
Three-dimensional model of the sarcoplasmic membrane system in the rat diaphragm.
Reprinted with permission from H. Schmalbruch, *Advan. Anat., Embryol. Cell. Biol.* 43:1, 1970.

respiratory capacity, a high glycogenolytic capacity and a high myosin ATPase activity; the fast-twitch red fibers have a high respiratory capacity, a high glycogenolytic capacity, and high myosin ATPase activity. The slow-twitch red fibers have a high respiratory capacity, a low glycogenolytic capacity and low myosin ATPase activity. (See Clin. Corr. 21.1.)

CLIN. CORR. **21.1**
DUCHENNE'S MUSCULAR DYSTROPHY

A number of components in the sarcoplasm of skeletal muscle are found to be present in lower than normal levels in patients with Duchenne's muscular dystrophy: glycolytic enzymes, creatine phosphokinase, adenosine monophosphate deaminase, ATP, phosphocreatine, K^+, and myoglobin. Most of the biochemical abnormalities have been interpreted as a failure to retain sarcoplasmic substances within the muscle fiber and *cannot* be stated to be specific inborn errors of metabolism.

Contractile Proteins

Myosin is the most abundant protein in skeletal muscle, accounting for 60–70% of total protein, and is the major protein of the thick filaments. Myosin is composed of two identical heavy chains about 200,000 mol wt, and four light chains of about 20,000 mol wt. Myosin has a double-headed globular region joined to a double stranded α-helical rod. The schematic representation of the myosin molecule is shown in Figure 21.4. The globular portions of the myosin have ATPase activity and will combine with

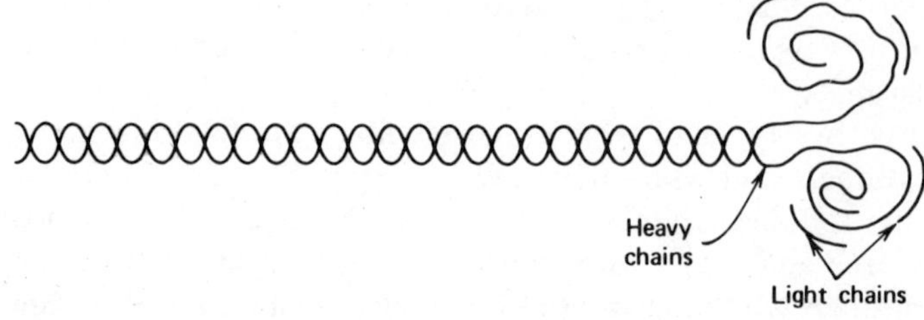

Figure 21.4
Schematic representation of the myosin molecule.
The myosin molecule consists of two heavy chains and four light chains. Each heavy chain has a molecular weight of ~200,000 and has an α-helical portion and a globular head. The four light chains are bound to the two globular heads of the heavy chains.

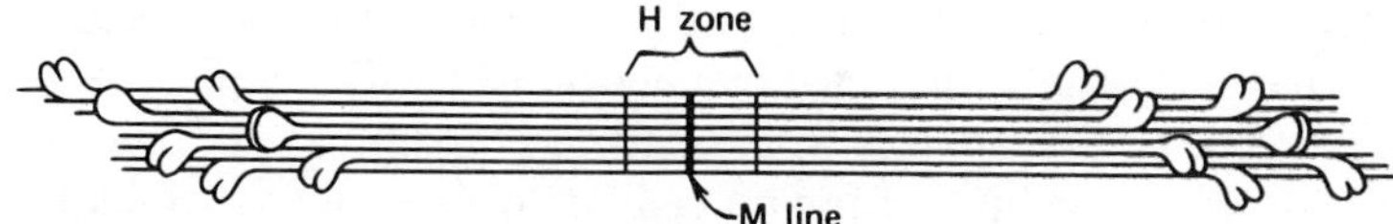

Figure 21.5
Schematic representation of a thick filament formed by the assembly of the individual myosin molecules (Figure 21.4).
The α-helical ends of the myosin heavy chains are joined together in a staggered array, with the result that every fourth pair of myosin heads are aligned with each other. The midpoint of the aggregate of the individual molecules (~400) is the M line, which is in the middle of the H zone.

actin. Two of the light chains (one on each globular head) are identical and may be removed with no loss of ATPase activity. The other two light chains are not identical and are required for both the ATPase activity and the actin-binding properties of the myosin. A region of the α-helical portion of the myosin molecule adjacent to the globular portion of the molecule is thought to be flexible. A thick myofilament consists of about 400 rods of myosin bound together as depicted in Figure 21.5. The midpoint at which the myosin fibers meet tail to tail is at the M line.

About 20–25% of total muscle protein is the globular protein, called actin. The globular form is called G-actin and consists of a single peptide chain. F-actin consists of a double-stranded helix of G-actin monomers forming a thin filament. A complex called actomyosin is formed when a solution of actin is mixed with a solution of myosin. Strands of actomyosin will contract in the presence of ATP.

Tropomyosin is a rod-shaped molecule found associated with the actin filaments. Tropomyosin consists of two similar α-helical peptide chains coiled around each other in a head-to-tail assembly and is attached noncovalently to the chains of F-actin. Troponin, a spherical molecule bound to the actin filaments, consists of three different subunits. A calcium-binding protein (TN–C) has two high and two low affinity Ca^{2+} binding sites and is the only subunit of troponin to bind Ca^{2+}. The binding of Ca^{2+} to TN–C induces a conformational change in this subunit which is recognized by both tropomyosin and actin. The inhibitory protein (TN–I) inhibits the interaction of actin with myosin and also inhibits ATPase activity. Tropomyosin-binding subunit (TN–T) mediates the binding of the TN–I and TN–C subunits to the actin–tropomyosin complex and also the binding of TN–I to TN–C. The probable relative positions of tropomyosin, troponin, and actin in the thin filament of muscle is shown in Figure 21.6. α-Actinin is associated with the Z line in the sarcomere and is the

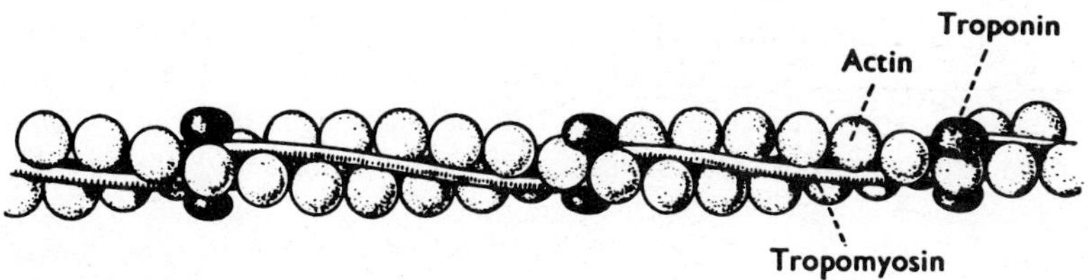

Figure 21.6
Probable relative positions of tropomyosin, troponin, and actin in the thin filament of muscle.
Tropomyosin lies in the groove of the helix, which is undisturbed. There is one tropomyosin and one troponin for each seven pairs of G-actin.
Reprinted with permission from H. E. Huxley, Regulation of muscle function by tropomyosin–troponin, in Y. Hatefi and L. Djavadi-Ohaniance (eds.), *The structural basis of membrane function*. New York: Academic Press, 1976, p. 319.

TABLE 21.1 Molecular Weights of Skeletal Muscle Contractile Proteins

Myosin	500,000
Heavy chain	200,000
Light chain	20,000
Actin monomer (G-actin)	42,000
Tropomyosin	70,000
Troponin	76,000
TN–C subunit	18,000
TN–I subunit	23,000
TN–T subunit	37,000
α-Actinin	200,000
C-protein	150,000
β-Actinin	60,000
M-protein	100,000

protein that binds the actin filaments to the Z line. γ-Component is found associated with the troponin molecule and has now been identified as creatine kinase. C-protein is isolated from preparations of myosin and is involved in the assembly of the myosin molecules into the thick filament. β-actinin is found associated with the F-actin molecules and is a length-determining factor for the assembly of the thin filaments. M-protein is detected in the M line. The approximate molecular weights of skeletal muscle contractile proteins are summarized in Table 21.1.

Mechanism of Muscle Contraction

Skeletal muscle contracts as the length of the sarcomeres decreases, which is accomplished by the actin filaments sliding past the myosin filaments into the H zone of the sarcomere. The sliding filament model of muscle contraction involves the ATPase activity of the globular portion of the myosin molecule, its ability to bind to actin, and the manner in which the molecules of myosin aggregate to form the thick filament. The fact that the globular heads of the myosin molecules in the filament are pointed in opposite directions on either side of the M line of the sarcomere means there is a directionality to the contractile force. The binding of the heads of myosin with the actin, forming cross-linkages, will exert a force on one side of the A band that is opposite to that on the other side of the A band. In other words, the thick filament is bipolar. The thin filaments also exhibit polarity. Myosin is degraded by proteolytic enzymes into particles called light and heavy meromyosin. The heavy meromyosin particles are the globular heads of the myosin molecule connected to short segments of the α-helical portions of the myosin molecule, and light meromyosin particles are the remaining segments of the α-helical chains. Particles of heavy meromyosin combine with actin filaments with the globular myosin heads attached to the actin filaments, and with the α-helical portions of the heavy meromyosin orientated in the same angular direction from the actin filament. Actin filaments, connected to the Z line, combine with heavy meromyosin with the α-helical portions orientated in opposite angular directions on either side of the Z line. Figure 21.7 is a schematic representation of the attachment of heavy meromyosin to actin filaments isolated free (*a*) and to actin filaments attached to the Z line (*b*). Since the thin filaments on one side of the Z line have the same orientation, whereas those on the opposite side have the reverse polarity, the force generated by the myosin heads binding to the actin is in the direction of the M line on either half of the sarcomere.

(*a*)

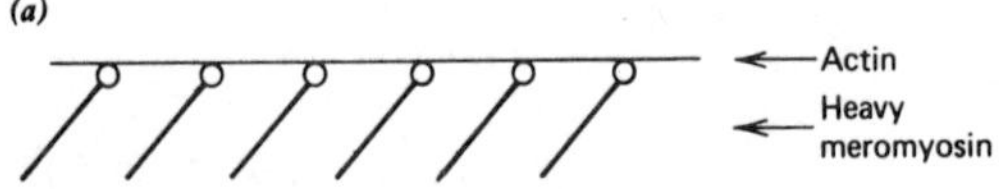

(*b*)

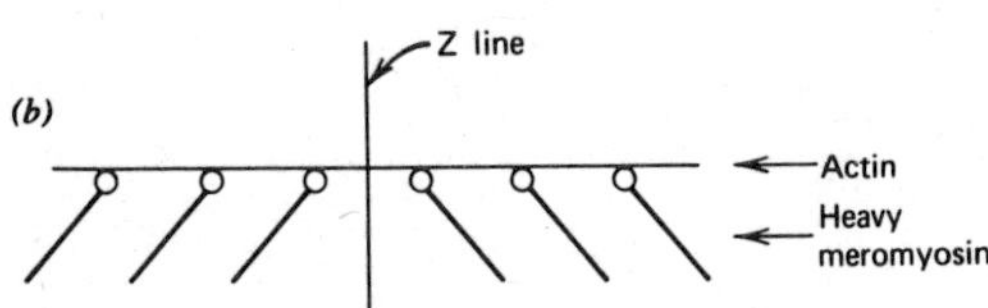

Figure 21.7
Schematic representation of the attachment of heavy meromyosin to actin filaments isolated free (a) and attached to the Z line (b).
Particles of heavy meromyosin combine with actin filaments with the α-helical portions of the particles orientated in the same angular direction from the actin filament (a). Heavy meromyosin particles combine with actin filaments attached to the Z line with the α-helical portions orientated in opposite angular directions on either side of the Z line (b).

The events thought to take place in skeletal muscle contraction are as follows. An electrical impulse from the motor nerve is transmitted to the muscle at the neuromuscular junction. This potential is transferred across the synapse by the release of acetylcholine. The impulse spreads over the entire sarcolemma, which then becomes depolarized. A potential difference of about 60 mV exists between the inside and outside of the resting muscle cell (positive outside). As the impulse spreads over the sarcolemma, the potential disappears as Na^+ enters the sarcoplasm, followed by a loss of K^+ to the exterior of the cell. The Na^+ channel in the sarcolemma consists of a large molecular weight glycoprotein. The transverse tubule becomes depolarized and transmits the impulse to all of the myofibrils within the fiber. The depolarization impulse is transmitted to the adjacent terminal cisternal portions of the sarcoplasmic reticulum from the transverse tubules in some unknown manner (possibly by a Ca^{2+}, Mg^{2+}-ATPase—an ATP energized Ca^{2+} pump). This Ca^{2+}, Mg^{2+}-ATPase is not the same protein that is found in the longitudinal sarcoplasmic reticulum. The longitudinal sarcoplasmic reticulum rapidly discharges

Ca^{2+} into the sarcoplasm. The efflux of Ca^{2+} involves a phosphorylation of ATPase and an opening of Ca^{2+} channels. The Ca^{2+} concentration in the sarcoplasm is increased by a factor of at least 10-fold. The Ca^{2+} binds to the TN–C subunit of troponin, which undergoes a conformational change that is recognized by tropomyosin. The tropomyosin moves, relative to the actin, which exposes the myosin binding sites on the actin. The actomyosin complex thus formed increases the ATPase activity. The ATP–myosin complex is converted to an ADP, P_i-myosin complex. Actin combines with the ADP-P_i-myosin complex to form actomyosin with the release of ADP and P_i. During this cycle, force is generated only while the myosin cross bridge is attached to the actin site. The net result of the series of reactions is the splitting of ATP to ADP with the accompanying release of energy that powers the muscle contraction. A major feature of this contraction cycle is that actin has a high affinity for myosin and for the myosin, ADP, P_i-complex but has a low affinity for the myosin–ATP complex. Therefore, actin alternately binds to myosin and is released from myosin as ATP is hydrolyzed. In simple terms, the energy produced by the hydrolysis of ATP is used for the attachment and detachment of actin and myosin to force the actin filaments into the H zone. The force-generating step involves a conformation change in the flexible region of the myosin molecule near the globular head. Relaxation of the skeletal muscle occurs with cessation of the nervous impulse and the return of the sarcolemma and transverse tubule to its original polarized state. That is, Na^+ and K^+ are pumped out and into the cell, respectively, by the Na^+-K^+-ATPase. The permeability of the sarcoplasmic reticulum to Ca^{2+} decreases, and Ca^{2+} is rapidly transported from the sarcoplasm into the sarcoplasmic reticulum by the ATP-dependent Ca^{2+} pump. (See Clin. Corr. 21.2.)

CLIN. CORR. 21.2
LOW SERUM POTASSIUM

Episodic muscle weakness may occur in situations of low serum potassium levels. The disorder may be due to a membrane abnormality since the sarcoplasmic reticulum in paralyzed muscle is inexcitable to electrical stimulation. The possibility exists that there is impaired release of Ca^{2+} from the sarcoplasmic reticulum.

Role of the Sarcoplasmic Reticulum in Muscle Contraction and Relaxation

The importance of the sarcoplasmic reticulum in furnishing Ca^{2+} for muscle contraction is well established. Sarcoplasmic reticulum from skeletal muscle is readily isolated and will reseal into closed vesicles that can be isolated by differential centrifugation. The vesicles retain the ability to pump Ca^{2+} coupled to ATP hydrolysis by the Ca^{2+}-dependent ATPase.

Four proteins are found in skeletal muscle sarcoplasmic reticulum. The major protein is a Ca^{2+},Mg^{2+}-dependent ATPase, which has two high affinity Ca^{2+}-binding sites. This ATPase forms a phosphoprotein intermediate when hydrolyzing ATP and its decomposition is activated by Mg^{2+} as follows:

$$\text{E} + \text{ATP} \xrightarrow{Ca^{2+}} \text{E—P} + \text{ADP}$$

$$\text{E—P} + H_2O \xrightarrow{Mg^{2+}} \text{E} + P_i$$

A low molecular weight (~12,000) proteolipid, rich in arginine and glutamic acid and covalently bonded to two fatty acids, may be involved in mediating the transport of Ca^{2+} across the lipid bilayer. Two calcium-binding proteins have been isolated that have no enzymatic activity. An acidic protein called calsequestrin has one-half of the amino acid residues as glutamic acid and aspartic acid. Another Ca^{2+}-binding protein is less acidic than the calsequestrin and binds about one-half as much Ca^{2+} as calsequestrin.

A tentative model for the Ca^{2+}-transport system in sarcoplasmic reticulum involves the ATPase and proteolipid as intrinsic membrane proteins. The ATPase has both polar and nonpolar domains and the nonpolar portion (possibly containing a Ca^{2+} ionophore site) is in the lipid bilayer. The polar portion is on the surface of the membrane and is the site for the hydrolysis of ATP. The interaction between the site of ATP hydrolysis and the portion of the ATPase buried in the lipid bilayer probably controls the transport of Ca^{2+} across the membrane. The Ca^{2+}-binding proteins are extrinsic proteins bound to the membrane inside the lumen of the sarcoplasmic reticulum and act to store the Ca^{2+} in the sarcoplasmic reticulum.

Energy Source for Muscle Contraction

The energy for muscle contraction is derived from the hydrolysis of ATP. The concentration of ATP in resting muscle will supply sufficient energy for contraction for only a fraction of a second, and the energy demands of intense contraction will exceed the capacity of the muscle to generate sufficient ATP by the metabolism of its various metabolic fuels. Mammalian muscle contains a reserve store of "high-energy phosphate" in the form of phosphocreatine. Resting muscle contains six times as much phosphocreatine as ATP, and whenever the expenditure of ATP exceeds its production, the phosphocreatine stores are used in an attempt to replenish the ATP supplies. The production of ATP from metabolic fuels is greatly increased in muscle as the work load is increased. The phosphocreatine stores are therefore considered a "backup system" for the generation of ATP. The phosphorylation of ADP by phosphocreatine is catalyzed by the enzyme creatine kinase as in Figure 21.8. After the completion of a contraction, or when the generation of ATP (by oxidation of metabolic fuel) in the muscle exceeds its utilization, phosphocreatine kinase catalyzes the phosphorylation of creatine by ATP. A second "backup system" for the generation of additional ATP is the enzyme myokinase that catalyzes the following reaction:

$$2ADP \longrightarrow ATP + AMP$$

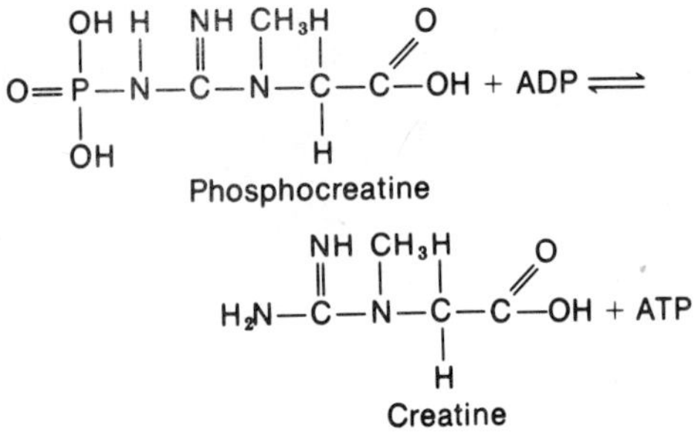

$$O{=}P(OH)_2{-}NH{-}C({=}NH){-}N(CH_3){-}CH_2{-}COOH + ADP \rightleftharpoons H_2N{-}C({=}NH){-}N(CH_3){-}CH_2{-}COOH + ATP$$

Phosphocreatine ⇌ Creatine

Figure 21.8
Reaction catalyzed by creatine kinase (creatine phosphotransferase).

Metabolic Fuel of Skeletal Muscle

After a Meal

Table 21.2 summarizes the metabolic fuels of skeletal muscle in various physiological states. The metabolic fuel of skeletal muscle after a meal is glucose, at which time the glucose and insulin levels in the blood are high, and free fatty acid levels are low. Glucose is metabolized principally by the glycolytic pathway, with less than 2% of glucose metabolized by the hexose monophosphate shunt. The glycolytic pathway and the citric acid cycle operate at ~10 and 5–10% of their maximal capacities, respectively, in resting muscle. Most of the products of glycolysis are completely oxidized in noncontracting muscle with only small amounts of lactate entering the blood. The rates of glucose uptake and lactate production by the skeletal muscle are stimulated in anoxia. The stimulation of the glycolytic rate in anoxia is considered to be via the activation of phosphofructokinase, pyruvate kinase, and glyceraldehyde 3-phosphate dehydrogenase. Both ATP and phosphocreatine inhibit all three enzymes and the amounts of both compounds are low in anoxic muscle. The enzyme phosphofructokinase is also activated by a variety of compounds that are found in increased amounts in skeletal muscle after anoxia. During contraction of skeletal muscle in the presence of excess glucose, the consumption of glucose and oxygen increases about 20-fold. (See Clin. Corr. 21.3.) The stimulation of the glycolytic rate is by the same mechanisms that occur in anoxia. The citric acid cycle activity is stimulated through the increased activity of isocitrate dehydrogenase. In prolonged heavy contraction, the rate of the production of pyruvate by glycolysis exceeds its rate of metab-

TABLE 21.2 The Metabolic Fuels of Skeletal Muscle in Various Physiological States

After a meal	Glucose
Fasting (short term)	Free fatty acids leucine, isoleucine, and valine
Fasting (long term)	Free fatty acids acetoacetate β-OH butyrate
Anoxia	Glucose, own glycogen
Tetanic contraction	Own glycogen

CLIN. CORR. **21.3**
LUFT'S SYNDROME

In Luft's syndrome, muscle weakness is caused by the mitochondria lacking respiratory control. There is poor coupling of oxidative phosphorylation to mitochondrial respiration.

olism by the citric acid cycle, and large amounts of lactate are transported to the liver. The fate of the lactate remaining in muscle after contraction has ceased has long been a matter of controversy, but it is now believed that fast-twitch red and white types of skeletal muscle can synthesize glycogen from lactate. In contrast, the slow-twitch red type of muscle has a limited capacity of glycogen synthesis because of a lack of fructose 1,6-bisphosphatase.

In Fasting

The metabolic fuel of skeletal muscle in the early stages of fasting is free fatty acids mobilized from adipose tissue which are the preferred fuel, since the presence of the free fatty acids actually suppresses the uptake and oxidation of glucose. Fatty acids are estimated to supply at least 50–60% of the energy needs of skeletal muscle both at rest and in contraction, and, unlike glucose, the utilization of fatty acids by muscle does not require insulin. The metabolic fuel of skeletal muscle in later stages of fasting are the ketone bodies, acetoacetate and β-hydroxybutyrate, synthesized by the liver. The utilization of fatty acids and ketone bodies by muscle spares glucose for tissues that utilize glucose as their principal metabolic fuel. Muscle cannot utilize fatty acids or ketone bodies for energy in anoxia. (See Clin. Corr. 21.4.)

The branched-chain amino acids (leucine, isoleucine, and valine) are degraded in fasting to yield energy in extrahepatic tissues, notably skeletal and cardiac muscle. Both fatty acids and epinephrine stimulate the oxidation of the branched-chain amino acids by skeletal muscle. The action of epinephrine on skeletal muscle branched-chain amino acid utilization, however, may be indirect via its lipolytic action on adipose tissue. The first step in the catabolism of the branched-chain amino acids is transamination with α-ketoglutarate to form glutamate and transamination of the glutamate with pyruvate to form alanine, which is released from muscle in fasting to be converted to glucose in liver and kidney (page 300).

CLIN. CORR. **21.4**
DEFIENCY OF CARNITINE PALMITYL TRANSFERASE

In a deficiency of the enzyme, carnitine palmityl transferase, there is an accumulation of lipid in the muscle cells. The muscle cells are unable to transport fatty acids across the inner mitochondrial membrane to be oxidized for energy.

In Tetanic Contraction

The total amount of glycogen in skeletal muscle far exceeds the amount of glycogen in liver. The amount of glycogen per unit weight in muscle is ~10% of the amount found in liver. However, the mass of skeletal muscle is over 25 times greater than liver. The glycogen in muscle is used as its energy source in anoxia and tetanic contraction. Muscle glycogen, in contrast to liver glycogen, is not depleted in fasting and is not broken down to glucose for use by other tissues because muscle lacks the enzyme glucose 6-phosphatase. At least two different systems have been found for the activation of the phosphorylase in muscle. Epinephrine activates the phosphorylase system in muscle (see page 315), but the time required for this mechanism is about 3 min, which is relatively slow in comparison to the activation of the phosphorylase induced by contraction. Contraction-induced phosphorylase activation is by the stimulatory action of Ca^{2+} on both activated and nonactivated phosphorylase kinase.

The Purine Nucleotide Cycle of Skeletal Muscle

A sequence of three enzyme reactions in skeletal muscle has been named the purine nucleotide cycle. The reactions catalyzed by these enzymes are shown in Figure 21.9. The enzymes of the purine nucleotide cycle in conjunction with the enzymes glutamate-oxaloacetate aminotransferase, fumarase, and malate dehydrogenase catalyze the net reaction,

$$\text{Glutamate} + NAD^+ + GTP + 2H_2O \longrightarrow \alpha\text{-ketoglutarate} + NADH + GDP + NH_4^+ + P_i$$

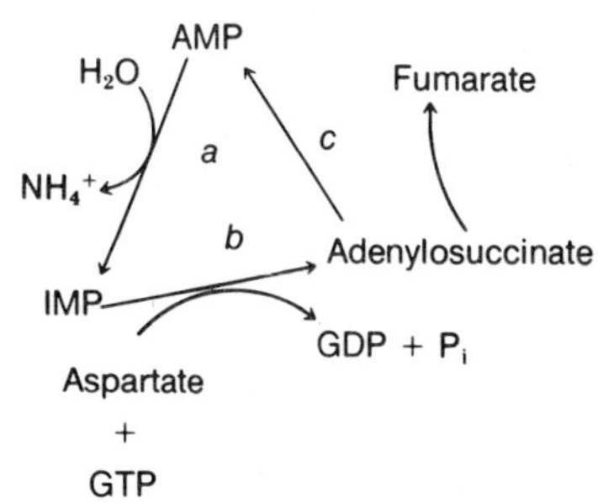

Figure 21.9
Reactions catalyzed by enzymes in the purine nucleotide cycle.
(a) Adenylate deaminase; (b) adenylosuccinate synthetase; (c) adenylosuccinase.

The cycle is the pathway for NH_4^+ formation in contracting skeletal muscle, for production of fumarate needed to replenish the intermediates of the citric acid cycle, and the oxidation of malate to oxaloacetate—a source of NADH for energy production by the respiratory chain. The cycle also controls the levels of adenine nucleotides, inorganic phosphate, and NH_4^+, all of which have an effect on the rate of glycolysis at the level of phosphofructokinase.

Effect of Training Exercise on Skeletal Muscle Metabolism

The size and the number of skeletal muscle mitochondria are greater in exercise-trained muscle than in nontrained muscle. Trained skeletal muscle has greater amounts of the enzymes used in fatty acid oxidation and ketone body utilization than does nontrained muscle and therefore has a greater capacity to utilize noncarbohydrates for energy. Thus the onset of the depletion of body carbohydrate, accumulation of lactate, and exhaustion during prolonged muscle contraction is delayed longer in trained muscle than in nontrained muscle.

21.2 CARDIAC MUSCLE

Cardiac muscle is striated and resembles red skeletal muscle but the fibers are more irregular in shape and have a variegated appearance caused by the branching of the fibers. Despite the branching, cardiac muscle fibers do not form an irregular network but are arranged in layers within which they tend to be parallel. The contractile proteins described for skeletal muscle are present in cardiac muscle and perform the same functions, even though the composition and the structures of the contractile proteins from the two tissues are not identical. Cardiac sarcoplasmic reticulum contains at least four of the major proteins found in the sarcoplasmic reticulum of skeletal muscle.

Regulation of Cardiac Muscle Contraction

The force and the rate of the contraction of the heart are stimulated by epinephrine, which stimulates the production of cAMP and leads to an increase in a protein kinase activity. A number of heart proteins are phosphorylated by the protein kinase, and all are thought to be involved in the control of cardiac contraction. The events in modulating cardiac contraction are suggested to be as follows. Epinephrine stimulates the formation of cAMP, which stimulates the cAMP-dependent protein kinase. This protein kinase catalyzes the phosphorylation of the myofibrillar proteins TN–I and myosin light chain, a sarcoplasmic membrane protein, phospholamban, and sarcolemma membrane proteins. Phospholamban is a protein that regulates active Ca^{2+} transport in cardiac sarcoplasmic reticulum. Calcium concentration in the sarcoplasm is increased by entry of Ca^{2+} through the sarcolemma and Ca^{2+} release from the sarcoplasmic reticulum. The entry of Ca^{2+} through the sarcolemma is accompanied by an efflux of Na^+, in other words a Ca^{2+}, Na^+-exchange system that involves three Na^+ ions per one Ca^{2+} ion. Negatively charged phospholipids (phosphatidylinositol and phosphatidylserine) are involved in the exchange process. After the contraction, the Ca^{2+} is sequestered back into the sarcoplasmic reticulum to be released at the time of the next contraction.

Metabolic Fuel of Cardiac Muscle

A summary of the metabolic fuel of heart in various physiological states is shown in Table 21.3. A variety of metabolic fuels can be utilized by the heart, but 60–90% of the total oxidative metabolism is accounted for by oxidation of fatty acids. Glucose, pyruvate, and lactate are utilized preferentially after a meal when the levels of free fatty acids in the blood are low. Both pyruvate and lactate inhibit the uptake and oxidation of free fatty acids and of the branched-chain amino acids isoleucine, leucine, and valine. Fatty acids mobilized from thc adipose tissue are the major metabolic fuel of heart in fasting; a lipoprotein lipase of heart is stimulated in fasting, increasing the hydrolysis of plasma triacylglycerols. The free fatty acids that enter the cells have to migrate through the cytoplasm in order to reach the outer membrane of the mitochondria where the fatty acids are activated into acyl CoA. A protein, molecular weight of 12,000, is found in the cytoplasm that transports the fatty acid across the cytoplasm. In addition, branched-chain amino acids are utilized for energy in short-term fasting. In long-term fasting, the ketone bodies become the preferred fuel for the heart. The utilization of glucose and endogenous glycogen by the heart increases about 10- to 20-fold in anoxia because the fatty acids and ketone bodies cannot be utilized for energy. Acetate, present in blood in significant amounts only after the ingestion of alcohol, is also used as an energy source for heart.

The inhibition of the utilization of glucose and pyruvate by fatty acids and the ketone bodies involves a number of control mechanisms. Fatty acids inhibit glucose entry into the cell and also inhibit the enzyme phosphofructokinase.

The rate-controlling reaction in the utilization of the branched-chain amino acids by heart is that of α-keto acid dehydrogenase, which utilizes as its substrates the α-keto acids from leucine, isoleucine, and valine. The inhibition of branched-chain amino acid utilization by pyruvate in the heart is attributed to a pyruvate inhibition of the α-keto acid dehydrogenase. The mechanism of the inhibition of fatty acid oxidation by pyruvate is not known.

The stimulation of glucose utilization in anoxia is by the mechanisms described for skeletal muscle. Three mechanisms have been suggested for the stimulation of the breakdown of heart glycogen in anoxia: an epinephrine activation of the phosphorylase system, an inhibition of phosphorylase phosphatase by the large amounts of AMP formed in anoxia, and by an activation of phosphorylase b by AMP. The supplies of cardiac glycogen are replenished after the anoxia by an activation of the glycogen synthase, which is activated (dephosphorylated) by a specific phosphatase.

TABLE 21.3 The Metabolic Fuels of Heart in Various Physiological States

Physiological State	*Metabolic Fuel*
After a meal	Glucose, lactate, pyruvate
Fasting (short-term)	Free fatty acids, triacylglycerols, leucine, isoleucine, valine
Fasting (long-term)	Acetoacetate, β-hydroxybutyrate
Anoxia	Glucose, glycogen
Heavy work load	Free fatty acids
Alcohol ingestion	Acetate

An increase in the level of mechanical work by the heart is accompanied by an increase in the oxygen consumption, with a concomitant increase in oxidative phosphorylation and a shift from glucose to palmitate utilization. The carnitine acyltransferase system plays a role in the coupling of the rate of fatty acid uptake by cardiac mitochondria to the activity of the citric acid cycle. In other words, the rate of translocation of acyl units across the inner mitochondrial membrane limits the rate of long-chain fatty acylcarnitine oxidation.

21.3 ADIPOSE TISSUE

''Body fat'' has been considered for many years to be connective tissue filled at random with droplets of fat, but when examined by electron microscopy, it becomes apparent that it is a specialized mammalian tissue. Adipose tissue has a central role in the energy metabolism of the entire animal. The end products of the digestion of dietary fat, carbohydrate, and protein are converted to triacylglycerols by the adipose tissue after a meal. The adipose tissue triacylglycerols are hydrolyzed to glycerol and fatty acids when the animal is in the fasting state, and the fatty acids are used by a variety of other tissues as their metabolic fuel. The triacylglycerols of adipose tissue are the major store of metabolic fuel for the body and are also a major source of energy for heat production. The oxidation of fatty acids yields more than twice the energy of carbohydrate or protein per unit weight. The deposition of the triacylglycerol in specialized cells eliminates the need for extensive storage of carbohydrate or triacylglycerol in the other tissues that might interfere with their function. The amount of adipose tissue triacylglycerol in the normal average adult human is about 25 lb (~11 kg), sufficient to maintain life for 40 days.

Distribution and Chemical Composition of Adipose Tissue

Adipose tissue is widely distributed in the body, in muscles, under the skin, around blood vessels, and in the abdominal cavity. White adipose tissue is involved in energy storage, while brown adipose tissue is involved in heat production of the animal. White adipose tissue cells are spherical, containing a large vacuole of triacylglycerol that occupies almost the entire cell, and the cytoplasm surrounding the lipid vacuole is no more than a thin film, invisible by light microscopy. A small number of mitochondria are found in the cytoplasm along the periphery of the central lipid droplet, a flattened Golgi apparatus is near the nucleus, and an endoplasmic reticulum is present.

Triacylglycerol constitutes ~80% of the wet weight of human white adipose tissue and ~99% of the lipid. Approximately 20 fatty acids are found in human adipose triacylglycerol, the principal ones being oleic acid (45%), palmitic acid (20%), linoleic acid (10%), stearic acid (6%), and myristic acid (4%).

Brown adipose tissue cells are polygonal, and their cytoplasm is more abundant and granular than that in white adipose cells. The cells contain a number of small lipid droplets, and the nucleus is not flattened as it is in the white adipose cells. The cytoplasm contains small amounts of endoplasmic reticulum and Golgi apparatus and numerous mitochondria. Both white and brown adipose tissue are well supplied with blood capillaries and are well innervated.

Adipose Tissue Metabolism after Feeding

Triacylglycerol in the chylomicrons and in the very low density lipoproteins (VLDL) are hydrolyzed to fatty acids and glycerol by lipoprotein lipase (LPL), which is secreted from the adipocyte and becomes associated with blood capillary walls, where it exerts its catalytic effect on the triacylglycerols. Insulin activates lipoprotein lipase, thus the enzyme is considered a control step in the assimilation of fatty acids from triacylglycerols into adipose tissue. Lipoprotein lipase requires activation by apolipoprotein C-II, which is a component of the surface film of chylomicrons and VLDL. Phospholipids are also required for the activation of LPL by apoprotein C-II. Insulin stimulates the rate of LPL synthesis and its secretion from the adipocyte to its endothelial site of activity. The fatty acids enter the adipocyte, with the help of a protein transporter in the membrane of the adipocyte and are esterified to form triacylglycerols, and the glycerol is transported to the liver, where it is used for the synthesis of glucose. The lipoprotein particles remaining after the action of lipoprotein lipase on VLDL are low density lipoproteins (LDL).

Most of the glucose taken up by adipose tissue after a meal is used for the synthesis of triacylglycerols with small amounts converted to glycogen. This is in contrast to the fate of glucose in liver, where most of the glucose is used for the synthesis of glycogen. The glycogen that is formed in adipose tissue after feeding is later converted to triacylglycerol.

Glycogen synthesis is stimulated after feeding and insulin is involved in some manner. The rate of entry of glucose into the adipocyte is a rate-determining step in glucose metabolism and is in some way stimulated by insulin. The uptake of glucose by adipocytes is by a process of passive mediated transport, and two glycoproteins have been isolated from adipocyte plasma membranes that may be the membrane transporter. Various proteolytic enzymes have an insulin-like activity on glucose uptake in the adipocyte. Several compounds that have sulfhydryl groups inhibit adipocyte glucose transport activity, and treatment of adipocytes with such compounds abolishes the ability of insulin to stimulate adipocyte glucose transport. Insulin has been reported to inhibit the phosphorylation of adipocyte membrane proteins by ATP; thus it has been suggested that insulin interaction with its plasma membrane receptor generates a second messenger, which regulates the phosphorylation of membrane proteins. A number of oxidants, including H_2O_2, stimulate glucose transport in the adipocyte, and the stimulation of H_2O_2 production in adipocytes by insulin has been found. The insulin effect on glucose transport may be, in part, an indirect effect mediated by its stimulation of intracellular rates of glucose utilization. Insulin also may induce glucose transport through a rapid and reversible translocation of glucose transport proteins from a large intracellular pool associated with the Golgi-enriched low density microsomal membrane fraction to the plasma membrane. The mechanism of the stimulation of adipocyte glucose transport by insulin is the subject of intensive investigation in a number of laboratories.

The phosphorylation of glucose to glucose 6-phosphate by adipocyte hexokinase is under the control of insulin and/or glucose. Adipose tissue from fed rats has higher hexokinase activities than adipose tissue from fasted rats. Approximately 23% of glucose taken up by the adipocyte is metabolized by the hexose monophosphate shunt, and the remainder is metabolized by glycolysis. The insulin-induced stimulation of both glucose transport into the adipocyte and of hexokinase activities produces an increase in the glycolytic rate after feeding. The reasons for the increase in glucose metabolism by the hexose monophosphate shunt after feeding are not known. The increase in the glycolytic rate and in hexose monophosphate shunt metabolism after feeding increases the production of NADH and NADPH, respectively. Some of the NADH is used for the reduction of dihydroxyacetone phosphate to glycerol 3-phosphate, and the remaining NADH is transhydrogenated to form NADPH (NADH + $NADP^+ \rightarrow NAD^+$ + NADPH), a reaction that is necessary in adipose tissue because the hexose monophosphate pathway produces only 65% of the NADPH needed for fatty acid synthesis. The pyruvate is converted to both the glycerol and the fatty acid moieties of the triacylglycerols that are synthesized in adipose tissue.

Most of the citrate formed from oxaloacetate and acetyl CoA leaves the mitochondria for conversion to acetyl CoA and oxaloacetate. The transfer of citrate out of the mitochondria means there is a loss of oxaloacetate from the mitochondria. The mitochondria are impermeable to cytosolic oxaloacetate and the major mechanism for the replenishment of mitochondrial oxaloacetate is the carboxylation of pyruvate.

The increase in the rate of glycolysis and the hexose monophosphate shunt after feeding are accompanied by increases in the activities of pyruvate dehydrogenase, acetyl CoA carboxylase, citrate cleavage enzyme, and fatty acid synthetase. The response of the key control enzymes in fatty acid synthesis in adipose tissue after feeding are listed in Table 21.4.

TABLE 21.4 The Enzyme Activities in Adipose Tissue That Are Increased in the Fed State as Compared to the Fasting State

Hexokinase
Glucose 6-phosphate dehydrogenase
Acetyl CoA carboxylase
Pyruvate dehydrogenase
Citrate cleavage enzyme
Fatty acid synthetase
Glycerol 3-phosphate acyltransferase

The stoichiometry of the conversion of glucose to palmitic acid has been calculated—taking into consideration all of the pathways involved: glycolysis; hexose monophosphate shunt; transhydrogenation; conversion of pyruvate to oxaloacetate, acetyl CoA, and citrate; cleavage of citrate to oxaloacetate and acetyl CoA; and synthesis of palmitate from acetyl CoA. The balanced overall reaction is

$$4.5\ \text{Glucose} + 4O_2 + 9\text{ADP} + 9P_i + 8H^+ \longrightarrow \text{palmitate} + 11CO_2 + 9\text{ATP} + 20H_2O$$

There is no involvement of the entire citric acid cycle in the conversion of glucose to fatty acid, and this is accomplished with a net production of ATP and of large amounts of water and CO_2.

Adipose tissue will remove amino acids from the blood for conversion to fatty acids and proteins. Insulin stimulates the transport of amino acids into the adipocytes and has a direct stimulatory effect on adipocyte protein synthesis. The amino acids differ in their ability to form fatty acids; that is, leucine is, by far, the most potent precursor of all the amino acids for fatty acid synthesis, presumably because the end products of leucine catabolism are acetoacetate and acetyl CoA. Acetoacetate also is incorporated into fatty acids in adipose tissue. Adipose tissue converts leucine to cholesterol at a significant rate and this synthesis is stimulated by insulin.

Triacylglycerol Synthesis in Adipose Tissue

The fatty acids derived from dietary triacylglycerol or synthesized from glucose and amino acids are used for triacylglycerol synthesis in the adipocyte. Three pathways of triacylglycerol synthesis in adipose tissue are (1) glycerol 3-phosphate pathway; (2) monoacylglycerol pathway; and (3) dihydroxyacetone phosphate pathway; the major pathway is the glycerol 3-phosphate pathway.

The activity of the enzyme catalyzing the acylation of glycerol 3-phosphate (acyl CoA: sn-glycerol 3-phosphate acyltransferase) is increased after feeding. There is a tendency for saturated fatty acids to occupy position 1 and unsaturated fatty acids to occupy position 2 of the triacylglycerol. A more random pattern of fatty acid distribution is found in position 3 of the triacylglycerol, with some preference for long-chain fatty acids. The acyltransferase has specificity with respect to the acyl CoA used in the esterification, and a protein "specifier factor" has been found that interacts with acyltransferase(s) to direct the acylation of palmitate to position 2 of the glycerol moiety.

Adipose Tissue Metabolism in Fasting

The triacylglycerols of adipose tissue are hydrolyzed in the fasting state to glycerol and free fatty acids by a hormone-sensitive lipase, which is activated by at least seven hormones (glucagon, epinephrine, ACTH, growth hormone, thyroxine, secretin, and the glucocorticoids). Glucagon and epinephrine are the principal lipolytic hormones in short-term fasting, whereas the glucocorticoids are the lipolytic hormones after long-term fasting or starvation. The hormone-sensitive lipase may be a misnomer because this enzyme preparation has at least three other substrates, including cholesterol esters. All of the above listed hormones, except the glucocorticoids, activate an adenylate cyclase in the adipocyte membrane, and the cAMP activates a protein kinase, which phosphorylates the lipase to its active phosphorylated form. The action of epinephrine on adipose tissue hormone-sensitive lipase is by a process that does not involve cAMP. The mechanism of the activation of adipose triacylglycerol hydrolysis by the glucocorticoids is not known, but does not involve cAMP. Insulin and/or glucose inhibit fatty acid mobilization from adipose tissue, therefore the mobilization of fatty acids from adipose tissue occurs mainly, if not entirely, in the fasting state. The mechanism of the inhibition of triacylglycerol hydrolysis by insulin is not certain and may be by an activation of phosphodiesterase which would increase the rate of destruction of cAMP. The prostaglandins also have an antilipolytic effect, in that they reduce the stimulation of lipolysis by epinephrine, glucagon, and so on, and inhibit the induction of cAMP formation by the lipolytic hormones.

Role of Adipose Tissue in Heat Production

Heat production by brown adipose tissue occurs primarily in newborn animals (including humans) and in adult hibernating animals, while white adipose tissue is involved in heat production in adults, especially during exposure to low temperatures. The production of heat by adipose tissue is instigated by epinephrine, which activates the hydrolysis of adipose tissue triacylglycerols to fatty acids and glycerol. The released fatty acids have a dual role in heat production, in that they are the source of reducing equivalents (by β-oxidation) for energy production by the respiratory chain of enzymes *and* also act as uncouplers of oxidative phosphorylation. That is, in the presence of the fatty acids, the energy produced by electron transport in the respiratory chain of enzymes is released as heat instead of being converted to ATP. It seems possible that the uncoupling of oxidative phosphorylation could well be accomplished by the fatty acids acting as carrier molecules for protons (proton ionophores) into the mitochondria, thus discharging the proton gradient established by the respiratory chain and bypassing the ATPase involved in the generation of ATP. Also, evidence has been found for a 32,000 dalton uncoupling protein in brown adipose tissue associated with a proton channel that would bypass the ATPase system.

21.4 LIVER

The liver has a dual afferent blood supply: the portal vein, carrying blood which has passed through the capillary beds of the alimentary tract, spleen, and pancreas; and the hepatic artery. A schematic representation of a liver lobule is shown in Figure 21.10. Blood from the branches of these two vessels mixes in passing through the sinusoidal capillaries of the liver lobules. The sinusoids drain into the central veins of the lobules, which are branches of the hepatic veins. The structural unit of liver is the lobule, consisting mainly of parenchymal cells. Branches of the afferent blood vessels together with the bile ducts (portal triad) run along the edges of each lobule. Sinusoidal capillaries pass through the parenchyma from the periphery of the lobule to the central vein. Minute bile capillaries run between the parenchymal cells, anastomose at the periphery of the lobule, and enter the bile ducts. The parenchyma is best defined as a continuous mass of cells perforated by a network of tunnels in which the sinusoids run, and the cells are called parenchymal cells. The walls of the sinusoids are lined by large cells with bulging nuclei, which are named Kupffer cells, and are functional phagocytes. The parenchymal cells, or hepatocytes, are the principal functioning cells of the liver.

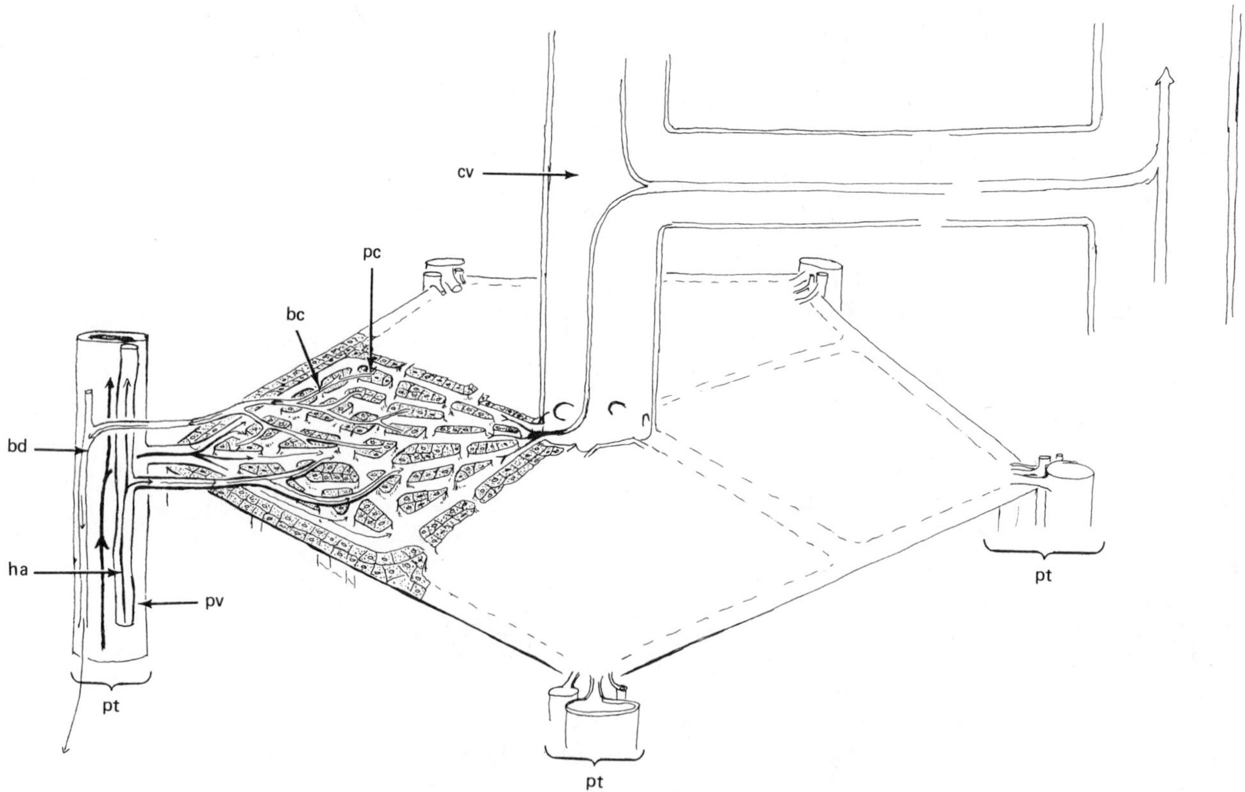

Figure 21.10
A schematic representation of a liver lobule.
Pt is the portal triad, consisting of the portal vein (pv), hepatic artery (ha), and the bile duct (bd). The central vein (cv) drains the lobule, the blood ultimately returning to the vena cava. The bile duct capillaries (bc) leave the parenchymal cells (pc), anastamose at the periphery of the lobule and enter the bile ducts.
Drawing by Dr. Donald W. Robertson, Department of Anatomy, College of Medicine, University of Minnesota.

General Functions of the Liver

The general functions of liver are: to furnish the metabolic fuel for a variety of tissues in fasting, that is, glucose, amino acids, and ketone bodies; to remove from the blood metabolic waste products generated by other tissues for conversion to other metabolic fuels or for conversion to compounds for excretion via the urine or feces; to synthesize compounds used as structural components for other tissues; and to detoxify biologically active compounds such as drugs, hormones, and poisons.

Metabolism

Carbohydrate Metabolism

A major function of liver is blood glucose homeostasis. The liver converts glucose in the blood to glycogen after a meal and converts glycogen to glucose in fasting, and also converts amino acids and lactate to glucose in fasting.

The liver parenchymal cells, in contrast to adipose and muscle cells, do not have a permeability barrier to glucose; thus insulin has no effect on uptake. Insulin, however, does stimulate the metabolism of glucose by liver, by stimulating glucokinase activity. The major fate of glucose 6-phosphate in liver is conversion to glycogen. Small amounts of glucose 6-phosphate enter the glycolytic pathway and the hexose monophosphate shunt pathway for the purposes of fatty acid synthesis.

The breakdown of glycogen to glucose 1-phosphate by phosphorylase in fasting is under the control of glucagon and epinephrine. The effects of epinephrine on liver glycogen breakdown are mediated by α-adrenergic receptors and involve a *cAMP-independent* mechanism, *and* also by β-adrenergic receptors involving the production of cAMP. The stimulation of α-adrenergic receptors is postulated to mobilize intracellular Ca^{2+}, which in turn stimulates phosphorylase kinase. The effect of glucagon on liver glycogen breakdown is mediated by β-adrenergic receptors and involves cAMP-dependent mechanisms.

The amount of glycogen stored in liver is not sufficient to maintain blood glucose levels in fasting, especially for periods of time exceeding about 12 h after a meal. The major source of blood glucose in fasting is gluconeogenesis in liver, which, like glycogen breakdown, is stimulated by the hormones glucagon and epinephrine. The process of gluconeogenesis is stimulated by the glucocorticoids in prolonged fasting and is inhibited by insulin after a meal.

Approximately 15% of the glucose metabolized by the liver is by the hexose monophosphate shunt for the purpose of generating NADPH for fatty acid and cholesterol synthesis. The rate of metabolism of glucose by the hexose monophosphate shunt is increased in feeding and decreased in fasting. The activity of glucose 6-phosphate dehydrogenase increases about fivefold after feeding, and the increase in enzyme activity has been accounted for by increases in the amount of enzyme. In addition, a high activity form of glucose 6-phosphate dehydrogenase has been found in livers of fed rats.

The amount of glucose metabolized by glycolysis in liver is small when compared to other tissues such as muscle or brain. The main reason for glycolysis in liver is to furnish pyruvate for conversion to acetyl CoA to be used for fatty acid synthesis. The rate of glycolysis in liver is increased in feeding and decreased in fasting, like the hexose monophosphate shunt.

Fructose is also metabolized by the liver and two pathological conditions occur leading to fructosuria (Clin. Corr. 21.5).

CLIN. CORR. 21.5
ESSENTIAL FRUCTOSURIA

Hepatic fructokinase is deficient in patients with fructosuria, and these individuals lack the ability to phosphorylate fructose to fructose 1-phosphate. Patients with hereditary fructose intolerance lack the hepatic enzyme fructose 1-phosphate aldolase.

Lipid Metabolism

The liver of fed animals synthesizes fatty acids from glucose and incorporates fatty acids into triacylglycerols and phospholipids. The primary fate of the fatty acids synthesized by the liver are the very low density lipoproteins (VLDL). The liver from fasted animals will utilize free fatty acids, released from adipose tissue, as its metabolic fuel. The carnitine transport system for entry of fatty acids into the mitochondria (page 375) is the control step that results in fatty acid oxidation by mitochondria in fasting or incorporation of fatty acids into phospholipids in feeding. (See Clin. Corr. 21.6.)

CLIN. CORR. **21.6**
DIABETIC KETOACIDOSIS

In diabetic ketoacidosis, there is an accumulation of triacylglycerol in the liver. This is probably the result of a decrease in the release of triacylgylcerol as VLDL from the liver into the circulation.

Protein Metabolism

Rapid protein synthesis and breakdown occurs in liver. The only serum proteins not synthesized in liver are the γ-globulins. Plasma proteins are constantly removed from circulation by liver and hydrolyzed to amino acids for utilization by extrahepatic tissues. The proteins enter the cells by a process of pinocytosis and are degraded to amino acids by intracellular cathepsins and other proteolytic enzymes in the lysosomes. The half-life of both serum and liver protein is about 10 days in contrast to the half-life of muscle protein which is much slower, that is, about 180 days.

The Metabolic Fuel of Liver

The energy demands of the liver are large and are supplied by lactate after a meal and fatty acids released from adipose tissue in fasting. Liver does have a lipoprotein lipase activity, and therefore plasma triacylglycerols are a possible source of fatty acids for the organ in fasting. (See Clin. Corr. 21.7.)

CLIN. CORR. **21.7**
LACTIC ACIDOSIS

Lactic acidosis is commonly found in people with diseases involving circulatory collapse or hypoxia. This also occurs in infants with defects in certain enzymes in glycolysis, gluconeogenesis, or citric acid cycle pathways; that is, glucose 6-phosphatase, fructose 1, 6-bisphosphatase, pyruvate carboxylase, or pyruvate dehydrogenase.

Biotransformation Reactions

A variety of biotransformation reactions occur solely or mainly in liver which are essential for the maintenance of the entire animal. A number of these transformations have been described in detail and will only be listed here: synthesis of bile acids from cholesterol (page 410); reduction of adrenal steroids and conjugation with glucuronic acid or sulfuric acid (page 572); conjugation of androgens and estrogens with glucuronic acid or sulfuric acid (page 572); formation of bile pigments (page 933); synthesis of purine and pyrimidine bases (page 496); formation of uric acid (page 503); and synthesis of urea (page 443).

The liver detoxifies a variety of drugs and poisons by a number of reactions: oxidation, reduction, hydrolysis, conjugation, and methylation.

Oxidative Reactions

Enzymes catalyzing oxidation reactions are present in the microsomal fractions of liver. The enzymes require NADPH and molecular oxygen and catalyze a variety of oxidative reactions commonly called "hydroxylation" reactions. Included in these reactions are aromatic and aliphatic hydroxylation, *N*-, *O*-, and *S*-dealkylations, sulfoxidation, *N*-oxidation, and epoxidation. The microsomal oxidase systems, commonly known as mixed function oxidases or monooxygenase systems, are comprised of three components: cytochrome P_{450}, a flavoprotein reductase, and a phospholipid. The system is referred to as the P_{450} system. A list of some compounds oxidized by this enzyme system is shown in Table 21.5. The general reaction catalyzed by mixed function oxidases is as follows:

$$RH + O_2 + NADPH + H^+ \longrightarrow ROH + H_2O + NADP^+$$

TABLE 21.5 Compounds That Are Oxidized by Mixed Function Oxidase in Liver

3-Methyl-4-aminoazobenzene
Aminopyrine
Biphenyl
Aniline
Dichloromethane
Benzo(α)pyrene
7-Ethoxycoumarin
N-Methyl-*p*-chloroaniline
Ethylmorphine

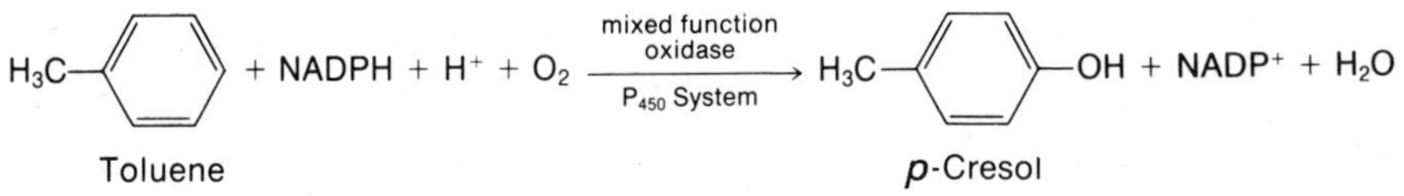

$$H_3C-C_6H_4 + NADPH + H^+ + O_2 \xrightarrow[P_{450}\ \text{System}]{\text{mixed function oxidase}} H_3C-C_6H_4-OH + NADP^+ + H_2O$$

Toluene — *p*-Cresol

Figure 21.11
Oxidation of toluene by the mixed function oxidase.

A specific example is the oxidation of toluene (Figure 21.11). The administration of a large number of compounds induces mixed function oxidase activity. The compounds noted for their induction of this enzyme activity are the barbiturates and polycyclic hydrocarbons such as benzo(*a*)pyrene.

The oxidation of alcohols to aldehydes or ketones and of aldehydes to carboxylic acids is catalyzed by two groups of enzymes in liver that do not involve the cytochrome P_{450} system: (1) The pyridine nucleotide-linked oxidoreductases utilize NAD^+ and are located in both the cytosol and mitochondria; (2) aldehyde oxidase, a metalloflavoprotein. The best known example of NAD^+-linked oxidoreductases is liver alcohol dehydrogenase, which catalyzes the reaction presented in Figure 21.12. Liver aldehyde dehydrogenase catalyzes the oxidation of a number of aldehydes, that is, formaldehyde and acetaldehyde, to their acids (Figure 21.13). Aldehyde oxidase is a metalloflavoprotein containing iron, molybdenum, and FAD, and catalyzes the oxidation of benzaldehyde to benzoic acid (Figure 21.14.).

$$CH_3CH_2OH + NAD^+ \xrightarrow[\text{dehydrogenase}]{\text{alcohol}} CH_3C(=O)-H + NADH + H^+$$

Ethanol — Acetaldehyde

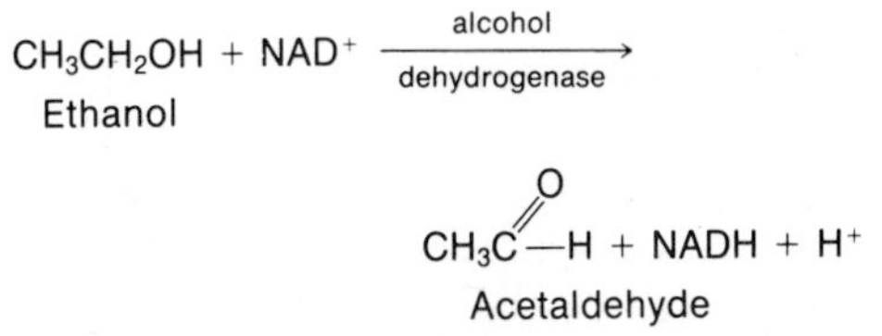

Figure 21.12
Reaction catalyzed by alcohol dehydrogenase.

$$CH_3C(=O)-H + NAD^+ + H_2O \xrightarrow[\text{dehydrogenase}]{\text{aldehyde}} CH_3C(=O)-OH + NADH + H^+$$

Acetaldehyde — Acetic acid

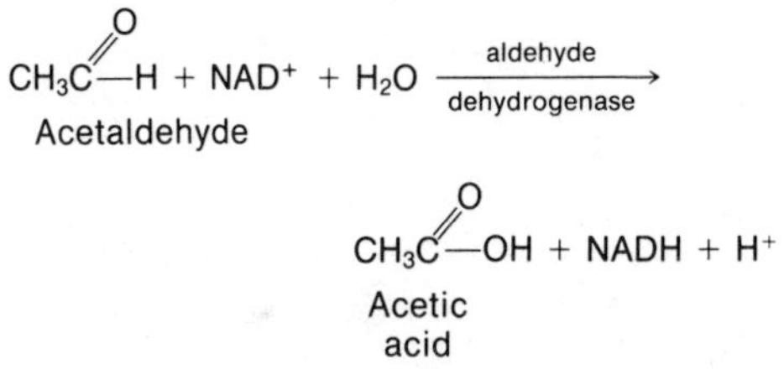

Figure 21.13
Reaction catalyzed by aldehyde dehydrogenase.

Reductive Reactions

Hepatic microsomal enzymes reduce both azo and nitro compounds by the addition of hydrogen. Azoreductase activity and nitroreductase are attributable to both NADPH–cytochrome P_{450} reductase and cytochrome P_{450}. The reduction of nitrobenzene to aniline is as shown in Figure 21.15. Aldehydes and ketones are reduced to alcohols by liver aldehyde or ketone reductases and the coenzyme NADPH (Figure 21.16). Liver ketone reductase reduces aromatic ketones to their alcohols.

$$C_6H_5-C(=O)H + FAD + H_2O \xrightarrow[\text{oxidase}]{\text{aldehyde}} C_6H_5-C(=O)OH + FADH_2$$

Benzaldehyde — Benzoic acid

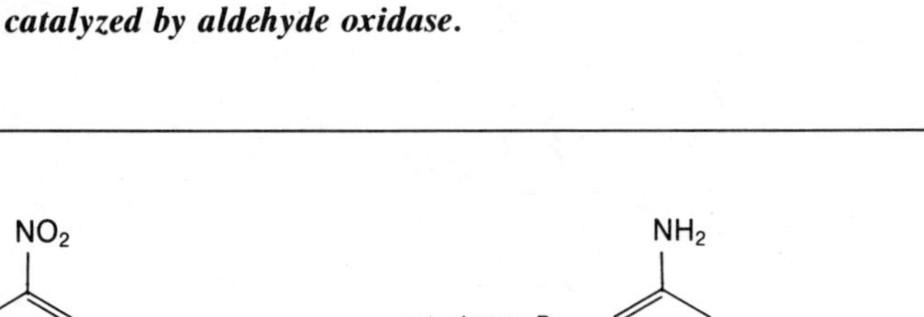

Figure 21.14
Reaction catalyzed by aldehyde oxidase.

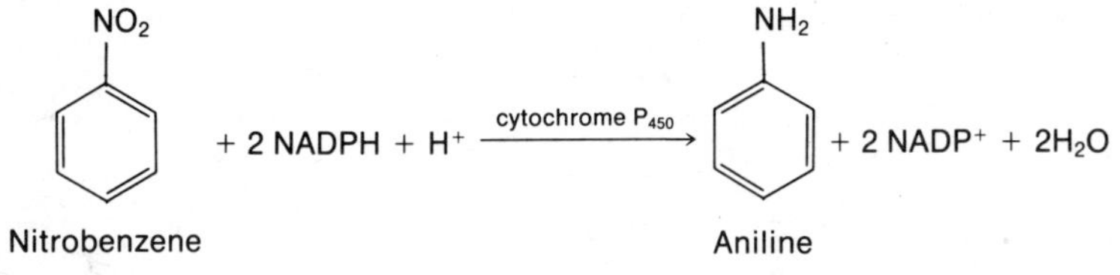

$$C_6H_5-NO_2 + 2\ NADPH + H^+ \xrightarrow{\text{cytochrome } P_{450}} C_6H_5-NH_2 + 2\ NADP^+ + 2H_2O$$

Nitrobenzene — Aniline

Figure 21.15
Reduction of nitrobenzene.

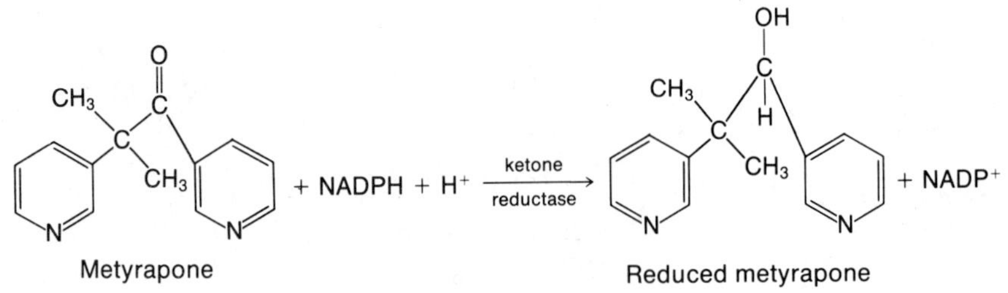

Metyrapone + NADPH + H^+ →(ketone reductase) Reduced metyrapone + $NADP^+$

Figure 21.16
Reduction of metyrapone.

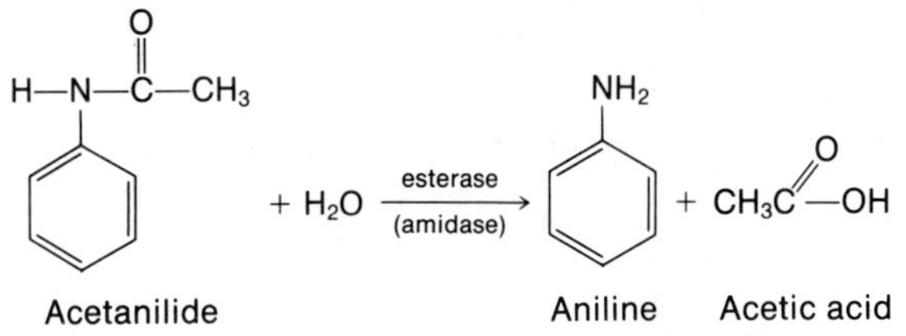

Acetanilide + H_2O →(esterase (amidase)) Aniline + Acetic acid

Figure 21.17
Esterase activity of liver.

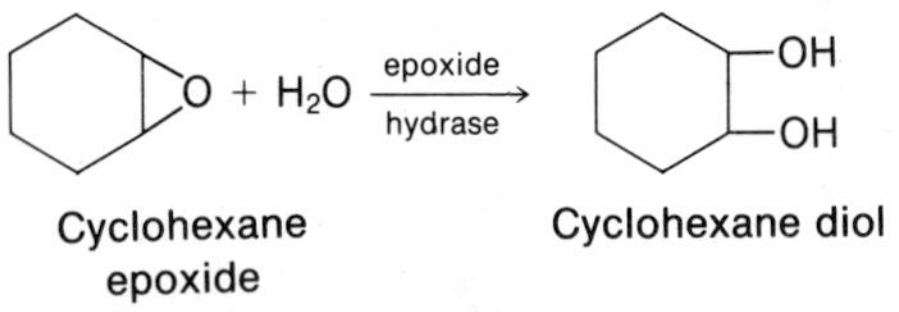

Cyclohexane epoxide + H_2O →(epoxide hydrase) Cyclohexane diol

Figure 21.18
Epoxide hydrase activity of liver.

Hydrolytic Reactions

Esterases are found in a variety of tissues, for example, plasma, brain, intestinal mucosa, erythrocyte, and muscle. An esterase is found in human liver that catalyzes the hydrolysis of acetanilide, procaine, xylocaine, and simple aliphatic esters. Epoxide hydrase catalyzes the conversion of epoxides to diols (Figures 21.17 and 21.18).

Conjugation Reactions

A large number of compounds are detoxified by conjugation reactions in the liver. Four types of compounds are conjugated with *glucuronic acid:* hydroxyl (both phenolic and alcoholic), carboxyl, sulfhydryl, and amino. The enzyme catalyzing the formation of glucuronides, called UDP-glucuronyltransferase, is found in the endoplasmic reticulum and a reaction it catalyzes is shown in Figure 21.19. Glucuronide formation is a major pathway of drug metabolism. Compounds bearing alcoholic or phenolic hydroxyl groups are conjugated with *sulfate,* catalyzed by the enzyme sulfotransferase, found in the cytosol of the liver cell. The sulfate group is transferred from 3′-phosphoadenosine 5′-phosphosulfate (PAPS) to yield the sulfate half ester as presented in Figure 21.20. *Glutathione* conjugates are thioethers formed by the combination of alkyl and aryl halides, epox-

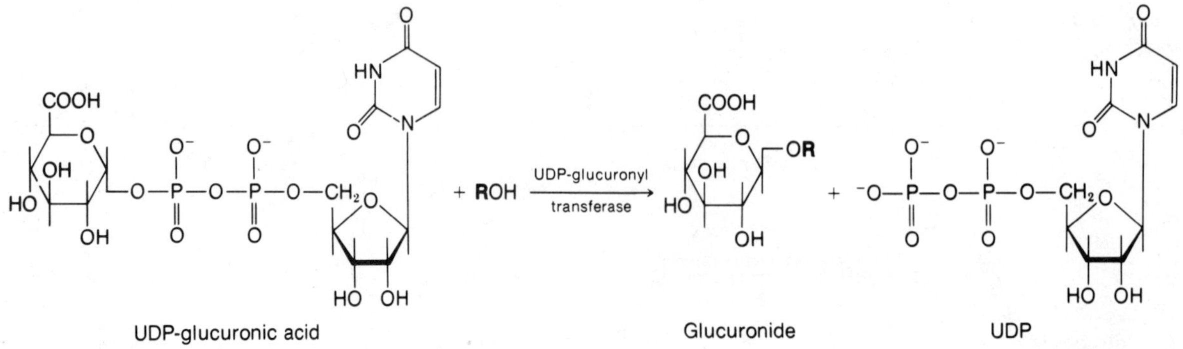

UDP-glucuronic acid + ROH →(UDP-glucuronyl transferase) Glucuronide + UDP

Figure 21.19
Conjugation of alcohols with glucuronic acid in liver.

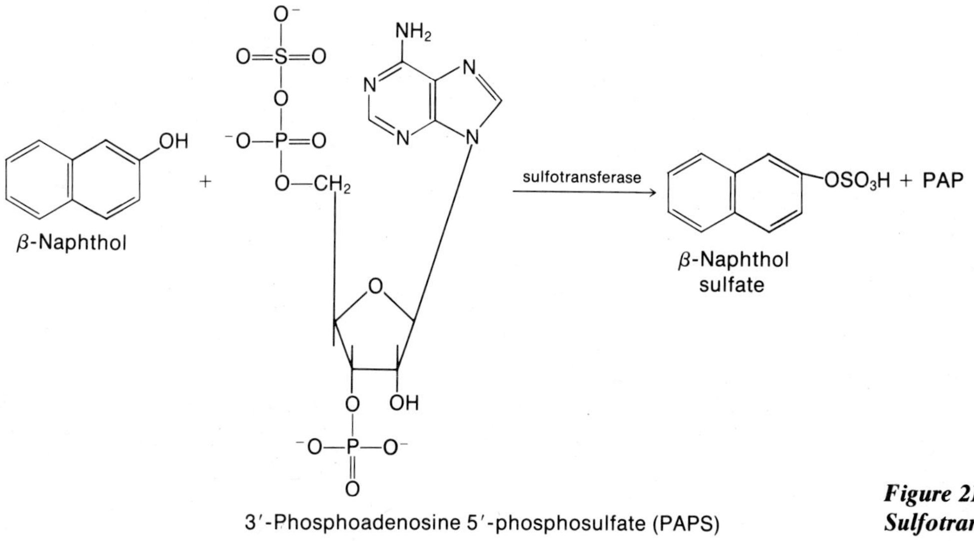

β-Naphthol + 3′-Phosphoadenosine 5′-phosphosulfate (PAPS) $\xrightarrow{\text{sulfotransferase}}$ β-Naphthol sulfate ($-OSO_3H$) + PAP

Figure 21.20
Sulfotransferase activity.

ides, and alkenes with glutathione. The enzymes that catalyze these reactions are glutathione transferases, found in the cytosol. Some glutathione conjugates are converted to mercapturic acids, catalyzed by *S*- and *N*-acetyltransferases. The formation of mercapturic acids is shown in Figure 21.21. *Glutamine* is conjugated with phenylacetic acid, and a number of drugs. The conjugation of phenylacetic acid with glutamine is as

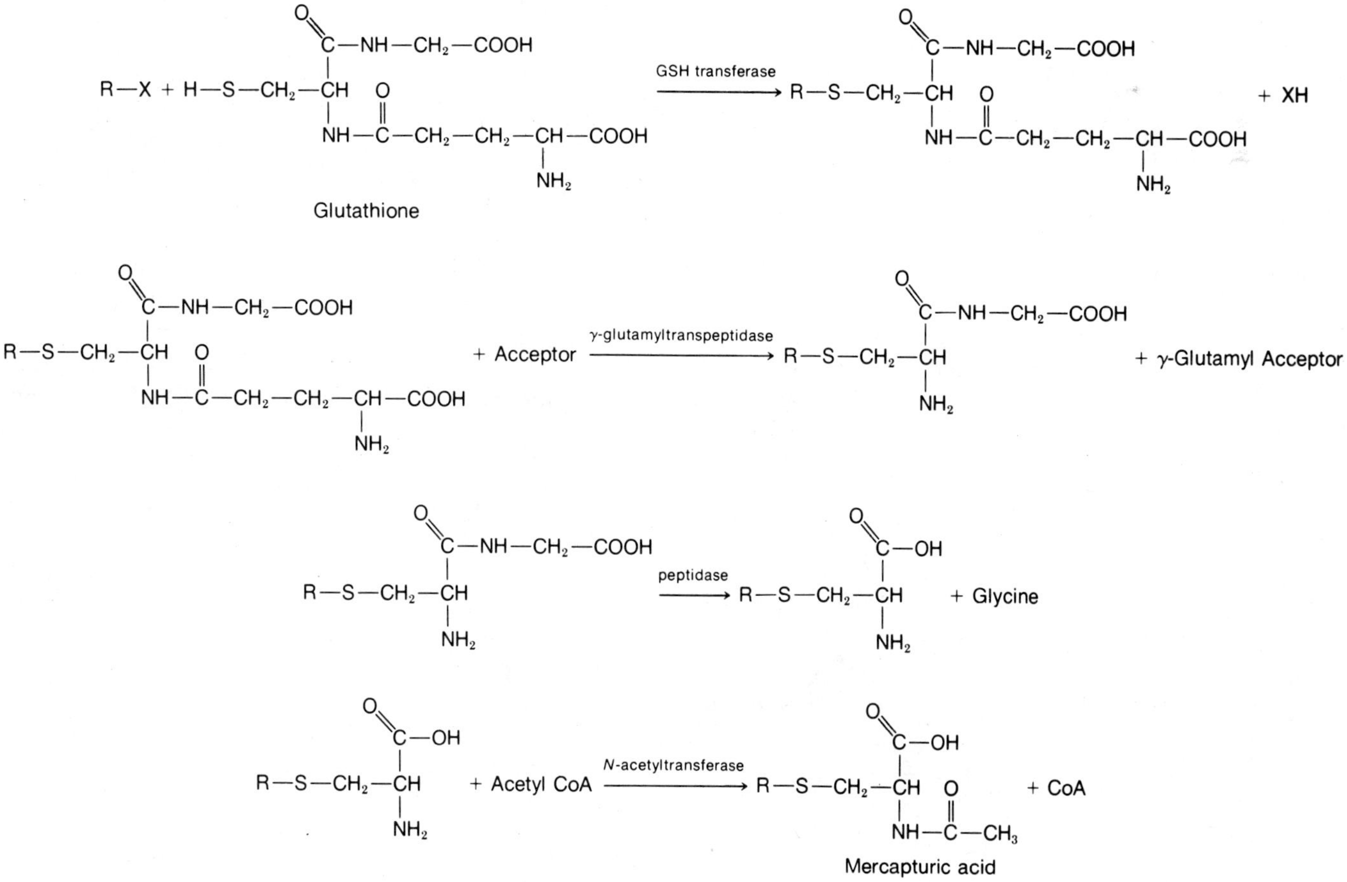

$$R{-}X + H{-}S{-}CH_2{-}CH(C(=O){-}NH{-}CH_2{-}COOH){-}NH{-}C(=O){-}CH_2{-}CH_2{-}CH(NH_2){-}COOH \xrightarrow{\text{GSH transferase}} R{-}S{-}CH_2{-}CH(C(=O){-}NH{-}CH_2{-}COOH){-}NH{-}C(=O){-}CH_2{-}CH_2{-}CH(NH_2){-}COOH + XH$$

Glutathione

$$R{-}S{-}CH_2{-}CH(C(=O){-}NH{-}CH_2{-}COOH){-}NH{-}C(=O){-}CH_2{-}CH_2{-}CH(NH_2){-}COOH + \text{Acceptor} \xrightarrow{\gamma\text{-glutamyltranspeptidase}} R{-}S{-}CH_2{-}CH(NH_2){-}C(=O){-}NH{-}CH_2{-}COOH + \gamma\text{-Glutamyl Acceptor}$$

$$R{-}S{-}CH_2{-}CH(NH_2){-}C(=O){-}NH{-}CH_2{-}COOH \xrightarrow{\text{peptidase}} R{-}S{-}CH_2{-}CH(NH_2){-}C(=O){-}OH + \text{Glycine}$$

$$R{-}S{-}CH_2{-}CH(NH_2){-}C(=O){-}OH + \text{Acetyl CoA} \xrightarrow{N\text{-acetyltransferase}} R{-}S{-}CH_2{-}CH(NH{-}C(=O){-}CH_3){-}C(=O){-}OH + \text{CoA}$$

Mercapturic acid

Figure 21.21
Pathway for the formation of mercapturic acids in liver.

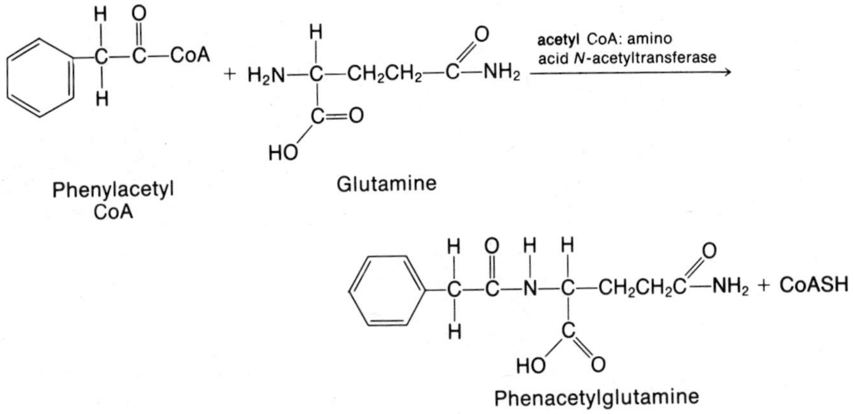

Figure 21.22
Conjugation of phenylacetate with glutamine.

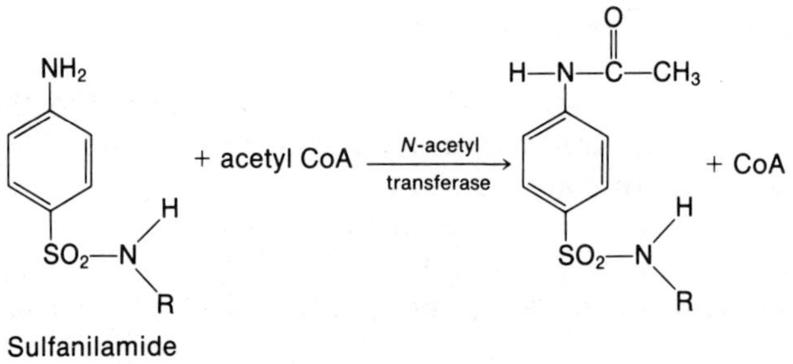

Phenylacetyl CoA + Glutamine $\xrightarrow{\text{acetyl CoA: amino acid } N\text{-acetyltransferase}}$ Phenacetylglutamine + CoASH

Figure 21.23
Acetylation of sulfanilamide.

Sulfanilamide + acetyl CoA $\xrightarrow{N\text{-acetyl transferase}}$ (acetylated sulfanilamide) + CoA

shown in Figure 21.22. Sulfanilamide is detoxified by forming the conjugate with *acetic acid* (Figure 21.23). Benzoic acid is conjugated with *glycine* to form hippuric acid (Figure 21.24). The enzyme that catalyzes the conjugation of glycine or glutamine is called acyl CoA: amino acid N-acyltransferase.

Methylation Reactions

A variety of methyltransferases that utilize *S*-adenosylmethionine as the methyl donor are found in liver. For example, catechol-*O*-methyltransferase catalyzes the *O*-methylation of norepinephrine, dopamine, epinephrine, dopa, and a number of drugs. The methylation of hydrogen sulfide is shown in Figure 21.25.

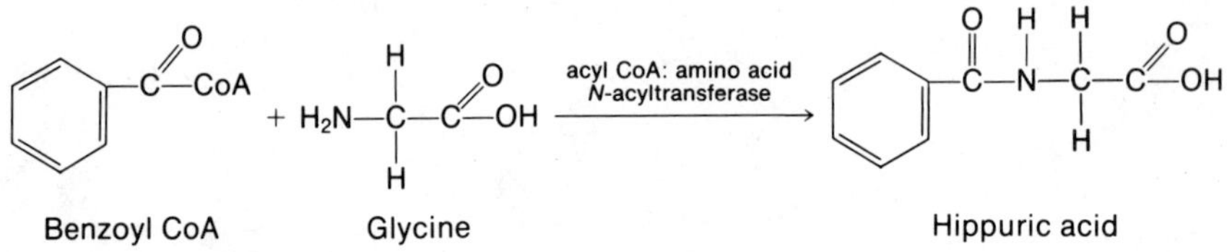

Figure 21.24
Formation of hippuric acid.

Benzoyl CoA + Glycine $\xrightarrow{\text{acyl CoA: amino acid } N\text{-acyltransferase}}$ Hippuric acid

$$H_2S + S\text{-adenosylmethionine} \xrightarrow{\text{thiol } S\text{-methyltransferase}} CH_3SH + S\text{-adenosylhomocysteine}$$

Figure 21.25
Formation of methylmercaptan from H_2S.

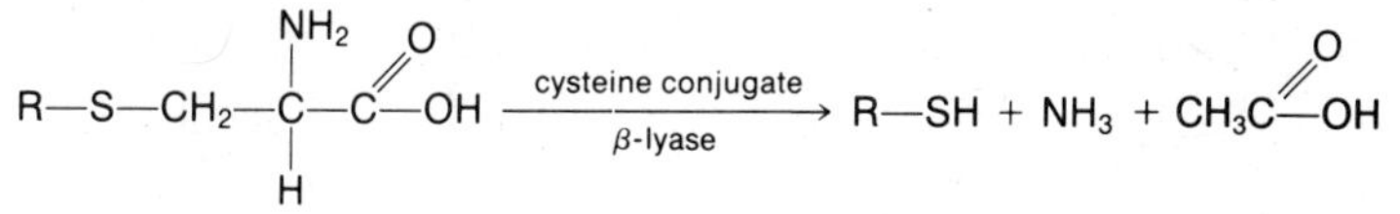

$$R{-}S{-}CH_2{-}\underset{H}{\overset{NH_2}{C}}{-}\overset{O}{C}{-}OH \xrightarrow[\beta\text{-lyase}]{\text{cysteine conjugate}} R{-}SH + NH_3 + CH_3\overset{O}{C}{-}OH$$

Figure 21.26
Action of cysteine conjugate β-lyase.

$$Ar{-}N(OH){-}\overset{O}{C}{-}R \xrightarrow[\text{acyltransferase}]{\text{arylhydroxamic acid}} Ar{-}NH{-}OH + R{-}\overset{O}{C}{-}\text{acyltransferase}$$

Arylhydroxamic acid → Arylhydroxylamine

$$\downarrow$$

$$Ar{-}NH{-}O{-}\overset{O}{C}{-}R$$

Acyloxyarylamine

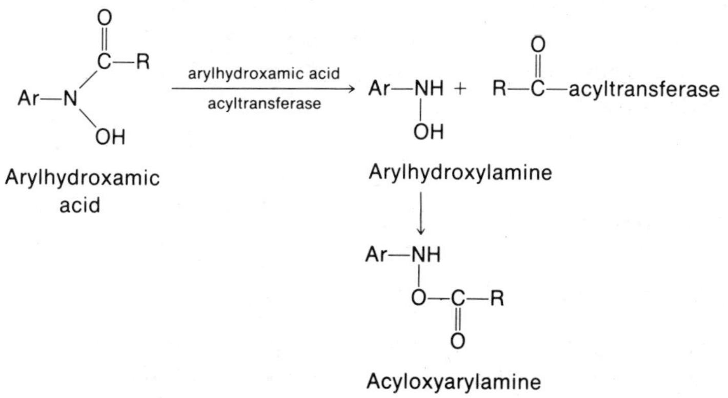

Figure 21.27
Activity of arylhydroxamic acid acyltransferase.

Other Detoxication Reactions

The cysteine conjugate of various aromatic drugs may be cleaved to the sulfhydryl derivative of the drug, NH_3, and acetic acid by the enzyme cysteine conjugate β-lyase (Figure 21.26). Arylhydroxamic acids are converted to arylhydroxylamines by arylhydroxamic acid acyltransferase. The arylhydroxylamine is then converted to *N*-acyloxyarylamine (Figure 21.27). The thioltransferases catalyze the transfer of a thiol residue to glutathione that results in the conversion of disulfides to sulfhydryl compounds (Figure 21.28). The glyoxylase system consists of two enzymes, glyoxylase I and glyoxylase II, which converts 2-oxyaldehydes to β-hydroxy acids (Figure 21.29). Cyanide is detoxified by sodium thiosulfate and the enzyme rhodanese (Figure 21.30).

$$X{-}S{-}S{-}Y + GSH \xrightarrow{\text{thioltransferase}} X{-}S{-}S{-}G + YSH$$

$$X{-}S{-}S{-}G + GSH \longrightarrow G{-}S{-}S{-}G + XSH$$

$$G{-}S{-}S{-}G + NADPH + H^+ \xrightarrow{\text{reductases}} 2GSH + NADP^+$$

Figure 21.28
Action of thioltransferase.
GSH, glutathione; examples of X-S-S-Y, coenzyme A, cysteine, homocysteine, and pantetheine.

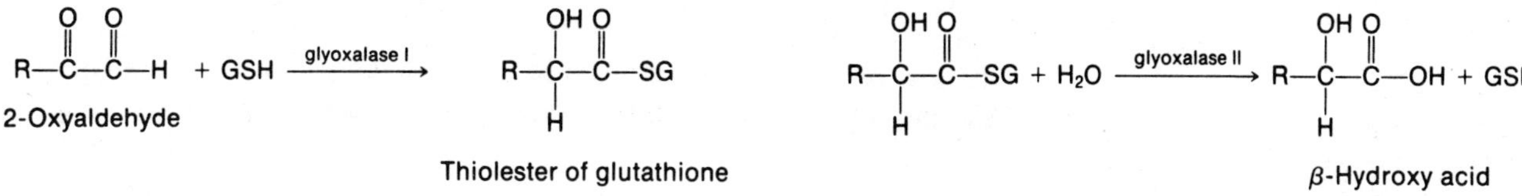

$$R{-}\overset{O}{C}{-}\overset{O}{C}{-}H + GSH \xrightarrow{\text{glyoxalase I}} R{-}\underset{H}{\overset{OH}{C}}{-}\overset{O}{C}{-}SG$$

2-Oxyaldehyde → Thiolester of glutathione

$$R{-}\underset{H}{\overset{OH}{C}}{-}\overset{O}{C}{-}SG + H_2O \xrightarrow{\text{glyoxalase II}} R{-}\underset{H}{\overset{OH}{C}}{-}\overset{O}{C}{-}OH + GSH$$

β-Hydroxy acid

Figure 21.29
Reactions catalyzed by glyoxalase I and II.
Some substrates for glyoxalase I are glyoxal, hydroxypyruvaldehyde, phenylglyoxal, and methylglyoxal.

$$Na_2S_2O_3 + NaCN \xrightarrow{\text{rhodanese}} NaSCN + Na_2SO_3$$

Sodium thiosulfate, Sodium cyanide, Sodium thiocyanate, Sodium sulfite

Figure 21.30
Reaction catalyzed by rhodanese.

The products of the various detoxication reactions described (i.e., oxidation, reduction, hydrolysis, conjugation, and methylation) are, for the most part, less toxic than the reactants. However, some products of the reactions, especially the oxidative reactions, are thought to be carcinogens and mutagens.

21.5 KIDNEY

Metabolic Fuel of Kidney

Mammalian kidney can utilize as its metabolic fuel palmitate, lactate, glutamine, glucose, citrate, glycerol, and ketone bodies. Palmitate furnishes 60–80% of the metabolic fuel of the intact kidney, with lactate as the second most important, and their utilization depends on the relative availabilities of the substrates in the plasma. Palmitate inhibits the utilization of lactate for energy but does not inhibit the kidney's uptake of lactate. The fate of lactate in kidney in the presence of large amounts of palmitate is conversion to glucose by gluconeogenesis. The relative contributions of citrate, glutamine, and glycerol to the energy requirements of the intact kidney have as yet to be determined. The utilization of the ketone bodies by kidney occurs in long-term fasting or starvation and in diabetes. The utilization of glucose by kidney is small and occurs mainly after feeding and in anoxia and has been calculated to account for only 2–6% of the oxygen consumption of the intact kidney. The major source of energy for the production of ATP by kidney is aerobic oxidation of substrates, with glycolysis producing only 4% of the ATP. The major utilization of ATP in kidney is for the reabsorption of NaCl by the kidney tubules.

Kidney cortex and kidney medulla differ greatly in their metabolism. Kidney cortex has a high rate of oxygen consumption, large amounts of citric acid cycle and respiratory chain enzymes, and a respiratory quotient of ~0.75 indicating that fatty acids are the principal metabolic fuel. The utilization of palmitate, lactate, ketone bodies, and glucose by oxidative metabolism in the intact kidney for energy occurs mainly in the kidney cortex. The oxidation of fatty acids by the cortex is accompanied by ketone body formation, as it is in liver. The medulla of the kidney has, on the other hand, high concentrations of the enzymes of glycolysis, low levels of the citric acid cycle and respiratory chain enzymes, low rates of oxygen consumption, and low energy requirements, which arise primarily from glycolysis. The source of the glucose for glycolysis by kidney medulla is plasma glucose after a meal and, perhaps, glucose synthesized in the cortex in fasting. The high respiratory metabolism of the cortex and the low respiratory metabolism of the medulla is correlated with a Po_2 of 80 mm, and 5–10 mm, in the cortex and medulla, respectively.

Gluconeogenesis in Kidney Cortex

The gluconeogenic enzymes present in kidney cortex can synthesize glucose more rapidly per unit weight than can liver, but the kidney does not contribute significant amounts of glucose to the bloodstream. The suggestion has been made that the glucose synthesized in the cortex is used by the medulla as its metabolic fuel. The rate of gluconeogenesis in kidney cortex, as in liver, is higher in fasting than in the fed state, and the hormones epinephrine and glucagon are implicated in the activation of the process via cAMP. The fatty acids and ketone bodies supply the energy for kidney cortex gluconeogenesis in fasting.

The rates of gluconeogenesis in kidney cortex are also stimulated in acidosis, but the physiological significance of this response to acidosis is not clear. The gluconeogenic enzyme phospho*enol*pyruvate carboxykinase is a major factor in the regulation of kidney cortex gluconeogenesis and the induction of its activities in acidosis is accompanied by increases in the amounts of enzyme present and by increases in the amount of messenger RNA for the enzyme. The nature of the signal acting on the renal cell to produce induction of the mRNA for PEP carboxykinase after acidosis is not known.

Ammonia Production in Kidney

Increased production and excretion of ammonium ion is a major component of the kidney's homeostatic response to onset of metabolic acidosis. Renal ammoniagenesis is initiated primarily by a phosphate-dependent glutaminase, which is localized within the inner membrane or matrix compartment of the mitochondria (Figure 21.31). There is a specific carrier system for transport of glutamine across the inner mitochondrial membrane. Several factors have been proposed for the stimulation of renal ammonia formation in acidosis, including an increase in the activity or amount of the major ammonia-producing enzyme, glutaminase, or an increased utilization of the end product inhibitor of this enzyme, glutamate. Also, evidence has been presented for the formation of ammonia by the kidney is by the purine nucleotide cycle.

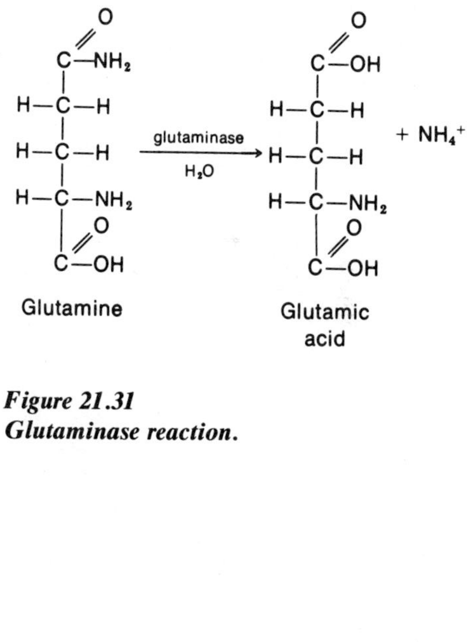

$$\underset{\text{Glutamine}}{H_2N{-}C(=O){-}CH_2{-}CH_2{-}CH(NH_2){-}COOH} \xrightarrow[H_2O]{\text{glutaminase}} \underset{\text{Glutamic acid}}{HO{-}C(=O){-}CH_2{-}CH_2{-}CH(NH_2){-}COOH} + NH_4^+$$

Figure 21.31
Glutaminase reaction.

Kidney Tubular Transport Mechanisms

Kidney is one of the richest sources of Na^+,K^+-stimulated ATPase, which is present in very high concentrations in the thick ascending loop of Henle and the proximal tubules. The reaction catalyzed by this enzyme is as follows:

$$\underset{\text{(inside)}}{3Na^+} + \underset{\text{(outside)}}{2K^+} + ATP + H_2O \longrightarrow \underset{\text{(outside)}}{3Na^+} + \underset{\text{(inside)}}{2K^+} + P_i + ADP + H^+$$

The function of this enzyme is the active transport of K^+ and Na^+ into and outside of the cell, respectively. A Na^+,K^+-ATPase has been isolated in a pure form from renal medulla and when constituted into phospholipid vesicles will exhibit active K^+ transport coupled to active Na^+ countertransport in a 2 : 3 ratio. The primary overall ion movement occurring in the ascending loop of Henle is, however, reabsorption of NaCl from the lumen of this tubular structure into peritubular space. An ATP-dependent (active transport) of NaCl has been identified in the ascending loop of Henle. The transport of both D-glucose and amino acids (in the proximal region of the nephron) from the glomerular filtrate into the blood is energy-dependent, saturable, and is cotransported with Na^+. The coupling of sugar and amino acid transport to the reabsorption of Na^+ may well be responsible for 10–15% of the total uptake in the proximal tubule.

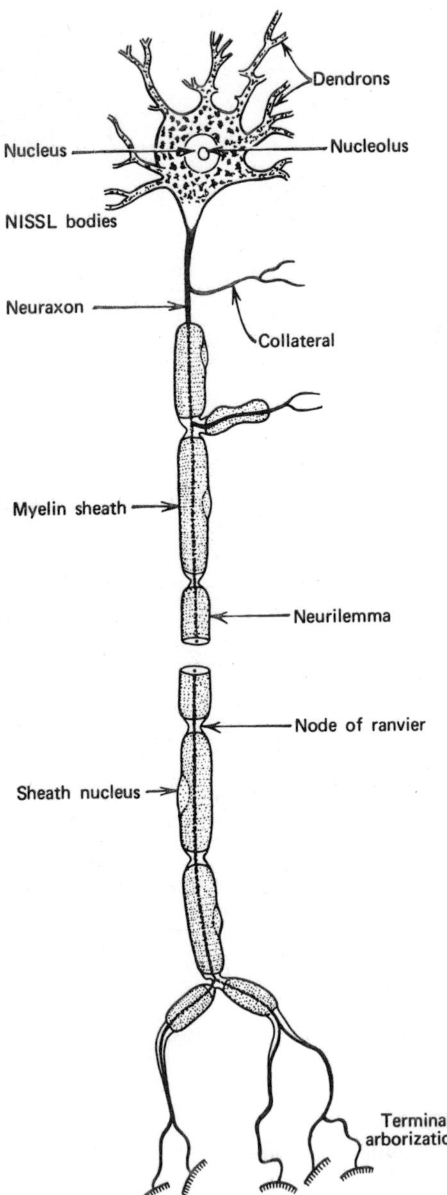

Figure 21.32
A motor nerve cell and investing membranes.

CLIN. CORR. **21.8**
MULTIPLE SCLEROSIS

Multiple sclerosis is a disease in which destruction of myelin occurs and in the absence of myelin the axon fails to conduct the nervous impulses. The cause of the destruction of myelin is uncertain, but a genetic predisposition for the disease occurs.

21.6 NERVOUS TISSUE

Structure

The major feature of the nerve cell, or neuron, is the presence of greatly elongated cytoplasmic processes, which in certain instances extend almost the length of the body. By means of these processes the neuron contacts other neurons, receptors, glands, or muscles and thus transmits or receives stimuli from them. There are about 10 billion neurons in the nervous system and most of these are located in the brain. All neurons have a cell body (a swollen portion of the neuron), which contains abundant cytoplasm and a large centrally located nucleus. The long extension of the cytoplasm that conducts the impulse away from the cell body is called the axon. The short arborized extensions of the cytoplasm that receive impulses from other neurons are called dendrites or dendrons. A motor nerve cell and investing membranes is depicted in Figure 21.32. The neurons are protected by various types of tubular structures (e.g., myelin sheath) that are not considered part of the neurons (see Clin. Corr. 21.8). A schematic diagram of an axon surrounded by myelin sheath is shown in Figure 21.33. The composition of the cytoplasm in the cell body and the extensions of the cytoplasm are not identical. The cytoplasm of the cell body contains Golgi apparatus, mitochondria, pigment granules (melanin and lipofuscin), droplets of lipid and glycogen, neurofibrils, neurotubules, neurofilaments, microfilaments, and rough endoplasmic reticulum referred to as Nissl bodies. The cytoplasm of the axon contains neurofibrils and mitochondria, but no rough endoplasmic reticulum, pigment, or lipid droplets. The absence of rough endoplasmic reticulum in the terminal regions of the dendrites and in the axons means that proteins are synthesized in the cell body and are transported down the interior of the axons and dendrites. Gray matter is the core of the nervous system containing cell bodies of the neurons, their dendrites, and the proximal portions of the axons. The zone mainly devoid of cell bodies and heavily invested with myelin is white matter.

Nervous Tissue Lipids

A major characteristic of nervous tissue is its high content of lipids and their unique structures. The lipid content of myelin, white matter, and gray matter is 80, 60, and 40% dry weight, respectively. Cholesterol is the second most abundant compound in the brain, and about 25% of the body cholesterol is present in nervous tissue. Cholesterol is synthesized in the brain and has an extremely low rate of breakdown or turnover number. Water is the most abundant substance in brain, and the water content, expressed as percent of fresh weight of myelin, white matter, and gray matter are 40, 70, and 80% of fresh weight, respectively.

Most of the fatty acids in the brain lipids are synthesized in situ, with only small amounts incorporated from the diet. The fatty acid composition of brain lipids is relatively constant, but variations have been reported with alterations in the lipid composition of the diet. Many of the fatty acids in nervous tissue are unsaturated and have long carbon chains, as many as 24 carbons. The palmitate, synthesized in the cytoplasm, is elongated by a mitochondrial and microsomal system, utilizing acetyl and malonyl CoA, respectively. Two enzymatic systems have been identified that desaturate the carbon chains between carbons 6-7 and 9-10, respectively.

The lipids found in nervous tissue and the locations where they are found in relatively high concentrations are shown in Table 21.6. All of the lipids listed in this table are found in varying amounts in white and gray

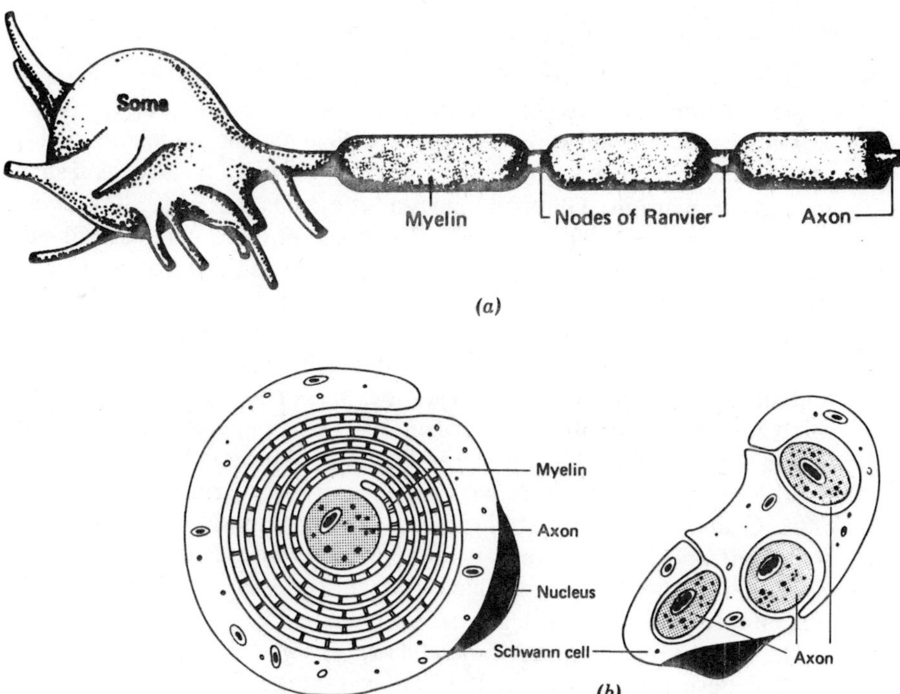

Figure 21.33
Schematic diagram of axon surrounded by myelin sheath and the unmyelinated node of Ranvier.
The drawing in the lower panel (b) is a cross section of a nerve, showing the axon surrounded by the myelin sheath composed of layers of Schwann cell plasma membrane.
Reprinted with permission from R. F. Schmidt, ed., *Fundamentals of Neurophysiology*. New York: Springer-Verlag, 1975, p. 8.

matter and in myelin. The pathways for the synthesis and breakdown of the lipids found in nervous and other mammalian tissues are described in Chapter 10. The source of choline and serine for brain phosphatidylcholine and phosphatidylserine synthesis, respectively, is exogenous (i.e., abstracted from the blood). The source of ethanolamine and inositol for the synthesis of phosphatidylethanolamine and phosphatidylinositols, respectively, are both endogenous and exogenous. No metabolic role has been assigned to any of the lipids found in nervous tissue other than that of cellular membrane functions.

TABLE 21.6 The Cerebral Lipids and the Sites in Which They Are Found in Relatively High Concentrations

Cholesterol	Approximately equal in myelin, white matter and gray matter
Phosphatidylcholine	Gray matter
Phosphatidylethanolamine	Gray matter
Phosphatidylserine	Gray matter
Phosphatidylinositol	Myelin (peripheral nerve)
Sphingomyelin	Approximately equal in myelin, white matter, and gray matter
Cerebrosides	Myelin and white matter
Sulfatides	Myelin and white matter
Gangliosides	Gray matter
Plasmalogens	Myelin and white matter
Polyglycerophosphatides	Gray matter

Nervous Tissue Proteins

Attempts to relate the structure of major proteins found in nervous tissue to the function of nervous tissue have not as yet been successful. Three proteins have received some attention. A highly acidic protein termed S-100 is found in large amounts in the glial cells of the brain and also in small amounts in neurons. Glutamic and aspartic acids comprise 30% of the amino acid residues. The protein is comprised of three-nonidentical subunits, and it has a high affinity for Ca^{2+}. The molecule undergoes a conformational change in the presence of Ca^{2+}. Another protein, given the designation of 14-3-2 from its sequence of elution in chromatographic steps is found mainly in neurons, and is also a highly acidic protein. The 14-3-2 protein has now been identified as an isozyme of enolase, the enzyme that catalyzes the reversible conversion of 2-phosphoglycerate to phospho*enol*pyruvate. Both the S-100 and the 14-3-2 protein have been implicated by some investigators with the memory or learning ability of the brain. Calmodulin, necessary for the Ca^{2+} activation of the enzyme cyclic nucleotide phosphodiesterase, is found in brain, and also is thought to be involved in the Ca^{2+}-dependent release of acetylcholine and norepinephrine from their vesicular stores (Figure 21.34).

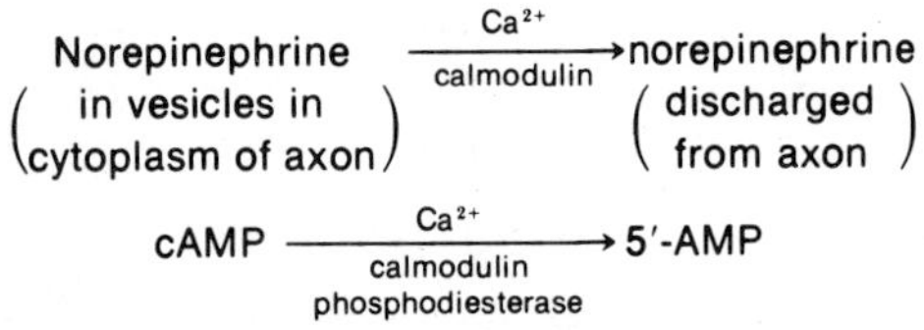

Figure 21.34
Calmodulin-mediated reactions in brain.

The major proteins found in myelin are a *basic protein* and a *proteolipid*. The basic protein, mol wt 18,000, constitutes 30% of the total protein in myelin, and contains a large number of arginine residues. Proteolipid is a combination of a protein, composed of a large number of hydrophobic amino acids, with a lipid. Proteolipid and the basic protein are located on the outer and inner surfaces of the myelin membrane, respectively, and abut each other within the lipid bilayer.

Neurotubules, Neurofilaments, and Microfilaments

The slender tubules transversing the cytoplasm from one dendrite to another or to an axon have a role in axonal flow or secretion. Microtubules are also probably involved in cell differentiation, intracellular transmission of signals from cell surface receptors and the transport of effectors of enzymes. Microtubules are involved in the regulation of cholesterol synthesis in glial cells. Tubulin is the major protein found in microtubules and comprises about 14% of the total protein in mammalian brain. Tubulin consists of two similar but not identical subunits, mol wt 55,000–60,000. Purified tubulin in the presence of Mg^{2+} and GTP will reconstitute or reassemble into microtubules. A diagrammatic view of the arrangement of the subunits of tubulin in the microtubules is shown in Figure 21.35. The most prominent protein in the neurofilaments is a component chemically distinct from tubulin. The neurofibrils are bundles of neurofilaments, and the protein of the microfilaments is actin.

The brain proteins have a rapid turnover rate relative to other body proteins. A mean half-life of brain proteins has been calculated to be about 85 h, or about 20 times faster than total body protein. Liver or serum proteins turnover at one-third the rate of brain protein. Data on the constant amounts of protein in brain, the high proteolytic activity of nervous tissue, the rapid incorporation of glucose into brain proteins, and the small arterial–venous differences of nitrogenous compounds for brain have been interpreted to indicate a retention and reutilization of NH_4^+ by nervous tissue, which is unique to this tissue. Some urea cycle enzymes are present in nerve tissue for, as yet, some unexplained reason. A retention and reutilization of the breakdown products of brain phospholipids is also known to occur. The retention and reutilization of brain protein and lipid seems logical in view of the relative impermeability of the "blood–brain barrier" and the resistance of the brain to deterioration during long-term starvation. The brain is the only tissue in which the breakdown

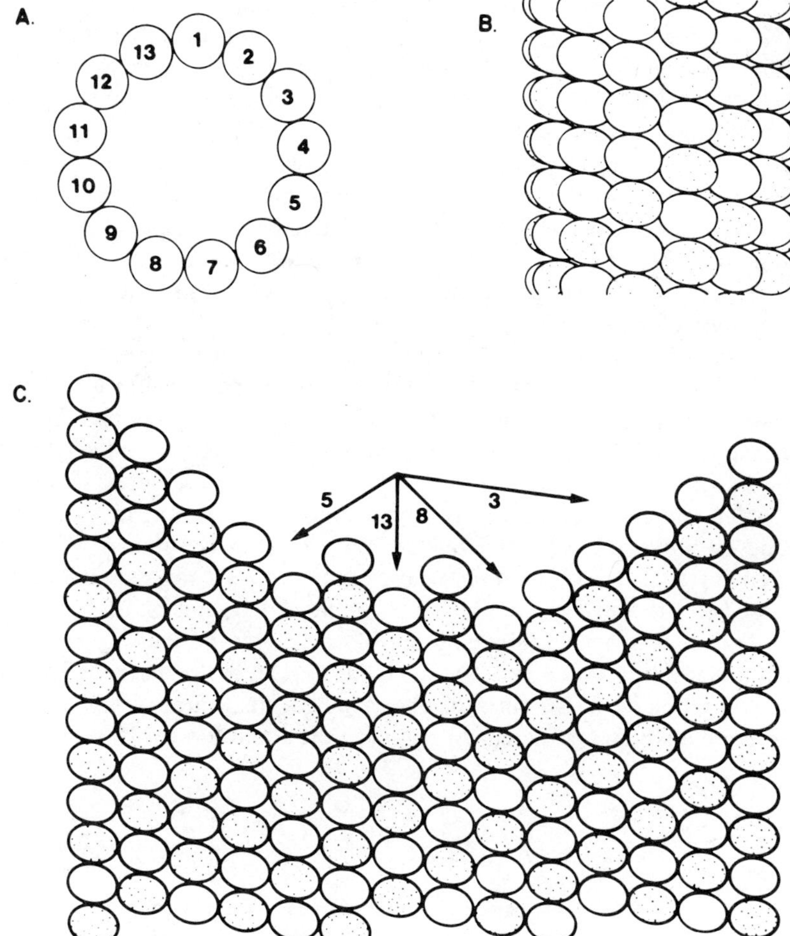

Figure 21.35
Arrangement of subunits of tubulin in microtubules.
(a) Diagramatic view of a cytoplasmic microtubule cross section showing arrangement of 13 protofilaments. (b) Diagrammatic view of cytoplasmic microtubules surface lattice showing the probable arrangement of the heterodimers. Stippled and clear ellipsoids represent α and β subunits. Monomers alternate along the three-start helix and within a protofilament. (c) Representation of the helix net as seen in an unfolded microtubule. Numbers at the bottom of the sheet correspond to the numbered protofilaments in (a). Each ellipsoid represents a monomer, stippled and clear ellipsoids representing the α and β subunits, respectively. Subunits alternate along the three-start helix, and the α–β dimer axis is parallel to the axis of the protofilament. The arrow designated 3 shows the angle of the three-start helix, arrow 13 the angle of the 13-start helix. α–β dimers can be viewed as arranged in a five-start right-handed helix (arrow 5) or an eight-start left-handed helix (arrow 8).

Reproduced with permission from J. A. Snyder and J. R. McIntosh, *Annu. Rev. Biochem.*, 45:706, 1976. Copyright 1976 by Annual Reviews, Inc.

products of protein and phospholipid are extensively reutilized. The other individual tissues in the body release their cellular breakdown products to the circulatory system for processing by the liver.

The Action Potential

Na^+ and K^+ are involved in the excitation and transmission of nerve impulses. The potential difference between the interior and exterior of a resting nerve cell is about 60 mV with the inside negative and the outside positive. The potential difference is mainly the result of the large amounts of K^+ on the inside of the cell relative to the outside of the cell. The sodium concentration outside the cell is larger than it is inside the cell, but this difference does not contribute greatly to the resting potential. The K^+ and Na^+ gradients are maintained by the Na^+,K^+-ATPase, whose action is to pump Na^+ out and K^+ into the cell. If a nerve cell is stimulated, the membrane becomes depolarized, and the potential difference between the inside and outside of the cell changes, and this change is called the action potential. The first event that produces the action potential is the increase of the permeability of the membrane to Na^+. A small amount of Na^+ enters the cell and the inside of the cell becomes positive with respect to the outside; that is, the membrane potential difference is about +35 mV. The potential difference is restored to the resting potential (~60 mV) by an increase in the permeability of the membrane to K^+, which leaks out because of the high intracellular K^+ concentration, and by a decrease in the rate of Na^+ entering the cell. These events are depicted in Figure 21.36. The potential difference is thus restored to its resting level, but the gradients between the intracellular and extracellular K^+ and Na^+ concentrations are not those found in the resting nerve. The ionic gradients are restored to the resting level by the pumping of Na^+ out and K^+ into the cell. This occurs with an expenditure of energy by the Na^+,K^+-ATPase. Mammalian brain, like heart and skeletal muscle, has a reservoir of high energy phosphate in the form of phosphocreatine, which along with the ATP supplies being sufficient to maintain the brain for 15–20 s.

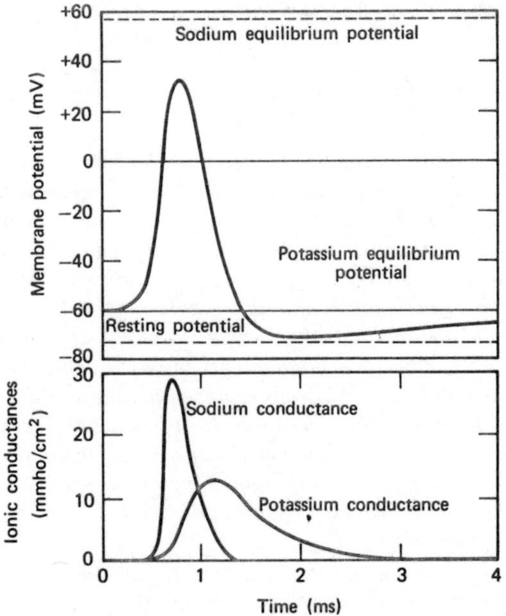

Figure 21.36
Schematic diagram of the potential changes in a neuron during stimulation.
Nerve impulse can be represented by changes in the voltage across the axon membrane (a) or by changes in the conductance of the membrane to sodium and potassium ions (b), both on a time scale of milliseconds. During the impulse the inside of the membrane becomes briefly positive with respect to the outside. After impulse has passed, the resting voltage is restored.

Redrawn from R. D. Keynes, Ion channels in the nerve cell membrane, *Sci. Am.* 240:126, 1979.

Metabolism of Nervous Tissue

Carbohydrate

The brain of a well-nourished animal utilizes large amounts of glucose as its only metabolic fuel. Glucose is nearly completely oxidized to CO_2 and H_2O: actually 25% of the oxygen consumption of the adult body is due to glucose oxidation in brain. The oxygen consumption of brain in infants and children up to 4 years of age accounts for about 50% of the body's oxygen consumption. The glycolytic pathway functions at about 20% of its capacity, but the citric acid cycle works at near maximum rate in a well-oxygenated brain, and small amounts of lactate leave the brain. The greatest danger to the survival of brain is anoxia against which it is protected by the Pasteur effect (page 283). The rate of glycolysis and lactate formation have been found to increase five- and eight-fold, respectively, after only 1 min of anoxia. The increase of the glycolytic flux in anoxia is attributed to the activation of hexokinase and phosphofructokinase. The details of the activation of these enzymes are not well understood. Phosphofructokinase is also activated by NH_4^+, which is found to be increased in anoxia as compared to normal brain.

A deficiency of glucose in blood also endangers the survival of the brain. Endogenous glycogen can supply a small amount of glucose in the absence of blood glucose, but cerebral failure occurs quickly after sudden removal of glucose from the blood. Cerebral failure does not occur, however, with long-term hypoglycemia, such as in starvation, because ketone bodies replace glucose as the brain's metabolic fuel. The utilization of ketone bodies by human brain is thought to be due to their increased concentration in blood in starvation and not to an increase in the permeability of the blood–brain barrier or to an increase in the activities of the enzymes involved in ketone body utilization.

The control of the synthesis and breakdown of brain glycogen seems to be similar to that observed in other tissues. Both the D and I forms of glycogen synthase are found in brain along with the enzymes that catalyze their interconversions. Insulin is thought to have no effect on brain glucose or glycogen metabolism, probably because insulin does not cross the blood–brain barrier. About 3–5% of glucose metabolized in brain is by the hexose monophosphate shunt, to furnish NADPH for fatty acid and cholesterol synthesis.

Amino Acids

Human brain contains about eight times the amount of free amino acids found in plasma and the concentration of the individual amino acids is also different. The concentration of aspartic and glutamic acids are about 300-fold greater in brain than in plasma, and γ-aminobutyric acid, *N*-acetylaspartic acid, and glutamine are present in high amounts. The rates of uptake of amino acids from the blood by the brain are low relative to other tissues. Three different carrier systems for uptake into brain from blood for neutral, basic, and acidic amino acids have been found. The ability of the brain to retain and reutilize nitrogen for amino acid synthesis may well be one reason for the low uptake of blood amino acids into the brain.

Brain levels of NH_4^+ increase in anoxia or ischemia and the source of the NH_4^+ is believed to be the purine nucleotide cycle enzymes (page 815). The immediate fate of the NH_4^+ is its use in the conversion of glutamic acid to glutamine. The utilization of the nitrogen from glutamate and glutamine for the synthesis of the nonessential amino acids from glucose in the brain seems likely. Some of the glutamate is decarboxylated to form the neurotransmitter γ-aminobutyric acid (GABA). GABA,

γ-Aminobutyrate + α-Ketoglutarate → (transaminase) → Succinic semialdehyde + Glutamate

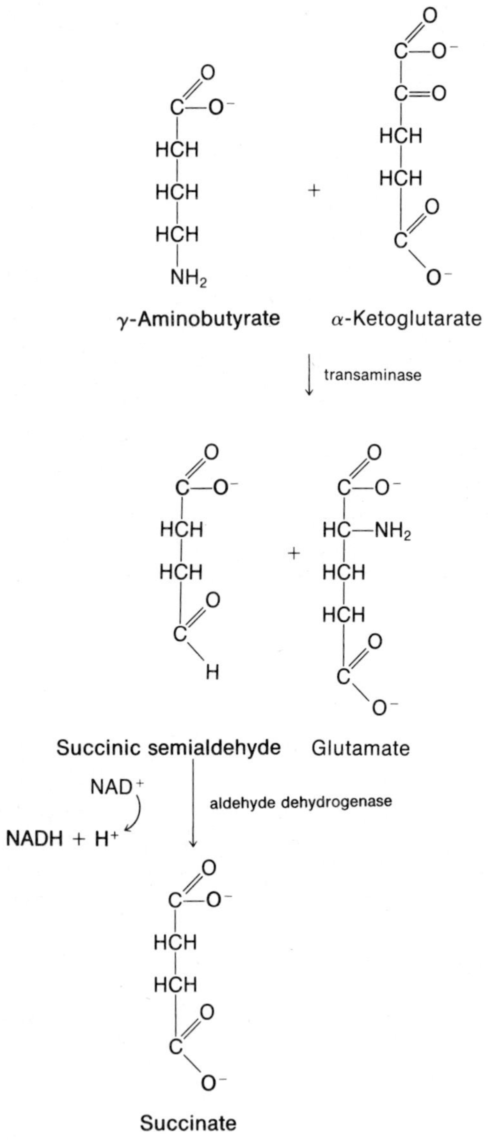

Succinic semialdehyde → (aldehyde dehydrogenase; NAD^+ → $NADH + H^+$) → Succinate

Figure 21.37
γ-Aminobutyric acid (GABA) shunt.

unlike other neurotransmitters that are resequestered into vesicles after their release, is catabolized via a process called the GABA shunt. The GABA shunt is outlined in Figure 21.37.

Glutamate and aspartate stimulate cAMP formation from ATP in brain. Two cAMP-dependent protein kinases are found in brain, which stimulate the phosphorylation of proteins in the synaptic membranes. The phosphorylation of synaptic membrane proteins is involved in the transmission of the nervous impulse across the synaptic junction. Biogenic amines (histamine and catecholamine) and a variety of stimuli known to cause depolarization of neural cell membrane also produce marked increases in the levels of cAMP in brain. The elevation of cAMP elicited by glutamate and aspartate is a unique characteristic of the brain cAMP system, in that all other amino acids are ineffective, and brain is the only tissue that responds to the two amino acids. Therefore, glutamate (and/or aspartate) may well be mediator(s) of the phosphorylation of synaptic membrane proteins and thus the transmission of the nervous impulse across the synaptic membrane.

A variety of inborn errors of metabolism have been described in which severe mental retardation occurs involving the amino acids arginine, phenylalanine, tyrosine, histidine, valine, and so on. The reasons for the mental retardation in these diseases are not known. Phenylpyruvic acid (a metabolite that accumulates in phenylketonuria) has been reported to inhibit glycolysis and pyruvate carboxylase, inhibit fatty acid synthetase, disrupt protein synthesis by converting brain polyribosomes to monoribosomes, and inhibit ketone body utilization. Any one of these effects would seem to be severe enough to interfere with normal brain development and function.

Nucleic Acid Metabolism

The large nerve cells are the most active producers of nucleic acids of any tissue cell. In spite of this, brain cannot synthesize pyrimidine bases but will incorporate uridine, probably made in the liver, into UMP. Brain can synthesize purine bases, but the purine salvage pathway (page 501) is probably the major pathway for the synthesis of GMP, IMP, and AMP.

21.7 LUNG

Glucose is readily metabolized by the lung for energy and for the synthesis of glycogen and lipids. Lung contains a glucose transport system that is stimulated by insulin, leading to increased glucose utilization and lactate production. Fatty acids and ketone bodies are utilized to a very limited extent by isolated lung mitochondria, lung slices, or by perfused lung. The production of lactate accounts for about 60% of the glucose metabolized by lung, and thus this tissue is a major source of blood lactate. The low oxidative capacity of lung is logical in view of its principal role in supplying oxygen to the body. About 10 and 4% of the glucose utilized by lung is by the citric acid cycle and the hexose monophosphate shunt, respectively.

The dependence of lung on insulin for glucose utilization would implicate the necessity for an alternate source of energy in fasting and diabetes. Lung lipoprotein lipase activities and the utilization of fatty acids by lung is increased in fasting. The increase in the oxidative metabolism of lung in fasting and diabetes would decrease the availability of O_2 for the rest of the body. In fact, if the energy for lung metabolism were to arise from the oxidative metabolism of fatty acids, the oxygen consumption of

TABLE 21.7 Composition of Canine Surfactant

	Percent of Total
Lipid	91
Saturated phosphatidylcholine	45
Unsaturated phosphatidylcholine	27
Phosphatidylglycerol	5
Phosphatidylethanolamine	2
Triacylglycerol	3
Cholesterol	7
Other lipid	2
Protein	7
Carbohydrate	2

SOURCE: From R. J. King, Utilization of alveolar epithelial Type II cells for the study of pulmonary surfactant. *Fed. Proc.* 38(12):2637, 1979.

lung would amount to 10% of the total oxygen consumption of a human, an increase of two- to five-fold over normal.

The role of the hexose monophosphate shunt in lung is to furnish NADPH to be used for the synthesis of fatty acids in surfactant and for the protection against injury to the lung tissue by oxidizing agents, for example, O_2, ozone, and superoxide radical (O_2^-).

The Surfactant System of Lung

The surface tension of alveolar fluid is lower than that of most biological fluids and the integrity of the alveolar space is maintained by the low surface tension of its fluid. A surfactant present in the alveolar fluid is responsible for the low surface tension and the absence or deficiency of this material is the cause of a variety of respiratory diseases. Surfactant is 80–90% lipid, of which phospholipids comprise ~80% of the total lipids. The lipid composition depends on the species of mammal; canine surfactant is shown in Table 21.7. The contributions of the dihydroxyacetone phosphate and the α-glycerol phosphate pathways are about equal for the synthesis of phosphatidylcholine and phosphatidylglycerol. The fatty acids of the surfactant are synthesized in the lung cells. The synthesis of phosphatidylcholine appears to be under the control of glucocorticoids (and perhaps thyroxine) in fetal lung, and this control involves hormone receptor systems. In addition, a protein phosphorylation mechanism, involving cAMP, is a factor in the control of phosphatidylcholine synthesis. Glycoproteins have been isolated from the trachea and bronchi of lung tissue (see Clin. Corrs. 21.9 and 21.10).

CLIN. CORR. 21.9 NEONATAL HYALINE MEMBRANE DISEASE (HMD)

Neonatal collapse of the lung alveoli is responsible for most cases of the respiratory distress syndrome of infants. This disease occurs most frequently in premature infants and in infants of diabetic mothers. Approximately 30% of infants born of women with severe diabetes die of this disease. The number of infant deaths with HMD is estimated to be 25,000 a year in the United States. The cause of HMD is a deficiency in the amount of surfactant in the alveoli of the lungs. The consequences of alveolar collapse are reduced functional residual capacity, lungs become stiff, hypoxia, and respiratory acidosis. Factors predisposing to the development of HMD are prematurity, perinatal asphyxia, diabetic mothers, cesarean section, and heredity.

CLIN. CORR. 21.10 ALVEOLAR PROTEINOSIS (AP)

Alveolar proteinosis may be a response of the lung to a variety of noxious agents, for example, dust. The epithelial cells of the alveoli secrete a proteinaceous material, which fills the alveoli and some of the respiratory bronchioles. The amorphous material consists of three glycoproteins, one of which has been characterized. The excess glycoprotein in the alveoli and bronchioles produces a reduction in vital capacity and in pulmonary compliance, and a reduction in O_2 uptake.

21.8 EYE

Lens

The lens is a colorless, transparent, avascular tissue, composed entirely of epithelial cells. The anterior surface consists of a layer of intact cells, whereas the rest of the lens consists of cells that progressively lose their mitochondria and nuclei—the loss of which is a factor that contributes to the transparency of the tissue. Mammalian lens consists of ~35% protein and four different proteins have been identified, called α, β, and γ crystallins, and albuminoid protein. The source of the lens nutrients and the route whereby its waste products are removed is a slowly circulating aqueous humor that bathes its front surface. The lens uses ATP for maintenance, growth and repair, protein synthesis, and transport of cations (via a Na^+,K^+-ATPase) and amino acids. If ATP production by the lens is disrupted, the lens quickly accumulates Na^+ and H_2O and becomes opaque. Most of the energy for ATP synthesis in the lens arises by glycolysis, and the percent glucose metabolized by the glycolysis and hexose monophosphate shunt is ~85 and 10%, respectively. Approximately 3% of the glucose is completely oxidized to CO_2 and H_2O by the citric acid cycle. The control of glycolysis by the lens is critical in that excess lactate damages the lens. The control step in lens glycolysis is hexokinase, which is inhibited by glucose 6-phosphate. A decrease in the amount of ATP in lens activates phosphofructokinase, which in turn produces a decrease in the amount of glucose 6-phosphate. The enzyme hexokinase is stimulated in the presence of low amounts of glucose 6-phosphate. An increase in ATP in lens has the opposite, or an inhibitory effect, on hexokinase. The NADH from glycolysis in the lens is used mainly for the conversion of pyruvate to lactate, with small amounts of reducing equivalents transferred to the few mitochondria present in lens by the α-glycerophosphate shuttle (page 239).

The only disease of lens is the formation of cataracts, an opacity of the lens. Cataracts can be produced experimentally in animals fed large amounts of galactose or sorbitol and are found in humans with hyper- and hypoglycemia and with deficiencies in galactokinase. Low glucose levels in the blood result in a decrease of lens hexokinase, thereby decreasing the glycolytic rate, and high amounts of glucose in the lens activate the sorbitol pathway leading to fructose formation. The sorbitol pathway is shown in Figure 21.38.

Glucose (HC=O, HCOH, HOCH, HCOH, HCOH, H) —NADPH → $NADP^+$→ Sorbitol (H, HCOH, HCOH, HOCH, HCOH, HCOH, HCOH, H) —NAD^+ → NADH + H^+→ Fructose (H, HCOH, C=O, HOCH, HCOH, HCOH, HCOH, H)

Figure 21.38
Sorbitol pathway.

Retina

The retina contains the enzymes of glycolysis, citric acid cycle, and hexose monophosphate shunt. Retina has one of the most active glycolytic systems and the highest rate of oxygen consumption of any tissue. The retina contains large amounts of lactate and lactate dehydrogenase, which means that the rate of pyruvate production by glycolysis exceeds its rate of oxidative metabolism even under conditions of adequate oxygen supplies. Low activities of the α-glycerophosphate shuttle enzymes (page 239) and the large amounts of lactate in retina are interpreted to indicate that NADH produced in glycolysis is used for the reduction of excess pyruvate to lactate. Lactate dehydrogenase of retina utilizes NADP as well as NAD; thus the production of NADPH from lactate is a factor in controlling the rate of glucose metabolism by the hexose monophosphate shunt. The production of NADPH by the hexose monophosphate shunt is a critical reaction for retina, in view of its utilization in the *rhodopsin cycle*.

Role of Rhodopsin in Vision

The retina contains two types of receptor cells: cones, which are specialized for color and detail vision in bright light, and rods, which are specialized for vision in dim light. Light waves striking these receptors produce chemical changes that give rise to nerve impulses that are transmitted to the brain. Vision in the rod cells is dependent upon a photosensitive pigment called rhodopsin or *visual purple*. Rhodopsin is a complex of the protein opsin and the 11-*cis* isomer of the aldehyde form of vitamin A (page 964), called retinal. Opsin is a hydrophobic protein (mol wt ~28,000). When light strikes rhodopsin in the retina, the 11-*cis*-retinal is isomerized to the all-*trans* form yielding an all-*trans*-retinal-protein complex called prelumirhodopsin (Figure 21.39). This is the only step in the rhodopsin cycle that is affected by light and is the step that is believed to be associated with the nerve impulse. The prelumirhodopsin undergoes a series of reactions, resulting in the cleavage of the complex to all-*trans*-retinal and the protein opsin.

The visual transduction process in retinal rods has been shown to involve a change in the cellular level of cyclic GMP. A photon is captured by rhodopsin, located in the disk membranes of the rod outer segment. This signal is then transmitted to a membrane-bound phosphodiesterase via a guanine nucleotide-binding regulatory protein called transducin. The activated phosphodiesterase rapidly converts cellular cyclic GMP to 5′-GMP. This process of phototransduction is coupled to a release of Ca^{2+} ions from the disk that inhibits the transport of Na^+ ions into the rod's outer segments. The synapse of the rod cell is stimulated by the hyperpolarized membrane, and the nerve impulse is transmitted to the brain.

Rhodopsin must be reconstituted for continued vision in dim light. The all-*trans* isomer of retinal is converted to 11-*cis*-retinal by the action of retinal isomerase in the retina, and the 11-*cis*-retinal combines with opsin to form rhodopsin. The all-*trans*-retinal has an alternate fate; it is reduced to all-*trans*-retinol by alcohol dehydrogenase and NADPH in the retina. The all-*trans*-retinol is transported to the liver as its fatty acid esters, and both the all-*trans*-retinol and its esters are isomerized to 11-*cis*-retinol and oxidized to 11-*cis*-retinal to be used for the formation of rhodopsin. The isomerization of all-*trans*-retinol to 11-*cis*-retinol occurs in the liver. The transport of retinol in the plasma is as a complex with retinol-binding protein.

Cornea

The cornea, like the lens, is an avascular organ and mainly derives its nutrients from the aqueous humor. The cornea also derives nourishment from limbus, tears, and from blood vessels in the lids. The single cell layer of the corneal endothelium contains the transport systems for nutrients and water. The constant removal of water from the cornea is necessary for the maintenance of an anhydrous and transparent cornea. The cornea

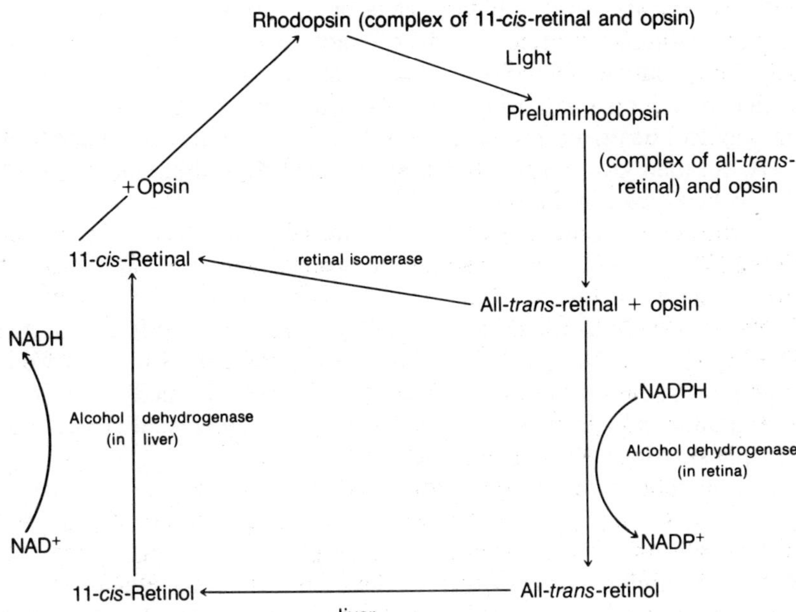

Figure 21.39
Rhodopsin cycle.

is a multilayered tissue and the major layer is the stroma, which consists of laminae of collagen parallel to its surface, with fibroblasts sandwiched between them.

Approximately 65% of the glucose used by the cornea is by the hexose monophate shunt—a higher percentage than reported for any mammalian tissue. Thirty percent of the glucose is metabolized by the glycolytic pathway. The cornea also has a high rate of oxygen consumption, which seems reasonable, since it is the portion of the eye exposed to the atmosphere. The end product of glucose metabolism in the stroma is lactate, which is transferred to the epithelium for oxidation to CO_2 and water by the citric acid cycle enzymes. A deficiency of oxygen supply to the cornea, such as might occur with improperly fitted contact lenses, will result in damage to the stroma because of an accumulation of lactic acid.

21.9 BLOOD CELLS

Reticulocytes

The immature red blood cell, called the reticulocyte, contains the subcellular organelles of all mammalian cells except a nucleus. The reticulocyte has the enzymes involved in glycolysis; hexose monophosphate shunt, citric acid cycle; oxidative phosphorylation; synthesis of hemoglobin, cholesterol, phospholipids, triacylglycerols, and purine nucleotides. Reticulocytes have higher rates of glucose consumption than erythrocytes and when their cellular respiration is inhibited, both the glucose uptake and lactate production are greatly increased. The Pasteur effect in reticulocytes is greater than that reported for any other mammalian tissue. The increase in glycolytic rate of reticulocytes in a lack of oxygen is attributed to an activation of phosphofructokinase and pyruvate kinase. Reticulocytes also have the enzyme glycerol kinase and can utilize glycerol for the production of energy.

Erythrocytes

The mitochondria, Golgi apparatus, and RNA are extruded from the reticulocyte in the process of maturation to form the erythrocyte, and therefore erythrocytes are not living cells in the strict sense. Erythrocytes cannot synthesize protein, glycogen, lipids (from glucose), and are incapable of oxidative metabolism. They utilize glucose by glycolysis and the hexose monophosphate shunt exclusively and have the ability to synthesize nucleotides from preformed purines. The erythrocytes survive and carry out their O_2 transport function for ~120 days after which time they are broken down in the body.

Erythrocytes contain ~35% solid material, which is almost all hemoglobin (32%). The remaining solids are proteins and lipids which form the stromal network for the support of the hemoglobin. The membrane is freely permeable to water, urea, creatinine, HCO_3^-, Cl^-, OH^-, and H^+. The transport of Na^+, K^+, Ca^{2+}, and perhaps P_i across the erythrocyte membrane requires energy. The phosphate content of the RBC is 50–100 fold greater than in plasma, most of which is organic, that is, hexose phosphate, triose phosphates, and ATP.

The metabolic fuel of the erythrocyte is glucose which is utilized by glycolysis. The energy of ATP is necessary to preserve the integrity of the erythrocyte and is also used for the membrane transport of ions. The NADH from glycolysis is used for the conversion of methemoglobin to hemoglobin and for the conversion of pyruvate to lactate. Methemoglobin is slowly but continually formed from hemoglobin in the body and is not capable of O_2 transport. Hexokinase, phosphofructokinase, pyruvate kinase, glyceraldehyde 3-phosphate dehydrogenase, and phosphoglycerate kinase are implicated in the control of glycolysis in the erythrocyte. A number of hemolytic anemias have been identified due to deficiencies of these enzymes (see Clin. Corr. 21.11). A major factor in the activation of glycolysis in the erythrocyte is inorganic phosphate. Approximately 10% of the glucose metabolized by the erythrocyte is by the hexose monophosphate shunt. The ratio of the oxidized to reduced form of glutathione, present in large amounts in the erythrocyte, the amount of H_2O_2 in the RBC, and a breakdown product of NADP, 2′-phosphoadenosine diphosphate ribose, are implicated in the control of the shunt. The glutathione and H_2O_2 may well control the rate of shunt metabolism by altering the ratio of the reduced to oxidized form of NADP.

A carrier protein is involved in the transport of glucose in erythrocytes and constitutes a major component of the erythrocyte membrane. The movement of glucose does not require energy or insulin. The number of sugar carrier sites per red cell has been calculated to be 6×10^4. An interchange of cholesterol and phospholipids between plasma and the erythrocyte membrane occurs, which is a nonenzymatic process that is a factor in maintaining the integrity of the membrane.

A cAMP-dependent protein kinase is found in human erythrocyte membrane. The function of this enzyme is unknown, since cAMP does not greatly stimulate protein kinase activity in the erythrocyte, and the erythrocyte has low levels of cAMP.

The modulating effect of 2,3-diphosphoglycerate on the combination of O_2 with hemoglobin is discussed on page 887.

CLIN. CORR. 21.11
HEMOLYTIC ANEMIA

Pyruvate kinase deficiency hemolytic anemia is the result of a deficiency of this enzyme in the erythrocyte and is an inherited disease. A second commonly found hemolytic anemia is the result of a deficiency of glucose 6-phosphate dehydrogenase in the RBC. Other hereditary hemolytic anemias are the result of deficiencies in the RBC of each of the following enzymes: hexokinase, glucosephosphate isomerase, phosphofructokinase, triosephosphate isomerase, 2,3-diphosphoglycerate mutase, phosphoglycerate kinase, 6-phosphogluconate dehydrogenase, glutathione reductase, glutathione peroxidase, and glutathione synthetase.

Leukocytes

The polymorphonuclear leukocytes possess the enzymes of the hexose monophosphate shunt, glycolysis, citric acid cycle, and respiratory chain. Only a small amount of glucose is oxidized via citric acid cycle and respiration. The primary purpose of the leukocyte is phagocytosis, which is the engulfing and destruction of particulate matter. Phagocytosis is an energy-requiring process associated with an increase in the rate of glycolysis and a great increase in the metabolism of glucose by the hexose monophosphate shunt. The role of the hexose monophosphate shunt is to produce H_2O_2 (from superoxide, O_2^-), which is used in the phagocytic process. Large amounts of glycogen are found in leukocytes, which is the energy source for phagocytosis in the absence of exogenous glucose. If glucose is present, glycogen breakdown does not occur in phagocytosis.

Platelets

Platelets consist of 50% protein and 15% lipid and contain a contractile protein (similar to actomyosin of muscle) and ATP. The actomyosin and ATP are involved in the clot retraction mediated by the platelets. Platelets contain substances required in blood clotting and enzymes characteristic of lysosomes and mitochondria.

The major energy source for platelets arises from the glycolysis of glucose. Under aerobic conditions, glycolysis accounts for one-half the ATP formed. Platelets increase the rate of glucose utilization and lactate production in a deficiency of oxygen. Approximately 20% of the glucose metabolized by platelets is by the hexose monophosphate shunt. The aggregation of platelets in blood clotting is an energy-requiring process, which is derived from its breakdown of glycogen and glycolysis.

Clotting of Blood

The formation of a blood clot is an important defense mechanism in the body. Blood clotting is a complicated process involving many events that are delicately controlled to prevent the formation of a clot unless it becomes necessary to retard the loss of blood from the body. The formation of a blood clot within an intact blood vessel results in an occlusion of the vessel, called a thrombosis. The formation of thrombi is a frequent cause of disability and death, that is, stroke and heart attack. The body has three major physiological responses to stop the loss of blood after an injury: (1) vasoconstriction of the blood vessel to reduce the flow of the blood to the injured site quickly; (2) an aggregation of the platelets in the blood at the site of injury to form a clump or a temporary clot; (3) the formation of a stronger and more permanent clot that will stop the blood loss until the injury is repaired. This so-called permanent clot is composed of the protein *fibrin*. The formation of fibrin from its precursor form in the blood, *fibrinogen,* is the result of a large number of events, some of which are triggered by chemical events associated with the clumping of the platelets to form the soft or temporary clot. Clumping of the platelets also releases chemicals that trigger the vasoconstriction of the blood vessels.

The events that result in the formation of fibrin, the permanent clot, can be triggered in vitro by two independent mechanisms or pathways. The *extrinsic pathway* is so called because factors not present in the blood are involved in triggering the clotting process. The *intrinsic pathway* is so called because all the components necessary for triggering the series of blood clotting events are present in the blood. Although the extrinsic and intrinsic pathways do not have a common origin, their final steps *are identical.* These are

1. Prothrombin → Thrombin

2. $$\underset{\text{(Soluble)}}{\text{Fibrinogen}} \xrightarrow{\text{thrombin}} \underset{\text{(Insoluble)}}{\text{Fibrin (the clot)}}$$

These two steps were discovered 40–50 years ago. The various other factors involved in the clotting pathways were discovered in large part because they were inactive or absent in patients with bleeding disorders, called hemophilias. The blood clotting factors and their functions are listed in Table 21.8.

Conversion of Fibrinogen to Fibrin

The fibrinogen concentration in human plasma is relatively high (about 0.3 g/dl), enough for fibrinogen to form a band visible in electrophoretic patterns of human plasma. Fibrinogen is synthesized in the liver and has a half-life in the blood of about 4 days. In patients with severe liver disease there may be a decreased plasma level of fibrinogen. In simple terms, the steps involved in the conversion of soluble fibrinogen to insoluble fibrin are (1) hydrolysis of Arg-Gly peptide bonds in fibrinogen to give fibrin monomers and fibrinopeptides A and B, respectively; (2) the fibrin monomers combine with one another both end to end and side by side to form a precipitate, called the *soft clot;* (3) the *soft clot* is converted to a *hard clot* by converting the end-to-end aggregation sites of the fibrin monomers into covalent bonds.

The details in the conversion of fibrinogen to fibrin are a beautiful example of the contributions of the knowledge of protein structure to its function in the body. Fibrinogen is a molecule that consists of three pairs of nonidentical peptide chains, α_2, β_2, and γ_2. The three pairs of chains are linked head to head and side by side by disulfide bonds near the amino

TABLE 21.8 Blood Coagulation Factors and Their Function

Name	*Factor Number*	*Site of Synthesis*	*General Functions and Properties*
Fibrinogen	I	Liver	Present in plasma and converted to the clot (fibrin) by *thrombin* (II_a). This involves proteolysis of the fibrinogen (I) by thrombin and a transglutaminase activity of the active form of *fibrin stabilizing factor* (FSF_a or XIII_a).
Prothrombin	II	Liver	Converted to the active form thrombin by proteolysis by the activated Stuart factor (X_a). Thrombin *also* activates proaccelerin (V), proconvertin (VII), antihemophilic factor (VIII_c), fibrin-stabilizing factor (XIII), and protein C–all by a proteolysis.
Thromboplastin (tissue factor)	III	Most tissues (lung, placenta, brain, and saliva)	This factor, in the presence of phospholipid, triggers the extrinsic clotting system. It is an accessory protein that aids in the activation of proconvertin (VII).
Calcium ions	IV		Binds vitamin K-dependent clotting factors prothrombin (II), proconvertin (VII), Christmas factor (IX), and Stuart factor (X) to phospholipid membranes. Also required for the activation of the fibrin-stabilizing factor (XIII).
Proaccelerin	V	Liver and endothelial cells (?)	The active form of proaccelerin does not have protease activity. It is a specific accessory protein for the activation of prothrombin (II) by the Stuart factor (X), Ca^{2+}, phospholipid complex in the intrinsic system. It accelerates the action of the Stuart factor (X).
Proconvertin	VII	Liver	The *active* form activates the Stuart factor (X) in the *extrinsic* pathway by proteolysis and can also activate the Christmas factor (IX).
Antihemophilic factor	VIIIc	Liver	Accessory protein that enhances the activation of the Christmas factor (IX) (not a proteolysis reaction).
Christmas factor	IX	Liver	The active form activates the Stuart factor (X) by a proteolytic reaction (in the *intrinsic* system).
Stuart factor	X	Liver	The active form converts prothrombin (I) to thrombin (II_a) by proteolysis.
Plasma thromboplastin antecedent	XI	Liver	The active form activates the Christmas factor by a proteolysis.
Hageman factor	XII	?	Upon contact of this factor with collagen, etc., *or* by the action of kallikrein, it becomes activated and triggers the intrinsic pathway. The activated Hageman factor (XII_a) is a proteolytic enzyme that activates plasma thromboplastin antecedent.
Fibrin-stablizing factor	XIII	?	The activated factor that has transglutaminase activity necessary for formation of the hard fibrin clot. It is activated by thrombin (II_a).
Protein C	XIV	Liver	Is converted to a protease by thrombin and *inactivates* proaccelerin V and antihemophilic factor (VIII_c). Protein C, thus has *anticoagulant* activity.
Protein S		Liver (?)	Protein S is not a serine protease. It acts as an accessory protein for the action of protein C. It activates protein C.
Prekallikrien		Liver	The precursor of the protease kallikrein, which acts to activate the Hageman factor (XII) in the contact phase of the intrinsic system.
High-molecular-weight kininogen		Liver	An accessory factor for the activation of the Hageman factor (XII) and plasma thromboplastin antecedent.
Von Willebrand factor	VIII antigen	Megacaryocytes and endothelium	A protein required for proper platelet adhesion and aggregation. This factor also serves as a "carrier" for the smaller factor VIII_c.

terminals to form an enlarged structure called the disulfide knot. (Fig. 21.40). The carboxyl terminals of the 3 pairs of chains are enlarged globular structures. Fibrin monomers are formed by the cleavage at Arg-Gly peptide bonds of the α and β chains near their amino terminal ends to form the fibrinopeptides A and B, which are 16 and 14 residues long, respectively. The removal of these four fibrinopeptides leaves the major portion of the molecule of the fibrinogen, called the fibrin monomer. The

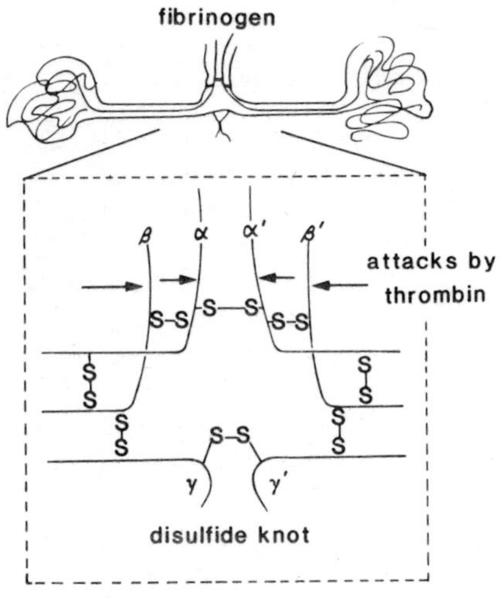

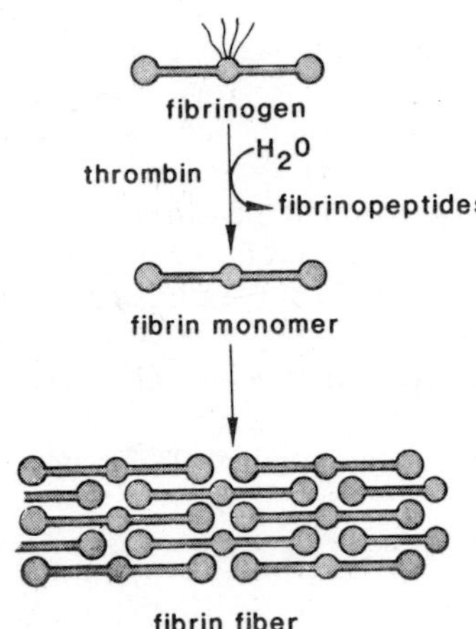

Figure 21.40
Diagrammatic representation of the fibrinogen molecule and its conversion to the soft clot of fibrin.

monomers associate side by side and end to end to form a fibrin fiber called a soft clot.

The joining of the fibrin monomers by covalent bonds is catalyzed by *fibrin-stabilizing factor* (FSF or Factor XIII), which is present in plasma and platelets. FSF is present in an inactive dimeric form, which like fibrinogen is hydrolyzed by thrombin to yield 2FSF peptides and smaller peptides. The 2FSF peptides are bound together and are inactive until they are separated by binding of Ca^{2+} to form an active form of FSF (called FSFa or XIII_a). The active form of FSF has transglutaminase activity which catalyzes the formation of covalent cross-links between the fibrin monomers. These cross-links are formed between the γ-amide group of glutamine and the ε-amino group of lysine. The soft clot thus becomes an insoluble hard clot of fibrin monomers linked end to end by a covalent linkage and side by side by disulfide bonds.

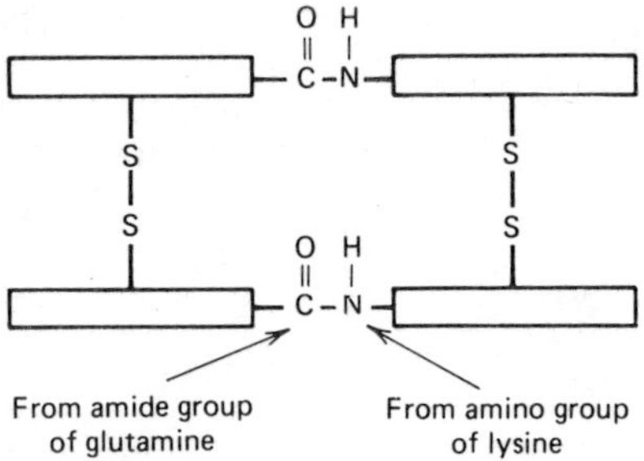

The cross-links give rigidity to the clot which is a lattice in which many of the cellular components of the blood are entrapped. The red color of the clot is due to the presence of red cells; when fibrinogen is converted to fibrin by prothrombin in the absence of red blood cells, the clot is colorless. Upon standing, the contractile proteins in the trapped cells cause the clot to retract into a hard pellet from which most of the serum has been extruded.

Conversion of Prothrombin to Thrombin

Fibrinogen is always present in blood, but the blood will not clot because the factor that converts fibrinogen to fibrin, called *thrombin,* is in its inactive form, called *prothrombin*. Prothrombin is converted to thrombin by an activated form of the *Stuart factor* (X_a). The activated Stuart factor is a serine endopeptidase attacking an arg-X bond. This is a limited proteolysis of prothrombin. *Two* reactions are catalyzed by the activated Stuart factor (X_a): (1) removal of an amino-terminal segment from the prothrombin and (2) opening of a disulfide loop. Both of these reactions are by the cleavage of an arg-X bond. The thrombin molecule consist of two peptide chains joined together by a disulfide bond. The Stuart factor (X_a) activation of prothrombin is an extremely slow reaction. For X_a to display physiological activity, the following must be present: Ca^{2+}, phospholipid, and another clotting factor, *proaccelerin* (V) in its active form (V_a). Proaccelerin is not an enzyme; it is a protein that facilitates the action of the Stuart factor on prothrombin. The phospholipids required for the action of the Stuart factor are phosphatidylserine and phosphatidylinositol that come from the platelets.

Intrinsic Pathway of Blood Clotting

The activation of this route for rapid coagulation is, in simple terms, by the *Hageman factor* (XII). It has been known for many years that contact of blood with glass (i.e., the glass test tube) activates the clotting process. This simple observation has been the basis for the discovery of the steps involved in the intrinsic pathway. The Hageman factor becomes activated when it is absorbed to glass surfaces or collagen or to platelet membranes. The activation is a proteolytic cleavage of Factor XII to XII_a, and the glass surface somehow makes the susceptible bond more available to a variety of proteases. When the Hageman factor becomes activated, it becomes a proteolytic enzyme and, in turn, starts the cascade of events in the intrinsic pathway. The relevance of the intrinsic pathway to in vivo coagulation has been disputed, since human patients lacking Hageman factor (XII) or plasma thromboplastin antecedent (XI) show no bleeding tendencies. The function of the extrinsic pathway for in vivo coagulation is well documented.

The activated Hageman factor (XII_a) activates the next factor in the cascade, called *plasma thromboplastin antecedent* (XI) by proteolysis. This activation is a complex step. It involves *prekallikrein,* a precursor of an active serine protease and high molecular weight *kininogen* (HM_wK) a precursor of *bradykinin*—an accessory factor. Specifically, activated Hageman factor (XII_a) converts prekallikrein to kallikrein, which *also* acts to convert XII to activated Hageman factor (XII_a) by a proteolysis. There are thus two routes for the activation of the Hageman factor—contact phase and proteolysis by kallikrein. The XI_a now activates the next factor, the *Christmas factor* (IX) by proteolysis—a reaction that requires Ca^{2+}. The Christmas factor is also activated by a proteolytic reaction catalyzed by active *proconvertin* (VII_a), which requires acidic phospholipids and Ca^{2+} ions.

The next step is the activation of the *Stuart factor* (X) by the activated Christmas factor (IX_a). The activated form of the Stuart factor converts prothrombin to thrombin. The activation of the Stuart factor is accomplished in two different ways, one operating in the intrinsic pathway and the other operating in the extrinsic pathway. In the intrinsic pathway, the Stuart factor is activated by the activated Christmas factor (IX_a), which requires the presence of Ca^{2+}, phospholipid, and the antihemophilic factor also known as blood coagulation factor VIII_c. Complete factor VIII is actually a complex of proteins that also includes Von Willebrand factor (factor VIII-related antigen), which is involved in coagulation at the level of platelet aggregation and serves as a "carrier" for the smaller factor VIII_c.

Extrinsic Pathway

The extrinsic pathway of clotting is much more rapid than the intrinsic pathway. This can occur in ~10 s when the tissue factors of the extrinsic pathway are added. The Stuart factor (X) becomes activated by a different mechanism than that in the intrinsic pathway. The activation requires Ca^{2+} and phospholipid (as in the intrinsic pathway), but the protease is *proconvertin* (VII) and a tissue factor called *thromboplastin* (III). Proconvertin is converted to its active form (VII_a) on proteolysis by thrombin (II_a), activated Stuart factor (X_a), or activated Hageman factor (XII_a). When a tissue is damaged, thromboplastin (III) and phospholipid are released, which stimulates the activity of the active form of proconvertin to generate thrombin.

Vitamin K and its Role in the Clotting Process

Prothrombin (II), proconvertin (VII), Christmas factor (IX), and the Stuart factor (X) are factors synthesized in the liver, require Ca^{2+} for their action, and also require the presence of vitamin K in the diet for their synthesis. The administration of antimetabolites of vitamin K, such as dicumarol, inhibit the synthesis of the coagulation factors. These clotting factors from animals fed dicumarol do not bind Ca^{2+} as well as the factors from animals fed a normal diet. Severe generalized liver disease results in deficient synthesis of coagulation factors (II, V, VII, IX, and X), which is expressed as a prolonged prothrombin time. The calcium-binding sites of the normal blood clotting factors are γ-carboxyglutamate residues in the amino terminal ends of each factor:

−OOC H COO− −OOC H COO−
C C
CH2 CH2 O
H
CH N C
N C C
H H
O

Prothrombin from animals fed dicumarol contain glutamate residues at these positions:

H H
H—C—COOH H—C—COOH
CH2 CH2 O
H
N C N C
H H C C
H H
O

Vitamin K is now known to be a cofactor for a carboxylase that catalyzes the carboxylation of glutamyl residues in the vitamin K dependent clotting factors:

H—N H—N
H H
HOOC—C—CH2—CH + 2CO2 HOOC—C—CH2—CH
H + O2 HOOC
C=O C=O
reduced form of vitamin K

The reduced form of vitamin K has the structure

OH
CH3
R
OH

In summary, the carboxylation of the γ-glutamyl groups in the Ca^{2+} requiring clotting factors by carboxylase, an enzyme that has reduced vitamin K as a cofactor, makes the clotting factors able to bind Ca^{2+}, which is necessary for their physiological activity.

CLIN. CORR. **21.12** DISORDERS OF PLATELET–VESSEL WALL INTERACTION: HEMORRHAGIC DISEASES

Several hemorrhagic diseases are caused by defects in either the platelets, plasma, or vessel walls. The Bernard-Soulier syndrome is an example of a defect in platelet adhesion and aggregration in which there is a deficiency or abnormality in a platelet membrane glycoprotein, or factor V or IX. There are diseases reported in which there is a deficiency of contact-promoting proteins secreted by platelets, a deficiency of platelet-secreted ADP and serotonin, and an abnormality of platelet arachidonic acid metabolism. Defects in platelet procoagulant activity have been reported in which Factor V_a binding sites on the platelet surface are deficient leading to decreased platelet adhesion.

Deficiency of platelet contact-promoting proteins is caused by a deficiency of the Von Willebrand factor, which is referred to as Von Willebrand's disease. Defects in the vessel wall are the results of genetic disorders of connective tissue, that is, collagen, elastin, or fibronectin.

Role of the Platelets in Clotting

The blood platelets, or thrombocytes, are made in the bone marrow. In the absence of platelets the blood will not clot properly. Platelets will adhere to collagen fibers and vasculature that become exposed upon injury to the tissue (Clin. Corr. 21.12). When the platelets adhere to the collagen and vasculature they shrink and excrete their contents such as ADP, serotonin, a thromboxane A_2, and platelet factor IV—a heparin-binding protein. The ADP that is released makes the platelets stick to one another to form the platelet plug. The shrinking of the platelets also exposes lipoproteins, which have a role in blood clotting, that is, conversion of prothrombin to thrombin and the activation of the Stuart factor. The platelets are also one source of proaccelerin (V) and antihemophilia factor (VIII_c) that is necessary for the formation of thrombin and the activated Stuart factor (X_a). The platelets are also the source of the fibrin-stabilizing factor (FSF) (XIII).

Inhibition of Blood Clotting

The clotting of blood can be prevented by the addition of substances that combine or tie up Ca^{2+} ions; that is, oxalate, fluoride, EDTA (a Ca^{2+} chelating agent), or citrate. When a sample of blood plasma is required for certain clinical tests, one of the above reagents is mixed with the blood.

Dicumarol is administered to patients to reduce the clotting ability over long periods of time. The effect of this treatment is to inhibit the formation of the carboxylated forms of prothrombin (II), proconvertin (VII), Christmas factor (IX), and Stuart factor (X). Excessive dicumarol treatment produces a bleeding disorder similar to hemophilia. This latter property has found extensive use for dicumarol as a rodenticide.

Heparin is the major clinical anticoagulant used in maintaining normal hemostasis. It can be administered to patients during and after surgery to retard blood clotting. Heparin is a high molecular weight polysaccharide containing large amounts of sulfonated sugars. It is found in many tissues, especially in the mast cells in the endothelium of the blood vessels. When heparin is administered or released from the mast cells, heparin combines with antithrombin III, which becomes activated and combines with and inhibits many active factors (prothrombin, Christmas factor, Stuart factor, plasma thromboplastin antecedent, Hageman factor, and kallikrein).

Another anticoagulant in the blood is protein C. Thrombin activates protein C by proteolysis, and protein C then hydrolyzes and *inactivates* proaccelerin (V) and the antihemophilia factor (VIII_c).

Lysis of Clots

The body contains a system that will dissolve or lyse the clot a few days after its formation. *Plasminogen* is synthesized in the kidney and is converted to its active form, *plasmin,* by plasminogen activators, which are proteases found in many tissues: (1) *urokinase* is a plasminogen activator that is synthesized in the kidney and that can be isolated from human urine; (2) activated Hageman factor in conjunction with prekallikrein and high molecular weight kininogen also activates plasminogen to plasmin.

Plasminogen is also converted to plasmin by *streptokinase,* a protein from β-hemolytic streptococci. Streptokinase is not a proteolytic enzyme but combines with plasminogen, altering its conformation, so that the altered plasminogen has no enzyme activity.

Both streptokinase and urokinase can cause bleeding to excess and are now considered to be "pharmacological agents"; that is, both will react with plasminogen in the *absence* of a clot. The probable compound for the physiological lysis of the clot is called *tissue plasminogen activator* (TPA) found in the vascular endothelium. TPA will act on plasminogen *only* if a clot is present.

The plasmin, however formed, hydrolyzes several bonds in fibrin to produce low molecular weight proteins thus breaking up the clot.

Summary and Miscellaneous Statements

With both the intrinsic and extrinsic pathways, the initial events that trigger the first reaction are amplified many fold with each succeeding stage. Blood clotting is a series of reactions called a *cascade*. Activated clotting factors are removed from the circulation by the hepatocytes of the liver. The nonactivated factors are not removed by the liver. Several activated clotting factors can either activate or inhibit other clotting factors. Thrombin (II_a) activates proaccelerin (V), antihemophilic factor (VIII_c), and fibrin-stabilizing factor (FSF). Thrombin (II_a) also activates protein C, which breaks down the active forms of proaccelerin and antihemophilic factor (VIII_c).

Most of the reactions in the cascade are proteolytic. Most, if not all, of the proteolytic enzymes are serine proteases; that is, serine is the active site of the proteolytic enzymes. Five factors serve as cofactors: proaccelerin (V) (for the Stuart factor X); antihemophilia factor VIII_c (for the Christmas factor (IX)); thromboplastin or tissue factor (III) (for proconvertin VII_a); protein S (for protein C), and HMW kininogen (for the Hageman factor (XII) *and* plasma thromboplastin antecedent (XI). Four vita-

TABLE 21.9 Some Additional Details in the Cascade Scheme Shown in Figure 21.42

Reaction A. The activation occurs when Factor XII becomes absorbed to certain ionic surfaces. Glass or kaolin function in vitro. Also the activated Hageman factor (XII_a) converts prekallikrein to kallikrein, which also converts inactive Hageman factor (XII) to the activated form (XII_a) by a proteolysis.

Reaction B. The activation of plasma thromboplastin antecendent (XI) also involves prekallikrein and high molecular weight kininogen–a precursor of bradykinin.

Reaction C. The activation of the Christmas factor (IX) *also* involves the proteolysis by the active form of proconvertin (VIIa). Proconvertin (VII) can be converted to its active form by thrombin (II), activated Stuart factor (X_a), or activated Hageman factor (XII_a). Acidic phospholipids and Ca^{2+} ions enhance the activation of the Christmas factor (IX).

Reaction D. The activation of the Stuart factor (X) by the active Christmas factor (IX_a) requires Ca^{2+}, phospholipid, and the activated form of the antihemophilic (factor VIII_c). The antihemophilic factor VIIIc is an accessory protein for this reaction–it is not a proteolytic enzyme.

Reaction E. The conversion of prothrombin (II) to thrombin (II_a) by the activated Stuart factor (X_a) also requires phospholipids, Ca^{2+} ions, and activated proaccelerin (V_a). Activated proaccelerin is an accessory protein in this reaction–it is not a proteolytic enzyme. Proaccelerin is activated by the proteolytic action of thrombin.

Reaction F. The soft clot is converted to a hard clot by the active form of the fibrin-stabilizing factor (FSF_a). The inactive form of FSF is converted to its active form by thrombin and Ca^{2+} ions.

Reaction G. This reaction, which triggers the extrinsic pathway by a proteolysis, is by the active form of proconvertin (VII_a). Proconvertin (VII) can be converted to its active form after tissue damage by the release of thromboplastin (a tissue factor) and phospholipid. These latter factors also enhance the ability to convert prothrombin to thrombin (II_a). Thrombin then feeds back by converting proconvertin into its active form (VII_a).

Reaction H. The activation of the Stuart factor (X) by the active form of proconvertin (VII_a) also requires thromboplastin, the tissue factor (III), plus phospholipids and Ca^{2+} ions.

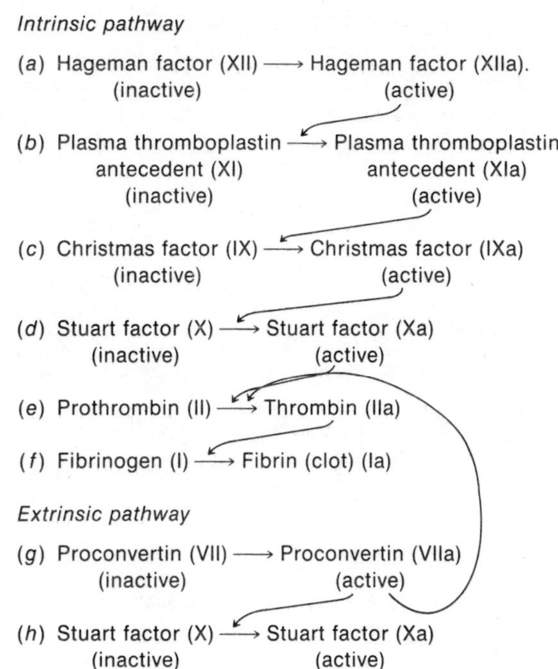

Figure 21.41
Simplified diagram of the intrinsic and extrinsic pathways of blood coagulation. *All of the reactions are proteolytic.*

min K dependent factors require Ca^{2+} (that is prothrombin (II), proconvertin (VII), Christmas factor (IX), and the Stuart factor (X)) for their function, and are synthesized in the liver.

A simplified diagram of the intrinsic and extrinsic pathways of blood clotting is shown in Figure 21.41. The additional details involved in the clotting pathways are summarized in Table 21.9.

21.10 SKIN

The skin consists of the outer epidermis and its supporting inner connective tissue, the dermis. The epidermis gives rise to hair, nails, and sweat glands, and all are considered to be continuous with the epidermis. The dermis contains the blood vessels, nerves, and lymphatics of the skin, and encloses the hair follicles, sebaceous glands, and sweat glands. Human skin accounts for about 10% of the body weight.

Approximately 80% of the glucose is metabolized by glycolysis, and about one-half of the remaining glucose is metabolized by the hexose monophosphate shunt and the other half is converted to glycogen. Glucose utilization by the epidermis is an insulin-dependent process. The source of energy for skin is glycolysis, and skin contributes large amounts of lactate to the blood. The concentration of lactate in skin is about 10-fold greater than in plasma. The control of glycolysis in skin is considered to be by the enzyme phosphofructokinase. The epidermis is capable of synthesizing DNA, RNA, and protein—which is principally keratin—a highly insoluble protein. A surface film of lipid containing large amounts of cholesterol and squalene is synthesized in the epidermal cells. The conversion of 7-dehydrocholesterol to cholecalciferol occurs in the epidermis.

Pigment-forming cells called melanocytes occur in the epidermis and are responsible for the formation and secretion of melanin. The first step in melanin synthesis is the hydroxylation of tyrosine to form 3,4-dihydroxyphenylalanine (dopa) which then undergoes a series of reactions to form compounds that polymerize to form melanin.

BIBLIOGRAPHY

General

Stanbury, J. B., Wyngaarden, J. B., Fredrickson, D. S., Goldstein, J. L., and Brown, M. S. *The metabolic basis of inherited disease,* 3rd ed. New York: McGraw-Hill, 1983.

Skeletal Muscle

Ebashi, S. Excitation-contraction coupling. *Annu. Rev. Physiol.* 38:293, 1976.

Huxley, H. E. The mechanism of muscular contraction. *Sci. Am.* 213(6):18, 1965.

Huxley, H. E. Regulation of muscle function by tropomyosin-troponin, in Y. Hatefi and L. Djavadi-Ohaniance (eds.), *The structural basis of membrane function.* New York: Academic, 1976, p. 313.

MacLennan, D. H. Resolution of the calcium transport system of the sarcoplasmic reticulum. *Can. J. Biochem.* 53:251, 1972.

Taylor, E. W. Chemistry of muscle contraction. *Annu. Rev. Biochem.* 41:577, 1972.

Adipose Tissue

Kinsell, L. W. (ed.). Adipose tissue as an organ. Proceedings of the Duel Conference of Lipids. Springfield, Ill.: C. C. Thomas, 1962.

Liver

Brodie, B. B., and Gillette, J. R. (eds.). *Concepts in biochemical pharmacology,* part 2: *Handbook of experimental pharmacology.* New York: Springer-Verlag, 1971.

Jakoby, W. B. (ed.). *Enzymatic basis of detoxication,* vols I & II. New York: Academic Press, 1980.

Jenner, P, and Testa, B. (eds.). *Concepts in drug metabolism,* Parts A and B. New York: Marcel Dekker, 1980.

Kidney

Anders, M. W. Metabolism of drugs by kidney. *Kidney Int.* 18:636, 1980.

Cohen, J. J., and Barac-Nieto, M. Renal metabolism of substrates in relation to renal function. *Handbook of physiology,* Section 8, 909, 1973.

Nervous Tissue

Albers, R. W., Agranoff, B. W., Katzman, R., and Siegel, G. J. (eds.). *Basic neurochemistry.* Boston: Little, Brown, 1972.

McIlwain, H., and Bachelard, H. S. *Biochemistry and the nervous system,* 4th ed. London: Churchill-Livingston, 1973.

Morell, P., and Norton, W. T. Myelin. *Sci. Am.* 242:88, 1980.

Lung

Cohen, A. B. Lung metabolism: cells. *Fed. Proc.* 38:2635, 1979.

Heinemann, H. O. Metabolism of lung. *Fed. Proc.* 32:1955, 1973.

Eye

Rathbun, W. B. Biochemistry of the lens and cataractogenesis: current concepts, in R. L. Peiffer (ed.), *Veterinary clinics of North America: small animal practice.* Philadelphia: W. B. Saunders, 110, 377, 1980.

Blood Cells

Harris, J. W., and Kellermeyer, R. W. *The red cell.* Cambridge, Mass.: The Commonwealth Fund, Harvard University Press, 1972.

QUESTIONS

C. N. ANGSTADT AND J. BAGGOTT

*Q*uestion *T*ypes are described inside the front cover.

1. (QT2) Contraction of skeletal muscle involves:
   1. an uptake of Ca^{2+} into the sarcoplasmic reticulum by the ATP-dependent Ca^{2+} pump.
   2. hydrolysis of ATP to provide energy for the movement of actin filaments into the H zone.
   3. pumping K^+ into and Na^+ out of the muscle cell by Na^+–K^+-ATPase.
   4. Ca^{2+}-dependent conformational changes in proteins that allow myosin and actin to bind to each other.
2. (QT3) A. Amount of phosphocreatine stored in resting muscle.
   B. Amount of ATP stored in resting muscle.
3. (QT2) Which of the following correctly denote(s) major energy sources in skeletal muscle?
   1. Glucose from muscle glycogen during anoxia
   2. Glucose from the blood following a meal
   3. Free fatty acids during early fasting
   4. Ketone bodies during long-term fasting
4. (QT1) Cardiac muscle:
   A. uses fatty acids in preference to glucose after a meal.
   B. shows a significant increase in utilization of glycogen in anoxia compared to a normal state.
   C. unlike skeletal muscle, cannot use branched-chain amino acids.
   D. increases its utilization of ketone bodies following alcohol ingestion.
   E. regulates its utilization of fatty acids by phosphorylation–dephosphorylation of key enzymes.
5. (QT2) In adipose tissue, insulin stimulates:
   1. amino acid uptake.
   2. lipoprotein lipase.
   3. glycogen synthesis.
   4. hexokinase activity.

6. (QT2) In adipose tissue, fatty acids:
   1. of various chain lengths and degrees of saturation are randomly attached to the three positions of the glycerol moiety.
   2. support thermogenesis both by serving as an oxidizable substrate and by uncoupling oxidative phosphorylation.
   3. may be synthesized from glucose only if a noncarbohydrate energy source is available.
   4. are mobilized when a "hormone sensitive lipase" is activated.
7. (QT2) Among the reactions the liver commonly uses to detoxify a variety of drugs is/are:
   1. cytochrome P_{450}-dependent oxidations.
   2. conjugation with glucuronic acid.
   3. NAD^+-linked oxidation of alcohols and aldehydes.
   4. NADPH-linked cleavage of aromatic rings.
8. (QT1) S-Adenosylmethionine:
   A. is the activated form of sulfate for conjugation reactions.
   B. forms glutathione conjugates in the presence of S-acetyl transferases.
   C. is reduced by a mixed function oxidase.
   D. is one of the substrates for the enzyme catechol-O-methyltransferase.
   E. is involved in the formation of mercapturic acids.
9. (QT2) In the brain during anoxia:
   1. the rate of glycolysis increases.
   2. a shift of fuel from all glucose to glucose–ketone bodies occurs.
   3. NH_4^+ levels increase.
   4. the GABA shunt is stimulated.
10. (QT1) The tissue with the highest rate of oxygen consumption is the:
    A. retina.
    B. cornea.
    C. kidney cortex.
    D. kidney medulla.
    E. lung.
11. (QT1) A high rate of oxygen consumption by a tissue correlates best with a high:
    A. oxygen availability.
    B. lactate production from glucose.
    C. glycolytic rate.
    D. rate of the hexose monophosphate pathway.
    E. none of the above.
12. (QT2) Glycolysis is the major source of energy for the:
    1. epidermis.
    2. erythrocyte.
    3. kidney medulla.
    4. lens.

    A. Kidney
    B. Skin
    C. Cornea
    D. Lung
    E. Reticulocyte

13. (QT5) Na^+ -dependent contransport of glucose.
14. (QT5) Synthesizes a surfactant to minimize surface tension.
15. (QT5) Site of the reaction, 7-dehydrocholesterol → cholecalciferol.

ANSWERS

1. C 2, 4 correct. 1, 3: The movement of Na^+, K^+, and Ca^{2+} is very important, but these are the directions of movement during relaxation rather than contraction. 2: Ca^{2+} binding to the TN–C subunit of troponin initiates these changes; activation of Ca^{2+}, Mg^{2+}–ATPase catalyzes the hydrolysis of ATP (p. 812).
2. A There is six times as much phosphocreatine as ATP in the resting muscle; thus, phosphocreatine provides a backup system for generation of ATP through the creatine kinase reaction (p. 814).
3. E All four correct. 1: Anaerobic metabolism requires glucose, and glycogen phosphorylase can be stimulated by either epinephrine or Ca^{2+}. 2: Following a meal, in blood, both glucose and insulin concentrations are high and free fatty acids are low. 3: Glucose and insulin are low, but the fasting activates the mobilization of fatty acids from adipose tissue. 4: The activities of the liver in long-term fasting lead to ketone body production (p. 814).
4. B A: Free fatty acids are very low following a meal. B: This is true for all tissues, since anything requiring participation of the citric acid cycle would be inhibited by the lack of oxygen. C: Branched-chain amino acids are used in fasting. D: Alcohol ingestion generates acetate, not ketone bodies. E: Fatty acid oxidation is controlled primarily by their rate of transport into mitochondria via the carnitine acyltransferase system (p. 816).
5. E All are correct (p. 819).
6. C 2 and 4 true. 1: The positions of glycerol are acylated nonrandomly (p. 820). 2: Uncoupling permits oxidation to be independent of ATP synthesis (p. 821). 3: Glucose can provide the material and the energy for fatty acid synthesis (p. 819). 4: Glucagon activates the lipase in a cAMP-dependent process (p. 821).
7. A 1, 2, 3 correct. 1: A number of drugs induce the mixed-function oxidase system. 2: Conjugation occurs with glutamine and sulfate, as well as with glucuronic acid. 3: Ethanol, for example, is metabolized by this type of mechanism. 4: One thing that does not occur very frequently is cleavage of aromatic rings. NADP-linked systems do reduce nitro groups (p. 824).
8. D *S*-Adenosylmethionine is the primary donor of methyl groups. A: The activated form of sulfate is a *phosphorylated* adenosine derivative (Section 21.4).
9. B 1, 3 correct. 1, 2, 3: Anoxia will inhibit the citric acid cycle and anything that requires it (e.g., ketone body utilization). The only source of energy is glycolysis, which is stimulated by many things including NH_4^+ (coming perhaps from the purine nucleotide cycle) 4: The GABA shunt is the catabolic path for the neurotransmitter, GABA, and involves the citric acid cycle (p. 837).
10. A Cornea and kidney cortex also have high rates of oxygen consumption (p. 840).
11. E The highest rate of oxygen consumption are found in tissues with the highest continuous requirements for energy. The retina has the highest rate of any tissue (p. 840). Oxygen availability is necessary, but well-oxygenated tissues such as the lung and blood use little oxygen (pp. 838, 842). The retina has an enormous glycolytic rate, but so does active white muscle. White muscle, however, cannot sustain maximal activity for long. The renal cortex also has a high oxygen consumption, but it preferentially uses fatty acids (p. 830).
12. E All are correct. 1: Epidermis is not well perfused with blood (p. 851). 2, 4: Erythrocytes lack mitochondria (p. 842), and the lens has few (p. 839). 3: The kidney medulla has a specialized circulatory system (vasa rectae), which keeps oxygen out and NaCl in (p. 830).
13. A It is used to reabsorb glucose from the glomerular filtrate. Other tissues take up glucose from the blood by passive mediated diffusion (p. 831).
14. D The surfactant keeps surface tension from collapsing the alveoli of the lungs (p. 839).
15. B This is why dietary vitamin D can be replaced by adequate exposure to sunlight (p. 851).

22

Iron and Heme Metabolism

MARILYN S. WELLS and WILLIAM M. AWAD, JR.

22.1 IRON METABOLISM: OVERVIEW

Iron is closely involved in the metabolism of oxygen; the properties of this metal permit the transportation and participation of oxygen in a variety of biochemical processes. The common oxidation states are either ferrous (Fe^{2+}) or ferric (Fe^{3+}); higher oxidation levels occur as short-lived intermediates in certain redox processes. Iron has an affinity for electronegative atoms such as oxygen, nitrogen, and sulfur, which provide the electrons that form the bonds with iron. These can be of very high affinity when favorably oriented on macromolecules. In forming complexes, no bonding electrons are derived from iron. There is an added complexity to the structure of iron: The nonbonding electrons in the outer shell of the metal (the incompletely filled 3*d* orbitals) can exist in two states. Where bonding interactions with iron are weak, the outer nonbonding electrons will avoid pairing and distribute throughout the 3*d* orbitals. Where bonding electrons interact strongly with iron, however, there will be pairing of the outer nonbonding electrons, favoring lower energy 3*d* orbitals. These two different distributions for each oxidation state of iron can be determined by electron spin resonance measurements. Dispersion of 3*d* electrons to all orbitals leads to the high-spin state, whereas restriction of 3*d* electrons to lower energy orbitals, because of electron pairing, leads to a low-spin state. Some iron–protein complexes reveal changes in spin state without changes in oxidation during chemical events (e.g., binding and release of oxygen by hemoglobin).

At neutral and alkaline pH ranges, the redox potential for iron in aqueous solutions favors the ferric state; at acid pH values, the equilibrium favors the ferrous state. In the ferric state iron slowly forms large polynuclear complexes with hydroxide ion, water, and other anions that may be present. These complexes can become so large as to exceed their solubility products, leading to their aggregation and precipitation with pathological consequences.

Iron can bind to and influence the structure and function of a variety of macromolecules, with deleterious results to the organism. To protect against such reactions, several iron-binding proteins function specifically to store and transport iron. These proteins have both a very high affinity for the metal and, in the normal physiological state, also have incompletely filled iron-binding sites. The interaction of iron with its ligands has been well characterized in some proteins (e.g., hemoglobin and myoglobin), whereas for others (e.g., transferrin) it is presently in the process of being defined. The major area of ignorance in the biochemistry of iron lies in the in vivo transfer processes of iron from one macromolecule to another. Several mechanisms have been proposed to explain the process of iron transfer. Two of these have been supported by excellent model studies but with varying degrees of relevance to the physiological state. The proposed processes are the following: First, the redox change of iron has been an attractive mechanism because it is supported by selective in vitro studies and because in many cases macromolecules show a very selective affinity for ferric ions, binding ferrous ions poorly. Thus reduction of iron would permit ferrous ions to dissociate, and reoxidation would allow the iron to redistribute to appropriate macromolecules. Redox mechanisms have only been defined in a very few settings, some of which will be described below. An alternative hypothesis involves chelation of ferric ions by specific small molecules with high affinities for iron; this mechanism has been supported also by selective in vitro studies. The chelation mechanism suffers from the lack of a demonstrably specific in vivo chelator. A third mechanism, which has been studied less well, is the possibility that conformational transitions are imposed upon iron-containing proteins leading to the loosening of ligands to iron. Because the redox potential strongly favors ferric ion at almost all tissue sites and because Fe^{3+} binds so strongly to liganding groups, the probability is that there are cooperating mechanisms regulating the transfer of iron.

22.2 IRON-CONTAINING PROTEINS

Iron binds to proteins either by incorporation into a protoporphyrin IX ring (see below) or by interaction with other protein ligands. Ferrous- and ferric-protoporphyrin IX complexes are designated heme and hematin, respectively. Heme-containing proteins include those that transport (e.g., hemoglobin) and store (e.g., myoglobin) oxygen; and certain enzymes that contain heme as part of their prosthetic groups (e.g., tryptophan pyrrolase). Discussions on structure–function relationships of heme proteins are presented in Chapters 6 and 23.

Nonheme proteins include transferrin, ferritin, a variety of redox enzymes that contain iron at the active site, and iron–sulfur proteins. A significant body of information has been acquired that relates to the structure–function relationships of some of these molecules.

Transferrin

The protein in serum involved in the transport of iron is *transferrin, a* β_1-glycoprotein synthesized in the liver, consisting of a single polypeptide chain of 78,000 daltons with two iron-binding sites. Sequence studies indicate that the protein is a product of gene duplication derived from a putative ancestral gene coding for a protein binding only one atom of iron. Although several other metals bind to transferrin, the highest affinity is for ferric ion. Transferrin does not bind ferrous ion. The binding of each ferric ion is absolutely dependent upon the coordinate binding of an anion, which in the physiological state is carbonate as indicated below:

$$\text{Transferrin} + Fe^{3+} + CO_3^{2-} \longrightarrow \text{transferrin} \cdot CO_3^{2-}$$

$$\text{Transferrin} \cdot Fe^{3+} \cdot CO_3^{2-} + Fe^{3+} + CO_3^{2-} \longrightarrow \text{transferrin} \cdot 2(Fe^{3+} \cdot CO_3^{2-})$$

In experimental settings, other organic polyanions can substitute for carbonate. Estimates of the association constants for the binding of ferric ion to transferrins from different species range from 10^{19} to 10^{31} M^{-1}, indicating for practical purposes that wherever there is excess transferrin free ferric ions will not be found. There is no evidence for cooperativity in the binding of iron at the two sites. In the normal physiological state, approximately one-ninth of all transferrin molecules are saturated with iron at both sites; four-ninths of transferrin molecules have iron at either site; and four-ninths of circulating transferrin are free of iron. The two iron-binding sites, although homologous, are not completely identical; they show some differences in sequences and also some differences in affinities for other metals (especially the lanthanides). Transferrin binds to specific cell surface receptors that mediate the internalization of the protein. The receptor is a transmembrane protein consisting of two monomers, each of 90,000 daltons; it favors the diferric form of transferrin. The transferrin–receptor complex is internalized with release of the iron from the complex followed by return of the receptor–apotransferrin complex to the cell surface where the apotransferrin is released to be reutilized in the plasma. The concentration of the receptor in different tissues reflects directly the need of a tissue for iron; thus the highest values are seen in red blood cells synthesizing hemoglobin and in the placenta.

Lactoferrin

Milk contains iron that is bound almost exclusively to a protein that is closely homologous to transferrin with two sites binding the metal. The iron content of lactoferrin varies, but the protein is never saturated. Surprisingly, the role of lactoferrin in facilitating the transfer of iron to intestinal receptor sites in the infant has not been carefully examined. Rather, major studies on the function of lactoferrin have been directed toward its antimicrobial effect, protecting the newborn from gastrointestinal infections. Microorganisms require iron for replication and function. The presence of incompletely saturated lactoferrin results in the rapid binding of any free iron leading to the inhibition of microbial growth by preventing a sufficient amount of iron from entering these microorganisms. Other microbes, such as *Escherichia coli*, which release competitive iron chelators, are able to proliferate despite the presence of lactoferrin, since the chelators transfer the iron specifically to the microorganism.

Ferritin

Ferritin is the major protein involved in the storage of iron. The protein consists of an outer polypeptide shell 130 Å in diameter with a central ferric-hydroxide-phosphate core 60 Å across. The apoprotein has a molecular weight of 440,000 and consists of 24 subunits of 18,500 daltons each. The ratio of iron to polypeptide is not constant, since the protein has the ability to gain and release iron according to physiological needs. With a capacity of 4,500 iron atoms, the molecule contains usually less than 3,000. Channels from the surface permit the accumulation and release of iron. When iron is in excess, the storage capacity of newly synthesized apoferritin may be exceeded. This leads to iron deposition adjacent to ferritin spheres. Histologically such amorphous iron deposition is called *hemosiderin*.

Ferritin facilitates the oxidation of ferrous ions to the ferric state. Ferritins derived from different tissues of the same species demonstrate differences in electrophoretic mobility in a fashion analogous to the differences noted with isoenzymes. In some tissues ferritin spheres form latticelike arrays, which are identifiable by electron microscopy. The synthesis of ferritin is inhibited by apoferritin molecules. Release of apoferritin molecules from the polyribosome occurs only after the incorporation of iron. Thus the association of apoferritin with polyribosomes leads to inhibition of synthesis of the protein when it is not needed (see Figure 22.1).

Other Nonheme Iron-Containing Proteins

Many iron-containing proteins are involved in enzymatic processes, most of which are related to oxidation mechanisms. The structural features of the ligands binding the iron are not well known, except for a few components involved in mitochondrial electron transport. These latter proteins are characterized by iron being bonded, with one exception, only to sulfur atoms. Four major types of such proteins are known (see Figure 22.2). The smallest (e.g., nebredoxin) found only in microorganisms, consists of a small polypeptide chain with a molecular weight of about 6,000, contain-

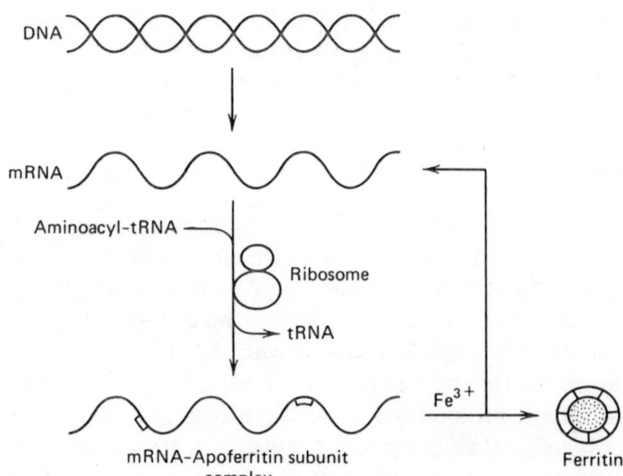

Figure 22.1
Translational control of apoferritin synthesis.

ing one iron atom bound to four cysteine residues. A second type consists of ferredoxins found in both plants and animal tissues where two iron atoms are found, each liganding to two separate cysteine residues and sharing two sulfide anions. The most complicated of the iron–sulfur proteins are the bacterial ferredoxins, which contain four atoms of iron, each of which is linked to single separate cysteine residues but also shares three sulfide anions with neighboring iron molecules to form a cube like structure. In some anerobic bacteria a family of ferredoxins may contain two of the third type of iron–sulfur groups per macromolecule. The most recently described ferredoxins (type IV) contain structures with three atoms of iron each linked to two separate cysteine residues and each sharing two sulfide anions, forming a planar ring. In one example of this ferredoxin type, an exception of iron atoms being liganded only to sulfur atoms was found where the sulfur of a cysteinyl residue was substituted by a solvent oxygen atom. The redox potential afforded by these different ferredoxins varies widely and is in part dependent upon the environment of the surrounding polypeptide chain that envelops these iron–sulfur groups. In nebredoxin the iron undergoes ferric–ferrous conversion during electron transport. With the plant and animal ferredoxins (type II iron–sulfur proteins) both irons are in the ferric form in the oxidized state; upon reduction only one iron goes to the ferrous state. In the bacterial ferredoxin (type III iron–sulfur protein) the oxidized state can be either $2Fe^{3+} \cdot 2Fe^{2+}$ or $3Fe^{3+} \cdot Fe^{2+}$, with corresponding reduced forms of $Fe^{3+} \cdot 3Fe^{2+}$ or $2Fe^{3+} \cdot 2Fe^{2+}$.

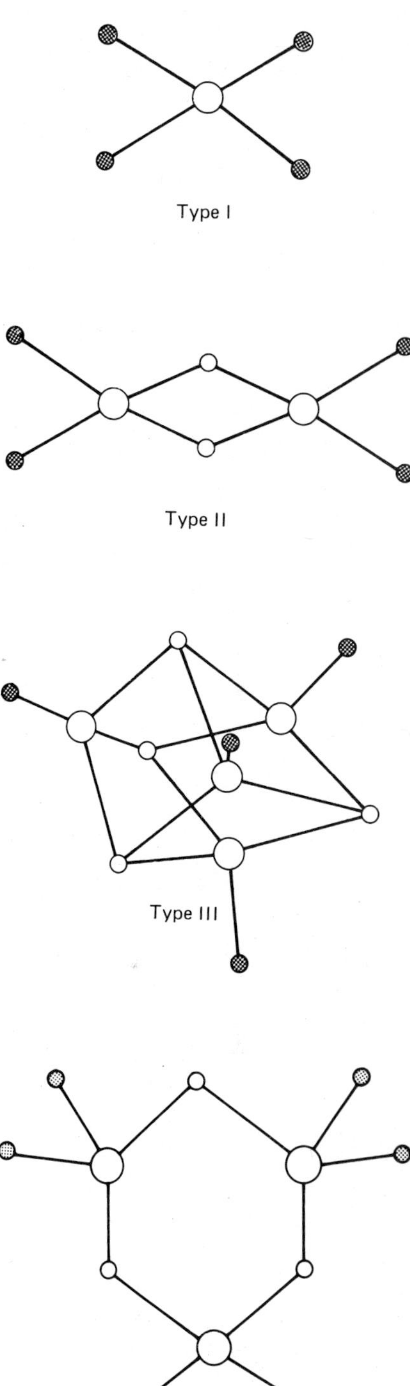

Figure 22.2
Structure of ferredoxins.
The large open circles represent the iron atoms; the small open circles represent the inorganic sulfur atoms; and the small stippled circles represent the cysteinyl sulfur atoms derived from the polypeptide chain. Variation in type IV ferredoxins can occur where one of the cysteinyl residues can be substituted by a solvent oxygen atom of a hydroxyl group.

22.3 INTESTINAL ABSORPTION OF IRON

The high affinity of iron for both specific and nonspecific macromolecules leads to the absence of significant formation of free iron salts, and thus this metal is not lost via usual excretory routes. Rather, excretion of iron occurs only through the normal sloughing of tissues that are not reutilized (e.g., epidermis and gastrointestinal mucosal cells). In the healthy adult male the loss is about 1 mg/day. In premenopausal women, the normal physiological events of menses and parturition substantially augment iron loss. A wide variation of such loss exists, depending upon the amounts of menstrual flow and the multiplicity of births. In the extremes of the latter settings, a premenopausal woman may require an amount of iron that is four to five times that needed in an adult male for prolonged periods of time. The postmenopausal woman who is not iron deficient has an iron requirement similar to that of the adult male. Children and patients with blood loss naturally have increased iron requirement.

Cooking of food facilitates the breakdown of ligands attached to iron, increasing the availability of the metal in the gut. The low pH of stomach contents permits the reduction of ferric ion to the ferrous state, facilitating dissociation from ligands. The latter requires the presence of an accompanying reductant, which is usually achieved by adding ascorbate to the diet. The absence of a normally functioning stomach reduces substantially the amount of iron that is absorbed. Some iron-containing compounds bind the metal so tightly that it is not available for assimilation. Contrary to popular belief, spinach is a poor source of iron because of an earlier erroneous record of the iron content and because some of the iron is bound to phytate (inositol hexaphosphate), which is resistant to the chemical actions of the gastrointestinal tract. Specific protein cofactors derived from the stomach or pancreas have been suggested as being facilitators of iron absorption in the small intestine.

The major site of absorption of iron is in the small intestine, with the largest amount being absorbed in the duodenum and a gradient of lesser absorption occurring in the more distal portions of the small intestine. The metal enters the mucosal cell either as the free ion or as heme; in the latter case the metal is split off from the porphyrin ring in the mucosal cytoplasm. The large amount of bicarbonate secreted by the pancreas neutralizes the acidic material delivered by the stomach and thus favors the oxidation of ferrous ion to the ferric state. The major barrier to the absorption of iron is not at the lumenal surface of the duodenal mucosal cell. Whatever the requirements of the host are, in the face of an adequate delivery of iron to the lumen, a substantial amount of iron will enter the mucosal cell. Regulation of iron transfer occurs between the mucosal cell and the capillary bed (see Figure 22.3). In the normal state, certain processes define the amount of iron that will be transferred. Where there is iron deficiency, the amount of transfer increases; where there is iron overload in the host, the amount transferred is curtailed substantially. One mechanism that has been demonstrated to regulate this transfer of iron across the mucosal–capillary interface is the synthesis of apoferritin by the mucosal cell. In situations in which little iron is required by the host, a large amount of apoferritin is synthesized to trap the iron within the mucosal cell and prevent transfer to the capillary bed. As the cells turn over (within a week), their contents are extruded into the intestinal lumen without absorption occurring. In situations in which there is iron deficiency, virtually no apoferritin is synthesized so as not to compete against the transfer of iron to the deficient host. There are other as yet

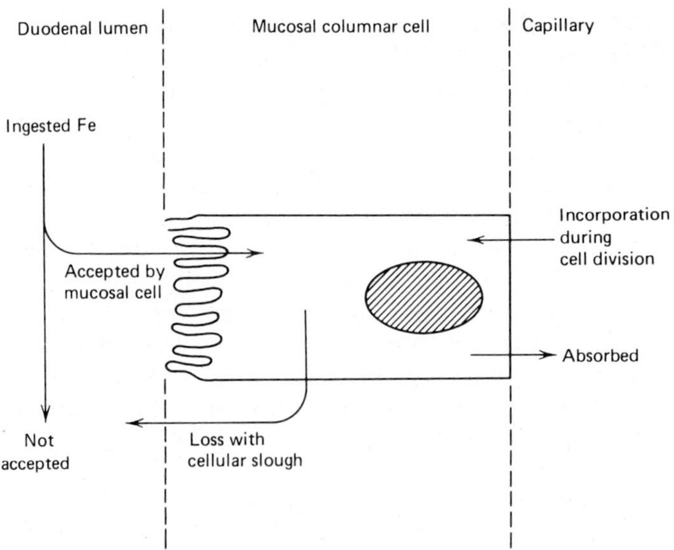

Figure 22.3
Intestinal mucosal regulation of iron absorption.
The flux of iron in the duodenal mucosal cell is indicated. A fraction of the iron that is potentially acceptable is transferred from the intestinal lumen into the epithelial cell. A large portion of ingested iron is not absorbed, in part because it is not presented in a readily acceptable form. Some iron is retained within the cell, bound by apoferritin to form ferritin. This iron is sloughed into the intestinal lumen with the normal turnover of the cell. A portion of the iron within the mucosal cell is absorbed and transferred to the capillary bed to be incorporated into transferrin. During cell division, which occurs at the bases of the intestinal crypts, iron is incorporated for cellular requirements. These fluxes change dramatically in iron-depleted or iron-excess states.

undefined positive mechanisms that increase the rate of iron absorption in the iron-deficient state. Iron transferred to the capillaries is trapped exclusively by transferrin.

22.4 IRON DISTRIBUTION AND KINETICS

A normal 70-kg male has 3–4 g of iron, of which only 0.1% (3.5 mg) is in the plasma. Approximately 2.5 g are in hemoglobin. Table 22.1 lists the distribution of iron in the human.

Normally about 33% of the sites on transferrin contain iron. Iron picked up from the intestine is delivered primarily to the marrow for incorporation into the hemoglobin of red blood cells. The mobilization of iron from the mucosa and from storage sites involves in part the reduction of iron to the ferrous state and its reoxidation to the ferric form. The reduction mechanisms have not been well described. On the other hand, conversion of the ferrous ion back to the ferric state is regulated by serum enzymes called ferroxidases as indicated below:

$$Fe^{2+} + \text{ferroxidase} \longrightarrow Fe^{3+} + \text{reduced ferroxidase}$$

Ferroxidase I is also known as ceruloplasmin. A deficiency of this protein is associated with Wilson's disease, in which there is progressive hepatic failure and degeneration of the basal ganglia associated with a characteristic copper deposition in the cornea (Kayser-Fleischer rings). The ferroxidase activity of ceruloplasmin is not clinically important, since there is no evidence for significant impairment of mobilization of iron in Wilson's disease. Another serum protein, ferroxidase II, appears to be the major serum component that oxidizes ferrous ions. If an inappropriately large amount of iron is administered by injection to a subject who is not iron deficient, this iron will be transported to the liver for storage in the form of ferritin.

In any disease process in which iron loss exceeds iron repletion, a sequence of physiological responses occurs. The initial events are without symptoms to the subject and involve depletion of iron stores without compromise of any physiological function. This depletion will be manifested by a reduction or absence of iron stores in the liver and in the bone marrow and also by a decrease in the content of the very small amount of ferritin that is normally present in plasma. Serum ferritin levels reflect slow release from storage sites during the normal cellular turnover that occurs in the liver; measurements are made by radioimmune assays. Serum ferritin is mostly apoferritin in form, containing very little iron. During this early phase, the level and percentage saturation of serum transferrin is not distinctly abnormal. As the iron deficiency progresses, the level of hemoglobin begins to fall and morphological changes appear in the red blood cells. Concurrently, the serum iron falls with a rise in the level of total serum transferrin, the latter reflecting a physiological adaptation in an attempt to absorb more iron from the gastrointestinal tract. At this state of iron depletion a very sensitive index is the percentage saturation of serum transferrin with iron (normal range 21–50%). At this point the patient usually comes to medical attention, and the diagnosis of iron deficiency is made. In countries in which iron deficiency is severe without available corrective medical measures, a third and severe stage of iron deficiency can occur, where there begins a depletion of iron-containing enzymes leading to very pronounced metabolic effects (see Clin. Corr. 22.1).

TABLE 22.1 Approximate Iron Distribution: 70–kg Man

	g	%
Hemoglobin	2.5	68
Myoglobin	0.15	4
Transferrin	0.003	0.1
Ferritin, tissue	1.0	27
Ferritin, serum	0.0001	0.004
Enzymes	0.02	0.6
Total	3.7	100

CLIN. CORR. 22.1 IRON-DEFICIENCY ANEMIA

Microscopic examination of a blood smear in patients with iron-deficiency anemia usually reveals the characteristic findings of microcytic (small in size) and hypochromic (underpigmented) red blood cells. A bone marrow aspiration will reveal no storage iron to be present and serum ferritin values are virtually zero. The serum transferrin value (expressed as the total iron-binding capacity) will be elevated (upper limits of normal: 410 μg/dl) with a serum iron saturation of less than 16%. Common causes for iron deficiency include excessive menstrual flow, multiple births, and gastrointestinal bleeding that may be occult. The common causes of gastrointestinal bleeding include medication (especially aspirin or cortisone-like drugs), hiatal hernia, peptic ulcer disease, gastritis associated with chronic alcoholism, and gastrointestinal tumors. The management of such patients must include both a careful examination for the cause and source of bleeding and supplementation with iron. The latter is usually provided in the form of oral ferrous sulfate tablets; occasionally intravenous iron therapy may be required. Where the iron deficiency is severe, transfusion with packed red blood cells may also be indicated.

CLIN. CORR. **22.2**
HEMOCHROMATOSIS AND IRON-FORTIFIED DIET

For many years there was a controversy as to whether food should be fortified with iron because of the prevalence of iron-deficiency anemia, especially among premenopausal women. Proponents suggested that if at least 50 mg of iron were incorporated per pound of enriched flour, dietary iron deficiency would be reduced markedly. Objecters stated that the possibility of toxicity from excess iron absorption through iron fortification was too great, and thus such a measure should not be sponsored. Recent studies have indicated that the gene prevalence for hemochromatosis is extraordinarily high, about 10% in the general population. Since the disease is expressed primarily in the homozygous state, about 0.25% of all individuals are at risk for hemochromatosis. A study in Sweden measured the serum iron and iron-binding capacity in a group of 347 subjects. No women showed evidence of iron overload. However, 5% of the men had persistent elevation of serum iron values, with 2% of the men having increased iron stores consonant with the distribution found in early stages of hemochromatosis. Of relevance is the fact that Sweden has had mandated iron fortification for the past 30 years, and approximately 42% of the average daily intake of iron is derived from fortified sources. This study points out the danger of iron-fortified diets. In other settings, however, such as in countries in which iron deficiency is widespread, fortification may still be the most appropriate measure.

Iron overload can occur in patients so that the iron content of the body can be elevated to values as high as 100 g. This may happen for a variety of reasons. Some patients have a recessive heritable disorder associated with a marked inappropriate increase in iron absorption. In such cases the serum transferrin can be almost completely saturated with iron. This state, which is know as idiopathic hemochromatosis, is more commonly seen in men because women with the abnormal gene are protected somewhat by menstrual and childbearing events. The accumulation of iron in the liver, the pancreas, and the heart can lead to cirrhosis and liver tumors, diabetes mellitus, and cardiac failure, respectively. The treatment for these patients is periodic withdrawals of large amounts of blood where the iron is in the contained hemoglobin. Another group of patients has severe anemias, among the most common of which are the thalassemias, a group of herditary hemolytic anemias. In these cases the subjects require transfusions throughout their lives, leading to the accumulation of large amounts of iron derived from the transfused blood. Clearly bleeding would be an inappropriate measure in these cases; rather, the patients are treated by the administration of iron chelators, such as desferrioxamine, which leads to the excretion of large amounts of complexed iron in the urine. Rarely, a third group of patients will acquire excess iron because they ingest large amounts of both iron and ethanol, the latter promoting iron absorption. In these cases excess stored iron can be removed by bleeding (see Clin. Corr. 22.2).

22.5 HEME BIOSYNTHESIS

Heme is produced in virtually all mammalian tissues. Its synthesis is most pronounced in the bone marrow and liver because of the requirements for incorporation into hemoglobin and the cytochromes, respectively. As depicted in Figure 22.4, heme is largely a planar molecule. It consists of one ferrous ion and a tetrapyrrole ring, protoporphyrin IX. The diameter of the iron atom is a little too large to be accommodated within the plane of the porphyrin ring, and thus the metal puckers out to one side as it coordinates with the apical nitrogens of the four pyrrole groups. Heme is one of the most stable of compounds, reflecting its strong resonance features.

Figure 22.5 depicts the pathway for heme biosynthesis. The following are the important aspects to be noted. First, the initial and last three enzymatic steps are catalyzed by enzymes that are in the mitochondrion,

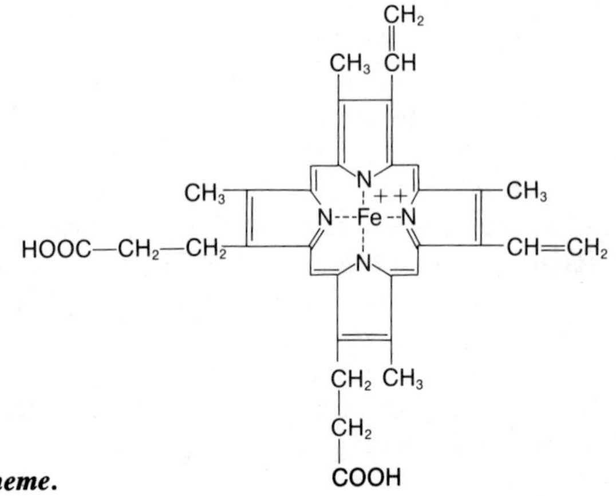

Figure 22.4
Structure of heme.

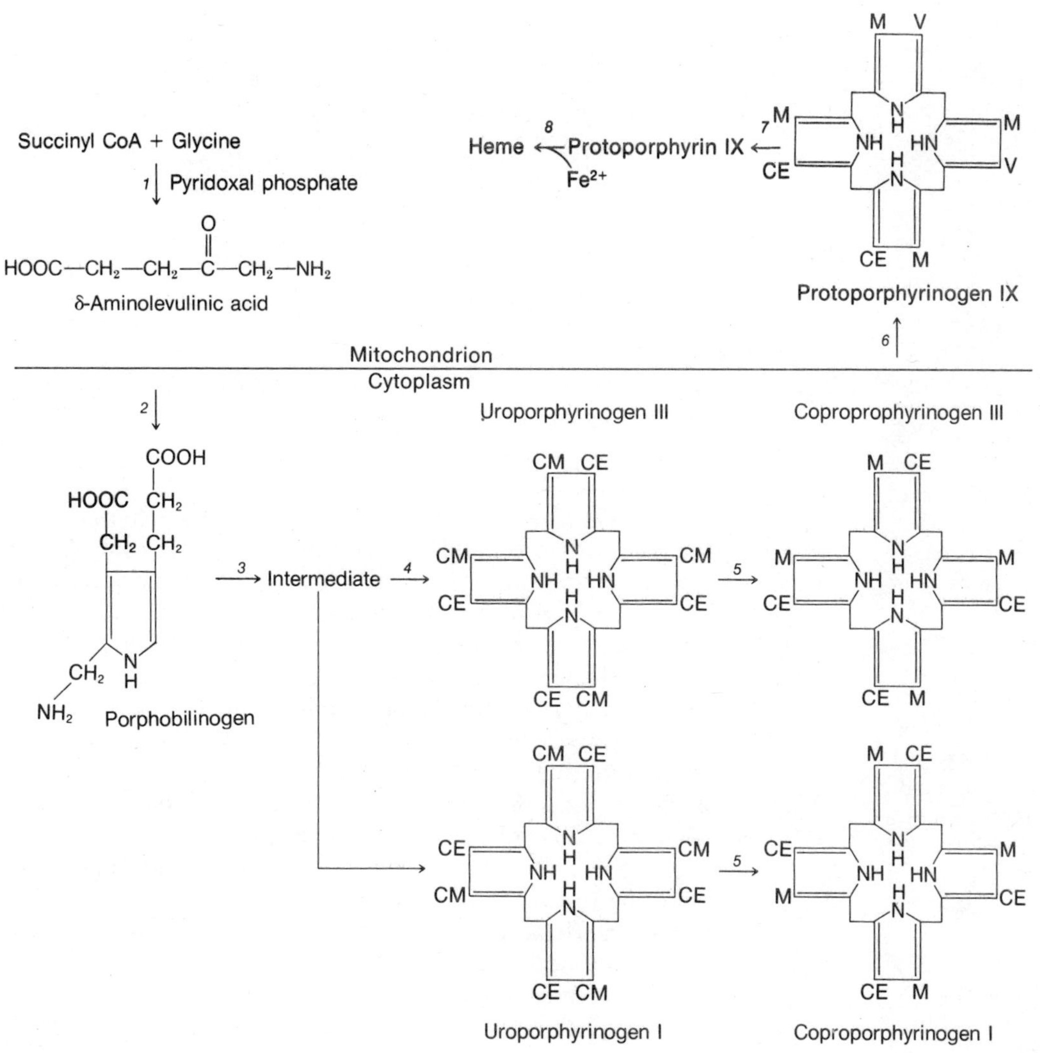

Figure 22.5
Pathway for heme biosynthesis.
The numbers indicate the enzymes involved in each of the biochemical steps according to the following code: 1, Ala synthase; 2, Ala dehydratase; 3, uroporphyrinogen I synthase; 4, uroporphyrinogen III cosynthase; 5, uroporphyrinogen decarboxylase; 6, coproporphyrinogen III oxidase; 7, protoporphyrinogen IX oxidase; 8, ferrochelatase. The pyrrole ligands are indicated by the following abbreviations: CE, β-carboxyethyl (propionic); CM, carboxymethyl (acetic); M, methyl; V, vinyl.

whereas the intermediate steps take place in the cytoplasm. This is important in considering the regulation by heme of the first biosynthetic step; this aspect is discussed below. Second, the organic portion of heme is derived totally from eight residues each of glycine and succinyl CoA. Third, the reactions occurring on the side groups attached to the tetrapyrrole ring involve the colorless intermediates known as porphyrinogens. The latter compounds, though exhibiting resonance features within each pyrrole ring, do not demonstrate resonance between the pyrrole groups. As a consequence, the porphyrinogens are unstable and can be readily oxidized, especially in the presence of light, by nonenzymatic means to their stable porphyrin products. In the latter cases resonance between pyrrole groups is established by oxidation of the four methylene bridges.

CLIN. CORR. **22.3**
ACUTE INTERMITTENT PORPHYRIA

A 40-year-old single, white woman appears in the emergency room in an agitated state, weeping, and complaining of severe abdominal pain. She states that she has been constipated for several days and has noted a feeling of marked weakness in her arms and legs and that "things do not appear to be quite right." Physical examination reveals a slightly rapid heart rate (110/min) and moderate hypertension (blood pressure of 160/110 mmHg). The only other significant findings are two well-healed abdominal operative scars. When queried, she relates that there have been earlier episodes of severe abdominal pain, in fact more severe than what she is presently experiencing. Exploratory abdominal operations undertaken on two of those past occasions revealed no abnormalities.

The usual laboratory tests are obtained and appear to be largely within normal limits. None of her neurological complaints appear to be well documented or to have any localized anatomical focus. The decision is made that the present symptoms are largely psychiatric in origin and have a functional rather than an organic basis. Because of her agitated state, the decision is made to sedate the patient with 60 mg of phenobarbital; a consultant psychiatrist agrees by telephone to see the patient in about 4 h. During the ensuing interval, the emergency room staff notices marked deterioration in the patient's status: Generalized weakness rapidly appears, progressing to a compromise of respiratory function. This ominous development leads to immediate incorporation of a ventilatory assistance regimen, with transfer of the patient to the intensive care unit for close physiological monitoring. Despite these measures the patient's condition deteriorates further and she dies 48 h later. A short time before death a urine sample of the patient is found to have a markedly elevated level of porphobilinogen.

This patient had acute intermittent porphyria, a disease of incompletely understood derangement of heme biosynthesis. This entity must be considered in any postpubertal patient who develops unexplained neurological, abdominal, or psychiatric symptoms. There is a dominant pattern of inheritance associated with an overproduction of the porphyrin precur-

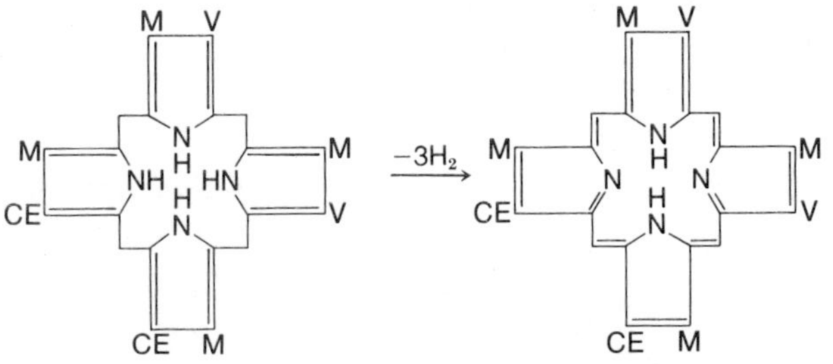

Figure 22.6
Action of protoporphyrinogen IX oxidase, an example of the conversion of a porphyrinogen to a porphyrin.

Figure 22.6 depicts the enzymatic conversion of protoporphyrinogen to protoporphyrin by this oxidation mechanism. This is the only known porphyrinogen oxidation that is enzyme regulated in humans; all other porphyrinogen-to-porphyrin conversions are nonenzymatic and catalyzed by light rather than catalyzed by specific enzymes. Fourth, once the tetrapyrrole ring is formed, the order of the R groups as one goes clockwise around the tetrapyrrole ring defines which of the four possible types of uro- or coproporphyrinogens are being synthesized. These latter compounds have two different substituents, one each for every pyrrole group. Going clockwise around the ring, the substituents can be arranged as ABABABAB (where A is one substituent and B the other), forming a type I porphyrinogen, or the arrangement can be ABABABBA, forming a type III porphyrinogen. In principle, two other arrangements can occur to form porphyrinogens II and IV, and these can be synthesized chemically; however, they do not occur naturally. In protoporphyrinogen and protoporphyrin there are three types of substituents, and the classification becomes more complicated; type IX is the only form that is synthesized naturally.

Derangements of porphyrin metabolism are known clinically as the porphyrias. This family of diseases is of great interest because it has revealed how complicated the regulation of heme biosynthesis is. The clinical presentations of the different porphyrias provide a fascinating exposition of biochemical regulatory abnormalities and their relationship to pathophysiological processes. Table 22.2 lists the details of the different porphyrias (see Clin. Corr. 22.3).

Enzymes in Heme Biosynthesis

Aminolevulinic Acid Synthase

Aminolevulinic acid synthase controls the rate-limiting step of heme synthesis in all tissues studied. The synthesis of the enzyme is not directed by mitochondrial DNA but occurs rather in the cytoplasm, being directed by mRNA derived from the nucleus. The enzyme is incorporated into the matrix of the mitochondrion. Succinyl CoA is one of the substrates and is found only in the mitochondrion. This protein has been purified to homogeneity from rat liver mitochondria. The enzyme consists of a dimer of subunits of 60,000 Da each; 50% inhibition of activity occurs in the presence of 5 μM hemin, and virtually complete inhibition is noted at a 20 μM concentration. The enzyme has a short biological half-life (about 60 min). Both the synthesis and the activity of the enzyme are subject to regulation by a variety of substances. The enzymatic reaction involves the condensation of a glycine residue with a residue of succinyl CoA (Figure 22.7).

TABLE 22.2 Derangements in Porphyrin Metabolism

Disease State	*Genetics*	*Tissue*	*Enzyme*	*Activity*	*Organ Pathology*
Acute intermittent porphyria	Dominant	Liver	1. ALA synthase 2. Uroporphyrinogen I synthase 3. Δ^4-5α-Reductase	Increase Decrease Decrease	Nervous system
Hereditary coproporphyria	Dominant	Liver	1. ALA synthase 2. Coproporphyrinogen oxidase	Increase Decrease	Nervous system; skin
Variegate porphyria	Dominant	Liver	1. ALA synthase 2. Protoporphyrinogen oxidase	Increase Decrease	Nervous system; skin
Porphyria cutanea tarda	Dominant	Liver	1. Uroporphyrinogen decarboxylase	Decrease	Skin, induced by liver disease
Hereditary protoporphyria	Dominant	Marrow	1. Ferrochelatase	Decrease	Gallstones, liver disease, skin
Erythropoietic porphyria	Recessive	Marrow	1. Uroporphyrinogen III cosynthase	Decrease	Skin and appendages; reticuloendothelial system
Lead poisoning	None	All tissues	1. ALA dehydrase 2. Ferrochelatase	Decrease Decrease	Nervous system; blood; others

The reaction has an absolute requirement for pyridoxal phosphate; the latter interacts with the nitrogen of glycine to form a Schiff's base. This generates a carbanion intermediate on the α-carbon of the glycine, allowing a nucleophilic attack and condensation with the succinyl group from succinyl CoA. The bound intermediate, α-amino-β-ketoadipic acid, decarboxylates to form the released product, aminolevulinic acid (ALA). Pyridoxal deficiency and drugs competing with pyridoxal phosphate lead to a decrease in enzyme activity.

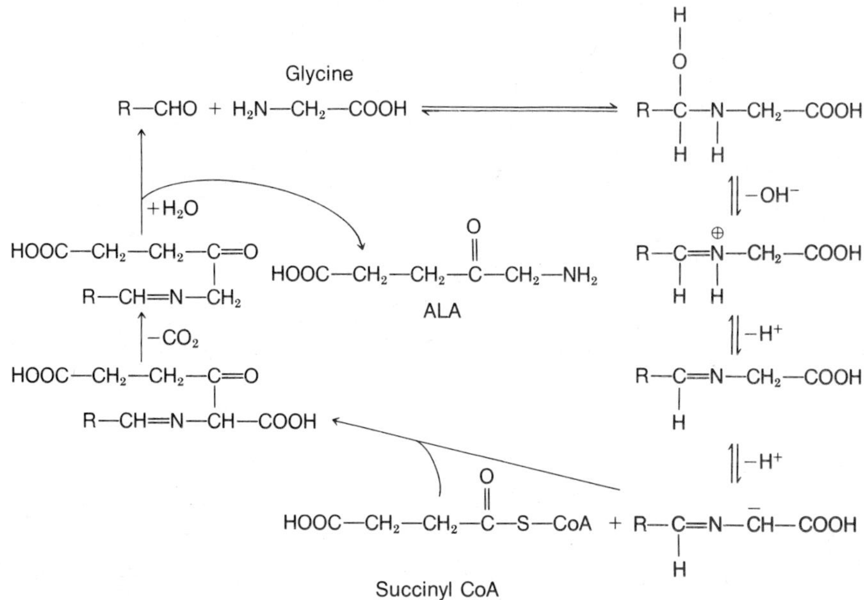

Figure 22.7
Synthesis of δ-aminolevulinic acid.

sors, ALA and porphobilinogen. Several associated well-defined enzyme abnormalities are noted in the cases that have been studied carefully. These include (1) a marked increase in ALA synthase, (2) a reduction by half of the activity of uroporphyrinogen I synthase, and (3) a reduction of one-half of the activity of steroid Δ^4-5α-reductase. The change in content of the second enzyme is consonant with a dominant expression. The change in content of the third enzyme is acquired and not apparently a heritable expression of the disease. It is believed that a decrease in uroporphyrinogen I synthase leads to a minor decrement in the content of heme in the liver. The low concentration of heme leads to a failure both to repress the synthesis and to inhibit the activity of ALA synthase. Since this disease is almost never manifested before puberty, it is thought that only with the induction of Δ^4-5β-reductase at adolescence does the disease become florid. Since these patients do not have a sufficient amount of Δ^4-5α-reductase, it is assumed that the observed increase in the 5β steroids is due to a shunting of Δ^4 steroids into the 5β-reductase pathway.

Pathophysiologically, the disease poses a great riddle: The derangement of porphyrin metabolism is confined to the liver, which anatomically appears normal, whereas the pathological findings are restricted to the nervous system. In the present case, involvement of (1) the brain led to the agitated and confused state, (2) the autonomic system led to the hypertension, increased heart rate, constipation, and abdominal pain, and (3) the peripheral nervous system and spinal cord led to the weakness and sensory disturbances. The conclusion generated by these observations is that there must be some hepatic product that circulates to reach and affect neural tissue. However, experimentally, no known intermediate of heme biosynthesis can cause the pathology noted in acute intermittent porphyria. In the present case there should have been a greater suspicion of the possibility of porphyria early in the patient's presentation. The analysis for porphobilinogen in the urine is a relatively simple test. The treatment would have been glucose infusion, the exclusion of any drugs that could cause elevation of ALA synthase, and, if her disease failed to respond satisfactorily despite these measures, the administration of intravenous hematin to inhibit the synthesis and activity of ALA synthase.

Acute intermittent porphyria is of historic political interest. The disease has been diagnosed in two descendants of

ALA Dehydrase

ALA dehydrase is a soluble cytosol component with a molecular weight of 280,000 and consisting of eight subunits, of which only four interact with the substrate. This protein also interacts with the substrate to form a Schiff's base, but in this case the ε-amino group of a lysine residue binds to the ketonic carbon of the substrate molecule (Figure 22.8). Two molecules of ALA condense asymmetrically to form porphobilinogen. ALA dehydrase is a sulfhydryl enzyme and is very sensitive to inhibition by heavy metals. A characteristic finding of lead poisoning is the elevation of ALA in the absence of an elevation of porphobilinogen.

Uroporphyrinogen I Synthase

The synthesis of uroporphyrinogen I involves head-to-tail condensations of four porphobilinogen residues, with the deamination of the primary amino groups. This enzyme has been purified from red blood cells and consists of a multiplicity of isoenzymes. The mechanism and the nature of the intermediates in this multistep reaction are not known. Normally this enzyme is closely associated with a second protein, *uroporphyrinogen III cosynthase*. The latter protein does not have any apparent enzymatic function but acts as a specifier protein, directing uroporphyrinogen I synthase to form the III isomer rather than the I isomer of uroporphyrinogen. The nature of the complicated biochemical steps modulated by the cosynthase is as yet undetermined. A rare recessive disease, erythropoietic porphyria, associated with marked light sensitization is due to an abnormality of red blood cell uroporphyrinogen III cosynthase. Here large amounts of the type I isomers of uroporphyrinogen and coproporphyrinogen are synthesized.

Uroporphyrinogen Decarboxylase

This enzyme acts on the side chains of the uroporphyrinogens to form the coproporphyrinogens. The protein catalyzes the conversion of both I and

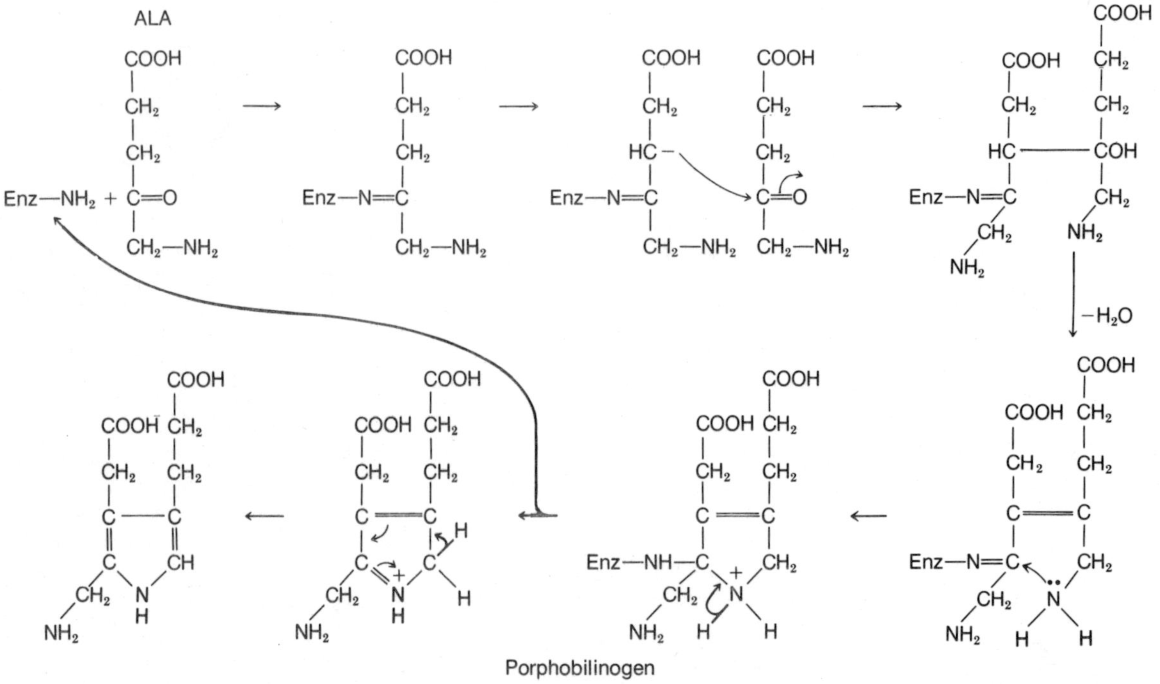

Figure 22.8
Synthesis of porphobilinogen.

III isomers of uroporphyrinogen to the respective coproporphyrinogen isomers. Uroporphyrinogen decarboxylase is inhibited by iron salts. Clinically the most common cause of porphyrin derangement is associated with patients who have a single gene abnormality for this enzyme, leading to 50% depression of the enzyme's activity. This disease, which shows cutaneous manifestations primarily with sensitivity to light, is known as porphyria cutanea tarda. The condition is not expressed unless patients either take drugs that cause an increase in porphyrin synthesis or drink large amounts of alcohol, leading to the accumulation of iron, which then acts to inhibit further the activity of uroporphyrinogen decarboxylase.

George III, suggesting that the latter's deranged personality preceding and during the American Revolution could possibly be ascribed to an affliction with this illness.

Coproporphyrinogen Oxidase

This mitochondrial enzyme is specific for the type III isomer of coproporphyrinogen, not acting on the type I isomer. Coproporphyrinogen III enters the mitochondrion and is converted to protoporphyrinogen IX. The mechanism of action is not understood. A dominant disease associated with a deficiency of this enzyme leads to a form of hereditary hepatic porphyria, known as hereditary coproporphyria.

Protoporphyrinogen Oxidase

This mitochondrial enzyme generates a product, protoporphyrin IX, which, in contrast to the other heme precursors, is very water-insoluble. Excess amounts of protoporphyrin IX that are not converted to heme are excreted by the biliary system into the intestinal tract. A dominant disease, variegate porphyria, is due to a deficiency of protoporphyrinogen oxidase.

Ferrochelatase

Ferrochelatase inserts ferrous iron into protoporphyrin IX in the final step of the synthesis of heme. Reducing substances are required for its activity. The protein is sensitive to the effects of heavy metals (especially lead) and, of course, to iron deprivation. In these latter instances, zinc instead of iron is incorporated to form a zinc–protoporphyrin IX complex. In contrast to heme, the zinc–protoporphyrin IX complex is brilliantly fluorescent and easily detectable in small amounts.

Regulation of Heme Biosynthesis

ALA synthase controls the rate-limiting step of heme synthesis in all tissues. Succinyl CoA and glycine are substrates for a variety of reactions. The modulation of the activity of ALA synthase determines the quantity of the substrates that will be shunted into heme biosynthesis. Heme (and also hematin) acts both as a repressor of the synthesis of ALA synthase and as an inhibitor of its activity. Since heme resembles neither the substrates nor the product of the enzyme's action, it is probable that the latter inhibition occurs at an allosteric site. Almost 100 different drugs and metabolites can cause induction of ALA synthase; for example, a 40-fold increase is noted in the rat after treatment with 3,5-dicarbethoxy-1,4-dihydrocollidine. The effect of pharmacological agents has led to the important clinical feature where in some patients with certain kinds of porphyria have had exacerbations of their condition following the inappropriate administration of certain drugs (e.g., barbiturates). ALA dehydrase is also inhibited by heme; but this is of little physiological consequence, since the maximal activity of the total amount of ALA dehydrase present is about 80-fold greater than that of ALA synthase, and thus heme-inhibitory effects are reflected first in the activity of ALA synthase.

Glucose or one of its proximal metabolites serves to inhibit heme biosynthesis in a mechanism that is not yet defined. This is of clinical relevance, since some patients manifest their porphyric state for the first time when placed on a very low caloric (and therefore glucose) intake. Other regulators of porphyrin metabolism include certain steroids. Steroid hormones (e.g., oral contraceptive pills) with a double bond in ring A between carbon atoms 4 and 5 can be reduced by two different reductases. The product of 5α-reduction has little effect on heme biosynthesis; however, the product of 5β-reduction serves as a stimulus for the synthesis of ALA synthase. The observation that 5β-reductase appears at puberty suggests this to be the reason why, in a few of the porphyrias, manifestations are not present at an earlier stage.

22.6 HEME CATABOLISM

The catabolism of heme-containing proteins presents two requirements to the mammalian host: (1) the development of a means of processing the hydrophobic products of porphyrin ring cleavage and (2) the retention and mobilization of the contained iron so that it may be reutilized.

Red blood cells have a life span of approximately 120 days. Senescent cells are recognized by their membrane changes and removed and engulfed by the reticuloendothelial system at extravascular sites. The globin chains denature, releasing heme into the cytoplasm. The globin is degraded to its constituent amino acids, which are reutilized for general metabolic needs.

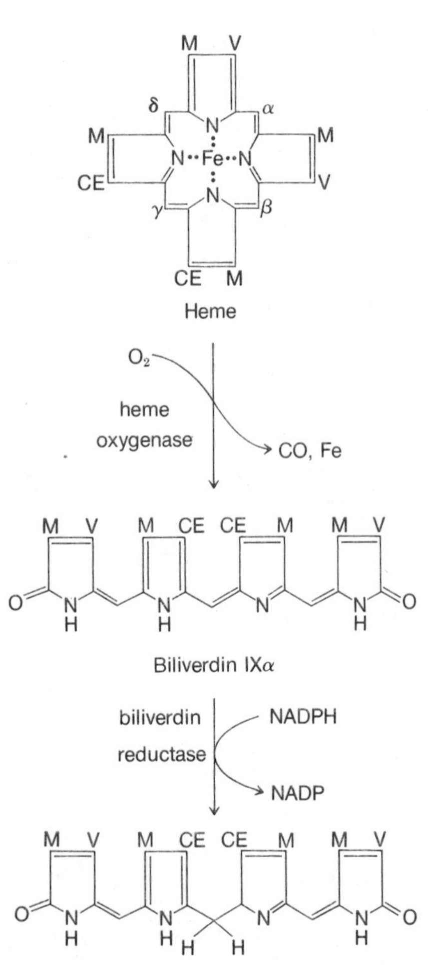

Figure 22.9
Formation of bilirubin from heme.
The Greek letters indicate the labeling of the methene carbons in heme.

Figure 22.9 depicts the sequence of events of heme catabolism. Heme is degraded primarily by a microsomal enzyme system in reticuloendothelial cells that requires molecular oxygen and NADPH. Cytochrome c serves as the major vehicle for regenerating the NADPH utilized in the reaction. Heme oxygenase is substrate inducible. The enzyme specifically catalyzes the cleavage of the α-methene bridge, which joins the two pyrrole residues containing the vinyl substituents. The α-methene carbon is converted quantitatively to carbon monoxide. The only endogenous source of carbon monoxide in man is the α-methene carbon. A fraction of the carbon monoxide is released via the respiratory tract. Thus the measurement of carbon monoxide in an exhaled breath provides an index to the quantity of heme that is degraded in an individual. The oxygen present in the carbon monoxide and in the newly derivatized lactam rings are generated entirely from molecular oxygen. The stoichiometry of the reaction requires 3 mol oxygen for each ring cleavage. Heme oxygenase will only use heme as a substrate with the iron possibly participating in the cleavage mechanism. Thus, free protoporphyrin III(IX) is not a substrate. The linear tetrapyrrole biliverdin IX is the product formed by the action of heme oxygenase. Biliverdin IX is reduced by biliverdin reductase to bilirubin IX.

Bilirubin Metabolism

Bilirubin is derived not only from senescent red cells but also from the turnover of other heme-containing proteins, such as the cytochromes. Studies with labeled glycine as a precursor have revealed that an early-labeled bilirubin, with a peak amount present within 1–3 h, appears a very short time after a pulsed administration of the labeled precursor. A larger amount of bilirubin appears much later at about 120 days, reflecting the turnover of heme in red blood cells. Early-labeled bilirubin can be divided into two parts: an early-early part, which reflects the turnover of heme proteins in the liver, and a late-early part, which consists of both the turnover of heme-containing hepatic proteins and the turnover of bone marrow heme, which is either poorly incorporated or easily released from red blood cells. The latter is a measurement of ineffective erythropoiesis and can be very pronounced in disease states such as pernicious anemia (see Chapter 26) and the thalassemias.

Bilirubin is poorly soluble in aqueous solutions at physiological pH values. When transported in plasma, it is bound to serum albumin with an association constant greater than $10^6\ M^{-1}$. Albumin contains one such high affinity site and another with a lesser affinity. At the normal albumin concentration of 4 g/dl, ~70 mg of bilirubin/dl plasma can be bound on the two sites. However, bilirubin toxicity (kernicterus), which is manifested by the transfer of bilirubin to membrane lipids, commonly occurs at concentrations greater than 25 mg/dl. This suggests that the weak affinity of the second site does not allow it to serve effectively in the transport of bilirubin. Bilirubin on serum albumin is rapidly cleared by the liver, where there is a free bidirectional flux of the tetrapyrrole across the sinusoidal–hepatocyte interface. Once in the hepatocyte, bilirubin is bound to several cytosolic proteins, of which only one has been well characterized. The latter component, ligandin, is a small basic component making up to 6% of the total cytoplasmic protein of rat liver. Ligandin has been purified to homogeneity from rat liver and characterized as having two subunits of molecular weights 22,000 and 27,000. Each subunit contains glutathione *S*-epoxide transferase activity, a function important in detoxification mechanisms of aryl groups. The stoichiometry of binding is one bilirubin residue per complete ligandin molecule. The functional role of ligandin and other hepatic bilirubin-binding proteins remains to be defined.

Once in the hepatocyte the β-carboxyethyl side chains of bilirubin are conjugated to form a diglucuronide (see Figure 22.10). The reaction mechanism includes the utilization of uridine diphosphoglucose, which is oxidized by a dehydrogenase to uridine diphosphoglucuronate. The latter serves as a glucuronate donor to bilirubin; different specific transferases

UDP-Glucose + 2NAD$^+$ ⟶ UDP-glucuronate + 2NADH + 2H$^+$

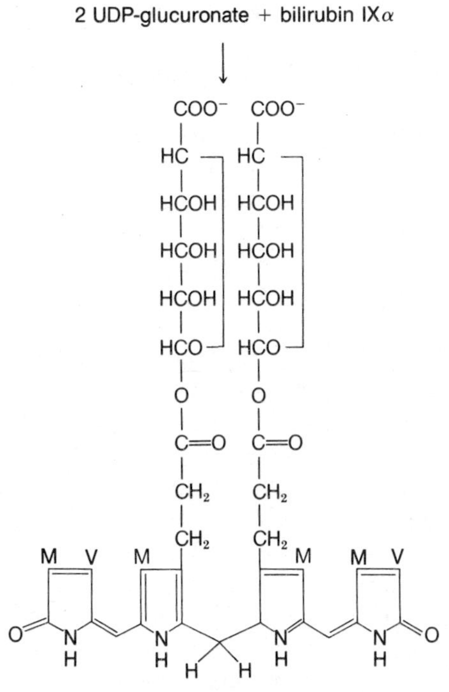

2 UDP-glucuronate + bilirubin IXα

↓

COO⁻ COO⁻ / HC HC / HCOH HCOH / HCOH HCOH / HCOH HCOH / HCO HCO / O O / C=O C=O / CH$_2$ CH$_2$ / CH$_2$ CH$_2$ / M V M M M V / O N H N H H H N H N H O

Bilirubin IXα diglucuronide

Figure 22.10
Biosynthesis of bilirubin diglucuronide.

form sequentially the mono- and diglucuronide adducts of bilirubin. The mammalian liver contains several uridine diphosphoglucuronyl transferases, each of which is substrate specific for the acceptor molecule. In normal bile the diglucuronide is the major form of excreted bilirubin, with only small amounts present of the monoglucuronide or other glycosidic adducts. Bilirubin diglucuronide is much more water-soluble than free bilirubin, and thus the transferase facilitates the excretion of the bilirubin into bile. Bilirubin diglucuronide is poorly absorbed by the intestinal mucosa. The glucuronide residues are released in the terminal ileum and large intestine by bacterial hydrolases; the released free bilirubin is reduced to the colorless linear tetrapyrroles known as urobilinogens. Urobilinogens can be oxidized to colored products known as urobilins, which are excreted in the feces. A small fraction of urobilinogen can be reabsorbed by the terminal ileum and large intestine to be removed by hepatic cells and resecreted in bile. When urobilinogen is reabsorbed in large amounts in certain disease states, the kidney serves as a major excretory site.

In the normal state plasma bilirubin concentrations are 0.3–1 mg/dl, and this is almost all in the unconjugated state. In the clinical setting conjugated bilirubin is expressed as direct bilirubin because it can be coupled readily with diazonium salts to yield azo dyes; this is the direct van den Bergh reaction. Unconjugated bilirubin is bound noncovalently to albumin and will not react until it is released by the addition of an organic solvent such as ethanol. The reaction with diazonium salts yielding the azo dye after the addition of ethanol is the indirect van den Bergh reaction, and this measures the indirect or the unconjugated bilirubin. Unconjugated bilirubin binds so tightly to serum albumin and lipid that it does not diffuse freely in plasma and therefore does not lead to an elevation of bilirubin in the urine. Unconjugated bilirubin has a high affinity for membrane lipids, which leads to the impairment of cell membrane function, especially in the nervous system. In contrast, conjugated bilirubin is relatively water-soluble, and elevations of this bilirubin form lead to high urinary concentrations with the characteristic deep yellow-brown color. The deposition of conjugated and unconjugated bilirubin in skin and the sclera gives the yellow to yellow-green color seen in patients with jaundice.

Recently a third form of plasma bilirubin has been described. This occurs only with hepatocellular disease in which a fraction of the bilirubin binds so tightly that it is not released from serum albumin by the usual techniques and is thought to be linked covalently to the protein. In some cases up to 90% of total bilirubin can be in this newly discovered covalently bound form.

The normal liver has a very large capacity to conjugate and mobilize the bilirubin that is delivered. As a consequence hyperbilirubinemia due to excess heme destruction, as in hemolytic diseases, rarely leads to bilirubin levels that exceed 5 mg/dl, except in situations in which functional derangement of the liver is present (see Clin. Corr. 22.4). Thus, marked elevation of unconjugated bilirubin reflects primarily a variety of hepatic diseases, including those that are heritable and those that are acquired. A severe jaundice in infants occurs with the Crigler-Najjar syndrome, in which there exists a homozygous complete functional deficiency of the specific uridine diphosphoglucuronyltransferase for bilirubin. A low-grade mild hyperbilirubinemia known as Gilbert's syndrome occurs in adults. Although this disease has not been as well characterized as the Crigler-Najjar syndrome, one of the findings is a moderate reduction in bilirubin–uridine diphosphoglucuronyl-transferase activity.

CLIN. CORR. 22.4 NEONATAL ISOIMMUNE HEMOLYSIS

Rh-negative women pregnant with Rh-positive fetuses will develop antibodies to Rh factors. These antibodies will cross the placenta to hemolyze fetal red blood cells. Usually this is not of clinical relevance until about the third Rh-positive pregnancy, in which the mother has had antigenic challenges with earlier babies. Antenatal studies will reveal rising maternal levels of IgG antibodies against Rh-positive red blood cells, indicating that the fetus is Rh positive. Before birth, placental transfer of fetal bilirubin occurs with excretion through the maternal liver. Because hepatic enzymes of bilirubin metabolism can be poorly expressed in the newborn, infants may not be able after delivery to process the large amounts of bilirubin that can be generated. At birth these infants usually appear unremarkable; however, the unconjugated bilirubin in the umbilical cord blood is elevated up to 4 mg/dl due to the hemolysis initiated by maternal antibodies. During the next 2 days the serum bilirubin rises, reflecting continuing isoimmune hemolysis, leading to jaundice, hepatosplenomegaly, ascites, and edema. If untreated, signs of central nervous system damage can occur, with the appearance of lethargy, hypotonia, spasticity, and respiratory difficulty, constituting the syndrome of kernicterus. Treatment involves exchange transfusion with whole blood, which is serologically compatible with both the infant's blood and maternal serum. The latter requirement is necessary to prevent hemolysis of the transfused cells. Additional treatment includes external phototherapy, which facilitates the metabolism of bilirubin.

Elevations of conjugated bilirubin level in plasma are attributable to liver and/or biliary tract disease. In simple uncomplicated biliary tract obstruction, the major component of the elevated serum bilirubin is the diglucuronide form, which is released by the liver into the vascular compartment. Biliary tract disease may be extrahepatic or intrahepatic, the latter involving the canaliculi and biliary ductules. The Dubin-Johnson syndrome is an autosomal recessive disease involving a defect in the biliary secretory mechanism of the liver. The excretion through the biliary tract of a variety of (but not all) organic anions is affected. The retention of undefined pigments in the liver in this disorder leads to a characteristic gray-black color of this organ. This pigment is apparently not derived from heme. A second heritable disorder associated with elevated levels of plasma conjugated bilirubin is Rotor syndrome. In this poorly defined disease no hepatic pigmentation occurs, and the associated finding of increased secretion of urinary coproporphyrins I and III is found. The relationship of the derangement of coproporphyrin metabolism to bilirubin metabolism is not well understood.

Intravascular Hemolysis

In certain diseases destruction of red blood cells occurs in the intravascular compartment rather than in the extravascular reticuloendothelial cells. In the former case the appearance of free hemoglobin and heme in the plasma potentially could lead to the excretion of these substances through the kidney with a substantial loss of iron. To prevent this occurrence specific plasma proteins are involved in scavenging mechanisms. Transferrin binds free iron and thus permits the reutilization of the metal. Free hemoglobin in the plasma leads to the following sequence of events. After oxygenation in the pulmonary capillaries, plasma oxyhemoglobin dissociates into $\alpha\beta$ dimers, which are bound to a family of circulating plasma proteins, the haptoglobins, having a high affinity for the oxyhemoglobin dimer. Since deoxyhemoglobin does not dissociate into dimers in physiological settings, it is not bound by haptoglobin. The stoichiometry of binding is two $\alpha\beta$-oxyhemoglobin dimers per haptoglobin molecule. Interesting studies have been made with rabbit antihuman-hemoglobin antibodies on the haptoglobin–hemoglobin interaction. Human haptoglobin interacts with a variety of hemoglobins from different species. The binding of human haptoglobin with human hemoglobin is not affected by the binding of rabbit antihuman-hemoglobin antibody. These studies suggest that haptoglobin binds to sites on hemoglobin, which are highly conserved in evolution and therefore are not sufficiently antigenic to generate antibodies. The most likely site for the molecular interaction of hemoglobin and haptoglobin is the interface of the α and β chains of the tetramer that dissociates to yield $\alpha\beta$ dimers. Sequence determinators have indicated that these contact regions are highly conserved in evolution.

The haptoglobins are α_2-globulins. Made in the liver, they consist of two pairs of polypeptide chains (α being the lighter and β the heavier). The genes for the α and β chains are linked so that a single mRNA is synthesized, generating a single polypeptide chain that is cleaved to form the two different chains. The β chains are glycopeptides of 39,000 daltons and are invariant in structure; α chains are of several kinds. The shorter α^{1S} chains each consist of 84 residues, with a molecular weight of about 9,000 varying only in the residues at positions 54 where Glu is found in α^{1S} and Lys in α^{1F}. The α^2 chain consists of an incomplete fusion-duplication of the genes for α^{1F} and α^{1S}, leading to a polypeptide chain with 143 residues. The haptoglobin peptide chains are joined by disulfide bonds between the α and β chains and between the two α chains. Thus, in contrast to the immunoglobulins, the disulfide bond between similar chains occurs with the smaller polypeptides. Because the α^2 chain has one more half-cystine residue than the α^1 chain, haptoglobins with the formula $\alpha_2^2B_2$ can polymerize into larger aggregates through disulfide linkages. There may not by symmetry in any single haptoglobin molecule, thus haptoglobins are known that have the following molecular structures: $\alpha^{1F}\alpha^{1S}\beta_2$ or $\alpha^{1F}\alpha^2\beta_2$. These variations in the structure of haptoglobins have been useful in analyses of population genetics.

The interaction of haptoglobin with hemoglobin leads to a complex that is too large to be filtered through the renal glomerulus. Free hemoglobin (appearing in the renal tubules and in the urine) will occur during intravascular hemolysis only when the binding capacity of circulating haptoglobin has been exceeded. Haptoglobin delivers hemoglobin to the reticuloendothelial cells. The heme in free hemoglobin is relatively resistent to the action of heme oxygenase, whereas the heme residues in an $\alpha\beta$ dimer of hemoglobin when attached to haptoglobin are very susceptible. This enhancement of oxygenase activity is especially pronounced with the $\alpha_2^2\beta_2$ haptoglobin type. It has been suggested that the high proportion of haptoglobin $\alpha_2^2\beta_2$ in certain populations (such as southeast Asia) where there is a high incidence of hemolytic disease is a reflection of a selective genetic advantage. The measurement of serum haptoglobin is used clinically as an indication of the degree of intravascular hemolysis. Patients who have significant intravascular hemolysis will have low or absent levels of haptoglobin because of the removal of haptoglobin–hemoglobin complexes by the reticuloendothelial system. Haptoglobin levels can also be low in severe extravascular hemolysis, in which the large load of hemoglobin in the reticuloendothelial system leads to the transfer of free hemoglobin into plasma.

Free heme and hematin appearing in plasma are bound by a β-globulin, hemopexin, which has a molecular weight of 57,000. One heme residue binds per hemopexin molecule. Hemopexin transfers heme to liver, where further metabolism by heme oxygenase occurs. In the normal state, hemopexin contains very little bound heme, whereas in intravascular hemolysis, the hemopexin is almost completely saturated by heme and is cleared with a half-life of about 7 h. In the latter instance excess heme will bind to albumin, with newly synthesized hemopexin serving as a mediator for the transfer of the heme from albumin to the liver. Hemopexin also binds free protoporphyrin.

BIBLIOGRAPHY

Bothwell, T. H., Charlton, R. W., and Motulsky, A. G. Idiopathic hemochromatosis, in J. B. Stanbury, J. B. Wyngaarden, D. S. Fredrickson, J. L. Goldstein, and M. S. Brown (eds.). *The metabolic basis of inherited disease*. New York: McGraw-Hill, 1983, p. 1269.

Gordon, E. R., Sommerer, U., and Goresky, C. A. The hepatic microsomal formation of bilirubin diglucuronide. *J. Biol. Chem.* 258:15028, 1983.

Huebers, H. A., and Finch, C. A. Transferrin: Physiologic behavior and clinical implications. *Blood* 64; 763, 1984.

Kappas, A., Sassa, S., and Anderson, K. E. The porphyrias, in J. B. Standbury, J. B. Wyngaarden, D. S. Fredrickson, J. L. Goldstein, and M. S. Brown (eds.). *The metabolic basis of inherited disease*. New York: McGraw-Hill, 1983, p. 1301.

Lustbader, J. W., Arcoleo, J. P., Birken, S., and Greer, J. Hemoglobin-binding site on haptoglobin probed by selective proteolysis *J. Biol. Chem.* 258:1227, 1983.

Maeda, N., Yang, F., Barnett, D. R., Bowman, B. H., and Smithies, O. Duplication within the haptoglobin Hp^2 gene. *Nature* 309:131, 1984.

McDonagh, A. F., Palma, L. A., and Lightner, D. A. Blue light and bilirubin excretion. *Science* 208:145, (1980).

Weiss, J. S., Gautam, A., Lauff, J. J., Sundberg, M. W., Jatlow, P. Boyer, J., and Seligson, D. The clinical importance of a protein-bound fraction of serum bilirubin in patients with hyperbilirubinemia. *New Engl. J. Med* 309:147, 1983.

Wolkoff, A. W., Chowdhury, J. R., and Arias, I. M. Hereditary jaundice and disorders of bilirubin metabolism, in J. B. Stanbury, J. B.

Wyngaarden, D. S. Fredrickson, J. L. Goldstein, and M. S. Brown (eds.). *The metabolic basis of inherited disease*. New York: McGraw-Hill, 1983, p. 1385.

Yamashiro, D. J., Tycko, B. Fluss, S. R., and Maxfield, F. R. Segregation of transferrin to a mildly acidic (pH 6.5) para-golgi compartment in the recycling pathway. *Cell* 37:789, 1984.

QUESTIONS

C. N. ANGSTADT and J. BAGGOTT

*Q*uestion *T*ypes are described inside the front cover.

A. Ferritin
B. Ferredoxin
C. Hemosiderin
D. Lactoferrin
E. Transferrin

1. (QT5) A type of protein in which iron is specifically bound to sulfur
2. (QT5) Exhibits an antimicrobial effect in the intestinal tract of newborns because of its ability to bind iron
3. (QT5) Delivers iron to tissues by binding to specific cell surface receptors
4. (QT2) In the intestinal absorption of iron:
   1. the presence of ascorbate enhances the availability of the iron.
   2. the regulation of uptake occurs between the lumen and the mucosal cells.
   3. the amount of apoferritin synthesized in the mucosal cell is related inversely to the need for iron by the host.
   4. iron bound tightly to a ligand, such as phytate, is more readily absorbed than free iron.
5. (QT1) Iron overload:
   A. cannot occur because very efficient excretory mechanisms are available.
   B. occurs in a deficiency of ferroxidase I (ceruloplasmin).
   C. would be accompanied by an increase in total serum transferrin.
   D. might be caused by the ingestion of large amounts of iron along with alcohol (ethanol).
   E. might be caused by the ingestion of iron chelators.
6. (QT2) The biosynthesis of heme requires:
   1. succinyl CoA.
   2. glycine.
   3. ferrous ion.
   4. propionic acid.
7. (QT1) Uroporphyrin III:
   A. is an intermediate in the biosynthesis of heme.
   B. does not contain a tetrapyrrole ring.
   C. differs from coproporphyrin III in the substituents around the ring.
   D. is formed from uroporphyrinogen III by an oxidase.
   E. formation is the primary control step in heme synthesis.
8. (QT2) Aminolevulinic acid synthase:
   1. requires pyridoxal phosphate for activity.
   2. is allosterically inhibited by heme.
   3. synthesis can be induced by a variety of drugs.
   4. is synthesized in the cytoplasm but catalyzes a mitochondrial reaction.
9. (QT1) Lead poisoning would be expected to result in an elevated level of:
   A. aminolevulinic acid.
   B. porphobilinogen.
   C. protoporphyrin I.
   D. heme.
   E. bilirubin.
10. (QT1) Ferrochelatase:
   A. is an iron-chelating compound.
   B. releases iron from heme in the degradation of hemoglobin.
   C. binds iron to sulfide ions and cysteine residues.
   D. is inhibited by heavy metals.
   E. is involved in the cytoplasmic portion of heme synthesis.
11. (QT2) Heme oxygenase:
   1. can oxidize the methene bridge between any two pyrrole rings of heme.
   2. requires molecular oxygen.
   3. produces bilirubin.
   4. produces carbon monoxide.

A. Direct bilirubin
B. Indirect bilirubin
C. Both
D. Neither

12. (QT4) Deposited in skin and sclera in jaundice
13. (QT4) Hepatic disease leads to major elevation of blood level of ___
14. (QT4) Biliary obstruction leads to major elevation of blood level of ___
15. (QT4) Acute intermittent porphyria is accompanied by increased blood level of ___
16. (QT1) Haptoglobin binds:
   A. a globin monomer.
   B. an oxyhemoglobin molecule.
   C. $\alpha\beta$-oxyhemoglobin dimers.
   D. a deoxyhemoglobin molecule.
   E. $\alpha\beta$-deoxyhemoglobin dimers.
17. (QT2) Haptoglobin:
   1. helps prevent loss of iron following intravascular red blood cell destruction.
   2. levels in serum are elevated in severe intravascular hemolysis.
   3. facilitates the action of heme oxygenase.
   4. binds heme and hematin as well as hemoglobin.

ANSWERS

1. B Animal ferredoxins, also known as nonheme iron-containing proteins, have two irons bound to two cysteine residues and sharing two sulfide ions (p. 859).
2. D As long as lactoferrin is not saturated, its avid binding of iron diminishes the amount available for growth of microorganisms (p. 857).
3. E Uptake of iron by tissues is governed by availability of receptors that internalize transferrin. Internalization is followed by release of the iron and recycling of the apotransferrin to the plasma (p. 857). Ferritin and hemosiderin (p. 858) are storage forms of iron.
4. B I, 3 correct. 1, 4: Ascorbate facilitates reduction to the ferrous state and, therefore, dissociation from ligands and absorption. 2: Substantial iron enters the mucosal cell regardless of need, but the amount transferred to the capillary beds is controlled. 3: Iron bound to apoferritin is trapped in mucosal cells and not transferred to the host (p. 860).
5. D Ethanol enhances the absorption of iron and could lead to an

overload. A: The high affinity of many macromolecules for iron prevents efficient excretion B: Deficiency of this enzyme has no significant effect on iron metabolism. C: Transferrin increases in iron deficiency to improve absorption. E: This is the treatment for iron overload when bleeding is inappropriate (p. 859, 861).

6. A 1, 2, 3 correct. 1, 2, 4: The organic portion of heme comes totally from glycine and succinyl CoA; the propionic acid side chain comes from the succinate. 3: The final step of heme synthesis is the insertion of the ferrous ion (p. 862, Figure 22.5).

7. C A, B, D: The tetrapyrrole porphyrins (except for protoporphyrin IX) are not intermediates but end products formed from the porphyrinogens nonenzymatically. E: The synthesis of aminolevulinic acid is the rate-limiting step (p. 864, Figure 22.5).

8. E All four correct. 1: The mechanism involves a Schiff's base with glycine. 2, 3: Heme both allosterically inhibits and suppresses synthesis of the enzyme, which can also be induced in response to need (many drug detoxications are cytochrome P_{450} dependent). 4: The gene for this enzyme is on nuclear DNA (p. 864).

9. A A–D: Lead inhibits ALA dehydrase so it inhibits synthesis of porphobilinogen and subsequent compounds. Heme certainly would not be elevated, because lead also inhibits ferrochelatase. E: Bilirubin is a breakdown product of heme, not an intermediate in synthesis (p. 866).

10. D This enzyme, in the mitochondria, catalyzes the last step of heme synthesis, the insertion of Fe^{+2}, and is sensitive to the effects of heavy metals (p. 867).

11. C 2, 4 correct. 1: The enzyme is specific for the methene between the two rings containing the vinyl groups (α-methene bridge). 2–4: It uses O_2 and the products are biliverdin and CO; the measurement of CO in the breath is an index of heme degradation (p. 868).

12. C Both conjugated (direct) and unconjugated (indirect) bilirubin are deposited (p. 870).

13. B Since the liver is responsible for conjugating bilirubin, hepatic disease leads to the elevation of unconjugated (indirect) bilirubin in blood (p. 870).

14. A Conjugated (direct) bilirubin is excreted in the bile (p. 871).

15. D Bilirubin reflects heme catabolism; porphyrias reflect a derangement of synthesis (Clin. Corr. 22.3).

16. C Haptoglobin binds dimers, two per haptoglobin molecule, specifically the oxyhemoglobin dimers since deoxyhemoglobin does not dissociate to dimers physiologically (p. 871).

17. B 1, 3 correct. 1, 2: Haptoglobin is part of the scavenging mechanism to prevent urinary loss of heme and hemoglobin from intravascular degradation of red blood cells. Since the complex is taken up by the reticuloendothelial system, the haptoglobin levels in serum are low. It also prevents clogging of the glomerular filter by hemoglobin. 3: Heme residues in the dimers bound to haptoglobin are more susceptible than free heme to oxidation by heme oxygenase. 4: Heme and hematin are bound by a β-globulin, while haptoglobin is an α-globulin (p. 872).

23

Gas Transport and pH Regulation

JAMES BAGGOTT

23.1 INTRODUCTION TO GAS TRANSPORT

Large organisms, especially terrestrial ones, require a relatively tough, impermeable outer covering to help ward off dust, twigs, nonisotonic fluids like rain and seawater, and other elements of the environment that would be harmful to living cells. One of the consequences of being large and having an impermeable covering is that individual cells of the organism cannot exchange gases directly with the atmosphere. Instead there must exist a specialized exchange surface, such as a lung or a gill, and a system to circulate the gases (and other materials, such as nutrients and waste products) in a manner that will meet the needs of every living cell in the body.

The existence of a system for the transport of gases from the atmosphere to cells deep within the body is not merely necessary, it has definite advantages. Oxygen is a good oxidizing agent, and at its partial pressure in the atmosphere, about 160 mmHg or 21.3 kPa [1 mmHg = 1 torr = 0.133 kPa (kilopascal)], it would oxidize and inactivate many of the components of the cells, such as essential sulfhydryl groups of enzymes. By the time oxygen gets through the transport system of the body its partial pressure is reduced to a much less damaging 20 mmHg (2.67 kPa) or less. In contrast, carbon dioxide is relatively concentrated in the body and becomes diluted in transit to the atmosphere. In the tissues, where it is produced, its partial pressure is 46 mmHg (6.13 kPa) or more. In the lungs it is 40 mmHg (5.33 kPa), and in the atmosphere only 0.2 mmHg (0.03 kPa), less abundant than the rare gas argon. Its relatively high concentration in the body permits it to be used as one component of a physiologically important buffering system, a system that is particularly useful because, upon demand, the concentration of carbon dioxide in the extracellular fluid can be varied over a rather wide range. This is discussed in more detail later in this chapter.

Oxygen and carbon dioxide are carried between the lungs and the other tissues by the blood. In the blood some of each gas is present in simple physical solution, but mostly each is involved in some sort of interaction with hemoglobin, the major protein of the red blood cell. There is a reciprocal relation between hemoglobin's affinity for oxygen and carbon dioxide, so that the relatively high level of oxygen in the lungs aids the release of carbon dioxide, which is to be expired, and the high carbon dioxide level in other tissues aids the release of oxygen for use by those tissues. Thus, a description of the physiological transport of oxygen and carbon dioxide is the story of the interaction of these two compounds with hemoglobin.

23.2 NEED FOR A CARRIER OF OXYGEN IN THE BLOOD

An oxygen carrier is needed in the blood simply because oxygen is not soluble enough in blood plasma to meet the body's needs. At 38°C 1 liter of plasma will dissolve only 2.3 ml oxygen. Whole blood, with its hemoglobin, has a much greater oxygen capacity. One liter of blood normally contains about 150 g hemoglobin, and each gram of hemoglobin can combine with 1.34 ml oxygen. Thus the hemoglobin in 1 liter of blood can carry 200 ml of oxygen, 87 times as much as plasma alone would carry. Without an oxygen carrier, the blood would have to circulate 87 times as fast to provide the same capacity to deliver oxygen. As it is, the blood makes a complete circuit of the body in 60 s under resting conditions, and in the aorta it flows at the rate of about 18.6 m/s. An 87-fold faster flow would require a fabulous high-pressure pump, would produce tremendously turbulent flow and high shear forces in the plasma, would result in uncontrollable bleeding from wounds, and would not even allow the blood enough time in the lungs to take up oxygen. The availability of a carrier not only permits us to avoid these impracticalities, but also gives us a way of controlling oxygen delivery, since the oxygen affinity of the carrier is responsive to changing physiological conditions.

The respiratory system includes the trachea, in the neck, which bifurcates in the thorax into right and left bronchi, as shown schematically in Figure 23.1. The bronchi continue to bifurcate into smaller and smaller passages, ending with tiny bronchioles, which open into microscopic gas-filled sacs called alveoli. It is in the alveoli that gas exchange takes place with the alveolar capillary blood.

As we inhale and exhale, the alveoli do not appreciably change in size. Rather, it is the airways that change in length and diameter as the air is pumped into and out of the lungs. Gas exchange between the airways and the alveoli then proceeds simply by diffusion. These anatomical and physiological facts have two important consequences. In the first place, since the alveoli are at the ends of long tubes that constitute a large dead space, and the gases in the alveoli are not completely replaced by fresh air with each breath, the gas composition of the alveolar air differs from that of the atmosphere, as shown in Table 23.1. Oxygen is lower in the alveoli because it is removed by the blood. Carbon dioxide is higher because it is added. Since we do not usually breathe air that is saturated with water vapor at 38°C, water vapor is generally added in the airways. The level of nitrogen is lower in the alveoli, not because it is taken up by the body, but simply because it is diluted by the carbon dioxide and water vapor.

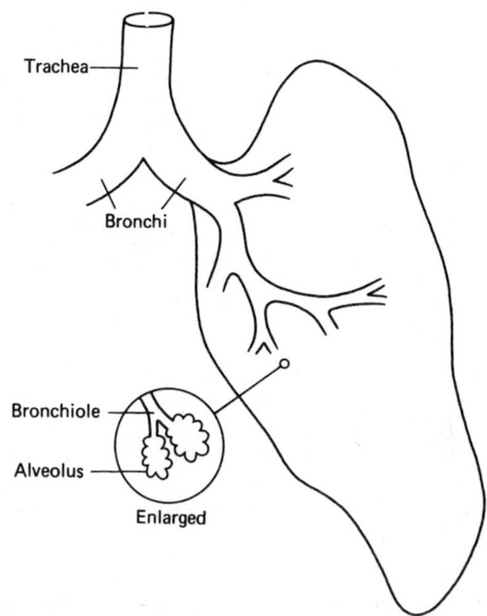

Figure 23.1
Diagram showing the respiratory tract.

TABLE 23.1 Partial Pressures of Important Gases Given in mmHg (kPa)

Gas	*In the Atmosphere*		*In the Alveoli of the Lungs*	
O_2	159	(21.2)	100	(13.3)
N_2	601	(80.1)	573	(76.4)
CO_2	0.2	(0.027)	40	(5.33)
H_2O	0	(0)	47	(6.27)
Total	760	(101)	760	(101)

A second consequence of the existence of alveoli of essentially constant size is that the blood that flows through the pulmonary capillaries during expiration, as well as the blood that flows through during inspiration, can exchange gases. This would not be possible if the alveoli collapsed during expiration and contained no gases, in which case the composition of the blood gases would fluctuate widely, depending on whether the blood passed through the lungs during an inspiratory or expiratory phase of the breathing cycle.

We have seen that an oxygen carrier is necessary. Clearly this carrier would have to be able to bind oxygen at an oxygen tension of about 100 mmHg (13.3 kPa), the partial pressure of oxygen in the alveoli. The carrier must also be able to release oxygen to the extrapulmonary tissues. The oxygen tension in the capillary bed of an active muscle is about 20 mmHg (2.67 kPa). In resting muscle it is higher, but during extreme activity it is lower. These oxygen tensions represent the usual limits within which an oxygen carrier must work. An efficient carrier would be nearly fully saturated in the lungs, but should be able to give up most of this to a working muscle.

Let us first see whether a carrier that binds oxygen in a simple equilibrium represented by

$$\text{Oxygen} + \text{carrier} \rightleftharpoons \text{oxygen} \cdot \text{carrier}$$

would be satisfactory. For this type of carrier the dissociation constant would be given by the simple expression

$$K_d = \frac{[\text{oxygen}][\text{carrier}]}{[\text{oxygen} \cdot \text{carrier}]}$$

and the saturation curve would be a rectangular hyperbola. This model would be valid even for a carrier with several oxygen-binding sites per molecule (which we know is the case for hemoglobin) as long as each site

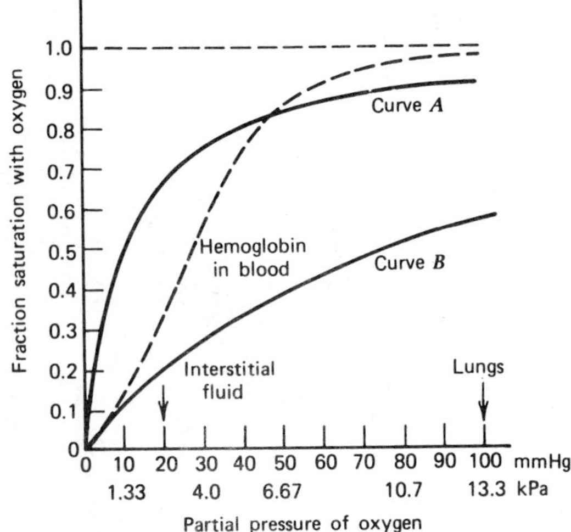

Figure 23.2
Oxygen saturation curves for two hypothetical oxygen carriers and for hemoglobin.
Curve A: Hypothetical carrier with hyperbolic saturation curve (a simple carrier) 90% saturated in the lungs and 66% saturated at the partial pressure found in interstitial fluid. Curve B: Hypothetical carrier with hyperbolic saturation curve (another simple carrier), 56% saturated in the lungs and 20% saturated at the partial pressure found in interstitial fluid. Dotted curve: Hemoglobin in whole blood.

were independent and not influenced by the presence or absence of oxygen at adjacent sites.

If such a carrier had a dissociation constant that permitted 90% saturation in the lungs, then, as shown in Figure 23.2A, at a partial pressure of 20 mmHg (2.67 kPa) it would still be 66% saturated, and would have delivered only 27% of its oxygen load. This would not be very efficient.

What about some other simple carrier, one that bound oxygen less tightly and therefore released most of it at low partial pressure, so that the carrier was, say, only 20% saturated at 20 mmHg (2.67 kPa)? Again, as shown in Figure 23.2B, it would be relatively inefficient; in the lungs this carrier could fill only 56% of its maximum oxygen capacity, and would deliver only 36% of what it could carry. It appears then that the mere fivefold change in oxygen tension between the lungs and the unloading site is not compatible with efficient operation of a simple carrier. Simple carriers are not sensitive enough to respond massively to a signal as small as a fivefold change.

Figure 23.2 also shows the oxygen-binding curve of hemoglobin in normal blood. The curve is sigmoid, not hyperbolic, and it cannot be described by a simple equilibrium expression. Hemoglobin, however, is a very good physiological oxygen carrier. It is 98% saturated in the lungs and only about 33% saturated in the working muscle. Under these conditions it delivers about 65% of the oxygen it can carry.

It can be seen in Figure 23.2 that hemoglobin is 50% saturated with oxygen at a partial pressure of 27 mmHg (3.60 kPa). The partial pressure corresponding to 50% saturation is called the P_{50}. P_{50} is the most common way of expressing hemoglobin's oxygen affinity. In analogy with K_m for enzymes, a relatively high P_{50} corresponds to a relatively low oxygen affinity.

It is important to notice that the steep part of hemoglobin's saturation curve lies in the range of oxygen tensions that prevail in the extrapulmonary tissues of the body. This means that relatively small decreases in oxygen tension in these tissues will result in large increases in oxygen delivery. Furthermore, small shifts of the curve to the left or right will also strongly influence oxygen delivery. In Sections 23.3, 23.5, and 23.6 we see how physiological signals effect such shifts and result in enhanced delivery under conditions of increased oxygen demand. Small decreases of oxygen tension in the lungs, however, such as occur at moderately high altitudes, do not seriously compromise hemoglobin's ability to bind oxygen. This will be true as long as the alveolar partial pressure of oxygen remains in a range that corresponds to the relatively flat region of hemoglobin's oxygen dissociation curve (Clin. Corr. 23.1).

Finally, we can see from Figure 23.2 that the binding of oxygen by hemoglobin is cooperative. At very low oxygen tension the hemoglobin curve tends to follow the hyperbolic curve which represents relatively weak oxygen binding, but at higher tensions it actually rises above the hyperbolic curve that represents tight binding. Thus it can be said that hemoglobin binds oxygen weakly at low oxygen tension and tightly at high tension. The binding of the first oxygen to each hemoglobin molecule somehow enhances the binding of subsequent oxygens.

Hemoglobin's ability to bind oxygen cooperatively is reflected in its Hill coefficient, which has a value of about 2.7. (The Hill equation is derived and interpreted on page 83.) Since the maximum value of the Hill coefficient for a system at equilibrium is equal to the number of cooperating binding sites, a value of 2.7 means that hemoglobin, with its four oxygen-binding sites, is more cooperative than would be possible for a system with only two cooperating binding sites, but it is not as cooperative as it could be.

CLIN. CORR. 23.1 CYANOSIS

Cyanosis is a condition in which a patient's skin or mucous membrane appears gray or (in severe cases) purple-magenta. It is due to an abnormally high concentration of deoxyhemoglobin below the surface, which is responsible for the observed color. The familiar blue of superficial veins is due to their deoxyhemoglobin content and is a normal manifestation of this color effect.

Cyanosis is most commonly caused by diseases of the cardiac or pulmonary systems, resulting in inadequate oxygenation of the blood. It can also be caused by certain hemoglobin abnormalities. Severely anemic individuals cannot become cyanotic; they do not have enough hemoglobin in their blood for the characteristic color of its deoxy form to be apparent.

23.3 HEMOGLOBIN AND ALLOSTERISM: EFFECT OF 2,3-DIPHOSPHOGLYCERATE

Hemoglobin's binding of oxygen was the original example of a homotropic effect (cooperativity and allosterism are discussed in Chapter 4), but hemoglobin also exhibits a heterotropic effect of great physiological significance. This involves its interaction with 2,3-diphosphoglycerate (DPG). Figure 23.3 shows the structure of DPG; it is closely related to the glycolytic intermediate, 1,3-diphosphoglycerate, from which it is in fact biosynthesized.

O   O⁻
 \\ /
  C
  |
H—C—O—PO_3^{2-}
  |
H—C—O—PO_3^{2-}
  |
  H

Figure 23.3
2,3-Diphosphoglycerate (DPG).

It had been known for many years that hemoglobin in the red cell bound oxygen less tightly than purified hemoglobin could (Figure 23.4). It had also been known that the red cell contained high levels of DPG, nearly equimolar with hemoglobin. Finally the appropriate experiment was done to demonstrate the relationships between these two facts. It was shown that the addition of DPG to purified hemoglobin produced a shift to the right of its oxygen-binding curve, bringing it into congruence with the curve observed in whole blood. Other organic polyphosphates, such as ATP and inositol pentaphosphate, also have this effect. Inositol pentaphosphate is the physiological effector in birds, where it replaces DPG, and ATP plays a similar role in some fish.

Monod's model of allosterism explains heterotropic interaction. Applying this model to hemoglobin, in the deoxy conformation (the T state) a cavity large enough to admit DPG exists between the β chains of hemoglobin. This cavity is lined with positively charged groups, and firmly binds one molecule of the negatively charged DPG. In the oxy conformation (the R state) this cavity is smaller, and it no longer accommodates DPG as easily. The result is that the binding of DPG to oxyhemoglobin is much weaker. Since DPG binds preferentially to the T state, the presence of DPG shifts the R–T equilibrium in favor of the T state; the deoxyhemoglobin conformation is thus stabilized over the oxyhemoglobin conformation (Figure 23.5). For oxygen to overcome this and bind to hemoglobin, a higher concentration of oxygen is required. Oxygen tension in the lungs is sufficiently high under most conditions to saturate hemoglobin almost completely, even when DPG levels are high. The physiological effect of DPG, therefore, can be expected to be upon release of oxygen in the extrapulmonary tissues, where oxygen tensions are low.

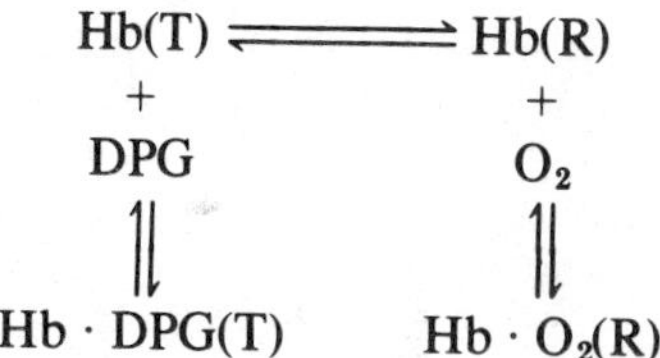

Figure 23.5
Schematic representation of equilibria among DPG, O_2 and the T and R states of hemoglobin.

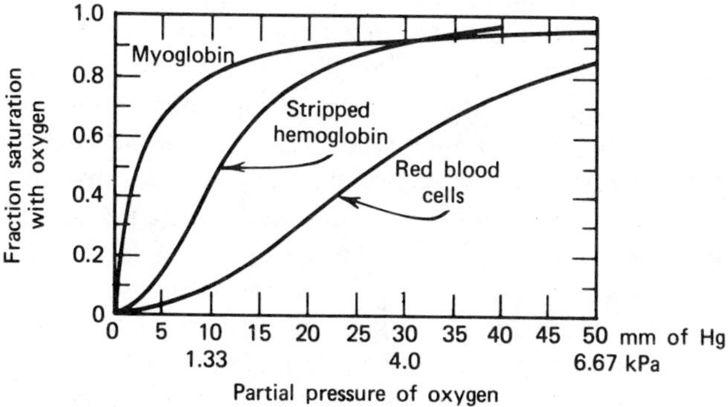

Figure 23.4
Oxygen dissociation curves for myoglobin, for hemoglobin that has been stripped of CO_2 and organic phosphates, and for whole red blood cells.

Data from Brenna, O., et al. *Advan. Exp. Biol. Med.* 28:19, 1972. Adapted from R. W. McGilvery. *Biochemistry: A functional approach,* 2nd ed. Philadelphia: W. B. Saunders, 1979, p. 236.

CLIN. CORR. 23.2 CHEMICALLY MODIFIED HEMOGLOBINS: METHEMOGLOBIN AND SULFHEMOGLOBIN

Methemoglobin is a form of hemoglobin in which the iron is oxidized from the iron (II) state to the iron (III) state. A tendency for methemoglobin to be present in excess of its normal level of about 1% may be due to a hereditary defect of the globin chain or to exposure to oxidizing drugs or chemicals.

Sulfhemoglobin is a species that forms when a sulfur atom is incorporated into the porphyrin ring of hemoglobin. Exposure to certain drugs or to soluble sulfides produces it. Sulfhemoglobin is green.

Hemoglobin subunits containing these modified hemes do not bind oxygen, but they change the oxygen-binding characteristics of the normal subunits in hybrid hemoglobin molecules containing some normal subunits and one or more modified subunits. The accompanying figure shows the oxygen-binding curve of normal HbA, 15% methemoglobin and 12% sulfhemoglobin. The presence of methemoglobin shifts the curve to the left, impairing the delivery of the decreased amount of bound oxygen. In contrast, the sulfhemoglobin curve is shifted to the right, a DPG-like effect. As a result, oxygen delivery is enhanced, partially compensating for the inability of the sulfur-modified hemes to bind oxygen.

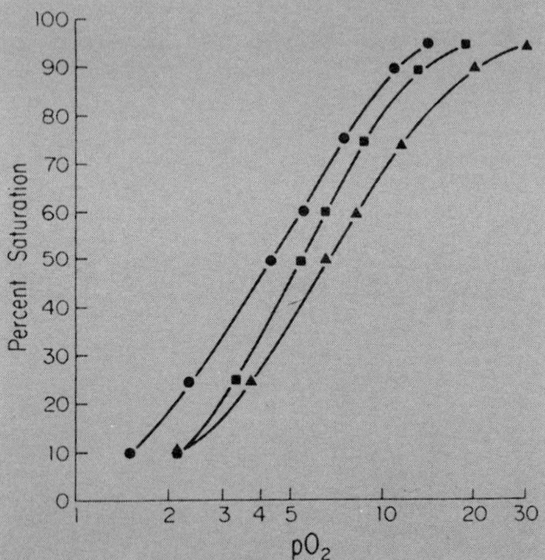

Oxygenation curves of unmodified hemoglobin A (squares) of a 15% oxidized hemolysate (circles) and of a hemolysate containing 12% sulfhemoglobin (triangles) in 0.1 M phosphate, pH 7.35, at 20°. Data from Park, C. M., and Nagel, R. L. *N. Engl. J. Med.* 310:1579, 1984.

The significance of a high DPG concentration is that the efficiency of oxygen delivery is increased. Levels of DPG in the red cell rise in conditions associated with tissue hypoxia, such as various anemias, cardiopulmonary insufficiency, and high altitude. These high levels of DPG enhance the formation of deoxyhemoglobin at low partial pressures of oxygen; hemoglobin then delivers more of its oxygen to the tissues. This effect can result in a substantial increase in the amount of oxygen delivered because the venous blood returning to the heart of a normal individual is (at rest) at least 60% saturated with oxygen. Much of this oxygen can dissociate in the peripheral tissues if the DPG concentration rises.

The DPG mechanism works very well as a compensation for tissue hypoxia as long as the partial pressure of oxygen in the lungs remains high enough that oxygen binding in the lungs is not compromised. Since, however, the effect of DPG is to shift the oxygen-binding curve to the right, the mechanism will not compensate for tissue hypoxia when the partial pressure of oxygen in the lungs falls too low. Then the increased efficiency of oxygen unloading in the tissues is counterbalanced by a decrease in the efficiency of loading in the lungs. This may be a factor in determining the maximum altitude at which people choose to establish permanent dwellings, which is about 18,000 feet (~5500m). There is evidence that a better adaptation to extremely low ambient partial pressures of oxygen would be a shift of the curve to the left.

23.4 OTHER HEMOGLOBINS

Although hemoglobin A is the major form of hemoglobin in adults and in children over 7 months of age, accounting for about 90% of their total hemoglobin, it is not the only normal hemoglobin species. Normal adults also have 2–3% of hemoglobin A_2, which is composed of two α chains like those in hemoglobin A and two δ chains. It is represented as $\alpha_2\delta_2$. The δ chains are distinct from the β chains, and are under independent genetic control. Hemoglobin A_2 does not appear to be particularly important in normal individuals.

Several species of modified hemoglobin A also occur normally. These are designated A_{Ia1}, A_{Ia2}, A_{Ib}, and A_{Ic}. They are adducts of hemoglobin with various sugars, such as glucose, glucose 6-phosphate, and fructose 1,6-bisphosphate. The quantitatively most significant of these is hemoglobin A_{Ic}. It arises from the covalent binding of a glucose residue to the N-terminal of the β chain. The reaction is not enzyme-catalyzed, and its rate depends on the concentration of glucose. As a result, hemoglobin A_{Ic} forms more rapidly in uncontrolled diabetics and can comprise up to 12% of their total hemoglobin under some circumstances. Hemoglobin A_{Ic} levels or total glycosylated hemoglobin levels have become a useful measure of how well diabetes has been controlled during the days and weeks before the measurement is taken; measurement of blood glucose only indicates how well it is under control at the time the blood sample is tàken. Chemical modification of hemoglobin A can also occur due to interaction with drugs or environmental pollutants (Clin. Corr. 23.2).

Fetal hemoglobin, hemoglobin F, is the major hemoglobin component of the newborn. It contains two γ chains in place of the β chains, and is represented as $\alpha_2\gamma_2$. Shortly before birth γ chain synthesis diminishes and β chain synthesis is initiated, and by the age of 7 months well over 90% of the infant's hemoglobin is hemoglobin A.

Hemoglobin F is adapted to the environment of the fetus, who must get oxygen from the maternal blood, a source that is far poorer than the atmosphere. In order to compete with the maternal hemoglobin for oxygen, fetal hemoglobin must bind oxygen more tightly; its oxygen-binding curve is thus shifted to the left relative to hemoglobin A. This is accomplished not through an intrinsic difference in the oxygen affinities of these hemoglobins but through a difference in the influence of DPG upon them. In hemoglobin F two of the groups that line the DPG-binding cavity have neutral side chains instead of the positively charged ones that occur in hemoglobin A. Consequently, hemoglobin F binds DPG less tightly and thus binds oxygen more tightly than hemoglobin A does. Furthermore, about 15–20% of the hemoglobin F is acetylated at the N-terminals; this is referred to as hemoglobin F_1. Hemoglobin F_1 does not bind DPG, and its affinity for oxygen is not affected at all by DPG. The postnatal change from hemoglobin F to hemoglobin A, combined with a rise in red cell DPG that peaks 3 months after birth, results in a gradual shift to the right of the infant's oxygen-binding curve (Figure 23.6). The result is greater delivery of oxygen to the tissues at this age than at birth, in spite of a 30% decrease in the infant's total hemoglobin concentration.

There are many inherited anomalies of hemoglobin synthesis in which there is formation of a structurally abnormal hemoglobin; these are called hemoglobinopathies. They may involve the substitution of one amino acid in one type of polypeptide chain for some other amino acid or they may involve absence of one or more amino acid residues of a polypeptide chain. In some cases the change is clinically insignificant, but in others it causes serious disease (Clin. Corr. 23.3).

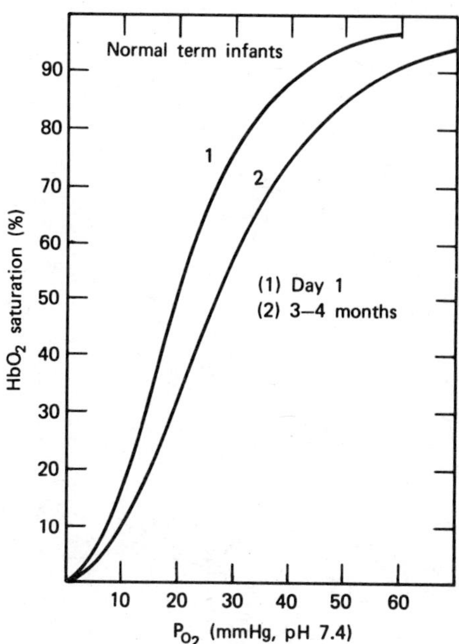

Figure 23.6
Oxygen dissociation curves after birth.
Adapted from Oski, F. A., and Delivoria-Papadopoulos, M. *J. Pediatr.* 77:941, 1970.

23.5 PHYSICAL FACTORS THAT AFFECT OXYGEN BINDING

Effect of Temperature

Temperature has a significant effect on oxygen binding by hemoglobin, as shown in Figure 23.7. At below-normal temperatures the binding is tighter, resulting in a leftward shift of the curve; at higher temperatures the binding becomes weaker, and the curve is shifted to the right.

The effect of elevated temperature is like that of high levels of DPG, in that both enhance unloading of oxygen. The temperature effect is physiologically useful, as it makes additional oxygen available to support the high metabolic rate found in fever or in exercising muscle with its elevated temperature. The relative insensitivity to temperature of oxygen binding at high partial pressure of oxygen minimizes compromise of oxygen uptake in the lungs under these conditions.

The tighter binding of oxygen that occurs in hypothermic conditions is not consequential in hypothermia induced for surgical purposes. The decreased oxygen utilization by the body and increased solubility of oxygen in plasma at lower temperatures, as well as the increased solubility of carbon dioxide, which acidifies the blood, compensate for hemoglobin's diminished ability to release oxygen.

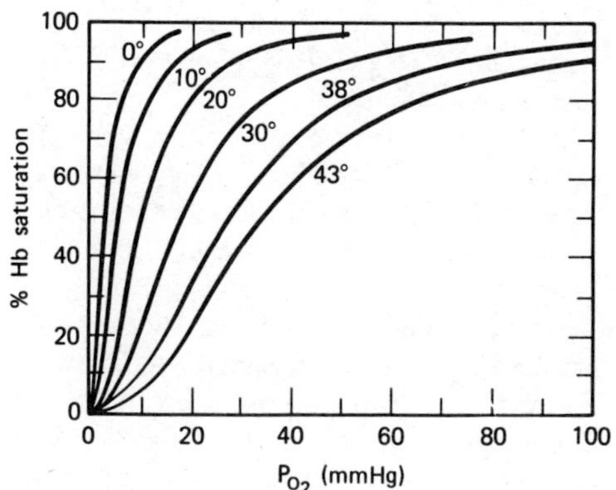

Figure 23.7
Oxygen dissociation curve for whole blood at various temperatures.
From Lambertson, Christian J. in *Medical physiology,* 11th ed., Philip Bard (ed.). St. Louis, Mo.: Mosby, 1961, p. 596.

Effect of pH

Hydrogen ion concentration influences hemoglobin's oxygen binding. As shown in Figure 23.8, low pH shifts the curve to the right, enhancing oxygen delivery, whereas high pH shifts the curve to the left. It is customary to express oxygen binding by hemoglobin as a function of the pH of the plasma because it is this value, not the pH within the erythrocyte, that is usually measured. The pH of the erythrocyte cell sap is lower than the plasma pH, but since these two fluids are in equilibrium, changes in the one reflect changes in the other.

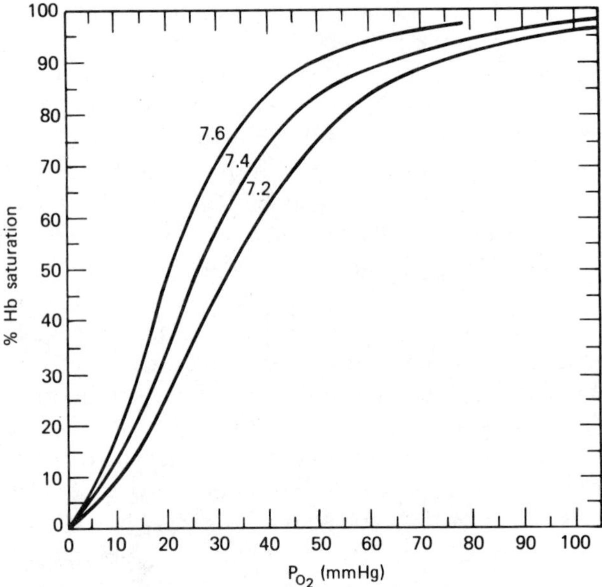

Figure 23.8
Oxygen dissociation curve for whole blood at various values of plasma pH.
Adapted from Lambertson, Christian J. in *Medical physiology,* 11th ed., Phillip Bard (ed.). St. Louis, Mo.: Mosby, 1961, p. 596.

The influence of pH upon oxygen binding is physiologically significant, since a decrease in pH is often associated with increased oxygen demand. An increased metabolic rate results in increased production of carbon dioxide and, as in muscular exercise, lactic acid. Lactic acid is also produced by hypoxic tissue. These acids produced by metabolism help release oxygen to support that metabolism.

The increase in acidity of hemoglobin as it binds oxygen is known as the *Bohr effect;* an equivalent statement is that the Bohr effect is the increase in basicity of hemoglobin as it releases oxygen. The effect may be expressed by the equation

$$HHb + O_2 \rightleftharpoons HbO_2 + H^+$$

Clearly this equation gives the same information as Figure 23.8, namely, that increases in hydrogen ion concentration will favor the formation of free oxygen from oxyhemoglobin, and conversely, that oxygenation of hemoglobin will lower the pH of the solution.

CLIN. CORR. **23.3**
HEMOGLOBINS WITH ABNORMAL OXYGEN AFFINITY

Some abnormal hemoglobins have an altered affinity for oxygen. If oxygen affinity is increased (P_{50} decreased), oxygen delivery to the tissues will be diminished unless some sort of compensation occurs. Typically the body responds by producing more erythrocytes (polycythemia) and more hemoglobin. Hb Rainier is an abnormal hemoglobin in which the P_{50} is 12.9 mmHg, far below the normal value of 27 mmHg.

In the accompanying figure the oxygen content in volume percent (ml oxygen/100 ml blood) is plotted *vs* partial pressure of oxygen, both for normal blood (curve *a*) and for the blood of a patient with Hb Rainier (curve *b*). Obviously the patient's blood carries more oxygen; this is because it contains 19.5 g Hb/100 ml instead of the usual 15g/100 ml.

Since the partial pressure of oxygen in mixed venous blood is about 40 mmHg, the volume of oxygen the blood of each individual can deliver may be obtained from the graph by subtracting the oxygen content of the blood at 40 mmHg from its oxygen content at 100 mmHg. As shown in the figure, the blood of the patient with Hb Rainier delivers nearly as much oxygen as normal blood does, although Hb Rainier delivers a significantly smaller fraction of the total amount it carries. Evidently polycythemia is an effective compensation for this condition, at least in the resting state.

23.6 CARBON DIOXIDE TRANSPORT

The carbon dioxide we produce is excreted by the lungs, to which it must be transported by the blood. Carbon dioxide transport is closely tied to hemoglobin and to the problem of maintaining a constant pH in the blood, a problem which will be discussed subsequently.

Carbon dioxide is present in the blood in three major forms, as dissolved CO_2, as HCO_3^- (formed by ionization of H_2CO_3 produced when CO_2 reacts with water), and as carbamino groups (formed when CO_2 reacts with amino groups of protein). Each of these is present both in arterial blood and in venous blood, as shown in the top three lines of Table 23.2. Net transport to the lungs for excretion is represented by the concentration difference between arterial and venous blood, shown in the last column. Notice that for each form of carbon dioxide the arterial–venous difference is only a small fraction of the total amount present; venous blood contains only about 10% more total carbon dioxide (total carbon dioxide is the sum of HCO_3^-, dissolved CO_2 and carbamino hemoglobin) than arterial blood does.

Carbon dioxide, after it enters the bloodstream for transport, generates hydrogen ions in the blood. Most come from bicarbonate ion formation, which occurs in the following manner.

Carbon dioxide entering the blood diffuses into the erythrocytes. The erythrocyte membrane, like most other biological membranes, is freely permeable to dissolved CO_2. Within the erythrocytes most of the carbon dioxide is acted upon by the intracellular enzyme, carbonic anhydrase, which catalyzes the reaction

$$CO_2 + H_2O \underset{\text{anhydrase}}{\overset{\text{carbonic}}{\rightleftharpoons}} H_2CO_3$$

This reaction will proceed in the absence of a catalyst, as is well known to all who drink carbonated beverages. Without the catalyst, however, it is too slow to meet the body's needs, taking over 100 s to reach equilibrium. Recall that at rest the blood makes a complete circuit of the body in only 60 s. Carbonic anhydrase is a very active enzyme, having a turnover number of the order of 10^6, and inside the erythrocytes the reaction reaches equilibrium within 1 s, less than the time spent by the blood in the capillary bed. The enzyme is zinc-requiring, and accounts for a portion of our dietary requirement for this metal.

The ionization of carbonic acid, $H_2CO_3 \rightleftharpoons H^+ + HCO_3^-$, is a rapid, spontaneous reaction. It results in the production of equivalent amounts of H^+ and HCO_3^-. Since, as shown in the last column of line 2 in Table 23.2, 1.69 meq of bicarbonate was added to each liter of blood by this process, 1.69 meq of H^+ must also have been generated per liter of blood. The addition of this much acid, over 10^{-3} equiv H^+, to a liter of water

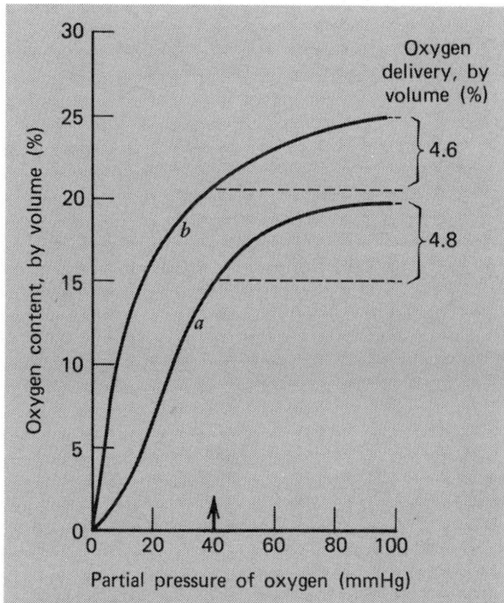

Curve a shows the oxygen dissociation curve of normal blood with a hemoglobin of 15 g/dl, P_{50} 27 mmHg, n 2.8, at pH 7.4, 37°C. Curve b shows that of blood from a patient with Hb Rainier, having a hemoglobin of 19.5 g/dl, P_{50} 12.9 mmHg, n 1.2, at the same pH and temperature. (1 mmHg ≈ 133.3 Pa.) On the right is shown the oxygen delivery. The compensatory polycythemia and hyperbolic curve of Hb Rainier results in practically normal arterial and venous oxygen tensions. Arrow indicates normal mixed venous oxygen tension.

From Bellingham, A. J. Hemoglobins with altered oxygen affinity. *Brit. Med. Bull.*, 32:234, 1976.

TABLE 23.2 Properties of Blood of Humans at Rest[a]

	Arterial			*Venous*			Δ		
	Serum	*Cells*	*Blood*	*Serum*	*Cells*	*Blood*	*Serum*	*Cells*	*Blood*
Hb carbamino groups meq/liter blood		*1.13*	*1.13*		*1.42*	*1.42*		*+0.29*	*+0.29*
HCO_3^-, meq/liter blood	13.83	*5.73*	*19.56*	14.84	*6.41*	*21.25*	+1.01	*+0.68*	*+1.69*
Dissolved CO_2, meq/liter blood	0.71	0.48	1.19	0.82	0.56	1.38	+0.11	+0.08	+0.19
Total CO_2, meq/liter blood	14.54	7.34	21.88	15.66	8.39	24.05	+1.12	+1.05	+2.17
Free O_2, mmol/liter blood			0.10			0.04			−0.06
Bound O_2, mmol/liter blood			8.60			6.01			−2.59
Total O_2, mmol/liter blood			8.70			6.05			−2.65
pO_2, mmHg			88.0			37.2			−50.8
pCO_2, mmHg			41.0			47.5			+6.5
pH	7.40	7.19		7.37	7.17		−0.03	−0.02	
Volume, cc/liter blood	551.7	448.3	1000	548.9	451.1	1000	−2.8	+2.8	0.0
H_2O, cc/liter blood	517.5	322.8	840.0	514.7	325.6	840.0	−2.8	+2.8	0.0
Cl^-, meq/liter blood	57.71	24.30	82.01	56.84	25.17	82.01	−0.88	+0.88	0.0

SOURCE: From Baggott, J. The contribution of carbamate to physiological carbon dioxide transport. *Trends Biochem. Sci.* 3:N207, 1978, with permission of the publisher.

[a] Hemoglobin, 9 mM; serum protein, 39.8 g/liter of blood; respiratory quotient, 0.82.

would give a final pH below 3. Since the pH of venous plasma has an average value of 7.37, clearly most of the H^+ generated during HCO_3^- production must be consumed by buffer action and/or other processes. This is discussed below.

Because of the compartmentalization of carbonic anhydrase, essentially all of the conversion of CO_2 to carbonic acid, and ultimately to HCO_3^-, occurs inside the erythrocyte. Negligible amounts of CO_2 react nonenzymatically in the plasma. This means that virtually all of the increase in HCO_3^- in venous as compared to arterial blood comes from intraerythrocyte HCO_3^- generation. To be sure, most of this diffuses into the plasma, so that venous plasma HCO_3^- is higher than arterial, but the erythrocyte was the site of its formation.

It has been observed that in the presence of carbonic anhydrase inhibitors, such as acetazolamide or cyanide, blood will still take up a certain amount of carbon dioxide rapidly. This is due to the reaction of carbon dioxide with amino groups of proteins within the erythrocyte to form carbamino groups.

$$R{-}NH_2 + CO_2 \rightleftharpoons R{-}NH{-}C(=O){-}O^- + H^+$$

Most reactions occur with the amino groups of hemoglobin. Deoxyhemoglobin forms carbamino hemoglobin more readily than oxyhemoglobin does, and oxygenation causes the release of CO_2 that had been bound in carbamino hemoglobin.

Carbamino hemoglobin formation occurs only with uncharged aliphatic amino groups, not with the charged form, $R{-}NH_3^+$. The pH within the erythrocyte is normally about 7.2, somewhat more acidic than the plasma. Since amino groups of proteins have p*K*s well to the alkaline side of 7.2, they will be mostly in the charged (undissociated acid) form. Removal of some of the uncharged form via carbamino group formation will shift the equilibrium, generating more uncharged amino groups and an equivalent amount of H^+.

$$R - NH_3^+ \rightleftharpoons R - NH_2 + H^+$$

Clearly the formation of a carbamino group is, like HCO_3^- formation, a process that generates H^+. But, while each CO_2 that forms HCO_3^- generates an equivalent amount of H^+, each equivalent of CO_2 that forms carbamino groups can be expected to produce somewhere between 1 and 2 equiv of H^+. The amount is somewhat less than 2 because the reestablishment of equilibrium shown in the above equation does not replace *every* $R{-}NH_2$ consumed during carbamino group formation; it merely reestablishes the previous *ratio* of $R{-}NH_2 : R{-}NH_3^+$. The lower the p*K* of the amino group, the closer the acid generation will be to 1 H^+ per CO_2 reacted.

The fact that only uncharged groups can form carbamino groups severely limits the number that can potentially participate in this reaction. Typical amino groups, such as the ε-amino groups in the side chains of lysyl residues, have p*K*s ~9.5–10.5. If the p*K* were 10.2, then at an intracellular pH of 7.2 only one ε-amino group in a thousand would be uncharged and able to react with carbon dioxide. The α-amino groups at the N-terminals of proteins, however, have much lower p*K*s, in the range of 7.6–8.4. This is because of the electron-withdrawing effect of the nearby oxygen of the peptide linkage. For an amino group with pK = 8.2, 1 out of every 10 molecules would be uncharged inside the cell and able to react with CO_2. A lower pK (or a higher intracellular pH) would result in an even greater availability of the group. Because of their lower pKs the α-amino groups at the N-terminals of hemoglobin's polypeptide chains are the principal sites of carbamino group formation. If all four N-terminal amino groups of hemoglobin are blocked chemically by reaction with cyanate, carbamino formation does not occur.

The N-terminal amino groups of the β chains form part of the binding site of DPG. Since the N-terminals cannot bind DPG and simultaneously form carbamino groups, a competition arises. CO_2 diminishes the effect for DPG and, conversely, DPG diminishes the ability of hemoglobin to form carbamino hemoglobin. Ignorance of the latter interaction led to a major overestimation of the role of carbamino hemoglobin in carbon dioxide transport. Prior to the discovery of the DPG effect, careful measurements were made of the capacity of purified hemoglobin (no DPG present) to form carbamino hemoglobin. The results were assumed to be applicable to hemoglobin in the erythrocyte, leading to the erroneous conclusion that carbamino hemoglobin accounted for 25–30% or more of carbon dioxide transport. It now appears that 13–15% of carbon dioxide transport is via carbamino hemoglobin.

Hemoglobin, in addition to being the primary oxygen carrier and a transporter of carbon dioxide in the covalently bound form of a carbamino group, also plays the major role in handling the hydrogen ions produced in carbon dioxide transport. It does this by buffering and by a second mechanism, which is discussed below. Hemoglobin's buffering power is due to its ionizable groups with pKs in the neighborhood of the intracellular pH of the erythrocyte. These include the four α-amino groups of the N-terminal amino acids and the imidazole side chains of the histidine residues. Hemoglobin has 38 histidines per tetramer; these therefore provide the bulk of hemoglobin's buffering ability.

In whole blood, buffering absorbs about 60% of the acid generated in normal carbon dioxide transport. Although hemoglobin is by far the most important nonbicarbonate buffer in blood, the organic phosphates in the erythrocytes, the plasma proteins, and so on, also make a significant contribution. Buffering by these compounds accounts for about 10% of the acid, leaving ~50% of acid control specifically attributable to buffering by hemoglobin. These buffer systems minimize the change in pH that occurs when acid or base is added, but do not altogether prevent that change. A small difference in pH between arterial and venous blood is therefore observed.

The remainder of the acid arising from carbon dioxide is absorbed by hemoglobin via a mechanism that has nothing to do with buffering. Recall that when hemoglobin became oxygenated it became a stronger acid and released H^+ (the Bohr effect). In the capillaries, where oxygen is released, the opposite occurs:

$$HbO_2 + H^+ \rightleftharpoons HHb + O_2$$

Simultaneously, carbon dioxide enters the capillaries and is hydrated:

$$CO_2 + H_2O \rightleftharpoons H^+ + HCO_3^-$$

Addition of these two equations gives

$$HbO_2 + CO_2 + H_2O \rightleftharpoons HHb + HCO_3^- + O_2$$

revealing that to some extent this system can take up H^+ arising from carbon dioxide, and can do so with no change in H^+ concentration (that is, with no change in pH). Hemoglobin's ability to do this, through the operation of the Bohr effect, is referred to as the *isohydric carriage* of CO_2. As already pointed out, there is a small A-V difference in plasma pH. This is because the isohydric mechanism cannot handle all the acid generated during normal CO_2 transport; if it could, no such difference would occur. Figure 23.9 is a schematic representation of oxygen transport and the isohydric mechanism, showing what happens in the lungs and in the other tissues.

Estimates of the importance of the isohydric mechanism in handling normal respiratory acid production have changed upward and downward over the years. The older, erroneous estimates arose out of a lack of knowledge of the multiple interactions in which hemoglobin participates. The earliest experiments, titrations of purified oxyhemoglobin and purified deoxyhemoglobin, revealed that oxygenation of hemoglobin resulted in release of an average of 0.7 H^+ for every O_2 bound. This figure still appears in textbooks, and much is made of it. Authors point out that with a Bohr effect of this magnitude the isohydric mechanism alone could handle all of the acid produced by the metabolic oxidation of fat (RQ of fat is 0.7), and buffering would be unnecessary. Unfortunately the experimental basis for this interpretation is physiologically unrealistic; the titrations were done in the total absence of carbon dioxide, which we now know binds to some of the Bohr groups, forming carbamino groups and diminishing the effect. When experiments are carried out in the presence of physiological amounts of carbon dioxide, there is a drastic diminution of the Bohr effect, so much so that at pH 7.45 the isohydric mechanism could handle only the amount of acid arising from carbamino group formation. This work, however, was carried out prior to our appreciation of the competition between DPG and carbon dioxide for the same region of the hemoglobin molecule. Finally, in 1971, careful titra-

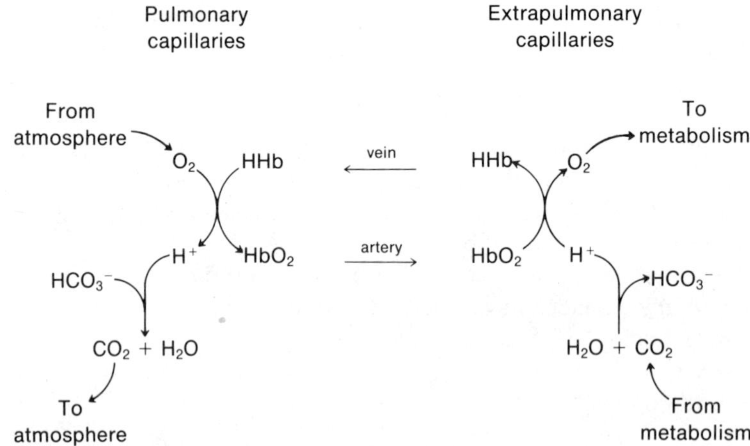

Figure 23.9
Schematic representation of oxygen transport and the isohydric carriage of CO_2 by hemoglobin.
In the lungs (left) O_2 from the atmosphere reacts with deoxyhemoglobin, forming oxyhemoglobin and H^+. The H^+ combines with the HCO_3^- to form H_2O and CO_2. The CO_2 is exhaled. Oxyhemoglobin is carried to extrapulmonary tissues (right), where it dissociates in response to low P_{O_2}. The O_2 is used by metabolic processes, and CO_2 is produced. CO_2 combines with H_2O to give HCO_3^- and H^+. H^+ can then react with deoxyhemoglobin to give HHb, which returns to the lungs, and the cycle repeats.

tions of whole blood under presumably physiological conditions were carried out, yielding a value of 0.31 H^+ released per O_2 bound. This value is the basis of the present assertion that the isohydric mechanism accounts for about 40% of the acid generated during normal carbon dioxide transport. The quantitative contributions of various mechanisms to the handling of acid arising during carbon dioxide transport are summarized in Table 23.3. The major role of hemoglobin in handling this acid is obvious.

TABLE 23.3 Distribution of the H^+ Generated During Normal Carbon Dioxide Transport

Buffering	
By hemoglobin	50%
By other buffers	10%
Isohydric mechanism (hemoglobin)	40%

We have seen that essentially all of HCO_3^- formation is intracellular, catalyzed by carbonic anhydrase, and that the vast bulk of the H^+ generated by CO_2 is handled within the erythrocyte. These two observations bear upon the final distribution of HCO_3^- between plasma and the erythrocyte.

Intracellular formation of HCO_3^- increases its intracellular concentration. Since the erythrocyte membrane is freely permeable to HCO_3^- (as well as to certain other ions, including Cl^- and H^+), HCO_3^- will diffuse out of the erythrocyte, increasing the plasma HCO_3^- concentration. Electrical neutrality must be maintained across the membrane as this happens. Maintenance of neutrality can be accomplished in principle either by having a positively charged ion accompany HCO_3^- out of the cell or by having some other negatively charged ion enter the cell in exchange for the HCO_3^-. Since the distribution of the major cations, Na^+ and K^+, is under strict control, it is the latter mechanism that is seen, and the ion that is exchanged for HCO_3^- is Cl^-. Thus as HCO_3^- is formed in red cells during their passage through the capillary bed, it moves out into the plasma and Cl^- comes in to replace it. The increase in intracellular $[Cl^-]$ is shown in the last line of Table 23.2. In the lungs, where all events of the peripheral tissue capillary beds are reversed, HCO_3^- migrates into the cells to be converted to CO_2 for exhalation, and Cl^- returns to the plasma. The exchange of Cl^- and HCO_3^- between the plasma and the erythrocyte is called the *chloride shift*.

The intracellular buffering of H^+ from carbon dioxide causes the cells to swell, giving venous blood a slightly (0.6%) higher hematocrit than arterial blood. (The hematocrit is the volume percent of red cells in the blood.) This occurs because the charge on the hemoglobin molecule becomes more positive with every H^+ that binds to it. Each bound positive charge requires an accompanying negative charge to maintain neutrality. Thus as a result of buffering there is a net accumulation of HCO_3^- or Cl^- inside the erythrocyte. An increase in the osmotic pressure of the intracellular fluid results from this increase in concentration of particles. As a consequence water migrates into the cells, causing them to swell slightly. Typically, an arterial hematocrit might be 44.8 and a venous hematocrit, 45.1, as shown in Table 23.2 by the line labeled "volume, cc/liter blood."

23.7 INTERRELATIONSHIPS AMONG HEMOGLOBIN, OXYGEN, CARBON DIOXIDE, HYDROGEN ION, AND 2,3-DIPHOSPHOGLYCERATE

By now it should be clear that multiple interrelationships of physiological significance exist among the ligands of hemoglobin. These interrelationships may be summarized schematically as follows:

$$\text{HHb}\begin{array}{l}\diagup \text{DPG}\\ \diagdown \text{CO}_2\end{array} + O_2 \rightleftharpoons HbO_2 + CO_2 + DPG + H^+$$

This equation shows that changes in the concentration of H^+, DPG, or CO_2 have similar effects on oxygen binding. The equation will help you remember the effect of changes in any one of these variables upon hemoglobin's oxygen affinity.

DPG levels in the red cell are controlled by product inhibition of its synthesis and by pH. Hypoxia results in increased levels of deoxyhemoglobin on a time-averaged basis. Since deoxyhemoglobin binds DPG more tightly, in hypoxia there is less free DPG to inhibit its own synthesis, and so DPG levels will rise due to increased synthesis. The effect of pH is that high pH increases DPG synthesis and low pH decreases DPG synthesis. Since changes in DPG levels take many hours to become complete, this means that the immediate effect of a decrease in blood pH is to enhance oxygen delivery by the Bohr effect. If the acidosis is sustained (most causes of chronic metabolic acidosis are not associated with a need for enhanced oxygen delivery), diminished DPG synthesis leads to a decrease in intracellular DPG concentration, and hemoglobin's oxygen affinity returns toward normal. Thus we have a system that can respond appropriately to acute conditions, such as vigorous exercise, but which when faced with a prolonged abnormality of pH readjusts to restore normal (and presumably optimal) oxygen delivery.

23.8 INTRODUCTION TO pH REGULATION

When we considered carbon dioxide transport we noted the large amount of H^+ generated by this process, and we considered the ways in which the blood pH was kept under control. Control of blood pH is important because changes in blood pH will cause changes in intracellular pH, which in turn may profoundly alter metabolism. Protein conformation is affected by pH, as is enzyme activity. In addition, the equilibria of important reactions that consume or generate hydrogen ions, such as any of the oxidation–reduction reactions involving pyridine nucleotides, will be shifted by changes in pH.

The normal arterial plasma pH is 7.40 ± 0.05; the pH range compatible with life is about 7.8–6.8. Intracellular pH varies with the type of cell. The pH of the erythrocyte is nearly 7.2, whereas most other cells are lower, $\sim$7.0. Values as low as 6.0 have been reported for skeletal muscle.

It is fortunate for both diagnosis and treatment of diseases that the acid–base status of the intracellular fluid influences and is influenced by the acid–base status of the blood. Blood is readily available for analysis, and when alteration of body pH becomes necessary, intravenous administration of acidifying or alkalinizing agents is efficacious.

23.9 BUFFER SYSTEMS OF PLASMA, INTERSTITIAL FLUID, AND CELLS

Each body water compartment is defined spatially by one or more differentially permeable membranes. Each type of compartment contains characteristic kinds and concentrations of solutes, some of which are buffers at physiological pHs. Although the solutes in the cytoplasm of each type of cell are different, most cells are similar enough that they are considered together for purposes of acid–base balance. Thus there are, from this point of view, three major body water components: plasma, which is contained in the circulatory system; interstitial fluid, the fluid that bathes the cells; and intracellular fluid.

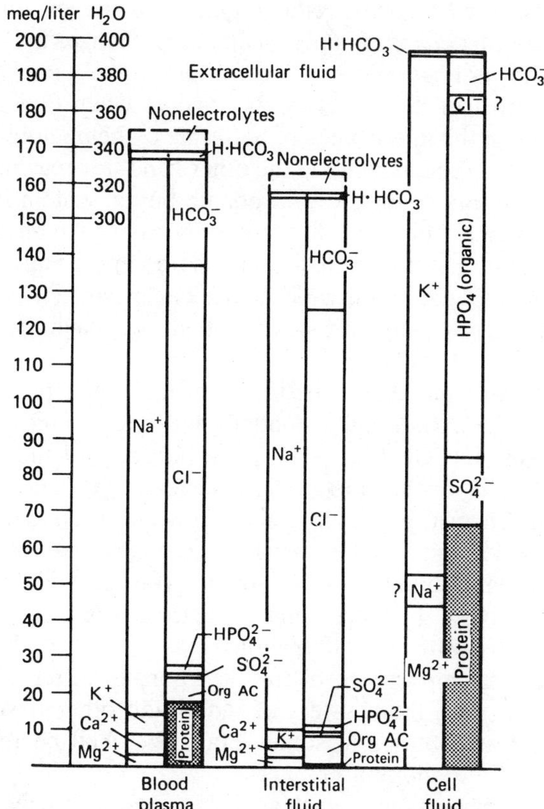

Figure 23.10
Diagram showing chief chemical constituents of the three fluid compartments. *Height of left half of each column indicates total concentration of cations; that of right half, concentration of anions. Both are expressed in meq/liter of water. Note that chloride and sodium values in cell fluid are questioned. It is probable that at least in muscle, intracellular phase contains some sodium but no chloride.*
Modified from Gamble. From Gregersen, Magnus I. in *Medical physiology*, 11th ed., Philip Bard, ed. St. Louis, Mo.: Mosby, 1961, p. 307.

The compositions of these fluids are given in Figure 23.10. In plasma the major cation is Na^+; small amounts of K^+, Ca^{2+}, and Mg^{2+} are also present. The two dominant anions are HCO_3^- and Cl^-. Smaller amounts of protein, phosphate, and SO_4^{2-} are also found, along with a mixture of organic anions (amino acids, etc.), each of which would be insignificant if taken separately. The sum of the anions equals, of course, the sum of the cations. It is apparent at a glance that the composition of interstitial fluid is very similar. The major difference is that interstitial fluid contains much less protein than plasma contains (the capillaries are not normally permeable to the plasma proteins) and, correspondingly, a lower cation concentration. Plasma and interstitial fluid taken together are called the extracellular fluid, and low molecular weight components equilibrate fairly rapidly between the two. H^+, for example, equilibrates between the plasma and interstitial fluid within about half an hour. The composition of intracellular fluid is strikingly different. K^+ is the major cation, while organic phosphates (ATP, DPG, glycolytic intermediates, etc.) and protein are the major anions.

As a result of the differences among these fluid compartments, each fluid makes a different contribution to buffering. The major buffer of the extracellular fluid, for example, is the HCO_3^-/CO_2 system. Since the pK of the HCO_3^-/CO_2 system is 6.1 (Table 23.4 lists the major physiological

TABLE 23.4 Acid Dissociation Constants of Major Physiological Buffers

Buffer System	pK
HCO_3^-/CO_2	6.1
Phosphate	
$HPO_4^{2-}/H_2PO_4^-$	6.7–7.2
Organic phosphate esters	6.5–7.6
Protein	
Histidine side chains	5.6–7.0
N-terminal amino groups	7.6–8.4

buffers and their p*K*s), extracellular fluid at a pH of 7.4 is not very effective in resisting changes in pH arising from changes in P_{CO_2}. Intracellular fluid, with its high levels of protein and organic phosphates, is responsible for most of the buffering that occurs when P_{CO_2} changes. We have already seen the importance of buffering by hemoglobin and organic phosphates within the red cell. On the other hand, for reasons that will be explained in Section 23.10, the bicarbonate buffer system is quite effective in controlling pH changes due to causes other than changes in P_{CO_2}. Extracellular fluid and intracellular fluid share almost equally in the buffering of strong organic or inorganic acids. The plasma $[HCO_3^-]$ is, therefore, an excellent indicator of the whole body's capacity to handle additional loads of these acids.

Since acid–base imbalance arising from metabolic production of organic acids is a common and potentially life-threatening condition, and since the plasma $[HCO_3^-]$ is such a good indicator of the whole body's capacity to handle further metabolic acid loads, plasma $[HCO_3^-]$ is of major clinical concern. It is hydrogen ion concentration that must be kept within acceptable limits, but measuring pH alone is like walking on thin ice while observing merely whether or not you are still on the surface. Knowledge of $[HCO_3^-]$ tells you how close to the breaking point the ice is and how deep the water is underneath.

Because of the importance of the bicarbonate buffer system and its interaction with the other buffers of blood and other tissues, we shall consider blood as a buffer in some detail. We shall begin with a brief consideration of a model buffer.

Every buffer consists of a weak acid, HA, and its conjugate base, A^-. Examples of conjugate base/weak acid pairs include acetate$^-$/acetic acid, NH_3/NH_4^+ and $HPO_4^{2-}/H_2PO_4^-$. Notice that the weak acid may be neutral, positively charged or negatively charged, and that its conjugate base must (since a H^+ has been lost) have one less positive charge (or one more negative charge) than the weak acid.

The degree of ionization of a weak acid depends on the concentration of free hydrogen ions. This may be expressed in the form of the Henderson–Hasselbalch equation (derived on page 11) as follows:

$$\mathrm{pH} = \mathrm{p}K + \log \frac{[\text{conjugate base}]}{[\text{acid}]}$$

This is a mathematical rearrangement of the fundamental equilibrium equation. It states that there is a direct relationship between the pH and the ratio of [conjugate base] : [acid]. It is important to realize that this *ratio,* not the absolute concentration of any particular species, is the factor that is related to pH. Use of this equation will help you to understand the operation of and to predict the effects of various alterations upon acid–base balance in the body.

Blood plasma is a mixed buffer system; in the plasma the major buffers are HCO_3^-/CO_2, $HPO_4^{2-}/H_2PO_4^-$ and protein/Hprotein. The pH is the same throughout the plasma, so each of these buffer pairs distributes independently according to its own Henderson–Hasselbalch equation:

$$\begin{aligned} \mathrm{pH} &= \mathrm{p}K_1 + \log \frac{[HCO_3^-]}{[CO_2]} \\ &= \mathrm{p}K_2 + \log \frac{[HPO_4^{2-}]}{[H_2PO_4^-]} \\ &= \mathrm{p}K_3 + \log \frac{[\text{protein}^-]}{[\text{Hprotein}]} \end{aligned}$$

Because each pK is different the [conjugate base]/[acid] ratio is also different for each buffer pair. Notice, though, if the ratio is known for any given buffer pair, one automatically has information about the others (assuming the pKs are known).

23.10 THE CARBON DIOXIDE–BICARBONATE BUFFER SYSTEM

As we have seen, the major buffer of the plasma (and of the interstitial fluid as well) is the bicarbonate buffer system. The bicarbonate system has two peculiar properties that make its operation unlike that of typical buffers. We shall examine this important buffer in some detail, since a firm understanding of it is the key to a grasp of acid–base balance.

In the first place, the component which we consider to be the acid in this buffer system is CO_2, which is not truly an acid, but an acid anhydride. It reacts with water to form carbonic acid, which is indeed a typical weak acid.

$$CO_2 + H_2O \rightleftharpoons H_2CO_3$$

Carbonic acid then rapidly ionizes to give H^+ and HCO_3^-.

$$H_2CO_3 \rightleftharpoons H^+ + HCO_3^-$$

If these two equations are added, H_2CO_3 cancels out, and the sum is

$$CO_2 + H_2O \rightleftharpoons H^+ + HCO_3^-$$

Elimination of H_2CO_3 from formal consideration is realistic since, not only does it simplify matters, but H_2CO_3 is, in fact, quantitatively insignificant. Because the equilibrium of the reaction,

$$CO_2 + H_2O \rightleftharpoons H_2CO_3$$

lies far to the left, H_2CO_3 is present only to the extent of 1/200 of the concentration of dissolved CO_2. Since the concentration of water is virtually constant, it need not be included in the equilibrium expression for the reaction, and one may write:

$$K = \frac{[H^+][HCO_3^-]}{[CO_2]}$$

The value of K is 7.95×10^{-7}.

The concentration of a gas in a solution is proportional to the partial pressure of the gas. Thus we measure the partial pressure of CO_2 (P_{CO_2}). P_{CO_2} is then multiplied by a conversion factor, α, to get the millimolar concentration of dissolved CO_2.

$$\alpha P_{CO_2} = \text{meq/liter}$$

α has a value of 0.03 meq/liter mmHg (or 0.225 meq/liter kPa) at 37°C. The equilibrium expression thus becomes

$$K = \frac{[H^+][HCO_3^-]}{0.03 \cdot P_{CO_2}}$$

and the Henderson–Hasselbalch equation for this buffer system becomes

$$\mathrm{pH} = 6.1 + \log \frac{[\mathrm{HCO_3^-}]}{0.03 \cdot P\mathrm{CO_2}}$$

with $[HCO_3^-]$ expressed in units of meq/liter (Clin. Corr. 23.4).

We said earlier that the bicarbonate buffer system, with a pK of 6.1, was not effective against carbonic acid in the pH range of 7.8–6.8, but that it was effective against noncarbonic acids. The usual rules of chemical equilibrium dictate that a buffer is not very useful in a pH range more than about one unit beyond its pK. Thus what needs to be explained is how the bicarbonate system can be effective against noncarbonic acids; its failure to buffer carbonic acid is expected. The manner in which it buffers noncarbonic acids in a pH range far from its pK is the second unusual property of this buffer system. Notice that the explanation of this property in the following paragraph involves the flow of materials in a living system, and so departs from mere equilibrium considerations.

Consider first a typical buffer, consisting of a mixture of a weak acid and its conjugate base. When a strong acid is added, most of the added H^+ combines with the conjugate base. As a result, [weak acid] increases and simultaneously [conjugate base] diminishes. The ratio of [conjugate base]/[weak acid] therefore changes, and so does the pH. Of course, the pH changes much less than if there were no buffer present. Now imagine a system in which the weak acid, as it is generated by the reaction of the added strong acid with the conjugate base, is somehow removed so that while [conjugate base] diminishes, [weak acid] remains nearly constant. In this case the ratio of [conjugate base]/[weak acid] would change much less for a given addition of strong acid, and the pH would also change much less. This is exactly what happens with the bicarbonate buffer system in the body. As strong acid is added, $[HCO_3^-]$ diminishes, and CO_2 is formed. But the excess CO_2 is exhaled, so that the ratio of $[HCO_3^-]/\alpha P\mathrm{CO_2}$ does not change so dramatically. In like manner, if strong base is added to the body, it will be neutralized by carbonic acid, but CO_2 will be replaced by metabolism, and again, the ratio of $[HCO_3^-]/P\mathrm{CO_2}$ will not change as much as would be expected. The bicarbonate buffer system in the body is thus an *open* system in which the $P\mathrm{CO_2}$ term is adjusted to meet the body's needs. If respiration should be unable to accomplish this adjustment, then $P\mathrm{CO_2}$ would change strikingly, and the bicarbonate system would be relatively ineffective, in keeping with the prediction of chemical equilibrium.

A graphical representation of the Henderson–Hasselbalch equation for the bicarbonate buffer system is a valuable aid to learning and understanding how this system reflects the acid–base status of the body. One of the most common of these representations is the pH–bicarbonate diagram, shown in Figure 23.11. $[HCO_3^-]$ of up to 40 meq/liter is shown on the ordinate; this is adequate to deal with most situations. Similarly, since plasma pH does not exceed 7.8 or (except transiently) fall below 7.0 in living patients, the abscissa of the graph is limited to the range of 7.0–7.8. The normal plasma $[HCO_3^-]$, 24 meq/liter, and the normal plasma pH, 7.4, are indicated. The third variable, CO_2, can be shown on a two-dimensional graph by assigning a fixed value to $P\mathrm{CO_2}$ and then showing, *for that value,* the relationship between pH and $[HCO_3^-]$. Figure 23.11 shows that relationship when $P\mathrm{CO_2}$ has its normal physiological value of 40 mmHg (5.33 kPa). The line is called the 40 mmHg (5.33 kPa) isobar. Whenever $P\mathrm{CO_2}$ is 40 mmHg (5.33 kPa), pH and $[HCO_3^-]$ must be somewhere on that line.

In a similar manner we can plot isobars for various abnormal values of

CLIN. CORR. 23.4 THE CASE OF THE VARIABLE CONSTANT

In clinical laboratories plasma pH and $P\mathrm{CO_2}$ are commonly measured with suitable electrodes, and plasma $[HCO_3^-]$ is then calculated from the Henderson–Hasselbalch equation using pK = 6.1. Although this procedure is generally satisfactory, there have been several reports of severely erroneous results in patients whose acid–base status was changing rapidly.[1] Clinicians who are attuned to this phenomenon urge that *direct* measurements of all three variables be made in acutely ill patients.

The clinical literature discusses this problem in terms of departure of the value of pK from 6.1. Studies of model systems suggest that this interpretation is incorrect; pK does change with ionic strength, temperature, and so on, and so does α, but not enough to account for the magnitude of the clinical observations.

Astute commentators have speculated that the real basis of the phenomenon is disequilibrium. The detailed nature of the putative disequilibrium has not yet been established, but it is probably related to the difference in pH across the erythrocyte membrane. Normally the pH of the erythrocyte is about 7.2, and the plasma pH is 7.4. If the plasma pH changes rapidly in an acute illness, the pH of the erythrocyte will also change, but the rate of change within the erythrocyte is not known. If the change within the erythrocyte lags sufficiently behind the change in the plasma, the system would indeed be in gross disequilibrium, and equilibrium calculations would not apply.

[1] See Hood, I. and Campbell, E. J. M. *N. Engl. J. Med.* 306:864, 1982.

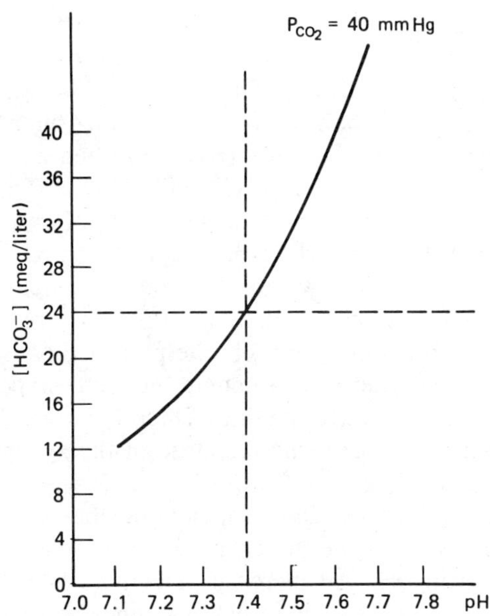

Figure 23.11
pH–Bicarbonate diagram, including the 40 mmHg (5.33 kPa) CO_2 isobar, and showing the normal values of plasma pH and bicarbonate ion concentration.

P_{CO_2}. These curves are shown in Figure 23.12. The range of values given covers those found in patients. Any point on the graph gives the values of the three variables of the Henderson–Hasselbalch equation for the bicarbonate system at that point. Since all you need to locate a point are any two of these variables, the third can be read directly from the graph.

Let us now see how the bicarbonate buffer system behaves when it is in the presence of other buffers, as it is in whole blood. First, let us acidify the system by increasing the concentration of the acid-producing component, CO_2. For every CO_2 that reacts with water to produce a H^+, one HCO_3^- will also form. Most of the H^+, however, will be buffered by protein and phosphate. As a result, [HCO_3^-] will rise much more than [H^+]. Similarly, if acid is removed from this system by decreasing P_{CO_2}, [HCO_3^-] will decrease. [H^+] will not decrease by an equivalent amount, though, because the other buffers will dissociate to resist the pH change. The results of these processes as they occur in whole blood, with its

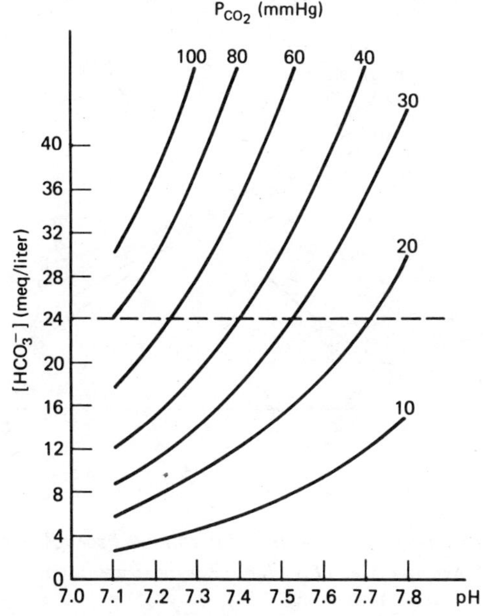

Figure 23.12
pH–Bicarbonate diagram, showing CO_2 isobars from 10 mmHg to 100 mmHg.

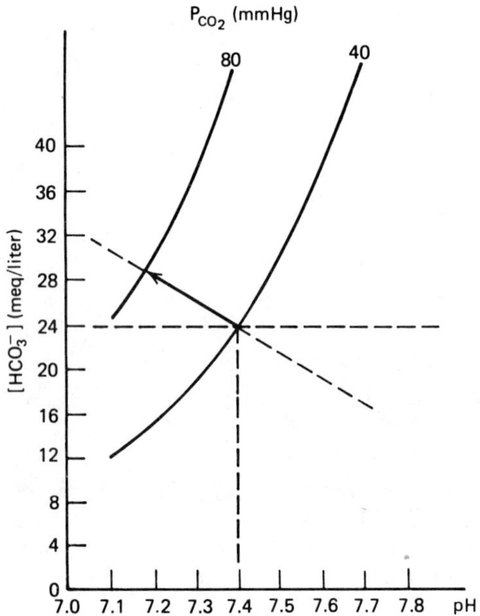

Figure 23.13
The buffering line of blood.
This pH–bicarbonate diagram shows the changes in pH and $[HCO_3^-]$ that occur in whole blood in vitro when P_{CO_2} is changed. Notice that the relationship between pH and $[HCO_3^-]$ is described by a straight line with a nonzero slope.

various intracellular and extracellular buffers, are shown in Figure 23.13. Let us start at the point that represents the normal values: pH of 7.4, $[HCO_3^-]$ of 24 meq/liter and P_{CO_2} of 40 mmHg (5.33 kPa). As P_{CO_2} rises to 80 mmHg (10.7 kPa), bicarbonate goes up to 28 meq/liter, an increase of 4 meq/liter. This means H_2CO_3 must have increased by 4 meq/liter, and that it immediately ionized, giving H^+ and HCO_3^-. The pH, however, drops to 7.18; this represents an increase in $[H^+]$ of only 26×10^{-6} meq/liter. The other 3.999974 meq/liter of H^+ produced by the ionization of carbonic acid were taken up by the phosphate, hemoglobin, plasma protein, and other buffer systems. If P_{CO_2} were to decrease, the opposite would occur. Thus by altering P_{CO_2} in the presence of HCO_3^- and other buffers, a line is generated with a definite nonzero slope. For the blood system, this is called the buffering line of blood. Notice that if P_{CO_2} is the only variable that is changed, the response of the system is confined to movements along this line.

The slope of the buffering line depends on the concentration of the nonbicarbonate buffers. If they were more concentrated, they would better resist changes in pH. An increase in P_{CO_2} to 80 mmHg (10.7 kPa) would then cause a smaller drop in pH, and since the more concentrated buffers would react with more hydrogen ions (produced by the ionization of carbonic acid), $[HCO_3^-]$ would rise higher. Thus the slope of the buffering line would be steeper.

Hemoglobin is quantitatively the second most important blood buffer, exceeded only by the bicarbonate buffer system. Since hemoglobin concentration in the blood can fluctuate widely in various disease states, it is the most important physiological determinant of the slope of the blood buffer line. Figure 23.14 shows how the slope of the blood buffer line varies with hemoglobin concentration.

Having now seen how the bicarbonate buffer system in blood responds to changes in P_{CO_2} and how this response is modified by changing the hemoglobin concentration, let us examine the response of blood to the addition of noncarbonic acids such as HCl, acetoacetic acid, and so on. We shall continue to analyze the situation in terms of the pH–bicarbonate diagram. The starting point will again be the normal state: pH = 7.4, $[HCO_3^-]$ = 24 meq/liter, and P_{CO_2} = 40 mmHg (5.33 kPa). As acid is

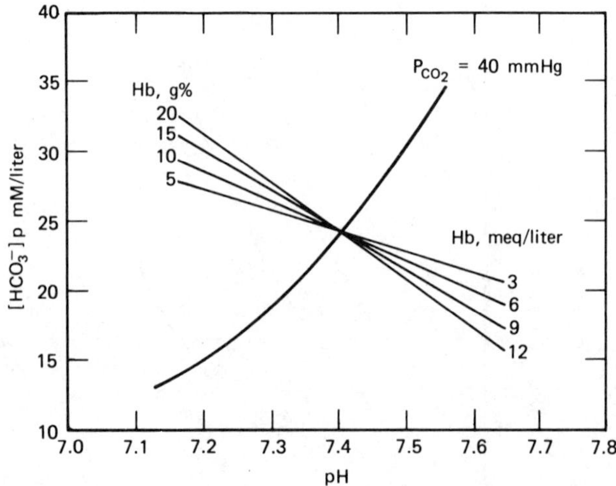

Figure 23.14
Slope of the buffering line of blood as it varies with hemoglobin concentration.
From Davenport, H. W. *The ABC of acid–base chemistry,* 6th ed. revised. Chicago, Ill.: University of Chicago Press, 1974, p. 55.

added, it will react with all the blood buffers, and the concentrations of all of the conjugate bases will decrease. Since the bicarbonate system is the major blood buffer, the decrease in [HCO_3^-] will be substantial. If P_{CO_2} is held constant at 40 mmHg (5.33 kPa) as a noncarbonic acid is added, the changes in the system can be represented by a point sliding down the 40 mmHg (5.33 kPa) isobar, as shown in Figure 23.15. If alkali is added to the blood, all the undissociated acids of the various buffer systems will participate in neutralizing it. Again, if this occurs at a fixed P_{CO_2} of 40 mmHg (5.33 kPa), the changes in the system will be represented by a point sliding up the 40 mmHg (5.33 kPa) isobar. Notice that, just as changes in P_{CO_2} were represented by points confined to the blood buffer line, changes due to the addition of acid or base at a fixed P_{CO_2} are represented by points confined to the CO_2 isobar.

The effects on blood of changing P_{CO_2} or of adding acid or alkali, as we have just described, are realistic qualitative models of what can happen in certain disease states. In the following sections we shall see how these changes occur in the body and how the body compensates for them.

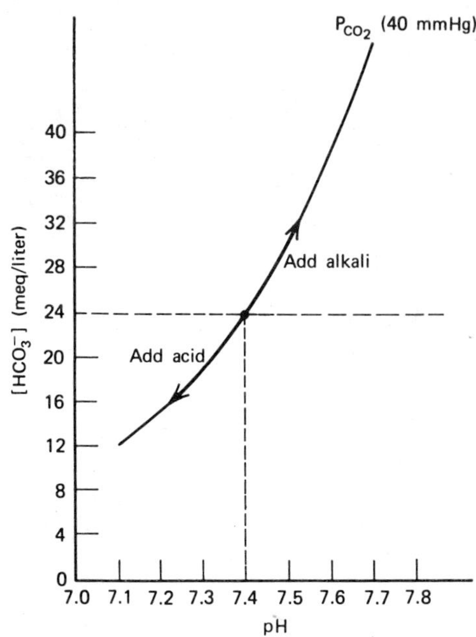

Figure 23.15
Effect of adding noncarbonic acid or alkali to whole blood with P_{CO_2} fixed at 40 mmHg.

23.11 ACID–BASE BALANCE AND ITS MAINTENANCE

It should come as no surprise that mechanisms exist whereby the body normally rids itself of excess acid or alkali. The physiological implication is that if a patient is in a state of continuing acidosis (excess acid or deficiency of alkali in the body) or alkalosis (excess alkali or deficiency of acid in the body), there must be a continuing cause of the imbalance. In such a situation the body's first task is to somehow compensate so plasma pH does not exceed the limits compatible with life. Assistance from the physician is sometimes necessary. The body's second task is to eliminate the primary cause of the imbalance, that is, to cure the disease, so that a normal acid–base status can be reestablished. Again, intervention by the physician may be needed.

All individuals, in sickness or in health, produce large amounts of acids every day. The major acid is CO_2. The amount of CO_2 produced depends on the individual's caloric expenditure; CO_2 production ranges from 12,500 meq/day to nearly 50,000 meq/day, and in an average young adult male, about 22,000 meq CO_2 are produced daily. This acid is volatile, and is normally excreted by the lungs. Inability of the lungs to perform this task adequately leads to respiratory acidosis or alkalosis. Respiratory acidosis is the result of hypoventilation of the alveoli, so that CO_2 accumulates in the body. Alveolar hypoventilation occurs when the depth or rate of respiration is diminished. Obstruction of the airway, neuromuscular disorders and diseases of the central nervous system are common causes of acute respiratory acidosis. Chronic respiratory acidosis is seen in patients with chronic obstructive lung disease, such as emphysema. Obviously, since the common element in all these conditions is increased alveolar P_{CO_2}, inhalation of a gas mixture with a high P_{CO_2} could also cause respiratory acidosis.

Respiratory alkalosis, on the other hand, arises from decreased alveolar P_{CO_2}. Hyperventilation due to anxiety is probably the most common cause. Central nervous system injury involving the respiratory center, salicylate poisoning, fever, and artificial ventilation are other causes. At high altitude, due to the decrease in total atmospheric pressure, alveolar P_{CO_2} also falls, producing chronic respiratory alkalosis.

Various amounts of nonvolatile acids are also produced by the body. The diet and the physiological state of the individual determine the kinds and amounts of these acids. Oxidation of sulfur-containing amino acids produces H^+ and SO_4^{2-}, the equivalent of sulfuric acid. Hydrolysis of phosphate esters is equivalent to the formation of phosphoric acid. The contribution of these processes depends on the amount of acid precursors ingested; for an individual consuming an average American diet there is a net daily acid production of about 60 meq.

Metabolism normally produces certain amounts of lactic acid, acetoacetic acid and β-hydroxybutyric acid. In some physiological or pathological states these are produced in excess, and accumulation of the excess causes acidosis. When an ammonium salt of a strong acid, such as ammonium chloride, or when arginine hydrochloride or lysine hydrochloride is administered, it is converted to urea, and the corresponding strong acid (hydrochloric acid in these examples) is synthesized. Ingestion of salicylates, methyl alcohol, or ethylene glycol results in production of strong organic acids. Accumulation of any of these nonvolatile acids leads to metabolic acidosis.

While it is obvious that excess acid production can cause acidosis, the same net effect can arise from abnormal loss of base, as could be predicted from the Henderson–Hasselbalch equation for the bicarbonate buffer system. Renal tubular acidosis is a condition in which this occurs. Abnormal amounts of HCO_3^- escape from the blood into the urine, leaving the body acidotic (Clin. Corr. 23.5). A more common cause of bicarbonate depletion is severe diarrhea. In this chapter it will be assumed that kidney function is normal.

CLIN. CORR. 23.5
THE ROLE OF BONE IN ACID–BASE HOMEOSTASIS

The average adult skeleton contains 50,000 meq of Ca^{2+} in the form of salts that are alkaline relative to the pH of plasma. In chronic acidosis this large reservoir of base is drawn upon to help control the plasma pH. Thus people with chronic kidney disease and severely impaired renal acid excretion do not experience a continuous decline in plasma pH and $[HCO_3^-]$. Rather, the pH and $[HCO_3^-]$ stabilize at some below-normal level. The resulting change in bone composition is not inconsequential, and clinical and roentgenologic evidence of rickets or osteomalacia often appear. Bone healing has been shown in these patients after prolonged administration of alkali in the form of sodium bicarbonate or citrate sufficient to restore plasma $[HCO_3^-]$.

Mammals do not synthesize alkaline compounds from neutral starting materials. Metabolic alkalosis therefore arises from intake of excess alkali or abnormal loss of acid. An alkali commonly taken by many people is sodium bicarbonate. A less obvious source of alkali is the salt of any metabolizable organic acid. Sodium lactate is often administered to combat acidosis; normal metabolism converts it to sodium bicarbonate. The net reaction is as follows:

$$Na^+ + CH_3CHOHCOO^- + 3O_2 \rightleftharpoons Na^+ + HCO_3^- + 2CO_2 + 2H_2O$$

Most fruits and vegetables have a net alkalinizing effect on the body for this reason. They contain a mixture of organic acids (which are metabolized to CO_2 and H_2O, and therefore have no long term effect on acid–base balance), and salts of organic acids, which give rise to bicarbonate. Abnormal loss of acid, as can occur with prolonged vomiting or gastric lavage, causes alkalosis. Alkalosis may also be produced by rapid loss of body water, as in diuresis, which may temporarily increase $[HCO_3^-]$ in the plasma and extracellular fluid.

Role of the Kidney

Excess nonvolatile acid and excess bicarbonate are excreted by the kidney. As a result the pH of the urine varies as a function of the body's need to excrete these materials. For an individual on a typical American diet urine pH is ~6, indicating a net acidification as compared to plasma. This is consistent with our knowledge that the typical diet results in a net production of acid. The pH of the urine can range from a lower limit of 4.4 up to 8.0.

A typical daily urine volume is about 1.2 liters. At the minimum urine pH of 4.4, $[H^+]$ is only 4×10^{-2} meq/liter, and it would take 1,250 liters of urine to excrete 50 meq of acid as free hydrogen ions. Clearly most of the acid we excrete must be in some form other than H^+. A form that can be excreted in a reasonable concentration, such as $H_2PO_4^-$ or NH_4^+, is needed.

Let us now see how the kidney accomplishes the excretion of acid or base. Figure 23.16 shows the fundamental functioning unit of the kidney, the nephron. Each human kidney contains at least a million of these. They serve first to filter the blood and then to modify the filtrate into urine.

Filtration occurs in the glomerulus, which consists of a tuft of capillaries enclosed by an epithelial envelope called the glomerular capsule (formerly Bowman's capsule). Water and low molecular weight solutes, such as inorganic ions, urea, sugars, and amino acids (but not normally substances with molecular weights above 70,000, such as plasma proteins), escape from these capillaries and collect in the capsular space. This ultrafiltrate of plasma then passes through the proximal convoluted tubule, where most of the water and solutes are reabsorbed. The tubule fluid continues through the loop of the nephron (loop of Henle) and through the distal convoluted tubule, where further reabsorption of some solutes or secretion of others occurs. The tubule fluid then passes into the collecting tubule, where concentration can occur if necessary. The fluid may now be called urine; it contains 1% or less of the water and solutes of the original glomerular filtrate.

The kidney regulates acid–base balance by controlling bicarbonate reabsorption and by secreting acid. Both of these processes depend on formation of H^+ and HCO_3^- from CO_2 and water within the tubule cells, shown in Figure 23.17A. The H^+ formed in this reaction is then actively secreted into the tubule fluid in exchange for a Na^+. Sodium uptake by the tubule cell is partly passive, with Na^+ flowing down the electrochemical

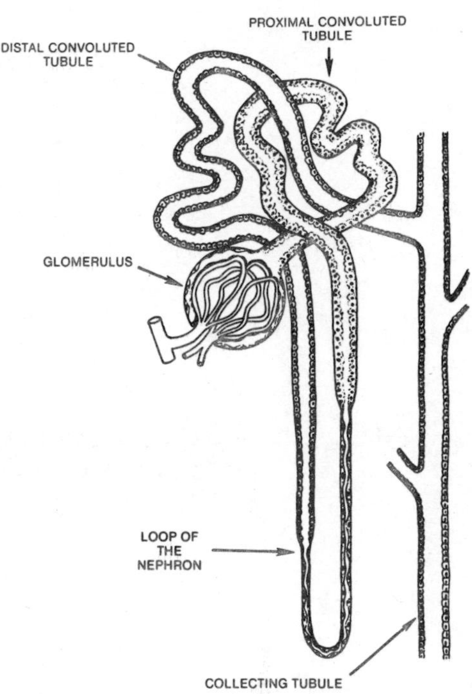

Figure 23.16
Diagram showing the essential features of a typical nephron in the human kidney.

Reprinted with permission from Smith, Homer W. *The physiology of the kidney*, London: Oxford University Press, 1937, p. 6.

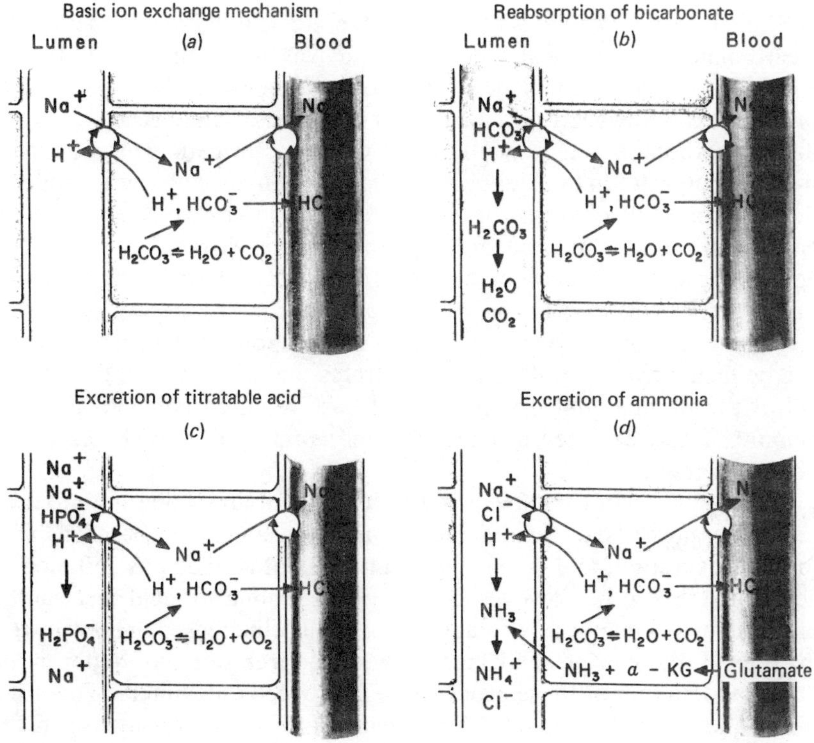

Figure 23.17
Role of the exchange of tubular cell H^+ ions in tubular fluid in renal regulation of acid–base balance.

(A) Basic ion exchange mechanism. (B) Reabsorption of bicarbonate. (C) Excretion of titratable acid. (D) Excretion of ammonia.

Adapted from Pitts, R. E., Role of ammonia production and excretion in regulation of acid–base balance, *New Engl. J. Med.*, 284:32, 1971, with permission of the publisher.

gradient, and partly active, via a Na^+, H^+-antiport system. At this point sodium has been reabsorbed in exchange for H^+, and sodium bicarbonate has been generated within the tubule cell. The sodium bicarbonate is then pumped out of the cell into the interstitial fluid, which equilibrates with the plasma.

The H^+ which has been secreted into the tubule fluid can now experience one of three fates. First, it can react with a HCO_3^-, as shown in Figure 23.17B, to form CO_2 and water. The overall net effect of this process is to move sodium bicarbonate from the tubule fluid back into the interstitial fluid. The name given to this is reabsorption of sodium bicarbonate.

As reabsorption of sodium bicarbonate proceeds, the tubule fluid becomes depleted of HCO_3^-, and the pH drops from its initial value, which was identical to the pH of the plasma from which it was derived. As HCO_3^- becomes less available and the pH comes closer to the pK of the $HPO_4^{2-}/H_2PO_4^-$ buffer system, more and more of the H^+ will be taken up by this buffer. Buffering is the second fate the H^+ can experience, and it is represented in Figure 23.17C. $H_2PO_4^-$ is not readily reabsorbed by the kidney. It passes out in the urine, and its loss represents net excretion of H^+.

Although phosphate is normally the most important buffer in the urine, other ions can become significant. For example, in diabetic ketoacidosis plasma levels of acetoacetate and β-hydroxybutyrate are elevated. They pass through the glomerular filter, and appear in the tubule fluid. Since acetoacetic acid has a pK = 3.6 and β-hydroxybutyric acid has a pK = 4.7, as the urine pH approaches its minimum of 4.4 these will begin to serve as buffers.

The effect of buffering is not only to excrete acid, but to regenerate the bicarbonate that was lost when the acid was first neutralized. Let us consider a situation in which the metabolic defect of a diabetic patient has produced β-hydroxybutyric acid. This acid immediately reacts with sodium bicarbonate and is neutralized, with the formation of sodium β-hydroxybutyrate. In the kidney, then, sodium β-hydroxybutyrate appears in the filtrate, is converted back to β-hydroxybutyric acid, which is excreted, and sodium bicarbonate is returned to the extracellular fluid. Net acid excretion and bicarbonate regeneration occur no matter what anion in the tubule fluid acts as the H^+ acceptor.

The amount of acid excreted as the acid component of a urinary buffer can be measured easily. One merely titrates the urine back to the normal pH of the plasma, 7.4. The amount of base required is identical to the amount of acid excreted in this form, and is referred to as the *titratable acidity* of the urine.

The formation of titratable acidity accounts for about one-third to one-half of our normal daily acid excretion. It is thus an important mechanism for acid excretion, and is capable of putting out as much as 250 meq of acid daily. There is, however, a limit to the amount of acid that can be excreted in this manner. Titratable acidity can be increased only by lowering the pH of the urine or by increasing the concentration of buffer in the urine, and neither of these processes can proceed indefinitely. The urine pH cannot go below about 4.4; evidently the sodium-for-hydrogen exchange mechanism is incapable of pumping H^+ out of the tubule cells against more than a thousandfold concentration gradient. Buffer excretion is limited not only by the solubility of the buffer, but by limitations to the supply of the buffer ion and of the cations that are necessarily part of the important buffer systems. If, for example, a 600 meq/day acid load were excreted as NaH_2PO_4, the body would be totally depleted of sodium in less than a week.

The third fate the H^+ can experience in the tubule fluid is neutralization by NH_3. The tubule cells can deamidate glutamine, forming glutamate and NH_3, as shown in Figure 23.17D. At a normal intracellular pH about 1% of the ammonia in the cell will be in the uncharged form. This form diffuses rapidly through the cell membrane and appears in the tubule lumen, where it then reacts with the H^+, and forms NH_4^+. NH_4^+ cannot easily diffuse through the cell membrane. Therefore, since the tubule fluid is more acid than the intracellular fluid, extraction of NH_3 from the cell into the tubule fluid occurs. Elimination of NH_4^+ in the urine contributes to net acid excretion.

NH_4^+ is normally a major urinary acid. Typically, one-half to two-thirds of our daily acid load is excreted as NH_4^+. For three reasons it becomes even more important in acidosis. In the first place, since the pK of NH_4^+ is 9.3, acid can be excreted in this form without lowering the pH of the urine, whereas formation of titratable acidity requires a decrease in urine pH. Second, enormous amounts of acid can be excreted in this form. Ammonia is readily available from amino acids, and in prolonged acidosis the NH_4^+ excretion system becomes activated. This activation, however, takes several days; it does not begin to adapt until after 2–3 days, and the process is not complete until 5–6 days after the onset of acidosis. Once complete, though, amounts of acid in excess of 500 meq can be excreted daily as NH_4^+. The third role of NH_4^+ in acidosis is that it spares the body's stores of Na^+ and K^+. Excretion of titratable acid, such as $H_2PO_4^-$, and of the anions of strong acids, such as acetoacetate, requires simultaneous excretion of a cation to maintain electrical neutrality. At the onset of acidosis this role is filled by Na^+, and as the body's Na^+ stores become depleted, K^+ excretion rises. If NH_4^+ did not then become available even a moderate acidosis could quickly become fatal.

Total acid excretion, the *total acidity* of the urine, is the sum of the titratable acidity and NH_4^+. Strictly speaking, one should subtract from this sum the urinary HCO_3^-, but this correction is seldom made in practice. Obviously, in severe metabolic acidosis, where the total acid excretion would be of greatest interest, the urine would be so acidic that HCO_3^- would be nil.

In alkalosis the role of the kidney is simply to allow HCO_3^- to escape. Metabolic alkalosis is therefore seldom long-lasting unless alkali is continuously administered or HCO_3^- elimination is somehow prevented. HCO_3^- elimination may be restricted if the kidney receives a strong signal to conserve Na^+ at a time when there is a deficiency of an easily reabsorbable anion, such as Cl^-, to be reabsorbed with it. Some diuretics cause this. The first renal response is to put out K^+ in exchange for Na^+ from the tubule fluid, and when K^+ stores are depleted, H^+ is exchanged for Na^+. This results in the production of an acidic urine by an alkalotic patient. If NaCl is administered, alkalosis associated with volume and Cl^- depletion may correct itself.

23.12 COMPENSATORY MECHANISMS

We have defined four primary types of acid–base imbalances and we have seen their chemical causes. Respiratory acidosis arises from an increased plasma P_{CO_2}. Respiratory alkalosis is caused by a decreased plasma P_{CO_2}. In metabolic acidosis addition of strong organic or inorganic acid (or loss of HCO_3^-) results in a decreased plasma $[HCO_3^-]$. Conversely, in metabolic alkalosis loss of acid from the body or ingestion of alkali raises the plasma $[HCO_3^-]$. Recall that in an acute respiratory acid–base im-

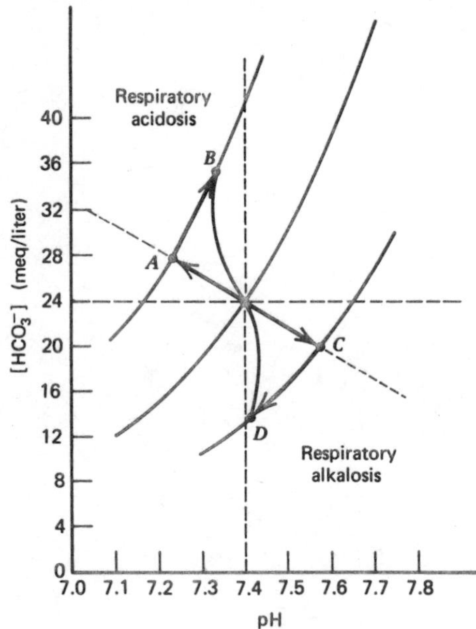

Figure 23.18
pH–Bicarbonate diagram showing compensation for respiratory acidosis (normal state to point B) and for respiratory alkalosis (normal state to point D).

balance, as long as the body has not attempted to compensate, the pH will be abnormal, and the $[HCO_3^-]$ will be somewhere on the buffer line. In an acute metabolic acid–base imbalance, if the patient has made no attempt to compensate, the pH will be abnormal and the $[HCO_3^-]$ will be somewhere on the 40mm isobar.

When the plasma pH deviates from the normal range, various compensatory mechanisms begin to operate. The general principle of compensation is that, since an abnormal condition has directly altered one of the terms of the $[HCO_3^-]/[CO_2]$ ratio, the plasma pH can be readjusted back toward normal by a compensatory alteration of the other term. For example, if a diabetic patient becomes acidotic due to excess production of ketone bodies, plasma $[HCO_3^-]$ will decrease. Compensation would involve decreasing the plasma $[CO_2]$ so that the $[HCO_3^-]/[CO_2]$ ratio, and therefore the pH, is readjusted back toward normal. Notice that compensation does not involve a return of $[HCO_3^-]$ and $[CO_2]$ toward normal. Rather, compensation is a secondary alteration in one of these, an alteration that has the effect of counteracting the primary alteration in the other. The result is that the plasma pH is readjusted toward normal. That this is necessarily so is evident from the Henderson–Hasselbalch equation.

$$\text{pH} = 6.1 + \log \frac{[HCO_3^-]}{0.03\ P_{CO_2}}$$

If $[HCO_3^-]$ changes, the only way to restore the original $[HCO_3^-]/[CO_2]$ ratio is to change P_{CO_2} in the *same direction*. If the primary change is in P_{CO_2}, the original ratio can be restored only by altering $[HCO_3^-]$ in the *same direction*.

Although some compensatory mechanisms begin to operate rapidly and produce their effects rapidly, others are slower. Several stages of compensation may therefore be seen. First is the acute stage, before any significant degree of compensation could possibly occur. After the acid–base imbalance has been in effect for a period of time the patient may become *compensated*. This means the compensatory mechanisms have come into play in a normal manner, as expected on the basis of experience with other individuals with an acid–base imbalance of similar type and degree. The "compensated state" does not necessarily imply that the plasma pH is within the normal range. Alternatively, the patient may show no sign of compensation, even though compensation is expected. This state is referred to as *uncompensated;* it arises because compensation cannot occur due to some other abnormality. Finally, there is an intermediate state where compensation is occurring but is not yet as complete as it should be. This is the *partially compensated* state. Factors which limit the compensatory processes will be discussed at the end of this section.

Let us now follow the course of acute onset of each type of acid–base imbalance and of the compensatory process. Each of these will be schematically illustrated in a pH–bicarbonate diagram.

Imagine an individual in normal acid–base balance who goes into an acute respiratory acidosis as a result of breathing a gas mixture containing a high level of CO_2. As P_{CO_2} rises, plasma pH will drop and $[HCO_3^-]$ will rise. (If a decrease in pH and a simultaneous rise in $[HCO_3^-]$ suddenly seems anomalous, turn back to Figure 23.13 and the text on page 894, and review the blood buffer line.) The point describing his or her condition will follow the buffer line to point A, as shown in Figure 23.18. Eventually a new steady-state P_{CO_2} will be established in the alveoli and in the blood, and no further change in P_{CO_2} will occur. The abnormal condition has

CLIN. CORR. **23.6**
ACUTE RESPIRATORY ALKALOSIS

An anesthetized surgical patient with a uretheral catheter in place was hyperventilated as an adjunct to the general anesthesia. Prior to hyperventilation normal values of plasma P_{CO_2} and pH were obtained. Alveolar ventilation was then increased mechanically, and a new steady state was reached, in which the plasma P_{CO_2} was 25 mmHg, and the pH was 7.55. Plasma HCO_3^- was not directly measured, but interpolation from a pH–bicarbonate diagram (e.g., Figure 23.12) or calculation from the Henderson–Hasselbalch equation reveals that the plasma $[HCO_3^-]$

fixed this patient on an abnormally high CO_2 isobar. If the condition is returned to normal, he can drop back to the 40 mmHg (5.33kPa) isobar and all will be well, but until that time all compensatory processes are confined to the higher CO_2 isobar. Compensation will, of course, consist of renal excretion of H^+. Since this is a bicarbonate-producing process, $[HCO_3^-]$ should rise, even though it is already above normal. This could have been predicted from the pH–HCO_3^- diagram with no knowledge of the renal mechanism of compensation. Since it is assumed that the individual is fixed on the high CO_2 isobar by the abnormal condition, the only way the pH can possibly be adjusted toward normal is by sliding up the isobar to point *B* in Figure 23.16. This movement is necessarily linked to an increase in $[HCO_3^-]$. Thus the correct analysis of this compensation could be made either from an understanding of the nature of the compensatory mechanism or from an appreciation of the physical chemistry of the bicarbonate buffer system as expressed in the pH–HCO_3^- diagram.

Although the path we have described, up the buffer line to point *A* and then up the isobar to point *B*, is a real possibility, it is likely that a respiratory acidosis would develop gradually, with compensation occurring simultaneously. The points describing this progress would fall on the curved line from the normal state to point *B*.

In sudden onset respiratory alkalosis P_{CO_2} drops rapidly. The pH rises and $[HCO_3^-]$ falls, following the buffer line to point *C* in Figure 23.18. Clinical Correlation 23.6 describes a case of acute respiratory alkalosis. As with respiratory acidosis, unless the cause of the decreased alveolar P_{CO_2} is removed, the patient is fixed on an abnormal CO_2 isobar. Compensation consists of renal excretion of HCO_3^-; plasma $[HCO_3^-]$ diminishes (at a fixed, subnormal P_{CO_2}), and the plasma pH decreases toward normal. This is described in Figure 23.18 by movement along the isobar from point *C* to point *D*. With a gradual onset of respiratory alkalosis, the bicarbonate buffer system would follow points along the curved line from the normal state to point *D*.

In metabolic acidosis two mechanisms are usually available for dealing with the excess acid. The kidneys increase their H^+ excretion, but this takes time, and is not adequate to return $[HCO_3^-]$ and the pH to normal. The other mechanism, which begins to operate almost instantly, is respiratory compensation. The acidosis stimulates the respiratory system to hyperventilate, decreasing the P_{CO_2}. Thus, if onset of a primary metabolic acidosis is represented in Figure 23.19 by a fall in plasma $[HCO_3^-]$ along the 40 mmHg (5.33 kPa) isobar from the normal state to point *E*, the compensatory decrease in P_{CO_2}, and the concomitant rise in pH will be along the line from *E* to *F*. Notice that this line is parallel to the buffer line, and so compensation for a metabolic acidosis involves not only the expected decrease in P_{CO_2}, but also a *further* small decrease in $[HCO_3^-]$. This is due to the same factor that causes the buffer line itself to have a slope: titration of the nonbicarbonate buffers. The inevitability and the magnitude of the further decrease in $[HCO_3^-]$ can be seen clearly in the pH–bicarbonate diagram.

The principles governing compensation for metabolic alkalosis are like those for metabolic acidosis, but everything is operating in the opposite direction. In metabolic alkalosis the primary defect is an increase in plasma $[HCO_3^-]$; it rises from the normal state to point *G* in Figure 23.19. The immediate physiological response is hypoventilation, followed by increased renal excretion of HCO_3^-. As a result of the hypoventilation the P_{CO_2} increases along the line from *G* to *H*, and a further small rise in $[HCO_3^-]$ occurs.

The respiratory response to a metabolic acid–base imbalance is rapid; an acute metabolic imbalance will not generally be seen outside the exper-

decreased to 21.2 meq/liter. Analysis of the urine showed negligible loss of HCO_3^- through the kidneys. It can be concluded that the decrease in $[HCO_3^-]$ was due to titration of bicarbonate by the acid components of the body's buffer systems. The point representing the patient's new steady-state condition clearly must be on the buffering line that represents whole-body buffering. (Since the buffers of the whole body are not identical in type or concentration to the blood buffers, the buffer line for the whole body will be analogous, but not identical, to the blood buffer line.)

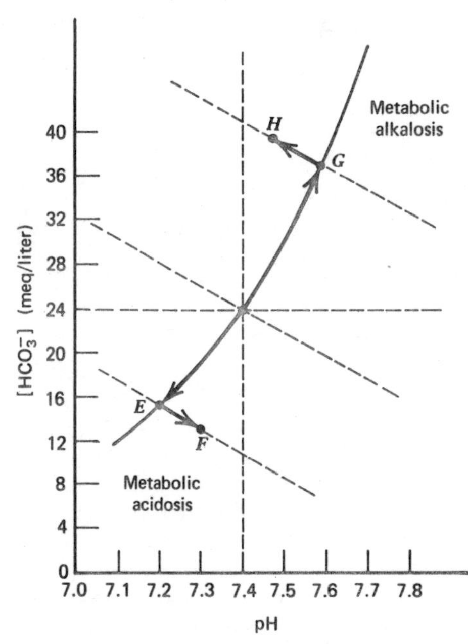

Figure 23.19
pH–Bicarbonate diagram showing compensation for metabolic acidosis (normal state to point F) and for metabolic alkalosis (normal state to point H).

CLIN. CORR. 23.7
CHRONIC RESPIRATORY ACIDOSIS

H.W. was admitted to the hospital with marked dyspnea, cyanosis, and signs of mental confusion. As his acute problems were relieved by appropriate treatment, his symptoms disappeared except for a continuing dyspnea. Blood gas analysis performed 8 days later yielded the following data: pH, 7.32; P_{CO_2}, 70 mmHg; $[HCO_3^-]$, 34.9 meq/liter. This is a typical compensation for this degree of chronic respiratory acidosis.

Another patient, C.Q., with chronic obstructive lung disease was found to have arterial plasma pH, 7.40; $[HCO_3^-]$, 35.9 meq/liter; and P_{CO_2}, 60 mmHg. A return of the plasma pH to 7.4 is not expected in the average patient with respiratory acidosis. Some patients are capable of attaining it, although it becomes less and less likely as P_{CO_2} increases. In this case, with a P_{CO_2} of 60 mmHg, a plasma pH of 7.4 lies outside the 95% probability range. Close questioning of the patient revealed that he had surreptitiously been taking a relative's thiazide diuretic, which superimposed a metabolic alkalosis upon respiratory acidosis.

imental laboratory. Indeed, if a physician sees a patient whose plasma pH, $[HCO_3^-]$, and P_{CO_2} are consistent with an acute metabolic imbalance, he concludes that the patient's compensatory mechanisms are impaired and that the patient cannot compensate. The patient would be suffering a mixed respiratory and metabolic acidosis or a mixed respiratory and metabolic alkalosis. Obviously, if a patient had a primary acidosis of one type (respiratory or metabolic) and a primary alkalosis of the other, both caused by independent diseases, the effects of the two on plasma pH would tend to cancel. But even if the pH were within the normal range due to such a circumstance, $[HCO_3^-]$ and P_{CO_2} would be abnormal.

How complete can the process of compensation be? Can the body totally compensate (bring the pH back to the normal range) for any imbalance? Generally, the answer is no. The organs used in compensation, the lungs and the kidneys, do not exist exclusively to deal with acid–base imbalance. There is a limit to how much one can hyperventilate; it is simply impossible to move air into and out of the lungs at an indefinitely high rate for an indefinitely long time. Also, one cannot suspend respiration merely to raise P_{CO_2} to some desired level. The kidney, too, has limits. As the P_{CO_2} rises above 70 mmHg (9.33 kpa) in respiratory acidosis, renal mechanisms for reabsorbing HCO_3^- fail to keep pace, and further increases in plasma $[HCO_3^-]$ are only about what could be expected from titration of the nonbicarbonate buffers (Clin. Corr. 23.7). In respiratory alkalosis renal excretion of excess HCO_3^- can, with time, be sufficient to return the plasma pH to within the normal range. Individuals who dwell at high altitude are typically in compensated respiratory alkalosis with their plasma pH within the normal range. For the other types of acid–base imbalance the exact degree of compensation expected of a patient with a given clinical picture is well worked out, but a detailed discussion is beyond the scope of this chapter. Suffice it to say that if a patient is compensating, but not as well as expected, this is taken to mean that the patient cannot compensate appropriately and must therefore have a mixed acid–base disturbance.

23.13 ALTERNATIVE MEASURES OF ACID–BASE IMBALANCE

Modern clinical laboratories generally report plasma bicarbonate concentration, and the value is used by the physician just as we have used it here. Some laboratories however, report total plasma CO_2. Total plasma CO_2 as reported by the clinical laboratory is the sum of bicarbonate and dissolved CO_2, and so is always slightly higher than $[HCO_3^-]$. At pH 7.4, for example, the ratio of $[HCO_3^-]$ to $[CO_2]$ is 20 : 1 (dissolved CO_2 is only 1/21 of the total CO_2); if $[HCO_3^-]$ is 24 meq/liter, $[CO_2]$ is 1.2 meq/liter and total CO_2 is 25.2 meq/liter. At pH 7.1, HCO_3^- is still 10 times as concentrated as dissolved CO_2. Because the major contributor to total CO_2 is HCO_3^-, total CO_2 is often used in the same manner as bicarbonate to make clinical judgments. Strictly speaking, total CO_2 also includes carbamino groups, but current clinical laboratory practice is to ignore carbamino groups when making a blood gas and pH report. If, however, they were included in a total CO_2 measurement it would not change the interpretation of the measurement, since carbamino groups, like dissolved CO_2, represent only a small fraction of the total CO_2.

The clinical importance of bicarbonate as a gauge of the whole body's ability to buffer further loads of metabolic acid (see Clin. Corr. 23.8) has given rise to several ways of expressing what the $[HCO_3^-]$ would be if

CLIN. CORR. 23.8
SALICYLATE POISONING

Salicylates are the most common cause of poisoning in children. A typical pathway of salicylate intoxication is plotted in the accompanying figure. The first effect of salicylate overdose is stimulation of the respiratory center, resulting in respiratory alkalosis. Renal compensation occurs, lowering the plasma [HCO_3^-]. A second, delayed effect of salicylate may then appear, metabolic acidosis. Since [HCO_3^-] had been lowered by the previous compensatory process the victim is at a particular disadvantage in dealing with the metabolic acidosis. In addition, but not shown in the graph, respiratory stimulation sometimes persists after the acidosis has run its course. Rational management of salicylate intoxication requires knowledge of the plasma pH and the plasma [HCO_3^-] or its equivalent throughout the course of the condition.

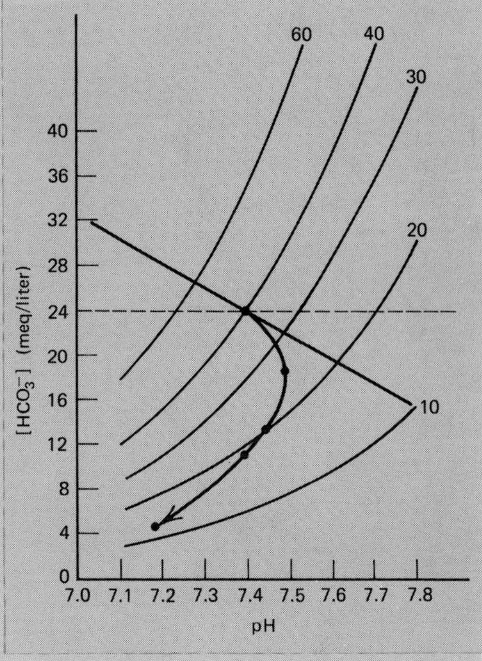

Data replotted from Singer, R. B. The acid–base disturbance in salicylate intoxication. *Medicine*, 33:1–13, 1954.

there were no respiratory component or respiratory compensation involved in a patient's condition. The *base excess* is one of the more common of these expressions. It is defined as the amount of acid that would have to be added to the blood to titrate it to pH 7.4 at a P_{CO_2} of 40 mmHg (5.33 kPa) at 37°C. Since the titration is carried out at the normal P_{CO_2}, only the metabolic contribution to acid–base imbalance (primary metabolic imbalance *and* nonrespiratory compensatory processes) would be measured. If a blood sample were acidic under the conditions of the titration, alkali would have to be added instead of acid, and the base excess would be negative.

The concept and the quantitation of base excess are most easily understood if we refer to the pH–bicarbonate diagram. In our discussion of the blood buffer line we saw how increasing the P_{CO_2} in blood, where other buffers are present, would result in a rise in [HCO_3^-] and a virtually identical decrease in the concentration of other buffer bases. This was because equivalent amounts of the other buffer bases were consumed as they buffered carbonic acid. Since virtually all the carbonic acid formed was buffered, for every HCO_3^- formed one conjugate base of some other system was consumed. In this situation the *total* base in the blood is not measurably changed; only the distribution of HCO_3^- and nonbicarbonate buffer conjugate base is changed. Thus, as long as one remains on the blood buffer line, [HCO_3^-] can change but total base will not. There will be no positive or negative base excess.

If, however, renal activity, diet or some metabolic process adds or removes HCO_3^-, then a positive or negative base excess will be seen. The patient's status will no longer be described by a point on the buffer line, and the base excess will be the difference between the observed plasma [HCO_3^-] and the [HCO_3^-] on the buffer line at the same pH. This is shown in Figure 23.20. In order to calculate this difference, the position of the buffer line (which can be determined from knowledge of the slope and the

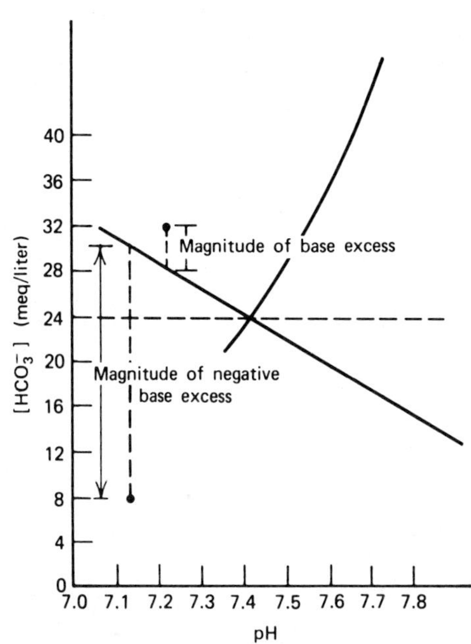

Figure 23.20
Calculation of base excess for a point above the blood buffer line, and calculation of negative base excess for a point below the blood buffer line.
The base excess is 32–28 = 4 meq/liter. The negative base excess is 30–8 = 22 meq/liter.

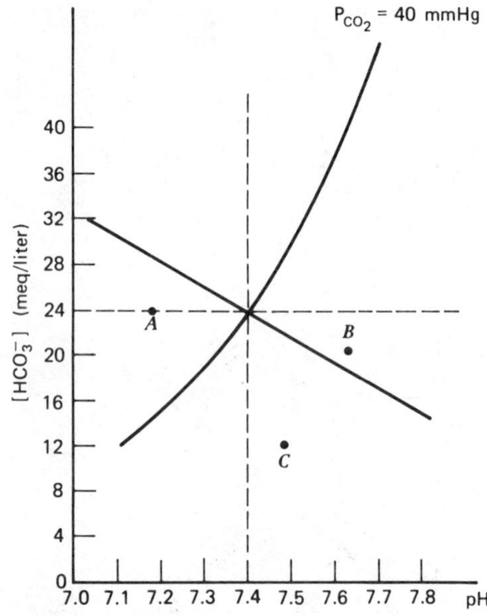

Figure 23.21
Examples showing the sign of the base excess at various points.
At points A and C there is a negative base excess. At point B the base excess is positive.

point representing the normal state) must be known. In the clinical laboratory it can be estimated by measuring the hemoglobin concentration and assuming that it is the major nonbicarbonate buffer.

The buffer line, then, is the dividing line between positive and negative base excess. Any point above it is in the region of positive base excess, and any point below it is in the region of negative base excess. This gives rise to situations that may seem peculiar at first. In Figure 23.21 the [HCO_3^-] at point *A* is normal, but the patient has a negative base excess. A positive or negative base excess occurs as a result of *compensation* for a respiratory acid–base imbalance or *directly* from a metabolic one. Respiratory compensation for a metabolic acid–base imbalance, since it involves movement along a line parallel to the buffer line (Figure 23.19), would cause no further change in the value of the base excess. Clin. Corr. 23.9 involves consideration of base excess.

23.14 THE SIGNIFICANCE OF Na^+ AND Cl^- IN ACID–BASE IMBALANCE

An important concept in diagnosing certain acid–base disorders is the so-called anion gap. Most clinical laboratories routinely measure plasma Na^+, K^+, Cl^-, and HCO_3^-. A glance back at the graph in Figure 23.10 will confirm that in the plasma of a normal individual the sum of Na^+ and K^+ is greater than the sum of Cl^- and HCO_3^-. This difference is called *A*, the anion gap; it represents the other plasma anions (Figure 23.10), which are not routinely measured. It is calculated as follows:

$$A = (Na^+ + K^+) - (Cl^- + HCO_3^-)$$

The normal value of *A* is in the range of 12–16 meq/liter. In some clinical laboratories K^+ is not measured; then the normal value is 8–12 meq/liter. The gap is changed only by conditions that change the sum of the cations or the sum of the anions, or by conditions that change both sums by different amounts. Thus administration or depletion of sodium bicarbonate would not change the anion gap because [Na^+] and [HCO_3^-] would be affected equally. Metabolic acidosis due to HCl or NH_4Cl administration would also leave the anion gap unaffected; here [HCO_3^-] would decrease, but [Cl^-] would increase by an equivalent amount, and the sum of [HCO_3^-] plus [Cl^-] would be unchanged. In contrast, diabetic ketoacidosis or methanol poisoning involves production of strong organic acids which react with HCO_3^-, decreasing its concentration. But since the [HCO_3^-] is replaced by some organic anion, the sum of [HCO_3^-] plus [Cl^-] decreases, and the anion gap increases.

The anion gap is most commonly used to establish a differential diagnosis for metabolic acidosis. In a metabolic acidosis with an increased anion gap, H^+ must have been added to the body with some anion other than chloride. Metabolic acidosis without an increased anion gap must be due either to accumulation of H^+ with chloride or to a decrease in the concentration of sodium bicarbonate. Thus, on the basis of the anion gap, certain diseases can be ruled out, while certain others would have to be considered. This information can be especially important in dealing with patients who cannot give good histories.

The electrolytes of the body fluids interact with each other in a multitude of ways. One of the most important of these involves the capacity for K^+ and H^+ to substitute for one another under certain circumstances. This can occur in the cell, where as we have seen, K^+ is the major cation.

CLIN. CORR. 23.9 EVALUATION OF CLINICAL ACID–BASE DATA

In a 1972 study of total parenteral nutrition of infants, it was found that infants who received amino acids in the form of a hydrolyzate of the protein fibrin maintained normal acid–base balance. In contrast, infants receiving two different mixtures of synthetic amino acids, FreAmine and Neoaminosol, became acidotic. Both synthetic mixtures contained adequate amounts of all the essential amino acids, but neither contained aspartate or glutamate. The fibrin hydrolyzate contained all of the common amino acids.

The accompanying figure shows the blood acid–base data from these infants. Notice that the normal values for infants, given by the dotted lines, are not quite the same as normal values for adults. (A child is not a small adult.)

The blood pH data show that the infants receiving synthetic mixtures were clearly acidotic. The low $[HCO_3^-]$ of the Neoaminosol group immediately suggests a metabolic acidosis, and the P_{CO_2} and base excess data are compatible with this interpretation. The FreAmine group, however, shows nearly normal $[HCO_3^-]$, and all of these infants have elevated P_{CO_2}'s. The P_{CO_2}'s indicate respiratory acidosis, but a simple respiratory acidosis should be associated with a slightly elevated $[HCO_3^-]$. The absence of this finding in most of the infants indicates that the acidosis must also have a metabolic component. This is confirmed by the observation that all the infants receiving FreAmine have a significant negative base excess.

The infants with mixed acid–base disturbances did, in fact, have pneumonia or respiratory distress syndrome. The metabolic acidosis, which all the infants receiving synthetic mixtures experienced, was due to synthesis of aspartic acid and glutamic acid from a neutral starting material (presumably glucose). Subsequent incorporation of these acids into body protein imposed a net acid load upon the body. Addition of aspartate and/or glutamate to the synthetic mixtures was proposed as a solution of the problem.

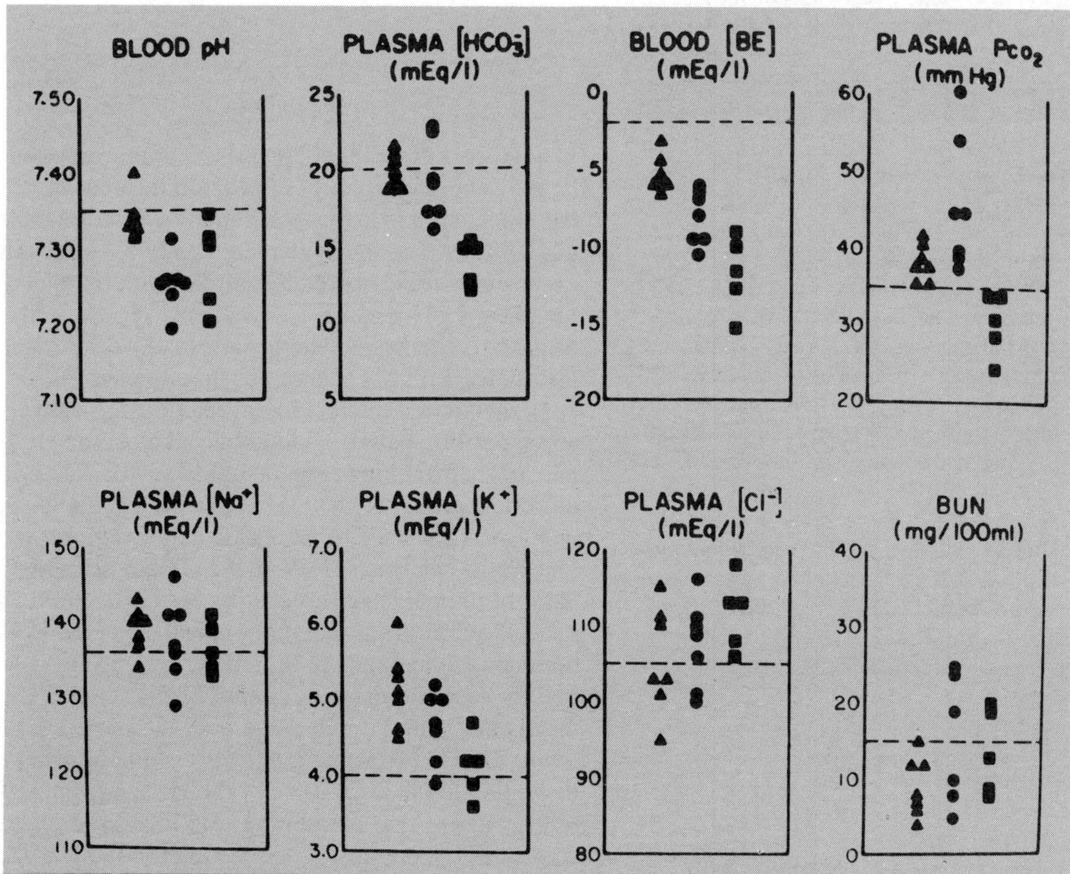

Blood acid–base data of patients receiving fibrin hydrolysate (▲) and of those receiving synthetic L-amino acid mixtures, FreAmine (●) and Neoaminosal (■). *Values are those observed at the time of the lowest blood base excess. Dashed lines represent accepted normal values for infants.*

Adapted from Heird, W. C. *New Engl. J. Med.*, 287:943, 1972.

CLIN. CORR. **23.10**
METABOLIC ALKALOSIS

Prolonged gastric lavage produces a metabolic alkalosis which is a good experimental model of the metabolic alkalosis that results from repeated vomiting. The following table gives plasma and urine acid–base and electrolyte data from a healthy volunteer on a low sodium diet who, after a control period, was subjected to gastric lavage for two days. After a 5-day recovery period he was placed on a low-potassium diet and given a sodium (130 meq/day) and chloride (121 meq/day) supplement.

During the control period the data are within normal limits. After gastric lavage which selectively removed HCl (Na^+, K^+, and water lost with the gastric juice were restored), an uncomplicated metabolic alkalosis developed. Notice that the subject excreted an alkaline urine, containing a substantial amount of HCO_3^-. Na^+ excretion increased, depleting the body's Na^+ stores. Plasma P_{CO_2} was not measured, but plotting the values of pH and HCO_3^- on a pH–bicarbonate diagram (e.g., Figure 23.12) allows one to interpolate a value of about 47 mmHg. Clearly, respiratory compensation was occurring. Plasma [K^+] was decreased. Plasma [Cl^-] decreased, but no more than would be expected on the basis of the changes in [Na^+], [K^+], and [HCO_3^-].

When the subject was placed on a low-potassium diet the alkalosis grew worse, and plasma [HCO_3^-] rose. Additional compensatory hypoventilation evidently prevented a further rise in plasma pH. Notice, though, that the urine became acid, in spite of the increased severity of the alkalosis. Na^+ was conserved, not in exchange for K^+, but in exchange for H^+. After several days of Na^+ and Cl^- administration, however, the subject was able to restore the depleted Cl^-, excrete the excess HCO_3^- and repair the acid–base imbalance with no other treatment.

	Control	*After Lavage*	*Low KCl*	*After NaCl*
PLASMA				
pH	7.4	7.50	7.48	7.41
HCO_3^- (meq/liter)	29.3	35.3	38.1	26.1
Na^+ (meq/liter)	138	134	141	144
K^+ (meq/liter)	4.2	3.2	2.9	3.2
Cl^- (meq/liter)	101	88	85	108
URINE				
pH	6.12	7.48	5.70	7.19
HCO_3^- (meq/day)	3	51	1	17
NH_4^+ (meq/day)	22	4	36	14
Titratable acidity (meq/day)	10	0	14	1
Total acidity (meq/day)	29	−47	49	−2
Na^+ (meq/day)	2	28	1	95

SOURCE: Data from Kassirer, J. P., and Schwartz, W. B. *Am. J. Med.*, 40:10, 1966.

In acidosis intracellular [H^+] rises, and it replaces some of the intracellular K^+. The displaced K^+ appears in the plasma, and in time is excreted by the kidneys. This leaves the patient with normal plasma [K^+] (normokalemia), but with seriously depleted body K^+ stores (hypokalia). Subsequent excessively rapid correction of the acidosis may then reverse events. As the plasma pH rises, K^+ flows back into the cells, and plasma [K^+] may decline to the point where muscular weakness sets in, and respiratory insufficiency may become life-threatening.

In the kidney the reciprocal relationship between K^+ and H^+ results in an association between metabolic alkalosis and hypokalemia. If hypokalemia arises from long-term insufficiency of dietary potassium or long-term diuretic therapy, K^+ levels in the cells will diminish, and intracellular [H^+] will rise. This leads to increased acid excretion, acidic urine and an alkaline arterial plasma pH. We have already seen how in an alkalotic individual a hormonal signal to absorb Na^+ can lead to K^+ loss and then to an exacerbation of the metabolic alkalosis (page 899). The association also operates in the opposite direction, with alkalosis leading to hypokalemia. In this case increased amounts of $Na^+ + HCO_3^-$ are presented to the distal convoluted tubule, where all K^+ secretion normally takes place (all *filtered* K^+ is reabsorbed; K^+ loss is due to distal tubular secretion). The distal tubule takes up some of the Na^+, but since HCO_3^- does not readily follow across that membrane, the increased Na^+ uptake is linked to increased K^+ secretion. K^+ excretion is complicated, with its control under the influence of a variety of hormones and other factors. The end result, however, is that metabolic alkalosis and hypokalemia go hand in hand, so much so that in some circles the term, "hypokalemic alkalosis," is used synonymously with metabolic alkalosis. Clinical Correlation 23.10 discusses a case of experimental metabolic alkalosis in which this occurred.

BIBLIOGRAPHY

Gas Transport

Bunn, H. F., Gabbay, K. H., and Gallop, P. M. The glycosylation of hemoglobin: relevance to diabetes mellitus. *Science* 200:21, 1978.

Kilmartin, J. V. Interaction of haemoglobin with protons, CO_2 and 2,3-diphosphoglycerate. *Br. Med. Bull.* 32:209, 1976.

Perutz, M. F., and Lehmann, H. Molecular pathology of human haemoglobin. *Nature* 219:902, 1968.

pH Regulation

Davenport, H. W. *The ABC of acid–base chemistry,* 6th ed. Chicago: University of Chicago Press, 1974.

Gabow, P. A., Kaehny, W. D., Fennessey, P. V., et al. Diagnostic importance of an increased serum anion gap. *N. Engl. J. Med.* 303:854, 1980.

Gamble, J. L., Jr., and Bettice, J. A. Acid–base relationships in the different body compartments: the basis for a simplified diagnostic approach. *Johns Hopkins Med. J.* 140:213, 1977.

Masoro, E. J., and Siegel, P. D. *Acid–base regulation: its physiology, pathophysiology and the interpretation of blood–gas analysis,* 2nd ed. Philadelphia: W. B. Saunders, 1977.

Siggaard-Andersen, O. *The acid–base status of the blood,* 4th ed. Baltimore: Williams & Wilkins, 1974.

QUESTIONS

J. BAGGOTT AND C. N. ANGSTADT

*Q*uestion *T*ypes are described inside the front cover.

1. (QT1) During a breathing cycle:
   A. the alveolar gases are completely exchanged for atmospheric gases.
   B. gas exchange between the alveoli and the capillary blood can occur at all times.
   C. gas exchange with the capillary blood occurs at the surface of all the airways.
   D. there is net uptake of nitrogen by the blood.
   E. atmospheric water vapor is taken up by the lungs.
2. (QT1) From an oxygen saturation curve for normal blood we can determine that:
   A. P_{50} is in the Po_2 range found in extrapulmonary tissues.
   B. oxygen binding is hyperbolic.
   C. an oxygen carrier is necessary.
   D. tighter oxygen binding occurs at lower Po_2.
   E. shifts of the curve to the left or right would have little effect on oxygen delivery.

A. α-Chains of hemoglobin   C. Both
B. β-Chains of hemoglobin   D. Neither

3. (QT4) Found in HbA, HbA_2, and HbF.
4. (QT4) Modified in the formation of HbA_{1c}.
5. (QT2) At a Po_2 of 30 mmHg hemoglobin's percent saturation will:
   1. increase with increasing temperature.
   2. decrease with decreasing pH.
   3. increase with increasing Pco_2.
   4. decrease with increasing 2,3-diphosphoglycerate concentration.
6. (QT2) Significant contributor(s) to the total carbon dioxide of whole blood include(s):
   1. bicarbonate ion.
   2. dissolved carbon dioxide (CO_2).
   3. carbamino hemoglobin.
   4. carbonic acid (H_2CO_3).
7. (QT1) 2,3-Diphosphoglycerate (DPG):
   A. is absent from the normal erythrocyte.
   B. is a homotropic effector for hemoglobin.
   C. binds more tightly to HbF than to HbA.
   D. synthesis increases when hemoglobin's $T \rightleftharpoons R$ equilibrium is shifted in favor of the T state.
   E. synthesis decreases when the erythrocyte pH rises.
8. (QT3) A. Contribution of intracellular buffers to minimizing changes in pH due to increased Pco_2.
   B. Contribution of extracellular buffers to minimizing changes in pH due to increased Pco_2.
9. (QT3) A. Increase in blood $[H^+]$ when Pco_2 rises from 40 to 75 mmHg.
   B. Increase in blood $[HCO_3^-]$ when Pco_2 rises from 40 to 75 mmHg.
10. (QT3) A. Ability of normal blood to buffer excess CO_2.
    B. Ability of anemic blood to buffer excess CO_2.
11. (QT2) Which of the following produce(s) H^+?
    1. Formation of bicarbonate ion from CO_2 and water.
    2. Formation of carbamino hemoglobin from CO_2 and hemoglobin.
    3. Binding of oxygen by hemoglobin.
    4. Oxidation of sulfur-containing amino acids.
12. (QT2) A substantial fraction of the urinary acid of a normal individual consists of:
    1. H_2CO_3.
    2. NH_4^+.
    3. acetoacetate.
    4. $H_2PO_4^-$.
13. (QT1) In a patient with diabetic ketoacidosis of long duration:
    A. the major urinary acid is $H_2PO_4^-$.
    B. hemoglobin's oxygen dissociation curve would be shifted to the right.
    C. the distribution of hemoglobin species would be the same as in a normal individual.
    D. 1 mol of bicarbonate is regenerated for every mole of $H_2PO_4^-$ formed in the renal tubule.
    E. hypoventilation would be expected.
14. (QT2) The following laboratory data are obtained from a patient: Pco_2 = 60 mmHg, HCO_3^- = 27 meq/liter, pH = 7.28; these values define a point on the patient's blood buffer line.
    1. The patient has an acute condition.
    2. The condition would lead to production of an alkaline urine.
    3. Of the blood buffers, hemoglobin is the most important in resisting this pH change.
    4. Increasing the alveolar Pco_2 could restore the plasma $[HCO_3^-$ to normal.
15. (QT2) During the process of compensation for a metabolic acid–base imbalance, which of the following would become increasingly abnormal?
    1. Plasma pH
    2. Blood Pco_2
    3. Base excess
    4. Plasma HCO_3^-

16. (QT2) In respiratory alkalosis:
    1. the acute state is associated with an abnormally low plasma [HCO_3^-].
    2. the mechanism of compensation causes an increase in the plasma [HCO_3^-].
    3. the plasma pH may be within the normal range in the fully compensated state.
    4. in the partially compensated state, there will be a negative base excess equal to the difference between 24 meq/liter and the actual plasma [HCO_3^-].
17. (QT2) Hypokalemia can be expected to:
    1. occur if the plasma pH is rapidly raised.
    2. lead to increased urine acidity.
    3. be associated with a high plasma [HCO_3^-].
    4. raise the value of the anion gap.

ANSWERS

1. B A and C: The alveoli, where gas exchange with the blood occurs, are of constant size, and exchange gases with the airways by diffusion. D,E: Water vapor and CO_2 are added to the alveolar gases by the lung tissue, diluting the nitrogen (p. 877).
2. A Po_2 of tissues is typically in the neighborhood of 20 mm in active muscle and is higher in less active situations. The normal P_{50}, 27 mm, is in this range (p. 878). B and D: The curve is sigmoid, with tighter binding at higher Po_2 (p. 878, Figure 23.2). C: If O_2 were soluble enough in plasma, no carrier would be necessary (p. 876). E: Shifts profoundly affect delivery (p. 878).
3. A It is the non-α chain that differs (p. 880).
4. B The β chains are nonenzymatically glycosylated (p. 880).
5. C 2 and 4 true. 1, 2, 3: High temperature, low pH (and therefore high Pco_2) favor dissociation; that is, decreased saturation (p. 881). 4: High DPG has the same effect (p. 879).
6. A 1, 2, and 3 true. Carbonic acid is present in very small amounts; the equilibrium strongly favors CO_2 and H_2O (p. 883, Table 23.2; see also p. 891).
7. D A and B: DPG is a normal component of the red cell, where it serves as a heterotropic effector of HbA (p. 879, Figure 23.4). C: It binds weakly or not at all to the HbF (p. 881). D and E: DPG binds to the T state, relieving product inhibition of DPG synthesis; DPG synthesis is also inhibited by low pH (p. 888).
8. A The bicarbonate system is a major extracellular buffer; with a pK of 6.1 it is ineffective toward CO_2. The intracellular buffers (phosphates and protein) are, however, effective (p. 889, Table 23.4).
9. B $CO_2 + H_2O \rightleftharpoons H^+ + HCO_3^-$, but most of the H^+ is taken up by buffers (p. 894).
10. A Hemoglobin is the major nonbicarbonate blood buffer (see question 8) (p. 894, Figure 23.14).
11. E 1, 2, 3, and 4 true. 1 and 2 are reactions whose products include H^+ (p. 891). 3 is the Bohr effect (p. 882). 4 is a major source of acid in the typical American diet (p. 896).
12. C 2 and 4 true. 1: The level of H_2CO_3 is very low. 3: Acetoacetate is not an acid; acetoacetic acid would appear only in some kinds of severe acidosis. 2 and 4 are true (p. 898–899).
13. D A: After adaption to acidosis NH_4^+ excretion rises enormously (p. 899). B: True only in acidosis of short duration; decreasing DPG in prolonged acidosis tends to restore the normal position (p. 888). C: Large amounts of HbA_{1c} would be expected (p. 880). D: See p. 898. E: Hyperventilation, to expel CO_2, would be expected (p. 901).
14. B 1 and 3 true. 1: High Pco_2, low pH, point on the blood buffer line define an acute respiratory acidosis. 2: An acid urine would be produced in compensation. 3: The nonbicarbonate buffers would be most important, and hemoglobin is the major one in blood. 4: Increasing Pco_2 would exacerbate the condition (p. 900).
15. C 2 and 4 true. 1: Plasma pH would be restored. 2: Pco_2 would fall below normal in acidosis or rise above normal in alkalosis. 3: Base excess would be unchanged. 4: Bicarbonate would decrease in acidosis or increase in alkalosis (p. 901, Figures 23.19 and 23.20).
16. B 1 and 3 true. 1 and 2: See p. 900, Figure 23.18. 3: This is the only acid–base abnormality in which compensation is expected to restore the plasma pH to 7.4 (p. 902). 4: There is a negative base excess equivalent to the difference between the patient's [HCO_3^-] and the [HCO_3^-] of the point on the blood buffer line at the same pH, a point that will be less than 24 meq/liter (p. 903, Figure 23.20).
17. A 1, 2, and 3 true. 1, 2, and 3: See p. 906. 4: Decreasing K^+ would lower the anion gap, but only by a small amount (p. 904).

24

Digestion and Absorption of Basic Nutritional Constituents

ULRICH HOPFER

24.1 OVERVIEW

Historical Aspects

Secretion of digestive fluids and digestion of food were some of the earliest biochemical events to be investigated at the beginning of the era of modern science. Major milestones were the discovery of hydrochloric acid secretion by the stomach and enzymatic hydrolysis of protein and starch by gastric juice and saliva, respectively. The discovery of gastric HCl production goes back to the American physician William Beaumont (1785–1853). In 1822 he treated a patient with a stomach wound. The patient recovered from the wound, but retained a gastric fistula (abnormal opening through the skin). Beaumont seized the opportunity to obtain and study gastric juice at different times during and after meals. Chemical analysis revealed, to the surprise of chemists and biologists, the presence of the inorganic acid HCl. This discovery established the principle of unique secretions into the gastrointestinal tract, which are elaborated by specialized glands.

Soon thereafter, the principle of enzymatic breakdown of food was recognized. Theodor Schwann, a German anatomist and physiologist (1810–1882), noticed in 1836 the ability of gastric juice to degrade albumin in the presence of dilute acid. He recognized that a new principle was involved and coined for it the word *pepsin* from the Greek *pepsis,* meaning digestion. Today the process of secretion of digestive fluids, digestion of food, and absorption of nutrients and of electrolytes can be described in considerable detail.

The basic nutrients fall into the classes of proteins, carbohydrates, and fats. Many different types of food can satisfy the nutritional needs of humans, even though they differ in the ratios of proteins to carbohydrates and to fats and in the ratio of digestible to nondigestible materials. Unprocessed plant products are especially rich in fibrous material that can be neither digested by human enzymes nor easily degraded by intestinal bacteria. The fibers are mostly carbohydrates, such as cellulose (β-1,4-glucan) or pectins (mixtures of methyl esterified polygalacturic acid, polygalactose, and polyarabinose). High-fiber diets enjoy a certain popularity nowadays because of a postulated preventive effect on development of colonic cancer.

Table 24.1 describes average contributions of different food classes to the diet of North Americans. The intake of individuals may substantially deviate from the average, as food consumption depends mainly on availability and individual tastes. The ability to utilize a wide variety of food is possible because of the great adaptability and digestive reserve capacity of the gastrointestinal tract.

TABLE 24.1 Contribution of Major Food Groups to Nutrient Supplies in the United States

	Total Consumption (g)	*Dairy Products, Except Butter (%)*	*Meat, Poultry, Fish (%)*	*Eggs (%)*	*Fruits, Nuts, Vegetables (%)*	*Flour, Cereal (%)*	*Sugar, Sweeteners (%)*	*Fats, Oils (%)*
Protein	100	22	42	6	12	18	0	0
Carbohydrate	381	7	0.1	0.1	19	36	37	0
Fat	155	13	35	3	4	1	0	42

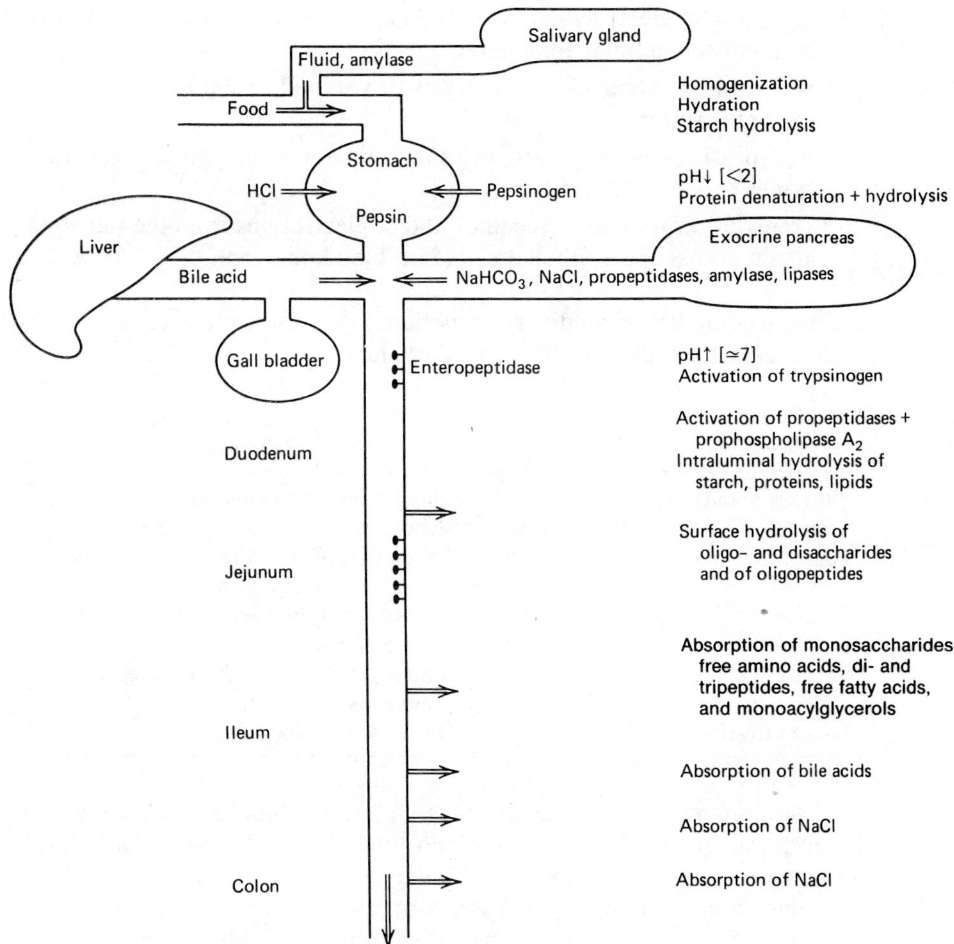

Figure 24.1
Gastrointestinal organs and their functions.

Knowledge of the nature of proteins and carbohydrates in the diet is important from a clinical point of view. Certain proteins and carbohydrates, although good nutrients for most humans, cannot be properly digested by some individuals and produce gastrointestinal ailments. Omission of the offending material and change to another diet can eliminate the gastrointestinal problems for these individuals. Examples of food constituents that can be the cause of gastrointestinal disorders are gluten, one of the protein fractions of wheat, and lactose, the disaccharide in milk.

Gastrointestinal Organs and Functions

The bulk of ingested nutrients consists of large polymers that have to be broken down to monomers before they can be absorbed and made available to all cells of the body. The complete process from food intake to absorption of nutrients into the blood consists of a complicated sequence of events, of which, at the minimum, the following steps are discernible (see Figure 24.1):

1. Mechanical homogenization of food and mixing of ingested solids with fluids secreted by the glands of the gastrointestinal tract
2. Secretion of digestive enzymes that hydrolyze macromolecules to oligomers, dimers, or monomers

3. Secretion of electrolytes, acid, or base to provide an appropriate environment for optimal enzymatic digestion
4. Secretion of bile acids as detergents to solubilize lipids and facilitate their absorption
5. Hydrolysis of nutrient oligomers and dimers by intestinal surface enzymes
6. Transport of nutrient molecules and of electrolytes from the intestinal lumen across the epithelial cells into blood or lymph

To accomplish these diverse functions, the gastrointestinal tract contains specialized glands and surface epithelia:

Organ	*Major Function in Digestion and Absorption*
Salivary glands	Elaboration of fluid and digestive enzymes
Stomach	Elaboration of HCl and proteases
Pancreas	Elaboration of $NaHCO_3$ and enzymes for intraluminal digestion
Liver	Elaboration of bile acids
Gallbladder	Storage of bile
Small intestine	Terminal digestion of food, absorption of nutrients and electrolytes
Large intestine	Absorption of electrolytes

The pancreas and the small intestine are essential for digestion and absorption of all basic nutrients. Fortunately, both organs have large reserve capacities. For example, maldigestion due to pancreatic failure becomes a problem only when the pancreatic secretion rate of digestive enzymes drops below one-tenth of the normal rate. The secretion of the liver (bile) is important for efficient lipid absorption, which depends on the presence of bile acids. In contrast, gastric digestion of food is nonessential for adequate nutrition, and loss of this function can be compensated for by the pancreas and the small intestine. Yet normal gastric digestion greatly increases the smoothness and efficiency of the total digestive process. The stomach aids in the digestion through its reservoir function, its churning ability, and initiation of protein hydrolysis, which, although small, is important for stimulation of pancreatic and gallbladder output. The peptides and amino acids liberated in the stomach serve as stimuli for the coordinated release of pancreatic juice and bile into the lumen of the small intestine, thereby ensuring efficient digestion of food.

24.2 DIGESTION: GENERAL CONSIDERATIONS

Site of Digestion

Since Schwann's discovery of gastric pepsin, it has been recognized that most of the breakdown of food is catalyzed by soluble enzymes and occurs within the lumen of the stomach or small intestine. However, the pancreas, not the stomach, is the major organ that synthesizes and secretes the large amounts of enzymes needed to digest the food. Secreted enzymes amount to at least 30 g protein/day in a healthy adult. The pancreatic enzymes together with bile are poured into the lumen of the second (descending) part of the duodenum, so that the bulk of the intraluminal digestion occurs distal to this site in the small intestine. However, pancreatic enzymes cannot completely digest all the nutrients to forms that can be absorbed. Even after exhaustive contact with pancreatic enzymes, a substantial portion of the carbohydrates and amino acids are present as dimers and oligomers that depend for final digestion on small intestinal surface or intracellular enzymes.

The importance of the small intestinal surface enzymes for digestion has been fully recognized only within the last 25 years. Its digestive functions could be examined and appreciated after methods for isolation and purification of the luminal (brush border) plasma membrane of enterocytes (intestinal epithelial cells) were developed. This membrane contains on its luminal side many di- and oligosaccharidases, amino- and dipeptidases, as well as esterases (Table 24.2). Many of these enzymes protrude up to 100 Å into the intestinal lumen, attached to the plasma membrane by an anchoring polypeptide that itself has no role in the hydrolytic activity. The substrates for these enzymes are the nutrient oligomers and dimers that result from the pancreatic digestion of food. The surface enzymes are glycoproteins that are relatively stable toward digestion by pancreatic proteases or toward detergents.

A third site of digestion is the cytoplasm of enterocytes. Intracellular digestion is of some importance for the hydrolysis of di- and tripeptides, which can be absorbed across the luminal plasma membrane.

Exocrine Secretion of Enzymes

The salivary glands, the gastric mucosa, and the pancreas contain specialized cells for the synthesis, packaging, and release of enzymes into the lumen of the gastrointestinal tract. (See Figure 24.2.) This secretion is termed "exocrine" because of its direction toward the lumen.

Proteins destined for secretion are synthesised on the polysomes of the rough endoplasmic reticulum, which is particularly abundant in exocrine cells. The NH_2-terminal amino acid sequence of nascent secretory proteins contains a signal sequence that results in anchorage of ribosomes to

TABLE 24.2 Digestive Enzymes of the Small Intestinal Surface

Enzyme (Common Name)	*Substrate*
Maltase	Maltose
Sucrase	Sucrose
Isomaltase	α-Limit dextrin
Glucoamylase	Amylose
Trehalase	Trehalose
β-Glucosidase	Glucosylceramide
Lactase	Lactose
Leucine aminopeptidase	Peptides with NH_2-terminal neutral amino acids
γ-Glutamyltransferase	Glutathione + amino acid
Enteropeptidase	Trypsinogen
Alkaline phosphatase	Organic phosphates

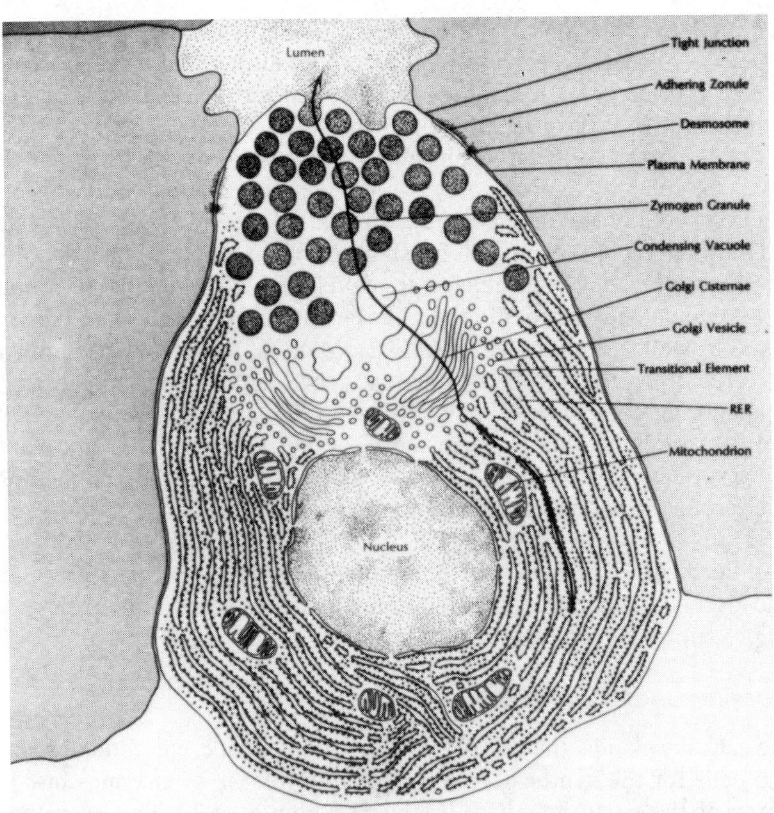

Figure 24.2
Exocrine secretion of digestive enzymes.
Reproduced with permission from Jamieson, J. D. Membrane and secretion, in *Cell Membranes: biochemistry, cell biology and pathology,* G. Weissmann and R. Claiborne (eds.). New York: HP Publishing Co., Inc., 1975. Figure by B. Tagawa.

the membrane and release of the peptide chain into the cisternal space of the endoplasmic reticulum. The amino acids forming the signal sequence may be clipped off during further processing. Secretory proteins are then transported from the endoplasmic reticulum to the Golgi complex in small membrane-bound vesicles. In the endoplasmic reticulum and also the Golgi apparatus, glycosylation may occur. Subsequent to processing, the secretory proteins are packaged into larger vesicles $\sim 1\ \mu m$ in diameter, which serve as storage forms until the stimulus for secretion is received. Proteases and phospholipase A are produced and stored as inactive precursors, also termed proenzymes or zymogens. Therefore the storage vesicles are also called *zymogen granules*. These zymogen granules are bounded by a typical cellular membrane with trilaminar appearance in conventional electron microscopy. When an appropriate stimulus for secretion is received by the cell, the granules move to the luminal plasma membrane, where their membranes fuse with the plasma membrane releasing the contents into the lumen. The process of fusion of granule membrane with plasma membrane and of release of secretory proteins is termed *exocytosis*. Activation of proenzymes occurs only after they are released from the cells.

Regulation of Secretion

The processes involved in the secretion of enzymes and electrolytes are regulated and coordinated. Elaboration of electrolytes and fluids simulta-

TABLE 24.3 Physiological Secretagogues

Organ	*Secretion*	*Secretagogue*
Salivary gland	NaCl, amylase	Acetylcholine (catecholamines?)
Stomach	HCl, pepsinogen	Acetylcholine, histamine, gastrin
Pancreas	NaCl, enzymes	Acetylcholine, cholecystokinin (secretin)
	$NaHCO_3$,NaCl	Secretin

neously with that of enzymes is required to flush any discharged digestive enzymes out of the gland into the gastrointestinal lumen. Coordination of macromolecule and electrolyte secretion must occur mainly at the tissue level because only a minor portion of secreted electrolytes appears to originate from those cells that produce zymogen granules. The physiological regulation of secretion occurs through *secretagogues* that interact with receptors on the contraluminal surface of the exocrine cells (Table 24.3). Neurotransmitters, hormones, pharmacological agents, and certain bacterial toxins can be secretagogues. Different exocrine cells, for example, in different glands, usually possess different sets of receptors. Interaction of the secretagogues with the receptors sets off a chain of events that ends with fusion of intracellular membrane-bounded granules with the plasma membrane and release of the granular material into the extracellular space. Two major types of intracellular messengers for signal transmission to the granules have been identified (Figure 24.3): (1) cleavage of inositol-1,4,5-trisphosphate from phosphatidyl-inositol-bisphosphate, which triggers Ca^{2+} release into the cytosol, and (2) activation of

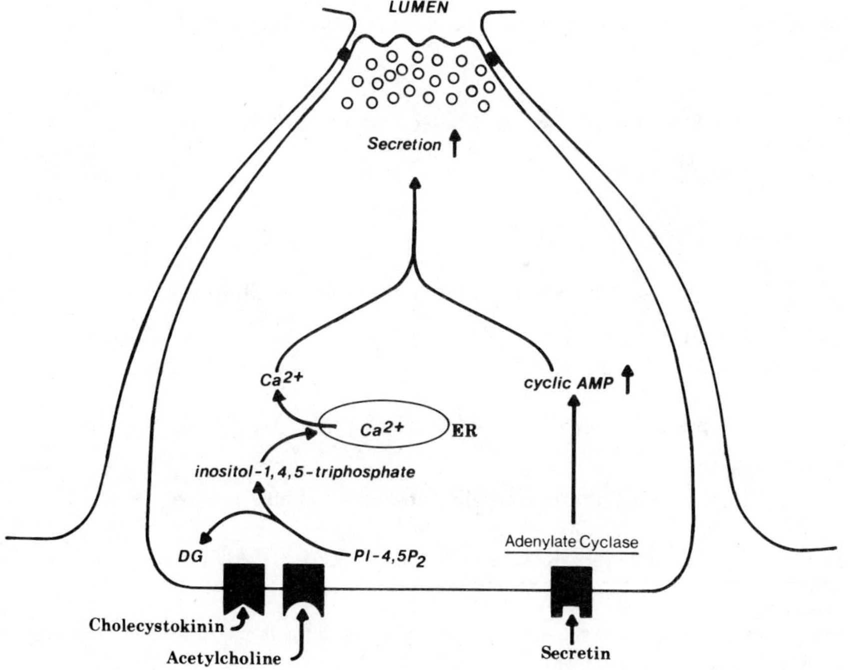

Figure 24.3
Cellular regulation of exocrine secretion in the pancreas.
Abbreviations: PI-4,5P$_2$ = phosphatidyl inositol-4,5-bisphosphate; DG = diacylglycerol; ER = endoplasmic reticulum.
Adapted from Gardner, J. D. *Annu. Rev. Physiol.* 41:63, 1979. Copyright 1979 by Annual Reviews Inc.

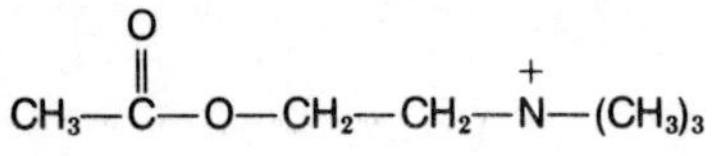

Figure 24.4
Acetylcholine.

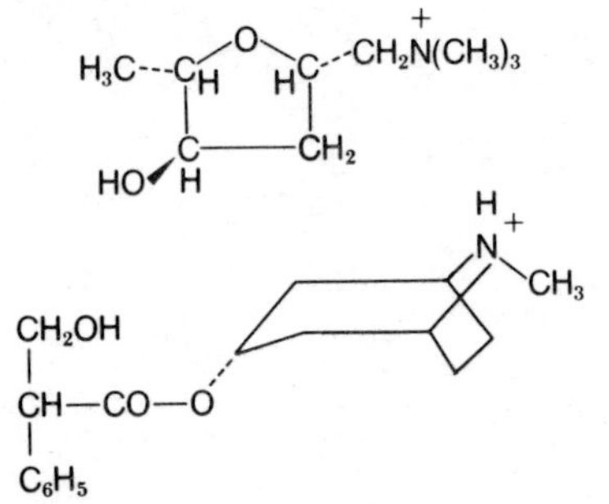

Figure 24.5
L(+)-Muscarine (top) and atropine (bottom).

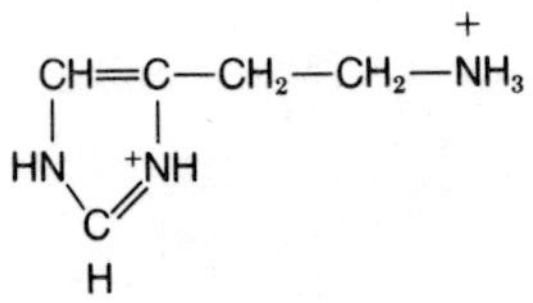

Figure 24.6
Histamine.

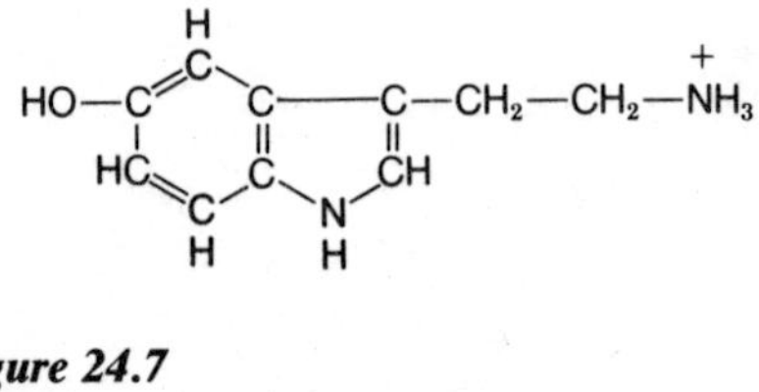

Figure 24.7
5-OH-Tryptamine (serotonin).

adenylate cyclase resulting in elevated cAMP levels. Secretagogues appear to activate either one of these pathways. Some bacterial toxins (e.g., cholera toxin) bypass the receptors and become secretagogues by directly activating one of the mechanisms for generating intracellular messengers.

Acetylcholine (Figure 24.4) elicits salivary, gastric, and pancreatic enzyme and electrolyte secretion. It appears to be the major neurotransmitter for stimulating secretion, with input from the central nervous system in salivary and gastric glands, or via local reflexes in gastric glands and the pancreas. The acetylcholine receptor of exocrine cells is of the *muscarinic* type, that is, it can be blocked by atropine (Figure 24.5). Most people have experienced the effect of atropine because it is used by dentists to "dry up" the mouth for dental work.

A second class of secretagogues consists of certain biogenic amines. For example, histamine (Figure 24.6) is a potent stimulator of HCl secretion. It interacts with a gastric-specific histamine receptor (=H_2 receptor) on the contraluminal plasma membrane of parietal cells. The cellular origin of histamine involved in the regulation of HCl secretion is not exactly known. Histamine as well as analogs, which act as antagonists at the H_2 receptor, are used medically to increase or decrease HCl output. 5-Hydroxytryptamine (serotonin) is another biogenic amine that is present in relatively high amounts in the gastrointestinal tract (Figure 24.7). It probably is involved in stimulation of NaCl secretion by the small intestinal mucosa.

A third class of secretagogues consists of peptide hormones. The gastrointestinal tract is rich in specialized epithelial cells, containing a large number of different biologically active amines or peptides. The peptides are localized in granules, usually close to the contraluminal pole of these cells and probably are released into the interstitial space. Hence these epithelial cells are classified as endocrine cells. Of particular importance are the peptides gastrin, cholecystokinin (pancreozymin), and secretin (Table 24.4).

Gastrin occurs predominantly as either a large peptide of 34 amino acids (G-34) or a smaller one of 17 residues (G-17) from the COOH-terminus of G-34. The functional portion of gastrin resides mainly in the last five amino acids of the COOH-terminus. Thus pentagastrin, an artificial pentapeptide containing only the last five amino acids, can be used specifically to stimulate gastric HCl and pepsin secretion. Gastrin as well

TABLE 24.4 Structure of Gastrin, Cholecystokinin, and Secretin

Human gastrin G-34-II[a] **G-17-II**

↓ ↓

[b]Glp-Leu-Gly-Pro-Gln-Gly-His-Pro-Ser-Leu-Val-Ala-Asp-Pro-Ser-Lys-Lys-Gln
|
[c]NH_2-Phe-Asp-Met-Trp-Gly-Tyr(SO_3H)-Ala-$(Glu)_5$-Leu-Trp-Pro-Gly

Cholecystokinin

Lys-Ala-Pro-Ser-Gly-Arg-Val-Ser-Met-Ile-Lys-Asn-Leu-Gln-Ser-Leu-Asp-Pro
|
[c]NH_2-Phe-Asp-Met-Trp-Gly-Met-Tyr(SO_3H)-Asp-Arg-Asp-Ser-Ile-Arg-His-Ser

Porcine secretin

His-Ser-Asp-Gly-Thr-Phe-Thr-Ser-Glu-Leu-Ser-Arg-Leu-Arg-Asp
|
[c]NH_2-Val-Leu-Gly-Gln-Leu-Leu-Arg-Gln-Leu-Arg-Ala-Ser

[a] Gastrin I is not sulfated.
[b] Glp = pyrrolidino carboxylic acid, derived from Glu through internal amide formation;
[c] —NH_2 = amide of carboxy terminal amino acid.

as cholecystokinin have an interesting chemical feature, a sulfated tyrosine, which considerably enhances the potency of both hormones.

Cholecystokinin and pancreozymin denote the same peptide. The different names stem from the times when only the functions were known and before the peptides had been purified. As implied by the names, the peptide stimulates gallbladder contraction and secretion of pancreatic enzymes. The peptide is secreted by epithelial endocrine cells of the small intestine, particularly in the duodenum. Cholecystokinin secretion is stimulated by luminal amino acids and peptides, usually derived from gastric proteolysis, by fatty acids and an acid pH. Cholecystokinin and gastrin are thought to be related in an evolutionary sense, as both share an identical amino acid sequence at the COOH-terminus.

Secretin is a polypeptide of 27 amino acids. This peptide is secreted by yet other endocrine cells of the small intestine. Its secretion is stimulated particularly by luminal pH <5. The major biological activity of secretin is stimulation of secretion of pancreatic juice rich in $NaHCO_3$. Pancreatic $NaHCO_3$ is essential for the neutralization of gastric HCl in the duodenum. Secretin also enhances pancreatic enzyme release, acting synergistically with cholecystokinin.

24.3 EPITHELIAL TRANSPORT

General Considerations

Solute movement across an epithelial cell layer is determined by the properties of epithelial cells, particularly their plasma membranes, as well as by the intercellular tight junctional complexes. (See Figure 24.8.) The tight junctions extend in a belt-like manner around the perimeter of each epithelial cell and connect neighboring cells. Therefore, the tight junctions constitute part of the barrier between the two extracellular spaces on either side of the epithelium, that is, the lumen of the gastrointestinal tract and the intercellular (interstitial) space on the other side. The tight junction also marks the boundary between the luminal and the contraluminal region of the plasma membrane of epithelial cells.

Two potentially parallel pathways for solute transport across epithelial cell layers can be distinguished: through the cells (transcellular) and through the tight junctions between cells (paracellular) (Figure 24.8). The transcellular route in turn consists mainly of two barriers in series, which are formed by the luminal and by the contraluminal plasma membrane. Because of this combination of different barriers in parallel (cellular and paracellular pathways) and in series (luminal and contraluminal plasma membranes), biochemical and biophysical information on all three barri-

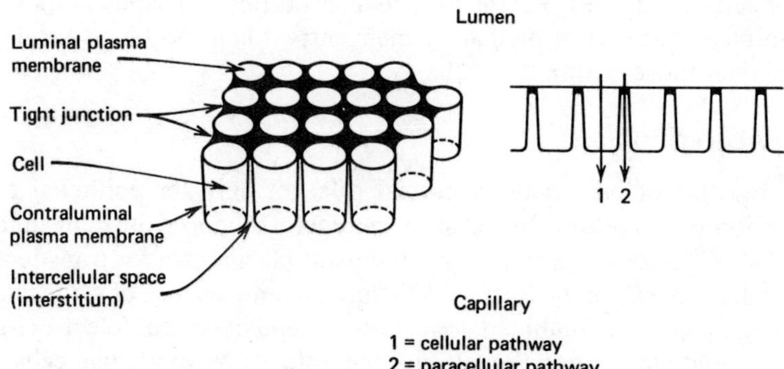

Figure 24.8
Pathways for transport across epithelia.

TABLE 24.5 Characteristic Differences Between Luminal and Contraluminal Plasma Membrane of Small Intestinal Epithelial Cells

	Luminal	*Contraluminal*
Morphological appearance	Microvilli in ordered arrangement (= brush border)	Few microvilli
Enzymes	Di- and oligosaccharidases Aminopeptidase Dipeptidases γ-Glutamyltransferase Alkaline phosphatase	Na^+,K^+-ATPase Adenylate cyclase
Transport systems	Na^+-monosaccharide cotransport Na^+-neutral amino acid cotransport Na^+-bile acid cotransport	Facilitated monosaccharide transport Facilitated neutral amino acid transport

ers as well as their mutual influence is required for understanding the overall transport properties of the epithelium.

One of the main functions of epithelial cells in the gastrointestinal tract is active transport of nutrients, electrolytes, and vitamins from one side of the epithelium to the other. The cellular basis for this vectorial solute movement must lie in the different properties of the luminal and contraluminal regions of the plasma membrane. The small intestinal cells provide a prominent example of the differentiation and specialization of the two types of membrane. They differ in morphological appearance, enzymatic composition, chemical composition, and transport functions (Table 24.5). The luminal membrane in contact with the nutrients in the chyme (the semifluid mass of partially digested food) is specialized for terminal digestion of nutrients through its digestive enzymes and for nutrient absorption through transport systems that accomplish concentrative uptake. Such transport systems are well known for monosaccharides, amino acids, peptides, and electrolytes. In contrast, the contraluminal plasma membrane, which is in contact with the intercellular fluid, capillaries, and lymph, has properties similar to the plasma membrane of most cells. It possesses receptors for hormonal or neuronal regulation of cellular functions, a Na^+,K^+-ATPase for removal of Na^+ from the cell and transport systems for the entry of nutrients for its own consumption. In addition, the contraluminal plasma membrane contains the transport systems necessary for exit of the nutrients derived from the lumen so that the digested food can become available to all cells of the body. Some of the transport systems in the contraluminal plasma membrane may fulfill both the function of catalyzing exit when the intracellular nutrient concentration is high after a meal and that of mediating their entry when the blood levels are higher than those within the cell.

NaCl Absorption

The transport of Na^+ plays a crucial role not only for epithelial NaCl absorption or secretion, but also in the energization of nutrient uptake. The Na^+,K^+-ATPase provides the dominant mechanism for transduction of chemical energy in the form of ATP into osmotic energy of a concentration (chemical) or a combined concentration and electrical (electrochemical) ion gradient across the plasma membrane. In epithelial cells this

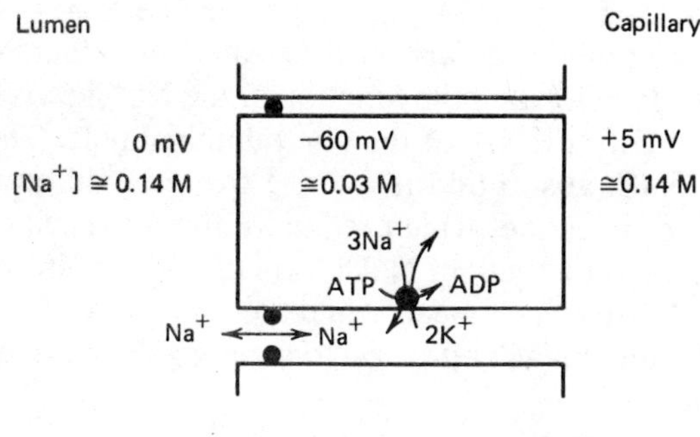

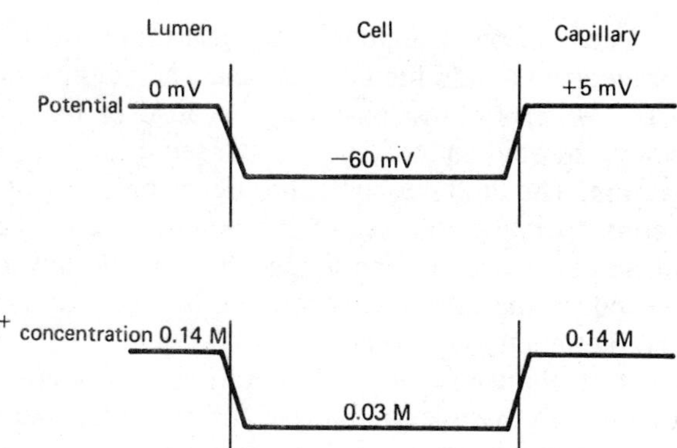

Figure 24.9
Na^+ concentrations and electrical potentials in enterocytes.

enzyme is located exclusively in the contraluminal plasma membrane (Figure 24.9). The best estimate for the stoichiometry of the Na^+,K^+-ATPase reaction appears to be 1 mol ATP coupled to the outward pumping of 3 mol Na^+ and the simultaneous inward pumping of 2 mol K^+. Because three cations leave the cell, while only two enter it during each cycle of ATP hydrolysis, the cytoplasm becomes electrically negative with respect to the extracellular fluid. In small intestinal epithelial cells, intracellular Na^+ concentrations are about 30 mM, that is, fivefold lower than in plasma, and the cytoplasm has a potential of about −60 mV relative to the extracellular solution.

Transepithelial NaCl movements are produced by the combined actions of the Na^+,K^+-ATPase and additional transport systems in the plasma membrane, which allow the entry of Na^+ or Cl^- into the cell. If NaCl can enter the cell passively at the luminal pole, NaCl absorption results from the combined action of NaCl entry and of Na^+,K^+-ATPase activity; if NaCl can enter the cell passively through the contraluminal plasma membrane, NaCl is secreted into the lumen.

At least two different types of transport systems for "passive" Na^+ movements across the luminal plasma membrane can be distinguished on the basis of inhibitors, hormonal regulation, and electrical fluxes associated with Na^+ transport. The epithelial cells of the lower portion of the large intestine possess a transport system that allows the uncoupled entry of Na^+ into the cell down its electrochemical gradient (Figure 24.10). This Na^+ flux is associated with an electrical current, that is, electrogenic, and can be inhibited by the drug amiloride at micromolar concentrations (Figure 24.11). The presence of this transport system, and hence NaCl absorption, is regulated by mineralocorticoid hormones. In contrast, epithelial cells of the upper portion of the intestine possess a transport system in the brush border membrane, which catalyzes an electrically silent Na^+-

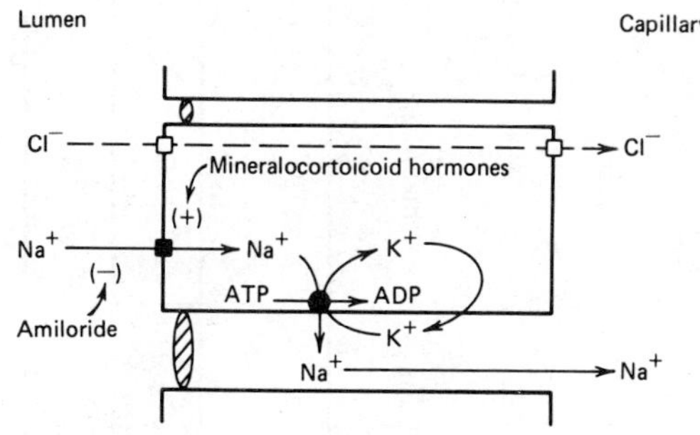

Figure 24.10
Model for epithelial NaCl transport in the lower intestine.

Figure 24.11
Amiloride.

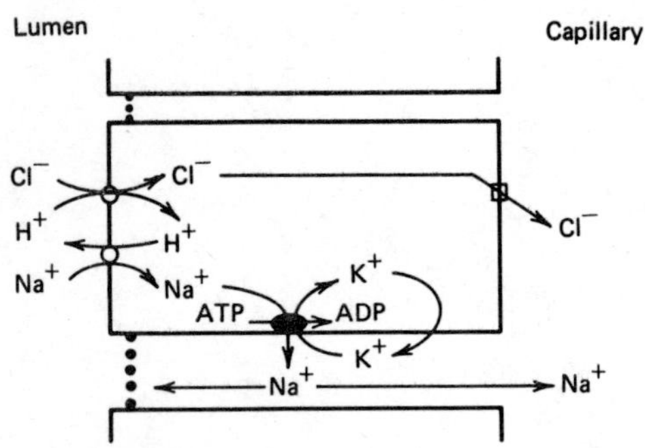

Figure 24.12
Model for epithelial NaCl transport in the upper intestine.

H^+ exchange. (See Figure 24.12.) The exchange is not affected by low concentrations of amiloride and not regulated by mineralocorticoids. In both cases, the location and the function of the Na^+,K^+-ATPase is identical, namely, to extrude Na^+ at the contraluminal pole. The necessity for two types of NaCl absorption may arise from the different functions of upper and lower intestine, which require different regulation. The upper intestine reabsorbs the bulk of NaCl from diet and from secretions of the exocrine glands after each meal, while the lower intestine participates in the fine regulation of NaCl retention, depending on the overall electrolyte balance of the body.

NaCl and $NaHCO_3$ Secretion

The epithelial cells of most regions of the gastrointestinal tract have the potential for electrolyte and fluid secretions. The major secreted ions are Na^+ and Cl^-. Water follows passively because of the osmotic forces exerted by any secreted solute. Thus, NaCl secretion secondarily results in fluid secretion. The fluid may be either hypertonic or isotonic, depending on the contact time of the secreted fluid with the epithelium and the tissue permeability to water. The longer the contact and the greater the water permeability, the closer the secreted fluid gets to osmotic equilibrium, that is, isotonicity (see Figure 24.13).

The cellular mechanisms for NaCl secretion are not completely established, but appear to involve the Na^+, K^+-ATPase located in the contraluminal plasma membrane of epithelial cells (Figure 24.14). The enzyme is implicated because cardiac glycosides, inhibitors of this enzyme, abolish

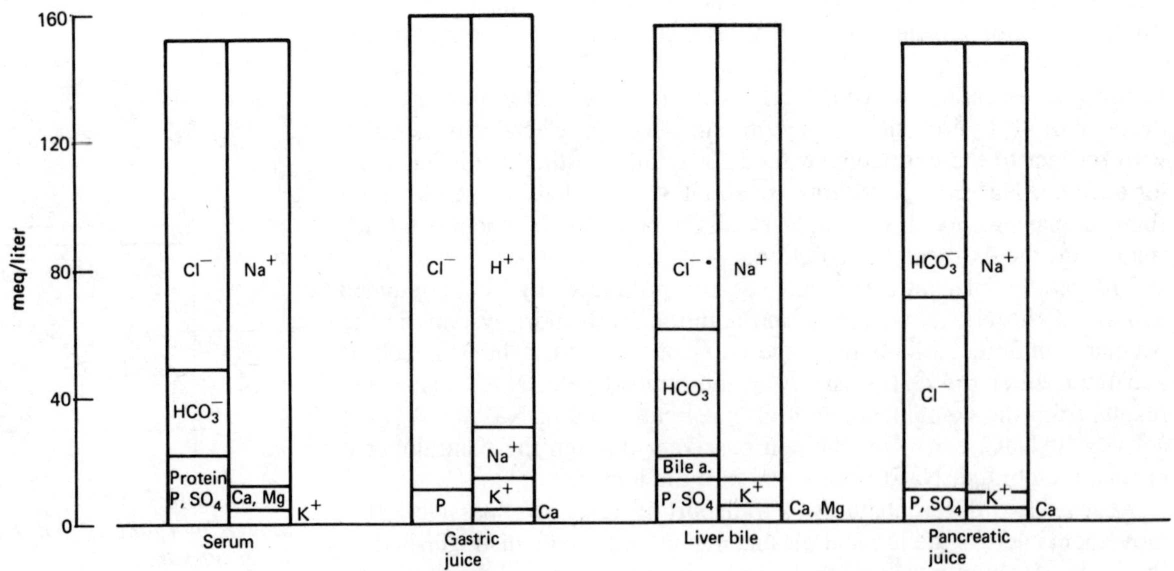

Figure 24.13
Ionic composition of secretions of the gastrointestinal tract.
Serum is included to facilitate comparison between fluids. Note the high H^+ concentration in gastric juice ($pH < 1$) and the high HCO_3^- concentration in pancreatic juice. P, organic and inorganic phosphate, SO_4, inorganic and organic sulfate, Ca, calcium, Mg, magnesium, bile a., bile acids.

Adapted from *Biological Handbooks, Blood and Other Body Fluids,* Federation of American Societies for Experimental Biology, 1961.

salt secretion. However, the involvement of Na^+,K^+-ATPase does not provide a straightforward explanation for a NaCl movement from the capillary side to the lumen because the enzyme extrudes Na^+ from the cell toward the capillary side. Thus the active step of Na^+ transport across one of the plasma membranes has a direction opposite to that of overall transepithelial NaCl movements. The current model of NaCl secretions, which is supported mainly by data from the salt-secreting glands of birds and sharks, has solved the paradox by an electrical coupling of Cl^- secretion across the luminal plasma membrane and Na^+ movements via the paracellular route illustrated in Figure 24.14.

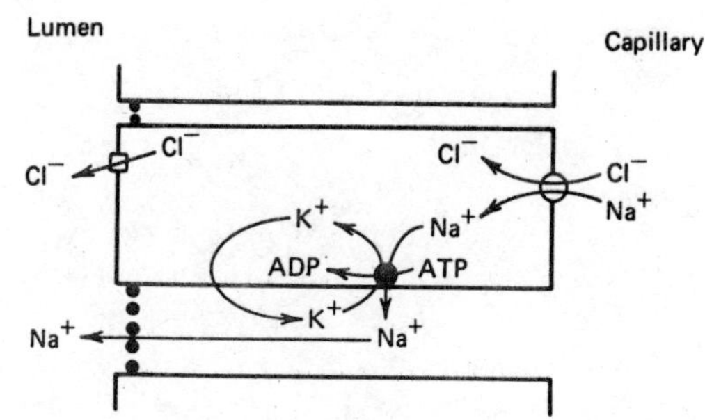

Figure 24.14
Model for epithelial NaCl secretion.

In the pancreas a fluid rich in Na^+ and Cl^- is secreted by acinar cells. This fluid provides the vehicle for the movement of digestive enzymes from the acini, where they are released, to the lumen of the duodenum. The fluid is modified in the ducts by the additional secretion of $NaHCO_3$. The HCO_3^- concentration in the final pancreatic juice can reach concentrations of up to 120 mM.

The permeability of the tight junction to water, Na^+, or other ions modifies active transepithelial solute movements. For example, a high permeability to Na^+ allows this cation to equilibriate between extracellular solutions of the luminal and of the intercellular compartments. Different regions of the gastrointestinal tract differ not only with respect to the transport systems that determine the passive membrane permeability (see above for amiloride-sensitive and amiloride-insensitive Na^+ entry), but also with respect to the permeability characteristics of the tight junction.

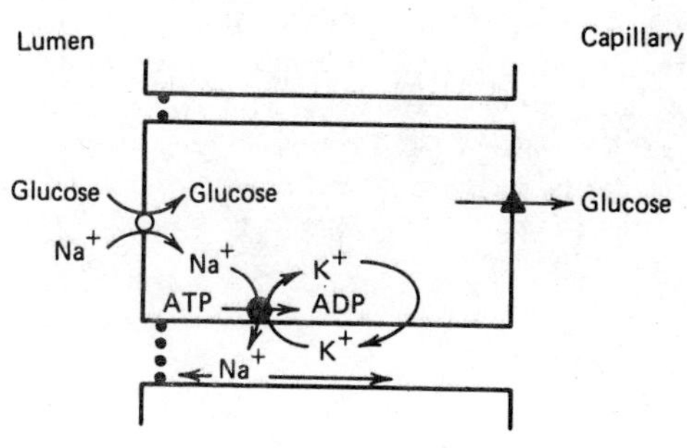

Glucose concentration profile normal to epithelial plane

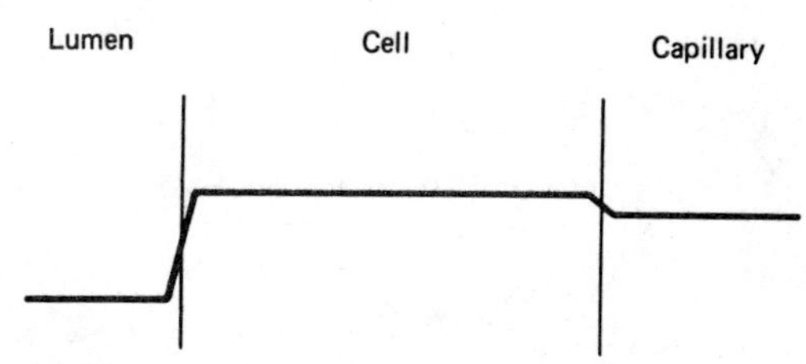

Figure 24.15
Model for epithelial glucose transport.

Energization of Nutrient Transport

Many solutes are absorbed across the intestinal epithelium against a concentration gradient. The energy for this "active" transport is directly derived from the Na^+ concentration gradient or the electrical potential across the luminal plasma membrane, rather than from the chemical energy of a covalent bond change, such as ATP hydrolysis. Glucose transport provides an example of uphill solute transport that is driven directly by the electrochemical Na^+ gradient and only indirectly by ATP (Figure 24.15).

In vivo, glucose is absorbed from the lumen into the blood against a concentration gradient. This vectorial transport is the combined result of several separate membrane events: (1) the asymmetric insertion of different transport systems for glucose into the luminal and the contraluminal plasma membrane, (2) the coupling of Na^+ and glucose transport across the luminal membrane, and (3) the ATP-dependent Na^+ transport out of the cell at the contraluminal pole. (See Figure 24.16).

$$
\begin{array}{rl}
3Na^+_{\text{lumen}} + 3Glc_{\text{lumen}} & \rightleftharpoons 3Na^+_{\text{cell}} + 3Glc_{\text{cell}} \\
3Na^+_{\text{cell}} + 2K^+_{\text{interstitium}} + ATP_{\text{cell}} & \longrightarrow 3Na^+_{\text{interstitium}} + 2K^+_{\text{cell}} + ADP_{\text{cell}} + P_{\text{cell}} \\
2K^+_{\text{cell}} & \rightleftharpoons 2K^+_{\text{interstitium}} \\
3Na^+_{\text{interstitium}} & \rightleftharpoons 3Na^+_{\text{lumen}} \\
3Glc_{\text{cell}} & \rightleftharpoons 3Glc_{\text{interstitium}} \\
\hline
\text{Sum:} \quad 3Glc_{\text{lumen}} + ATP_{\text{cell}} & \longrightarrow 3Glc_{\text{interstitium}} + ADP_{\text{cell}} + P_{\text{cell}}
\end{array}
$$

Figure 24.16
Transepithelial glucose transport as translocation reactions across the plasma membranes and the tight junction.

CLIN. CORR. 24.1
ELECTROLYTE REPLACEMENT THERAPY IN CHOLERA

Huge, life-threatening intestinal electrolyte and fluid secretion (diarrhea) occurs in patients with cholera, an intestinal infection by *Vibrio cholerae*. The secretory state is a result of cholera enterotoxin produced by the bacterium. The toxin activates adenylate cyclase (see page 596) and thereby turns on electrolyte secretion. Modern, oral treatment of cholera takes advantage of the presence of Na^+-glucose cotransport in the intestine, which is not regulated by cAMP and remains intact in this disease. In this case, the presence of glucose allows uptake of Na^+ to replenish body NaCl. Composition of solution for oral treatment of cholera patients is glucose 110 mM, Na^+ 99 mM, Cl^- 74 mM, HCO_3^- 39 mM, and K^+ 4 mM.

Carpenter, C. C. J., in Secretory diarrhea, edited by M. Field, J. S. Fordtran, and S. G. Schulz, *Am. Phys. Soc.* Bethesda, MD, pp. 67–83, 1980.

The luminal plasma membrane contains a transport system that facilitates a tightly coupled movement of Na^+ and D-glucose (or structurally similar sugars). The transport system mediates glucose and Na^+ transport equally well in both directions. However, because of the higher Na^+ concentration in the lumen and the negative potential within the cell, the observed direction is from lumen to cell, even if the cellular glucose concentration is higher than the luminal one. In other words, downhill Na^+ movement normally supports concentrative glucose transport. Concentration ratios of up to 20-fold between intracellular and extracellular glucose have been observed in vitro under conditions of blocked efflux of cellular glucose. In some situations Na^+ uptake via this route is actually more important than glucose uptake (Clin. Corr. 24.1).

The contraluminal plasma membrane contains another type of transport system for glucose, which allows glucose to exit. This transport system facilitates equilibration of glucose across the membrane, whereby the direction of the net flux is determined by the glucose concentration gradient. The two glucose transport systems in the luminal and contraluminal plasma membrane share glucose as substrate, but otherwise differ considerably in terms of Na^+ as cosubstrate, specificity for other sugars, sensitivity to inhibitors, or biological regulation. Since both Na^+-glucose cotransport and the simple, facilitated diffusion are not inherently directional, "active" transepithelial glucose transport can be maintained under steady-state conditions only if the Na^+, K^+-ATPase continues to move Na^+ out of the cell. Thus the active glucose transport is indirectly dependent on a supply of ATP and an active Na^+, K^+-ATPase.

The advantage of an electrochemical Na^+ gradient serving as intermediate in the energization is that the Na^+, K^+-ATPase can drive many different nutrients. The only requirement is presence of a transport system catalyzing cotransport of the nutrient with Na^+.

Secretion of HCl

The parietal (oxyntic) cells of gastric glands are capable of secreting HCl into the gastric lumen. Luminal H^+ concentrations of up to 0.14 M (pH 0.8) have been observed. (See Figure 24.17.) As the plasma pH = 7.4, the parietal cell transports protons against a concentration gradient of $10^{6.6}$. The free energy required for HCl secretion under these conditions is minimally 9.1 kcal/mol HCl (=38 J/mol HCl), as calculated from

$$\Delta G' = RT\,2.3 \log 10^{6.6} \qquad (RT = 0.6 \text{ kcal/mol at } 37°\text{C})$$

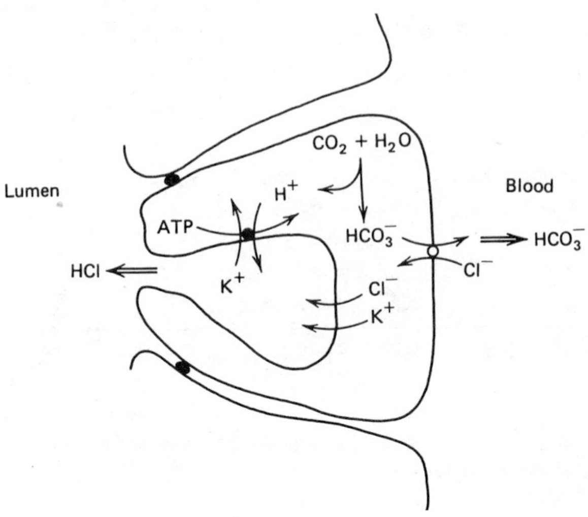

Figure 24.17
Model for secretion of hydrochloric acid.

A K^+-activated ATPase is intimately involved in the mechanism of active HCl secretion. This enzyme is unique to the parietal cell and is found only in the luminal region of the plasma membrane. It couples the hydrolysis of ATP to an electrically silent obligatory exchange of K^+ for H^+, secreting H^+ and taking K^+ into the cell. The stoichiometry appears to be 1 mol of transported H^+ and K^+ for each mol of ATP:

$$\text{ATP}_{\text{cell}} + \text{H}^+{}_{\text{cell}} + \text{K}^+{}_{\text{lumen}} \rightleftharpoons \text{ADP}_{\text{cell}} + \text{Pi}_{\text{cell}} + \text{H}^+{}_{\text{lumen}} + \text{K}^+{}_{\text{cell}}$$

In the steady state, HCl can be elaborated by this mechanism only if the luminal membrane is permeable to K^+ and Cl^- and the contraluminal plasma membrane catalyzes an exchange of Cl^- for HCO_3^-. The exchange of Cl^- for HCO_3^- is essential to refurnish the cell with Cl^- and to prevent accumulation of base within the cell. Thus, under steady-state conditions, secretion of HCl into the gastric lumen is coupled to movement of HCO_3^- into the plasma.

24.4 DIGESTION AND ABSORPTION OF PROTEINS

General Considerations

The total daily protein load to be digested consists of ~70–100 g dietary proteins and 35–200 g of endogenous proteins from digestive enzymes and sloughed-off cells. Digestion and absorption of proteins are very efficient processes in healthy humans, since only about 1–2 g of nitrogen are lost through feces each day, which is equivalent to 6–12 g of protein.

With the exception of a short period after birth, oligo- and polypeptides (proteins) are not absorbed intact in appreciable quantities by the intestine. Proteins are broken down by hydrolases with specificity for the peptide bond, that is, by peptidases. This class of enzymes is divided into endopeptidases (proteases) which attack internal bonds and liberate large peptide fragments, and exopeptidases which cleave off one amino acid at a time from either the COOH- or the NH_2-terminus. Thus exopeptidases are further subdivided into carboxy- and aminopeptidases. Endopeptidases are important for an initial breakdown of long polypeptides into smaller products, which can then be attacked more efficiently by the exopeptidases. The final products are free amino acids and di- and tripeptides, which are absorbed by the epithelial cells. (See Figure 24.18).

The process of protein digestion can be divided into a gastric, a pancreatic, and an intestinal phase, depending on the source of peptidases.

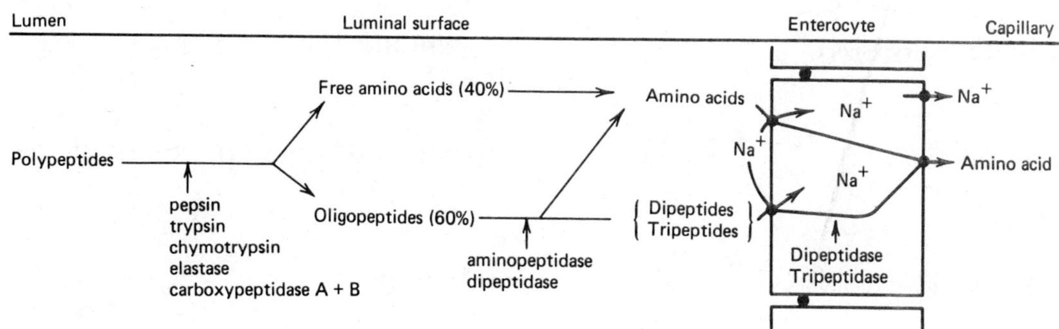

Figure 24.18
Digestion and absorption of proteins.

Gastric Digestion of Proteins

Gastric juice is characterized by the presence of HCl and therefore a low pH <2 as well as the presence of proteases of the pepsin family. The acid serves to kill off microorganisms and also to denature proteins. Denaturation makes proteins more susceptible to hydrolysis by proteases. Pepsins are unique in that they are acid-stable; in fact, they are active at acid but not at neutral pH. The catalytic mechanism that is effective for peptide hydrolysis at the acid pH depends on two carboxylic groups at the active site of the enzymes. Pepsin A, the major gastric protease, prefers peptide bonds formed by the amino group of aromatic acids (Phe, Tyr). (See Table 24.6)

Active pepsin is generated from the proenzyme pepsinogen by the removal of 44 amino acids from the NH_2-terminus (pig enzyme). Cleavage of the peptide bond between residues 44 and 45 of pepsinogen can occur as either an intramolecular reaction (autoactivation) below pH 5 or by active pepsin (autocatalysis). The liberated peptide from the NH_2-terminus remains bound to pepsin and acts as "pepsin inhibitor" above pH 2. This inhibition is released either by a drop of the pH below 2 or further degradation of the peptide by pepsin. Thus, once favorable conditions are reached, pepsinogen is converted to pepsin by autoactivation and subsequent autocatalysis at an exponential rate.

The major products of pepsin action are large peptide fragments and some free amino acids. The importance of gastric protein digestion does not lie so much in its contribution to the breakdown of ingested macromolecules, but rather in the generation of peptides and amino acids that act as stimulants for cholecystokinin release in the duodenum. The gastric peptides therefore are instrumental in the initiation of the pancreatic phase of protein digestion.

TABLE 24.6 Gastric and Pancreatic Peptidases

Enzyme	*Proenzyme*	*Activator*	*Reaction Catalyzed*	
Carboxyl Proteases				
Pepsin A	Pepsinogen A	Autoactivation, pepsin	CO↓—NHCH(R)CO↓—NHCH(R′)CO	R = tyr, phe, leu
Serine Proteases				
Trypsin	Trypsinogen	Enteropeptidase, trypsin	CO—NHCH(R)CO↓—NHCH(R′)CO	R = arg, lys
Chymotrypsin	Chymotrypsinogen	Trypsin	CO—NHCH(R)CO↓—NHCH(R′)CO	R = tyr, trp, phe, met, leu
Elastase	Proelastase	Trypsin	CO—NHCH(R)CO↓—NHCH(R′)CO	R = ala, gly, ser
Zn-Peptidases				
Carboxypeptidase A	Procarboxypeptidase A	Trypsin	CO↓—NHCH(R)CO_2^-	R = val, leu, ile, ala
Carboxypeptidase B	Procarboxypeptidase B	Trypsin	CO↓—NHCH(R)CO_2^-	R = arg, lys

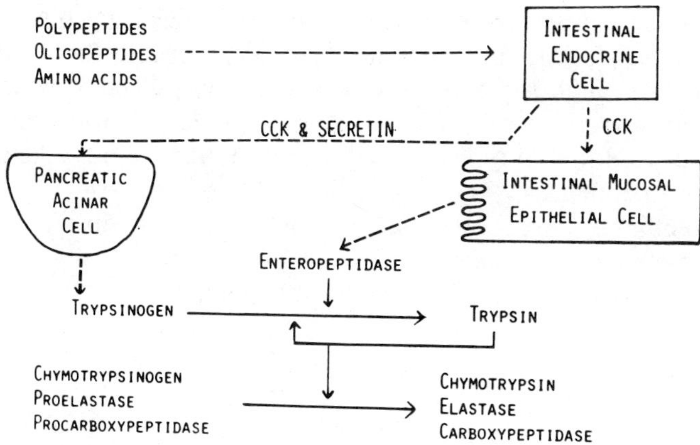

Figure 24.19
Secretion and activation of pancreatic enzymes.
Abbreviation: CCK = cholecystokinin.

Reproduced with permission from Freeman, H. J., and Kim, Y. S. *Annu. Rev. Med.* 29:102, 1978. Copyright 1978 by Annual Reviews Inc.

Pancreatic Digestion of Proteins

The pancreatic juice is rich in proenzymes of endopeptidases and carboxypeptidases. (See Figure 24.19.) These proenzymes are activated only after they reach the lumen of the small intestine. The key to activation is enteropeptidase (old name: enterokinase), a protease produced by duodenal epithelial cells. Enteropeptidase activates pancreatic trypsinogen to trypsin by scission of a hexapeptide from the NH_2-terminus. Trypsin in turn autocatalytically activates more trypsinogen to trypsin and also acts on the other proenzymes, thus liberating the endopeptidases chymotrypsin and elastase and the carboxypeptidases A and B. Since trypsin plays a pivotal role among pancreatic enzymes in the activation process, pancreatic juice normally contains a small molecular weight peptide that acts as a trypsin inhibitor and neutralizes any trypsin formed prematurely within the pancreatic cells or pancreatic ducts.

Trypsin, chymotrypsin, and elastase are endopeptidases with different substrate specificity as shown in Table 24.6. They are all active only at neutral pH and depend on pancreatic $NaHCO_3$ for neutralization of gastric HCl. The mechanism of catalysis of all three enzymes involves an essential ''serine'' residue. Thus reagents that interact with serine and modify it, inactivate the enzymes. A prominent example of such a reagent is diisopropylphosphofluoridate (Figure 24.20).

Figure 24.20
Inactivation of "serine"-proteases by diisopropylphosphofluoridate.

The polypeptides generated from ingested proteins by the action of gastric and pancreatic endopeptidases are degraded further within the small intestinal lumen by carboxypeptidase A and B. The pancreatic carboxypeptidases are metalloenzymes that require Zn^{2+} for activity and thus possess a different type of catalytic mechanism than the ''carboxy'' or ''serine'' peptidases.

The combined action of pancreatic peptidases results in the formation of free amino acids and small peptides of 2–8 residues. Peptides account for about 60% of the amino nitrogen at this point.

Small Intestinal Digestion of Proteins

Since pancreatic juice does not contain appreciable aminopeptidase activity, final digestion of di- and oligopeptides depends on small intestinal enzymes. The luminal surface of intestinal epithelial cells is particularly rich in aminopeptidase activity, but also contains dipeptidases. The end products of the cell surface digestion are free amino acids and di- and tripeptides. Amino acids and small peptides are then absorbed by the epithelial cells via specific amino acid or peptide transport systems. The di- and tripeptides are generally hydrolyzed within the cytoplasmic compartment before they leave the cell. The cytoplasmic dipeptidases explain why practically only free amino acids are found in the portal blood after a meal. The virtual absence of peptides had previously been taken as evidence that luminal protein digestion had to proceed all the way to free amino acids before absorption could occur. However, it is now established that a large portion of dietary amino nitrogen is absorbed in the form of small peptides with subsequent intracellular hydrolysis. Exception to this general rule are di- and tripeptides containing proline and hydroxyproline or unusual amino acids, such as β-alanine in carnosine (β-alanylhistidine) or anserine [β-alanyl(1-methyl)histidine] after ingestion of chicken meat. These peptides are not good substrates for the intestinal cytoplasmic dipeptidases and therefore are available for transport out of the cell into the portal blood.

CLIN. CORR. **24.2**
NEUTRAL AMINO ACIDURIA (HARTNUP DISEASE)

Transport functions, like enzymatic functions, are subject to modification by mutations. An example of a genetic lesion in epithelial amino acid transport is Hartnup disease, named after the family in which the disease entity resulting from the defect was first recognized. The disease is characterized by the inability of renal and intestinal epithelial cells to absorb neutral amino acids from the lumen. In the kidney, in which plasma amino acids reach the lumen of the proximal tubule through the ultrafiltrate, the inability to reabsorb amino acids manifests itself as excretion of amino acids in the urine (amino aciduria). The intestinal defect results in malabsorption of free amino acids from the diet. Therefore the clinical symptoms of patients with this disease are mainly those due to essential amino acid and nicotinamide deficiencies. The latter is explained by a deficiency of tryptophan, which serves as precursor for nicotinamide. Investigations of patients with Hartnup disease revealed existence of different intestinal transport systems for free amino acids and di- or tripeptides. The genetic lesion does not affect transport of peptides which remains as a pathway for absorption of protein digestion products.

Absorption of Free Amino Acids

The small intestine has a high capacity to absorb free amino acids. Most L-amino acids can be transported across the epithelium against a concentration gradient, although the need for concentrative transport in vivo is not obvious, since luminal concentrations are usually higher than the plasma levels of 0.1–0.2 mM. Amino acid transport in the small intestine has all the characteristics of carrier-mediated transport, such as discrimination between D- and L-amino acids and energy- and temperature-dependence. Additionally, genetic defects are known to occur in humans (Clin. Corr. 24.2).

On the basis of genetics and of transport experiments, at least six brush-border specific transport systems for the uptake of L-amino acids from the luminal solution can be distinguished:

1. For neutral amino acids with short or polar side chains (Ser, Thr, Ala).
2. For neutral amino acids with aromatic or hydrophobic side chains (Phe, Tyr, Met, Val, Leu, Ile).
3. For imino acids (Pro, Hyp).
4. For β-amino acids (β-Ala, taurine).
5. For basic amino acids and cystine (Lys, Arg, Cys-Cys).
6. For acidic amino acids (Asp, Glu).

The mechanism for concentrative transepithelial transport of L-amino acids appear to be similar to those discussed for D-glucose (see Figure 24.15). Na^+-dependent transport systems have been identified in the luminal (brush border) membrane and Na^+-independent ones in the contraluminal plasma membrane of small intestinal epithelial cells. Similarly, as for active glucose transport, the energy for concentrative amino acid transport appears to be derived directly from the electrochemical Na^+ gradient and only indirectly from ATP. The amino acids are not chemically modified during membrane transport, although they may be metabolized within the cytoplasmic compartment.

Absorption of Intact Proteins

The fetal and neonatal small intestines can absorb intact proteins. The uptake occurs by endocytosis, that is, the internalization of small vesicles of plasma membrane, which contain ingested macromolecules. The process is also termed "pinocytosis" because of the small size of vesicles. The small intestinal pinocytosis of protein is thought to be important for the transfer of maternal antibodies (γ-globulins) to the offspring, particularly in rodents. The pinocytotic uptake of proteins is not important for nutrition, and its magnitude usually declines after birth. Persistance of low levels of this process beyond the neonatal period may, however, be responsible for absorption of sufficient quantities of macromolecules to induce antibody formation.

24.5 DIGESTION AND ABSORPTION OF CARBOHYDRATES

Digestion of Carbohydrates

Dietary carbohydrates provide a major portion of the daily caloric requirement. From a digestive point of view, it is important to distinguish between mono-, di-, and polysaccharides. (See Table 24.7.) Monosaccharides need not be hydrolyzed prior to absorption. Disaccharides require the small intestinal surface enzymes for break-down into monosaccharides, while polysaccharides depend additionally on pancreatic amylase for degradation. (See Figure 24.21.)

Starch is a major nutrient. It is a plant polysaccharide with a molecular weight of more than 100,000. It consists of a mixture of linear chains of glucose molecules linked by α-1,4-glucosidic bonds (amylose) and of branched chains with branch points made up by α-1,6 linkages (amylopectin). The ratio of 1,4- to 1,6-glucosidic bonds is about 20 : 1. Glycogen is

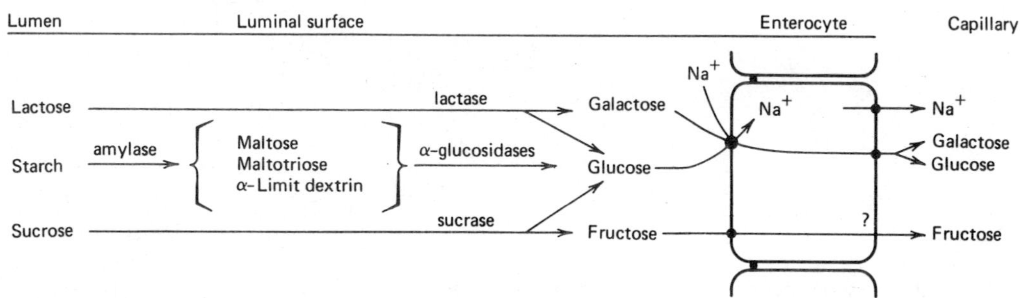

Figure 24.21
Digestion and absorption of carbohydrates.

TABLE 24.7 Dietary Carbohydrates

Carbohydrate	*Typical Source*	*Structure*	
Amylopectin	Potatoes, rice, corn, bread	α-Glc(1 ⟶ 4)$_n$Glc with α-Glc(1 ⟶ 6) branches	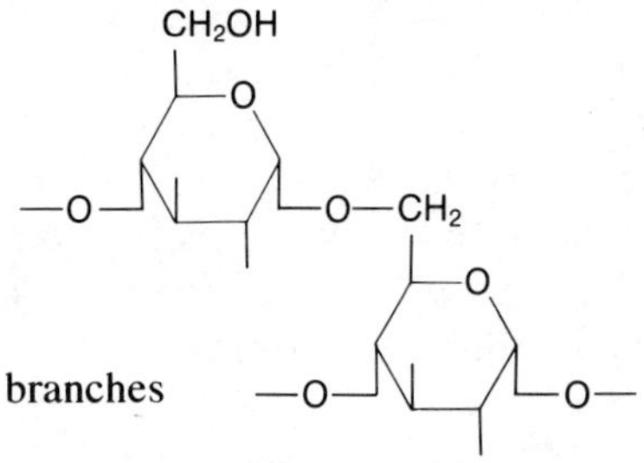
Amylose	Potatoes, rice, corn, bread	α-Glc(1 ⟶ 4)$_n$Glc	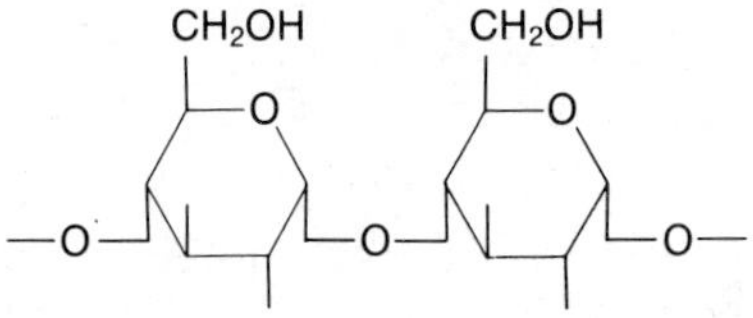
Sucrose	Table sugar, desserts	α-Glc(1 ⟶ 2)β-Fru	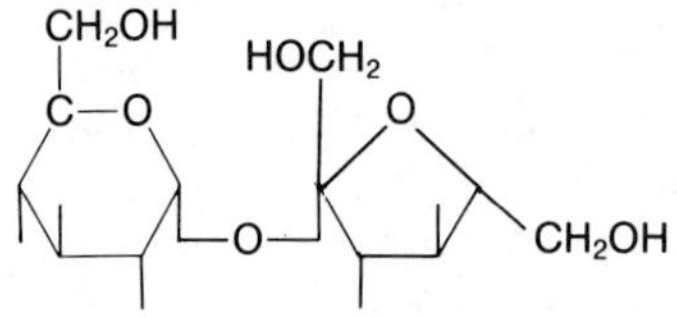
Trehalose	Young mushrooms	α-Glc(1 ⟶ 1)α-Glc	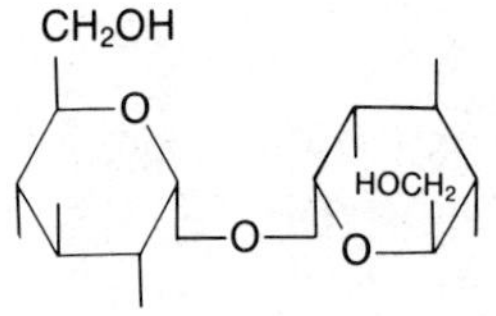
Lactose	Milk, milk products	β-Gal(1 ⟶ 4)Glc	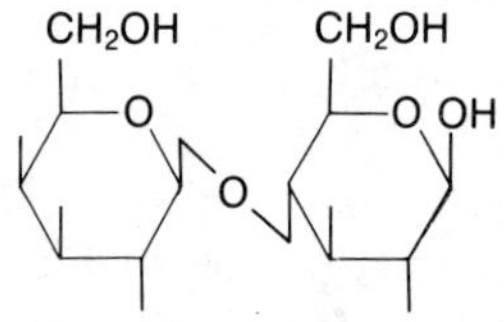
Fructose	Fruit, honey	Fru	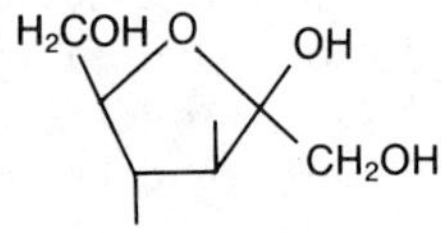
Glucose	Fruit, honey, grape	Glc	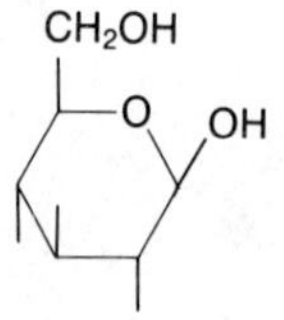
Raffinose	Leguminous seeds	α-Gal(1 ⟶ 6)α-Glc (1 ⟶ 2)β-Fru	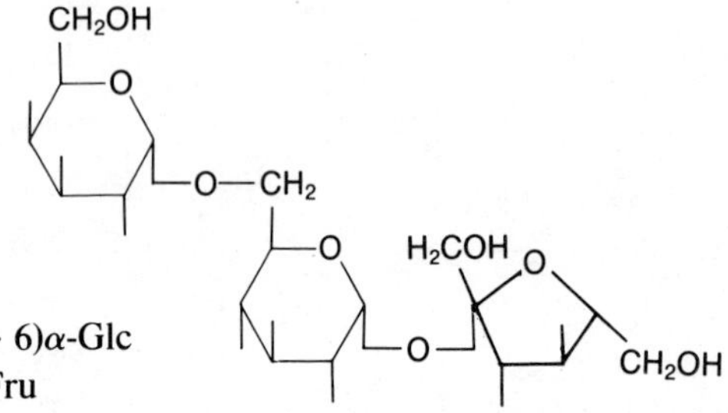

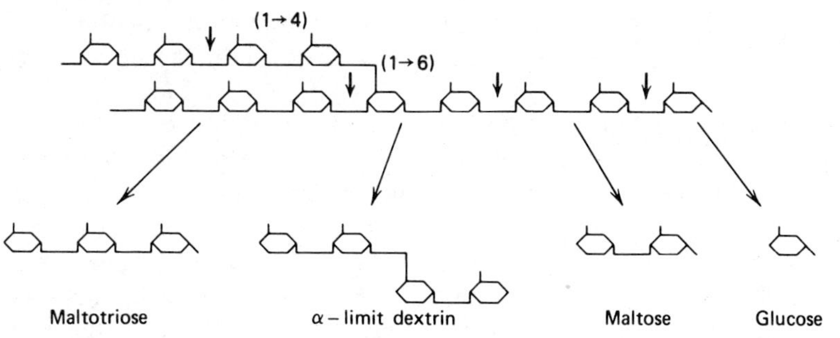

Figure 24.22
Digestion of amylopectin by salivary and pancreatic α-amylase.

an animal polysaccharide similar in structure to amylopectin. The two compounds differ in terms of the number of branch points, which occur more frequently in glycogen.

Hydrated starch and glycogen are attacked by the endosaccharidase α-amylase, which is present in saliva and pancreatic juice. (See Figure 24.22.) Hydration of the polysaccharides occurs during heating, which is essential for efficient digestion of starch. Amylase has specificity for internal α-1,4-glucosidic bonds; α-1,6 bonds are not attacked, nor are α-1,4 bonds of glucose units that serve as branch points. The pancreatic isoenzyme is secreted in large excess relative to starch intake and is more important than the salivary enzyme from a digestive point of view. The products of the digestion by α-amylase are mainly the disaccharide maltose [α-Glc(1 → 4)Glc], the trisaccharide maltotriose [α-Glc(1 → 4)α-Glc(1 → 4)Glc], and so-called α-limit dextrins containing on the average eight glucose units with one or more α-1,6-glucosidic bonds.

Final hydrolysis of di- and oligosaccharides to monosaccharides is carried out by surface enzymes of the small intestinal epithelial cells (Table 24.8). Most of the surface oligosaccharidases are exoenzymes which clip off one monosaccharide at a time from the nonreducing end. The capacity of the α-glucosidases normally is much greater than that needed for completion of the digestion of starch. Similarly, there usually is excess capacity for sucrose (table sugar) hydrolysis relative to dietary intake. In con-

TABLE 24.8 Di- and Oligosaccharidases of the Luminal Plasma Membrane in the Small Intestine

Enzyme	*Specificity*	*Natural Substrate*	*Product*
exo-1,4-α-Glucosidase (glucoamylase)	α-(1 ⟶ 4)Glucose	Amylose	Glucose
Oligo-1,6-glucosidase (isomaltase)	α-(1 ⟶ 6)Glucose	Isomaltose, α-dextrin	Glucose
α-Glucosidase (maltase)	α-(1 ⟶ 4)Glucose	Maltose, maltotriose	Glucose
Sucrose-α-Glucosidase (sucrase)	α-Glucose	Sucrose	Glucose, fructose
α,α-Trehalase	α-(1 ⟶ 1)Glucose	Trehalose	Glucose
β-Glucosidase	β-Glucose	Glucosyl-ceramide	Glucose, ceramide
β-Galactosidase (lactase)	β-Galactose	Lactose	Glucose, galactose

CLIN. CORR. 24.3
DISACCHARIDASE DEFICIENCY

Intestinal disaccharidase deficiencies are encountered relatively frequently in humans. Deficiency can be present in either a single enzyme or several enzymes for a variety of reasons (genetic defect, physiological decline with age, or as result of "injuries" to the mucosa). Of the disaccharidases, lactase is the most common enzyme with an absolute or relative deficiency, which is experienced as milk intolerance. The consequences of an inability to hydrolyze lactose in the upper small intestine are (1) inability to absorb lactose, and (2) bacterial fermentation of ingested lactose in the lower small intestine. Bacterial fermentation results in the production of gas (distension of gut, flatulence) and osmotically active solutes that draw water into the intestinal lumen (diarrhea).

trast, β-galactosidase (lactase) can be rate-limiting in humans for hydrolysis and utilization of lactose, the major milk carbohydrate (Clin. Corr. 24.3).

Di-, oligo-, and polysaccharides that are not hydrolyzed by α-amylase and/or intestinal surface enzymes cannot be absorbed; therefore they reach the lower tract of the intestine, which from the lower ileum on contains bacteria. Bacteria can utilize many of the remaining carbohydrates because they possess many more types of saccharidases than humans. The monosaccharides that are released as a result of bacterial enzymes are predominantly anaerobically metabolized by the bacteria themselves, resulting in such degradation products as short-chain fatty acids, lactate, hydrogen gas (H_2), methane (CH_4), and carbon dioxide (CO_2). These compounds can cause fluid secretion, increased intestinal motility, and cramps, either because of increased intraluminal osmotic pressure, distension of the gut, or because of a direct irritant effect of the bacterial degradation products on the intestinal mucosa.

The well-known problem of flatulence after ingestion of leguminous seeds (beans, peas, soya) can be traced to oligosaccharides, which cannot be hydrolyzed by human intestinal enzymes. The leguminous seeds contain modified sucrose to which one or more galactose moieties are linked. The glycosidic bonds of galactose are in the α-configuration, which can only be split by bacterial enzymes. The simplest sugar of this family is raffinose [α-Gal(1 → 6)α-Glc (1 → 2) β-Fru] (see Table 24.7).

Trehalose [α-Glc(1 → 1)α-Glc] is a disaccharide that occurs in young mushrooms. The digestion of this sugar requires a special disaccharidase, trehalase.

Absorption of Monosaccharides

The major monosaccharides that result from the digestion of di- and polysaccharide hydrolysis are D-glucose, D-galactose, and D-fructose. Absorption of these and other minor monosaccharides are carrier-mediated processes that exhibit such features as substrate specificity, stereospecificity, saturation kinetics, and inhibition by specific inhibitors.

At least two types of transport systems are known to catalyze the uptake of monosaccharides from the lumen into the cell: (1) a Na^+-monosaccharide cotransport system with high specificity for D-glucose and D-galactose, which catalyzes "active" sugar transport; (2) a Na^+-*independent*, facilitated-diffusion type of monosaccharide transport system with specificity for D-fructose. Additionally, a Na^+-*independent* monosaccharide transport system with specificity for D-glucose and D-galactose is present in the contraluminal plasma membrane. It mediates exit of monosaccharides from epithelial cells (for principles of transepithelial glucose transport (see page 921). The properties of the two transport systems accepting D-glucose have been investigated in some detail and are compared in Table 24.9. Although overlapping, the substrate specificity and

TABLE 24.9 Characteristics of Glucose Transport Systems in the Plasma Membranes of Enterocytes

Characteristic	*Luminal*	*Contraluminal*
Effect of Na^+	Cotransport with Na^+	None
Good substrates	D-Glc, D-Gal, α-methyl-D-Glc	D-Glc, D-Gal, D-Man, 2-deoxy-D-Glc
Inhibition by	Phlorizin	Cytochalasin B

the sensitivity to inhibitors suggest that different proteins are involved in Na^+-*dependent* and Na^+-*independent* glucose transport. The Na^+-glucose cotransport is inhibited by low concentrations of the plant glycoside phlorizin (Figure 24.23), whereas the Na^+-independent one is inhibited by cytochalasin B, (Figure 24.24), which is structurally unrelated to sugars. The inhibitors are not specific for the intestine, but are apparently effective throughout the body, that is, phlorizin also inhibits the Na^+-*dependent* glucose transport of renal proximal tubules and cytochalasin B inhibits the Na^+-*independent* one of erythrocytes, adipose tissue, muscle cells, and other tissues.

Figure 24.23
Phlorizin (phloretin-2′-β-glucoside).

24.6 DIGESTION AND ABSORPTION OF LIPIDS

General Considerations

An adult man ingests about 60–100 g of fat per day. Triacylglycerols constitute more than 90% of the dietary fat. The rest is made up of phospholipids, cholesterol, cholesterol esters, and free fatty acids. Additionally, 1–2 g cholesterol and 4–5 g phosphatidylcholine (lecithin) are secreted into the small intestinal lumen as constituents of bile.

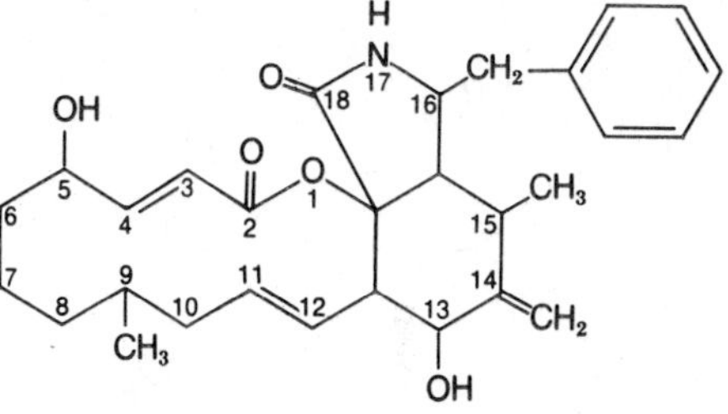

Figure 24.24
Cytochalasin B.

Lipids are defined by their good solubility in organic solvents. Conversely, they are sparingly or not at all soluble in aqueous solutions. The poor water solubility presents problems for digestion because the substrates are not easily accessible to the digestive enzymes in the aqueous phase. Additionally, even if ingested lipids are hydrolyzed into simple constituents, the products tend to aggregate to larger complexes that make poor contact with the cell surface and therefore are not easily absorbed. These problems are overcome by (1) increases in the interfacial area between the aqueous and the lipid phase, and (2) "solubilization" of the hydrolysis products with detergents. Thus changes in the physical state of lipids are intimately connected to chemical changes during digestion and absorption. (See Figure 24.25.)

At least five different phases can be distinguished:

1. Hydrolysis of triacylglycerols to free fatty acids and monoacylglycerols
2. Solubilization of free fatty acids and monoacylglycerols by detergents (bile acids) and transportation from the intestinal lumen toward the cell surface
3. Uptake of free fatty acids and monoacylglycerols into the cell and resynthesis to triacylglycerols

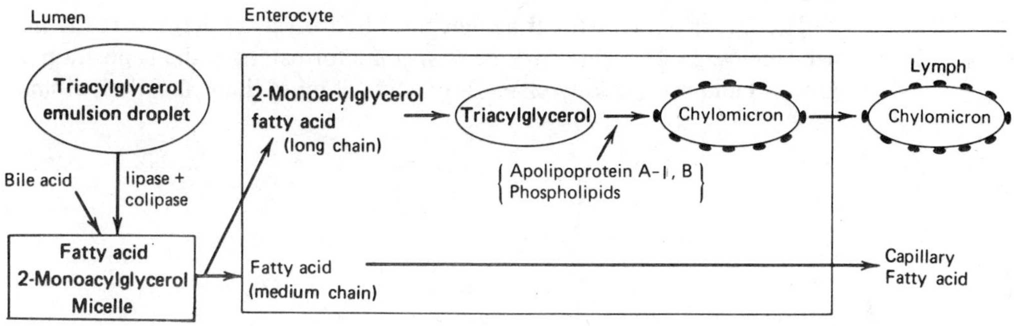

Figure 24.25
Digestion and absorption of lipids.

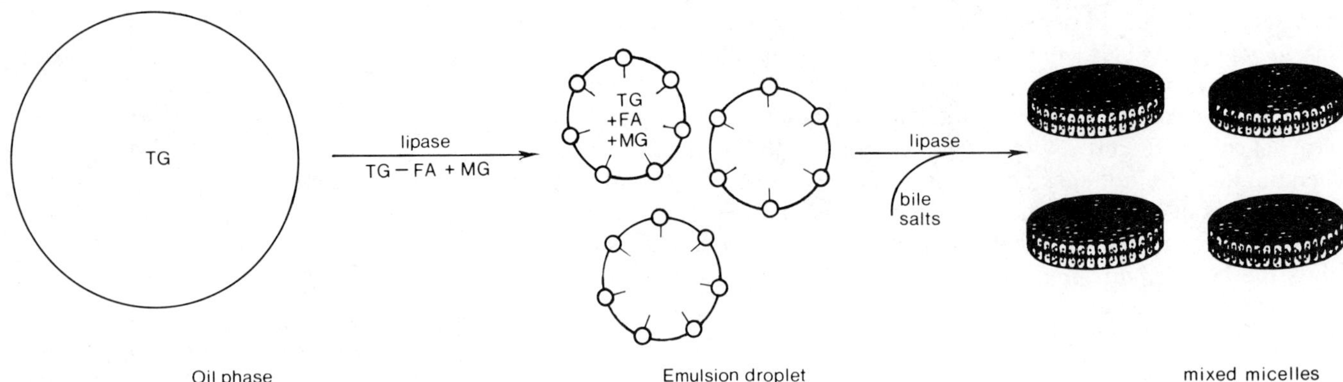

Figure 24.26
Changes in physical state during triacylglycerol digestion.
Abbreviations: TG = triacylglycerol, FA = fatty acid, MG = monoacylglycerol.

4. Packaging of newly synthesized triacylglycerols into special lipid-rich globules, called chylomicrons
5. Exocytosis of chylomicrons from cells and release into lymph

Digestion of Lipids

The digestion of lipids is initiated in the stomach by an acid-stable lipase, most of which is thought to originate from glands at the back of the tongue. However, the rate of hydrolysis is slow because the ingested triacylglycerols form a separate lipid phase with a limited water-lipid interface. The lipase adsorbs to that interface and converts triacylglycerols into fatty acids and monoacylglycerols. (See Figure 24.26.) The importance of the initial hydrolysis is that some of the water-immiscible triacylglycerols are converted to products that possess both polar and nonpolar groups. Such products spontaneously adsorb to water–lipid interfaces, and therefore are said to be *surfactive*. In effect, surfactants confer a hydrophilic surface to lipid droplets and thereby provide a stable interface with the aqueous environment. Among the dietary lipids, free fatty acids, monoacylglycerols, and phospholipids are the major surfactants. One of the effects of the action of lipase is then a release of surfactive molecules and through these an increase in interfacial area between lipid and water. At constant volume of the lipid phase, any increase in interfacial area produces dispersion of the lipid phase into smaller droplets (emulsification). Thus lipase autocatalytically enhances the availability of more triacylglycerol substrate through an increase in interfacial area.

The major enzyme for triacylglycerol hydrolysis is the pancreatic lipase (Figure 24.27). This enzyme is specific for esters in the α-position of glycerol and prefers long-chain fatty acids of more than 10 carbon atoms.

$$\underset{\text{Triacylglycerol}}{\begin{array}{c} H_2C{-}O{-}\overset{O}{\overset{\|}{C}}{-}R_1 \\ R_2{-}\overset{O}{\overset{\|}{C}}{-}O{-}CH \\ H_2C{-}O{-}\overset{O}{\overset{\|}{C}}{-}R_3 \end{array}} + 2H_2O \longrightarrow \begin{array}{c} R_1{-}CO_2H \\ + \\ R_2{-}CO_2{-}CH(CH_2{-}OH)_2 \\ + \\ R_3{-}CO_2H \end{array}$$

$\longrightarrow$ 2 fatty acids and 1 monoacylglycerol

R = hydrocarbon chain

Figure 24.27
Mechanism of action of lipase.

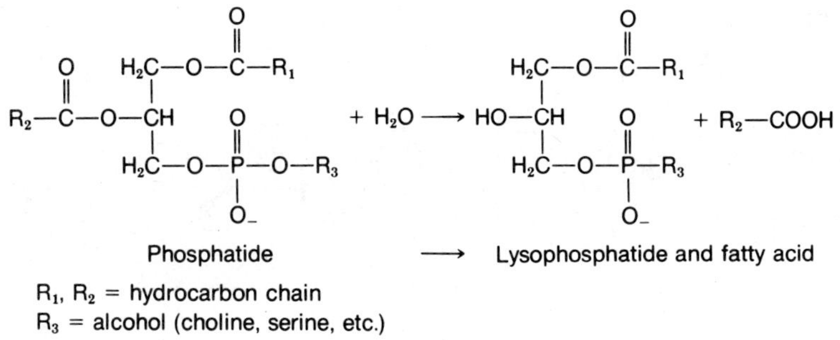

Figure 24.28
Mechanism of action of phospholipase A_2.

Triacylglycerol hydrolysis by the pancreatic enzyme also occurs at the water–lipid interface of emulsion droplets. The products are free fatty acids and β-monoacylglycerols. The purified form of the enzyme is strongly inhibited by the bile acids that normally are present in the small intestine during lipid digestion. The problem of inhibition is overcome by the addition of colipase, a small protein with a molecular weight of 12,000 dalton. Colipase binds to both the water-lipid interface and to lipase, thereby anchoring and activating the enzyme. It is secreted by the pancreas as procolipase and depends on tryptic removal of a NH_2-terminal decapeptide for full activity.

In addition to lipase, pancreatic juice contains another less specific lipid esterase. This enzyme acts on cholesterol esters, monoglycerides, or other lipid esters, such as esters of vitamin A with carboxylic acids. In contrast to triacylglycerol lipase, the less specific lipid esterase requires bile acids for activity.

Phospholipids are broken down by specific phospholipases. Pancreatic secretions are especially rich in the proenzyme for phospholipase A_2 (Figure 24.28). As other pancreatic proenzymes, this one, too, is activated by trypsin. Phospholipase A requires bile acids for activity.

Role of Bile Acids in Lipid Absorption

Bile acids are biological detergents that are synthesized by the liver and secreted with the bile into the duodenum. At physiological pH values, the acids are present as anions, which exhibit the detergent properties. Therefore, the terms bile acids and bile salts are often used interchangeably (Figure 24.29).

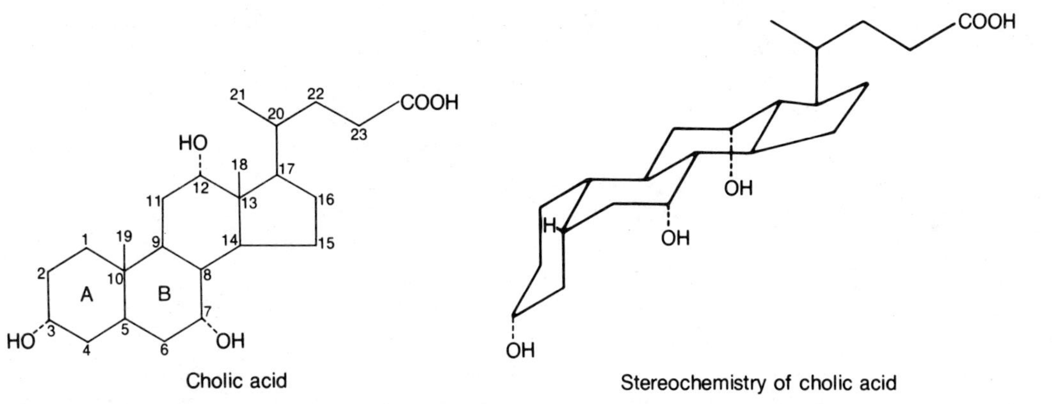

Figure 24.29
Cholic acid, a bile acid.

TABLE 24.10 The Effect of Conjugation on the Acidity of Cholic, Deoxycholic, and Chenodeoxycholic Acid

Bile Acid	*Ionized Group*	pK_a
Unconjugated bile acids	—CO_2^- of cholestanoic acid	≃5
Glycoconjugates	—CO_2^- of glycine	≃3.7
Tauroconjugates	—SO_3^- of taurine	≃1.5

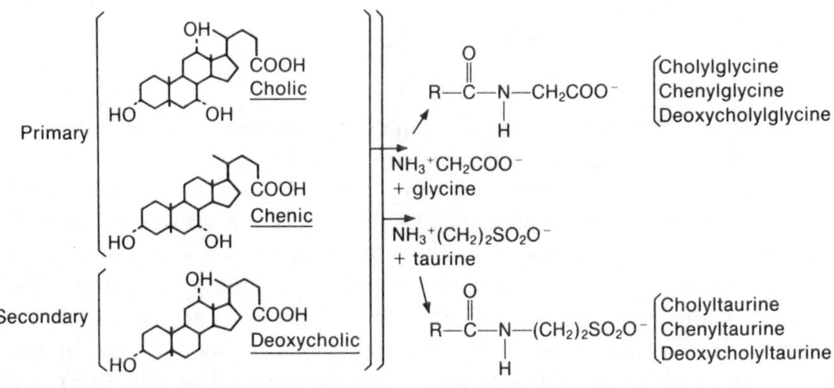

SOURCE: Reproduced with permission from Hofmann, A. F. *Handbook of Physiology* 5:2508, 1968.

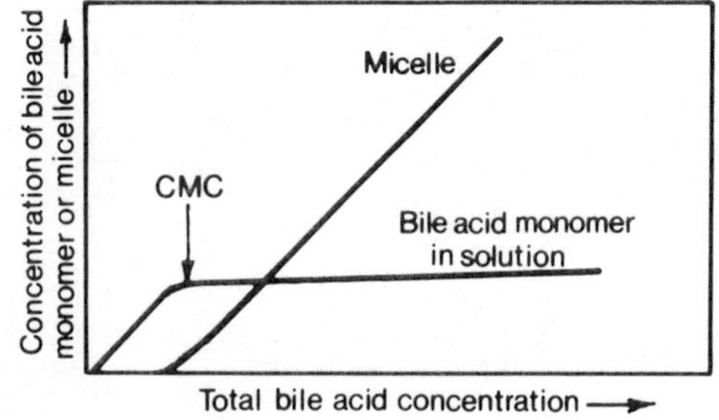

Figure 24.30
Solubility properties of bile acids in aqueous solutions.
Abbreviation: CMC = critical micellar concentration.

Bile acids at pH values above the p*K* (see Table 24.10) reversibly form aggregates at concentrations above 2–5 mM. These aggregates are called "micelles," and the minimal concentration necessary for micelle formation is the *critical micellar concentration* (Figure 24.30). Micelles are distinct from lipid emulsion droplets in that micelles are smaller than emulsion droplets, and bile acid molecules in the micelle are in equilibrium with free bile acid in solution.

The arrangements of bile acids in micelles is such that the hydrophobic portions of the molecule are removed from contact with water, while hydrophilic groups remain exposed to the water molecules. The hydrophobic region of bile acids is formed by one surface of the fused ring system, while the carboxylate or sulfonate ion and the hydroxyl groups on the other side of the ring system are hydrophilic. Since the major driving forces for micelle formation are the removal of apolar, hydrophobic groups from and the interaction of polar groups with water molecules, the distribution of polar and apolar regions places some constraints on the stereochemical arrangements of bile acid molecules within a micelle. Four bile acid molecules are sufficient to form a very simple micelle as shown in Figure 24.31.

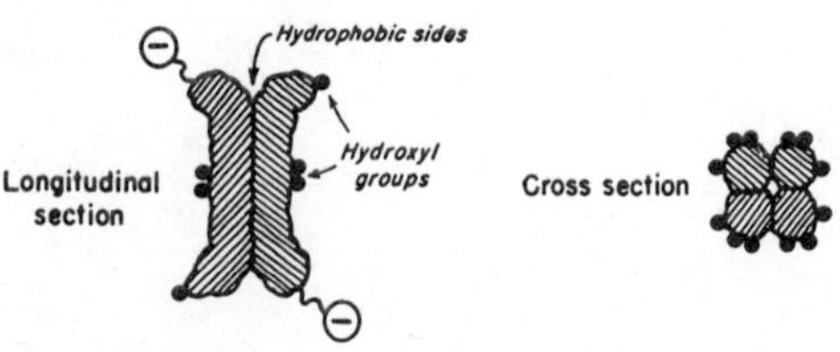

Figure 24.31
Diagrammatic representation of a Na⁺ cholate micelle.
Reproduced with permission from Small, D. M. *Biochim. Biophys. Acta* 176:178, 1969.

Bile salt micelles can solubilize other lipids, such as phospholipids and fatty acids. These mixed micelles have disklike shapes whereby the phospholipids and fatty acids form a bilayer and the bile acids occupy the edge positions, rendering the edge of the disk hydrophilic (Figure 24.32). Within the mixed phospholipid-bile acid micelles, other water-insoluble lipids, such as cholesterol, can be accommodated and thereby "solubilized" (for potential problems see Clin. Corr. 24.4).

During triacylglycerol digestion, free fatty acids and monoacylglycerols are released at the surface of fat emulsion droplets. In contrast to triacylglycerols, which are water-insoluble, free fatty acids and monoacylglycerols are slightly water-soluble, and molecules at the surface equilibrate with those in solution. The latter in turn become incorporated into bile acid micelles. Thus the products of triacylglycerol hydrolysis are

CLIN. CORR. 24.4 CHOLESTEROL STONES

Liver secretes, as constituents of bile, phospholipids and cholesterol together with bile acids. Because of the limited solubility of cholesterol, its secretion in bile can result in cholesterol stone formation in the gallbladder. Stone formation is a relatively frequent complication of bile secretion in humans.

Cholesterol molecules are practically insoluble in aqueous solutions. However, they can be incorporated into mixed phospholipid–bile acid micelles up to a mole ratio of 1 : 1 for cholesterol to phospholipids and thereby "solubilized." (See accompanying figure). Moreover, the liver can produce supersaturated bile with a higher ratio than 1 : 1 of cholesterol to phospholipid. This excess cholesterol has a tendency to crystallize out. Such bile with excess cholesterol is considered *lithogenic,* that is, stoneforming. The crystal formation usually occurs in the gallbladder, rather than the hepatic bile ducts, because contact times between bile and any crystallization nuclei are greater in the gallbladder. The tendency to secrete bile supersaturated with respect to cholesterol is inherited and found more frequently in females than in males, often associated with obesity. Supersaturation also appears to be a function of the size and nature of the bile acid pool as well as the secretion rate.

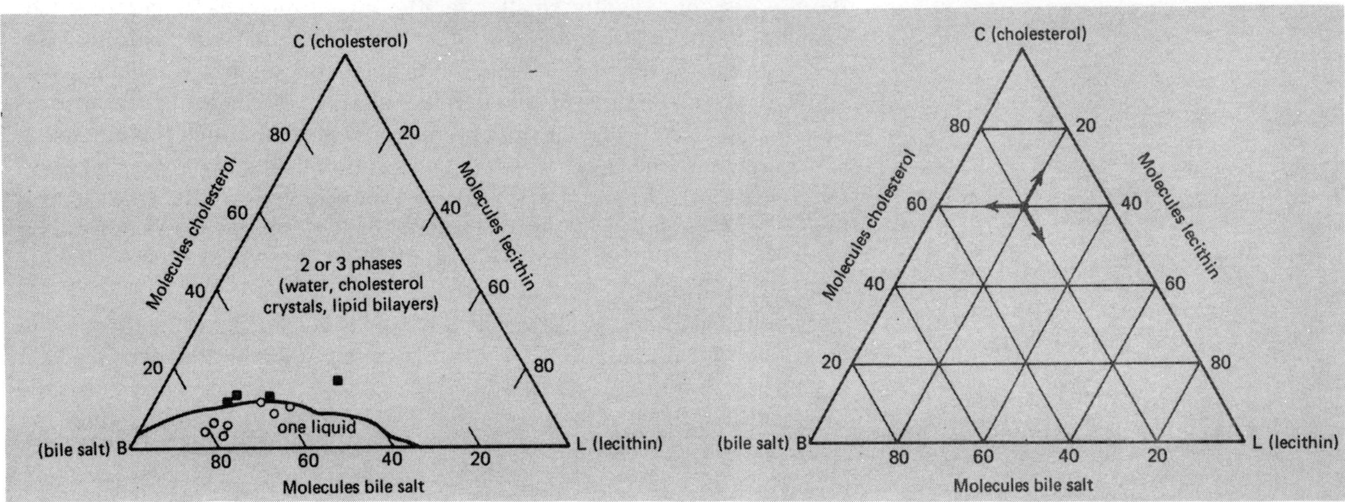

Diagram of the physical states of mixtures of 90% water and 10% lipid.
The 10% lipid is made up of bile acids, lecithin, and cholesterol, and the triangle represents all possible ratios of the three lipid constituents. Each point within the triangle corresponds to a particular composition of the three components, which can be read off the graph as indicated; each point on one of the sides corresponds to a particular composition of just two components. The left triangle contains the composition of gallbladder bile samples from patients without stones (○) and with cholesterol stones (■). Lithogenic bile has a composition that falls outside the "one liquid" area in the lower left corner.

Redrawn from Hofmann, A. F., and Small, D. M. *Annu. Rev. Med.* 18:362, 1967. Copyright 1967 by Annual Reviews Inc.

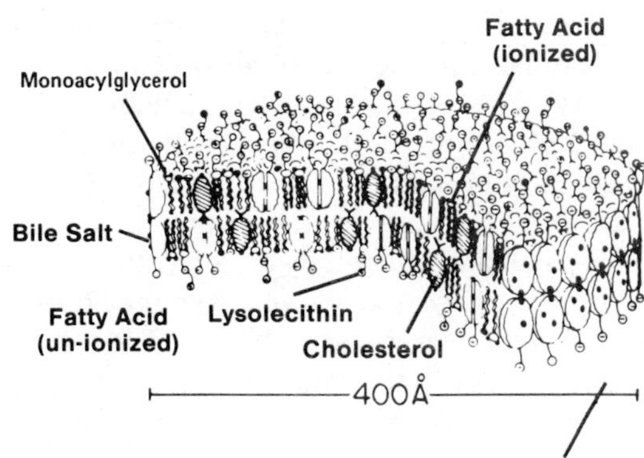

Figure 24.32
Proposed structure of the intestinal mixed micelle.
The bilayered disk has a band of bile salt at its periphery and other, more hydrophobic components (fatty acids, monoacylglycerol, phospholipids, and cholesterol) protected within its interior.

Reproduced with permission from M. C. Carey, In *The liver: Biology and pathology,* Arias, A. M., Popper, H., Schacter, D., et al. (eds). New York: Raven Press, 1982.

continuously transferred from emulsion droplets to the micelles (Figure 24.26).

Micelles provide the major vehicle for moving lipids from the intestinal lumen to the cell surface where absorption occurs. Because the fluid layer next to the cell surface is poorly mixed, the major transport mechanism for solute flux across this "unstirred" fluid layer is diffusion down the concentration gradient. With this type of transport mechanism, the delivery rate of nutrients at the cell surface is proportional to their concentration difference between luminal bulk phase and cell surface. Obviously, the unstirred fluid layer presents problems for sparingly soluble or insoluble nutrients, in that reasonable delivery rates cannot be achieved. Bile acid micelles overcome this problem for lipids by increasing their effective concentration in the unstirred layer. The increase in transport rate is nearly proportional to the increase in effective concentration and can be 1,000-fold over that of individually solubilized fatty acids, in accordance with the different solubility of fatty acids as micelles or as individual molecules. This relationship between flux and effective concentration holds because the diffusion constant, another parameter that determines the flux, is only slightly smaller for the mixed micelles as compared to micellar constituents free in solution. Thus, efficient lipid absorption depends on the presence of sufficient bile acids to "solubilize" the ingested and hydrolyzed lipids in micelles. In the absence of bile acids, the absorption of triacylglycerols does not completely stop, although the efficiency is drastically reduced. The residual absorption depends on the slight water solubility of the free fatty acids and monoacylglycerols. Unabsorbed lipids reach the lower intestine where a small part can be metabolized by bacteria. The bulk of unabsorbed lipids, however, is excreted with the stool (=steatorrhea).

Micelles also serve as transport vehicles through the unstirred fluid layers for those lipids that are even less water-soluble than fatty acids, such as cholesterol or the vitamins A and K. For these compounds, bile acid secretion is absolutely essential for absorption by the intestine.

Absorption of Lipids

The uptake of lipids by the epithelial cells of the small intestine can be explained on the basis of diffusion through the plasma membrane. Absorption is virtually complete for fatty acids and monoacylglycerols, which are slightly water-soluble. It is less efficient for water insoluble lipids. For example, only 30–40% of the dietary cholesterol is absorbed.

Within the intestinal cell, the fate of absorbed fatty acids depends on chain length. Fatty acids of medium chain length (6–10 carbon atoms) pass through the cell into the portal blood without modification. In contrast, the long-chain fatty acids (>12 carbon atoms) become bound to a soluble, fatty acid-binding protein in the cytoplasm and are transported to the endoplasmic reticulum, where they are resynthesized into triacylglycerols. Glycerol for this process is derived from the absorbed 2-monoacylglycerols and, to a minor degree, from glucose. The resynthesized triacylglycerols form lipid globules to which surface-active phospholipids and special proteins, termed *apolipoproteins,* adsorb. The lipid globules migrate within membrane-bound vesicles through the Golgi to the basolateral plasma membrane. They are finally released into the intracellular space by fusion of the vesicles with the basolateral plasma membrane. The final size can be several micrometers in diameter. Because the lipid globules can be so large and because they leave the intestine via lymph vessels, they are called *chylomicrons* (chyle = milky lymph that is present in the intestinal lymph vessels, lacteals, and the thoracic duct after a lipid meal; the work chyle is derived from the Greek *chylos*, which means juice). The intestinal apolipoproteins can be distinguished from those of the liver by antisera. The major intestinal ones are called A-I and B apolipoproteins. Apolipoprotein B is essential for chylomicron release from enterocytes (see Clin. Corr. 24.5).

While medium-chain fatty acids from the diet reach the liver directly with the portal blood, the long-chain fatty acids bypass the liver by being released in the form of chylomicrons into the lymphatics. The intestinal lymph vessels drain into the large body veins via the thoracic duct. Blood from the large veins first reaches the lungs and then the capillaries of the peripheral tissues, including adipose tissue and muscle, before it comes into contact with the liver. Fat and muscle cells in particular take up large amounts of dietary lipids for storage or metabolism. The bypass of the liver may have evolved to protect this organ from a lipid overload after a meal.

The differential handling of medium- and long-chain fatty acids by intestinal cells can be specifically exploited to provide the liver with fatty acids, which constitute high-caloric nutrients. Short- and medium-chain fatty acids are not very palatable; however, triacylglycerols synthesized from these fatty acids are quite palatable and can be used as part of the diet. In the small intestine the triacylglycerols would be hydrolyzed by pancreatic lipase and thus provide fatty acids that reach and can be utilized by the liver.

CLIN. CORR. 24.5
ABETALIPOPROTEINEMIA

Abetalipoproteinemia is a human, recessively inherited disorder. It is characterized by the absence of all lipoproteins containing apobetalipoprotein, i.e., chylomicrons, very low density lipoproteins (VLDL), and low density lipoprotein (LDL). The clinical features are triacylglycerol malabsorption and deficiency of fat-soluble vitamins, notably vitamin E. The defect is absent synthesis of apobetalipoprotein rather than inhibition of release from cells. In the enterocyte, lipid droplets accumulate at the apex of the cells.

24.7 BILE ACID METABOLISM

All bile acids are synthesized initially within the liver from cholesterol, but they can be modified by bacterial enzymes during passage through the intestinal lumen. The primary bile acids synthesized by the liver are cholic and chenodeoxycholic acid. The secondary bile acids are derived from the primary bile acids by bacterial dehydroxylation in position 7 of the ring structure, resulting in deoxycholate and lithocholate, respectively (Figure 24.33).

Primary and secondary bile acids are reabsorbed by the intestine into the portal blood, taken up by the liver, and then resecreted into bile. Within the liver, primary as well as secondary bile acids are linked to either glycine or taurine via an isopeptide bond. These derivatives are called glyco- and tauro- conjugates, respectively, and constitute the forms that are secreted into bile. With the conjugation, the carboxyl group of the unconjugated acid is replaced by an even more polar group. The p*K* values of the carboxyl group of glycine and of the sulfonyl group of taurine are lower than that of unconjugated bile acids, so that conjugated bile acids remain ionized over a wider pH range (Table 24.10). The conjugation is partially reversed within the intestinal lumen by hydrolysis of the isopeptide bond.

The total amount of conjugated and unconjugated bile acids secreted per day by the liver is 16–70 g for an adult. As the total body pool is only 3–4 g, bile acids have to recirculate 5–14 times each day between the intestinal lumen and the liver. Reabsorption of bile acids is important to conserve the pool, and the lower ileum contains a specialized Na^+-bile acid cotransport system for concentrative reuptake. Thus during a meal bile acids from the gallbladder and the liver are released into the lumen of the upper small intestine, pass with the chyme down the small intestinal lumen, are reabsorbed by the epithelium of the lower small intestine into the portal blood, and then extracted from the portal blood by the liver

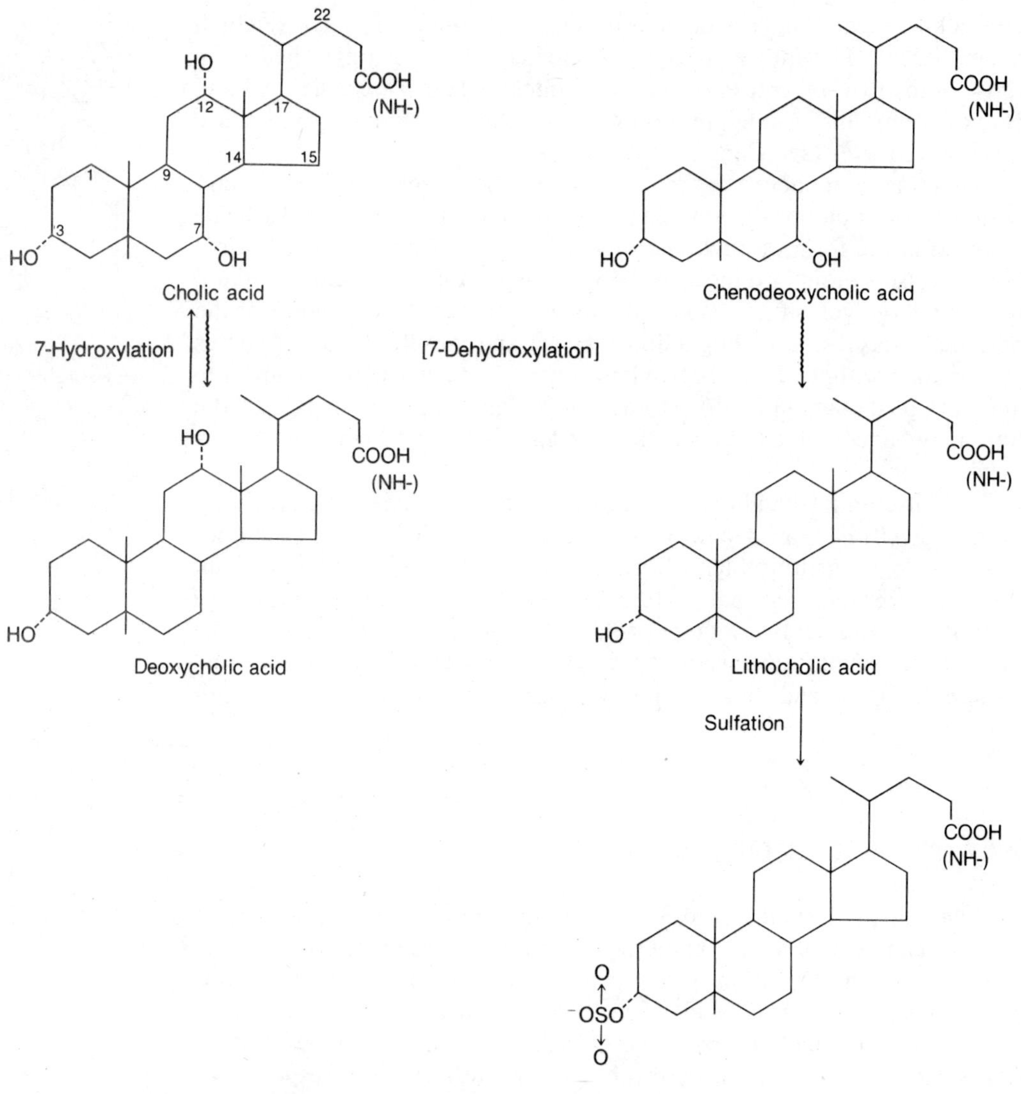

Figure 24.33
Bile acid metabolism in the rat.
Straight arrows indicate reactions catalyzed by liver enzymes; bent arrows indicate those of bacterial enzymes within the intestinal lumen. (NH −), glycine or taurine conjugate of the bile acids.

parenchymal cells during passage through the liver. The process of secretion and reuptake is referred to as the enterohepatic circulation (Figure 24.34). The reabsorption of bile acids by the intestine is quite efficient as only about 0.5 g bile acids escape reuptake each day and are secreted with the feces. Serum levels of bile acids normally vary with the rate of reabsorption and therefore are highest during a meal when the enterohepatic circulation is most active.

Cholate, deoxycholate, chenodeoxycholate, and their conjugates continously participate in the enterohepatic circulation. In contrast, most of the lithocholic acid that is produced by the action of bacterial enzymes within the intestine is sulfated during the next passage through the liver. The sulfate ester of lithocholic acid is not a substrate for the bile acid transport system in the ileum, and therefore is excreted with the feces.

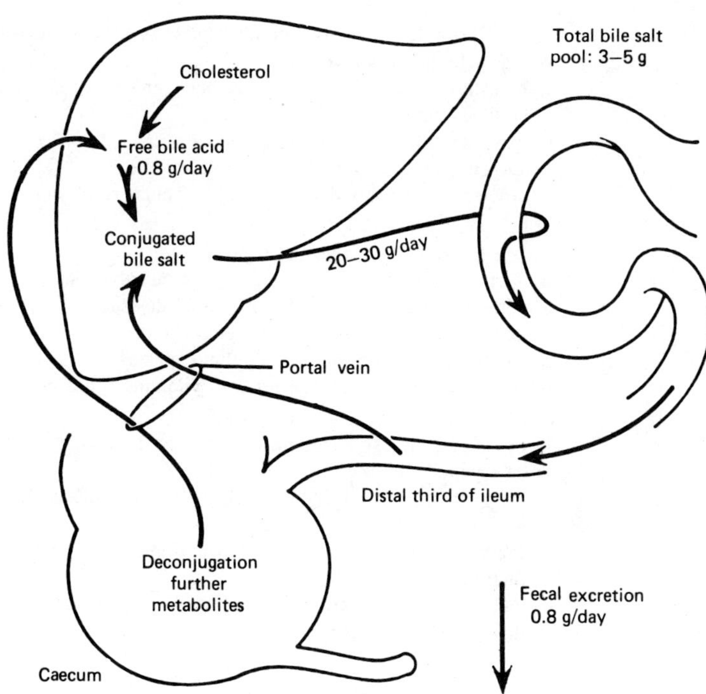

Figure 24.34
Enterohepatic circulation of bile acids.
Redrawn from Clark, M. L., and Harries, J. T. in *Intestinal absorption in man,* I. McColl and G. E. Sladen (eds.). New York: Academic Press, 1975, p. 195.

BIBLIOGRAPHY

Cristofaro, E., Mottu, F., and Wuhrmann, J. J. Involvement of the raffinose family of oligosaccharides in flatulence, in H. L. Sipple and K. W. McNutt (eds.), *Sugars in nutrition.* New York: Academic, 1974, p. 314.

Harris, T. J. (ed.). Familial inherited abnormalities. *Clin. Gastroenterol.* 11:17–72, 87–140, 1982.

Johnson, L. R. *Physiology of the gastrointestinal tract,* vols. 1 and 2. New York: Raven Press, 1981.

McColl, I., and Sladen, G. E. (eds.), *Intestinal absorption in man.* New York: Academic, 1975.

Peters, T. J. (ed.). Biochemical mechanisms in gastroenterology. *Clin. Gastroenterol.* 10:521–582, 671–706, 1981.

Porter, R., and Collins, G. M. *Brush border membranes.* Ciba Foundation Symposium, vol. 95. London: Pitman, 1983.

Sleisenger, M. H. (ed.). Malabsorption and nutritional support. *Clin. Gastroenterol.* 12:323–378, 1983.

QUESTIONS

J. BAGGOTT AND C. N. ANGSTADT

*Q*uestion *T*ypes are described inside the front cover.

1. (QT1) All of the following are involved in the digestive process *except:*
   A. mouth.
   B. pancreas.
   C. liver.
   D. spleen.
   E. gallbladder.
2. (QT2) Active forms of enzymes that digest food may normally be found:
   1. in soluble form in the lumen of the stomach.
   2. attached to the luminal surface of the plasma membrane of intestinal epithelial cells.
   3. dissolved in the cytoplasm of intestinal epithelial cells.
   4. in zymogen granules of pancreatic exocrine cells.
3. (QT1) Histamine stimulates secretion of:
   A. amylase by the salivary glands.
   B. HCl by the stomach.
   C. gastrin by the stomach.
   D. hydrolytic enzymes by the pancreas.
   E. $NaHCO_3$ by the pancreas.
4. (QT2) The luminal and contraluminal membranes of small intestinal epithelial cells:
   1. both contain systems for mediated transport of Na^+.
   2. differ in that only the contraluminal membrane contains Na^+, K^+-ATPase.
   3. both contain active transport systems that involve Na^+.
   4. both contain an energy-linked (active) transport system for glucose.

A. Stomach C. Both
B. Small intestine D. Neither

5. (QT4) Site of protein digestion.
6. (QT4) Site of ATP-linked K^+, H^+-antiport.
7. (QT4) Site of chymotrypsinogen synthesis.
8. (QT4) Site of amino acid absorption.
9. (QT1) Oral administration of large amounts of tyrosine could be expected to interfere with the intestinal absorption of:
   A. leucine.
   B. lysine.
   C. glycine.
   D. aspartate.
   E. none of the above.

A. Starch C. Both
B. Protein D. Neither

10. (QT4) Digestion is more efficient after heating.
11. (QT4) Absorbed intact (i.e., without digestion) under certain conditions.

A. Autoactivation C. Both
B. Autocatalysis D. Neither

12. (QT4) Pepsin.
13. (QT4) Gastric lipase.
14. (QT2) During the digestion and absorption of triacylglycerols:
    1. a gastric lipase is involved.
    2. hydrolysis occurs at the interface between lipid droplets and the aqueous phase.
    3. most of the triacylglycerol hydrolysis is carried out by a pancreatic enzyme that is inhibited by bile acids.
    4. efficiency is greatly decreased if bile acids are absent.
15. (QT1) Micelles:
    A. are the same as emulsion droplets.
    B. form from bile acids at all bile acid concentrations.
    C. although they are formed during lipid digestion, do not significantly enhance utilization of dietary lipid.
    D. always consist of only a single lipid species.
    E. are essential for the absorption of vitamins A and K.
16. (QT2) In the metabolism of bile acids:
    1. the liver synthesizes the primary bile acids, cholic, and chenodeoxycholic acids.
    2. primary and secondary bile acids may be conjugated to glycine or taurine.
    3. secondary bile acids are formed from primary bile acids by intestinal bacteria.
    4. daily bile acid secretion by the liver is approximately equal to daily bile acid synthesis.

ANSWERS

1. D A: Mechanical homogenization, fluid, amylase. B: HCO_3^-, amylase, proteases, lipase. C: Bile acid synthesis. E: Bile acid storage (p. 912).
2. A 1, 2, and 3 true. Zymogen granules contain inactive proenzymes or zymogens, which are not activated until after release from the cell (p. 912-913).
3. B Stimulation of H_2 receptors of the stomach causes HCl secretion (p. 915).
4. A 1, 2, and 3 true. Only the luminal surface contains a system for the Na^+-linked active uptake of glucose. Only the contraluminal surface contains the Na^+, K^+-ATPase. The contraluminal surface contains a transport system for glucose that equilibrates glucose across the membranes (p. 918, Table 24.5, p. 921).
5. C Pepsin acts in the stomach, producing peptides that stimulate cholecystokinin release, which activates the pancreatic stage of protein digestion. Pancreatic proteases and intestinal peptidases are active in the small intestine (p. 923).
6. A This system is responsible for H^+ secretion into the gastric lumen by the oxyntic cells (p. 922). See also item 4 above.
7. D Chymotrypsinogen is synthesized in the pancreas and is activated by trypsin in the small intestine (p. 924, Table 24.6).
8. B The small intestine has at least six amino acid transport systems (p. 926).
9. A Tyrosine shares a transport system with Val, Leu, Met, Phe, and Ile (p. 926).
10. C α-Amylase attacks hydrated starch more readily than unhydrated; heating hydrates the starch granules (p. 929). Proteolytic enzymes act more readily on denatured proteins (p. 924).
11. B Some proteins are absorbed intact a short time after birth. Persistence of this phenomenon may be involved in some food allergies (p. 927). In some species, for example the cow, this mechanism is of critical importance in transferring maternal antibodies to the newborn.
12. C Autoactivation is an intramolecular reaction, which produces active pepsin. The pepsin thus formed can, by autocatalysis, activate other pepsinogen molecules (p. 924).
13. B Gastric lipase digestion products are surface active agents, which stabilize small emulsion droplets. This process enlarges the surface area upon which the lipase acts. Thus the process is autocatalytic (p. 932).
14. E 1, 2, 3, and 4 true. 1 and 2: See item 13. 3: Pancreatic lipase carries out the bulk of lipid digestion but is inhibited by bile acids (p. 932). 4: A protein, colipase, prevents bile acid inhibition in vivo (p. 933). Bile acid micelles are the major vehicles for moving lipids to the cell surface, where absorption occurs (p. 936).
15. E A: Micelles are of molecular dimensions and are highly ordered structures; emulsion droplets are much larger and are random (p. 932, Figure 24.26; p. 935, Figure 24.32). B: Micelle formation occurs only above the critical micellar concentration (CMC); below that concentration the components are in simple solution (p. 934, Figure 24.31). C: See item 14. D: Micelles may consist of only one component, or they may be mixed (p. 934). E: The lipid soluble vitamins must be dissolved in mixed micelles as a prerequisite for absorption (p. 936).
16. A 1, 2, and 3 true. The primary bile acids are synthesized in the liver. In the intestine they may be modified to form the secondary bile acids. Both are reabsorbed and recirculated (enterohepatic circulation). Both are conjugated to glycine or taurine. Only a small fraction of the bile acid escapes reuptake; this must be replaced by synthesis (p. 937).

25

Principles of Nutrition I: Macronutrients

STEPHEN G. CHANEY

25.1 OVERVIEW

Nutrition is best defined as the utilization of foods by living organisms. Since the process of food utilization is clearly biochemical in nature, the major thrust of the next two chapters is a discussion of basic nutritional concepts in biochemical terms. However, simply understanding basic nutritional concepts is no longer sufficient. Nutrition appears to attract more than its share of controversy in our society, and a thorough understanding of nutrition almost demands an understanding of the issues behind these controversies. Thus these chapters also explore the biochemical basis for some of the most important nutritional controversies.

Why so much controversy in the first place? Part of the problem is simply that nonscientists feel competent to be "experts." After all, nutrition is concerned with food, and everyone knows about food. However, part of the problem is scientific. Both food itself and our utilization of it are very complex. Quite often the ideal scientific approach of examining only one variable at a time—isolated from other, related variables—is simply inadequate to handle this complex situation. Human variability raises further problems. In studying biochemistry the tendency is to study metabolic pathways and control systems as if they were universal, yet probably no two people utilize nutrients in exactly the same manner. Finally, it is important to realize that much of our knowledge of nutrient requirements and functions comes from animal studies. The days of using prison "volunteers" for nutritional experimentation are behind us. Yet animals are seldom completely adequate models for human beings, since their biochemical makeup almost always differs (the ability of most other animals to synthesize their own ascorbic acid is just one illustration). Thus, despite the best efforts of many reputable scientists, some of the most important nutritional questions have yet to be answered.

The study of human nutrition can logically be divided into three areas: undernutrition, overnutrition, and ideal nutrition. With respect to undernutrition, the primary concern in this country is not with the nutritional deficiency diseases, which are now quite rare. Today more attention is directed toward potential, or subclinical, nutritional deficiencies. Overnutrition, on the other hand, is a particularly serious problem in developed countries. Current estimates suggest that between 15 and 30% of the United States population are obese, and obesity is known to have a number of serious health consequences. Finally, there is increasing interest today in the concept of ideal, or optimal nutrition. This is a concept that has meaning only in an affluent society. Only when the food supply becomes abundant enough so that deficiency diseases are a rarity does it become possible to consider the long-range effects of nutrients on health. This is probably the most exciting and least understood area of nutrition today.

25.2 ENERGY METABOLISM

By now you should be well acquainted with the energy requirements of the body. Much of the foods we eat are converted to ATP and other high energy compounds, which are in turn utilized by the body to drive biosynthetic pathways, generate nerve impulses, and power muscle contraction. We generally describe the energy content of foods in terms of calories. Technically speaking, we are actually referring to the kilocalories of heat energy released by combustion of that food in the body. Some nutritionists today prefer to use the term kilojoule (a measure of mechanical energy), but since the American public is likely to be counting calories rather than joules in the foreseeable future, we will restrict ourselves to that term here. Experimentally, calories can be measured as the heat given off when a food substance is completely burned in a bomb calorimeter, although the caloric value of foods in our body is slightly less due to incomplete digestion and metabolism. The actual caloric values of protein, fat, carbohydrate, and alcohol are roughly 4, 9, 4, and 7 kcal/g, respectively. Given these data and the composition of the food, it is simple to calculate the caloric content (input) of the foods we eat. Calculating caloric content of foods does not appear to be a major problem in this country. Millions of Americans appear to be able to do that with ease. The problem lies in balancing caloric input with caloric output. Where do these calories go?

In practical terms, there are four principal factors that affect individual energy expenditure: surface area (which is related to height and weight), age, sex, and activity levels. (1) The effects of surface area are thought to be simply related to the rate of heat loss by the body—the greater the surface area, the greater the rate of heat loss. While it may seem surprising, a lean individual actually has a greater surface area, and thus a greater energy requirement, than an obese individual of the same weight. (2) Age, on the other hand, may reflect two factors: growth and lean muscle mass. In infants and children more energy expenditure is required for rapid growth, and this is reflected in a higher basal metabolic rate (rate of energy utilization in the resting state). In adults (even lean adults), muscle tissue is gradually replaced with fat and water during the aging process, resulting in a 2% decrease in basal metabolic rate (BMR) per decade of adult life. (3) As for sex, women tend to have a lower BMR than men due to a smaller percentage of lean muscle mass and the effects of female hormones on metabolism. (4) The effect of activity levels on energy requirements is obvious. However, most of us overemphasize the immediate, as opposed to the long-term, effects of exercise. For example, one would need to jog for over an hour to burn up the calories found in one piece of apple pie.

Yet, the effect of a regular exercise program on energy expenditure can be quite beneficial. Regular exercise will increase basal metabolic rate, allowing one to burn up calories more rapidly 24 h a day. A regular exercise program should be designed to increase lean muscle mass and should be repeated 4 or 5 days a week but need not be aerobic exercise to have an effect on basal metabolic rate. For an elderly or infirm individual, even a daily walk will, with time, help to increase basal metabolic rate.

The effect of all of these variables on energy requirements can be readily calculated and, assuming light activity levels (a safe assumption for most Americans) and ideal body weight (not such a safe assumption), are presented in tabular form as Recommended Dietary Allowances (RDAs) for energy in Table 25.1.

The RDAs are average values and are widely quoted, but they tell us little about the energy needs of individuals. For example, body composition is known to affect energy requirements, since lean muscle tissue has a higher basal metabolic rate than adipose tissue. This, in part, explains the higher energy requirements of athletes and the lower energy requirements of obese individuals. Hormone levels are important also, since thyroxine, sex hormones, growth hormones, and, to a lesser extent, epinephrine and cortisol are known to increase BMR. The effects of epinephrine and cortisol probably explain in part why severe stress and major trauma significantly increase energy requirements. Fever also causes a significant increase in energy requirements. Finally, energy intake itself has an inverse relationship to expenditure in that during periods of starvation or semistarvation BMR can decrease up to 50%. This is of great survival value in cases of genuine starvation, but not much help to the person who wishes to lose weight on a calorie-restricted diet. Clearly then, the above tables are only general guidelines and may bear little resemblance to the energy needs of a given individual—a fact that is important to remember in the treatment of obesity.

TABLE 25.1 Recommended Dietary Allowances for Energy Intake[a]

Age (yr) and Sex Group	*Weight (lb)*	*Height (in.)*	*Energy Needs (kcal)*
Infants			
0.0–0.5	13	24	kg × 115
0.5–1.0	20	28	kg × 105
Children			
1–3	29	35	1,300
4–6	44	44	1,700
7–10	62	52	2,400
Males			
11–14	99	62	2,700
15–18	145	69	2,800
19–22	154	70	2,900
23–50	154	70	2,700
51–75	154	70	2,400
76+	154	70	2,050
Females			
11–14	101	62	2,200
15–18	120	64	2,100
19–22	120	64	2,100
23–50	120	64	2,000
51–75	120	64	1,800
76+	120	64	1,600
Pregnancy			+300
Lactation			+500

SOURCE: From Recommended Dietary Allowances, Revised 1980, Food and Nutrition Board, National Academy of Sciences–National Research Council, Washington, D.C.

[a] The data in this table have been assembled from the observed median heights and weights of children, together with desirable weights for adults for mean heights of men (70 in.) and women (64 in.) between the ages of 18 and 34 years as surveyed in the U.S. population (DHEW/NCHS data).

25.3 PROTEIN METABOLISM

Normal Fate of Dietary Protein

Protein carries a certain mystique as a "body-building" food. While it is true that protein is an essential structural component of all cells, protein is equally important for maintaining the output of essential secretions such as digestive enzymes and peptide hormones. Protein is also needed to synthesize the plasma proteins, which are essential for maintaining osmotic balance, transporting substances through the blood, and maintaining immunity. However, the average adult in this country consumes far more protein than needed to carry out these essential functions. The excess protein is simply treated as a source of energy, with the glucogenic amino acids being converted to glucose and the ketogenic amino acids being converted to fatty acids and keto acids. Both kinds of amino acids will of course eventually be converted to triacylglycerol in the adipose tissue if fat and carbohydrate supplies are already adequate to meet energy requirements. Thus for most of us the only body-building obtained from high-protein diets is adipose tissue.

It has always been popular to say that the body has no storage depot for protein, and thus adequate dietary protein must be supplied with every meal. However, in actuality, this is not quite accurate. While there is no separate class of "storage" protein, there is a certain percentage of protein that undergoes a constant process of breakdown and resynthesis. In the fasting state the breakdown of this store of body protein is enhanced, and the resulting amino acids are utilized for glucose production, the synthesis of nonprotein nitrogenous compounds, and for the synthesis of the essential secretory and plasma proteins described above (see also Chapter 14). Even in the fed state, some of these amino acids will be utilized for energy production and as biosynthetic precursors. Thus the turnover of body protein is a normal process—and an essential feature of what is called nitrogen balance.

Nitrogen balance (Figure 25.1) is simply a comparison between the intake of nitrogen (chiefly in the form of protein) and the excretion of nitrogen (chiefly in the form of undigested protein in the feces and urea

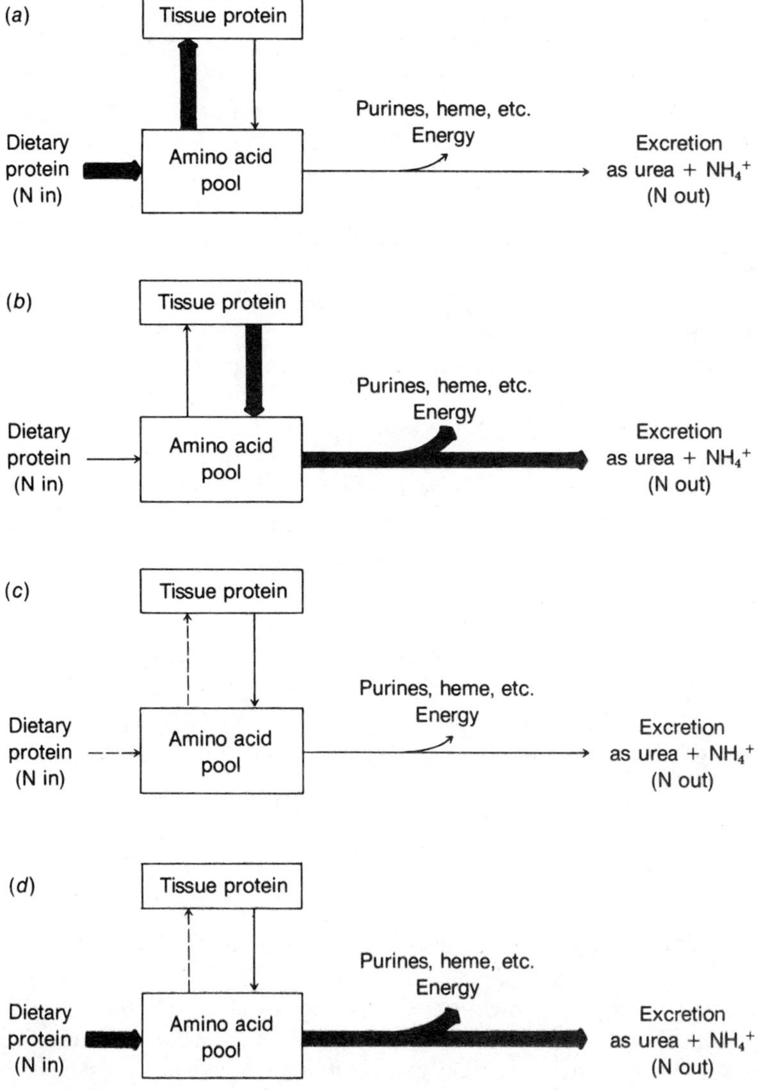

Figure 25.1
Factors affecting nitrogen balance.
Schematic representations of the metabolic interrelationship involved in determining nitrogen balance. (a) *Positive nitrogen balance (growth, pregnancy, lactation, recovery from metabolic stress).* (b) *Negative nitrogen balance (metabolic stress).* (c) *Negative nitrogen balance (inadequate dietary protein).* (d) *Negative nitrogen balance (lack of an essential amino acid). Each figure represents the nitrogen balance resulting from a particular set of metabolic conditions. The dominant pathways in each situation are indicated by heavy arrows.*

and ammonia in the urine). The normal adult will be in nitrogen equilibrium, with losses just balanced by intake. Negative nitrogen balance can result from an inadequate dietary intake of protein, since the amino acids utilized for energy and biosynthetic reactions are not replaced. However, it also is observed in injury when there is net destruction of tissue and in major trauma or illness in which the body's adaptive response causes increased catabolism of body protein stores (see Chapter 14). A positive nitrogen balance will be observed whenever there is a net increase in the body protein stores, such as in a growing child, a pregnant woman, or a convalescing adult.

Essential Amino Acids

In addition to the amount of protein in the diet, several other factors must be considered. One of these is the complement of essential amino acids present in the protein. The essential amino acids are those amino acids that cannot be synthesized by the body (Chapter 12). If just one of these essential amino acids is missing from the diet, the body cannot synthesize new protein to replace the protein lost due to normal turnover, and a negative nitrogen balance results (Figure 25.1). Obviously then, the complement of essential amino acids in any given protein will determine how well it can be used by our body. The "usefulness" of protein is most often expressed in terms of either net protein utilization (NPU) or biological value (BV).

NPU determinations involve putting subjects on a protein-free diet until their stores of labile protein are depleted. Then the protein to be tested is fed to the subject and the amount of this protein actually retained is calculated by subtracting the amount of nitrogen lost from the amount of nitrogen originally fed to the subject. The percentage of this dietary protein retained is defined as NPU:

$$\text{NPU} = \frac{\text{N retained}}{\text{N intake}} \times 100 = \frac{\text{N intake} - \text{N output}}{\text{N intake}} \times 100$$

The NPU score of a particular protein is a measure of two things: how well that protein is digested, and how well it is utilized once the amino acids have been absorbed into the system (which is primarily dependent on its complement of essential amino acids). BV, on the other hand, is defined as the percentage of the absorbed nitrogen that is retained by the body:

$$\text{BV} = \frac{\text{N retained}}{\text{N absorbed}} \times 100$$

Experimentally, this is determined by correcting nitrogen input for the undigested protein that appears in the feces. BV is solely a measure of how well the protein is used once it has been digested and absorbed.

Generally most animal proteins contain all of the essential amino acids in about the quantities needed by the human body and thus have high NPU scores (see Table 25.2). Vegetable proteins, on the other hand, often lack one or more essential amino acids and may, in some cases, be more difficult to digest. This is reflected by lower NPU scores. Even so, vegetarian diets can provide adequate protein provided (1) enough extra protein is consumed to provide sufficient quantities of the essential amino acids and/or (2) two or more proteins are consumed together, which complement one another in amino acid content. For example, if corn (which is deficient in lysine) is combined with legumes (which are defi-

CLIN. CORR. 25.1 VEGETARIAN DIETS AND PROTEIN-ENERGY REQUIREMENTS

One of the most important problems of a purely vegetarian diet (as opposed to a lacto-ovo vegetarian diet), is the difficulty in obtaining sufficient calories and protein. The potential caloric deficit results from the fact that the caloric densities of fruits and vegetables are much less than the meats they replace (30–50 cal/100 g vs 150–300 cal/100 g). The protein problem is generally threefold: (1) most plant products are much lower in protein (1–2 g protein/100 g vs 15–20 g/100 g); (2) most plant protein is of low biological value; and (3) some plant proteins are incompletely digested. Actually, any reasonably well-designed vegetarian diet will usually provide enough calories and protein for the average adult. In fact, the reduced caloric intake may well be of benefit because strict vegetarians do tend to be lighter than their nonvegetarian counterparts.

However, whereas an adult male may require about 0.8 g protein and 38 cal/kg body weight, a young child may require 2–3 times that amount. Similarly a pregnant woman needs an additional 30 g of protein and 300 cal/day and a lactating woman an extra 20 g of protein and 500 cal. Thus both young children and pregnant and lactating women run a risk of protein-energy malnutrition.

However, it is possible to provide sufficient calories and protein even for these high-risk groups provided the diet is adequately planned. There are three principles that can be followed to design a calorie/protein-sufficient vegetarian diet for

TABLE 25.2 Net Protein Utilization Values of Some Common Foods

Protein	*NPU in Young Children*
Human milk	95
Whole hen egg	87
Cow's milk	81
Polished rice	63
Peanuts	57
Soybean flour	54
Whole wheat	49
Maize	36

SOURCE: Taken from p67 of WHO Technical Report #522.

cient in methionine but rich in lysine), the NPU for the mixture approaches that of animal protein. The adequacy of vegetarian diets with respect to protein and calories is discussed more fully in Clin. Corr. 25.1, and the need for high quality protein in low protein diets in renal disease is discussed in Clin. Corr. 25.2.

Effect of Carbohydrate and Fat: Protein Sparing

Another factor that must be considered in determining protein requirements is the dietary intake of fat and carbohydrate. If these components are present in insufficient quantities, some of the dietary protein must be used for energy generation and is unavailable for building and replacing tissue. Thus, as the energy (calorie) content of the diet from carbohydrate and fat increases, the need for protein decreases. This is referred to as protein sparing. Carbohydrate is somewhat more efficient at protein sparing than fat—presumably because carbohydrate can be used as an energy source by almost all tissues, whereas fat cannot.

Normal Adult Protein Requirements

Assuming adequate calorie intake and a NPU of 75% on the mixed protein in the average American diet, the recommended protein intake is 0.8 g/(kg body wt · day). This amounts to about 56 g protein/day for a 70-kg (154-lb) man and about 44 g/day for a 55-kg (120-lb) woman. These recommendations would obviously need to be increased on a vegetarian diet if the overall NPU was less than 75%.

Other Factors Affecting Protein Requirements

Because dietary protein is essential for the synthesis of new body tissue, as well as for maintenance and repair, the need for protein increases markedly during periods of rapid growth. Such growth occurs during pregnancy, infancy, childhood, and adolescence. These recommendations are summarized in Table 25.3.

Once growth requirements have been considered, age does not seem to have much effect on protein requirements. If anything, the protein requirement may slightly decrease with age. However, older people need and generally consume fewer calories, so high-quality protein should provide a larger percentage of their total calories. Furthermore, some older people may have special protein requirements due to malabsorption problems.

One of the myths of popular nutrition is that athletes need more protein. While there may be a slight increase in protein requirements during periods of intensive body-building, vigorous exercise alone increases only energy needs—not protein requirements. Thus the protein needs of the athlete are only slightly greater than those of the nonathlete and can be met from a well-balanced diet that provides for the additional caloric needs.

Illness, major trauma, and surgery all cause a major catabolic response in the body. Both energy and protein needs are very large, and the body responds by increasing production of glucagon, glucocorticoids, and epinephrine (see Chapter 14). In these situations the breakdown of body protein is greatly accelerated and a negative nitrogen balance results unless protein intake is increased (Figure 25.1). Although this increased protein requirement is of little significance in short-term illness, it can be vitally important in the recovery of hospitalized patients as discussed in the next section (see also Clin. Corr. 25.3 and 25.4).

young children: (1) whenever possible, include eggs and milk in the diet. They are both excellent sources of calories and high-quality protein. (2) Include liberal amounts of those vegetable foods with high caloric density in the diet. These include nuts, grains, dried beans, and dried fruits. (3) Include liberal amounts of high-protein vegetable foods, at the same meal, which have complementary amino acid patterns.

CLIN. CORR. 25.2 LOW-PROTEIN DIETS AND RENAL DISEASE

Chronic renal failure is characterized by the buildup of the end products of protein catabolism (mainly urea). Since these toxic end products are responsible for many of the symptoms associated with renal failure, some degree of dietary protein restriction is usually necessary. The amount of protein restriction is dependent on the severity of the disease. It is easy to maintain patients in nitrogen equilibrium for prolonged periods on diets containing as little as 40 g protein/day if the diet is calorically sufficient. However, diets containing less than 40 g/day pose problems. Protein turnover continues and one is forced to walk a tightrope between providing enough protein to avoid negative nitrogen balance, but little enough to avoid buildup of waste products.

The strategy employed in such diets is twofold: (1) to provide a minimum of protein, primarily protein of high BV, and (2) to provide the rest of the daily calories in the form of carbohydrates and fats. The goal is to provide just enough essential amino acids to maintain positive nitrogen balance. In turn, the body should be able to synthesize the nonessential amino acids from other nitrogen-containing metabolites. Enough carbohydrate and fat are provided so that essentially all of the dietary protein can be spared from energy metabolism. With this type of diet, it is possible to maintain a patient on 20 g protein/day for considerable periods.

However, such diets are extremely monotonous and difficult to follow. A typical 20-g protein diet (the Giovannetti diet) is shown below:

1. One egg *plus* $\frac{3}{4}$ cup milk or 1 additional egg or 1 oz meat.

2. One-half pound of deglutenized (low protein) wheat bread; all other breads and cereals must be avoided—this includes almost all baked goods.
3. A limited amount of certain low-protein, low-potassium fruits and vegetables.
4. Sugars and fats to fill the rest of the needed calories; however, cakes, pies, and cookies would need to be avoided.

The palatability of this diet would be improved considerably, if it were possible to include more foods containing protein of low BV (vegetables, cereals, bread) for variety, yet still maintain a positive nitrogen balance. One approach to this problem is to use dietary supplements (usually in liquid or tablet form), which provide only the essential amino acids. Another approach is to use supplements containing the carbon skeletons (in the form of α-hydroxy and α-keto acids) of the essential amino acids. Both of these techniques do work, but their feasibility is limited by cost and poor taste.

TABLE 25.3 Recommended Dietary Allowances for Protein

Age (yr) and Sex Group	*Weight*		*Height*		*Protein*
	kg	*lb*	*cm*	*in.*	*(g)*
Infants					
0.0–0.5	6	13	60	24	kg × 2.2
0.5–1.0	9	20	71	28	kg × 2.0
Children					
1–3	13	29	90	35	23
4–6	20	44	112	44	30
7–10	28	62	132	52	34
Males					
11–14	45	99	157	62	45
15–18	66	145	176	69	56
19–22	70	154	177	70	56
23–50	70	154	178	70	56
51+	70	154	178	70	56
Females					
11–14	46	101	157	62	46
15–18	55	120	163	64	46
19–22	55	120	163	64	44
23–50	55	120	163	64	44
51+	55	120	163	64	44
Pregnancy					+30
Lactation					+20

SOURCE: From Recommended Dietary Allowances, Revised 1980, Food and Nutrition Board, National Academy of Sciences, National Research Council, Washington, D.C.

CLIN. CORR. 25.3 HYPERALIMENTATION AND THE CATABOLIC RESPONSE

The normal metabolic response to infection, trauma, and surgery is a complex and carefully balanced catabolic state. As discussed in the text, epinephrine, glucagon, and cortisol are released, greatly accelerating the rates of lipolysis, proteolysis, and gluconeogenesis. The net result is an increased supply of fatty acids and glucose to meet the increased energy demands of such major stress. The high serum glucose does, in turn, result in some elevation of circulating insulin levels. However, the moderate increase in insulin is more than counterbalanced by the increased levels of epinephrine and other hormones, which render some tissues less responsive to insulin. Skeletal muscle, for example, uses very little of the serum glucose, but continues instead to rely on free fatty acids and its own catabolized protein as a primary source of energy. It also continues to export amino acids (primarily alanine) for use elsewhere in the body, resulting in a very rapid depletion of body protein stores.

For many years, hospitalized patients routinely received intravenous solutions of glucose and water, on the assumption that the primary goal of the catabolic response was to provide adequate glucose levels during periods of stress. The idea was to provide an exogenous source of glucose for increased energy needs and at the same time to raise insulin levels sufficiently to exert a sparing effect on wastage of skeletal protein. In fact, intravenous glucose was very effective at reaching those objectives. Circulating insulin levels do become high enough so that muscle relies primarily on glucose as an energy source, and the overall negative nitrogen balance can be minimized.

However, recent studies have shown that the normal catabolism of skeletal protein during stress situations plays another very important role. It provides the amino acid building blocks for the synthesis of albumin, transferrin, and other essential secretory proteins and for the production of immunocompetent cells and their secretory products. High insulin levels, of course, prevent this response. Patients maintained on 5% glucose while experiencing catabolic stress from surgery,

25.4 PROTEIN-ENERGY MALNUTRITION

The most common form of malnutrition in the world today is protein energy malnutrition (PEM). In the developing countries inadequate intake of protein and energy is all too common, and it is usually the infants and young children who suffer most. While the actual symptoms of protein-energy insufficiency vary widely from case to case, it is common to classify all cases as either marasmus or kwashiorkor. Marasmus is usually defined as inadequate intake of both protein and energy and kwashiorkor as inadequate intake of protein in the presence of adequate energy intake. More often than not, the diets associated with marasmus and kwashiorkor may be similar with the kwashiorkor being precipitated by conditions of increased protein demand. The marasmic infant will have a thin, wasted appearance and will be small for his/her age. If the PEM continues long enough the child will be permanently stunted in both physical and mental development. In kwashiorkor, on the other hand, the child will often have a deceptively plump appearance due to edema. Other telltale symptoms associated with kwashiorkor are dry, brittle hair, diarrhea, dermatitis of various forms, and retarded growth. Perhaps the most devastating result of both marasmus and kwashiorkor is reduced ability of the afflicted individuals to fight off infection. They have a reduced number of T lymphocytes (and thus diminished cell-mediated immune response) as well as defects in the generation of phagocytic cells and production of immunoglobulins, interferon, and other components of the immune system. Many of these individuals die from secondary infections, rather than from the starvation itself. In the United States, classical marasmus and kwashiorkor are exceedingly rare, but milder forms of protein-energy malnutrition are seen.

The most common form of PEM seen in the United States today occurs in the hospital setting. A typical course of events is as follows: The patient is not eating well for several weeks or months prior to entering the hospital due to chronic or debilitating illness. He/she enters the hospital with major trauma, severe infection, or for major surgery, all of which cause a large negative nitrogen balance. This is often compounded by difficulties in feeding the patient or by the necessity of fasting the patient in preparation for surgery or diagnostic tests. The net result is PEM as measured by low levels of serum albumin and other serum proteins or by decreased cellular immunity tests. Recent studies have shown that hospitalized patients with demonstrable PEM have delayed wound healing, decreased resistance to infection, increased mortality, and increased length of hospitalization. Currently most major hospitals have instituted programs to monitor the nutritional status of their hospitalized patients and to intervene where necessary to maintain a positive nitrogen and energy balance (Clin. Corr. 25.4).

25.5 EXCESS PROTEIN-ENERGY INTAKE

Much has been said in recent years about the large quantities of protein that the average American consumes. Certainly most of us do consume far more than needed to maintain positive nitrogen balance. The average American currently consumes 99 g of protein, 68% of it from animal sources. However, most studies seem to show that a healthy adult can consume that quantity of protein with no apparent harm. Concern has been raised about the possible effect of high-protein intake on calcium requirements. Some studies suggest that high-protein intakes increase

trauma, or infection show a very rapid fall in both serum albumin and immunocompetence and can often develop kwashiorkor-like symptoms in as little as 1–2 weeks. This can delay wound healing and predispose the patient to potentially fatal infections.

Current hospital practice is to use 5% glucose solutions in short-term situations when the patient has been previously well nourished. However, when the patient enters the hospital in a malnourished state or when oral intake is not possible for a long time, protein supplementation is usually included.

CLIN. CORR. 25.4 PROVIDING ADEQUATE PROTEIN AND CALORIES FOR THE HOSPITALIZED PATIENT

A highly catabolic hospitalized patient may require 35–45 kcal/kg/day and 2–3 g protein/kg/day. A patient with severe burns may require even more. The physician has a number of options available to provide this postoperative patient with sufficient calories and protein to insure optimal recovery. When the patient is simply unable to take in enough food, it may be adequate to supplement the diet with high-calorie/high-protein preparations (which are usually mixtures of homogenized corn starch, egg, milk protein, and flavorings). When the patient is unable to take in solid food or unable to digest complex mixtures of foods adequately, elemental diets are usually administered via a nasogastric tube. Elemental diets consist of small peptides or purified amino acids, glucose and dextrins, some fat, vitamins, and electrolytes. These diets are generally low residue and can be used in patients with lower gastrointestinal tract disturbances. They are also very efficiently digested and absorbed in the absence of pancreatic enzymes or bile salts. These diets are sometimes sufficient to meet most of the short-term caloric and protein needs of a moderately catabolic patient.

However, when the patient is severely catabolic or unable to digest and absorb foods normally, parenteral (intravenous) nutrition is necessary. The least invasive method is to use a peripheral, slow-flow vein in a manner similar to any other IV infusion. The main limitation of this method is hypertonicity. If the infusion fluid is too hypertonic, there is endothelial cell damage and thrombosis. However, a solution of 5% glucose and 4.25% purified amino acids can be used safely. This solution will usually provide enough protein to maintain positive nitrogen balance, but will rarely provide enough calories for long-term maintenance of a catabolic patient.

The most aggressive nutritional therapy is total parenteral nutrition. Usually an indwelling catheter is inserted into a large fast-flow vessel such as the superior vena cava, so that the very hypertonic infusion fluid can be rapidly diluted. This allows solutions of up to 60% glucose and 4.25% amino acids to be used, providing sufficient protein and most of the calories for long-term maintenance. Intravenous lipid infusion is often added to boost calories and provide essential fatty acids. All of these methods can be used to prevent or minimize the negative nitrogen balance associated with surgery and trauma. The actual choice of method depends on the patient's condition. As a general rule it is preferrable to use the least invasive technique.

CLIN. CORR. 25.5 FAD DIETS: HIGH PROTEIN–HIGH FAT

Among the most popular fad diets over the years have been the high-protein/high-fat diets such as the Stillman and Atkins diets. Actually, at first glance, it seems as if these diets should have a sound metabolic basis. The basic premise is that if one severely restricts carbohydrate intake, it is possible to eat large amounts of high-protein/high-fat foods because the body will not be able to utilize the fat efficiently. This hypothesis is primarily based on the fact, made abundantly clear in any biochemistry textbook, that glucose is needed to replenish the intermediates of the citric acid cycle. Thus, in the absence of glucose, fat should simply be converted to ketone bodies and be disposed of.

There are several problems with this oversimplified hypothesis. First, it ignores the fact that many amino acids can be readily converted to citric acid cycle intermediates. Second, many tissues in the body are perfectly able to use ketone bodies for energy generation. Third, the loss of ketones in the urine cannot possibly lead to any significant caloric deficit. Maximum ketone excretion is about 20 g (100 kcal)/day.

urinary loss of calcium and thus may accelerate the bone demineralization associated with the aging process. However, this issue is far from settled.

Perhaps the more serious nutritional problem in this country is excessive energy consumption. In fact, obesity is the most frequent nutritional disorder in the United States. A discussion of the treatment of obesity is clearly beyond the scope of this chapter, but it is worthwhile to consider some of the metabolic consequences of obesity. One striking clinical feature of overweight individuals is a marked elevation of serum free fatty acids, cholesterol, and triacylglycerols irrespective of the dietary intake of fat. Why is this? Obesity is obviously associated with an increased number and/or size of adipose cells. Furthermore, these cells contain fewer insulin receptors and thus respond more poorly to insulin, resulting in increased activity of the hormone-sensitive lipase. The increased lipase activity along with the increased mass of adipose tissue is probably sufficient to explain the increase in circulating free fatty acids. These excess fatty acids are, of course, carried to the liver, where they are broken down to acetyl CoA, which is a precursor for both triacylglycerol and cholesterol synthesis. The excess triacylglycerol and cholesterol are released as very low density lipoprotein particles, leading to higher circulating levels of both triacylglycerol and cholesterol (for more detail on these metabolic interconversions, see Chapters 10 and 14).

A second striking finding in obese individuals is higher fasting blood sugar levels and decreased glucose tolerance. Fully 80% of adult onset diabetics are overweight. Again the culprit appears to be the decrease in insulin receptors, since many adult onset diabetics have higher than normal insulin levels. Because of these metabolic changes, obesity is one of the primary risk factors in coronary heart disease, hypertension, and diabetes. This is nutritionally significant because all of these metabolic changes are reversible. Quite often reduction to ideal weight is the single most important mode of nutritional therapy. Furthermore, when the individual is at ideal body weight, the composition of the diet becomes a less important consideration in maintaining normal serum lipid and glucose levels.

Any discussion of weight reduction regimens should include a mention of one other metabolic consequence of obesity. Aldosterone levels are also elevated, leading to increased retention of both sodium and water. Furthermore, in some cases aldosterone levels increase even more when the obese person begins dieting. Thus, in effect as the fat stores are metabolized, they are converted to water (which is denser than the fat), and the water may be largely retained. In fact, some individuals may actually observe short-term weight gain on certain diets, even though the diet is working perfectly well in terms of breaking down their adipose tissue. This metabolic fact of life can be psychologically devastating to the dieters, who expect to see quick results for all their sacrifice. This is one major reason for the popularity of the low-carbohydrate diets, which decrease water retention (Clin. Corr. 25.5).

25.6 CARBOHYDRATES

The chief metabolic role of carbohydrates in the diet is for energy production. Any carbohydrate in excess of that needed for energy is converted to glycogen and triacylglycerol for long-term storage. The human body can adapt to a wide range of carbohydrate levels in the diet. Diets high in carbohydrate result in higher steady-state levels of glucokinase and some of the enzymes involved in the hexose monophosphate shunt and triacylglycerol synthesis. Diets low in carbohydrate result in higher steady-state levels of some of the enzymes involved in gluconeogenesis, fatty acid oxidation, and amino acid catabolism. Glycogen stores can also be affected by the carbohydrate content of the diet (Clin. Corr. 25.6). Very low levels of carbohydrate result in a permanent state of ketosis similar to that seen during starvation. The mechanism has been discussed earlier (Chapter 14). If continued over a period of time this ketosis may, in some instances, be detrimental to the patient's health. Most Americans, of course, do consume more than adequate carbohydrate levels.

The most common nutritional problems involving carbohydrates are seen in those individuals with various carbohydrate intolerances. The most common form of carbohydrate intolerance is diabetes mellitus, which is caused either by lack of insulin production or lack of insulin receptors. This causes an intolerance to glucose and those simple sugars that can be readily converted to glucose. The dietary treatment of diabetes usually involves limiting intake of most simple sugars, and increasing intake of those carbohydrates that are better tolerated, mostly complex carbohydrates, but including some simple sugars such as fructose and sorbitol. Lactase insufficiency is also a common disorder of carbohydrate metabolism affecting over 30 million people in the United States alone. It is most prevalent among blacks, Asians, orientals, and South Americans. Without the enzyme lactase, the lactose is not significantly hydrolyzed or absorbed. It remains in the intestine where it acts osmotically to draw water into the gut and serves as a substrate for conversion to lactic acid, CO_2, and H_2S by intestinal bacteria. The end result is bloating, flatulence, and diarrhea—all of which can be avoided simply by eliminating milk and milk products from the diet.

25.7 FATS

Triacylglycerols, or fats, can be directly utilized by many tissues of the body as an energy source and, as phospholipids, are an important part of membrane structure. Any excess fat in the diet can be stored as triacylglycerol only. As with carbohydrate, the human body can adapt to a wide range of fat intakes. However, some problems can develop at the extremes (either high or low) of fat consumption. At the low end, essential fatty acid (EFA) deficiencies may become a problem. The fatty acids linoleic, linolenic, and arachidonic acid cannot be made by the body and thus are essential components of the diet. These EFA are needed for maintaining the function and integrity of membrane structure, for fat metabolism and transport, and for synthesis of prostaglandins. The most characteristic symptom of essential fatty acid deficiency is a scaly dermatitis. EFA deficiency is very rare in the United States, being seen primarily in low-birth weight infants fed on artificial formulas lacking EFA and in hospitalized patients maintained on total parenteral nutrition for long periods of time.

At the other end of the scale, there is legitimate concern that excess fat in the diet does cause elevation of serum lipids and thus an increased risk of heart disease. Most experts agree with that general conclusion. Unfortunately, there is no firm consensus as to how much is too much. However, our understanding of metabolism tells us that any fat consumed in excess of energy needs (except for the small amount needed for membrane formation) has nowhere to go but our adipose tissue, and obesity is correlated with an increased risk of heart disease, diabetes, and stroke.

It is important, however, to realize that this diet does appear to "work" for many patients. The apparent success of the diet is related primarily to two factors. In the first place, any low carbohydrate diet results in a significant initial water loss. This is primarily due to depletion of glycogen reserves, since 3 g of water is bound for every 1 g of glycogen. It is this rapid initial weight loss which makes the diet so appealing. Second, while a high-protein/high-fat diet sounds appealing initially, it is relatively unpalatable and expensive, leading ultimately to decreased caloric intake.

This diet is also not without its health risks. The high-fat content may contribute to atherosclerosis and heart disease. The lack of fruits and vegetables may lead to vitamin deficiencies. Finally, ketosis should be avoided by pregnant women (ketone bodies can be harmful to the developing fetal brain), and high-protein intakes should be avoided by anyone with a history of kidney disease.

CLIN. CORR. 25.6 CARBOHYDRATE LOADING AND ATHLETIC ENDURANCE

Much of the folklore concerning special nutritional requirements or specific nutritional regimes for athletes has little basis in fact. However, one common practice, that of carbohydrate loading, does appear to have biochemical backing. This technique is frequently used by track athletes to increase endurance. It consists of a 4- to 5-day period of heavy exercise while on a low carbohydrate diet, followed by 1–3 days of light exercise while on a high carbohydrate diet. The initial low carbohydrate-high energy demand period appears to cause a depletion of muscle glycogen stores. Apparently, the subsequent change to a high carbohydrate diet causes a slight rebound effect with the production of higher than normal levels of insulin and growth hormone. Under these conditions glycogen storage is favored and glycogen stores can reach almost twice the normal amounts. This does significantly increase endurance, since exercising muscle relies preferentially on glucose for its energy needs and glycogen is the most readily metabolizable form of glucose.

Test subjects on a high-fat and high-protein diet had less than 1.6 g of glycogen/100 g of muscle and could perform a standardized work load for only 60 min. When the same subjects then consumed a high carbohydrate diet for 3 days, their glycogen stores increased to 4 g/100 g of muscle and the same workload could be performed for up to 4 h. With time, of course, the body would adapt to the high carbohydrate intake, with insulin and glycogen levels returning to normal. No potential side effects of this dietary regimen are known at present, although the diet will also cause retention of water in the muscle.

25.8 FIBER

Dietary fiber is defined as those components of food that cannot be broken down by human digestive enzymes. It is incorrect, however, to assume that fiber is indigestible. Some fibers are, in fact, at least partially broken down by intestinal bacteria. Our knowledge of the role of fiber in human metabolism has expanded significantly in the past decade. Our current understanding of the metabolic roles of dietary fiber is based on three important observations: (1) there are several different types of dietary fiber, (2) they each have different chemical and physical properties, and (3) they each have different effects on human metabolism, which can be understood, in part, from their unique properties.

The major types of fiber and their properties are summarized in Table 25.4. Cellulose and hemicellulose increase stool bulk and decrease transit time. These are the types of fiber that should most properly be associated with the effects of fiber on regularity. They also decrease intracolonic pressure and appear to play a beneficial role with respect to diverticular diseases. By diluting out potential carcinogens and speeding their transit through the colon, they may also play a role in reducing the risk of colon cancer. Lignins, on the other hand, play a slightly different role. In addition to their bulk enhancing properties, they adsorb organic substances such as cholesterol and appear to play a role in the cholesterol-lowering effects of fiber. The mucilaginous fibers, such as pectin and gums, have a very different mode of action. They tend to form viscous gels in the stomach and intestine and slow the rate of gastric emptying, thus slowing the rate of absorption of many nutrients. The most important clinical role of these fibers is to slow the rate at which carbohydrates are digested and absorbed. Thus, both the rise in blood sugar and the subsequent rise in insulin levels are significantly decreased if these fibers are ingested along with carbohydrate-containing foods. These fibers also help to lower serum cholesterol levels in most people. Whether this is due to their effect on insulin levels (insulin stimulates cholesterol synthesis and export) or to other metabolic effects (perhaps caused by end products of partial bacterial digestion) is not known at present.

TABLE 25.4 Major Types of Fiber and Their Properties

Type of Fiber	*Major Source in Diet*	*Chemical Properties*	*Physiological Effects*
Cellulose	Unrefined cereals Bran Whole wheat	Nondigestible Water insoluble Absorbs water	↑ Stool bulk ↓ Intestinal transit time ↓ Intracolonic pressure
Hemicellulose	Unrefined cereals Bran Whole wheat	Partially digestible Usually water insoluble Absorbs water	↑ Stool bulk ↓ Intestinal transit time ↓ Intracolonic pressure
Lignin	Vegetables	Nondigestible Water insoluble Adsorbs organic substances	↑ Stool bulk Binds cholesterol Binds carcinogens (?)
Pectin	Fruits	Digestible Water soluble Mucilaginous	↓ Rate of gastric emptying ↓ Rate of sugar uptake ↓ Serum cholesterol
Gums	Legumes	Digestible Water soluble Mucilaginous	↓ Rate of gastric emptying ↓ Rate of sugar uptake ↓ Serum cholesterol

25.9 COMPOSITION OF MACRONUTRIENTS IN THE DIET

From the foregoing discussion it is apparent that there are relatively few instances of macronutrient deficiencies in the American diet. Thus, much of the interest in recent years has focused more on the question of whether there is an ideal diet composition consistent with good health. It would be easy to pass off such discussions as purely academic, yet our understanding of these issues could well be vital. Heart disease, stroke, and cancer kill many Americans each year, and if some experts are even partially correct, many of these deaths could be preventable with prudent diet. So it is only fitting that we now turn to the question of diet composition.

Lipid Composition of the Diet

Most of the current discussion centers around two key issues: (1) Can serum cholesterol and triacylglycerol levels be controlled by diet? (2) Does lowering serum and triacylglycerol levels protect against heart disease? The controversies centered around dietary control of cholesterol levels illustrate perfectly the trap one falls into by trying to look too closely at each individual component of the diet instead of the diet as a whole. For example, there are at least four components that can be identified as having an effect on serum cholesterol: cholesterol itself, polyunsaturated fatty acids (PUFA), saturated fatty acids (SFA), and fiber. It would appear obvious that the more cholesterol one eats, the higher the serum cholesterol would be. However, cholesterol synthesis is tightly regulated via a feedback control at the hydroxymethylglutaryl CoA reductase step, so decreases in dietary cholesterol have relatively little effect on serum cholesterol levels (Chapter 10). One can obtain a more significant reduction in cholesterol and triacylglycerol levels by increasing the ratio of PUFA/SFA in the diet (see Clin. Corr. 25.7). While the effects of PUFA are more dramatic, the biochemistry of their action is still uncertain. Finally, some plant fibers, especially lignins and gums from vegetables and pectin from fruits appear to decrease the absorption of cholesterol and bile acids.

Actually there is very little disagreement with respect to these data. The question is, what can be done with the information? Much of the disagreement arises from the tendency to look at each dietary factor in isolation. For example, it is indeed debatable whether it is worthwhile placing a patient on a highly restrictive 300-mg cholesterol diet (1 egg = 250 mg of cholesterol) if his serum cholesterol is lowered only 5–10%. Likewise, changing the PUFA/SFA ratio from 0.3 (the current value) to 1.0 would either require a radical change in the diet by elimination of foods containing saturated fat (largely meats and fats) or an addition of large amounts of rather unpalatable polyunsaturated fats to the diet. For many Americans this would be unrealistic. Fiber is another good example. One could expect, at the most, a 5% decrease in serum cholesterol by adding any reasonable amount of fiber to the diet. (Very few people would eat the 10 apples per day needed to lower serum cholesterol by 15%.) Are we to conclude then that any dietary means of controlling cholesterol levels are useless? Only if each element of the diet is examined in isolation. For example, recent studies have shown that vegetarians, who have lower blood cholesterol levels plus higher PUFA/SFA ratios and higher fiber intakes, may average 25–30% lower cholesterol levels than their nonvegetarian counterparts. Perhaps, more to the point, diet modification of the type acceptable to the average American has been shown to cause a

CLIN. CORR. 25.7 POLYUNSATURATED FATTY ACIDS AND RISK FACTORS FOR HEART DISEASE

The recent NIH study confirming that reduction of elevated serum cholesterol levels can reduce the risk of heart disease has rekindled interest in the effects of diet on serum cholesterol levels and other risk factors for heart disease. We have known for years that one of the most important dietary factors regulating serum cholesterol levels is the ratio of polyunsaturated fats (PUFAs) to saturated fats (SFAs) in the diet.

One of the most interesting recent developments is the discovery that different types of polyunsaturated fatty acids have different effects on lipid metabolism and on other risk factors for heart disease. As discussed in Chapter 9, there are two families of polyunsaturated essential fatty acids—the ω-6, or linoleic family (18:2(9,12)), and the ω-3, or linolenic family (18:3(9,12,15)).

Recent clinical studies have shown that the ω-6 PUFAs (chief dietary source = linoleic acid from plants and vegetable oils) primarily decrease serum cholesterol levels, with only modest effects on triacylglycerol levels. The ω-3 PUFAs (chief dietary source = eicosapentaenoic acid from certain ocean fish and fish oils), on the other hand, are two to five times more effective in lowering serum cholesterol levels and cause significant decreases in triacylglycerol levels as well. The biochemical mechanism behind these different effects on serum lipid levels is unknown at present.

The ω-3 PUFAs have yet another unique physiological effect that may decrease the risk of heart disease—they decrease platelet aggregation. The mechanism of this effect is a little more clear. Arachidonic acid (ω-6 family) is known to be a precursor of thromboxane A_2 (TXA_2), which is a potent proaggregating agent, and prostaglandin I_2 (PGI_2), which is a weak antiaggregating agent (see Chapter 10). The ω-3 PUFAs are thought to act by one of two mechanisms: (1) Eicosapentaenoic acid (ω-3 family) may be converted to thromboxane A_3 (TXA_3), which is only weakly proaggregating, and prostaglandin I_3 (PGI_3), which is strongly antiaggregating. Thus, the balance between proaggregation and antiaggregation would be shifted toward a more antiaggregating condition as the ω-3 PUFAs displace ω-6 PUFAs as a source of precursors to the thromboxanes and prostaglandins. (2) The ω-3 PUFAs may also act by simply inhibiting the conversion of arachidonic acid to TXA_2. Although these two mechanisms are considered the most likely at present, other pathways may very well be involved as well.

The unique potential of eicosapentaenoic acid and other ω-3 PUFAs in reducing the risk of heart disease is currently being tested in numerous clinical trials. Although the results of these clinical studies may affect dietary recommendations in the future, it is well to keep in mind that no long-term clinical studies of the ω-3 PUFAs have yet been carried out. No major health organization is yet recommending that we attempt to replace ω-6 with ω-3 PUFAs in the American diet.

10–15% decrease in cholesterol levels in long-term studies. The second question has been much more difficult to answer. However, a recent 7-year clinical trial sponsored by the National Institutes of Health has proved conclusively that lowering serum cholesterol levels will reduce the risk of heart disease in men. However, it is important to keep in mind that serum cholesterol is just one of many risk factors. Unfortunately, some nutritionists, and certainly the popular press, have overemphasized the importance of this one risk factor in preventing heart disease.

Carbohydrate

Much of the current dispute in the area of carbohydrates centers around the amount of refined carbohydrate in the diet. It is possible to see in the popular press simple sugars (primarily sucrose) blamed for almost every ill from tooth decay to heart disease and diabetes. In the case of tooth decay, these assertions are clearly correct. In the case of heart disease and diabetes, however, the linkage is more obscure.

It is evident that much of the excess dietary carbohydrate in the American diet is converted to triacylglycerol in the liver, exported as VLDL, and stored in the adipose tissue. It is even somewhat logical to assume that simple sugars, which are absorbed and metabolized very rapidly, might cause a slightly greater elevation of triacylglycerols than complex carbohydrates. In actual studies with human volunteers, an isocaloric switch from a diet high in starch to one high in simple sugars, does cause a transient rise in triacylglycerol levels. However, over a period of 2–3 months, adaptation occurs and the triacylglycerol levels return to normal. Thus, for most individuals, there is no evidence that simple sugars can cause a permanent elevation of serum triacylglycerols.

The situation with respect to diabetes is probably even less direct. Whereas restriction of simple sugars is often desirable in a patient who already has diabetes, there is little direct evidence that an excess of simple sugars in the diet is a direct cause of diabetes.

In fact, recent studies show less than expected correlation between the type of carbohydrate ingested and the subsequent rise in serum glucose levels (Table 25.5). Ice cream, for example, causes a much smaller increase in serum glucose levels than either potatoes or whole wheat bread. It turns out that other components of food—such as protein, fat, and certain types of fiber—are much more important than the type of carbohydrate present in determining how rapidly glucose will enter the bloodstream. However, foods rich in simple sugars do have a very high caloric density, contributing to overeating and obesity. Obesity, as discussed earlier, does have a direct relationship to heart disease and diabetes. This may go a long way toward explaining some of the epidemiologic studies linking consumption of simple sugars with heart disease and diabetes.

Protein

Much concern has also been voiced recently about the type of protein in the American diet. Epidemiologic data and animal studies suggest that consumption of animal protein is associated with increased incidence of heart disease and various forms of cancer. One would assume that it is probably not the animal protein itself that is involved, but the associated fat and cholesterol. What sort of protein should we consume? Although the present diet may not be optimal, a strictly vegetarian diet is not without some health risks of its own, unless the individual is nutritionally very well informed (Clin. Corr. 25.1). Perhaps a middle road is best. Clearly there are no known health dangers associated with a mixed diet that is lower in animal protein than the current American standard.

TABLE 25.5 Glycemic Index[a] of Some Selected Foods[b]

Grain & cereal products		Root vegetables	
Bread (white)	69 ± 5	Beets	64 ± 16
Bread (whole wheat)	72 ± 6	Carrots	92 ± 20
Rice (white)	72 ± 9	Potato (white)	70 ± 8
Sponge cake	46 ± 6	Potato (sweet)	48 ± 6
Breakfast cereals		Dried legumes	
All bran	51 ± 5	Beans (kidney)	29 ± 8
Cornflakes	80 ± 6	Beans (soy)	15 ± 5
Oatmeal	49 ± 8	Peas (blackeye)	33 ± 4
Shredded wheat	67 ± 10		
Vegetables		Fruits	
Sweet corn	59 ± 11	Apple (golden delicious)	39 ± 3
Frozen peas	51 ± 6	Banana	62 ± 9
		Oranges	40 ± 3
Dairy products		Sugars	
Ice cream	36 ± 8	Fructose	20 ± 5
Milk (whole)	34 ± 6	Glucose	100
Yogurt	36 ± 4	Honey	87 ± 8
		Sucrose	59 ± 10

[a] Glycemic index is defined as the area under the blood glucose response curve for each food expressed as a percentage of the area after taking the same amount of carbohydrate as glucose (means of 5–10 individuals).
[b] Data from Jenkins, David A. et al. Glycemic index of foods: A physiological basis for carbohydrate exchange. *Am. J. Clin. Nutr.* 34:362, 1981.

Fiber

Because of our current knowledge about the effects of fiber on human metabolism, most suggestions for a prudent diet recommend an increase in dietary fiber. The main question is: "How much is enough?" Because of possible effects of high fiber diets on nutrient absorption, it would seem wise not to increase fiber levels too much. The current fiber content of the American diet is about 25 g per day.

Most experts feel that an increase to 35–40 g would be safe and beneficial. Since we know that different fibers have different metabolic roles, this increase in fiber intake should come from a wide variety of fiber sources—including fresh fruits and vegetables as well as the more popular cereal sources of fiber (which are primarily cellulose and hemicellulose).

Recommendations

In the midst of all of this controversy, it would seem to be premature to make specific recommendations with respect to the ideal dietary composition for the American public. Yet that is just what several private and government groups have done in recent years. This movement was spearheaded by the Senate Select Committee on Human Nutrition which first published its *Dietary Goals for the United States* in 1977. The Senate Select Committee recommended that the American public reduce consumption of total calories, total fat, saturated fat, cholesterol, simple sugars, and salt to "ideal" goals more compatible with good health (Figure 25.2). In recent years the USDA, the American Heart Association, the American Diabetes Association, and the National Research Council all have published similar recommendations (Table 25.6). These recommendations have become popularly known as the "prudent diet." How valid is the scientific basis of the recommendations for a prudent diet? Is there evidence that a prudent diet will improve the health of the general public? These remain very controversial questions at present.

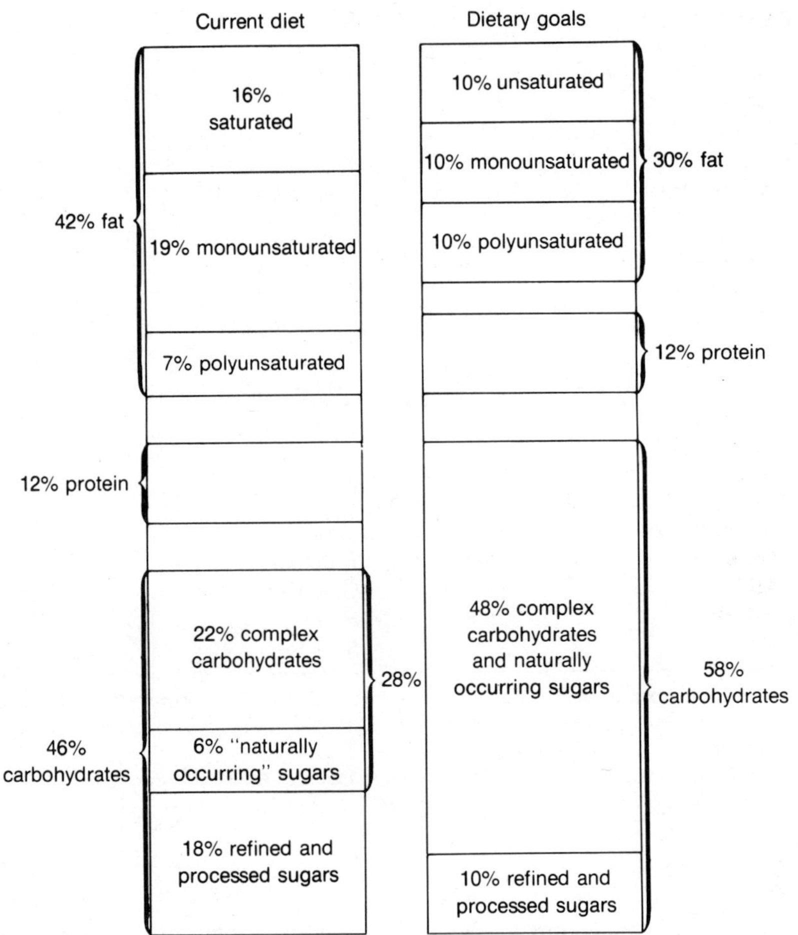

Figure 25.2
United States dietary goals.
Graphical comparison of the composition of the current U.S. diet and the dietary goals for the U.S. population suggested by the Senate Select Committee on Human Nutrition.
From *Dietary goals for the United States, 2nd ed.,* U. S. Government Printing Office, Washington, D.C., 1977.

The most important argument against such recommendations is that we presently do not have enough information to set concrete goals. We might be creating some problems while solving others. For example, the goals of reducing total fat and saturated fat in the diet are best met by replacing animal protein with vegetable protein. This, in turn, might reduce the amount of available iron and vitamin B_{12} in the diet. It is also quite clear that the same set of guidelines do not apply for every individual. For example, exercise is known to raise serum HDL cholesterol and obesity is known to elevate cholesterol, triacylglycerols, and reduce glucose tolerance. Thus the very active individual who maintains ideal body weight can likely tolerate higher fat and sugar intakes than an obese individual.

On the "pro" side, however, it clearly can be argued that all of the dietary recommendations are in the right direction for reducing nutritional risk factors in the general population. Furthermore, similar diets have been consumed by our ancestors and by people in other countries with no apparent harm. Whatever the outcome of this debate in the years ahead, it will undoubtedly shape much of our ideas concerning the role of nutrition in medicine.

TABLE 25.6 Dietary Recommendations of Major Health Organizations

	Recommendations of			
	American Heart Association[a] *(to ↓ risk of heart disease)*	*National Research Council*[b] *(to ↓ risk of cancer)*	*Senate Select Committee on Human Nutrition*[c]	*USDA*[d]
Obesity	Reduce to ideal weight	—	Reduce to ideal weight	Maintain ideal weight
Fat	30% of total calories	30% of total calories	30% of total calories	Avoid too much fat
Type of fat	10% SFA	—	10% SFA	Avoid too much saturated fat
	10% MUFA	—	10% MUFA	
	10% PUFA	—	10% PUFA	
Cholesterol	<300 mg/day	—	300 mg/day	Avoid too much cholesterol
Carbohydrate	55% of total calories	—	58% of total calories	
Type of carbohydrate	Increase complex carbohydrates and fiber-rich foods	Fiber may be beneficial	48% complex CHO 10% simple sugars	Eat foods with starch and fiber Avoid too much sugar
Vitamins/minerals	—	Include fruits, vegetables, and whole grain foods in diet	—	Eat a variety of foods
Alcohol	If you drink alcohol, do so in moderation	Avoid excess alcohol consumption	— —	If you drink do so in moderation
Sodium		Avoid salt-cured foods	Limit sodium intake to 5 g/day	Avoid too much sodium

[a] American Heart Association. Recommendations for treatment of hyperlipidemia in adults. American Heart Association, 1984.
[b] National Research Council. Diet, nutrition, and cancer. National Academy Press, 1982.
[c] Senate Select Committee on Human Nutrition. Dietary goals for the United States, 2nd ed. U.S. Government Printing Office, 1977.
[d] U.S. Department of Agriculture. Nutrition and your health. U.S. Government Printing Office, 1980.

BIBLIOGRAPHY

Starvation

Cahill, G. F., Jr. Starvation in man. *N. Engl. J. Med.* 282:668, 1970.

Young, V. R., and Scrinshaw, N. S. The physiology of starvation. *Sci. Am.* 225:214, 1971.

Occurrence and Management of Protein Energy Malnutrition

Bistrian, B. R. Interaction of nutrition and infection in the hospital setting. *Am. J. Clin. Nutr.* 30:1228, 1977.

Bistrian, B. R., and Blackburn, G. L. Assessment of protein-calorie malnutrition in the hospitalized patient in H. Schneider, C. Anderson, and D. Coursin (eds.). *Nutritional support of medical practice,* 2nd. ed. New York: Harper and Row, 1983, p. 128.

Flatt, J. P., and Blackburn, G. L. The metabolic fuel regulatory system: implications for protein sparing therapies during caloric deprivation and disease. *Am. J. Clin. Nutr.* 27:175, 1974.

Hopkins, B. S., Bistrian, B. R., and Blackburn, G. L. Protein-calorie management in the hospitalized patient, in H. Schneider, C. Anderson, and D. Coursin (eds.). *Nutritional support of medical practice,* 2nd ed. New York: Harper and Row, 1983, p. 140.

Metabolic Consequences of Overweight

Bray, G. A. *The obese patient*. Philadelphia, W. B. Saunders, 1976.

Glass, A. R., Burman, K. D., Dahms, W. T., and Boehm, T. M. Endocrine function in human obesity. *Metabolism.* 30:89, 1981.

Gordon, E. S. Metabolic aspects of obesity. *Adv. Metab. Disord.* 4:229, 1970.

Van Itallie, T. B. Obesity: Adverse effects of health and longevity. *Am. J. Clin. Nutr.* 32:2723, 1979.

Vegetarian Diets

Lappé, F. M. *Diet for a small planet.* New York: Ballantine, 1975.

Raper, N. R., and Hill, M. M. Vegetarian diets. *Nutr. Rev.* 32:29, 1974.

Fad Weight Loss Diets

Berland, T. *Rating the diets.* New York: Signet, 1979.

Macronutrient Composition and Health

Glueck, C. J., and Conner, W. E. Diet-coronary heart disease relationships. *Am. J. Clin. Nutr.* 31:727, 1978.

Harris, W. S., Connor, W. E., and McMurray, M. P. The comparative reductions of the plasma lipids and lipoproteins by dietary polyunsaturated fatty acids: Salmon oil versus vegetable oils. *Metabolism* 32:179, 1983.

Jenkins, D. J. A., Wolener, T. M. S., Taylor, R. H., Barker, H., Fielden, H., Baldwin, J. M., Bowling, A. C., Newman, H. C., Jenkins, A. L., and Goff, D. V. Glycemic index of foods: A physiological basis for carbohydrate exchange. *Am. J. Clin. Nutr.* 34:362, 1981.

Nagakawa, Y., Orimo, H., Harasawa, M., Morita, I., Yashiro, K., and Murota, S. Effect of eicosapentaenoic acid on the platelet aggregation and composition of fatty acid in man. *Atherosclerosis* 47:71, 1983.

National Heart, Lung, and Blood Institute. The lipid research clinics' coronary primary prevention trial results: I. Reduction in the incidence of coronary heart disease; II. The relationship of reduction in incidence of coronary heart disease to cholesterol lowering. *JAMA* 251:351, 365, 1984.

Reiser, R. Oversimplification of diet-coronary heart disease relationships. *Am. J. Clin. Nutr.* 31:865, 1978.

Spiller, G. A., and Kay, R. M. *Medical aspects of dietary fiber.* New York: Plenum Book Co., 1980.

Dietary Recommendations

American Heart Association. *Recommendations for treatment of hyperlipidemia in adults.* Dallas: American Heart Association, 1984.

Food and Nutrition Board of the National Academy of Sciences. *Towards healthful diets.* Washington, D.C.: U.S. Government Printing Office, 1980.

Harper, A. E. Arguments against U.S. dietary goals. *J. Nutr. Ed.* 9:154, 1977.

Hegsted, D. M. Dietary goals—a progressive view. *Am. J. Clin. Nutr.* 31:1504, 1978.

National Research Council. *Diet, nutrition and cancer.* Washington: National Academy Press, 1982.

Senate Select Committee on Human Nutrition. *Dietary goals for the United States, 2nd ed.,* Stock no. 052-070-04376-8. Washington, D.C.: U.S. Government Printing Office, 1977.

U.S. Department of Agriculture. *Nutrition and your health, dietary guidelines for Americans,* Stock no. 017-001-00416-2. Washington, D.C.: U.S. Government Printing Office, 1980.

QUESTIONS

C. N. ANGSTADT AND J. BAGGOTT

*Q*uestion *T*ypes are described inside the front cover.

1. (QT2) Of two people with approximately the same weight, the one with the higher basal energy requirement would most likely be:
   1. taller.
   2. female if the other were male.
   3. younger.
   4. under less stress.
2. (QT3) A. Basal metabolic rate of a person consuming about 2,000 kcal/day
   B. Basal metabolic rate of a person consuming about 600 kcal/day
3. (QT1) The primary effect of the consumption of excess protein beyond the body's immediate needs will be:
   A. excretion of the excess as protein in the urine.
   B. an increase in the "storage pool" of protein.
   C. an increased synthesis of muscle protein.
   D. an enhancement in the amount of circulating plasma proteins.
   E. an increase in the amount of adipose tissue.
4. (QT1) Which of the following individuals would most likely be in nitrogen equilibrium?
   A. A normal, adult male
   B. A normal, pregnant female
   C. A growing child
   D. An adult male recovering from surgery
   E. A normal female on a very low protein diet
5. (QT1) If an individual consumes 100 g of a test protein that is 15% nitrogen and excretes 2 g of nitrogen in the feces and 6 g in the urine, the biological value (BV) of that protein is about:
   A. 47.
   B. 54.
   C. 80.
   D. 92.
   E. 94.
6. (QT1) Vegetarian diets:
   A. cannot meet the body's requirements for all of the essential amino acids.
   B. generally contain proteins with higher NPU scores than do diets containing meat.
   C. are adequate as long as two different vegetables are consumed in the same meal.
   D. would require less total protein to meet the requirement for all of the essential amino acids.
   E. require that proteins consumed have essential amino acid contents that complement each other.
7. (QT2) In which of the following circumstances would a protein intake of 0.8 g protein/kg body weight/day probably not be adequate?
   1. the NPU of the protein consumed was less than 75.
   2. infancy.
   3. severe burn.
   4. about 85–90% of total calories from carbohydrate and fat.
8. (QT1) Kwashiorkor is:
   A. the most common form of protein–calorie malnutrition in the United States.
   B. characterized by a thin, wasted appearance.
   C. an inadequate intake of food of any kind.
   D. an adequate intake of total calories but a specific deficiency of protein.
   E. an adequate intake of total protein but a deficiency of the essential amino acids.
9. (QT2) An excessive intake of calories:
   1. is likely to have adverse metabolic consequences if continued for a long period of time.
   2. leads to metabolic changes that are usually irreversible.
   3. is frequently associated with an increased number or size of adipose cells.
   4. is frequently associated with an increased number of insulin receptors.
10. (QT1) A diet very low in carbohydrate:
    A. would cause weight loss because there would be no way to replenish citric acid cycle intermediates.
    B. would result in no significant metabolic changes.

C. could lead to a chronic ketosis.
D. would lead to water retention.
E. would be the diet of choice for a diabetic.

11. (QT1) Lactase insufficiency:
A. is a more serious disease than diabetes mellitus.
B. has no clinical symptoms.
C. causes an intolerance to glucose.
D. causes an intolerance to milk and milk products.
E. affects utilization of milk by the liver.

12. (QT1) Dietary fat:
A. is usually present, although there is no specific need for it.
B. if present in excess, can be stored as either glycogen or adipose tissue triacylglycerol.
C. should include linoleic and linolenic acids.
D. should increase on an endurance training program in order to increase the body's energy stores.
E. if present in excess, does not usually lead to health problems.

13. (QT1) Which one of the following dietary regimens would be most effective in lowering serum cholesterol?
A. Restrict dietary cholesterol
B. Increase the ratio of polyunsaturated to saturated fatty acids
C. Increase fiber content
D. Restrict cholesterol and increase fiber
E. Restrict cholesterol, increase PUFA/SFA, increase fiber

14. (QT1) Most nutrition experts currently agree that an excessive consumption of sugar causes:
A. tooth decay.
B. diabetes.
C. heart disease.
D. permanently elevated triacylglycerol levels.
E. all of the above.

A. 10% of total calories
B. 12% of total calories
C. 30% of total calories
D. 48% of total calories
E. 58% of total calories

The dietary goal recommended by the Senate Select Committee on Human Nutrition for:

15. (QT5) Polyunsaturated fatty acids is
16. (QT5) Complex carbohydrates and naturally occurring sugars is _____

17. (QT1) A complete replacement of animal protein in the diet by vegetable protein:
A. would be expected to have no effect at all on the overall diet.
B. would reduce the total amount of food consumed for the same number of calories.
C. might reduce the total amount of iron and vitamin B_{12} available.
D. would be satisfactory regardless of the nature of the vegetable protein used.
E. could not satisfy protein requirements.

ANSWERS

1. B 1, 3 correct. 1: A taller person with the same weight would have a greater surface area. 2: Males have higher energy requirements than females. 3: Energy requirements decrease with age. 4: Stress, probably because of the effects of epinephrine and cortisol, increase energy requirements (Section 25.2).

2. A The BMR can decrease up to 50% on an inadequate caloric intake (p. 943).

3. E A: Protein is not found in normal urine except in very small amounts. The excess nitrogen is excreted as NH_4^+ and urea, whereas the excess carbon skeletons of the amino acids are used as energy sources. B–D: There is no discrete storage form of protein, and although some muscle and structural protein is expendable, there is no evidence that increased intake leads to generalized increased protein synthesis. E: Excess protein is treated like any other excess energy source and stored (minus the nitrogen) eventually as adipose tissue fat (Section 25.3).

4. A B–D: Although normal, pregnancy is also a period of growth, requiring positive balance as does a period of convalescence. E: Inadequate protein intake leads to negative balance (Section 25.3).

5. B $$BV = \frac{N_{intake} - N_{output}}{N_{absorbed}} \times 100 = \frac{(100 \times 0.15) - (2 + 6)}{(100 \times 0.15) - 2} \times 100$$
A: This is NPU because the intake is not corrected for loss in the feces. (Note: It is necessary to convert grams of protein to grams of nitrogen by multiplying by the percentage content of nitrogen) (p. 946).

6. E A–E: It is possible to have adequate protein intake on a vegetarian diet provided enough is consumed (protein content is generally low) and there is a *mixture* of proteins that supplies all of the essential amino acids since individual proteins are frequently deficient in one or more (Clin. Corr. 25.1, p. 946).

7. A 1, 2, 3 correct. 1: The standard protein requirement is based on an NPU of mixed protein of 75%. 2, 3: Periods of rapid growth require extra protein, as does major trauma. 4: This level of calories from carbohydrate and fat is more than adequate for protein sparing (Section 25.3).

8. D A: The most common protein–calorie malnutrition occurs in severely ill, hospitalized patients who would be more likely to have generalized malnutrition. B, C: These are the characteristics of marasmus. E: This would lead to negative nitrogen balance but does not have a specific name (Section 25.4).

9. B 1, 3 correct. 1: Excess caloric intake will lead to obesity if continued long enough. 2: Fortunately most of the changes accompanying obesity can be reversed if weight is lost. 3, 4: Many of the adverse effects of obesity are associated with an increased number of adipocytes that are deficient in insulin receptors (Section 25.5).

10. C A: This is a popular myth but untrue because many amino acids are glucogenic. B, C, E: The liver adapts by increasing gluconeogenesis, fatty acid oxidation, and ketone body production, contraindicated for a diabetic. D: Low carbohydrate leads to a depletion of glycogen with its stored water, accounting for rapid initial weight loss on this kind of diet (Section 25.6, Clin. Corr. 25.5).

11. D B, D, E. Lactase insufficiency is an inability to digest the sugar in milk products, causing intestinal symptoms, but is easily treated by eliminating milk products from the diet. A, C. Diabetes, caused by inadequate insulin or insulin receptors, inhibits appropriate utilization of glucose (Section 25.6).

12. C A, C: Linoleic and linolenic acids are essential fatty acids and so must be present in the diet. B, D: Excess carbohydrate can be stored as fat but the reverse is not true. D: Carbohydrate loading has been shown to increase endurance. E: High-fat diets are associated with many health risks (Section 25.7, Clin. Corr. 25.6).

13. E Any of the measures alone would decrease serum cholesterol slightly, but to achieve a reduction of more than 15% requires all three (Section 25.9).

14. A This is the only direct linkage shown. B, C: There may be an association with these conditions but not a direct cause–effect relationship. D: Transient elevations may occur on an isocaloric switch from a high starch to a high simple sugar diet but not a permanent elevation (Section 25.9).
15. A See Figure 25.2, p. 956.
16. D See Figure 25.2, p. 956.
17. C A, C: This would reduce the amount of fat, especially saturated fat, but could also reduce the amount of necessary nutrients that come primarily from animal sources. B: The protein content of vegetables is quite low, so much larger amounts of vegetables would have to be consumed. D, E: It is possible to satisfy requirements for all of the essential amino acids completely if vegetables with complementary amino acid patterns, in proper amounts, are consumed (Section 25.9, Clin. Corr. 25.1).

26

Principles of Nutrition II: Micronutrients

STEPHEN G. CHANEY

The micronutrients play a vital role in human metabolism, since they are involved in almost every biochemical reaction and pathway known to man. However, the biochemistry of these nutrients is of little interest unless we also know if dietary deficiencies are likely. Alarming reports of nutritional deficiencies in the American diet continually appear in the popular press. Is there any truth to these reports? On the one hand, the American diet is undoubtedly the best that it ever has been. Our current food supply provides us with an abundant variety of foods all year long—a luxury not available in the "good old days"—and deficiency diseases have become medical curiosities. On the other hand, our diet is far from optimal. The old adage that we get everything we need from a balanced diet is true only if in fact we eat a balanced diet. Unfortunately, most Americans do not know how to select a balanced diet. Foods of high caloric density and low nutrient density (often referred to elsewhere as empty calories or junk food) are an abundant and popular part of the American diet, and our nutritional status suffers because of these food choices. Obviously then, neither alarm nor complacency are fully justified. We need to know how to best evaluate the adequacy of our diet.

26.1 ASSESSMENT OF MALNUTRITION

Why does one see so many reports of vitamin and mineral deficiencies in the popular press when deficiency diseases are so rare? In many cases these reports result from the misinterpretation of valid scientific data. One needs to be aware that there are three increasingly stringent criteria for measuring malnutrition. (1) Dietary intake studies, which are usually based on a 24-hour recall, are the least stringent. In the first place, 24-hour recalls almost always tend to overestimate the number of people with deficient diets. Also, poor dietary intake alone is usually not a problem in this country unless the situation is compounded by increased need. Thus dietary surveys are not indicative of malnutrition by themselves, although they are often quoted as if they were. (2) Biochemical assays, either direct or indirect, are a more useful indicator of the nutritional status of an individual. At their best, they can indicate subclinical nutritional deficiencies, which can be treated before actual deficiency diseases develop. However, all biochemical assays are not equally valid—an unfortunate fact that is not sufficiently recognized. Furthermore, changes in biochemical parameters due to stress need to be interpreted with caution. The distribution of many nutrients in the body changes dramatically in a stress situation such as illness, injury, and pregnancy. A drop in level of one nutrient in one tissue compartment (usually blood) need not signal a deficiency or an increased requirement. It could simply reflect a normal metabolic adjustment to stress. (3) The most stringent criterion is, of course, the appearance of clinical symptoms. However, it would be desirable to intervene long before clinical symptoms became apparent.

The question remains: When should dietary surveys or biochemical assays be interpreted to indicate the necessity of nutritional intervention? At what level should we become concerned? Obviously, the situation is complex and controversial, but the following general guidelines are probably useful. Dietary surveys are seldom a valid indication of general malnutrition unless the average intake for a population group falls significantly below the standard (usually two-thirds of the Recommended Dietary Allowance) for the nutrients. However, by looking at the percentage of people within a population group who have suboptimal intake, it is possible to identify high-risk population groups that should be monitored more closely. This is the real value of dietary surveys. Biochemical assays, on the other hand, can definitely identify subclinical cases of malnutrition where nutritional intervention is desirable provided (a) the assay has been shown to be reliable, (b) the deficiency can be verified by a second assay, and (c) there is no unusual stress situation that may alter micronutrient distribution. In evaluating nutritional claims and counterclaims it is well to keep an open, but skeptical, mind and evaluate each issue on the basis of its scientific merit. While serious vitamin and mineral deficiencies are rare in this country, mild to moderate deficiencies can be found in many population groups. It is important for the clinician to be aware of these population groups at risk and their most probable symptoms.

26.2 RECOMMENDED DIETARY ALLOWANCES

One hears a lot about the Recommended Dietary Allowances (RDA). What are they and how are they determined? Briefly, the Recommended Dietary Allowances are the levels of intake of essential nutrients considered in the judgement of the Food and Nutrition Board of the National Research Council on the basis of available scientific knowledge, to be adequate to meet the known nutritional need of practically all healthy persons. Optimally, the RDAs are based on daily intake sufficient to prevent the appearance of nutritional deficiency in 95% of the population. This determination is relatively easy to make for those nutrients associated with dramatic deficiency diseases, such as vitamin C and scurvy. In other instances more indirect measures must be used, such as tissue saturation or extrapolation from animal studies. In some cases, such as vitamin E, in which no deficiency symptoms are known to occur in the general population, the RDA is simply defined as the normal level of intake in the American diet. Obviously, there is no one set of criteria that can be used for all micronutrients, and there is always some uncertainty and debate as to the correct criteria. Furthermore, the criteria are constantly changed by new research. The Food and Nutrition Board meets every 5 years (most recently in 1984) to consider currently available information and update their recommendations.

The RDAs serve as a useful general guide in evaluating the adequacy of diets and (as the USRDA) the nutritional value of foods. However, the RDAs have several limitations that should be kept in mind. Some of the most important limitations are as follows: (1) The RDAs represent an ideal average intake for groups of people and are best used for evaluating nutritional status of population groups. The RDAs are not meant to be standards or requirements for individuals. Some individuals would have no problem with intakes below the RDA, whereas others might develop deficiencies on intakes above the RDA. (2) The RDAs were designed to meet the needs of healthy people and do not take into account any special needs arising from infections, metabolic disorders, or chronic diseases. (3) Since present knowledge of nutritional needs is incomplete, there may be unrecognized nutritional needs. To provide for these needs, the RDAs should be met from as varied a selection of foods as possible. No single food can be considered complete, even if it meets the RDA for all known nutrients. This is an important consideration, especially in light of the current practice of fortifying foods of otherwise low nutritional value. (4) The RDAs make no effort to define the "optimal" level of any nutrient, since optimal levels are almost impossible to define on the basis of current scientific information.

26.3 FAT-SOLUBLE VITAMINS

Vitamin A

The active forms of vitamin A are retinol, retinaldehyde, and retinoic acid. These substances are synthesized by plants as the more complex carotenoids (Figure 26.1), which are cleaved to retinol by most animals and stored in the liver as retinol palmitate. Liver, egg yolk, butter, and whole milk are good sources of the preformed retinol. Dark green and yellow vegetables are generally good sources of the carotenoids. The conversion of carotenoids to retinol is rarely 100%, so that the vitamin A potency of various foods is expressed in terms of retinol equivalents (1 RE is equal to 1 μg retinol, 6 μg β-carotene, and 12 μg of other carotenoids). β-Carotene and other carotenoids are the major sources of vitamin A in the American diet. These carotenoids are first cleaved to retinol and then converted to other vitamin A metabolites in the body (Figure 26.1).

Vitamin A plays a number of important roles in the body. However, only in recent years has its biochemistry become well understood (Figure

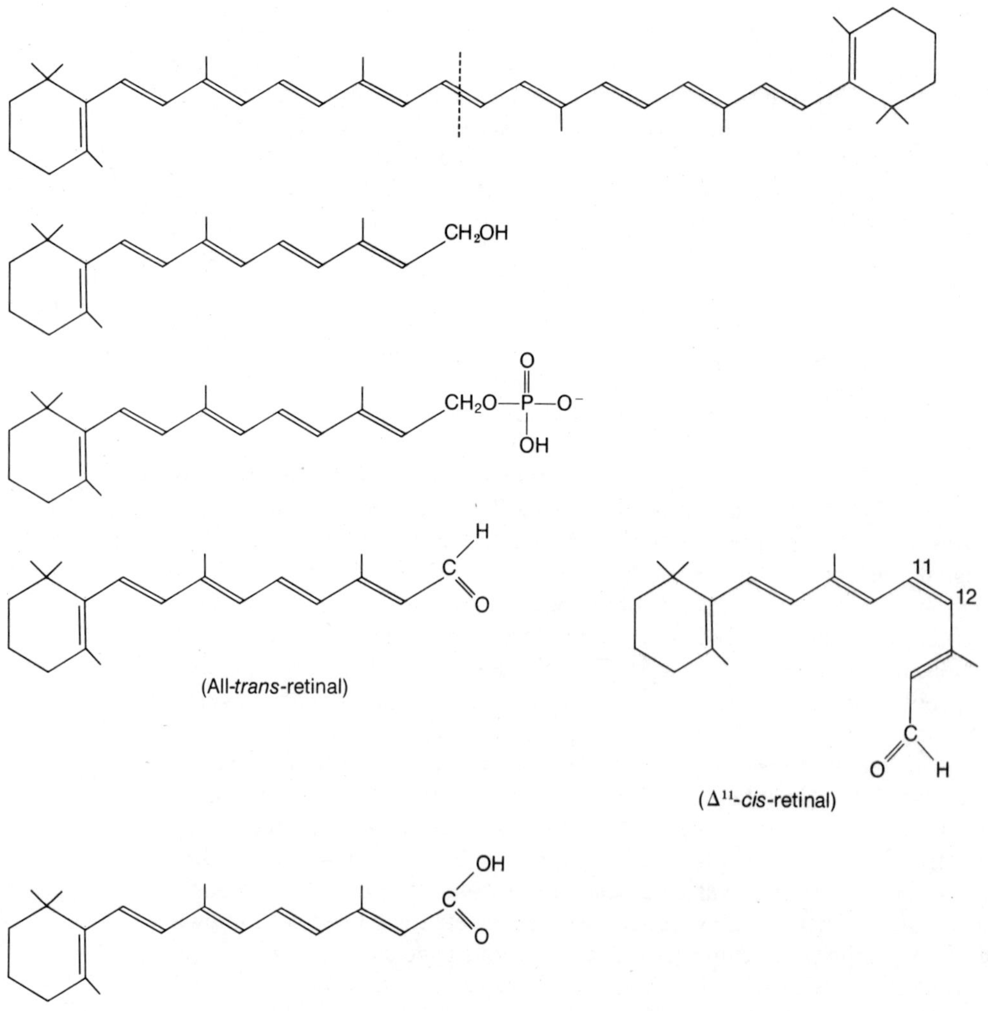

Carotenoids (β-carotene)

Retinol (vitamin A) CH_2OH

Retinyl phosphate $CH_2O-P(=O)(OH)-O^-$

Retinal

(All-*trans*-retinal)

11 12

(Δ^{11}-*cis*-retinal)

Retinoic acid

Figure 26.1
Structures of vitamin A and related compounds.

26.2). β-Carotene and some of the other carotenoids have recently been shown to play an important role as antioxidants. At the low oxygen tensions prevalent in the body, β-carotene is an extremely effective antioxidant and might be expected to reduce the risk of those cancers initiated by free radicals and other strong oxidants. Several retrospective clinical studies have suggested that adequate dietary β-carotene may play an important role in reducing the risk of lung cancer—especially in people who smoke. The National Institutes of Health is currently testing this hypothesis in a multicenter prospective study.

Retinol can be converted to retinyl phosphate in the body. The retinyl phosphate appears to serve as a glycosyl donor in the synthesis of some glycoproteins and mucopolysaccharides in much the same manner as dolichol phosphate, which has been discussed previously (Chapter 8, Figure 26.3). It appears that retinyl phosphate is essential for the synthesis of certain glycoproteins needed for normal growth regulation and for mucous secretion.

Both retinol and retinoic acid bind to specific cytosolic receptor proteins, which transport these compounds into the nucleus where they bind to chromatin and affect the synthesis of specific proteins involved in the regulation of cell growth and differentiation. Thus, both retinol and retinoic acid can be considered to be acting like the steroid hormones regulating growth and differentiation.

Finally, in the Δ^{11}-*cis*-retinal form, vitamin A becomes reversibly associated with the visual pigments. When light strikes the retina, a number of complex biochemical changes take place, resulting in the generation of a nerve impulse, conversion of the retinal to the all-*trans* form, and its dissociation from the visual pigment. Regeneration of more visual pigment requires isomerization back to the Δ^{11}-*cis* form which occurs in the liver (p. 841) and to some extent in the eye (Figure 26.4).

Based on what is now known about the biochemical mechanisms of vitamin A action, its biological effects are easier to understand. For example, vitamin A is known to be required for the maintenance of healthy epithelial tissue. We now know that retinol and/or retinoic acid are required to prevent the synthesis of high molecular weight forms of keratin and that retinyl phosphate is required for the synthesis of mucopolysaccharides (an important component of the mucus secreted by many epithelial tissues). The lack of mucus secretion leads to a drying of these cells, and the excess keratin synthesis leaves a horny keratinized surface in place of the normal moist and pliable epithelium. It has also been ob-

β-carotene (*antioxidant*)
↓
retinol (*steroid hormone*) → retinyl (*glycoprotein synthesis*)
↓
retinal (*visual cycle*)
↓
retinoic acid (*steroid hormone*)

Figure 26.2
Vitamin A metabolism and function.

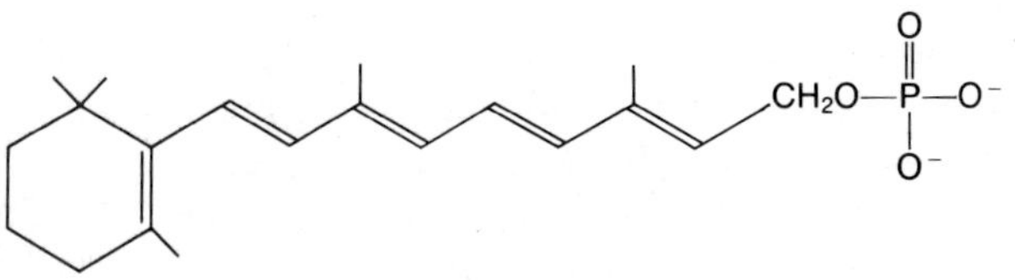

$CH_2O{-}P(=O)(O^-){-}O^-$

Retinol phosphate

$CH_2CH(CH_3)CH_2CH_2OP(=O)(O^-){-}OH$ (16–20)

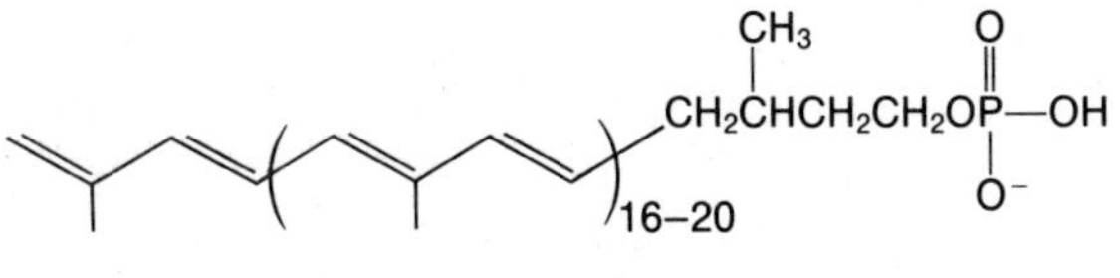

Dolichol phosphate

Figure 26.3
Carriers involved in glycoprotein synthesis.

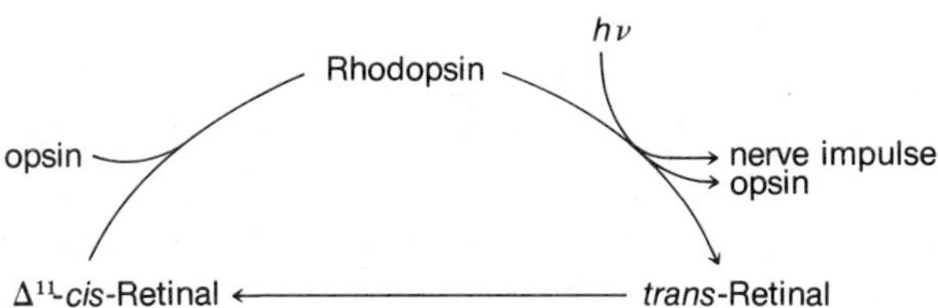

Figure 26.4
The role of vitamin A in vision.

served that vitamin A deficiency can lead to anemia caused by impaired mobilization of iron from the liver. Recent studies have shown that retinol and/or retinoic acid are required for the synthesis of the iron transport protein transferrin.

Finally, vitamin A-deficient animals have been shown to be more susceptible to both infections and cancer. The decreased resistance to infections is thought to be due to the keratinization of the mucosal cells lining the respiratory, gastrointestinal, and genitourinary tract. Under these conditions fissures readily develop in the mucosal membranes, allowing microorganisms to enter. Some evidence suggests that vitamin A deficiency may impair the immune system as well. The protective effect of vitamin A against many forms of cancer probably results from the antioxidant potential of β-carotene and the effects of retinol and retinoic acid in regulating cell growth.

Since vitamin A is stored by the liver, deficiencies of this vitamin can develop only over prolonged periods of inadequate uptake. Mild vitamin A deficiencies are characterized by follicular hyperkeratosis (rough keratinized skin resembling "goosebumps"), anemia (biochemically equivalent to iron deficiency anemia, but in the presence of adequate iron intake), and increased susceptibility to infection and cancer. Night blindness is also an early symptom of vitamin A deficiency. Severe vitamin A deficiency leads to a progressive keratinization of the cornea of the eye known as xerophthalmia in its most advanced stages. In the final stages, infection usually sets in, with resulting hemorrhaging of the eye and permanent loss of vision.

The severe symptoms of vitamin A deficiency are generally seen only in developing countries. In this country even mild vitamin A deficiencies are rare, but the potential for deficiencies does exist. For most people (unless they happen to eat liver) the dark green and yellow vegetables are the most important dietary source of vitamin A. Unfortunately, these are the foods most often missing from the American diet. Nationwide, dietary surveys indicate that between 40 and 60% of the population consumes less than two-thirds of the RDA for vitamin A. The important question is how significant these dietary surveys are. While plasma vitamin A levels are low in a significant number of individuals, the clinical symptoms of vitamin A deficiency are rare. Follicular hyperkeratosis is occasionally seen and is the most characteristic symptom of vitamin A deficiency. The anemia and decreased resistance to infection and cancer are obviously of great concern in preventive medicine, but they are too nonspecific to be useful indicators of vitamin A status. Night blindness is seldom seen in the general population. While clinically detectable vitamin A deficiency is rare in the general population, it is a fairly common consequence of severe liver damage or diseases that cause fat malabsorption (Clin. Corr. 26.1).

Since vitamin A does accumulate in the liver, large amounts of this vitamin over prolonged periods of time can be toxic. Doses of 15,000 to 50,000 RE per day over a period of months or years will prove to be toxic for many children and adults. The usual symptoms include bone pain, scaly dermatitis, enlargement of liver and spleen, nausea, and diarrhea. It

CLIN. CORR. **26.1**
NUTRITIONAL CONSIDERATIONS FOR MALABSORPTION SYNDROMES

Patients with malabsorption diseases have a high risk of malnutrition. As an example, let us examine the nutritional consequences of two diseases with malabsorption components. Cystic fibrosis (CF) involves a generalized dysfunction of the exocrine glands that leads to formation of a viscid mucus, which progressively plugs their ducts. Obstruction of the bronchi and bronchioles leads to pulmonary infections, which are usually the direct cause of death. However, in many cases the exocrine glands of the pancreas are also affected, leading to a deficiency of pancreatic enzymes and often a partial obstruction of the common bile duct.

The deficiency (or partial deficiency) of pancreatic lipase and bile salts leads to severe malabsorption of fat and fat-soluble vitamins. Calcium tends to form insoluble salts with the long-chain fatty acids, which accumulate in the intestine. While these are the most severe problems, some starches and proteins are also trapped in the fatty bolus of partially digested foods. This physical entrapment, along with the deficiencies of pancreatic amylase and pancreatic proteases, can lead to protein-calorie malnutrition as well. In the treatment of these patients, the total fat content of the diet is decreased drastically and medium-chain triacylglycerols (MCTs) are

is, of course, virtually impossible to ingest toxic amounts of vitamin A from normal foods unless one eats polar bear liver (6,000 RE/g) regularly. Most instances of vitamin A toxicity are due to the use of massive doses of this vitamin to treat acne or prevent colds. Fortunately, this practice, while once common, is now relatively rare because of increased public awareness of vitamin A toxicity.

Whereas vitamin A itself is too toxic to be used therapeutically, certain synthetic retinoids (chemical derivatives of retinoic acid) are being used or tested because they often have lower toxicity. For example, 13-*cis*-retinoic acid is widely used in the treatment of acne and etretinate (an aromatic analog of all *trans*-retinoic acid) is being used in the treatment of psoriasis and related disorders (Figure 26.5). Other synthetic retinoids have been shown to be useful in the prevention and treatment of cancers in laboratory animals.

Vitamin D

Technically, vitamin D could be considered a hormone rather than a vitamin. Cholecalciferol (D_3) is produced in the skin by ultraviolet irradiation of 7-dehydrocholesterol, a normal metabolite of cholesterol (see Figure 26.6). Thus, as long as the body is exposed to adequate sunlight, there is little or no dietary requirement for vitamin D. The best dietary sources of vitamin D_3 are saltwater fish (especially salmon, sardines, and herring), liver, and egg yolk. Milk, butter, and other foods are routinely fortified with ergocalciferol (D_2) prepared by irradiating ergosterol from yeast. Vitamin D potency is measured in terms of cholecalciferol units (1 μg cholecalciferol or ergocalciferol).

Both cholecalciferol and ergocalciferol are metabolized identically. They are carried to the liver where the 25-hydroxy derivative is formed. 25-(OH)D is the major circulating derivative of vitamin D, and it is in turn converted into the biologically active 1α,25-dihydroxy derivative in the kidney (Clin. Corr. 26.2). The synthesis of 1,25-$(OH)_2$D is stimulated by low calcium and/or low phosphate levels. High calcium levels cause synthesis of an inactive 24,25-$(OH)_2$D derivative.

1,25-$(OH)_2$D acts in concert with parathyroid hormone (PTH), which is also produced in response to low serum calcium. There is some evi-

used as a partial replacement, since they can be absorbed directly through the intestinal mucosa in the absence of bile salts and pancreatic lipase. Total calories in a CF diet are increased to compensate for the inefficiency of digestion, with most of the calories coming from simple sugars. Protein nutrition is improved with a high protein diet [3–5 g/(kg/day)] and pancreatic extracts taken orally with meals. Water-soluble forms of vitamins A, D, E, and K are usually included in the diet as well.

Gluten-sensitive enteropathy (celiac disease or nontropical sprue) appears to be caused by an immunologic reaction to the protein gluten found in wheat, rye, oats, and barley. This immunologic reaction causes atrophy of the intestinal villi, resulting in loss of surface area, brush border enzymes, and transport proteins. Secondary to the villous atrophy is impaired production of the hormone cholecystokinin-pancreozymin, which normally stimulates pancreatic enzyme and bile salt secretion. Thus many of the nutritional problems are the same as with the CF patient. However, the loss of brush border enzymes and transport proteins leads to more severe protein and calcium malabsorption, and may cause malabsorption of folic acid and iron. A gluten-free diet is, of course, the most important therapy for such a patient. Nutritional problems, when they do arise, can be handled in much the same manner as for the CF patient. Calcium, iron, and folic acid supplements may also be important.

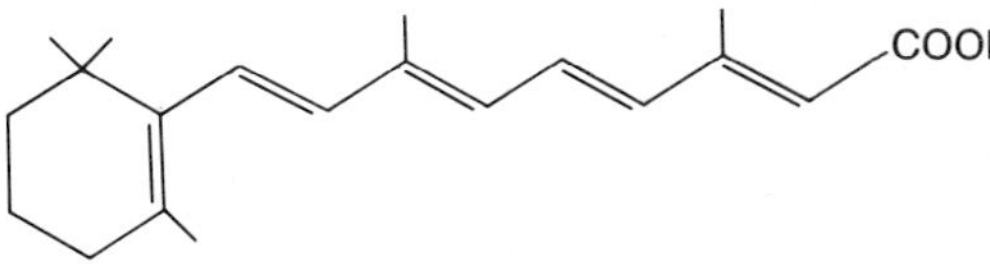

COOH

All-*trans*-Retinoic acid

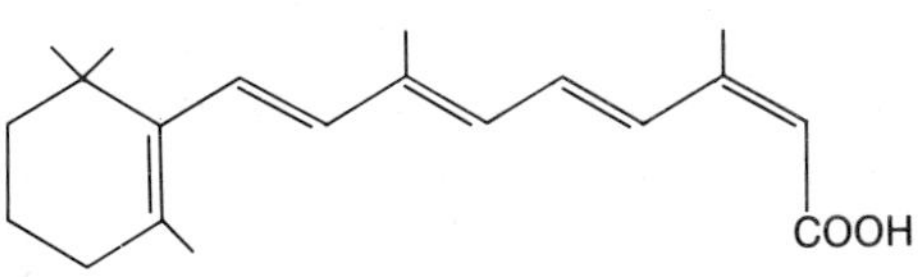

COOH

13-*cis*-Retinoic acid

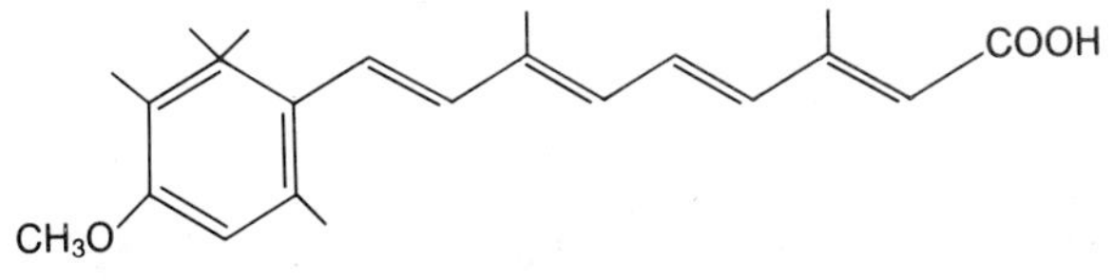

COOH

CH_3O

Etretinate

Figure 26.5
Therapeutically useful analogs of vitamin A.

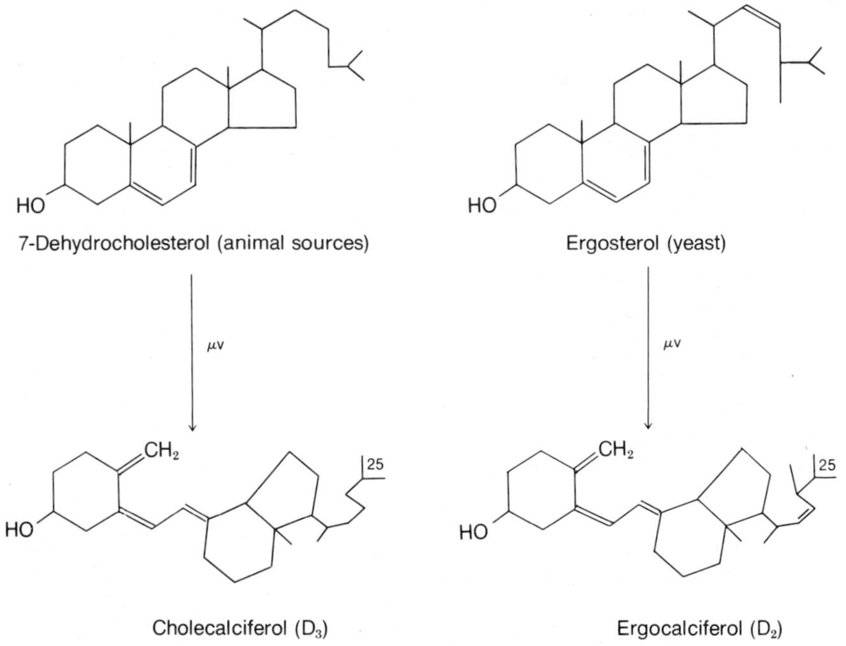

Figure 26.6
Structures of vitamin D and related compounds.

dence that the 1α-hydroxylase activity in the kidney is at least partially controlled by PTH. Once formed, the 1,25-$(OH)_2D$ acts alone as a typical steroid hormone in intestinal mucosal cells, where it induces synthesis of a protein required for calcium transport. In the bone 1,25-$(OH)_2D$ and PTH can act synergistically to promote bone resorption (demineralization) by stimulating osteoblast formation and activity. Finally, PTH inhibits calcium excretion in the kidney. The overall response of calcium metabolism to several different physiological situations is summarized in Figure 26.7. The response to low serum calcium levels is characterized by elevation of PTH and 1,25-$(OH)_2D$, which act to enhance calcium absorption and bone resorption and inhibit calcium excretion (Figure 26.7*a*). High serum calcium levels cause production of the hormone calcitonin, which inhibits bone resorption and enhances calcium excretion. Furthermore, the high levels of serum calcium and phosphate increase the rate of bone mineralization (Figure 26.7*b*). Thus bone serves as a very important reservoir of the calcium and phosphate needed to maintain homeostasis of serum levels. When vitamin D and dietary calcium are adequate, no net loss of bone calcium occurs. However, when dietary calcium is low, PTH and 1,25-$(OH)_2D$ will cause net demineralization of bone to maintain normal serum calcium levels. Vitamin D deficiency also causes net demineralization of bone due to elevation of PTH (Figure 26.7*c*).

The most common symptoms of vitamin D deficiency are rickets in young children and osteomalacia in adults. Rickets is characterized by continued formation of osteoid matrix and cartilage, which are improperly mineralized resulting in soft, pliable bones. In the adult demineralization of preexisting bone takes place, causing the bone to become softer and more susceptible to fracture. This osteomalacia is easily distinguishable from the more common osteoporosis, by the fact that the osteoid matrix remains intact in the former, but not in the latter.

Because of fortification of dairy products with vitamin D, dietary deficiencies are very rare. The cases of dietary vitamin D deficiency that do occur are most often seen in low-income groups, the elderly (who often

CLIN. CORR. **26.2**
RENAL OSTEODYSTROPHY

In patients with chronic renal failure, a complicated chain of events is set in motion, leading to a condition known as renal osteodystrophy. The renal failure results in an inability to produce the 1,25-$(OH)_2D$, and thus bone calcium becomes the only important source of serum calcium. The situation is complicated further by increased renal retention of phosphate and a resulting hyperphosphatemia. The serum phosphate levels are often high enough to cause metastatic calcifications, which tends to lower serum calcium levels further (the solubility product of calcium phosphate in the serum is very low and a high serum level of one component necessarily causes a decreased concentration of the other). The hyperphosphatemia and hypocalcemia stimulate parathyroid hormone secretion, and the resulting hyperparathyroidism further accelerates the rate of bone loss. One ends up with both bone loss and metastatic calcifications. In this case, simple administration of high doses of vitamin D or its active metabolites would not be sufficient, since the combination of hyperphosphatemia and hypercalcemia would only lead to more extensive metastatic calcification. The readjustment of serum calcium levels by high calcium diets and/or vitamin D supplementation must be accompanied by phosphate reduction therapies. The most common technique is to use phosphate binding antacids that bind phosphate and make it unavailable for absorption (usually those containing aluminum hydroxide).

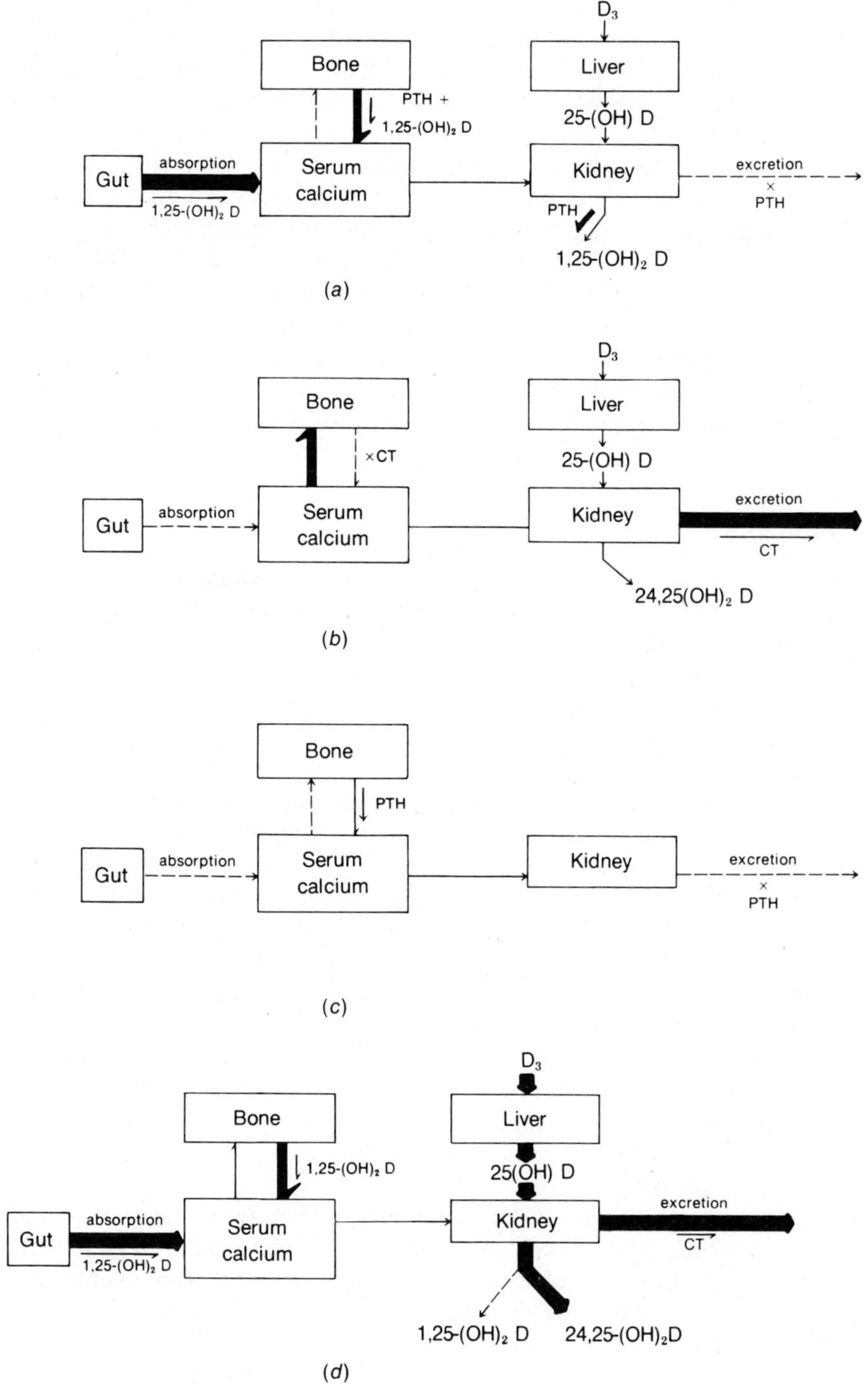

Figure 26.7
Vitamin D and calcium homeostasis.
(a) Low serum calcium. (b) High serum calcium. (c) Low vitamin D. (d) Excess vitamin D. The dominant pathways of calcium metabolism under each set of metabolic conditions are shown with heavy arrows. The effect of various hormones on these pathways is shown by → (stimulation) or X (repression). PTH = parathyroid hormone; CT = calcitonin; D = cholecalciferol; 25-(OH)D = 25-hydroxycholecalciferol; 1,25-$(OH)_2D$ = 1 α,25-dihydroxycholecalciferol.

also have minimal exposure to sunlight), strict vegetarians (especially if their diet is also low in calcium and high in fiber), and chronic alcoholics. Most cases of vitamin D deficiency, however, are a result of diseases causing fat malabsorption or severe liver and kidney disease (Clin. Corr. 26.1). Certain drugs also interfere with vitamin D metabolism. Anticonvulsant drugs inhibit the 25-hydroxylation reaction in the liver (Clin. Corr. 26.3). Corticosteroids also have a similar effect and have been shown to cause bone demineralization when used for long periods of time.

Vitamin D can also be toxic in doses 10-100 times the RDA. The mechanism of vitamin D toxicity is summarized in Figure 26.7*d*. Enhanced calcium absorption and bone resorption cause hypercalcemia, which can lead to metastatic calcifications. The enhanced bone resorption also causes bone demineralization similar to that seen in vitamin D deficiency. Finally, the high serum calcium leads directly to hypercalciuria which predisposes the patient to formation of renal stones.

CLIN. CORR. 26.3 ANTICONVULSANT DRUGS AND VITAMIN REQUIREMENTS

Anticonvulsant drugs such as phenobarbital or diphenylhydantoin (DPH) present an excellent example of the type of drug–nutrient interactions which are of concern to the physician. Anticonvulsants inhibit the conversion of vitamin D to the 25-hydroxy derivative in the liver and enhance its conversion to inactive metabolites. Whereas children and adults on these drugs seldom develop rickets or osteomalacia, as many as 65% of those on long-term therapy will have abnormally low serum calcium and phosphorus and abnormally high serum alkaline phosphatase. Some bone loss is usually observed in these cases. The amount of supplemental vitamin D required to correct this problem is one-and-a-half to two-and-a-half times the RDA. Anticonvulsants also tend to increase needs for vitamin K, leading to an increased incidence of hemorrhagic disease in infants born to mothers on anticonvulsants.

Anticonvulsants also seem to increase the need for folic acid and B_6. Low serum folate levels are seen in 75% of patients on anticonvulsants and megaloblastic anemia may occur in as many as 50% without supplementation. By biochemical parameters, 30–60% of the children on anticonvulsants exhibit some form of B_6 deficiency. However, clinical symptoms of B_6 deficiency are rarely seen. From 1 to 5 mg of folic acid and 10 mg of vitamin B_6 appear to be sufficient for most patients on anticonvulsants. Since folates may speed up the metabolism of some anticonvulsants, it is important that excess folic acid not be given.

Vitamin E

For many years vitamin E was described as the "vitamin in search of a disease." While vitamin E deficiency diseases are still virtually unknown, its metabolic role in the body has become better understood in recent years. Vitamin E occurs in the diet as a mixture of several closely related compounds, called tocopherols. α-Tocopherol is considered the most potent of these and is used as the measure of vitamin E potency (1-tocopherol equiv = 1 mg α-tocopherol).

First and foremost, vitamin E appears to play an important role as a naturally occurring antioxidant. Due to its lipophilic structure it tends to accumulate in circulating lipoproteins, cellular membranes, and fat deposits, where it reacts very readily with molecular oxygen and free radicals. It acts as a scavenger for these compounds, protecting unsaturated fatty acids (especially those in the membranes) from peroxidation reactions. Vitamin E appears to play a role in cellular respiration, either by stabilizing coenzyme Q or by helping transfer electrons to coenzyme Q. It also appears to enhance heme synthesis by increasing the levels of δ-aminolevulinic acid (ALA) synthetase and ALA dehydratase. Most of these vitamin E effects are thought to be an indirect effect of its antioxidant potential, rather than its actual participation as a coenzyme in any biochemical reactions.

Symptoms of vitamin E deficiency vary widely from one animal species to another. In various animals vitamin E deficiencies can be associated with sterility, muscular dystrophy, central nervous system changes, and megaloblastic anemia. In humans, however, the symptoms are usually limited to increased fragility of the red blood cell membrane (presumably due to peroxidation of membrane components), although neurological symptoms have been reported following prolonged vitamin E deficiency associated with malabsorption diseases.

Premature infants fed on formulas low in vitamin E often develop a form of hemolytic anemia that can be corrected by vitamin E supplementation. Adults suffering from fat malabsorption show a decreased red blood cell survival time, but seldom develop anemia itself. Hence, vitamin E supplementation is often necessary with premature infants and in cases of fat malabsorption (Clin. Corr. 26.4).

Studies on the recommended levels of vitamin E in the diet have been hampered by the difficulty in producing severe vitamin E deficiency in man. In general it is assumed that the vitamin E levels in the American diet are sufficient, since no major vitamin E deficiency diseases have been found. However, vitamin E requirements do increase as the intake of

CLIN. CORR. 26.4 THERAPEUTIC USES OF VITAMIN E

Although vitamin E deficiencies are rare, use of vitamin E in therapeutic doses is fairly widespread. In the past vitamin E has been promoted as a cure for so many diseases that it approached the status of a patent medicine. Many of the most heavily promoted uses of vitamin E were completely unfounded. Unfortunately, the resulting controversy has obscured more recent discoveries about vitamin E. For example, since vitamin E deficiency leads

polyunsaturated fatty acids (PUFA) increases. While the recent emphasis on high PUFA diets to reduce serum cholesterol may be of benefit in controlling heart disease, the propensity of PUFA to form free radicals on exposure to oxygen may lead to an increased cancer risk. Thus it appears only prudent to increase vitamin E intake along with high PUFA diets.

As a fat-soluble vitamin, E has the potential for toxicity. However, it does appear to be the least toxic of the fat-soluble vitamins. No instances of toxicity have been reported at doses of 600 mg/day or less. A few scattered reports of malaise and easy fatigability have been reported at doses of 800 mg/day.

Vitamin K

Vitamin K is found naturally as K_1 (phytylmenaquinone) in green vegetables and K_2 (multiprenylmenaquinone), which is synthesized by intestinal bacteria. The body is also able to convert synthetically prepared menaquinone (Menadione) and a number of water-soluble analogs to a biologically active form of vitamin K (see Figure 26.8). Dietary requirements are measured in terms of micrograms of vitamin K_1 with the RDA for adults being in the range of 70–140 μg per day.

Vitamin K_1 has been shown to be required for the conversion of several clotting factors and prothrombin to the active state. The mechanism of this action has been most clearly delineated for prothrombin. Prothrombin is synthesized in an inactive precursor form called preprothrombin. Conversion to the active form requires a vitamin-K-dependent carboxylation

to red cell hemolysis, supplemental vitamin E has been used in the treatment of various hemolytic anemias. Vitamin E has been shown to increase red cell survival in some patients with glucose-6-phosphate dehydrogenase deficiency, β-thalassemia, and sickle cell anemia. However, in most cases the increase is rather small, and it is too early to know whether vitamin E supplementation will reduce the severity of the symptoms in a significant number of these patients. Animal studies have suggested that 200 mg of vitamin E/day may protect lung tissue from the damaging effects of ozone in high smog areas. A regimen of 300–400 mg of vitamin E/day for a period of at least 3–6 months has been shown to alleviate a condition known as intermittent claudication (leg pain when walking due to poor circulation). Similar levels of vitamin E inhibit one of the oxidation steps leading to synthesis of certain prostaglandins. This appears to cause mild antiinflammatory properties and a prolongation of clotting time. The latter property has recently raised some interest, since vitamin E inhibits the same cyclooxygenase reaction as aspirin. However, no studies have been carried out to show whether vitamin E can be effective in reducing the recurrence of myocardial infarctions. This brings us to perhaps the most controversial use of vitamin E—to prevent and/or treat heart disease. None of the double-blind studies carried out to date have shown any significant effect of vitamin E. However, most of these studies have been criticized on the basis of the relatively low doses and short time periods involved.

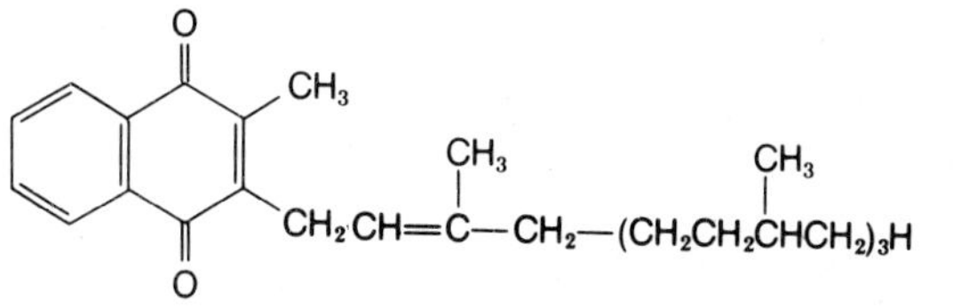

O
CH₃
CH₃ CH₃
$CH_2CH{=}C{-}CH_2{-}(CH_2CH_2CHCH_2)_3H$
O
Vitamin K_1 (green leafy vegetables)

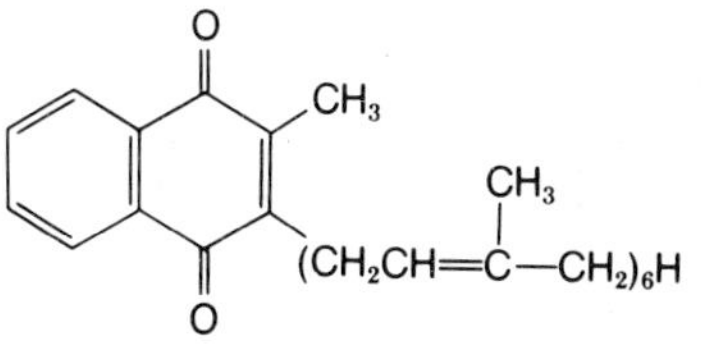

O
CH₃
CH₃
$(CH_2CH{=}C{-}CH_2)_6H$
O
Vitamin K_2 (intestinal bacteria)

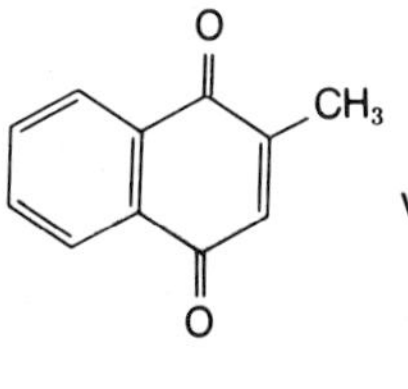

O
CH₃
O
Vitamin K_3 (menadione)

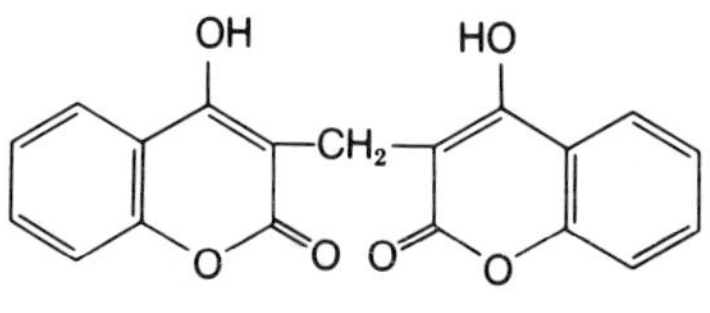

OH HO
CH_2
O O O O
Dicumarol

Figure 26.8
Structures of vitamin K and related compounds.

$$\underset{\text{Precursor}}{\mathrm{COOH{-}CH_2{-}CH_2{-}}} \xrightarrow[\text{Vitamin K}]{HCO_3^-} \underset{\text{Prothrombin}}{\mathrm{(HOOC)(COOH)CH{-}CH_2{-}}}$$

Figure 26.9
Function of vitamin K.

of specific glutamic acid residues to γ-carboxyglutamic acid (Figure 26.9). The γ-carboxyglutamic acid residues are good chelators and allow prothrombin to bind calcium. The prothrombin–Ca^{2+} complex in turn binds to the phospholipid membrane, where proteolytic conversion to thrombin can occur in vivo. The mechanism of the carboxylation reaction has not been fully clarified, but appears to involve the intermediate formation of a 2,3-epoxide derivative of vitamin K. Dicumarol, a naturally occurring anticoagulant, may inhibit the reductase which converts the epoxide back to the active vitamin.

The only known symptom of vitamin K deficiency in man is increased coagulation time. Since vitamin K is relatively abundant in the diet and synthesized in the intestine, deficiencies are very rare. The most common deficiency is seen in newborn infants (Clin. Corr. 26.5). Vitamin K deficiency is also seen in patients with obstructive jaundice and other diseases leading to severe fat malabsorption (Clin. Corr. 26.1) and patients on long-term antibiotic therapy (which may destroy vitamin K-synthesizing organisms in the intestine). Finally, vitamin K deficiency is occasionally seen in the elderly, who are prone to poor liver function (reducing preprothrombin synthesis) and fat malabsorption. Certainly vitamin K deficiency should be suspected in any patient demonstrating easy bruising and prolonged clotting time.

CLIN. CORR. 26.5 NUTRITIONAL CONSIDERATIONS IN THE NEWBORN

Newborn infants are at special nutritional risk. In the first place, this is a period of very rapid growth, and needs for many nutrients are high. Some micronutrients (such as vitamins E and K) do not cross the placental membrane well and tissue stores are low in the newborn infant. The gastrointestinal tract may not be fully developed, leading to malabsorption problems (particularly with respect to the fat-soluble vitamins). The gastrointestinal tract is also sterile at birth, and the intestinal flora that normally provide significant amounts of certain vitamins take several days to become established. If the infant is born prematurely, the nutritional risk is slightly greater, since the gastrointestinal tract will be less well developed and the tissue stores will be less.

The most serious nutritional complication of newborns appears to be hemorrhagic disease of the newborn. Newborn infants, especially premature infants, have low tissue stores of vitamin K and lack the intestinal flora necessary to synthesize the vitamin. Breast milk is also a relatively poor source of vitamin K. Approximately one out of 400 live births shows some signs of hemorrhagic disease. One milligram of the vitamin at birth is usually sufficient to prevent hemorrhagic disease.

Iron is another potential problem. Most newborn infants are born with sufficient reserves of iron to last 3–4 months (although premature infants are born with smaller reserves). Since iron is present in low amounts in both cow's milk and breast milk, iron supplementation is usually begun at a relatively early age by the introduction of iron-fortified cereal. Vitamin D levels are also considered to be somewhat low in breast milk and supplementation with vitamin D is usually recommended. However, some recent studies have suggested that the iron in breast milk is present in a form that is particularly well utilized by the infant and that earlier studies probably underestimated the amount of vitamin D available in breast milk. Other vitamins and minerals appear to be present in adequate amounts in breast milk as long as the mother is getting a good diet.

Artificial formulas present some special nutritional concerns. The high temperatures necessary for pasteurization and sterilization can lead to loss of vitamins C and B_6. There have been instances in which B_6-deficient formulas have reached the marketplace, resulting in a severe enough deficiency to cause convulsions. Vitamin E levels have occasionally caused problems in artificial formulas. While breast milk contains sufficient quantities of vitamin E, cow's milk does not. This can occasionally lead to a hemolytic anemia, especially if the formulas are also high in iron and PUFA. However, manufacturers appear to have learned from past mistakes, and most artificial formulas do contain enough of these three nutrients now. Recent studies have suggested that in situations in which infants must be maintained on assisted ventilation with high oxygen concentrations, supplemental vitamin E may reduce the risk of bronchopulmonary dysplasia and retrolental fibroplasia—two possible side effects of oxygen therapy.

In summary, most infants are provided with supplemental vitamin K at birth to prevent hemorrhagic disease. Breast-fed infants are usually provided with supplemental vitamin D, with iron being introduced along with solid foods. Bottle-fed infants are provided with supplemental iron. The formulas themselves are usually fortified with vitamins D, E, C, and B_6. If infants must be maintained on oxygen, supplemental vitamin E may be beneficial.

26.4 WATER-SOLUBLE VITAMINS

The water-soluble vitamins differ from the fat-soluble vitamins in several important aspects. In the first place, most of these compounds are readily excreted once their concentration surpasses the renal threshold. Thus toxicities are rare. It is popular to speak of these vitamins as "not being stored by the body." While that is not quite accurate, the metabolic stores are quite labile and depletion can often occur in a matter of weeks or months. Since the water-soluble vitamins are coenzymes for many common biochemical reactions, it is often possible to assay vitamin status by measuring one or more enzyme activities in isolated red blood cells. These assays are especially useful if one measures both the endogenous enzyme activity and the stimulated activity following addition of the active coenzyme derived from that vitamin.

Most of the water-soluble vitamins are converted to coenzymes, which are utilized either in the pathways for energy generation or hematopoiesis. Deficiencies of the energy-releasing vitamins produce a number of overlapping symptoms. In many cases the vitamins participate in so many biochemical reactions that it is impossible to pinpoint the exact biochemical cause of any given symptom. However, it is possible to generalize that because of the central role these vitamins play in energy metabolism, deficiencies show up first in rapidly growing tissues. Typical symptoms include dermatitis, glossitis (swelling and reddening of the tongue), cheilitis at the corners of the lips, and diarrhea. In many cases nervous tissue is also involved due to its high energy demand or specific effects of the vitamin. Some of the common neurological symptoms include peripheral neuropathy (tingling of nerves at the extremities), depression, mental confusion, lack of motor coordination, and malaise. In some cases demyelination and degeneration of nervous tissues also take place. These deficiency symptoms are so common and overlapping that they can be considered as properties of the energy-releasing vitamins as a class, rather than being specific for any one.

26.5 ENERGY-RELEASING WATER-SOLUBLE VITAMINS

Thiamine (Vitamin B_1)

Thiamine (Figure 26.10) is rapidly converted to the coenzyme thiamine pyrophosphate (TPP), which is required for the key reactions catalyzed by pyruvate and α-ketoglutarate dehydrogenases (see Figure 26.11). Thus, the cellular capacity for energy generation is severely compromised in thiamine deficiency. TPP is also required for the transketolase of the pentose phosphate pathway. While the pentose phosphate pathway is not quantitatively important in terms of energy generation, it is the sole source of ribose for the synthesis of nucleic acid precursors and the major source of NADPH for fatty acid biosynthesis and other biosynthetic pathways. The red blood cell transketolase is also the enzyme most commonly used for measuring thiamine status in the body. Finally, TPP appears to play an important role in the transmission of nerve impulses. TPP (or a related metabolite, thiamine triphosphate) is localized in peripheral nerve membranes. It appears to be required for acetylcholine synthesis and may also be required for ion translocation reactions in stimulated neural tissue.

Thiamine

Figure 26.10
Structure of thiamine.

Although the biochemical reactions involving TPP are fairly well characterized, it is not clear how these biochemical lesions result in the symptoms of thiamine deficiency. The pyruvate dehydrogenase and transketolase reactions are the most sensitive to thiamine levels. Thus thiamine

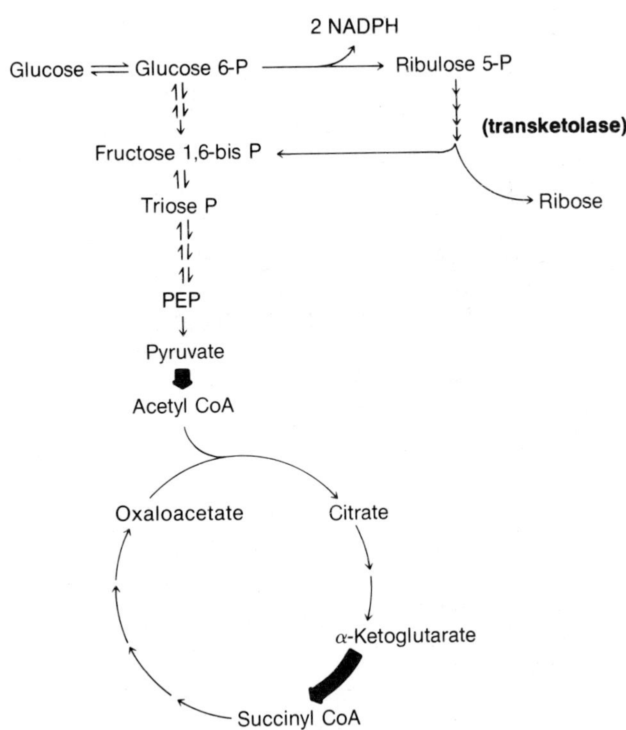

Figure 26.11
Summary of important reactions involving thiamine pyrophosphate.
The reactions involving thiamine pyrophosphate are indicated in boldface type.

deficiency appears to selectively inhibit carbohydrate metabolism, causing a buildup of pyruvate. The cells may be directly affected by the lack of available energy and NADPH or may be poisoned by the accumulated pyruvate. Other symptoms of thiamine deficiency involve the neural tissue and probably result from the direct role of TPP in nerve transmission.

Loss of appetite, constipation, and nausea are among the earliest symptoms of thiamine deficiency. Mental depression, peripheral neuropathy, irritability, and fatigue are other early symptoms and probably directly relate to the role of thiamine in maintaining healthy nervous tissue. These symptoms of thiamine deficiency are most often seen in the elderly and low-income groups on restricted diets. Symptoms of moderately severe thiamine deficiency include mental confusion, ataxia (unsteady gait while walking and general inability to achieve fine control of motor functions), and ophthalmoplegia (loss of eye coordination). This set of symptoms is usually referred to as Wernicke–Korsakoff syndrome and is most commonly seen in chronic alcoholics (Clin. Corr. 26.6). Severe thiamine deficiency is known as *beriberi*. Dry beriberi is characterized primarily by advanced neuromuscular symptoms, including atrophy and weakness of the muscles. When these symptoms are coupled with edema, the disease is referred to as wet beriberi. Both forms of beriberi can be associated with an unusual type of heart failure characterized by high cardiac output. Beriberi is found primarily in populations relying exclusively on polished rice for food, although cardiac failure is sometimes seen in alcoholics as well.

The thiamine requirement is proportional to the caloric content of the diet and will be in the range of 1.4–1.5 mg per day for the normal adult. This requirement should be raised somewhat if carbohydrate intake is excessive or if the metabolic rate is elevated (due to fever, trauma, preg-

CLIN. CORR. **26.6** NUTRITIONAL CONSIDERATIONS IN THE ALCOHOLIC

Chronic alcoholics run considerable risk of nutritional deficiencies. The most common problems are neurologic symptoms associated with thiamine or pyridoxine deficiencies and hematological problems associated with folate or pyridoxine deficiencies. The deficiencies seen with alcoholics are not necessarily due to poor diet alone, although it is often a strong contributing factor. Alcohol causes pathological alterations of the gastrointestinal tract, which often directly interfere with absorption of certain nutrients. Also the liver is often the most important site of activation and storage of many vitamins.

nancy, of lactation). Coffee and tea both contain substances that destroy thiamine, but this is not a problem for individuals consuming normal amounts of these beverages. The routine enrichment of cereals has assured that most Americans have an adequate intake of thiamine on a normal mixed diet.

Riboflavin

Riboflavin is converted to the coenzymes flavin adenine dinucleotide (FAD) and flavin mononucleotide (FMN), both of which are involved in a wide variety of redox reactions. The flavin coenzymes are essential for energy production and cellular respiration. The most characteristic symptoms of riboflavin deficiency are angular cheilitis, glossitis, and scaly dermatitis (especially around the nasolabial folds and scrotal areas). The

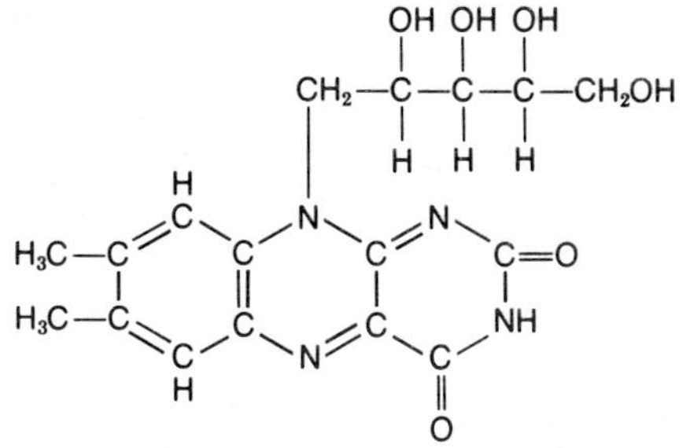

Riboflavin [7,8-dimethyl-10-(1′-D-ribityl)isoalloxazine]

best flavin-requiring enzyme for assaying riboflavin status appears to be erythrocyte glutathione reductase. The recommended riboflavin intake is 1.2–1.7 mg/day for the normal adult. Foods rich in riboflavin include milk, meat, eggs, and cereal products. Riboflavin deficiencies are quite rare in this country. When riboflavin deficiency does occur, it is usually seen in chronic alcoholics.

Niacin

Niacin is not a vitamin in the strictest sense of the word, since the body is capable of making some niacin from tryptophan. However, the conversion of tryptophan to niacin is relatively inefficient (60 mg of tryptophan are required for the production of 1 mg of niacin) and occurs only after all of the body requirements for tryptophan (protein synthesis and energy production) have been met. Since the synthesis of niacin requires thiamine, pyridoxine, and riboflavin, it is also very inefficient on a marginal diet. Thus, in practical terms, most people require dietary sources of both tryptophan and niacin. Niacin (nicotinic acid) and niacinamide (nicotinamide) are both converted to the ubiquitous oxidation-reduction coenzymes NAD^+ and $NADP^+$ in the body.

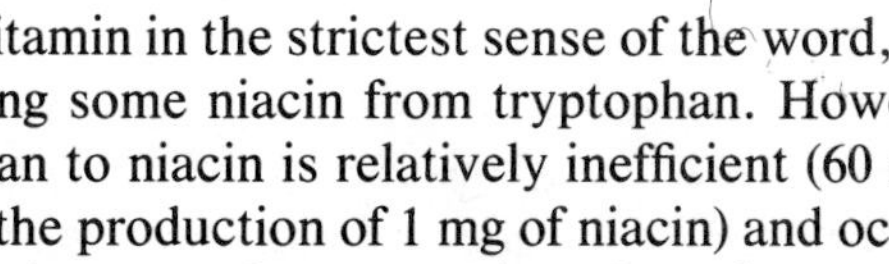

Niacin   Niacinamide

Borderline deficiencies of niacin are first seen as a glossitis of the tongue, somewhat similar to riboflavin deficiency. Pronounced deficiencies lead to pellagra which is characterized by the three Ds: dermatitis, diarrhea, and dementia. The dermatitis is characteristic in that it is usually seen only in skin areas exposed to sunlight and is symmetric. The neurologic symptoms are associated with actual degeneration of nervous tissue.

The severe liver damage associated with chronic alcoholism appears to interfere directly with storage and activation of certain nutrients.

Up to 40% of hospitalized alcoholics are estimated to have megaloblastic erythropoiesis due to folate deficiency. Alcohol appears to directly interfere with folate absorption and alcoholic cirrhosis impairs both activation and storage of this nutrient. Another 30% of hospitalized alcoholics have sideroblastic anemia or identifiable sideroblasts in erythroid marrow cells characteristic of pyridoxine deficiency. The problem here appears to be an impaired utilization of pyridoxine with much of the pyridoxine being degraded rather than being activated to pyridoxal phosphate. Some alcoholics also develop a peripheral neuropathy which responds to pyridoxine supplementation.

The most common nutritionally related neurological disorder is the Wernicke–Korsakoff Syndrome. The symptoms include mental disturbances, ataxia (unsteady gait and lack of fine motor coordination), and uncoordinated eye movements. Congestive heart failure similar to that seen with beriberi is also seen in a small number of these patients. While this syndrome may only account for 1–3% of alcohol-related neurologic disorders, the response to supplemental thiamine is so dramatic that it is usually worth consideration. The thiamine deficiency appears to arise from impaired absorption and utilization.

While those are the most common nutritional deficiencies associated with alcoholism, deficiencies of almost any of the water-soluble vitamins can occur and cases of alcoholic scurvy and pellagra are occasionally reported. Alcoholic patients do have decreased bone density and an increased incidence of osteoporosis. This probably relates to the lack of the 25-hydroxylation step in the liver as well as an increased rate of metabolism of vitamin D to inactive products by an activated cytochrome P_{450} system. Dietary calcium intake is also often poor. In fact, alcoholics generally have decreased serum levels of zinc, calcium, and magnesium due to poor dietary intake and increased urinary losses. Iron-deficiency anemia is very rare unless there is gastrointestinal bleeding or chronic infection. In fact, excess iron is a more common problem with alcoholics. Many alcoholic beverages contain relatively high iron levels, and alcohol appears to enhance iron absorption.

Because of food fortification, pellagra is a medical curiosity in the developed world. Today it is primarily seen in alcoholics, patients with severe malabsorption problems, and elderly on very restricted diets. Pregnancy, lactation, and chronic illness lead to increased needs for niacin, but a varied diet will usually provide sufficient amounts.

Since tryptophan can be converted to niacin, and niacin itself can exist in a free or bound form, the calculation of available niacin for any given food is not a simple matter. For this reason, niacin requirements are expressed in terms of niacin equivalents (1 niacin equiv = 1 mg free niacin). The current recommendation of the Food and Nutrition Board for a normal adult is 13–19 niacin equivalents (N.E.) per day. The richest food sources of niacin are meats, peanuts and other legumes, and enriched cereals.

When nicotinic acid (but not nicotinamide) is used in pharmacologic doses (2–4 g/day) it appears to cause a number of metabolic effects in the body not related to its normal function as a vitamin. For example, vasodilation (flushing) is a very immediate reaction. Over the longer term there is a decreased mobilization of fatty acids from adipose tissue, a marked decrease in circulating cholesterol and lipoproteins (especially LDL), and an elevation of serum glucose and uric acid. These effects can be explained in part by an effect of nicotinic acid on cAMP levels. While the cholesterol lowering effects of nicotinic acid may be desirable in certain controlled clinical situations, there are potential side effects of pharmacologic doses of this vitamin. The reduced mobilization of fatty acids from adipose tissue causes depletion of the glycogen and fat reserves in skeletal and cardiac muscle. The tendency toward elevated glucose and uric acid could cause problems if someone is borderline for diabetes or gout. Finally, continued use of nicotinic acid in those doses is sometimes associated with elevated levels of serum enzymes suggestive of liver damage.

Pyridoxine (Vitamin B_6)

Pyridoxine, pyridoxal, and pyridoxamine are all naturally occurring forms of vitamin B_6 (see Figure 26.12). All three forms are efficiently converted by the body to pyridoxal phosphate which is required for the synthesis, catabolism, and interconversion of amino acids. The role of pyridoxal phosphate in amino acid metabolism has been discussed previously (Chapter 11) and will not be considered here. While pyridoxal phosphate-dependent reactions are legion, there are a few instances in which the biochemical lesion seem to be directly associated with the symptoms of B_6 deficiency. Some of these more important reactions are summarized in Figure 26.13. Obviously, pyridoxal phosphate is essential for energy production from amino acids and can be considered an energy-releasing vitamin. Thus some of the symptoms of severe B_6 deficiency are similar to those of the other energy-releasing vitamins. Pyridoxal phosphate is also required for the synthesis of the neurotransmitters serotonin and norepinephrine and appears to be required for the synthesis of the sphingolipids necessary for myelin formation. These effects are thought to explain the irritability, nervousness, and depression seen with mild deficiencies and the peripheral neuropathy and convulsions observed with severe deficiencies. Pyridoxal phosphate is required for the synthesis of δ-aminolevulinic acid, a precursor of heme. B_6 deficiencies occasionally cause sideroblastic anemia, which is characteristically a microcytic anemia seen in the presence of high serum iron. Pyridoxal phosphate is also an essential component of the enzyme glycogen phosphorylase. It is covalently linked to a lysine residue and stabilizes the enzyme. This role of B_6 appears to explain the decreased glucose tolerance associated with defi-

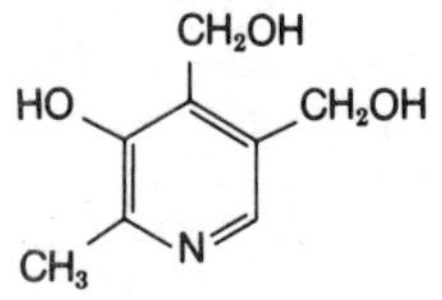

Pyridoxine

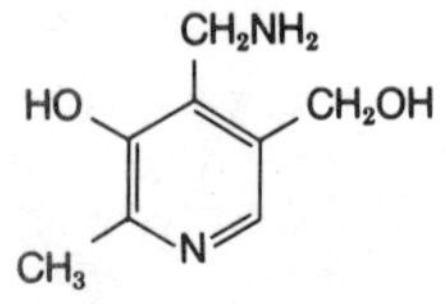

Pyridoxamine

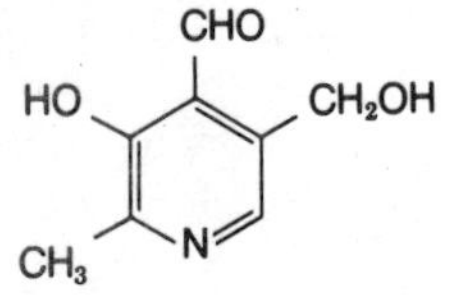

Pyridoxal

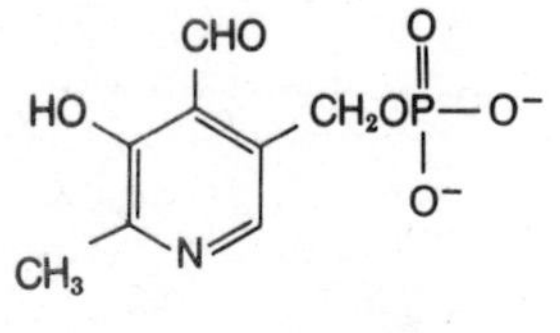

Pyridoxal phosphate

Figure 26.12
Structures of vitamin B_6.

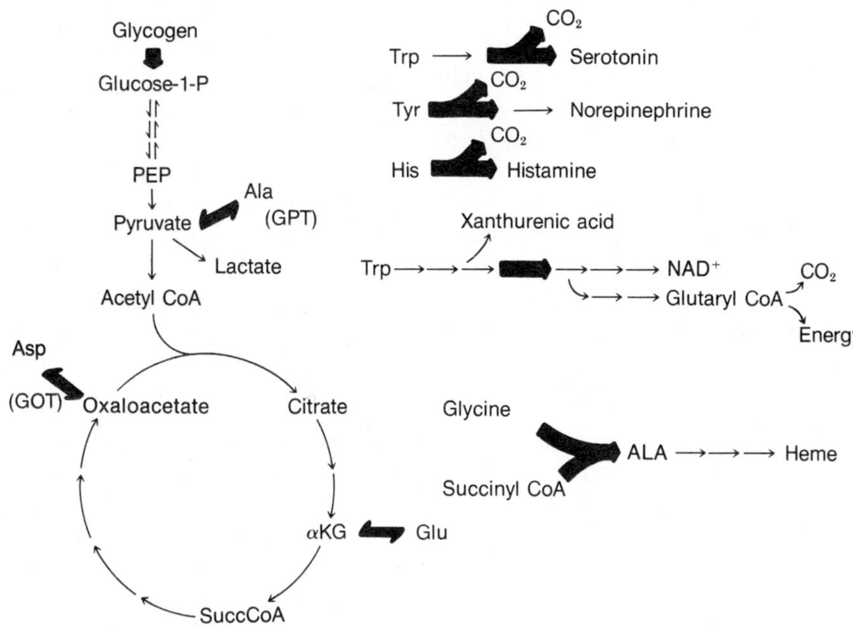

Figure 26.13
Some important metabolic roles of pyridoxal phosphate.
The reactions requiring pyridoxal phosphate are indicated with heavy arrows. ALA = δ-aminolevulinic acid; αKG = α ketoglutarate; GPT = glutamate pyruvate transaminase; and GOT = glutamate oxaloacetate transaminase.

ciency, although B_6 may have some direct effects on insulin metabolism as well. Finally, pyridoxal phosphate is one of the cofactors required for the conversion of tryptophan to NAD. While this may not be directly related to the symptomatology of B_6 deficiency, a tryptophan load test is one of the most sensitive indicators of vitamin B_6 status (Clin. Corr. 26.7).

The amount of B_6 required in the diet is roughly proportional to the protein content of the diet. Assuming that the average American consumes close to 100 g of protein per day, the RDA for vitamin B_6 has been set at 1.8–2.2 mg/day for a normal adult. This requirement is increased during pregnancy and lactation and may increase somewhat with age as well. Vitamin B_6 is fairly widespread in foods, but meat, vegetables, whole-grain cereals, and egg yolks are among the richest sources.

The evaluation of B_6 nutritional status has become a controversial topic in recent years. Both questions of dietary intake and nutritional needs have been clouded by the issue of how to adequately measure B_6 status. Some of this controversy is discussed in Clin. Corr. 26.7. In terms of dietary intake, it has usually been assumed that the average American diet is adequate in B_6 and it is not routinely added to flour and other fortified foods. However, recent nutritional surveys have cast doubt on that assumption. In several instances, a significant fraction of the survey population was found to consume less than two-thirds of the RDA for B_6. Although dietary intake of B_6 may be marginal for many individuals, this does not necessarily cause adverse affects unless coupled with increased demand. Pregnancy, for example, is usually considered to increase the needs for B_6. The usual recommendation is for an additional intake of 0.5 mg/day.

There are, however, a few instances where B_6 deficiencies are clear-cut and noncontroversial. For example, newborn infants rapidly develop symptoms of hyperirritability and convulsive seizures when fed milk or

CLIN. CORR. 26.7
VITAMIN B_6 REQUIREMENTS FOR USERS OF ORAL CONTRACEPTIVES

The controversy over vitamin B_6 requirements for users of oral contraceptives best illustrates the potential problems associated with biochemical assays. For years, one of the most common assays for vitamin B_6 status has been the tryptophan load assay. This assay is based on the observation that when tissue pyridoxal phosphate levels are low, the normal catabolism of tryptophan is impaired and most of the tryptophan is catabolized by a minor pathway leading to synthesis of xanthurenic acid (see Figure 26.12). Under many conditions, the amount of xanthurenic acid recovered in a 24-h urine sample following ingestion of a fixed amount of tryptophan is a valid indicator of vitamin B_6 status. When the tryptophan load test was used to assess the vitamin B_6 status of oral contraceptive users, however, alarming reports started appearing in the literature. Not only did oral contraceptive use increase the excretion of xanthurenic acid considerably, but the amount of pyridoxine hydrochloride needed to return xanthurenic acid excretion to normal was 20 mg/day. This amounts to 10 times the RDA and almost 20 times the level required to maintain normal B_6 status in control groups. As might be expected, this observation received much popular attention in spite of the fact that most classical symptoms of vitamin B_6 deficiency were not observed in oral contraceptive users.

More recent studies using other measures of vitamin B_6 have painted a slightly different picture. For example, erythrocyte glutamate pyruvate transaminase (EGPT) and erythrocyte glutamate oxaloacetate transaminase (EGOT) are both pyridoxal phosphate-containing enzymes.

One can also assess vitamin B_6 status by measuring the endogenous activity of these enzymes and the degree of stimulation by added pyridoxal phosphate. These types of assays show a much smaller difference between nonusers and users of oral contraceptives. The minimum level of pyridoxine hydrochloride needed to maintain normal vitamin B_6 status as measured by these assays was only 2.0 mg/day, which is equal to the RDA and only 1.5 to 2.0 times greater than that needed by nonusers.

Why the large discrepancy? For one thing, it must be kept in mind that for an assay such as the tryptophan load test, what you are actually measuring is the activity of a pyridoxal phosphate containing enzyme *relative* to the overall need for that enzyme. Any unrelated event which increased the overall rate of tryptophan catabolism via the pathway in Figure 26.13 would increase the need for the pyridoxal phosphate-dependent enzyme(s) in that pathway. For example, it is known that the levels of tryptophan dioxygenase (the first enzyme of the pathway) can be increased by hormonal stimulus (for example by hydrocortisone). It seems probable then, that the estrogens in oral contraceptives may also increase the levels of tryptophan dioxygenase, thus creating an increased pyridoxine requirement for tryptophan metabolism without necessarily affecting pyridoxine requirements for other metabolic processes in the body.

Does this mean that vitamin B_6 status is of no concern to users of oral contraceptives? Oral contraceptives do appear to increase vitamin B_6 requirements slightly. Several dietary surveys have shown that a significant percentage of women in the 18- to 24-year age group consume diets containing less than 1.3 mg pyridoxine/day. If these women are also using oral contraceptives, they are at some increased risk for developing a borderline deficiency. Furthermore, there are documented cases of depression and decreased glucose tolerance in oral contraceptive users, which responded to pyridoxine supplementation. While these cases are the exception rather than the rule, they do demonstrate that poor diet plus the slightly increased demands for B_6 due to oral contraceptive use can be expected to lead to deficiencies in some individuals. While the tryptophan load test was clearly misleading in a quantitative sense, it did alert the medical community to a previously unsuspected nutritional risk.

formulas containing less than 50 μg of vitamin B_6/liter. Also the drug isoniazid (isonicotinic acid hydrazide), which is commonly used in the treatment of tuberculosis reacts with pyridoxal or pyridoxal phosphate to form a hydrazone derivative, which inhibits pyridoxal phosphate-containing enzymes. Patients on long-term isoniazid treatment develop a peripheral neuoropathy, which responds well to B_6 therapy. Finally, penicillamine (β-dimethylcysteine), which is used in the treatment of patients with Wilson's disease, cystinuria, and rheumatoid arthritis, reacts with pyridoxal phosphate to form an inactive thiazolidine derivative. Patients treated with penicillamine occasionally develop convulsions, which can be prevented by B_6 supplementation.

Other Energy-Releasing Vitamins

Pantothenic acid is an essential component of coenzyme A (CoA) and of phosphopantetheine of fatty acid synthase and thus is required for the metabolism of all fat, protein, and carbohydrate via the citric acid cycle. In short, more than 70 enzymes have been described to date which utilize CoA or ACP derivatives. In view of the importance of these reactions, one would expect pantothenic acid deficiencies to be a serious concern in man. However, this does not appear to be the case and the reasons are essentially twofold: (1) pantothenic acid is very widespread in natural foods, probably reflecting its widespread metabolic role, and (2) most symptoms of panthothenic acid deficiency are vague and mimic those of other B vitamin deficiencies.

Biotin is the prosthetic group for a number of carboxylation reactions, the most notable being pyruvate carboxylase (needed for synthesis of oxaloacetate for gluconeogenesis and replenishment of the citric acid cycle) and acetyl CoA carboxylase (fatty acid biosynthesis). Biotin is found in peanuts, chocolate, and eggs and is usually synthesized in adequate amounts by intestinal bacteria. Biotin deficiency is generally seen only following long-term antibiotic therapy or excessive consumption of raw egg white. The raw egg white contains a protein, avidin, which binds biotin in a nondigestible form. However, in humans raw egg white must comprise 30% of the caloric intake (approximately 20 egg whites/day) to precipitate a biotin deficiency.

26.6 HEMATOPOIETIC WATER-SOLUBLE VITAMINS

Folic Acid (Folacin)

The simplest form of folic acid is pteroylmonoglutamic acid. However, folic acid usually occurs as polyglutamate derivatives with from 2 to 7 glutamic acid residues (Figure 26.14). These compounds are taken up by intestinal mucosal cells and the extra glutamate residues removed by conjugase, a lysosomal enzyme. The free folic acid is then reduced to tetrahydrofolate by the enzyme dihydrofolate reductase and circulated in the plasma primarily as the free N^5-methyl derivative of tetrahydrofolate (Figure 26.14). However, inside the cells, tetrahydrofolates are found primarily as polyglutamate derivatives, and these appear to be the biologically most potent forms. Folic acid is also stored as a polyglutamate derivative of tetrahydrofolate in the liver.

Various 1-carbon tetrahydrofolate derivatives are used in biosynthetic reactions (see Figure 26.15). They are required, for example, in the synthesis of choline, serine, glycine, methionine, purines, and dTMP. Since adequate amounts of choline and the amino acids can usually be obtained

Folic acid

N^5-Methyltetrahydrofolate

Figure 26.14
Structure of folic acid and N^5-methyl tetrahydrofolate.

from the diet, the participation of folates in purine and dTMP synthesis appears to be metabolically most significant. However, under some conditions the folate-dependent conversion of homocysteine to methionine can make a significant contribution to the available pool. Methionine, of course, is converted to *S*-adenosylmethionine, which is also used in a number of biologically important methylation reactions.

The most pronounced effect of folate deficiency is inhibition of DNA synthesis due to decreased availability of purines and dTMP. This leads to an arrest of cells in S phase and a characteristic "megaloblastic" change

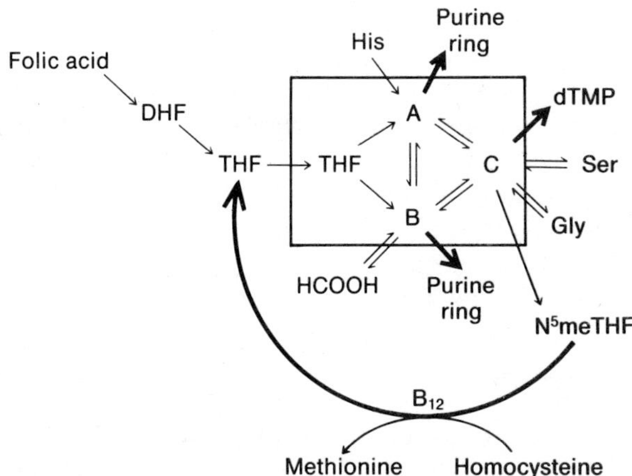

Figure 26.15
Metabolic roles of folic acid and vitamin B_{12} in one carbon metabolism.
The metabolic interconversions of folic acid and its derivatives are indicated with light arrows. Pathways relying exclusively on folate are shown with heavy arrows. The important B_{12}-dependent reaction converting N^5-methyltetrahydrofolate back to tetrahydrofolate is also shown with a heavy arrow. The box encloses the "pool" of 1-carbon derivatives of tetrahydrofolate. DHF = dihydrofolate; THF = tetrahydrofolate; A = 5,10-methenylTHF; B = 10-formylTHF; C = 5,10-methyleneTHF; N^5meTHF = N^5-methylTHF; and dTMP = deoxythymidylic acid.

in the size and shape of the nuclei of rapidly dividing cells. The block in DNA synthesis also slows down the maturation of red blood cells, causing production of abnormally large "macrocytic" red blood cells with fragile membranes. Thus a macrocytic anemia associated with megaloblastic changes in the bone marrow is fairly characteristic of folate deficiency.

There is some uncertainty over the incidence of folate deficiencies. Folates occur very widely in foods, especially meats and a variety of different forms, which are utilized with varying efficiencies. Assessment of nutritional status is further complicated by the use of different biochemical methods to measure folate levels. For example, the serum folate level decreases rapidly on restricted folate intake and is a very sensitive indicator of folate status. However, levels of polyglutamate folate derivatives in red blood cells decrease much more slowly and are more indicative of tissue levels. Symptoms of folate deficiency appear only as tissue folates are depleted. There can be many causes of folate deficiency, including inadequate intake, impaired absorption, increased demand, and impaired metabolism. Some dietary surveys have suggested that inadequate intake may be more common than previously supposed. However, as with most other vitamins, inadequate intake may not be sufficient to trigger symptoms of folate deficiency in the absence of increased requirements or decreased utilization.

Perhaps the most common example of increased need occurs during pregnancy and lactation. As the number of rapidly dividing cells in the body increases, the need for folic acid increases. This situation is complicated by decreased absorption and increased clearance of folate during pregnancy. By the third trimester the folic acid requirement has almost doubled. In the United States almost 20-25% of otherwise normal pregnancies are associated with low serum folate levels, but actual megaloblastic anemia is rare and is usually seen only after multiple pregnancies. Normal diets seldom supply the 800 μg of folate needed during pregnancy, so many physicians routinely recommend supplementation. Folate deficiency is common in alcoholics (Clin. Corr. 26.6). Folate deficiencies are also seen in a number of malabsorption diseases and are occasionally seen in the elderly, due to a combination of poor dietary habits and poor absorption.

There are a number of drugs that also directly interfere with folate metabolism. Anticonvulsants and oral contraceptives interfere with folate absorption (Clin. Corr. 26.3). Oral contraceptives and estrogens also appear to interfere with folate metabolism in their target tissue. Long-term use of any of these drugs can lead to folate deficiencies unless adequate supplementation is provided. For example, 20% of patients using oral contraceptives develop megaloblastic changes in the cervicovaginal epithelium, and 20-30% show low serum folate levels.

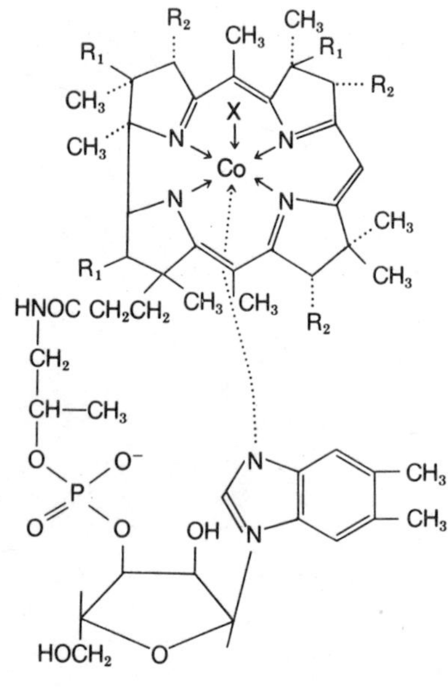

Figure 26.16
Structure of vitamin B_{12} *(cobalamine).*

Vitamin B_{12} (Cobalamine)

Pernicious anemia, a megaloblastic anemia associated with neurological deterioration, was invariably fatal until 1926 when liver extracts were shown to be curative. Subsequent work showed the need for both an extrinsic factor present in liver and an intrinsic factor produced by the body. Vitamin B_{12} was the extrinsic factor. Chemically, vitamin B_{12} consists of cobalt in a coordination state of six, coordinated in 4 positions by a tetrapyrrol (or corrin) ring, in one position by a benzimidazole nitrogen, and in the sixth position by one of several different ligands (Figure 26.16). The crystalline forms of B_{12} used in supplementation are usually hydroxycobalamine or cyanocobalamine. In foods B_{12} usually occurs bound to protein in the methyl or 5′-deoxyadenosyl forms. To be utilized the B_{12}

must first be removed from the protein by acid hydrolysis in the stomach or trypsin digestion in the intestine. It then must combine with "intrinsic factor," a protein secreted by the stomach, which carries it to the ileum for absorption.

In man there are two major symptoms of B_{12} deficiency (hematopoietic and neurological), and only two biochemical reactions in which B_{12} is known to participate (Figure 26.17). Thus it is very tempting to speculate on exact cause and effect mechanisms. The methyl derivative of B_{12} is required for the conversion of homocysteine to methionine and the 5-deoxyadenosyl derivative is required for the methylmalonyl CoA mutase reaction (methylmalonyl CoA → succinyl CoA), which is a key step in the catabolism of some branched-chain amino acids. The neurologic disorders seen in B_{12} deficiency are due to progressive demyelination of nervous tissue. It has been proposed that the methylmalonyl CoA which accumulates interferes with myelin sheath formation in two ways. (1) Methylmalonyl CoA is a competitive inhibitor of malonyl CoA in fatty acid biosynthesis. Since the myelin sheath is continually turning over, any severe inhibition of fatty acid biosynthesis will lead to its eventual degeneration. (2) In the residual fatty acid synthesis that does occur, methylmalonyl CoA can substitute for malonyl CoA in the reaction sequence, leading to branched-chain fatty acids, which might disrupt normal membrane structure. There is some evidence supporting both mechanisms.

The megaloblastic anemia associated with B_{12} deficiency is thought to be due to the effect of B_{12} on folate metabolism. The B_{12}-dependent homocysteine to methionine conversion (homocysteine + N^5-methyl THF → methionine + THF) appears to be the only major pathway by which N^5-methyltetrahydrofolate can return to the tetrahydrofolate pool (Figure 26.15). Thus in B_{12} deficiency there is a buildup of N^5-methyltetrahydrofolate and a deficiency of the tetrahydrofolate derivatives needed for purine and dTMP biosynthesis. Essentially all of the folate becomes "trapped" as the N^5-methyl derivative. B_{12} also may be required for uptake of folate by cells and for its conversion to the biologically more active polygluta-

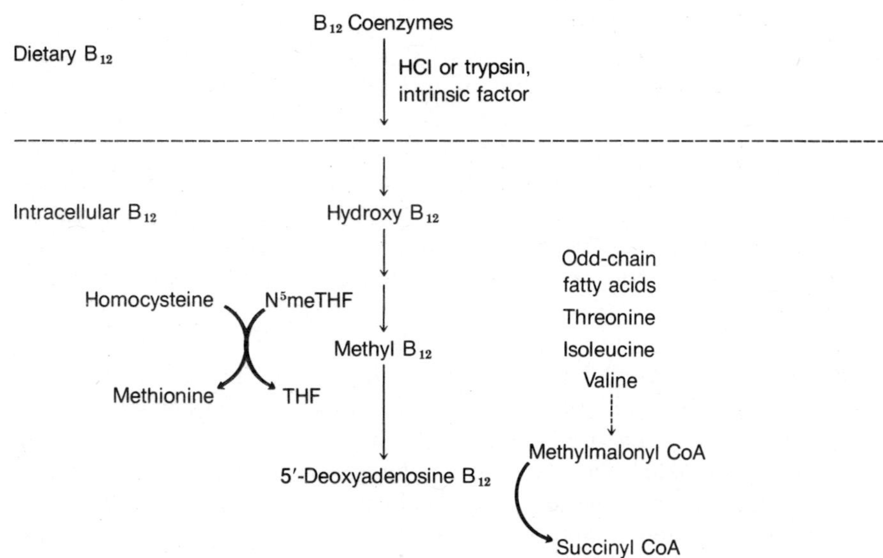

Figure 26.17
Metabolism of vitamin B_{12}.
The metabolic interconversions of B_{12} are indicated with light arrows, and B_{12} requiring reactions are indicated with heavy arrows. Other related pathways are indicated with dashed arrows.

mate forms. High levels of supplemental folate can overcome the megaloblastic anemia associated with B_{12} deficiencies but not the neurological problems. Hence caution must be utilized in using folate to treat megaloblastic anemia.

Vitamin B_{12} is widespread in foods of animal origin, especially meats. Furthermore, the liver stores up to a 6-year supply of vitamin B_{12}. Thus, deficiencies of B_{12} are extremely rare. They are occasionally seen in older people due to insufficient production of intrinsic factor and/or HCl in the stomach. B_{12} deficiency can also be seen in patients with severe malabsorption diseases and in long-term vegetarians.

26.7 OTHER WATER-SOLUBLE VITAMINS

Ascorbic Acid

Vitamin C or ascorbic acid is a 6-carbon compound closely related to glucose. Its main biological role appears to be as a reducing agent in a number of important hydroxylation reactions in the body. For example, there is clear evidence that ascorbic acid is required for the hydroxylation of lysine and proline in protocollagen. Without the hydroxylation of these amino acids, the protocollagen is unable to properly cross-link into normal collagen fibrils. Thus vitamin C is obviously important for maintenance of normal connective tissue and for wound healing, since the connective tissue is normally laid down first. Vitamin C is also necessary for bone formation, since bone tissue has an organic matrix containing collagen as well as the inorganic, calcified portion. Finally, collagen appears to be a component of the ground substance surrounding capillary walls, so vitamin C deficiency is associated with capillary fragility.

Since vitamin C is concentrated in the adrenal gland—especially in periods of stress—it has also been postulated to be required for the hydroxylation reactions involved in the synthesis of some corticosteroids. Ascorbic acid has other important properties as a reducing agent, which appear to be nonenzymatic. For example, it aids in the absorption of iron by reducing it to the ferrous state in the stomach. It spares vitamin A, vitamin E, and some B vitamins by protecting them from oxidation. Also, it enhances the utilization of folic acid, either by aiding the conversion of folate to tetrahydrofolate or the formation of polyglutamate derivatives of tetrahydrofolate. Finally, vitamin C appears to be a biologically important antioxidant. The National Research Council has recently concluded that adequate amounts (RDA levels) of antioxidants such as β-carotene and vitamin C in the diet reduce the risk of cancer. The data for other naturally occurring antioxidants such as vitamin E and selenium are not yet conclusive.

Most of the symptoms of vitamin C deficiency can be directly related to its metabolic roles. Symptoms of mild vitamin C deficiency include easy bruising and the formation of petechiae (small, pinpoint hemorrhages in the skin) due to increased capillary fragility. Mild vitamin C deficiencies are also associated with decreased immunocompetence. Scurvy itself is associated with decreased wound healing, osteoporosis, hemorrhaging, and anemia. The osteoporosis results from the inability to maintain the organic matrix of the bone, followed by demineralization. The anemia results from the extensive hemorrhaging coupled with defects in iron absorption and folate metabolism.

Since vitamin C is readily absorbed, vitamin C deficiencies almost invariably result from poor diet and/or increased need. There is some uncertainty over the need for vitamin C in periods of stress. In severe stress or trauma there is a rapid drop in serum vitamin C levels. In these situations most of the body's supply of vitamin C is mobilized to the adrenals and/or the area of the wound. These facts are clear but their interpretation is variable. Does this represent an increased demand for vitamin C, or merely a normal redistribution of vitamin C to those areas where it is needed most? Do the lowered serum levels of vitamin C impair its functions in other tissues in the body? The current consensus appears to be that the lowered serum vitamin C levels do indicate an increased demand, but there is little agreement as to how much.

A similar situation exists with respect to the effect of various drugs on vitamin C status. Smoking has been shown to cause lower serum levels of vitamin C. Aspirin appears to block uptake of vitamin C by white blood cells. Oral contraceptives and corticosteroids also lower serum levels of vitamin C. While there is no universal agreement as to the seriousness of these effects on vitamin C requirements, the possibility of marginal vitamin C deficiencies should be considered with any patient using these drugs over a long period of time, especially if dietary intake is less than optimal.

Of course, the most controversial question surrounding vitamin C is its use in megadoses to prevent and cure the common cold. Ever since this use of vitamin C was first popularized by Linus Pauling in 1970, the issue has generated considerable controversy. However, some reliable double-blind studies appear to have substantiated the claim in part. The number of colds experienced by vitamin C-supplemented groups appeared to be about the same as for control groups, but the severity and duration of the colds were significantly decreased. Thus, while vitamin C does not appear to be useful in preventing the common cold, it does appear to moderate its symptoms. There is no clear indication at present as to how much vitamin C is required to achieve this effect. In the original experiment the control group was eating a balanced diet providing at least 45 mg of ascorbic acid/day while the experimental group was receiving 1–4 g/day. Subsequent experiments have suggested that considerably less vitamin C may achieve the same result. The mechanism by which vitamin C ameliorates the symptoms of the common cold is not known. It has been suggested that vitamin C is required for normal leukocyte function or for synthesis and release of histamine during stress situations.

While megadoses of vitamin C are probably no more harmful than the widely used over-the-counter cold medications, there are some potential side effects of high vitamin C intake which should be considered. For example, oxalate is a major metabolite of ascorbic acid. Thus, high ascorbate intakes could theoretically lead to the formation of oxalate kidney stones in predisposed individuals. Pregnant mothers taking megadoses of vitamin C may give birth to infants with abnormally high vitamin C requirements. Earlier suggestions that megadoses of vitamin C interfered with B_{12} metabolism have proved to be erroneous.

26.8 MACROMINERALS

Calcium

Calcium is the most abundant mineral in the body. Most of this calcium is in the bone, but the small amount of calcium outside of the bone functions in a number of essential processes. It is required for many enzymes, mediates some hormonal responses, and is essential for blood coagulation. It is also essential for muscle contractility and normal neuromuscular irritability. In fact only a relatively narrow range of serum calcium levels is compatible with life. Since maintenance of constant serum calcium levels is so vital, an elaborate homeostatic control system has evolved. Part of this was discussed earlier in the section on vitamin D metabolism. Low serum calcium stimulates formation of 1,25-dihydroxycholecalciferol, which enhances calcium absorption. If dietary calcium intake is insufficient to maintain serum calcium, 1,25-dihydroxycholecalciferol and parathyroid hormone stimulate bone resorption. Long-term dietary calcium insufficiency, therefore, almost always results in net loss of calcium from the bones.

However, dietary calcium requirements are very difficult to determine due to the existence of other factors that affect availability of calcium. One important factor is vitamin D. As discussed in Chapter 25, excess protein in the diet may upset calcium balance by causing more rapid excretion of calcium. Finally, exercise increases the efficiency of calcium utilization for bone formation. Thus calcium balance studies carried out on Peruvian Indians, who have extensive exposure to sunlight, get extensive exercise, and subsist on low protein vegetarian diets, indicate a need for only 300–400 mg calcium/day. However, calcium balance studies carried out in this country consistently show higher requirements and the RDA has been set at 800 mg/day, with an additional 400 mg allowance for pregnant and lactating women.

The chief symptoms of calcium deficiency are similar to those of vitamin D deficiency, but other symptoms such as muscle cramps are possible with marginal deficiencies. Dietary surveys indicate that a significant portion of certain population groups in this country do not have adequate calcium intake—especially low-income children and adult females. This is of particular concern because these are the population groups with particularly high needs for calcium. For this reason, the U.S. Congress has established the WIC (Women and Infant Children) Program to assure adequate protein, calcium, and iron for indigent families with pregnant/lactating mothers or young infants.

Dietary surveys also show that 34–47% of the over-60 population consume less than one-half the RDA for calcium. This is also the age group most at risk of developing osteoporosis, which is characterized by loss of the organic matrix as well as progressive demineralization of the bone. The causes of osteoporosis are multifactorial and largely unknown, but it appears likely that part of the problem has to do with calcium metabolism. The ability to convert vitamin D to the active 1,25-dihydroxy metabolite decreases with age, especially in women past menopause. Recent studies suggest that postmenopausal women may need up to 1,200 mg calcium/day just to maintain calcium balance. There have been suggestions that this elevated calcium requirement, coupled with years of inadequate dietary intake, may be a factor in osteoporosis. There is no definitive evidence that high dietary calcium alone can prevent osteoporosis, but recent studies suggest that supplemental calcium in the range of 1 g/day will significantly slow the rate of bone loss. Recent studies have also suggested that inadequate intake of calcium may result in elevated blood pressure. Although this hypothesis cannot be considered to have been conclusively demonstrated, it is of great concern because most low sodium diets (which are recommended for patients with high blood pressure) severely limit dairy products (which are the main dietary source of calcium for Americans).

Other Macrominerals

Phosphorous is a universal constituent of living cells and, for that reason, is almost always present in adequate amounts in the diet. Hypophosphatemia is one symptom of vitamin D deficiency. It can also occasionally be seen following excessive use of antacids containing aluminum hydroxide or calcium carbonate, which form insoluble precipitates with phosphate. Uncontrolled metabolic acidosis can lead to excessive phosphate loss in the urine. The initial symptom of hypophosphatemia is muscle weakness, but eventually a form of rickets can develop.

Magnesium is also ubiquitous in living tissue. It is required for many enzyme activities and for neuromuscular transmission. Magnesium deficiency is most often observed in conditions of alcoholism, use of certain diuretics, and metabolic acidosis. The main symptoms of magnesium deficiency are weakness, tremors and cardiac arrhythmia. The discovery that heart muscle from patients with myocardial infarctions was low in magnesium lead to the hypothesis that magnesium deficiency might be a predisposing condition for various forms of heart disease. However, there is no evidence that the patients with heart disease had, in fact, consumed diets inadequate in magnesium. The possibility must be considered that a redistribution of tissue magnesium takes place as a result of the heart attack. There is some evidence that supplemental magnesium may help prevent the formation of calcium oxalate stones in the kidney.

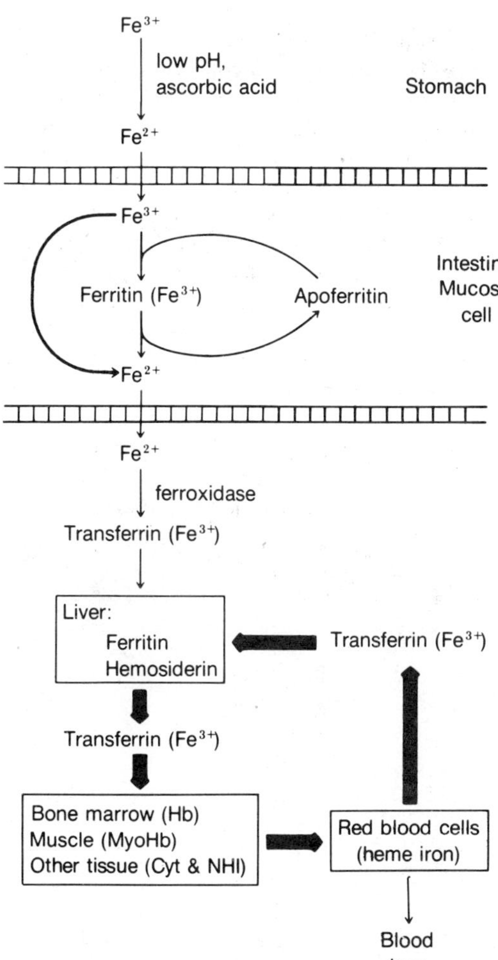

Figure 26.18
Overview of iron metabolism.
This figure reviews some of the features of iron metabolism discussed previously in Chapter 22. The heavy arrows indicate that most of the body's iron is efficiently reutilized by the pathway shown. Hb = hemoglobin; MyoHb = myoglobin; Cyt = cytochromes; and NHI = nonheme iron.

26.9 TRACE MINERALS

Iron

Iron metabolism is unique in that it operates largely as a closed system, with iron stores being efficiently reutilized by the body. Not only are iron losses normally minimal (<1 mg/day), but iron absorption is also minimal under the best of conditions. Iron usually occurs in foods in the ferric form bound to protein or organic acids. Before absorption can occur, the iron must be split from these carriers (a process that is facilitated by the acid secretions of the stomach) and reduced to the ferrous form (a process that is enhanced by ascorbic acid). Only 10% of the iron in an average mixed diet is usually absorbed, but the efficiency of absorption can be increased to 30% by severe iron deficiency. Iron absorption and metabolism have been discussed previously (Chapter 22) and are summarized in Figure 26.18.

Iron, of course, plays a number of important roles in the body. As a component of hemoglobin and myoglobin, it is required for O_2 and CO_2 transport. As a component of cytochromes and nonheme iron proteins, it is required for oxidative phosphorylation. As a component of the essential lysosomal enzyme myeloperoxidase, it is required for proper phagocytosis and killing of bacteria by neutrophils. The best-known symptom of iron deficiency is a microcytic hypochromic anemia. Iron deficiency is also associated with decreased immunocompetence.

Assuming a 10% efficiency of absorption, the Food and Nutrition Board has set a recommended dietary allowance of 10 mg/day for a normal adult male and 18 mg/day for a menstruating female (see Clin. Corr. 26.8). For pregnant and lactating females these allowances are raised to 30–60 mg/day. While 10 mg of iron can easily be obtained from a normal diet, 18 mg is marginal at best and 30–60 mg can almost never be obtained. The best dietary sources of iron are meats, dried legumes, dried fruits, and enriched cereal products.

Iron-deficiency anemia has long been considered the most prevalent nutritional disorder in the United States. Young children need enough iron to allow for a continuing increase in blood volume, as do pregnant females. Menstruating females lose iron through blood loss and lactating

CLIN. CORR. 26.8
CALCULATION OF AVAILABLE IRON

The RDA for iron for women of childbearing age has been estimated at 18 mg/day, based on a need for 1.5–2.0 mg of absorbed iron and an average absorption of 10%. While this calculation makes meal planning simple enough, it has long been regarded as unsatisfactory. In the first place, the 10% average figure is not very useful, since absorption can vary from 2% to over 30% depending on the food source and need. Also, while it is almost impossible to design a diet containing more than 6 mg iron/1,000 kcal, most American women of childbearing age do not suffer from iron-deficiency anemia. A much more useful method is to calculate the actual amount of dietary iron available for absorption in the diet and compare that to the 1.5–2.0 mg needed. This calculation is based on the following data: (1) 40% of the iron in meat is heme iron and the efficiency of absorption of heme iron is 23%. This does not depend on the presence of other food factors. (2) All other dietary iron is nonheme iron, and absorption is dependent on other food, primarily the presence of ascorbic acid and/or some factor(s) present in meat.

Availability	*Consumed in Presence of*	*% Absorption*
Low	<1 oz of meat or <25 mg vit. C	3
Medium	1–3 oz of meat or 25–75 mg vit. C	5
High	>3 oz of meat or >75 mg vit. C or >1 oz of meat + >25 mg vit. C	8

Calculations based on this information provide a much more rational estimate of dietary iron. For example, 3 oz of soybeans contain more than twice as much iron as a 3-oz steak. However, at 3% availability, they would provide only 60% as much available iron. By improving availability to 5%, one could obtain an equivalent amount of iron. This adds a new dimension to planning vegetarian diets. Not only should the protein be balanced and include ample quantities of iron rich sources (dried beans and dried fruit), but these meals should be consumed along with fresh fruit and/or fruit juices to provide enough vitamin C for efficient absorption. While 75 mg of vitamin C might be difficult to obtain in a single meal without supplementation, many fresh fruits and fruit juices will supply 50 mg.

These calculations are also useful in planning diets containing iron-fortified food products. While $FeSO_4$ taken on an empty stomach is well absorbed, it is very poorly utilized in some fortified foods. A

females through production of lactoferrin. Thus iron deficiency anemia is primarily a problem for these population groups. This is reflected in dietary surveys, which indicate that 95% or more of children and menstruating females are not obtaining adequate iron in their diet. It is also reflected in biochemical measurements of a 10–25% incidence of iron deficiency anemia in this same group. Iron-deficiency anemia is also occasionally a problem with the elderly due to poor dietary intake and increased frequency of achlorhydria.

Because of the widespread nature of iron-deficiency anemia, government programs of nutritional intervention such as the WIC program have emphasized iron rich foods. There has also been discussion of more extensive iron fortification of foods. There is concern among some nutritionists that iron deficiency has been overemphasized (Clin. Corr. 26.9). Since iron excretion is very limited, it is possible to build up toxic levels of iron. The excess iron leads to a condition called hemochromatosis in which iron deposits are found in abnormally high levels in many tissues. This can lead to liver, pancreatic, and cardiac dysfunction as well as pigmentation of the skin. This condition is usually only seen in hemolytic anemias and liver disease, but some concern has been voiced that if iron fortification of foods were to become more widespread, iron overload could become more prevalent.

Iodine

Dietary iodine is very efficiently absorbed and transported to the thyroid gland, where it is stored and used for the synthesis of the thyroid hormones triiodothyronine and thyroxine. These hormones play a major role in regulating the basal metabolic rate of the adult and the growth and development of the child. Saltwater fish are the best natural food sources of iodine and in earlier years population groups living in inland areas suffered from the endemic deficiency disease goiter. The most characteristic symptom of goiter is the enlargement of the thyroid gland to the point where a large nodule is visible on the neck. Since iodine has been routinely added to table salt, goiter has become relatively rare in this country. However, in some inland areas, mild forms of goiter may still be seen in up to 5% of the population. It is a possible deficiency for individuals not using iodinized salt provided that they have no other dietary sources of iodine.

Zinc

Zinc absorption appears to be dependent on a transport protein, metallothionein. Over 20 zinc metalloenzymes have been described to date, including both RNA and DNA polymerases. Zinc deficiencies in children are usually marked by poor growth and impairment of sexual development. In both children and adults zinc deficiencies result in poor wound healing. Zinc is also present in gustin, a salivary polypeptide that appears to be necessary for normal development of taste buds. Thus zinc deficiencies also lead to decreased taste acuity.

The few dietary surveys that have been carried out in this country have indicated that zinc intake may be marginal for some individuals. However, few symptoms of zinc deficiency other than decreased taste acuity can be demonstrated in those individuals. Severe zinc deficiency is seen primarily in alcoholics (especially if they have developed cirrhosis), patients with chronic renal disease or severe malabsorption diseases and occasionally in patients on long-term total parenteral nutrition (TPN). The most characteristic early symptom of zinc deficient patients on TPN is dermatitis. Zinc is occasionally used therapeutically to promote wound healing and may be of some use in treating gastric ulcers.

Copper

Copper absorption may also be dependent on the protein metallothionein, since excess intake of either copper or zinc interferes with the absorption of the other. Copper is contained in a number of important metalloenzymes, including cytochrome c oxidase, dopamine β-hydroxylase, superoxide dismutase, and lysyl oxidase. The lysyl oxidase is necessary for the conversion of certain lysine residues in collagen and elastin to allysine, which is needed for cross-linking. Some of the symptoms of copper deficiency in man are leukopenia, demineralization of bones, anemia, fragility of large arteries, and demyelination of neural tissue. The anemia appears to be due to a defect in iron metabolism. The copper-containing enzyme ferroxidase is necessary for conversion of iron from the Fe^{2+} state (in which form it is absorbed) to the Fe^{3+} state (in which form it can bind to the plasma protein transferrin). The bone demineralization and blood vessel fragility can be directly traced to defects in collagen and elastin formation. The causes of the other symptoms are not known.

Copper balance studies carried out with human volunteers seem to indicate a minimum requirement of 1.5 to 2.0 mg/day. Thus the RDA has been set at 2–3 mg/day. Most dietary surveys find that the average American diet provides only 1 mg at $\leq$2000 cal/day. At present, this remains a puzzling problem. No symptoms of copper deficiency have been identified in the general public. It is not known whether there exist widespread marginal copper deficiencies, or whether the copper balance studies are inaccurate. Recognizable symptoms of copper deficiency are usually seen only in two relatively rare hereditary diseases, Menke's syndrome and Wilson's disease. Menke's syndrome is associated with a defect in copper transport across cell membranes. Wilson's disease is associated with abnormal accumulation of copper in the tissues and can be treated with the naturally occurring copper chelating agent penicillamine.

Chromium

Chromium probably functions in the body primarily as a component of glucose tolerance factor (GTF), a naturally occurring substance that appears to be a coordination complex between chromium, nicotinic acid, and the amino acids glycine, glutamate, and cysteine. GTF potentiates the effects of insulin, presumably by facilitating its binding to cell receptor sites. The chief symptom of chromium deficiency is impaired glucose tolerance, a result of the decreased insulin effectiveness.

The frequency of occurrence of chromium deficiency is virtually unknown at present. The RDA for chromium has been set at 50–200 μg for a normal adult. The best current estimate is that the average consumption of chromium is around 60 μg/day in the United States. Unfortunately, the range of intakes is very wide (5–100 μg) even for individuals otherwise consuming balanced diets. Those most likely to have marginal or low intakes of chromium are individuals on low caloric intakes or consuming large amounts of processed foods. Some concern has been voiced that many Americans may be marginally deficient in chromium. However, it is difficult to assess the extent of this problem, if it exists, until better chromium analyses of food become available.

The situation is further confused by individual differences in chromium absorption and utilization. For some individuals, GTF appears to be a hormone-like substance in that they can utilize dietary chromium salts, niacin, and amino acids to synthesize GTF. However, other individuals

serving of iron-fortified cereal with milk alone may actually provide very little available iron. Practical application of this type of information has helped solve problems such as the mysterious occurrence of iron-deficiency anemia in some children on "well-balanced" school meal programs. As long as an iron-fortified cereal was served at 8:00 and orange juice at 10:00, there was a continued incidence of iron deficiency. By moving the orange juice to 8:00 with the cereal, the problem virtually disappeared.

CLIN. CORR. 26.9 IRON-DEFICIENCY ANEMIA

While iron-deficiency anemia is one of the most common nutritional deficiencies in this country, there is some confusion as to just how common it is. The problem is that iron depletion occurs very gradually over a period of time. Iron deficiency occurs in at least three distinct stages. (1) In the first stage, iron stores (usually as ferritin and hemosiderin) are depleted. (2) Once the iron stores have been depleted, serum iron levels fall and the total iron binding capacity (TIBC) of transferrin increases. (3) In the final stage, hemoglobin levels fall and anemia becomes evident. Mean corpuscular hemoglobin concentration (MCHC) is the best indicator of anemia, while percentage saturation of transferrin (which can be calculated from transferrin and TIBC determinations) is the best measure of simple iron deficiency. Obviously, the percentage saturation of transferrin is a more sensitive indicator of iron deficiency than MCHC. However, both determinations have sometimes been used interchangeably to estimate the incidence of iron-deficiency anemia.

That leads us to the second question. Just how serious are the symptoms associated with iron-deficiency anemia? In most cases the symptoms are mild enough that the anemia is seldom the reason that the patient comes to see the physician. The most common symptoms are mild fatigue, weakness, and anorexia. In most cases, these symptoms seem to be associated with depleted tissue stores rather than the anemia itself. As the iron-deficiency anemia becomes more severe, the fingernails become thin and flat with a characteristic spoon-shaped appearance (koilonychia). If these were the only possible symptoms of iron-deficiency anemia, a major public health effort to prevent iron deficiency would not appear to be warranted. However, there is the possibility that iron deficiency may lead to an increased susceptibility to infection. When tissue stores of iron are depleted, there is an impairment of cell mediated immunity and phagocytic activity. Unfortunately, it is very difficult to correlate accurately the frequency of infection with the iron status in a human population. However, since that possibility clearly exists and a significant portion of the population evidences iron deficiency by one measure or another, several major public health measures have been undertaken to improve iron availability in the diet. These include iron fortification of flour and, more recently, the Women and Infant Children (WIC) Program.

CLIN. CORR. **26.10**
NUTRITIONAL CONSIDERATIONS FOR VEGETARIANS

A vegetarian diet poses certain problems in terms of micronutrient intake that need to be recognized in designing a well-balanced diet. Vitamin B_{12} is of special concern, since it is found only in foods of animal origin. B_{12} should be obtained from fortified foods (such as some brands of soybean milk) or in tablet form. However, surprisingly few vegetarians ever develop pernicious anemia, perhaps because an adult who has previously eaten meat will have a 6- to 10-year store of B_{12} in his liver.

Iron is another problem. The best vegetable sources of iron are dried beans, dried fruits, whole grain or enriched cereals, and green leafy vegetables. Vegetarian diets can provide adequate amounts of iron provided that these foods are regularly selected and consumed with vitamin C-rich foods to promote iron absorption (Clin. Corr. 26.8). However, iron supplementation is usually recommended for children and menstruating females.

When milk and dairy products are absent from the diet, certain other problems must be considered as well. Normally, dietary vitamin D is obtained primarily from fortified milk. While some butters and margarines are fortified with vitamin D, they are seldom consumed in sufficient quantities to supply significant amounts of vitamin D. Although adults can usually obtain sufficient vitamin D from exposure

utilize chromium salts very poorly and appear to need preformed GTF in the diet. Unfortunately, very little is known about the requirements for preformed GTF in the general population or about its distribution in natural foods. While it is clear that most diabetics do not respond significantly to either chromium or GTF, there are well documented cases in which GTF has been useful in treating cases of diabetes.

Selenium

Selenium appears to function primarily in the metalloenzyme glutathione peroxidase, which destroys peroxides in the cytosol. Since the effect of vitamin E on peroxide formation is limited primarily to the membrane, both selenium and vitamin E appear to be necessary for efficient scavenging of peroxides. Selenium is one of the few nutrients not removed by the milling of flour and is usually thought to be present in adequate amounts in the diet. The selenium levels are very low in the soil in certain parts of the country, however, and foods raised in these regions will be low in selenium. Fortunately, this effect is minimized by the current food distribution system, which assures that the foods marketed in any one area are derived from a number of different geographical regions.

Other Trace Minerals

Manganese is a component of pyruvate carboxylase and probably other metalloenzymes as well. Molybdenum is a component of xanthine oxidase. Deficiencies of both of these trace minerals are virtually unknown in man. Fluoride is known to strengthen bones and teeth and is usually added to drinking water.

26.10 THE AMERICAN DIET: FACT AND FALLACY

Much has been said lately about the supposed deterioration of the American diet. How serious a problem is this? Clearly Americans are eating much more processed food than their ancestors. These foods differ from simpler foods in that they have a higher caloric density and a lower nutrient density than the foods they replace. However, these foods are almost uniformly enriched with iron, thiamine, riboflavin, and niacin. In many cases they are even fortified (usually as much for sales promotion as for nutritional reasons) with as many as 11–15 vitamins and minerals. Unfortunately, it is simply not practical to replace all of the nutrients lost, especially the trace minerals. Imitation foods present a special problem in that they are usually incomplete in more subtle ways. For example, the imitation cheese and imitation milkshakes that are widely sold in this country usually do contain the protein and calcium one would expect of the food they replace, but often do not contain the riboflavin, which one should also obtain from these items. Fast food restaurants have also been much maligned in recent years. Some of the criticism has been undeserved, but fast food meals do tend to be high in calories and fat and low in certain vitamins and trace minerals. For example, the standard fast food meal provides over 50% of the calories the average adult needs for the entire day, while providing <5% of the vitamin A and <30% of biotin, folic acid, and pantothenic acid. Unfortunately, much of the controversy in recent years has centered around whether these trends are "good" or "bad." This simply obscures the issue at hand. Clearly it is possible to obtain a balanced diet which includes processed, imitation, and fast foods if one compensates by selecting foods for the other meals which are low in caloric density and rich in nutrients. However, without such compensation the "balanced diet" becomes a myth. Unfortunately, few nutritionists have taken the initiative in pointing out that such a compensation is both necessary and possible if one wishes to consume a balanced diet.

What do dietary surveys tell us about the adequacy of the American diet? The most comprehensive dietary survey presently available is the 1977–1978 Nationwide Food Consumption Survey carried out by the USDA, which analyzed 3-day dietary reports from 38,000 Americans. It showed that many Americans may be consuming suboptimal amounts of iron, calcium, vitamin A, vitamin B_6, and vitamin C. Less extensive dietary surveys have suggested that a significant fraction of the population might have inadequate intakes of folic acid and certain trace minerals. How are these data to be interpreted? In every instance, biochemical measurements show significantly fewer individuals with marginal nutritional status, and clinical symptoms of these deficiencies are rare indeed. Thus a physician need not be alarmed by these reports of potential dietary deficiencies, but should be aware of them when dealing with patients with increased nutrient requirements. Clearly these dietary surveys tell us that the diet provides no surplus of most nutrients. Thus, if a patient has increased nutritional needs, it is unlikely that he or she will meet these needs from diet alone without proper nutritional counseling and/or dietary supplementation.

to sunlight, dietary sources are oten necessary during periods of growth and for adults with little exposure to sunlight. Vegetarians may need to obtain their vitamin D from fortified foods such as cereals, certain soybean milks, or in tablet form. Riboflavin is found in a number of vegetable sources such as green leafy vegetables, enriched breads, and wheat germ. However, since none of these sources supply more than 10% of the RDA in normal serving sizes, fortified cereals or vitamin supplements may become an important source of this nutrient. The important sources of calcium for vegetarians include soybeans, soybean milk, almonds, and green leafy vegetables. Those green leafy vegetables without oxalic acid (mustard, turnip, and dandelion greens, collards, kale, romaine, and loose leaf lettuce) are particularly good sources of calcium. However, none of these sources is equivalent to cow's milk in calcium content, so calcium supplements are usually recommended during periods of rapid growth.

26.11 ASSESSMENT OF NUTRITIONAL STATUS IN CLINICAL PRACTICE

Having surveyed the major micronutrients and their biochemical roles, it might seem that the process of evaluating the nutritional status of an individual patient would be an overwhelming task. It is perhaps best to recognize that there are three factors that can add to nutritional deficiencies: poor diet, malabsorption, and increased nutrient need. Only when two or three components overlap in the same person (Figure 26.19) do the risks of symptomatic deficiencies become significant. For example, infants and young children have increased needs for iron, calcium, and protein. Dietary surveys show that many of them consume diets inadequate in iron and some consume diets that are low in calcium. Protein is seldom a problem unless the children are being raised as strict vegetarians (see Clin. Corr. 26.10). Thus, the chief nutritional concerns for most children are iron and calcium. Teenagers tend to consume diets low in calcium, magnesium, vitamin A, vitamin B_6, and vitamin C. Of all of these nutrients, their needs are particularly high for calcium and magnesium during the teenage years, so these are the nutrients of greatest concern. Young women are likely to consume diets low in iron, calcium, magnesium, vitamin B_6, folic acid, and zinc—and all of these nutrients are needed in greater amounts during pregnancy and lactation. Older women often consume diets low in calcium, yet they may have a particularly high need for calcium to prevent rapid bone loss. Finally, the elderly in general tend to have poor nutrient intake due to restricted income, loss of appetite, and loss of the ability to prepare a wide variety of foods. They are also more prone to suffer from malabsorption problems and to use multiple prescription drugs that increase nutrient needs (Table 26.1).

Illness and metabolic stress often cause increased demand or decreased utilization of certain nutrients. For example, diseases leading to fat malabsorption cause a particular problem with absorption of calcium and the fat-soluble vitamins. Other malabsorption diseases can result in

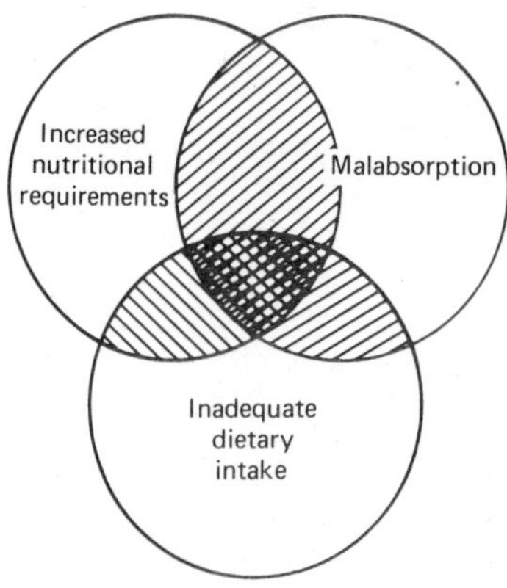

Figure 26.19
Factors affecting individual nutritional status.
Schematic representation of three important risk factors in determining nutritional status. A person in the white area would have very low risk of any nutritional deficiency, whereas people in the crosshatched areas would be much more likely to experience some symptoms of nutritional deficiencies.

TABLE 26.1 Drug-Nutrient Interactions

Drug	*Potential Nutrient Deficiencies*
Alcohol	Folic acid Vitamin B_6
Anticonvulsants	Thiamine Vitamin D Folic acid Vitamin K
Corticosteroids	Vitamin D and calcium Zinc Potassium Vitamin B_6 Vitamin C
Diuretics	Potassium Zinc Calcium, magnesium
Oral contraceptives and estrogens	Vitamin B_6 Folic acid Vitamins C and B_{12}

deficiencies of many nutrients depending on the particular malabsorption disease. Liver and kidney disease can prevent activation of vitamin D and storage or utilization of many other nutrients including vitamin A, vitamin B_{12}, and folic acid. Severe illness or trauma increase the need for calories, protein, and possibly some micronutrients such as vitamin C and certain B vitamins. Long-term use of many drugs in the treatment of chronic disease states can affect the need for certain micronutrients. Some of these are summarized in Table 26.1.

Who then is at nutritional risk? Obviously, this depends on many factors. Nutritional counselling will be an important part of the treatment for infants, young children, and pregnant/lactating females. A brief analysis of a dietary history and further nutritional counseling will also be important when dealing with many other high-risk patients.

BIBLIOGRAPHY

Recommended Dietary Allowances

Food and Nutrition Board of the National Academy of Sciences. *Recommended dietary allowances,* 9th ed. Washington, D.C.: National Academy of Sciences, 1980.

Micronutrients—General Information

The Nutrition Foundation. *Present knowledge in nutrition.* 5th ed. New York, 1985.

Vitale, J. J. *Vitamins.* Kalamazoo, Mich.: Upjohn Co., 1976.

Vitamin A

Goodman, D. S. Vitamin A and retinoids in health and disease. *New Engl. J. Med.* 310:1023, 1984.

Shekelle, R. B., Liu, S., Raynor, W. J., Leper, M., Melija, C., Rossof, A. H. Dietary vitamin A and risk of cancer in the western electric study. *Lancet* II:1185, 1981.

Wolf, Multiple uses of vitamin A. *Phys. Rev.* 64:873, 1984.

Vitamin E

Bieri, J. G., Coresh, L., and Hubbard, V. S. Medical uses of vitamin E. *New Engl. J. Med.* 308:1063, 1983.

Howitt, M. K. Vitamin E: A reexamination. *Am. J. Clin. Nutr.* 29:569, 1976.

Vitamin C

Anderson, T. W. Large scale trials of vitamin C. *Ann. N.Y. Acad. Sci.* 258:498, 1975.

Calcium

Heaney, R. P., Recker, R. R., and Saville, P. D. Calcium balance studies and calcium requirements in middle-aged women. *Am. J. Clin. Nutr.* 30:1603, 1977.

McCarron, D. A. Calcium and magnesium nutrition in human hypertension. *Ann. Intern. Med.* 98:800, 1983.

Chromium

Saner, G. *Chromium in nutrition and disease*. New York: Alan R. Liss, 1980.

Vitamin B_6 and Oral Contraceptives

Bossé, T. R., and Donald, E. A. The vitamin B_6 requirement in oral contraceptive uses. *Am. J. Clin. Nutr.* 32:1015, 1979.

Kirksey, A., Keaton, K., Abernathy, R. P., and Greger, J. L. Vitamin B_6 nutritional status of a group of female adolescents. *Am. J. Clin. Nutr.* 31:946, 1978.

Dietary Iron

Manson, E. R., Hallberg, L., Layrisse, M., Hegsted, D. M., Cook, J. D., Mertz, W., and Finch, C. A. Estimation of dietary iron. *Am. J. Clin. Nutr.* 31:134, 1978.

Nutrition and Alcoholism

Shaw, S., and Lieber, C. S. Alcoholism, in H. Schneider, C. Anderson, and D. Coursin (eds.), *Nutritional support of medical practice*. New York: Harper & Row, 1977, p. 202.

Drug-Nutrient Interaction

Theuer, R. D., and Vitale, J. J. Drug and nutrient interactions, in H. Schneider, C. Anderson, and D. Coursin (eds.), *Nutritional support of medical practice*. New York: Harper & Row, 1977, p. 297.

Dietary Surveys

Pao, E. M., and Mickle, S. J. Problem nutrients in the United States, *Food Technology*. 35:58, 1981.

U.S. Department of Health, Education, and Welfare. *Highlights of the ten-state nutrition survey,* DHEW Publication No. 72-8134. Washington, D.C.: U.S. Government Printing Office, 1972.

U.S. Department of Health, Education, and Welfare. *Preliminary findings of the first health and nutrition examination survey, 1971–1973,* DHEW Publication No. (HRA) 74-1219-1. Washington, D.C.: U.S. Government Printing Office, 1974.

QUESTIONS

J. BAGGOTT AND C. N. ANGSTADT

*Q*uestion *T*ypes are described inside front cover.

A. Visual cycle   C. Both
B. Cholecalciferol metabolism   D. Neither

1. (QT4) Liver plays a critical role.
2. (QT4) Kidney plays a critical role.

A. Fat-soluble vitamins   C. Both
B. Water-soluble vitamins   D. Neither

3. (QT4) Excess intake may be harmful.
4. (QT4) Can generally be synthesized in adequate amounts by healthy people who have an adequate protein–calorie intake.

A. Vitamin A
B. Biotin
C. Niacin
D. Vitamin D
E. Vitamin B_{12} (cobalamine)

5. (QT5) Requirement may be totally supplied by intestinal bacteria.
6. (QT5) Precursor is synthesized by green plants.
7. (QT5) Tryptophan is a precursor.
8. (QT5) Deficiency may be seen in long-term adherence to a strict vegetarian diet.
9. (QT5) Not required in the diet if adequate exposure of the body to sunlight occurs.
10. (QT2) Ascorbic acid:
    1. is a biological reducing agent.
    2. is absorbed with difficulty.
    3. aids in iron absorption.
    4. is harmless in high doses.
11. (QT2) In assessing the adequacy of a person's diet:
    1. intake of the RDA (recommended dietary allowance) of every nutrient assures adequacy.
    2. a drop in the plasma concentration of a nutrient is a clear sign of an increased requirement.
    3. a 24-hr dietary intake history provides an adequate basis for making a judgment.
    4. currently administered medications must be considered.

A. Calcium
B. Iron
C. Iodine
D. Copper
E. Selenium

12. (QT5) Absorption is inhibited by excess dietary zinc.
13. (QT5) Risk of nutritional deficiency is high in postmenopausal women.
14. (QT5) Risk of nutritional deficiency is high in young children.
15. (QT5) Unsupplemented diets of populations living in inland areas may be deficient.
16. (QT5) Essential component of glutathione peroxidase.

ANSWERS

1. C Liver converts *trans*-retinol to 11-*cis*-retinol (p. 965) and cholecalciferol to 25-hydroxycholecalciferol (p. 967).
2. B Here 25-hydroxycholecalciferol is converted to the active 1,25-dihydroxy compound (p. 967).
3. C High levels of vitamin A (p. 966) and vitamin D (p. 970) are toxic. Water-soluble vitamins are often regarded as harmless, but this is not true of niacin (p. 976).
4. D Although some niacin is synthesized in the body (p. 975), no

vitamin can be synthesized in adequate amounts. Vitamin D is not properly a vitamin, but a hormone (p. 967).

5. B See p. 978.
6. A β-Carotene, from green plants, is converted to vitamin A (p. 964).
7. C See p. 975.
8. E This vitamin is from animal sources (p. 982).
9. D 7-Dehydrocholesterol, a normal metabolite of cholesterol, is converted to cholecalciferol by ultraviolet irradiation (p. 967).
10. B 1 and 3 true. 1, 2, and 3: See p. 982. 4: See p. 983.
11. D Only 4 true. 1: RDAs meet the known needs of most healthy people. Individual variations may occur, and certainly stress or disease can be expected to change some requirements. 2: This could reflect a mere redistribution, not deficiency. 3: Can you be sure that any 24-h diet history is either accurate or representative of the individual's typical diet? (p. 962) 4: Isoniazid and penicillamine affect B_6 (p. 978), and antacids interfere with phosphate absorption (p. 985).
12. D See p. 987.
13. A This seems to be related to decreased vitamin D metabolism (p. 984).
14. B Rapid growth in children causes high demands for iron (p. 985).
15. C The problem is rare in the United States due to the common use of iodized salt (p. 986).
16. E See p. 988.

APPENDIX

Review of Organic Chemistry

CAROL N. ANGSTADT

FUNCTIONAL GROUPS

Alcohols

The general formula of alcohols is R—OH, where R equals an alkyl or aryl group. They are classified as *primary, secondary,* or *tertiary,* according to whether the hydroxyl (—OH)-bearing carbon is bonded to no carbon or one, two, or three other carbons:

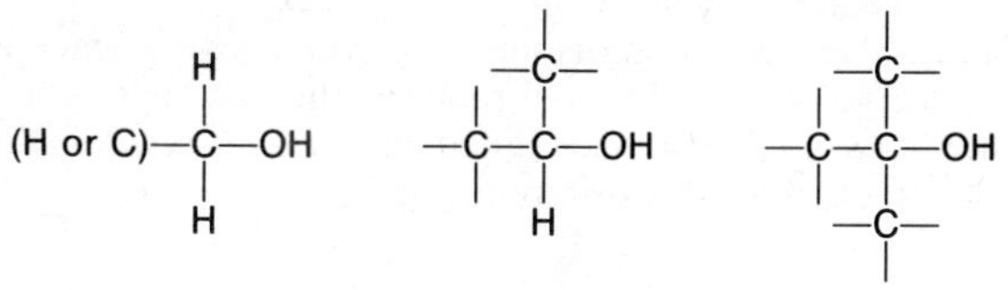

(H or C)—CH₂—OH	(C)₂CH—OH	(C)₃C—OH
Primary	Secondary	Tertiary

Aldehydes and Ketones

Aldehydes and ketones contain a carbonyl group —C(=O)—; *aldehydes* are R—C(=O)—H, and a *ketone* has two alkyl groups at the carbonyl group R—C(=O)—R′.

Acids and Acid Anhydrides

Carboxylic acids contain the functional group R—C(=O)—OH (R—COOH). Dicarboxylic and tricarboxylic acids contain two or three carboxyl groups. A carboxylic acid ionizes in water to a negatively charged carboxylate ion:

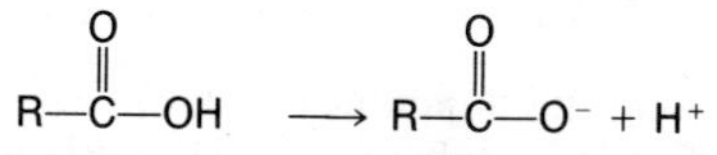

$$\text{R—C(=O)—OH} \longrightarrow \text{R—C(=O)—O}^- + \text{H}^+$$

Carboxylic acid   Carboxylate ion

Names of carboxylic acids usually end in "-ic" and the carboxylate ion in "-ate." *Acid anhydrides* are formed when two molecules of acid react with loss of a molecule of water. An acid anhydride may form between two organic acids, two inorganic acids, or an organic and an inorganic acid:

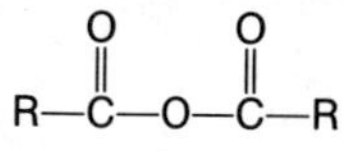

R—C(=O)—O—C(=O)—R

Organic anhydride

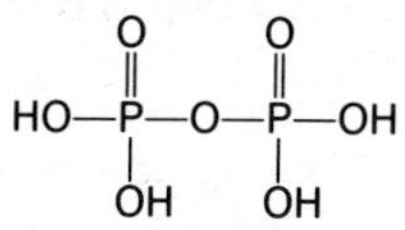

HO—P(=O)(OH)—O—P(=O)(OH)—OH

Inorganic anhydride

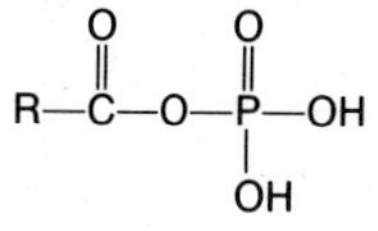

R—C(=O)—O—P(=O)(OH)—OH

Organic–inorganic anhydride

Esters

Esters form in the reaction between a carboxylic acid and an alcohol:

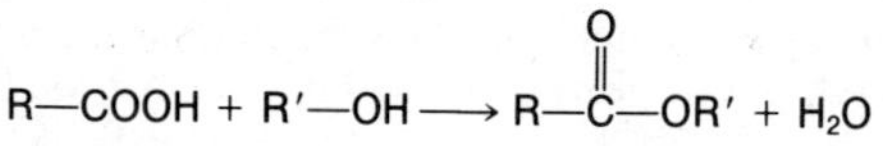

$$\text{R—COOH} + \text{R}'\text{—OH} \longrightarrow \text{R—C(=O)—OR}' + H_2O$$

Esters may form between an inorganic acid and an organic alcohol, for example, glucose 6-phosphate (see carbohydrates).

Hemiacetals, Acetals, and Lactones

A reaction between an aldehyde and an alcohol gives a *hemiacetal,* which may react with another molecule of alcohol to form an *acetal:*

$$R\text{—}CHO \xrightarrow{R'\text{—}OH} R\text{—}\overset{\overset{\displaystyle OH}{|}}{\underset{\underset{\displaystyle H}{|}}{C}}\text{—}OR' \xrightarrow{R''\text{—}OH} R\text{—}\overset{\overset{\displaystyle OR''}{|}}{\underset{\underset{\displaystyle H}{|}}{C}}\text{—}OR'$$

Hemiacetal Acetal

Lactones are cyclic esters formed when an acid and an alcohol group on the same molecule react and usually requires that a five- or six-membered ring be formed.

Unsaturated Compounds

Unsaturated compounds are those containing one or more carbon–carbon multiple bonds, for example, a double bond: $\text{—}C{=}C\text{—}$.

Amines and Amides

Amines, $R\text{—}NH_2$, are organic derivatives of NH_3 and are classified as *primary, secondary,* or *tertiary,* depending upon the number of alkyl groups bonded to the nitrogen. When a fourth substituent is bonded to the nitrogen, the species is positively charged and called a *quaternary ammonium ion:*

$R\text{—}NH_2$	$R\text{—}\underset{\underset{\displaystyle H}{|}}{N}\text{—}R'$	$R\text{—}\underset{\underset{\displaystyle R''}{|}}{N}\text{—}R'$	$R\text{—}\overset{\overset{\displaystyle H\ (\text{or } R''')}{|}}{\underset{\underset{\displaystyle R''}{|}}{N^+}}\text{—}R'$
Primary amine	Secondary amine	Tertiary amine	Quaternary ammonium ion

Amides contain the functional group $\text{—}\overset{\overset{\displaystyle O}{\|}}{C}\text{—}\overset{\overset{\displaystyle H}{|}}{N}\text{—}X$; X can be H (simple) or R (*N* substituted). The carbonyl group is from an acid, and the *N* is from an amine. If both functional groups are from amino acids (see Amino acids), the amide bond is referred to as a *peptide bond.*

TYPES OF REACTIONS

Nucleophilic Substitutions at an Acyl Carbon

$$\begin{matrix} R \\ \ \ \diagdown \\ \ \ \ C{=}\ddot{O}{:} \\ \ \ \diagup \\ L \end{matrix} + {:}X\text{—}H \rightleftharpoons R\text{—}\overset{\overset{\displaystyle XH^+}{|}}{\underset{\underset{\displaystyle L}{/}}{C}}\text{—}\ddot{\underset{..}{O}}{:} \rightleftharpoons R\text{—}\overset{\overset{\displaystyle X}{|}}{\underset{\underset{\displaystyle LH^+}{/}}{C}}\text{—}\ddot{\underset{..}{O}}{:}^- \longrightarrow \begin{matrix} X \\ \ \ \diagdown \\ \ \ \ C{=}O \\ \ \ \diagup \\ R \end{matrix} + LH$$

New compound Leaving group

If the acyl carbon is on a carboxylic group, the leaving group is water. Nucleophilic substitution on carboxylic acids usually requires a catalyst or conversion to a more reactive intermediate; biologically this occurs via enzyme catalysis. X—H may be an alcohol (R—OH), ammonia, amine ($R\text{—}NH_2$), or another acyl compound. Types of nucleophilic substitutions include *esterification, peptide bond formation,* and *acid anhydride* formation.

Hydrolysis and Phosphorolysis Reactions

Hydrolysis is the cleavage of a bond by water:

$$R\text{—}\overset{\overset{\displaystyle O}{\|}}{C}\text{—}OR' + H_2O \rightleftharpoons R\text{—}\overset{\overset{\displaystyle O}{\|}}{C}\text{—}OH + R'\text{—}OH$$

Hydrolysis is often catalyzed by either acid or base. *Phosphorolysis* is the cleavage of a bond by inorganic phosphate:

$$\text{Glucose-glucose} + HO\text{—}\overset{\overset{\displaystyle O}{\|}}{\underset{\underset{\displaystyle O^-}{|}}{P}}\text{—}O^- \longrightarrow \text{Glucose 1-phosphate} + \text{glucose}$$

Oxidation–Reduction Reactions

Oxidation is the loss of electrons; *reduction* is the gain of electrons. Examples of oxidation are as follows:

1. $Fe^{+2} + \text{acceptor} \longrightarrow Fe^{+3} + \text{acceptor} \cdot e^-$
2. $\text{S(ubstrate)} + O_2 + DH_2 \longrightarrow S\text{—}OH + H_2O + D$
3. $S\text{—}H_2 + \text{acceptor} \longrightarrow S + \text{acceptor} \cdot H_2$

Some of the group changes that occur on oxidation–reduction are

1. $\text{—}CH_2OH \rightleftharpoons \text{—}\overset{\overset{\displaystyle H}{|}}{C}{=}O$
2. $\rangle C\text{—}OH \rightleftharpoons \rangle C{=}O$
3. $\text{—}\overset{\overset{\displaystyle H}{|}}{C}{=}O \rightleftharpoons \text{—}\overset{\overset{\displaystyle O}{\|}}{C}\text{—}OH$
4. $\text{—}CH_2NH_2 \rightleftharpoons \text{—}\overset{\overset{\displaystyle H}{|}}{C}{=}O + NH_3$
5. $\text{—}CH_2\text{—}CH_2\text{—} \rightleftharpoons \text{—}CH{=}CH\text{—}$

STEREOCHEMISTRY

Stereoisomers are compounds with the same molecular formulas and order of attachment of constituent atoms but with different arrangements of these atoms in space.

Enantiomers are stereoisomers in which one isomer is the mirror image of the other and requires the presence of a chiral atom. A *chiral carbon* (also called an *asymmetric atom*) is one that is attached to four different groups:

$$D\blacktriangleright\overset{\overset{\displaystyle B}{\vdots}}{\underset{\underset{\displaystyle E}{|}}{C}}\blacktriangleleft A \qquad A\blacktriangleright\overset{\overset{\displaystyle B}{\vdots}}{\underset{\underset{\displaystyle E}{|}}{C}}\blacktriangleleft D$$

Enantiomers

Enantiomers will be distinguished from each other by the designations *R* and *S* or *D* and *L*. The maximum number of stereoisomers possible is 2^n, where *n* is the number of chiral carbons. A molecule with more than one chiral center will be an achiral molecule if it has a point or plane of symmetry.

Diastereomers are stereoisomers that are not mirror images of each other and need not contain chiral atoms. *Epimers* are diastereomers that contain more than one chiral carbon and differ in configuration about *only one* asymmetric carbon.

Anomers are a special form of carbohydrate epimers in which the difference is specifically about the anomeric carbon (see page 996). Diastereomers can also occur with molecules in which there is restricted rotation about carbon–carbon bonds.

Double bonds exhibit *cis-trans isomerism*. The double bond is in the *cis*-configuration if two similar groups are on the same side and is *trans* if two similar groups are on opposite sides. Fused ring systems, such as those found in steroids (see page 1000), also exhibit *cis–trans* isomerism.

a / b	a / b
Trans-rings	*Cis*-rings

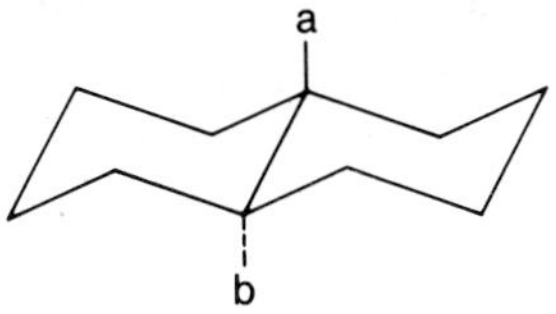

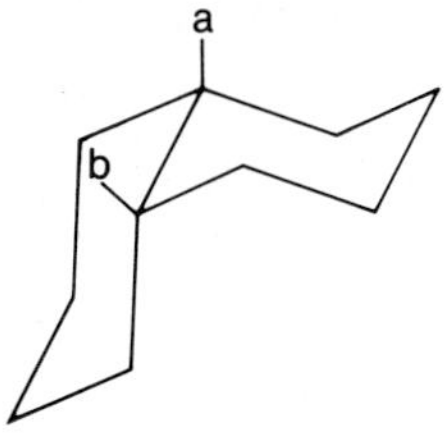

TYPES OF FORCES INVOLVED IN MACROMOLECULAR STRUCTURES

A *hydrogen bond* is a dipole–dipole attraction between a hydrogen atom attached to an electronegative atom and a nonbonding electron pair on another electronegative atom:

$$\underset{\delta^-}{:\ddot{X}}\!-\!\underset{\delta^+}{H} \;\cdots\cdots\; \underset{\delta^-}{:\ddot{X}}\!-\!\underset{\delta^+}{H}$$

Hydrogen bonds of importance in macromolecular structures occur between two nitrogens, two oxygens, or an oxygen and a nitrogen.

A *hydrophobic interaction* is the association of nonpolar groups in a polar medium. *van der Waals* forces consist of dipole and induced dipole interactions between two nonpolar groups. A nonpolar residue dissolved in water induces a highly ordered, thermodynamically unfavorable, solvation shell. Interaction of nonpolar residues with each other, with the exclusion of water, increases the entropy of the system and is thermodynamically favorable.

Ionic (*electrostatic*) *interactions* between charged groups can be attractive if the charges are of opposite signs or repulsive if they are of the same sign. The strength of an electrostatic interaction in the interior of a protein molecule may be high. Most charged groups on the surface of a protein molecule interact with water rather than with each other.

A *disulfide bond* (—S—S—) is a covalent bond formed by the oxidation of two sulfhydryl (—SH) groups.

CARBOHYDRATES

Carbohydrates are polyhydroxy aldehydes or ketones or derivatives of these. Monosaccharides ("simple sugars") are those carbohydrates that cannot be hydrolyzed into simpler compounds. The name of a type of monosaccharide includes the type of function, a Greek prefix indicating the number of carbons, and the ending -ose; for example, *aldohexose* is a six-carbon aldehyde and *ketopentose* a five-carbon ketone. Monosaccharides may react with each other to form larger molecules. With fewer than eight monosaccharides, either a Greek prefix indicating the number or the general term *oligosaccharide* may be used. *Polysaccharide* refers to a polymer with more than eight monosaccharides. Oligo- and polysaccharides may be either homologous or mixed.

Most *monosaccharides* are asymmetric, an important consideration since enzymes usually work on only one isomeric form. The simplest carbohydrates are glyceraldehyde and dihydroxyacetone whose structures, shown as Fischer projections, are as follows:

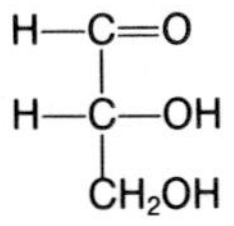

D-Glyceraldehyde	L-Glyceraldehyde	Dihydroxyacetone
H—C=O	H—C=O	CH_2OH
H—C—OH	HO—C—H	C=O
CH_2OH	CH_2OH	CH_2OH

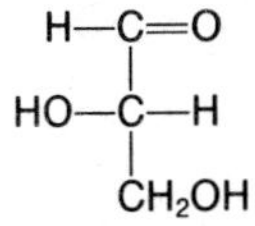

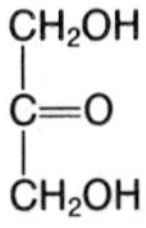

D-Glyceraldehyde may also be written as follows:

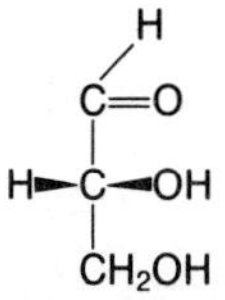

H
C=O
H▸C◂OH
CH_2OH

In the Cahn–Ingold–Prelog system, the designations are R (rectus; right) and S (sinister, left).

The configuration of monosaccharides is determined by the stereochemistry at the asymmetric carbon furthest from the carbonyl carbon (number 1 for an aldehyde; lowest possible number for a ketone). Based on the *position* of the —OH on the highest number asymmetric carbon, a monosaccharide is D if the —OH projects to the *right* and L if it projects to the *left*. D and L monosaccharides with the same name are *enantiomers*, and the substituents on all asymmetric carbons are reversed as in

D-Glucose	L-Glucose
H—C=O	H—C=O
H—C—OH	HO—C—H
HO—C—H	H—C—OH
H—C—OH	HO—C—H
H—C—OH	HO—C—H
CH_2OH	CH_2OH

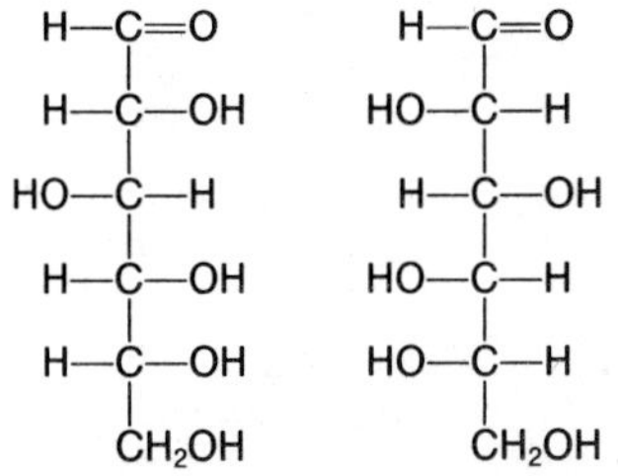

Epimers (e.g., glucose and mannose) are stereoisomers that differ in the configuration about *only one* asymmetric carbon. The relationship of —OH's to *each other* determines the specific monosaccharide. Three aldohexoses and three pentoses of importance are

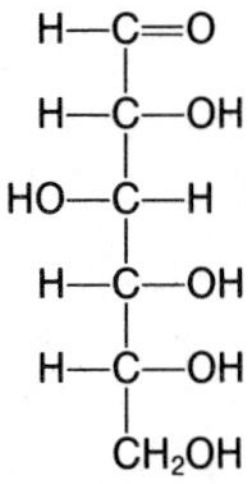

D-Glucose	D-Mannose	D-Galactose
H—C=O	H—C=O	H—C=O
H—C—OH	HO—C—H	H—C—OH
HO—C—H	HO—C—H	HO—C—H
H—C—OH	H—C—OH	HO—C—H
H—C—OH	H—C—OH	H—C—OH
CH_2OH	CH_2OH	CH_2OH

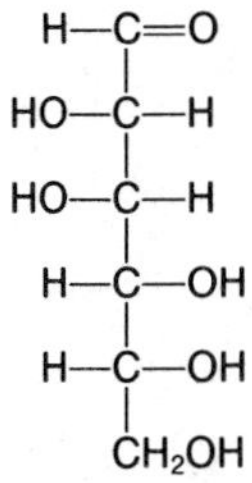

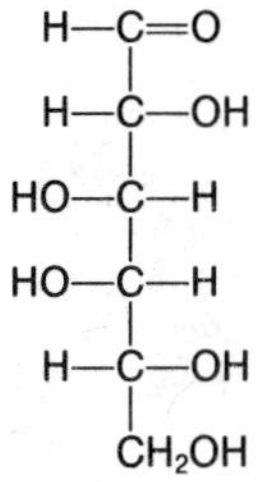

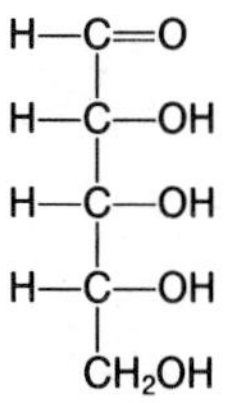

D-Ribose	D-Ribulose	D-Xylulose
H—C=O	CH_2OH	CH_2OH
H—C—OH	C=O	C=O
H—C—OH	H—C—OH	HO—C—H
H—C—OH	H—C—OH	H—C—OH
CH_2OH	CH_2OH	CH_2OH

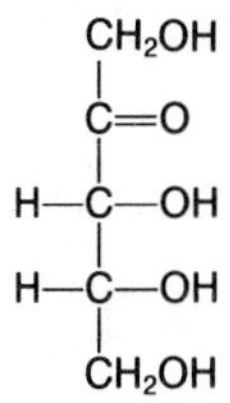

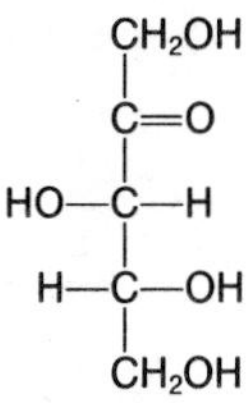

Fructose, a ketohexose, differs from glucose only on carbons 1 and 2:

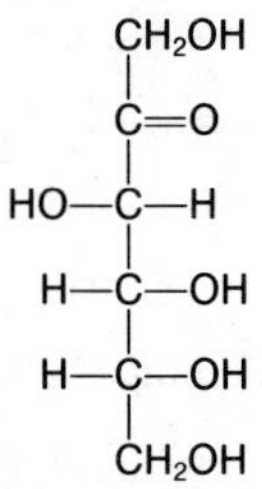

CH_2OH–C=O–HO–C–H–H–C–OH–H–C–OH–CH_2OH

Five- and six-carbon monosaccharides form *cyclic hemiacetals* and *hemiketals* in solution (see page 994). A new asymmetric carbon is generated so two isomeric forms are possible:

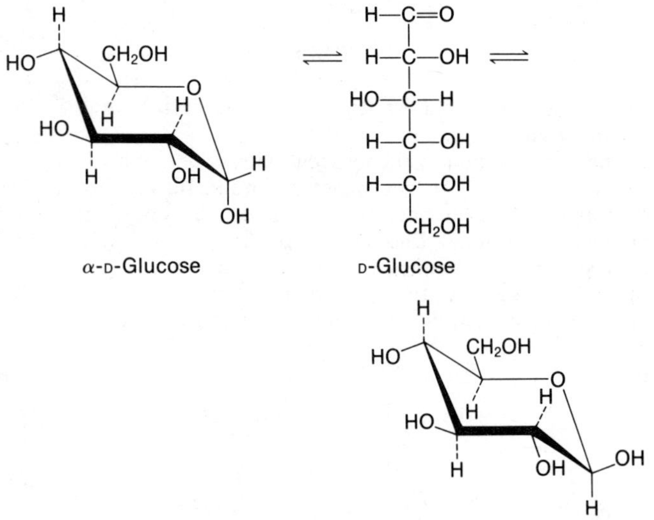

α-D-Glucose ⇌ D-Glucose ⇌

β-D-Glucose

Both five (furanose)- and six-membered (pyranose) ring structures are possible, although pyranose rings are more common. A furanose ring is written as follows:

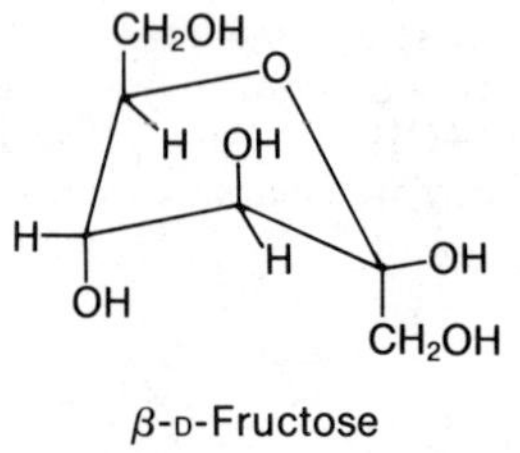

β-D-Fructose

The isomer is designated α- if the —OH and the —CH_2OH on the two carbons linked by the oxygen are *trans* to each other and β- if they are *cis*. The hemiacetal or hemiketal forms may also be written as modified *Fischer projection formulas:* α if —OH on the acetal or ketal carbon projects to the same side as the ring and β if on the opposite side:

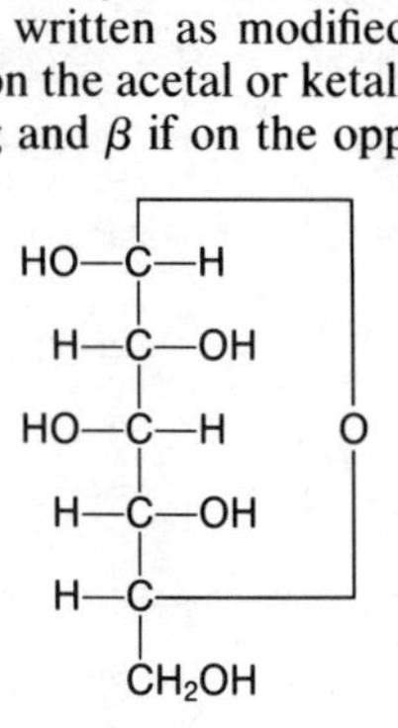

β-D-Glucose 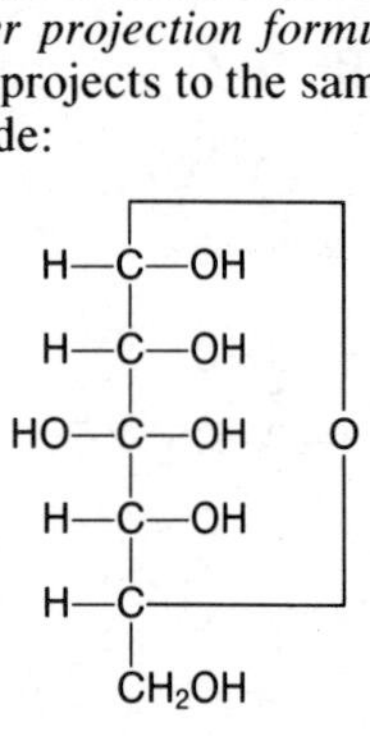 α-D-Glucose

Haworth structures are used most commonly:

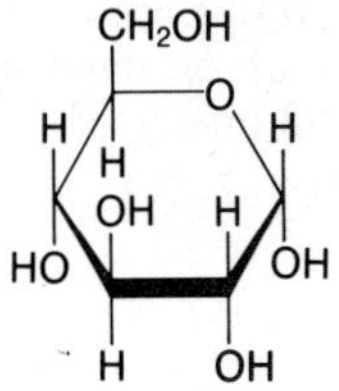

α-D-Glucose 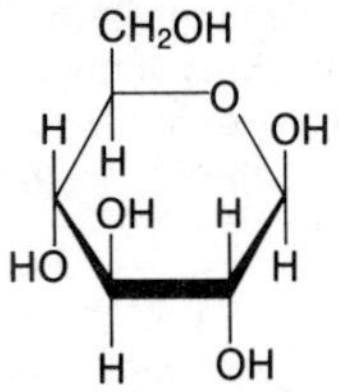 β-D-Glucose

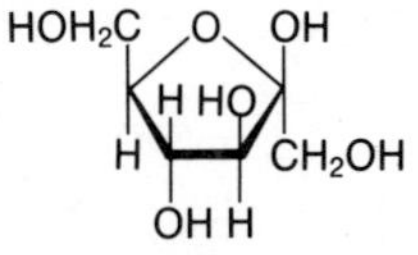

β-D-Fructose

The ring is perpendicular to the plane of the paper with the oxygen written to the back (upper) right, carbon 1 to the right, and substituents above or below the plane of the ring. The —OH at the acetal or ketal carbon is below in the α-isomer and above in the β. Anything written to the right in the Fischer projection is written down in the Haworth structure.

In this text, a *modified Haworth structure,* not indicating ring orientation, will be used:

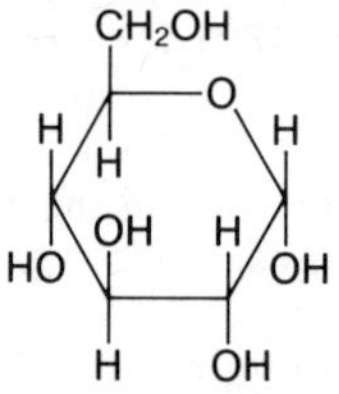

α-D-Glucose

The α- and β-forms of the same monosaccharide are special forms of epimers called *anomers,* differing only in the configuration about the anomeric (acetal or ketal) carbon. Monosaccharides exist in solution primarily as a mixture of the hemiacetals (or hemiketals) but react chemically as aldehydes or ketones. *Mutarotation* is the equilibration of α- and β-forms through the free aldehyde or ketone. Substitution of the H of the anomeric —OH prevents mutarotation and fixes the configuration in either the α- or β-form.

Monosaccharide Derivatives

A *deoxymonosaccharide* is one in which an —OH has been replaced by —H. In biological systems, this occurs at carbon 2 unless otherwise indicated. An *amino monosaccharide* is one in which an —OH has been replaced by —NH_2, again at carbon 2 unless otherwise specified. The amino group of an amino sugar may be *acetylated:*

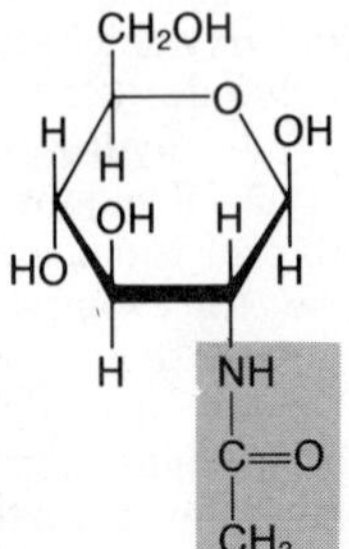

β-*N*-Acetylglucosamine

An aldehyde is reduced to a primary and a ketone to a secondary *monosaccharide alcohol* (*alditol*). Alcohols are named with the base name of the sugar plus the ending *-itol* or with a trivial name (glucitol = sorbitol). Monosaccharides that differ around only two of the first three carbons yield the same alditol. D-Glyceraldehyde and dihydroxyacetone give glycerol:

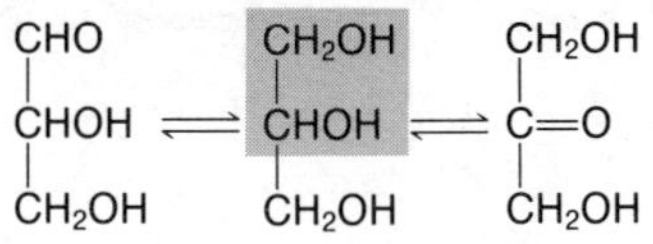

CHO / CHOH / CH_2OH ⇌ CH_2OH / CHOH / CH_2OH ⇌ CH_2OH / C=O / CH_2OH

D-Glucose and D-fructose give D-sorbitol; D-fructose and D-mannose give D-mannitol. Oxidation of the terminal —CH_2OH, but not of the —CHO, yields a *-uronic acid*, a *monosaccharide acid:*

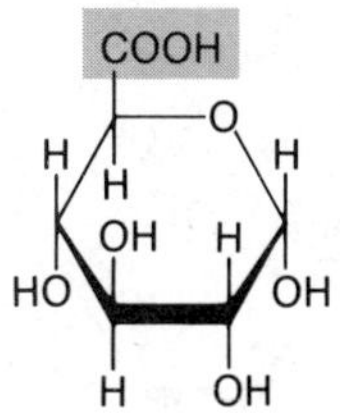

D-Glucuronic acid

Oxidation of the —CHO, but not the —CH_2OH, gives an *-onic acid:*

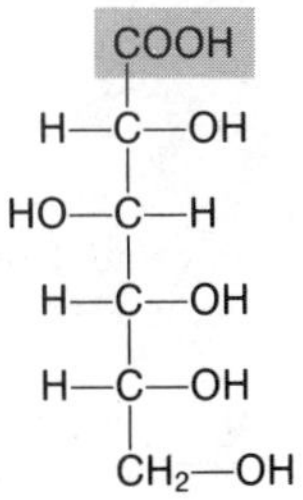

D-Gluconic acid

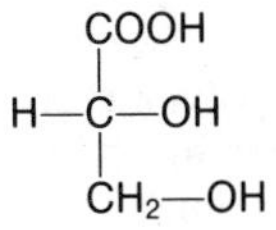

D-Glyceric acid

Oxidation of both the —CHO and —CH_2OH gives an *-aric acid:*

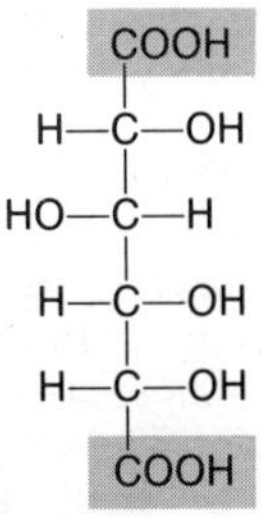

D-Glucaric acid

Ketones do not form acids. Both -onic and -uronic acids can react with an —OH in the same molecule to form a *lactone* (see page 994):

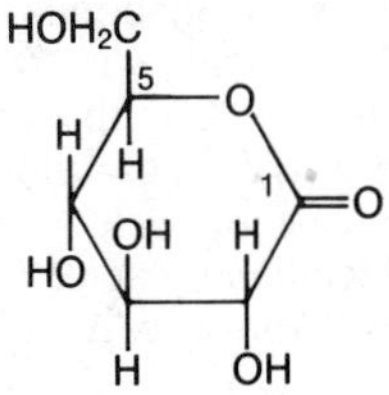

Gluconolactone

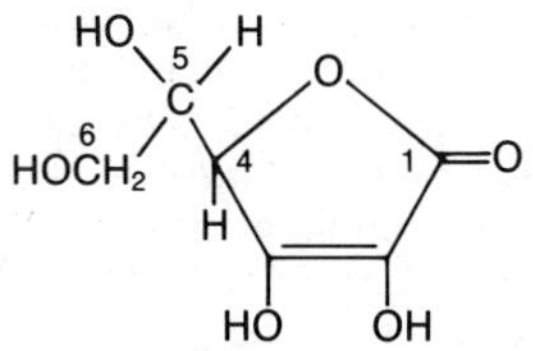

L-Ascorbic acid (Derivative of L-gulose)

Reactions of Monosaccharides

The most common *esters* of monosaccharides are phosphate esters at carbons 1 and/or 6 (see page 993):

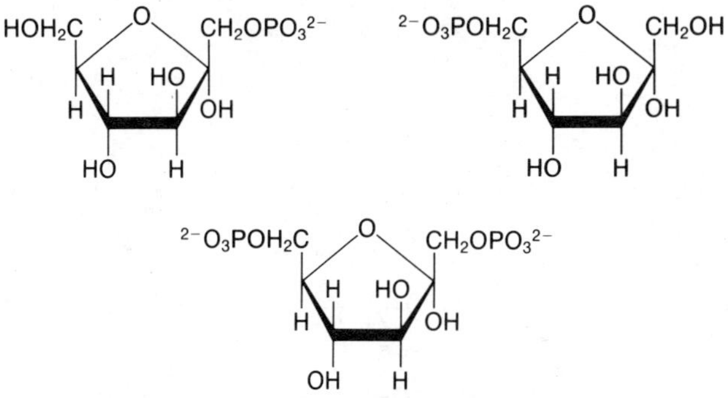

To be a *reducing sugar,* mutarotation must be possible. In alkali, enediols form that may migrate to 2,3 and 3,4 positions:

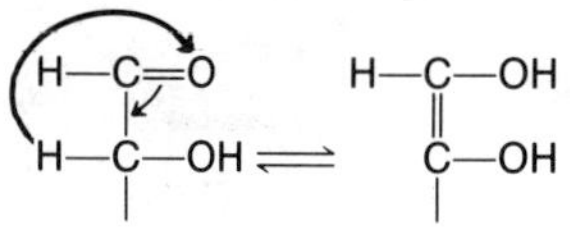

H—C=O / H—C—OH ⇌ H—C—OH / C—OH

Enediols may be oxidized by O_2, Cu^{2+}, Ag^+, Hg^{2+}. Reducing ability is more important in the laboratory than physiologically. A hemiacetal or hemiketal may react with the —OH of another monosaccharide to form a disaccharide (*acetal, glycoside*) (see page 998):

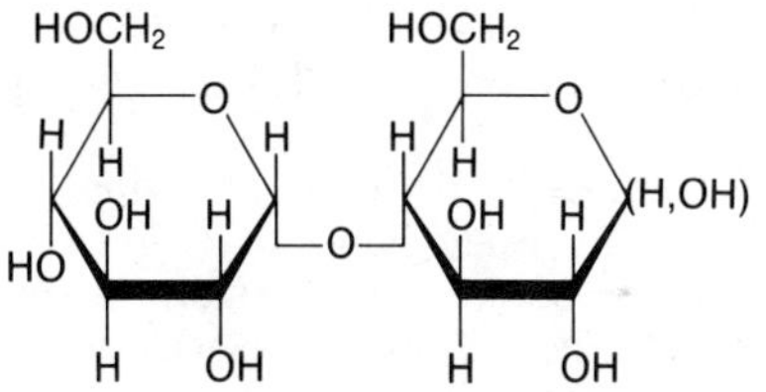

(H,OH) (May be either α or β)

α-1,4-Glycosidic linkage

One monosaccharide still has a free anomeric carbon and can react further. Reaction of the anomeric —OH may be with any —OH on the other monosaccharide, including the anomeric one. The anomeric —OH that has reacted is fixed as either α- or β- and cannot mutarotate or reduce. If the glycosidic bond is not between two anomeric carbons, one of the units will still be free to mutarotate and reduce.

Oligo- and Polysaccharides

Disaccharides have two monosaccharides, either the same or different, in glycosidic linkage. If the glycosidic linkage is between the two anomeric carbons, the disaccharide is nonreducing:

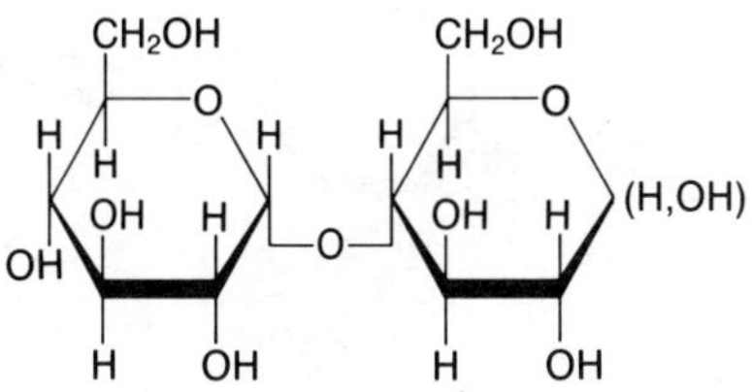

Maltose

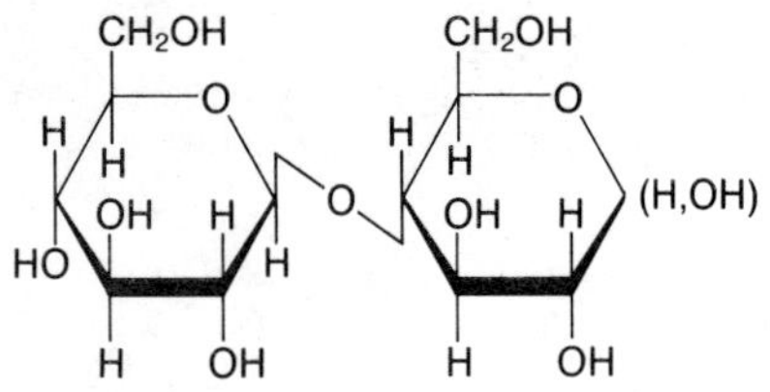

Cellobiose

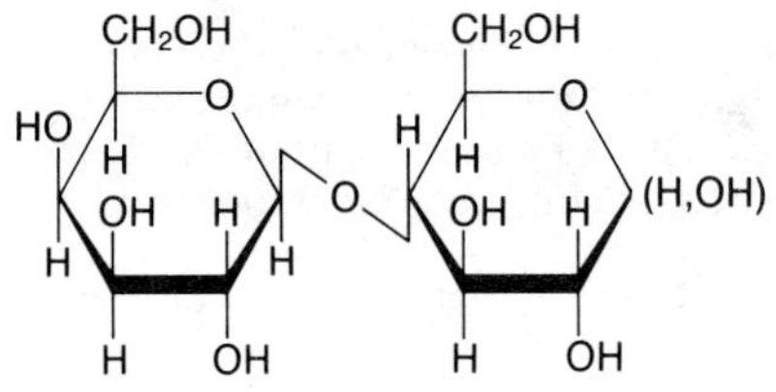

Lactose

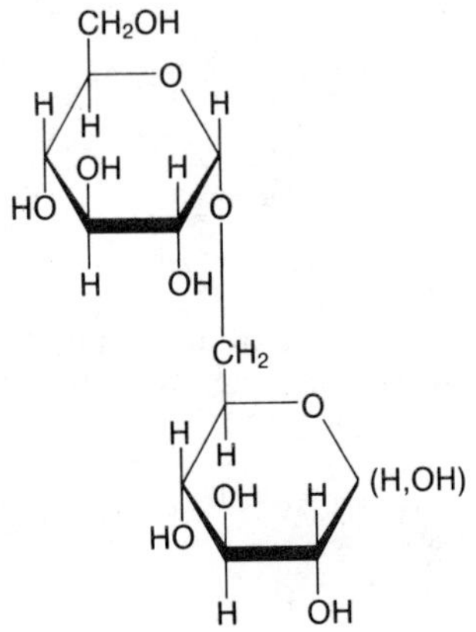

Isomaltose

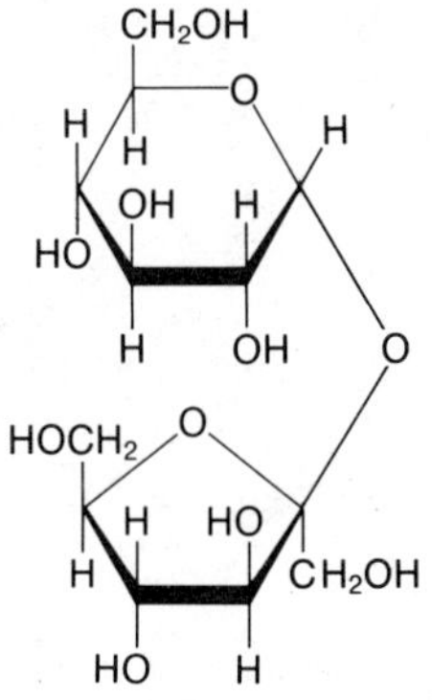

Sucrose

Maltose = 4-*O*-(α-D-glucopyranosyl)D-glucopyranose; reducing
Isomaltose = 6-*O*-(α-D-glucopyranosyl)D-glucopyranose; reducing
Cellobiose = 4-*O*-(β-D-glucopyranosyl)D-glucopyranose; reducing
Lactose = 4-*O*-(β-D-galactopyranosyl)D-glucopyranose; reducing
Sucrose = α-D-glucopyranosyl-β-D-fructofuranoside; nonreducing

As many as thousands of monosaccharides, either the same or different, may be joined by glycosidic bonds to form *polysaccharides*. The anomeric carbon of one unit is usually joined to carbon 4 or 6 of the next unit. The ends of a polysaccharide are not identical (reducing end = free anomeric carbon; nonreducing = anomeric carbon linked to next unit; branched polysaccharide = more than one nonreducing end). The most common carbohydrates are homopolymers of glucose; for example, starch, glycogen, and cellulose. Plant starch is a mixture of *amylose,* a linear polymer of maltose units, and *amylopectin,* branches of repeating maltose units (glucose–glucose in α-1-4 linkages) joined via isomaltose linkages. *Glycogen,* the storage form of carbohydrate in animals, is similar to amylopectin, but the branches are shorter and occur more frequently. *Cellulose,* in plant cell walls, is a linear polymer of repeating cellobioses (glucose–glucose in β-1-4 linkages). *Mucopolysaccharides* contain amine sugars, free and acetylated, -uronic acids, sulfate esters, and sialic acids in addition to the simple monosaccharides. *N-Acetylneuraminic acid,* a sialic acid, is

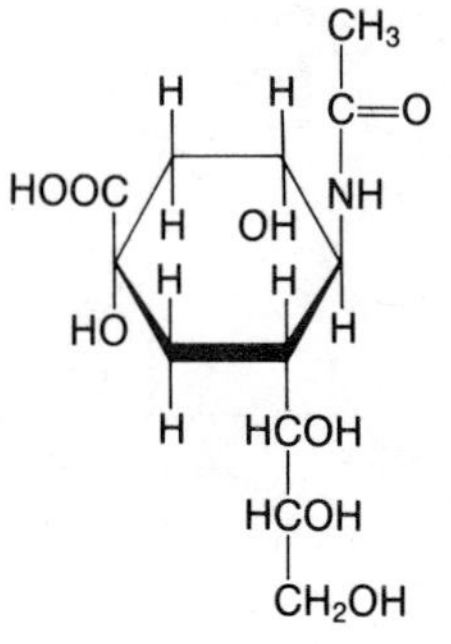

LIPIDS

Lipids are a diverse group of chemicals related primarily because they are insoluble in water, soluble in nonpolar solvents, and found in animal and plant tissues.

Saponifiable lipids yield salts of fatty acids upon alkaline hydrolysis. *Acylglycerols* = glycerol + fatty acid(s); *phosphoacylglycerols* = glycerol + fatty acids + HPO_4^{2-} + alcohol; *sphingolipids* = sphingosine + fatty acid + polar group (phosphorylalcohol or carbohydrate); *waxes* = long-chain alcohol + fatty acid. *Nonsaponifiable lipids* (*terpenes, steroids, prostaglandins,* and related compounds) are not usually subject to hydrolysis. *Amphipathic* lipids have both a polar "head" group and a nonpolar "tail." Amphipathic molecules can stabilize emulsions and are responsible for the lipid bilayer structure of membranes.

Fatty acids are monocarboxylic acids with a short (<6 carbons), medium (8–14 carbons), or long (>14 carbons) aliphatic chain. Biologically important ones are usually linear molecules with an even number of carbons (16–20). Fatty acids are numbered using either Arabic numbers (—COOH is 1) or the Greek alphabet (—COOH is not given a symbol; adjacent carbons are α, β, γ, etc.). *Saturated fatty acids* have the general formula $CH_3(CH_2)nCOOH$. (*Palmitic acid* = C_{16}; *stearic acid* = C_{18}). They tend to be extended chains and solid at room temperature unless the chain is short. Both trivial and systematic (prefix indicating number of carbons + *anoic acid*) names are used. $CH_3(CH_2)_{14}COOH$ = palmitic acid or hexadecanoic acid.

Unsaturated fatty acids have one or more double bonds. Most naturally occurring fatty acids have *cis* double bonds and are usually liquid at room temperature. Fatty acids with *trans* double bonds tend to have higher melting points. A double bond is indicated by Δ^n, where *n* is the number of the first carbon of the bond. *Palmitoleic* = Δ^9-hexadecenoic acid; *oleic* = Δ^9-octadecenoic acid; *linoleic* = $\Delta^{9,12}$-octadecadienoic acid; *linolenic* = $\Delta^{9,12,15}$-octadecatrienoic acid; *arachidonic* = $\Delta^{5,8,11,14}$-eicosatetraenoic acid.

Since the p*K*s of fatty acids are about 4–5, in physiological solutions, they exist primarily in the ionized form, called salts or "soaps." Long-chain fatty acids are insoluble in water, but soaps form micelles. Fatty acids form esters with alcohols and thioesters with coenzyme A.

Biochemically significant reactions of unsaturated fatty acids are

1. *Reduction:* —CH═CH— + XH_2 ⟶ —CH_2CH_2— + X
2. *Addition of water:* —CH═CH— + H_2O ⟶ —CH(OH)—CH_2—
3. *Oxidation:* R—CH═CH—R′ ⟶ R—CHO + R′CHO

Prostaglandins, thromboxanes, and *leukotrienes* are derivatives of 20-carbon, polyunsaturated fatty acids, especially arachidonic acid. *Prostaglandins* have the general structure:

PGE$_2$

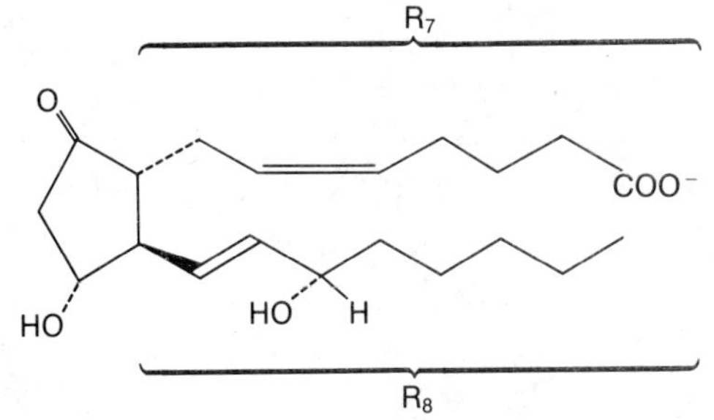

R_7 COO$^-$ HO HO H R_8

The series differ from each other in the substituents on the ring and whether carbon 15 contains an —OH or —O · OH. The subscript indicates the number of double bonds in the side chains. Substituents indicated by ——(β) are above the plane of the ring; ---(α) below:

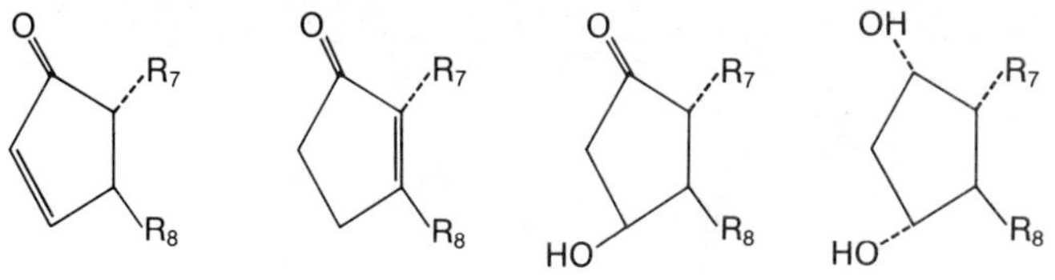

PGA PGB PGE PGF

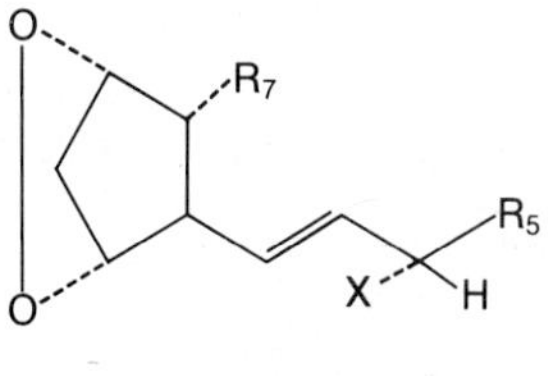

PGG(X = —OH); PGH(X = —OOH) 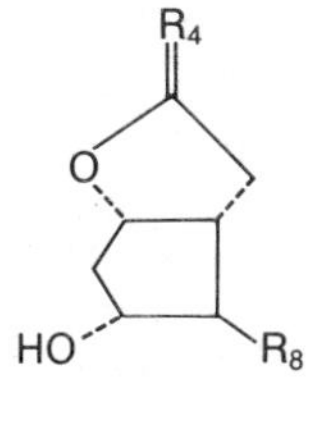 PGI

Thromboxanes have an oxygen incorporated to form a six-membered ring:

TXA$_2$

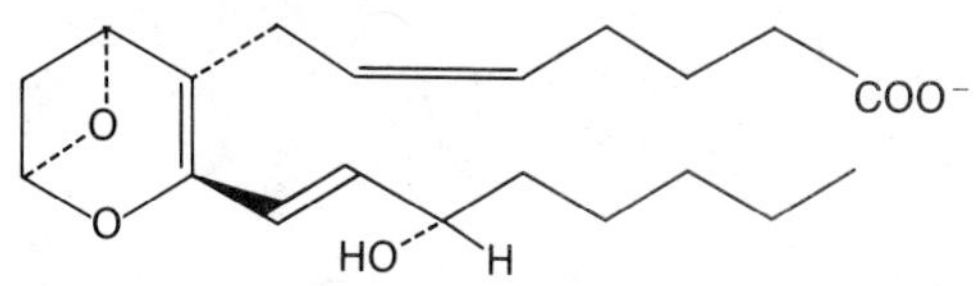

Leukotrienes are substituted derivatives of arachidonic acid in which no internal ring has formed: R is variable:

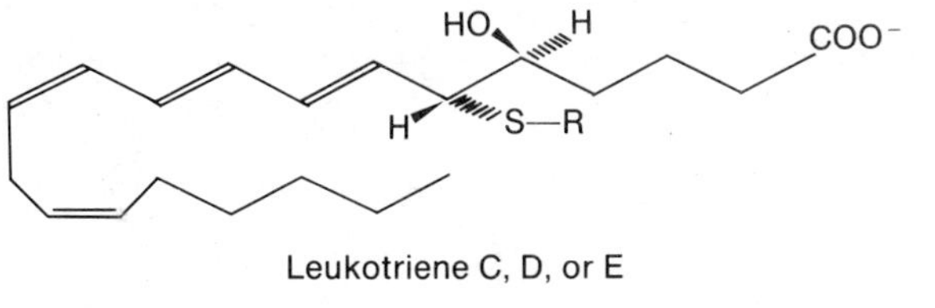

Leukotriene C, D, or E

Acylglycerols are compounds in which one or more of the three hydroxyl groups of glycerol is esterified. In *triacylglycerols* (triglycerides) all three hydroxyl groups are esterified to fatty acids. At least two of the three R groups are usually different. If R_1 is not equal to R_3, the molecule is asymmetric and of the L configuration:

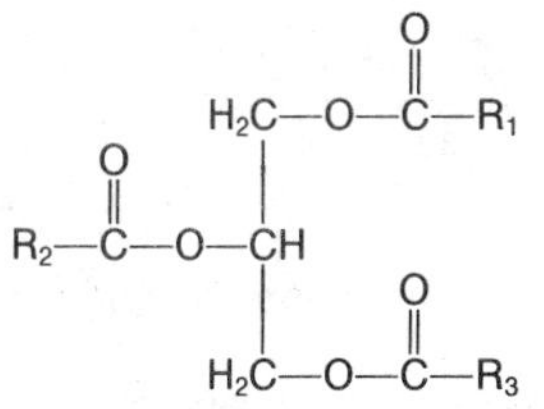

The properties of the triacylglycerols are determined by those of the fatty acids they contain with *oils* liquids at room temperature (preponderance of short-chain and/or *cis*-unsaturated fatty acids) and *fats* solid (preponderance of long-chain and/or saturated).

Triacylglycerols are hydrophobic and do not form stable micelles. They may be hydrolyzed to glycerol and three fatty acids by strong alkali or enzymes (lipases). *Mono-* (usually with the fatty acid in the β (2) position) and *diacylglycerols* also exist in small amounts as metabolic intermediates. Mono- and diacylglycerols are slightly more polar than triacylglycerols. *Phosphoacylglycerols* are derivatives of L-α-glycerolphosphate (L-glycerol 3-phosphate):

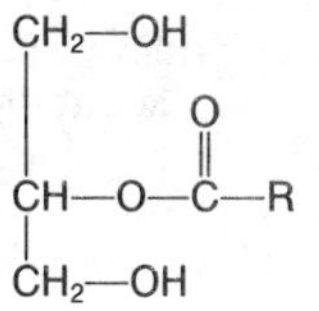

β-Monoacylglycerol L-α-Glycerolphosphate

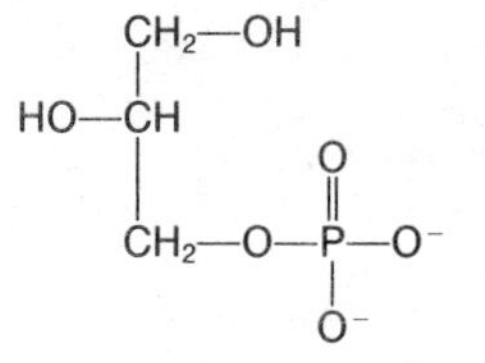

The parent compound, *phosphatidic acid* (two —OH groups of L-α-glycerolphosphate esterified to fatty acids), has its phosphate esterified to an alcohol (X—OH) to form several series of phosphoacylglycerols. These are amphipathic molecules, but the net charge at pH 7.4 depends upon the nature of X—OH. In *plasmalogens,* the —OH on carbon 1 is in *ether,* rather than ester, linkage to an alkyl group. If *one* fatty acid (usually β) has been hydrolyzed from a phosphoacylglycerol, the compound is a *lyso*-compound; for example, lysophosphatidylcholine (lysolecithin):

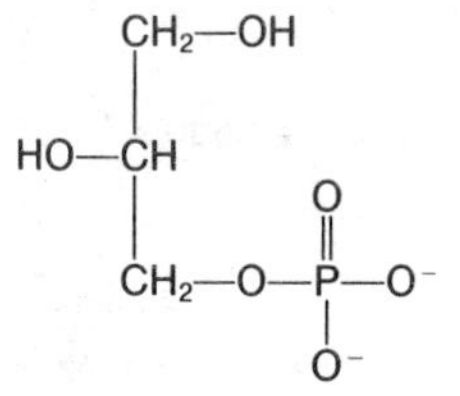

L-α-Glycerol phosphate 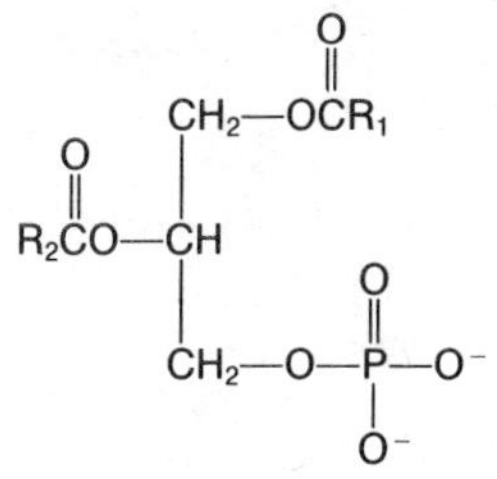 A phosphatidate

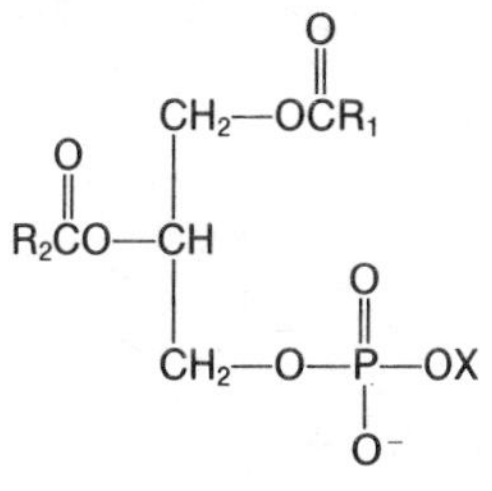

A phosphoacylglycerol 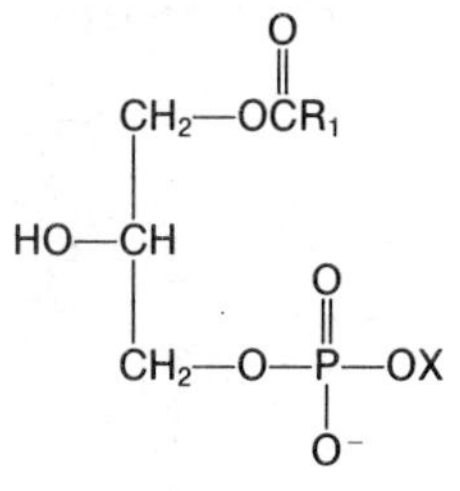 A lysocompound

Sphingolipids are complex lipids based on the 18 carbon, unsaturated alcohol, sphingosine. In *ceramides,* a long-chain fatty acid is in amide linkage to sphingosine:

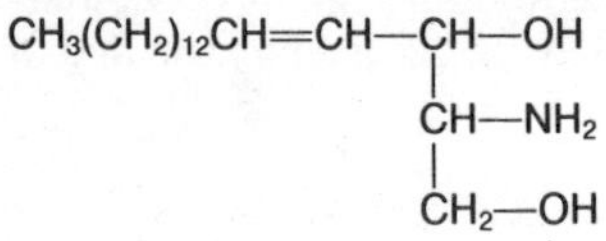

```
CH3(CH2)12CH=CH—CH—OH        CH3(CH2)12CH=CH—CH—OH
                |                            |
                CH—NH2                       |   H  O
                |                            |   |  ||
                CH2—OH                       HC—N—C—R
                                             |
                                             CH2—OH
```

Sphingosine 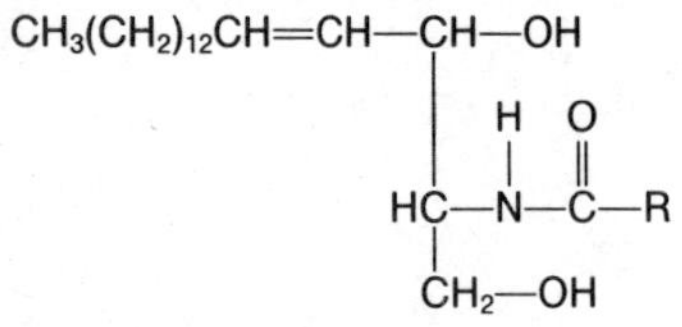 A ceramide

Sphingomyelins, the most common sphingolipids, are a family of compounds in which the primary hydroxyl group of a ceramide is esterified to phosphorylcholine (phosphorylethanolamine):

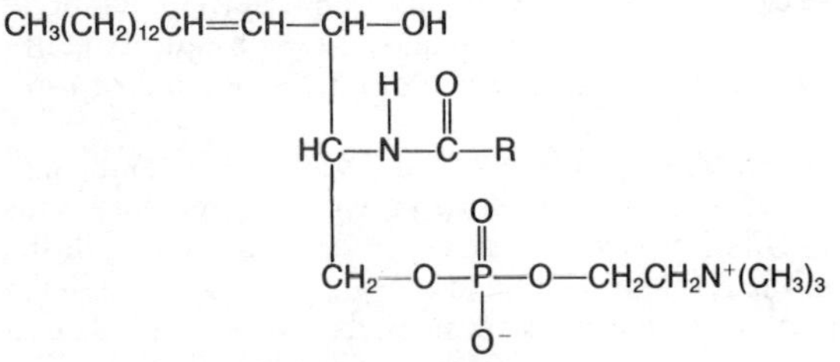

```
CH3(CH2)12CH=CH—CH—OH
                |
                |   H  O
                |   |  ||
                HC—N—C—R
                |
                |        O
                |        ||
                CH2—O—P—O—CH2CH2N+(CH3)3
                         |
                         O−
```

They are amphipathic molecules, existing as zwitterions at pH 7.4 and the only sphingolipids that contain phosphorus. *Glycosphingolipids* do not contain phosphorus but contain carbohydrate in glycosidic linkage to the primary alcohol of a ceramide. They are amphipathic and either neutral or acidic if the carbohydrate moiety contains an acidic group. *Cerebrosides* have a single glucose or galactose linked to a ceramide. *Sulfatides* are galactosylceramides esterified with sulfate at carbon 3 of the galactose:

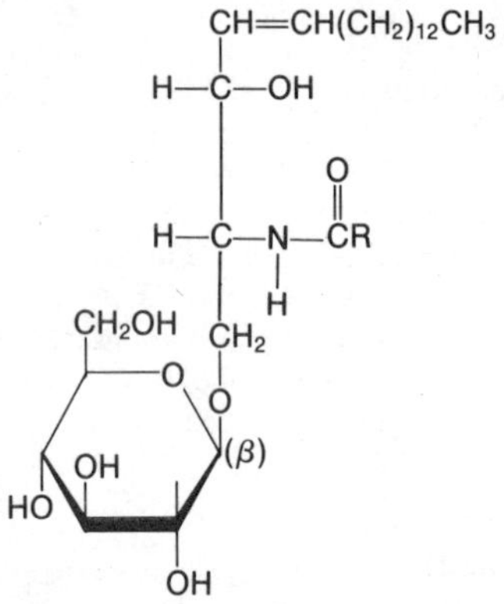

Glucosylceramide (glucocerebroside)

Globosides (*ceramide oligosaccharides*) are ceramides with two or more neutral monosaccharides, whereas *gangliosides* have an oligosaccharide containing one or more sialic acids.

Steroids are derivatives of cyclopentanoperhydrophenanthrene. The steroid nucleus is a rather rigid, essentially planar structure with substituents above or in the plane of the rings designated β (solid line) and those below called α (dashed line):

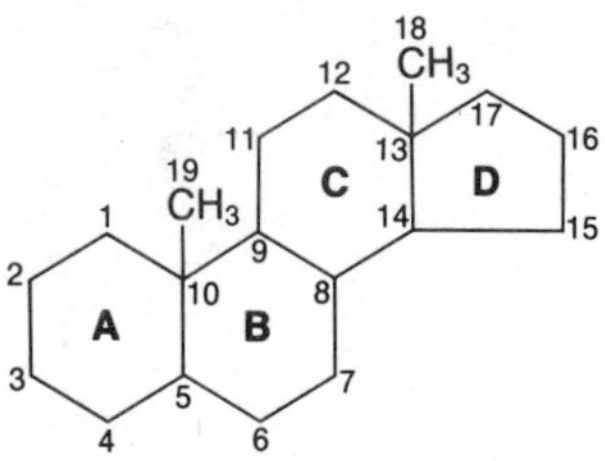

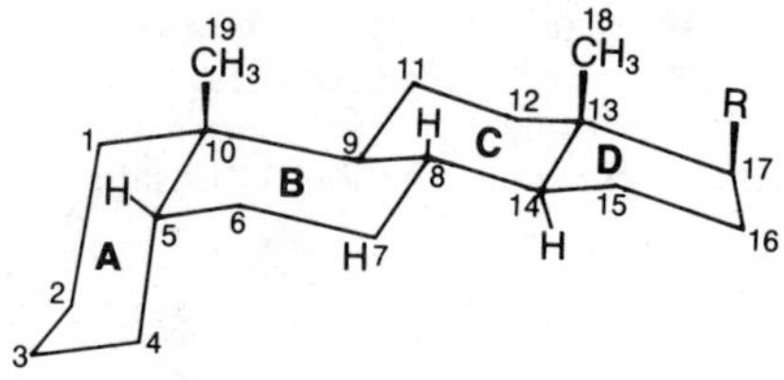

A and B rings *cis,* rest *trans*

Most steroids in humans have methyl groups at positions 10 and 13 and frequently a side chain at position 17. Sterols contain one or more —OH groups, free or esterified to a fatty acid. Most steroids are nonpolar. In a liposome or cell membrane, *cholesterol* orients with the —OH toward any polar groups; cholesterol esters do not. *Bile acids* (e.g., cholic acid) have a polar side chain and so are amphipathic:

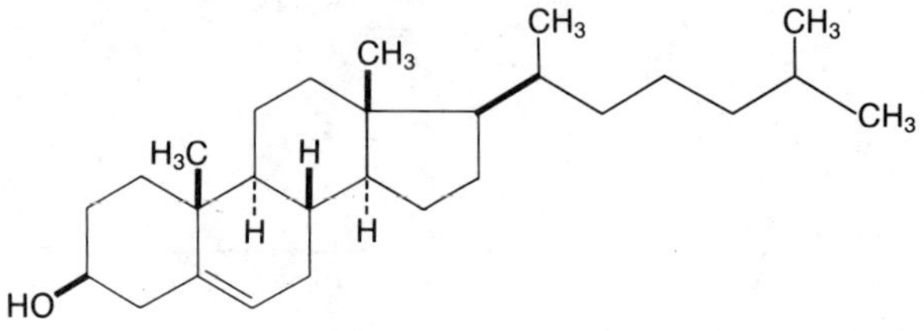

Cholesterol

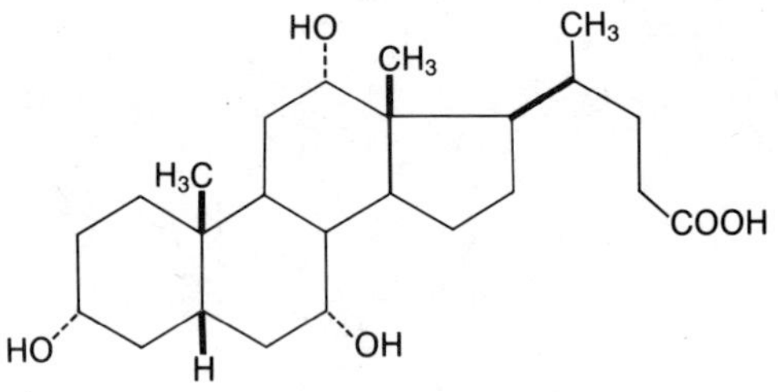

Cholic acid

Steroid hormones are oxygenated steroids of 18–21 carbons. *Estrogens* have 18 carbons, an aromatic ring A, and no methyl at carbon-10. *Androgens* have 19 carbons and no side chain at carbon-17. *Glucocorticoids* and *mineralocorticoids* have 21 carbons including a two-carbon, oxygenated side chain at carbon-17. *Vitamin D_3* (*cholecalciferol*) is not a sterol but is derived from 7-dehydrocholesterol in humans:

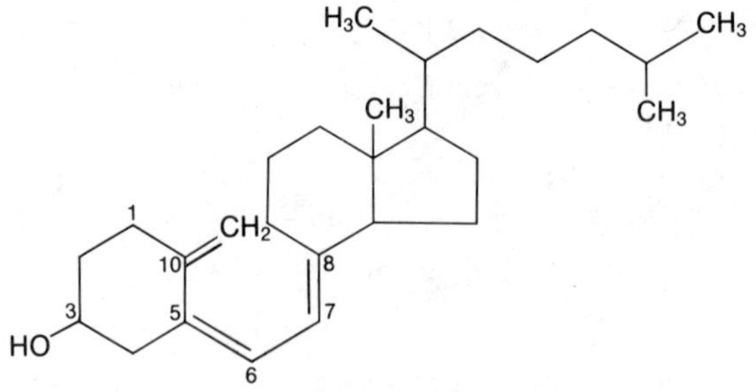

Cholecalciferol

Terpenes are polymers of two or more isoprene units. *Isoprene* is

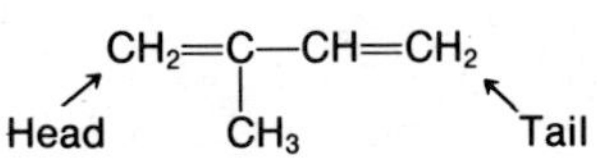

```
   CH2=C—CH=CH2
   ↗    |       ↖
Head   CH3       Tail
```

Terpenes may be linear or cyclic, with the isoprenes usually linked head to tail and most double bonds *trans* (but may be *cis* as in vitamin A). *Squalene,* the precursor of cholesterol, is a linear terpene of six isoprene units. Fat-soluble *vitamins* (A, D, E, K) contain polymers of isoprene units:

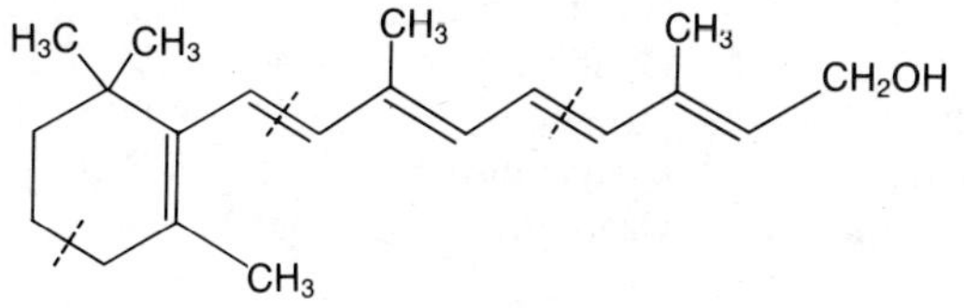

Vitamin A

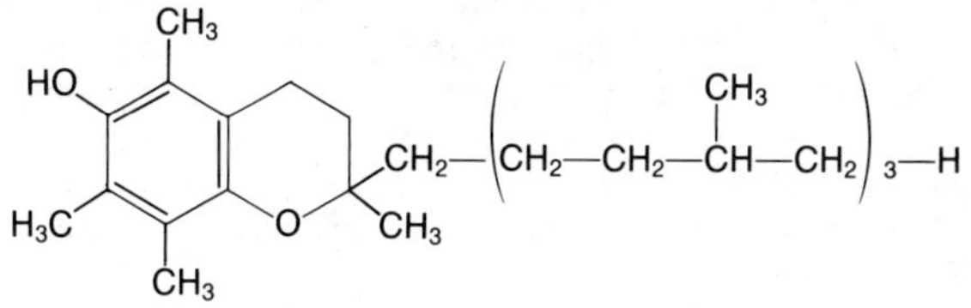

Vitamin E (α-tocopherol)

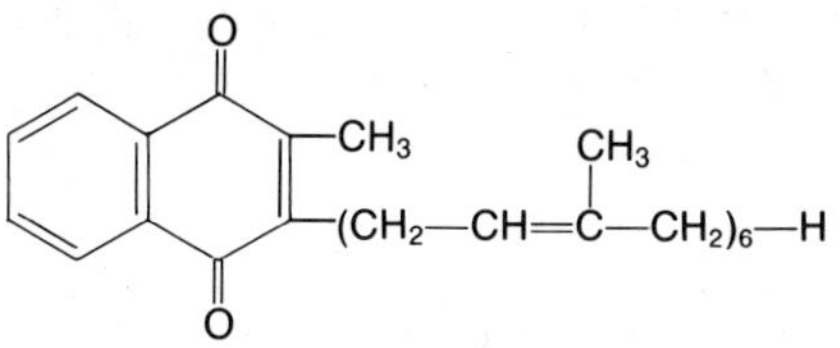

Vitamin K_2

AMINO ACIDS

Amino acids contain both an *amino* ($-NH_2$) and a *carboxylic acid* ($-COOH$) group. Biologically important amino acids are usually α-amino acids with the formula

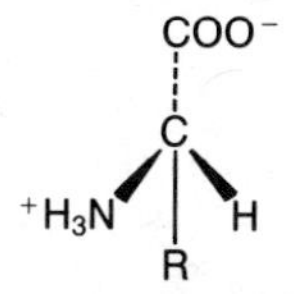

L-α-Amino acid

The amino group, with an unshared pair of electrons, is basic, with a pK_a of about 9.5, and exists primarily as $-NH_3^+$ at pH's near neutrality. The carboxylic acid group (pK about 2.3) exists primarily as a carboxylate ion. If R is anything but —H, the molecule is asymmetric with most naturally occurring ones of the L-configuration (same relative configuration as L-glyceraldehyde; see page 995).

The *polarity* of amino acids is influenced by their side chains (R groups) (see page 30 for complete structures). *Nonpolar* amino acids include those with large, aliphatic, aromatic and imino groups. [aliphatic = Val, Leu, Ile; aromatic = Phe, Trp, (unionized) Tyr); imino = Pro]. *Intermediate polarity* amino acids are those with small alkyl chains, nonionizable hydroxyl or sulfur groups, and amides (alkyl = Gly, Ala; hydroxyl = Ser, Thr; sulfur = Met; amides = Gln, Asn). Amino acids with ionizable side chains are *polar*. The pKs of the side groups of arginine, lysine, glutamate, and aspartate are such that these are nearly always charged at physiological pH, whereas the side groups of histidine (pK = 6.0) and cysteine (pK = 8.3) exist as both charged and uncharged species at pH 7.4 (acidic = Glu, Asp, Cys; basic = Lys, Arg, His).

All amino acids are at least *dibasic acids* because of the presence of both the α-amino and α-carboxyl groups, the ionic state being a function of pH. The presence of another ionizable group will give a tribasic acid, as shown for cysteine.

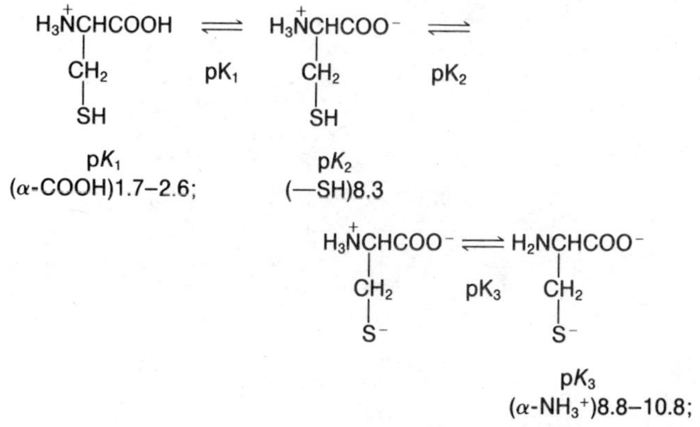

pK_1 (α-COOH)1.7–2.6; pK_2 (—SH)8.3

pK_3 (α-NH_3^+)8.8–10.8;

The *zwitterionic* form is the form in which the *net* charge is zero. The *isoelectric point* is the average of the two pK's involved in the formation of the zwitterionic form. In the above example this would be the average of $pK_1 + pK_2$.

PURINES AND PYRIMIDINES

Purines and *pyrimidines,* often called *bases,* are nitrogen-containing heterocyclic compounds with the structures

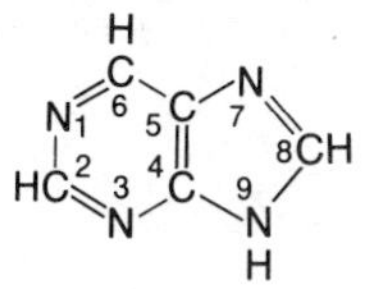

Purine

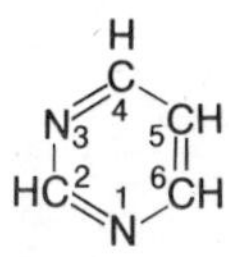

Pyrimidine

Major bases found in nucleic acids and as cellular nucleotides are

Purines		*Pyrimidines*	
Adenine:	6-amino	Cytosine:	2-oxy, 4-amino
Guanine:	2-amino, 6-oxy	Uracil:	2,4-dioxy
		Thymine:	2,4-dioxy, 5-methyl

Other important bases found primarily as intermediates of synthesis and/or degradation are

Hypoxanthine:	6-oxy	Orotic acid:	2,4-dioxy, 6-carboxy
Xanthine:	2,6-dioxy		

Oxygenated purines and pyrimidines exist as *tautomeric* structures with the keto form predominating and involved in hydrogen bonding between bases in nucleic acids:

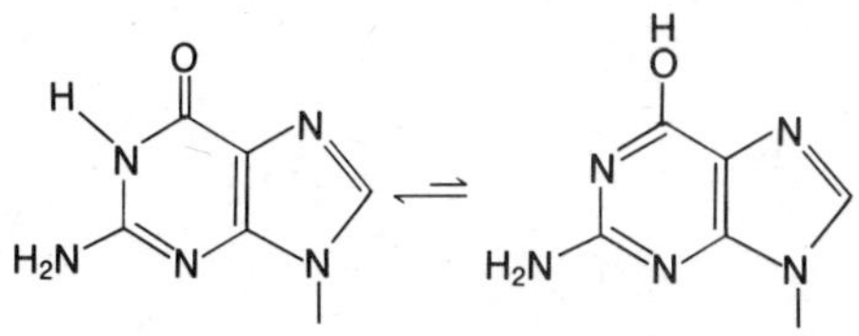

Keto

Enol

Nucleosides have either β-D-ribose or β-D-2-deoxyribose in an *N*-glycosidic linkage between carbon 1 of the sugar and nitrogen 9 (purine) or nitrogen 1 (pyrimidine).

Nucleotides have one or more phosphate groups esterified to the sugar. Phosphates, if more than one is present, are usually attached to each other via phosphoanhydride bonds. Monophosphates may be designated as either the base monophosphate or as an *-ylic acid* (AMP: adenylic acid):

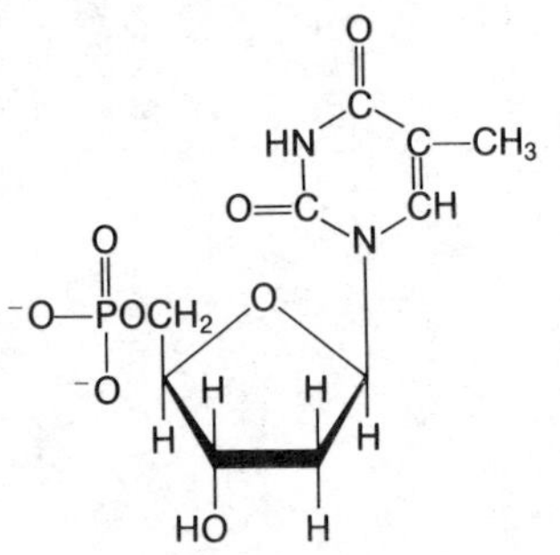

A pyrimidine nucleotide

By conventional rules of *nomenclature,* the atoms of the base are numbered 1–9 in purines or 1–6 in pyrimidines and the carbons of the sugar 1′–5′. A nucleotide with an unmodified name indicates that the sugar is ribose and the phosphate(s) is/are attached at carbon 5′ of the sugar. Deoxy forms are indicated by the prefix d (dAMP = deoxyadenylic acid). If the phosphate is esterified at any position other than 5′, it must be so designated [3′-AMP; 3′-5′-AMP (cyclic AMP; cAMP)]. The nucleosides and nucleotides (ribose form) are named as follows:

Base	*Nucleoside*	*Nucleotide*
Adenine	Adenosine	AMP, ADP, ATP
Guanine	Guanosine	GMP, GDP, GTP
Hypoxanthine	Inosine	IMP
Xanthine	Xanthosine	XMP
Cytosine	Cytidine	CMP, CDP, CTP
Uracil	Uridine	UMP, UDP, UTP
Thymine	dThymidine	dTMP, dTTP
Orotic acid	Orotidine	OMP

Minor (*modified*) bases and nucleosides also exist in nucleic acids (see page 700 for a list). *Methylated* bases have a methyl group on an amino group (*N*-methyl guanine), a ring atom (1-methyl adenine) or on a hydroxyl group of the sugar (2′-*O*-methyl adenosine). *Dihydrouracil* has the 5–6 double bond saturated. In *pseudouridine,* the ribose is attached to carbon 5 rather than to nitrogen 1.

In *polynucleotides* (*nucleic acids*), the mononucleotides are joined by phosphodiester bonds between the 3′-hydroxyl of one sugar (ribose or deoxyribose) and the 5′-hydroxyl of the next (see page 629 for the structure).

Index

NORMAL CLINICAL VALUES: BLOOD*

INORGANIC SUBSTANCES

Ammonia	80–110 μg/dl
Calcium	8.5–10.5 mg/dl
Carbon dioxide	24–30 meq/liter
Chloride	100–106 meq/liter
Copper	100–200 μg/dl
Iron	50–150 μg/dl
Lead	50 μg/dl or less
Magnesium	1.5–2.0 meq/liter
P_{CO_2}	35–40 mmHg 4.7–6.0 kPa
pH	7.35–7.45
Phosphorus	3.0–4.5 mg/dl
P_{O_2}	75–100 mmHg 10.0–13.3 kPa
Potassium	3.5–5.0 meq/liter
Sodium	135–145 meq/liter
Sulfate	2.9–3.5 mg/dl

ORGANIC MOLECULES

Acetoacetate	0.3–2.0 mg/dl
Ascorbic acid	0.4–1.5 mg/dl
Bilirubin	
Direct	0.4 mg/dl
Indirect	0.6 mg/dl
Carotenoids	0.8–4.0 μg/ml
Creatinine	0.6–1.5 mg/dl
Glucose	70–110 mg/dl
Lactic acid	0.6–1.8 meq/liter
Lipids	
Total	450–1000 mg/dl
Cholesterol	120–220 mg/dl
Phospholipids	9–16 mg/dl as lipid P
Total fatty acids	190–420 mg/dl
Triglycerides	40–150 mg/dl
Phenylalanine	0–2 mg/dl
Pyruvic acid	0–0.11 meq/liter
Urea nitrogen (BUN)	8–25 mg/dl
Uric acid	3.0–7.0 mg/dl
Vitamin A	0.15–0.6 μg/ml

PROTEINS

Total	6.0–8.4 g/dl
Albumin	3.5–5.0 g/dl
Ceruloplasmin	27–37 mg/dl
Globulin	2.3–3.5 g/dl
Insulin	6–20 μU/ml

ENZYMES

Aldolase	1.3–8.2 mU/ml
Amylase	4–25 U/ml
Cholinesterase	0.5 pH U or more/h
Creatine phosphokinase (CPK)	5–55 mU/ml
Lactic dehydrogenase	60–120 U/ml
Lipase	2 U/ml or less
Nucleotidase	0.3–3.2 Bodansky U
Phosphatase (acid)	0.1–0.63 Sigma U/ml
Phosphatase (alkaline)	13–39 IU/liter
Transaminase (SGOT)	10–40 U/ml

PHYSICAL PROPERTIES

Blood pressure	120/80 mmHg
Blood volume	8.5–9.0% of body weight in kg
Iron binding capacity	250–410 μg/dl
Osmolality	285–295 mOsm/kg H_2O
Hematocrit	37–52%

NORMAL CLINICAL VALUES: URINE*

Acetoacetate (acetone)	0
Amylase	24–76 U/ml
Calcium	150 mg/d or less
Copper	0–100 μg/d
Coproporphyrin	50–250 μg/d
Creatine	under 100 mg/d
Creatinine	15–25 mg/kg body weight/d
5-Hydroxyindoleacetic acid	2–9 mg/d
Lead	120 μg/d or less
Phosphorus (inorganic)	varies; average 1 g/d
Porphobilinogen	0
Protein (quantitative)	less than 150 mg/d
Sugar	0
Titratable acidity	20–40 meq/d
Urobilinogen	up to 1.0 Ehrlich U
Uroporphyrin	0

* Selected values are taken from normal reference laboratory values in use at the Massachusetts General Hospital and published in the *New England Journal of Medicine* **302:**37, 1980. The reader is referred to the complete list of reference laboratory values in the literature citation for references to methods and units. dl, deciliters (100 ml); d, day.